Animal Physiology

ANIMAL PHYSIOLOGY

An environmental perspective

Patrick J. Butler
University of Birmingham, UK

J. Anne Brown
University of Exeter, and Aquatonics Ltd, UK

D. George Stephenson
La Trobe University, Australia

John R. Speakman
University of Aberdeen, UK, and Chinese Academy of Sciences, PR China

Great Clarendon Street, Oxford, OX2 6DP,
United Kingdom

Oxford University Press is a department of the University of Oxford.
It furthers the University's objective of excellence in research, scholarship,
and education by publishing worldwide. Oxford is a registered trade mark of
Oxford University Press in the UK and in certain other countries

Impression: 1

Published in the United States of America by Oxford University Press
198 Madison Avenue, New York, NY 10016, United States of America

British Library Cataloguing in Publication Data
Data available

Library of Congress Control Number: 2017964396

ISBN 978–0–19–965545–8

Printed in Great Britain by
Bell & Bain Ltd., Glasgow

Overview of the book

Contents

Preface

Wild animals survive in a variety of complex environments; they are exposed to predictable and unpredictable changes in their particular environment on a daily or seasonal basis. Environmental animal physiology is an integrative discipline which brings together elements of anatomy, evolutionary biology, earth sciences, ecology, and aspects of physics, chemistry, biochemistry and genetics to explore the physiological basis of how animals interact with and survive in their natural environment. We live through times when almost all natural environments are undergoing relatively rapid change, and many of these changes, such as the pollution of air and water, removal of natural food sources, environment fragmentation, and (according to most experts) climate change, are the result of human activity. We aim to show how an understanding of the physiology of animals in their natural habitats helps us to understand not only how and why animals evolved the way they did, but how we can act to protect at least some of them from the extreme effects of the changes affecting their environments. In short, we show how environmental physiology can inform conservation.

In **Part 1**, we set the foundation for the topics covered in the remainder of the book by introducing a range of fundamental processes that are essential to life. We consider the diversity of habitats on Earth in which animals live, and examine animal groups and their evolutionary relationships. An animal's DNA is fundamental to its physiological functioning; we learn how cells translate the information captured in DNA into the functional outcomes that underpin life. We also explore the impacts of historic environmental change on animals as well as the potential future impacts of current trends that may be uppermost in your minds.

The efficient use of energy is of paramount importance to animals, particularly when resources are scarce. We explore the different feeding strategies used by animals to obtain the energy they require to carry out all the essential functions of life and learn how animals convert the chemical energy in food molecules into the energy they need to power all body functions.

Our emphasis in this book is always on the whole animal, but to appreciate how animals function we need to know about the cells that form the basic structural and functional units of animals, and how these cells are organized. We look at the general properties of animal cells, and how animals maintain a suitable internal environment in which their cells are protected from external influences. We then examine those fundamental principles governing the main exchanges between the cells within animals and between an animal and its environment.

In **Parts 2 to 4** of the book we discover how different organ systems—respiratory and circulatory systems, excretory organs and endocrine systems—enable animals to interact with their environment, and how environmental temperature profoundly affects the physiology of animals.

In **Part 5**, we explore how the sensory and nervous systems provide animals with information on their internal as well as their external environment and how they, together with the endocrine system, are involved in the control and co-ordination of muscles, reproduction, salt and water balance and the cardio-respiratory systems.

We also examine how different animal species have evolved to spend the whole or part of their lives in extreme environmental conditions, such as exceptionally high or low temperatures, high altitude or extreme salinities and, in the case of some air-breathing animals, under water. We also reveal how other animals are able to perform exceptional feats of endurance such as long-distance, non-stop migrations. Where appropriate, we discuss the use of molecular technologies to advance our understanding of how animals have evolved and become adapted to their different environments.

Audience and approach

The book is written for biology undergraduates who have a basic grounding in maths and science, and are now ready to study animal physiology in some detail. Our presentation of animal physiology is based on several features that pervade the book:

- **A readable and engaging text to keep students' interest and attention, by layering of information.** We focus our main narrative on key concepts, using boxes to expand on particular topics or themes, thus enabling students to grasp difficult concepts in a progressive, layered way.
- **A readily accessible account of the physical and chemical principles** underpinning animal physiology. The physical and chemical principles underpinning the functioning of the organs and physiological systems are often

skimmed over by students using existing textbooks, which can lead to a superficial or confused understanding. To address this, we take particular care to provide students with a robust yet readily accessible and engaging introduction to the physical and chemical properties of the environments in which animals live.

- **The use of carefully chosen examples** to illustrate how different groups of animals have evolved different solutions to deal with the environmental problems they face.
- **The exploration of animal physiology in the context of global issues.** We consider topics such as climate change and pollution from the point of view of their physiological effects on animals, exploring how our understanding of such topics can be translated into new approaches to conservation.

Learning features: a note for students

Animal Physiology uses a number of learning features to make your learning as effective as possible:

- **Boxes** expand on a range of topics covered in the text, giving you the opportunity to go into them in more detail.
- **Case studies throughout the text** include recent investigations using the most up-to-date technology to further our understanding of how animals respond to and deal with environmental events and changing habitats. Some case studies also describe examples of biomimicry to illustrate how our understanding of animals' adaptations to their environment is being practically applied to problems in novel ways, in some cases to address environmental issues that humans are facing.
- **Experimental panels reinforce the nature of animal physiology as an experimental science and the overall process of scientific enquiry;** they explore the variety of methodologies employed in contemporary research, set out how investigators interpret the data gathered, and review the conclusions drawn and implications of their findings.
- **Full colour, high-quality annotated illustrations of physiological processes** help you to gain a clearer understanding of important physiological structures and processes by aiding visualization.
- **Further information sources at the end of each section** encourage you to take your learning one step further by consulting carefully chosen review articles and other appropriate sources.
- **Cross references throughout each chapter** help you to make connections between the topics discussed throughout the book, and encourage you to view the subject as a coherent, unified whole.
- **Key terms** are emboldened when first introduced; those key terms that also appear in the accompanying online glossary are coloured.
- **Key concepts** and ideas within each chapter are summarized in the bulleted list at the end of each chapter to enable you to check your grasp of the main concepts discussed.
- **Study questions at the end of each chapter** provide opportunities for you to refine your understanding, and include questions that require the manipulation and/or interpretation of data, giving you the opportunity to develop this important skill.
- **An appendix** provides a concise summary of the units used in animal physiology, as a single point of reference, and to help you avoid confusion and understand the relationships between units.

Online support

Oxford University Press offers a comprehensive ancillary package for lecturers and students using *Animal Physiology – An environmental perspective*. Go to www.oup.com/uk/butler to find out more.

For students:

- **Original articles:** a list of original articles consulted during the writing of each chapter so that you can explore the original research for yourself.
- **Additional case studies and experimental approach panels** to augment those in the printed book.
- **Answers to numerical questions:** full solutions to numerical questions so that you can verify your working.

For registered adopters of the text:

- **Digital image library:** Includes electronic files in JPEG format of every illustration, photo, graph and table from the text.

Acknowledgements

We thank Jonathan Crowe, who has fully supported us throughout the process of writing this book, right from its initial conception. He has tirelessly committed himself to the project and its demands, even when he might sensibly have passed on the editorial role. We have been fortunate in receiving his invaluable help with careful and perceptive editing of the chapters.

We must acknowledge the very helpful and often extensive feedback from our reviewers, and colleagues; without this feedback a textbook with the breadth of *Animal Physiology – An environmental perspective* would not be possible. Our list of reviewers does not do justice to the fact that several of them gave us feedback on multiple chapters. The time they committed to this, and the suggestions and constructive comments they provided, have helped enormously in shaping the book and its scope. So, for their time and input we thank:

Ann Linda Baldwin, PhD, University of Arizona
Dr Debbie Bartlett, University of Greenwich
Holly Bates, Trent University
Chris Brown, University of Worcester
Sheena Cotter, University of Lincoln
Heidi Engelhardt, University of Waterloo
Stephanie M. Gardner, Purdue University
Fritz Geiser, University of New England
Katie Gilmour, Bishop's University
Malcolm S. Gordon, University of California, Los Angeles
Dr Caron Inouye, California State University, East Bay
Kenneth Long, California Lutheran University
Martin Luck, University of Nottingham
Terri J. Maness, Louisiana Tech University
Dr Tom Moon, University of Ottawa
Dr Bob Newby, Central Queensland University
Benjamin L. Predmore, University of South Florida
Andrew Pye, University of Exeter
Robert Roer, University of North Carolina Wilmington
Jim Staples, University of Western Ontario
William Velhagen, New York University
Alan Waterfall, University of Nottingham
Tim Whalley, University of Stirling
Dr Nia Whiteley, Bangor University
Joe Williams, Ohio State University
Brian Wilson, Acadia University, Wolfville
Professor Philip Withers, University of Western Australia
Kathryn Yuill, University of Plymouth

We are also grateful for the input of student reviewers, which has enabled us to improve the illustrations and clarify explanations of difficult topics. We thank the following friends and colleagues who provided unpublished materials for us to include:
Rexford Ahima, Peter Arkwright, Nyambayar Batabaya, Michael Berenbrink, Randy Davis, Tom Kunz, Paul Ponganis, John Rensberger, Jon Russ and Chloe Rutter.

In particular, we thank Dona Boggs-Aitken for her input to the text by providing case study examples of biomimicry.

Our greatest thanks go to our families, in particular to Vivien, Phil and Gabriela, for their patience, support and encouragement, without which this book would not have been written.

Abbreviations

ABO Air-breathing organ
ACBP Acyl-CoA binding protein
ACC Antarctic Circumpolar Current
ACE Angiotensin converting enzyme
ADH Antidiuretic hormone
ADL Aerobic dive limit
AFP Antifreeze peptides
AFGP Antifreeze glycopeptide / glycoproteins
AKH Adipokinetic hormone
ALD Anterior latissimus dorsi
AMP Adenosine monophosphate
ANP Atrial natriuretic peptide
ANS Autonomic nervous system
BAT Brown adipose tissue
BCF Body and/or caudal fin
BMR Basal metabolic rate
BNP B-type natriuretic peptide
CA Carbonic anhydrase
CFTR Cystic fibrosis transmembrane conductance regulator
CG Chorionic gonadotropin
CGRP Calcitonin gene-related peptides
CHH Crustacean hyperglycaemic hormone
CNP C-type natriuretic peptide
CNS Central nervous system
COP Colloid osmotic pressure
COT Cost of transport
CPG Central pattern generator
CRH Corticotrophin releasing hormone
CS Citrate synthase
CSA Cross-sectional area
DGC Discontinuous gas-exchange cycles
DRIP Drosophila integral protein
DVN Dorsal vagal motor nucleus
ECF Extracellular fluid
EPOC Excess post-exercise oxygen consumption
EPSP Excitatory postsynaptic potential
ETH Ecdysis triggering hormone
EWL Evaporative water loss
FATP Fatty acid transport protein
FG Fast-twitch glycolytic
FMR Field metabolic rate
FOG Fast-twitch, oxidative glycolytic
FSH Follicle-stimulating hormone
FW Fresh water
GAS General adaptation syndrome
GFR Glomerular filtration rate
GH Growth hormone
GHK Goldman–Hodgkin–Katz
GIH Gonad-inhibiting hormone
GnIH Gonadotrophin inhibiting hormone
GnRH Gonadotrophin releasing hormone
GPCR G-protein coupled receptor
HAN High-altitude native
HIF Heat increment of feeding
HPA Hypothalamus–pituitary–adrenal
HPI Hypothalamus–pituitary–interrenal
HRE Hormone response element
HSP Heat shock (stress) proteins
ICF Intracellular fluid
IUCN International Union for Conservation of Nature
IPSP Inhibitory postsynaptic potential
ITM Intermediate-term memory
JH Juvenile hormone
LAN Low-altitude natives
LCFA Long-chain fatty acids
LH Luteinizing hormone
LP Lateral pyloric
LTD Long-term depression
LTM Long-term memory
LTP Long-term potentiation
LWS Long-wave sensitive
MET Mechanoelectrical transduction
MHC Myosin heavy chain
MIH Moult-inhibiting hormone
MLCK Myosin light chain kinase
MLCP Myosin light chain phosphatase
MMR Maximum metabolic rate
MPF Median and/or paired fin
MR Metabolic rate
MRLC Myosin regulatory light chain
NA Nucleus ambiguus
NAD Nicotinamide adenine dinucleotide
NMR Nuclear magnetic resonance
NST Non-shivering thermogenesis
NTS Nucleus of the solitary tract
OCLTT Oxygen- and capacity-limited thermal tolerance
OEC Oxygen equilibrium curve
PD Pyloric dilator
PETH Pre-ecdysis triggering hormone
PNS Peripheral nervous system
POA Pre-optic area
PSP Postsynaptic potential
PTH Parathyroid hormone
PUFA Polyunsaturated fatty acids

RER	Respiratory exchange ratio
REWL	Respiratory evaporative water loss
RMR	Resting metabolic rate
RMT	Relative medullary thickness
ROS	Reactive oxygen species
RQ	Respiratory quotient
RVLM	Rostral ventrolateral medulla
RyR	Ryanodine receptor
SCFA	Short-chain fatty acids
SCN	Suprachiasmatic nuclei
SDA	Specific dynamic action
SFO	Subfornical organ
SGLT	Sodium-glucose-transporters
SMR	Standard metabolic rate
SNGFR	Single nephron glomerular filtration rates
SO	Slow-tonic oxidative
STPD	Standard temperature and pressure dry
SW	Seawater
TCA	Tricarboxylic acid
TEF	Thermic effect of food
TEWL	Total evaporative water loss
TNF	Tumour necrosis factor
TNZ	Thermoneutral zone
TRH	Thyrotrophin releasing hormone
TRP	Transient receptor potential
TSH	Thyroid-stimulating hormone
UT	Urea transport proteins; urea transporters
VEGF	Vascular endothelial growth factor
VIP	Vasoactive intestinal peptide
VLDL	Very-low-density lipoproteins
VNP	Ventricular natriuretic peptide
WVD	Water vapour density

Coral polyps with their emerging tentacles that capture microscopic organisms suspended in the water. The energy input from these organisms supplements energy obtained from the photosynthetic activity of the symbiotic zooxanthellae that give the corals their characteristic colour.
Image source: FLPA/Alamy Stock Photo

Part 1

Animals and their environment

Part 1 reviews the fundamental aspects of biology that we need in order to understand animal physiology from an environmental perspective, which we go on to explore throughout the rest of the book. In biological terms an animal is any organism classified as a member of the kingdom Animalia, from comb jellies to humans.

First, in **Chapter 1**, we consider the diversity of habitats on Earth in which animals thrive, and examine animal groups and their evolutionary relationships. An animal's DNA is fundamental to its physiological functioning; we learn how cells translate the information captured in DNA into the functional outcomes that underpin life. We also explore the impacts of historic environmental change on animals, as well as the potential future impacts of current trends that may be uppermost in your minds.

Animals need energy to stay alive—to maintain their complex level of organization, and to drive the myriad of physiological processes explored throughout this book. In **Chapter 2** we explore the different feeding strategies used by animals to obtain this energy. Here, we discover just how effective suspension feeding can be for aquatic animals, from mussels to blue whales, including the coral polyps (shown here). We learn how animals convert the chemical energy in food molecules into the energy they need to power all body functions. The efficient use of energy is of paramount importance to animals, particularly when resources are scarce.

Our emphasis in this book is always on the whole animal, but to appreciate how animals function we need to know about the cells that form the basic structural and functional units of animals, and how these cells are organized. In **Chapter 3**, we look at the general properties of animal cells, and how animals maintain a suitable internal environment in which their cells are protected from external influences. We then examine those fundamental principles governing the main exchanges between the cells within animals and between an animal and its environment.

1

The diversity of animals and their interactions with natural environments

Life on Earth has existed for the past 3.8 billion years or more. Around 1.5 billion years ago the first animals evolved from a single-celled marine flagellate and have ultimately given rise to up to one billion species[1] so far. All animals are subjected to the fundamental biological law of survival: to live long enough to reproduce and so pass on their genetic material to a new generation. But such survival is an unceasing battle against predators, infections and parasites. To survive, animals must compete with others to obtain the essential nutrients and water for bodily functions. They must also cope with environmental conditions and their changes. The long-term outcome is that no animal species survives indefinitely. The average lifespan of a species from its first appearance is estimated at about 10 million years, although there are also a small number of species that seem to have survived for hundreds of millions of years[2].

Currently living animal species are estimated to represent only ~1 per cent of all animal species that have ever existed, though the actual number of living species is greater now than at any point in the past 500 million years, at about 8–10 million. However, the May 2019 UN report from the Intergovernmental Science-Policy Platform on Biodiversity and Ecosystem Services estimates the current extinction rate of species to be 100 to 1000 fold greater than over the past 10 million years, and it is far outpacing the origination rate of new species. This trend, caused primarily by human activity, can only be reversed through urgent transformative action by the international community, without which many scientists believe that the Earth will experience another **mass extinction event**.

New species arise through an evolutionary[3] process that gives rise to new varieties of organisms that are able to exploit available ecological niches in the environment. A huge variety of environments exist, each with its own particular physical and chemical characteristics—from the dry hot deserts to the lightless depths of the oceans. This enormous range of environments has contributed to the vast diversity of animals that have evolved to live in them—in so doing achieving the remarkable feat of being able to make a home in what may seem the most inhospitable of surroundings.

Throughout this book we explore how the physiological systems of animals have evolved to ensure they can survive and reproduce in their particular environment, how physiological systems enable animals to operate in changing conditions without compromising their well-being, and how common systems have evolved in different ways to allow different species to occupy vastly contrasting environmental niches. We also discover how the well-being of animals is constrained by their environment, and the challenges many species face as climate change pushes physiological systems beyond their normal limits. Our understanding of these constraints may help us to conserve species that might otherwise be at risk.

We begin by setting out the scope of environmental animal physiology before discussing the range of environments and the challenges they pose for the animals that occupy them. These environments are dynamic systems that change on various timescales—from the natural tidal cycles usually of approximately 12 hours that we observe along coastlines, to diurnal, seasonal and to longer-term climatic changes which we, as humans, may influence. We consider some of these changes and how they may compromise physiological functioning, forcing some populations of some species to redistribute or face extinction.

We also discuss the diverse range of biological characteristics that animals exhibit, setting the scene for when we explore the functioning of different physiological systems in later chapters. Finally, in the last section of this chapter we consider how animal populations adapt to changes in the environment, how new species come to exist through natural

[1] An animal species (plural species, abbreviation spp.) is generally defined as a group of similar individuals capable of exchanging genes to produce fertile offspring in the wild.

[2] For example, hagfish (a group of jawless fish) have hardly changed for more than 300 million years.

[3] Evolution refers to the change in the heritable characteristics (traits) of a population through successive generations as we discuss in Section 1.5.3.

selection and how rapid environmental change—defined as a rapid change or disturbance of the environment caused by natural ecological processes or by human interference—is placing animals under stress, increasing their chance of extinction once a major catastrophe occurs.

1.1 What is environmental animal physiology?

Environmental animal physiology investigates how animals function in strikingly different environments—from deep seas to sea shores; from the top of mountains to lakes or deserts. Virtually every aspect of any environment that we could consider impinges on an animal's physiology: its existence, its ability to obtain food and energy, its dispersion or migration, and its reproduction. This is what environmental animal physiology is all about.

Environments are rarely constant, particularly on land, and to survive animals must cope with whatever changes they meet. Environmental physiology explores how animals respond to environmental changes, whether these occur on a daily basis, seasonally or as long-term trends that could affect populations.

In order to understand how animals function, we need to know the *mechanisms* by which organs, such as the heart, kidneys and muscles, function[4] and how these are controlled by nerves[5] and hormones[6]. Physiologists ask questions such as: How do animals sense and react to their environment[7]? How do nerves conduct signals at up to 120 metres per second, and how do muscles[8] function to allow movement and behaviour? How do the kidneys[9] excrete wastes and regulate body fluids, and how do environmental conditions affect their functioning?

Environmental animal physiology is an integrative discipline; Figure 1.1 shows how it brings together elements of anatomy, evolutionary biology, Earth sciences, ecology, and aspects of physics, chemistry, biochemistry and genetics. First, to understand how animals function, we need to know about the structure of their organs and tissues (their anatomy). Structure and function are intrinsically linked: a car with square wheels—an inappropriate structure—would not function very well! When it comes to physiological systems, then, we need to know the structure of organs, like the heart and kidneys, and the cells that make up the organs and tissues to understand their functioning. Physiologists also

[4] Sections 3.2 and 3.3 discuss the organization of cells, organs and tissues in animals.
[5] Chapter 16 discusses how the nervous system functions.
[6] Hormones are the topic of Chapter 19 and discussed at many points throughout the book, for example in Chapter 20 (Reproduction).
[7] Animal senses are the focus of Chapter 17.
[8] We examine muscle function and movement in Chapter 18.
[9] We discuss kidney excretion in Chapter 7.

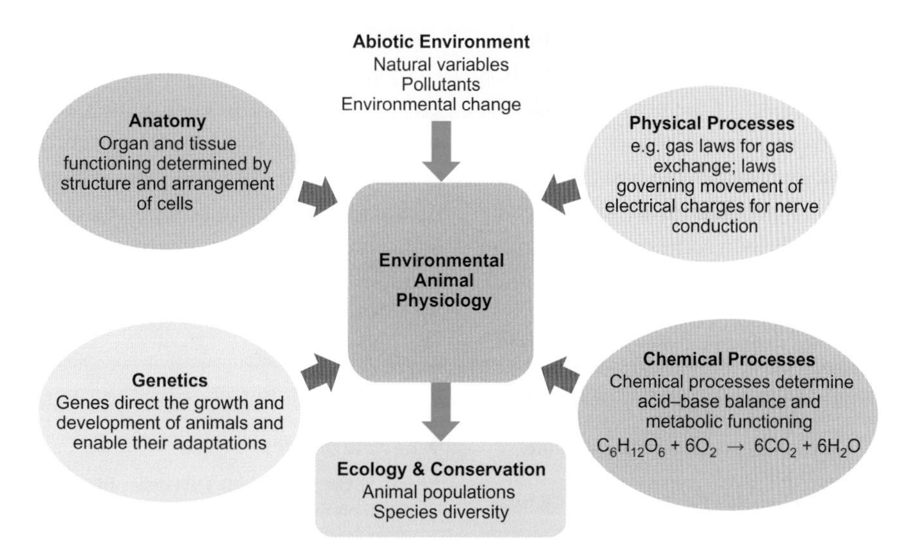

Figure 1.1 Environmental animal physiology is an integrative discipline which informs ecology and conservation

Abiotic variables, including natural variables and those resulting from human activities affect physiological functioning. Understanding physiological function requires information on the anatomy of organs and tissues. Genes determine the expression of enzyme and other proteins that determine tissue functioning, which is also governed by physical laws and chemical principles; the equation shows the chemical oxidation of glucose, which releases energy from carbohydrate foodstuffs. Environmental physiology informs ecology and conservation by providing mechanistic explanations of ecological changes and the impacts of environmental change or pollutants on animal species and populations, which guide conservation.

study genes to make predictions about how particular differences evolved[10], and to gain insights into how animals cope with non-biological (**abiotic**) variables in their environment such as temperature[11], oxygen[12] and availability of water[13].

Much of our understanding of animal physiology is based on laboratory studies, often using strains of laboratory animals that have a history of artificial selection. Such investigations can only tell us so much about how animals cope in their natural environments. To understand environmental physiology, laboratory studies need augmenting with studies of animals in their natural environment. Field physiology is an essential part of building an understanding of how animals function in the wild. For instance, some animals undertake spectacular migrations that have long fascinated physiologists. These migrations raise questions. For example, do animals prepare for their migration? If so, to what extent? In seeking answers, physiologists discovered that birds, such as barnacle geese (*Branta leucopsis*[14]), have a high level of physical fitness when they commence migration, despite not showing any increase in flight activity beforehand[15].

Environmental changes are of great concern and may adversely affect the well-being of individual animals. Loss of a suitable habitat due to its destruction, degradation or fragmentation is a major threat to wildlife, and changes to the abiotic conditions in environments, such as changes in temperature or water availability, are additional threats. The physiology of animals determines whether they will cope or not.

Pollution presents a further environmental challenge and we need to consider the physiology of animals to understand the impacts of particular pollutants on animals. It is possible to use **biomarkers** to detect the physiological state of animals in advance of damage by pollutants. Thus, environmental animal physiology provides the basis for understanding why and how animal populations are influenced by environmental pollutants[16], which in some cases can explain why populations decline.

[10] For example, in Figure 13.13 we examine the evolution of the capillaries in the eyes and swim bladders of teleost fish, while in Section 9.3.3 we examine the evolution of some key enzymes that result in thermal adaptation.

[11] Part 3 examines animals in relation to environmental temperature.

[12] Part 4 examines gas exchange by animals.

[13] Part 2 examines the adaptive mechanisms of animals in different habitats that achieve water balance.

[14] The scientific names of organisms are italicized. The genus name is capitalized and is followed by the name of the species, which is not capitalized.

[15] Section 15.2.3 discusses physiological aspects of migration of barnacle geese.

[16] For example, Case Study 5.2 explores why some fish are sensitive to nitrite pollution but others protected from such effects, while in Section 18.6.2 we examine the effect of low environmental pH on the swimming ability of some species of freshwater fish. Box 7.3 discusses ammonia toxicity.

A striking example of population decline whose cause was revealed by physiological study is the dramatic decline in vulture populations throughout India that occurred in the 1990s[17]. Once investigated from a physiological perspective it became clear that the vultures had consumed pharmaceutical contaminants in their diet of animal flesh. Their inability to excrete nitrogenous waste[18] led to its accumulation in major body organs, and death due to kidney and liver failure. This devastating effect was difficult to predict in advance, but once recognized conservation and recovery programmes for vultures began. Attention has since turned to the protection of other raptors that could be similarly at risk. This is an example in which environmental animal physiology guided conservation (sometimes called conservation physiology).

The example of vulture populations in India is one in which environmental physiology identified the mechanism *after* damage had occurred, and which helped to inform conservation strategies. In a similar way, detecting the damaging effects of acid rain caused by emissions of sulfur and nitrogen oxides on aquatic animals, which were identified by many physiological investigations in the 1970s and 1980s, has led to measures to curb these emissions and thus conserve animal populations.

Environmental physiology can also provide mechanistic explanations that help to predict the effects of environmental change in *advance* and to identify those species and populations that are most vulnerable. For example, in Case Study 1.1 we consider studies that predict the effects of ocean acidification (a decrease in pH) on marine animals.

By understanding physiological functioning and the tolerances of animals to environmental factors we can start to understand why some animals occur in particular habitats, while others do not. And with an appreciation of how pollutants affect physiological processes we can begin to understand how changes in environmental conditions may alter suitable habitats into unsuitable ones for particular species, and thus affect the abundance of animal populations and their distributions.

› *Review articles*

Bograd SJ, Block BA, Costa DP, Godley BL (2010). Biologging technologies: new tools for conservation. Introduction. Endangered Species Research 10: 1–7.

Costa DP, Sinervo B (2004). Field physiology: physiological insights from animals in nature. Annual Review of Physiology 66: 209–238.

[17] Case Study 7.2 explores the physiological reasons for the decline of vultures and the encouraging beginnings of recovery achieved since then.

[18] We discuss uric acid excretion in Section 7.4.3.

Case Study 1.1 Ocean acidification

About a quarter of the carbon dioxide (CO_2) emitted into the atmosphere is estimated to be absorbed by the oceans. Consequently, the global increases in atmospheric concentrations are increasing the concentrations of CO_2 in the oceans, causing **ocean acidification**. Ocean acidification means that there is an *increase* in H^+ concentrations of seawater, which can be measured as a *decrease* in the pH[1], and a decrease in the ocean concentrations of carbonate ion (CO_3^{2-}). The reactions involved in these chemical changes are summarized in Figure A.

How has ocean pH changed historically, and what are the future forecasts? Between 1750 and the early 1990s, the average ocean pH is estimated to have declined by 0.1 pH units. Recordings over the last 10–20 years indicate a faster current rate of decline of ocean pH, as shown by the examples in Figure B. If the changes in pH continue at the current rates, further pH decreases of 0.1 pH units could occur within about 50–60 years.

Ocean acidification has direct implications for marine organisms, particularly those with shells (such as molluscs) or skeletons (like corals) that consist of mineral forms of calcium carbonate ($CaCO_3$). Surface seawater is usually supersaturated with calcium carbonate ($CaCO_3$) and hence solid forms of calcium carbonate are usually stable, including the calcareous shells and skeletons of living marine organisms, calcareous rocks, broken shells and mineral particles in sea sediments. However, ocean acidification causes a reduction in seawater concentrations of CO_3^{2-}, as shown in Figure A. If under-saturation of $CaCO_3$ is reached, then we can expect that dissolution of mineral forms of $CaCO_3$, including the shells and skeletons of organisms (into Ca^{2+} and CO_3^{2-}), will occur.

There is already evidence of dissolution of $CaCO_3$ in the shells of some marine species due to the more acidic environment—for example, the shells of various species of sea butterflies in polar and sub-polar regions, where seawater is under-saturated with respect to $CaCO_3$. Sea butterflies are a major group of small molluscs in the zooplankton, the animal component of **plankton**. Since sea butterflies are important food for carnivorous predators in the zooplankton, as well as polar fishes and baleen whales[2], the impacts of declining ocean pH on sea butterflies could have far-reaching ecological effects.

Some marine calcifiers (such as corals and mussels) can increase the rate at which they form their shell or skeleton, by raising the pH of the fluid at the site of $CaCO_3$ crystallization[3], and might maintain the rate of calcification even when water chemistry appears unsuitable. However, such maintenance of calcification requires energy-dependent transport processes, which may affect the energy balance and adversely affect growth. It is thought that by the middle of this century the levels of CO_2 and pH may be such that there is no net growth of coral reefs.

Evidence of the effects of acidification on growth has emerged from studies of California mussels (*Mytilus californianus*). The growth of the motile larval stages of the mussels declined when they were reared in seawater of reduced pH or seawater

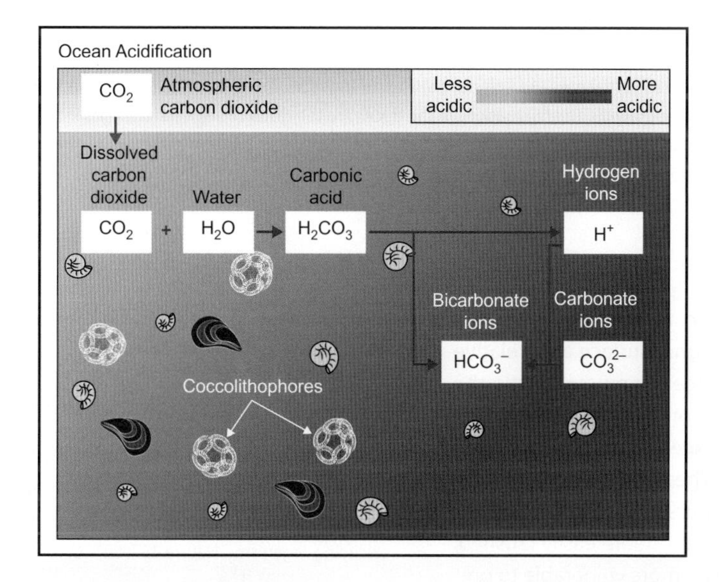

Figure A Chemical reactions in ocean acidification

Atmospheric carbon dioxide is absorbed by the oceans (and other water bodies such as rivers and lakes). Some of the extra carbon dioxide remains as dissolved carbon dioxide, but some reacts with water and is converted into carbonic acid, which dissociates to bicarbonate and protons (H^+). When CO_2 increases, the protons released from the dissociation of carbonic acid decrease the pH and bind to CO_3^{2-}, which reduces the seawater concentrations of CO_3^{2-} and increases the concentration of bicarbonate, as shown by the back arrow from CO_3^{2-} to HCO_3^-. The organisms illustrated (various molluscs, and coccolithophores, which are a group of single-celled calcifying phytoplankton—the plant component of plankton) depend on the formation of aragonite or calcite, which are calcium carbonate-based minerals, in structures, such as their shells, or the calcified plates surrounding the coccolithophores. For some species the process of calcification has been predicted to be negatively affected by ocean acidification, but other species appear unaffected, and some may show evolutionary adaptation.

Source: adapted from: http://www.oceanacidification.org.uk/.

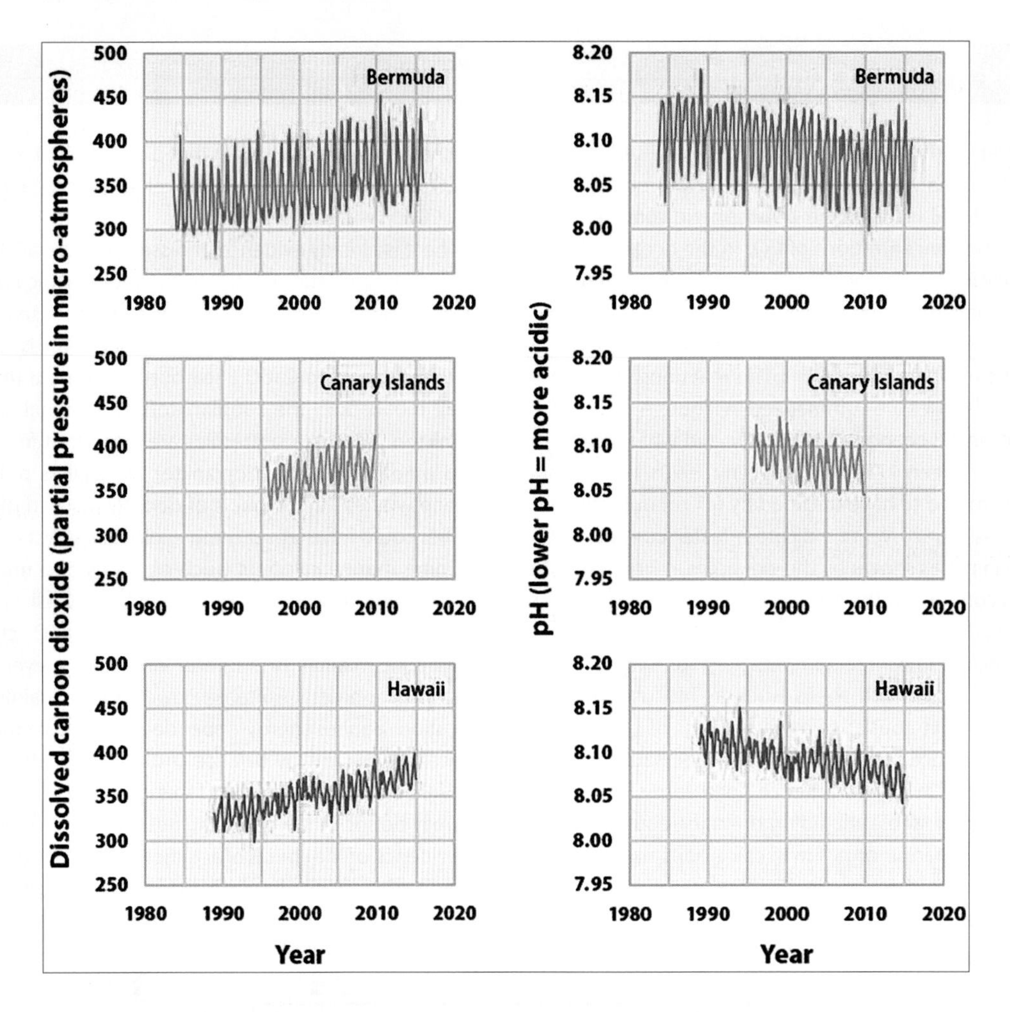

Figure B Ocean carbon dioxide levels and acidity between 1983 and 2015 at three locations

Changes are shown for carbon dioxide levels, given as partial pressure (left panels) and pH (right panels) at two observation stations in the North Atlantic Ocean (Canary Islands and Bermuda) and one in the Pacific (Hawaii). The fluctuations in the traces are the result of seasonal changes in temperature. The overlying trend is an increase in carbon dioxide and reduction in pH at all three locations.

Source: https://www.epa.gov/climate-indicators/climate-change-indicators-ocean-acidity.

containing CO_2 at the levels predicted to occur by 2100. The smaller mussels have a 40 per cent higher surface area relative to their mass than larger mussels[4]. Once the mussels settle the smaller ones are at a higher risk of mortality because of their higher rates of desiccation and/or excessive heat gain, when they are exposed between the tides. The researchers also found that the smaller shells of California mussels reared in seawater with a higher level of CO_2 are thinner and more easily crushed, so newly settled mussels are likely to be more vulnerable to predation by the juvenile crabs (also 'carrying' a shell) that are more resistant to the effects of acidification.

› Find out more

Gaylord B, Hill TM, Sanford E, Lenz EA, Jacobs LA, Sato KN, Russell AD, Hettinger A (2011). Functional impacts of ocean acidification in an ecologically critical foundation species. Journal of Experimental Biology 214: 2586–2594.

Gazeau F, Quiblier C, Jansen JM, Gattuso J-P, Middelburg JJ, Heip CHR (2007). Impact of elevated CO_2 on shellfish calcification. Geophysical Research Letters 34(7): L07603.

Hofmann GE, Evans TG, Kelly MW, Padilla-Gamiño JL, Blanchette CA, et al. (2014). Exploring local adaptation and the ocean acidification seascape—studies in the California Current large marine ecosystem. Biogeosciences 11: 1053–1064.

Orr JC, Fabry VJ, Aumont O, Bopp L, Doney SC, Feely RA, et al. (2005). Anthropogenic ocean acidification over the twenty-first century and its impact on calcifying organisms. Nature 437: 681–686.

[1] A decrease in pH is equivalent to increased acidity; pH units are on a logarithmic scale of H^+ concentrations, as we explain in Box 13.2, such that a decrease of 0.1 pH unit corresponds to a 26 per cent increase in H^+ concentration.

[2] The feeding of baleen whales is discussed in Case Study 2.1.

[3] Shell formation in molluscs is discussed in Section 21.2.5.

[4] The relationship between size of an animal and its surface area/volume (mass) is examined in Box 2.3. This relationship can drive exchanges across body surfaces (e.g. heat uptake or water loss).

1.2 Natural environments: where and under what conditions do animals live?

The environments in which communities of animals and plants live are called **ecosystems**. Animals (and plants) in ecosystems interact with the surrounding non-biological abiotic environment. Within ecosystems, animals live in **habitats**, which supply all their needs: food, minerals and salts[19], water, oxygen and suitable temperatures.

[19] Salts are neutral compounds which dissociate in water into positively charged cations and negatively charged anions.

At a global level, similar ecosystems found on different landmasses or in different oceans are called **biomes**. A biome is characterized by its physical, chemical and climatic characteristics, and the interacting communities of animals and plants it supports. There is a fundamental division of biomes into two main strands: aquatic biomes and terrestrial (land) biomes.

1.2.1 Aquatic biomes and habitats

The Earth has been called the 'blue planet' because of the prevalence of water, which covers approximately 71 per cent of its surface, primarily in the oceans. Figure 1.2A, an image from outer space, shows the two largest oceans: the Pacific Ocean, which holds 49 per cent of the Earth's water, and the Atlantic Ocean, holding 23 per cent of Earth's water. The plan drawing in Figure 1.2B also shows

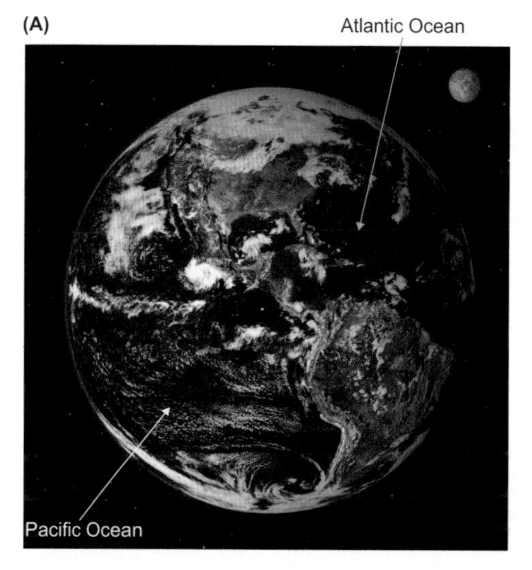

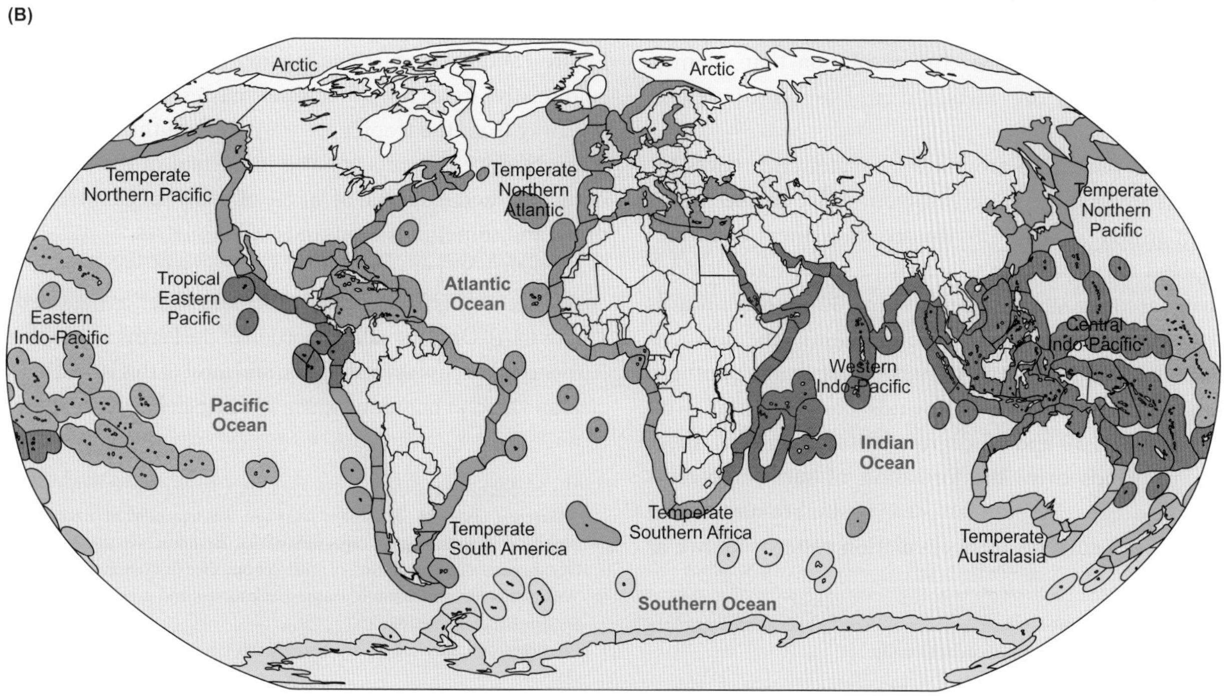

Figure 1.2 The 'Blue Planet' Earth from space and its marine coastal ecoregions

(A) True-colour image showing North and South America as they appear from 35,000 km above the Earth. The image is the result of a combination of data from two satellites. (B) Global map of marine coastal and shelf ecoregions within colour-coded and labelled provinces identified by the team effort of more than 1000 biologists and biogeographers as regions that have some cohesion on an evolutionary timescale and hold endemic species. The ecoregions occur in areas shallower than 200 m and extend to 370 km offshore.

Sources: (A) http://dvidshub.net/image/843476/earth-space-image-day; (B) Spalding MD et al (2007). Marine ecoregions of the world: A bioregionalization of coastal and shelf areas. Bioscience 57: 573-583.

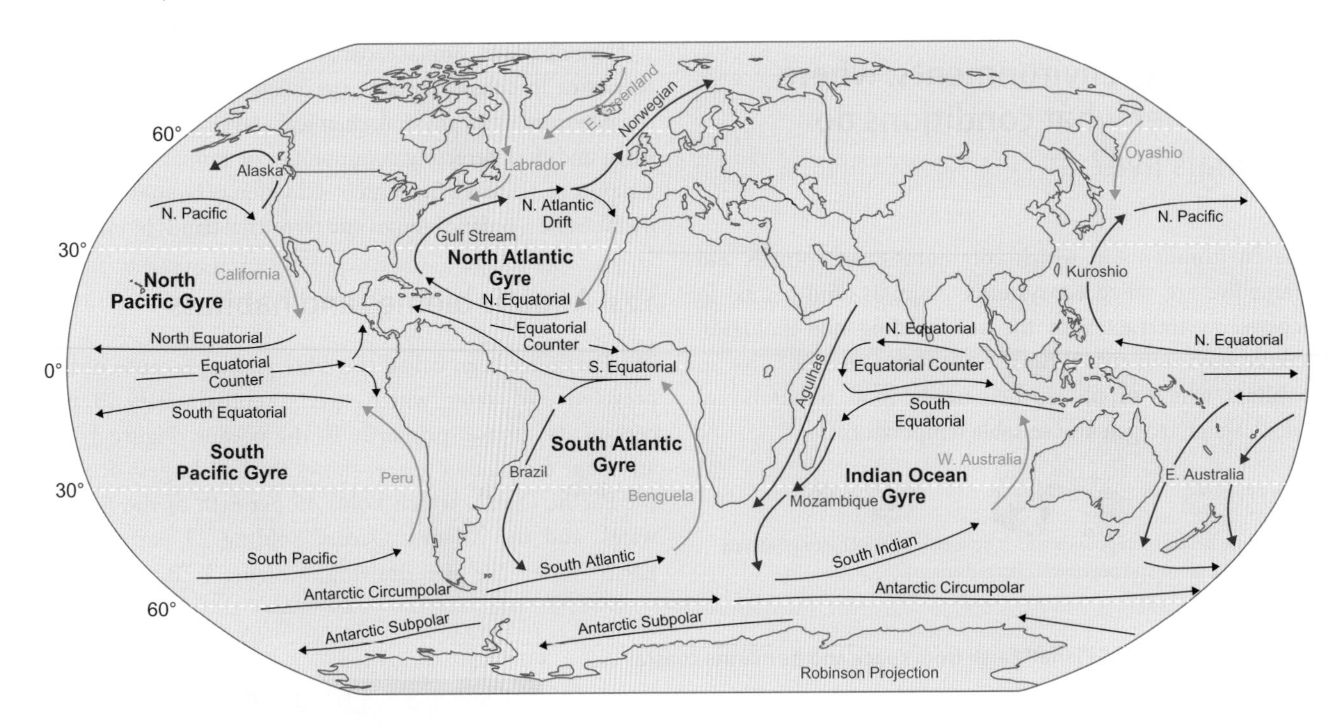

Figure 1.3 Surface ocean currents

Warm currents (red arrows) and cold currents (blue arrows) are defined as 'warm' or 'cold' by comparison to the ambient temperature, not the actual temperature of the current. Black arrows indicate currents predominantly travelling in an eastward or westward direction.

Source: adapted from Pidwirny M (2006). Surface and Subsurface Ocean Currents: Ocean Current Map. Fundamentals of Physical Geography, 2nd Edition. http://www.physicalgeography.net/fundamentals/.

the Indian Ocean, which holds 20 per cent of Earth's water, and the Southern Ocean, which holds about 5 per cent of the Earth's water.

Some marine animals cover vast distances and depths within the oceans, but most marine organisms occur in coastal areas. At a finer level of resolution than biomes, coastal **ecoregions** are identified and are illustrated in Figure 1.2B.

Warm air rises from around the equator and moves towards the poles in an easterly direction due to the Earth's rotation around its axis. This massive movement of air creates global wind patterns which include the trade winds[20] (blowing at lower latitudes below about 30 degrees towards the equator from north-east to south-west in the northern hemisphere and south-east to north-west in the southern hemisphere) and the westerly winds (blowing from west to east between 30 and 60 degrees latitude). In turn, these prevailing winds drive the formation of surface currents within the oceans. Because the ocean basins are surrounded by land, ocean currents form circulation patterns known as gyres. Figure 1.3 shows the five major ocean gyres. In the Antarctic, water flows around the Earth in the great circumpolar currents[21].

The surface ocean currents help to redistribute heat around the globe in the upper layers of the oceans. Notice in Figure 1.3 that the currents moving towards the tropics are cold relative to surrounding water masses, while those moving away from the tropics are warmer than the surrounding water. Surface currents are also important in the migration of turtles[22] and dispersal of larvae of many species over hundreds and sometimes thousands of kilometres.

More than 97 per cent of the total volume of global water occurs in the oceans and seas, and other salty water, as indicated in Figure 1.4(i). Only a small proportion of the Earth's water (about 2.5 per cent) occurs as **fresh water**, and most of this is held in glaciers and ice caps or ground water. Warmer global temperatures[23]

[20] Trade winds are named as such because of their historical use by sailors crossing the world's oceans in order to trade.

[21] The establishment of the massive Antarctic Circumpolar Current around Antarctica, about 25–22 million years ago, was a significant driver of the evolution of physiological capacities of marine animals in the Southern Ocean, such as the evolution of antifreezes, as we discuss in Section 9.2.4.

[22] We discuss the migration of leatherback turtles and green turtles in Sections 10.3.2 and 18.6.2.

[23] Predicted climate changes are discussed further in Section 1.5.4.

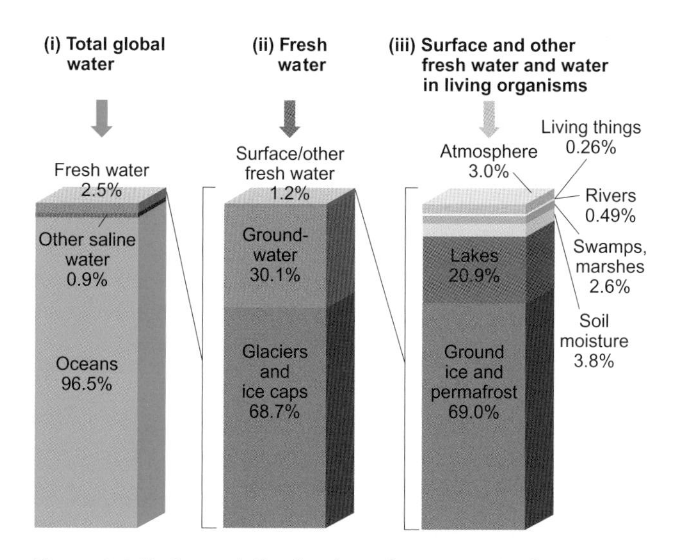

Figure 1.4 Estimated distribution of water on Earth

The distribution of Earth's water shown as: i) proportions of total water; ii) distribution of fresh water; and iii) surface water and other fresh water (numbers are rounded so do not add up to exactly 100%).

Source: data from Gleick PH (ed). Water in Crisis: A Guide to the World's Fresh Water Resources. Oxford: Oxford University Press, 1993.

are likely to reduce the amount of fresh water in the ice caps and glaciers and hence are likely to contribute to the predicted rise in seawater levels. Although these predictions are complex and estimates differ, recent forecasts predict that the amount of melting of ice sheets and glaciers by 2100 (based on a mid-range level of increase in temperatures) could add almost 37 cm to ocean levels. With other factors, overall sea levels could rise more than 50 cm by 2100.

Only a very small part of the water on Earth is present in surface fresh water and in living organisms, which together represent about 1.2 per cent of Earth's fresh water as Figure 1.4(ii) illustrates. The water in living organisms represents only about 0.003 per cent of the total fresh water on Earth ($0.26/100 \times 1.2$).

While the total mass of water on Earth is fairly constant, its partitioning into the atmosphere, fresh water, ice and permafrost, the oceans and living organisms varies depending on climatic variables, such as temperature, which influence the Earth's water cycle due to physical processes such as evaporation (driving more water vapour into the atmosphere at higher temperatures), precipitation, and run-off from land into rivers and the oceans.

Just 0.008 per cent of Earth's water occurs in freshwater habitats such as rivers and freshwater lakes, swamps and marshes. Even though this is a tiny fraction of Earth's total water it still amounts to about 105,000 km^3 (about 10^{14} m^3)—roughly equivalent to the volume that would be contained in a lake of 100 m depth (0.1 km) covering an area of 1000 km by 1000 km. This volume of water is sufficient to provide the bulk of the drinking water needed by all land animals and creates a diversity of freshwater habitats. However, most freshwater habitats and supplies of drinking water are under constant threat of pollution, which can adversely affect water quality[24].

Abiotic conditions in aquatic habitats

In all aquatic habitats, two major sub-types of habitats (and hence two major descriptors for animals) are identified, based on their locations either on the bottom or in the water column. Figure 1.5 illustrates these locations in marine and estuarine habitats and the key environmental variables that differ in these locations:

- **Benthic** animals (or benthos) live on sediments at the lowest level of a body of water such as the bottom of the ocean or a lake or on the coasts and in estuaries through which water flows from the land into the sea. The area of sediment—mud, sand, cobbles and boulders—uncovered during each tidal cycle[25] is called the **intertidal zone**. Many animals occupy the intertidal zone and live on or in the sediment where they are exposed to changing environmental conditions (air/water temperature, oxygen availability and a risk of dehydration) with the ebb and flow of each tide.
- **Pelagic** animals live in the water column and open waters, i.e. away from the benthic zone. Figure 1.5 illustrates the five pelagic zones in the oceans.
  - Most pelagic animals occupy the **epipelagic zone** through which sunlight penetrates to a depth of about 200 m, enabling plants and algae to photosynthesize, which generates oxygen that animals use for **aerobic metabolism**[26].
  - The deeper **mesopelagic zone** (middle layer) extends to around 1,000 metres. Although some light penetrates into the mesopelagic (or twilight zone), it is insufficient for photosynthesis. Oxygen levels usually decline with increasing depth, depending on the rate of oxygen usage and the influence of water currents.
  - Beneath the mesopelagic, in the **bathypelagic zone**, which extends to depths of around 4000 m, and in the abyssopelagic zone below, there is no light except for that emitted by bioluminescent organisms. Animals living here consume material falling from above.

[24] We examine several impacts of pollution on fresh water in later chapters. For example, Case Study 5.2 examines nitrite pollution and Experimental Panel 20.1 examines the impacts of endocrine disruptor chemicals.

[25] The tidal cycle—the rise and fall of sea levels—is caused by the combined effects of gravitational forces exerted by the moon and the sun, and rotation of the Earth.

[26] We discuss aerobic metabolism in Section 2.4.2.

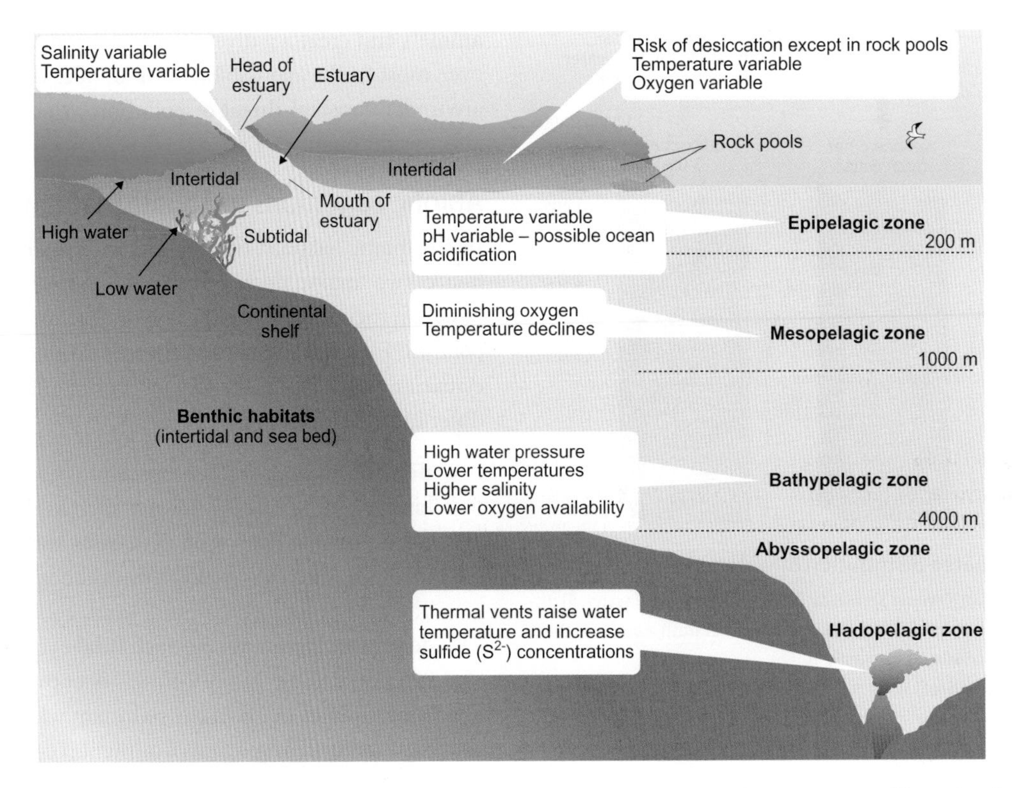

Figure 1.5 Marine and estuarine habitats showing the important environmental variables experienced by animals in various locations

- The **hadopelagic** zone (or trench zone) occurs in the deepest trenches of the ocean floors, where **hydrothermal vents** form fissures in the Earth's crust from which heated water and gases emerge. Hydrothermal vents occur predominantly on the boundaries of Earth's plates at depths of 2000 m to 7700 m. Biological diversity is higher around the hydrothermal vents than in the rest of the deep sea because the chemicals ejected from the vents provide food for thick mats of bacteria that form the base of food chains.

With increasing depth, water pressure increases due to the weight of the water above it: for every 10 m of descent, the pressure increases by approximately 1 atmosphere (101.3 kPa)[27]. So an animal going from the water surface, where the pressure is 1 atmosphere, to a depth of 10 m experiences a doubling of the external pressure exerted on them. Animals that breathe air and that dive to great depths need to cope with the mechanical effects of changing external pressures on the body, and the resulting effects of pressure on lung volumes and gas pressures. These animals need to use oxygen effectively through their dive until they next surface for a breath. Recent advances in technology have enabled scientists to investigate how aquatic birds and mammals manage their oxygen stores during natural voluntary dives[28].

Solar radiation warms surface waters—to a greater extent at the equator than at the poles. Hence, temperature often decreases with increasing depth in a water body, resulting in **thermal stratification** which influences the behaviour and physiological functioning of animals[29]. Where surface layers are warmed, a layer of water known as a **thermocline** may occur in which water temperature decreases abruptly with depth, more so than above or below the thermocline, as shown for a typical ocean in Figure 1.6.

In much of the world's oceans, a thermocline occurs between a depth of 100 or 200 m to a depth of 1000 m, but the exact depths vary with latitude and season. The thermocline is semi-permanent in the tropics, but more variable in temperate regions and is often deepest in summer. By contrast, the thermocline in colder polar regions is shallow and may be absent. Thermoclines can also be observed in lakes in which warming of the surface layers in summer results in a less dense surface layer (called the epilimnion) on top

[27] Appendix 1 explains the units for measuring pressure.

[28] Section 15.3.3 examines physiological adaptations of diving birds and mammals.

[29] Section 9.1.1 discusses thermal preferences and thermal selection of aquatic ectotherms.

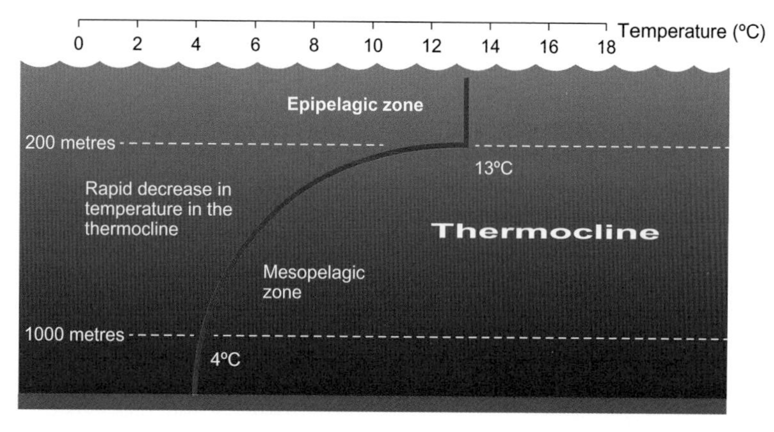

Figure 1.6 Typical ocean thermocline

The red line shows a typical temperature profile for seawater. The thermocline is recognized by the rapid decrease in temperature from the warmer mixed upper epipelagic zone to the lower temperatures of the deeper mesopelagic zone. Below 1000 m to a depth of about 4000 m seawater temperature remains relatively constant. Below 4000 m the temperature of the seawater ranges from near freezing to just above the freezing point as depth increases.

Source: adapted from National Oceanic and Atmospheric Administration (NOAA); https://oceanservice.noaa.gov/facts/thermocline.html.

of colder denser water (the hypolimnion) separated by a thermocline.

In open seas, surface water temperatures are relatively stable within a geographical location on both a daily basis and seasonally in comparison to the more variable conditions on land. Surface ocean water temperatures vary from about −2.0°C at the poles to 33°C closer to the equator. In contrast, land surface temperatures vary from −25°C to up to +65°C, as recorded in several parts of Australia, Asia and Africa.

Sea surface temperatures are continuously monitored by ships and stationary or drifting buoys as part of the effort to understand the impacts of climate change on global temperatures. More recently, we have gained further insights into sea temperatures from data collected from marine mammals with attached satellite-linked tags. The data collected are used to calculate global average sea surface temperatures and are compared to historic data.

The earliest data for sea surface temperatures collected in the 1800s are now considered unreliable. Consequently, more recent historic data are used for comparison with ongoing measurements to calculate the temperature anomaly (i.e. the increase since a particular period). For instance, the global sea surface temperatures in 2014 were 1.13°C higher than in the period between 1971 and 2000.

The increases in sea temperatures are most pronounced in climate change 'hotspots'. For example, sea surface temperatures in southeast Australia[30] have increased by 2–3°C since the mid-twentieth century. Such changes have inevitable implications for the distribution and functioning of marine organisms, some of which may not cope with continuing increases in temperatures[31].

Most aquatic habitats contain dissolved substances (carbon dioxide, salts, nitrogen and phosphorous compounds) which support photosynthesis if there is sufficient light penetration. Photosynthesis provides the food[32] and the oxygen necessary for animal life, which in turn produces carbon dioxide[33]. A balance is reached between carbon dioxide consumption (photosynthesis) and production by the aerobic metabolism of animals. The increasing amount of atmospheric carbon dioxide disturbs this balance and has implications for chemical conditions in aquatic habitats. Recent studies show that absorption of carbon dioxide by oceans causes **ocean acidification** (an increase in hydrogen ion concentration which translates into a reduced water pH)[34]. We consider some of the data showing these trends and the potentially damaging effects on marine ecosystems in Case Study 1.1.

Salt composition of aquatic habitats

Table 1.1 compares the concentrations of the main ions resulting from the dissociation of the dissolved salts in ocean seawater and the lower salt concentrations in fresh water. The striking differences have important effects on the aquatic animals that occupy marine or freshwater habitats,

[30] Case Study 9.1 examines some of the impacts of the changing sea surface temperatures in another hot spot.

[31] Chapter 9 examines the effects of temperature on the functioning of marine animals.

[32] Section 2.3.1 discusses food sources of animals and their digestion.

[33] Section 11.1.2 discusses factors influencing the amounts of oxygen and carbon dioxide in aquatic habitats.

[34] Water pH (which is related to the logarithm of the inverse hydrogen ion concentration) is explained in Box 13.2.

1

Table 1.1 Composition of ocean seawater and fresh water in rivers

Average concentrations of main ions in the Atlantic Ocean (seawater) and fresh water in rivers are tabulated. Seawater is primarily a sodium chloride solution, while the main ions in river water are calcium and bicarbonate. River water varies between continents; the range of average concentrations of ions for rivers in North America, South America, Europe, Asia and Australia are given in brackets below average values.

Ion	Atlantic seawater mmol L^{-1}	Fresh water mmol L^{-1}
Sodium (Na^+)	468	0.27 (0.13–0.48)
Chloride (Cl^-)	546	0.22 (0.14–0.34)
Magnesium (Mg^{2+})	53	0.17 (0.06–0.23)
Calcium (Ca^{2+})	10.3	0.38 (0.09–0.78)
Sulfate (SO_4^{2-})	28	0.12 (0.03–0.25)
Potassium (K^+)	10.2	0.06 (0.04–0.06)
Bicarbonate (HCO_3^-)	2.3	0.96 (0.51–1.56)

Sources: River water—Livingston DA (1963); Seawater data: Summerhayes CP, Thorpe SA (1996); sulfate and bicarbonate from Rankin JC, Davenport JA (1981).

because they need to maintain body fluid parameters within fairly narrow limits[35].

Seawater contains high concentration of sodium chloride and other salts such as magnesium sulfate and calcium carbonate. These concentrations result from erosion of submerged continental shelves together with the input of salts from rivers entering the seas over a period of around 3.4 billion years, and some input from volcanic and hydrothermal vents. The much lower concentrations of ions in fresh water mainly result from the gradual weathering of rocks by rainwater, which releases minerals and salts that form soils through which water percolates into rivers and lakes. Hence the variablity in water composition of rivers and lakes illustrated in Table 1.1 is related to the geology of the river catchment. In some locations, salts are carried up into air currents from the oceans before being deposited in rain and subsequently percolating into freshwater habitats.

The total content of salts in natural waters determines its **salinity**. Salinity was originally measured by drying down samples of water, and was expressed as the amount of solid material (in grams) per kg of solution, or parts per thousand. Nowadays salinometers measure the conduction of electricity and give readings as practical salinity units (abbreviated to psu)[36].

The average salinity of open oceans is 35 psu (although it ranges from 33 to 37 psu)[37]. Fresh water is not strictly defined in terms of its salinity but is usually considered to have a salinity of less than 1 psu[38]. The salinity of the freshwater rivers and lakes depends on the river catchment but is never zero because it is never entirely devoid of salts. However, the overall concentration of salts in fresh water is invariably low (1–5 mmol L^{-1}), as indicated by Table 1.1.

The higher salinity of surface seawater due to evaporation increases its density compared to the density of lower layers, giving it a tendency to sink to greater depths. Sinking of dense seawater takes some dissolved oxygen to the deeper layers of the ocean.

The input of river water and run-off from land reduces the salinity of estuaries and coastal seawater and creates areas of intermediate salinities—**brackish water**—in estuaries, sea lochs, fjords, salt marshes and mangrove swamps. These are particularly challenging habitats for animals that must tolerate the wide variations in environmental salinity resulting from freshwater input at the head of the estuary, sea loch or fjord, and seawater input on each tide at the mouth of the estuary. Similarly, large flows of fresh water into enclosed areas or bays, such as the Hudson Bay in North America and the Baltic Sea in northern Europe, create brackish environments.

Figure 1.7A examines the average salinity of the different regions of the Baltic Sea in which only a low exchange of seawater occurs along the narrow shallow passage which connects the Baltic Sea to the North Sea. Large freshwater inputs enter the Baltic, mainly via the Gulf of Bothnia, the Gulf of Finland and the Gulf of Riga. The fresh water input, combined with general shallowness of the Baltic Sea (which has an average depth of about 55 metres) results in a decrease in surface salinity values from about 30 psu in the Skagerrak and 22 psu in the Kattegat to about 5 psu within a few hundred kilometres to the north and east; the most northerly part of the Baltic reaches almost freshwater conditions.

Salinity differences have a major effect on the community of animals found in aquatic environments due to

[35] Chapter 5 examines body fluid balance of aquatic organisms.

[36] In most natural waters, values in parts per thousand (ppt or ‰) are virtually identical to values in practical salinity units, which is a dimensionless scale. Hence, salinity is often simply stated as a number but as a reminder we will use psu to indicate practical salinity units.

[37] For organisms, osmotic concentrations (= osmolality) rather than salinity as such determine water movements, by osmosis into or out of the organism, as we discuss in Section 4.1.1; ocean seawater has an osmolality of about 1000 mOsm kg^{-1}.

[38] Some authorities use <0.5 psu rather than <1 psu as the cut-off point for considering a water body as fresh water.

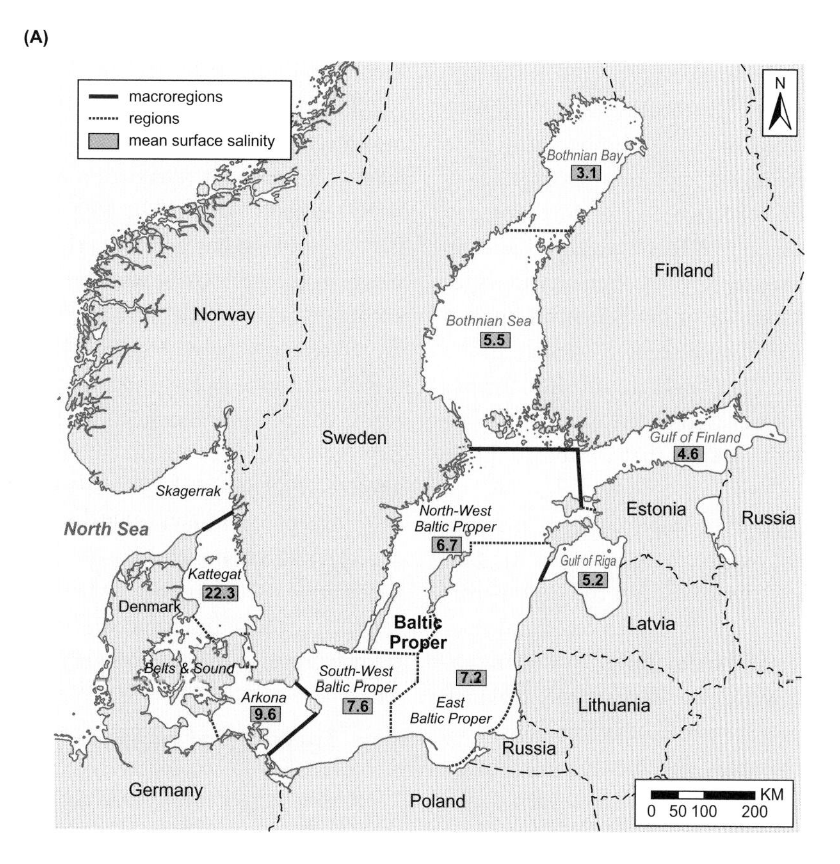

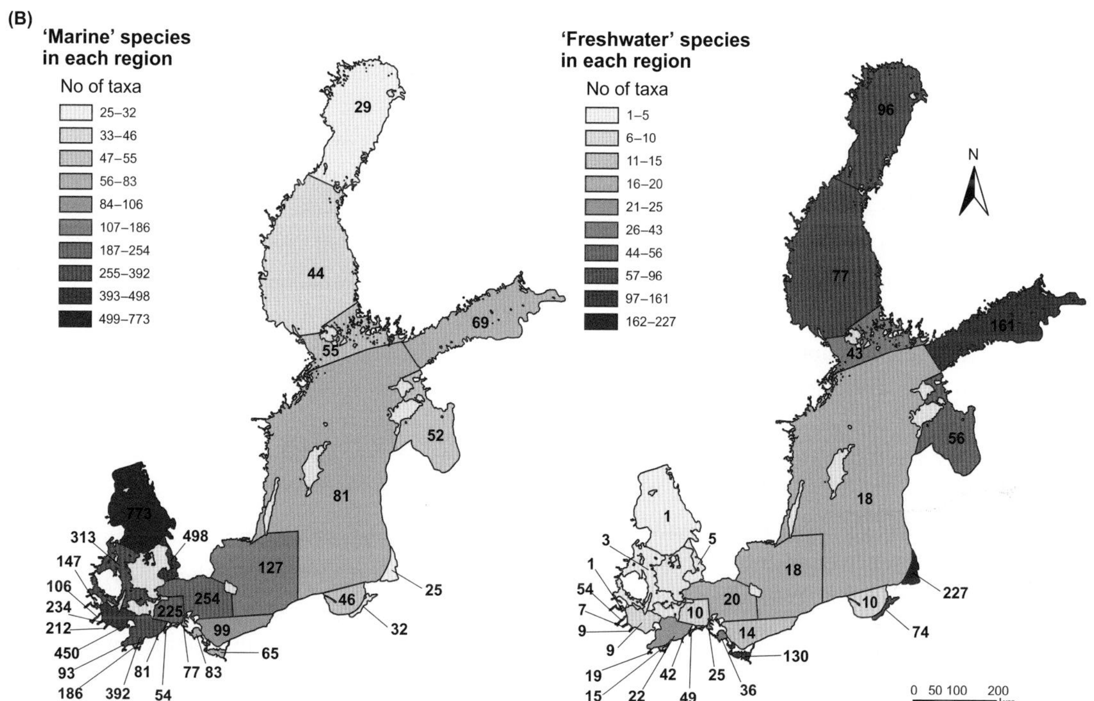

Figure 1.7 Mean surface salinity of the regions of the Baltic Sea and diversity of invertebrate macrobenthos

(A) Average salinities in the 10 regions of the Baltic Sea (data in green boxes) vary from 3.1 psu in the most northern region (the Bothnian Bay) to 22.3 psu in the Kattegat, which lies between Denmark and Sweden. Only a low exchange of seawater occurs along this narrow shallow passage which connects the Baltic Sea to the North Sea.
(B) Regional distribution of macrobenthic invertebrates that are 'marine' or 'freshwater' species. Macrobenthic animals are by definition big enough to be visible by the naked eye (larger than 1 mm and at least 0.5 mm in their smallest dimension). Greater species diversity of marine species occurs in the south (more saline) than north, while higher species diversity of freshwater species occurs in the north and east where salinities are lowest.

Source: Ojaveer H et al (2010). Status of biodiversity in the Baltic Sea. PLoS ONE 5(9): e12467.

their different saline tolerances[39]. Figure 1.7B shows the numbers of macrobenthic invertebrate species in the different parts of the Baltic Sea. In the Kattegat, predominantely marine species occur and the number of marine species decreases eastwards and northwards as salinity declines. Notice in Figure 1.7B that in the northern and eastern coastal waters there are at least two times and up to nine times more freshwater species than marine species. The combined total number of macrobenthic species is, however, much higher in the Kattegat (774) than in the entire Baltic (Sea) Proper (244) and diminishes further north and in the Gulf of Riga (108) to the east. This is indicative of the low number of species that are able to thrive in brackish conditions.

An additional abiotic factor of importance in the Baltic Sea is the availability of oxygen. Large parts of the Baltic are shallow, particularly the Gulfs, but the central part of the Baltic Sea and parts of the Bothnian Sea are deeper (100–300 m); here, a permanent **halocline** at 70–100 m separates the upper less saline water from deeper water in which salinity reaches 9 to 13 psu. The permanent halocline and a seasonal thermocline (in summer) together impede vertical mixing of the water column, which results in oxygen depletion in the deeper areas. Benthic animals living here must tolerate brackish water and cope with low concentrations of oxygen to survive.

Hypersaline environments

Where circulation of ocean water is restricted—such as in lagoons—and temperatures are high, evaporation may create **hypersaline** environments, with salinities above those of ocean seawater. In tropical lagoons, such as the Caimanero lagoon in Mexico, for example, salinity increases by about 2 psu each day in summer to reach 300 psu—about eight times that of the oceans[40]. The same principle applies to rock pools between the tides (so over a shorter timescale). Animals living in rock pools need to tolerate these changing conditions, not just in salinity, but also in temperature and available oxygen in the pool.

Evaporation of water in the Dead Sea in the Middle East, coupled with limited fresh water input from the River Jordan, has raised salinity above 300 psu. The Dead Sea is effectively a large saline lake as there is now no connection to an open sea and freshwater input from a single source. This hypersaline environment does not support any aquatic animals, although its pink tinge is due to the presence of salt-loving bacteria.

Like the Dead Sea, the Great Salt Lake in Utah, USA, is also a remnant of an ancient sea, now forming a terminal lake with no outlet rivers running to the ocean. A railway built across the lake in the 1900s divides the lake into two bays (Gunnison Bay and Gilbert Bay) shown in the satellite image in Figure 1.8A. A wooden bridge used to allow water exchange between the bays but was replaced in 1959 by a more solid causeway. Figure 1.8B shows the changing salinities since the railway construction: Gunnison Bay to the north has become much saltier than the southerly Gilbert Bay, which receives almost all of the freshwater input. Few animals can survive the extreme salinities of the Great Salt Lake, except for the larvae of two species of brine flies (*Ephydra* spp.) and brine shrimps (also known as sea monkeys, *Artemia franciscana*), which thrive in huge numbers because they can regulate the composition of their body fluids in extremely hypersaline environments[41].

1.2.2 Terrestrial biomes and habitats

Various schemes have been devised to classify terrestrial biomes. Figure 1.9 illustrates a scheme showing 14 terrestrial biomes, which includes tropical and subtropical forests, grasslands and shrublands, mangroves, tundra and deserts. At a finer level of resolution, we can recognize ecoregions within biomes. As an example, Figure 1.10 illustrates the broad ecoregions of North America.

The most important factor distinguishing terrestrial biomes and ecoregions is their climate[42], which results from differences in rainfall and temperature at different latitudes and elevations. Rainfall and its distribution throughout the year affect the humidity of biomes, allowing the distinction of humid, semi-humid, semi-arid or arid biomes. A similar scheme classifies biomes as **hydric** (wet), **mesic** (moist) and **xeric** (very dry).

There is no exact definition of a desert but most classifications encompass two main factors:

- Rainfall (precipitation): the total volume of annual precipitation, and the number of days when precipitation occurs. Annual precipitation in deserts is less than the water loss to the atmosphere by evaporation and transpiration of water (by plants).

[39] The physiological functioning of animals in different salinities and their saline tolerances are explored in Chapter 5.

[40] A salinity of eight times ocean seawater equals an osmotic concentration of about 8000 mOsm kg^{-1} (osmotic concentrations are explained in Section 4.1.1). This may lead to large water loss from the body of animals by osmosis (Section 3.1.2) across permeable external surfaces.

[41] We discuss the composition of intracellular and extracellular fluids in Section 3.2.5 and 4.1.1. In Section 5.3.1 we examine the capacity of brine shrimps to regulate the composition of their body fluids.

[42] Climates are commonly characterized by average and typical ranges of temperature, precipitation and evaporation over periods of about 30 years.

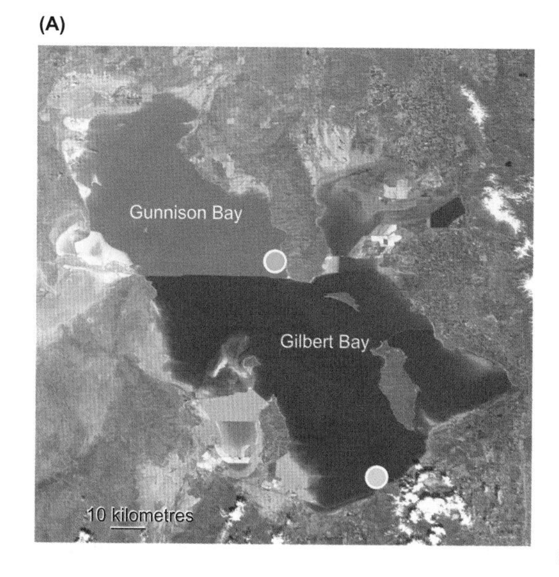

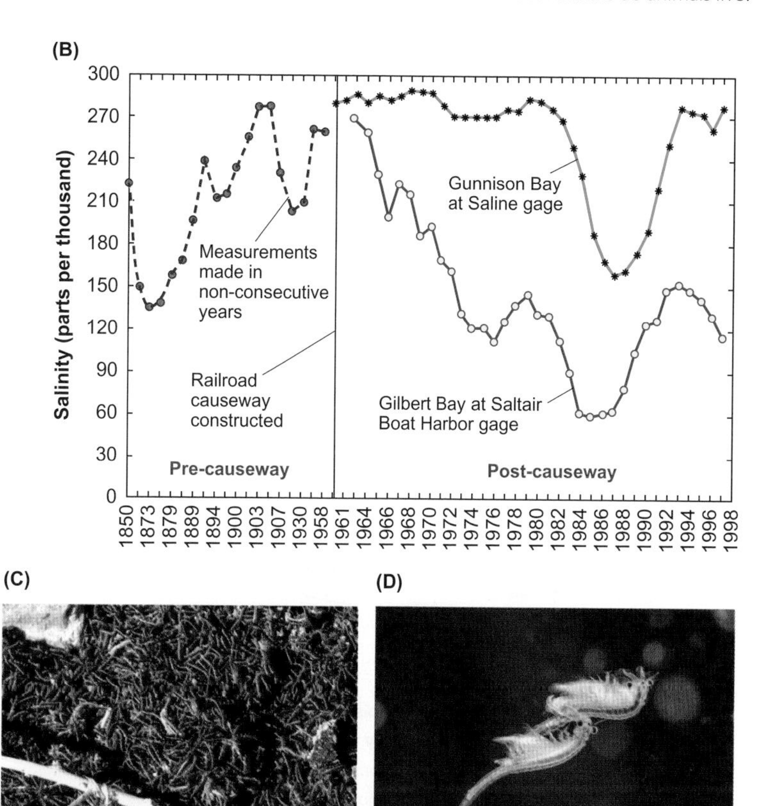

Figure 1.8 Salinity of Great Salt Lake, Utah, United States

The image (A) taken from the International Space Station shows the striking separation of the more saline Gunnison Bay north of the railway crossing and Gilbert Bay to the south which is lower in salinity. The pink colouration of Gunnison Bay is due to red pigments from the high density of salt-tolerant bacteria. Before the construction of the railway causeway water samples were collected from Gilbert Bay, only from the location shown by the blue dot. After the construction of the railway causeway in 1959, two stations, shown by blue and green dots on the image, were used for regular monitoring of salinities. Salinity of the Great Salt Lake between 1850 and 1998 is shown in (B), which demonstrates the increasing salinity differences that emerged after construction of the railway causeway. Image (C) shows the castings of the larvae of brine flies (*Ephydra hians* and *Ephydra cinerea*) that pile up on the shores of the Great Salt Lakes. (D) illustrates the brine shrimps (sea monkeys: *Artemia franciscana*), which thrive in the Great Salt Lake, examined under a microscope. Brine shrimps are a few millimetres in length. The male has enlarged second antennae that are used to clasp the female during mating, as shown in the photograph.

Sources: (A) http://eoimages.gsfc.nasa; (B) adapted from http://ut.water.usgs.gov; (C) http://learn.genetics.utah.edu; (D) Brine shrimp: blickwinkel/Alamy Stock Photo.

- Temperature: deserts are often classified as 'hot' or 'cold' deserts. In hot deserts temperatures can reach 50°C during the day in summer, which increases the rate of evaporation. Winter temperatures may, however, decline to 0°C or less, depending on the geographical location (latitude and within a continental mass). Most hot deserts occur at low latitudes, as illustrated by Figure 1.9, such as the Sahara in North Africa and Arabian desert in the Arabian Peninsula, and the deserts of the southwest of the United States and Mexico and Australia. Cold (temperate) deserts[43] occur at higher latitudes; an example is the Gobi desert in Asia (parts of China and Mongolia).

[43] The Antarctic and Artic land masses are also sometimes considered as cold deserts on the basis of their low annual precipitation (as snow), although this classification is not encompassed in Figure 1.9.

Taking account of rainfall and temperature, deserts encompass:

- Extremely arid hot deserts, where any rainfall is infrequent and highly unpredictable and temperatures high. Periods of at least 12 consecutive months without rainfall at all are typical. The best example is the The Rub' al Khali or 'Empty Quarter', and is the largest contiguous sand desert in the world extending to about 650,000 km^2 that forms part of the larger Arabian desert in the southern third of the Arabian Peninsula.
- Arid hot deserts in which there is some rainfall, but usually less than 250 mm annually. This level of rainfall is typical of many areas of the American deserts.
- Semi-arid grasslands (the steppes), where mean annual precipitation is 250–500 mm but rates of evaporation and

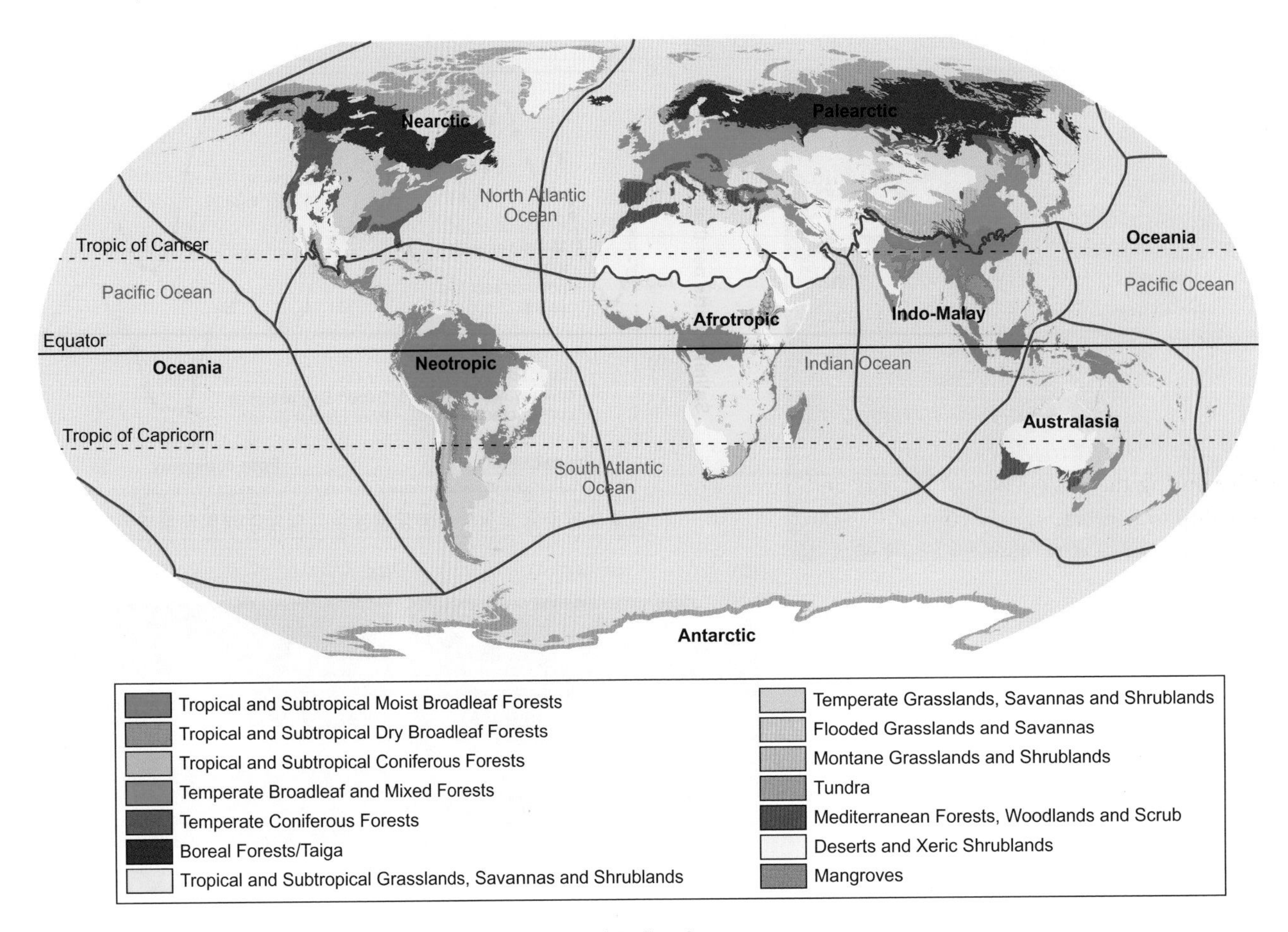

Figure 1.9 Fourteen terrestrial biomes in eight biogeographical realms

Note that Oceania appears on both sides of the illustration, but is actually continuous.

Source: adapted from Olson DM et al (2001). Terrestrial ecoregions of the world: a new map of life on earth. Bioscience 51: 933-938.

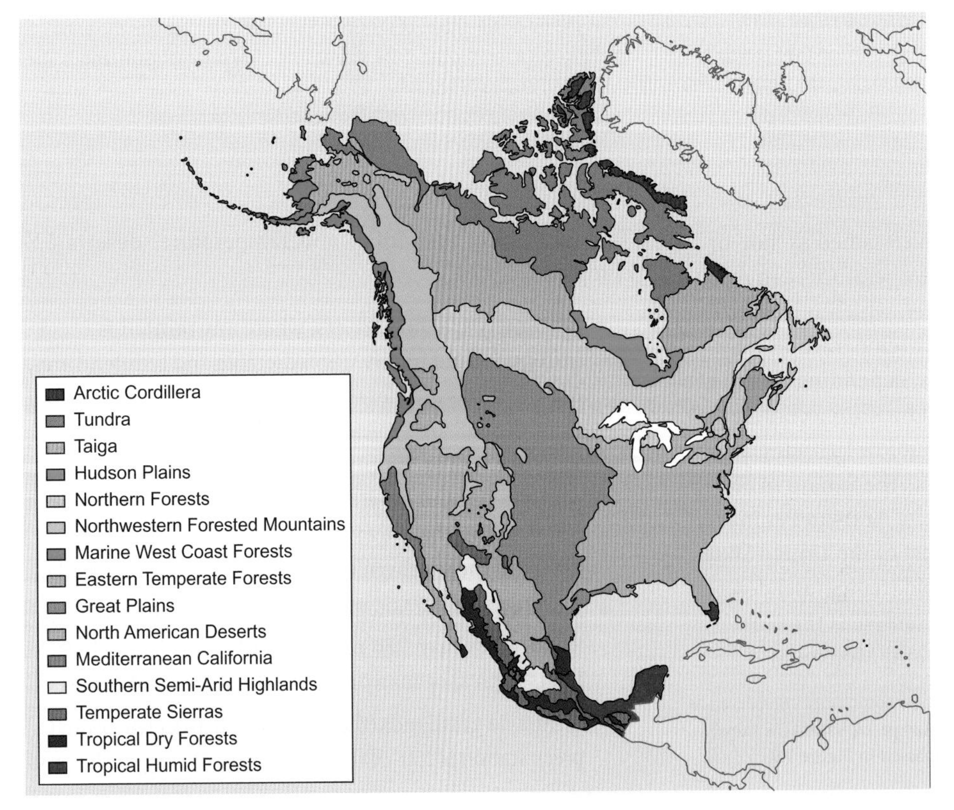

Figure 1.10 Ecoregions of North America

The map shows the 15 broad (Level 1) ecoregions of North America developed by the Commission for Environmental Cooperation Working Group in 1997.

Source: Commission for Environmental Cooperation Working Group; Ecoregions of North America – Encyclopaedia of Earth. https://editors.eol.org/eoearth/wiki/Ecoregions_of_North_America_(CEC).

transpiration are high. Typically, there are large temperature differences between summer highs of up to 45°C and low winter temperatures of –50°C or lower, and temperature differences between day and night of 30°C.

The lack of water is the major determining factor of the animal species present in deserts, their physiological traits (characteristics) and lifestyles. To survive in such habitats animals must exhibit physiological and behavioural adaptations to produce water during metabolism, obtain natural sources of water and conserve it. For example, some of the beetles living in hot deserts have evolved a novel ability to collect fog water, while others take up water from the atmosphere[44]. Desert mammals in hot deserts are often inactive during the hottest part of the day and have evolved an ability to generate highly concentrated urine[45]. One of the mammals most profoundly adapted to live in the extremely hot deserts of the Arabian Peninsula is the Arabian oryx (*Oryx leucoryx*) and we learn about its particular adaptations for survival in such a hostile environment in later chapters[46].

Latitude and elevation results in global temperature zonation, which distinguishes subtropical and tropical biomes with mean annual temperatures of around 28°C, tundra with sub-zero mean annual temperatures, and temperate biomes with temperature varying over the year. The occurrence of particular animal species in each biome and the ecosystems within a biome is, to a large extent, the result of their physiological functioning, especially their ability to maintain a constant body temperature, or not, and associated behaviours[47]. Changing patterns of activity on an annual basis may also occur and can involve a resting phase when conditions are inhospitable (**diapause, hibernation** or **aestivation**)[48].

Within any habitat, the external environment experienced by an animal is locally based and the microclimate needs to be considered. Some animals generate their own microclimate. For example, land and aquatic animals breathe out carbon dioxide and breathe in oxygen, which influences the chemistry of their immediate environment. Burrowing animals occupy a different microclimate (temperature, humidity[49], oxygen and carbon dioxide concentrations) than animals living on the Earth's surface and can reduce their water losses by increasing the water vapour density (**humidity**) of the air in their burrow.

› Review articles

Jones P (2016). The reliability of global and hemispheric surface temperature records. Advances in Atmospheric Sciences 33: 269–282.

Land PE, Shutler JD, Findlay HS, Girard-Ardhuin F, Sabia R, et al. (2015). Salinity from space unlocks satellite-based assessment of ocean acidification. Environmental Science and Technology 49: 1987–1994.

Ojaveer H, Jaanus A, MacKenzie BR, Martin G, Olenin S, et al. (2010). Status of biodiversity in the Baltic Sea. PLOS ONE 5: e12467.

Olson DM, Dinerstein E, Wikramanayake ED, Burgess ND, Powell GVN, Underwood EC, D'Amico JA, Itoua I, Strand HE, Morrison JC, Louks CJ, Allnutt TF, Ricketts TH, Kura Y, Lamoreux JF, Wettengel WW, Hedao P, Kassem KR (2001). Terrestrial ecoregions of the world: a new map of life on earth. BioScience 51: 933–938.

Simpson SD, Jennings S, Johnson MP, Blanchard JL, Schön, P-J, Sims DW, Genner MJ (2011). Continental shelf-wide response of a fish assemblage to rapid warming of the sea. Current Biology 21(18): 1565–1570.

Simpson SD, Harrison HB, Claereboudt MR, Planes S (2014). Long-distance dispersal via ocean currents connects Omani clownfish populations throughout entire species range. PLOS ONE 9: e107610.

Spalding MD, Fox HE, Allen GR, Davidson N, Ferdaña ZA, Finlayson M, Halpern BS, Jorge MA, Lombana A, Lourie SA, Martin KD, McManus E, Molnar J, Recchia CA, Robertson J (2007). Marine ecoregions of the world: A bioregionalization of coastal and shelf areas. BioScience 57: 573–583.

Useful websites

National Oceanic and Atmospheric Administration—https://www.noaa.gov/

US Geological Survey—https://www.usgs.gov/

United States Environmental Protection Agency—https://www.epa.gov/

1.3 How animal groups are related to each other

Animal diversity is measured in terms of number and abundance of species, or genera[50], where a species is a group of animals that can interbreed in nature to produce fertile offspring[51] and a genus (plural genera) designates a group of species that exhibit similar characteristics and are assumed to be closely related. All animals evolved from a common ancestor.

We can learn more about the evolutionary development and diversification of a species[52] by exploring how its genetic

[44] The uptake of atmospheric water in some insects and arachnids is discussed in Section 6.2.3.

[45] We explore the physiological adaptations of mammals in xeric habitats in Sections 6.1.4 and 7.2.3.

[46] The Arabian oryx is a particular focus in Sections 6.1.2, 6.1.4 and 6.2.1 in which we examine the aspects of physiological functioning that maintain its water balance.

[47] Chapters 9 and 10 examine temperature influences on animals in detail.

[48] Diapause is common among temperate insects, as we discuss in Section 9.2. We discuss the physiology of hibernation in Section 10.2.8.

[49] Humidity is expressed in gram of water vapour per m^3 of air. We discuss the influence of burrows on an animal's microclimate in Sections 6.1.1 and 15.3.2.

[50] In a scientific name, the genus name is capitalized and italicized and is followed by the species' name, which is also italicized but is not capitalized—for example, *Homo sapiens* for humans. A third italicized Latin name following the name of a species designates a subspecies, i.e. a subgroup of that species. For example the dog (*Canis lupus familiaris*) is a subspecies of the grey wolf (*Canis lupus*).

[51] However, note that there are situations where hybrids capable of reproduction occur in nature particularly for closely related species—for example, species of water frogs of the *Pelophylax* genus.

[52] New species arise through an evolutionary process called speciation, which we discuss in Section 1.5.4.

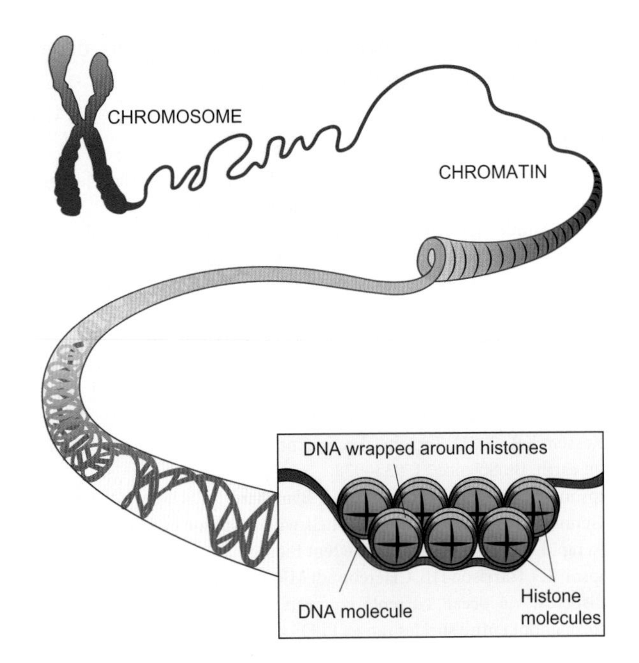

Figure 1.11 Organization of the genetic material in chromosomes

The thread-like DNA molecules with their double-helix configuration are wrapped around proteins called histones for compaction and regulation of the genes located on them. The very long DNA molecule wrapped around the histone molecules form the chromatin from which chromosomes are made.
Source: after National Institutes of Health http://commonfund.nih.gov/epigenomics/figure.aspx.

material is modified as it is passed from parent to offspring. The evolution of a genetically related group of organisms is known as **phylogeny** (from Greek *phylon* = tribe, clan and *genetikós* = origin, source, birth). The term phylogeny also refers to the pattern of evolutionary relationships among organisms.

1.3.1 The genetic material of an animal contains the information that determines its general makeup

The complete genetic material of an animal is referred to as its **genome** and consists of DNA (deoxyribonucleic acid) molecules in their well-known double-helix configuration, packaged in chromosomes that are predominantly located within the nucleus (nuclear genome), but also in mitochondria (mitochondrial genome)[53]. As shown in Figure 1.11, chromosomes are made of chromatin, which consists of the threadlike, negatively charged DNA molecule firmly coiled around a family of small, positively charged proteins[54] called histones. The chromosomes in the nucleus generally exist in pairs (homologous chromosomes) and are only visible during cell division[55], when the DNA molecules in the chromatin are more tightly packed. In each pair, one chromosome is from the mother and the other from the father.

The two strands of the DNA molecule in the double-helix configuration consist of nucleotide subunits as shown in Figure 1.12A–C. Each nucleotide is made of one of four nitrogen-containing nucleobases (A for adenine, G for guanine, C for cytosine and T for thymine), one five-carbon sugar (deoxyribose) and one phosphate group (P). The backbone of the two DNA strands consists of sugar molecules linked together by negatively charged phosphate groups, while the nucleobases are positioned perpendicular to the axis of the double helix, as illustrated in Figures 1.12B and 1.12C. Hydrogen bonds[56] between complementary nucleobases on the two strands (A–T and G–C) hold the two strands together and confer high stability to the double helix.

It is the sequence of nucleotides (each with one of the four types of nucleobases: A, G, T, C) on the DNA strands that carries the genetic information in each individual cell of a particular animal. Genes are portions of DNA that either:

- encode a functional molecule of ribonucleic acid (RNA)[57], or
- encode a protein product, or
- regulate the expression of a functional RNA molecule, or
- regulate the expression of a functional protein molecule.

The **genotype** of an organism generally refers to the DNA-coded, inheritable information, while its **phenotype** refers to its observable features. The phenotype is predominantly determined by the proteins expressed, which define the structure and function of individual cells, tissues and organs in the body. Since genes dictate which proteins are expressed at a particular time, the genotype effectively controls the phenotype.

Figure 1.13 summarizes the flow of information from the DNA to proteins in animal cells. First, the information

[53] The mitochondria are cellular organelles that play a major role in energy metabolism, as we discuss in Section 2.4.2.

[54] Proteins are chains of amino acids, as we discuss in Box 2.1.

[55] Cell division in mitosis is outlined in Box 3.2.

[56] Hydrogen bonds are formed when one hydrogen atom attached to a strongly electronegative atom, such as oxygen or nitrogen, exists in the vicinity of another electronegative atom with a lone pair of electrons. Hydrogen bonds are weaker than true covalent and ionic bonds.

[57] RNA molecules consist of one single strand of nucleotides. Like DNA, each RNA nucleotide subunit is made of one of four nitrogen-containing nucleobases A, G, C and uracil (U) instead of T, one five-carbon sugar (ribose instead of deoxyribose in DNA nucleotides) and one phosphate group.

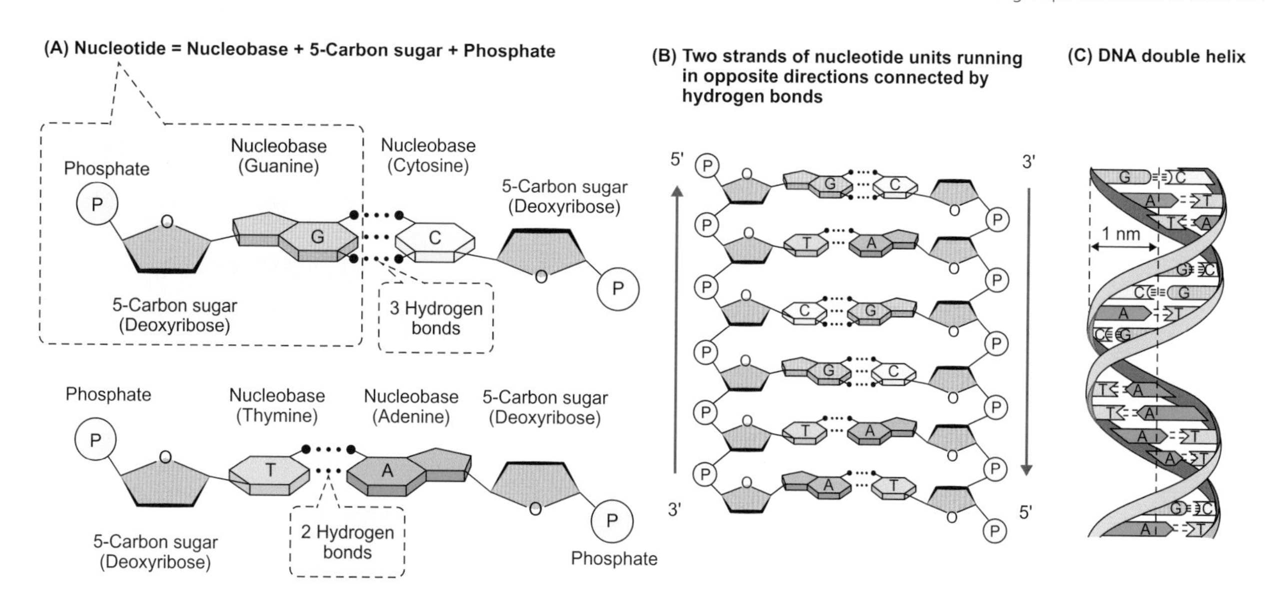

Figure 1.12 Deoxyribonucleic acid (DNA) structure

(A) DNA is made of nucleotide subunits containing one of four nucleobases: adenine (A), cytosine (C), guanine (G) and thymine (T). C and G pair together via three hydrogen bonds, while A and T pair together via two hydrogen bonds.

(B) The nucleotides polymerize to form strands which have the backbone made of the phosphate (P) and the sugar groups (deoxyribose). The association of two strands with complementary nucleobases is stabilized by hydrogen bonds to form a DNA molecule. The 5′ and 3′ mean 'five prime' and 'three prime', which indicate the carbon numbers in the DNA's sugar backbone. The 5′ carbon has a phosphate group attached to it and the 3′ carbon a hydroxyl group. The asymmetry gives a DNA strand a 'direction' indicated by arrows.

(C) The three-dimensional structure of a DNA molecule is in the form of a double helix with a radius of 1 nm.

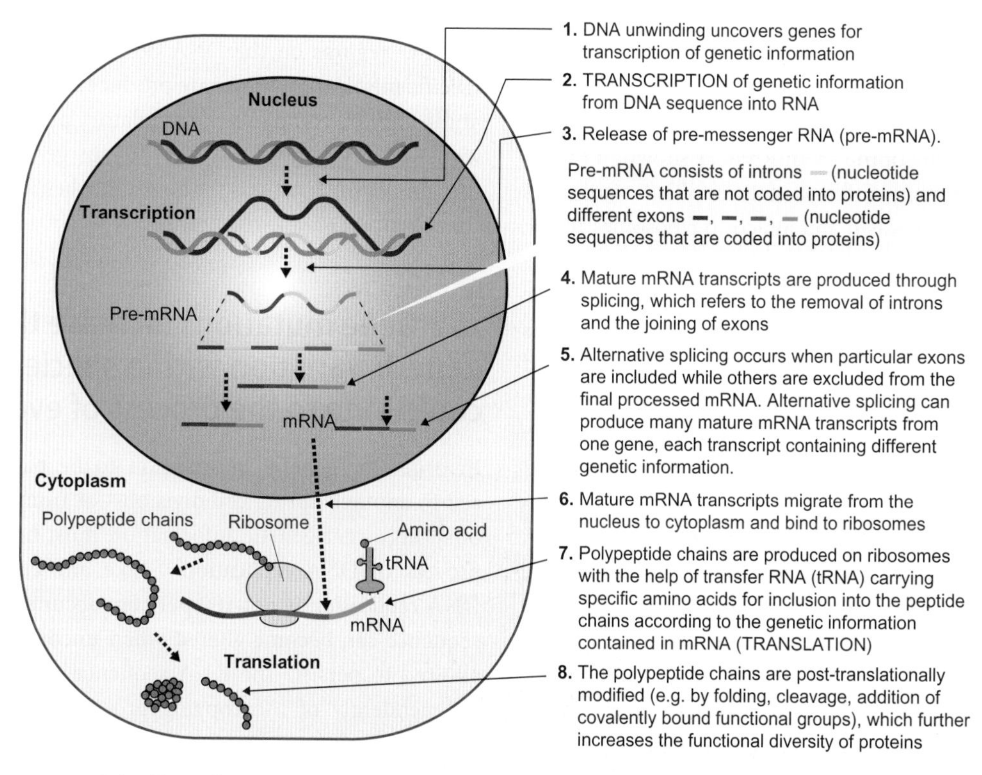

Figure 1.13 Summary of the flow of genetic information from DNA to proteins in animal cells

Genetic information carried by DNA is transcribed in RNA in the nucleus and then is translated into different proteins in the cytoplasm. Alternative splicing contributes to protein diversity since one single gene transcript (pre-mRNA) can undergo hundreds of different splicing patterns and will therefore code for hundreds of different proteins. The functional diversity of proteins is further enhanced by post-translational modifications.

Table 1.2 The standard genetic code (written in RNA nucleobases) showing the start and stop codons, as well as the correlation between codons and the naturally occurring amino acids

1st base	2nd base								3rd base
	U		C		A		G		
U	UUU	Phenylalanine	UCU	Serine	UAU	Tyrosine	UGU	Cysteine	U
	UUC		UCC		UAC		UGC		C
	UUA	Leucine	UCA		UAA	Stop codons	UGA	Stop codon	A
	UUG		UCG		UAG		UGG	Tryptophan	G
C	CUU	Leucine	CCU	Proline	CAU	Histidine	CGU	Arginine	U
	CUC		CCC		CAC		CGC		C
	CUA		CCA		CAA	Glutamine	CGA		A
	CUG		CCG		CAG		CGG		G
A	AUU	Isoleucine	ACU	Threonine	AAU	Asparagine	AGU	Serine	U
	AUC		ACC		AAC		AGC		C
	AUA		ACA		AAA	Lysine	AGA	Arginine	A
	AUG	Methionine or Start codon*	ACG		AAG		AGG		G
G	GUU	Valine	GCU	Alanine	GAU	Aspartic acid	GGU	Glycine	U
	GUC		GCC		GAC		GGC		C
	GUA		GCA		GAA	Glutamic acid	GGA		A
	GUG		GCG		GAG		GGG		G

* The translation into a protein begins at the first AUG codon in the coding region of an mRNA

contained in a gene is transcribed into a pre-messenger RNA (pre-mRNA) consisting of complementary nucleotide sequences called **exons** that are later coded into proteins and nucleotide sequences called introns, which are not coded into proteins. Generally, the protein-coding region of the genome is called the **exome**.The complementary RNA nucleobases to the A, G, T and C bases on DNA are uracil (U), C, A and G, respectively.

Alternative splicing of the pre-mRNA, as shown in Figure 1.13, can produce many different mature mRNA transcripts that migrate into cytoplasm and attach to ribosomes in the cytoplasm, where proteins are assembled according to the genetic information captured in the mRNA. The base sequence in the mRNA transcript is 'read' by the ribosome in units called codons. Each codon comprises three nucleobases and encodes a specific amino acid that is added by the ribosome at a specific position in the emerging polypeptide chain. Each amino acid is carried to the ribosome by a specific transfer RNA molecule (tRNA). The codon on the mRNA pairs with the complementary sequence of three nucleobases on a tRNA molecule, called the anticodon, and the amino acid is added to the polypeptide chain emerging from the ribosome. The correlation between a codon and a specific amino acid is the basis of the genetic code, which is common to all animals. This process in which ribosomes produce polypeptide chains in which the amino acid sequence is determined by the sequence of nucleobases in the mRNA is called **translation**.

With four different RNA nucleobases (U, C, G and A) it is possible to make $4^3 = 64$ unique codon combinations. As shown in Table 1.2 each combination encodes a specific amino acid[58] (but one amino acid can be encoded by up to six different codons) and specific combinations also encode the beginning and the end of the polypeptide chain (protein molecule) that is synthesized on ribosomes. The polypeptide chains are post-translationally modified to produce specific active proteins.

Other regions on the DNA determine when and where specific mRNA transcripts are produced according to a multitude of more complex gene regulatory processes. Thus, the sequence of nucleotides in the various genes in DNA also contains the instructions for how and when specific proteins are produced.

1.3.2 Gene mutations increase the genetic variation within species and contribute to the process of evolution

In order for genetic information to be passed from one cell to another during cell division—or from a parent to its offspring—that genetic information must be copied in the process of DNA replication[59]. Like any copying process, DNA replication is not 100 per cent accurate and the DNA sequence can become altered when uncorrected errors are allowed to persist. The DNA sequence can also be altered independently of the copying process by random events

[58] The list of amino acids from which proteins are made is shown in Box 2.1, Table A.

[59] We discuss chromosomes and DNA replication in Box 3.2.

caused by environmental factors such as ionising radiation[60]. A permanent alteration in the DNA sequence of a particular gene is called a gene mutation.

Some inherited mutations can impart certain advantages to the respective individual. For example, if a particular mutation in a gene permits the animal to become better camouflaged and therefore less conspicuous to both prey and predators, then the chance of that individual surviving long enough to reproduce increases compared with other individuals in the population. As a result, in time, the respective mutation is naturally selected and spreads more widely within the population. In contrast, gene mutations that are detrimental to the survival of an individual are eliminated: if the individual can not survive long enough to reproduce that gene mutation will fail to be passed on to the next generation.

Many mutations are neutral (neither harmful nor beneficial) but can become important for a population when environmental conditions change. For example, some mutations may endow an animal with an ability to resist desiccation better than the rest of the population. If access to water is plentiful, such mutations are of no particular advantage or disadvantage, but if water availability decreases, then the mutation becomes essential for the survival of the population.

Mutations in a specific gene give rise to different forms of that gene, called **alleles**. An allele is a heritable **genetic variant** of a particular DNA sequence in the genome. For example, differences in eye colour and blood type between humans are determined by different alleles of the genes that determine eye colour and blood type. Different alleles of the same gene appear at the same place on a chromosome. An organism is considered homozygous for a particular gene when the same allele is present in the same location on both chromosomes of a homologous pair, and heterozygous if different alleles are present in the same location on homologous chromosomes.

About 20 to 30 per cent of genes in a species may exist in two or more allelic states such that the number of possible genetic combinations may exceed the total number of individuals within a species and a particular genetic variant (= specific DNA sequence of nucleotides) may never occur, or re-occur. For example if 40 genes exist in two allelic states, then the number of different combinations of these allelic states is greater than 1 trillion (2^{40} which equals 1.1×10^{12}). The variation in the DNA sequence in the genome of individual members of a population is called **genetic variation**. The number of gene variants (alleles) in a population fluctuate randomly over time due to the chance that particular alleles disappear because individuals carrying those alleles die before they reproduce.

[60] Ionizing radiation refers to electromagnetic radiation that can remove electrons from atoms.

Genetic drift refers to the random change over time (through successive generations) in the frequency of alleles in a gene pool. This random process affects small populations to a greater extent than large populations because small populations have only a few copies of particular alleles such that the loss of only a few individuals could remove those alleles from the genetic pool. By comparison, the number of individuals in large populations with particular alleles is high and hence the chance of losing alleles from the genetic pool through genetic drift is low.

The genetic variation within a population plays a central role in the process of evolution through **natural selection** that we discuss in Section 1.5.3. In this process, heritable traits that increase the survival rate and reproductive capacity become more common in successive generations of the population, until eventually the respective traits spread to the entire population and become population-level characteristics.

1.3.3 Genomic information helps to infer and construct phylogenies

Powerful comparisons can be made to reveal similarities in gene sequences between the genomes of different species using bioinformatic analyses, in which each nucleotide is identified by a letter and a codon is equivalent to a word. Comprehensive genomic studies comparing the genomes of various forms of life, support the idea that all forms of life on Earth evolved from one **universal ancestor**.

Carl Woese, an American microbiologist and biophysicist, pioneered the study of differences in the gene of a ribosomal RNA (16S rRNA), which has a very slow rate of mutation. In collaboration with George Fox, Woese introduced a system of classification of life forms on Earth, which recognizes the fundamental divides between three **domains** (initially called kingdoms): Bacteria, Archaea (**archaeans**) and Eukaryota (or **eukaryotes**). This system, which was refined based primarily on molecular differences observed in other genes, is shown in Figure 1.14.

Eukaryotes consist of cells that have a nucleus containing most of their genetic material (DNA) and have membrane-enclosed organelles[61]. In evolutionary terms the eukaryotes are closer to archaeans than to bacteria, as shown in Figure 1.14. Both archaeans and bacteria lack nuclear membranes and are called **prokaryotes**[62]. Eukaryotes include a diverse group of unicellular organisms, such as protozoa (amoebae,

[61] Figure 3.8 examines cellular organelles.

[62] Archaeans display different biochemistry from bacteria and generally include microorganisms that grow best at temperatures between +80 and +100°C (hyperthermophile), where most other organisms would die. Indeed, it has been suggested that life may have originated around hydrothermal vents.

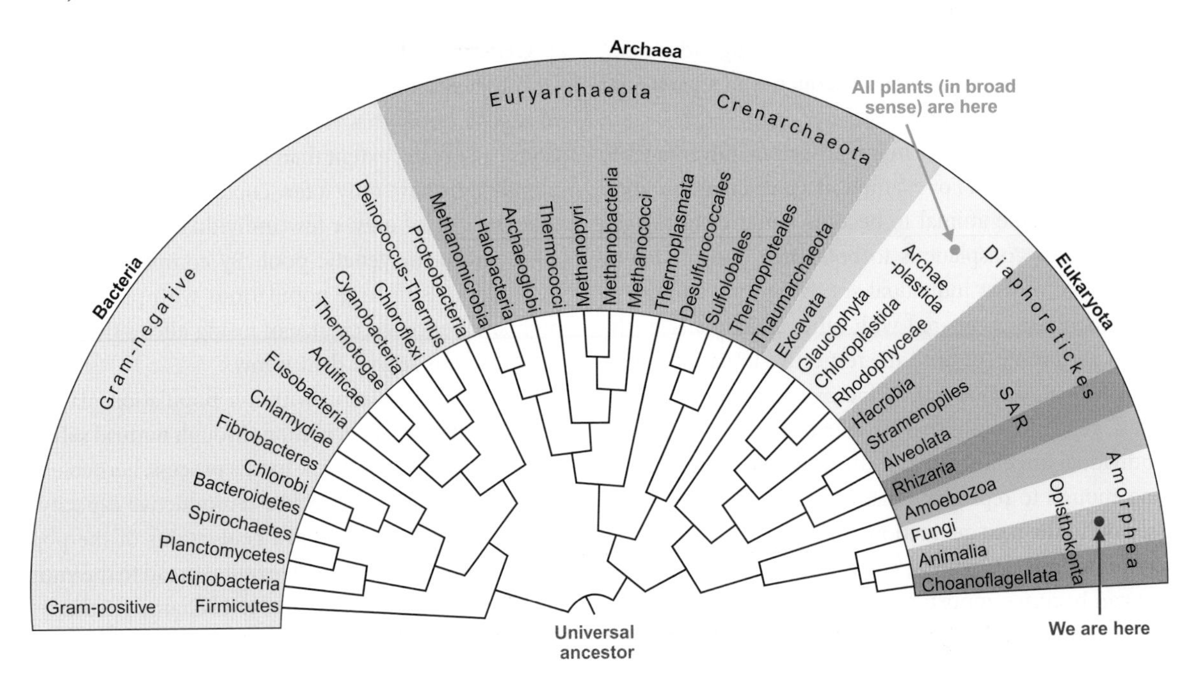

Figure 1.14 The 'Tree of Life', showing the classification of organisms at the highest ranks into three domains: Bacteria, Archaea and Eukaryota

The classification is based on sequenced genomes as at the end of 2012. All organisms evolved from the Universal Ancestor, believed to be a single-celled organism surrounded by a lipid bilayer membrane, which had a water-based cytoplasm with low sodium and high potassium ion concentrations in which hundreds of enzyme-catalysed reactions took place. The Universal Ancestor used ATP as energy currency, contained a ring-shaped coil of double-stranded DNA composed of four nucleotides and used a genetic code composed of three-nucleotide codons which was used to make proteins via RNA intermediates using 20 amino acids. Note that, according to this classification, animals are closer to fungi than to any other forms of life shown on the diagram, except for choanoflagellates.

Source: Adl SM, Simpson AGB et al (2012). The revised classification of eukaryotes. Journal of Eukaryotic Microbiology 59: 429–493 and Ciccarelli FD et al (2006). Toward automatic reconstruction of a highly resolved tree of life. Science 311: 1283–1287.

flagellates and ciliates), unicellular algae and fungi, as well as multicellular organisms that include multicellular algae and fungi, plants, slime moulds and the entire animal kingdom (**Animalia**).

In terms of individual organisms, eukaryotes represent only a very small proportion of all living organisms. Indeed, the bacteria in the human gut (which are prokaryotes) outnumber the cells in the human body 10 to 1! However, due to their much larger mass, the total biomass of all eukaryotes in the world is estimated to be about the same as that of prokaryotes.

The diagram shown in Figure 1.14 is called a **phylogenetic tree**. In recent years, the reconstruction of the animal phylogenetic tree has progressed significantly based primarily on increased knowledge of the molecular differences of whole genomes or specific genes, but also taking into consideration the fossil record and observations made with more traditional methods based on developmental patterns and morphological and physiological differences. Figure 1.15 indicates the phylogenetic relationships between the major groups of animals, which we consider in the next section. Groups joined together in the tree are considered to have descended from a common ancestor, even though they may look quite different.

A **clade** is a grouping of organisms that includes a common ancestor and all its living and extinct descendants. Animals as a whole, for example, form a clade, but so do all mammals. A branching diagram showing the phylogenetic relationships among organisms is called a cladogram[63]. Each node on the cladogram indicates the evolutionary divergence from a common ancestor. The main difference between a cladogram and a phylogenetic tree is that the length of the branches in some phylogenetic trees may be interpreted as an estimate of evolutionary time.

Overall, phylogenetic studies strongly support the propositions that:

- Animals are a **monophyletic group** (i.e. a group with a single ancestral lineage).
- The common ancestor of all animals was a unicellular flagellate.
- The closest living relatives of animals are the choanoflagellates, which are free-living unicellular and colonial eukaryotes.

[63] Cladogram is derived from the Greek, *clados* meaning 'branch'. An example of the use of cladograms to show the evolution of physiological processes is shown in Figure 13.13.

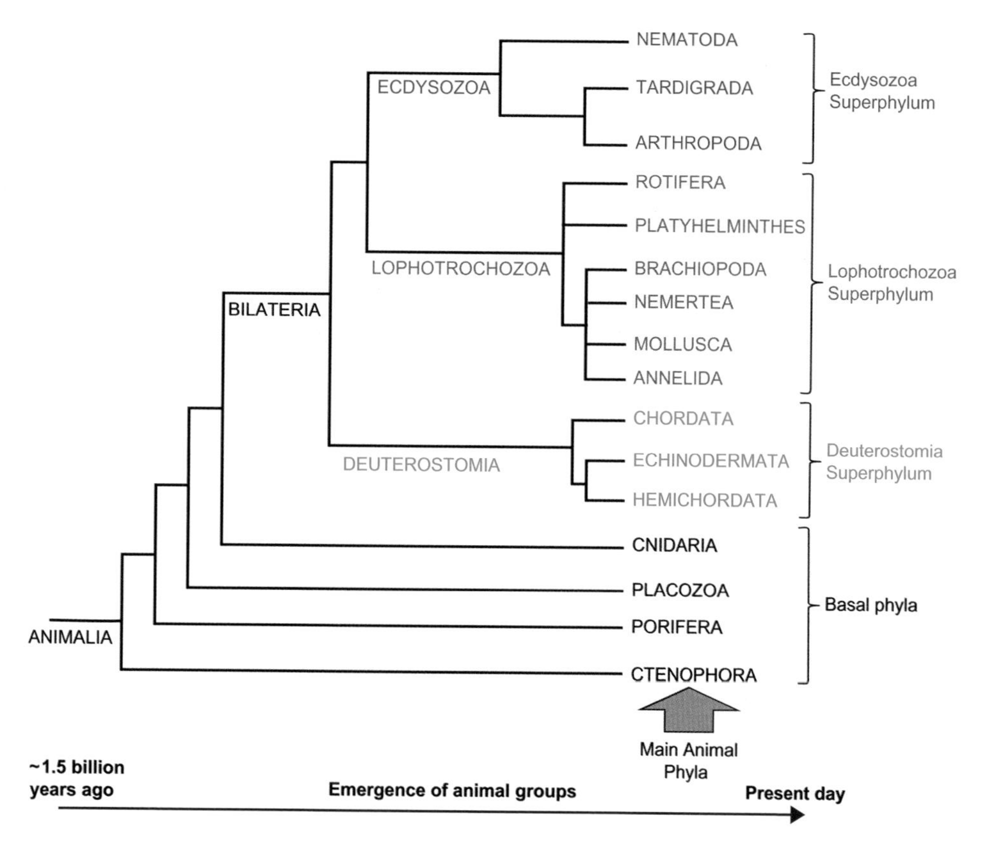

Figure 1.15 Phylogenetic tree of main groups of animals based primarily on genetic analyses

The position of some groups changed considerably between 2001 and 2013 due to the increased body of available genetic information. Note that the length of the branches in this phylogenetic tree gives only a coarse indication of the relative evolutionary time.

Source: adapted from Dunn CW et al (2014). Animal phylogeny and its evolutionary implications. Annual Review of Ecology, Evolution, and Systematics 45: 371-395. and Moroz LL et al (2014). The ctenophore genome and the evolutionary origins of neural systems. Nature 510: 109-114.

Phylogenetic information is now incorporated in comparative physiological studies to take account of the fact that animal groupings, such as different species, are not independent of each other. Since the various animal groups are differentially related to each other, according to their position on the phylogenetic tree, the level of relatedness between groups needs to be taken into account when performing statistical analyses on such physiological data. Therefore, commonly used statistical analyses that assume independence between data obtained on different taxa should not be used.

› *Review articles*

Adl SM, Simpson AGB et al. (2012). The revised classification of eukaryotes. Journal of Eukaryotic Microbiology 59: 429–493.

Ciccarelli FD, Doerks T, Von Mering C, Creevey CJ, Snel B, Bork P (2006). Toward automatic reconstruction of a highly resolved tree of life. Science 311: 1283–1287.

Dunn CW, Giribet G, Edgecombe GD, Hejnol A (2014). Animal phylogeny and its evolutionary implications. Annual Review of Ecology, Evolution, and Systematics 45: 371–395.

Garland T Jr, Bennett AF, Rezende EL (2005). Phylogenetic approaches in comparative physiology. Journal of Experimental Biology 208: 3015–3035.

Moroz LL, Kocot KM, Citarella MR, Dosung S, et al. (2014). The ctenophore genome and the evolutionary origins of neural systems. Nature 510: 109–114.

1.4 Animal diversity

Biodiversity within a defined space refers to the number of different species and the variations within these species. We may examine the biodiversity in an ecoregion or a biome, for example—or we can consider the biodiversity of the entire planet (which accommodates some 8.7 million species according to one estimate). Biodiversity is not equally distributed across the planet. In particular:

- Marine coastal habitats typically have a high diversity of animal life probably due to the origin of life in the sea around 1.5 billion years ago, and the extensive period of evolution of marine animals since then.
- The coastlines have the highest diversity of marine habitats and hence higher biodiversity than open seas.
- Biodiversity is higher in warmer tropical waters than at the poles.
- Similarly on land and in fresh water, higher biodiversity occurs in tropical areas than temperate areas (e.g. in tropical rain forests compared to temperate forests).
- Biodiversity on land is higher in areas of high water availability than areas of low water availability (e.g. tropical rain forests compared to deserts).

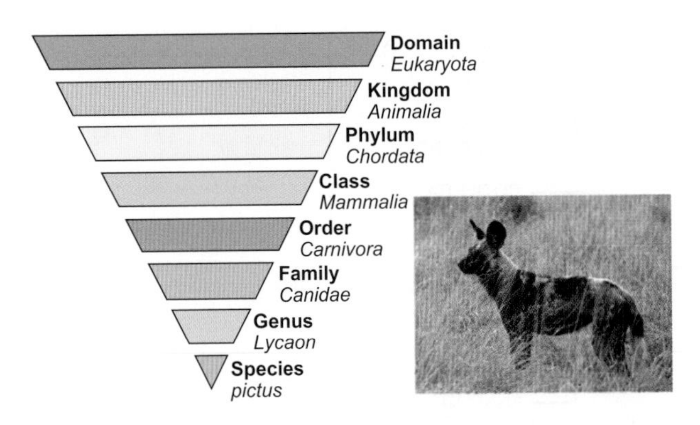

Figure 1.16 Taxonomic hierachy showing taxonomic ranks

The species and its genus identifies a particular organism—in this example the African wild dog (*Lycaon pictus*). At the next higher taxonomic rank, African wild dogs are members of the family Canidae (which includes domestic dogs, wolves and all foxes). Canidae are one of the families within the order Carnivora (which includes cats, lions and bears, as well as dogs and other members of the Canidae). At the next taxonomic level, Carnivora are an order with the class Mammalia, which is within the phylum Chordata. Chordata are a part of the Animalia within the domain Eukaryota.

Image: African wild dog or painted wolf in Botswana, © Phil Smith, Aquatonics Ltd.

The exact causes of these patterns in biodiversity are a topic of debate. They are likely to result from interacting factors, including the relative environmental stability (current and historical) and availability of energy (through the Sun's radiation[64]) which supports a higher biomass of individuals at the tropics.

To understand animal diversity we need to understand the relationships between different groups (taxa, plural of **taxon**) of animals. Animals are traditionally classified in a hierarchical system of taxonomic levels, which Figure 1.16 illustrates. A single animal can be classified as a species, a genus, placed within a family, in an order, at the next level within a class, and finally placed in a particular **phylum**, which is the most fundamental level of animal classification.

Animals within a phylum share particular anatomical features and characteristics that differ from those of animals in other phyla (plural of phylum). Many aspects of animal morphology determine their physiological functioning, as we learn throughout this book, so the organization of animals in phyla or sub-divisions within phyla is widely used by physiologists studying particular animal groups.

Early biologists decided how animals were related by identifying morphological and developmental differences. Some of the traditional morphology-based taxonomies have been confirmed by modern genetic analysis of gene sequences that investigate the evolutionary relationships between organisms. However, analysis of DNA has also led to many changes in the perceived relationships of animals and even the possibility of new phyla. A good example is a group of about 350 species of marine worms that were long considered to be members of the phylum Platyhelminthes[65] (flatworms) based on their flatworm-like morphology. Gene sequences from these simple worms indicated that they represent an independent evolutionary branch and led to their classification within a separate phylum: Acoelomorpha.

As molecular data accumulate further changes in animal classification are inevitable, particularly among the rarer phyla for which we currently have less information, but it is generally agreed that there are 33 or so animal phyla. Looking back at Figure 1.15 we see a current consensus phylogenetic tree of the major animal groups based mainly on genetic analyses. Figure 1.15 also identifies the three so-called **superphyla** that we consider in the next sections.

1.4.1 The basal phyla

Genetic data indicate that four basal phyla branched off early in evolution, as Figure 1.15 shows. While all animals are multicellular, the organization of cells early in the evolution of animals was simpler. The basal phyla do not have the organized tissues and organs that we see in more complex animals[66]. They have no kidneys, liver or muscle tissue, for example.

The basal phyla are:

- Ctenophora, which encompasses about 100–150 existing species of marine comb jellies. Comb jellies can be recognized by the characteristic combs of cilia along their side, as shown in Figure 1.17A, which they use for locomotion. Comb jellies vary enormously in size from a few millimetres to over a metre. As their common name suggests, the comb jellies consist mainly of a mass of jelly sandwiched between an outer cell-layer and a lining of their internal cavity.
- Porifera (the sponges; an example of which is shown in Figure 1.17B) have a jelly-like mass of material between two layers of cells. About 5000–10,000 or so living species of sponges occur today, mainly in seawater. Many unspecialized cells in sponges can transform into other cell types and migrate between the main cell layers and central material or reform sponges when they are broken up into pieces. Pores and channels across the body wall of sponges allow water to pass through bringing in food and oxygen and removing waste.
- Placozoa (literally meaning flat animals), which consist of an outer layer of cells enclosing loosely arranged cells.

[64] Section 8.4.1 discusses solar radiation and its impacts on animals.

[65] Platyhelminthes are described in Section 1.4.2.

[66] Section 3.3.2 discusses the organization of tissues and organs of animals.

Figure 1.17 Animals in three of the basal phyla lacking tissues

Comb jellies and sponges vary in size depending on the species from a few millimetres to over a metre. *Trichoplax* is small (about 1 mm across) and flattened.

Images: comb jelly, © Richard Hermann/Minden Pictures/Getty Images; sponge, Pbsouthwood/Wikimedia Commons; placozoan, *Trichoplax*, Bernd Schierwater/Wikimedia Commons.

Only one single species is recognized: *Trichoplax adhaerens*, which is shown in Figure 1.17C. However the genetic diversity of *Trichoplax* individuals suggests there may be multiple species in this group. *Trichoplax* feeds by engulfing particles of detritus much like single-celled protozoans such as *Amoeba* (which in current phylogenetic analyses is not included among the Animalia).

- The Cnidaria (cnidarians) is also considered a basal phylum but its members (jellyfish, sea anemones, hydroids and corals) have the beginnings of tissue layering, which Figure 1.18 illustrates. An innermost cell layer, the gastrodermis, lines the **gastrovascular cavity** where digestion occurs. Uptake of oxygen by diffusion across the gastrodermis supplements the oxygen uptake across the outer **epidermis**[67]. Between the epithelial cells of the epidermis, particularly in the tentacles, there are specialized cells called **cnidocytes**, which give the cnidarians their name. Each cnidocyte holds a cnidocyst, the most common form of which are nematocysts, which function like barbed harpoons in defence from predators and in capturing prey, as Figure 1.18C illustrates.

Figures 1.18A and B also illustrate the non-cellular **mesoglea** separating the epidermis and gastrodermis of cnidarians. The mesoglea has a lower density than water, which explains why jellyfish, which have a thick layer of mesoglea, float and are much more mobile than the attached polyps of sea anemones or sessile corals.

A further characteristic of cnidarians (compared to groups that evolved later) is their radial symmetry: if you imagine cutting across a sea anemone like a pie, each piece will look roughly the same. There is no right- or left-hand side to a sea anemone, but there is a top and bottom. A single opening acts as both the mouth and anus of the polyp or medusa. Sea anemones attach to a suitable rock at their circular base, but the pedal disc and the polyp can to some extent move around on its base.

In contrast to the mobility of anemones and jellyfish, coral polyps are completely sessile. Coral reefs of hard (scleractinian) corals are highly diverse ecosystems, as illustrated in Figure 1.19A. The health of coral reefs is critically dependent on the symbiotic relationship of the living coral polyps with photosynthetic zooxanthellae (microscopic unicellular dinoflagellates)[68] living in the cells of the gastrodermis. Hence, most corals only thrive at depths of up to 50 metres in clear (nutrient poor) waters, and coral diversity diminishes at greater depths. In tropical regions, where there is a high coral diversity, the warm, less dense surface seawater creates a stable thermocline. Plants and algae in the surface layers use the available nutrients and maintain the high clarity of the water. However, upwelling of cooler water containing nutrients or nutrient input from coastal run-off may cause algal blooms, which reduce the amount of light reaching the zooxanthellae for photosynthesis.

The bulk of a coral reef consists of interconnected old coral skeletons of the mineral aragonite, a form of calcium carbonate ($CaCO_3$) that corals produce from calcium (Ca^{2+}) and carbonate (CO_3^{2-}) ions in seawater, as illustrated in Figure 1.19C. Hence, coral reefs are an important component of the global carbon balance, although they are not as an effective sink for atmospheric carbon dioxide as the long-lived trees in forests. In Case Study 1.1, we consider the possible effects of ocean acidification due to the rising amounts of carbon dioxide in the atmosphere on corals and reef growth.

[67] Methods of gaseous exchange in water are discussed in Section 12.2.

[68] Unhealthy coral polyps eject their zooxanthellae and the coral reef loses colour. In extreme cases coral bleaching occurs and the corals become stark white. Rising sea temperatures and increased exposure to ultraviolet radiation are widely believed to cause stress of the polyps and result in the coral bleaching that may lead to their death.

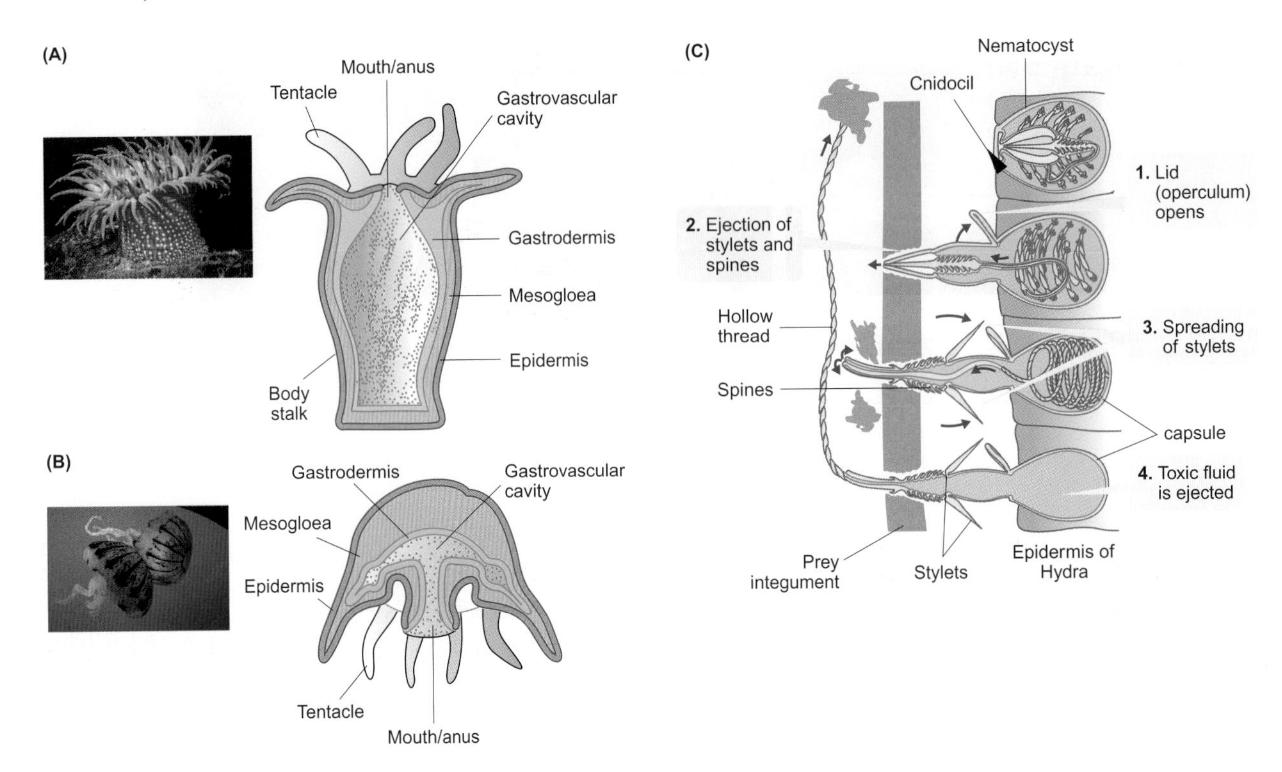

Figure 1.18 Schematic drawings of cnidarians in cross section

Two forms of cnidarians occur: (A) relatively sessile polyps such as marine sea anemones and freshwater *Hydra* (B) mobile floating medusae. Many cnidarians alternate between these forms during their life cycle. The drawings show the outer epidermis, inner gastrodermis and between these a layer of mesogloea. The epidermis of cnidarians contains cnidocytes holding cnidocysts. The most common form are nematocysts as shown for *Hydra* in (C). When the external surface is stimulated by touch or chemicals the capsule of the nematocysts swells due to water influx by osmosis and rising pressure. Within microseconds or less the nematocyst discharges, as shown in (C) beginning with ejection of the stylets and spines, and is hurled out at any nearby prey organisms, penetrating the integument of the prey organisms and holding them in place while toxic substances contained in the capsule of the nematocysts are injected into the prey. The prey is immobilized and then consumed.

Sources: (A), (B): adapted from: http://biology.unm.edu/jpeg. Images: http://pnwscuba.smugmug.com/; http://www.ucmp.berkeley.edu/ (C): adapted from Ruppert EE, Barnes RD (1994). Invertebrate Zoology, Saunders College Publishing.

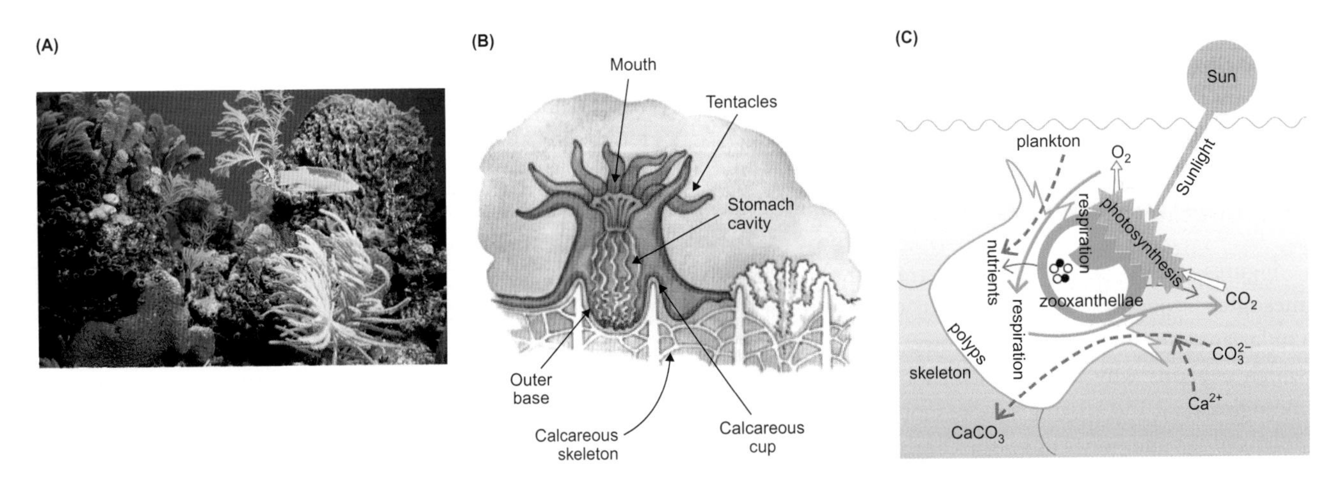

Figure 1.19 How coral reefs form

(A) Example of diverse healthy coral reef.

(B) Structure of living coral polyp and its relationship with the calcium carbonate ($CaCO_3$)-based skeleton which is secreted from the basal plates of the calcareous cups in which each live polyp sits at the surface of the coral. Secretion of plate after plate can increase healthy reefs by 0.3 to 2 cm per year for large corals, or even up to 10 cm per year for branching corals.

(C) Symbiotic relationship between coral polyps and photosynthetic zooxanthellae. The zooxanthellae provide most of the nutrients and energy needs of the corals. These nutrients are supplemented by catching planktonic organisms. Waste products from the polyps provide nitrogen, phosphorus and sulfur, and carbon dioxide, which the symbiotic zooxanthellae use in photosynthesis. The zooxanthellae need light penetration through clear surface seawater for their photosynthesis, which generates oxygen (O_2) used by the polyps for respiration. The $CaCO_3$ skeleton is produced from (Ca^{2+}) and carbonate (CO_3^{2-}) ions in seawater.

Sources: A, Spanish hogfish at coral reef, https://www.noaa.gov/; B, https://www.ausmepa.org.au/; C, http://www.okinawa-coral.co.jp.

1.4.2 Bilateral animals

Early in the evolution of animals, bilaterally symmetrical animals, which have similar right and left halves, evolved, and most animals living today are bilaterally symmetrical. All of these animals—be they a worm, fly, crab, or human—have the same network of genes that control the orientation of the body—the formation of right and left side, head to tail orientation, and dorsal and ventral orientation, and the associated development of organs and nervous systems. The so-called developmental toolkit of genes that achieves the body plan includes the well-known Hox genes[69], which do not occur in animals in the basal phyla.

Bilateral animals (Bilateria) gave rise to three superphyla: the Deuterostomia (**deuterostomes**), Lophotrochozoa (**lophotrochozoans**) and Ecdysozoa (**ecdysozoans**). In lophotrochozoans and ecdysozoans, the **blastopore**, which is the first opening of the developing embryo, forms the mouth. Older evolutionary schemes grouped the lophotrochozoans and ecdysozoans as protostomes (meaning first mouth). In contrast, the defining morphological characteristic of the deuterostomes (meaning second mouth) is that in their embryos the blastopore forms the anus and a second opening becomes the mouth.

Lophotrochozoans

Looking back at Figure 1.15, notice that Lophotrochozoa includes several phyla: Platyhelminthes (**platyhelminths**, flatworms), Nemertea (unsegmented nemertean worms), Annelida (**annelids**, segmented worms and worms that have lost segmentation), Rotifera (rotifers), Mollusca (**molluscs**) and Brachiopoda (brachipods). All lophotrochozoans show a form of embryonic cleavage in which cell division occurs obliquely to the main axis (spiral cleavage), which results in a spiral arrangement of their cells. However, each of the phyla have characteristic features that determine their physiological functioning.

The platyhelminths (flatworms), of which there are about 20,000 species, differ from other bilateral animals by their lack of an internal enclosed fluid cavity[70] or any specialized organs for the uptake of oxygen. The nemertean worms (or ribbon worms) are similarly unsegmented worms represented by about 1200 species, some of which are small but some of which extend many metres in length. Like the flatworms, nemertean worms lack large fluid-filled body cavities so they can stretch and deform. Both flatworms and ribbon worms rely on their flat body surface[71] for gaseous exchange (uptake of oxygen and excretion of carbon dioxide). Some are free living in intertidal, marine and freshwater habitats, and feed on living or dead animal tissues, but the flukes and tapeworms are parasitic in the gut of animals[72] and absorb digested food through their body surface.

Annelids are a diverse group of segmented worms and derived groups of worms represented by more than 22,000 species that live in a wide range of moist or aquatic habitats, which enables them to take up oxygen and excrete carbon dioxide via their thin body surfaces, although some annelids have external gills used for exchange of oxygen and carbon dioxide. The majority of annelids are marine polychaetes (Polychaeta), most of which are characterized by outgrowths called parapodia[73] with bristle-like chaetae (or setae) projecting from them. The parapodia are used for locomotion, gas exchange and for transporting ions across the body wall[74].

In contrast to the polychaetes, parapodia are absent and there are few chaetae in oligochaetes (Oligochaeta), such as earthworms[75] and leeches (Hirudinea). Oligochaetes and leeches are hermaphrodites, i.e. each individual has both male and female sex characteristics. Leeches have lost the walls separating internal segments so they can distend and stretch after feeding. A characteristic feature of leeches is their sucker at each end of the body, which they use to move: they use each sucker in turn as a temporary anchor point and alternately contract circular muscles that extend the body and longitudinal muscles that shorten the body.

The phylum Mollusca is diverse and about 93,000 living species have been described. Figure 1.20A shows the basic body plan of the hypothesized form of ancestral molluscs. This body plan has been modified during the evolution of vastly different molluscan forms, although the basic plan remains.

Most molluscs are **gastropods**, including all snails, limpets, slugs and nudibranchs (sea slugs)[76]. Most of the gastropods are aquatic species, usually marine/estuarine species. The only terrestrial molluscs (slugs and land snails) are gastropods that evolved a lung which they use for oxygen uptake on land[77]. Gastropods feed using a **radula**, a tongue-like

[69] Hox genes are highly conserved among animals; the hox genes and their products, hox proteins, control the head to tail axis of development, and the development of legs or wings on particular body segments.

[70] See Figure 4.2 for further discussion of body cavities.

[71] Section 11.2.1 discusses diffusion of gases across animals with flat body surfaces.

[72] Flukes and tapeworms as adults are usually parasites of vertebrates but in the larval stage occupy intermediate hosts that can be invertebrates or vertebrates.

[73] Figure 20.4 illustrates a polychaete with pronounced parapodia.

[74] Section 5.3.1 examines osmoregulation in some polychaetes and Section 12.3.1 discusses gas exchange across the body wall of aquatic animals.

[75] We discuss earthworms in several parts of the book. For example, Figure 20.14A illustrates their reproduction; Figure 7.31 illustrates their excretory processes, Section 14.3 discusses their circulatory system and Figure 18.31 illustrates their locomotory mechanism.

[76] See Figure 20.7B for an example of a nudibranch.

[77] Section 12.3.3 discusses the lungs of molluscs and their use in air-breathing.

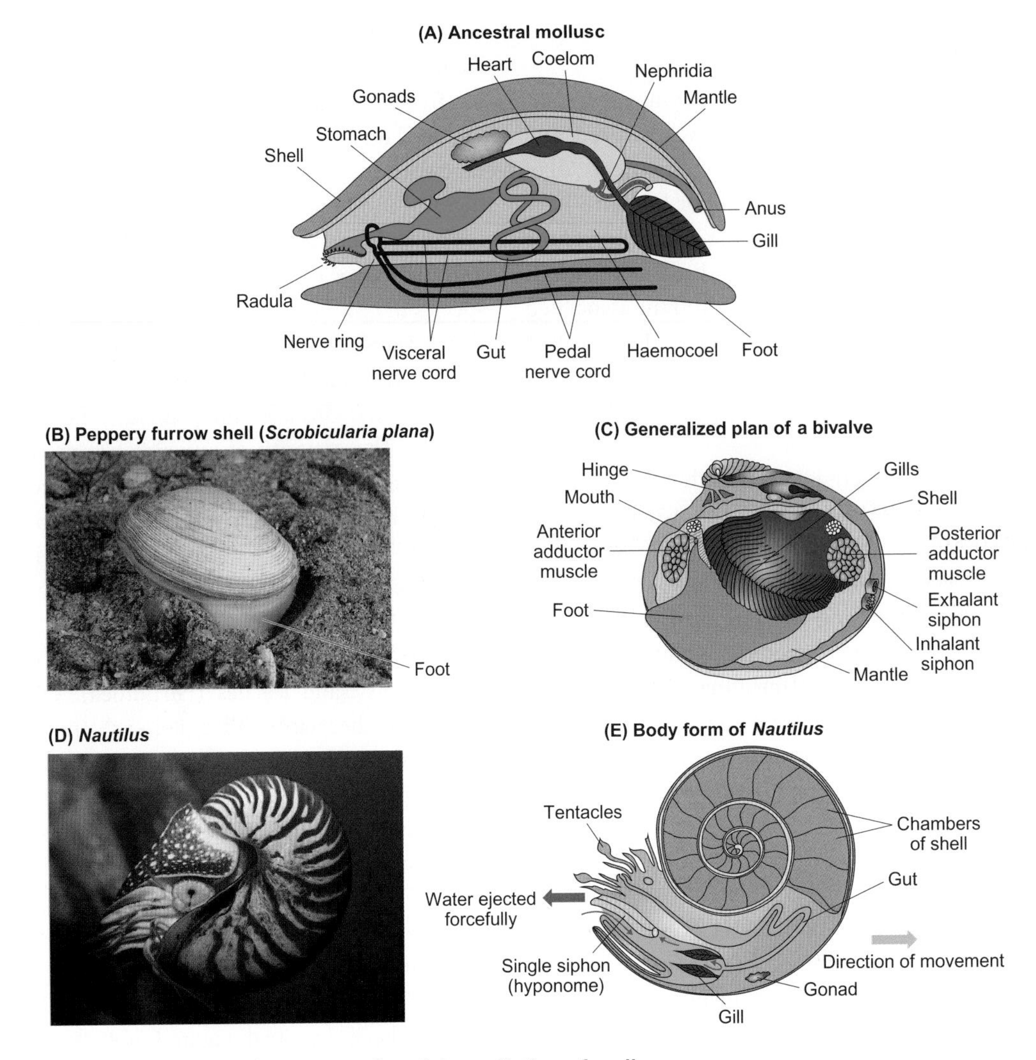

Figure 1.20 Molluscan body form and two examples of the radiation of molluscs

(A) Characteristic features of an ancestral mollusc that gave rise to the different classes of molluscs. Two of these are illustrated.

(B) Bivalve mollusc—the peppery furrow shell (*Scrobicularia plana*) using its foot to start burrowing. The peppery furrow shell varies in diameter reaching up to 6.5 cm and burrows to depths of up to 20 cm.

(C) The general body form of a bivalve.

(D) *Nautilus*—the only member of the cephalopods that has an external shell. *Nautilus* vary in size; most *Nautilus* species reach up to 20 cm in shell diameter.

(E) General body form of *Nautilus* showing the ejection of water used for respiration and locomotion.

Sources: B, Peppery furrow shell (*Scrobicularia plana*): age fotostock / Alamy Stock Photo; C, http://www.marlin.ac.uk/; D, Nautilus - http://palaeo.gly.bris.ac.uk/; E, Clarkson ENK (1998). Invertebrate Palaentology and Evolution, 4th edition. Oxford: Blackwell Science Ltd.

structure with rows of microscopic teeth that they use for scraping food, such as algae, from hard surfaces and for rasping at vegetation[78]. Some carnivorous snails (gastropods) bore through the shells of other molluscs using their radula.

Bivalve molluscs, an example of which is shown in Figure 1.20B, evolved a hinged shell consisting of two halves. Most bivalves are fairly sedentary, but many use their modified foot, which Figures 1.20B and 1.20C illustrate, for burrowing[79]. Bivalves obtain oxygen using large gills (**ctenidia**), which they also use to filter food from the water passing over their gills[80].

An outer shell occurs in most molluscs (except the pulmonate slugs with lungs), sea slugs and almost all **cephalopods** in which the outer shell has been lost. The outer shell is secreted from the mantle[81], which forms the dorsal body wall

[78] Figure 2.11 shows the organization of the feeding apparatus in gastropod molluscs.

[79] Section 14.3.1 discusses the role of the circulatory system in the functioning of the foot in bivalve molluscs.

[80] We discuss filter feeding in general in Section 2.3.1 and specifically in the bivalve zebra mussel in Case Study 5.1.

[81] Section 21.2.5 discusses the formation of the molluscan shell.

and protrudes beyond the visceral mass, as shown in Figures 1.20A and 1.20C. The mantle covers the visceral mass (digestive components, and organs for reproduction and excretion) and encloses the gills, which are used for gas exchange[82].

In cephalopods (squid, octopus and cuttlefish), the most striking features are the tentacles, and the lack of an external shell, except in the six species of *Nautilus* that retain a coiled external shell, as shown in Figures 1.20D and 1.20E. Cephalopods have a well developed brain, particularly species of *Octopus*, and high visual capacity[83]. All cephalopods have a well developed capacity for movement. This movement is achieved by ejecting water through a single siphon (funnel or hyponome) located below the tentacles, which can be turned to different positions to direct locomotion. Figure 1.20E illustrates the entry of water around the siphon into the mantle cavity; water passes over the gills before being forced out under pressure through the siphon, propelling *Nautilus* backwards.

The Brachiopoda (brachipods), also called lamp shells, show a superficial similarity to bivalved molluscs as they have two half shells, but they form a distinct phylum as reinforced by genetic analyses, which show that brachiopods are not closely related to molluscs. All brachiopods are marine filter feeders that occur mostly in cold waters in polar locations or at depths where temperatures decline. Profound changes in the assemblage of brachiopod fossils are seen on a geological timescale. Only about 300 species currently exist, but at their peak they were a highly diverse group of abundant animals. Brachiopod fossils, of which there are >12,000 species, are used as indicators of ancient seawater temperatures and how these temperatures have changed as a consequence of climate change over geological time, with losses occurring after temperature increases. After the mass extinction of animals about 250 million years ago (the Permian–Triassic mass extinction[84]) the diversity of brachiopods only partially recovered.

Ecdysozoans

Ecdysozoa was as a term introduced in 1997 to describe the superphylum (clade) that includes Nematoda (nematodes: non-segmented roundworms), Tardigrada (tardigrades, commonly called water bears) and Arthropoda (**arthropods**). Nematodes are a surprising relative of tardigrades and arthropods as their morphological appearance, as roundworms, does not suggest an evolutionary relationship to the jointed arthropods or the eight-legged tardigrades. However, analysis of gene sequence data reveals the origin of nematodes, tardigrades and arthropods from a common ancestor. The key characteristic of all Ecdysozoa is that they all shed their outer **cuticle** (their exoskeleton) when they grow by the process of **ecdysis** (moulting). 'Ecdysozoa' literally means 'moulting animals'.

In nematodes the pressure of the internal body fluids pushes against the cuticle and results in the circular cross-section from which their common name of roundworm stems. Nematodes are amazingly diverse; about 25,000 species have been identified, but it is thought that a million species of nematodes may exist, as nematodes occur in virtually every conceivable habitat, and most of them have not been described. Some nematodes are parasitic; these include the common roundworms, hookworms and pinworms.

Tardigrades are ubiquitous in a wide diversity of terrestrial and aquatic habitats ranging from the deep sea, where they can survive the high pressures, to the highest mountains. More than 1000 species of tardigrades exist throughout the world, although we would find it hard to spot them with the naked eye because of their small size (just 0.1 to at most 1.5 mm in length as adults). In a physiological sense, tardigrades are most remarkable for their resistance to extreme environmental conditions that would kill most other organisms. They can withstand extremely low temperatures—as low as −200°C for a few days—and survive at temperatures as high as 150°C, so they can live in hot springs. Tardigrades are able to survive without food or water for many years in a state of **anhydrobiosis**[85] after which they can rehydrate and reproduce.

The Arthropoda is the most diverse phylum on Earth, with estimates varying from a little over 1 million to up to 10 million living species, many of which are insects. The key morphological features of arthropods are their body segmentation, jointed appendages and an external calcium-rich exoskeleton. The waterproofing of the exoskeleton allows many arthropods to live on land without suffering from dehydration[86]. The arthropod exoskeleton protects the animal and provides support for muscle attachment[87], which allows their appendages to be used for walking on land, swimming and flying. However, the exoskeleton also presents a potential problem: as the animal grows, it becomes too large for its 'skin'. To grow any more, it is therefore necessary to shed the exoskeleton. We can imagine that such a dramatic event requires a multiplicity of physiological control processes. We discuss the control of moulting in Chapter 19 and the management of calcium stores during moulting in Chapter 21[88].

[82] Section 12.2.2 discusses gaseous exchange via the gills of molluscs.

[83] We discuss *Octopus* brain and vision in Sections 16.5.1 and 17.2.1.

[84] Figure 1.23 illustrates the five mass extinction events over past 542 million years.

[85] Anhydrobiosis is a state of extreme dehydration and reduced metabolism; in this state, the tardigrades contain very little water, as we discuss in Section 4.1. In Case Study 8.1 we discuss how tardigrades cope with anhydrobiosis and how this ability has inspired methods for preservation of biological materials.

[86] Waterproofing of insects is discussed in Section 6.1.3.

[87] Section 18.5.1 describes how muscles attach to the exoskeleton.

[88] We discuss moulting in crustaceans and insects in Sections 19.5.1 and 19.5.3, respectively and the management of calcium stores during crustacean moulting in Section 21.2.5.

The Arthropoda diversified into four subphyla:

- Chelicerata (chelicerates) which includes **arachnids** (spiders, mites and scorpions) that occupy a diverse range of mostly (but not exclusively) terrestrial habitats.
- Myriapoda (myriapods: millipedes and centipedes), all of which are terrestrial species.
- Crustacea (**crustaceans**): crabs, lobsters, shrimps and sessile barnacles. Most crustaceans are aquatic species but some are semi-terrestrial or terrestrial.
- Hexapoda (**hexapods**) mostly insects and insect-like animals with six thoracic legs of many different types living mainly in terrestrial habitats and some in fresh water. The most diverse groups of insects are the Coleoptera (**coleopterans**: beetles and weevils), Lepidoptera (**lepidopterans**: butterflies and moths), true flies (Diptera, **dipterans**) and bees, wasps and ants (Hymenoptera, **hymenopterans**).

Wing development of insects allows flight[89], which is considered to be an important aspect of their diversification.

Deuterostomes

Deuterostomes include the following phyla:

- Echinodermata (**echinoderms**: starfish, sea urchins, sea stars and sea cucumbers), which are entirely marine species.
- Hemichordata (hemichordates, which include acorn worms). There are only about 120 known species of hemichordates, all of which are marine benthic (bottom-living) animals. The hemichordates are the closest phylogenetic relatives of the chordates among the invertebrates.
- Chordata (**chordates**) representing about 50,000 species, including *Homo sapiens*.

[89] We discuss insect flight in Section 18.6.3.

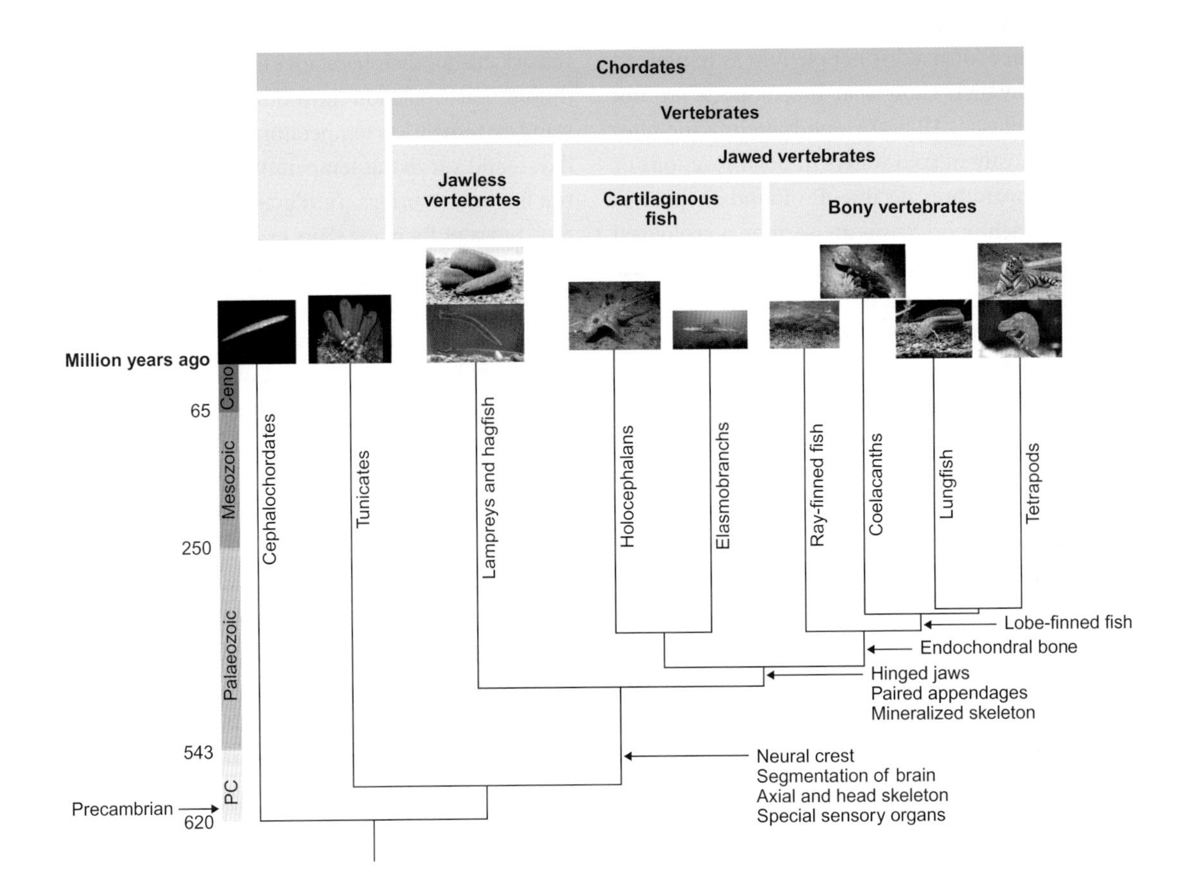

Figure 1.21 Chordate phylogeny showing some of the major steps during evolution

Evolutionary time is indicated on the left side of the illustration, which distinguishes the Palaeozoic, Mesozoic and Cenozoic (Ceno) eras. The neural crest is a transient embryonic structure in vertebrates that gives rise to neural and non-neural tissue. Lampreys and hagfish are jawless vertebrates. Jawed vertebrates (gnathostomes) divided into two major groups: cartilaginous fish with largely cartilaginous endoskeletons and bony vertebrates, about 450 Mya. The holocephalans and elasmobranchs diverged about 420 Mya. Endochondral bone forms within cartilage and involves partial or complete replacement of cartilage by calcification.

Sources: Vekatesh B et al (2014). Elephant shark genome provides unique insights into gnathostome evolution. Nature 505: 174-179. Images – Tunicate, sea squirt: http://a-z-animals.com/; Lamprey: Tilt Hunt/Wikimedia Commons; Elasmobranch, spiny dogfish: © Doug Perrine/Nature Picture Library; Cephalochordate (*Branchiostoma lanceolatum*): Hans Hillewaer/Wikimedia Commons; Coelacanth: http://marinebio.org/; Tetrapods: © Phil Smith, Aquatonics Ltd; Hagfish: https://www.wired.com/wp-content/uploads/2014/05/hagfish-getty-01a.jpg; Holocephalan, spotted ratfish: http://vignette2.wikia.nocookie.net/pugetsound/images/0/02/Spotted_ratfish.jpg; Marbled lungfish: https://primitivefishes.files.wordpress.com; Ray-finned fish, rainbow trout: Jack Perks/Alamy Stock Photo.

Figure 1.15 includes no mention of the widely used terms **invertebrates** (animals without a vertebral column) and **vertebrates** (animals with a vertebral column), although we use these terms throughout this book. Vertebrates actually form a group within the Chordata, as shown in Figure 1.21. Members of all the other subphyla are invertebrates, including two groups of non-vertebrate chordates—the Tunicata (tunicates; sometimes called Urochordata) and Cephalochordata. More than one million different species of invertebrate animals have been described, compared to about 66,000 described vertebrate species, although most invertebrates are much smaller than most vertebrates.

A unique characteristic of chordates is their possession of a stiff but flexible rod called the **notochord** (hence the term chordates), at least at some point in their life cycle. The notochord is made out of a cartilage-like material and forms a rigid structure for muscle attachment that is sufficiently flexible to allow an elongated animal to bend somewhat. Further chordate characteristics (again at some point in their life cycle) are a dorsal hollow neural tube (nerve cord), repeated muscle blocks (myotomes) on either side of their body and pharyngeal slits (or holes) between the outside of the body and the pharynx.

Tunicates

The tunicates encompass about 2000 species, all of which are marine (mostly sea squirts). They are mostly sessile and have an outer 'tunic' containing cellulose and two siphons through which they draw in and expel water through the body wall. Looking at the photograph in Figure 1.21 you might think tunicates are relatives of the sponges rather than close relatives of the vertebrates. However, the chordate features of tunicates become apparent at the larval stage. Thus, tunicate larvae swim using muscle fibres attached to a notochord, much like amphibian tadpoles; have a small brain connected to a nerve cord; and possess a single, simple eye[90]. All these chordate features are lost when the larva settles and develops to form a sessile adult.

The final chordate characteristic appears in the adult tunicate: bands of pharyngeal slits appear in the walls of the large pharynx that fills most of the interior. Water flowing into the tunicate, via an upper siphon, passes through bands of slits into the surrounding water-filled cavity and it is then expelled through a ventral siphon. Food particles in the water are caught up by a mucous net and pass into the oesophagus connected to the lower end of the pharynx[91].

Cephalochordates

The other invertebrate group of chordates, the cephalochordates (lancelets or amphioxi (singular amphioxus))[92], consists of about 30 marine species. In contrast to tunicates, lancelets retain a notochord and a nerve cord throughout their life, both of which extend along the whole body. The anterior end of the nerve cord is only slightly enlarged, indicating a lack of a brain. Segmental muscles attached to the notochord allow uncoordinated swimming, much like larval tunicates. In fact, the lancelets spend most of their life buried in the marine sediments, obtaining particles of food from the water by trapping them on a film of mucus covering their gill slits (pharyngeal slits).

Vertebrates

Two of the earliest vertebrate groups to have evolved are the lampreys and hagfish, which retain the notochord as adults. About 40 living species of lampreys and more than 60 species of hagfish exist but recent finds of several new species suggest more may exist.

Lampreys and hagfish are grouped together as **cyclostomes** ('round mouths') based on both morphological and molecular data. The cyclostomes are also known as **agnathans** or jawless fish, which distinguishes them from all other jawed vertebrates, (gnathostomes), as illustrated in Figure 1.21. Lampreys and hagfish have a cartilaginous skull but lack the mineralized vertebrae that are a morphological characteristic of other vertebrates, an observation which has contributed to a debate about their inclusion with the vertebrates. However, the lampreys have cartilaginous dorsal skeletal elements of the vertebrae so are argued to have lost the ventral elements. Segmentally arranged cartilaginous elements have also been identified recently along the ventral side of the notochord in hagfish.

Vertebrate embryos have notochords, but in most vertebrates the notochord develops into the discs between each pair of vertebrae. The vertebrates evolved skeletons, with paired appendages, and muscles that allow extensive and complex movements such as running, swimming and flight[93]. Development of a structured brain (protected by a skeleton) and sensory organs allow vertebrates to coordinate and control their movements and to sense their environment and respond accordingly.

The sharks, skates and rays (most of which are marine) are able to produce dermal bone (e.g. in teeth and spines) but do not replace their cartilaginous skeleton with a bony endoskeleton, seemingly because they lack the genes encoding calcium-binding phosphoproteins. The sharks, skates and rays are **elasmobranchs**, which are one of the two groups of cartilaginous fish (Class Chondrichthyes), which diverged about 420 million years ago, as illustrated in Figure 1.21, and of which there are currently 1100 described species. The other group, the holocephalans (or chimaeras) flourished 350 million years ago but are now represented by only 33 species (such as ratfish and ghost sharks).

In contrast to cartilaginous fish, two sister groups of fish, the **lobe-finned fish** (Sarcopterygii) and **ray-finned fish** (Actinopterygii, actinopterygians), evolved **endochondral bone** in which calcification occurs within cartilage, which is

[90] We discuss simple eyes in Section 17.2.1.

[91] We discuss filter feeding in Section 2.3.1.

[92] Most amphioxi are of the genus *Branchiostoma*.

[93] Section 18.6 explores locomotion in some detail.

partially or entirely destroyed as bone forms. The lobe-finned fish have fleshy pectoral and pelvic fins which articulate with the pectoral and pelvic girdles via single bones.

Tetrapods (amphibians, turtles, lizards and snakes, crocodiles, birds and mammals) evolved from a lobe-finned vertebrate. One group of lobe-finned fish is represented by coelacanths and another by lungfish (which can breathe air using lungs[94]) as shown in Figure 1.21. Only two species of coelacanths and six species of lungfish exist today. Figure 1.21 includes an image of a marbled lungfish (*Protopterus aethiopicus*) and a coelacanth (*Latimeria chalumnae*) in its natural deep ocean habitat.

The coelacanths have drawn enormous interest from biologists as they were only known from 300–400 million year old fossils and were thought to have gone extinct until living coelacanths were discovered off the South African coast in 1938, and then later in the Indian Ocean around the Comoro Islands, and in Indonesia.

Charles Darwin referred to lungfish as almost living fossils and the term 'living fossil' was widely adopted when the living coelacanths were discovered in 1938. Of course the term 'living fossil' is a contradiction in terms, as all fossils are dead, but the idea behind it is that the living species shares morphological characteristics with fossil ancestors that have remained unchanged for millions of years. Comparison of protein-coding gene sequences of lungfish and coelacanths and other vertebrates suggests slower rates of change (based on the nucleobase substitutions) in lungfish and coelacanths than in mammals, bony fish and most (but not all) tetrapods, explaining at least in part how these species managed to survive almost unchanged morphologically for 300–400 million years.

The lungfish and coelacanths are of particular interest to biologists because they can give us clues as to the particular molecular and structural characteristics that enabled the transition from an aquatic to terrestrial life. Molecular studies of nuclear genes have produced overwhelming evidence that the closest living relatives of land vertebrates was an ancestral lungfish, as indicated in Figure 1.21, not a coelacanth.

The evolution of a gas-filled **swim bladder** in ray-finned fishes allows them to adjust their buoyancy[95] and explore a wide range of habitats with minimal use of energy. At least partially because of this, the bony **teleost fish** diversified such that they now represent almost 50 per cent of all living vertebrate species (almost 27,000 species) that occur in almost every aquatic habitat.

Throughout this book we deal with birds separately from reptiles, even though genetic data indicate that birds are more closely related to crocodilians (crocodiles and alligators) than crocodiles are to turtles, lizards and snakes. As such, from a purely phylogenetic perspective, birds should be grouped with reptiles. However, the birds which evolved about 150 million years ago from a group of dinosaurs are physiologically quite different from living reptiles. For instance, all birds regulate their body temperature by internal heat generation (they are **endotherms**), while living reptiles are **ectotherms**[96].

Mammals evolved from a group of reptiles called synapsids, which arose about 300 million years ago. All mammals are recognizable by three unique characteristics:

- Their three middle ear bones[97], two of which are derived from the lower jaw bones of mammalian ancestors.
- The suckling by female mammals of newborn offspring by **lactation**[98], secreting milk produced by modified sweat glands or mammary glands.
- Their possession of body hair (at some stage in development, even in whales and dolphins). The evolution of body hair forms an important aspect of body temperature regulation of many mammals, which (like birds) regulate relatively higher body temperatures than most environmental temperatures[99], and has implications for the functioning of all tissues and organs.

According to a survey of the Earth's mammals by the International Union for Conservation of Nature (IUCN) in 2008, there are about 5400 species of extant (still existing) mammals, but almost a quarter of these are globally threatened, primarily due to habitat loss. Although the number of mammalian species is about half the number of bird species, mammals show greater size diversity—from the bumblebee bat (*Craseonycteris thonglongyai*), which is 30 mm long and weighs about 2 g, to the blue whales (*Balaenoptera musculus*) that grow to up to 30 metres in length and 180,000 kg in weight.

Mammals have evolved to exploit almost every type of habitat. Most mammals are terrestrial, but some are fossorial (they live underground) and some, such as bats, are adapted for life in trees and for flight. About 84 species, including baleen and toothed whales, and porpoises, seals, walruses and sea lions live in the oceans or along coasts. Mammals' different lifestyles encompass the ability to run and jump, swim, burrow and in some cases fly (or glide).

The earliest mammals, like their reptile ancestors, reproduced using eggs. These egg-laying mammals, called monotremes, were originally widespread, but only the duck-billed platypus and four species of echidnas exist today in Australia and New Guinea. Except for these monotremes, mammals

[94] All living lungfish, the Australian lungfish *Neoceratodus forsteri*, South American lungfish *Lepidosiren paradoxa*; and four African species of the genus *Protopterus* are freshwater species. Section 12.3.3 discusses air-breathing by lungfish.

[95] We examine the functioning of swim bladders in Case Study 13.1.

[96] Chapter 10 discusses endothermy; Chapter 9 discusses ectothermy.

[97] Section 17.4.1 discusses the function of the middle ear bones in mammals.

[98] Control of lactation by prolactin is discussed in Section 19.2.2.

[99] We discuss endothermy and body temperature regulation in Chapter 10.

are characterized by giving birth to live young, although at markedly different stages of development. In marsupials, the young are born at a very early stage of development and are then provided with nourishment from their mother, in a pouch[100]. There are currently about 240 species—the opossums, koalas, kangaroos, wombats and wallabies, which are mostly endemic to Australia and associated islands.

Placental mammals (about 3800 species) represent about 70 per cent of mammalian species. In placental mammals the foetus develops to a more advanced stage before birth because the **placenta**[101] provides oxygen and nutrients to the developing foetus until birth.

The original schemes for the classification of mammals relied on studies of fossil morphology and the morphology and development of living species. In contrast, modern genetic studies investigate the similarities and differences in DNA sequences of species to build up a phylogenetic tree, and have revised our views of the phylogeny of mammals. Figure 1.22 shows the proposed phylogeny of mammals based on DNA sequences. This cladogram should not be seen as rigid, but rather as the current consensus and a working hypothesis, which will need frequent re-evaluation as new data emerge.

The molecular data indicate that placental mammals form four major phylogenetic groups (monophyletic clades). These clades map onto broad geographical areas, which suggests that the initial diversification of placental mammals took place before the continents separated.

- Afrotheria evolved in Africa, early in mammalian radiation. This grouping includes elephants, hyraxes, manatees (sea cows) and dugong, the aardvark (*Orycteropus afer*) and tenrecs.
- Xenarthra, which evolved in the Americas, encompasses sloths, anteaters and armadillos.
- Euarchontoglires includes humans and other primates (great apes and monkeys), placing us in the mammal grouping that includes flying lemurs (Order Dermoptera), tree shrews (Order Scandenta), rabbits, hares and pikas (Order Lagomorpha), and rats and mice (Order Rodentia).
- Laurasiatheria encompasses the placental mammals that evolved in Laurasia, which over geological time formed Europe and Asia. The clade Laurasiatheria includes cats, dogs and wolves (Order Carnivora), bats (Order Chiroptera), cows, camels, deer, horses (previously classified as members of the order Artiodactyla or even-toed ungulates) and whales. Thus, genetic studies revealed that the cetaceans (whales, dolphins and porpoises) evolved from an artiodactyl ancestor.

[100] Figure 20.34 shows an example, the full term foetus and a suckling newborn of tammar wallaby (*Macropus eugenii*).

[101] Figure 20.3 shows the structure of the human placenta. We discuss different types of placentae in Section 20.1.3.

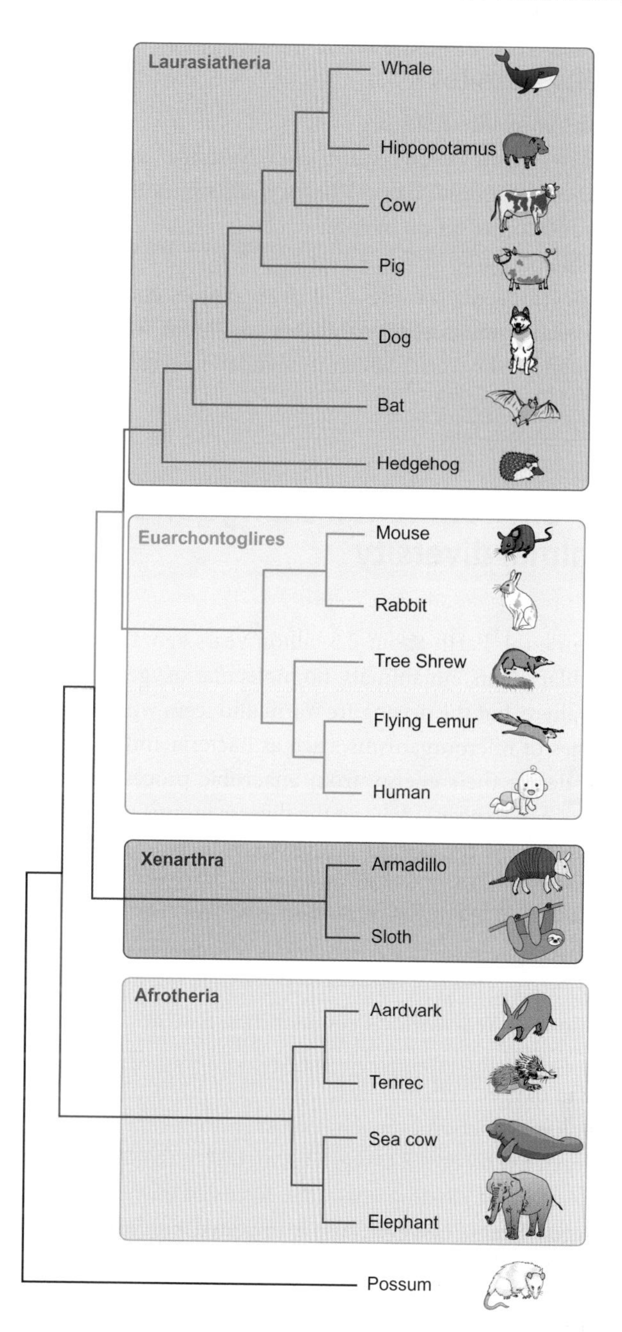

Figure 1.22 Mammalian phylogeny

Monotremes and marsupials (represented by possum in the diagram) evolved before the placental mammals. The four major clades of placental mammals are distinguished by different colours: Afrotheria (orange), is the earliest branching lineage, followed by Xenarthra (purple), the Euarchontoglires (green), which includes humans and other primates, and Laurasiatheria (blue).
Source: Bromham L. An introduction to molecular evolution and phylogenetics. 2nd Edition. Oxford: Oxford University Press, 2016.

› *Review articles*

Gaston KJ (2000). Global patterns in biodiversity. Nature 405: 220–227.

Goreau TF, Goreau NI, Goreau TJ (1979). Corals and coral reefs. Scientific American 241: 124–137.

Nelson DR (2002). Current status of the Tardigrada: Evolution and Ecology. Integrative and Comparative Biology 42: 652–659.

Springer MS, Stanhope MJ, Madsen O, de Jong WW (2004). Molecules consolidate the placental mammal tree. Trends in Ecology and Evolution 19: 430–438.

1

Useful websites

https://animaldiversity.org

This website is an online encyclopaedia: a searchable database of animal natural history, distribution, classification and conservation biology.

https://eol.org/

Encyclopaedia of life is a free online encyclopaedia compiled from existing databases and input from experts throughout the world, which is intended to eventually cover every living species.

1.5 Environmental change and animal diversity

Imagine the Earth about 2.5 billion years ago: there are no vascular plants, no animals, no molecular oxygen in the atmosphere, but the oceans are warm and teem with countless species of microorganisms, such as bacteria and archaeans, that derive their energy from anaerobic processes (in the absence of oxygen). Among the diverse groups of anaerobic organisms in the oceans are photosynthetic cyanobacteria (blue-green algae), which produce molecular oxygen from water as a waste product of photosynthesis.

The appearance of cyanobacteria was directly responsible for initiating huge environmental changes over hundreds of millions of years, which resulted in a dramatic rise in the concentration of molecular oxygen in the atmosphere[102]. The rise in oxygen concentration caused one of the most significant life extinction events in Earth's history, called the Great Oxygen Event, or the Oxygen Catastrophe, that drove to extinction most existing anaerobic forms of life[103]. The Great Oxygen Event, however, also stimulated biodiversification as some organisms evolved that could use oxygen in their metabolic processes. These organisms could obtain much more energy from the oxidation of foodstuff in the presence of oxygen than in its absence[104].

The increased efficiency of energy production from foodstuff allowed microscopic organisms to grow and reproduce faster and eventually evolve into various aerobic forms of life, including vascular plants and animals that were much larger and more complex than their anaerobic relatives. This discussion underlies the strength and depth of interactions between life on Earth and the abiotic environment of Earth and shows that the evolution of specific life forms like cyanobacteria induced massive environmental changes, which, in turn, dramatically influenced the composition and diversity of life forms on Earth.

Another important outcome of the presence of molecular oxygen in the atmosphere was the formation of the **ozone (O_3) layer** under the influence of ultraviolet radiation from the sun[105]. The ozone layer acts as a shield that protects the Earth's surface from damaging ultraviolet (UV) radiation; its formation allowed life to move out of oceans onto land[106].

In the last 500 million years there were five major mass extinction events in which 50 to 96 per cent of the then existing species disappeared. Figure 1.23 shows the number of existing genera of marine animals over the past 542 million years. Although we do not know the exact cause that triggered each of these catastrophic events, there is evidence to suggest that impact by large asteroids and massive volcanic explosions were responsible for sudden environmental changes that would have led to mass extinction events.

Despite their catastrophic nature, each of these events also paved the way for the evolution of new forms of life. For example, dinosaurs first appeared after the Permian–Triassic mass extinction event about 250 million years ago; the Cretaceous mass extinction event that killed the dinosaurs (except for some theropod dinosaurs that evolved into birds) when it took place about 65 million years ago allowed mammals to diversify rapidly and evolve. The net outcome so far is that animal diversity is greater now than it has ever been; Figure 1.23 uses marine animal life as an example to illustrate this point.

1.5.1 Environmental factors can alter the pattern in which genes are switched on and off

When genes are switched either 'on' or 'off' they affect the characteristics of cells or organisms—that is, their phenotype. Some changes in the phenotype are reversible, while others are irreversible. Also some changes affect only one individual cell or organism, while others are heritable, and affect future generations of cells and/or organisms.

[102] We discuss the changes of oxygen concentration in the Earth's atmosphere in Section 11.1.1.

[103] Oxygen can be toxic to cells through the formation of oxygen free radicals, such as superoxide (O_2^-), which damage cell membranes and impair cellular metabolism. Anaerobic organisms lack enzymes that convert oxygen free radicals to harmless molecules. We also discuss in Section 11.2.2 how the much lower oxygen partial pressure in cells than in the atmosphere reduces the formation of oxygen free radicals.

[104] We discuss in Section 2.4 how much more energy is obtained by aerobic than by anaerobic metabolism.

[105] Some oxygen molecules split into two oxygen atoms under the influence of UV radiation high in the atmosphere. One of these atoms then combines with one molecule of oxygen to form an ozone molecule (O_3).

[106] Note that water absorbs damaging UV radiation as we discuss in Section 17.2 and therefore protects aquatic organisms from UV radiation.

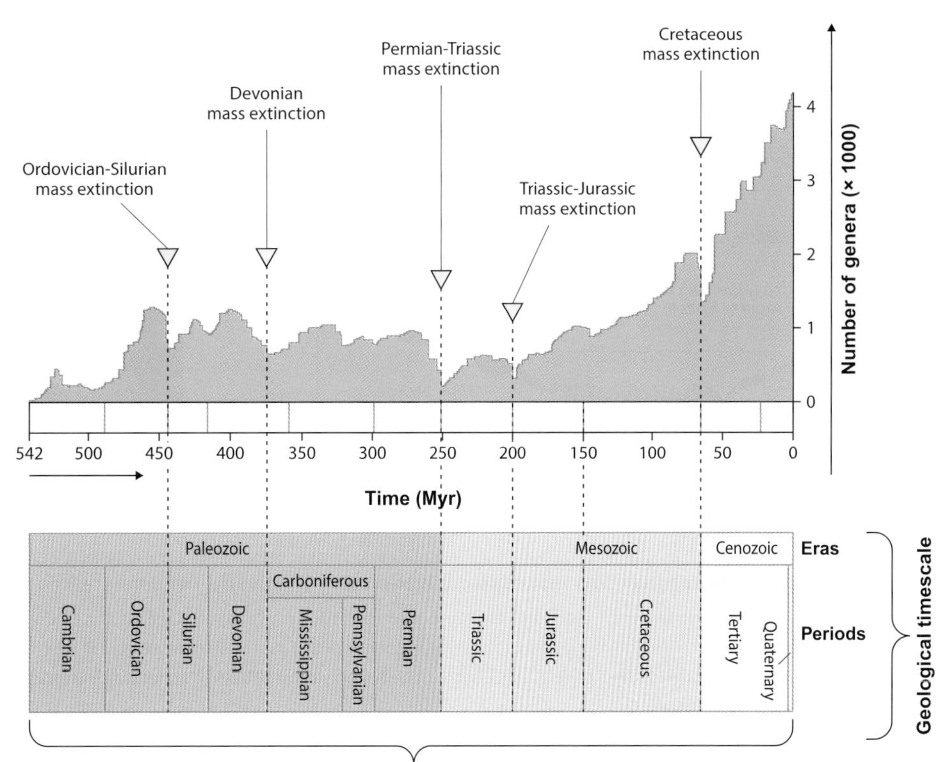

Figure 1.23 Diversity of marine animal genera, defined as distinct genera alive at any given time over the past 542 million years (Myr) corresponding to the current eon

The past 542 Myr cover the Phanaerozoic eon on the geological timescale. The 'big five' mass extinction events are indicated by arrow heads. The Permian–Triassic mass extinction was the biggest, causing a massive 96% extinction of all species.

Source: adapted from Rohde RA, RA Muller (2005). Cycles in fossil diversity, Nature 434: 208-210.

The phenotype of cells or organisms is not only determined by the genotype, but also by environmental influences. For example, external factors, such as diet, stress level, physical activity, climate, pollution and lifestyle, can all alter the pattern of gene expression in specific cells of an animal.

Phenotypic flexibility (or plasticity) generally refers to the ability of one genotype to produce more than one phenotype under different environmental conditions[107].

The study of phenotypic changes caused by modifications in the pattern in which genes are switched on or off without there having been an alteration to the underlying genotype is called **epigenetics** (from Greek *epi* = outside of, above and *genetikós* = origin, source, birth). In its precise usage, epigenetics is concerned only with heritable changes in gene expression occurring without the DNA sequence changing[108]. Similarly, the term **epigenetic trait** refers to a stable heritable phenotypic characteristic resulting from changes in a chromosome that exclude changes to its DNA sequence.

Figure 1.24A shows two mechanisms of epigenetic change that can be inherited via chromosome modification, without changes in the DNA sequences:

- DNA methylation (addition of a methyl group, $-CH_3$, to some cytosine nucleobases in the genome) alters the pattern of activation or inhibition of specific genes; and
- Histone modification changes the way DNA is wrapped around histones, causing DNA's three-dimensional structure to be altered, and hence affecting how readily genes in that portion of DNA can be expressed.

Other types of epigenetic changes involve non-coding RNAs[109] which control the splicing pattern of pre-messenger RNAs (Figure 1.13) or the stability of messenger RNA transcripts.

An example of an epigenetic change is the process of cell differentiation in animal embryos: a single fertilized egg divides into daughter cells that eventually differentiate into all the different types of cells in the organism such as nerve cells, muscle cells, skin cells, bone cells, liver cells, etc[110]. All cells in an organism have the same genome, but exogenous factors cause some genes to be switched on while others are inhibited at a particular time, which causes cells to differentiate into specific cell types. When these fully differentiated cells divide further, their progeny retains these particular

[107] There are many examples of phenotypic flexibility in this book. For example, we discuss phenotypic flexibility in gut function in Section 2.3.4, muscle response to training in Section 15.2.2, responsiveness to stress in Sections 19.3.4 and amphibian metamorphosis in response to temperature in Section 19.4.2.

[108] In its broader usage, epigenetics is also concerned with non-heritable changes.

[109] A non-coding RNA is an RNA molecule that is not translated into a protein. Non-coding RNAs include tRNAs, ribosomal RNAs, which are an integral part of ribosomes, microRNAs (miRNAs), which bind to mRNA transcripts and induce their degradation, and small nuclear RNA (snRNA) involved in pre-mRNA splicing.

[110] We discuss the major general types of cells in animals in Section 3.2.2.

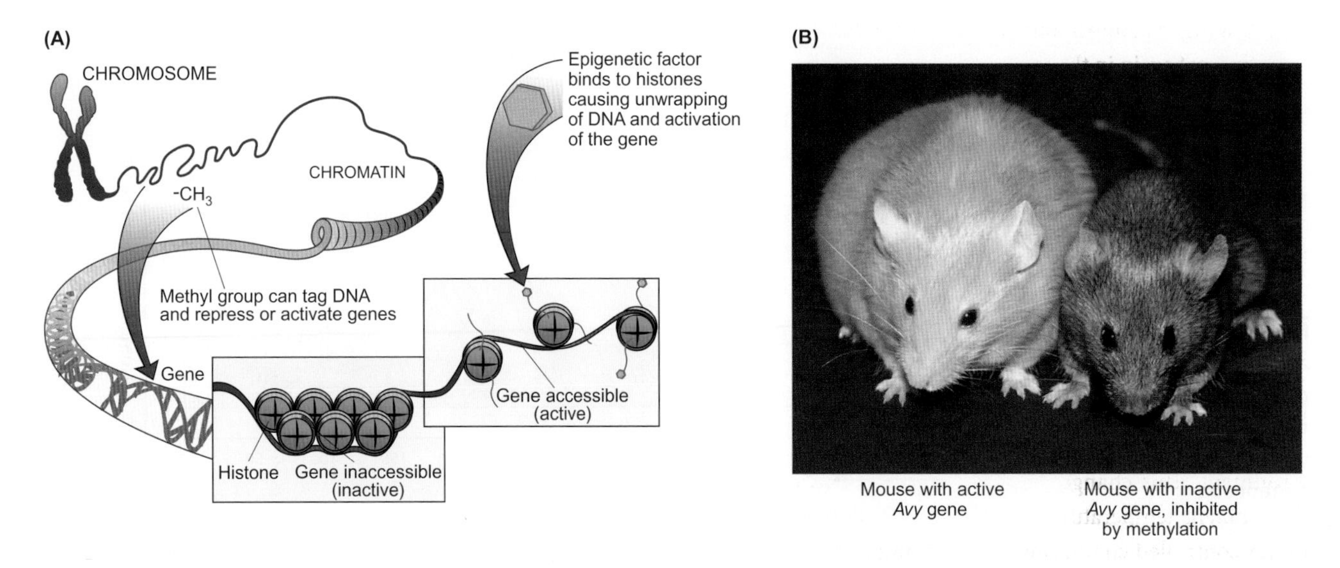

Figure 1.24 Epigenetic mechanisms that cause activation and inactivation of genes, without changing the DNA sequence

(A) Methylation of specific cytosines on the DNA chain can cause gene repression or activation. Similarly, the binding of epigenetic factors to the histone molecules alters the extent to which the DNA is wrapped around the histones causing activation or repression of specific genes.

(B) Photograph of two genetically identical mice but which differ in the methylation pattern at a specific gene locus (*Avy* gene) that impacts hair colouration, weight and predisposition to diabetes.

Sources: A: US National Institutes of Health. http://commonfund.nih.gov/epigenomics/figure.aspx B: photograph by Randy Jirtle and Dana Dolinoy / *Wikimedia Commons*

characteristics, indicating that the respective changes were inherited. Many cancers are the result of epigenetic changes to particular cells that are induced by external factors, which promote uncontrolled cell division, forming invasive tumors that eventually kill the animal.

At the level of the whole organism a variant epigenetic state must arise in the reproductive cells (gametes) and be maintained for one or more generations for it to have a heritable effect. The environment has been shown to induce chromosome modifications and changes in inherited characteristics. However, it is not easy to prove that the inherited characteristics are actually caused by chromosome modifications in the gametes—such as a change in DNA methylation or histone modification—rather than by other mechanisms.

It is likely that many heritable phenotypic differences between closely related species are mainly caused by epigenetic variants. For example, comparative analyses of DNA methylation pattern in humans and primates (chimpanzees, bonobos, gorillas and orangutans) which share 97 to 99 per cent of their protein-coding DNA, showed that hundreds of genes are differently methylated in each group, suggesting that epigenetic differences, rather than differences in DNA (and protein sequences), may be at least partly responsible for the different phenotypes observed.

A direct demonstration that a heritable phenotype change can be induced by changing the level of gene methylation was shown in mice. As illustrated in Figure 1.24B, the de-methylation of the mouse gene *Avy*, which is normally inhibited by methylation, produces a striking heritable change in phenotype: the mouse becomes obese with a characteristic yellow coat.

Epigenetic inheritance is an area of great interest to evolutionary biologists, although tracking epigenetic states over evolutionary timescales is extremely challenging.

1.5.2 Acclimatization, acclimation and evolutionary adaptation

Animals respond to changes in external conditions generally by counteracting the negative effects and enhancing the positive effects that changes have on an animal's well-being. **Acclimatization** refers to the process of adjustment of an individual animal's morphology, physiology and behaviour to a gradual change in the natural environment, which enables the animal to maintain or even enhance its performance. Acclimatization occurs over a relatively short period of time compared with the animal's lifespan. An important characteristic of acclimatization is that it is reversible if and when environmental conditions reverse.

All animals acclimatize to those environmental changes brought about by the change in seasons. For example:

- Some mammals shed their heavy and generally darker winter coats in spring, which reduces the chance of overheating in summer and increases mobility. They grow the coat again in the autumn, in preparation for winter[111].

- Humans decrease the concentration of salt in the sweat under hot conditions to reduce its loss through this pathway[112].
- Fish and aquatic invertebrates acclimatize to changes in water temperature and salinity[113] and a reduction in oxygen availability (hypoxia) within certain ranges.

While acclimatization refers to an animal's response following natural changes in the environment, **acclimation** refers to the response by an animal to artificial changes in the environment. This change is usually in one variable, such as the ambient temperature[114] or the oxygen concentration[115], under controlled conditions in a laboratory.

Unlike acclimatization and acclimation, which refer to reversible adjustments occurring during an individual animal's lifespan in response to reversible environmental changes, **evolutionary adaptation**[116] refers to changes within a population over many generations in response to persistent changes in the environment. Evolutionary adaptation has a genetic component to it. For example, the accumulation of mutations in the genome of an individual animal may impart a specific advantage to that individual animal over other individuals in the population, increasing its fitness (enhanced survival rate and reproductive capacity) by making it better suited to live and reproduce in that environment. As time progresses, there will be more descendants from that particular individual than from other individuals in the population, making the respective mutations more prevalent in the population.

Not all characters displayed by a group of animals are evolutionary adaptations for performing a particular function. For example, some neutral mutations that are neither beneficial nor deleterious may become fixed in the genome of a population by chance and turn out to be beneficial when environmental conditions change. Alternatively some characters may be evolutionary adaptations for performing a completely different function; under new conditions, such characters become important for different functions that are essential for the survival of the population. Such characters that did not evolve by natural selection for that specific function, but were already present in the population for other causes, are called **exaptations**[117].

An example of exaptation is the evolutionary adaptation of an existing body structure for a new function, such as the use of feathers for bird flight. Feathers are thought to have evolved initially for heat regulation or display behaviour in a group of terrestrial dinosaurs called theropods, from which birds evolved. Terrestrial theropods covered in feathers lived millions of years before flight evolved. Since feathers did not evolve by natural selection to enable bird flight, feathers as such are not an adaptation for bird flight, but an exaptation.

Generally, if a trait displayed by one animal group also occurs in one or more phylogenetically related animal groups living in different types of habitat, then that trait is likely an exaptation.

1.5.3 Mechanisms of evolutionary change

Evolution is defined as the change in heritable traits of a population through successive generations. Heritable traits can not only be genetic and epigenetic in nature (as we already discussed), but can also be ecologically or behaviourally adaptive. For example, feeding of individuals on different types of food may be an heritable ecological trait within a population. Examples of heritable behavioural traits in animals include behaviours for optimizing foraging techniques that are observed in rats and chimpanzees over many generations, the development of song patterns in birds, and the development of distinct routes that fish populations travel to specific sites for spawning or schooling on coral reefs.

Heritable traits within a population can be altered by four mechanisms:

- Gene mutations, which are permanent alterations in the DNA sequence of particular genes.
- Genetic drift, which causes variation in the frequency of alleles (gene variants) in the population from one generation to another generation simply due to chance.
- Gene migration (or gene flow), in which new individuals with different heritable traits join a population.
- Natural selection, in which a heritable trait increases the fitness of the population, i.e. its survival rate and reproductive capacity, such that the trait becomes more common in successive generations of the population, until it eventually spreads through the entire population and becomes a population-level characteristic.

[111] We discuss this aspect of thermoregulation in endotherms in Section 10.2.7.
[112] We discuss the control of salt loss in Section 21.1.1.
[113] Section 5.3 discusses acclimatization to salinity.
[114] Section 9.3 discusses the acclimation of fish to water temperature.
[115] Section 15.3.1 discusses the responses of aquatic organisms to hypoxia.
[116] 'Evolutionary adaptation' should be preferentially used instead of simply 'adaptation' to avoid confusion since the term 'adaptation' is also used to indicate physiological adaptation (without a genetic component) in sensory physiology as we discuss in Section 17.1.
[117] Exaptations are also called 'pre-adaptations' by some authors, a term that we try to avoid because it conveys a sense of purpose that is contrary to evolutionary change driven by natural selection.

The scientific theory of evolution by natural selection was formulated by Charles Darwin in his book *On the Origin of Species*, published in 1859. At that time little was known about the genes and chromosomes that form the basis of heredity. In Darwin's view, successive generations of a population are replaced by offspring from parents that are better adapted to survive and reproduce in a particular environment in which natural selection takes place. We now know that natural selection can only operate if there are heritable intra-specific variations, i.e. when some individuals have different heritable characteristics from others.

When environmental conditions change, larger populations possessing a more diverse set of functional traits have a greater chance of including genotypes that persist through particular fluctuations in environmental conditions. Natural selection then drives the evolutionary adaptation of the surviving genotypes to the changed environment. In this way, new phenotypes emerge in the surviving population. The significance of the concept of evolution based on natural selection lies in the proposition that evolution simply happens without plan or purpose and that every organism, as part of a population, plays a concrete role in it—however tiny it may be—without being aware of that role.

The effects of evolution are limited by different constraints

It is worth noting that although evolution through natural selection is a powerful mechanism for changing population characteristics leading to the emergence of new species, the effects of evolution are limited by different types of constraints:

- Physical constraints limit the body plan and size of organisms. For example, 10 m long cockroaches cannot exist because their exoskeleton would be too heavy to carry.
- Morphogenetic constraints allow only a limited number of ways in which an animal can develop. For example, limb formation in vertebrates can only occur in a limited number of ways; also loss of specific functional genes due to mutations limit the direction in which development can happen.
- Evolutionary constraints where the same genes affect multiple traits limit the types of evolutionary adaptations. Specific mutations in those genes may be beneficial with respect to some traits but detrimental with respect to others.

Some animal species which have changed little over long periods of time, such as the coelacanths and lungfish, might be considered to have reached an evolutionary dead end and have limited capacity to adapt to new environmental conditions. This may explain why such animals are generally restricted to narrow ecological niches.

Evolution can lead to the occurrence of similar traits in different groups of animals

When a similar trait or character is present in two or more species but was not present in their last common ancestor, then we talk about **homoplasy** (meaning same form). Homoplasy can occur as a result of **convergent evolution**, **parallel evolution** or **evolutionary reversal**.

Convergent evolution occurs when similar features evolve independently by *different pathways* in different groups of animals. Classical examples of convergent evolution include the evolution of powered flight in insects, birds and bats[118], the evolution of streamlined bodies in marine mammals and fish, lift-based swimming in turtles, tunas, penguins and dolphins[119] and the camera eyes of cephalopods (squid and octopus) and vertebrates[120].

Parallel evolution occurs when similar or identical features evolve independently by *similar pathways* in different groups of animals. For example, placental and marsupial mammals followed independent evolutionary pathways on different land masses after the geological separation of the continents, about 100 million years ago, but evolved into some surprisingly similar phenotypes. Eurasian wolves (*Canis lupus*) are similar to Tasmanian wolves (*Thylacinus cynocephalus*, roughly meaning pouched dog with a wolf's head); common moles (Talpidae family) are similar to marsupial moles (Notoryctidae family) and flying squirrels (Sciuridae family) are similar to (marsupial) sugar gliders (*Petaurus breviceps*).

Evolutionary reversal or reversal homoplasy occurs when there is a return from a more recent character to a former character/condition. An example of reversal homoplasy is the loss of limbs in legless amphibians, legless lizards and snakes.

1.5.4 Environmental factors play a key role in speciation

A key question in evolutionary biology is: how do species arise? Different species are by definition not capable of exchanging genes to produce fertile offspring. Therefore, evolutionary biologists generally believe that at an early stage in speciation two populations of the same species are prevented from interbreeding such that the two populations become reproductively isolated. Environmental factors play a key role in **reproductive isolation** between populations of the same species, which can occur in several ways. These include:

- Formation of a physical barrier due to geographical change (mountain range or island formation), which splits a population into two or more isolated groups.

[118] We discuss powered flight in Section 18.6.3.
[119] We discuss lift-based swimming in Section 18.6.2.
[120] We discuss vision in Section 17.2.1.

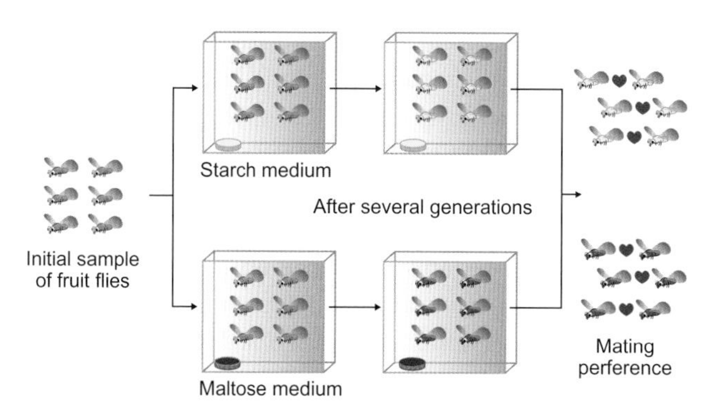

Figure 1.25 Reproductive isolation is an important step in the process of speciation

Example of reproductive isolation of two populations achieved by feeding different diets to one population of *Drosophila* flies.

Source: results from Dodd DMB (1989). Reproductive isolation as a consequence of adaptive divergence in *Drosophila pseudoobscura*. Evolution 43: 1308-1311. Drawing from http://evolution.berkeley.edu/.

- Penetration of a relatively small group of a population into an isolated area where it becomes separated from the main group but can still survive.
- The presence of two separate habitats with distinct characteristics in an area occupied by one population. The two groups drift apart and eventually become reproductively isolated.

Once reproductively isolated, two groups originating from the same animal population are subjected to different selective pressures and different occurrences of gene mutations and epigenetic modifications. Hence, the descendants of the two isolated groups become genetically distinct and eventually cannot exchange genes to produce fertile offspring; they therefore become different species.

Figure 1.25 illustrates an example of how reproductive isolation can be achieved between two populations of the same species. In this experiment, a population of fruit flies (*Drosophila melanogaster*) was split into two groups, with one fed on starch and the other on maltose[121]. When the flies were brought together after about 35 generations of isolation, the starch-fed flies preferred to mate with other starch-fed flies and maltose-fed flies preferred to mate with other maltose-fed flies: reproductive isolation had occurred. Reproductive isolation has been demonstrated for different species of fruit flies and with different types of food.

We now know that the reproductive isolation in the experiment shown in Figure 1.25 was not a consequence of genetic or epigenetic influences, but was due to the presence of different communities of microorganisms in the two types of food ingested by the flies. Different communities of microorganisms in the insects' gut led to the production of different odours, which act as pheromones[122] and are recognized by flies eating the same food. Thus, in this case, the reproductive isolation was essentially via the different types of food. This is therefore not a speciation event, but a prerequisite for a future speciation event.

The hawthorn fly (*Rhagoletis pomonella*) appears to be undergoing the beginnings of speciation in North America following the introduction of the domestic apple tree to America at the beginning of the 17th century. The change is driven by environmental factors, which induce an earlier fruiting cycle of the apple (*Malus* spp.) than of the hawthorn (*Crataegus* spp.) trees. Since that time, a population of hawthorn flies feeding exclusively on apples has evolved. This population does not interbreed with the hawthorn-feeding population and differs in several important ways from the original population. For example, the apple-feeding population grows and matures faster and is genetically differentiated from the original hawthorn-feeding population, indicating genetic divergence between the two populations.

Moreover, the population of apple-feeding flies acts as an agent for the evolution of new species of parasitic wasps. Parasitic wasps of these flies lay eggs into the body of the fly maggot, with the wasp larvae ultimately killing the fly maggot. The emergence of new species of the parasitic wasps is also driven by environmental factors: apple maggots can burrow deeper into apples than into smaller hawthorn fruit, making them better able to escape wasps, which cannot penetrate deep enough with their egg-laying ovipositor to reach the apple maggots. This example supports the view that speciation can create new opportunities for speciation of animals at a higher trophic level[123].

1.5.5 What are the possible future impacts of environmental change?

We live in a period of rapid environmental change driven directly or indirectly by human impacts causing loss of habitat through activities such as:

- Deforestation, particularly in tropical regions where about 50 per cent of the Earth's animal species live;
- Habitat fragmentation, the breaking up of a habitat into smaller parcels (called patches) that no longer interconnect;
- Pollution of land and waterways, which can make them uninhabitable;
- Infestation with invasive species which compete with native species and may destroy habitats;

[121] As we discuss in Box 2.1, maltose is a disaccharide and starch is a polysaccharide.

[122] We discuss pheromones in Section 17.3.1.

[123] We discuss trophic levels in Section 2.3; parasites are one trophic level higher than the hosts.

- Climate change, which manifests itself as a significant rise in Earth's average surface temperature and changes in the weather patterns across the globe.

Climate change is associated with a 30 per cent increase in average global carbon dioxide concentration in the atmosphere over the past 60 years[124] most likely due to human activities. According to the report of the Intergovernmental Panel on Climate Change (IPPC)[125] between 1880 and 2012 the Earth's averaged temperature over all land and ocean surfaces increased by 0.85°C, which is the fastest average rate of temperature change over the past 1000 years.

The Earth's rise in average surface temperature is argued to have caused a number of events including:

- ice loss from land-bound glaciers and ice packs
- rising sea levels
- rising sea temperature[126]
- increased incidence of extreme land weather events, such as storms, floods, drought and fires, causing shifts in vegetation and affecting the sources of fresh water available to terrestrial animals.

Such changes in the environment could exceed the ability of many animal species to migrate, acclimatize or adapt to the new conditions, causing increased rates of extinction.

Animal biodiversity is particularly sensitive to loss of habitat, which causes a reduction in the size of animal populations, and hence a decrease of intra-specific genetic variation. These changes reduce the ability to adapt to future changes in the environment, which increases the probability of extinction of animal populations. Indeed, it has been estimated that the extinction rate of species is now 100 to 1000 fold greater than before the appearance of the modern humans about 200,000 years ago.

We are clearly living through a significant evolutionary event that has a negative influence on the Earth's faunal population. Some have labelled this event as the 6th mass extinction. Conservation strategies based on a deeper understanding of how particular groups of animals function under specific conditions will help reduce the extinction rate and markedly lessen further loss of biodiversity[127]. Once a species becomes extinct it is forever, because the long chains of events responsible for its evolution can never be repeated.

› *Review articles*

Bambach RK (2006). Phanerozoic biodiversity mass extinctions. Annual Review of Earth and Planetary Sciences 34: 127–155.

Fahrig E (2003). Effects of habitat fragmentation on biodiversity. Annual Review of Ecology, Evolution, and Systematics 34: 487–515.

Kovalchuk I (2012). Transgenerational epigenetic inheritance in animals. Frontiers in Genetics 3: article 76.

Newman M (1997). A model of mass extinction. Journal of Theoretical Biology 189: 235–252.

Pimm SL, Jenkins CN, Abell R, Brooks TM, Gittleman JL, Joppa LN, Raven PH, Roberts CM, Sexton JO (2014). The biodiversity of species and their rates of extinction, distribution, and protection. Science 344: 987.

Rohde RA, Muller RA (2005). Cycles in fossil diversity. Nature 434: 208–210.

Solomon S, Plattner G-K, Knutti R, Friedlingstein P (2009). Irreversible climate change due to carbon dioxide emissions. Proceedings of the National Academy of Sciences of the United States of America 106: 1704–1709.

[124] We discuss the change in atmospheric carbon dioxide concentration in Section 11.1.1.

[125] The IPPC was formed in 1988 by the United Nations Environment Programme and the World Meteorological Organization to assess the state of scientific knowledge about the human role in climate change.

[126] We discuss rising sea temperatures in Section 1.2.1.

[127] Case Study 7.2 explores the physiological reasons for the decline of vulture population and recent conservation efforts to restore populations. We discuss in Sections 17.3.1, 17.5 and 17.7.1 how pollution interferes with sensory systems of animals, and in Case Study 10.1 we discuss how climate change may be affecting the timing of migration in some species of birds.

Checklist of key concepts

What is environmental animal physiology?

- Environmental animal physiology:
  - Investigates how animals function in different environments.
  - Provides mechanistic explanations that help predict the impacts of environmental change on the most vulnerable species.
  - Provides the basis for understanding the optimal conditions for animals.
  - Gives an understanding of the effects of environmental pollutants on animal life.

Natural environments: where and under what conditions do animals live?

- Terrestrial and aquatic **biomes** are characterized by the **abiotic** conditions and the communities of organisms present.
- Latitude distinguishes subtropical, tropical, tundra and temperate biomes.
- Two similar schemes are used to classify terrestrial biomes, ecoregions and habitats based on their climate:
  - **hydric** (wet), **mesic** (moist) and **xeric** (very dry)
  - humid, semi-humid, semi-arid and arid.

- Sea surface temperatures are relatively stable compared to those on land; since 1900 the average sea surface temperatures have risen, affecting the functioning of marine organisms.
- In aquatic habitats:
  - **benthic** animals live on the bottom sediments or in the **intertidal zone** where environmental conditions change with the ebb and flow of each tide.
  - **pelagic** animals live in the water column, mainly in the **epipelagic zone**.
- Aquatic habitats contain dissolved oxygen and nutrients that feed the plants and algae, which themselves provide food and oxygen for animals.
- **Fresh water** is usually considered to have a total salt content (**salinity**) of < 1 practical salinity unit (psu), whereas surface ocean **seawater** has salinity of 33–37 psu.
- **Brackish water** habitats of variable salinity have a major effect on the community of animals present because saline tolerances vary.
- The higher salinity of surface seawater causes it to sink, carrying dissolved oxygen into deeper water.
- At greater depths animals must cope with increasing water pressures and usually declining oxygen availability.
- Solar radiation may result in **thermal stratification** of water bodies.
- Predicted increases in atmospheric carbon dioxide causing **ocean acidification** are likely to have damaging effects on some marine animals.

How animal groups are related to each other

- **Phylogeny** refers to the inferred evolution of a genetically related group of organisms.
- A **species** is a group of animals that can interbreed in nature to produce fertile offspring.
- A **genus** is a group of species that are genetically closely related and exhibit similar characteristics.
- All life on Earth evolved from a **universal ancestor**; all animals evolved from a unicellular choanoflagellate.
- Animal **phylogenetic trees** show the inferred evolutionary relationships among different groups of animals.
- An ancestor together with all descendants (living or extinct) form a **clade**.
- The genetic material of organisms can be modified through gene mutations; **alleles** are variations of particular genes.
- The flow of information from DNA to proteins involves **transcription** of genetic information carried by DNA into mRNA in the nucleus followed by **translation** into different proteins in the cytoplasm.
- The functional diversity of proteins is enhanced by alternative splicing and post-translational modifications.
- **Genetic drift** refers to the random change in frequency of gene variants (alleles) through successive generations in the population.

Animal diversity

- **Biodiversity** refers to the number of different species and the variations within these species on the entire planet, in a biome, an ecoregion or (more commonly) a smaller area.
- Biodiversity is unequally distributed across the planet:
  - Biodiversity on land is highest in areas of high water availability than in areas of low water availability.
  - Marine habitat biodiversity is higher on the coastlines than open seas.
  - Biodiversity is higher in low latitude areas than at higher latitudes.
- All members of the **Animalia** are multicellular, but members of the four basal phyla do not have the organs and tissues that occur in members of other phyla.
- Bilaterally symmetrical animals evolved in members of three superphyla: **Lophotrochozoa** (Mollusca, Annelida, Platyhelminthes and Rotifera), **Ecdysozoa** (Arthropoda, Tardigrada and Nematoda) and **Deuterostomia** (Echinodermata, Hemichordata and **Chordata**, which includes the vertebrates).
- **Arthropods** are characterized by their body segmentation, jointed appendages and rigid waterproofed exoskeleton.
- The phylum Arthropoda is the most diverse phylum on Earth and includes hexapods, myriapods, crustaceans and chelicerates.
- All chordates have a **notochord** at some point in their life cycle.
- **Vertebrates** are distinguished from **invertebrates** by their vertebral column and internal skeleton.
- Two groups of fish (**bony fish**) evolved bony skeletons: **lobe-finned fish** (Sarcopterygii) and **ray-finned fish** (Actinopterygii).
- Ray-finned fish gave rise to a diverse range of bony teleosts that now represent almost 50 per cent of living vertebrate species.
- An ancestral lobe-finned lungfish gave rise to all **tetrapods** (amphibians, reptiles, birds and mammals).
- Mammals, except for the egg-laying monotremes, give birth to live young and have three unique physiological characteristics: three middle ear bones, lactation and suckling of newborns, and body hair that is important in temperature regulation.
- The phylogenetic groups of placental mammals suggest diversification of mammals before continents separated.

Environmental change and animal diversity

- Environmental change is both:
  - a major factor causing extinction of species
  - a major driver in the origination of new species.
- **Acclimatization** refers to reversible adjustments occurring during an individual animal's lifespan in response to reversible changes in the natural environment.
- **Acclimation** refers to an animal's response to an artificial change in the environment (in a laboratory).
- **Evolutionary adaptations** are changes over many generations within a population driven by natural selection in response to persistent environmental change.

1

- **Exaptations** are characters already present in a population that acquire functions for which they were not originally selected for.
- Intra-specific genetic, epigenetic, ecological and behavioural variations that manifest as heritable traits confer differential survival rate and reproductive capacity to populations.
- A decline in intra-specific variation reduces a population's ability to adapt to future environmental changes, which increases the probability of its extinction.
- **Natural selection** is the process by which characteristics that increase the fitness of a population become more common in successive generations of the population.
- **Evolution** is the change in heritable traits of a population through successive generations and happens without plan or purpose through gene mutations, genetic drift, migration and natural selection.
- The effects of evolution are limited by different constraints.
- Convergent evolution, parallel evolution and evolutionary reversal are different forms of **homoplasy**.
- **Reproductive isolation** of two populations of the same species is a necessary early step in the evolution of new species (**speciation**).
- Rapid environmental change driven directly or indirectly by human activities causes habitat loss and an increase in the rate of animal species extinction to unprecedented levels.

Study questions

1. Discuss the most important environmental variables for animals living in: (i) intertidal habitats; (ii) the epipelagic zone; (iii) around hydrothermal vents; (iv) the bathypelagic zone. (Hint: Section 1.2.1)
2. How is Earth's water distributed? What proportions of Earth's water occur in freshwater and marine habitats? (Hint: Section 1.2.1)
3. Explain how atmospheric carbon dioxide interacts chemically with seawater. What are the potential implications of this interaction for marine organisms over this century? (Hint: Case Study 1.1)
4. Discuss the trends in surface seawater temperatures since 1900 and their potential implications for marine organisms over the rest of the 21st century. (Hints: Section 1.2.1 and Section 9.3)
5. What do we mean by biodiversity? Outline the main global patterns of diversity and discuss some of the contributory factors thought to account for these patterns. (Hint: Section 1.4)
6. Give two examples of hypersaline environments, and discuss the animal diversity found in these environments. (Hint: Section 1.2.1)
7. Outline two schemes used to classify terrestrial biomes and habitats. (Hint: Section 1.2.2)
8. The genetic code is universal; what does this suggest about the origin of life? (Hint: Section 1.3.1)
9. Explain how genetic information passes from DNA to protein synthesis in animals. (Hint: Sections 1.3.1)
10. Explain in broad terms what a phylogenetic tree is and how phylogeny generally helps with the classification of organisms. (Hint: Section 1.3.3)
11. Explain why high coral diversity is associated with particular parts of the world. Why are corals important in the global carbon balance? (Hint: Section 1.4.1)
12. Discuss the key characteristics and diversity of animals in the following phyla: (i) Arthropoda; (ii) Mollusca; (iii) Annelida; (iv) Chordata. (Hint: Section 1.4.2)
13. What is the major difference between epigenetic- and genetic-based traits? Give two examples of epigenetic changes. (Hint: Section 1.5.1)
14. What are histones and what role can they play in epigenetic changes? (Hint: Section 1.5.1)
15. Explain the difference between acclimatization and evolutionary adaptation. (Hint: Section 1.5.2)
16. Explain what is meant by heritable intra-specific variation and the role it plays in evolution. (Hint: Sections 1.5.3)
17. Discuss why extinction of a species is forever. (Hint: Section 1.5.5)
18. Explain how environmental change associated with human activities increases the rate of species extinction. (Hint: Section 1.5.5)

Bibliography

Bromham L (2016). An Introduction to Molecular Evolution and Phylogenetics. 2nd Edition. Oxford: Oxford University Press.

Darwin C (1859). The Origin of Species. https://ebooks.adelaide.edu.au/d/darwin/charles/d22o/index.html.

Dawkins R (2006). The Selfish Gene, 3rd Edition. Oxford: Oxford University Press.

Hall BK, Hallgrimsson B (2014). Strickberger's Evolution, 5th Edition. Burlington, MA: Jones and Bartlett Learning.

Holland P (2011). The Animal Kingdom. A Very Short Introduction. Oxford: Oxford University Press.

Miller GT, Spoolman SE (2016). Environmental Science. 15th Edition. Boston, MA: Cengage Learning.

Minelli A (2009). Perspectives in Animal Phylogeny and Evolution. Oxford: Oxford University Press.

Telford MJ, Littlewood DTJ eds. (2009). Animal Evolution: Genomes, Fossils, and Trees. Oxford: Oxford University Press.

2

Energy metabolism: generating energy from food

Animals need a continuous supply of energy to stay alive. Energy is required for essentially all biological processes, including the synthesis of new molecules and structures for growth, reproduction, tissue maintenance and regeneration; circulation of the blood; salt and water balance; excretion; sensory perception; and coordination of body activities. Hence, energy is a vital resource that animals must obtain from their environment.

This chapter focuses on how animals meet their bodies' demands for energy and on the common processes involved in the flow of energy, a process that starts with the acquisition of food and sees the transfer of energy from food to molecules of adenosine triphosphate (ATP)[1], the common currency for energy transfer and utilization in animal cells.

2.1 What is energy?

For animals, energy is the currency of life. Like any other currency, it can be acquired and spent over a relatively short period of time; alternatively, it can be stored and used at a later date, as we see in birds or mammals before migration or hibernation. There are different forms of energy, such as mechanical energy associated with motion, chemical energy associated with bonds between atoms, electrical energy associated with charged particles and thermal energy associated with internal movement of molecules. Energy can be transformed from one form to another, but cannot be created or destroyed. This fact is described by the first law of thermodynamics[2], also known as the law of energy conservation. For example, the mechanical energy produced by muscles is obtained from the chemical energy stored in ATP molecules (which is itself transferred to ATP from foodstuffs).

The level of order (or disorder) in a system is described by its entropy[3]. An ordered state has low entropy and a disordered state has high entropy. According to the second law of thermodynamics, the entropy of an isolated system—i.e. a system that cannot exchange energy or matter with its environment—has a tendency to increase: in other words, an isolated system tends to become disordered.

However, all living animals are open systems rather than isolated systems for which an unchecked increase in disorder is not compatible with the successful maintenance of life. As such, living organisms use a continuous input of energy and materials from nutrients in their environment to oppose or even reduce their entropy (disorder) to maintain (or even enhance) their level of order—for example, during growth and development.

Even though the entropy of animals, as open systems, does not increase, they do not violate the second law of thermodynamics: the entropy of the universe as a whole—of which animals are part—continues to rise inexorably.

The thermal energy of a substance determines how fast its molecules move, vibrate or rotate; the faster the motion, the higher the thermal energy and the higher the temperature of that substance, as illustrated in Figure 2.1. This random movement of molecules is called thermal motion. The temperature at which substances have no thermal energy and at which the molecules do not move is called **absolute zero** (0 kelvin[4], K) and corresponds to –273°C[5].

A change in thermal energy is associated with the transfer of heat. Figure 2.1 shows that heat always flows from regions of higher temperature to regions of lower temperature. This is because of the difference in thermal motion between the two regions: more molecules in the higher-temperature region, which are faster-moving, move into the region of lower

[1] We discuss the key role played by ATP in energy transfer and usage in Section 2.2.1.

[2] Thermodynamics is the branch of physical science that deals with the relations between all forms of energy.

[3] Entropy is a thermodynamic parameter which is measured in $J\ K^{-1}$ as shown in Appendix 1.

[4] Appendix 1 contains information on the various scales and units used for temperature measurements. For example, 20°C corresponds to an absolute temperature of 273 + 20 K = 293 K.

[5] More precisely–273.15°C.

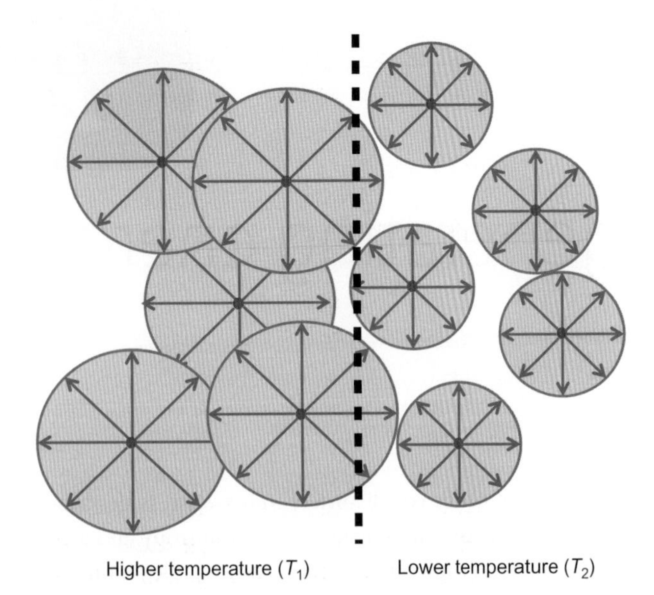

Figure 2.1 Diagrammatic representation of heat conduction from higher to lower temperature

Molecules, shown as red dots, are surrounded by arrows pointing in all directions to indicate that they move randomly. The circle around each molecule represents the space within which the molecule can be found after a short period and the size of the circle around the molecule indicates that the molecule can travel further at the higher temperature (T_1) than at the lower temperature (T_2). At higher temperatures, the molecules move faster and the thermal energy is greater. The dotted line can be crossed within the given time by two molecules from left to right, but only by one molecule from right to left. When this happens, there is a net transfer of thermal energy across the dotted line from left to right, i.e. from a region of higher temperature to a region of lower temperature.

temperature than move in the opposite direction. As a result, thermal energy is transferred from a region of higher temperature to a region of lower temperature.

The unit for energy in the international system of units (SI)[6] described in Appendix 1 is the joule (J). Another widely used unit for energy, commonly added to packets or tins containing processed food, is the calorie (cal). The energy content of food is normally expressed in kcal (1000 cal) or Cal. One calorie (cal) is the amount of energy necessary to increase the temperature of 1 g water by 1°C (from 14.5 to 15.5°C) and is equivalent to 4.1840 J.

2.1.1 What forms of energy do animals use?

The forms of energy that are used by animals are:

- Chemical energy
- Mechanical energy
- Electrical energy
- Radiation energy
- Thermal energy

All animals extract chemical energy stored in nutrients by changing the chemical bonds between atoms in food substances.

Animals use mechanical energy for movement and locomotion, for lifting or carrying items, for pumping blood around the body and for the filtering of body fluids associated with excretion. Mechanical energy in the form of soundwaves produced by animals themselves is also used for communication between them.

Electrical energy occurs when positive electrical charges carried by ions, such as protons (H^+), sodium (Na^+) and potassium (K^+), are separated from negative charges by biological membranes[7]. As such, electrical energy drives the synthesis of ATP, the transport mechanisms that regulate the composition of the body fluids[8], and the functioning of the nervous system and sensory organs[9].

Radiation energy (usually called radiant energy) in the form of electromagnetic waves is important for temperature regulation, particularly in some species of ectotherms[10]. Solar radiation is used by a number of species to raise their body temperature before embarking on various activities such as feeding, avoiding a predator, finding a mate. Radiation energy also plays the predominant role in animal vision[11].

Thermal energy in animals flows in the form of heat between different parts of their body or between the animal's body and its environment, always from a region of higher temperature to a region of lower temperature. Changes in thermal energy are also associated with various reactions in the body that increase or decrease the random movement of molecules. Body temperature rises if the heat gain is greater than the heat loss, but temperature decreases when the heat gain is less than the heat loss. So the temperature of an organism depends on the balance between the rate of heat gain and the rate of heat loss[12].

Chemical and electrical forms of energy are generally regarded as high-grade forms of energy, because they can be used for powering physiological processes. In contrast, thermal energy is considered low-grade energy because its

[6] SI is an abbreviation of *Le Système Internationale d'Unités* (in French), commonly known as the metric system.

[7] Electrical energy carried by ions is discussed in Box 16.1.
[8] The processes involved in body fluid regulation are discussed in Part 2.
[9] We discuss in Chapters 16 and 17 how the nervous system and sensory organs function.
[10] Ectotherms are animals that mainly use external sources of energy to regulate their body temperature as we discuss in Chapter 9.
[11] Vision is discussed in section 17.2.1.
[12] Part 3 discusses heat balance in animals.

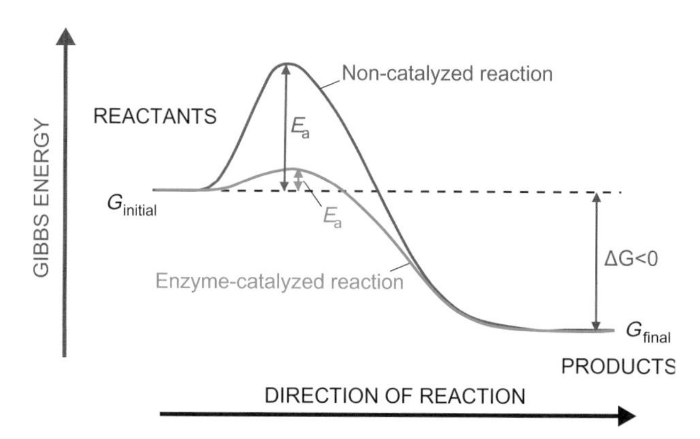

Figure 2.2 The Gibbs energy change ($\Delta G = G_{final} - G_{initial}$) in an exergonic chemical reaction

The chemical bonds that are broken in the reactant molecules are weaker (and therefore have a higher Gibbs energy) than the new bonds that are formed in the products of the reaction. For the reaction to progress, an activation energy (E_a) barrier needs to be crossed before energy is released. An enzyme that is specific for a particular reaction greatly increases the velocity of the reaction by lowering the activation energy, but it does not alter the total amount of energy released (ΔG).

effective conversion to other forms of energy depends on large differences in temperature (tens to hundreds of °C) between different parts of the system—a level of difference that simply does not exist in animals. Nevertheless, heat production is physiologically important for controlling body temperature, which influences the velocity of biochemical reactions[13].

2.1.2 Biochemical reactions produce or absorb energy

When biochemical reactions take place, energy is either released to or absorbed from the environment, depending on the relative energies of the chemical bonds that form and break during the course of the reaction. The amount of energy released or absorbed in processes that occur at constant temperature and pressure (as is the case for most biochemical and physiological reactions in animals) is called the change in **Gibbs energy (ΔG)**[14]. Animals extract energy from nutrient molecules in many reaction steps by gradually transferring the chemical energy contained in their food either directly or indirectly to ATP. Significantly, the change in Gibbs energy (ΔG) is the same irrespective of whether the reaction takes place in one step outside an animal's body, or in many steps within the animal's body.

Importantly, ΔG gives an indication of whether a reaction is feasible or not. ΔG is negative when the final state of a process or reaction has less Gibbs energy than the initial state. A reaction with a negative ΔG is feasible without energy input from outside—that is, it tends to happen spontaneously. This is because the system moves towards a more stable state, which is characterized by less Gibbs energy. This type of process/reaction is **exergonic** (from Greek *exo* meaning outside and *ergon* meaning work) and energy released from exergonic reactions can be used to initiate other reactions, form new bonds, move molecules against a difference in concentration or perform mechanical work. However, it is always the case that some of the released chemical energy cannot be used to perform work and is instead released as heat. If this energy is not lost to the outside, it spreads out as heat within the body and may be used by an animal to help regulate temperature.

Figure 2.2 illustrates the change in Gibbs energy during a theoretical exergonic reaction. Notice how, for the reaction to proceed, the reactants need to pass through an intermediate state characterized by a higher Gibbs energy than that of the initial state (e.g. two reactant molecules need to collide head on). Only then can the reaction progress, forming products with a lower Gibbs energy than the initial reactants. The difference in Gibbs energy between the initial state and the intermediate state is called the **activation energy (E_a)**, as illustrated in Figure 2.2[15].

Biochemical reactions are catalysed by enzymes, which act to reduce the activation energy of a given reaction, as shown in Figure 2.2. For example, an enzyme molecule may bind two reactant molecules close to each other thereby facilitating their reaction without them needing to collide. When the activation energy is reduced, the velocity of reaction increases without a change in ΔG.

If ΔG for a reaction is positive—that is, the final state has a greater Gibbs energy than the initial state—the reaction or process cannot spontaneously proceed without energy input from outside. We say the process is **endergonic** (from Greek *endo* meaning within) because it absorbs energy in order to happen. Protein synthesis and ion transport by ATP-driven ion pumps[16] are examples of endergonic reactions.

A different pair of terms, **exothermic** and **endothermic**[17], refer to the release and absorption of energy in the form of heat. For example, evaporation and melting are endothermic processes because heat is absorbed from the environment as the reactions proceed. Conversely, the oxidation of foodstuffs is exothermic: heat is released.

[13] The effect of temperature on biochemical reactions is discussed in section 8.2.1.

[14] Willard Gibbs (1839-1903) was an American scientist who introduced the concept of energy for processes that occur at constant temperature and pressure. Gibbs energy is also known as free enthalpy.

[15] Activation energy and its importance in biological processes is discussed in Section 8.1.

[16] ATP-driven ion pumps are essential for cell function as we discuss in Section 3.2.5.

[17] An endothermic process should not be confused with endotherms.

2

2.1.3 Equilibria and steady states

When a system is at **thermodynamic equilibrium**, there is no tendency for spontaneous change and the macroscopic properties of the system do not alter with time. At thermodynamic equilibrium the system is simultaneously in thermal, mechanical, chemical, electrical and radiative equilibrium.

Animals cannot be at thermodynamic equilibrium. However, they can be in *one* kind of equilibrium at a given time. For example, a snail can be in thermal equilibrium, when the temperature is uniform throughout its body and is same as the ambient temperature. However, a living snail can never reach chemical equilibrium: this would imply that all of the different molecules of which it is composed were uniformly distributed throughout its body!

No energy input is needed to maintain a particular kind of equilibrium. However, in order to survive, animals need to maintain stable conditions that are far from equilibrium as we learn throughout the book. Such stable conditions, called steady states, are achieved by physiological processes that continuously use significant amounts of energy. For example, all animal cells continuously use considerable amounts of energy to ensure the composition of their internal environment, which is different from that of their extracellular environment, remains stable. These differences are essential for cells to function, as we discuss in Section 3.2.5.

Further reading for Section 2.1

› Review articles

Dyson FJ (1954). What is heat? Scientific American 191: 58–63.

Schneider ED, Kay JJ (1995). Order from disorder: The thermodynamics of complexity in biology. In: *What is Life? The Next Fifty Years: Speculation on the Future of Biology*. Eds Murphy MP, O'Neill LAJ. Cambridge: Cambridge University Press, pp. 161–173.

Crowe J, Bradshaw T (2014). Chapter 14 in Chemistry for the Biosciences. The Essential Concepts. 3rd Ed. Oxford: Oxford University Press

2.2 Metabolism, energy metabolism and metabolic rates

Metabolism consists of all the reactions within an animal which allow it to stay alive and respond to changes in its environment. Metabolic reactions are involved in a number of different processes:

- breakdown of food particles into smaller molecules in the alimentary canal to facilitate their absorption across the gut epithelium[18],
- uptake of substances into cells, transport between cells, and storage,
- synthesis of proteins such as enzymes, and structural and transport proteins,
- removal of waste products from cells and formation and excretion of nitrogenous waste[19].

Intermediary metabolism refers to the whole set of biochemical reactions that take place within cells and include those involved in the synthesis of cellular components and ATP production.

Metabolism can be subdivided into **catabolism** and **anabolism**.

- Catabolism encompasses the metabolic processes whereby larger molecules such as food particles are broken down into simpler molecules in enzyme-catalysed reactions. Catabolic reactions are mostly exergonic; some of the chemical energy released is utilized by cells and the rest is dissipated as heat.
- Anabolism comprises the metabolic processes responsible for the synthesis of new, more complex molecules from simple precursor molecules. Such processes include protein synthesis from amino acids or the assembly of energy storage compounds such as glycogen from glucose[20]. Most anabolic reactions are endergonic, requiring chemical energy to proceed.

2.2.1 Metabolic reactions result in energy flow

The energy changes associated with the various biochemical reactions that take place in an animal's body and with the interactions between an animal and the environment are collectively known as **energy metabolism**.

Adenosine triphosphate (ATP) is the common currency for energy transfer and utilization in animal cells

The molecular unit for energy transfer in all animal cells is the ATP molecule. As shown in Figure 2.3A, ATP consists of adenine attached to ribose and three phosphate (P_i) groups. The bond between the last two phosphate groups can be relatively easily broken by ATP-ase enzymes. These enzymes use one molecule of water to hydrolyse an ATP molecule into one ADP (**adenosine diphosphate**) molecule

[18] We discuss the structure and function of epithelia (plural of epithelium) in Section 4.2.

[19] Section 7.4 discusses the nitrogenous excretion of animals.

[20] We discuss carbohydrates and proteins in Box 2.1.

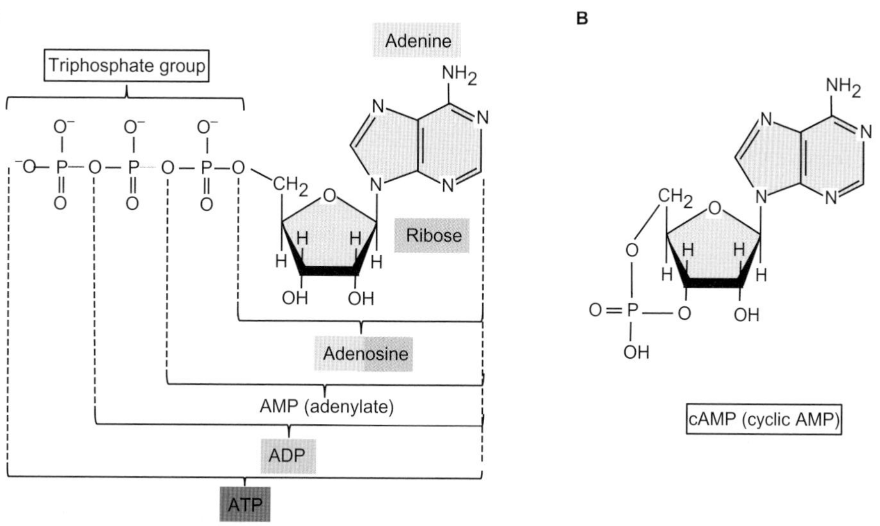

Figure 2.3 General structure of the ATP, ADP, AMP and cyclic AMP molecules

(A) The ATP molecule consists of adenine (a purine base) attached to ribose (a 5-carbon sugar) and three phosphate (P_i) groups attached to each other. The bond between the second and the third phosphate groups (shown in red) can be relatively easily broken with the formation of one ADP (adenosine diphosphate) molecule and one inorganic phosphate molecule (P_i) and release of about 50–55 kJ mol^{-1} Gibbs energy under physiological conditions. The bond between the second and the first phosphate (shown in pink) can also be broken with formation of AMP and inorganic phosphate and releases a similar amount of Gibbs energy. (B) The structure of cyclic AMP (cAMP).

and one inorganic phosphate molecule (P_i), as indicated in Figure 2.3A. The Gibbs energy liberated by hydrolysing ATP to ADP and P_i is about −50 to −55 kJ mol^{-1} in a resting cell.

ATP:

- drives many different types of endergonic metabolic reactions, such as transport of molecules and ions against differences in concentration, protein synthesis and animal movement,
- is used by kinase enzymes to attach a phosphate group (P_i, PO_4^{3-}) to specific chemical groups on proteins, carbohydrates and lipids; this type of reaction, called phosphorylation, plays an important role in the regulation of protein function such as enzyme activity,
- acts as a signalling molecule in various processes involving the nervous system[21] and urine formation[22],
- is used to produce cyclic AMP (cAMP, shown in Figure 2.3B), which is an important second messenger that regulates many cellular processes[23],
- is involved in DNA replication.

ADP can be further hydrolysed to AMP (adenosine monophosphate) and P_i, as indicated in Figure 2.3A, with the further liberation of energy. Processes that either consume or produce ATP are sensitive to changes in the ratio between AMP and ATP concentrations (the AMP/ATP ratio). Cells use the AMP/ATP ratio to sense how much energy is available to them and generally regulate their function so that the AMP/ATP ratio is maintained within strict limits.

The process of oxidation liberates energy from foodstuffs

Animals usually obtain the energy they need to survive from the oxidation of organic compounds ingested in foodstuffs. Energy is predominantly obtained by oxidizing carbohydrates and triacylglycerols (fats) in the process of **aerobic respiration**, which occurs in the mitochondria[24], where most ATP is produced. Box 2.1 summarizes the biochemical characteristics of carbohydrates and lipids (which include triacylglycerols).

In strict chemical terms, oxidation refers to the loss of electrons, while reduction refers to the gain of electrons. In less strict terms, oxidation is defined as a gain of oxygen and reduction as a loss of oxygen. For example, the formation of water by burning hydrogen (H_2) in the presence of oxygen (O_2) involves the oxidation of two hydrogen atoms, each of which loses one electron, and the reduction of one oxygen atom, which gains two electrons to form one water molecule (H_2O). Alternatively, we can say that hydrogen is oxidized when water is formed because it gains oxygen.

Animals which live in environments lacking oxygen, such as intestinal parasites and animals that live around

[21] We discuss how ATP can act as a transmitter molecule in chemical synapses in Section 16.3.4.

[22] We discuss urine formation in kidneys in Section 7.4.

[23] We discuss how cAMP is used as a second messenger in section 19.1.4 and Figure 19.6.

[24] Mitochondria are cellular organelles found in large number in the cells of most animals, as we discuss in Section 2.4.

hydrothermal vents on the ocean floor, can still oxidize food in the absence of oxygen (i.e. there is still a loss of electrons) by respiring anaerobically, using processes which we discuss later in this chapter. **Anaerobic respiration** may also be used during relatively short bursts of high-intensity exercise[25].

[25] Burst exercise is discussed in section 15.2.4.

When carbohydrates and fatty acids are oxidized in the presence of oxygen (O_2), carbon dioxide (CO_2) and water (H_2O) molecules are the end products.[26] During such

[26] In addition to the energy production, the water produced by aerobic metabolism, which is known as **metabolic water**, is important for the water balance of some animals, particularly when water supplies on land are restricted, as we discuss in Section 6.2.1; oxidation of fat is the primary source for metabolic water production.

Box 2.1 Carbohydrates, lipids and proteins are essential biomolecules for animal function

Carbohydrates

Carbohydrates are produced in green plants from carbon dioxide and water by photosynthesis, with sunlight providing the necessary energy for the reaction to proceed. It is difficult to overstate the importance of carbohydrates to animal life and energy metabolism: most plant material consumed as food is made out of carbohydrates in the form of starches, sugars or fibres from the cell wall of plants.

Carbohydrates occur in four chemical forms:

- **monosaccharides**—simplest forms of carbohydrate molecules, such as glucose, fructose and galactose shown in Figure A and ribose shown in Figure 2.3;
- **disaccharides**—two monosaccharide molecules linked together, as in sucrose, lactose, and trehalose shown in Figure B;
- **oligosaccharides**—comprising three to nine monosaccharides linked together. Figure C shows raffinose, which consists of three monosaccharides (galactose, glucose and fructose) and is found in whole grains, beans, cabbage and other vegetables;
- **polysaccharides**—comprising large numbers of monosaccharides linked together. Examples include **starches, glycogen** and **cellulose**, which are all polymers of repeating molecules of glucose.

When in solution, monosaccharides occur mainly as cyclic structures; Figure A shows the cyclic forms of glucose, fructose and galactose, which are the main building blocks of naturally occurring carbohydrates in food. Although glucose, fructose and galactose have the same general formula, ($C_6(H_2O)_6$), the spatial arrangements of the hydroxyl (–OH) groups differ, resulting in right-handed D forms (from the Latin '*dexter*' meaning right side) and the left-handed L forms (from the Latin '*laevus*' meaning left side). Naturally occurring carbohydrates are D forms; L-type forms cannot normally be metabolized by animals.

The position of side groups varies with respect to the C-ring, as shown in Figure A, resulting in α- and β-configurations, which confer different physical and chemical properties to the individual molecules.

Two monosaccharides link together by a **glycosidic bond**, as shown in Figure B, to form disaccharides. These disaccharides include:

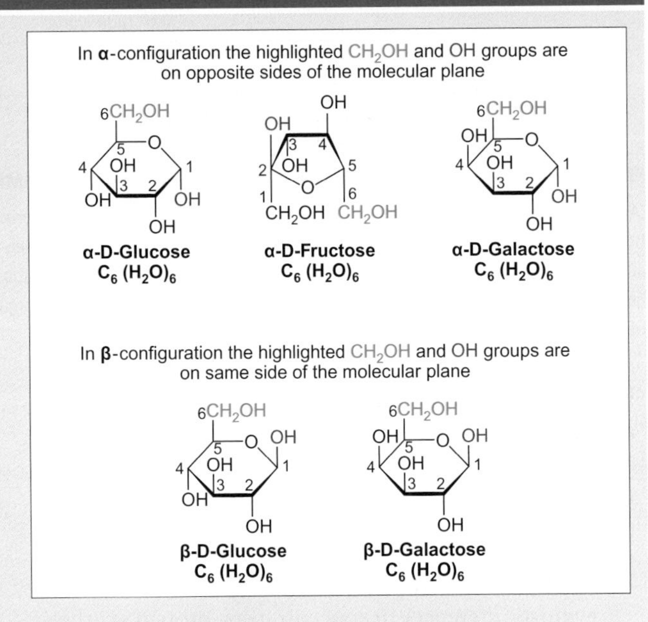

Figure A The α- and β-forms of the cyclic structures of naturally occurring monosaccharide molecules

Examples of naturally occurring monosaccharide molecules in different α- and β-configurations; the position of the carbon atoms in the respective molecule is indicated by numbers. The α- and β-configurations of monosaccharides in solution are not fixed, as one molecule of monosaccharide can change from one type to the other type of configuration. All shown monosaccharides have the same general formula ($C_6(H_2O)_6$), but, individually, they have distinct chemical properties due to differences in their 3D structure.

- sucrose (table sugar), consisting of one molecule of α-D-glucose and one molecule of β-D-fructose joined by an α-1, β-2 glycosidic bond;
- lactose (milk sugar, found in the milk of all female mammals), the disaccharide of β-D-galactose and β-D-glucose, joined by a β -1,4 glycosidic bond;
- trehalose, the disaccharide of two α-D-glucose molecules joined by a α-1,2 glycosidic bond is the main form of carbohydrate in the haemolymph of insects.

Oligosaccharides are also linked together by glycosidic bonds, as shown in Figure C for raffinose.

Amylose is a starch consisting of 300 to 3000 glucose units connected in a linear fashion by α-1,4-disaccharide bonds

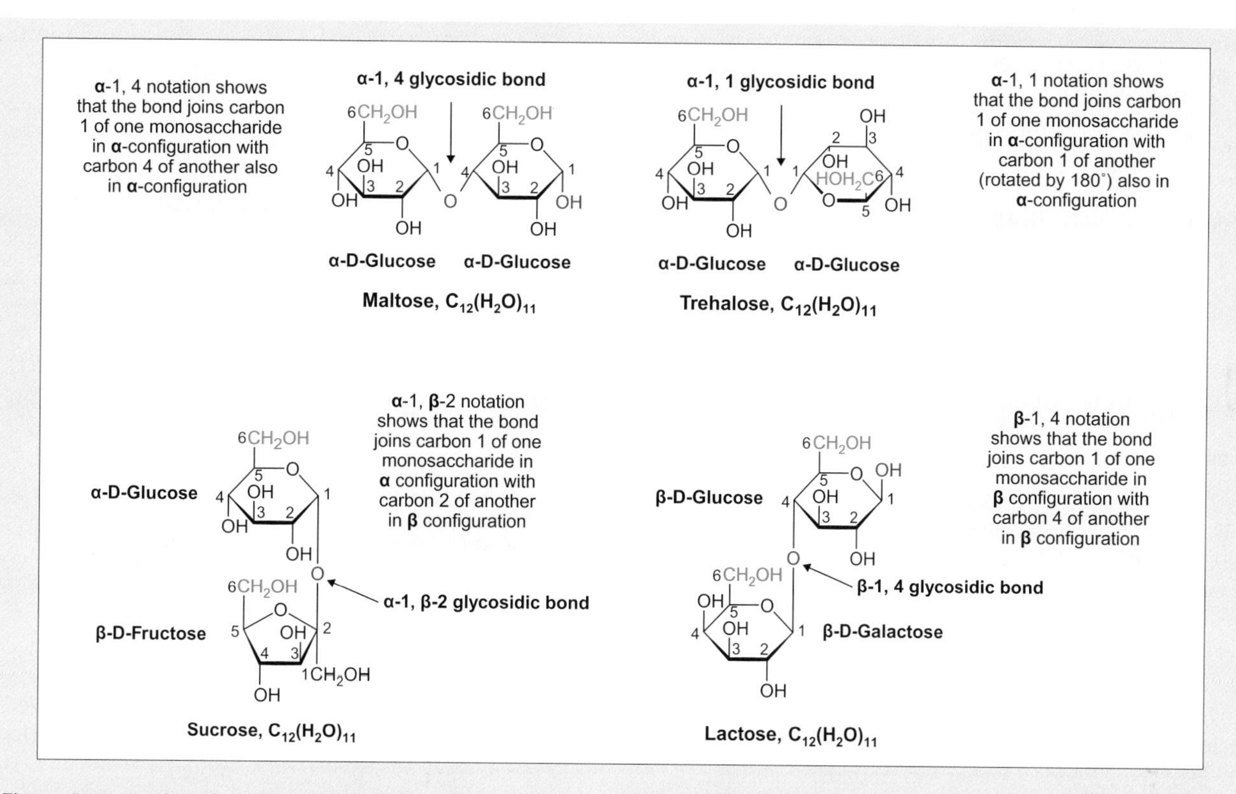

Figure B Examples of disaccharides of importance to animals with same general formula $C_{12}(H_2O)_{11}$, but different glycosidic bonds and distinct chemical properties

The formation of glycosidic bonds between two monosaccharides involves the removal of one molecule of water; such bonds are possible only between specific configurations (α or β) and carbons of the respective monosaccharides, fixing the monosaccharides into specific configurations.

shown in Figure B. **Amylopectin** is a more complex starch which contains up to 30,000 glucose units linked by a combination of α-1,4 and α-1,6-disaccharide (glycosidic) bonds. An example of a α-1,6-glycosidic bond is shown in Figure C. Starchy foods include all grain products, potatoes, beans, lentils, beets, pumpkin, parsnips and carrots.

Figure C Molecular structure of the oligosaccharide raffinose consisting of three monosaccharide units (α-D-galactose, α-D-glucose and β-D-fructose)

Cellulose is a rigid linear polymer of up to 5000 glucose units connected by β-1,4 disaccharide bonds. Cellulose is the main structural support material in plants and represents more than 50 per cent of the total organic matter in the world. Most animals do not possess enzymes that break the β-glycosidic bonds in cellulose.

Glycogen is the major form of carbohydrate stored in animals. It has a similar chemical structure to amylopectin, with branches arising about every 10 glucose units. Glycogen exists in the form of granules developed around a protein **glycogenin** at its core and can have millions of glucose units in its molecule.

Chitin is a structural polysaccharide, from which the exoskeleton of arthropods is made. It has a linear structure similar to cellulose and is made of nitrogen-modified glucose units, linked by β-1,4 glycosidic bonds.

Lipids

Lipids are organic compounds that, like carbohydrates, consist of carbon, hydrogen and oxygen. Lipids are generally insoluble in water and occur in food as molecular aggregates held together by hydrophobic (water-hating) interactions. Lipids have an important role in energy metabolism, energy storage, signal communication and formation of cellular membranes[1].

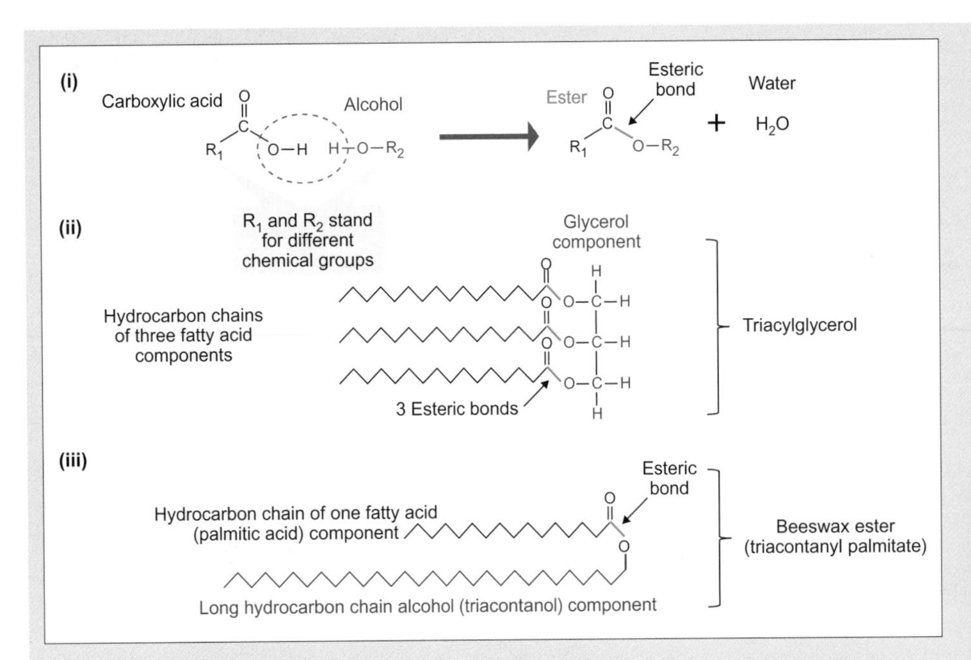

Figure D Structural basis of triacylglycerols and waxes

(i) Formation of esteric bonds between carboxyl and hydroxyl groups; (ii) structural representation of triacylglycerols; (iii) structural representation of a wax molecule (beeswax ester).

Five of the most important types of lipids in animals are **fatty acids, triacylglycerols** (or triglycerides), **glycerolphosholipids** (or phosphoglycerides), **waxes** and **steroids**.

Fatty acid molecules have a carboxyl group (–COOH) at one end called the head, connected to a long hydrocarbon chain called the tail. Saturated fatty acids have single bonds between all pairs of carbons, while unsaturated fatty acids have one or more double bonds between carbon atoms.

As we discuss in Box 9.2, the composition of a fatty acid is indicated by a pair of numbers separated by a colon: the first number indicates the number of carbons present; the second number indicates the number of double bonds. Using this notation, the most common naturally occurring saturated fatty acids are the **18:0** and the **16:0** fatty acids, commonly known as **stearic** (also known as octadecanoic) and **palmitic** (hexadecanoic) fatty acids, respectively. The most common unsaturated fatty acid is **oleic** or **18:1 (9)** (9-octodecanoic) fatty acid. The number in parentheses indicates that the double bond is between carbons 9 and 10 (when counting from the carboxyl group).

Fatty acids occur rarely as free molecules in the body of animals because their carboxylic group is normally linked either to an alcohol group via a so-called **esteric bond**, as shown in Figure D (i), or to an amino group.

Triacylglycerols are commonly called fats or oils depending on their state at room temperature: fats are solid at room temperature, whereas oils are liquid. Triacylgycerols are derived from two compounds, **fatty acids** and **glycerol**, as shown in Figure D (ii). Triacylglycerols are the molecules of choice for energy storage in animals as the hydrophobic nature of fats allows them to aggregate as body fat that contains very little water. Complete oxidation of 1 g of tissue rich in fat (triacylglycerols) yields several fold more useable chemical energy than oxidation of 1g of tissue rich in proteins or carbohydrates, as indicated in Table 2.1.

Glycerophospholipids are a very important class of so-called **phospholipids**, shown in Figure 3.10. Phospholipids are the major components of all biological membranes.

Waxes are esters of fatty acids and long chain alcohols called fatty alcohols, as shown in Figure D (iii). Waxes are particularly prevalent in marine organisms.

Steroids are a class of lipids derived from cholesterol, which is the precursor of steroid hormones[2], including the androgens, oestrogens, progestins, glucocorticoids and mineralocorticoids.

Proteins

Proteins constitute more than 50 per cent of the dry material in animal cells and are directly involved in the structure and function of all living organisms. More specifically, proteins:

- act as enzymes, which catalyse diverse metabolic reactions;
- form complex cellular structures;
- transport molecules between different cellular compartments and between different parts of an animal's body;
- convert chemical energy into mechanical energy;
- respond to specific stimuli;
- replicate the genetic material;
- transport respiratory gases in blood.

Proteins consist primarily of carbon, hydrogen, nitrogen and oxygen and are produced in all animal cells by linking together **amino acids** in a sequence dictated by information encoded in the animal's genes, as indicated in Figure 1.13. A protein molecule consists of at least 20–30 amino acids[3] and can reach a molecular mass up to 10^6 dalton (D).

Twenty amino acids are encoded by DNA; these 'standard' amino acids are listed in Table A. Each of these amino acids has both the amino group ($-NH_2$) and the carboxylic group (–COOH) attached to the first (α-) carbon atom that generally

Table A Naturally occurring amino acids in proteins coded by DNA

Amino acid	Three letter code	Amino acid	Three letter code	Amino acid	Three letter code
alanine	ala	glycine	gly	proline	pro
arginine	arg	histidine*	his	serine	ser
asparagine	asn	isoleucine*	ile	threonine*	thr
aspartic acid	asp	leucine*	leu	tryptophan*	trp
cysteine	cys	lysine*	lys	tyrosine	tyr
glutamine	gln	methionine*	met	valine*	val
glutamic acid	glu	phenylalanine*	phe		

The asterisks indicate the nine amino acids called 'essential amino acids' because they cannot be synthesized by the human cells, and therefore are essential in the diet of humans. The essential amino acids can vary between different species. For example, rats also require arginine in their diet, in addition to those required by humans. Source: Berg JM et al (2011). Biochemistry, International edition. New York: WH Freeman.

also bears one hydrogen atom, as illustrated in Figure E. The fourth bond of the carbon is attached to a chemical group called the side chain, denoted R. There are 20 different amino acid side chains, one for each of the 20 standard amino acids found in animal proteins. The side chain can feature a number of active groups including nitrogen-containing or carboxylic groups, hydroxyl groups (–OH), or sulfur-containing groups (–SH).

All amino acids (except for glycine whose 'side chain' comprises a single hydrogen atom) can have L-type or D-type spatial configurations as explained in Figure E. All amino acids found in proteins have the L-type configuration shown in Figure E[4].

The amino acids in proteins are linked together by **peptide bonds** to form **polypeptide chains**. The peptide bonds are formed during an endergonic reaction (requiring energy from the outside) called a condensation reaction during which the carboxylic group of one amino acid reacts with the amino group of another amino acid to release one molecule of water as shown in Figure F.

Proteins are made of one continuous polypeptide chain or from a cluster of several polypeptide chains. The sequence of amino acids making up a protein's polypeptide chain (or chains) is known as its **primary structure.**

In solution, under physiological conditions, the carboxylic groups (–COOH) at one end of the polypeptide chains and on the side chains of the amino acids lose their protons to become negatively charged ($-COO^-$). In contrast, the amino groups ($-NH_2$) at the other end of the polypeptide chains and on the side chains of amino acids gain protons to become positively charged ($-NH_3^+$). At physiological pH, neutral amino acids carry an equal number of positive ($-NH_3^+$) and negative charges ($-COO^-$); cationic amino acids arginine, lysine and histidine have a net positive charge (because they have extra amino groups on the side chain); and anionic amino acids glutamic and aspartic acids carry a net negative charge (because they have an extra carboxylic group on the side chain).

The attractive electrostatic forces that operate between the (positively charged) protons attached to the amino groups and the negatively charged $-COO^-$ groups give rise to so-called **hydrogen bonds**. Hydrogen bonds also form between the carbonyl (C=O) and the amino (NH) groups along the backbone of the polypeptide chains, shaping local segments on the polypeptide chains into particular configurations as shown in Figure G. These particular configurations form what we call the **secondary structure** of a protein. The most common secondary structures caused by hydrogen bonds along the backbone of polypeptide chains are the so-called **α-helices** and **β-sheets** shown in Figure G.

The **tertiary structure** of a polypeptide chain refers to its overall three-dimensional shape. For example, Figure H shows the three-dimensional structure of the enzyme amylase in human saliva, which plays a key role in the digestion of starches and consists of a single polypeptide chain. The tertiary structure is typically stabilized by strong interactions between adjacent segments of a polypeptide chain—for example, the formation of disulfide bridges (–S–S– bonds) between cysteine residues on adjacent segments—that cause different segments of the polypeptide chain to become covalently bonded together. These types of interaction cause folds, bends and loops in the polypeptide chain.

The highest level of structural organization in proteins, the **quaternary structure**, refers to the way in which multiple subunits of a protein cluster together to form either a globular structure like haemoglobin, as we discuss in Section 13.1.1, an ionic channel structure as we discuss in Chapter 16 (Figure 16.23), or a fibrous structure like the **collagen** or the **myosin filament** in muscle, as discussed in Section 18.1.

The amino acid side chains in a polypeptide chain can also be chemically modified, which alters the interactions between different segments of the polypeptide chain, the tertiary structure and functional properties of the protein. The polypeptide chains of proteins can also become bonded to non-peptide groups, called **prosthetic groups**, like sugars

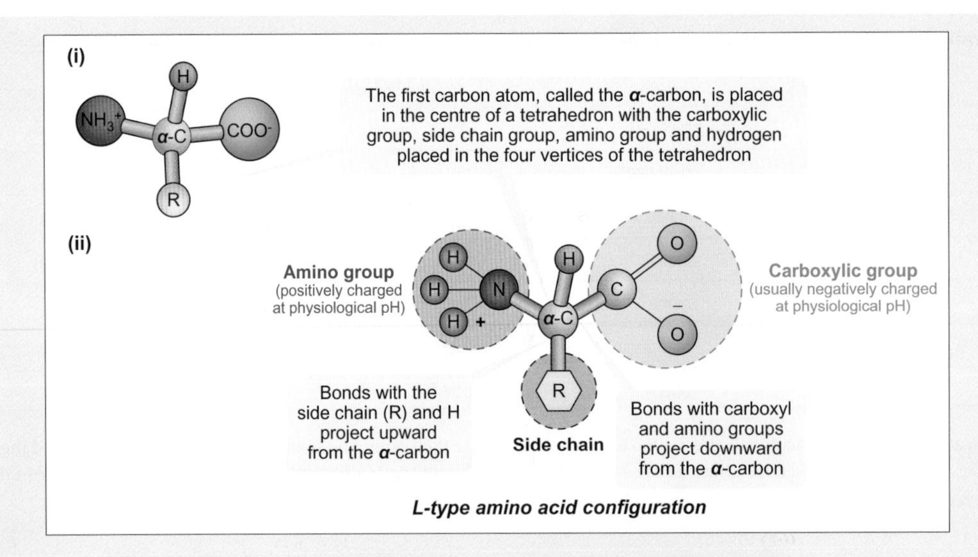

Figure E Generic structure of the α-amino acid molecules in the L-type spatial configuration

(i) In this spatial representation of the amino acids, the bonds of the α-carbon with the side chain R and H project upward and bonds between the α-carbon and the carboxylic, and amino groups project downward. All amino acids (except glycine) have two spatial configurations, the L-type configuration, shown above and the D-type configuration, where the position of the side chain (R) is reversed with the position of H. Since glycine has two hydrogen atoms at the α-carbon (i.e. side chain R in glycine is H), reversing the side chain R (i.e. H) with the other hydrogen atom does not change the spatial configuration of the molecule. Thus, glycine has effectively only one spatial configuration because the L- and D-type spatial configurations overlap. (ii) Other bonds between atoms in amino acids are represented by black lines. There is a double bond between the carbon in the carboxylic group and one of the oxygen atoms in this group. You can recognize the L-type configuration by imagining that you are tilting the tetrahedron toward you to bring the hydrogen atom closest to you and looking along the bond of the hydrogen atom with the α-carbon. In the L-type configuration you get the word CORN if you read the groups in a clockwise direction (CO from the carboxylic group, R from the side chain and N from the amino group).

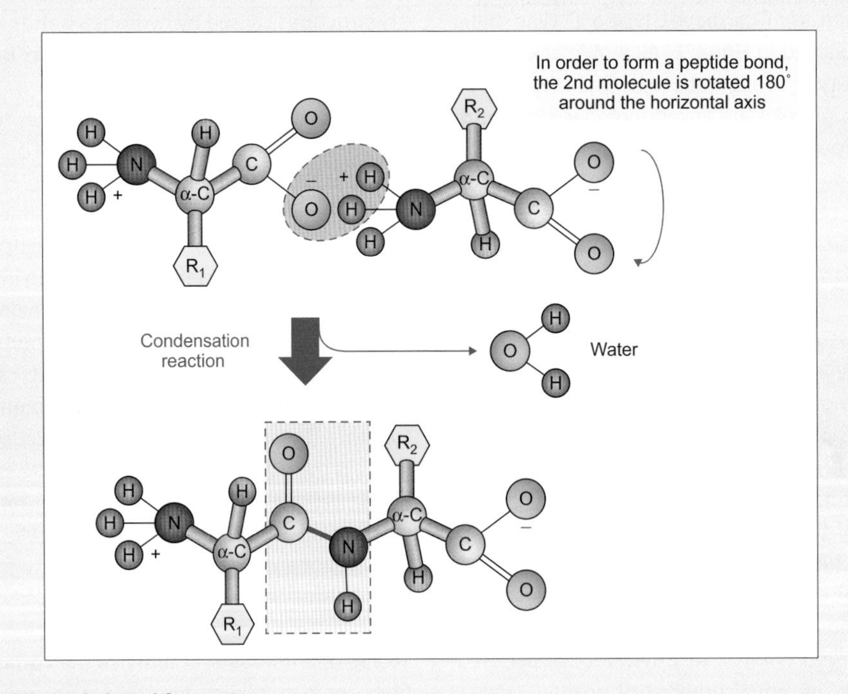

Figure F Diagram of peptide bond formation

Peptide bonds are formed from a condensation reaction between the carboxylic group of one amino acid and the amino group of another amino acid with the release of one molecule of water (H_2O). The reverse reaction, in which one molecule of water is used to break the peptide bond, is called hydrolysis.

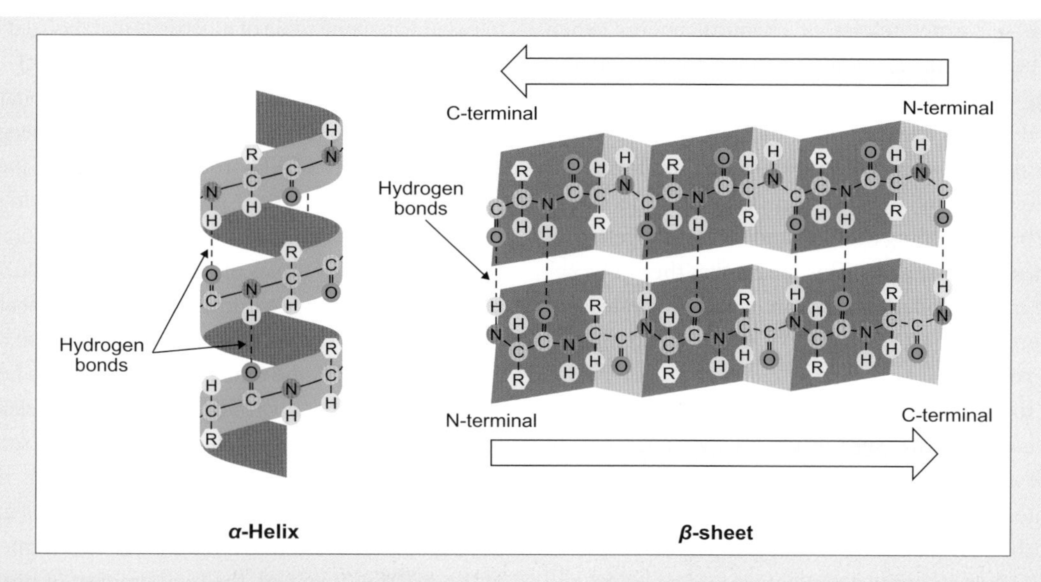

Figure G Two major ways in which backbone peptides interact to determine the secondary structure of proteins

Some amino acids (methionine, alanine, leucine, glutamic acid and lysine) have the propensity to form right-handed helices, called α-helices, while others (valine, isoleucine, threonine, phenylalanine, tyrosine and tryptophan) form pleated sheets, called β-sheets. In β-sheets, the polypeptide chains are anti-parallel to each other. α-Helices are 3–10 amino residue long with 3.6 amino acids per turn. β-Sheets are 4–40 amino residue long and are less common than α-helices. β-Sheets are conventionally represented by thick arrows pointing towards the C-terminal. Haemoglobin is about 75% α-helices, while β-sheets are common in fibrous proteins found in silk, feathers and hair.

and nucleic acids, which change the properties of the entire molecule.

A most useful data bank for exploring in detail the 3D shapes of proteins is at http://www.rcsb.org/pdb/home/home.do

[1] Biological membranes are discussed in Section 3.2.3.

[2] We discuss cholesterol and steroid hormones in Section 19.1.

[3] A linear chain of less than 20–30 amino acid residues is normally called a peptide.

[4] This is in contrast to monosaccharides which occur in nature in the D-configuration.

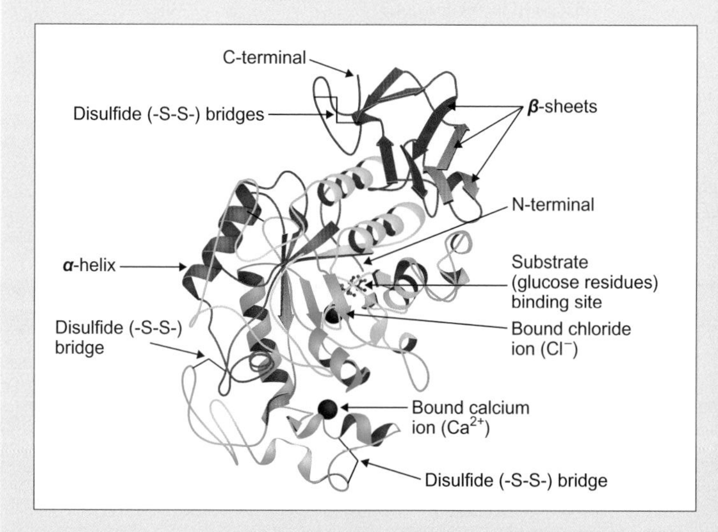

Figure H Diagram of the three dimensional structure of the human salivary amylase.

Salivary amylase is an enzyme which plays a key role in the digestion of starch. The enzyme is made of one single peptide chain of 496 amino-acid residues and displays typical secondary-structural elements (α – helices and β – sheets). Several disulphide (-S-S-) bridges which contribute to the overall fold (tertiary structure) of the protein are shown in thin black lines. The enzyme is active only when calcium (Ca^{2+}) and chloride ions (Cl^-) are bound to it at the sites indicated on the diagram.

Source: Ramasubbu N et al (1996). Structure of human salivary alpha-amylase at 1.6 A resolution: implications for its role in the oral cavity. Acta Crystallographica, Sect.D 52: 435-446. DOI: 10.2210/pdb1SMD/pdb.

reactions we see a net release of chemical energy when new bonds between the C, H and O atoms are formed and a net uptake of energy when previous bonds are broken. The chemical energy released when the new bonds holding carbon dioxide and water molecules together are formed is markedly greater overall than the chemical energy necessary to break the bonds holding the C, H and O atoms together in carbohydrates and fatty acids. This means that the Gibbs energy change in the oxidation reaction of carbohydrates and fatty acids (producing CO_2 and H_2O) is negative and that the reaction is exergonic.

In order to calculate the energy released from the oxidation of a carbohydrate, such as glucose, we need first to consider the amount of energy needed to break down the molecule into its component elements in their most stable states, i.e. carbon (C), molecular hydrogen (H_2) and molecular oxygen (O_2) at a chosen temperature and pressure and then subtract this value from the total amount of energy that is released when C is oxidized to CO_2 and H_2 is oxidized to H_2O at that chosen temperature and pressure.

As shown in Figure 2.4, the breakdown of a glucose molecule ($C_6H_{12}O_6$) into 6C, 6H_2 and 3O_2 at a temperature of 25°C (298 K) and pressure of 100 kPa (750 mm Hg)[27] requires an energy (ΔG) of **+911 kJ mol^{-1}**. The energy released by the formation of six molecules of carbon dioxide (ΔG = **–2367 kJ mol^{-1}**) and six water molecules (ΔG = **–1422 kJ mol^{-1}**) at same temperature (25°C) and pressure (100 kPa) amounts to **–3789 kJ mol^{-1}**. Therefore, the amount of energy released when one mole of glucose is metabolized to CO_2 and H_2O at 25°C and 100 kPa is **–2878 kJ (–3789 kJ + 911 kJ)**.

The synthesis of ATP from ADP and phosphate is an endergonic reaction (it uses energy); the Gibbs energy change is about +50 kJ mol^{-1} of ATP. Therefore, in theory, the energy released by the oxidation of 1 mole of glucose could produce up to 57 moles of ATP (2878 kJ/50 kJ = 57.6). In practice, however, only 27 to 29 moles of ATP[28] can be produced for each mole of glucose that is oxidized. This means that about 50 per cent of the Gibbs energy released by the oxidation of glucose is transferred to ATP; the remainder is released as heat.

Eventually all energy transferred to ATP is released as heat if the animal is at steady-state and does not perform external work or change its composition (e.g. by growth, digestion of food or excretion). Since the amount of Gibbs energy does not depend on the nature and number of intermediary steps of the reaction involved, the total amount of heat produced by an animal corresponds to the *total* amount of Gibbs energy utilized to keep the animal at steady state, after corrections for external work and changes in body composition. This statement is also valid for situations where animals oxidize food in the absence of oxygen by anaerobic respiration. The amount of energy used can be measured by monitoring the amount of heat produced by an animal at constant temperature and pressure with a piece of equipment known as a **calorimeter**.

The average amount of energy used daily by adult animals at steady state can also be determined over periods of

[27] Appendix 1 explains the international system of SI units and provides information on conversions to alternative units.

[28] The number of ATP molecules produced from 1 molecule of glucose is discussed in Section 2.4.3.

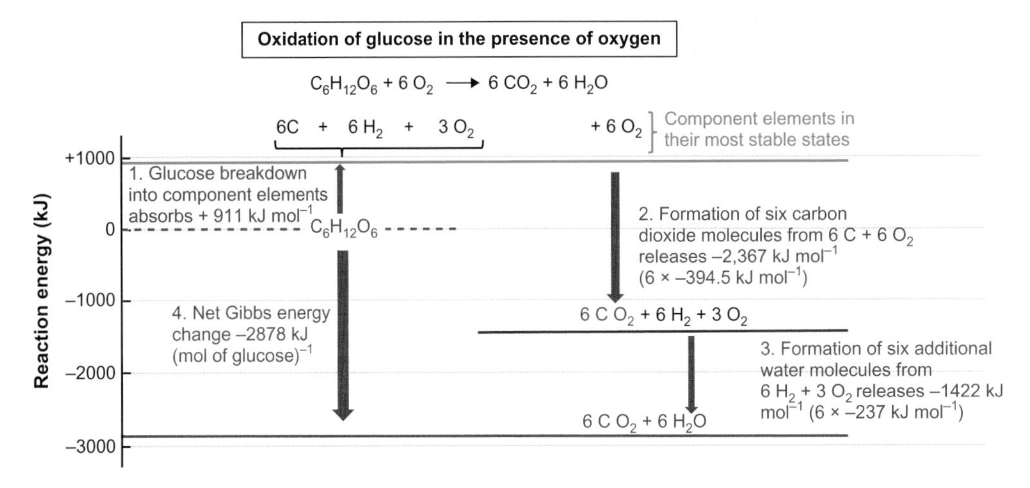

Figure 2.4 Reaction energy balance for the oxidation of glucose in the presence of oxygen at 25°C (298 K) and 100 kPa (750 mmHg) pressure

The net Gibbs energy change is the balance between the energy required to break down the glucose molecule ($\Delta G > 0$) into the most stable form of its component elements (C, H_2, O_2) and the energy released ($\Delta G < 0$) by the formation of the six molecules of carbon dioxide and the six water molecules from the respective component elements (C, O_2 for carbon dioxide; and H_2, O_2 for water).

several days to weeks by measuring the difference between the energy value of all ingested food and the energy value of all excreta. However, the most versatile, and therefore most widely used, method for determining the amount of energy used by animals is based on measurements of the amount of oxygen consumed, as discussed in Box 2.2.

This method can only be used with animals whose energy needs are fulfilled primarily from aerobic respiration. It is based on the widely accepted fact that roughly the same amount of energy is released when the common types of foodstuff—carbohydrates, lipids and proteins—are oxidized in the presence of a given amount of molecular oxygen. Indeed, according to Table 2.1, the amount of energy released when 1 mmol O_2 is used to oxidize carbohydrates, lipids and proteins varies only slightly, between 0.43 and 0.47 kJ (mmol O_2)$^{-1}$. For simplicity, we use an average value of 0.45 kJ released per mmol O_2 consumed or the conventional values given in Table 2.1 for individual types of foodstuff. The possibility of some inaccuracy must be borne in mind because relatively large errors (of up to 38 per cent) were recorded when direct measurements of energy utilization by an animal were made by measuring heat production using a calorimeter compared with indirect measurements of energy utilization by converting oxygen consumption into its equivalent energy.

Data in Table 2.1 show that the oxidation of 1 g of dehydrated lipid releases more than double the amount produced from the oxidation of 1 g of dehydrated carbohydrate or protein. Table 2.1 also shows that natural foods rich in carbohydrates and proteins contain more water than lipids in adipose tissue, causing an even larger discrepancy between the amount of energy produced by the oxidation of 1 g of fat-containing foods vs 1 g of carbohydrate- and protein-containing foods. Thus, fats have about 7.5 times greater energy density than carbohydrates and proteins in the body and are generally used by all animals as a form of energy storage[29].

2.2.2 Respiratory quotient

A useful indicator to help determine what type of foodstuff animals use at a particular time for providing their energy needs is the **respiratory quotient** or **RQ**. The RQ is defined as the ratio between the rate of production of carbon dioxide ($\dot{M}CO_2$, mol min^{-1}) in the tissues and the rate of consumption of oxygen ($\dot{M}O_2$, mol min^{-1}) when foodstuffs are oxidized in the presence of oxygen[30]:

$$RQ = \dot{M}CO_2 / \dot{M}O_2 \qquad \text{Equation 2.1}$$

However, it is impossible to measure directly the exchange of CO_2 and O_2 at the cellular level in whole animals, so the **respiratory exchange ratio (RER)** is determined from the exchange of CO_2 and O_2 between the animal and its environment. The RER only represents the RQ when the animal is in a steady state, i.e. when both the exchange rates of CO_2 and O_2 between the animal and the environment and the body temperature are stable[31]. Values of RQ for carbohydrates, lipids, and proteins are given in Table 2.1.

The usefulness of RQ becomes apparent if we consider how RQ changes when different types of foodstuff molecules are oxidized. Figure 2.4 shows how the oxidation of one glucose molecule consumes six molecules of oxygen and produces six molecules of carbon dioxide:

$$C_6H_{12}O_6 + 6\,O_2 \rightarrow 6\,CO_2 + 6\,H_2O \qquad \text{Equation 2.2}$$

More generally, when one carbohydrate molecule with the general formula ($C_m(H_2O)_n$) is oxidized, one molecule of CO_2 is produced for each molecule of O_2 consumed.

$$C_m\left(H_2O\right)_n + m\,O_2 \rightarrow m\,CO_2 + n\,H_2O \qquad \text{Equation 2.2a}$$

Consequently, when carbohydrates are oxidized, the RQ is 1.0.

By contrast, when lipids are oxidized, the number of CO_2 molecules produced is considerably smaller than the number of O_2 molecules consumed. This is because lipid molecules contain a smaller proportion of oxygen atoms in their molecules than carbohydrates. For example, the complete oxidation of a molecule of stearic acid ($C_{18}H_{36}O_2$) requires 26 molecules of O_2 and produces 18 molecules of CO_2.

$$C_{18}H_{36}O_2 + 26\,O_2 \rightarrow 18\,CO_2 + 18\,H_2O \qquad \text{Equation 2.3}$$

Consequently, the RQ value for stearic acid is 0.69 (18 mol CO_2/26 mol O_2). On average, RQ = 0.71 when energy is obtained from the oxidation of lipids.

Since animals generally extract most of their energy requirements from the oxidation of carbohydrates and lipids, an RQ value close to 1.0 indicates predominantly carbohydrate oxidation, while an RQ value close to 0.7 indicates predominantly lipid oxidation. The RQ associated with protein oxidation varies depending on whether the dominant form in which nitrogen is excreted is ammonia, urea, or uric acid, as shown in Table 2.1[32]. Rainbow trout (*Oncorhynchus mykiss*), for example, excrete about 20 per cent of nitrogen waste as urea and the rest as ammonia. Thus, the RQ for protein metabolism in this species is 0.94. For a mixed diet in general, the RQ value is around 0.8.

[29] We discuss the energy density of fat and protein in an animal carcass in Section 2.3.

[30] *M* indicates mol and the dot over the *M* indicates rate (mol/time). These are explained in more detail in Section 11.1.2.

[31] The relationship between RQ and RER is discussed further in Section 12.3.3.

[32] The generation of nitrogenous waste products from proteins is discussed in Section 7.4.

2

Box 2.2 Animal respirometry

Animal respirometry refers to the quantitative measurement of gas exchange (O_2 and CO_2) between an animal and its environment with a piece of apparatus called a **respirometer**. Respirometry is used to measure the metabolic rate (i.e. the rate of oxygen consumption) and the **respiratory exchange ratio (RER**, i.e. the ratio between rate of oxygen consumption and rate of carbon dioxide production) of animals.

Modern respirometers use sensitive gas analysers to measure with high levels of accuracy and precision the concentration of oxygen and carbon dioxide in the air and oxygen in water. These devices enable the continuous measurement of metabolic rate. There are two types of respirometers, **open** and **closed**. In a closed respirometer, no water or air flows through the respirometer chamber and the decrease in the concentration of oxygen over a given amount of time is used to calculate the rate of oxygen consumption. For air, the CO_2 has to be absorbed to prevent its build-up in the chamber.

In open respirometry, air or water flows through an animal chamber as shown in Figure A (i) and (ii), or air flows through a mask worn by the animal as shown in Figure A (iii). Mask-based respirometry permits measurements in birds or bats flying in a wind tunnel.

To calculate the rates of oxygen consumption and carbon dioxide production in air breathers (and rate of oxygen consumption in water breathers), it is necessary to know the flow rates and fractional concentrations of the two gases into and out of the animal chamber.

For the measurement of rate of oxygen consumption in air, the carbon dioxide and water vapour are removed from the air before it enters the oxygen analyser and after it leaves the animal chamber. Assuming that the flow rate of air (FR) is measured and adjusted to standard temperature and pressure for dry air (STPD) conditions[1] as air enters the chamber, and $F_{in}O_2$ and $F_{out}O_2$ are the fractional concentrations of O_2 measured by the oxygen analyser when air enters and exits the chamber, respectively, then the rate of oxygen consumption ($\dot{M}O_2$) is obtained from the following expression:

Air flow rate at STPD as it enters the chamber

Fractional O_2 concentration of air entering the chamber

Fractional O_2 concentration of air exiting the chamber

Rate of oxygen consumption at STPD

$$\dot{M}O_2 = FR\ (F_{in}O_2 - F_{out}O_2)/(1 - F_{out}O_2)$$

As an example, if a 25 g mouse is at rest in the animal chamber and the flow of air entering the chamber, *FR*, is 4 mL s^{-1}, with $F_{in}O_2 = 0.2095$ and $F_{out}O_2 = 0.2075$, then $\dot{M}O_2 = 0.01$ mL O_2 s^{-1} = 0.00045 mmol (0.45 µmol) O_2 s^{-1}. Assuming a conversion factor of 0.45 kJ mmol^{-1} O_2 as we discuss in Section 2.2.1, then the metabolic rate is 0.202 J s^{-1}.

Similar measurements and calculations can be performed with aquatic animals as shown in Figure A(ii). If the rate of CO_2 production ($\dot{M}CO_2$) is measured in air-breathing animals, RER can be calculated. However, because CO_2 is very soluble in water[2], it is very difficult to make accurate measurements of $\dot{M}CO_2$ in aquatic animals unless the water entering the respirometer is decarbonated and has a low background level of total CO_2.

[1] We discuss in Section 11.1.2 the relationship between volume, pressure, temperature and molecular mass of gases. 1 mL O_2 at STPD = 0.045 mmol O_2.

[2] The properties of CO_2 in air and water are discussed in Section 11.1.2.

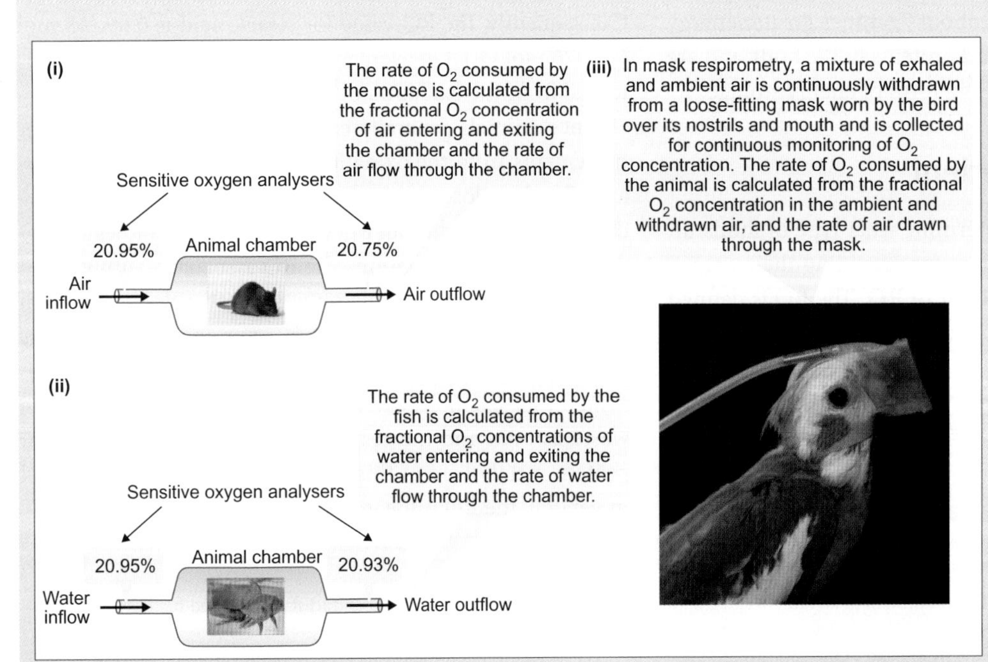

Figure A General principles used in open-flow respirometer systems for measuring oxygen consumption in animals

(i) Respirometer system used with air breathing animals, (ii) respirometer system used with aquatic animals, and (iii) mask-respirometer system used to measure rates of oxygen consumption in birds. Cockatiel (*Nymphicus hollandicus*) wearing a respirometry mask.

Source in (iii) Morris CR, Nelson FE, Askew GN (2010). The metabolic power requirements of flight and estimations of flight muscle efficiency in the cockatiel (*Nymphicus hollandicus*). Journal of Experimental Biology 213: 2788-2796.

Table 2.1 Energy density, amount of energy released and respiratory quotient when foodstuffs are fully oxidized

Foodstuff	Glycogen (mainly in liver but also in muscles of vertebrates)	Lipids (in adipose tissue)	Protein (mainly in skeletal muscle and gastrointestinal tract)
Energy produced per mmol O_2 consumed [kJ (mmol O_2)$^{-1}$]	0.47	0.44	0.43
Energy density when dehydrated (kJ g^{-1})	17.5	39.6	17.8
Water content when in the body (per cent)	75	10	70
Energy density when hydrated in the body (kJ g^{-1})	4.4	35.6	5.3
Amount of O_2 consumed for the oxidation of 1 g of dehydrated foodstuff (mmol)	37.5	89.3	43
Respiratory quotient (rate of CO_2 production/rate of O_2 consumption)	1.0	0.71	0.97 (if nitrogen is excreted as ammonia) 0.81 (if nitrogen excreted as urea) 0.74 (if nitrogen excreted as uric acid)

All values are given for standard temperature (273 K) and pressure (101 kPa). 1 L O_2 at standard temperature and pressure (STPD) corresponds to 44.6 mmol and 1.43 g O_2.Note that data in row 5 are for common dehydrated foodstuff—carbohydrates, lipids and proteins—rather than for specific tissues in the body.

Sources: after Flatt J-P (1995). Use and storage of carbohydrates and fat. American Journal of Clinical Nutrition 61 (suppl): 952S-959S; Lauff RF, Wood CM (1996). Respiratory gas exchange, nitrogenous waste excretion, and fuel usage during aerobic swimming in juvenile rainbow trout. Journal of Comparative Physiology B 166: 501-509. Schmidt-Nielsen K (1997). Animal Physiology. 5th edn. Cambridge: Cambridge University Press. Jenni L, Jenni-Eiermann S (1998). Fuel supply and metabolic constraints in migrating birds. Journal of Avian Biology 29: 521-528.

When preparing for hibernation[33], migration[34] or reproduction[35] animals convert carbohydrates into fat. In such cases, the RQ value can exceed 1.0 because the conversion of carbohydrates to fat releases oxygen, thereby reducing the amount of oxygen needed from the environment.

Based on all of the information above, it is possible to use the measured RQ value to calculate the contributions of the carbohydrate, lipid and protein oxidation to the overall energy metabolism. For example, RQ measurements first showed that penguins fasting during egg incubation initially metabolize fat almost exclusively and then shift to protein as their time on the nest increases[36].

2.2.3 Metabolic rates

Metabolic rate refers to the amount of energy used by an animal per unit time. It is usually presented in kJ per day or watts[37] ($W = J\ s^{-1}$). Knowing the value of the metabolic rate for animals undergoing specific activities is of great practical importance because it allows us to determine the type and amount of food animals (humans included) need to consume under specific conditions in order to maintain an energy balance.

This knowledge is important for managing animal populations in the wild (such as mitigating the effects of climate change on some species), in planning expeditions, space travel, for the food industry and animal production and aquaculture. Knowledge of the metabolic rate in different populations is also important for understanding evolution since the traits of individuals that can convert excess energy into increased reproductive potential will have a greater chance to be selected for through the process of natural selection.

The metabolic rate of animals is usually measured as the rate of oxygen consumption by **respirometry** using the methods outlined in Box 2.2.

As a baseline, it is necessary to know the minimum rate of energy utilization needed to maintain all processes necessary to sustain the life of the animal. Knowing this value, we can then measure the increased energy cost associated with specific physiological activities such as digestion and absorption of nutrients, reproductive functions, locomotion, and keeping warm in the cold and cool in warm weather.

The baseline metabolic rate is measured when an animal is:

- in its non-reproductive state,
- awake in its resting state,
- post-absorptive (all food intake has been digested and absorbed) and
- equilibrated at a specified ambient temperature.

In ectotherms, which include all animals other than birds and mammals, the minimum metabolic rate is called the **standard metabolic rate (SMR)**. The SMR varies depending on the temperature of the environment.

Birds and mammals are called endotherms because they use internal sources of energy to regulate their body temperature above that of their environment. The minimum metabolic rate in endotherms is called the **basal metabolic rate (BMR)**. The BMR is measured in the animal's

[33] We discuss the laying down of fat in preparation for hibernation in Sections 9.4.3 and 10.2.7.

[34] Migration in birds is discussed in sections 10.2.7 and 15.2.3.

[35] Section 20.1.2 discusses the build-up of endogenous reserves of protein and lipid in birds during reproduction.

[36] We discuss the reproductive fast in king and emperor penguins in Case Study 2.2.

[37] The watt (W) is the SI unit for power as shown in Appendix 1.

thermoneutral zone[38], which is the range of ambient temperatures where the rate of heat production is at its minimum and the animal does not require additional amounts of energy to control its body temperature.

It is not always easy to tell when an animal has completed food digestion and absorption. For example, as we discuss in Sections 2.3.4 and 15.1.1, food digestion and absorption in some carnivores such as pythons may take several days or longer depending on the ambient temperature. Similarly, herbivores[39] graze more or less continuously throughout the day, so it is not easy to know when they are in a post-absorptive state. In such cases, the lowest value of the metabolic rate during the daily cycle is measured as the baseline instead of BMR. This is known as the **resting metabolic rate (RMR)**.

Thus, in principle, mammals and birds have only one value for their BMR (or RMR). However, in practice, the BMR (or RMR) depends on the proportion of lean body mass of an animal, which in some species can change markedly throughout the year. All other animal groups have values of SMR (or RMR) that vary according to environmental temperature.

2.2.4 Scaling of metabolic rate with body mass

BMR and SMR have been used to understand the energetic costs associated with the maintenance of body functions in wild animals of different sizes and belonging to different taxonomic groups. For example, we might ask: do BMR and SMR vary directly with body mass?

Figure 2.5A shows the relationship between BMR and body mass (M_b) for mammals. The values are plotted on logarithmic axes so that we can include in a single graph animals whose body masses differ by over one million fold, from Etruscan shrews (*Suncus etruscus*) weighing about 2.4 g to Asian elephants (*Elephas maximus*) weighing around 3.7 tonnes (3,670,000 g). The equation describing this relationship is called the **scaling equation** of BMR with M_b.

$$BMR = a M_b^b \qquad \text{Equation 2.4}$$

where the factor a and exponent b are called the **coefficient** and the **scaling factor**, respectively.

If the scaling factor b equals 1, then metabolic rate varies in direct proportion to body mass and the relationship is said to be **isometric** because we expect metabolic rate to be proportional to the total volume of cells in the body (which, in turn, is directly related to the body volume and mass, as discussed in Box 2.3). However, if metabolic rate does not vary in direct proportion to body mass, i.e b is different from 1, then the relationship is said to be **allometric**.

Equation 2.4 changes to a linear expression if log(BMR) values are plotted against the corresponding log(M_b) values as shown in Figure 2.5B (log(BMR) = log a + b log(M_b)). As shown in Figure 2.5, the scaling equation of BMR with M_b is clearly not isometric because the scaling factor b is less than 1.0. Thus, metabolic rate scales allometrically

[38] The thermoneutral zone of endotherms is discussed in Section 10.1.

[39] Herbivores, which eat food of plant origin, also have large communities of microbes in their gut. Herbivores therefore pose the additional problem that these communities are considered as part of their body.

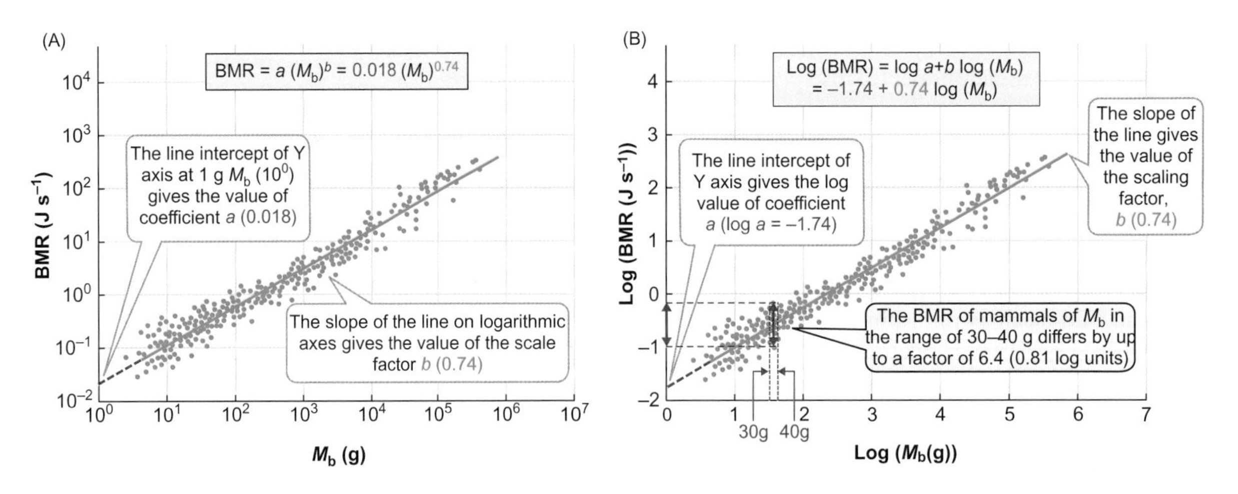

Figure 2.5 Relationship between basal metabolic rate (BMR) and body mass (M_b) in 626 species of mammals

(A) BMR is expressed as a function of M_b using double logarithmic scales. The solid line is the plot of the allometric equation BMR = a $(M_b)^b$ with coefficient a = 0.018 J s^{-1} and scaling factor b = 0.74. (B) log (BMR) is expressed as a function of log (M_b) using linear scales. The solid line is the plot of the allometric equation log (BMR) = log a + b log (M_b) with coefficient a = 0.018 J s^{-1} and scaling factor b = 0.74. Notice that the BMR data points for mammals of similar mass can differ by several fold (> 0.8 log units). Adapted from Savage VM et al (2004). The predominance of quarter-power scaling in biology. Functional Ecology 18: 257-282.

Box 2.3 Isometric vs allometric scaling

In geometrically similar objects (i.e. objects that have the same form and change their linear dimensions in proportion), the surface **area** and the **volume** are always proportional to one particular dimension squared and cubed, respectively. This is shown in Figure A, using the cube as an example, as all cubes are similar objects.

More generally, for all similar objects, the surface *Area* is proportional to $(Length)^2$:

$$Area \propto (Length)^2 \qquad \text{Equation A}$$

and the *Volume* is proportional to the *Length* cubed:

$$Volume \propto (Length)^3 \qquad \text{Equation B}$$

Consequently, the *Length* increases proportionally to the square root of the *Area*:

$$Length \propto \sqrt{(Area)} = (Area)^{1/2} \qquad \text{Equation C}$$

or to the cube root of the *Volume*:

$$Length \propto \sqrt[3]{Volume} = Volume^{1/3} \qquad \text{Equation D}$$

Since the *Area* increases proportionally to $Length^2$, it also follows that the *Area* increases proportionally to the cube root of the *Volume*, squared:

$$Area \propto Volume^{2/3} \qquad \text{Equation E}$$

Isometric scaling (from Greek *isos* = same, *metria* = measure) refers to the occurrence of proportional relationships involving anatomical, physiological or behavioural measurements that follow these simple rules. For example, the relationship between length of a leg and body mass, M_b (which is proportional to the volume) in a growing animal is described by the following relationship:

$$Leg\ length = a\,M_b^{1/3} \qquad \text{Equation F}$$

where a is a constant known as the **proportionality coefficient**.

We then say that the leg length increases **isometrically** with the body mass, because it follows the relationship given in Equation D. For example, frogs are known to grow in size isometrically a few weeks following metamorphosis. Also, the size of the shell of snails increases isometrically with their body mass.

Isometric scaling generally sees all volume-based properties changing in proportion to body mass (M_b) or body length cubed (Equation B); all surface-based properties changing in proportion to $M_b^{2/3}$ (Equation E) or body-length squared (Equation A); and all length-based properties changing in proportion to $M_b^{1/3}$ (Equation F) or to body length.

Allometric scaling (in Greek *allo* means variable) refers to the occurrence of relationships that differ from the isometric type because animals of different sizes are not geometrically similar, or do not maintain geometric similarity as they grow. Nevertheless, when expressing a particular physiological or morphological parameter as a function of body size, the experimental data points can be well described by an **allometric equation** characterized by two variable parameters: the proportionality coefficient a and the **scaling factor** b

$$y = a\,x^b \qquad \text{Equation G}$$

Importantly, the scaling factor in the allometric equation (Equation G) is not prescribed as in isometric scaling but can

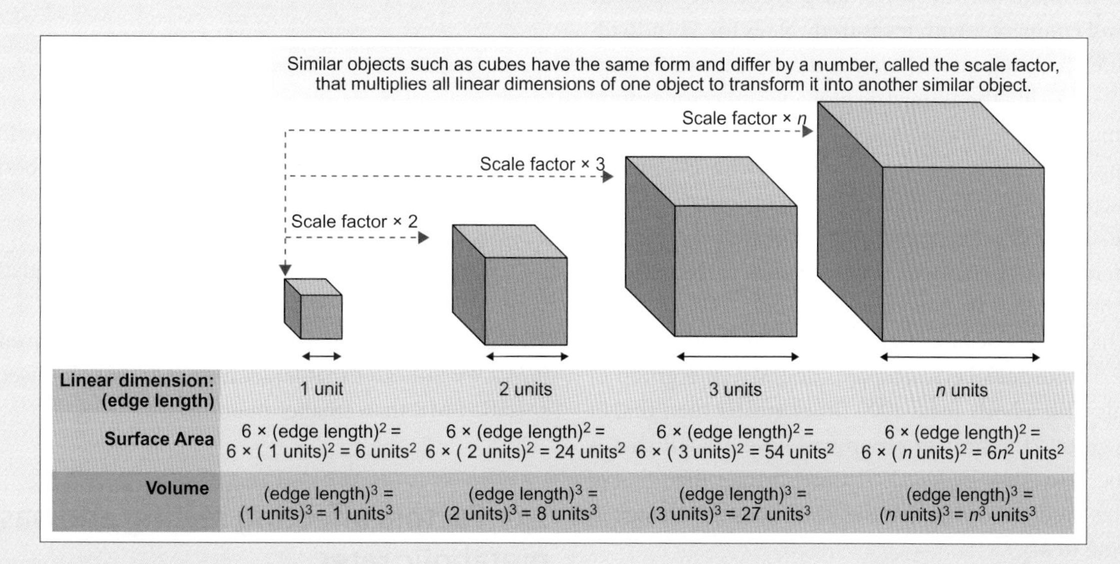

Linear dimension: (edge length)	1 unit	2 units	3 units	*n* units
Surface Area	6 × (edge length)² = 6 × (1 units)² = 6 units²	6 × (edge length)² = 6 × (2 units)² = 24 units²	6 × (edge length)² = 6 × (3 units)² = 54 units²	6 × (edge length)² = 6 × (*n* units)² = 6*n*² units²
Volume	(edge length)³ = (1 units)³ = 1 units³	(edge length)³ = (2 units)³ = 8 units³	(edge length)³ = (3 units)³ = 27 units³	(edge length)³ = (*n* units)³ = *n*³ units³

Figure A Similar objects scale isometrically.

In isometric scaling, surface area and volume of different objects are proportional to one particular dimension squared and cubed, respectively. In the case of cubes, *surface area* = 6 × *(edge length)*² *and volume* = *(edge length)*³.

take any value that is different from the prescribed value(s) for isometric scaling. Hence, b must be: different from 1 when scaling a volume-based property to M_b, different from 2/3 when scaling a surface-based property to M_b, and different from 1/3 when scaling a length-based property to M_b.

The value of the scaling factor b can be readily obtained by plotting the independent (x) and the dependent (y) variables in Equation G on double logarithmic (log–log) axes—that is plotting log y against log x. This is because if we take the logarithm of both sides of the allometric equation, then:

$$\log y = \log (a\,x^b) = \log a + b \log x \qquad \text{Equation H}$$

Substitution of *log y* with Y and *log x* with X changes Equation H into a typical linear equation:

$$Y = \log a + bX \qquad \text{Equation I}$$

The slope of the line described by Equation I is the scaling factor b. The intercept on the Y axis (when $X = 0$) is log a. Since log x was substituted with X, it means that $X = 0$ when $x = 1$.

with body mass. The two forms of representation of the scaling equation of BMR with M_b shown in Figure 2.5 are entirely equivalent to each other and both are in common use. Here we use the form shown in equation 2.4. While the allometric equation fits the overall BMR data for a wide range of body masses, there is a relatively large variation in BMR of mammals of similar mass. For example, Figure 2.5B highlights that the range of BMR of mammals with a body mass of 30–40 g can differ by up to a factor of 6.4. The different body composition of mammals of the same mass can be a contributing factor because the RMR of lean tissue is several fold greater than that of adipose tissue.

The relatively large variation in BMR in mammals of the same mass precludes the use of the scaling equation to predict the BMR value for a *particular* mammal of a certain body mass. For example, we can use an allometric plot of the BMR of almost 150 species of murids (rats, mice and voles) to estimate the BMR of a 33 g animal in the same group, and compare it with measured values for 11 individual spinifex hopping mice (*Notomys alexis*) with an average mass of 33 g. Using the allometric plot gives an error around the estimated value, which is about 20 times greater than that around the mean measured value for the 11 spinifex mice. Such a high level of uncertainty points out the limitation of using an allometric plot to demonstrate any significant difference between two mean estimates corresponding to body mass values that are relatively close.

So, what does a scaling factor of 0.70 to 0.74[40] (0.74 in Figure 2.5) indicate? Let's compare the BMR of a group of mammals with a body mass 2000 times greater than that of another group. The heavier group will have a BMR that is only $2000^{0.70}$ to $2000^{0.74}$ (205–277) times greater than that of the lighter group, showing that the BMR does not change in proportion to M_b.

The relationship between average mass-specific BMR (= BMR/M_b) and M_b for mammals is displayed in Figure 2.6, where the experimental data points were fitted by the allometric Equation 2.5, obtained by dividing both sides of Equation 2.4 by the body mass:

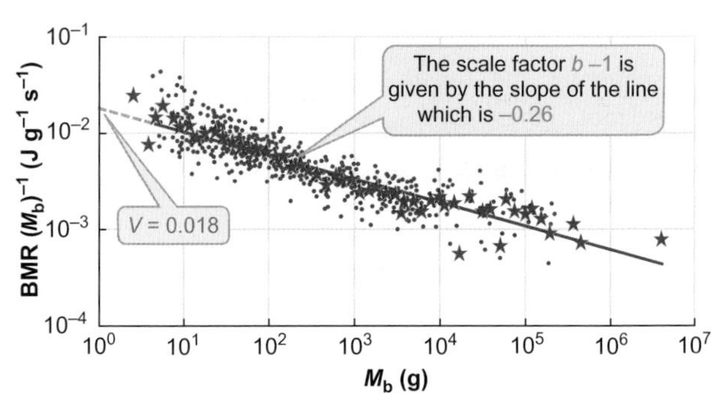

Figure 2.6 Plot of the mass-specific basal metabolic rate *BMR* $(M_b)^{-1}$ against body mass (M_b) in mammals

The mass-specific BMR data points for 626 species of mammals are shown as small points on the graph. The red stars are average mass-specific data falling within one of 10 intervals per 10-fold change in body mass. This grouping method limits the number of points to a maximum of 10 for each 10-fold change in body mass and greatly reduces the six-fold range in BMR data that is apparent in Figure 2.5. The solid line is the plot of the allometric equation **BMR** $(M_b)^{-1} = a\,(M_b)^{(b-1)}$ with coefficient $a = 0.018$ J s^{-1} and scaling factor $b - 1 = -0.26$.

Modified from Savage VM et al (2007). Scaling of number, size, and metabolic rate of cells with body size in mammals. Proceedings of the National Academy of Sciences USA 104: 4718-4723.

$$\text{BMR}/M_b = a(M_b)^{b-1} \qquad \text{Equation 2.5}$$

The negative scaling factor $b-1 = -0.26$ (derived from 0.74 – 1), indicates that larger mammals have a lower mass-specific BMR than smaller mammals.

2.2.5 Factors affecting resting and basal metabolic rates

How universal is the relationship between BMR (or SMR) and M_b? Does it extend to animal groups other than mammals? Figure 2.7 shows the scaling of BMR with M_b for placental mammals; two distinct groups of birds,

[40] Note that there is disagreement among researchers about the actual value of the scaling factor for mammals in the range 0.70 to 0.74.

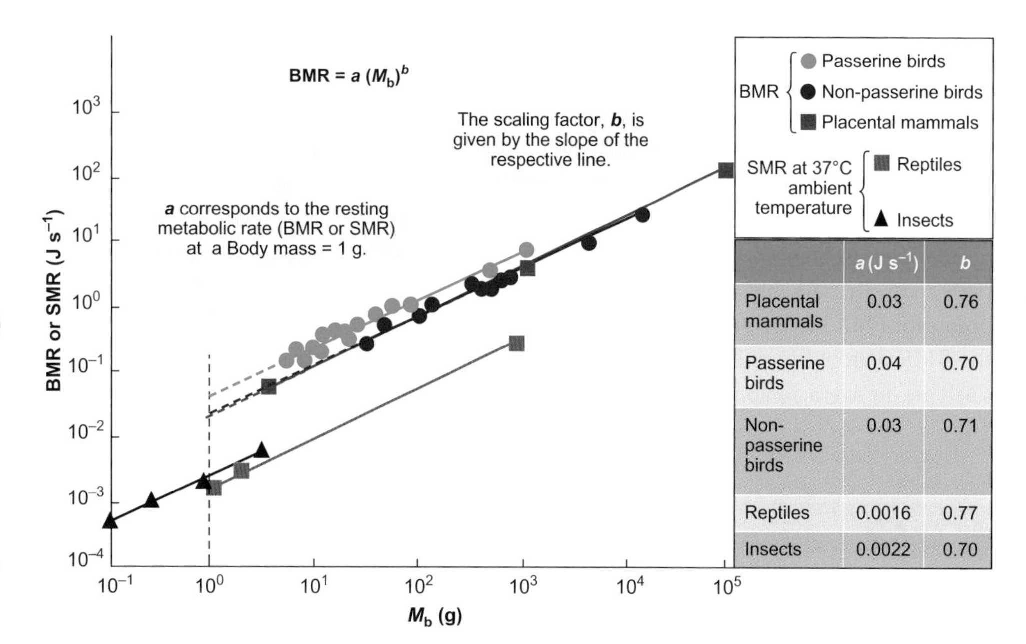

	a (J s⁻¹)	b
Placental mammals	0.03	0.76
Passerine birds	0.04	0.70
Non-passerine birds	0.03	0.71
Reptiles	0.0016	0.77
Insects	0.0022	0.70

Figure 2.7 Scaling of basal or resting metabolic rates (BMR or SMR) with body mass (M_b) in birds, mammals, reptiles and insects. (BMR for birds and mammals and SMR for reptiles and insects at 37°C)
The straight lines fitting the data points plotted on double logarithmic axes are derived from the scaling equation ((BMR or SMR) = $a\,M_b^{\,b}$. The parameters a and b are given for each group in the table.
From Gavrilov VM (2013). Origin and development of homoiothermy: A case study of avian energetics. Advances in Bioscience and Biotechnology 4: 1-17.

passerines[41] (or perching birds) and all other birds (non-passerines); reptiles; and insects. The SMRs of reptiles and insects are for animals at 37°C to enable comparisons with placental mammals, which have a body temperature close to 37°C.

Figure 2.7 shows that resting metabolic rates (BMR or SMR) of taxonomically diverse animal groups are allometric functions of body mass. While the values for the coefficient a (y-intercept) differ, the lines for the different groups are almost parallel; that is, they have similar slopes (or scaling factors, (b)) that lie between 0.70 and 0.77.

There is disagreement among researchers about whether the scaling factor b has a universal value. A scaling factor close to 0.67 would be consistent with a link between metabolic rate and surface area in animals of different sizes[42]. Nevertheless, the important fact is that small animals within a taxonomic group have a larger mass-specific metabolic rate than large animals in the same group, showing that the metabolic rate does not change in proportion to the mass (or volume) of animals.

Figure 2.7 shows that the BMR of placental mammals is about 15–20-fold (1.2–1.3 logarithmic units) greater than the SMR of reptiles and insects of similar body mass at similar body temperatures. This difference indicates that body temperature per se is not the major determinant of the difference between RMR of endotherms and ectotherms. Indeed, the high level of complex activities and work intensities in endotherms is only possible due to the presence of high energy, oxygen-consuming circulatory and nervous systems, which greatly elevate the level of the BMR in endotherms and, therefore, the food requirements of these animals. Thus, the major driver of the evolution of endothermy is thought to be the increased ability to perform more complex, energy-consuming activities, rather than a simple rise in the velocity of biochemical reactions.

While the main determinants of RMRs have been related to body size, taxonomy and phylogeny within a taxonomic class, other factors have been shown to reduce the variation of parameters in the allometric equation when animals are grouped based on differences in these factors. Such factors include climate, biogeography and habitat, diet, reproductive strategy and life history. As an example, small desert mammals have a BMR that is up to 24 per cent smaller than that of other mammals of similar size living in environments with moderate moisture, called mesic habitats[43]. Similar differences have been reported between desert and non-desert birds.[44]

2.2.6 Field metabolic rate

The most ecologically relevant indicator of energy utilization in animals is the **field metabolic rate (FMR)**, which refers to the average daily metabolic rate of a free-living animal in its natural environment. The ability of animals to utilize energy efficiently in their environment is critical to their reproductive success.

[41] Passerine birds are characterized by the arrangement of their toes (three toes point forward and one toe is directed backward), which allows them to perch. The passerines comprise more than half of all bird species and include the songbirds.

[42] As we discuss in Box 2.3 (Equation E) a quantity that is proportional to the area of an object would scale with a factor of 2/3 = 0.67 for objects of different sizes.

[43] Section 1.2.2 discusses the classification of terrestrial habitats based on water availability.

[44] Desert and non-desert birds are discussed in Section 10.2.3.

2

FMR includes the energy costs for maintaining an animal's basic functions (BMR or SMR) and reproductive status, as well as the energy used for everything else—for example, foraging, feeding, digesting food, absorbing nutrients, and avoiding predation. Box 2.4 describes three methods used to measure metabolic rate in wild animals that are free to roam.

As shown in Figure 2.8, FMR is an allometric function of body mass. Remarkably, the regression line fitted to the data points for FMR has an almost identical slope to that fitted to the BMR data but is shifted to higher values. In mammals, FMR is, on average, about three times greater than BMR, irrespective of body mass (based on the shift in the relationship shown in Figure 2.8 of about 0.5 logarithmic units, which is equal to 3). This conclusion is important because it allows us to estimate FMR of wild mammals; we would expect it to be, on average, about three times their BMR.

The FMR of birds and reptiles is, on average, about 3.5–4 times greater than their respective BMRs and SMRs. Since the SMR of reptiles is about 20 times smaller than the BMR in endotherms of similar body mass, and reptiles experience longer periods of inactivity than endotherms of similar mass, it follows that on an annual basis, birds and mammals

Box 2.4 How is field metabolic rate estimated?

The main techniques for estimating the field metabolic rate (FMR) in terrestrial animals are the **doubly labelled water (DLW)** technique, the **heart rate** method and the **overall dynamic body acceleration (ODBA)** technique.

The DLW technique described in Figure A measures the average rate of CO_2 production by an animal ($\dot{M}CO_2$) after its body water has been first enriched with **stable** or **radioactive isotopes**[1] of hydrogen and oxygen. A fluid sample is taken before the animal is released in the field and a final fluid sample is taken after the animal has been recaptured. Assuming an average respiratory exchange ratio (RER) of 0.8, as discussed in **Section 2.2.2**, we can calculate the rate of oxygen consumption ($\dot{M}O_2$), since $\dot{M}O_2 = \dot{M}CO_2/RER = \dot{M}CO_2/0.8 = 1.25\dot{M}CO_2$. The average estimated metabolic rate during the experimental period is then calculated based on the energy released per mol of O_2 consumed (0.45 kJ energy released per mmol O_2 consumed).

The heart rate method uses the heart rate as an indicator of the rate of energy consumption. Heart rate can be continuously monitored using implanted heart rate monitoring devices and data loggers that store heart rate data for many months. After the implantation of the monitoring devices, the animal is released in the field and is later recaptured to download the data. The heart rate method relies on the existence of a predictable relationship between heart rate and rate of oxygen consumption[2] over a range of activity levels for the species being studied. The changes in heart rate due to specific activities in the field can then be converted into the associated changes in the rate of energy utilization.

The ODBA technique uses accelerometers attached to the body of animals in the field. The accelerometers can make measurements with high time resolution of muscle activity-related movements. An acceleration index termed the overall dynamic body acceleration (ODBA) has been used as a proxy in animal ecology studies to estimate the amount of energy expenditure in wild animals that are active. The major limitation

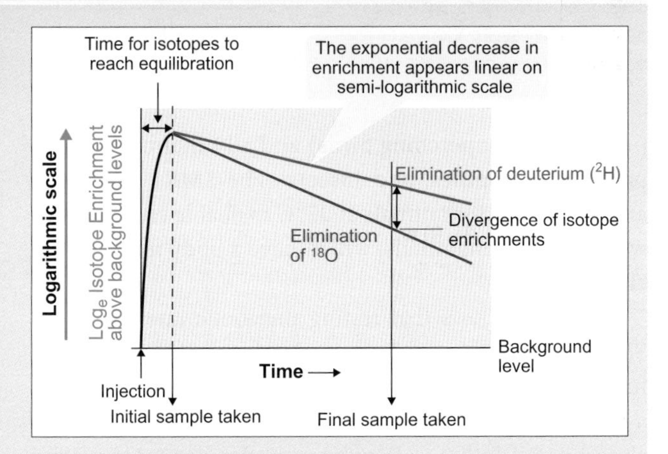

Figure A Schematic diagram for the measurement of CO_2 production using the doubly isotopically labelled water (DLW) technique

Following injection, the isotopes of hydrogen (^{2}H) and oxygen (^{18}O) distribute throughout the animal's body fluids and reach equilibrium after one or more hours depending on the size of the animal. Thereafter, the isotopes are gradually eliminated from the body and their concentrations decrease exponentially towards their background levels. The oxygen isotope is eliminated as both CO_2 and H_2O, while the hydrogen isotope is eliminated only as H_2O. The amount of CO_2 that was produced (and lost) (mol) in the time interval between samples is calculated from the divergence of the two isotope enrichments over that time interval. In our example, using semi-logarithmic scales, 50% of the enriched ^{18}O and 30% of enriched ^{2}H were eliminated in the time interval between samples. The divergence of the two isotope enrichments is therefore 20% and the amount of CO_2 produced corresponds to 20% of total body water of the animal. Since the total amount of water in an animal is normally a known fraction of body mass, the average rate of CO_2 production in the animal ($\dot{M}CO_2$) is calculated by dividing the total amount of CO_2 produced over the time interval between samples.

Source: Modified from Butler PJ et al (2004). Measuring metabolic rate in the field: the pros and cons of the doubly labelled water and heart rate methods. Functional Ecology 18: 168-183.

[1] An isotope is one of two or more forms of the same atom (element) that contain equal numbers of protons but different numbers of neutrons in their nuclei. The isotopes differ in relative atomic mass but not in chemical properties.

[2] The relationship between heart rate and rate of oxygen consumption is given in Section 15.1.

of the method is that a significant amount of energy expenditure in animals is associated with internal processes such as ion transport across membranes, synthesis of organic molecules and digestion, which cannot be measured with accelerometers.

All three methods for estimating FMR should be appropriately calibrated for each species of animal used against measured $\dot{M}CO_2$ or $\dot{M}O_2$ under conditions that mimic as closely as possible those to be encountered by the animals in their natural environment. Also, any changes in body mass could affect these relationships, particularly for the ODBA method[3].

[3] Force = mass × acceleration, and work is done (energy expended) when a force moves.

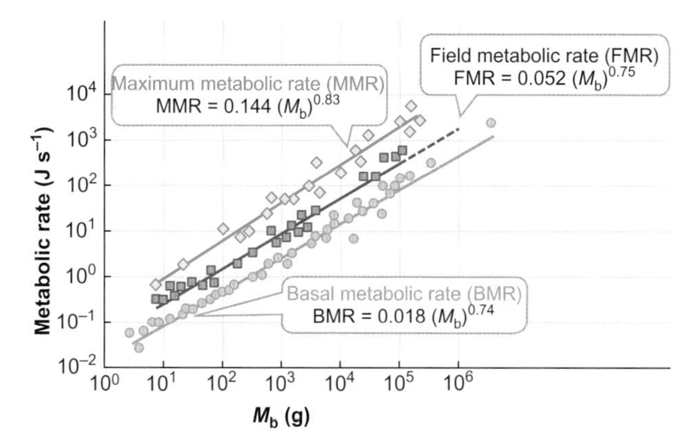

Figure 2.8 Scaling of maximum (MMR), field (FMR) and basal metabolic rates (BMR) with body mass (M_b) in mammals

The values for MMR and BMR were obtained in the laboratory using respirometry. The grouping method described in Figure 2.6 was used for the three sets of data to make separation of the three plots clearer.

From Savage VM et al (2004). The predominance of quarter-power scaling in biology. Functional Ecology 18: 257-282.

spend about 30 times more energy to function than a reptile or other ectothermic vertebrates of similar body mass. This puts endotherms at a great disadvantage in terms of their energy requirements. However, the high energy-intensive endothermic way of life is used to support more complex nervous, circulatory, respiratory, digestive and musculoskeletal systems, which in turn sustain a high level of complex activities. These activities enable endotherms to obtain and process food at much higher rates than ectotherms of similar size. In this way, the high energy-cost disadvantage is turned into a major ecological advantage, ensuring the biological success of birds and mammals, provided they can acquire the necessary amount of food.

The daily level of energy expenditure in humans is, on average, considerably less than the 3 × BMR which occurs in other species of mammals in the wild. There are two main reasons for this:

- wild mammals lack the ease of access to food that most humans have,
- wild mammals generally do not have the shelter from the environment available to most humans.

Nevertheless, there are situations in which the daily level of physical activity in humans exceeds 3 × BMR. For example, professional cyclists during gruelling competitions or explorers hauling sleds to the Poles need 4.5–4.7 times their BMR to fuel their activities.

The maximum sustainable rate of energy expenditure in some species of wild birds is close to four times BMR. For example, female macaroni penguins (*Eudyptes chrysolophus*) have a sustained metabolic rate of 3.5 to 4 × their resting FMR during the brood and crèche periods of their reproductive cycle.

The metabolic rate associated with the maximum level of activity in animals, the **maximum metabolic rate (MMR)**, can be induced by intense exercise in most species of animals, but can only be sustained for relatively short durations (in the order of many minutes rather than hours). Figure 2.8 shows that MMR of mammals is also an allometric function of body mass and, on average, is about 10–20 fold greater than the BMR. The scaling factor b for MMR in mammals is larger than that for BMR (0.83 compared to 0.74) and the logarithmic value of coefficient a is also higher by 0.9 units for the MMR than BMR. Thus, if we divide the allometric equation for MMR ($= 0.144\,(M_b)^{0.83}$) by the allometric equation for BMR ($= 0.018\,(M_b)^{0.74}$) we obtain another allometric equation that gives the ratio between MMR and BMR at different body masses:

$$\text{MMR/BMR} = 8.0\,(M_b)^{0.83-0.74} = 8.0\,(M_b)^{0.09} \qquad \text{Equation 2.6}$$

The MMR/BMR ratio is close to 10 for small mammals around 10 g:

$$8 \times 10^{0.09} = 8 \times 1.23 = 9.8$$

and close to 22 for large mammals around 100 kg (10^5 g):

$$8 \times 10^{5 \times 0.09} = 8.0 \times 2.82 = 22.6$$

Thus, larger mammals have, on average, a greater capacity for increasing their metabolic rate above the resting state than smaller mammals. Elite athletes can also increase their metabolic rate to 20 times BMR, but race horses (*Equus ferus caballus*)[45] and pronghorn antelope (*Antilocapra americana*) are able to increase their metabolic rate by a staggering factor of 50 times BMR for several minutes during running. Such high performance is essential for the antelope[46] to escape from predators such as cheetahs.

[45] Race horses are artificially selected for their high performance, as we discuss in Case Study 15.1.

[46] The high metabolic rate of exercising pronghorn antelope is also discussed in Section 15.1.3.

Further reading for Section 2.2

› *Review articles*

Butler PJ, Green JA, Boyd IL, Speakman JR (2004). Measuring metabolic rate in the field: the pros and cons of the doubly labelled water and heart rate methods. Functional Ecology 18: 168–183.

Frappell PB, Butler PJ (2004). Minimal metabolic rate, what it is, its usefulness, and its relationship to the evolution of endothermy: a brief synopsis. Physiological and Biochemical Zoology 77: 865–868.

Hulbert AJ, Else PL (2000). Mechanisms underlying the cost of living in animals. Annual Reviews in Physiology 62: 207–235.

Hulbert AJ, Else PL (2004). Basal metabolic rate: history, composition, regulation, and usefulness. Physiological and Biochemical Zoology 77: 869–876.

Nagy KA, Girard IA, Brown TK (1999). Energetics of free-ranging mammals, reptiles, and birds. Annual Reviews in Nutrition 19: 247–277.

Nagy KA (2005). Field metabolic rate and body size. Journal of Experimental Biology 208: 1621–1625.

Savage VM, Gillooly JF, Woodruff WH, West GB, Allen AP, Enquist BJ, Brown JH (2004). The predominance of quarter-power scaling in biology. Functional Ecology 18: 257–282.

White CR, Seymour RS (2005). Allometric scaling of mammalian metabolism. Journal of Experimental Biology 208: 1611–1619.

White CR, Kearney MR (2014). Metabolic scaling in animals: methods, empirical results and theoretical explanations. Comparative Physiology 4: 231–256

2.3 Energy intake from food

The vast majority of energy necessary for an animal's survival is obtained from the oxidation of organic compounds in food. Therefore, the energy metabolism of animals is inexorably linked to food intake. The energy flow in the **food chain** involves photosynthetic cyanobacteria, algae and plants, called **primary producers**, which trap energy from the sun to produce carbohydrate molecules such as sugars. **Primary consumers** feed on organic products made by primary producers.

In aquatic environments, the phytoplankton, representing microscopic plants (1–50 μm) suspended in the body of water, are consumed by zooplankton (protozoa, which are unicellular organisms, small animals and immature stages of larger animals < 200 μm), which can be regarded as primary consumers. In terrestrial habitats, plants are eaten by herbivores and omnivores. A variable number of other levels of consumers (called **trophic levels**) can exist along the food chain in aquatic and terrestrial habitats. Examples of simple food chains in a marine and a terrestrial habitat are shown in Figure 2.9.

The energy transfer efficiency between successive trophic levels is, on average, 10 per cent, but can range from 2 per cent to 50 per cent. On average, therefore, only 10 per cent of the total energy at one trophic level is passed to the next trophic level. As an example, imagine a habitat where the primary producers generate 100 units of energy. The next trophic level would use this energy and generate 10 units of energy that can be used at the next energy level, and so on, until the total energy of organisms at the 5th trophic level is only a tiny fraction ($10^{-4} \cdot 100$ units = 0.01 units) of the energy generated by primary producers.

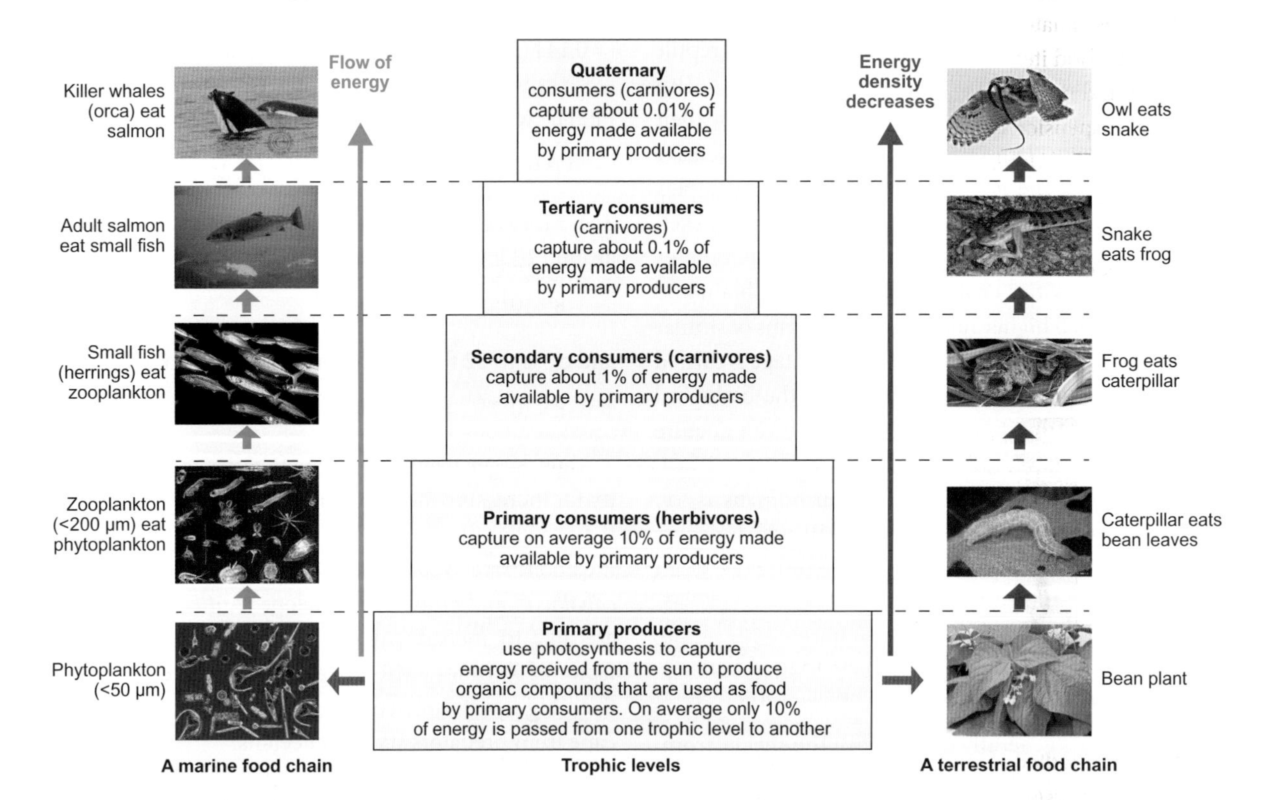

Figure 2.9 Simple food chains with five trophic levels in marine and terrestrial environments

Sources: Marine food chain: Phytoplankton: http://marinephytoplankton.org; Salmon: Hans-Petter Fjeld/Wikimedia Commons Caterpillar © Phil Bendle; Snake eats frog (c) Stewart Macdonald; Bean leaves, http://www.koosacupuncture.com

2.3.1 Types of food and feeding methods

Different foodstuffs comprise varying proportions of solids and liquids, both of which may contain the three major types of foodstuff, carbohydrates, fats and proteins, also in various proportions. The major characteristics of these foodstuffs are discussed in Box 2.1 and Table 2.1.

Different foods also vary in terms of their energy density—that is, the amount of energy contained per unit mass. The energy density of plant material varies from about 2.5 kJ g^{-1} in leaves to about 15 kJ g^{-1} in some seeds such as acacia (*Acacia* spp.) seeds. By contrast, the energy density of animal carcasses depends on the ratio of protein to fat. Because of its high water content, muscle has an energy density of about 5.3 kJ g^{-1} and stored glycogen an energy density of about 4.4 kJ g^{-1}, as shown in Table 2.1. However, the energy density of adipose tissue in the body is about 35.6 kJ g^{-1}. For example, the energy density of the edible carcass of a white-tailed deer (*Odocoileus virginianus*) consisting of 17.5 per cent body fat is just over 10 kJ g^{-1}.

Animals use a diversity of feeding strategies to acquire the foodstuff they need for their survival and reproduction.

Aquatic animals are often suspension feeders

Many species of aquatic animals feed directly on the **plankton** (singular plankter from Greek meaning wanderer) suspended in the water column. All these animals—called **suspension feeders**—use anatomical structures that act as strainers to remove small food items from the water. Many aquatic invertebrates, including some crustaceans, sponges and bivalve molluscs are suspension feeders. Sessile suspension feeders, such as mussels (*Mytilus* spp.) and oysters (*Ostrea* spp.), pump water over their gills[47] not only for gas exchange, but also for feeding, as shown in Figure 2.10A. Microscopic particles, including small (5–50 μm) phyto- and zooplankton, are filtered by the gills and transported to the digestive tract for ingestion. Many suspension feeders are able to filter out particles that are of no or low nutritional value before they reach the mouth, and reject them. In bivalve molluscs this sorting process is performed by the labial palps: prominent triangular flaps hanging on either side of the mouth.

[47] Ventilation of the gills of bivalve molluscs is discussed in section 12.2.2.

Given the low energetic efficiency between successive trophic levels, it is no surprise that some of the largest animals feed directly on the plankton and organisms smaller than several centimetres suspended in the water column. Thus, both whale sharks (*Rhincodon typus*)—the largest living non-cetacean animals, growing to more than 12 m in length and weighing more than 20,000 kg—and blue whales (*Balaenoptera musculus*)—by far the largest living animals with weights exceeding 130,000 kg—are suspension feeders.

Blue whales filter their food using a **baleen**, which is shown in Figure 2.10B. The baleen consists of hundreds of plates hanging down from the upper jaw, which form a comb through which large quantities of water are pushed. When it feeds, a baleen whale opens its mouth widely, taking in large volumes of seawater together with small organisms such as krill (*Euphausia superba,* which is only about 6 mm long), copepods and little fish. The partial closure of its mouth forces water through the baleen. Organisms that are unable to move through the baleen are trapped and then brushed by the tongue down the throat of the whale.

Thus, the largest animal on Earth consumes some of the smallest. As such, it harvests organisms low down the food chain, many of which are themselves suspension feeders. Blue whales have access to a much more abundant source of energy than if they consumed organisms from higher trophic levels such as large fish, marine birds or other species of marine mammals.

(A)

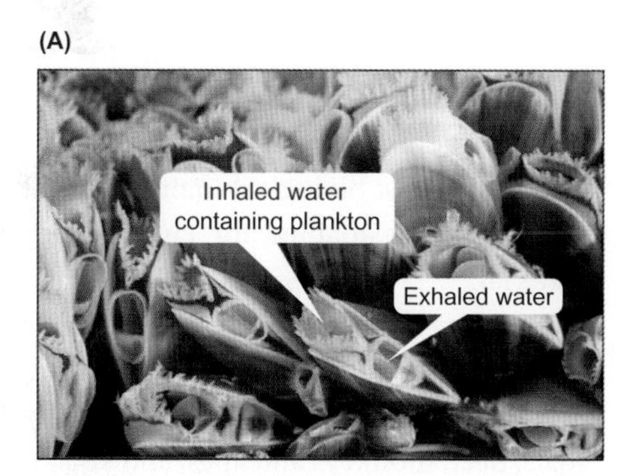

(B)

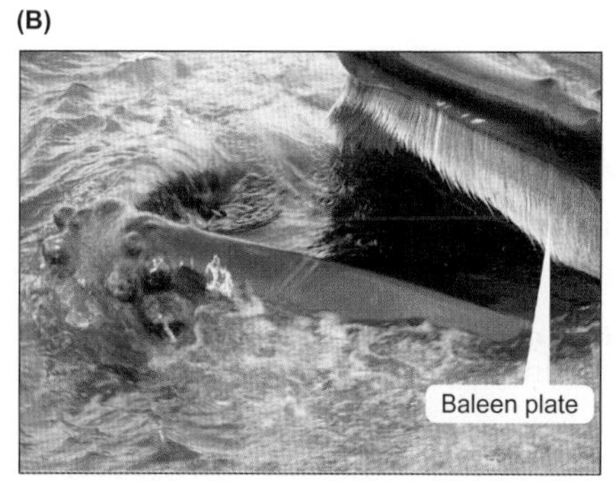

Figure 2.10 Filter feeding in bivalve molluscs (*Mytilus edulis*) and baleen whales (*Balaenoptera musculus*)

(A) Image of bivalve molluscs (size 2–10 cm) feeding on plankton. Water continuously enters through the inhalent siphon and is expelled through the exhalent siphon. (B) Image of a blue whale (11–30 m long) feeding on small organisms (few centimetres in size, like krill). Source A: © Alexander Semenov; B: http://citadel.sjfc.edu/students-virtual/a/arw00081/public_html/webdesign/mobilesite/whalesite.html#five.

Figure 2.11 Feeding apparatus in gastropods

(Ai) Photograph of a common garden snail (*Cornu aspersum*) and (Aii) diagram of the feeding apparatus in gastropods, which consists of a tongue covered in tiny teeth called a radula. The radula is moved back and forth by muscles dislodging food particles from the substrate that are ingested. The radula is supported by a cartilaginous projection called odontophore. (Bi) Photograph of a freshwater snail (*Marstonia comalensis*) and (Bii) Photomicrograph taken in a scanning electron microscope (EM) of the central teeth on the radula of the snail. (Ci) Photograph of limpets (*Patella vulgata*) which grow to about 60 mm diameter and (Cii) scanning EM micrograph of one of the 1620 radular teeth of a limpet.

From Ai: Jon Sullivan/Wikimedia Commons; Aii Benjamin de Bivort/Wikimedia Commons Bi and Bii: Hershler R, Liu H-P (2011). Redescription of *Marstonia comalensis* (Pilsbry & Ferriss, 1906), a poorly known and possibly threatened freshwater gastropod from the Edwards Plateau region (Texas). ZooKeys 77: 1-16. Ci: Tango22/Wikimedia Commons and Cii: Barber AH, Lu D, Pugno NM (2015). Extreme strength observed in limpet teeth. Journal of the Royal Society Interface 12: 2014.1326.

We discuss in Case Study 2.1 (see Online Resources) how the filtering apparatus of baleen whales has inspired the development of industrial self-cleaning filtering devices.

Many animals graze or browse on plants and algae

Grasses, other plants and algae eaten by grazers or browsers offer little resistance to being eaten, but have poor nutrient content, although the water content of plant material can be important to the water balance of land animals[48]. Thus, grazers, such as herbivorous mammals, tend to feed more or less continuously and consume relatively large quantities of food compared with animals that feed on fruits, seeds and animal material.

Plant material has to be thoroughly macerated to prepare it for digestion. The radula in gastropod molluscs, described in Figure 2.11, is well adapted for grazing. The very hard and sharp teeth on the radula enable gastropods to scrape algae and other material off hard surfaces and destroy plant cell walls. The radula teeth of limpets are made of the strongest known natural material, comparable to the strongest commercial carbon fibres: tightly packed, very thin iron-based mineral fibres laced with protein. The force that such fibres can withstand without breaking is considerably greater than that for spider silk, which, until recently, was considered to be the strongest biological material. Vertebrate grazers also have dentition adapted for cutting and macerating plant material.

Browsing animals feed on foliage from shrubs and trees in terrestrial habitats or on submerged algae, plants and corals in aquatic habitats. Typical browsers on land include giraffes, goats, deer, monkeys and apes, and koalas. Aquatic browsers include some species of fish, sea urchins and marine mammals such as dugongs and manatees. Elephants both browse and graze because their diet consists not only of tall grasses, but also of shrubs and tree foliage.

[48] Water balance in land animals is discussed in Section 6.2.1

Animals feeding on liquids

Nectar feeders, such as insects (bees, moths and butterflies) and some birds (hummingbirds, sunbirds and honeyeaters), have evolved special adaptations for accessing the sugar-(carbohydrate-)rich nectar produced by plants. The insects use modified mouthparts that form a tubular **proboscis** which draws nectar from flowers by capillary action, while birds have highly elongated beaks and extensible tongues that are either tubular or grooved with brush-like filaments at the tip to collect the nectar. Insect sap feeders, like aphids, have the mouthparts modified to act as a stylet, which is used to pierce the plant wall and suck out the sap.

Blood-feeding (hematophagous) animals use blood from fish, birds or mammals as their main source of food. Blood is full of nutrients, such as protein, and can be taken without great effort. Blood feeding evolved separately in different groups of animals including annelids like leeches; insects like mosquitoes (such as *Culex* spp.), bed bugs (*Cimex* spp.) and kissing bugs (*Rhodnius* spp.)[49]; fish like toothpick fish (*Vandellia cirrhosa*); and mammals like vampire bats (such as *Desmodus rotundus*).

Common specialized features of blood-feeding animals are their ability to puncture or cut through the skin of their victim, often without the victim noticing (due to injection of chemicals with anaesthetic action) and the release of anticoagulants that prevent the victim's blood from clotting. Indeed, the early development of anticoagulant medicines was based on substances found in the saliva of leeches.

Animals feeding on fruit and seeds

Fruit eaters feed on fruits that are rich in sugars (carbohydrates); seeds contained in fruits generally pass undigested in the faeces of fruit-eating birds and many mammals including monkeys and apes, fruit bats, bears and elephants.

Seed eaters have evolved strategies to remove the indigestible protective covering of seeds to gain access to the nutritious mass inside. Seeds are the most nutritious parts of a plant. Some birds, like most parrots, have strong sharp beaks which they use while keeping the seed in one of their claws to remove, crack and skilfully discard the seed shell before ingesting the seed's content. Squirrels also use their teeth while holding a nut in their front paws to crack and remove the nutshell to gain access to the nut's content. However, most seed-eating birds rely on their muscular **gizzard**, which contains small stones, to macerate the hard seeds into a paste.

Predators and scavengers

Predation involves the capture, killing and eating of other animals and is a strategy adopted by a diverse group of animals. Since prey animals normally offer considerable resistance to being eaten, predators use a wide variety of techniques for capturing their prey:

- predators like spiders and jellyfish use snares,
- wolves hunt in packs, while lions collaborate to ambush prey,
- solitary cats stalk their prey,
- praying mantises and snakes employ camouflage so that the prey does not recognize the predator until it is too late,
- great white sharks (*Carcharodon carcharias*), the Earth's largest predatory fish at up to 8 m long and weighing up to 3000 kg, and killer whales or orcas (*Orcinus orca*), the largest member of the dolphin family, use brute force.

Once captured, the prey is killed and eaten by tearing, shredding and chewing, or is simply swallowed whole—as is the case with insects eaten by vertebrates, or with various animals eaten by some species of snakes such as Burmese pythons[50].

Many predators rely on acute senses, such as hearing, vision or smell[51], to locate and capture their quarry. In response, many prey animals have evolved defences against predation which include toxins, hard shells or spines; the ability to defend themselves and fight back; mimicking unrelated poisonous animals; or fast reflexes to help them rapidly move away.

Scavengers feed on a range of dead, often rotting, organic material. Some—such as blowflies, wasps, vultures, crows, hyenas, coyotes and racoons—feed on dead animals; some—such as ants, termites, snails, millipedes and earthworms—feed on leaf litter; others—such as dung beetles and flies—feed on manure. Thus, scavengers fulfil an important role in all ecosystems by maintaining the 'hygiene' of the environment.

2.3.2 The passage of food through the digestive tract

Once ingested, food is subjected to further mechanical and chemical processing in the **gastrointestinal (GI) tract** (gut), which usually consists of a tube with chambers forming different parts of the gut. The gut is lined by an epithelial layer consisting of tightly packed cells; this layer forms a barrier between the internal environment of the animal and the lumen of the digestive tract, which is connected to the external environment.

The simplest guts occur in cnidarians (jellyfishes, sea anemones and corals) and free-living flat worms that have a simple sac-like gut with a single opening through

[49] Blood feeding of kissing bugs is discussed in Section 21.1.4.

[50] We discuss feeding by Burmese pythons in Section 15.2.1.

[51] The senses are discussed in Chapter 17.

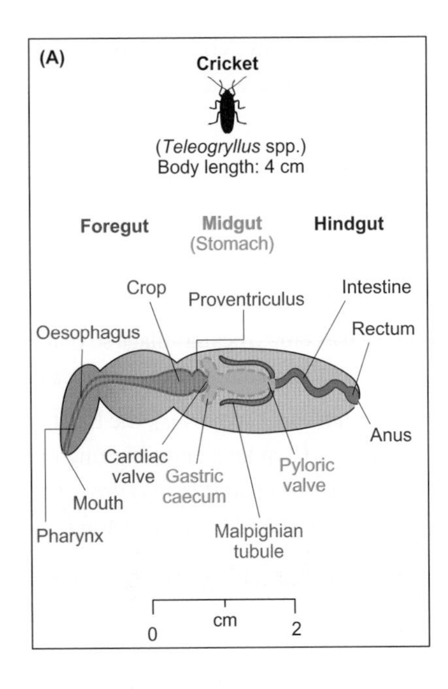

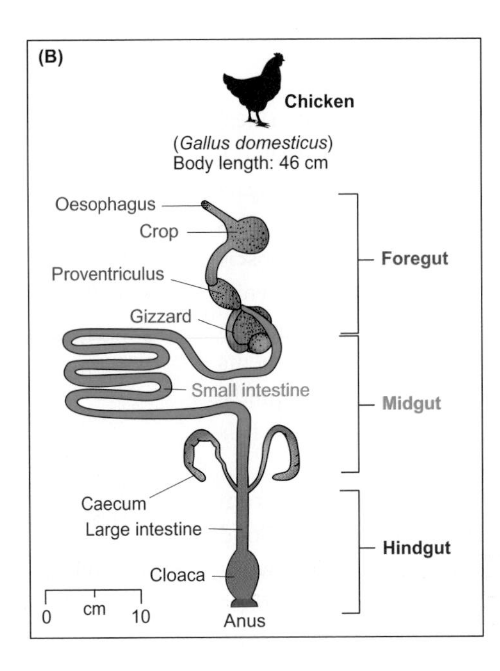

Figure 2.12 The digestive tract of insects and birds

(A) Morphological components of the digestive tract of crickets. The foregut and the hindgut of insects are lined with chitin, the same material that makes up much of the exoskeleton of the insect. When an insect moults (sheds its exoskeleton) it also sheds the internal lining of the fore- and hindguts. Malpighian tubules are the excretory organs in insects. (B) Morphological components of the digestive tract of the domestic chicken. Fermentation of ingested plant material in birds takes place in the caeca.

Source B: Stevens CE, Hume ID (1998). Contributions of microbes in vertebrate gastrointestinal tract to production and conservation of nutrients. Physiological Reviews 78: 393-427.

which food enters and waste products are released to the environment[52].

A simple tubular, one-way digestive tract with a mouth at one end and an anus at the other end evolved in roundworms (nematodes). The one-way digestive tract offers the great advantage of allowing regional specialization and independent intake of food and elimination of waste products.

More complex developments of a one-way digestive tract evolved in annelids, arthropods, molluscs, echinoderms and chordates[53] (which include the vertebrates). In these animals, the food is propelled through the gut by **peristalsis**, in which waves of contraction and relaxation of the smooth muscle around the tubular gut propel its contents in one direction, independent of the body movements of the animal. The evolution of an active mechanism of gut content propulsion allowed intermittent feeding and specialization of the gut to store food temporarily in enlarged chambers, such as the **crop** of annelids, arthropods and most species of birds, and the **stomach** of most vertebrates. For comparative purposes, the digestive tract in these animals can be divided into three sections: the **foregut**, **midgut** and **hindgut**.

In **insects**, the foregut comprises the **buccal cavity** (mouth), the **pharynx**, the **oesophagus**, the crop and the **proventriculus**, as shown in Figure 2.12A. The foregut is responsible for storage, grinding of solid food, and transport of the food to the midgut, where food is broken down into small molecules by enzymes and absorbed across the epithelium into the insect's body fluids. At the end of the midgut, the **pyloric valve** regulates the movement of material from the midgut to the hindgut.

Urine produced in the **malpighian tubules**[54] also enters the hindgut at the junction with the midgut and mixes with undigested material entering from the midgut. Absorption of water, salts and other beneficial substances takes place in the hindgut before a mixture of faeces and urine is excreted through the anus.

The foregut in birds refers to the section of the gastrointestinal tract from the buccal cavity (mouth) up to and including the gizzard, as shown in Figure 2.12B. The intestine in birds corresponds to the midgut, and the hindgut comprises the ceaca (singular—caecum), the large intestine (also called the colon) and the cloaca.

The food ingested by birds enters the oesophagus which in many species forms a pouch, called a crop, which primarily regulates the passage of food to the acid-secreting segment of the stomach, called the **proventriculus**. The acid-treated food then enters the gizzard, considered the 'muscular stomach', where it is macerated into a paste-like **chyme**, which passes into the small intestine where the food particles are broken down into molecules that can be absorbed. All ingested material that is not absorbed is moved to the large intestine (which is very short in birds) and then to the cloaca where it mixes with urine from the kidney before it is excreted through the anus as a mixture of faeces and urine[55].

[52] Section 1.3 discusses the two-cell layered Cnidaria and Figure 1.12 illustrates the gastrovascular cavity of sea anemones and jellyfish.

[53] Figure 1.14 illustrates probable chordate phylogeny.

[54] We discuss urine formation in insects and the handling of water and salts by the hindgut in Section 7.3.4.

[55] Post-renal salt and water absorption that occurs in the cloaca and large intestine (colon) of birds is particularly important for species living in arid habitats, as we discuss in Section 6.1.4.

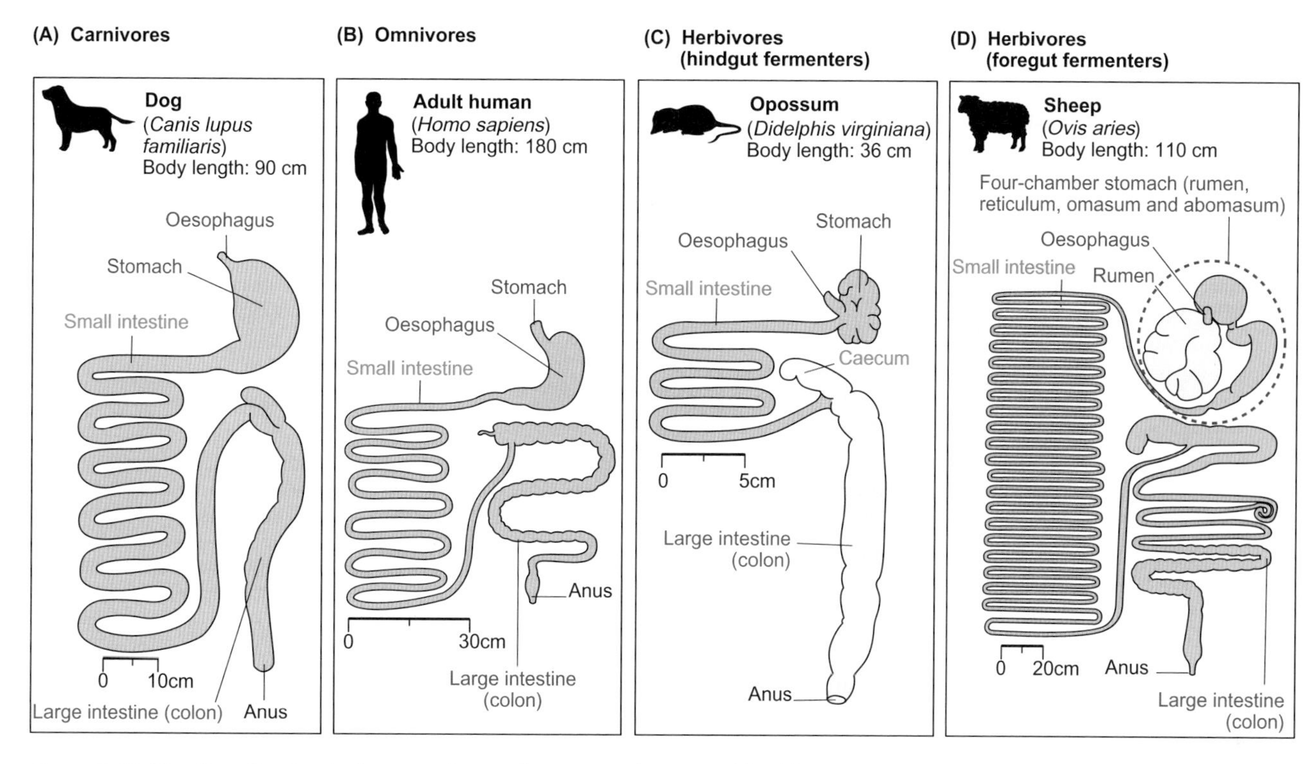

Figure 2.13 The digestive tract of mammals: carnivores, omnivores and herbivores

(A) Carnivores have simple stomachs and short digestive tract. (B) Omnivores have a relatively simple stomach and longer large intestines than carnivores. (C, D) Herbivores have digestive tracts that are specialized in parts to support communities of mixed fermenting unicellular microorganisms. In hindgut fermenters (C), fermentation occurs in the caecum and in an enlarged colon, whereas in foregut fermenters (D) fermentation occurs in specialized chambers in the stomach; for example, the rumen in ruminants acts as a fermentation vat. Sites of the digestive tract where fermentation takes place are highlighted in yellow. Notice that the diagrams are not drawn to the same scale, see scale bars.

Source: Stevens CE, Hume ID (1998). Contributions of microbes in vertebrate gastrointestinal tract to production and conservation of nutrients. Physiological Reviews 78: 393-427.

Birds usually have a pair of pouches connected to the junction between the small intestine and the large intestine called caeca, as shown in Figure 2.12B, where fermentation takes place[56].

The foregut of mammals refers to the section of the gastrointestinal tract from the mouth up to and including the stomach. Food is transferred from the mouth directly into the stomach via the oesophagus. The acidic content of the stomach (chyme) is neutralized as it enters the small intestine, which is generally long and coiled, as shown in Figure 2.13.

Food particles are broken down into simple molecules in the small intestine and are then absorbed across the intestine's epithelium. The large intestine or colon continues from the small intestine and material that has not been absorbed is discharged through the anus as faeces. At the junction between the small intestine and the large intestine, there is a pouch called the caecum, which together with the large intestine forms the hindgut in mammals.

In humans, a 5–10 cm long pouch, the **appendix**, extends from the caecum. It has recently been proposed that the appendix may act as a sanctuary for beneficial bacteria, particularly during a bout of diarrhoea.

The GI tracts of different mammals have very different morphologies and dimensions which are related to their diets. Figure 2.13 illustrates the main differences:

- **Carnivores**, like the dog, eat food of animal origin and have a relatively short colon.
- **Omnivores**, like humans, eat food of both plant and animal origin; omnivores have a relatively longer colon than carnivores.
- **Herbivores** eat food of plant origin and have digestive tracts that are specialized to support communities of mixed fermenting unicellular microorganisms in parts of their digestive tract.
  - **Hindgut fermenters**, like opossums, horses and rabbits, ferment the plant material in the caecum and in an enlarged colon.
  - **Foregut fermenters**, like sheep and other ruminants such as cattle and deer, have four specialized chambers in the stomach to provide suitable environments for microbial communities where fermentation takes place.

[56] Passerine birds feeding on seeds and insects have rudimentary caeca.

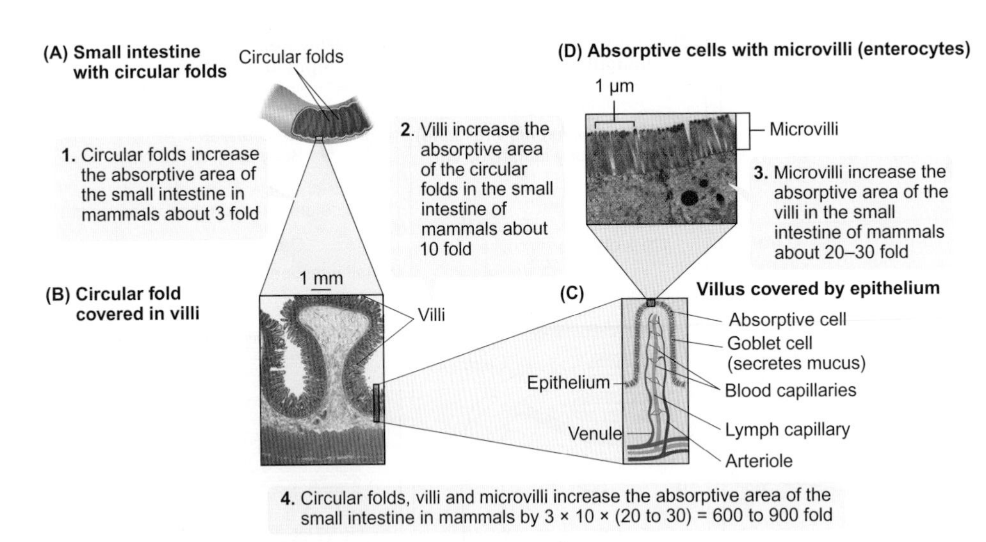

Figure 2.14 The absorptive area of the small intestine is increased several hundred fold by the presence of circular folds, villi and microvilli

Sources: B Stacey and Sternberg (eds) (2007). Histology for Pathologists. Third edition. Lipincott Williams & Wilkins. http://flylib.com/books/en/2.953.1.30/1/; D Mukherjee TM, Williams AW (1967). A comparative study of the ultrastructure of microvilli in the epithelium of small and large intestine of mice. Journal of Cell Biology 34: 447-461.

2.3.3 The digestion and absorption of different foodstuffs

Food can only be absorbed across the gut epithelium after its breakdown into small molecules during the process of hydrolysis, more commonly known as **digestion**. To be absorbed, the molecules of digested foodstuff must move from the gut lumen across the epithelial cells into the interstitial (tissue) fluid from where they enter the circulatory system. The rate at which absorption occurs depends on the rates of nutrient transport across a given area of intestine and the potential surface area available. The potential absorptive surface area is increased substantially in the intestine of many species of vertebrates by the presence of circular folds, villi and microvilli on absorptive epithelial cells, called enterocytes, as shown in Figure 2.14. The potential surface area for absorption is greater in endotherms than in ectotherms of similar mass.

Digestion and absorption of carbohydrates

Carbohydrates[57] in food can only be absorbed across the gut epithelium after enzymes (**glycosidases**) secreted into the GI tract or tethered to the surface of the gut epithelium break them down into monosaccharides such as glucose, fructose and galactose.

Amylases[58] break down starches by hydrolysing α-1,4-glycosidic bonds between the glucose subunits of carbohydrate molecules. However, amylases cannot hydrolyse terminal α-1,4-glycosidic bonds. Most animals secrete amylases; the most important amylase in vertebrates is secreted by the pancreas[59], which breaks down starches to maltose[60], but salivary glands in some species, including humans, also secrete amylase.

Disaccharides[61], such as maltose, sucrose and lactose, are hydrolysed to monosaccharides under the action of enzymes on the surface of the intestinal epithelium (microvilli):

- maltose is hydrolysed into two glucose molecules by the enzyme **maltase**,
- sucrose is cleaved into glucose and fructose molecules by the enzyme **sucrase**,
- **lactase** in mammals breaks down the lactose in milk into glucose and galactose.

The transport of monosaccharides across the gut epithelium is enabled by two families of transporters[62]: the **sodium-glucose-transporters** (SGLTs), which permit the absorption of glucose (and other monosaccharides) from the intestinal lumen into enterocytes against a concentration difference, and the **facilitative glucose carriers** (GLUTs), through which glucose and/or other monosaccharides are transported down a concentration difference across membranes. The transport mechanism of monosaccharides through the

[57] We discuss the molecular structure of carbohydrates in Box 2.1.

[58] We consider the molecular structure of amylase in Box 2.1 (Figure H).

[59] The pancreas is a digestive and endocrine organ of vertebrates that produces an alkaline solution containing digestive enzymes that assist the breakdown of carbohydrates, lipids and proteins in the small intestine. The pancreas also produces several hormones that circulate in the blood, such as insulin, that control the concentration of blood glucose, as we discuss in Section 2.3.5.

[60] Maltose is a disaccharide, as we discuss in Box 2.1.

[61] We discuss the molecular structure of disaccharides in Box 2.1.

[62] We discuss transporter proteins in Section 3.2.4.

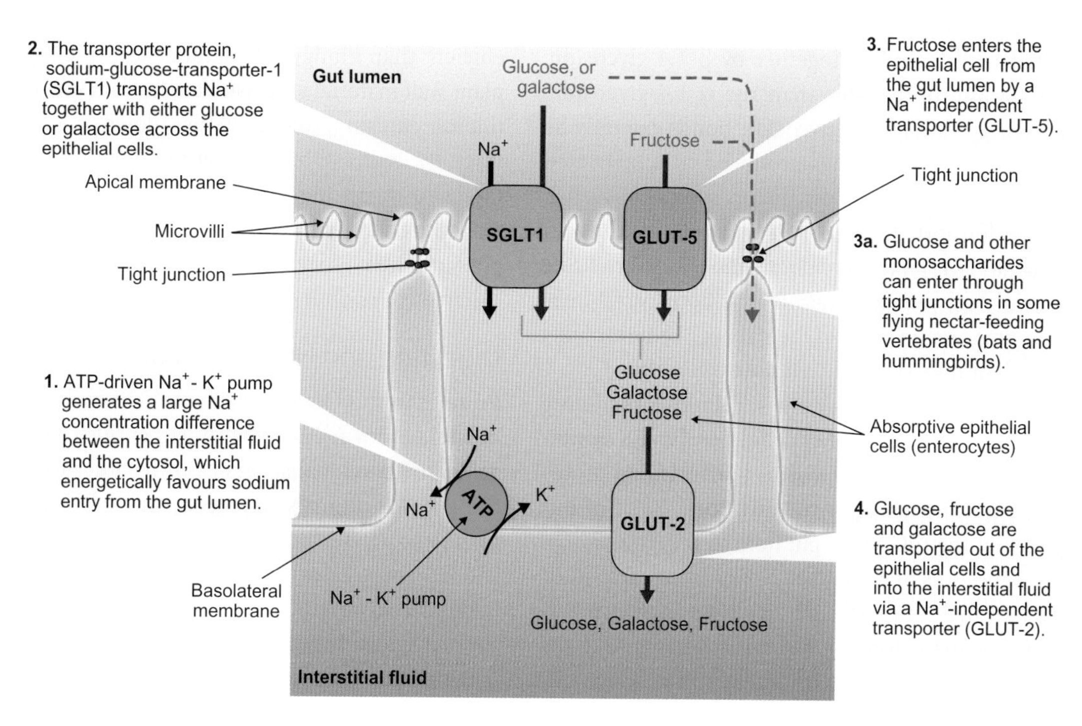

Figure 2.15 Absorption of monosaccharides across gut epithelium (based on studies on mammals)

intestinal wall has been most intensely studied in mammals and Figure 2.15 illustrates the major features of this transport mechanism:

- Glucose and galactose are transported into enterocytes against a concentration difference by a member of the SGLT family of transporters, called SGLT1. This process is an example of a **secondary active transport**[63] mechanism driven by sodium pumps (Na^+, K^+-ATPase)[64] in the basal borders of the cells.
- Fructose enters the cell from the gut by passive transport down a difference in concentration via a member of the GLUT family of transporters, GLUT-5.
- Glucose, fructose and galactose are transported passively out of the epithelial cells and into the interstitial fluid via another member of the GLUT family, GLUT-2.
- The monosaccharides enter the circulatory system from the interstitial fluid by simple diffusion and travel first to specific organs for processing and storage before continuing to the heart. For example, in vertebrates, substances absorbed in the small intestine travel first to the liver[65] for processing[66] via a system of veins forming the so-called **portal system**.

Recent studies have indicated that small flying vertebrates, such as hummingbirds and nectar-feeding bats can also absorb glucose (and other monosaccharides) by diffusion through tight junctions[67] between epithelial cells (paracellular pathway). The tight junctions in this case act as a sieve allowing high capacity absorption of glucose to support flight, which is energetically very demanding[68] in hummingbirds and small bats. The paracellular pathway for glucose transport may also occur in other vertebrate species.

In insects and other invertebrates, glucose is rapidly converted to **trehalose** (a disaccharide of glucose molecules[69]) in **fat body** cells on the haemolymph side of the gut. Trehalose is the major form in which carbohydrates are transported in the insects' haemolymph.

Cellulose digestion

Cellulose is an important structural component of the primary cell wall of green plants and many forms of algae.

[63] We discuss secondary active transport in Section 3.2.4.
[64] Section 4.2.2 discusses the functioning of Na^+,K^+-ATPase.
[65] The liver or its functional equivalent is an essential organ of vertebrates and some invertebrates. It plays a major role in carbohydrate metabolism, detoxification, protein synthesis and fat digestion.
[66] The liver cells in vertebrates have the capacity to convert fructose and galactose into residues of glucose that are then stored in the form of glycogen or glycogen-like material, which are polymers of glucose.
[67] We discuss tight junctions in Sections 3.2.3 and in Figure 4.10.
[68] We discuss the energy cost for flying in Section 18.6.1.
[69] We consider the molecular structure of trehalose in Box 2.1.

Cellulose consists of a linear chain of up to many thousands of glucose units linked together by β-1,4 glycosidic bonds,[70] which are more stable than the α-1,4 glycosidic bonds found in starches. β-1,4 glycosidic bonds can only be hydrolysed by **cellulase** enzymes produced by a restricted number of invertebrates including insects such as silverfish (*Ctenolepisma lineata*), termites (*Reticulitermes speratus*) and cockroaches (*Cryptocercus punctulatus*). No vertebrate produces cellulases, yet many herbivores, both vertebrates and invertebrates, obtain most of their energy requirements by digesting plant material with cellulose cell walls.

In general, animals which can digest cellulose, including those that produce some forms of cellulases, have a symbiotic relationship with microorganisms (bacteria and protozoa[71]) in their gut, which form the **gut flora**. Most of our knowledge about the role of the gut flora in the digestion of plant wall compounds like cellulose comes from early studies on domesticated animals, such as cattle, in which fermentation occurs in the foregut. The microorganisms in the foregut live in an anaerobic environment and produce cellulase enzymes that break down cellulose into glucose and also generate a cocktail of short-chain fatty acids (SCFA), such as acetic acid (CH_3COOH) and propionic acid (CH_3–CH_2–COOH), as well as methane, a more potent greenhouse gas than CO_2.

The energy obtained from the degradation of glucose to SCFAs may be partly used by the microorganisms themselves. The remainder of the SFCAs produced by the microorganisms are absorbed by diffusion across the gut epithelium into the blood of the host animal. Once absorbed, these SFCAs can be readily oxidized in the presence of oxygen, or converted into fats or carbohydrates.

Similar reactions in the presence of microorganisms occur also in the hindguts of animals such as horses, rabbits, rodents, elephants and rhinoceroses—all hindgut fermenters. Indeed, more recent studies have shown that the hindgut of all terrestrial vertebrates contains bacteria that perform a similar role to those in foregut fermenters.

Food passes through the gut in hindgut fermenters about twice as fast as in foregut fermenters. Hindgut fermenters use the products of fermentation mainly for their own benefit. In contrast, the microorganisms of foregut fermenters utilize a large proportion of the SCFAs themselves. As a result, animals using hindgut fermentation grow faster than foregut fermenters. Hindgut fermenters can also expel their gut flora during a period of dormancy like hibernation, which reduces the energetic cost of maintaining live gut flora.

Foregut fermenters, however, are about 50 per cent more efficient than hindgut fermenters at extracting energy from plant wall material and also benefit from the digestion of the microorganism protein further down the gastrointestinal tract. By contrast, hindgut fermenters need to re-ingest the faeces to digest the microorganisms and benefit optimally from the energy content of the food. Several species of hindgut fermenters, such as rabbits and guinea pigs, practice this behaviour, which is called **coprophagy**.

Digestion and absorption of proteins

Proteins in food items, which we discuss in Box 2.1, need to be broken down into very short peptide chains (2–4 amino acids) or individual amino acids in order to be absorbed in the gut. All animals have enzymes called **peptidases** that hydrolyse peptide bonds.

Some peptidases, called **endopeptidases,** break peptide bonds *between* particular amino acids, splitting peptides into shorter chains. The most well-known digestive endopeptidases in vertebrates are **pepsin**, **trypsin** and **chymotrypsin**.

- Pepsin is activated in the acidic environment of the stomach of vertebrates. Generally, invertebrates do not produce pepsin in their digestive system, but have **cathepsins**[72], which are activated at low pH.
- Trypsin appeared early in the evolution of animals and is produced in the digestive system of all animal phyla. Trypsin and chymotrypsin are produced by the pancreas of vertebrates in an inactive form and are activated in the gut in a slightly alkaline environment.

Those peptidases called **exopeptidases** hydrolyse peptide bonds at *the end* of a polypeptide chain in proteins: aminopeptidases catalyse the cleavage of the amino acid at the amino terminal and carboxypeptidases catalyse the cleavage of the amino acid at the carboxy (C)-terminal of the peptide chain. Exopeptidases are widely distributed throughout the animal kingdom.

Under the action of peptidases, the proteins are digested to short chains and single amino acids that cross the gut epithelium in two stages, as shown in Figure 2.16.

- Dipeptides and tripeptides are rapidly absorbed with protons (H^+) via a transporter protein called PepT-1. Indeed, it is now believed that more than 50 per cent of the ingested protein is absorbed as short peptides, at least in mammals. Inside the epithelial cells, these short peptides are broken down into single amino acids.

[70] We illustrate the different types of glycosidic bonds in Box 2.1.

[71] Bacteria are unicellular microorganisms that have cell walls but lack organelles and an organized nucleus. Protozoa are unicellular eukaryotes, which have organelles and a nucleus.

[72] Vertebrates also have cathepsins localized in cellular organelles called lysosomes, plasma membrane and cell nucleus, which are discussed in Section 3.2.1. Cathepsins are also implicated in cellular turnover such as bone resorption.

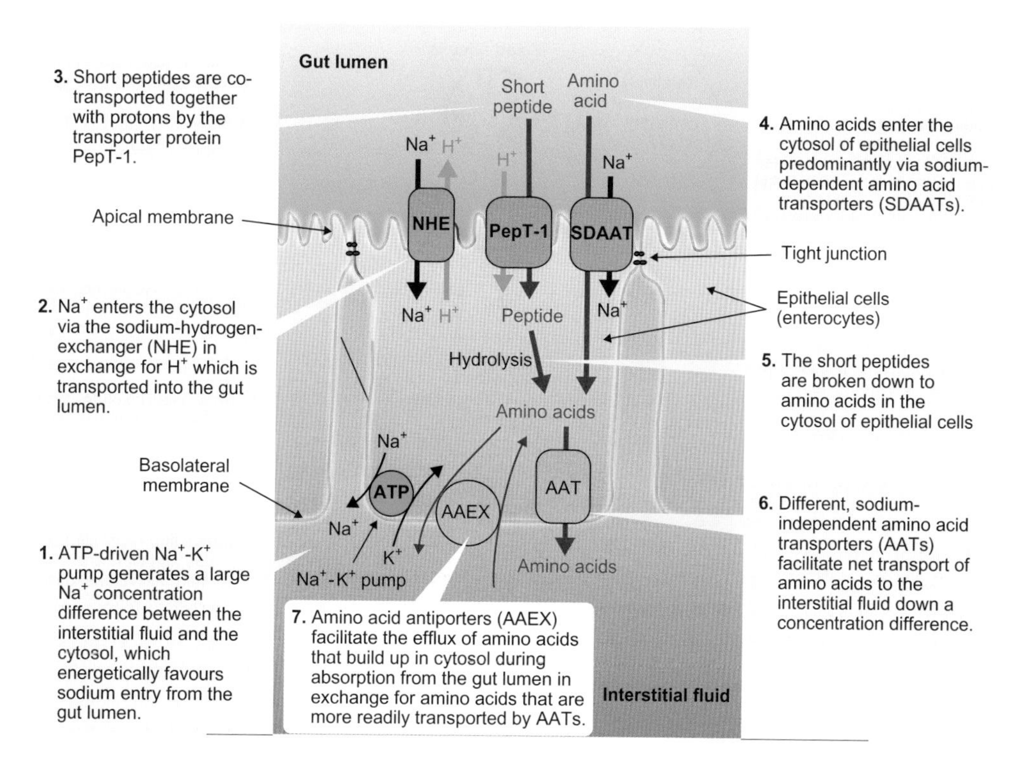

Figure 2.16 Absorption of small peptide and individual amino acids across gut epithelium

Amino acid antiporters in the basolateral membrane of enterocytes (AAEX) exchange amino acids in the cytosol for amino acids in the interstitial fluid. The AAEX antiporters work together with AAT transporters to release into the interstitial fluid those amino acids that are not transported by AATs; these include cationic amino acids, encompassing many of the so-called essential amino acids described in Box 2.1. Conversely, the AAEX antiporters also ensure that valuable amino acids are not depleted in the cytosol of enterocytes, as they would transport amino acids that are lacking in the cytosol in exchange for amino acids that are produced in enterocytes.

- Absorption of individual amino acid molecules from the gut fluids generally involves co-transport with sodium ions (Na^+).
- On the other side of the epithelial cells, passive transporters on the basolateral membrane, which include different types of amino acid transporters and antiporters (Figure 2.16), export amino acids into the interstitial fluid[73] from where they enter the circulation and are transported to all cells in the body of the animal.
- Most cells can synthesize non-essential amino acids such as alanine and glutamine from metabolic intermediates and use Na^+ co-transporters to maintain certain levels of non-essential amino acids. Antiporters in the cellular membranes facilitate exchange of non-essential amino acids for the essential amino acids needed to synthesize specific proteins as determined by the genes that are active in each individual cell at a particular time.

Digestion and absorption of lipids

Lipids[74] are a major source of stored energy in food and occur primarily as **triacylglycerols** (or triglycerides), which are also known as fats and oils. Triacylglycerols are esters of fatty acids and glycerol. Other types of lipids include glycerolphospholipids and cholesterol, which are important components of cellular membranes (as we discuss in Section 3.2.3), steroids and waxes. Waxes are esters of fatty acids with longer chain (fatty) alcohols (Fig Diii in Box 2.1) and are particularly prevalent in marine organisms. It has been estimated that at least half of the photosynthetic production of organic matter exists at some stage as wax esters of polyunsaturated fatty acids and polyunsaturated fatty alcohols. Most marine animals including crustaceans, fish, marine birds and some mammals store waxes.

The most important products of lipid digestion are the monoacylglycerols (or monoglycerols) and fatty acids. Before absorption, triacylglycerols are broken down into monoacylglycerols and fatty acids by enzymes called **lipases,**

[73] Transport of amino acids into the interstitial fluid is also discussed in Section 13.2.1.

[74] We discuss different categories of lipids in Box 2.1.

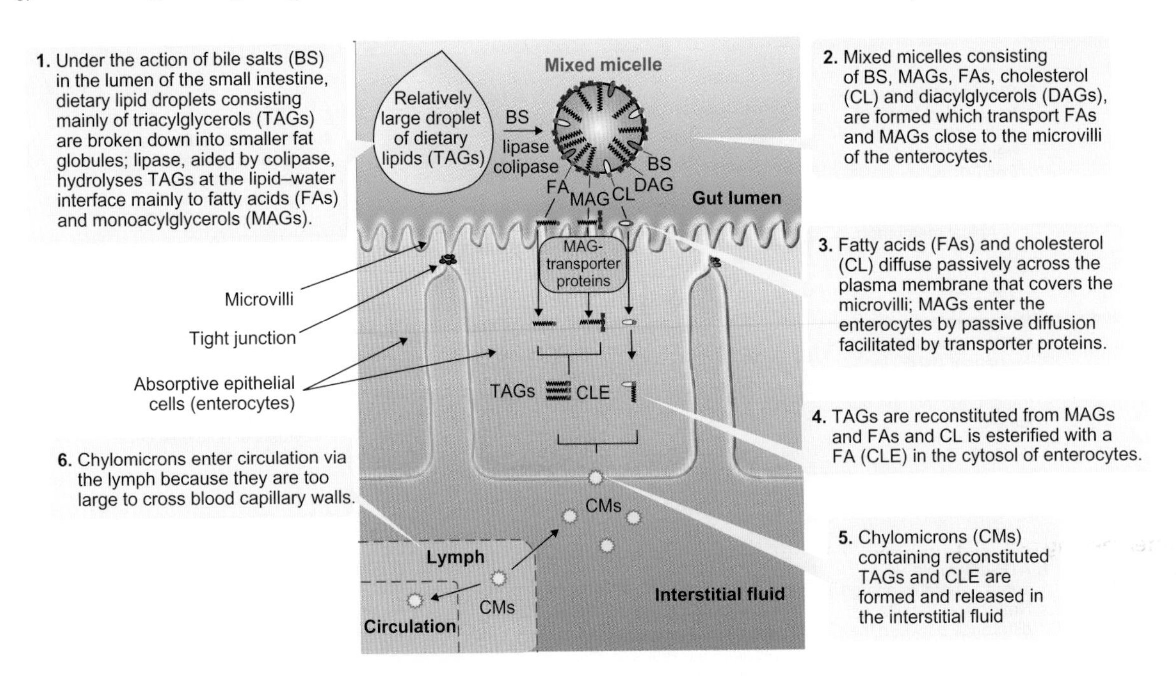

Figure 2.17 Absorption of lipids across gut epithelium (based on studies of mammals)

while waxes are hydrolysed into fatty acids and fatty alcohols by wax lipases. Since lipases are water soluble, while lipids are not, lipases can only act at the interface between lipids and digestive fluids. Attachment of lipases to the water–lipid interface is facilitated by small proteins called **colipases**.

In order to increase the fluid–lipid interface, animals use emulsifiers to break down the larger masses of lipids into small droplets. In vertebrates, bile salts secreted from the liver act as emulsifiers.

The released fatty acids, monoacylglycerols, cholesterol and alcohols can readily cross the cellular membrane of the epithelial cells[75] as shown in Figure 2.17. After entering the epithelial cells, triacylglycerols are re-constituted from monoacylglycerols and fatty acids and the cholesterol is esterified with a fatty acid. The resulting triacylglycerols are released to the interstitial fluid in combination with esterified cholesterol and specific proteins as small droplets of fat, called **chylomicrons** in mammals (generally <1 μm diameter) and **protomicrons** in birds (mean diameter about 150 nm). Protomicrons can enter the bloodstream across capillary walls, but chylomicrons are too large to cross capillary walls. Instead, they enter the lymph, which drains into the circulatory system, which in turn transports them to cells[76].

[75] We discuss the transport of molecules across cellular membranes in Section 3.2.1.

[76] We discuss the transport of lipids in the circulation and into the mitochondria of cells in Section 13.2.2.

2.3.4 Digestion, absorption and distribution of nutrients influence metabolic rate

Following the ingestion of food by animals, a significant amount of heat is produced. This heat production is associated with acid and enzyme secretion, digestion, absorption, distribution and storage of food material within the animal's body. For example, considerable heat is produced when peptide and glycosidic bonds are broken during digestion. Similarly, the absorption and transport of nutrient molecules across the gut epithelium is associated with an increased energetic cost and the production of heat.

The increase in metabolic rate above resting level associated with feeding is known as **specific dynamic action (SDA)**, **heat increment of feeding (HIF)** or **thermic effect of food (TEF)**. The magnitude and the duration of SDA depends on the animal species, the environmental conditions (including the ability to dissipate excess heat) and the amount and type of food consumed. For example, the SDA in humans and rodents lasts only a couple of hours but lasts about 12 hours or more in herbivores like cows because of the time it takes to digest the cellulose in the plant material they eat. In ectotherms, such as fish and reptiles, the duration of SDA is also dependent on the ambient temperature, with the duration increasing as the ambient temperature—and hence rates of digestion and absorption—decrease.

A particularly large SDA effect is observed in Burmese pythons (*Python bivittatus*) after fasting periods between

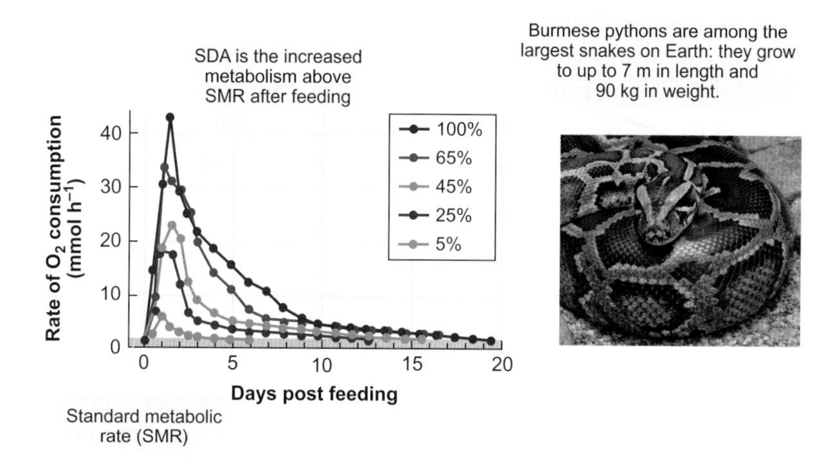

Figure 2.18 The specific dynamic action (SDA, or heat increment of feeding) in Burmese pythons (*Python bivittatus*) kept at 30°C after feeding on different meal sizes consisting of rodents

Meal size is expressed as percentage of snake body mass, which, on average, was around 700 g. The SDA is expressed as rate of oxygen consumption above standard metabolic rate (SMR). Average SDA coefficient, representing the ratio between the energy cost of SDA and the amount of energy ingested with the meal, was 0.32 ± 0.01, indicating that about one third of the meal's energy content was used by the snake in processes associated with the swallowing, acid and enzyme secretion, digestion, nutrient absorption and transport within the body, and protein synthesis.

Modified from Secor SM (2008). Digestive physiology of the Burmese python: broad regulation of integrated performance. Journal of Experimental Biology 211: 3767-3774
Photo: http://www.listenandlearn.org/

meals. Figure 2.18 shows the time course of the metabolic rate (measured as rate of oxygen consumption) in pythons kept at 30°C after ingesting meals of different sizes[77]. The pythons had been fasting for one month before experimentation. After pythons were fed, the metabolic rate increased steadily, reaching a peak within 36 h post-feeding, which was up to 40 fold greater than the SMR at 30°C for the larger meal sizes. The metabolic rate then decreased steadily to the resting level (SMR at 30°C) after 5–20 days, depending on the meal size. The magnitude of the SDA increased with the meal size and on average was about 32 per cent of the energy content of the meal (known as the SDA coefficient).

During fasting periods between meals, pythons exhibit a progressive reduction in the size of the endothelial cells (enterocytes) in the intestine, which results in a significant saving of energy. However, within 6 h post-feeding about 2400 genes in the intestine undergo differential expression and the mass of the intestine increases by about 70 per cent 12–24 h after ingestion, driven in part by a fourfold increase in the length of the microvilli. The increase in the mass of the small intestine is primarily the result of an increase in size (hypertrophy) of the enterocytes and secondarily of an increase in cell number (hyperplasia). The very large SDA effect observed in pythons reflects the sizeable amount of energy required for producing such significant changes in the gut morphology and function.

The reversible changes in gut structure and function observed in pythons are examples of **phenotypic flexibility**[78] and are common in all vertebrates when there are abnormally long periods of fasting (for that species) followed by subsequent feeding.

2.3.5 Energy storage, balance, hunger and satiety

Naturally occurring food contains carbohydrates, lipids, proteins and water in various proportions, as indicated at the beginning of this section and in Table 2.1. There is also an energetic cost associated with the digestion and processing of food: about 5 per cent of the total energy content for fats, 10 per cent for carbohydrates and 20 per cent for proteins. Therefore, the amount of food that needs to be ingested in order to provide the useable energy required to cover an animal's energy requirements varies greatly. For example, after digestion and absorption, the amount of energy that is useable to an animal is about:

- 34 kJ after ingesting 1 g adipose tissue from a carcass,
- 20 kJ after consuming 1 g seed kernels rich in fats,
- 3.5–6 kJ after taking in 1 g of meat, depending on source and fat content and
- 1–4 kJ after eating 1 g fresh fruit, which is predominantly carbohydrate and water.

[77] The cardiovascular responses to SDA in pythons are discussed in Section 15.2.1.

[78] Phenotypic flexibility is the ability of an organism to reversibly modify its phenotype. This is further discussed in Section 15.2.2.

Thus, an animal would need to consume about 9–34 times more fruit than fat, 5–20 more fruit than seeds, 6–10 fold more meat than fat or about 1–6 fold more fruit than meat to obtain the energy required for its survival.

Energy balance and storage

Animals achieve energy balance when the rate of energy intake in food equals the rate of energy expenditure. If the rate of energy intake exceeds its rate of expenditure over a prolonged period, the excess energy is stored as fat. Conversely, if the rate of energy expenditure exceeds the rate of energy intake on a prolonged basis, then the deficit leads to depletion of energy stores in the body, loss of body mass and eventual starvation.

As we discuss in Section 2.2.1, fat is the most efficient form of storing energy because it contains little water and has the greatest energy density. For example, 100 kJ of useable metabolic energy is stored in about 3 g adipose tissue, 18 g muscle or 25 g glycogen[79], such that from an energy storage point of view, 1 g fat stores six times and eight times more energy than 1 g muscle or 1 g glycogen, respectively. Indeed, all animals store energy mainly in the form of neutral fats (triacylglycerols). Many hibernating animals store energy as fat deposits before hibernation commences then slowly deplete these stores during hibernation[80].

King and emperor penguins (*Aptenodytes patagonicus* and *A. forsteri*) are among those species that fast for very long periods of time during the breeding season as a natural part of their annual cycle. The penguins deposit fat before courtship and mating and use it gradually during the breeding season. Adult king penguins naturally fast for about one month during their breeding cycle, which occurs during the Antarctic summer, whereas male emperor penguins fast during the Antarctic winter for about four months. The female emperor penguin lays one egg at a nest site which may be over 100 km from the sea. The egg is incubated solely by the fasting male emperor penguin, which is exposed to temperatures as low as –50°C and wind speeds of up to 200 km h^{-1}. We discuss the reproductive fast in king and emperor penguins in Case Study 2.2.

[79] Glycogen is the main form in which carbohydrates are stored in the body, as we discuss in Box 2.1.

[80] The storage and use of fat stores before and during hibernation is discussed in Section 10.2.7.

Case Study 2.2 The reproductive fast in king (*Aptenodytes patagonicus*) and emperor (*A. forsteri*) penguins

After laying her egg, a female emperor penguin goes to the sea for feeding, while the male penguin takes care of the egg and fasts. The female only returns at about the time the egg has hatched, taking over the care of the chick from the male penguin, which returns to sea for feeding. After an average of 115 days of fasting, the male penguin loses about 40 per cent of its initial body mass, primarily from fat stores.

Data on body mass and metabolic rate during fasting have been obtained from king penguins during the Antarctic summer, but similar changes occur in male emperor penguins during their winter fast. Both body mass and basal metabolic rate fall steeply at the onset of fasting; this is **phase I of fasting**. These initially steep falls are followed by more steady reductions during the majority of the fast (**phase II of fasting**), as shown in Figure A. Blood glucose levels are maintained close to their normal value. Both body mass and metabolic rate fall by similar proportions during phase II, so the mass-specific metabolic rate remains constant. At the end of phase II, the rate of loss of body mass increases quite substantially, shown in Figure A(i); this is **phase III of fasting**.

At the beginning of the fasting period, a 38 kg emperor penguin has approximately 11 kg of dry lipids and 6 kg of dry protein, as shown in Figure B(i). When the bird reaches a body mass of about 24 kg, it has around 2 kg of dry lipid remaining but approximately 5 kg of dry protein. The remainder of the loss in body mass could be accounted for by the loss of water associated with the adipose tissue and protein in the body[1] which is metabolized by the bird. From the data in Figure B(i), it has been estimated that during phase II, fat is the major source of energy (96 per cent). Although protein is largely spared during phase II, it is still metabolized, as shown in Figure B(ii), albeit at a slower rate than during the non-fasting period.

After the depletion of the relatively small glycogen stores, glucose for the central nervous system[2] may be mainly derived from amino acids. During phase III, the proportion of energy obtained from proteins increases substantially as the animal reduces its rate of fat metabolism. It is this increased usage of protein, which is less energy dense than fat, that leads to the increased rate of mass loss during phase III.

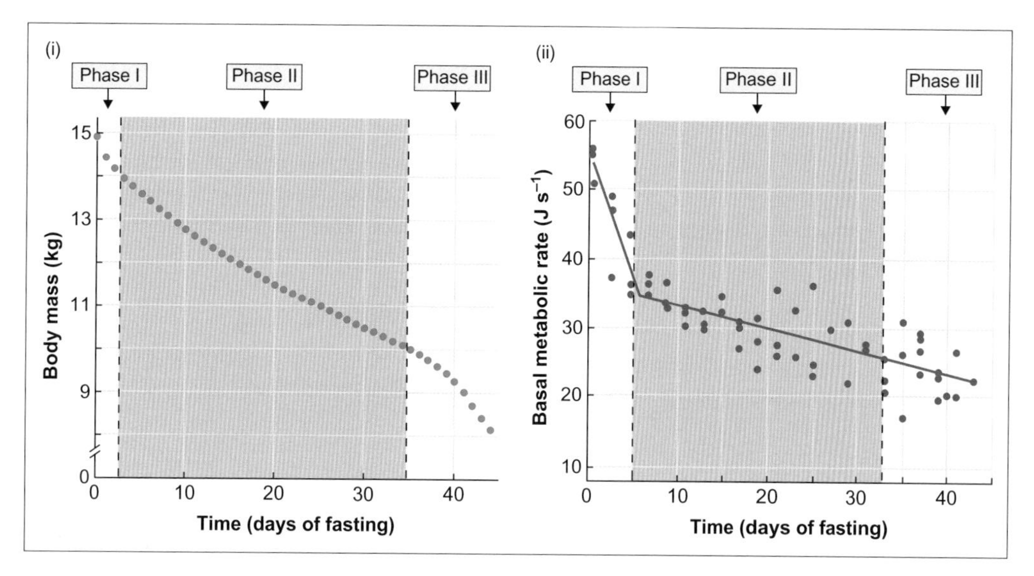

Figure A Changes in body mass and metabolic rate during the reproductive fast of king penguins (*Aptenodytes patagonicus*)

(i) Body mass showing the three different phases of the fasting period in king penguins. (ii) Metabolic rate decreases substantially during phase I but slows down during phases II and III.

Note: different groups of king penguins were used in the body mass and BMR experiments. Source: Cherel Y et al (1988). Fasting in king penguin. I. Hormonal and metabolic changes during breeding. American Journal of Physiology 254: R170-R177.

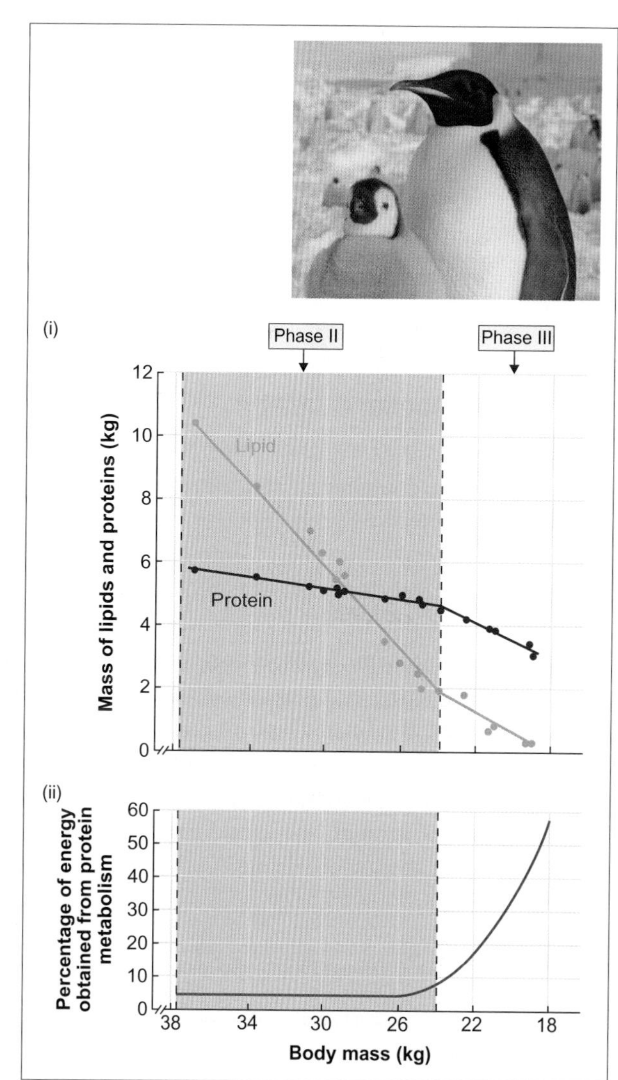

Figure B Roles of lipid and protein during the reproductive fast of emperor penguins (*Aptenodytes forsteri*)

(i) Masses of dry lipid and dry protein in penguins in relation to body mass; (ii) the proportion of metabolic rate derived from proteins during the fast. There is a substantial increase during phase III. In (ii), data were obtained from 17 birds. Note, it is not possible to record phase I for emperor penguins in the field, probably because they have started fasting before capture.

Source: (i) Groscolas R, Robin J-P (2001). Long-term fasting and re-feeding in penguins. Comparative Biochemistry and Physiology 128A: 645-655 (ii) Robin J-P et al (1998). Behavioral changes in fasting emperor penguins: evidence for a 'refeeding signal' linked to a metabolic shift. American Journal of Physiology 274: R746 -R753. Photo, with permission from Paul Ponganis.

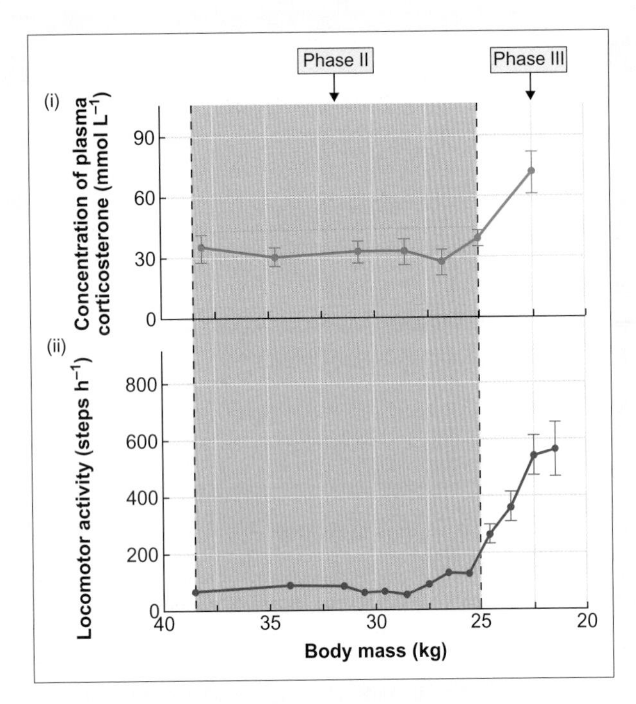

Figure C Plasma corticosterone and locomotor activity in emperor penguins (*Aptenodytes forsteri*) in relation to body mass during a winter fast

Both corticosterone and locomotor activity increase when body mass falls below about 25 kg. Mean values ± SE of mean from six penguins.

Source: Robin J-P et al (1998). Behavioral changes in fasting emperor penguins: evidence for a 'refeeding signal' linked to a metabolic shift. American Journal of Physiology 274: R746 -R753.

When an incubating male emperor penguin reaches a body mass of about 25 kg, its plasma corticosterone[3] and locomotor activity increase, as shown in Figure C(i) and C(ii). Eventually the male penguin abandons the egg/newly hatched chick if the female has not returned. The fat that remains enables the male to walk back to the sea to re-feed. The increase in plasma corticosterone could play an important role in initiating the re-feeding behaviour.

› Find out more

Groscolas R, Robin J-P (2001). Long-term fasting and re-feeding in penguins. Comparative Biochemistry and Physiology 128A: 645–655.

[1] The energy densities of lipids and protein, both in their dry state and when in the body, are given in Table 2.1.

[2] The nervous system in vertebrates is discussed in Section 16.1.2 and Box 16.1.

[3] The synthesis and structure of the steroid hormone corticosterone is illustrated in Figure 19.3 and discussed in Section 19.1.1.

Control of hunger and satiety

Animals have complex regulatory mechanisms to maintain energy balance. Mammals and other vertebrates have hormones that regulate energy input by adjusting hunger signals for food intake, and energy output by adjusting the proportion of energy used for ATP production, fat storage, glucose storage in the form of glycogen and heat loss. The regulatory mechanism for energy balance has been most intensely studied in mammals in which two peptide hormones[81] called **ghrelin** and **leptin**[82] interact.

Ghrelin is produced by specialized cells in the stomach wall. When the stomach is empty, ghrelin is released. It then circulates through the blood and binds to specific receptors in the hypothalamus[83], which gives rise to the sensation of hunger. The secretion of ghrelin stops when the stomach is stretched following food intake, which contributes to the termination of feeding. Ghrelin also increases gastric acid secretion and the motility of the gastrointestinal tract, which prepares the body for food intake. Therefore, ghrelin was dubbed the 'hunger hormone'. However, more recent studies in mice suggest that ghrelin does not alter the amount of food intake, but actually promotes the expansion of the fat tissue to increase an animal's energy stores.

Leptin, on the other hand, is the product of the *Ob* gene[84], which is expressed in fat cells. More leptin is produced when the fat level (that is, energy store) rises following feeding. Leptin circulates in the blood and binds to leptin receptors, which are located in the same cells as the ghrelin receptors in the hypothalamus. When the concentration of leptin in

[81] We discuss the mode of cell signalling of peptide hormones in Section 19.1.4.

[82] Evidence for the presence of hormones that are functional homologues to leptin and ghrelin have been found not only in vertebrates such as fish, but also in insects.

[83] Hypothalamus is part of the brain in vertebrates as we discuss in Section 16.1.2.

[84] Mutations in the *Ob* gene in mice causes obesity.

the blood rises above a set level, the number of activated leptin receptors increases; this increase contributes to a decrease in appetite. Leptin is therefore often called the 'satiety hormone'. At high concentrations, leptin also binds to receptors on non-adipose cells, stimulating their rate of energy expenditure, which dissipates as heat. The outcome is a decrease in the body's energy stores.

The location of the ghrelin and leptin receptors ensures that the two competing signals—one that indicates that the level of the fat energy stores has been exceeded (the leptin signal) and one that indicates that the stomach is empty (the ghrelin signal)—are received by the same cells in the hypothalamus. These cells respond by inducing a sensation that falls at an appropriate point between satiety and hunger, to ensure that energy balance is maintained.

Other hormones[85] released from the small intestine, particularly in the presence of luminal lipids, collectively act as a brake to reduce gastric motility and inhibit the emptying of the stomach. In this sense they can be viewed as intestinal 'satiety hormones'. Their action delays the passage of the **chyme** through the small intestine and maximizes absorption of nutrients across the epithelium of the small intestine.

An important element of energy metabolism is the hormone-dependent regulation of nutrient levels in the blood[86]. For example, the pancreas releases **insulin** in response to an increase in blood glucose (and some amino acids) from digested food, and releases **glucagon** when the blood glucose concentration is too low. Insulin promotes absorption of glucose from the blood by skeletal muscle to be stored as glycogen, and by adipose cells to be stored as fat for later utilization rather than for short-term energy usage. In contrast, glucagon causes the liver to convert stored glycogen into glucose, which is released into the bloodstream. Thus, insulin and glucagon are part of a feedback system that maintains a stable blood glucose concentration. Sections 19.2.3 and 19.3.2 also discuss the role of other hormones, such as melatonin and glucocorticoid hormones, in the regulation of the energy metabolism.

› *Review articles*

Bayne BL (1998). The physiology of suspension feeding by bivalve molluscs: an introduction to the Plymouth 'TROPHEE' workshop. Journal of Experimental Marine Biology and Ecology 219: 1–19.

Benson AA, Lee RF, Nevenzel JC (1972). Wax esters: major marine metabolic energy sources. Biochemical Society Symposia 35: 175–187.

Cordain L, Miller JB, Eaton SB, Mann N, Holt SHA, Speth JD (2000). Plant-animal subsistence ratios and macronutrient energy estimations in worldwide hunter-gatherer diets. American Journal of Clinical Nutrition 71: 682–692.

Flatt J-P (1995). Use and storage of carbohydrates and fat. American Journal of Clinical Nutrition 61 (suppl.): 952S–959S.

Friedman JM (2002). The function of leptin in nutrition, weight and physiology. Nutrition Reviews 60: S1–S14.

Karasov WH (2017). Integrative physiology of transcellular and paracellular intestinal absorption. Journal of Experimental Biology 220: 2495–2501.

Karasov WH, Hume ID (1997). Vertebrate gastrointestinal system. In Dantzler WH, Ed. Handbook of Physiology, Section 13: Comparative Physiology Vol. 1, pp 409–480.

McCue MD (2006). Specific dynamic action: A century of investigation. Comparative Biochemistry and Physiology, Part A 144: 381–394.

Mu H, Høy CE (2004). The digestion of dietary triacylglycerols. Progress in Lipid Research 43: 105–133.

Muhlia-Almazán A, Sánchez-Paz A, Carreño FL (2008). Invertebrate trypsins: a review. Journal of Comparative Physiology B 178: 655–672.

Place AR (1992). Comparative aspects of lipid digestion and absorption: physiological correlates of wax ester digestion. American Journal of Physiology 263: R464–R471.

Post DM (2002). The long and short of food chain length. Trends in Ecology and Evolution 17: 269–277.

Stevens CE, Hume ID (1998). Contributions of microbes in vertebrate gastrointestinal tract to production and conservation of nutrients. Physiological Reviews 78: 393–427.

Watanabe H, Tokuda G (2001). Animal cellulases. Cellular and Molecular Life Sciences 58: 1167–1178.

Wang T, Hung CCY, Randall DJ (2006). The comparative physiology of food deprivation: from feast to famine. Annual Reviews of Physiology 68: 223–251.

2.4 Intermediary metabolism involved in extracting energy from foodstuff

Intermediary metabolism comprises the whole set of life-supporting biochemical reactions within cells which are all necessary to animal life and are intrinsically linked to the flow of energy. These reactions enable animal cells to extract energy from nutritive molecules in a form that they can use to synthesize new molecular structures and perform diverse functions.

The constraints imposed by the limited range of temperatures within which animal life is possible and by the stringent conditions in the cellular environment have led to the development of sets of **metabolic pathways** involving enzyme-catalysed reactions that convert specific reactants to particular end products. The metabolic pathways that appeared in early life forms have been preserved with little change during evolution in all forms of life. The currency unit for energy transfer in all animal cells, the ATP molecule, can be produced either in the presence of oxygen (**aerobic metabolism**) or in the absence of oxygen (**anaerobic metabolism**). The first stage of both processes is **glycolysis**.

2.4.1 Glycolysis

Glycolysis consists of a set of enzyme-catalysed reactions, shown diagrammatically in Figure 2.19, in which glucose

[85] Cholecystokinin (CCK), neurotensin, peptide YY (PYY) and glucagon-like peptides (GLP).

[86] Figure 19.27 and the relevant section of text covers some of the interactive effects of hormones on plasma glucose, fatty acids and amino acids.

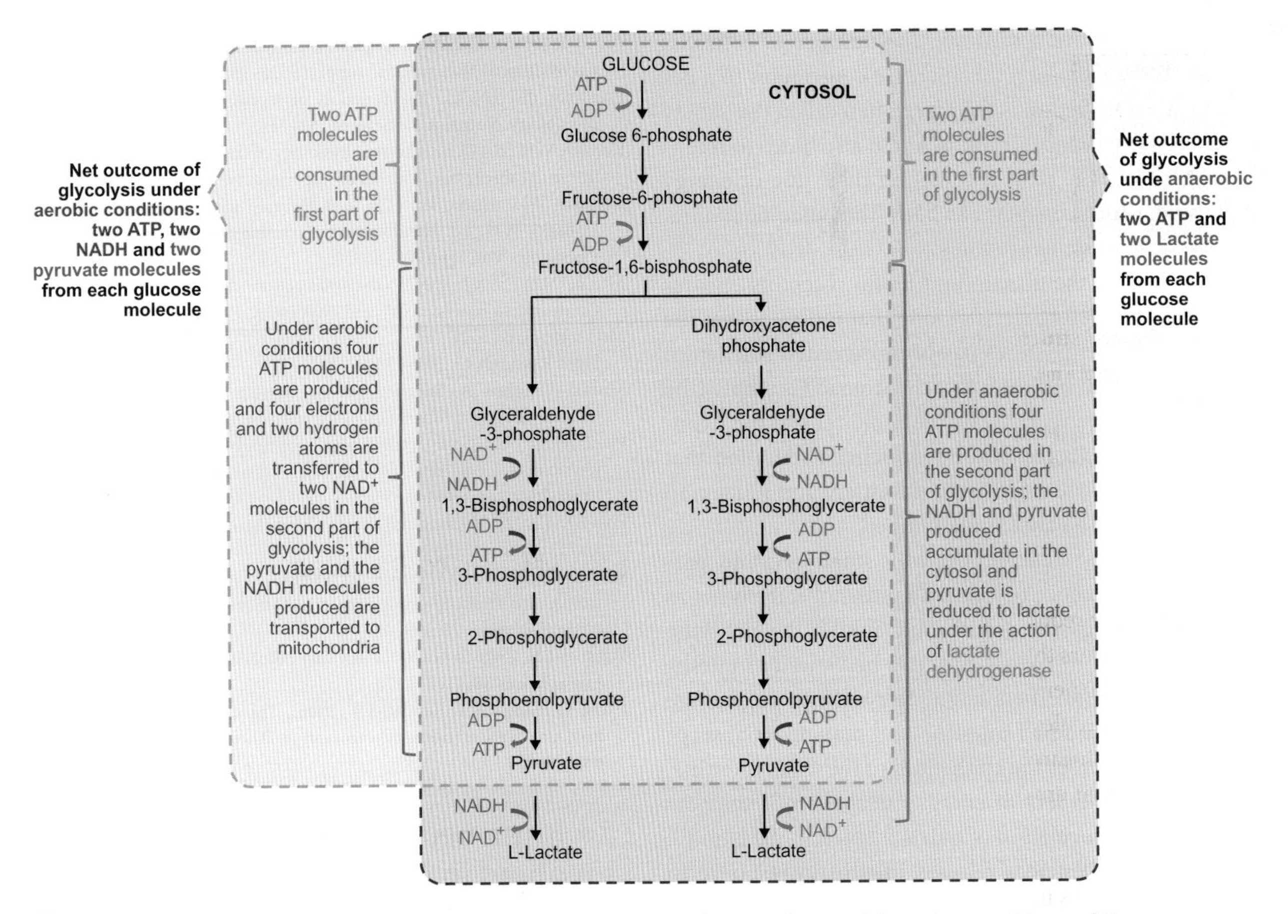

Figure 2.19 Glycolysis: the enzymatic breakdown of glucose in cytoplasm under aerobic and anaerobic conditions

Fructose-1,6-bisphosphate is broken down into one molecule of glyceraldehyde-3-phosphate and one molecule of dihydroxyacetone phosphate. The latter is then converted into glyceraldehyde-3-phosphate. Note that L-lactate is also produced whenever mitochondrial activity is not keeping pace with the amounts of pyruvate and NADH generated. For most animals, lactate is not a dead-end product as it can be exported to other cells where it can be reconverted to pyruvate and further oxidized in their mitochondria with production of ATP, when sufficient oxygen is available.

is broken down in the cell cytoplasm into two molecules of pyruvic acid (pyruvate). Glycolysis cannot proceed without the input of energy from two molecules of ATP in the initial reaction steps. Then, in later steps, four molecules of ATP are generated from four molecules of ADP and four phosphate ions, and two molecules of the oxidized coenzyme[87] nicotinamide adenine dinucleotide (NAD^+) are reduced to NADH. Thus, the net outcome of glycolysis is the production from one molecule of glucose of two molecules of pyruvate, two molecules of ATP and two molecules of NADH, which act as electron carriers.

In the presence of oxygen, the pyruvate produced by glycolysis is oxidized in the mitochondria, the organelles found in large numbers in the cells of most animals in which most ATP is produced. Figure 2.20 illustrates the double membrane of the mitochondria; the inner membrane is highly folded to form **cristae** that are packed with the membrane proteins involved in cellular respiration. The space enclosed by the inner membrane contains the **mitochondrial matrix** which is crammed with enzymes involved in the oxidation of molecules derived from foodstuff.

Anaerobic glycolysis

Under anaerobic conditions, the NADH produced in the second part of glycolysis (during which glyceraldehyde-3-phosphate is oxidized to 1,3-bisphosphoglycerate as shown in Figure 2.19) accumulates in the cytosol because it cannot be oxidized in the mitochondria. At higher NADH concentrations, the two pyruvate molecules are reduced to two molecules of L-lactate with the oxidation of two NADH molecules to two NAD^+ molecules, as illustrated in Figure 2.19. Thus, anaerobic glycolysis results in the conversion of one molecule of glucose into two molecules of lactate, with

[87] Coenzymes are molecules that accept specific functional groups of atoms or electrons and the associated energy from donor molecules in enzyme-catalysed reactions and then transfer these to acceptor molecules. Just like enzymes, coenzymes are continuously recycled without being consumed.

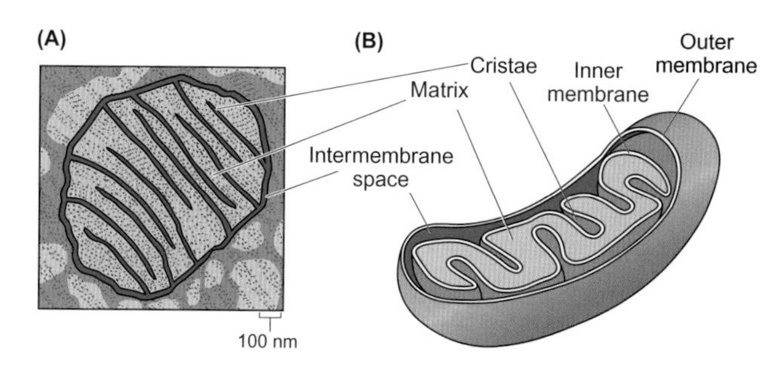

Figure 2.20 General structure of mitochondria

(A) Electron micrograph of a mitochondrion; (B) diagrammatic representation of a mitochondrion.

Source A: Alberts B, Johnson A, Lewis J et al (2002). Molecular Biology of the Cell. 4th edition. New York: Garland Science.

no net oxidation or reduction. In this process, the liberation of some energy occurs through the rearrangement of chemical bonds. About 50 per cent of this energy is used to regenerate two ATP molecules; the rest is dissipated as heat.

Since life first appeared in an environment lacking oxygen[88], it is likely that anaerobic glycolysis played a key role in providing metabolic energy to the various life forms that evolved on Earth over a period of about two billion years. As oxygen concentrations in the atmosphere started to rise[89], so animals evolved more efficient, oxygen-dependent metabolic pathways for extracting metabolic energy from foodstuff. Nevertheless, anaerobic glycolysis may have been retained to provide precursor molecules for **aerobic pathways** that extract energy from foodstuff and as a residual source of metabolic energy when oxygen is in short supply. For example, parasites in the gut of animals and mud dwellers that have no or little access to oxygen generate a large part of their energy by anaerobic glycolysis. Intense muscle activity also causes rapid depletion of oxygen in the muscle; under such conditions some ATP in muscle is regenerated by anaerobic glycolysis[90]. Note that the lactate produced during intense exercise in muscle is not a dead-end product because it can be exported to better oxygenated neighbouring cells or to the heart where it is converted back to pyruvate and oxidized in the mitochondria to produce more ATP.

Other sugar molecules, such as fructose and galactose, and glycerol can enter glycolysis after these molecules are converted by appropriate enzymes to one of the intermediary products of glycolysis shown in Figure 2.19. For example, galactose is converted to glucose-6-phosphate in four steps.

2.4.2 Aerobic metabolism

Three coenzyme molecules play a central role in the aerobic pathways of energy extraction from foodstuff, which occur in mitochondria: coenzyme A, which transfers the acetyl group between donor and acceptor molecules, and two electron transport molecules, NADH and $FADH_2$ (the reduced form of coenzyme flavin adenine dinucleotide). Under aerobic conditions, pyruvate is converted in the mitochondrial matrix to acetyl-coenzyme A in a reaction known as the **oxidative decarboxylation** of pyruvate. This reaction is depicted in Figure 2.21. During this process, the acetyl group is transferred from pyruvate to coenzyme A, one molecule of CO_2 is produced and one NAD^+ is reduced to NADH. The acetyl-coenzyme A undergoes further oxidation in the tricarboxylic (TCA) cycle, which we discuss later in this section. In addition to carbohydrates, fatty acids and proteins also undergo aerobic metabolism with the production of acetyl-coenzyme A, NADH and $FADH_2$.

[88] We discuss in Section 1.5 how animals evolved.

[89] Atmospheric changes in oxygen concentration are discussed in Section 11.1.1.

[90] Anaerobic metabolism during intense exercise is discussed in Section 15.2.4.

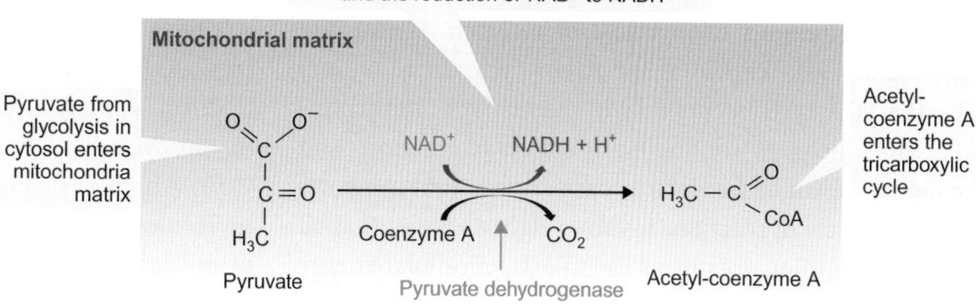

Figure 2.21 The oxidative decarboxylation of pyruvate in the mitochondrial matrix

The reaction can take place only under aerobic conditions when the NADH and acetyl-coenzyme A are recycled back to NAD^+ and coenzyme A. As two molecules of pyruvate are produced for each molecule of glucose, two molecules of acetyl-CoA, two molecules of NADH and two molecules of CO_2 are produced per molecule of glucose during this stage of aerobic metabolism.

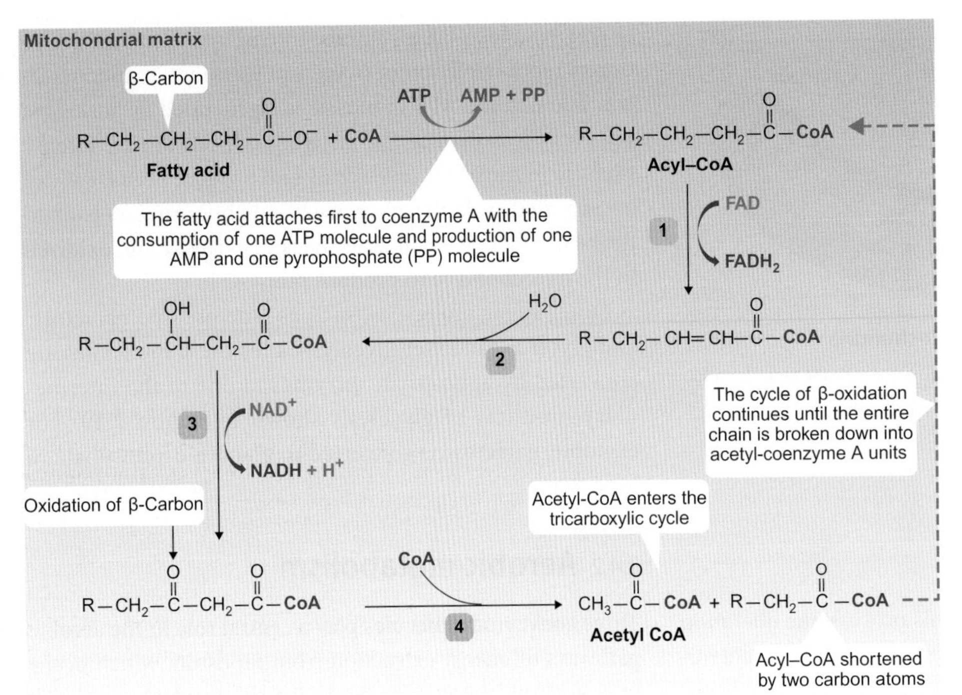

Figure 2.22 The β–oxidation cycle of fatty acids
Each cycle of β-oxidation, which liberates one acetyl-CoA unit, occurs in a sequence of four reactions: 1, dehydrogenation by FAD; 2, hydration; 3, oxidation by NAD^+ which is reduced to NADH and releases one proton; 4, cleavage into an acetyl-CoA molecule which enters the tricarboxylic cycle, and an acyl-CoA molecule which continues the cycle of β-oxidation. Note that ATP is converted to AMP (not ADP) in the activation step before fatty acids enter the first cycle of β- oxidation. Energetically this is equivalent with the use of 2 ATP molecules that are converted to 2 ADP molecules.

β-oxidation of fatty acids

Figure 2.22 summarizes the oxidation of fatty acids through a set of reactions known as beta (β)-oxidation. This set of reactions produces the three major precursor molecules required for the aerobic pathways: acetyl-coenzyme A, NADH and $FADH_2$. In this process the β-carbon of the fatty acid (i.e. the carbon next to the carboxyl group (–COOH)) undergoes oxidation to a carbonyl (–C=O) group and results in successive removal of two carbons from the fatty acid carbon chain. This reaction takes place in the mitochondrial matrix and generates from the two carbons removed from the fatty acid one acetyl-coenzyme A (which enters the tricarboxylic acid cycle), one NADH and one $FADH_2$ molecule, which are used in the aerobic section of ATP production.

The tricarboxylic acid cycle

The tricarboxylic acid (TCA) cycle or citric acid cycle (named after the first reactant in the cycle) is a universal metabolic pathway present in all aerobic animals. This pathway was identified as a cycle by Hans Adolf Krebs, who received the Nobel Prize for Physiology or Medicine in 1953 for his contributions to this field of research. Hence, the TCA cycle is also known as the Krebs cycle. In all animal cells, the TCA cycle occurs only in the matrix of mitochondria, which helps to confine the products of reactions and the reduced form of the electron carriers to a tight compartment. The cycle of reactions involves eight enzymes that all oxygen-consuming organisms use to extract cellular energy from carbohydrates, fats and proteins.

Figure 2.23 shows the major steps in the TCA cycle. The acetyl group of acetyl-CoA produced by the decarboxylation of pyruvate from glycolysis and by β-oxidation of fatty acids is transferred to a four-carbon acceptor compound (oxaloacetate) to form citrate (a six-carbon compound with three carboxyl groups). The citrate then undergoes a series of reactions causing:

- release of two CO_2 molecules,
- transfer of electrons and associated energy to four coenzyme electron carriers: three NAD^+ molecules are reduced to form NADH and one FAD molecule is reduced to form $FADH_2$,
- formation of one guanosine triphosphate molecule (GTP) from one molecule of guanosine diphosphate (GDP) and one inorganic phosphate (P_i), followed by conversion of GTP to GDP and ADP to ATP,
- regeneration of an oxaloacetate molecule, which can accept another acetyl group from another acetyl-CoA molecule to repeat the cycle.

Under physiological conditions, the TCA cycle is mainly regulated by the accumulation of the reduced form of the electron carrier NADH, which inhibits several enzymes in the cycle, and by substrate availability, such as acetyl-CoA, which is determined by the availability of foodstuff molecules. The NADH concentration depends on its rate of utilization in other reactions, particularly during oxidative phosphorylation, which we discuss in the next section.

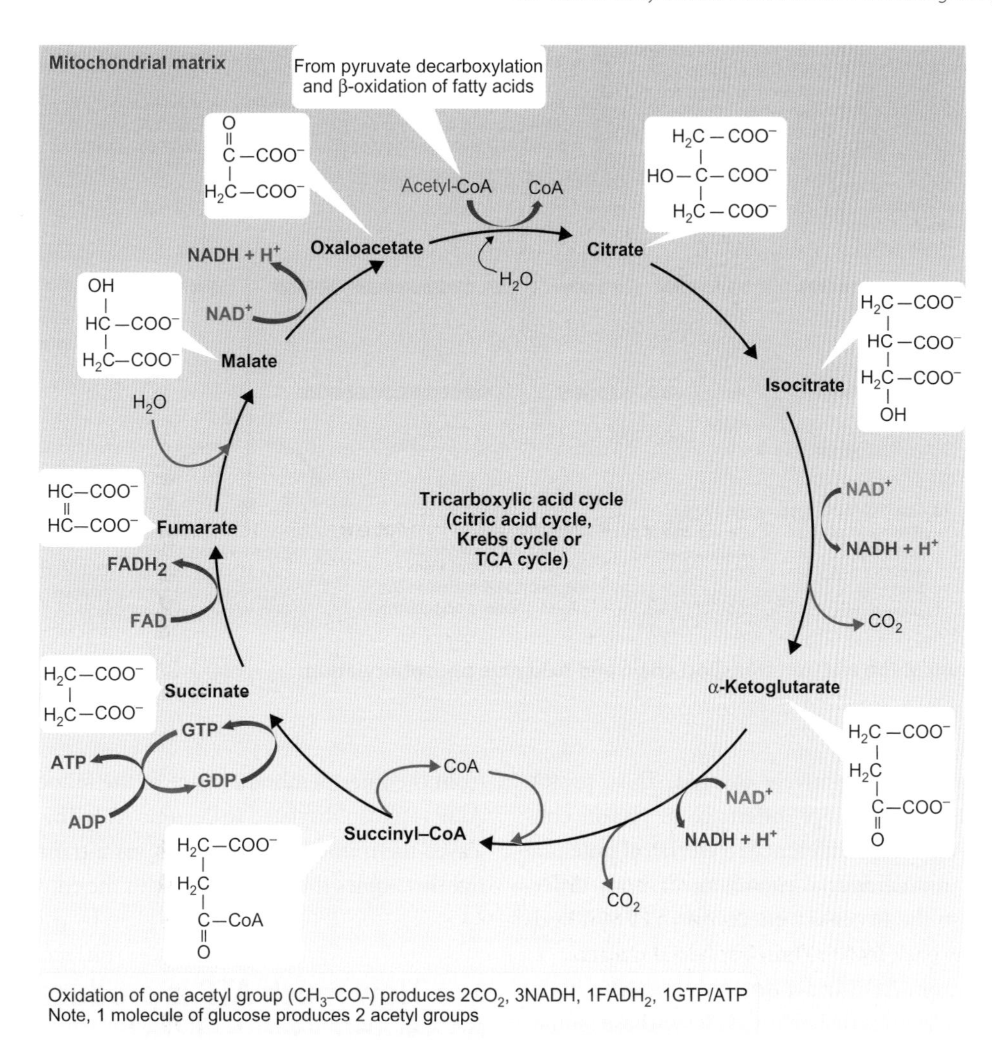

Figure 2.23 The tricarboxylic cycle (TCA cycle, citric acid cycle or Krebs cycle)

This regulatory mechanism prevents the wastage of large amounts of metabolic energy by overproducing NADH.

Amino acids can also be used for energy production after the removal of the amino group (which in vertebrates takes place in the liver and kidney), with the remainder of the molecule entering glycolysis, β-oxidation or the TCA cycle.

Electron transport and oxidative phosphorylation

The final segment of **aerobic respiration** couples the oxidation of the electron carriers NADH and $FADH_2$ to the phosphorylation of ADP to become ATP. As such, this process is called **oxidative phosphorylation**. The sequence of events is shown in Figure 2.24.

The electrons released from NADH and $FADH_2$ pass through **the electron transport chain**, which consists of four protein complexes (I–IV) located in the inner mitochondrial membrane. When the electrons reach their final destination on complex IV, they are transferred onto oxygen molecules together with protons, resulting in the production of water molecules as shown in Figure 2.24.

The energy produced from the flow of electrons through complexes I to IV drives the movement of protons (H^+) across the inner mitochondrial membrane from the mitochondrial matrix into the small intermembrane space between the two mitochondrial membranes. This creates a very large difference in proton concentration between the intermembrane space, which has a high H^+ concentration (and hence a low pH[91]), and the mitochondrial matrix, which has a relatively low H^+ concentration (and a higher pH).

Moreover, the positive charge carried by protons when they are driven across the inner mitochondrial membrane makes the intermembrane space more positively charged

[91] The relationship between hydrogen ion concentration and pH is discussed in Section 13.3.

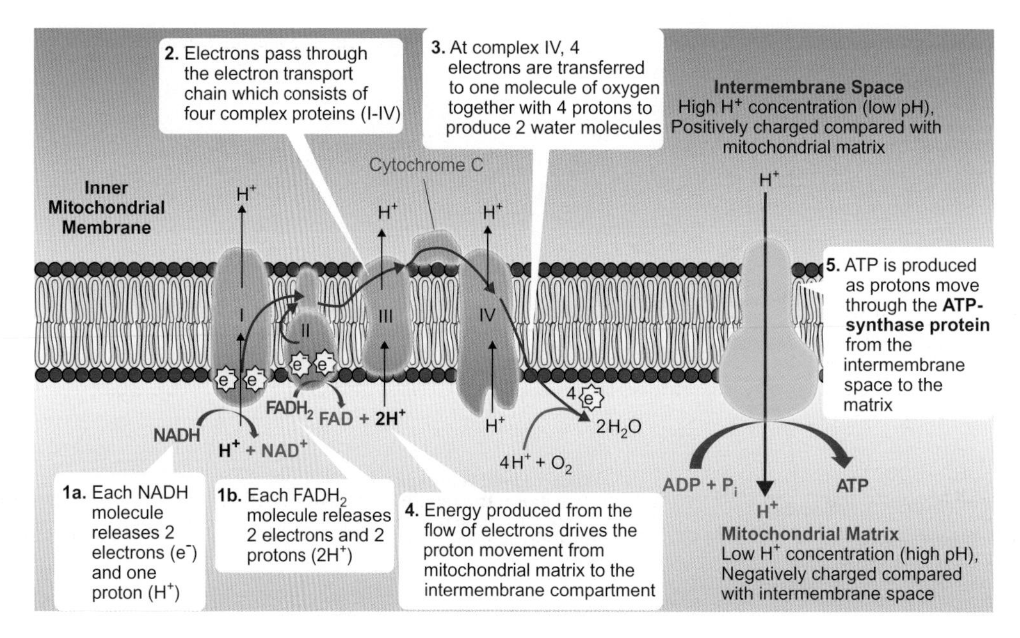

Figure 2.24 Diagram of the electron transport chain and oxidative phosphorylation

and the matrix relatively negatively charged. Thus, proton pumping generates both a large difference in proton concentration and a very large difference in electrical potential across the inner mitochondrial membrane[92]. Both differences contribute to the necessary energy for ATP synthesis when protons flow back into the mitochondrial matrix.

ATP synthesis from ADP and inorganic phosphate in the mitochondrial matrix is carried out by **ATP synthase** (sometimes called F_1F_0 ATPase). ATP synthase is also located in the inner mitochondrial membrane as shown in Figure 2.24. When protons pass through ATP synthase, moving down a difference in both concentration and electrical potential, from the intermembrane space to the mitochondrial matrix, ADP is phosphorylated to become ATP. The synthesized ATP leaves the mitochondrial matrix via a transport system that tightly couples the exit of ATP with the entry of ADP.

[92] We discuss how an electrical potential difference across the membrane affects the movement of electrically charged particles across cell membranes in Section 3.1.3 and Box 16.3.

2.4.3 How much ATP is produced by aerobic metabolism?

Figure 2.25 shows the balance sheet for ATP production from one molecule of glucose. The maximum theoretical yields for ATP production from the oxidation of NADH and $FADH_2$ molecules present in the mitochondrial matrix are traditionally considered three and two ATP molecules respectively. However, there are two possible pathways by which the two NADH molecules generated in the cytosol at

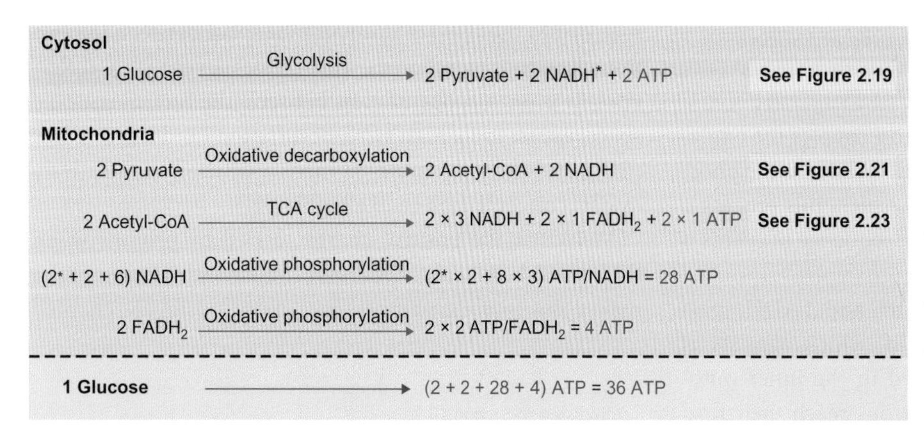

Figure 2.25 Overall absolute maximum ATP production from the oxidation of one molecule of glucose under aerobic conditions

In these calculations, we used the maximum theoretical yields for the ATP production from NADH (3 ATP) and $FADH_2$ (2 ATP) except for the ATP production from NADH produced in the cytosol during glycolysis (*), where the conversion yield may be 2, depending on which pathway is used. Note, these values are not reached in animals because of inefficiencies.

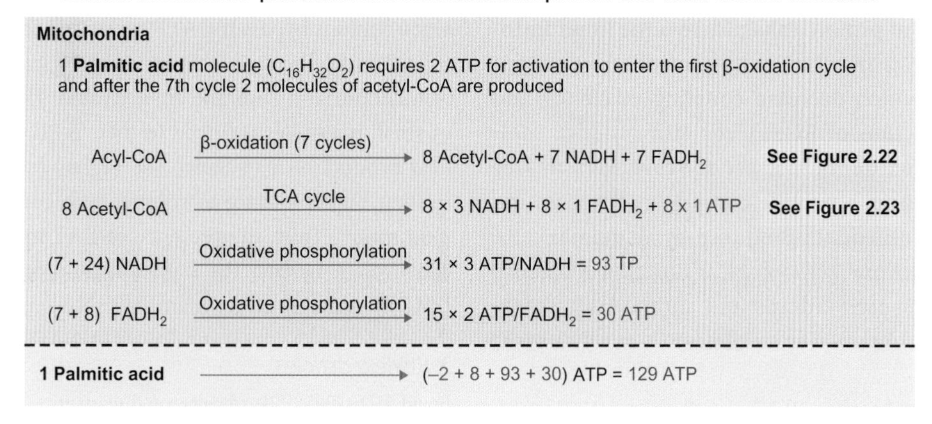

Figure 2.26 Overall absolute maximum ATP production from one molecule of palmitic acid under aerobic conditions

In these calculations we used the maximum theoretical yields for the ATP production from NADH (three ATP) and $FADH_2$ (two ATP). Note, these values are not reached in animals because of inefficiencies.

the end of aerobic glycolysis can be transported across the mitochondrial membranes to the mitochondrial matrix. By one pathway, the maximum ATP yield from NADH molecules is reduced from three to two, whereas it remains at three by the other pathway.

Using these values, the absolute maximum number of ATP molecules produced from the oxidation of one molecule of glucose in the presence of oxygen is either 36 or 38 ATP molecules, depending on which pathway is used to transport the two NADH molecules from the cytosol, as depicted in Figure 2.25. However, theoretical analyses and experimental observations suggest that the efficiency for ATP production, called **coupling efficiency**, is not 100 per cent and that a maximum number of 27 or 29 ATP molecules may actually be produced from the complete oxidation of one molecule of glucose. As we discuss in Sections 2.1.2 and 2.2.1, the Gibbs energy change for the production of 1 mol ATP from 1 mol ADP and 1 mol phosphate is about $+50\ \text{kJ}\cdot\text{mol}^{-1}$ ATP under physiological conditions. Then, the net Gibbs energy transferred to produce 27–29 mol ATP is about +1400 kJ ($50\ \text{kJ}\cdot\text{mol}^{-1} \times 28$ mol ATP = 1400 kJ). This amounts to about 50 per cent of the total Gibbs energy released from the oxidation of 1 mol glucose (2878 kJ) shown in Figure 2.4.

What is more, some protons leak back across the inner mitochondrial membrane without passing through the ATP-synthase; this leakage reduces ATP production even further. Such leakage of protons converts electrical energy into heat and may constitute approximately 20 per cent of BMR in a range of vertebrates and invertebrates.

By comparison, under anaerobic conditions, only two molecules of ATP are obtained per molecule of glucose. In this case the efficiency of anaerobic respiration in extracting chemical energy from glucose to produce ATP is only 3.5 per cent (2 mol ATP × 50 kJ mol^{-1}/2878 kJ = 3.5 per cent)[93].

When fatty acids and glucose are fully oxidized in the presence of oxygen, about twice as many ATP molecules are produced per g of fatty acid than per g of glucose. As illustrated in Figure 2.26, the full oxidation of one molecule of palmitic acid—the most common fatty acid in animals and plants—generates a maximum of 129 ATP molecules. However, as with the oxidation of glucose, theoretical and experimental evidence suggests that the actual maximum number of ATP molecules produced from the oxidation of one molecule of palmitic acid is about 96 ATP. If we now consider that 1 mol (i.e. 180 g) glucose yields about 28 ATP molecules and that 1 mol (i.e. 256 g) palmitic acid yields about 96 ATP molecules, then the maximum yield of ATP production is 0.156 ATP moles/g glucose and 0.375 ATP moles/g palmitic acid. This indicates that the ATP yield per g is 2.4 times greater for palmitic acid than for glucose.

Oxidative phosphorylation is vital if animals are to extract the vast amount of energy they need to survive, but it also results in the production of chemically reactive molecules of oxygen called **reactive oxygen species (ROS)**[94] such as **superoxide** (an oxygen molecule with an extra, unpaired electron, O_2^-). The production of ROS leads to the formation of other chemically reactive molecules that contain unpaired electrons, called **free radicals**.

The presence of free radicals can have damaging effects on cells because they chemically react with and modify

[93] Note, however, that the actual efficiency of the anaerobic glycolysis for producing 2 mol ATP at a cost of 2 × 50 kJ for each mol of glucose that is broken down into 2 mol lactic acid with the liberation of about 200 kJ (mol glucose)$^{-1}$ Gibbs energy, is much higher, and close to 50 per cent (2 × 50 kJ/200 kJ =50 per cent).

[94] We discuss the production of reactive oxygen species in Section 11.2.2.

cellular components, such as cell membranes, proteins and DNA. The control of proton leakage across the inner mitochondrial membrane may be tightly controlled by uncoupling proteins[95], which decrease coupling efficiency and reduce the production of ROS. Cells also produce molecules called **antioxidants** that act as scavengers of free radicals. **Glutathione,** a peptide comprising three amino acids, is a major antioxidant produced by all animal cells. Importantly, for normal physiological function a balance must exist between free radicals and antioxidants (some which can also be obtained from diet, such as vitamins C and E).

As we discuss in Section 10.2.5, small mammals in particular have a highly vascularized brown-coloured fat tissue packed with mitochondria, called brown adipose tissue (BAT), which is specialized for heat production. Heat is produced when protons bypass the ATP synthase and leak back into the mitochondrial matrix when uncoupling proteins in the inner mitochondrial membrane are activated. The presence of BAT in small mammals is linked to their need for a higher rate of heat production to balance the higher rate of heat loss due to their higher surface to volume ratio compared with larger mammals[96]. Thus, the presence of BAT in smaller mammals can explain in part the higher mass- (or volume-) specific BMR than that in larger mammals shown in Figure 2.6.

› *Review articles*

Brand MD (2005). The efficiency and plasticity of mitochondrial energy transduction. Biochemical Society Transactions 33: 897–904.

Divakaruni AS, Brand MD (2011). The regulation and physiology of mitochondrial proton leak. Physiology 26: 192–205.

[95] Uncoupling proteins are discussed in Section 10.2.5.

[96] Equation E in Box 2.3, shows that $Area \propto Volume^{2/3}$, so $Surface\ area/Volume \propto Volume^{-1/3}$, indicating that the surface to volume ratio decreases as the size (volume) of the animal increases.

Checklist of key concepts

General principles of energy transfer in animals

- All animals need a regular input of energy from the environment to maintain their complex level of organization and to function.
- Animals obtain most of the energy necessary for their survival in the form of **chemical energy** associated with changes in chemical bonds between atoms in their foodstuff.
- Chemical energy can be converted in the body of animals into other forms of energy including electrical, mechanical and thermal energy, as necessary.
- Reactions in the body can proceed spontaneously only if the energy content of the products of reaction (in the form of Gibbs energy) is lower than the energy content of the reactants.
- Non-spontaneous reactions do not happen without energy input from outside.
- All animals use **ATP** as the **common form of energy currency** to power reactions in the body that would otherwise not be energetically feasible: energy is generally released when ATP is hydrolysed to ADP.
- Equilibria are stable conditions that do not require energy input; physiological steady states are stable conditions that require continuous energy input.

Metabolism, energy metabolism and metabolic rates

- **Metabolism** encompasses all reactions within the body of an animal which allow it to stay alive and respond to changes in its environment.
- Metabolism can be broadly subdivided into **catabolism** and **anabolism**:
  - Catabolism refers to metabolic processes that involve the breakdown of larger compounds into smaller molecules—generally with the release of energy that is used by cells.
  - Anabolism refers to metabolic processes that involve more complex molecular structures being synthesized from simpler molecules, using energy obtained from catabolic processes.
- **Energy metabolism** concerns the flow of energy between different reactions in the body of the animal and between the animal and its environment.
- Most animals extract the bulk of their energy from the oxidation of carbohydrates and fats in the presence of molecular oxygen—a process called **aerobic respiration**.
- Oxidation of carbohydrates in the absence of oxygen can also provide energy through **anaerobic respiration.**
- The most common method for estimating the amount of energy used by animals is based on measurements of the amount of oxygen consumed.
- The amount of energy that can be released by aerobic respiration from 1 g of fat-containing food is several fold greater than from 1 g of carbohydrate- or protein-containing food.
- **The respiratory quotient (RQ)** is the ratio between the rate of CO_2 production and rate of O_2 consumption at the cellular level.
- The RQ provides insight into the type of foodstuff used by animals to cover their energy requirements.
- **The respiratory exchange ratio (RER)** is recorded in whole animals and equates to RQ when the animal is in a steady state.
- **Basal metabolic rate (BMR)** refers to the minimum rate of energy utilization required to sustain the life of endotherms.

- **Standard metabolic rate (SMR)** refers to baseline metabolic rate required to sustain the life of ectotherms at a particular temperature.
- Smaller-bodied animals have, on average, larger mass-specific BMRs or SMRs than larger-bodied animals of the same taxonomic groups.
- The BMR of endotherms is 15–20 fold greater than the equivalent SMR of ectotherms of similar mass and at similar body temperature.
- **Field metabolic rate (FMR)** is the average daily metabolic rate of free-living animals in the wild.
- FMR is on average about three-fold greater than the BMR (SMR).
- The **maximum metabolic rate (MMR)** is about 10 × BMR in smaller mammals, but MMR in larger mammals exceeds BMR by a factor of 20 or more.
- On an annual basis, endotherms expend about 30 times more energy than ectotherms of similar mass.

Energy intake in the form of food

- The vast amount of energy necessary for an animal's survival is obtained from food.
- Animals adopt diverse feeding strategies to acquire food, including filter feeding, grazing on plant materials, browsing, nectar and sap feeding, blood feeding, fruit and seed feeding, predation and scavenging.
- One-way digestive tracts with muscular walls allow intermittent feeding independent of an animal's body movements.
- The digestive tract is lined by an epithelium, which separates the lumen from the internal environment of the animal, and is generally divided into three sections: foregut, midgut and hindgut.
- Animals have complex regulatory mechanisms involving hormones to ensure energy balance. In mammals we talk about a **'hunger hormone' (ghrelin)** and **'satiety' hormones (leptin and intestinal hormones).**

Intermediary metabolism involved in energy extraction from foodstuff

- ATP is the molecular unit for energy transfer in all animal cells.
- Cells use the ratio between ATP/AMP concentrations to sense how much energy is available to them to function and maintain the energetic status within strict limits.
- **Coenzymes NAD^+, FAD** and **coenzyme A** play crucial roles in transferring energy from the oxidation of foodstuff to ATP.
- **Glycolysis** results in the production of ATP in the cytoplasm both under aerobic and anaerobic conditions.
- Under aerobic conditions, glycolysis also transfers energy from glucose to NAD^+, FAD and pyruvate, which can be used to produce much more ATP in mitochondria.
- **The tricarboxylic acid (TCA) cycle** in the mitochondrial matrix extracts energy from the oxidation of the acetyl group from pyruvate or fatty acid molecules with the help of coenzyme A.
- **The electron transport chain** and **oxidative phosphorylation** are the final segments of aerobic respiration in which energy is transferred from the oxidation of NADH and $FADH_2$ to ATP with the transfer of electrons to molecular oxygen to produce water in the presence of protons.

Study questions

* Answers to these numerical questions are available online. Go to www.oup.com/uk/butler

1.* According to Figure 2.4 the amount of Gibbs energy liberated at 25°C and 100 kPa from the oxidation of 1 mol (180 g) glucose is −2878 kJ. Using the information provided in Figure 2.4:
   i) calculate the amount of Gibbs energy released from the oxidation of 1 mol (46 g) ethanol (C_2H_5OH) in an animal at 25°C and 100 kPa considering that each mol of ethanol produces 2 mol of CO_2 and 3 mol H_2O, and that the breakdown of ethanol into its component elements in their stable states absorbs 174 kJ mol^{-1},
   ii) compare the amount of Gibbs energy produced at 25°C and 100 kPa from the aerobic oxidation of 1 g ethanol to that produced from the oxidation of 1 g glucose.
2.* Calculate the amount (in g) of water produced from the aerobic oxidation of:
   i) 18 g of glucose and
   ii) 18 g of stearic acid.

   (Hint. Look at Equations 2.2 and 2.3 and consider that the molecular mass of glucose, stearic acid and water are 180 g, 284 g and 18 g, respectively.)
3. Describe the meaning of the following terms: basal metabolic rate, standard metabolic, field metabolic rate and maximum metabolic rate. (Hint: See Sections 2.2.3 and 2.2.6.)
4.* Estimate the steady state RER value of a mammal, which provides 60 per cent of its energy needs from the oxidation of carbohydrates and 40 per cent from the oxidation of fats, knowing that the RQ values for providing the energy needs from carbohydrates and fats are 1.0 and 0.70, respectively. (Hint: Assume that about the same amount of oxygen is consumed to generate a certain amount of energy from either carbohydrates or fats, as shown in Table 2.1.)
5.* The resting metabolic rate (RMR) of teleosts fish in their natural environment is described by the following allometric equation:

   RMR ($J\ s^{-1}$) = 0.00055 × (body mass in g)$^{0.8}$

2

i) Calculate the RMR values corresponding to fish of 50 g, 250 g and 1000 g, and describe how the respective mass-specific BMR values compare.

ii) How does the RMR of a 1 kg teleost fish compare with the BMR of an average 1 kg mammal, considering that the allometric equation for mammals is BMR ($J\ s^{-1}$) = 0.018 × (body mass in g)$^{0.74}$ as shown in Figure 2.5A.

6. Justify the basis for assessing the amount of energy utilization by an animal under aerobic conditions from measurements of heat production or the amount of oxygen consumption by the animal. (Hint: Look at Table 2.1 and Sections 2.2.1 and 2.2.3.)

7. Explain what is meant by the term specific dynamic action (SDA). (Hint: Section 2.3.4.)

8. Describe what glycosidic, peptide and esteric bonds are and in what type of foods such bonds are encountered. (Hint: Box 2.1.)

9. Compare and contrast three feeding methods employed by animals. (Hint: Section 2.3.1.)

10. Estimate the absolute maximum number of ATP molecules that could be produced from the β-oxidation of stearic acid ($C_{18}H_{36}O_2$) assuming that three and two ATP molecules are produced for each molecule of NADH and $FADH_2$ oxidized, respectively. (Hint: Section 2.4.3 and Figure 2.26.)

11. Highlight the major steps involved in the chemical digestion and absorption of carbohydrates, proteins and lipids. (Hint: Section 2.3.3.)

12. Explain how animals digest cellulose. (Hint: Section 2.3.3.)

13. Outline the general mechanism for achieving energy balance in mammals. (Hint: Section 2.3.5.)

Bibliography

Alberts B, Johnson A, Lewis J, et al (2002). Molecular biology of the cell. 4th ed. New York: Garland Science.

Atkinson DE (1977). Cellular energy metabolism and its regulation. Cambridge, MA: Academic Press.

Berg JM, Tymoczko JL, Stryer L (2011). Biochemistry, International edition. New York: WH Freeman, p. 287.

Crowe J, Bradshaw T (2014). Chemistry for the biosciences. The essential concepts, 3rd ed. Oxford: Oxford University Press.

Garrett RH, Grisham CM (2012). Biochemistry, 5th ed. Pacific Grove, CA: Brooks Cole Publishing.

Schneider ED, Sagan D (2006). Into the cool: energy flow, thermodynamics and life. Chicago, IL: University of Chicago Press.

Speakman JR (1997). Doubly labelled water: theory and practice. London: Chapman and Hall.

Stevens CE, Hume ID (2004). Comparative physiology of the vertebrate digestive system. Cambridge: Cambridge University Press.

3

Cells and organisms, and their interactions with their environment

Cells are the basic structural and functional units of all free-living organisms. In animals, cells are organized into functional groups called tissues, which are themselves organized into functional groupings called **organs**. Several organs work together in an **organ system**. In order to maintain their integrity, animals need a constant supply of energy and nutrients, and waste products and heat need to be continually removed.

Generally, animals satisfy the needs of their cells by providing a stable internal environment[1], which protects their cells from direct external influences. The ability of living systems to regulate their internal environment and maintain a relatively stable set of conditions is known as **homeostasis**[2] (from the Greek *hómos* meaning similar and *stásis* meaning standing still). However, different animal groups need different sets of stable conditions for their cells to function, as we learn in later chapters of this book. For example, the maintenance of a stable body temperature is usually critical for mammals and birds, but less so for the arthropods and the vast majority of the so-called cold-blooded animals, or ectotherms[3].

In this chapter, we examine the physical principles governing the movement of molecules, ions and heat in biological systems, the general properties of animal cells and their interactions with the extracellular environment. We learn how animals receive a constant flow of information about their internal and external environments and how their integrated responses to this information produces and maintains a stable internal environment within which their cells can function.

[1] The concept of a stable internal environment was introduced by the French physiologist Claude Bernard, who was one of the great 19th century physiologists.

[2] The American physiologist Walter Cannon coined the term *homeostasis* in the 1920s.

[3] Ectotherms are discussed at length in Chapters 8 and 9.

3.1 Physical principles govern the flow of heat and the movement of ions and molecules in animals

Animals and their environments are part of the physical world so some understanding of physical principles is essential to appreciate how their cells work and how animals interact with their environment. Physical principles determine heat transfer between animals and their environment; they also govern the movement of all molecules, gases and ions across membranes, between cells, and between an animal and its environment.

3.1.1 Temperature differences determine heat conduction

We discuss in Section 2.1 (Fig 2.1) that heat always flows from a region of higher temperature into a region of lower temperature. This applies to all animals, whether we consider heat transfer between parts of the animal's body, or the transfer of heat to and from the environment across an animal's skin. For example, Arabian oryx (*Oryx leucoryx*) lie on the ground and use their front hooves to dig out a shallow depression in cooler sand in which they lie, as illustrated in Figure 3.1A. Lying in cooler sand maximizes the conduction of heat to the ground after heat becomes stored in their body during the hot summer days and therefore helps the oryx to cool down[4].

Heat conduction is governed by fundamental physical principles that apply whether heat is conducted along a metal bar or through the insulating barrier layer of an animal's skin. Figure 3.1B illustrates the flow of heat from a warmer compartment (at temperature T_1) to a colder compartment (at temperature T_2) separated by an insulating layer. These

[4] Heat storage during the day allows Arabian oryx to conserve water, as we discuss in Section 6.1.2. We discuss heat storage in other large mammals in Section 10.2.2.

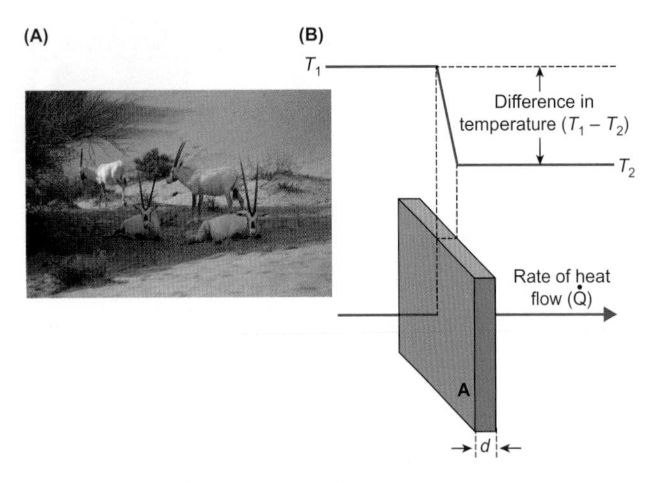

Figure 3.1 Heat conduction across a barrier

(A) Arabian oryx (*Oryx leucoryx*), two of which are lying in a shallow depression in the sand, in the shade of a tree. The contact with cooler soil maximizes heat conduction for cooling. Source: © Werner Layer/ Juniors Bildarchiv
(B) The rate of heat flow, measured as the amount of heat crossing the barrier per unit time is directly proportional to the area of the barrier (A), and the difference in temperature across the barrier (T_1–T_2), and inversely proportional to the barrier thickness (*d*). The thermal properties of the barrier determine its thermal conductivity.

principles are expressed by Equation 3.1, which is called **Fourier's law of heat conduction**[5] (after its originator).

Thermal conductivity (J s⁻¹ m⁻¹ °C⁻¹); Difference in temperature (°C)

$$\dot{Q} = k\frac{(T_1 - T_2)A}{d} \qquad \text{Equation 3.1}$$

Rate of heat flow (J s⁻¹); Barrier thickness (m); Surface Area (m²)

Equation 3.1 shows that several factors determine the rate of heat flow ($\dot{Q}$) (i.e. the amount of heat that flows across the barrier per unit time):

- The rate of heat flow is directly proportional to the difference in temperature ($T_1 - T_2$) between the two compartments. (The Arabian oryx in Figure 3.1A enhances the rate of heat loss by increasing this temperature difference.)
- The rate of heat flow is directly proportional to the surface area of the insulating layer (*A*), which animals often adjust to control their rates of heating or cooling.
- The rate of heat flow is inversely proportional to the thickness of the barrier/insulating layer (*d*); adjustments to barrier thickness are important in altering the rate of heat flow across the body surface of many animals[6].
- The thermal properties of the insulating barrier layer determine the ease of heat conduction—its **thermal conductivity** (*k*).

[5] Heat conduction is discussed in more detail in Section 8.4.2.

[6] Seasonal adjustments of the insulating layer of **endothermic** animals are discussed in Section 10.2.7.

3.1.2 Differences in concentration determine rates of diffusion of substances

In a similar way to the transfer of heat from an area of high temperature to one of lower temperature, the movement of molecules of uncharged substances occurs from a region of higher concentration to a region of lower concentration, due to the random motion of individual molecules, as illustrated by Figure 3.2.

In Figure 3.2 the concentration of a substance is greater on the left (C_1) than on the right (C_2) of an imaginary interface, shown as a dotted line, and temperatures are uniform. Hence, within a short time interval more molecules cross the dotted line from left to right than from right to left. Consequently, there is always a net movement of molecules from the region of higher concentration to the region of lower concentration. This process is **diffusion** and is described quantitatively by **Fick's law of diffusion**[7].

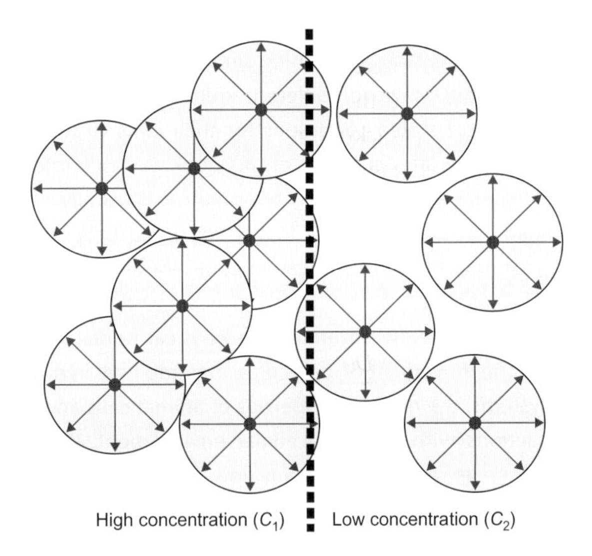

Figure 3.2 Diffusion of an uncharged substance in a homogenous medium from a higher concentration (C_1) to a lower concentration (C_2)

Each molecule is represented by a blue dot and the random movement of the molecule is represented by arrows pointing in all directions. The circle around each molecule represents the space which can be randomly occupied by that molecule after a particular time. In this example, there are three molecules (blue dots) that can cross the dotted line from left to right but only one molecule that can cross the dotted line from right to left in our chosen time interval.

[7] This law takes its name from the German physiologist Adolf Fick, who formulated it in the 1800s.

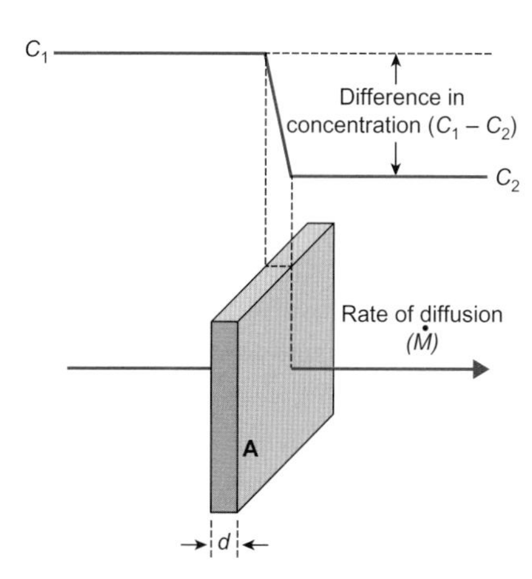

Figure 3.3 Fick's law of diffusion of a substance through a permeable barrier of thickness *d*

The rate of diffusion of a substance measured as the amount crossing the permeable barrier per unit time ($\dot{M}$) is directly proportional to the area of the barrier (A), the difference in concentration of the substance across the barrier (C_1–C_2), and inversely proportional to the thickness of the barrier (d). The proportionality constant, known as the diffusion coefficient, depends on the properties of the substance and the barrier.

For diffusion of molecules across a barrier, as illustrated in Figure 3.3, Fick's law of diffusion states that the rate of diffusion of a substance across a surface area is proportional to the difference in concentration of the substance across the barrier ($C_1 - C_2$) and the surface area (A), and is inversely proportional to the thickness of the barrier (d). Equation 3.2 expresses these relationships.

Diffusion coefficient (property of the substance and the barrier)(cm s⁻¹)

Difference in concentration (mol L⁻¹)

Surface area (cm²)

$$\dot{M} = \frac{D(C_1 - C_2)A}{d} \qquad \text{Equation 3.2}$$

Rate of diffusion of a substance (mmol s⁻¹)

Thickness of barrier (cm)

In Equation 3.2, D is the **diffusion coefficient**, which varies according to the particular molecule being considered, the properties of the barrier and the medium containing the molecule of interest. For biological systems, diffusion coefficients are usually expressed in $cm^2\ s^{-1}$ (rather than the SI unit of $m^2\ s^{-1}$, as a metre is a too long a distance to be meaningful). Concentration differences are usually measured in $mol\ L^{-1}$ (rather than the SI unit[8] of $mol\ m^{-3}$; one litre = 1000 cm^3 or 0.001 m^3). Adopting the commonly used unit in Equation 3.2, barrier thickness is in cm, surface area in cm^2, and the rate of diffusion of a substance is calculated in millimol per second[9] ($mmol\ s^{-1}$)[10].

The amount of a substance diffusing across a unit area per unit time ($\dot{M}/A$)[11] is known as its **flux** (J), which is usually measured in $mol\ cm^{-2}\ s^{-1}$ when area is expressed in cm^2 (or $mol\ m^{-2}\ s^{-1}$ if the SI unit for area (m^2) is used).

Notice the similarities between Figure 3.1 and 3.3 and hence between Equation 3.1 and Equation 3.2. Indeed, Fick's law of diffusion was based on Fourier's law of heat conduction.

The diffusion coefficient relative to the thickness of a barrier (D/d) is known as the **permeability constant** of the barrier for the particular substance (P_s). The permeability constant (given in $cm\ s^{-1}$ or $m\ s^{-1}$) describes the ease with which a substance can pass across the barrier[12]. Seen from the opposite perspective, the inverse of permeability ($1/P_s$) expresses the **resistance** of a barrier to the passage of a substance (in $s\ cm^{-1}$ or $s\ m^{-1}$). A good example of the importance of this concept biologically is the investigation of skin resistance among land animals, which reveals the evolution of skins that limit water losses in some animal groups[13].

Fick's law of diffusion across a barrier can be expressed in a simpler way, as shown in Equation 3.3, which we obtain by dividing both sides of Equation 3.2 by A (since $\dot{M}/A$ equals the flux (J) and D/d is the permeability constant (P_s) as explained above):

Difference in concentration for substance S across the barrier

$$J = P_s(C_1 - C_2) \qquad \text{Equation 3.3}$$

Net flux of substance S across the barrier ($\dot{M}/A$)

Barrier permeatability for substance S (D/d)

Fick's law of diffusion also governs the movement of gases, such as oxygen and carbon dioxide, across biological barriers.

[8] Appendix 1 explains the international system of SI units and provides information on conversions to alternative units.

[9] $(cm^2\ s^{-1}) \times (mol\ L^{-1}) \times (cm^2) \times (cm)^{-1} = cm^3\ s^{-1}\ mol\ L^{-1} = mol\ s^{-1}\ cm^3$ $(1000\ cm^3)^{-1} = 0.001\ mol\ s^{-1} = mmol\ s^{-1}$.

[10] Alternatively, entering values in SI units in Equation 3.2: D ($m^2\ s^{-1}$), concentration difference ($mol\ m^{-3}$) surface area (m^2), thickness (m) gives the rate of diffusion of a substance in $mol\ s^{-1}$.

[11] The presence of a dot above a quantity signifies a rate, e.g. $\dot{M}$ is the amount of substance per second.

[12] For cell membranes, the concept of permeability applies equally to uncharged solutes that can dissolve in the membrane and those that must pass through channels in the membrane, as we discuss in Section 3.2.4 and 4.2.2.

[13] Table 6.2 in Section 6.1.3 examines skin resistance to evaporative water loss in land vertebrates.

3

However, in this case the equation is expressed slightly differently, because gases can move between media with different properties—for example, between air and aqueous solution. The difference in **partial pressures** of gases drives gaseous diffusion. In air, the partial pressure of a gas is the pressure the gas would exert if it were alone in the volume of air at the particular temperature. When in solution, the partial pressure of a gas is equivalent to the partial pressure of the gas in the air surrounding that solution when at equilibrium[14]. Thus, according to Fick's equation for gases, a gas moves from a region of higher partial pressure (P_1) to a region of lower partial pressure (P_2).

Using the concept of a difference in partial pressure for the diffusion of a gas across a permeable barrier, the rate of gas diffusion (i.e. the amount of gas that diffuses per unit time ($\dot{M}$) across a barrier of thickness d and surface area A) is given by Equation 3.4 in which K is **Krogh's diffusion constant**[15] for the particular gas under consideration. Note the similar form of Equations 3.2, 3.3 and 3.4: these equations are based on common principles.

Krogh's diffusion constant of the gas across the barrier ($mmol\ cm^{-1}\ s^{-1}\ Pa^{-1}$)

Difference in partial pressure of gas across barrier (Pa)

$$\dot{M} = \frac{K(P_1 - P_2)A}{d} \qquad \text{Equation 3.4}$$

Surface area (cm^2)

Rate of gas diffusion ($mmol\ s^{-1}$)

Thickness of the barrier (cm)

Equation 3.4 is particularly important for determining the rate of diffusion of oxygen and carbon dioxide across biological membranes and between animals and their environment. The principles described by this equation form the basis of oxygen and carbon dioxide diffusion across the gas-exchange surfaces in the lungs of birds and mammals, across the skin of amphibians, and across the gills and other respiratory surfaces of fish and invertebrates[16]. The principles of gas diffusion are also important for determining the rate of evaporative water loss (loss of water vapour) by land animals[17].

14 We learn more about the physical properties of gases, such as oxygen and carbon dioxide, in air and in water in Section 11.1.2, and the effects of temperature on the partial pressure of gases in Box 11.1.

15 Named after August Krogh, a Danish physiologist who was a Nobel Prize winner in 1920. Section 11.2.1 considers the factors that affect Krogh's diffusion constants of gases in more detail.

16 Chapter 12 discusses gas exchange between animals and their environment.

17 Equation 6.1 in Section 6.1.1 describes the factors determining evaporative water loss is a modification of Fick's law of diffusion.

Osmosis

Water, like any other molecule, moves from a region of higher water concentration to one with a lower water concentration following the principles outlined by Equations 3.2 and 3.3. The diffusion of water across a **semi-permeable** barrier—one that is permeable to water molecules but not to all of the substances dissolved in it (called **solutes**)—occurs by the process of **osmosis**. Osmosis is an important process in physiology because it drives the passive movement of water across biological membranes, whether this is between cells and the extracellular fluid, between body fluid compartments, or across the external surfaces of aquatic animals[18].

Looking at Figure 3.4, notice that the concentration of water decreases when solutes are dissolved in it, simply because the solutes occupy a volume which otherwise would be occupied by water molecules. When a semi-permeable barrier is placed between a compartment containing a water-based (aqueous) solution with a high concentration of solutes and a solution containing a low concentration of solutes, as in Figure 3.4, osmosis causes a net flow of water across the semi-permeable barrier into the more concentrated solution. Osmosis continues until equilibrium is established, either by equalization of solute concentrations or by the build-up of sufficient hydrostatic pressure in the compartment containing the higher concentration of solutes. The hydrostatic pressure drives water flow in the opposite direction to osmosis, as shown in Figure 3.4, and at equilibrium balances net water passage by osmosis.

Hydrostatic pressure acts as a significant driving force for fluid movement in plants, but such effects are less prevalent among animals, in which osmosis generally has the dominant effect on fluid movements across membranes. However, the hydraulic force present in the circulatory systems of animals due to blood pressure may drive fluid movement through the capillary walls. This process is crucial to the formation of urine by the kidney of most vertebrates and some invertebrates[19].

Figure 3.5 illustrates the definition of the **osmotic pressure** of a solution. The osmotic pressure of a solution corresponds to the pressure that needs to be exerted on that solution to prevent the flow of water across a semi-permeable barrier from a compartment that contains pure water.

It is important to remember that a higher osmotic pressure is associated with a higher concentration of dissolved solutes and, therefore, with lower water concentration and lower water

18 The importance of osmosis is emphasized in Part 2 of this book when discussing water movement across epithelial cell membranes (Section 4.2.3), cell volume regulation (Section 4.3), the water balance of aquatic animals (Chapter 5) and kidney function (Chapter 7).

19 Section 7.1.2 discusses the various forces driving fluid movements in the vertebrate kidney, and these forces are discussed for capillaries in general in Section 14.1.5.

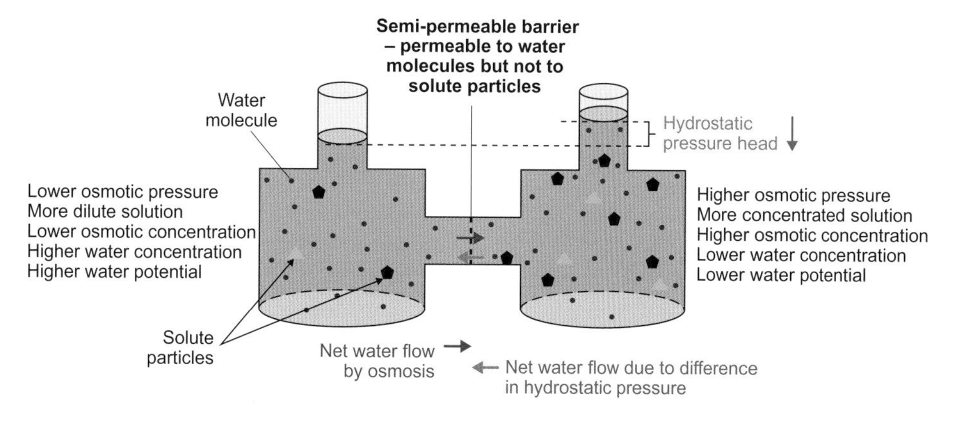

Figure 3.4 Water movement by osmosis

Net water flow by osmosis (red arrows) across a semi-permeable barrier always occurs from a solution of lower osmotic pressure to a solution of higher osmotic pressure, or from a more dilute solution to a more concentrated solution, or from a lower osmotic concentration solution to a higher osmotic concentration solution. This means that water always flows from a solution with higher water concentration to a lower water concentration (or from a solution with higher water potential—a higher potential energy of water per unit volume—to a solution of lower water potential).

The net water flow by osmosis results in the solution moving higher up the column from the right-hand chamber, which creates a hydrostatic force (blue arrow, driving water movement from right to left compartments) that opposes osmosis. Water movement from the left to the right continues until an equilibrium is reached when the net water flow from right to left due to the difference in hydrostatic pressure is sufficient to equal the net water flow by osmosis.

potential. Consequently, as Figure 3.4 shows, there is always a net flow of water across a semi-permeable barrier from a solution with a lower osmotic concentration[20] to a more osmotically concentrated solution, or from a solution of lower osmotic pressure to a solution of higher osmotic pressure—that is, unless osmosis is prevented by pressure being applied on the more concentrated solution, as shown in Figures 3.4 and 3.5.

3.1.3 The movement of charged solutes across biological barriers is determined simultaneously by concentration and electrical potential differences

Molecules of solutes that are electrically neutral cross biological barriers according to Equations 3.2 and 3.3. However, the movement of charged solutes, such as ions, across biological barriers is determined by the combined effects of concentration and charge differences. Ions are formed in aqueous solution when salts (including common salt, NaCl) dissolve in water and dissociate into positively charged **cations** (e.g. sodium ion, Na^+) and negatively charged **anions** (e.g. chloride ion, Cl^-).

If anions and cations are distributed unevenly across a barrier that is **selectively permeable**[21] to the respective ions, then an **electrical potential difference**, or voltage develops across the barrier, just like between the poles of an electrical battery[22]. This voltage influences the movement

[20] We discuss the measurement of osmotic concentrations in Section 4.1.1.

[21] For inorganic ions (e.g. potassium (K^+), sodium (Na^+), chloride (Cl^-)), the selective permeability of biological barriers (such as the cell membrane) depends on the number and properties of ion channels per unit of cross-sectional area that are permeable to the respective ions, which we discuss further in Sections 3.2.3 and 4.2.2.

[22] Box 16.3 explains how electrical potentials differences develop across barriers.

Figure 3.5 Osmotic pressure

The osmotic pressure of a solution is defined as the hydrostatic pressure that needs to be applied on that solution to prevent net inflow of water across a semi-permeable barrier, which separates that solution from a compartment containing pure water. Osmotic pressure (like any other pressure measurement) is expressed in pascals
(see Appendix 1 for details of SI units).

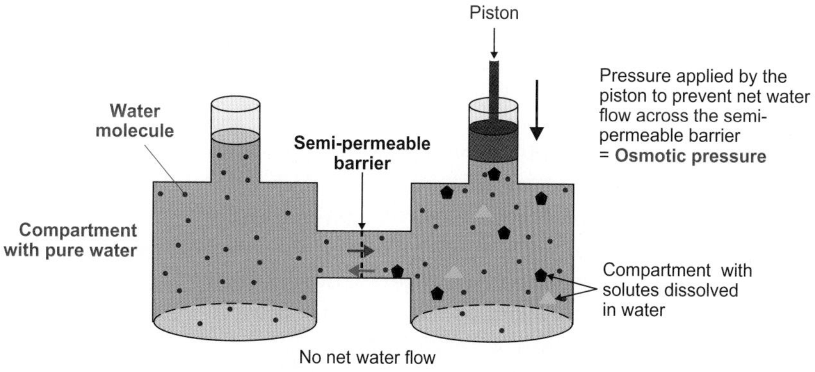

of ions across the barrier. In such situations, we need to consider not only the effect of the different concentrations of ions across the barrier, but also the effect of different electrical forces acting on the ions. Box 3.1 explores the terms and principles concerning the movement of electrical charges.

For example, on the more positive side of a barrier, the repulsive electrostatic forces push positively charged cations across the barrier[23], while attractive electrostatic forces acting on negatively charged anions impede them from crossing the barrier. On the more negative side of the barrier, repulsive electrostatic forces push negatively charged anions across the barrier, but attractive forces impede cations from crossing.

[23] Electrical charges of the same sign repel each other, while electrical charges of opposite sign attract each other.

Box 3.1 The movement of electrical charges: terms and principles

Charges on atoms or molecules (ions). Ions carry net electrical charges due to loss or gain of one or more electrons. The SI unit for electrical charge is the coulomb (C) and one electron carries a charge of 1.6×10^{-19} C. In practical terms, one coulomb (C) is the amount of positive charge carried by 10.36 μmol of monovalent cations, like K^+ or Na^+, or the amount of negative charge carried by 10.36 μmol monovalent anions like Cl^-. The total charge carried by 1 mol of monovalent cations equals 96,485 coulombs and is called **Faraday's number** (*F*).

Capacitance is the ability to store electrical charge. The SI unit for capacitance is the farad (F), named after Michael Faraday (not to be confused with Faraday's number, *F*, explained above). A 1 F capacitor charged with 1 C of electrical charge has an electrical potential difference (voltage) of 1 V (i.e. $1\ F = 1\ C \times (1\ V)^{-1}$ or $1\ C = 1\ F \times 1\ V$).

Coulomb's law describes the interactions between electrically charged particles; it states that charges of opposite sign attract each other, while charges of the same sign repel each other. The force of interaction is proportional to the product of the charges carried by the particles and inversely proportional to the square of the distance between particles.

Electric current refers to the number of coulombs carried per second. The SI unit for electric current is the ampere (A) and $1\ A = 1\ C\ s^{-1}$.

Electric field is the electric force per unit charge, measured in newtons (N) per coulomb ($N\ C^{-1}$) or volts per metre ($V\ m^{-1}$). These units are equivalent as $1\ N\ C^{-1} = 1\ V\ m^{-1}$. The direction of the electric field is given by the direction of the electrical force exerted on a positive charge.

Electric potential energy is the potential energy measured in joules (J) that results from Coulomb forces and is associated with the configuration of a particular set of electrical charges in a defined system like that in Figure 3.7. Electric potential energy should not be confused with the term **electrical potential**, which is the electric potential energy per unit charge (J/C) and is measured in volts ($1\ V = 1\ J\ C^{-1}$). The quantity that is physically meaningful is the *difference* in electrical potential, which represents the work which would have to be done to move one unit of charge from one point to another, which is commonly referred to as voltage. For example when the voltage of a battery is given as 1.5 V, it means that the electrical potential difference between the positive and the negative poles of the battery is 1.5 V.

Electrical conductance is the ease with which an electric current, such as that carried by ions, flows along a certain path. It is defined as the ratio between current intensity and voltage (i.e. difference in electrical potential, explained above). The SI unit for conductance is the siemens (S) and $1\ S = 1\ A\ V^{-1}$. **Electric conductivity** (or the specific conductance of a material) is the ability of that material to conduct an electric current. The ability of one unit volume of a material to conduct an electric current is measured in siemens (S) per metre ($S\ m^{-1}$).

Electrical resistance is the reciprocal of electrical conductance and describes the extent to which a material opposes the flow of an electric current, defined as the ratio between voltage (difference in electrical potential) and current intensity. The SI unit for resistance is the ohm (Ω), where $1\ \Omega = 1\ V\ A^{-1}$. **Electric resistivity** (or the specific electrical resistance) of a material is the reciprocal of electric conductivity and refers to how strongly a material (including gases and solutions) opposes the flow of electric current. It is normally measured in Ω m.

Electrochemical potential energy (measured in joules, J) is the combination of chemical potential energy resulting from the concentration of an ion and the electric potential energy resulting from the charge carried by the ion and the electrical potential at the site where the ion is located. **Electrochemical potential** refers to the electrochemical potential energy per mol measured in $J\ mol^{-1}$. As for the electrical potential, the quantity that is physically meaningful is the *difference* in electrochemical potential, which represents the work which would have to be done to move one mol of ions from one point to another.

Equilibrium potential refers to the electrical potential difference that balances out the ion concentration difference across a barrier that is permeable to that ion. The equilibrium potential can be viewed as a conversion of the chemical potential difference associated with an ion's concentration difference into an electrical potential difference. Equilibrium potential is measured in volts (V) and is given by the Nernst equation (Equation 3.5).

Ohm's law describes the relationship between electric current intensity through a conductor (*I*), electrical potential difference measured across the conductor (voltage, *V*) and electrical resistance of the conductor (*R*): $I = V/R$. Notice the similarities between Ohm's law, Fick's law of diffusion (Equation 3.2) and Fourier's law of heat conduction (Equation 3.1).

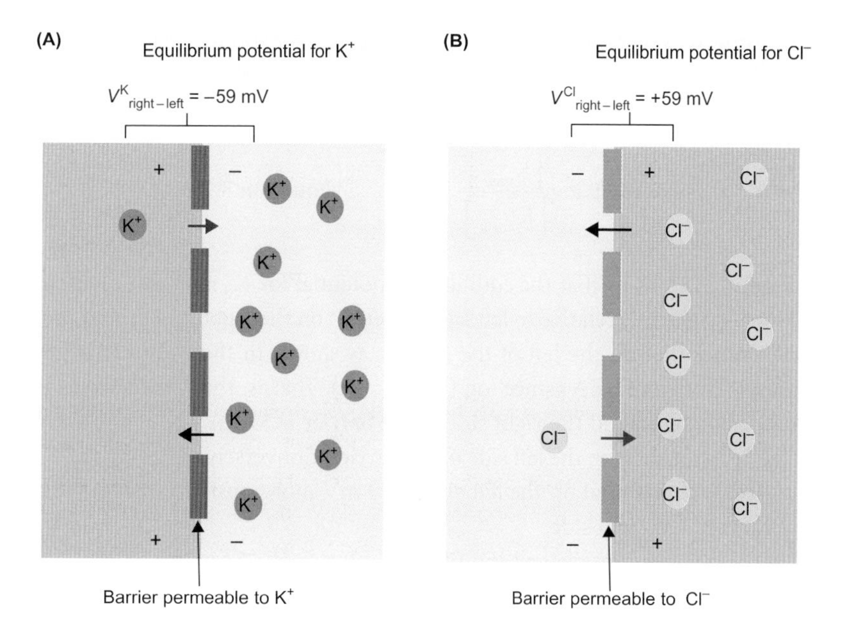

Figure 3.6 Diagrammatic representation of the equilibrium potential for potassium (K^+) and chloride (Cl^-) ions

In these examples there is a 10-fold concentration difference for (A) K^+ and (B) Cl^- between the two compartments. At the equilibrium potential, the electrical forces acting on the respective ions balance the concentration difference such that there is no net movement of the respective ions across the barrier (the net ion movement by diffusion from right to left, shown by black arrows is balanced by the net movement of the respective ions from left to right under the influence of electrostatic forces, shown by red arrows). The values for the equilibrium potentials are calculated using Nernst equations 3.7 (for K^+) and Equation 3.8 (for Cl^-).

The net movement of an ion driven by the concentration difference across a permeable barrier stops at equilibrium potential

Let us consider the simple case illustrated in Figure 3.6A, where a barrier that is permeable only to potassium ions (K^+) separates two compartments, and the right compartment has a 10-fold greater K^+ concentration than the left compartment. The difference in potassium ion concentration on the two sides of the barrier drives the diffusion of potassium ions across the barrier from right to left. However, if the left compartment is more positively charged than the right compartment, a repulsive electrostatic force is exerted on the potassium ions in the left compartment, which drives them in the opposite direction—from left to right across the barrier.

The electrical potential difference across the barrier (measured in volts, V) that balances out the effect of the concentration difference on the movement of potassium ion across the barrier is called the **equilibrium potential** for K^+. The effect of the electrical force opposing diffusion, which is driven by the concentration difference, is similar to the effect of hydrostatic pressure opposing osmosis, as we discuss in relation to Figure 3.4.

Figure 3.6B illustrates the same principle for chloride (Cl^-) ions when the concentration is 10 times greater on the right side than on the left side of a selectively permeable barrier for chloride. Notice, however, that the left compartment in this case needs to be negatively charged to drive the negatively charged chloride ion in the opposite direction to its diffusion as driven by the concentration difference.

The first quantification of equilibrium potentials was made by the German physicist Walther Nernst, who developed the **Nernst equation** in the nineteenth century; this is shown in Equation 3.5. The Nernst equation calculates the electrical potential difference (in volts, V) that balances the concentration difference of a particular ion distributed across a selectively permeable barrier.

$$V^{ion}_{right\text{-}left} = \frac{RT}{zF}\,ln\frac{[ion_{left}]}{[ion_{right}]} \qquad \text{Equation 3.5}$$

Equilibrium potential for a specific ion = electric potential difference between the right and the left side of the barrier where there is no net flux of the ion across the barrier (V)

Gas constant (8.31 J mol^{-1} K^{-1})

Absolute temperature (K)

The elementary number of positive or negative electrical charges carried by the ion

Faraday's number (96,485 C)

Natural logarithm of the ratio between the ion concentration on the left and that on the right side of the barrier

Specifically, the Nernst equation gives the electrical potential difference across a selectively-permeable barrier ($V_{right\text{-}left}$) such that the particular ion concentrations on the right side $[ion_{right}]$ and left side $[ion_{left}]$ of the barrier/membrane are in equilibrium (i.e. there is no net movement of the ion across the barrier).

If we use logarithm to base 10 (*log*) instead of the natural logarithm (*ln*), then the *log* is multiplied by the factor 2.3 (= *ln* 10) and becomes:

$$V^{ion}_{right\text{-}left} = 2.3\frac{RT}{zF}\log\frac{[ion_{left}]}{[ion_{right}]} \qquad \text{Equation 3.6}$$

The value of 2.3 $RT \div zF$ for positive monovalent cations, such as potassium (K^+), for which $z = +1$, at 25°C (298 K) is +59 mV[24] so:

$$V^{K^+}_{\text{right-left}}(\text{mV}) = 59 \log \frac{[K^+_{\text{left}}]}{[K^+_{\text{right}}]} \qquad \text{Equation 3.7}$$

Equation 3.7 tells us that the equilibrium potential for K^+ is −59 mV (right side relative to left side) when K^+ on the right is ten times that on the left of the barrier, as shown in the example in Figure 3.6A (since log (1/10) = −1). That is, the electrical potential on the right side of the barrier is 59 mV more negative than on the left side of the barrier. Conversely, the electrical potential on the left side is 59 mV more positive than on the right side.

For a negative monovalent ion like the chloride (Cl^-) for which $z = -1$, the equilibrium potential at 25°C (298 K) is given by the following expression:

$$V^{Cl^-}_{\text{right-left}}(\text{mV}) = -59 \log \frac{[Cl^-_{\text{left}}]}{[Cl^-_{\text{right}}]} \qquad \text{Equation 3.8}$$

In this case, when the concentration of Cl^- is 10 times greater on the right than on the left side of the barrier, as shown in Figure 3.6B, the equilibrium potential for Cl^- is $-59 \log \frac{1}{10} = -59 \times (-1) = +59\,\text{mV}$. That is, the electrical potential on the right side of the barrier is 59 mV more positive than on the left side of the barrier. Conversely, the electrical potential on the left side is 59 mV more negative than on the right side.

We have discovered that a potential difference (right–left) of only −59 mV or +59 mV across a selectively permeable barrier is sufficient to balance out a 10-fold difference (right–left) in the concentration of monovalent cations or anions, respectively, across the barrier.

Predicting membrane potentials for a combination of ions and barrier permeabilities to ions: the Goldman-Hodgkin-Katz equation

The **Goldman-Hodgkin-Katz (GHK) equation**[25] predicts the difference in the electrical potential across a barrier that is permeable to various degrees to any set of monovalent ions. The equation allows calculation of the voltage across a barrier at which a *steady state* occurs—when there is no *net* charge movement (no current flow) across the barrier. The GHK equation is widely used because it allows calculation of the electrical potential difference that establishes across a barrier for any set of (monovalent) ion concentrations and specified (relative) ion permeabilities.

The main monovalent ions in organisms are sodium (Na^+), potassium (K^+) and chloride (Cl^-). If the permeabilities of the barrier to these ions are designated as P_{Na}, P_K and P_{Cl}, and the concentrations of the ions are shown in square brackets, then the Goldman-Hodgkin-Katz equation is expressed as Equation 3.9:

Electrical potential difference between the right and the left side of a barrier that is permeable to K⁺, Na⁺ and Cl⁻ when there is no net charge carried across the barrier.

Sum of cation (Na⁺, K⁺) concentrations on the left multiplied by respective membrane permeability and anion (Cl⁻) concentration on the right multipled by its membrane permeability.

$$V^{\text{GHK}}_{\text{right-left}} = 2.3\frac{RT}{F}\log\frac{P_K[K^+_{\text{left}}] + P_{Na}[Na^+_{\text{left}}] + P_{Cl}[Cl^-_{\text{right}}]}{P_K[K^+_{\text{right}}] + P_{Na}[Na^+_{\text{right}}] + P_{Cl}[Cl^-_{\text{left}}]} \qquad \text{Equation 3.9}$$

The value of 2.3*RT*/*F*=59 mV at 25°C (see text related to Equation 3.8).

Logarithm base 10 of ratio between two sums.

Sum of cation (Na⁺, K⁺) concentrations on the right multipled by respective membrane permeability and anion (Cl⁻) concentration on the left multiplied by its membrane permeability.

Notice that the numerator includes K^+ and Na^+ concentrations on the left side of the barrier plus Cl^- concentration on the right while the denominator includes K^+ and Na^+ concentrations on the right side of the barrier together with Cl^- concentrations on the left. This is because of the negative charge of Cl^-, compared to the positive charge carried by Na^+ and K^+.

Figure 3.7 shows an example in which the electrical potential difference between the right and the left side of the barrier stabilizes at −32.9 mV. Notice that even though there is no *net* movement of electrical charge across the barrier, there *are* net movements of individual ions across the barrier. For example, from Equation 3.7 we know that an electrical potential difference $V_{\text{right-left}}$ of −59 mV is necessary to balance out the 10-fold difference in K^+ concentration between the right and the left side of a barrier. However, according to the GHK equation applied to the situation in Figure 3.7, the $V_{\text{right-left}}$ is −32.9 mV. Therefore, there will be a net flux of K^+ across the barrier, from the higher concentration (right side) to the lower concentration (left side).

Furthermore, the equilibrium potential for Na^+ needed to balance out the 10-fold higher concentration on the left side than on the right side of the barrier is a $V_{\text{right-left}}$ of +59 mV. Therefore, when $V_{\text{right-left}}$ is −32.9 mV, there will be a net flow of Na^+ from left to right aided by both the concentration difference and the electrical potential difference.

[24] The value of +59 mV is calculated from $2.3 \times 8.31\ \text{J mol}^{-1}\text{K}^{-1} \times 298\ \text{K} / ((+1) \times 96{,}485\ \text{C mol}^{-1}) = 0.059\ \text{J C}^{-1} = 59\ \text{mV}$.

[25] Box 16.3 discusses the general principles on which the GHK equation is based. The equation is named after the American David Goldman and the English Nobel laureates Alan Hodgkin and Bernard Katz who developed the equation in the 1940s.

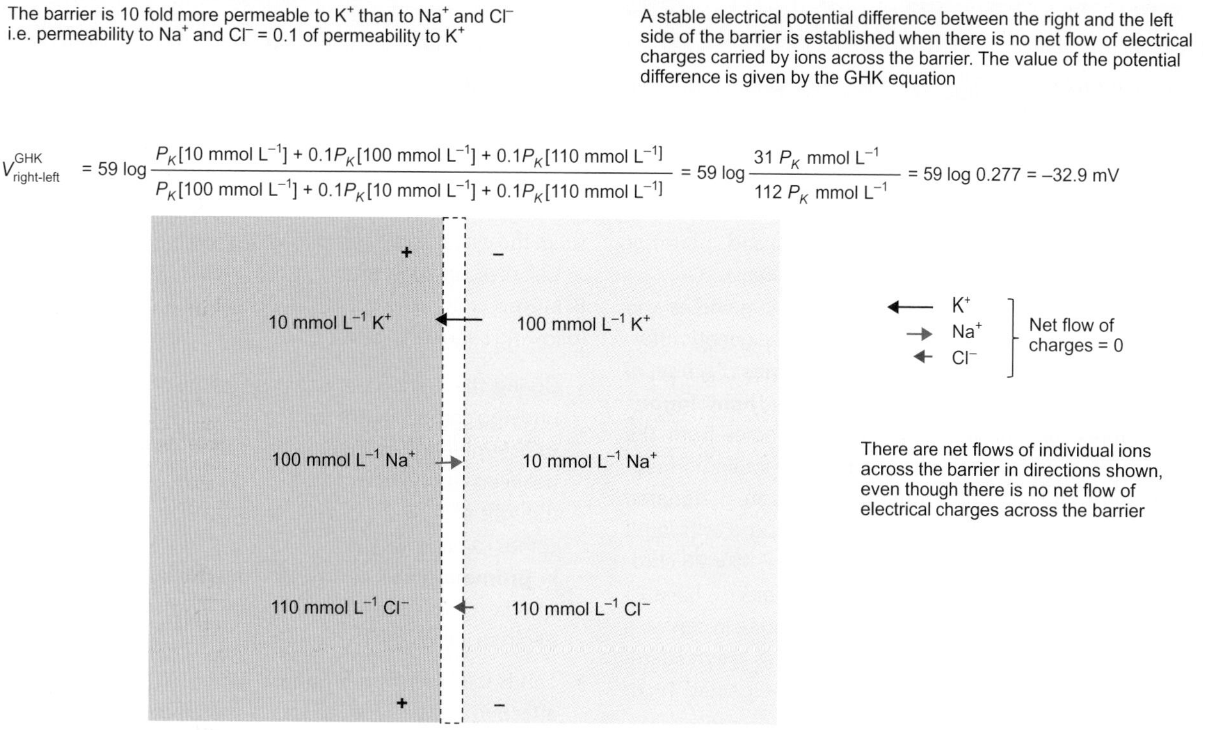

The barrier is 10 fold more permeable to K^+ than to Na^+ and Cl^- i.e. permeability to Na^+ and Cl^- = 0.1 of permeability to K^+

A stable electrical potential difference between the right and the left side of the barrier is established when there is no net flow of electrical charges carried by ions across the barrier. The value of the potential difference is given by the GHK equation

$$V^{GHK}_{right\text{-}left} = 59 \log \frac{P_K[10\ \text{mmol L}^{-1}] + 0.1P_K[100\ \text{mmol L}^{-1}] + 0.1P_K[110\ \text{mmol L}^{-1}]}{P_K[100\ \text{mmol L}^{-1}] + 0.1P_K[10\ \text{mmol L}^{-1}] + 0.1P_K[110\ \text{mmol L}^{-1}]} = 59 \log \frac{31\ P_K\ \text{mmol L}^{-1}}{112\ P_K\ \text{mmol L}^{-1}} = 59 \log 0.277 = -32.9\ \text{mV}$$

+ | −

10 mmol L^{-1} K^+ ← 100 mmol L^{-1} K^+

100 mmol L^{-1} Na^+ → 10 mmol L^{-1} Na^+

110 mmol L^{-1} Cl^- ← 110 mmol L^{-1} Cl^-

+ | −

← K^+, → Na^+, ← Cl^- : Net flow of charges = 0

There are net flows of individual ions across the barrier in directions shown, even though there is no net flow of electrical charges across the barrier

Figure 3.7 Example of establishment of an electrical potential difference according to the Goldman-Hodgkin-Katz (GHK) equation

The barrier separating the two compartments containing different concentrations of potassium (K^+) and sodium (Na^+) ions has a 10 times greater permeability for K^+ than for Na^+ and Cl^-, i.e. the relative permeability to Na^+ and Cl^- is 0.1 of K^+ permeability. Relative permeabilities are entered into the GHK equation shown and calculate an electrical potential difference that is established between the right and left compartment at 25°C = 298 K of −32.9 mV. Note that a different electrical potential is established for the same distribution of ions across the barrier if the relative permeabilities change. Also note that the ion concentrations in the two compartments change little for as long as the net amount of Na^+, K^+ and Cl^- ions flowing across the barrier is considerably smaller than the total amount of respective ions in the two compartments.

Finally, the equilibrium potential for Cl^- is zero, because the concentration of Cl^- is the same on both sides of the barrier. However, because $V_{right\text{-}left} = -32.9$ mV, the electrical forces will cause a net Cl^- flux across the barrier from right to left.

Overall, the electrical charges carried by the K^+, Na^+ and Cl^- fluxes balance out, as shown in Figure 3.7. In most physiological situations in which the GHK equation is used, the K^+, Na^+ and Cl^- fluxes are very small and there are no significant changes to the ion concentrations across the barrier[26].

In conclusion, the net movement of ions across biological barriers is governed not only by concentration differences, but also by the electric forces that act on ions when there is an electrical potential difference across the barrier. The combined chemical potential energy of one mol of ions (which is higher at higher concentrations) and electrostatic potential energy of one mol of ions (which is the product of the electrical charge carried by one mol of ions and the electrical potential at the site where the ions are located) is known as the **electrochemical potential**. According to general physical and chemical principles, ions move from a higher electrochemical potential to a lower electrochemical potential (i.e. ion movements are driven by electrochemical potential differences) in a similar way to the movement of electrically neutral molecules from a higher to a lower concentration.

We learn more about ion movements in Chapters 4, 5 and 16 when we find out how they are involved in the ionic regulation of aquatic animals and in the generation, transmission and integration of nerve impulses.

[26] We discuss ion fluxes across neuron membranes in Section 16.2.

3.2 General properties of animal cells

All animal cells arise from other cells through the process of cell division outlined in Box 3.2. Cells are bounded by a surface membrane known as the **plasma membrane**, which acts as the interface and boundary between the cell's content and its external (**extracellular**) environment.

3

Box 3.2 Cell division

Animal cells undergo division by either the process of **mitosis** or **meiosis**. Cell division by mitosis is responsible for animal growth and development, tissue regeneration and replacement of damaged cells. Mitosis occurs through the entire life of an animal. In contrast, cell division by meiosis is only responsible for the production of **gametes** (egg cells and sperm) or **gametogenesis** for sexual reproduction of animals[1].

All cells in the body of an animal, except the gametes and their immediate precursor cells, contain the same genetic information, which is encoded in a set of **chromosomes**[2]. Each set of chromosomes is made up of a series of pairs called **homologous chromosomes**; one of each pair typically comes from the mother, the other from the father. Cells that contain homologous chromosomes are called **diploid** cells and are designated $2n$. For example, the cells of the domestic dog (*Canis lupus familiaris*) have 39 pairs of chromosomes or $2 \times 39 = 78$ chromosomes, while the cells in our body (*Homo sapiens*) have 23 pairs of chromosomes or $2 \times 23 = 46$ chromosomes. In contrast, cells that contain only one copy of chromosomes like the gamete cells are known as **haploid** cells and are designated $1n$, or simply n.

Cell division, whether by mitosis or meiosis, is always preceded by a stage during which the DNA is replicated such that the nucleus of the dividing diploid animal cell already contains a complete set of **replicated chromosomes** and two **centrosomes**[3]. When the process of cell division begins, the replicated chromosomes in the nucleus undergo compaction, which results in the recognizable four-arm structure of a chromosome shown in Figure A. Notice that each replicated chromosome contains two identical elements called **chromatids** attached to each other at the **centromere**; when the sister chromatids segregate they become two separate daughter chromosomes. Thus, before the process of cell division begins, the number of chromosomes in the nucleus of the dividing cell is effectively twice that in the otherwise normally functioning diploid cell. During cell division, the nucleus divides first and this is followed by the process of **cytokinesis** by which the dividing cell splits into two daughter cells. The two daughter cells share between them the cytoplasmic components of the mother cell.

Cell division by mitosis is schematically illustrated in Figure B. Mitosis is the division of the nucleus into two nuclei in the following sequential phases:

- During the first phase, called **prophase**, the chromosomes undergo compaction in the nucleus and **microtubules** begin to assemble in the cytoplasm to form the **mitotic spindle** between the two centrosomes. The mitotic spindle may be thought of as a set of cellular tracks along which chromosomes can be pulled.
- In **prometaphase**, the nuclear membrane breaks up; the spindle enters the nuclear region, and the chromosomes begin to interact with the spindle.
- This is followed by **metaphase**, when all chromosomes are attached at the level of their centromeres to the mitotic spindle and line up in the middle of the spindle along a plane called the equatorial or **metaphase plate**.
- In **anaphase**, the sister chromatids segregate and become separate daughter chromosomes. The spindle elongates, the sister chromosomes are pulled to opposite ends of the cell, and the microtubules begin to degrade. Anaphase results in the precise division of the genetic information into two identical pools, each containing a complete complement of chromosomes.
- Finally, in **telophase**, a new nuclear membrane forms around each set of chromosomes, the spindle disintegrates and the chromosomes, which until now were in a compacted form, begin to spread out, or decondense.

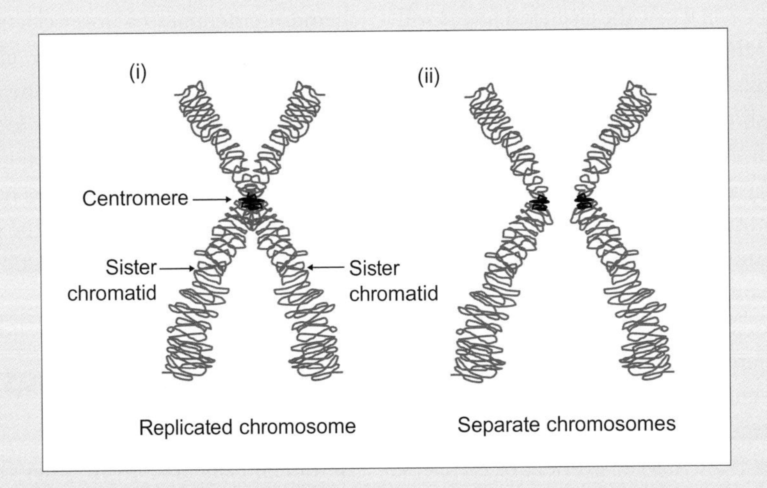

Figure A
(i) Diagram of a condensed replicated chromosome as it appears during metaphase. One replicated chromosome consists of two identical sister chromatids attached to each other at the centromere. (ii) Segregation of the sister chromatids gives rise to two separate chromosomes.

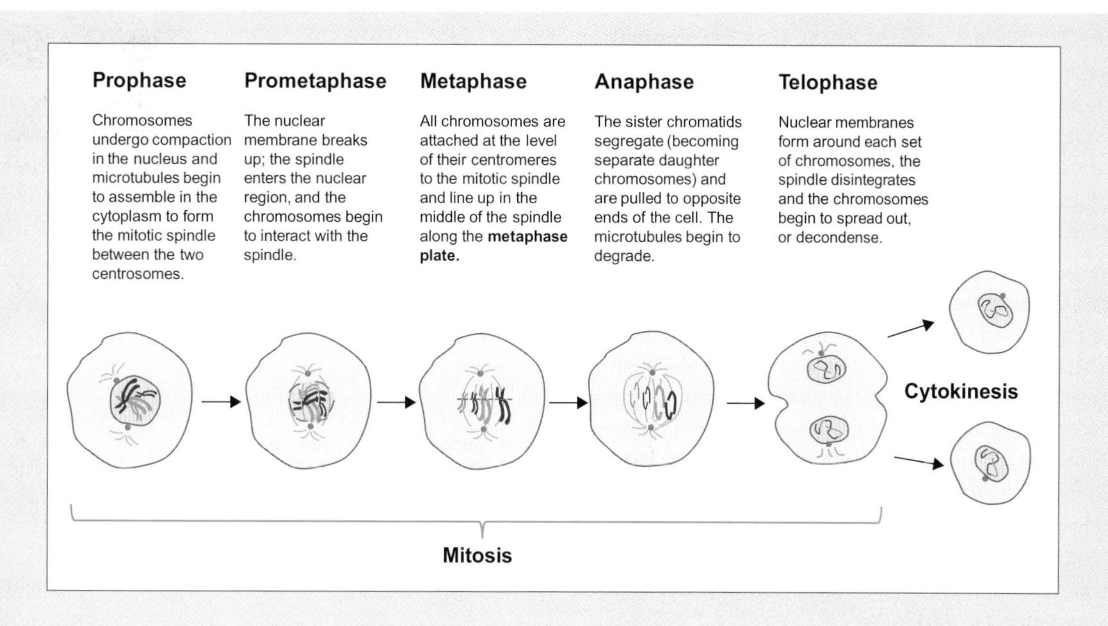

Figure B

Cell division by mitosis produces two cells that are genetically identical to each other and to the mother cell. There are two pairs of homologous chromosomes in this cell: one pair of larger chromosomes and one pair of smaller chromosomes. In each pair of homologous chromosomes, one is from the mother (red) and the other from the father (green).

Telophase is immediately followed by cytokinesis whereby the dividing cell splits into two daughter cells. The two daughter cells contain identical genetic information to each other and to their mother cell and share between them the cytoplasmic components of the mother cell in approximately equal proportions.

Cell division by meiosis involves many of the mechanisms that operate during cell division by mitosis. Just like mitosis, meiosis begins with one diploid cell containing two copies of each chromosome. Unlike mitosis, however, meiosis produces four haploid cells that contain only one copy of each chromosome as shown in Figure C.

Meiosis involves two cell divisions in succession: meiosis I, which consists of prophase I, metaphase I, anaphase I and telophase I; and meiosis II, which consists of prophase II, metaphase II, anaphase II and telophase II. Like mitosis, meiosis I begins with the diploid mother cell containing a set of replicated chromosomes, where each chromosome is made up of two identical sister chromatids. The cell then divides into two daughter cells. However, unlike the daughter cells resulting from mitosis, the daughter cells resulting from the first cell division of meiosis are not genetically identical to each other or to the mother cell. Moreover, as these two daughter cells divide further during meiosis II, they produce four genetically diverse haploid cells.

How is meiosis responsible for producing this genetic diversity?

1. During prophase I, homologous chromosomes (one from the mother and one from the father) come into close proximity and become paired, as shown in Figure D.

2. The pairing of homologous chromosomes allows exchange of genetic material between sections of the maternal and paternal chromosomes as also shown in

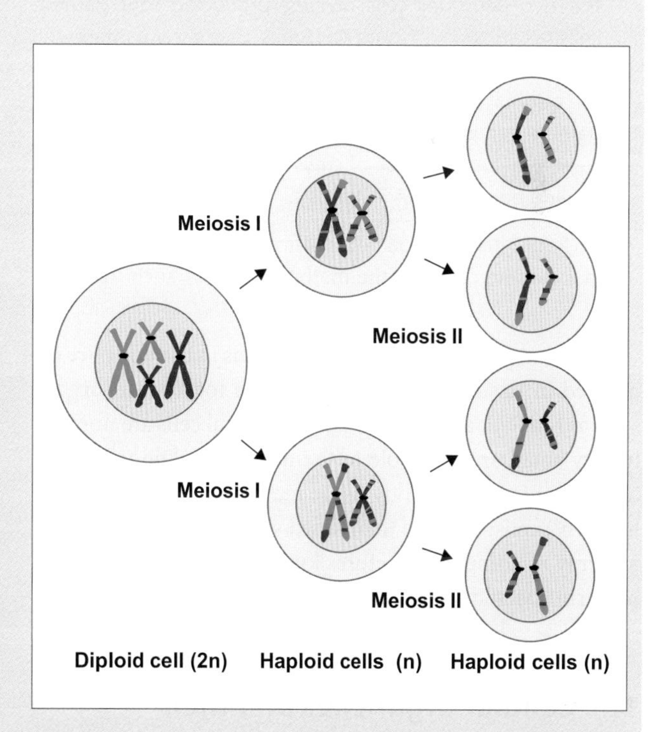

Figure C

Cell division by meiosis involves cell divisions in succession in meiosis I and meiosis II, which produces four haploid cells that are genetically different from each other.

3

Figure D. This process, known as **DNA recombination**, leads to **DNA cross-overs** between the two homologous chromosomes and plays a major role in DNA repair. Notice in Figure D that because of DNA cross-overs, the two chromatids on each of the two homologous chromosomes are no longer identical to each other.

3. The sister chromatids on each chromosome do not segregate during anaphase I to become separate chromosomes. Instead, homologous chromosomes in a pair are pulled apart and randomly migrate to opposite poles of the spindle to produce two pools, each with only half the original number of chromosomes. After the nuclear membrane forms around each set of (recombined) chromosomes, during telophase I, the mother cell divides into two haploid daughter cells, each daughter cell receiving only half of the chromosomes, as shown in Figure C. Notice that each chromosome still consists of a pair of sister chromatids and that the two sister chromatids on each chromosome are no longer identical to each other.

4. There is no chromosome replication before the second process of division (meiosis II) begins. During anaphase II, the two chromatids of the homologue chromosomes in the two daughter cells separate just like in the process of mitosis to become separate chromosomes. These chromosomes are then randomly divided into two pools in each of the two daughter cells and a nuclear membrane forms around each pool during telophase II. Thereafter, each of the two daughter cells divides, producing four gamete cells, as shown in Figure C. Notice that each gamete cell contains only half the number of chromosomes that were in the original mother cell and that each chromosome in the gamete cell is a unique mixture of paternal and maternal DNA as a consequence of cross-overs between the maternal and paternal DNA in meiosis.

Upon fertilization[4] one gamete cell from the father (sperm) fuses with a gamete cell from the mother (egg cell), such that each gamete contributes half of the set of chromosomes that make up the complete diploid genome of the new offspring. Notice that each chromosome in the offspring is now different from the homologous chromosomes in the body of either the mother or the father, thus giving rise to increased genetic variation in sexually reproducing animals on which natural selection can act[5].

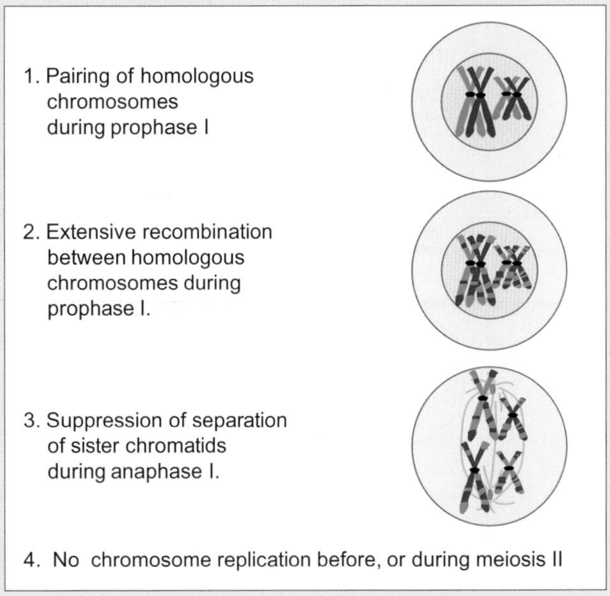

1. Pairing of homologous chromosomes during prophase I

2. Extensive recombination between homologous chromosomes during prophase I.

3. Suppression of separation of sister chromatids during anaphase I.

4. No chromosome replication before, or during meiosis II

Figure D Distinguishing characteristics of meiosis
Note that unequal DNA cross-overs between homologous chromosomes can lead to gene duplication on one chromosome and gene deletion on its pair.

1 Box 20.1 outlines gametogenesis.
2 The chromosome is an organized structure of coiled DNA and protein as we discuss in Section 1.3.1.
3 The centrosome is a cytoskeletal organelle described in Section 3.2.1.
4 We examine the process of fertilization in Section 20.5.
5 We discuss the importance of genetic variation in the process of evolution by natural selection in Section 1.5.4.

Cells that perform particular functions in the body are organized in **tissues**, and tissues function together in **organs** such as the liver and kidney. The individual cells are normally very small and invisible to the naked eye. Did you know that 1 litre (L) of soft body tissue may contain up to one trillion (10^9) cells? Despite their small size, animal cells contain an array of complex structures known as **organelles**, which perform specific functions in the life of each cell.

3.2.1 Cellular organelles and their functions

Intracellular membranes that have a similar structure to the plasma membrane (see Section 3.2.3) of cells bound many organelles. Figure 3.8 depicts a number of the membrane-bound organelles normally present in animal cells.

- The **nucleus** is a relatively large organelle containing most of the genetic material in a cell. The nucleus is surrounded by a **nuclear envelope**, which is a double membrane with pores that allow substances to pass across it. At the nuclear pores, the inner and the outer membranes of the nuclear envelope fuse. The nucleus also contains the **nucleolus**, where the synthesis and assembly of **ribosomes** takes place. Ribosomes are organelles located outside the nucleus that synthesize the proteins needed by the cells, as shown in Figure 1.13.
- The **mitochondria** provide a continuous supply of useable energy to cells in the form of adenosine triphosphate (ATP)[27].

27 Section 2.4.2 discusses ATP production in mitochondria in some detail.

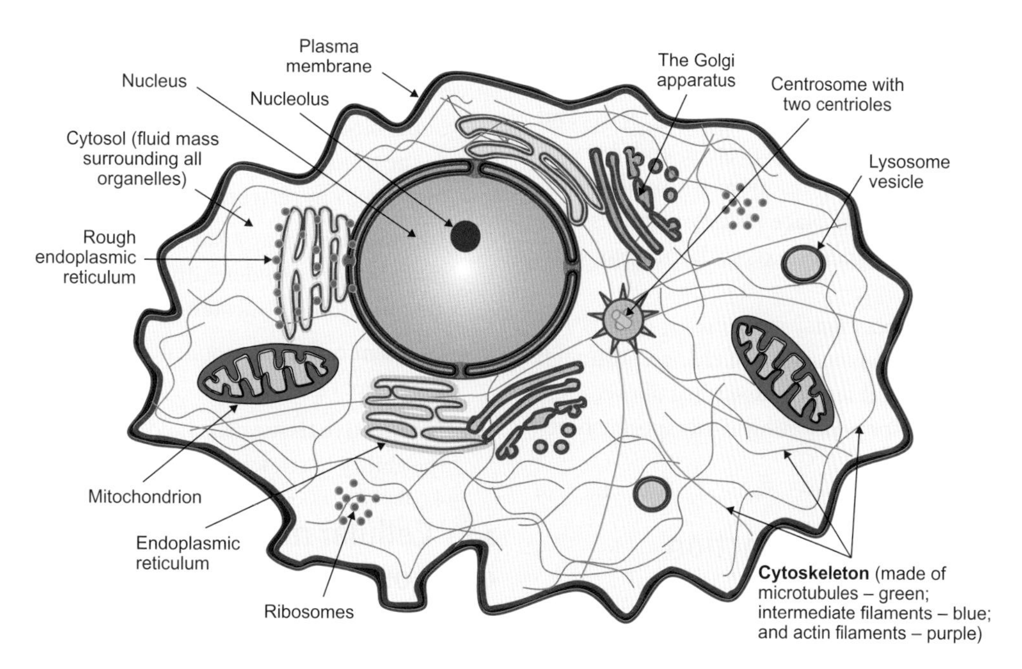

Figure 3.8 Diagrammatic representation of cellular organelles in an animal cell

- The **endoplasmic reticulum** is a network of interconnected membrane-bound compartments. The endoplasmic reticulum is involved in the production and storage of protein and lipid substances essential to cell and tissue function. It also acts as a store for calcium ions (Ca^{2+}) that play a key role in cell signalling[28]. The membrane of the endoplasmic reticulum is continuous with the outer layer of the nuclear envelope, as shown in Figure 3.8.
- The **Golgi apparatus**[29] consists of a stack of flat compartments connected to the endoplasmic reticulum. Within these compartments, molecules destined for secretion, such as hormones, are processed and packaged[30].
- Vesicles contain packages of molecules. Molecules leaving the Golgi apparatus are packaged in vesicles. Vesicles known as **lysosomes** contain enzymes that break down damaged molecules and molecules no longer needed by cells.

Figure 3.8 also shows organelles that are not surrounded by membranes, such as the ribosomes. Sometimes the ribosomes are associated with the endoplasmic reticulum to form a structure known as the **rough endoplasmic reticulum**.

[28] We examine the role of calcium in muscle contraction in Chapter 18 and in some other cell signalling processes in Section 19.1.4.

[29] The Golgi apparatus is named after the Italian cytologist Camillo Golgi, who discovered it at the end of the 19th century.

[30] We explore the secretion of hormones in Section 19.1.1.

The **cytoskeleton** is also not surrounded by a membrane. The cytoskeleton acts as a scaffold and determines to a large extent the spatial organization of the cell. The cytoskeleton gives mechanical strength and also plays a major role in the movement of vesicles and other organelles within the cell.

The cytoskeleton is made of three main types of filamentous structures: **microtubules**, **intermediate filaments** and **actin filaments**. The microtubules are long, relatively thick, hollow cylinders made of the protein tubulin; they play a crucial role in the spatial arrangement of the cell and also in intracellular transport. All microtubules are attached at one end to the **centrosome** (or the cell centre) located close to the nucleus. Looking at Figure 3.8, notice that the centrosome contains two small cylindrical structures perpendicular to each other; these are called **centrioles**. The centrioles are composed of short and very thin microtubules and are involved in the formation of the spindle apparatus responsible for the segregation of chromosomes during cell division, as shown in Box 3.2. The intermediate filaments are rope-like protein structures that extend across the cell, conferring mechanical strength to it. The actin filaments are the thinnest filaments of the cytoskeleton; they are flexible helical polymeric structures made of the protein actin. Actin filaments are distributed across the cell in a variety of configurations and are involved in cell shape changes and various forms of cell motility[31].

[31] The actin filaments play a key role in muscle contraction as we discuss in Section 18.1.2.

The liquid in cells is called the **cytosol** (intracellular fluid); it forms a fluid matrix around the organelles. The cytosol contains many macromolecules, ions and water-soluble molecules, including protein and enzyme complexes that regulate cellular reactions within the cytosol, while many reactions take place in the organelles. The cell cytosol and cellular organelles, except the nucleus, constitute the **cytoplasm**.

Importantly, the organizational structure of cells is not rigidly set but is in a continuous state of change. Although most cells share the general features described above, most cells become specialized for performing highly specific functions in the bodies of animals.

3.2.2 Cell types and their function

Almost all cells in the body of an animal (except gamete cells and their precursor cells, and the red blood cells of mammals, which lack a nucleus) contain the same genetic information. However, major functional differences arise among animal cells during their development as a consequence of different elements of genetic information being expressed in different cells: the genes of cells located in different parts of the body switch on and off in characteristic temporal patterns that cause cells to differentiate into different cell types[32]. Based on their main function, we can recognize the following cell types:

- **Epithelial cells** form barriers located at the surface of the animal at the interface with its external environment or that line internal body cavities such as the gut and the kidney tubules[33].
- Storage cells, such as adipose (fat) cells, store energy in the form of **lipids**[34].
- Cells, such as liver and kidney cells of vertebrates and fat body cells of insects, are involved in the metabolism, detoxification or modification and excretion of various substances.
- **Germ cells** are essential for gamete production (gametogenesis) for sexual reproduction[35].
- **Excitable cells**, such as nerve and muscle cells, respond rapidly to suitable stimuli with an electrical change (excitation) at the plasma membrane[36].
- **Secretory cells**, such as endocrine cells, release chemicals (hormones) with specific functions[37].
- Some blood cells are involved in gas transport such as the transport of oxygen and carbon dioxide by red blood cells in vertebrates[38].
- **Immune cells** play a key role in protecting an animal's integrity by defending the body against pathogens and foreign substances.
- **Pigment cells** are responsible for the colouration of the animal's surface.
- **Connective tissue cells** hold tissues together and have the ability to stretch and contract passively.

[32] This is an example of epigenetic change as we discuss in Section 1.5.2.

[33] Chapters 4 to 7 discuss the functioning of many types of epithelial cells in ion and water transport processes that are essential for regulating their internal environments.

[34] For an outline of lipids, see Box 2.1.

[35] We discuss sexual reproduction in Chapter 20.

[36] We learn how excitable cells work in Chapters 16 and 18; how they are involved in sense organs in Chapter 17, and their functioning in controlling physiological processes in Chapters 19 to 22.

3.2.3 Cellular membranes, their composition and properties

Arguably, the most important structural elements of animal cells are the membranes. Cell membranes are indispensable to life as we know it because they demarcate the boundaries of individual cells and of the many cellular organelles within them and thereby allow the compartment they enclose (the cell or organelle) to have a different fluid composition from the surrounding environment.

Cellular membranes are very thin structures, just 7–10 nm thick. Look at Figure 3.9 and notice that cell membranes are made of two main components:

- A lipid component, mainly the **lipid bilayer** that occupies up to 99 per cent of the membrane surface in plasma membranes and results in properties that are common to all cell membranes.
- Protein molecules embedded in the lipid bilayer, which are responsible for the specific functions of particular membranes. Some proteins pass through the lipid bilayer while some occur on one side of the lipid bilayer or the other. Some proteins bond to other protein molecules embedded in the lipid bilayer and link to the cell's cytoskeleton.

The lipid component of cell membranes

The lipid bilayer is made of **amphipathic** molecules, i.e. molecules with both a **hydrophobic** (water-hating) region and a **hydrophilic** (water-loving) region. **Phospholipids** are the major amphipathic lipid molecules in the cell membranes of animals. Other amphipathic molecules in the plasma membranes are **glycolipids** (in which a sugar molecule is covalently bound to the lipid) and **cholesterol** located between neighbouring phospholipid molecules.

[37] Chapter 19 explores hormone function.

[38] Section 13.1 discusses gas transport in blood cells.

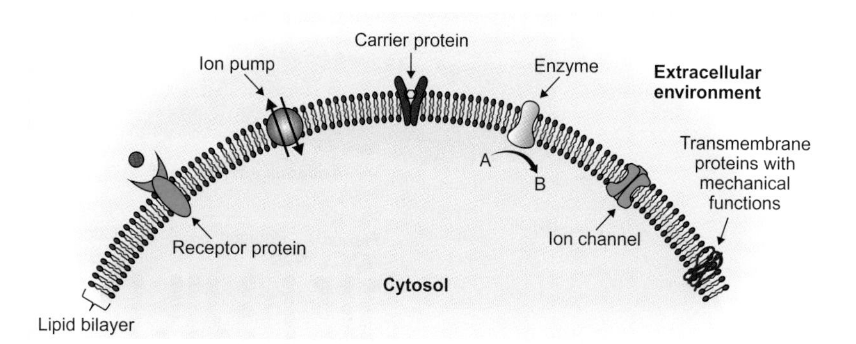

Figure 3.9 Diagram showing the lipid and protein components of plasma membranes

The lipid component (lipid bilayer) gives a common structural basis and a set of common properties to all cellular membranes, while the protein components are responsible for imparting specific functions to membranes such as acting as receptors for specific molecules (receptor proteins) and facilitating transport of specific solutes across the membrane via ion pumps, carrier proteins and ion channels.

Figure 3.10 shows one example of a phospholipid molecule, **phosphatidylcholine** in this case. Phospholipid molecules consist of one glycerol molecule, which is generally bound covalently to two fatty acid molecules (forming a so-called **diacylglycerol**) and to a negatively charged phosphate group that is covalently bound to a simple hydrophilic organic group (choline in Figure 3.10). The two hydrocarbon chains of the fatty acid molecules form the hydrophobic tail of the phospholipid molecule, while the region containing the negatively charged phosphate residue and the hydrophilic organic group forms the hydrophilic head of the phospholipid molecule.

Looking at Figure 3.10, notice how the hydrophilic head of the phospholipid molecule associates with water molecules. Water molecules are dipolar, that is they have centres of positive charge (on the hydrogen atoms) and a centre of negative charge (on the oxygen atom), as shown in Figure 3.10B. As such, they interact through electrostatic forces with electrically charged regions on phospholipid molecules; the interaction occurs because regions of molecules of opposite charge attract each other while regions of molecules of the same charge repel each other.

Hydrophobic regions of the phospholipids do not interact with water molecules but can interact with each other. Therefore, in an aqueous environment, the phospholipids of cell membranes aggregate in such a way that the hydrophobic tails are shielded from the aqueous surroundings, while their hydrophilic heads are exposed to the surrounding water molecules, which gives the characteristic lipid bilayer seen in all biological membranes and illustrated in Figures 3.11, 3.12 and 3.13. All cell membranes show a fundamental property of self-sealing or re-sealing when disrupted in an aqueous solution, which Figure 3.11 illustrates.

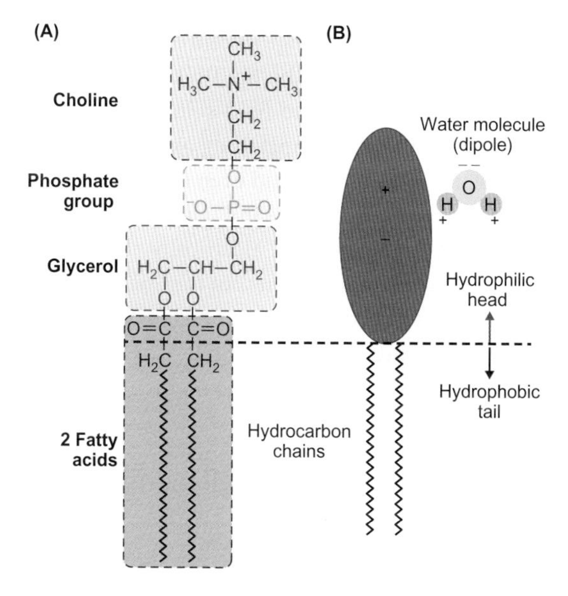

Figure 3.10 Diagrammatic representation of the phospholipid phosphatidylcholine

(A) The chemical formula of phosphatidylcholine; (B) generalized representation of the phospholipid molecule with its hydrophobic tail and hydrophilic head, which interacts with water molecules.

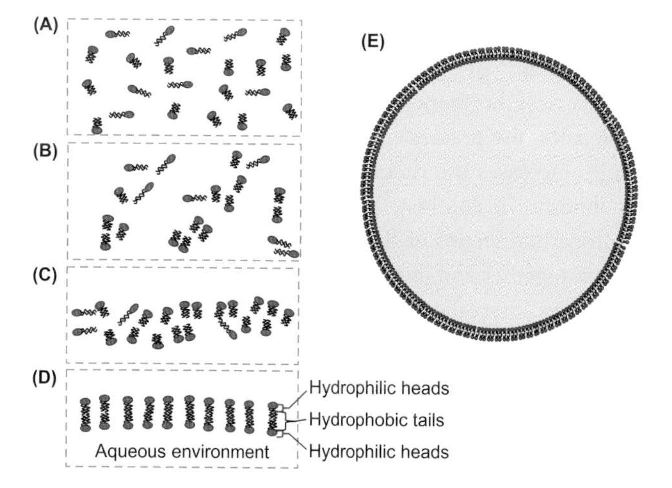

Figure 3.11 Spontaneous aggregation of phospholipid molecules to form a lipid bilayer

(A) to (D) illustrate the progressive aggregation of phospholipid molecules in aqueous solutions which forms a lipid bilayer in which the hydrophilic heads in both monolayers associate with the aqueous environment, while the tails create an inner hydrophobic environment between the monolayers. In aqueous solutions the bilayer closes on itself to form a closed compartment (membrane vesicles) as illustrated in (E).

3

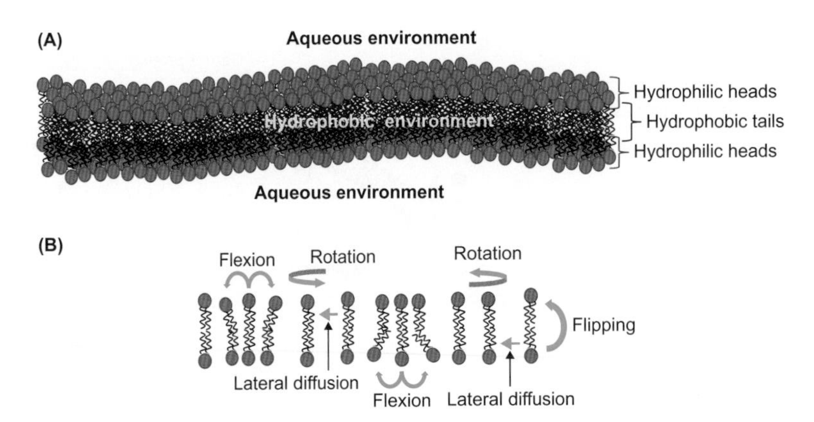

Figure 3.12 The lipid bilayers and movements within the bilayer

(A) The arrangement of amphipathic molecules in the lipid bilayer; (B) individual molecules in the bilayer can flex, rotate, diffuse in their own monolayer and even flip from one monolayer to the other.

The ability of cellular membranes to seal and form vesicles when disrupted is used as an important tool for studying the properties and functioning of membranes from particular tissues.

Within the lipid bilayer, the individual amphipathic molecules are not rigid and immobile. They can move laterally within their monolayer, or rotate and flex, and even flip from one of the layers to the other, as depicted in Figure 3.12. This behaviour means that the lipid bilayer acts like a two-dimensional fluid.

The fluidity of cell membranes is influenced by their chemical composition, in particular the amount of cholesterol and the type of fatty acids in the hydrophobic tails of the phospholipids. Cholesterol is a two-way regulator of membrane fluidity. At higher temperatures when the membrane fluidity rises due to increased movement of the phospholipid molecules, the presence of relatively rigid cholesterol molecules increases the rigidity of the membrane and reduces its fluidity. In contrast, at lower temperatures, when the hydrocarbon chains of the phospholipid molecules tend to cluster together and increase membrane rigidity, cholesterol molecules intercalate between the phospholipid molecules preventing them from clustering and oppose the decrease of membrane fluidity.

Changes in the proportion of longer chain fatty acids that are unsaturated (i.e. contain double or triple carbon–carbon bonds[39]) within cell membranes also play an important role in maintaining membrane fluidity in response to changing temperature of cells and tissues, or in parts of the body where temperature is lower[40]. These effects occur because unsaturated carbon bonds cause kinks in the long carbon chains, which increase the volume occupied by the hydrocarbon chains, so the phospholipids become less closely packed, making the overall structure less rigid.

For example, the bone marrow in the toes of the reindeer (*Rangifer tarandus*), where the temperature is low, contains more unsaturated fatty acids than the bone marrow at the head of the femur, where temperature is higher. The reindeer (or caribou) is an Arctic and Subarctic dweller, which often walks on ice. In order to minimize heat loss through their hoofs, caribou maintain the temperature of the lower part of the legs closer to that of the substrate on which it walks using a countercurrent heat-exchange system[41].

An important feature of the lipid bilayer shown in Figure 3.13 is its permeability to lipid-soluble substances and uncharged small molecules, including oxygen, carbon dioxide, alcohol, ammonia, urea, and even water molecules to a certain degree. However, notice in Figure 3.13 that the lipid bilayer prevents the passage of electrically charged solutes. Nevertheless, the extreme thinness of the lipid bilayer allows ions of opposite electrical charge on each side of the bilayer to attract each other strongly without crossing the membrane, leading to their accumulation on the two sides of the lipid bilayer.

The protein component of cell membranes

The protein component of cellular membranes consists of molecular assemblies embedded in the lipid bilayer. The

[39] Box 9.2 shows fatty acid structures, including chain length and degree of unsaturation/saturation, and gives some examples.

[40] We learn much more about changes in membrane fatty acid composition and their effects on membrane fluidity in Section 9.3.2 for animals that do not regulate their body temperature (**ectotherms**).

[41] We discuss countercurrent heat exchangers in Section 3.3.1.

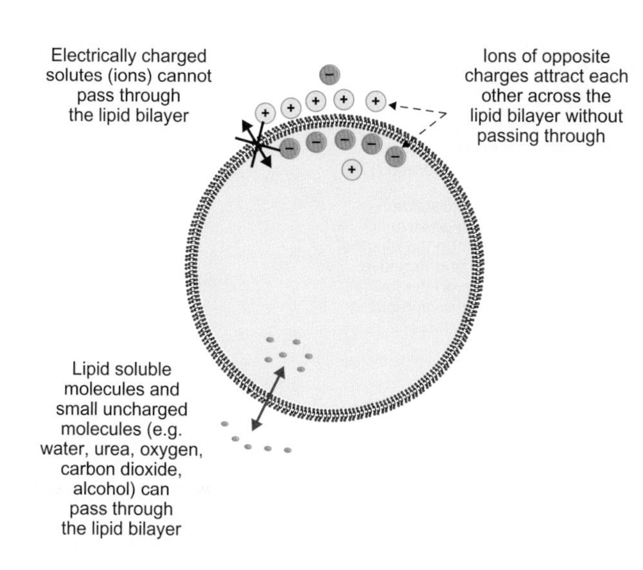

Figure 3.13 The lipid bilayer is permeable to lipid soluble substances and small uncharged molecules but not to electrically charged solutes

The lipid bilayer separates ions carrying positive and negative charges.

membrane proteins can either float in the lipid bilayer while aggregated with other lipid molecules to form so-called **lipid rafts** or can be anchored to the cytoskeleton. Looking back at Figure 3.9 notice that membrane proteins have four important functions:

- Transmembrane proteins span the membrane and are responsible for making physical connections between the cell's cytoskeleton and the extracellular matrix or form cell junctions. Structurally, the transmembrane proteins involved in forming **tight junctions** (claudins, occludin and junctional adhesion proteins) interact with cytoplasmic proteins connected to the actin-based cytoskeleton to form a complex and continuous network of protein strands that block the movement of substances between epithelial cells. However, small molecules can sometimes pass through the pathways between epithelial cells and the tight junctions[42].
- Some membrane proteins act as **receptors** for specific compounds, such as hormones that have signalling functions.
- Other membrane proteins act as **enzymes** that catalyse certain cellular reactions.
- Many of the membrane proteins are **transport proteins**, such as ion pumps and channels, that transport specific solutes across the membrane, and are responsible for the asymmetric distribution of substances across the cell membranes.

[42] Figure 4.10 illustrates a tight junction between epithelial cells, the functioning of which influences the permeability to water and solutes in many tissues, as we discuss further in Section 4.2.1.

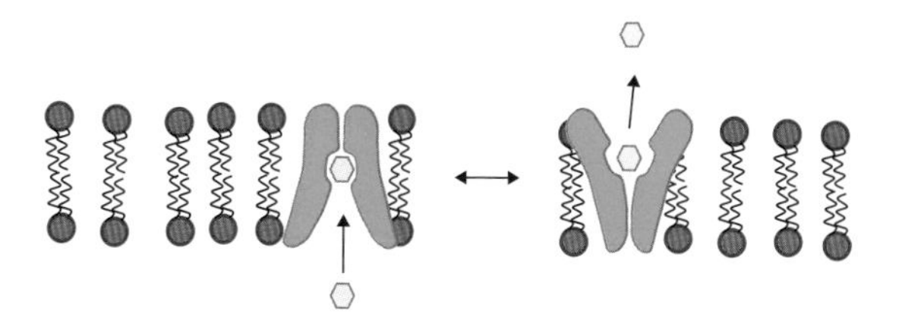

Figure 3.14 Diagrammatic representation of the functioning of a carrier protein

The carrier protein flips between two states exposing its binding site for a specific molecule either on one side and or on the other side of the membrane.

How do transport proteins operate?

One group of transport membrane proteins, known as **carrier proteins**, have binding sites for specific water-soluble molecules which otherwise would not pass through the lipid bilayer, and thereby facilitate the passage of these substances via the plasma membrane. Upon binding the substance, the carrier protein changes from one conformation to another. Figure 3.14 shows how the binding site for a specific molecule is exposed to one side of the membrane when the carrier is in one conformation and to the other side of the membrane when conformation changes to a different state. The change from one molecular conformation (shape) to the other facilitates the movement of the substance across the membrane.

Many transport proteins form **channels** in the membrane that facilitate the passage of specific water-soluble molecules from one side of the membrane to the other side. The membrane permeability for these water-soluble molecules depends directly on the number of channels that are open.

Most membrane channels are gated, which means that the proteins forming the channel can undergo structural change such that they either physically block passage through the channel (like a gate may block a footpath) or swing open to allow free passage through the channel. The functional state of the channels—whether or not they are open or closed—is under physiological control.

Figure 3.15 shows how some channels are sensitive to changes in the voltage (electrical potential difference) across the membrane: these are **voltage-gated channels**. Such channels play a crucial role in the generation of **action potentials** in nervous tissue[43].

Other channels open when a specific molecule (known as a ligand) binds to one or several sites on the channel. These **ligand-gated channels**, depicted in Figure 3.16, play an

[43] We learn how action potentials are generated and how they propagate in Section 16.2.

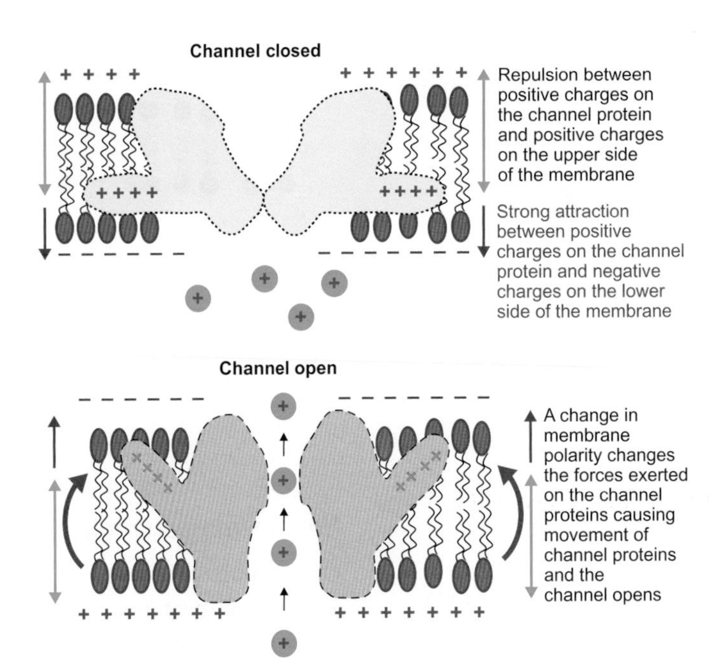

Figure 3.15 Diagrammatic representation of a voltage-gated ion channel

The channel closes and opens in response to electrical forces exerted between fixed charges on the channel proteins and the electrical polarity (+ or −) of the membrane.

important role in transmitting signals between cells and in some sensory cells[44].

There are also **mechano-sensitive channels**, which are sensitive to stress (stretch) applied to the lipid bilayer. Such channels, shown diagrammatically in Figure 3.17, play a central role in mechanoreception and regulation of cell volume[45].

Other channels are sensitive to changes in pH or Ca^{2+} concentrations. Many channels change properties when phosphate ions (PO_4^{3-}) bind covalently to specific sites (we say that these channels become phosphorylated). Such channels are important in ion movements and play an important role in learning and memory.

All cells have a membrane potential, which can be negative, positive or (rarely) zero

The presence of specific channels in the plasma membrane, which confer different membrane permeability to Na^+, K^+ and Cl^- ions, and the uneven distribution of these ions across the membrane, result in the development of an electrical potential difference across the plasma membrane as illustrated in the earlier Figure 3.7. The electrical potential difference between

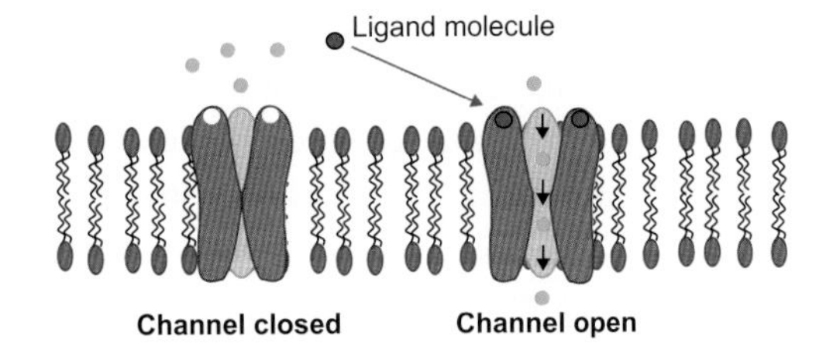

Figure 3.16 Diagrammatic representation of a ligand-gated channel

These channels open when specific (ligand) molecules bind to specific sites on the channel proteins.

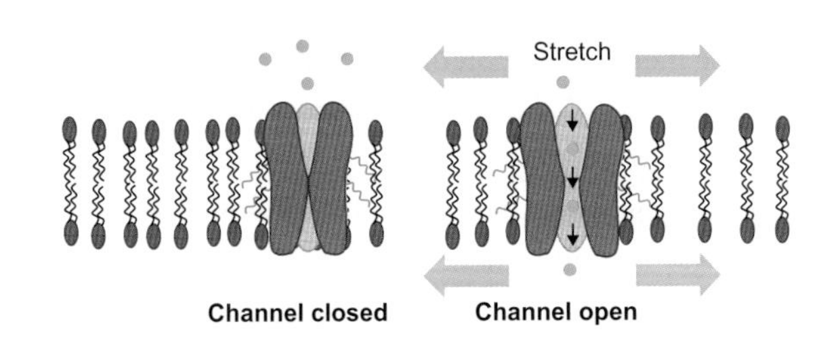

Figure 3.17 Diagrammatic representation of a mechano-sensitive channel

Some mechano-sensitive channels open when the lipid bilayer is stretched. These channels are sensitive to forces acting on the lipid bilayer due to strong hydrophobic interactions between the channel proteins (via structures represented here in blue) and the hydrophobic core of the lipid bilayer.

[44] The functioning of ligand-gated channels in signal transmission between neurons is discussed in Section 16.3.

[45] Section 4.3 (Figure 4.21) discusses this type of channel when we examine cell volume regulation in more detail.

the inside of the cell (V_{in}) and the outside the cell, in the extracellular environment, (V_{out}) is called the **membrane potential**[46] V_M (= V_{in} − V_{out}) and corresponds numerically to the amount of energy needed (measured in joules, J) to move one unit of positive charge (1 coulomb, C) across the membrane[47] that is V = J/C.

All cells have a membrane potential whose value depends upon the physiological state of the cell. Small changes in the membrane potential of cells, in either direction, have major effects on the movement of ions across the plasma membrane and the functioning of cells in neural, muscular and osmoregulatory tissues, which we discuss in Chapters 5, 16, 17 and 18.

Notice in Figure 3.18 that membrane potential (V_M) can be measured by making electrical contact with the outside and inside of the cell and measuring the difference between the electrical potential inside the cell (V_{in}) and outside the cell (V_{out}), just like measuring the voltage difference between the poles of an electrical battery. The electrical charge 'stored' by a plasma membrane (Q_M, measured in coulombs, C) is proportional to the membrane potential (V_M, measured in volts, V). The proportionality constant between the electrical charge stored per unit area of membrane (Q_M measured in C m^{-2}) and the membrane potential V_M is called **membrane capacitance** C_M, which is measured in farads[48] per m^2 (F m^{-2}), giving the following relationship:

$$Q_M = C_M V_M \qquad \text{Equation 3.10}$$

The common structural basis of the lipid bilayer confers on all cellular membranes a capacitance (C_M) of about 10 mF m^{-2}. Furthermore, since C_M is constant for a particular cell, it follows from Equation 3.10 that membrane charge and membrane potential are proportional to each other. Therefore, these two terms—membrane charge and membrane potential—are often used interchangeably. Thus, stating that a cell is more negatively charged inside than outside is the same as saying: (i) the cell has a negative membrane potential, (ii) the electrical potential of the cytosol is lower than the electrical potential of the extracellular environment, or (iii) there is a positive electrical potential difference between extracellular environment and the cytosol.

Using Equation 3.10 and a value for C_M of 10 mF m^{-2} we can calculate the electrical charge 'stored' per unit area

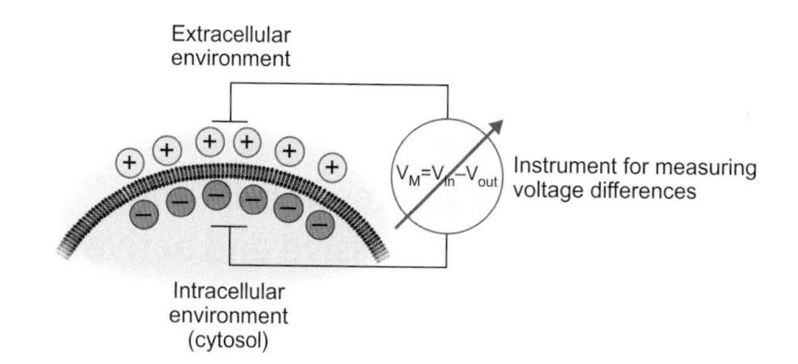

Figure 3.18 Measurement of the membrane potential

The membrane potential measured in volts (V) equals the difference in electrical potential (electric potential energy per unit charge). Membrane potential (V_M) is the difference between the electrical potential inside the cell (V_{in}) and outside the cell, in the extracellular environment (V_{out}) i.e. $V_M = V_{in} - V_{out}$

of membrane (Q_M) for a known membrane potential. For example, if the potential difference between the inside and the outside of the membrane V_M is −50 mV (which is typical of many cells at rest) then Q_M = 10 (mF m^{-2}) × 0.05 (V) = 0.5 mC m^{-2} (as 1 C = 1 F × 1V, as explained earlier in Box 3.1).

How many moles of monovalent ions, such as sodium (Na^+) or potassium (K^+), does this represent and how does this compare to the total amount of monovalent cations in cells?

The electrical charge carried by 1 mol of monovalent ions is given by the Faraday number (F = 96,485 C)[49], so in the example above, the 'storage' of 0.5 mC m^{-2} for a membrane potential of −50 mV, represents the storage of positive charges outside the membrane equal to 0.5 (mC m^{-2})/96,485 (C mol^{-1}) = 5.18 nmol per m^2 of membrane. Similarly, 0.5 mC m^{-2} of negative charges stored on the inside of the membrane corresponds to 5.18 nmol monovalent anions *per square metre* of membrane.

Cells obviously have a tiny surface area of plasma membrane compared to one square metre so, for instance, a vertebrate spherical cell of 4 μm diameter has a surface area of about 50 μm^2 (calculated from 4 π (cell radius)^2 = 4 π 2^2 μm^2 = 50 μm^2) Therefore, such a cell has only 2.6 × 10^{-19} mol[50] of monovalent cations separated from the same amount of monovalent anions when the membrane potential is −50 mV and this amount of monovalent ions represents only

[46] The mechanisms responsible for generation, maintenance and alteration of the cellular membrane potential are discussed in detail in Section 16.2, where we learn about the basis of electrical activity in neurons.

[47] The volt and coulomb are named after two influential physicists: Alessandro Volta (an Italian physicist) and Charles-Augustin de Coulomb (a French physicist) respectively.

[48] The farad is named after the 19th century English physicist Michael Faraday, who discovered electromagnetism.

[49] Look also at Box 3.1, which contains more information on electrical charges.

[50] If you check this for yourself, you will need to be careful with the units and remember that 1 μm equals 10^{-6} of a metre and 1 nmol is 10^{-9} of a mol.

0.00005 of the total amount of K^+ in animal cells[51], which is very small indeed.

3.2.4 Transport of solutes in cells and across membranes: passive and active transport

Despite their thinness, cellular membranes act as significant barriers against the free movement of solutes—typically they slow down (retard) the movement of molecules across them. The factor by which the movement of a particular substance is slowed down when it crosses a membrane is known as the retardation factor. Depending on the nature of the solute and the state of the particular membrane, the retardation factor can be between 50 to 10^{11} or even greater. You may be surprised to know that even lipid-soluble molecules passing through the lipid bilayer can be slowed down 50 to 10,000-fold, depending on the particular molecule. For example, the diffusion coefficient of the alcohol ethanol is reduced about 60-fold when it passes through the plasma membrane of red blood cells compared with its diffusion in water.

In many instances, the retardation factor is under physiological control, meaning that the membrane can become more or less permeable to particular molecules depending on the functional state of the cell. This difference in permeability can be achieved by altering the number of channels that are open, or the number of carriers that are available to transport particular molecules.

The most important physical process driving the movement of solutes at a cellular level is diffusion from regions of higher to lower concentration[52]. Diffusion within cells distributes most substances uniformly in the cytosol over distances of about 10 μm (equivalent to the size of most cells) in only a few seconds. However, diffusion is very ineffective over larger distances. For instance, it would take many months to distribute the same substance uniformly over a distance of 10 cm (about the size of a coffee cup). For this reason, diffusion plays a major role in determining the upper limit of the size of most cells: beyond these size limits, diffusion would be too ineffective for maintenance of life processes.

Passive transport across membranes

Passive transport in general refers to the movement of substances from a high concentration to a lower concentration (if the substances are not electrically charged), or from a higher electrochemical potential[53] to a lower electrochemical potential (if the substances are electrically charged) without the cell expending energy to facilitate this movement. As shown in Figure 3.19, the passage of substances by diffusion through the lipid bilayer and via channels takes place by passive transport. Movement of some substances by

[51] The concentration of K^+ in a vertebrate cell is about 150 mmol L^{-1} = 150 mol m^{-3}; the volume of the spherical cell $4/3\ \pi\ 2^3\ \mu m^3 = 33.5\ \mu m^3 = 33.5 \times 10^{-18}\ m^3$ and the total amount of K^+ in the spherical cell = concentration × volume = $150\ mol\ m^{-3} \times 33.5 \times 10^{-18}\ m^3 = 5.0 \times 10^{-15}$ mol. Hence, 2.6×10^{-19} mol represents only 0.00005 of the total amount of K^+ (2.6×10^{-19} mol/5.0×10^{-15} mol).

[52] Section 3.1.2 describes the principles of simple diffusion from high to low concentrations.

[53] The term electrochemical potential is defined in Box 3.1.

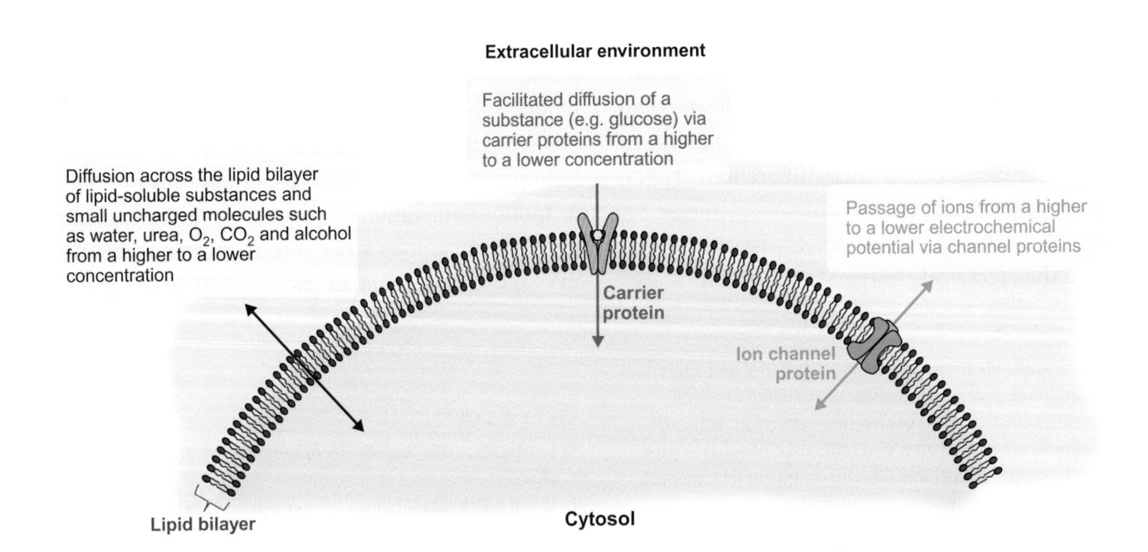

Figure 3.19 Passive movement of substances across membranes

carrier proteins also takes place by passive transport; such carrier proteins are known as **passive carriers**. Examples include the family of **facilitative glucose carriers (GLUTs)** that operate in most cells. The movement of substances by passive carriers is also known as **facilitated diffusion**.

Active transport across membranes

Active transport refers to the situation in which cellular energy is used to drive the movement of substances by carrier proteins across a membrane from a lower concentration to a higher concentration (if substances are not charged) or from a lower electrochemical potential to a higher electrochemical potential (if substances are charged). Such carrier proteins are known as **active carriers**.

Figure 3.20 shows how we can distinguish between **primary** (or direct) and **secondary** (or indirect) **active carriers**. Primary or direct active transport occurs when carriers use energy directly, in the form of ATP, to transport a substance across the membrane against an electrochemical potential difference, from a lower electrochemical potential to a higher electrochemical potential, as illustrated by Figure 3.20A. Such carriers are known as **ion pumps**. For example, the ubiquitous **sodium/potassium (Na^+/K^+) pump** uses energy released from the hydrolysis of one ATP molecule to remove three sodium ions from the cytosol and bring two potassium ions into the cell during each pump cycle[54].

The Na^+/K^+ pump is responsible for generating a very large electrochemical potential difference across the membrane for Na^+. This large difference in potential energy is used by several carriers to simultaneously transport specific substances from a lower to a higher electrochemical potential coupled to the movement of Na^+ from a higher electrochemical potential to a lower electrochemical potential. This is an example of secondary (or indirect) active transport, which Figure 3.20B illustrates.

Carriers that transport two coupled molecules (e.g. Na^+ and an amino acid such as alanine) in the same direction

[54] Figure 4.13 examines the pumping cycle of Na^+/K^+ pumps (Na^+, K^+-ATPase) more closely.

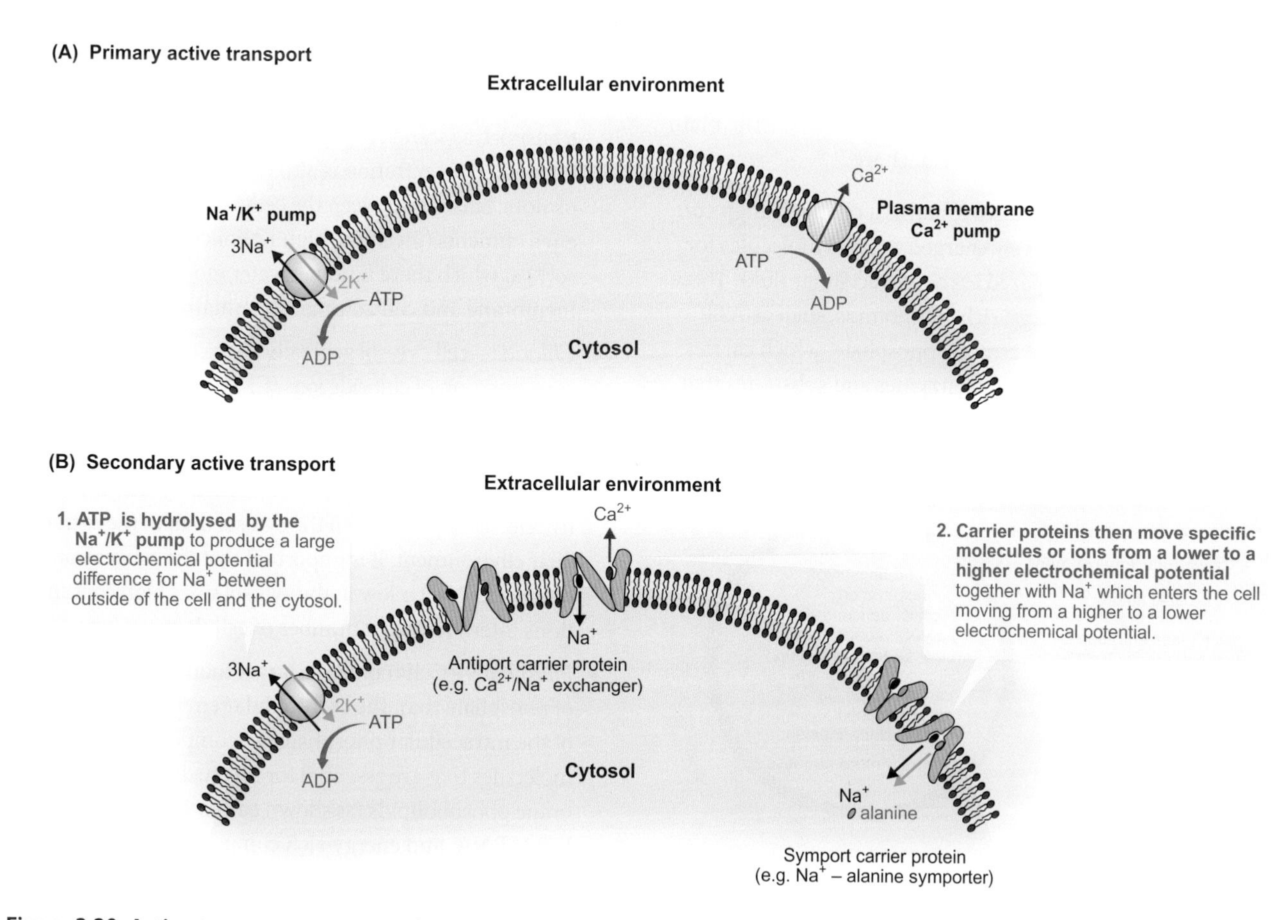

Figure 3.20 Active transport across membranes

(A) Primary active transport involves use of ATP by carrier proteins such as the Na^+/K^+ pump and the plasma membrane Ca^{2+} pump to move ions from a lower to a higher electrochemical potential.

(B) Secondary active transport across membranes: the Na^+/K^+ pump generates a very high electrochemical potential difference between the extracellular environment and the cytosol which is then used by carrier proteins to move various molecules and ions from a lower to a higher electrochemical potential.

are known as **symport carriers**, while carriers that transport two coupled molecules in opposite directions are known as **antiport carriers**. Both are shown in Figure 3.20. Antiport carriers are also known as **exchangers** because they effectively exchange one solute, which enters the cytosol, for another molecule or ion, which leaves the cytosol. For example, removal of Ca^{2+} from the cytosol occurs in exchange for Na^+ via Ca^{2+}/Na^+-exchangers.

From this discussion, it is apparent that cell membranes act both as selective barriers to diffusion and as passageways for the movement of specific molecules against their concentration and/or electrochemical potential differences with energy being needed to move substances from a lower to higher concentration, or from a lower to higher electrochemical potential. We look next at the net effect of these transport processes in cellular membranes on the composition of the cytosol.

3.2.5 Distribution of substances across plasma membranes

Several important differences in the ionic composition of the cytosol and the extracellular environment apply to most cells of most species regardless of their environments[55]. The main differences summarized in Figure 3.21 are:

- *Organic anions*—Cytosol contains a relatively high concentration of negatively charged organic molecules (organic anions) compared to the extracellular fluid. These anions include energy-rich compounds, such as ATP (e.g. $Mg\text{-}ATP^{2-}$) and creatine phosphate, which carries two negative charges, and enzymes and substrates that participate in chemical reactions in the cytosol.

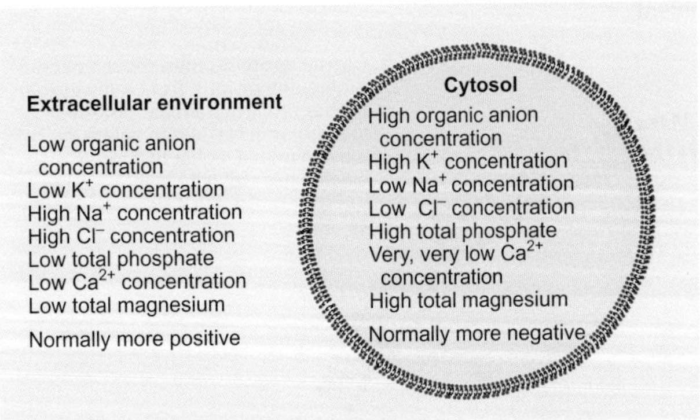

Extracellular environment
Low organic anion concentration
Low K^+ concentration
High Na^+ concentration
High Cl^- concentration
Low total phosphate
Low Ca^{2+} concentration
Low total magnesium
Normally more positive

Cytosol
High organic anion concentration
High K^+ concentration
Low Na^+ concentration
Low Cl^- concentration
High total phosphate
Very, very low Ca^{2+} concentration
High total magnesium
Normally more negative

Figure 3.21 The asymmetric distribution of substances between the cytosol and the extracellular environment

- *Potassium*—Cells contain K^+ at 8 to 50 times the concentration of the extracellular environment. The large K^+ concentration difference and normally higher permeability of the plasma membrane to K^+ compared with its permeability to Na^+ generates a negative membrane potential difference between the cytosol and the extracellular environment, as we considered earlier in Figure 3.7. A high potassium concentration in the cytosol is also necessary for protein synthesis.
- *Sodium*—The cell cytosol contains Na^+ concentrations that are 7–8 times lower than in the extracellular environment. These concentration differences result from the continuous operation of Na^+/K^+ pumps located in the plasma membrane that use energy released by the hydrolysis of ATP to pump out the Na^+ that diffuses slowly into the cells, while simultaneously maintaining high K^+ concentrations in the cytosol. In resting mammals, the Na^+/K^+ pumps in their plasma membranes use about 30 per cent of the average energy expenditure of cells. The considerable build-up of electrochemical potential associated with the uneven distribution of Na^+ across the plasma membrane acts as a source of energy to transport various molecules from a lower electrochemical potential to a higher electrochemical potential by secondary active transport, as shown in Figure 3.20B. The low intracellular Na^+ concentration is also essential to maintain the osmotic balance between the cellular and extracellular environments (given the high K^+ concentrations in cytosol[56]) at which there is no net water movement across the membrane and cell volume is maintained[57].
- *Chloride*—cell cytosol generally has a considerably lower concentration of chloride ions (Cl^-) than extracellular fluids. The uneven distribution of Cl^- across the plasma membrane arises mainly from the high concentration of organic anions in the cytosol combined with a more negative electrical potential in the cytosol than in the extracellular environment. It is important that the concentration of Cl^- (like Na^+) is low in the cytosol since high concentrations interfere with a number of cellular reactions.
- *Phosphate*—Cells contain a larger total concentration of phosphate than the extracellular environment. Most of the intracellular phosphate is a component of organic molecules that are essential for life, such as the membrane phospholipids (as shown earlier in Figure 3.10A), DNA, RNA, and energy-rich compounds containing phosphate such as ATP and ADP. Phosphorus as a free phosphate ion (PO_4^{3-}) is called **inorganic phosphate**,

[55] These common differences provide strong evidence that animals (and other forms of life) evolved from a common ancestor, as we discuss in Section 1.3.

[56] As an example, Figure 4.5 shows ion concentrations in the cytosol (intracellular fluid) and extracellular fluid of muscle cells.

[57] We examine cell volume regulation more fully in Section 4.3.

usually designated as P_i because phosphate in solutions occurs in various ionic forms, depending on pH (the concentration of hydrogen ion). In extracellular fluid and at physiological pH, phosphate occurs mainly as a mixture of hydrogen phosphate (HPO_4^{2-}) and dihydrogen phosphate ($H_2PO_4^-$).

- *Calcium*—The concentration of calcium ions (Ca^{2+}) in the cytosol is 10,000 to 100,000 times less than in either the extracellular environment or the endoplasmic reticulum (where Ca^{2+} is stored). The extremely low Ca^{2+} concentrations in cytosol are achieved through the action of Ca^{2+} pumps in the plasma membrane that utilize ATP, and the presence of Na^+/Ca^{2+} exchangers that remove Ca^{2+} from the cytosol while moving Na^+ into the cytosol from a higher to a lower electrochemical potential. In addition, Ca^{2+} pumps in the endoplasmic reticulum are important for transporting cytosolic Ca^{2+} into stores from where it can be released, when required, via alternative transport proteins.
- The huge difference between Ca^{2+} concentrations of cytosol and the extracellular environment (or endoplasmic reticulum/mitochondria) is essential in enabling Ca^{2+} to play a major role in regulating cell function: relatively small amounts of Ca^{2+} entering the cytosol produce very large changes in the cytosolic Ca^{2+} concentrations. Such events activate many cell processes, including mitochondrial function, secretion, growth, division, ciliary movement and are particularly important for muscle contraction[58]. Calcium overload, on the other hand, plays a key role in inducing cell death, known as **apoptosis**.
- *Magnesium*—The total and ionized concentrations of Mg^{2+} in the cytosol are normally high. This is beneficial for cells because Mg^{2+} is a cofactor in many reactions. For example, Mg^{2+} binds to ATP to form an Mg-ATP complex, which is the true substrate for many reactions that use the energy liberated from the hydrolysis of ATP. Mg^{2+} is also the main divalent cation (i.e. with two positive electrical charges) in the cytosol and plays an important role in stabilizing cellular membranes, most likely by stabilizing the structure of the membrane proteins.

Understanding the difference in ionic composition between the cytosol and the extracellular environment is of great practical importance. Based on such knowledge it is possible to prepare physiological saline solutions to mimic the ionic composition of the extracellular fluids of an animal when partially replacing body fluids lost from an animal's body during surgery, during the treatment of dehydration, or when studying the properties of isolated cells and tissues under controlled conditions. Experiments performed in physiological solutions are known as ***in vitro*** experiments (from the Latin meaning 'within glass'), to distinguish them from experiments conducted ***in vivo*** (from the Latin meaning 'in living'), i.e. in the living animal.

For cells to function correctly their internal environment must be maintained, not only with respect to the various ionic species mentioned here, but also with respect to many other parameters including ATP concentration, pH, organelle structure, cell volume, and factors involved in gene regulation. The process of maintaining this constant cellular environment is called **cellular homeostasis**. For this, cells need energy, which they mostly extract from the energy in the chemical bonds of carbohydrates and lipids[59], which is then converted into an energy 'currency' (ATP) that can be used to drive most cellular processes.

Cellular homeostasis does not mean that physiological conditions are set at fixed (rigid) values; rather, they adopt a range of values depending on the type of cell, animal species and the physiological state of the animal. For example, the intracellular K^+ concentration in cold-submerged, hibernating frogs (*Rana temporaria*) decreases by up to 30 per cent in skeletal muscle, but not in the heart muscle. There is also a 30 per cent reduction in the extracellular Na^+ concentration in hibernating frogs but this is not the case for hibernating ground squirrels (*Citellus lateralis*).

› Review articles

Berridge MJ, Bootman MD, Roderick H (2003). Calcium signalling: dynamics, homeostasis and remodelling. Nature Reviews Molecular Cell Biology 4: 517–529.

van Meer G, Voelker DR, Feigenson GW (2008). Membrane lipids: where they are and how they behave. Nature Reviews Molecular Cell Biology 9: 112–124.

Orrenius S, Zhivotovsky B, Nicotera P (2003). Regulation of cell death: the calcium–apoptosis link. Nature Reviews Molecular Cell Biology 4: 552–565.

Pollard TD, Cooper JA (2009). Actin, a central player in cell shape and movement. Science 326: 1208–1212.

Pizzo P, Drago I, Filadi R, Pozzan T (2012). Mitochondrial Ca^{2+} homeostasis: mechanism, role, and tissue specificities. Pflügers Archiv—European Journal of Physiology 464: 3–17.

3.3 Interactions of animals with their environments

In the previous section, we explored the interactions between cells and the extracellular environment. Now we move to a higher level of biological organization and examine the interactions between animals with their external environment. These interactions allow them to maintain a stable internal environment in which their cells can function, while being protected from variable external influences.

[58] We discuss in detail the regulation of muscle contraction by calcium ions in Section 18.1.

[59] We discuss in detail how ATP is produced in animal cells in Section 2.4.

The maintenance of a relatively stable internal environment is known as **animal homeostasis** (to distinguish it from cellular homeostasis, which refers to maintenance of a relatively stable *cellular* environment). Animal homeostasis generally maintains physiological stability. Just like cellular homeostasis, however, animal homeostasis does not refer to firmly fixed values for physiological characteristics, but to ranges. For example, the body temperature of healthy humans is not rigidly fixed at 37°C but varies between about 36.7 and 37.3°C, depending on the time of the day and other factors[60].

More dynamic changes in physiological parameters in response to changing and particularly challenging environmental changes have been described as **allostasis** (from the Greek *allo* meaning variable and *stasis* meaning standing still), although some physiologists consider allostasis to be a natural extension of homeostasis for more dynamic conditions. However, a key aspect of the concept of allostasis is the proactive ability to change physiological variables in *advance* of demand, while homeostasis is generally seen as a series of reactive responses to maintain internal stability. As such, the anticipatory nature of allostasis is usually associated with brain functioning and often hormonal changes in animals with complex brain function (birds and mammals) and has more recently been considered for fish.

First, we examine the basic principles that govern the physical processes that facilitate interactions between animals and their environments before considering the complex physiological systems animals employ to maintain internal stability.

3.3.1 Physical processes facilitating animal interactions with their environment

In order to grow, reproduce and survive, animals need a continuous supply of energy and nutrients[61], and information about their environments[62]. The passive physical processes of diffusion of molecules and heat conduction at the interfaces between animals and their environment facilitate interactions between them. In addition, the maintenance of a stable internal environment requires waste products to be removed by specific tissues and organs such as the kidneys[63].

In this section we discuss further physical processes—**convection**, **radiation** and **evaporation**—that facilitate the exchange of heat and water between animals and their environment, and how the loss of heat or water can sometimes be minimized by countercurrent exchange mechanisms.

Convection

Convection is a physical process during which a bulk flow of a fluid (gas or liquid) accompanies the transfer of heat or matter. For instance, as we discuss in Section 1.2.1, warm air rises from around the equator and generates global wind patterns which, in turn, drive the formation of ocean currents. More generally, when thermal energy expands a fluid, the density of the fluid declines, which drives its flow in **free convection**.

Among animals, movements of the fluid medium in which they live—air or water—as a result of free convection can increase heat loss if their surface temperatures exceed those of the surrounding fluids (as in birds and mammals). The temperature difference warms the air or water close to the bird's or mammal's body (by conduction and radiation), causing air or water to rise, which drives the movement of cooler air or water toward the animal's surfaces; this in turn leads to more heat loss. If the fluid surrounding the animal moves, or the animal moves in its environment, then additional convective heat losses will occur in a slightly different process of **forced convection**[64].

We see forced convection in the circulatory systems of animals, where the pumping action of the heart moves the blood. In this case, convection of blood distributes heat, gases, nutrients and various solutes within an animal's body. Another good example of forced convection occurs when water is forced over the respiratory surfaces of aquatic animals or air is moved over the respiratory surfaces of land animals. In these cases, forced convection refreshes the respiratory medium (water or air), which enhances the uptake of oxygen and excretion of carbon dioxide. Forced convection occurs during the panting of mammals, birds and some reptiles for regulation of body temperature[65]. In aquatic animals, forced convection facilitates solute exchange and heat transfer between the animal and its environment.

Radiation

The term 'radiation' describes the transfer of energy through space by means of electromagnetic waves in the same way as the Sun transmits light and heat to Earth. The electromagnetic energy emitted by an object is proportional to the absolute temperature of the object raised to the power of four (T^4) and has a higher peak frequency and intensity the hotter the surface of the object[66].

Thermal radiation is generated when the thermal energy of an object is converted into electromagnetic energy and occurs when the body has a temperature above 0 K (–273°C). Animals with a surface temperature of 17–42°C (290 to 315 K) emit electromagnetic radiation mainly in the far infrared

[60] Section 10.2 examines variations in body temperatures of mammals and Section 10.2.6 discusses the control systems regulating body temperatures.

[61] Chapter 2 examines the energy needs and provision for animals.

[62] For a detailed discussion of animal senses, look at Chapter 17.

[63] Chapter 7 discusses kidney function.

[64] We discuss free and forced convection further in Section 8.4.3.

[65] We discuss panting in Section 10.2.2.

[66] Section 8.4.1 discusses these physical relationships in some detail.

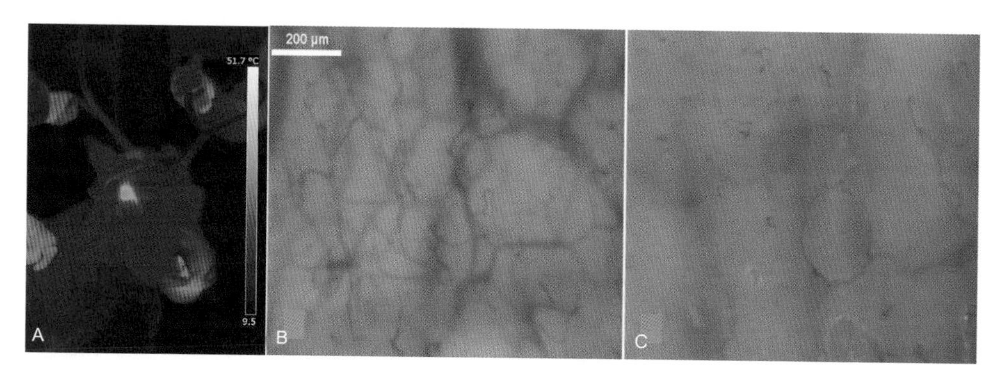

Figure 3.22 Infrared image of a reindeer's head

(A) The temperature scale indicates surface temperature of the nose is at about 24°C after the reindeer has exercised (on a test treadmill). The higher surface temperatures in the nose and eyes compared to the rest of the body, which is well insulated, allows the reindeer to lose heat by radiation and is thought to be due to the richly vascularized microcirculation (shown in (B)) compared to that of humans (C), which is demonstrated by the video microscopy images.

Source: Ince C et al (2012). Why Rudolph's nose is red: observational study. British Medical Journal 345: e8311.

part of the spectrum, which we cannot see, although some animals can[67].

The infrared radiation of animals can be detected by infrared (thermal imaging) cameras, which create a false colour image related to the extent of the radiation. Hotter objects emit higher intensity infrared radiation so thermal imaging cameras help us understand which parts of an animal are radiating most heat. Figure 3.22 uses a thermal image of a reindeer to illustrate this principle; reindeer have a well-vascularized 'red nose', which allows heat loss by radiation after exercise. We learn about these processes and the thermal relations of animals in Chapters 9 and 10.

The Sun has a surface temperature of around 5800 K and peak radiation intensity in the visible part of the spectrum, as well as sizeable components in the ultraviolet and infrared parts of the spectrum. Of course, life depends on the radiation from the Sun reaching the Earth: even small fluctuations can have major consequences for the climate and weather patterns and hence for animal life on the planet. Fortunately, most of the **ultraviolet radiation** from the sun (that is damaging to life) does not reach the surface of the Earth because of the presence of the ozone layer in the atmosphere and scattering[68].

In order to regulate the heat received directly from the thermal radiation of the Sun, animals:

- move to maximize uptake of thermal radiation or protect themselves from the direct radiation from the Sun;
- change position with respect to the direction of the Sun's rays;
- change the level of absorption of thermal radiation at their surfaces by changing colour or changing the flow of blood close to their surfaces, for example.

Evaporation

Conversion of liquid water into water vapour (evaporation) takes place when it absorbs energy (in the form of heat) from animal surfaces (which consequently experience heat *loss*). Not surprisingly, such heat loss can form a significant part of the control of body temperature[69] as each gram of water evaporated under standard conditions[70] results in the loss of 2257 J. (This value—the amount of energy required to evaporate one gram of water under standard conditions is called the **latent heat of water vaporization**. 'Latent heat' refers to the heat associated with the change in phase—in this case liquid to gas.)

The amount of energy lost as heat during the evaporation of just 1 g of water is sufficient to decrease the temperature of 100 g of water by about 5.4°C. The water content of most animals is very high, so evaporation of 1 g of water from the surface of an animal weighing about 100 g can decrease its body temperature by about 5°C. Hence, many terrestrial animals regulate their body temperature by evaporating water from various moist surfaces (such as their respiratory surfaces). However, such water losses could have significant effects on fluid balance, so these fluids need replacing to avoid the animal suffering dehydration[71].

Countercurrent exchange

When two fluids running in parallel, but in opposite (countercurrent) directions, are separated by a thin layer through which transfer can occur, heat, solutes, water or gases can

[67] We discuss the spectral range of electromagnetic radiation to which animals are sensitive in Section 17.2.1.

[68] We discuss the formation of ozone in Section 1.5 and show the transmittance of particular wavelengths through the atmosphere in Figure 17.9.

[69] Evaporative water loss by terrestrial animals as part of their body temperature regulation is examined in Section 10.2.2.

[70] Values are normally quoted for standard temperature 273 K and pressure (1.01 × 105 Pa = 1 atmosphere). The slightly different values quoted for latent heat of evaporation by different sources are likely to be due to quoting for slightly different conditions.

[71] Section 6.1 discusses evaporative water loss in land animals and the ability of some species to limit these losses.

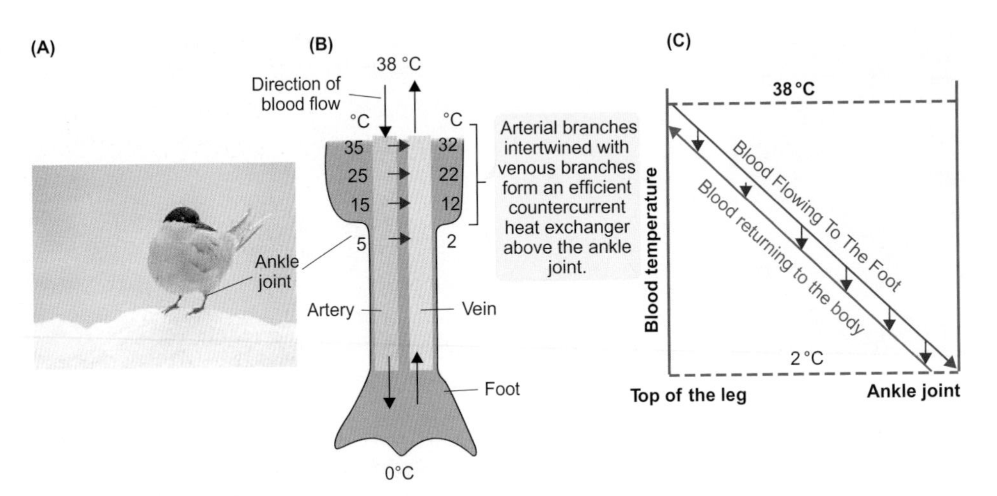

Figure 3.23 Countercurrent heat exchanger in the legs of the arctic tern (*Sterna paradisaea*) standing on ice

(A) Tern standing on ice; (B) heat is transferred from the arterial blood flowing to the tern's foot, to the venous blood returning to the body such that the blood is cold (0°C) when it reaches the foot and warm when is returned to the body; (C) the flow of blood in opposite directions through blood vessels close to each other allows for a steady temperature difference to be established above the ankle joint along the tern's leg. This countercurrent heat exchanger is responsible for a very effective retention of heat by the bird with little loss to the external environment.

Source: A: Copyright Rich Reid

be efficiently exchanged between the fluids (gas or liquid) without use of additional energy—the system operates passively. **Countercurrent-exchange** systems occur between animals and their external environment and are also important *within* animals[72] where they allow exchanges between blood flowing in opposite directions through vessels lying close to each other.

A good example of countercurrent exchange is the heat exchange that occurs in the legs and feet of birds. Despite standing on ice or snow, or in icy water, countercurrent heat exchange avoids them losing heat through their legs and feet. This is because heat moves from the blood flowing to the feet into the blood returning to the body in a countercurrent exchange, as illustrated in the schematic diagram in Figure 3.23. The blood is relatively cold when it reaches the feet, but its temperature is kept at just above freezing point by adjusting the flow through the parallel blood vessels.

Since blood flowing *to* the feet is a bit warmer than blood *leaving* the feet, at any given point in the parallel vessels, as shown in Figure 3.23B, heat flows into the blood leaving the feet, which reduces heat loss to the environment. Countercurrent heat exchange leads to a stable temperature difference between the blood flowing to the feet and the blood returning to the body, which Figure 3.23C illustrates. A similar mechanism operates in the legs of the reindeers (and many other species), allowing them to walk on ice[73].

Countercurrent multiplier systems

In a countercurrent exchange system, no cellular energy is expended to maintain differences between the fluids moving in opposite directions. However, if cellular energy *is* expended, and a fluid moves in opposite directions through the arms of a loop or hairpin-like structure, then very large differences can be established between the start and the turning point of the hairpin structure, by a **countercurrent multiplier system**. Countercurrent multipliers were first identified in the kidneys of mammals in which they enable the conservation of water and excretion of concentrated urine when an animal is dehydrated[74].

3.3.2 Physiological systems maintain animal stability

In order to achieve animal homeostasis, tissues and organ systems are necessary. The following physiological systems are responsible for regulating specific parameters:

72 Figure 7.23A examines the countercurrent exchange system, which exchanges ions and water in parallel blood vessels in the mammalian kidney. Other examples of countercurrent exchangers are discussed in Sections 10.2.4 and 12.2.2.

73 Section 3.2.3 discusses further temperature adaptations of the fatty acid composition in reindeer legs.

74 We learn about the production of concentrated urine by mammals and birds in Section 7.2.3. We also look at another example, in the swim bladder of bony fish, in Case Study 13.1.

- a **gas-exchange (respiratory) system** for uptake of O_2 from the environment and excretion of the CO_2 continuously produced by cells;
- an **excretory system** to remove toxic waste products;
- a **digestive system** to extract nutrients (amino acids, sugars and fat molecules) from food;
- **sensory and integrative (neural and hormonal) systems** to sense, process and respond to signals from the external and internal environments to keep the animal alive;
- a **reproductive system** to maintain (or expand) populations in communities of animals within ecosystems;
- a **locomotory system** to facilitate movement in search of food, water, or shelter, to find a partner for reproduction, and to escape predators;
- larger animals have also developed a **circulatory system** to provide cells with oxygen and metabolic substrates, facilitate communication between various tissues and organ systems, and remove waste products and heat produced by cells;
- an **immune system**, which is responsible for protecting the identity (self) of an animal by neutralizing potentially harmful foreign agents (pathogens) that may enter the body and cause disease.

The function of each of these physiological systems is controlled and coordinated in each animal to enable it to survive, mature and reproduce, while interacting with its external environment.

3.3.3 Control of physiological systems

For animals to survive they must control their internal environment using negative and positive feedback loops. A feedback loop is a mechanism in which the output from a system feeds information back into the system, which then modifies the system's output[75]. **Negative feedback** occurs when the output from the system feeds back and reduces the output from the system[76]. In contrast, **positive feedback** occurs when output from the system feeds back and increases the output from the system.

An everyday example of a negative feedback loop is a heater controlled by a thermostat. If the temperature of the room is below the set temperature on the thermostat, the heater produces heat. However, if the temperature in the room rises above the set temperature, the negative feedback loop in the thermostat switches off the heater: the output from the heater (the heat) feeds back into the system (the heater) to reduce its output and the heater switches off.

The presence of a negative feedback loop thus dampens changes of a particular variable thereby tending to keep the system in a stable state with respect to that variable. Many such negative feedback loops operate in animals to maintain overall control of their internal environment such as specific ion concentrations and osmolality.

In contrast, positive feedback loops amplify changes to a particular variable, moving the system away from its equilibrium state for the variable. The generation of nerve impulses involves a positive feedback mechanism that leads to the explosive generation of a nerve impulse[77]. Positive feedback loops are also important in some hormonal control mechanisms, the best example of which is the normal control of ovulation of vertebrates[78].

Neural and hormonal control mechanisms

Generally, nervous and hormonal (endocrine) systems integrate and coordinate the activity of the different types of cells, tissues and physiological systems present in the body of animals[79].

The nervous system is specialized for the rapid transmission of signals in the body, often over long distances via nerve cells or **neurons**. The neurons play a crucial role in the reception, transmission, integration and processing of information about the internal and external environment and in an animal's responses to changes in the external and internal conditions. Output signals from the nervous system activate **effector cells** to generate a coordinated response to the particular situation. Neurons connect to each other through specialized junctions called **synapses**[80] and provide the physical basis for many of the negative feedback loops that function in animals to maintain homeostasis.

The nervous system of animals that display bilateral symmetry[81] can be generally subdivided into:

- a **central nervous system** (CNS) that *coordinates* the activities of all parts of the body;
- a **peripheral nervous system** (PNS) located outside the CNS, which connects the CNS with various parts of the body to enable the animal to respond specifically to various stimuli.

[75] Figure 19.5 illustrates feedback loops in hormonal (endocrine) systems.

[76] Figure 19.17 illustrates the general principles of negative feedback loops controlling an endocrine axis involving a series of hormones.

[77] We examine the generation of nerve impulses in Section 16.2.5.

[78] We learn about the positive feedback loops that control ovulation in Section 20.4.1.

[79] We examine the functions of the nervous and endocrine systems in Chapters 16 and 19, respectively.

[80] We discuss how synapses function in Section 16.3.

[81] We discuss the evolution of bilateral symmetry, as compared to the radial symmetry of cnidarians (sea anemones, corals and jellyfish) in Section 1.4.2.

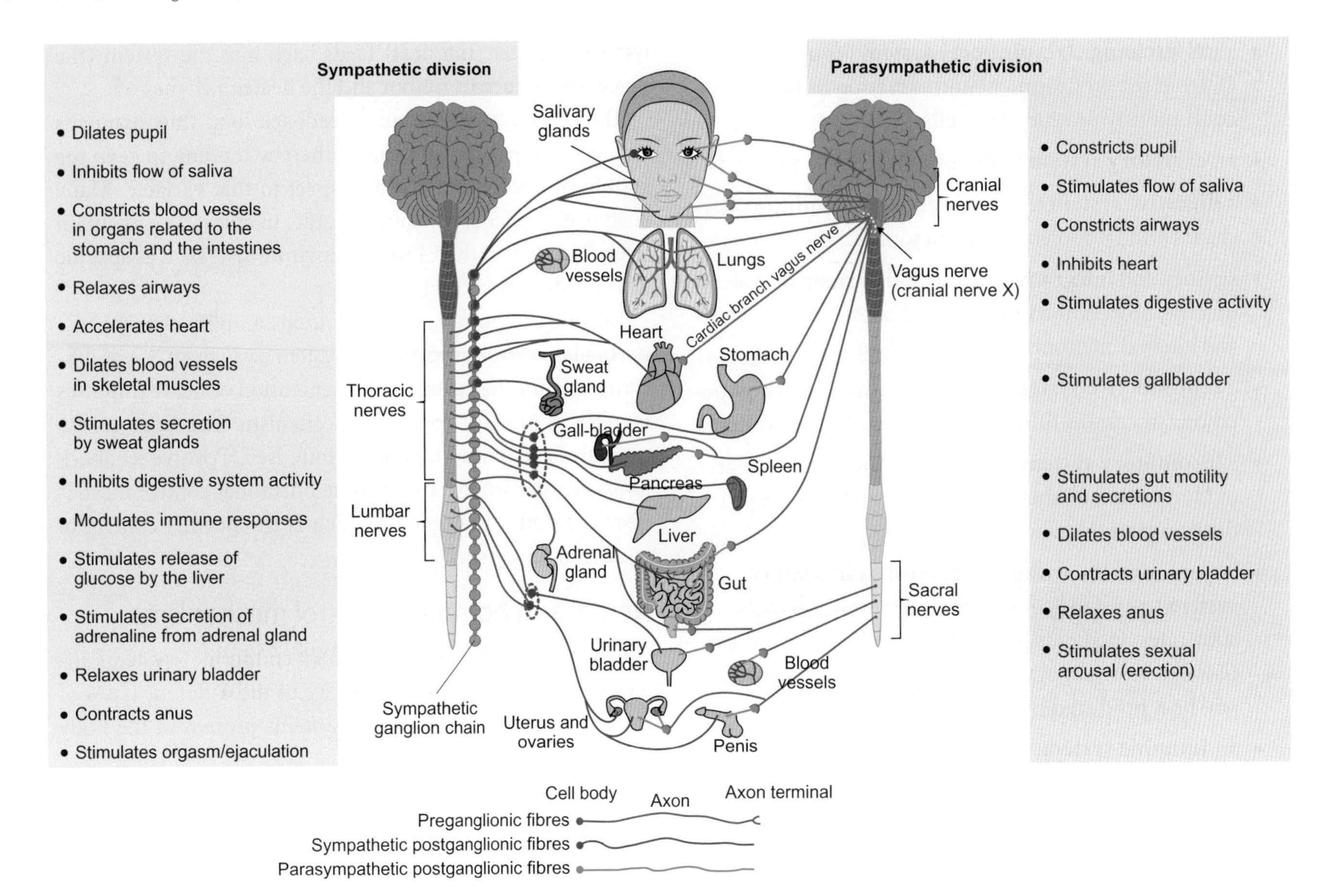

Figure 3.24 Control of internal organ function in mammals by the sympathetic and parasympathetic divisions of the autonomic nervous system

Both the sympathetic and the parasympathetic divisions of the autonomic nervous system have left and right sides, but only one side of each is shown in the diagram: sympathetic nerves and ganglia are shown on the left and parasympathetic nerves and ganglia are shown on the right. The actions of each division are listed.

The part of the peripheral nervous system that is involved in controlling tissue and organ function allowing them to maintain homeostasis is the **visceral** or **autonomic nervous system**. The autonomic nervous system of vertebrates comprises two parts: the **parasympathetic division** and the **sympathetic division**. Figure 3.24 shows the innervation of the internal organs in mammals by each division. The parasympathetic division is responsible for maintaining normal body functions in the animal during rest while the sympathetic division mobilizes body systems during activity or when animals are stressed[82].

Animals often adjust their behaviour in response to their senses[83], in response to changes in internal variables, and when they are stressed. In such circumstances, higher centres in the brain are usually involved. Figure 3.25 shows the different parts of the vertebrate brain, which are involved in control of the functioning of vertebrates, and which we learn more of in many of the later chapters in this book.

The higher centres of the brain also influence the functioning of hormone (endocrine) systems, which as Figure 3.26 illustrates typically use the circulatory system to convey **hormones** from one part of the body to another, in the bloodstream[84]. Hormone secretion generally occurs over a longer time than neural signals, although some hormones—**neurohormones**—are in fact synthesized and secreted by neurons. Hormones and neurohormones travel in the bloodstream or extracellular fluids for some distance before binding to specific receptors on the so-called **target cells** and resulting in modification to the

[82] We discuss the effects of stress on animals and their responses in Section 19.3.

[83] We discuss how animals sense their environment in Chapter 17.

[84] Endocrine systems and their actions are the topic of Chapter 19. We also discuss hormone actions on physiological systems at many other points in this book.

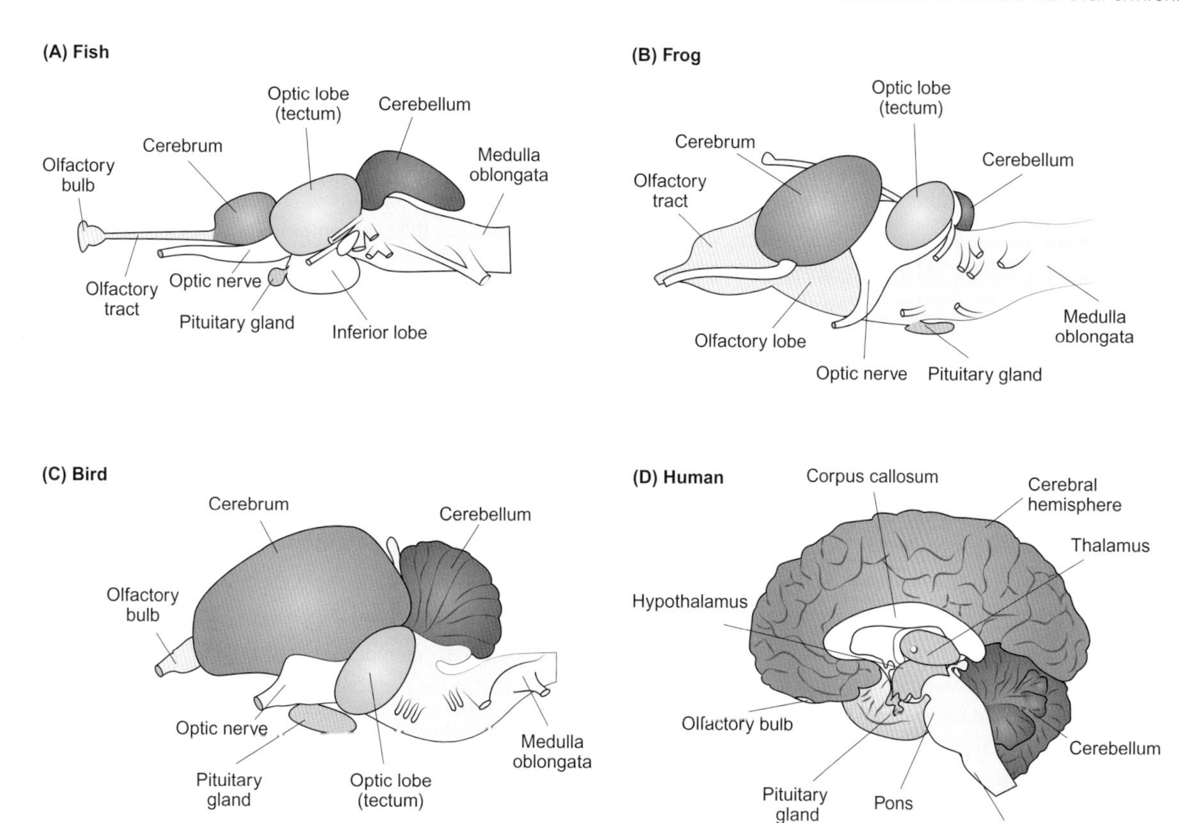

Figure 3.25 Lateral view of the brain and its main structures in representative animals of four groups of vertebrates

(A) Fish; (B) frog; (C) bird; and (D) mammal (human). The human brain has been sectioned transversally along the line of bilateral symmetry.

Source: modified from Eckert. Animal Physiology, 5th Edition. New York: W H Freeman & Company. 2002, which was adapted from Romer AS. The vertebrate body. Philadelphia: Saunders. 1955.

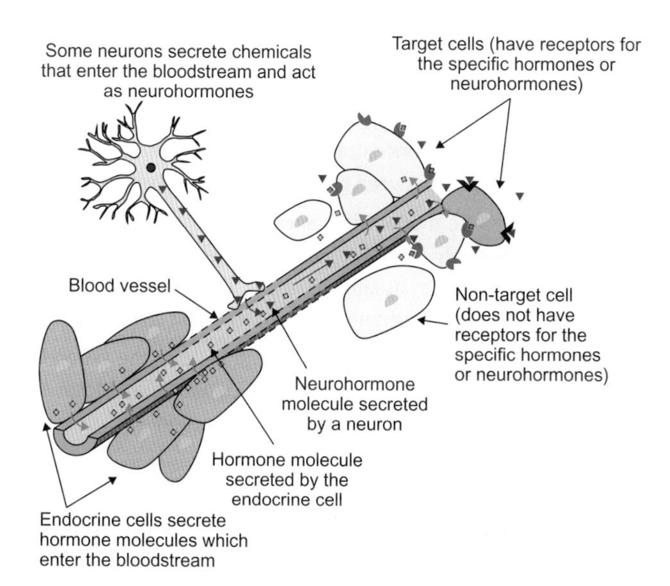

Figure 3.26 Schematic diagram showing the basis of how the endocrine system works

functioning of these cells[85]. Thus, hormone systems and the central and peripheral nervous system act in close cooperation to coordinate body functions and the behaviour of animals—they control their internal environment, and their responses to changes in the external environment, metabolism, feeding, growth, physical development and maturation, and reproductive processes, and thereby their populations.

› *Review articles*

Day TA (2005). Defining stress as a prelude to mapping its neurocircuitry: no help from allostasis. Progress in Neuro-Psychopharmacology and Biological Psychiatry 29: 1195–1200.

McEwen BS, Wingfield JC (2003). The concept of allostasis in biology and medicine. Hormones and Behavior 43: 2–15.

Schulkin J (2010). Social allostasis: anticipatory regulation of the internal milieu. Frontiers in Evolutionary Neuroscience 2: 111.

[85] Many hormones act over relatively short distances, for example within an organ/tissue, as we discuss in Section 19.1.

Checklist of key concepts

The flow of heat and the movement of ions and molecules within animals and between them and their environments:

- The flow of heat, uncharged solutes and gases across biological barriers occurs from regions of higher temperature/concentration or partial pressure (gases) to regions of lower temperature/concentration or partial pressure.
- The rate of transfer across a barrier is directly proportional to the temperature/concentration or partial pressure difference between the two sides and the barrier surface area and is inversely proportional to barrier thickness.
- The ease with which substances can pass across a barrier is expressed by **permeability constants**.
  - The proportionality constant for heat transfer is called the **thermal conductivity** and depends on the thermal properties of the barrier.
  - The proportionality constant for diffusion is called the **diffusion coefficient**, and depends additionally on the properties of the barrier, on the type of molecule and the fluid medium.
- **Osmosis** drives passive diffusion of water across semi-permeable membranes from relatively dilute into more concentrated solutions.
- The **osmotic pressure** of a solution is the pressure needed to prevent the flow of pure water into the solution across a semi-permeable membrane.
- An **electrical potential difference** or voltage across a barrier influences the movement of charged solutes (ions) across the barrier.
- The movement of ions across biological barriers is driven by **electrochemical potential differences**.
  - Ions move from regions of higher electrochemical potential to regions of lower electrochemical potential.
- The **equilibrium potential** for an ion across a barrier corresponds to the electrical potential difference that balances the effect of the concentration difference on the movement of the ion across the barrier; the equilibrium potential is calculated using the **Nernst equation**.
- The **Goldman-Hodgkin-Katz equation** calculates the voltage that becomes established across a selectively permeable barrier for given distributions of permeant monovalent ions across the barrier.

General properties of animal cells

- Cells contain membrane-bound and non-membrane-bound **organelles**.
- Cell membranes consist of a lipid component called the **lipid bilayer** and a protein component.
- The lipid bilayer consists mainly of phospholipids with **hydrophobic heads** and **hydrophilic tails** (=**amphipathic**).
- The lipid bilayer is permeable to lipid-soluble substances and small uncharged molecules but impedes passage of charged solutes.
- The chemical composition of the lipid bilayer determines the membrane's fluidity.
- The **plasma membranes** of cells act both as selective barriers to diffusion and as passageways for the movement of specific molecules against their concentration and/or electrochemical potential differences.
- The plasma membranes contain carrier proteins, channels, and ion pumps that facilitate the passage of water-soluble molecules.
- Most membrane channels are **gated**: they open or close in response to voltage differences across the membrane, the binding of specific ligands to channel sites, or stress (stretch) of the lipid bilayer.
- The uneven distribution of ions across the plasma membrane and different membrane permeability to various ions results in the development of **electrical potential differences** across the membrane.
- Most cells have a negative membrane potential difference between the cytosol and the extracellular environment.
- Changes in membrane permeability to specific ions cause changes in the membrane potential that have large effects on cell functioning.
- **Passive transport** across membranes refers to movement of uncharged solutes by diffusion from a higher concentration to a lower concentration, and movement of charged solutes from a higher **electrochemical potential** to a lower electrochemical potential.
- **Active transport** uses energy, directly or indirectly, from ATP to drive the passage of uncharged solutes across membranes from lower to higher concentrations, or the passage of ions from a lower to a higher electrochemical potential.
- Carriers that simultaneously transport molecules in the same direction are **symport** carriers, while carriers transporting molecules in the opposite directions are **antiport** carriers (**exchangers**).
- Cytosol typically contains a relatively high concentration of organic anions, K^+, phosphate and magnesium compared to its surrounding extracellular fluid, but relatively low concentrations of Na^+ and Cl^- and very low concentration of ionized Ca (Ca^{2+}).
- Low intracellular Na^+ concentration is important for maintaining the osmotic balance between cytosol and fluids outside the cells.
- The large K^+ concentration difference across the plasma membrane and relatively high permeability of plasma membranes to K^+ compared to Na^+ usually generates a negative membrane potential (cytosol more negative than extracellular environment).
- The large difference between Ca^{2+} concentrations in the extracellular environment (and endoplasmic reticulum) and cytosol is essential for regulating many aspects of cell function.

Interactions of animals with their environments

- Bulk flow of a gas or liquid transfers heat or matter by **convection**.
- Animals control the absorption of thermal radiation from the Sun's radiation by changing orientation, colour changes and/or changing the flow of blood close to their surfaces.

- **Evaporation** converts liquid water into water vapour, and takes place with absorption of energy (heat), so evaporation from animal surfaces reduces body heat.
- The **countercurrent** flow of fluids provides an efficient arrangement for the passive exchange of heat, solutes, water or gases between an animal and its external environment, and within animals.
- **Animal homeostasis** refers to the maintenance of a relatively stable internal environment by animals using **negative and positive feedback loops** to control physiological systems.
- The control of tissue and organ function that underlies homeostasis is mediated by the **autonomic nervous system**.
- Dynamic changes in physiological parameters in response to challenges are described as **allostasis** and allow anticipatory responses in advance of demand.
- Neural and hormonal systems coordinate body functions, behaviour, and responses to external changes.

Study questions

***Answers to these numerical questions are available online. Go to www.oup.com/uk/butler**

1. Describe the function of the main cellular organelles in animal cells. (Hint: Section 3.2.1.)
2. Highlight the major features of a phospholipid molecule that make it suitable as the major component of cellular membranes. What major features of cellular membranes are associated with the lipid component? (Hint: Section 3.2.3.)
3. Illustrate the main route by which lipid soluble molecules pass through cell membranes. (Hint: Section 3.2.3.)
4. Explain what major features of cellular membranes are associated with the protein component. (Hint: Section 3.2.3.)
5. Give four examples of major differences in the ion composition of cytosol and the extracellular fluid of animal cells and explain the basis of these differences. (Hint: Section 3.2.4.)
6. Why do cells have a low intracellular Na^+ concentration? (Hint: Section 3.2.5.)
7.* The flux of glycerol entering some cells was measured at 0.02 mmol m^{-2} s^{-1} in the presence of 10 mmol L^{-1} extracellular glycerol. Calculate the value of the membrane permeability for glycerol in these cells assuming the intracellular glycerol concentration was zero. Would you expect the rate of glycerol entry (influx) to change if the membrane potential changes from −60 to −30 mV? (Hints: Section 3.1.2; use Equation 3.3 and consider whether the glycerol molecule carries an electrical charge.)
8. Explain the difference between direct and indirect active transport. (Hint: Section 3.2.4.)
9.* Calculate the equilibrium potential for Na^+ expressed as $V^{Na}_{right\text{-}left}$ at 7°C when the Na^+ concentration on the left side of a permeable barrier is 100 mmol L^{-1} and that on the right side of the barrier is 50 mmol L^{-1}. Explain what happens to the equilibrium potential for Na^+ when the concentration of Na^+ increases by a factor of two on both sides of the barrier. (Hints: Section 3.1.3 and Equation 3.6.)
10.* Calculate the difference in electrical potential across a barrier that is permeable to K^+, Cl^- and Na^+ using the Goldman-Hodgkin-Katz equation for the following conditions:
    - temperature, 17°C;
    - K^+ concentrations: 10 mmol L^{-1} on left and 100 mmol L^{-1} on the right of the barrier;
    - Na^+ concentrations: 10 mmol L^{-1} on right and 100 mmol L^{-1} on the left of the barrier;
    - Cl^- concentrations: 110 mmol L^{-1} on left and right of the barrier (to balance out the electrical charges on the cations);
    - barrier permeability for Na^+ and Cl^- is 100-fold smaller than the barrier permeability for K^+ (P_{Na}, $P_{Cl} = 0.01\ P_K$). (Hint: use Equation 3.9 in Section 3.1.3.)

Bibliography

Aidley DJ (1998). Physiology of Excitable Cells. 4th Ed. Cambridge: Cambridge University Press.

Alberts B, Bray D, Lewis J, Raff M, Roberts K, Watson JD (2007). Molecular Biology of the Cell. 5th Ed. New York: Garland Science.

Cooper GM, Hausman R (2000). The Cell: A Molecular Approach. 5th Ed. Southerland, MA: Sinauer.

Katz B (1966). Nerve, Muscle and Synapse. New York: McGraw-Hill.

Randall D, Burggren W, French K (2002) Eckert Animal Physiology: Mechanisms and Adaptations. 5th ed. New York: W.H. Freeman and Co.

Widmaier EP, Raff H, Strang KT (2016). Vander's Human Physiology: The Mechanisms of Body Functions. 14th Ed. New York: McGraw-Hill Publishing.

Salmon leaping up falls during upstream migration to spawn.

Part 2

Water and salts

Overview

In this Part we explore how animals maintain the correct body fluid composition and volume in environments in which salt and water availability varies greatly.

We start in **Chapter 4** by examining the underlying processes that ultimately control the volume and composition of body fluids. Here we learn how cell volume is controlled and explore the properties of cells that influence how water and ions move within and between them.

Building on these basic principles, we explore how animals function in their particular environments; we look at animals living in aquatic environments in **Chapter 5** and those living on land in **Chapter 6**.

In these chapters, we discover the influence of drinking, diet and metabolism on the water and salt balance of animals—and how different animals have evolved ingenious adaptations both to maximize the uptake of water from their surroundings and to minimize water loss. We also learn about the evolution of the transport processes in particular organs and tissues that allow the ionic composition and volume of the body fluids to be regulated.

In **Chapter 7**, we take a close look at the functioning of kidneys in vertebrates and their invertebrate equivalents, and learn about the importance of these organs in regulating the volume and salt composition of the extracellular fluids of many animals.

Throughout these four chapters, we will meet examples of animals that flourish in vastly different environments, either on land or in water. Some are adapted to freshwater rivers, others for life in oceans, where the problems they face in terms of regulating salt and water are very different. Migrating salmon (pictured here) move between these two environments during their lifetime, and we will learn about the profound adjustments to the mechanisms of salt and water balance they must make to undertake such migrations.

4

Body fluid regulation: principles and processes

Life evolved in water, and all cellular processes involve water either directly or indirectly. The cells of animals do not form a continuous mass but have spaces between them. These spaces are filled with fluids—body fluids—consisting of water containing a variety of solutes. Different fluid-filled spaces have evolved in animals, but in each case, the fluid *within* the cells is distinct from fluid *outside* the cells. In this chapter, we discuss how the composition of fluid outside the cells affects cell volume and we examine the repertoire of mechanisms for cell volume regulation that are fundamental to the way cells are protected in all animals.

Epithelial sheets of cells form barriers between the body fluids, and between the internal fluids and the outside world. The properties of these barriers and the way they function are crucial in protecting the performance of cells by determining the passage of water and solutes between the body fluids, and between the body fluid and the outside world. During the course of this chapter, we examine the fundamental principles and transport processes at work in these epithelia. These processes determine the capacity of animals to live in environments with very different water and salt availability. Later chapters of the book explore how these processes underpin the functioning of animal tissues and organs such as gills, salt glands and kidneys; we explore the mechanisms by which aquatic animals regulate their body fluid volume and composition, and how some animals have adapted to cope with desiccating conditions on land.

4.1 Animal body fluids

The water content of animals generally accounts for 60 to 95 per cent of their body mass. Soft-bodied invertebrates are at the upper end of the range, whereas animals with a higher proportion of mineralized skeleton, such as arthropods and vertebrates, are at the lower end. Only a few rarities, such as tardigrades (water bears) and a few nematodes[1], insects and rotifers can survive a state of anhydrobiosis in which they contain almost no water—a state that allows these organisms to survive extreme drought or extremes of temperature.

Tardigrades live in moist habitats but enter anhydrobiosis if external water dries up. They can subsequently recover from a state of anhydrobiosis even after being sent into space. In anhydrobiosis, metabolism is undetectable, which emphasizes the importance of water in cells as the medium for metabolic processes.

The fluid inside the cells is called **intracellular fluid** (ICF). The cell membrane separates the intracellular fluid from fluid outside the cells—the **extracellular fluid** (ECF). The majority of the extracellular fluid fills the spaces between the cells; this is **interstitial fluid** (sometimes simply called tissue fluid). These major divisions of fluid compartments are illustrated in Figure 4.1 for vertebrates, in which extracellular fluid occurs both as interstitial fluid surrounding the cells and as the fluid component of blood (the plasma) in blood vessels.

In most of the blood capillaries of vertebrates, an exchange of water and ions occurs across the capillary endothelium[2]. Among most invertebrates, the vascular system is usually more open than among the vertebrates[3], so fluid flows from vessels into the fluid-filled space surrounding tissues and organs.

Most animals have fluid-filled body cavities between their outer covering (the epidermis) and their gut cavity. In some animal groups, such as arthropods and molluscs, a primary body cavity is derived from the remnant of the fluid-filled **blastocoel** that forms during the blastula stage of development of embryos[4]. In arthropods and molluscs, the primary

[1] We discuss the characteristics of tardigrades and nematodes (roundworms) in Section 1.4.2. In Case Study 8.1, we learn how understanding anhydrobiosis helped in the development of methods for preservation of biological materials.

[2] Capillaries of different tissues vary in the structure of their endothelial lining (the endothelium). Pores in the endothelium that allow exchange of water and ions are discussed in Section 14.1.5.

[3] We discuss open and closed circulatory systems again in Section 14.1.1.

[4] Figure 20.32 shows a blastula and its blastocoel.

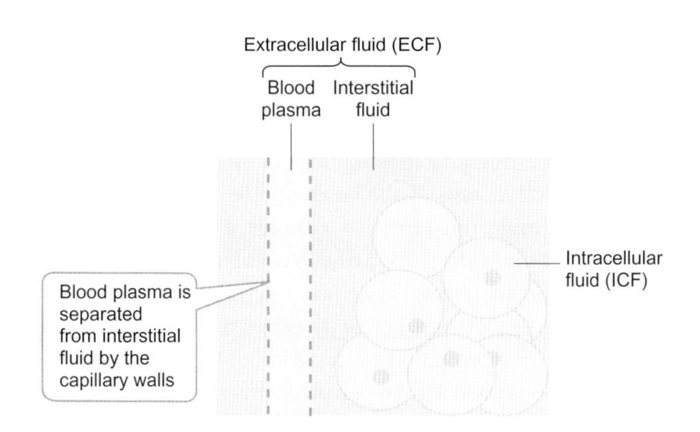

Figure 4.1 Distribution of body fluids in intracellular fluid (ICF) and extracellular fluid (ECF) compartments of vertebrates

Blood plasma (the fluid component of the blood), and interstitial fluid between the cells make up the bulk of the ECF of vertebrates.

body cavity is known as the **haemocoel** and contains haemolymph (the fluid equivalent of blood) that bathes the tissues. In many animal groups, a secondary body cavity develops to form the main body cavity, the **coelom**.

Different arrangements of body layers and body cavities mean that three body plans are recognized: acoelomate, pseudocoelomate and coelomate. Figure 4.2 illustrates these general body plans.

In the acoelomate body plan of flatworms, illustrated in Figure 4.2A, there is no fluid cavity in the mesoderm. This arrangement relies on diffusion alone for gas exchange and cell nourishment, which therefore restricts the size (volume) of the animals. The acoelomate body plan of flatworms was originally thought to be the primitive condition, but molecular evidence has given researchers reason to question this idea, and suggests that platyhelminthes may, in the course of their evolution, have actually lost the fluid-filled coelom that occurs in other animals.

The coelomic cavity that occurs in most multicellular animals (molluscs, annelids, arthropods and chordates—including vertebrates[5]) lies within the mesoderm, between the inner endoderm, which gives rise to the digestive tract, and the outer ectoderm, which forms the body wall. This arrangement, which Figure 4.2C illustrates, results in a surrounding layer of mesoderm, distinguishing a true coelom from the pseudocoelom of nematode worms illustrated in Figure 4.2B. The fluid-filled circulatory system and lymphatic systems, which lie within the coelom of vertebrates, separate circulating blood and lymph from interstitial fluid but allow exchange of gases, water, and other small molecules across the vessel walls[6].

Figure 4.3 shows the distribution of body fluids between the intracellular and extracellular compartments of selected species, as a percentage of total body water. Among the examples shown, sea hares have an unusually high proportion of ECF, probably because of their mode of locomotion, using a fluid-filled foot. In most vertebrate groups, 20 to 30 per cent of the total body water is extracellular fluid (the combination of plasma and interstitial fluid), and the circulatory system usually holds about six per cent of the body water in blood plasma. Teleost fish are an exception and have a lower relative blood volume, so their plasma volume is less than three per cent of body water.

[5] Figures 1.15 and 1.21 illustrate the phylogenetic organization of animal groups.

[6] Section 14.1.5 discusses these exchange processes in detail.

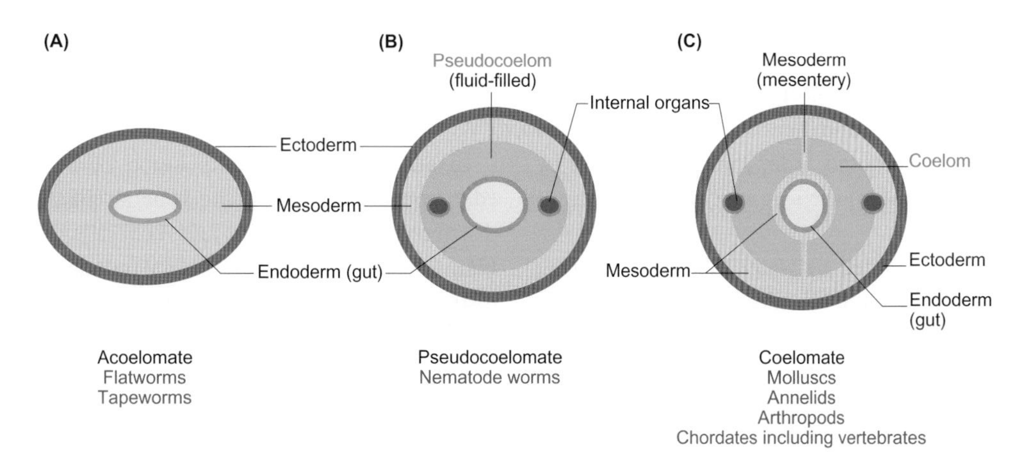

Figure 4.2 Body plans of animals

(A) Acoelomate; (B) pseudocoelomate with fluid-filled pseudocoelom; (C) coelomate, with fluid-filled coelom. The phylogenetic organization of animals and their diversity are discussed in Sections 1.3 and 1.4. The coelom is quite small and surrounds the heart as the pericardial cavity in molluscs and has almost disappeared in arthropods. The movement of fluid in the coelomic cavity is discussed in Section 14.1.1.

4

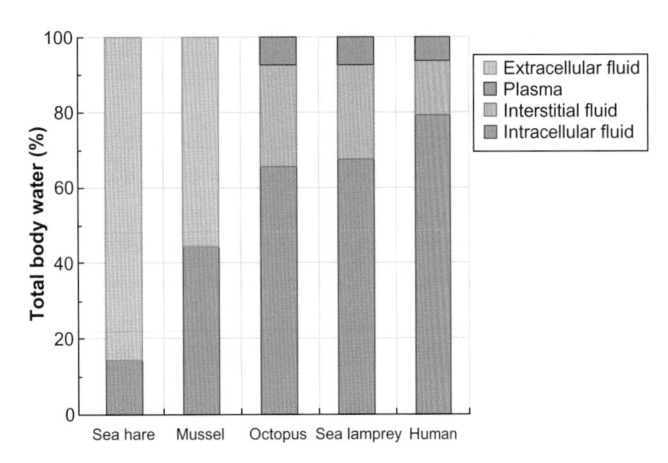

Figure 4.3 Distribution of body water in the intracellular compartment and the extracellular compartment of selected species

Extracellular fluid of octopus, sea lamprey (*Petromyzon marinus*) and humans, which have closed circulatory systems, is distributed between the vascular system and the interstitial compartment. In contrast, in molluscs, which have an open circulation, distinct separation of a blood compartment is not possible; data are shown for the proportion of body fluid in the extracellular and intracellular compartments of two molluscs – sea hares (*Aplysia* sp.) and freshwater pearl mussels (*Margaritifera margaritifera*).

Sources of data: Potts WTW and Parry G. Osmotic and Ionic Regulation in Animals. Oxford: Pergamon Press, 1964.

4.1.1 Intracellular and extracellular fluids differ in composition

The solvent for all body fluids is water, in which **solutes**, such as glucose, some hormones, gases, such as carbon dioxide, and salts, such as sodium chloride and potassium chloride, readily dissolve. The salts are **electrolytes**, because their interaction with water results in ionization, whereby they dissociate into separate oppositely charged ions. Charged ions in solution interact, which influences the **effective concentration**, or the activity of the ions, as we discuss in Box 4.1, and hence the concentrations of each ion that can undergo independent transport by the mechanisms involved in the salt and water balance of animals.

Charged ions are hydrated by water molecules. During hydration, a shell of water molecules (a hydration shell) forms around the ion, stabilized by electrostatic interactions. Figure 4.4 is a conceptual diagram of this process. The shell of water molecules attracts further water molecules, forming a second, less strongly attracted hydration shell. The hydration of ions in solution means that they function as larger molecules than the ion alone, a characteristic that becomes important in determining their passage across biological membranes.

Box 4.1 Electrolytes in solution

Electrolytes are salts that dissociate into cations and anions in solution. These ions interact; ions of opposite charges attract each other, ions of similar charge repel. These interactions determine the **effective concentration** or chemical activity (often shortened to activity) of each ion compared to that of an uncharged solute (a non-electrolyte). The **activity coefficient** (a factor of below 1) expresses the fraction of equivalent ions compared to those for an ideal solute in an ideal solution in which there are no chemical interactions. We can calculate the predicted effective concentration (activity) of a particular ion, by multiplication of the measured concentration of an ion by the expected activity coefficient of the ion in solution. The accuracy of such calculations is implicitly dependent on the accuracy of the activity coefficient, which varies depending on the overall concentration of ions in the particular solution (ionic strength) and temperature.

Activity coefficients of about 0.65 are typical of monovalent ions in the body fluids of marine invertebrates, while the activity coefficient for sodium (Na^+) is around 0.75 in the extracellular fluids of many vertebrates which contain Na^+ at around 150 mmol L^{-1}. Hence, in vertebrates, the effective Na^+ concentration (Na^+ activity) is around $0.75 \times 150 = 112.5$ mmol L^{-1}.

For a precise understanding of transport processes across cell membranes such as the movement of ions through ion channels, it is necessary to know the effective concentrations of ions present in intracellular and extracellular fluids. For example, the equilibrium potential given by the Nernst equation (Equation 3.5) depends on the ratio between effective concentrations of an ion across a membrane. However, since the activity coefficient of a particular ion has about the same value on both sides of a biological membrane, it means that the activity coefficients cancel out and that therefore, for all practical purposes, the ratio between effective concentrations is the same as the ratio between normal concentrations. For simplicity, therefore, in discussing the salt and water balance of animals (and other physiological processes in this book), we generally refer to the measured (dissolved) ion concentrations, rather than effective concentrations.

In all animals, the composition of ECF differs markedly from that of ICF[7]. The general characteristics of ion distribution between ICF and ECF cells are illustrated in Figure 4.5, for mammalian skeletal muscle. Note the unequal distribution of sodium ions (Na^+), potassium ions (K^+) and chloride ions (Cl^-) between ICF and ECF depicted in this diagram.

[7] Figure 3.21 examines the general differences in solute concentrations of ICF and ECF. The exact concentrations of ions in ECF and ICF differ between organisms and cell types. We examine some differences in composition of extracellular fluid in Chapter 5, such as the high urea concentrations in sharks and rays shown in Table 5.4.

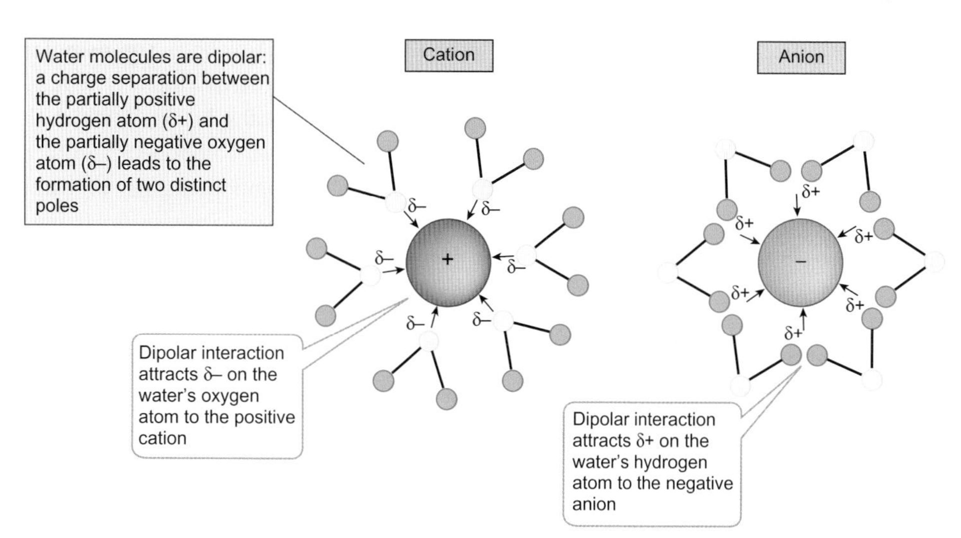

Figure 4.4 Hydration of ions by water molecules

Charged ions interact with the partial charge on water molecules. The positive charge of cations interacts with the partial negative charge (δ–) on the oxygen atom (yellow) of the water molecule. The negative charge of anions interacts with the partial positive charge (δ+) of hydrogen atoms (green) in water. This forms a shell of water molecules around the ion.

Source: Crowe J, Bradshaw T. Chemistry for the Biosciences, 3rd Edition. Oxford: Oxford University Press, 2014.

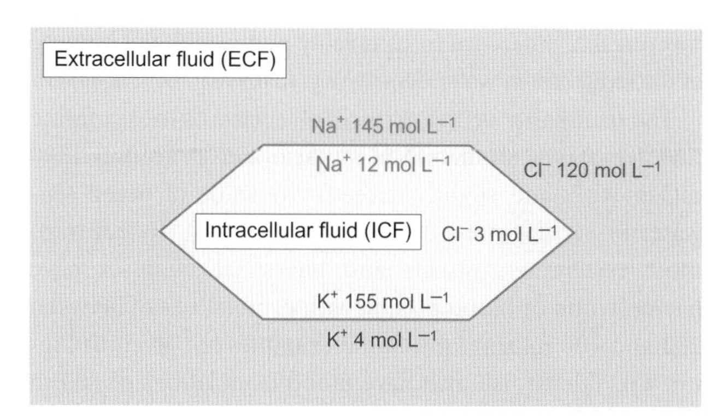

Figure 4.5 Potassium (K^+), sodium (Na^+) and chloride (Cl^-) concentrations in intracellular fluid of mammalian skeletal muscle cells and the surrounding extracellular (interstitial) fluid

The difference in composition between intracellular and extracellular fluids is because of active transport of Na^+ and K^+ and the electric charge across the membrane, which we discuss in Section 3.1.3.

In all animals, Na^+ is the major cation in the ECF, while Cl^- is the major anion. Among most vertebrates, the Na^+ concentration of ECF is around 120 to 210 mmol L^{-1} while Cl^- is slightly less at around 80 to 195 mmol L^{-1}. The Na^+ and Cl^- concentrations of extracellular fluids are higher among marine invertebrates at around 450 mmol L^{-1} for Na^+ and 550 mmol L^{-1} for Cl^-—similar to the seawater they occupy. In contrast to the high concentrations of Na^+ in extracellular fluids, the concentrations of Na^+ are relatively low in the ICF.

The major cation in ICF is K^+, which occurs at 100 to 200 mmol L^{-1}, while K^+ concentration of ECF is maintained at concentrations of a few mmol L^{-1}, as shown in Figure 4.5. The high intracellular concentration of K^+ means that cells can initially buffer the K^+ influx into ECF after consumption of food containing high concentrations of K^+ (such as meat and fruits), until the excess K^+ is excreted by the kidneys[8]. An influx of K^+ into the ECF results in initial transfer of K^+ into cells, due to the activity of the **sodium pump (Na^+, K^+-ATPase)**[9].

The normal distribution of K^+ between ICF and ECF is essential for maintaining normal cell function, particularly of excitable tissues such as nerve and muscle cells. In these cells, the difference in K^+ concentrations of ECF and ICF is partially responsible for maintaining the normal resting membrane potential and hence the excitability of the cells[10]. Subnormal concentrations of K^+ in extracellular fluids including blood (hypokalaemia[11]) result in an increase in the membrane potential of nerve and muscle cells making them less sensitive to electrical stimuli. In contrast, hyperkalaemia (elevated concentrations of K^+ in blood plasma and extracellular fluids) depolarizes cells, so the excitable cells become more excitable.

The combination of solutes in biological fluids determines the osmotic concentration

Chapter 3 introduces the concept of **osmosis**. The pressure needed to stop the water movement seen during osmosis is the **osmotic pressure**[12]. However, osmotic pressures are rarely measured when studying the regulation of salt and water balance (**osmoregulation**) of animals. Rather, animal osmoregulation is best understood from measurements of the **osmotic concentrations** of body fluids, resulting from the total concentration of solute molecules in solution, and (for aquatic animals) the osmotic concentration of the water in which they live.

8 We examine K^+ secretion by the kidney of vertebrates in Section 7.2.4.

9 We discuss the functioning of Na^+, K^+-ATPase (the sodium pump) in Section 4.2.2.

10 Section 3.2.3 explains membrane potentials, while Chapters 16 and 18 discuss nerve and muscle function and their excitability.

11 Hypokalaemia may result from an excessive loss of K^+ either from the kidneys or more usually from the gastrointestinal tract because of chronic diarrhoea or vomiting. Hyperkalaemia can result from a failure to excrete the excess potassium taken in each day in food.

12 Section 3.1.2 discusses the concept of osmotic pressures and osmosis.

Osmotic concentrations are fundamental to understanding body fluid regulation of animals. The osmotic concentration of body fluids determines the movement of water, by osmosis, which occurs whenever differences in the osmotic concentrations (and, hence, water concentrations) occur in adjacent body fluid compartments. Thus, osmotic concentrations determine the movements of water between the intracellular and extracellular fluids. Osmotic concentrations also determine water flow by osmosis across the external permeable surfaces of aquatic animals.

Osmotic concentration is one of a number of properties of solutions, known collectively as **colligative properties**, which depend on the total number of particles dissolved in a volume of solvent, rather than the type of particles. The colligative properties of solutions are:

- osmotic concentration
- freezing point depression
- melting point depression
- boiling point elevation
- vapour pressure depression.

The chemical type or individual molecular masses of the dissolved particles are irrelevant to osmotic concentrations, and, by definition, all colligative properties vary in direct proportion to each other. Consequently, the measurement of one colligative property gives a measure of the others.

Values for osmotic concentrations of body fluids are generally obtained by using a commercial osmometer to measure freezing point depression in just a few microlitres of fluid. To do this, a small thermistor[13] is inserted into the fluid sample, and the sample is supercooled, i.e. cooled to below its freezing point but without freezing. The supercooled sample is then frozen by introducing an ice crystal and vigorously agitating it. Cooling stops and the temperature rises gradually until the sample begins to melt, at which point temperature stabilizes until the whole sample melts[14]. The machine determines the temperature at which the sample melts (the melting point), which is the same as freezing point. Pure water freezes at 0°C and the presence of solutes in a solution lowers its freezing point below 0°C. For example, the freezing point of seawater is −1.86°C. The depression in freezing point to below 0°C is related directly to the osmotic concentration of the sample.

Osmometers measure osmotic concentrations as **osmolality** (osmoles per kg of solvent abbreviated as osmol kg^{-1} or Osm kg^{-1}), that is, the osmotic concentration relative to a *mass* of solvent (in our case, water). Seawater freezing at −1.86°C has an osmotic concentration of about 1 Osm kg^{-1}. The **osmolarity** of a solution is related to its molarity (mole L^{-1}). A 1 molar solution (1 mol L^{-1}) of an ideal solution, in which the properties are linearly related to each component and there are no interactions between the components, has an osmolarity of 1 Osmolar (1 Osm L^{-1}). Osmolarity is similar, but not identical to the osmolality of a solution; for osmolarity, osmotic concentration is expressed relative to a *volume* of solution (osmoles per litre, abbreviated as osm L^{-1} or Osm L^{-1}).

In reality, the osmolarity and osmolality of physiological solutions may not be significantly different in animals whose body fluids are at 0 to 25°C, since one litre of physiological solutions weighs very close to 1 kg. Therefore, for practical purposes, Osm kg^{-1} is approximately equal to Osm L^{-1} for physiological solutions. Only in mammals and birds, which regulate body temperatures at 37 to 44°C[15], is there a very small but measureable difference in osmolality and osmolarity because of the reduction in water density (e.g. to 0.990 g mL^{-1} at 44°C).

The osmolality of the extracellular fluid of most marine invertebrates is similar to ocean seawater at around 1000 mOsm kg^{-1}, and slightly exceeds the value of ocean seawater in cartilaginous fish (sharks and rays). By contrast, most vertebrates, insects and terrestrial molluscs have haemolymph or blood plasma with osmolalities of between 200 and 400 mOsm kg^{-1}. Some examples are given in Figure 4.6. Teleost fish also have blood osmolalities of 300 to 400 mOsm kg^{-1}. In contrast, terrestrial crustaceans, such as woodlice and land crabs, have higher haemolymph osmolalities (500 to 900 mOsm kg^{-1}). These differences have important implications for the mechanisms of body fluid regulation that we discuss in detail in Chapters 5 to 7.

The osmolality of body fluids results from the combination of non-dissociable molecules, such as glucose, and the effective concentration[16] of the inorganic ions in these fluids. Non-dissociable molecules contribute to the osmotic concentration of a solution in a manner that is directly proportional to their molar concentrations. By contrast, inorganic electrolytes dissociate into two or more ions that *each* contribute to the osmotic concentration of a solution.

If completely dissociated in the water, a strong electrolyte, such as potassium chloride (KCl) or sodium chloride (NaCl), forms two dissolved ions: K^+ and Cl^- or Na^+ and Cl^-, respectively, while calcium chloride ($CaCl_2$) forms three dissolved ions: Ca^{2+} and two Cl^-. However, this ideal behaviour only occurs if the individual ions do not interact in the solution, which rarely applies in biological solutions. The non-ideal behaviour of NaCl in physiological solutions means that there are fewer than two free ions per mole of NaCl so

[13] A thermistor is a temperature sensitive resistor, i.e. the measured resistance is dependent on temperature.

[14] Figure 9.20 shows an example of a cooling curve and freezing point depression of a sample of body fluids.

[15] We discuss body temperatures and their regulation in endotherms in Chapter 10.

[16] Box 4.1 explains effective concentrations (ion activities).

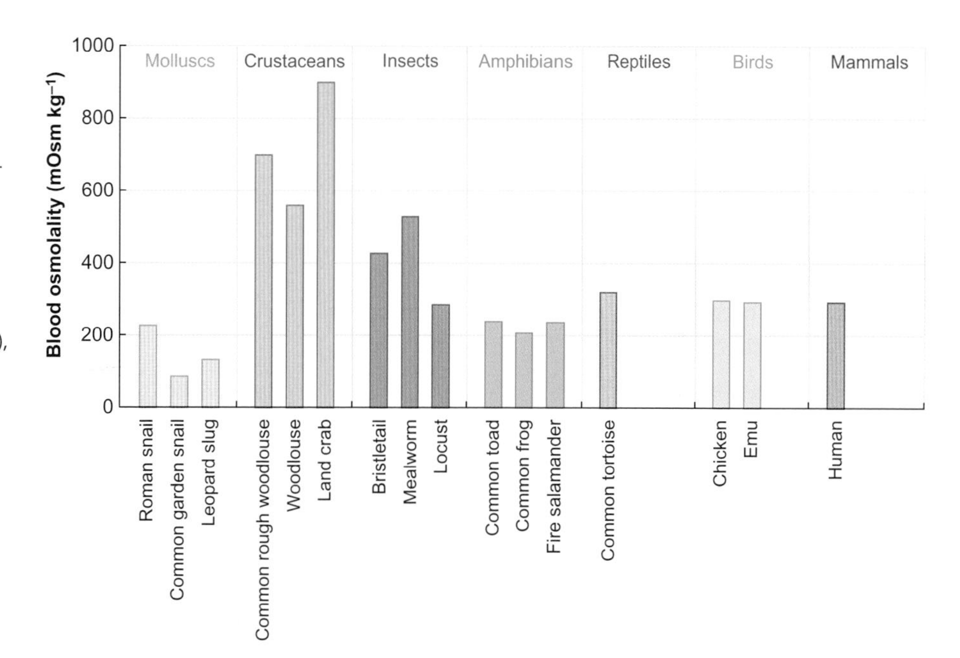

Figure 4.6 Blood osmolality in selected terrestrial animals

Data show blood osmolality of Roman snail (*Helix pomatia*), common garden snail (*Cornu aspersum*), leopard slug (*Limax maximus*), common rough woodlouse (*Porcellio scaber*), woodlouse (*Oniscus* sp.), blackback land crab (*Gecarcinus lateralis*), bristletail (*Petrobius* sp.), meal worm (*Tenebrio molitor*), locust (*Locusta* sp.), common toad (*Bufo bufo*), common frog (*Rana temporaria*), fire salamander (*Salamandra salamandra*), common tortoise (*Testudo graeca*), chicken (*Gallus gallus domesticus*), emu (*Dromaius novaehollandiae*) and humans.

Data sources: Potts WTW, Parry G. Osmotic and Ionic Regulation in Animals. Oxford: Pergamon Press, 1964; and Skadhauge E, Maloney SK, Dawson TJ (1991). Osmotic adaptation of the emu (*Dromaius novaehollandiae*). Journal of Comparative Physiology B 161: 173-178.

the contribution to osmolality is less than twice the molarity of NaCl.

Specifically, at 25°C, it has been determined that NaCl in a solution of 150 mmol L^{-1} functions as if it dissociates into 1.86 ions rather than two ions: the proportion of dissociated ions is 1.86/2 (0.93). This factor is the osmotic coefficient (Φ)[17] and values for many other strong electrolytes are of a similar order of magnitude.

Knowing the osmotic coefficient of an electrolyte allows us to predict the osmotic concentration of salt solutions used for intravascular infusion or tissue incubation. Several electrolytes are often mixed together to prepare physiological solutions that mimic extracellular fluid, in which case each electrolyte contributes to the total osmotic concentration in an additive way.

Taking a single electrolyte as an example, what is the calculated osmolarity of 135 mmol L^{-1} NaCl at 25°C?

The predicted osmolarity equals the molarity of the NaCl solution, multiplied by 2 (as dissociation produces two ions) multiplied by the osmotic coefficient (proportion of NaCl that dissociates into two ions).

$$\text{At } 25^{\circ}\text{C osmolarity} = 135 \times 2 \times 0.93 \text{ mOsm L}^{-1}$$
$$= 251.1 \text{ mOsm L}^{-1}$$

The calculated value is close to the osmolarity of the blood plasma of amphibians in which Na^+ and Cl^- are the major ions, so intravenous injection of an amphibian with test compounds in 135 mmol L^{-1} NaCl avoids disruption of the osmotic balance of their extracellular fluids.

Charged macromolecules in body fluids influence ion distribution

Cells (and blood plasma) contain charged macromolecules, such as proteins and organic phosphates, that do not pass (passively) through the cell membrane. Therefore, these charged macromolecules are known as impermeant solutes. The majority of these charged macromolecules are negatively charged (i.e. impermeant anions) at physiological pHs.

In the early 1900s, an American physicist, Josiah Willard Gibbs, and Frederick Donnan, a British chemist, recognized that the presence of impermeant solutes in body fluids could influence the distribution of diffusible charged solutes like K^+ and Cl^- ions. If this occurs, *in the absence of other interfering influences*, there will be an unequal distribution of ions and a charge difference (a difference in electrical potential[18]) between the two compartments. The compartment containing the impermeant anion will become more negatively charged and will have a higher osmotic concentration than the other compartment. This Gibbs–Donnan effect applies not only to cells, but also to the distribution of permeable ions across any barrier that is not permeable to larger molecules, such as the wall of capillaries. The Gibbs–Donnan effect across such a barrier meets two conditions:

- The concentration of *diffusible* cations multiplied by the concentration of *diffusible* anions is identical on the outside (o) and inside (i) of the barrier. Considering just the molar concentrations of the major cations $[K^+]$ and $[Na^+]$ and the molar concentration of the major anion $[Cl^-]$ in

[17] The osmotic coefficient is related to (but not identical to) the activity coefficient discussed in Box 4.1.

[18] Charge differences and electrical potential are discussed in more detail in Section 3.1.3.

body fluids[19], we can express this equilibrium of diffusible ions across a membrane separating extracellular body fluids using Equation 4.1:

$$([Na^+]_i + [K^+]_i) \times [Cl^-]_i = ([Na^+]_o + [K^+]_o) \times [Cl^-]_o \quad \text{Equation 4.1}$$

- On each side of the barrier, the total number of positively charged solutes equals the total number of negatively charged solutes. So, in the compartment without impermeant anions (o):

$$[Na^+]_o + [K^+]_o = [Cl^-]_o \quad \text{Equation 4.2}$$

- The presence of impermeant macromolecules (A^-) in the other compartment (i) affects the distribution of the charged ions at equilibrium. Again, considering just the Na^+, K^+ and Cl^- and the impermeant anion in this compartment, which we assume has a single charge (but, in reality, impermeant anions could have several charged groups), it follows:

$$[Na^+]_i + [K^+]_i = [Cl^-]_i + [A^-]_i \quad \text{Equation 4.3}$$

For aquatic animals, Gibbs–Donnan effects may affect the distribution of ions between the extracellular fluids (which contain proteins) and the external environment when there is a selectively permeable[20] interface—a phenomenon that could explain some of the differences in the concentration of ions between seawater and the body fluids of marine organisms.

It is difficult to determine Gibbs–Donnan effects for complex biological solutions using equations, but there is a simple solution to the problem, which has been used to examine the Gibbs–Donnan effect for a range of marine organisms. To adopt this solution, which Figure 4.7 illustrates, extracellular fluids of each animal are placed in a cellophane bag that is permeable to water molecules and inorganic ions but not to larger molecules (a **semi-permeable membrane**) and suspended in seawater. When equilibrium in the distribution of charged ions is reached, the bag stays fully expanded because the solution in the bag has a greater osmotic concentration than the solution outside the bag and therefore water is drawn in, keeping the bag expanded.

The measured concentration of each ion at equilibrium indicates the Gibbs–Donnan effect on the particular ion. If the equilibrium concentration differs from the actual concentration in the extracellular fluids of the particular animal, then the existence of an active transport process in ion regulation is implied[21].

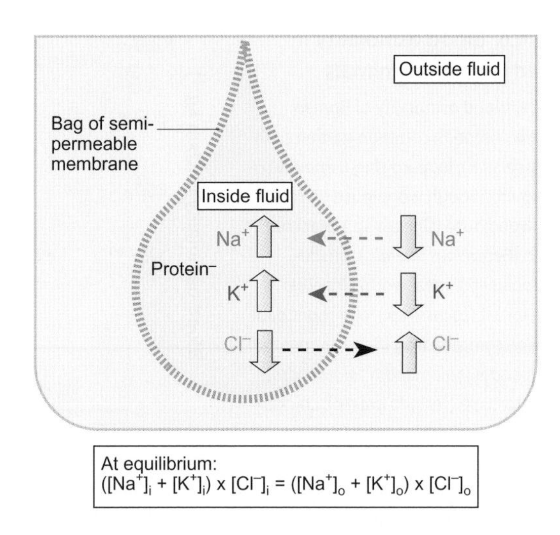

Figure 4.7 Gibbs–Donnan equilibrium

The presence of negatively charged impermeant protein molecules affects the distribution of charged ions across semi-permeable membranes that are permeable to water and inorganic ions and impermeable to proteins. The example shown considers the major diffusible ions: sodium (Na^+), potassium (K^+) and chloride (Cl^-). Diffusion occurs in both directions across the membrane. The dashed arrows crossing the membrane show the net effects, and the up and down arrows indicate the changes at equilibrium: ↑ equals an increase, ↓ equals a decrease in the final equilibrium concentrations [ ] of Na^+, K^+ and Cl^- on the inside (i) relative to outside (o) of the semi-permeable membrane. At equilibrium the product of the sum of the concentrations of the diffusible cations (Na^+, K^+) and anions (Cl^-) is equal on both sides of the membrane, as shown in the equation.

Figure 4.7 illustrates the principle of the dialysis bag method for Na^+, K^+ and Cl^-. Diffusion of inorganic ions occurs but not of the impermeant protein molecules. At equilibrium, diffusible cations, such as K^+ and Na^+, have a greater concentration in the bag because of the presence of the negatively charged impermeant protein molecules; by contrast, anions like Cl^- have a lower concentration in the bag, in agreement with the Gibbs–Donnan effect. This uneven distribution of diffusible ions and the higher osmotic concentration of the fluid in the compartment containing the impermeant anions, which tends to draw in water by osmosis[22], are key features of the Gibbs–Donnan effect.

In theory, the Gibbs–Donnan effect could influence the ionic concentrations of intracellular fluids. In reality, however, the distribution of ions across cell membranes is determined mainly by the presence of ion transporters and ion pumps, some of which we examine in the next section[23].

[19] When the Gibbs–Donnan equation is applied to cells, Na^+ may be removed from the equation, because as we discuss in Section 3.2.5, the cellular membranes of most resting cells have a very low permeability for sodium ions.

[20] We also discuss the passage of ions across selectively-permeable membranes in Section 3.1.3.

[21] We examine the ionic composition of marine animals in some detail in Section 5.1 where we consider whether the composition results from Gibbs–Donnan effects or active transport mechanisms.

[22] Volume regulation of cells is discussed in Section 4.3.

[23] Section 3.2.4 gives an overview of the basic mechanisms for transport of solutes in cells and across membranes.

› *Review articles*

Balavoine G (1998). Are platyhelminthes coelomates without a coelom? An argument based on the evolution of *Hox* genes. American Zoologist 38: 843–858.

Hoffman EK, Lambert IH, Pedersen SF (2009). Physiology of cell volume regulation in vertebrates. Physiological Reviews 89: 193–277.

Rieger RM, Purschke G (2005). The coelom and the origin of the annelid body plan. In: Bartolomaeus T and Purschke (eds.). Morphology, Molecules, Evolution and Phylogeny in Polychaeta and Related Taxa. Developments in Hydrobiology 179: Dordrecht: Springer.

Savitz D, Sidel VW, Solomon AK (1964). Osmotic properties of human red cells. Journal of General Physiology 48: 79–81.

Thorson TB (1961). The partitioning of body water in osteichthyes: phylogenetic and ecological implications in aquatic vertebrates. The Biological Bulletin 120: 238–254.

4.2 Transepithelial transport

Animals achieve solute and water balance by regulating the rate of solute and water transport across sheets of epithelial cells, which act as barriers between their extracellular fluids and the outside world. These epithelia cover their outer surfaces, including the gills, and line the cavities of organs, such as the intestine, and tubular tissues, such as the kidney tubules and (in some species) their salt glands. These tubular structures transport solutes and water between a fluid cavity and the extracellular fluid compartment.

Solutes and water cross epithelia by two routes, which Figure 4.8 depicts: via the **paracellular pathway** *between* the epithelial cells and via a transcellular route *through* the cells themselves.

4.2.1 Paracellular transport by osmoregulatory epithelia

In the epithelia of vertebrates, tunicates (sea squirts) and some of the epithelia of arthropods, body fluid regulation involves passive transport[24] of solutes and water molecules via paracellular pathways. For example, paracellular transport occurs in the epithelia forming the walls of kidney tubules, and the epithelia covering the gills of fish. In each case, the passive paracellular transport has certain fundamental characteristics:

- no energy requirement
- always occurs in the direction of equilibrium
- depends upon transepithelial differences in concentration
- for ions, is influenced by electrical charge (voltage) differences across the epithelium.

For specific ions under consideration, the difference in concentration across an epithelium, together with the charge difference, are described as the difference in **electrochemical potential**[25] across the epithelium and has a major effect on the diffusion of charged ions via the paracellular pathways.

[24] Passive transport and active transport are outlined in Section 3.2.4.

[25] Section 3.1 discusses the effects of **electrochemical potential differences** on charged ions.

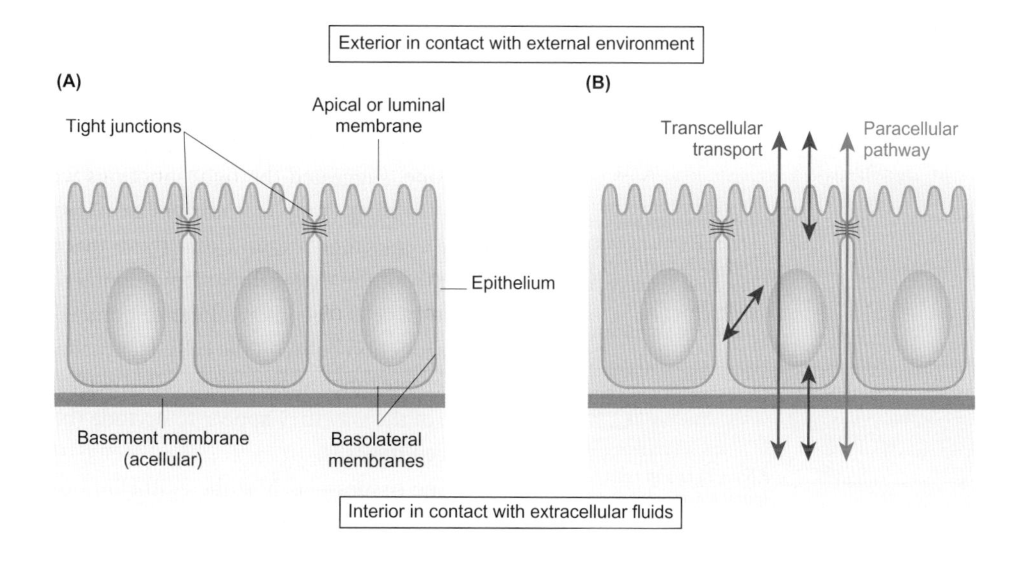

Figure 4.8 Characteristic features of an epithelium and routes for solute and water transport across the epithelium

(A) The epithelial cells sit on a basement membrane that faces the interior and is in contact with extracellular fluids. Cells are in contact at 'tight junctions' that delimit the apical/luminal membrane from the basolateral membranes of the epithelial cells.

(B) Transport of water and solutes across epithelia can occur by a transcellular route (**blue arrow**), across the apical or luminal epithelium (in tubular epithelia) and across the basolateral membranes, in either direction (**red arrows**), or via an extracellular route following a paracellular pathway (**green arrow**): via the 'tight junctions' between the cells.

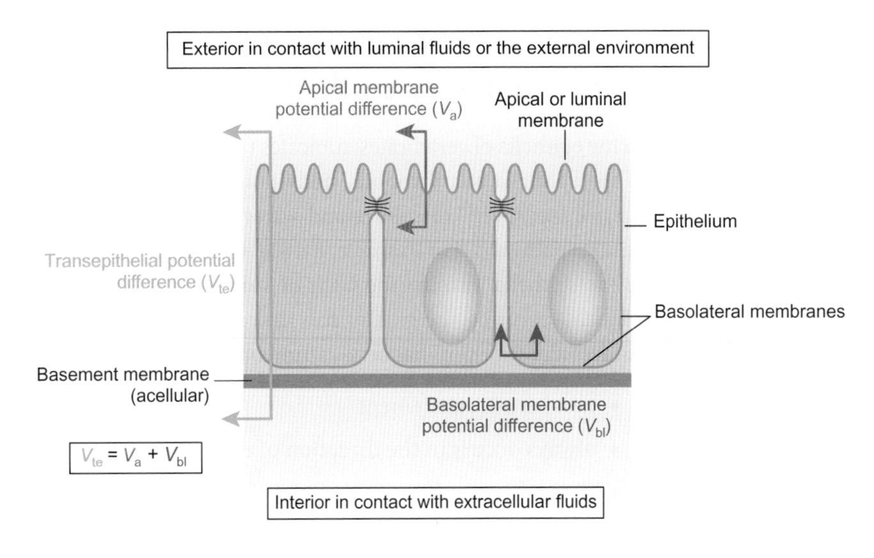

Figure 4.9 Epithelium showing locations for measuring the charge differences (potential difference in mV) across the epithelium and its component membranes

The transepithelial potential difference is measured by placing microelectrodes in the extracellular fluids and in the apical/luminal fluids. Transmembrane potential differences are measured as the voltage difference between a microelectrode in the cell and a second microelectrode in the extracellular fluids (for the potential difference across the basolateral membrane) or luminal/external fluids (for the potential difference across the apical membrane). The transepithelial potential difference (V_{te}) equals the sum of the potential differences across the apical/luminal (V_a) and basolateral membranes (V_{bl}); i.e. $V_{te} = V_a + V_{bl}$.

Electrophysiological methods are used to investigate transepithelial potential differences (charge effects), as shown in Figure 4.9[26].

For paracellular transport, solutes and water molecules must pass through the so-called 'tight junctions' that occur at the apical or luminal junctions between the adjacent epithelial cells, as indicated in Figure 4.8. Figure 4.10 illustrates the three-dimensional arrangement of tight junctions with close apposition of lateral membranes from adjacent cells (often called 'kissing points'). The transmembrane proteins that form the tight junctions interact with cytoplasmic proteins that connect to the cell cytoskeleton[27], and form a ring of tight junctions with adjacent cells, like a tight band. In a similar way, some tissues, particularly those of invertebrates, have ladder-like septate junctions encircling each cell in which the rungs of the ladder forming the junction cross the paracellular space between the cells.

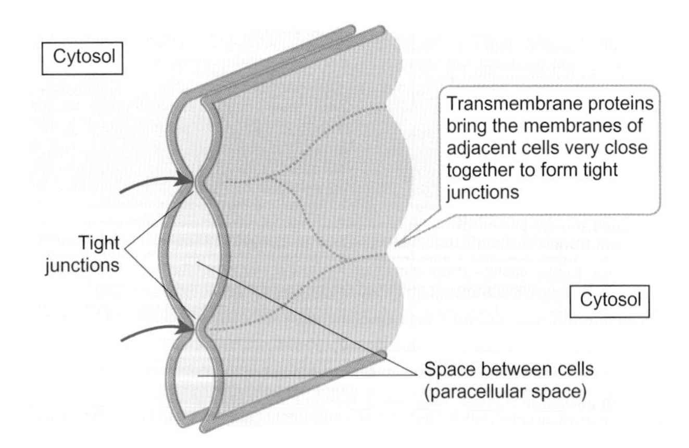

Figure 4.10 Diagrammatic representation of tight junctions between epithelial cells

In an apical/luminal location in transporting epithelia, the tight junctions have a dual role as:

- 'Fences' that delimit the membrane proteins in the apical/luminal membrane of the cell from membrane proteins in basolateral membranes.
- 'Gates' that mediate the passage of solute and water between cells.

Initially, studies using macromolecules, such as colloidal lanthanum, which can be visualized in the electron microscope, suggested the tight junctions were tightly closed gates forming complete barriers to paracellular diffusion. These experiments found that these macromolecules only move from the extracellular fluids along the space between epithelial cells (the intercellular or paracellular space) as far as the junction between the cells. Hence, both tight junctions and septate junctions are often called occluding junctions because of the way they appear to occlude (block) the pathway between the cells.

Some tissues—such as the epithelium of the mammalian urinary bladder—do have truly tight junctions at the luminal side of the epithelium. These tight junctions are important in holding urine in the bladder and can maintain large osmotic differences across the epithelium. However, it became clear in the 1960s and early 1970s that tight junctions do not always prevent the passage of smaller molecules or water between epithelial cells. Indeed, the 'tight' junctions of many epithelia (such as the gall bladder epithelium of the

[26] Figure 16.15 in Section 16.2 explains how potential differences across membranes (membrane potentials) are measured.
[27] The cell cytoskeleton is discussed in Section 3.2.1.

mudpuppy (*Necturus maculosus*)[28], proximal tubules of the kidney[29] and epithelia in the gills of teleost fish, when acclimated to seawater[30]) are in reality quite leaky. Leaky junctions allow various amounts of ions and water to pass along the paracellular pathway, and are in fact the main determinant of its **permeability** to solutes and water.

Claudins are integral membrane proteins that determine the permeability of tight junctions

The discovery of **claudins** (from the Latin *claudere* meaning to close) in 1988 was a major breakthrough in beginning to understand how tight junctions function. Many different types of claudins exist: 26 different claudin genes occur in humans and even more occur in teleost fish. More than 60 different genes encoding claudins have been reported in 16 species of teleost fish, many of which seem to be fish-specific.

Why do fish have so many claudins? One idea is that claudins in the tight junctions of the epithelia of osmoregulatory organs are linked to the adaptation and acclimation of fish to environments with very different salt contents. In particular, more than 40 claudin genes have been identified in the gills of teleosts where the claudins are hypothesized to determine the tightness or leakiness of the gill tissue and, hence, the loss or gain of water and ions across the gills[31].

Analogous proteins to the claudins of vertebrates occur in the septate junctions of fruit flies (*Drosophila*) in which at least two different claudin proteins have been identified and called sinuous and mega. These claudins occur in epithelia such as the lining of the tracheal system that forms the network of tubes used for gas exchange, and the lining of the hindgut and the salivary gland, although the exact functioning of these proteins is unclear.

Vertebrate claudins are much more fully understood and are functionally divided into those acting as barriers to particular solutes and those that increase the leakiness of the tight junctions, functioning more like pores. In addition, some claudins seem to restrict or assist the passage of specific solutes. Consequently, the patterns of claudin expression in tissues can have a major role in determining the permeability of particular paracellular pathways. Some claudins, such as claudin-1 in mammalian epidermis, operate as barriers to macromolecules, solutes and water, while at least three claudins form pores for cations, and two others form pores for anions.

Figure 4.11A shows how claudins insert in the cell membrane as four transmembrane loops with two extracellular loops. Claudins from adjacent cells interact to form the tight junction, as shown diagrammatically in Figure 4.11B. Interaction of two different claudins on opposing membranes may form an ion pore. Amino acids in the first extracellular loop, which line the pore, may determine the permeability of the paracellular pathway, although we need details of the tertiary structure to gain a better understanding of the exact functioning of claudins in tight junctions.

One of the best studied is claudin-2, which occurs particularly in relatively leaky epithelial tissues, where water

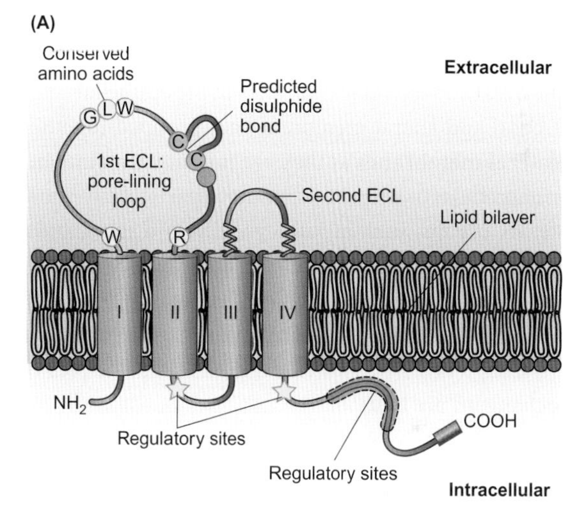

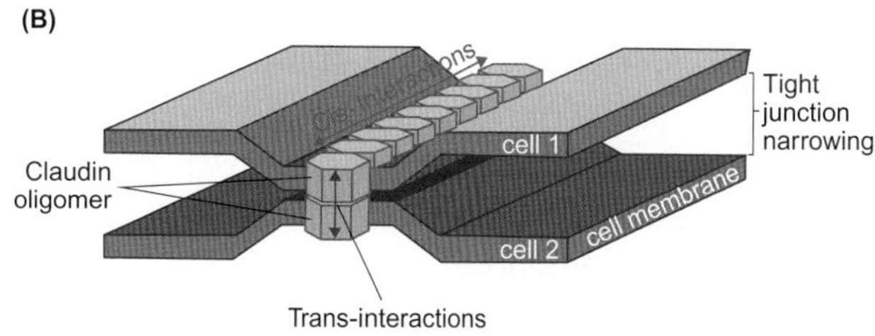

Figure 4.11 Claudin structure

(A) Model of claudin protein secondary structure based on amino acid sequence. The molecule contains four transmembrane domains (I to IV) and two extracellular loops (ECL). The first ECL contains a conserved triplet of amino acids necessary for biological function that may ensure proper folding allowing interaction with the claudin in the adjacent cell. Two cysteines are also likely to stabilize the molecule via a disulfide bond.

(B) Interactions of claudin monomers at tight junction. Diagram shows interaction of identical claudin monomers on lateral cell membranes of cell 1 and 2 (trans-interactions – red arrow) and adjacent monomers (cis-interaction – blue arrow). Interaction of different claudins can also occur.

Sources: A: adapted from Günzel D, Yu ASL (2013). Claudins and the modulation of tight junction permeability. Physiological Reviews 93: 525-569; B: Krause G et al (2008). Structure and function of claudins. Biochimica et Biophysica Acta 1778: 631-645.

[28] Work on *Necturus maculosus* gall bladder by Jared Diamond and Eberhard Frömter (reviewed by Diamond (1977) as listed in the further reading for this section), was pivotal in identifying the paracellular pathway for ion passage across this epithelium.

[29] We discuss proximal tubular functioning in Section 7.2.

[30] Section 5.1.3 examines the gill epithelium of fish, which in marine fish allows passage of Na^+ via the paracellular pathways.

[31] We consider claudins in fish gills in more detail in Sections 5.1.3, 5.2.2 and 5.3.3.

and ion fluxes occur. The clearest example is the occurrence of claudin-2 in the proximal tubule of the mammalian kidney, where it acts as a cation permeable pore with an estimated radius of 30–40 nm. This pore allows Na^+ (and to a lesser extent Cl^-), as well as water, to be reabsorbed across the highly permeable tubular epithelium by the paracellular route[32]. Such passive reabsorption saves energy that would otherwise be used in active transcellular transport.

4.2.2 Transcellular transport of solutes by osmoregulatory epithelia

Transcellular transport of solutes involves passage across the cells, in either direction, as shown in Figure 4.8B. Solute uptake into the extracellular fluids of an animal (such as ion uptake across the gills of fish, or ion reabsorption from kidney tubules), involves initial transport across the **apical** (or **luminal) membrane**, i.e. the outer surface of the cells, followed by passage through the cytoplasm, and then transport across the basal and lateral membranes of the epithelial cells. Usually, basal and lateral membranes are grouped together as **basolateral membranes** as they are functionally similar.

The transcellular transport of solutes (particularly ions) and water by organs, such as kidneys, gills and salt glands, is fundamental to the regulation of salt and water balance by multicellular animals, and uses a complex array of transport mechanisms. We examine the overall processes in particular tissues in Chapters 5–7. In all cases, the basic principles of such mechanisms are that:

- Transporter proteins may act as channels across the apical or basolateral membranes of the epithelium of osmoregulatory tissues. These channels allow passive transport under favourable conditions, by which millions of ions can move through the channels each second. Specific transport proteins allow particular ions and other solutes across membranes, and some allow water transport.

  In each case, the fundamental characteristics of this passive transport are similar to those that apply to passive transport across the entire epithelia. Hence, the passive transport has no energy requirement, and occurs in the direction of equilibrium. This means that passive transport across apical or basolateral membranes depends upon transmembrane differences in concentrations, and for ions is influenced by electrical charge (voltage) differences across the particular membrane.
- Ion pumps in osmoregulatory tissues transport particular ions across the apical and/or basolateral membranes by active transport. The key characteristics of active transport by these tissues are:
  - use of adenosine triphosphate (**ATP**) as the energy source[33]
  - transport against concentration and electrical differences
  - the creation of electrochemical differences that drive secondary active transport[34] by other membrane proteins.

It is not possible to discuss in this chapter all of the complex array of transport mechanisms in osmoregulatory tissues. Instead, we focus in this sub-section on some of the transport mechanisms that we meet most frequently in later chapters to provide the background needed to understand general principles.

Sodium pumps in basolateral membranes of osmoregulatory tissues

The most commonly occurring ion pump in osmoregulatory tissues is the sodium/potassium pump (often simply called the sodium pump). The sodium pump is actually an enzyme, **Na^+, K^+-ATPase**, which functions as an ion transporter. Na^+, K^+-ATPases located in the membrane are essential for the maintenance of a membrane potential difference[35]. However, Na^+, K^+-ATPase activity in osmoregulatory tissues occurs at much greater levels than necessary simply to maintain the concentration differences of ions between intracellular and extracellular fluids (low cellular Na^+; high cellular K^+).

Particularly high levels of Na^+, K^+-ATPase occur in those organs that are specialized for sodium pumping, such as the mammalian kidneys, in which Na^+, K^+-ATPase is partially responsible for at least 50 per cent of the sodium reabsorption by kidney tubules. Hence, organs such as kidneys have a high rate of metabolism and a rich supply of mitochondria[36]. There is a significant relationship between the Na^+, K^+-ATPase activity and sodium transport by different segments of the kidney tubules[37].

In the rectal glands of sharks, Na^+, K^+-ATPase activity is 10 times greater than even the highest levels in mammalian tissues. Hence, shark rectal glands are widely used by biochemists studying the function of Na^+, K^+-ATPase.

[32] We discuss proximal tubular functioning in more detail in Section 7.2.1.

[33] ATP is the common currency for energy transfer and utilization in animal cells, as we discuss in Chapter 2.

[34] Section 3.2.4 examines the concept of secondary active transport.

[35] The role of Na^+, K^+-ATPase in maintaining membrane potential is discussed in Section 3.2.4.

[36] Figure 7.12 illustrates these features of kidney tubules.

[37] Section 7.2 examines Na^+ reabsorption by kidneys and the participation of Na^+, K^+-ATPase and other transporters in this process.

Polarized distribution of Na^+, K^+-ATPase in salt-transporting epithelia

The Na^+, K^+-ATPase in the membranes of salt-transporting epithelial cells is not distributed at random: no Na^+, K^+-ATPase is found in apical membranes, but it occurs in basolateral membranes that often extend as dense invaginations into the cell cytoplasm. This basolateral location of Na^+, K^+-ATPase is a key factor in determining the ion transporting properties of osmoregulatory epithelia.

Why is a basolateral location for Na^+, K^+-ATPase in ionoregulatory epithelia important in ion regulation? The basolateral—as opposed to apical—location facilitates the movement of Na^+ from the cytosol to extracellular fluid, lowering cellular concentrations of Na^+. As a result, directional transfer of Na^+ across the epithelium can occur if Na^+ uptake occurs via the apical membrane. Such uptake may be from the external media (for aquatic animals) or from other fluids, such as in the kidney tubules.

In addition, the pumping process of basolateral Na^+, K^+-ATPase in epithelial cells drives secondary active transporters. The best-known secondary active transporters in osmoregulatory tissues are **Na^+, K^+, $2Cl^-$ co-transporters** found in the basolateral membranes of chloride secretory tissues including the rectal gland of elasmobranchs, salt glands of reptiles and birds, and gills of marine teleost fish[38]. In a similar way, basolateral Na^+, K^+-ATPase drives Na^+, glucose co-transporters in the apical (luminal) membranes of the kidney tubules, providing the driving force for glucose reabsorption. In addition, the **electrogenic** (charge) effects of Na^+, K^+-ATPase can drive Cl^- transfer from intracellular fluids into extracellular fluids.

The pumping cycle of Na^+, K^+-ATPase

To understand how Na^+, K^+-ATPase drives such processes, we need to look more closely at its pumping cycle. Na^+, K^+-ATPase is one of the P-ATPases—a classification that tells us about the way it functions in the pumping cycle. The activation of P-ATPases occurs when a phosphate group attaches to the molecule in a process known as **phosphorylation**. All P-ATPases transport specific ions such as Ca^{2+} (calcium pumps) or Na^+ and K^+ (sodium–potassium pumps) across membranes.

P-type ATPases are also referred to as E1–E2 ATPases because they change from one enzyme (E) conformation (E1) to a new conformation (E2) during their pumping cycle. To understand this cycle, we need to look more closely at the Na^+, K^+-ATPase molecule shown in Figure 4.12.

[38] Na^+, K^+, $2Cl^-$ co-transporters linked to Na^+, K^+-ATPase pumping by salt excreting glands and the gills of marine teleosts are illustrated in Figures 5.14 and 5.7, respectively.

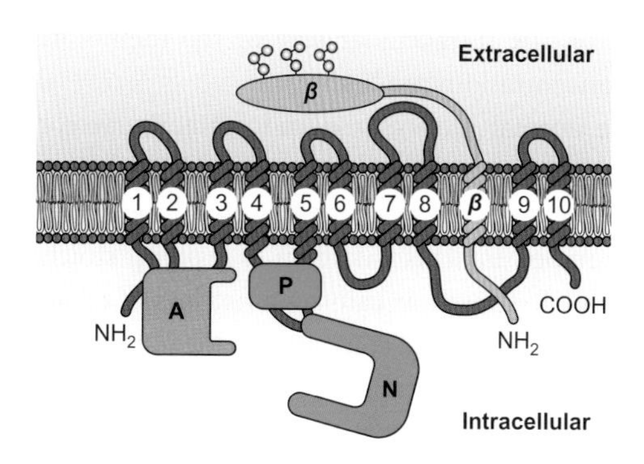

Figure 4.12 Diagram of Na^+, K^+-ATPase in the lipid membrane

The main part of Na^+, K^+-ATPase—the α-subunit—has 10 transmembrane segments (1–10), which are shown in an unfolded position, but actually form a bundle within the membrane. The intracellular domains of the α-subunit are the functional 'engine' of Na^+, K^+-ATPase. These are: (i) the nucleotide domain (N; green) made up of the main part of the large intracellular loop, which forms a kind of pocket for ATP binding, (ii) the phosphorylation domain (P; blue) made up of the parts of the large intracellular loop between the 4th and 5th transmembrane segments, (iii) the actuator domain (A; orange), formed by the NH_2-terminal segment preceding the 1st transmembrane segment and the intracellular loop between the 2nd and 3rd transmembrane segments, which results in dephosphorylation. The β-subunit of the Na^+, K^+-ATPase molecule has a single transmembrane segment, a short intracellular NH_2 terminus, and a large extracellular domain.

Source: adapted from Horisberger J-D. (2004). Recent insights into the structure and mechanism of the sodium pump. Physiology 19: 377-387.

Na^+, K^+-ATPase has two main components: the α and β subunits. The relatively small β-subunit occurs predominantly on the extracellular side of the molecule and is important for the folding of the α-subunit and its insertion in the membrane. The larger α-subunit extends through the lipid bilayer into the cytoplasm, as shown in Figure 4.12. Usually two α-subunits and two β-subunits form each functional transmembrane unit.

The α-subunits are of most importance for the ion pumping cycle, which is illustrated in Figure 4.13. The essential feature of Na^+, K^+-ATPase functioning is the continual switching between Na^+ and K^+ binding, resulting from the changes in the enzyme's molecular shape. The α-subunits contain the catalytic part of the molecule, which releases the energy for ion transport from ATP. Other important functional parts of the α-subunit, which are illustrated in Figure 4.12, are the nucleotide binding domain, which binds ATP, and a P-domain to which a phosphate group from ATP is attached (the domain is phosphorylated) after the entry of Na^+ from the cytosol (which results in conformational changes to the molecule).

Phosphorylated Na^+, K^+-ATPase (the E1 conformation) has a lower affinity for Na^+, which is released on the extracellular side of the membrane (where Na^+ concentrations are higher). The E2 conformation has a high affinity for K^+ from the extracellular fluids. Binding of K^+ triggers dephosphorylation, and lowers the affinity for K^+ ions, which are released on the intracellular side of the membrane, again where concentrations are higher.

At the normal intracellular and extracellular concentrations of Na^+, K^+ and high intracellular ATP relative to ADP, each normal cycle pumps three Na^+ out of the cell while two K^+ are pumped in. This is known as a pump stoichiometry of 3:2 (Na:K). As well as causing a net loss of osmotic concentration inside the cell compared to the outside, this pump stoichiometry means that the sodium pump is electrogenic, i.e. it generates a net flow of charge and results in a net electronegativity within the cells relative to the extracellular fluid.

A specific chemical inhibitor of Na^+, K^+-ATPase, ouabain, also binds with high affinity to the α subunit from the extracellular side of the membrane, and prevents the conformational changes that are an essential part of ion pumping. This action means that ouabain is often used in experimental studies of the ionoregulatory processes of animals as a means of revealing the involvement of Na^+, K^+-ATPase.

Ouabain binds to Na^+, K^+-ATPase in a 1:1 molar ratio, so radiolabelled ouabain can be used to investigate the number of sodium pumps per cell. Experiments using radiolabelled ouabain show tightly packed sodium pumps in osmoregulatory tissues such as the mammalian kidney, where up to 40–50 million pumps per cell occur in some parts of the tubule.

Ouabain occurs naturally in circulation, and at low concentrations stimulates heart function. This action results from the inhibition of Na^+ extrusion by Na^+, K^+-ATPase, which leads to an increase in intracellular Na^+ concentrations in the pumping cells of the heart—the **cardiac myocytes**. An elevated intracellular Na^+ concentration in these cells inhibits another ion transport mechanism that occurs via the sodium/calcium (Na^+/Ca^{2+}) ion exchange proteins in the cardiac myocytes, which pump Ca^{2+} out of the cells in exchange for Na^+. Hence, ouabain treatment leads to Ca^{2+} accumulation by cardiac cells; when released in response to an action potential, the increased Ca^{2+} causes a greater contractile force[39]. At higher doses, ouabain becomes toxic and causes breakdown of normal cardiovascular and respiratory function.

The toxic effects of ouabain have long been used by tribes in Africa who coat the arrows they use for hunting with ouabain from the bark and sap of poison-arrow trees (*Acokanthera* sp.). Amazingly, the African crested rat (*Lophiomys imhausi*) seems to indulge in a similar process. This rat gnaws and chews the roots and bark of poison-arrow trees, but is insensitive to the toxic effects of ouabain in the chewed mixture. The rat then carefully slavers this mixture onto specialized hairs on its flanks. These hairs are hollow, so they rapidly absorb the poisonous mix, which dries to forms a defence mechanism. Any attack on the rat by a potential predator stimulates the rat to change its posture and expose the pronounced area of poisoned hairs along each flank, drawing the predator's attention to an area where a bite will deliver toxic ouabain.

Proton pumps

Not all of the pumps using ATP as an energy source undergo phosphorylation and dephosphorylation during ion transport (like Na^+, K^+-ATPase, in Figure 4.13), as **proton pumps** demonstrate.

[39] We discuss the functioning of cardiac muscle in detail in Section 18.3.

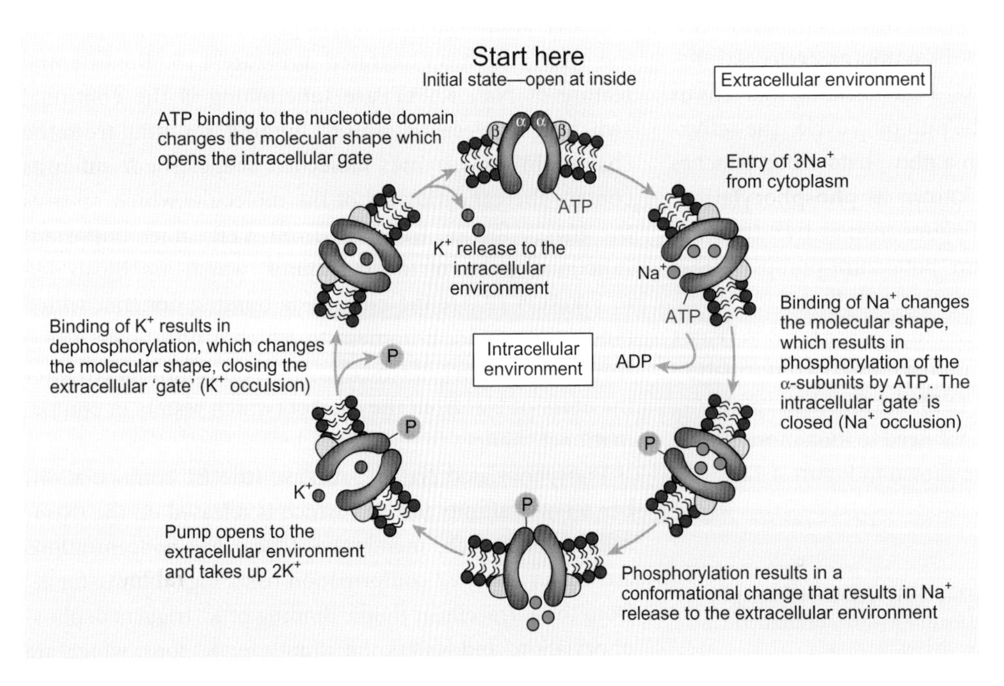

Figure 4.13 Schematic diagram of the functional cycle of Na^+, K^+-ATPase

The cycle is continuous, but best understood by starting at the top centre of the diagram ('start here').

Adapted from Becker WM et al. The World of the Cell. 4th Edition. Benjamin/Cummings Pub Co, 2000; and Horisberger J-D (2004). Recent insights into the structure and mechanism of the sodium pump. Physiology 19: 377-387.

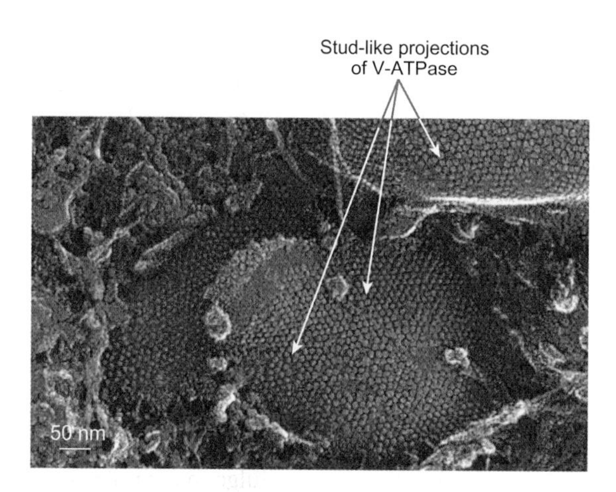

Figure 4.14 Proton pumps (V-ATPase) in toad urinary bladder

Freeze-etching reveals hundreds of stud-like projections in each epithelial cell. These studs of about 10 nm in diameter correspond to the cytoplasmic, internal part of V-ATPase.

Source: Reproduced from Brown D et al (2009). Regulation of the V-ATPase in kidney epithelial cells: dual role in acid base homeostasis and vesicle trafficking. Journal of Experimental Biology 212: 1762-1772.

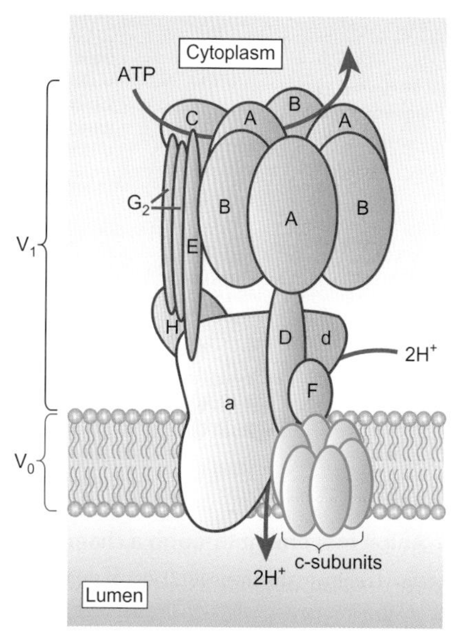

Figure 4.15 Schematic representation of the subunits in V-ATPase

The V_0 transmembrane sector transports H^+ across lipid membranes. This part of V-ATPase is composed of an **a** sub unit, **d** subunit, and six **c** subunits (4 **c**, 1 **c′** and 1 **c″**) connected to the V_1 cytoplasmic component of V-ATPase, which is made up of eight types of subunits (A–H). Subunit A catalyses ATP hydrolysis forming ADP and inorganic phosphate (P_i).

Source: Brown D et al (2009). Regulation of the V-ATPase in kidney epithelial cells: dual role in acid base homeostasis and vesicle trafficking. Journal of Experimental Biology 212: 1762-1772.

Proton pumps transport protons (H^+) into vesicles, vacuoles and lysosomes, hence their alternative names of V-type (vacuolar) ATPases or V-type H^+-ATPases often shortened to **V-ATPases**. In protozoans, V-ATPases play an important role in the formation of fluid for excretion by the contractile vacuole, which controls cell volume. However, V-ATPases are not all restricted to vesicles and vacuoles. V-type-ATPase proton pumps also occur at high densities in acid-excreting epithelial cells. Such cells occur in the **osteoclasts**[40] of bone, the **Malpighian tubules** of insects, the gill epithelia of some fish and crustaceans, in kidney tubules, and in the epithelial cells lining the amphibian bladder. Figure 4.14 illustrates such cells in the amphibian bladder.

All V-ATPases are complex proteins, made up of many subunits, as shown in Figure 4.15. Part of a large cytoplasmic component, (V_1) is involved in ATP hydrolysis, which provides the energy to transport protons against differences in H^+ concentration of 100 times or more. The transmembrane component (V_o) is responsible for proton transport.

Proton transport by V-ATPase in animal cells is regulated in various ways. These include changes in the amount of V-ATPase subunits in the cells, association and dissociation of the V_o and V_1 parts of the V-ATPase, and intracellular trafficking and recycling of vesicles containing V-ATPase between membranes.

Proton pumping creates a build-up of protons on the apical (luminal) side of membranes, which is usually accompanied by either:

- Passive transport via ion channels of an equal number of anions, such as Cl^-, in the same direction as the pumping of protons. We see this during the secretion of hydrochloric acid by the epithelium of the stomach, for example.
- Transport of an equal number of cations such as K^+ or Na^+, in the opposite direction to H^+ ions—for example, in the apical border of the gill epithelium of freshwater fish[41].

Ion channels in ionoregulatory processes

There are hundreds of different types of ion channels in the epithelia of animals, and we are only just beginning to understand the exact route and the reasons for the passage of particular ions through a given channel (i.e. its ion selectivity). Selectivity generally seems to result from a combination of factors, including the charges on the amino acid residues making up the channel. In some cases, a narrow bottleneck in the channel functions as a physical selectivity filter. Figure 4.16 illustrates the key stages in the passage of ions across a membrane through an ion channel.

One general feature of the epithelial cells of osmoregulatory and excretory tissues is that they often have particular types of K^+ channels in the basolateral membranes

[40] In osteoclasts, V-ATPase is a key participant in the process of bone resorption; we explore bone turnover in Figure 21.31.

[41] Section 5.2.2 discusses Na^+ uptake across the gill epithelia of freshwater fish.

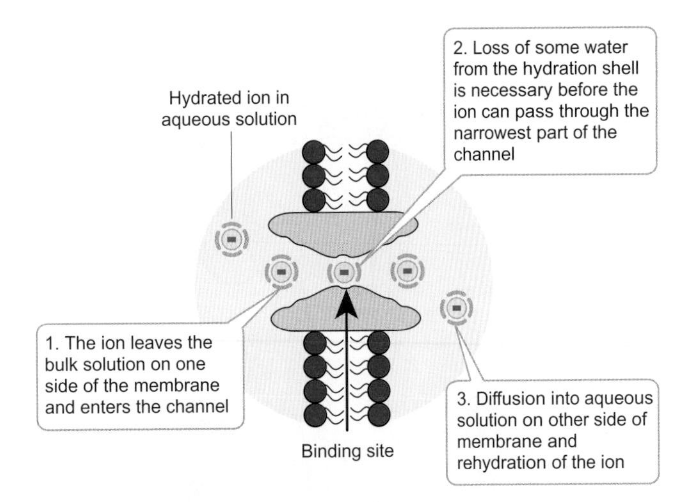

Figure 4.16 Anion movement through a channel

Three stages occur: (1) entry to the channel; (2) loss of some of the hydration shell (dehydration) to allow passage through the narrowest part of the channel, often in single file or with only a couple of ions passing through the narrowest part of the channel at any one time. The narrowest part of the channel may prevent passage of some ions and thus determine channel selectivity. Binding sites may channel ions in particular directions or impede passage of some ions; (3) diffusion of the rehydrated ions.

Source: adapted from Liu X et al (2003). CFTR: What's it like inside the pore? Journal of Experimental Zoology 300A: 69-75.

known as **leak channels**. These K^+ channels allow leakage out of K^+ that has been transported into the cells via Na^+, K^+-ATPase. This leakage of K^+ allows the continual operation of Na^+, K^+-ATPase and the resultant driving of ion transport processes.

Chloride channels

Table 4.1 lists some of the many osmoregulatory processes that involve chloride reabsorption or secretion via **chloride channels**. Several distinct categories of membrane channels permit Cl^- transport:

- Calcium-sensitive chloride channels (CLCAs) are activated in response to increased intracellular concentrations of calcium ions (Ca^{2+}).

Table 4.1 Examples of osmoregulatory processes which employ chloride channels

Tissue	Osmoregulatory transport process
Gills of marine bony fish	Excretion of Cl^- across apical membrane of gill epithelial cells
Salt glands and elasmobranch rectal gland	Cl^- secretion by apical membrane of salt glands
Intestine	Uptake of Cl^- via basolateral membrane of intestine of marine fish
Kidney tubules	Cl^- reabsorption across basolateral membranes
Insect Malpighian tubules	Cl^- secretion into excreta of insects

- Chloride channels functioning as dimers of two similar subunits, each with a pathway for Cl^- conductance. These Cl^- channels, known as CLC proteins, occur in many tissues, including kidney tubules, where they facilitate Cl^- reabsorption via the basolateral membranes[42].
- Cystic fibrosis transmembrane conductance regulator (CFTR) proteins—so named because mutations of the CFTR gene in humans cause cystic fibrosis. CFTR proteins occur in the apical border of several fluid-secreting epithelia, including sweat and pancreatic ducts, the small intestine and the apical membranes of the epithelial cells lining the air passages of the lung.

In the mammalian lung, CFTR is important for the normal hydration of the mucous lining of the airways. The lack of anion secretion in cystic fibrosis reduces fluid secretion and at least partially accounts for mucus build up in the lung. In non-mammalian vertebrates, CFTR-like proteins are involved in ion regulation. For instance, chloride excretion across the apical membranes of ion-excretory cells in the gill epithelia of marine fish occurs via CFTR-like proteins[43].

Figure 4.17 shows the characteristic five domains of a CFTR protein and its gate-like operation, which allows anions to move across the lipid bilayer of epithelial cells. The anion selectivity of CFTR is not strong: the channel allows a range of other anions to pass through, including HCO_3^-, halide anions like iodide and bromide, water and urea, but Cl^- passes through most easily.

Urea transport proteins

Some channel proteins in cell membranes transport urea and are essential components of osmoregulatory tissues such as the kidney. Two main sub-types of urea transport proteins (UTs) occur: UT-A (in epithelial cells) and UT-B (in vascular endothelial cells and erythrocyte cell membranes). Six sub-types of UT-A proteins are recognizable in mammals, three of which occur at high levels in the kidney (UT-A1, UT-A2 and UT-A3), and UT-B occurs in the renal vasculature, so the kidney is rich in urea transport proteins[44].

Although the mammalian kidney excretes urea, **urea transporters** are essential to the retention of urea by the kidney as part of the mechanism for generating osmotically concentrated urine whenever there is a need to conserve

[42] Figure 7.16 illustrates this process in the loop of Henle of the mammalian nephron.

[43] Section 5.1.3 discusses the involvement of chloride channels in ion-regulation of marine teleost fish.

[44] We discuss the urea transporter proteins in various components of the kidney in Section 7.2.3, and their hormonal regulation is discussed in Section 21.1.4.

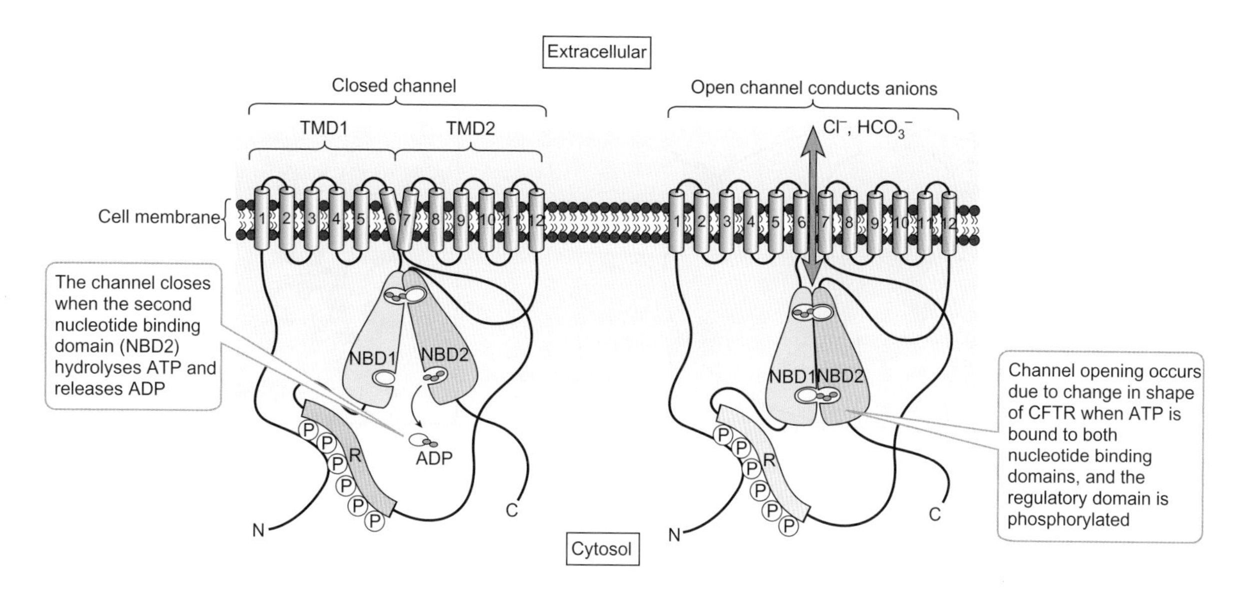

Figure 4.17 Structure and function of chloride channel formed by cystic fibrosis transmembrane conductance regulator (CFTR) protein

The CFTR protein is composed of 12 transmembrane (TM) helices forming two transmembrane domains (TMD1 and TMD2). The 6th helix of TMD1 extends into the cytosol where it is connected to the first nucleotide binding domain (NBD1) (which binds ATP), then the regulatory domain (R) with regulated phosphorylation sites (P). The regulatory domain is connected to the second transmembrane domain (TMD2), and finally a second nucleotide (ATP) binding site (NBD2). ATP binding and hydrolysis by the cytoplasmic NBDs is coupled with the opening of an anion channel regulated by the phosphorylation of the R domain. The conformational changes necessary for channel activity are poorly understood. The diagram shows a simplified 2D arrangement of TMDs, but 3D models suggest several TM helices line the pore for ion passage across the membrane. Passage of chloride ions (or other anions) is predicted to occur after loss of at least some of the water of hydration of the ions with ions passing through a central pore that has a narrow central region formed by the 3D organization of transmembrane domains.
Source: adapted from Berridge MJ (2012). Cell Signalling Biology. Module 3 Ion Channels http://www.cellsignallingbiology.co.uk/.

water. Urea transport proteins also occur in the kidneys of cartilaginous fish living in seawater, in which they are essential for retaining urea as a major contributor to their high plasma osmolality[45].

4.2.3 Water passage across epithelia

More than 90 per cent of the water in cells is free to move in and out as only a small amount of intracellular water is bound to macromolecules (such as proteins, carbohydrates and nucleic acids) which restrict its movement out of the cells.

Water molecules are small enough (being about 0.3 nm diameter) to diffuse slowly through the lipid membranes of most cells[46]. However, the rate of water flux (volume per unit area per min) often exceeds the rate at which simple diffusion through the lipid bilayer could occur. Higher water fluxes result from the occurrence of **water channels** (pores) that permit higher rates of water passage.

Aquaporins enhance water fluxes

Convincing evidence for the existence of water channels called **aquaporins** was obtained in the late 1980s by the renowned work of Peter Agre[47] and his team. They discovered an abundant protein in the plasma membranes of erythrocytes. This protein was subsequently called aquaporin 1 (AQP1), and since then more than 200 aquaporins have been identified in plants, bacteria, invertebrates and vertebrates.

In mammals, 13 aquaporins have been identified, although the functional importance of all of these is not fully understood. The AQP0 protein functions as a water channel in the lens of the eye and is important in maintaining its transparency. By contrast, AQP1, AQP2 and AQP3 function as water channels in the tubular epithelia of kidneys[48].

Aquaporins 3, 7, 9 and 10 are also permeable to some small, uncharged solutes, such as urea and glycerol, so these mammalian proteins are often classified as aquaglyceroporins. Glycerol and urea are larger molecules than water, but their passage through the **aquaglyceroporins** is faster than for water, which demonstrates the selective nature of the transport through these two types of channel. However, the biological significance of aquaglyceroporins is uncertain.

[45] We examine the osmoregulation of cartilaginous fish, including urea retention in more detail in Section 5.1.4.

[46] We discuss the factors affecting diffusion in Section 3.1.2.

[47] Peter Agre was awarded a Nobel Prize in Chemistry in 2003 for his discovery of aquaporins.

[48] We examine the role of aquaporins in the kidney in Sections 7.2.1 and 21.1.4.

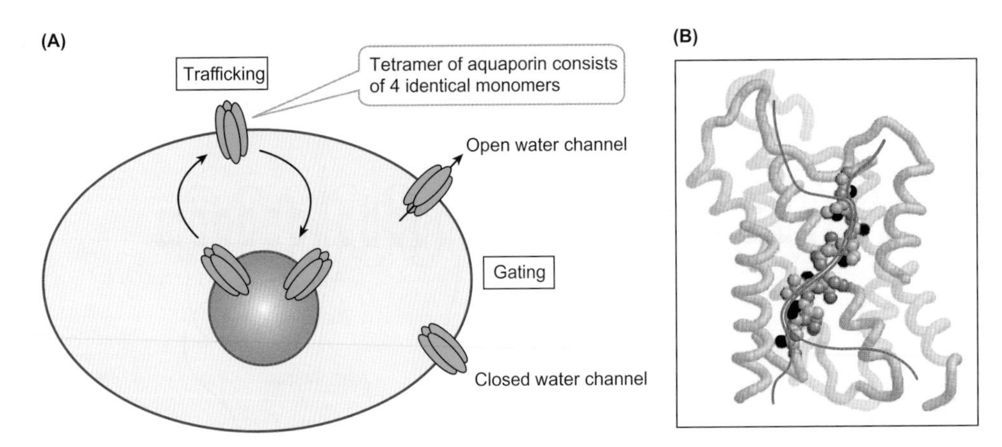

Figure 4.18 Aquaporin regulation and translocation

(A) In animals, aquaporins appear to be regulated primarily by trafficking to the plasma membrane. A decrease in water flux occurs when aquaporins are internalized into vesicles. The alternative mechanism of 'gating', as a result of structural changes that close or open the channel, is the dominant mechanism in regulation of aquaporins among plants. (B) Three-dimensional arrangement of a monomer of AQP2 showing the water channel (orange lines), which is wide at both ends and with a long curved central pore.

Sources: A: Sjöhamn J, Hedfalk K (2014). Unraveling aquaporin interaction partners. Biochimica et Biophysica Acta 1840: 1614-1623; B: Shapiguzov AY (2004). Aquaporins: structure, systematics, and regulatory features. Russian Journal of Plant Physiology 51: 127-137.

Further work indicates that in some locations, aquaporins are important in facilitating the passage of gases: carbon dioxide and ammonia, although these findings are controversial. Nevertheless, there is convincing evidence for excretion of carbon dioxide and ammonia, as well as the passage of water, via an AQP1 sub-type in the epithelium covering the yolk sac of zebrafish larvae (*Danio rerio*)[49].

Aquaporins are usually tetramers: they consist of four identical subunits, as shown in Figure 4.18A. Each of the four monomers consists of six transmembrane α-helices connected by alternate intracellular and extracellular loops. In each monomer, the transmembrane domains form a single water channel. The arrangement of the four monomers forms a larger central pore that may have additional functionality. For instance, passage of carbon dioxide and ammonia may occur predominantly via the central pore.

Detailed structural studies indicate that each water channel broadens into a funnel at either side of an elongated pore, which passes through the cell membrane, as shown in Figure 4.18B. Consequently, the structure is often likened to an 'hourglass'. To navigate the aquaporin, water molecules must pass through the central narrow part of the pathway. The specificity of aquaporins for water passage is thought to result from hydrogen bonding between water molecules and the specific amino acids at the narrowest part of the channel. In AQP1 the narrow part is about 3 nm in diameter, just large enough for water molecules to pass through in single file.

Despite having to pass through the narrowest part in single file, up to around 10^9 molecules of water can pass through each AQP1 water channel every second so when cell membranes contain water channels they are highly water permeable. In contrast, cell membranes with few aquaporins exhibit low water permeability, an effect that is clearly demonstrated by studies of African clawed frog (*Xenopus*) oocytes (eggs).

Frog oocytes do not normally have aquaporins in their plasma membrane, but if mRNA for AQP1 is injected into the egg, the resulting expression of AQP1 in the plasma membrane increases its osmotic water permeability to 200×10^{-4} cm s^{-1}, which is about a third of the rate at which water diffuses in unconstrained solutions. Figure 4.19 shows the dramatic effect of this manipulation. When placed in water, frog oocytes with AQP1 in their plasma membrane rapidly swell and burst within a few minutes due to the osmotic water influx across the plasma membrane.

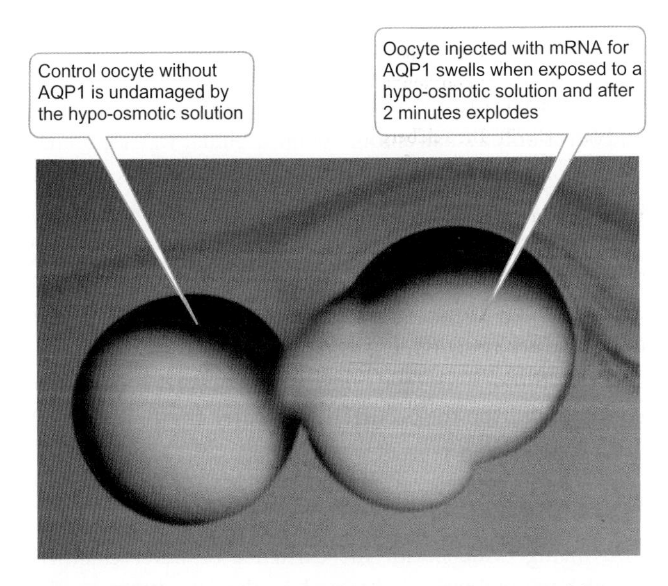

Figure 4.19 Expression of aquaporin in African clawed frog (*Xenopus laevis*) oocytes shows that it is a water channel

Source: Agre P (2000). Aquaporin water channels in kidney. Homer W Smith Award Lecture. Journal of American Society of Nephrology 11: 764-777.

[49] These cells also excrete ammonia via another membrane transporter (Rhesus protein), which we discuss in Section 7.4.1.

A crucial aspect of the functioning of aquaporins is that despite water moving readily through them the passage of ions and other charged solutes does not occur. Therefore, water passage through aquaporins in the plasma membrane does not affect the electrical charge difference[50] between intracellular and extracellular fluids.

Aquaporins may either be permanently present in cell membranes (i.e. constitutively expressed), or their presence may be regulated such that osmotic water fluxes are controlled. Although structural changes causing opening or closure (gating) of water channels have been proposed for some mammalian tissues, the gating of aquaporins has only been demonstrated in plants and microbes; its occurrence in animal tissues remains controversial. In animals, water transport via aquaporins is usually regulated by the phosphorylation of aquaporin proteins and their translocation and subsequent insertion into the apical or basolateral borders of the epithelia, as indicated in Figure 4.18A. Hence, animals primarily control transcellular water flow as part of their osmoregulatory mechanisms by modifying the distribution and abundance of aquaporins in particular membranes[51]. Water passage via aquaporins is also a significant component of cell volume regulation, which we explore in the next section.

› *Review articles*

Agre P (2000). Aquaporin water channels in kidney. Homer W. Smith Award Lecture. Journal of American Society of Nephrology 11: 764–777.

Balkovetz DF (2006). Claudins at the gate: determinants of renal epithelial tight junction paracellular permeability. American Journal of Physiology Renal Physiology 290: 572–579.

Breton S, Brown F (2007). New insights into the regulation of V-ATPase-dependent proton secretion. American Journal of Physiology Renal Physiology 292: F1–10.

Day RE, Kitchen P, Owen DS, Bland C, Marshall L, Conner AC, Bill RM, Conner MT (2014). Human aquaporins: Regulators of transcellular water flow. Biochimica et Biophysica Acta 1840: 1492–1506.

Diamond JM (1977). Twenty-first Bowdich lecture. The epithelial junction: bridge, gate, and fence. Physiologist 20: 10–18.

Fenton RA, Knepper MA (2007). Urea and renal function in the 21st century: insights from knockout mice. Journal of American Society of Nephrology 18: 679–688.

Frömter E (1979). The Feldberg Lecture 1976. Solute transport across epithelia: what can we learn from micropuncture studies in kidney tubules? Journal of Physiology 288: 1–31.

Günzel D, Yu ASL (2013). Claudins and the modulation of tight junction permeability. Physiological Reviews 93: 525–569.

Horisberger J-D (2004). Recent insights into the structure and mechanism of the sodium pump. Physiology 19: 377–387.

Kolosov D, Bui P, Chasiotis H, Kelly SP (2013). Claudins in teleost fishes. Tissue Barriers 1: e25391.

Linsdell P (2014). Functional architecture of the CFTR chloride channel. Molecular Membrane Biology 31: 1–16.

Marshall WS (2002). Na^+, Cl^-, Ca^{2+} and Zn^{2+} transport by fish gills: retrospective review and prospective synthesis. Journal of Experimental Zoology 293: 264–283.

Sjöhamn J, Hedfalk K (2014). Unraveling aquaporin interaction partners. Biochimica et Biophysica Acta 1840: 1614–1623.

[50] Section 3.2.3 discusses the electrical charge differences across the plasma membrane of cells.

[51] We discuss the regulation of AQP2 in mammalian kidney tubules in Section 21.1.4.

4.3 Regulation of cell volume

The cells of animals in which the composition of the extracellular fluid is tightly regulated are bathed in a stable environment and are generally protected from the likelihood of frequent cell shrinkage or swelling. Despite this tight regulation, the cells of some mammalian tissues experience frequent changes in the osmolality of the extracellular fluid around them. For instance, intestinal cells experience dilution of the extracellular fluid after water intake and absorption, which threatens the cells with water influx by osmosis, resulting in volume expansion. In contrast, cells in the tubules and vessels deep in the kidney (in the inner medulla) are exposed to the high osmotic concentrations in the surrounding interstitial fluid whenever concentrated urine is being produced[52]. Such conditions threaten the cells with possible shrinkage due to an osmotic water efflux.

The osmolality of the extracellular fluids of vertebrates such as fish and amphibians are generally less tightly regulated than in mammals, and variations in extracellular fluid osmolality are even greater among invertebrates living in environments that vary in osmolality[53]. The cells in these animals face possible shrinkage or swelling on a regular basis, whenever there are changes in the extracellular fluid composition.

Our discussion of water movements across epithelia so far has focused on the osmolality of extracellular fluid. However, when examining cell volumes the concept of **tonicity** (rather than osmolality) is used. The tonicity of a solution (which has no units) reflects the concentration of solutes that *cannot* pass across the plasma membrane, compared to the non-penetrating solutes in the cytosol. Figure 4.20 illustrates this concept and the outcome on cell volume for cells placed in **isotonic**, **hypotonic** or **hypertonic** solutions.

- An isotonic solution has no effect on the volume of the cells. An isotonic solution has the same concentration of non-penetrating solutes as the cells. (This does not mean that the cells and the external solutions in Figure 4.20A have the same osmotic concentration, as osmotic concentrations include penetrating and non-penetrating solutes).
- When cells are bathed in a hypotonic solution (the Greek *hupo* means under), with a relatively low concentration of non-penetrating solutes compared to the cytosol, water enters the cells and they swell and become leaky, sometimes eventually bursting, as illustrated in Figure 4.20B. The swelling of cells in hypotonic solutions occurs because cells typically have a water permeability that is several orders of

[52] Section 7.2.3 discusses the production of hyperosmotic urine, which relies on the high osmotic concentration of the interstitial fluid in driving water absorption from the tubular fluid.

[53] We discuss the changes in extracellular fluid osmolality of invertebrates living in variable salinities in Section 5.3.1.

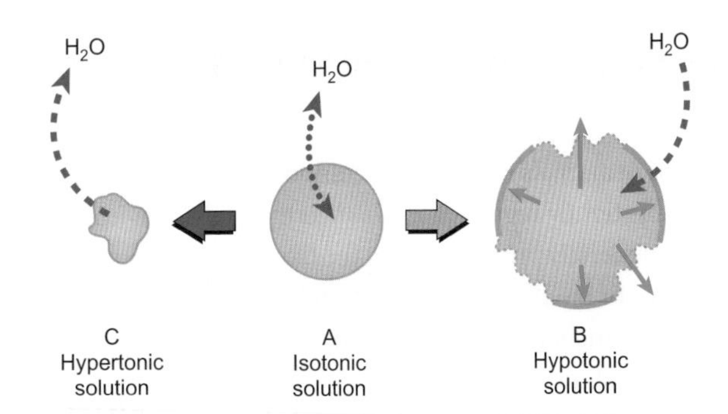

Figure 4.20 Rapid responses of animal cells to changing tonicity of external solutions

In an isotonic solution (A) the small amount of water diffusing across the cell wall, in both directions, does not affect cell volume. When transferred to a hypotonic solution (B), the cell swells. If the solution is sufficiently hypotonic the large water influx may burst the cell. Transfer to a hypertonic solution (C) results in shrinkage and distortion of the membrane (known as crenulation).

magnitude higher than their permeability to ions (Na^+, K^+, and Cl^-). Water inflow occurs by simple diffusion or more rapidly through the aquaporins in the cell membrane, by osmosis. In theory, in the absence of cell volume regulation, volume expansion of the cells continues until the internal and external fluids are in balance.

- When cells are bathed in a hypertonic solution (the Greek *hyper* means over or above), the cells shrink due to the outward movement of water, which is illustrated in Figure 4.20C. Cell shrinkage concentrates the cell contents, which has damaging effects on cell function—for example, by leading to a crowding of macromolecules. Crowding of intracellular regulatory proteins, nucleic acids and polysaccharides impedes chemical reactions, such as enzyme functions. Hence, mechanisms for regulating cell volume are a fundamental part of ensuring the proper functioning of cells.

In practice, compensatory processes begin almost immediately following a change in cell volume, although if synthesis of new proteins and intracellular solutes are necessary, adjustments continue over hours or days. So we might ask—how do cells sense volume changes so that regulatory responses are initiated? Three types of event appear to act as potential signals:

- the changing cell content of specific ions and total ionic strength
- macromolecular crowding of intracellular contents
- mechanical changes in the lipid bilayer that act as mechanosensors, as illustrated in Figure 4.21, which activate transport processes, such as mechanosensory ion channels.

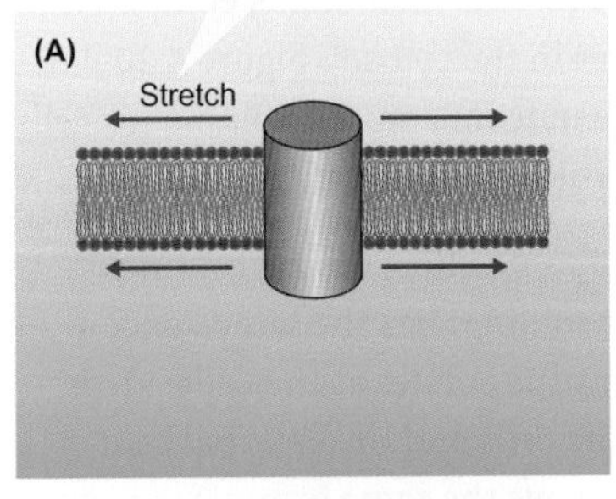

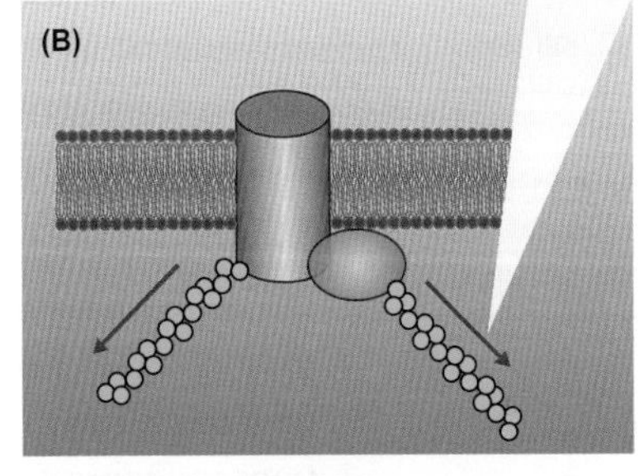

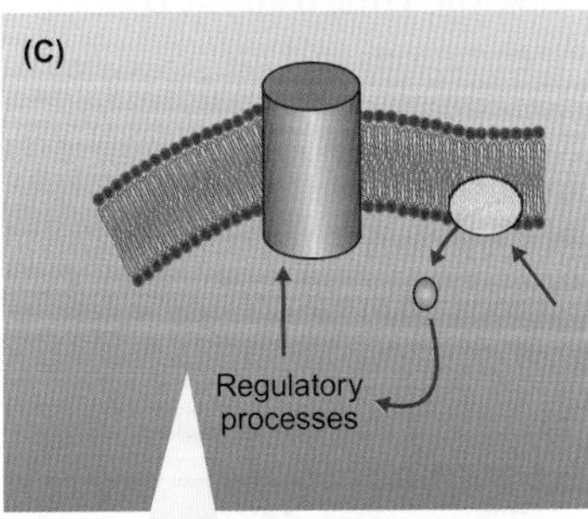

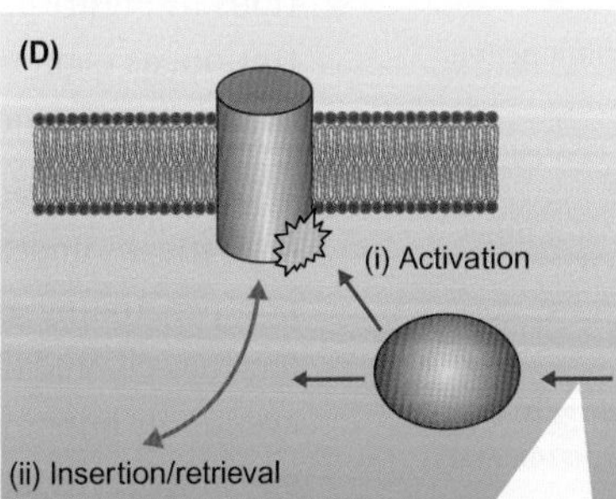

Figure 4.21 Four possible mechanisms for mechanosensing that activate membrane proteins after cell volume expansion

The brown cylinder is a volume-sensitive membrane protein.

Source: adapted from Hoffman EK et al (2009). Physiology of cell volume regulation in vertebrates. Physiological Reviews 80: 193-277.

4.3.1 Responses to cell shrinkage or swelling

A general principle underpinning the regulation of cell volume is the rapid efflux or influx of intracellular osmolytes when the cell volume exceeds or falls below a set point[54].

- An increase in the cell content of osmolytes occurs after cell shrinkage *below* the set point for volume, and results in an influx of water by osmosis. These responses constitute a **regulatory volume increase**.
- A decrease in the cell content of osmolytes occurs if cell volume increases *above* the set point for volume, and results in an efflux of water by osmosis. These responses constitute a **regulatory volume decrease**.

Regulatory volume increase in response to shrinkage of vertebrate cells

Figure 4.22 outlines the key elements in a regulatory volume increase although the participation of these transport mechanisms varies between cell types and animal species:

- Active uptake of organic osmolytes, such as taurine, occurs in the cells of some organs such as the brain, kidney and heart.
- Activation of volume-sensitive proteins in the plasma membrane exchange Na^+ with H^+ in a 1:1 ratio. The Na^+/H^+ exchange mechanism (NHE) was first identified in the 1960s in red blood cells of three-toed salamanders (*Amphiuma tridactylum*). NHEs that respond to cell shrinkage have since been identified in most vertebrate cells studied. In addition to the activation of NHE by cell shrinkage, cell swelling inhibits NHEs.
- Activation of Na^+, K^+, $2Cl^-$ co-transporters, which were first identified in the red blood cells (erythrocytes) of birds,[55] were soon after recognized as a general response to cell shrinkage in many tissues of vertebrates. However, Na^+, K^+, $2Cl^-$ co-transporters are not always a component of regulatory volume increase. For example, there is a notable

[54] Section 3.3.3 explains the principle of the control of physiological systems.

[55] The erythrocytes of birds and almost all vertebrates except mammals are nuclear cells with mitochondria. As such, nuclear erythrocytes have a greater capacity to generate the ATP that powers Na^+, K^+-ATPase and hence the secondary active transport processes such as Na^+, K^+, $2Cl^-$ co-transporters or Na^+/H^+ exchange.

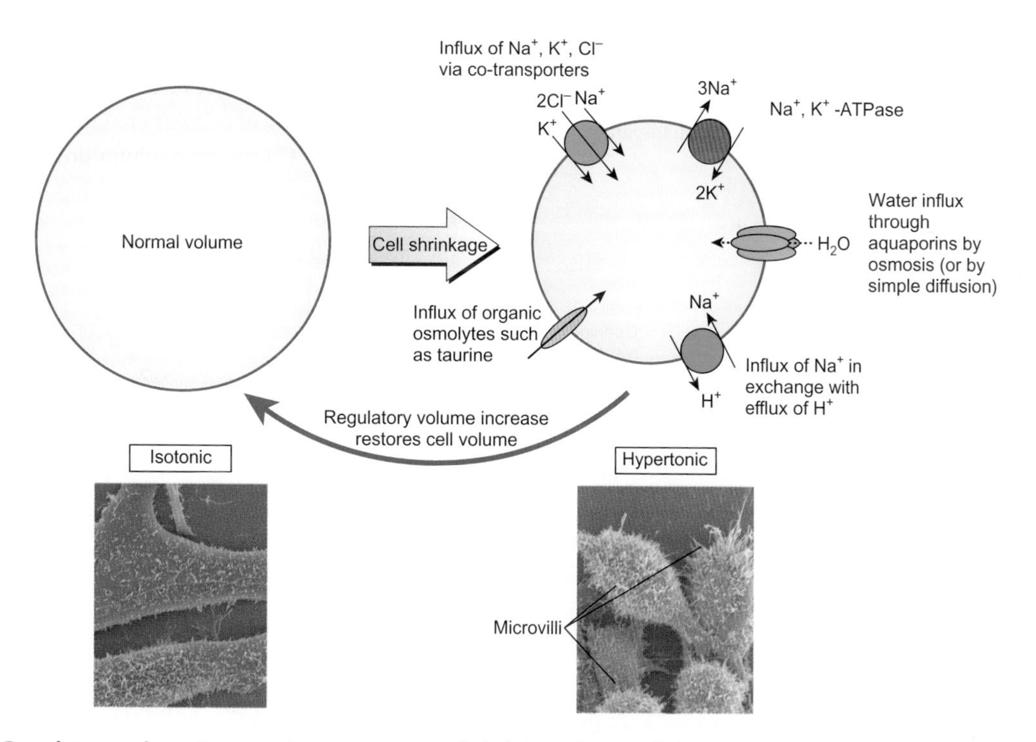

Figure 4.22 Regulatory volume increase in response to cell shrinkage in vertebrates

Cell shrinkage stimulates regulatory volume increase by ion influxes through Na^+/H^+ exchangers and Na^+, K^+, $2Cl^-$ co-transporters (energized by Na^+, K^+-ATPase), which increase the ionic content of the cells. In some tissues influx of other osmolytes, particularly taurine, occurs as part of the mechanism for cell volume regulation. These osmolytes draw in water by osmosis, which restores cell volume.

The two scanning electron microscopy images show cells (Ehrlich Lettre Ascites cells) under isotonic conditions (normal volume) and after 5 min in a hypertonic solution. Note the marked increase in microvilli (projections of the plasma membrane) in the cells in the hypertonic solution, which are typical of shrinking cells in which the lipid bilayer becomes extensively folded. Images taken at similar (unspecified) magnifications.

Source of images: Hoffman EK et al (2009). Physiology of cell volume regulation in vertebrates. Physiological Reviews 80: 193-277.

absence of such a component in the regulation of the volume of the red blood cells of teleost fish and amphibians. In these animal groups, regulation of red blood cell volume relies on the Na^+/H^+ exchange mechanism, together with an anion exchange mechanism, usually Cl^-/HCO_3^- exchange.

While our focus here is on cell volume regulation, it is important to be aware that Na^+/H^+ exchange proteins (NHEs) and Na^+, K^+, $2Cl^-$ co-transporters are involved in a variety of other physiological processes. In particular, these transporters are important components of many of the epithelial ion transport mechanisms in kidneys, salt glands, gills and intestine that we meet in subsequent chapters. These transporters are secondary active transporters[56] usually energized by the sodium pumping of Na^+, K^+-ATPase, which maintains the electrochemical difference for passive Na^+ entry into the cells.

Regulatory volume decrease in response to swelling of vertebrate cells

Figure 4.23 shows several events that occur in response to cell swelling. These events result in the loss of intracellular osmolytes, which drives water loss from the cells by osmosis, and leads to a decrease in cell volume.

- Opening of K^+ channels is a major feature of regulatory volume decrease. There are many different types of K^+ channels in cell membranes, several of which are volume-sensitive, and appear to be directly activated by stretch. Opening of K^+ channels allows an increase in K^+ efflux through the cell membrane (K^+ **conductance**). The flow of K^+ during this efflux occurs from the higher concentrations in intracellular fluids to the extracellular fluid.
- An increase in K^+ conductance increases membrane potential (i.e. causes hyperpolarization—a greater negative charge inside the cells), and hence increases Cl^- efflux via open anion channels.
- K^+, $2Cl^-$ co-transporters occur in many cell types, and are activated by cell swelling.
- Release of taurine is a further component of the rapid regulatory volume decrease of some cells such as those in the brain and kidney.

[56] Section 4.2.2 discusses some of the secondary active transport mechanisms driven by Na^+, K^+-ATPase.

Different categories of cellular osmolytes are manipulated in cell volume regulation

Figures 4.22 and 4.23 show that two kinds of cellular osmolytes are manipulated during regulatory volume responses. These are:

(i) ionic osmolytes—mostly K^+ and Cl^-;

(ii) organic osmolytes, mostly **compatible osmolytes** (solutes), so-called because they are compatible with the functioning of cellular enzymes and other cellular proteins. Compatible solutes do not interfere with

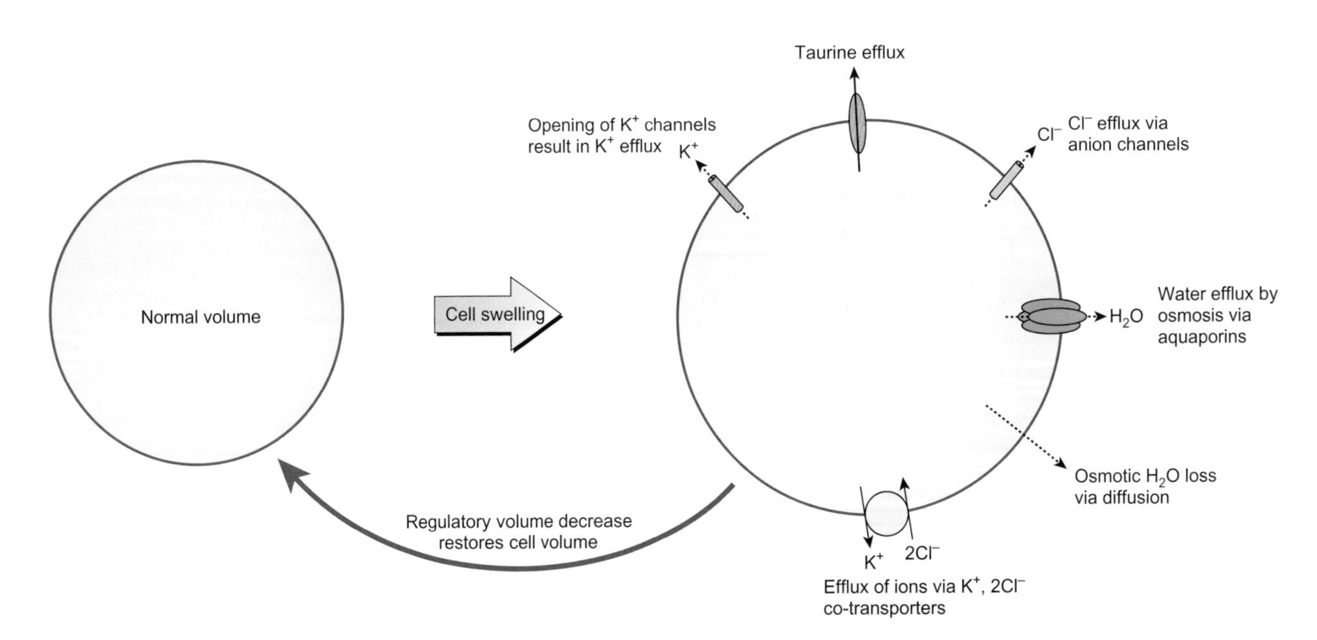

Figure 4.23 Regulatory volume decrease in response to cell swelling in vertebrate cells

Cell volume expansion (swelling) leads to a regulatory volume decrease by rapid elimination of K^+ via K^+ channels and K^+, $2Cl^-$ co-transporters, efflux of Cl^- through ion channels in some cells, and activation of taurine efflux in some cells, such as those in the brain. These processes result in osmotic water efflux by simple diffusion and more rapidly via aquaporins.

cell functioning in the way that high concentrations of inorganic ions do when they bind to proteins and cause unfolding.

Four main types of organic osmolytes can be modified by animal cells during volume regulation:

- polyhydric alcohols, such as sorbitol, which are synthesized from glucose and myo-inositol
- amino acids, particularly taurine, but also alanine, glycine, arginine, serine, glutamic acid, proline and aspartic acid
- amino acid derivatives such as the methylamine, glycerophosphorylcholine and ammonium compounds, such as glycine betaine
- Urea, which destabilizes proteins (though this destabilization is counteracted in the presence of methylamines)[57].

The type of osmolyte manipulated during volume regulation by particular cell types depends primarily on the osmotic concentration of the extracellular fluids in which the cells exist. As a general principle, cells surrounded by extracellular fluid with an osmotic concentration of 300 to 400 mOsm kg^{-1} usually rely mainly on adjustments to the intracellular concentration of ions (which account for 60–70 per cent of the intracellular osmolytes). This situation applies to most animals living in fresh water, most terrestrial vertebrates, and insects.

However, in some vertebrate tissues, organic osmolytes are most important in rapid cell volume regulation. Taurine regulation is particularly important in the regulation of the cell volume of neural tissue and other excitable tissues. In the mammalian heart, taurine accounts for up to 50 per cent of the free amino acid pool in the cells. In the brain, where Na^+ and K^+ differences across the cell membrane are critical for signal transmission between neurons[58], the use of ions for cell volume regulation would be problematic. In brain tissue, a high blood Na^+ concentration initially results in some tissue accumulation of Na^+, K^+ and Cl^-, but then some recovery occurs. As such, ion accumulation is insufficient to explain cell volume regulation. Instead, the majority of the solutes that accumulate in cells of the mammalian cerebrum (part of the forebrain) are amino acids and their derivatives. The taurine content of mice brains increases by about 50 per cent during prolonged increases in the Na^+ concentration of blood plasma, which would otherwise cause cell shrinkage.

The accumulation of compatible organic solutes is a general feature of cells that are surrounded by extracellular fluids with a high osmotic concentration as occurs in the inner medulla of the mammalian kidneys, for example. Figure 4.24 shows some results for rabbits. The osmotic concentration of extracellular (interstitial) fluid increases during dehydration as a key part of the mechanism enabling production of osmotically concentrated urine[59]. Hence, the concentrations of sodium and urea in the inner medulla are much higher when low volumes of urine are excreted (during an antidiuresis) than during a diuresis (when dilute urine is excreted). Such changes mean there is a risk of changes in

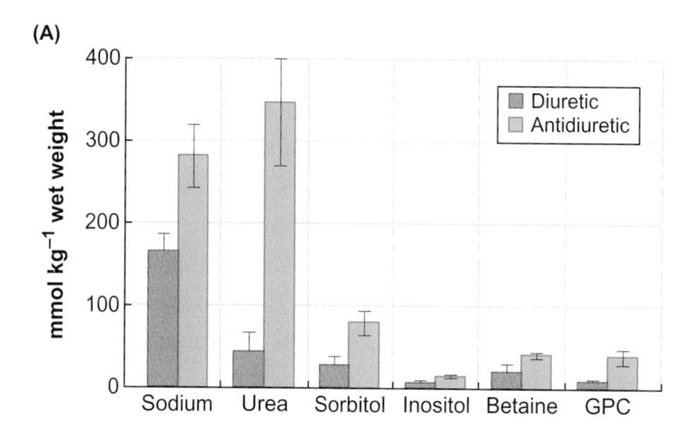

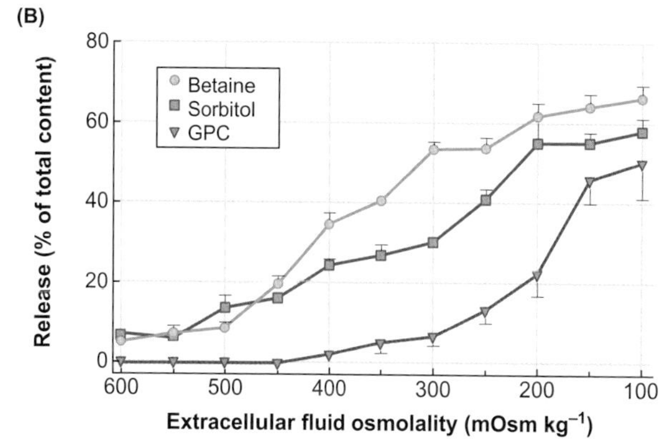

Figure 4.24 Counteracting and compatible solutes in renal medulla of mammals

(A) Osmolyte content of the inner medulla of the rabbit during dehydration, resulting in an antidiuresis (low rate of urine excretion) and during a diuresis. During dehydration, extracellular fluid concentrations of urea and sodium chloride in the kidney medulla increase, as part of the mechanisms to generate hyperosmotic urine. Cell volume regulation is achieved by increases in glycerophosphorylcholine (GPC), sorbitol, inositol and betaine in the medullary tissue. (B) Progressive release of organic osmolytes (sorbitol, betaine, GPC) from inner medullary cells of the rabbit kidney in relation to a decreasing osmolality of extracellular fluid. Release of the organic osmolytes occurs at different thresholds; release of GPC is delayed relative to release of betaine and sorbitol. Sources: (A) Garcia-Perez A, Burg MB (1990). Importance of organic osmolytes for osmoregulation by renal medullary cells. Hypertension 16: 595-602; (B) Kinne RKH (1993). The role of organic osmolytes in osmoregulation: from bacteria to mammals. Journal of Experimental Zoology 265: 346-355.

[57] The accumulation of urea in cartilaginous fish, together with the influence of methylamines, such as trimethylamine oxide (TMAO), is discussed in more detail in Section 5.1.4.

[58] We discuss the differences in Na^+ and K^+ across cell membranes and their effect on membrane potentials in Sections 3.2.3 and in greater detail in Section 16.2.

[59] See Section 7.2.3 for a description of the mechanisms responsible for producing hyperosmotic urine in mammals.

cell volume. Over several days of dehydration, during which concentrated urine is excreted, a gradual accumulation of organic osmolytes occurs in the renal tubular cells. Note in Figure 4.24A the higher concentrations of sorbitol, inositol and glycerophosphorylcholine (GPC) in the medullary tissue of dehydrated (antidiuretic) rabbits.

Accumulation of inositol occurs due to the activity of the Na^+-inositol co-transporters in the basolateral membranes of the renal medullary cells. Sorbitol accumulates as a result of its synthesis from glucose (catalysed by aldose reductase): during dehydration, the activity of aldose reductase increases. Taurine accumulation also occurs due to influx via a specific taurine transporter protein (Tau-T), while Na^+, Cl^-/betaine co-transporters stimulate betaine influx.

In contrast, rapid rehydration of a dehydrated mammal leads to a return of the interstitial osmolality to around 300 mOsm kg^{-1}, which triggers a rapid release of the organic osmolytes by kidney tubules in the medulla, as shown in Figure 4.24B.

The cells of invertebrates living in marine habitats or brackish waters are routinely bathed in extracellular fluid that with an almost identical osmolality (iso-osmotic or **isosmotic**) to the seawater/brackish water in which the animals live[60]. Similarly, marine cartilaginous fishes (sharks and rays) have extracellular fluids that are close to isosmotic with seawater (actually slightly hyperosmotic[61]). In these marine and brackish water animals, organic osmolytes are the dominant cellular osmolytes, rather than ions; ionic concentrations could not be raised sufficiently for the intracellular fluid to reach a level that is isotonic with extracellular fluid without interfering with cell function. Hence, these animals primarily modify the organic osmolyte content of their cells to regulate cell volume.

The cells of invertebrates living in marine and brackish habitats have a high amino acid content (a large amino acid pool), which may reach 800 mmol L^{-1} when they are in seawater. These animals adjust the cell content of amino acids in response to cell shrinkage and cell swelling using the processes summarized in Figure 4.25, although not all of the processes shown in the diagrams necessarily occur in any one species or animal group. For example, in response to cell swelling, molluscs emphasize amino acid fluxes across the cell membrane followed by their metabolism and excretion of ammonia[62], while crustaceans emphasize amino acid catabolism and an increase in protein synthesis.

Cell shrinkage results in protein degradation, increased uptake of amino acids from extracellular fluid into the cells and/or increased amino acid synthesis, so the cell content of amino acids increases. These processes then draw water into the cells by osmosis. Although Figure 4.25 suggests that cell volume regulation results from a general adjustment of the pool of free amino acids, the specific amino acids used for cell volume regulation actually vary between species and cell type, although it is not clear why this is the case.

[60] The composition of the extracellular fluid of marine invertebrates is discussed in Section 5.1.1.

[61] Osmoregulation of cartilaginous fish is explored in Section 5.1.4.

[62] We discuss ammonia excretion in aquatic animals in Section 7.4.1.

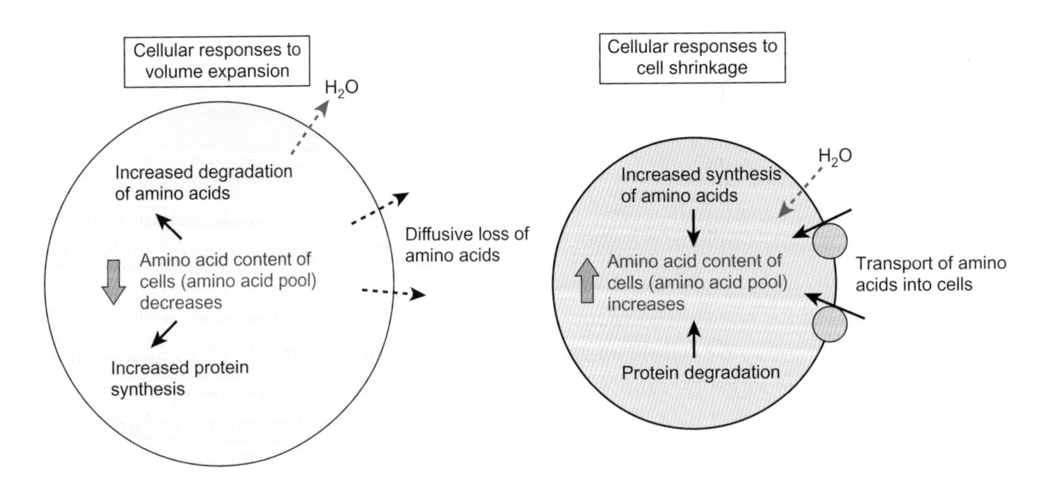

Figure 4.25 Manipulation of the cell content of amino acids for cell volume regulation

Cellular volume expansion is corrected by increased degradation of amino acids, increased use of amino acids in protein synthesis and increased cell membrane permeability to specific amino acids allowing them to diffuse out of the cell, which leads to water loss by osmosis. Cell shrinkage results in corrective responses that increase the content of specific amino acids in the cells by transport of amino acids across the cell membrane, reducing membrane permeability to amino acids, synthesis of some amino acids and stimulation of protein degradation. The increase in cellular osmolytes draws in water by osmosis.

› Review articles

Hoffman EK, Lambert IH, Pedersen SF (2009). Physiology of cell volume regulation in vertebrates. Physiological Reviews 89: 193–277.

Forster RP, Goldstein L (1979). Amino acids and cell volume regulation. The Yale Journal of Biology and Medicine 52: 497–515.

Garcia-Perez A, Burg MB (1991). Renal medullary organic osmolytes. Physiological Reviews 71: 1081–1115.

Kinne RKH (1993). The role of organic osmolytes in osmoregulation from bacteria to mammals. Journal of Experimental Zoology 265: 346–355.

Pierce SK (1982). Invertebrate cell volume control mechanisms: a coordinated use of intracellular amino acids and inorganic ions as osmotic solute. The Biological Bulletin 163: 405–419.

Yancey PH (2005). Organic osmolytes as compatible, metabolic and counteracting cytoprotectants in high osmolarity and other stresses. Journal of Experimental Biology 208: 2819–2830.

Checklist of key concepts

Animal body fluids

- The bulk of the body mass of animals is water, distributed between (i) **intracellular fluid** (ICF) inside the cells and (ii) **extracellular fluid** (ECF) outside the cells.
- Extracellular and intracellular fluids differ markedly in composition:
  - sodium (Na^+) is the major cation in ECF;
  - potassium (K^+) is the major cation in ICF.
- Ionization of strong **electrolytes** forms oppositely charged ions that are transported independently during body fluid regulation.
- The total number of dissolved particles of solute in a solution determines the **osmotic concentration (osmolality)** of the solution.
- In aquatic animals, the osmotic concentration of fluids on either side of their **permeable** external epithelia influences the rate of water flow by osmosis.
- Charged macromolecules that are unable to pass via membranes influence the distribution of diffusible ions.
- Active transport processes can be revealed by comparing the predicted *Gibbs–Donnan* effect with the actual distribution of ions.

Transepithelial transport

- Body fluid regulation depends on solute and water transport across epithelial cells via:
  - **paracellular pathways** between the cells passing through **'tight' junctions** at the apical border;
  - a **transcellular route** across apical and basolateral membranes, passing through the cell cytoplasm.
- 'Tight' junctions may actually be leaky due to the properties of the *claudins* that form the junctions.
- Solutes and water move through the paracellular pathways by *passive transport*.
- Epithelial transport of ions and water via the transcellular route occurs by both passive transport and *active transport* mechanisms.
- The sodium pump, **Na^+, K^+-ATPase**, is a P-ATPase that plays a major role in osmoregulation.
- **Proton pumps** on the apical membranes of epithelia drive transport of other ions via ion channels by creating an **electrochemical potential difference**.
- **Urea transporters** in the cell membranes of some tissues, such as kidney tubules, enable the passive transport of urea.
- Many types of ion channels, such as **chloride channels**, occur in the epithelial cells of osmoregulatory and excretory tissues.
- Cell membranes become highly water permeable if they contain water channels called **aquaporins**.

Regulation of cell volume

- Cell volume regulation is required for normal cell function in animals.
- Changes in cell volume are described in relation to the **tonicity** of the fluids on either side of the plasma membrane:
  - If cell volume is stable, the external solution is **isotonic** to ICF.
  - Cells swell in **hypotonic** solutions and may burst due to volume expansion.
  - Cells shrink in **hypertonic** solutions due to the outward flow of water.
- Changes in cell volume stimulate rapid compensatory responses:
  - **Regulatory volume increase** in response to cell shrinkage involves an increase in the osmolyte content of the cells, which draws in water by osmosis.
  - **Regulatory volume decrease** in response to cell swelling reduces the cell content of osmolytes, which results in an efflux of water by osmosis.
- Cell volume regulatory responses result from changes in the cell content of inorganic osmolytes and/or organic osmolytes that are typically **compatible solutes** i.e. do not interfere with the functioning of enzymes and other cellular proteins.

Study questions

*** Answers to these numerical questions are available online. Go to www.oup.com/uk/butler.**

1. Which ions make the major contribution to plasma osmolality? What are the major ions in intracellular fluid? (Hint: Section 4.1.1; see also Figure 3.21.)

2.* Calculate the osmolarity (mOsm L^{-1}) of the following solutions at 25°C, using values for the osmotic coefficients (at 25°C) of solutions of sodium chloride (NaCl) and potassium chloride (KCl) of 0.93 and 0.92, respectively:

(a) a 160 mmol L^{-1} solution of NaCl;

(b) a solution containing NaCl at 50 mmol L^{-1} and KCl at 10 mmol L^{-1}.

(Hint: Section 4.1.1.)

3. What do we mean by a colligative property of a solution? Give three examples of colligative properties. How are colligative properties used to assess the osmotic concentrations of a solution? (Hint: Section 4.1.1.)
4. What are the main influences on paracellular transport in osmoregulatory epithelia? (Hint: Section 4.2.1.)
5. In an experiment, some liver cells were placed in a hypertonic solution of sucrose (a non-permeable osmolyte). Predict what happened to these cells over the following 4 hours and explain these events from your knowledge of regulatory volume responses. (Hint: Section 4.3.1.)
6. Frog eggs do not swell when they are in pond water. Explain what happens when mRNA for aquaporin is injected into the eggs. (Hint: Section 4.2.3.)
7. Why are Na^+, K^+, $2Cl^-$ co-transporters called secondary active transporters? How does this transport process differ from direct active transport? (Hints: Sections 4.2.2 and 4.3.1.)
8. Red blood cells are permeable to urea. What will happen when red blood cells are added to an isosmotic solution of urea? (Hint: Section 4.3.)
9. Explain how Na^+, K^+-ATPase transports sodium and potassium across the cell membrane. What effect does this process have on the electrochemical differences between intracellular and extracellular fluids? (Hint: Section 4.2.2.)
10. Explain how ion channels influence the passage of specific named ions across cell membranes. (Hint: Section 4.2.2.)
11. What do we mean by the term compatible solute? Give three examples and explain how these solutes are employed in specific examples of cell volume regulation. (Hint: Section 4.3.1.)

Bibliography

Berridge MJ (2012). Cell Signalling Biology. London: Portland Press.

Boron WF, Boulpaep EM (2012). Medical Physiology. 3rd Edition. Philadelphia, PA: Elsevier.

Bradley TJ (2009). Animal Osmoregulation. Oxford Animal Biology Series. Oxford: Oxford University Press.

Yang B, Sands JM (Editors) (2014). Urea Transporters. Subcellular Biochemistry 73. Dordrecht: Springer Science.

5

Osmotic and ionic regulation in aquatic animals

The osmotic and ionic concentrations of aquatic environments vary enormously. The oceans and seas have high concentrations of sodium chloride and other ions, such as magnesium and sulfate, while rivers and lakes are relatively devoid of ions, except for a few hypersaline lakes[1]. Such environmental differences and their variabilities over time have profound implications for the animals that live in these aquatic habitats and in brackish habitats of intermediate **salinity**[2]. In this chapter, we examine the ways that animals have adapted to the contrasting osmotic and ionic nature of marine and freshwater habitats.

Some aquatic animals are particularly tolerant of changes in environmental salinity. We investigate how they respond to such changes and discover that while some animals conform to any external changes in salinity, others regulate the composition of their body fluids. This ability allows some animals to regulate their body fluid composition during their migrations between rivers and the seas, or vice versa, which they undertake for periods of growth or reproduction.

5.1 Marine animals

The highest diversity of animal life occurs in marine habitats, probably because of the origin of life in ancient seas and the extensive period over which animals in these seas have been able to evolve[3]. Hence, all animal phyla have marine representatives. Many of these animals (but not all) are in an osmotic equilibrium with seawater so their extracellular fluids are described as **iso-osmotic** to the environment (from the Greek word for equal = *iso*); iso-osmotic is usually shortened to **isosmotic**.

[1] We examine the ionic composition of aquatic habitats, including hypersaline lakes, in Section 1.2.1.

[2] Section 1.2.1 discusses the measurement of salinity as a measure of the salt content of a solution.

[3] We outline the diversity of animals in Section 1.4.

5.1.1 High sodium chloride concentrations in extracellular fluids of isosmotic marine invertebrates

The two major ions in seawater are sodium (Na^+) and chloride (Cl^-) and these ions occur at similar concentrations in the extracellular fluids of most marine invertebrates. Table 5.1 gives some examples. Therefore, most marine invertebrates are isosmotic to seawater (or close to isosmotic), with an osmolality of around 1000 mOsm kg^{-1}.

Table 5.1 emphasizes the similar Na^+ and Cl^- concentrations in the extracellular fluids of many invertebrate marine animals to those in seawater, but shows some differences in the concentrations of other ions. Do these differences indicate active regulation of these ions? To answer this question, we first consider the effects of charged macromolecules in the extracellular fluids, such as proteins, that do not pass easily through the semi-permeable membranes forming the interface between extracellular fluids and seawater. These macromolecules influence the distribution of diffusible ions.

Proteins are generally negatively charged (anions) at the pH of body fluids. Their presence in extracellular fluid therefore results in a slightly higher concentration of positively charged ions (cations), such as Na^+, and slightly lower levels of anions, such as sulfate (SO_4^{2-}), in the extracellular fluids of marine invertebrates than occur in seawater[4]. However, the differences observed in many species of marine invertebrates are larger than the relatively small effects of impermeant anions, such as proteins, and in some cases, the difference is in the opposite direction to that predicted. For example, Table 5.1 shows the haemolymph concentrations of magnesium (Mg^{2+}) in Norway lobsters (*Nephrops norvegicus*) are less than the Mg^{2+} concentrations in seawater (not more, as the effects of impermeant anions would predict). The explanations of ionic differences are not always clear, but active transport mechanisms are implied.

[4] This unequal distribution is known as the Gibbs–Donnan equilibrium and is discussed in more detail in Section 4.1.1.

Table 5.1 Concentrations of major ions (mmol L^{-1}) and protein (mg mL^{-1}) in extracellular fluids of selected marine invertebrates compared to the composition of seawater

The tabulated values are for blood, except for sea mouse in which the values are for coelomic fluid extracted from the fluid-filled body cavity (coelom)[1].

Species	Sodium (Na^+)	Chloride (Cl^-)	Potassium (K^+)	Calcium (Ca^{2+})	Magnesium (Mg^{2+})	Sulfate (SO_4^{2-})	Protein
Atlantic seawater	468	546	10.2	10.3	53	28.2	–
Sea mouse (*Aphrodite*)	476	557	10.5	10.5	54.6	26.5	0.2
Norway lobster (*Nephrops norvegicus*)	541	552	7.8	12	9.3	19.8	33
European spider crab (*Maja squinado*)	488	554	12.4	13.6	44.1	14.5	–
Sea urchin (*Echinus* sp.)	474	557	10.1	10.6	53.5	28.7	0.3
Squid (*Loligo* sp.)	456	578	22.2	10.6	55.4	8.1	150
Blue mussel (*Mytilus edulis*)	474	553	12.0	11.9	52.6	28.9	1.6

[1] Section 4.1 discusses the fluid-filled body cavities of animals, including the coelom.

Source of data: Potts WTW, Parry G. Osmotic and Ionic Regulation in Animals. Oxford: Pergamon Press, 1964. Seawater composition from Summerhayes CP, Thorpe SA. Oceanography. An Illustrated Guide. Manson Publishing Ltd. CRC Press, 1996.

A possible link between the haemolymph concentrations of Mg^{2+} and the activity of crustaceans has been a source of speculation for many years after Mg^{2+} concentrations in the haemolymph of species of crustaceans were measured alongside their behaviour. American and Norway lobsters, which move extremely rapidly, have the lowest concentrations of Mg^{2+}, as shown by the data in Figure 5.1. By contrast, slower-moving spider crabs (*Maja squinado*) have much higher haemolymph concentrations of Mg^{2+}.

The effect of haemolymph concentrations of Mg^{2+} on the walking activity of crustaceans can be explored by keeping them in artificial seawater containing different concentrations of Mg^{2+}; at higher concentrations of Mg^{2+} in seawater the haemolymph concentrations of Mg^{2+} increase. Figure 5.2 shows the results of studies of Antarctic spider crabs (*Eurypodius latreillei*). The spider crabs are more active when Mg^{2+} concentrations of haemolymph are less than those in seawater. Such differences may reflect the narcotic actions of Mg^{2+} as addition of magnesium chloride to seawater induces a state of relaxation (a kind of narcosis) in marine invertebrates. An alternative explanation is that differences in Mg^{2+} concentrations in the haemolymph result in parallel

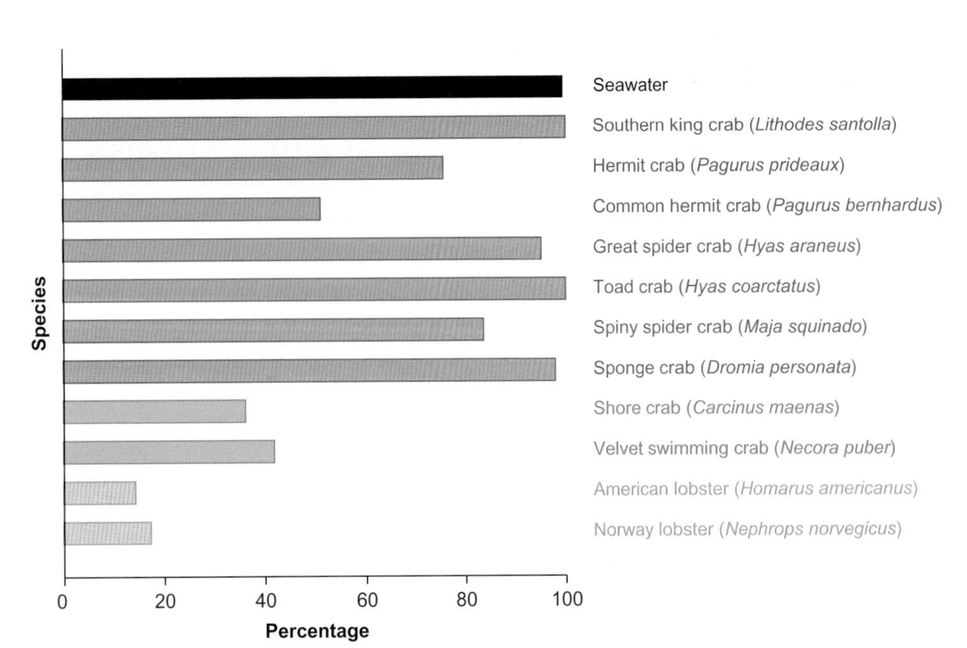

Figure 5.1 Haemolymph concentrations of magnesium (Mg^{2+}) in decapod crustaceans, as a percentage of the Mg^{2+} concentration of seawater

Species shown in blue have high relative Mg^{2+} concentrations and are least active. Species in green are more active and have intermediate Mg^{2+} concentrations. The Mg^{2+} concentrations of the haemolymph of two species of lobsters, shown in orange, are the most active of the species compared.

Source: adapted from Robertson JD (1953). Further studies on ionic regulation in marine invertebrates. Journal of Experimental Biology 30: 277-296.

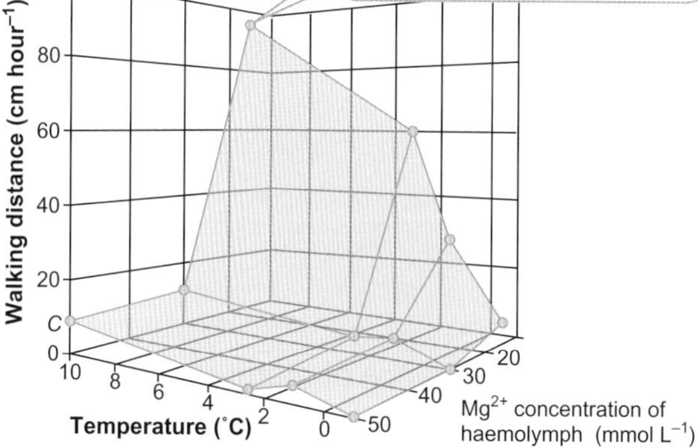

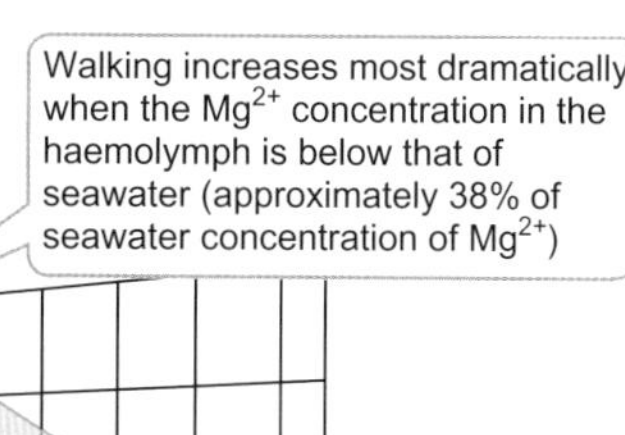

Figure 5.2 Walking activity of Antarctic spider crabs (*Eurypodius latreillei*) acclimated to seawater containing different concentrations of magnesium (Mg^{2+}) at different temperatures

Although temperature has small effects on walking activity, the effects of Mg^{2+} concentrations in the haemolymph are much more pronounced.

Source: Frederich M et al (2000). Haemolymph Mg^{2+} regulation in decapod crustaceans: physiological correlates and ecological consequences in polar areas. Journal of Experimental Biology 203: 1383-1393. Antartctic Spider crab image: ©Aquarius.

changes in cytosolic concentrations of Ca^{2+} that have direct effects on muscle performance[5].

Calcium (Ca^{2+}) concentrations in haemolymph appear to modulate the effects of Mg^{2+} in crustaceans. High haemolymph concentrations of Ca^{2+} result in excitability of crabs and could counteract the narcotic actions of Mg^{2+}. In general, a Ca^{2+}/Mg^{2+} concentration ratio of 0.4 to 2 occurs in the haemolymph of active decapods, while less active species have lower Ca^{2+} concentrations relative to Mg^{2+}, giving Ca^{2+}/Mg^{2+} concentration ratios of 0.2 to 0.3.

It is thought that Mg^{2+} reduces the ability of crabs and lobsters to increase their heart rate, which is essential for effective circulation and respiratory function during activity[6]. Hence, although decapod crustaceans are isosmotic in a marine habitat, because of high concentrations of Na^+ and Cl^-, they must regulate their haemolymph concentrations of Mg^{2+} to lower levels than those in seawater in order to stay active and hunt effectively. They do this by excreting Mg^{2+} in their urine[7].

We might ask whether a relationship between blood Mg^{2+} and activity occurs more broadly among marine invertebrates. Looking again at Table 5.1, notice that squids (*Loligo*) have a high Mg^{2+} concentration in their haemolymph; however, squids are very active, like other cephalopods. *Loligo* has a higher concentration of potassium (K^+) in its extracellular fluid than in seawater, and it has been argued that K^+ stimulates the neuromuscular system of squids. However, it appears that the main factor determining physical activity of squids is not their ionic concentrations but the structural development that allows them to propel themselves by ejecting a jet of water.

5.1.2 Hagfish are isosmotic to seawater

Hagfish are marine jawless fish (agnathans)[8], which are the only animal group aside from invertebrates with concentrations of Na^+ and Cl^- in their extracellular fluid that are similar to those in seawater. These high Na^+ and Cl^- concentrations explain why hagfish are isosmotic to seawater.

Although Na^+ and Cl^- concentrations of the plasma (extracellular fluids) of hagfish are similar to those of seawater, the concentrations of several other ions are less than those in seawater, as indicated by the data in Table 5.2. These differences suggest that hagfish actively regulate several types of ions.

The simple nephrons of the hagfish kidney[9] do not absorb water, yet the urine concentrations of several ions exceed plasma concentrations. The data in Table 5.2 indicate that K^+, Mg^{2+}, SO_4^{2-} and PO_4^{2-} are secreted into the kidney tubules[10] for excretion in the urine.

The plasma concentrations of Ca^{2+} in hagfish are less than the Ca^{2+} concentrations of seawater, but there is no evidence from the data in Table 5.2 that net secretion by the kidney occurs, as urine/plasma concentration ratios for Ca^{2+} are below 1.0. Two alternative routes for Ca^{2+} excretion occur in hagfish:

- Hagfish may regulate Ca^{2+} concentrations of the extracellular fluids by excreting Ca^{2+} in bile. The Ca^{2+} concentration in the bile of hagfish is 12× more than in their blood plasma.

[5] We discuss the effects of cytosolic Mg^{2+} on Ca^{2+} release from the sarcoplasmic reticulum and the impacts on muscle function in Sections 18.1.6 and 18.2.2.

[6] We learn more about circulation during exercise in Section 15.2.2.

[7] Section 7.3.3 examines the excretion of Mg^{2+} by the antennal glands of crustaceans.

[8] The morphological characteristics of the agnathans are outlined in Section 1.4.2.

[9] For a comparison of the simple nephron of hagfish to those of other species, see Figure 7.11.

[10] Section 7.2.4 examines some of the secretory processes in kidneys.

Table 5.2 Ionic composition of blood plasma, urine and mucus of the Pacific hagfish (*Eptatretus stoutii*)

Green font indicates blood plasma concentrations of ions that are less than those of the seawater in which the animals were held. Red font indicates urine concentrations and urine/plasma concentration ratios of ions that are secreted by the kidney tubules and excreted in urine.

Ion (mmol L^{-1})	Seawater	Blood plasma	Mucus	Urine	Urine/plasma concentration ratio
Sodium (Na^+)	496	570	95	533	0.97
Chloride (Cl^-)	543	547	–	548	1.0
Potassium (K^+)	10	7.0	207	11	1.57
Calcium (Ca^{2+})	11	4.5	21	3.6	0.8
Magnesium (Mg^{2+})	52	12	77	14.7	1.23
Sulfate (SO_4^{2-})	26	0.9	–	7.3	8.59
Phosphate (PO_4^{2-})	–	2.3	–	8.9	3.91

Sources: Munz FW, McFarland WN (1964). Regulatory function of a primitive vertebrate kidney. Comparative Biochemistry and Physiology 13: 381-400.

- Secretion of mucus will have a significant influence on Ca^{2+} and Mg^{2+} balance since hagfish produce copious volumes of mucus, which contain concentrations of Ca^{2+} and Mg^{2+} exceeding those in seawater, as shown in Table 5.2. Mucus has an intracellular origin and hence has a low Na^+ concentration, but high K^+ concentration[11].

5.1.3 Hypo-osmoregulation in marine vertebrates

Most marine vertebrates (but not elasmobranchs: sharks and rays) regulate the osmolality of their extracellular fluid to below the osmolality of seawater. These animals are described as **hypo-osmotic** to the seawater (often shortened to **hyposmotic**; 'hypo' is from the Greek *hupo*, meaning under). Marine teleosts, and reptiles, birds and mammals living in marine habitats, have plasma osmolalities of between 350 and 400 mOsm kg^{-1}, containing Na^+ and Cl^- at 150 to 170 mmol L^{-1}, which is about a third of the osmolality and Na^+ and Cl^- concentrations of ocean seawater.

Most biologists consider that the ancestors of present-day marine vertebrates that are hyposmotic to seawater evolved the lower concentrations of sodium chloride in their extracellular fluids when they entered osmotically dilute freshwater habitats, which reduced their energetic requirements for ion regulation. During the course of evolution, representatives of most vertebrate groups, some crustaceans, and a few insect larvae re-entered more saline habitats, but have retained the relatively low osmolality of the body fluids of their ancestors.

[11] Figure 3.21 and Figure 4.5 show the typical distribution of ions in cells and extracellular fluid.

Water and salt balance of marine teleosts

The maintenance of lower ionic and osmotic concentrations in body fluid (which defines **hypo-osmoregulation**) in an aquatic environment is challenging for marine teleosts. Hypo-osmoregulation has profound effects on the movement of water and ions across the semi-permeable epithelia that are in contact with their surroundings, the gills in particular.

Figure 5.3 illustrates the main challenges and the processes involved in the hypo-osmoregulation of marine teleosts:

- The higher osmolality of seawater compared to extracellular fluids results in water loss by osmosis, mainly across the gills, at a rate determined by[12]:
  - the difference in osmotic concentrations;
  - the surface area over which osmosis occurs (usually expressed in cm^2);
  - the water permeability of the barrier, i.e. the ease with which water crosses the barrier (cm^3 per cm^2 area of barrier per hour, per unit of osmotic concentration difference).
- Only low volumes of urine are excreted by marine teleosts[13] (compared to freshwater teleosts), but nevertheless this adds to the water loss that needs to be balanced by water intake.
- The passive influx of particular ions, via the gills, depends on (i) the gill permeabilities to ions, and (ii) the difference in concentrations of particular ions and the electrical charge difference across the gill membranes, which together determine the

[12] Equations 3.2 and 3.3 describe the influence of the factors determining water loss.

[13] We examine kidney function of marine teleosts in more detail in Sections 7.1.3 and 7.2.4.

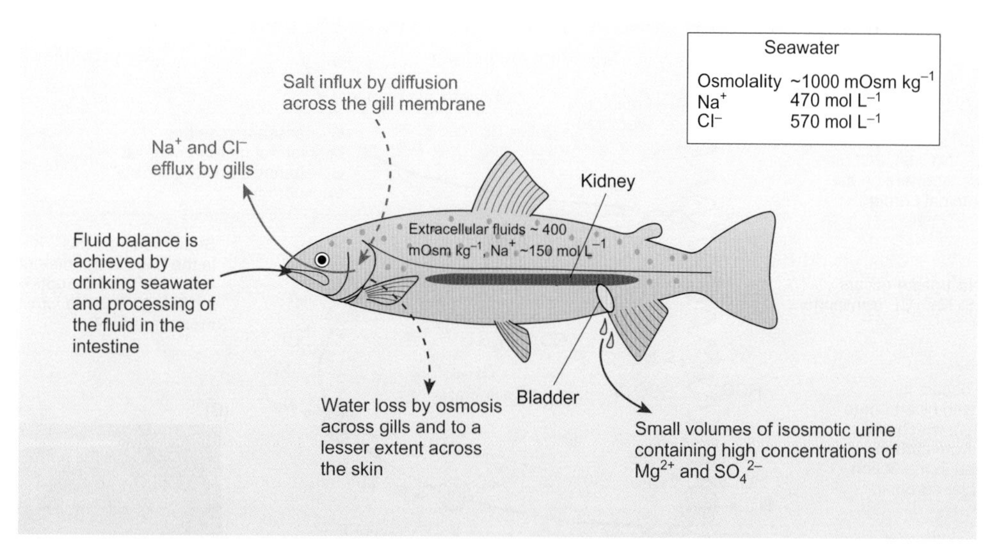

Figure 5.3 Osmoregulatory processes in marine teleosts

Processes involving water loss or replacement are shown in blue. Processes involved in loss of NaCl and uptake are shown in brown. Water losses via osmosis across the gills and in the urine are balanced by fluid intake by drinking and processing of this fluid in the intestine. Influx of salts across the gills and in the intestine are balanced by the net efflux across the gills. Passive influx via the gills is indicated by dotted lines; active efflux is indicated by a solid line.

difference in electrochemical potential across the epithelial cells[14].

- Na^+ and Cl^- and other ions such as Mg^{2+} and SO_4^{2-} are present at higher concentrations in seawater than in the extracellular fluids of marine teleosts.
- Excretion of Na^+ and Cl^- via the gills is essential in balancing the continual influx of these ions across the gills and gastrointestinal tract.
- The kidneys of marine teleosts excrete high concentrations of Mg^{2+}, SO_4^{2-} and Cl^- in the urine, which are important in compensating for the ion influxes[15].

Marine teleosts maintain water balance by drinking seawater

Marine fish achieve water balance by drinking large volumes of seawater to compensate for water losses[16]. Drinking seawater to maintain water balance may sound surprising, as the dangers for humans of consuming seawater are well known: we are unable to excrete the excess salts that we would acquire by doing so. In sharp contrast, marine teleosts have evolved special mechanisms for excreting salt across their gills, which we consider later.

Drinking by marine teleosts is stimulated by a swallowing reflex in response to their detection of high concentration of Cl^- in seawater that enters the buccal cavity during gill ventilation for gas exchange. The high concentration of Cl^- in the intestinal fluids then initiates a negative feedback mechanism, which reduces further drinking. The distension of the stomach also exerts a negative feedback effect. In addition, drinking rates are controlled by an area of the hindbrain, the area postrema, which responds to neural input from receptors that are sensitive to the volume of extracellular fluids and hormones[17].

Drinking is the first step in obtaining water, which leads to the absorption of 70–90 per cent of the ingested fluids in the intestine. Initially, partial desalination of the fluid occurs in the oesophagus, as salts move passively into the extracellular fluids, which reduces the osmotic concentration of the fluid entering the stomach.

In the stomach, water moves *into* the ingested water, by osmosis, and Na^+ and Cl^- move by passive diffusion from the fluid in the stomach into the extracellular fluids. Fluid in the stomach of unfed animals reaches an osmotic concentration of approximately 500 mOsm kg^{-1}. When fluids enter the intestine, the osmotic concentration of the intestinal fluid

[14] Section 3.1.3 explains the effects of differences in ion concentrations and electrical charge on ion passage across biological membranes. Selective permeability of biological membranes depends on the number of ion channels per unit of cross-sectional area and the proportion of open channels, as discussed in Sections 3.2.3 and 4.2.2.

[15] We discuss the formation of urine in detail in Chapter 7; Section 7.2.4 explores the tubular secretion of Mg^{2+} and SO_4^{2-}.

[16] Marine crustaceans that hypo-osmoregulate, such as brine shrimps, similarly maintain fluid balance by drinking and processing of fluid by the gut.

[17] Section 21.1.3 discusses the control of drinking and fluid absorption from the intestine of marine teleosts, in particular its hormonal regulation.

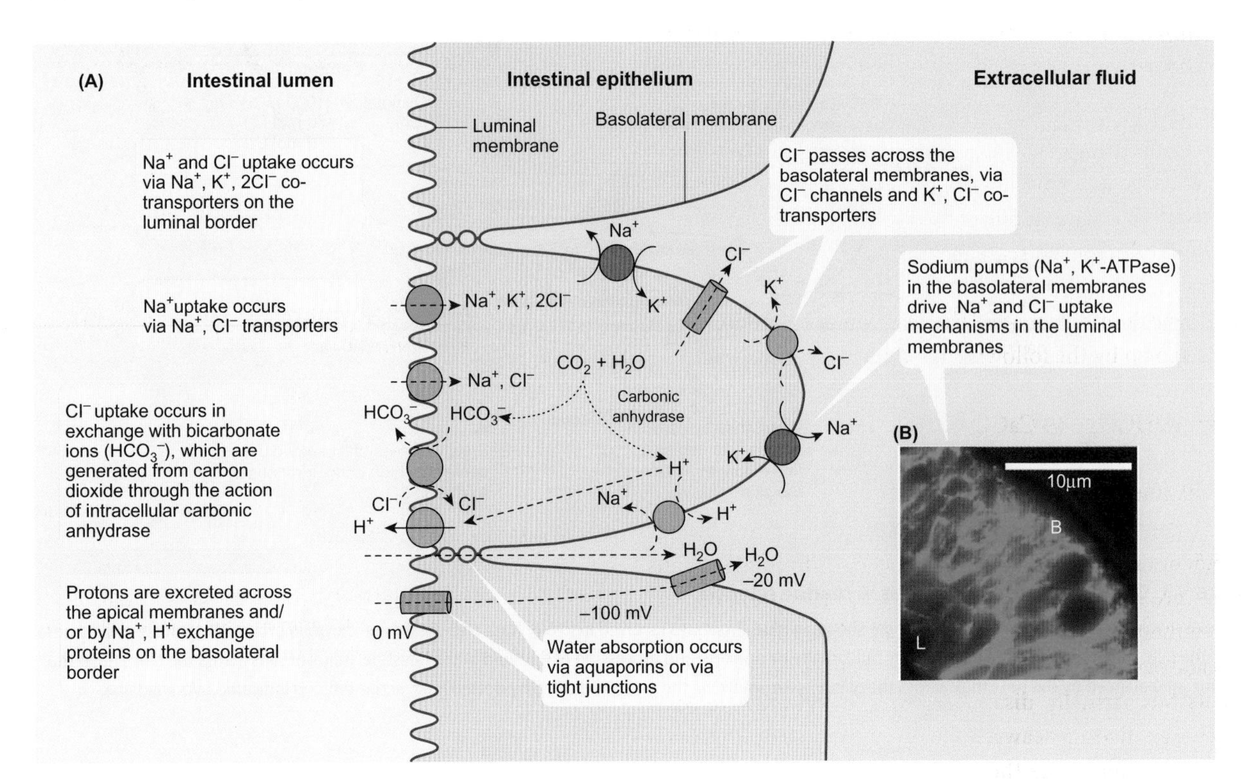

Figure 5.4 Transport processes in the marine teleost intestine

(A) Basolateral sodium pumps (Na^+, K^+-ATPase) drive Na^+, Cl^- and K^+ uptake on the luminal border, which in turn drives water absorption. Chloride uptake also occurs in exchange with bicarbonate ions generated by hydration of endogenous carbon dioxide through the action of intracellular carbonic anhydrase, or that enter the cells via a Na^+, HCO_3^- co-transporter on the basolateral membranes (not shown). Dotted lines and arrows indicate passive processes; solid lines and arrows indicate active processes. (B) High levels of Na^+, K^+-ATPase (green immunofluorescence) localized in the basolateral membranes of the intestine of the Gulf toadfish (*Opsanus beta*) using an antibody to Na^+, K^+-ATPase. The basolateral membranes penetrate the cells from the basal membrane (B), but do not reach the luminal side of the cells (L).

Sources: A: Marshall WS, Grosell M (2006). Ion transport, osmoregulation, and acid-base balance. Chapter 6. pp 179-230, In: The Physiology of Fishes. 3rd Edition (Editors: Evans DH, Clairbourne JB) CRC Press, Boca Raton, USA. B: Tresguerres M et al (2010). Modulation of NaCl absorption by [HCO_3^-] in the marine teleost intestine is mediated by soluble adenylyl cyclase. American Journal of Physiology Regulative and Integrative Comparative Physiology 299: R62–R71.

is initially reduced further, so that it becomes close to the osmolality of the extracellular fluid.

Despite the similar osmotic concentrations of fluid in the intestine and extracellular fluids, water absorption occurs because of its coupling to the absorption of NaCl[18]. A complex array of ion transporter proteins in the basolateral and apical (luminal) membranes of the intestinal epithelium result in absorption of NaCl that draws water from the luminal fluid into the extracellular fluids. The interlinked processes are still not fully understood, but the main elements are summarized in Figure 5.4, which illustrates the following features:

- Active sodium pumps (Na^+, K^+-ATPases)[19] on the basolateral membranes of the intestinal epithelial cells transport Na^+ into the intercellular spaces between the epithelial cells. These sodium pumps maintain the electrochemical differences that drive absorption of Na^+ and Cl^- from fluid in the lumen of the intestine.
- Absorption of Na^+ and some of the Cl^- across the luminal membrane of the intestine occurs via two co-transporter proteins. These are Na^+, K^+, $2Cl^-$ co-transporters and Na^+, Cl^- co-transporters. K^+ recycles via ion channels in the luminal membrane (not shown in Figure 5.4) and re-enters the intestinal fluid, which allows continued operation of Na^+, K^+, $2Cl^-$ co-transporters.
- The passage of Cl^- from the epithelial cells into the lateral spaces between the epithelial cells occurs via two routes: Cl^- channels and K^+, Cl^- co-transporters.
- These ion transport processes create a high osmotic concentration in the fluid in the lateral interspaces between the cells, which draws water from the intestinal lumen by osmosis.
- The passage of water occurs via the **paracellular pathway** or via **aquaporins**[20] (water channels) in the cell membranes. Aquaporins occur at high levels in the intestinal epithelium and rectal epithelium of marine fish.

[18] Similar processes of water absorption linked to salt reabsorption occur in the proximal tubules of the kidney, as we learn in Section 7.2.1.

[19] The sodium pump (Na^+, K^+-ATPase) carries three Na^+ out of the cell and two K^+ into the cells, as illustrated in Figure 4.13 in Section 4.2.2.

[20] Section 4.2.3 discusses aquaporins (water channels) and their importance in determining epithelial water permeability.

Secretion of bicarbonate facilitates water absorption from the intestine of marine teleosts

Secretion of bicarbonate ions (HCO_3^-) into the intestinal lumen (in exchange with Cl^-, as shown in Figure 5.4) results in an alkaline intestinal fluid with a pH of about 9 (HCO_3^- is a base)[21], containing 50–100 mmol L^{-1} of bicarbonate. In this fluid, bicarbonate reacts with calcium (Ca^{2+}) and magnesium (Mg^{2+}) ions to form a complex of insoluble calcium and magnesium carbonates. The reaction of bicarbonate and Ca^{2+} is shown by the following chemical equation:

$$Ca^{2+} + 2HCO_3^- \leftrightarrow CaCO_3 + CO_2 + H_2O \quad \text{Equation 5.1}$$

The calcium–magnesium carbonates either form white tubes coated with mucus, as shown by the example in Figure 5.5, or are excreted within the faeces. Although mucous tubes were first described in the 1960s, little attention was paid to them until it was noticed that white tubes or white deposits occur in the intestine or rectum of all marine teleosts but virtually disappear when **euryhaline** teleosts (which can live in seawater or fresh water)[22] are moved into fresh water. Since then, the notion that the precipitation of calcium and magnesium carbonates in marine fish aids water absorption by removing ions from solution has gained much support. This process is thought to facilitate up to 50–70 per cent of the water absorption from the intestine of marine teleosts.

To see the benefits, Box 5.1 (see Online Resources) compares the theoretical position if there was no carbonate precipitation to the situation in which precipitation of a complex of calcium and magnesium carbonates reduces the total concentration of dissolved ions. Precipitation of carbonates reduces the osmolality of the intestinal fluid, by 70–105 mOsm kg^{-1}, which promotes additional water absorption from the intestine (above that linked to NaCl absorption), by osmosis.

The source and routes by which HCO_3^- enters the intestinal fluids remains a topic of debate, and pathways seem to vary between species, and possibly also along the intestinal tract. Two alternatives are suggested:

- Basolateral entry of HCO_3^- into intestinal cells via Na^+, HCO_3^- co-transporters (although this has not been included in Figure 5.4), followed by excretion via the Cl^-, HCO_3^- exchangers on the luminal membrane.
- Hydration of endogenous carbon dioxide (CO_2) in the cytosol of the intestinal epithelium, catalysed by carbonic anhydrase, followed by excretion of HCO_3^- via the luminal membrane, as indicated in Figure 5.4. This mechanism requires excretion of protons (H^+) to prevent cell acidification and reversal of the reaction. Basolateral exchange of Na^+ with H^+ and apical H^+ pumps are suggested to occur, as indicated in Figure 5.4, but the exact mechanism(s) for proton excretion is uncertain and may vary between species[23].

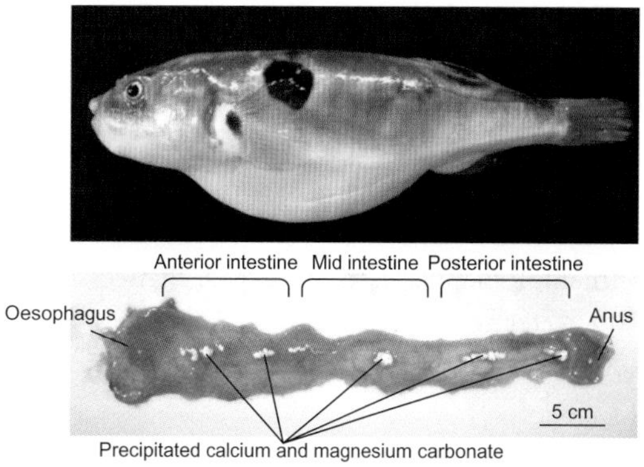

Figure 5.5 Carbonate precipitates in the intestine of obscure pufferfish (*Takifugu obscurus*) living in seawater

Precipitates of calcium and magnesium carbonates occur along the entire intestine; this species has no stomach.

Source: upper image: http://mitofish.aori.u-tokyo.ac.jp. lower image: Kurita Y et al (2008). Identification of intestinal bicarbonate transporters involved in formation of carbonate precipitates to stimulate water absorption in marine teleost fish. American Journal of Physiology Regulatory Integrative and Comparative Physiology 294: R1402-1412.

Bicarbonate secretion in the intestine of marine teleosts is also thought to be important in Ca^{2+} balance, given that the formation of calcium carbonates results in calcium excretion rather than absorption from the intestinal fluids. An idea that has emerged is that excretion of Ca^{2+} helps to prevent an excessive concentration of Ca^{2+} in the plasma, thereby helping to protect marine fish from the formation of kidney stones.

Recent work indicates that bicarbonate secretion also occurs in the intestine of jawless lampreys and cartilaginous fish, such as sturgeons and at least some sharks. These findings, in conjunction with the virtually ubiquitous excretion of carbonates by marine teleosts, and a consideration of fish biomass on a global scale, suggest that carbonate excretion by marine fish makes a significant contribution to the inorganic marine carbon cycle[24]. Conservative estimates of fish

[21] Box 13.2 explains the pH scale and acids and bases.

[22] Section 5.3.3 examines osmoregulation of euryhaline fish.

[23] Section 4.2.2 discusses proton (H^+) pumps (V-type H^+-ATPase). Excess protons entering extracellular fluids are ultimately excreted by ion transporting cells in the gills of marine teleosts.

[24] The ocean carbon cycle includes the reaction of calcium with bicarbonate to form insoluble carbonates.

biomass coupled with the rates of carbonate excretion by marine fish suggest that fish account for 3–15 per cent of global carbonate production, while estimates that are more liberal place the figure at 9–45 per cent.

The contribution of fish to the inorganic marine carbon cycle by their excretion of carbonates could be affected by environmental change. Two envisaged changes are an increase in water temperature through climate change, and ocean acidification[25] due to increasing ocean concentrations of CO_2:

- Increased water temperatures increase the metabolic rates of ectothermic fish by a factor of about 1.8 for each 10°C rise in water temperature[26]. An increase in metabolic rate increases production of CO_2, which may result in increased carbonate excretion by individual fish. However, fish biomass at a community level has been predicted to decrease with increasing temperatures, for a constant rate of primary production, which could offset the effects of individual fish on the total rate of carbonate production.
- Ocean acidification as the result of increasing ocean concentrations of CO_2 could cause elevated CO_2 levels in the blood of fish and hence stimulate their bicarbonate excretion[27]. Such changes mean that fish could become even more significant in the ocean carbon cycle.

Marine teleosts excrete excess salts via their gills

Marine teleosts face an influx of NaCl in two main locations:

- The gills, across which there is a diffusive influx of salts as indicated in Figure 5.3.
- Via the gastrointestinal tract (i) as a result of drinking to obtain water from seawater by the processes outlined in Figure 5.4A and (ii) from food items, particularly if their diet consists mainly of marine invertebrates. A kilogram of marine invertebrates contains about 143 millimoles of NaCl (based on 30 per cent of their body mass as extracellular fluids with a NaCl concentration of about 475 mmol L^{-1}). By contrast, one kilogram of teleost fish contains about 45 millimoles of NaCl (based on 30 per cent of the mass as extracellular fluid with a NaCl concentration of about 150 mmol L^{-1}).

To maintain salt balance, the excess must be excreted. While some Na^+ and Cl^- are excreted in the urine, this process cannot eliminate the excess as the maximum osmotic concentration of urine is isosmotic to extracellular fluids. Urine excreted by marine teleosts contains high concentrations of Mg^{2+} and SO_4^{2-}, which limits the Na^+ and Cl^- concentrations in the isosmotic urine. Therefore, the kidneys of teleosts cannot possibly excrete the Na^+ and Cl^- gained by diffusive influx across the gills and from the gastrointestinal tract. Instead, marine teleosts rely on their gills[28] for this purpose.

The process by which the gills of marine teleosts excrete salts was first studied in the 1930s by Ancel Keys while working with August Krogh in Copenhagen. Their work showed active Cl^- excretion by the gills of eels held in seawater. They identified specialized cells in the gills and called them chloride secreting cells, which was later shortened to **chloride cells**. We now know that these cells are responsible for the transport of several ions, not just chloride, so **ionocytes** is a better name for them.

The ionocytes represent only a small fraction of the epithelial cells in the gills (<10 per cent) with the rest of the cells being flat pavement cells. A high density of ionocytes occurs on the yolk sac skin of the larval stages of marine teleosts, where they are responsible for ion excretion. As the larvae develop, ionocytes begin to appear in the developing gills.

The ionocytes have characteristic features, which Figure 5.6 illustrates:

- The apical membrane of the ionocytes is concave at an apical pit or apical crypt where the surface epithelium of the cell is exposed to seawater. These pits are visible amongst the carpet of pavement cells.
- Beneath the apical surface, the cytoplasm contains many small tubules and vesicles that shuttle toward and fuse with the apical membrane, inserting transporter proteins into the apical membrane.
- The basal and lateral borders of the cells are invaginated and form an extensive interconnected tubular system. The tubular system is closely associated with mitochondria that fuel ion transport, so these cells are also known as **mitochondria-rich cells**.
- Accessory cells, with fewer mitochondria and a less dense tubular system than the mitochondria-rich cells, share the apical crypt, as illustrated in Figure 5.7.

So how do the mitochondria-rich ionocytes of the gills of marine teleosts operate? Figure 5.7 illustrates the key processes. A good place to start is with the Na^+, K^+-ATPase that occurs in high amounts in the basolateral membrane and the

[25] Case Study 1.1 examines ocean acidification and some of the predicted effects on marine organisms.

[26] Chapter 9 discusses the limited control of body temperature by ectotherms, although they can use behavioural thermoregulation.

[27] Section 13.3 discusses these processes, which are involved in the maintenance of acid–base balance.

[28] The gills of fish are complex multipurpose structures, as we learn in later chapters, used simultaneously for ion regulation, gas exchange (Section 12.2.2), acid–base balance (Section 13.3.2) and nitrogenous excretion (Section 7.4.1).

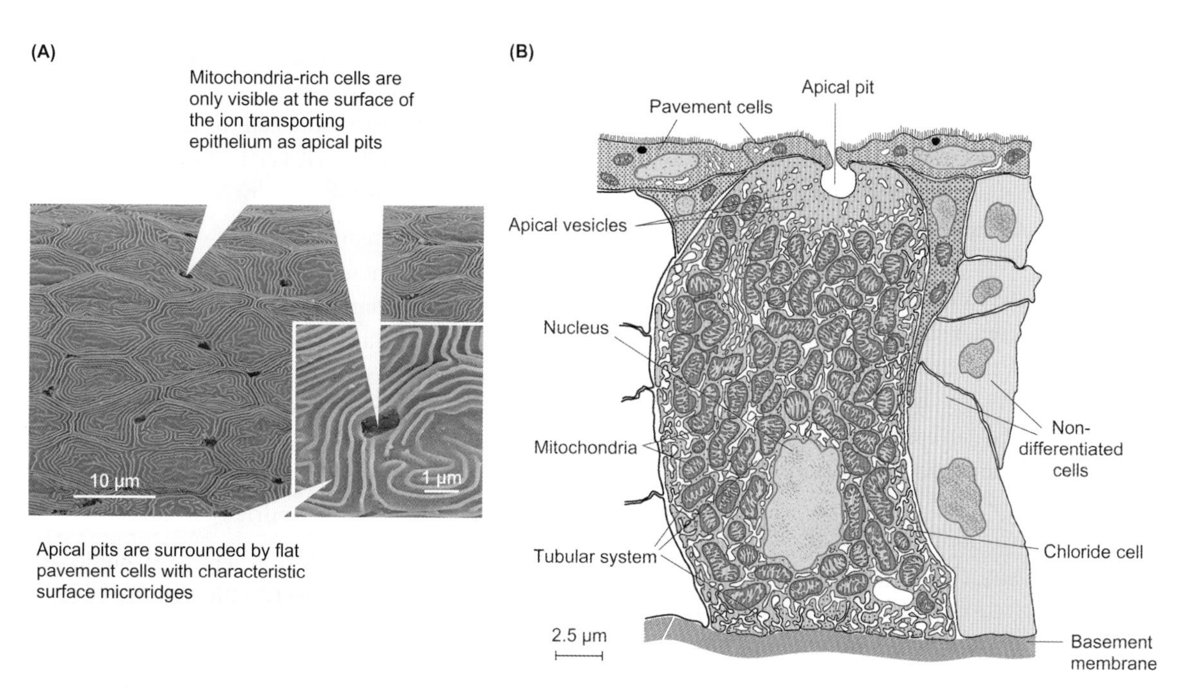

Figure 5.6 Mitochondria-rich ionocytes (or chloride cells) of a coastal species of killifish (*Fundulus heteroclitus*) living in seawater

(A) Scanning electron micrograph of the gill surface showing apical pits of mitochondria-rich cells in the gill epithelium. (B) Diagram of mitochondria-rich cell from the opercular epithelium, which is characterized by an apical pit, a dense population of mitochondria and a branching tubular system which connects with the basal and lateral plasma membranes (basolateral membranes).

Sources: A: Katoh F, Kaneko T (2003). Short-term transformation and long-term replacement of branchial chloride cells in killifish transferred from seawater to freshwater, revealed by morphofunctional observations and a newly established 'time-differential double fluorescent staining' technique. Journal of Experimental Biology 206: 4113–4123. B: adapted from Degnan KJ, Karnacky KJ, Zadunaisky JA (1977). Active chloride transport in the in vitro opercular skin of a teleost (*Fundulus heteroclitus*), a gill- like epithelium rich in chloride cells. Journal of Physiology 271:155-191.

extensive tubular system of these cells, and which essentially powers chloride secretion by the cells. The pumping of $3Na^+$ out of the cells and $2K^+$ into the cells by Na^+, K^+-ATPase using ATP (Figure 5.7, process 1) creates an electrochemical potential difference (low Na^+ concentrations and negative charge in the cells) that enables Na^+ to be carried into the cells by Na^+, K^+, $2Cl^-$ co-transporters in the basolateral membranes (Figure 5.7, process 3). These co-transporters

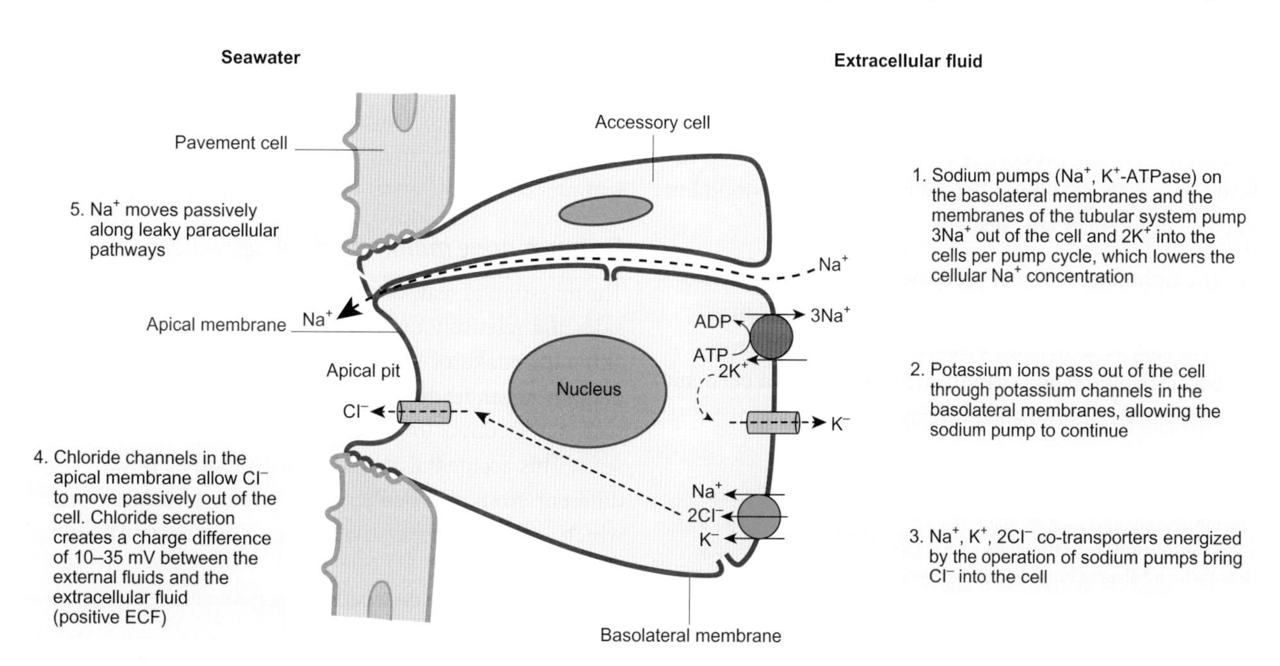

Figure 5.7 Model of mechanisms resulting in sodium chloride excretion by mitochondria-rich cells of a marine teleost

Ion transport mechanisms occur simultaneously, but the process is easier to understand if it is examined as a series of numbered steps.

Source: adapted from Marshall WS (2002). Na^+, Cl^-, Ca^{2+} and Zn^{2+} transport by fish gills: retrospective review and prospective synthesis. Journal of Experimental Zoology 293: 264-283.

occur in many chloride secretory organs, including rectal glands of elasmobranch fish and salt glands of birds and reptiles, which we examine later.

An accumulation of K^+ ions in the cells is avoided by the presence of K^+ channels in the basolateral membranes, which allows diffusion back into the extracellular fluid (Figure 5.7, process 2), and continued recycling by Na^+, K^+-ATPase.

Entry of Cl^- into the mitochondria-rich cells results in an electrochemical difference that favours Cl^- passage across the apical border of the cells, through chloride channels (Figure 5.7, process 4). The dominant chloride channel in the apical border of the gill epithelium of marine teleosts is a channel known as the cystic fibrosis transmembrane conductance regulator (CFTR)[29].

The excretion of Cl^- ions by the cells is an electrogenic process—i.e. it creates charge differences across the gill epithelium. The secretion of Cl^- creates more negatively charged conditions outside the apical membrane than in the extracellular fluids, or more positively charged conditions on the basolateral side of the cells (as seen from the opposite perspective), of +10 to +35 mV relative to the seawater.

The difference in electrical charge across the gill epithelium is usually sufficient for it to result in the passive diffusion of Na^+ across the epithelium, with Na^+ attracted by the more negative outer face of the apical side of the epithelium and repelled by the more positive side of the epithelium. The actual driving force for transepithelial movement of charged ions is the difference between the actual transepithelial charge and the epithelial potential difference at which equilibrium occurs (the equilibrium potential)[30].

Na^+ excretion occurs by diffusion along the paracellular pathways between mitochondria-rich cells and adjacent accessory cells (Figure 5.7, process 5). In contrast, Na^+ generally does not pass readily between accessory cells and the pavement cells, or between the pavement cells because of differences in the properties of the tight junctions[31] at the apical border. These differences are not fully understood but some key differences are apparent:

- The tight junctions between the mitochondria-rich cells and accessory cells of the gill epithelia of marine teleosts are relatively shallow, with fewer cross-connecting strands than the deeper junctions between accessory cells and adjacent pavement cells, or between pavement cells. In some species, such as killifish (*Fundulus heteroclitus*), cytoplasmic extensions (or 'arms') of accessory cells interdigitate with adjacent mitochondria-rich cells making the leaky routes for paracellular excretion of Na^+ more prominent.
- Particular proteins—**claudins**—function as pores within the tight junction complex. The heterogeneity in sub-types of claudins in particular gill cells is not fully understood, but two types (claudin-10d and claudin-10e) appear most important in the leaky tight junctions between mitochondria-rich and accessory cells of marine fish. These claudins are absent from cultures of pavement cells and appear to be restricted to the junctions between mitochondria-rich and accessory cells.

[29] Figure 4.17 gives details on the operation of the cystic fibrosis transmembrane conductance regulator (CFTR).

[30] The **equilibrium potential** for an ion is the electrical potential difference which balances out the ion concentration difference across the barrier, and is quantified by the **Nernst equation**, as we discuss in Section 3.1.3.

[31] The structural arrangement of tight junctions, which are composed of transmembrane proteins linked to the cell cytoskeleton, are discussed in Section 4.2.1.

Hypo-osmoregulation of marine reptiles, seabirds and mammals

Marine mammals, reptiles and seabirds have extracellular fluids with an osmolality of about 400 mOsm kg^{-1}, which makes them highly hyposmotic to seawater (like marine teleosts), but their water losses are generally low compared with marine teleosts. This difference is because:

- An evolutionary stage on land means that seabirds, and marine reptiles and mammals have relatively impermeable skins[32].
- Air breathing avoids an intimate contact of seawater with the internal respiratory surfaces of marine reptiles, seabirds and marine mammals, but small amounts of water loss occur during breathing[33], and in the urine.

The amounts of lost water may be low enough for the intake of water in food (**preformed water**) and production of water during metabolism (**metabolic water**)[34] to balance water loss. If this occurs, then there would be no need to drink seawater to achieve water balance.

Do any marine mammals drink seawater?

The ingestion of seawater by marine mammals is usually considered a relatively unimportant part of their water budget, although intake of small amounts of seawater occurs accidentally, when feeding. For example, the accidental intake of seawater accounts for up to 9 per cent of the water influx in seals. This accidental 'drinking' with food is, however, quite different from the deliberate drinking of larger volumes, which appears to be relatively rare among marine mammals.

[32] We learn more about the water resistance of the integument of terrestrial animals in Section 6.1.3.

[33] Section 6.1.2 discusses the respiratory water loss of land vertebrates.

[34] We discuss preformed and metabolic water in Sections 6.2.1 and 6.2.2, respectively.

Despite the overall picture, a few marine mammals do drink seawater in some circumstances—for instance, when extra water intake is necessary during thermal stress when venturing onto land. Adult males, juveniles and pups of Antarctic fur seals (*Arctocephalus gazella*) have been observed drinking both seawater and fresh water while free ranging on shore. The pups drink most frequently, maybe because they are left for up to 15 days between the maternal visits when fluid intake occurs as milk. The juveniles and pups seem to prefer to drink fresh water from seeps and streams, although they have been seen drinking salt water from rock pools on 25–30 per cent of their transits to and from the sea.

Feeding clearly has an important impact on the water balance of marine mammals by providing preformed water and metabolic water. In the absence of such sources of water from food and metabolism, starved animals start to drink seawater, but at relatively low rates. Fasting short-finned pilot whales (*Globicephala macrorhynchus*), and three species of dolphins, common dolphins (*Delphinus delphis*), Pacific white-sided dolphins (*Lagenorhynchus obliquidens*) and Pacific bottlenose dolphins (*Tursiops truncatus gilli*), have been observed drinking seawater at 4.5 to 13 mL kg^{-1} body mass day^{-1} during fasting.

Ultimately, seawater can only provide net pure water intake, devoid of the solutes it contains, if an animal can excrete the NaCl it acquires and which allows water to be absorbed by osmosis from the gastrointestinal tract. Marine teleosts use their gills to excrete the extra salts. Many seabirds and marine reptiles can excrete excess salts using **salt glands**[35]. Marine mammals do not have these options and can only excrete the excess salts in their urine.

One litre of seawater contains 468 mmol of Na^+ (and slightly more Cl^-), so every litre of seawater absorbed from the gut results in an intake of about 468 mmol Na^+. Hence, a net water gain from seawater can only occur in marine mammals that are capable of producing urine with a NaCl concentration of more than that of seawater. For many mammalian species, the maximum concentration of NaCl in the urine is too low for a net water gain, which is evident from the data for pilot whales, Weddell seals (*Leptonychotes weddellii*), Cape fur seals (*Arctocephalus pusillus*), California sea lions (*Zalophus californianus*) and bottlenose dolphins in Table 5.3.

In Weddell seals for example, the values in Table 5.3 indicate that the minimal urine volume to excrete the Na^+ absorbed from one litre of seawater is 1.42 L (calculated by dividing the 468 mmol Na^+ acquired from one litre of seawater by the maximum Na^+ concentration in urine, in this case 330 mmol L^{-1}). This calculation reveals that a greater volume of water is needed to excrete the Na^+ than the volume obtained by drinking seawater. In other words, a net negative water balance of 0.42 L occurs per litre of seawater consumed by Weddell seals.

[35] The functioning of salt glands is explored in Section 5.1.5.

Table 5.3 Maximum sodium and chloride concentrations and osmolality of urine of selected marine mammals compared to seawater

Species	Urinary Na^+ (mmol L^{-1})	Urinary Cl^- (mmol L^{-1})	Urine osmolality (mOsm kg^{-1})
Sea otter (*Enhydra lutris*)	505	555	2130
Harbour seal (*Phoca vitulina*)	523	508	2050
Bottlenose dolphin (*Tursiops truncatus*)	460	632	1815
California sea lion (*Zalophus californianus*)	442	608	–
Cape fur seal (*Arctocephalus pusillus*)	368	567	2364
Weddell seal (*Leptonychotes weddellii*)	330	–	1760
Short-finned pilot whale (*Globicephala macrorhynchus*)	263	–	–
Seawater	468	546	~1040

Source of data: Ortiz RM (2001). Osmoregulation in marine mammals. Journal of Experimental Biology 204: 1831-1844. For seawater composition: Summerhayes CP, Thorpe SA. Oceanography. An illustrated guide. Manson Publishing Ltd. CRC Press, 1996.

The maximum concentrations of Na^+ in the urine of two marine mammals, sea otters (*Enhydra lutris*) and harbour seals (*Phoca vitulina*), are sufficiently high to enable water devoid of Na^+ to be obtained after drinking seawater. In sea otters, the maximum concentration of Cl^- in urine, given in Table 5.3, is also sufficient to excrete the Cl^- obtained by drinking seawater. This probably explains why sea otters are one of the few marine mammals known to consume seawater on a regular basis, at up to 124 mL kg^{-1} day^{-1}. Consumption of one litre of seawater results in intake of 468 mmol Na^+ assuming its complete absorption in the gastrointestinal tract. Their excretion of urine containing Na^+ at up to 505 mmol L^{-1} means that the intake of Na^+ can be excreted in 0.927 L (927 mL) of urine (calculated by dividing the 468 mmol Na^+ acquired from one litre of seawater by the maximum Na^+ concentration in urine, in this case 505 mmol L^{-1}). Hence, from each litre (1000 mL) of seawater consumed, sea otters can obtain 73 mL (1000–927 mL) of pure water.

Drinking seawater by marine and coastal birds and reptiles

Birds and reptiles (like land mammals) normally drink fresh water to stay in water balance[36]. However, birds held in captivity will drink brackish water as an alternative to fresh

[36] We examine drinking by land animals in Section 6.2.5.

(A)

(B)

(C)

Figure 5.8 Oesophagus of the leatherback turtle (*Dermochelys coriacea*)

Leatherback turtles feed mainly on jellyfish and when available large colonial tunicates (pyrosomes) as shown in (A). Large backward pointing spiny projections line the oesophagus (B) and hold the jellyfish or pieces of tunicate in the oesophagus when water is ejected, but these spines also 'capture' plastic bags (C).

Sources: A: National Geographic Image Collection/Alamy Stock Photo; B: http://www.fisheries.noaa.gov; C: http://www.seaturtle.org/ .

water, and some species will drink seawater. We do not know, however, whether birds living in marine and coastal habitats *choose* to drink seawater to maintain their water balance. If they do, they will also take in salts. The excretion of excess salts should not be a problem, because marine and coastal birds can excrete a highly concentrated NaCl solution from specialized salt glands. Like birds, some marine reptiles have well-developed salt glands which enable them to excrete the salt intake from drinking seawater or in their food.

Perhaps the best example of a marine reptile that maintains water balance by drinking seawater is the Galápagos marine iguana (*Amblyrhynchus cristatus*). About 45 per cent of the water contributing to the water balance of marine iguanas is from seawater, with a further 50 per cent coming from the water content of food and 5 per cent from metabolism (metabolic water).

The drinking of seawater to maintain water balance is also particularly important in newly hatched leatherback turtles (*Dermochelys coriacea*), which can spend several days digging vertically up from the deep nest. During their emergence, the metabolism of internalized yolk liberates water, but the hatchlings still lose about 10 per cent of their body mass, mostly as water, before they reach the sea. By drinking seawater, they quickly re-establish water balance. However, their need to drink seawater diminishes with age, partly because of the reduced integumentary permeability that limits water losses and their consumption of jellyfish, which contain large amounts of fluid.

Adult leatherback turtles avoid ingesting seawater and eject any that makes its way into the oesophagus while eating jellyfish (which constitutes the bulk of their diet) or large colonial tunicates (pyrosomes), as shown in Figure 5.8A. Spiny projections (papillae) lining the throat and oesophagus, which are shown in Figure 5.8B, hold the jellyfish in place when the oesophagus contracts and forces water backward, out of the mouth.

Unfortunately, leatherback turtles, which are critically endangered worldwide, mistake plastic bags and other floating pieces of plastic for jellyfish. Once ingested, these bags and pieces of plastic are almost impossible to regurgitate: they too get caught on the spines lining the throat and oesophagus, as shown in Figure 5.8C. A thorough examination of historic data from post-mortem examinations of leatherback turtles indicates that between 1968, when plastic in the gastrointestinal tract was first mentioned, and 2007, 37 per cent of autopsied leatherback turtles had plastic in their gastrointestinal tracts, with plastic bags being the most common item. Clearly such items obstruct normal food processing and have implications for turtles' health. In this analysis, the amount and location of the plastic was considered sufficient to have blocked the gut of 8.7 per cent of the turtles examined, making it a probable cause of their deaths.

Sea snakes avoid drinking seawater, even if they become dehydrated. Amphibious sea snakes (sea kraits, *Laticauda* spp.) forage in marine environments, but digest and reproduce on land, and need fresh water to rehydrate after periods in seawater. In the absence of such water, they dehydrate—and their

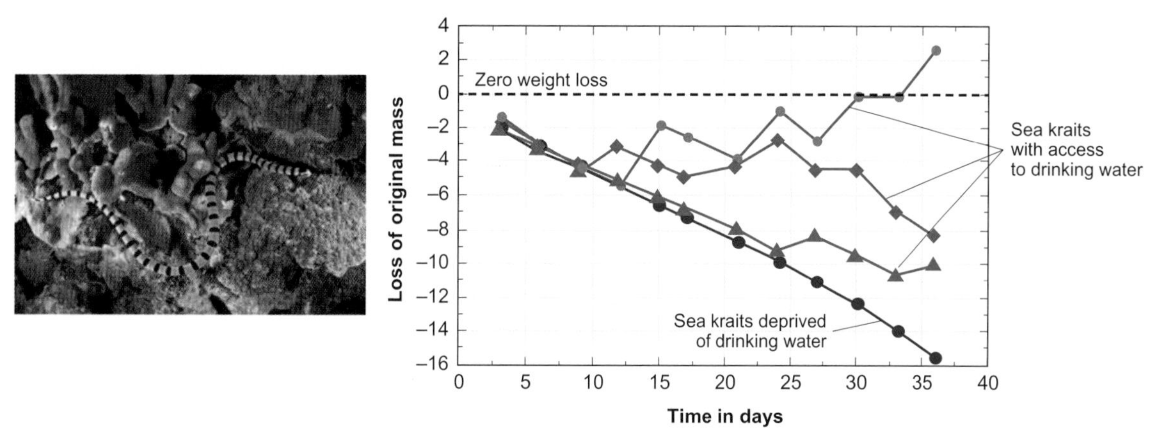

Figure 5.9 Effects of drinking on body mass of black-banded sea kraits (*Laticauda semifasciata*)

When deprived of access to fresh water, the body mass of the sea kraits declines. Three of the group of seven sea kraites that were given access to fresh water from day 9, for 1 hour every 3 days, lost less body mass than the sea kraites deprived of drinking water, or gained water.

Source: Lillywhite HB et al (2008). Sea snakes (*Laticauda* spp.) require fresh drinking water: implication for the distribution and persistence of populations. Physiological and Biochemical Zoology 81: 785–796. Sea krait image: Colin Marshall/www.flpa-images.co.uk

body mass progressively declines—as shown by Figure 5.9, due to water loss primarily across the skin. Sea kraits feed mainly on teleost fish (which are hyposmotic to seawater) so they also obtain water in their diet and during metabolism. However, their need for low-salinity drinking water possibly explains the predominance of sea kraits in coastal areas with sources of fresh water nearby, and the positive correlation between annual precipitation in the Indo-West Pacific tropical region and the diversity of sea snakes. This correlation led to a suggestion that, after rainfall, sea snakes drink from the freshwater lens that occurs before rainwater mixes with the more saline deeper water.

5.1.4 Organic osmolytes are used as osmotic ballast in some marine vertebrates

A few types of marine vertebrates are unusual among vertebrates in being isosmotic or slightly **hyperosmotic** to seawater mainly because their body fluids contain high concentrations of organic solutes, primarily urea and trimethylamine oxide (TMAO). Several disparate groups of animals demonstrate this so-called **ureosmotic** strategy of osmoregulation:

- Coelacanths (*Latimeria*), of which only two living species are known to exist[37].
- The crab-eating frog (*Fejervarya cancrivora*), which is the only known amphibian species that can live permanently in brackish waters in the mangrove forests of Southeast Asia, and forages in seawater. Crab-eating frogs can also live in fresh water. When in saline water, crab-eating frogs rapidly synthesize sufficient urea to cause slight hyperosmoregulation.
- Cartilaginous fish (Chondrichthyes): the elasmobranchs (sharks, skates and rays) and holocephalans (chimaeras)[38].

Isosmotic and slightly hyperosmotic vertebrates do not have particularly high NaCl concentrations in their body fluids, unlike the marine invertebrates (examples given back in Table 5.1) or hagfish (data given back in Table 5.2). Note in Table 5.4 that the extracellular fluids of the isosmotic and slightly hyperosmotic vertebrates have a lower Na^+ concentration than in seawater (Cl^- concentrations are similarly low), i.e. they are hypo-ionic just like marine teleost fish.

When urea retention results in slight hyperosmoregulation, as in elasmobranch fish, some water inflow occurs by osmosis. Until recently, this water influx was thought to be sufficient to balance the low volume of water excreted in urine and the saline solution secreted by their **rectal gland**. Consequently, it was thought that elasmobranchs do not need to drink seawater. However, experiments have shown that drinking of seawater supplements the osmotic water influx as and when necessary, and drinking is therefore included in Figure 5.10.

Urea as osmotic ballast: problems and solutions

To maintain high concentrations of urea in the body fluids requires urea synthesis. Cartilaginous fish produce urea by the **ornithine–urea cycle**[39] in the liver and to some extent in

[37] The phylogenetic tree in Figure 1.21 illustrates the evolutionary steps that gave rise to coelacanths that were abundant in the Devonian period; we discuss coelacanths further in Section 1.4.2.

[38] Section 1.4.2 gives further information on the diversity of cartilaginous fish.

[39] Figure 7.43 outlines the synthesis of urea by the ornithine–urea cycle. For most vertebrates, urea is a nitrogenous waste substance and is excreted.

Table 5.4 Osmolality and concentrations of major osmolytes in blood plasma of coelacanths, sharks and crab-eating frogs

Blood plasma osmolality is raised to approximately isosmotic (coelacanth) or slightly hyperosmotic with the external environment by retention of urea and trimethylamine oxide (TMAO), together with some elevation of plasma concentrations of sodium chloride compared to those present in the blood plasma of other vertebrates. The osmolality of the surrounding seawater in each case is also listed; the crab-eating frogs were living in seawater of 800 mOsm kg^{-1} (around 80 per cent of ocean seawater).

Species	Solute concentration (mmol L^{-1})			Osmolality (mOsm kg^{-1})	Osmolality of seawater (mOsm kg^{-1})
	Na^+	Urea	TMAO		
Coelacanth (*Latimeria chalumnae*)	197	377	122	932	1035
Spiny dogfish (*Squalus acanthias*)	286	351	71	1018	930
Bull shark (*Carcharhinus leucas*)	289	370	NM	1067	1047
Crab-eating frog (*Fejervarya cancrivora*)	252	350	NM	830	800

Osmotic concentrations are tabulated as osmolality (mOsm kg^{-1}) but were reported as osmolarity (mOsm L^{-1}). NM: not measured

Sources: Coelacanths: Griffith RW et al (1974). Serum composition of the coelacanth *Latimera chalumnae*. Journal of Experimental Zoology 187: 87-102. Spiny dogfish: Burger JW, Hess WN (1960). Function of the rectal gland in the spiny dogfish. Science 131: 670-671. Bull sharks: Pillans RD, Franklin CE (2004). Plasma osmolyte concentrations and rectal gland mass of bull sharks *Carcharhinus leucas*, captured along a salinity gradient. Comparative Biochemistry and Physiology Part A 138: 363-371. Crab-eating frogs: Gordon MS, Schmidt-Nielsen K, Kelly HM (1961). Osmotic regulation in the crab-eating frog (*Rana cancrivora*). Journal of Experimental Biology 38: 659-678.

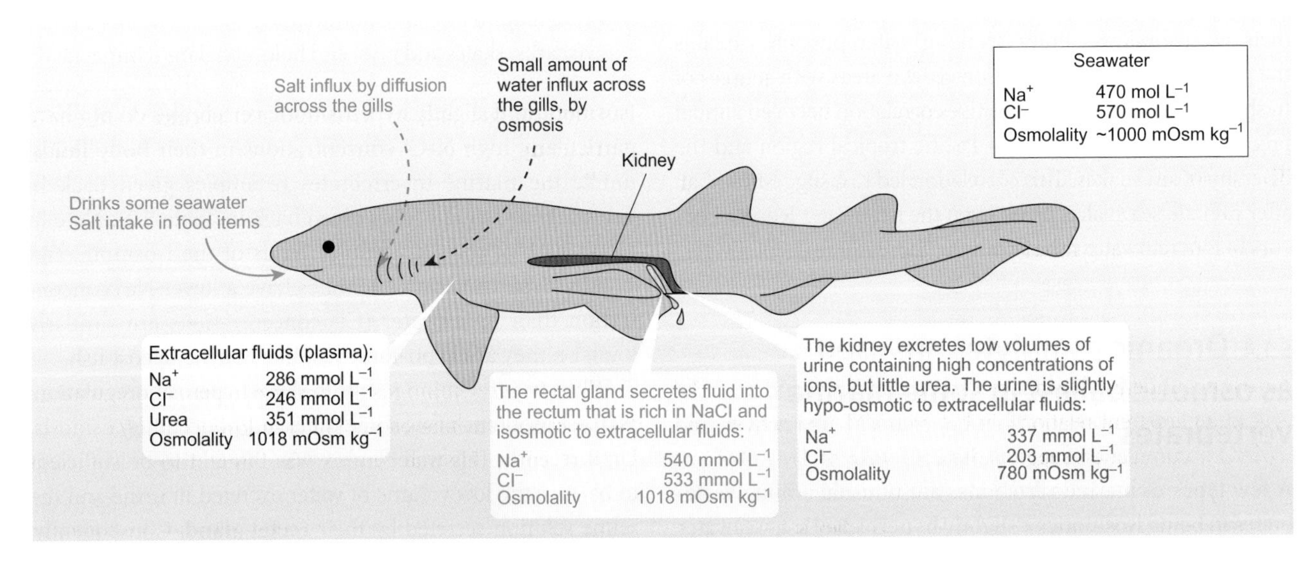

Figure 5.10 Osmoregulation of marine elasmobranchs

A small amount of hyperosmoregulation is achieved by retention of urea and trimethylamine oxide (TMAO) and relatively high concentrations of sodium chloride compared to marine teleost fish. The slight hyperosmoregulation results in some osmotic water influx. Further water intake occurs by drinking seawater, and its processing by the gastrointestinal tract. Water is lost in urine produced by the kidney and in rectal fluids secreted by the rectal gland. The ions gained across the gills, in food and by drinking are excreted by the rectal gland and kidney.

Source of data: Burger JW, Hess WN (1960). Function of the rectal gland in the spiny dogfish. Science 131: 670-671 for data for extracellular fluid, rectal gland fluids and urine for spiny dogfish (*Squalus acanthias*), which are typical of marine elasmobranchs. Passive processes are indicated by dotted lines; processes involving use of energy are indicated by solid lines.

other tissues such as muscle. Synthesis of urea is energetically costly, but costs seem to be balanced by the savings in energy that would otherwise be required for hypo-osmoregulation[40].

The high concentrations of urea in extracellular fluids are associated with similarly high concentrations of urea in the intracellular fluids, and hence high osmotic concentrations, which are important for cell volume regulation[41]. The high concentrations of urea in intracellular fluids have potential

[40] Energy is required for hypo-osmoregulation, for example in ion transport processes in the gills and gastrointestinal tract, which we discuss in Section 5.1.3.

[41] Cell volume regulation is discussed in Section 4.3.

implications for the proper functioning of polypeptide-based cellular enzymes: most polypeptides denature in solutions containing high concentrations of urea. Indeed, urea at a concentration of 8 mol L^{-1} is commonly used in biochemistry to disrupt non-covalent bonds. The concentration of urea in animals using urea as an osmotic ballast is much less than this, but we might still expect the tertiary structure of proteins and the subunit association of enzymes to be affected at the urea concentrations of 0.3 to 0.4 moles L^{-1} (300 to 400 mmol L^{-1}).

Two solutions to this apparent problem occur:

- A few enzymes are functionally modified. One of the best examples of a functionally modified enzyme is **lactate dehydrogenase** (LDH)[42]. LDH comprises four subunits; four potential combinations of the A and B subunits occur in different forms of LDH (LDH isozymes) that differ in their sensitivity to urea. Elasmobranchs' tissues typically express LDH isozymes that are most resistant to urea.
- Stabilizing effects of other organic compounds counteract the destabilizing effects of urea. These substances are called **counteracting solutes**. For the vast majority of elasmobranch proteins, the stabilizing effects of counteracting solutes, principally methylamines, counteract the effects of urea. Trimethylamine oxide (TMAO) is the most common stabilizing methylamine in urea-retaining animals.

The question of whether elasmobranchs obtain TMAO in the diet or synthesize it from trimethylamine has intrigued physiologists. The synthesis of TMAO requires the enzyme trimethylamine oxidase (TMAoxi), which only occurs sporadically among elasmobranchs. For example, high TMAoxi activity occurs in the liver of nurse sharks (*Ginglymostoma cirratum*) but is undetectable in several species of skates and rays. The lack of an apparent relationship between TMAoxi activity and TMAO accumulation suggests dietary intake of TMAO is important for many species. What happens if the intake of TMAO in food items declines in these species? Studies suggest those marine elasmobranchs that lack TMAoxi can compensate for a dietary shortfall of TMAO by synthesizing alternative methylamines that stabilize enzymes, such as betaine and sarcosine.

The evolution of different forms of enzymes, and protein stabilization by counteracting solutes, such as TMAO, means that the tissues and organs of elasmobranchs, such as the heart, are adapted to function properly only in the presence of high concentrations of urea.

Cartilaginous fish limit their urea losses

Cartilaginous fish to a large extent retain urea, which limits the need for continual high levels of urea synthesis. Retention of urea occurs in two ways (although urea loss is not zero):

- restriction of urea loss via the gills;
- reduced excretion of urea in the urine by reabsorbing filtered urea in the kidney tubules.

The low permeability of the gill epithelium to urea limits the loss of urea from the extracellular fluids into seawater. If elasmobranch gills were as permeable to urea as those of teleost fish they would lose urea at about 10 mmol kg^{-1} h^{-1}, which would require 70 per cent of the body's urea to be replaced every day. In fact, the gills of marine elasmobranchs are about 80 times less permeable to urea than the gills of marine teleosts.

Restriction of urea passage across the basolateral membranes and into the cells limits the passive efflux of urea across the apical border of the branchial epithelium. The low permeability to urea of the basolateral membranes of the gill epithelial cells results from increased packing of **phospholipids** and a high molar ratio of the lipid cholesterol relative to phospholipids in these membranes (3.7 to 1 in shark gills compared to the 1 to 1 ratio in other aquatic animals)[43].

There is also evidence that additional mechanisms transport urea back into the extracellular fluids from the epithelial cells, across the basolateral membranes. Despite these restrictions, the bulk of urea loss still occurs via the gills and accounts for the nitrogenous excretion of cartilaginous fish[44]. Excretion of urea in the urine accounts for only 4–20 per cent of the total urea loss of cartilaginous fish.

Studies by Homer Smith, in the 1930s provided the first evidence that elasmobranchs do not excrete large amounts of urea in their urine (despite inevitably filtering urea into their kidney tubules as urea is a relatively small molecule)[45]. Spiny dogfish (*Squalus acanthias*), for example, reabsorb more than 90 per cent of the filtered urea in their kidney tubules.

Ever since Homer Smith's findings, there has been enormous interest in discovering how the kidneys of cartilaginous fish reabsorb urea. Given the understanding of urea transporters[46] and their influence on urea permeability of mammalian kidney tubules[47], significant effort has been invested in identifying urea transporters in the kidneys of cartilaginous fish. Urea transporters were first identified in tissues of the spiny dogfish, in which highest levels occur in

[42] LDH is a major enzyme in anaerobic glycolysis as illustrated in Figure 2.19.

[43] Section 3.2.3 explains the structure of biological membranes and their characteristic phospholipid bilayers. Box 2.1 discusses the types of lipids in organisms.

[44] We discuss urea excretion including that of cartilaginous fish in Section 7.4.2.

[45] Urea has a molecular mass of 60 Da, whose structure is shown in Figure 7.43, so it is readily filtered by the glomeruli into the kidney tubules, as we discuss for small molecules in general in Section 7.1.1.

[46] Section 4.2.2 discusses urea transporter proteins.

[47] We discuss urea handling by mammalian kidneys in Section 7.2.3.

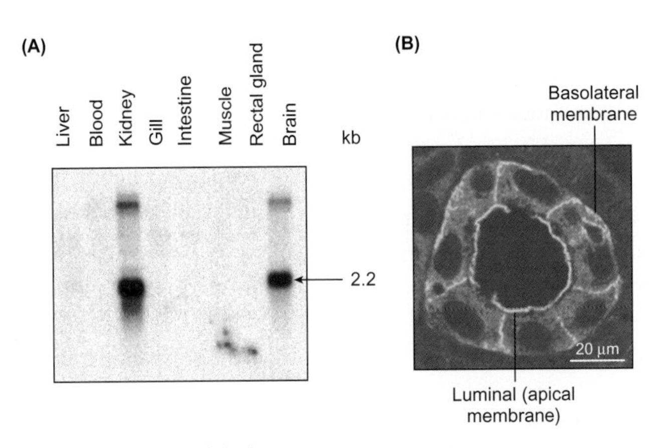

Figure 5.11 Shark urea transporter in elasmobranch kidneys

(A) Northern blot of mRNA in tissues of spiny dogfish (*Squalus acanthias*) probed with radiolabelled (^{32}P) full-length cDNA for the shark urea transporter. Dark bands of 2.2 kb identify urea transporters in kidney and brain tissue. (B) Localization of urea transporters in the collecting tubule of the kidney of the banded houndshark (*Triakis scyllium*), by immunohistochemistry using an antibody to urea transporter protein visualized with a fluorescent second antibody. The signal is most intense in the luminal (apical) plasma membranes, but there is also some expression in the basolateral membranes.

Sources: A: Smith CP, Wright PA (1999). Molecular characterization of an elasmobranch urea transporter. American Journal of Physiology 276 (Regulatory Integrative and Comparative Physiology 45): R622-626. B: Hyodo S et al (2014). Morphological and functional characteristics of the kidney of cartilaginous fishes: with special reference to urea reabsorption. American Journal of Physiology Regulatory and Integrative Comparative Physiology 307: R1381-1395.

the kidney, as shown by Figure 5.11A. Subsequently, urea transporters showing a high level of homology to the urea transporter UT-A2 of mammalian kidney tubules have been identified in the kidneys of several elasmobranch species, including Atlantic stingrays (*Dasyatis sabina*) and little skates (*Leucoraja erinacea*).

The complex renal tubules of marine cartilaginous fish (elasmobranchs and chimaeras) are several centimetres in length and run back and forth as countercurrent loops between a sinus zone and the densely packed bundle zone, where five tubules from the four countercurrent loops[48] run alongside each other. This anatomical arrangement has long been thought to be essential for urea reabsorption by the kidney, but the precise mechanism is uncertain.

Localization of urea transporters to particular nephron segments is necessary to work out how cartilaginous fish reabsorb urea. Recent investigations of banded houndsharks (*Triakis scyllium*) and bull sharks (*Carcharhinus leucas*) found urea transporters only in the collecting tubule. Figure 5.11B shows that expression of urea transporters is highest in the luminal membranes. These urea transporters could allow passive transport of urea into the cells, as the first step in urea reabsorption.

Urea transporters also occur in the basolateral membranes of the collecting ducts, although at lower levels of expression than the luminal membranes, as shown in Figure 5.11B. However, we might expect the high concentrations of urea in extracellular fluids to oppose the passive transport of urea across the basolateral membranes. The exact transport processes leading to urea reabsorption are uncertain, but are thought to involve active transport mechanisms for sodium and chloride, and the passage of water via water channels (aquaporins), such that low concentrations of urea occur around the collecting duct, which facilitates the passive reabsorption of urea.

[48] Figure 7.11 gives an impression of the complexity of nephrons in the kidneys of cartilaginous fish.

5.1.5 Salt glands and rectal glands secrete salts for excretion by hypo-ionic marine vertebrates

Hypo-ionic regulation by cartilaginous marine fish, reptiles and birds, living in marine and coastal habitats, requires excretion of the excess NaCl they acquire in their diet, or by drinking seawater, and, for cartilaginous fish, by passive influx across the gill epithelium. The excess salts are excreted using salt-secreting glands.

Cartilaginous fish and coelacanths have a salt-secreting rectal gland consisting of many blind-ended interconnected tubules, each lined by a layer of secretory epithelial cells. Fluid secreted into the lumen of the tubules runs toward the centre of the gland where the tubules connect in a central duct. The central duct carries the secreted fluid into the posterior end of the intestine, close to the rectum.

There are striking similarities between the secretory functioning of the rectal glands of cartilaginous fish and coelacanths and the salt glands of marine reptiles (marine turtles, estuarine crocodiles, sea snakes and marine lizards/iguanas) and coastal birds. Figure 5.12 shows the different locations of salt glands in present-day and reptiles and birds.

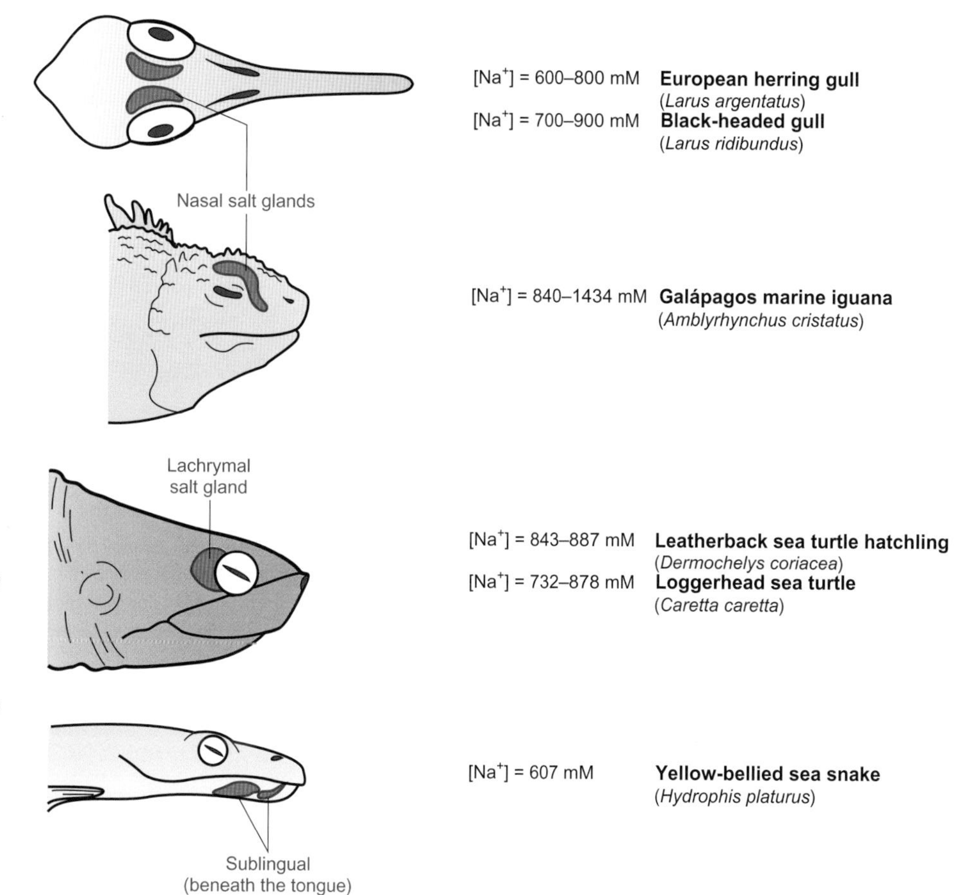

Figure 5.12 Salt glands in some species of bird and reptile

Drawings show location of salt glands in selected avian and reptilian species. The data to the right of each drawing show the sodium concentration $[Na^+]$ of fluid (mM; mmol L^{-1}) collected from the salt glands after salt loading of each species.

Data obtained from: Dunson WA, 1969; Reina RD, Jones TT, Spotila JR, 2002; Schmidt-Nielsen K, 1960; Schmidt-Nielsen K and Fange R, 1958.

Salt glands are thought to have evolved in the late Paleozoic (about 300 million years ago) from a common ancestor. Direct fossil evidence for soft tissues is often elusive, but the salt glands of reptiles and birds often affect the size and shape of the surrounding skull, which gives permanent evidence of salt glands in fossilized skulls. For example, fossils of Jurassic marine crocodiles (*Geosaurus* spp.) dating to about 140 million years ago have allowed reconstruction of the salt glands that lay beneath the skull. Also, an exceptionally well-preserved specimen of the oldest known sea turtle (*Santanachelys gaffneyi*), from the Early Cretaceous (about 110 million years ago), found in eastern Brazil, shows that this turtle had large orbital spaces for huge salt glands that would have surrounded its eyes. These salt glands were modified lachrymal (tear) or post-orbital salt glands. Present-day sea turtles and terrapins secrete salty tears from the lachrymal salt glands, as indicated in Figure 5.12.

Salt glands can be very large, as represented by the lachrymal salt glands of leatherback turtles in Figure 5.13A. However, lachrymal salt glands are not the only type of salt gland to have evolved in reptiles. Marine iguanas have nasal salt glands from which they sneeze or spray out salts. Contrary to the popular myth, crocodiles do produce tears, but 'crocodile tears' are to lubricate their eyes and have little to do with salt regulation. Instead, salt-tolerant estuarine crocodiles have salt glands opening onto the surface of the tongue. Figure 5.13B shows the densely packed tubules in lobules of one of the 30–40 lingual salt glands of estuarine crocodiles (*Crocodylus porosus*) that live in Northern Australia, eastern India and parts of South East Asia.

Sea snakes, like crocodiles, secrete salt into the oral cavity. Some species use premaxillary salt glands in the anterior roof of the mouth, while others have sublingual salt glands (beneath the tongue) that empty into a canal in the tongue sheath. The sea snake expels the salty secretion into the seawater when it puts its tongue out.

For reptiles, the salt glands are the only means of excreting excess salt because reptiles cannot excrete urine that is hyperosmotic to their extracellular fluids (due to their lack of countercurrent loops in the kidney nephrons). By contrast, birds excrete urine that is, on average, about twice the osmotic concentration of plasma. However, their urine enters the **cloaca** and hindgut (colon), where reabsorption of salts and water occurs before excretion, so the urine is not an effective means of excreting excess salts[49]. All birds have at least rudi

[49] We learn more about the production of hyperosmotic urine by birds in Section 7.2.3; Figure 6.33 in Section 6.1.4 illustrates the processing of water and ions in the colon and cloaca of birds.

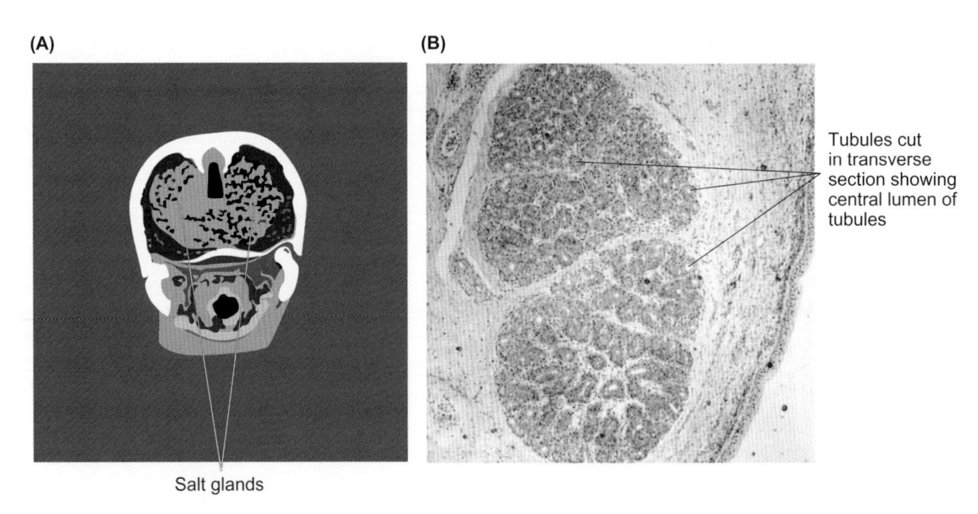

Figure 5.13 Salt glands of reptiles investigated by histology and computer tomography

(A) Drawing of a transverse section through the head of a leatherback turtle (*Dermochelys coriacea*) showing the enormous lachrymal salt glands. The red area shows a layer of blubber around the glands, which insulates them and maintains high rates of ion transport by the salt-transporting mechanisms. The leatherback turtle has an elevated core temperature of ~25°C relative to the cold waters in which it forages, due to its **gigantothermia** (we discuss gigantothermy in Section 10.3.2). (B) Histological section of one of the lingual salt glands of the estuarine crocodile (*Crocodylus porosus*). The gland is one of 30–40 salt glands on the tongue, each of which are densely packed with blind-ending tubules in well-vascularized lobules.

Sources: A: drawing based on image in Davenport J et al (2009). Fat head: an analysis of head and neck insulation in the leatherback turtle (*Dermochelys coriacea*). Journal of Experimental Biology 212: 2753-2759. B ©Craig Franklin, University of Queensland

mentary nasal salt glands located in shallow depressions of their skull, but the size and activity of these glands depends on their intake of sodium chloride. When active, the nasal glands secrete a salty fluid, which flows into two main ducts that run through the beak and empty at the nostrils.

Composition of salt and rectal gland secretions

To be able to excrete an excess of salts from the body fluids, animals must excrete fluid that contains higher concentrations of Na^+ and Cl^- than those of their extracellular fluids. Earlier, Figure 5.10 gave some data for the composition of rectal gland fluid. Notice that the fluid is isosmotic to plasma (extracellular fluid) but consists entirely of high concentrations of Na^+ and Cl^-, (at about twice those in plasma). Consequently, rectal gland secretions are an effective means of excreting excess NaCl.

Salt gland secretions of birds and reptiles are hyperosmotic to their extracellular fluids, but, as yet, we do not understand how these salt glands generate hyperosmotic salt solutions. Figure 5.12 gives data for the Na^+ concentrations of fluid secreted by the salt glands of some birds and reptiles. At 500–1430 mmol L^{-1}, the Na^+ and Cl^- concentrations of the secreted fluids exceeds the 300–400 mmol L^{-1} found in extracellular fluids.

The salt gland secretions produced by most species contain slightly lower concentrations of Cl^- than Na^+ and only a small amount of potassium (K^+; about 25 mmol L^{-1}). However, Galápagos marine iguanas (*Amblyrhynchus cristatus*) are a notable exception: their nasal glands excrete fluid containing K^+ at up to 235 mmol L^{-1}. This is an important means of getting rid of the excess K^+ they acquire by eating marine algae that contain high levels of K^+, and which the salt glands and kidneys[50] excrete.

Mechanism of salt secretion by salt and rectal glands

All salt glands and the rectal gland appear to work by similar mechanisms, which show a striking similarity to the mechanisms for Na^+ and Cl^- excretion by the mitochondria-rich cells of marine teleosts shown in Figure 5.7. However, there is one crucial difference: salt efflux across the gills of marine teleosts enters seawater directly, while salt glands secrete salty fluid into tubules in which it is either conveyed to the exterior or is sneezed out.

Figure 5.14 summarizes the processes in salt and rectal gland secretion. The key points are:

- In all cases, the epithelial cells of the secretory tubules are rich in mitochondria and contain large amounts of Na^+, K^+-ATPase located in membrane infoldings. The salt glands of birds have basolateral infoldings, but lateral infoldings dominate the reptilian salt glands. However, any functional importance of these differences is not understood.

[50] Section 7.2.4 discusses K^+ secretion by vertebrate kidneys that results in excretion in the urine.

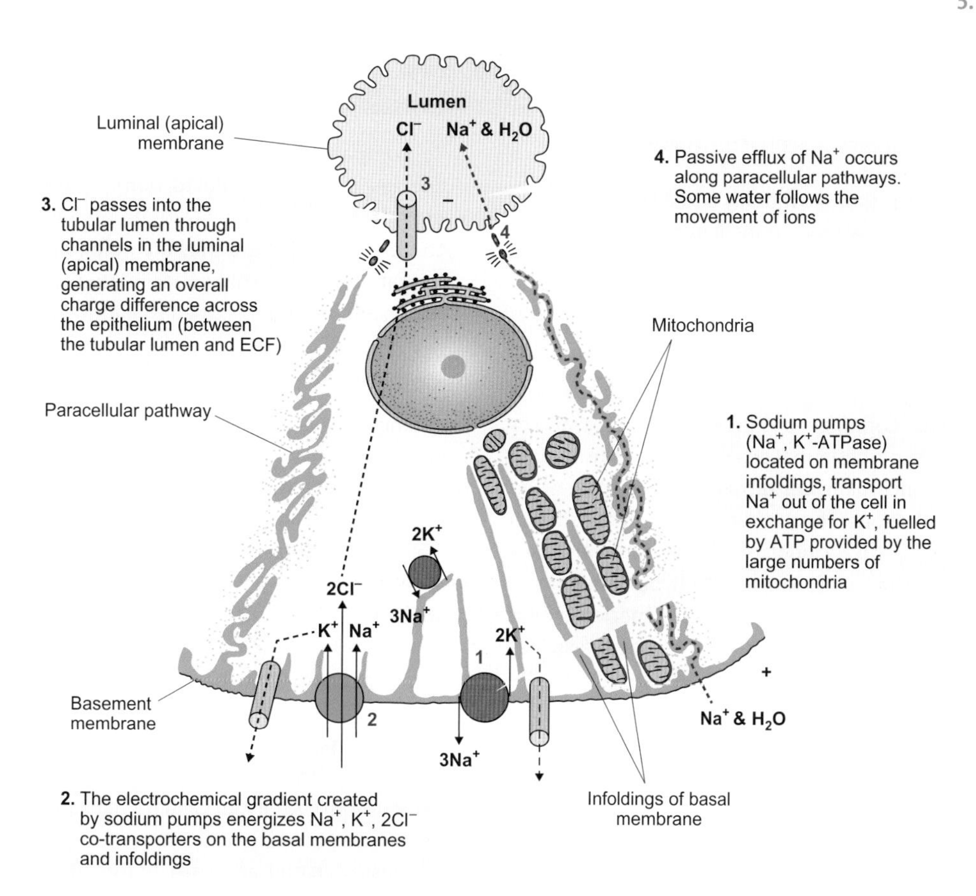

Figure 5.14 Secretory mechanisms in salt and rectal glands

- Na^+, K^+-ATPase plays a critical role in driving salt secretion by affecting electrical charge across the basolateral membrane and driving Na^+, K^+ $2Cl^-$ co-transporters on the basolateral membranes, which bring Cl^- into the cells.
- Ultimately, the passage of Cl^- through channels in the luminal (apical) membrane is the rate-limiting step in determining the rate of fluid secretion.
- In contrast to the transcellular passage of Cl^- ions, Na^+ secretion occurs via the paracellular pathways between the epithelial cells of the tubules. Water also passes into the secreted fluid via the paracellular pathways.

The activity of Na^+, K^+-ATPase makes salt secretion by the salt glands energetically costly, but secretion is intermittent and matches the needs arising from salt intake and the resultant rise in plasma concentrations of sodium chloride. The tighter the regulation of plasma concentrations of Na^+ and Cl^-, the greater the energetic demand will be in driving salt secretion. In Experimental Panel 5.1 we explore some studies of sea snakes that despite having salt glands appear to save energy by tolerating higher extracellular fluid concentrations of sodium chloride.

Environmental factors influence salt and rectal gland function

In marine elasmobranchs, influx of Na^+ and Cl^- occurs continuously across the gills and stimulates the steady excretion of salts by the rectal gland. The rate of fluid secretion strongly correlates with the rate at which ions are delivered to the rectal gland—that is, the rate at which blood flows to the gland in the rectal artery, which divides into capillaries surrounding each tubule[51]. An additional intake of salts after a salty meal increases Na^+, K^+-ATPase activity of the rectal gland, which increases salt secretion. The signal for the increasing salt secretion in these circumstances may be an increase in Na^+ and/or Cl^- concentrations of extracellular fluids or the resultant increase in plasma volume caused by the redistribution of fluids between intracellular and extracellular fluid compartments.

In birds and reptiles, the salt glands vary in mass and secretory activity. These variations appear related to the salt intake of individuals, influenced by the habitats they frequent and their diet. For example, the nasal glands of species of bird that rarely visit the seashore, such as snipe (*Capella gallinago*),

[51] Rectal gland secretion is under nervous and hormonal control, which are discussed in Section 21.1.2.

Experimental Panel 5.1 Hypernatraemia among sea snakes

Background

Marine birds and reptiles that secondarily adopted a marine life evolved salt glands. These salt-secreting glands usually maintain the salt concentrations of extracellular fluids *within narrow limits*, but for marine sea snakes this concept has been challenged.

Species of sea snakes with functional salt glands appear unable to maintain water and salt balance without access to fresh water for drinking[1]. Dehydration of amphibious sea snakes results in a decrease in body mass during periods in seawater. These studies led to an alternative hypothesis: that sea snakes have developed a physiological tolerance of **hypernatraemia**[2]—an increase in plasma NaCl concentrations. Thus, in addition to reduced skin permeability to sodium, and evolution of salt glands, some snakes may have evolved a tolerance of some level of hypernatraemia.

All sea snakes have salt glands, but different species exhibit different lifestyles. Some sea snakes are 'true' marine sea snakes that never leave seawater. By contrast, sea kraits (*Laticauda* spp.) are amphibious: they spend prolonged periods foraging in coral reef systems for fish on which they prey, but return to land for digestion, mating and egg laying. We might therefore expect sea kraits that can restore normal plasma sodium when they return to land and find fresh water to drink to be more tolerant of hypernatraemia than the fully marine true sea snakes.

Methods

1. To test the hypothesis that sea kraits have high tolerance of hypernatraemia, blood samples were collected from free-ranging sea kraits and the concentration of Na^+ in plasma was measured.
2. To compare the extent of hypernatraemia in a range of species, data were collated from published studies reporting plasma Na^+ concentrations of sea snakes acclimated to fresh water (where they are able to drink ad libitum and hydrate fully), sea snakes acclimated to (100 per cent) seawater, and free-ranging sea snakes in seawater.

Results and discussion

Free-ranging sea kraits and true sea snakes captured in seawater were found to have higher plasma Na^+ concentrations than the typical 140–150 mmol L^{-1} in reptiles (and other tetrapods in marine or terrestrial habitats). The mean plasma Na^+ concentration of free-ranging *Laticauda saintgironsi* was 181 mmol L^{-1}, while free-ranging *Laticauda laticaudata* had a mean plasma Na^+ concentration of 189 mmol L^{-1}, as shown in Figure A(i). These findings indicate that free-ranging sea kraits display a high tolerance to hypernatraemia, despite having a functional salt gland.

Figure A includes data for plasma Na^+ concentrations in three amphibious species of sea kraits, and four fully marine snakes. The data show that when these snakes are acclimated to fresh water in the laboratory, plasma Na^+ concentrations are restored to within the normal range.

In contrast, hypernatraemia occurs in both amphibious sea kraits and fully marine sea snakes when they are free ranging or held in 100 per cent seawater. Unexpectedly, the true sea snakes, which have highly developed salt glands, show the highest tolerance of hypernatraemia, with plasma Na^+ concentrations exceeding 200 mmol L^{-1} in three of the species studied.

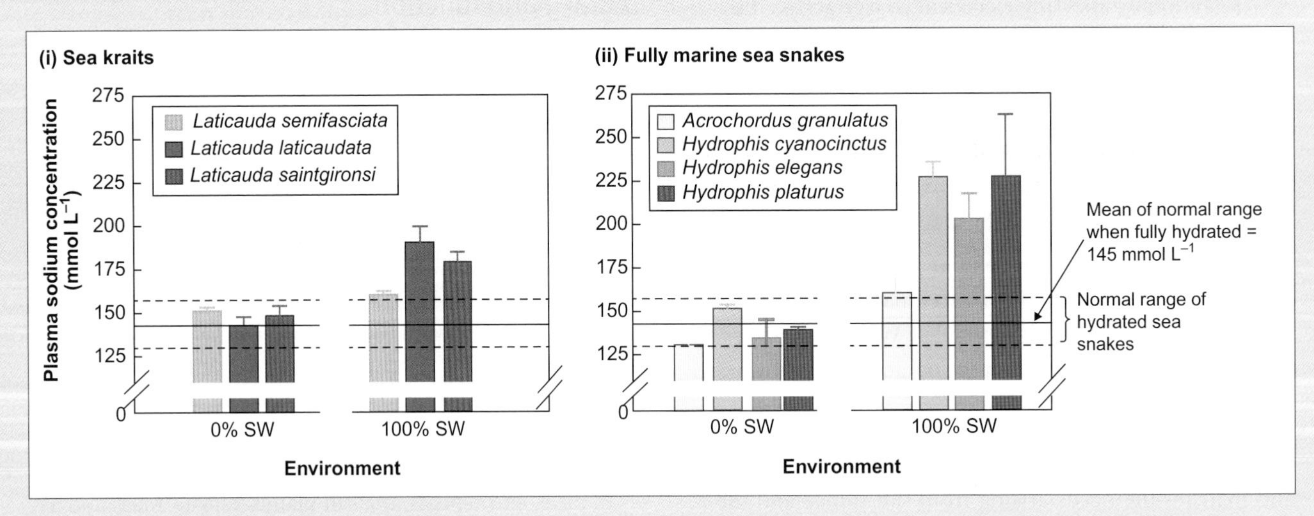

Figure A Plasma sodium concentrations of various species of sea snakes when acclimated to fresh water (0% seawater, SW) and 100% seawater

The left-hand panel (i) shows data for three species of amphibious sea kraits; the right-hand panel (ii) shows data for three species of fully marine true sea snake (bar-bellied sea snakes: *Hydrophis elegans*, blue-banded sea snakes: *Hydrophis cyanocinctus* and yellow-bellied sea snakes: *Hydrophis platurus*) and fully marine little file snakes (*Acrochordus granulatus*). All species have functional salt glands. Lines across panels show the range and mean values of plasma Na^+ concentrations of fully hydrated sea snakes. Data columns show means ± standard deviations.

Source: adapted from Brischoux F, Kornilev YV (2014). Hypernatremia in dice snakes (*Natrix tessellata*) from a coastal population: implications for osmoregulation in marine snake prototypes. PLOS ONE 9(3) e92617.

Conclusions and implications

We can conclude that the salt glands of sea snakes only become effective in reducing plasma Na^+ concentrations at relatively high plasma concentrations of sodium chloride, which are significantly above the normal Na^+ concentrations of reptiles. The thresholds for activation of the salt glands appear to be highest among the true sea snakes.

It is uncertain whether the hypernatraemia of sea snakes has implications for physiological performances. However, the data in Figure A suggest that any such effects are likely to be greatest in true sea snakes that show the highest level of hypernatraemia.

Hypernatraemia by sea snakes will save energy because salt secretion by the lingual salt glands involves active transport processes. By contrast, tighter salt regulation would impose energy demands that might not be sustainable on a day-to-day basis, and could potentially adversely affect other energy-demanding processes, such as swimming and prey capture.

› Find out more

Brischoux F, Kornilev YV (2014). Hypernatremia in dice snakes (*Natrix tessellata*) from a coastal population: implications for osmoregulation in marine snake prototypes. PLOS ONE 9: e92617.

Brischoux F, Briand MJ, Billy G, Bonnet X (2013). Variations of natremia in sea kraits (*Laticauda* spp.) kept in seawater and fresh water. Comparative Biochemistry and Physiology Part A 166: 333–337.

Brischoux F, Tingley R, Shine R, Lillywhite HB (2012). Salinity influences the distribution of marine snakes: implications for evolutionary transitions to marine life. Ecography 35: 001–010.

[1] We discuss drinking of fresh water by sea snakes in Section 5.1.3.

[2] Natrium refers to salt, while -aemia refers to the blood.

5

are relatively small, with a mass of 0.1 to 0.3 mg g body mass^{-1}. In contrast, the nasal glands of shore birds that feed on crabs and worms, such as European herring gulls (*Larus argentatus*), reach 1 mg g body mass^{-1} and contain more tubules.

Among birds, a high intake of salts either from the diet or from the drinking of saline water has been shown to increase the size of the nasal gland, the number of cells within the gland, and the Na^+, K^+-ATPase activity of the gland, leading to an increase in the rate of salt secretion. However, even well-developed salt glands are relatively inactive until a sufficient excess of salts or an increase in plasma volume stimulates salt secretion.

Many of the experimental studies of salt/rectal gland secretion have involved the administration of known amounts of salt (a salt load) by intravenous injection: beyond a certain amount, salt loads typically trigger secretion by the salt/rectal gland of an animal. Even for birds that do not naturally experience salt loads (such as ducks), the number of cells in the salt glands increases by two to three-fold within 48 hours of a salt load, and pre-existing inactive cells become secretory cells. These cells exhibit an increase in basolateral membrane infoldings, an increased number of mitochondria, and elevated Na^+, K^+-ATPase activity.

There have been fewer studies of salt secretion by the salt glands of reptiles than birds, but similar effects of salt loads are apparent in reptiles and stimulate salt secretion if sufficient. Figure 5.15 shows the rapid increase in fluid secretion after administering salt loads to hatchling leatherback sea turtles and juvenile green sea turtles (*Chelonia mydas*).

Many vertebrates can detect the salt content of their food using their taste buds[52], which could allow them to select food items according to their salt balance. The functioning of the salt glands appears to be an important factor in influencing the food selection of coastal birds that feed on intertidal and estuarine invertebrates of variable salt contents. This conclusion is the result of recent studies of red knots (*Calidris canutus*). Red knots with relatively small salt glands (after prolonged access to fresh water) were found to prefer mud snails (*Peringia ulvae*) with a low salt content, rather than saltier snails; such behaviour limits the salt loading. However, the preference for snails with a lower salt content disappeared when the salt glands enlarged (after a long period of feeding on a salty diet without access to fresh water). These studies indicate the importance of behaviour in the osmoregulation of free-ranging animals that have a choice of food items from variable salinity environments.

[52] Section 17.3.2 discusses taste receptors.

› *Review articles*

Chasiotis H, Kolosov D, Bui K, Kelly SP (2012). Tight junctions, tight junction proteins and paracellular permeability across gill epithelium of fishes: A review. Respiratory Physiology & Neurobiology 184: 269–281.

Evans DH, Piermarini PM, Choe KP (2005). The multifunctional fish gill: dominant site of gas exchange, osmoregulation, acid-base regulation and excretion of nitrogenous waste. Physiological Reviews 85: 97–177.

Grosell M (2011). Review. Intestinal anion exchange in marine teleosts is involved in osmoregulation and contributes to the oceanic inorganic carbon cycle. Acta Physiologica 202: 421–434.

Hildebrandt J-P (2001). Coping with excess salt: adaptive functions of extrarenal osmoregulatory organs in vertebrates. Zoology 104: 209–210.

Hyodo S, Kakumura K, Takagi W, Hasegawa K, Yamaguchi Y (2014). Morphological and functional characteristics of the kidney of cartilaginous fishes: with special reference to urea reabsorption. American Journal of Physiology—Regulatory, Integrative and Comparative Physiology 307: R1381–1395.

Larsen EH, Deaton LE, Onken H, O'Donnell M, Grosell M, Dantzler WH, Weihrauch D (2014). Osmoregulation and excretion. Comprehensive Physiology 4: 405–573.

Ortiz RM (2001). Review. Osmoregulation in marine mammals. Journal of Experimental Biology 204: 1831–1844.

Wilson RW, Wilson JM, Grosell M. (2002). Review. Intestinal bicarbonate secretion by marine teleost fish—why and how? Biochimica et Biophysica Acta – Biomembranes 1566: 182–193.

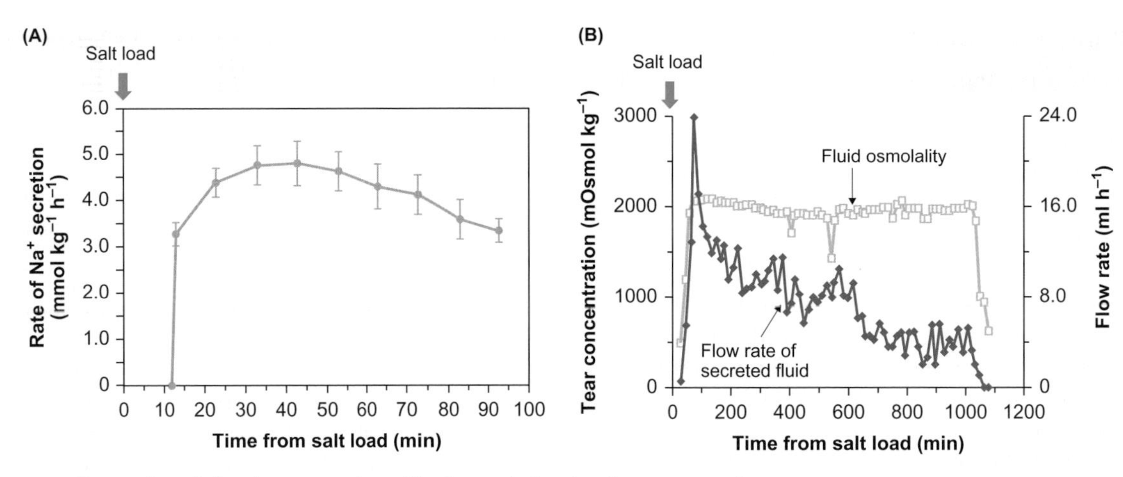

Figure 5.15 Effects of a salt load on secretion of lachrymal glands of two sea turtles

(A) The rate of sodium (Na^+) secretion by lachrymal salt glands of hatchling leatherback turtles (*Dermochelys coriacea*) increases rapidly after an intravenous salt load (27 mmol NaCl kg body mass^{-1}) at time 0 min and reaches a peak rate of secretion within about 30 min. (B) Effect of a salt load (20 mmol NaCl kg body mass^{-1}) on green sea turtles (*Chelonia mydas*) in which 4–5 mmol NaCl kg body mass^{-1} is necessary to switch on secretion by the salt gland. After sufficient salt loading at 0 min, the volume flow rate of secreted fluid (filled symbols) rapidly increases, and then diminishes. The osmolality of the fluid secreted by the salt glands at its plateau and the duration of response are related to the amount of salt administered. After the salt load of 20 mmol NaCl kg body mass^{-1} fluid osmolality plateaus at about 2000 mOsm kg^{-1}.

Sources: A: Reina RD et al (2002). Salt and water regulation by the leatherback sea turtle *Dermochelys coriacea*. Journal of Experimental Biology 205: 1853-1860. Data points are means ± standard errors for 12 turtles; B: Nicolson SW, Lutz PL (1989). Salt gland function in the green sea turtle *Chelonia mydas*. Journal of Experimental Biology 144: 171-184.

5.2 Animals living in freshwater habitats

On an evolutionary timescale, animals ventured from marine habitats into a range of freshwater habitats, probably initially venturing into estuaries and brackish water habitats, such as salt marshes, from where they moved into rivers, ponds and lakes[53]. As they did, the osmotic concentration of their extracellular fluids decreased. This would have reduced the energy costs of maintaining the higher salt concentrations of their marine ancestors.

Freshwater habitats contain few ions compared to seawater or brackish water, and osmotic conformation to freshwater habitats (as found in many animals living in seawater) is not an option: the cells of such organisms simply cannot function in such osmotically dilute fluids. Instead, animals that live in freshwater habitats must hyperosmoregulate. That is, they regulate the osmolality and NaCl concentrations of their extracellular fluids at concentrations exceeding those in the surrounding fresh water. The extent of their **hyperosmoregulation** varies according to the animal group and species in question, as illustrated by the examples in Table 5.5.

The lowest osmolalities and Na^+ concentrations of extracellular fluids occur in the freshwater bivalves and are strikingly low in zebra mussels (*Dreissena polymorpha*). An understanding of the ion balancing mechanisms and the need to maintain the low extracellular fluid concentrations of K^+ of zebra mussels has led to the development of novel methods for controlling this invasive pest (while conserving indigenous species) in some water bodies of North America and Europe, which we discuss in Case Study 5.1.

[53] Section 1.2.1 discusses freshwater habitats.

Table 5.5 Extracellular fluid osmolality and Na^+ concentration of selected examples of invertebrates living in fresh water

Average Na^+ concentration in fresh water (river water) is shown for comparison.

Species	Osmolality mOsm kg^{-1}	Na^+ mmol L^{-1}
Fresh water	<10	0.27
Zebra mussel (*Dreissena polymorpha*)	41	18
Green hydra (*Hydra viridissima*)	40–50	–
Swan mussel (*Anodonta cygnea*)	44	16
Wandering snail (*Radix balthica*)	126	–
Water flea (*Daphnia magna*)	136	–
Yellow fever mosquito (larva) (*Aedes aegypti*)	266	–
Freshwater shrimp (*Palaemonetes antennarius*)	403	210
European crayfish (*Astacus astacus*)	436	212

Sources: river water; Livingston DA (1963). Chemical composition of Rivers and Lakes. Chapter G in: Data of Geochemistry. Sixth edition, Fleischer M (ed). United States Department of the Interior, Geological Survey. Invertebrate examples: Potts WTW, Parry G. Osmotic and Ionic Regulation in Animals. Oxford: Pergamon Press, 1964; Byrne RA, Dietz TH (2006). Ionic and acid-base consequences of exposure to increased salinity in the zebra mussel, *Dreissena polymorpha*. Biological Bulletin 211: 66–75.

Case Study 5.1 Zebra mussels: controlling an invasive species using potassium

Zebra mussels (*Dreissena polymorpha*) are generally considered to be one of the world's most ecologically and economically important pests. Zebra mussels originated in Eastern Europe (the Black Sea and Caspian Sea areas), probably from an estuarine ancestor. Once outside their native range, in North American and European fresh waters, populations expanded rapidly and soon became of concern because of their effect on native species. In North America, zebra mussels rapidly spread throughout the Great Lakes and into streams and rivers in at least 25 US states.

Unlike other freshwater bivalves, zebra mussels attach to solid substrates using byssal threads, like marine mussels. They can cover the entire substrate, attaching both to each other (as shown Figure A) and to indigenous species, with which they compete for the phytoplankton on which they feed. Zebra mussels also attach to water intake grills, the inside of pipes for water intake by industrial facilities, such as power stations and water treatment works, and irrigation pipelines. By stacking on top of each other, water flow through the pipes is drastically reduced and may even become virtually blocked. There is, therefore, an enormous interest in developing effective control methods both from an ecological and economic perspective.

Various methods have been attempted in an effort to control zebra mussel populations, the most promising of which was derived from an understanding of their ionoregulatory physiology. Zebra mussels have epithelia with an exceptionally high permeability to ions, which may explain the particularly low osmotic concentration of their haemolymph (shown in Table 5.5).

The high permeability to ions of zebra mussels seems to be their Achilles' heel in terms of providing the means of reducing invasive populations. If potassium salts (usually potassium chloride, KCl) are added to the fresh water they inhabit, rapid influx of K^+ increases the haemolymph concentrations. Cell functioning depends on the correct ratios of K^+ and Na^+ across the plasma membrane[1] and is adversely affected by the K^+ influx.

Figure A Zebra mussels

An invasive pest in North America and Europe

Source: Gordon MacSkimming/Fotostock

Zebra mussels use K^+ for cell volume regulation[2], rather than the amino acids used by marine mussels and other marine invertebrates. Hence, the plasma membrane of zebra mussel cells readily transport K^+ via channels and/or Na^+, K^+, $2Cl^-$ co-transporter proteins. High extracellular concentrations of K^+ therefore result in K^+ influx into the cells of zebra mussels, which causes swelling of the tissues. When potassium salts are added to the water, swelling of the gill tissue of zebra mussels can reach the point at which gas exchange is impeded and the mussels ultimately asphyxiate. By contrast, indigenous aquatic molluscs and fish are much less sensitive to external K^+ concentrations, presumably because of their lower permeability to potassium.

Addition of KCl to the water in sufficient amounts to exert toxic effects on zebra mussels is not the best approach from an environmental stance and for flowing systems is probably unrealistic, although it does work in contained systems such as lakes or reservoirs. However, the impacts of K^+ on zebra mussels inspired a 'Trojan horse' design of BioBullets® to deliver lethal particles of KCl to these suspension feeders[3].

BioBullets consist of KCl encapsulated in a palatable outer fat layer. The manufacturers (BioBullets Ltd) use different materials within the particle to obtain optimal buoyancy, dispersion and palatability to the target animal. Figure B(i) shows some of the microscopic particles, with a mean diameter of about 100 µm. These microparticles must be ingested within a few hours of application, after which time they break up, releasing small amounts of KCl into the water. The effects on water concentrations of KCl are insignificant, so BioBullets are considered to have a minimal environmental impact on other animal groups.

Zebra mussels are extremely efficient suspension feeders, being capable of removing particles between about 1 to 200 µm and processing 1 to 2 L of water every day. Hence, after adding BioBullets to a blocked pipeline, the zebra mussels filter out the BioBullets, which effectively concentrates the KCl within them. Figure B(ii) illustrates a zebra mussel that has captured some BioBullets from the water column. Some of the encapsulated KCl particles are transported along the gills towards the mouth from where they are ingested.

BioBullets have been seen within the gut and it is likely that these ingested particles are at the smaller end of the size range as zebra mussels preferentially ingest particles of 15 to 40 µm. Consumption of the toxic microparticles then results in death of the zebra mussels, with mortality of 60–65 per cent recorded after a single application of BioBullets.

› Find out more

Aldridge DC, Elliott P, Moggridge GD (2006). Microencapsulated BioBullets for the control of biofouling zebra mussels. Environmental Science & Technology 40: 975–979.

Costa R, Aldridge DC, Moggridge GD (2011). Preparation and evaluation of biocide-loaded particles to control the biofouling zebra mussel,

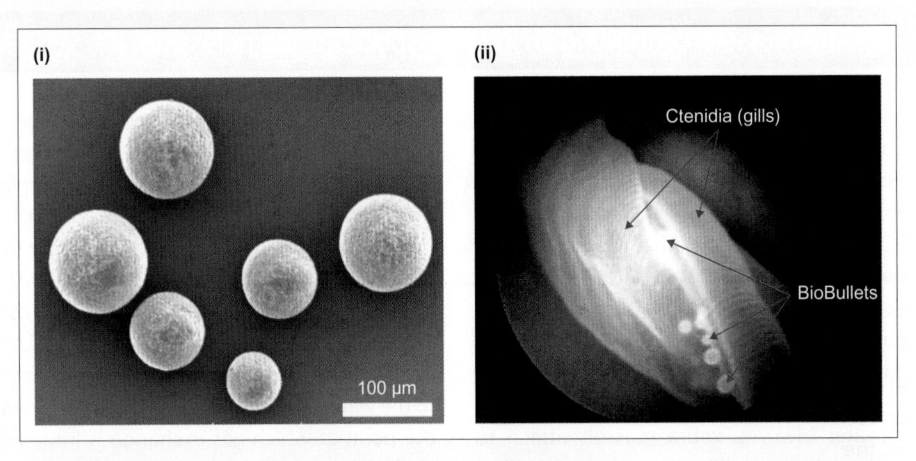

Figure B BioBullets and their filtration by zebra mussels (*Dreissena polymorpha*)

(i) Scanning electron micrograph shows range of sizes of microscopic particles; (ii) BioBullets of various sizes along the gills of a live zebra mussel. The photograph was obtained by video recording using an endoscope.

Source: Aldridge DC et al (2006). Microencapsulated BioBullets for the control of biofouling zebra mussels. Environmental Science & Technology 40: 975-979.

Dreissena polymorpha. Chemical Engineering Research and Design 89: 2322–2329.

Sousa R, Novais A, Costa R, Strayer DL (2014). Invasive bivalves in fresh waters: impacts from individuals to ecosystems and possible control strategies. Hydrobiologia 735: 233–251.

[1] Section 3.1.3 discusses the importance of ion permeabilities of plasma membranes.

[2] Section 4.3 examines cell volume regulation of animals.

[3] Section 2.3.1 discusses suspension (filter) feeding as a means of nutrient intake.

Looking back at Table 5.5, note that molluscs, such as wandering pond snails (*Radix balthica*), insect larvae such as mosquito larvae (e.g. *Aedes aegypti*), and some crustaceans, such as water fleas (*Daphnia magna*), form an intermediate grouping with extracellular fluid osmolalities of 120 to 260 mOsm kg^{-1}. These values are still only half the osmotic concentration of the body fluids of other crustaceans such as freshwater shrimps (*Palaemonetes* spp.) and European crayfish (*Astacus astacus*).

Vertebrates living in freshwater habitats regulate blood osmolality at 200 to 350 mOsm kg^{-1}, as shown by examples in Figure 5.16. The Na^+ and Cl^- concentrations of extracellular fluids (blood plasma) are regulated at 90 to 180 mM, depending on the species. Figure 5.16 includes data for Amazon stingrays (*Potomotrygon* sp.), which live entirely in fresh water. River stingrays (of which about 25 species are described, but more suspected to exist) have long fascinated physiologists, because they abandoned the urea retention of marine elasmobranchs illustrated in Figure 5.10. While marine elasmobranchs have plasma urea concentrations of above 300 mmol L^{-1}, the plasma urea concentration of Amazon stingrays is 1.2 mmol L^{-1}, and their plasma

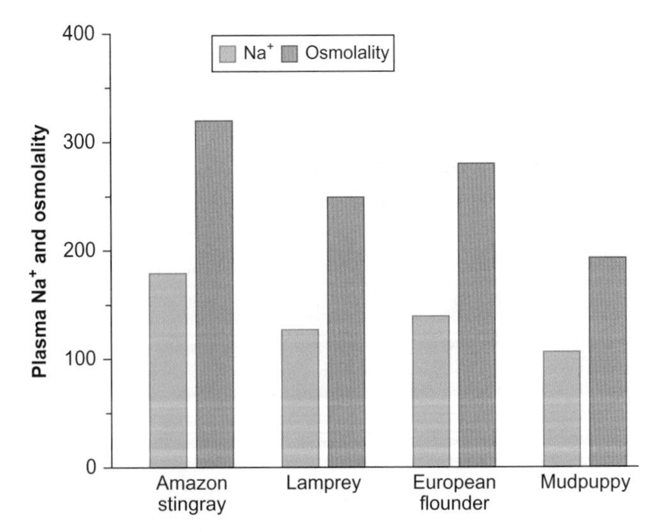

Figure 5.16 Plasma Na^+ concentrations (mmol L^{-1}) and plasma osmolality (mOsm kg^{-1}) of selected vertebrates held in fresh water

Data are shown for Amazon stingray (*Potomotrygon* sp.), river lamprey (*Lampetra fluviatilis*), European flounder (*Platichthys flesus*) and mudpuppy (*Necturus maculosus*).

Data from Garland HO et al, 1975; Rankin JC et al, 1980; Bond H et al, 2002; Wood CM et al, 2002.

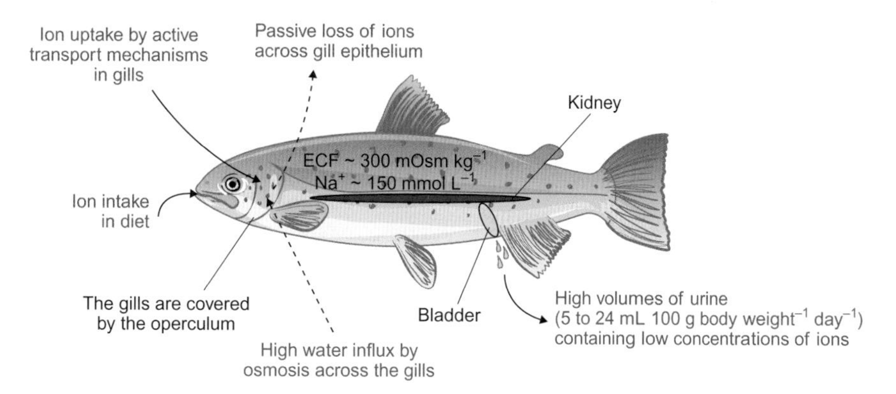

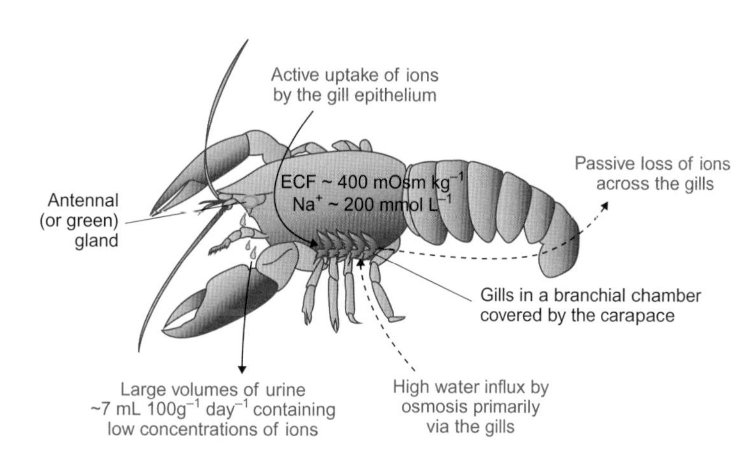

Figure 5.17 Salt and water exchanges between the environment and the extracellular fluid (ECF) of freshwater teleosts and crayfish

Blue font indicates processes determining water balance; urine volumes are given as mL per 100g body mass per day. Brown font indicates processes determining salt balance. Passive processes are shown by dashed line and arrows; processes involving energy expenditure are shown by solid lines and arrows.

osmolality is similar to that of freshwater teleosts. Freshwater stingrays have no need for the salt secreting rectal gland that occurs in marine elasmobranchs[54], and the rectal gland is much reduced in size and is non-functional.

All animals in fresh water need to breathe and many aquatic animals use gills or their skin to take up oxygen from the water[55]. The surfaces used for gas exchange by aquatic animals have particular characteristics—high surface area, thinness, and gaseous permeability[56]—that inevitably result in high permeabilities to water and ions. Hence, all freshwater animals face two major challenges, which Figure 5.17 illustrates:

- Passive influx of water occurs by osmosis across the permeable epithelia in intimate contact with the more osmotically dilute environment—the gills of fish and crustaceans, and **integument** (skin and underlying connective tissue of amphibians, freshwater snakes and turtles, or cuticle of insect larvae).
- Freshwater habitats have low ion concentrations, which result in an efflux of Na^+ and Cl^- from the extracellular fluids across ion permeable epithelial membranes (gills or skin), by diffusion. The rates of passive diffusion of ions across permeable epithelia are related to the combined effects of charge and concentration differences (electrochemical differences) and ion permeabilities[57]. In fresh water, the very low ion concentrations have an overriding effect on ion losses. Smaller ion losses also occur in urine.

These two challenges mean that two principles apply to all freshwater animals:

- to remain in water balance freshwater animals must excrete the excess water influx;
- to maintain salt balance and hyperosmoregulate, freshwater animals must take up Na^+ and Cl^- from the relatively dilute water in which they live.

[54] Figure 5.14 outlines the mechanism of Na^+ and Cl^- secretion by the rectal glands of marine elasmobranchs.

[55] We discuss aquatic respiration in Section 12.2.

[56] Gaseous exchange surfaces are discussed in Section 11.2.

[57] Section 3.1.3 discusses the effects of charge, ion concentrations and membrane permeability on the passage of ion via membranes.

5.2.1 Water balance of freshwater animals

According to physical principles[58], the rate of water influx into a freshwater animal depends on:

- the difference in osmotic concentrations between fresh water and the extracellular fluids of the animal, which results in water influx by osmosis;
- the area of permeable surfaces across which osmosis occurs;
- the water permeability of these surfaces.

From these principles, we can deduce that a reduction in the osmotic concentration of the extracellular fluids reduces the driving force for water influx, which may reduce the osmotic water influx (depending on other factors). Vertebrate animals do not reduce the osmolality of extracellular fluids to much below 200 mOsm kg^{-1}, as shown by the examples in Figure 5.16. Only a few soft-bodied bivalves appear to have reduced their extracellular fluid osmolalities significantly. Even in these animals, however, high rates of urine production occur (for example, about 30 per cent of their body weight per day in swan mussels, *Anodonta cygnea*), which indicates that the surface epithelia are highly water permeable and that the high water influx is subsequently excreted.

The influence of the differences in osmotic concentration on water fluxes are clearly demonstrated for the water permeable skin of amphibians, by *in vitro* study of isolated pieces of skin of known area. Such investigations show that the rate of water transfer across the skin varies according to the species and habitat of amphibians.

In the presence of a constant difference in osmotic concentrations (about 200 to 220 mOsm kg^{-1}) water passes across the skin of totally aquatic frogs (such as African clawed frogs, *Xenopus laevis*) and the tadpoles of amphibious species at 2 to 8 μL cm^{-2} skin h^{-1}. In amphibious frogs, water passage across the skin occurs at a somewhat higher rate of 5 to 13 μL cm^{-2} h^{-1}. By contrast, water influx at 19 to 28 μL cm^{-2} h^{-1} occurs via the skin of more terrestrial species of frogs and toads, such as cane toads (*Rhinella marina*) and common toads (*Bufo bufo*). From these measurements, it is apparent that the skin of aquatic amphibians is less water permeable than that of terrestrial amphibians, which reduces water influxes, but never to zero.

The water permeability of amphibian skin is governed by the properties of the junctions between the epithelial cells, which could allow transepithelial water fluxes, and the presence of aquaporins (water channels) that facilitate water passage across the plasma membranes of the epithelium[59]. The low water permeability of the skin of totally aquatic frogs (*Xenopus laevis*) appears to result from the lack of aquaporins in their skin.

In a similar way to amphibians, crustaceans that are confined to freshwater habitats, such as European crayfish (*Astacus astacus*), have integuments with a relatively low permeability to water compared to the integument of crustaceans confined to marine or high-salinity habitats. Among euryhaline[60] crustaceans (such as species of prawns), which can live in different salinities, water permeability is *higher* when they are in fresh water and low salinities than when they acclimatize to higher salinities. Although these differences are poorly understood from a mechanistic perspective, it is clear that lower water permeabilities reduce water fluxes.

The integument of terrestrial insects (the cuticle) has low water permeability because of the waxy substances found in the outermost part of the cuticle[61]. Insect larvae that live in freshwater habitats would stick on the surface at the air–water interface if they were completely covered with a waxy cuticle that repels water. In fact, mosquito larvae do have a small waxy area at the tip of the respiratory siphon that is a kind of breathing tube for gas exchange. The waxy area sticks the breathing tube to the water–air interface, but the rest of the cuticle is water permeable. Those insect larvae that rely on their gills for respiration[62], such as the mayflies and dragonflies, are vulnerable to osmotic water influx across the gills, in a similar way to crustaceans and fish. Similarly, water influx occurs across the cuticle of insect larvae that rely on cutaneous gas exchange.

To remain in fluid balance and avoid dilution of the ions in extracellular fluids due to water influxes, freshwater animals must excrete large volumes of dilute urine each day. Generally, freshwater teleost fish excrete 5 to 24 mL 100 g body $mass^{-1}$ day^{-1}, while amphibians living in fresh water typically excrete 30 to 40 mL 100 g^{-1} day^{-1}—in other words, about a third of their body mass each day. To put this in perspective, if humans generated a similar volume of urine, this would amount to about 23 L day^{-1} (assuming body mass of 70 kg), which would mean having to empty our bladders about 60 times each day (every 15 min while we are awake).

Daily urine production by freshwater fish and some aquatic invertebrates are temperature dependent, with higher rates of urine output at higher temperatures. An important factor driving an additional water influx and excretion is

[58] Equations 3.2 and 3.3 show the mathematical relationships that determine water fluxes. These principles apply equally to all aquatic animals, whatever the salinity of the environment.

[59] Aquaporins are discussed in Section 4.2.3. We examine the hormonal control of aquaporins in the skin of amphibians in Case Study 21.1.

[60] We examine euryhalinity in more detail in Section 5.3.

[61] The insect integument (cuticle) is illustrated in Figure 6.17.

[62] Figure 12.39 illustrates the respiratory gills of these animals.

the influence of temperature on oxygen uptake for oxidative metabolism[63]. At higher temperatures, an increase in gill ventilation and/or an elevated surface area is required for gaseous exchange, which not only increases oxygen uptake, but also increases the water influx across the gills. The extra water influx results in a higher rate of urine excretion which maintains fluid balance. For example, the freshwater fish, white suckers (*Catostomus commersonii*), excrete urine at a rate of 29 ml kg body mass^{-1} day^{-1} when water temperature is 2°C, but urine output increases to 106 mL kg body mass^{-1} day^{-1} when the water temperature rises to 18°C. These differences give a Q_{10} value (increase for a 10°C temperature interval)[64] of 2.2 for urine excretion.

The high urine volumes excreted every day by freshwater animals result from the continual high rates of **primary urine** formation by the **glomeruli** of the vertebrate kidney, the **antennal glands** of crustaceans, **nephridia** of freshwater invertebrates such as leeches, or the **Malpighian tubules** of insect larvae[65]. The high rates of primary urine formation are coupled with low rates of fluid absorption within the tubules forming the excretory organs or the urinary bladder, into which some animals pass their urine for storage before its excretion.

Despite the high rates of **glomerular filtration** by the kidneys of teleost fish living in fresh water, fewer than half of the glomeruli in the kidney are typically filtering in resting animals at any one time, so there is plenty of spare capacity to increase filtration, and hence increase urine output when necessary[66]. Such circumstances may arise frequently in natural conditions as the permeability of the gills to water is sensitive to the fluctuating concentrations of adrenaline and/or noradrenaline in the blood circulation, and concentrations of these hormones may increase in response to a multitude of environmental events[67].

5.2.2 Salt balance of freshwater animals

Freshwater animals reabsorb sodium and chloride from the excretory fluid that passes through the tubular systems of their kidneys (vertebrates) or nephridia (invertebrates), which produce osmotically dilute urine, and excrete the water influx that enters their body fluids. In this way, freshwater animals generate urine with an osmolality of about 10 per cent of their extracellular fluids[68]. Nevertheless, some loss of ions in the urine is inevitable and adds to passive ion losses across permeable body surfaces.

Freshwater teleosts reduce paracellular permeability of ion-transporting epithelia

Early ultrastructural studies of the gills and opercular epithelium of teleost fish noted distinct differences in the cell-to-cell junctions of species living in fresh water compared to those of marine teleosts[69]. The tight junctions between the ionocytes and adjacent epithelial cells of the gills of freshwater teleosts are often deep and tight, and have more numerous cross-connecting strands than the shallower junctions between the gill epithelial cells of marine teleosts. These differences are thought to explain the lower permeability to ions of freshwater teleosts than marine teleosts. Furthermore, the tight junctions become even deeper during the acclimation of freshwater species such as goldfish (*Carassius auratus*) to ion poor water, which reduces the passive ion loss via the gills. To understand such differences requires information on the constituent proteins, **occludin** and **claudins** that determine the functioning of the tight junctions[70].

Numerous claudins and occludins have been identified in the epithelia of teleost fish, but the factors determining the expression of most of these proteins are poorly understood. A study of goldfish showed that mRNA for occludin and six types of claudins increased during their acclimation to ion poor water. The greatest consistency in studies so far is that claudin-30 appears to have an important role in forming a permeability barrier in the ion-transporting epithelia of freshwater fish.

Zebrafish (*Danio rerio*) are a favourite species for recent investigations because of the possibility of manipulating the genome of one-cell embryos to create larvae with particular translational gene knockdowns (morpholinos)[71]. Figure 5.18 examines some of the results of studies of zebrafish morpholinos. In normal larvae, the outer epithelial layer of skin overlying the **yolk sac** is rich in ionocytes that take up ions; these cells appear in the gills later in development. Claudin-b (an orthologue of claudin-30) occurs in cell-to-cell junctions between the yolk sac skin cells, as illustrated in Figure 5.18A, where it appears to reduce the solute permeability of the paracellular pathways between the skin cells. The importance

[63] We discuss aerobic metabolism in Section 2.2.

[64] Section 8.1 discusses Q_{10} in more detail.

[65] Urine formation among vertebrates and invertebrates is the topic of Chapter 7.

[66] The variability in glomerular filtration rates of teleost fish due to alterations in the population of filtering glomeruli are the main controller of urine output, which Section 7.1.3 explores in more detail.

[67] Section 19.3 discusses adrenaline and noradrenaline (also known as epinephrine and norepinephrine) and responses to stress.

[68] We explore the generation of dilute urine by particular animal groups in Sections 7.2.2 (vertebrates), 7.3.1 (freshwater rotifers) and 7.3.3 (crustacean antennal glands).

[69] Leaky tight junctions at the apico-lateral borders of the paracellular pathways between ionocytes and the accessory cells of marine fish are examined in Section 5.1.3.

[70] Tight junction proteins are discussed in Section 4.2.1.

[71] Morpholinos are produced by microinjection of antisense oligonucleotides that bind to translation start sites for particular genes.

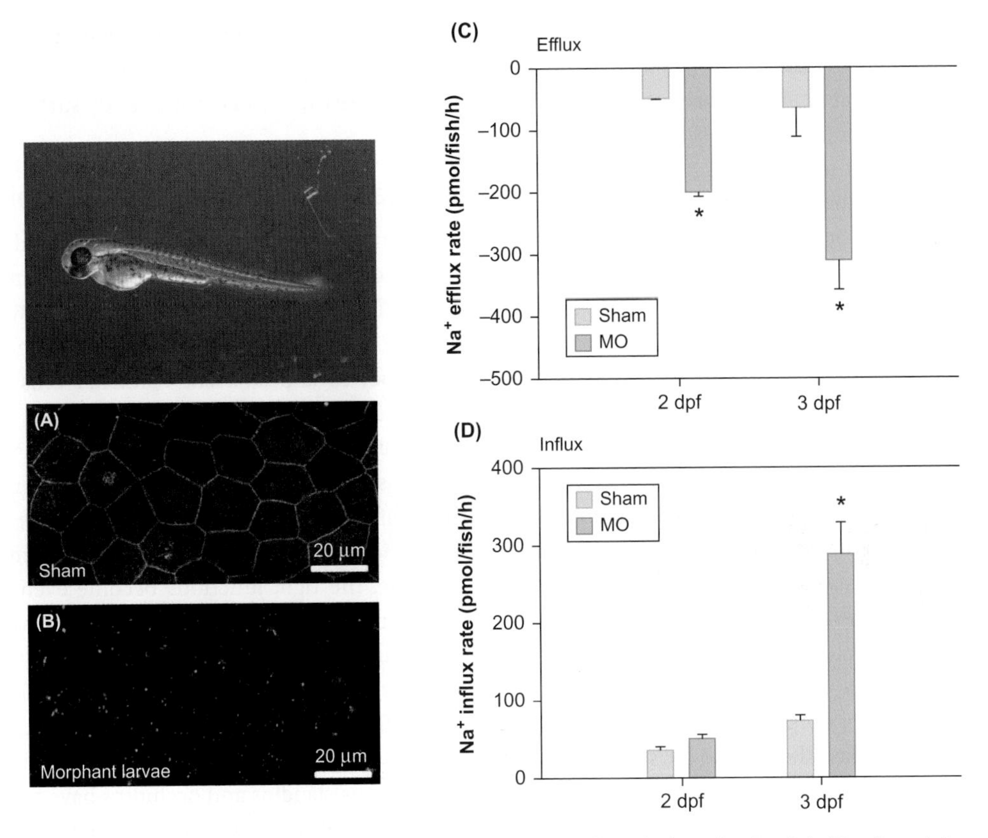

Figure 5.18 Tight junction protein regulates epithelial permeability to sodium in larval zebrafish (*Danio rerio*)

(A) and (B) show images of yolk-sac skin viewed by fluorescent confocal microscopy after incubation with an antibody to zebrafish claudin-b (an orthologue of claudin-30d). (A) Control—sham-injected fish shows clear cell-to-cell contacts on cell borders delineated by the antibody reaction with the claudin protein. (B) Morphant larvae in which the one cell stage embryo stage was subjected to knockdown of the gene for claudin-b, showing substantial reduction in expression of the claudin. (C) The knockdown of the claudin (morphants – MO) significantly increases the rate of passive efflux of Na^+ compared to control larvae at 2 or 3 days post-fertilization (dpf). (D) By 3 days post-fertilization, the larvae begin to partially compensate for the increased efflux by increasing Na^+ influx.* $P<0.05$.

Source: Kwong RWM, Perry SF (2013). The tight junction protein claudin-b regulates epithelial permeability and sodium handling in larval zebrafish, *Danio rerio*. American Journal of Physiology Regulatory Integrative and Comparative Physiology 304: R504–R513. Zebrafish image: © Steven J Baskauf

of claudin-b in restricting passive efflux of Na^+ is illustrated by the data in Figure 5.19C, which shows that increased Na^+ efflux occurs after claudin-b knockdown.

Overall, freshwater teleosts maintain low ion permeability that limits passive ion losses. There is also evidence that the gill permeability to ions can be further reduced in environmental conditions in which ion acquisition becomes more challenging. For example, an increase in occludin expression in the gill epithelium occurs in goldfish (*Carassius auratus*) during their acclimation to ion-poor water. Similar increases in occludin in the gill epithelia of freshwater fish occur in response to reduced ion intake in food. Such responses probably involve hormonal control mechanisms. Although not fully understood, there is evidence that the hormones **prolactin** and **cortisol**[72] reduce Na^+ paracellular permeability of the ion-transporting epithelia of teleost fish, via effects on occludin and specific claudins, which results in the tightening of the cell-to-cell junctions and a reduction in the rate of passive ion loss.

Active ion uptake by freshwater animals

The first evidence of ion uptake in freshwater animals emerged in the pioneering experiments of August Krogh, in the 1930s. His elegant experiments showed that frogs and goldfish take up ions from solutions containing as little as 0.01 mmol L^{-1} of sodium (Na^+) or chloride (Cl^-). In these experiments, he placed each animal in water that was devoid of either Na^+ (by replacing sodium chloride by choline chloride) or Cl^- (by replacing sodium chloride by sodium sulfate). These manipulations led to a reduction in the concentration of either Na^+ or Cl^- in the animal's extracellular fluids, depending on the missing ion. Subsequent uptake of the depleted ion (Na^+ or Cl^-) when the animal was placed in fresh water containing these ions resulted in declining Na^+ or Cl^- concentrations in the surrounding fresh water.

[72] We discuss cortisol in Section 19.3.1 and prolactin in Section 19.2.2. Further actions of prolactin on teleost gills influence calcium balance, which we discuss in Section 21.2.2.

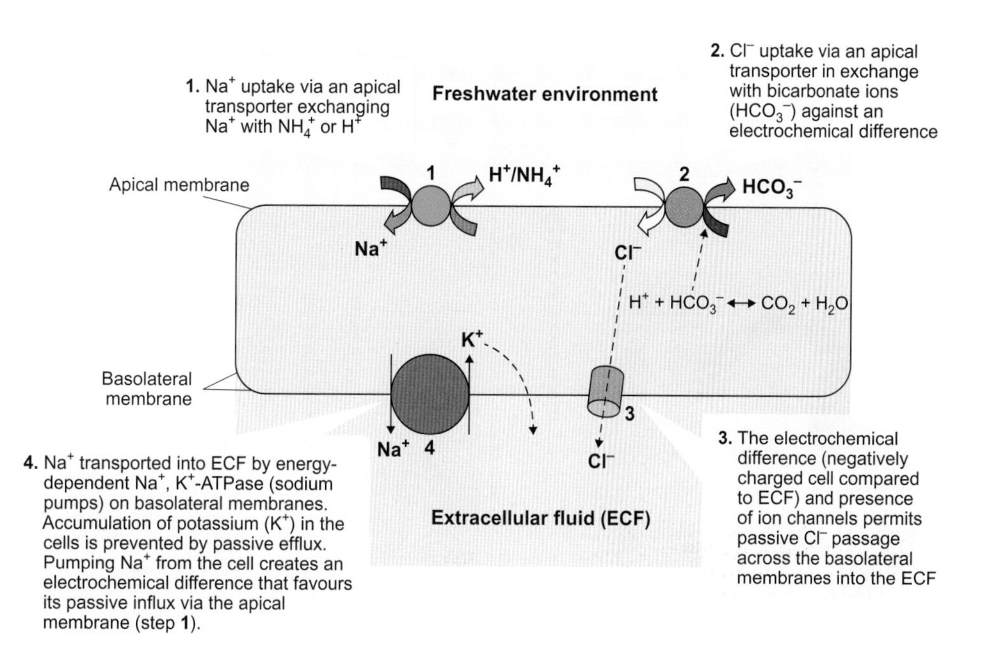

Figure 5.19 Early model of ion uptake by freshwater teleost gill

This simple model suggests two independent membrane transporter proteins, one taking up sodium ions (Na^+) and one taking up chloride ions (Cl^-) from the environment. On the basolateral membranes, passage of Cl^- into the extracellular fluid (ECF) occurs passively via ion channels, while Na^+ passage is active (energy dependent).

Crucially, these experiments showed an important principle that applies to all freshwater animals: *the uptake mechanisms for Na^+ and Cl^- uptake work independently*. Frogs can take up Na^+ from solutions of sodium chloride or sodium carbonate, and take up Cl^- from a solution of sodium chloride, ammonium chloride, or choline chloride[73]. August Krogh did not make any electrical measurements, but suggested an exchange of each ion for a similarly charged ion maintains electrical neutrality. Exchange of Cl^- for bicarbonate ions (HCO_3^-) and Na^+ for ammonium ions (NH_4^+) or H^+ was proposed. Figure 5.19 shows these exchanges in a simple model of the gill processes in freshwater teleosts.

The site for active uptake of ions from fresh water varies among animal groups. Some animals use their skin (or parts of it) for ion uptake; this applies among freshwater worms, leeches, frogs and tadpoles, and fish larvae that take up ions via the outer layer of the epidermis covering the yolk sac.

Some insect larvae, such as those of mosquitoes and chironomids (a type of dipteran) take up ions via prominent **anal papillae** at their posterior end, which Figure 5.20A illustrates. The anal papillae consist of a thin external cuticle lined internally by a layer of epithelial cells surrounding an internal lumen that is continuous with the internal body of the animal. The importance of the anal papillae in maintaining ion balance in these larvae is illustrated by the effects of their removal, which results in declining NaCl content of larvae living in ion-poor fresh water.

The anal papillae of mosquito larvae reared in progressively lower concentrations of NaCl show an increase in the number of mitochondria in the epithelium, suggesting an increase in the active uptake of ions. Recent work on larvae of harlequin flies (*Chironomus riparius*) similarly indicates an increase in ion uptake by the anal papillae of larvae moved into ion-poor water but that this process occurs in two phases. Initial increases in the rate of ion fluxes occur within existing tissue, while over a longer time the size of the anal papillae increases, as shown in Figure 5.20B; the 1.8-fold increase in the surface area of the papillae, approximately doubles Na^+ and Cl^- influxes. Studies with specific inhibitors of various transport proteins suggest that Na^+ uptake by the midge larvae involves exchange with protons (H^+), while Cl^- uptake occurs via exchange with bicarbonate ions (HCO_3^-), much like the model for fish illustrated in Figure 5.19. However, the transport mechanisms of most invertebrates remain unexplored.

Many freshwater animals take up ions across their multifunctional gills[74]. This applies to insect larvae like mayfly and dragonfly larvae, freshwater crustaceans, and teleost fish. Krogh's studies introduced a membrane to separate the anterior and posterior ends of the fish, identifying for the first

[73] Hormones, in particular aldosterone, which we discuss in Section 21.1.1, stimulate uptake of Na^+ via the skin of frogs.

[74] The multifunctional role of gills is discussed at many points in the book: ionoregulation (discussed in this Chapter), gas exchange (Section 12.2.2), acid–base balance (Section 13.3.2), and nitrogenous excretion (Section 7.4.1).

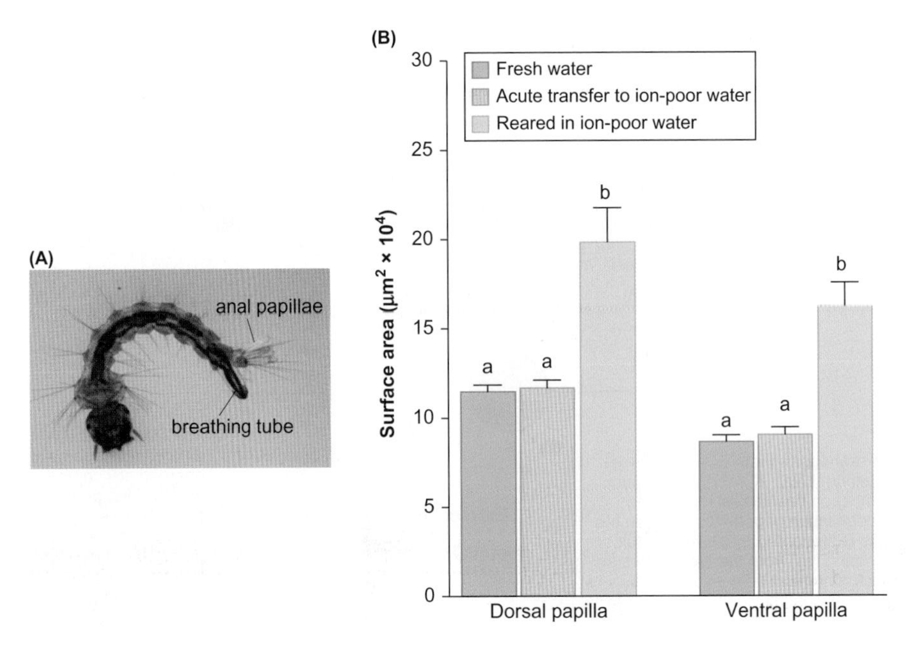

Figure 5.20 Anal papillae in insect larvae for ion uptake

(A) Larva of the tiger mosquito (*Aedes albopictus*) showing four anal papillae located around the anus close to the breathing tube. One is green because it has taken up a virus labelled with a green fluorescent protein. This procedure can be used to introduce insect toxins into recombinant densoviruses (=small single-stranded DNA without a viral envelope) which offers a route for rapid targeting and killing mosquitoes without chemical control. (B) The surface area of dorsal and ventral papillae of the larvae of harlequin flies (non-biting midges) *Chironomus riparius* held in fresh water, within 48 h of transfer to ion-poor water and after longer-term (~30 days) rearing in ion-poor water. Different letters on bars indicate a significant increase in the surface area of dorsal and ventral papillae after rearing the larvae in ion-poor water.

Sources: A: Gu J-B et al (2010). A recombinant AeDNA containing the insect-specific toxin, BmK IT1, displayed an increasing pathogenicity on *Aedes albopictus*. American Journal of Tropical Medicine and Hygiene 83: 614–623; B: Nguyen H, Donini A (2010). Larvae of the midge *Chironomus riparius* possess two distinct mechanisms for ionoregulation in response to ion-poor conditions. American Journal of Physiology Regulatory and Integrative Comparative Physiology 299: R762–R773.

time that ion uptake occurs at the anterior end of the teleost fish. In the 80 years since these studies, there has been intense investigation of the mechanisms for Na^+, Cl^- and Ca^{2+} uptake across the gills of freshwater fish, and via the epithelium that covers the inside of the operculum.

The main cells responsible for active uptake of ions are the **mitochondria-rich cells** (**ionocytes**), of which multiple types occur. Looking at Figure 5.21, note the high density of mitochondria-rich cells on the trailing edge of the gill filament and toward the trailing edge between the lamellae. The mitochondria-rich cells usually make up less than 10 per cent of the surface area of the gill epithelium and are surrounded by many **pavement cells**, which Figure 5.21C illustrates, although striking changes in the number and location of these cells can occur in response to environmental conditions. For example, Figure 5.22 shows the proliferation of mitochondria-rich cells along the gill lamellae of rainbow trout that occurs when they are held in ion-deficient soft water and which increases ion uptake. Fish use their gill lamellae for oxygen uptake and carbon dioxide excretion[75] so the increasing thickness of the lamellae shown in Figures 5.22B compared to Figure 5.22A could have adverse implications for oxygen uptake by the gills; however, an increase in gill ventilation appears to compensate for the structural changes.

Sodium uptake by freshwater teleosts

Over the years there has been intense debate about the mechanism of Na^+ uptake by freshwater fish. Recent localization of transport proteins, by use of specific antibodies, suggests that one reason for the debate is that Na^+ uptake mechanisms differ between species, with at least three major mechanisms occurring:

1. Apical Na^+/H^+ exchange[76] in which the uptake of Na^+ occurs in exchange for protons, which are excreted, as illustrated in Figure 5.23.
2. Proton pump (V-type H^+-ATPase)-assisted Na^+ uptake via apical Na^+ channels driven by the electrical charge difference that results from the activity of the proton pumps[77], which is illustrated in Figure 5.24.
3. Na^+ uptake via **Na^+, Cl^- co-transporters** or **Na^+, K^+, $2Cl^-$ co-transporters** (often abbreviated to NCC and NKCC, respectively).

[75] We discuss gaseous exchange via fish gills and the physical factors affecting this process in Section 12.2.2.

[76] Na^+/H^+ exchange proteins are often abbreviated to NHE of which several sub-types exist.

[77] Section 4.2.4 explores proton pumps, specifically V-type H^+-ATPase in some detail.

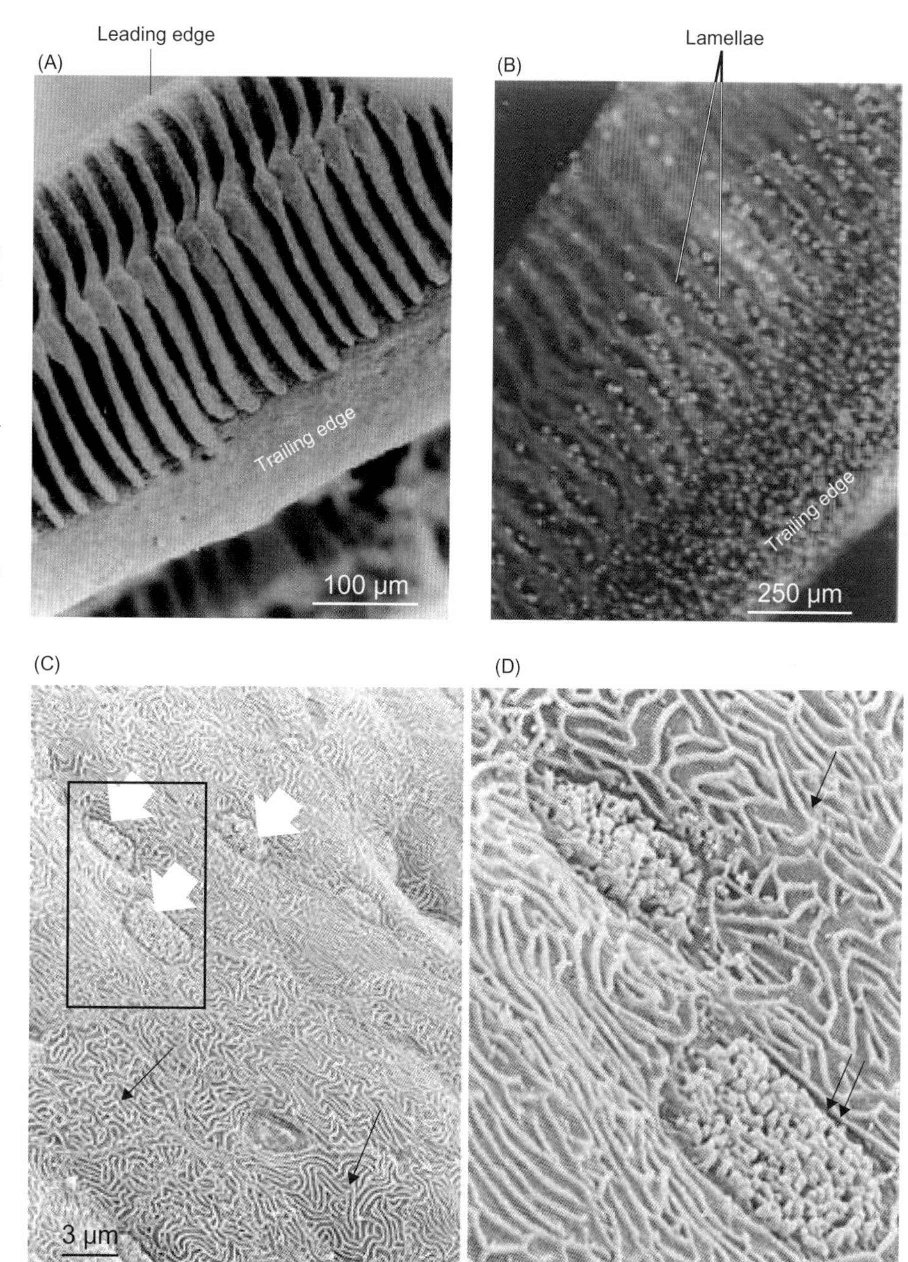

Figure 5.21 Mitochondria-rich cells in gill epithelium of teleost fish living in fresh water

(A) Scanning electron micrograph of the gill filament of Mozambique tilapia (*Oreochromis mossambicus*) showing the many leaf-like lamellae running between the leading and trailing edges of each gill filament. Water flows between the lamellae from the leading to trailing edges. (B) Tilapia gill filament in which live mitochondria of mitochondria-rich cells are stained cells (orange) with DASPMI (dimethyl-amino-styryl-methyl-pyridinium-iodine). The image was obtained with a confocal laser scanning microscope. Note that the mitochondria-rich cells occur predominantly between the lamellae towards the trailing edge and along the flattened trailing edge of the filament, which is the area usually examined in the scanning electron microscope. (C & D) Surface view of the gill epithelium of a freshwater Japanese eel (*Anguilla japonica*) viewed in the scanning electron microscope. Large flat pavement cells with extensive surface microridges (thin black single arrows) account for most (> 90%) of the gill epithelial surface area. Smaller domed mitochondria-rich cells (indicated by white arrows) occur between the pavement cells. Two of these cells, within the box on (C), are illustrated in (D); these cells have prominent microvilli on their apical surface (double arrows), but do not have the apical pits of the ionocytes of marine teleosts shown in Figure 5.6A.

Sources: A & B: Van der Heijden AJH et al (1997). Mitochondria-rich cells in gills of tilapia (*Oreochromis mossambicus*) adapted to fresh water or sea water: quantification by confocal laser scanning microscopy. Journal of Experimental Biology 200: 55-64. C & D: Reproduced from Wong CKC, Chan DKO (1999). Isolation of viable cell types from the gill epithelium of Japanese eel Anguilla japonica. American Journal of Physiology 276 (Regulatory Integrative Comparative Physiology 45: R363–R372.

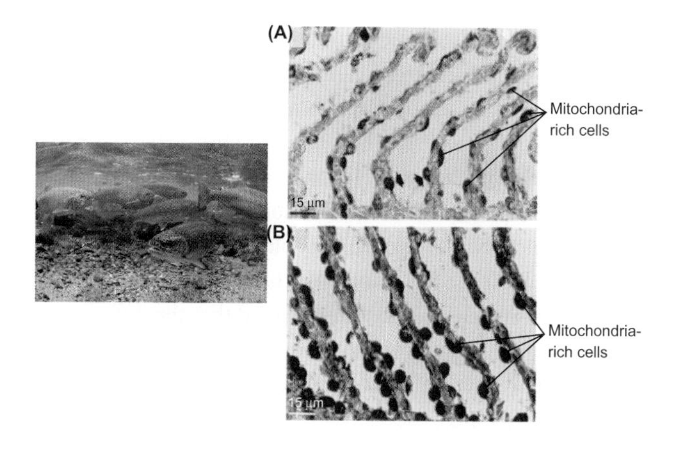

Figure 5.22 Mitochondria-rich ionocytes in gill lamellae of rainbow trout (*Oncorhynchus mykiss*)

To localize ionocytes in the lamellae of trout living in: (A) hard water (high Ca^{2+} concentration: 0.4 mmol L^{-1}) and (B) ion-deficient soft water (low Ca^{2+} concentration: 0.05 mmol L^{-1}) isolated gill arches were treated with zinc iodide in osmium tetraoxide, which stains the high density of membranes in the mitochondria of these cells an intense black. Note that the ionocytes increase in size and number in trout living in ion-deficient soft water.

Source: Reproduced from Perry SF (1998). Relationships between branchial chloride cells and gas transfer in freshwater fish. Comparative Biochemistry and Physiology Part A: Molecular & Integrative Physiology 119: 9-16. Rainbow trout image: Jack Perks/Alamy Stock Photo

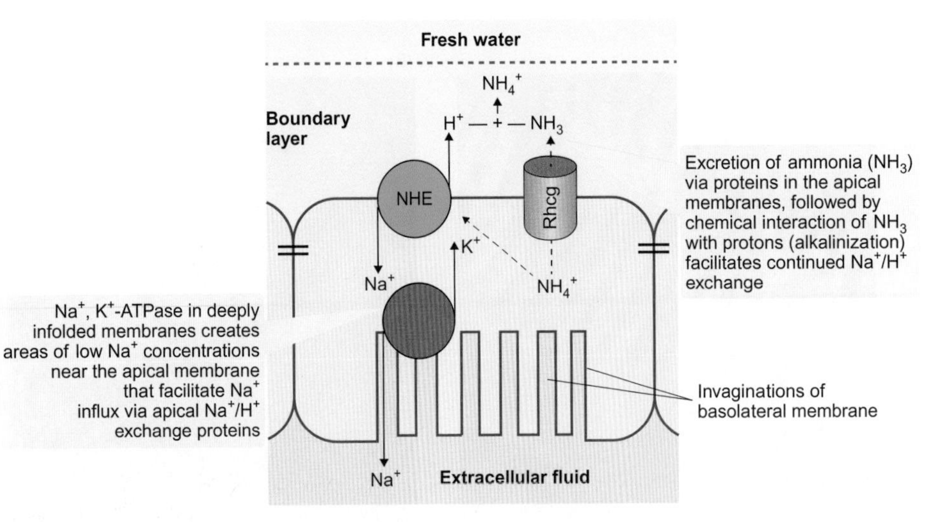

Figure 5.23 Model of uptake of Na^+ via Na^+/H^+ exchanger (NHE) proteins by ion-transporting cells of the gills of freshwater teleost fish

Rhcg is a type of ammonia transporting Rhesus protein.

Source: adapted from Kumai Y, Perry SF (2012). Review. Mechanisms and regulation of Na^+ uptake by freshwater fish. Respiratory Physiology & Neurobiology 184: 249– 256.

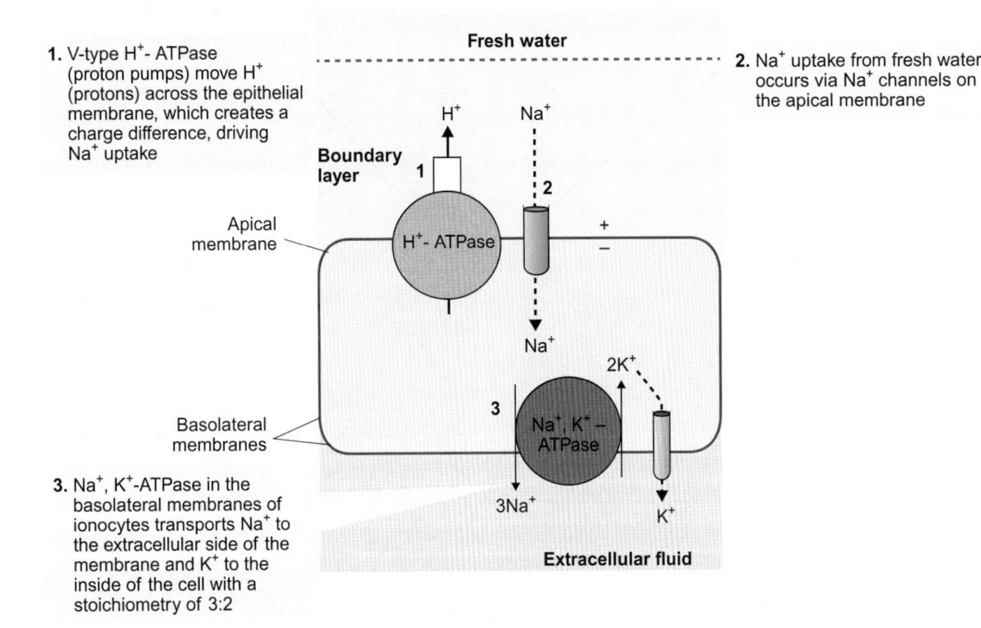

Figure 5.24 H^+-ATPase assisted Na^+ uptake by gill ionocytes of freshwater teleosts

Apical Na^+/H^+ exchanger proteins have been identified in mitochondria-rich ionocytes of several fish species, including zebrafish, goldfish and rainbow trout. Furthermore, the expression of Na^+/H^+ exchangers, at the mRNA or protein level, increases under environmental conditions in which the uptake of Na^+ or excretion of H^+ is stimulated—for example, in zebrafish exposed to ion-poor soft water (with low Na^+ and Ca^{2+} concentrations). These and other results suggest that apical Na^+/H^+ exchangers play an important part in Na^+ uptake by some freshwater fish, at least in some circumstances. Nevertheless, it has been argued that the differences in H^+ and Na^+ concentrations across the apical membranes of the gill epithelial cells are likely to be insufficient to drive the Na^+/H^+ exchange.

Figure 5.23 offers two feasible explanations of the apparent paradox and proposes how Na^+/H^+ exchange could facilitate Na^+ uptake via gill ionocytes of freshwater teleosts:

- The activity of Na^+, K^+-ATPase in the folded basolateral membrane of ion-transporting epithelial cells may create a low Na^+-microenvironment close to the apical membrane that facilitates Na^+ influx via Na^+/H^+ exchange proteins in the apical membranes, even from dilute fresh water. However, Na^+, K^+-ATPase activity in some ion-transporting cells of some species of freshwater fish is not particularly high.
- In some species, alternative processes may link Na^+/H^+ exchange to ammonia efflux from the gills, as

shown in Figure 5.23, in which Na^+/H^+ exchange proteins form a type of functional **metabolon** with transport proteins that conduct ammonia (NH_3)[78] via the apical membranes of ionocytes. The H^+ and NH_3 interact to form NH_4^+, resulting in alkalization of the external boundary layer, as shown in Figure 5.23. These processes have been found to be important in Na^+ uptake by zebrafish larvae, when they are held in acidic or soft water environments, and in medakas (*Oryzias latipes*). However, the extent to which the functional metabolon of the ammonia transporters (**Rhesus proteins**) and Na^+/H^+ exchange proteins occurs among species of freshwater fish is uncertain.

Figure 5.24 illustrates an alternative model of Na^+ uptake involving H^+-ATPase assistance, which began to emerge in studies of rainbow trout in the 1990s. A key feature is the presence of V-type H^+-ATPases (proton pumps[79]) in gill ionocytes. In rainbow trout, proton pumps also occur in the pavement cells, but this is not true of all species. By moving H^+ across the apical membrane of the gill epithelium, the inside of the cells becomes more negatively charged and drives Na^+ uptake through Na^+ channels in the apical membrane, which Figure 5.24 illustrates. Finally, basolateral Na^+, K^+–ATPase removes the Na^+ brought into the cell. Some ionocytes in the lamellae of the rainbow trout contain both V-type H^+-ATPase and Na^+, K^+-ATPase, and this 'co-localization' provides evidence for the model of events shown in Figure 5.24.

Studies using specific inhibitors of V-type H^+-ATPase, such as bafilomycin, provide convincing evidence for the importance of proton pumps in the Na^+ uptake of many species of freshwater teleosts, although some species appear to lack apical proton pumps in their gill epithelium. For instance, bafilomycin prevents Na^+ uptake in carp (*Cyprinus carpio*), zebrafish and goldfish. In zebrafish larvae, in which a subset of gill ionocytes express H^+-ATPase, gene knockdown of V-type H^+-ATPase results in a reduction in whole body Na^+ content, although the Na^+ channels are ultimately the limiting factor for Na^+ uptake.

The third alternative mechanism for Na^+ uptake is via Na^+, Cl^- or Na^+, K^+, $2Cl^-$ co-transporter proteins that occur in the ionocytes of several freshwater teleosts. For example, zebrafish and Mozambique tilapia (*Oreochromis mossambicus*) express Na^+, Cl^- co-transporters in the apical membranes of mitochondria-rich cells in their gills. The expression of these co-transporters increases during the acclimation of tilapia to water containing low concentrations of Cl^-, which suggests the Na^+, Cl^- co-transporters are a significant component in ion regulation of tilapia.

The relative importance of Na^+, Cl^- co-transporter proteins in Na^+ and Cl^- uptake is uncertain. In knockdown experiments on zebrafish embryos, the lack of Na^+, Cl^- co-transporters reduced Cl^- uptake and resulted in a decline in whole-body Cl^- content; however, whole-body Na^+ content increased, presumably via compensatory mechanisms. The multiple mechanisms for Na^+ uptake may enable freshwater fish to cope with fluctuating conditions that may affect their mechanisms for Na^+ uptake in different ways.

Basolateral Na^+, K^+-ATPase plays an important part in the extrusion of the Na^+ brought into the cells, by whatever means (Na^+ channels, Na^+/H^+ exchange or Na^+, Cl^- co-transporters), and maintains low cellular concentrations of Na^+. Therefore, Na^+, K^+-ATPase is included in each of the diagrams in Figures 5.19, 5.23 and 5.24. Structural differences in the subunits of Na^+, K^+-ATPase[80] in freshwater fish are generally considered to promote Na^+ extrusion from the cells and hence Na^+ uptake from the environment.

Chloride uptake by freshwater animals

Most (but not all[81]) teleost fish living in fresh water rely on the active uptake of Cl^- across their gill epithelium. The uptake of Cl^- may occur via the apical mechanisms for Na^+ uptake discussed in the previous section: that is, via Na^+, Cl^- or Na^+, K^+, $2Cl^-$ co-transporter proteins. However, for many species the main uptake has been shown to be via Cl^-/HCO_3^- exchange proteins, as suggested in the early model of gill transport processes illustrated earlier in Figure 5.19.

An important piece of evidence for Cl^-/HCO_3^- exchange is that stimulation of HCO_3^- excretion by intravenous infusion of HCO_3^- also results in an increase in Cl^- uptake, and increases the area of mitochondria-rich cells in the gill epithelium. In another study, uptake of Cl^- was reduced in trout treated with inhibitors of carbonic anhydrase, which reduces the availability of HCO_3^- for Cl^-/HCO_3^- exchange by the gill epithelium.

The process of Cl^- uptake by gill ionocytes appears to involve two types of Cl^-/HCO_3^- exchange proteins. These membrane proteins are part of a large family of solute carrier proteins (SLCs) that transport a multiplicity of solutes and that form many sub-families. An anion exchanger in the

[78] We discuss ammonia transporters in more detail in Section 7.4.1.

[79] Section 4.2.2 discusses proton pumps more generally.

[80] The basic structural arrangement of the subunits of Na^+, K^+-ATPase is illustrated in Figure 4.12; Figure 4.13 illustrates its pumping cycle.

[81] More information on groups, such as eels, that do not appear to rely on Cl^- uptake via the gills is given in Case Study 5.2.

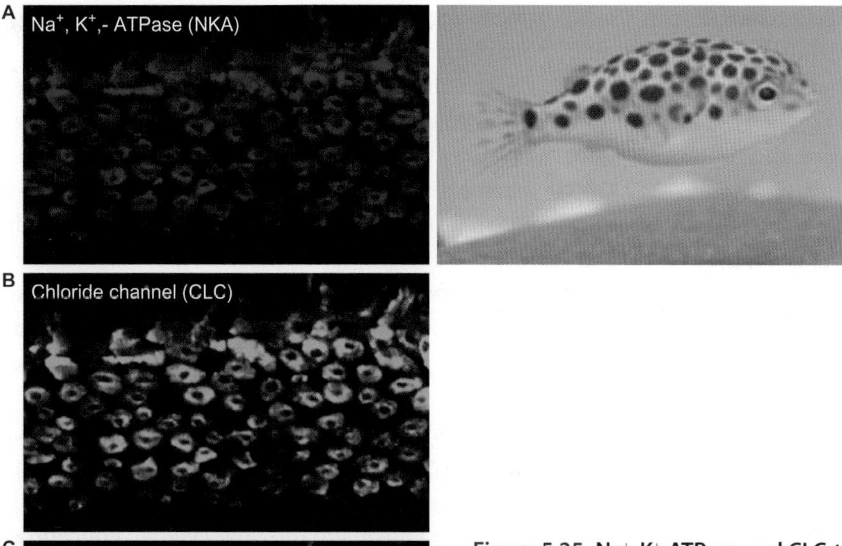

Figure 5.25 Na$^+$, K$^+$-ATPase and CLC-type chloride channels in gill filaments of spotted green pufferfish (*Dichotomyctere nigroviridis*) acclimated to fresh water

Antibodies to Na$^+$, K$^+$-ATPase and CLC were used separately and in combination, and fluorescent staining of bound antibody was examined by confocal microscopy. The results show co-localization of Na$^+$, K$^+$-ATPase in (A) red fluorescence, and CLC in (B) green fluorescence, in the same mitochondria-rich cells. (C) Shows the merged images, which clearly illustrates co-localization.

Source: Tang C-H et al (2010). Chloride channel ClC-3 in gills of the euryhaline teleost, *Tetraodon nigroviridis*: expression, localization and the possible role of chloride absorption. Journal of Experimental Biology 213: 683-693. Photo: © K H Liu

bicarbonate group of solute transporters (SLC4) occurs in the gills of some species of teleosts, while three members of the SLC26 group of multifunctional anion exchangers have been identified in zebrafish embryos.

Gene knockdown experiments of zebrafish provide the clearest evidence for the involvement of SLC26 proteins in Cl^- uptake: when transcription is inhibited by knocking down the three SLC26 genes, Cl^- uptake is reduced. An additional piece of evidence suggesting the importance of SLC26 transporters comes from studies of adult zebrafish, which respond to reduced water concentrations of Cl^- and increased concentrations of HCO_3^- by increasing the expression of mRNA for the SLC26 transporters in mitochondria-rich cells in the gills. These studies provide strong evidence for Cl^-/HCO_3^- exchange being responsible for the Cl^- uptake of freshwater teleosts, but it is not yet understood how such a process takes place against the unfavourable electrochemical difference (bearing in mind a negative charge occurs inside the cells relative to the outside).

Transfer of Cl^- across the basolateral membranes is also poorly understood, but has long been thought to occur via chloride channels, as shown in Figure 5.19. A member of the group of the CLC type of Cl^- channels[82] occurs in high levels in the gills of freshwater teleosts. Figure 5.25 shows some results from studies in which co-localization of CLC channels occurs with Na$^+$, K$^+$-ATPase in mitochondria-rich cells in the gill epithelium. The spotted green pufferfish examined in these studies tolerates a wide range of salinities and in doing so adjusts the abundance of chloride channels according to environmental salinity[83].

Ion transporter mechanisms may predispose animals to the possible uptake of ionic pollutants if these pollutants can interact with the ion transporters and substitute or compete with the normally transported ion. This principle is thought to explain the uptake of toxic levels of nitrite in freshwater animals that take up Cl^- using Cl^-/HCO_3^- exchange mechanisms. Case Study 5.2 examines this problem and what lies behind the different sensitivities of aquatic animals to nitrite.

Salt intake in the diet of freshwater animals

Many experiments on freshwater animals have starved them to avoid contamination of the experimental system by faeces, and to avoid the stimulatory effects of feeding on metabolic and enzymatic processes. Hence, the influence of salts in the food consumed by freshwater animals is often overlooked. However, a theoretical analysis of the likely intake of Na$^+$ in the diet of freshwater salmonids compared to intake via the

[82] Section 4.2.2 discusses types of chloride channels that are involved in osmoregulation. CLC proteins are dimers of two similar subunits that function as pathways for Cl^- conductance.

[83] The euryhalinity of pufferfish is considered further in Section 5.3.3.

Case Study 5.2 Nitrite toxicity is linked to osmoregulatory processes

Nitrite (NO_2^-) concentrations in unpolluted surface waters are usually very low (1–2 μg/l = 0.02–0.04 μmol L^{-1}). However, the concentrations of NO_2^- can increase if there is a large amount of decaying organic matter, or contamination by sewage or other effluents containing ammonia. High ammonia concentrations can overload bacterial oxidation of NO_2^- to nitrate (NO_3), which is part of the two-step pathway for nitrification of ammonia to nitrate illustrated in Figure A, and which forms part of the nitrogen cycle[1]. In natural habitats, polluting effluents can lead to water concentrations of NO_2^- that are 0.02–0.4 mmol L^{-1} or more, i.e. 1000 to 10,000 times higher than the concentrations in unpolluted water.

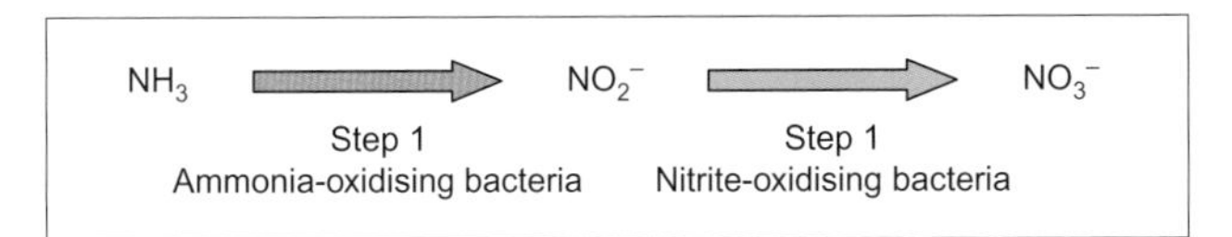

Figure A Steps in the two-step process of nitrification

The classical view is that nitrification of ammonia requires: first ammonia-oxidizing bacteria, such as *Nitrosomonas* sp. and then nitrite-oxidizing bacteria, such as *Nitrobacter* in soil and marine environments and *Nitrospira* sp. The recent identification of bacteria that are complete ammonia oxidizers (Comammox), such as *Nitrospira* spp. in freshwater and terrestrial environments, which possess all of the enzymes for complete nitrification of ammonia via nitrite challenges the long-held view of an obligatory division of the steps in nitrification between different bacterial species.

Aquatic animals themselves add to the ammonia load in their environment, as they excrete ammonia[2]. Hence, nitrite accumulates in the environment if step 1 of nitrification exceeds step 2—for instance, in intensive freshwater aquaculture systems. This means that there is a significant risk of nitrite reaching a concentration above which toxic effects are seen. (The no observable effect concentration (NOEC) for freshwater habitats for salmonid fish has been estimated as 0.01 mg L^{-1}.)

Aquatic animals exposed to NO_2^- in fresh water accumulate up to 10 times the environmental concentration of NO_2^- in their blood. The nitrite oxidizes iron in haemoglobin and forms methaemoglobin, which cannot bind with oxygen. Consequently, severe internal **hypoxaemia** may occur.

There is compelling evidence that NO_2^- mainly enters aquatic animals via the epithelial transporter sites that are responsible for Cl^- uptake, although the competition between Cl^- and NO_2^- is poorly understood mechanistically. Nevertheless, competition between NO_2^- and Cl^- explains why some aquatic animals are vulnerable to NO_2^- and others not.

Animals living in brackish waters or in seawater are not susceptible to NO_2^- uptake and accumulation because NO_2^- uptake is inhibited by the large number of Cl^- ions in saline water. Experiments with different levels of Cl^- added to the water indicate that a molar ratio of Cl^- relative to NO_2^- of 8:1 to 17:1 (depending on the species) completely alleviates NO_2^- toxicity. If we consider that a molar ratio of about 10:1 for Cl^- relative to NO_2^- completely protects a given species then Cl^- at a concentration of 1.8 mmol L^{-1} will block the toxicant action of 0.18 mmol L^{-1} nitrite (which can occur in a polluted environment). This is quite a low concentration of Cl^- ions, when considering environmental levels.

Seawater contains approximately 450 mmol L^{-1} chloride, so 0.4 per cent seawater (1.8 mmol L^{-1}) would contain sufficient Cl^- to protect animals against 0.18 mmol L^{-1} NO_2^-. This means that nitrite toxicity is highly improbable for animals living in estuaries and salt marshes or in seawater. Looked at from a different perspective, a NO_2^- concentration of above 0.45 mmol L^{-1} (20.7 mg L^{-1}) would be needed to overcome the protective effect of even 1 per cent seawater (approximately 4.5 mmol L^{-1} Cl^-). However, the risks for freshwater animals are much greater.

Many (though not all) freshwater fish take up Cl^- across the apical membranes of their gills using a Cl^-/HCO_3^- exchange mechanism, as illustrated by Figure 5.19. A similar transport mechanism occurs in crustacean gills. Uptake of some NO_2^- in place of Cl^- by these freshwater animals means that the plasma or haemolymph concentrations of Cl^- can be significantly reduced when they are exposed to nitrite.

In such situations, the ion-transporting cells often proliferate as a compensatory response. However, not all freshwater animals are equally sensitive to nitrite. The data in Figure B show that channel catfish (*Ictalurus punctatus*) are sensitive to NO_2^-. The channel catfish is an important species in aquaculture where NO_2^- concentrations may increase.

Animals that lack Cl^-/HCO_3^- exchange mechanisms, such as freshwater planarians, are less sensitive to NO_2^-. Among freshwater fish, sensitivity to NO_2^- is 1000 times greater in the most sensitive species than the least sensitive species. Eels (*Anguilla rostrata* and *Anguilla anguilla*) show low sensitivity to NO_2^-, which seems to be because of the minimal rate of Cl^- uptake across their gills. Eels must rely instead on Cl^- intake in their diet.

› Find out more

Kroupova H, Machova J, Piackova V, Blahova J, Dobsikova R et al (2008). Effects of subchronic nitrite exposure on rainbow trout (*Oncorhynchus mykiss*). Ecotoxicology and Environmental Safety 71: 813–820.

Tomasso JR, Grosell M (2005). Physiological basis for large differences to nitrite among freshwater and freshwater-acclimated euryhaline fishes. Environmental Science & Technology 39: 98–102.

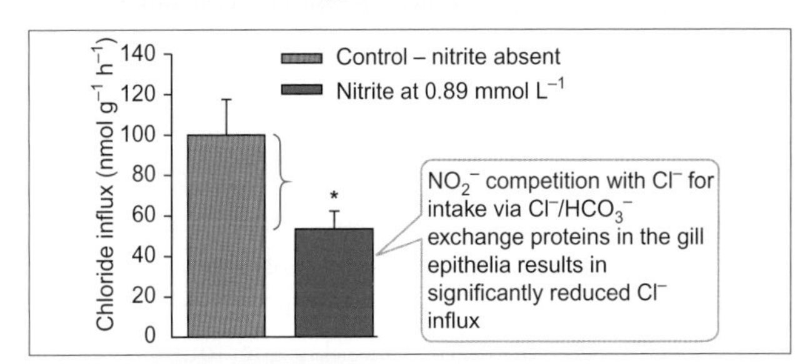

Figure B Effects of exposure to nitrite on chloride influx of juvenile channel catfish (*Ictalurus punctatus*)

Chloride influx was measured using $^{36}Cl^-$ as a radiolabelled tracer, monitoring its disappearance from the water and appearance in the fish. Values are means ± standard errors *n* = 5 per group; *P<0.05. Source: Tomasso JR, Grosell M (2005). Physiological basis for large differences in resistance to nitrite among freshwater and freshwater-acclimated euryhaline fishes. Environmental Science & Technology 39: 98-102.

[1] The other part of the nitrogen cycle is denitrification (reduction) that converts nitrate to nitrite and nitrogen gas.

[2] We find out more about ammonia excretion in Section 7.4.1.

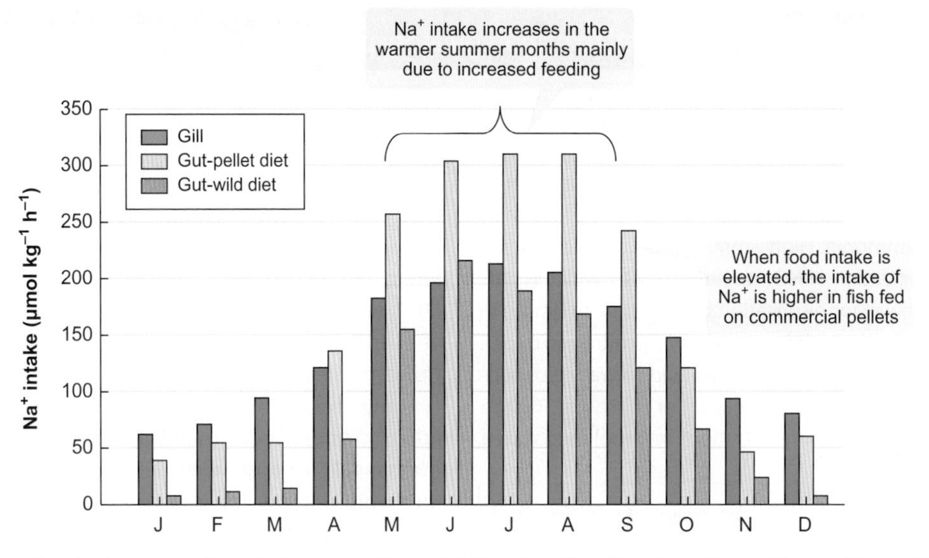

Figure 5.26 Seasonal variation in predicted dietary and branchial intake of sodium by wild and farmed juvenile salmonid fish living in fresh water

The predicted Na^+ influxes are based on data for branchial Na^+ influxes of rainbow trout (*Oncorhynchus mykiss*) held at different water temperatures and consuming either a diet of invertebrates (as for wild fish) or commercial pellets (as in farmed fish) in the northern hemisphere. Data show Na^+ intake via the gills and gut. For gut intake of Na^+, the intake is shown for fish fed on commercial food pellets and fish fed a wild diet of invertebrates.

Source: Wood CM, Bucking C (2010). The role of feeding in salt and water balance, in: Fish Physiology, Volume 30: The Multifunctional Gut (Eds: Grosell M, Farrell AP, Brauner CJ). Cambridge MA: Academic Press.

gills, which is summarized by Figure 5.26, suggests dietary salt intake can be significant.

In temperate regions, reduced feeding of aquatic ectotherms occurs when water temperatures decline[84], which reduces the dietary intake of Na^+. Figure 5.26 indicates the dominance of Na^+ uptake via the gills of freshwater teleosts in winter. In contrast, when feeding increases in the warmer summer months, gut absorption of Na^+ is predicted to equal or exceed uptake via the gills, although intake by both routes increases.

The graph in Figure 5.26 also shows profound differences between captive fish fed on commercial pellets and wild fish-eating invertebrates: captive fish consume much more salt in their diet than wild fish. Dried pellet food contains around 220 mmol Na^+ per kg, which is similar to the Na^+ content of dried food organisms such as freshwater shrimps, water hoglouse (*Asellus aquaticus*) and the larvae of species of stoneflies and caddis flies. However, it is important to remember that wild food organisms contain about 75 per cent water, so a diet of wet invertebrates contains about 55 mmol Na^+ per kg.

The predictions of Figure 5.26 lay the foundation for experimental studies. From such studies, it has become evident that the diet provides most of a fish's needs for K^+, 50 per cent of its intake of inorganic phosphate, 15–30 per cent of Ca^{2+} and 50 per cent of Mg^{2+} requirements. The balance is obtained from fresh water by uptake via the gills or skin. For freshwater fish, the dietary intake of Na^+ and Cl^- varies between 0 and 60 per cent of the total intake and varies seasonally because of the seasonal variations in food intake. Typically, a significant increase in plasma concentrations of Na^+ and/or Cl^- occurs within a few hours of eating a meal containing 18 mmol NaCl kg^{-1} or more.

We have seen that food intake can play a significant role in the salt balance of aquatic animals, but food shortage or even starvation does not usually significantly reduce the plasma concentrations of Na^+ and Cl^- of most species. This is because most species can compensate for the shortfall by taking up extra Na^+ and Cl^- via their gills or skin. However, dietary intake of NaCl becomes more important in species that lack particular mechanisms for ion uptake via the gills. Species, such as eels, that cannot take up Cl^- via their gills (or skin) are more dependent on their diet for Cl^- balance.

Ion uptake in the diet is also likely to be particularly important if the uptake of ions by the gills becomes compromised for example by exposure to various toxicants. For example, fish living in the ion-poor low pH waters of Scandinavia, the Canadian Shield and the Amazon basin exhibit reduced ion influx via the gills coupled with an increase in passive ion losses. In these conditions, the intake of ions in food is particularly important, and feeding may increase. Experimental studies of rainbow trout exposed in the laboratory to pH 5.2 ion-poor soft water (sub-lethal conditions)

[84] Metabolic rate is reduced by low environmental temperatures (except for endotherms: birds and mammals), as we discuss in Section 2.2.3.

found they eat more than rainbow trout kept at a neutral pH. In a similar way, dietary NaCl helps fish to survive exposure to elevated concentrations of dissolved copper. There is no evidence that these fish are able to select saltier food. However, the increased loss of Na^+ across the gills and in the urine of fish exposed to copper are more effectively replaced when they can obtain Na^+ in their diet.

› *Review articles*

Chasiotis H, Kolosov D, Bui K, Kelly SP (2012). Tight junctions, tight junction proteins and paracellular permeability across gill epithelium of fishes: A review. Respiratory Physiology & Neurobiology 184: 269–281.

Kumai Y, Perry S (2012). Review. Mechanisms and regulation of Na^+ uptake by freshwater fish. Respiratory Physiology & Neurobiology 184: 249–256.

Larsen EH, Deaton LE, Onken H, O'Donnell M, Grosell M, Dantzler WH, Weihrauch D (2014). Osmoregulation and excretion. Comprehensive Physiology 4: 405–573.

Perry SF, Shahsavarani A, Georgalis T, Bayaa M, Furimsky M, Thomas SLY (2003). Channels, pumps and exchangers in the gill and kidney of freshwater fishes: Their role in ionic and acid-base regulation. Journal of Experimental Zoology 300A: 53–62.

Wood CM, Bucking C (2010). The role of feeding in salt and water balance. In: Fish Physiology, The Multifunctional Gut of Fish: Volume 30 (Eds: Grosell M, Farrell AP, Brauner CJ). Chapter 5, pp. 165–212. Cambridge, MA: Academic Press.

5.3 Osmoregulation in changing salinities

Most aquatic animals can survive only a narrow range of salinities (osmotic concentrations) and live in seawater or fresh water—they are **stenohaline** (from Greek for narrow: *stenos* and salt: *halos*). Some, however, are **euryhaline** (from Greek: *eurys* meaning wide), meaning that they tolerate variable salinities. Although euryhaline animals are less common than stenohaline ones, there are euryhaline representatives among aquatic insect larvae, polychaete worms, crustaceans (crabs, prawns), fish, amphibians and reptiles. Some live in rock pools in intertidal areas, some in estuaries in which the input of river water reduces the salinity to a level between that of seawater and fresh water[85], and also the salinity varies with the tidal cycle, which creates particularly challenging conditions for the animals living in such habitats.

Species that can tolerate estuarine conditions can do well as these are often areas of high primary productivity and reduced competition. Therefore, in areas such as the Chesapeake Bay on the Atlantic coast, which is the largest estuary in the US, there are large populations of species of crustaceans, molluscs and fish that are tolerant of variable salinities. These populations form a crucial food source for migrant wading birds.

[85] Section 1.2.1 discusses the variability of salinity in natural habitats.

There is no exact cut-off between euryhaline and stenohaline animals, so qualifying terms, such as 'extremely' and 'slightly', are necessary to describe individual species as anything from extremely stenohaline to slightly euryhaline or extremely euryhaline. A capacity for saline tolerance has been linked to the spread of invasive species, particularly invertebrates and fish that hitch a ride in the ballast tank of ships. Saline tolerant species from the Ponto-Caspian area (Black, Azov and Caspian Seas) are particularly problematic, and their colonization success has been attributed to their tolerance of a wide range of temperatures and salinities.

5.3.1 Euryhaline osmoconformers and osmoregulators

When aquatic animals are exposed to water of a range of osmotic concentrations, two major sub-categories of euryhaline are recognizable. Figure 5.27 illustrates the distinguishing characteristics of perfect **osmoconformers** and perfect **osmoregulators**.

- Osmoconformers are species that conform to the osmotic changes in environmental osmolality. These aquatic animals are isosmotic to their environment. Hence, euryhaline osmoconformers tolerate a wide range of changes in the osmolality of their extracellular fluids.
- Osmoregulators are species that regulate the osmotic composition of the extracellular fluids such that their cells are bathed in fluids of a relatively stable osmolality even when environmental osmolality alters. Perfect osmoregulators hyperosmoregulate at external osmolalities below the isosmotic point (when body fluid osmolality equals external osmolality), while hypo-osmoregulation occurs when external osmolalities exceed the isosmotic point.

Euryhaline osmoconformers follow the changes in external osmolality

Most marine invertebrates are stenohaline osmoconformers with a limited ability to tolerate the changes in body fluid osmolality that follow changes in external osmolalities. By contrast, some marine and estuarine invertebrates are more euryhaline, and are able to survive in a wider range of salinities despite their osmoconformation.

Good examples of euryhaline osmoconformers occur among molluscs, including oysters and mussels, and

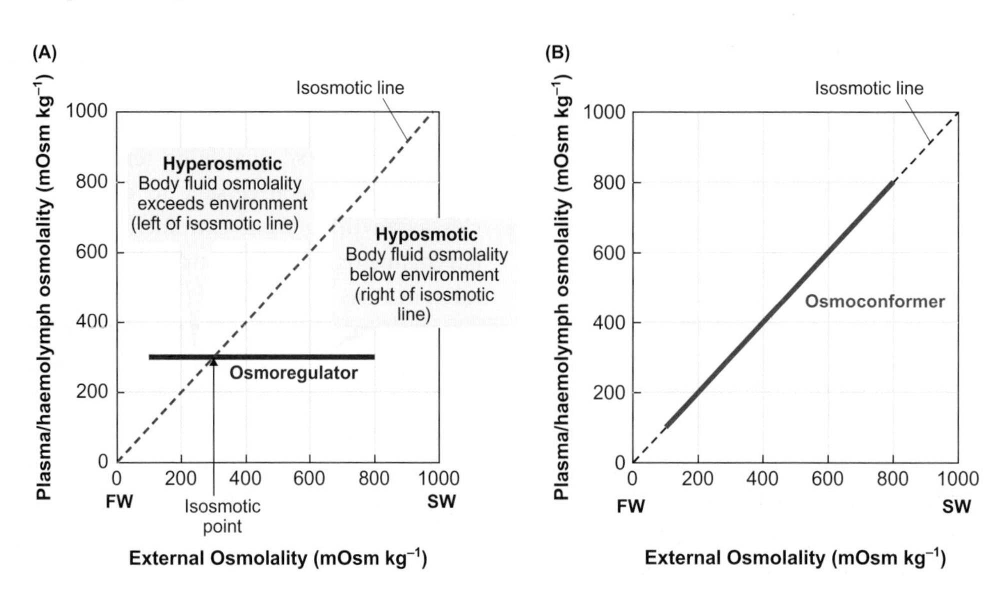

Figure 5.27 Principle of osmoregulation and osmoconformation in theoretical euryhaline animals that are perfect osmoregulators (blue line) or perfect osmoconformers (red line)

(A) The perfect osmoregulator has a constant osmolality over a range of external osmolalities. The graph shows external osmolalities between fresh water (FW) rivers and lakes, and seawater (SW). Osmoregulators show hyperosmoregulation at external osmolalities that are less than the isosmotic point at which the body fluid osmolality equals that of the external environment. Hypo-osmoregulation occurs when external osmolalities exceed the isosmotic point.
(B) The osmolality of the body fluids of a perfect osmoconformer follows the isosmotic line, i.e. identical changes occur in body fluid osmolality to the changes in the osmolality of the external medium.

polychaete worms such as ragworms (*Hediste diversicolor*). Figure 5.28 shows the relationship between the osmolality of the environment and the osmolality of extracellular fluids for ragworms, which only deviate from osmoconformation when external osmolality is less than about 600 mOsm kg^{-1} (about 60 per cent of ocean seawater).

Osmoconformers exposed to changing external osmotic concentrations face challenges to their extracellular fluid volume and cell volumes. If external salinity increases, an osmoconforming animal loses water by osmosis; if external salinity decreases, an uptake of water via permeable surfaces results in volume expansion.

Euryhaline osmoconformers within their tolerance range are able to regulate extracellular fluid volume and cell volumes. For example, lugworms (*Arenicola marina*), which are a common polychaete in intertidal areas with variable salinities, are able to volume regulate completely when transferred from pure seawater to about 60 per cent seawater, but if they

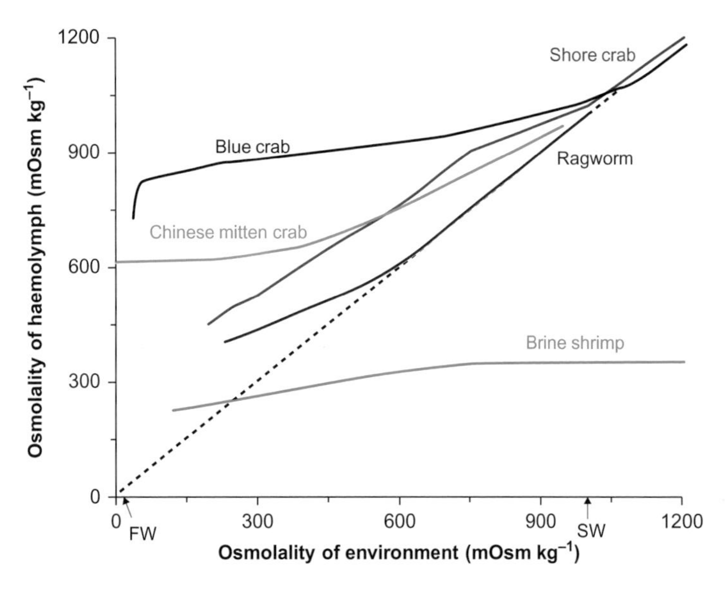

Figure 5.28 Change in osmolality of the haemolymph (blood) of estuarine invertebrate animals during their exposure to different external osmolalities

The dashed line is the line of equality between the osmolality of the environment and haemolymph (blood), i.e. the isosmotic line. Coloured lines show changing haemolymph osmolality of brine shrimps (*Artemia* sp.), shore crabs (*Carcinus maenas*), polychaete ragworms (*Hediste diversicolor*), Chinese mitten crabs (*Eriocheir sinensis*) and blue crabs (*Callinectes sapidus*) when exposed to different external osmolalities. Approximate osmolalities for fresh water (FW) and ocean seawater (SW) are indicated. Except for Chinese mitten crabs these species cannot tolerate freshwater habitats, so the lines stop before reaching the osmolality of fresh water.

are moved to 30 per cent seawater they retain about 10 per cent of the water intake.

An increase in external osmolality resulting in a similar increase in osmolality of the extracellular fluids (osmoconformation) requires an increase in the cell content of osmolytes to achieve cell volume regulation, while a decrease in external osmolality requires a decrease in the cell content of osmolytes[86]. The particular osmolytes that are regulated vary between species but commonly include the following amino acids: glycine, alanine, proline, serine, taurine and glycine betaine. As an example, Figure 5.29 shows data for the amino acid content of cells from the body wall of polychaete worms (*Hediste japonica*) and demonstrates their ability to adjust the cell content of amino acids (alanine and glycine, in particular), in response to changing environmental salinity[87].

The release of cellular amino acids into the extracellular fluids when euryhaline osmoconformers enter lower salinities is generally followed by deamination of the amino acids (removal of $-NH_2$) and excretion of ammonia[88] (but the carbon skeletons are presumably retained). The reduced intracellular concentrations of amino acids in bivalve molluscs exposed to reduced salinities also results from an increased use of amino acids in protein synthesis. In contrast, intracellular free amino acid content of mussel tissues increases when they are exposed to higher salinities, due to an increase in the transport of amino acids into the cells and breakdown (catabolism) of intracellular proteins.

The pattern of changes in individual amino acids varies, which indicates independent regulatory mechanisms. In bivalve molluscs, cytoplasmic levels of alanine rise very quickly after transfer to a higher salinity, and other amino acids, such as glycine and taurine, accumulate over time. A similar pattern of rising levels of alanine, in advance of glycine, occurs in polychaete worms, and is illustrated by the data in Figure 5.29.

In both polychaetes and molluscs, cell volume regulation occurs more rapidly and more completely when external osmolality decreases than when external osmolality increases. This is probably because the release of amino acids from cells when animals are exposed to low salinities is an easier process than synthesizing extra amino acids and transporting them into cells or breakdown of intracellular proteins. For example, when the heart (ventricle) of the ribbed mussel *Geukensia demissus*, initially held in normal seawater, is exposed to 50 per cent seawater, an efflux of taurine, glycine and alanine from the heart cells occurs over the next 2 h. The rapid efflux of amino acids may result from an increase in the permeability of the plasma membrane to amino acids when an influx of water into the cells increases the tension on the cell membrane.

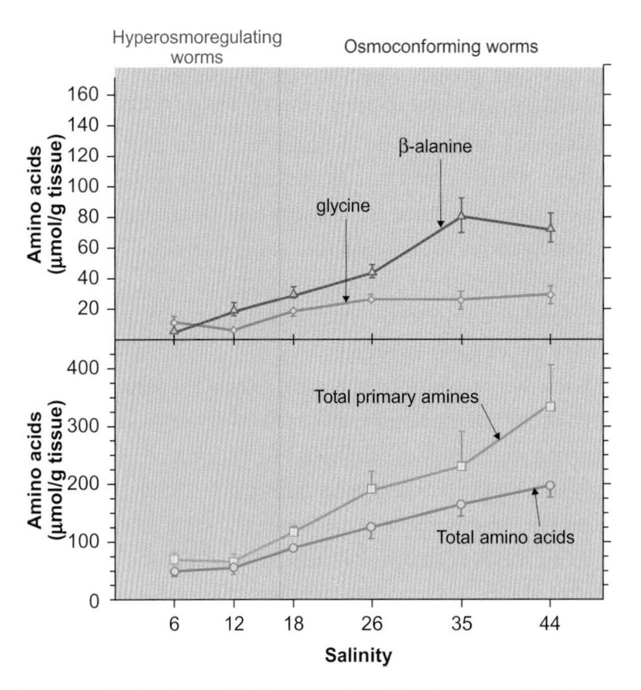

Figure 5.29 Amino acids in the cells of the body wall of the polychaete worm, an Asian ragworm, *Hediste japonica*, held in water of different salinities

The worm osmoconforms between salinities of 16 and 44 psu (area of blue overlay), but at lower salinities (pink overlay) shows some hyperosmoregulation, as other polychaetes such as ragworms (see Figure 5.28). When salinity increases, the rise in extracellular fluid osmolality is accompanied by an increase in amino acids within the cells. Note the earlier and greater rise in alanine than glycine, which suggests independent regulation of these amino acids.

Data are means ± standard deviations for 4–9 measurements.

Source: adapted from Hoeger U, Abe H (2004). β-Alanine and other free amino acids during salinity adaptation of the polychaete *Nereis japonica*. Comparative Biochemistry and Physiology Part A 137: 161–171.

Euryhaline osmoregulators control the osmolality of their body fluids

Animals that mainly live in fresh water typically hyperosmoregulate when they are in fresh water or when experiencing low external salinities (brackish water). Above the isosmotic point (when body fluid osmolality equals the external osmolality), however, they osmoconform (i.e. are isosmotic). This pattern of osmoregulation is called **hyper-isosmotic regulation** and also occurs in many marine invertebrates that osmoconform in seawater and slightly lower salinities, but begin to hyperosmoregulate as salinities decline.

Shore crabs (*Carcinus maenas*) that live on European coasts and in the more saline reaches of estuaries illustrate the hyper-isosmotic pattern of osmoregulation, which is shown by the red line in Figure 5.28. By contrast, Chinese

[86] The principles involved in cell volume regulation are examined in more detail in Section 4.3.

[87] Figure 4.25 illustrates the general processes in adjusting cell content of amino acids for cell volume regulation.

[88] Section 7.4.1 discusses ammonia excretion by animals in aquatic environments.

mitten crabs (*Eriocheir sinensis*) have more stable blood osmolality and greater tolerance of low salinities (osmolalities) than most crabs, and can tolerate both fresh water and seawater, as indicated by the pale blue line in Figure 5.28. Chinese mitten crabs are an invasive pest in Europe, and their salinity tolerance is an important factor in competing with indigenous shore crabs that cannot tolerate fresh water. Chinese mitten crabs move further up estuaries, beyond the tidal limit and into fresh water, but must return to the outer estuary or seawater to breed.

Blue crabs (*Callinectes sapidus*), which are common along much of the Atlantic and Gulf coasts of America, are also tolerant of a wide range of salinities, but must breed in higher salinity waters: the larvae of blue crabs cannot tolerate salinities of less than about 22 psu[89]. However, the euryhalinity of blue crabs increases with development, and adult blue crabs have a better ability to hyperosmoregulate than most marine and estuarine crustaceans. Adult blue crabs are hyper-isosmotic regulators and maintain a relatively stable haemolymph osmolality over much of their salinity range, as illustrated by the black line in Figure 5.28.

In contrast to a hyper-isosmotic pattern of regulation, some animals demonstrate an alternative pattern of **hyper-hyposmotic regulation**, in which the hyperosmoregulation occurs when they are in relatively low salinities but alters to hypo-osmoregulation at higher salinities (above an isosmotic point). This pattern comes closer to the theoretical perfect osmoregulation of Figure 5.27A.

Hyper-hyposomotic regulation occurs in all migratory species of agnathan and teleost fish, and in some crustaceans, such as the brine shrimps (*Artemia* spp.), as illustrated by the green line in Figure 5.28. The ability of brine shrimps to hypo-osmoregulate allows them to thrive in hypersaline lakes, such as the Great Salt Lake in the US in salinities of up to six times higher than ocean seawater[90].

5.3.2 Euryhaline sharks and rays

Elasmobranchs (sharks and rays) live mainly in marine environments, but some species are euryhaline. Two species known to undertake long migrations from coastal areas through brackish water and into fresh water are Atlantic stingrays (*Dasyatis sabina*) and bull sharks (*Carcharhinus leucas*). Tagging studies have shown that some bull sharks migrate between Lake Nicaragua and the Caribbean Sea, and some also migrate along the Brisbane River in Australia.

Euryhaline species of elasmobranchs, like their marine relatives, remain tied to urea retention as a fundamental element of their hyperosmoregulation, whether they are in seawater or fresh water. Figure 5.30 illustrates the hyperosmoregulation of bull sharks. When captured at various locations along the Brisbane River, Australia, these bull sharks are found to reduce their plasma osmolality, primarily by reducing plasma concentrations of urea, as they move into progressively lower salinities. A higher level of hyperosmoregulation in lower salinities increases osmotic water influx, so the bull sharks need to excrete a greater volume of water.

Urine production of Atlantic stingrays increases by a factor of eight when they are held in brackish water (440 mOsm kg^{-1}) compared to the volume they excrete when held in higher salinity water (850 mOsm kg^{-1}), closer to ocean seawater. In fresh water, Atlantic stingrays excrete 15 times more urine per hour than stingrays in seawater.

High volumes of urine carry greater amounts of urea, but this is not the only explanation for the declining plasma concentrations of urea when elasmobranchs enter water with lower salinity. Reduced synthesis of urea may also occur. Furthermore, some species reduce urea transporters in the kidney tubules, as shown by the data in Figure 5.31; the declining urea retention by the kidney increases urea excretion.

Sharks, skates and rays have less need for a salt-excreting rectal gland[91] when they are living in waters with lower Na^+ and Cl^- concentrations than seawater. The rectal gland has atrophied in bull sharks living in the Rio San Juan and Lake Nicaragua, which are unlikely to venture from the lake and down the vast river system into saline waters (although some do). However, bull sharks that undertake frequent migrations along estuaries need to be able to switch on salt secretion by the rectal gland when they enter more salty environments. This may explain why bull sharks captured in fresh water in the Brisbane River (Australia), 70 km upstream, have rectal glands similar in size relative to their body mass to the rectal glands of bull sharks captured closer to the sea.

Spending short periods in lower salinity water probably has more influence on the secretory activity of the rectal gland than its size. Studies on Atlantic stingrays support this idea, as the Na^+, K^+-ATPase activity of the rectal gland declines when the stingrays are in lower salinities. If we look back at Figure 5.14 we see the involvement of Na^+, K^+-ATPase in salt secretion by the rectal glands.

Although Na^+, K^+-ATPase activity of the rectal gland of Atlantic stingrays declines when they enter low salinity water, the Na^+, K^+-ATPase activity of their gill tissue increases. The higher Na^+ (208 mmol L^{-1}) and Cl^- (203 mmol L^{-1}) concentrations in the extracellular

[89] Open oceans have an average salinity of 35 practical salinity units (psu) (ranging from 33–37), as we discuss in Section 1.2.1.

[90] Section 1.2.1 and Figure 1.8 provide more information on the salinity of the Great Salt Lake.

[91] Section 5.1.5 discusses the role of the rectal gland in salt secretion by marine elasmobranchs.

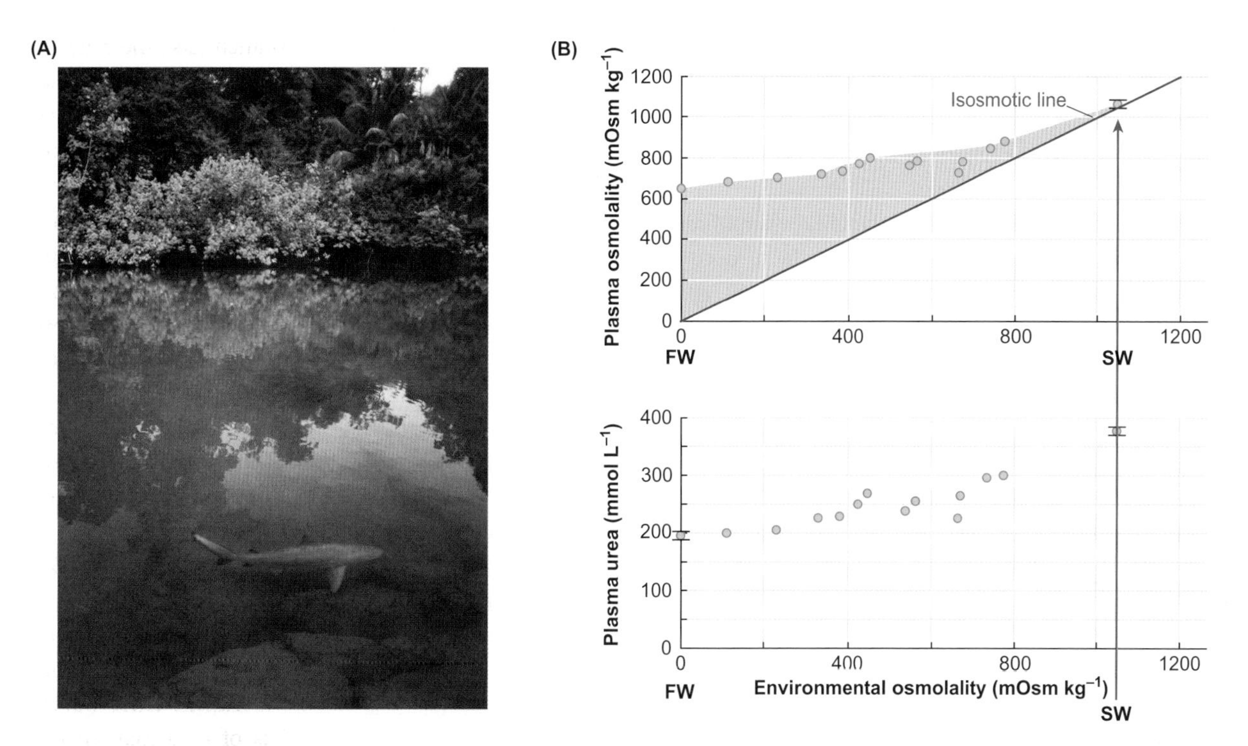

Figure 5.30 Bull sharks (*Carcharhinus leucas*) migrating up rivers reduce the plasma concentration of urea and plasma osmolality

(A) Bull shark entering a river. (B) Plasma osmolality and urea concentrations of bull sharks captured in environments of different osmolalities along the Brisbane River, Australia. Plasma osmolality is always above that of the environment (i.e. bull sharks always hyperosmoregulate). For comparison, the red line indicates a state of isosmocity between the environment in which they captured and their extracellular fluids, which never exists. Although plasma osmolality declines when bull sharks leave seawater (SW) and enter lower salinity waters, plasma osmolality never drops below 640 mOsm kg^{-1} even in fresh water (FW). The shaded area emphasizes the progressive increase in the level of hyperosmoregulation as external osmolality declines. Most of the decrease in plasma osmolality is due to the decrease in the concentration of urea, although the plasma concentration of urea is 192 mmol L^{-1} even in bull sharks in fresh water.

Source: adapted from Pillans RD, Franklin CE (2004). Plasma osmolyte concentrations and rectal gland mass of bull sharks *Carcharhinus leucas*, captured along a salinity gradient. Comparative Biochemistry and Physiology Part A 138: 363-371. Photograph: © Adrian Hepworth/www.photoshot.com

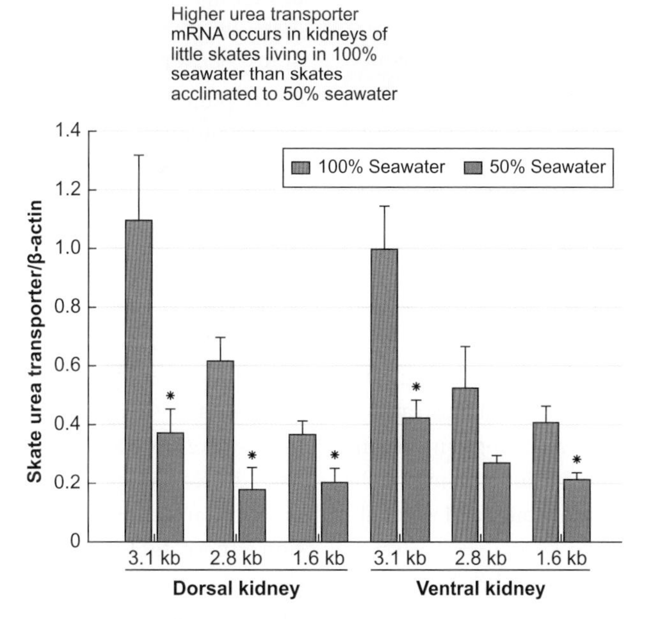

Figure 5.31 Urea transporter mRNA in kidneys of little skate (*Leucoraja erinacea*) living in 100% or 50% seawater

Expression of three different-sized bands (3.1, 2.8 and 1.6 kb) were measured in ventral and dorsal sections of the kidney. Data for expression of the urea transporter are shown relative to the expression of a housekeeping gene (β-actin mRNA used as a control gene). Values are means ± standard errors. Asterisks indicate statistical significance between the four fish in each experimental group.

Source: Morgan RL et al (2003). Regulation of a renal urea transporter with reduced salinity in a marine elasmobranch, *Raja erinacea*. Journal of Experimental Biology 206: 3285-3292.

fluids of bull sharks than in fresh water result in salt loss via the gills, as in freshwater teleosts (illustrated by the schematic in Figure 5.17). To maintain salt balance an uptake of these ions (or intake in the diet) is essential; this process involves Na^+, K^+-ATPase.

The uptake of Na^+ and Cl^- by euryhaline elasmobranchs when living in low salinity waters is thought to occur via two distinct populations of ionocytes in the gill epithelium. Chloride uptake appears to take place via a Cl^-/anion exchange protein known as **Pendrin**, driven by a basolateral V-type H^+-ATPase, while sodium uptake occurs via different cells, probably via apical Na^+/H^+ exchange (as in freshwater teleosts and agnathans). Uptake of Na^+ into the cells is followed by pumping of Na^+ into the extracellular fluids across the basolateral borders of the cells via Na^+, K^+-ATPase. This process mirrors that shown for freshwater teleosts back in Figure 5.23.

5.3.3 Euryhaline teleosts

The most remarkable examples of euryhalinity occur among teleost fish, such as species of salmon and eels, and the jawless lampreys that migrate over long distances between freshwater and seawater habitats at various stages in their life cycles. Two types of life cycle migrations occur:

- **Anadromous** migrations (from the Greek *ana* meaning up and *dromos* meaning running). Salmon species and lampreys are anadromous fish migrating from the seas and oceans up rivers to spawn.
- **Catadromous** migrations (from the Greek *kata* meaning down, and *dromos* running) when fish move down rivers into the sea to spawn. Eel species are remarkable catadromous species. European eels (*Anguilla anguilla*) migrate from rivers and estuaries some 6000 km (3700 miles) to spawning grounds in the Sargasso sea (south of Bermuda) that overlap with those of the American eel (*Anguilla rostrata*).

When migratory teleosts are in fresh water, they hyperosmoregulate, and function like freshwater teleosts, as illustrated earlier in Figure 5.17. When they are in seawater they hypo-osmoregulate like the marine teleost illustrated in Figure 5.3. In fact, only relatively small changes in plasma osmolality occur because of the altered functioning of the gills and kidney and changes in drinking rates and water absorption in the gastrointestinal tract.

Euryhaline teleosts in fresh water or brackish water excrete copious volumes of hyposmotic urine and drink minimal amounts[92], absorbing only small quantities of water from the intestine. By contrast, in seawater, euryhaline teleosts excrete small volumes of isosmotic urine, and drinking increases. It has been speculated that claudins maintain the low paracellular permeability of the intestinal epithelium to divalent ions (Mg^{2+} and SO_4^{2-}) of euryhaline teleosts living in seawater in which the functioning of the intestinal epithelium becomes tuned to the acquisition of water. An increase in Na^+, K^+-ATPase in the intestine of the migrating silver European eels (*Anguilla anguilla*) is a good example of the activation of the transport processes that drive water absorption from the intestine (by the interlinked processes that we examined back in Figure 5.4). Acclimation to seawater results in enhanced bicarbonate secretion by the intestinal epithelium and an increase in intestinal aquaporins enhances water withdrawal from the intestine.

Of all the ionoregulatory tissues of euryhaline teleosts, the gills are the most studied. Environmental salinity has profound effects on the way that cells making up the gill epithelium link together at their apical borders. The deep cell junctions that occur at the apical border of the gill epithelium of fish acclimated to fresh water and which maintain the relatively low gill permeability to ions are replaced by leaky junctions which allow Na^+ excretion when euryhaline teleosts enter hyperosmotic environments (such as seawater).

Recent work indicates the importance of tight junction proteins in adjusting the paracellular permeability of the gill epithelium to ions in euryhaline teleosts. Although relatively few species have been studied and the exact changes are uncertain, two claudins (claudin-10d and -10e) appear to be important in acclimation to seawater based on their increased abundance and co-localization with ionocytes rich in Na^+, K^+-ATPase in species such as the euryhaline spotted green pufferfish (*Tetraodon nigroviridis*). An increase in these claudins is thought to allow Na^+ excretion via the paracellular pathway when the fish are living in hyperosmotic environments or during the catadromous migration to sea of species such as the Atlantic salmon (*Salmo salar*).

The gills excrete Na^+ and Cl^- when euryhaline fish are in hyperosmotic environments like seawater, using the transport processes examined back in Figure 5.7. In contrast, when they enter hyposmotic environments (like fresh water) a diverse array of adjustments occur that see the gills switching to take up Na^+, Cl^- and Ca^{2+} instead[93]. Recent studies have developed specific antibodies to investigate transport proteins, and molecular probes to investigate gene expression. These tools are being used to explore the specific changes in the gills (and other tissues and organs) of euryhaline fish. As an example, Figure 5.32 examines the changes in chloride channels in the gill ionocytes of spotted green

[92] We discuss kidney function of euryhaline teleosts further in Section 7.1.3, aspects of the hormonal regulation of kidney function in Sections 21.1.4 and 21.1.2, and hormonal control of drinking in Section 21.1.3.

[93] The mechanisms for Na^+ and Cl^- uptake are discussed in Section 5.2.2; Section 21.2.2 discusses calcium uptake by the gills of teleosts living in fresh water and its hormonal regulation.

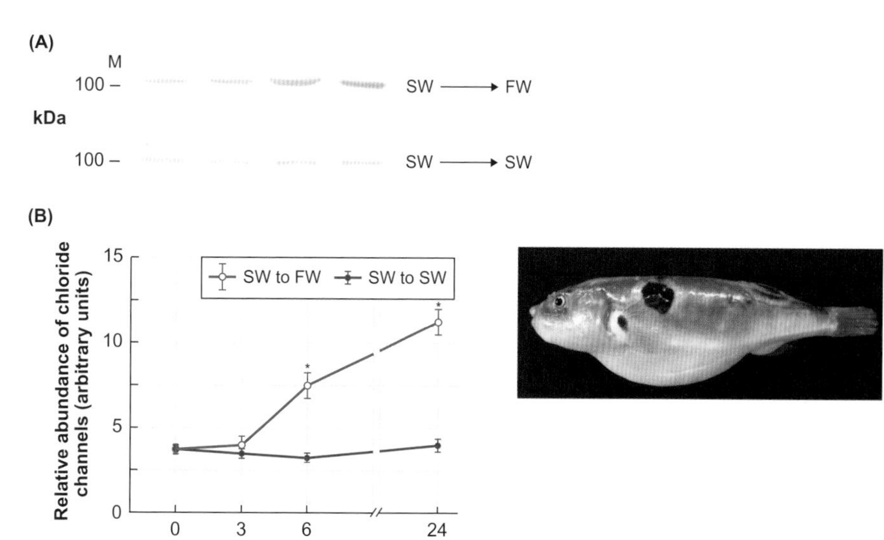

Figure 5.32 Abundance of chloride channels in basolateral membranes of ionocytes in gill filaments of spotted green pufferfish (*Dichotomyctere nigroviridis*) increases following transfer from seawater (SW) to fresh water (FW)

(A) Immunoblots of a single band at 102 kDa representing chloride channels (CLC-3 protein) in gill tissue. Gill samples were taken at 0 h (i.e. before transfer) and at 3, 6 and 24 h after transfer of pufferfish from SW to FW, and from control pufferfish transferred from SW to SW. Transfer from SW to FW increased the amount of CLC protein. (B) Relative abundance of chloride channel protein at various time points after transfer from SW to FW shows an increase by almost three times, compared to controls (SW to SW transfer).

Source: Tang C-H, Hwang L-Y, Lee, T-H (2010). Chloride channel ClC-3 in gills of the euryhaline teleost, *Tetraodon nigroviridis*: expression, localization and the possible role of chloride absorption. Journal of Experimental Biology 213: 683-693.

pufferfish. Looking back at Figure 5.25 reminds us of the co-localization of chloride channels and Na^+, K^+-ATPase in the ionocytes of spotted green pufferfish. The abundance of chloride channels in pufferfish acclimated to seawater increases rapidly (within a few hours) when the pufferfish are transferred to fresh water, as shown in Figure 5.32; in fresh water Cl^- uptake (rather than excretion) becomes essential.

Na^+, K^+-ATPase is a key component of the ion transport processes of gill ionocytes regardless of whether fish are living in fresh water and actively taking up ions, or are in seawater where active ion efflux occurs. The ionocytes concerned with ion uptake or ion excretion can be investigated by using antibodies to Na^+, K^+-ATPase, as demonstrated in Figure 5.33. Note that Na^+, K^+-ATPase occurs in ionocytes in the lamellar epithelium of Atlantic salmon smolts (the stage that migrates to seawater) when they are in fresh water, but these cells disappear when the salmon smolts are in seawater. This observation has led to the hypothesis that lamellar ionocytes are responsible for ion uptake when the salmon is in fresh water, but this role becomes unnecessary in seawater. Larger ionocytes in the filament epithelium at the base of the lamellae, shown in Figure 5.33B, are more prominent in salmon acclimated to seawater than those in fresh water; these ionocytes are primarily responsible for NaCl excretion via the gills when in seawater. Several studies have demonstrated that growth hormone and cortisol[94] trigger similar increases in the size and number of ionocytes, although their interactions and exact mechanisms of action at a cellular level are not fully understood.

While the functional division of ionocytes shown in Figure 5.33 occurs in the migratory stage of salmonids (the smolt), and in eels, it is not a strong feature of those non-migratory euryhaline fish that retain an ability to acclimate to different

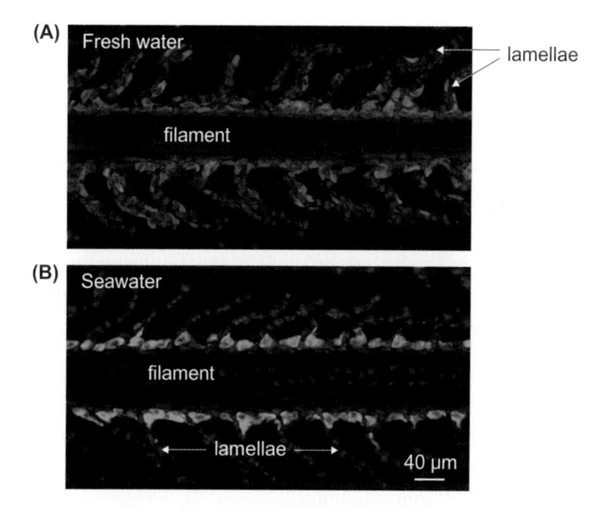

Figure 5.33 Ionocytes of Atlantic salmon (*Salmo salar*) smolts in fresh water (A) and after transfer to seawater (B)

Each image shows the gill filament running left to right with lamellae either side of the gill filament perpendicular to it. Binding of an antibody to the α subunit (the main catalytic unit) of Na^+, K^+-ATPase shows as green; cell nuclei are counterstained magenta. Na^+, K^+-ATPase is located in the extensive tubular system and hence immunoreactivity occurs through most of the cell. In salmon in fresh water, Na^+, K^+-ATPase positive ionocytes occur in the filament epithelium at the base of the lamellae, and in the lamellar epithelium. In salmon in seawater, the Na^+, K^+-ATPase positive ionocytes in the lamellae disappear, while those in the filament epithelium increase in size and show stronger staining.

Source: Hiroi J, McCormick SD (2012). New insights into gill ionocytes and ion transporter function in euryhaline and diadromous fish. Respiratory Physiology and Neurobiology 184: 257-268.

[94] Section 19.2.2 discusses the peptide hormone growth hormone which is synthesized and secreted by the anterior pituitary, while Section 19.3.1 discusses the steroid hormone cortisol.

salinities throughout their life. Some non-migratory euryhaline species, such as spotted green pufferfish, lack lamellar ionocytes containing high levels of Na^+, K^+-ATPase, regardless of whether they are in fresh water or seawater.

In salmon and trout, the organization of Na^+, K^+-ATPase-rich ionocytes in the lamellae or filament epithelium of the gills and the development of different cell types has been linked to the presence of different isoforms of Na^+, K^+-ATPase[95]. These were first demonstrated in rainbow trout (*Oncorhynchus mykiss*) in which a freshwater isoform of Na^+, K^+-ATPase increases when the trout acclimate to fresh water, whereas a seawater isoform of Na^+, K^+-ATPase dominates in the gills when the trout acclimate to seawater.

Atlantic salmon undergo a **parr-smolt transformation** prior to seaward migration in which differences in Na^+, K^+-ATPase isoforms and their localities develop. The freshwater isoform of Na^+, K^+-ATPase occurs in the lamellar and filament ionocytes when salmon smolts are in fresh water. Acclimation to seawater results in a loss of the freshwater isoform and its replacement by the seawater isoform in the larger ionocytes of the gill filament shown in Figure 5.33B, which, probably to a large extent, mediates the tolerance of salinity by smolts that migrate downstream to the sea. Interestingly, smolts remaining in fresh water have a large number of ionocytes that express both isoforms, which suggests that there is a gradual shift of gill functioning and cell types between that shown in Figure 5.33A and Figure 5.33B. Once exposed to seawater, however, almost no ionocytes express both isoforms—the seawater isoform takes over.

Overall, it seems that the freshwater and seawater isoforms of Na^+, K^+-ATPase in salmon, trout, and other euryhaline teleosts are regulated according to environmental salinity. It seems probable that these isoforms evolved in multiple unrelated lines (demonstrating convergent evolution) but we need to know about the functioning of these isoforms of Na^+, K^+-ATPase and their structural differences in a broader range of species to be sure.

[95] Na^+, K^+-ATPase isoforms differ in the α-subunit, which is the major catalytic part of the molecule. Section 4.2.2 outlines the structure of the enzyme in more detail.

› Review articles

Evans DH (2008). Teleost fish osmoregulation: what have we learned since August Krogh, Homer Smith and Ancel Keys. American Journal of Physiology Integrative Comparative Physiology 295: R704–713.

Evans DH, Piermarini PM, Choe KP (2005). The multifunctional fish gill: dominant site of gas exchange, osmoregulation, acid-base regulation and excretion of nitrogenous waste. Physiological Reviews 85: 97–177.

Hazon N, Wells A, Pillans RD, Good JP, Anderson WG, Franklin CE (2003). Urea based osmoregulation and endocrine control in elasmobranch fish with special reference to euryhalinity. Comparative Biochemistry and Physiology Part B 136: 685–700.

Hiroi J, McCormick SD (2012). New insights into gill ionocytes and ion transporter function in euryhaline and diadromous fish. Respiratory Physiology and Neurobiology 184: 257–268.

Larsen EH, Deaton LE, Onken H, O'Donnell M, Grosell M, Dantzler WH, Weihrauch D (2014). Osmoregulation and excretion. Comprehensive Physiology 4: 405–573.

Checklist of key concepts

- The osmotic and ionic concentrations of aquatic environments and their variability have profound implications for the salt and water balance of animals living in these habitats.

Isosmotic marine animals

- The extracellular fluids of most marine invertebrates and agnathan hagfish are **isosmotic** to seawater (or close to isosmotic) due to high Na^+ and Cl^- concentrations.
- Some marine invertebrates regulate the concentration of particular ions in extracellular fluids.

Most marine vertebrates hypo-osmoregulate

- Most marine vertebrates (except cartilaginous fish) have extracellular fluid osmolalities that are almost a third of seawater osmolality.
- Hypo-osmoregulation results in water being lost at a rate determined by the permeability and surface area of permeable epithelia, and the osmotic concentration difference between extracellular fluids and seawater.
- Marine teleosts balance water loss by drinking seawater and absorbing much of the ingested fluid, a process facilitated by salt absorption, bicarbonate secretion and precipitation of carbonates in the intestine.
- The excretion of carbonates contributes to the inorganic marine carbon cycle.
- Animals that drink seawater need to be able to excrete the salts acquired from the seawater in the gastrointestinal tract in order to obtain pure water.
- Marine teleosts excrete Na^+ and Cl^- via **mitochondria-rich cells** in the gills and/or parts of the skin.
- **Preformed water** in food and **metabolic water** are usually sufficient to offset water loss by sea birds, mammals, and some marine reptiles.
- In most marine mammals, the maximum concentrations of NaCl in urine are too low to allow them to obtain pure water by drinking seawater.

Marine vertebrates can be hyperosmotic or isosmotic but hypo-ionic to seawater

- Some marine vertebrates use organic **osmolytes** as osmotic ballast and are **hyperosmotic** or **isosmotic** to seawater.

- Some proteins and enzymes of urea-retaining vertebrates are adapted to the presence of urea but most are stabilized by solutes, such as TMAO, that counteract the destabilizing effects of urea.
- After urea is synthesized, it is retained by its loss, via the gills or in urine, being restricted.
- **Salt glands** and **rectal glands** excrete excess salts by similar mechanisms to the gill ionocytes of marine teleosts. Such secretion is energetically costly due to the activity of Na^+, K^+-ATPase.
- The size and secretory activity of the nasal glands of **birds** is related to their habitat and the intake of sodium chloride.
- Even well-developed salt glands are relatively inactive until a sufficient excess of salts or an increase in plasma volume stimulates salt secretion.

Hyperosmoregulation and ionoregulation of animals living in fresh water

- Freshwater animals face the challenges of:
  - water gain by osmosis across permeable epithelia in contact with the osmotically dilute environment,
  - Na^+ and Cl^- efflux by diffusion.
- The rate of water influx in freshwater animals is determined by:
  - the area of permeable epithelia,
  - their permeability,
  - the difference in osmotic concentrations of the extracellular fluids of an animal and its aquatic environment.
- The water permeability of epithelia is governed by the properties of the **tight junctions** between cells and the **aquaporins** that facilitate the passage of water across the plasma membranes.
- Freshwater animals typically produce large volumes of dilute urine to maintain fluid balance.
- Specialized cells in locations such as skin, gills and anal papillae are responsible for the ion uptake that maintains salt balance of freshwater animals.
- Freshwater fish possess three mechanisms for apical Na^+ uptake via ionocytes:
  - **Na^+/H^+ exchangers**,
  - **proton pumps** driving Na^+ uptake via apical Na^+ channels,
  - Na^+ uptake via **Na^+, Cl^- co-transporters** or **Na^+, K^+, $2Cl^-$ co-transporters**.
- Chloride uptake by freshwater teleosts mainly occurs independently of Na^+ uptake, mostly via Cl^-/HCO_3^- exchange proteins in gill ionocytes.
- The Cl^-/HCO_3^- exchange proteins in the epithelia of freshwater teleosts makes them vulnerable to the toxic effects of nitrite (NO_2^-).
- Dietary salts are often a significant but variable component of the ion intake of aquatic animals, but are influenced by seasonal variations in food intake.
- Ion uptake in the diet is particularly important if uptake via the gills is compromised by environmental conditions.
- Species of fish that cannot take up Cl^- via their gills or skin must be highly dependent on dietary intake.

Osmoregulation in changing salinities—euryhaline species

- Most aquatic animals are **stenohaline**: they tolerate a narrow range of salinities. **Euryhaline** animals have a wider salinity tolerance.
- **Euryhaline osmoconformers** follow the changes in external osmolality.
- Euryhaline osmoconformers regulate their cell volume by adjusting the cytoplasmic pool of organic osmolytes, especially amino acids.
- The osmolality of the body fluids of **euryhaline osmoregulators** is fairly stably maintained.
  - **Hyper-isosmotic regulation** occurs in animals living mainly in fresh water, where they hyperosmoregulate, but osmoconformation occurs at salinities exceeding their **isosmotic point**.
  - **Hyper-hyposmotic regulation** occurs in animals that hyperosmoregulate in relatively low salinities but hypo-osmoregulate at salinities above the isosmotic point.
- **Euryhaline elasmobranchs** are always slightly hyperosmotic, but respond to low environmental salinity:
  - by reducing urea synthesis and increasing urea excretion;
  - by taking up Na^+ and Cl^- via the gills.
- Rectal gland secretion switches on in more salty environments.
- **Migratory teleosts** hyperosmoregulate in fresh water and hypo-osmoregulate in seawater by virtue of multiple changes in functioning of gills, kidney, intestine and control of drinking.

Study questions

* Answers to these numerical questions are available online. Go to www.oup.com/uk/butler

1. What do you understand by the terms osmoregulation and osmoconformation? Why might osmoregulation be more common amongst vertebrate animals than invertebrates? (Hints: Sections 5.1.1, 5.1.3, 5.2; See also Section 4.1.1.)
2. Explain the terms isosmotic, hyperosmotic and hyposmotic, giving examples of animals that fit each description. (Hints: These terms are fundamental to understanding the way animals function in aquatic environments; examples occur in all sections.)
3. What are tight junction proteins? Explain when and how the tight junctions can influence the salt and water balance of teleost fish. (Hints: Sections 5.1.3, 5.2.2 and 5.3.3; see also Section 4.2.1.)

4. What internal and external factors affect the exchange of water and ions between body fluids and aquatic environments? (Hints: Sections 5.1.3, 5.1.4, 5.2.1, 5.2.2, 5.3.2, 5.3.3; Case Study 5.2.)

5.* The Weddell seal excretes urine with a maximum sodium concentration of 330 mmol L^{-1}. If the seal drinks 500 mL seawater containing sodium at 468 mmol L^{-1} and absorbs all the sodium and water, would it be in negative or positive water balance because of drinking seawater? Calculate the net volume (mL) of water gained or lost by drinking 500 mL seawater. (Hints: Section 5.1.3; Table 5.3.)

6. How is urea retention involved in the osmoregulation of particular types of marine vertebrates? What are the advantages and disadvantages of urea retention when living in seawater and how are the possible problems overcome? How do euryhaline examples of the same animal groups alter these processes? (Hints: Sections 5.1.4 and 5.3.2.)

7. What are the common features of the gill ionocytes of marine teleosts, the epithelial cells of elasmobranch rectal glands and the epithelial cells of the nasal glands of birds? Draw a diagram that illustrates the operation of the interlinked ion transport processes in these tissues. (Hints: Sections 5.1.3 and 5.1.5.)

8.* A marine flounder drinks seawater to maintain water balance. The seawater contains Ca^{2+} at 10 mmol L^{-1} and Mg^{2+} at 53 mmol L^{-1}, none of which is absorbed in the gastrointestinal tract.

   (i) If 80 per cent of the volume of seawater ingested is absorbed, what are the theoretical concentrations of Ca^{2+} and Mg^{2+} in the rectal fluid that the flounder excretes?

   (ii) Experiments found that the rectal fluid contained Ca^{2+} at 2 mmol L^{-1}, Mg^{2+} at 160 mmol L^{-1} and 110 mmol L^{-1} of bicarbonate/carbonate. What percentage of the Ca^{2+} and Mg^{2+} passing along the gut of the flounder are removed from solution by gastrointestinal processing?

   (iii) What is the effect on the osmolality of the intestinal fluids, assuming (an approximation) that 1 mol of the ions equals 1 osmol?

   (Hints: Section 5.1.3; Box 5.1 in Online Resources.)

9. The diet of some aquatic animals helps in their ionic regulation, while in other animals the diet creates osmotic/ionic demands. Explain these ideas giving examples. (Hints: Sections 5.1.3, 5.1.4, 5.1.5 and 5.2.2; Case Panel 5.2).

10. Name three sites where ion uptake occurs in named freshwater species. What are the proposed mechanisms for Na^+ and Cl^- uptake by freshwater teleosts? (Hint: Section 5.2.)

11. What changes in salt and water movements occur when a salmon or an eel migrates from fresh water into seawater? (Hints: Sections 5.1.3, 5.2.2 and 5.3.3.)

12. How do euryhaline invertebrates that osmoconform regulate their cell volume? (Hints: Section 5.3.1; see also Section 4.3.1.)

13. Exotic (non-indigenous) aquatic species that become invasive when they enter new habitats are of worldwide concern. How would you go about investigating whether a non-indigenous species of amphipod (a crustacean) found in a coastal survey could have arrived in the ballast saline water of ships, and whether it might disperse up estuaries or into freshwater habitats? (Hint: Section 5.3.1.).

Bibliography

Evans DH, Clairbourne JB (Editors) (2006). The Physiology of Fishes. 3rd Edition. Boca Raton, FL: CRC Press.

Gallardo B, Aldridge DC (2013). Review of the ecological impact and invasion potential of Ponto Caspian invaders in Great Britain. Cambridge: Cambridge Environmental Consulting Cambridge (UK). www.nonnativespecies.org

Grosell M, Farrell AP, Brauner CJ (Editors) (2010). The Multifunction Gut of Fish Fish Physiology: Volume 30. Cambridge, MA: Academic Press.

Rankin JC, Jensen FB (Editors) (1993). Fish Ecophysiology. London: Chapman and Hall.

6

Water balance of land animals

When animals evolved to live on land, they faced a problem not experienced in aquatic environments—that of dehydration. Animals, such as soft-bodied annelid worms and slugs, and land crabs and woodlice, occupy terrestrial habitats but are restricted to damp or humid habitats where the risk of dehydration is reduced. Only animals of three phyla—the arthropods (including insects and arachnids, centipedes and millipedes), some molluscs (land snails), and the vertebrates[1]—have evolved a greater level of independence from watery habitats such that they can fully exploit terrestrial habitats.

Among the vertebrates, the fish with lobed fins living in the Late Devonian period could clamber onto land, if only for brief periods. Fossils of the lobe-finned fish *Tiktaalik roseae*, dated to 380 million years ago, show that this species had front fins which could bend at a wrist joint and bones that articulated with a strong shoulder bone. These and other features of *Tiktaalik* suggest it and other transitional species could carry themselves onto land, and that this led to the evolution of the tetrapods (amphibians, reptiles, birds and mammals) that fully exploit land habitats. On land, they would face the challenge of maintaining water balance.

In this chapter, we explore the major routes for water loss from land animals and the evolutionary adaptations[2] that limit these water losses. We discuss how these animals obtain sufficient water to balance their water losses. We also investigate some of the special adaptations of animals that successfully occupy dry habitats, where water is in short supply. In many cases, the adaptations that reduce their water losses provide striking examples of homoplasy in which similar adaptations have arisen in different lineages by convergent evolution[3].

[1] Vertebrates are members of the phylum Chordata, as we discuss in Section 1.4.2 and illustrate in Figure 1.21.

[2] Sections 1.5.2 and 1.5.3 discuss the concept of non-reversible evolutionary adaptations, as distinct from more rapid and reversible acclimatization of animals to their natural environments.

[3] We examine the mechanisms behind evolutionary convergence and divergence in Section 1.5.3.

6.1 Water loss from animals living on land

Figure 6.1 summarizes the main routes for water loss by land animals, in their urine and faeces, and by evaporation. Land animals differ in their relative water losses via these routes. Soft-bodied animals usually occupy relatively moist habitats, but in a drying environment lose water by evaporation across their skins at much faster rates than animals living in **xeric** habitats[4]. Animals that live in habitats with an intermediate level of water availability—**mesic** habitats—lose water at a rate between those of animals living in humid (**hydric**) habitats and xeric habitats. Such differences are the result of the adaptations that have evolved among animals. In all cases, similar physical processes drive evaporative water losses.

Figure 6.1 Routes of water loss in land animals

Green iguana (*Iguana iguana*). Image: © Katrina Hopkins.

[4] Section 1.2.2 outlines the classification of terrestrial habitats in relation to water availability.

6.1.1 Physical processes determine evaporative water loss

Evaporation requires heat input[5], which changes water from a liquid (the condensed state of water) into a gas (**water vapour**). Hence, evaporation is a form of gaseous diffusion in which water vapour diffuses from a solution (sol) that has a high partial pressure (P) of water vapour (P_{sol}) into air that has a lower partial pressure of water vapour (P_{air})[6].

What do we mean by the water vapour pressure of a solution (P_{sol}) from which evaporation occurs? The easiest way to understand the concept of the water vapour pressure of a solution is to imagine a sealed container holding a volume of pure water with air above the solution, as shown in Figure 6.2. At the air–water interface, water molecules in the solution exchange with those in the air, and vice versa. The partial pressure of water vapour in air (P_{air}) is the portion of the total atmospheric pressure that results from the water vapour contained in the air. However, unlike other gases in air (such as oxygen, nitrogen or carbon dioxide), there is a limit to the amount of water vapour that air holds before saturation occurs, at the **saturation water vapour pressure**. The vapour pressure of air saturated with water vapour above a volume of *pure water* equals P_w.

Temperature influences the vapour pressure of saturated air, as illustrated in Figure 6.3, such that the concentration of water vapour in air (the water vapour density) exceeds 50 g m^{-3} at 40°C and is about five times more than in air saturated at 10°C. From the opposite perspective, reducing the temperature of saturated air results in condensation of water to form a mist or fog of minute droplets of water in the air.

Note that the water vapour pressure of an aqueous solution (P_{sol}) is smaller than that of pure water (P_w) because the solute molecules at the surface of the solution impede water molecules escaping into the vapour phase, but do not hinder their return. The rate of evaporation is determined by the difference between the partial pressure of water vapour in the solution from which evaporation occurs (P_{sol}) and the partial pressure of water vapour in air (P_{air}) (i.e. $P_{sol} - P_{air}$), or the difference in water vapour density (usually expressed in g cm^{-3} or g m^{-3}). As this difference declines, so does the rate of evaporation.

Rate of evaporative water loss from animals

Evaporative water loss from animals occurs across external surfaces (the **integument**—i.e. the skin or cuticle) including the invagination of these surfaces into respiratory organs. Hence, the total rate of evaporation from animals is influenced by both the difference in partial pressures of water vapour (of water vapour densities) and the characteristics of the external surfaces: its area, thickness and water **permeability**.

Equation 6.1[7] expresses the combined influences of these parameters on the rate of evaporative water loss (expressed in cm^3 s^{-1}, which equals g s^{-1} or mL s^{-1}). In this equation, A is the area over which evaporation occurs, and d is the distance, which

[5] See Section 3.3.1 for an explanation of the latent heat of vaporization. This heat input is important in evaporative cooling for dissipating excess heat produced during metabolism, which we discuss in Sections 8.4.4, 9.1.3, 10.2 2 and 10.2.3.

[6] Section 3.1.2 outlines the physical principles of diffusion; for further explanation of partial pressures of gases and gaseous diffusion (as it applies to respiratory gases) see Section 11.1.

[7] Equation 6.1 is based on Fick's Law of Diffusion, as modified for gases, which we discuss in Section 3.1.2 and is similar to Equations 3.1 and 3.4.

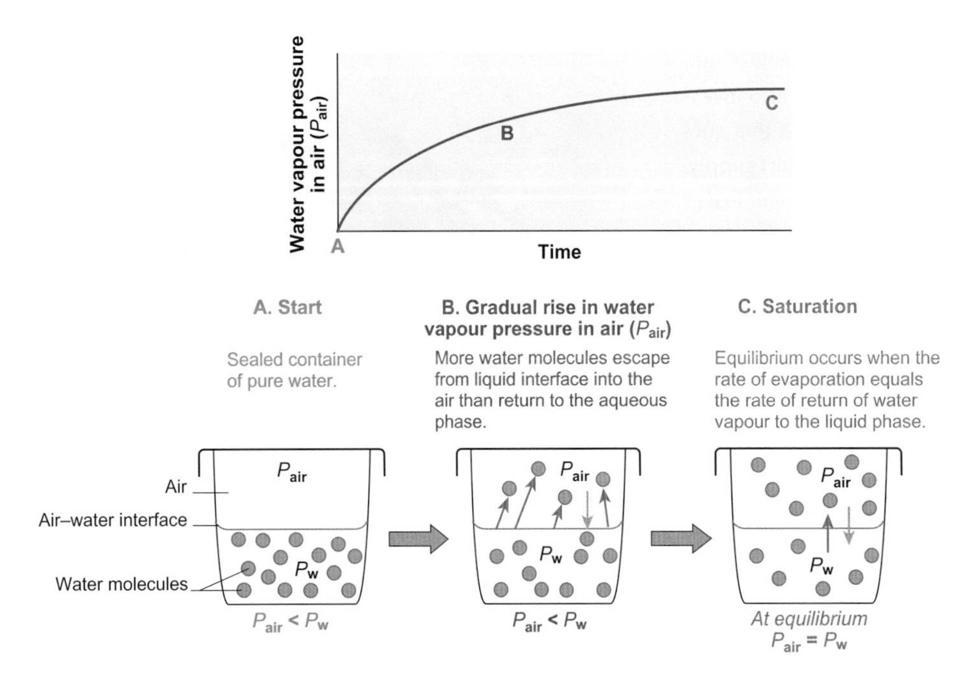

Figure 6.2 Evaporation of pure water in a closed container

Graph shows rising water vapour pressure in air (P_{air}) over time. The gradual processes of evaporation (B) result in an increase in the partial pressure of water vapour in air (P_{air}) until an equilibrium is established between the solution and water vapour in air. At equilibrium (C), the air is saturated with water vapour, and P_{air} equals the partial pressure of water vapour in the liquid phase (P_w).

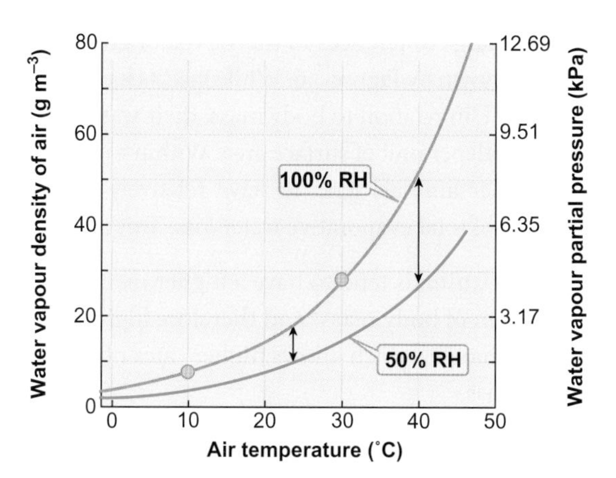

Figure 6.3 Relationship between water vapour pressures and water vapour concentrations (water vapour densities) in air with changing air temperatures and at two relative humidities

Orange circles show that two environments saturated with water vapour (100% relative humidity; RH) have differing water vapour partial pressures at different air temperatures. The difference in water vapour pressure and water vapour densities between two environments with different relative humidities (100% and 50% RH) is small at low temperatures (represented by the double-ended arrows) but increases as temperature rises.

may separate the solution from air. For land animals, the proportionality factor k, which represents the permeability of the integument, is the most important variable that animals can adjust to limit the loss of water by evaporation across the integument.

Proportionality factor ($cm^2\ s^{-1}\ kPa^{-1}$)

Area of integument across which evaporation occurs (cm^2)

Difference in partial pressure of water vapour between solution and air (kPa)

$$EWL = \frac{kA(P_{sol} - P_{air})}{d}$$

Evaporative water loss ($cm^3\ s^{-1}$)

Distance separating the solution from air (cm)

Equation 6.1

Air saturated with water vapour impedes evaporation because $P_{sol} - P_{air}$ is zero or less, so physiologists sometimes use the concept of **saturation deficit** (or vapour deficit) as a measure of the ease with which evaporation occurs, or the drying capacity of air. The saturation deficit is the difference between the saturation water vapour pressure, at the temperature of the air, and the actual (measured) water vapour pressure.

Relative humidity is a similar concept to saturation deficit in terms of assessing the drying capacity of air. Relative humidity expresses absolute humidity (the water vapour density) as a percentage of the humidity of fully saturated air at the same temperature. In other words, percent relative humidity at a particular temperature = absolute humidity/maximum humidity × 100. When air is fully saturated with water vapour it has a relative humidity of 100 per cent. Taking an example, at an air temperature of 20°C absolute humidity (water vapour density) of fully saturated air is 17.4 g m^{-3}, as shown in Figure 6.3. So if the water vapour density is say 10 g m^{-3}, at the same temperature (20°C), then the relative humidity will be 57.5 per cent (10/17.4 × 100).

At high air temperatures, the elevation in saturation water vapour density reduces relative humidity, which facilitates evaporation. However, relative humidity is not an exact measure of the water content of air and can mislead our assessment of the driving force for passive evaporative water losses if there are changes in temperatures. Figure 6.3 shows the relationships between air temperature and the partial pressure of water vapour in air at 100 per cent and 50 per cent relative humidity, which have important implications:

- Two environments with the same relative humidity do not have the same partial pressure of water vapour in air (or water vapour densities) if temperature differs.
- The water vapour pressures in air can be the same at different relative humidities if temperatures differ.
- The difference in the partial pressures of water vapour in air (water vapour densities) of two environments where there are different relative humidities (100 per cent and 50 per cent relative humidity) increases as temperature rises.

Figure 6.4 illustrates the importance of considering environmental vapour pressures when examining habitat variability and diffusional movement of water vapour. In the example shown, water vapour moves into the cooler burrow because of the lower water vapour pressure in the burrow than in the air at the ground surface.

The length and structure of the burrows of scorpions vary among related species according to habitat conditions. For example, the black rock scorpion (*Urodacus manicatus*), a mesic species that lives in Australia, occupies relatively shallow burrows under rocks, whereas the inland robust scorpion (*Urodacus yaschenkoi*), which is a xeric species, excavates tortuous burrows that reach up to 100 cm deep. In deep burrows, the declining temperatures reduce water vapour pressures, drawing in water vapour, which buffers the scorpions against high evaporative water losses[8].

The rate of total evaporative water loss (TEWL) of animals can be measured by the gravimetric method, which

[8] Case Study 6.1 discusses the impacts of burrowing on the water balance of amphibians.

6

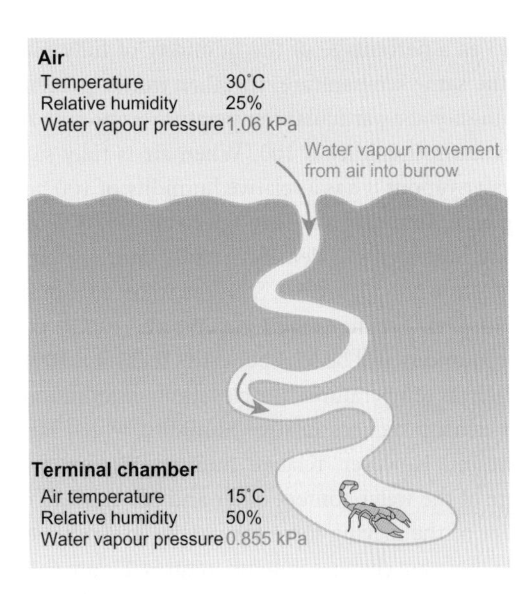

Figure 6.4 Water vapour movement into deep, cooled burrow of a desert scorpion

The difference in water vapour pressures—with higher external vapour pressure than in the cooler burrow—results in the inward diffusion of water vapour.

monitors changes in whole body mass, as outlined in Box 6.1. As an example, Figure 6.5 shows some data obtained by the gravimetric method for European pigeon ticks (*Argas reflexus*). The data in Figure 6.5B show the negative linear relationship between the TEWL of the ticks, as indicated by the per cent decrease in body mass per hour, and the water density of the surrounding air.

There are large differences in the TEWL of groups of land animals, as shown by Figure 6.6. While insects have a relatively high TEWL in relation to body mass, their water losses are relatively small per unit of surface area. Within a phylogenetic group, smaller animals tend to have relatively high mass-related rates of total evaporative water loss. This is because:

- Smaller animals tend to have a higher metabolic rate per gram of body mass[9] and therefore higher rates of gas exchange, which allows higher rates of respiratory water loss.
- The surface area of the integument of smaller animals tends to be higher relative to the body volume[10], which allows more evaporative water loss.

We might expect that natural selection will have resulted in lower TEWL in animals of a particular phylogenetic group living in xeric (arid) environments than those living in mesic environments. Figure 6.7 supports this hypothesis for mammals; mammals living in arid habitats have slightly lower TEWL than mammals living in a mesic habitat when body mass and phylogenetic relationships are taken into account. To explain such differences, and to compare different groups of animals

[9] Section 2.2.4 examines the relationship between metabolic rate (as oxygen consumption) and body mass.

[10] Figure A in Box 2.3 shows how the increasing size of a similar shaped object results in a smaller surface area relative to its volume.

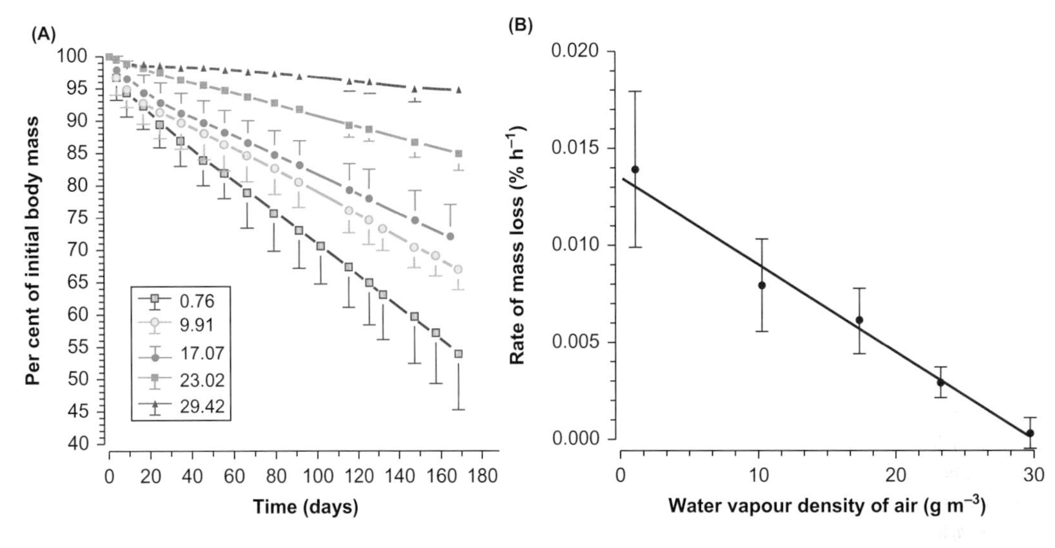

Figure 6.5 Evaporative water loss in European pigeon ticks (*Argas reflexus*)

(A) Percentage loss of wet body mass increases linearly over time and is related to water vapour density of the air. The nymphs of the soft ticks in these experiments were kept at a constant temperature (30°C) in darkness at different water vapour densities (0.76–29.4 g m^{-3}) and starved, so no water was gained from food. The loss of body mass is primarily indicative of evaporative water loss (and the very small loss of body mass through metabolism – measured as less than 0.005% of dry mass h^{-1}). (B) The rate of loss of body mass shows a negative linear relationship to the water vapour density of the air, with high rates of water loss (indicated by loss of body mass) when water vapour density is low, diminishing to very low rates of water loss when water vapour densities are high.

Source: adapted from data in Dautel H (1999). Water loss and metabolic water in starving *Argas reflexus* nymphs (Acari: Argasidae). Journal of Insect Physiology 45: 55-63.

Box 6.1 Measurement of evaporative water losses in animals

Evaporative water losses of terrestrial animals are often measured by the gravimetric method, which measures the loss of mass from the whole body over a period of time. The loss of body mass equals the total evaporative water loss (TEWL) plus any loss of fluid in faeces or in urine, but is balanced by any intake of water in food or by drinking. When possible, allowance is also made for metabolic water production[1]. Bringing these components together gives the following relationship:

TEWL = loss of body mass (g) – water lost in urine (g) – water lost in faeces (g) + water acquired in food or metabolism (g) + water intake by drinking (g)

The animal will also be losing mass as carbon (in the form of CO_2), due to oxidation of food substances, which can either be measured by monitoring gas exchange or be estimated from metabolic rate. However, this factor is often ignored, particularly in animals with low metabolic rates.

Gravimetric determination of water loss is susceptible to some potential errors:

- Repeated weighing involves handling, which may in itself affect water loss as a result of stress. Continuous weighing on a minute-to-minute basis can avoid such problems.
- Care needs to be taken to minimize unstirred **boundary layers**[2], which cause changes in vapour pressure close to the animal, by using flow-through systems in which air is continually moved into and out of the chamber containing the animal, as shown in Figure 6.7.

As an alternative to the gravimetric method, instantaneous evaporative water losses can be measured using electronic sensors that are sensitive to water vapour, and can be combined with flow-through respirometry to measure oxygen consumption and CO_2 release by animals held in an enclosed space, with a continual flow of air into and out of the space as in Figure 6.8. The rate of loss of water vapour is measured by the difference in water vapour content of the air delivered into the space and air leaving the space multiplied by the rate of air flow through the system. This method is extremely sensitive and allows measurements to be made even in animals as small as fruit flies (*Drosophila*).

[1] Metabolic water production is discussed in Section 6.2.2.

[2] Boundary layers are explained in Section 8.4.3.

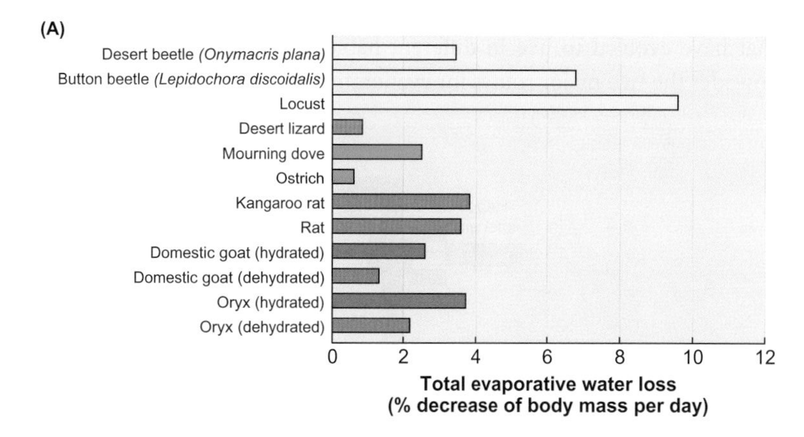

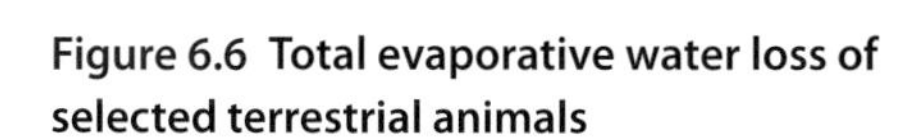
Figure 6.6 Total evaporative water loss of selected terrestrial animals

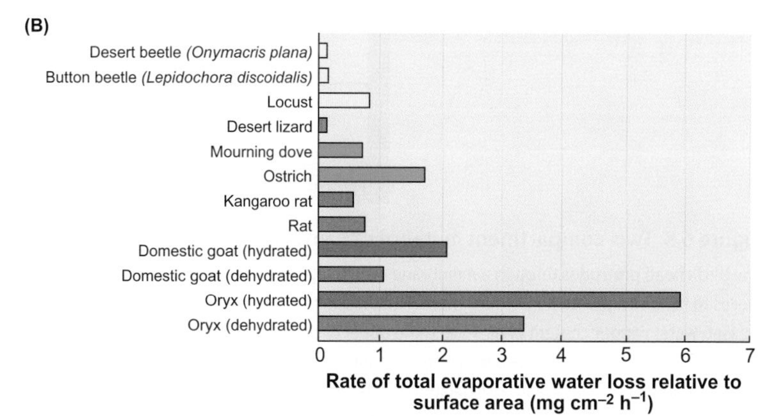

Rates of evaporative water loss are shown as % decrease in body mass per day (A) and relative to surface area, expressed as mg cm^{-2} h^{-1} (B). Insects (yellow bars) show the highest rates of water loss relative to their body mass, but when water loss is expressed per cm^2 of surface area, mammals (domestic goat, *Capra hircus* and East African oryx, *Oryx beisa*) have relatively high rates of evaporative water loss (brown and purple bars), while lower rates of water loss occur in insects, particularly those species living in arid habitats (two species of Namib Desert beetles) and in desert reptiles represented here by a lizard the desert iguana (*Dipsosaurus dorsalis*).

Source for data: Louw GN. Physiological Animal Ecology. Harlow: Longman Scientific & Technical, 1993.

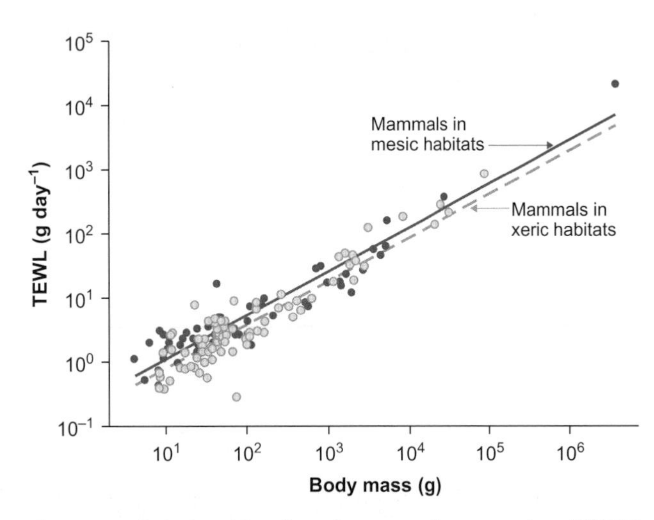

Figure 6.7 Relationship of total evaporative water loss (TEWL) and body mass for 136 mammals separated by habitat

Eighty of the mammals were identified as living in xeric (arid) habitats, while 56 live in mesic habitats. The mammals ranged in body mass (M_b) from 4 g to 3500 g (the Asian elephant). Species living in habitats where water is in short supply have significantly lower TEWL than those in mesic habitats, after accounting for differences in body mass. Data are plotted on double logarithmic scales and the blue solid and orange dashed lines represent the allometric equation of TEWL to body mass (TEWL = a $(M_b)^b$) for each group. (Section 2.2.4 explains the principles of allometric and other scaling equations.) When all the mammals were incorporated into a single analysis (not shown), TEWL was related to body mass with the value for b (the slope of the relationship) equal to 0.69.

Source: Van Sant MJ et al (2012). A phylogenetic approach to total evaporative water loss in mammals. Physiological and Biochemical Zoology 85: 526-532.

that have evolved to live in different habitats, we need to consider the two major routes for evaporative loss:

- respiratory water loss from the moist surfaces of respiratory systems;
- cutaneous water loss via the skin or cuticle, which can, in fact, be the respiratory surface of some land animals.

Classical experiments by Arthur Ramsay in the 1930s paved the way for understanding the relative rates of water loss across the body surfaces and the respiratory water loss of insects. To prevent respiratory water loss (and measure water loss via the cuticle) he blocked the **spiracles** from the respiratory system of dead cockroaches (*Periplaneta americana*) with wax. To prevent water loss across the cuticle (and measure respiratory water loss), Ramsay coated the body surface with shellac (a resin produced by some insects) that dries as an impervious layer. While such methods work for insects, alternative approaches are necessary for other animal groups.

Figure 6.8 shows a system used to separate cutaneous and respiratory water losses of birds, although the small amount of cutaneous EWL from the head is included in the measurements of respiratory water losses. Respiratory masks have also been developed that exclude cutaneous EWL from the head in measurements of respiratory EWL[11]. However, masking or restraint of animals may well stress the animal under investigation. Hence, methods that provide direct measurements of evaporative water loss are preferable. For instance, hummingbirds will take food from a specific location where respired air can be monitored while they are hovering and feeding. Figure 6.9

[11] Box 2.2 Figure A(iii) shows a photograph of a bird fitted with a respiratory mask.

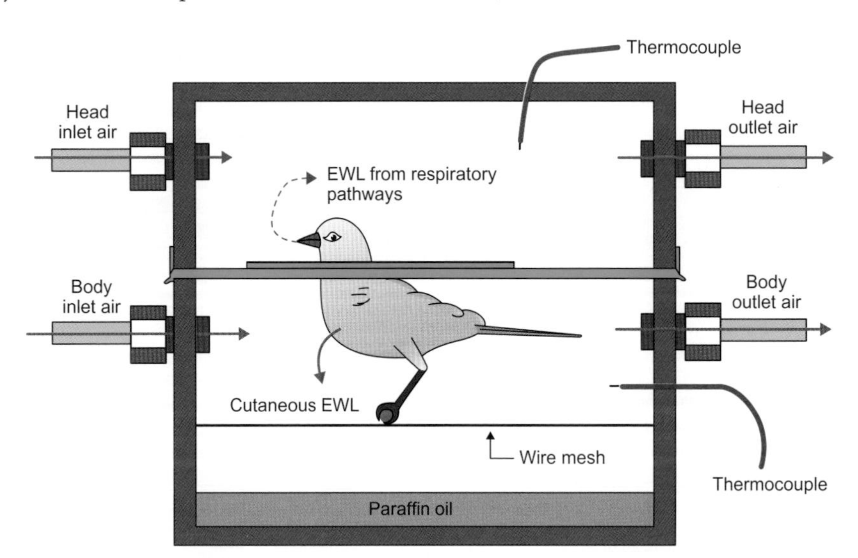

Figure 6.8 Two-compartment metabolic chamber partitioning respiratory and cutaneous evaporative water loss (EWL)

The bird's head protrudes through a membrane, which separates upper and lower chambers. The lower half of the chamber contains a perch on a mesh to allow faeces to fall through into paraffin oil. There is a flow of air through the upper and lower chambers. The EWL from the respiratory pathways is measured as the increased water vapour content of air leaving the upper chamber compared to air entering this chamber. Cutaneous water loss (except from the small area of head) is calculated as the difference between water vapour in the air leaving the lower chamber and air entering the lower chamber. The arrangement also allows respiratory water loss to be measured in relation to oxygen consumption if the oxygen content of the air entering and leaving the upper chamber is also measured.

Source: adapted from Wolf BO, Walsberg GE (1996). Respiratory and cutaneous evaporative water loss at high environmental temperatures in a small bird. Journal of Experimental Biology 199: 451-457.

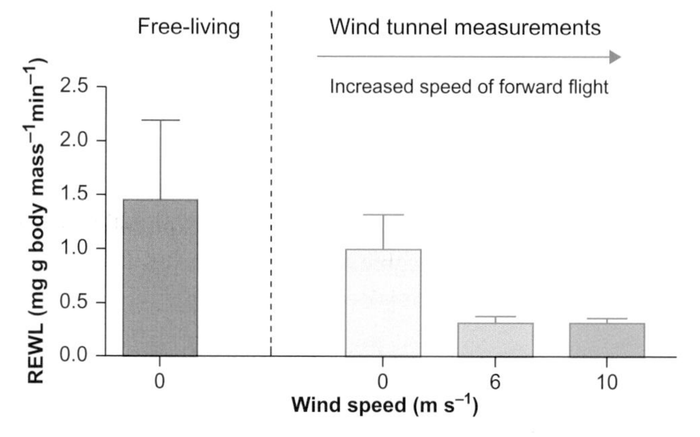

Figure 6.9 Respiratory evaporative water loss of rufous hummingbirds (*Selasphorus rufus*)

The hummingbirds were trained to take food from a location where their respiratory evaporative water loss (REWL) was monitored during natural hovering (pink column of data), and during hovering in a wind tunner (wind speed zero), and during forward flight in a wind tunnel. The REWL is lower during the less energy-demanding forward flight than during hovering.

Values are mean ± standard deviation.

Source: Powers DR et al (2012). Respiratory evaporative water loss during hovering and forward flight in hummingbirds. Comparative Biochemistry and Physiology Part A 161: 275–285. Image: Dean E Biggins, US Fish & Wildlife Service/Wikimedia Commons.

shows some of the results of such studies in which respiratory EWL is calculated based on the difference between the water content of expired air and ambient inspired air.

6.1.2 Evaporative water loss from respiratory organs

Most land animals have internal respiratory structures such as the vertebrate lung or the tracheae of insects. During inspiration of unsaturated air, evaporation from the moist surfaces adds water vapour to the inspired air until it becomes saturated with water vapour (at the body temperature of the animal). Hence, water vapour is lost with each expired breath.

The amount of EWL from the respiratory system, expressed in Equation 6.2, depends on:

- the rate of oxygen consumption (metabolic rate, mol of oxygen per unit time);
- the water loss for each mole of O_2 consumed.

$$\underset{(\text{mL h}^{-1})}{\text{Rate of respiratory EWL}} = \underset{(\text{mol O}_2\ \text{h}^{-1})}{\text{Rate of O}_2\ \text{consumption}} \times \underset{(\text{mL mol O}_2^{-1})}{\text{H}_2\text{O loss per mole O}_2\ \text{uptake}}$$

Equation 6.2

An important principle shown by Equation 6.2 is that respiratory water loss will increase if for any reason the rate of oxygen consumption increases, for example during physical activity. Figure 6.9 shows an example of the increased respiratory EWL that occurs in hummingbirds during hovering compared to the lower respiratory EWL during more energy efficient forward flight.

The respiratory EWL of those groups of land animals in which body temperatures vary by large amounts throughout the day (**ectotherms**)[12] is influenced by the associated changes in metabolic rate. Although the body temperature of reptiles varies throughout the day, they exhibit behavioural thermoregulation[13]. When investigated at a constant environmental temperature of 23°C, the respiratory EWL of reptiles generally accounts for 20–40 per cent of their TEWL, as shown by the data in Table 6.1. However, the respiratory EWL of reptiles can reach 50–60 per cent of TEWL at air

[12] Ectothermy is the focus of Chapter 9.

[13] Section 9.1 discusses the behavioural (and physiological) thermoregulation of reptiles.

Table 6.1 Total evaporative water loss (TEWL) and losses via respiratory and cutaneous routes in selected species of terrestrial reptiles

Species	Habitat	TEWL (g 100 g body mass^{-1} day^{-1})	Respiratory water loss % TEWL	Cutaneous water loss % TEWL
Desert tortoise	Desert	0.2	24	76
Chuckwalla	Desert	0.3	34	66
Sonoran gopher snake	Desert	0.9	36	64
Green iguana	Tropical forest	0.8	28	72
Common box turtle	Temperate forest	0.9	24	76

The desert tortoise (*Gopherus agassizii*) chuckwalla (*Sauromalus ater*) and Sonoran gopher snake (*Pituophis catenifer affinis*) are species native to the Mojave and Sonoran deserts in Southwest USA and Northwest Mexico. The green iguana (*Iguana iguana*), shown in Figure 6.1, is native to the Caribbean and Central and South America. The common box turtle (*Terrapene carolina*) is native to Eastern USA and Mexico. However, for comparability, all animals were studied when resting in dry air at 23°C.

Sources: Schmidt-Nielsen K (1969). The neglected interface: The biology of water as a liquid gas system. Quarterly Reviews of Biophysics 2: 283-304 and Schmidt-Nielsen K. Animal Physiology. Adaptation and Environment. 5th Edition. Cambridge: Cambridge University Press, 1997.

temperatures of 40°C, when cooling by evaporative water loss becomes necessary.

Land vertebrates can reduce respiratory evaporative water loss by cooling the nasal cavities

Mammals and birds generally have higher rates of metabolism (oxygen consumption) than reptiles because of their maintenance of higher and stable body temperatures (endothermy[14]). However, evidence to suggest that land vertebrates can reduce their respiratory water losses by cooling their nasal cavity first emerged in the 1960s when Knut Schmidt-Nielsen and his collaborators discovered that air is often exhaled at a temperature below body temperature. They argued that cooling of the exhaled air results from a temporal separation of the countercurrent flow of inhaled and exhaled air and heat exchange[15] in two stages:

- During inspiration of unsaturated air, evaporation from the nasal membranes progressively adds water vapour to the inspired air. Due to the latent heat of vaporization required for this evaporation, the epithelium of the nasal passages is cooled—to a greater extent closer to the outside where more water evaporates, and progressively less so as the inhaled air moves through the passages. Nevertheless, the higher temperature of the nasal passages warms up the inhaled air such that the temperature of inhaled air gradually approaches the body temperature of the animal.
- Air leaving the lungs during expiration is saturated with water vapour at the body temperature of the animal. As the air passes through the cooled nasal passages, heat is transferred from the expired air to the nasal passages. Cooling of the expired air reduces the saturated water vapour pressure of the air, which results in condensation of water onto the nasal surfaces and the release of the vaporization heat, which further warms up the nasal surfaces. Thus, heat and water that might be lost during expiration are retrieved in the nasal passages.

The effectiveness of water retrieval from expired air will depend mainly on the degree of cooling of the expired air in the nasal passages, which is determined by the:

- surface area involved in heat exchange—the greater the surface area the better the heat exchange;
- distance between the air stream and cooled surfaces—a shorter distance improves heat exchange;
- velocity of the flow of air—a slower flow of air allows more time for heat exchange.

The nasal passages of birds and mammals have coiled bony or cartilaginous projections called respiratory turbinates[16] covered with moist epithelium that increase the surface area for heat exchange; these are crucial for effective heat exchange and cooling of the nasal passages in animals that

[14] We discuss endothermy in detail in Chapter 10.

[15] Countercurrent systems separated in space (rather than time as in the nasal countercurrent system) are common in animals and used for exchange of heat, gases or solutes, as outlined in Section 3.3.1.

[16] We discuss respiratory turbinates as a means of heat conservation in birds and mammals in Section 10.2.7. Figure 10.25 illustrates the respiratory turbinates of several species.

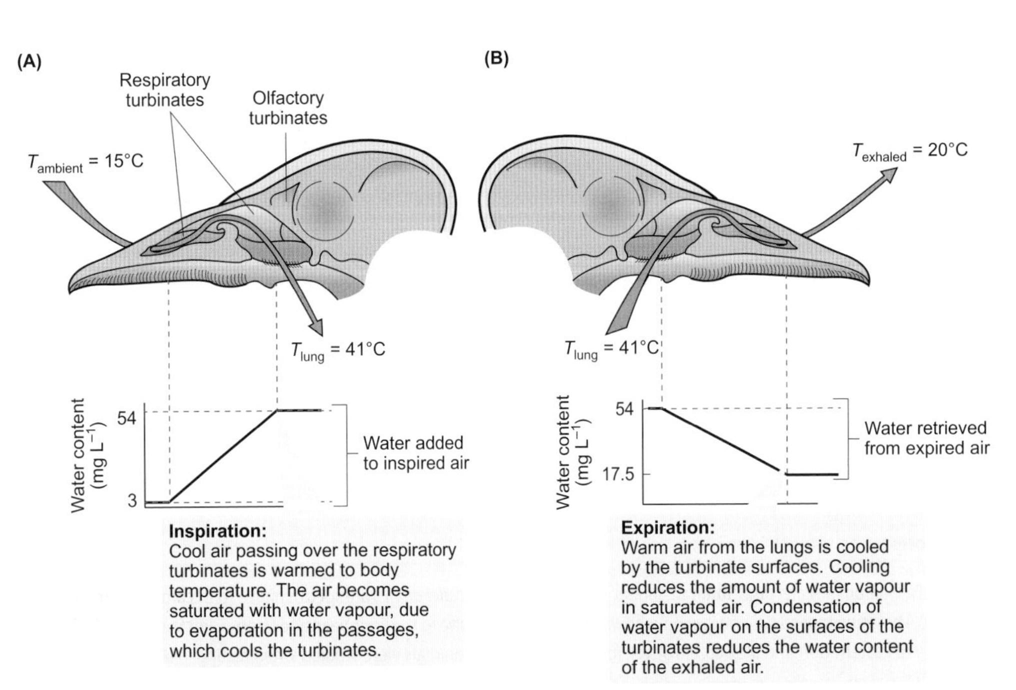

Figure 6.10 Predicted retrieval of respiratory evaporative water loss by cooling in respiratory turbinates of birds

(A) Inspired air at 15°C ($T_{ambient}$) reaches a body temperature of 41°C by heat exchange processes. The graph beneath shows water added to inspired air. (B) During expiration air starts at a body (lung) temperature 41°C but cools to 20°C ($T_{exhaled}$) during passage through the turbinates. The graph beneath shows water retrieved from expired air.

Source: Hillenius WJ, Ruben JA (2004). The evolution of endothermy in terrestrial vertebrates: who? when? why? Physiological and Biochemical Zoology 77: 1019–1042.

maintain high rates of lung ventilation. Figure 6.10 is a model of the effect of cooling on water retrieval by the turbinates from expired air of a bird. The cooled air remains saturated with water vapour, but contains less water vapour than it would if it was at body temperature.

Respiratory turbinates are absent in most reptiles, so it is often assumed that little water vapour is retrieved in the nasal passages of reptiles. However, several studies suggest that nasal cooling can allow retrieval of some of the water vapour in the respiratory tract of reptiles.

Early studies of desert iguana (*Dipsosaurus dorsalis*) reported cooling of exhaled air by about 0.5°C below body temperature, when body temperature is close to ambient air temperature, and greater nasal cooling when the body temperatures exceeds ambient temperatures, while basking in the sun. For example, at a body temperature of 42°C (the average body temperature in active wild desert iguanas) and an ambient temperature of 30°C and 30 per cent relative humidity, the exhaled air is cooled to 35°C (7°C below body temperature). Figure 6.11 shows evidence of cooling of expired air in the deep nasal passages of the South American rattlesnake (*Crotalus durissus*) by up to 6°C. These data indicate that evaporation and cooling of the respiratory walls could retrieve significant amounts of water from expired air.

How much water can animals save by cooling the expired air? We can use the graphs in Figure 6.3 to answer this question. To see how this works:

- assume that the water density of inhaled air is 20 g H_2O per m^3 (=20 mg L^{-1});
- then assuming a body temperature of 38°C, the graph in Figure 6.3 shows us that saturated air in the lungs at body temperature will contain 46 g H_2O per m^3 (46 mg L^{-1});
- therefore, the amount of water vapour added to inhaled air before it reaches saturation at body temperature is 26 mg H_2O L^{-1} (i.e. 46 minus 20);
- if exhaled air is then cooled to 35°C in the nasal passages and condensation occurs, the data in the

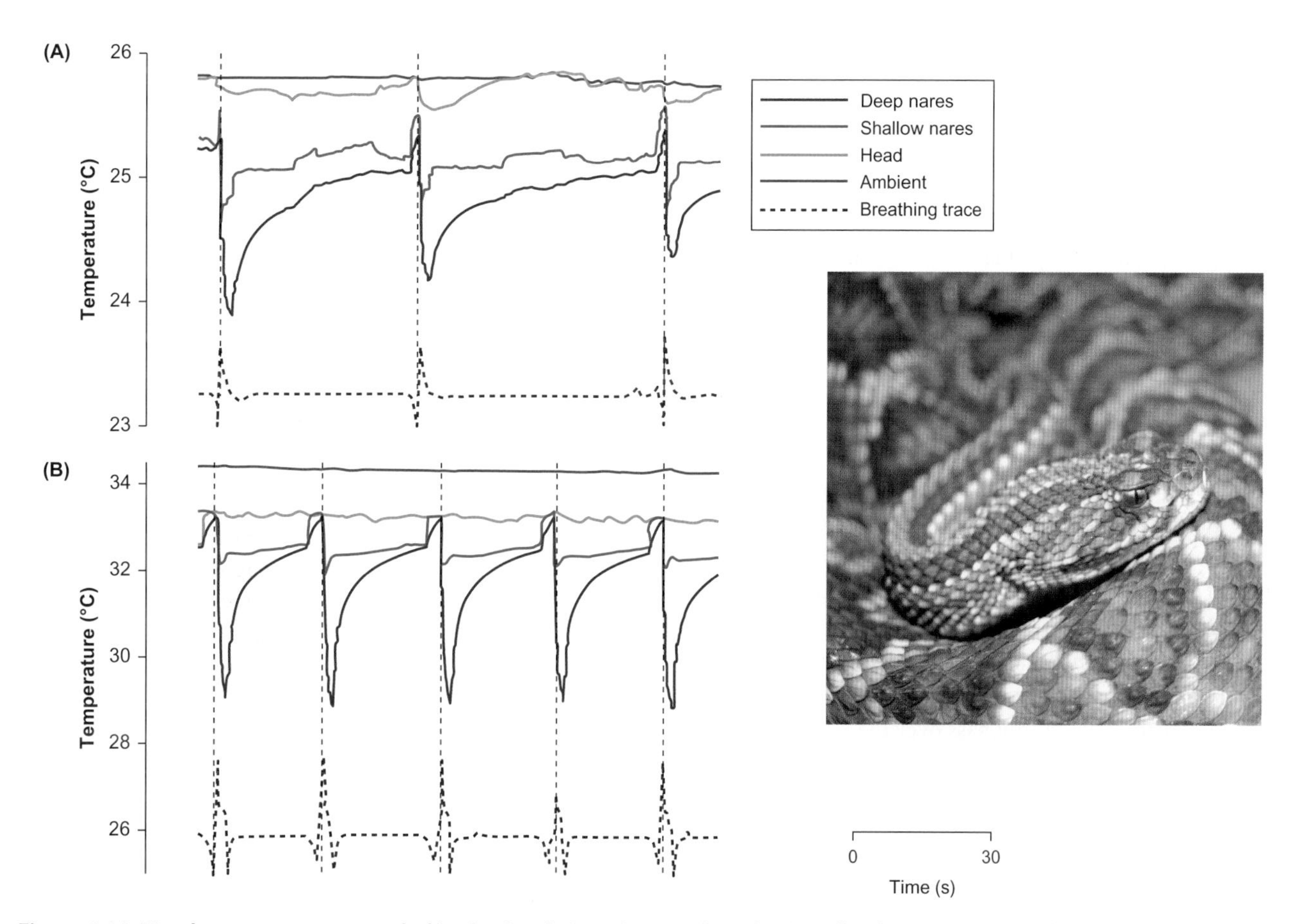

Figure 6.11 Nasal temperatures recorded in the South American rattlesnake *Crotalus durissus* at two ambient temperatures

Ambient temperature is shown by red lines. Body temperatures equal ambient temperature in these ectotherms. The breathing trace (dashed blue line) shows inspiration upwards and expiration downwards. Dashed vertical lines indicate the onset of each inspiration. Note that air temperatures recorded in the deep nares (deeper part of the nostrils) (purple) change dramatically during inspiration and expiration. The temperature scales differ in (A) and (B), but careful examination of the graphs shows a greater change in the temperatures recorded in the deep nares when ambient temperature is above 34°C (B) than at close to 26°C (A).

Source: Tattersall GJ et al (2006). Respiratory cooling and thermoregulatory coupling in reptiles. Respiration Physiology and Neurobiology 154: 302-318. Image: © Patrick JEAN / muséum d'histoire naturelle de Nantes / Wikimedia Commons.

graph back in Figure 6.3 indicate that expired saturated air will contain approximately 40 g H_2O per m^3 (=40 mg H_2O L^{-1});

- the water vapour added to inspired air and actually lost during expiration in this example is 20 mg H_2O L^{-1} (40 minus 20), whereas the potential loss in the absence of nasal cooling would be 26 mg H_2O L^{-1} (46 minus 20);
- the difference between the actual loss of water vapour (20 mg H_2O L^{-1}) and the potential loss in the absence of nasal cooling in this example (26 mg H_2O L^{-1}) is 6 mg H_2O L^{-1}, which represents a 23 per cent saving of the potential water loss.

These theoretical calculations indicate considerable water savings could be achieved by the condensation of water vapour on nasal membranes.

One way of assessing the *actual* retrieval of water vapour by the nasal passages is to compare evaporative water losses during normal breathing and after bypassing the nasal passages (breathing through the mouth). Experiments on four large bird species (Japanese quails, *Coturnix japonica*; pigeons, *Columba livia*; European herring gulls, *Larus argentatus* and American crows, *Corvus brachyrhynchos*) show an average reduction in respiratory EWL of 55–71 per cent in these species during normal breathing in which air passes across the respiratory turbinates, compared to breathing when bypassing the nasal passages. Expression of these values in relation to the total volume of water that the birds process each day indicates an overall 10–12 per cent saving. Although this may seem a small overall water saving, it is important to remember that this process requires no metabolic energy. In its absence, animals would need to use energy to search for additional drinking water to maintain their water balance.

We might expect water retrieval from expired air to be most important in species living in deserts. However, among birds there seems to be little relationship between the availability of water in their natural habitat and their nasal cooling and retrieval of water vapour by the respiratory turbinates. In fact, for at least one small desert bird, the desert lark (*Ammomanes deserti*), occlusion of the external nares (nostrils) such that expired air cannot pass over the respiratory turbinates appears to have no effect on the TEWL.

Humans have relatively wide nasal passages and the small surface area for heat exchange these passages provide only cools exhaled air by a few degrees below their core body temperature. Consequently, the saving of water by condensation in the nasal passages of humans is relatively small. By contrast, the nasal passages are complex and narrow in small rodents, such as kangaroo rats (*Dipodomys* sp.), in which expired air is at 11°C below the body temperature of 38°C at an air temperature of 30°C and 25 per cent relative humidity. This cooling allows more than half of the potential respiratory EWL to be condensed, which may be seen as adaptive for a species that lives in arid habitats. However, a similar level of nasal cooling occurs in lab rats and there is no evidence of a correlation between cooling and the aridity of natural habitats.

Dromedary camels (*Camelus dromedarius*) have evolved two mechanisms for the retrieval of water vapour from expired air at night:

- At night, condensation of water occurs in the cooled nasal passages, because the expired air is up to 10°C cooler than the body temperature of around 36°C, as shown in the lower panel of Figure 6.12A. Condensation at night retrieves up to 46 per cent of the water vapour in expired air, in the elaborate respiratory turbinates of camels that have a surface area of over 1000 cm^2 (compared with 160–180 cm^2 in humans). During the higher daytime temperatures[17], camels partially lose heat by exhaling warm air with a correspondingly high-water content; there is minimal cooling of the expired air, as shown in Figure 6.12A, so condensation is much less significant than at night.
- In dehydrated camels, expired air becomes unsaturated at night and in the early morning, as Figure 6.12B illustrates, but this does not occur during the daytime. Desaturation of exhaled air, together with the condensation due to cooling of the expired air increases the retrieval of EWL to up to 60 per cent of the water vapour that would otherwise be lost during breathing. The desaturation of expired air is believed to result from the hygroscopic properties of dried mucus and salts covering the nasal membranes of dehydrated camels at night. Hygroscopic materials absorb water in a reversible way, such that absorption during exhalation is followed by desorption during the inhalation of dry air.

Reduced or intermittent ventilation may limit evaporative water loss

Some birds and mammals that live in arid environments can reduce their rate of oxygen consumption during long periods of water restriction and hence reduce their respiratory EWL. Such reductions occur in ostriches (*Struthio camelus*), dromedary camels and Arabian oryx (*Oryx leucoryx*). In the Arabian oryx, a 50 per cent reduction in metabolic rate (measured in field conditions) occurs in the hot summer months (when water intake is reduced by almost a third) compared with the cooler months. A reduction in

[17] In the higher ambient temperatures during the day camels prioritize protection of the brain from overheating; inhaled air warms the turbinates by heat exchange and cools the blood, which we discuss further in Section 10.2.4.

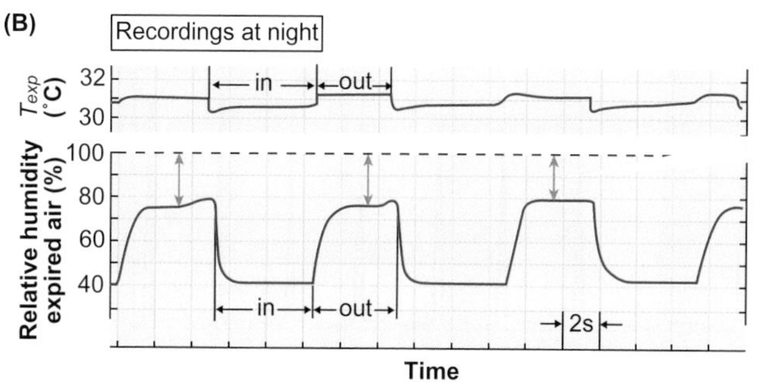

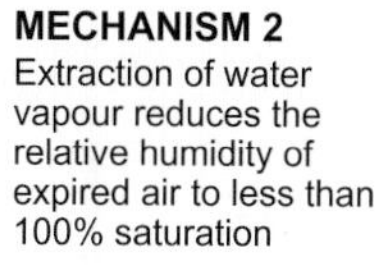

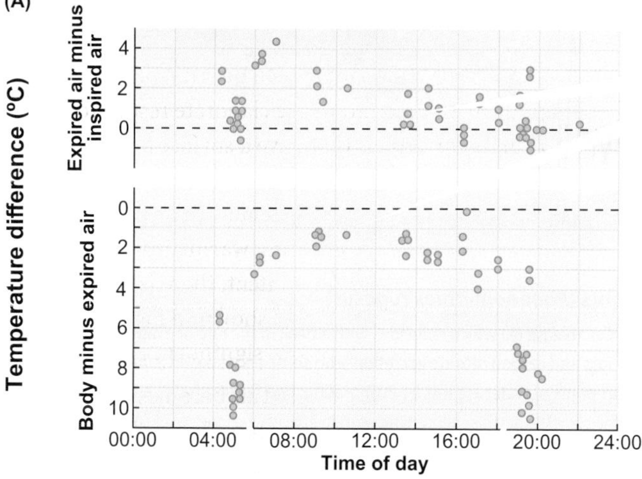

Minimal cooling of expired air occurs during daytime; the temperature of expired air is close to body temperature and similar to or above that of inspired air

MECHANISM 1

At night up to 10°C difference exists between the body (rectal) temperature and the temperature of expired air, which results in condensation of water vapour in the nasal passages

Figure 6.12 Dromedary camels (*Camelus dromedarius*) retrieve water vapour from exhaled air at night by two mechanisms

(A) Difference between the temperature of expired and inspired air (upper panel) and difference between the body (rectal) temperature and temperature of expired air (lower panel) throughout the day. The cooling of expired air at night results in the condensation of water vapour in the nasal passages (mechanism 1). (B) Recordings at night time of the temperature and relative humidity of expired air from dehydrated camels. Temperatures of expired air (T_{exp}, upper panel) are below the body temperatures at night, which is about 36°C, and result in condensation in the nasal passages (mechanism 1) together with extraction of water vapour (mechanism 2).

Source: Schmidt-Nielsen K, Schroter RC, Shkolnik A (1981). Desaturation of exhaled air in camels. Proceedings of the Royal Society of London. Series B, Biological Sciences 211: No 1184. Image of camel in Oman © J.A. Brown.

metabolic rate will reduce EWL from the pulmonary system and will hence reduce TEWL.

Intermittent ventilation occurs in many species of amphibians and reptiles[18]. For example, sporadic and low breathing rates occur in desert reptiles, such as chuckwallas (*Sauromalus ater*), which only take two to three breaths each hour. However, it is not certain whether these intermittent and low breathing rates are a mechanism triggered in response to dehydration to reduce water loss.

Many insects and insect-like animals (Hexapoda) exhibit **discontinuous gas-exchange cycles** (DGCs)[19] in which the spiracles are only wide open in the short O-phase[20]. When the spiracles are fully closed, or fluttering between slightly open and closed, evaporative water loss is restricted but the animal can still take up oxygen from the tracheal system.

By keeping the spiracles closed or in the flutter phase for long periods, water loss is confined largely to the short O-phase, as shown in the example in Figure 6.13. The vapour pressure in the tracheal system is saturated whichever pattern of gas exchange occurs, but discontinuous ventilation leads to CO_2 accumulation, allowing faster efflux of CO_2 during the short O-phase. Hence, respiratory EWL relative to CO_2 excretion should be reduced during discontinuous gas exchange. This idea led to the **hygric hypothesis** in which the water saving was invoked as an adaptive explanation of the evolution of DGCs.

Functional studies of several insect species, such as the beetles *Protaetia cretica* examined in Experimental Panel 6.1, see Online Resources, and the larvae of the butterfly propertius duskywing (*Erynnis propertius*), support the hygric hypothesis. Propertius duskywing caterpillars switch between continuous and discontinuous ventilation when quiescent; their respiratory water loss is lower during DGCs

[18] Section 12.3.3 discusses intermittent ventilation of amphibians and reptiles.

[19] Hexapoda are the most diverse of the four groups of arthropods, as we outline in Section 1.4.2. Discontinuous gas-exchange cycles occur among many types of insects: beetles, cockroaches, moths, almost all ants and some bees, locusts, crickets, grasshoppers, some true bugs and true flies, and in other arthropods respiring via tracheal systems such as ticks and centipedes.

[20] Section 12.4.1 gives further information on the characteristics of the different phases of discontinuous gas-exchange cycles of insects.

6

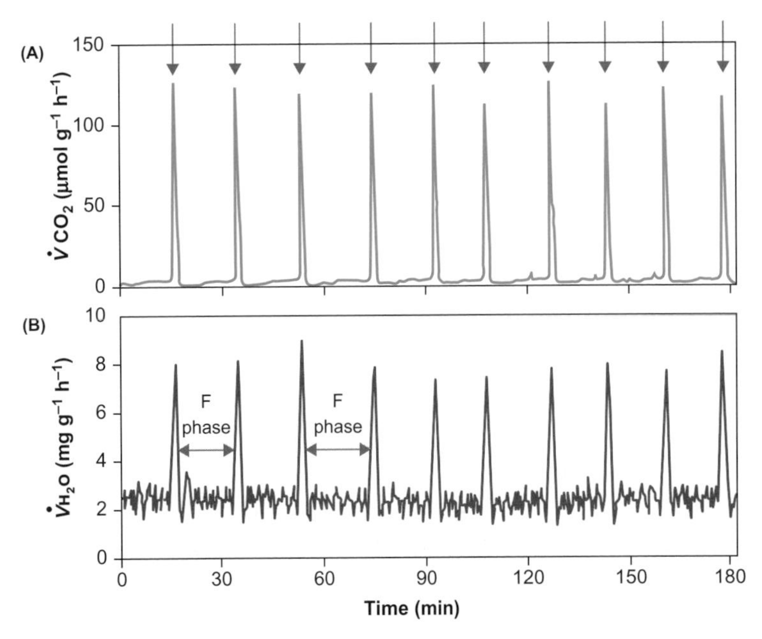

Figure 6.13 Discontinuous ventilation of female rough harvester ants (*Pogonomyrmex rugosus*)

At each opening of the spiracles (O-phase) indicated by blue arrows there is a burst of CO_2 release (A) which coincides with a peak in the rate of evaporative water loss ($\dot{V}H_2O$), shown in (B). Between each opening there is a relatively low rate of water loss and carbon dioxide excretion (and some air intake) in a flutter phase (F-phase) when the spiracles flutter between slightly open and then close so very small amounts of water are lost by evaporation; two flutter phases are labelled on (B).

Source: Lighton JRB et al (1993). Spiracular control of respiratory water loss in female alates of the harvester ant *Pogonomyrex rugosus*. Journal of Experimental Biology 179: 233–244. Image: © Alex Wild/www.alexanderwild.com

than during continuous ventilation (4.4 per cent and 12.8 per cent of total EWL, respectively).

An ecological advantage of using DGCs as a means of water saving is likely to be most apparent if water availability is compromised, for whatever reasons. The female rough harvester ants (*Pogonomyrmex rugosus*) shown in Figure 6.13 must survive severe water shortage after insemination, when they take in no food or water for several weeks. Reducing their water loss at this time should increase their chance of survival, a part of which may involve their discontinuous ventilation.

From measurements of the rates of water loss in insects using DGCs or in its absence, we can calculate how DGCs influence the rate of water loss (and possibly survival times) of insects deprived of water. For example, in propertius duskywing caterpillars at 0 per cent relative humidity, such calculations indicate it would take 19 days for a 30 per cent reduction in the water content of caterpillars using DGCs to be reached; whereas if using continuous gas exchange they would reach this level in about half the time (10.6 days).

A recent study of three grasshopper species living in habitats of differing aridities in Israel found a correlation between habitat aridity and the prevalence of DGCs (percentage occurrence of DGCs among many individuals). The grasshoppers *Tmethis pulchripennis* live in an arid desert habitat where there is very low annual rainfall and show the highest prevalence of DGCs (61 per cent). This species of grasshopper also had the lowest respiratory EWL relative to metabolic rate of the three species and made water savings by its use of discontinuous rather than continuous gas exchange. By comparison, two mesic species (*Ocneropsis* spp.) living in habitats with greater rainfall, show 52 per cent and 19 per cent prevalence of DGCs but without any evidence that this achieves significant water saving. From these studies of grasshoppers, we might envisage that DGCs generally have greatest benefits for insects living in the most arid environments.

However, DGCs are not restricted to insects living in dry habitats, and not all insects that occupy xeric habitats exhibit DGCs or continue to do so when they are dehydrated. It seems that multiple independent evolutionary origins of DGC occurred and that DGC may have more than one adaptive function among insects[21].

Thermoregulation of endotherms can influence respiratory water loss

Regulation of the body temperature by endothermic birds and mammals that live at high ambient temperatures relies on evaporative cooling[22], which involves either respiratory or cutaneous EWL. Figure 6.14 shows the typical changes in respiratory and cutaneous EWL with air temperature of passerines, which is the largest grouping of birds encompassing more than half of the species of birds. Note in Figure 6.14 that between air temperatures of 30 and 36°C, roughly

21 We discuss the ideas and hypotheses regarding the evolution of DGCs in more detail in Section 12.4.1.

22 Section 10.2.2 examines the evaporative cooling of endotherms in detail.

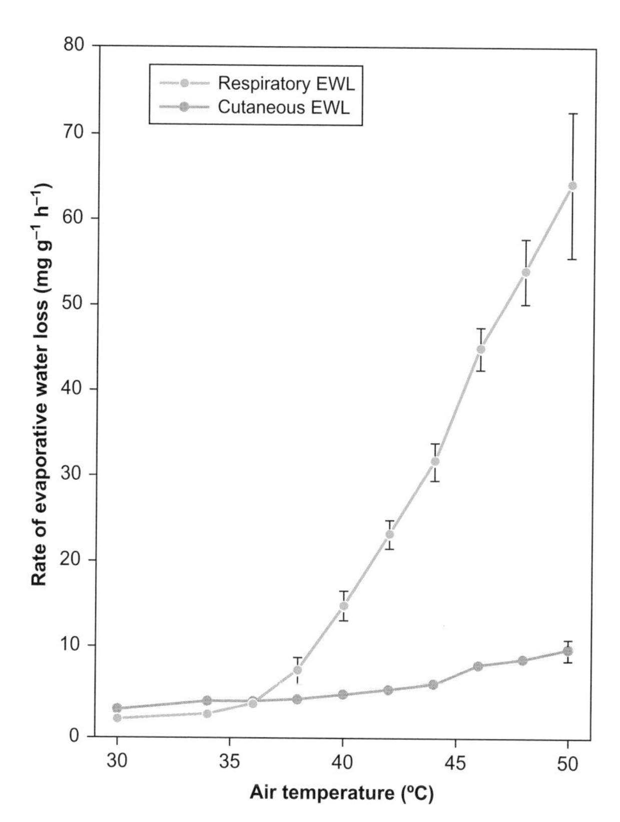

Figure 6.14 Respiratory and cutaneous evaporative water loss (EWL) in a small desert bird, the verdin (*Auriparus flaviceps*), in relation to air temperatures

Source: Wolf BO, Walsberg GE (1996). Respiratory and cutaneous evaporative water loss at high environmental temperatures in a small bird. Journal of Experimental Biology 199: 451-457. Image: Patrick Coin/Wikimedia Commons

equal respiratory EWL and cutaneous EWL occur in verdins (*Auriparus flaviceps*). Above 36°C, there is a greater dependency on respiratory evaporation for heat dissipation in passerines. However, some bird groups, such as various pigeons and doves, make greater use of their skin for evaporative cooling than respiratory evaporation, particularly in extreme heat stress when air temperatures exceed 40°C.

Relaxation of body temperature regulation in animals living in xeric habitats could, in theory, save some of the water that would otherwise be lost in evaporative cooling. This process would involve storage of some of the heat gained during foraging followed by its later dissipation in the shade by radiation, convection and conduction[23].

Daily storage of a proportion of the heat load, with offloading at night, gives rise to daily variations in body temperature of 2°C or more (**temporal heterothermy**). This phenomenon has been suggested to be a mechanism to conserve water in species of large ruminants, including dromedary camels. However, the fluctuations in body temperature that had been seen in captive animals were much less when free-ranging animals were studied[24]. There are two notable exceptions, in which heterothermy occurs in the wild; these are Arabian oryx and Arabian sand gazelles (*Gazella marica*). Both of these species live in hyper-arid conditions.

The Arabian oryx is a vulnerable species, which became extinct in the wild in 1972, but has since been successfully reintroduced into protected areas in Oman, Saudi Arabia, Israel, United Arab Emirates and Jordan, with further introductions envisaged. All of the 1000 or so Arabian oryx in the wild currently live in protected areas in the deserts of the Arabian Peninsula.

Figure 6.15 shows the results of the first study of six free-ranging Arabian oryx that were implanted with temperature-sensitive radiotransmitters and allowed to roam freely over the following three years. The graphs in Figure 6.15 show that the body temperatures of these Arabian oryx varied daily by an average variation of 4.1°C during the summer months, and by a lesser average of 1.5°C during winter months. Later studies confirmed these results and observed an amplitude in body temperature of up to 7.7°C in some individuals during hot dry conditions, which exceeds that recorded for any other mammal.

So how much water can Arabian oryx save by their daily heat storage? We can get some idea of an answer to this question by examining the water saving seen when body temperature rises to a maximum of 40.5°C, which occurs during summer, as shown by the data in Figure 6.15. A body temperature of 40.5°C is 2.1 degrees above the average body

[23] The principles of radiation, conduction and convection are outlined in Section 8.4.

[24] Experimental Panel 10.1 examines some of the experiments that monitored body temperature fluctuations of African free-ranging mammals.

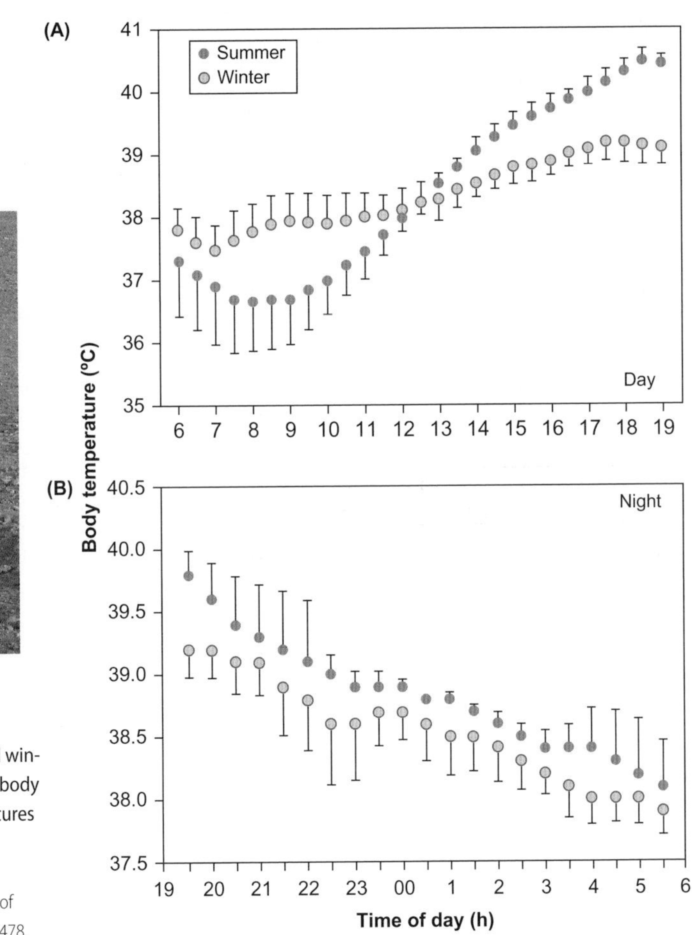

Figure 6.15 Heterothermy in Arabian oryx (*Oryx leucoryx*) during summer and winter

Daily rhythm of body temperatures in six free-ranging oryx in summer and winter. (A) Shows body temperatures during the day. The daytime variation in body temperature was greater in summer than winter. (B) Shows body temperatures during the night. Heat stored during the day is offloaded overnight.

Data are means ± standard deviations for 59 days and 12 nights between May and September. Source: Ostrowski S et al (2003). Heterothermy and the water economy of free-living Arabian oryx (*Oryx leucoryx*). Journal of Experimental Biology 206: 1471-1478. Image: © J.A. Brown

temperature of 38.4°C. Hence, estimated daily heat storage for an oryx with average body mass of 92.9 kg is:

$$\text{Heat storage} = \text{specific heat of tissues } (3.48\text{ kJ kg}^{-1}\,°\text{C}) \times \text{daily temperature difference } (2.1°\text{C day}^{-1}) \times \text{average body mass } (92.9\text{ kg}) = 679\text{ kJ day}^{-1} \quad \text{Equation 6.3}$$

The water saving by heat storage (measured in litres) can be calculated by dividing the energy saving in kJ day^{-1} (679) by the latent heat of vaporization (which at 38°C is 2404 kJ L^{-1}). Hence, storage of 679 kJ day^{-1} saves 679/2404 = 0.28 L of water per day from being lost via evaporative cooling.

Even this amount is a significant volume of water to find in the water budget of an animal living in a hyper-arid environment, in which there is no drinking water available. However, actual water savings are likely to be much greater. The reason for a greater saving is that overnight body temperatures are lower. Hence, the average daily amplitude of body temperatures is 4°C to 5°C in dry conditions and may reach up to 7.7°C in some Arabian oryx. Substituting 7.7°C into the calculation shown in Equation 6.3 gives a value for heat storage of 2489 kJ day^{-1}, which saves about a litre of body water each day from being used for temperature regulation[25].

The heterothermy of Arabian oryx is the likely explanation for their relatively low total evaporative water losses compared with other ungulates, which is shown by the data in Figure 6.16. Notice in this graph that the difference between the Arabian oryx and other species is greatest when the oryx have a restricted water intake.

The Arabian oryx may be at or close to its limit for maintaining water balance. Air temperatures typically reach above 40°C during the daytime of the hot dry season but reach 47°C in the Arabian deserts of the Empty Quarter (Rub' al Khali, which is most of the southern third of the Arabian Peninsula). They may even reach highs of above 50°C. Can Arabian oryx

[25] By allowing body temperature to rise during daytime the Arabian oryx reduces the rate of heat gain, which is proportional with the difference between ambient and body temperatures, as we discuss in Section 8.4. Figure 3.1A shows how Arabian oryx can lose heat at a faster rate to the sandy substrate after creating a depression in which they lie. They also lose heat into the cooler air overnight.

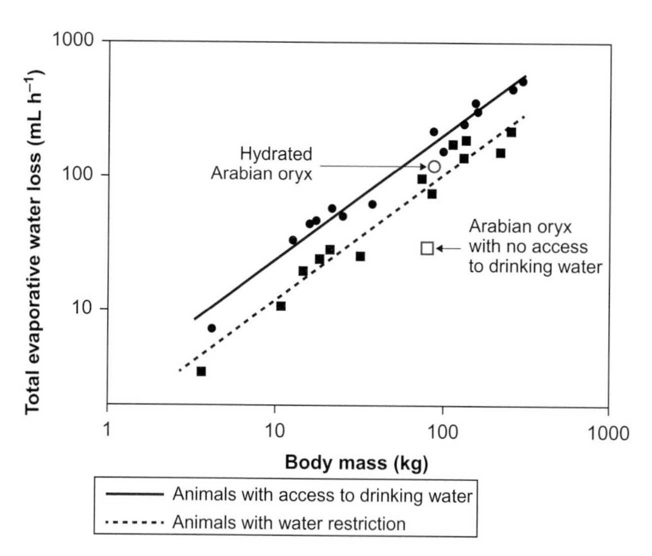

Figure 6.16 Evaporative water loss of ungulates

Double logarithmic plot of the rate of total evaporative water loss (EWL) against body mass for hydrated ungulates with access to drinking water (solid blue line) and ungulates from arid zones with water restriction (dashed blue line). The line relationships were obtained by least squares regression. The red symbols indicate EWL for Arabian oryx, which do not normally drink but do so if water is available. The red circle shows EWL for hydrated oryx given access to food and 4.5 L of water per day. The red square shows EWL for Arabian oryx after 5 months of gradual water restriction, by 60–70% after 3 weeks, when water provided equalled the water intake of free-living oryx from food in summer. The EWL of Arabian oryx is 54% and 35% below mass-predicted values (based on regression line equations) for other ungulates.

Source: adapted from Ostrowski S et al (2006). Physiological acclimation of a desert antelope, Arabian oryx (*Oryx leucoryx*) to long-term food and water restriction. Journal of Comparative Physiology B 176: 191-201.

cope with the further increases in ambient temperatures that are predicted by climate models? This question is a serious concern for conservationists protecting the Arabian oryx. Predicted increases in average temperature for the Arabian Peninsula by 2100 are at least 1.5°C for the lowest predicted concentrations of greenhouse gases, 4°C at higher predicted levels, and possibly 5–7°C for the worst-case model[26]. These increases are coupled with greater predicted increases in night-time temperatures than during the day, which will reduce the capacity for oryx to lose stored heat overnight.

6.1.3 Evaporative water loss across the integument

One of the most pronounced features of land animals, particularly those species living in arid environments, is their ability to restrict evaporative water losses via their integument due to waterproofing mechanisms. The waterproofing mechanisms exhibited by land animals appear to be primarily an inherited characteristic, as evident from the data for skin **resistances** (*R*) to evaporative water loss of vertebrate groups given in Table 6.2.

[26] The 2014 Intergovernmental Panel of Climate Change (IPCC) used four models to predict global and regional effects based on greenhouse gas concentrations, called Representative Concentration Pathways (RCPs): RCP2.6, RCP4.5, RCP6, and RCP8.5, which consider the range of radiative force in 2100 relative to pre-industrial values at +2.6, +4.5, +6.0 and +8.5 W/m^2, respectively.

Table 6.2 Skin resistance to evaporative water loss in selected species of land vertebrates

Taxon	Species	Habitat	Resistance (s cm^{-1})
Amphibians	Cane toad (*Rhinella marina*)	Mesic	1.7
	American green tree frog (*Hyla cinerea*)	Mesic arboreal	1.2–3.1
	Couch's spadefoot toad (*Scaphiopus couchi*)	Xeric burrowing	5
	Waxy monkey leaf (tree) frog (*Phyllomedusa sauvagii*)	Xeric arboreal	206–300
	Orange-legged leaf frog (*Phyllomedusa hypochondrialis*)	Xeric arboreal	364
Reptiles	Common box turtle (*Terrapene carolina*)	Mesic	78
	Green iguana (*Iguana iguana*)	Mesic	370
	Palestine viper (*Daboia palaestinae*)	Mesic	706–878
	Common chukwalla (*Sauromalus ater*)	Xeric	1360
Birds	Rock dove (*Columba livia*)	Widespread	45–130
	Chicken (*Gallus gallus domesticus*)	Mesic; domestic	88–101
	Japanese quail (*Coturnix japonica*)	Xeric	78–297
Mammals	Human stratum corneum (*Homo sapiens*)	Widespread	377
	House mouse (*Mus musculus*)	Widespread	116–201

Habitats are classified as mesic (with a moderate supply of moisture) or xeric (arid) when there is a severe lack of available water[1]. For amphibian species, burrowing and arboreal (tree-climbing) species are identified. A range of values is given where data are available from several studies of the same species at different relative humidities or different temperatures. Such natural variables have large effects on skin resistance particularly among birds.

Sources: Lillywhite HB (2006) Review - Water relations of tetrapod integument. Journal of Experimental Biology 209: 202-226; and Wyoga ML, Kersten CA (2013). Effects of water vapour density on cutaneous resistance to evaporative water loss and body temperature in green tree frogs (*Hyla cinerea*). Physiological and Biochemical Zoology 86: 559-566.

[1] We discuss the classification of terrestrial habitats in Section 1.2.2.

6

Estimates of total skin resistance (expressed in s cm^{-1}) are usually obtained by measuring the rate of EWL from a fluid-filled test chamber covered by a known area (A) of skin which is exposed for a measured period of time to a measured difference in water vapour density (WVD), using Equation 6.4:

$$R = A \times \frac{\text{WVD(sol)} - \text{WVD(air)}}{\text{Cutaneous EWL}}$$

Total resistance (s cm^{-1}); Difference in saturated water vapour densities at skin temparature and measured in air (g cm^{-3}); Area of skin (cm^2); EWL across the skin (g s^{-1})

Equation 6.4

Different animal groups use different mechanisms for waterproofing their integument. Among these, the insects provide some of the best examples of waterproofing, which enable some species to survive in even the most xeric habitats.

Waterproofing of insect cuticles

The waterproofing of the insect cuticle is mediated by a matrix of waxy substances in the outermost part of the cuticle, which covers the whole body. The cuticle also lines the trachea, tracheoles, foregut and hindgut.

Figure 6.17 shows the basic multilayered structure of the insect cuticle. The cuticle is secreted by the underlying **epidermis**, which is included in Figure 6.17A. Each secreted layer gradually moves outward, away from the epidermis. Initially, these layers are flexible, and the external dimensions of an insect only become fixed when the exocuticle becomes hardened (tanned or sclerotized) by cross-linking of proteins.

The **epicuticle** provides the waterproofing of the cuticle. The epicuticle consists of several separate layers, as shown in Figure 6.17B. The inner epicuticle, which consists of layers of lipoproteins, is topped by a thinner outer epicuticle; above this is a waterproofing wax layer. Some species, such as kissing bugs (*Rhodnius* sp.), also have an outermost thin layer of 'cement', as shown in Figure 6.17B.

Early experiments in the 1930s and 1940s began to reveal the water-resistant properties that limit water loss via the insect cuticle when a droplet of water placed on the surface of cockroaches was found to stay there indefinitely. In contrast, a droplet of water placed on a sheet of glass at room temperature rapidly evaporates. Under the microscope, a greasy film was seen on the surface of the droplet of water placed on the cockroaches, which prevents evaporation. These experiments gave the first clues to the existence of an external greasy barrier to water loss in cockroaches. The wax layer is soft and relatively mobile in cockroaches (but not all species), so it formed a greasy layer on top of the water droplet.

All insect species have hydrocarbons in their wax layer, but these vary in chain length from 12 to 50 carbons atoms. The closer packing of saturated long-chain hydrocarbons[27] means they are the most effective material for waterproofing the cuticle. Figure 6.18 shows the structures of the chemical compounds found in epicuticular lipids of insects, of which

[27] Lipid packing of hydrocarbons is illustrated in Figure 9.41.

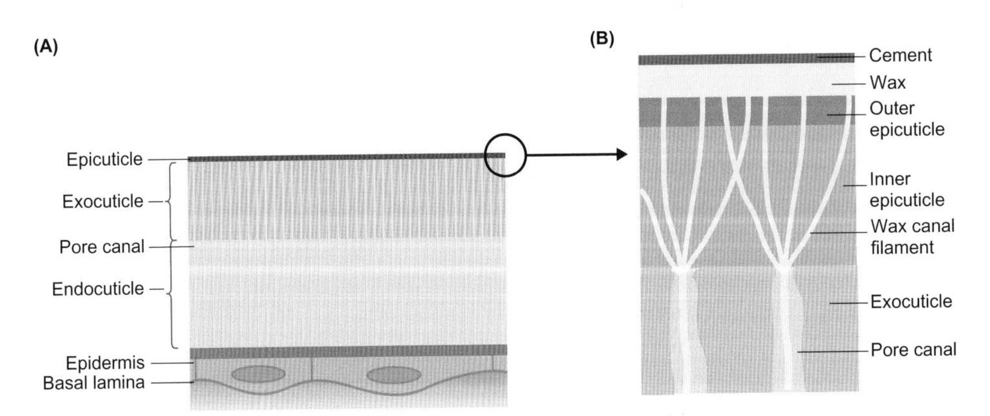

Figure 6.17 Schematic structure of the multilayered insect integument

(A) Section through the entire integument. Beneath the outermost epicuticle there is a hardened exocuticle and beneath this an inner undifferentiated endocuticle. (B) Section through the epicuticle of an insect such as the kissing bug (*Rhodnius* sp.) in which there is a cement layer, overlying the layer of wax. Pore canals pass through the cuticle. The pore canals contain tubular filaments (wax canal filaments) that convey waterproofing lipids to the surface of the cuticle.

Source: Chapman RF (1998). The Insects. Structure and Function, 4th Edition. Cambridge: Cambridge University Press.

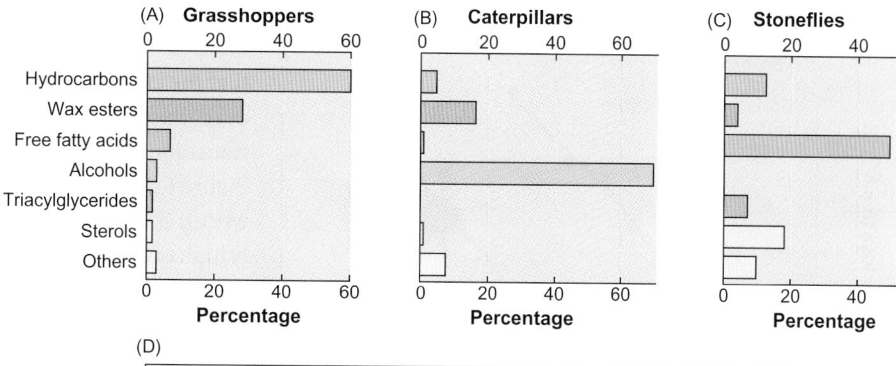

(D)

	Chain lengths
Hydrocarbons $CH_3-(CH_2)_X-CH_3$	12–37
Wax esters $CH_3-(CH_2)_X-O-C(=O)-(CH_2)_X-CH_3$	–
Free fatty acids $CH_3-(CH_2)_X-C(=O)-OH$	10–22
Alcohols $CH_3-(CH_2)_X-CH_2-OH$	12–34
Triacylglycerides CH_2-O-fatty acid / $CH-O-$fatty acid / CH_2-O-fatty acid	–
Sterols (HO)	–

Figure 6.18 Main groups of compounds in epicuticular wax as a per cent of total in selected species of insects

There is a mix of waterproofing compounds in the epicuticular waxes of insects. In grasshoppers (A) hydrocarbons are the dominant waterproofing compound. In caterpillars (B) and other larvae of butterflies, moths and beetles, straight-chain alcohols dominate. In adult stoneflies (C) free fatty acids are the dominant waterproofing compound in the epicuticular wax. The structures of the main types of compounds and their chain lengths (number of C atoms) are illustrated in (D).

Source: Chapman RF (1998). The Insects. Structure and Function, 4th Edition. Cambridge: Cambridge University Press.

the saturated alkanes, unsaturated alkenes and methyl-branched chains are the most common.

These hydrocarbons are synthesized from fatty acids by oenocytes (named for their usual amber colour: *oeno* means wine in Greek). The oenocytes are large specialized cells rich in endoplasmic reticulum and mitochondria that occur beneath or attached to the epidermis or associated with the fat body. The aliphatic alcohols (shown in Figure 6.18) may form esters[28], although these are often a relatively minor component of the waterproofing agents, as indicated in Figure 6.18. In some species, such as stoneflies, free fatty acids are the dominant hydrocarbon in the wax layer.

Some species of tenebrionid beetles that live in the African Namib Desert (such as *Stenocarpa gracilipes*) and the Sonoran Desert in North America (such as *Cryptoglossa verrucosa*) produce an unusual filamentous wax bloom in response to high temperatures and low humidity. These wax blooms are thought to have two main functions: (i) reduction in water loss by increasing the diffusion distance for loss of water vapour, via tortuous pathways between the long filaments of wax, and (ii) reduction in heat uptake, by increasing reflectance of solar radiation[29].

Temperature effects on insect waterproofing

Most insects and arachnids show a marked decrease in the effectiveness of the physical barrier to water loss as temperature increases such that there is a decrease in the waterproofing properties of the cuticular lipids at a particular transition temperature. The best way to reveal such changes is by examining evaporative water loss (EWL) at a range of cuticular temperatures, as shown in the example in Figure 6.19. Notice the marked increase in the rate of water loss by cockroaches when the cuticular temperature reaches about 30°C. Cockroaches can live at temperatures higher than 30°C but only remain in water balance if they increase water acquisition from food or by metabolism[30].

[28] Figure D in Box 2.1 illustrates the formation of esteric bonds in lipids.

[29] Section 9.1.3 discusses changes in reflectance as a means of regulating heat uptake.

[30] Sections 6.2.1 and 6.2.2 discuss water intake in food and by metabolism.

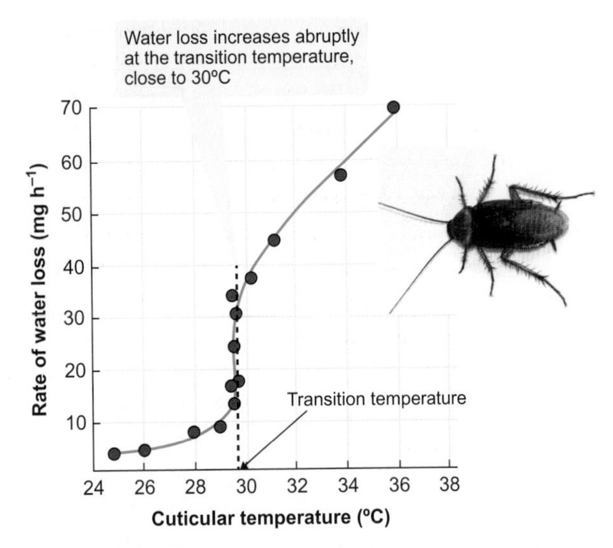

Figure 6.19 Evaporative water loss across the cuticle of American cockroaches (*Periplaneta americana*)

Water loss was measured by changes in body mass after blocking the spiracles with wax to prevent respiratory water loss.

Source: adapted from Beament JWL (1958). The effect of temperature on the waterproofing mechanism of an insect. Journal of Experimental Biology 35: 494-519. Image: © Peter J Bryant.

Transition temperatures are usually higher than the maximum temperatures experienced naturally, which tells us that the cuticle usually provides an excellent barrier to EWL over the natural temperature range of a species. However, transition temperatures differ between species, as shown in Figure 6.20, and between life stages. Kissing bugs and mealworms (larvae of the beetle *Tenebrio molitor*) have particularly high transition temperatures, which indicates that the cuticle of these species provides an excellent barrier to EWL even beyond their usual temperature range.

In marked contrast, some soil-living species, such as wireworms (larvae of *Agriotes* sp.), have poorly defined transition temperatures; in these species any increase in temperature has a significant effect on EWL. Does this mean that wireworms and other soil larvae have a poor capacity to waterproof their cuticle? In reality, immediately after moulting[31], wireworms are resistant to cuticular water loss. Movement

[31] Section 19.5.3 discusses the process of moulting by insects.

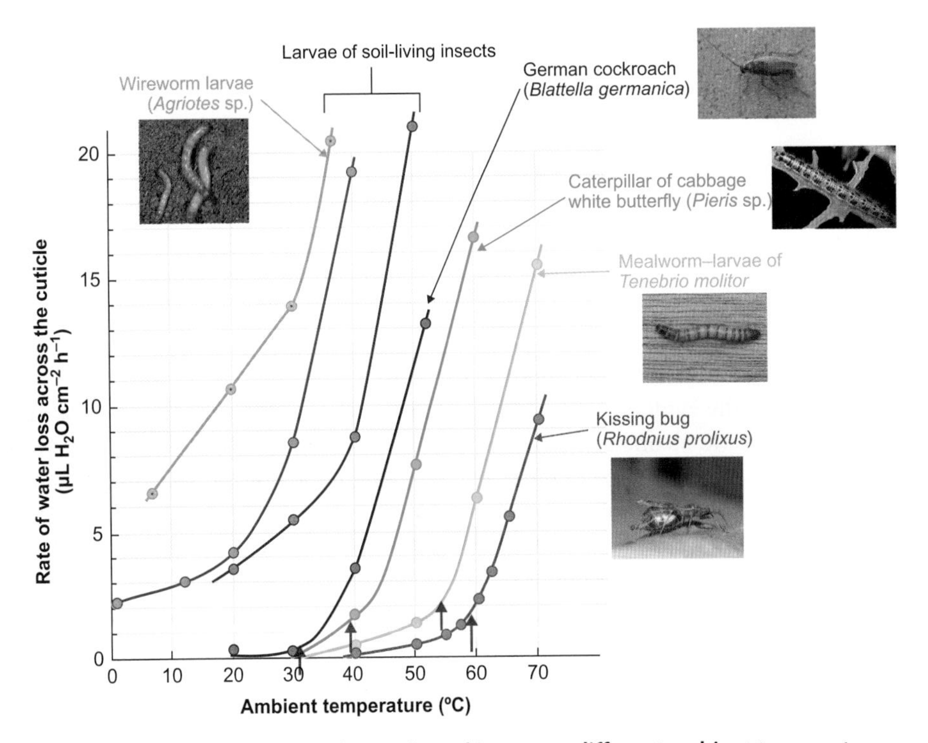

Figure 6.20 Rate of cuticular evaporation of water from selected insects at different ambient temperatures

All measurements were made on animals held in dry air. Each point is a mean value for 4–8 individuals. The transition temperature, at which water loss across the cuticle increases more rapidly, differs between species (shown by red arrows). The high transition temperatures of kissing bugs (about 60°C) and mealworms (55°C) are above their natural temperature ranges, which indicates a high resistance to the effects of temperature on cuticular water loss. For soil-living larvae transition temperatures are often indistinguishable.

Source: adapted from Wigglesworth VB (1945). Transpiration through the cuticle of insects. Journal of Experimental Biology 21: 97-114. Images: wireworm larvae http://www.efagalicia; German cockroach © Luis Miguel Bugallo; caterpillar (cabbage white butterfly) Soebe/Wikimedia Commons; mealworm Rasbak/Wikimedia Commons; kissing bug http://www.inctem.bioqmed.ufrj.br.

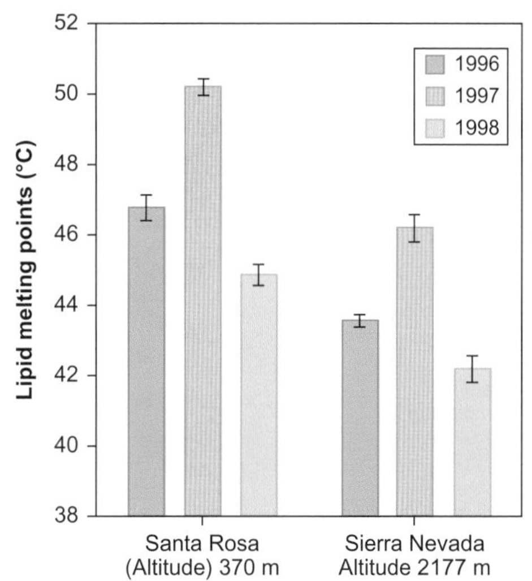

Figure 6.21 Melting points of cuticular lipids in two populations of californian grasshoppers (*Melanoplus sanguinipes*)

Lipid melting points were consistently higher (in three consecutive years of study) for grasshoppers from the Santa Rosa population living at lower altitude (370 m) than grasshoppers from the Sierra Nevada population (at 2177 m).

Source: Rourke BC (2000). Geographic and altitudinal variation in water balance and metabolic rate in a California grasshopper, *Melanoplus sanguinipes*. Journal of Experimental Biology 203: 2699-2712. Image: MostlyDross/flickr.

of the wireworms through the soil particles abrades the cuticle, and, with time, scratches and other damage to the waxy waterproof layer accumulate. However, this does not usually pose a problem for wireworms and other insect larvae living in moist soils. The comparisons of water losses shown in Figure 6.20 are for animals held in a dry atmosphere (such that conditions are equivalent between species), but water loss in moist soils is generally limited by the high water vapour pressure in the pores in the soil. In these conditions, uptake of water vapour from the most soil generally balances water losses[32].

Like wireworms, waterproofed species, such as cockroaches, can suffer damage by abrasions that increase EWL, but rapid secretion of lipids normally repairs the damage. Thus, the main difference between well-waterproofed insect species and more water permeable soil-living larvae seems to result from the inability to repair the damage to their cuticle caused by the continuous abrasion by soil particles.

What happens to the lipid matrix of the epicuticle when temperature increases? The answer seems to be that individual components start to melt[33] changing the lipid matrix from a crystalline gel state to a fluid state, in which bigger gaps occur between the molecules, which allows a higher rate of EWL. The point at which 50 per cent of the lipids melt is, by convention, taken as the cuticular melting point.

Cuticular melting points provide a precise way of comparing different animals and revealing the effects of their cuticular lipids on EWL. In cockroaches, the cuticular melting point and transition temperatures are similar, at close to 30°C. In contrast, most of the cuticular lipids in kissing bugs remain in a gel state until 75°C when 50 per cent are melted. Hence, these bugs can cope with extremely high temperatures without suffering high cutaneous rates of EWL.

We might ask whether variations in cuticular lipid melting points influence the water losses exhibited by populations of a species living in different habitats. To begin to investigate this question, populations of migratory grasshoppers (*Melanoplus sanguinipes*) living at different altitudes in California have been studied. Figure 6.21 shows some of the results. Grasshoppers living in the hotter climate, at a lower altitude, have cuticular lipids with a higher melting point than those of grasshoppers living at a higher altitude. Such changes in cuticular lipids translate into effects on water loss. When grasshoppers from the different locations are exposed to the same temperatures, grasshoppers from the higher altitude population have higher rates of water loss than grasshoppers from lower altitudes.

Cutaneous water loss by amphibians

The skin of amphibians consists of a thin epidermis[34] overlying the dermis, as shown in Figure 6.22. The outermost layer of the epidermis of amphibians—the **stratum corneum** of dead cells—is a thin layer (1 μm thick). Evaporative water loss typically occurs from the continually replenished moist layer of mucus on the surface of the skin, at a rate equivalent to a free water surface, and therefore usually accounts for about 99 per cent of their total evaporative water losses. However, many amphibians are essentially aquatic or semi-aquatic and live in damp environments with a high ambient water vapour density that reduces evaporative water loss, as illustrated by Equation 6.1.

[32] Section 6.2.4 discusses the uptake of water vapour by soil organisms and burrowing animals.

[33] Box 9.2 gives some information on the melting points of fatty acids, which are influenced by chain length, whether the molecule is saturated, and its degree of unsaturation.

[34] Amphibians exchange oxygen and carbon dioxide across the thin epidermis, which we discuss in Section 12.2.1, and take up water and ions via the skin when they are in water or in contact with a moist substrate.

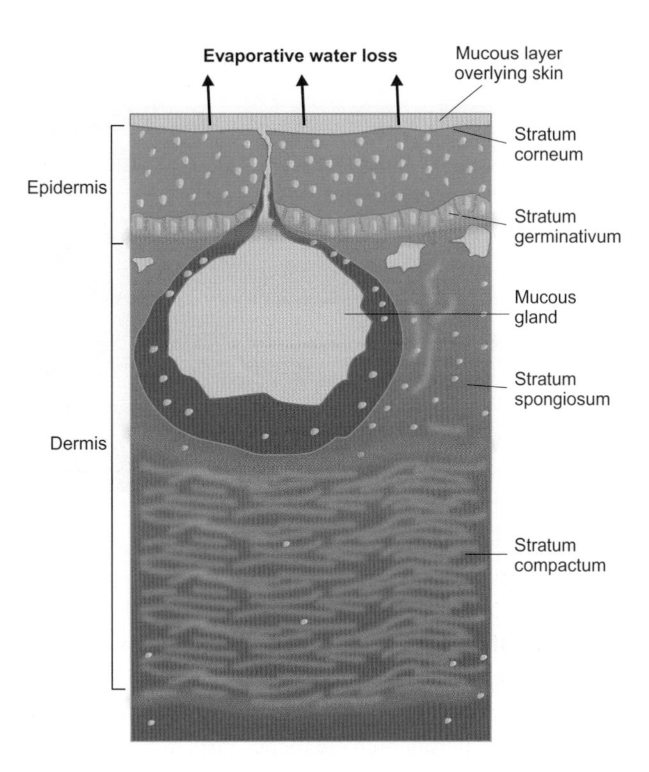

Figure 6.22 Schematic drawing of the skin of a typical amphibian

The thin layer of dead cells forming the stratum corneum overlies living cells of the epidermis, which are continuously generated by the divisions of cells in the stratum germinativum (also called stratum basale as it is the basal layer of the epidermis). Mucus is secreted by dermal mucous glands. The moist layer of mucus facilitates evaporative water loss in a drying atmosphere.

Source: adapted from Lillywhite HB (2006). Review - Water relations of tetrapod integument. Journal of Experimental Biology 209: 202-226.

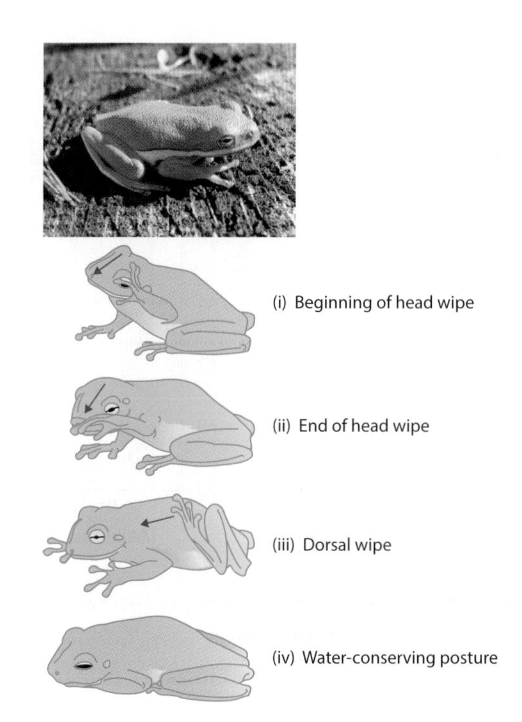

Figure 6.23 Common wiping movements and water-conserving posture of Florida tree frogs (*Hyla* sp.)

Drawings show the beginning (i) and end of a head wipe (ii), and a dorsal wipe (iii). The red arrows on the drawings indicate the direction of these wiping movements. These movements distribute lipid secretions which reduce water losses across the skin. After wiping, most frogs display the water-conserving posture (iv).

Source: adapted from Barbeau TR, Lillywhite HB. (2005). Body wiping behaviours associated with cutaneous lipids in hylid tree frogs of Florida. Journal of Experimental Biology 208: 2147-2156. Image: © Greg Schechter/flickr.

In contrast to the majority of amphibian species, some are more terrestrial, **fossorial** (adapted to digging), **arboreal** (living in or among trees) or **scansorial** (climbing). Arboreal and scansorial species are more independent of water because of their mechanisms for waterproofing the skin, which reduce evaporative water loss.

Waterproofing of tree frogs

Nearly all tree frogs have some cutaneous resistance to evaporative water loss. The first evolutionary step in reducing evaporative water loss across the skin may have been simply to secrete lipids along with mucus by species that experience seasonal periods of water shortage. The next step may have been the evolution of specific dermal glands that secrete waterproofing substances onto the skin, which occurs in tree frogs in Florida, South America, Australia and India.

Secretion of lipids onto the skin occurs in response to tactile stimulation of their skin by the front limbs using wiping behaviours, such as those illustrated in Figure 6.23. Many tree frogs increase the frequency with which they wipe themselves if temperature increases, and in response to dehydration. The physiological triggers of such changes are uncertain, but recent studies suggest that American green tree frogs (*Hyla cinerea*) increase the frequency and intensity of lipid secretions in response to low humidity.

After wiping, the frogs usually adopt a water-conserving posture (typical of dehydrated amphibians) that helps restrict water loss. In this position—crouching toward the substrate with their legs tucked in—the ventral surface is close to the substrate. If the substrate is damp, water uptake across the ventral skin can occur by osmosis[35].

The chemical nature of the waterproofing agents secreted by tree frogs is poorly understood, but the presence of lipids in the secretions is shown by their staining with Sudan Black B. Lipids are also detected in granular glands in the dermis of the skin. Intermittent discharge of lipids from cutaneous glands probably accounts for the variations in skin resistances that occur at different water vapour densities, which Figure 6.24 illustrates.

[35] Case Study 21.1 discusses the hormonal control of water uptake across the pelvic patch of the skin of dehydrated amphibians.

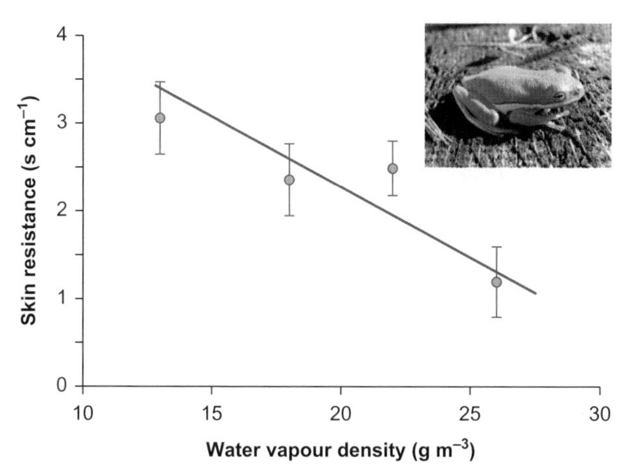

Figure 6.24 Cutaneous resistance to water loss of American green tree frogs (*Hyla cinerea*) exposed to different water vapour densities of air

Skin resistance in the frogs is negatively and linearly related to water vapour density, which suggests that the frogs secrete lipids in response to a lowering of water vapour density. The air was at 30°C moving at a velocity of 100 cm s^{-1} in a wind tunnel.

Values are means ± standard errors.

Sources: adapted from Wygoda ML, Kersten CA (2013). Effects of water vapour density on cutaneous resistance to evaporative water loss and body temperature in green tree frogs (*Hyla cinerea*). Physiological and Biochemical Zoology 86: 559-566. Image: © Greg Schechter/flickr.

The wiping behaviour of frogs differs between species and is most complex in the more waterproofed *Phyllomedusa* species: the waxy monkey leaf (tree) frog and the orange-legged leaf frog, which live in arid and semi-arid regions of Central and South America. These regions have high ambient temperatures, a long dry season and only short periods of low rainfall. The wiping manoeuvres of these leaf frogs control both the thickness and distribution of the lipid layer on all four limbs and the dorsal and ventral body surfaces. These leaf frogs secrete a waxy layer that is up to 100 molecules thick, making it comparable to the epicuticular wax layer of insects and arachnids, and thus is an example of convergent evolution, in which similar traits evolved in different lineages[36].

The skins of the leaf frogs are highly resistant to EWL, as shown by the data in Table 6.2. After wiping, these frogs usually become immobile and adopt a resting position with their legs beneath them, head down and eyes closed. This behaviour is thought to reduce the cracking of the protective wax layer and possible disturbance of the lipid protection.

Burrowing and cocoon formation reduces water loss of some amphibians

Some amphibian species living in deserts escape dehydration by burrowing underground during the day into moist cooler conditions and are only active at night. These animals rely on the water content of the soil to maintain their water balance by cutaneous water uptake[37]. As an example, Figure 6.25 shows how soil moisture increases with depth in the sand dunes of Western Australia into which sandhill frogs (*Arenophryne rotunda*) burrow.

During the dry season, some species of fossorial (burrowing) frogs enter a resting period of aestivation[38] in moist soil in which they remain dormant, only emerging after rainfall to breed. The reduction of metabolism during aestivation means that

[36] Section 1.5.3 discusses evolutionary convergence (homoplasy).

[37] Higher water content enhances water uptake from the moist substrate across their skin, which we discuss in Section 6.2.3.

[38] Aestivation is a prolonged state of low rates of metabolism similar to hibernation, which we discuss in Section 9.2.2

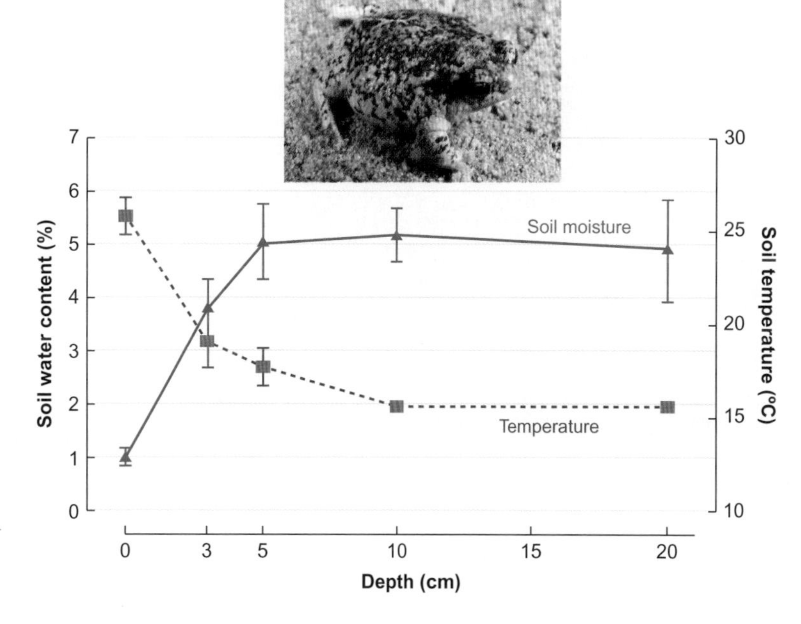

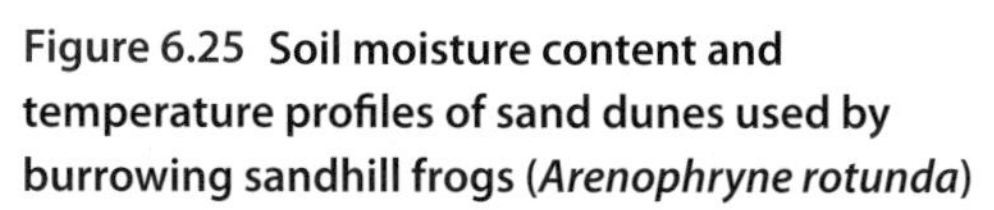

Figure 6.25 Soil moisture content and temperature profiles of sand dunes used by burrowing sandhill frogs (*Arenophryne rotunda*)

Burrowing sandhill frogs occur in restricted areas of Western Australia. The temperature of the sand dunes in which they burrow during the day decreases by about 10°C between the surface and 10 cm depth, where temperature stabilizes. Soil moisture content (determined by weight difference after drying) increases from the surface to 5 cm depth and then stabilizes at around 5%.

Source: adapted from Cartledge VA et al (2006). Water relations of the burrowing sandhill frog, *Arenophryne rotunda* (Myobatrachidae). Journal of Comparative Physiology B: 176: 295-302. Image: © Northern Sandhill Frog, B. Maryan, Western Australian Museum.

a minimal amount of water is produced by metabolism. Evaporative water losses from the pulmonary system and via the skin are initially balanced by water absorption from urine stored in the urinary bladder. After depletion of this supply of fluid, further water loss increases the osmolality of the body fluids.

Studies of spadefoot toads (*Scaphiopus* sp.) show that their burrows are 6–8 cm deep during their active period, but they burrow deeper during aestivation. At the start of the dry season, they are in burrows at a depth of 20 cm, but they reach depths of 90 cm as the season progresses, only returning to the surface as the wet season approaches. At greater depth, the reduction in temperature and the higher water content of the soil reduce the driving force for cutaneous EWL and facilitate water uptake.

During aestivation, some frogs form a cocoon, which seals them from the environment. Case Study 6.1 explores the water balance of some of these cocoon-forming frogs.

Case Study 6.1 Aestivation of burrowing frogs in cocoons

Many species of amphibians live in dry desert conditions where there are annual droughts. During the dry season they need to minimize evaporative water losses and generally survive by aestivation in underground burrows. Among the burrowing frogs, nine separate genera secrete cocoons in which they aestivate; indeed, they may remain in their cocoon for several years. The cocoon covers almost the entire body surface, as shown in Figure A, leaving only the nares (nostrils) open to allow continued exchange of gases in the lungs, although at a depressed rate.

Cocoon formation occurs in some North American, Central and South American and African frog species, but many of the cocoon-forming species occur in Australia, including probably all species in the genera *Cyclorana* and *Neobatrachus*. The Australian *Cyclorana* species are commonly known as water-holding frogs because of the large amount of water they hold in their urinary bladder. During aestivation, the water in the bladder is withdrawn and urine increases in concentration to reach that of the frogs' haemolymph, which also increases in concentration.

Figure A Water-holding frog (*Cyclorana* sp.) in its outer cocoon

The surface soil has been removed to expose the aestivating frog.

Source: http://ourriverourfuture.org/

At the start of aestivation, *Cyclorana platycephala* stores dilute urine in a bladder that holds 57 per cent of its standard body mass (mass without the bladder filled). This volume has been estimated to allow aestivation for 2–3 years given the rate of water loss seen during aestivation.

Based on laboratory studies, it is widely accepted that the cocoons help to reduce evaporative water loss during aestivation, although studies of the water balance of frogs in the wild or the effects of cocoons formed naturally by burrowing frogs are rare: finding such frogs once they have burrowed is difficult. However, this difficulty has been overcome during studies of Australian northern burrowing frogs (*Neobatrachus aquilonius*) and other species thanks to assistance from Aboriginal elders with a lifetime's knowledge of burrowing frogs, and an ability to locate them by surface features.

Cocoons are made up of layers of sloughed skin interspersed with other substances. Figure B shows this layering in the cocoons of aestivating frogs. The continual sloughing of skin to add to the cocoon occurs throughout aestivation, so the number of layers present in a cocoon indicates the time since it was first formed. Six *Neobatrachus aquilonius* excavated from their burrows had cocoons consisting of 81 to 229 layers—the highest number recorded for any cocooning species. The addition of layers at a rate of 0.2 to 0.35 layers each day means that these frogs had spent at least 6 months and up to more than 3 years in their cocoons.

In *Cyclorana* species, the occurrence of two types of lipids, cerebrosides and ceramides, in their cocoons suggests that these lipids are secreted between the layers of sloughed skin where they act as the main substance that reduces water passage through the cocoon. The same types of lipids occur in the skin of birds and mammals but are not present in the epidermis of *Cyclorana* spp. It is thought that these lipids must occur beneath the epidermis, in dermal glandular cells, and are secreted into the cocoon.

So, to what extent do cocoons reduce cutaneous water loss? Figure C shows area-specific rates of cutaneous water loss in each of five species of aestivating water-holding frogs. Measurements were made when the frogs were in their cocoons, and after removal of the cocoon. In all five species, removal of the

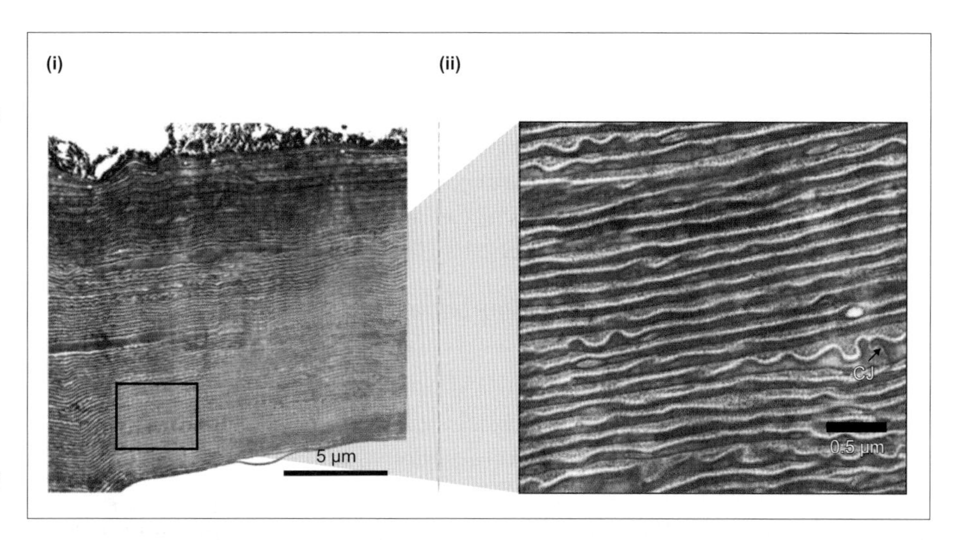

Figure B Electron micrographs through the cocoon after removal from Australian northern burrowing frogs (*Neobatrachus aquilonius*) excavated from a claypan

(i) Entire wall of a cocoon; (ii) higher magnification of boxed area showing cell layers with interspersed material, and one of the cell junctions (CJ).

Source: Cartledge VA et al (2006). Water balance of field-excavated aestivating Australian desert frogs, the cocoon-forming *Neobatrachus aquilonius* and the non-cocooning *Notaden nichollsi* (Amphibia: Myobatrachidae). Journal of Experimental Biology 209: 3309-3321.

cocoon results in a very large increase in cutaneous water loss. These studies show that the cocoons reduce water loss by an average of 86 per cent for the five species. The calculated total resistance to water loss of cocooned frogs varied from 40 to 119 s cm^{-1} compared to a range of 2.3 to 4.4 s cm^{-1} after removal of the cocoon. Thus, cocoons restrict evaporative water loss.

Studies of *Neobatrachus aquilonius* have shown that cocoon formation is not obligatory for species capable of producing them. Instead, it seems that the frogs produce cocoons in response to dehydrating soil conditions. This conclusion is based on the fact that the frogs excavate burrows in different types of soils. Those frogs buried in claypans (containing 13.5 per cent clay, 84.6 per cent sand and a small amount of silt) were found to form cocoons, but those frogs buried in a more loamy sand (6.3 per cent clay, 86.6 per cent sand and 6.3 per cent silt) had not formed cocoons. Examination of the water balance between the frogs and water in the soil offers an explanation.

The claypans contained only half of the moisture content necessary to balance the osmotic pressure of the plasma of the burrowing frogs[1], so water loss would occur. In such conditions, having buried themselves into the claypans before moisture levels decreased, cocoon formation by the frogs would reduce the rate of water loss. But these frogs may not form cocoons until they are in negative water balance: an earlier formation of a cocoon would restrict water uptake from the soil.

In the more sandy sites, where the burrowing frogs did not have cocoons, the moisture content of the soils was closer to being in balance with the plasma osmolality of the frogs, but again it indicated a negative water potential that would draw water from the frogs. The normal plasma osmolality in frogs at this location (195 mOsm kg^{-1} compared to >300 mOsm kg^{-1} in frogs from the claypans) indicated an insignificant level of dehydration at the time, perhaps due to the use of water in the bladder to balance cutaneous water losses. Further drying or more prolonged aestivation could induce eventual cocoon formation.

Figure C Cutaneous water loss for five species of *Cyclorana* frogs when in a cocoon or without a cocoon

Frogs without cocoons were rehydrated before measurements were made. All measurements were made by placing the frogs in a flow-through chamber with vapour pressure measurements of inflowing and outflowing air. Respiratory water loss was minimal, so measurement of the difference in humidity of inflowing and outflowing air gave a value for the cutaneous water loss. Numbers in brackets are numbers for each species.

Source: Sadowski-Fugitt LM et al (2012). Cocoon and epidermis of Australian *Cyclorana* frogs differ in composition of lipid classes that affect water loss. Physiological and Biochemical Zoology 85: 40-50.

› Find out more

Booth DT (2006). Effect of soil type on burrowing behavior and cocoon formation in the green-striped burrowing frog, *Cyclorana alboguttata*. Canadian Journal of Zoology 84: 832–838.

Cartledge VA, Withers PC, McMaster KA, Thompson GG, Bradshaw SD (2006). Water balance of field-excavated aestivating Australian desert frogs, the cocoon-forming *Neobatrachus aquilonius* and the non-cocooning *Notaden nichollsi* (Amphibia: Myobatrachidae). Journal of Experimental Biology 209: 3309–3321.

Sadowski-Fugitt LM, Tracy CR, Christian KA, Williams JB (2012). Cocoon and epidermis of Australian *Cyclorana* frogs differ in composition of lipid classes that affect water loss. Physiological and Biochemical Zoology: Ecological and Evolutionary Approaches 85: 40–50.

[1] To make this assessment it is necessary to convert plasma osmolality to the equivalent water pressure in kPa.

6

Internal barriers to water loss via the skins of mammals, birds and reptiles

In contrast to the external waterproofing of the skins of some frogs, the reptiles, birds and mammals have evolved internal mechanisms for waterproofing their skins that are an important component of their adaptation to a terrestrial life.

The stratum corneum is the permeability barrier of mammals

Figure 6.26 illustrates the layers (strata) of the mammalian skin. The outermost layer of the epidermis, the stratum corneum (a 'horny layer'), consists largely of **corneocytes**, which are dead cells that lack nuclei or cell organelles. In humans, the stratum corneum is 6 to 20 μm thick (a quarter or less of the thickness of a piece of paper), except for a few locations, such as the soles of the feet, where we have a thicker protective layer of corneocytes. The outermost cells of the stratum corneum consists of fully cornified **keratinocytes** (cells with a high **keratin** content) that are sloughed off and continuously replaced from below by the division of cells in the lowest layer of the epithelium: the stratum basale. Division of cells in the stratum basale pushes the cells above toward the surface of the skin.

The stratum corneum was for many years thought to have little functional activity, but it became clear in the 1960s that, in fact, the stratum corneum forms a vital barrier between the external environment and internal tissues. This barrier function is essential in limiting the transepithelial loss of water and also in preventing cutaneous infections. The function of the stratum corneum as a barrier to water loss can be demonstrated by stripping away corneocytes using adhesive strips (tape stripping), which progressively increases the transepithelial water loss with each stripping.

In mammals, the barrier function of the stratum corneum has been likened to 'bricks and mortar,' in which the corneocytes are the bricks, as illustrated by the schematic in Figure 6.27. The mortar between these bricks consists of a matrix of lipids that impedes the passage of water between the interdigitating corneocytes. These lipids are synthesized from precursors extruded from the lamellar bodies within the keratinocytes of the stratum granulosum. After synthesis, the lipids form intercellular lamellar lipid bilayers between the interdigitating corneocytes, as illustrated in Figure 6.27.

The intercellular lamellar lipids of the mammalian stratum corneum comprise three types of lipid: the ceramides, which are illustrated in Figure 6.28, and which may account for up to 50 per cent of the lipid content of the stratum corneum; cholesterol (about 25 per cent by mass); and various fatty acids (10–20 per cent by mass).

The mammalian skin normally shows a rapid ability to respond to an increase in transepidermal evaporative water loss by releasing stores of lipids from the lamellar bodies and replenishment of the lipid matrix in the stratum corneum. However, in some of the disease states that affect the skin, a persistently diminished ceramide content results in high cutaneous EWL.

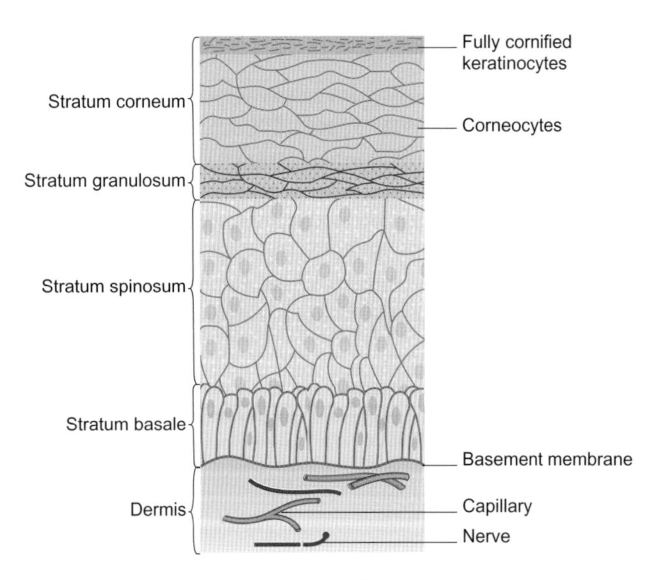

Figure 6.26 Layers in epidermis of mammalian skin

The skin consists of an outer epidermis overlying the dermis, and beneath this there is subcutaneous fat tissue. Formation of the keratinocytes commences in the stratum basale (also known as the stratum germinativum) and forms cells that move outward. Keratinization begins in the stratum spinosum and continues as cells move through the stratum granulosum. The outer layer—the stratum corneum—consists of corneocytes that lack nuclei or cellular organelles. Loss of the fully cornified keratinocytes from the outer surface of the stratum corneum occurs at a rate that is balanced by formation of new cells by the stratum basale.

Cutaneous water loss and lipids in the stratum corneum of birds

The surface specific rates of cutaneous water loss by species of birds living in arid areas are less than for similar species living in mesic conditions. For instance, larks from arid zones of Arabia (greater hoopoe larks (*Alaemon alaudipes*) and Dunn's larks (*Eremalauda dunni*)) have rates of cutaneous EWL that are about 25 per cent less than those of larks from mesic areas (Eurasian skylarks (*Alauda arvensis*) and woodlarks (*Lullula arborea*)) when measured at the same temperature (25°C). These differences in EWL suggest that natural selection has resulted in genetic differences in the properties of the stratum corneum such that water conservation occurs in species living in arid locations.

In addition, there is evidence that acclimatory responses of cutaneous EWL occur in response to periods of water shortage in some arid species. For instance, zebra finches (*Taeniopygia guttata*) reduce the rate of cutaneous EWL by 50 per cent within 16 h of having drinking water withdrawn, and further reductions occur if water deprivation continues. Replenishment of their water supply then leads to a return to the normal higher levels of water loss. Whether such responses will be sufficient to cope with the predicted increases in global temperatures is doubtful.

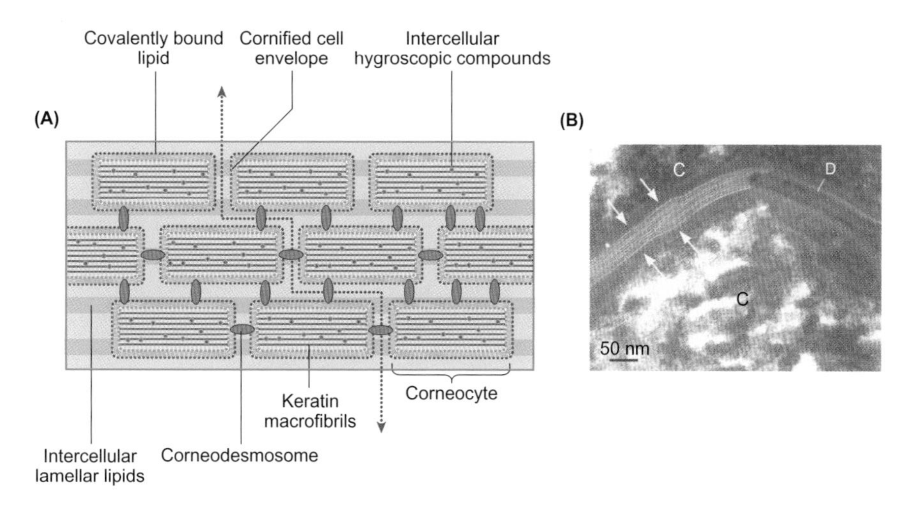

Figure 6.27 Bricks and mortar of the mammalian stratum corneum

(A) Schematic showing the bricks—the corneocytes consisting primarily of keratin macrofibrils and protected by a cornified outer layer (the cell envelope), which are held together in a stable arrangement by cell junctions called corneodesmosomes. The corneodesmosomes act as 'rivets' in both the plane of each layer of corneocytes and between layers, until they are degraded by proteolytic enzymes prior to sloughing of the corneocytes. Within the corneocytes a mixture of hygroscopic compounds maintains skin hydration. The mortar between the bricks consists of layers of lipids (ceramides, cholesterol and fatty acids). The route for water passage between the interdigitating corneocytes (inwards or outwards) is indicated by the dotted blue line and arrows. (B) Electron micrograph of human stratum corneum, showing lipid bilayer (between the arrows), and a corneodesmosome (D), between adjacent corneocytes (C).

Sources: A: adapted from Harding CR (2004). The stratum corneum: structure and function in health and disease. Dermatologic Therapy 17: 6-15. B: Wertz PW (2000). Lipids and barrier function of the skin. Acta Dermato-Venereologica, Supplement 208: 7-11.

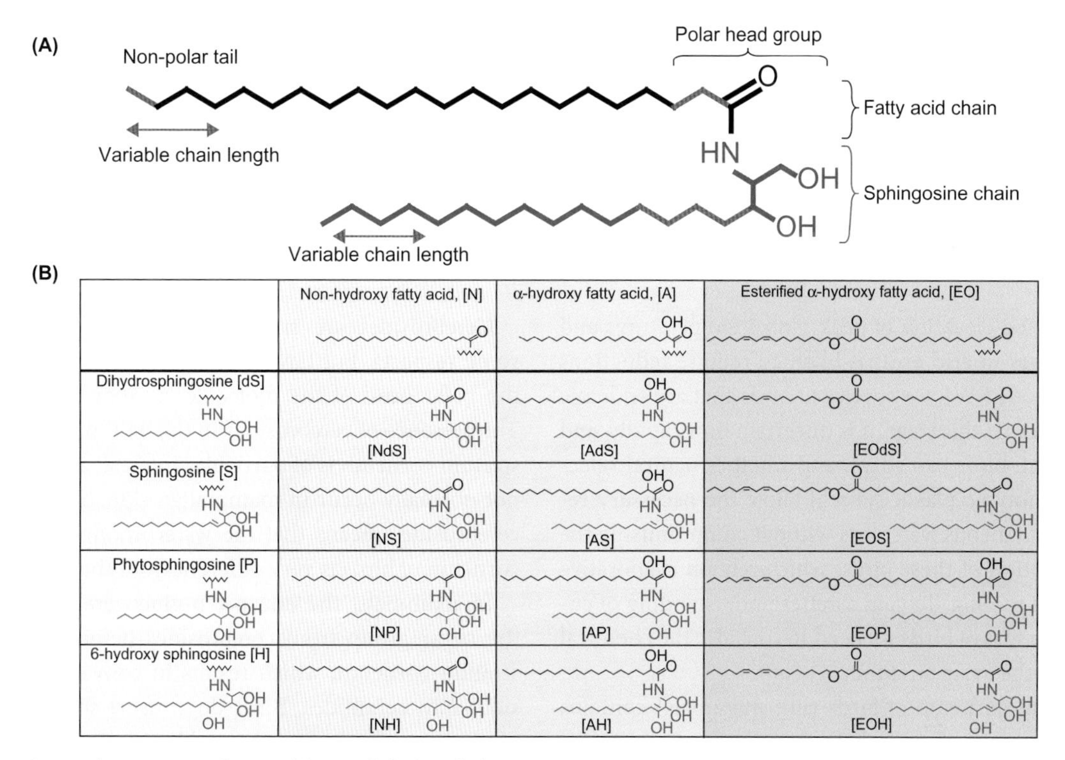

Figure 6.28 General structure of ceramides and their subclasses

(A) General structure of ceramides showing non-polar tail and polar head. All ceramides consist of a fatty acid chain, linked to a variable length (usually 16 to 20 carbon atoms) sphingosine chain, so ceramides vary widely in total carbon chain length. (B) Linkage of one of four sphingosine bases to a non-hydroxy fatty acid is shown by the general structures in the pale brown panel and forms ceramides denoted as [NdS], [NS], [NP], [NH]. Ceramides formed from hydroxy-fatty acids (blue panel) each have an additional hydroxy group in the polar head and are denoted as [AdS], [AS], [AP], [AH]. Ceramides formed from an esterified fatty acid (with O groups in the hydrocarbon chain) are shown in the green panel and denoted as [EOdS], [EOS], [EOP], [EOH]. All 12 ceramide subclasses occur in the stratum corneum of humans, while mice skin contains nine subclasses and lacks the three dihydrosphingosine [dS]-linked subclasses of ceramide.

Source: van Smeden J et al (2014). The important role of stratum corneum lipids for cutaneous barrier function. Biochimica et Biophysica Acta 1841: 295-313.

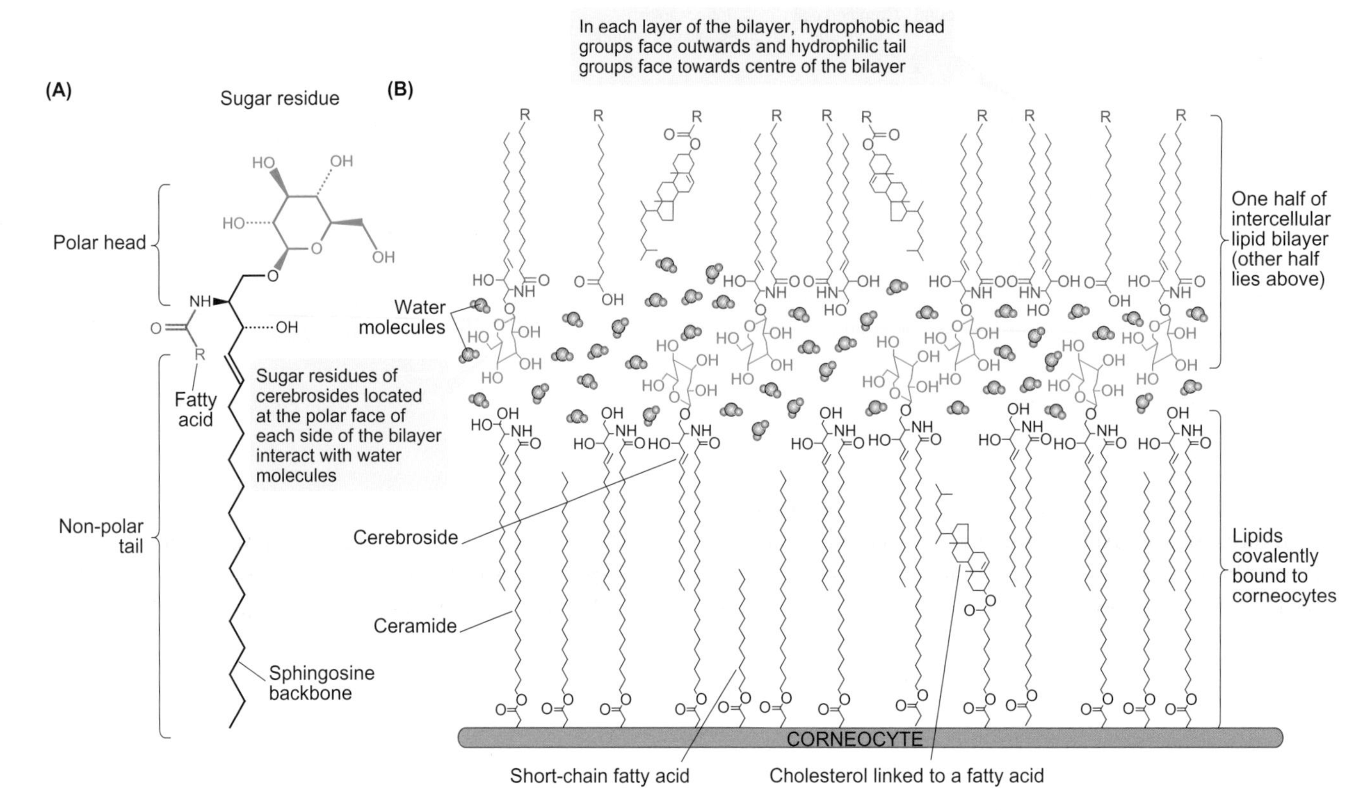

Figure 6.29 Cerebroside structure and a model of their arrangement in the stratum corneum of birds

(A) Birds have cerebrosides in the outer layer of their epidermis (stratum corneum). Like the ceramides shown in Figure 6.28, cerebrosides consist of a polar head group linked to a fatty acid hydrocarbon chain (simplified to R in this diagram) via an amide (NH) bond, and a non-polar hydrocarbon tail formed from a sphingoid base. The marked difference between ceramides and cerebrosides lies in the linkage of cerebrosides to a single sugar residue (shown in green). (B) Model of the interface of covalently bound and intercellular lipids showing proposed relationship between bound lipids (black) on the surface of a corneocyte and intercellular lipids (blue) in the avian stratum corneum. In this model, interaction of water with hydroxyl groups sequesters water within the stratum corneum outside the bilayer of lipids so water penetration of the stratum corneum is inhibited and cutaneous water loss reduced. Short-chain fatty acids and cholesterol esters also bind to the corneocytes and inhibit water passage between corneocytes.

Source: B: Champagne AM et al (2012). Lipid composition of the stratum corneum and cutaneous water loss in birds along an aridity gradient. Journal of Experimental Biology 215: 4299-4307.

The predicted elevation of peak global temperatures and water shortages in arid areas is likely to require reductions in the cutaneous EWL for species of birds that are endemic to arid regions. At this stage, it is uncertain how rapidly, and to what extent, birds can adapt, and whether natural selection (or phenotypic plasticity) will allow the necessary reductions in cutaneous water loss without compromising the thermoregulation of these birds, which rely on evaporative water loss for cooling. To gain a better understanding of cutaneous water loss of birds we need to consider the chemical properties of the avian stratum corneum.

The stratum corneum of birds (like mammals) contains high concentrations of lipids that are synthesized in the stratum corneum after the release of precursors and enzymes from multigranular bodies formed deeper in the skin. In birds, free fatty acids and esters of fatty acids often account for more than 50 per cent of the total lipids, which are accompanied by triacylglycerides (up to 25 per cent of total lipids), and generally smaller amounts of cholesterol/cholesterol esters, ceramides and cerebrosides.

Cerebrosides are structurally similar to ceramides in some respects, but have a hexose sugar group attached to the polar head of the molecule, as shown in Figure 6.29. The presence of cerebrosides in the lipid matrix in the avian stratum corneum is of particular interest, as cerebrosides do not normally occur in mammalian skin. The occurrence of cerebrosides means that the waterproofing of the stratum corneum of birds is very different from that of mammals.

In mammals, the enzyme β-glucocerebrosidase cleaves the sugar group from cerebrosides before they reach the stratum corneum, which results in conversion of cerebrosides into ceramides. Birds do convert some (but not all) of the cerebrosides in their skin to ceramides, using β-glucocerebrosidase. Hence, large quantities of cerebrosides occur in the avian stratum corneum.

A genetic defect in β-glucocerebrosidase in humans causes cerebrosides to occur in the stratum corneum, such that the rate of cutaneous EWL increases. This increase is thought to be caused by the bulkiness of the cerebrosides, due to the sugar group, and because the sugar group disrupts lipid packing, allowing a higher rate of penetration of water vapour. Similarly, in gene knockout mice that are deficient in β-glucocerebrosidase, water loss via the skin increases by 10 to 50 times relative to control mice that possess a functioning β-glucocerebrosidase gene. On this basis, we might expect that the presence of cerebrosides in the stratum corneum of birds would result in a high rate of EWL.

At first it was thought that cerebrosides in the avian stratum corneum could explain the generally high rates of cutaneous evaporative water of birds, in which cutaneous water loss generally exceeds 50 per cent of total EWL. However, contrary to expectations, the amount of cerebrosides in the skin of birds does not seem to be the explanation for the high rates of cutaneous water loss. In general, desert species of birds have a higher proportion of cerebrosides in their stratum corneum than mesic species. Among a single species (house sparrows, *Passer domesticus*) populations living in arid areas of Saudi Arabia have *higher* amounts of cerebrosides than populations of temperate house sparrows living in Ohio, USA, yet the birds in Arabia have a 25 per cent *lower* rate of cutaneous water loss.

The cerebroside content of the stratum corneum of birds living in habitats exhibiting a range of aridities has been found to correlate negatively with the rate of cutaneous water loss, as shown by Figure 6.30. These results suggest that higher rates of cutaneous water loss occur in bird species with lower amounts of cerebrosides in the stratum corneum—exactly the opposite of the original prediction. These findings suggest that lipids in the stratum corneum of birds may be organized very differently from those in mammals in which cerebrosides increase water loss.

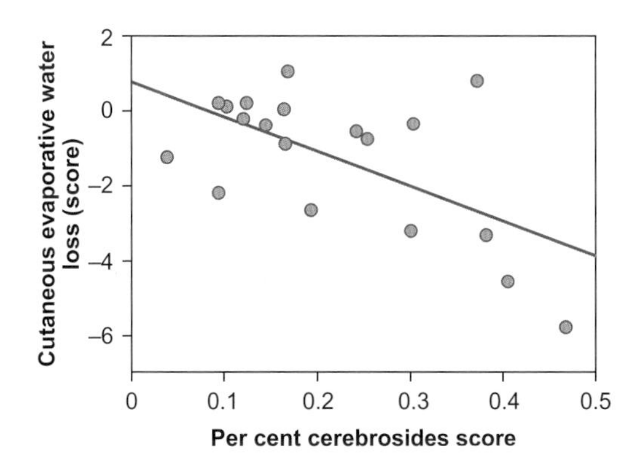

Figure 6.30 Relationship between cerebrosides in avian stratum corneum and cutaneous evaporative water loss of 20 species of birds in habitats of different aridities

Birds with a high per cent score of cerebrosides in their stratum corneum have lower rates of cutaneous evaporative water loss (lower scores). The cutaneous water loss of these birds was lower in species from environments with lower water availability (greater aridity). The data shown were obtained by principal components analysis and are phylogenetically independent scores (i.e. they take account of phylogenetic relationships, as we discuss in Section 1.3.3).

Source: Champagne AM et al (2012). Lipid composition of the stratum corneum and cutaneous water loss in birds along an aridity gradient. Journal of Experimental Biology 215: 4299-4307.

Exactly how lipids are arranged in the avian stratum corneum is uncertain, but new models of lipid organization in the stratum corneum of birds have emerged. The model in Figure 6.29B may look complex, but the overall concepts are relatively straightforward. The model focuses on the interaction of the two compartments of lipid molecules in the stratum corneum: (i) lipids covalently bound to the surface of the protein envelope surrounding corneocytes that provide the scaffold for the organization of (ii) the intercellular lipids.

The key point that Figure 6.29B illustrates is that the arrangement of lipids facilitates the interaction of water molecules with the hydroxyl groups (OH^-) of the cerebroside sugar groups, without disrupting the lipid layers between corneocytes (in which hydrophilic head groups of the bilayer face outward and the hydrophobic tails face inward). Further work is needed to test the model, and to work out exactly how cerebrosides form hydrogen bonds with those water molecules that are believed to form a barrier to the diffusion of water vapour across the stratum corneum. Sequestering of water may ultimately form aggregates of water molecules that move more slowly through the stratum corneum, so reducing cutaneous water loss.

Lipid layers in the epithelium of reptilian skin

Among reptiles (like birds), cutaneous water loss is generally more than half of the total evaporative water loss. However, several studies have shown a negative relationship between and cutaneous EWL and habitat aridity. For example, a study of Puerto Rican crested anoles (lizards, *Anolis cristatellus*) from xeric and mesic habitats showed that populations of lizards from xeric habitats lose water at a slower rate than populations from mesic habitats. These differences are sustained under a range of experimental conditions, which suggests that the lizards are genetically adapted to their habitats. The differences in cutaneous EWL appear to be due to differences in the amount and types of lipids in the skin, with greater amounts in xeric and terrestrial species than the lizards living in aquatic environments.

The organization of the layers of the reptilian epidermis varies between different types of reptiles, but we have most information for snakes and lizards. The outermost layer is the β-layer—so-called because of the presence of β-keratin in this layer. The β-keratin is important in strengthening the skin and results in the tougher skin of reptiles than occurs in mammals[39] but has little effect on skin waterproofing. Below the β-layer are layers of cells in the mesos layer. Below the mesos layer, there is a flexible α-layer, containing α-keratin.

Of the various layers in the epidermis, the mesos layer is critical for the waterproofing of snakes and lizards. This

[39] Two forms of keratin occur: α-keratins, which form flexible filaments, and β-keratins, which fold into less flexible sheets. The flexible skins of amphibians and mammals contain only the more flexible α-keratin, whereas the tougher skins of reptiles and birds (and scales) also contain β-keratin.

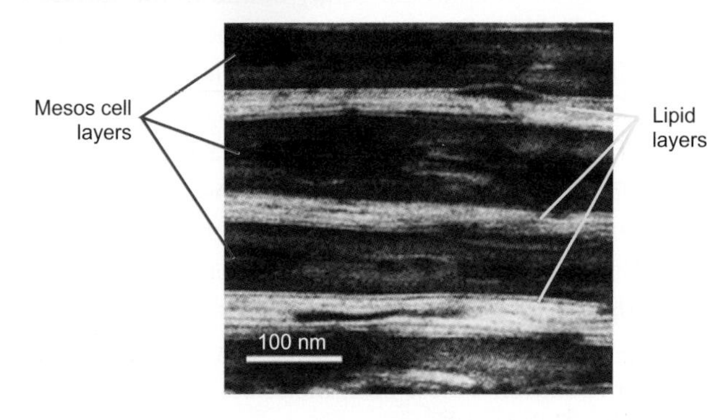

Figure 6.31 Lipid waterproofing of snake skin

The electron micrograph shows (lipid layers) between mesos cell layers of the epidermis of the grass snake (*Natrix natrix*). These lipid layers form the permeability barrier of the skin.

Source: Lillywhite HB (2006). Review - Water relations of tetrapod integument. Journal of Experimental Biology 209: 202-226. Image of grass snake © Karl Larsaeus

conclusion is supported by tape-stripping experiments that remove layers of the epidermis, one at a time. Stripping of the β-layer has little or no effect on water loss, but there is a rapid increase in cutaneous EWL when the stripping starts to remove some of the mesos layers. Recovery of the normal rate of CWL occurs gradually as the mesos tissue begins to be replaced.

The mesos layer shows some similarities to the stratum corneum of mammals. Layers of lipids between the mesos cells (keratinocytes) form the barrier to cutaneous water loss. Layering of mesos cells and lipids is particularly striking in snakes, as shown in Figure 6.31.

Snakes and lizards do not continuously slough off individual dead corneocytes, but instead shed their entire epidermal layer periodically, replacing it by a new one formed from the layers beneath. During this ecdysis, their protection against evaporative water loss declines. For example, the skin resistance of green iguanas declines from 370 s cm^{-1} to 108 s cm^{-1} after their skin is shed.

Snakes and lizards can adjust the number of layers making up the mesos layer after a moult and hence alter the water resistance of their new skin. Figure 6.32 shows the doubling in thickness of the mesos layer of California kingsnakes (*Lampropeltis californiae*) after they first shed their skin. Newly hatched kingsnakes possess about three lipid layers, as shown in Figure 6.32B, but 6–7 layers develop in the new epidermal layers, as illustrated by the electron micrograph in Figure 6.32C.

Such events probably explain the behaviour of kingsnakes (and other species of snakes), which switch from reclusive behaviour to dispersive behaviour after their first moult. Immediately after hatching, the kingsnakes tend to hide in damp locations where they moult several days later. After the first moult the skin's resistance to water loss increases from about 440 s cm^{-1} to almost 900 s cm^{-1}, so cutaneous EWL declines to about a half of its value in newly hatched snakes. A further increase in the skin's resistance to EWL occurs after the second moult of kingsnakes, such that it reaches more than 1200 s cm^{-1}, and it is possible that the maximum resistance is even higher in arid conditions.

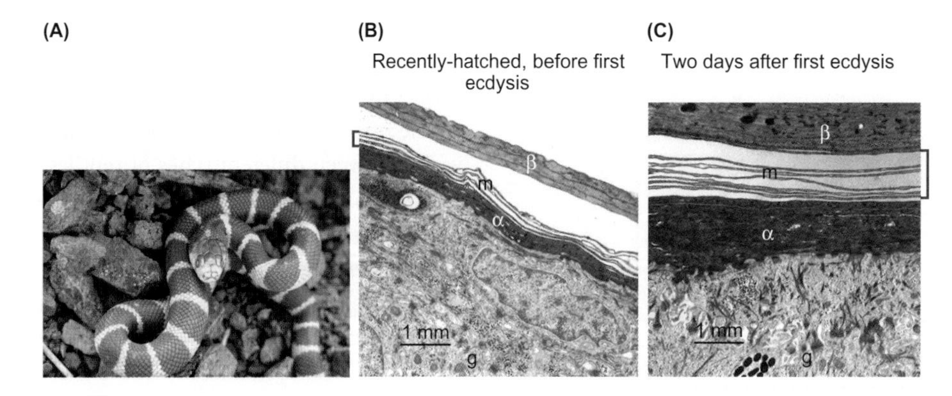

Figure 6.32 Increased mesos layer thickness in epidermis of Californian kingsnakes (*Lampropeltis californiae*) after first ecdysis

(A) The young kingsnake in the photograph is grey in colour, which occurs when the snakes are close to the time when the skin is shed. (B) and (C) are electron micrographs showing the ultrastructure of the epidermis of recently hatched kingsnakes. (B) is taken from a kingsnake just after hatching before its first moult (ecdysis). (C) is taken after the first moult and shows the increase in thickness of each of the three layers of the epidermis: the outer β-keratin layer (β), the mesos layer (m), and the α-keratin layer (α) that lies above the stratum germinativum (g). Separation of the β-keratin and mesos layers and the gaps in the mesos layers are artefacts from the tissue preparation. The main water permeability barrier in the epidermis of reptiles is the lipid layers in the mesos layer (indicated by red brackets), which doubles in the number of layers and overall thickness after the first moult.

Source: Tu MC et al (2002). Postnatal ecdysis establishes the permeability barrier in snake skin: new insights into barrier lipid structures. Journal of Experimental Biology 205: 3019–3030. Image: Calibas/Wikimedia Commons

6.1.4 Excretory water loss

In addition to evaporative water loss, land animals lose water in their urine and faeces, as indicated earlier in Figure 6.1. The ability of land animals to adjust the amount of water their kidneys excrete in urine is a key element of their adaptations to terrestrial life. Animals can reduce the volume of water loss in their urine in two fundamentally different ways:

- reabsorption of water from the urine;
- excretion of precipitated nitrogenous waste substances, in only small amounts of fluid.

Increased urine concentration saves water

Animals, such as annelid worms, slugs and amphibians, that normally occupy damp environments usually excrete urine that is more dilute than their body fluids, much like animals living in fresh water. However, if these animals become dehydrated, they reduce the amount of water lost in the urine by:

- reducing the rate at which the excretory organs form the primary urine from the extracellular fluids[40];
- increasing water reabsorption from the urine by osmosis, driven by solute reabsorption in the tubular components of excretory organs. This isosmotic water reabsorption[41] results in excretion of a reduced volume of urine, but since water reabsorption follows solute reabsorption, the maximum concentration of the urine is typically equal to their blood plasma or extracellular fluids.

Similar limitations to the ability to concentrate the urine occur in reptiles, even though they often occupy arid environments. Hence, reptiles are incapable of generating urine with a higher osmotic concentration than their extracellular fluids, as indicated by the data in Table 6.3.

[40] Sections 7.1 and 7.3 discuss the mechanisms by which animals adjust their rate of formation of primary urine, and we discuss its hormonal control in Section 21.1.4.

[41] Isosmotic volume reabsorption by kidney tubules is examined in Section 7.2.1.

6

Table 6.3 Maximum urine concentration of selected species of land reptiles, birds and mammals

Species	Maximum urine osmotic concentration (mOsm kg^{-1})	Urine/plasma osmotic concentration ratio
REPTILES		
Western bearded dragon (*Pogona minor*)	288	0.76
Claypan dragon (*Ctenophorus salinarum*)	295	0.85
Mojave desert tortoise (*Gopherus agassizii*)	320	0.90
Texas horned lizard (*Phrynosoma cornutum*)	327	1.0
BIRDS		
Emu (*Dromaius novaehollandiae*)	506	1.5
Chicken or domestic fowl (*Gallus gallus domesticus*)	538	1.6
Crested pigeon (*Ocyphaps lophotes*)	655	1.7
Ostrich (*Struthio camelus*)	890	2.6
Savannah sparrow (*Passerculus sandwichensis beldingi*)	2020	5.8
MAMMALS		
Human (*Homo sapiens*)	1430	4.2
Naked mole rat (*Heterocephalus glaber*)	1521	4.6
Cape ground squirrel (*Xerus inauris*)	1934	7.5
Arabian oryx (*Oryx leucoryx*)	2363	7.5
Dromedary camel (*Camelus dromedarius*)	2800	8
Merriam's kangaroo rat (*Dipodomys merriami*)	6382	16
Fat sand rat (*Psammomys obesus*)	6340	17
Spinifex hopping mouse (*Notomys alexis*)	9370	25

Values were obtained by water deprivation, during periods of seasonal dehydration (desert tortoises), after salt-loading or after administration of antidiuretic hormones[2]. The extent of an animal's ability to produce hyperosmotic urine is indicated by the maximum urine osmotic concentration divided by the plasma osmotic concentration at the time (the urine/plasma osmotic concentration ratio).

Sources: Beuchat CA 1990, 1996; Dantzler WH 1989, 2016; Dantzler WH, Bradshaw SD 2009; Dawson TJ et al 1991; Ford SS, Bradshaw SD 2006; Laverty G, Skadhauge E 2008; Louw GN 1993; Marsh AC et al 1978; Nagy KA, Medica PA 1986; Ostrowski S et al 2006; Urison NT, Buffenstein R 1994.

[2] Section 21.1.4 discusses the actions of antidiuretic hormones on the vertebrate kidney.

In contrast, mammals and birds can produce urine that is more osmotically concentrated than plasma (i.e. hyperosmotic to plasma). By producing hyperosmotic urine, birds and mammals can use their kidneys to obtain **osmotically free water**—i.e. water free of the solutes and ions (osmolytes) contained in the initial filtrate formed in the kidney from plasma.

The values in Table 6.3 show that dehydrated humans can reduce the loss of water in the urine by excreting urine with an osmotic concentration four times greater than that of plasma. But just how good are humans at producing hyperosmotic urine compared with other land animals? Looking at Table 6.3, we find that mammalian species living in arid environments can produce urine with osmolalities up to 25 times those of plasma. Desert rodents have the most pronounced ability to form concentrated urine. For example, Merriam's kangaroo rats (*Dipodomys merriami*) and spinifex hopping mice (*Notomys alexis*) can achieve urine/plasma osmolality ratios of 16–25. As with mammals, the bird species with the best ability to produce hyperosmotic urine (and thus to conserve water) are those living in arid environments[42].

Not all desert animals have the same ability to produce hyperosmotic urine. Naked mole rats (*Heterocephalus glaber*), which live in arid zones of Africa, are only slightly better at concentrating their urine than humans. The naked mole rat has little protection from evaporative water loss across its naked skin, and limited access to drinking water. However, its permanent underground existence, in a microhabitat with a high humidity, limits its actual evaporative water loss to amounts that naked mole rats can balance by water acquisition in food and by metabolism[43].

Urinary excretion of poorly soluble nitrogenous wastes saves water

Most land animals excrete the nitrogenous wastes that result from protein catabolism in their urine. Soluble nitrogenous wastes, in the form of ammonia and urea, require excretion of water in which these nitrogenous substances dissolve. However, some land animals excrete nitrogenous wastes, such as **uric acid**[44], which is secreted (as urate) into the urine. Urates have low solubility and precipitate when reabsorption of water occurs.

Precipitation of urates from urine reduces the osmotic concentration of the urine, which is determined by the substances dissolved in it. Precipitation of solutes from the urine allows further water reabsorption (by osmosis); as such, it makes an important contribution to the reduction of excretory water losses by disparate groups of land animals. Precipitation of nitrogenous wastes occurs in birds and reptiles, insects and arachnids, land snails, and some of the waterproofed species of frogs that occupy xeric habitats[45].

In birds, reptiles and insects, urine generated by the kidney (or **Malpighian tubules** in insects) enters the **cloaca** or hindgut where mixing of urine and faeces usually occurs. Reabsorption of water from urine in the hindgut may further precipitate uric acid and, hence, allow the absorption of more water before excretion of a semi-solid mix of faeces and nitrogenous wastes.

Many birds use reverse peristalsis to move the mixture of urine and gut fluids backward along the gut into the **coprodeum**, **colon** and the intestinal caecae. The hyperosmotic urine may actually draw water into the coprodeum, as shown in Figure 6.33A. However, this movement of water is more than compensated for by water absorption from the colon, which is driven by ion pumping by the colonic epithelium.

Post-renal salt and water absorption is hormonally regulated[46] and is particularly important for species of birds living in arid habitats. Water absorption in these species is enhanced by the high absorptive surface area of the coprodeum and colon, which results from the regularly-spaced tall folds (villi) of the mucosal surface. In zebra finches (*Taeniopygia guttata*) and emus (*Dromaius novaehollandiae*) the tall villi are covered by shorter villi, which further increases the absorptive surface area[47]. Figure 6.34 illustrates the elaborate arrangement of villi and microvilli in the colon of the emu. Such villi do not occur in the colon (large intestine) of mammals, or in all species of birds. However, the villi and microvilli are particularly well developed in the colon of those species of birds living in arid habitats. The high absorptive surface area of the colon and coprodeum of emus and zebra finches enables them to absorb a higher proportion of water from the urine prior to its excretion.

The ostrich is unusual among birds in having a strong sphincter between the coprodeum and the terminal part of its colon. This sphincter allows temporal separation of the excretion of its urine and faeces, which Figure 6.33B illustrates. This separation thus preserves the concentrating effects of the kidneys (urine/plasma osmotic concentration ratio = maximum of 2.7). During defaecation, the rectal-coprodeum sphincter is relaxed, and urine is stored in the coprodeum, in a similar way to urine storage by a bladder. When the

[42] Differences in the kidney structures of birds and mammals that determine maximum urine osmolalities are discussed in Section 7.2.3.

[43] Water acquisition in food and by metabolism is discussed in Sections 6.2.1 and 6.2.2.

[44] We discuss the use of uric acid as nitrogenous waste more fully in Section 7.4.3.

[45] Figure 7.47 shows the distribution of forms of nitrogenous wastes in various amphibian species.

[46] We discuss the regulation of post-renal transport by aldosterone in relation to salt and water balance in Section 21.1.1.

[47] The increase in surface area by villi and microvilli has similarities to that illustrated in Figure 2.14 for the small intestine of mammals and birds, where it facilitates the absorption of nutrients.

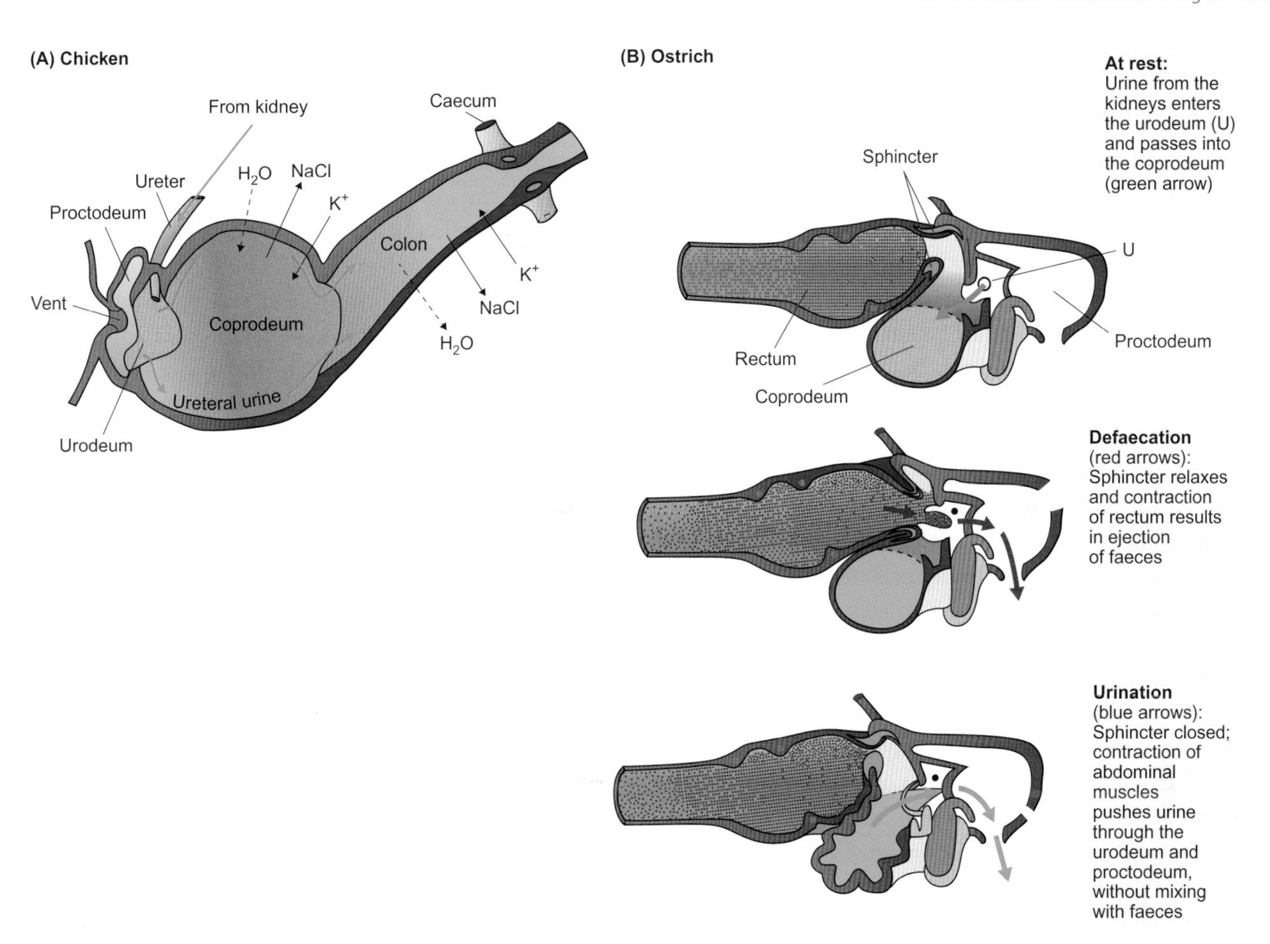

Figure 6.33 Fluid processing in the cloaca and colon of chickens (*Gallus gallus domesticus*) and ostriches (*Struthio camelus*)

The cloaca is an expanded tubular structure made up of proctodeum, urodeum and coprodeum, with a common opening, called the vent, for digestive, reproductive and urinary systems. (A) In chickens, urine enters the urodeum from the ureters and may flow backward (blue arrows) into the posterior digestive tract—the coprodeum, colon and caecae—where epithelial transport processes modify the composition of the mixture of faeces and urine before excretion. Water movement into the coprodeum dilutes the fluid, NaCl may be withdrawn and K^+ is added to the fluid. In the colon and caecae, water reabsorption is linked to ion transport processes controlled hormonally in relation to salt intake. (B) In ostriches, a sphincter separates faeces in the rectum from urine stored in the bladder-like coprodeum. The diagrams illustrate separation of urine excretion from defaecation by the operation of the sphincter.

Sources: A: adapted from Choshniak I, Munck BG, Skadhauge E (1977). Sodium chloride transport across the chicken coprodeum. Basic characteriscitcs and dependence on sodium chloride intake. Journal of Physiology 271: 489-504; B: Laverty G, Skadhauge E (2008). Adaptive strategies for post-renal handling of urine in birds. Comparative Biochemistry and Physiology Part A 149: 246-254.

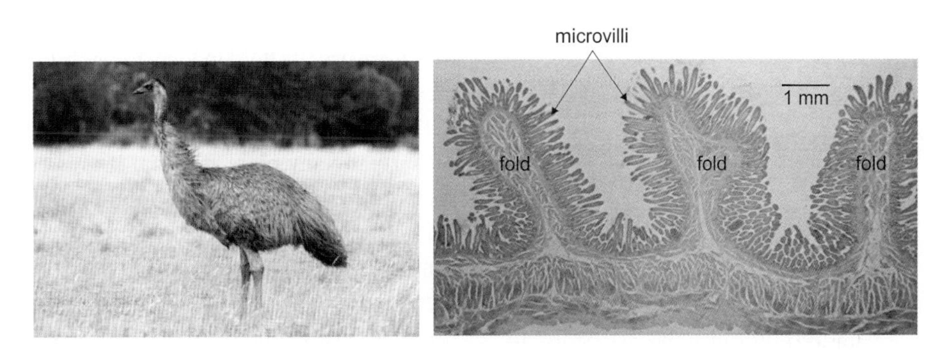

Figure 6.34 The colon of the emu

The emu colon has tall folds, each bearing numerous villi, which results in an extensive surface area for absorptive processes.

Source: Johnson OW, Skadhauge E (1975). Structural-functional correlations in the kidneys and observations of colon and cloacal morphology in certain Australian birds. Journal of Anatomy 120: 495-505. Emu image: benjamint444, https://en.wikipedia.org/wiki/GNU_free_documentation_license.

rectal-coprodeum sphincter is closed, urine enters the **urodeum** from the ureters, and then the coprodeum. For urination, contraction of the abdominal muscles pushes urine out without it being mixed with the faeces.

In insects, urine generated by their Malpighian tubules is at most equal in osmolality to their haemolymph[48], but further processing of the urine in the hindgut can generate excreta in which the fluid component becomes hyperosmotic to their haemolymph by 500 to 5500 mOsm kg^{-1}. For example, desert locusts, cockroaches and blowflies produce excreta with an osmotic concentration up to five times higher than that of their haemolymph. Production of hyperosmotic excreta depends on ion transport processes in specialized thickened areas of the rectum known as the rectal pads[49].

6

Reabsorption of water in the hindgut and rectum ultimately determines the amount of water lost in faeces. In mammals, faecal water content is usually below 50 per cent (by mass) but varies from less than 20 per cent to more than 80 per cent among desert rodents. For example, spinifex hopping mice excrete faeces containing 19 per cent water by mass, even when they are fully hydrated. In these mice, no further water absorption occurs after 24 h of dehydration, but some species living in arid conditions, such as the Arabian oryx, reduce the water content of their faeces when food and water are restricted.

› *Review articles*

Braun EJ (2003). Regulation of renal and lower gastrointestinal function: role in fluid and electrolyte balance. Comparative Biochemistry and Physiology Part A 136: 499–505.

Contreras HL, Heinrich EC, Bradley TJ (2014). Hypotheses regarding the discontinuous gas exchange (DGC) of insects. Current Opinion in Insect Science 4: 48–53.

Gibbs AG (1998). The role of lipid physical properties in lipid barriers. American Zoologist 38: 268–279.

Laverty G, Skadhauge E (2008). Adaptive strategies for post-renal handling of urine in birds. Comparative Biochemistry and Physiology, Part A 149: 246–254.

Lillywhite HB (2006). Review—Water relations of tetrapod integument. Journal of Experimental Biology 209: 202–226.

Williams JB, Muñoz-Garcia A, Champagne A (2012). Climate change and cutaneous water loss of birds. Journal of Experimental Biology 215: 1053–1060.

6.2 Balancing water loss

Animals balance their water losses by acquiring **preformed water** in food items, by producing **metabolic water** during metabolism and by water uptake from the environment, either by drinking or by absorption of water across the integument.

6.2.1 Preformed water in food

Foods items contain variable amounts of water known as preformed water. Fruits, meat and some plant material contain 60 per cent to 90 per cent water (by mass). Consequently, their consumption makes an important contribution to water balance. By contrast, lipids, for example in the adipose tissue of vertebrates, contain much less water (about 10 per cent by mass). The preformed water occurs either as bulk water that flows out if food items are squeezed or is loosely adsorbed but released during digestion[50].

Some herbivores may adjust their selection of food items to optimize the intake of preformed water in relation to their needs for water balance. Large herbivores, such as desert goats, acquire most of their water needs from their food by selecting species of plants that provide enough water, even if this is at the expense of the nutritional content of the food items.

One of the animal groups that has been well studied in relation to preformed water in their diet are locusts, and in particular the migratory locust (*Locusta migratoria*). Migratory locusts have the widest distribution of any locust species, being found in Europe, Africa, Arabia, Asia, Australia and New Zealand. Most migratory flights are over short distances, but migratory locusts sometimes become gregarious and form huge swarms that migrate for hundreds of kilometres, covering several kilometres each day.

Wild locusts actively select food items with high water content. Hence, of the water taken in by locusts eating their normal diet of luscious leafy materials, 98.8 per cent comes from preformed water. Migratory locusts prefer grasses and eat almost their own body mass in food each day. Hence, a 1-gram locust eating leafy material that contains 75 per cent water would obtain 750 μL of preformed water (= 75 mg water per 100 g body mass per day). As shown in Figure 6.35A, this intake of water is more than enough to balance the total water losses in the faeces/urine, via the cuticle, and the respiratory loss via the spiracles. When eating such foods, excess water can be lost either in the faeces, which are usually fluid in locusts feeding on lush foods, or by increasing ventilatory losses.

Despite their preference for grasses, migratory locusts will eat virtually any plant material if grasses are in short supply, and therefore swarms of migratory locusts can severely damage crops and tree stems. They even resort to the consumption of much drier food items if that becomes necessary. Figure 6.35B illustrates what happens in the extreme, if the locust's diet consists of only dry food items.

[48] Section 7.3.4 discusses urine production by insects.

[49] Figure 7.38 examines the functioning of rectal pads.

[50] Even 'dry' food contains water molecules chemically bound to the food particles, but bound water in food particles is difficult to remove and hence unimportant in water balance.

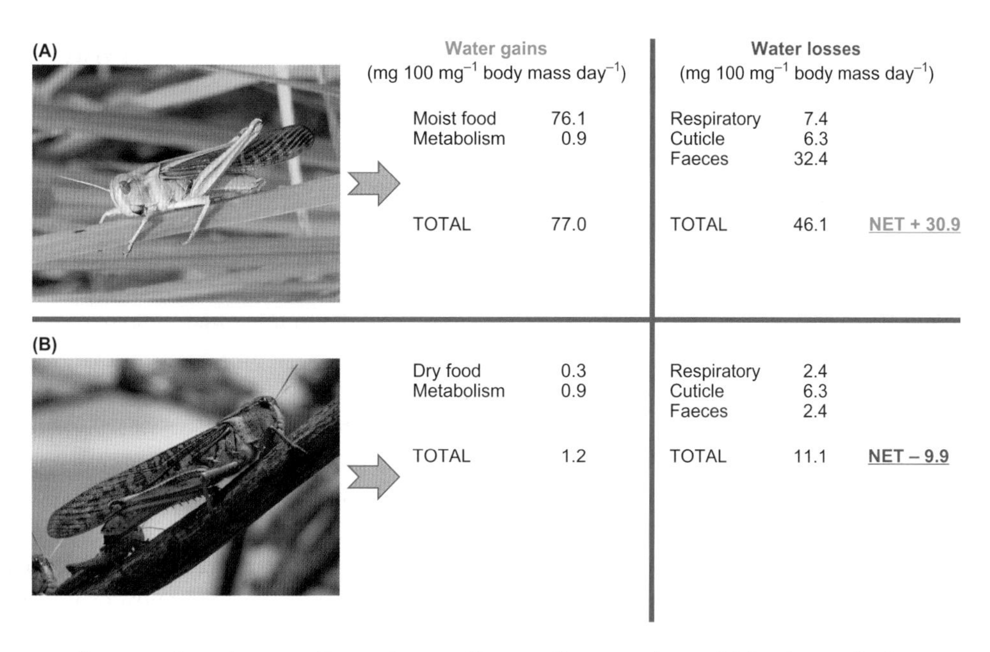

Figure 6.35 Sources of water gain and routes of water losses of locusts (*Locusta migratoria*) feeding on fresh green grass or dry food

(A) In normal circumstances *Locusta migratoria* feed on leafy material. The water content of the diet is the major source of water gain, and loss of water in the faeces is the major route for water loss; evaporative water loss via the respiratory system and across the cuticle are much smaller. A positive water balance is maintained with excess water excreted as necessary. (B) If locusts consume dry foods, a negative water balance occurs. In such circumstances, locusts drink water if it is available to restore water balance.

Data from Chapman RF. The Insects. Structure and Function, 4th Edition. Cambridge: Cambridge University Press, 1998. Images: A: Christiaan Kooyman/Wikimedia Commons B: Jonathan Hornung/Wikimedia Commons

In such circumstances, the locusts are in an overall negative water balance (based on food intake and its metabolism, and water losses). Hence, such locusts need to drink water to restore their water balance. This balance sheet probably explains why locusts and grasshoppers studied in the laboratory cease feeding if they are given dry food without the provision of an adequate supply of drinking water to maintain water balance. In such experiments, the amount of dry food consumed is proportional to water intake, with maximal intake of dry food only occurring when there is provision of unlimited water.

Similar principles apply to other land animals, although drinking water may be unavailable to animals living in arid habitats, so there is a risk of a negative water balance. In extreme circumstances, when food items contain little water, animals may reduce their feeding, which reduces the water loss associated with the excretion of faeces and nitrogenous wastes[51].

In arid habitats, dry plant materials tend to absorb moisture when the water vapour density of air rises. In a similar way, seeds that contain much less water than leaves, at 5–10 per cent water by mass, absorb water when taken into the burrows of seed-eating animals; this water is subsequently acquired when the seeds are eaten. Water vapour may also condense onto plants as dew if the saturation pressure on the plant surfaces falls below the saturation pressure at air temperatures. Dew occurs in the Namib Desert on about 50 days per year, while morning fogs deposit water droplets onto plant surfaces on an additional 60–70 days. Because of the absorption of water vapour, the water content of desert grasses typically peaks during the night and early morning, when atmospheric humidity increases compared to the lower humidity at midday, which is illustrated in Figure 6.36. Many herbivores, such as beisa oryx (*Oryx beisa*) and Grant's gazelles (*Nanger granti*), survive droughts in arid conditions in North-East Africa by feeding overnight or at dawn, taking advantage of the higher water content of food items at these times.

Arabian oryx obtain about 80 per cent of the water they need from the water content of dwarf shrubs, desert grasses and acacia leaves. While this vegetation seems dry, it may contain 50–60 per cent water on a mass basis at dawn. However, water intake does not reach the amounts seen among herbivores living in less arid conditions. Indeed, Arabian oryx have one of the lowest rates of water influx in food among ungulates for their body mass, at about 44 per cent lower in spring, and 77 per cent lower in summer.

[51] Section 6.1.4 briefly examines excretory water loss, while Section 7.4 examines nitrogenous excretion and the associated water loss.

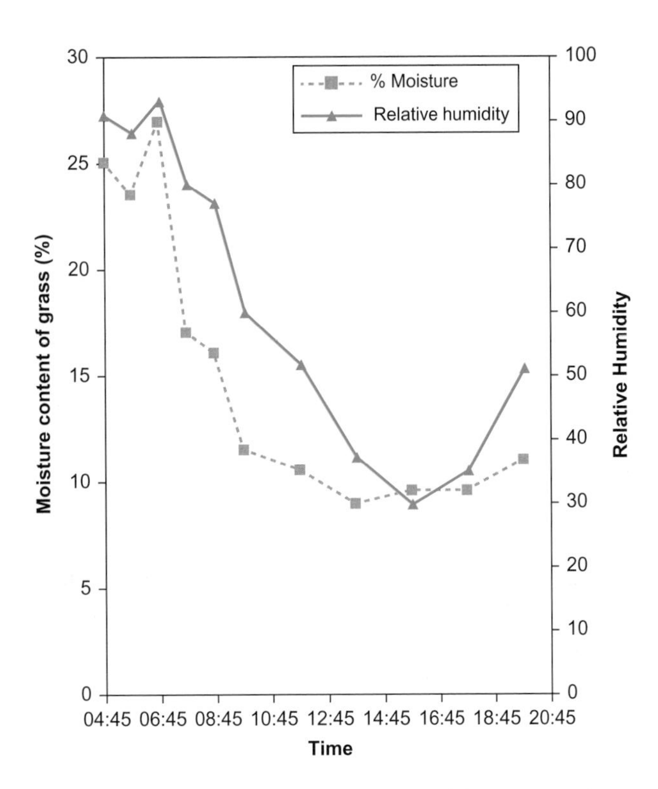

Figure 6.36 Changing moisture content of desert grass due to daily changes in relative humidity

The moisture content of the desert grass (*Stipagrostis uniplumis*) peaks at 27% when atmospheric humidity rises to a peak of 95% relative humidity in the early morning (6:45); thereafter, atmospheric humidity declines and the moisture content of the grass follows a similar but slightly delayed pattern.

Source: Hentschel JR, Seely MK (2008). Ecophysiology of atmospheric moisture in the Namib Desert. Atmospheric Research 87: 362-368.

6.2.2 Obtaining water by metabolism

Metabolism of carbohydrates, protein and lipids (fats) yields metabolic water. During aerobic metabolism, the oxidation of carbohydrates, lipids and proteins[52] results in reduction of oxygen to form metabolic water, as shown for glucose metabolism in Equation 6.5.

$$C_6H_{12}O_6 + 6O_2 \rightarrow 6CO_2 + 6H_2O \qquad \text{Equation 6.5}$$

The amount of water generated by metabolism depends on the amount of hydrogen present in the particular foodstuffs or the type of energy store that is metabolized and which forms water. The stoichiometry of Equation 6.5 (that is, the number of molecules of each substance present in the equation relative to each other) tells us that one mole of glucose (which has a mass of 180 g) generates six moles of water (a total mass of 108 g: each mole of water has a mass of 18 g). This means that each gram of completely oxidized glucose yields 0.6 g (i.e. 0.6 mL) of water. This amount is the same in all animals as it is fixed by the chemical processes.

Table 6.4 compares the amount of metabolic water formed by oxidation of carbohydrate (glycogen), lipid and protein. Glycogen generates slightly less water than the amount we calculated for glucose because of the slightly lower H content.

Lipid (fat) metabolism generates the highest amount of water per gram of substrate metabolized (although the exact amount depends on the degree of saturation of the fats[53]). However, Table 6.4 shows that metabolism of lipids provides more than twice the energy as the metabolism of protein or carbohydrate, when expressed per gram of dry matter (as **energy density** g^{-1} dry matter), so animals need to metabolize less fat than carbohydrates for a particular metabolic rate. These differences are even more pronounced on a wet weight basis since natural foods rich in carbohydrates and proteins contain greater amounts of preformed water than lipids[54].

The final column of Table 6.4 shows the net metabolic water production for each kilojoule (kJ) of energy for each

[52] Section 2.4.2 discusses aerobic metabolism.

[53] Boxes 2.1 and 9.2 discuss lipids and their structure.

[54] Table 2.1 shows that fat stores in animals have about 7.5 times greater energy density than carbohydrates and proteins and are used by all animals as a form of energy storage.

Table 6.4 Metabolic water production during oxidation of carbohydrate, lipid and protein

	Water production per gram dry matter oxidized (ml H_2O g^{-1})	Energy density (kJ g^{-1} dry matter)	Metabolic water production per energy expended (ml H_2O kJ^{-1})
Carbohydrate (glycogen[3])	0.56	17.5	0.032
Lipid in adipose tissue	1.07	39.6	0.027
Protein (urea excretion)	0.39	17.8	0.022
Protein (uric acid excretion)	0.50	17.8	0.028

Metabolic water production per energy expended (ml H_2O kJ^{-1}) is calculated by the division of the data for metabolic H_2O per g dry matter oxidized (column 1) by energy density (column 2).

Sources: data from Schmidt-Nielsen K. Animal Physiology. Adaptation and Environment. 5th Edition. Cambridge: Cambridge University Press, 1997 and Jenni L, Jenni-Eiermann S (1998). Fuel supply and metabolic constraints in migrating birds. Journal of Avian Biology 29: 521-528.

[3] Glycogen is the main form of carbohydrate storage in the liver and muscles of vertebrates, as we discuss in Box 2.1.

of the main types of foodstuffs on a *dry weight* basis. The differences are small, but metabolism of carbohydrates yields slightly larger amounts of metabolic water per kJ of energy than other metabolic substrates.

The amount of water gained by oxidation of protein depends on the chemical form(s) of nitrogenous waste produced and excreted[55]. These wastes contain different amounts of hydrogen that is excreted rather than being oxidized to form water. Each mole of urea ($(H_2N)_2CO$) contains two hydrogen atoms per atom of N excreted, while a mole of uric acid ($C_5H_4N_4O_3$) has one hydrogen atom per N atom. Hence, metabolism of protein followed by uric acid excretion generates more water than metabolism of protein and urea excretion. As a result, mammals (which predominately excrete urea) increase their possibility of water stress if feeding on high protein diets in a dehydrating environment, where insufficient drinking water is available.

In most ectothermic animals, their low rates of metabolism result in relatively small amounts of metabolic water in relation to overall water balance. For example, migratory locusts feeding on their normal leafy diet, such as those shown back in Figure 6.35A, produce 0.9 mg water per 100 g body mass per day from metabolism, which accounts for less than 2 per cent of the daily water they require to maintain water balance.

Locusts do not increase their metabolic water production when water is in short supply because this would need extra supplies of food and the additional oxidative metabolism would add to their oxygen requirement. This additional need for oxygen would then result in increased evaporative water losses from the respiratory system and would offset any benefits[56]. These interacting factors can be expressed in a water-balance index of metabolic water production relative to evaporative water loss:

$$\text{Water balance index} = \frac{\text{Metabolic water production}}{\text{Evaporative water loss}}$$

Equation 6.6

If the water balance index is less than 1.0, the water gained by increasing metabolism will be less than the water lost by evaporation. For the locusts back in Figure 6.35, the values for the water balance index are:

$0.9/13.7 = 0.066$ when consuming wet food
$0.9/8.7 = 0.103$ when eating dry food.

These values confirm the relative unimportance of metabolic water in their water balance.

[55] Section 7.4 discusses the nitrogenous wastes of different animal groups.

[56] Equation 6.2 shows the relationships between evaporative water loss and oxygen consumption.

Although most animals gain only relatively small amounts of metabolic water, metabolic water is critical if some animals are to survive. Perhaps the best examples are desert rodents, such as kangaroo rats (*Dipodomys* spp.) and pocket mice (*Perognathus* spp.), in which high rates of metabolism are important for their water balance. Except for lactating females, kangaroo rats never drink water.

Early laboratory studies of kangaroo rats by Knut and Bodil Schmidt-Nielsen suggested that these animals obtain 90 per cent of their water needs by oxidative metabolism of a carbohydrate-rich diet of dry seeds and dry plants, which contains only small amounts of preformed water. Ambient temperatures have an important influence on water balance because of the way body temperature is regulated through evaporative water loss. While feeding on seeds, a water balance index of 1.0 (based on Equation 6.6) occurs at ambient temperatures of between 20 to 25°C, depending on the species of kangaroo rat. However, at higher ambient temperatures, an increase in evaporative water loss is necessary to maintain core body temperatures, which reduces the water balance index to below 1.0.

Kangaroo rats, like other desert rodents, may solve the problem of a lower water balance index at high ambient temperatures by staying in cool underground burrows during the heat of the day to reduce evaporative water losses. In addition to being cooler, underground burrows may have a higher water vapour density than at the surface, which reduces pulmonary evaporative water loss[57]. In extreme circumstances, however, this may not be a viable option. Studies of kangaroo rats living in the central Sonoran Desert (Arizona, USA) indicate that burrowing alone is not sufficient to allow rats on a diet of seeds to maintain their water balance during the summer months.

Monitoring of kangaroo rats living in the central Sonoran Desert by using temperature transmitting collars indicates that they spend most of their time in relatively shallow burrows (<1 metre deep) where ambient temperatures are above 35°C for almost two months of the summer. The use of night cameras has revealed that the kangaroo rats leave their burrows soon after sundown and concentrate feeding into a short period of about an hour, when air temperatures at the surface are still 35–39°C. This unusual decision to leave the burrows before temperature declines to the overnight minimum may be explained by the equally unusual finding that they consume any available insects along with succulent green vegetation, rather than merely eating seeds and dry vegetation as traditionally thought.

Figure 6.37 examines the daily gains and losses of water for kangaroo rats held in the laboratory at 35°C (the temperature of their burrows) while feeding on just seeds. The diet provides about 54 g metabolic water per 100 g seeds, which accounts

[57] Equation 6.1 provides an explanation of this effect.

6

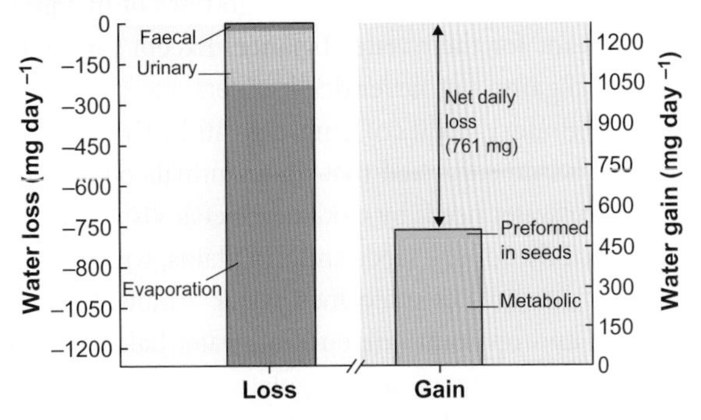

Figure 6.37 Water losses and water gain in Merriam's kangaroo rats (*Dipodomys merriami*) feeding on dry seeds at an ambient temperature of 35°C

Water gained from preformed water in food plus metabolic water does not balance the total water loss by evaporation from the respiratory tract, in faeces and in urine. The daily shortfall is about 760 µL (760 mg) in a 35 g animal, which equals about 2.2% of its body mass per day.

Source: Tracey RL, Walsberg GE (2002). Kangaroo rats revisited: re-evaluating a classic case of desert survival. Oecologia 133: 449-457. Image: Bcexp/WikimediaCommons/CC-BY-SA-3.0.

for almost all of the water gain; at 35°C, however, it is insufficient to balance the water losses by evaporation and the additional smaller water losses in the urine and faeces. The outcome is a significant water deficit equivalent to 2.2 per cent of body mass per day. Since mammals cannot tolerate overall water loss of more than 12–14 per cent of body mass, the kangaroo rats would survive less than a week at this level of water deficit.

It seems that consumption of insects and succulent vegetation, which have higher preformed water than seeds, is essential for kangaroo rats to maintain water balance during the harsh desert summer. Green vegetation is thought to be the greatest consistent contributor to water balance and insects a more ephemeral food source, with likely seasonal variations, but the extent of such variations in dietary items on a seasonal basis is uncertain.

The importance of metabolic water is not restricted to desert rodents with high metabolic rates: it is also important in arthropods that occupy dry environments with low (or no) water availability. Ticks, mites and scorpions have a low metabolic rate but still generate an amount of metabolic water which, when coupled with their low evaporative water losses, ensures survival in dehydrating conditions. For example, the European pigeon tick survives between annual breeding seasons (and even for several years) using metabolic water as their main input of water. Long-term survival is limited not by evaporative water loss but by exhaustion of the metabolic reserves that are needed to produce metabolic water.

6.2.3 Water uptake via the integument

Some land animals that live in damp environments, or that seek out such habitats when they are dehydrated, can compensate for water losses by taking up water across their integument, in a process known as **contact-rehydration**. Contact-rehydration involves similar behavioural responses—a water absorption response—in evolutionarily diverse groups, from slugs to amphibians. Typically, animals move onto a moist surface and adopt a relatively flattened position, maximizing the surface area of the integument in physical contact with a moist substrate such as damp vegetation, damp soil, or temporary puddles and pools of water. Even desert snails (*Xerocrassa seetzeni*) use contact-rehydration, having searched out water that has condensed onto shrubs overnight.

The results of laboratory studies of leopard slugs (*Limax maximus*) illustrated in Figure 6.38 show that contact rehydration initiates rapid water absorption across the foot, at a rate of 8 µl cm^{-2} min^{-1}. As such, 10 minutes is generally long enough to rehydrate. In the wild, leopard slugs tend to seek out shallow pools of water in the late evening and overnight in order to rehydrate. However, slugs in the laboratory tolerate large changes in body mass due to water loss (in the order of 60–70 per cent) before initiating contact rehydration.

Amphibians are more sensitive than slugs to the loss of body water. A 10 per cent loss of body mass (through water loss) is sufficient to stimulate most frogs and toads to exhibit behaviours resulting in water absorption. During water absorption, the hind legs splay so that an area of skin known as the pelvic patch is pressed close to moist/wet surfaces. The pelvic patch is well-vascularized skin in the ventral and pelvic areas. Although the pelvic patch represents only about 10 per cent of the surface area of the skin of frogs and toads, it is responsible for the majority of the animal's water uptake.

The rate of the water absorption by amphibians is determined by a combination of:

- The difference in osmotic concentrations between the external fluid on the skin and internal body fluids. Dehydration increases the osmolality of extracellular fluids—that is, it lowers the water concentration in the extracellular fluid compared with the higher water concentration (lower osmotic concentration) of pools of fresh water—which increases the driving force for water absorption during contact rehydration.
- The water permeability of the integument, which determines the ease with which water crosses the

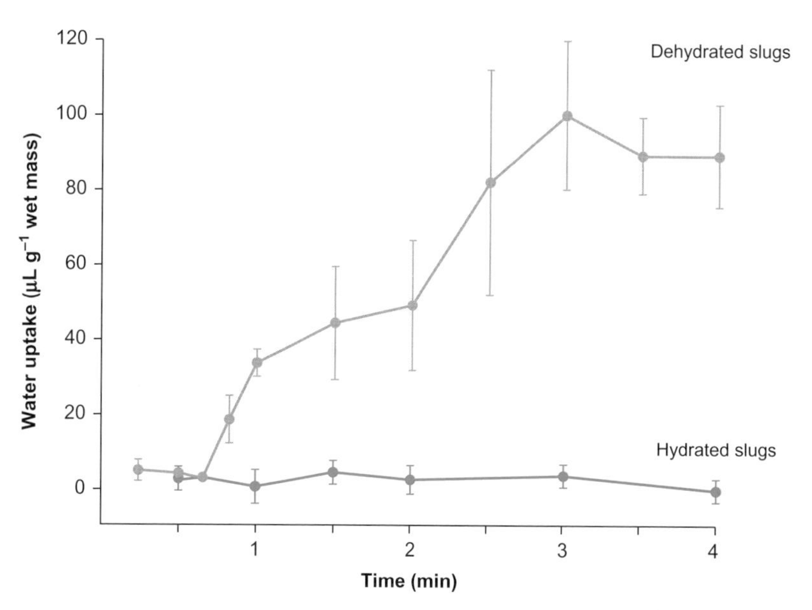

Figure 6.38 Contact rehydration of leopard slugs (*Limax maximus*)

Water uptake was measured by weighing the slugs before and after various times on a wet pad, which they voluntarily moved toward after 60–70% dehydration. Measurements were also made in fully hydrated slugs placed on a wet pad. The surface area in contact with the pads was similar in both groups.

Each point is mean ± standard deviation from 5–9 slugs and shows a high level of individual variability. Source: Prior DJ (1984). Analysis of contact-rehydration in terrestrial gastropods: osmotic control of drinking behaviour. Journal of Experimental Biology 111: 63-73. Image: Michal Manas/Wikimedia Commons

integument. In amphibians, the water permeability of the ventral pelvic patch, which is responsible for most of the water uptake, is increased by the hormone **arginine vasotocin**[58]. Arginine vasotocin stimulates an increase in skin permeability by the insertion of water channels (aquaporins)[59]—in the skin epithelial cells.

In contrast to amphibians, the integument of most reptiles has low water permeability and typically a relatively high water resistance, as shown by the examples back in Table 6.2. The high resistance limits both evaporative water loss and water uptake for rehydration, so land reptiles cannot rehydrate by uptake of water via their integument.

Soil characteristics influence water balance for burrowers with water-permeable skins

The characteristics of the soil in which soil-dwelling organisms live—and particularly its particle size—determine the movement of water through it (its hydraulic conductivity). Water moves more easily (the soil has a higher hydraulic conductivity) in sandy soils, which possess a high proportion of larger particles (63–2000 μm in size), than through the finer particles of soils that have a high proportion of clay (<2 μm) and silt (2–63 μm)[60]. Hence, it is easier for an animal with a highly permeable skin to absorb water from sandy soil than from a clay soil.

Whatever the type of soil, water uptake can only occur if the water concentration in the soil is greater than that in the extracellular fluids of the burrowing animal. Such physical factors appear to influence the behaviour of burrowing frogs, as shown by one of the few studies of simultaneous behavioural patterns and water balance of sandhill frogs (*Arenophryne rotunda*). These frogs live completely independently of free water bodies. They are active at night when they lose water by evaporation, but burrow during the daytime. Sandhill frogs taken from these burrows were found to have plasma osmolalities of around 290 mOsm kg^{-1}, which is similar to normal values for well-hydrated amphibians and indicates that sandhill frogs are able to rehydrate in their burrows.

To understand how rehydration of sandhill frogs occurs in their burrow, we need information on the soil moisture content around the burrow, and its relationship to water vapour pressure so that these values can be related to the water concentration of the body fluids of sandhill frogs. Such analyses revealed that the burrow soil had relatively high moisture content (4 per cent) after recent (but infrequent) rain. (Look back at Figure 6.25 for data on soil moisture content around the burrows of sandhill frogs.) This moisture content is higher than strictly necessary for the daily rehydration of sandhill frogs: if the soil were to be dried out to 2 per cent moisture there would still be enough water vapour in the soil for equilibrium to be reached with the body fluids of the frogs.

[58] Case Study 21.1 examines hormonal regulation of the permeability of the ventral skin by arginine vasotocin in dehydrated amphibians.

[59] Aquaporins are discussed in Section 4.2.3.

[60] These sizes are based on the international scale for classification of soils.

Like amphibians, some soil insects, such as springtails, have cuticles with a relatively low resistance to evaporative water loss (in contrast to most insects). This means they could lose water vapour across their relatively permeable cuticle; equally, however, they could gain water via the same route if they are in damp conditions. Actual water movements will depend on the osmotic concentration of their body fluids and how this relates to the humidity of soil pores.

The osmolality of the haemolymph of deeper-burrowing springtails (*Protaphorura tricampata*) living in damp soils is about 220 mOsm kg^{-1}. This osmolality would be in equilibrium with a soil pore relative humidity of 99.6 per cent at 20°C, as indicated by the data in Figure 6.39A. However, what happens when low levels of rainfall reduce the soil moisture content and create conditions in which the springtails could lose significant amounts of water? Do springtails simply lose water into the soil pores and eventually die?

Field measurements indicate that springtails survive even when the relative humidity of pore waters reaches 97.5 per cent. To put these conditions in perspective, the permanent wilt point for plants occurs at about 98.9 per cent relative humidity. Deep-burrowing springtails such as *Protaphorura tricampata* avoid dehydration even in persistent droughts. Indeed, they maintain stable water content by increasing the osmotic concentration of their body fluids during soil water deficit, as shown in Figure 6.39B. The gradual increase in the osmolality of the body fluids of the springtails is the result of the accumulation of amino acids (primarily alanine and proline), the carbohydrate trehalose, and free fatty acids. By their accumulation of sufficient osmolytes during a drought to raise the osmotic concentrations in their body fluids, these highly permeable springtails allow a continual influx of water.

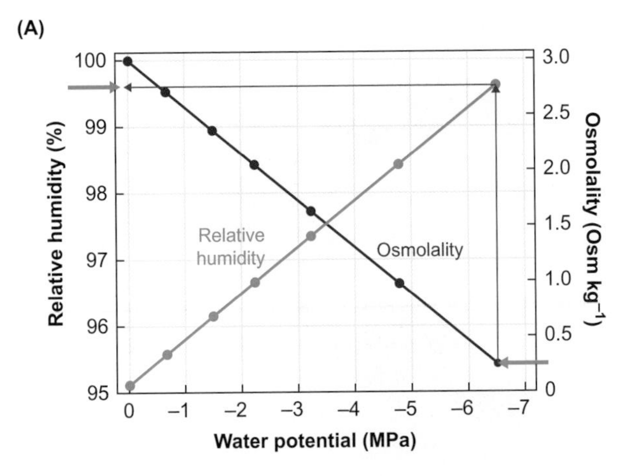

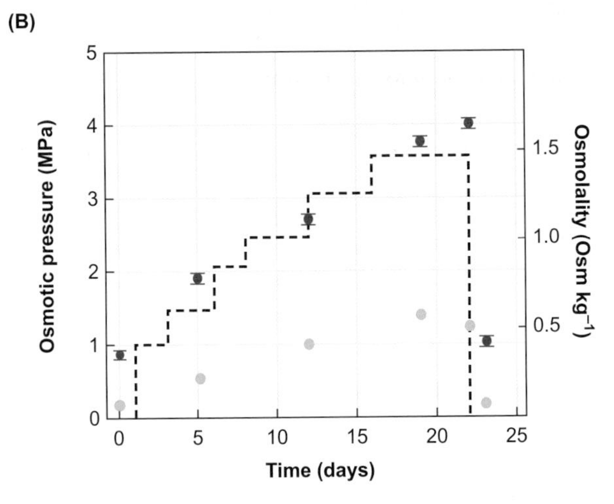

Figure 6.39 Soil characteristics and water balance of burrowing springtails (*Protaphorura tricampata*)

(A) Relationships between soil pore relative humidity and water potential, and the osmolality of body fluids of a soil organism that is in equilibrium with the surrounding atmosphere, at 20°C. Reading off the graph, an organism with an osmolality of 0.22 Osm kg^{-1} (i.e. 220 mOsm kg^{-1}, shown by the green arrow on the right-hand Y axis), would be in equilibrium with soil pore relative humidity of 99.6%, at 20°C. (B) The combined contribution of compatible osmolytes (free amino acids and trehalose; green circles) to the osmotic pressure (red circles) of the body fluids of springtails (*Protaphorura tricampata*) increases over a period of stepped increase in soil dehydration (dashed line). Equivalent osmolality values are shown by the right-hand Y axis. Addition of water to the soil at day 22 results in a decrease in the osmotic pressure (osmolality) of the springtails within 1 day.

Source: Holmstrup M, Bayley M (2013). *Protaphorura tricampata*, a eudaphic and highly permeable springtail that can sustain activity by osmoregulation during extreme drought. Journal of Insect Physiology 59: 1104-1110.

6.2.4 Absorption of atmospheric water vapour by arthropods

Water uptake from unsaturated air by some land arthropods is one of the most fascinating aspects of their water balance. By taking up water vapour from the atmosphere, some arthropods are able to exploit water that is abundant as a gas even when liquid sources of water, and water in food, are absent. How and exactly where each species takes up water vapour is often still a mystery for many species, although some general principles have emerged since water vapour absorption was first reported during the study of mealworms (the larvae of *Tenebrio molitor* beetles) in the 1930s.

The initial studies weighed starved mealworms after holding them in air of known humidities. Despite the initial doubt about the validity of these results, it is now clear that water vapour absorption among arthropods has evolved on multiple occasions in diverse groups. More than 70 species of unrelated arthropod groups of insects, mites and crustacean isopods (woodlice species) are now known to be capable of absorbing atmospheric water vapour.

The uptake of vapour is particularly prevalent among small species that need to compensate for high evaporative losses from a large surface area relative to their volume, and for species that live in dry habitats and eat food containing very small amounts of water (or none at all). This description includes pest species, such as larvae of clothes moths, and species of beetles that are pests, which live in dried food products; these include mealworms, tobacco beetles (*Lasioderma serricorne*), and mites such as flour mites (*Acarus siro*) and dust mites (*Dermatophagoides farinae*). Similarly, species of ticks and fleas can maintain their water balance by exploiting water vapour even when years pass between meals.

Each species that is capable of water vapour uptake can do so if relative humidity is above a certain value called the **critical equilibrium humidity**, which is characteristic for the species. Critical equilibrium humidities range from 95 per cent to 43 per cent, the lowest value of which occurs in tobacco beetles. For many species (although not all) the rate of uptake of water is independent of relative humidity, provided humidity exceeds their critical equilibrium humidity.

As an example, Figure 6.40 shows data for pre-pupae of a rat flea *Xenopsylla brasiliensis*, which gains about 14 per cent of its body mass by uptake of water vapour in the 3–4 days after it ceases to feed and forms a cocoon. Note in the graph that weight gain for the first few days after cocoon formation is independent of humidity between 90 per cent and 50 per cent relative humidities, and this uptake occurs irrespective of the humidity they previously experienced. At 50 per cent relative humidity nearly all the fleas survive through to emergence from the pupae; at lower humidities, however, the absorption of water vapour ceases such that, at 40 per cent relative humidity, nearly all fleas die, having lost weight (because of evaporative water losses) as shown by the data in Figure 6.40.

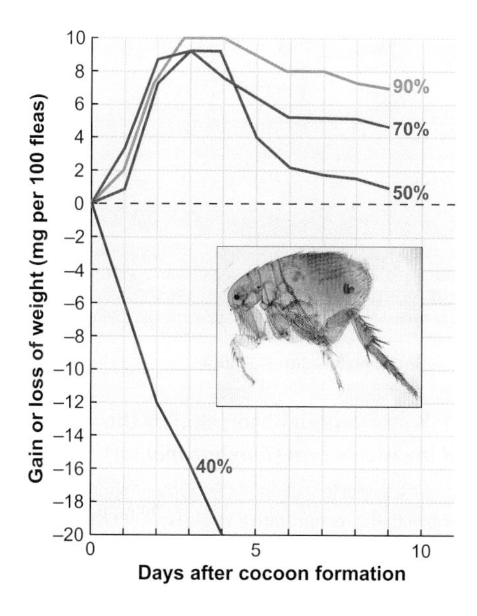

Figure 6.40 Changes in weight resulting from absorption of water vapour in pre-pupal stage of fleas (*Xenopsylla brasiliensis*)

Water uptake in the pre-pupae (first 3–4 days after cocoon formation) determines the survival of fleas until they emerge from their cocoons. Water vapour absorption in the first 3–4 days results in water gain provided the relative humidity (RH) is 50 per cent or greater. Pupae do not absorb water vapour, so when the pre-pupae moult into pupae, inside the cocoon, their loss of water by evaporation is not compensated for by water uptake, so pupal weight gradually decreases. In the pre-pupae held at 40 per cent RH, water vapour uptake does not occur and water loss results in death within a few days.

Source: adapted from Edney EB (1947). Laboratory studies on the bionomics of the rat fleas, *Xenopsylla brasiliensis*, Baker, and *X. cheopis*, Roths. II. Water relations during the cocoon period. Bulletin of Entomological Research 38: 263-280; Edney EB (1977). Water Balance in Land Arthropods. Springer-Verlag. Image of an adult of a related flea species (*Xenopsylla cheopis*): http://www.cdc.gov

For many arthropod species, the mechanism or location of the water vapour absorption is unknown but water uptake is thought to result from active transport processes driven by the organism in question because it ceases when the animal dies and is inhibited by anaesthesia. It is also clear that uptake of water vapour from sub-saturated air cannot be entirely passive and be driven solely by the overall difference in water concentrations externally and internally. In fact, the water concentrations within animals and in air under the conditions in which water absorption occurs favour passive water loss (by evaporation), not absorption.

The solutes in body fluids reduce the concentration of water molecules in internal fluids compared to pure water, and hence reduce the vapour pressure of the solution. Therefore, the value for the relative humidity of air that is in equilibrium with the body fluids is reduced compared to pure water, but only slightly—to just below 100 per cent (saturated)[61]. Body fluids of land arthropods at an osmolality of 300 mOsm kg^{-1} are in equilibrium with an atmosphere of 99.5 per cent relative humidity, but these effects are much too small to explain water vapour uptake from air at 95 per cent to 43 per cent relative humidity as is seen among arthropods.

Rather than relying on entirely on passive processes, the uptake of water vapour seems to involve active processes. The idea is that solutes accumulate in an area that becomes sufficiently hyperosmotic (and hence has a low enough water concentration) for water vapour to move from air with a high water density (high water vapour concentration) into the area of lower water density (low water vapour concentration).

Rectal absorption of water vapour

Evidence that the absorption of water vapour occurs in the rectum of some species came first from mealworms (*Tenebrio*). When the anus of the mealworms was blocked with wax, the mealworms could not balance their evaporative water loss by water vapour absorption. By contrast, mealworms with an open anus gain weight if the atmosphere has a relative humidity of above about 88 per cent. Similar experiments in other species support the idea that rectal or anal absorption of water vapour occurs in some species such as firebrats (*Thermobia domestica*).

In firebrats, the rectum is divided into two chambers, the most posterior of which forms an anal sac that opens to the

[61] Pure water is in equilibrium with saturated air (with a relative humidity of 100 per cent), as Figure 6.2 illustrates.

exterior when the anal valves are open. In hydrated animals, these valves are closed, but dehydrated firebrats, exposed to relative humidities of above 45 per cent, open and close their anal valves every 1–2 s, thereby ventilating the anal sac with air. Absorption of water vapour from the air is believed to occur in the anal sac, and structural features of the sac are a strong indication that active transport processes drive the absorption of water vapour.

In particular, the epithelium of the rectum of firebrats has one of the highest tissue concentrations of mitochondria. The elongated tubular mitochondria are packed between deep infolds of the apical membrane in a so-called apical complex. There is strong evidence that the apical complex is necessary for the maintenance of water balance by water vapour absorption, as the absorption ceases when the apical complex is lost during cuticular moulting[62], and recommences when the complex reforms. However, we do not understand exactly how the apical complex is linked to uptake of water vapour by firebrats.

The processes involved in absorption of atmospheric water vapour from the rectum by mealworms are more fully understood than those of firebrats. Mealworms have an unusual structural arrangement of their excretory Malpighian tubules[63] and rectum, which form a **cryptonephridial complex**. Figure 6.41 illustrates this complex, in which the Malpighian tubules cover the posterior two-thirds of the rectum. A perinephric membrane forms a protective barrier, which isolates most of the length of the Malpighian tubules from the haemolymph. The low water permeability of the perinephric membrane prevents water influx from the haemolymph into the concentrated fluid that forms within the Malpighian tubules and in the perinephric space around the rectum and tubules.

The high osmotic concentrations of the fluid in the Malpighian tubules and within the perinephric space, which reaches up to 6800 mOsm kg^{-1} in dehydrated mealworms, results primarily from the high concentrations of potassium (K^+) and chloride (Cl^-) ions. Potassium reaches 1800 to 2700 mmol L^{-1} in mealworms kept in dry conditions as a result of active secretion of K^+, while Cl^- moves passively across the Malpighian tubules. The exact transporting location for production of the hyperosmotic fluid is uncertain, but it is thought to be unlikely that the thin-walled blisters and leptophragma cells (shown in Figure 6.41A) are responsible for the active K^+ secretion as suggested by early studies.

The highly concentrated fluids in the cryptonephridial complex are essential for drawing water across the cuticle of the rectum, by osmosis, from the mixture of faeces and urine (passed from the Malpighian tubules into the midgut). This process creates virtually dry faeces. When the mealworms suck humid air into the posterior rectum through the anus, by a muscular pumping action, water vapour first moves into the

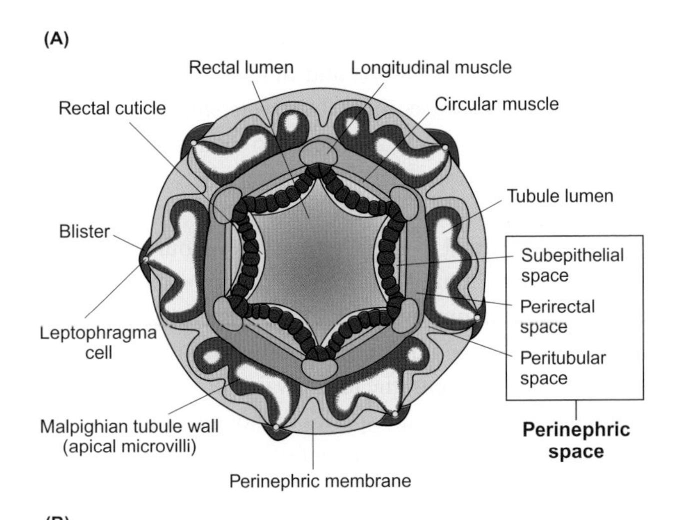

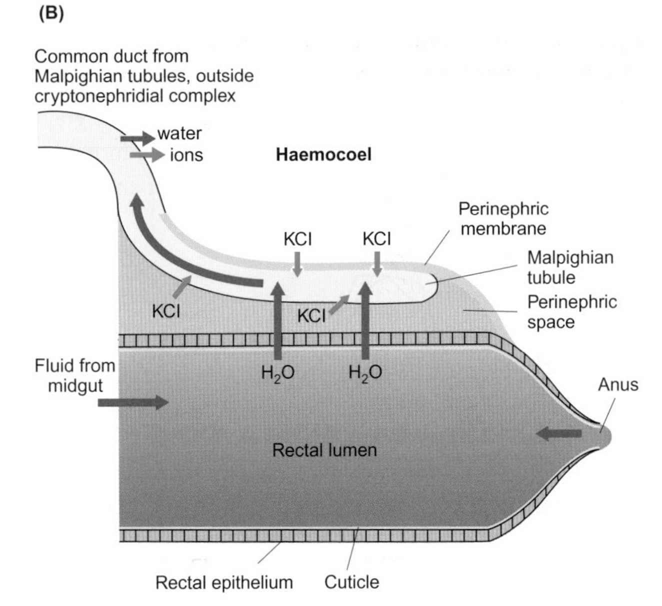

Figure 6.41 Water vapour absorption by the cryptonephridial complex of the mealworm (*Tenebrio molitor*)

(A) Schematic of a transverse section of the cryptonephridial complex of mealworms showing the rectum with a single layer of cells forming the rectal epithelium. The rectum is surrounded by circular and longitudinal muscles. Outside of these muscle layers there are six Malpighian tubules that secrete urine. The Malpighian tubules are held in a perinephric membrane. The blister area is a thin part of the perinephric membrane, which is associated with a leptophragma cell. (B) The Malpighian tubules, which are held alongside the rectum beneath the perinephric membrane, join anteriorly and enter the midgut. Hence, the excreta are a mixture of urine and faeces. During water vapour absorption (blue arrows), the rectum is open and air is pulled into the rectum. Water vapour in the air moves from the rectum into the highly concentrated fluids in the perinephric space and Malpighian tubules.

Sources: A: O'Donnell MJ, Machin J (1991). Ion activities and electrochemical gradients in the mealworm rectal complex. Journal of Experimental Biology 155: 375-402.
B: adapted from Bradley TJ (2009). Animal Osmoregulation. Oxford Biology Series. Oxford: Oxford University Press.

[62] The cuticle (shown in Figure 6.17) lines the rectum, as illustrated in Figure 6.41, for *Tenebrio*.

[63] Section 7.3.4 discusses urine formation by the Malpighian tubules of insects.

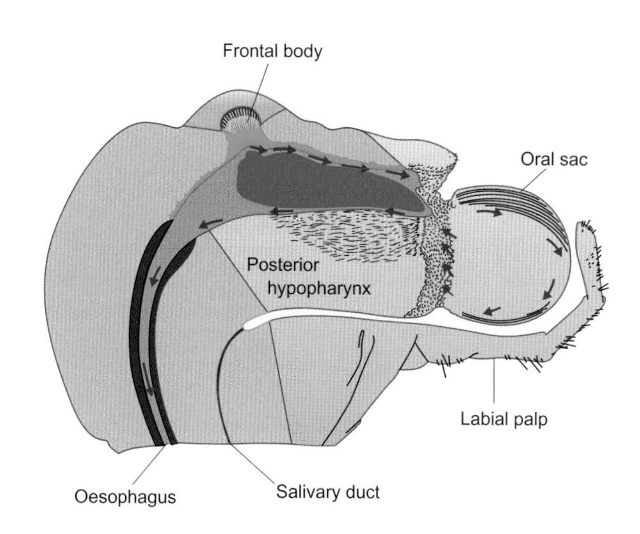

Figure 6.42 Head of the desert cockroach (*Arenivaga investigata*) showing bladder-like oral sacs used in water vapour absorption

The head has been cut in cross-section, so that one half is visible, and parts removed to expose the hypopharynx. Water condenses on hairs on the two oral sacs (one shown). Fluid formed in the frontal bodies (one shown) flows to the two oral sacs, via the route indicated by red arrows, and is conveyed to the oesophagus.
Source: O'Donnell MJ (1982). Hydrophilic cuticle - the basis for water vapour absorption by the desert burrowing cockroach, *Arenivaga investigata*. Journal of Experimental Biology 99: 43-60.

perinephric fluids where there is a lower water concentration and then into the Malpighian tubules. Ultimately, absorbed water enters the haemolymph when the Malpighian tubules leave the confines of the cryptonephridial complex and water and ion exchanges with haemolymph occur.

Oral sites for absorption of water vapour

Oral sites of solute secretion linked to water vapour absorption have been identified in several arthropod species, in many cases by applying wax to the suspected areas. The idea behind this manipulation is that wax layers prevent the secretion from drawing in water vapour. If water absorption is blocked, then water imbalance should occur. Refinement of this technique has localized the secretions of some species to specific parts of the salivary glands—for example, parts near to the mouth in lone star ticks (*Amblyomma americanum*).

In desert cockroaches (*Arenivaga investigata*) water vapour absorption involves two oral sacs (or bladders) which the cockroaches inflate and protrude from their mouth, as illustrated in Figure 6.42. In this case, water vapour absorption seems to result from the condensation of water vapour onto hydrophilic hairs on the sacs. One piece of evidence in support of this idea is that the surface temperature of the sacs is about 0.2°C higher than that of the surrounding mouthparts, which is thought to result from the release of heat during condensation[64]. A pulse of fluid secreted by the frontal bodies, which passes over the oral sacs, is thought to result in the release of water from the hairs. This water is then carried into the oesophagus, along the route shown in Figure 6.42.

6.2.5 Drinking water

Drinking water is an obvious source of water influx; as such, it may seem odd to have left our discussion of this form of water gain until now. The reason for doing so is that drinking is a behavioural response. Many invertebrates and vertebrates drink water, although drinking is rare in amphibians, except for a few species of arboreal frogs that drink from water that has collected in leaves or from other sources[65].

Provided that sufficient drinking water is available in open sources, such as water holes and ponds, animals typically adjust their drinking to maintain water balance[66]. Hence, the drinking rates of animals depend on water losses and the amount of water acquired through metabolism and food intake. Recall from Figure 6.35 that locusts feeding on dry food must drink in order to maintain water balance, but if they are consuming grass containing large quantities of preformed water then they do not drink.

Some large mammals living in dry savannas obtain sufficient water in their diet and from metabolism, and reduce water losses sufficiently to enable them to live almost independently of drinking water. Among herbivores, species that do not rely on drinking water often co-exist with species that are more reliant on drinking to maintain water balance. For example, in the African Serengeti, common elands (*Taurotragus oryx*) and Grant's gazelles (*Nanger granti*) do not need a daily intake of water. These species cope with 1 to 3 weeks of water deprivation before dehydration drives them to seek drinking water. In contrast, blue wildebeest (*Connochaetes taurinus*) and waterbucks (*Kobus ellipsiprymnus*) need a regular intake of drinking water, at least every few days. The availability of water largely drives the extensive annual migrations of these water-dependent species. When streams dry up in the long dry season, wildebeest and zebras need to migrate toward less arid regions.

African and Asian elephants (*Loxodonta africana* and *Elephas maximus*) not only rely on a supply of drinking water, but also require water for evaporative cooling to such an extent that the management of surface water supplies has been suggested as a way of managing populations in those reserves where elephants are overabundant.

Dromedary camels (*Camelus dromedarius*) have a striking ability to survive for weeks without drinking water, but cannot do so indefinitely. Rather, they drink once water

[64] There is heat loss during evaporation, which we discuss in Section 3.3.1; the opposite process—heat release—occurs during condensation.

[65] Although amphibians do not drink by mouth, they do take up water via their skin, and this is sometimes called cutaneous drinking.

[66] Section 21.2.3 examines the hormonal and neural control of drinking behaviour by vertebrates.

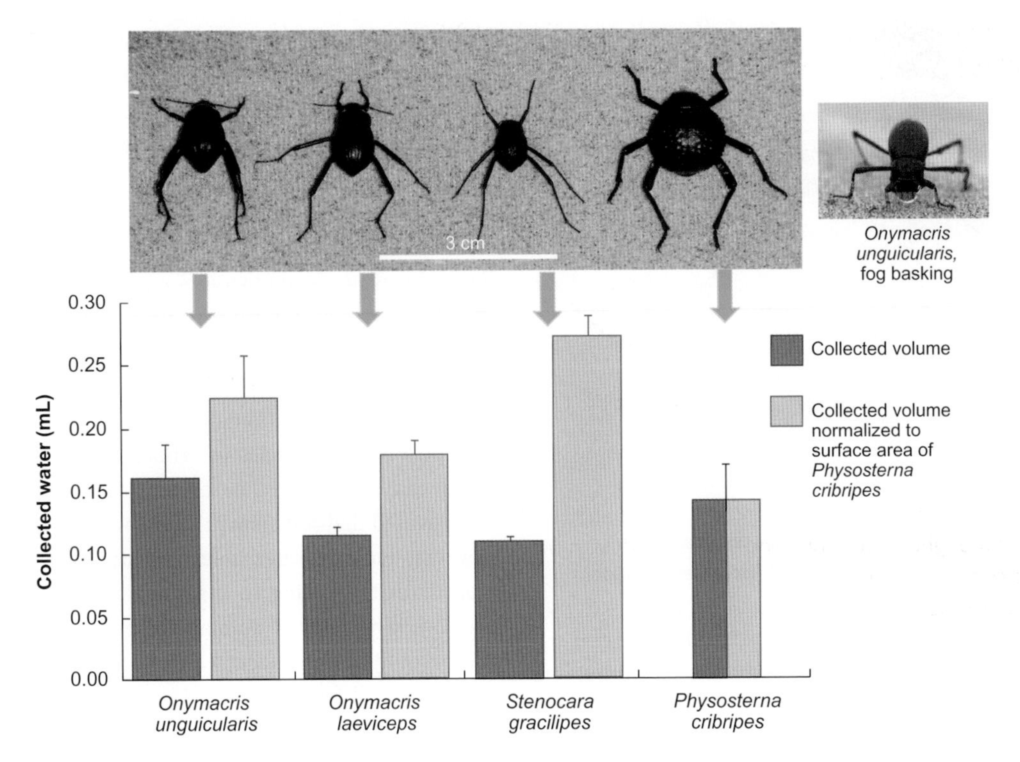

Figure 6.43 Fog harvesting efficiency of four species of tenebrionid Namib Desert beetles

Purple columns in the graph show water collected in 2 hours by the dorsal body surface of each of four species of Namib beetles when they are held in a fog. The blue columns give data for water collection by a standardized surface area—the surface area of the largest species, the desert toktokkie (*Physosterna cribripes*). Image of single beetle shows the head-stander beetle *Onymacris unguicularis* while fog basking.

Source: Nørgaard T, Dacke M (2010). Fog-basking behaviour and water collection efficiency in Namib Desert darkling beetles. Frontiers in Zoology 7: 23.

is available, when up to a third of their body mass can be consumed in a few minutes. With this high level of water intake, there is a danger of the blood and other extracellular fluids being rapidly diluted as water moves out of the gut. Camels avoid this problem by absorbing water from the gut slowly, over a 24-hour period.

It used to be thought that camels carry a large volume of water in their rumen to use between episodes of drinking, but we now know that camels hold no more water in their rumen than other ruminants. Camels drink to rehydrate rather than for water storage. The key to their long period of survival between drinks lies mainly in their ability to conserve water by producing concentrated urine and by minimizing evaporative water loss during ventilation[67]. Camels can also tolerate high levels of water loss—of 30 per cent to 40 per cent—whereas most mammals, including humans, only tolerate losses amounting to 10 per cent to 15 per cent of their total body water content.

A somewhat unusual source of drinking water is used by tenebrionid beetles living in the hyperarid Namib Desert, where fogs stretch 100 km from the coast several times each month. Most beetles living in the dunes drink water that condenses from the fog onto vegetation or on sand ridges that the beetles sometimes create by digging a trough across the path of incoming fog. A few species use a third strategy for passive collection of drinking water by **fog basking**. To fog bask, the beetle places itself on the top of a ridge on the dunes and positions itself so that as much of its dorsal surface as possible intercepts the fog. Fog water then condenses on the cooler forewings (elytra) of the beetles. By standing on its head during fog basking, the condensed water runs down into its mouth, as shown in the image of *Onymacris unguicularis* in Figure 6.43.

The graph in Figure 6.43 shows the rates of water collection by four species of Namib beetles placed in a fog-generating system. In these experiments, *Onymacris unguicularis* and *Stenocara gracilipes* had the highest rates of water collection. While *Onymacris unguicularis* is a natural fog-basking species, *Stenocara gracilipes* probably is not, but seems to have the right surface characteristics to give the highest rates of water collection. The exact reasons are unclear, but it has been suggested that water, which has condensed on bumps on its dorsal surface, is channelled via hydrophobic areas between the bumps toward the mouth. Case Study 6.2 discusses how some of the Namib Desert beetles, and other animals, have inspired the development and manufacture of various water collection devices for use by humans.

[67] The minimization of respiratory evaporative water loss by camels is discussed in Section 6.1.2; conservation of water in the kidney is discussed in Section 6.1.4.

Case Study 6.2 Biologically inspired approaches to harvesting water from air

Some animals living in arid regions have evolved materials that encourage condensation of water from the air because of their chemical and surface structural characteristics. By investigating such features, materials scientists have developed, and continue to refine, their own designs of biomimetic materials for harvesting water from air, whether from fog, overnight dew or air conditioning systems.

Beetle mimics

The hardened front wing cases of some Namib Desert beetles[1] have been found to have hydrophilic bumps on a hydrophobic (waxy) surface, as shown in Figure A. This arrangement has been argued to enhance the collection of tiny water droplets out of fog. These tiny droplets are thought to grow into bigger water drops as more droplets coalesce out of the fog, then they fall off the bumps and slide down the hydrophobic channels between them until they reach the beetle's mouth for it to drink. British investigators tested their hypothesis that this structural design enhances the harvesting of water from fog in a simple experiment using glass beads (0.6 mm spheres) of a similar size to the bumps on the beetle's back, which they embedded into a wax coating on glass microscope slides.

Four sets of experimental slides were used for this study. On two sets of slides, glass beads with unwaxed (hydrophilic) tops were used. These beads were either randomly arranged, with an average spacing of 0.5 mm, or placed in an orderly pattern at 0.6 mm intervals. The other two sets of slides were either covered in smooth hydrophobic wax or left uncoated as smooth hydrophilic clean glass. The slides were inclined at a 45° angle, sprayed with a fine mist, and the amounts of water collected at the base of each slide were compared.

The greatest and most consistent amount of water was collected from the slides with hydrophilic bumps on the hydrophobic background: double the amount of water was collected from the slide with the regular array of hydrophilic beads than from the smooth wax surface. Slightly less was collected when the array of beads was randomly arranged than for a regular arrangement. By comparison, the amount collected from the hydrophilic glass slide was very variable as it ran down different routes, with six of the ten trials collecting nothing.

Several approaches to the manufacture of successful 'beetle mimics' have now been taken, with variations in the materials used as well as the size and arrangement of the 'bumps'. The comparison of a variety of patterns and materials has shown that the optimal design for water collection from air is very much like that of the beetle. The optimal diameter of hydrophilic spots is 500 μm, with 1 mm spacing; for comparison, the bumps on the beetle shown in Figure A are about 500 μm in diameter and 500 μm to 1.5 mm apart.

Investigators in Australia have recently developed a mimic of the beetle's design with a material that is easier to make and potentially less expensive than those used previously. This material incorporates hydrophilic bumps on a hydrophobic background of thin polymer films as shown in Figure B. An advantage of this material is that it can be produced as large sheets of sufficient size to rest on the sloping roofs of homes or tents to collect water by condensation.

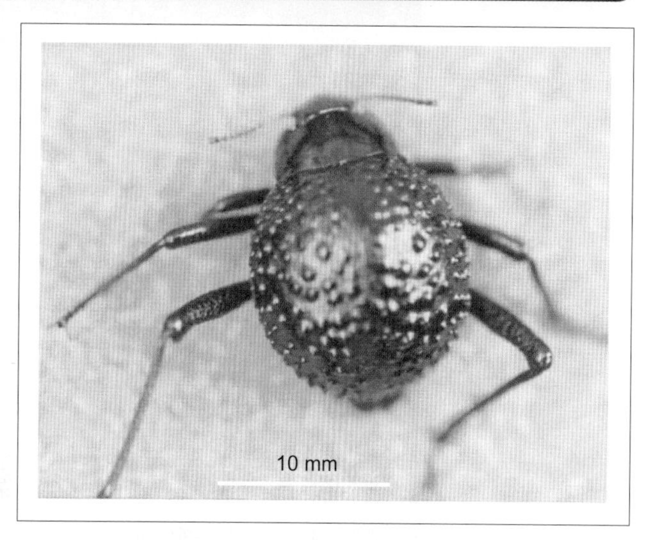

Figure A Namibian desert beetle

Source: Parker AR, Lawrence CR (2001). Water capture by a desert beetle. Nature: 414: 33-34.

Spider web mimics

Spider silk is made of hydrophilic protein fibres and thus tends to attract water, as shown in Figure C(i), with cribellate spiders forming a web like that shown in Figure C(ii). They use a comb-like structure, called a cribellum, to separate the silk fibres emerging from their spinnerets into many much finer fibres that are particularly effective at collecting water from humid air.

Chinese scientists have used their understanding of the directional water collecting properties of the silk in the webs of cribellate spiders to create bio-inspired artificial water collection devices. They analysed how the structure of the silk fibres changes in wetness such that each fibre develops a periodic pattern that alternates between dense 'spindle-knots' made of randomly aggregated hydrophilic nanofibrils, and adjoining narrower 'joints', made of orderly aligned nanofibrils that give it its directional water collection capability. This pattern is illustrated in Figure C(iii). The joints act as condensation sites that attract

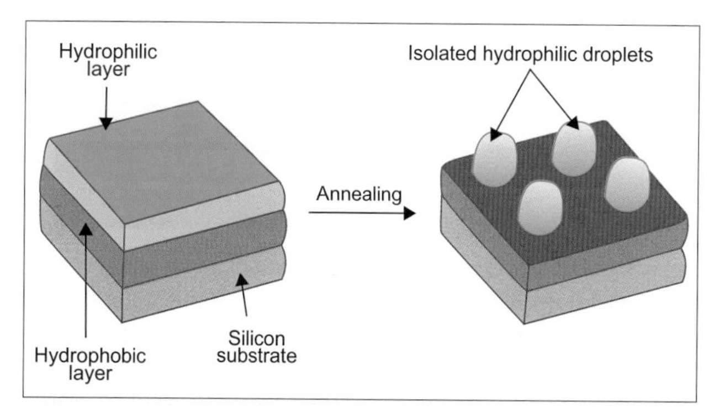

Figure B Making a beetle mimic material for water collection

Source: Thickett SC et al (2011). Biomimetic surface coatings for atmospheric water capture prepared by dewetting of polymer films. Advanced Materials 23: 3718-3722.

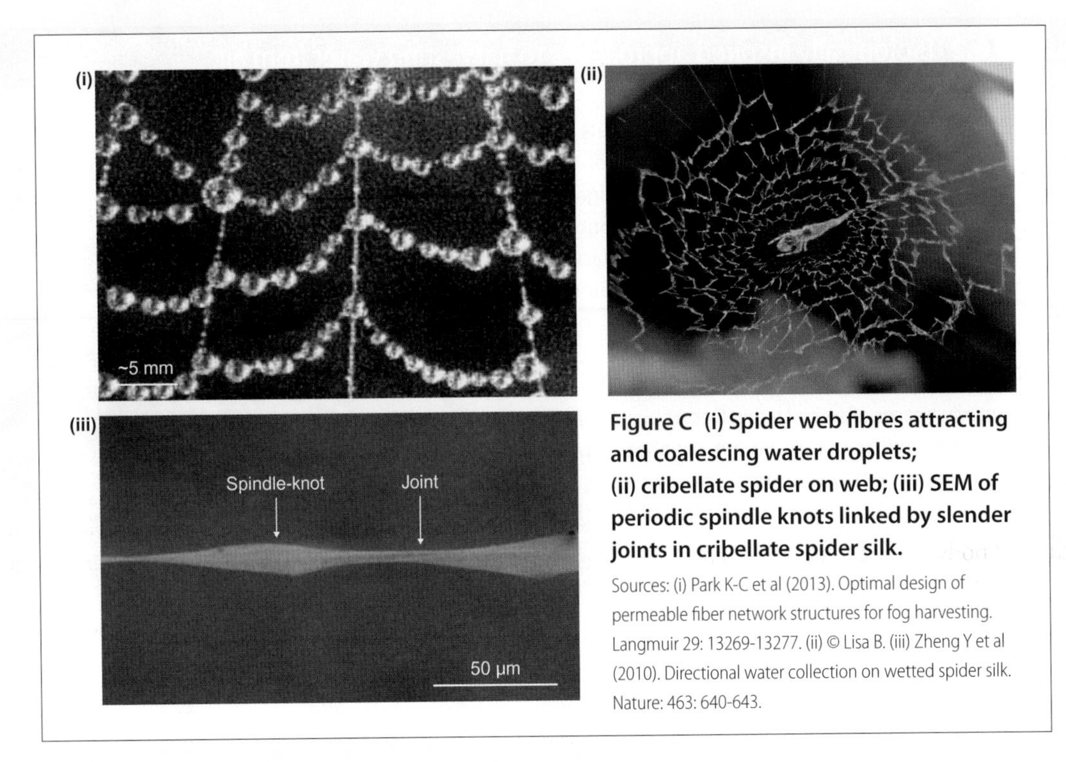

Figure C (i) Spider web fibres attracting and coalescing water droplets; (ii) cribellate spider on web; (iii) SEM of periodic spindle knots linked by slender joints in cribellate spider silk.

Sources: (i) Park K-C et al (2013). Optimal design of permeable fiber network structures for fog harvesting. Langmuir 29: 13269-13277. (ii) © Lisa B. (iii) Zheng Y et al (2010). Directional water collection on wetted spider silk. Nature: 463: 640-643.

and then move tiny micrometre-sized droplets to the spindle knots, which act as drop-enlarging sites.

The Canadian non-profit organization *FogQuest* has sponsored the use of vertical nets to collect water from fog in Central and South America, and Africa. In one village in Guatamala, 30 large fog-collection nets collect 6000 L of fresh clean water each day. However, while effective, these relatively large mesh polyolefine nets, which are illustrated in Figure D(i), capture only about 2 per cent of the available water.

By studying the designs and material wettability of both plant and animal water collecting structures, materials scientists are developing improvements to both the fibre surface wettability and configurations of meshes used in harvesting fog water. For example, the latest design from a group at the Massachusetts Institute of Technology has been tested in the field, in cooperation with FogQuest. Using extremely fine stainless steel filaments, with chemically enhanced surface wettability, in a fine mesh whose dimensions both reduce clogging on the one hand and loss to wind on the other, the new fog net, which is shown in Figure D(ii) and (iii), extracts 10 per cent of the available water from the air—a five-fold improvement over the nets currently in use.

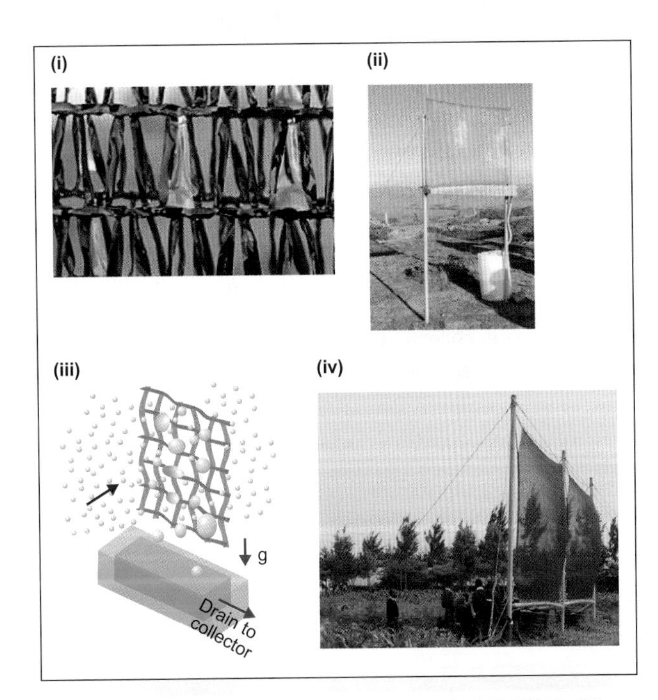

Figure D (i) Large polyolefine fog nets originally used by FogQuest; (ii) newer fog catching fine mesh designed by the group at MIT; (iii) diagram of mesh collecting water droplets from fog laden air flow (blue arrow) that fall, under the influence of gravity (g), into the water storage container; (iv) Tanzanian school children collecting clean fresh water from a fog net system.

Sources: (i) Fogquest (www.fogquest.org); (ii) © 2013, American Chemical Society (iii) adapted from Park KC et al (2013). Optimal design of permeable fiber network structures for fog harvesting. Langmuir 29: 13269-13277. (iv) © Dona Aitken

› Find out more

Bai H, Ju J, Sun R, Chen Y, Zheng Y, Jiang L (2011). Controlled fabrication and water collection ability of bioinspired artificial spider silks. Advanced Materials 23: 3708–3711.

Parker AR, Lawrence CR (2001). Water capture by a desert beetle. Nature 414: 33–34.

Park K-C, Chhatre SS, Srinivasan S, Cohen RE, McKinley GH (2013). Optimal design of permeable fiber network structures for fog harvesting. Langmuir 29: 13269–13277.

Thickett SC, Neto C, Harris AT (2011). Biomimetic surface coatings for atmospheric water capture prepared by dewetting of polymer films. Advanced Materials 23: 3718–3722.

Zheng Y, Bai H, Huang Z, Tian X, Nie F-Q, Zhao Y, Zhai J, Jiang L (2010). Directional water collection on wetted spider silk. Nature 463: 640–643.

[1] Originally described as *Stenocara* sp. but later argued to be *Physostema* sp.

6.2.6 Water turnover

The overall water influx rates of free-ranging wild animals can be measured after injecting a known amount of isotopically labelled water (usually deuterated water, ^{2}HHO, also known as heavy water) into the body fluids of a captured animal. The labelled water is diluted by the body fluids into which it disperses. Then, the rate of water influx balancing water losses (water turnover) determines the rate of decline of the concentration of the isotope in the body water.

The very high rates of water turnover of most amphibians result in errors that preclude the use of this approach, but the labelled water method can be used in birds, reptiles and mammals. Mass-specific water influx is higher in smaller species than in larger species if the relationship between water influx and body mass is allometric[68] and follows Equation 6.7 in which the scaling factor (b) is 0.6 to 0.8:

$$\text{Total rate of water turnover} - a \times (\text{body mass})^b$$

Equation 6.7

[68] Box 2.3 explains the principles of allometric relationships and scaling factors.

As an example, Figure 6.44 shows the linear relationships between log (water influx rate) and log (body mass) for marsupials occupying arid habitats and marsupials in non-arid habitats. Those species living in arid habitats have water influx rates that are about 35 per cent less than those of marsupials living in non-arid habitats. This tells us that adaptation of marsupials to arid environments is associated with lower water loss and hence reduced need for water intake—in other words, reduced water turnover.

Similar reductions in water requirements occur in placental mammals and birds in arid environments compared with similar-sized species living in non-arid environments. These reductions indicate convergent evolution—another example of evolutionary homoplasy in separate lineages. In birds of 10 to 500 g, in arid environments, the water influx rates are, on average, 47 per cent lower than similarly sized non-desert species. In small placental mammals (of about 100 g), water influx rates of desert species are about 50 per cent less than those of non-desert species of similar size, but the difference decreases with greater body size.

One way of assessing water influx in free-ranging animals is to calculate their **water economy index**: their water influx rate relative to the rate of energy metabolism (mL per kJ

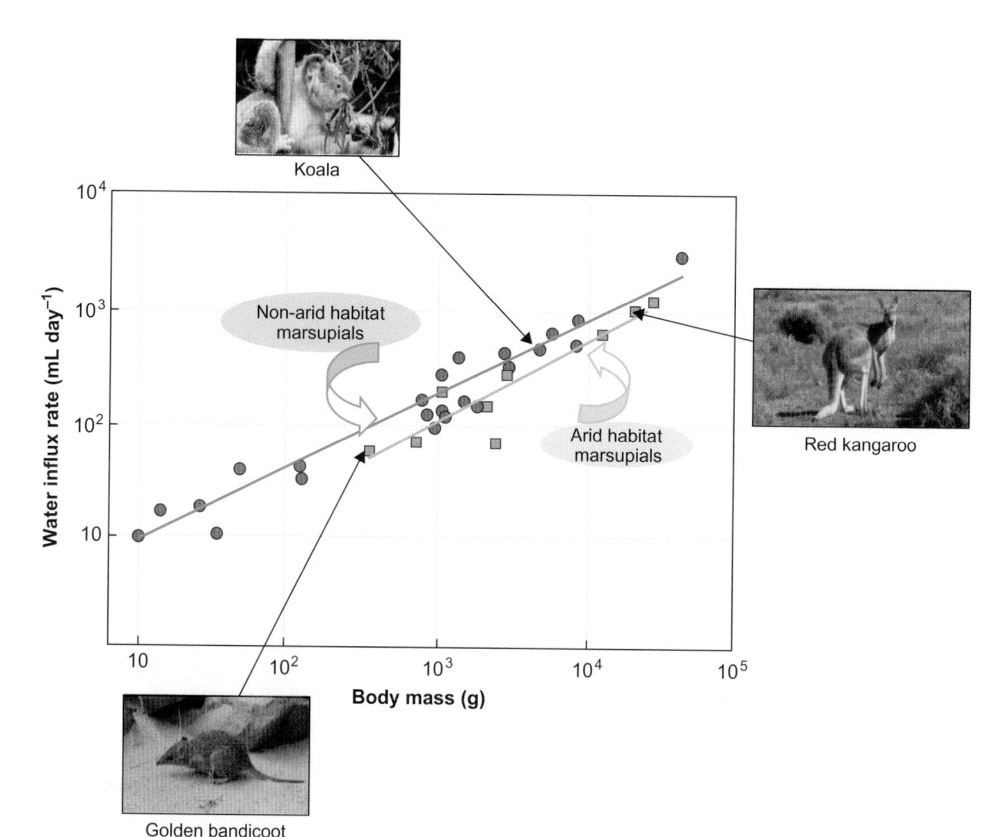

Figure 6.44 Relationships of total water influx rates for free-ranging marsupials living in arid or other habitats

Double logarithmic plot of water influx rate and body mass shows a lower water influx rate in marsupials of similar mass in arid habitats compared with those in non-arid environments, but the slopes of the lines are not significantly different. The allometric relationship for marsupials in non-arid habitats: water influx rate (mL day^{-1}) = 1.87 × (g body mass)$^{0.637}$; for marsupials in arid habitats: water influx rate (mL day^{-1}) = 0.777 × (g body mass)$^{0.693}$.

Source: Nagy KA, Bradshaw SD (2000). Scaling of energy and water fluxes in free-living arid-zone Australian marsupials. Journal of Mammalogy 81: 962-970. Images: kangaroo http://museumvictoria.com.au; golden bandicoot © Ross Knowles; koala Arnaud Gaillard/Wikimedia Commons.

energy metabolized), with both measured using stable isotopes[69] as follows:

$$\text{Water economy index (mL kJ}^{-1}\text{)} = \frac{\text{Rate of water intake (mL day}^{-1}\text{)}}{\text{Rate of energy metabolism (kJ day}^{-1}\text{)}} \quad \text{Equation 6.8}$$

The amount of water gained in food (as preformed water and metabolic water) depends on the food type, as indicated in Table 6.4. In the absence of drinking, the highest water economy indices occur in herbivorous animals eating plant materials that contain large amounts of preformed water. Hence, in herbivorous animals, water economy indices are typically 0.15 to 0.27 mL water intake per kJ metabolized. Carnivores have lower water economy indices of 0.07 to 0.17 mL kJ^{-1}, while animals eating dry seeds have water economy indices of around 0.04 mL kJ^{-1}.

The values for the water economy index of carnivores, herbivores and seedeaters can reveal the extent to which particular animals rely on additional intake of drinking water for their water turnover. For example, if the calculated water economy index for an animal is more than 0.27 mL kJ^{-1}—the upper end of the range for herbivorous animals—then the act of drinking adds to the water intake (or, more rarely, there is uptake of water from the atmosphere).

[69] Stable isotopes of elements are energetically stable and do not decay.

› Reviews

Hentschel JR, Seely MK (2008). Ecophysiology of atmospheric moisture in the Namib Desert. Atmospheric Research 87: 362–368.

Laverty G, Skadhauge E (2008). Adaptive strategies for post-renal handling of urine in birds. Comparative Biochemistry and Physiology Part A 149: 246–254.

Schmidt Nielsen K, Schmidt Nielsen B (1952). Water metabolism of desert mammals. Physiological Reviews 32: 135–166.

Checklist of key concepts

Water losses by animals living on land

- **Evaporative water loss** (EWL) is driven by the difference in partial pressures of water vapour in the solution from which evaporation occurs (P_{sol}) and in the air surrounding an animal (P_{air}).
- **Burrowing** reduces EWL of some animals when there is a higher water vapour pressure of air in the burrow than at the surface.
- EWL from the respiratory system occurs in direct proportion to the rate of oxygen consumption, and hence increases during activity.
- A reduction in oxygen consumption reduces EWL of some birds and mammals living in arid environments.
- **Condensation** of evaporative water in the cooled nasal cavities and respiratory turbinates of vertebrates retrieves water vapour from expired air, and is a well-developed adaptation of mammals living in arid environments.
- Some insects living in arid environments conserve water by using **discontinuous ventilation**, which restricts respiratory EWL to the short period when the spiracles are open.
- Some large mammals living in extremely arid habitats save significant amounts of water that would otherwise be lost through evaporative cooling by relaxing their regulation of body temperature and storing heat during the day, which results in **heterothermy**.
- An integument with a low permeability to diffusion of water vapour, which equals a high water resistance, restricts water loss.
- Resistance of the integument of land animals to EWL results from either:
  - **secretion of external lipids and waxy substances**, as in arboreal frogs and insects, or
  - the presence of **internal lipid barriers**, as in mammals, birds and reptiles.
- Animal species that use their skin for respiration avoid high cutaneous water losses by behavioural adaptations that include seeking out humid habitats, restricted daytime activity, burrowing into moist soil, and aestivating underground in cocoons.
- Animals with permeable skins that live in damp environments are often in positive water balance and excrete urine with a lower osmolality than their body fluids to eliminate excess water.
- If these animals become dehydrated, their urine concentration increases toward, but not above, that of the extracellular fluids.
- Mammals, birds and insects can maintain water balance by generating urine that is more osmotically concentrated than plasma (**hyperosmotic urine**), which is best developed in species adapted to live in arid environments.
- Excretion of poorly soluble nitrogenous wastes, such as uric acid, allows extra water absorption when the nitrogenous wastes precipitate in the hindgut or bladder.

Balancing water loss

- Animals balance water losses by:
  - acquiring **preformed water** in food items,
  - generating water during metabolism, and/or
  - water uptake from the environment.
- If food has a low water content, some animals reduce feeding, which reduces their loss of water during excretion of nitrogenous waste.
- **Metabolic water** is generated in variable amounts by oxidative metabolism of carbohydrates, lipids and proteins:
  - Fat metabolism generates the highest amount of water per gram of substrate.

- Relative to energy production, metabolism of carbohydrates yields the highest amount of water per kJ of energy produced, and protein (with urea as the nitrogenous waste) the least.

- An increase in metabolism increases generation of metabolic water, but this gain may be offset by EWL from the respiratory system.
- A **water-balance index** of above 1 indicates net water gain.
- The survival of some **xeric** species is reliant on metabolic water.
- Some land animals with highly permeable integuments can readily compensate for water losses by **contact-rehydration**, which results in water uptake by osmosis.
- The water concentration of soils must exceed that of extracellular fluid for an animal to acquire water by uptake from the soil.
- The **hydraulic conductivity** of soils influences the rate of water uptake across the skin of soil-dwelling organisms.
  - Water uptake occurs more rapidly in sandy soils that have a high hydraulic conductivity than in soils containing a high proportion of clay or silt.
- Many terrestrial arthropods maintain water balance in the absence of liquid sources by absorbing water vapour from air with a relatively high water density into an area of low water density.
- Drinking water is an important source of water influx for many species, but not all.
- Some animals drink water that has condensed on vegetation, and some desert beetles can employ their own body to intercept fog water.
- In free-ranging animals, calculation of a **water economy index** provides a useful measure for comparison of species and the extent to which they employ drinking water in water turnover.

Study questions

*** Answers to these numerical questions are available online. Go to www.oup.com/uk/butler.**

1. Describe the insect cuticle. How can the properties of insect cuticles explain the differences in the rate of water loss by different species? From your understanding of these processes, why and to what extent are abrasive powders or soaps (which dissolve fats and lipids) likely to be effective in controlling some insect pests? (Hint: Section 6.1.3.)
2. How are water balance, respiration and temperature regulation of mammals and birds linked in the process of determining water losses? Your answer should give specific animal examples. (Hints: Sections 6.1.1 and 6.1.2.)
3.* Over a 24-hour period, a 38 g mammal in a gravimetric experiment gave the following values for various components of its water budget.

   Change in body mass = –2 mg

   Preformed water in food = 100 mg

   Metabolic water = 480 mg

   Weight of urine = 160 mg

   Water content of faeces = 10 per cent by mass

   Mass of faeces = 200 mg

   During the experiment, the animal was not given any drinking water. Use the values provided to estimate the total evaporative water (TEWL) loss of the animal. What other factor could be considered to obtain a more accurate value? Would inclusion of this additional factor increase or decrease the value for TEWL? (Hint: Box 6.1.)
4. Outline the structure of the skin of a typical frog. How does this structure influence water loss? Outline two ways that some frog species reduce their cutaneous water loss. (Hints: Section 6.1.3; Case Study 6.1; Case Study 21.1 also gives relevant information regarding hormonal control processes.)
5. Where is the main water permeability barrier to be found in the skin of mammals? How does this layer achieve its function? Discuss the evidence that the water permeability barrier of birds, and of reptiles differ from the barrier in the skin of mammals. (Hint: Section 6.1.3.)
6. What mechanisms do Arabian oryx living in the hyper-arid Arabian deserts employ to control their water balance? (Hints: Sections 6.1.2, 6.1.4, 6.2.1, 6.2.2.)
7.* A kangaroo rat has a body temperature of 38°C. It breathes in air at 30°C, which contains water vapour at 12.3 mg L^{-1}. The air expired by the kangaroo rat is at 25°C. Based on these data, what proportion of the water that could potentially have been lost in respiration is saved by cooling of the exhaled air? (Hint: Section 6.1.2; refer to Figure 6.3; note that 1 m^3 = 1000 L.)
8. Outline the sources of water that allow animals to achieve water gain, giving examples of particular species that rely heavily on each of these water sources. (Hint: Section 6.2.)
9. Predict how and why changes in house heating and ventilation can affect the water balance of dust mites. (Hints: Sections 6.1.1, 6.1.2, 6.1.3, 6.2.2, 6.2.4.)
10. Explain with examples why certain herbivores can live in the absence of drinking water while others are reliant on a regular intake of drinking water to maintain their water balance. (Hints: Sections 6.2.1, 6.2.2, 6.2.5.)
11. What types of nitrogenous wastes are excreted by (i) birds and (ii) mammals? How do these differences influence water balance? (Hints: Section 6.1.4; see also Section 7.4 for additional information on nitrogenous excretion.)
12. Give three reasons why water vapour is an important aspect of the water balance of particular arthropod species. Does their uptake of water vapour imply active water transport? You discover beetle larvae living among a container of rice. How do you think these larvae might be maintaining their water balance? What experiments could you carry out to check your ideas? (Hint: Section 6.2.4.)

Bibliography

Bradley TJ (2009). Animal Osmoregulation. Oxford Animal Biology Series. Oxford: Oxford University Press.

Chown SL, Nicolson S (2004). Insect Physiological Ecology: Mechanisms and Patterns. Chapter 4: Water balance physiology. Oxford: Oxford University Press.

Dantzler WH (2016). Comparative Physiology of the Vertebrate Kidney. 2nd Ed. American Physiological Society. New York: Springer.

Dodd CR (2009). Amphibian Ecology and Conservation. A Handbook of Techniques. Oxford: Oxford University Press. (Chapter 20: Lillywhite HB Physiological ecology: field methods and perspective.)

Evans DH (2009). Osmotic and Ionic Regulation: Cells and Animals. Boca Raton, FL: CRC Press.

Hadley NF (1994). Water Relations of Terrestrial Arthropods. San Diego, CA Academic Press.

Hillman SS, Withers PC, Drewes RC, Hillyard SD (2009). Ecological and Environmental Physiology of Amphibians. Oxford: Oxford University Press.

Louw G (1993). Physiological Animal Ecology. Harlow: Longman Scientific & Technical.

McNab BK (2002). The Physiological Ecology of Vertebrates: A View From Energetics. Ithaca, NY: Comstock Publishing Associates, Cornell University Press.

Nolan V Jr, Thompson CF (Editors) (2001). Current Ornithology. Volume 16. New York: Kluwer Academics/Plenum Publishers. (Chapter 2: Williams JB, Tieleman. Physiological ecology and behavior of desert birds.)

Schmidt-Nielsen K (1997). Animal Physiology. Adaptation and Environment. 5th Ed. Cambridge: Cambridge University Press.

Stocker TF, Qin D, Plattner G-K, Tignor M, Allen SK, Boschung J, Nauels A, Xia Y, Bex V and Midgley PM (Editors) (2013). Climate Change 2013: The Physical Science Basis. Contribution of Working Group I to the Fifth Assessment Report of the Intergovernmental Panel on Climate Change. Cambridge and New York: Cambridge University Press, Cambridge. (Chapter 12: Collins et al. Long-term climate change: projections, commitments and irreversibility. pp. 1029–1136.)

7

Kidneys and excretion

To stay alive and function properly, cells operate in a regulated environment, within the extracellular fluids. Regulating the composition and volume of this fluid environment is the job of the kidneys of vertebrates, the nephridia of invertebrates, and the Malpighian tubules in most insects, and some arachnids and myriapods.

In mammals, the kidneys are solely responsible for regulating the solute and water composition of extracellular fluids. In non-mammalian vertebrates and invertebrates, other organs, such as gills, salt glands and the gastrointestinal tract, also participate in regulating the composition and volume of extracellular fluids, as discussed in Chapters 5 and 6. However, if the kidneys or Malpighian tubules do not function properly in any species, an inevitable disruption of the volume and composition of the extracellular fluids occurs, which can have dire consequences. Disruption of the correct balance of ions influences the transmission of the action potentials necessary for nerves and muscles to function[1]; the accumulation of potassium disrupts the normal heart rhythm[2] and can even result in cardiac arrest, while volume expansion may lead to life-threateningly high blood pressures.

Most vertebrate groups also rely on their kidneys for excretion of their nitrogenous waste. Poor kidney function results in such waste accumulating, potentially to toxic concentrations, which can ultimately be fatal. To understand how healthy animals avoid this problem we examine how nitrogenous wastes are processed and excreted in their urine. We also explore alternative routes for nitrogenous excretion, such as the gills of fish and aquatic invertebrates.

The start of our journey to understand the functioning of vertebrate kidneys and the nephridia or Malpighian tubules of invertebrates is the initial step of forming the **primary urine**. We then examine how the composition of the primary urine is adjusted before its excretion as the final (or definitive) urine. Some animal groups, such as birds, reptiles and insects, pass their urine into the posterior gut where further modifications occur before excretion of a mix of urine and faeces.

1 Section 16.2.4 explains the formation of action potentials. Chapter 18 discusses muscle function.

2 We discuss the pumping cycle of the heart, and the generation of the cardiac rhythm in Sections 14.1.2 and 22.3.1, respectively.

7.1 Production of the primary urine

Animals produce primary urine in one of two ways: by secretion or by **ultrafiltration**. Figure 7.1 illustrates these alternatives.

Secretory processes, as in Figure 7.1A, generate the primary urine in the Malpighian tubules of insects and in the kidney tubules of a small number of species of marine teleost fish. In each case, these tubules have a poor arterial blood supply and solutes and water that form the primary urine are secreted into the tubules from extracellular fluid outside the tubules.

The secretion of solutes and ions draws water into the tubules, by osmosis, so the primary urine may become nearly isosmotic to the extracellular fluids. Some ions enter the tubular fluid passively, while others enter because of active processes. Subsequent reabsorption of many of the ions in the tubular fluid usually occurs, as shown in the lower part of Figure 7.1A, which explains the description of such kidneys as secretion–reabsorption organs.

Rather than using secretion–reabsorption devices, most vertebrates and some invertebrates, such as molluscs and decapod crustaceans (lobsters and crabs), generate primary urine by ultrafiltration of their blood plasma (vertebrates) or **haemolymph**[3] (invertebrates), as shown in Figure 7.1B. The process of ultrafiltration is driven by hydrostatic pressure differences across a filtering membrane, which results in the bulk passage of water containing molecules of filterable solutes (ions and organic compounds such as glucose, urea, and amino acids). Macromolecules, such as plasma proteins and any solutes and ions bound to plasma proteins, remain mostly unfiltered.

Ultrafiltration in vertebrate kidneys occurs in the **glomerulus** of the **nephron**. Each glomerulus consists of an interconnected cluster of branching capillaries, like that shown in Figure 7.2. The complexity of the glomerulus varies among different vertebrate groups and is generally greater in mammals than in the glomeruli of non-mammalian species.

3 The haemolymph is the functionally equivalent to the blood of animals with closed circulatory systems, as we discuss in Section 14.1.1.

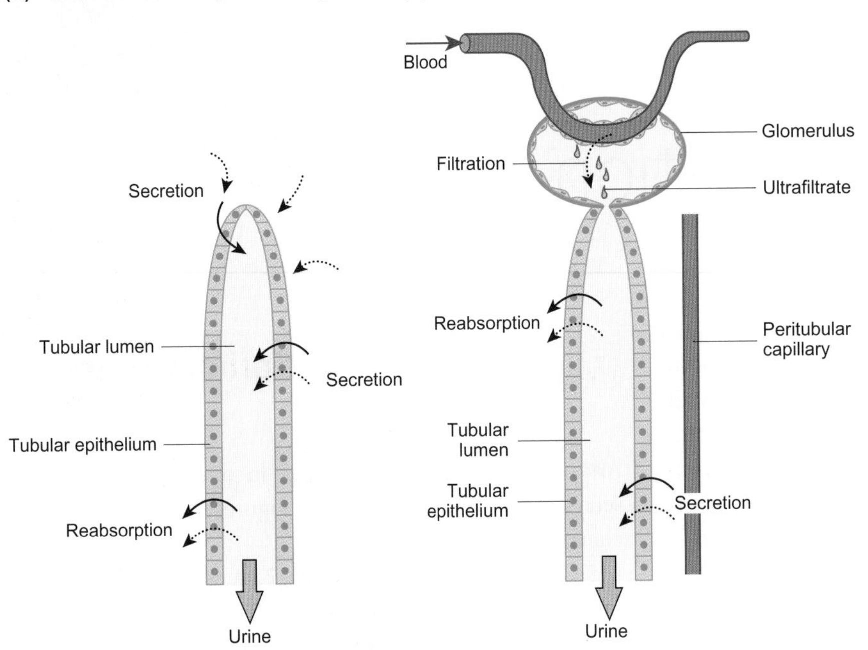

Figure 7.1 Primary urine formation

(A) Secretion–reabsorption kidneys make primary urine by secretion from the surrounding extracellular fluids into the lumen of the sealed tubules. Reabsorption processes by the tubular epithelium, later in the tubules, remove some of the secreted solutes/ions.
(B) Filtration–reabsorption kidneys make primary urine by ultrafiltration of blood delivered to a filtering unit (the glomerulus in the vertebrate kidney). The blood supply is shown as a single loop, for simplicity, but is actually more complex. Reabsorption involves active and passive processes. Secretory processes driven by active and passive processes in the tubular epithelium may add to the filtrate. Solid lines and arrows indicate active (energy-dependent) transport processes. Dotted lines and arrows indicate passive processes.

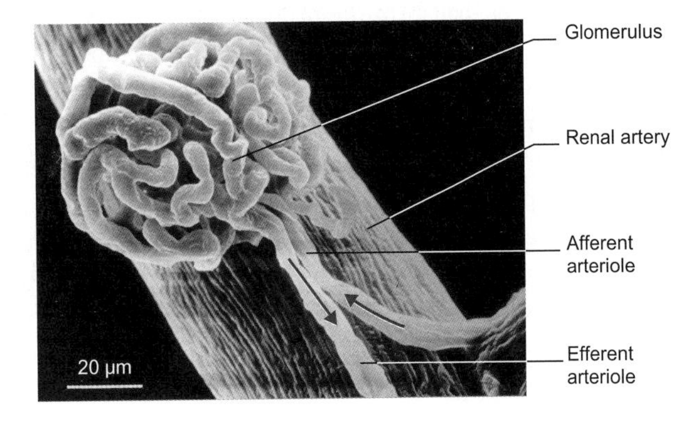

Figure 7.2 The glomerulus: the ultrafiltration device of the vertebrate kidney

Photomicrograph of a resin cast of a glomerulus from a rainbow trout (*Oncorhynchus mykiss*) viewed in the scanning electron microscope after injection of a resin into the arterial supply to the kidney. The glomerulus is lying on a renal artery, which enters a renal arteriole before giving rise to afferent arterioles. Red arrows show direction of blood flow via the afferent arteriole into the glomerulus and out via the efferent arteriole.

Source: Brown JA (1985). Renal microvasculature of the rainbow trout *Salmo gairdneri*: scanning electron microscopy of corrosion casts of glomeruli. The Anatomical Record 213: 505-513.

The glomerular capillaries arise from an **afferent**[4] glomerular arteriole, which branches from the renal arteries that supply the kidney. The glomerular capillaries rejoin to form an **efferent** glomerular arteriole that carries blood away from the glomerulus. Each glomerulus is held in a cup-shaped **Bowman's capsule**[5], which collects the filtrate that passes into the tubular part of the nephron.

In a few vertebrate species, including representatives from all the major groups, some glomeruli (the plural of glomerulus) are close to the surface of the kidney. Consequently, it is possible to study ultrafiltration by extracting small amounts of primary urine from the Bowman's capsule of these superficial glomeruli, using micropuncture. Box 7.1 outlines this experimental approach, which has played a fundamental role in helping us understand how kidneys work.

The classical studies of Alfred Richards collected samples of ultrafiltrate (primary urine) from the Bowman's capsules of frogs. His studies showed that the concentrations of small solutes in the filtrate, such as sodium and chloride ions, glucose and urea, are almost identical to the concentrations in blood plasma. The Gibbs–Donnan effect[6] of charged macromolecules (proteins) held back in the plasma accounts for the very small differences in ion concentrations between the plasma and filtrate. The concentration of plasma protein in the ultrafiltrate is extremely low (or even zero). Hence, solutes that are bound to plasma proteins, including a proportion of some of the plasma electrolytes, such as calcium[7], are also non-filterable.

After forming primary urine by ultrafiltration, the kidneys and nephridia reabsorb any filtered proteins and variable amounts of ions and water. Consequently, we call

4 Afferent means going toward; efferent means directed away from.
5 Named after William Bowman, more than 100 years ago.

6 Section 4.1.1 discusses the Gibbs–Donnan effect on the distribution of ions.
7 In mammals, almost 50 per cent of calcium is bound to plasma albumins and plasma globulins.

Box 7.1 Microtechniques to study the function of individual nephrons

Renal tubular micropuncture to investigate tubular function

The development of micropuncture methods to study nephron function by Alfred N. Richards and his colleagues in the 1920s was one of the greatest advances in renal physiology during the 20th century. Their pioneering and painstaking work established methods that are still used today to discover how the tubular segments function. Micropuncture involves the insertion of a glass micropipette with a sharp tip a few μm wide into the lumen of a superficial nephron or nephridial tubule. The pipette is initially filled with a mineral oil, such as liquid paraffin (pre-saturated with water to avoid extracting water from the fluid sample), to avoid evaporation of the small volume of fluid collected. Once a tubule is punctured, slight suction overcomes the resistance of the tip of the micropipette and tubular fluid flows readily up the pipette.

Micropuncture to measure single nephron glomerular filtration rates (SNGFR)

SNGFR can be measured by two micropuncture methods:

- Collection of all the fluid filtered into Bowman's capsule, after blocking the flow of fluid along the tubule using paraffin oil ejected from the micropipette.
- Collection of tubular fluid from a section of renal tubule after injection of a small amount of paraffin oil into the tubule; if all the fluid is removed, the oil will not move along the tubule. The volume collected is generally less than the volume filtered because of water absorption by the renal tubule. Water absorption can be allowed for in calculating SNGFR by use of a substance such as inulin[1] that is not reabsorbed or secreted by renal tubules. Inulin is freely filtered and increases in concentration in the tubular fluid (TF) compared to plasma because of water reabsorption. Hence:

Inulin concentration in tubular fluid (g mL^{-1})

$$\text{SNGFR} = \text{Volume collected in micropipette (nL min}^{-1}) \times \frac{[\text{TF}]_{\text{in}}}{[\text{P}]_{\text{in}}}$$

Inulin concentration in plasma (g mL^{-1})

Micropuncture to study glomerular haemodynamics

To assess glomerular haemodynamics, a micropipette filled with NaCl (2M) is inserted into the glomerular capillaries and/or Bowman's capsule. The pressure in these locations results in a small amount of fluid entering the pipette tip. This fluid movement into the tip of the pipette causes an increase in the resistance of an electrical circuit formed by connecting with the animal's blood system. The increased resistance automatically triggers the operation of a pump that pushes NaCl down the micropipette, until electrical resistance returns to normal. Thus, the plasma/NaCl interface is held in a constant position when the pressure in the pipette is identical to that at the puncture site (so it is called a servo-null apparatus). The pressure monitoring device has a fast response time so can detect changing pressures in the glomerular capillaries during each heartbeat. These changes in pressure are essential to confirm the correct location of the pipette within the glomerular capillaries.

Tubular microperfusion

Microperfusion techniques were developed in the 1930s (again by Alfred Richards and his colleague Arthur Walker) to study tubular function in amphibians after filtration was stopped by microinjecting mercury from a micropipette into the neck of the nephron, or a Bowman's capsule visible on the surface of the kidney. After filtration was stopped, a fluid (the perfusate) was continuously pumped into the nephron beyond the mercury. This fluid was then collected downstream of the injection site, with a second micropipette. The tricky part in the procedure is making sure that the flow of perfusate mimics normal flow rates. Figure A shows a slightly different approach, known as stationary microperfusion, which avoids this problem.

Stationary microperfusion involves the insertion of a large droplet of paraffin oil (which is preferable to mercury) from a

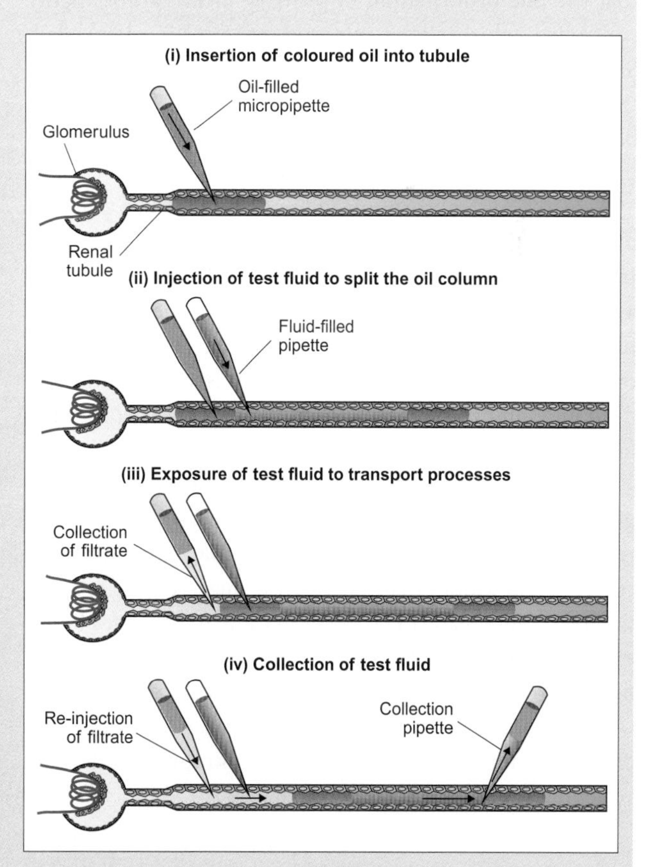

Figure A Stationary microperfusion

The process involves the series of steps (i) to (iv).

micropipette; the oil is used to isolate a section of tubule for investigation, as shown in Figure A(i). A second micropipette is used to insert a volume of fluid with a known composition to split the oil, as in Figure A(ii). Held between oil in the tubule, normal transport processes continue, while newly formed filtrate is collected in the first pipette, so as to hold the test fluid in place, which is shown in Figure A(iii). The filtrate is re-injected into the tubule after a certain time period, which pushes the test fluid down the tubule to allow collection via a third pipette, as shown in Figure A(iv).

A similar approach can be used *in vitro* for studying isolated sections of kidney tubules or nephridia. In this case, one end of the tubule is perfused continuously and the fluid is collected at the other end. Work *in vitro* has the advantage that manipulation of the composition of the bathing media—for example, by inclusion of chemical inhibitors—allows ideas about ion transport mechanisms to be tested. By analysing the composition and volume of collected fluid, tubular microperfusion allows us to characterize solute transport mechanisms, to measure the rates of absorption or secretion of fluids, and to investigate unidirectional or net solute fluxes.

[1] Box 7.2 outlines the use of inulin to monitor kidney filtration rates.

these kidneys and nephridia filtration–reabsorption devices. Human kidneys create around 180 litres of filtrate per day but excrete only about 1 litre of this as urine; this means that the kidney tubules reabsorb 99 per cent of the filtered water, along with the bulk of the solutes.

Given the two very different ways to make primary urine, we might ask: what are their relative advantages and disadvantages? A significant advantage of secretory kidneys is that the rate of formation of primary urine can be highly controlled, which allows intermittent operation if needed. Such control could be a significant advantage for animals that have highly variable body fluid volumes. For instance, blood-feeding medicinal leeches, such as *Hirudo medicinalis*, switch on secretory processes after feeding[8].

A secretion–reabsorption kidney could reduce primary urine formation when water conservation is necessary. In an evolutionary sense, this explains the occurrence of **aglomerular** kidneys (kidneys without glomeruli), which function by secretion–reabsorption in at least 30 species of marine teleost fish[9]. However, this is an evolutionary dead end, and most marine (and freshwater) teleosts have filtration–reabsorption kidneys.

Filtration followed by reabsorption is energetically costly given that the vast majority of the filtered solutes are also reabsorbed. However, the major advantage of filtration–reabsorption kidneys is that small changes in the proportion of a solute reabsorbed result in dramatic changes in net excretion. Hence, filtration–reabsorption kidneys can exert fine control over the composition of extracellular fluids.

We know most about vertebrate kidneys, so the next parts of this chapter examine the process of ultrafiltration among vertebrates in some detail.

[8] We explore nephridial function of medicinal leeches in Section 7.3.2.

[9] Water conservation by marine teleosts, which excrete only small volumes of urine, is discussed in Section 5.1.3.

7.1.1 Structural basis of ultrafiltration in glomerular capillaries

Figure 7.3 shows the three-layered extracellular route for ultrafiltration across glomerular capillaries[10]. First, the ultrafiltrate passes through pores (fenestrae) in the **endothelium** of the capillaries. Then the ultrafiltrate flows through the **basement membrane**, which seems to act like a hydrated gel with continuously collapsing and reforming water-filled channels. Fibres of collagen (chains of amino acids) in the basement membrane form a network that restricts the movement of large macromolecules. Negatively charged molecules (glycoproteins) occur in the basement membrane and impede filtration of negatively charged proteinaceous molecules and protein-bound solutes, but the extent of this effect is controversial and probably differs between species.

Finally, fluid flows through the epithelial layer of the glomerular capillaries, which consists of **podocytes** ('foot cells', from the Greek word *podo* for foot) that extend as primary processes along the length of the capillary and which give rise to secondary processes or pedicels (foot processes). The pedicels from one podocyte interdigitate with those from another podocyte, forming filtration slits in the gaps between. Slit diaphragms at the base of the filtration slits have many tiny pores in a zip-fastener-like arrangement. The size of the openings (in mammals approximately 14 nm by 4 nm) allows the passage of bulk water and small solutes, but impedes the passage of large proteins.

Ultimately, whether each solute is filtered (or not) depends on the combination of its size, shape and charge. Plasma albumin, with an ellipsoid shape of an effective molecular radius of 3.5 nm, and flexibility, could be just small enough, in theory, to penetrate through the pores in the slit diaphragms.

[10] Fluid and solute exchange across capillary walls is discussed more generally in Section 14.1.5.

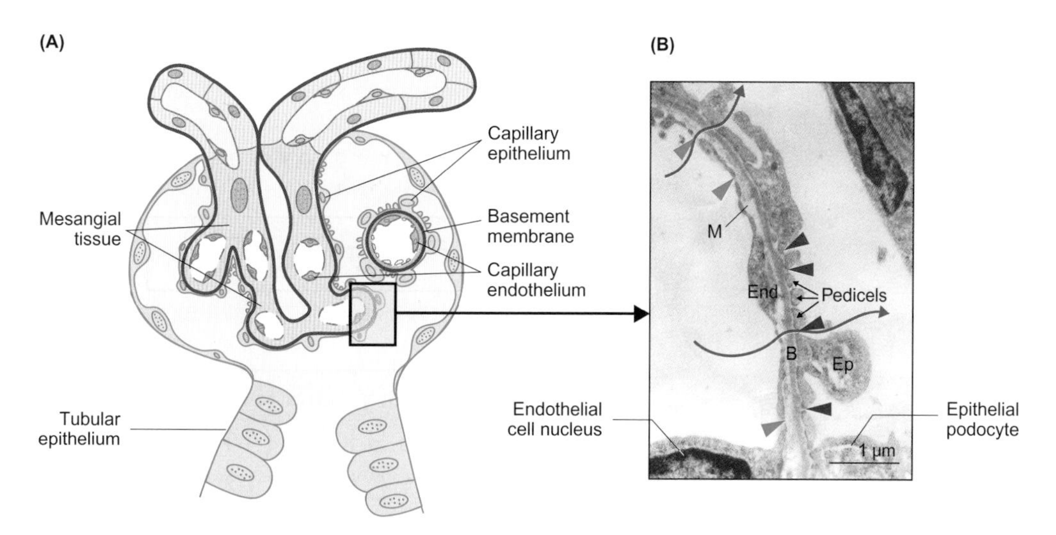

Figure 7.3 Three tissue layers of glomerular capillaries of the rainbow trout (*Oncorhynchus mykiss*)

(A) Simplified drawing of cross–section through a glomerulus. Contractile mesangial tissue occurs between the capillary loops. The glomerular capillary wall consists of three distinct tissue layers: capillary endothelium (purple), basement membrane (black) and capillary epithelium (pale green). The square marks an area examined at higher magnification, by transmission electron microscopy. (B) Electron micrograph showing the capillary endothelial layer (End) with fenestrae (green arrowheads), the basement membrane (B) and the capillary epithelial layer (Ep), including epithelial podocytes and pedicels. M = mesangial matrix penetrating between the capillary endothelium and basement membrane. Slit diaphragms (indicated by red arrowheads) occur between each pair of pedicels. Blue arrows show the extracellular route for the flow of water and filterable solutes across the glomerular capillaries.

Source: A: adapted from Brown JA, Rankin JC, Yokota SD. Glomerular haemodynamics and filtration in single nephrons of non-mammalian vertebrates. In: New Insights in Vertebrate Kidney Function (Eds. Brown JA et al). pp 1-44. Cambridge: Cambridge University Press, 1993. B: Brown JA, Taylor SM, Gray CJ (1983). Glomerular ultrastructure of the trout, *Salmo gairdneri*. Cell Tissue Research 230: 205-218.

At physiological pH, however, plasma proteins are negatively charged, which impedes their passage through the capillary wall. For example, filtration of less than 1 per cent of plasma albumin occurs in the glomeruli of the amphibian mudpuppy (*Necturus maculosus*) and healthy rats filter less than 0.1 per cent of the plasma albumin. An increase of plasma proteins in the urine (**proteinuria**) occurs in some renal diseases due to an increase in the permeability of the glomerular barrier to proteins.

7.1.2 Haemodynamics of glomerular filtration

Figure 7.4 shows the haemodynamic forces that account for ultrafiltration by the glomerular capillaries. Blood pressure in the glomerular capillaries (P_{GC}) is the hydrostatic force driving filtration. Blood pressure in the afferent arterioles is generally higher than in other arterioles because of the high resistance of the narrow bored efferent arteriole through which blood leaves the glomerulus[11].

Filtration creates a hydrostatic backpressure in the Bowman's capsule (P_{BC}) because of the resistance to the flow of fluid along the narrow lumen of the nephron; this backpressure impedes filtration to some extent. In addition, ultrafiltration, by definition, results in an accumulation of plasma proteins and other macromolecules in the glomerular capillaries. These molecules create a transmembrane difference in the **colloid osmotic pressure**[12], which tends to pull water back into the capillaries.

A few vertebrate species have some glomeruli visible at the kidney surface (some fish, a few amphibian species and a few mammalian species). In these species, the hydrostatic pressure in glomerular capillaries (P_{GC}) and the backpressure in Bowman's capsule (P_{BC}) can be measured using the micropuncture techniques outlined in Box 7.1, and samples of blood plasma and ultrafiltrate collected to determine the difference in colloid osmotic pressures (P_{COP}). Figure 7.4 includes such data for the Munich Wistar strain of laboratory rats. In these rats, a net effective filtration pressure—the pressure for ultrafiltration—of approximately 2 kPa occurs in the superficial glomeruli.

The analysis of filtration haemodynamics in Figure 7.4 does not consider the fact that the filtration of fluid from the blood plasma during its passage along the capillaries increases the colloid osmotic pressure in the glomerular capillaries. On entering the glomerulus, P_{COP} in the Munich Wistar rats is ~2.3 kPa, but by the time blood leaves the glomerulus in the efferent arteriole, P_{COP} has increased to ~4.4 kPa. In reality, P_{COP} increases in a non-linear fashion

[11] Section 14.2.1 outlines the relationships between fluid pressure, resistance and flow.

[12] The colloid osmotic pressure (COP) due to plasma proteins is known as **oncotic pressure**; it is only about 1 per cent of the absolute osmotic pressure, so the filtrate is still considered isosmotic to plasma.

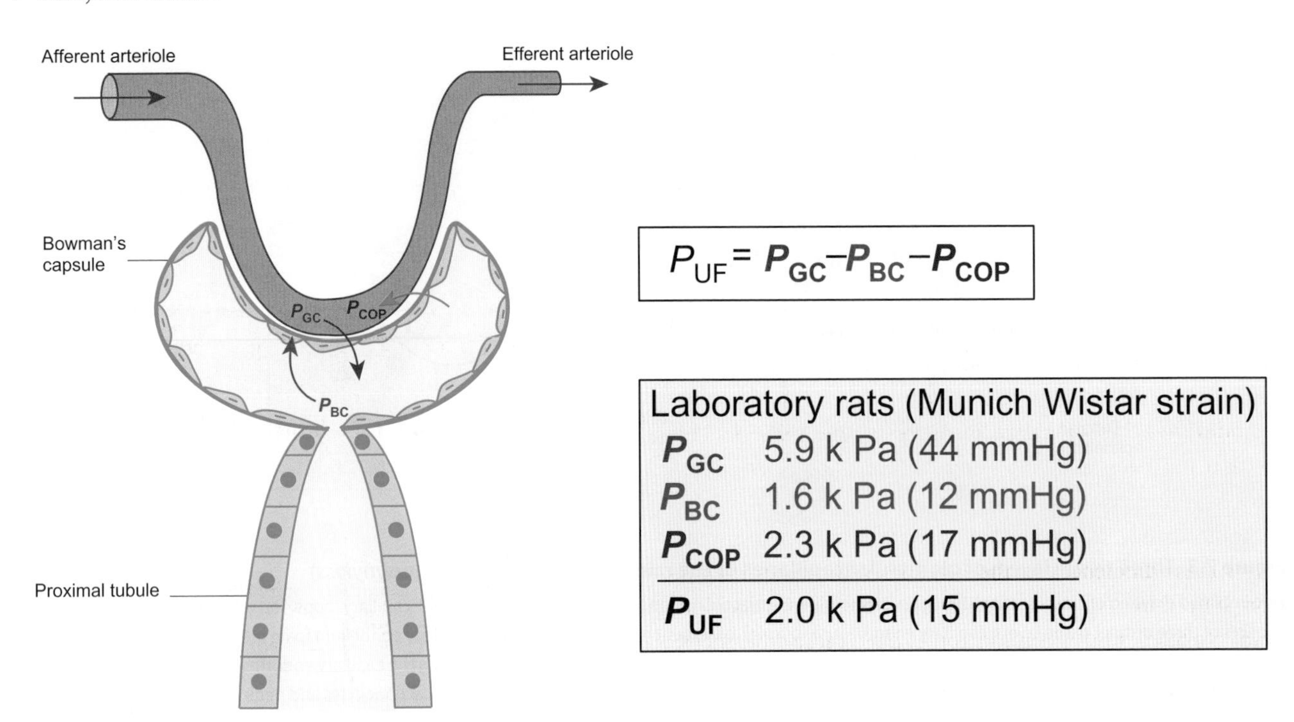

Figure 7.4 Haemodynamic forces result in ultrafiltration in the vertebrate glomerulus

The network of glomerular capillaries is simplified to a single loop. The listed values for pressure measurements are the approximate values for the Munich Wistar strain of rats and show a net positive pressure for ultrafiltration (P_{UF}). This effective filtration pressure is determined by the balance between the hydrostatic pressure in the glomerular capillaries (P_{GC}) acting for filtration and the combined effect of the forces acting against filtration. The factors acting against filtration are the backpressure in Bowman's capsule (P_{BC}) and the difference in colloid osmotic pressures of plasma in the capillaries and the filtrate (P_{COP}).

Data source: Maddox DA et al (1975). Determinants of glomerular filtration in experimental glomerulonephritis in the rat. The Journal of Clinical Investigation 55: 305-318.

along the glomerular capillaries; initially a rapid rise occurs, which gradually slows as the rate of ultrafiltration decreases.

Looking again at the data in Figure 7.4, we can work out that a P_{COP} of 4.4 kPa in the efferent arterioles of the rats added to the backpressure in Bowman's capsule (P_{BC}) of 1.6 kPa (total 6.0 kPa opposing filtration) means that the effective filtration pressure in these rats must reach zero somewhere within the glomerular capillary knot. In other words, a state of filtration pressure equilibrium occurs. Filtration pressure equilibrium occurs in several mammalian species, such as common squirrel monkeys (*Saimiri sciureus*) and rabbits, and among various non-mammalian species, such as an amphibian salamander, the two-toed amphiuma (*Amphiuma means*).

The effective force for filtration (P_{UF}) remains above zero throughout the glomerulus of river lampreys (*Lampetra fluviatilis*) living in freshwater rivers, commencing at ~0.72 kPa (5.4 mmHg) when blood enters the glomerulus. However, river lampreys living in brackish water (at the equivalent of 20–30 per cent ocean seawater) have a lower arterial blood pressure, which reduces P_{GC} such that P_{UF} is only 0.093 kPa (0.7 mmHg) when blood enters the glomerulus; filtration pressure equilibrium some way along the glomerular capillaries is inevitable. Such changes in response to changing environmental conditions are central to the adjustment of the rate of filtration by the individual glomeruli, by which lampreys maintain their water balance. The changes in individual filtration rates drive changes in the rates of urine excretion that match the rates of water influx (or efflux) by osmosis, and which are determined by the difference in osmotic concentrations between internal body fluids and the aquatic environment[13].

Filtration rates of individual glomeruli

The filtration rates of individual nephrons are known as single nephron glomerular filtration rates (SNGFRs). The classical method to measure SNGFRs involves micropuncture and timed collection of tubular fluid from superficial segments of the nephron. Renal tubules generally reabsorb some of the filtered water, so the volume collected is less than the volume initially filtered. Box 7.1 explains how a suitable marker substance that is freely filtered, but which remains entirely in the tubular fluid after filtration, can be used to take account of water absorption from the tubular fluid and, hence, to calculate SNGFRs. In most species, however, micropuncture is not an option for measuring SNGFRs because superficial kidney tubules (or glomeruli) are rare.

[13] Lampreys hyperosmoregulate when they are in fresh water or brackish water (with body fluids more osmotically concentrated than external fluids), but hypo-osmoregulate in seawater, in a similar way to euryhaline teleosts, which we discuss in Section 5.3.3.

An indirect approach to measure SNGFRs involves the intravascular injection of sodium ferrocyanide. Ferrocyanide ions are freely filtered by the glomeruli. After allowing 20–30 seconds of filtration, the kidney is snap-frozen, which holds the filtered ferrocyanide in the nephrons. A chemical reaction between the ferrocyanide and ferric chloride in pieces of kidney kept at subzero temperatures converts ferrocyanide in the tubules into a visible blue precipitate, Prussian blue. The nephrons are then painstakingly microdissected from the pieces of softened kidney tissue. The length of tubule containing Prussian blue (formed from filtered ferrocyanide) gives a qualitative measure of the SNGFR. This method has been adapted using radiolabelled sodium ferrocyanide to obtain quantitative values.

Studies using micropuncture and the ferrocyanide technique reveal large differences in the SNGFRs with values ranging from 0 nL min^{-1} (in some nephrons in non-mammalian kidneys)[14] to 52 nL min^{-1}. Three factors explain these differences:

- The permeability (leakiness) to water of the glomerular capillaries: typically, glomerular capillaries are about 100 times more permeable to water than other capillary beds, because of the structural features shown back in Figure 7.3.
- Effective filtration pressure (P_{UF}) varies among animal groups. Mammals and birds have relatively high arterial blood pressures, which result in relatively high values for P_{UF} compared with non-mammalian vertebrates, and may explain why mammals have the highest SNGFRs at 30–52 nL min^{-1}. However, the SNGFRs of bird kidneys are typically much less, at 6–16 nL min^{-1}, which indicates that other variables also play an important part in determining their SNGFRs.
- The size of a glomerulus determines its filtering surface area and complexities such as the interconnections between the capillary loops add to the filtering surface area. In the kidneys of mammals and birds, glomerular size and SNGFRs are directly related.

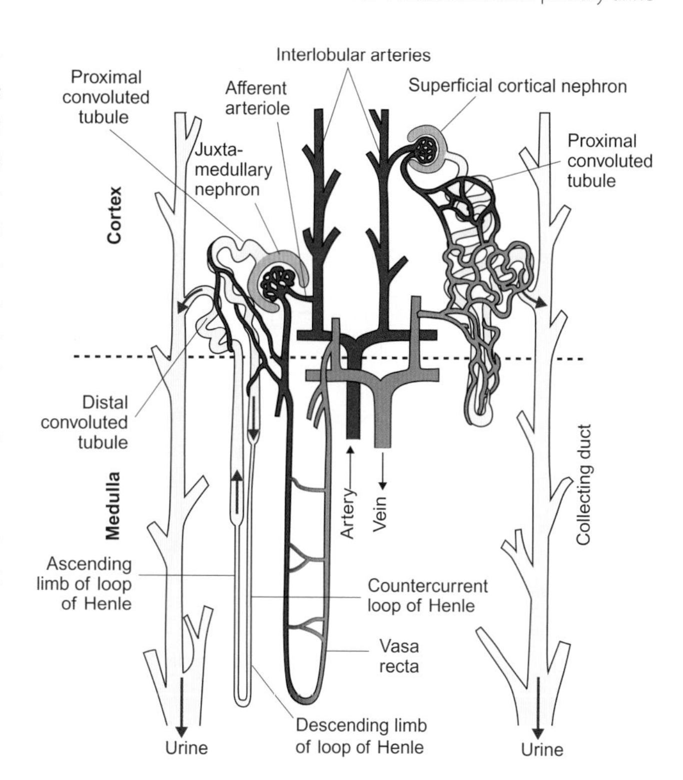

Figure 7.5 Superficial cortical and juxtamedullary nephrons of the mammalian kidney and their vascular supply

The kidney nephrons consist of a glomerulus which empties into the proximal convoluted tubule, then the countercurrent loop of Henle and finally the distal convoluted tubule before emptying into the collecting duct. Purple arrows show the direction of flow of the tubular fluid. The arterial supply to the mammalian kidney is via the renal artery which divides into interlobular arteries. Afferent arterioles arise from the interlobular arteries. The efferent arteriolar pathways differ in their arrangement depending on the location of the nephrons within the kidney. In superficial cortical nephrons, the efferent arterioles enter a capillary network around the same entire nephron. In deeper juxtamedullary nephrons, most of the efferent blood circulation is organised around the loop of Henle in straight bundles of vessels: the vasa recta.

Source: adapted from Schmidt-Nielsen K. Animal Physiology. Adaptation and Environment. 5th Edition. Cambridge: Cambridge University Press, 1997.

Figure 7.5 illustrates the arrangement of the nephrons in the mammalian kidney. Superficial cortical nephrons near to the surface of the kidney cortex have relatively small glomeruli compared with the glomeruli of juxtamedullary nephrons that occur close to (juxta) the medulla of the kidney, and which have long loops of Henle. Hence, the cortical nephrons have a relatively small filtering surface area and lower SNGFR than the deeper juxtamedullary nephrons. In a study of the Munich Wistar strain of laboratory rats, the mean SNGFRs of cortical nephrons was 36 nL min^{-1}, while the larger glomeruli of juxtamedullary nephrons had a mean SNGFR of 52 nL min^{-1}.

Cortical and medullary regions of the kidney are also broadly recognizable in birds, as illustrated in Figure 7.6, but the boundary between the two areas is less distinct than in mammals. Most of the glomeruli in the avian kidney are in the cortical region where the smallest glomeruli have just a single capillary loop with no cross-branching connections; these are the simplest of any vertebrate glomeruli. The cortical nephrons in the avian kidney have no countercurrent loop. The afferent arterioles supplying these loopless nephrons arise from four intralobular arteries running as two pairs on each side of each kidney lobule, as shown in Figure 7.6. The *intra*lobular arteries arise from an *inter*lobular artery supplying each lobule, which in turn arise from the renal artery.

[14] Section 7.1.3 discusses the switching of filtration on and off (glomerular intermittency).

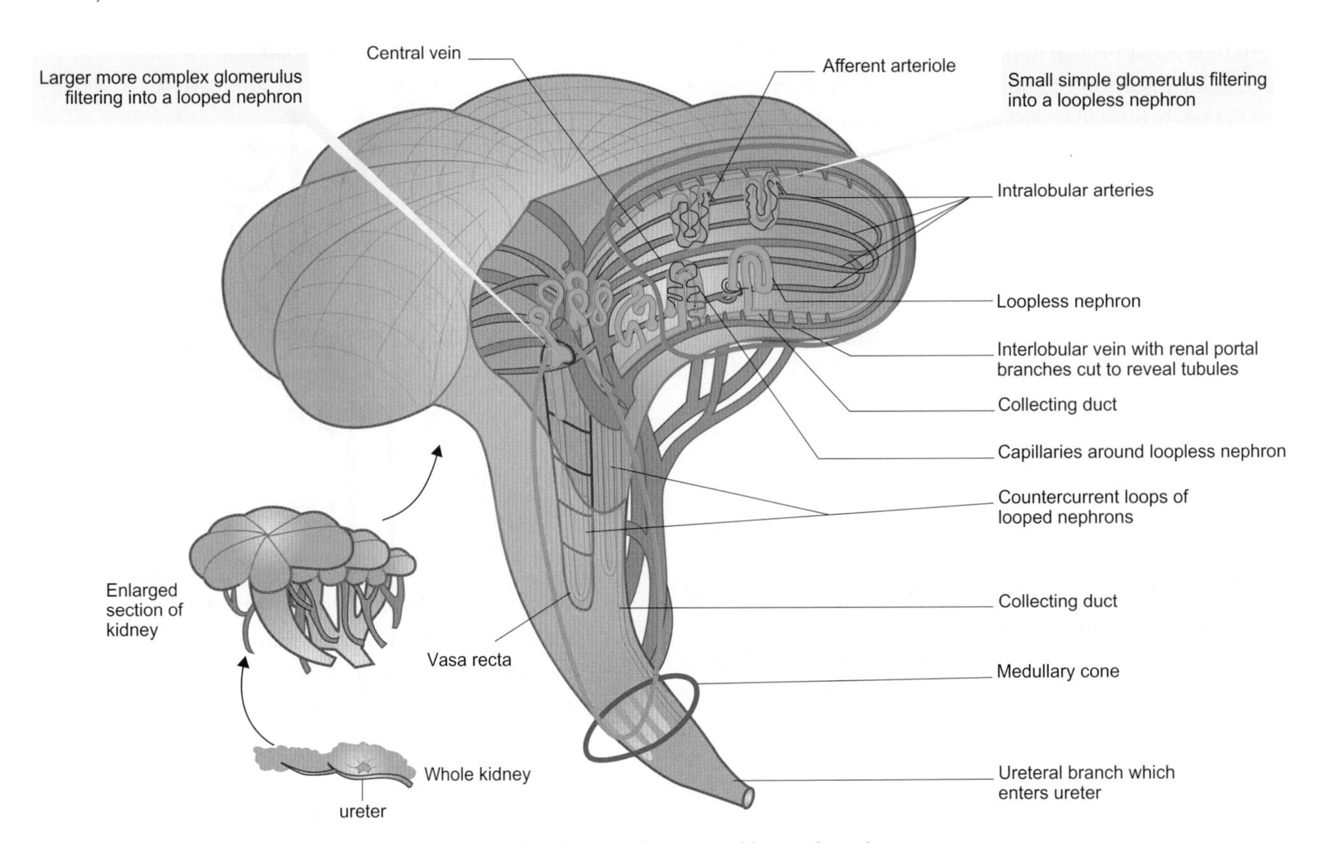

Figure 7.6 Organization of the avian kidney with loopless nephrons and looped nephrons

The elongated kidney of birds is compressed dorsoventrally and made up of lobules joined together like petals on a flower. Cortical and medullary regions are indicated by the areas encircled in green (cortical) and orange (medullary), although there is a gradation between two extremes. (i) The smallest and simplest glomeruli occur in the outer (cortical) parts of each kidney lobule, arranged on both dorsal and ventral surfaces. These small glomeruli filter into short, loopless nephrons consisting of a proximal and distal segment and empty into the collecting ducts that run around the outer edges of the kidney. Urine flowing down the collecting ducts passes through the medullary cone and from there into the ureter. (ii) The largest glomeruli are situated towards the centre of the kidney, and filter into looped nephrons. The countercurrent loops are surrounded by vasa recta capillaries that arise from the efferent arterioles. The vasa recta run parallel to the countercurrent loop.

Source: adapted from Braun EJ. Renal function in birds. In: New Insights in Vertebrate Kidney Function (Eds. Brown JA et al). Cambridge: Cambridge University Press, 1993.

The medullary region of the avian kidney contains a few larger glomeruli supplied by afferent arterioles that arise from the interlobular arteries. These larger glomeruli filter into nephrons that have countercurrent loops, as shown in Figure 7.6. The relative proportions of looped and loopless nephrons have been studied in detail in a few species of birds. For example, about 10 per cent of the nephrons of Gambel's quails (*Callipepla gambelii*), and 30 per cent of those in European starlings (*Sturnus vulgaris*) have loops; the remaining nephrons are loopless.

The different glomerular sizes of the looped and loopless nephrons in the avian kidneys mean that the glomerular surface areas can vary widely. In keeping with these differences in surface area, the SNGFRs of the larger medullary glomeruli of well-hydrated Gambel's quails are more than double those of the smaller cortical glomeruli (in one study: 15.8 nL min^{-1} compared to 6.4 nL min^{-1}).

It is difficult to measure the filtering surface area of glomeruli, or its permeability, so these factors are usually combined in an ultrafiltration coefficient (K_f) which represents the product of the total filtering surface area of the glomerulus and the hydraulic conductivity of the filtering capillaries. The ultrafiltration coefficient is expressed in nL min^{-1} kPa^{-1} (volume filtered per unit time per kPa of net filtration pressure). This means that SNGFR will be determined by the ultrafiltration coefficient (nL min^{-1} kPa^{-1}) multiplied by the effective filtration pressure (kPa), as described by Equation 7.1, which combines all the main factors that influence SNGFR:

Single nephron glomerular filtration rate (nL min^{-1})

Ultrafiltration coefficient for entire glomerulus (nL min^{-1} kPa^{-1})

$$SNGFR = K_f \times P_{UF} \qquad \text{Equation 7.1}$$

Effective filtration pressure (kPa)

Equation 7.1 tells us that the measurement of SNGFR and the haemodynamic forces that determine the effective filtration pressure (P_{UF}) enable calculation of K_f. Such studies show that K_f of glomeruli varies between species. For example, K_f (nL min^{-1} kPa^{-1}) is approximately 13 in river lampreys, 36 in rats and 57 in the salamander *Amphiuma means*.

Values for K_f are not as fixed as once thought. Variations in the filtering surface area of the glomerular capillaries and their hydraulic conductivity will influence K_f and hence SNGFRs, and can occur as a result of at least two events:

- The contraction of elements in the mesangial cells, which occur between the capillary loops (as shown earlier in Figure 7.3A) can reduce blood flow through parts of the glomerulus, and hence reduce the surface area involved in ultrafiltration.
- Structural changes in the capillary wall can influence capillary hydraulic conductivity. For example, a reduction in the number of slit diaphragms (shown back in Figure 7.3B) reduces hydraulic conductivity. Such changes occur during the acclimation of rainbow trout (*Oncorhynchus mykiss*) to seawater where they are thought to be an important component in reducing the number of filtering glomeruli as part of the mechanisms for reducing urine excretion for water conservation[15].

Among teleost fish, glomerular size tends to be larger among freshwater species than marine species. Furthermore, long-term acclimation to seawater of some euryhaline species of teleosts (that can acclimate to a range of salinities)[16] results in a decrease in glomerular size. However, despite these apparent size relationships, the limited information available does not suggest that SNGFRs decline. Instead, there is strong evidence that alterations to the filtering populations of glomeruli are the main means of controlling the total kidney GFR in teleost fish. Such changes play a major role in adjusting kidney GFR of most non-mammalian vertebrates, as we learn in the next section.

7.1.3 Glomerular filtration rates

Within a kidney, the sum total of all the individual nephron filtration rates makes up the total kidney glomerular filtration rate (GFR). The overall kidney GFR can be measured using marker substances that pass without restriction across the capillary walls of the glomeruli in the bulk water flow. To be a suitable marker for GFR, the substance must then pass along the tubule without any reabsorption, and without any of the substance being added by secretion. If such conditions apply, then all the filtered material is excreted (but no more). Box 7.2 explains these concepts in more detail.

Inulin[17] is the 'gold standard' for measuring GFR. Inulins have a molecular weight of between about 500 and 6000 Da.

[15] Figure 7.7 examines the filtering population of glomeruli in fresh water and seawater-acclimated trout. We explore osmoregulation of marine and freshwater teleosts more broadly in Section 5.1.3 and 5.2 respectively.

[16] We explore osmoregulation in euryhaline teleosts in Section 5.3.3.

[17] Inulins are polysaccharides stored in some plant tubers such as dahlias and onions.

Box 7.2 Measurement of glomerular filtration rates (GFR) and clearance techniques

The measurement of total kidney GFR, using substances such as inulin, employs a technique known as **clearance measurement**. Calculation of clearances calculates the *volume* of plasma that is completely emptied (cleared) of a substance per unit time, to account for the amount excreted per unit time, and is usually expressed in mL min^{-1} or mL h^{-1}.

The renal clearance of a solute is a theoretical concept since plasma is never totally cleared of any solute during the passage of plasma through the glomerular capillaries. Nevertheless, clearance values are extremely valuable in understanding kidney function. The theoretical volume of plasma cleared of a solute (S) is given by the amount of S excreted per unit time divided by its plasma concentration, which is expressed in Equation A:

$$\text{Clearance of S (mL min}^{-1}) = \frac{[U]_s \times \dot{V}}{[P]_s} \qquad \text{Equation A}$$

Urine concentration of S (mg mL^{-1})
Volume of urine per unit time (mL min^{-1})
Plasma concentration of S (mg mL^{-1})

Clearance measurements to monitor GFR

The rate of clearance (volume cleared per unit time) of substances that fulfil certain criteria provide a measure of glomerular filtration rate (GFR). Ideal GFR markers should be:

- unimpeded from glomerular filtration;
- not reabsorbed from the renal tubules;
- not secreted into the renal tubules.

Figure A illustrates these concepts and shows how solutes, such as inulin, that fulfil the above criteria for GFR markers, are processed by the kidney. The amount that is filtered per unit time ultimately becomes the amount excreted per unit time (with just a slight delay in passing along the kidney tubules). These relationships are expressed mathematically in Equation B:

Amount filtered per unit time = Amount excreted per unit time

$$GFR \times [F]_{in} = \dot{V} \times [U]_{in} \qquad \text{Equation B}$$

Glomerular filtration rate (mL min^{-1})
Concentration of inulin (in) in the glomerular filtrate (F) (mg L^{-1})
Volume of urine per unit time (mL min^{-1})
Concentration of inulin in urine (mg L^{-1})

Equation B can be rearranged to give:

$$GFR = \frac{[U]_{in}}{[F]_{in}} \times \dot{V} \qquad \text{Equation C}$$

Samples of filtrate are not easily obtained, but for freely filtered GFR markers, such as inulin, the concentration in the filtrate equals the concentration in the plasma. Hence, Equation C becomes:

$$GFR = \frac{[U]_{in}}{[P]_{in}} \times \dot{V} \qquad \text{Equation D}$$

Notice that Equation D is identical for inulin to Equation A, the general equation for calculating clearances. In other words, GFR measurements are clearance measurements (i.e. the volume of plasma cleared per unit time to account for the excretion of a GFR marker per unit time equals the volume filtered per unit time (GFR)).

Clearances of solutes relative to GFR give information on tubular function

After filtration, the renal tubules usually reabsorb some of a filtered solute and secretory processes may add to the amount that is filtered. The net effect of these processes on a particular substance can be assessed by calculating its clearance compared to that of a GFR marker.

A relative clearance (compared to GFR) greater than 1 indicates net tubular secretion. The more secretion that occurs, the higher the relative clearance value becomes. A relative clearance value of less than 1 indicates either: (i) the substance has not been freely filtered (because it is too large a molecule or because it binds to unfiltered plasma proteins), or (ii) net reabsorption of the substance occurs in the renal tubules, which reduces the amount in the urine.

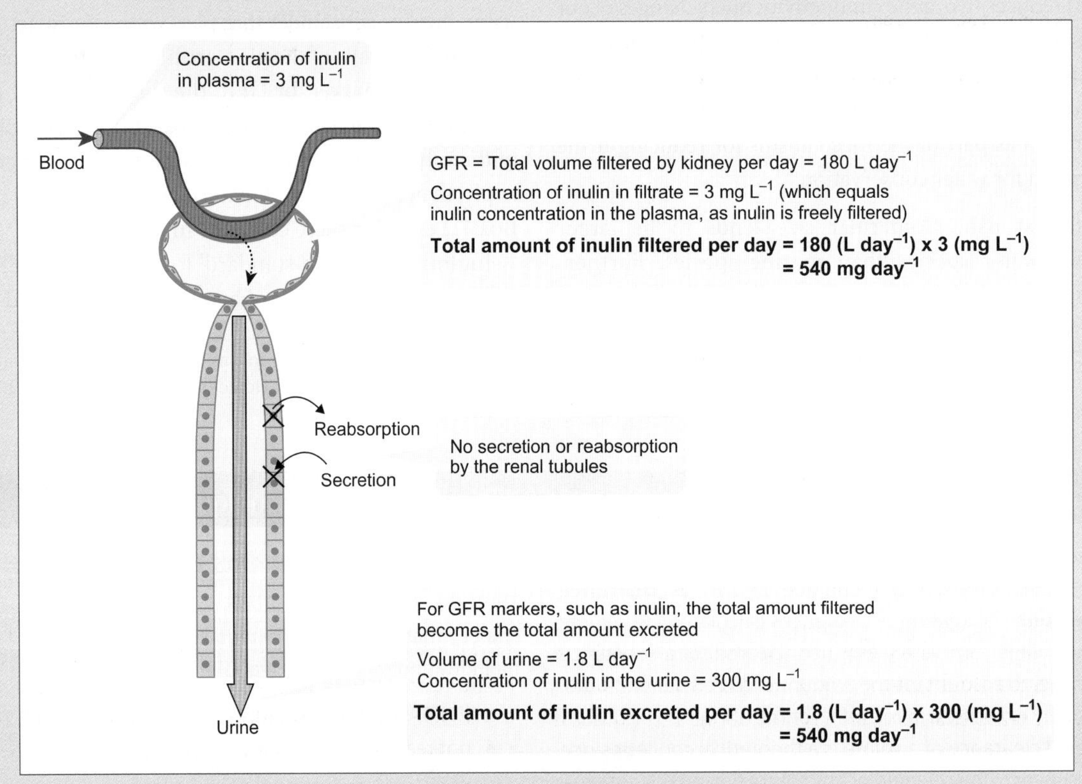

Figure A Concepts relating to measurement of glomerular filtration rate

Once in the circulation, glomerular filtration of inulin is unrestricted. Renal tubular micropuncture in species as diverse as amphibian mudpuppies (*Necturus maculosus*) and rats shows reabsorption of less than 1 per cent of the filtered inulin. However, inulin is not naturally present in the circulation, so veterinary and clinical studies rarely use inulin to monitor GFR.

An acceptable alternative to inulin is creatinine, which is always present in the circulation as it is a product of muscle turnover. Creatinine is a small molecule (molecular mass 113 Da) so it is freely filtered by the glomeruli. However, in many species (such as dogs and humans), the renal tubules reabsorb some creatinine, but generally secrete more than is reabsorbed so measurements of creatinine clearance[18] usually overestimate GFR by 10–40 per cent. Despite this, plasma concentrations of creatinine are a good index of healthy kidney function, and are routinely measured to assess GFR for clinical purposes. A decrease in GFR increases the plasma concentration of creatinine compared with that expected for an individual's age, sex and body mass.

[18] Box 7.2 explains the concept of clearance measurements.

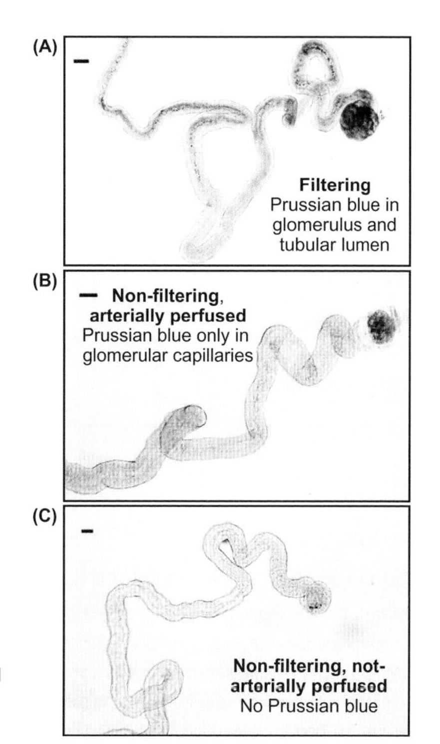

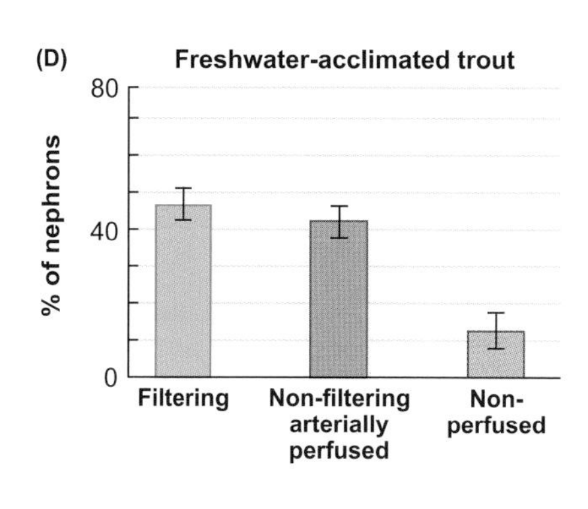

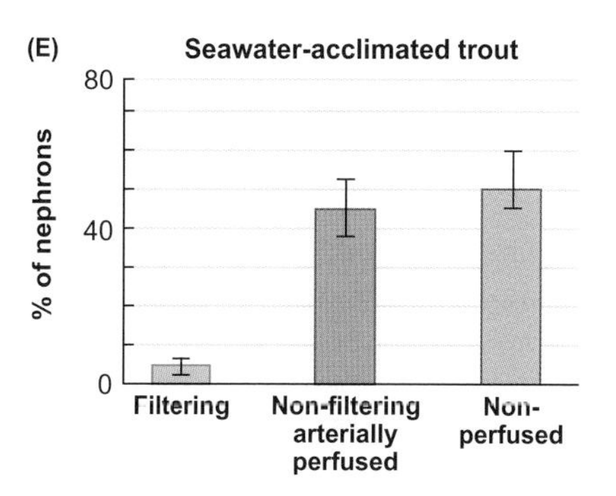

Figure 7.7 Glomerular intermittency occurs in the kidney of rainbow trout (*Oncorhynchus mykiss*)

The ferrocyanide technique (outlined in Section 7.1.2) reveals three glomerular states based on the location of Prussian blue precipitate: (A) Filtering—with Prussian blue precipitate in the glomerulus and tubular lumen; (B) Non-filtering but arterially perfused—with Prussian blue precipitate in the glomerulus only and no precipitate in the tubule; and (C) Non-perfused: non-filtering nephrons, not-arterially perfused—with no Prussian blue precipitate in the glomerulus or tubule. Graphs show the higher proportion of filtering nephrons in kidneys of (D) freshwater-acclimated trout than (E) seawater-acclimated trout. Bars on A, B & C = 100 μm

Source: Images: © J.A.Brown; D & E: adapted from Brown JA et al (1980). Angiotensin and single nephron glomerular function in the trout *Salmo gairdneri*. American Journal of Physiology Regulatory Integrative and Comparative Physiology (8): 239: R509-514.

Highly variable GFRs drive changes in urine excretion rates in non-mammalian vertebrates

In non-mammalian vertebrates (fish, amphibians, reptiles and birds), kidney GFR is highly variable and generally results in proportional changes in the volume of urine produced per unit time. In these animal groups, GFR is influenced by internal fluctuations in blood pressure and environmental circumstances that affect the state of hydration and salt balance.

Non-mammalian species living in terrestrial habitats tend to have relatively high GFRs after water intake or expansion of their extracellular fluid volume, but lower GFRs during dehydration. For example, an average GFR of about 170 mL/kg/h in European starlings decreases by 60 per cent during dehydration to about 70 mL/kg/h. For aquatic non-mammalian species that acclimate to different salinities, GFR is usually higher when the animals acclimate to fresh water, in which there is an excess water influx, than when they acclimate to brackish or seawater[19].

Alteration in GFR could, in theory, result from changes in the number of filtering glomeruli or the rate at which individual glomeruli filter (SNGFRs). River lampreys are a rare example in which SNGFR correlates well with GFR, and which suggests that changes in individual SNGFRs drive adjustments of total kidney GFR in river lampreys. These events appear to result from the altered glomerular haemodynamics. For example, blood pressure in the glomerular capillaries (P_{GC}) declines when lampreys acclimate to brackish water, which reduces the effective filtration pressure (P_{UF}) compared to that of freshwater lampreys, and reduces SNGFR.

Studies of lesser spotted dogfish (*Scyliorhinus canicula*) show changes in SNGFR occur in cartilaginous sharks. The exact events in the glomerulus are uncertain, but when blood pressure and renal blood flow increase during treatment with adrenaline[20], increased SNGFRs at least partially account for the increase in total GFR and urine output. However, alongside these changes of SNGFR, dogfish have some capacity to adjust the number of filtering glomeruli.

Changes in the population of filtering glomeruli (**glomerular intermittency**) of teleost fish, amphibians and reptiles almost exclusively drive the changes in GFR that are essential in maintaining the water balance of these animal groups. Figure 7.7 illustrates some of the clearest evidence for such changes in the filtering population of glomeruli of rainbow trout. In rainbow trout acclimated to fresh water about 45 per cent of the glomeruli are filtering, which leaves plenty of spare capacity to increase kidney GFR by recruiting extra glomeruli, if necessary. On the other hand, acclimation to seawater (when GFR and urine output decline) results in a 10-fold reduction in the filtering population of glomeruli. Figure 7.7D & E also illustrate changes in the pattern of vascular perfusion of the glomerular capillaries that result in an increased proportion

[19] Chapter 5 discusses the osmoregulation of aquatic animals.

[20] Adrenaline is also known as epinephrine.

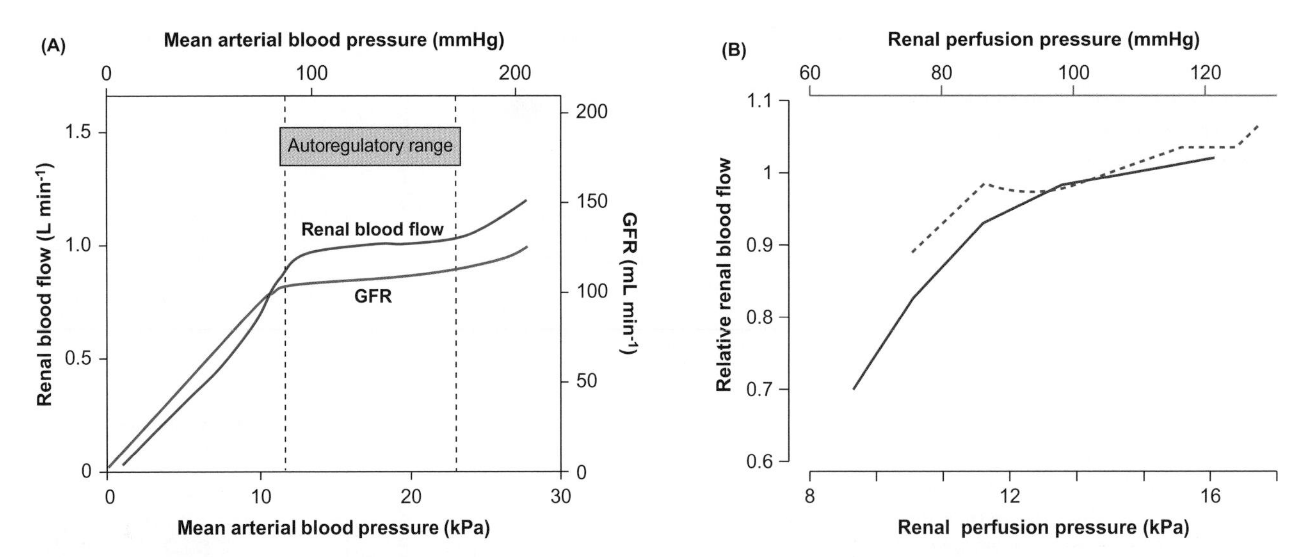

Figure 7.8 Autoregulation of glomerular filtration rates (GFR) and renal blood flow (RBF) in mammals

(A) Stable GFR (blue line) and renal blood flow (red line) occur across the autoregulatory range of arterial blood pressures in large mammals. These data are for perfused kidneys in which no external hormones or neural control can influence kidney function. (B) Pressure-flow relationship data from experiments on isolated rat kidneys show a less discrete autoregulatory range. A clamp on the aorta was used to manipulate renal perfusion pressure (RPP). The solid purple line shows results with sequential reduction of RPP on relative renal blood flow (RBF). The red dashed line shows results from random pressure manipulations. The values for RBF are normalized relative to the value at a reference pressure (typically 13.3 kPa; 100 mmHg), hence values of relative RBF below 1 indicate a decrease in RBF relative to the reference pressure.

Sources: A: adapted from Lote CJ. Principles of Renal Physiology, 5th Edition. Springer, 2012. B: Cupples WA, Braam B (2007). Assessment of renal autoregulation. American Journal of Physiology, Renal Physiology 292: F1105–F1123.

of non-arterially perfused glomeruli in seawater-acclimated trout, which inevitably decreases kidney GFR.

Among birds, the cortical glomeruli that filter into loopless nephrons appear to function intermittently, much like the glomeruli of fish, amphibians or reptiles. For example, over-hydration of Gambel's quails' results in an increase in the percentage of filtering loopless nephrons from 71 per cent to 100 per cent. During dehydration or after a salt load, the loopless nephrons of Gambel's quails stop filtering, in part due to hormonal regulation, but the looped nephrons continue to filter[21]. However, changes in SNGFRs also occur; the over-hydration of Gambel's quails results in a doubling of SNGFR in both loopless and looped nephrons.

Most of the species that show glomerular intermittency have a renal portal supply delivering venous blood to their kidneys. This venous system could provide nourishment to the renal tubules in the absence of arterial blood flow to many of the glomeruli. In the avian kidney, in which a muscular valve controls blood flow through the renal portal system, the portal blood flow is thought to influence renal haemodynamics and affect glomerular filtration. It is possible that opening and closing of the valve controlling renal portal flow is coordinated with the intermittent filtration by glomeruli. The renal portal supply of the avian kidney enters each lobule via an interlobular vein, which runs around the periphery of the lobule, as shown in Figure 7.6. Branches from the interlobular vein pass through blood sinuses around the tubules (peritubular sinuses) which also receive blood from the efferent arterioles of the cortical glomeruli, and then enter the central vein, before leaving the kidney.

[21] In Section 7.2.3, we discover how the different types of nephrons in bird kidneys enable production of hyperosmotic urine during dehydration or after a high salt intake. We learn in Section 21.1.4 how arginine vasotocin (AVT) influences the filtering population of glomeruli in birds and redistributes filtration toward the looped nephrons.

Autoregulation of GFR by mammals and birds

In mammals and birds, GFR is generally much more stable than in fish, amphibians and reptiles. In mammals, GFR is relatively independent of changes in blood pressure. Figure 7.8 illustrates the stability of GFR between mean arterial blood pressures of about 10 to 24 kPa (75 to 180 mmHg). A stable GFR is associated with stable renal blood flow, also shown in Figure 7.8, and is an example of autoregulation. In other words, the stability results from the kidney itself and occurs independently of hormonal or neural control from outside the kidney. Hence, autoregulation of mammalian kidneys continues after severing the renal nerves and occurs in isolated kidneys studied *in vitro*. In a similar way, the kidneys of birds show relatively stable GFRs when arterial blood pressures vary between 8 and 16 kPa (60 to 120 mmHg).

Autoregulation does not mean that arterial blood pressure never affects the GFR of mammals and birds. In fact, GFR can be quite variable, partly because normal mean arterial blood pressures are close to the bottom end of the autoregulatory range. Nevertheless, autoregulation does mean that GFR remains relatively stable if blood pressure increases within the autoregulatory range.

For many years, the autoregulation of renal blood flow among mammals has been considered essential for the stabilization of renal function and to prevent fluctuations in the pressure in the

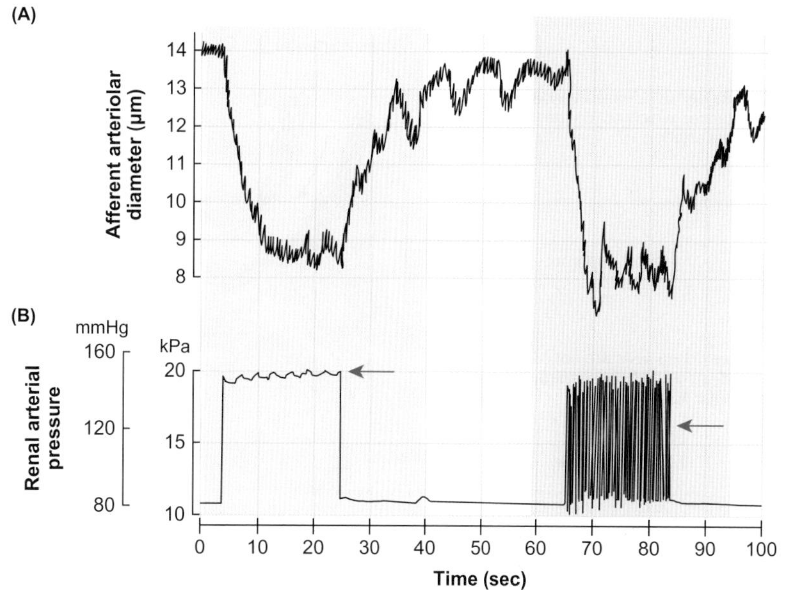

Figure 7.9 The myogenic response in kidney autoregulation

Trace (A) shows the change in afferent arteriolar diameter of an isolated rat kidney in response to the perturbations of perfusion pressure shown in (B). To isolate the myogenic response from the tubuloglomerular feedback mechanism, rats with 'hydronephrotic kidneys' that lack renal tubules were used. In the first part of the traces (blue panel), a pressure *increase* of about 8 kPa (60 mmHg), within the autoregulatory range, causes a marked and sustained *decrease* in afferent arteriolar diameter (vasoconstriction). In the second part of the trace (pink panel) a train of pulses of increased blood pressure over the same pressure range caused an equal and also sustained vasoconstriction (narrowing) of the afferent arteriole. The average pressure over the period of application (indicated by the two blue arrows) differed in the two scenarios but peak pressures were similar (almost 20 kPa; 150 mmHg). The results indicate a response to peak systolic blood pressures rather than mean pressures.

Source: Loutzenhiser R et al (2006). Renal autoregulation: new perspectives regarding the protective and regulatory roles of the underlying mechanisms. American Journal of Physiology Regulation Integrative and Comparative Physiology 290: R1153–R1167.

glomerular capillaries (P_{GC}) and hence GFR. If autoregulation did not occur, an increase in blood pressure seen during exercise, for example, would increase the filtered load to the tubules and could increase salt and volume loss. However, this may not be the only explanation. Autoregulation protects the glomerular capillaries from damage induced by high blood pressures. The glomerular capillaries are prone to physical damage at high systemic blood pressures and there is a marked association between poor renal autoregulation and hypertensive renal disease leading to protein filtration and excretion.

The dual regulation of GFR and renal blood flow when there is an increase in mean systemic blood pressure indicates that autoregulation in such circumstances results from increased pre-glomerular vascular resistance[22], which reduces blood flow to the glomeruli. The afferent arterioles are the primary site of autoregulation, with the interlobular arteries that lie upstream of the afferent arterioles being involved to some extent. Although still not fully understood, autoregulation results from at least two separate and interacting mechanisms: a rapid renal myogenic response and a slower tubuloglomerular feedback (TGF) mechanism.

The myogenic response involves vasoconstriction of the smooth muscle of the afferent arteriole when the transmural pressure (tension) on the vessel wall increases, due to stretching of the wall because of an increase in blood pressure. The myogenic response is a fast-responding system: the delay after an increase in pressure to the start of vasoconstriction by smooth muscle fibres in the wall of the afferent arteriole is only approximately 0.3 s. Figure 7.9 shows the decrease in afferent arteriolar diameter **(vasoconstriction)** that occurs when arterial pressure increases in isolated perfused rat kidneys. These and other data suggest rapid responses to *peak* arterial pressures **(systolic blood pressure)** during the cycle of cardiac events[23].

The myogenic vasoconstrictor response to increased blood pressure increases resistance to blood flow, which regulates glomerular capillary pressure (P_{GC}). Thus, both renal blood flow and GFR stay constant even if systemic arterial blood pressure increases. On the other hand, if arterial blood pressure decreases (within the autoregulatory range), automatic relaxation of the afferent arteriole occurs and again maintains both renal blood flow and P_{GC}. Thus, these responses control glomerular blood flow by changing the bore of the afferent arteriole (the resistance to blood flow). Birds seem to rely entirely on such local responses in autoregulation of GFR and blood flow, whereas mammals have further interactive mechanisms.

In mammals, slower tubuloglomerular feedback (TGF) also influences afferent arteriolar tone. The relative contribution of the myogenic response and TGF is controversial, although studies suggest that TGF accounts for 20–50 per cent of the autoregulatory response. The two mechanisms appear to act in concert to achieve precise minute-by-minute regulation of GFR and the delivery of salt (NaCl) to a special part of the renal tubule at the distal end of the thick ascending loop of Henle. This region consists of closely packed tall (columnar) cells that form the macula densa.

Figure 7.10 shows the location of the macula densa in the distal nephron, between the afferent and efferent arterioles of the glomerulus of the same nephron. An increase in salt delivery to the macula densa stimulates the release of the vasoconstrictor substances such as adenosine, which results in vasoconstriction of the afferent arteriole to the glomerulus of the same nephron and may also act on nearby nephrons.

[22] The effects of vascular resistance on blood flow are discussed in more detail in Section 14.2.1.

[23] The pumping cycle of the heart and the resulting fluctuations in arterial pressure are discussed in Section 14.1.2.

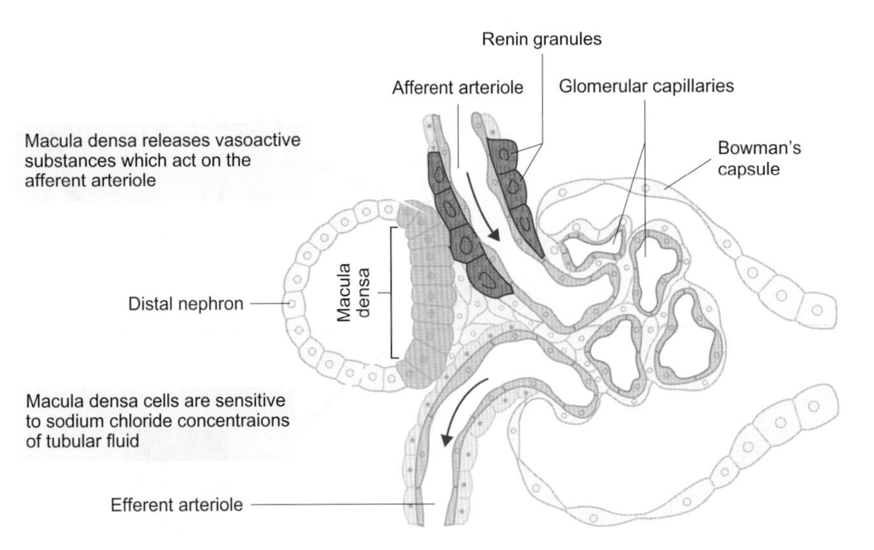

Figure 7.10 Macula densa of distal nephron

The macula densa is a region of about 15–25 closely packed tall (columnar) epithelial cells in the distal nephron, at the distal end of the ascending limb of the loop of Henle. The macula densa lies close to the glomerulus between the afferent and efferent arterioles. Red arrows show direction of blood flow. The macula densa cells detect the sodium chloride delivered to this point in the nephron and release vasoactive substances that control the vascular tone of the afferent arteriole, such that SNGFR and glomerular blood flow are autoregulated. The macula densa is also one of the controllers of renin release from granules in cells in the wall of the afferent arteriole by which high concentrations of sodium chloride inhibit renin release.

In addition, a high delivery of chloride ions to the macula densa suppresses the release of an enzyme, renin, from the granular cells of the afferent arteriole, which are illustrated in Figure 7.10, and hence reduces the generation of the hormone **angiotensin II**[24]. Angiotensin II stimulates vasoconstriction of both afferent and efferent renal arterioles, so the control of renin release by the macula densa acts as an independent mechanism to modulate TGF.

Although autoregulation of GFR usually occurs in mammals (and to a lesser extent in birds), we should bear in mind that the data in Figures 7.8 and 7.9 are for isolated perfused kidneys. In whole animals, with intact neural and hormonal systems, the autoregulatory mechanisms can be overridden, particularly when blood volume or blood pressure decreases. In such circumstances, activation of sympathetic nerves that innervate the afferent arterioles initiates vasoconstriction as part of the reflex responses to regulate blood pressure[25]—exactly the opposite response to the vasodilation that we might predict if autoregulation occurred. The vasoconstriction of afferent arterioles caused by activation of sympathetic nerves results in a reduction in GFR, which is important in helping to restore blood volume.

› *Review articles*

Aukland K (2001). Odd E. Hanssen and the Hanssen method for measurement of single-nephron glomerular filtration rate. American Journal of Physiology Renal Physiology 281: F407–F413.

Cupples WA, Braam B (2007). Assessment of renal autoregulation. American Journal of Physiology Renal Physiology 292: F1105–F1123.

Deen WM, Lazzara MJ, Myers BD (2001). Structural determinants of glomerular permeability. American Journal of Physiology Renal Physiology 281: F579–596.

Loutzenhiser R, Griffin K, Williamson G, Bidani A (2006). Renal autoregulation: new perspectives regarding the protective and regulatory roles of the underlying mechanisms. American Journal of Physiology Regulatory, Integrative and Comparative Physiology 290: R1153–R1167.

Richards AN (1938).The Croonian lecture: Processes of urine formation. Proceedings of Royal Society London B 126: 398–432.

Schweda F (2015). Salt feedback on the renin-angiotensin-aldosterone system. Pflügers Archives – European Journal of Physiology 467: 565–576.

7.2 Kidney tubules and their functions

The tubular arrangement of the nephrons of the major groups of animals varies tremendously. Figure 7.11 is a schematic illustration of the segments in the nephrons of hagfish, lampreys, teleost and elasmobranch fish, and amphibians and reptiles. The simplest nephrons occur in hagfish in which each of the large glomeruli filters fluid into a short neck segment that form right and left rows of just 30–35 nephrons. These nephrons empty into right and left ducts that are structurally similar to the **proximal**[26] **segment** of the nephrons of all other vertebrates.

The proximal segment often forms distinguishable subsections before entering the intermediate and/or **distal segment** (named because of its distance from the glomerulus). The distal segment is thought to have evolved in early vertebrates living in fresh water. All present-day vertebrates except marine teleosts have a distal segment, which is responsible for producing osmotically dilute urine, which we discuss in Section 7.2.2. Marine teleosts do not produce osmotically dilute urine and instead their kidneys conserve water[27].

Figure 7.12 illustrates the key feature of the proximal and distal segments:

- All proximal segments have a well-developed **brush border** that increases the luminal surface area for

[24] The control of renin release is complex and affected by several factors; we discuss this complexity and the renin-angiotensin system in Section 21.1.1.

[25] We discuss the barostatic (baroreceptor) reflex of vertebrates more extensively in Section 22.4.2. Figure 3.24 outlines the mammalian sympathetic nervous system.

[26] Proximal is derived from the Latin proximus, meaning nearest; the proximal segment is the segment closest to the glomerulus.

[27] Figure 5.3 illustrates the osmoregulatory processes of marine teleosts that continually lose water by osmosis across their permeable external surfaces due to the higher osmolality of the surrounding seawater compared to their extracellular fluids.

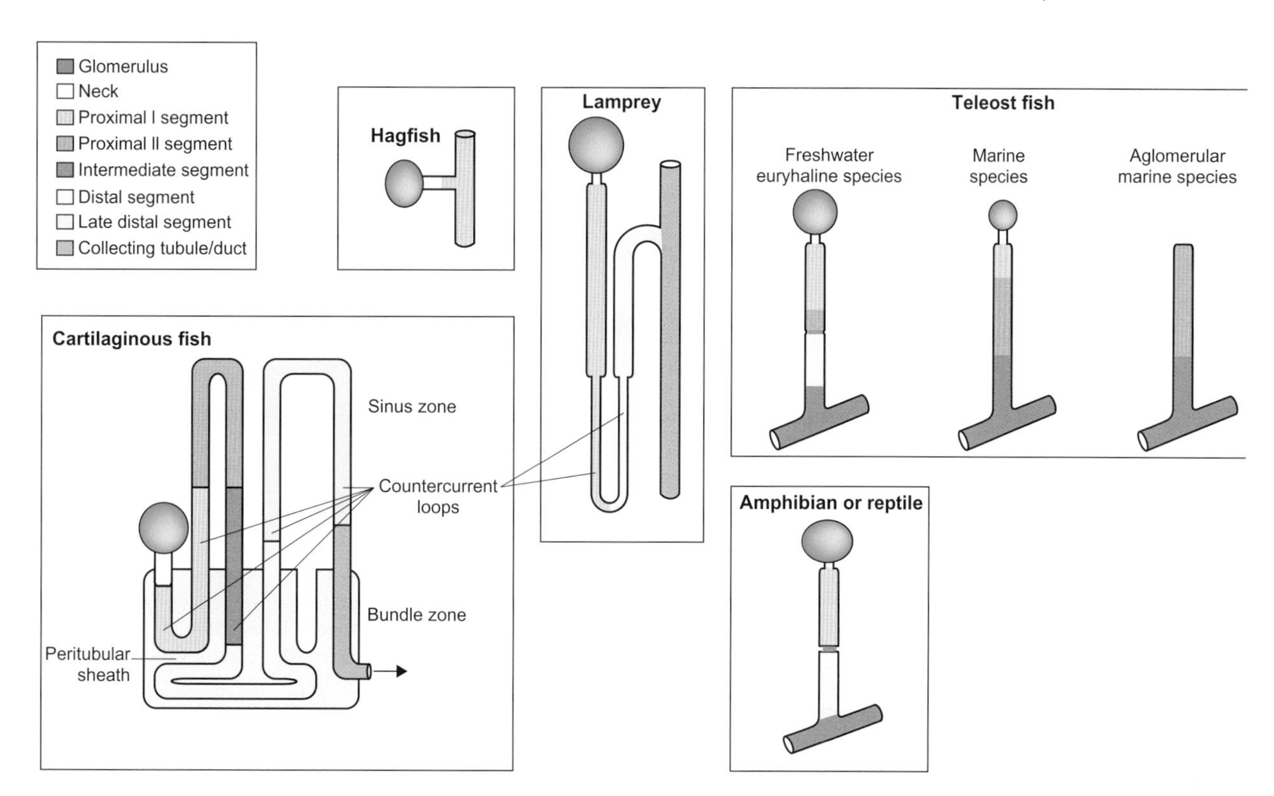

Figure 7.11 Schematic diagrams of nephron segmentation of hagfish, lamprey, cartilaginous fish, teleost fish, amphibians and reptiles

Marine teleosts lack a distal segment in their nephrons. Countercurrent loops occur in the nephrons of lampreys and cartilaginous fish (elasmobranchs and holocephalans) that have the most complex nephrons in which several countercurrent loops are enclosed by a peritubular sheath, in a bundle zone. The terminology for nephron segments of cartilaginous fish varies between species and researchers.

The diagram shown is derived from Dantzler WH. Comparative Physiology of the Vertebrate Kidney. Springer, 1989 (for hagfish, teleosts, amphibians); Hyodo S et al (2014) American Journal of Physiology Regulatory, Integrative and Comparative Physiology 307: R1381-1394 (for cartilaginous fish)

reabsorption of solutes and water. In mammals, the proximal segment is tallest in the first subsection of the proximal tubule, which Figure 7.12A illustrates. The distal segment lacks the brush border, as shown in Figure 7.12B.

- Infoldings of the basal membrane of the tubular epithelium are associated with many mitochondria that fuel active transport processes. These infoldings are most pronounced in the distal segment where they are associated with many large mitochondria.

In some animal groups, new segments appear between the proximal and distal segments, such as the parallel system of loops in the nephron of cartilaginous fish. A single countercurrent loop occurs in the nephrons of lampreys, which is included in Figure 7.11, and in some nephrons in bird kidneys (see Figure 7.6). The countercurrent loops that characterize mammalian nephrons, which Figure 7.5 illustrates, are crucial for producing urine that is hyperosmotic to plasma, which we explore in Section 7.2.3.

The renal tubules are the site of a vast array of reabsorptive and secretory processes that determine the volume of urine excreted and its composition. It is impossible for us to consider how the kidney deals with every one of the solutes in urine in a single chapter. In the subsections that follow, however, we focus on some of the major processes in the renal tubules that regulate extracellular fluid volume and osmolality by adjusting the volume of urine excreted, and the concentrations of

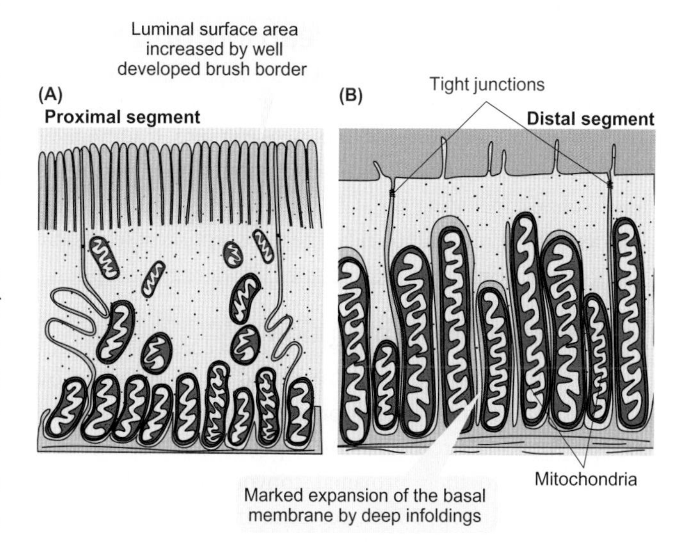

Figure 7.12 Diagrams of the ultrastructure of a single cell from proximal segment (first section) and distal segment of the mammalian renal tubule

Source: Forster R P (1973). Comparative vertebrate physiology and renal concepts. In: Handbook of Physiology, Section. 8: Renal Physiology. Chapter 8 pp 161-184. (Eds. Orloff J, Berliner RW). American Physiological Society, Bethesda.

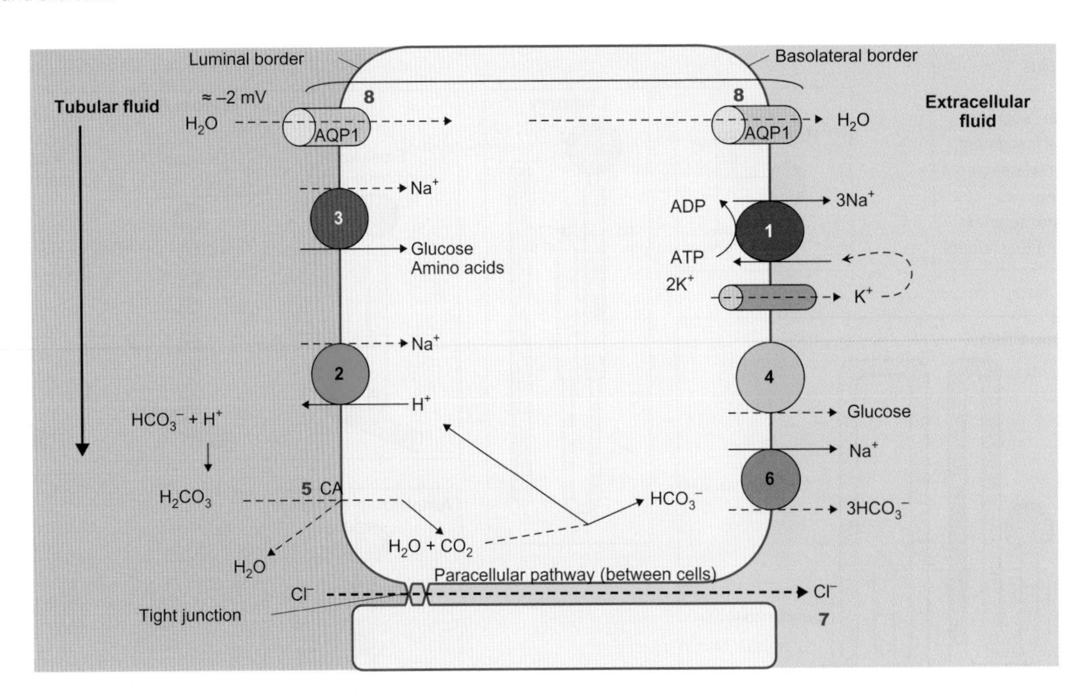

Figure 7.13 Model of main solute and water transport processes of early proximal convoluted segment of the mammalian nephron

1 Na^+, K^+-ATPase in the basolateral membranes results in a transepithelial potential of about −2 mV (lumen negative) and drives luminal Na^+ uptake by two mechanisms: Na^+/H^+ antiport exchange on the luminal border of the tubular epithelium **2**, and **3** cotransport of Na^+ with glucose or amino acids on the luminal border of the tubular epithelium. **4** Glucose passes out of the basolateral borders of the cells via a facilitated diffusion transporter. **5** The Na^+/H^+ antiport exchange on the luminal border (**2**) results in absorption of virtually all the filtered bicarbonate. This process involves carbonic anhydrase (CA) in the luminal brush border that results in CO_2 influx and formation of cellular bicarbonate that leaves the cells via **6**: basolateral Na^+/HCO_3^- cotransporters. **7** The proximal tubule allows passive chloride passage via the paracellular pathways driven by the electrochemical difference. **8** Aquaporin (AQP1) on both luminal and basolateral borders allow continual passive water reabsorption by osmosis.

sodium (Na^+) and chloride (Cl^-) in the final urine. We also explore how the renal tubules of selected animal groups handle potassium (K^+), magnesium (Mg^{2+}) and sulfate (SO_4^{2-})[28], and how the renal systems of mammals handle urea.

7.2.1 Solute and water reabsorption by the proximal tubule

The proximal segments of the nephron are traditionally thought to be responsible for reabsorption of the bulk of filtered solutes, including Na^+, Cl^-, Ca^{2+}, K^+ and organic solutes such as glucose and amino acids. The reabsorption of solutes by the proximal segment is accompanied by water reabsorption.

Sodium chloride reabsorption

The percentage of the filtered Na^+ and Cl^- reabsorbed by the proximal segment varies considerably among vertebrate groups. Mammals reabsorb 60–70 per cent of the Na^+ and Cl^- filtered into their proximal convoluted tubules while birds reabsorb about 50–60 per cent. In contrast, amphibians and most reptiles reabsorb less Na^+ and Cl^- in the proximal segments of their nephrons, although still about 20–45 per cent of the filtered quantity. We do not know the exact amounts of Na^+ or Cl^- reabsorbed by the proximal tubule of most elasmobranchs or teleosts.

Lampreys only reabsorb 10 per cent of the filtered Na^+ and Cl^- in the proximal segment of the nephron, but this low proportion does not appear to reflect a particularly low *rate* of reabsorption. River lampreys, amphibians, most reptiles and birds reabsorb Na^+ at similar rates: about 30 pmol min^{-1} mm^{-1} tubule. However, reabsorption of Na^+ by mammalian proximal tubules occurs six times faster, which accounts for their reabsorption of a large proportion of the filtered load of Na^+ and Cl^- by the end of the proximal tubule. In mammals, Na^+ reabsorption occurs most rapidly in the earliest part of the proximal tubule and diminishes as the number of mitochondria and the membrane surface area (luminal and basolateral) decrease along the length of the proximal tubule.

So how does the reabsorption of Na^+ and Cl^- occur? Several interlinked processes occur, many of which are summarized in the schematic for the early proximal segment of mammals shown in Figure 7.13. This diagram shows several aspects of the reabsorptive process that apply more generally among vertebrates:

- Sodium pumps (**Na^+, K^+-ATPase**)[29] in the basolateral membranes (step 1 in Figure 7.13) drive Na^+ uptake from tubular fluid.

[28] We also explore the renal handling of calcium (Ca^{2+}) and its hormonal regulation in Section 21.2.

[29] Figure 4.13 in Section 4.2.2 outlines the functioning of the Na^+, K^+-ATPase (the sodium pump).

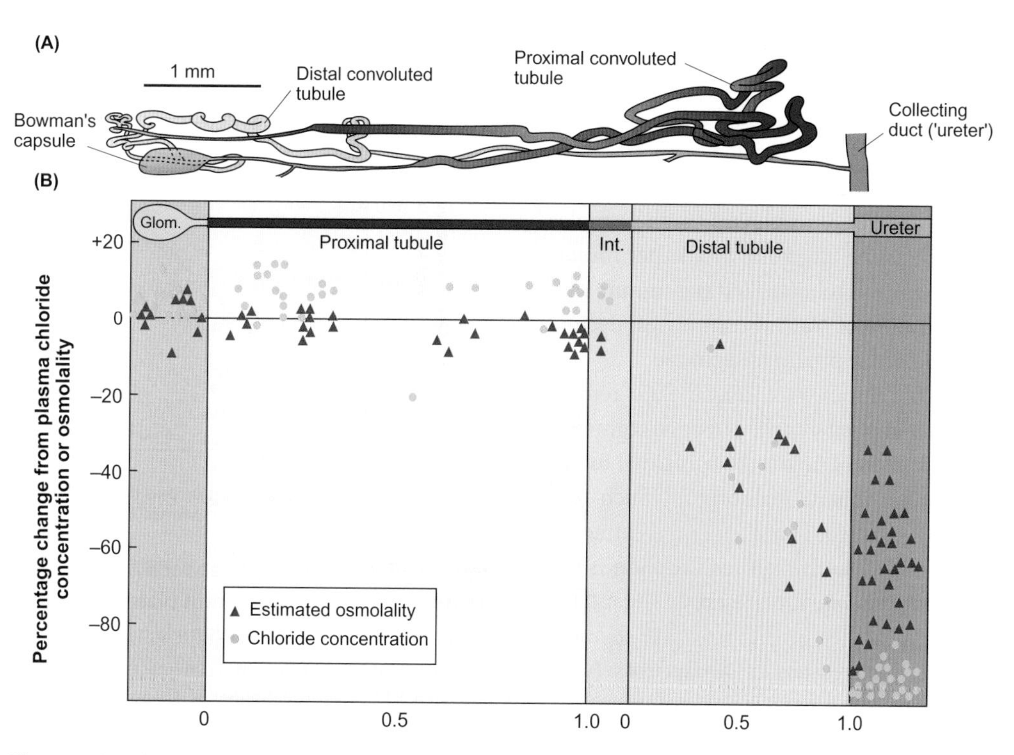

Figure 7.14 Changes in tubular fluid composition along the nephrons of mudpuppies (*Necturus maculosus*)

(A) Layout of an entire nephron as it appears within the kidney. The large nephrons with a systematic arrangement enable collection of fluid from known positions which can later be measured off against the length off the proximal and distal segments of the nephron. (B) Relative chloride concentrations (green circles) and estimated total osmotic concentration of the tubular fluid (red triangles) compared to plasma. A value of 0 indicates tubular fluid equal in total chloride concentration or osmolality to plasma. Samples from the glomerulus (Glom.) are close to 0. In the proximal and intermediate (Int.) segments tubular fluid remains close in composition to plasma. In the distal tubule NaCl reabsorption results in a reduction in the osmotic concentration and chloride concentration of the tubular fluid.

Sources: A: Richards AN (1938). The Croonian lecture: Processes of urine formation. Proceedings of Royal Society, London B 126: 398-432. B: Walker AM et al (1936). The total molecular concentration and the chloride concentration of fluid from different segments of the renal tubule of Amphibia. American Journal of Physiology 118: 121-129.

- Na^+ uptake from tubular fluid on the luminal border of the proximal tubular epithelium occurs mostly via Na^+/H^+ exchange (transporter 2 in Figure 7.13) and to a lesser extent via Na^+ cotransport with glucose and amino acids (transporter 3 in Figure 7.13).
- The Na^+/H^+ exchange effectively results in reabsorption of most of the filtered bicarbonate[30], as illustrated in Figure 7.13, since H^+ combines with bicarbonate ions (HCO_3^-) in the tubular fluid and forms CO_2 that enters the cell. In mammalian kidneys, Na^+/H^+ exchange in proximal tubules is subject to neural and hormonal control, by sympathetic innervation of the kidneys, and the hormone angiotensin II[31], respectively.

The pathways for Cl^- reabsorption by the proximal tubule vary between species and among subsections of the proximal tubule. In the early proximal segment of mammalian nephrons (the focus of Figure 7.13) the transepithelial electrochemical difference[32] allows passive transport of Cl^- along the paracellular pathway between the epithelial cells. The initial step is passage through the **claudin** proteins of '**tight junctions**'[33] at the luminal border of the epithelium. A cellular route for Cl^- reabsorption may also occur in some animal groups, such as amphibians, in which Cl^-/anion exchange proteins occur in the luminal membranes. The exchange process allows Cl^- to pass into the cells, after which they pass through ion channels in the basolateral membranes.

We might expect that reabsorption of NaCl by the proximal segment would lower the osmotic concentration[34] of the tubular fluid. However, micropuncture and microperfusion experiments (using the techniques outlined in Box 7.1) show little or no change in the osmolality of the tubular fluid as it passes along the proximal tubule. As an example, Figure 7.14 shows some data obtained from the micropuncture studies of nephrons of amphibian mudpuppies. Note in Figure 7.14 that the osmotic concentration of the tubular fluid remains similar to that of the initial filtrate (i.e. similar to plasma osmolality) along the entire proximal tubule.

The results in Figure 7.14 tell us that the proximal tubule (segment) is highly water permeable and allows water to

[30] Bicarbonate reabsorption is important for acid–base balance, which we learn about in Section 13.3.

[31] Figure 21.3 outlines the stages in forming angiotensin II by the renin-angiotensin system.

[32] Electrochemical differences are explained in Section 3.1.3.

[33] Section 4.2.1 discusses tight junctions and claudins in some detail.

[34] Osmotic concentrations (usually measured as osmolality in mOsm kg^{-1}) are explained in Section 4.1.1.

follow the reabsorption of osmolytes (Na^+, Cl^- and other solutes). While the *concentration* of Na^+ or Cl^- (mmol L^{-1}) and the osmotic concentration of fluid (mOsm kg^{-1}) in the proximal tubule remain unchanged, reabsorption of water reduces the volume of the fluid as it moves through the proximal tubule. Hence, the total *amount* of osmolytes (milli Osmoles) or NaCl (milli moles) passing along the tubule per minute, as expressed by the volume of tubular fluid per minute multiplied by the concentration of NaCl or osmolytes, declines dramatically by the end of the proximal segment.

Since the osmotic concentration of tubular fluid remains stable and similar to that of plasma in the proximal segment of all vertebrates, the absorption of fluid by the proximal tubules is considered as isosmotic volume reabsorption. Such reabsorption cannot alter the osmolality of the plasma/extracellular fluids but isosmotic NaCl and water reabsorption is very important in maintaining the *volume* of the extracellular fluids.

The high water permeability of the proximal segments of nephrons is due to the presence of **aquaporins** (water channels)[35] in the luminal and basolateral membranes. In Figure 7.13, the passage of water into the cells occurs through AQP1 in the luminal membrane. The water permeability of the proximal segment is sufficiently high that differences in osmotic concentrations in the fluid-filled intercellular spaces of as little as 2–5 mOsm kg^{-1} draw water from the cells via AQP1 in the basolateral membranes.

The importance of AQP1 in water reabsorption by the proximal tubule is apparent in *AQP1* gene 'knockout' mice that are deficient in AQP1. In these mice, the proximal segments reabsorb about 20 per cent less water than those of mice with a functioning *AQP1* gene. An additional route for water reabsorption may also exist, as revealed by other gene knockout studies: 25 per cent reduction in water reabsorption by the proximal segments occurs in mice deficient in claudin-2, which occurs in tight junctions between the epithelial cells of the proximal tubule. These studies support the idea that water reabsorption in some species can occur both through the cells (a transcellular route) and between the cells (a paracellular route)[36].

Saturable carriers reabsorb glucose

Reabsorption of all or most of the filtered glucose by the kidney tubules is a fundamental process in all vertebrates, which is mediated primarily by high capacity carrier proteins[37] in the luminal border of the proximal tubular epithelium. In the mammalian kidney, there are two subtypes of **sodium-glucose-transporters** (referred to as SGLT: S sodium, G glucose, L linked, T transporters). High-capacity SGLT2 occurs in the early proximal tubule segments of the mammalian kidney,

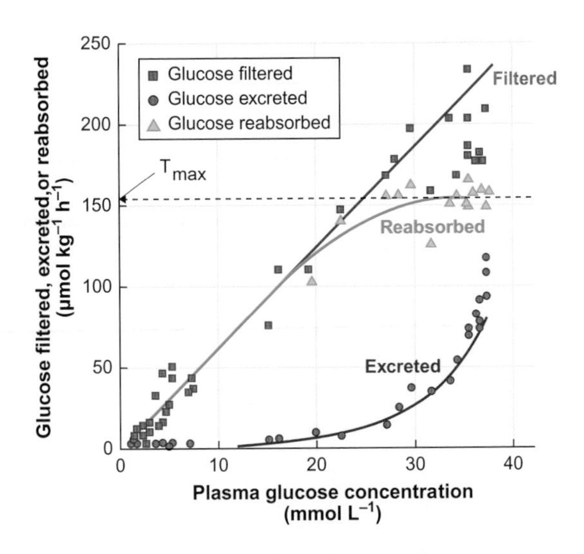

Figure 7.15 Relationships between filtered, excreted and reabsorbed glucose at different plasma concentrations in the rainbow trout (*Oncorhynchus mykiss*)

At normal plasma concentrations, all filtered glucose is reabsorbed and none is excreted. If plasma concentrations of glucose are increased, the carriers begin to saturate and glucose begins 'spilling over' into the urine and is excreted. Excreted glucose rises further if plasma concentrations of glucose increase further. Reabsorption reaches a plateau of maximal tubular reabsorption (T_{max}) when all the glucose carriers are occupied.

Source: Bucking C, Wood CM (2005). Renal regulation of plasma glucose in the freshwater rainbow trout. Journal of Experimental Biology 208: 2731-2739.

while lower capacity (but high affinity) SGLT1 occurs in later segments of the proximal tubule and absorbs glucose at low concentrations. Only a few studies have identified SGLTs in non-mammalian vertebrates for example, in little skates (*Leucoraja erinacea*), an SGLT1 has been identified which extends through large parts of the nephron and into the collecting duct.

The general principle in the operation of SGLTs is that active transport of sodium by Na^+ K^+-ATPase on the basolateral border (shown in Figure 7.13) drives the cotransport of glucose with Na^+ from the tubular fluid, across the luminal border[38]. The rising concentration of glucose in the epithelial cells leads to diffusion across the basolateral border (step 4 in Figure 7.13). The paracellular pathway (between the cells) probably allows a small amount of glucose to leak back into the tubular fluid, but reabsorption usually occurs later when the fluid flows through the distal nephron.

In healthy kidneys, the carrier capacity of the nephron is usually sufficient to reabsorb all filtered glucose, although the carriers can become saturated if the plasma concentrations of glucose increase sufficiently. Figure 7.15 illustrates the effects of increasing plasma concentrations of glucose:

- Initially, reabsorption keeps pace with the increasing amount of filtered glucose.

[35] We discuss the importance of aquaporins in determining epithelial water permeability more generally in Section 4.2.3.

[36] Figure 4.8 depicts these alternative routes for transepithelial transport.

[37] The concept of passive (non-energy dependent) transport by membrane carrier proteins in discussed in Section 3.2.4.

[38] This process also occurs in the absorption of monosaccharides across the gut epithelium, as illustrated in Figure 2.15.

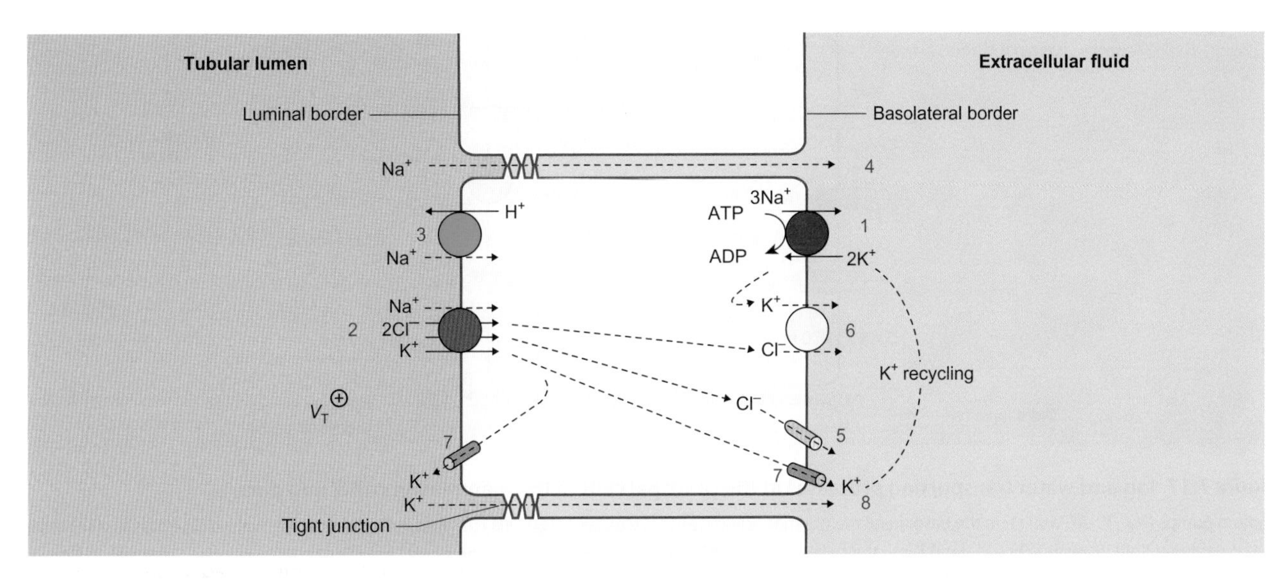

Figure 7.16 Model of the ion transport processes in the early distal tubules of non-mammalian vertebrates and the thick ascending limb of the loop of Henle of long-looped avian nephrons and mammalian nephrons

Sodium pumps (Na^+, K^+-ATPase) (1) occur in the basolateral membranes. These pumps reduce intracellular Na^+ concentrations driving influx on the luminal border, via a cotransporter (2) which transports each Na^+ with one K^+ and $2Cl^-$, and via Na^+ /H^+ exchangers (3). In mammals, transepithelial Na^+ reabsorption also occurs along the paracellular pathways between the cells (4), driven by the lumen positive transepithelial voltage (V_T^+). Chloride ions leave the cells through basolateral channels (5) and are also transported across the basolateral borders via carriers in which there is coupled transport of Cl^- and K^+ (6). Potassium ions exit from the cells via channels in the luminal and basolateral membranes (7) and also pass via 'tight junctions' and along the paracellular pathways (between the cells) (8), which allows continual recycling of K^+.

Source: adapted from Dantzler WH (2003). Regulation of renal proximal and distal tubule transport: sodium, chloride and organic anions. Comparative Biochemistry and Physiology Part A: Molecular & Integrative Physiology 136: 453-478.

- Reabsorption begins to saturate when the filtered amount of glucose begins to exceed the reabsorptive capacity of the kidney.
- If plasma glucose concentrations increase further, the filtered amount reaches the maximum reabsorptive capacity of the carriers, which is known as the **tubular maximum** (T_{max}) for glucose. Such saturation of glucose reabsorption occurs in unmanaged diabetes mellitus, when plasma concentrations of glucose exceed the normal range. Thus, glucose in the urine is an indicator of diabetes mellitus in mammals.

7.2.2 Distal nephron production of osmotically dilute (hypo-osmotic) urine

Animals with an excess of water in their body fluids, such as amphibians living in freshwater ponds, must excrete this water in order to regulate their body fluid volume and composition. They do this by reabsorbing solutes (mainly sodium and chloride), without the associated water, from tubular fluid passing through the distal convoluted tubule and collecting ducts. This process creates urine with a lower osmolality than plasma (i.e. is hypo-osmotic to plasma). Fish, amphibians and reptiles can generate urine with an osmolality that is 10 per cent of their plasma osmolality (i.e. with a urine/plasma osmolality ratio of 0.1). In a similar way, mammals can generate urine with a urine/plasma osmolality ratio of around 0.3.

Micropuncture and microperfusion studies of the distal nephron provide detailed information on the mechanisms leading to urinary dilution in some species, such as the amphibian mudpuppy. We found earlier, in Figure 7.14, that tubular fluid is isosmotic to plasma along the entire proximal tubule of mudpuppies. Notice in Figure 7.14 that once the tubular fluid enters the distal tubule its osmolality decreases and continues to do so along the entire length of the distal nephron.

In mammals, solute reabsorption without the associated water by the distal nephron retrieves about 25 per cent of filtered Na^+. Figure 7.16 illustrates the processes involved in transcellular reabsorption of Na^+ and Cl^- in the ascending limb of the loop of Henle. Similar processes occur in the early distal segment of the nephrons of many vertebrates. A key feature is the basolateral Na^+, K^+-ATPase (sodium pumps) that occur in the prominent deep infoldings of the basolateral membranes (shown earlier in Figure 7.12B). The sodium pumps drive ion uptake from the tubular fluid through Na^+/H^+ exchange proteins and Na^+, K^+, $2Cl^-$ cotransporters. About half of the sodium reabsorption by the distal nephron of mammals occurs in this transcellular manner and about half occurs along the paracellular pathway. We do not know whether the paracellular pathways in the distal nephron operate in NaCl reabsorption by other vertebrates, or how important this route may be.

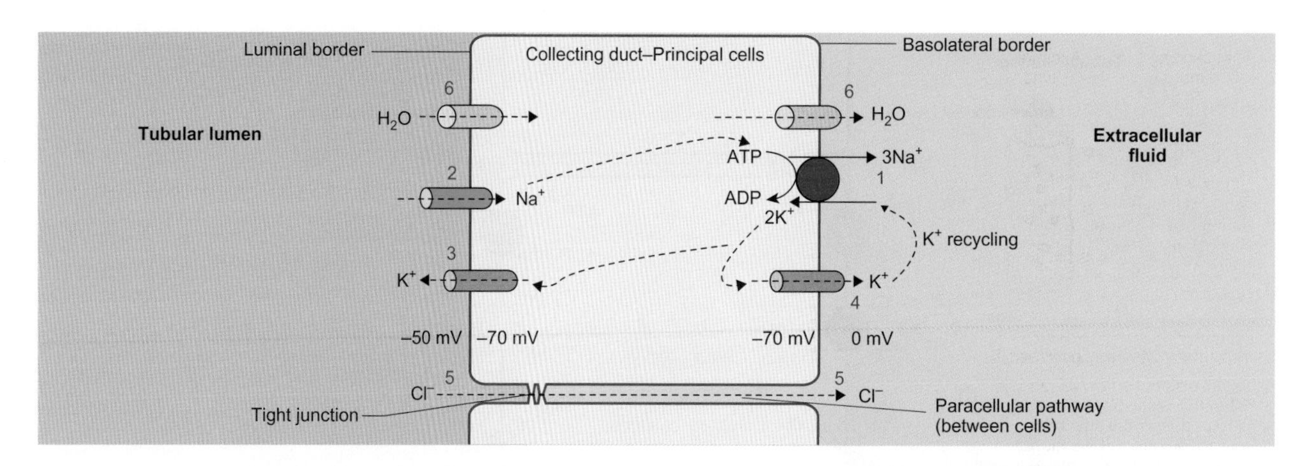

Figure 7.17 Ion and water transporting processes in the principal cells of the mammalian collecting duct

Sodium pumps (Na^+, K^+-ATPase) (1) in the basolateral membranes drive the passive uptake of Na^+ via channels in the luminal membrane (2). Potassium secretion into the tubular fluid involves active K^+ uptake into the cells, via Na^+, K^+-ATPase in the basolateral membranes (1) followed by diffusion via K^+ channels in the luminal membranes (3). Some backflux of K^+ occurs via ion channels in the basolateral membranes (4). Absorption of Cl^- ions occurs passively via 'tight junctions' and along the paracellular pathways, between the cells, due to the electrochemical difference across the cells (5). Water permeability of the collecting duct epithelium is determined by the hormonal regulation of aquaporins in the luminal and basolateral membranes (6), which we discuss further in Section 21.1.4.

In mammals, the passage of tubular fluid through later parts of the distal nephron—particularly the collecting duct that receives fluid from several nephrons—is the most significant stage in the potential production of hypo-osmotic urine, by solute reabsorption *without* water as a result of the hormonal control of water permeability of the tubular epithelium[39]. Notice in Figure 7.17, that Na^+, K^+-ATPase in the basolateral membrane once again establishes the electrochemical difference driving Na^+ reabsorption from the tubular fluid. In the principal cells (the dominant large cells) of the collecting duct, Na^+ uptake across the luminal border of the cells is via Na^+ channels rather than Na^+/H^+ exchangers or Na^+, K^+, $2Cl^-$ cotransport mechanisms that occur in earlier parts of the distal nephron. Although collecting ducts reabsorb only about 1 per cent of the filtered Na^+ this process and its hormonal regulation[40] is essential in balancing Na^+ excretion according to the Na^+ intake in the diet. At very low rates of Na^+ intake, Na^+ concentrations of urine can reach extremely low levels (<1 mmol L^{-1}).

Production of dilute urine by fish nephrons

In comparison to mammals and amphibians, we have a poor understanding of how other vertebrate groups produce dilute urine, although recent studies have begun to identify Na^+ transporter proteins in the distal convoluted tubule and collecting ducts of fish. These transporters include Na^+, K^+, $2Cl^-$ cotransporters in the distal tubule and Na^+, Cl^- cotransporters in the collecting duct of freshwater fish, but not sodium channels, which are so important in mammalian collecting ducts.

Studies of the expression of Na^+ transporter proteins in the euryhaline obscure puffer fish (*Takifugu obscurus*) give some information about their relative importance in producing dilute urine. Transfer of this pufferfish from fresh water to seawater (where they no longer need to produce dilute urine to maintain salt and water balance[41]) results in a 2.7-fold decrease in the expression of Na^+, Cl^- cotransporters, which become internalized in the cytoplasm, where they are likely to be non-functional. This finding suggests that Na^+, Cl^- cotransporters are involved in producing the osmotically dilute urine of freshwater fish but become less important once the pufferfish enters seawater. In fish (and amphibians) living in fresh water, ion absorption continues in the urinary bladder to produce urine that is more osmotically dilute than the urine generated by the distal tubule and collecting ducts.

7.2.3 Hyperosmotic urine production by mammals and birds

The ability of mammals and birds to excrete urine that is hyperosmotic to plasma, i.e. with a higher osmolality than plasma, is vital to life on land particularly when water is in short supply. When the intake of water is low and the osmolality of the blood plasma increases, mammals and birds excrete urine that is hyperosmotic to their extracellular fluids. The maximum osmotic concentration of the urine generated by individual species correlates with the aridity

[39] We discuss the hormonal control of water permeability of the distal nephron in Section 21.1.4.

[40] Aldosterone stimulates salt reabsorption by the distal nephron, which we discuss in more detail Section 21.1.1.

[41] A comparison of Figures 5.17 and 5.3 illustrates the changing demands for osmoregulation by teleosts when they live in fresh water and seawater, respectively.

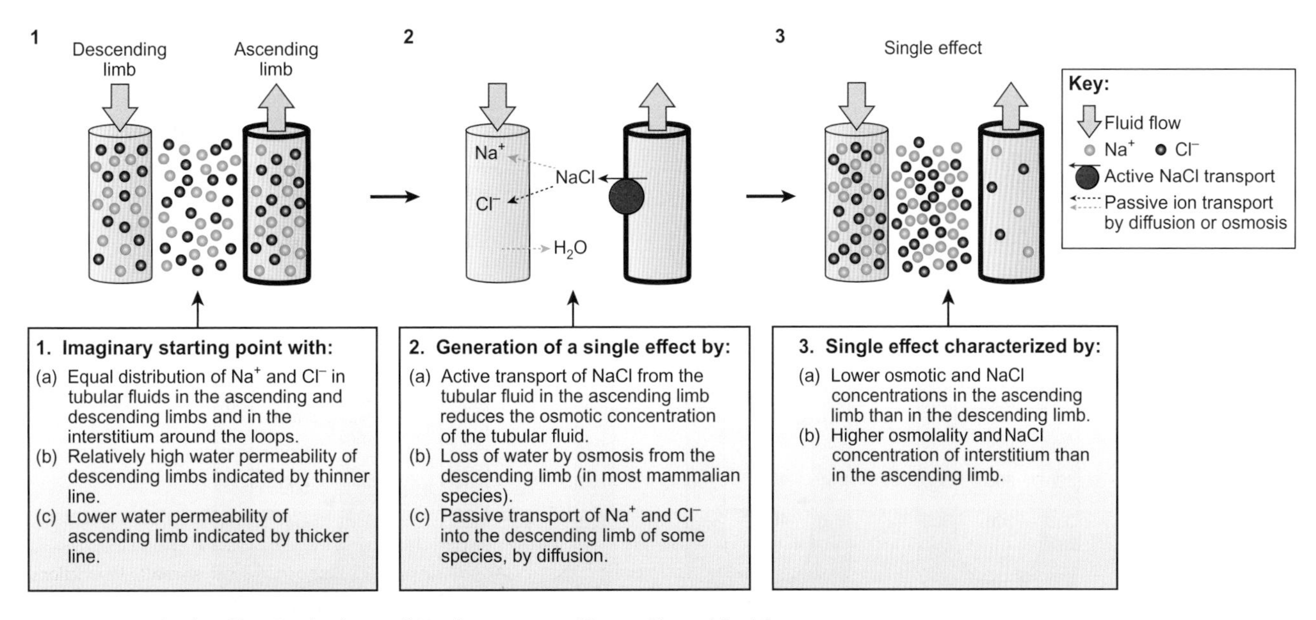

Figure 7.18 Single effect in the loop of Henle generated by sodium chloride pumps

of their habitat, which is illustrated by the examples back in Table 6.3.

To produce hyperosmotic urine, an animal must reabsorb water from the kidney tubules without the organic solutes and ions (osmolytes) that were contained in the fluid filtered by the glomeruli. Hence, the absorbed water is **osmotically free water**. The ability of mammals to produce hyperosmotic urine has long been considered to result from the striking organization of the kidney tubules and vasculature of the mammalian kidney.

Countercurrent multiplication by the mammalian loop of Henle

The 1950s saw a major conceptual advance in thinking when Werner Kuhn put forward a **countercurrent multiplication** model to explain the production of hyperosmotic urine by mammals. This model proposes that concentrated urine results from the multiplication of the 'single effect' between the descending and ascending limbs seen at any level of the countercurrent loop of Henle.

Figure 7.18 illustrates the single effect in which active transport[42] carries Na^+ and Cl^- out of the ascending limb into the interstitium between the ascending and descending limbs. In this model, the relatively low water permeability of the ascending limb inhibits water movement from the tubular fluid. Hence, pumping of ions out of the ascending limb reduces the NaCl concentration (and osmotic concentration) of tubular fluid in the ascending limb. On the other hand, the descending limb of the loop has a relatively high-water permeability, which allows water to pass from the descending limb into the interstitium, by osmosis. Taken together, these processes create a difference in the concentration of osmolytes and NaCl between the tubular fluid in the ascending and descending limbs of the loop of Henle, with lower concentrations in the ascending limb than in the descending limb.

The difference in concentration *across the loop* is called a single effect to distinguish it from the multiplication effect *along the loop* that results in a greater end-to-end difference in osmotic concentration. The multiplication effect results from the flow of fluid through the countercurrent limbs of the loop of Henle, which Figure 7.19 illustrates as a series of steps. In this way, countercurrent multiplication creates increasingly concentrated interstitial fluids between the cortex and the medulla, through which the collecting ducts run, in parallel to the loops of Henle, as shown in Figure 7.20.

The fluid leaving the ascending limb of a nephron to enter the distal convoluted tubule is either isosmotic or most usually hypo-osmotic to plasma (as in Figure 7.20), but is never hyperosmotic. This apparent paradox is solved by what happens more distally in the kidney tubules. Production of hyperosmotic urine (or more hypo-osmotic urine) occurs in the distal convoluted tubule and collecting duct so that animals either:

- excrete urine that is more dilute than body fluids, after salt reabsorption by the distal convoluted tubule and collecting duct without water following the salt reabsorption (as outlined in Section 7.2.2);
- excrete hyperosmotic urine due to the multiple actions of **arginine vasopressin (AVP)** on the distal nephron, illustrated by Figure 7.20. AVP is also called **antidiuretic hormone (ADH)** due to its antidiuretic actions; AVP increases the water permeability of the late distal tubule

[42] Active pumping of ions in countercurrent multipliers distinguishes them from the passive processes of countercurrent heat exchanges, which we discuss in Section 3.3.1.

[43] We explore the control of circulating concentrations of AVP and consider its actions in more detail in Section 21.1.4.

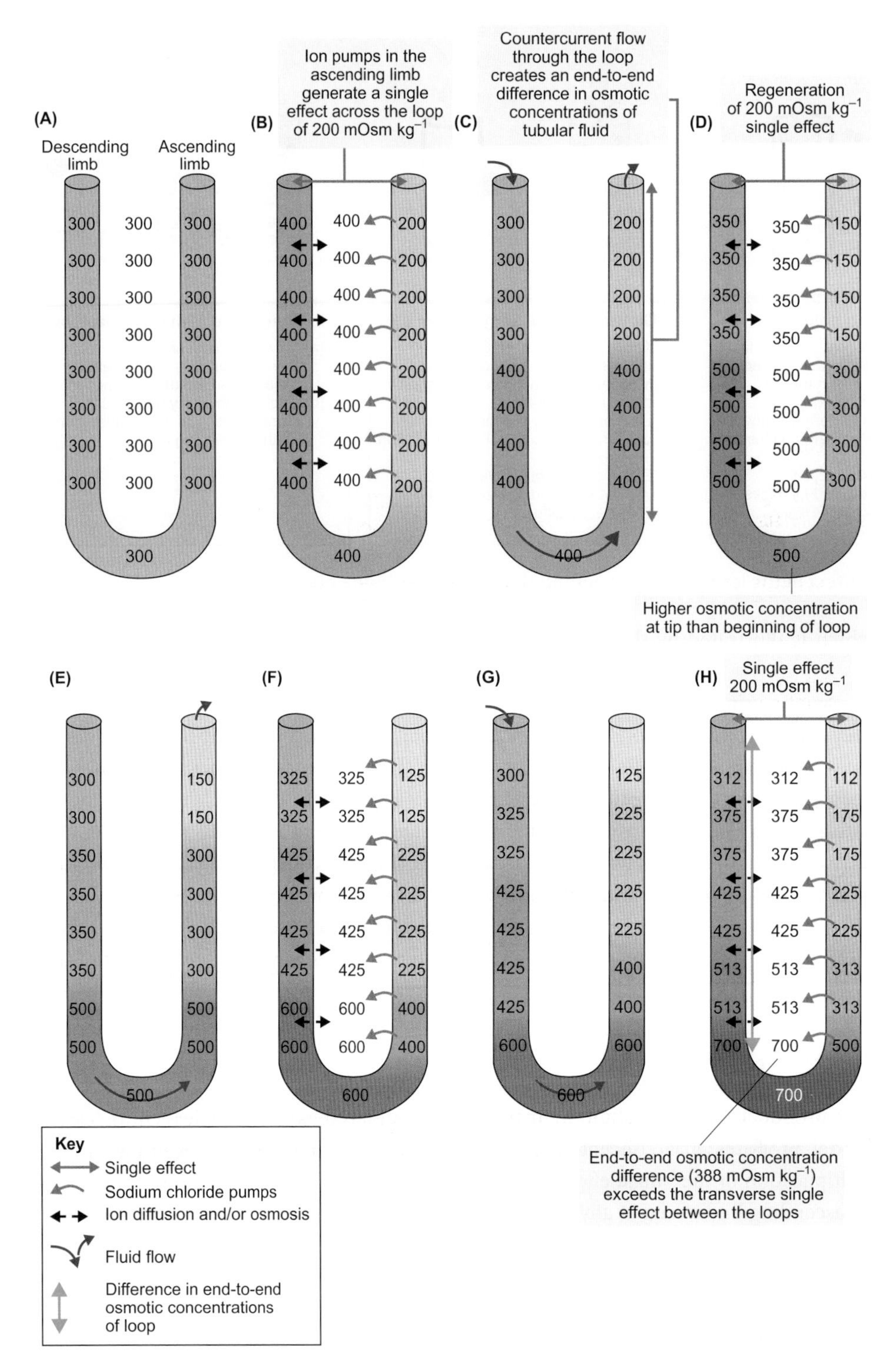

Figure 7.19 Conceptual illustration of countercurrent multiplication model for loop of Henle

Countercurrent multiplication shown as a series of alternating steps (A to H). Numbers are osmotic concentrations (mOsm kg^{-1}) of the fluid along descending and ascending limbs, and in the interstitium around the loops. Shaded colouring indicates relative osmotic concentrations at each stage. The single effect between loops is 200 mOsm kg^{-1}. Countercurrent flow multiplies the end-to-end difference in cortex to medulla osmotic concentrations. The steps shown are: (A) The model starts with both limbs full of fluid at 300 mOsm kg^{-1}; fluid leaving the proximal segment and entering the loop is isosmotic to plasma. (B) Ion pumping by the ascending limb carries NaCl into the interstitium. Tubular fluid in the descending limb comes into osmotic balance with the interstitium, which leads to an identical 'single effect' between the two limbs (200 mOsm kg^{-1}) at all levels of the loop. (C) Countercurrent flow of fluid from the descending limb into the ascending limb increases the end-to-end difference in osmotic concentrations. (D) Pumps recreate the single effect. (E) to (G) continues the alternate steps. The amount of fluid movement through the loop is reduced in (E) compared to (C) (which could occur if water moves from the descending limb, by osmosis), and further decreases in (G) compared to (E). In (F) to (H) the end-to-end osmotic (top to loop tip) concentration difference exceeds the transverse single effect between the loops and is greatest in (H, green arrow).

Source: adapted from Pitts RF. Physiology of the Kidney and Body Fluids. 2nd Edition. Chicago: Year Book Medical Publishers Inc., 1972.

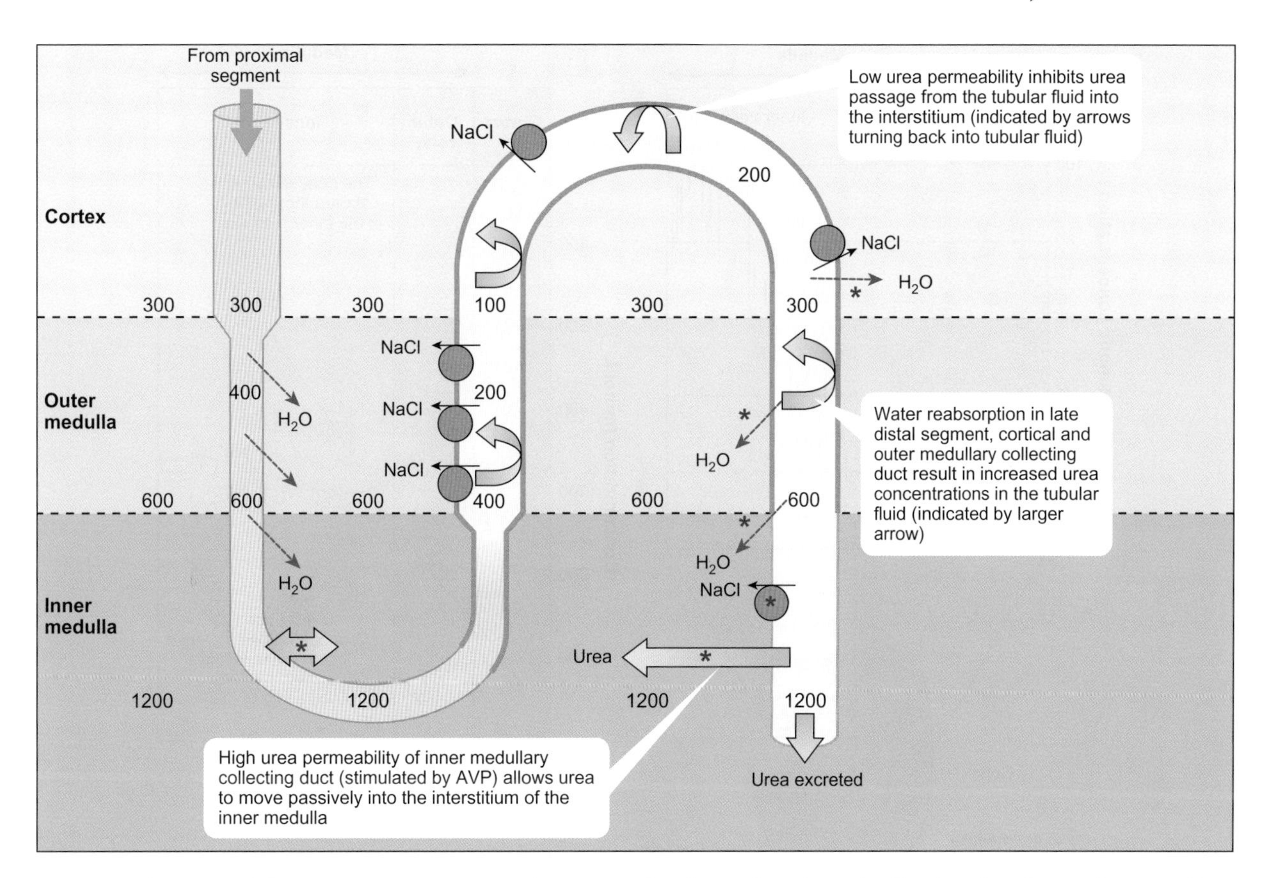

Figure 7.20 Schematic of the transport processes in the mammalian loop of Henle and collecting duct for production of low volumes of hyperosmotic urine

The diagram depicts a long loop of Henle with its tip in the inner medulla. Numbers indicate osmotic concentrations (mOsm kg^{-1}) in tubular fluid and kidney interstitial fluid (between the tubules). Green shading indicates the gradually increasing osmotic concentration of the interstitium between the cortex and the medulla. The blue dashed lines indicate passive osmotic efflux of water from the descending limb of the loop of Henle; the ascending limb has a low water permeability. Green arrows indicate urea movements. Tubular segments with low urea permeability are outlined with a thickened orange line. AVP stimulates the processes shown by red asterisks, which lead to production of hyperosmotic urine: (i) stimulation of sodium chloride reabsorption in the collecting duct; (ii) increased tubular water permeability of the distal nephron which increases water reabsorption; (iii) increased urea permeability of the inner medullary collecting duct; (iv) increased urea permeability in thin descending limb of the loop of Henle.

and cortical collecting ducts, which results in water being reabsorbed from the collecting duct during its passage through the increasingly concentrated interstitium. At high concentrations of AVP, tubular fluid can reach osmotic equilibrium with the interstitial fluids[43].

The first experimental evidence for countercurrent multiplication by the loops of Henle to create an increasingly concentrated interstitium between the cortex and medulla came from studies of rat kidneys. The kidneys were first frozen *in situ* and then cut into a series of slices through the cortex and then the medulla on a cold microscope stage.

Fluid osmolality determines the temperature at which frozen fluid thaws: fluid with a higher osmolality thaws at a relatively low temperature[44]. In frozen slices of rat kidney, the slices through the cortex thaw at a similar temperature to samples of blood plasma, which indicates that the interstitial and tubular fluids in the cortex are isosmotic to plasma. Slices from the outer zone of the medulla of rat kidneys thaw at a lower temperature than slices of cortex, which indicates a higher fluid osmolality of the interstitial and tubular fluids within the outer medulla. However, Figure 7.21A shows that the most striking increase in osmolality occurs in the inner medulla.

Na^+ and Cl^- are important osmolytes in the outer medulla

The Na^+ and Cl^- concentrations gradually increase with depth through the outer medulla, as shown by Figure 7.21B. This increase in ion concentrations results from the transport processes in the thick ascending segment of the loop of Henle, explored back in Figure 7.16. In some species, AVP increases NaCl transport out of the thick ascending limb by stimulating Na^+, K^+, $2Cl^-$ cotransporter proteins, which increases the ability to generate hyperosmotic urine.

[44] Section 9.2.3 and Figure 9.20 discuss the relationship between fluid osmolality and freezing point depression (which generally equals melting point).

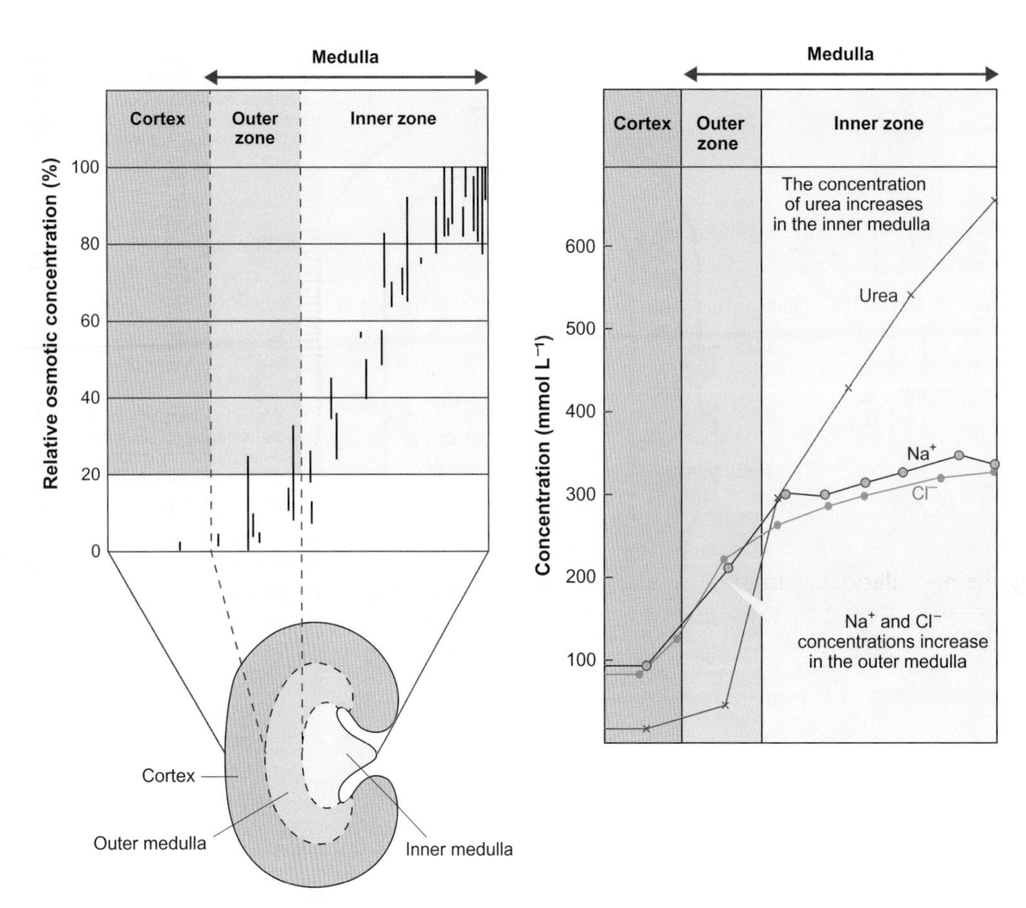

Figure 7.21 Osmotic concentration and solute composition of fluid from slices of mammalian kidneys

(A) Osmotic concentration of fluid from slices taken through the kidney of dehydrated laboratory rats. Fluid in slices from the cortex has an osmolality equal to that of plasma, which is shown as zero. Maximal concentration is plotted as 100%. Osmolality steadily increases through the medulla from the outer zone, which contains the first part of the loops of Henle, and through the inner zone, which contains the deeper parts of the loops, the surrounding interstitium and vasculature. The bars at each depth in the kidney (on the *x* axis) show the range of values recorded. (B) Mean concentration of sodium (Na^+), chloride (Cl^-) and urea (mmol L^{-1}) from cortex and medullary slices of the kidneys of dehydrated dogs.

Source: Pitts RF. Physiology of the Kidney and Body Fluids. 2nd Edition Chicago: Year Book Medical Publishers Inc., 1972.

Urea is an important osmolyte in the interstitium of the inner medulla

Figure 7.21B shows no significant further increase in Na^+ or Cl^- in the inner medulla of the dog. In keeping with this finding, there is no evidence of active NaCl transport from the thin part of the ascending limb in the inner medulla of many of the species examined, such as rats, rabbits and golden hamsters. In these species, the thin ascending limb has high Na^+ and Cl^- permeability so ion fluxes across this part of the nephron are probably entirely passive.

Rather than NaCl, urea is the important osmolyte in the interstitium of the inner medulla that drives water absorption and production of hyperosmotic urine in many species (but not all). Notice the increase in urea concentrations in the inner medulla of the kidneys of dehydrated dogs, shown in Figure 7.21B. Urea concentrations can reach 1 molar in the inner medulla of dog kidneys.

[45] Section 4.2.2 discusses the subtypes of urea transporters. We explore AVP actions on the kidney in more detail in Figure 21.23 (Section 21.1.4).

The rise in urea concentrations in the medullary interstitium during dehydration results from an increased presence of **urea transporters** (UT-A1 and UT-A3) in the inner medullary collecting duct, driven by the increasing concentration of circulating AVP[45]. In contrast, the tubular epithelia of the distal tubule, cortical collecting duct and outer medullary collecting duct have relatively low urea permeabilities. Therefore, water reabsorption during fluid passage through the late distal tubule, cortical collecting duct and outer medullary collecting duct (also stimulated by AVP) results in a gradual increase in the urea concentration of the tubular fluid. These processes create the conditions for passive movement of urea into the inner medullary interstitium via the urea transporters. Urea in the medullary interstitial fluid enters capillaries and is trapped by the countercurrent exchange system, which we examine shortly, such that during dehydration, urea recycles within the kidney and enables continued generation of hyperosmotic urine. This model of events is supported by the evidence from studies of knockout mice lacking the genes for UT-A1 and UT-A3. These knockout mice excrete greater amounts of urea and have a defective ability to produce concentrated urine when drinking water is restricted.

There is evidence for another type of urea transporter, UT-A2 in mammalian kidney tubules. The UT-A2 transporters occur in the thin descending limbs of short loops of Henle in the outer medulla, and during prolonged dehydration, UT-A2 is expressed in the thin descending limbs of longer loops of Henle in the inner medulla, as shown in Figure 7.20. However, the significance of UT-A2 transporters is uncertain, as gene knockout mice lacking UT-A2 have only a mild deficit in their ability to produce concentrated urine. One suggestion is that the UT-A2 is most important in maintaining high urea concentrations in the medulla when urea production declines because of low protein intake[46].

Not all species appear to rely on urea as their means of concentrating the medullary interstitium. For example, fat sand rats (*Psammomys obesus*) that live in North African and Middle Eastern arid habitats do not accumulate urea in the inner medulla, yet sand rats can produce hyperosmotic urine with urine osmolalities of up to about 6400 mOsm kg^{-1}. Fat sand rats appear to rely on NaCl in the interstitium of the inner medulla to generate hyperosmotic urine.

Countercurrent exchange and blood flow in the vasa recta of mammalian kidneys

Blood supply to the medulla of the mammalian kidney occurs via the **vasa recta**[47], which are long capillaries that arise from the efferent drainage of the juxtamedullary glomeruli, as shown earlier in Figure 7.5. The vasa recta run in bundles as countercurrent loops that lie close to the loops of Henle and collecting ducts. This parallel arrangement of structures creates a striated appearance to the medulla that is particularly striking in mammalian species that live in arid environments. Notice in Figure 7.22 that striations run through the medulla and into the papilla of spinifex hopping mice (*Notomys alexis*), which are widespread in arid areas of central and western Australia. These mice can produce extremely hyperosmotic urine, at over 9300 mOsm kg^{-1}.

Capillaries are permeable to water and small solutes, so the countercurrent looping of the vasa recta helps to maintain the increase in osmotic concentrations of the interstitium between the cortex and medulla, while supplying the medullary tissue with oxygen. Figure 7.23A illustrates how blood flowing toward the medulla gains Na^+ and Cl^- by diffusion and loses water by osmosis. On return to the cortex, the opposite processes occur: blood plasma in the vasa recta loses salts and gains water in the passive process of **countercurrent exchange**[48], by diffusion and osmosis.

[46] Protein intake determines nitrogenous excretion and hence production of urea in mammals, which we discuss in Section 7.4.2.

[47] Derived from the Latin *vasa* meaning vessels and *recta* meaning straight.

[48] Section 3.3.1 outlines the concept of countercurrent exchange by passive diffusion.

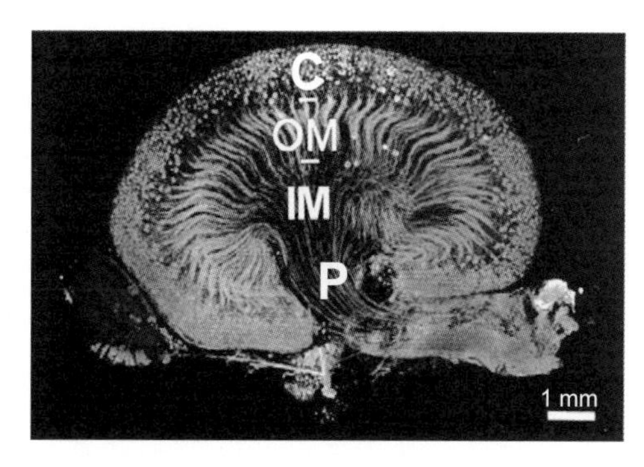

Figure 7.22 Cross-sectional anatomy of the kidney of the spinifex hopping mouse (*Notomys alexis*)

The kidney shown was sectioned across the cortex (C), outer medulla (OM), inner medulla (IM) and papilla (P) after intravascular injection of Microfil to form a cast of the microvasculature. The dots in the cortex are glomeruli. In species living in arid conditions, a single striated papilla penetrates the renal pelvis. The striations running into the papilla are the vasa recta which run parallel to the collecting ducts. Urine collecting in the papilla enters the ureter and is excreted.

Source: Gordge L, Roberts JR (2008). Kidney function in the Spinifex hopping mouse, *Notomys alexis*. Comparative Biochemistry and Physiology 150: 90-101.

The descending vasa recta are also highly permeable to urea due to the presence of urea transporter proteins (UT-B type), which helps to trap urea in the inner medulla. The balance is not always perfect, so some solutes do escape from the kidney in blood that has a slightly higher osmolality than when it entered the medulla. However, the loss of solutes is far less than if blood flowed straight through the medulla, in the theoretical arrangement shown in Figure 7.23B, which would reduce the difference in osmotic concentrations between the cortex and medulla.

The effectiveness of the countercurrent exchange process in the vasa recta depends on the structural arrangements and the rate of blood delivery through these capillaries. Slower flow allows longer for plasma in the vasa recta to approach the osmolality of the adjacent interstitial fluid. By contrast, higher rates of blood flow carry a larger proportion of solutes out of the kidney, which reduces the driving force for water withdrawal from the collecting ducts. After denervation of the kidneys, the rate of blood flow through the vasa recta increases, which suggests that neural activity participates in achieving low rates of blood flow to the medulla. However, the concentration of circulating arginine vasopressin (AVP) is usually the major influence on renal blood flow rates. AVP has a pronounced vasoconstrictor action, which reduces renal blood flow during dehydration.

Dehydrated mammals have higher blood concentrations of AVP[49], which increases water reabsorption from the distal parts of the nephron, as illustrated in Figure 7.20. The additional water reabsorbed by late distal tubules and cortical collecting ducts readily enters the network of blood capillaries in the renal cortex. In the medulla, the additional water reabsorbed from the medullary collecting ducts during

[49] Figure 21.20 outlines the control of AVP release.

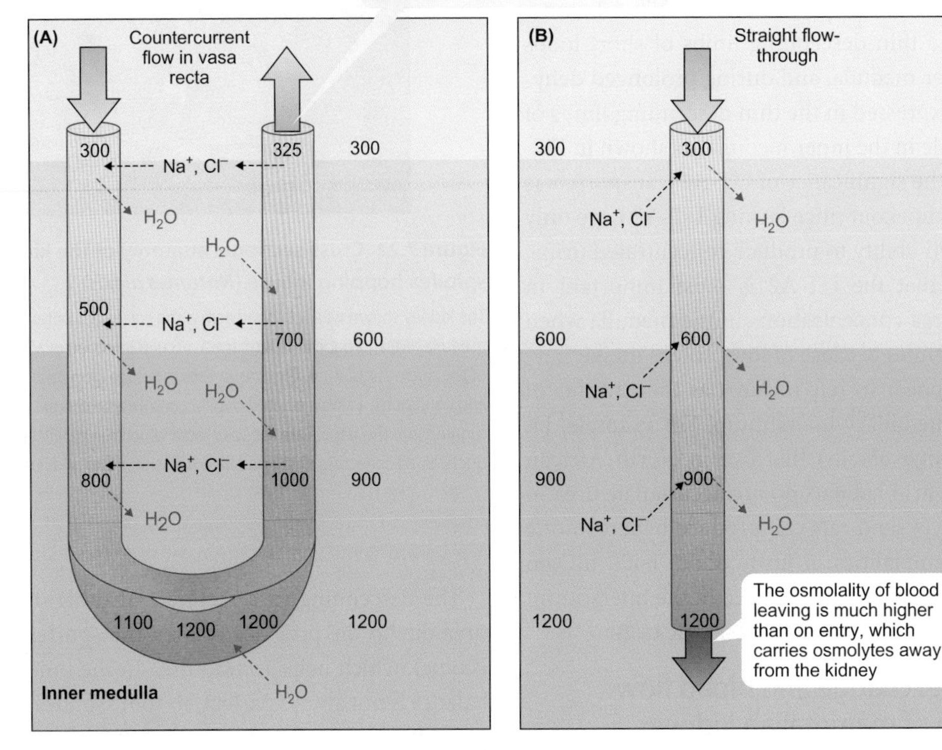

Figure 7.23 Countercurrent exchange of water, Na^+ and Cl^- by vasa recta in the renal medulla retains the cortical-medullary osmotic gradient

The numbers give osmolality (mOsm kg^{-1}) of blood plasma and interstitial fluids for (A) passive countercurrent exchange system in vasa recta and (B) a vessel flowing straight through the kidney from cortex to medulla. Countercurrent exchange of Na^+ and Cl^- in the vasa recta maintains the increasing osmotic concentration between the cortex and inner medulla. Blood plasma leaving the loop, in the cortex, is only slightly higher in osmolality than on entry to the vasa recta. The number of pathways connecting descending and ascending limbs of the vasa recta differs between species. These connecting pathways are often rare and restricted to the tip, as shown in (A), so blood entering the inner medulla travels to the tip of the loop. Water is drawn into the tip because of an increase in colloid osmotic pressure resulting from the higher concentration of plasma proteins due to water passage out of the fluid in the descending limb. In contrast, flow-through vessels (B) would carry large quantities of solutes away in blood plasma with a high osmolality (similar to that of the interstitium). Actual values in both (A) and (B) depend on the rates of blood flow.

dehydration needs to be removed by the vasa recta in order to maintain the high osmolality of the medulla interstitium, which is essential for continued water reabsorption from the collecting ducts. A key component in the descending vasa recta (which Figure 7.23A illustrates) is the passage of water out of the capillaries. This movement of water increases the concentration of plasma proteins in vasa recta blood, creating a higher colloid osmotic pressure force for the absorption of water[50], which facilitates the uptake of the reabsorbed water and its removal from the kidney.

During dehydration, the rate of blood flow through the vasa recta is reduced both by the vasoconstrictor action of AVP, and by the additional outward movement of water in the descending vasa recta by osmosis. Video microscopy of the kidneys of dehydrated rats shows that these processes result in blood flow through the descending vasa recta at a rate that is about half of that in the ascending vasa recta.

[50] Section 14.1.5 examines the haemodynamic forces on capillaries.

Loop development of mammalian nephrons

From the description of the countercurrent multiplier model back in Figure 7.19, we would predict that the osmotic concentration of fluid at the tip of the loops would increase as the length of the loop of Henle increases, and result in a higher maximum osmotic concentration of urine. Actual data reveal a more complex picture. Small desert rodents whose loops of Henle are, in absolute terms, quite short (7–14 mm) produce the most hyperosmotic urine, sometimes exceeding 8000 mOsm kg^{-1}. Larger mammals, such as horses, with larger kidneys and hence longer countercurrent loops (37 mm), produce urine with a maximum osmolality of 1770 mOsm kg^{-1}.

It seems that at least two important factors determine the production of concentrated urine:

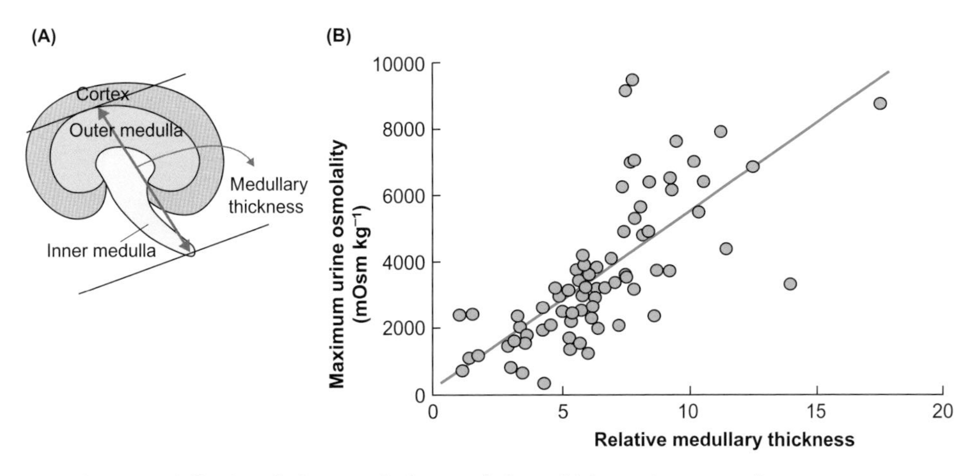

Figure 7.24 Maximum urine osmolality in relation to relative medullary thickness in mammals

(A) Medullary thickness is the distance between the border between the cortex and the outer medulla and the tip of the inner medulla. (B) Relationship between maximum urine osmolality of different mammals and the relative medullary thickness: medullary thickness/(kidney length × width × depth)$^{0.33}$.

Source: B: Beuchat CA (1996). Structure and concentrating ability of the mammalian kidney: correlations with habitat. American Journal of Physiology 271 (Regulatory Integrative Comparative Physiology 40): R157-179.

7

- the proportion of longer-looped nephrons generating the most hyperosmotic urine;
- the relative length of the longest loops of Henle for a given kidney size (or volume).

Proportion of long- and short-looped nephrons

Mammals living in habitats in which they have a ready availability of water or which consume large quantities of water in their food have few or no long-looped nephrons. Good examples are the semi-aquatic moles (*Desmana moschata*), European beavers (*Castor fiber*), West Indian manatees (*Trichechus manatus*) and hippopotami (*Hippopotamus amphibius*). However, there are notable exceptions, suggesting that other factors are important.

For example, domestic dogs (*Canis lumpus familiaris*) have only long-looped nephrons in their kidneys, yet excrete only modestly concentrated urine (maximum osmolality ~2600 mOsm kg^{-1}), while some small desert mammals have fewer than 35 per cent long-looped nephrons but excrete much more hyperosmotic urine. Merriam's kangaroo rats (*Dipodomys merriami*)[51] and fat sand rats live in arid environments and can generate urine with an osmolality exceeding 6000 mOsm kg^{-1} but their kidneys comprise only 27 per cent and 34 per cent long-looped nephrons, respectively.

Relative medullary thickness as a measure of longest loops of Henle

Long-looped nephrons begin as a juxtamedullary glomerulus at the border of the cortex and medulla, as shown earlier in Figure 7.5, but the tip of their countercurrent loops turn at different points in the inner medulla. Those nephrons with the longest loops are most important in generating the highest osmotic concentration of the deeper medullary interstitium, and hence in generating highly hyperosmotic urine. Figure 7.24A illustrates one way of assessing the maximum length of the looped nephrons, by measuring the thickness of the medulla of a kidney—the distance from the beginning of the medulla to the tip of the longest loop in the inner medulla.

The relative medullary thickness (RMT) takes account of kidney size; RMT equals the medullary thickness (mm) divided by kidney size, itself expressed as the cube root of the product of three linear dimensions (length, width and thickness, in cm). Equation 7.2 shows this relationship:

$$\text{Relative medullary thickness} = \frac{\text{Medullary thickness}}{(\text{Kidney length} \times \text{width} \times \text{depth})^{0.33}}$$

Equation 7.2

Data for more than 70 mammalian species, in Figure 7.24B, show that maximum urine osmolality correlates with RMT, although there is a high level of unexplained variability, much of which may result from the structural heterogeneity in the organization of the collecting ducts and the renal vasculature, and the arrangement of the renal pelvis.

Large mammals have a smaller RMT than smaller mammals, as shown in Figure 7.25. However, notice that mammals living in arid habitats, and which survive without regular access to water, have a higher RMT for a given body size than mammals spending their lives in close association with water. Species living in habitats in which they have access to water within a day or two (mesic species) lie between those that live in arid and aquatic habitats.

[51] We learn about the water balance of kangaroo rats more broadly, including their behaviour and their water acquisition in food and by metabolism, in Section 6.2.

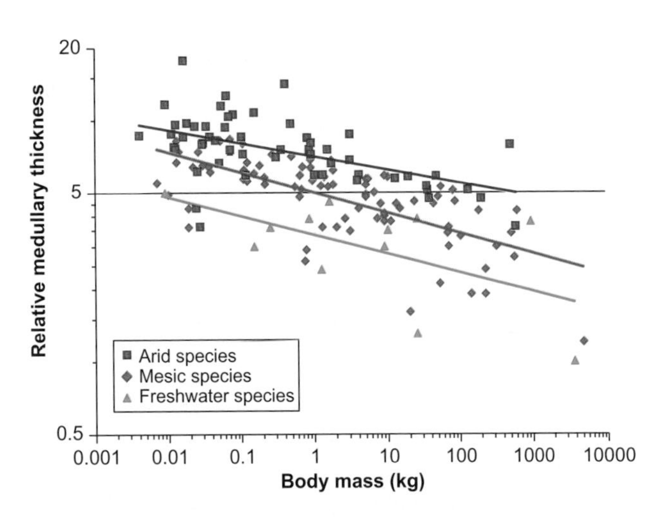

Figure 7.25 Relationships between relative medullary thickness and body mass of mammals in relation to habitat

Scaling of relative medullary thickness (RMT) with body mass (kg) for mammals is separated according to their habitats (arid, mesic or freshwater). In all cases RMT decreased with body mass. For a given body mass, the highest RMTs occur in mammals living in arid habitats, while RMTs of mammals living in mesic habitats are intermediate between those of mammals living in arid habitats and mammals living in freshwater habitats.

Data taken from Beuchat CA (1996). Structure and concentrating ability of the mammalian kidney: correlations with habitat. American Journal of Physiology 271 (Regulatory Integrative Comparative Physiology 40): R157-179.

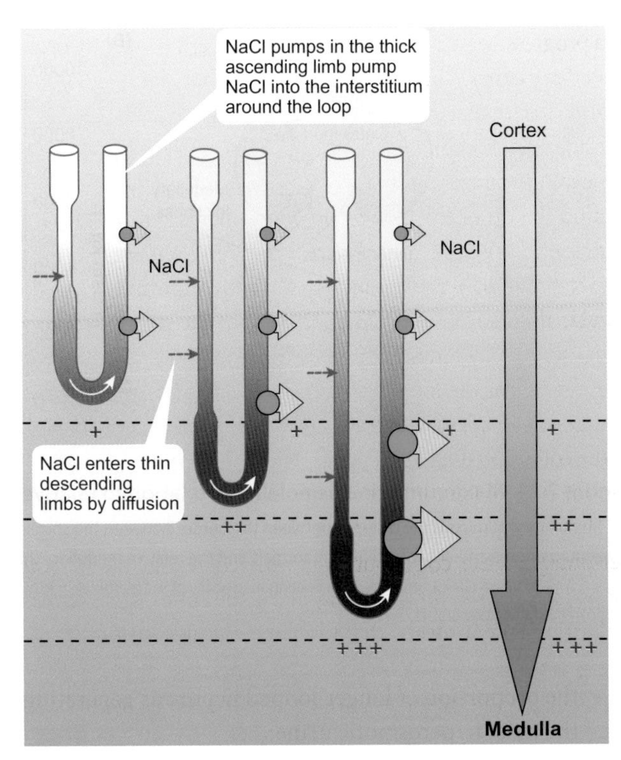

Figure 7.26 Proposed cascade mechanism for NaCl recycling for formation of hyperosmotic urine in birds

Transport of NaCl out of ascending limb (blue circle and arrows) increases the osmotic concentration of the interstitium. NaCl diffuses into the adjacent descending limb. The countercurrent flow creates more concentrated fluid at the tip of the loop (as in mammals). Relative shading within the tubules indicates relative osmotic concentrations. A cascade of increasing length of the loop results in an increasingly concentrated interstitium around the loop (+ to ++, to +++). As loops lengthen the increasing osmotic concentration in the tubular fluid results in an increase in the effectiveness of the NaCl pumps, indicated by larger circles and arrows, which increases the difference in osmotic concentrations between the cortex and medulla (shaded green arrow).

Source: adapted from Nishimura H, Fan Z (2003). Regulation of water movement across vertebrate renal tubules. Comparative Biochemistry and Physiology Part A. Molecular & Integrative Physiology 136: 479–498.

Production of hyperosmotic urine by birds

Birds can excrete urine with a higher osmotic concentration than their plasma, but, on average, urine cannot exceed about twice the concentration of plasma. Those birds living in arid or salty environments mirror mammals in having the best ability to excrete hyperosmotic urine. However, even birds living in arid conditions usually exhibit a maximum urine/plasma osmolality ratio of 1.5 to 2.7, compared with the values of four and above that are seen in land mammals. The savannah sparrow (*Passerculus sandwichensis beldingi*) is a rare exception among birds in being able to excrete urine with a urine/plasma osmolality ratio greater than five.

The poorer ability of birds to produce osmotically concentrated urine compared to mammals is likely to result from the absence of a loop of Henle in 70–90 per cent of the nephrons in bird kidneys. Earlier, Figure 7.6 showed the arrangement of loopless nephrons on both dorsal and ventral sides of the outer (cortical) area of each kidney lobule. Only a relatively small number of nephrons have long countercurrent loops dipping into the medulla. A gradual transition from short loops to longer loops results in tapered medullary cones holding countercurrent loops that run in parallel both to one another and to the associated capillaries (vasa recta) and collecting ducts. This arrangement, and the evidence from studies of the transport characteristics of the loops, provides a working model of how birds concentrate their urine.

Unlike mammals, birds cannot use urea as part of the concentrating mechanism because they produce uric acid rather than urea as their prime nitrogenous waste[52]. Birds appear to rely entirely on NaCl to increase the osmotic concentration of the interstitium and draw out water from the collecting duct. The thick ascending limb of the looped bird nephrons, like those of mammals, pumps NaCl into the interstitium, as illustrated in Figure 7.26.

In birds, movement of water from the lumen of the thin descending limb seems to play little or no role in concentrating the tubular fluid as it flows toward the tip of the loop (unlike most mammals). However, the thin descending limb is equally permeable to Na^+ and Cl^-, which allows diffusion from the interstitium, and recycling of these ions into thick ascending limbs and back again. Figure 7.26 shows the arrangement of a cascade of nephrons in which a short-looped nephron occurs next to a longer-looped nephron, next to longer ones, and so on, resulting

[52] We learn about the nitrogenous waste products and their synthesis in Section 7.4.

in a progressive increase in loop length toward the centre of the medullary cones. This system results in higher NaCl concentrations in the tubular fluid at the tip of the longest loops, not unlike the situation in mammals. There is a greater difference between the osmotic concentration of the cortical interstitium and that around the tip of the loop as the loops increase in length. Experimental evidence from Japanese quails (*Coturnix japonica*) shows that an increase in sodium chloride concentrations in the tubular fluid stimulates NaCl efflux from the thick ascending limb so the concentration mechanism progressively increases in effectiveness as the loop length increases. The processes shown in Figure 7.26 allow withdrawal of water from the tubular fluid in the collecting duct[53] as it passes through the increasingly concentrated medullary interstitium, as seen in mammals.

The collecting ducts of loopless nephrons are on the periphery of the cone, but join with collecting ducts from looped nephrons about half way along the cone. This anatomical arrangement means that tubular fluid produced by loopless nephrons can become concentrated later, in the collecting ducts of the medullary cones. This model of events may explain why the urine concentrating ability of birds correlates with the size of the kidney relative to body mass. The relatively larger kidneys of species producing more hyperosmotic urine tend to have a greater proportion of medullary tissue and a relatively large number of longer medullary cones. In other words, there is more emphasis on longer-looped nephrons within the kidney of these species.

Birds pass the urine from the collecting ducts, via the ureters, and into the **cloaca** where post-renal absorption of salt and water occurs[54]. The cloaca of birds is an expanded tubular structure into which both the ureters and the hindgut empty, and from which birds excrete a mixture of urine and faeces. Many birds use reverse peristalsis to move the mixture of urine and gut fluids backward into the colon where further modification of the composition and volume of the fluid mixture of urine and faecal matter occurs. Nevertheless, there is evidence that maximal urine osmolality of bird species, and structural modifications of the kidney in different species, influence the ability of birds to cope with life in arid habitats, much like mammals.

7.2.4 Tubular secretion by vertebrate nephrons

A significant component of tubular function involves secretion. The secretion of potassium (K^+) by vertebrate nephrons is essential in regulating potassium balance[55]. Secretion of sodium chloride drives fluid secretion. In those marine teleost fish that lack glomeruli (the aglomerular species), fluid secretion is the only way to form the primary urine, but secretory processes also occur in marine teleosts that have glomeruli, in which the secretion of magnesium (Mg^{2+}) and sulfate (SO_4^{2-}) are essential to divalent ion balance.

Secretion of potassium excretes excess intake

The kidneys are the organs principally responsible for the long-term maintenance of K^+ balance in response to the variations of K^+ intake in the diet and K^+ loss in the urine and faeces. Insectivorous and carnivorous animals take in K^+ in the cells of the animal meats they consume, while some plants and fruits contain very large amounts of K^+ (but low amounts of sodium). The dietary intake of K^+ usually exceeds requirements and results in net K^+ secretion by the nephron. An increase in K^+ intake stimulates K^+ secretion by the kidney tubules, while a decrease in K^+ intake reduces K^+ secretion.

The kidneys initially filter K^+ from the plasma in the glomeruli, but reabsorption of 80 per cent or more of the filtered K^+ occurs in the proximal part of the nephrons; in mammalian kidneys, some K^+ reabsorption also occurs in the loop of Henle, as shown by the schematic back in Figure 7.16. Hence, most K^+ excretion results from the secretion of K^+ by the tubular cells of the distal parts of the nephron. In mammalian kidneys, K^+ secretion starts in the early distal convoluted tubule and progressively increases along the distal nephron and into the collecting duct.

The process of K^+ secretion is passive and mainly results from passage via K^+ channels in the luminal membrane of the tubular epithelial cells. Luminal (apical) membranes are an unusual location for K^+ channels, which typically occur in the basolateral borders of epithelial cells where they allow recycling of K^+ pumped into the cells by Na^+, K^+-ATPase.

Earlier, Figure 7.17 illustrates the two-step process of K^+ secretion by the principal cells of mammalian collecting ducts. The Na^+, K^+-ATPase in the membranes of the basolateral border of the epithelial cells transports K^+ into the cells (process 1 in Figure 7.17) sustained by a continual supply of K^+ in the blood plasma delivered to the kidney. From the cells, passive diffusion of K^+ across the luminal membrane occurs via K^+ channels (process 3 in Figure 7.17) due to a favourable electrochemical difference, which is the main determinant of K^+ secretion.

The rate at which passive secretion of K^+ occurs in vertebrate nephrons is determined by:

- the number of K^+ channels in the luminal membrane (i.e. its K^+ permeability);
- the electrochemical difference for K^+ across the luminal membrane driving the passive transport via

[53] Hormonal control of the kidney of birds by **arginine vasotocin (AVT)** redistributes filtration toward the looped nephrons producing hyperosmotic urine; AVT also appears to act preferentially on water conservation by these looped nephrons, which we discuss in Section 21.1.4.

[54] Post-renal salt and water absorption in the cloaca (particularly the coprodeum), and colon, of birds is particularly important for species of birds living in arid habitats as we discuss in Section 6.1.4. Hormonal regulation of post-renal processing of salts is also discussed in Section 21.1.1.

[55] K^+ balance is essential in determining nerve and muscle function as we outline in Section 4.1.1.

> K^+ channels. The electrochemical difference results from the combination of the potential (voltage) difference by which a more electronegative lumen enhances K^+ passage via the K^+ channels, and the concentration difference, in which an increase in cellular K^+ concentrations or a decrease in tubular fluid concentration of K^+ enhance secretion via the K^+ channels.

The hormone **aldosterone**[56] is an important regulator of the factors listed above and hence stimulates K^+ secretion by the distal nephron when there is an excess of K^+ in the extracellular fluid. Firstly, aldosterone increases the number of open K^+ channels in the luminal membrane, which enhances the rate of K^+ secretion. Aldosterone also increases the number of open Na^+ channels in the luminal membrane, which increases Na^+ reabsorption and therefore increases the electrochemical gradient that favours K^+ secretion. Finally, aldosterone stimulates basolateral Na^+-K^+-ATPase, which increases cellular K^+ concentrations, which again favours secretion.

Potassium secretion increases the tubular fluid concentration of K^+, which tends to inhibit the rate of further secretion. Hence, an increase in the rate of fluid (and sodium) delivery to the distal nephron, which dilutes the tubular fluid and reduces the build up of K^+ concentrations, enhances the rate of K^+ secretion. An increase in the rate of Na^+ delivery to the distal nephron stimulates Na^+ reabsorption from the tubular fluid, which makes the lumen more negative, and again enhances the rate of K^+ secretion.

Two types of K^+ channels occur in the luminal membranes of the principal cells in the mammalian collecting duct. The dominant route for K^+ secretion is thought to be via the renal outer medullary potassium channels (ROMK channels), which have a high probability of being open, but low K^+ conductance. By comparison, a second type of K^+ channel with higher conductance, called maxi-K^+ channels, become active under high tubular flow conditions.

In contrast to the situation for mammals, we have a poor understanding of K^+ secretion by the kidneys of non-mammalian vertebrates except for a few amphibian species, in particular tiger salamanders (*Ambystoma tigrinum*). As newborns, these salamanders live in water and use gills for gas exchange. The influx of K^+ across their skin (and possibly gills) makes it easy to alter their plasma concentrations of K^+ and study renal K^+ secretion.

Exposure of tiger salamanders to water containing 50 mmol L^{-1} KCl for 14 days, which increases K^+ influx, results in a dramatic increase in K^+ excretion. In such conditions, there is an almost 10-fold increase in the density of K^+ channels in the luminal membranes of the initial collecting tubule. These K^+ channels have the characteristics of maxi-K^+ channels and only become highly conductive when extracellular fluid concentrations of K^+ increase, which provides a mechanistic basis for salamanders (and probably other amphibians) to secrete K^+ when necessary in order to regulate K^+ balance.

Marine fish kidneys secrete magnesium and sulfate to maintain low plasma concentrations

The conditions facing marine fish are different from those facing terrestrial vertebrates. For those vertebrates living on land, (Mg^{2+}) and sulfate (SO_4^{2-}) are often in short supply such that their availability is determined by dietary intake. Excretion by the kidneys balances the intake in the diet, with the bulk of the filtered Mg^{2+} and SO_4^{2-} reabsorbed by the renal tubules. However, seawater contains Mg^{2+} at around 53 mmol L^{-1} and SO_4^{2-} at around 28 mmol L^{-1}, while plasma concentrations are 0.5–2.5 mmol L^{-1}. Marine teleosts drink seawater to maintain fluid balance, so an influx of Mg^{2+} and SO_4^{2-} may occur in the intestine[57] and additional influx is also possible across the gills. The maintenance of low plasma concentrations of these divalent ions is only possible because of secretion by the kidney and the excretion of urine containing Mg^{2+} at 80 to 150 mmol L^{-1} and SO_4^{2-} at 40 to 120 mmol L^{-1}.

Analysis of fluid secreted by the isolated perfused[58] proximal tubules of marine fish shows the tubular fluid in this segment contains Mg^{2+} at 22 mmol L^{-1}, which indicates its important role in Mg^{2+} secretion. Injection of the stable isotope of magnesium (^{26}Mg) into fish, followed by imaging of its location, has shown ^{26}Mg located in the cells of the proximal tubule, which indicates a transcellular pathway of Mg^{2+} secretion.

Figure 7.27 illustrates a proposed model for Mg^{2+} secretion by fish nephrons. Notice in this model Mg^{2+} passively enters the proximal tubular cells through ion channels in the basolateral border because of the negative intracellular voltage and low intracellular concentration of Mg^{2+}. At the luminal border of the cells, the electrochemical difference suggests that secondary active transport of Mg^{2+} occurs, possibly via a Mg^{2+}/Na^+ exchanger, energized by the basolateral Na^+, K^+-ATPase, or by luminal exchange of Mg^{2+} with protons (H^+). The Mg^{2+}/Na^+ exchange may be via Mg^{2+} transporters that are one of the large family of solute carriers (Slc): Slc41a1, which occurs in vacuoles in the luminal (apical) cytoplasm of the proximal tubules of pufferfish. One of the pieces of evidence is that expression of mRNAs encoding this Mg^{2+} transporter increase by five-fold in the kidney of euryhaline obscure pufferfish during its acclimation

[56] Aldosterone is a steroid hormone produced by the adrenocortical tissue, which we discuss in Section 19.3.1. We discuss the actions of aldosterone on the distal nephron further in Section 21.1.1.

[57] We discuss drinking of marine teleosts in Section 5.1.3. Most Mg^{2+} in drinking water is not absorbed from the intestine of marine teleosts; a large amount of Mg^{2+} and Ca^{2+} precipitate as carbonates, as part of the mechanisms for water retrieval, and the carbonates are excreted rectally.

[58] Box 7.1 outlines the technique of tubular perfusion for studying individual nephrons.

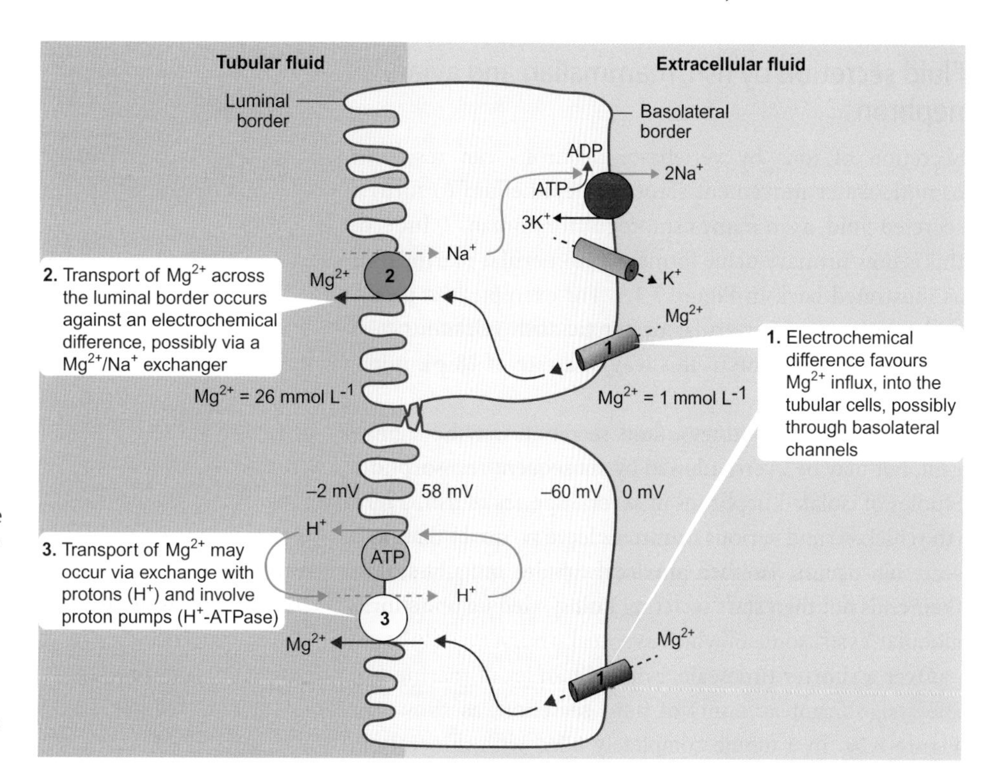

Figure 7.27 Model for transport processes involved in secretion of magnesium in proximal tubules of marine teleosts

Numbered processes are shown on the basolateral and luminal (brush) border of the proximal tubular cells. Dashed lines through ion channels indicate passive transport.

Source: adapted from Beyenbach KW (2004). Kidneys sans glomeruli. American Journal of Physiology Renal Physiology 286: F811-827.

to seawater, when excretion of Mg^{2+} becomes essential for survival.

Secretion of sulfate by the proximal tubules of marine fish appears to be more complex and involve the concerted action of a number of possible processes, which Figure 7.28 illustrates:

- movement of SO_4^{2-} into the epithelial cells via the basolateral membranes in exchange with hydroxyl ions (OH^-);
- a central role for **carbonic anhydrase** in maintaining the exchange with OH^- on the basolateral border;
- luminal exchange of sulfate with bicarbonate ions (HCO_3^-) and/or Cl^-.

We explore these ideas further in Experimental Panel 7.1, available online, by examining the results of studies of sulfate secretion by the proximal tubules of Japanese eels (*Anguilla japonica*) and winter flounders (*Pseudopleuronectes americanus*).

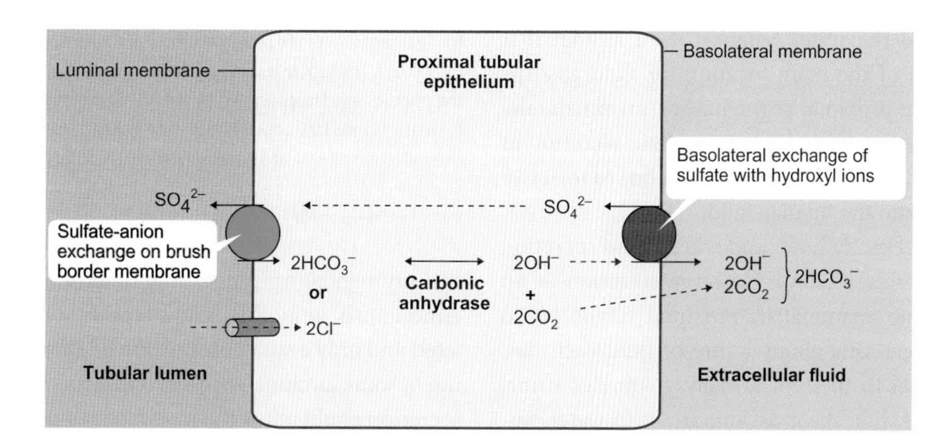

Figure 7.28 Model of transport mechanisms resulting in sulfate secretion in the proximal tubule of marine teleosts

On the basolateral membranes, the influx of sulfate ions (SO_4^{2-}) is suggested to result from an exchange with OH^-. This exchange is maintained by the formation of OH^- as a result of carbonic anhydrase activity. Influx of SO_4^{2-} allows its efflux via an electrically neutral exchange with HCO_3^- and/or Cl^- in luminal membranes. Carbonic anhydrase plays a central role in maintaining these exchanges by de-hydroxylation of HCO_3^- forming CO_2 and OH^-. Carbon dioxide diffuses via the basolateral membrane and interacts with OH^-, reforming HCO_3^- in the extracellular fluids.

Fluid secretion by fish, mammalian and avian nephrons

Secretion of ions by vertebrate nephrons can result in osmotic water movement across the epithelium to form a secreted fluid, as in many osmoregulatory organs[59]. Indeed, this is how primary urine forms in aglomerular marine fish, as illustrated back in Figure 7.1A. For example, the proximal tubules of aglomerular oyster toadfish (*Opsanus tau*) secrete primary urine in this way at a rate of 28 pL min^{-1} mm^{-1} tubule.

Even in filtering kidneys, fluid secretion may be significant, but may be overshadowed by subsequent reabsorption. Studies of isolated nephrons of several species of fish, a bird (the chicken) and various mammals leave no doubt that fluid secretion occurs. Isolated proximal tubules initially seal at their ends but then start secreting fluids, swelling up to form globular 'cysts', some of which eventually burst.

Over a shorter timescale, isolated tubules of marine fish show significant amounts of fluid secretion, as shown in Figure 7.29. In a tubule completely filled with mineral oil, the secreted fluid rapidly breaks up the oil, and gradually pushes it along the tubule, expelling all the oil after about an hour. Experiments like these show that fluid secretion occurs in the proximal segments of fish, amphibians, birds and mammals, and the collecting duct of the inner medulla of mammalian kidneys.

The second proximal segment of various glomerular species of fish has been found to secrete fluid at a rate of up to 54 pL min^{-1} mm^{-1} tubule. From these measurements, the rate of secretion for whole nephrons is estimated to be about 100 pL min^{-1}. As single nephron glomerular filtration rates are 350 to 650 pL min^{-1}, fluid secretion is not insignificant and may be almost a third of the fluid passing along the nephron.

Fluid secretion will have the greatest potential significance in relation to the water balance of an animal if it occurs downstream of the main location for fluid absorption, i.e. beyond the proximal segment. For mammals, the collecting duct is in the very best position; fluid secretion in the inner medullary collecting duct has been linked to active chloride transport into the tubular fluid.

We do not know exactly how (and when) fluid secretion is important in filtering kidneys. However, secretion of up to 160 pL min^{-1} per mammalian proximal tubule could result in humans secreting about 1 litre of fluid each day. This volume is equal to the typical daily volume of urine. Fluid secretion is likely to be most important in water balance when there are low filtration rates. For example, renal blood flow during the dives of marine mammals falls outside the autoregulatory range, so glomerular filtration declines. Among fish, amphibians and reptiles, GFR is not autoregulated and only a small proportion of glomeruli may be filtering in some circumstances, as we learned in Figure 7.7. Fluid secretion could allow the non-filtering nephrons to contribute to excretion.

[59] For example, fluid secretion by salt glands, which we examine in Figure 5.14 in Section 5.1.5.

(A)

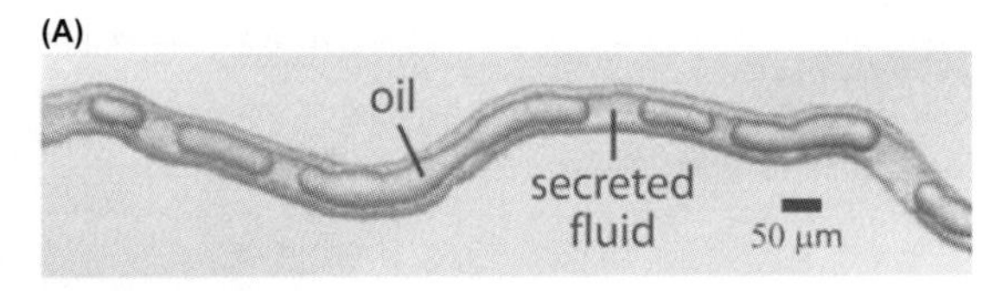

(B)

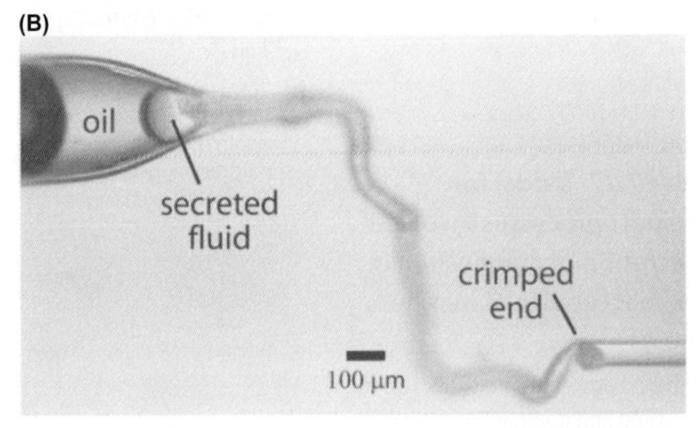

(C)

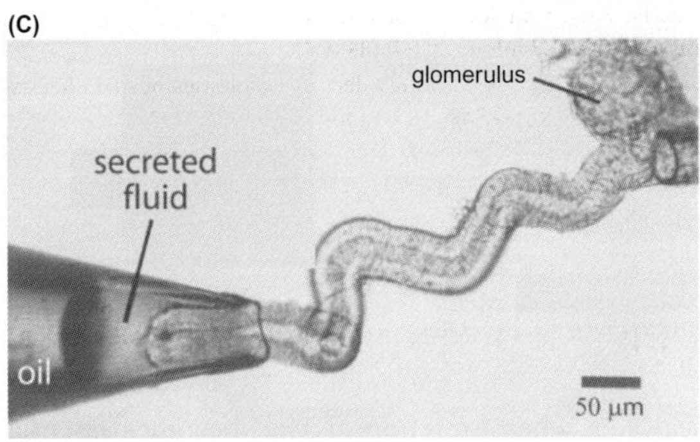

Figure 7.29 Fluid secretion in isolated renal proximal tubules of winter flounders (*Pseudopleuronectes americanus*) in seawater and killifish (*Fundulus heteroclitus*) adapted to fresh water

(A) Isolated flounder tubule held between micropipettes (not shown). The tubule was first filled with mineral oil. Secretion of fluid breaks up the oil. (B) Flounder proximal tubule crimped closed with a turn in a holding pipette at the right-hand side of the image. The open end of the tubule is held in a collection pipette on the left. The photographic focus shows secreted fluid accumulating at an average rate of 37 pL min^{-1} mm^{-1} tubule in the collection pipette, in front of light mineral oil to avoid evaporation. (C) Fluid secretion by the first segment of the proximal tubule of a killifish acclimated to fresh water for 35 days. The tubule is crimped closed near its glomerulus, on the right of the photograph. Fluid secreted by the first part of the proximal tubule can leave the tubule by its open end in the left-hand pipette. Fluid accumulated in the collection pipette at an average rate of 34 pL min^{-1} mm^{-1} tubule.

Source: reproduced from Beyenbach KW (2004). Kidneys sans glomeruli. American Journal of Physiology Renal Physiology 286: F811-827.

› *Review articles*

Bankir L, de Rouffignac C de (1985). Urinary concentrating ability: insights from comparative anatomy. American Journal of Physiology 249 (Regulatory Integrative Comparative Physiology 18): R643–666.

Beyenbach KW (2004). Kidneys sans glomeruli. American Journal of Physiology Renal Physiology 286: F811–827.
Dantzler WH (2003). Regulation of renal proximal and distal tubule transport: sodium, chloride and organic anions. Comparative Biochemistry and Physiology Part A: Molecular & Integrative Physiology 136: 453–478.
Dantzler WH (2005). Commentary. Challenges and intriguing problems in comparative renal physiology. Journal of Experimental Biology 208: 587–594.
Houllier P (2014). Mechanism and regulation of renal magnesium transport. Annual Review of Physiology 76: 411–430.
Palmer BF (2015). Regulation of potassium homeostasis. Clinical Journal of American Society of Nephrology 10: 1050–1060.
Palmer LG and Schnermann J (2015). Integrated control of Na transport along the nephron. Clinical Journal of the American Society of Nephrology 10: 676–687.
Pannabecker TL (2013). Comparative physiology and architecture associated with the mammalian urine concentrating mechanism: role of inner medullary water and urea transport pathways in the rodent medulla. American Journal of Physiology Regulatory and Integrative Comparative Physiology 304: R488–R503.
Richards AN (1938). The Croonian lecture: Processes of urine formation. Proceedings of Royal Society, London B 126: 398–432.
Sands JM (2004). Micropuncture: unlocking the secrets of renal function. American Journal of Physiology Renal Physiology 287: F866–F867.

7.3 Invertebrate nephridia

We generally understand much less about the nephridia of invertebrates than the kidneys of vertebrates, but we do have some information about the functioning of the **protonephridia** of rotifers, the **metanephridia** of earthworms and leeches, the **antennal glands** of crustaceans, renal organs of molluscs and the **Malpighian tubules** of insects, which we examine in the following subsections.

7.3.1 Protonephridia of rotifers

Protonephridia occur in a wide range of invertebrate groups, including flatworms, rotifers and some annelids and molluscs. These excretory organs consist of one or more tubules ending in one or more terminal organs each containing a tuft of cilia or flagellae that flicker (like a flame) when examined under the microscope, so they are known as flame cells or flame bulbs. Figure 7.30 shows an example of the protonephridia of rotifers (*Asplanchna priodonta*).

The beating of the flagellae in the terminal organs may generate a sufficient reduction in pressure to draw fluid into the terminal organ from the body cavity via the membrane surrounding the terminal cell of the terminal organ. Hence, protonephridia may function by ultrafiltration. The ultrastructure of the outer surface consists of filtration slits and an overlying matrix, which supports the idea that ultrafiltration forms the primary urine, but there are no direct measurements of the forces for filtration equivalent to those available for vertebrates[60].

The rotifer *Asplanchna priodonta* is one of the few species in which functional investigation of the protonephridia is feasible—something made possible because of its:

- size—it is one of the biggest rotifers, measuring 400–1500 μm in length and 200–400 μm in breadth;
- transparency, which means that body organs are visible, as shown in the image in Figure 7.30A;
- bladder, which enables collection of samples of urine.

[60] We learn about the haemodynamic forces that drive ultrafiltration in the vertebrate glomerulus in Section 7.1.2.

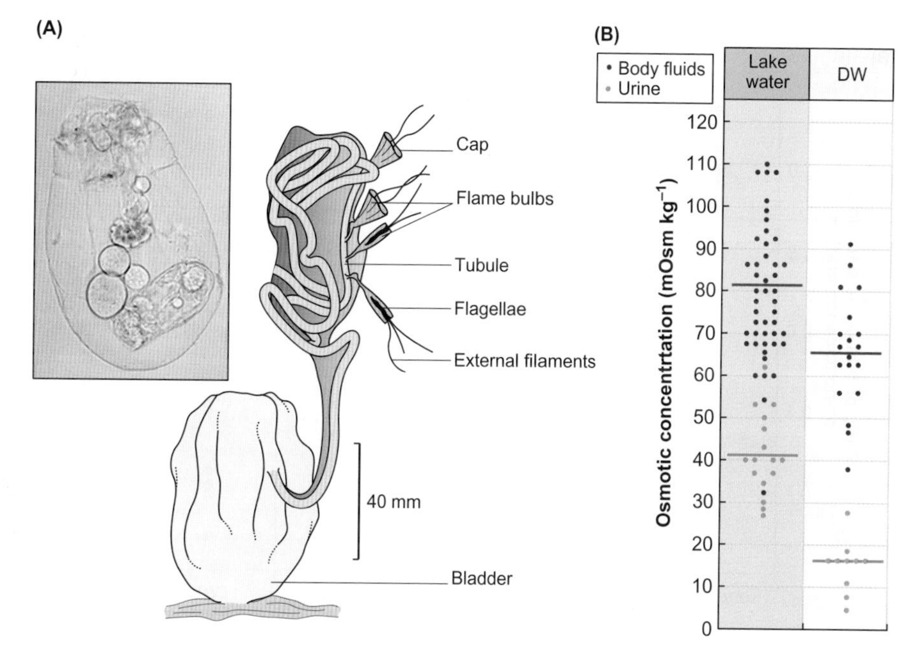

Figure 7.30 Protonephridial functioning of the freshwater rotifer (*Asplanchna priodonta*)

(A) Protonephridia showing flame bulbs. External filaments attach the cap of the flame bulbs to the body wall. The flame bulbs filter primary urine into the tubule and the fluid passes into a bladder. Contraction of the bladder results in excretion of urine. The image of *Asplanchna* shows an embryo inside. (B) Osmolality of body fluids and excreted urine of individual rotifers are indicated by purple and orange symbols, respectively. Average values are indicated by straight lines for each data set. In rotifers held in lake water (osmolality 18 mOsm kg^{-1}) the urine osmolality exceeds that of rotifers held in distilled water (DW).

Source: adapted from Braun G et al (1966). Studies on the ultrastructure and function of a primitive excretory organ, the protonephridium of the rotifer *Asplanchna priodonta*. Pflügers Archives. 289: 141-154. Image: VIM VAN EGMOND/SCIENCE PHOTO LIBRARY

Figure 7.30B shows some data for the osmolality of body fluids and urine of *Asplanchna* held in different environmental conditions. When the rotifers are held in lake water, the difference between the external osmolality (18 mOsm kg^{-1}) and body fluid osmolality (average 81 mOsm kg^{-1}) results in a constant inflow of water across the integument, by osmosis. This water inflow is excreted in urine that is hypo-osmotic to the body fluids, at an average osmolality of 42 mOsm kg^{-1}, which indicates reabsorption of osmolytes from the tubular fluid.

Measurement of sodium concentrations in body fluids and urine indicates a decrease from 17 mmol L^{-1} (in the filtered extracellular fluid) to 13 mmol L^{-1} by the time the fluid reaches the bladder. The tubule also reabsorbs about 30 per cent of the filtered volume[61]. Thus, *Asplanchna* creates osmotically dilute urine in its protonephridia, in a somewhat similar way to vertebrate nephrons, and thereby excretes the excessive water influx. Furthermore, *Asplanchna* can alter the composition of the urine in response to environmental conditions.

When placed in ion-free (distilled) water, which increases water influx, *Asplanchna* increases urine production by almost 30 per cent within 30 minutes. At the same time, enhanced tubular reabsorption of Na^+ reduces the Na^+ concentration of the urine and, as shown in Figure 7.30B, urine osmolality decreases and stabilizes at an average of about 15 mOsm kg^{-1}.

[61] These data are based on studies in which radiolabelled inulin was injected into the rotifers, to monitor water movements based on measurement of the urine/plasma inulin concentration ratio, following the principles outlined in Box 7.2.

7.3.2 Metanephridia of earthworm and medicinal leech

Metanephridia occur in animals with a distinct body cavity between the gut and the body wall (a coelom[62]). The metanephridia open into the coelom through a funnel-shaped nephrostome, which drains fluid from the coelom into the attached tubule where its composition is modified before excretion. As filtration of fluid from blood vessels forms the coelomic fluid, excretion via metanephridia is essentially a filtration–reabsorption process. However, the functioning of the metanephridia of most invertebrate species is poorly understood, except for a few species such as earthworms (*Lumbricus terrestris*) and the European medicinal leech *Hirudo medicinalis*.

In earthworms, active reabsorption of sodium ions and passive reabsorption of chloride ions from the primary urine occurs as the fluid passes along the subsections of the segmentally arranged pairs of metanephridia. The low permeability of the metanephridia to water prevents it following the absorption of solutes. As a result, fluid osmolality decreases. Figure 7.31 shows the declining osmolality, which reaches about 20 per cent of the osmolality of the blood by the distal end of the nephridia.

In contrast to the earthworm, the segmentally arranged pairs of nephridia of leeches *secrete* the primary urine. The rate of fluid secretion by the nephridia determines the volume of urine excreted. Figure 7.32A shows a model of

[62] Coelomic and acoelomic animals are distinguished in Figure 4.2.

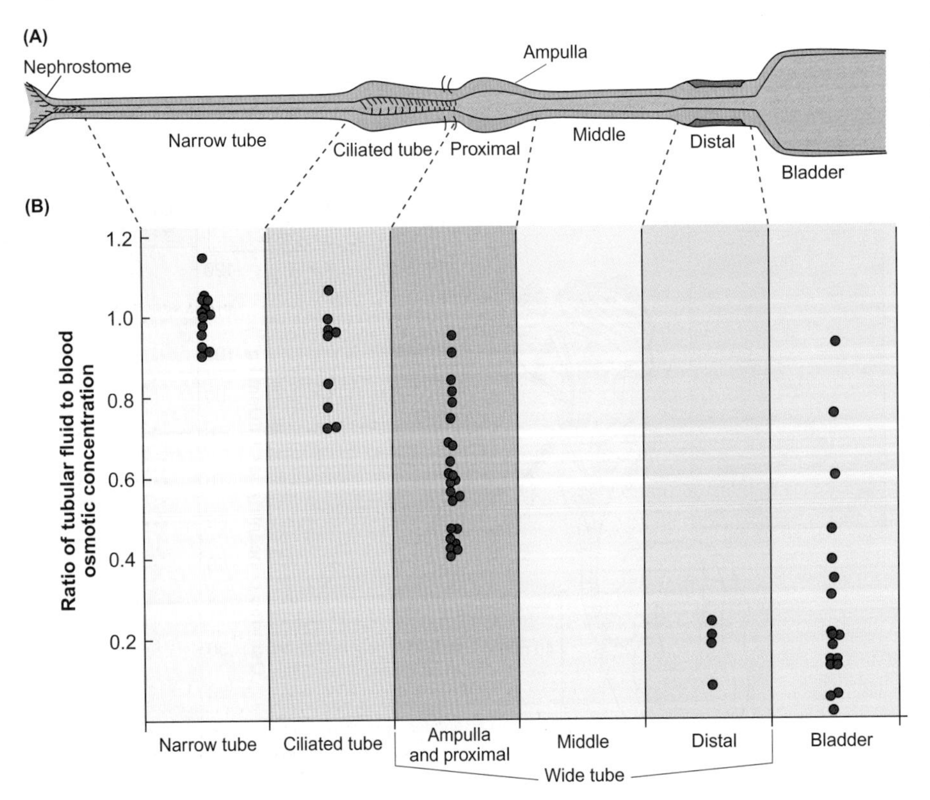

Figure 7.31 Metanephridia of the common earthworm (*Lumbricus terrestris*) reabsorb osmolytes

(A) A diagram of an uncoiled metanephridium showing the subsections. Each segmentally-arranged metanephridium is normally coiled, and needs unravelling to reveal the nephrostome connected to proximal and distal segments of a tube which ends in a bladder that stores urine before its excretion. (B)The tubular fluid/blood osmotic concentration along the metanephridia. At a value of 1.0 there are equal osmotic concentrations. Values below 1.0 indicate reabsorption of osmolytes. Progressive reabsorption occurs until the osmotic concentration of the tubular fluid relative to the blood is about 0.2 in the distal section. Values for samples of fluid collected from the bladder are highly variable, which suggests that fluid movement and/or addition of solutes occurs in the bladder.

Source: Riegel JA. Comparative Physiology of Renal Excretion. New York: Hafner Publishing Company, 1972.

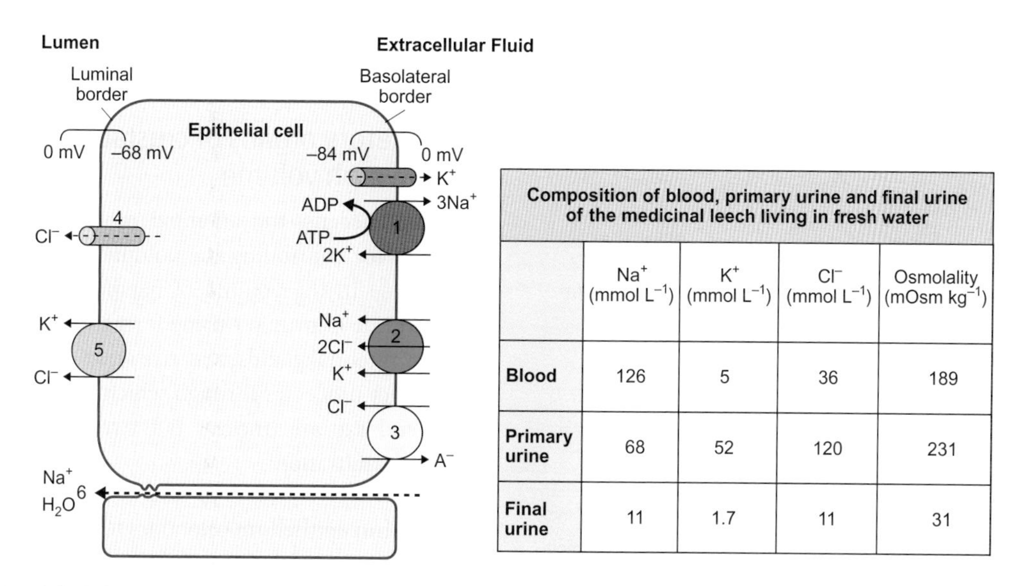

Composition of blood, primary urine and final urine of the medicinal leech living in fresh water

	Na^+ (mmol L^{-1})	K^+ (mmol L^{-1})	Cl^- (mmol L^{-1})	Osmolality (mOsm kg^{-1})
Blood	126	5	36	189
Primary urine	68	52	120	231
Final urine	11	1.7	11	31

Figure 7.32 Model of the mechanism of primary urine production by the European medicinal leech (*Hirudo medicinalis*) and composition of blood, primary urine and final urine of leeches living in fresh water

(1) Basolateral Na^+, K^+-ATPase energizes (2) Na^+, K^+, $2Cl^-$ co-transporters on the basal border of the cells that transport Cl^- ions into the cells. There is also evidence of (3) Cl^-/anion exchangers on the basal border. Secretion of Cl^- through the luminal border of the epithelium occurs via (4) Cl^- channels and (5) K^+, Cl^- co-transporters. (6) Sodium (Na^+) passage across the epithelium follows its electrochemical difference, along the paracellular route, and water passes by osmosis across the leaky epithelium. The table of data shows the urine composition of the primary urine and the final urine of medicinal leeches when they are living in fresh water.

Source: adapted from Zerbst-Boroffka I et al (1997). Chloride secretion drives urine formation in leech nephridia. The Journal of Experimental Biology 200: 2217–2227.

the secretory transport processes in the metanephridia of medicinal leeches. Note that the secretion of K^+ and Cl^- ions (via co-transport proteins and ion channels in the luminal membrane) drive Na^+ and water passage through the paracellular pathways. The outcome is primary urine containing high concentrations of K^+, at about 10 times those of the leech's blood, and Cl^- concentration at about 3.5 times those in the blood; in contrast, Na^+ concentrations of the primary urine are about half those of the blood, as shown in the tabulated data in Figure 7.32.

For most of their life, European medicinal leeches live in fresh water, where they take up water and lose salt across the integument, and excrete a volume of urine equal to their own body mass each day. Salt loss in the urine is minimized by reabsorbing 85 per cent of the Na^+ and 97 per cent of the K^+ secreted in the primary urine, to produce final urine with a low osmolality, which the data in Figure 7.32 illustrate.

During feeding, medicinal leeches consume large volumes of blood (up to 10 times their body mass). The blood consumed is hyperosmotic to the leeches' own blood and coelomic fluid: the blood of freshwater vertebrates typically has an osmolality of ~300 mOsm kg^{-1}, while the blood of leeches is less than 200 mOsm kg^{-1}. After feeding, the increased extracellular fluid volume stimulates an increase in urine volume—a **diuresis**[63].

The feeding diuresis in leeches is a direct result of an increase in primary urine production, which is the only means by which leeches regulate urine output. The diuresis results from opening of the paracellular pathways which increases Na^+ concentrations in the urine by 15-fold after feeding, coupled with an increase in transcellular KCl transport by four-fold[64]. These processes drive an increase in water passage via the leakier paracellular pathways.

The change from K^+-enriched toward Na^+-enriched primary urine is the first stage in the natriuresis that occurs after feeding. The second step is a decrease in reabsorption of NaCl from later parts of the nephridia. The NaCl reabsorption is proportional to the NaCl concentrations in the blood of the leech, with a decrease from 85 per cent reabsorption of the secreted amount when leeches are in fresh water to just 3 per cent reabsorption during the diuresis after feeding. The combined effects of additional fluid secretion (to deal with the volume intake) and reduced NaCl reabsorption can increase the rate of NaCl excretion by as much as 80 times after feeding, which eliminates the excess salt intake.

[63] We discuss diuretic factors controlling urine output of animals in Section 21.1.4, but they are poorly understood in leeches.

[64] Blood-sucking insects similarly switch to production of Na^+-enriched primary urine after feeding, but by different mechanisms. We discuss these mechanisms in Section 7.3.4.

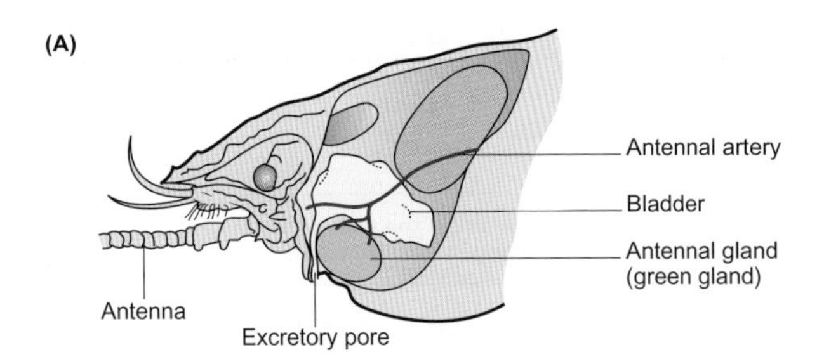

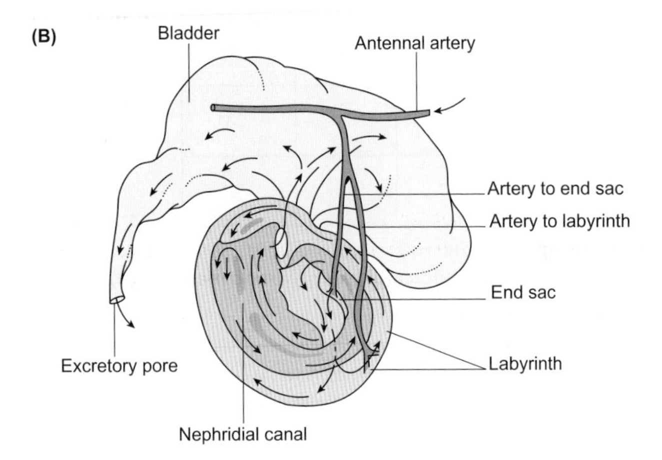

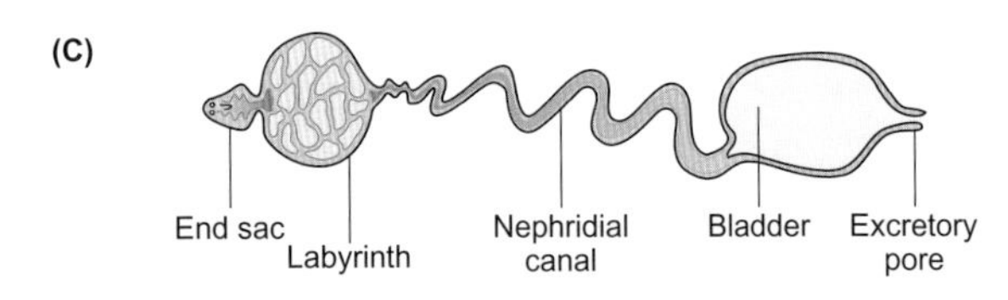

Figure 7.33 Antennal gland of white clawed crayfish (*Austropotamobius pallipes*)

(A) Location of one of the pair of antennal glands in the head of a crayfish. The antennal gland opens through an excretory pore at the base of the antenna. (B) Antennal gland removed from the head. Arrows show direction of fluid flow, starting by fluid filtration in the end sac. Fluid then enters interconnected pathways in the labyrinth and flows into the nephridial canal. Finally, the fluid flows into a bladder, before excretion through the excretory pore. (C) Unravelled antennal gland showing its subsections.

Source: Potts WTW, Parry G. Osmotic and Ionic Regulation in Animals. Oxford: Pergamon Press, 1964.

7.3.3 Antennal glands of crustaceans

The excretory organs of crustaceans are located in their head, as paired globular structures, illustrated in Figure 7.33, which are, in fact, tubular structures opening as an excretory pore (nephropore) at the base of the second antennae or within the mouthparts, at the base of the second maxillae. Hence, these organs are antennal glands or maxillary glands. Their greenish colour explains their alternative name of green glands.

Once unravelled, subsections of the antennal/maxillary glands become apparent. The unravelled antennal glands of decapods like crayfish, illustrated in Figure 7.33C, start as end sacs (or coelomosacs), pass through a proximal part or labyrinth and then a nephridial tubule (canal), and end in a bladder, which opens via an excretory pore.

The excretory organs of crustaceans regulate haemolymph volume

In all crustaceans, the antennal or maxillary glands are important for controlling the volume of the haemolymph. An increase in the volume of haemolymph results in an increased rate of primary urine formation and its excretion from the antennal glands, and *vice versa*.

The exact processes through which primary urine forms in the end sac are unclear, but two anatomical features suggest ultrafiltration:

- The thin epithelium of podocytes and pedicels covering the wall of the end sacs is very similar in structure to the glomerular barrier of vertebrates[65].
- An arterial supply to the antennal glands divides into many small vessels and ends as blood sinuses on the end sacs.

Although poorly understood compared with the vertebrate glomerulus, blood pressure close to the end sacs appears to provide a hydrostatic force for ultrafiltration. Presumably, changes in blood volume affect blood pressures in the circulation[66] and hence determine ultrafiltration, thereby affecting urine output.

Modification of urine composition by decapod crustaceans

Antennal glands of many (but not all) decapod crustaceans that live in fresh water or low salinity habitats excrete osmotically dilute urine, containing less than 20 mmol L^{-1} Na^+ and Cl^-. In these species, several lines of evidence suggest sodium pumps (Na^+, K^+-ATPase) are a key component in the active Na^+ reabsorption from the primary urine:

- In species of decapod crustaceans that live exclusively or mainly in fresh water, cells in the distal tubular part of the antennal gland exhibit high immunoreactivity for Na^+, K^+-ATPase. These cells have a well-developed system of basal membrane infoldings associated with mitochondria[67].
- Marine decapod crustacean species consistently have lower Na^+, K^+-ATPase activity in their antennal glands. The antennal gland of marine species either entirely lack a distal nephridial tubule or when

[65] Look back at Figure 7.3 for a reminder of the arrangement of epithelial cells in the filtration barrier of vertebrates.

[66] Section 14.3.3 discusses the circulatory system of crustaceans.

[67] Look back at Figure 7.12 to find similar features in the distal segment of vertebrate nephrons.

present it has poorly developed basal membrane infoldings and few mitochondria. These species produce isosmotic urine with similar Na^+ and Cl^- concentrations to those of their haemolymph, so their antennal glands do not regulate the haemolymph concentrations of NaCl.

Marine and estuarine crustaceans are exposed to high external concentrations of magnesium (Mg^{2+}) and sulfate (SO_4^{2-}); secretion of these ions by the antennal glands is necessary to maintain body fluid composition[68]. As an example, acclimation of giant tiger prawns (*Penaeus monodon*) to a salinity 1.3 times greater than that of ocean seawater results in the excretion of urine containing Mg^{2+} at a concentration 13.5 times higher than in their haemolymph. Their antennal glands secrete the vast majority (95 per cent) of this Mg^{2+}. By contrast, acclimation of giant tiger prawns to a reduced salinity (equivalent to approximately 15 per cent ocean seawater) results in a decline in the urine concentration of Mg^{2+} to 2.3 times that of haemolymph.

The antennal glands of crustaceans secrete or reabsorb calcium ions (Ca^{2+}) depending on environmental factors that influence Ca^{2+} balance (such as food availability, growth, external salinity). Within the glands, a series of membrane proteins transport Ca^{2+} via a transcellular route, and keep cytosolic concentrations stable. These transporter proteins include epithelial Ca^{2+} channels that mediate passive uptake of Ca^{2+} from the primary urine, via the luminal border, plasma membrane Ca^{2+}-ATPase, which functions as a calcium pump on the basolateral border of the cells, and Na^+/Ca^{2+} exchange proteins for extrusion of Ca^{2+} across the basolateral membranes.

Freshwater crayfish reabsorb Ca^{2+} from the primary urine, although to differing extents depending on the stage in the moult cycle[69]. Reabsorption of Ca^{2+} has been found to increase significantly (by as much as twice its initial value) in the post-moult stage of the red swamp crayfish *Procambarus clarkii*, which is native to northern Mexico and southern parts of US, but which has become an invasive pest species in some areas where it was introduced.

The increased Ca^{2+} reabsorption from primary urine during post-moult presumably compensates for the dilution of the haemolymph resulting from the absorption of fresh water prior to **ecdysis** (shedding of the exoskeleton). Hence, we can anticipate that changes in the expression of Ca^{2+} transport proteins occur in the antennal gland and in a general sense facilitate Ca^{2+} absorption in relation to requirements. Figure 7.34 provides some evidence for this idea and shows the increase in abundance of mRNA for Ca^{2+} channels (that allow Ca^{2+} transport into the cells) and two transport proteins involved in Ca^{2+} export via the basolateral membranes, specifically localized to the area of the nephridial canal, when red swamp crayfish are in pre- and post-moult stages.

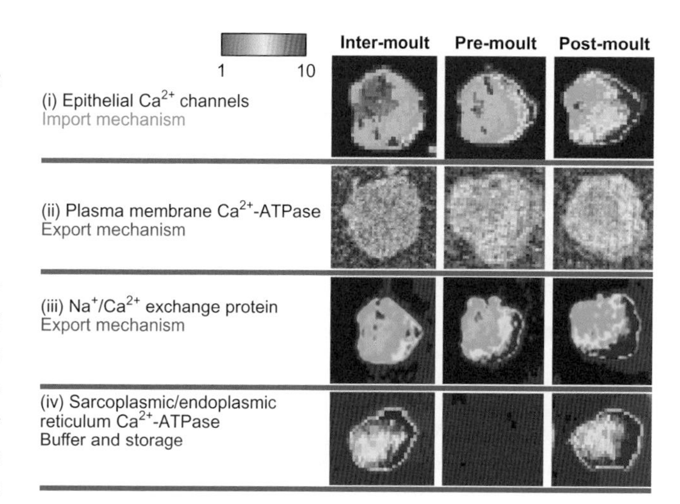

Figure 7.34 Cross-sections of antennal glands of freshwater red swamp crayfish (*Procambarus clarkii*) showing expression of mRNAs for proteins that are critical for calcium reabsorption

Images show sections of antennal glands, approximately 500 μm diameter, from crayfish at three stages in the moult cycle (inter-moult, pre-moult and post-moult) after *in situ* hybridization using molecular probes to examine the localization and abundance of mRNAs for Ca^{2+} transport and cytosolic Ca^{2+} storage. The colours indicates relative levels of expression (blue = 1, low expression, grading through green, and yellow to red = 10, high expression on an arbitrary scale): (i) Epithelial Ca^{2+} channels for absorption of Ca^{2+} (import) from the tubular fluid. (ii) Plasma membrane Ca^{2+}-ATPase for export of cytosolic Ca^{2+} into extracellular fluids on the basolateral border of the epithelial cells. (iii) Na^+/Ca^{2+} exchange proteins for Ca^{2+} export into extracellular fluids via the basolateral membranes. (iv) Sarcoplasmic/endoplasmic reticulum Ca^{2+}-ATPase for buffering Ca^{2+} concentrations in the cytosol, by storage in intracellular organelles. The highest levels of all mRNA—indicated by yellow/red signal—are on the right-hand edge of the images, where the nephridial tubule (canal) occurs. Over this area, mRNA abundance of each of the proteins for Ca^{2+} transport is higher during pre-moult than inter-moult and higher again in post-moult.

Source: Wheatly MG et al (2007). Paradox of epithelial cell calcium homeostasis during vectorial transfer in crayfish kidney. General and Comparative Endocrinology 152: 267–272.

7.3.4 Excretion by insect Malpighian tubules

The Malpighian tubules[70] of insects arise as blind ending tubules from the junction of the midgut and hindgut (ileum and rectum), as shown in Figure 7.35. These are an insect's equivalent of the tubules in vertebrate kidneys, although they do not form a distinct organ but instead lie freely in the body cavity. The number of Malpighian tubules varies

[68] Table 5.1 includes the ionic composition of seawater. Similarly, secretion of Mg^{2+} and SO_4^{2-} occurs in the nephrons of marine teleosts, which we explore in Section 7.2.4.

[69] All crustaceans moult as they grow and need to calcify their new exoskeleton. We discuss storage of Ca^{2+} and its processing in the moult cycle more fully in Section 21.2.5.

[70] Malpighian tubules were named after the Italian Marcelo Malpighi who first described them in 1600.

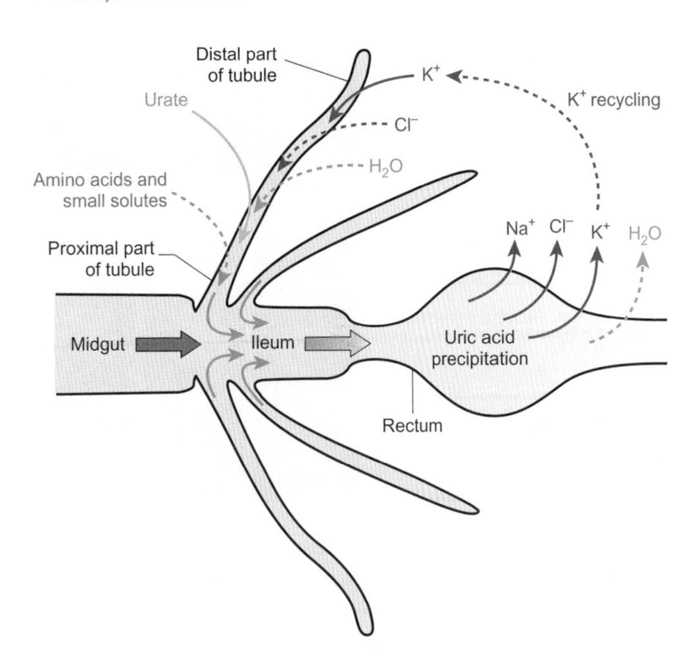

Figure 7.35 Secretion of primary urine by insect Malpighian tubules of fruit flies (*Drosophila*)

Solid lines show active transport processes, dashed lines show passive processes. Malpighian tubules develop as outgrowths from the hind gut, so the proximal part is closest to the hindgut and the blind end is the distal part of the tubule (the opposite of the naming of the vertebrate nephron). In *Drosophila* a third part—the main segment—occurs between proximal and distal segments (not shown). Formation of primary urine is by active secretion of K^+ in most insect species, which draws in Cl^- ions, and water (by osmosis). Active and passive mechanisms transport organic solutes such as urate and amino acids into the tubular fluid. In the rectum (or in the second part of the tubule of some species) K^+ absorption occurs, which enables K^+ recycling and continued fluid secretion. Reabsorption of K^+ and other ions leads to passive water reabsorption, which results in uric acid precipitation in the rectum allowing extra water reabsorption.

between species. For example, grasshoppers and locusts have several hundred, while fruit flies (*Drosophila* spp.) have just two pairs of Malpighian tubules: a posterior pair projects backward on either side of the hindgut and the anterior pair projects forward into the thoracic region.

Each Malpighian tubule consists of a single layer of epithelial cells on a basement membrane. In some species, the secretory cells of the Malpighian tubules consist entirely of large principal cells; these cells are packed with mitochondria and have prominent microvilli on the luminal membrane. However, several species of different insect orders also have smaller cells intercalated between the principal cells, as shown for the Malpighian tubules of *Drosophila* in Figure 7.36. These small cells have a star-like appearance, hence their name: stellate cells.

Malpighian tubules secrete the primary urine

Secretion of primary urine by Malpighian tubules usually involves secretion of K^+ and Cl^-, as indicated in Figure 7.35. Sodium secretion also occurs in some species, particularly those insects that feed on vertebrate blood, which is rich in sodium chloride. The secretion of ions draws in water by osmosis, because of the high water permeability of the tubules. Indeed, insect Malpighian tubules secrete fluid at the highest rate of any known epithelium—as high as their own cell volume every 10 seconds in *Drosophila*. A difference in

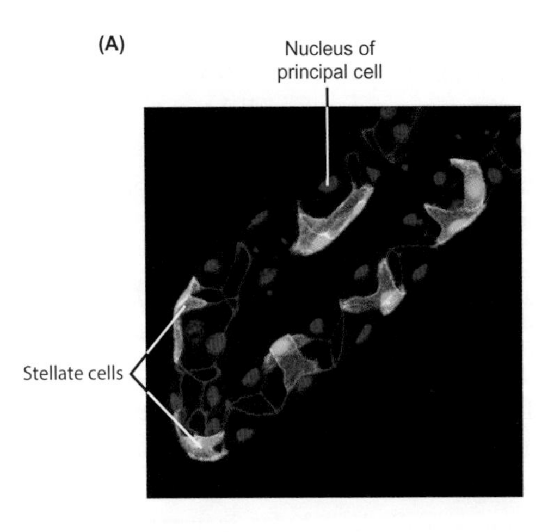

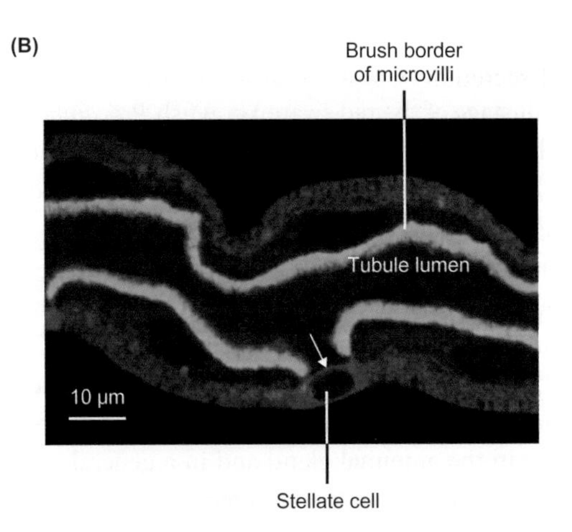

Figure 7.36 Cell types in Malpighian tubules of *Drosophila* and location of V-type H⁺-ATPase (proton pumps)

(**A**) Photomicrograph of a Malpighian tubule showing stellate cells labelled with green fluorescent protein. The septate junctions between the cells are labelled red and nuclei are blue. (The tubules are approximately 35 μm across). (**B**) Immunostaining of tubules after exposure to antisera to V-type H^+-ATPase (proton pumps). The green staining identifies proton pumps in the luminal brush border of the principal cells, which is absent from stellate cells as shown by the gap in the luminal staining (white arrow).

Source: A: Dow JAT, Romero MF (2010). *Drosophila* provides rapid modeling of renal development, function, and disease. American Journal of Physiology Renal Physiology 299: F1237-1244; B: Du J et al (2006). The SzA mutations of the B subunit of the *Drosophila* vacuolar H^+ ATPase identify conserved residues essential for function in fly and yeast. Journal of Cell Science 119: 2542–2551

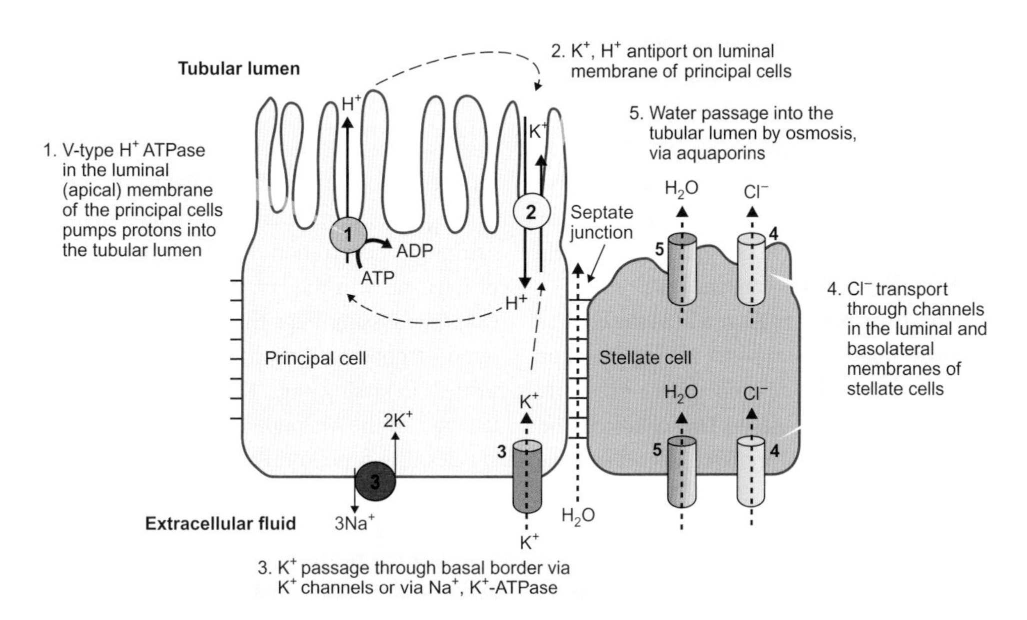

Figure 7.37 Model of secretory processes in Malpighian tubule (main segment) of fruit flies (*Drosophila*)

Annotations 1–3 illustrate the transport steps driving K^+ secretion by principal cells. Annotations 4 and 5 illustrate routes for passive flux of Cl^- via channels on the basal and luminal borders of the stellate cells, and transcellular passage of water across the epithelium. Some water passage may also occur via septate junctions between the cells. Sources: Based on model proposed by Dow JAT, Romero MF (2010). *Drosophila* provides rapid modeling of renal development, function, and disease. American Journal of Physiology Renal Physiology 299: F1237-1244.

osmotic concentration of just a few mOsm kg^{-1} is sufficient to cause water to pass into the tubular lumen. Hence, secretion of primary urine acts as the means of regulating the volume of the haemolymph[71] but not its osmolality as primary urine is almost isosmotic to the insect's haemolymph.

Various other solutes present in haemolymph (such as amino acids) also slowly enter the tubular fluid, apparently via the paracellular route between the epithelial cells. Ions, such as magnesium (Mg^{2+}), inorganic phosphate and uric acid/urate—the main nitrogenous waste of insects[72]—are also transported into the Malpighian tubules. For example, pronounced secretion of Mg^{2+} and SO_4^{2-} occurs in saline-tolerant mosquitoes within a few hours of their exposure to hyperosmotic saline. (We learn more about these mosquitoes later in Case Study 7.1).

Work to characterize the transport processes occurring in Malpighian tubules has mainly concentrated on *Aedes* mosquitoes and *Drosophila*. Although the details are not completely resolved, it *is* clear that Malpighian tubules are unusual in using a proton pump (V-type H^+-ATPase)[73] on the luminal border to drive the secretion of electrolytes, rather than using Na^+, K^+-ATPase (on the basolateral border) as in the metanephridia of leeches examined earlier in Figure 7.32.

Figure 7.36B shows the clear location of proton pumps on the luminal brush border of the principal cells of the Malpighian tubules, but their absence from stellate cells. A high density of proton pumps occurs in the luminal membranes of Malpighian tubules of many insect species, such as ants, locusts, kissing bugs (*Rhodnius* spp.), the beetle *Tenebrio molitor*, and in the main and proximal parts of the Malpighian tubules of fruit flies (*Drosophila* spp.).

Experimental studies of Malpighian tubules have led to the conclusion that proton pumps drive the secretion of K^+ (and in some cases Na^+) in exchange for H^+, as shown in Figure 7.37. Notice in this schematic that the supply of K^+ ions in *Drosophila* occurs via ion channels in the basal epithelium, or via Na^+, K^+-ATPase.

For dipteran species that have both principal cells and stellate cells in their Malpighian tubules, we need to consider the possible role of the stellate cells in the production of primary urine. The small size of these cells has made investigations difficult, to the extent that stellate cells seemed to have little involvement in excretion for many years. However, patch clamping[74] of the luminal cell membrane, which allows the study of ion channels, has revealed the fundamental importance of stellate cells in determining chloride fluxes. Such chloride fluxes may occur via ion channels, especially as recent studies indicate a high level of expression of genes encoding chloride channels in Malpighian tubules.

71 We discuss the hormonal control of the rate of excretion of urine by insects in Section 21.1.4.

72 Production of nitrogenous wastes is discussed in Section 7.4.

73 V-type (H^+)-ATPase is discussed further in Section 4.2.2.

74 Box 16.4 outlines the principle of studying ion channels by monitoring the current resulting from the passage of ions through membrane channels when the membrane potential is clamped at one value (voltage clamping); for patch clamping a patch of membrane is attached to the tip of a glass pipette to enable study of single ion channels.

Figure 7.37 includes chloride channels on both luminal and basolateral borders of the stellate cells, which would allow transcellular fluxes of Cl^-, but the exact pathway remains controversial and probably varies between species, particularly as some species appear to lack stellate cells and rely entirely on principal cells to produce the primary urine. For instance, in *Aedes* mosquitoes, a basolateral Cl^-/HCO_3^- exchanger in stellate cells seems likely to transport Cl^- into the cells.

The secretion of ions drives water passage across the epithelium of the Malpighian tubules. Does water transport occur by a paracellular route—as thought for many years—or via the cells? The discovery of aquaporin genes in *Drosophila*, in the mid-1990s, raised the likelihood that transcellular water passage occurs via water channels[75]. An aquaporin gene known as *DRIP* (Drosophila integral protein) was subsequently found to encode an aquaporin localized specifically to the luminal membrane of stellate cells. Similarly, aquaporin-1 (AQP-1) occurs in stellate cells of the mosquitoes *Anopheles gambiae*.

We should remember that insects are diverse, and not all species possess stellate cells. Consequently, aquaporins may occur in the principal cells of some species, and a paracellular route could still apply in some species. We do not know how much of the transepithelial passage of water by Malpighian tubules goes through the cells, and how much moves between the cells.

[75] Section 4.2.3 discusses aquaporins in some detail.

Modification of the primary urine of insects in Malpighian tubules and hindgut

Insects cannot cope with the loss of all the ions and solutes secreted into the primary urine, so recovery is necessary. In some species, particularly those that excrete urine with a lower osmotic concentration than their haemolymph, reabsorption of ions starts in the last part of the Malpighian tubules. For example, kissing bugs that consume large volumes of blood hypo-osmotic to their haemolymph excrete relatively dilute urine[76]; in kissing bugs, K^+ and Cl^- reabsorption occurs in the late part of the Malpighian tubules. However, for most insect species, the functioning of the hindgut determines the final osmotic and ionic composition of the definitive urine.

Freshwater insect larvae, such as mosquitoes, reabsorb K^+, Na^+ and Cl^- in the hindgut, but not water, so the excreted rectal fluid is typically hypo-osmotic to their haemolymph. For example, the rectal fluid of the larvae of *Aedes aegypti* mosquitoes has an osmolality about one tenth that of the haemolymph and contains Na^+ at 4 mmol L^{-1}, whereas the Na^+ concentration of haemolymph is 87 mmol L^{-1}. In contrast, saline-tolerant species of mosquitoes generate hyper-osmotic excreta in the final part of their rectum; we explore the processess involved in Case Study 7.1.

[76] Figure 21.18 gives the composition of the tubular fluid in the upper (distal) part of the Malpighian tubules of *Rhodnius* compared to the excreted fluid, after reabsorption of K^+ and Cl^-.

Case Study 7.1 Excretory function in larvae of saline tolerant mosquitoes

More than 3500 species of mosquitoes exist, about 95 per cent of which have aquatic larvae that are saline-intolerant. The larvae of these species are restricted either to freshwater habitats, or water with an osmolality well below that of their haemolymph (~300 mOsm kg^{-1}, and osmotically equivalent to about 30 per cent ocean seawater). Figure A illustrates the results of some saline tolerance tests and shows the contrasting survival of several saline-intolerant and saline-tolerant species of mosquitoes.

Only around 5 per cent of mosquito species survive in the salty habitats found in saline marshes, mangrove swamps, hypersaline ponds or mineral springs, in which the osmolality of the external fluids exceeds that of the body fluids. Since mosquitoes act as vectors of malaria and other disease agents, the saline tolerance of these species has implications for their distribution and habitat use that could influence disease transmission.

The mosquito species that are tolerant of saline are dispersed across the phylogenetic tree in different mosquito subfamilies that span 10 genera, which suggests that saltwater tolerance evolved repeatedly, and independently, during the diversification of mosquitoes.

Different osmoregulatory strategies have evolved among the saline-tolerant species of mosquitoes. The larvae of *Culex* and *Culiseta* species osmoconform, i.e. they stay in osmotic balance with the external medium; when exposed to saline they accumulate organic compatible solutes, such as proline and trehalose (a sugar), in their haemolymph[1].

In contrast, the larvae of saline-tolerant species of *Ochlerotatus, Aedes* and *Anopheles* can control their haemolymph osmolality, whether they are in fresh water or in saline with osmotic concentrations of up to or beyond 1000 mOsm kg^{-1} (similar to seawater); these saline-tolerant species are osmoregulators and use quite different mechanisms to osmoregulate.

Examination of the rectal anatomy of 10 species of *Ochlerotatus* and *Aedes* identifies a clear relationship between the saline tolerance of some species and the anatomy of their rectum. The saline-tolerant species have a rectum consisting of two parts, as shown in Figure B, while the obligate freshwater species have a single rectal segment. The epithelium of the posterior segment of the rectum of saline-tolerant species has a deeply folded apical membrane and many associated mito-

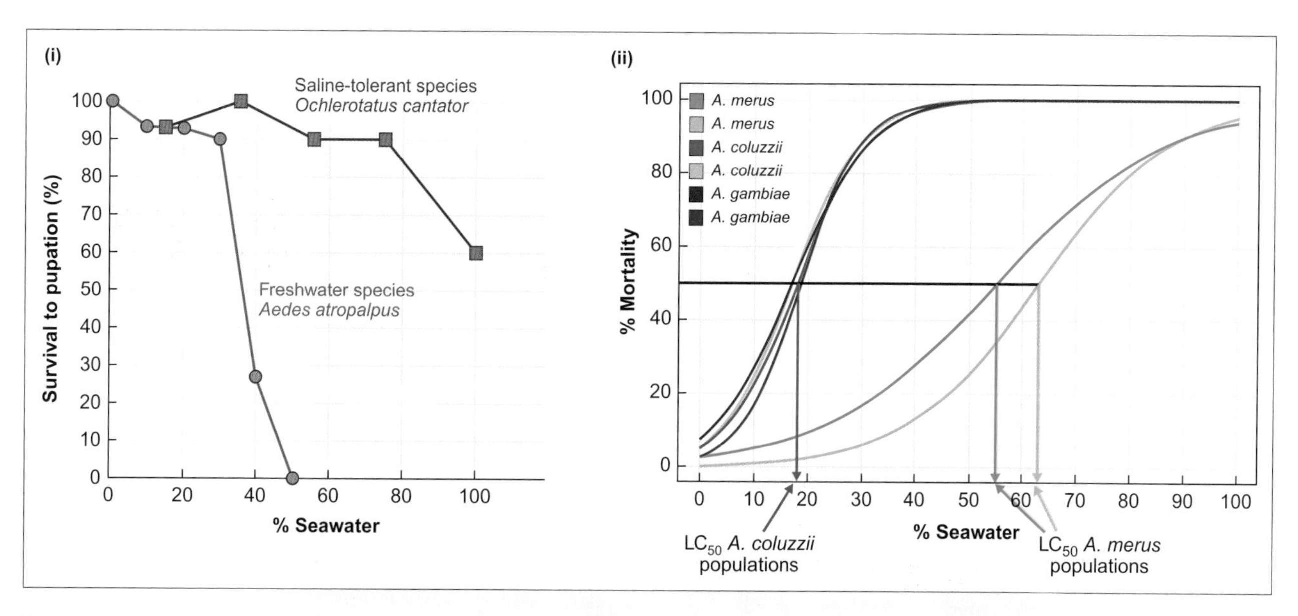

Figure A Survival to the pupal stage of development of mosquito larvae reared in various salinities

(i) Data show the lower survival to pupation of a freshwater species of mosquito larvae (American rock pool mosquitoes *Aedes atropalpus*) in various % ocean seawater compared to the survival of a saline-tolerant species brown saltmarsh mosquito *Ochlerotatus cantator*. (ii) Curves show percentage mortality of three *Anopheles* species of mosquitoes (two populations of each species) fitted to mortality data during chronic exposure to salinity when newly hatched larvae were reared to pupae in various % seawater solutions. *Anopheles merus* (green lines) has a much higher level of saline tolerance than the obligate freshwater species (*Anopheles coluzzii* and *Anopheles gambiae*), as illustrated by comparison of the LC_{50} (lethal concentrations for 50% of the population) for *Anopheles merus* populations (55% and 63% seawater) and *Anopheles coluzzii* (18% seawater); these values are indicated on the graph.

Source: A: Albers MA, Bradley TJ (2011). On the evolution of saline tolerance in the larvae of mosquitoes in the genus *Ochlerotatus*. Physiological and Biochemical Zoology 84: 258-267. B: adapted from White BJ et al (2013). Dose and developmental responses of *Anopheles merus* larvae to salinity. Journal of Experimental Biology 216: 3433-3441.

chondria that are linked to the active transport of ions. The posterior rectum, which is relatively water impermeable, secretes Na^+, K^+, Cl^- and Mg^{2+}, as indicated in Figure B. The secretion of ions generates excreta with fluid osmolality of up to 3000 mOsm kg^{-1} before excretion.

Figure B also shows the processes that occur in the Malpighian tubules and anterior rectum and which create hyposmotic urine when the mosquito larvae are in fresh water or water with osmolality below that of their haemolymph.

All *Anopheles* species have a single rectal compartment regardless of their salt tolerance, but the special feature of the rectum of *Anopheles* larvae is that it consists of two cell types, as shown in Figure C. Cells in the dorsal anterior rectum (DAR) form approximately 25 per cent of the rectum. The other cells form the ventral anterior and posterior parts of the rectum, and are thought to be used in ion excretion or reabsorption, depending upon the localization of key transporter proteins.

The saline tolerance of one species, *Anopheles merus*, which exploits coastal areas in East Africa, has been examined in some detail; some of the results are outlined in Figure C, in which salt

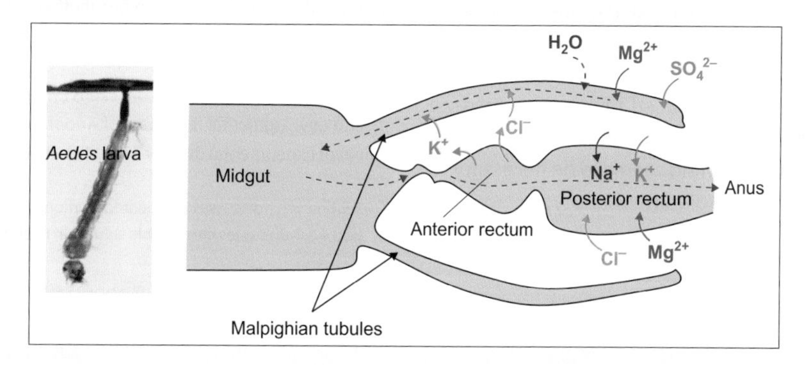

Figure B Diagram of the excretory system of larva of saline-tolerant black salt marsh mosquitoes

Active secretion of magnesium (Mg^{2+}) and sulfate (SO_4^{2-}) occurs in the first part of the Malpighian tubules, followed by potassium (K^+) and chloride (Cl^-) secretion in the later part of the tubule. Water follows the secretion of ions in the Malpighian tubules resulting in an isosmotic urine, which flows into the midgut and then into the rectum. In the anterior rectum, ions are reabsorbed and recycled (as in terrestrial insects), which slightly reduces the osmolality of the fluid. When in hyperosmotic saline, secretion of Na^+, K^+, Mg^{2+} and Cl^- into the relatively water impermeable posterior rectum creates hyperosmotic excreta.

Source: adapted from Bradley TJ. Animal Osmoregulation. Oxford: Oxford University Press, 2009. Image: larva of *Aedes* sp. mosquito; © RC Russell.

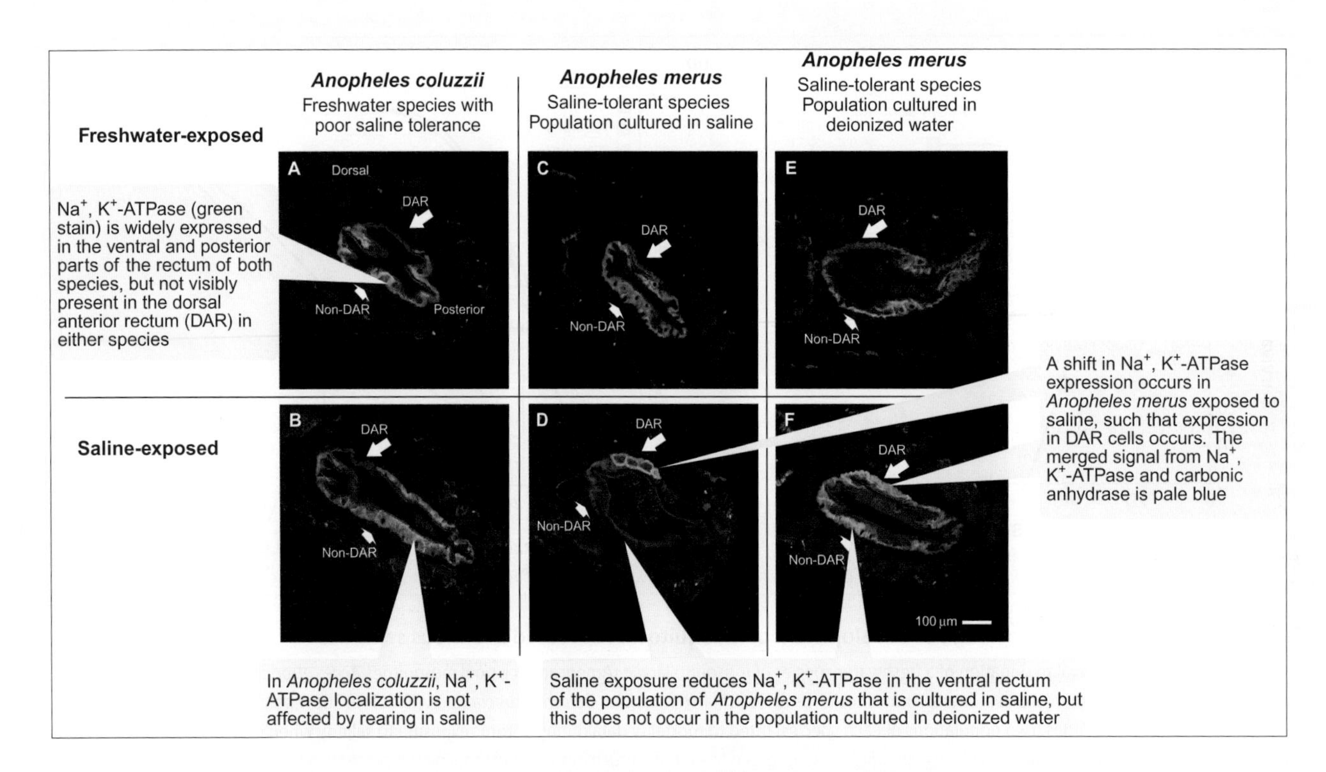

Figure C Rectal localization of major ion transport proteins in larvae of *Anopheles* mosquitoes

Images show cross sections of the rectum of two species of mosquito larvae when reared in fresh water (top three panels: A, C, E) and when reared in saline solutions (bottom three panels: B, D, F), at approximately 20 per cent saline (weight by volume) for *Anopheles coluzzi* (a relatively saline-intolerant species) and using 50 per cent seawater for *Anopheles merus* (a saline-tolerant species). Two populations of *Anopheles merus* were studied: one in which the mosquitoes were cultured in saline equivalent to about 50 per cent seawater since collection in the early 1990s (panels C, D), and one cultured in deionized water (panels E, F). Antisera were used to (i) delineate the lumen of the rectum by localizing Na^+/H^+ exchangers (red stain), (ii) localize carbonic anhydrase (blue stain) that occurs only in the dorsal anterior rectum (DAR), and (iii) localize Na^+, K^+-ATPase (green stain). All micrographs are at the same magnification.

Source: White BJ et al (2013). Dose and developmental responses of *Anopheles merus* larvae to salinity. Journal of Experimental Biology 216: 3433-3441.

tolerance (and culture in 50 per cent seawater) is associated with a shift in the localization of Na^+, K^+-ATPase compared to rearing in fresh water. Exactly how this process is mechanistically linked to salt secretion is not yet understood.

Further studies on *Anopheles merus* demonstrate that inheritance of saline tolerance interacts with its phenotypic expression. Complete expression of saltwater tolerance requires early exposure of the larvae to saline during a relatively brief developmental window. When saline exposure is delayed by more than 24 h, even saline-tolerant species have reduced tolerance compared to that of larvae exposed to saline as hatchlings. Thus, an initial saline exposure within 24 h of hatching seems to be necessary to allow full expression of the saline-tolerant phenotype.

› Find out more

Albers MA, Bradley TJ (2011). On the evolution of saline tolerance in the larvae of mosquitoes in the genus *Ochlerotatus*. Physiological and Biochemical Zoology: Ecological and Evolutionary Approaches 84: 258–267.

Smith KE, Raymond SL, Valenti ML, Smith PJS, Linser PJ (2010). Physiological and pharmacological characterization of the larval *Anopheles albimanus* rectum supports a change in protein distribution and/or function in varying salinities. Comparative Biochemistry and Physiology A Molecular and Integrative Physiology 157: 55–62.

White BJ, Kundert PN, Turissini DA, Ekeris LV, Linser PJ, Besansky NJ (2013). Dose and developmental responses of *Anopheles merus* larvae to salinity. Journal of Experimental Biology 216: 3433–3441.

[1] Section 5.1.1 discusses osmoconformation of marine invertebrates, and Section 4.3.1 discusses compatible solutes in relation to cell volume regulation.

In terrestrial insects, the hindgut usually consists of a narrower anterior part—the ileum—and a broader rectum, as shown back in Figure 7.35. The ileum and rectum retrieve variable amounts of water, and solutes such as glucose, amino acids and salts. For example, reabsorption of K^+ and Cl^- occurs in the ileum and rectum of desert locusts (*Schistocerca gregaria*). In the ileum, water may follow the ions by osmosis, reducing the volume of urine, such that the fluid remains isosmotic. When desert locusts are feeding on foliage with high water content, the rectum is relatively impermeable to

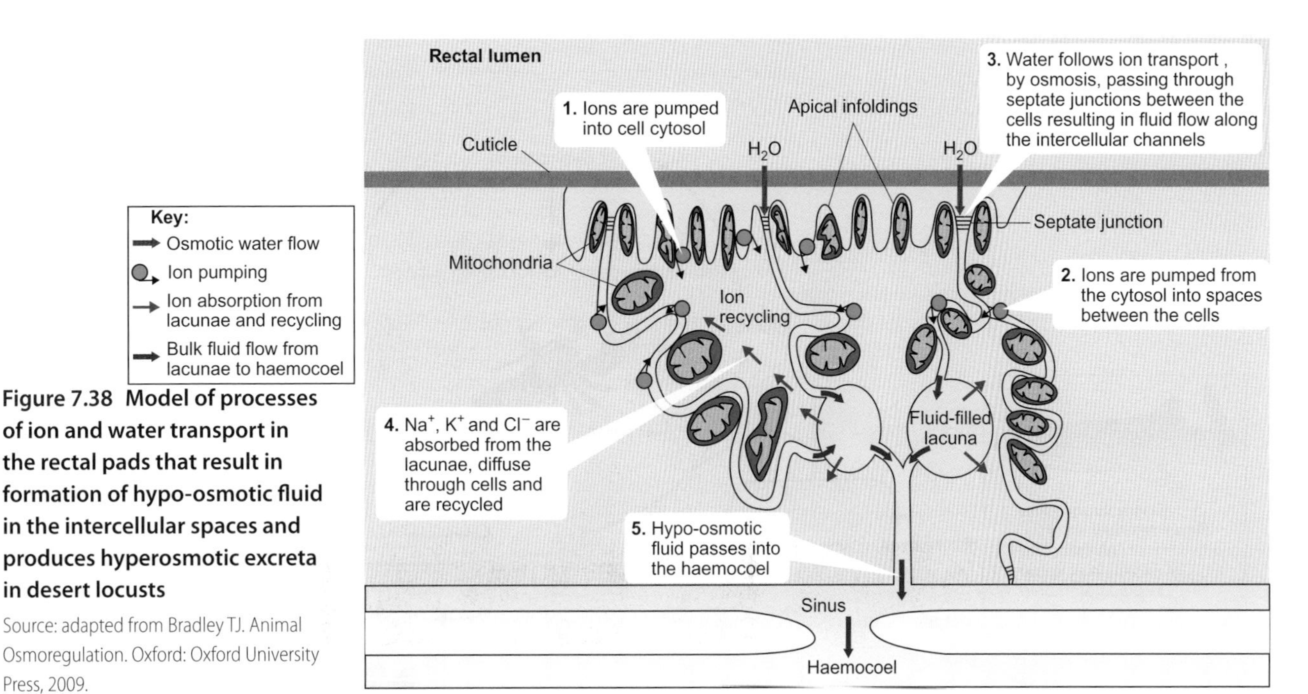

Figure 7.38 Model of processes of ion and water transport in the rectal pads that result in formation of hypo-osmotic fluid in the intercellular spaces and produces hyperosmotic excreta in desert locusts

Source: adapted from Bradley TJ. Animal Osmoregulation. Oxford: Oxford University Press, 2009.

water so ion absorption from the rectal fluid generates excretory fluids that are hypo-osmotic to haemolymph. However, when the locusts feed on relatively dry food material, the excreta become hyperosmotic to haemolymph[77].

Desert locusts, cockroaches and blowfly can produce excreta up to five times more concentrated than their haemolymph due to the transport processes in the thickened areas of columnar epithelium forming the rectal pads arranged radially around the rectum. Figure 7.38 is a diagrammatic representation of the structures and processes in the rectal pads that extract hypo-osmotic fluid from the rectum and thereby generate hyperosmotic excreta:

- Mitochondria associated with dense infoldings of the apical membrane fuel active transport of ions from the rectal fluids into the cytosol.
- Highly convoluted intercellular channels associated with many mitochondria energize active ion transport from the cytosol into intercellular channels, which draws in water from the rectal lumen (by osmosis).
- Fluid flows from the luminal side of the intercellular channels toward the basal side of the cells where the channels open into large fluid spaces (lacunae).
- Reabsorption of Na^+, K^+ and Cl^- from the lacunae and the ducts into which they drain produces fluid that is hypo-osmotic to haemolymph (approximately 60 mmol L^{-1} Na^+ and 110 mmol L^{-1} Cl^- compared with 145 mmol L^{-1} Na^+ and 160 mmol L^{-1} Cl^- in haemolymph).
- This hypo-osmotic fluid drains into the sinus below the epithelial cells and enters the haemocoel.
- The continual recycling of ions absorbed from the lacunae and lower part of channels back into the upper parts of the cells and channels allows the continued withdrawal of water from the rectum, which generates increasingly hyperosmotic excreta.

The structural adaptations of the rectum and transport processes enable conservation of water from both urine and faeces. By contrast, mammals cannot raise the osmolality of the fluid component of their faeces beyond that of the plasma.

An additional factor that contributes to the conservation of water in the hindgut of insects results from their excretion of uric acid as nitrogenous waste. The absorption of water by the hindgut increases the concentration of uric acid, resulting in its precipitation. In removing solute by precipitation, further water absorption by osmosis can occur. We examine uric acid excretion in more detail in the next section, in which we explore the various forms of nitrogenous excretion among animal groups.

› *Review articles*

Beyenbach KW, Skaer H, Dow JAT (2010). The developmental, molecular, and transport biology of Malpighian tubules. Annual Review of Entomology 55: 351–7.

Dow JAT (2012). The versatile stellate cell—More than just a space-filler. Journal of Insect Physiology 58: 467–472.

Dow JAT, Davies SA (2003). Integrative physiology and functional genomics of epithelial function in a genetic model organism. Physiological Reviews 83: 687–729.

[77] We compare the water budget of locusts feeding on moist food and dry food in Figure 6.35.

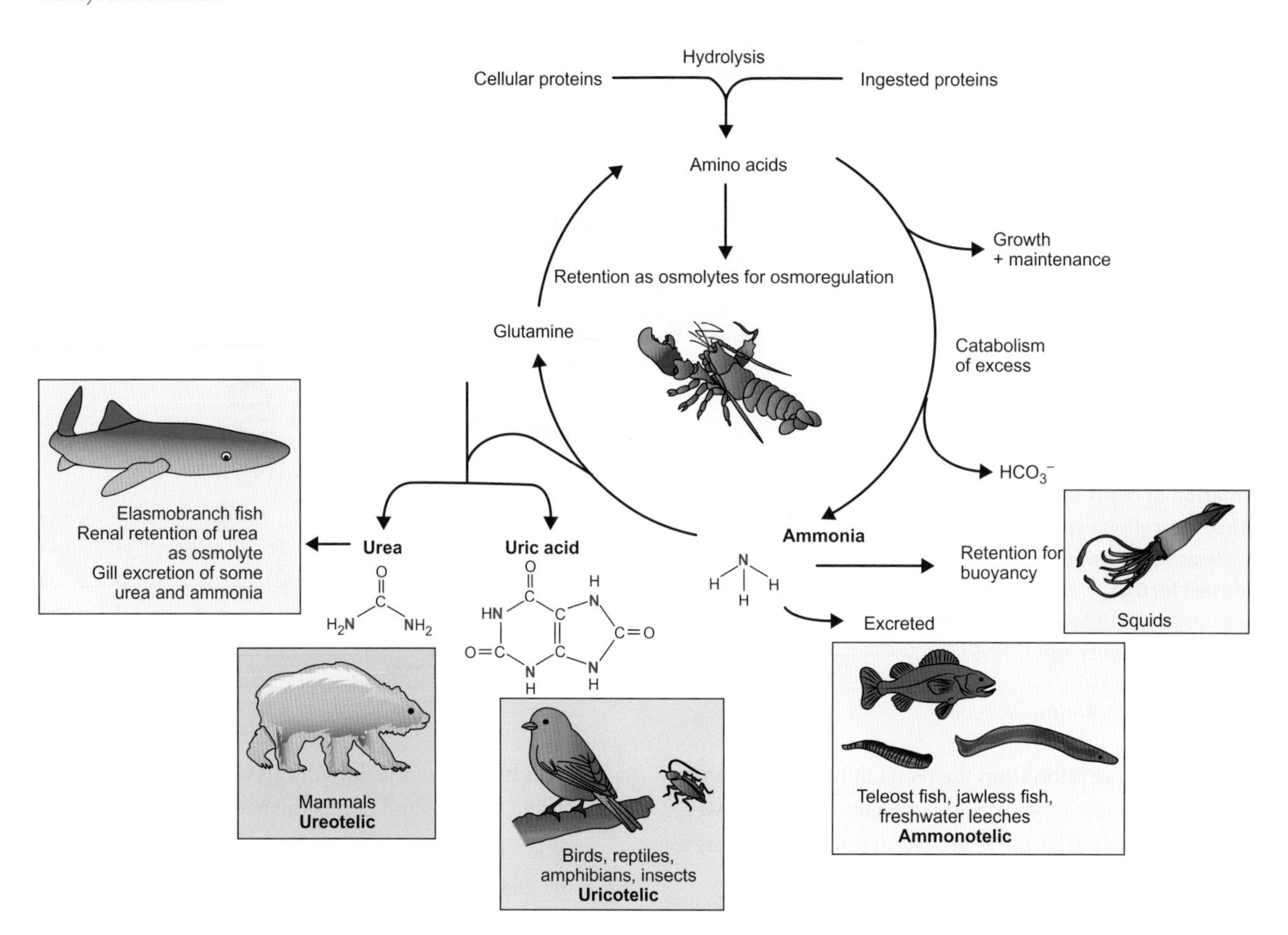

Figure 7.39 Overview of nitrogen metabolism and nitrogenous waste products excreted by animals

Aquatic animals are usually ammonotelic, but elasmobranch fish (sharks, skates and rays) convert ammonia to urea, which is used to increase plasma osmolality (as we discuss in Section 5.1.4). High plasma concentrations of urea in elasmobranchs lead to loss at the gills, along with excretion of smaller amounts of ammonia. Mammals are ureotelic—they excrete urea in the urine. Birds, reptiles and most insects are uricotelic—they excrete uric acid from the kidney or Malpighian tubules in a mix of faeces and urine. Amino acids are retained by marine invertebrates as important osmolytes in their osmoconformation and cell volume regulation, which we discuss in Section 5.3.1. A few specialized invertebrates, such as squids, hold ammonia in buoyancy chambers where it replaces heavier ions such as Ca^{2+}, Mg^{2+} and SO_4^{2-}. Freshwater leeches (one of the few freshwater invertebrates in which nitrogenous excretion has been studied) primarily excrete ammonia.

Source: adapted from schematic of Wright PA (1995). Nitrogen Excretion: Three end products, many physiological roles. Journal of Experimental Biology 198: 273-281.

Dow JAT, Romero MF (2010). *Drosophila* provides rapid modeling of renal development, function, and disease. American Journal of Physiology Renal Physiology 299: F1237–1244.

Freire CA, Onken H, McNamara JC (2008). A structure–function analysis of ion transport in crustacean gills and excretory organs. Comparative Biochemistry and Physiology, Part A 151: 272–304.

Wheatly MG, Gao Y, Gillen CM (2007). Paradox of epithelial cell calcium homeostasis during vectorial transfer in crayfish kidney. General and Comparative Endocrinology 152: 267–272.

7.4 Nitrogenous excretion

Figure 7.39 provides an overview of nitrogen metabolism and excretion in animals. A small part of the nitrogenous excretion (about 5 per cent) results from the metabolism of nucleic acids, while the majority of the nitrogenous waste results from the catabolism of the excess amino acids obtained from proteins in the diet and protein turnover. Nitrogenous excretion is therefore highly dependent on the diet. Herbivores have a relatively low intake of protein and excrete less nitrogen than carnivores whose high protein diet typically provides more amino acids than needed for growth and tissue repair, and in which the excretion of nitrogenous waste generally increases after feeding.

Based on the particular nitrogenous substance(s) excreted as shown in Figure 7.39 we can distinguish three main groups of animals:

- **Ammonotelic** animals, which by definition excrete more than 50 per cent of their nitrogenous waste as ammonia resulting from the deamination of the amino group ($-NH_2$) of amino acids and addition of an H atom (chemical reduction) in the liver and other tissues. The production of ammonia does not require ATP, so it is an ideal nitrogenous waste, as long as the toxic effects of ammonia outlined in Box 7.3 are avoided by rapid excretion and dilution. Each molecule of ammonia (NH_3) contains a single N atom.

Box 7.3 Ammonia tolerance and toxicity

The sensitivity of animals to ammonia varies between species. Some insects tolerate high environmental and internal concentrations. For example, Australian sheep blowflies (*Lucilia cuprina*) tolerate haemolymph concentrations exceeding 10 mmol L^{-1}. Among species less tolerant of ammonia, aquatic species are generally more tolerant of ammonia in their blood than terrestrial species. Teleost fish usually tolerate total ammonia concentrations (NH_3 plus NH_4^+) in plasma of 0.5 to 1 mmol L^{-1}, but at higher levels deaths may occur. Haemolymph concentrations of total ammonia in aquatic crabs are usually 0.05 to 0.4 mmol L^{-1}; at higher concentrations, toxic effects occur.

Toxicity of ammonia is not usually due to a single effect, but instead results from the combination of effects that occur in different tissues. Some of the most significant effects are:

- Ammonia (NH_3) formed by glutamate deamination is released in the mitochondria, where rapid combination with protons may disrupt the proton gradient across the inner mitochondrial membrane and impede **oxidative phosphorylation**[1]. A reduction in ATP will have diverse effects on tissue function.
- Ammonium ions (NH_4^+) can substitute for K^+ in various transporter proteins, probably because NH_4^+ and K^+ have an identical hydrodynamic radius[2]. Thus, ammonium ions can enter cells via K^+ channels, thereby depolarizing the membrane potential. This results in a multiplicity of subsequent effects, such as effects on neurotransmission. Substitution of NH_4^+ for K^+ in the sodium pump, Na^+, K^+-ATPase[3], can deplete intracellular K^+. Substitution of NH_4^+ for K^+ in Na^+, K^+, 2 Cl^- cotransporters can disrupt ion balance.
- In the mammalian brain, glutamine synthetase produces glutamine from NH_4^+ and glutamate. Glutamine is often considered a means of ammonia detoxification in the brain, particularly in the **astrocytes**[4] (which provide metabolic support to the neurons) where glutamine synthetase is mainly located. However, recent studies suggest that elevated concentrations of ammonia in the mammalian brain that increase glutamine synthesis actually result in toxic action on the brain, rather than alleviating toxicity. This is known as the Trojan horse hypothesis, in which glutamine—the Trojan horse—enters the mitochondria of the cells via a carrier; hydrolysis of glutamine in brain mitochondria then unloads NH_3 and induces the production of reactive oxygen species[5] that cause mitochondrial dysfunction. The subsequent decline in ATP production may cause a failure to regulate cell volume[6] (which depends on Na^+, K^+-ATPase) and hence explain the swelling of brain astrocytes in patients with liver failure, in which ammonia concentrations increase because of the reduced synthesis of urea from ammonia.

[1] Figure 2.24 illustrates the proton gradient across the inner mitochondrial membrane and oxidative phosphorylation.

[2] The hydrodynamic radius of ions is determined by the formation of hydration shells as we examined in Figure 4.4.

[3] The operation of the sodium pump (Na^+, K^+-ATPase) which pumps K^+ into cells in exchange for Na^+ is discussed in Section 4.2.2.

[4] Figure 16.3 illustrates an astrocyte; their functioning is discussed in Section 16.1.

[5] Reactive oxygen species and their damaging effects on cell function are discussed in Section 2.4.3.

[6] Cell volume regulation is discussed in Section 4.3.

- **Ureotelic** animals synthesize and excrete most (usually defined as more than 50 per cent) of their nitrogenous waste as urea. Note in Figure 7.39 that each urea has two N atoms per molecule.
- **Uricotelic** animals synthesize uric acid (or related urate salts) as their main nitrogenous waste. Note that each molecule of uric acid contains four N atoms.

Water availability plays a pivotal role in determining the patterns of nitrogenous excretion, which do not follow clear phylogenetic groupings. As such, aquatic animals (teleost fish, amphibian tadpoles, aquatic insect larvae and freshwater leeches) mostly excrete ammonia across their gills or skin. By contrast, terrestrial animals (adult amphibians, mammals) mostly excrete urea in their urine. However, excretion of uric acid is associated with the least water and uric acid is the major nitrogenous product in birds, most reptiles, most insects, and a few amphibians, as indicated in Figure 7.39. In addition, changes in water availability during the development of some species result in changes in the type of nitrogenous product excreted, as we learn in the following sections.

7.4.1 Ammonia excretion

Ammonia (NH_3) is a gas at physiological temperatures, and some terrestrial invertebrates (land snails, isopods and some land crabs) excrete gaseous ammonia into the air. However, most ammonotelic animals live in aquatic habitats and excrete NH_3 into the surrounding water. The water generally dilutes and disperses the excreted ammonia, such that passive diffusion can continue to be driven by the concentration differences (partial pressure differences) across epithelia in contact with the external water.

Ammonia dissolved in water acts as a weak base, whereby it rapidly associates with protons (H^+) from the water (H_2O, which acts as an acid) to form the ammonium ion (NH_4^+) and OH^-. The ammonium ion is a weak acid (it will donate a proton, albeit not that readily, and it reacts with bases, for example OH^-, in water to reform NH_3 and H_2O). Equation 7.3 combines these two reactions. Hence, in any fluid there is a pH-dependent equilibrium between NH_3 and NH_4^+, which form a conjugate acid–base pair.

$$NH_3 + H_2O \rightleftharpoons NH_4^+ + OH^- \qquad \text{Equation 7.3}$$

The total ammonia in any aqueous solution is the sum of NH_3 and NH_4^+. Temperature, and to a lesser extent salinity (in natural environments) affect the proportions of NH_3 and NH_4^+, but pH has the greatest effect. At a pH of 9.5 (the value for the acid dissociation constant (pK_a) of NH_4^+ at 15°C), there are equal proportions of NH_4^+ and NH_3 present at equilibrium[78]. If the pH is below the pK_a, the equilibrium shifts to the right, increasing the proportion of NH_4^+ molecules present relative to NH_3. At a pH above the pK_a (when the concentration of H^+ is low), the equilibrium shifts to the left, generating a greater proportion of NH_3.

Physiological solutions such as body fluids generally have a pH between 7.4 and 8.0 (i.e. below the pK_a of NH_4^+) so the equilibrium of Equation 7.3 shifts to the right, and NH_4^+ dominates. The small proportion of NH_3 in body fluids readily crosses cell membranes, driven by the differences in partial pressures[79], which enables excretion of NH_3 by aquatic animals.

Ammonia excretion often occurs via membrane proteins that are part of the Rhesus (Rh) complex of proteins, so-named because of their relationship in a molecular sense to the Rhesus blood group antigen of humans. However, the Rh proteins that transport ammonia are functionally quite different and have nothing to do with blood groups or immunity. The Rh proteins transporting NH_3 allow passage in either direction depending on the difference in concentration on either side of the membrane. These proteins evolved among early invertebrates; they occur in the skin of a freshwater platyhelminth (*Schmidtea mediterranea*), a soil nematode (*Caenorhabditis elegans*), freshwater ribbon leeches (*Erpobdella obscura*), the anal papillae of mosquito larvae (*Aedes aegypti*), skin of African clawed frogs (*Xenopus laevis*), and the gills of crabs and teleost fish.

In some aquatic habitats, ammonia concentrations are too high for excretion by passive diffusion to occur routinely. For example, total ammonia may reach 2.5 mmol L^{-1} within a few centimetres of the surface of marine sediments. Consequently, animals may supplement the passive efflux of NH_3 by excretion of NH_4^+. By contrast to NH_3 permeability, the permeability of cell membranes to NH_4^+ is generally low, but interaction with active transport proteins may enable NH_4^+ excretion in some species, as we learn in the following sections.

Ammonia excretion by fish

The nitrogenous waste of most teleosts (and lampreys) is 80 to 90 per cent ammonia, with small amounts of urea accounting for the remainder. Ammonia excretion by teleost fish has been a controversial topic of research since 1929, when the classic experiments of Homer Smith first showed the nitrogenous excretion of fish to be primarily ammonia, excreted via their gills.

Figure 7.40 illustrates a current model of the key processes for ammonia excretion by freshwater teleosts. In this model, diffusion of NH_3 across the gill epithelia (process 1), due to the difference in partial pressures, is a major route for excretion. Passage of NH_3 through Rh proteins also occurs in the basolateral and apical membranes of the epithelial cells. These transporters are denoted as Rhag, Rhbg, and Rhcg (g indicates glycosylation; a, b and c are different types). Notice in Figure 7.40 that Rhbg occurs in the basolateral membranes, while Rhcg is in the apical membranes. The third type (Rhag) occurs in erythrocyte cell membranes where it may facilitate the transfer of ammonia into the extracellular fluids in gill tissue.

The first clear evidence for a significant role of the Rh proteins in the day-to-day ammonia excretion came from studies of larval zebrafish (*Danio rerio*), in which knockdown of *Rhag*, *Rhbg* and *Rhcg* genes inhibited ammonia excretion. Recent studies suggest that aquaporins[80] may also facilitate ammonia excretion by some epithelia. In particular, knockdown of a particular subtype of aquaporin reduced ammonia efflux via the epithelium covering the yolk sac of zebrafish larvae, by approximately 30 per cent.

High external concentrations of NH_3 may adversely affect the concentration differences that drive passive diffusive loss of NH_3, and may even cause ammonia influx from the environment into the extracellular fluids. However, there is evidence that incorporation of extra Rh proteins into the cell membranes of the gill epithelia, in such conditions, facilitates NH_3 excretion. For example, in rainbow trout, elevated external concentrations of ammonia result in an increase in the mRNA transcripts of the gene for apical Rhcg.

Continuous excretion of ammonia by fish appears to rely on bonding of the excreted NH_3 to protons (H^+) in the unstirred boundary layer (close to the gills) to form NH_4^+. This process is known as 'acid trapping', and is included in Figure 7.40. The idea behind acid trapping is that it maintains a sufficient concentration difference for continued diffusion of NH_3 and its passage through membrane channels.

Creation of an acidic environment in the boundary layer of the gill is thought to depend on multiple loosely linked transport processes. Proton pumps occur in the apical membranes of the gills of freshwater teleosts, as illustrated in Figure 7.40, in which they may drive Na^+/H^+ exchange[81]; as such, ammonia excretion is loosely (but not directly) linked to Na^+ uptake in a type of functional **metabolon**. An additional source of protons is the hydration of carbon dioxide

[78] We discuss acids and bases, and the concept of dissociation constants in more detail in Section 13.3.

[79] Ammonia can pass across cell membranes like other gases (for example O_2 and CO_2); we discuss the principles of gas transport driven by differences in partial pressure in Chapter 11.

[80] Aquaporins (water channels) are best known in relation to water passage across epithelial membranes, as we discuss in Section 4.2.3.

[81] Section 5.2.2 discusses ion transport mechanisms in the gills of freshwater teleosts.

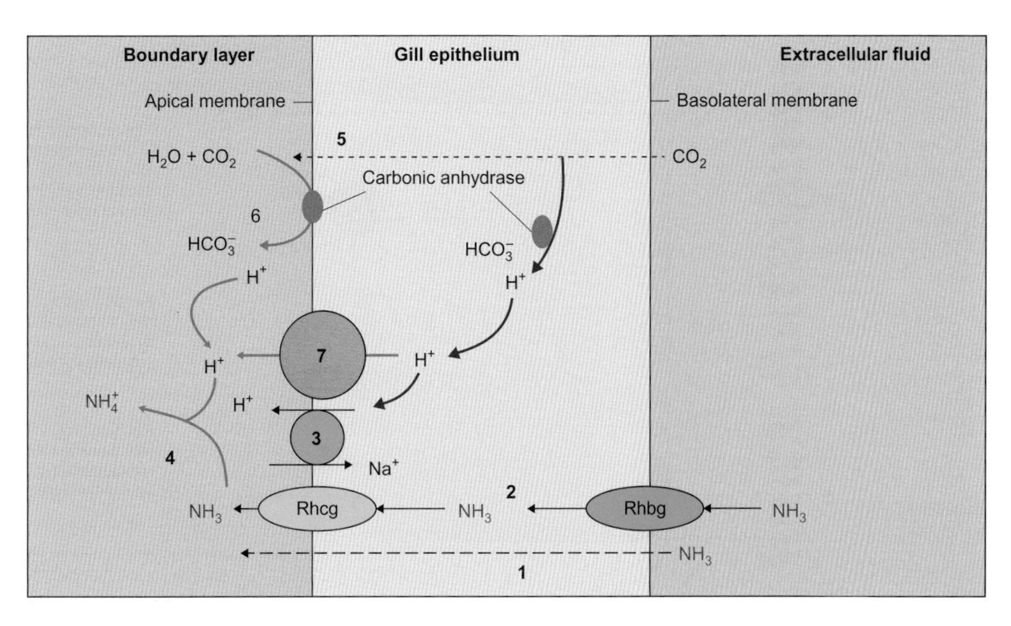

Figure 7.40 Model of ammonia (NH_3) transport and alkalinization in the boundary layer of the gill epithelium of freshwater teleost fish

Ammonia (NH_3) passage occurs across the gills by (1) simple diffusion across the epithelium (dashed line) and (2) via two types of Rhesus proteins that act as ammonia transporters in the gill epithelial cells: Rhbg in the basolateral membranes and Rhcg in the apical membrane. Apical passage of NH_3 appears to be coupled to Na^+ uptake across the apical membrane via Na^+/H^+ exchange proteins (3) and Na^+ channels (not shown; Section 5.2.2 discusses the ion transporting mechanisms of the gill that map onto the processes shown in this figure). In the boundary layer of water, excreted ammonia bonds with protons (H^+) to form NH_4^+ (4). Protons in the boundary layer result from (5) diffusion of CO_2 across the gill followed by (6) its hydration in the boundary layer catalysed by carbonic anhydrase, to form H^+ and bicarbonate ions (HCO_3^-), and (7) transport of protons by proton pumps in the apical membrane of the gill. Formation of NH_4^+ maintains a difference in the partial pressure of NH_3 across the gill membranes, which allows continued excretion of NH_3 by simple diffusion and via the Rhesus proteins.

Source: adapted from Wright PA, Wood CM (2009). A new paradigm for ammonia excretion in aquatic animals: role of Rhesus (Rh) glycoproteins. Journal of Experimental Biology 212: 2303–2312.

in the boundary layer or in the gill epithelium, both of which are included in Figure 7.40.

Some studies suggest that some marine teleosts can excrete NH_4^+ across their gills. The idea is that the substitution of NH_4^+ for K^+ in the basolateral sodium pumps (Na^+, K^+-ATPase) carries NH_4^+ into the cells. The excretion of NH_4^+ from the gill cells may involve apical Na^+/H^+ exchange proteins, with NH_4^+ acting as an alternative exchange ion to H^+. This is an old idea, but remains controversial, and species variations seem to occur. However, air-breathing giant mudskippers (*Periophthalmodon schlosseri*), which live in burrows in mangrove swamps, seem to make use of sodium pumps (Na^+, K^+-ATPase) to excrete ammonia, based on the inhibition of ammonia excretion by ouabain, which inhibits Na^+, K^+-ATPase. A similar process seems to occur in crustaceans, which we look at next.

Ammonia excretion by crustaceans

As in teleost fish, ammonia excretion across the gills of crustaceans involves a series of processes, which may vary between species. A model of the processes in shore crabs (*Carcinus maenas*), shown in Figure 7.41, includes transcellular diffusion of NH_3, paracellular passage of NH_3 or NH_4^+, and basolateral influx of NH_4^+ via sodium pumps (Na^+, K^+-ATPase) and K^+ channels. Transport of ammonia across the cells is thought to occur in cytoplasmic vesicles that fuse with the apical membrane and excrete NH_4^+ from where it diffuses across the cuticle.

Intertidal animals, such as crabs, routinely experience periods in air (between tides) when they cannot effectively excrete ammonia across their gills. In such circumstances, ammonia-transporting Rh proteins in the basolateral membranes of branchial cells could transport ammonia back into the extracellular fluids. Such a mechanism is included in Figure 7.41.

Some crab species deliberately venture further onto land and for longer periods, which raises a question: How do these species solve the apparent problem of excreting nitrogenous wastes when they are out of water for long periods?

Some species of crabs can store nitrogenous waste as alternative products, such as the amino acid glutamine or urate[82], which they re-convert to ammonia before excretion. For example, an air-breathing species of semi-terrestrial crab (freshwater crab, *Austrothelphusa transversa*) native to arid parts of inland Australia survives long periods of drought in a deep burrow only emerging when the rains start, at which point it functions as an aquatic (freshwater) crab. While in its burrow, nitrogenous waste is stored as urate. The need to re-immerse in water to excrete nitrogenous waste as ammonia,

[82] Equation 7.4 explains the formation of urates from uric acid.

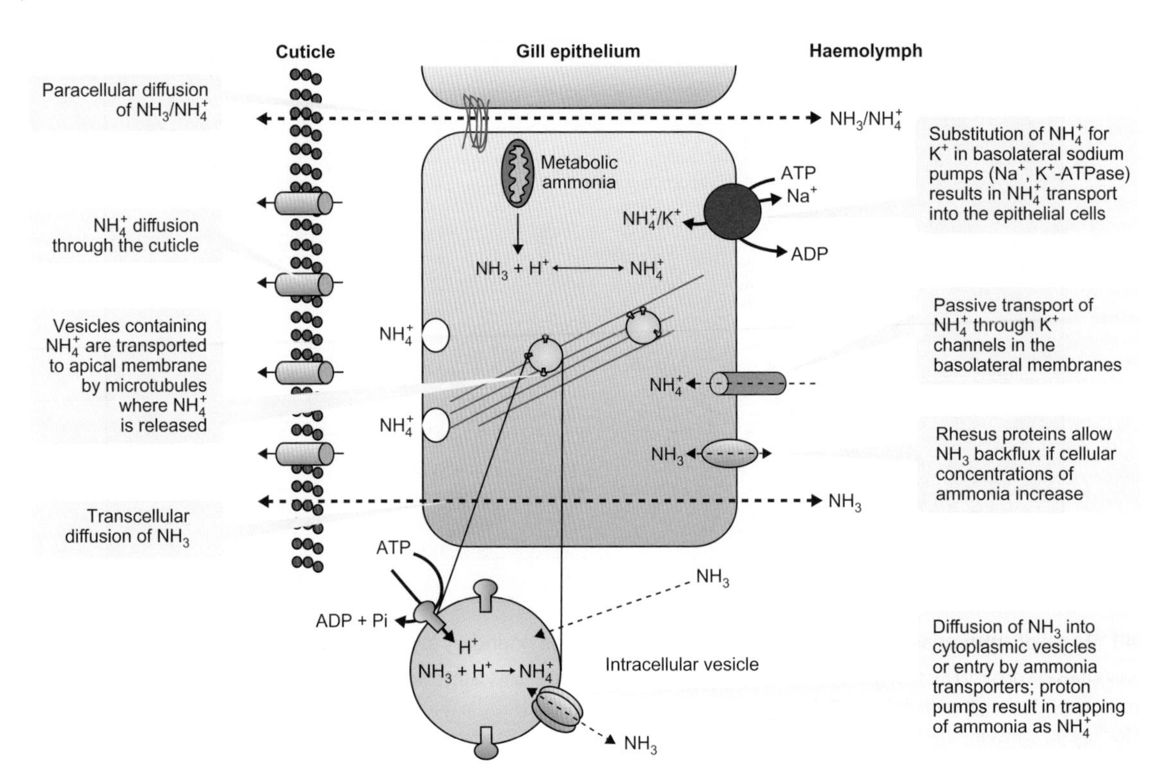

Figure 7.41 Model of ammonia excretion across gill epithelium and cuticle of shore crabs, also known as European green crabs *Carcinus maenas* Source: adapted from Weihrauch D et al (2004). Ammonia excretion in aquatic and terrestrial crabs. Journal of Experimental Biology 207: 4491-4504.

by branchial mechanisms, may ultimately limit the duration that semi-terrestrial crabs spend away from water.

Nitrogenous excretion is potentially most problematic for those land crabs that spend most of their time on land, and only rarely immerse themselves. These terrestrial crabs have evolved a neat solution: they excrete ammonia into a chamber of urine that covers their gills. Reabsorption of salts in the urine occurs in this chamber, in exchange for NH_4^+ excretion across the gill membranes. In Christmas Island red crabs (*Gecarcoidea natalis*), branchial exchange of Na^+ in the urine for NH_4^+ increases the concentration of ammonia in the chamber from 0.36 mmol L^{-1} to as much as 10.8 mmol L^{-1}. In Atlantic ghost crabs (*Ocypode quadrata*), it seems that NH_3 excretion occurs across the gills, and bonding with protons forms NH_4^+ in the acidic urine (pH ~5.4).

We can imagine a drawback to relying on urine production to provide the fluid into which land crabs excrete their nitrogenous waste. What happens when urine production declines dramatically in the dry season? It seems that in such conditions, land crabs store nitrogenous waste as urates or purines, such as guanine[83], in connective tissue distributed throughout their bodies.

Some insects excrete ammonia

Insects generally excrete uric acid as their main nitrogenous excretory product. However, studies from the 1970s onward have found that insect species deal with nitrogenous waste in a variety of ways, including the excretion of ammonia and urea. It is perhaps not surprising that the particular excretory product seems to depend on the habitat and feeding habits of particular species.

Early studies detected ammonia excretion in the hindgut of the larvae of the grey flesh flies (*Sarcophaga bullata*), a species that ingests rotting meat, and which therefore has to deal with large amounts of ammonia, but the mechanism of ammonia excretion in flesh flies is not understood.

Ammonia excretion by insects predominantly occurs among those species or life stages that live in water and which typically have a higher tolerance of ammonia than other animal groups. For example, haemolymph concentrations of up to 5 mmol L^{-1} are tolerated by aquatic blackfly larvae and ammonia up to 1.5 mmol L^{-1} is tolerated by mosquito larvae (*Aedes aegypti*) compared to a tolerance of <0.4 mmol L^{-1} by mammals.

Mosquito larvae excrete ammonia across their anal papillae[84]. The photograph in Figure 7.42A shows one of these finger-like outpushings, which are covered with a single layer of epithelium in contact with the surrounding water. The lumen of each anal papillae is continuous with the haemocoel. Figure 7.42A also shows two ion-selective electrodes recording NH_4^+ and H^+. Notice in Figure 7.42B

[83] The chemical formula of guanine is $C_5H_5N_5O$.

[84] Anal papillae are also used for salt uptake, which we discuss in Section 5.2.2.

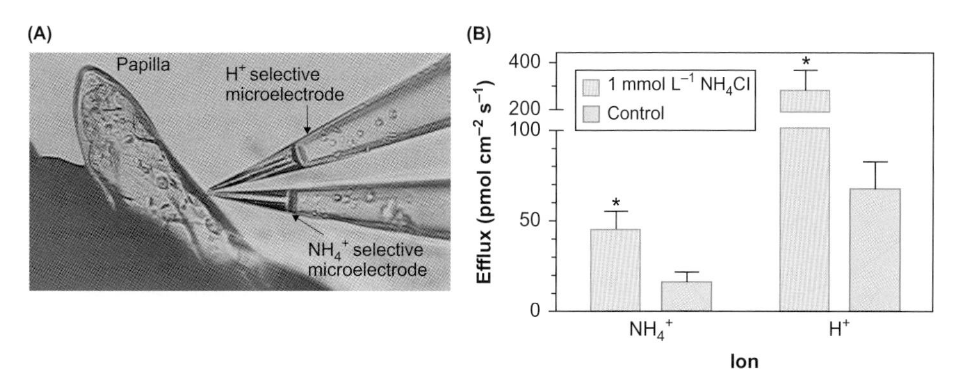

Figure 7.42 Ammonia and proton effluxes from the anal papillae of the larvae of yellow fever mosquitoes *Aedes aegypti*

(A) Digital image of an anal papilla with two closely placed microelectrodes detecting H^+ and NH_4^+ close to the surface of the papilla. (B) Effluxes of NH_4^+ and H^+ recorded from larvae exposed to an increase in ammonia (1 mmol L^{-1} NH_4Cl) for 3 days (orange columns) are significantly higher (*, $P<0.05$) than those of control larvae, exposed to 1 mmol L^{-1} NaCl (green columns).

Source: Weihrauch D, Donini A, O'Donnell MJ (2012). Ammonia transport by terrestrial and aquatic insects. Journal of Insect Physiology 58: 473–487.

that when the mosquito larvae are exposed to high environmental concentrations of ammonia (as NH_4Cl) both NH_4^+ and H^+ effluxes increase. We are beginning to understand the mechanisms behind these effluxes, which appear to involve several types of ammonia transporter proteins and show some similarities to the processes in crustacean and fish gills.

The Rh proteins in the basal membrane of the epithelium of the anal papillae of mosquito larvae are thought to allow bi-directional transport of NH_3 between the haemolymph and cytoplasm and in the reverse direction. Mosquito larvae are often naturally exposed to high external concentrations of ammonia, and in such circumstances a decrease in the Rh proteins in the basal membrane may act as a protective mechanism to impede ammonia influx. An additional type of ammonia transporter in the basal membrane is thought to facilitate NH_4^+ transport into the epithelial cytosol of the anal papillae. From the cytosol, NH_3 passage through the apical membrane of the epithelium, via Rh proteins, facilitates ammonia efflux. The apical Rh proteins are thought to work in conjunction with apical proton pumps and Na^+/H^+ exchange proteins that create an acidic boundary layer and result in formation of NH_4^+, in a similar way to that shown in Figure 7.40 for fish gills, thereby maintaining NH_3 excretion.

7.4.2 Urea excretion

Urea has a lower toxicity than ammonia, which allows toleration of higher blood concentrations than ammonia, for example, 0.05–0.2 mmol L^{-1} in mammals and 0.9–2.5 mmol L^{-1} in birds, and excretion at high concentrations in small amounts of water. However, the lower toxicity, and hence less requirement for water to dilute urea, needs to be balanced against the energy cost of synthesizing urea. Figure 7.43 shows the **ornithine-urea cycle** (or ornithine cycle), which mammals, turtles and tortoises, and adult amphibians use to synthesize urea; notice that every molecule of urea produced via this cycle uses three ATP molecules.

The final step in the ornithine-urea cycle requires the enzyme arginase, and its presence is often taken to indicate that an animal can synthesize urea. However, urea formation actually requires five enzymes, as illustrated by the series of steps in Figure 7.43. The genes for these enzymes are widely believed to have occurred in early vertebrates, but we cannot know whether all were expressed.

Urea excretion in fish

Marine cartilaginous fish[85] and the less well-studied coelacanths (lobe-finned fish) are unusual among ureotelic animals because they synthesize urea as an essential ingredient for their osmoregulation and hence reabsorb urea in the kidney. Although urea excretion via the gills of elasmobranchs is restricted[86], more than 90 per cent of the nitrogenous excretion of elasmobranchs occurs as urea, via the gills.

Instead of the increase in nitrogenous excretion typically seen after feeding in many types of animals, cartilaginous fish need to conserve nitrogen and ureogenesis is a high priority. Cartilaginous fish produce urea by the ornithine-urea cycle, as in other ureotelic vertebrates, although the enzyme in the initial synthetic step is slightly different: it is carbamoyl phosphatase (CPS) III, which preferentially uses glutamine as opposed to ammonia as the nitrogen donor. The ammonia produced by metabolism is converted to glutamine before entering the ornithine-urea cycle.

[85] Figure 1.21 illustrates the evolution of two groups of cartilaginous fish: the elasmobranchs (sharks, skates and rays) and the holocephalans (or chimaeras).

[86] Section 5.1.4 discusses urea retention by elasmobranchs at the gills and in the kidney, which results in their slight hyperosmoregulation, and maintains water influx for excretory processes.

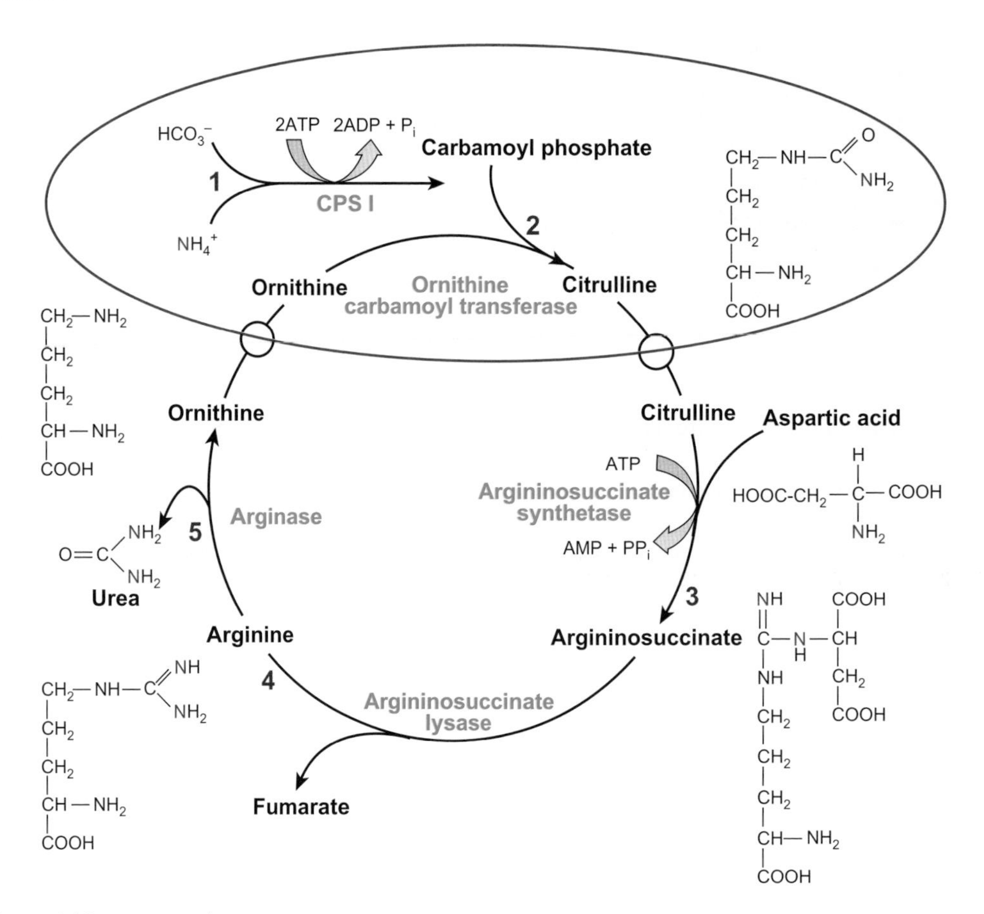

Figure 7.43 The ornithine-urea cycle

The cycle involves reactions within the mitochondria (red ellipse), in the cytosol of the liver and to a lesser extent in the kidney, and in some animal groups (e.g. cartilaginous fish) in muscle tissue. The numbered sequence of reactions is as follows: (1) Combination of ammonia (which provides one of the N atoms in urea) with CO_2 (from bicarbonate, HCO_3^-) in mitochondria to form carbamoyl phosphate using the enzyme, carbomyl phosphate synthetase I (CPS I). This reaction needs two ATP molecules. (2) Carbamoyl phosphate forms citrulline (an amino acid) from ornithine, using the enzyme ornithine carbamoyl transferase. Citrulline is transported to the cytosol and converted into argininosuccinate (3) by addition of aspartic acid (which provides the second nitrogen (N) atom in urea); the reaction requires the enzyme, argininosuccinate synthetase, and is driven by hydrolysis of ATP to AMP and PP_i with both high energy bonds in ATP ultimately cleaved. (4) Arginine and fumarate are produced from argininosuccinate, using argininosuccinate lysase. (5) Arginase cleaves urea from arginine producing cytosolic ornithine, which is transported to the mitochondrial matrix to perpetuate the cycle.

Maintenance of the high concentrations of urea in cartilaginous fish requires food intake (to supply the nitrogen). For example, studies suggest that spiny dogfish (*Squalus acanthias*) need to feed every 5 to 6 days to maintain plasma concentrations of urea, and more frequently if they are to grow. Hence, plasma concentrations of urea decrease by almost 20 per cent after about a month of starvation. After feeding, the activity of ornithine-urea cycle enzymes increases and the amino acids obtained from the food are channelled into urea production, which restores plasma concentrations of urea.

Fascinating recent studies have shown that the gills of sharks are functionally adapted to take up ammonia from seawater, rather than excreting ammonia via the gills like bony fishes and other aquatic animals. The measured rates of ammonia uptake suggest that the nitrogen obtained in this way is equivalent to about a third of the intake by feeding. For sharks, that feed infrequently, such uptake could be an important nitrogen supplement allowing them to maintain plasma levels of urea.

Most teleost fish are generally ammonotelic and do not possess a full complement of active enzymes for the ornithine-urea cycle, but teleosts may synthesize urea from arginine or employ the **uricolytic pathway**[87], to synthesize urea when necessary. Several species of teleost are capable of switching to urea excretion if environmental conditions, such as high ammonia concentrations, impede passage across the gills, or there is a lack of water to dilute the ammonia. These fish are **facultatively ureotelic.**

Air-breathing fish[88] can excrete urea if necessary. For example, walking catfish (*Clarias batrachus*) usually excrete 35–40 per cent of total nitrogen as urea produced via the

[87] The uricolytic pathway is illustrated in Figure 7.45.

[88] We discuss air-breathing among fish in Section 12.3.3.

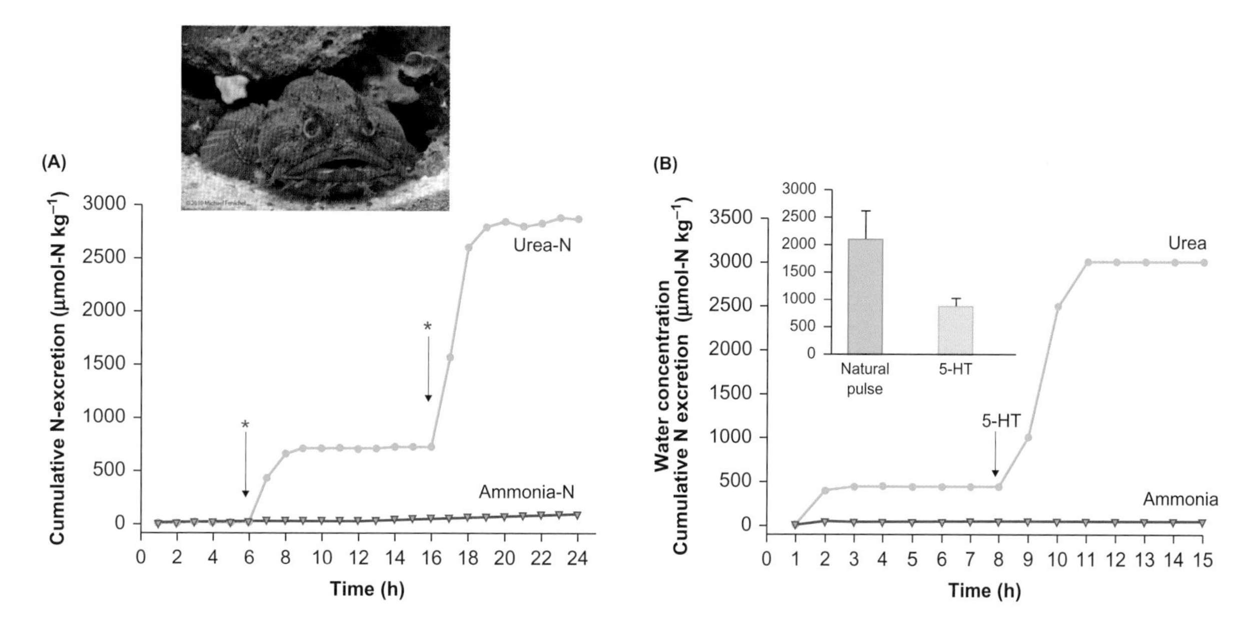

Figure 7.44 Urea excretion in pulses by Gulf toadfish (*Opsanus beta*)

(A) Almost the entire day's excretion of nitrogenous waste is in one or two pulses of urea (red asterisk and arrows). Ammonia excretion only accounts for a small proportion of nitrogenous excretion. (B) Pulses of urea excretion are induced by injection of serotonin (5-HT; 3×10^{-6} mol L^{-1}). The inset histogram compares the urea excretion in natural pulses and urea pulses induced by serotonin.

Source: Wood CM et al (2003). Pulsatile urea excretion in the gulf toadfish: mechanisms and controls. Comparative Biochemistry & Physiology Part B 136: 667-684. Image of toadfish: © Michael Fenichel

ornithine-urea cycle. Lungfish, such as African lungfish (*Protopterus* spp.), excrete mainly ammonia when they are in water, but change to urea synthesis if the pond dries up, when they enter a resting state of **aestivation** in a cocoon in the drying mud and retain urea until water returns. During aestivation, the lungfish have low metabolic rates. Once back in water, urea rapidly diffuses across the gills and the lungfish resume ammonia excretion.

In addition to achieving nitrogenous excretion, urea excretion may function as a form of chemical signalling in at least some species of fish. Gulf and oyster toadfish are perhaps the best examples. In non-stressful conditions in the laboratory, toadfish are ammonotelic, but their possession of a complete set of ornithine-urea cycle enzymes means they are capable of synthesizing urea.

When toadfish are crowded, or if they become exposed to high ammonia concentrations in the surrounding water, urea excretion increases to the point that it can account for 75 per cent or more of their nitrogenous excretion. Figure 7.44A shows an example of the dramatic pulses of urea excretion that occur 1–3 times a day. An injection of urea does not stimulate a pulse of excretion, so urea excretion does not seem to occur when blood concentration reaches a particular threshold concentration. Rather, toadfish seem to make behavioural decisions as to when to excrete pulses of urea. Exactly why toadfish excrete pulses of urea is uncertain, but individual toadfish are territorial and the pulses of urea appear to have a role in social communication.

Figure 7.44B shows that injection of serotonin (5-hydroxytryptamine, 5-HT) induces the excretion of a pulse of urea. Serotonin (5-HT) is stored in neuroepithelial cells in the gills; its release is thought to increase the permeability of the gill to urea, by 30–40 times, which results in the release of a pulse of urea.

At least one species of teleost fish—the Lake Magadi tilapia (*Alcolapia grahami*)—is **obligatory ureotelic**; it is obliged to excrete urea as its nitrogenous waste. Lake Magadi is a strongly alkaline soda lake (pH 10), so a high proportion of ammonia exists as un-ionized NH_3, which impedes passive diffusion of ammonia across the gills. Conversion of ammonia into urea, in a fully functional ornithine-urea cycle in the liver, allows the tilapia to excrete urea, primarily via the gills. The urea permeability of the gills of Lake Magadi tilapia is about five times more than that of toadfish during their pulses of urea excretion. However, the mechanisms for urea excretion are remarkably similar, with both species using urea transporter proteins (UT-A type[89]).

Even among the insects, which are usually considered uricotelic, some species can synthesize urea. Insects lack one or more genes encoding the enzymes of the ornithine-urea cycle, but urea can be generated by hydrolysis of arginine by arginase. An alternative uricolytic pathway illustrated in Figure 7.45 is also apparent in some insects, for example yellow fever mosquitoes (*Aedes aegypti*) in which the genes

[89] Section 4.2.2 discusses the types of urea transporters.

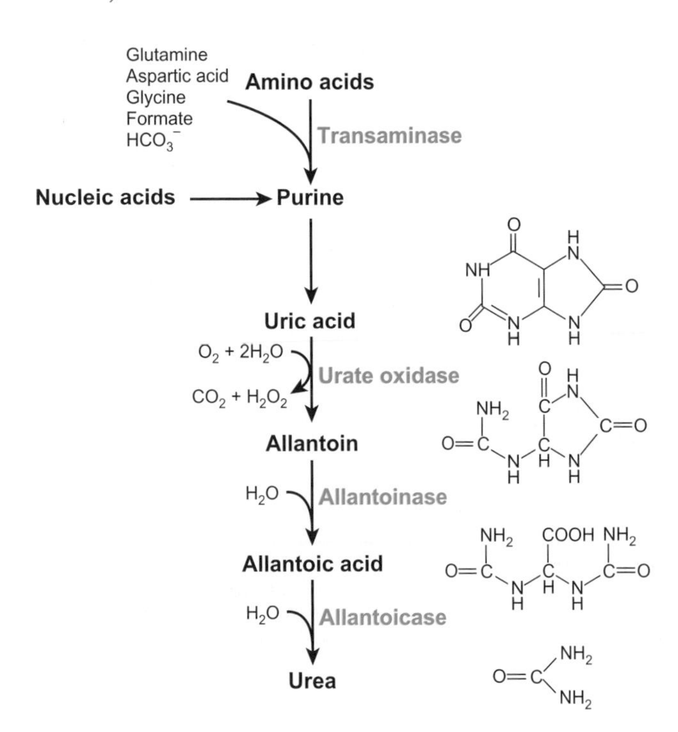

Figure 7.45 Uricolytic pathway for production of uric acid and urea

The uricolytic pathway produces urea from uric acid and involves the enzymes urate oxidase (uricase), allantoinase and allantoicase. The full pathway forms uric acid, allantoin, allantoic acid and urea. Uric acid is formed from a purine ring that is synthesized from complex interactions of amino acids or by nucleic acid metabolism.

for the three key enzymes in the pathway (urate oxidase, allantoinase and allantoicase) occur.

Figure 7.46 shows data for the excretion of large amounts of urea by female mosquitoes accompanied by much smaller amounts of allantoin and allantoic acid. These mosquitoes also excrete uric acid but at about half the amounts of urea. Uric acid excretion declines to insignificant amounts after the female mosquitoes feed on blood when their rate of urine excretion increases to regulate body fluid volume. Figure 7.46 shows that the female mosquitoes activate the uricolytic pathway and increase excretion of urea after feeding.

Urea excretion by amphibians

Most amphibian larvae develop in water and hence are ammonotelic. Species that remain in water as adults remain ammonotelic throughout their life, as shown by two examples in Figure 7.47, but most amphibians change to urea excretion during metamorphosis when they begin to express the enzymes of the ornithine-urea cycle that enables urea synthesis. Figure 7.47 shows the decrease in the per cent excretion of ammonia, and the increasing excretion of urea in terrestrial species of amphibians after metamorphosis.

The unusual breeding habits of some amphibians living in more extreme habitats appear to explain their adoption of urea excretion at an earlier stage of development, as larvae. For example, tadpoles of shovel-nosed chamber frogs (*Leptodactylus bufonius*), which develop in foam nests in a mud chamber on land, excrete predominantly urea. The omission of an ammonotelic stage in these tadpoles avoids ammonia toxicity. Similarly, tadpoles of waxy monkey leaf (tree) frogs (*Phyllomedusa sauvagii*) excrete more than 50 per cent of their nitrogenous waste as urea. After metamorphosis, these tree frogs move a step further in water conservation[90] by becoming uricotelic, as shown by the data for this species in Figure 7.47.

Some amphibians, such as crab-eating frogs (*Fejervarya cancrivora*) can survive in saline environments, when the concentration of urea in their plasma increases, which maintains some water intake, in a similar way to cartilaginous fish[91]. After

[90] Section 6.1.3 discusses the waterproofing of waxy monkey leaf (tree) frog.

[91] We discuss the retention of urea as osmotic ballast in some aquatic vertebrates living in saline/marine habitats in Section 5.1.4.

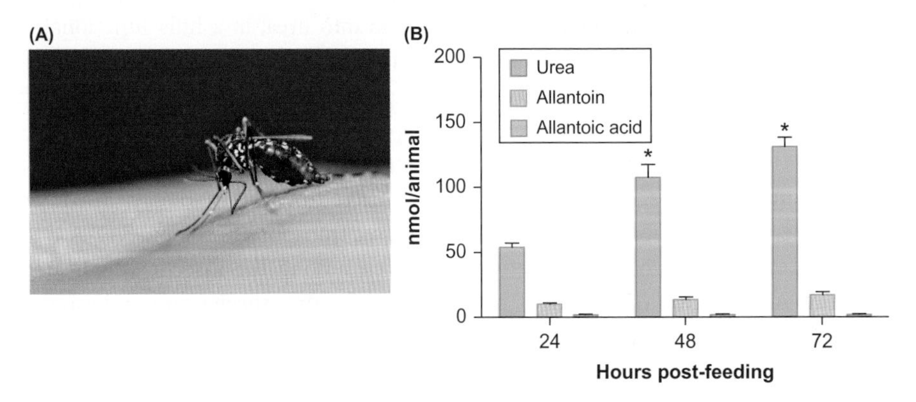

Figure 7.46 Urea excretion in urine and faeces of female yellow fever mosquitoes (*Aedes aegypti*) after a blood meal

(A) Feeding female mosquito (*Aedes aegypti*). (B) After a blood meal, the mosquitoes mainly excrete urea, together with much small amounts of allantoin and allantoic acid. Data are means ± standard errors (n=5); asterisks indicate a significant increase in urea excretion 48 and 72 h after feeding compared to urea excretion 24 h after feeding; $P<0.05$. Sources: A: Feeding mosquito: mrfiza/shutterstock.com; B: Scaraffia PY et al (2008). Discovery of an alternate metabolic pathway for urea synthesis in adult *Aedes aegypti* mosquitoes. Proceedings of the National Academy of Sciences of United States of America 105: 518–523.

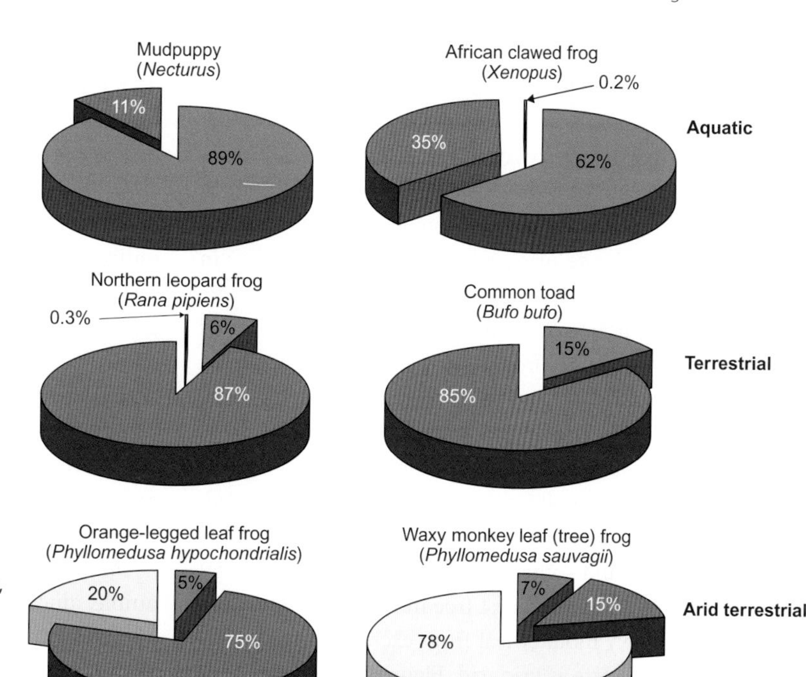

Figure 7.47 Ammonotelic, ureotelic and uricotelic amphibians in relation to habitat

The pie diagrams show the per cent excretion of ammonia, urea and uric acid/urates in two amphibians from each of three habitat groups: fully aquatic (throughout the life cycle), terrestrial and arid terrestrial.

Source: Data from Loveridge JP (1993). Nitrogenous excretion in the Amphibia. In: New Insights in Vertebrate Kidney Function (Eds. Brown JA, Balment RJ, Rankin JC), pp 135-144. Cambridge, Cambridge University Press.

several days in brackish water, an increase in CPS I (carbomyl phosphate synthetase I) in the liver results in increased synthesis of urea.

7.4.3 Uric acid excretion

In physiological solutions, such as blood plasma, uric acid is mainly present as sodium and potassium urate. The acid dissociation constant (pK_a) of the reaction shown in Equation 7.4 is about 5.75; in other words, at pH 5.75 there is 50 per cent uric acid and 50 per cent urate. At normal plasma or haemolymph pH of 7.4–7.8, about 98 per cent is urate (which lacks the H^+ associated with N-9 in uric acid).

$$\text{Uric acid} \rightleftharpoons \text{Urate}^- + \text{H}^+ \qquad \text{Equation 7.4}$$

Nucleic acid metabolism generates purines[92] that result in the formation of a small amount of uric acid, but most uric acid is synthesized from amino acids in a process involving about 15 catalysed reactions. In addition, glycine, aspartic acid (providing one N atom), glutamine (providing two N atoms), formate and bicarbonate are required, as indicated back in Figure 7.45. In vertebrates, these reactions occur mainly in the liver and to a lesser extent in kidney tissue, while insects generate uric acid mainly in the fat body.

The synthesis of uric acid requires a high level of energy input: at least 15 ATP molecules per molecule synthesized. Given this, we might ask why birds, most reptiles, most terrestrial insects and land snails are uricotelic (i.e. predominantly excrete uric acid).

The main advantages of excreting uric acid/urate compared to urea or ammonia as a means of excreting nitrogen are that uric acid is relatively non-toxic to animals and requires minimal water for its excretion. Uric acid and its urate salts are poorly soluble[93] so reabsorption of water from the excreted fluids of uricotelic animals pushes the concentration above the solubility limit, resulting in precipitation of uric acid/urates. This precipitation reduces the osmotic concentration (which is determined by the dissolved substances) of the urine or faecal fluid, allowing additional water absorption by osmosis. Such absorption of water plays an important part in the water balance of uricotelic terrestrial animals.

All animals that reproduce using **cleidoic (= closed) eggs** are uricotelic; these include birds, reptiles, insects and land snails. In cleidoic eggs, only gases (including water vapour) exchange with the environment and excretory products remain in the egg, but only a small volume of water is available for association with the nitrogenous wastes.

Bird embryos can synthesize ammonia, urea and uric acid but the proportions of these nitrogenous waste substances

[92] Section 1.3.1 discusses the structure of deoxyribonucleic acid (DNA) in which two of the base pairs (adenine and guanine) are purines.

[93] Uric acid solubility at 20°C is 0.06g L^{-1} (0.357 mmol L^{-1}) compared to urea solubility of 1079 g L^{-1} (17.96 mol L^{-1}).

changes during development. Excretion of ammonia throughout development would lead to toxic concentrations in the egg, while excretion of urea would increase the osmotic pressure of fluids surrounding the embryo over the period of development, and would cause damage to it. Uric acid is relatively non-toxic and production rises early in development and soon reaches its solubility limit. Urate/uric acid precipitates accumulate in the allantois as a semi-solid, non-toxic material.

Adult amphibians usually excrete urea in their urine as their nitrogenous waste, but tree frogs have highly developed mechanisms to conserve water[94], including changes in their handling of nitrogenous waste. To conserve water, some species of frogs resort to storing nitrogen in their skin, in forms such as guanine[95], and only excrete nitrogen as urea when water is available. However, the routine excretion of uric acid is a better way to conserve water, and some frogs have evolved this strategy. For example, southern foam nest tree frogs (*Chiromantis xerampelina*) and waxy monkey leaf frogs excrete a large part of their nitrogenous waste as uric acid. Figure 7.47 indicates that almost 80 per cent of the nitrogenous waste of waxy monkey tree frogs is uric acid. The ability of these tree frogs to excrete uric acid instead of urea reduces their daily requirement for water in excreting urine by more than 90 per cent.

Reptiles and birds show some of the clearest relationships between water availability and nitrogenous excretory patterns. Crocodiles living in fresh water, which have plenty of drinking water available to them, excrete mainly ammonia, although they retain a capacity to synthesize uric acid. Although we might envisage the excretion of ammonia to be an adaptation to a semi-aquatic life, with plenty of available fresh water, crocodiles living in seawater also excrete uric acid.

Studies of the urine composition of turtles and tortoises indicate a mixture of nitrogenous wastes that shows a close relationship between the water availability in their habitat and the proportions of ammonia, urea and uric acid in their urine. The urine of those species living in almost wholly aquatic habitats contains only small amounts of uric acid (accounting for <5 per cent of the total N in the urine), while uric acid accounts for >50 per cent of the nitrogenous waste in urine of species that live in very dry areas such as the Indian star tortoises (*Geochelone elegans*).

Birds that habitually have fresh water available—for example, ducks or domestic hens—excrete more ammonia and less uric acid than birds coping with less water. Similarly, some hummingbirds feeding on nectar and fruit-eating birds (with a low protein (N) intake but high water intake) adjust ammonia excretion accordingly: they are facultatively ammonotelic. In contrast, carnivorous birds that consume large amounts of amino acids synthesize large amounts of uric acid and excrete up to 90 per cent of their nitrogenous waste as urate. Case Study 7.2 examines this process and the devastating consequences of its failure in vultures in the Indian sub-continent after consuming carcasses containing veterinary drugs that interfere with uric acid excretion. Recognition of these toxic actions is driving changes in their use in an effort to conserve these threatened species.

[94] Section 6.1.3 discusses water conservation in tree frogs.

[95] Guanine also occurs in nucleic acids (DNA and RNA), as we discuss in Section 1.3.1.

Case Study 7.2 Excretion of uric acid and diclofenac toxicity in vultures

Two decades ago, vulture populations across the Indian subcontinent (India, Pakistan and Nepal) began to collapse catastrophically. Three endemic species of vultures, white-rumped (*Gyps bengalensis*), Indian (*Gyps indicus*) and slender-billed (*Gyps tenuirostris*), showed rapid and sustained declines in total population size that were among the most severe of any bird species (as illustrated for the white-rumped vulture in Figure A). The population decreases of more than 97 per cent since the early 1990s resulted in the International Union for the Conservation of Nature (IUCN) placing these species on their list of Critically Endangered species.

As well as having direct implications for conservation, the declining populations of vultures have implications for public health, since vultures play a critical role in the disposal of the huge number of cattle carcasses (by feeding on them), which reduces disease transmission among domestic and wild ungulates.

Despite the first evidence of population decline in the 1990s, it was not until 2004 that information about its cause began to emerge, when a study of wild vultures identified a link between deaths and the dietary intake of diclofenac by vultures. As such, diclofenac is now the best-known and clearest instance of a veterinary or human drug resulting in an adverse response in free-ranging wildlife populations.

Diclofenac is a non-steroidal anti-inflammatory drug (NSAID) used for clinical and veterinary purposes to treat inflammation and/or pain associated with diseases (e.g. arthritis) or injury. Such treatments leave accumulated diclofenac in the tissues. When vultures scavenge on the carcasses of dead livestock (mainly cattle), they consume diclofenac in the carcasses of treated animals.

The post-mortem examination of more than 250 specimens of wild white-rumped vultures found that 85 per cent of them had severe visceral gout. Visceral gout is characterized by extensive deposition of urate salts and uric acid crystals on the surface and within many of the visceral organs, including heart, liver, spleen and lung[1]. Figure B shows an example of visceral gout, in this case affecting the liver of the African white-backed vulture (*Gyps africanus*). Uric acid deposition results in tissue damage to

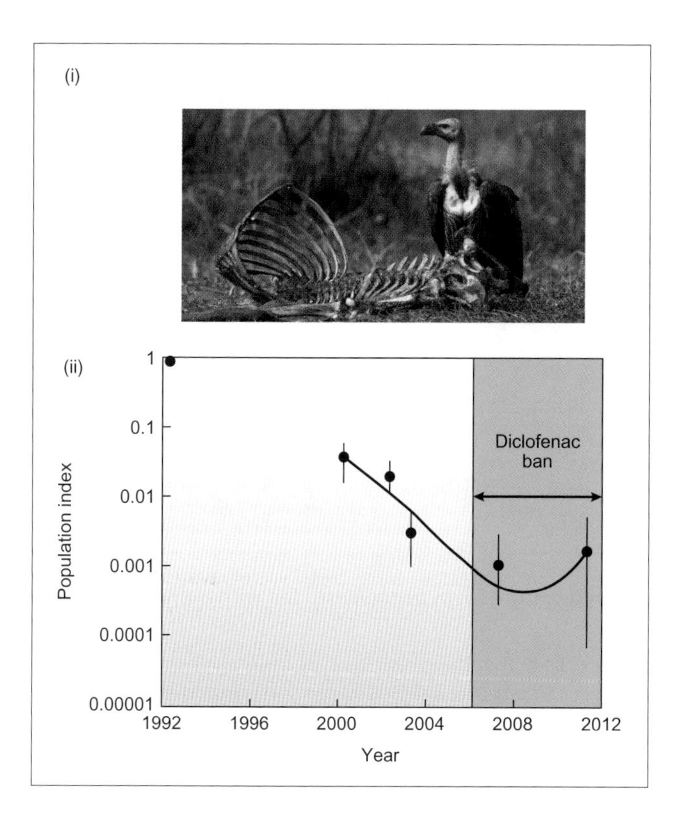

Figure A Relative population of the white-rumped vulture (*Gyps bengalensis*)

(i) Vulture beside a stripped carcass; (ii) population data from repeat surveys of a large number of road transects in India between 2000 and 2012, expressed relative to the population in 1992. The toxic effects of diclofenac, a non-steroidal anti-inflammatory drug used to treat inflammation and pain, which is consumed in carcasses, are blamed for the declining population of vultures. Diclofenac was banned in India, Pakistan and Nepal in 2006.

Vertical lines show estimated 95% confidence limits.

Source: Balmford A (2013). Pollution, Politics, and Vultures. Science 339: 653-654. Vulture image (c) Mike Lane/Biosphoto

all major organs; if sufficient diclofenac is consumed, death occurs within 2–3 days. Studies suggest that all *Gyps* species are potentially vulnerable to diclofenac toxicity.

Birds usually excrete uric acid/urates, so an accumulation indicates a breakdown in the normal excretory processes. A rise in plasma concentrations of uric acid in animals given diclofenac-contaminated food at known doses confirms this idea. Notice in Figure C that there is a positive relationship between the dose of diclofenac administered and the concentration of plasma uric acid soon after.

The increase in plasma concentrations of uric acid results from a failing ability of the proximal tubules to secrete urate ions, such that its excretion declines. There is evidence of severe kidney damage, particularly of the proximal segment of the nephrons. Figure D shows an example of such damage in an white-rumped vulture (a captive vulture that could not be released) fed with buffalo meat containing diclofenac.

In contrast to the toxic effects of diclofenac on *Gyps* vultures, diclofenac intake by vulture species in North and South America, at concentrations of up to more than 100 times the estimated median lethal dose for *Gyps* vultures, does not result in overt signs

Figure B Post-mortem examination of an African white-backed vulture (*Gyps africanus*) after experimental oral dosing with diclofenac at 0.8 mg kg^{-1}

Visceral gout indicated by white arrow is visible on the liver surface.

Source: Swan G et al (2006). Toxicity of diclofenac to *Gyps* vultures. Biology Letters 2: 279–282.

of toxicity (visceral gout, renal necrosis, elevation in plasma uric acid). Turkey vultures (*Cathartes aura*) clear the diclofenac within a few days and any residues are undetectable in liver or kidney. Such a marked difference between Old World and New World species is likely to reflect differences in the actions of diclofenac on biochemical pathways that could vary between individual species.

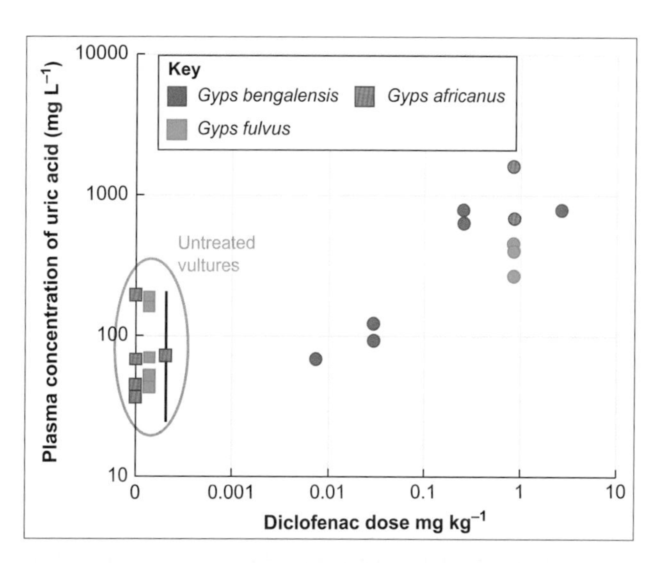

Figure C Uric acid concentration in the plasma of three species of *Gyps* vultures, 24 h after treatment with various doses of diclofenac and in untreated vultures

Plasma uric acid levels after diclofenac treatment were correlated with the dose administered when data for the three species were combined and for *Gyps bengalensis* alone, although three *Gyps bengalensis* that received low doses of diclofenac had uric acid concentrations within the normal range for wild/untreated *Gyps africanus* and the Griffon vulture *Gyps fulvus*. The data include a mean with 2.5th to 97.5th percentiles (indicated by vertical lines) for 14 untreated *Gyps africanus*.

Source: Swan G et al (2006). Toxicity of diclofenac to *Gyps* vultures. Biology Letters 2: 279–282.

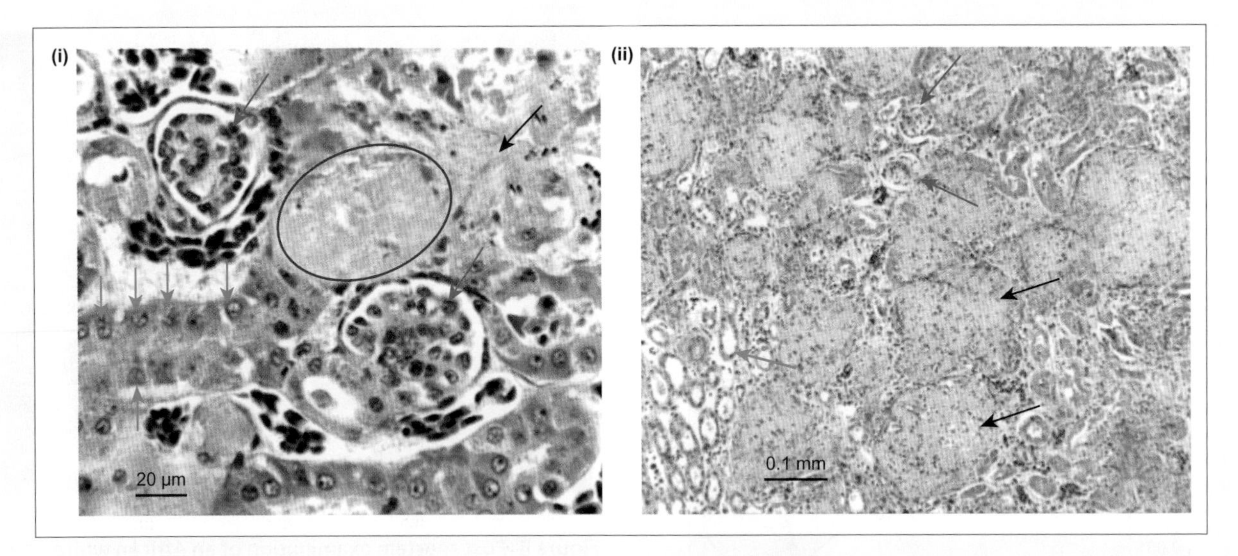

Figure D Histological sections of the kidney from a captive white-rumped vulture (*Gyps bengalensis*) fed with buffalo meat containing diclofenac

(i) Kidney of white-rumped vulture after feeding on meat containing 6.4 mg diclofenac per kg body mass. The red ring shows an area of non-functional proximal convoluted tubule, adjacent to an area of viable but swollen cells with prominent nuclei (green arrows). Delicate lightly stained urate crystals cross tubular and vascular boundaries (black arrow). The glomeruli (blue arrows) are normal in histological appearance. (ii) Kidney from a wild white-rumped vulture found dead in Punjab Province, Pakistan, showing numerous large aggregates of urate material and cell debris (black arrows) instead of the normal renal tubular architecture. The epithelium of a group of cross sections of distal convoluted tubule (green arrow) shows viable tubular epithelia; the lumen of these tubules contains cell debris. The glomeruli (blue arrows) are normal in appearance (the smaller size of glomeruli in panel (ii) compared to (i) is because the magnification is lower).

Source Meteyer CU et al (2005). Pathology and proposed pathophysiology of diclofenac poisoning in free-living and experimentally exposed oriental white-backed vultures (*Gyps bengalensis*). Journal of Wildlife Diseases 41: 707–716.

Based on the evidence that diclofenac in domestic livestock was the major cause of the declining vulture populations in the Indian subcontinent, the governments of India, Pakistan, and Nepal banned the veterinary use of diclofenac in 2006. Bangladesh introduced a similar ban in 2010. However, diclofenac intended for human is still available for potential (and likely) use on livestock, so a search for alternative less/non-toxic NSAIDs began.

Meloxicam appears to be a good alternative to diclofenac. In experimental studies, six *Gyps* species showed no ill effects after dosing with meloxicam rather than diclofenac, and continued to excrete uric acid effectively. In a series of studies testing the safety of meloxicam the serum uric acid concentrations remained within the normal limits in *Gyps bengalensis* and *Gyps indicus* held captive in India and unable to be released.

The reasons behind the marked difference in meloxicam and diclofenac toxicity to vultures and their differing sensitivities are not clear. To understand these differences we need a better understanding of their mechanisms of action at a cellular level in vultures—and for that matter in other raptors (birds of prey, feeding on other animals). Recent studies suggest that the steppe eagles (*Aquila nipalensis*) which feed on dumped carcasses in India are also subject to diclofenac toxicity, which widens the diversity of raptors under threat from the misuse of pharmacological agents. Nevertheless, the recognition of the toxic actions is driving changes in their use in an effort to conserve these threatened species.

› Find out more

Balmford A (2013). Pollution, politics, and vultures. Science 339: 653–654.

Harris RJ (2013). The conservation of Accipitridae vultures of Nepal: a review. Journal of Threatened Taxa 5: 3603–3619.

Meteyer CU, Rideout BA, Gilbert M, Shivaprasad HL, Oaks JL (2005). Pathology and proposed pathophysiology of diclofenac poisoning in free-living and experimentally exposed oriental white-backed vultures (*Gyps bengalensis*). Journal of Wildlife Diseases 41: 707–716.

Oaks JL, Gilbert M, Virani MZ, Watson RT, Meteyer CU, Rideout BA, Shivaprasad HL, Ahmed S, Chaudhry MJI, Arshad M, Mahmood S, Ali A, Khan AA (2004). Diclofenac residues as the cause of vulture population decline in Pakistan. Nature 427: 630–633.

Rattner BA, Whitehead MA, Gasper G, Meteyer CU, Link WA, Taggart MA, Meharg A, Pattee OH, Pain DJ (2008). Apparent tolerance of turkey vultures (*Cathartes aura*) to the non-steroidal anti-inflammatory drug diclofenac. Environmental Toxicology and Chemistry 27: 2341–2345.

Sharma AK, Saini M, Singh SD, Prakash V, Das A, Dasan RB, Pandey S, Bohara D, Galligan TH, Green RE, Knopp D, Cuthbert RJ (2014). Diclofenac is toxic to the steppe eagle *Aquila nipalensis*: widening the diversity of raptors threatened by NSAID misuse in South Asia. Bird Conservation International 24: 282–286.

Swan GE, Cuthbert R, Quevedo M, Green RE, Pain DJ, Bartels P, Cunningham AA, Duncan N, Meharg AA, Oaks JL, Parry-Jones J, Shultz S, Taggart MA Verdoorn G, Wolter K (2006). Toxicity of diclofenac to *Gyps* vultures. Biology Letters 2: 279–282.

Swan G, Naidoo V, Cuthbert R, Green RE, Pain DJ, Swarup D, Prakash V, Taggart M, Bekker L, Das D, Diekmann J, Diekmann M, Killian E, Meharg A, Patra RC, Saini M, Wolter K (2006). Removing the threat of diclofenac to critically endangered Asian vultures. PLoS Biology 4(3): e66.

[1] A less severe form of gout occurs in humans in which deposition of needle-like crystals of uric acid occurs in joints, skin, or in other tissues.

The bulk of the excretion of uric acid/urates results from active secretion of urate by the proximal segment of the kidney nephrons in which urate specific transporters and multipurpose transporters of organic anions occur. Studies of perfused proximal tubules from snakes suggest that the secretory process is usually 60–70 per cent saturated and plasma concentrations have relatively little effect on urate secretion rates, although there is a poor understanding of the transport mechanisms involved in urate secretion.

Water absorption in the distal nephron and cloaca of birds and reptiles, and in the rectum of insects, increases the concentration of urate/uric acid in the excretory fluids. When concentrations rise, the urate/uric acid comes out of solution because of low solubility. In birds, which excrete hyperosmotic urine, we might predict the precipitation of uric acid/urates in the renal tubules, as concentrations increase. How, then, do birds avoid precipitated material blocking their renal tubules? The answer seems to be that the interaction of urate and uric acid with low concentrations of proteins in the urine forms a colloidal suspension of spheres of 3 to 13 μm in diameter small enough to pass along the renal tubules and avoids their coalescence into larger particles. After post-renal modification of the urine, in the cloaca and colon[96], the characteristic white sludge of urine and faeces is excreted. This is an ideal way of excreting excess nitrogen when water is a precious commodity.

› *Review articles*

Dantzler WH (2005). Commentary. Challenges and intriguing problems in comparative renal physiology. Journal of Experimental Biology 208: 587–594.

Weihrauch D, Morris S, Towle DW (2004). Review. Ammonia excretion in aquatic and terrestrial crabs. Journal of Experimental Biology 207: 4491–4504.

Weihrauch D, Donini A, O'Donnell MJ (2012). Ammonia transport by terrestrial and aquatic insects. Journal of Insect Physiology 58: 473–487.

Wilkie MP (2002). Ammonia excretion and urea handling by fish gills: present understanding and future research challenges. Journal of Experimental Zoology 293: 284–301.

Wright PA (1995). Review. Nitrogen excretion: Three end products, many physiological roles. Journal of Experimental Biology 198: 273–281.

Wright PA, Wood, CM, (2009). A new paradigm for ammonia excretion in aquatic animals: role of Rhesus (Rh) glycoproteins. Journal of Experimental Biology 212: 2303–2312.

[96] We discuss post-renal modification of urine composition by birds in Sections 6.1.4 and 21.1.1

Checklist of key concepts

Animals produce the primary urine by secretion or ultrafiltration

- The size and electrical charge of solutes determines their ultrafiltration by the vertebrate **glomerulus** via an extracellular route across the capillary endothelium, basement membrane and slit diaphragms in the capillary epithelium.
- An **effective filtration pressure** for ultrafiltration by vertebrate glomeruli results from the hydrostatic force (blood pressure) in the capillaries acting against the backpressure in Bowman's capsule.
- The **colloid osmotic pressure** difference due to accumulation of plasma proteins in the capillaries acts against filtration.
- **Single nephron glomerular filtration rates** (**SNGFRs**) are determined by the product of the effective filtration pressure (kPa) and the **ultrafiltration coefficient** (nL min^{-1} kPa^{-1}).
- Total kidney **glomerular filtration rate** (**GFR**) results from the sum total of the individual SNGFRs in a kidney.
- The measurement of GFR using markers such as inulin calculates the volume of plasma completely cleared (emptied) of the substance per unit time, which is a **clearance measurement**.
- Net reabsorption or secretion of a substance by the renal tubules is given by its relative clearance compared to the clearance of a GFR marker.
- Marker compounds to monitor GFR should ideally be (i) unimpeded from filtration, (ii) not reabsorbed by the renal tubules and (iii) not secreted into the renal tubules.
- In most non-mammalian vertebrates, the filtering population of glomeruli determines overall GFR and is a major influence on urine output and hence fluid balance.
- Almost all species demonstrating glomerular intermittency have a renal portal (venous) supply that could nourish the renal tubules deprived of an arterial blood supply.
- **Autoregulation** of kidney GFR and renal blood flows occurs in mammals and birds.
  - When blood pressure increases, pre-glomerular vascular resistance may increase by (i) a rapid **myogenic response** increasing the resistance to blood flow, (ii) in mammals, a slower **tubuloglomerular feedback** mechanism that regulates sodium chloride delivery to the distal nephron.

The proximal tubule reabsorbs the bulk of filtered solutes and water and also secretes ions and fluid

- In the proximal tubule, sodium pumps (**Na^+, K^+-ATPase**) in the basolateral membranes of the epithelium drive Na^+ reabsorption via the luminal border.

- The high water permeability of proximal tubules results in water reabsorption following the solute reabsorption, which is important for maintaining the volume of extracellular fluids.
- Glucose carriers reabsorb glucose from the proximal tubule, but may become saturated if plasma concentrations rise to the point at which the filtered load reaches its **tubular maximum**.
- Fluid secretion can account for almost 30 per cent of the fluid passing along the nephrons of teleost fish and could allow non-filtering nephrons to contribute to excretion.
- Proximal tubular secretion of magnesium and sulfate maintains the low plasma concentrations of these ions in marine fish.

The distal nephron of vertebrates modifies urine composition and volume

- Animals with excess water in their body fluids excrete osmotically dilute urine by reabsorbing Na^+ and Cl^- from distal parts of the nephron.
- Production of urine that is hyperosmotic to plasma, by mammals and birds, is vital to their water conservation for life on land.
- The maximum osmotic concentration of urine of individual species correlates with the **aridity** of their habitat.
- Production of hyperosmotic urine requires absorption of **osmotically free water** from the tubular fluid.
- The **countercurrent multiplication** model proposes the generation of hyperosmotic urine by multiplication of the difference in osmotic concentrations between descending and ascending limbs of the countercurrent loop.
- Urea is the most important osmolyte in the inner medulla of some mammalian species.
- Production of hyperosmotic urine in mammals results from the actions of **arginine vasopressin** (**antidiuretic hormone**), which increases the osmotic water permeability of the late distal nephron and collecting duct, and its other actions.
- Countercurrent blood flow in the vasa recta of mammalian kidneys and **countercurrent exchange** diffusion can maintain the increasing concentration of osmolytes between the cortex and medulla.
- Mammals living in arid habitats generally have a better ability to generate hyperosmotic urine because of:
  - a higher proportion of longer-looped nephrons;
  - increased **relative medullary thickness** for a given body size than mammals spending their lives in close association with water.

$$\text{Relative medullary thickness} = \frac{\text{Medullary thickness}}{(\text{Kidney length} \times \text{width} \times \text{depth})^{0.33}}$$

- In the avian kidney, hyperosmotic urine is probably generated by **sodium chloride recycling** in a cascade of short- to long-looped nephrons but further water can be extracted when urine enters the cloaca and colon.
- Secretion of potassium ions (K^+) by the distal nephron regulates K^+ balance in accord with dietary intake.
- Fluid secretion in distal parts of the mammalian nephron may be most important in water balance when GFR is low.

Invertebrate excretory organs regulate the volume and composition of extracellular fluids

- Insect **Malpighian tubules** secrete primary urine at the highest rate of any known epithelium; secretion of various ions draws water into the tubules by osmosis.
- **Proton pumps** on the luminal border of the Malpighian tubules are a key component of the secretory processes of several insect species.
- Post-renal processing of urine and faecal matter, in the rectum, modifies urine composition and volume, and in some species generates hyperosmotic excretory fluid.
- Some invertebrate species living in dilute salinities produce osmotically dilute urine by reabsorbing ions from distal elements of their excretory organs.
- Secretory processes in the antennal glands of marine and estuarine crustaceans are essential in maintaining body fluid concentrations of magnesium, sulfate and calcium.

Nitrogenous excretion

- Nitrogenous wastes (N-waste) mainly result from metabolism of excess amino acids, so rates of N-excretion depend on an animal's diet and increase after feeding in carnivores.
- The chemical form of N-waste(s) is determined mainly by water availability in an animal's habitat, rather than phylogeny.
- More than 50 per cent of the N-waste of most aquatic animals is ammonia; they are said to be are **ammonotelic**. Such animals rely on dilution to avoid ammonia toxicity.
- The form of ammonia in solution is pH dependent:
  - equal amounts of NH_4^+ and NH_3 occur at pH 9.5 at 15°C;
  - below pH 9.5 NH_4^+ exceeds NH_3;
  - above pH 9.5 NH_3 exceeds NH_4^+.
- Production of alternative N-wastes for temporary storage allows some aquatic animals (such as land crabs) to travel onto land for short periods.
- Terrestrial animals are usually **ureotelic** (urea excreting), or **uricotelic** (synthesize uric acid as their main N-waste).
- In aquatic environments, high external concentrations of ammonia can have multiple toxic effects.
- Some fish living in high ammonia conditions always excrete urea (= **obligatory ureotelic**); others switch to urea excretion when ammonia increases (= **facultative ureotelic**).
- Marine cartilaginous fish synthesize urea, which is essential for their high plasma osmolality.
- The main advantage of synthesizing uric acid as N-waste, despite the high demand for ATP, is the very small amount of associated water loss.
- Urates have low solubility; their precipitation in excretory fluids allows extra water absorption by osmosis.
- Recognition of the toxic actions of veterinary medicines that impede uric acid excretion by scavenging birds (raptors) has resulted in measures to conserve these threatened species.

Study questions

* Answers to these numerical questions are available online. Go to www.oup.com/uk/butler

1. Explain the two main processes involved in the production of primary urine. What are the relative advantages and disadvantages of these two alternatives? What experiments could you carry out to determine the process an animal is using? (Hint: Section 7.1.)
2. Kidney clearance techniques, micropuncture, *in situ* tubular microperfusion, and use of isolated tubules are all techniques for investigating renal function. Explain how each of these techniques could enable investigation of the actions of an environmental toxicant. (Hints: Boxes 7.1 and 7.2; Sections 7.1, 7.2; Case Study 7.2.)
3. Why does inulin clearance give a good measure of kidney GFR? Under what circumstances would the monitoring of creatinine be a preferable approach for assessing glomerular function? (Hints: Box 7.2; Section 7.1.3.)
4.* A micropuncture study of the kidney of an animal provided the following mean values:
   a. SNGFR = 20 nL min^{-1}
   b. Blood pressure in the glomerular capillaries = 6.7 kPa
   c. Hydrostatic pressure in Bowman's capsule = 1.6 kPa
   d. Colloid osmotic pressure difference between a sample of renal arterial blood plasma and fluid in Bowman's capsule = 3.3 kPa.

   Using these values estimate the ultrafiltration coefficient (K_f) of the glomeruli of the animal. (Hint: Section 7.1.2.)
5. Give two examples of environmental circumstances in which the extracellular fluid volumes could *decline* for (a) a teleost fish and (b) a mammal. Draw up a list of the likely renal responses of each species. (Hints: Sections 7.1.3 and 7.2.)
6. Describe (with diagrams) the anatomical arrangement of the nephrons and the vasculature of the kidneys of birds and mammals. Which aspects of these arrangements are important for their production of hyperosmotic urine? (Hints: Sections 7.1.2 and 7.2.3.)
7.* In a small mammal, a blood sample was collected 1 hour after injection of inulin into the circulatory system. The volume of urine excreted over the next 15 hours measured 10.17 mL. Analysis of the sodium (Na^+) and inulin concentrations in the blood plasma and urine samples gave the following values:

   Plasma Na^+ concentration = 145 mmol L^{-1}
   Plasma inulin concentration = 1.5 mg L^{-1}
   Urine inulin concentration = 20 mg L^{-1}
   Urine Na^+ concentration = 9.7 mmol L^{-1}

   Use these data to estimate:
   (i) Glomerular filtration rate of the animal (mL min^{-1}).
   (ii) The urine/plasma concentration ratio for inulin. What does this tell us about the ability of this animal to conserve water?
   (iii) The sodium clearance (mL h^{-1}).

   (Hints: Section 7.1.3; Box 7.2; see also Table 6.3.)
8. Give two examples of active secretory processes in nephrons. How do such secretions occur in teleost fish? (Hints: Section 7.2.4; Experimental panel 7.1, see Online Resources.)
9.* The dimensions of the kidney for a 70 kg human are: length 12 cm, width 6 cm, depth (thickness) 3 cm. If the distance between the beginning of the medulla and the tip of the loops in the papilla of the kidney is 19 mm, what is the relative medullary thickness? (Hints: Section 7.2.3; Figure 7.25.)
10. Insects do not have kidneys. How do insects control their extracellular fluid volumes and composition? (Hints: Section 7.3.4; Case Study 7.1.)
11. Why do terrestrial animals mainly excrete urea or uric acid, while aquatic animals usually excrete ammonia? Give two examples of situations in which this typical pattern of nitrogenous excretion does *not* apply. (Hint: Section 7.4.)
12. Discuss the strategies that crabs have adopted for excreting nitrogenous waste. How do these strategies differ between crabs occupying different habitats? (Hint: Section 7.4.1.)

Bibliography

Bradley TJ (2009). Animal Osmoregulation. Oxford: Oxford University Press.

Brown JA, Balment RJ, Rankin JC Editors (1993). New Insights in Vertebrate Kidney Function. Cambridge: Cambridge University Press.

Chapman RF (1998). The Insects. Structure and Function. 4th Edition. Cambridge: Cambridge University Press.

Dantzler WH (2016). Comparative Physiology of the Vertebrate Kidney. 2nd Edition. New York: Springer.

Lote CJ (2012). Principles of Renal Physiology. 5th Edition. New York: Springer.

Riegel JA (1972). Comparative Physiology of Renal Excretion. New York: Hafner Pub Co.

Wright P, Anderson P (Editors) (2001). Nitrogen Excretion. Fish Physiology Volume 20. Academic Press.

Yang B, Sands JM (Editors) (2014). Urea Transporters. Subcellular Biochemistry, Volume 73. Dordrecht: Springer.

Frozen wood frog (*Rana sylvatica*).
Image courtesy Jan Storey.

Part 3

Temperature

In this Part we explore the effects of temperature on biological functions, how animals cope with variation in their ambient temperature, and how they have adapted to these changes in ways that ensure their survival. We start in **Chapter 8** with a discussion of the chemical principles that reveal why temperature has such profound effects on biological systems and introduce the principles of heat exchange. In **Chapters 9 and 10** we consider in more detail how different animal groups are adapted in different ways to survive and function within the confines set by these principles.

Some groups of animals can only raise their body temperature above that of the environment by obtaining heat from outside their body; by contrast, members of other groups can achieve the same effect by generating sufficient heat from within their bodies. We find that the problem of temperature regulation is very different for animals that live and breathe in water compared with those living on land (or at least those breathing air). We discover how the solutions to the problem of thermoregulation that have evolved reflect this fundamental dichotomy.

In these three chapters we encounter a range of remarkable animals, from frogs that can allow their bodies to freeze completely to animals that use heat production from their own bodies to kill their enemies.

8

Temperature and the principles of heat exchange

The term 'thermoregulation' describes the strategies adopted by animals to manage their heat balance so that their body temperature remains within a relatively narrow range. Thermoregulation in birds and mammals is mainly achieved by physiological mechanisms, including the internal generation of heat. In the other groups of animals, behavioural mechanisms play a much more dominant role, with the behaviour being determined by the temperature of the environment and, for terrestrial animals, by the radiant heat from the sun.

To understand why all animals need to control their body temperature to a greater or lesser extent, it is necessary to consider the effects of temperature on biological functions. In this respect, biological processes are similar to chemical reactions, which are also temperature dependent. We therefore start this chapter by considering temperature dependence as a chemical phenomenon. We then expand this approach to look at how temperature affects biological processes, and conclude by examining the general principles that govern heat flow between animals and their environment.

An understanding of these principles is necessary to appreciate the effects of temperature on biological systems and to understand how different groups of animals manage their heat budgets.

8.1 The effect of temperature on chemical reactions

Molecules in a gas or in solution undergo random motion[1] and are continually colliding with each other. The energy within the gas or solution is related to the rate at which its component molecules move and is called **kinetic energy**. The speed of chemical reactions depends on temperature because the molecules involved need to collide with each other before they will react. As temperature increases, the kinetic energy of the reacting molecules increases. As a result, they move faster and collide more frequently increasing the probability that they will react with each other.

Central to understanding the effect of temperature on chemical reactions is the concept of **activation energy**. Molecules not only need to collide to react, but they must do so at or above a certain level of kinetic energy—the activation energy. The frequency distribution of kinetic energies of molecules within a gas or solution is known as the **Maxwell–Boltzmann distribution**, which is illustrated in Figure 8.1A. This distribution is positively skewed[2] and its exact shape depends on the temperature of the chemicals.

The energy that molecules must have before they will react with each other when they collide lies to the right of the Maxwell–Boltzmann distribution. As temperature increases, the Maxwell–Boltzmann distribution shifts towards the right, as illustrated in Figure 8.1B. When this shift occurs, a greater proportion of molecules will possess sufficient energy to be above the activation energy and will therefore react with each other.

The rate of product formation in a chemical reaction is called the **rate or velocity of reaction**, v. the rate of reaction is directly proportional to the concentration of the reactants and the proportionality constant is called the **rate constant**. The rate constant is dependent on the temperature at which the reaction takes place. The units used for the rate constant vary depending on the type of reaction as discussed in Box 8.1 (see Online Resources).

An example of the temperature dependence of a chemical reaction is shown in Figure 8.2A. Note that this relationship is described by an exponential curve when plotted on a

1 Random movement of molecules is discussed in Sections 2.1 and 3.1.2.

2 Positively skewed means that the tail on the right-hand side of a distribution is longer than that on the left.

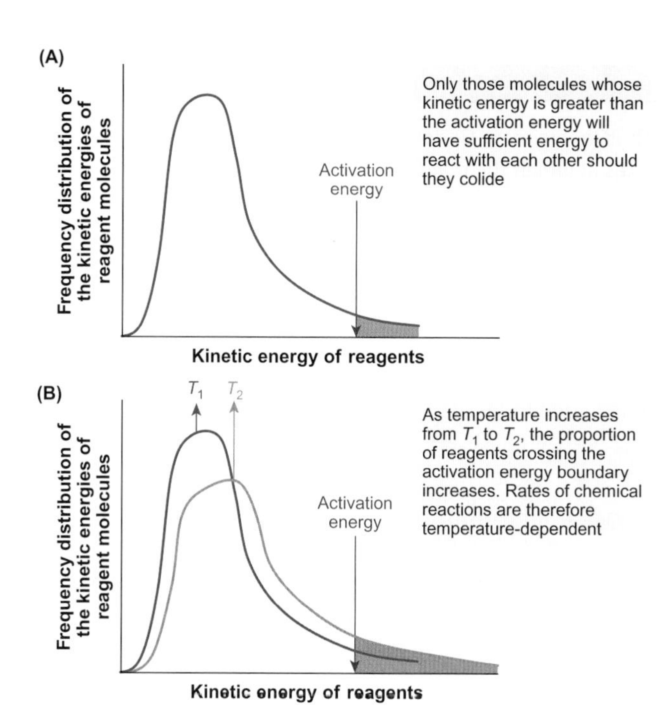

Figure 8.1 Hypothetical plots of Maxwell–Boltzmann distribution and activation energy

(A) In order to react, the reagents must first cross the activation energy boundary. The frequency distribution of the kinetic energies of the reagent molecules follows a skewed distribution (Maxwell–Boltzmann distribution) that is dependent on temperature. (B) At higher temperatures, the Maxwell–Boltzmann distribution shifts to the right.

linear scale. As temperature rises, the rate constant increases and the rate of reaction will also increase in proportion to the rate constant if the concentrations of the reactants do not change. By convention, the relationship is characterized by constructing what is called an **Arrhenius plot**, which is a transformation of the exponential curve into a straight line, shown in Figure 8.2B. To obtain this plot, we take the log to base e ($\log_e$, also known as the natural logarithm, ln) of the reaction rate, and plot it against the reciprocal of absolute temperature (in kelvin, K)[3]. The resultant relationship is linear over the range within which reaction rates increase with temperature because of the $\log_e$ conversion. This temperature dependence of reactions is generally characterized by a value called the $\boldsymbol{Q_{10}}$ (Q_{10} is discussed further in Box 8.2).

The Q_{10} is the extent to which the reaction rate changes with a 10°C change in the temperature. The Q_{10} relationship can be explained by the increased number of molecules whose kinetic energy is above the activation energy when temperature increases, as explained in Figure 8.1. If the relationship represented by the Arrhenius plot is linear over a range of temperature, say from 5 to 45°C, then Q_{10} will be constant within that temperature range. In other words, if the temperature of a reaction with a Q_{10} of 2 increases from 5 to 15°C the reaction rate will double, and it will also double if the temperature increases from 35 to 45°C. However, if the relationship represented by the Arrhenius plot deviates from a straight line, then the Q_{10} values will differ between different temperature ranges.

As a general rule, many chemical reactions have a Q_{10} value around 2.0 (from about 1.7 to 2.5) so, if a biological process has a similar Q_{10} value, it is said to show the 'expected' dependency on temperature. A Q_{10} value well outside this range suggests that something is happening beyond the effects of activation energy on chemical events alone. We encounter several such effects in the other chapters in this Part of the book.

[3] Kelvin is the SI unit for absolute temperature and 0°C = 273 K or, more accurately 273.15 K, as discussed in Section 2.1.

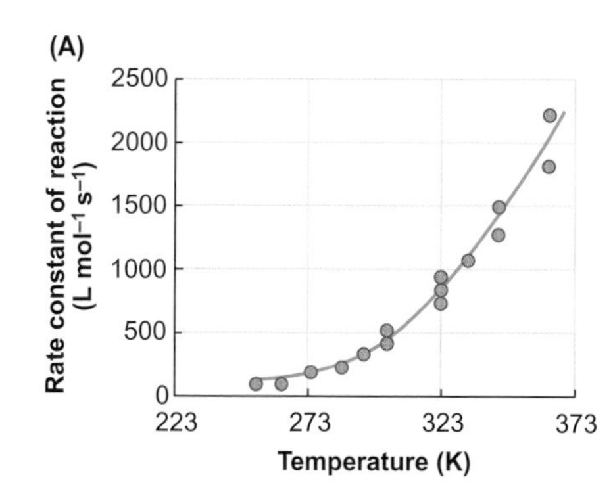

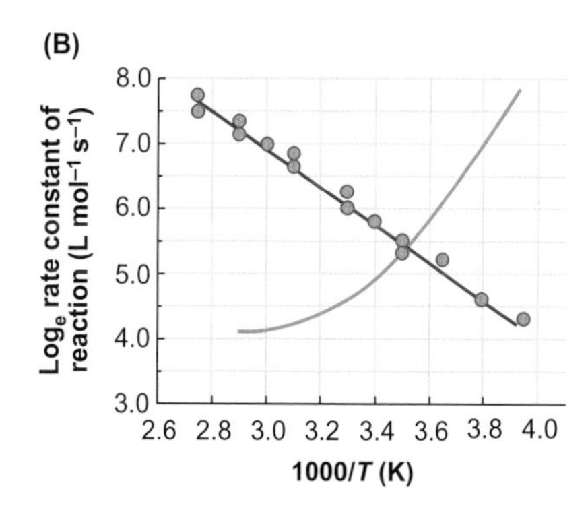

Figure 8.2 Dependency of a chemical reaction (polymerization of n-butyl methacrylate by free radicals) on temperature

(A) The rate constant of the reaction increases as the temperature rises; since the rate of reaction is proportional to the rate constant, it means that the rate of reaction increases as the temperature rises. (B) The equivalent Arrhenius plot (red line) of the reaction shown in A (blue line) is obtained by plotting $\log_e$ of the reaction rate against 1000/absolute temperature (T in kelvin). The slope of an Arrhenius plot reflects the activation energy of the reaction.

Data from Beuermann S and Buback M (1998). Critically-evaluated propagation rate coefficients in free radical polymerisations – II. Alkyl methacrylates. Pure and Applied Chemistry 70: 1415-1418.

Box 8.2 Q_{10} and temperature effects

The Q_{10} of a reaction characterizes how dependent the rate of reaction is on temperature. Specifically, it expresses by how much the rate of reaction increases for a standard increase in temperature of 10°C. Larger values of Q_{10} imply a very profound effect of temperature on the rates of reaction, while a Q_{10} value of 0 would indicate complete independence of the reaction from temperature. So, with a Q_{10} of 2, the rate of a reaction doubles for every 10°C increase, while a Q_{10} of 3 tells us that the rate triples with every 10°C increase.

The changes in rate for every 10°C increase in temperature, give rise to an exponential relationship, as shown in Figure A. Exponential relationships can be expressed as follows:

$$y = ab^x \qquad \text{Equation A}$$

The y variable depends on both a constant, b, raised to the power of the x variable and a second constant, a.

If we measure the rate of a reaction (V_1 and V_2) at two different temperatures (T_1 and T_2), and replace b with the symbol Q_{10}, we obtain:

$$V_2 = V_1 Q_{10}^{(T_2 - T_1/10)} \qquad \text{Equation B}$$

It is more convenient to use the logarithmic form of Equation A:

$$\log y = \log a + x \log b \qquad \text{Equation C}$$

which becomes:

$$\log V_2 = \log V_1 + \log Q_{10} \times (T_2 - T_1)/10 \qquad \text{Equation D}$$

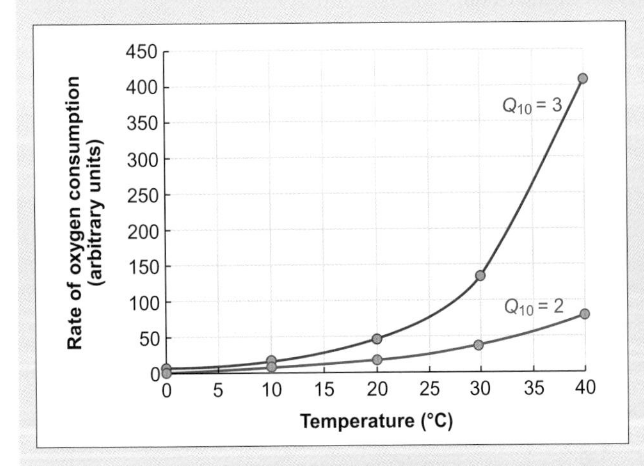

Figure A Theoretical plot of the effect of temperature on rate of a reaction at two different values of Q_{10}

Thus, to estimate Q_{10} for a reaction, we need to measure the rates at which the reaction proceeds (V_1 and V_2) and the two temperatures at which these measurements are made (T_1 and T_2 respectively). We can then rearrange Equation D to give log Q_{10}:

$$\log Q_{10} = (\log V_2 - \log V_1)\,(10/(T_2 - T_1)) \qquad \text{Equation E}$$

Which means that:

$$Q_{10} = \left(\frac{V_2}{V_1}\right)^{\left(\frac{10}{T_2 - T_1}\right)} \qquad \text{Equation F}$$

It is not necessary to obtain rates of a process exactly 10°C apart in order to calculate Q_{10}. However, the temperatures should be sufficiently different from each other to enable an accurate calculation of Q_{10}. Also, the Q_{10} of a biological process may vary at different temperature intervals within the same species.

For example: rates of oxygen consumption in lesser spotted dogfish (*Scyliorhinus canicula*) acclimated to three different temperatures are:

15.4 μmol kg^{-1} min^{-1} at 7°C

20.7 μmol kg^{-1} min^{-1} at12°C

32.8 μmol kg^{-1} min^{-1} at 17°C

What are the Q_{10} values at each of the two temperature intervals and for the full 10°C range?

Using Equation F for the interval 7°C to 12°C, we have:

$$Q_{10} = \left(\frac{20.7\ \mu\text{mol kg}^{-1}\ \text{min}^{-1}}{15.4\ \mu\text{mol kg}^{-1}\ \text{min}^{-1}}\right)^{\left(\frac{10}{12° - 7°}\right)}$$

Therefore: $Q_{10} = (1.34)^2 = \mathbf{1.80}$

Similarly, for the interval from 12°C to 17°C we have:

$$Q_{10} = \left(\frac{32.8}{20.7}\right)^2 = (1.58)^2 = \mathbf{2.50}$$

The overall Q_{10} from 7°C to 17°C is $\left(\frac{32.8}{15.4}\right) = \mathbf{2.13}$.

8.2 The effect of temperature on biological processes

Virtually all biological processes show some sort of dependence on temperature. This includes cellular processes, processes at organ level, physiological reactions at the level of the entire animal and even behaviour. In this section we explore in detail some of the similarities and differences between the rates of biological reactions and the simpler chemical reactions described in the previous section.

8.2.1 The effect of temperature on the rate of biological processes

The effects of temperature on rates of biological processes are generally more complex than the simple linear Arrhenius plots that occur for most chemical reactions. In particular, biochemical reactions (in test tubes) often show a transition or breakpoint at which the rate of reaction starts to get slower as temperature increases above a critical level. This effect is particularly obvious in the reaction characterized in Figure 8.3. The main reason for this phenomenon is that biochemical reactions are **catalysed** by enzymes. Enzymes accelerate reaction rates by effectively lowering the activation energy of a given reaction. But why is the activity of enzymes affected by temperature?

Most enzymes are proteins that work because they have particular three-dimensional (tertiary)[4] structures that assist reagents in colliding in the correct orientations. However, proteins are susceptible to disruption by temperature changes: as temperature increases, the tertiary structure of enzymes starts to 'unravel' causing a reduced capacity to catalyse reactions successfully. Hence, above the breakpoint in the Arrhenius curve, an increase in temperature has a negative effect on reaction rates because enzymatic catalysis is inhibited.

In hibernating mammals, such as Richardson's ground squirrel (*Urocitellus richardsonii* formerly, *Spermophilus richardsonii*)[5], whose body temperature is low, the activity of an enzyme can differ from that when the animal is at its normal body temperature. This difference can be illustrated by constructing Arrhenius plots as the slope of such a plot signifies the activation energy of the enzyme. The activation energy of an enzymic reaction occurring during the transition from normal body temperature to the low temperature of hibernation is about 33 per cent greater in muscle from hibernating squirrels than from that of non-hibernating animals. This increase in activation energy, together with other data, indicates that the total activity of the enzyme is reduced when the squirrels enter hibernation.

8.2.2 Effects of temperature on mortality and the temperature limits to life

The effects of temperature on biology are not limited to the impact it has on the rates at which physiological processes and behaviour proceed. At some point, the effects of temperature start to become seriously detrimental, either when

[4] The structure of proteins is discussed in Box 2.5.

[5] Hibernation is discussed in Sections 10.2.7 and 15.4.

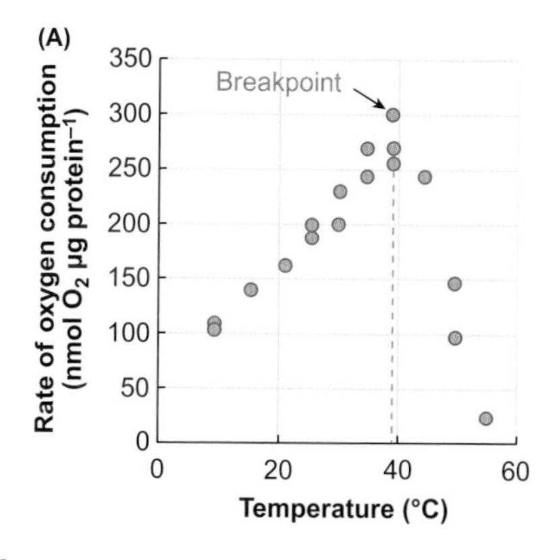

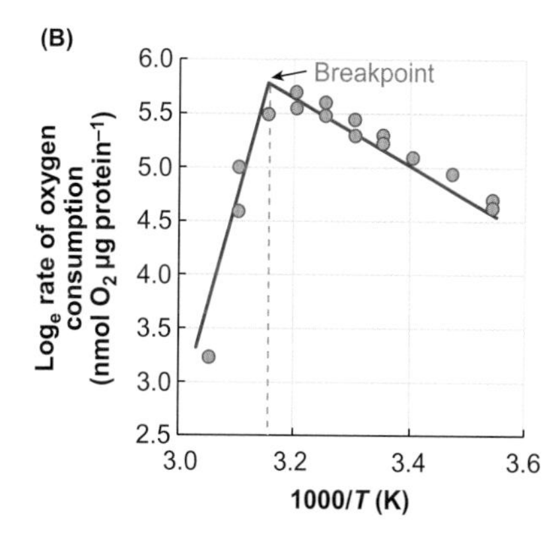

Figure 8.3 Effect of temperature on rate of oxygen consumption by mitochondria of abalone (*Haliotis* spp.)

A Plot of rate of oxygen consumption against temperature (°C) showing the break-point at about 39°C (purple dotted line). **B** The equivalent Arrhenius plot of the data in A, which also shows the breakpoint.

Based on Dahlhoff E and Somero GN (1993). Effects of temperature on mitochondria from abalone (genus *Haliotus*): adaptive plasticity and its limits. Journal of Experimental Biology 185: 151-168.

temperature increases or when it falls, and the animals die. The point at which the temperature reaches a fatal level varies between individuals, with the quoted figure normally being the temperature at which 50 per cent of the animals die over a fixed period. This is an analogous concept to the lethal dose of a toxic compound that kills 50 per cent of the animals exposed to it (the lethal dose 50 or LD_{50}). The lethal temperature at which half the animals die is abbreviated to **LT_{50}**.

To establish the LT_{50} of a species, groups of animals are exposed to a range of temperatures. The numbers that die in each group are recorded and then plotted against the temperature to which the group was exposed. A hypothetical example when temperature is increased is shown in Figure 8.4.

Several points can be taken from Figure 8.4:

- There is a temperature below which all the animals survive for the set time period.
- There is a temperature above which every animal in the experiment dies.
- Between these two temperatures, the relationship between mortality and temperature is described by a curve.
- The LT_{50} is the point at which 50 per cent mortality occurs, which can be read directly from the graph.

In general, the more prolonged an exposure time, the lower the LT_{50}. Animals can survive brief exposures to adverse temperatures, but the temperature may become lethal if they are exposed for more prolonged periods. For example, freshwater crayfish (*Austropotamobius pallipes*) have been extensively used in studies of lethal temperatures. If crayfish are kept at 10°C and then transferred to 32°C they all die within 60 minutes of transfer. After 18 minutes, only 10 per cent of the animals are dead so, 90 per cent of crayfish could survive exposure to the 'lethal' temperature of 32°C if the exposure was limited to 15 minutes. As such, LT_{50} measurements only make sense if the exact time of exposure of the animals to the different temperatures is specified.

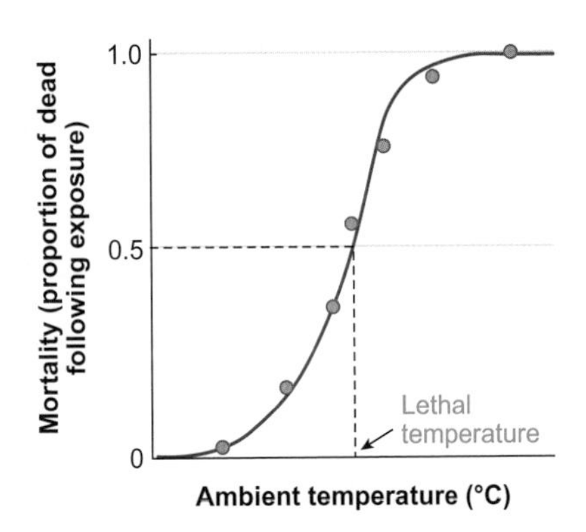

Figure 8.4 Hypothetical mortality plot for groups of animals exposed to different temperatures for a fixed time period

The number of animals that die (mortality) in each group increases in relationship to temperature in a sigmoidal fashion. The lethal temperature is defined as the temperature at which 50% of the animals die (LD_{50}).

Different organisms tolerate variations in body temperature to different extents. The highest temperatures at which organisms have been reported are in the deep ocean vents. Some microorganisms can live around black smoker vents[6] at over 250°C. In these conditions, the water is prevented from boiling because of the extreme pressures. At the surface (where pressure is much lower) bacteria from hot springs are known routinely to experience temperatures of 60–80°C. Some species of animals, such as a polychaete annelid (*Paralvinella sulfincola*) can live at temperatures around 50–55°C, the highest demonstrated for a marine metazoan[7].

At the other extreme, many organisms die when their tissues freeze at sub-zero temperatures. However, some species can survive in such conditions[8], with some (such as tardigrades—phylum Tardigrada) having the capacity to survive for several hours at the temperature of liquid helium (−269°C). Tardigrades are able to survive such temperatures because they become desiccated, a process known as **anhydrobiosis** (which means life without water). The principles of anhydrobiosis have been applied to the storage of biological materials at room temperature, as outlined in Case Study 8.1.

Temperature limits to life in individual species

Although the temperature limits to animal life as a whole are wide, the capacity of any single species to survive over a large temperature range is relatively limited (with a few rare exceptions like the tardigrades mentioned above). That said, the actual limits vary enormously between different species. To illustrate this let us contrast two groups of small mammals and two species of small fish.

Small insectivorous bats weigh 5–20 g and generally regulate their body temperatures to be around 37°C, which is the typical level for a eutherian mammal. The upper lethal temperature for these bats is typically 42–46°C. However, most species of bats are very tolerant to the cooling of their body and will survive body temperatures of about 4°C. Indeed, some can survive periods during which their

[6] Black smokers are forms of hydrothermal vents which emit superheated fluids through fissures in the Earth's surface. Hydrothermal vents are briefly discussed in section 1.2.1.

[7] A metazoan is a multicellular animal.

[8] Survival at subzero temperatures is discussed in Sections 9.2.3 and 9.2.4.

Case Study 8.1 Nature inspires an energy-saving approach to the preservation of biological materials

Tardigrades, nematodes, rotifers and brine shrimps live in extremely variable and often unpredictable habitats. Some such animals must be capable of surviving almost total desiccation in an ametabolic state, known as anhydrobiosis, for long periods of time. They return to a metabolic state simply by becoming rehydrated[1].

Research into anhydrobiosis in animals like tardigrades has revealed that the process replaces water with a molecule that can protect structure but prevent function. Important elements of anhydrobiosis include changing the permeability characteristics of the animals' surface material and essentially replacing the water molecules within their tissues with specific sugars such as the disaccharide trehalose. These sugars protect phospholipids and proteins in cell membranes from desiccation damage.

Trehalose is found in very high concentrations in anhydrobiotic organisms; it can comprise from 20 to 50 per cent of their dry weight. This sugar replaces the primary water attached by hydrogen bonds to macromolecules, as shown in Figure A(i). It also contributes to a process known as **vitrification**, or **glassification**. Glassification involves the production of an intracellular organic glass that stabilizes the cells' contents and halts deleterious interactions between intracellular components. Dehydration raises the temperature at which glassification occurs (*Tg*) from colder to more normal ambient temperatures. At such temperatures, the high concentration of sugars increase their viscosity so much that they enter a glass-like state—in other words, they form a homogeneous non-crystalline solid. The analogy has been made that it is like 'shrink-wrapping' the molecules in a protective glass-like coating.

Research biologists in universities, biotechnology and pharmaceutical industries, and technicians in criminology laboratories have to store biological materials for long periods of time, typically by using super-cold freezers that hold the samples at −20 to −80°C or in liquid nitrogen at around −200°C. The widespread use of freezers in laboratories consumes huge amounts of energy, estimated recently to cost $30 billion annually worldwide. Furthermore, the requirement for the refrigeration of vaccines limits their distribution in places lacking refrigera-

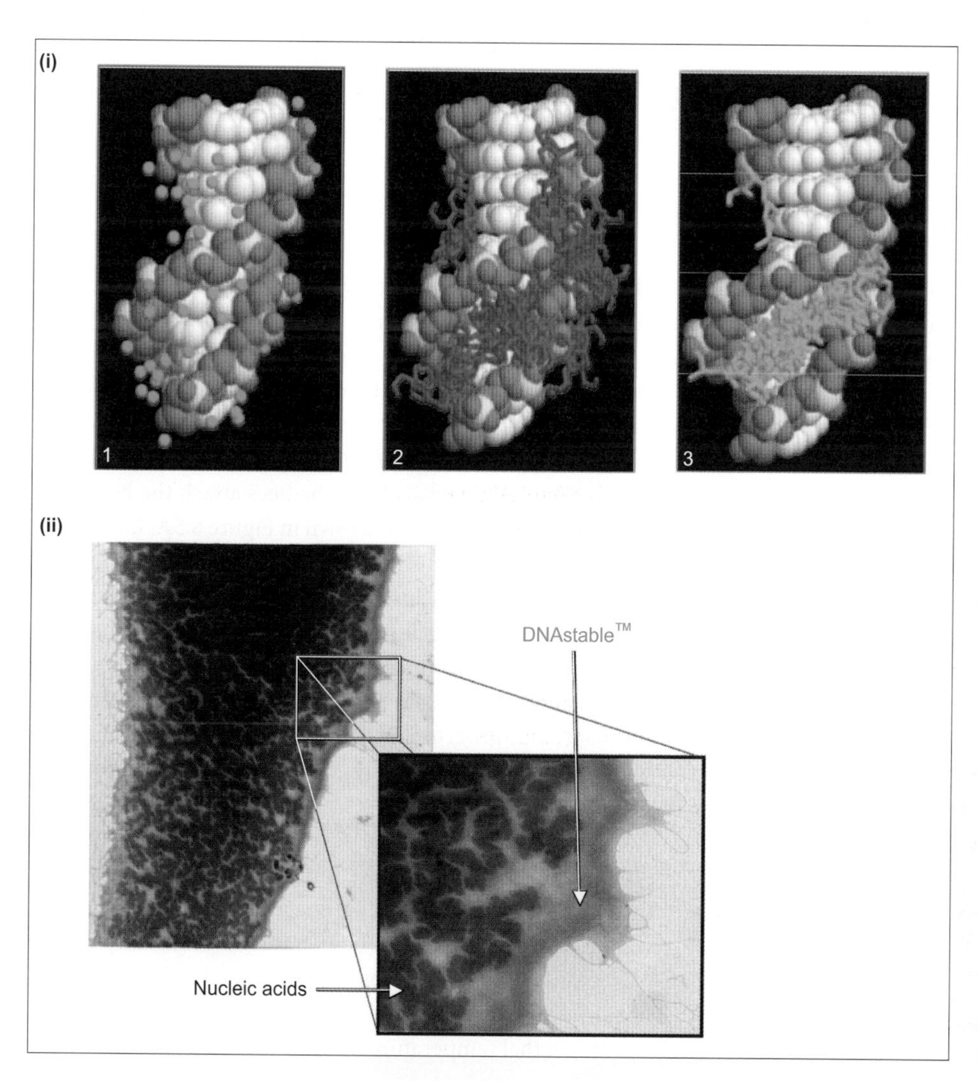

Figure A Synthetic polymers enable the preservation of biological materials at room temperature

(i) 1: Model of normally hydrated nucleic acid (turquoise balls represent water molecules). 2: Natural system with trehalose (blue) replacing the coating of water molecules. 3: A synthetic polymer, DNAstable™ (green), fitting in to the same groove in the nucleic acid molecule as the trehalose does in the natural system. (ii) Electron micrograph of a nucleic acid sample showing the thermostable layer of DNAstable™ that forms around the nucleic acid molecules. This layer helps to stabilize the molecules and protect them from desiccation damage.

Reproduced from: Rozieres SD et al (2007). Room temperature storage of Biological Samples. PAGS. Blow N (2009). Biobanking: freezer burn. Nature Methods 6: 173-178

tors. The long-term exposure to freezing temperatures and potentially multiple freeze–thaw cycles can also degrade precious biological samples.

To address the need for long-term storage of biological materials at room temperature, several companies, such as the Biomatrica company (San Diego, California), GenVault (Carlsbad, California) and Nova Laboratories Ltd (Leicester, UK) have developed approaches to the storage of biological materials at ambient temperature that use dehydration and glassification processes inspired by the study of organisms capable of anhydrobiosis.

The role played by trehalose in anhydrobiotic organisms—that of replacing the water molecules bound to other molecules in anhydrobiotic organisms—is performed instead by a patented synthetic polymer, the biomimetic product DNAstable™ by Biomatrica, shown in Figure A(i). Upon dehydration, DNAstable™ becomes a thermostable, dissolvable glass that 'packages' every single molecule in the sample, as illustrated in Figure A(ii).

To store a biological tissue sample, it is applied to a DNAstable™—lined test tube or microsample well and dehydrated. The sample remains protected from damage by UV or heat for long periods of time at room temperature; when needed for analysis it is simply rehydrated. The rehydration dissolves and washes away the protective coating, and water molecules are restored to their normal positions around the molecules, which then regain their functional characteristics.

Independent studies of the products of both Biomatrica and GenVault have shown them to be at least as effective at safely storing DNA and RNA samples at room temperature as freezing to –20°C. Indeed, in some cases they have proven to be even better at preserving function upon rehydration than after the standard freeze–thawing processes. Biomatrica's products can also be used for long term storage of blood samples.

The Biomatrica and Nova Laboratories have found that the dehydration and glassification process with sugars or mixed synthetic polymers can also be used for the storage of vaccines and medications at room temperature. This process greatly increases the ability to get these life-saving biochemicals to remote regions of less developed countries, where reliable refrigeration is rare.

› Find out more

Blow N (2009) Biobanking: freezer burn. Nature Methods 6: 173–177.

Clegg JS (2001) Cryptobiosis—a peculiar state of biological organization. Comparative Biochemistry and Physiology B 128: 613–624.

Crowe, JH, Carpenter JF, Crowe LM (1998). The role of vitrification in anhydrobiosis. Annual Review of Physiology 60: 73–103.

Stern V (2010). Dry out, put away. The Scientist 24: 19–20.

Wan E, Akana M, Pons J, Chen J, Musone S, Kwok P-Y, Liao W (2010). Green technologies for room temperature nucleic acid storage. Current Issues in Molecular Biology 12: 135–142.

Rozieres SD, Ohgi S, Faix PH, Muller-Cohn, J, Muller R (2007). Room temperature storage of Biological Samples. PAGS pdf (http://www.biomatrica.com/media/posters/PAGS%202007%20-%20Room%20Temperature%20Storage% 20of%20Biological%20Samples.pdf)

[1] Section 4.1 also discusses anhydrobiosis in tardigrades.

body is super-cooled[9] to around –5°C during hibernation. Contrast the wide thermal tolerance of these bats with that of a group of small shrews. These shrews also weigh between 5–20 g, feed on insects and regulate their body temperatures to be around 38°C. However, if a shrew heats up to 42°C or if it cools to about 33°C it dies. The range of permissible body temperatures for the shrews is only about 10°C, compared with a range about four times greater for the insectivorous bats.

Similar comparisons are possible with many other groups of animals. For example, Antarctic icefish routinely live under and in cracks in the Antarctic ice at sea temperatures of –1.9°C[10]. However, they cannot tolerate cooling beyond –4°C, and are generally dead when temperatures rise much above 6°C. In contrast, bullfish (*Ictiobus* sp.) survive at any temperature between 4 and 34°C.

We use different terms to reflect the different capacities to survive ambient temperature variations exhibited by animals: **eurythermal** (Greek: *eurys*—wide, broad, *thermé*—heat) animals have a wide temperature tolerance and **stenothermal** (Greek, *stenos*—narrow) animals can tolerate only narrow temperature limits.

A remarkable example of how animals can utilize their different tolerances of ambient temperature is provided by Japanese honeybees (*Apis cerana japonica*), which are predated by Japanese giant hornets (*Vespa mandarinia japonica*). When a hornet attacks a hive, the bees attack the hornet by forming a ball around it, as shown in Figure 8.5A. It had been thought that the bees sting the hornet to kill it, but we now know that the small cluster of bees generates so much heat that the centre of the cluster, where the hornet is, can rise to a temperature of around 47°C, as indicated in Figure 8.5C. This is above the lethal temperature of the hornet, which is 44–46°C, but lower than the lethal temperature of the bees (48–50°C). The bees literally cook the hornet to death!

Upper lethal temperature limits

The fact that the temperature limits to life as a whole exceed by far the limits for individual species suggests that the processes driving the effect of temperature on mortality do not reflect some uniform factor common to all animals. Several ideas have been put forward to explain the patterns we observe in lethal temperature limits.

[9] Super-cooling is discussed in Section 9.2.4.

[10] The temperature of the surface seawater of Antarctica is relatively stable. It may warm to 0°C in summer, but during winter it is at its freezing point of –1.8 to –1.9°C.

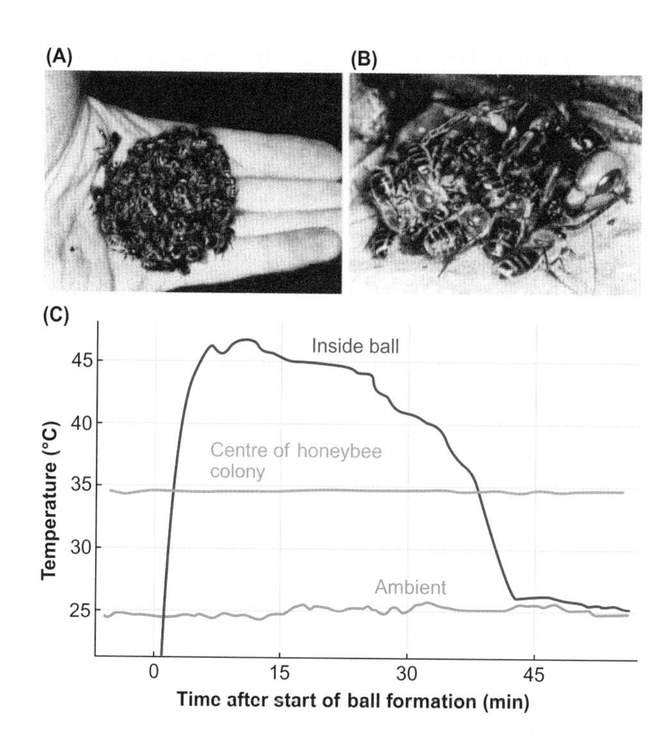

Figure 8.5 Japanese honeybees (*Apis cerana japonica*) using heat to kill a Japanese giant hornet (*Vespa mandarinia japonica*)

(A) Defensive ball of about 400 honeybees, which show no inclination to sting. (B) Giant hornet killed by heat from the ball of honeybees. (C) Temperatures measured from inside a ball of honeybees, within a colony of honeybees and in the surrounding environment (ambient).

Reproduced from Ono M et al (1995). Unusual thermal defence by a honey bee against mass attack by hornets. Nature 377: 334-336. and Ono M et al (1987). Heat production by balling in the Japanese honeybee, *Apis cerana japonica* as a defensive behavior against the hornet, *Vespa simillima xanthoptera* (Hymenoptera: Vespidae). Experientia 43: 1031-1034.

Effects of temperature on protein structure

As we note in Section 8.2.1, most enzyme catalysts are proteins, the tertiary structures of which are adversely affected at temperatures between 40 and 50°C[11]. The range of 40 to 50°C corresponds to the range of temperatures at which breakpoints in Arrhenius plots[12] often occur and at which many terrestrial vertebrates and invertebrates die. One idea then, is that animals die because of direct temperature effects on proteins (denaturation and coagulation) that compromise their capacity to catalyse reactions and are reflected in the breakpoints in the Arrhenius curves for particular reactions.

On the face of it there appears to be a problem with this idea. We know that basic metabolic processes are preserved across all organisms that utilize oxygen as a metabolic substrate[13]. For example, conversion of oxaloacetate and acetyl-coenzyme A to citrate is catalysed by citrate synthase in all animals (and plants) that utilize the tricarboxylic acid (TCA) cycle (also known as the Krebs cycle)[14]. How is it possible then, that some of these enzymes become dysfunctional in ice fish exposed to 6°C, but the same enzymes function perfectly well in mammals at body temperatures of 35–40°C, and in birds at 40–45°C?

Two main possibilities might provide an answer to this question:

- The first is that some animals may have unique metabolic pathways that are critically sensitive to temperature. Species of ice fish, for example, may die at 5–6°C because they have a metabolic process that fails at these temperatures. Such a process is not present in mammals and birds that remain unaffected. While this possibility might explain deviations between species, the diversity of the critical temperature limits across all species would require enormous variability in metabolic processes for it to be correct in all instances. Such large variability is inconsistent with the evidence that metabolic processes are essentially the same in all animals.
- In the last decade, our knowledge of the protein sequences of some enzymes (in particular lactate dehydrogenase—LDH) from a diverse range of organisms has improved, and a second possible answer to the question posed above is starting to emerge.

Subtle changes in the amino acid sequences of enzymes can have profound effects on their temperature sensitivity and stability. It is possible therefore, that an enzyme catalysing aspects of metabolism in ice fish becomes denatured or coagulated at 5–6°C whereas the same enzyme in eurythermal fish living at higher temperatures only becomes dysfunctional at 30–35°C[15], due to as little a difference as a single to a few amino acids.

An additional feature that could specifically make ice fish more vulnerable to small increases in temperature is their inability to increase their production of **heat shock proteins** (HSP). Heat shock proteins are protein chaperones which reduce and repair damage to cellular proteins that arises from many forms of stress, most notably high temperature[16]. However, the increased production of HSP in response to heat stress is lacking in ice fish and many other species of Antarctic marine ectotherms[17]. This failure to respond to heat stress means that such species have reduced abilities to minimize damage to proteins following an increase in temperature. This may be one reason for their extreme stenothermality and may limit their ability to cope with climate (temperature) change.

[11] Section 8.2.1 discusses the effect of temperature on the structure of proteins.
[12] The breakpoint of an Arrhenius plot is shown in Figure 8.3.
[13] Metabolic pathways are discussed in Section 2.4.
[14] We discuss the tricarboxylic acid cycle in Section 2.4.2.
[15] The specific example of LDH is discussed in section 9.3.2.
[16] Heat shock (stress) proteins are discussed in Section 9.3.1 and Box 9.1.
[17] We define the term ectotherm in Section 8.5.

Different Q_{10} values along a sequence of different reactions

High temperature may lead to mortality before proteins become denatured and completely inactive. This is evident from the fact that the breakpoints in the Arrhenius plots for given enzyme-linked reactions (reflecting the point at which enzymes become inactivated) often lie well above the temperatures at which the animals perish. Different biological reactions have different Q_{10} values, so a serious metabolic disruption might occur because different reactions in a sequence are affected by the temperatures to different extents.

We can illustrate this potential effect by considering a simple metabolic pathway, where substrate X is converted to substrate Y, which is itself converted to substrate Z. The enzymatic catalysis of the conversion of X to Y may have a Q_{10} of 1.7, but that of Y to Z may have a Q_{10} of 3:

$$X \rightarrow Y \qquad Y \rightarrow Z$$
$$Q_{10} = 1.7 \qquad Q_{10} = 3$$

In these circumstances, the consumption of substrate by the conversion of Y to Z as temperature increases will greatly exceed that of X to Y, and therefore the middle element of the pathway (Y) will potentially become seriously depleted[18]. If Y is involved in other reactions, the repercussions of this single rate disparity could be widespread. The potential for catastrophic breakdown increases as the number of links in the pathway gets larger, and particularly if they are at different values for Q_{10}.

Temperature effects on the lipid bilayer

Another problem with a change in temperature is the effect it has on the lipid bilayers that comprise the cellular substructures and surface of the cell[19]. Lipid bilayers have embedded in them a whole host of proteins, some of which act as signal receptors[20]. Enzyme complexes are also often bound to membranes. Membrane fluidity is a key feature that affects how these membrane proteins are able to interact.

Lipids comprise a major proportion of the cellular membrane, with the fluidity of these lipids being sensitive to temperature[21]. This sensitivity is apparent whenever we place a pat of butter into a fridge; at 20°C the butter is soft and easily manipulated, but at 4°C it is hard.

The more fluid the lipids in the membrane become, the less ordered they are. The fluidity of the lipids in a membrane can be estimated by measuring the polarization of fluorescence from a molecule such as 1,6-diphenyl-1,3,5-hexatriene (DPH) which has been experimentally positioned within the membrane, when it is illuminated with polarized light. A low value of polarization means that the lipids are relatively fluid, whereas a high value of polarization means that fluidity is lower.

Plots of polarization as an indicator of fluidity for membranes from four species of fish, a bird and a mammal, as a function of temperature are shown in Figure 8.6. Note that, although the curve shifts to the right in species with higher body temperature, the polarizations measured at the different acclimation temperatures[22] of different species of fish or at the body temperatures of birds and mammals, are similar. Thus, the intrinsic influence of temperature on membrane fluidity is offset by the rightward shift of the curves in animals with higher body temperature. This is known as **homeoviscous acclimation**[23].

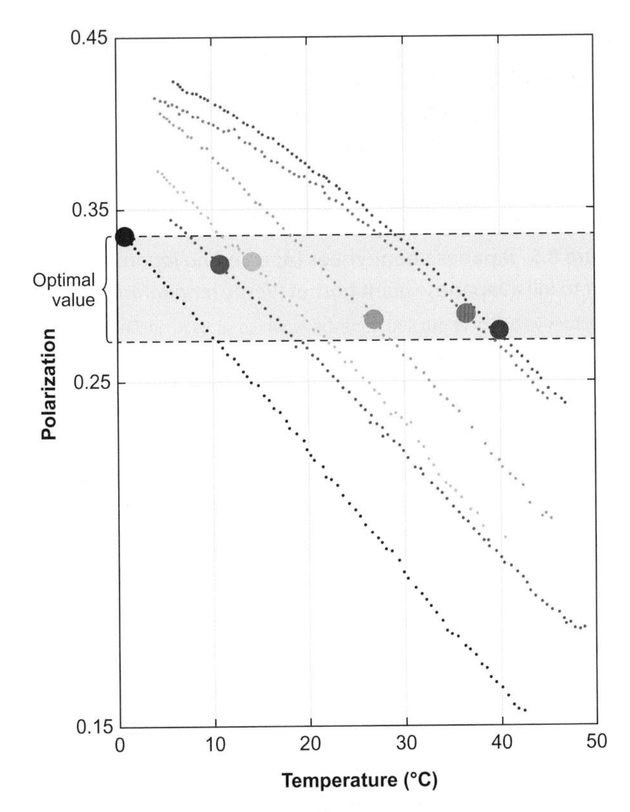

Figure 8.6 Membrane lipid structure of brain synapses of four species of fish and two of mammals as a function of temperature

Lower values of polarization of light indicate increasing disorder of the cell membranes—their fluidity increases. The acclimation temperatures of the fish and body temperature of the mammals are shown by the coloured dots: ● the Antarctic fish (*Notothenia neglecta*), ● rainbow trout (*Oncorhynchus mykiss*), ● perch (*Perca fluviatilis*), ● convict cichlid (*Cichlasoma nigrofasciatum*), ● hamster (*Mesocricetus auratus*), ● starling (*Sturnus vulgaris*). Note that fluidity of the cell membranes is similar when measured at the acclimation temperature of the fish and at the body temperature of the birds and mammals.

Source: adapted from Cossins, AR and Bowler K. Temperature Biology of Animals. London and New York: Chapman and Hall, 1987.

[18] When one particular substrate becomes depleted, it represents a rate limiting step in the overall reaction.

[19] Lipid bilayers are discussed in Section 3.2.3.

[20] Section 19.1.4 discusses hormone receptors.

[21] Some characteristics of fatty acids are discussed in Box 9.1.

[22] Acclimation temperature is one to which animals are exposed in an experiment. Acclimation, acclimatization and adaptation are discussed in Section 1.5.1.

[23] Homeoviscous acclimation is discussed further in Section 9.3.2.

Lower lethal temperature limits

Complex reasons lie behind the death of animals as their body temperatures fall. Ice formation in tissues causes widespread tissue damage that is often lethal. Many organisms can tolerate such events, while many others die long before their body temperature falls low enough for ice formation[24].

Thermal denaturation of enzymes at low temperatures is unlikely, but the influence of temperature on reactions in pathway sequences could be profound. Once a rate of reaction becomes critically low, vital processes may stop and this may be the ultimate cause of death from cold, particularly in birds and mammals. For example, once axon conductance stops, any breathing motions will halt, and heart failure will occur. Lack of oxygen in tissues (**hypoxaemia**[25]) may then be the immediate result of declining body temperature.

Because metabolic rate may be severely depressed at low body temperature by normal Q_{10} effects, the time taken to die may be quite long, even if ventilation and circulation have ceased. This is why occasional lucky individuals survive after having fallen through ice on frozen ponds. Such individuals may appear dead on recovery because they have stopped breathing and their hearts have apparently stopped (but are actually beating slowly and weakly), but can be slowly warmed and often make a complete recovery.

› *Review articles*

Fisher, CR, Takai K, Le Bris N (2007). Hydrothermal vent ecosystems. Oceanography 20: 14–23.

Mayer MP, Bukau B (2005). Hsp70 chaperones: cellular functions and molecular mechanism. Cellular and Molecular Life Sciences 62: 670–684.

Rebecchi L, Altiero T, Guidetti R (2007). Anhydrobiosis: the extreme limit of desiccation tolerance. Invertebrate Survival Journal 4: 65–81.

Somero GN (2004). Adaptation of enzymes to temperature: searching for the basic 'strategies'. Comparative Biochemistry and Physiology B 139: 321–333.

Somero GN (2010). The physiology of climate change: how potentials for acclimatization and genetic adaptation will determine 'winners' and 'losers'. Journal of Experimental Biology 213: 912–920.

8.3 Environmental temperature variation

The data on ranges of environmental temperature tolerance show that unless animals obtain some stability in their body temperatures, their bodily functions (and often their lives) will be at the mercy of environmental fluctuations in temperature. This does not pose much of a problem if the environment is relatively stable, but just how stable (or, indeed, variable) is environmental temperature, and what problems confront different organisms as a consequence of any variability?

Table 8.1 **Specific heat capacities of different materials**

	J kg^{-1} K^{-1}
Water	4182
Cotton	1298
Air	1002
Glass	837
Granite	816
Copper	383

Source: Cossins, AR and Bowler K. Temperature Biology of Animals. London and New York: Chapman and Hall, 1987.

Earth comprises two major habitat types: the oceans (and large bodies of fresh water) and the land[26]. The mass specific heat capacity of water is about 4–5 times greater than that of air and granite, respectively, as demonstrated in Table 8.1. This difference plays a major role in environmental temperature stability. When the heat of the sun reaches Earth, its effect on habitat temperatures depends on whether it falls on a body of water or on land. Much more heat is required to change the temperature of a large volume of water compared to a similar volume of air[27], so bodies of water are well buffered against temperature change. In the deep oceans, for example, ambient temperatures average 4.3°C and vary by less than 0.1°C over a year.

For animals that live in the ocean, stability of body temperature is generally not an issue, particularly if they are small, because they exist at the temperature of the medium in which they live (except marine birds and mammals). These animals have not evolved mechanisms to respond to rapid changes in temperature because they do not experience such changes in their natural environments. If we plot the biological functions of these organisms as a function of temperature, they follow the classic linear Arrhenius plot with Q_{10} values around 2.0–2.5 for the range of temperatures in their environment.

Figure 8.7 shows that to experience a change in surface temperature of 10°C in the Pacific Ocean in March, an organism would need to shift across the planet surface by a distance of around 2200 km. The main problem marine organisms face is the much more prolonged annual change in temperature that occurs as the seasons progress. The surface temperatures of the ocean can change by 10–15°C during the course of a year, depending on latitude.

Contrast the temperature stability in the ocean with what happens on the land. Figure 8.8 shows the ambient temperature measured around the entrance to the burrow of a small mammal, the plateau pika (*Ochotona curzoniae*), on the grasslands of the Qinghai-Tibetan plateau. What is significant

[24] Freeze tolerance and freeze avoidance are discussed in Sections 9.2.3 and 9.2.4.

[25] Hypoxaemia is discussed in section 15.3.1.

[26] Different habitats are discussed in more detail in Section 1.2.

[27] Note that 1 kg of air at 20°C occupies a volume of 830 L, whereas 1 kg of water occupies a volume of approximately 1 L, so the thermal capacity of water is almost 3500 times greater than that of air on a volume specific basis.

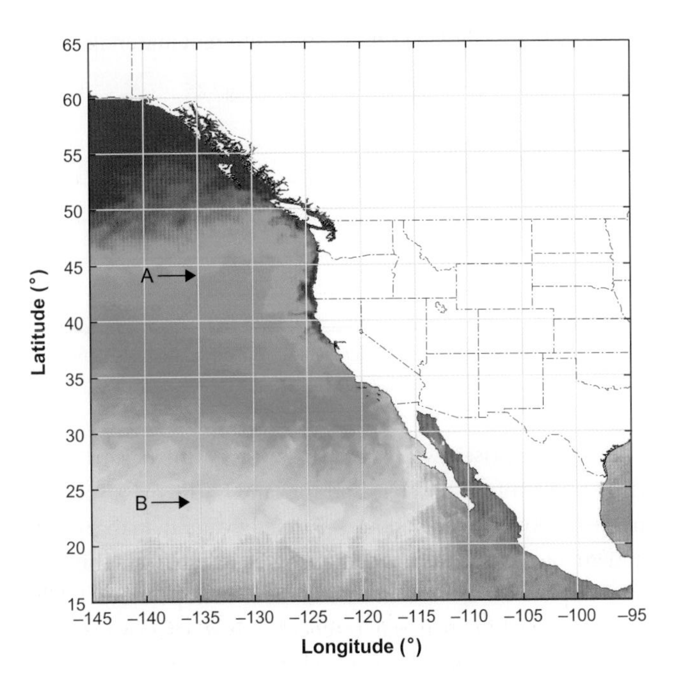

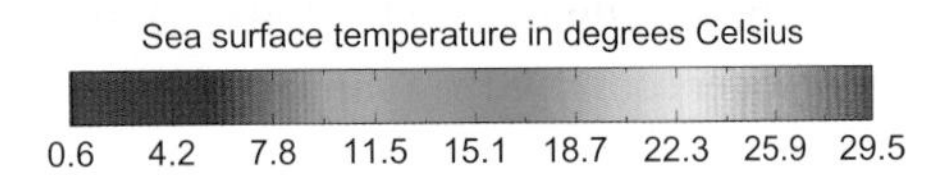

Figure 8.7 Surface temperature of the Pacific Ocean (°C) off the coast of the USA recorded on 22 March 2014

The arrows denote two points that are 10°C different (point A is at approximately 11.5°C and point B is at 21.5°C. The distance between these points is 20° of latitude (equivalent to 2220 km).

Reproduced from: NOAA http://www.ospo.noaa.gov/.

Figure 8.8 Ambient temperatures measured every 3 minutes around a burrow of the plateau pika (*Ochotona curzoniae*) on the Qinghai-Tibetan plateau (37° N, 100° E), at an altitude of 3200 m during four summer days (2008)

(A) Three temperature loggers were located within 0.6 m of each other. The upper photo shows details of the location. Logger A was tucked under the edge of a rock where it was not exposed to the sun, logger B was exposed on the surface and logger C was inserted 30 cm down an adjacent burrow. The lower photograph shows the general location of the burrow and loggers on the grasslands of the plateau. (B) Despite the apparent featureless nature of the grasslands the loggers reveal large differences in microclimate over very small distances. During sunny days (4–6 July inclusive) the surface temperature was routinely 10°C and up to 30°C hotter than that in the burrow only 50–60 cm away. Grey bars on the temperature plot indicate periods of darkness.

Unpublished data: Speakman JR, Wang DH, Chi QS and Zhang YM.

about these simple measurements is how they contrast with the temperature changes in the ocean, shown in Figure 8.7. The probes around the burrow were less than 1 m apart, yet the temperature difference between them was greater than that recorded over 2200 km of open ocean.

The above examples demonstrate that thermoregulation is generally a much more significant process for animals living on land than for those living in water. Indeed, if we consider the range of ambient temperatures found in most terrestrial habitats, it is apparent that, unless they exert some control over their body temperatures, most animals would be dead within a few days, and the rest would experience large fluctuations in their biological functions. To understand how animals manage their heat budgets to control their body temperatures, we need first to consider the general processes of energy exchange.

8.4 General processes of heat exchange

Heat can be exchanged by four mechanisms: radiation, conduction, convection and evaporation. Animals are able to maintain a degree of balance between how much heat enters and is generated by the body and how much heat leaves it. They achieve this balance by manipulating the flow of energy along the four different routes for heat exchange we have just mentioned—radiation, conduction, convection and evaporation. Let us now consider each of these routes in turn.

8.4.1 Radiation

All objects emit **electromagnetic radiation**. The nature and amount of this radiation is dramatically affected by the surface temperature of the object and whether it is a good or a poor emitter of radiation. In general, the amount of emitted radiation energy is described by Equation 8.1:

Amount of emitted radiation ($W\ m^{-2}$)

Boltzmann constant ($5.67 \times 10^{-8}\ W\ m^{-2}\ K^{-4}$)

Emissivity

$$\dot{Q}_{rad} = B\varepsilon T^4 \quad \text{Equation 8.1}$$

Surface temperature in kelvin (K) raised to the fourth power

The dot over the Q signifies a rate.

Radiation energy is exchanged between an object and its surroundings according to Equation 8.2, which describes the rate of net transfer of radiation energy between them:

$$\dot{Q}_{rad} = B\varepsilon\left(T_1^4 - T_2^4\right) \quad \text{Equation 8.2}$$

Table 8.2 **Emissivities for some common materials**

Skin (human)	0.98
Water	0.95
Wood	0.9
Concrete	0.8
Aluminium	0.2
Copper	0.03

Source: Cossins, AR and Bowler K. Temperature Biology of Animals. London and New York: Chapman and Hall, 1987.

Where ε and T_1 are the emissivity and temperature of the object, respectively and T_2 is the temperature of the surroundings.

The **emissivity** of an object is given a value between 0 and 1.0 to explain how good an emitter of radiation the surface of the object is compared to a perfect emitter (for which emissivity is 1.0). A perfect emitter is called a **black-body** emitter. A black body also completely absorbs radiation in all wavelengths and reflects nothing. So, a body that is a perfect emitter at a given wavelength is also a perfect absorber of radiation at that wavelength. Confusingly, we do not necessarily perceive a black body as being black, as the concept applies to all wavelengths. For example, human skin radiates like a black body, irrespective of its visible colour, light or dark. The skin reflects shorter wavelengths in the visible spectrum, but almost perfectly absorbs and radiates longer wavelengths in the infrared, which are outside the visible spectrum[28]. Some typical values of emissivity for common objects are listed in Table 8.2.

Apart from some unusual objects at the surface of Earth, such as molten larva, surface temperatures of Earth range between the extremes of 183 to 340 K (i.e. about –90 to +70°C). The temperatures in this range are substantially colder than the temperature of the sun, which has a surface temperature of about 5800 K. We know the surface temperature of the sun, despite the fact that no man-made object has been sufficiently close to record it[29]. We can measure its emitted radiation ($\dot{Q}_{rad}$), take its emissivity as 1 and then calculate its surface temperature using Equation 8.1.

Although the sun is only 17 times hotter than objects at the surface of Earth in terms of absolute temperature, the fact that surface temperature is raised to the fourth power in Equation 8.1 means that the amount of radiation the sun emits per square metre of its surface is about 64,000 kW m^{-2}. This radiation spreads out in all directions and after travelling a distance of almost 150,000,000 km, the amount of the sun's radiation that reaches the earth is around 1.37 kW m^{-2} when

[28] We discuss the visible spectrum in Section 17.2.1.

[29] Solar probe plus will approach to within 5.9 million km of the surface of the sun.

8

the surface is perpendicular to the direction of the sun's rays. This incoming radiation from the sun is called **insolation**.

At any given surface temperature, the wavelengths of electromagnetic radiation emitted by an object follow a positively skewed distribution, as shown in Figure 8.9A. As an object heats up, the peak of this distribution increases in intensity and shifts to shorter and shorter wavelengths. This shift explains the change in colour we observe, from red to yellow to white, as the temperature of an object, such as an iron rod, increases. In other words, the wavelength of maximum emission energy is inversely proportional to the absolute temperature of an object. This is known as **Wien's displacement law**, after the German physicist Wilhelm Wien, who was awarded the 1911 Nobel prize in Physics for his work on the radiation of heat.

The wavelength at which peak emission intensity occurs can be calculated as shown in Equation 8.3:

Wavelength of maximum emission intensity (μm) — Wien's constant (μm K)

$$\lambda_{(max)} = \frac{2897}{T} \qquad \text{Equation 8.3}$$

Surface temperature of object (K)

Figure 8.9B shows a plot of this relationship and further illustrates the shift in peak radiation intensity of an object with changes in its surface temperature.

At the temperature of the sun's surface, the wavelengths at which peak radiation intensity occur are short (around 500 nm), so this radiation is often referred to as **short-wave radiation**. The peak wavelengths of light that our eyes respond to range from around 400 to 700 nm—the visible spectrum—and correspond to the peak wavelengths emitted by the sun. Figure 8.9A shows the distribution of wavelengths emitted by a black body with a surface temperature close to that of the sun. Shorter wavelengths are in the ultraviolet (UV) range, which are too short for our eyes to detect[30]. Once an object's surface temperature cools to below about 4000 K, the wavelengths at which its peak emission intensity occur are outside the visible spectrum and are in the infrared (IR) range. This radiation is the radiation emitted by the Earth and is sometimes referred to as long-wave radiation; its wavelengths are too long for our eyes to detect[30].

Why then are objects with lower surface temperatures still visible to us? The answer is that although we no longer see the emitted radiation, we can still see *reflected* shortwave radiation that is in the visible spectrum. The capacity of an

[30] Some other species are able to detect ultraviolet radiation and some can detect infrared radiation, as we discuss in Section 17.2.1.

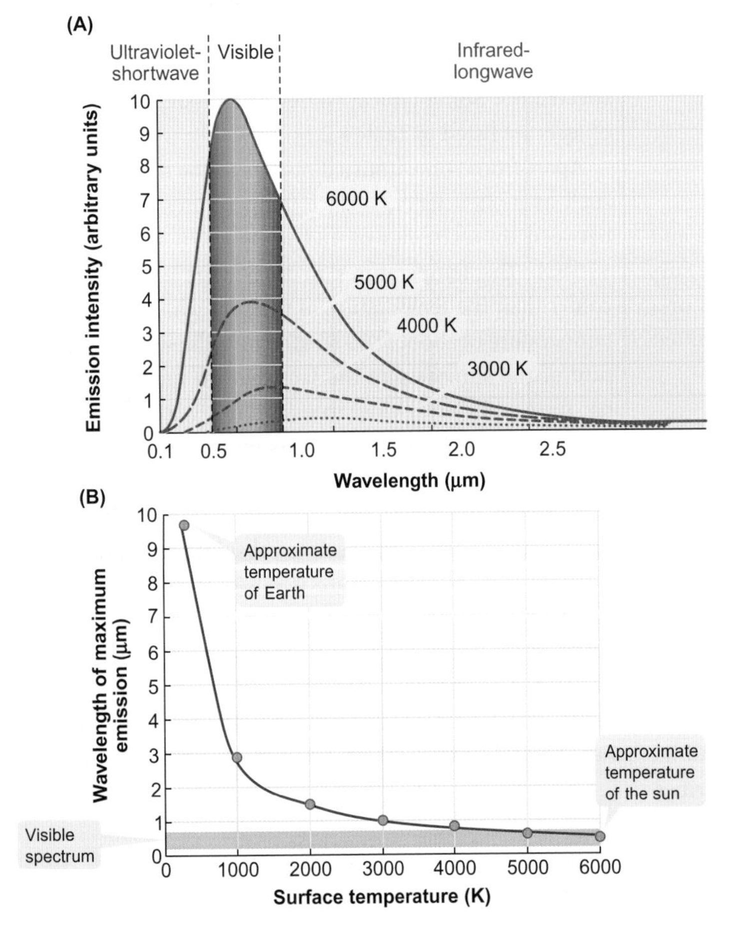

Figure 8.9 Weins displacement law

(A) The wavelength of emitted radiation from an object decreases as the temperature of the object increases. (B) For most animals there are only two objects of significance regarding radiation exchange—the sun at around 5800 K and earth-borne objects at around 250–350 K. The former emits shortwave radiation (in the visible spectrum) and the latter emits long-wave radiation (in the infrared).

Source: A: Adapted from: HyperPhysics (http://hyperphysics.phy-astr.gsu.edu) (c) C R Nave.

object to reflect short-wave radiation is an important parameter because it defines how much radiation that object will absorb when placed in sunlight. In fact, the colours of the objects we observe in the world around us relate to the wavelengths of light that are reflected from the objects, that is, those wavelengths that are not absorbed.

The capacity of an object to reflect short-wave radiation is known as **albedo** (Latin for white). Albedo is defined on a scale from 0 to 1.0, where 1.0 is a perfect reflector and 0.0 is a perfect absorber. Objects with low albedo tend to be dark coloured and those with high albedo tend to be light coloured. So, our perception of the darkness of an object is related to its albedo, not to its emissivity.

We can appreciate the significant roles that albedo (reflectivity) and emissivity (and therefore absorption) play in determining heat balance by performing a simple experiment. Put on a black cotton T-shirt on a hot sunny day, and go out and sit in the sun. Then, after a while, change into a white cotton T-shirt. Both shirts have the same emissivity as they are made from the same material (cotton cloth has an emissivity of about 0.8), but the black one has a low albedo. So, when exposed to direct sunlight, the white T-shirt reflects much of the incident energy, keeping you cool, whereas the black one absorbs more, making you hot. A more natural example of this phenomenon is the zebra, which we discuss below.

The experiment with the T-shirts raises the question: What is the significance of indigenous people from temperate climates having less pigmented and hence lighter, high albedo skins compared with those who originate in tropical areas? We might expect those from cooler climates to have darker skins to absorb more radiation and keep them warm. This question is discussed in Case Study 8.2.

Case Study 8.2 Why is skin lighter in people from higher latitudes?

Figure A shows that indigenous people from higher latitudes have small amounts of melanin (melanism) in their skin compared to those from tropical and sub-tropical regions. The most likely explanation for this difference is that darker skins absorb some of the incoming radiation (particularly the very short-wave UV) that could otherwise have a damaging effect on DNA in the cells of the epidermis, while lighter skins allow the absorption of UV radiation for the synthesis of vitamin D.

If unchecked, solar radiation can lead to skin cancers. The nature of the protection conferred by melanin has become most apparent in tropical areas where there are large immigrant white populations, such as Australia. In white Australians of European origin, death rates due to melanoma during 1985–1994 were in the region of 2.5/100,000 person years for women and 5/100,000 person years in men. In the dark-skinned aboriginal tribes, melanoma is virtually unknown.

If dark skin protects against UV-mediated DNA damage, and also provides a thermoregulatory advantage, why did populations that migrated into Northern latitudes evolve to lose their black pigmentation? The answer may lie in the fact that absorbed radiation has some benefits as well as disadvantages. The most notable advantage relates to the synthesis of vitamin D_3 (cholecalciferol) in the basal layers of the skin, a process driven by UV light with wavelengths around 295 to 300 nm.

The level of melanism therefore, strikes an optimal balance between blocking out UV light to prevent DNA damage and an increased risk of skin cancer, and allowing sufficient UV light to penetrate the basal skin levels to synthesize vitamin D_3, which plays a key role in calcium homeostasis[1]. If *Homo sapiens* originated in tropical regions with high solar radiation and spread to higher latitudes from there, then the initial presence of a dark skin would have given protection against incoming UV

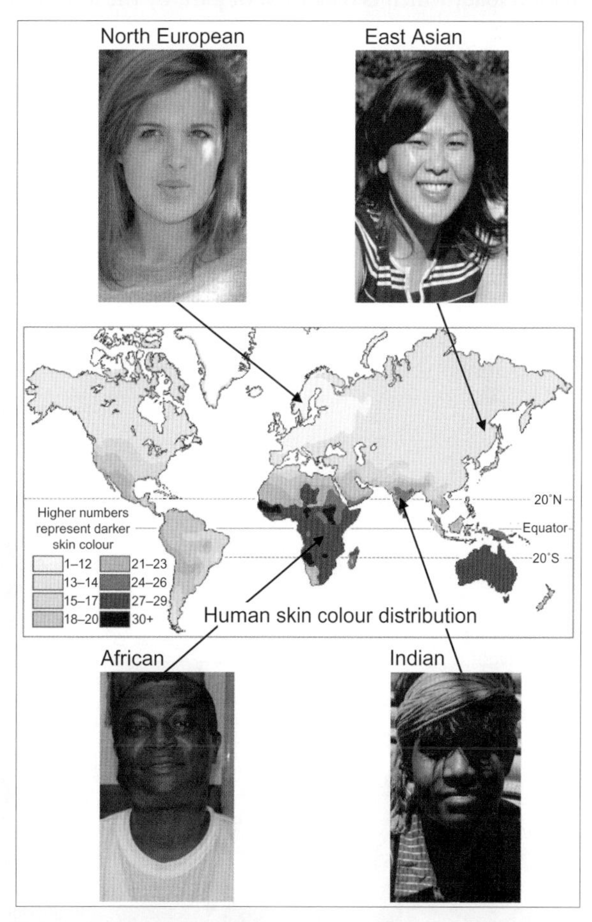

Figure A Skin pigmentation of humans from different geographical regions

Indigenous peoples from tropical regions where solar radiation is most intense have darker skins than those from temperate regions.

Source of map: from an original image © 1959 Renato Biasutti.

radiation, while still enabling the synthesis of vitamin D_3. During subsequent migration of *H. sapiens* to higher latitudes with reduced solar radiation, the evolution of a reduction in melanism would have enabled the penetration of sufficient UV radiation for vitamin D_3 synthesis while maintaining a relatively low risk of skin cancer.

› Find out more

Giles GG, Armstrong BK, Burton RC, Staples MP, Thursfield VJ (1996). Has mortality from melanoma stopped rising in Australia? Analysis of trends between 1931 and 1994. BMJ 312: 1121–1125.

Stringer C (2003). Out of Ethiopia. Nature 423: 692–695.

[1] Vitamin D_3 synthesis and the actions of vitamin D_3 metabolites on calcium balance are discussed in section 21.2.1

The effect of surface colouration on radiation uptake is, in fact, even more complex than the example we have considered so far. This is because radiation absorption also depends on the surface structure to some extent, which influences the degree of penetration of the radiation[31].

Terrestrial animals are exposed to a complex set of radiation inputs, as shown in Figure 8.10. For example, all objects surrounding an animal emit long-wave radiation and reflect short-wave radiation onto it. The different sources of radiation that impinge on an animal make up its total incoming radiation load, which is balanced, in part, by the long-wave radiation it emits. Almost all animals are good radiation emitters because they consist mostly of water which is a good emitter, as shown in Table 8.2. There is little scope to alter the proportion of water in an organism, but albedo (short-wave reflection) can be altered readily, as illustrated in Figure 8.11, and sometimes over relatively short timescales.

Although we cannot see long-wave radiation directly, we can detect it using devices known as thermographs. These machines convert long-wave radiation into short-wave radiation that we can see, enabling us to visualize regional variations in the levels of emitted radiation. If surface emissivity differences are minor, this measurement will approximately relate to the surface temperature of the object in question.

[31] Degree of penetration of solar radiation is discussed for fur and feathers in Section 10.2.3.

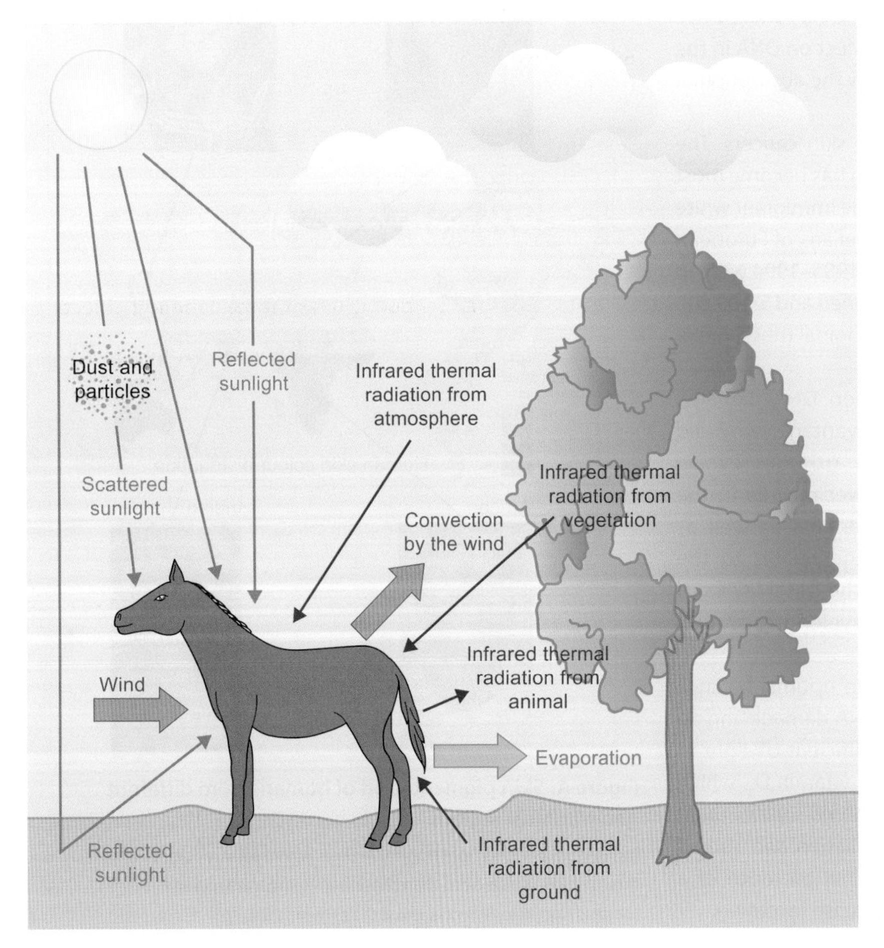

Figure 8.10 Heat exchange between animals and their environment

Arrows indicate the direction of energy flow. All sunlight and radiation from the atmosphere is short-wave radiation. As well as impinging directly on the animal, the short-wave (ultraviolet and visible) radiation from the sun is reflected onto the animal from surrounding objects. Radiation from vegetation, the ground and the animal is long wave (infrared) radiation. There may also be some conduction between the animal and the ground, although probably not much for a standing horse.

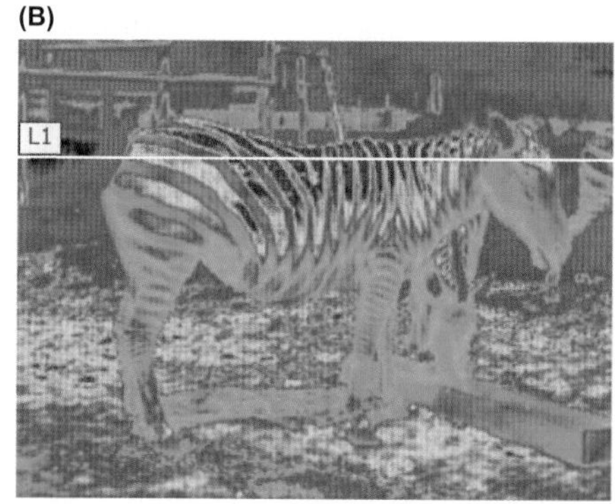

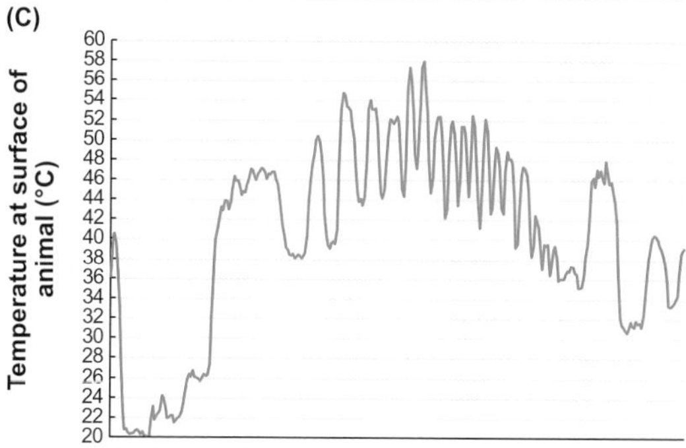

Figure 8.11 Visible and infrared images of a Grant's zebra (*Equus burchelli*)

(A) The animal is photographed using visible light and the solar light is reflected off its surface. (B) Image of infrared emitted radiation. The intensity of this radiation has been computer processed to give a false colour image and calculations have been made using equation 8.1 to translate these intensities into surface temperatures, assuming a surface emissivity of 0.95. This allows us to visualize the temperatures on the surface of the animal without touching it. (C) The temperature profile along the line (L1) in (B). The images and data show how the surface of the zebra body is hotter on the black stripes than on the white stripes because of absorbance of solar radiation by the black stripes.

Source: McCafferty DJ (2007). The value of infrared thermography for research on mammals: previous applications and future directions. Mammal Review 37: 207-223.

With some computation, these surface temperatures can be quantified, as shown in Figure 8.11. While the function of zebras' stripes is controversial[32], the photograph of a zebra (*Equus burchelli*) shown in Figure 8.11 illustrates how surface albedo influences the absorption of solar radiation and thereby affects the surface temperature of an animal. The surface temperatures on the black stripes of a zebra are higher than the surface temperatures on the white stripes, because more solar radiation is absorbed on the low albedo black surface. This is a natural example of the T-shirt experiment we discuss above.

8.4.2 Conduction

Generally, metals feel colder to us than non-metallic materials at the same temperature. For example, a metal chair tends to feel cooler than a wooden one in the same room. So why does the metal feel colder? The reason is that metals are good conductors of heat energy, as illustrated in Table 8.3.

Whenever two objects are in physical contact, heat from the hotter object, say a hand, flows to the cooler object, a chair. The rate of heat flow depends on the temperature difference between the objects, the area of contact between the objects (the cross-sectional area), the distance over which the heat is being conducted (its path distance) and a constant, the thermal conductivity of the object through which the heat is being conducted. The thermal conductivities of several substances are given in Table 8.3, where it can be seen that metals have greater thermal conductivities than wood.

Heat also flows within the *same* object when different regions are at different temperatures, such as a metal rod which is being heated at one end. Generally, the rate of heat exchanged by conduction, $\dot{Q}_{cond}$, is defined by Fourier's law[33]:

Table 8.3 Thermal conductivities of some common objects

	Conductivity ($W\ m^{-1}\ K^{-1}$)
Copper	385
Stainless steel	16
Wood	1.4
Glass	0.8
Water	0.6
Muscle	0.42
Adipose tissue (fat)	0.2
Feathers	0.16
Fur	0.04
Air (still)	0.026

Source: Cossins AR and Bowler K. Temperature Biology of Animals. London and New York: Chapman and Hall, 1987. Withers PC. Comparative Animal Physiology. New York: Saunders College Publishing, 1992.

[32] Recent analysis suggests that the stripes may function to reduce the likelihood of the animals being attacked by biting flies.

[33] Fourier's law and its similarity to Fick's law for the passive diffusion of gases and ions are discussed in more detail in Sections 3.1.1 and 3.1.2.

Thermal conductivity (J s⁻¹ cm⁻¹ °C⁻¹) | Area of contact or cross-sectional area (cm²) | Difference in temperature (°C) | Rate of heat conduction (J s⁻¹) | Path distance (cm)

$$\dot{Q}_{cond} = \frac{kA(T_1 - T_2)}{d} \quad \text{Equation 8.4}$$

When our skin is in contact with two different materials which are both at the same temperature, but lower than the temperature of the skin, the material with higher conductivity feels colder. This is because more heat flows from the skin to the material of higher conductivity than to the material of lower conductivity, hence the temperature of the skin drops more when it is in contact with the material of higher conductivity than when it is contact with the material of lower conductivity.

The term **thermal conductance** is often used as a measure of the capacity of an object to conduct heat. This term combines the thermal conductivity (k) of the material the object is made from, with the cross-sectional area (A) and path distance (d).

$$\text{Thermal conductance } (C) = \frac{kA}{d} \quad \text{Equation 8.5}$$

The value of the thermal conductance indicates the ease with which a material conducts heat. The opposite of this, the resistance of a material to conducting heat, is known as **insulation**:

$$\text{Insulation } (R) = \frac{d}{kA} \quad \text{Equation 8.6}$$

So, the information in Table 8.3 indicates that adipose tissue (fat) has a lower thermal conductance, and hence greater insulation, than muscle and can be used to reduce heat flow. Consequently, fat has evolved as a form of thermal insulation in aquatic mammals, such as seals, whales and dolphins[34], all of which have body temperatures above ambient and live in water, which has a high thermal conductivity.

Table 8.3 shows that water has a thermal conductivity that is 23 times greater than that of still air. We perceive this difference when swimming in an indoor pool even if the air and water temperatures are the same and several degrees (approximately 10°C) below body temperature; upon entering the water, it feels colder than the surrounding air. The water drains more heat from the body than air at the same temperature because of its higher thermal conductivity. This higher rate of heat loss causes a greater drop in skin temperature.

The extent of both radiation and conduction depend on the difference in temperature between an object and its surrounding environment. If the environment is cooler than the object, heat will flow from the object to the environment. Equation 8.2 indicates that for radiation, the two temperatures are raised to the power of 4. However, over the relatively small temperature differences in absolute temperature that exist between a bird or a mammal and its environment, the rate of radiation heat transfer is effectively proportional to $T_b - T_{amb}$. It is possible, therefore, to combine and simplify Equations 8.2 and 8.4 to produce the following:

Rate of heat loss (J s⁻¹, or W) | Overall thermal conductance of animal (W °C⁻¹) | Difference between body and ambient temperatures (°C)

$$\dot{Q} = C(T_b - T_{amb}) \quad \text{Equation 8.7}$$

Equation 8.7 is known as the Newtonian model of heat balance and applies, in principle, to birds and mammals.

In Equation 8.7, heat losses by evaporation and convection have been ignored, and all of the constants in Equations 8.2 and 8.4 have been combined to give the one constant, C, the overall thermal conductance of the animal. Animals can vary their thermal conductance in response to changes in environmental temperature: in cold ambient temperatures, birds and mammals can decrease their thermal conductance to *reduce* the rate of heat loss, while in warm temperatures they can increase their thermal conductance to *increase* the rate of heat loss. In order for a bird or mammal to maintain a constant body temperature, its metabolic rate would need to match the rate of heat loss[35].

The last thing to note in Table 8.3 is the fact that fur and feathers have a greater conductivity than still air. The key word here is 'still', which is critical when we consider convection in the next section.

8.4.3 Convection

Animals live immersed in a fluid medium, air or water. If this fluid is absolutely stationary, heat flows from the body by conduction and radiation only, but fluids surrounding animals are seldom stationary. Imagine an endothermic animal (bird or mammal) standing motionless in still air at 10°C with its body at say 38°C. Its surface temperature will be around 30°C or so (depending on the extent of its surface insulation). Because the animal is hotter than the air, heat

[34] Insulation in aquatic birds and mammals is discussed in Section 10.2.8.

[35] The Newtonian model of heat loss for birds and mammals is discussed further in Section 10.1.

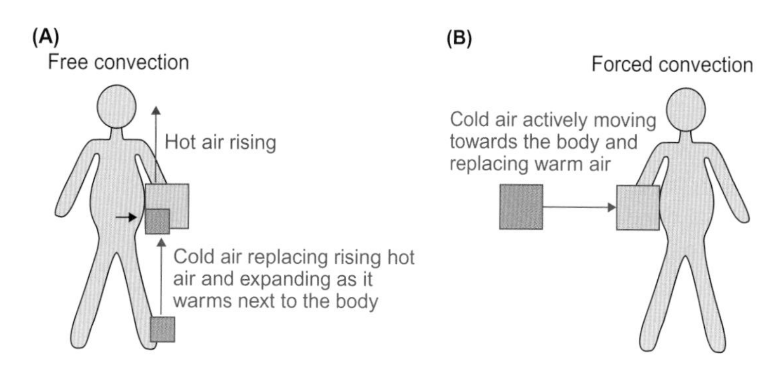

Figure 8.12 Convective heat loss by a stationary body, which is warmer than the surrounding air

In (A) there is no movement of the air around the body. Any small volume of air adjacent to the body such as the small blue square, will be heated by conduction from the body surface. As it heats it will expand (becoming the larger, red square) and it will rise. The warm air will be replaced by cooler air from below. This is known as free convection. In (B) the air is moving and heated volumes of air adjacent to the body are directly replaced by gross air movements. This is known as forced convection.

flows from the animal into the air immediately surrounding it. When this happens, the air expands and rises[36] as shown in Figure 8.12A.

The heated air breaks away from the surface of the animal and is replaced by cooler air from below. The unavoidable air movement that occurs when a hot object is placed in a cooler fluid, even when the object is completely stationary, is called **free convection** and increases the rate of heat flow that occurs by conduction alone.

The scenario we have just described only applies routinely to birds and mammals, as most other species of animals are at the same temperature as their surroundings at some time during the daily cycle, whereby no free convection occurs.

If the animal moves, or if the fluid it is immersed in moves independently of the animal, then additional convective heat losses occur, as illustrated in Figure 8.12B. The process is called forced convection. The rate of heat loss by **forced convection** depends on:

- the temperature difference between the object and its surroundings,
- the thermal conductivity of the fluid medium,
- the shape and size of the object,
- its orientation relative to the flow,
- the rate at which the fluid is flowing.

An important concept in understanding the nature of convective heat loss is that of a **boundary layer**[37]. The boundary layer is a layer of still air (or water) that is against the surface of an animal (or other object), as illustrated in Figure 8.13A. Table 8.3 shows that still air is a very good thermal insulator because of its poor thermal conductivity. So if the layer of still air against the surface of an animal is stripped away (by a high rate of fluid flow over the animal, for example), the rate of heat flow from the animal increases.

Figure 8.13B shows how the insulating effect of a boundary layer is reduced as it is progressively depleted. Increases in fluid movements from an initially low overall velocity have a greater impact on the rate of heat loss than an equivalent increase from an initially higher fluid velocity. In other words, the increase in rate of heat loss is not linearly related to absolute fluid velocity, but rather to the relative increase in velocity. Specifically, forced convective heat loss is approximately related to the square root of fluid velocity.

Of particular importance to the rate of heat loss from an animal is whether the fluid flow over the object is turbulent or laminar. The difference between **turbulent** and **laminar flows**[38] is that turbulent flows disrupt the boundary layer much more than laminar flows. Laminar flows are typical of slow fluid speeds and become turbulent at higher speeds, but low and high are relative to the size and shape of the object

[36] The heated air rises because it has a lower density than the cooler surrounding air.

[37] Boundary layers are discussed in Section 12.2.1.

[38] Turbulent and laminar flows in tubes are discussed in Section 14.2.4.

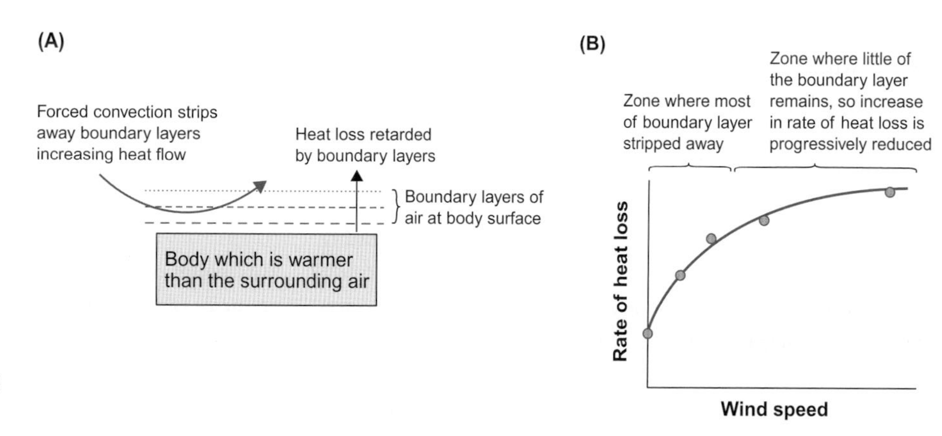

Figure 8.13 Effects of forced convection on rate of heat loss in air

(A) Rate of heat loss is normally retarded by boundary layers of still air but can be increased by movement of the air. (B) Air movements strip away the boundary layers thus elevating rate of heat loss, but there is a diminishing effect as wind speed increases.

being considered. Relatively large, unstreamlined animals create more turbulence than relatively small streamlined animals.

Weather reports in temperate zones often mention something called **wind chill**. Wind chill is related to the velocity of the wind and makes an endothermic animal feel colder than if it was surrounded by still air at the same temperature. Wind chill is calculated by making some assumptions about the nature of convective heat flow from humans in relation to wind velocity. Thinking about animals more generally, an estimate of the rate of heat loss by forced convection assumes that it is related to the square root of wind velocity. We can use this estimate to determine the drop in ambient temperature that would produce an equivalent rate of heat loss if wind velocity was 0 (that is with no wind chill). For example, at an air temperature of 0°C, a wind speed of 12 m s^{-1} will actually make the temperature feel like the equivalent of −8°C in still air.

8.4.4 Evaporation

The flow of heat by convection, conduction and radiation only occurs along a temperature difference from hot to cold, as illustrated in Equations 8.2 and 8.4 for radiation and conduction. So animals that find themselves in environments in which ambient temperature *exceeds* their body temperatures are unable to avoid heating up unless they can use mechanisms that do not rely on heat transfer along a temperature difference. Fortunately, there is such a mechanism: evaporation.

A certain amount of energy input is required to transform a gram of water into vapour at a given temperature. This amount of energy is called the latent heat of vaporization and it varies slightly with temperature. At 25°C, the latent heat of vaporization of water is 2444 J g^{-1}, whereas at 40°C it is 2408 J g^{-1}. So, if water evaporates from the surface of a body, it takes energy (in the form of heat) with it. Animals can use this phenomenon to get rid of excess heat against a difference in temperature. This effect is true whatever the source of the water.

Animals exploit the effect of evaporation in two different ways—namely by using a source of water that is external to their body, or one that is internal. For example, humans and animals, such as elephants, may pour or spray water over their head on hot days, as shown in Figure 8.14. In these instances, water is applied from sources outside the bodies. Alternatively, and more generally, if water is not readily available outside their bodies, terrestrial animals can exploit the fact that their bodies consist of 60–70 per cent water: they can supply the water from within their body by sweating, by panting, by licking their skin and even by defecating moist faeces on to their feet, as happens with some species of birds. However, there are limits to how long an animal can sustain the use of internal water without drinking before it becomes severely dehydrated[39].

Figure 8.14 An elephant spraying itself in Chitwan National Park, Nepal

Source of image: Ali Abas/Alamy Stock Photo.

8.4.5 The heat balance equation and temperature regulation

In order for the body temperature of an animal to remain constant, the rate of heat loss must be equal to the rate of heat gain. However, body temperature is not always constant, as we discuss in the next two chapters. If the rate of heat loss is greater than the rate of heat gain, body temperature falls. Conversely, if the rate of heat gain is greater than the rate of heat loss, heat is stored in the body and the body temperature rises.

Heat gain by an animal is not always from metabolism; it may be by conduction if external temperature exceeds internal temperature, or by radiation from external sources, such as the sun. Evaporation almost always results in heat loss except in unusual circumstances such as when a relatively cool body is in contact with warmer, moist air. All of these possible routes of heat gain or loss can be incorporated into a single **heat balance equation**:

Rate of metabolic heat production; Rate of radiation heat exchange; Rate of conductive and convective heat exchange; Rate of heat storage in the body; Rate of evaporative heat loss

$$\dot{Q}_{met} = \pm\dot{Q}_{rad} \pm\dot{Q}_{con} \pm\dot{Q}_{stor} - \dot{Q}_{evap} \quad \text{Equation 8.8}$$

This equation forms the basis for the discussions on temperature regulation in Chapters 9 and 10.

[39] Water loss by sweating or panting is discussed in Section 6.1, and evaporative heat loss in birds and mammals is discussed in Section 10.2.2.

› Review articles

McCafferty DJ (2007). The value of infrared thermography for research on mammals: previous applications and future directions. Mammal Review 37: 207–233.

Osczevski R, Bluestien M (2005). The new wind chill equivalent temperature chart. Bulletin of the American Meteorological Society 86: 1453–1458.

Speakman JR, Król E (2010). Maximal heat dissipation capacity and hyperthermia risk: neglected key factors in the ecology of endotherms. Journal of Animal Ecology 79: 726–746.

8.5 How do we describe different thermoregulatory strategies?

In Chapters 9 and 10, we explore the various strategies adopted by animals to regulate their body temperature. To be able to discuss these strategies clearly however, we need to be aware of some specific terminology. We end this chapter by reviewing this terminology in preparation for the chapters that follow.

Scientists who carried out early research into the responses of animals to temperature noted that some groups of animals seemed to maintain their body temperatures at constant levels almost continuously, while other groups showed considerable variability in their body temperatures, often closely matching the surrounding temperature. These two strategies were termed **homeothermy** (Greek, *homoios*—similar, *thermé*—heat) and **poikilothermy** (Greek, *poikilos*—varied). Birds and mammals are generally homeotherms, and other vertebrate groups and the invertebrates are poikilotherms. But these definitions and their taxonomic affiliations run into problems as soon as the thermoregulation of animals is studied more closely.

Many reptiles may actually maintain a stable body temperature for protracted periods, particularly if they have access to a source of radiation, such as the sun[40], while mammals and birds show small changes in body temperatures over the daily cycle; they regulate their body temperature within a narrow range. To reflect the fact that some species of birds and mammals go through daily or seasonal variations in body temperature which exceed those seen during homeothermy, a new term was introduced: **heterothermy** (Greek, *heteros*—other).

[40] Behavioural thermoregulation in reptiles is discussed in Section 9.1.1.

Examples of daily heterothermy are found in some species of mammals living in hot, arid parts of the world, while hibernation is an example of seasonal heterothermy[41]. **Regional heterothermy** occurs in some species of birds and mammals that allow the temperature of parts of their body to change with respect to core temperature. Some species that live in cold environments allow peripheral structures, such as limbs with little or no insulation, to fall below core temperature[42].

Other discrepancies in this terminology are still apparent. For example, many species of fish have body temperatures that approximate to the environmental temperature (like poikilotherms). Consequently, where environmental temperature is relatively constant, most species of fish are, in effect, homeotherms.

The main problem with these terms is that they rely on the observed variations in body temperature, but do not adequately reflect the underlying thermoregulatory processes that generate the variability. Hence, two animals may have relatively stable body temperatures—like a fish and a bird—but the reasons for the stability may be completely different. To overcome this problem, new terms were introduced to reflect the source of heat that is used for thermoregulation. Animals that obtain the heat required to maintain their body temperature above environmental temperature from sources outside their body are called ectotherms[43] (Greek, *ektos*—outside) or exotherms, while animals deriving such heat internally are called endotherms (Greek, *endon*—within). Most birds and mammals, a handful of reptiles, some fish and a few invertebates are endotherms. All other animals are ectotherms.

All of these different thermoregulatory behaviours and physiological strategies are discussed in more detail in the following two chapters.

› Review articles

Glossary of terms for thermal physiology, 3rd edition 2001 Japanese Journal of Physiology 51: 245–280.

[41] Examples of daily and seasonal heterothermy are discussed in further detail in Sections 10.2.2 and 10.2.7.

[42] Examples of regional heterothermy are discussed in further detail in Section 10.2.8.

[43] It should be noted that all animals, including ectotherms, derive some heat from biochemical reactions inside their body, as we discuss in Section 2.1.2, but most ectotherms require additional heat, from outside their bodies, in order to maintain their body temperature above that of their environment

Checklist of key concepts

- The **kinetic energy** of two reactants has to be above a certain level, the **activation energy**, before they will react with each other.
- If the temperature of a reactant increases, a greater proportion of its molecules will have an energy that exceeds the activation energy; they will react more readily.

- The rate of a chemical reaction is directly proportional to the concentration of the reactant molecules and the proportionality constant is called the **rate constant**.
- As temperature rises, the rate constant increases and so the rate of reaction increases. This relationship can be represented by an **Arrhenius plot**.
- The extent to which a rate of reaction changes with a 10°C change in temperature is called the Q_{10}.
- Q_{10} is not necessarily constant over a given temperature range.

How temperature affects biological processes

- In many biological processes there is a **breakpoint** in an Arrhenius plot at which the rate of reaction slows as temperature continues to increase.
- If temperature is too high, or too low, animals die, but these lethal temperatures vary between species and between individuals of a species.
- The **lethal temperature** at which 50 per cent of the animals of a species die under experimental conditions is known as the $\mathbf{LT_{50}}$.
- **Eurytherms** can survive a wide range of environmental temperatures while **stenotherms** can only tolerate a narrow range.
- The upper lethal temperature limit may vary between species because of a number of factors, which include subtle changes in amino acid sequences of key enzymes and different Q_{10} values of the different reactions making up the overall reaction pathway.
- A process known as **homeoviscous acclimation** occurs in animals so that the fluidity of their cell membranes is similar even though they have different body temperatures.
- Terrestrial environments can have wildly different temperatures, both over distance and over time, whereas large bodies of water, such as the oceans, are much more stable.

The processes of heat exchange

- Animals can exchange heat with their environment in four different ways: **radiation, conduction, convection** and **evaporation**.
- All objects emit **electromagnetic radiation**, with the amount emitted depending on the **Boltzmann constant**, the **emissivity** of the object and the temperature of the object raised to the power of four.
- A perfect emitter and absorber of radiation is known as a **black body**.
- As an object heats up, the peak intensity of the radiation it emits shifts to shorter and shorter wavelengths. This is known as **Wien's displacement law**.
- The capacity of an object to reflect short-wave radiation is known as its **albedo**.
- The colour of an object is determined by the wavelength within the visible spectrum that is reflected by the object rather than being absorbed.
- The rate of heat conduction of an object depends on its **thermal conductance** and the difference in temperature between the object and its surroundings.
- There are two different types of convection:
  - **Free convection** occurs when a warm animal in a cooler fluid increases the temperature of the fluid next to its body, causing the fluid to rise and be replaced by cooler fluid.
  - **Forced convection** occurs when an animal moves or if the fluid surrounding the animal moves independently of the animal.
- A **boundary layer** of fluid next to the animal reduces the rate of heat lost by convection.
- **Wind chill** makes an animal feel colder than if it was surrounded by still air at the same temperature.
- If ambient temperature is above that of the animal, it can only lose heat by the **evaporation** of water from its body surface.
- Approximately 2.4 kJ are required to convert a gram of water into vapour with no change in temperature. This amount of heat is known as the **latent heat of vaporization**.
- Water from external sources can be applied to the body surface or an animal can **sweat**, **pant** and **lick** regions of its body or even defaecate on its feet in order to provide water for evaporation.
- In an animal with a stable body temperature, the loss of heat via all the above mechanisms is balanced by the metabolic production of heat and can be expressed in an overall **heat balance equation**.

Strategies of thermoregulation in animals

- **Ectotherms** use external supplies of heat to raise the temperature of their bodies above that of the ambient medium.
- **Endotherms**, which include birds, mammals and a few species of reptiles, fish and invertebrates, use internally generated heat to raise and maintain their body temperature above that of the environment.
- **Poikilotherms** exert little or no control over their body temperature such that it follows that of the surrounding environment.
- **Homeotherms** maintain their body temperature within a narrow range.
- Homeothermy occurs not only in the endothermic birds and mammals, but also in some ectothermic species which live in relatively stable thermal environments
- Species of birds and mammals that go through daily or seasonal variations in body temperature which are greater than those seen during homeothermy are called **heterotherms**.
- **Regional heterothermy** occurs when the temperature of the extremities of a bird or mammal, such as the legs, are allowed to fall below that the core temperature of the animal.

Study questions

*** Answers to these numerical questions are available online. Go to www.oup.com/uk/butler**

1. A seal is basking on a beach on a warm sunny day. It has positioned itself so that it is in 10 cm of shallow water and therefore only the lower part of its body is submerged. Outline all the ways in which the animal is exchanging heat with its environment. (Hint: Section 8.4.)

2.* A researcher measures the rate of oxygen uptake ($\dot{M}O_2$) of a fish to be 5.5 μmol O_2 g^{-1} h^{-1} at 20°C, and 4.2 μmol O_2 g^{-1} h^{-1} at 15°C. What is the Q_{10} of $\dot{M}O_2$? Is this Q_{10} unusual for what might be anticipated for a fish? Assume that the Q_{10} is constant over the range from 0 to 30°C. Can you predict what the $\dot{M}O_2$ of the fish will be at 5°C and 30°C? (Hint: equations in Box 8.2.)

3.* The table below shows some data on the oxygen consumption of an insect measured at ambient temperatures between 5 and 50°C. Using these data, produce an Arrhenius plot to show the relationship between temperature and rate of oxygen consumption. Interpret the plot you have produced. (Hint: Section 8.1 and Figure 8.3.)

Temperature (°C)	Rate of oxygen consumption (μmol O_2 g^{-1} h^{-1})
5	1.78
10	2.90
15	3.93
20	5.67
25	8.92
30	11.4
35	17.1
40	24.8
45	21.3
50	14.5

4. Compare and contrast the variability of ambient temperature in the oceans and on the land. What are the consequences of this variation for the animals that live in these different habitats? (Hint: Sections 8.2 and 8.3.)

5.* Two kangaroo rats (rat A and rat B) are standing in the desert at noon on a clear day. Rat B has a metabolic rate of 1 W and 45.7 per cent of this is dissipated in all directions as long-wave radiation. It is 1 m away from rat A. The radiation from the sun is 1.37 kW m^{-2} and 50 per cent of this is intercepted by the atmosphere. The receiving surface is the same in both cases. What provides more heat to rat A, the sun or rat B? The surface area of a sphere is $4\pi r^2$. (Hint: Section 8.4.1.)

6. Explain why a swimming pool feels cold when you jump into it, despite the fact a thermometer shows the water and the air above it to be at the same temperature, which is several degrees below body temperature. (Hint: Section 8.4.2.)

   A bird is sitting in the top of a tree exposed to the wind, which is at a lower temperature than that of the bird. Explain why the increase in heat loss from the bird when the wind increases from 0 to 1 ms^{-1} is greater than the increase in heat loss when the wind changes from 5 to 6 ms^{-1}. (Hint: Section 8.4.3, Figure 8.13B.)

7. Outline the meaning of the terms poikilothermy, homeothermy, heterothermy, endothermy and ectothermy. Why are the latter two terms now preferred as descriptors of thermoregulation strategies? (Hint: Section 8.5.)

Bibliography

Atkins P, de Paula J (2011). Physical Chemistry for the Life Sciences, 2nd edn. Oxford University Press.

Cossins AR, Bowler K (1987). Temperature Biology of Animals. Chapman and Hall, London and New York.

Holman JP, White PRS (1992). Heat Transfer 7th edn. McGraw-Hill, London and New York.

Schmidt-Nielsen K (1997). Animal Physiology. Adaptation and environment. 5th edn. Cambridge University Press. Cambridge.

9

Temperature regulation in ectotherms

Animals that need to obtain heat from sources outside their body for body temperatures to be raised are **ectotherms**[1]. In fact, almost all invertebrates and all amphibians, most fish and most reptiles are ectotherms.

Early studies suggested that the body temperature of ectotherms blindly tracks any change in external temperatures. Although broadly true of aquatic ectotherms if they remain in one location, this idea has turned out to be far from correct for ectotherms living on land. Terrestrial ectotherms can reduce the fluctuations of body temperature by using complex behaviours and in some cases exert physiological control on the rate of heat exchange with the environment.

Ectotherms adjust their cell and tissue function in response to changes in environmental temperatures, beginning to acclimate[2] to a temperature change in the laboratory within hours. In natural environments, thermal **acclimatization** occurs where a suite of variables often change at the same time, only one of which is temperature (for example, during changing seasons).

Both **acclimation** and acclimatization are examples of **phenotypic flexibility** and as such are reversible. In contrast, irreversible genetic **adaptation** of species or populations results from natural selection. Long-term changes in the thermal environment, such as those resulting from global climate changes, may drive genetic adaptation of populations and species over evolutionary time, but these are not changes at an individual level. Hence, if we move a tropical fish adapted to relatively high water temperatures from the tropics to polar waters the individual fish would not become a polar fish, no matter how long it remained there. In this chapter, we examine some of the adaptations of ectotherms, such as the adaptations of ants living in hot deserts that enable them to tolerate extremely high temperatures; at the other end of the temperature scale, we explore adaptations of animals living in freezing conditions.

9.1 Thermal relations of ectotherms with their environments

In order to raise their body temperatures, ectothermic animals need heat input from outside their body. While metabolism inevitably generates heat[3], it readily dissipates through an animal's tissues and body fluids. If the environment is cooler than the animal, the heat generated by metabolism will flow from the body surfaces into the environment, by **radiation** (for land animals), **conduction** and **convection**[4]. On the other hand, if an environment is warmer than the animal's body, heat will flow into the body tissues.

The rate at which ectotherms lose or gain heat across their body surfaces depends on whether they are on land or in water. Water conducts heat about 23 times faster than air[5], so, when in water, ectothermic animals readily lose the heat generated by metabolism. On the other hand, if water temperature increases, ectotherms readily gain heat.

An additional factor that determines heat exchange in water is that water holds less oxygen per unit volume than air. Consequently, water-breathers need to pass relatively more water across their gaseous exchange surfaces than air-breathers need to pass air in order to obtain an equivalent amount of oxygen. The actual difference between the amount of oxygen in water and air depends on the temperature and salinity of the water; oxygen is less soluble in warmer than cooler water and an increase in the salt content of the water slightly lowers oxygen solubility. For example, pure water at 0°C holds about 5 per cent of the oxygen held in air at the same temperature and seawater holds about 4 per cent of the oxygen held in air[6]. By contrast, pure water at 40°C holds less than 3 per cent of the oxygen in air at the same temperature while seawater holds less than 2.5 per cent of the oxygen held in air.

1 Ectotherm is derived from the Greek, *ektos* meaning outside, and *thermos*, meaning heat.

2 We discuss the general principles of acclimation, acclimatization and evolutionary genetic adaptation in Section 1.5.2.

3 Section 2.2.1 discusses the heat energy generation by metabolism.

4 Radiation conduction and convection are explained in relation to heat exchange in Sections 8.4.1, 8.4.2 and 8.4.3, respectively.

5 The thermal conductivity at 20°C of still air is 0.026 W m^{-1} K^{-1} while that of water is 0.6 W m^{-1} K^{-1}.

6 Table 11.2 gives data for the capacity of water and air to hold oxygen (oxygen capacitance) at various temperatures.

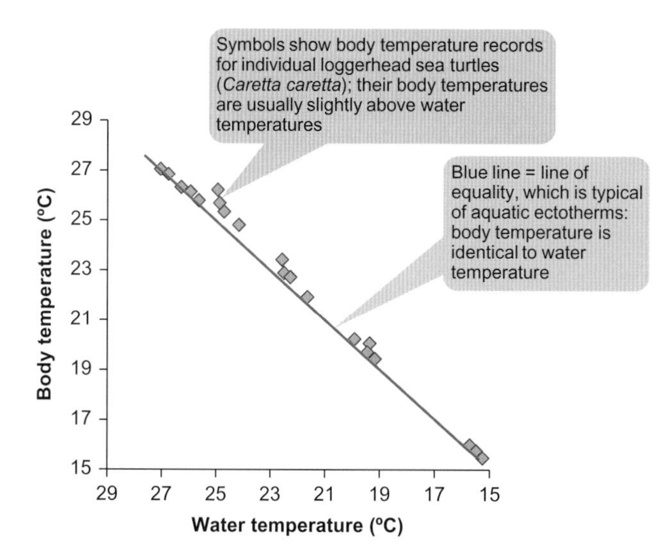

Figure 9.1 Relationship between body temperature and water temperature in aquatic ectotherms

Source: adapted from Hochscheid S, Bentivegna F, Speakman JR (2004). Long-term cold acclimation leads to high Q_{10} effects on oxygen consumption of loggerhead sea turtles *Caretta caretta*. Physiological and Biochemical Zoology 77: 209–222.

The difference in the amount of oxygen in water and air results in a much greater cooling or warming effect for water breathers than air breathers: about 920 times greater at 40°C in freshwater ectotherms. The factor of 920 is due to the combined effect of moving 40 times more water across the gas-exchange surface and the 23-times-higher **thermal conductivity** of water than air.

Because of the ready transfer of heat, the body temperature of aquatic ectotherms usually matches (or almost matches) the temperature of the water, following the isothermal line shown in Figure 9.1. Only a few exceptions occur among aquatic ectotherms. For example, body temperatures of loggerhead sea turtles (*Caretta caretta*) are 1–2°C above the temperature of the water in which they are swimming, as shown in Figure 9.1. The reason for the small difference between body temperature of the turtles and water temperature is that the carapace of turtles provides a barrier to heat loss so heat is lost less easily than in most aquatic ectotherms.

9.1.1 Behavioural control of body temperatures in ectotherms

Only animals fixed to one spot, like barnacles on intertidal rocks, are completely at the mercy of environmental temperatures. When covered with seawater, water temperature determines the body temperature of such animals; by contrast, when the tide goes out, air temperature and the sun's radiation determine their temperature. Few ectothermic animals lack control of their body temperature in this way, however. Instead, many ectothermic animals can change their behaviour to influence their body temperatures to some extent. Such **behavioural thermoregulation** is most complex among terrestrial ectotherms that can readily adjust the rate of heat uptake and heat loss by the way they position themselves in their environment. However, even aquatic ectotherms faced with life in a medium that readily conducts heat often show some level of behavioural thermoregulation.

Thermal selection by aquatic ectotherms

The simplest form of behavioural thermoregulation is thermal selection by aquatic animals that 'select' their thermal environment by moving faster when body temperature is further from their **preferred body temperature**. They do this by altering their turning frequency or their speed.

Thermal selection is the main way in which mobile aquatic ectotherms influence their body temperature. Although water temperatures are relatively stable on a day-to day basis, spatial variations occur. Given the opportunity, then, aquatic ectotherms select water at a temperature that matches the temperature to which they have acclimated (in a laboratory) or acclimatized (in the wild) and where physiological function is optimal[7].

Figure 9.2 shows an example of thermal selection by juvenile opaleye (*Girella nigricans*). Adult opaleyes live in shallow subtidal areas near the kelp beds on the coasts of California and the Mexican Baja California. After inshore reproduction, schools of young opaleyes move into the intertidal zone and occupy shallow pools. The temperature of these pools varies: in upper shore pools water temperature reaches up to 31°C in summer, but pool temperatures vary down the shore by more than 10°C. Field observations suggest that the juvenile opaleyes move between the intertidal pools, selecting those that coincide with their preferred temperature.

Temperature preferences of some ectotherms change seasonally. For example, tadpoles of American bullfrogs (*Lithobates catesbeianus*) and northern leopard frogs (*Lithobates pipiens*) placed in a thermal gradient accumulate in water at 23–24°C in April, but by June they accumulate in water at 27–28°C, and by July they select water at 30°C. In the wild, these temperature preferences mean that tadpoles tend to occupy the surface layers of shallow water warmed by solar irradiation. Their apparent increase in preferred temperature as summer progresses may be beneficial in reducing the development time before they metamorphose into the young frogs that are less dependent on the pools that may dry up.

Although thermal selection can often maintain relatively stable body temperatures in small pools, this is not possible when temperatures change seasonally. Even in the open ocean where temperatures are relatively stable, surface water temperatures may vary seasonally by 10°C. As such, annual migration over thousands of kilometres could be required

[7] We discuss acclimation in Section 9.3.2.

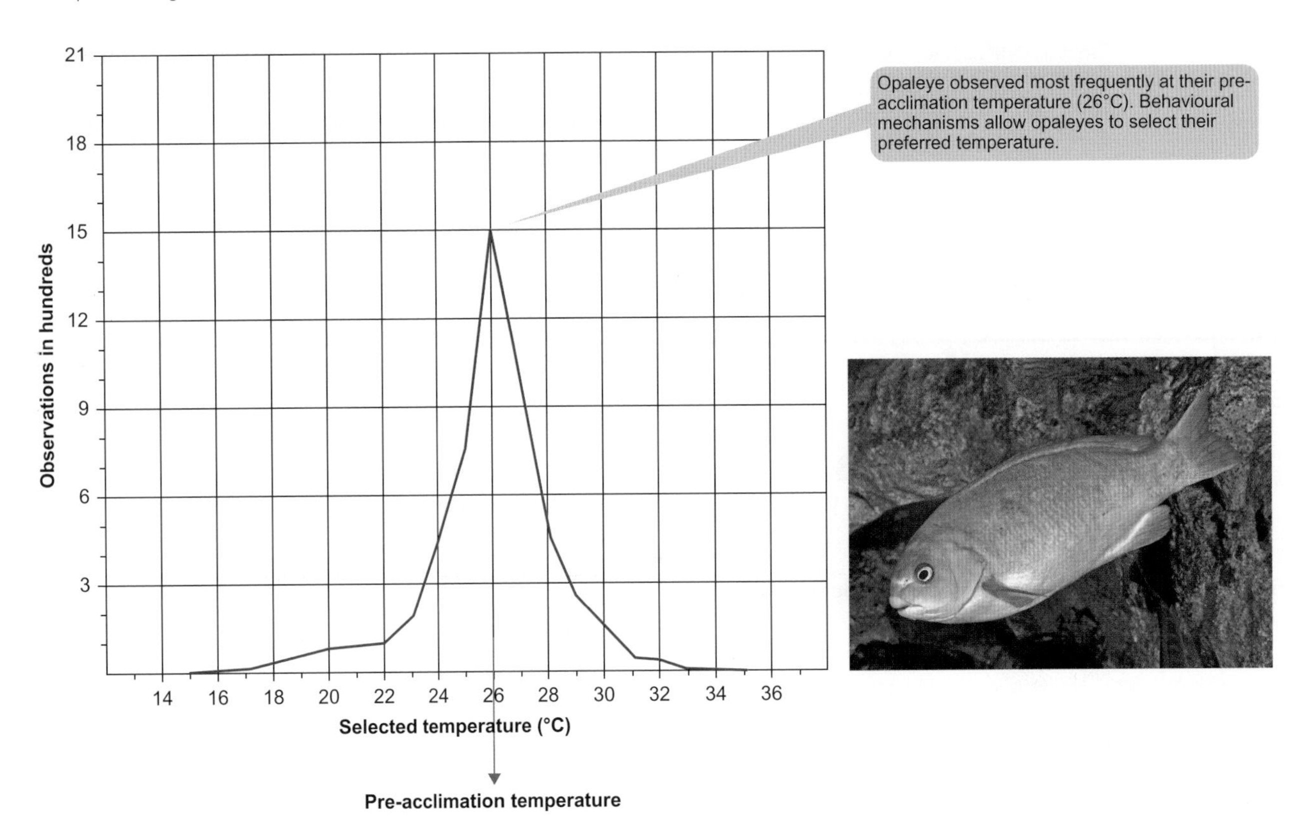

Figure 9.2 Temperature selection by juvenile opaleye (*Girella nigricans*)

The graph shows the number of observations at each temperature for thirty-nine opaleye exposed to a thermal gradient (15°C to 35°C) after acclimation to 26°C.

Source: Norris KS (1963). The functions of temperature in the ecology of the percoid fish *Girella nigricans* (Ayres). Ecological Monographs 33: 23–62. Image of opaleye: © D. Ross Robertson.

for an animal to maintain an approximately constant body temperature. Most marine ectotherms are incapable of such migrations so their body temperatures vary seasonally. However, some fish, such as the skipjack tuna (*Katsuwonus pelamis*) do undertake seasonal migrations linked to changes in water temperatures due to seasonal weather patterns. Figure 9.3 illustrates the migratory patterns that allow skipjack tunas to remain in water of 28–29°C.

Thermal selection is probably a factor in the changing distributions of species in response to the effects of climate change, which has resulted in rising sea temperatures[8]. Case Study 9.1 examines some evidence of the changing distributions of fish species.

Shuttle thermoregulation of reptiles

Among terrestrial ectotherms, behavioural thermoregulation often involves shuttling back and forth between basking in sunlight and cooling down in shade, a cool burrow or cool water. While basking, body temperatures gradually increase due to the uptake of heat from solar radiation (**insolation**)[9]. By contrast, when in shade, a burrow or cool water, the animal cools down because heat dissipates into cooler air or water.

The balance between the convection of heat to air (heat loss) and solar irradiation (heat gain) over time determines the net heat balance of terrestrial ectotherms that have little body surface area in contact with the ground. However, the temperature of the substrate on which an animal basks can result in additional heat gain (or heat loss) by conduction across the body surface in contact with the substrate.

Shuttle thermoregulation is particularly common among reptiles, but less frequent among amphibian species in which high rates of evaporative water loss across the skin result in cooling, which is likely to offset the heat gain by basking, but may cause potential problems for water balance[10]. This may explain why the species of frogs that do bask in sunlight often have more waterproofed skins.

The nature of shuttle regulation suggests that it is controlled by temperature receptors with two **set points**. In such a system, cooling in the shade would result in a decrease in body temperature until it reaches a lower set point, which switches on warming behaviour. Basking in sunlight then raises body temperature until it reaches the high set point, when cooling behaviour commences, and so on. The precision of such a regulatory system depends on the interval between the two set points.

[8] Section 1.2.1 discusses changing sea temperatures.

[9] Solar radiation is discussed in some detail in Section 8.4.1.

[10] Sections 6.1.1 and 6.1.3 discuss the factors that determine evaporative water loss across the skin.

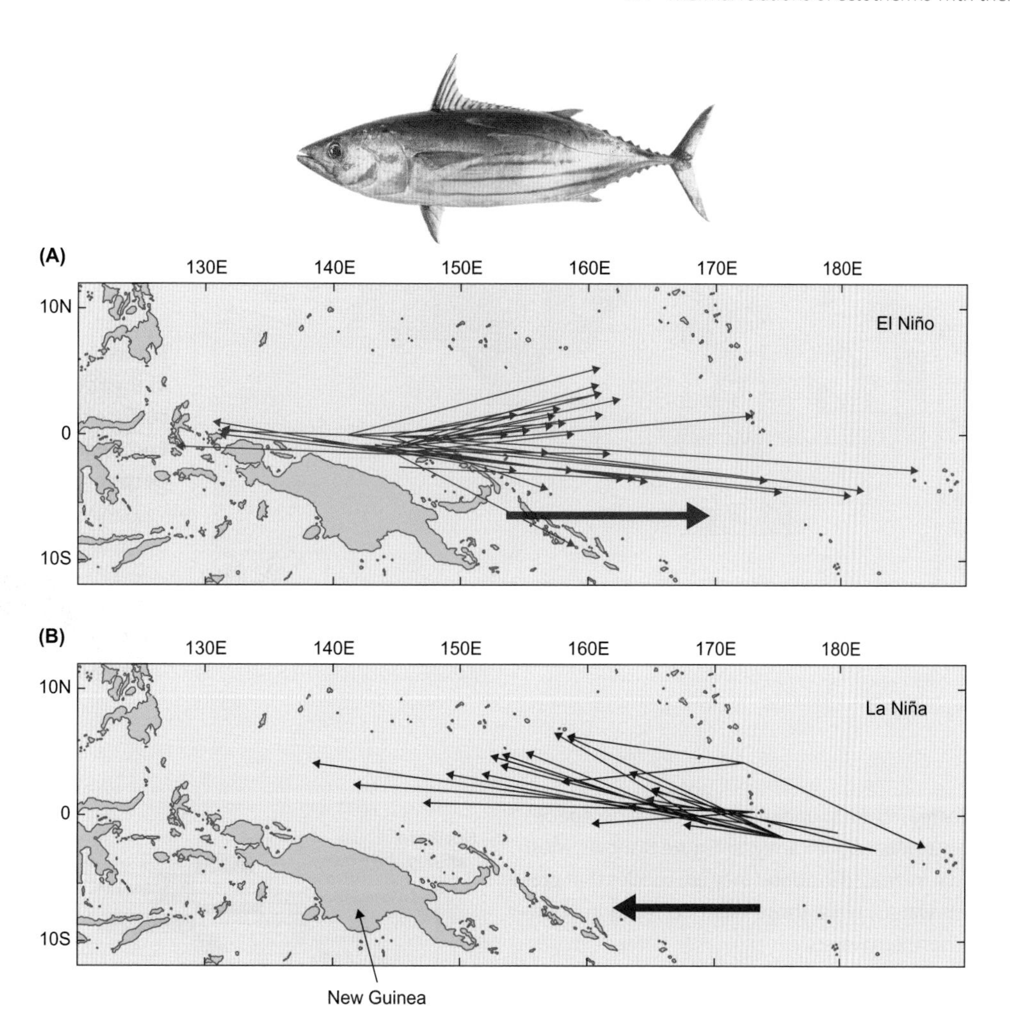

Figure 9.3 Movements of radio-tagged skipjack tuna (*Katsuwonus pelamis*)

Movements of tuna occur in response to changes in ocean temperatures resulting from the El Niño (warm phase) of the Southern Oscillation associated with a band of warm ocean water that develops in the central and eastern tropical Pacific and oscillates with a cooler phase (La Niña). The dominant movements of the tuna follow the movements of warm water, such that the highest densities occur in the area of water of >28–29°C. Each arrow shows the movement of an individual tuna after release. The dominant movements indicated by the large arrows follow the warm current, either away from New Guinea (red arrow in A) or toward New Guinea (blue arrow in B).

Source: Lehodey P et al (1997). El Niño Southern Oscillation and tuna in the western Pacific. Nature 389: 715–718. Tuna image: ©CSIRO.

Case Study 9.1 Impacts of rising sea temperatures on fish assemblages in the north-east Atlantic

Rising sea temperatures are likely to result in fish movement and redistribution given their mobility and ability to exhibit thermal selection. Rising sea temperatures may also affect abundances by directly affecting the survival of fish larvae and growth of juveniles, and by indirect effects on the food chain.

In this panel, we focus on the north-east Atlantic continental shelf region of Europe, where sea surface temperatures have risen more sharply than global averages[1]. Figure A shows the annual increase in sea surface temperatures of the European continental shelf between 1980 and 2008, during which there has been an average increase of 0.06°C per year; the mean sea surface temperature increased by up to 1.31°C over the 30 years. Hence, the North Sea is one of the 20 'hotspots' for major global warming, and a further rise of 1.8°C is predicted in the next 50 years.

Figure B summarizes some of the results of a comprehensive analysis of data including the results of more than 25,000 bottom trawl surveys of **demersal** species (which live and feed on or near the bottom) covering an area of about 1.2 million km^2 of seabed. The full analysis included in excess of 100 million individuals from 177 species of fish (or species groups). The analysis revealed that relative abundances of 72 per cent of the common fish species had responded to rising sea temperatures. Twenty-seven of the most

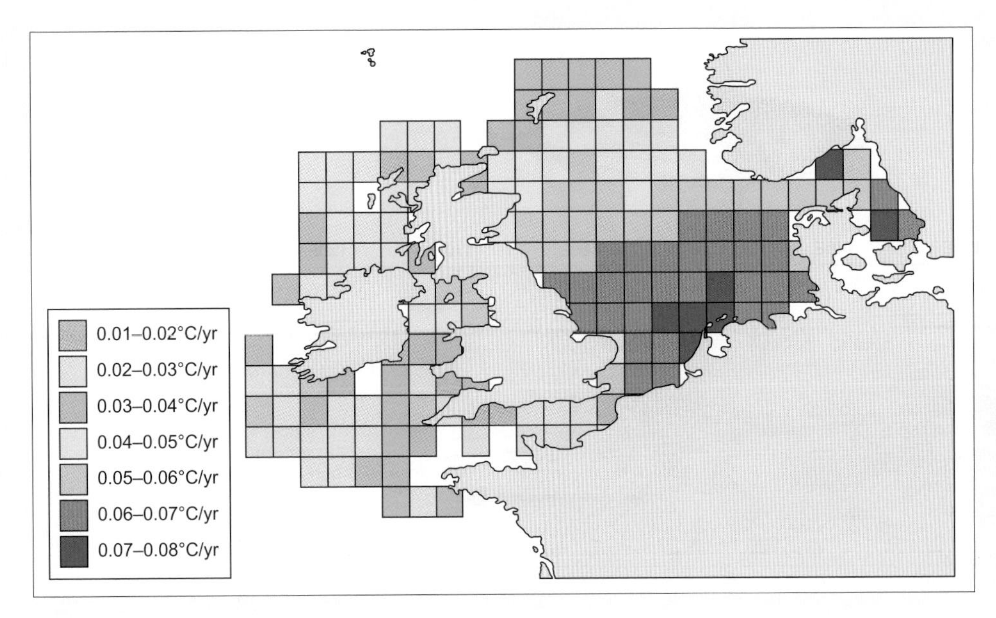

Figure A Spatial variation in the rate of increase in sea surface temperatures in the European continental shelf between 1980 and 2008

Source: Simpson SD et al (2011). Continental shelf-wide response of a fish assemblage to rapid warming of the sea. Current Biology 21: 1–6.

common species increased in abundance whereas nine species declined in abundance. Most significantly, as Figure B shows, most of the fish species that increased in abundance are species that prefer warmer southern waters, while those that decreased in abundance prefer colder northern waters.

More recently, attention turned to the distribution of **pelagic species** of marine fish that are not dependent on benthic habitats for food or shelter. As such, these species have more potential to show more rapid changes in abundance as a result of the rapid redistribution of adults in response to sea surface temperatures, and their planktonic larval stages and rapid life cycles.

Analysis of the results of almost 58,000 survey trawls across the European continental shelf between 1965 and 2012 shows a strong shift in the assemblage of pelagic species in the North Sea and Baltic Sea[2]. Figure C shows results for the six most common demersal species whose distributions have all altered. As a result, the cold-water assemblage of the 1960s has become a community of warmer water fish species. Specifically Atlantic herring (*Clupea harengus*) and European sprats (*Sprattus sprattus*), which were common in the 1960s to the 1980s, have been replaced by Atlantic mackerel (*Scomber scrombrus*), Atlantic horse mackerel (*Trachurus trachurus*), European anchovies (*Engraulis encrasicolus*) and European pilchards (*Sardina pilchardus*). These species accounted for almost 40 per cent of fish harvested in the north-east Atlantic between 2000 and 2011.

The changes in fish communities are considered to be one of the most dramatic consequences of climate changes that may influence the entire food web, from zooplanktonic larvae to top predators such as seals, dolphins and seabirds. The predicted increase in sea surface temperatures of the European continental

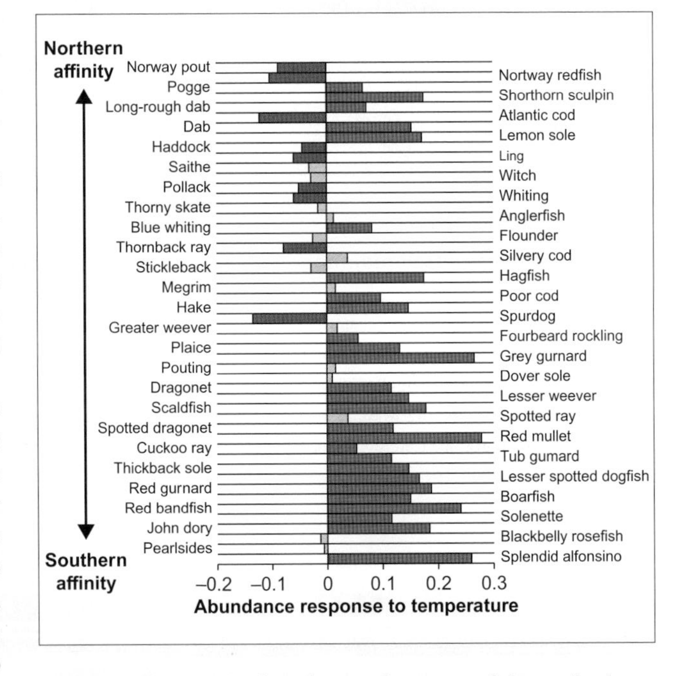

Figure B Changes in abundance of common fish species in northern Europe in response to increases in sea temperatures between 1980 and 2008

Species are ranked according to their northern/southern affinities based on information on species ranges and temperature preferences. Abundance responses are based on the correlations of the association of mean annual log abundance with temperature. Positive values indicate an increase in abundance in research surveys (species shown by red bars); negative values indicate a decrease in abundance (species shown by blue bars); species represented by grey bars showed smaller changes in abundance.

Source: Simpson SD et al (2011). Continental shelf-wide response of a fish assemblage to rapid warming of the sea. Current Biology 21: 1–6.

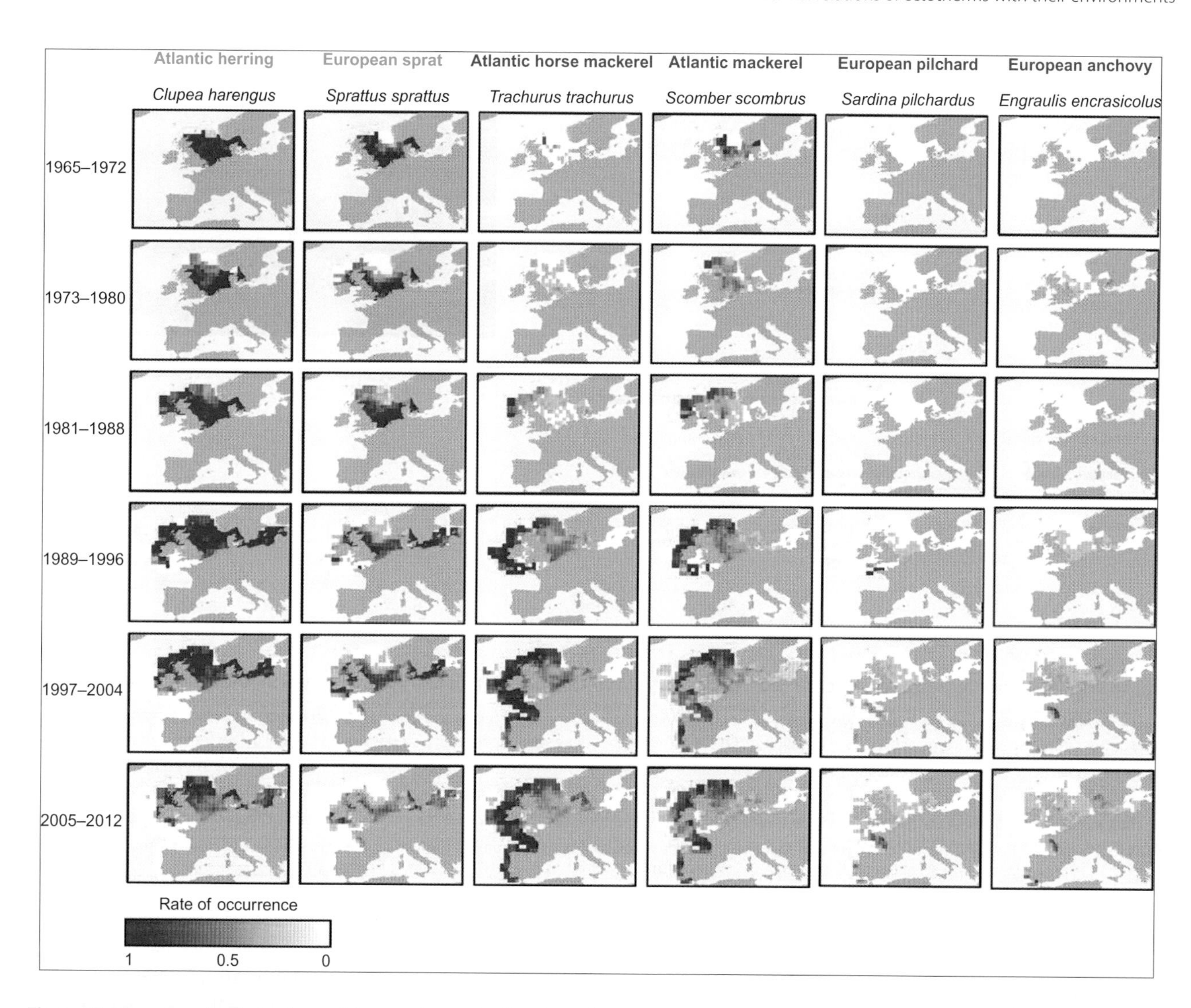

Figure C Warming shelf seas have driven sub-tropicalization of European pelagic fish communities

The boxes illustrate the changes of occurrence and distribution of the six most common pelagic fish species in the north-east Atlantic over the last four decades. Species are listed at the top of the diagram. Atlantic herring and European sprat are cold-water species (blue font), while the other four (pink font) are warm-water species. The maps illustrate the annual rate of occurrence for each species averaged for each of six time-periods (8 years per block) in colour-coded grid cells: red (high) to white (low); cells in light blue are where no data were included.

Source: Montero-Serra I et al (2015). Warming shelf seas drive the subtropicalization of European pelagic fish communities. Global Change Biology 21: 144–153.

shelf over the rest of this century remains within the limits of the thermal tolerances[3] of the fish species present. However, recent predictions of the abundance of demersal species based on a wide array of variables (including surface and near-bottom temperatures, and water depth) indicate that thermal conditions over the next 50 years are likely to constrain the fish species unless they can track the rate of warming by acclimation or genetic adaptation.

› Find out more

Montero-Serra I, Edwards M, Genner MJ (2015). Warming shelf seas drive the subtropicalization of European pelagic fish communities. Global Change Biology 21: 144–153.

Rutterford LA, Simpson SD, Jennings S, Johnson MP, Blanchard JL, Schön P-J, Sims DW, Tinker J, Genner MJ (2015). Future fish distributions constrained by depth in warming seas. Nature Climate Change Letters Online publication.

Simpson SD, Jennings S, Johnson MP, Blanchard JL, Schön P-J, Sims DW, Genner MJ (2011). Continental shelf-wide response of a fish assemblage to rapid warming of the sea. Current Biology 21: 1–6.

[1] Section 1.2.1 briefly examines global trends in sea surface temperatures; in August 2015, global average sea surface temperature across oceans was 0.78°C above the average for the 20th century.

[2] Section 1.2.1 discusses the Baltic Sea in more detail.

[3] Section 8.2.2 discusses the concept of lethal limits and thermal tolerances. Section 9.3.1 examines the concept of an optimal temperature range and physiological tolerance ranges of ectotherms.

9

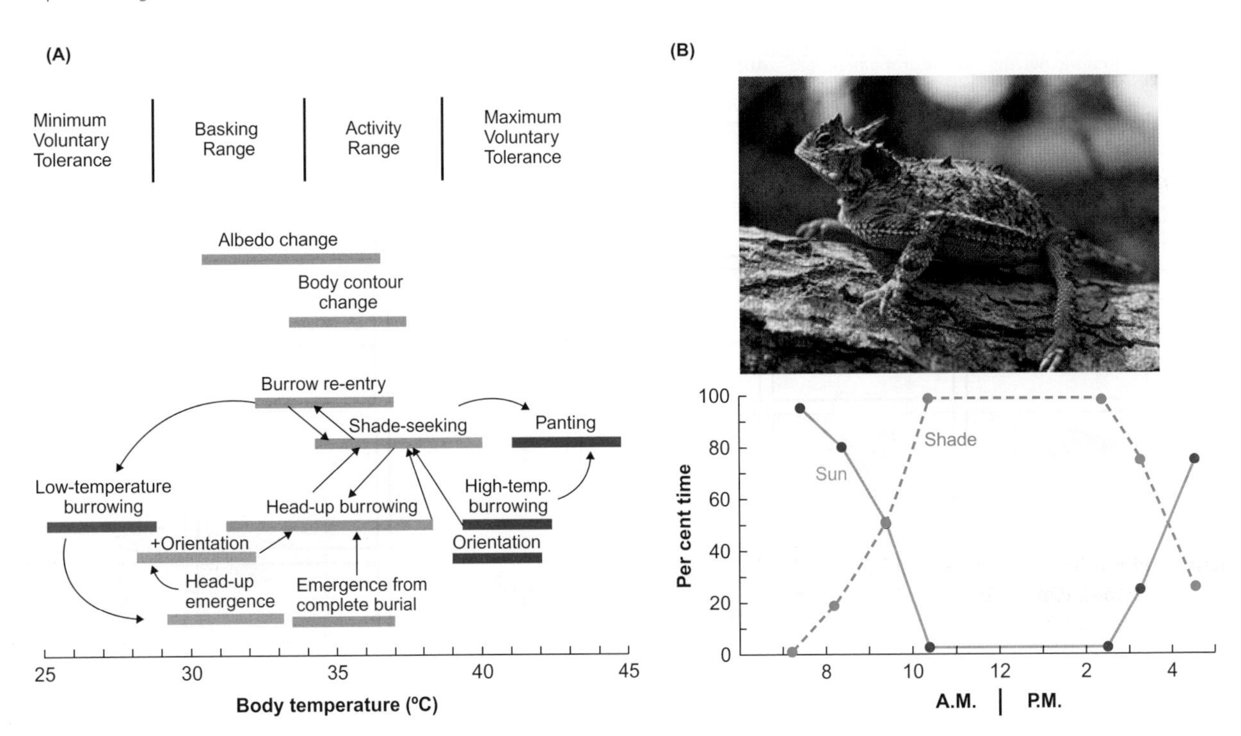

Figure 9.4 Behaviour patterns of coast horned lizards (*Phrynosoma coronatum*) in relation to body temperature and time of day

(A) Some of the complex patterns of temperature-related behaviours recorded for horned lizard. Arrows indicate changes in behaviour. Bars represent the range of body temperatures when each type of behaviour is exhibited. Colour scale: blue = below set temperature; pink = warming but still below set temperature; orange = basking and activity range; red = overheating and reaching tolerance limit. At low temperatures overnight, the animals occupy burrows. They emerge from the burrows and orientate to take up heat when body temperature is 28–32°C. Shade-seeking and re-entry into burrows occurs at body temperatures of 33–39°C. Orientation minimizes further heat uptake when body temperature is 39–42°C and the lizards re-enter their burrows. Panting, which results in heat loss by evaporative water loss occurs at 41–44.5°C. (B) Percentage of time that horned lizards spend basking in the sun (solid line) and hiding in shade (dashed line) throughout the day.

Source: adapted from Heath JE (1965). Temperature regulation and diurnal activity in horned lizards. Univ. California (Berkeley) Publications in Zoology 64: 97–136. Image of horned lizard: © Texas Parks and Wildlife Department.

A closer temperature interval results in more frequent shuttling but achieves tighter regulation of body temperature.

Coast horned lizards (*Phrynosoma coronatum*), which occur in the southern United States, are a particularly good example of a species using proportional heating in which behaviour is related to the difference between body temperature and the preferred body temperature, or the set points of temperature sensitive receptors. Figure 9.4 illustrates the complexity of the lizards' daily behaviour patterns.

Soon after sunrise, horned lizards leave their temporary burrows, which protect them from low temperatures at night, and warm themselves by basking in sunlight. During the day, they shuttle between sun and shade to maintain body temperatures. As air temperatures rise, the lizards spend more time in the shade and less in sunlight, such that between about 10:30 and 14:30 few animals bask in sunlight, as Figure 9.4B illustrates. As ambient temperature starts to decrease in the late afternoon, basking occurs again to defend the falling body temperature.

Between body temperatures of 29 and 35°C the horned lizards achieve proportional heating by adjusting their orientation relative to the sun according to the deviation of body temperature from the set point; maximum heat uptake occurs when they position themselves at right angles to the sun's rays. Horned lizards also use complex postural adjustments that vary the area of skin exposed to solar radiation such that when body temperature decreases the lizards expose proportionally more surface area by spreading their ribs and flattening the dorsal body surface. If body temperature reaches above 39°C, the lizards re-burrow by pushing their snout into the sand and moving it from side to side.

Body size influences behavioural thermoregulation of some reptiles

Larger animals have lower surface areas relative to their body mass (or volume) than smaller animals[11], which reduces both the rate of heat loss and heat gain relative to body mass and results in lower fluctuations of body temperature in larger rather than smaller animals. Such differences occur among free-ranging estuarine crocodiles (*Crocodylus porosus*), as illustrated in Figure 9.5A. The crocodiles shuttle between basking in sunlight and cooling in water or shade. Body temperature of small

[11] Section 10.2.1 examines the relationship between body size and surface area more closely. We discuss the concept of **gigantothermy** (a form of endothermy) in large reptiles in Section 10.3.2.

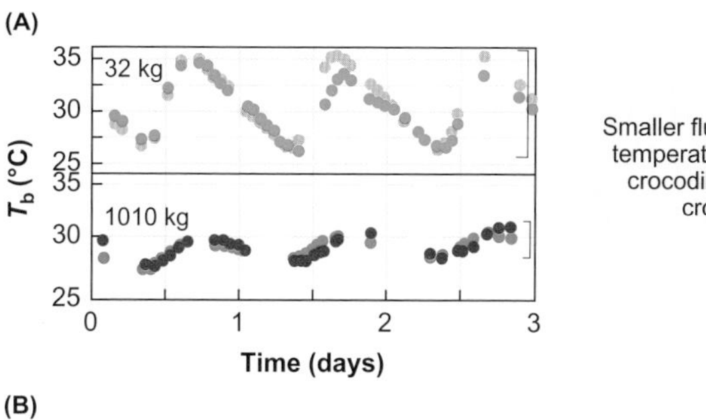

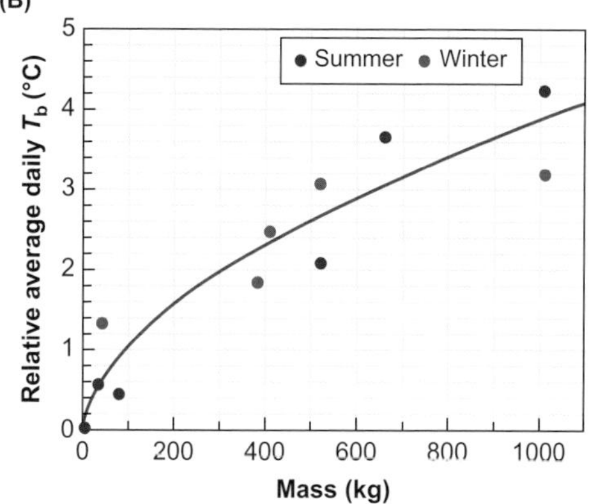

Figure 9.5 Body temperature (T_b) of crocodiles in relation to body mass

(A) Diurnal (daily) fluctuations in T_b of a small (32 kg) estuarine crocodile (*Crocodylus porosus*) (upper panel) are greater than those of a large (1010 kg) (lower panel) crocodile. Blue and red circles show measured values recorded by temperature sensitive radiotransmitters in the stomach of free-ranging crocodiles. Orange and green circles show values predicted from a mathematical model of heat flows based on body sizes and agree with measured values. (B) Relative average daily body temperature (T_b) of estuarine crocodiles increases with body mass. All values are expressed relative to those of the smallest species of freshwater crocodile (*Crocodylus johnsoni*).
Source: Seebacher F (2003). Dinosaur body temperatures: the occurrence of endothermy and ectothermy. Paleobiology 29: 105–122. Image of saltwater crocodile. http://animals.nationalgeographic.com.

9

crocodiles fluctuate over the daily cycle by almost 10°C, but by only 2–3°C in the large crocodiles (which is not much more than the 0.5°C daily variation of human body temperatures).

In addition to their body temperature being more stable, the average body temperature of large crocodiles is relatively high, as shown in Figure 9.5B. Taking small crocodiles weighing 2.6 kg in winter and 3.5 kg in summer as reference points the average daily body temperature was 3–4°C higher over a size range of 1000 kg. Conduction of heat in crocodiles occurs across the outer layer of exposed skin, muscle and fat and a thicker inner component of bone, tissues and fat. With an increase in size, the surface layers increase in thickness, which reduces the rate of heat loss. Modelling of these various components throughout the day predicts the higher average body temperature found in larger crocodiles. Figure 9.5A shows that modelling also predicts an almost identical diurnal pattern of fluctuations in body temperature to those measured for wild crocodiles.

A nocturnal lifestyle avoids high temperatures of hot deserts

Deserts typically receive high radiant energy input during the day. Many of the small ectotherms, which would receive a high thermal load if they were active during the day, adopt a nocturnal lifestyle and avoid the highest temperatures by burrowing.

Spiders are a good example of animals that retreat into deep burrows during the day. Figure 9.6 illustrates the importance of the burrow in protecting tarantulas from surface temperatures reaching up to 55°C. In Arizona deserts, *Aphonopelma* tarantulas do not move far from their burrows, except when males are looking for females in the mating season. The tarantulas remain at body temperatures of 27–30°C in the bottom of the burrows, even during the hottest part of the summer.

Like spiders, most scorpion species—and particularly those living in arid deserts—are nocturnal and occupy burrows or hide in rock crevices or under rocks during the day. Some desert scorpions such as *Hadrurus* species maintain their cycle of nocturnal activity and remain inactive during the day even when kept in constant darkness at a constant temperature and humidity. This observation suggests the scorpions have an endogenous mechanism that controls their activity cycle.

Thermophilic ants are active during the heat of the day

In marked contrast to the scorpions and tarantulas that avoid high daytime temperatures, some species of invertebrates are thermophilic. By definition, thermophilic species are active at above 45°C, which is a lethal temperature for non-thermophilic species. The most well known are the thermophilic ants that forage at soil or sand surface temperatures of 50–70°C. Thermophilic ants provide a striking example of independent parallel evolution in phylogenetically well-separated genera that occur in northern and southern hemispheres[12]:

[12] Parallel evolution of similar features in phylogenetically separated genera is a form of **homoplasy** in which a similar trait occurs in two or more species, but was not present in the last common ancestor, as we discuss in Section 1.5.3.

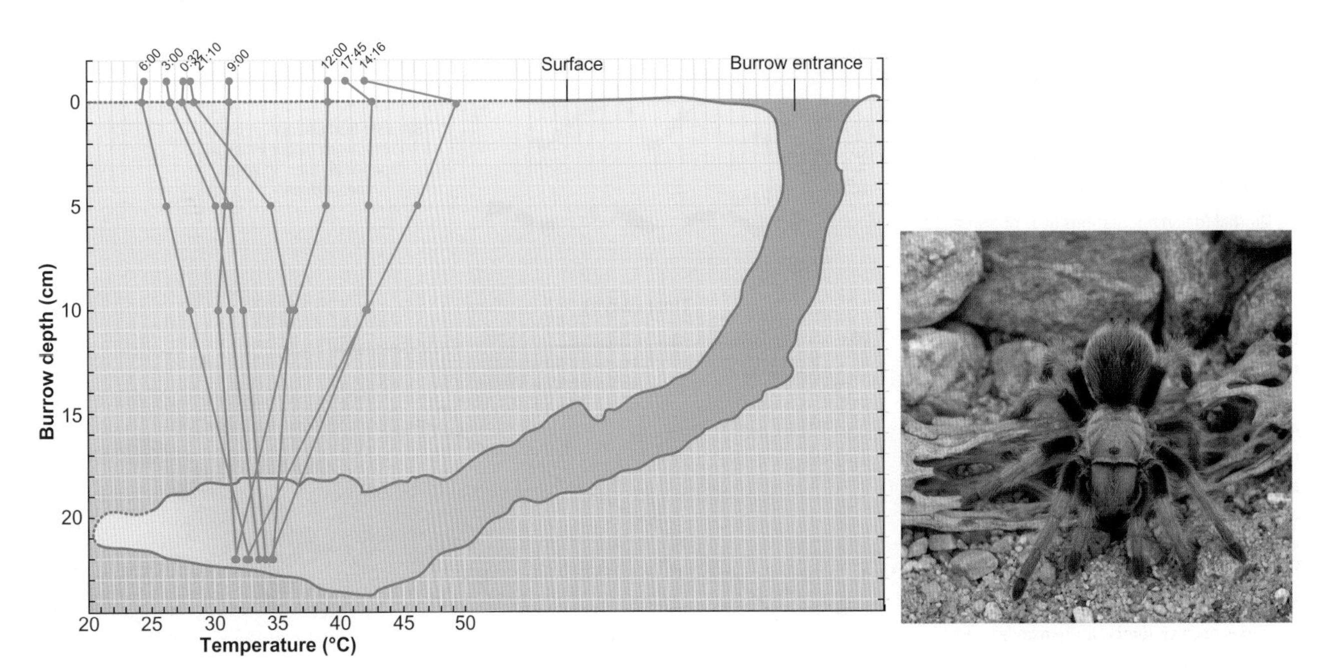

Figure 9.6 Temperature in the burrows of a North American tarantula (*Aphonopelma* sp.)

Temperatures were measured on the walls of burrows in south-eastern Arizona in July taking measurements at various depths (indicated by solid symbols) at various times of day. A 24-cm-deep burrow is shown in cross-section. Surface temperatures in the afternoon exceed the upper critical maximum for the tarantulas (~43°C). Retreat into the base of the burrow where temperatures are fairly stable throughout the day allows the tarantulas to maintain a body temperature of 32–35°C.

Sources: Seymour RS, Vinegar A (1973). Thermal relations, water loss and oxygen consumption of a North American tarantula. Comparative Biochemistry and Physiology 44A: 83–96. Image of Arizona (Mexican) blond tarantula: © Kim Taylor/naturepl.com

- In Australia, red honey ants (*Melophorus bagoti*) are active on desert soils with temperatures of 50–70°C.
- In the dry savannahs and deserts of southern and eastern Africa, such as the Namib Desert stretching along the coast of Angola, Namibia and South Africa, *Ocymyrmex* ants occupy habitats where sand surface temperatures reach above 60°C and the ants are active at up to 67°C.
- *Cataglyphis* species of thermophilic ants occur in semi-arid and arid habitats of the northern hemisphere: Mediterranean areas and the Sahara and Gobi Deserts. In the Sahara Desert, *Cataglyphis bicolor* ants forage on sands with surface temperature exceeding 70°C. In the cooler northerly parts of their distribution, *Cataglyphis* species even enter a resting phase (diapause)[13] during winter whereas non-thermophilic species of ants remain active at 25°C–30°C.

Despite the remarkable abilities of thermophilic ants in the wild, survival of these ants in laboratory conditions is less spectacular. Australian desert ants only survive about an hour at a laboratory air temperature of 55°C. At a similar temperature, Sahara Desert ants (*Cataglyphis* sp.) survive 10 to 25 min, while Namib Desert ants (*Ocymyrmex* sp.) are in a heat coma within 25 seconds. So how do these animals survive in the wild? Although the ants have different genetic lineages, they share characteristics that allow them to forage in extreme heat.

A major factor in their survival when out foraging is their behaviour. The ants run fast between refuges of shade cast by dry vegetation, or climb away from the hot sand onto dry stalks or pebbles for respites from the heat. During their respites, the thermal load taken up by the ant's body can dissipate by radiation, conduction, and convection into the cooler air[14]. While the high surface area of the ants relative to their body mass results in a high rate of thermal loading it also allows rapid dissipation of the thermal load.

The activity of the thermophilic ants increases as sand or soil surface temperatures increase during the day—but they take more respites of increasing length. Figure 9.7 shows some data for Namib Desert ants, which take infrequent respites (about once every 2 minutes) when air temperatures are below 50°C, but respites are increasingly frequent at above 50°C (shown by the red line on Figure 9.7A). At surface temperatures above 60°C, the ants spend more than 50 per cent of their time in respites between rapid searches for food lasting just 10 seconds on average. This behaviour is fortunate: laboratory studies show that this surface temperature would induce a heat coma after 20–25 seconds. This really is living life on the edge, but by running between thermal refuges, the ants only fleetingly expose themselves to the hot sand surface.

[13] We discuss insect diapause in Section 9.2.1.

[14] We discuss radiation, conduction and convection in relation to heat exchange in Section 8.4.

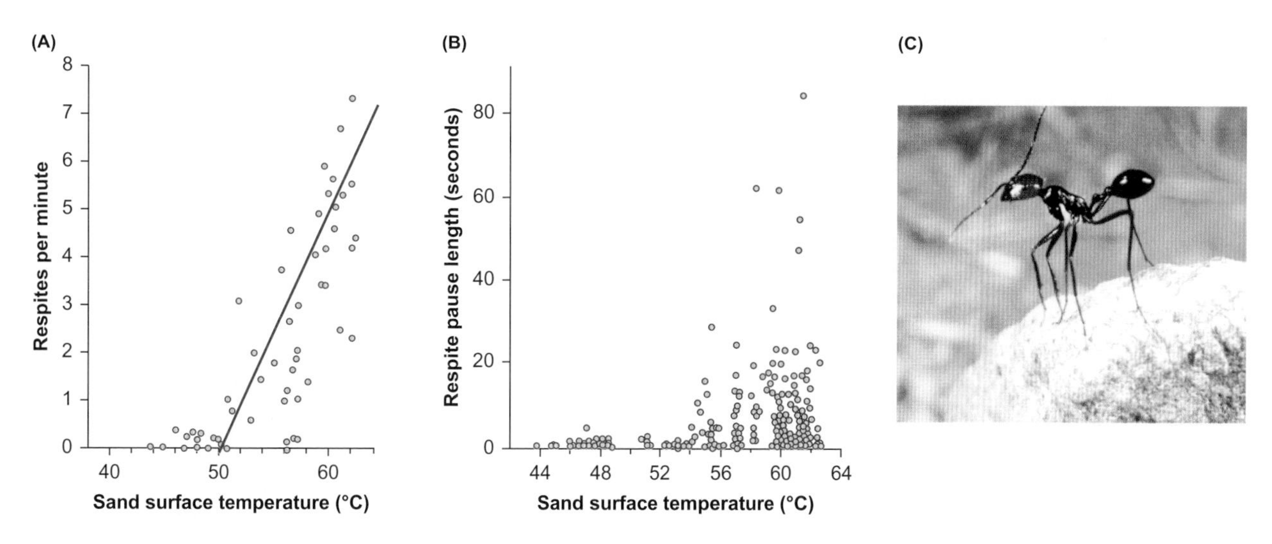

Figure 9.7 Frequency and duration of use of thermal refuges (respites) by desert ants

(A) Number of respites per minute of Namib Desert ants (*Ocymyrmex* sp.) in relation to sand temperature. The red line shows the pronounced increase in respites in refuges once air temperature rises above 50°C. (B) Duration of respites by Namib Desert ants in relation to sand temperature. (C) Worker ant of *Ocymyrmex velox* during respite (cooling off) behaviour in stilt posture.

Sources: A and B: Marsh AC (1985). Thermal responses and temperature tolerance in a diurnal desert ant, *Ocymyrmex barbiger*. Physiological Zoology 58: 629–636. C: Sommer S, Wehner R (2012). Leg allometry in ants: Extreme long-leggedness in thermophilic species. Arthropod Structure & Development 41: 71–77.

When thermal loads approach lethal levels, thermophilic ants often climb vegetation or retreat to higher ground and stretch their legs, as illustrated by the ant shown in Figure 9.7C. After a few moments in this stilt position the ant jumps down and returns to foraging. This behaviour suggests that the length of the legs of thermophilic desert ants could be under strong selective pressure. Recent comparison of leg lengths of small, medium and larger species of two genera (*Cataglyphis* and *Ocymyrmex*) of thermophilic ants with size-matched ants in closely related non-thermophilic genera living in moist habitats has found that the thermophilic desert ants do have longer legs than non-thermophilic species. Long legs may be advantageous in thermophilic species for two reasons:

- Air temperatures decrease with distance from the hot substrate.
- Long legs enable the ants to run faster—and a higher running speed will reduce the time taken to find food within a particular distance.

Within 4 mm from the surface, which is about the length of the long legs of desert ants, air temperatures decrease by 12°C (relative to a surface temperature of 60°C) during the hottest period of the day. The temperature reduction, to about 48°C, enables body temperature to fall below the critical thermal maxima[15] of thermophilic ants which vary between 51 and 57°C depending on the species. If an animal stands on surface features that raise them higher, temperatures decline further. *Ocymyrmex* ants routinely stand on small pebbles that can raise them to about 20 mm above the sand surface, where air temperatures are 10–17°C lower than on the surface below depending on the time of day.

Thermophilic ants in the Namib Desert move relatively slowly (at about 4 m min^{-1}) when sand temperature is 30°C, but as the data in Figure 9.8 illustrate, they increase their running speed with surface temperature, and by 60°C are running at 22 m min^{-1}. Other ants living in hot deserts run even faster. For example, two species of Sahara Desert ants (*Cataglyphis*) can run at 60 m min^{-1} (1 m s^{-1}) which for an animal of less than 1 cm equates to 100

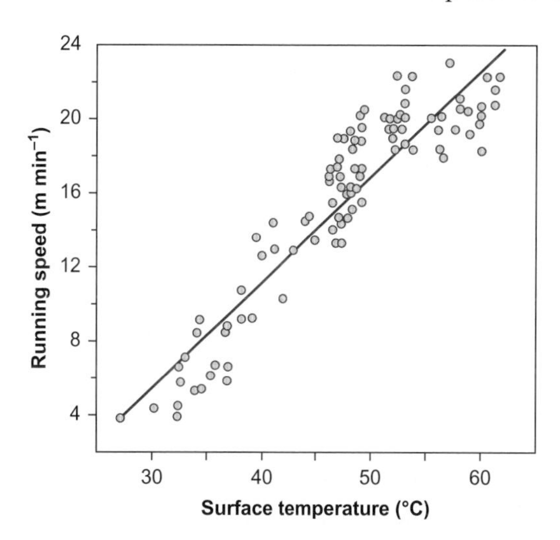

Figure 9.8 Running speeds of Namib Desert ants (*Ocymyrmex* sp.) in relation to sand surface temperature

Source: Marsh AC (1985). Microclimatic factors influencing foraging patterns and success of the thermophilic desert ant, *Ocymyrmex barbiger*. Insectes Sociaux, Paris 32: 286–296.

[15] Critical thermal maxima of ectotherms are determined by raising ambient temperatures from acclimation temperature, at a constant rate at which body temperatures follows ambient temperature, without a significant time lag. At the critical thermal maximum locomotion becomes disorganized; the animal cannot escape the conditions and death is imminent as temperatures continue to rise.

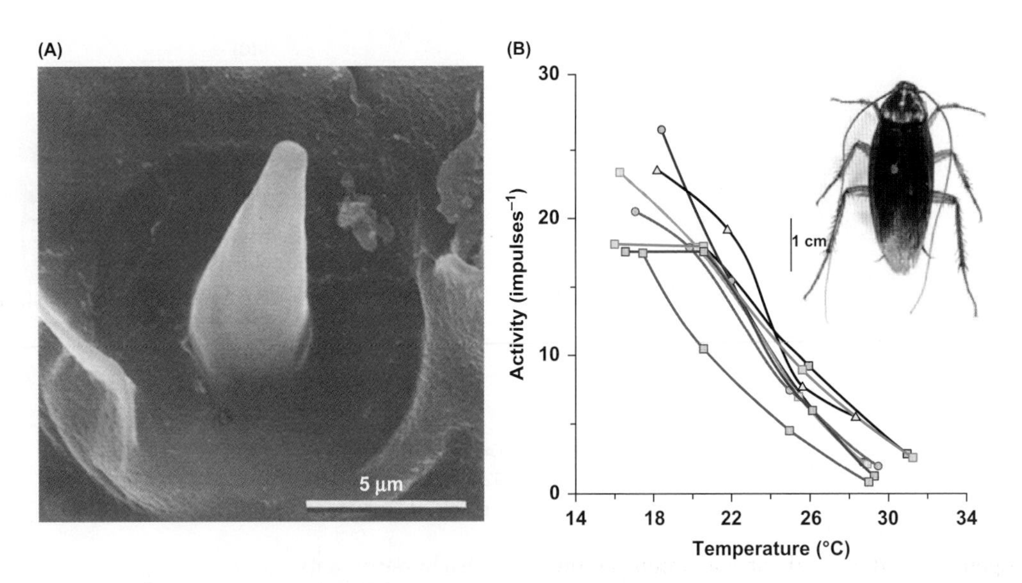

Figure 9.9 Cold receptors of the American cockroach (*Periplaneta americana*)

(A) Scanning electron micrograph of the thermohygro (humidity) sensillum housing on the cockroach antenna. (B) Steady-state activities recorded from seven cold receptors. Lower temperatures increase the number of impulses per second.

Source: Nishikawa M et al (1992). Response characteristics of two types of cold receptors on the antennae of the cockroach, *Periplaneta americana* L. Journal of Comparative Physiology A 171: 299–307. American cockroach image: http://www.landcareresearch.co.nz/__data/assets/image/0005/40991/AmericanCockroach.jpg.

body lengths per second. These faster running speeds reduce the period of exposure to the hot desert surface, which reduces heat uptake and the risk of dehydration resulting from the loss of water by evaporation[16].

The morphological adaptations and behavioural responses of thermophilic ants explain why they are able to be active when surface temperatures are so high. But we might ask—why not avoid the heat loading completely? It seems perverse to come out to forage when temperatures are already above 50°C and for activity to peak at the hottest time of the day, when temperatures are close to the critical thermal maximum of the ants. However, this unusual behaviour has two benefits:

- Thermophilic ants are able to forage for dead insects and other arthropods that have succumbed to heat or desiccation stress. Their tolerance of high heat loads enables them to avoid competition with less heat tolerant competitor species of ants.
- Foraging at high temperatures avoids meeting the desert lizards that prey on ants. Experiments have shown that the thermophilic ants fall prey to lizards within about 5 minutes when placed on the desert surface at lower temperatures than those at which they are routinely active.

Thus, thermophilic ants occupy a thermal niche defined by high predation pressure at lower temperatures and increased tolerance of high temperatures. In some species, a part of tolerance of high temperatures may be the continued synthesis of **heat-shock proteins**[17] that protect their tissues from thermal damage. Sahara Desert ants synthesize heat-shock proteins at temperatures below those they normally experience and continue this process at higher temperatures (45°C) than related non-thermophilic species, which stop synthesizing heat-shock proteins at 39°C.

9.1.2 Sensing the thermal environment

The ability of ectotherms to respond to their environment behaviourally suggests they can detect the temperature on their body surfaces or internally, by temperature sensing mechanisms (thermoreception[18]).

External thermoreceptors in ectotherms

Peripheral (external) **thermoreceptors** can convey immediate information to an animal about the thermal conditions on their body surface. Studies of external thermoceptors of invertebrates have concentrated on insects, mainly cockroaches, which have cold receptors on their antennae and legs. An example is shown in Figure 9.9, which also illustrates the higher frequency of impulses when external temperature decreases. The receptors respond to temperature change rather than being able to detect what the exact temperature is. Despite their apparent temperature responses, the involvement of thermoreceptors in the control

[16] Section 6.1.1 outlines the physical processes that determine the evaporative water loss of terrestrial animals.

[17] We examine heat shock (stress) proteins in Box 9.1 in Section 9.3.1.

[18] We discuss thermoreception (thermoception) in detail in Section 17.5; Section 10.2.6 discusses thermoreceptors and temperature control of birds and mammals.

of behavioural thermoregulation and thermal selection is uncertain because removal of structures bearing thermoceptors does not prevent thermal selection by cockroaches in laboratory tests.

Peripheral thermoreceptors also occur among ectothermic vertebrates (fish, amphibians and reptiles), but we have a poor understanding of how they operate, except for the pit organs of some snakes, which detect infrared thermal radiation (wavelength 5–30 μm) of their prey.

Pit organs evolved independently in several different snake families. Pythons and some boas have three or more pit organs called labial pits, which are located between the scales along the upper or lower lip. Pit vipers have a pair of larger facial pits one between each eye and nostril, as shown in the rattlesnake in Figure 9.10A. Infrared radiation passes through the opening of the facial pit and impinges on a membrane at the back of the pit that is packed with thermoreceptors, which transmit thermal signals to the brain[19].

In addition to prey detection, at least 12 species of pit vipers (mainly rattlesnakes) are able to use their facial pits to sense the thermal environment and stimulate thermal selection. Figure 9.10B examines some of the first results that demonstrated the ability of rattlesnakes to sense their thermal environment. The rattlesnakes were first subjected to a thermal load, and then placed in test systems in which they could either find a thermal refuge at 30°C and offload excess heat, or (depending in which direction they moved) enter a hot environment (at 40°C). If the snakes moved at random when placed in a Y maze they would end up at each side of the Y in half (0.5) of the trials. In fact, the snakes located the thermal refuge at the end of one arm of the Y in 0.75 of the trials. By contrast, when the pit organs were blocked the snakes *did* move at random and ended up in the thermal refuge in 0.5 of trials. More complex test arenas with four hide boxes or four semi-natural burrows (only one with a thermal refuge) confirmed the ability of the rattlesnakes to locate the thermal refuge, if their facial pits were open, as shown by the results illustrated in Figure 9.10B (ii) and (iii).

Such abilities enable these snakes to escape from extreme temperatures and locate thermal refuges such as rodent burrows and rock crevices. Indeed, sensing the thermal environment may have been the first function of facial pits, before the evolution of the pit organs enabled prey detection.

Internal thermoreceptors in ectotherms

Internal thermoreceptors convey information about tissue temperatures. Thermoreceptors within the central nervous system could trigger thermal selection behaviour and behavioural responses that reduce variations in body

(A)

(B)

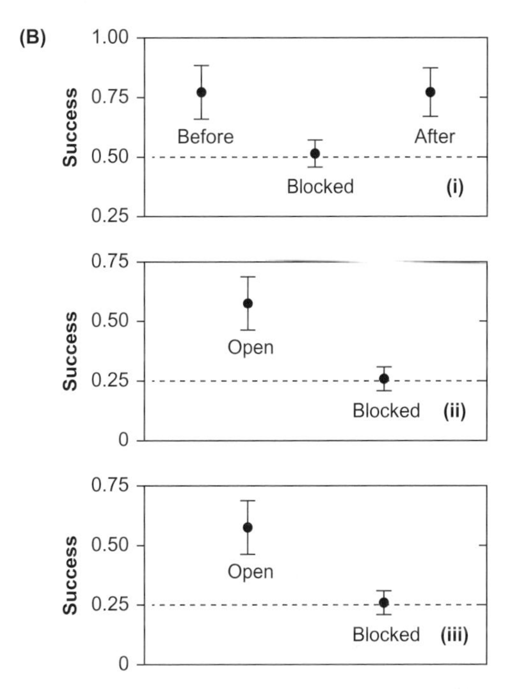

Figure 9.10 Facial pits of rattlesnakes

(A) Photograph of the head of a canebrake rattlesnake showing the location of one of the pair of facial heat sensing pits between the eyes and nostrils. (B) Results of choice experiments with western diamondback rattlesnakes (*Crotalus atrox*). The snakes were subjected to a heat stress (40°C) before being placed in the test arena to assess their ability to locate a thermal refuge at 30°C. The dashed lines represent the expected rate of success for random movements. Snakes were first tested with intact pits (results shown as 'before'), then after temporary blockage of the pits with a small polystyrene foam ball placed in the pit cavity and gluing a piece of aluminium foil over the opening of the pit to reflect incoming thermal radiation and retain the foam ball; the polystyrene insulated the pit membrane from heating. The polystyrene was easily removed after testing the effects of blockage. (i) Experiment in Y maze. Blockade of facial pits resulted in loss of directed movements. There was a 50% (0.5) chance of randomly finding the thermal refuge following blockade compared to approximately 75% success prior to blocking the pits and after unblocking the pits. (ii) Circular area with four hide boxes. There was 25% random chance (0.25) of finding the refuge location. With open pits the success rate was more than doubled. (iii) Circular arena with four burrows to simulate natural retreats. With open pits the success rate was more than double the 25% (0.25) random success rate.

Data points represent means and 95% confidence intervals for 12, 16 and 16 snakes in the three experiments, respectively.

Source: Krochmal AR, Bakken GS (2003). Thermoregulation is the pits: use of thermal radiation for retreat site selection by rattlesnakes. Journal of Experimental Biology 206: 2539–2545. Image of canebrake rattlesnake: © Edwin Giesbers / naturepl.com

[19] We consider the anatomical arrangement of pit organs and their nervous supply more fully in Section 17.2.1.

temperature. Such capacity appears to occur both among invertebrate and vertebrate ectotherms.

Thermosensitive neurons occur in the ganglia making up the central nervous system of cockroaches, locusts and crickets. Some neurons increase their firing rate in response to warming, while others respond to cooling, but we do not understand exactly how these responses determine their behavioural responses to temperature.

Among ectothermic vertebrates (fish, amphibians and reptiles), thermosensitive neurons occur at the base of the brain in the **hypothalamus**[20]. That these receptors are involved in behavioural thermoregulation becomes evident by their destruction. For example, destruction of the preoptic area of the hypothalamus, by passing an electric current through an electrode implanted in this tissue to create a hypothalamic lesion, disrupts the thermal selection of goldfish (*Carassius auratus*). Similarly, hypothalamic lesions disrupt the shuttling responses to environmental temperature of desert iguanas (*Dipsosaurus dorsalis*).

An alternative experimental approach to the destruction of the thermosensitive neurons is stimulation of these thermoreceptors. Such experiments have shown that cooling of the brainstem results in an animal remaining in a warm environment for longer so that body temperature increases more than usual. On the other hand, warming of the brainstem results in animals leaving a warm environment when body temperature is lower than usual.

Recent studies indicate additional tissue-specific thermosensitivity of the liver, heart and muscle tissue of some ectotherms such as the estuarine crocodile (*Crocodylus porosus*), in which these thermoreceptors influence shuttling behaviour between cooling in water and basking in the sun. The tissue thermoreceptors of reptiles function as a type of ion channel, which opens transiently in response to temperature; these are known as **transient receptor potential (TRP) channels** or thermoTRP channels[21]. TRPV1 and TRPM8 (the letters V and M indicate that these are from separate sub-families) are highly responsive to temperature. TRPV1 opens in response to heat, while TRPM8 is sensitive to cold.

9.1.3 Physiological regulation of heat exchange in ectotherms

Behavioural thermoregulation is critical to the control of body temperature of ectotherms, but some ectotherms couple their behaviour with physiological mechanisms that influence internal tissue temperatures by:

- changes in skin lightness (**albedo**), which determines the proportion of the incident solar radiation that is reflected from an animal's external surfaces;
- changes in blood flowing to the surface tissues, which influences heat exchange between the animal and its environment;
- respiratory cooling by evaporative water loss from respiratory passages, which numerous reptiles use to avoid overheating[22].

Changing albedo affects heat uptake of some ectotherms

Albedo[23] (lightness of colour) or reflectance is a measure of the reflectance of solar radiation from an animal's surface. Hence, albedo determines the proportion of incident radiation that is absorbed, which in turn determines the *rate of change* in body temperature. Among animals of similar size and posture in similar environmental conditions, those with a high albedo, which tend to be lighter, heat up more slowly than darker animals with a low albedo (reflectance). Albedo does not, however, affect the final equilibrium temperature reached if an animal remains in the same place indefinitely.

Some amphibians and reptiles adjust their rate of heat uptake by adjusting skin lightness. One of the best examples are species of chameleons, such as Namaqua chameleons (*Chamaeleo namaquensis*) and flap-necked chameleons (*Chamaeleo dilepis*), that become lighter at higher body temperatures, which reduces the rate of further warming during basking. Their lightness increases the reflectance of visible and infrared wavelengths, as shown in Figure 9.11.

Basking as a component of thermoregulatory behaviour is less common in amphibians than reptiles probably because of the high rates of evaporative water loss across the permeable skin of most amphibians. Those amphibians that do raise their body temperatures by basking in sunlight tend to have relatively lower water permeability of their skin[24].

Some tree frogs that bask in the sun have a unique pigment, pterorhodin, in the skin cells containing pigments. Pterorhodin reflects infrared radiation (which is not visible to humans) with peak reflectance at about 800 nm, which may allow them to bask for long periods without overheating.

[20] Thermosensitive neurons in central nervous system of mammals are also located predominantly in the hypothalamus as we discuss in Section 10.2.6.

[21] Section 17.5 discusses thermoreception (thermoception) including TRP channels in more detail.

[22] Respiratory cooling is not a strong feature of most terrestrial ectotherms and is more generally associated with endothermic birds and mammals, which we examine in Section 10.2.2; its impact on their water balance is explored in Section 6.1.2.

[23] Section 8.4.1 explains albedo and its measurement.

[24] Section 6.1.3 discusses evaporative water loss of amphibians and skin permeability in more detail.

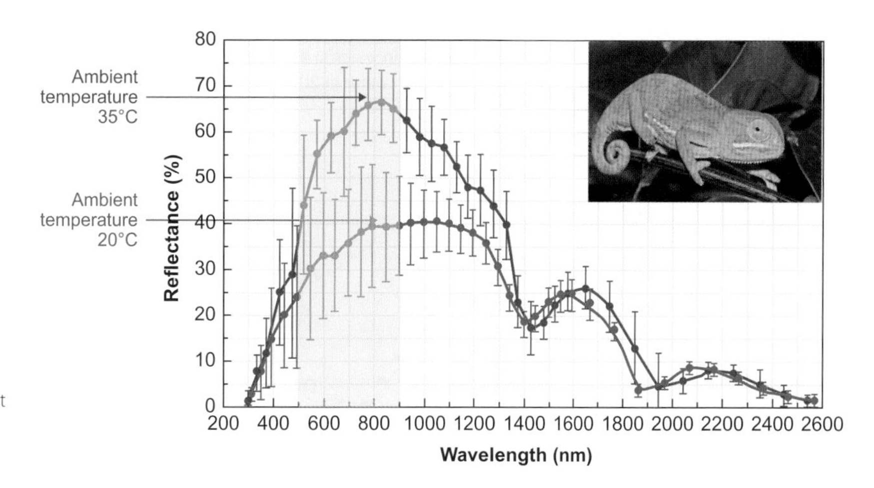

Figure 9.11 Reflectance of radiation of varying wavelengths by the flap-necked chameleon (*Chamaeleo dilepis*) at two ambient temperatures

Data are shown for the reflectance at an ambient temperature of 20°C (blue data points and line) and 35°C (red data points and line). The reflectance in the visible and infrared wavebands (500–900 nm; yellow overlay) increases from a peak of around 40% (blue arrow) to a peak of around 70% (red arrow) as it gets hotter.

Source: Walton BM, Bennett AF (1993). Temperature-dependent color change in Kenyan chameleons. Physiological Zoology 66: 270–287. Image of chameleon © Tyrone Ping/Flickr.

The pigment may also have a role in camouflage as the frogs rest on leaves that reflect similar infrared wavelengths.

In basking amphibians, colour changes that influence body temperature while basking could be of thermoregulatory importance in fine-tuning the heat gain. One species examined in some detail is a South American frog (Santa Barbara tree frog, *Bokermannohyla alvarengai*), which spends hours exposed on lichen-covered stones. Figure 9.12A shows one of these frogs with normal cryptic colouration. However, in Figure 9.12B note the striking colour change to a stark white of a frog exposed to sunlight. Temperature recording in individual frogs shows that cooler dark-coloured frogs placed in the sun become lighter in colour as they warm up. The lightening will reduce the rate of heat uptake as the body temperature increases, although the frogs become conspicuous to predators. As such, these frogs must balance the need for cryptic colouration, behavioural thermoregulation and water balance.

In contrast to animals that change colour, some species are permanently dark due to the permanent dispersion of granules of dark pigments of melanin in the melanophores of the skin, even when they are on a light background. Studies of insects generally support the idea that **melanism** reduces reflectance and is thermally beneficial for insects living in cold conditions.

Among vertebrates, one of the clearest demonstrations of the effect of melanism occurs in related species of lizards occupying geographically separate locations of South Africa: two melanistic species (black-girdled lizard, *Cordylus niger* and Oelofsen's girdled lizard,*Cordylus oelofseni*) and one non-melanistic species (Cape girdled lizard, *Cordylus cordylus*). The non-melanistic lizards, which live mostly in warmer inland areas at lower altitudes, reflect 15 per cent of the incident radiation, whereas the two melanistic species of lizards reflect only 5–7 per cent. These differences result in a faster rate of heating for the melanistic lizards. A faster heating rate and hence slightly higher body temperatures in the melanistic species (by about 2°C) could be beneficial in cold winter conditions, particularly in *Cordylus oelofseni*, which lives in cooler mountain locations.

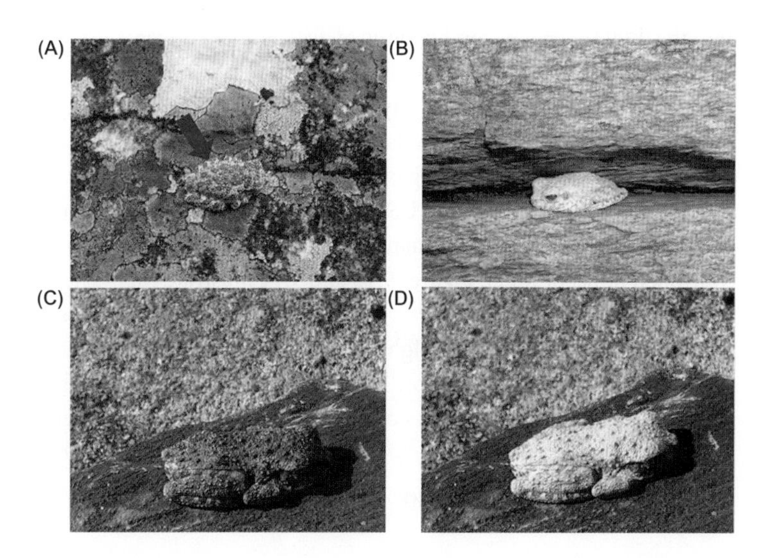

Figure 9.12 South American frogs *Bokermannohyla alvarengai* showing colour changes

Dramatic colour differences occur for camouflage as shown by the cryptically coloured frog in (A) while resting on lichen-covered rocks. The same frog is shown in (B) after its temperature increased after exposure to full sunlight. (C) and (D) show a frog in which body temperature was recorded; in (C) the frog had a body temperature of 18°C, while in (D) the frog is light coloured after 1 h in the sun when body temperature reached 28°C.

Source: Tattersall GJ, Eterovick PC, de Andrade DV (2006). Tribute to R. G. Boutilier: Skin colour and body temperature changes in basking *Bokermannohyla alvarengai* (Bokermann 1956). Journal of Experimental Biology 209: 1185–1196.

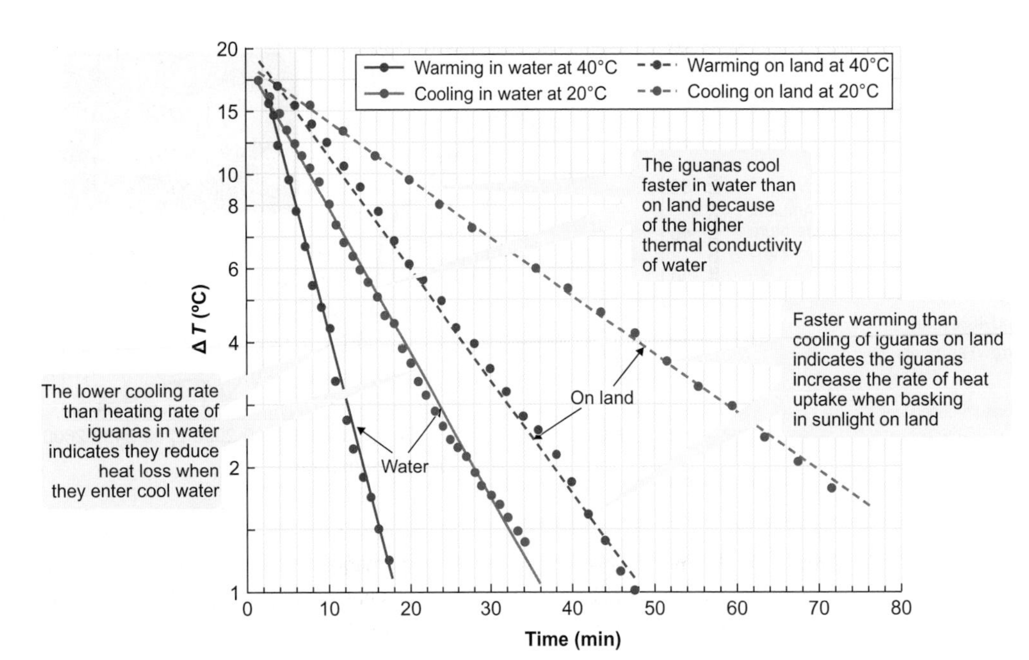

Figure 9.13 Heating (solid lines) and cooling (dashed lines) of Galápagos marine iguana (*Amblyrhynchus cristatus*) held in and out of water

The difference between ambient temperature and body temperature (ΔT°C), expressed as a positive value, is plotted against time, which allows comparison of the slopes of the four lines as a measure of the relative rate of heating or cooling in water and on land.

Source: adapted from Bartholomew GA, Lasiewski RC (1965). Heating and cooling rates, heart rate and simulated diving in the Galapagos marine iguana. Comparative Biochemistry and Physiology 16: 573–582.

Blood flow to surface tissues influences heat exchange in some ectotherms

Some vertebrate ectotherms control heat uptake and heat loss by regulating blood flow to surface tissues. The first clear evidence for such processes came from studies of Galápagos marine iguanas (*Amblyrhynchus cristatus*). These iguanas bask on lava rocks along the coasts of the Galápagos Islands, where they take up heat from solar radiation and by conduction of heat from the rocks across their skin, until they reach a body temperature of about 37°C. At this point, the iguanas dive into the water to either feed on algae in the splash zone or further offshore in large kelp beds.

To gain access to the kelp beds they dive to depths of 6–10 metres, where water temperatures are around 22–27°C. The lower water temperatures compared to body temperature and the high thermal conductivity of water and convection result in an inevitable loss of heat across their skin[25]. To examine whether iguanas have any control of heat loss during their feeding dives, George Bartholomew brought back a few iguanas from the Galápagos Islands to his laboratory at the University of California, USA, to investigate the rates of cooling and heating of iguanas held in temperature-controlled tanks of water. Figures 9.13 and 9.14 illustrate the results of these experiments.

The rates of heat loss are higher when the iguanas are in water than when they are on land, as can be expected because of the higher thermal conductivity of water than air. Concentrating on what happens in water, notice that the slope of the line for cooling in water (solid blue line) is shallower than the slope of the line for warming in water (solid red line). This indicates that iguanas reduce the loss of heat when they enter cool water. The calculated rate of cooling (°C per min) when there is a 10°C difference between the water temperature and body temperature is half the rate at which iguanas warm in water with a similar temperature difference. Reducing heat loss will allow the iguanas to extend the time spent foraging before they reach body temperatures at which they must surface and warm up.

Looking again at Figure 9.13, focus on what happens when the iguanas move back onto land. The relative slopes of the dashed red and blue lines indicate that warming on land occurs at about twice the cooling rate. These results imply that iguanas can control the rates of heat exchange across their surface tissues, reducing heat loss while cooling in water but increasing the rate of heat uptake while warming on land.

In reptiles, heart rate is a major determinant of the rate of blood flow through the arterial system, which delivers blood to the surface tissues where heat exchange occurs. Hence, heart rate has important effects on the rate of heat uptake or loss. An increase in body temperatures will increase heart rate due to Q_{10} effects[26] on heart function (and *vice versa*), which will increase blood flow to the body surface, and hence aid heat uptake during basking of reptiles in sunlight. However, examination of heart rates in relation to body

[25] The general principles governing rates of heat exchange are explained in Section 8.4.

[26] Box 8.2 discusses Q_{10} effects on physiological processes.

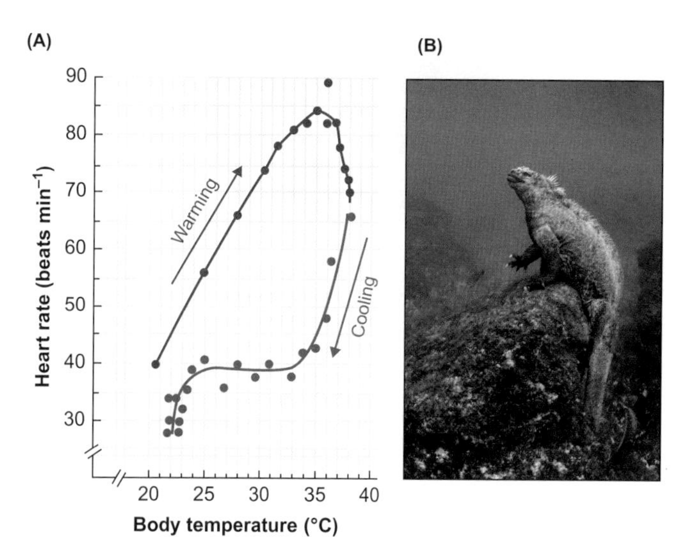

Figure 9.14 Change in heart rate of Galápagos marine iguana (*Amblyrhynchus cristatus*) submerged in water in relation to body temperature during warming and then cooling

When warming the rising body temperature increases heart rate. When subsequently cooled while feeding on algae the decline in heart rate is greater than expected from the changes in temperature alone, which is indicated by the steep decline. This difference results in a displacement of the warming and cooling lines on the graph such that at any given body temperature the heart rate of a submerged marine iguana is lower when it is cooling than when it is warming.

Sources: A: Bartholomew GA, Lasiewski RC (1965). Heating and cooling rates, heart rate and simulated diving in the Galapagos marine iguana. Comparative Biochemistry and Physiology 16: 573–582. B: © Danita Delimont / Alamy Stock Photo

temperature reveals that body temperature is not the only control of heart rate in reptiles. If this were true, the relationship between heart rate and body temperature would overlie each other when the reptile was warming and cooling.

Figure 9.14 shows the changing heart rates of Galápagos marine iguanas during warming and cooling are displaced. A **hysteresis** in heart rates is apparent, which means that, at any particular temperature, a lower heart rate occurs when the animal is cooling (reducing heat loss) than during warming. A hysteresis of heart rates also occurs in other reptilian species, loggerhead turtles (*Caretta caretta*) and estuarine crocodiles (*Crocodylus porosus*).

The results for estuarine crocodiles shown in Figure 9.15 indicate that the **autonomic nervous system**[27] is important for controlling heart rates during warming and cooling. Inhibition of the autonomic nervous system abolishes the hysteresis of heart rates during warming and cooling, as shown in Figure 9.15B, but this does not appear to occur in all species.

The heart rate hysteresis may have evolved with a closed circulatory system[28], in which the heart pumps blood around the

[27] We outline the autonomic nervous system in Section 3.3.3.
[28] Distinction of closed circulatory systems and their evolution from more open systems is discussed in Section 14.1.1.

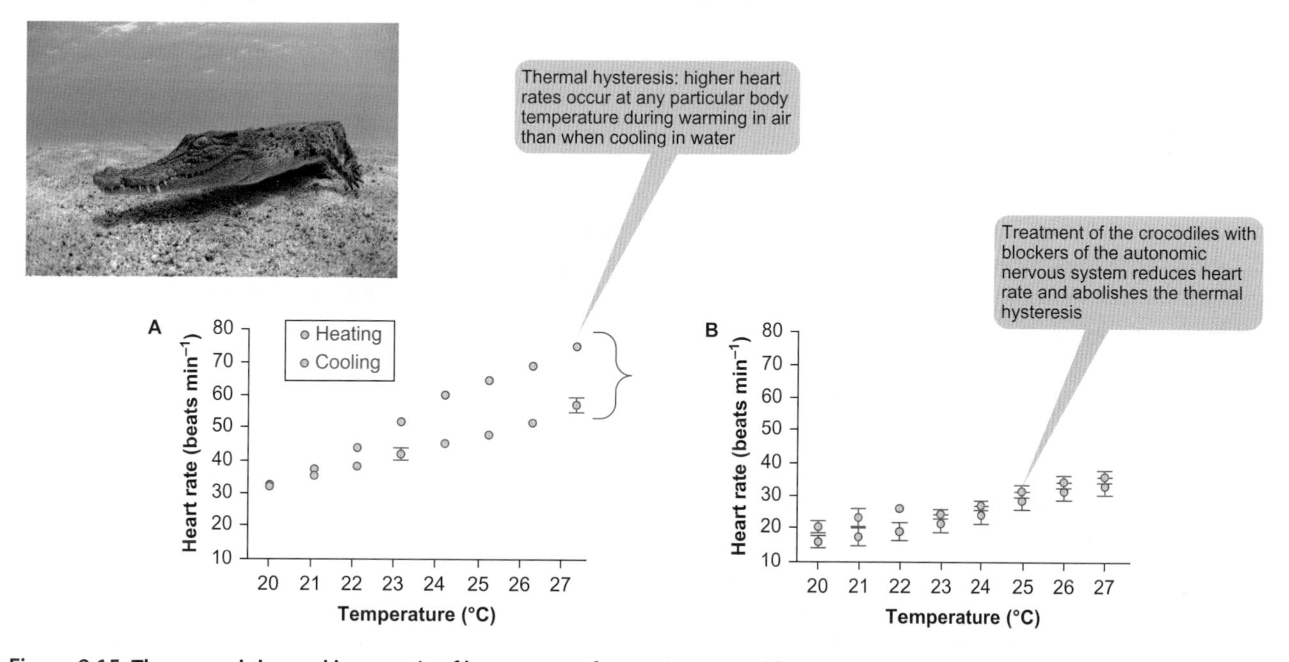

Figure 9.15 The normal thermal hysteresis of heart rates of estuarine crocodiles (*Crocodylus porosus*)

(A) Normal thermal hysteresis of heart rates, which is blocked in (B) by inhibitors of the autonomic nervous system. β-adrenergic receptors were blocked with propranalol; cholinergic receptors were blocked with atropine.

Data points show means ± standard errors (sometimes too small to be visible) for 13 crocodiles.

Source: Seebacher F, Franklin CE (2007). Redistribution of blood within the body is important for thermoregulation in an ectothermic vertebrate (*Crocodylus porosus*). Journal of Comparative Physiology B 177: 841–848. Image of estuarine crocodile: Helmut Corneli / Alamy Stock Photo

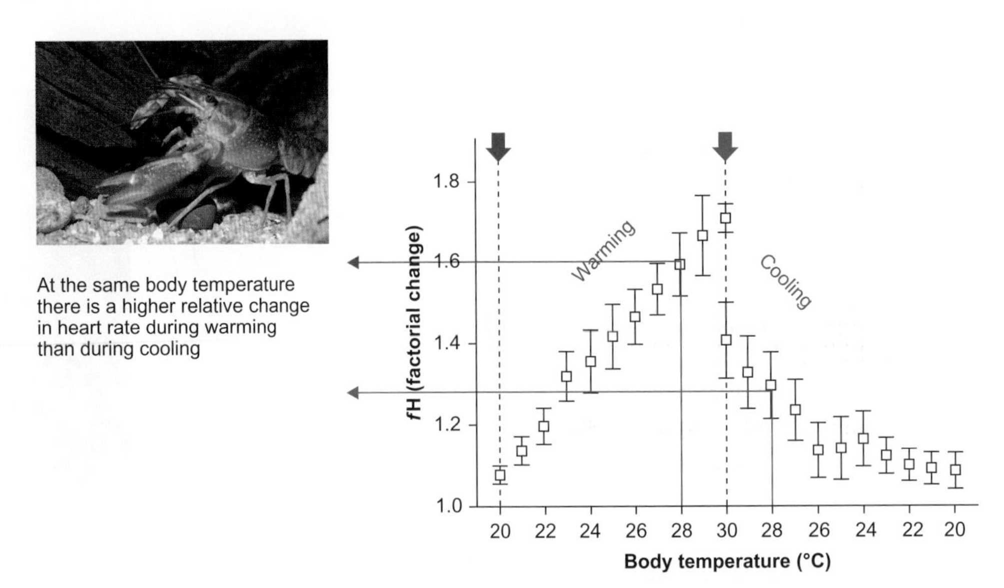

Figure 9.16 Relative changes in heart rates in an Australian freshwater crayfish, the common yabby (*Cherax destructor*) in response to changing body temperatures during warming and subsequent cooling

Heart rates were measured after warm water had been introduced to warm the crayfish (at red arrow) and during cooling (blue arrow) when warm water input was discontinued. The vertical dashed lines indicate the body temperature when hot water was introduced (left line) and removed (right line). The measured heart rates were divided by the resting heart rate (measured prior to the experiment) to calculate the relative change in heart rate (*f*H).

Data are means ± standard errors for eight crayfish.

Source: adapted from Goudkamp JE et al (2004). Physiological thermoregulation in a crustacean? Heart rate hysteresis in the freshwater crayfish *Cherax destructor*. Comparative Biochemistry and Physiology Part A 138: 399–403. Crayfish image: © Patrick Morris/ardea.com.

9

body, as it also occurs in at least one crustacean—an Australian freshwater crayfish, the common yabby (*Cherax destructor*). Figure 9.16 shows the thermal hysteresis in heart rates of these crayfish; when cool water is added, heart rates decrease almost immediately even though the change in body temperature is negligible.

In addition to the control of heart rates, estuarine crocodiles and turtles can control the distribution of blood between tissues during heating and cooling. Loggerhead sea turtles can control the blood flow to the flippers, which increases during warming and declines during cooling, thereby increasing warming rates and reducing cooling rates respectively.

Figure 9.17 illustrates the regulation of blood distribution in estuarine crocodiles, which was studied using microspheres (25 μm in diameter) injected into the vasculature to measure blood flows during warming and cooling. The rate of blood delivery to the tissues determines the relative capture of microspheres by each type of tissue or organ. From Figure 9.17, it is apparent that the crocodiles direct a greater proportion of blood to the dorsal skin (increasing heat uptake) and muscle during heating, while reducing blood flow to the tail and duodenum. In contrast, the duodenum receives a higher proportion of the blood during cooling while the dorsal skin receives less, which will reduce the rate of cooling.

Reptiles also cool by evaporative water loss from respiratory pathways

Respiratory cooling involves evaporative water loss from the upper airways, mouth (buccal cavity) and tongue during normal breathing, **panting**, or by gaping with an open mouth with the tongue hanging out.

One way of revealing the extent of respiratory cooling without disturbing an animal is to use thermal imaging to measure the temperature of various parts of the body. Figure 9.18 examines some thermal images and data from such images of rattlesnakes (*Crotalus durissus*). The temperature differences between lower nasal surface and head temperatures result from respiratory cooling and an ability to retain heat in the head when necessary. After feeding, respiratory cooling increases and offloads the heat from food digestion[29]. This increased cooling is evident from the increase in the difference between nasal and body temperatures shown in Figure 9.18A. Similar respiratory cooling occurs after activity as shown by the thermal image in Figure 9.18B(c) by comparison to the thermal image of a quiescent snake in Figure 9.18B(a).

[29] Section 2.3.4 discusses heat generation during digestion.

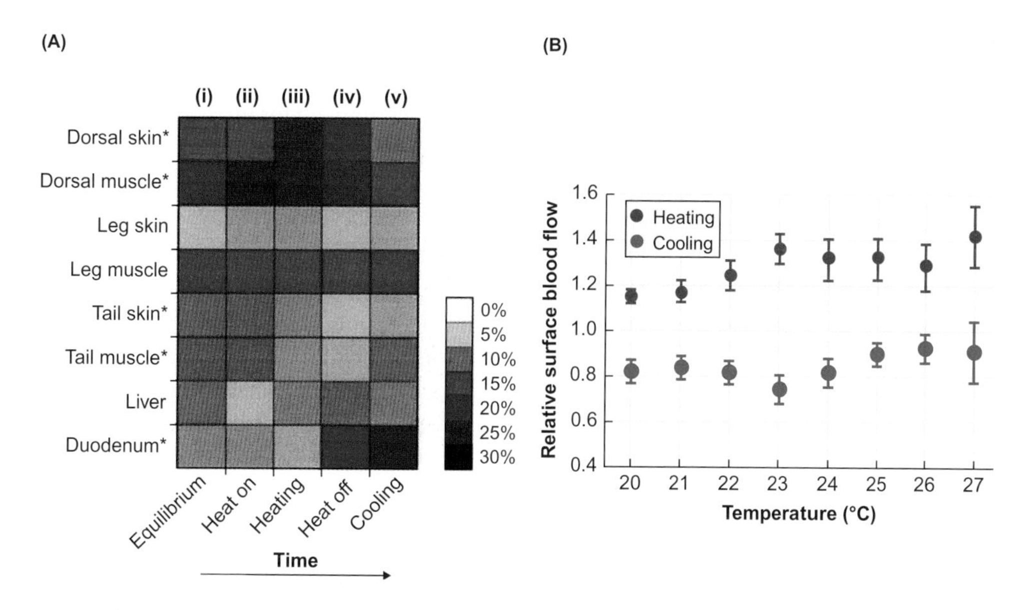

Figure 9.17 Experimental study of blood distribution in estuarine crocodiles (*Crocodylus porosus*)

(A) Colour grid shows percentage of cardiac output flowing to different regions of the body based on intensity of colour (according to the colour bar at the right of the figure). Data are shown for relative blood flow for the five consecutive phases of the experiments indicated beneath the grid: (i) Animals in thermal equilibrium with the environment (at body temperature of 20°C), (ii) 30–60 s after heat lamp switched on, (iii) mid-way through heating (at 24°C body temperature), (iv) 30–60 s after heat lamp turned off, (v) mid-way through cooling (body temperature at 24°C). During heating elevated blood flow to dorsal skin and muscle occurs accompanied by decreased blood flow to the duodenum and tail. * Tissues where blood flow differed significantly between thermal states. (B) Real-time measurements of surface blood flow measured using a Doppler surface flow probe attached to the dorsal surface of each crocodile. Results (shown as means ± standard errors) are expressed as a ratio of baseline measurements during thermal equilibrium. Ratios of above 1 indicate an increase in cutaneous blood flow during heating, while ratios below 1 indicate reduced cutaneous blood flow during cooling.

Source: Seebacher F, Franklin CE (2007). Redistribution of blood within the body is important for thermoregulation in an ectothermic vertebrate (*Crocodylus porosus*). Journal of Comparative Physiology B 177: 841–848.

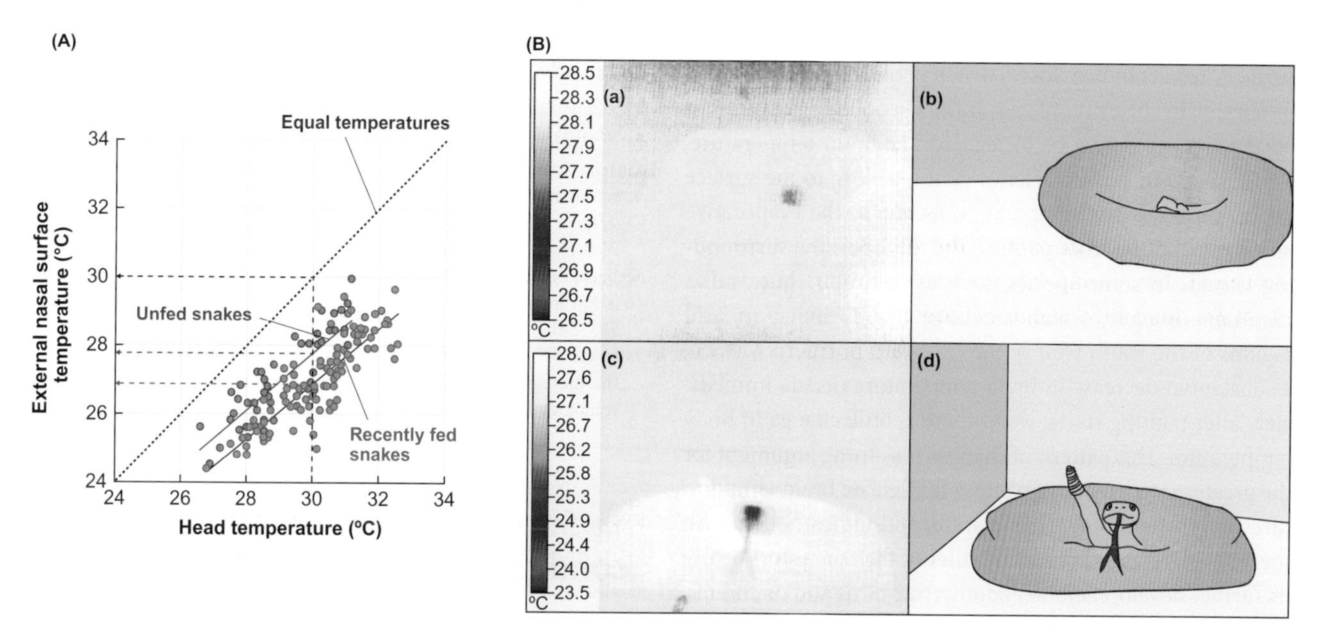

Figure 9.18 Effects of respiratory cooling by rattlesnakes (*Crotalus durissus*)

(A) Relationship between temperatures of head and external nasal temperature derived from thermal images of rattlesnakes in various conditions. Respiratory cooling routinely lowers nasal temperature compared to the head surface temperature of rattlesnakes by ~2°C to >3°C (reading off from the red and blue dashed line and arrows compared to black dashed line and arrow). The nasal temperatures decrease to a greater extent in recently-fed snakes (blue symbols) than unfed snakes (red symbols). (B) Evaporative cooling could protect the brain from overheating in reptiles. Colour coding for the two thermal images (a and c) is shown on the left of each thermal image. The outline drawings (b and d) show the two rattlesnakes are in contrasting states of activity. After a period of no breathing (a) and (b) head cooling is barely perceptible, and much less than during a high level of activity and tail rattling as shown in the thermal image (c) and illustrated in (d).

Source: Tattersall GJ et al (2006). Respiratory cooling and thermoregulatory coupling in reptiles. Respiratory Physiology and Neurobiology 154: 302–318.

Simultaneous measurement of core body temperatures and the temperature of the head of desert lizards has revealed that head temperatures during basking initially increase more rapidly than core body temperature. Hence, brain temperatures during warming can often be 2–4°C higher than body temperature. These differences appear to result from heat exchange between the internal jugular vein and the internal carotid artery, which lie close to one another and run in opposite directions. While basking, heat retention in the head occurs by countercurrent exchange[30] of heat between warm blood leaving the head, carried in the jugular vein, and cooler blood in the internal carotid en route to the head. As basking proceeds, the countercurrent exchange is disabled to allow heat to flow into the body while avoiding overheating of the brain. The uncoupling of heat exchange is associated with an increase in respiratory cooling.

Numerous reptiles, particularly those species living in arid environments, pant to reduce their body temperature once brain/body temperatures rise above a certain point: the panting threshold, which avoids heat stress. In panting, the breathing frequency increases and the breaths become shallow. Panting occurs via an open mouth often with a protruding tongue. Some species also move the floor of the mouth and throat during gaping and panting, which may increase evaporative cooling by increasing convective heat losses. Looking back at Figure 9.4, we find that coast horned lizards pant when body temperatures reach 41°C. The resulting evaporative water loss from the respiratory tract and mouth surfaces results in heat loss[31], which cools the animal.

The main priority of respiratory cooling (panting in particular) appears to be to regulate the brain temperature. Warm blood in carotid arteries running close to the surface of the buccal cavity (pharynx) cools due to the evaporative heat loss resulting from panting and this cools the surrounding tissues. In some species, such as common chuckwallas (*Sauromalus ater*), which are large lizards living in arid regions of the south west of the USA and northern Mexico, a substantial decrease in brain temperature occurs immediately after panting starts, when there is little change in body temperature. This pattern of changes is a strong argument for the greater importance of panting to regulate brain temperatures rather than body temperature, and indicates that such mechanisms evolved in early reptiles or their ancestors before its further development in endothermic birds and mammals.

Chuckwallas store fluid in their lymph sacs, which they use in respiratory cooling, keeping body temperature at about 1°C below ambient temperatures of up to 45°C; at the same time, however, they maintain brain temperature at 2.7°C below ambient for up to 8 hours. A simple experiment illustrates the importance of panting for cooling the brain in chuckwallas: taping their mouth shut to prevent panting removes the difference between brain temperatures and body temperatures.

The panting of heat-stressed reptiles appears to be controlled by neural systems that receive input from thermoreceptors in the skin, body organs and brain. In lizards, panting appears to be controlled in a proportional way relative to the deviation from a set-point temperature. As such, lizards spend more time panting and pant more continuously with a wider gape when ambient temperatures are high. However, there is a poor understanding of how the inputs from peripheral thermoreceptors in the skin and thermoreceptors in the central nervous systems are integrated to control panting. In some species, the brain thermoreceptors appear to be the most sensitive regulator of panting. For instance, the temperature threshold for activating panting in chuckwallas is lowest in brain tissue and highest in the skin, but high brain temperature alone does not stimulate panting, which only commences when body temperature exceeds 38°C.

Among desert lizards, further neural input conveys information about the state of hydration[32] and severe dehydration appears to increase the temperature threshold for increased panting. Panting for thermoregulation gains a higher priority when higher body temperatures necessitate it even if it has an adverse effect on water balance.

› *Review articles*

Clusella-Trullas S, van Wyk JH, Spotila JR (2007). Thermal melanism in ectotherms. Journal of Thermal Biology 32: 235–245.

Lagerspetz KYH, Vainio LA (2006). Thermal behaviour of crustaceans. Biological Reviews 81: 237–257.

Seebacher F, Franklin CE (2005). Physiological mechanisms of thermoregulation in reptiles: a review. Journal of Comparative Physiology B 175: 533–541.

Tattersall GJ, Cadena V, Skinner MC (2006). Respiratory cooling and thermoregulatory coupling in reptiles. Respiratory Physiology and Neurobiology 154: 302–318.

9.2 Surviving cold or subzero conditions

Ectotherms living in temperate areas (where low temperatures occur seasonally) may escape by migration to warmer climates before winter, returning the following year when temperatures increase. This behaviour is common among

[30] Section 3.3.1 outlines the principle of countercurrent exchange. We also discuss heat exchange for brain cooling in exercising mammals in Section 10.2.4.

[31] Evaporation requires heat input in changing water from a liquid (the condensed state of water) into a gas (water vapour), as we discuss in Section 3.3.1.

[32] Sections 21.1.3 and 21.1.4 discuss the osmoreceptors and volume receptors of vertebrates that convey information about their level of hydration.

insects, such as butterflies, that often use wind currents to assist the migration. Because of their short lifespan, the outward migration of insects usually involves adults, with the next generation undertaking the return trip.

Those ectotherms that spend the winter in the relatively cold conditions of polar or temperate zones, or at high altitudes, often enter a state of arrested development in which they overwinter. Among terrestrial arthropods living in seasonally variable habitats, a winter diapause is ubiquitous but the developmental stage that enters diapause varies between species. An alternative resting state—hibernation—occurs in some of the ectothermic terrestrial vertebrates (amphibian and reptiles).

9.2.1 Insect diapause

The characteristic feature of diapause is a reduction of metabolic rate (oxygen consumption) by about a factor of 10. While the rate of biochemical reactions declines, they do not cease entirely during diapause. In insect pupae, for example, the manufacture of RNAs and proteins continues, but at very low levels[33]. Hence, diapausing insects survive without feeding for several months by using their stores of lipids and carbohydrates as energy sources[34].

Early researchers thought conventional oxygen-dependent metabolism ceased in diapausing insects, but two lines of evidence question this idea. First, an accumulation of the products of **anaerobic metabolism**[35], such as lactate and pyruvate, does not occur during diapause. Second, pupae in diapause die after about 3 days of oxygen starvation, which suggests that the primary metabolic route is aerobic metabolism. However, respiratory poisons (such as cyanide and carbon monoxide) do not significantly inhibit the metabolism of many insect species when they are in diapause, which might seem surprising if aerobic metabolism does indeed continue. The likely explanation is that the effect of respiratory poisons is hard to detect when metabolic rates are very low. This idea is supported by the observation that some diapausing insects with relatively high metabolic rates during diapause, such as larvae of larch sawflies (*Pristiphora erichsonii*), are sensitive to cyanide poisoning.

Diapause is not simply a delay in development initiated by *current* conditions. Rather, insects monitor environmental cues and respond by preparing for diapause by storing lipids, proteins and carbohydrate, particularly in the fat body. These energy stores serve as fuels during the diapause and when development resumes at the end of diapause. Storage proteins synthesized by the fat body are also stored in the haemolymph of many insect species. During diapause itself, animals need to maintain the state of reduced metabolism, but continuously monitor environmental cues to allow them to reverse the diapause and emerge at the appropriate time.

Insects enter diapause at different developmental stages: some as eggs, pupae or larvae, but others (such as most diapausing beetles and some butterflies) as adults which enter a phase of **reproductive diapause**, in which the gonads remain undeveloped. One remarkable example of reproductive diapause occurs in adult monarch butterflies (*Danaus plexippus*). Case Study 9.2 examines their extensive migration to overwintering locations where they rest in diapause. An integral part of the survival of Monarch butterflies is the exceptional longevity of migrating adults, which live from August through to the following March.

Regulation of diapause

The major external cue that initiates preparations for diapause is the changing daylength or **photoperiod**. Typically, the shortening days in late summer result in a switch from the phase of active development, growth and reproduction into an induction phase for diapause. Insects usually seek out favourable microclimates in which to enter a winter diapause, typically seeking out small crevices, after tunnelling into rotting wood, or burrowing into soil, where they can avoid the lowest winter temperatures.

The photoperiod at which half the population of a species enters diapause is called the **critical photoperiod** and is specific for each insect species and its geographic locality (latitude in particular). The influence of locality suggests the existence of genetic adaptations of species that enable them to respond to the earlier winter at higher latitudes.

In one species—the pitcher-plant mosquito (*Wyeomyia smithii*)—altitude also influences the timing of diapause, which suggests that temperature, which declines with altitude, also influences the timing of diapause. Linked to their response to photoperiod and temperature, picture plant mosquitoes provide a rare example of **evolutionary adaptation**[36] to recent climate change. Rising temperatures over the recent past have lengthened the growing season for the picture plants, and the initiation of diapause by the mosquitoes living in the picture plants now takes place significantly later than 40 years ago, particularly in the more northerly populations where selection pressures are stronger and genetic variation greater.

Photoperiod cues initiate changes in the **hormonal systems** that initiate and maintain diapause. An important hormone is **diapause hormone**, which is a **neurohormone**

[33] Diapause is in some respects similar to the state of **torpor** in some endotherms, which we discuss in Section 10.2.7.

[34] Energy storage is discussed in detail in Section 2.3.5.

[35] Section 2.4 examines aerobic and anaerobic metabolism.

[36] We discuss further examples of thermal adaptations of ectotherms in Sections 9.2.3, 9.2.4, and 9.3.3.

Case Study 9.2 Spectacular migration of monarch butterflies

In summer, monarch butterflies (*Danaus plexippus*) are distributed throughout southern Canada and the northern part of the United States. These butterflies lay their eggs on milkweeds (species of *Asclepias*), which the caterpillars eat, but the milkweeds become unavailable in winter in northerly areas; this may partly explain the butterflies' southerly migration.

As temperatures start to fall at the end of summer, the butterflies become gregarious and start migrating south. Figure A shows the migratory routes of the two main populations of monarch butterflies. Those west of the Rocky Mountains migrate to a wintering site in southern California where they cluster in eucalyptus trees. The monarch butterflies to the east of the Rocky Mountains demonstrate a more spectacular migration of about 50 km each day, which takes them to the mountain tops of central Mexico—a distance of up to 4000 km. As many as 300 million individuals spend winter in Mexico, clustered in huge numbers in the branches and trunks of oyamel fir trees.

Monarch butterflies appear to be very sensitive to wind patterns and use winds when they can, or fly close to the ground if they encounter adverse winds. They may even temporarily halt the trip until more favourable conditions occur. They may also avoid weather problems by flying relatively high (up to 1200 metres), enabling them to benefit from favourable winds at higher altitudes, but must descend to feed on nectar to fuel the flight and to deposit a fat store in their abdomen, which is needed to survive the winter.

In their overwintering site, the butterflies congregate in large clusters in a **reproductive diapause**[1] until the spring, at which point they start to become more active and begin reproductive development so they are ready to mate as days lengthen and temperatures rise. The map in Figure B shows that, by the end of February, some of the monarch butterflies in Mexico have begun moving northward, but virtually none make it all the way north to their summer habitat. Instead, they must follow a two-step process. First, overwintering butterflies migrate to the southern states (Texas and Louisiana), where females arriving in March lay eggs that rapidly develop into new butterflies; it is these butterflies that complete the migration northward, arriving in the north-eastern US and Canada in mid-June, where they feed on the newly emerging milkweed.

A crucial factor in the migratory patterns of monarch butterflies is the longevity of the migrating butterflies than move south in autumn. While butterflies born in summer on average live less than two months, as shown in Figure C, the migrating butterflies survive from autumn through to the following March. What makes the difference between the summer and migrating butterflies? Analysis of the concentration of **juvenile hormone**[2] in the blood of these **seasonal morphs** has shown that monarch butterflies

9

(i)

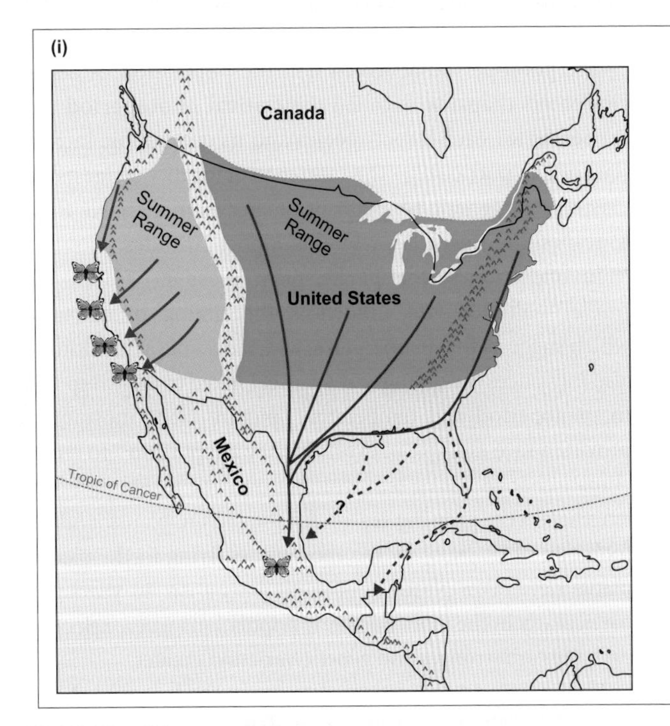

(ii)

Figure A Southerly migration of monarch butterflies

(i) Two migratory populations of the monarch butterfly occur in North America, distinguished here by red and purple lines. Summer ranges of populations are shown in the green bands across North America. The western population breeds west of the Rocky Mountains in spring and summer. In Autumn this population migrates to overwintering sites mainly along the California Coast. The much larger, eastern population breeds over several generations east of the Rocky Mountains and in the autumn migrates southwards to overwintering sites south of the Tropic of Cancer in central Mexico. (ii) Monarch butterflies resting on eucalyptus during their southern migration to California.

Sources: (i) Brower LP (1996). Monarch butterfly orientation: missing pieces of a magnificent puzzle. Journal of Experimental Biology 199: 93-103. (ii) © yhelfman/ Shutterstock..

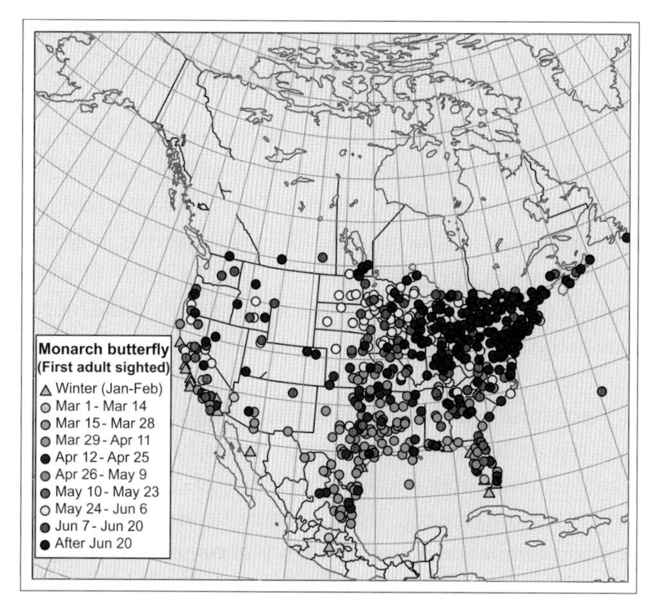

Figure B The journey north. First sightings of monarch butterflies between January/Feb and June 2014

Recorded sightings during southerly and northern migrations are available from https://www.learner.org/jnorth/monarch/News.html

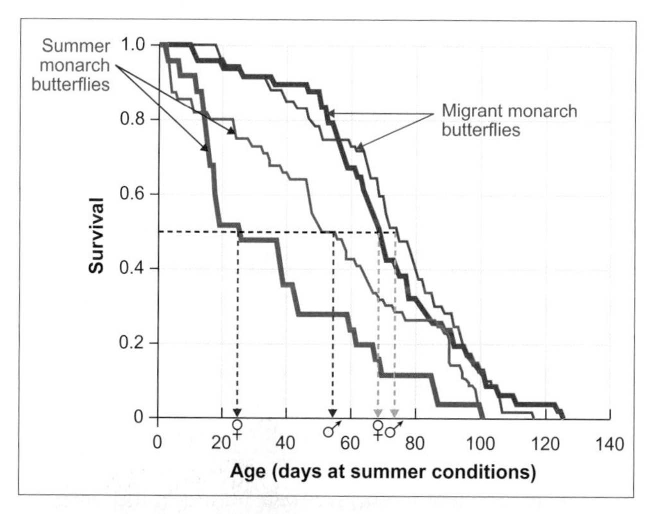

Figure C Survival of migrating monarch butterflies and butterflies collected in summer

The butterflies were held in identical (summer) conditions during the tests. Data for males are shown by thinner lines and data for females are shown by thicker lines. Migrants (red lines) survive longer than summer butterflies (blue lines) whether they are male or females. Reading off the graph at 0.5 (dashed lines) indicates the median life expectancy for summer butterflies is 25 days for females and 55 days for males. By comparison later in the year the migrant butterflies had a median life expectancy of about 69 days for females and 75 days for males.

Source: Herman WS, Tatar M (2001). Juvenile hormone regulation of longevity in the migratory monarch butterfly. Proceedings of the Royal Society London B 268: 2509–2514.

have 100-fold more juvenile hormone in their blood in summer than in autumn/winter.

Low concentrations of juvenile hormone are essential if the butterflies are to survive into spring, as is apparent from the data shown in Figure D. Removal of the glands that produce juvenile hormone (the corpora allata) enhances the longevity (survival)

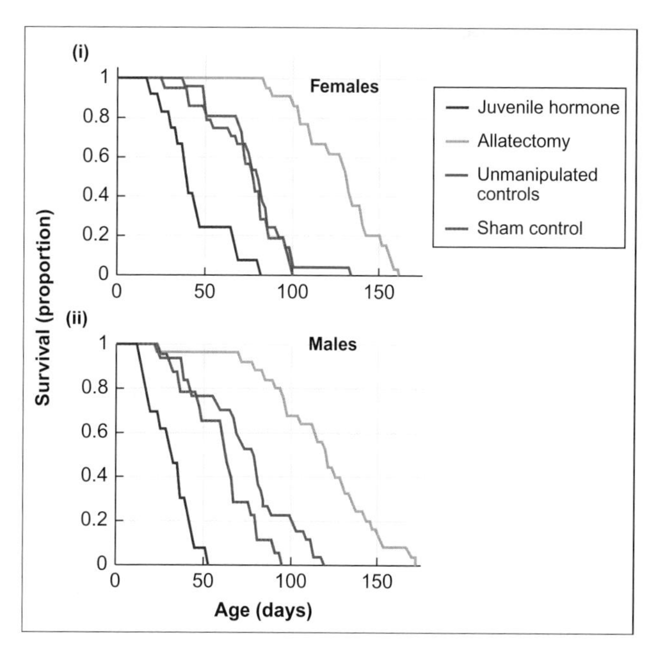

Figure D Effect of juvenile hormone on longevity (survival) of monarch butterflies held in summer conditions

External (topical) application of juvenile hormone reduced the survival of monarch butterflies compared to the controls, whereas removal of the corpora allata (allatectomy), where juvenile hormone is synthesised, almost doubled survival times compared to the controls. Control groups are unmanipulated butterflies, and sham operated butterflies in which all the steps involved in allatectomy were performed, except for removal of the corpora allata. Further controls (not shown) treated with acetone, because juvenile hormone was dissolved in acetone, were similar to unmanipulated controls.

Source: adapted from Herman WS, Tatar M (2001). Juvenile hormone regulation of longevity in the migratory monarch butterfly. Proceedings of the Royal Society London B 268: 2509–2514.

of the butterflies compared to controls with the glands left in place. On the other hand, treatment of butterflies with juvenile hormone reduces their longevity.

Genetic studies have sequenced the genome of monarch butterflies, enabling specific genes that control migration to be identified. These studies suggest that male and female butterflies reduce their levels of juvenile hormones in different ways during the diapause. It is thought that migrating males reduce the synthesis and release of juvenile hormone, while females appear to increase its turnover.

› Find out more

Brower LP (1996). Monarch butterfly orientation: missing pieces of a magnificent puzzle. Journal of Experimental Biology 199: 93–103.

Herman WS and Tatar M (2001). Juvenile hormone regulation of longevity in the migratory monarch butterfly. Proceedings of the Royal Society London B 268: 2509–2514.

[1] Diapause is a state of rest with a low metabolic rate in which the butterflies exhibit a low body temperature. In reproductive diapause there is no gonadal development.

[2] Section 19.5.2 discusses juvenile hormone and its production by the corpora allata of insects.

secreted by the sub-oesophageal ganglion[37]. Diapause hormone often acts as a stimulant of diapause in insect eggs.

One well-studied species is the domestic silkmoth (*Bombyx mori*), although initiation of diapause in this species is unusual as it is initiated so early in the life of female silkmoths that long summer days and high temperatures program her embryos to secrete diapause hormone during their late pupal stage, when their eggs are maturing in their ovaries. These eggs subsequently enter diapause.

Diapause hormone stimulates trehalase activity in the developing ovaries, which generates high levels of glycogen in the mature eggs. The conversion of glycogen to polyols, such as glycerol and sorbitol, then induces the eggs to enter diapause. By contrast, a decrease in sorbitol levels may act as a signal to end diapause. However, the eggs must also experience a period of cold exposure before their diapause ends.

Diapause hormone does not always stimulate the diapause of insects. For example, tiny amounts of diapause hormone (10^{-12} mol L^{-1}) are sufficient to terminate the pupal diapause of corn earworms (*Helicoverpa zea*). As corn earworms are an agricultural pest, desynchronization of their diapause by chemical agents could offer a way to control them by forcing them to be active (terminating their diapause) when climatic conditions are unfavourable. However, diapausing pupae have already gone underground so it would be difficult to administer such agents. Compounds applied to larvae to prevent their entry into the pupal diapause offers a better prospect of success.

Adult diapause in insects is usually (but not always) induced by the *reduced* secretion of **juvenile hormones**, by a pair of neurosecretory glands (the **corpora allata**) in the neck of most insects. The concentration of juvenile hormone in monarch butterflies must remain low to sustain the unusual longevity of adult butterflies and their overwinter survival during diapause, as the data in Case Study 9.2 demonstrate.

9.2.2 Winter hibernation of amphibians and reptiles

Many amphibians and reptiles living in temperate habitats survive through the cold winter months in a state of lethargy—a hibernation—in which they cease foraging and become quiescent in a **hibernaculum**. The lethargic state of overwintering ectotherms is not like the hibernation of endothermic mammals, which regulate body temperature at a lower set point[38]. Hibernating amphibians and reptiles cease their daily patterns of behavioural thermoregulation and allow their body temperatures to follow those of the environment.

Many species of frogs and turtles overwinter under ice in cold but unfrozen water where they remain submerged for several months. Painted turtles (*Chrysemys picta*), which are widespread in slow-moving freshwater habitats in North America, often bury themselves in almost a metre of sediment at the bottom of ponds or lakes, although some hibernate in the burrows of mammals and under vegetation in the bank side.

Temperature is the most important trigger of amphibian and reptile hibernation. Painted turtles hibernate when water temperatures decrease to below 15–18°C. The signal for them to emerge from hibernation also seems to be ambient temperature, rather than photoperiod, so if temperature increases the hibernating animal may awaken. The final emergence occurs when water temperature rises in spring and any ice has gone. As such, the period of hibernation is determined geographically: northern populations of painted turtles hibernate for 5 to 6 months continuously, while further south, where ambient temperatures are higher, some populations do not hibernate at all. The geographic range of frogs extends further north than that of painted turtles, so their hibernation periods can be longer.

Survival through hibernation

The key factors in surviving through the winter hibernation are: (i) an availability of sufficient energy stores for the period of starvation, and (ii) oxygen supply.

Frogs and turtles are air-breathers but while hibernating underwater can make use of oxygen in the water. Frogs have a well-developed capacity to take up oxygen from water across their skin[39]. The lipid stores of amphibians are usually more than sufficient to survive their period of hibernation. In European common frogs (*Rana temporaria*) for example, which hibernate in Finland for 7–8 months, 80 per cent of the energy used during hibernation comes from lipids. However, hibernating frogs use glycogen stores for anaerobic metabolism if conditions become anoxic (lacking oxygen). Hibernating frogs do not survive long in such conditions, probably because of the build up of lactate, which disturbs acid–base balance[40].

Freshwater turtles have a poorer ability than frogs for gas exchange by non-pulmonary routes, but their oxygen requirements decrease during hibernation, so the oxygen

[37] Figure 19.37 illustrates the endocrine and neural systems of insects.

[38] We discuss hibernation (and torpor) of endotherms in Section 10.2.7.

[39] We discuss gas exchange via the skin in water in Section 12.2.1.

[40] The generation of lactate during anaerobic metabolism is outlined in Section 2.4.1. We discuss the effects of metabolic substrates on acid–base balance in Section 13.3.

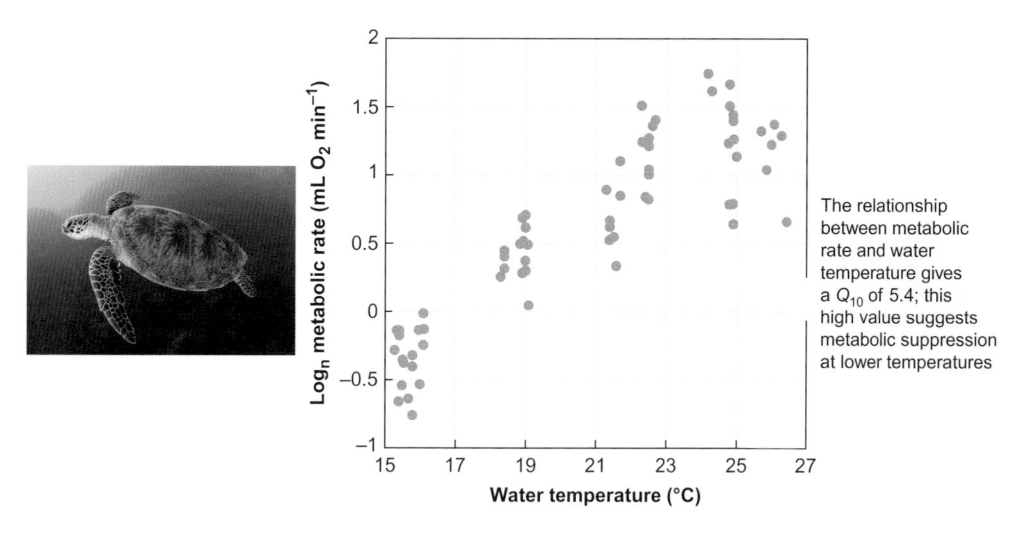

Figure 9.19 Effects of water temperature on metabolic rate of loggerhead sea turtles (*Caretta caretta*)

Source: Hochscheid S, Bentivegna F, Speakman JR (2004). Long-term cold acclimation leads to high Q_{10} effects on oxygen consumption of loggerhead sea turtles *Caretta caretta*. Physiological and Biochemical Zoology 77: 209–222. Image of sea turtle: © Turtle Island Restoration Network.

supply is sufficient for aerobic metabolism in several species. Part of the reason for a lower oxygen need during hibernation is that body temperature is much lower than when these ectotherms are active and are able to thermoregulate behaviourally by basking in the sun to raise their body temperature above ambient. An additional factor is the **metabolic suppression** which occurs during hibernation, and which reduces the metabolic rate beyond the effects of temperature alone.

In the active temperature range for ectotherms, the metabolic rate typically decreases 2-3-fold (Q_{10} effect)[41] for a decrease in temperature of 10°C. However, for both turtles and frogs, Q_{10} values are much higher during hibernation—for example, up to 8.5 in turtles. Some loggerhead sea turtles appear to rest/hibernate in a state of quiescence at the bottom of shallow seas when water temperatures decline. The linear relationship between the metabolic rate of loggerhead sea turtles and water temperatures, which Figure 9.19 shows, gives a Q_{10} of approximately 5.4 indicating metabolic suppression occurs at low water temperatures.

We do not fully understand the initiation of or regulation of the metabolic suppression at a cellular level during the hibernation of amphibians and reptiles; functionally, however, it is apparent that metabolic suppression will reduce the rate at which stored substrates are used. In one study, painted turtles used 70 per cent of the stored glycogen and almost 40 per cent of the stored lipids during hibernation, as well as 11 per cent of their protein. The fact that stores remained when the turtles emerged from hibernation, however, indicates that their energy supplies were sufficient. The glycogen stores in the liver of painted turtle are sufficient to support anaerobic metabolism at 3°C for up to 6 months, which exceeds the likely period of anoxia in the wild. By contrast, other turtles have a poorer capacity to survive in anoxic conditions. For example, pond sliders (*Trachemys scripta*) use 80 per cent of their glycogen within less than 6 weeks at 3°C.

9.2.3 Freeze-tolerant ectotherms

Many invertebrates, such as intertidal marine species (for example, barnacles and mussels), and the many overwintering cold-hardy species of insects—including butterflies and moths, bees and wasps, cockroaches, true flies, grasshoppers and beetles—that occur in temperate and polar regions can tolerate the freezing of their body fluids. Once frozen some freeze-tolerant species can cool to temperatures as low as −50°C to −80°C and still function normally after thawing.

Freeze tolerance is much rarer among vertebrate species than invertebrates, although some reptiles (turtles) and at least six species of frogs are freeze-tolerant. Of these amphibians the wood frog (*Rana sylvatica*) is the most studied. Wood frogs are common in forest habitats between the Appalachian Mountains and north to Alaska. Wood frogs often burrow into the forest floor for a period of hibernation, but only to a relatively shallow depth that is insufficient to avoid subzero conditions. At the southern margin of their range, wood frogs experience temperatures of −2°C to −4°C for a few days at a time, whereas subzero conditions occur at their northern limits for many weeks, and wood frogs in Alaska survive freezing temperatures to as low as −16°C to −18°C. The frozen wood frogs fully recover when thawed.

[41] Q_{10} and its calculation are explained in Box 8.2 and Section 8.1.

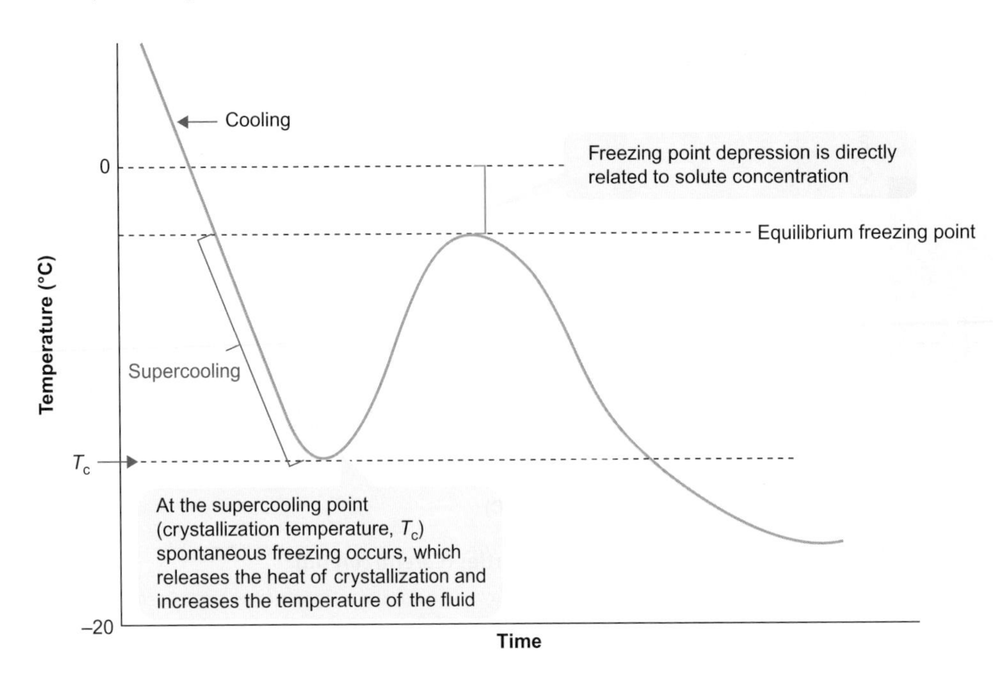

Figure 9.20 Cooling curve for the body fluids of an ectotherm

Cooling initially reduces the temperature of the fluid to below its freezing point when the fluid is supercooled, without freezing. At the supercooling point, spontaneous freezing releases heat which raises temperature; this process continues until all the water is frozen. Temperature settles at the equilibrium freezing point, which is below 0°C, before declining again if cooling continues.

Body fluids, like any aqueous solutions, may cool by several degrees below zero before freezing becomes inevitable at the **supercooling point** (or, more correctly, the crystallization temperature). The supercooling point of the body fluids of most freeze-tolerant species is about −5 to −10°C. Figure 9.20 illustrates the supercooling point on a typical cooling curve for a solution. Freezing at the supercooling point releases the heat of crystallization, which causes an increase in temperature. When all the water is frozen, the temperature settles at the equilibrium **freezing point**, which for pure water would be at 0°C, but the presence of solutes in any solution depresses its freezing point to below 0°C to an extent that is directly related to the concentration of solutes in the solution. Once frozen, a gradual increase in temperature leads to melting at the melting point; melting point and freezing point are normally identical in freeze-tolerant animals and most typical solutions.

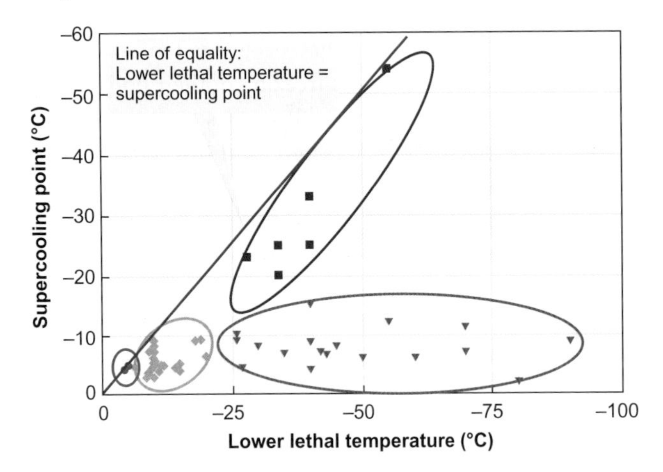

Figure 9.21 Relationship between the supercooling point (°C) and the lower lethal temperature for 53 freeze-tolerant species of insects

The insects are separated into four groupings: (1) partially freeze-tolerant (blue circles), (2) moderately freeze-tolerant (orange diamonds), (3) strongly freeze-tolerant (brown triangles) and (4) freeze-tolerant with a very low supercooling point (purple squares).

Source: adapted from Sinclair BJ (1999). Insect cold tolerance: how many kinds of frozen? European Journal of Entomology 96: 157–164.

Freeze tolerance is well developed in the Alaskan adult ground beetles (*Pterostichus brevicornis*), which survive for over a month in rotting wood with body temperatures as low as −70°C, while the diapausing larvae of a drosophilid fly (*Chymomyza costata*) survive to −80°C. However, not all species are equally freeze-tolerant, as shown in Figure 9.21. In this graph, four sub-groups of insects are identified based on their lower lethal temperature (the temperature at which they soon die when temperatures are gradually reduced) and the temperature at which their body fluids freeze (the supercooling point).

Partially freeze-tolerant insects (blue symbols in Figure 9.21) such as the larvae of crane flies (*Tipula* species) tolerate some ice formation, but have relatively high lethal temperatures and do not survive if ice formation continues to the equilibrium freezing point. These insects may represent an intermediate position between true freeze tolerance and freeze avoidance.

Moderately freeze-tolerant insect species (orange symbols in Figure 9.21) have relatively high supercooling points, but once frozen survive for reasonably long periods. They die if body temperatures decline to about 10°C below their supercooling point, but their freeze tolerance is usually sufficient to survive the minimal environmental temperatures experienced.

Strongly freeze-tolerant insect species (brown symbols in Figure 9.21) begin to freeze at relatively high supercooling points and survive to very low temperatures. They only die when temperatures are usually 20°C to 70°C below their supercooling point.

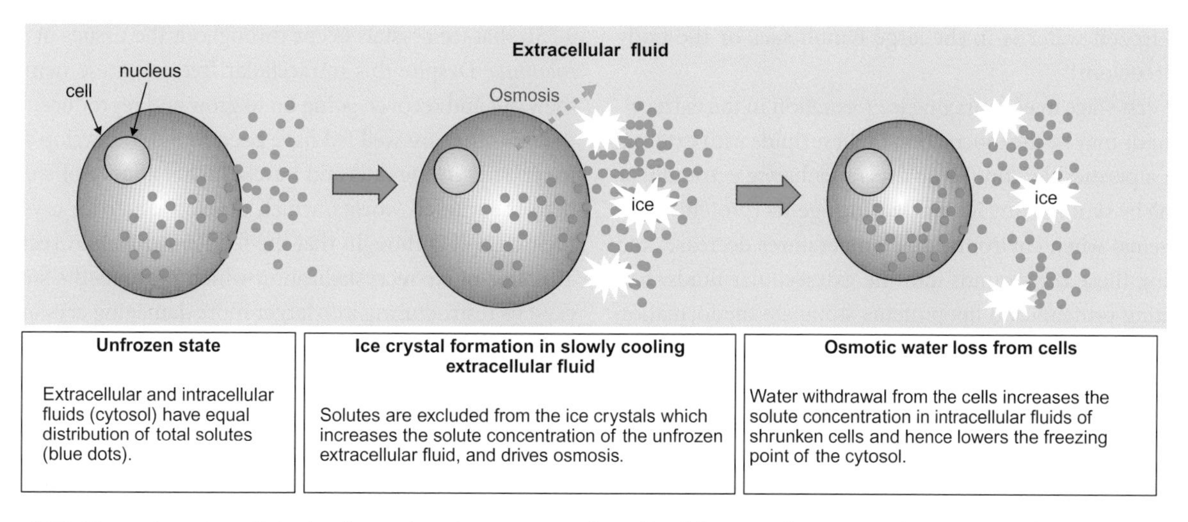

Figure 9.22 Stages in extracellular ice formation that protects the cells of freeze-tolerant ectothermic animals

Low supercooling points occur in some freeze-tolerant insect species (purple symbols in Figure 9.21) which may protect them from freezing, although their freeze tolerance is less striking than in many insects. In reality, it seems there is a continuum between various levels of freeze tolerance and freeze intolerance (freeze avoidance).

Freeze-tolerant species usually promote extracellular ice formation

The formation of ice crystals inside cells when environmental temperatures fall can be catastrophic because expanding ice ruptures cell membranes. Damage to the plasma membrane can cause leakage of potassium from the cells, which disrupts the normal ionic differences between intracellular and extracellular fluids[42]. Ectotherms living at subzero temperatures need to avoid these devastating events. Therefore, an important characteristic of freeze-tolerant species is that freezing generally occurs exclusively in the extracellular fluids.

Formation of ice in the extracellular fluids reduces the volume of remaining fluid, as illustrated in Figure 9.22, which increases the solute concentration of this fluid. The resulting difference in the osmotic concentration of extracellular and the intracellular fluids draws water from the cells by osmosis. As temperatures decline the process continues and the cells slowly dehydrate as more of the body water becomes frozen in extracellular fluids. Some water always remains unfrozen and bound to cellular proteins and other macromolecules in the cells. In this way freeze-tolerant animals may survive freezing of up to 65 per cent of their total body water, as illustrated by wood frogs in Figure 9.23, which remain in this frozen state for several weeks. In wood frogs, about 25 per cent

[42] Section 4.1.1 discusses the composition of intracellular and extracellular fluids.

Frozen wood frog

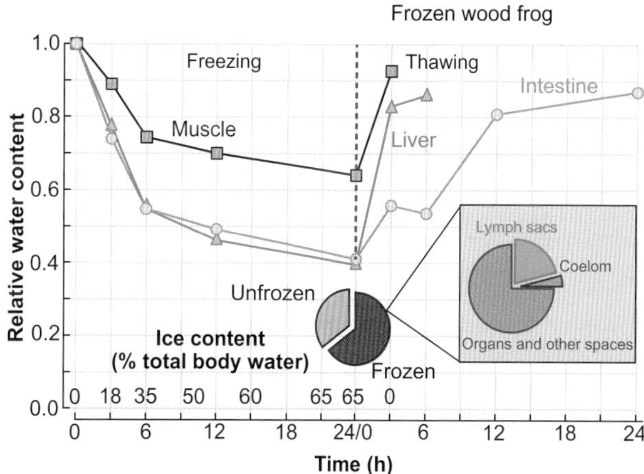

Figure 9.23 Water content of gastrocnemius muscle, liver and intestine of wood frogs (*Rana sylvatica*) during freezing and subsequent thawing

Temperature was reduced to −2.5°C in a 24 h freezing episode and then raised to 5°C, which resulted in thawing. Relative water content of tissues decreases during freezing while the ice content (as a % of body water) increases to up to 65% of total body water. During freezing, more than half of the water in some organs relocates to extracellular fluid spaces; about 25% of the ice is in the lymph sacs and the body cavity (coelom) as indicated in the boxed pie chart. Tissues rehydrate within several hours of thawing.

Sources: Lee RE Jr., Costanzo JP (1998). Biological ice nucleation and ice distribution in cold hardy ectothermic animals. Annual Reviews of Physiology 60: 55–72. Image of frozen wood frog: © Janet M Storey.

of the frozen water is in the large lymph sacs or the body cavity (coelom).

The first stage in encouraging ice formation in the extracellular fluids may be the inoculation of these fluids with external ice via a permeable skin. Some species enhance extracellular freezing by synthesizing **ice-nucleating agents** (proteins or lipoproteins) when environmental temperatures decrease, and secreting these compounds into the extracellular fluids. Ice-nucleating proteins and lipoproteins stimulate the formation of ice nuclei that act as foci for the addition of water molecules and the slow growth of ice crystals in the extracellular fluids.

In addition to specialized proteins, some insects appear to use inorganic molecules, such as calcium or potassium phosphate, as ice nucleators. Figure 9.24 shows some of the spherules of calcium phosphate that seem to encourage freezing of the excretory fluids in the Malpighian tubules[43] of larvae of goldenrod gallflies (*Eurosta solidaginis*). These larvae reach –40°C during their winter diapause in galls on stems of goldenrod in North America.

Intracellular freezing is usually fatal. However, at least one invertebrate, a nematode worm (*Panagrolaimus* sp.) that lives in the Antarctic, tolerates freezing of up to 82 per cent of its body water including intracellular freezing and survives at temperatures as low as –80°C. Its transparent appearance means that ice formation can be seen under a specialized cryomicroscope. Figure 9.25A shows this process. In nematodes, the cuticle acts as a barrier to ice nucleation, but ice propagates through body orifices, especially the excretory pore, and maybe the mouth and anus. At temperatures of –1°C to –4°C the ice in *Panagrolaimus* is confined to the extracellular compartment; at –5°C or lower, however, ice propagates through the whole body within 0.2 seconds. (In most freeze-tolerant animals freezing takes many hours or even days.)

A neat way of looking at the location of freezing is by freeze fracture, which examines cells once split open. Notice in Figure 9.25B that ice crystals occur throughout the tissues of *Panagrolaimus*. Despite this intracellular freezing these nematodes thaw out and recover, going on to grow and reproduce. *Panagrolaimus* that are well fed have been found to develop smaller ice crystals in the cells and have a higher chance of surviving than the starved worms, which develop larger ice crystals in their cells. It is thought that the fed *Panagrolaimus* secrete an inhibitor of ice recrystallization which prevents the small ice crystals restructuring into larger more damaging crystals.

Synthesis of cryoprotectants protects tissues from damage

Toleration of freezing relies partly on the addition of **cryoprotectants** to extracellular and intracellular fluids, which act in two main ways:

- Cryoprotectants in extracellular fluids reduce the freezing point of the fluid, which slows the rate of ice formation.
- The addition of cryoprotectants to intracellular fluids limits cell shrinkage as less water loss occurs before reaching an osmotic equilibrium.

Among invertebrates, several different cellular solutes act as cryoprotectants. Some insects use amino acids such as proline, others employ polyols such as sorbitol, and some accumulate glycerol[44]. Oligochaete worms employ glucose as a cryoprotectant, while nematode worms use another carbohydrate, trehalose[45], as a cryoprotectant.

The synthesis of cryoprotectants requires the activation of cryoprotectant-synthesizing enzymes and the activation of enzymes that degrade these substances when the

[43] Section 7.3.4 examines the function of Malpighian tubules in excretion and ionoregulation by insects.

[44] Glycerol at extremely high concentrations may also depress the supercooling point of some species and confer resistance to freezing, which we examine in Section 9.2.4.

[45] Section 2.3.3 discusses the use of the carbohydrate trehalose (shown in Box 2.1) in insect metabolism.

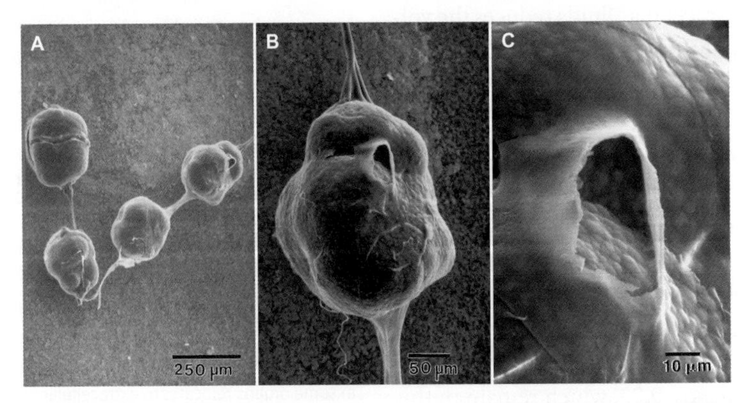

Figure 9.24 Scanning electron micrographs of spherules of calcium phosphate in the Malpighian tubules of larvae of goldenrod gall flies (*Eurosta solidaginis*) while in diapause.

(A) String of evenly-spaced spherules (each larva contains 25–45 spherules), which expand the diameter of the narrow Malpighian tubule. (B) Higher magnification of the spherule on the far right of A. (C) Further magnification shows the conglomerate appearance of numerous round particles.

Source: Mugnano JA et al (1996). Fat body cells and calcium phosphate spherules induce ice nucleation in the freeze tolerant larvae of the gall fly *Eurosta solidaginis* (Diptera, Tephritidae). Journal of Experimental Biology 199: 465–471.

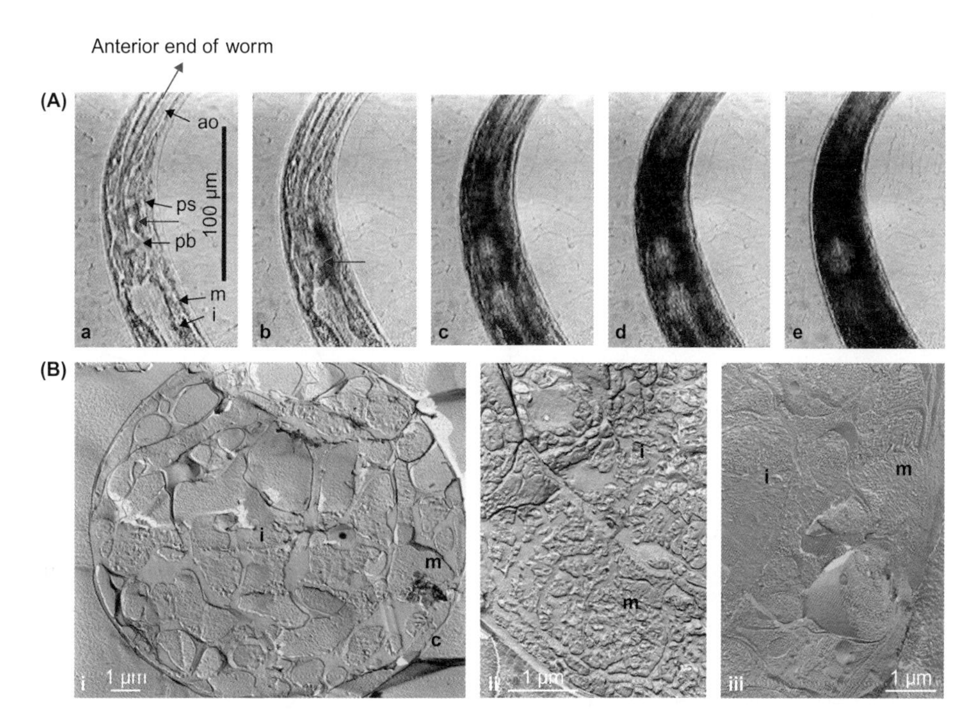

Figure 9.25 Extracellular and intracellular freezing in the freeze-tolerant Antarctic nematode *Panagrolaimus*

(A) Sequential images (a–e, at same magnification) show darkening of the body during freezing. Red arrows indicate the start of freezing in the space between the oesophagus and the outer body wall, close to the excretory pore. Freezing spreads through the extracellular fluids in the pseudocoel (ps) and then propagates into the intestine (i), anterior oesophagus (ao), muscle (m) and posterior bulb of the oesophagus (pb), within 0.2 seconds of the beginning of freezing. (B) Freeze-fracture replicas showing ice crystals in all parts of the body of frozen *Panagrolaimus*: (i) cross-section at a relatively low magnification; the higher magnification image in (ii) shows many small ice crystals; image (iii) is from a control, non-frozen specimen which shows the normal appearance after freeze fracture of muscle. c, cuticle; i, intestine; m, muscle cells.

Source: Wharton DA, Ferns DJ (1995). Survival of intracellular freezing by the Antarctic nematode *Panagrolaimus davidi*. Journal of Experimental Biology 198: 1381–1387.

temperature increases in spring. One of the species that has been well studied is the goldenrod gall fly (which we also examined in Figure 9.24 in relation to ice nucleation) whose larvae spend about 11 months of the year living inside the galls on stems of goldenrod. Figure 9.26A shows the increasing concentrations of glycerol and sorbitol in the larvae in autumn, which reaches a peak in winter. Figure 9.26B examines the increasing activity of key enzymes for synthesizing sorbitol. The graphs also show the increase in activity of enzymes that degrade the cryoprotectants in spring, which reduces the concentrations of cryoprotectants. Freeze tolerance is lost after the subsequent metamorphosis to pupae.

Most freeze-tolerant amphibians only start to produce cryoprotectants once ice formation starts. In wood frogs, glucose mobilized from glycogen stores in the liver is the main cryoprotectant. An increase in the **glucose transporters** in plasma membranes results in glucose accumulation in the cells of all tissues. Glucose accumulation in cells has additional importance: once a frog is partially frozen, the distribution of oxygen to the tissues is limited (if it happens at all). Metabolism consequently becomes exclusively anaerobic, fuelled by the glucose distributed to the tissues prior to freezing.

Cell protection and cell processes

Survival of the cells in freeze-tolerant animals is likely to rely on a multitude of processes that allow them to cope with cell shrinkage, increasing ionic concentrations, and the shortage of oxygen. The heart tissue of frozen wood frogs shows an increased expression of the gene encoding hypoxia-inducible factor 1, which is a protein involved in tolerating low oxygen availability (hypoxia)[46].

When freezing starts, expression of the gene for fibrinogen increases. Fibrinogen is an important factor in blood clotting and may be important in increasing clotting capacity to deal with any bleeding injuries caused by ice crystal damage. In addition to such specific responses, multiple changes in gene transcription, translation and post-translational modification of gene products[47] are highly likely and are thought to be important for halting

[46] Section 22.2.3 discusses hypoxia-inducible factor-1 (HIF-1) in more detail.

[47] Section 1.3.1 gives an overview of the flow of information from DNA (genes) to cell proteins.

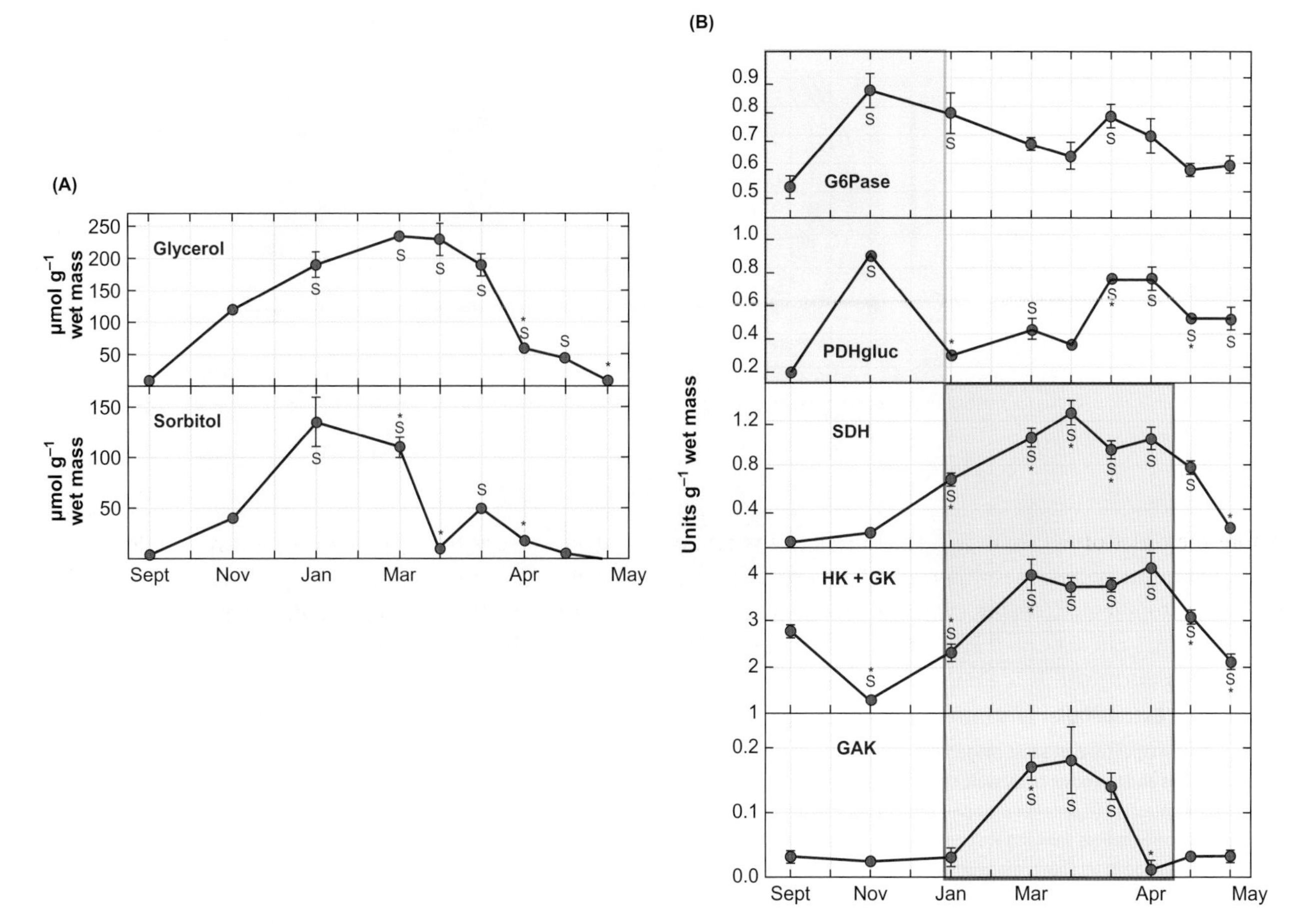

Figure 9.26 Seasonal changes in the concentrations of cryoprotectants (glycerol and sorbitol) and the enzymes involved in their production and degradation in the freeze tolerant larvae of the goldenrod gall fly (*Eurosta solidaginis*)

(A) Concentrations of glycerol and sorbitol rise to a peak in winter and decline in spring, at the end of diapause. (B) Sorbitol synthesis relies on increasing glucose-6-phosphatase (G6Pase) and polyol-dehydrogenase using glucose as the substrate (PDHgluc) (green overlay). Declining sorbitol in spring is due to an increase in sorbitol dehydrogenase (SDH) (orange overlay), and hexokinase (HK); the assay of HK included glucose kinase (GK). Increasing activity of glyceraldehyde kinase (GAK) in spring degrades glyceraldehyde that is produced from glycerol. Enzyme assays were at 25°C.

Values significantly different ($P<0.05$) from September values are indicated by s; and from previous values are indicated by asterisks.

Source: Storey KB (1997). Organic solutes in freezing tolerance. Comparative Biochemistry Physiology A 117A: 319–326.

the translation of unnecessary proteins during freeze tolerance and for reducing ATP use.

Recent studies suggest that such responses are likely to involve a newly discovered mechanism for rapid and reversible post-transcriptional regulation which involves multiple non-coding RNAs, including microRNAs (miRNAs). Study of miRNAs is a relatively new and expanding area of research, but already several miRNAs have been found to be dynamically regulated in freeze-tolerant wood frogs, intertidal snails (*Littorina littorea*) and cold-hardy insects. The responsiveness of particular miRNAs appears to be a common feature of both freeze tolerance and metabolic suppression, which supports the idea that miRNAs are important in prioritizing the use of ATP.

9.2.4 Freeze avoidance

Many cold-hardy terrestrial arthropods, some turtles, and many polar fish, avoid the formation of ice inside their bodies despite external water temperatures dropping below zero. Freeze avoidance involves one or both of two mechanisms:

- During supercooling, the body fluids of an animal remain unfrozen, as illustrated earlier by Figure 9.20. Supercooling is an unstable condition not controlled by an animal. If water temperature declines further body temperature follows the decline, and freezing becomes increasingly likely as body fluids approach the temperature of crystallization (or supercooling point).

- Antifreeze compounds **are synthesized**, which depress the freezing point of body fluids.

Supercooling of freeze-avoidant ectotherms

Many terrestrial invertebrates that avoid freezing during winter show a seasonal increase in their capacity to supercool. Figure 9.27 shows that the supercooling capacity of animals is size-related. Many insect eggs, and small arthropods (such as springtails, mites and aphids) weighing <100 μg (<10^{-4} g), can supercool to −25°C or below before reaching their temperature of crystallization. In contrast, larger species generally supercool to −5°C to −15°C, but the explanation of this size relationship is not certain.

A cessation of feeding and clearing of the alimentary tract before the onset of winter is thought to reduce the risk of freezing in supercooling ectotherms. This is probably because the stopping of feeding avoids the intake of bacteria and fungi that can act as ice nucleators to stimulate ice formation. Support for this idea has come from experimental studies of lady beetles (*Hippodamia convergens*) fed with ice-nucleating bacteria; the intake of the bacteria raises the crystallization temperature at which ice forms in the bodies of the beetles from −16°C to −3.5 °C. Also, in springtails, the lethal temperature for animals with full guts is −8°C but is reduced to −28°C in animals that have evacuated their gut contents.

Among vertebrates, the hatchlings of painted turtles have the highest capacity to supercool. Female painted turtles dig nests to a depth of about 10 cm in which they deposit clutches of eggs. The eggs hatch during late summer but the hatchlings stay in the nest until the following spring. In the nest the hatchlings can be exposed to temperatures at several degrees below zero in some locations, and possibly as low as −12°C for several weeks.

The temperature of crystallization (supercooling point) of hatchling painted turtles is typically −13°C to −17°C and can reach as low as −20°C. Hence, hatchlings can usually avoid freezing by supercooling. However, the characteristics of the soil in the nest chamber affects the risk of freezing. If the females lay eggs in nests made of soil with a high water content there is a higher risk of ice propagation across the skin of the hatchlings, and poorer survival rates (below 50 per cent) than in nests in loamy sand with a low water content (where 90 per cent survive). In such situations, the turtle hatchlings have been found to become freeze tolerant. Excavation of natural nests containing frozen turtles has shown that these turtles may survive after thawing, but we have a poor understanding of their freeze-tolerance mechanisms.

Marine teleost fish living toward the poles show only a relatively low degree of supercooling, by a few °C, compared to the much greater supercooling of hatchling painted turtles and many overwintering insects, yet the supercooling of these fish may be sufficient to avoid freezing, which Per F. Scholander first demonstrated in the 1950s. Scholander observed their survival in seawater at −1.5°C; when he put an ice cube in the water, thereby exposing the fish to ice crystals, they immediately froze, even though the water in which they were swimming was still liquid. Does this mean that these fish routinely face a risk of ice formation in their tissues? Their behaviour is the solution to this conundrum. Sea ice is largely a surface phenomenon, so the concentration of ice crystals to act as nucleating agents and initiate freezing of the fish diminishes with depth. Consequently, some fish species can avoid freezing when coastal temperatures drop in winter by moving into deeper waters and supercooling. For example, in northern waters, Atlantic cod (*Gadus morhua*) move offshore to a depth of 100–400 metres where they supercool.

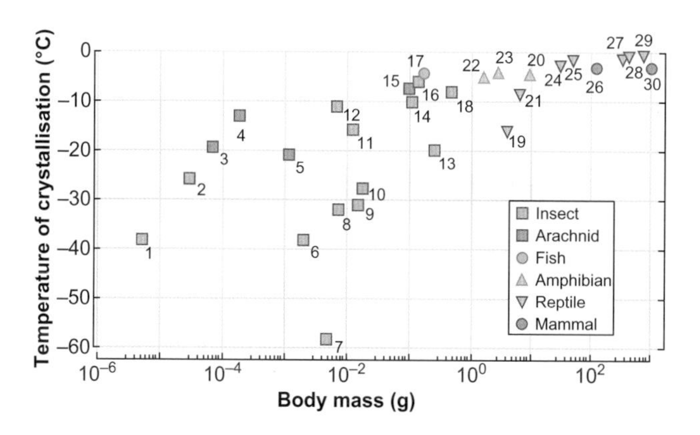

Figure 9.27 Effect of body size on supercooling ability

At their temperature of crystallization (=supercooling point) body fluids begin to freeze. Individuals with a smaller body mass cool to much lower temperatures before freezing, although there is a large amount of variability in the supercooling point of animals weighing less than a gram. Examples 1 and 2 are eggs of insect species; 5, 6, 7 and 9 are arthropod larvae, 17 are juvenile teleost fish, 19 are hatchlings of turtles; all other examples are adult animals.

Source: Lee RE Jr., Costanzo JP (1998). Biological ice nucleation and ice distribution in cold hardy ectothermic animals. Annual Reviews of Physiology 60: 55–72.

Freeze avoidance using antifreezes

Toward the South Pole, in the Southern Ocean (Antarctic Ocean), the temperature of the sea is often at its freezing point of −1.8 to −1.9°C[48]. Marine teleosts could theoretically live in a permanently supercooled state beneath the surface sea ice and platelet ice, but would need to avoid contact with ice crystals in the water to avoid freezing. However, many species of Antarctic fish live in surface waters where it would

[48] Seasonal variations in water temperature or with depth are limited because of the Antarctic circumpolar current and Antarctic polar front illustrated in Figure 9.30B.

be difficult to avoid freezing because of the ice crystals in the water column. So how do they survive? The answer lies in their evolution of antifreezes that enable them to avoid freezing. Animals living in freezing conditions may possess two types of antifreeze:

- Colligative[49] antifreezes lower the freezing point of body fluids due to the increase in the total concentration of solutes. Examples of colligative antifreezes are small solutes such as glucose and trehalose, glycerol, mannitol and sorbitol.
- Non-colligative antifreeze compounds have specialized chemical properties that suppress the growth of ice crystals in a solution, and thereby lower the freezing point relative to the melting point, which results in a **thermal hysteresis**.

Colligative antifreezes

The first observation that insects produce glycerol and sorbitol when they experience cold conditions and enter a diapause was by the Japanese researcher Harou Chino while working on eggs of silkmoths in 1957. Reflecting on his discovery about 30 years later, he explained that his aim was to try to explain why the glycogen content of the eggs declined during diapause. He stated that *'one day in 1956, I unintentionally tasted the concentrated aqueous extract from the diapausing eggs and, to my surprise, I found that it was very sweet!'* As a result of this 'unintentional' discovery, we now know that many overwintering insects accumulate colligative antifreezes to such an extent that they make up 15–25 per cent of their body mass.

Like insects in diapause, a few species of fish synthesize colligative antifreezes in response to a seasonal decline in temperature. The most studied example is the rainbow smelt (*Osmerus mordax*), which lives along the Atlantic coast of Canada. Rainbow smelt migrate into fresh water to spawn in spring, returning to seawater to spend the autumn and winter. When temperatures decline, the rainbow smelt begin to synthesize glycerol. Notice in Figure 9.28 the increasing glycerol concentration in blood plasma of rainbow smelt in the winter months. This colligative antifreeze enables them to spend winter under the ice near the shore.

One problem with relying on small solutes to depress the freezing point of body fluids is that large quantities are necessary to provide effective protection, but glycerol is one of a small number of solutes that can increase to unusually high levels without damaging cell function. In rainbow smelt, glycerol concentrations in the plasma can reach above 400 mmol L^{-1} in winter compared to concentrations below 1 mmol L^{-1} in spring to late summer. Large quantities of glycerol, along with the accumulation of smaller amounts of urea and trimethylamine oxide in the plasma and other tissues, reduce the freezing point of the extracellular fluids of rainbow smelt to about −1.9°C in winter.

Glycerol is a small molecule ($C_3H_8O_3$: molecular mass about 92 Da) so it is lost across body surfaces, such as the gills and skin, and lost in the urine[50]. Rainbow smelt

[49] All colligative properties (freezing point, melting point and osmolality) of solutions are determined by the total concentration of solutes present, not what they are, as we discuss in Section 4.1.1. Based on colligative properties alone freezing point equals melting point.

[50] We discuss the filtration of small solutes by the kidney in Section 7.1.

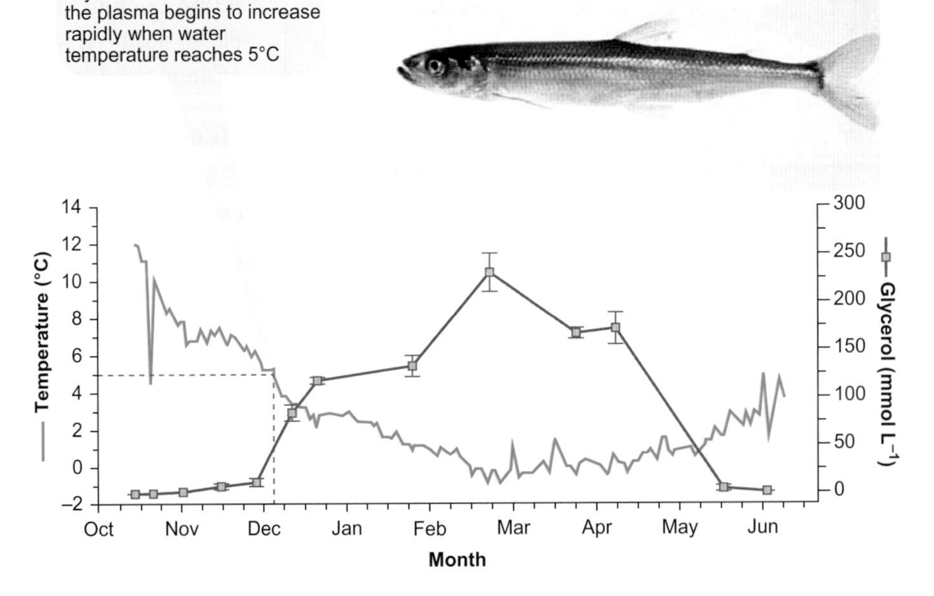

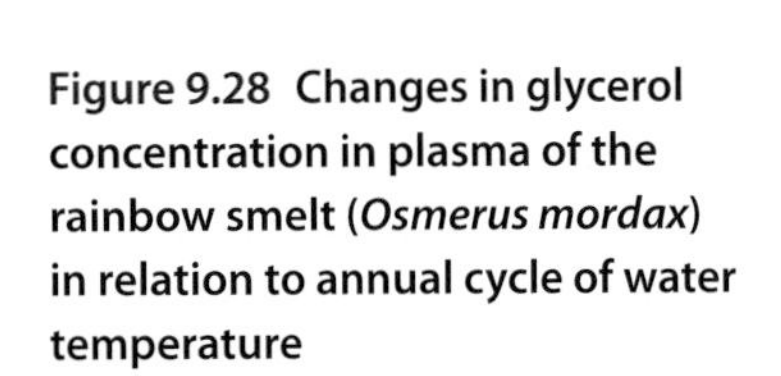

Figure 9.28 Changes in glycerol concentration in plasma of the rainbow smelt (*Osmerus mordax*) in relation to annual cycle of water temperature

Source: Driedzic WR, Ewart KV (2004). Control of glycerol production by rainbow smelt (*Osmerus mordax*) to provide freeze resistance and allow foraging at low winter temperatures. Comparative Biochemistry and Physiology Part B 139: 347–357. Image of rainbow smelt: http://fisheries.noaa.gov.

therefore need to produce glycerol continuously throughout winter to replace losses. The continual production of glycerol in winter could be a significant drain on resources. However, the presence of glycerol enables rainbow smelt to exploit large populations of protein-rich invertebrates living under the ice. The fact that these prey items provide ample protein, and hence amino acids that are used in glycerol production, provides a neat circularity in the story linking biochemical adaptation and acclimatization to the local environment.

How do rainbow smelt synthesize glycerol?

The high activity of aspartate amino transferase and alanine amino transferase seen in the liver of glycerol-producing fish, compared to the levels in the liver of species that do not generate glycerol, suggests glycerol synthesis from amino acids. Liver activity of a third enzyme, phosphoenolpyruvate carboxykinase (PEPCK), converts amino acids to glyceraldehyde-3-phosphate, which can then enter the pathways for glycerol synthesis shown in Figure 9.29. The second route in Figure 9.29 is thought to dominate because:

- There is strikingly high activity of glycerol 3-phosphate dehydrogenase (GPDH) in the liver of rainbow smelt in winter as indicated in Figure 9.29.
- Reversal of the winter events can be induced in laboratory experiments in which an *increase* in seawater temperature in winter *reduces* GPDH activity and results in a *decrease* in the concentration of blood glycerol.

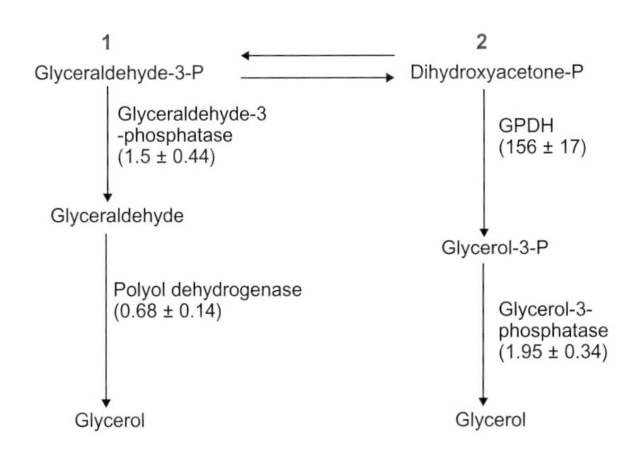

Figure 9.29 Pathways for glycerol production by hepatocytes of rainbow smelt (*Osmerus mordax*) in winter

1 Conversion from glyceraldehyde-3-phosphate (-3-P) catalysed by glyceraldehyde-3-phosphatase and polyol dehydrogenase. **2** Conversion from dihydroxyacetone phosphate (DHA-P), catalysed by glycerol-3-phosphate dehydrogenase (GPDH) and glycerol-3-phosphatase. Activities of these enzymes vary seasonally with water temperature. Numbers next to the enzymes indicate maximal enzyme activities measured *in vitro* (μmol min^{-1} g^{-1} wet tissue) ± standard deviations.

Source: adapted from Driedzic WR, Ewart KV (2004). Control of glycerol production by rainbow smelt (*Osmerus mordax*) to provide freeze resistance and allow foraging at low winter temperatures. Comparative Biochemistry and Physiology Part B 139: 347-357.

Most of the species of marine teleost fish living in the Arctic or the Southern Ocean[51] cannot synthesize significant amounts of glycerol. They have slightly higher concentrations of sodium chloride in their extracellular fluids than fish from milder temperate and tropical regions, but only at concentrations that have a minor colligative effect on the freezing point of their extracellular fluids (about −0.22°C). Increasing the concentration of other small solutes can depress the freezing point of the cytosol by an additional −0.08°C, to about −1.0°C. Nonetheless, this freezing point is well above the freezing point of the seawater in the Southern Ocean (−1.8 to −1.9°C). Additional protective measures are essential to avoid freezing in the icy waters, where temperatures are often at the freezing point of seawater. The most significant aspect of the freeze avoidance of polar teleosts, other than supercooling, is the incorporation into their body fluids of non-colligative antifreezes that inhibit the growth of ice crystals.

Non-colligative antifreezes

The defining characteristic of animals that synthesize proteins or glycoproteins as non-colligative antifreezes is the thermal hysteresis between the freezing point and melting point of their plasma/haemolymph (essentially a depression of freezing point relative to melting point). We explore this hysteresis further in Experimental Panel 9.1. The hysteresis results from adsorption of molecules of antifreeze to tiny ice crystals, which inhibits their growth, and is known as the adsorption–inhibition effect.

Many arthropods (mainly beetles) and several species of spiders, mites and centipedes combine supercooling as the means of avoiding freezing in winter with the synthesis of non-colligative antifreezes. However, our greatest understanding of non-colligative antifreezes is for marine teleost fish living in polar and sub-polar regions. In the late 1960s, Arthur de Vries identified the first antifreezes in the blood of Antarctic fish; this historic moment was vital in driving the many subsequent studies. We now know that many types of antifreeze proteins and glycoproteins occur in at least 11 different families of teleosts.

Table 9.1 lists the major categories of peptide and protein antifreezes and their occurrence in different fish groups and some arthropods. The amazing diversity of chemical forms of antifreezes and their distribution among animal taxa indicates independent **convergent evolution**[52] of antifreezes. Most of the northern species relying on antifreezes in winter

[51] Figure 1.2 gives an overview of the major oceans and marine coastal ecoregions of the Earth.

[52] Convergent evolution involves the evolution of similar features by different pathways in different groups of animal, as we discuss in Section1.5.3.

Experimental panel 9.1 The relative effectiveness of non-colligative antifreezes in animals living in subzero habitats

Background

Non-colligative antifreezes by definition reduce the freezing point of a solution but do not affect its melting point, which results in a *thermal hysteresis*. Hence, a thermal hysteresis of body fluid freezing points compared to melting points is a diagnostic characteristic for the presence of molecules of non-colligative antifreeze. In polar marine fishes, thermal hysteresis values reach 2°C, which result in freezing points below −2.7°C. This thermal hysteresis is sufficient to protect them from freezing in seawater that freezes at −1.86°C. By contrast, many terrestrial insect and spider species cope with much lower winter temperatures; those with non-colligative antifreezes typically exhibit 2–5°C thermal hysteresis and many reach 8–9°C thermal hysteresis in winter, despite levels of antifreeze compounds in their body fluids 10 times or more below those in fish. This raises the question: are insect/spider antifreeze peptides (AFP) more effective than those of most polar fish? Experiment 1 explores this question.

Although the AFPs (and antifreeze glycopeptides, AFGPs) produced by different animal species are chemically and structurally diverse, an overall concept of an adsorption-inhibition mechanism has arisen. This concept hypothesizes that once molecules of antifreeze adsorb on an ice crystal, growth of the crystal can only occur slowly between the molecules of bound proteins. It was further proposed that the effectiveness of insect AFPs results from their binding to all the faces of the hexagonal structure of ice crystals; experiment 2 examined this proposition.

Methods

Experiment 1 explored the effectiveness of AFPs by directly examining nanolitre samples under the binocular microscope in a specialized temperature-controlled base, which held the fluid sample to be tested in paraffin oil, to avoid evaporation. The sample was first frozen, and the temperature increased slowly to melt the ice until one tiny crystal was just visible at the melting point of the sample. On re-cooling, the ice crystal began to grow immediately if no antifreeze peptides were present. If, on the other hand, AFP was present, the ice crystal did not grow until the temperature reached the lower hysteretic freezing point[1]. The growth of the ice crystals was followed using video recording in the presence of realistic amounts of the purified AFP from two species that synthesize AFP seasonally: winter flounders (*Pseudopleuronectes americanus*) and larvae of eastern spruce budworms (*Choristoneura fumiferana*), which is one of the most destructive native pests of spruce and fir in the eastern United States and Canada.

Experiment 2 examined the binding of AFP to ice crystals by pre-labelling the AFP of spruce budworm with a green fluorescent protein and examining ice crystals by fluorescent microscopy.

Results

Figure A shows drawings from video microscopy recordings of the growth of ice crystals in extracellular fluid, and demonstrates that even when spruce budworm AFP levels are 20-fold less than those of winter flounder, the depression of freezing point is four-fold more. When ice growth does occur the ice crystals look very different in the two species. In the winter flounder, growth of ice occurs along one axis to form a spicule. In contrast, AFP of spruce budworm results in ice crystals that grow in multiple directions, as shown by the image in Figure A.

The results of experiment 2 showed AFP of spruce budworm adsorbs to all planes of the ice crystal, resulting in fluorescence of all planes, as shown in Figure B.

Discussion and implications

The multiple directions of slow growth of ice in spruce budworm are explained by structural studies of the AFP. Figure C shows the grid of threonine residues that occur on one side of the molecule in two lines. This positioning is due to the repeated occurrence of the T-X-T motif (where X is any amino acid and T is threonine) in each loop. The T-X-T motif has also been identified in the AFPs of mealworms (*Tenebrio molitor*) and a related species, *Den-*

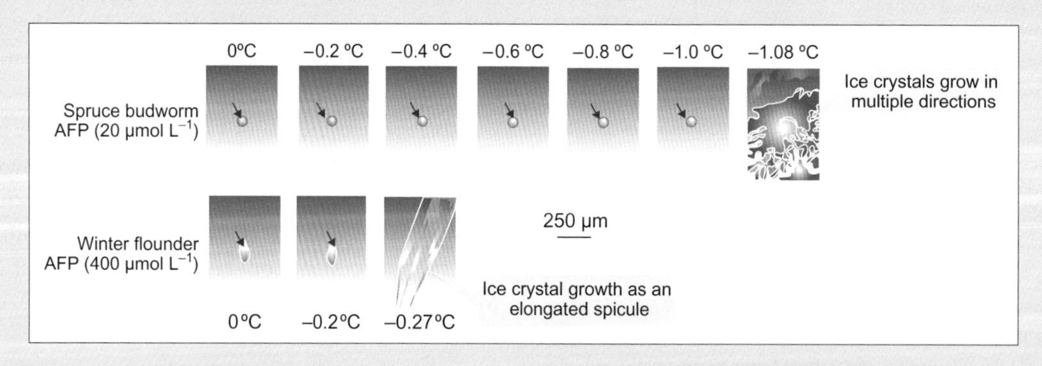

Figure A Drawings from video microscopy comparing the effectiveness of antifreeze peptides (AFP) from the eastern spruce budworm and the winter flounder

Drawings show an ice crystal (red arrows) at −0.2°C intervals of cooling in the presence of 20 μmol L^{-1} spruce budworm AFP and 400 μmol L^{-1} winter flounder AFP. There was no observable growth of ice crystals until −0.27°C for winter flounders and −1.08°C for spruce budworm.

Source: adapted from Graether SP et al (2000). β-Helix structure and ice-binding properties of a hyperactive antifreeze protein from an insect. Nature 406: 325-326.

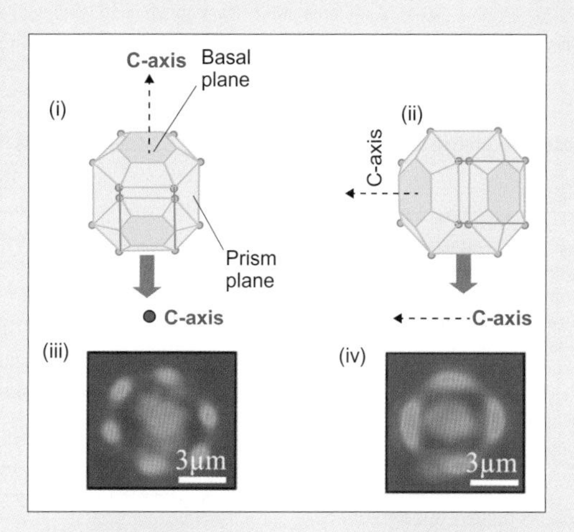

Figure B Binding of AFP of spruce budworm to ice crystals

Ice crystals are hexagonal structures, as shown in (i) and (ii), in which water molecules form layers of hexagonal plates. The top and bottom face (or plane) of each hexagonal layer are the basal planes. The drawing (i) shows a single hexagonal crystal of ice standing on its basal plane, while in drawing (ii) the ice crystal is on its side with basal planes at the left and right edges. Ice growth in the absence of AFP is primarily along the C-axis by stacking of the hexagonal layers. Images (iii) and (iv) show fluorescence images corresponding to the positions of the crystals in (i) and (ii). The fluorescence is the result of adsorption of fluorescently labelled AFP. Adsorption of ADP occurs on all planes of the ice crystals.

Source: Pertaya N et al (2008). Direct visualization of spruce budworm antifreeze protein interacting with ice crystals: basal plane affinity confers hyperactivity. Biophysical Journal 95: 333–341.

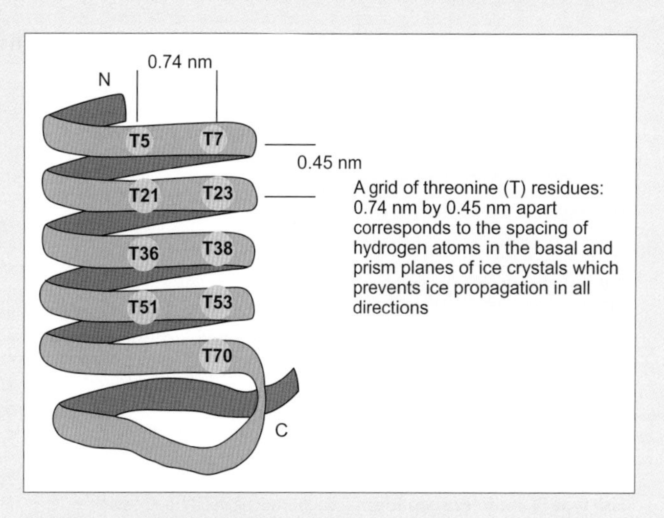

Figure C Schematic representation of antifreeze peptide (AFP) of eastern spruce budworms

The AFP of spruce budworms comprises 90 amino acids arranged in a helix. The locations of the ice-binding threonine residues are indicated by the numbered locations: T5, T7, etc.

Source: adapted from Graether SP et al (2000). β-Helix structure and ice-binding properties of a hyperactive antifreeze protein from an insect. Nature 406: 325-326.

droides canadensis. Hence, the T-X-T motif is probably a common ice-binding element of insect AFPs. Together, these motifs are believed to present a threonine-rich flat face that binds to ice crystals. The exposed OH^- (hydroxyl) groups on threonine residues are prime candidates for forming hydrogen bonds with ice crystals.

› Find out more

Graether SP, Kuiper MJ, Gagné SM, Walker VK, Jia Z, Sykes BD, Davies PL (2000). β-Helix structure and ice-binding properties of a hyperactive antifreeze protein from an insect. Nature 406: 325–328.

Pertaya N, Marshall CB, Celik Y, Davies, PL, Braslavsky I (2008). Direct visualization of spruce budworm antifreeze protein interacting with ice crystals: basal plane affinity confers hyperactivity. Biophysical Journal 95: 333–341.

[1] The difference between the freezing point and melting point gives a value for the thermal hysteresis due to the antifreezes.

to avoid freezing synthesize peptide antifreezes, but some, such as the Arctic cods *Boreogadus saida* synthesize antifreeze glycoproteins (AFGP).

The structure of AFGPs is almost identical among icefish living in the Antarctic Ocean and the Arctic cods, even though these groups of fish diverged over 40 million years ago. This suggests that independent evolution, on at least two occasions, gave rise to the genes encoding the helical antifreeze proteins that are made up of a highly variable number of repeats of the basic tripeptide sequence Alanine-Alanine/Proline-Threonine (Ala-Ala/Pro-Thr) with a disaccharide group attached to each threonine along one side of the helical molecule.

The analysis of gene sequences for antifreeze glycopeptides in Antarctic notothenioids provide strong evidence for the evolution of an AFGP gene in an ancestor of the icefish that led to the adaptive radiation, illustrated in Figure 9.30A. This adaptive radiation resulted in five families of notothenioids dominating the Antarctic fish fauna. These fish are thought to have evolved from the common ancestor after the separation of South America from Antarctica, and the subsequent establishment of a massive circular current around Antarctica—the Antarctic Circumpolar Current (ACC)—about 25–22 million years ago. Figure 9.30B illustrates the circular flow running clockwise around the entire continent in a current of 200 km to more than 1000 km width and extending to the ocean floor.

The establishment of the ACC resulted in thermal isolation of Antarctica, glaciation and freezing of the coastal waters, which is thought to have generated present-day conditions about 10–15 million years ago. The ACC would have been a significant driver for the evolution of the marine

Table 9.1 Classification of antifreeze proteins and peptides in fish and insects

The peptides type I to IV are traditionally numbered in order of their discovery.

Protein type	Species examples or fish group	Protein sequence and structure	Molecular mass (kDa)
Fish antifreeze glycoprotein	Arctic cod (*Boreogadus saida*) Antarctic notothenioids (icefish)	Helical protein Polymer—repeats of Ala-Ala/Pro-Thr with disaccharide attached to each Thr Arctic cod—occasional substitution of Thr by Arg	2.6 to 34
Fish type I antifreeze peptide	Flat fish: Winter flounders (*Pseudopleuronectes americanus*) Plaice and flounders (e.g. *Pleuronectes* sp.)	Small α-helix 11-residue repeats– Thr $X_{(10)}$	3 to 5
Fish type II antifreeze peptide	Sea ravens (*Hemitripterus* spp.) Atlantic herring (*Clupea harengus*) Rainbow smelt (*Osmerus mordax*)	Folded protein Non-repetitive; cysteine-rich	14 to 24
Fish type III antifreeze peptide	Ocean pout (*Zoarces americanus*)	Globular protein, one flat surface	7
Fish type IV antifreeze peptide	Longhorn sculpin (*Myxocephalus octodecimspinosis*)	Helical protein	12.3
Insect—beetle antifreeze peptide	Fire-coloured beetle larvae (*Dendroides canadensis*)	β-helix 12- or 13-residue repeats Thr and Cys rich	9
Insect—moth antifreeze peptide	Eastern spruce budworm (*Choristoneura fumiferana*)	Cys, Thr, Ser-rich Non-repetitive	9

Sources: Cheng C-H C (1998). Evolution of the diverse antifreeze proteins. Current opinions in genetics and development 8: 715–720; Cheng C-H C (2009). Freezing avoidance in polar fishes. In: Encyclopaedia of Life Support Systems (EOLSS). Eds. C Gerday and Glansdorff N. Extremophiles, Volume II pp 215–232.

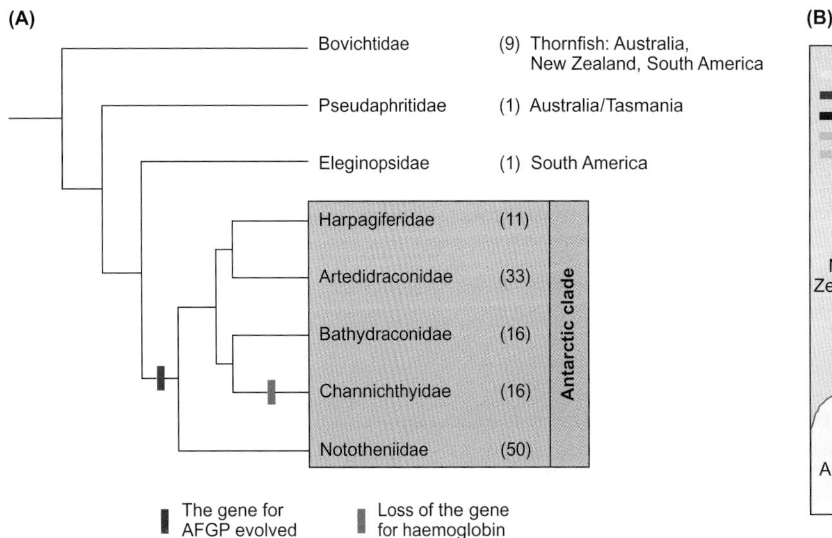

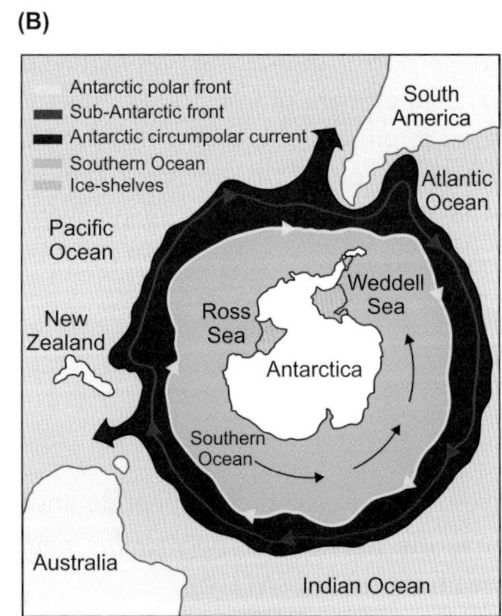

Figure 9.30 Evolution of antifreeze glycoproteins (AFGPs) in the suborder Notothenioidei (notothenioids)

(A) Phylogenetic relationships among notothenioids. The number of species in each family is shown in brackets. An Antarctic clade of five families (overlaid by the blue box) accounts for 126 of the total of 137 notothenioid species and represents ~46% of Antarctic fish species; in shallow bays and shelves, notothenioids account for 77% of species and 90% of fish biomass. The AFGPs evolved in an ancestral notothenioid, indicated by the red bar. Hence almost all fish in the Antarctic clade have AFGPs. Just a small number of non-Antarctic species that are members of the Nototheniidae family and that are not exposed to freezing conditions, have lost the AFGP gene or reduced the number of genes and AFGP synthesis (e.g. black cods of New Zealand and Patagonian toothfish (*Dissostichus eleginoides*) in South America). The top three families in the phylogenetic tree occur in non-freezing conditions in the coastal waters of Tasmania/Australia, New Zealand or South America, as indicated. After the evolution of antifreezes, channichthyid icefish lost the gene for haemoglobin, as indicated by the blue bar. (B) The Southern Ocean and its oceanographic current system surrounding Antarctica. The development of the Antarctic Circumpolar Current (ACC) was a significant driver in the evolution and specialisation of Antarctic fish. The ACC is the largest current on Earth in terms of length and transport capacity. It flows in a clockwise direction around Antarctica and creates the Antarctic Polar Front of the Southern Ocean, which impedes migration between Antarctic and sub-Antarctic regions.

Source: Beers JM, Jayasundara N (2015). Antarctic notothenioid fish: what are the future consequences of 'losses' and 'gains' acquired during long-term evolution at cold and stable temperatures? Journal of Experimental Biology 218: 1834–1845. The phylogenetic relationships are based on the results for mitochondrial DNA (Near & Cheng, 2008).

fauna of the Southern Ocean. The extinction of fish species that lacked antifreezes would have created an opportunity for species with antifreezes to colonize the freezing ice-laden habitats. In the absence of competition, these species could diversify over an evolutionary time span and now dominate the fish fauna of the Southern Ocean.

The liver of northern fish species synthesizes their antifreeze peptides, so the presumption for many years was that the liver was the main organ responsible for the synthesis of AFGPs. However, this idea was questioned since large amounts of liver total RNA were needed to demonstrate any expression. Sequencing projects have more recently been able to encompass tens of thousands of expression sequence tags (ESTs) and yet have found no evidence of the expression of AFGPs by the liver of Antarctic notothenioids. We now know that the AFGPs are secreted by the pancreatic tissue (which is diffuse in teleosts) and the anterior part of the stomach. The AFGPs secreted by the pancreas enter the intestine, via the pancreatic duct, and inhibit the growth of any ice crystals ingested with seawater. Intact AFGPs may also be reabsorbed from the intestine and enter the circulation.

Sequence analysis of the AFGP gene of notothenioid fish indicates that it arose by repeat duplication of a non-coding region of a functionally unrelated gene for a pancreatic trypsinogen-like protease. Of course, the repetitive Ala-Ala-Thr sequence in AFGP is quite short, and nucleotide sequences that code for this sequence probably occur elsewhere in the fish genome. So how do we know that this small segment of the trypsinogen gene gave rise to the same sequence in the AFGP gene? The answer is that the ends of the repeated sections of the AFGP gene are flanked by almost identical sequences to those in the trypsinogen gene, which Figure 9.31 illustrates. Notice in this diagram that the AFGP gene of the Arctic cod also has 21 repeats coding for Ala-Ala-Thr, but in the Arctic cod, the flanking regions have no similarity to the trypsinogen gene (or other known genes).

The evolution of the antifreeze gene in Antarctic notothenioid fish is an interesting insight into how adaptive processes can occur at a genetic level. The trypsinogen gene exhibits no native antifreeze activity in its protein product, and the antifreeze protein exhibits no trypsin activity, yet we have strong evidence that one is derived from the other. Further evidence for the origin of the AFGP gene from the trypsinogen gene arose from the identification of an intermediate (chimaeric) gene that encodes both isoforms of AFGP and a trypsinogen-like protease in the Antarctic toothfish (*Dissostichus mawsoni*). This gives us a rare view of the genesis of a new protein. Expansion of the small AFGP sequence and shedding of the protease sequence allowed Antarctic notothenioid fish to produce antifreeze and hence to flourish in the Southern Ocean.

Antarctic fish that avoid freezing by producing antifreezes must constantly maintain the concentrations of antifreeze peptides in their body fluids. Their evolution of **aglomerular** kidneys[53], which lack the glomeruli typical of the nephrons in most vertebrate kidneys, means that these Antarctic fish avoid filtration and excretion of their antifreeze glycopeptides. Aglomerular kidneys have not evolved in most fish living in

[53] Section 7.1 discusses vertebrate kidney structure and glomerular filtration.

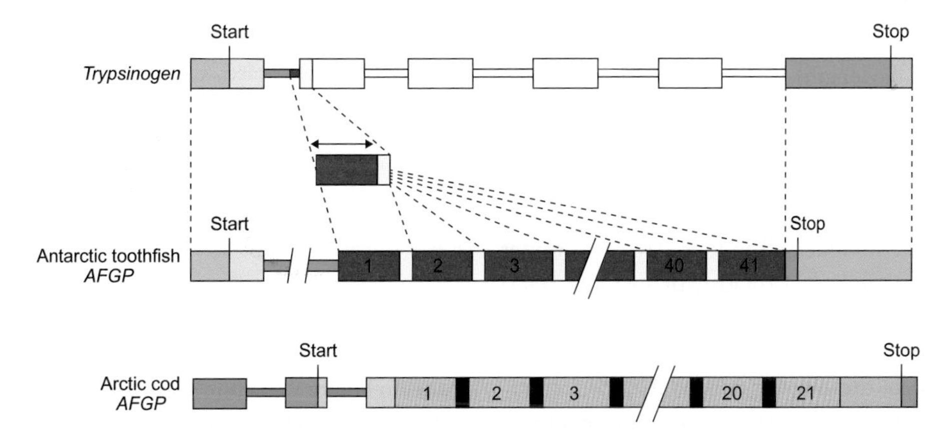

Figure 9.31 Gene structure of pancreatic trypsinogen and antifreeze glycoproteins of Antarctic toothfish (*Dissostichus mawsoni*) and Arctic cod (*Boreogadus saida*)

Regions of identical colours in different genes share sequence similarity. Large boxes represent the coding regions (exons), while thinner boxes represent non-coding regions (introns). The dashed lines indicate the regions of homology between the gene coding for trypsinogen and the antifreeze glycoprotein gene (*AFGP*) in the Antarctic toothfish; the white regions of the *trypsinogen* gene are absent in *AFGP*. The segment below the double-headed arrow shows expansion of a sequence element in the gene for trypsinogen that appears to have given rise to the repeating sequence, numbered 1, 2, etc to 41 in *AFGP*, and which give rise to the repeating tripeptide (Ala-Ala-Thr) residues of the antifreeze glycoprotein. The *AFPG* gene of Arctic cod has completely different repeating regions that do not map onto trypsinogen or any other known gene. The regions between the *AFGP* repeats (yellow or black spacers) are the presumed sites of post-translational cleavage. The diagrams are not drawn to scale.

Source: Logsdon JM Jr., Doolittle WF (1997). Origin of antifreeze protein genes: A cool tale in molecular evolution. Proceedings of the National Academy of Sciences of the United States of America 94: 3485–3487.

northern and Arctic waters, but filtration of the antifreeze peptides (and hence loss in the urine) is impeded because the antifreeze peptides are negatively charged at physiological pH, like most plasma proteins.

Large-scale genome-wide studies have recently revealed the extent of the adaptive radiation of Antarctic notothenioids to the constant cold conditions in the Southern Ocean[54]. The studies found 177 notothenioid gene families expressed at levels many times greater in Antarctic toothfish than in non-notothenioid warm-water species. In many cases, the increased expression results from the duplication of genes over geological time: 118 protein genes are duplicated in Antarctic toothfish, often to a dramatic extent (up to >300-fold). Combining these findings, more than 200 genes were identified as specifically augmented in Antarctic notothenioids and are considered to be functionally important for cold adaptation. The importance of each one of these genes is not clear, but adaptation of processes such as protein synthesis and folding, lipid metabolism and immunity has been identified[55].

Arctic and sub-Arctic teleost fish produce antifreezes seasonally

Seasonal acclimatization of northern polar and sub-polar fish involves the seasonal synthesis of antifreeze proteins and peptides (AFP), which results in rising and falling concentrations during the year, as Figure 9.32 illustrates for two species of flounders. From these data, it is apparent that the amounts of AFPs reach a peak when temperatures reach their minimum, but the rise in levels begins well before temperatures are sufficiently low for there to be a significant risk of freezing.

In flounders, and other species, the seasonal decrease in photoperiod (short daylength) is the major trigger for the production of AFP by the liver. In some species (such as Atlantic cod), however, cold temperatures stimulate the synthesis of glycopeptide antifreeze.

Figure 9.33 illustrates the series of steps regulating AFP production in winter flounders. The long days of summer stimulate the synthesis and release of **growth hormone** (GH) by the pituitary gland[56], and hence stimulate the secretion of insulin-like growth factor I (IGF-1) by the liver. IGF-1 inhibits the transcription of the gene for AFP so AFP concentrations remain low until the circulating concentration of GH naturally declines due to the decreasing daylength. The blue panel in Figure 9.33 shows how short daylengths remove the inhibition of transcription of the gene coding for AFP and lead to an increase in the amount of AFP in circulation.

While overall control seems to rely on photoperiod cues, Figure 9.33 also indicates how ambient water temperature appears to fine-tune the response. In cold conditions, production of the precursor for AFP occurs within the cell. A final cleavage step is then necessary to produce the mature AFP after export of pro-AFP from the cell. Regulation of the production of AFP is therefore possible at a number of levels and by changing the activity of enzymes that

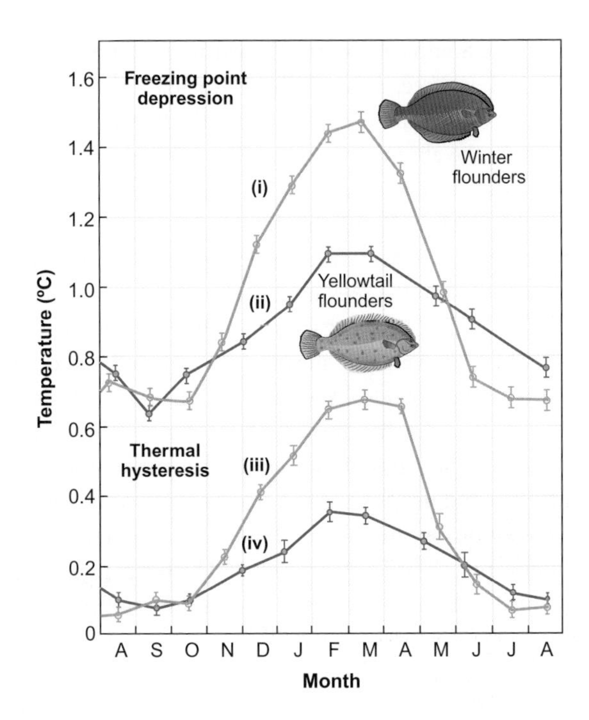

Figure 9.32 Antifreeze peptides are synthesized seasonally in northern species

Lines (i) and (ii) show seasonal changes in freezing point depression of blood plasma samples from winter flounders (*Pseudopleuronectes americanus*) and yellowtail flounders (*Limanda ferruginea*). A high freezing point depression equates to a lower freezing point (FPt), due to the synthesis of antifreeze peptide, which causes a thermal hysteresis (difference between FPt and melting point), as shown by lines (iii) and (iv) for winter flounders and yellowtail flounders, respectively. Freezing point depression (antifreeze synthesis) reaches a peak in January to March in both species of flounder, reaching higher levels in winter flounders than in yellowtail flounders, and then declines in spring and summer. The plasma FPt of –0.7°C in summer is due to the concentration of solutes, mainly Na^+ and Cl^-, as we discuss in Section 4.1.1. Note that an increase in freezing point depression and thermal hysteresis begins in November, which is in advance of any possibility of freezing. Mean water temperatures during these studies were 9.1°C in October; 5.9°C in November, 2.0°C in December, 0°C in January, –0.5°C in February, 0°C in March and April, 1°C in May, 5°C June, 11°C July, 13°C in August, 12°C in September.

Sources: Scott GK et al (1988). Differential amplification of antifreeze protein genes in the Pleuronectinae. Journal of Molecular Evolution 27: 29–35.

[54] We examine thermal adaptation further in Section 9.3.

[55] In addition, the notothenioid genome shows a loss of some traits in some species. The loss of haemoglobin in the ice fish is the best example; the implications on respiratory function are discussed in Section 13.1.2. The lack of a heat shock response to increased temperatures by Antarctic notothenioids may limit their ability to cope with climate change, as we discuss in Section 8.2.2.

[56] We discuss pituitary gland function in some detail in Section 19.2.

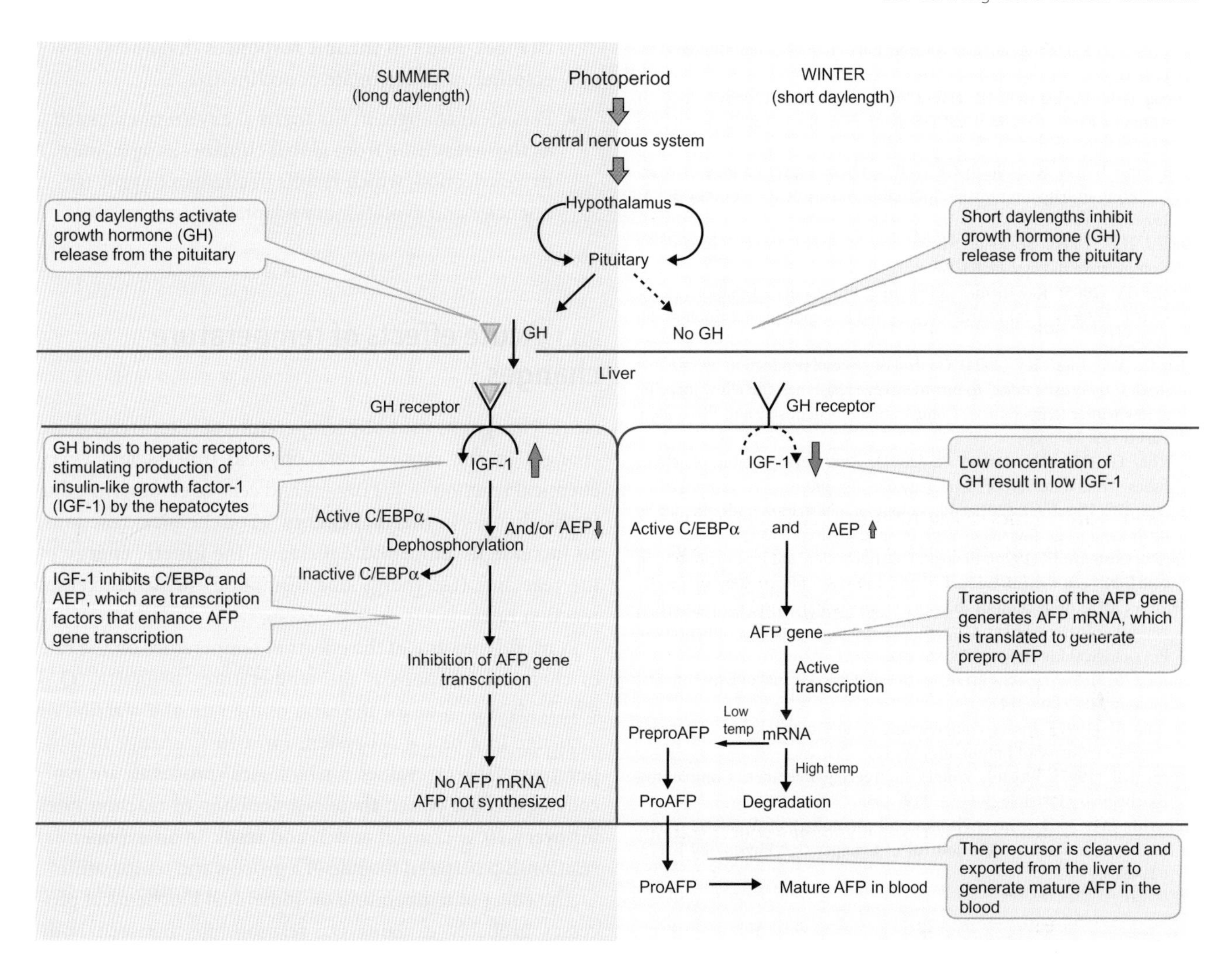

Figure 9.33 Regulation of antifreeze peptide (AFP) production by the liver of winter flounders (*Pseudopleuronectes americanus*)

The left side of the diagram (overlaid in pink) shows summer events, while on the right, overlaid in blue, are the events in winter when AFP production increases. AEP = antifreeze-enhancer binding protein. Section 19.2.2 discusses the hypothalamic–pituitary axis in more detail.

Source: Fletcher GL et al (2001). Antifreeze proteins of teleost fishes. Annual Review of Physiology 63: 359–90.

cleave the immature precursors. At higher temperatures, any AFP mRNA present is likely to be degraded without translation, so AFP is not produced.

Production of AFP also depends on the number of copies of the AFP gene present in the genome—a phenomenon known as gene dosage. The number of copies of AFP genes can be very large; for example, the winter flounder has approximately 100 copies of the gene coding for its most abundant AFP. Yellowtail flounder (*Limanda ferruginea*) generally has fewer gene copies (35). A look back at Figure 9.32 reminds us that winter flounder produce more AFP (and hence the plasma has a higher thermal hysteresis) than yellowtail flounder, which live in deeper water. Swimming at depth, below the ice, reduces the need for antifreeze peptides in yellowtail flounder, and supercooling is an alternative, less energy costly way of avoiding freezing. A mixture of supercooling and use of AFP probably occurs.

Variations in the number of copies of AFP genes are also apparent within a species. Yellowtail flounders occur along the eastern coast of North America. Around Newfoundland, in colder waters, the yellowtail flounders produce more AFP than those living further south, along the coast at New Brunswick. The Newfoundland population has 150 copies of the AFP gene compared to only 30–40 copies in the New Brunswick population. The higher gene dosage of the more northerly populations allows a rapid increase in antifreeze peptide production when necessary.

› *Review articles*

Beers JM, Jayasundara N (2015). Antarctic notothenioid fish: what are the future consequences of 'losses' and 'gains' acquired during long-term evolution at cold and stable temperatures? Journal of Experimental Biology 218: 1834–1845.

Biggar KK, Storey KB (2015). Insight into post-transcriptional gene regulation: stress-responsive microRNAs and their role in the environmental stress survival of tolerant animals. Journal of Experimental Biology 218: 1281–1289.

Cheng C-H C, Detrich III HW (2007). Molecular ecophysiology of Antarctic fishes. Philosophical Transactions of the Royal Society B 362: 2215–2232.

Constanzo JP, Lee RE, Ultsch GR (2008). Physiological ecology of overwintering in hatchling turtles. Journal of Experimental Zoology 309A: 297–379.

Danks HV (2004). Seasonal adaptations in Arctic insects. Integrative Comparative Biology 44: 85–94.

Danks HV (2005). Key themes in the study of seasonal adaptations in insects I. Patterns of cold hardiness. Applied Entomology and Zoology 40: 199–211.

Driedzic WR, Ewart KV (2004). Control of glycerol production by rainbow smelt (*Osmerus mordax*) to provide freeze resistance and allow foraging at low winter temperatures. Comparative Biochemistry and Physiology Part B 139: 347–357.

Fletcher GL, Hew CL, Davies PL (2001). Antifreeze proteins of teleost fishes. Annual Review of Physiology 63: 359–390.

Jackson DC, Ultsch GR (2010). Physiology of hibernation under the ice by turtles and frogs. Journal of Experimental Zoology 313A: 311–327.

Lee RE, Costanzo JP (1998). Biological ice nucleation and ice distribution in cold hardy ectothermic animals. Annual Review of Physiology 60: 55–72.

Nylin S (2013). Induction of diapause and seasonal morphs in butterflies and other insects: knowns, unknowns and the challenge of integration. Physiological Entomology 38: 96–104.

Peck L (2015). DeVries: the art of not freezing fish. Journal of Experimental Biology 218: 2146–2147.

Sinclair BJ (1999). Insect cold tolerance: How many kinds of frozen? European Journal of Entomology 96: 157–164.

Storey KB (1997). Organic solutes in freezing tolerance. Comparative Biochemistry Physiology 117A: 319–326.

Wharton DA (2003). The environmental physiology of Antarctic terrestrial nematodes: a review. Journal of Comparative Physiology B 173: 621–628.

9.3 Temperature change over three timeframes: implications for tissue functioning in ectotherms

The cells and tissues of ectotherms need to function over the range of temperatures resulting from the changes in body temperature that occur, even when behavioural thermoregulation[57] and physiological processes reduce the fluctuations. Changes in environmental temperature—and hence body temperature of ectotherms—occur over three timeframes:

- Acute (rapid) changes in the thermal environment result in **acute responses** by cells and tissues, within minutes to hours, and acute effects on physiological processes.
- Chronic changes in the thermal environment lasting several days, weeks or months, such as seasonal changes, result in **chronic responses** that often allow ectotherms to acclimate or acclimatize[58].
- Persistent changes in the thermal environment, such as those resulting from global climate changes, may drive selection, which results in changes in gene frequencies and evolutionary adaptations.

9.3.1 Acute effects of temperature changes

Acute changes in the body temperature of ectotherms drive changes in the rate of every physical process and every chemical reaction in the organs and cells of ectotherms, due to the changes in **thermal energy**[59]. For chemical reactions, an increase in temperature increases the **kinetic energy** of the reactants (i.e. energy due to motion), which increases reaction rates.

A useful measure of the effects of temperature on the rate of chemical reactions is the temperature coefficient (**Q_{10}**) which gives a value for the ratio of the rate of a reaction or a change in a biological system for a 10°C change in temperatures. The Q_{10} values for biological processes are typically between 2 (when there is a doubling of reaction rates for every 10°C change) and 3, and result in an exponential relationship between the rate of a process and temperature[60].

The effect of temperature on individual biochemical processes probably explains Q_{10} relationships between body temperature and the total rate of oxygen consumption (i.e. metabolic rate[61]) of resting animals. An example of this relationship is shown in Figure 9.34 for mummichogs (*Fundulus heteroclitus*), which are **eurythermal** fish—i.e. they tolerate a wide range of temperatures.

A simple way to show the effects of temperature on the physiological performance of ectotherms is to plot a **thermal performance curve** for a particular parameter against ambient temperature, as in Figure 9.35. In this drawing, peak performance corresponds to the thermal optima for the measured parameter, in this case oxygen concentration in the blood. The temperature range over which performance is optimal are highly likely to vary for different parameters and determining the most crucial parameter(s) for the overall performance of particular species is challenging, and is the subject of lively debate.

[57] Body temperatures of ectotherms are influenced by the behaviours we discuss in Section 9.1.1, and the physiological processes discussed in Section 9.1.3, which all influence heat exchange.

[58] The alternative possibilities of migration, dispersion into new areas, or entering a resting phase (diapause or hibernation) are discussed in Case Studies 9.1 and 9.2, and Sections 9.2.1 and 9.2.2.

[59] We outline the different forms of energy used by animals in Section 2.1.1.

[60] Box 8.2 explains Q_{10} calculations; Figure A in this box shows the exponential relationships for Q_{10} values of 2 and 3.

[61] More detailed discussion of metabolic rate can be found in Section 2.2.

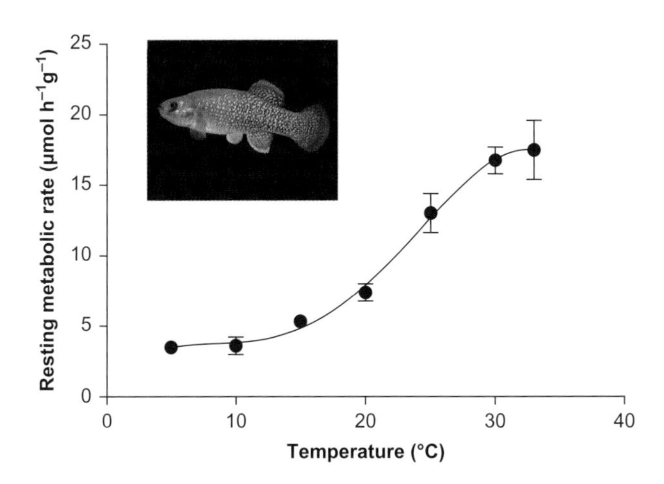

Figure 9.34 Routine metabolic rate (oxygen consumption) of mummichogs (*Fundulus heteroclitus*)

The graph shows the effects of water temperature (which equals body temperature) on routine metabolic rates, i.e. when fish are resting quietly but not necessarily at their minimum level of activity. Values are means ± standard errors.

Source: Schulte PM (2015). The effects of temperature on aerobic metabolism: towards a mechanistic understanding of the responses of ectotherms to a changing environment. Journal of Experimental Biology 218: 1856–1866. image of mummichog: © John Brill.

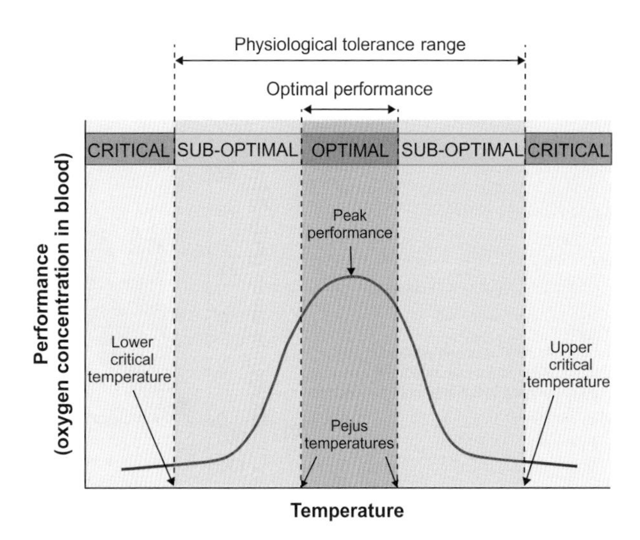

Figure 9.35 Performance curve—diagrammatic representation of oxygen- and capacity-linked thermal tolerance of aquatic ectotherms

The curve represents a thermal performance curve. Optimal performance occurs when oxygen circulated in the blood or in the haemolymph (invertebrates) is high and relatively stable. At either end of the optimal performance range, at the pejus temperatures, oxygen in the blood declines. Aerobic metabolism may continue within the physiological tolerance range (depending on levels of activity) until lower or upper critical temperatures when available oxygen can no longer sustain aerobic metabolism.

For aquatic ectotherms, increased water temperature increases the amounts of oxygen needed for metabolism yet the rising water temperature reduces the amount of dissolved oxygen in the water[62]. Hence, the capacity for oxygen uptake and transport is considered a major determinant of the thermal tolerance of aquatic ectotherms and is called the oxygen- and capacity-limited thermal tolerance (OCLTT) hypothesis.

In Figure 9.35, optimal performance occurs at high and relatively stable concentrations of oxygen in the blood. According to the OCLTT hypothesis, high availability of oxygen in the blood is necessary for resting tissue metabolism and for the maximum needs for foraging or to catch prey, grow, reproduce and to undertake competitive interactions[63]. This does not mean that animals necessarily exploit the maximum capacity of OCLTT everywhere in their natural range. They may occupy habitats outside the thermal optimal range perhaps selecting such habitats because of the availability of particular food items.

At the **pejus temperatures** (*pejus* in Latin means to grow worse), the availability of oxygen in the blood declines, although below-optimal oxygen could be sufficient for aerobic metabolism depending upon levels of activity. Only at the **upper critical temperature** or **lower critical temperature** is survival seriously compromised and only a short period of survival in these conditions is likely.

The performance curves of ectotherms are, in fact, usually asymmetric, with a skew to the left, rather than symmetrical, like the drawing in Figure 9.35. In practice, a slow-rising phase of variable slope reaches a plateau. When high temperatures reach the upper pejus temperature we typically see a steep decline that has been attributed to protein denaturation or changes in their functional state. Variation in the shape of thermal performance curves are evident and may partly result from species differences (genetic adaptations) or the particular trait or parameter studied (which may be unrelated to oxygen). In addition, thermal acclimation or acclimatization could affect performance curves.

These complications and variations do not negate the OCLTT hypothesis, however, which is supported by studies of many aquatic ectotherms (fish and marine invertebrates). That said, support for the applicability of these concepts in air-breathing ectotherms, such as insects, is limited, although interestingly their aquatic larvae do follow the OCLTT principles.

In Figure 9.36, we examine an example of the OCLTT principles by examining thermal performance curves of spider crabs (*Maja squinado*) in laboratory studies. Looking at Figure 9.36C, notice that heart rate increases steadily

[62] The solubility of oxygen in water declines as temperature increases, which we discuss in Section 9.1.

[63] The difference between the maximum sustainable rate of oxygen consumption and the resting rate of oxygen consumption are described as **aerobic scope**, which we discuss in Section 15.2.2. The thermal range for optimal performance of aquatic ectotherms matches the window of aerobic scope.

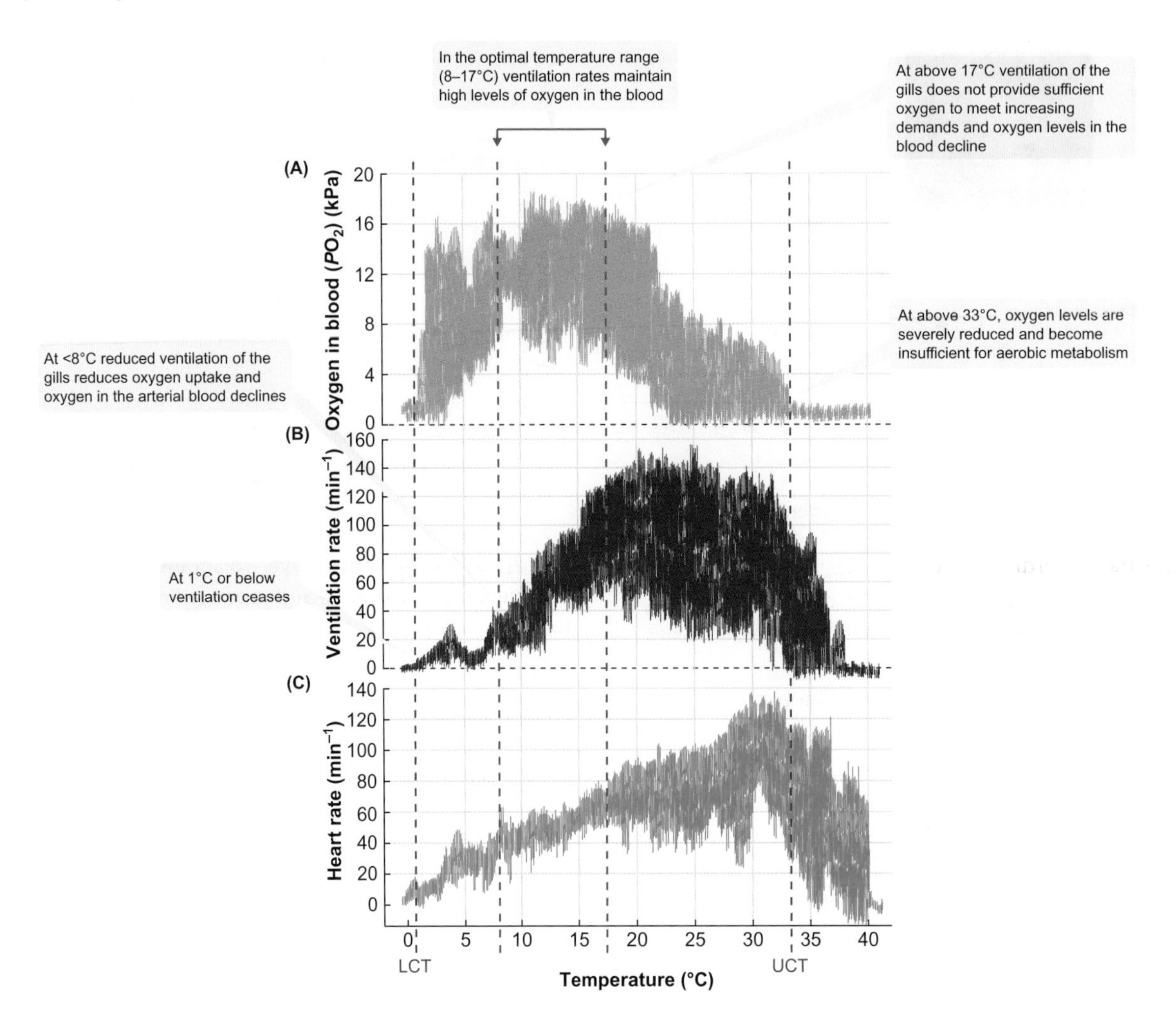

Figure 9.36 Effects of acute changes in water temperature on the ability of spider crab (*Maja squinado*) to match oxygen supply and demand

Spider crabs pre-acclimated to water at 12°C were exposed to decreasing temperatures (to 0°C) or increasing temperature (to 40°C) while continously recording the fluctuations in heart rate (C), ventilation rate (B) and oxygen levels in the blood (A, as partial pressure PO_2); Section 11.1.2 explains oxygen measurements in blood. LCT = Lower critical temperature; UCT = upper critical temperature.

Sources: Frederich M, Pörtner HO (2000). Oxygen limitation of thermal tolerance defined by cardiac and ventilatory performance in spider crab, *Maja squinado*. American Journal of Physiology 279: R1531–1538.

between 0°C and 30°C, but above 30°C, heart rate declines sharply. The involvement of the heart in the circulation of oxygen is not, however, the main factor in determining the thermal tolerance of spider crabs. Ventilation of their gills[64], which is illustrated by the data in Figure 9.36B, determines oxygen uptake and is the main determinant of the thermal tolerance of spider crabs.

The optimal temperature range for spider crabs based on the data in Figure 9.36A is 8°C to 17°C. Within this temperature range, the crabs have relatively stable and high oxygen levels in the arterial blood. Turning to the ventilation rates in Figure 9.36B, we find a reduced rate of ventilation below 8°C, which does not appear to provide sufficient oxygen to meet demands, so blood levels of oxygen decline. Between 17°C to 30°C, the demand for oxygen increases due to the increasing body temperature, but ventilation rates remain relatively constant, so oxygen levels in the blood again decline.

From the data in Figure 9.36, a lower critical temperature of 1°C and an upper critical temperature of 33°C occur when inadequate ventilation or a complete cessation of ventilation result in very low blood oxygen levels. The spider crabs then become increasingly reliant on anaerobic metabolism even though the water contains high levels of oxygen.

[64] Section 12.2.2 discusses oxygen uptake via the gills of crabs.

Although heart rate is not the limiting factor in determining the upper critical temperature range for spider crabs, cardiac function plays a major role in determining the capacity to supply oxygen in many species, including fish[65] and many marine invertebrates. A good example of how heart rate determines the thermal range of species is given by blue mussels (*Mytilus edulis*)[66] in which heart rates increase with body temperature up to a critical thermal maximum (CT_{max} or H_{crit} equivalent to upper critical temperature) at which point any further increase in temperature leads to a rapid decline in heart rate.

The CT_{max} for blue mussels alters during thermal acclimation. At 14°C, the average CT_{max} is 25.5°C, whereas acclimation to 21°C increases CT_{max} to 28.5°C. However, blue mussels cannot acclimate beyond 28.5°C, when the predicted CT_{max} is about 31°C. High mortalities occur at 32°C, probably due to the shortfall in cardiac function, which will contribute to setting the upper thermal limits for blue mussels.

Differences in heart function of different *Mytilus* species (genetic thermal adaptations) are particularly interesting as they provide an explanation of the recent changes in distribution of mussels on the west coast of the USA. The invasive species *Mytilus galloprovincialis*, which is of Mediterranean origin, is more heat tolerant than the native more cold-tolerant bay mussels *Mytilus trossulus*. After acclimation to 21°C, the average CT_{max} is 26.0°C in the native species compared to 30.7°C in the invasive species. The differences in thermal performance of heart function could explain the success of the invasive species in replacing the native species along much of the coastline of southern and central California, and indicate that *Mytilus galloprovincialis* is likely to expand its range further as sea temperatures rise.

Upper and lower critical temperatures may also influence the abundance of species of fish that cannot escape from such conditions. (The more mobile fish species may redistribute using thermal selection, as we discussed in Case Study 9.1). An example is the common eelpout (*Zoarces viviparus*), which occupies rocky shores and tide pools in coastal areas of northern Europe. Eelpouts generally prefer cool waters but do not relocate when temperatures exceed their upper critical temperature of about 22.5°C. Results of field studies summarized by the data in Figure 9.37 indicate that high summer temperatures affect the abundance of eelpouts the following year. Thermal conditions are also likely to set the southern distribution limit for eelpouts in the German Wadden Sea.

Heat shock and cold shock

Exposure of ectotherms to a rapid and severe change in temperature can damage many cell proteins leading to protein unfolding and a loss of the configuration (shape) necessary for their functioning. Heat shock or cold shock initiates a

[65] Experimental Panel 15.1 and Section 15.2.2 explore the importance of heart rate in OCLTT of migrating Pacific salmon during their upstream migration for spawning and the adverse effects of rising temperatures.

[66] We describe the cardiovascular system of molluscs (non-cephalopod molluscs) in Section 14.3.1.

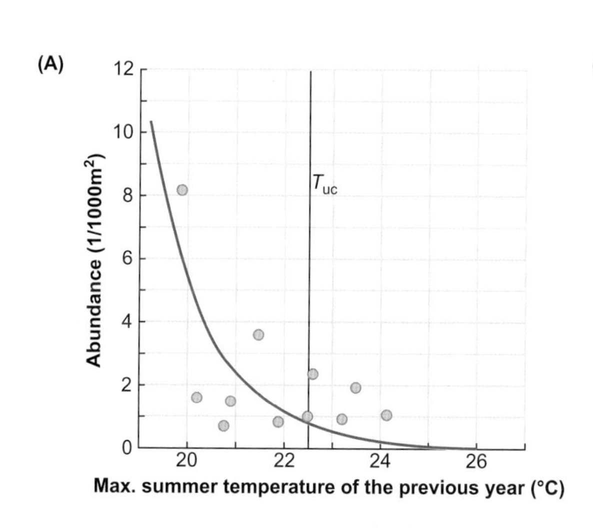

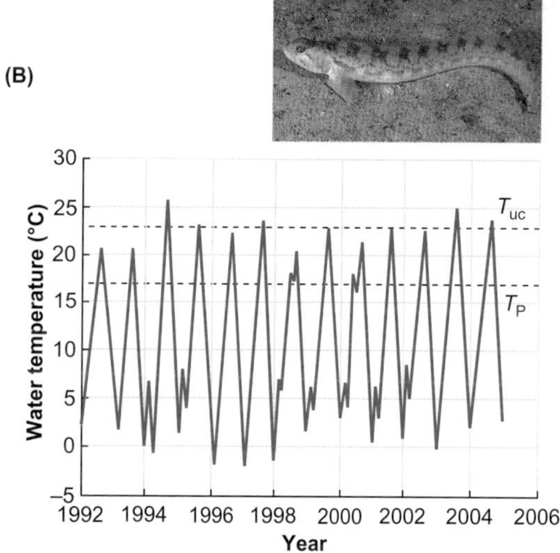

Figure 9.37 High temperatures reduce the abundance of eelpouts in the German Wadden Sea

(A) The abundance of viviparous eel pouts (*Zoarces viviparus*) in the Wadden Sea in spring relates to maximum (max) water temperatures in the previous year. T_{uc} = upper critical temperature. (B) Daily water temperatures in the Wadden Sea between 1992 and 2005. The data show that summer maxima exceed the pejus temperature of eelpouts (T_p) and in some years exceed their upper critical temperature (T_{uc}).

Source: Pörtner HO, Knust R (2007). Climate change affects marine fishes through the oxygen limitation of thermal tolerance. Science 315: 95–97. Image of eelpout: © Klas Malmbery.

9

rapid and ubiquitous cellular response, which activates the synthesis of **heat-shock proteins** that protect the cells from such damage.

Heat-shock proteins are specific proteins produced by the cells of all organisms (including bacteria and plants) in response to a variety of stressful conditions. These proteins were first recognized as a cellular response to temperature elevation (heat shock), hence their name. However, we now know that a wide variety of stressful but non-lethal conditions activates a response, which explains their alternative name: stress proteins. Box 9.1 gives further information on heat-shock (stress) proteins and the activation and regulation of this cellular response.

Cells synthesize stress proteins in response to severe cold stress, infection, chemical pollutants or other chemical shocks, exposure to ultraviolet (UV) light or hypoxia (oxygen deficiency). In fact, any stimulus that causes unfolding of tissue proteins is likely to stimulate the synthesis of stress proteins. The stress proteins restore the folding and repair reversibly damaged proteins, while any proteins beyond repair are degraded by proteolysis as a result of a complex chain of reactions involving the activation of another cell protein—**ubiquitin**.

Although stress proteins are typically associated with heat (or cold) shock, some invertebrates living high on the marine intertidal, such as black turban snails (*Tegula funebralis*), activate stress proteins and ubiquitin as a part of their normal daily response to exposure during the tidal cycle. This tells us that natural heat stress when the tide retreats can be sufficient to result in reversible and irreversible damage to the cell proteins on a daily basis. The energetic implications of activating stress protein responses, which uses ATP[67], may be a limiting factor for the success of intertidal species—particularly for the sessile species on the upper shore, which are exposed to the greatest thermal fluctuations but have the least time for feeding when covered by the incoming tide. High body temperatures will increase energy consumption (Q_{10} effects), and the additional energy demand due to synthesis of stress proteins may thus set the upper limits to thermal tolerances and the upper limits of distribution on the intertidal.

Among terrestrial invertebrates, an autumn/winter cold shock is important for inducing a rapid cold hardening response which enhances subsequent cold tolerance and enables survival in subzero conditions. Figure 9.38 illustrates the cold hardening of flesh flies (*Sarcophaga bullata*) exposed to a cold shock of 0°C for 2 h. This cold shock enables a large proportion (about 65 per cent) of the cold-hardened flies to survive when exposed to a more severe cold shock (−8°C for 2 h)[68], whereas no flesh flies survive such conditions without the cold-hardening stage. Similar differences between cold-hardened flies and non-hardened flies can be seen at a tissue and cell level, and are apparent by examination of the fat body.

The fat body is essential in insects for carbohydrate, lipid and protein metabolism, much like the liver of vertebrates. Only about 20 per cent of the fat body cells survive when flesh flies are exposed to the severe cold shock of −8°C without cold hardening, as illustrated in Figure 9.38. By contrast, 70 per cent of the fat body cells survive after cold hardening. The improved survival of these cells appears to be at least partly due to an increase in the proportion of unsaturated fatty acids in cell membranes. Box 9.2 examines the importance of fatty acid composition on the fluidity (melting point) of cell membranes.

9.3.2 Acclimation and acclimatization to temperature changes

Figure 9.39 illustrates the three types of response of ectotherms that result from changes in environmental temperature and influence the observed rate of a typical biological process (for example an enzyme reaction rate). In type I responses, illustrated by Figure 9.39A, there is a lack of adjustment in the rate of the physiological process or biochemical reaction. Hence, reaction rates increase with temperature following a Q_{10} relationship[69]. More usually, ectotherms show type II and type III responses, which result in similar rates of reaction at different acclimation temperatures.

Type II acclimation involves translational responses in which the relationship between temperature and the rate of reaction or the rate of a physiological variable moves (is translated) to a different position on the graph, as shown in Figure 9.39B, but without a change in the slope of the relationship. Comparing Figures 9.39A and 9.39B indicates that the rate of reaction, after a reduction in temperature and cold acclimation, is higher at the temperature of the cold-acclimated animals than in Figure 9.39A when no acclimation occurs.

Type III responses involve a rotational modification (change in slope) of the relationship between reaction rate and temperature, which Figure 9.39C shows. At the acclimation temperatures, the outcome is the same as for translational modifications: Q_{10} temperature effects on the rate of the reaction are compensated for. However, the outcome is not the same at all temperatures because of the change in slopes of the relationship.

[67] ATP production as the source of energy for physiological and biochemical processes is discussed in Section 2.4.

[68] Survival involves freeze avoidance by supercooling, which we discuss in section 9.2.4.

[69] Box 8.2 discuses Q_{10} type responses.

Box 9.1 Heat-shock (stress) proteins

Stress proteins function as **chaperone proteins**, which stabilize the folding of proteins into their correct shape (configuration). Such stabilization is essential given that the ability of a protein to function correctly is dependent upon its shape. Stress proteins help to refold any proteins damaged by cellular stresses; where damage is irreversible these proteins facilitate the removal of damaged proteins from the cells by proteolysis. Several families of heat (stress) proteins exist, each with characteristic molecular mass of 60, 70 or 90 kDa. Hence, these proteins are designated as heat-shock proteins (hsp): hsp60, hsp70 or hsp90. The amino acid sequences of these proteins are highly conserved, which indicates their common evolutionary origin and function.

Figure A illustrates a widely accepted model of the mechanism of activation of a heat shock (stress) response by the cells. This model is called the cellular thermometer model. In this scheme the heat-shock proteins determine the amount of free heat-shock factor 1 (HSF1) in the cell cytosol, which in turn activates the nuclear response that results in the increased expression of heat-shock proteins.

Figure A shows the stages in the activation process, beginning in the non-stressful state. In this state, the activation of a stress protein response is inhibited because of the association of hsp proteins with HSF1 protein monomers to form an inactive multi-chaperone complex in the cytosol (Figure A, point 1). This association inhibits migration of free HSF1 into the cell nucleus, which is necessary to activate a stress protein response.

During heat stress (or other cell stresses resulting in protein damage), weakening of hydrogen bonds and ionic interactions leads to an unfolding and dissociation of cellular proteins. Heat-shock proteins bind more easily to unfolded proteins than to HSF1 and the multichaperone complex dissociates, as shown by Figure A, point 2. The binding of the heat-shock proteins to unfolded proteins inhibits their aggregation and assists in protein refolding.

The release of HSF1 monomers from the protein complex allows HSF1 to diffuse into the cell nucleus (Figure A, point 3), where HSF1 forms trimers (point 4) that bind to a heat-shock element (HSE) of the genes for heat-shock proteins. Binding is necessary for these genes to be expressed (point 5). If an excess of hsps are produced beyond that necessary to bind to unfolded proteins, autoregulation of the heat-shock (stress) response sees the HSF trimers dissociate from the HSE and inhibit hsp production, as indicated in Figure A (point 6).

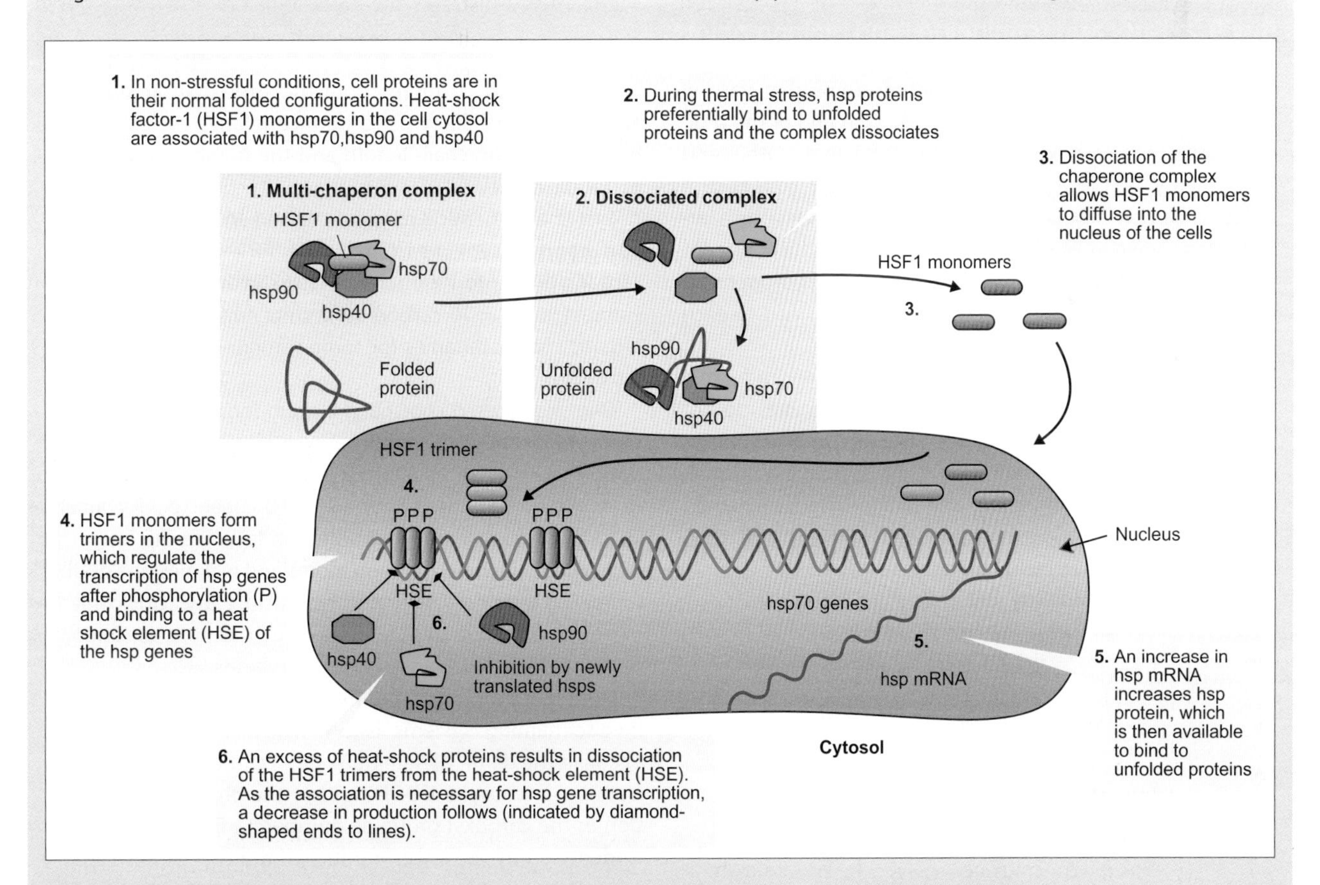

Figure A Model of activation of a heat-shock protein response during heat or other cellular stress

The blue box illustrates an unstressed cellular state when the heat (stress) protein response is inhibited. The numbered pink box and illustrations show the sequential stages in which the activation of a heat (stress) protein proceeds.

Source: Tomanek L, Somero GN (2002). Interspecific- and acclimation-induced variation in levels of heat-shock proteins 70 (hsp70) and 90 (hsp90) and heat-shock transcription factor-1 (HSF1) in congeneric marine snails (genus *Tegula*): implications for regulation of *hsp* gene expression. Journal of Experimental Biology 205: 677-685.

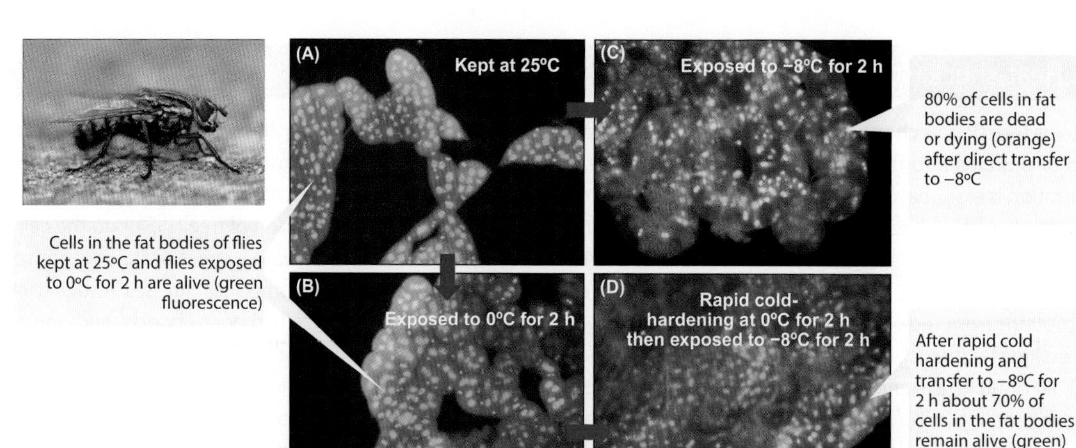

Figure 9.38 Rapid cold hardening enhances survival of fat body cells of grey flesh flies (*Sarcophaga bullata*)

The viability of the cells in the fat bodies were assessed using fluorescent dyes that distinguish live and dead (or dying) cells based on their colour. Cells fluorescing green are alive; red-orange cells are dead or injured. Red arrows indicate the experimental manipulations of the flies.

Source: Lee RE Jr. et al (2006). Rapid-cold hardening increases membrane fluidity and cold tolerance of insect cells. Cryobiology 52: 459–463. Fly image: © Paul Wheeler.

Box 9.2 Fatty acid terminology and melting points

Fatty acids consist of chains of carbon atoms that vary in length, from 4 to 30 (but usually 12–24). One end of the molecule contains a carboxyl group (COOH), while the other end has a methyl (CH_3) group. Saturated fatty acids consist of a chain of carbon atoms each with two hydrogen atoms attached as shown in Figure A(i). In unsaturated fatty acids, not all carbon atoms have the maximum number of hydrogens attached; some of the carbon atoms are instead joined to each other by double bonds, as shown in Figure A(ii).

Fatty acids have been given common names, and Table A includes some of these. Although common names do not convey any information about the structure and properties of fatty acids, they are often used if the systematic (formal) chemical names are complex. The systematic chemical names, however, tell us far more about the structural properties of a fatty acid and allow us to deduce relative melting points based on carbon chain length and the number and position of double bonds.

(i) Saturated Fatty acid

(ii) Unsaturated Fatty acid

Figure A Types of fatty acids

Systematic chemical names are based on the number of carbon atoms present in each molecule (for example etha- = two carbons; buta- = four carbons; hexa- = six carbons; octa- = eight carbons; deca- = 10 carbon atoms, etc.). Table A lists the systematic chemical names for some saturated and unsaturated fatty acids.

One widely used shorthand system for naming fatty acids gives the number of carbon atoms (C) and the number of double bonds (D), separated by a colon—that is, C:D. For example, the saturated fatty acid 18:0 (stearic acid) contains 18 carbon atoms and no double bonds, while the fatty acid 18:1 (oleic acid) has 18 carbon atoms and includes one double bond; it is a monounsaturated fatty acid. Fatty acids containing *more* than one double bond—such as 20:4 (arachidonic acid)—are called *poly*unsaturated fatty acids.

Double bonds occur in different places within fatty acids. Precise naming systems clarify where the double bonds occur. In the delta (Δ) naming system, carbons are numbered from the carboxyl (–COOH) end of the chain (with the carbon in the carboxyl group as number 1). Each double bond is identified by the lower number of the two carbons that it joins. The carbons with double bonds are then identified by a prefix, delta (Δ). For example, the 18:1 fatty acid (oleic acid), which has a single double bond between carbons 12 and 13 (numbered from the carboxyl end of the molecule), is identified as 18:1 Δ^{12}.

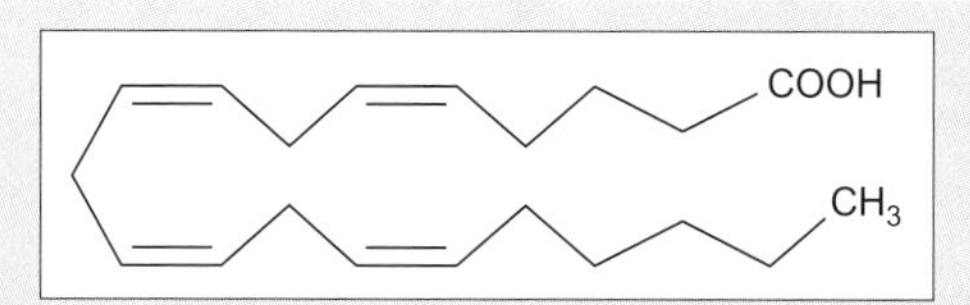

Figure B Structure of arachidonic acid

In the omega (Ω) classification of fatty acids, carbons are counted from the methyl (CH_3–) end of the fatty acid and Ω refers to the position of the double bond closest to the methyl end. For example, in oleic acid (18:1) the double bond is six carbon atoms from the methyl end of the molecule, so this is an omega-6 (Ω-6) fatty acid.

Both the delta and omega naming systems can also specify the location of multiple double bonds. For example, arachidonic acid (20:4), whose structure is shown in Figure B, has four double bonds, making it 20:4 Ω-6,9,12,15 in the omega system (or $\Delta^{5,8,11,14}$ in the delta system). In the omega system, only the position of the lowest double bond is generally stated. Arachidonic acid is therefore an Ω-6 fatty acid.

The desaturase enzymes involved in manufacturing unsaturated fatty acids are also classified by the Δ or Ω schemes to explain where double bond(s) are formed. For example, Δ^9-desaturase creates a double bond at the 9th position from the carboxyl end.

For ectothermic animals, the melting points of fatty acids are particularly important in determining membrane fluidity. In Table A, notice that:

- Increasing chain length increases the melting point of saturated fatty acids, as is evident by examining the data in red typeface for fatty acids of various C:Ds.
- The presence of just one double bond significantly decreases the melting point of a fatty acid as shown by comparison of values for saturated fatty acids and monounsaturated fatty acids with similar number of carbon atoms (data in green).
- Increasing the number of double bonds (degree of unsaturation) to form polyunsaturated fatty acids reduces melting points even further. The three examples given (data in blue) are fatty acids that are liquids at the normal body temperatures of almost all ectotherms.

Table A Melting points (°C) of some saturated and unsaturated fatty acids

C = number of carbon atoms, D = number of double bonds.

Saturated fatty acids				Monounsaturated fatty acids				Polyunsaturated fatty acids			
Common name	**Systematic name (-acid)**	**C:D**	**Melting point (°C)**	**Common name**	**Systematic name (-acid)**	**C:D**	**Melting point (°C)**	**Common name**	**Systematic name (-acid)**	**C:D**	**Melting point (°C)**
Butyric acid	Butanoic	4:0	−7.9								
Caproic acid	Hexanoic	6:0	−3.4								
Caprylic acid	Octanoic	8:0	16.5								
Capric acid	Decanoic	10:0	31.5								
Lauric acid	Dodecanoic	12:0	43.8								
Myristic acid	Tetradecanoic	14:0	53.9	Myristoleic acid	9-Tetradecanoic	14:1	−4.5				
Palmitic acid	Hexadecanoic	16:0	62.5								
Stearic acid	Octadecanoic	18:0	69.3	Oleic acid	9-Octadecanoic	18:1	16.3	Linoleic acid	9,12-Octadecadienoic	18:2	−6.9
Arachidic acid	Eicosanoic	20:0	75.4					α-Linolenic acid	9,12,15-Octadecatrienoic	18:3	−16.5
								Arachidonic acid	5,8,11,14-Eicosatetraenoic	20:4	−49.5

Source: PubChem (https://pubchem.ncbi.nlm.nih.gov)

Type II and III thermal acclimation responses result from one or more of the following:

- quantitative changes in the amount of enzyme activity;
- qualitative changes in the form of an enzyme (different **isozymes**);
- modulative changes in the environment in which enzymes function.

Let us now consider some examples of each type of acclimation response.

Quantitative changes in the activities of the enzymes involved in aerobic pathways often increase during cold

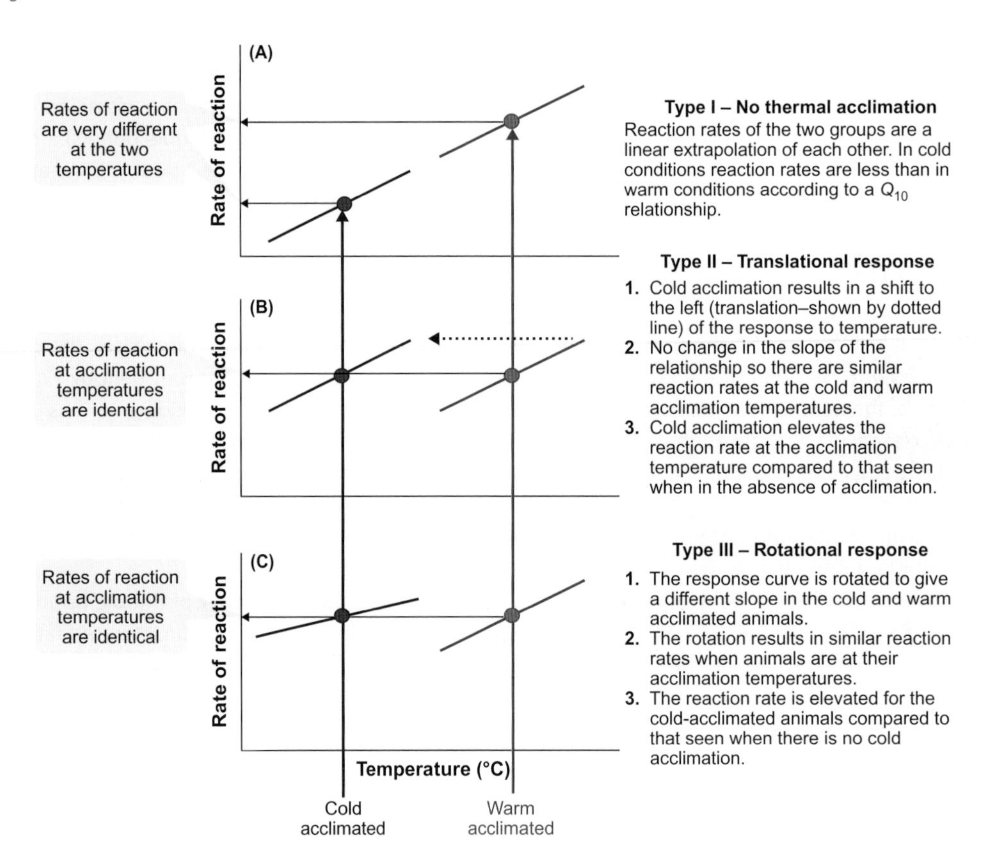

Figure 9.39 Three types of thermal acclimation

The three plots show the results of a theoretical laboratory experiment in which an ectotherm is exposed to warm (red) and cold (blue) conditions for a period of potential thermal acclimation. Then the rate of an enzyme reaction is measured at a range of temperatures. The relationships between reaction rate and temperature reveal which of the three types of acclimation has occurred. The blue lines, arrow and circles on each graph indicate data for animals exposed to cold conditions. The red lines, arrow and circles indicate data for animals exposed to warm conditions.

acclimation of ectotherms. For instance, the activity of cytochrome c oxidase (the last enzyme in the respiratory **electron transport chain** in mitochondria[70]) increases by almost 70 per cent in goldfish (*Carassius auratus*) after they are transferred from water at 25°C to water at 5°C and acclimate to the lower temperature.

Initial changes in enzyme activity can result from the activation of inactive enzyme molecules or the transfer of enzymes into the cytoplasm from a location where the enzyme was previously inactive (e.g. sequestered on a membrane). Slower responses involve increased translation of messenger RNA into proteins, at the ribosomes. For example, when water temperature declines, oyster toadfish (*Opsanus tau*) increase the activity of amino-acetyl transferase, which transfers amino acids to the ribosome during protein synthesis and increases the capacity for enzyme synthesis.

Changes in isozymes (qualitative responses) are often an important part of thermal acclimation. A well-studied example is lactate dehydrogenase (LDH)[71], of which there are five isozymes. Each comprises four subunits (monomers) of two types: M (for skeletal muscle) and H (for heart). (These two types are sometimes called A and B instead to reflect the *LDH-A* and *LDH-B* genes they are coded by). Hence, the five LDH isozymes are designated as M4 (where all monomers are M type), 3MH, 2M2H, M3H and 4H (where all monomers are H type), which are also identified as LDH5 to LDH1, respectively.

The differences in LDH isozymes depend on just a few changes in the amino acid sequence of the enzyme but result in marked differences in their sensitivity to temperature: the forms with the highest number of H monomers are more thermally stable than isozymes with more M monomers. Some (but not all) fish species (e.g. goldfish)

[70] Section 2.4.2 discusses the electron transport chain and oxidative phosphorylation.

[71] LDH catalyses conversion of pyruvate to lactate, in anaerobic glycolysis, which Figure 2.19 outlines.

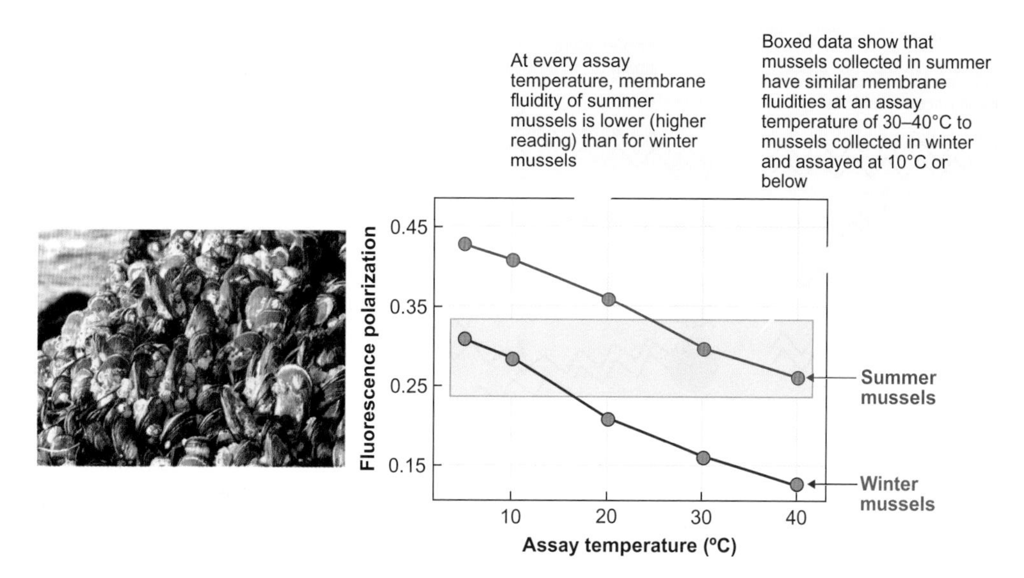

Figure 9.40 Effects of temperature on fluidity of gill membranes of California mussels (*Mytilus californianus*) in summer and winter

Membrane fluidity is inversely related to the polarization reading, i.e. a relatively high polarization reading results from a relatively low fluidity. Data are means for 12 mussels collected in summer and 20 mussels collected in winter.

Source: Williams EE, Somero G (1996). Seasonal, tidal cycle and microhabitat-related variation in membrane order of phospholipid vesicles from gills of the intertidal mussel, *Mytilis californianus*. Journal of Experimental Biology 199: 1587–1596. Image of mussels: Till Luckenbach/Helmholtz Centre for Environmental Research © UFZ.

change the proportion of the LDH isozymes at different acclimation temperatures.

While many species have well-developed mechanisms to acclimate to temperature changes in the laboratory and acclimatize in the natural environment, others do not. Marine invertebrates of the Southern Ocean (Antarctic) have a poor capacity to adjust to temperature changes. This is probably the result of their long evolutionary history (>15 million years) in stable conditions, which has resulted in genetic adaptation to function at permanently low temperatures. Such species are likely to be among the most sensitive to the projected rise of sea temperatures in the Southern Ocean[72]. Their only escape would be thermal adaptation (genetic modifications)[73] at a sufficiently fast rate to offset the effects of rising temperatures.

Compensation for temperature effects on membrane fluidity

The movement of phospholipid molecules within *isolated lipid bilayers*[74] of the cell membranes increases when temperature increases, whereas a decrease in temperature makes the membranes more ordered and gel-like. We might therefore grow bacteria (*E. coli*) at temperatures of between 15°C and 43°C and would then discover that membrane fluidity of the bacteria is very similar at different temperatures.

The compensation process was named **homeoviscous adaptation**; homeoviscous has the literal meaning: constant viscosity. Genetic *adaptation* probably did occur in the bacteria due to a selection process over many generations of rearing at the particular temperatures. In general, however, ectotherms employ **homeoviscous acclimation** (or acclimatization) to adjust membrane fluidity in response to changes in ambient temperatures, generally over several days or weeks. In some cases, there is evidence of more rapid changes in membrane fluidity.

Figure 9.40 shows an example of the seasonal homeoviscous acclimatization that is common among intertidal marine organisms. The California mussels (*Mytilus californianus*) in this study were exposed to natural solar radiation during each low tide. In summer the mussels warm to higher body temperatures than in winter. Figure 9.40 shows the direct effect of assay temperature on membrane fluidity: higher fluidity (lower polarization reading) is an inevitable effect of the higher assay temperatures, whatever the season. More importantly, at all assay temperatures the membranes of mussels collected in summer are less fluid (higher polarization reading) than those from mussels collected in winter.

How do animals modify membrane fluidity?

Animals modify membrane fluidity by changing the proportions of fatty acids in the membrane phospholipids. Earlier, Box 9.2 outlines the range of fatty acids in membranes and their melting points, the combination of which determines membrane fluidity. The saturated fatty acids and longer-chain

[72] Section 1.2.1 discusses rising sea temperatures.
[73] We discuss adaptations in Section 9.3.3.
[74] Figure 3.12 illustrates the lipid bilayer that constitutes the bulk of cell membranes.

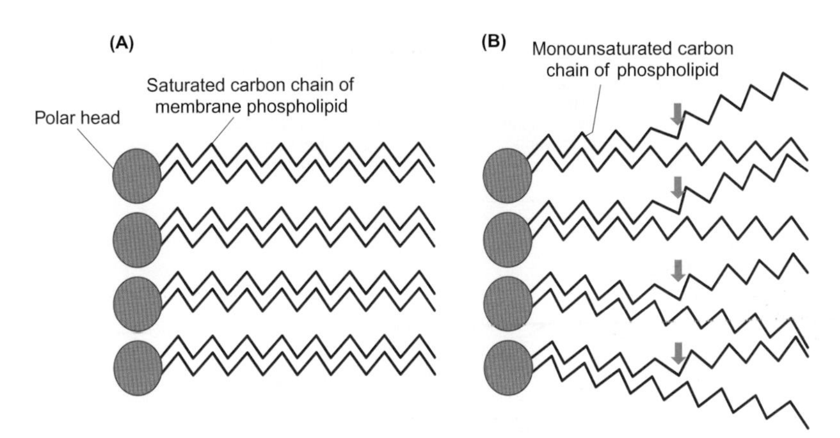

Figure 9.41 Diagram of phospholipids of lipid bilayer of membranes, showing effects of unsaturation

(A) Membrane phospholipids with saturated fatty acids stack closely together. (B) Membrane phospholipids with unsaturated fatty acids kink in the location of each double bond which makes them occupy more volume in the lipid bilayer. Diagram shows a simplified membrane with monounsaturated phospholipids with a single kink (at green arrows).

fatty acids have lower fluidities (higher melting points) than unsaturated fatty acids, which is evident from the examples given in Table A in Box 9.2.

Ectotherms often incorporate a higher proportion of unsaturated fatty acids into their membranes during acclimatization (or acclimation) to colder environments. The unsaturated fatty acids pack less easily, as illustrated in Figure 9.41B, and may therefore enhance cold tolerance by reducing the organization of the phospholipids and increasing membrane fluidity. In contrast, acclimation to warmer conditions involves incorporating into the membranes a higher proportion of saturated fatty acids that lack double bonds in the carbon chain and which therefore pack more closely, as Figure 9.41A illustrates. Close packing increases membrane stability and viscosity, i.e. reduces membrane fluidity.

To manipulate membrane fluidities, animals either manufacture the necessary fatty acids internally or in some cases obtain specific fatty acids, such as linoleic acid and α-linolenic acid, from plants in their diet. Some animals use the fatty acids from plants to synthesize the longer chain Ω-3 and Ω-6 polyunsaturated fatty acids necessary for reducing membrane fluidity during acclimatization (or acclimation) to colder environments.

An important component of thermal acclimatization or acclimation of membrane fluidity is the activity of the enzymes responsible for elongation, desaturation or saturation of lipids. Various desaturase enzymes remove hydrogen atoms from fatty acid molecules to create unsaturated fatty acids. Figure 9.42 shows a good example of cold acclimation by carp (*Cyprinus carpio*) in which Δ^9-desaturase is activated in the liver.

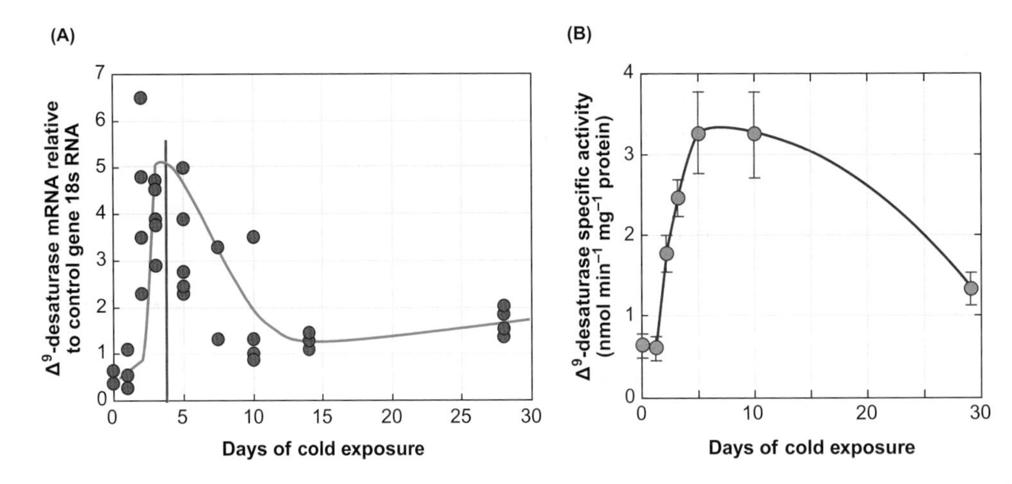

Figure 9.42 Fatty acid desaturase activity of the liver of common carp (*Cyprinus carpio*) increases during cold acclimation

The carp had been maintained at 30°C for at least 3 months before water temperature was cooled to 10°C, over 3 days, and held at this temperature. Panel (A) shows desaturase transcripts (mRNA) relative to a control 'housekeeping' gene (18s RNA) for individual carp. The green line shows the change in mean values. Increased gene expression is evident in some carp after 48 h but on average the peak occurs after 3–5 days. The peak effect subsides to a persistent level that is about 2–3 times that of warm-acclimated carp. These effects are similar to the increase in Δ^9-desaturase specific activity in carp liver during cold acclimation, shown in panel (B), which is followed by a decline to 2–3 times the activity of warm-acclimated trout.

Source: adapted from Tiku PE et al (1996). Cold-induced expression of Δ^9-desaturase in carp by transcriptional and posttranslational mechanisms. Science 271: 815–818.

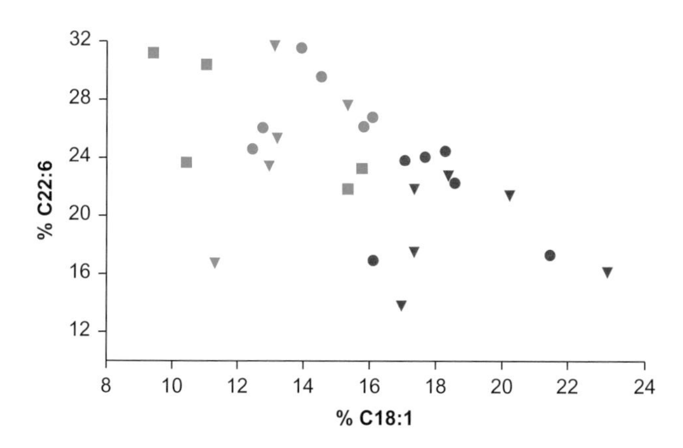

Figure 9.43 Percentages of oleic acid (C18:1) and docosahexaenoic acid (C22:6) in fatty acids of surviving and non-surviving alewives (*Alosa pseudoharengus*) exposed to a cold challenge in fresh water

The cold challenge involved reducing water temperature from the acclimation temperature of 20°C to 2°C, over 19 days. The survivors (green symbols) tended to have a higher percentage of C22:6 and less C18:1. Non-survivors (red symbols) had relatively high levels of monounsaturated C18:1 and lower levels of C22:6. There was no evidence of post mortem changes in the fatty acid composition of non-survivors before they were removed from the tanks as data from survivors left in their tanks for 24 h after sacrifice (green squares) grouped with the survivors. Survival was independent of whether the fish were fed a diet high in polyunsaturated fatty acids (*Artemia*: triangles) or containing lower levels of polyunsaturated fatty acids (*Daphnia*: circles).

Source: Snyder RJ, Hennessey TM (2003). Cold tolerance and homeoviscous adaptation in freshwater alewives (*Alosa pseudoharengus*). Fish Physiology and Biochemistry 29: 117–126.

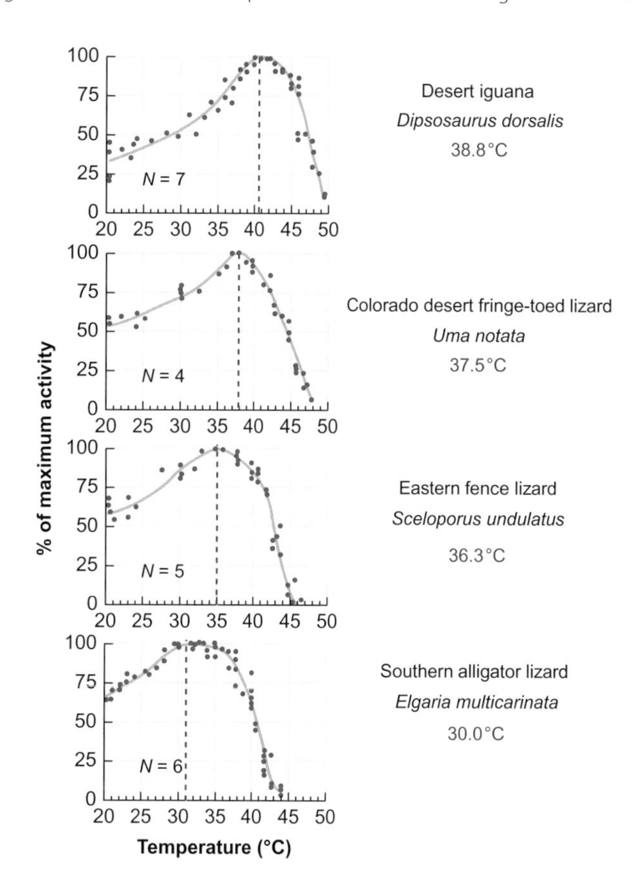

Figure 9.44 Influence of incubation temperature on enzyme activities of myosin-adenosine triphosphatase (myosin-ATPase) from skeletal muscles of four lizard species

Samples of limb and dorsal trunk musculature were used in these experiments. Enzyme activity is plotted as a % of maximal activity. The preferred temperatures of each species based on thermal gradient tests in the laboratory are given in red typeface under the names of each species. In each case, preferred temperatures are close to the temperature at which maximum enzyme activity occurs (indicated by dashed red lines). *N* = number of animals of each species.

Source: Licht P (1964). The temperature dependence of myosin-adenosinetriphosphatase and alkaline phosphatase in lizards. Comparative Biochemistry and Physiology 12: 331–340.

The initial increase in Δ^9-desaturase in carp occurs within 24–48 hours of a reduction in water temperature and results from activation of pre-existing latent desaturase. Then, over a period of 3–5 days, mRNA transcripts for Δ^9-desaturase increase, as shown in Figure 9.42A. This increase results in peak enzyme activity after about 5 days, as shown in Figure 9.42B. From this example, we have discovered a series of events that gives considerable flexibility in matching the response to needs according to the extent and period of cooling.

An ability to increase desaturase activity and generate unsaturated fatty acids seems to be critical to the survival of cold conditions for some species. A good example is a species of herring common in the Great Lakes of North America, alewives (*Alosa pseudoharengus*), which in fresh water are prone to massive winter die-offs. Alewives collected from the wild and given a cold challenge in the laboratory may die. Those that die have lower levels of polyunsaturated fatty acids, particularly docosahexaenoic acid (C22:6) and arachidonic acid (C20:4) than the surviving fish. Figure 9.43 shows the relationship between the monounsaturated fatty acid oleic acid (C18:1) and the polyunsaturated fatty acid docosahexaenoic acid (C22:6) of surviving alewives. Interestingly, these differences are independent of any change in the dietary intake of fatty acids so individual differences in desaturase activity probably determine the ability of alewives to convert oleic acid to docosahexaenoic acid and survive.

9.3.3 Thermal adaptation of enzymes and metabolism

Persistent temperature changes result in changes in gene frequencies and evolutionary adaptations that may by revealed by examining related species that occupy different thermal habitats. Figure 9.44 shows an example of the thermal adaptation of four species of lizards. In each species, myosin ATPase activity in the skeletal muscles[75] peaks at a tempera-

[75] The maximum velocity of shortening of a muscle is proportional to the myosin ATPase activity in the muscle, as we discuss in Section 18.5.2.

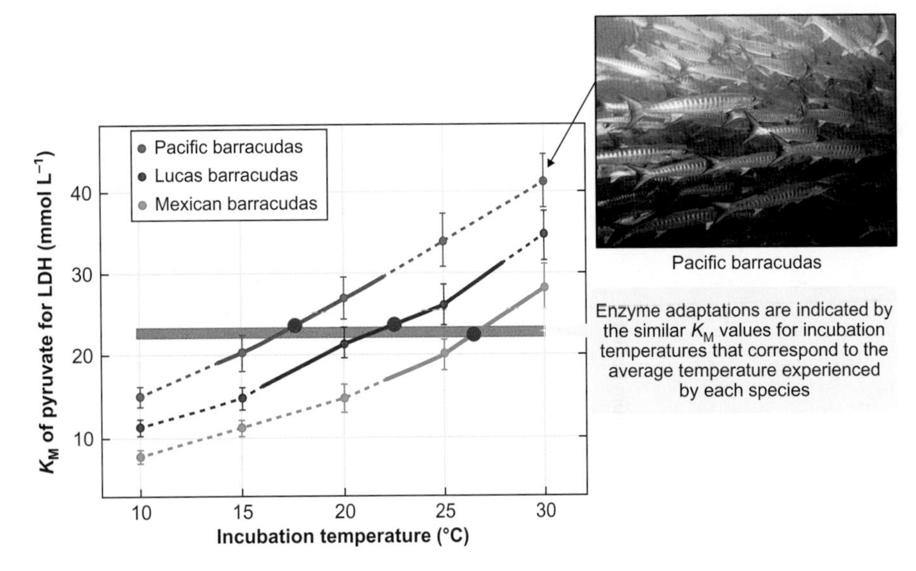

Figure 9.45 Enzyme adaptation of lactase dehydrogenase (LDH) from skeletal muscles of barracudas on the west coast of North America south to Panama

The thick part of each line indicates the range of temperatures they experience in their natural habitats. Red dots indicate average temperature of mid-point of geographic range of each species. The K_M of pyruvate for M4-LDH (LDH5) is strongly dependent on incubation temperature in all species, but values consistently differ between species and show enzyme adaptations that achieve temperature compensation.

Source: adapted from Graves JE, Somero GN (1982). Electrophoretic and functional enzyme evolution in four species of eastern Pacific barracudas from different thermal environments. Evolution 36: 97-106. Image of barracudas: http://marinebio.org.

ture that matches the temperature preference of the lizard. The maximal force of muscle contraction occurs when myosin-ATPase is at its peak, so the muscles of each species are adapted to result in maximum running speed at a temperature that matches their environment.

A more precise way to examine whether particular enzymes have thermally adapted over evolutionary timescales is to study the kinetic functioning of enzymes in congeneric species (species of the same genus) that occupy habitats with different thermal conditions. By studying congeneric species, their evolutionary divergence is small and differences in amino acid sequences are more likely to be the result of temperature adaptation. An exquisite example of this approach is the study of adaptation of the M4-LDH isozyme of lactate dehydrogenase in *Sphyraena* species of barracudas.

The M4-LDH isozyme occurs at high levels in white (glycolytic) muscles of teleost fish, where it is important in burst (anaerobic) swimming[76]. Barracudas are ferocious predators, which rely on bursts of high speed swimming to catch their prey, so evolutionary adaptation of the LDH isozymes of barracudas may well be a key component of their success in different habitats.

Figure 9.45 shows some of the evidence for thermal adaptation of LDH in the barracuda species that occur in different locations along the west coast of North America and south to Panama. Pacific barracudas (*Sphyraena argentea*) occur mainly in southern California. Lucas barracuda (*Sphyraena lucasana*) occur along the Baja California and south along the Mexican coast, while Mexican barracudas *Sphyraena ensis* occur further south and along the Panama coast. Figure 9.45 shows the range of temperatures experienced by each species.

[76] We explore muscle fibres and their use in swimming by fish in Sections 18.2.3 and 18.5.1.

As expected, an increase in incubation temperature increases the rate of LDH function whatever the species, as illustrated by the data in Figure 9.45. A good way to reveal thermal adaptation is to examine the kinetic properties of an enzyme *at the same incubation temperature* for all three species. Considering 25°C in Figure 9.45 we find the lowest K_M (highest affinity)[77] for muscle LDH in the Mexican barracudas living furthest south, while LDH of muscle from Pacific barracudas living furthest north (at lower temperatures) has the highest K_M (lowest affinity).

Now looking at the K_M values at the mid-range temperature at which each barracuda species lives (the red dots on Figure 9.45) we find quite consistent K_M values for the three species, which suggests an evolutionary adaptation has compensated for the persistent temperature differences in the habitats of the barracudas. A difference of just 3°C to 5°C appears to be sufficient to have resulted in the evolution of LDH variants in the three species of barracudas.

The next questions is by how much does a gene (and the amino acid sequence of the enzyme it encodes) need to change for adaptation to occur, and where in the molecule do structural changes occur? Structural studies have shown that the adaptations of LDH involve changes in just one to three amino acids outside the conserved active site of the enzyme. Comparing the structures of LDH in Pacific and Lucas barracudas reveals that alanine substitution for valine

[77] K_M (the Michaelis constant) is the substrate concentration at which the enzyme converts substrate into product at half its maximum rate, and is therefore related to the affinity of the substrate for enzyme molecules (i.e. the strength of binding between an enzyme and its substrate); roughly speaking K_M is inversely related to the affinity of the enzyme for its substrate. This inverse relationship means that a *low* K_M indicates *high* affinity and relatively small amounts of substrate saturate the enzyme so maximum reaction velocity occurs with relatively low substrate concentrations.

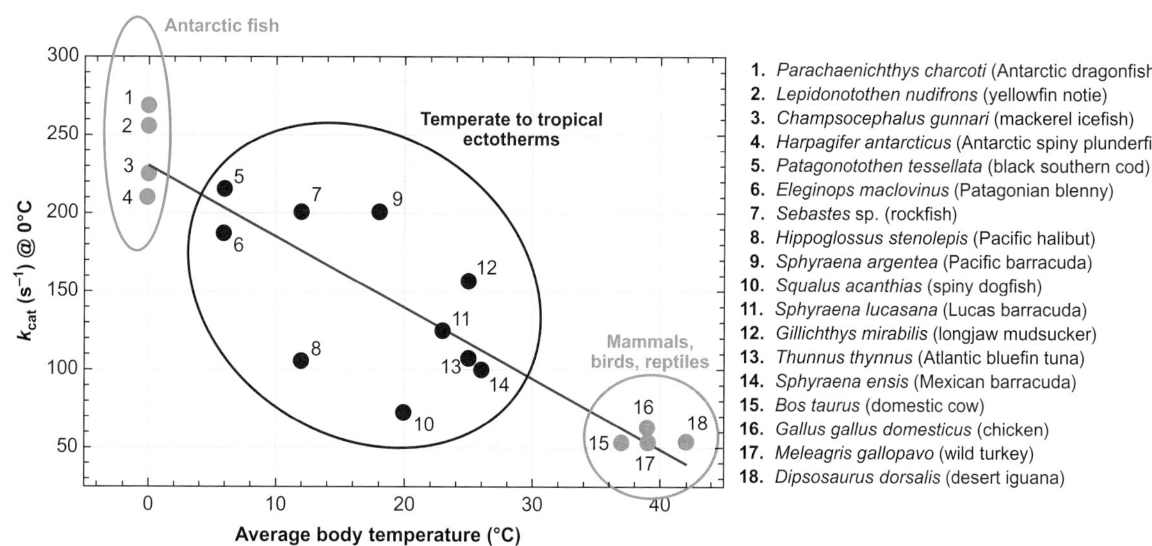

Figure 9.46 Catalytic rate constant of lactate dehydrogenase in muscle tissue measured at 0°C, for 18 vertebrate species, in relation to their average body temperature

LDH activity (M4-LDH = LDH5) correlated with body temperature among the vertebrate species studied. Species 5 and 6 are South American fish; species 7 to 13 are temperate fish; 14 is a tropical fish species; species 18 is an iguana; species 15 to 17 are endothermic mammals and birds (endothermy is discussed in detail in Chapter 10). Catalytic rate constant (k_{cat}) is expressed as enzymatic reactions catalysed per second.

Source: Pörtner H-O et al (2007). Thermal limits and adaptation in marine Antarctic ectotherms: an integrative view. Philosophical Transactions of the Royal Society B 362: 2233–2258.

occurs at position 61, and serine substitution for glycine occurs at position 68 in Pacific barracudas. In another example of LDH adaptation, just a single amino acid change (at position 8) in the LDH of the blackfin icefish (*Chaenocephalus aceratus*) living in the Southern Ocean explains its high K_M (low affinity) for the substrate pyruvate.

Another kinetic property of enzymes related to K_M is the **catalytic rate constant (k_{cat})**, which expresses the rate of product formation when an enzyme is saturated with substrate. The k_{cat} is therefore a measure of the maximum number of substrate molecules converted to a product per unit time. Figure 9.46 examines k_{cat} for M4-LDH in muscle tissue at a constant incubation temperature (0°C) in 18 different species of animals (including some endotherms). Notice from the graph that the rate of catalysis (k_{cat}) is four to five times higher in Antarctic species of fish than for animals with higher average body temperatures. This indicates that Antarctic fish have evolved LDHs that partially offset the effects of very low water temperature. The relationship in Figure 9.46 also shows more generally that enzyme adaptation of species that have evolved to live at colder temperatures results in higher catalytic rate constants than for the enzymes of warm-adapted species.

Thermal adaptation of enzymes often involves changes in their stability. Figure 9.47 shows a good example of the adaptation of ATPase extracted from the white (glycolytic) muscle of species of fish living in very different thermal habitats. Making measurements at 37°C in all cases reveals very different stabilities. The data in Figure 9.47 show that ATPase in muscle samples taken from Antarctic species when incubated at 37°C denatures within a few minutes, whereas myofibrillar ATPase from tropical fish and fish living in hot

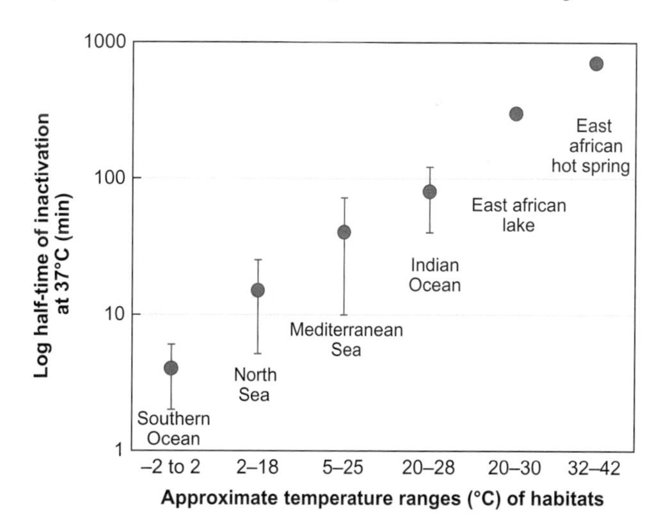

Figure 9.47 Myofibrillar ATPase has a longer half-life in species from warmer habitats than those living in colder habitats

Half-life at 37°C was used as a measure of enzyme denaturation or stability of ATPase extracted from white (glycolytic) muscle fibres of 27 species of bony fish from six thermally different marine habitats. The species represented are three species from the Southern (Antarctic) Ocean, six species from the North Sea, eight Mediterranean species, eight species from the Indian Ocean, one from an east african lake and one species from an east african hot spring. Temperature range for the species in each group are given on the x-axis.

Data are means and ranges or individual values from Johnson IA, Walesby NJ (1977). Molecular mechanisms of temperature adaptation in fish myofibrllar adenosine triphosphatases. Journal of Comparative Physiology 119: 195-206.

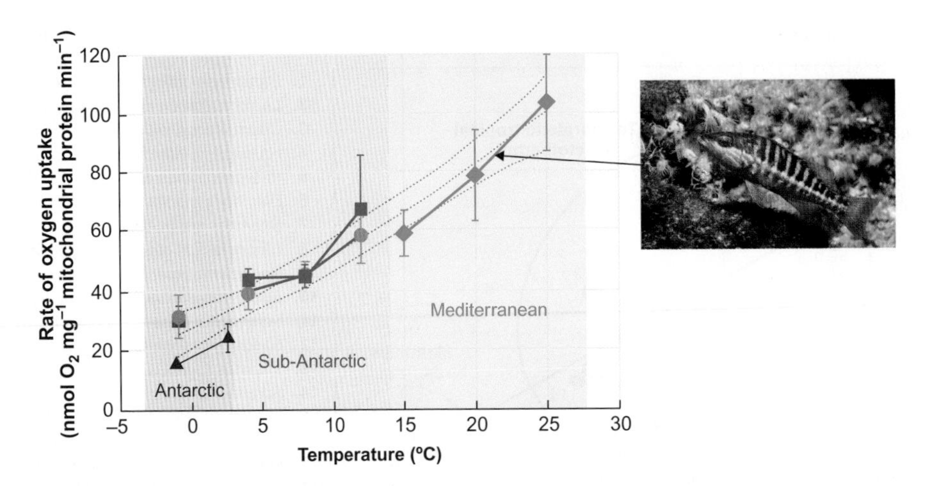

Figure 9.48 Rate of oxygen consumption of mitochondria extracted from red muscles of teleost fish species relative to the average temperature of their habitats

Mediterranean species: comber (*Serranus cabrilla*) (green diamonds); sub-Antarctic species (blue symbols): Patagonian blenny (*Eleginops maclovinus*) (circles) and Maori cod (*Notothenia angustata*) (squares); Antarctic species (purple symbols): dusky rockcod (*Trematomus newnesi*) (circle), yellowfin notie (*Lepidonotothen nudifrons*) (square) and black rockcod (*Notothenia coriiceps*) (triangles).

Data are means ± standard errors for 5–6 measurements. Dotted lines show the line fitted to all the data and its 95% confidence limits.

Source: Johnson IA et al (1998). Latitudinal variation in the abundance and oxidative capacities of muscle mitochondria in perciform fishes. Journal of Experimental Biology 201: 1–12. Image of comber: © Reinhard Dirschert/Alamy Stock Photo.

springs has evolved to function at higher temperatures and lasts for many hours at 37°C.

Given the compensatory adaptations seen for individual enzymes, we might ask whether ectotherms living in cold environments achieve temperature compensation of their overall metabolic rate. This concept is called **metabolic cold adaptation**.

Metabolic cold adaptation

August Krogh, the great Danish physiologist, speculated in 1916 that metabolic cold adaptation explains the abundant and diverse communities of ectotherms found in the cold oceans. Ever since Krogh's speculations the concept of metabolic cold adaptation has been one of the most controversial in physiological ecology and remains so.

If metabolic cold adaptation occurs then different ectothermic species, *when tested at the same temperature*, should show differences in metabolic rates, with species from cold climates having a higher rate of metabolism than those from warm climates, in a similar way to the enzyme adaptation we examined in Figure 9.45. When tested at their habitat temperatures, metabolic rates would be equal in species from different thermal habitats if complete metabolic compensation occurs. On the other hand, if there is no adaptation at all then the rates of metabolism should be equal when tested at the same temperature for species living at different latitudes (and habitat temperatures).

Much of the work on metabolic cold adaptation over the past 100 years has focused on species of fish or insects. The idea initially received substantial support, but early investigations suffered from technical problems, such as not allowing a sufficient amount of time after feeding before measuring the rate of oxygen consumption[78]. It is now accepted by most researchers that while there is some degree of metabolic compensation in some animals it is insufficient to overcome completely the effects of habitat temperatures on oxygen consumption. For example, polar fish generally have metabolic rates that are much less than those of tropical or temperate species when measured at their habitat temperature.

The high level of oxygen consumption by mitochondria shows little apparent temperature compensation and therefore is thought to be an important constraint on metabolic cold adaptation. Notice in Figure 9.48 that the rate of oxygen consumption by mitochondria extracted from the red (aerobic) muscle[79] of marine fish species living at different latitudes follows a similar relationship in Antarctic, sub-Antarctic and Mediterranean species. If temperature compensation occurs, we would expect to find different slopes of the relationship between the rate of oxygen uptake and water temperature in the three groupings.

These results suggest that the capacity of individual mitochondria to generate ATP in the muscle of polar fish could be limited, which would have important consequences for

[78] The effects of feeding on metabolic rate are discussed in Section 2.3.4.

[79] Section 18.2.3 gives details of muscle types and their function.

power generation. However, an increase in the number of mitochondria might overcome the problem. Typically, mitochondria occupy 30–60 per cent of the muscle volume in non-polar fish. In some Antarctic species, such as the Antarctic silverfish (*Pleuragramma antarcticum*), the muscle fibres are surrounded by a high density of mitochondria, which could compensate for the effect of the very low body temperature on mitochondrial function.

› *Review articles*

Cunnane SC (2003). Review. Problems with essential fatty acids: time for a new paradigm? Progress in Lipid Research 42: 544–568.

Danks HV (2005). Key themes in the study of seasonal adaptations in insects I. Patterns of cold hardiness. Applied Entomology & Zoology 40: 199–211.

Hazel JR (1995). Thermal adaptation in biological membranes: Is homeoviscous adaptation the explanation? Annual Review of Physiology 57: 19–42.

Pörtner HO, Peck L, Somero G (2007). Thermal limits and adaptation in marine Antarctic ectotherms: an integrative view. Philosophical Transactions of the Royal Society B 362: 2233–2258.

Pörtner HO, Farrell AP (2008). Physiology and climate change. Science 322: 690–692.

Pörtner HO (2010). Oxygen- and capacity-limitation of thermal tolerance: a matrix for integrating climate-related stressor effects in marine ecosystems. Journal of Experimental Biology 213: 881–893.

Schulte PM (2015). The effects of temperature on aerobic metabolism: towards a mechanistic understanding of the responses of ectotherms to a changing environment. Journal of Experimental Biology 218: 1856–1866.

Somero GN (2004). Review. Adaptation of enzymes to temperature: searching for basic 'strategies'. Comparative Biochemistry and Physiology, Part B 139: 321–333.

Somero GN (2010). The physiology of climate change: how potentials for acclimatization and genetic adaptation will determine 'winners' and 'losers'. Journal of Experimental Biology 213: 912–920.

Somero GN (2012). The physiology of global change: linking patterns to mechanisms. Annual Review of Marine Science 4: 39–61.

Tomanek L, Somero GN (2002). Interspecific- and acclimation-induced variation in levels of heat-shock proteins 70 (hsp70) and 90 (hsp90) and heat-shock transcription factor-1 (HSF1) in congeneric marine snails (genus *Tegula*): implications for regulation of *hsp* gene expression. Journal of Experimental Biology 205: 677–685.

Checklist of key concepts

- Almost all invertebrates, all amphibians, most fish and most reptiles are ectotherms.

Behaviour, temperature sensing and heat exchange of ectotherms

- Ectotherms lose the heat they generate by metabolism to cooler environments by radiation (land animals), conduction and convection, or gain heat from warmer environments by similar processes.
- The body temperature of almost all aquatic ectotherms matches water temperature because of the rapid heat exchange between the two.
- Ectotherms influence their body temperatures by **behavioural thermoregulation**.
- The simplest form of behavioural thermoregulation is **thermal selection** according to the **preferred body temperatures** of aquatic animals.
- Terrestrial ectotherms adjust their rate of heat uptake and heat loss by shuttle behaviour.
- The nocturnal lifestyle of small ectotherms, often accompanied by burrowing, avoids high daytime temperatures in hot deserts.
- Control of behavioural thermoregulation by ectotherms employs internal and external **thermoreceptors**.
- **Albedo** (skin lightness) influences the proportion of the incident solar radiation absorbed and the proportion reflected from the external surfaces of an animal and, hence, the rate of heat uptake, but does not affect the equilibrium temperature when an animal remains in the same place indefinitely.
- **Melanism** describes the permanent darkness resulting from pigment dispersion.
- Melanism in melanistic insects and lizards may be thermally beneficial as a consequence of reduced reflectance in cold conditions.
- Some reptiles influence heat exchange with their environment by controlling the rate of blood flow to surface tissues.
- **Thermal hysteresis** of changing heart rates during warming and cooling occurs in some reptiles and crustaceans; a greater increase in heart rate during warming maximizes heat uptake, while a greater *decrease* during cooling reduces heat loss.
- Among reptiles, brain temperatures typically exceed body temperatures by a few degrees due to heat exchange mechanisms.
- Evaporative cooling by panting and uncoupling respiratory cooling and heat exchange mechanisms avoids heat stress in reptiles once brain/body temperature rises above a particular threshold.

Ectotherms in cold and subzero conditions

- Some ectotherms living in temperate areas migrate to warmer climates before winter.
- Some animals cope with cold conditions by entering a resting phase called **diapause** or **hibernation**.
- **Diapause hormone** in insect eggs initiates and maintains the diapause.
- Adult diapause in insects is usually induced by declining concentrations of **juvenile hormone**.
- During diapause, metabolic rate decreases by about a factor of 10, with stored lipids and carbohydrates being used as energy sources.
- The **critical photoperiod** is the daylength at which half the population of a species enters diapause and is specific for each species and its geographic locality (latitude).

- Temperature is an additional influence on the timing of diapause in some species.
- In temperate habitats, low temperatures trigger the hibernation of amphibians and reptiles in which **metabolic suppression** reduces the rate of usage of energy stores and oxygen needs.
- Ectotherms living in subzero temperatures demonstrate a continuum between **freeze tolerance** and freeze intolerance (**freeze avoidance**).
- Freeze-tolerant species tolerate up 60–65 per cent of total body water within extracellular fluids (ECF) being frozen by:
  - Synthesizing and secreting ice-nucleating proteins/lipoproteins into the ECF, which encourages extracellular freezing.
  - Inoculation of ECF with ice via permeable body surfaces.
- **Cryoprotectants** protect the cells of freeze-tolerant species from damage e.g. by reducing cell shrinkage.
- Intracellular freezing is usually fatal, aside from exceptional cases.
- Freeze-intolerant (freeze-avoidant) animals living in subzero temperatures (below their freezing points) supercool, and/or synthesize antifreezes.
- The two main types of antifreezes in freeze-intolerant animals are:
  - **Colligative antifreezes**, small solutes such as glucose, sorbitol and glycerol.
  - **Non-colligative antifreezes** which suppress the growth of ice crystals and lower freezing point of body fluids.
- The seasonal decrease in photoperiod (shortening daylengths) switches on the synthesis of antifreeze proteins of some northern polar and sub-polar fish species.

Implications of temperature change for tissue functioning of ectotherms

- Acute changes in the body temperature drive rapid changes in the rate of physical and chemical processes in ectotherms (**Q_{10}** effects).
- Thermal performance curves show the effects of temperature on the physiological performance of ectotherms.
- According to the oxygen- and capacity-limited thermal tolerance (OCLTT) hypothesis, the capacity for oxygen uptake and transport is the major determinant of thermal tolerance of aquatic ectotherms.
  - **Optimal performance** occurs at high and relatively stable concentrations of blood oxygen.
  - Oxygen availability in the blood declines at the **pejus temperatures**, but may be sufficient for aerobic metabolism.
  - Temperatures beyond the **upper critical** or **lower critical temperatures** compromise survival and affect the abundance and distribution of species.
- OCLTT and linked parameters, such as maximum heart rates and ventilatory performance, allow prediction of the impacts of climate change.
- Heat shock, cold shock and other stressful stimuli cause protein unfolding in cells, which initiates the synthesis of **heat shock (stress) proteins** that restore the folding and repair of reversibly damaged proteins.
- Cold shock induces rapid **cold hardening** by many terrestrial invertebrates, which are then able to survive subzero conditions.
- Chronic changes in environmental temperatures can result in reversible changes in the rate of biological processes or reaction of ectotherms by:
  - **translational responses**, which compensate for Q_{10} effects by altering the relationship between the rate of a process/reaction and body temperature;
  - **rotational modification**, which changes the slope of the relationship between reaction rate and temperature and affects the outcome at different temperatures.
- Thermal acclimatization or acclimation of enzyme function may result from:
  - quantitative changes in enzyme activity;
  - qualitative changes of **isozymes**;
  - modulative changes in an enzyme's environment.
- In **homeoviscous acclimation** (or **acclimatization**) a change in the proportions of saturated and unsaturated fatty acids in the membrane phospholipids adjusts membrane fluidity in response to ambient temperatures:
  - A higher proportion of more closely packed saturated fatty acids increase membrane stability and reduce membrane fluidity during acclimation to warmer conditions.
  - A higher proportion of unsaturated fatty acids is incorporated into the cell membranes during cold acclimation.
- Persistent thermal changes, e.g. due to global climate changes, may result in changes in gene frequencies and evolutionary **adaptations** by driving selection.
- Thermal adaptation of enzymes, such as lactate dehydrogenase (LDH), indicate that adaptations may involve just a few amino acids that lie outside the substrate-binding site of an enzyme.
- Thermal adaptation of enzymes often involves changes in their stability.

Study questions

*** Answers to these numerical questions are available online. Go to www.oup.com/uk/butler**

1. How do aquatic ectotherms respond to differences in water temperature? Outline suitable experimental designs that could enable investigation of the behavioural thermoregulation of two named species. (Hint: Section 9.1.1.)
2. Why do terrestrial ectotherms often have body temperatures that differ from air temperatures? (Hints: Section 9.1; Section 8.4 discusses the principles of heat transfer.)

3. Explain how some terrestrial ectotherms are adapted to live in hot deserts. (Hint: Section 9.1.1; Box 9.1.)
4. What do we mean by albedo? How do albedo and melanism influence the thermoregulation of named ectotherms? (Hint: 9.1.3.)
5. Outline the characteristic features of the winter hibernation of named ectothermic vertebrates. What influences their survival over winter? (Hint: 9.2.2.)
6. What do we mean by the term diapause? What is the significance of this process in insects? (Hints: Section 9.2.1 and Case Study 9.2.)
7. Explain the behavioural and physiological processes that have evolved in freeze-tolerant animals. (Hint: Section 9.2.3.)
8. Why do teleost fish in cold Arctic and Antarctic waters face the potential risk of their bodies freezing, while this is less of a problem for marine invertebrates? (Hints: Sections 9.2.3, 9.2.4; Sections 5.1.1 and 5.1.3 give relevant information on the osmotic composition of body fluids.)
9. Outline the main types of antifreeze proteins that have evolved. How can their effectiveness be studied? (Hints: Section 9.2.4; Experimental Panel 9.1.)
10. Two types of thermal acclimation responses occur in ectotherms. How do these responses differ? (Hint: Section 9.3.2.)
11. Researcher A is overheard speaking to researcher B saying 'So after I transferred the frog to the cold temperature its levels of cytochrome B adapted to the changed conditions'. Explain the error in this statement. (Hints: Section 9.3; Section 1.5.2 also discusses relevant definitions.)
12. What is homeoviscous acclimatization? Explain what benefits this has in a named species. (Hint: Section 9.3.2.)
13. What does C20:4 Ω6, tell you about a fatty acid? How and why do ectotherms manipulate their fatty acid composition? (Hints: Section 9.3.2; Box 9.2.)
14. Outline an example of thermal adaptation of an enzyme in ectotherms. Do these responses result in adaptation of metabolic rates? (Hint: Section 9.3.3.)
15. How can an understanding of thermal tolerances, acclimatization and adaptation help in predicting the effects of climate change of ectotherms in different habitats and geographical locations? (Hints: Section 9.3 and Case Study 9.1.)
16.* The catalytic rate constant (k_{cat}) for lactate dehydrogenase (LDH) in muscle cells of a species of frog kept at 20°C when measured at 20°C was 150 per second. Based on this value calculate the following:
   (i) Assuming there is a similar quantity of LDH in the cells, what would you expect k_{cat} to be when measured at 0°C after the frogs were moved to these conditions and held there for 2 hour if the Q_{10} of the reaction rate is 2?
   (ii) Predict what would happen to k_{cat} measured at 0°C after keeping the frogs at this temperature for 3 months if complete compensation for temperature occurs. (Hint: Section 9.3.)

Bibliography

di Prisco G, Pisano E, Clarke A (1998) (Editors). Fishes of Antarctica. A Biological Overview. Springer Verlag.

Ernst CH, Lovich JE (2009). Turtles of the United States and Canada. Second Edition. Baltimore, MD: John Hopkins University Press.

Gerday C, Glansdorff N (2009) (Editors). Encyclopaedia of Life Support Systems (EOLSS) Theme 6.73 Extremophiles. Cheng C-H. C. Freezing avoidance in polar fishes. pp 215–232. Eolss Publishers UK.

Hochachka PW, Somero GN (2002). Biochemical Adaptation: Mechanisms and Process in Physiological Evolution. Oxford: Oxford University Press.

Nijhout HF (1994). Insect Hormones. Chapter 7. Diapause. Princetown University Press.

Rogers AD, Johnston NM, Murphy EJ, Clarke A (2012) (Editors). Antarctic Ecosystems: An Extreme Environment in a Changing World. Chapter 13 by Pörtner HO, Peck LS, Somero G N. Mechanisms defining thermal limits and adaptation in marine ectotherms: an integrative view. Chichester: John Wiley & Sons.

Rustan AC, Drevon CA (2005). Fatty Acids: Structures and Properties. In: Encyclopedia of Life Sciences. Wiley & Sons. els.net.

10

Temperature regulation in endotherms

When animals first invaded land, they encountered a medium with a lower thermal conductivity and a much lower thermal capacity, on a volume basis, than water[1]. That medium was air. The lower thermal capacity of air means that its temperature is more variable than that of water, which can pose problems for some species. However, it also means that the rate of heat flow from a warm body in air is much lower than it would be in water[2]. This lower rate of heat flow in air provides an environment in which the energy required to maintain a body at a temperature several degrees above that of its environment is much lower than it would be in water. Thus, air is a more suitable medium in which birds and mammals, with their high body temperatures, could evolve. Animals that regulate their body temperature above that of the environment by using external sources of heat are called ectotherms[3], whereas those that generate heat internally are known as endotherms (Greek, *endon*—within, *thermé*—heat), which we explore in this chapter. Endothermic animals—birds and mammals—are able to generate sufficient heat internally to balance heat losses and thereby maintain body temperature independent of environmental temperature.

Endothermy has evolved on a number of occasions in terrestrial animals[4], some of which have become secondarily aquatic. Living in water poses additional problems for sustaining internal heat production because of the high thermal conductivity of water relative to air.

The main animal groups alive today that are endothermic are the mammals and birds. There are also a few species of reptiles, fish and insects that are able to warm specific parts of their body, such as the locomotor muscles—a phenomenon known as **regional endothermy**.

In this chapter, we start by examining the theoretical amounts of heat that endothermic individuals must produce in order to maintain a constant body temperature. We then discuss the particular features of endothermy in the different taxonomic groups. The thermoregulatory status of extinct animals such as the dinosaurs remains controversial, so toward the end of this chapter we review the evidence for and against endothermy in dinosaurs.

1 The thermal conductivity of water is about 23 times that of air, and its volume-specific thermal capacity at 20°C, is almost 3500 times greater, as shown in Tables 8.1 and 8.3.

2 Principles of heat flow are discussed in Section 8.4.

3 We discuss ectothermy in Chapter 9.

4 Evolution of endothermy is briefly discussed in Section 2.2.5.

10.1 Heat requirements of endotherms

Imagine a cup of coffee with a temperature of 40°C, which is approximately the same temperature as a bird or mammal. In a room with an air temperature at 40°C, no heating of the coffee is needed to maintain its temperature because there would be no temperature difference to drive the loss of heat from the cup. If the room temperature is about 20°C, however, the coffee would continuously lose heat by long-wave radiation and, if in still air, by free convection to the air in the room and by conduction[5] to the table. If the air is moving, heat would also be lost by forced convection. Eventually the coffee would cool to room temperature. The cooling curve would follow an exponential approach to the ambient temperature as shown in Figure 10.1A.

In order to keep the coffee at 40°C, energy (as heat) has to be supplied at a sufficient rate to balance the initial rate of heat loss—for example by an electric heating coil submerged into the coffee. The lower the temperature of the air in the room, the greater difference in the temperature between the coffee and the room and, therefore, the greater the rate at which heat would need to be supplied to the coffee in order to maintain its temperature at 40°C. A plot of the rate of heat input required to keep the coffee temperature constant against room (ambient) temperature looks like the one depicted in Figure 10.1B.

The characteristics of the cup also affect rate of heat loss. The shape of the cup and the material from which it is made—particularly the thermal conductivity (and hence thermal conductance)[6] of that material—are the important factors. For example, a polystyrene cup will keep the coffee hot for longer than a ceramic mug. But why is this the case?

Expanded polystyrene is filled with small bubbles of still air, making its thermal conductance[7] lower than that of a ceramic mug. Consequently, expanded polystyrene increases

5 Routes for heat exchange are discussed in Section 8.4.

6 Thermal conductivity and thermal conductance are discussed in Section 8.4.2.

7 The inverse of thermal conductance is thermal resistance or insulation, so as thermal conductance decreases, thermal resistance (insulation) increases, as illustrated in Equations 8.5 and 8.6.

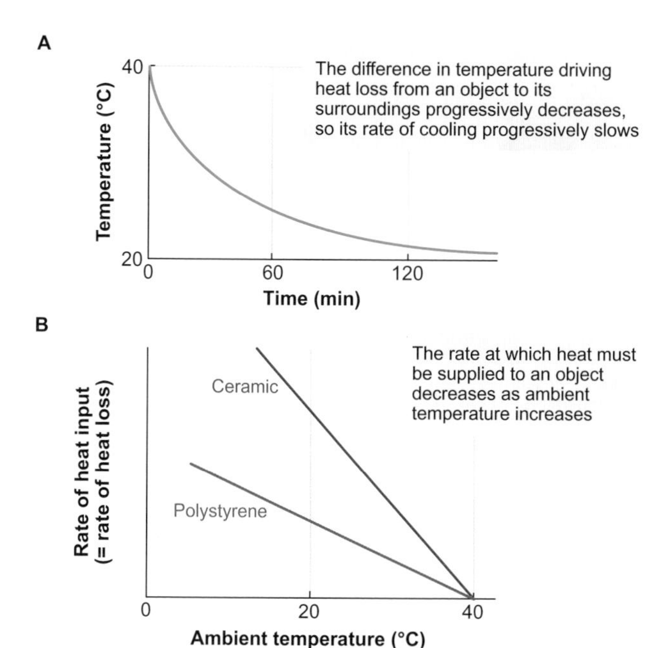

Figure 10.1 Hypothetical cooling characteristics for an object (e.g. a cup of coffee) above ambient temperature

(A) The object cools exponentially from its initial temperature (40°C) to the ambient temperature (20°C). (B) Rate of heat input required to maintain an object (a cup of coffee at 40°C) at a constant temperature above ambient temperature over a range of temperatures. Two lines are shown—one for coffee in a ceramic cup and another for coffee in a polystyrene cup. The slope of each line is the thermal conductance of the material from which the cup is made.

the resistance to heat loss and is, therefore, a good thermal insulator. This insulation means that the rate of heat flow is less from coffee in a polystyrene cup than from coffee in a ceramic mug when the temperature difference between the coffee and the air is the same. Therefore, the rate of heat input needed to balance the rate of heat loss is correspondingly lower. This is reflected in the shallower slope of the line in Figure 10.1B for the polystyrene cup. Overall, then, the slopes of the lines in Figure 10.1B are reflections of the rates of heat loss and the thermal conductances of the two materials.

The body of an endotherm can be compared to the cup of hot coffee. It has a core that it is maintained a temperature of about 40°C (usually around 38°C in many mammals and around 40°C in many birds), and usually has an external coat of fur or feathers, or a subcutaneous layer of fat, that provides insulation and reduces the rate of heat loss. In equatorial regions, air temperature is closer to body temperature than at higher latitudes (the higher the latitude, the lower the air temperature). So, the evolution of effective insulation enabled endothermic animals to move to higher latitudes.

Endothermic animals can vary their thermal conductance by simply tucking in their limbs to reduce the surface area that is exposed, or by making their hairs or feathers stand up on end (**piloerection**—Latin, *pilus* hair in mammals, and **ptiloerection**—Greek, *ptilon* feather in birds) to increase the insulating layer of air. Overall thermal conductance can also be reduced by the presence of countercurrent heat exchangers[8] in the extremities, such as those in the legs of birds and the tails and flippers of aquatic mammals. Whatever an animal does to control rate of heat loss, there are two limits to these manipulations: the point at which thermal conductance is at its maximum and the point at which it is at its minimum.

A living animal always produces a certain level of heat as a result of the cellular processes keeping it alive. In endotherms, this is called the **basal metabolic rate** (BMR)[9]. Figure 10.2 shows the lines of maximal and minimum conductance and BMR. The BMR line crosses the thermal conductance lines at two different ambient temperatures; it intersects the line of rate of heat loss at minimal thermal conductance at the lower critical temperature (T_{lc}), and crosses the line of rate of heat loss at maximal thermal

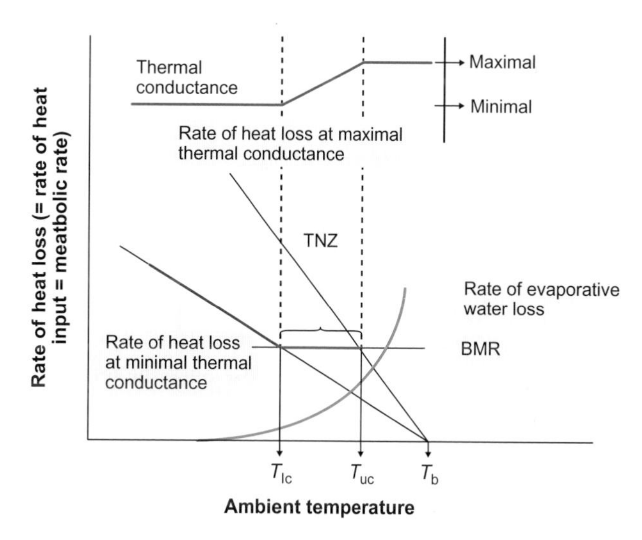

Figure 10.2 Theoretical rate of heat loss lines at maximal and minimal thermal conductance and basal metabolic rate (BMR) for an endotherm regulating its body temperature at T_b, in relation to ambient temperature

The point at which the basal metabolic rate crosses the thermal conductance lines defines the upper and lower critical temperatures (T_{uc} and T_{lc}, respectively) and between these is the thermoneutral zone (TNZ). Actual rate of metabolism for a typical endotherm is shown in red. The blue line above the graph shows changes in thermal conductance of an animal between its maximal and minimal values. Green line shows evaporative water loss. Note that the thermal conductance lines in the graph extrapolate back to body temperature (T_b).

[8] Countercurrent exchange is discussed in Sections 3.3.1 and 12.2.2 and Box 12.1. The heat exchangers in the tails and flippers of aquatic mammals are discussed further in Section 10.2.8.

[9] The internal source of heat in all animals and the difference between standard metabolic rate in ectotherms and basal metabolic rate in endotherms are discussed in Sections 2.2.1 and 2.2.3.

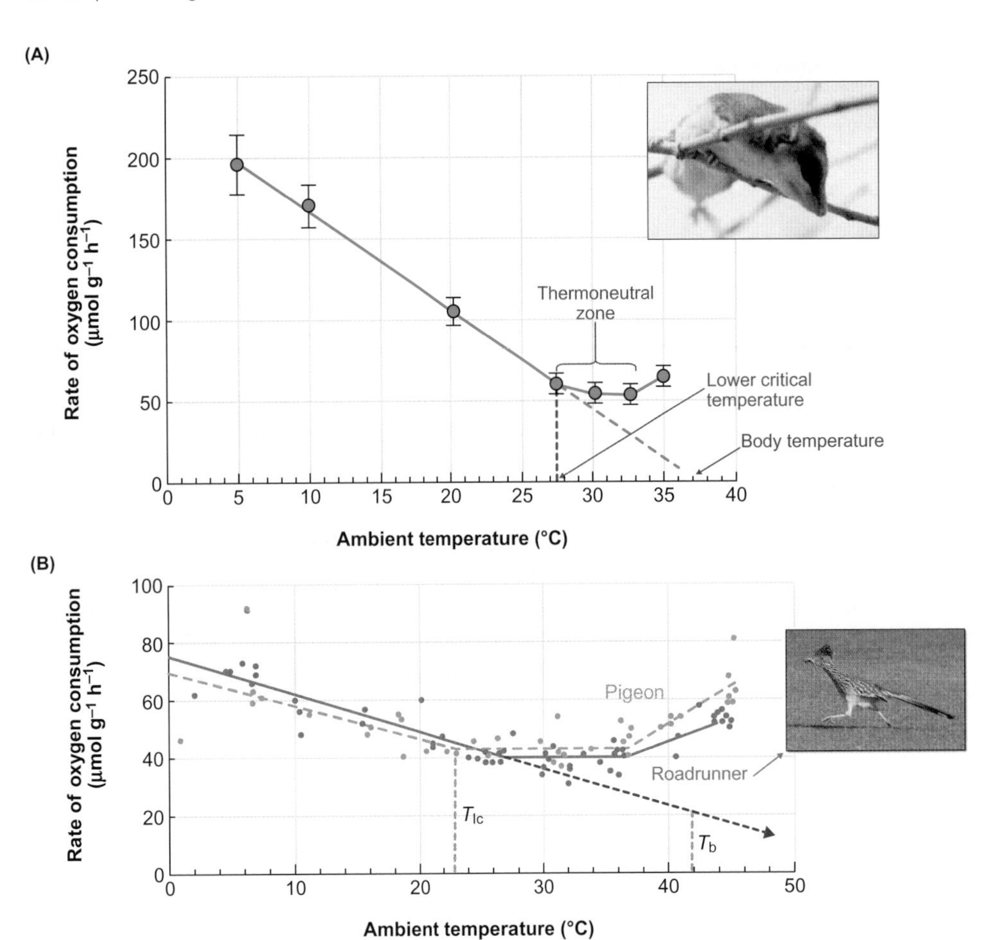

Figure 10.3 Metabolic rate (rate of oxygen consumption) plotted against ambient temperature for mammals and birds

(A) This graph shows the thermoneutral zone, lower critical temperature (T_{lc}) and body temperature (T_b) of the Chilean mouse opossum (*Thylamys elegans*) at normal body temperature. (B) In some species of birds such as common pigeons (*Columba livia*) and road runners (*Geococcyx californianus*) the thermal conductance line does not extrapolate back to body temperature.

Source: A: adapted from Bozinovic F et al (2005) Energetics, thermoregulation and torpor in the chilean mouse-opossum *Thylamys elegans* (Didelphidae). Revista Chilena de Historia Natural 78: 199–206. B: adapted from Calder, WA and Schmidt-Nielsen, K (1967). Temperature regulation and evaporation in the pigeon and the roadrunner. American Journal of Physiology. 213: 883-889. Opossum image courtesy of Yamil Hussein; roadrunner image courtesy of Lip Kee.

conductance at the upper critical temperature (T_{uc}). These are called 'critical' temperatures because once the temperature goes lower than T_{lc}, or higher than T_{uc}, the animals face a thermoregulatory problem.

An animal can modify its thermal conductance between the lower and upper critical temperatures, as represented by the blue line in Figure 10.2, so that the rate at which heat input is required to maintain a stable body temperature exactly matches BMR. Within this ambient temperature range there is no change in the rate of heat production, so this range is called the thermoneutral zone (TNZ). The pattern of variation in energy demands in relation to ambient temperature is known as the standard **Newtonian model**[10], after Sir Issac Newton, the famous physicist and mathematician who first derived the principles on which it is based, or the **Scholander–Irving model**, named after the two workers who, in 1950, were the first to apply Newton's principles to the study of animal energetics.

If ambient temperature rises above T_{uc}, the rate of heat production of an animal exceeds that needed to maintain body temperature, yet thermal conductance is at its maximum. The animal is producing too much heat but cannot increase conductance further to lose the excess heat, and cannot immediately lower its basal metabolic rate. We see how different animals cope with this problem in the next section.

When ambient temperature falls below the T_{lc}, the animal has the opposite problem; the rate at which heat is required to maintain body temperature is more than its BMR but its insulation is at its maximum as the animal cannot further reduce its thermal conductance, as illustrated in Figure 10.2. Below the T_{lc}, body temperature is maintained by increasing the rate of heat production above BMR to match the increased rate of heat loss as ambient temperature falls.

Like the ectotherms, endotherms can obtain additional heat from outside their bodies. However, mammals and birds can also generate the extra heat needed internally, whereas the majority of ectotherms cannot. Examples of the rate of heat production (presented as rate of oxygen consumption) in relation to ambient temperature for a typical mammal and for some species of birds are shown in Figure 10.3. In mammals and some species of birds, the thermal conductance line extrapolates back to body temperature, as shown in Figure 10.3A. By contrast, in many species of birds, such as pigeons (*Columba livia*) and roadrunners (*Geococcyx californianus*), there is a different pattern. Figure 10.3B shows

[10] The Newtonian model of heat balance is discussed in Section 8.4.2.

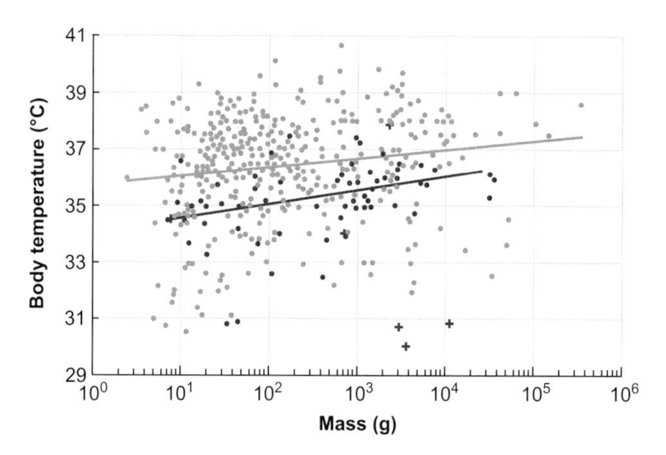

Figure 10.4 Body temperature in relation to body mass for mammals

Upper line is eutherians (placental mammals, ●) and the lower line metatherians (marsupials, ● and monotremes, +). Reproduced from White, CR and Seymour, RS (2003). Mammalian basal metabolic rate is proportional to body mass$^{2/3}$. Proceedings of the National Academy of Sciences of the United States of America 100: 4046-4049.

that the thermal conductance lines of these two species of birds do not extrapolate to body temperature on the x-axis but extend to a higher temperature.

A possible explanation for this difference is that birds such as pigeons and roadrunners do not continue to decrease their conductance (increase their insulation) to its minimum before increasing their rate of heat generation. Instead, their metabolic rate increases while conductance is still falling. The result is that the metabolic rate does not increase along a line of fixed thermal conductance, and hence the thermal conductance line does not extrapolate to body temperature.

10.2 How endotherms regulate their body temperature

Most mammals regulate their body temperatures at around 36–37°C, but eutherians (placental mammals) regulate at slightly higher values than the metatheria (marsupials) and monotremes (egg-laying mammals), as shown in Figure 10.4. On the other hand, birds generally maintain a higher body temperature than that of mammals, at around 38–40°C.

Regulated body temperature also varies with time of day. Figure 10.5A,B illustrates that for mammals and birds active during daylight (**diurnal**—Latin, *diurnalis* daily—animals) regulate their body temperature to be about 2–3°C lower at night than during the day. The opposite is true for animals that are active at night (**nocturnal**—Latin, *nocturnalis* of the night). These daily changes in body temperature may be absent in species that spend at least part of the year at high latitudes and when there is continuous daylight during summer, as shown in Figure 10.5C. Similarly, there is an absence of circadian activity in other species living at high latitudes, such as reindeer (*Rangifer tarandus*) and Svalbard ptarmigan (*Lagopus mutus hyperboreus*).

Body size (mass) seems to have an effect on the extent of the daily variation in body temperature. For example, in relatively large mammals, like humans, the difference between active and quiescent phases may be less than 1°C, but in small mammals, such as mice, the difference may be 2–3°C. This difference is related to the fact that larger animals tend to have a smaller surface area/volume ratio (sa/vol ratio) than smaller animals, and therefore lose heat more slowly when not active, although shape of the body can also influence daily temperature variations.

10.2.1 The importance of mass in thermoregulation

Big objects cool down more slowly than small objects, even if they are made of the same material because the ratio of their surface area to their volume is lower than in smaller objects.

Imagine a cube measuring 10 mm on each side. The total volume would be 10 × 10 × 10 mm = 1000 mm^3. The total surface area would be (six sides, each 10 mm × 10 mm) = 600 mm^2. If two similar cubes are placed together, they would have twice the volume (2000 mm^3), and twice the mass, but not twice the surface area. The surface area would now be 10 mm × 10 mm for two sides and 10 mm × 20 mm for four sides, giving a total of 1000 mm^2. In other words, the ratio of the surface area relative to the volume would get smaller, from 0.6 (600 mm^2/1000 mm^3) to 0.5 (1000 mm^2/2000 mm^3).

Heat is generated by the volume of a body, but is lost across the body's surface, so its lower surface area to volume ratio makes a larger object easier to keep warm. By contrast, smaller objects have a larger surface area to volume ratio and lose heat faster. These relationships do not apply to flat, thin shapes, however, as explained when we explore gas exchange in Section 11.2.1.

Examples of thermoregulatory curves for a large and a small bat are shown in Figure 10.6A. Note that the thermal conductance (slope of the line below the lower critical temperature) for the larger animal is higher than that of the smaller animal because of its greater surface area. This relationship changes if we make these plots in a slightly different way, by dividing all the rates of oxygen consumption by body mass, as illustrated in Figure 10.6B. The curve for the heavier animal now lies below that for the lighter animal, although the lower critical point temperatures (T_{lc}) remain in the same place. These plots show that the rate of oxygen consumption ($\dot{M}O_2$) *relative to body mass* (known as **mass-specific** $\dot{M}O_2$) required to keep a larger animal warm is lower than that for a smaller animal and that this difference is greater the colder it gets. However, as shown in Figure 10.6A, larger endotherms still need a greater overall rate of heat production (measured

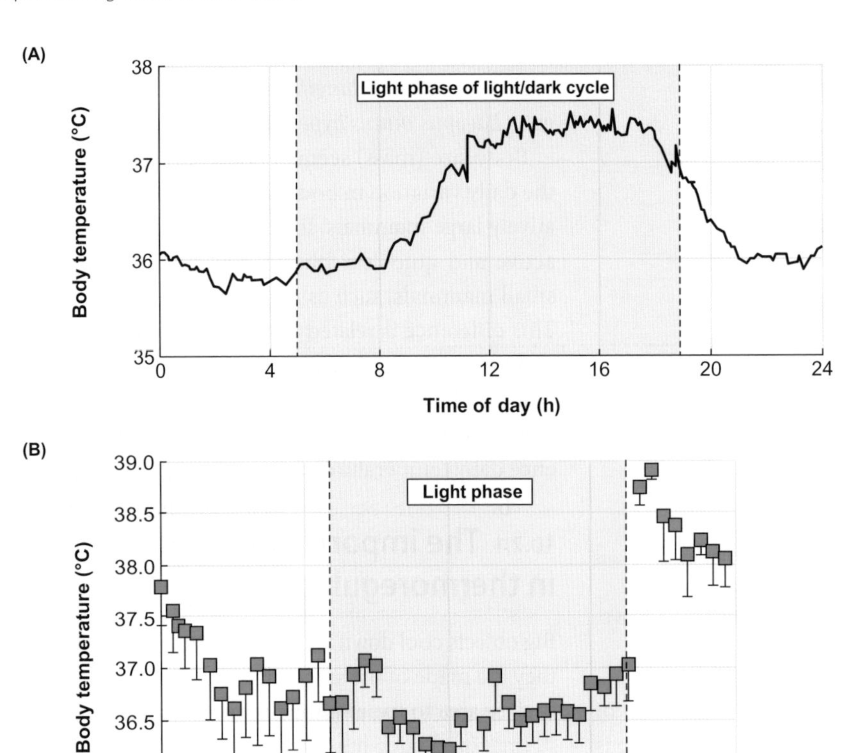

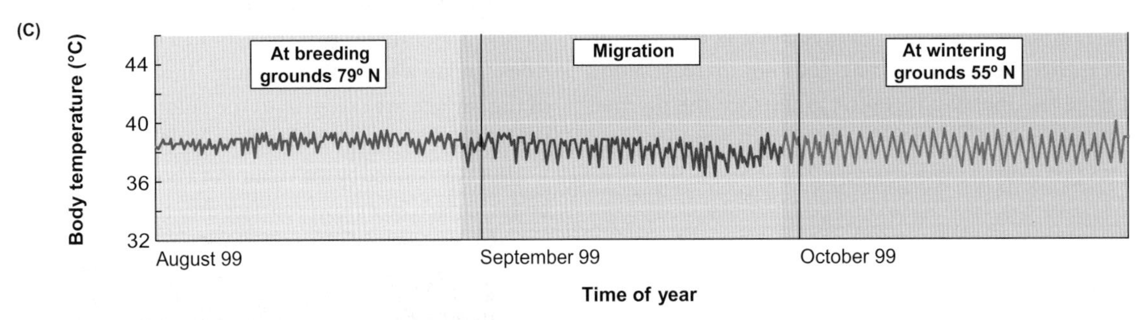

Figure 10.5 Diurnal patterns of variation in body temperature

(A) Typical daytime active mammal (Richard's ground squirrel, *Spermophilus richardsonii*). (B) Night time active mammal (mouse, *Mus musculus*). (C) Barnacle goose (*Branta leucopsis*) before, during and after its migration from its high Arctic breeding grounds, where there is constant daylight, to its wintering grounds in southern Scotland, where there is a daily light/dark cycle. Note the lack of a clear diurnal rhythm in body temperature at the breeding grounds compared with that at the wintering grounds.

Sources: A: Refinetti, R. 1999. Relationship between the daily rhythms of locomotor activity and body temperature in eight mammalian species. American Journal of Physiology 277: R1493- R1500 B: Reproduced from Speakman, JR (2008) The physiological cost of reproduction in small mammals. Philosophical Transactions of the Royal Society 363: 375-398 C: Butler and Woakes, unpublished data

as $\dot{M}O_2$) to keep themselves warm compared with smaller endotherms, whatever the temperature.

At any given environmental temperature, some animals in a community may be well below their T_{lc}, while others may be within their thermoneutral zone (TNZ). For example, at an ambient temperature of 20°C, a small shrew (*Sorex* sp., body mass approximately 10 g) would be about 15°C below its T_{lc} and would need to generate a substantial amount of heat above that generated by its BMR in order to keep itself warm. However, at the same ambient temperature, a 1 kg hedgehog (*Erinaceus* sp.) would be within its TNZ. Thus, in a thermoregulatory sense, 'cold' and 'hot' are relative to the ambient temperature when considered in relation to the animal's thermoregulatory characteristics, as illustrated earlier in Figure 10.2.

Endothermic animals may experience fluctuations of ambient temperature above their upper and below their lower critical temperatures for varying periods of time. How animals respond to such variations in ambient temperature depends at least partly on the duration of the changes.

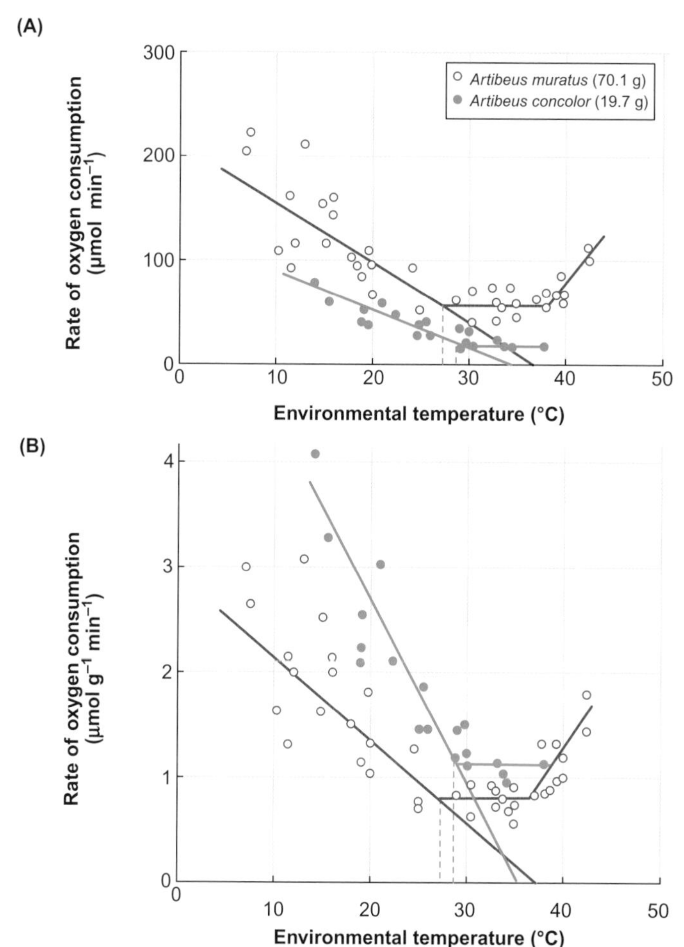

Figure 10.6 Thermoregulatory curves for a large (open symbols) and small (closed symbols) species of bat

(A) Whole animal metabolic rates; (B) Mass-specific metabolic rates. Note, the slopes of the lines below the lower critical temperatures (blue dashed lines) indicate the thermal conductances.

Modified from McNab, BK (2002) The Physiological Ecology of Vertebrates. A View from Energetics. Comstock Publishing Associates.

10.2.2 Thermoregulatory responses to periodic heat loads

High temperatures that temporarily exceed the upper critical temperature (T_{uc}) may be encountered in many tropical and desert environments. In these situations, animals have several options, including those that facilitate heat loss by increasing radiation, conduction, convection or evaporation[11]. Options include:

- increasing blood flow to the periphery,
- forced convection,
- increase rate of heat loss by evaporative cooling,
- storage of heat.

[11] Radiation, conduction, convection and evaporation are discussed in section 8.4.

Increasing blood flow to the periphery as a means of heat loss

Increasing blood flow to the peripheral tissues raises the surface temperature and leads to an immediate increase in the rate of heat loss by long-wave radiation, conduction and convection. Because birds and mammals generally have an insulative covering that results in heat retention, simply increasing the flow of blood to surface tissues will not necessarily elevate the rate of heat loss. Rather, they need to increase the flow of blood to specific parts of their bodies where the covering of insulation is thinnest in order to lose heat.

Thermal images[12] of the surface temperatures of mammals and birds at different ambient temperatures allow us to see the differences that result from the regional distribution of blood flow and how that can maximize the rate of heat loss. An example of such an image is shown in Figure 10.7B, which is

(A)

(B)

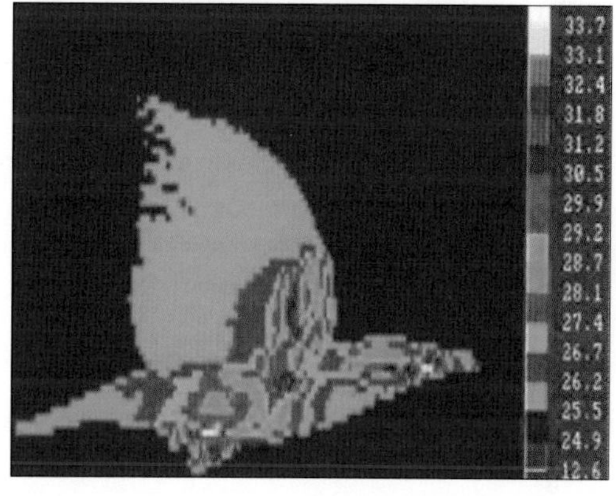

Figure 10.7 European starling (*Sturnus vulgaris*) flying in a wind tunnel

(A) The bird is photographed using visible light. (B) Thermal image of the same bird. Hotspots are clearly visible on the forewing, around the eyes and the lower segments of the legs. The thermal imaging camera detects infrared emitted radiation only.

Source: A: Speakman JR, unpublished; B: Reproduced from: Ward, S et al (1999). Heat transfer from starlings *Sturnus vulgaris* during flight Journal of Experimental Biology 202: 1589-1602.

[12] Section 8.4.1 and Figure 8.12 give more information on thermal images.

compared with an ordinary photograph of the same bird in Figure 10.7A. Such images reveal that many mammals and birds have areas of their bodies that are poorly insulated and that they can flood with blood, if necessary, to dissipate heat. These are called **thermal windows** because they maximize the ability of heat to flow from the body. The underside of the wings of birds (Figure 10.7B) and the tail of a rat are such examples.

Figure 10.8A depicts a cross-section of a rat's tail. At room temperature, the body temperature of the rat is at its normal value (around 37°C) and the rate of blood flow into the tail is usually quite low. When the body temperature of the rat is artificially raised above its normal value, however, the lateral veins in the tail vasodilate, as shown in Figure 10.8B, raising its surface temperature and increasing the rate of heat loss.

Heat loss by forced convection

As well as increasing their surface temperatures, animals can actively assist heat loss by increasing forced convection, such as by fanning themselves. African elephants (*Loxodonta africana*) use their large ears to fan themselves when the ambient temperature rises above the T_{uc}, in addition to increasing the rate of blood flow to their ears. However, using the ears as fans actually generates internal heat because of the muscular work involved. Hence, at some ambient temperature, the elevated rate of heat production will offset the elevated rate of heat loss arising from the increased convection.

Figure 10.9 shows that this offsetting of heat loss by heat production is increasingly likely as ambient temperature approaches body temperature and the temperature difference driving convective heat loss from the ears to the surrounding air becomes progressively smaller. Consequently, fanning is not a commonly observed strategy in small mammals, since their T_{uc} is closer to body temperature.

Evaporative heat loss

The most common way to lose heat when body temperature rises above the T_{uc} is by evaporation, which increases more rapidly above T_{uc}, as demonstrated earlier in Figure 10.2. Heat is required to change liquid water into a vapour at the same temperature[13], so when water evaporates from the

13 Sections 3.3.1 and 8.4.4 explain evaporation.

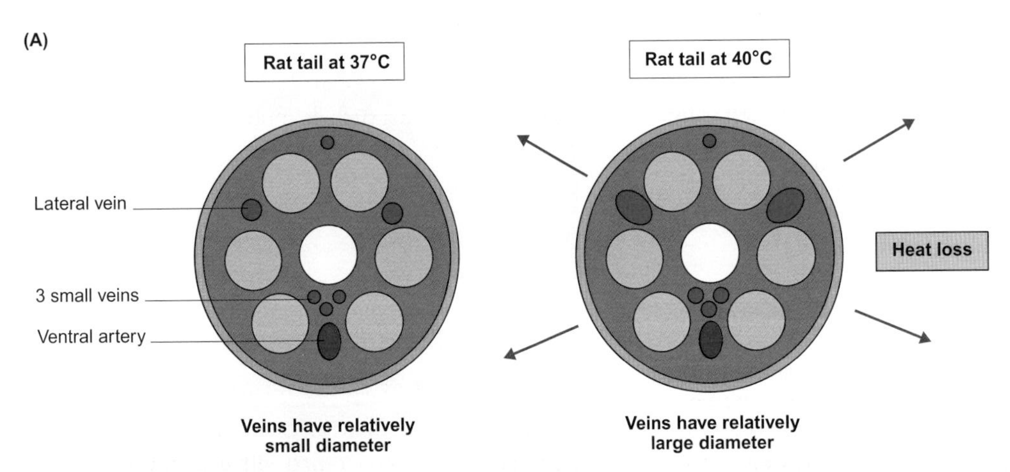

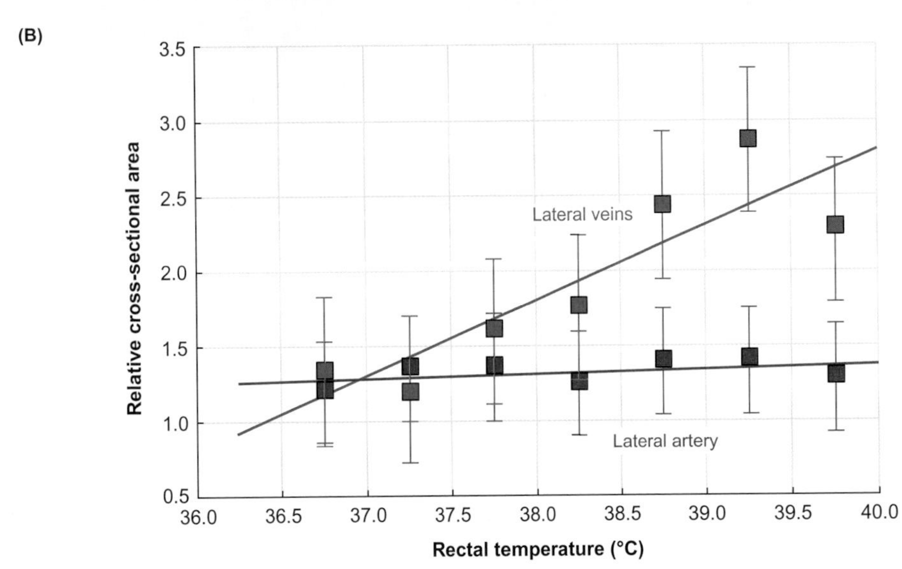

Figure 10.8 Heat loss across the tail of a rat

(A) Diagrammatic representation of cross-section of base of rat tail at body temperatures of 37 and 40°C based on NMR imaging. (B) Plot showing cross sectional areas of lateral veins and artery in the base of the tail in relation to rectal temperature.

Reproduced from Vanhoutte, G et al (2002). In-*vivo* non-invasive study of the thermoregulatory function of the blood vessels in the rat tail using magnetic resonance angiography. NMR in biomedicine 15: 263-269.

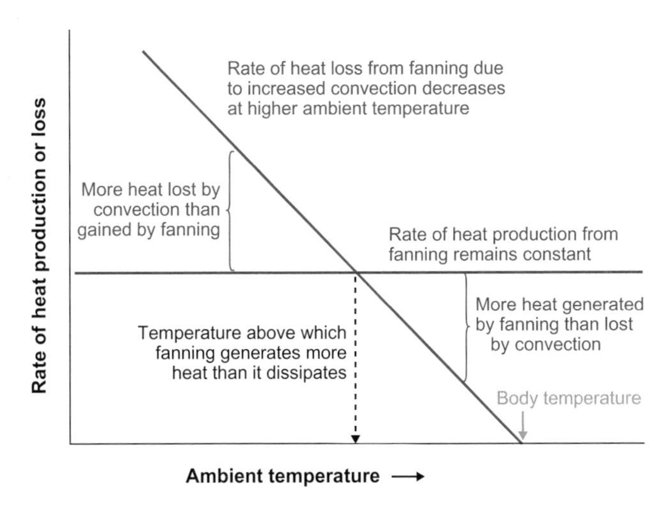

Figure 10.9 Effects of ambient temperatures on rates of heat loss and heat gain from fanning behaviour in animals such as the African elephant (*Loxodonta africana*)

Above a certain temperature, fanning becomes disadvantageous as the muscular activity generates more heat than the fanning dissipates. This situation is a result of the smaller difference in temperature between ambient air and the animal's body.

surface of an animal, heat is lost. The most effective way for an endotherm to make use of evaporation is to cover the body with water from the environment and let it evaporate. The elephant is again a good example, drawing up water in its trunk and spraying itself[14]. For most animals, however, hot temperatures coincide with a shortage of external water supplies; in these conditions, animals derive the evaporative water internally. As the internal supply of water is limited[15], however, its use in this way can only be for a limited duration.

Animals can evaporate water for cooling in several ways. Probably the most common is simply to lick the body surface with saliva, which rats do when their body temperature exceeds 38.5°C. This behaviour appears to be centrally controlled since it can be elicited by direct heating of the hypothalamus[16]. Two additional and more effective ways of evaporative cooling are sweating and panting.

Sweating involves the activity of specialized glands that are spread across the skin surface. Sweat glands secrete fluids (sweat) that are **hyposmotic**[17] to the blood plasma onto the body surface, where they evaporate. As sweat is not pure water, sweating involves an inevitable loss of solutes, such as the important sodium, potassium and chloride ions, as well as water. Animals that live routinely in hot environments produce sweat that is low in solutes to minimize such losses.

[14] Figure 8.14 shows a photograph of an elephant spraying itself.

[15] Water balance in land animals is discussed in Chapter 6.

[16] The hypothalamus is a region of the brain involved in temperature regulation as discussed in Sections 10.2.5 and 16.1.2.

[17] Hyposmotic fluids have a lower osmolality than the plasma, as explained in Section 5.1.3.

Birds and many species of mammals (e.g. dogs, pigs and rodents) lack large numbers of sweat glands but can lose large amounts of heat by evaporative cooling across the respiratory surfaces[18]. In doing so, the animals pant, and an immediate advantage of evaporative fluid loss while panting compared with sweating is that it does not involve loss of solutes.

During panting, mammals rapidly increase their ventilation rate. The actual rate normally matches the natural **resonant frequency** of the respiratory system[19]. Minimal energy is required to oscillate a system at its resonant frequency, as energy is first stored elastically and then released. This process minimizes the energy cost of panting and hence the amount of heat it generates.

Increasing ventilation rates could lead to an excessive loss of carbon dioxide, which would cause a **respiratory alkalosis**[20]. One way in which panting animals avoid such respiratory alkalosis is by reducing their **tidal volume**, i.e. they move a smaller volume of air backward and forward through the respiratory tract[21]. As a result, the majority of the increased air movement during panting is restricted to the upper airways where there is no gaseous exchange[22] but from where heat can still be lost by evaporation. During panting in dogs, there are increases in blood flow to both the mucosa of the nasal passages and to the tongue, indicating that both surfaces are important in evaporative heat loss.

Some species of birds, such as mute swans (*Cygnus olor*), have a similar pattern of panting to that in mammals, whereas others, such as flamingos (*Phoenicopterus rube*) and pigeons, have different patterns that combine high- and low-frequency ventilation. In flamingos, the low-frequency, high-volume components regularly interrupt the higher-frequency, low-volume components, whereas in pigeons, the high-frequency component is superimposed on the low-frequency component. The lower-frequency components are predominantly concerned with gas exchange; the higher-frequency ventilation is concerned with evaporative cooling.

Storage of heat

An additional strategy to cope with high temperatures and incoming heat load from solar radiation (insolation) is to store the incoming heat, allowing the body temperature to rise, although there is an upper limit to this[23]. Experimental Panel 10.1

[18] We discuss evaporative cooling via the respiratory system in some species of reptiles in Section 9.1.3.

[19] Resonant frequency of a system is determined by the elastic properties of the tissues and their size.

[20] See Sections 13.3.1 and 13.3.2 for an explanation of why an alkalosis would be a problem.

[21] Changes in ventilation pattern can be seen in panting dogs.

[22] Distribution of air within the respiratory systems of birds and mammals is discussed in Section 12.3.3.

[23] Upper lethal temperature limit is discussed in Section 8.2.2.

Experimental Panel 10.1 Measuring heat storage in large herbivores living in hot arid conditions

Background

The classical studies of Knut and Bodil Schmidt-Nielsen and colleagues in the 1950s measured the body temperature of camels (*Camelus dromedaries*) by inserting a mercury thermometer into the rectum of an animal to a depth of 120 mm and leaving it there for at least three minutes. Thus, the animals had to be in captivity, and were kept in a yard outside a laboratory in the full glare of the sun for most of the day. These conditions would simulate the hot dry summer conditions of their natural environment. Measurements were taken every hour, with the data showing that in summer and when deprived of water, the animals began the day with a low body temperature but allowed this to rise by over 6°C. They off-loaded this stored heat during the cooler nights by non-evaporative means, as shown in Figure A panel (i).

The extent of the heat storage and increase in body temperature during the day, known as adaptive heterothermy, was dependent on how much water the camels had to drink. If the animals were well hydrated, their body temperatures

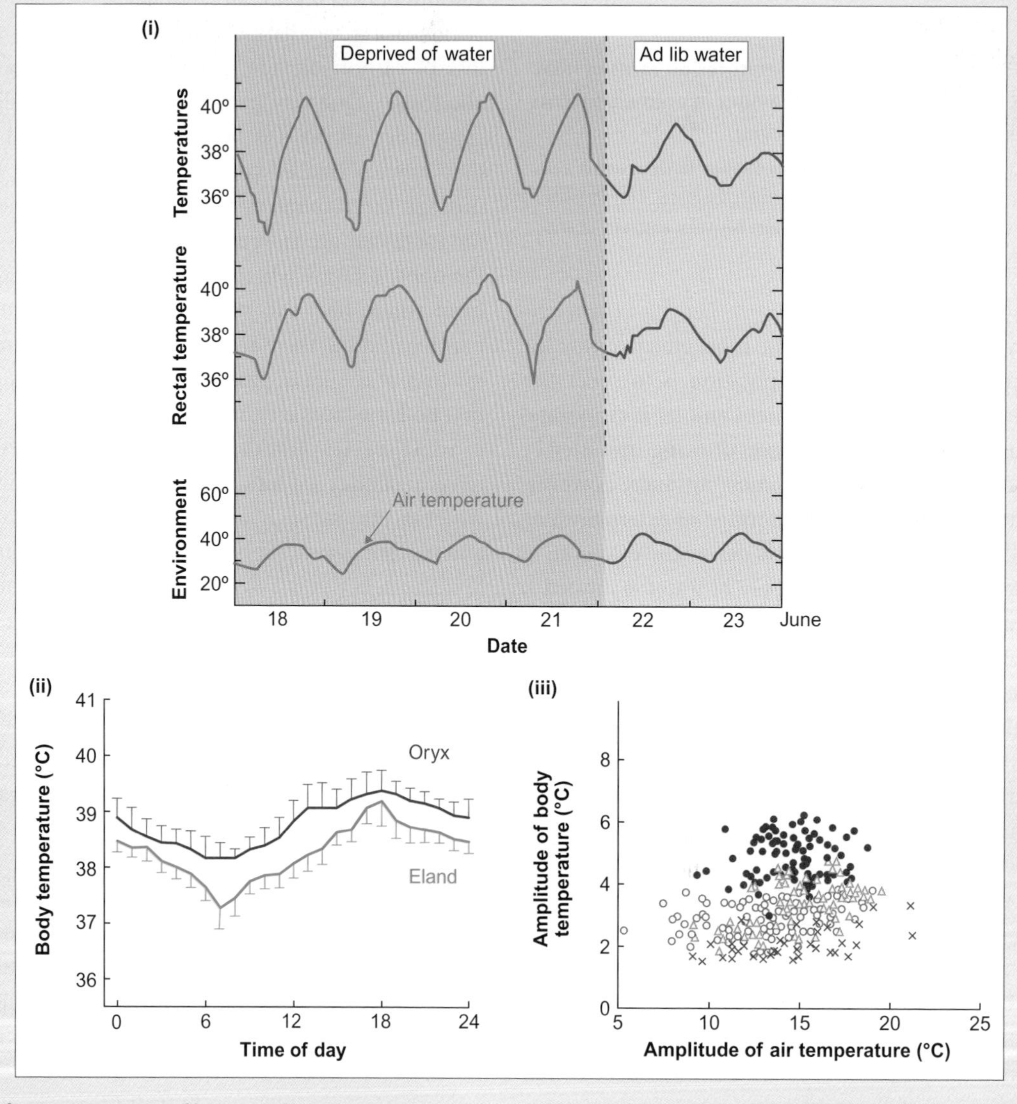

Figure A Body temperatures of large arid-zone mammals

(i) Body temperature (°C) in two camels (*Camelus dromedarius*) in captivity that have been allowed access to or deprived of water. Also shown is environmental temperature. When deprived of water, the animals allow their bodies to heat up substantially during the day and then shed this heat at night, instead of dissipating heat by evaporation. When well-watered, the camels regulate their body temperatures more closely by evaporative means. (ii) Mean ± SD of data from a South African oryx or gemsbok (*Oryx gazella*) and an eland (*Tragelaphus oryx*) free-ranging in their natural environment for 1–3 weeks. Note the lower amplitude in daily temperature excursions than in the captive camels deprived of water shown in (i). (iii) Graph showing no correlation between the amplitude of body temperature and amplitude of air temperature in free-ranging Arabian oryx (*O. leucoryx*) during both summer and winter when the conditions were: x, warm-wet; •, hot-dry; Δ, warm-dry; ○, cool-dry.

(i) adapted from Schmidt-Nielsen, K et al (1956). Body temperature of the camel and its relation to water economy. American Journal of Physiology 188: 103-112. (ii) reproduced from Fuller A et al (2004). The eland and the oryx revisited: body and brain temperatures of free-living animals. International Congress Series 1275: 275-282 (iii) reproduced from Hetem RS et al (2010). Variation in daily rhythm of body temperature of free-living Arabian oryx (*Oryx leucoryx*): does water limitation drive heterothermy? Journal of Comparative Physiology B 180: 1111-1119

Figure B South African oryx or gemstock (*Oryx gazella*) standing in the shade of a tree in the central Kalahari National Park, Botswana

Image: PW Arkwright – with permission.

only oscillated by about 2°C, and they used evaporation to regulate body temperature. This study on camels, together with similar experiments in the early 1970s on African mammals, including eland (*Tragelaphus oryx*) and oryx (*Oryx gazella*), led to the idea that large mammals in warm arid environments use adaptive heterothermy as a strategy for coping with short-term heat stress when there is limited access to water. Such a strategy would reduce the need to lose heat by evaporation and thus conserve water[1]. The question is, though: What do these animals do when they are free to roam in their natural environment?

Experimental approach

More recent investigations have been conducted by Duncan Mitchell and his colleagues who employed implanted electronic devices to store temperature data from free-ranging animals in their natural environment for periods of several weeks. The data storage devices were surgically implanted into the abdomens of anaesthetized animals under sterile conditions. The animals were allowed to recover fully before being released into the wild. The eland and South African oryx (or gemsbok) used in the studies were recaptured between 45 and 60 days after surgery, re-anaesthetized and the devices removed, as described above. Arabian oryx (*O. leucoryx*) were recaptured a year after surgery and the devices similarly removed. All animals were then released into their natural environment following full recovery.

Experiment 1

Figure A panel (ii) shows results from African eland and oryx. Notice in these graphs that the average change in body temperature was only around 2°C, even though the conditions were often ideal for the use of adaptive heterothermy: the days were warm to hot, the animals did not drink when exposed to a high ambient temperature, and the nights were cool. However, the large swings in body temperature that are the main characteristic of adaptive heterothermy did not occur.

Based on these data and on measures of the behaviour of the animals in the wild, it has been suggested that the animals are able to use behavioural means, such as seeking shade during the day (as shown in Figure B) and huddling together at night, to reduce heat gain or loss during these periods when free in their natural environment.

Experiment 2

Arabian oryx live in the very hot and dry Arabian desert. Following on from earlier studies by Stéphane Ostrowski and colleagues, Mitchell's group obtained data from free-ranging individuals which revealed average daily excursions in body temperature of almost 5°C (maximum 7.7°C) during the summer[2]. These data suggest that this species does employ heterothermy. On the other hand, a classic feature of adaptive heterothermy is a positive relationship between amplitude of the daily change in body temperature with amplitude of the daily change in air temperature. Figure A panel (iii) shows that no such relationship exists in Arabian oryx, implying that the degree of heterothermy is not driven by ambient temperature, but by other factors. High environmental temperatures and limited access to water may be the major explanation for heterothermy in this species. This idea is supported by the fact that, when in captivity with food and unlimited water available, a male oryx had a daily temperature fluctuation of 2.6°C throughout the year, despite being exposed to temperatures similar to those experienced by the free-ranging animals.

To illustrate the importance of behavioural modifications in dealing with hotter conditions, it has been shown that Arabian oryx move from a diurnal or crespuscular pattern of activity to more nocturnal activity without a reduction in total activity over a 24 h period. Such modifications of behaviour may ameliorate some of the effects of the predicted increase in summer temperature in the southwest of the Arabian Peninsula, although may not be sufficient to ensure long-term survival of the species[3].

Overall findings

In general, it seems that the daily temperature excursions recorded from large ungulates living in warm arid environments are similar to those that might be expected in any mammal, as illustrated in Figure 10.5. However, more data, including environmental and behavioural data, are required from a range of different species of free-ranging animals in their natural environment to obtain a more accurate picture of how large mammals cope with high temperatures and limited water supplies in the wild.

› Find out more

Fuller A, Maloney SK, Mitchell G, Mitchell D (2004). The eland and the oryx revisited: body and brain temperatures of free-living animals. International Congress Series 1275: 275–282.

Fuller A, Dawson T, Helmuth B, Hetem RS, Mitchell D, Maloney SK (2010). Physiological mechanisms in coping with climate change. Physiological and Biochemical Zoology 83: 713–720.

Hetem RS, Strauss WM, Fick LG, Maloney SK, Meyer LCR, Shobrak M, Fuller A, Mitchell D (2010). Variation in the daily rhythm of body temperature of free-living Arabian oryx (*Oryx leucoryx*): does water limitation drive heterothermy? Journal of Comparative Physiology B 180: 1111–1119.

Mitchell D, Maloney SK, Jessen C, Laburn HP, Kamerman PR, Mitchell G, Fuller A (2002). Adaptive heterothermy and selective brain cooling in arid-zone mammals. Comparative Biochemistry and Physiology B 131: 571–585.

Ostrowski S, Williams JB, Ismael K (2003). Heterothermy and the water economy of free-living Arabian oryx (*Oryx leucoryx*). Journal of Experimental Biology 206: 1471–1478.

Schmidt-Nielsen K, Schmidt-Nielsen B, Jarnum, SA, Houpt TR (1957). Body temperature of the camel and its relation to water economy. American Journal of Physiology 188: 103–112.

Taylor CR (1970a). Strategies of temperature regulation: effects on evaporation in East African ungulates. American Journal of Physiology 219: 1131–1135.

Taylor CR (1970b). Dehydration and heat: effects of temperature regulation of East African ungulates. American Journal of Physiology 219: 1136–1139.

[1] Evaporative water loss in mammals is discussed in Section 6.1

[2] More details of seasonal temperature variations and water balance in Arabian oryx are given in Section 6.1.2 and Figure 6.15.

[3] Predicted increases in environmental temperature in the Arabian peninsula by 2100 are discussed in Section 6.1.2.

examines some of the methods used to investigate the extent to which heat storage is used in captive and free-ranging animals.

10.2.3 Temperature regulation in hot environments

Heat stress in hot areas is much more of a concern for larger animals than smaller ones because larger animals have a relatively low surface area compared to their volume. It is important that the rate of heat flow away from a large animal is not unduly retarded by surface insulation. With this in mind, we see an inverse relationship between the size of mammals in hot climates and the thickness of their fur, or pelage, as shown in Figure 10.10 for antelope species. Indeed, larger animals (rhinoceros, hippopotamus and elephants) have lost their external insulation almost completely, which facilitates heat dissipation.

Restricting physical activity to the coolest part of the day is an obvious response to perpetually elevated ambient temperatures. This behaviour combines two strategies:

- maximizing the difference between body and environmental temperatures, and
- the avoidance of incoming solar radiation.

Most desert living mammals are nocturnal or **crepuscular**[24]. Hiding in a burrow during the day is a useful strategy to avoid solar radiation at the warmest part of the day, but this is less of an option for larger animals. However, larger animals can reduce exposure to incoming radiation by seeking shade. Indeed, many wild animals seek the shade of the occasional trees in hot areas, as discussed in Experimental Panel 10.1.

In many hot deserts there are no trees, so larger animals are forced to spend protracted periods exposed to high levels of solar radiation. We discuss, in Section 10.1, how fur and feathers increase insulation and reduce the loss of heat from an endotherm in a cold environment, but fur and feathers can also reduce the absorption of solar radiation and thus reduce the potential heat load falling on the animal. Red kangaroos (*Macropus rufus*) live in areas with high incident solar radiation but shelter in shade during hot summer days. In contrast, emu (*Dromaius novaehollaniae*) living in similar environments feed in the open during the day.

The kangaroos studied have a red-coloured relatively thin (9 mm) coat of fur, while emus have a dark grey, relatively thick (45 mm) coat of feathers. Although the coat of the kangaroos reflects more than twice as much solar radiation than that of emus, solar radiation penetrates to a proportionately shallower depth of the coat in emus (10 per cent) than in red kangaroos (16 per cent). The net effect of these factors is that, at low wind speeds, about 9 per cent of incident radiation reaches the skin of emus whereas 23 per cent reaches that of red kangaroos, as shown in Figure 10.11. At higher wind speeds, these proportions become even lower, so that at a wind speed of 10 m s^{-1}, almost no incident radiation reaches the skin of emu.

[24] Crepuscular animals are mainly active during twilight hours of dawn and dusk.

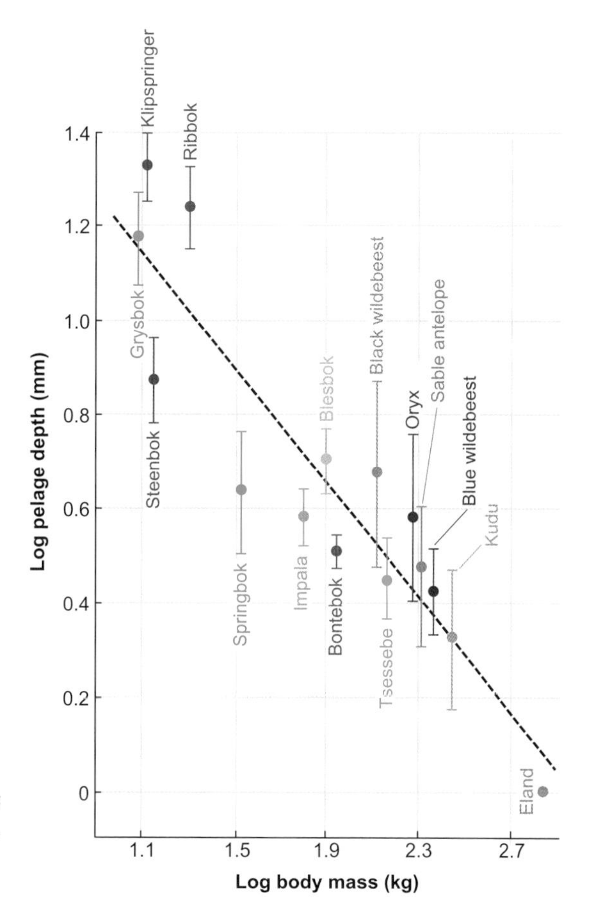

Figure 10.10 Fur (pelage) thickness of African antelopes in relation to body mass

Antelopes with larger body masses have thinner pelage.

Reproduced from: Hofmeyr, MD, Louw, GN (1987). Thermoregulation, pelagic conductance and renal function in the desert-adapted springbok, *Antidorcas marsupialis*. Journal of Arid Environments 13: 137-51. Images: grysbok: Masteraah/Wikimedia Commons; oryx: PW Arkwright; eland: Johnny Magnusson.

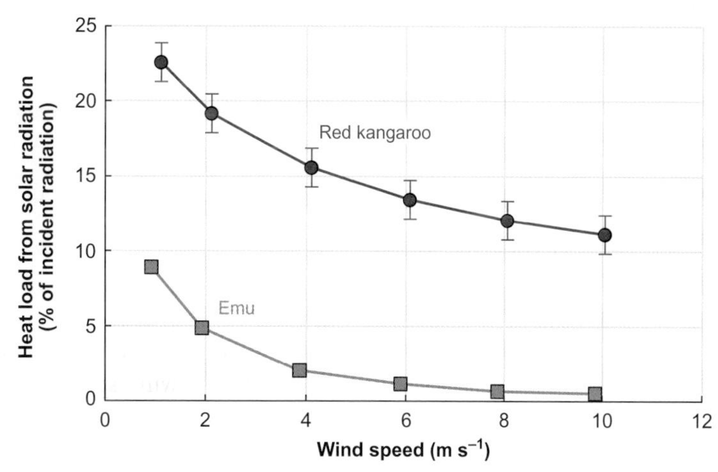

Figure 10.11 The proportion of solar radiation that reaches the skin of emus (*Dromaius novaehollandiae*) and red kangaroos (*Macropus rufus*) at different wind speeds.

At low wind speeds, a lower proportion of incident radiation reaches the skin of emu than that of red kangaroos and these proportions decrease at higher wind speeds

Modified from Dawson TJ and Maloney SK (2004). Fur versus feathers: the different roles of red kangaroo fur and emu feathers in thermoregulation in the Australian arid zone. Australian Mammology 26: 145-151. Images: emu: Fir0002/Flagstaffotos; kangaroo: courtesy of Drs.

A potentially major contribution to the way animals adapt to hot temperatures is to reduce heat production from basal metabolic rate (BMR); BMR is generally lower in mammals that live closer to the equator and which are therefore exposed to high environmental temperatures. Also, in general, birds living in deserts have a BMR that is 17–25 per cent lower than that of non-desert species.

10.2.4 Thermoregulation and exercise

Endothermic animals generally need to keep heat trapped inside their bodies, hence their surface insulation. However, during exercise this may not be the case. Indeed, the internal generation of heat during exercise may exceed that required to keep the body at a fixed temperature, which can be problematic.

When humans begin to exercise in cold weather they may need extra clothing to keep them warm. Soon after they have started the exercise, however, additional heat is generated by the active muscles. This additional heat exceeds the person's thermoregulatory needs, at which point the clothes needed to trap heat prior to exercise become a hindrance to heat loss. Fortunately, a human can remove layers of clothing. By contrast, other endothermic species are unable to shed their external insulation, fur or feathers, in the short term.

We discuss in Section 10.2.2 that many endotherms have thermal windows, which are less well insulated than other areas. During the heat stress of exercise, animals can offload excess heat from their thermal windows by increasing the rate at which they are perfused. Sled dogs, such as Siberian huskies in the Arctic, are a good example. These dogs must survive outside in the severe cold of the Arctic winter, and, consequently, have very thick fur on their backs. In the cold, they curl up into balls so that their less heavily furred bellies are concealed. However, they generate large amounts of heat when working that cannot be effectively dissipated because of their well-developed insulative covering. To combat this, these dogs frequently lie down with their lightly furred undersides pressed into cold snow, after a period of hard work, in order to offload the heat generated while exercising. Other animals, such as red kangaroos, lick themselves after exercising to promote heat loss by evaporation. Panting[25], during and/or after exercise is also often used as a means of losing excessive heat by evaporation.

Countercurrent heat exchangers and selective brain cooling

In addition to thermal windows, some animals have specific structures that allow them to protect vulnerable organs, such as the brain, from overheating when they exercise. Hot blood flowing into the brain from the body could cause considerable problems. A famous film clip of Gabriela Andersen-Scheiss at the end of the 1984 Los Angeles Olympics marathon shows Ms Andersen-Scheiss struggling with a lack of limb coordination and disorientation as she enters the stadium. Figure 10.12 shows her being lifted onto a stretcher after staggering across the finishing line after taking almost 6 minutes to complete the final 400 m circuit of the stadium. As a result, she finished in 37th place, despite entering the stadium in 2nd place. Lack of coordination and disorientation are classic symptoms of heat exhaustion during exercise caused by overheating in the brain. These effects are so well established that the pre-cooling of athletes' bodies prior to races where heat exhaustion might be a possibility has attracted much attention.

To avoid the effects of overheating the brain, many animals simply refuse to run when their body temperatures reach a critical point; in cheetah (*Acinonyx jubatus*) the critical temperature is 40.5°C, in brown rats (*Rattus norvegicus*) it is 39.4°C, in humans (at least those not trying to win an Olympic medal), it is 40°C. However, the data obtained for cheetahs and rats were from animals running on a treadmill. Data from cheetahs hunting in the wild and using implanted data storage devices indicate that they do not abandon a hunt, whether successful or not, because of an increase in body temperature (T_b) above the 24 h mean value of 38.3°C. The 'critical' temperature of 40.5°C was very rarely reached, and far less frequently than a cheetah abandoning its hunt.

Figure 10.12 Medical officers carry Gabriela Andersen-Scheiss to a stretcher after she staggered across the finish line in the 1984 Los Angeles Olympic marathon

Ms Andersen-Scheiss showed the classic signs of brain hyperthermia during the 5 minutes 44 seconds it took her to complete the final lap (400 m) of the race.

[25] Panting is discussed in Section 10.2.2.

Special structures called **respiratory turbinates**[26] are present in the nasal cavities of birds and mammals and enable the transfer of heat from the warm nasal venous blood to the incoming cooler airflow. This transfer results in a cooling of the blood. However, the brain is not supplied by this cooled venous blood (which is poorly oxygenated), but by warm arterial blood, so a mechanism is needed to transfer heat from the hot arterial blood coming from the body into the cool venous blood.

[26] Respiratory turbinates are discussed in Section 10.2.7 and Case study 10.2.

In some species of mammals (cats and even-toed ungulates—pigs, camels, goats, cattle, antelope, etc.), this heat transfer is achieved by the cool venous blood from the nasal turbinates draining into a large sinus or venous plexus at the base of the brain, as shown in Figure 10.13. The carotid arteries carrying warm oxygenated blood from the body to the brain divide into smaller arterioles and capillaries that form a network which passes through this sinus or intermingle with the venous plexus. Consequently, warm blood in the arterioles passes in the opposite direction to cool blood in the venous plexus.

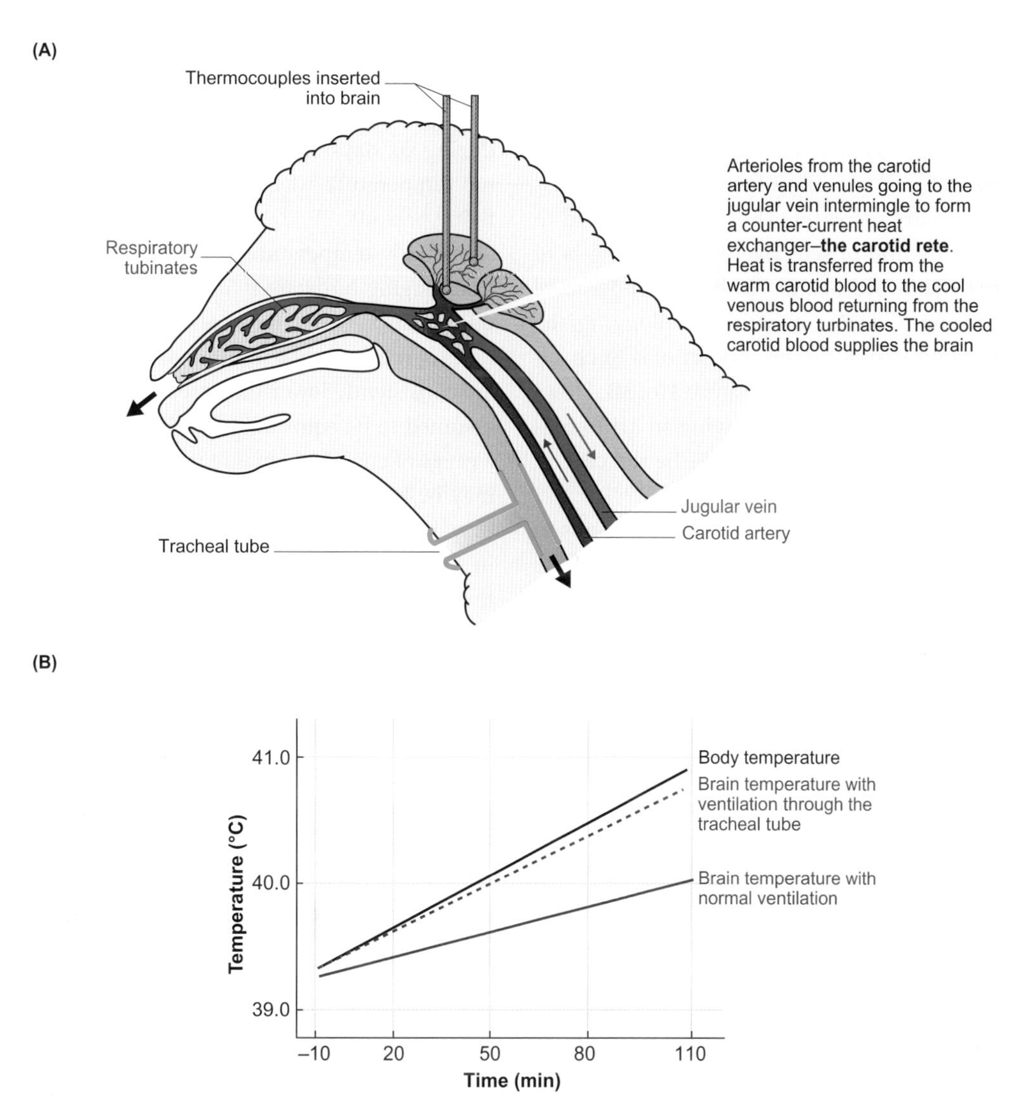

Figure 10.13 Regulation of brain temperature independent of the body temperature in a sheep

(A) Heat in venous blood entering the nasal cavity is offloaded in the turbinates. The cooled venous blood flows to a plexus of small veins, venules. Hot oxygenated blood from the body in arterioles from the carotid artery intermingle with the venules forming a countercurrent heat exchanger—a rete. Heat in the oxygenated arterial blood is transferred to the cooler venous blood in the rete. The venous blood flows into the jugular vein back to the body. (B) Data showing that as the body heats up during exercise (black line) the rete allows the animal to stop heat generated in the body from flowing into the brain and therefore enables the animal to maintain brain temperature relatively constant (solid red line). The importance of this mechanism for regulating brain temperature is shown by bypassing the turbinates and forcing ventilation to occur via a tube inserted to the trachea, as shown in (A). When the animal is placed on tracheal ventilation, the brain temperature does not remain much cooler than the body temperature (dashed red line).

Source: A: adapted from Laburn, HP et al (1988). Effects of tracheostomy breathing on brain and body temperatures in hyperthermic sheep. Journal of Physiology 406: 331-344. B: data from Laburn et al., 1988.

Such countercurrent exchange[27] complexes are often called **retia mirabilia** (singular **rete mirabile**, which means wonderful net) and describes the intricately intertwined net-like structure of the capillaries. In mammals, this complex is known as the carotid rete while the analogous structure in birds is associated with the eyes and is known as the ophthalmic rete.

Figure 10.13A illustrates the location of the carotid rete in sheep. In mammals with this arrangement, the temperature of the brain can be cooled by around 1°C below that of the rest of the body. This cooling is known as selective brain cooling.

Laboratory experiments on species of even-toed ungulates have demonstrated that selective cooling of the brain becomes apparent when the temperature of the blood rises above approximately 39°C; they have also shown that the difference between blood and brain temperatures increases as blood temperature increases. The experiments led to the idea that the function of selective brain cooling is to prevent the brain from overheating as body temperature increases (for example, during exercise), prolonging the time an animal could exercise in hot conditions. Such selective brain cooling is illustrated in Figure 10.13B. However, studies on free-ranging animals living in arid environments indicate that this may not always be the case in the wild.

In free-ranging animals, selective brain cooling seems to be seldom employed. Even when it is, the temperature of the brain is often less than 0.4°C lower than that of the blood. Arabian oryx, which live in the extremely hot and dry conditions of the Arabian desert, are an exception. Heat storage in Arabian and African oryx are discussed in Experimental Panel 10.1. However, there is no selective brain cooling during strenuous exercise in many large ungulates living in arid conditions, including African oryx, and brain temperature can exceed 41°C.

It has been suggested that when even-toed ungulates that live in arid regions (such as the Arabian desert) are under moderate heat stress, an increase in brain temperature will trigger selective brain cooling. This selective cooling will reduce the stimulus for evaporative cooling via the nasal passages, thus conserving water. However, if an animal undergoes intense activity, for example when being chased by a predator, the temperature of both the body and brain will increase. In this situation, an accompanying stress-related increase in the activity of the sympathetic nervous system[28] leads to constriction of selected blood vessels. As a result, the cooled venous blood leaving the nasal passages bypasses the carotid rete. Under such conditions, selective brain cooling no longer occurs and the brain temperature rises. This temperature rise triggers mechanisms of evaporative heat loss in order to rapidly reduce body temperature. In these circumstances, the animal's immediate survival is more important than water balance.

When a bird begins to fly from rest, its rate of heat production may increase by an order of magnitude in a matter of seconds. Measurements of abdominal (body) temperature of barnacle geese flying in wind tunnels indicate that the body temperature increases from 40.4°C to 41.3°C during the first minute or so of flight. However, the data in Figure 10.14 demonstrate that this may not be the case during free flights in the wild; in six wild barnacle geese during six days of their autumn (fall) migration, abdominal temperature was 39.0°C during both flying and non-flying periods.

As body temperature does not appear to increase when free-ranging birds are flying long distances, there is no requirement for selective cooling of the brain. The temperatures of the brain of a number of species of birds in captivity are consistently lower than cloacal temperatures (which are assumed to be equivalent to central core temperature) over a range of environmental temperatures. However, these data may be artefacts of experimental conditions in the laboratory. For example, in the one species that has been studied in the wild (ostrich, *Struthio camelus*), the temperature of the brain is rarely below that of the arterial blood. It would appear, therefore, that the ophthalmic rete in birds may not function in a similar fashion to the carotid rete in some species of mammals.

These observations of mammals and birds illustrate the importance of obtaining data from wild animals in their natural environment rather than relying solely on data from laboratory-based observations or experiments.

10.2.5 Thermoregulatory responses to cold conditions

When ambient temperature falls below the lower critical temperature, endotherms must elevate their rate of heat production to match the increased rate of heat loss if they are to maintain their normal body temperature. This is achieved by two fundamentally different mechanisms in mammals: **shivering thermogenesis**, and a mechanism that does not involve shivering called **non-shivering thermogenesis** or NST (Greek *thermé*—heat, *genesis*—origin). During short-term exposure to the cold, the rate of heat production increases almost immediately (within seconds to minutes). The relative contribution of NST to heat production is greater in smaller mammals than in larger ones, in animals

[27] The principles of countercurrent exchange are discussed in Section 3.3.1.

[28] See Sections 16.1.3 and 22.4.3 for discussion of the sympathetic nervous system and its role in the regional distribution of blood.

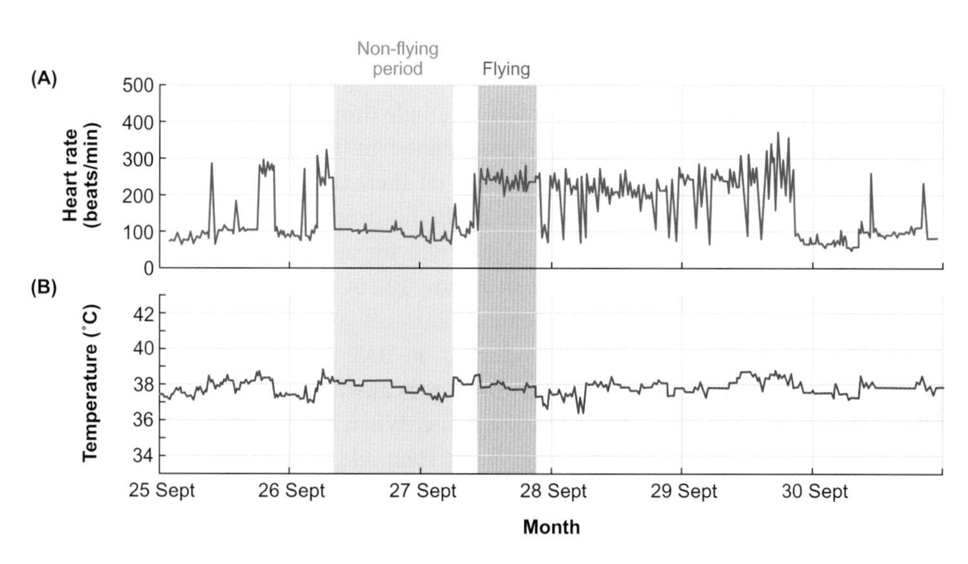

Birds stop several times during their migrations

There was no increase in abdominal temperature during migratory flights

Figure 10.14 Migration and body temperature of barnacle geese (*Branta leucopsis*)

The data presented were obtained from small implanted data storage devices. Heart rate and body temperature in a wild barnacle goose on its autumn (fall) migration from the island of Spitzbergen in the high Arctic to Caerlaverock in south-west Scotland (Figure 10.29B shows details of this migration). When heart rate is high, around 300 beats min^{-1}, the birds are flying.

Modified from Butler PJ and Woakes AJ (2001). Seasonal hypothermia in a large migrating bird: saving energy for fat deposition? Journal of Experimental Biology 204: 1361-1367

previously exposed to cold conditions, and also in younger compared to older mammals. Generating heat by shivering (shivering thermogenesis) seems to be the only mechanism present in adult birds, as we discuss later.

Non-shivering thermogenesis involves a change to the way in which the oxidation of metabolic substrates in mitochondria is harnessed by the cells. Normally, the mitochondria use the oxidation of these substrates to drive the phosphorylation of ADP to produce ATP, a process called oxidative phosphorylation. During oxidative phosphorylation, hydrogen ions (protons) are pumped outward across the inner mitochondrial membrane to generate a difference in proton concentration across the membrane. These protons move back across the membrane to the mitochondrial matrix via the enzyme complex ATP synthase[29]. During this process, the protons drive a molecular motor that generates ATP from ADP and inorganic phosphate. There are, however, many other routes that protons can take back into the mitochondrial matrix which bypass the ATP synthase. In particular, the mitochondrial membranes of animals have specific protein pores that can be opened to facilitate movement of protons back into the matrix.

David Nicholls of Dundee University and his colleagues were the first to discover one of these protein pores in the late 1970s. This protein pore became known as the **uncoupling protein (UCP)**, because it uncouples the movement of protons across the membrane from the synthesis of ATP. In some older literature this protein is called thermogenin to reflect its role in thermoregulation. More recently, it has been called UCP1 to distinguish it from additional uncoupling proteins that have since been described.

The role of UCP1 in non-shivering thermogenesis

UCP1 is a 32 kD protein that consists of three repeated segments each involving about 100 amino acid residues, to give a the total protein that is about 305–310 amino acids long. UCP1 is found mainly in a specialized tissue called **brown adipose tissue (BAT)**, which, in turn, is found only in eutherian (placental) mammals[30], and most predominantly in small or hibernating species. Figure 10.15 shows a northern blot analysing gene expression of UCP1 in different tissues from a small rodent, the field vole (*Microtus agrestis*). Notice how the mRNA for UCP1 is only present in brown adipose tissue. Brown adipose tissue is also important in temperature regulation in newborn humans, and has recently been shown to be functionally present in healthy adults[31].

[29] Oxidative phosphorylation is discussed in Section 2.4.2.

[30] UCP1 has recently been identified in adipose tissue of marsupials, suggesting that an ancestral form of BAT was present before the divergence of marsupials and eutherians.

[31] As brown adipose tissue consumes fatty acids and only produces heat, it might be possible to activate it using drugs as part of a treatment for obesity.

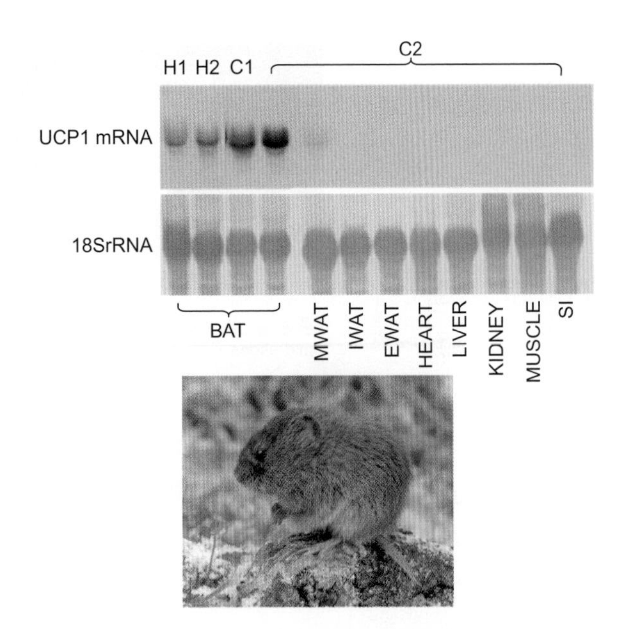

Figure 10.15 Northern blot showing tissue distribution of mRNA for uncoupling protein-1 (UCP1) in a small mammal (field vole *Microtus agrestis*) relative to ribosomal RNA (18S)

The gene for UCP1 is only expressed in brown adipose tissue (BAT). H1 and H2 refer to voles kept in the hot (21°C) and C1 and C2 refer to voles kept in the cold (8°C). Note the UCP1 gene in BAT is expressed more in the cold (spots are darker). IWAT, MWAT and EWAT refer to inter-scapular, mesenteric, and epidydimal white adipose tissue respectively. SI is the small intestine.

Król and Speakman, unpublished. Image: courtesy of José Ramón Pato Vicente.

Brown adipose tissue is histologically different from white adipose tissue and all other tissue types, as illustrated in Figure 10.16. First, BAT is very heavily vascularized. Second, the cells of BAT contain many small droplets of fat, which are shown in Figure 10.16B and C, rather than the characteristic single droplet of lipid present in a white adipocyte, as depicted in Figure 10.16A. Having many small droplets of fat increases the overall surface area to volume ratio of the fat droplets, which helps to mobilize the fat as a substrate for aerobic metabolism. Finally, as illustrated in Figure 10.16C, the tissue of BAT is packed with mitochondria.

Heat production in BAT is switched on in response to stimulation by the sympathetic nervous system, which innervates the tissue. When the nerve endings of the sympathetic nervous system release **noradrenaline**[32] it binds to β-3 adrenoceptors[33] at the surface of BAT and generates a signal that activates UCP1. A possible scheme for activation of UCP1 is shown in Figure 10.17.

But why do protons bypass ATP synthase when they enter the mitochondria of BAT?

[32] Also called norepinephrine, see Section 16.3.4.

[33] See Table 16.1 for a full list of adrenoceptor types in vertebrates.

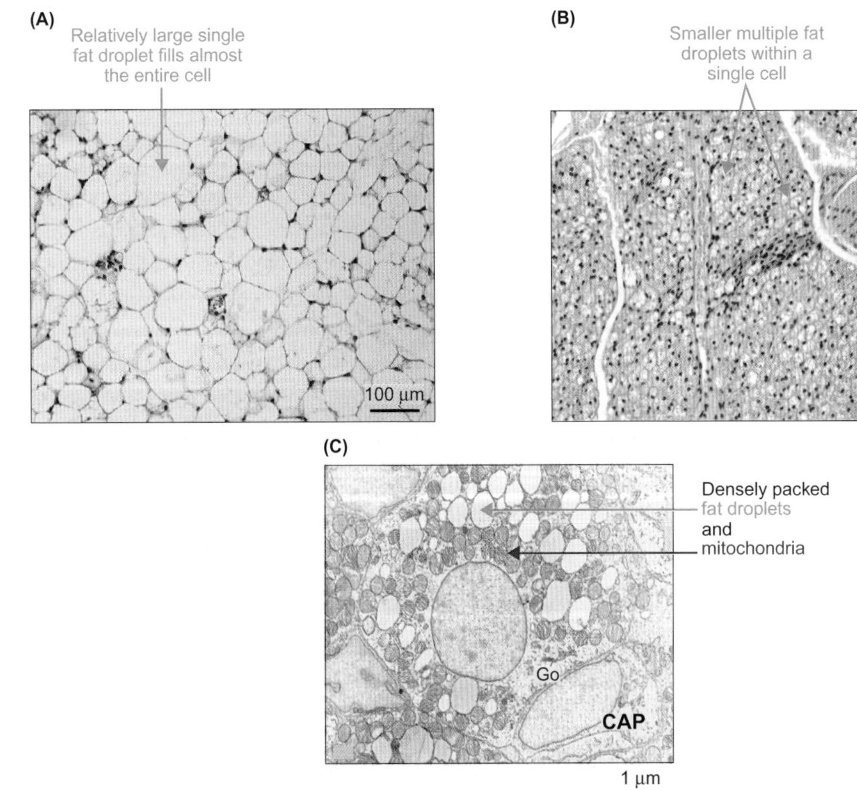

Figure 10.16 Histology of white and brown adipose tissues

(A) Cells of white adipose tissue are completely dominated by a single large lipid droplet. (B) Cells of brown adipose tissue contain many smaller lipid droplets and are packed with mitochondria. (C) Transmission electron micrograph of brown adipose tissue showing high density of mitochondria and fat droplets. Go, Golgi apparatus; CAP, capillary.

A, B: Rexford Ahima, with permission. C: Reproduced from: Cinti, S (2001). The adipose organ: morphological perspectives of adipose tissues. Proceedings of the Nutrition Society 60: 319-328

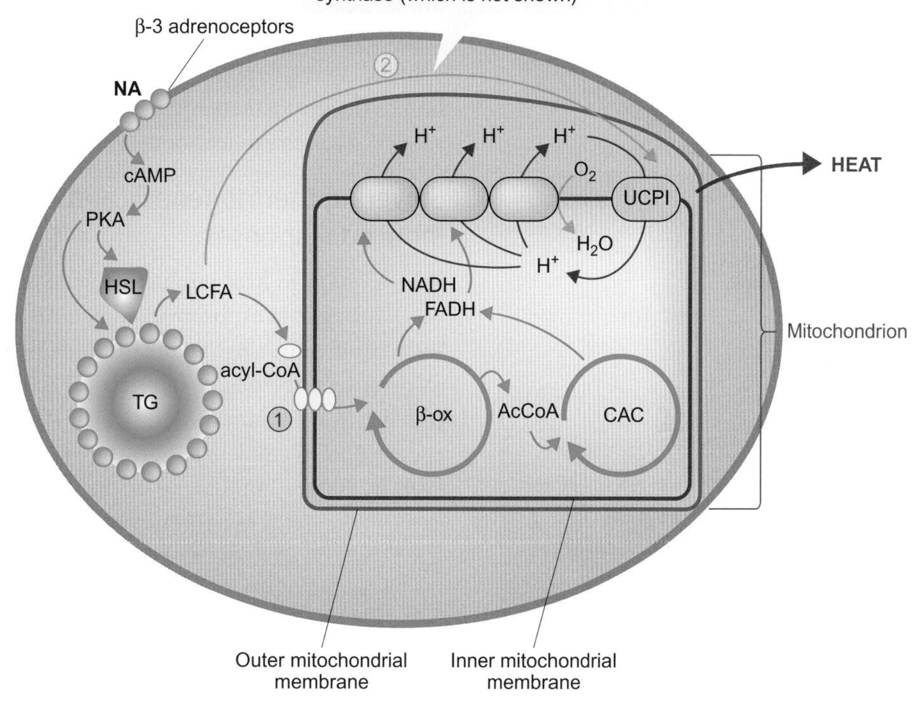

Figure 10.17 Proposed mechanism for stimulation of thermogenesis in brown adipocytes

Noradrenaline (NA; also called norepinephrine) stimulates a cascade of events involving cyclic AMP (cAMP), protein kinase A (PKA) and hormone-sensitive lipase (HSL), which lead to the release of fatty acids (FA) from triglyceride (TG) or fat droplets These FA are: (1) converted to acyl-CoA and transported across the mitochondrial membranes. Once inside a mitochondrion, the acyl-Co-A enters the β-oxidation cycle (β-ox) and the citric acid cycle (CAC) leading to the formation of the reduced electron carriers FADH and NADH, which are then oxidized along the electron transport chain (indicated by the series of grey boxes). Protons are pumped out of the mitochondrial matrix into the intermembrane mitochondrial space during this process (2) Long-chain fatty acids (LCFA) are part of a LCFA/H^+ symporter involved in the re-entry of the protons into the mitochondrion, against the difference in electrochemical potential via UCP-1 rather than via ATP synthase (which is not shown), thus generating heat instead of ATP.

Modified from Cannon B and Nedergaard J (2004). Brown adipose tissue: function and physiological significance. Physiological Reviews 84: 277-359.

Figure 10.18A shows how each of the three segments of the UCP1 protein includes two helices, which span the inner mitochondrial membrane (transmembrane helices); taken together, there are six transmembrane helices with three loops on one side of the membrane and three on the other. It is suggested that each pair of transmembrane helices form a pore or channel. Several factors can open or inhibit opening of the pore: long-chain fatty acids (LCFA) open it to allow the flow of protons down their electrochemical difference, while purine nucleotides, such as ATP/ADP and GTP/GDP, have an inhibitory effect.

The mechanism by which UCP1 transports protons across the inner mitochondrial membrane has been the subject of much debate. Recent experiments using patch clamp[34] techniques have provided evidence that the pore in UCP1 acts as a LCFA/H^+ symporter[35], translocating H^+ from the intermitochondrial membrane space to the mitochondrial matrix. This has been called the LCFA-shuttling model and is shown in Figure 10.18B, C and D. Exactly what removes the inhibition of UCP1 by purine nucleotides is unknown, although patch clamping experiments indicate that LCFAs may be involved.

When protons move through a pore in UCP1 they release their potential energy, not by synthesizing ATP, but directly as heat, as indicated in Figure 10.17. This is an example of a **futile cycle**[36], in which the only product is heat. It should be noted that about 50 per cent of the energy released from glucose is lost as heat, even when oxidation is coupled to the production of ATP[37].

Four additional UCPs have been described. UCP2 has a very wide tissue distribution, UCP3 is almost completely restricted to muscle with small amounts in brown adipose tissue, and UCP4 and 5 are restricted to the brain. There

[34] Patch-clamping is discussed in Section 16.2.4 and Figure 16.18.

[35] Symporters (symport carriers) are discussed in Section 3.2.4.

[36] Futile biochemical cycles are also discussed in Sections 10.3.1 and 10.3.2 for a specialized eye muscle in some species of fish and in muscles of some species of insects, respectively.

[37] The amount of ATP produced by the aerobic oxidation of glucose is discussed in Section 2.4.3.

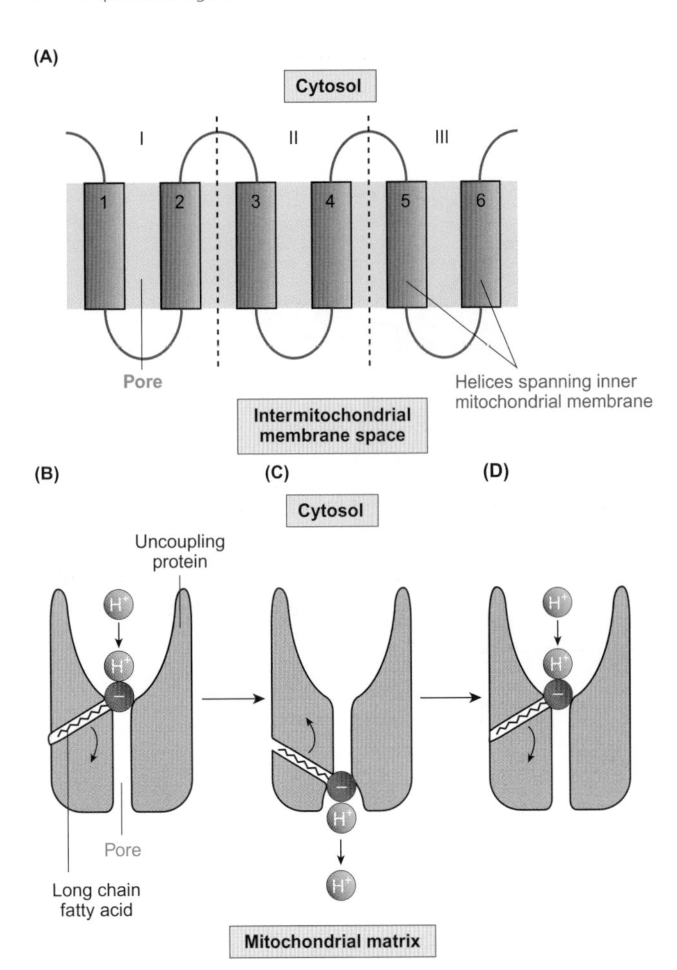

Figure 10.18 The long-chain fatty acid (LCFA) model of H^+ transport by UCP1

Note that UCP1 spans the inner mitochondrial membrane. One LFCA anion and one H^+ are transported with each cycle. (A) Diagram showing the three repeated segments, each containing two helices, and two loops which span the inner mitochondrial membrane. Each segment consists of about 100 amino acids. (B) The LFCA anion binds to the UCP1 at the bottom of a hypothetical cavity in the intermitochondrial. membrane space. Binding of H^+ to UCP1 only occurs after the LCFA binds to UCP1. (C) Conformational change in UCP1 leads to the translocation of LCFA and H^+ is released on the opposite side of the inner mitochondrial membrane. The LCFA remains associated with the UCP1. (D) The LCFA returns to its original position to begin another cycle.

Source: A: Modified from Klingenberg M and Echtay KS (2001). Uncoupling proteins: the issues from a biochemist point of view. Biochimica et Biophysica Acta 1504: 128-143; B, C, D: Reproduced from Fedorenko A et al (2012). Mechanism of fatty-acid-dependent UCP1 uncoupling in brown fat mitochondria. Cell, 151: 400-413.

has been considerable debate over the roles of these UCPs in the process of heat production. Studies involving the surgical removal of the brown adipose tissue have been unsuccessful as the tissue rapidly regenerates. In contrast, studies involving genetic engineering to produce animals that lack given UCPs indicate that the only uncoupling protein that plays any role in heat production during cold exposure is UCP1.

The mechanisms described above do not account for the full thermogenic output of BAT; thyroid hormones and maybe thyroid stimulation hormone[38], are also necessary. Their influence includes increasing the stimulatory effect of noradrenaline on thermogenesis and increasing expression of the UCP1 gene via cAMP. As well as a peripheral effect, thyroid hormone may also have a central effect by directly stimulating the hypothalamus[39], which leads to the stimulation of the sympathetic nerves that innervate BAT and activate UCP1 in the mitochondria[40].

Shivering thermogenesis and other cold-induced responses

The elevation in heat production (rate of oxygen consumption) during cold exposure in adult birds is probably derived solely from shivering. Figure 10.19 shows a linear relationship between the degree of electrical activity in the muscles (the electromyogram, EMG), which is indicative of muscle contraction, as ambient temperature falls below the T_{lc}. However, in 2001 a homologue of the mammalian uncoupling

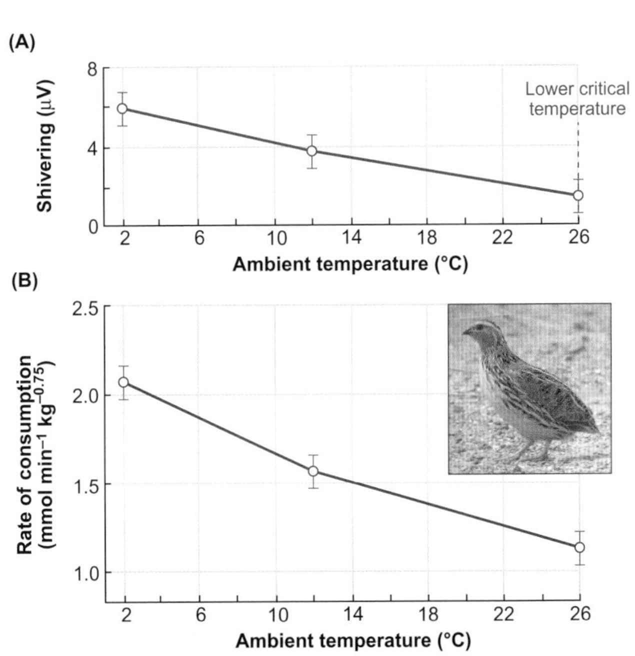

Figure 10.19 Thermoregulatory responses in birds

As ambient temperature falls below the lower critical temperature of 26°C in the Japanese quail (*Coturnix japonica*), both shivering (as indicated by electrical activity in the muscles) and rate of oxygen consumption increase linearly, which suggests that shivering is the only form of thermogenesis in adult birds. Note that rate of oxygen consumption is related to body mass raised to 0.75, which reflects the scaling of metabolic rate to body mass, as discussed in Section 2.2.4. Vertical lines, ± SE of mean values

Modified from: Hohtola E and Stevens ED (1986). The relationship of muscle electrical activity, tremor and heat production to shivering thermogenesis in Japanese quail. Journal of Experimental Biology 125: 119-135.

[38] Thyroid hormones and thyroid-stimulating hormone are discussed in Section 19.4.
[39] The hypothalamus is discussed in Section 16.1.2.
[40] The role of thyroid hormone in thermogenesis in BAT is discussed in Section 19.4.2.

protein was identified in the chicken genome. Figure 10.20A illustrates that the avian homologue, avUCP, is expressed exclusively in skeletal muscle (similar to mammalian UCP3) and that its expression increases by about 3–4-fold after exposure to the cold. What is more, Figure 10.20B shows that the rate of oxygen consumption in Muscovy ducklings (*Cairina moschata*) increases before there is an increase in EMG activity when the birds are exposed to the cold.

The difference between T_{lc} and threshold temperature for shivering is greater in birds that are raised in cold conditions. These data from Muscovy ducklings strongly link an increase in avUCP expression in the ducklings with non-shivering thermogenesis (NST). However, there is no evidence of mitochondrial uncoupling, so an unknown mechanism must be involved. There is no evidence for a role of avUCP in NST in adult birds; it has only been identified in young birds.

Behavioural thermoregulation in response to the cold

Some species use behavioural responses to reduce the rate of heat loss when in cold conditions. One way for an animal to minimize the rate of heat loss is by changing its microclimate, for example by building a nest to trap heat lost from the body. Inside the nest, the animal is exposed to a warmer temperature than would be the case outside. This effect can be magnified if the animals in question huddle together to trap the heat of several individuals, a behaviour exhibited by male emperor penguins (*Aptenodytes forsteri*) while incubating their eggs on the Antarctic ice during winter.

Huddling brings about benefits from two different perspectives. First, it modifies the microclimate; second, individuals sitting adjacent to one other do not lose as much heat across the surfaces that are in mutual contact, so they effectively behave like a single, larger individual in terms of the rate of heat loss. The relative contribution of these two mechanisms in emperor penguins is about one third for the microclimate and approximately two thirds for the reduction in body surfaces exposed to the cold. However, in groups of small field voles (*Microtus agrestis*) they appear to make about an equal contribution to the reduction of energy demands.

Another important behavioural response is for an animal to protect itself from exposure to the wind, particularly if it is a cold wind, as this can cause an increased rate of heat loss by forced convection. For example, incubating male emperor penguins in the middle of a huddle are protected from the

(A)

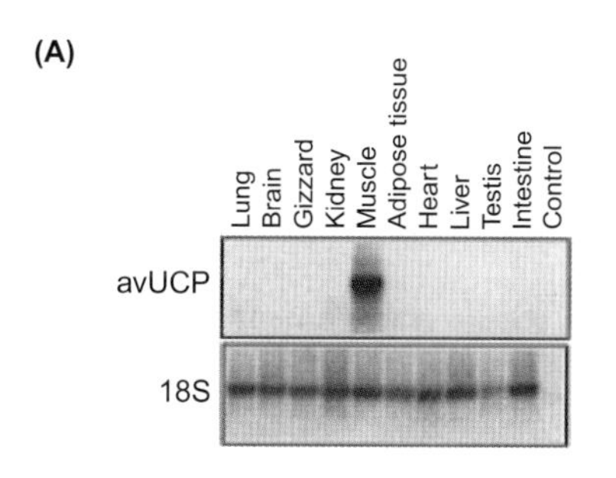

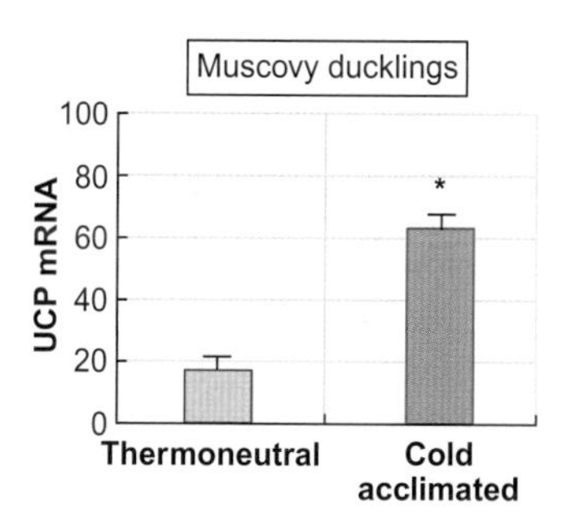

(B)

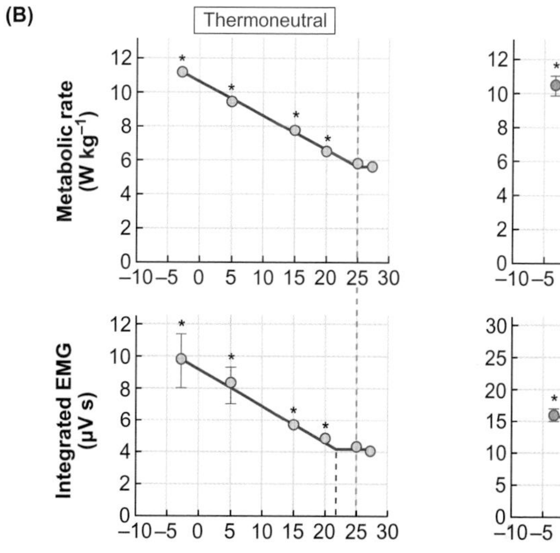

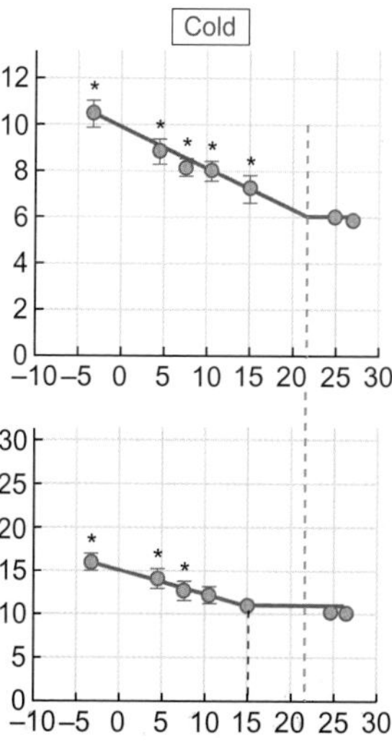

Figure 10.20 Avian uncoupling protein (avUCP)

(A) Avian uncoupling protein is localised to muscle in chicken (*Gallus gallus*) and its expression is increased in response to cold exposure in 5-week-old Muscovy ducklings (*Cairina moschata*). The tissue distribution of the 'housekeeping' ribosomal protein 18S gene expression is also shown. (B) Relationship between lower critical temperature (T_{lc}, green dashed line) and onset of shivering (black dashed line) in 5-week-old Muscovy ducklings raised at 25°C (thermoneutral) and 11°C (cold). Note that the difference in shivering threshold and T_{lc} is greater in the 'cold' birds. Vertical lines, ± SE of mean values of eight (thermoneutral) or five (cold) ducklings. *Significantly different from values in the thermoneutral zone, $P < 0.05$.

Reproduced from: (A) Raimbault S et al (2001). An uncoupling protein homologue putatively involved in facultative muscle thermogenesis in birds. Biochemical Journal 353: 441-444. (B) Teulier L et al (2010). Cold-acclimation-induced non-shivering thermogenesis in birds is associated with the upregulation of avian UCP but not with innate uncoupling or altered ATP efficiency. Journal of Experimental Biology 213: 2476-2482.

cold Antarctic wind. The forced convecton gives rise to wind chill, which is when the air feels colder than it would if it was still[41]. Conversely, if the sun shines on a cold day, an animal may be able to warm itself by basking, even if air temperature is below the animal's lower critical temperature.

We have now seen the variety of ways in which endotherms respond to both hot and cold temperatures outside their thermoneutral zone. We now need to explore the mechanisms involved in the activation of BAT and shivering to generate heat and in the sweating or panting responses when an animal begins to overheat.

10.2.6 Control systems that regulate responses to changing temperatures

The mechanisms used to control the body temperature of birds have not been so clearly identified as those in mammals, although the basic principles appear to be similar in the two groups. The following discussion concentrates mainly on mammals.

Endotherms regulate their body temperature using a combination of peripheral and central temperature sensors. Endothermic animals have receptors in their skins that are exclusively responsive to temperature. These receptors fire at a peak rate at a particular temperature. There are two distinct groups of peripheral temperature receptors, as shown in Figure 10.21. One set, the **cold receptors**, is located in or immediately beneath the epidermis and has a rather flat response curve over its active range. Firing rates begin to fall when temperatures are below about 20°C or above 34°C. In contrast, the second group of peripheral temperature receptors, the **warm receptors**, is located slightly deeper in the dermis and is silent below 30°C while firing reaches a peak at around 45°C. From the relative firing rates of these receptors, animals can assess the peripheral temperature in different parts of their bodies.

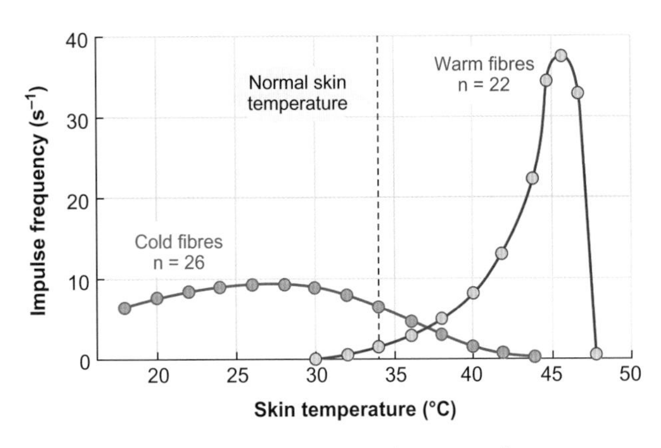

Figure 10.21 Temperature receptors in mammals

Average activity in nerve fibres from cold and warm skin receptors as a function of temperature. Cold-sensitive receptors respond only to cooling and warm-sensitive receptors respond to warming.

Reproduced from: Patapoutian A et al (2003). ThermoTRP channels and beyond: mechanisms of temperature sensation. Nature Reviews, Neuroscience 4: 529-539.

The peripheral temperature receptors possess a subclass of transient receptor potential (TRP) ion channels[42], known as thermoTRP channels. Activation of thermoTRP channels leads to an inward, nonselective cationic current and, therefore, depolarization of the resting potential of the receptor cell.

There are also **thermosensitive** neurons in the central nervous system and an important location in mammals is in the pre-optic area (POA) of the hypothalamus at the base of the brain, with some also found in the spinal cord. However, in birds, thermosensitive neurones are mainly located in the spinal cord and even distributed throughout the whole body. The characteristics of the thermosensitive cells in the spinal cord of mammals are as yet unclear. Most of those in the POA and skin of mammals are warm receptors, with fewer cold receptors. In fact, these two types of receptors do not have equally important roles; defence against both cold and heat are initiated by changes in activity of the warm-sensitive receptors. An increase in activity of warm-sensitive receptors in the POA causes increased activity of the warm-defence responses, while a decrease in their activity initiates cold-defence responses.

It has traditionally been thought that the temperature-control system is a single, central thermostat, which integrates incoming information from thermoreceptors and compares it to a set point temperature. Any deviation from the set point is detected (error signal) and the appropriate response set in train: if incoming information signals a temperature above the set point, the rate of heat acquisition is reduced (for example, by cessation of voluntary physical activity) and rate of heat loss is increased (insulation decreased); if the incoming signal denotes a temperature below the set point, rate of heat loss is reduced (insulation increased) and if the incoming temperature is below the T_{lc}, rate of heat production increases.

Recent experimental evidence indicates that a unified set point model is too simplistic for temperature control in mammals and that there are, in fact, multiple independent controllers acting via separate motor pathways. As a result, the use of the term '**set point**' has been challenged, given its association with a unified control system, with the term 'balance point' being proposed instead. However, the independent pathways do appear to act together in a coordinated manner when the balance point is reset, for example on a diurnal basis or during fever, so we will use the term 'set point' in this book.

A detailed analysis of the available data led to the conclusion that there are at least four independent neural outflow pathways from the POA, as depicted in Figure 10.22. All four

[41] Forced convection and wind chill are discussed in Section 8.4.3.

[42] TRP ion channels in thermoreceptors are discussed in Section 17.5.

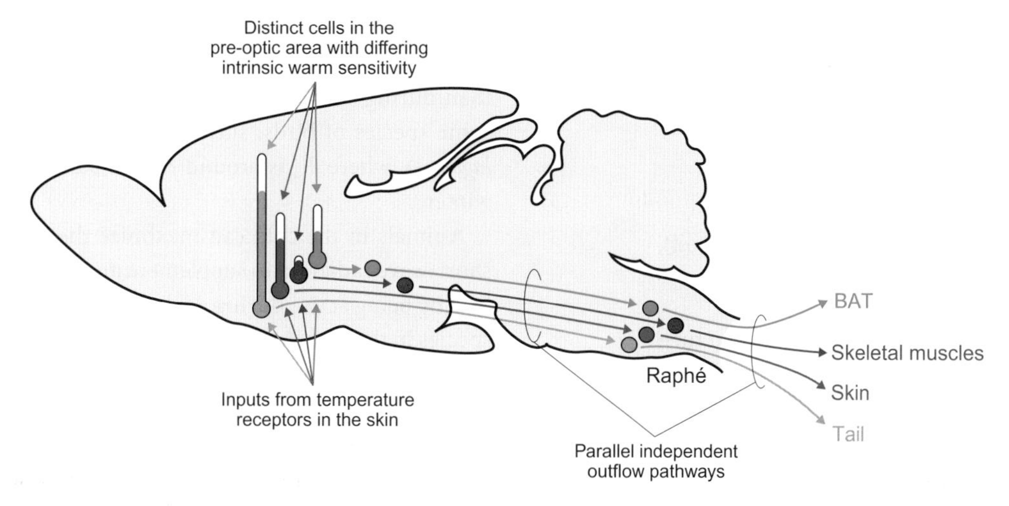

Figure 10.22 Temperature control based on a multiple, independent set of pathways

Diagrammatic representation of longitudinal section through the brain of a rat showing a model that consists of four distinct cold-defence pathways with inputs from temperature sensors in the pre-optic area of the hypothalamus and in the skin and with outflows to the blood vessels in the tail (of rats) and skin, to the skeletal muscles (the fusimotor output) and to the brown adipose tissue (BAT). The distinct groups of pre-optic neurons have different thermal sensitivity, which is indicated by the length of the 'thermometers'.

Reproduced from: McAllen RM et al (2010). Multiple thermoregulatory effectors with independent central controls. European Journal of Applied Physiology 109: 27-33

are activated by cooling the skin or core and are inhibited by a small region in a ventral region of the medulla oblongata[43] called the raphé. The temperature sensors in the hypothalamus have different thermal thresholds and the core and skin sensors have different relative responses to changes in temperature. The responsiveness to core temperature is greatest in the tail group of outflow fibres, followed by the skin, BAT and motor fibres to the hind limb muscles (fusimotor fibres). The response to skin temperature is the opposite way, with the fusimotor group being the most responsive and the tail group the least responsive. Exactly how the set point is encoded is uncertain, but it is clear that it is not rigidly fixed. This lack of rigidity is evident from the diurnal variations in body temperatures shown back in Figure 10.5.

Fever affects the thermoregulatory set point

There are other situations in which more significant changes in the regulated temperature occur for example, during hibernation or during fever. During fever, there is an increase in the set point, usually by 1–2°C, at which the mechanisms for heating and cooling are activated. Human subjects with fever who have a high body temperature, may 'feel' cold and switch on mechanisms, such as shivering, to heat themselves up.

Fevers are often initiated when the animal is infected by microorganisms, although the response does not appear to be directly stimulated by the microorganisms themselves. Rather, the white blood cells react to the microorganisms to produce a variety of compounds, such as prostaglandins and thromboxanes (collectively called pyrogens), which enter the circulation.

Pyrogens stimulate the release of prostaglandin E2 (PGE2), which interacts with neurons in the POA, and fever persists as long as PGE2 remains elevated in the POA. This persistence of a fever suggests that PGE2 is a major component of the temperature-control mechanism.

The significance of fever in mammals has been demonstrated by a number of correlative studies on humans and experiments on other species; the higher the temperature of ferrets (*Mustela* sp.) following infection with a flu virus in their nasal passages, the lower the incidence of live viruses in nasal washings. Elevated body temperature (fever) also occurs in ectotherms following infection or injected toxins and contributes to their survival. For example, when desert iguanas (*Dipsosaurus dorsalis*) are infected with a bacterium, they survive longer when maintained at higher temperatures (longer than 7 days at 42°C, just over 3 days at 34°C). Under normal conditions, high body temperature in lizards is achieved by behavioural means (they move to a higher temperature in a thermal gradient) rather than by endothermic means[44].

10.2.7 Thermoregulatory responses to seasonal decreases in temperature

Animals living in temperate and polar regions encounter annual shifts from summer temperatures, which often include periods in or around the animals' thermoneutral zone (TNZ), to winters during which temperatures could be significantly below their lower critical temperature (T_{lc}).

[43] The medulla oblongata in the brain is discussed in section 16.1.2

[44] Section 9.1 discusses aspects of temperature control in reptiles and other ectotherms.

10

Endotherms respond to a seasonal decrease in temperature in a number of different ways, which include:

- reducing their thermal conductance,
- reducing their body mass,
- migrating away from the cold area,
- surviving winter by storing energy (food),
- surviving winter by undergoing torpor.

We now explore each of these responses in turn.

Reducing thermal conductance during winter

A common response to a seasonal decrease in temperature in mammals is to increase the thickness of their fur by moulting from a summer to a winter coat. Thicker fur provides additional insulation, as shown in Figure 10.23, which reduces the rate of heat loss. However, fur growth in winter is related to body size: larger animals have thicker fur and increase their fur thickness in winter to a much greater extent than smaller animals. Note that this pattern is opposite to that seen in hot environments, where larger animals have thinner pelts, as shown back in Figure 10.10.

The consequence of increased fur thickness is that the relationship between ambient temperature and the rate of heat production required to maintain body temperature is different between winter and summer animals. During winter, the rate of oxygen consumption in snowshoe hares (*Lepus americanus*), is lower at a given environmental temperature

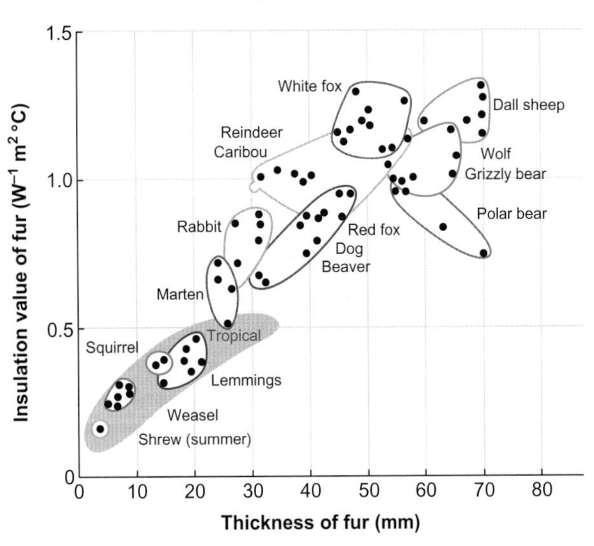

Figure 10.23 Plot of insulation value against fur thickness for the fur of a number of species of mammals

Smaller animals such as shrews and weasels have shorter fur than larger ones such as foxes and reindeer and, therefore, have poorer insulation. In fact, insulation of the fur of the smaller animals in this plot overlaps that for many tropical species, indicated by the pink-shaded area. For animals larger than foxes, there is no clear relationship between the thickness of their fur and its insulation value.

Modified from: Scholander PF et al (1950). Body insulation of some arctic and tropical mammals and birds. Biological Bulletin 99: 225-226.

than that during summer, as illustrated in Figure 10.24. Also, lower critical temperature (T_{lc}) is lower during winter (0°C) than during summer (11°C). A similar situation occurs in some species of birds, such as willow ptarmigan (*Lagopus lagopus*), where T_{lc} is around 8°C in summer, but −6°C in winter.

Animals in the cold can maximize their uptake of solar short-wave radiation to supplement any internally generated heat by being active during daylight. A greater proportion of species in the high Arctic are active during the daytime compared with those in desert regions. We might imagine that the uptake of solar radiation could be enhanced by reducing surface albedo (making the body surface darker). However, many species, such as snowshoe hares and polar bears (*Ursus maritimus*), seem to do exactly the opposite; they become white in winter, as shown in Figure 10.24. The structure of the fur in animals living in cold, snowy environments may not only give rise to the white colouration, which acts as an effective camouflage against the snow, it may also increase its insulative properties by producing the scattering of long-wave radiation back to the animal. For example, polar bears do not produce a strong thermal image against snow.

Mammals and birds ventilate their lungs more than other air-breathing vertebrates because of their higher resting metabolic rates. This expired air will be at or close to body temperature (approximately 35–40°C). Birds and mammals routinely decrease their effective thermal conductance by reclaiming heat in expired air during lung ventilation. This is particularly significant during cold conditions.

The potentially costly loss of heat during expiration has been drastically reduced by the evolution of the **respiratory turbinates** in the nasal passages of most species of mammals and birds. While olfactory turbinates for the sense of smell are present in all land vertebrates, respiratory turbinates are only present in mammals and birds. They consist of highly scrolled or branched bony or cartilaginous structures covered in a moist epithelium, as shown in Figure 10.25A.

The way in which they function is elegantly simple. During inspiration, air passes over the respiratory turbinates and is warmed close to body temperature. As a result, the turbinates are cooled, both by the temperature of the incoming air and by evaporative water loss. During expiration, the warm air leaving the respiratory system passes over the cooler surfaces of the turbinates in the respiratory passages and is cooled down, thereby conserving heat, as illustrated in Figure 10.25B. This mechanism can be thought of as a countercurrent system[45], with alternating flow occurring in a single tube, rather than continuous flow occurring in opposite directions in two adjacent sets of tubes, as in the gills

[45] See Section 3.3.1 for explanation of countercurrent exchange.

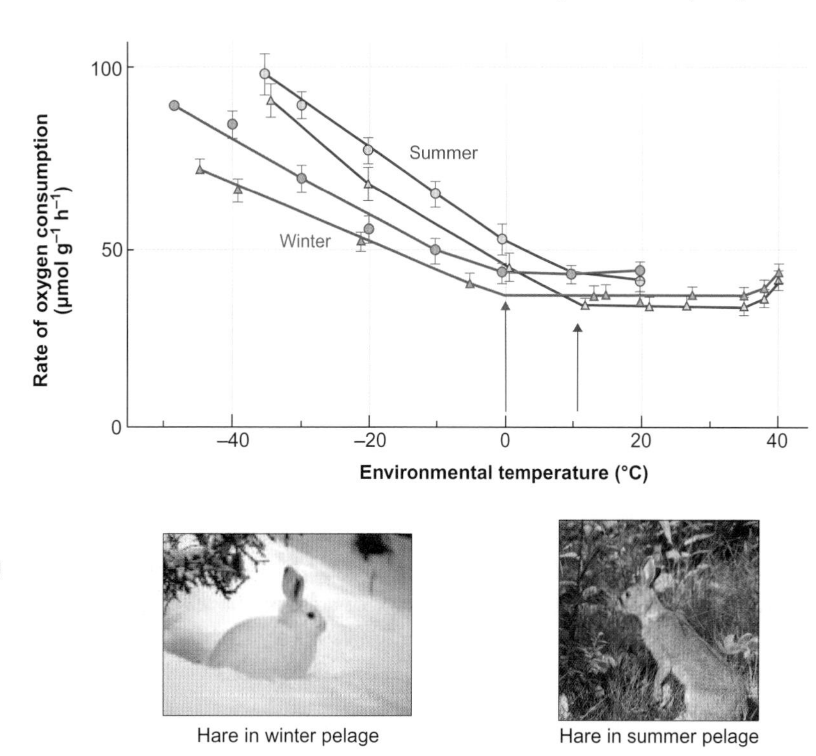

Figure 10.24 Relationship between rate of oxygen consumption and environmental temperature in snowshoe hares (*Lepus americanus*) during two summers (July, August) and two winters (January to March)

Note that in winter: (i) the lower critical temperature is lower (blue arrow) and (ii) the relationship between rate of oxygen consumption and environmental temperature is less steep, hence energy demands increase less as temperature falls during winter than by a similar extent during summer. Red arrow indicates lower critical temperature in summer. Symbols indicate mean values ± SEM. N = 5–9.

Modified from: Hart JS et al (1965). Seasonal acclimatization in varying hare (*Lepus americanus*). Canadian Journal of Zoology 43: 731-744. Images: hare in winter pelage © R. Bruce Gill; hare in summer pelage © Marc F O'Brien.

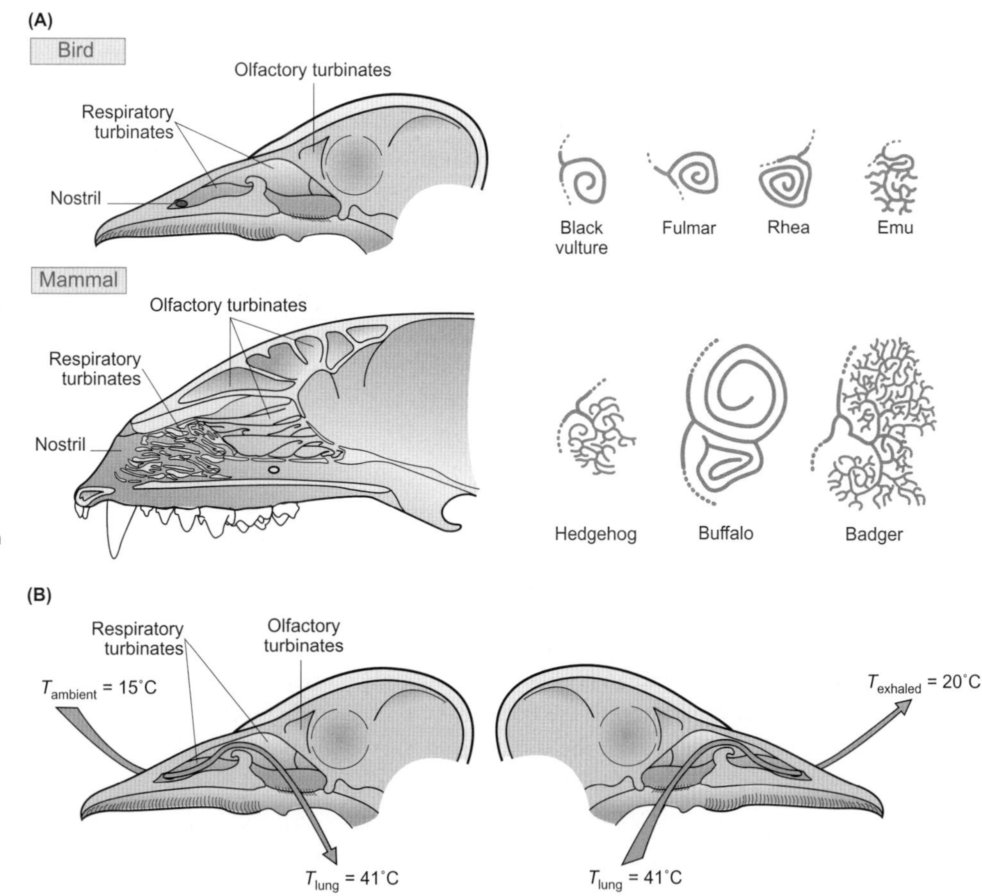

Figure 10.25 Conservation of heat during ventilation of the lungs of birds and mammals

(A) Nasal passages of a bird and mammal showing the olfactory and respiratory turbinates. Also shown are cross-sections of respiratory turbinates of some species of birds (top right) and of some species of mammals (bottom right). (B) Functioning of respiratory turbinates showing the countercurrent flow of air in a single tube during inhalation and exhalation. In the example shown, the turbinates are cooled by the incoming air (at 15°C) and heat in the air from the lungs (at 41°C) is transferred to the turbinates so that air leaving the nasal passages is at 20°C.

Reproduced from: Hillenius WJ and Ruben JA (2004). The evolution of endothermy in terrestrial vertebrates: who? when? why? Physiological and Biochemical Zoology 77: 1019-1042.

of many water-breathing animals, or through a rete, as in the heat exchangers in the brain of many species of mammals[46].

The effectiveness of the nasal turbinates has been demonstrated in birds ranging in mass from 120 to 800 g. The average temperature of the expired air is around 20°C and the daily saving in energy expenditure is around five per cent.

Reducing body mass during winter

Another response to the seasonal cold in winter is to get smaller. These changes often involve reducing the amounts of lean and fat body tissue, but in some exceptional cases include complete skeletal remodelling; the animals actually physically shrink themselves. The best known example of this is the seasonal change in the size of the skull of shrews (*Sorex* sp.), which was first described by the Polish zoologist August Dehnel in 1949. This lead to the seasonal and reversible changes in the skeleton and major organs that have been described for some species of mammals being called **Dehnel's phenomenon**. Inspection of the thermoregulatory response curves of small and large animals shown back in Figure 10.6, reveals the big advantage that being smaller provides; the total energy demand is reduced.

Many studies have helped us understand the stimuli for changes in body mass and pelage, notably those on Siberian hamsters (*Phodopus sungorus*). The primary stimulus triggering the response is the change in photoperiod. When captive hamsters are switched from a long day photoperiod (16 hours light, 8 hours dark) to a short day (8 hours light, 16 hours dark), they start to lose body mass after a period of about 2 weeks, losing around 30 per cent of their body mass after about 10 weeks. These changes occur independently of ambient temperature.

Getting smaller may reduce energy requirements, but if this reduction is more than offset by a decline in the rate of energy (food) intake it will not be advantageous. Whether or not getting smaller will be beneficial probably depends on the exact dietary intakes of the animals in question and the distribution of food resources during the winter, which we discuss later in this section.

There is one situation in which the effects of body size on energy supply and demand favours being larger; where there is periodically no food available at all and animals must rely completely on stored energy reserves. The energy density of fat[47] is 35.6 kJ g^{-1}, making it the most space-efficient way to store energy. It is not surprising, therefore, that animals store energy for long-term use predominantly as fat. If the amount of stored fat is in direct proportion to the body mass, Box 10.1 (available online) explains why the duration an animal survives without any access to food is much greater for

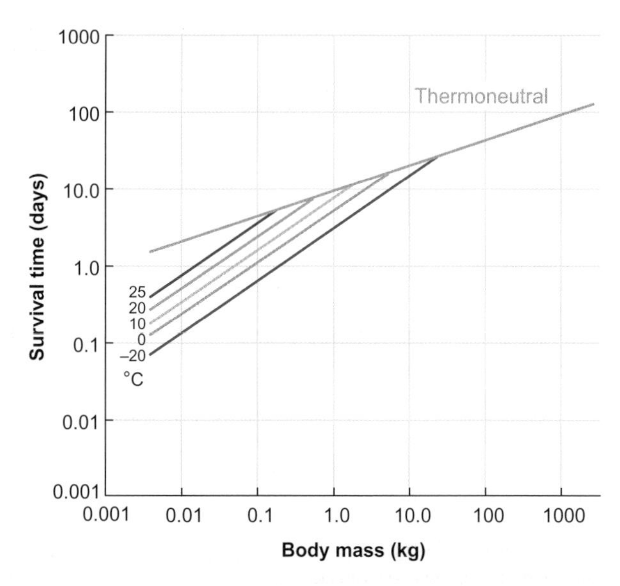

Figure 10.26 Fasting endurance (survival time) of animals of different body sizes (body mass) at thermoneutral temperatures and at different ambient temperatures

Survival time gets longer as animals get larger and this effect is greater at cold temperatures.

Reproduced from: Lindstedt SL and Boyce MS (1985). Seasonality, fasting, endurance and body size in mammals. American Naturalist 125: 873-878.

a larger compared with a smaller animal, even though the total energy demand of the larger animal is greater. The survival time is called an animal's 'fasting endurance', and its relationship to body mass is illustrated in Figure 10.26.

Whether or not it is better for an animal to be larger or smaller in the cold depends on its food supply. If food supply is only intermittently available, and animals need long fasting endurance, selection will favour being larger. But if food is always available at low levels then selection might favour being smaller.

In some circumstances, the supply of food in the winter may fall to such low levels that an animal would still not be in energy balance, even if it was to increase its surface insulation and reduce its body size. Under these circumstances animals need to do something more radical. One response to this situation is to migrate away from the area of restricted food supply.

Migrating to warmer climates during winter

For most mammals, particularly small ones, migration is not feasible because terrestrial locomotion is simply too slow to get them out of the danger area (and, with the exception of bats, flying is not an option for them). Migration is a much more feasible strategy for flying animals because their greater speed allows them to cover the required distances in less time and they can easily cross physical barriers.

An animal flying at 10 ms^{-1}, for example, could cover 100 km in just under 3 hours; unsurprisingly then, migration is a strategy used by many bat species. An example of such

[46] The rete in the brain of sheep is shown in Figure 10.13.

[47] Energy density of fat is discussed in Section 2.2.1.

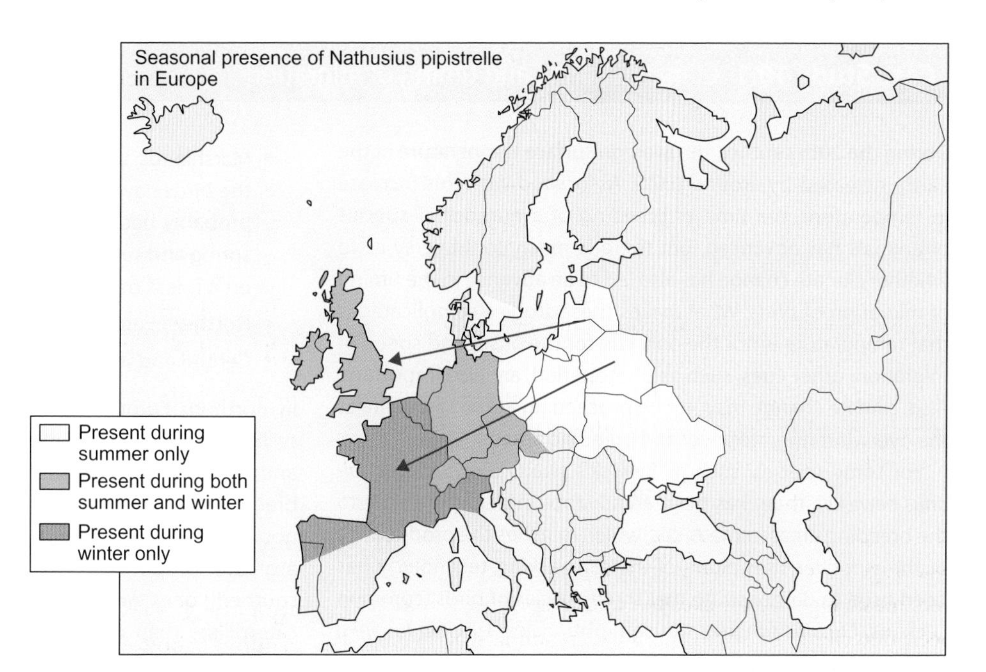

Figure 10.27 Migration routes of Nathusius' pipistrelle (*Pipistrellus nathusii*)

The bats migrate from the cold continental countries to milder coastal areas.

Modified from: www.nathusius.org.uk with permission.

a seasonal migrant is the small Nathusius' pipistrelle bat (*Pipistrellus nathusii*) which migrates from cold continental countries in winter to milder coastal areas. Figure 10.27 shows how these bats spend their summer in large numbers in the Eastern European states of Latvia, Lithuania and Poland, but appear in Belgium and France in winter, about 1500 km south-west of their summer habitat. A population probably also flies across the North Sea into Great Britain, because bats are often captured on North Sea oil exploration platforms during both autumn and spring.

The biannual migration of some species of birds is one of the wonders of the natural world. Many of these species encounter barriers such as oceans and mountain ranges during their migrations. For example, a race of bar-tailed godwits (*Limosa lapponica baueri*) have been tracked flying non-stop for 8 days, covering an average distance of over 10,000 km across the Pacific ocean from their breeding grounds in southwestern Alaska to overwinter (the austral summer) in New Zealand. Others, like bar-headed geese (*Anser indicus*), fly over the Himalayas during their migration between the high plateaux of central Asia, where they breed, and India, where they overwinter[48].

One species that has been studied in some detail is barnacle geese (*Branta leucopsis*), which breeds on the island of Spitzbergen in the high Arctic and migrates south in the autumn (fall) to southern Scotland, stopping at Bear Island and along the Norwegian coast, as illustrated in Figure 15.15. One of the issues that has emerged from such studies is the extent to which climate change may be affecting the behaviour of migrants, as we discuss in Case Study 10.1.

[48] Adaptations to flight at high altitude in bar-headed geese is discussed in Section 15.3.1.

Surviving winter by storing energy (food)

If an animal is unable to move from an area with insufficient food to get through the winter, it may store energy during the period when food is available and then use this throughout the winter. There are two ways to achieve this:

- The animal can store the food inside its body as fat, as we discuss earlier in this section.
- It can store the food externally in a food store or cache.

The storage of food outside the body allows much greater amounts of energy to be stored, because space is less limited and the store does not impede speed of travel or manoeuvrability, but it is not without problems. The store may become contaminated by mould, for example, and other animals may steal the food from the store. Nevertheless, many animals use this strategy despite the potential pitfalls.

A well-studied example of food storage is exhibited by North American red squirrels (*Tamiasciurus hudsonicus*), which do not hibernate. During autumn, red squirrels collect pine cones and deposit them into a store at the base of a tree. In good crop years a typical squirrel might collect over 50 kg of cones. The store is protected from theft because the squirrels are strongly territorial. However, this is not a foolproof strategy, as they cannot continuously attend the food pile, although theft by other squirrels is generally not a serious issue, particularly in years when cones are plentiful.

Case Study 10.1 Climate change and timing of migration

During the 20th century, the average surface temperature of the Earth increased by around 0.6°C. Associated with this increase in temperature, the time of breeding of a number of species of animals has advanced, but there is much controversy as to whether climate change has also led to an advance in the timing of migration of birds. What makes these studies complicated is that temperature is not the only cue for the onset and speed of migration; other cues, such as photoperiod, are also important. Also, climate change may not be proceeding at the same rate at the overwintering, stopover and breeding areas.

For some species, such as Bewick's swans (*Cygnus columbianus bewickii*), there has been an advance in departure time to the breeding areas in the Arctic, which matches the productivity of the stopover sites. State-of-the-art tracking technology has been used to demonstrate that three species of birds (common cuckoos, *Cuculus canorus*; thrush nightingales, *Luscinia luscinia*; red-backed shrikes, *Lanius collurio*) that migrate between Europe and Africa follow the greenness of the vegetation throughout their annual migrations. They achieve this by adjusting the timing and direction of their movements to coincide with the availability of the vegetation.

It would appear, however, that climate change has often created a mismatch between the timing of migration and the timing of maximum food availability at the stopover and/or breeding grounds. North American wood warblers (Family Parulidae) have not advanced their migration time, despite their main prey species, the eastern spruce budworm (*Choristoneura fumiferana*), having advanced its breeding. This means that the warblers arrive too late in their breeding grounds to take full advantage of the peak in caterpillars. The American robin (*Turdus migratorius*) now arrives 14 days earlier at its breeding area in the Rocky Mountains than it did 20 years ago, but there has been no advance in the date of snow melt. As a result, there is now an 18-day delay between the birds' first arrival and bare ground appearing. Such mismatches appear to be more apparent for long-distance migrants, as they are not able to predict the start of spring at their breeding grounds.

Long-distant migrants may not have changed the arrival dates at their breeding grounds to the same extent as short-distant migrants. Christiaan Both and his colleagues have analysed data collected during 1984 and 2004 from a number of species of insectivorous passerines, which breed in the Netherlands, in order to examine the idea that long-distant migrants are more vulnerable to climate change than residents and short-distant migrants.

They divided the data into birds that live in:

- Temperate forests where there is a short burst of mainly herbivorous insects that feed on young leaves of deciduous trees before the production of secondary plant compounds begins. It has been shown that breeding in forest birds is highly synchronized with this peak in food supply. What is more, the peak of forest insects has advanced in relation to earlier springs.
- Marshlands, which are dominated by *Phragmites* reed. Here the birds have more extended periods of food abundance, probably because the reed continues to grow throughout spring and summer and the biomass of the insects that feed on it is less peaked than in forests.
- Northern Europe.
- Central and western Europe.

In northern Europe, spring temperatures have not increased or have increased only mildly compared with those in central and western Europe at the time long-distant migrants arrive to breed, so the laying dates of resident and migratory birds have not advanced in northern Europe. Also, northern forests have a greater proportion of coniferous trees compared with more southerly ones, and conifers have a less peaked abundance of caterpillars than deciduous trees. This makes more northerly habitats less seasonal in this respect, compared with those in central and western Europe.

Figure A summarizes the data from the analyses. Figure A(i) shows that the behaviour of long-distant migrants has led to a greater decline in populations that breed in forest habitats where the food peak in spring is short (average decrease by 38 per cent), compared with those that breed in marshes, which have a more extended period of food production (average increase by 158 per cent). In contrast, populations of residents and short-distant migrants have not declined in either habitat, which suggests that the quality of the habitats has remained unchanged. Figure A(ii) shows that, within species, long-distant migrants decline more in forests than in marshes and Figure A(iii) shows that populations of long-distant migrants that breed in forests in western Europe and in northern Europe have declined more substantially in western Europe (average decrease 35 per cent), where springs have become much warmer, compared with those in northern Europe (average decrease 9 per cent). Taken together, these data suggest that:

- the quality of the breeding grounds is not a factor as resident birds and short-distant migrants are less affected than long-distant migrants at the same breeding sites;
- the decline in populations of the long-distant migrants is most likely related to climate change causing earlier springs in western and central Europe and the fact that these migrants do not respond to the changes.

Some larger mammals in the Arctic also migrate over large distances to respond to changes in food availability during winter. The best studied of these are the movements of caribou (reindeer: *Rangifer tarandus*) and polar bears (*Ursus maritimus*) in northern America. These movements appear to be responsive to seasonal alterations in predation risk for their offspring (caribou) or distributions and accessibility of their prey (polar bears) and do not reflect direct responses to seasonal changes in ambient temperature.

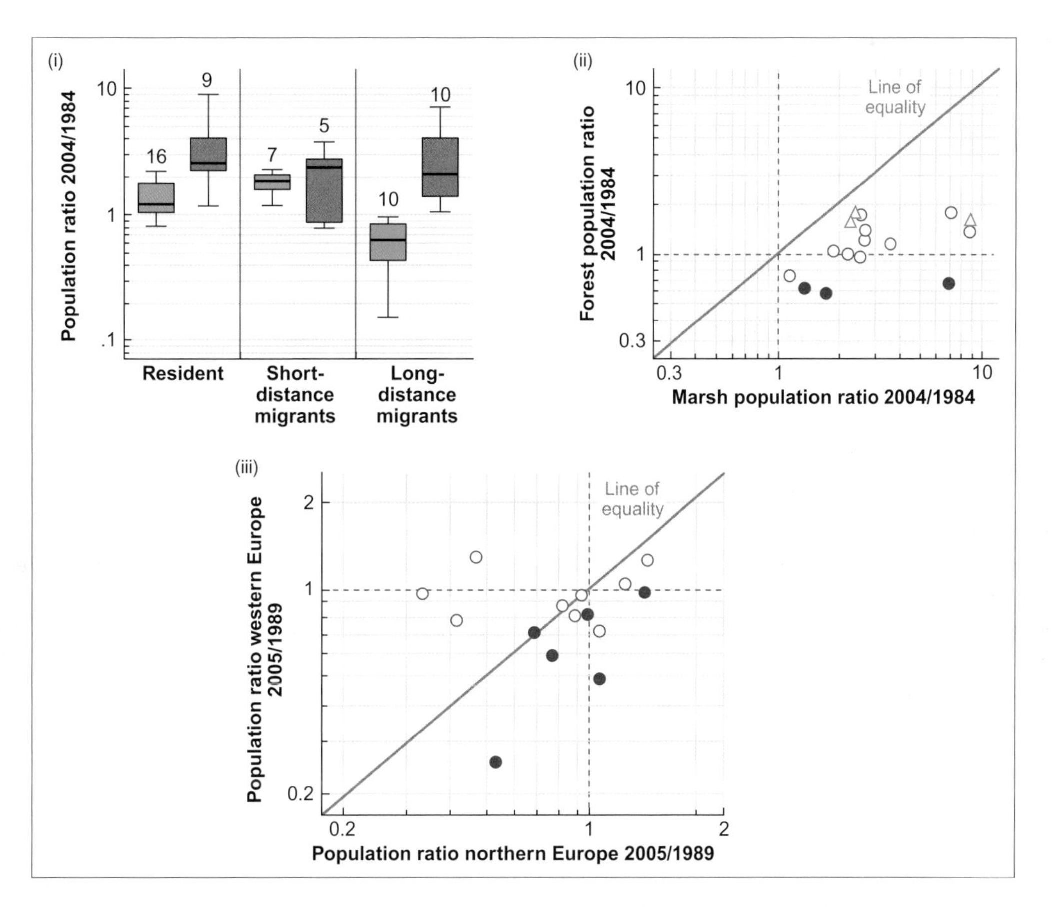

Figure A Factors influencing populations of migratory insectivorous birds between 1984 and 2004

(i) For residents and short-distant migrants there is no difference in the population ratio between forest-dwellers (green-shaded boxes) and those that live in marshes (brown-shaded boxes). However, there is a decline in the ratio in long-distant migrants that breed in forests compared with those that spend the spring in marshes. The boxes enclose the middle half of the data, the horizontal lines across the boxes represent the median values and the vertical lines indicate the range of the data. The numbers above the boxes are the numbers of birds. (ii) Within-species population trends in forests and marshes. Long-distance migrants (red solid circles) decline more in forests than in marshes, compared with residents (open triangles) and short distance migrants (open circles). (iii) Population trends of species that breed in forests in northern and western Europe. Long-distant migrants are depicted by red solid circles and short-distance migrants and residents by open circles. Note, that those species that fall below the line of equality decline more in western than in northern Europe, and that most of long-distant migrants fall below this line.

Reproduced from Both C et al (2010). Avian population consequences of climate change are most severe for long-distance migrants in seasonal habitats. Proceedings of the Royal Society B 277: 1259-1266.

› Find out more

Both C, van Turnhout CAM, Bijlsma RG, Siepel H, van Strien AJ, Foppen RPB (2010). Avian population consequences of climate change are most severe for long-distant migrants in seasonal habitats. Proceedings of the Royal Society of London B 277: 1259–1266.

Thorup K, Tøttrup AP, Willemoes M, Klaassen RHG, Strandberg R, Vega ML, Dasari HP, Araújo MB, Wikelski M, Rahbek C (2017). Resource tracking within and across continents in long-distant bird migrants. Science Advances 3: e1601360.

Visser ME, Both C (2005). Shifts in phenology due to global climate change: the need for a yardstick. Proceedings of the Royal Society of London B 272: 2561–2569.

Red squirrels build a nest in a tree above the food cache. During winter in the Yukon in western Canada, temperature falls to –20°C on average, and even as low as –40°C, with an extensive snow covering of between 300 to 600 mm. During this period the squirrels spend almost their entire day inside the nest.

Figure 10.28A shows how activity outside the nest is related to ambient temperature; the colder it gets, the less time the squirrels spend outside. This behaviour minimizes the squirrels' exposure to low ambient temperatures when their energy requirements would be excessive. Figure 10.28B shows the response of the squirrels to warming in winter; the warmer the ambient temperature the more active squirrels become so the greater their energy demands. This response is the opposite to that of the other species of mammals at higher temperatures and contrary to what would be

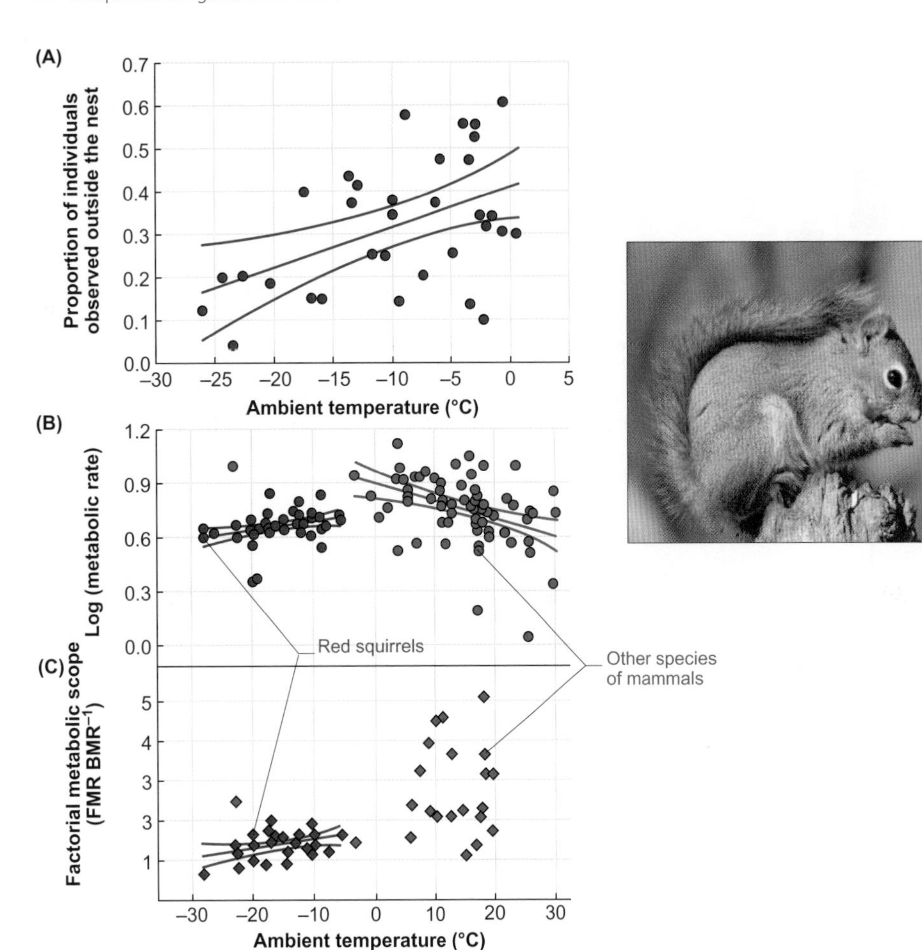

Figure 10.28 Activity patterns and energy expenditures of North American red squirrels (*Tamiasciurus hudsonicus*) in relation to ambient temperature during winter in the Yukon

(A) Proportion of individuals outside the nest. (B) Field metabolic rate (FMR) of red squirrels and other mammals in relation to ambient temperatures. (C) Factorial metabolic scope (FMR BMR^{-1}) of the red squirrels and other mammals. Note that in winter, FMR of red squirrels declines with decreasing temperature whereas in other species of mammals at higher environmental temperatures, the opposite is the case. BMR, basal metabolic rate. Lines represent the least squares regressions and 95% confidence intervals.

Reproduced from: Humphries MM et al (2005) Expenditure freeze: the metabolic response of small mammals to cold environments. Ecology Letters 8: 1326-1333. Image: © Mircea Costina.

10

predicted from the Newtonian model for the relationship between rate of energy expenditure and ambient temperature, as depicted earlier in Figure 10.2.

By storing their food in advance of winter, squirrels minimize the time they spend exposed to the cold conditions. In consequence, their total energy requirements are only slightly more than their resting energy demands in the thermoneutral zone (averaging about 1.6 times basal metabolic rate (BMR), as depicted in Figure 10.28C), despite ambient temperatures routinely being 30 to 40°C colder than the squirrels' lower critical temperature.

Surviving winter by undergoing torpor

A different approach to surviving the winter in areas where food is scarce or unavailable is for an animal to massively reduce its energy demands by reducing the level at which regulatory control is held over body temperature (i.e. the regulatory set point).

The reduction in energy demands and decline in body temperature to levels where an animal becomes moribund is known as **torpor** (Greek—sluggishness). In some species, torpor enables an animal to survive for several days or even for the entire winter on its stored fat reserves; this is called **hibernation**[49] (Latin, *hibernatio*—the action of passing the winter). Other species reduce their energy demands and drop body temperature on a daily basis, a phenomenon known as **daily torpor**. Animals that become torpid on a daily basis can vary its incidence and duration day-by-day in response to the availability of food, environmental temperature and other demands on their rate of energy expenditure.

Characteristics of hibernation

The only species of bird that remains torpid continuously for several days (10 or more) is the common poorwill (*Phalaenoptilus nuttallii*) and whether or not this should be considered as hibernation is a matter of debate. So, the remainder of this section will concentrate on mammals.

During hibernation, mammals generally reduce their body temperatures to values that are only just above ambient temperature, as shown in Figure 10.29A. By contrast, their body temperature can be several degrees above ambient

[49] Hibernation in ectotherms is discussed in Section 9.2.2.

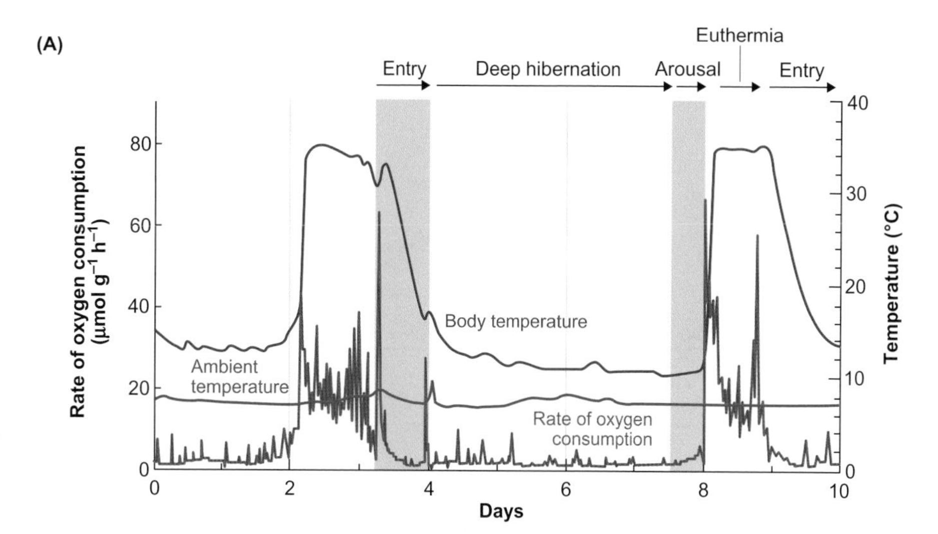

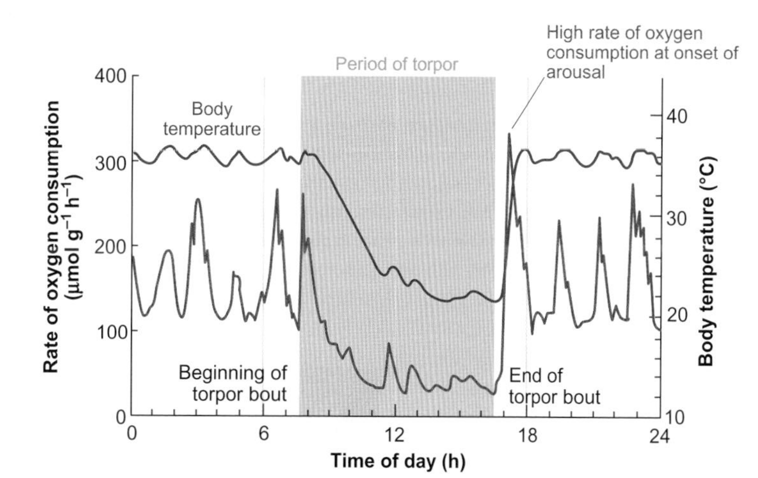

Figure 10.29 Body temperature (T_b) and rate of oxygen consumption (metabolic rate) before, during and after hibernation and a bout of daily torpor

(A) Hibernation bout in an Alpine marmot (*Marmota marmota*); (B) Daily torpor in a Siberian (Djungarian) hamster (*Phodopus sungorus*). As (B) is on a more expanded timescale, it can be seen that the decline in the rate of oxygen consumption occurs before the change in T_b at the beginning of torpor. It also begins to increase before the rise in T_b at the end torpor. Rate of oxygen consumption during the initial stages of arousal greatly exceeds that before the onset of hibernation/torpor.

Reproduced from: A: Heldmaier G et al (2004). Natural hypometabolism during hibernation and daily torpor in mammals. Respiratory Physiology & Neurobiology 141: 317-329. B: Heldmaier G et al (1999). Metabolic adjustments during daily torpor in the Djungarian hamster. American Journal of Physiology 276: E896-E906.

during daily torpor as shown in Figure 10.29B. By making sequential measures of both metabolic rate and body temperature, it has become clear that metabolic rate does not follow the fall in body temperature. In fact, it decreases beforehand, as shown for daily torpor in Figure 10.29B. Although not so obvious in Figure 10.29A because it is plotted on a more compressed timescale, the same occurs at the onset of hibernation. In other words, there is an active depression of metabolic rate to what is known as a **regulated hypometabolic state**.

A Q_{10} effect[50] on metabolic rate during torpor/hibernation may be of greater significance in some species than in others. This may depend on the pattern of torpor (it may be more important during daily torpor than during hibernation), body mass and body temperature during the torpid state. However, regulated hypometabolism during torpor is not always accompanied by a large reduction in body temperature, as we discuss later in this section.

[50] Q_{10} is explained in Section 8.1 and Box 8.2.

Figure 10.30 shows that during hibernation, metabolic rate falls by a greater proportion in smaller animals than in larger ones. Small hibernating mammals can save up to 98 per cent of the energy they expend when euthermic (have a normal body temperature). As a consequence, hibernation is most frequent among smaller mammals and only rarely occurs in larger species such as yellow-bellied marmots (*Marmota flaviventris*) and black bears (*Ursus americanus*). An average value for energy savings during hibernation is about 90 per cent. Energy savings during daily torpor are less than during hibernation, being about 40 per cent over a 24 h period. However, long-term energy budgets of Siberian (or Djungarian) hamsters (*Phodopus sungorus*) undergoing daily torpor reveal energy savings of up to 67 per cent compared to individuals that remain active.

It is important to recognize that hibernating animals have not abandoned thermoregulation, but have merely reduced the body temperature around which they regulate. This change in regulated body temperature can be easily demonstrated if the ambient temperature during hibernation is reduced to extremely low levels. Hibernators do not

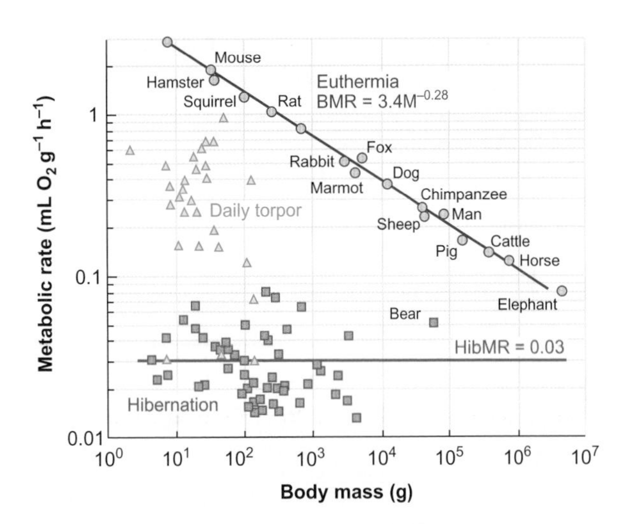

Figure 10.30 Metabolic rates (MR) of mammals of different body sizes during daily torpor and hibernation (open triangles and filled squares, respectively)

In order to illustrate the extent of metabolic depression during daily torpor and hibernation, data are also given for basal metabolic rate (BMR) of animals at normal body temperature (euthermia). Note, that MR during hibernation (HibMR) is not affected by body mass whereas basal metabolic rate during euthermia is, with larger animals having a lower mass-specific BMR than smaller animals. This means that the relative difference between BMR and HibMR is greater in smaller animals.

Reproduced from: Heldmaier G et al (2004). Natural hypometabolism during hibernation and daily torpor in mammals. Respiratory Physiology & Neurobiology 141: 317-329.

allow their body temperatures to continue falling, but at some point start to elevate their rate of heat production to defend their body temperatures at the new, lower set point (T_{set}). Figure 10.31 illustrates the difference in the value of T_{set} between hibernating and non-hibernating marmots. Such a downward shift in the T_{set} relative to normal body temperature (which is termed **euthermia**—Greek, *eus*—good, well, *thermos*—warm) is known as regulated hypothermia or anapyrexia (Greek, *ana*—reverse, *pyretos*—fever).

The major benefit of hibernation is the reduction in energy demands that allows the animals to survive, often without feeding, for the entire winter, relying only on the energy they have stored in their bodies or perhaps in a food store. There are, however, two fundamental disadvantages of entering this hibernating state:

- The animal must switch off many basic biological processes to reduce its rate of energy expenditure to a sufficiently low level.
- Hibernating animals almost completely lack any functional abilities—their muscles are extremely slow to contract, and so they are unable to run away or to defend themselves against dangers. This potentially makes them vulnerable to predation.

Figure 10.31 Response of metabolic rate (MR) to change in temperature of the hypothalamus in a euthermic and hibernating yellow-belied marmot (*Marmota flaviventris*)

Note that when in the hibernating condition, hypothalamic temperature fell to about 7°C before there was an increase in MR (this is the set point temperature or T_{set}), whereas in the euthermic animal the T_{set} was about 36.5°C. Also, note the much lower metabolic rate during hibernation.

Reproduced from: Florant GL and Heller HC (1977). CNS regulation of body temperature in euthermic and hibernating marmots (*Marmota flaviventris*). American Journal of Physiology 232: R203-R208. Image: Davefoc/Wikimedia Commons.

However, unlike the situation in inactive non-hibernating mammals, there is limited, if any, loss of muscle mass (atrophy) in hibernating bats, such as the greater tube-nosed bat (*Murina leucogaster*), and black bears,[51] despite their long periods of inactivity. Similarly, the inactivity of hibernation does not lead to a net loss of bone mass in black bears, as both the rates of bone formation and resorption decrease by a similar proportion. Although there is significant atrophy of leg muscles in golden mantled ground squirrels (*Callospermophilus* (formerly *Spermophilus*) *lateralis*), there are also significant increases in the activity of oxidative enzymes in the muscles, which could be important for heat generation by shivering during arousal.

The fact that animals are in a cost–benefit trade-off during hibernation has been demonstrated by manipulating the quantity of the food available to eastern chipmunks (*Tamias striatus*) during the entry phase to hibernation. The results are shown in Figure 10.32. The food cache available to the chipmunks was manipulated by providing the animals with supplementary food during the phase when they were building up the store. In response to the additional food, the animals did not reduce their body temperature by as much nor for such long periods. However, it is not only the calorific content of the food that determines depth and duration of hibernation.

Significance of unsaturated fatty acids during hibernation

At the low body temperatures reached during hibernation, normal body fat would become solid. Consequently, the proportion of polyunsaturated fatty acids (PUFAs[52]—which

[51] Hibernation in black bears is discussed later in this section.

[52] Unsaturated fatty acids are discussed in Box 9.2.

(A)

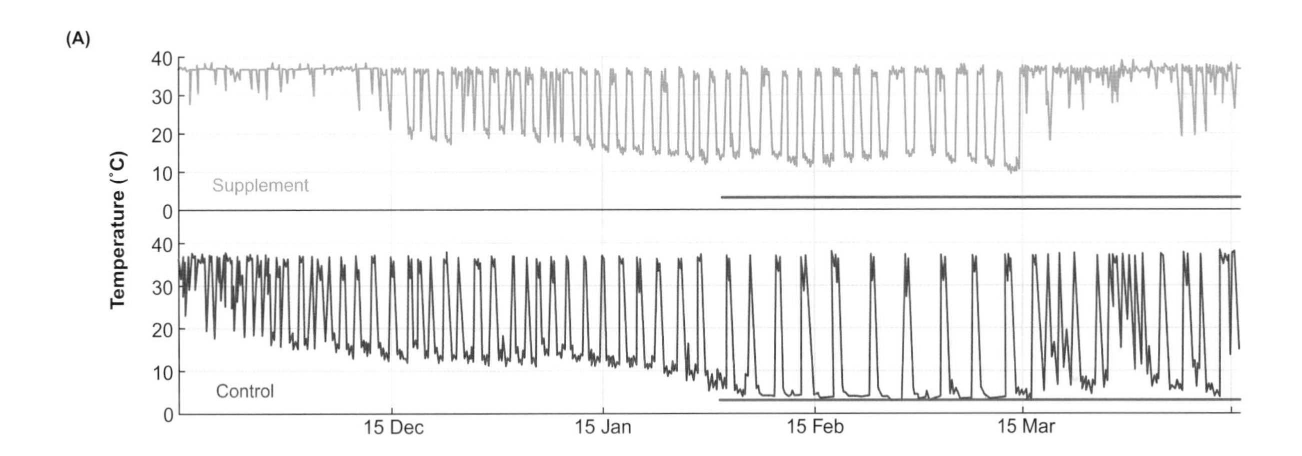

(B)

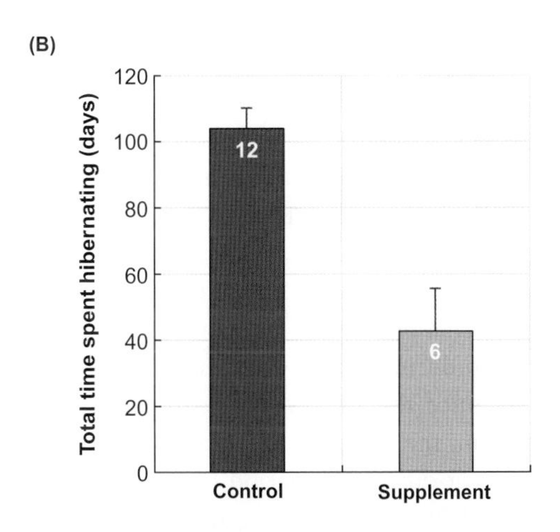

Figure 10.32 Effect of food supplements on patterns of hibernation in eastern chipmunks (*Tamais striatus*)

(A) Recordings of body temperature (T_b) of two individual hibernating chipmunks at a study site in Quebec, Canada. One had not received supplementary food, control, and one had received supplementary food. Note that the drop in T_b during the winter period (Feb–mid Mar) is greater in the control group. The lower line of each panel indicates the temperature of the soil. (B) Duration of the hibernating periods is greater in the control group. Vertical lines indicate + SEM and numbers of animals used are shown within the columns.

Reproduced from: Munro D et al (2005). Torpor patterns of hibernating eastern chipmunks *Tamais striatus* vary in response to the size and fatty acid composition of food. Journal of Animal Ecology 74: 692-700. Image: © Phil Armitage.

have low melting points) in the bodies of many hibernators is higher than in non-hibernators. For example, Belding's ground squirrels (*Urocitellus beldingi*) have almost twice the amount of PUFA in their white adipose tissue during autumn compared to non-hibernating rodents.

The fluidity of cell membranes is also important in order to preserve their permeability and allow for normal activity of transmembrane proteins[53]. As such, the ratio of unsaturated to saturated fatty acids for a variety of phospholipids increases in cellular and subcellular membranes of hibernators.

Most studies have found that high PUFA diets enhance the depth and duration of hibernation, but it has been suggested that the effects on hibernation are not the result of PUFAs in general, but rather the result of shifts in the ratio of n-6 PUFAs to n-3 PUFAs in membrane phospholipids.

Characteristics and possible function of arousals

In almost all the hibernating mammals studied so far, the animals drop their body temperatures (T_b) during the early phase of hibernation for relatively short periods during the daily cycle. Gradually, the reductions in T_b become greater until the animals enter a prolonged period of regulated hypothermia lasting several days, as shown in Figure 10.32A. During these periods the body temperature generally declines to within a few degrees of ambient.

In species from high latitudes, such as Arctic ground squirrels (*Urocitellus parryii*), soil temperature surrounding their **hibernaculum** (Latin for winter quarters) may reach several degrees below freezing. Consequently, T_b in these animals routinely falls to below 0°C, and to as low as −2.9°C. Because the duration of hibernation is so long, these animals spend the majority of their lives in a state where their body temperature is below freezing, although their plasma shows no evidence of having high solute concentrations nor of containing any antifreeze molecules[54].

Between the periods of torpor, the animals increase their body temperatures back almost to the temperature at which they regulate in the summer, as shown in Figure 10.32A. These periods are known as **arousals** and often last around 10–20 hours or so, before the animals then re-enter the torpid state.

During the arousal phase from hibernation, the animals have to heat up their bodies from temperatures that can be around freezing to between 30–35°C. Assuming the body of a ground squirrel has a specific heat of 3.4 kJ kg^{-1} $°C^{-1}$ (this is the generally accepted mean value for animal tissue), a 1 kg ground squirrel would need to generate 119 kJ of heat to raise its temperature by 35°C (1 kg × 3.4 kJ kg^{-1} $°C^{-1}$ × 35°C). During the arousal period, therefore, hibernators have very high rates of energy expenditure, as shown in Figure 10.29. It has been calculated that the arousal periods can consume 70 per cent of the total energy demands over the entire winter period, despite the fact that they may only last for three per cent of the total time.

Figure 10.33 shows that the source of heat during arousal in bats is located primarily in the region between the shoulder blades. This is also the location of the major deposit of brown adipose tissue (BAT) in hibernators, supporting the proposal that heat production during arousal comes entirely from BAT.

During winter, animals occasionally exhaust their reserves of white adipose tissue and do not survive the winter hibernation. It seems that their energy balance is on a knife edge. Yet they could easily survive the entire winter if they did not engage in arousal. For this reason, and because these arousal periods are observed in all hibernators that allow their bodies to drop to low temperatures, it is widely assumed that they serve an essential biological function. However, to date, we do not know exactly what that function is, although several functions have been proposed:

- to dispose of potentially toxic metabolic by-products that accumulate during hibernation,
- to feed and/or drink,
- to sleep,
- to repair the brain from damage that occurs during torpor.

Many species of hibernators, including ground squirrels (*Urocitellus* spp.) do not leave the hibernaculum to feed or drink during their arousal. On the other hand, experiments on Kuhl's pipistrelle bat (*Pipistrellus kuhlii*), have provided support for the hypothesis that at least some species undergo arousal to drink.

The electroencephalograms (EEGs) of ground squirrels during the hibernation/arousal cycle suggest that the animals go to sleep during the arousal period, as depicted in Figure 10.34. Since the EEG virtually disappears during deep hibernation, the animals may progressively accumulate a sleep deficit during the period of torpor, which can only be paid back by sleeping during arousal. However, this may not be the complete story. If the 'warming up for sleep' idea is correct, then if a newly aroused animal is temporarily prevented from sleeping, it should develop an even greater sleep deficit, and respond by increasing the period it subsequently spends in the arousal state and asleep, but this does

[53] Membrane fluidity is discussed in sections 3.2.3, 8.2.2 and 9.3.2.
[54] Antifreeze molecules are discussed in Section 9.2.4.

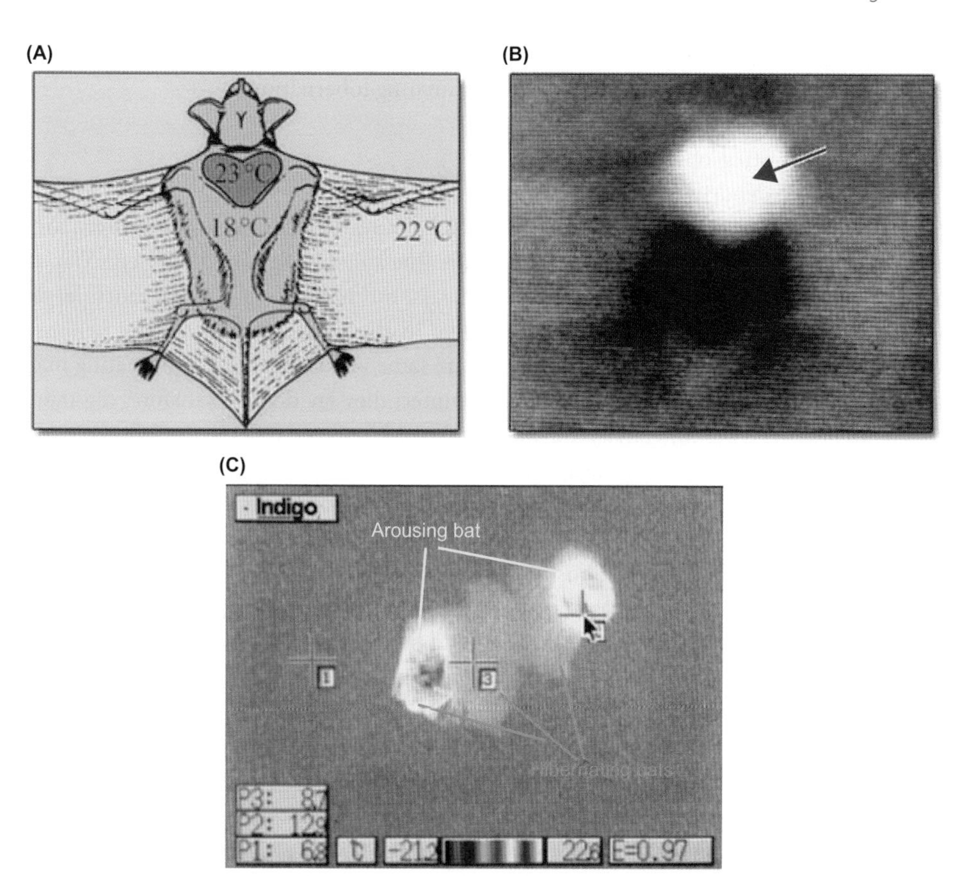

Figure 10.33 Localization and heat generation of brown adipose tissue (BAT) in bats

(A) Dorsal view of a bat showing the localisation of the interscapular BAT (red) and the major temperatures recorded by a thermographic scan. (B) Thermogram of the dorsal surface of an arousing bat. The higher the temperature, the brighter the image (indicated by the arrow). (C) This image shows four bats in hibernation two of which are in the process of arousal. The three targets 1, 2 and 3 refer to the sites where surface temperatures were measured. The measurements are indicated as P1, P2 and P3 at the lower left-hand corner. P1 is for the cave wall (temperature 6.8°C), P2 (with arrow) is for a bat during arousal (temperature 12.9°C) and P3 is for a still torpid bat (temperature 8.7°C).

Source: A and B reproduced from: Hayward JS and Lyman CP (1967). Nonshivering heat production during arousal from hibernation and evidence for the contribution of brown fat. In: Fischer KP et al (Eds), Mammalian Hibernation. American Elsevier: New York pp. 346–354. C: T.H. Kunz – with permission.

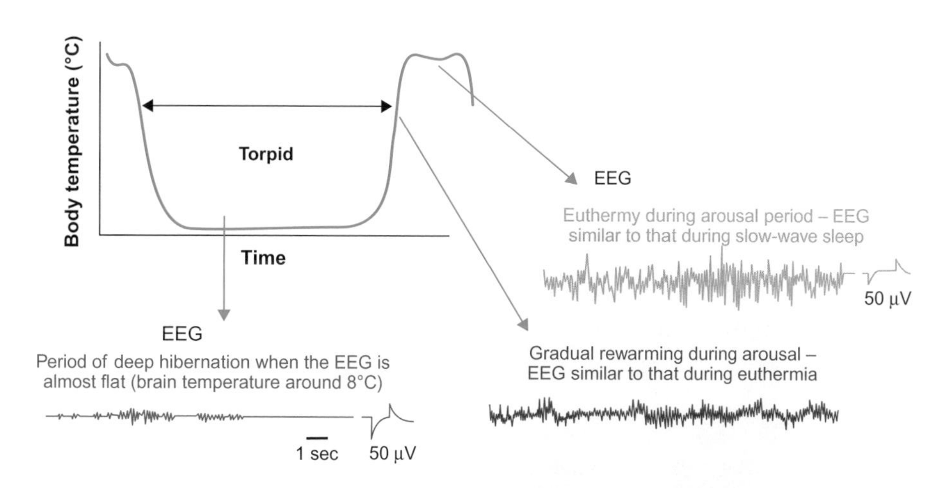

Figure 10.34 Electroencephalograms (EEGs) during deep hibernation and arousal of a Belding's ground squirrel (*Urocitellus beldingi*)

During hibernation with brain temperature at around 8°C there are only very shallow waves but during arousal, as the brain warms up, the EEG is initially at a higher frequency, which is similar to that when the animals are at normal body temperature, known as euthermia. However, as the arousal period progresses, the EEG activity becomes more like that during slow-wave sleep, suggesting that the animals go to sleep during the arousal period.

EEG traces modified from Walker JM et al (1977). Sleep and hibernation in ground squirrels (*Citellus spp*): electrophysiological studies. American Journal of Physiology 233: R213-R221.

not happen. Keeping aroused animals awake for a while does not increase the time they then spend asleep before re-entering torpor.

Perhaps one of the most interesting observations made, is that the synaptic buttons on dendritic spines[55] in the brain appear to regress during hibernation, a process that is reversed when the animals become aroused. These buttons are involved in connectivity and memory formation within the brain, implying that the brain gradually deteriorates during torpor. Indeed, some of the regions in which this deterioration occurs are concerned with spatial memory. Consequently, the reversals of the loss of connectivity during arousal are important for normal functioning of the animal following hibernation. There is also evidence that arousal reverses some important biochemical changes in the brain that accumulate with time during torpor, including the phosphorylation of a protein in the brain known as the tau protein, as shown in Figure 10.35.

Tau proteins stabilize microfilaments in the cytoskeleton[56] of cells and phosphorylation of tau leads to destabilization of microtubules in nerve cells and to the formation of neurofibrillary tangles. This destabilization is one of the damaging processes that occur during progression of Alzheimer's disease; Alzheimer's patients have very high levels of phosphorylation of the tau protein in their brains. By the end of a 10-day period of hibernation, the brains of ground squirrels are similar to those of Alzheimer's patients, particularly in the extent of tau phosphorylation and the existence of neurofibrillary tangles. Figure 10.35 also shows that ground squirrels dephosphorylate tau during arousal. They also eliminate the tangles. However, the formation of phosphorylated tau in hibernating animals is selective; it affects some neurons but not others, such as those involved in terminating hibernation.

[55] Dendritic spines are discussed in section 16.3.6

[56] The cytoskeleton is discussed in Section 3.2.1

Hibernation without arousals

Not all species that undergo long-term torpor exhibit arousals. The animals we have discussed so far are relatively small. Bears, on the other hand, are much larger—but those that live in the temperate and Arctic regions behave in much the same way as the small hibernating mammals during the winter; they lay down a fat store, dig themselves into a refuge, seal it off and emerge the following spring. The question of whether bears are torpid or awake in their dens is, therefore, of great interest.

Figure 10.36 shows that the changes in body temperature in black bears are radically different from those of other hibernators. Body temperature only falls to around 30°C or so and periodic arousals back to euthermic levels do not occur. This raises the question: are the bears hibernating or not? The answer depends on whether the definition of hibernation is based on changes in body temperature or on changes in metabolic rate.

The metabolic rates of hibernating animals (during the periods between arousals) are all very similar, as shown back in Figure 10.29. However, notice where black bears fit into this plot. Their metabolic rate during hibernation is 25 per cent of that in similar-sized euthermic animals, which is relatively less of a drop than that seen in other, smaller hibernators, but closer to that seen in species that undergo daily torpor. Also, the body temperature of the bear only declines to around 30°C because of its small surface area to volume ratio and thermal inertia. So, bears do exhibit hypometabolism just like other hibernators (although not to such a large extent), and they do not cool down as much because they are so big.

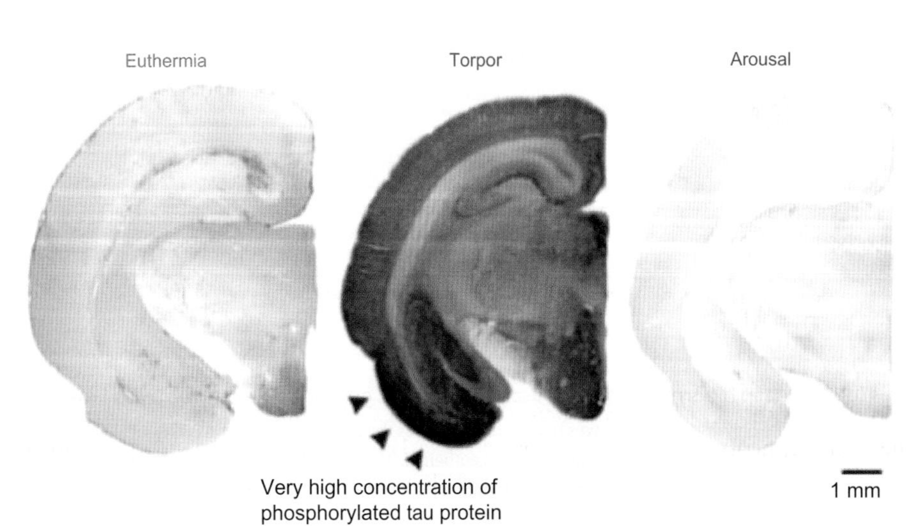

Figure 10.35 Brain slices from European ground squirrel, (*Spermophilus citellus*) showing increase in tau phosphorylation (darkening) during torpor

Immunohistochemical detection of phosphorylated tau by monoclonal antibody AT8 in brains from torpid squirrels and its reversal during subsequent arousal. Note, immunological methods involve the use of highly selective antibodies to detect the presence of specific antigens (proteins).

Reproduced from Arendt T et al (2003). Reversible paired helical filament-like phosphorylation of tau is an adaptive process associated with neuronal plasticity in hibernating animals. Journal of Neuroscience 23: 6972–6981.

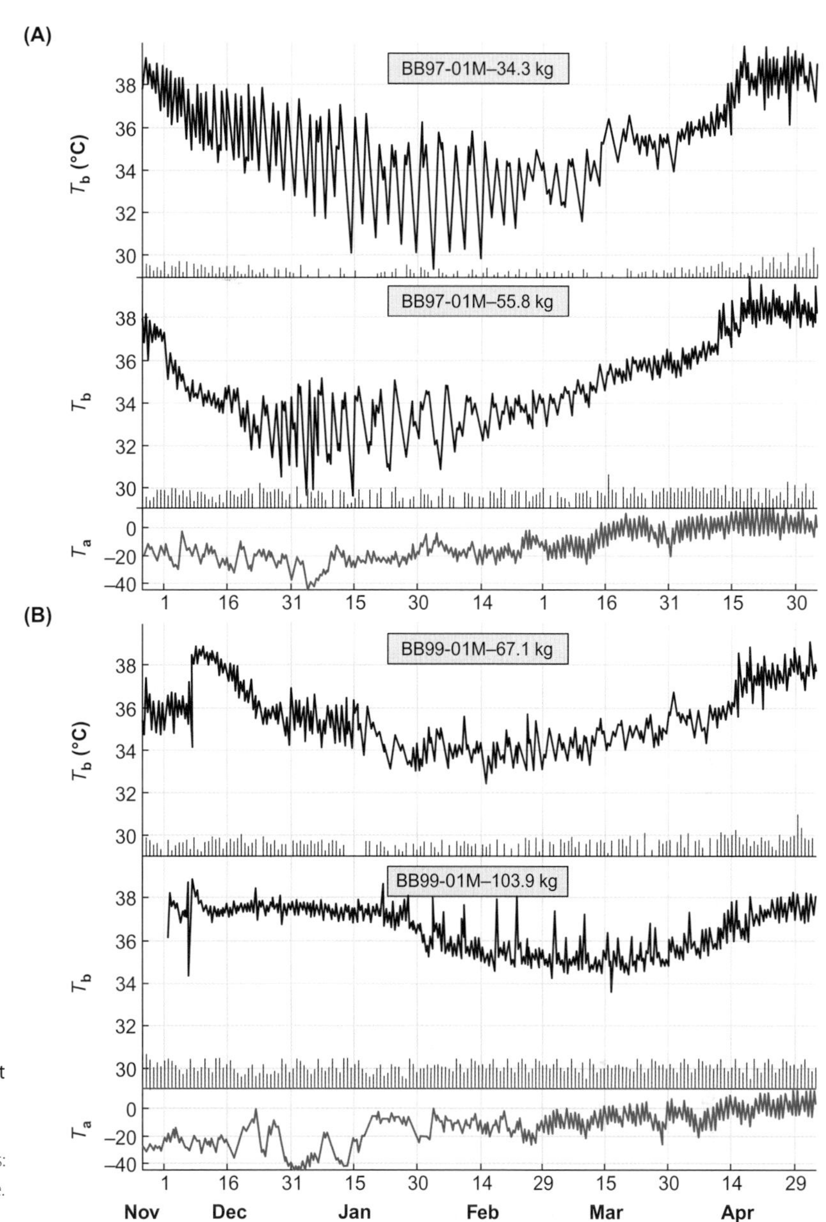

Figure 10.36 Body temperatures of black bears (*Ursus americanus*) throughout hibernation

The traces show body temperature (T_b, black lines), temperature outside the dens (T_a, blue lines) and movements of the bears (purple lines) recorded over 2 different years. (A) and (B) two bears in each of 1997 and 1999, respectively.

Reproduced from: Tøien Ø et al (2011) Hibernation in black bears: independence of metabolic suppression from body temperature. Science 331: 906-909.

These observations from black bears suggest that the periodic arousals observed in at least some hibernating species are only necessary if the body temperature falls to, and remains at, relatively low levels. Hence the process of arousal may not always be related to the decline in metabolic rate but to the temperature that the body reaches after the metabolic rate has been depressed. This still does not explain what arousal is for, but perhaps tells us something about the process that initiates it.

Some other species of hibernators do not always exhibit periodic arousals. At least three species of Madagascan lemurs undergo hibernation, with one of these, nocturnal fat-tailed dwarf lemurs (*Cheirogaleus medius*), showing two distinct patterns of hibernation depending on ambient temperature. The lemurs hibernate in tree holes during the cool dry season (April–October) when the availability of food and water is low. In most individuals living in a hibernaculum that is not well insulated, body temperature fluctuates between 10 and over 30°C, closely tracking the daily fluctuations in ambient temperature for many weeks or even months. Under these conditions, the lemurs do not undergo arousals from their hibernation. However, in those individuals living in well-insulated hibernacula, the daily fluctuations in both ambient and body temperatures are much less, at about 3.5°C and average body temperature is around 25°C. Under these conditions, the animals show regular but short (less than 6 h) arousals about once a week.

These data from *C. medius* in relatively warm conditions indicate that hibernation does not necessarily need to be accompanied by persistently low ambient temperature, nor by low body temperature, but is more likely to be related to the limited availability of food, which can be accommodated by a reduction in metabolic demands. The data also show that arousals can occur if body temperature does not regularly reach a sufficiently high level during the period of hibernation.

The mechanisms which control the entry into, the maintenance of, and arousal from hibernation are not well understood, although there have been many studies which have resulted in a number of hypotheses being proposed.

Control of hibernation

Torpor may be an extension of slow wave sleep (also called non-rapid eye movement—NREM—sleep), when the thermoregulatory set point (T_{set}) is lower than that in awake animals, although there is no evidence that torpor and sleep perform similar functions. Indeed, as shown in Figure 10.34, there is little or no electrical activity in the brain during deep hibernation, in contrast with that seen during sleep.

The suprachiasmatic nucleus (SCN) is situated in the anterior part of the hypothalamus and plays an important role in controlling circadian rhythms in mammals[57]; it may also be an integral component of the process of hibernation. One piece of evidence in support of the involvement of the SCN in hibernation is the fact that its metabolic rate is higher than that in other parts of the brain both during the entrance to hibernation and during hibernation itself. The SCN may be an integrating region, receiving stimuli from the retina, the endocrine systems and other areas and sending signals to the autonomic nervous system (ANS)[58] and some other areas to influence metabolic and thermoregulatory processes, although the evidence for this is controversial.

A component of the ANS, the parasympathetic nervous system, may control entry into hibernation. One of its actions is to decrease heart rate by as much as 50 per cent before body temperature falls significantly and the Q_{10} effects of temperature take over. The other component of the ANS, the sympathetic nervous system, initiates an increase in heart rate and arousal.

A central component of hibernation and torpor is the controlled reduction of basal metabolic rate. Hypothermia in non-hibernating animals causes many potentially lethal changes, including a fall in the concentration of ATP and a reduction in membrane potential, which can quickly become irreversible. However, none of these damaging changes occurs in hibernating or torpid animals.

During the hypometabolism and controlled hypothermia of hibernation or torpidity, there are large reductions in ATP turnover, changes in patterns of substrate use and changes in gene expression. Energetically expensive processes, such as protein synthesis, and ion transport are suppressed. For example, protein synthesis in the brain of hibernating golden-mantled ground squirrels is only 0.04 per cent of that in the brain of euthermic ground squirrels.

An important mechanism in metabolic suppression is the reversible phosphorylation of many enzymes. Phosphorylation suppresses the activities of many enzymes involved in carbohydrate metabolism and ion transport, such as Na^+, K^+-ATPase which is responsible for 5–40 per cent of total ATP turnover, depending on the type of cell. Thus, suppressing the activity of these enzymes causes a dramatic reduction in metabolic rate.

Figure 10.37 shows what happens to Na^+, K^+-ATPase activity during hibernation. The maximum Na^+, K^+-ATPase activities in skeletal muscle, kidney and liver of hibernating ground squirrels are between 40 to 60 per cent of those in euthermic squirrels. Subsequent treatment of extracts from the tissues of hibernating squirrels with alkaline phosphatase (which dephosphorylates the enzyme) fully restores enzyme activity. So it is envisaged that activity of Na^+, K^+-ATPase during hibernation is suppressed by phosphorylation, but that this is reversed (that is, dephosphorylation occurs) when the animal enters arousal[59] and the enzyme is needed for shivering thermogenesis.

Hibernation and climate change

We discuss in Case Study 10.1 how climate change may be affecting the long-term success of some species of long distance migrating birds. There is also emerging evidence that climate change is affecting the fitness of some species of hibernators. Columbian ground squirrels (*Urocitellus colombianus*) live in the Rocky Mountains of North America and, because of the short growing seasons of the plants in the regions where they live, spend 8–9 months of the year hibernating. Adult females spend about half of their 3–4 months of activity engaged in reproduction and after that, they and their offspring must accumulate sufficient fat stores to survive the forthcoming period of hibernation. Any reduction in the time available for these processes could be detrimental to individuals and to the population.

---

57 The role of the SCN in the circadian rhythm of mammals is discussed in Section 17.2.2.

58 The autonomic nervous system is discussed in Section 16.1.3.

59 Recall the phosphorylation of tau protein in the brain during hibernation and its subsequent dephosphorylation during arousal, as discussed earlier in this section and depicted in Figure 10.35.

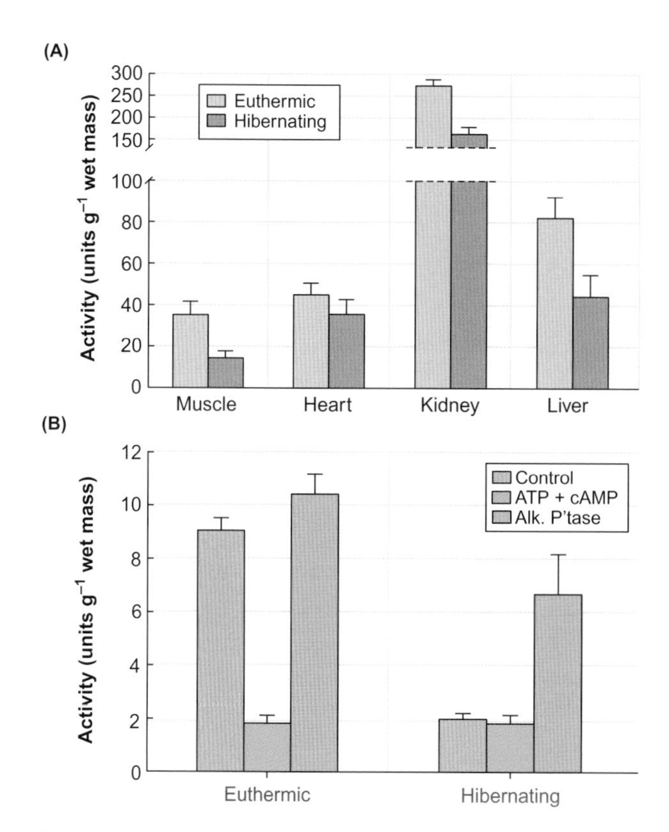

Figure 10.37 Effect of hibernation on activities of Na^+, K^+-ATPase in tissues of the golden-mantled ground squirrel (*Callospermophilus lateralis*)

(A) Activity of Na^+, K^+-ATPase in tissues from animals at normal body temperature (euthermic) and during hibernation. Hibernation causes a significant reduction in Na^+, K^+-ATPase activity in all tissues except the heart. (B) Phosphorylation of Na^+, K^+-ATPase in muscle extracts from euthermic animals with ATP and cAMP causes a reduction in activity of Na^+, K^+-ATPase to the level seen in muscle extracts from hibernating animals. Subsequent treatment with alkaline phosphatase (Alk. P'tase) restores activity of Na^+, K^+-ATPase in extracts from both euthermic and hibernating animals by dephosphorylating Na^+, K^+-ATPase.

Reproduced from Storey K and Storey JM (2004). Metabolic rate depression in animals: transcriptional and translational controls. Biological Reviews 79: 207-233.

Between 1992 and 2012, the average date of emergence of adult females from hibernation was delayed by around 0.5 days per year. Date of snowmelt was delayed by 2.8 days per year over the same period due to the increased likelihood of late season snowstorms, but there was no significant change in spring temperature. Females emerged later during years of delayed snowmelt and during those years there was a decline in individual fitness (overwinter survival and number of offspring produced in any one year that survived the winter). As a result of these changes in individuals, there has also been a decline in population growth rate during the study period. This study illustrates that the effects of climate change are not always the result of increasing temperature. Other climatic changes can also be affecting wild populations of animals.

Daily torpor

Daily torpor is common in a number of species of mammals and birds, including marsupials, bats, carnivores, rodents and hummingbirds. The pattern of daily torpor in birds is similar to that of mammals, which is shown in Figure 10.29B. No consistent pattern of daily change in body temperature is seen in birds, as illustrated by data from three species of hummingbirds shown in Figure 10.38A. However, Figure 10.38B shows that they regulate their body temperature at about 3–4°C above ambient temperature.

The use of daily torpor in birds and mammals appears to depend on food supply and ambient temperature. Torpor is more likely when food supply is low and temperatures are cold. This is consistent with torpor being a thermoregulatory strategy that allows small endotherms to survive periodic energy crises. This view is supported by laboratory studies which show that the depth of torpor (i.e. the fall in body temperature) is related to the level of energy deprivation. For example, if blue-naped mouse birds (*Urocolius macrourus*) are progressively deprived of food, the depth of their torpor increases over successive days, as illustrated in Figure 10.39.

It has also been suggested that some species of hummingbirds enter torpor at night just prior to migration, not because of a shortage of food, but to save on the energy cost of thermoregulation and conserve fat for their migratory flight the next day.

10.2.8 Thermoregulation in aquatic mammals

Living in an aquatic habitat increases the problem of heat loss for endotherms. Water has 23 times the thermal conductivity of air, so when mammals enter water, their rate of heat loss increases enormously. Yet many species of mammals, of widely different masses, occupy aquatic habitats, from the diminutive water shrew (*Neomys fodiens*) weighing about 10 g, to the largest animals on the planet, blue whales (*Balenoptera musculus*), which weigh up to 100,000 kg.

These animals cope with this situation by several different mechanisms which reduce the rate of heat loss:

- trapping a layer of air within their fur;
- replacing fur by a layer of subcutaneous fat (blubber);
- regional heterothermy in the periphery.

Trapping a layer of air within the fur to increase insulation

Looking back at Table 8.3 we examined the thermal conductivities of different materials. One perhaps surprising point is that the thermal conductivity of fur is greater than

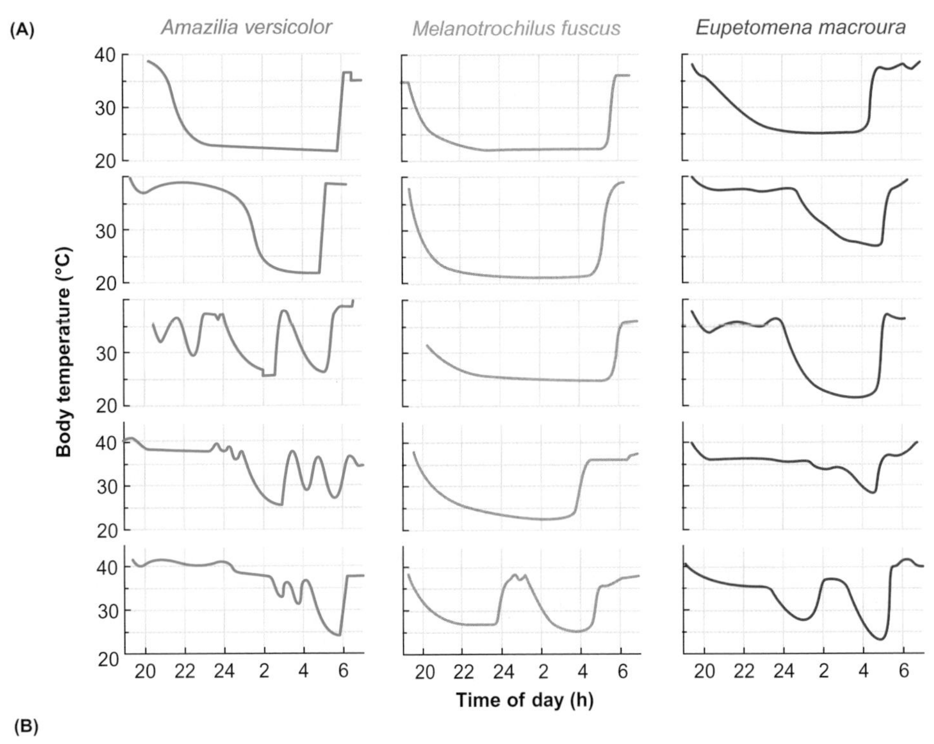

(B)

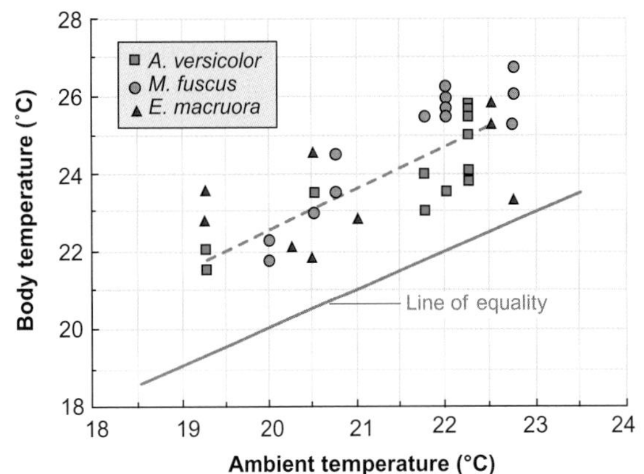

Figure 10.38 Torpor in hummingbirds

A Patterns of body temperature of five individuals from each of three species of hummingbirds monitored in the wild. **B** The relationship between body temperature (T_b) and ambient temperature during torpor of these three species showing the birds regulate T_b at about 3–4°C higher than ambient.

Reproduced from: Bech C et al (1997). Torpor in three species of Brazilian hummingbird under semi-natural conditions. The Condor, 99: 780-788.

that of still air. The insulative properties of fur are primarily the result of the layer of still air trapped within it. When fur becomes wetted, however, its insulative properties decline enormously. Look at the examples in Table 10.1, which compare the insulation of polar bear and beaver pelts in air and when saturated with water. The prevention of the ingress of water into the pelage retains a layer of still air and provides maximal insulation.

The fur of sea otters (*Enhydra lutris*) is specially adapted to maximize its ability to trap air close to the skin. Sea otter fur is the most dense of any mammal that has been examined so far, with about 1200 hairs (mm skin)$^{-2}$. This fur comprises a densely packed layer of thin under hairs and longer, hollow 'guard hairs'. The cuticle of the under hairs has a striking pattern of sharply sculpted fins with deep grooves between them, which trap air. To maintain this ability to trap air, the animals groom for up to 2 hours a day.

Trapping a layer of air next to the skin retains the thermal properties of the air, despite the animals being in water. But there are several disadvantages to having an air-filled pelage. First, its buoyancy makes the initial stages of diving to any depth difficult and, second, fur creates drag. Moreover, if the animals submerge to any depth, the air layer will compress and become less effective as an insulator, although the compression will decrease the buoyancy. Polar bears do not have fur that effectively traps air; consequently, the fur loses almost all of its insulative properties when wetted, as shown in Table 10.1. Nonetheless, a combination of subcutaneous fat and regional heterothermy (which we discuss next in this section) provide some insulation, but prolonged exposure to cold Arctic water can lead to excessive heat loss in these animals.

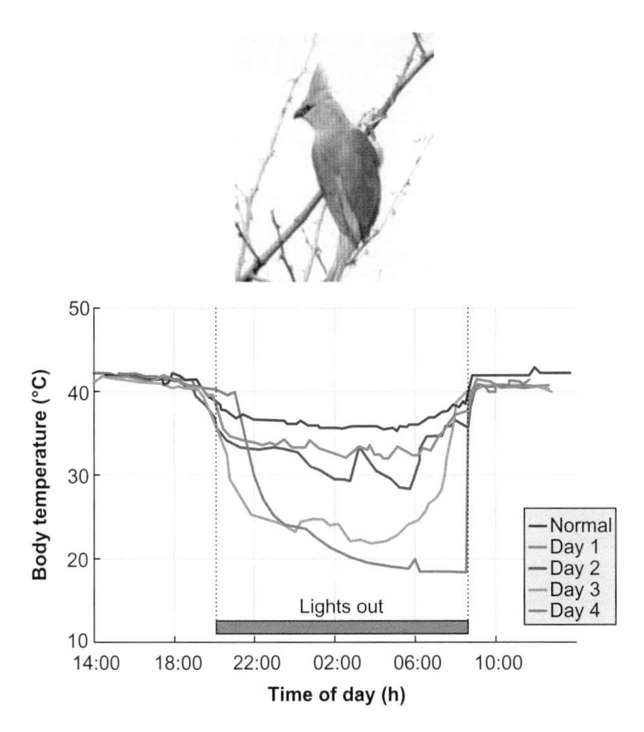

Figure 10.39 Body temperatures of blue-naped mouse birds (*Urocolius macrourus*) in captivity under successive days of food restriction

The extent and depths of the regulated hypothermia/torpor are greater as the duration of food restriction increases. Ambient temperature is 15°C.

Reproduced from: Schleucher E (2004). Torpor in birds: taxonomy, energetics and ecology. Physiological and Biochemical Zoology 77: 942-949. Image (c) Nick Ransdale.

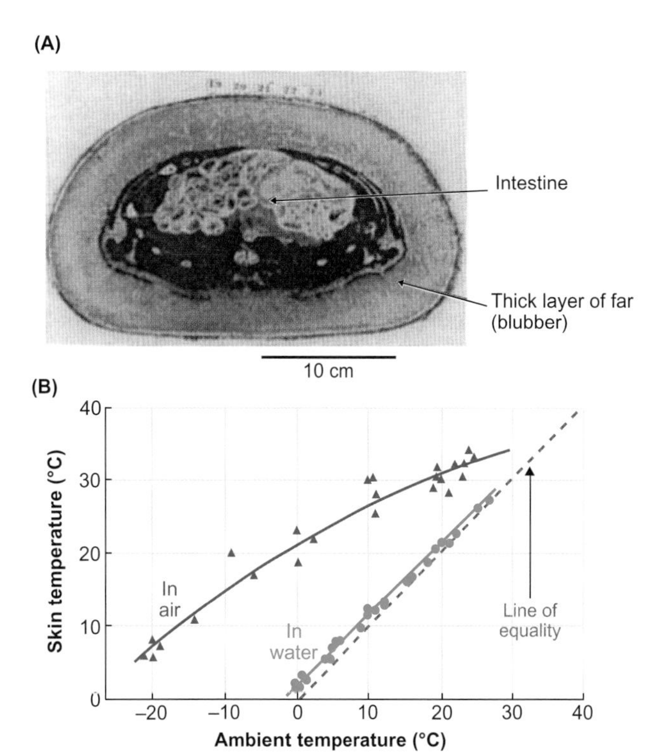

Figure 10.40 Insulation in marine mammals such as seals (*Phoca* spp.)

(A) Cross section of a seal showing the thick layer of blubber just beneath the skin; (B) Skin temperature of seals in water and in air. In water the skin temperature is almost equal to that of the surrounding water, which indicates that the skin is not well perfused with blood and that heat loss across it is minimal. However, in air, the surface of the skin in much hotter than the surrounding air, suggesting that the skin is well perfused with blood and that there is some heat loss across it.

Reproduced from (A) Scholander PF et al (1950) Body insulation of some arctic and tropical mammals and birds. Biological Bulletin 99: 225-236 (B) Hart JS et al (1959). The energetics of harbour seals in air and in water with special consideration of seasonal changes. Canadian Journal of Zoolog 37: 447-457.

Replacing fur by a layer of subcutaneous fat as a means of insulation

An alternative strategy that overcomes the difficulties presented by trapping a layer of air has evolved in many aquatic mammals; they have insulation on the inside of the skin, rather than on the outside, and lose their external hair completely. The lack of fur also allows the animal to have a smooth exterior with minimal drag; their locomotion has relatively low energetic cost.

Diving mammals have a deposit of fat under the skin, although this is not as effective an insulator as still air, as shown earlier in Table 8.3. Nonetheless, this strategy has evolved independently among the cetaceans (whales and porpoises), the pinnipeds (seals) and also in polar bears. Figure 10.40A shows how thick the layer of fatty blubber is beneath the skin of a seal. This layer of blubber occupies about 58 per cent of the cross sectional area of the body. Figure 10.40B shows that when in water, the skin temperature of a seal is only marginally above that of the water, suggesting that the layer of blubber acts as a very effective barrier (insulator) to heat flow: very little heat is lost from the surface of the animal. However, this insulation leads to potential difficulties when the animals generate excessive heat internally during exercise or when they enter warm water. Thermoregulatory problems could also arise if animals with blubber leave the water and spend time on land, as is routinely the case for many seals and fur seals.

When the seals are in air, their skin temperature is around 10–20°C above the air temperature, as revealed in Figure 10.40B. This difference indicates that the animals are able to bypass their blubber layer and flood the skin with blood,

Table 10.1 Insulation values of fur of beaver and polar bear when dry and when wet

	Insulation (W m^{-2} °C^{-1})	
Species	**Dry**	**Wet**
Beaver (*Castor canadensis*)	0.7	0.1
Polar bear (*Ursus marinus*)	0.92	0.05

Data from Scholander PF et al (1950). Body insulation of some arctic and tropical mammals and birds. Biological Bulletin 99: 225-226.

enabling them to avoid overheating when they are out of the water.

While bypassing the blubber layer is an effective mechanism to avoid hyperthermia in air, there is a potential problem when the animals dive, particularly to any depth. During active diving, there is a conflict between the thermoregulatory demands to perfuse the surface with blood in order to dissipate heat, and the diving response, which aims to reduce flow to the peripheral tissues[60]. Studies of bottlenose dolphins (*Tursiops truncatus*) show that these animals have very effective thermal windows on their flanks and dorsal fins (similar thermal windows are present in seals). However, heat flow to the thermal windows is reduced during a dive, which is consistent with vasoconstriction as part of an oxygen conservation response. The animals wait until they resurface before increasing blood flow to these areas to dissipate any heat stored during the dive.

Regional heterothermy in extremities to reduce rate of heat loss

When in cold water or air, the loss of heat across the extremities of the body of aquatic mammals is a potential problem. This is avoided by the use of countercurrent heat exchangers[61]. For example, in bottlenose dolphins, some of the extremities which are used as thermal windows can also be used to conserve heat in colder water. When in the wild, there are seasonal differences between the heat flux across the fluke and flippers in these dolphins. In winter they conserve heat, by way of a countercurrent exchange system whereas in summer, the countercurrent heat exchangers are bypassed and the animals use these extremities as thermal windows to dissipate heat. In summer, average heat flux (± standard error of the mean) across the pectoral flipper was 155 ± 15 W m^{-2}, whereas in winter it was 86 ± 34 W m^{-2}. Similarly, at the fluke, the comparable values were 126 ± 12 Wm^{-2} and 75 ± 33 W m^{-2}, respectively.

› *Review articles*

Baker MA (1979). A brain-cooling system in mammals. Scientific American 240: 130–139.
Brock JA, McAllen RR (2016). Spinal cord thermosensitivity: An afferent phenomenon? Temperature 3: 232–239.
Cannon B, Nedergaard J (2004). Brown adipose tissue: function and physiological significance. Physiological Review 84: 277–359.
Cannon B, Nedergaard J (2010). Thyroid hormones: igniting brown fat via the brain. Nature Medicine 16: 965–967.
Drew KL, Buck CL, Barnes BM, Christian SL, Rasley BT, Harris MB (2007). Central nervous system regulation of mammalian hibernation: implications for metabolic suppression and ischemia tolerance. Journal of Neurochemistry 102: 1713–1726.
Gilbert C, Blanc S, Le Maho Y, Ancel A. 2008. Energy saving processes in huddling emperor penguins: from experiments to theory. Journal of Experimental Biology 211: 1–8.
Klingenberg M (1999). Uncoupling protein—a useful energy dissipater. Journal of Bioenergetics and Biomembranes 31: 419–430.
McAllen RM, Tanaka M, Ootsuka Y, McKinley MJ (2010). Multiple thermoregulatory effectors with independent central controls. European Journal of Applied Physiology 109: 27–33.
Mitchell D, Maloney SK, Jessen C, Laburn HP, Kamerman PR, Mitchell G, Fuller A (2002). Adaptive heterothermy and selective brain cooling in arid-zone mammals. Comparative Biochemistry and Physiology Part B 131: 571–585.
Morrison SF, Nakamura K, Madden CJ (2008). Central control of thermogenesis in mammals. Experimental Physiology 93: 773–797.
Nedergaard J, Cannon B (1990). Mammalian hibernation. Philosophical Transactions of the Royal Society of London B 326: 669–686.
Nedergaard J, Matthias A, Golozoubova V, Jacobson A, Cannon B (1999). UCP1: the original uncoupling protein—perhaps the only one? New perspectives on UCP1, UCP2 and UCP3 in the light of the bioenergetics of the UCP1-ablated mice. Journal of Bioenergetics and Biomembranes 31: 475–491.
Nicholls DG, Rial E (1999). A history of the first uncoupling protein, UCP1. Journal of Bioenergetics and Biomembranes 31: 399–406.
Romanovsky AA (2007). Thermoregulation: some concepts have changed. Functional architecture of the thermoregulatory system. American Journal of Physiology 292: R37–R46.
Ruf T, Arnold W (2008). Effects of polyunsaturated fatty acids on hibernation and torpor: a review and hypothesis. American Journal of Physiology 294: R1044–1052.
Simon E, Pierau K-F, Taylor, KCM (1986). Central and peripheral thermal control of effectors in homeothermic temperature regulation. Physiological Reviews 66: 235–300.
Storey KB (2000). Turning down the fires of life: metabolic regulation of hibernation and estivation. In: Storey KB (ed.) Molecular Mechanisms of Metabolic Arrest. pp 1–21. Oxford: BIOS Scientific Publishers.
Storey KB (2010). Out cold: biochemical regulation of mammalian hibernation—a mini-review. Gerontology 56: 220–230.

10.3 Endothermic fish, reptiles and insects

Even though mammals and birds are the only groups of animals that are continuously endothermic, a few species of fish, reptiles and insects can raise and maintain temperature of all or part of their body above ambient for a sufficient period of time to enable specific behaviours, such as swimming, incubation or flight, to occur. Because the heat is usually only generated and retained in specific regions of the body, it is called regional endothermy.

10.3.1 Locomotion and non-shivering thermogenesis as sources of heat in endothermic fish

When fish swim from warm to cold water their body temperature falls and their ability to perform certain tasks, such as the generation of power from their muscles, is considerably reduced. Figure 10.41 shows that the generation of power

[60] Section 15.3.3 discusses the diving responses to conserve oxygen for the central nervous system and locomotor muscles.
[61] Countercurrent exchange is discussed in Section 3.3.1.

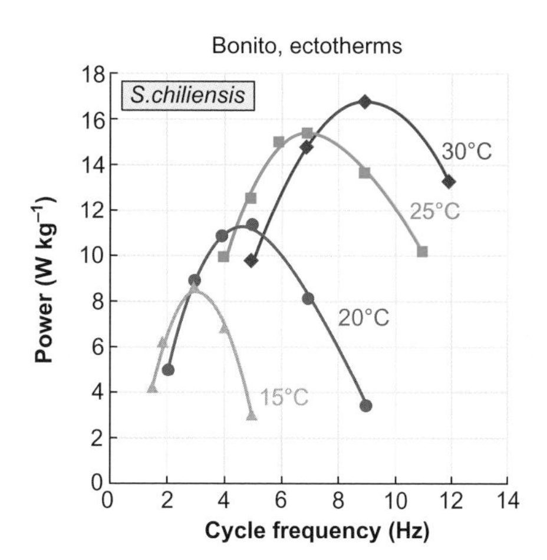

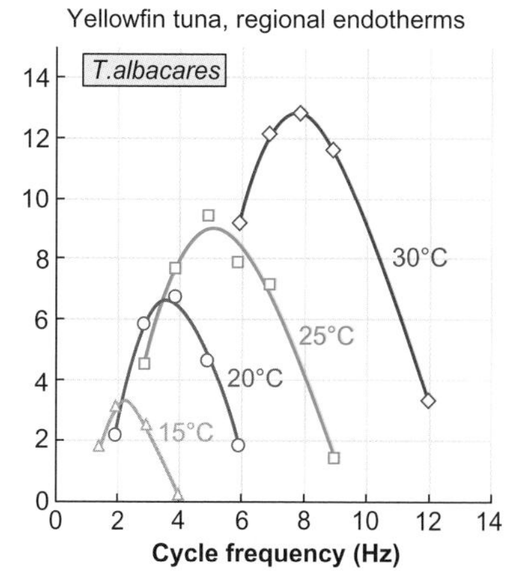

Figure 10.41 Power versus frequency curves for oxidative (slow) muscle from Eastern Pacific bonito (*Sarda chiliensis*) and yellowfin tuna (*Thunnus albacares*)

In both species, maximum power output of the muscles decreases with decreased temperature.

Source: adapted from Altringham JD and Block BA (1997). Why do tuna maintain elevated slow muscle temperatures? Power output of muscle isolated from endothermic and ectothermic fish. Journal of Experimental Biology 200: 2617-2627.

in muscles removed from two species of fish, Eastern Pacific bonito (*Sarda chiliensis*—ectotherms) and yellowfin tuna (*Thunnus albacores*—regional endotherms) is lower at lower temperatures. If a fish could keep its important tissues and organs warm when in cold water, it would have an advantage. This may seem to be impossible, as the temperature of blood passing through the gills equilibrates with that of the surrounding water. Nonetheless, a small number of fish species have evolved strategies to keep parts of their bodies warmer than the environment.

While the source of the heat varies, all endothermic fish make use of countercurrent heat exchangers situated between the endothermic tissue and the periphery in order to reduce the loss of heat across the gills. The venous blood leaving the tissue is warmed by the metabolic activity of the tissue and a proportion of the heat is then reclaimed by the colder arterial blood flowing to the tissue, rather than being lost. Some species of fish, such as billfish (marlin, sailfish and swordfish) and butterfly mackerel (*Gasterochisma melampus*) only elevate temperature in the eyes and brain and use a special heater organ associated with an eye muscle to do this. Other species—including mako, white sharks and tunas—are able to raise the temperature of the viscera and oxidative (red) muscle fibres, as well as the eyes and brain.

Tuna retain the heat generated by their red locomotor muscles by using heat exchangers (retia, singular: rete[62]) located centrally beneath the vertebral column, as shown in Figure 10.42A, or at the lateral sides of the red muscle blocks. Both locations are shown in Figure 10.42B. The heat exchangers serve primarily to maintain the temperature of red muscle that extends inward to a far greater extent than that in most fish, and reaches the vertebral column. The close proximity of the blood vessels to each other in the retia greatly facilitates the exchange of heat between the warm blood in the venules and the cooler arteriolar blood. Figure 10.42B shows the location of the major arteries and veins that are connected to the retia.

In billfish, the heat is generated by non-shivering thermogensis in modified cells of eye muscles that have lost their contractile elements. Instead, the cells have a profusion of mitochondria, occupying 60 per cent or more of the total volume, with t-tubules and sarcoplasmic reticulum running between the mitochondria, as shown in Figure 10.43A. The sarcoplasmic reticulum is the muscle cell equivalent of endoplasmic reticulum; it acts as an intracellular calcium store from which calcium ions are released to activate the contractile machinery and promote muscle contraction[63]. The sarcoplasmic reticulum in the skeletal muscle and heater organ cells contains high levels of two large protein complexes comprising multiple subunits:

- Ca^{2+}-ATPase that acts as a Ca^{2+} pump, transporting Ca^{2+} from the cytosol into the lumen of the cytoplasmic reticulum. This Ca^{2+}-ATPase is also called the sarco/endoplasmic-reticulum-Ca^{2+}-ATPase (SERCA).
- Ca^{2+} release channel that allows calcium to move out of the sarcoplasmic reticulum. This Ca^{2+} channel binds with high affinity to the alkaloid ryanodine and is, therefore, called the ryanodine receptor (RyR).

Figure 10.43B shows how these large protein complexes are anchored into the membrane of the sarcoplasmic reticulum. SERCA actively pumps Ca^{2+} into the sarcoplasmic reticulum, a process that is coupled to the hydrolysis of ATP to ADP and inorganic phosphate (Pi). The cells of the heater organ are under direct control of oculomotor neurones via

[62] See Section 10.2.3 for explanation of the term 'rete'.

[63] For more details of the sarcoplasmic reticulum and muscle contraction, see Section 18.1.4.

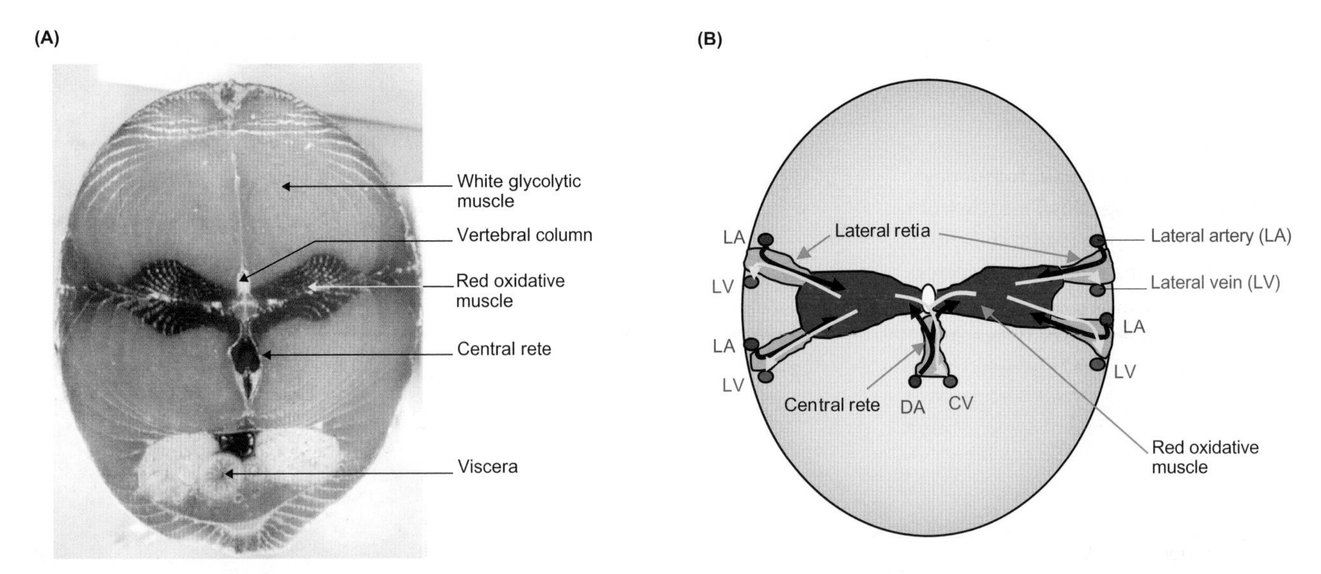

Figure 10.42 Counter-current heat exchangers in tuna

(A) Section through skipjack tuna (*Katsuwonus pelamis*) showing the red oxidative muscle block, within which the heat exchangers are located, and the white glycolytic muscle. In this species of tuna, the main heat exhanger is located centrally, below the vertebral column, whereas in others, such as Atlantic bluefin tuna (*Thunnus thynnus*), the heat exchangers are located laterally. (B) Diagram of section through a tuna showing the locations of the major blood vessels (arteries dark red and veins in blue), the internalized red muscle blocks (red) and the retia (pink). Cool oxygenated blood flows into the muscles via the retia from the arteries (black arrows). Warmed deoxygenated blood from the muscles (yellow arrows) returns to the veins via the retia exchanging heat as it does so. DA, dorsal aorta; CV, cardinal vein.

A: Reproduced from Stevens, ED (2011) The retia. In: Farrell AP (ed) Encyclopedia of Fish Physiology: From Genome to Environment, vol 2. pp 1119-1131. San Diego, Academic Press.

chemical synapses, which we discuss in Section 18.2.1. In response to stimulation by oculomotor neurons, the Ca^{2+}-release channels (RyRs) open by a mechanism that is similar to that in skeletal muscle which we discuss in Section 18.2.2.

Once the RyRs open, there is a flux of Ca^{2+} out of the sarcoplasmic reticulum and at the same time, Ca^{2+} is continuously pumped back via SERCA, generating heat from the hydrolysis of ATP. This is known as a futile cycle[64]. When the nervous stimulation ends, the ryanodine receptor closes. But what are the advantages to some species of fish of being able to keep areas of their body above ambient temperature?

Functional significance of regional endothermy in fish

Figure 10.41 demonstrates that power generation in muscles of fish is temperature-dependent, so it has been suggested that fish which are able to keep their locomotor muscles warm in a cold environment can produce greater power in the muscles. An additional advantage is the ability of their muscle to clear lactate more rapidly after anaerobic exertions (produced mostly in white, glycolytic muscle), because lactate clearance is temperature dependent. These abilities are advantageous when reduced environmental temperature leads to a reduction in performance of a fish's prey species, which lack regional endothermy. However, there are some downsides: in some endothermic fish, such as lamnid sharks and tuna, the temperature dependence of muscle performance has become so tuned to high temperatures that they are forced to swim continuously to keep their locomotory muscles warm[65].

Many fish and invertebrates show daily migration patterns whereby they move from the surface at night to much deeper water during the day. This is seen as a protection mechanism from the visual predators that are active in the upper surface layers during daytime. Diving to depth takes animals into cooler water below the thermocline. Regional endothermy allows some species of tuna, billfish and lamnid sharks to follow their prey to depths of 200 m or more during their daily cycle without losing the functions of their muscles, digestive tract, eyes or brain, because the temperatures of these organs are maintained above ambient during deep dives, as suggested by the data in Figure 10.44. There is, however, a potential problem by entering such cold water, particularly for species such as bigeye tuna (*Thunnus obesus*) which routinely dive to depths of 700 m, where water temperature can be as low as 3°C.

The blood that supplies the heart of this species does not pass through a heat exchanger, so the temperature of the heart is virtually the same as that of the environment. However, the heart is very temperature sensitive, with its rate of contraction

[64] Other examples of futile cycles are discussed in Sections 10.2.5 and 10.3.2.

[65] A consequence of the continual swimming in endothermic fish is that they ventilate their gills by ram ventilation, as discussed in Section 12.2.2.

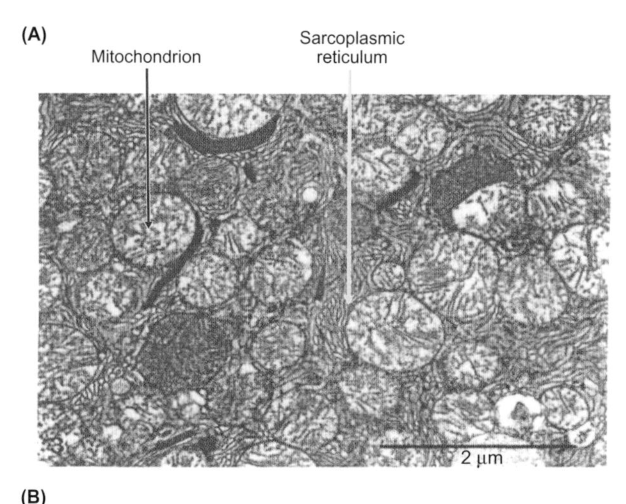

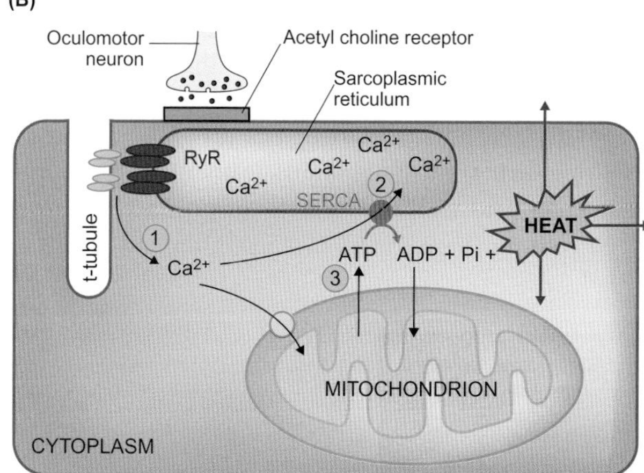

Figure 10.43 Heater organ of billfish

(A) Transmission electron micrograph of cell from the heater organ of Atlantic blue marlin (*Makaira nigricans*). The cell is packed with mitochondria and sarcoplasmic reticulum (SR). (B) Mechanism of heat production in heater organ cells. (1) Activity in the occulomotor nerve causes the ryanodine receptor (RyR) to open allowing Ca^{2+} to move from inside the sarcoplasmic reticulum to the cytoplasm. (2) The Ca^{2+} is then pumped back in to the sarcoplasmic reticulum via the ATP-dependent Ca^{2+} pump (SERCA), involving utilization of ATP and generation of heat. The futile pumping of calcium into the SR which leaks back out uses up energy with no net benefit apart from heat production. (3) ATP is regenerated in the mitochondria. This process is upregulated at higher cytoplasmic Ca^{2+}.

Sources: A: reproduced from Tullis A et al (1991). Activities of key metabolic enzymes in the heater organs of scombroid fish. Journal of Experimental Biology 161: 383-403. B: adapted from Morrissette JM et al (2003). Characterisation of ryanodine receptor and Ca^{2+}-ATPase isoforms in the thermogenic heater organ of blue marlin (*Makaira nigricans*). Journal of Experimental Biology 206: 805-812

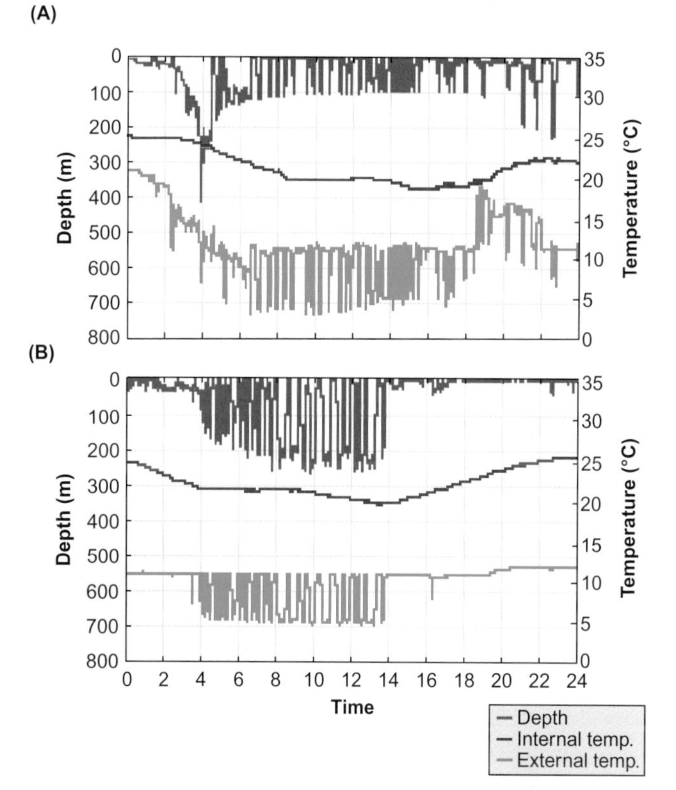

Figure 10.44 Body temperature of Atlantic bluefin tuna (*Thunnus thynnus*) in relation to external temperature

Behaviour and body temperature of two Atlantic bluefin tuna, in relation to time of day. Note that, despite large changes in external temperature (green line), body temperature (red line) varies little and is maintained at several degrees above ambient.

Reproduced from Blank JM et al (2004). *In situ* cardiac performance of Pacific blue fin tuna hearts in response to acute temperature change. J. Exp. Biol. 207, 881-890. Image: © istock/Whitepointer.

being slower at lower temperatures. This decrease in heart rate is compensated for in bigeye tuna by an increase in the force of contraction of ventricular muscles—but such compensation does not occur in yellowfin tuna. The lack of compensation in yellowfin tuna may explain why they rarely dive below 300 m where water temperature is between 17°C and 30°C.

The compensatory response in bigeye tuna during cold exposure suggests that there is an accompanying increase in cardiac stroke volume[66], which will tend to maintain cardiac output[67] despite the lower heart rate. Drugs that block the release and reuptake of Ca^{2+} into the sarcoplasmic reticulum completely abolish the initial increase in contractile force in ventricular muscle of bigeye tuna in response to the cold.

These data suggest that the hearts of bigeye tuna have evolved a mechanism to enhance the level of Ca^{2+} and thus increase the force of contraction[68] of cold ventricular muscle.

[66] Amount of blood pumped with each heartbeat.

[67] Amount of blood pumped by the heart per minute.

[68] We discuss the relationship between ionized Ca^{2+} concentration and force in Section 18.1.6.

10.3.2 Muscle contraction as a heat source in endothermic reptiles and insects

Some species of reptiles and insects also use muscle activity to generate heat, but only the larger marine turtles and female pythons incubating eggs are able to maintain body temperature significantly above ambient over an extended period.

Endothermy in marine turtles

For some species of reptiles, the aquatic environment is their primary habitat and they come onto land only infrequently. The most conspicuous group of reptiles that do this are the sea turtles. In general, marine turtles, like other reptiles, have effective physiological, anatomical and behavioural mechanisms by which they can regulate their body temperature above water temperature to a greater or lesser degree, mainly depending on their level of activity. However, their inability to withstand consistently cold water confines the distribution of most species of marine turtles to tropical and subtropical latitudes. Leatherback turtles (*Dermochelys coriacea*) are notable exceptions.

Leatherback turtles are the largest extant reptiles, with adults routinely weighing in excess of 900 kg and having a carapace length approaching 1.75 m. Unlike other sea turtles, leatherbacks may occasionally be found well outside of tropical and subtropical waters, and it remained unclear for a long time if these were simply occasional stray animals. The use of externally attached satellite transmitters to ocean-going animals has now enabled researchers to follow the movements of individual animals.

Leatherback turtles undergo an annual migration, shown in Figure 10.45, which takes them north at the end of the breeding season. By late summer and autumn (fall) the animals congregate off the coast of North America where they feed on populations of large jellyfish, such as *Cyanea capillata* and *Aurelia aurita*. They then migrate southward to tropical waters where they remain until the breeding season the following spring. A complete migration may be between 6000 and 12,000 km in distance.

During their annual migrations, leatherback turtles routinely enter cold temperate waters for protracted periods. In the cold northern waters (averaging around 10–12°C),

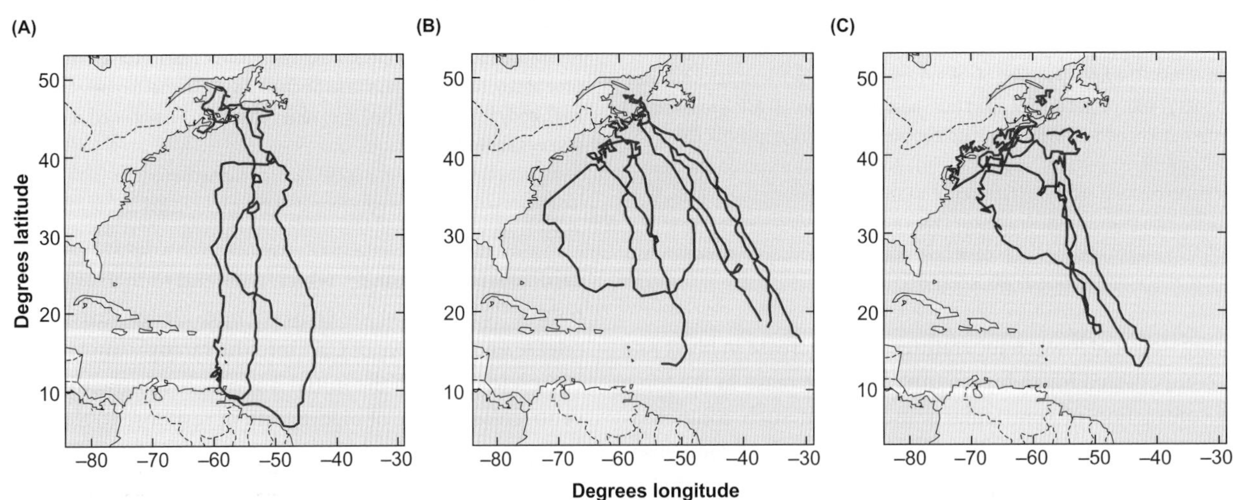

Figure 10.45 Tracks of leatherback turtles (*Dermochelys coriacea*) in the north Atlantic

The leatherback turtles were studied using attached satellite transmitters. (A) mature males, (B) mature females and (C) sub-adults. In all cases the animals were initially tagged off the north coast of Canada. The tracks indicate seasonal migrations between these cold temperate waters and warm tropical waters over the annual cycle. For those animals where the tags remained attached long enough, the animals returned to Canada the following summer/autumn.

Reproduced from James MC et al (2005). Behaviour of leatherback sea turtles, *Dermochelys coriacea*, during the migratory cycle. Proceedings of the Royal Society B 272: 1547-1555. Photo from http://www.nestonline.org/.

they spend most time in relatively shallow water and may be motionless at the surface for long periods. As they move southward into warmer waters they start to make deeper dives to over 200 m or so, taking them from surface waters at around 25°C into deeper, cooler waters. This behaviour may enable them to offload the metabolic heat generated by swimming.

One of the few studies on free-ranging leatherback turtles reveals that when in tropical waters, the turtles sustain a very stable body temperature, which averages around 29°C, despite spending considerable time diving into deep water at ambient temperatures down to 16°C. They also spend time in surface waters that average 26°C. Stomach temperature is less stable than body temperature, occasionally dropping rapidly and then recovering over the following 50–100 minutes. These periodic falls in temperature probably reflect feeding or drinking events when the animal ingests water/food that is at the environmental temperature.

Leatherback turtles achieve these temperature differences between their body and the surrounding water by possessing several adaptations that allow them to trap the metabolic heat generated during swimming. These adaptations include a thick layer of subcutaneous blubber, particularly round the head and neck, which is unique among reptiles. The animals also have countercurrent heat exchangers in their flippers and their massive size gives them an overall very favourable surface area to volume ratio that minimizes rate of heat loss. Hence, the term **gigantothermy** has been coined to define this form of endothermy.

Endothermy in snakes

Many species of terrestrial reptiles attend nests and show varying degrees of parental involvement in their offspring, but only the pythons invest their own body heat to regulate the incubation temperature of their clutches. Figure 10.46A shows a female ball python (*Python regius*)

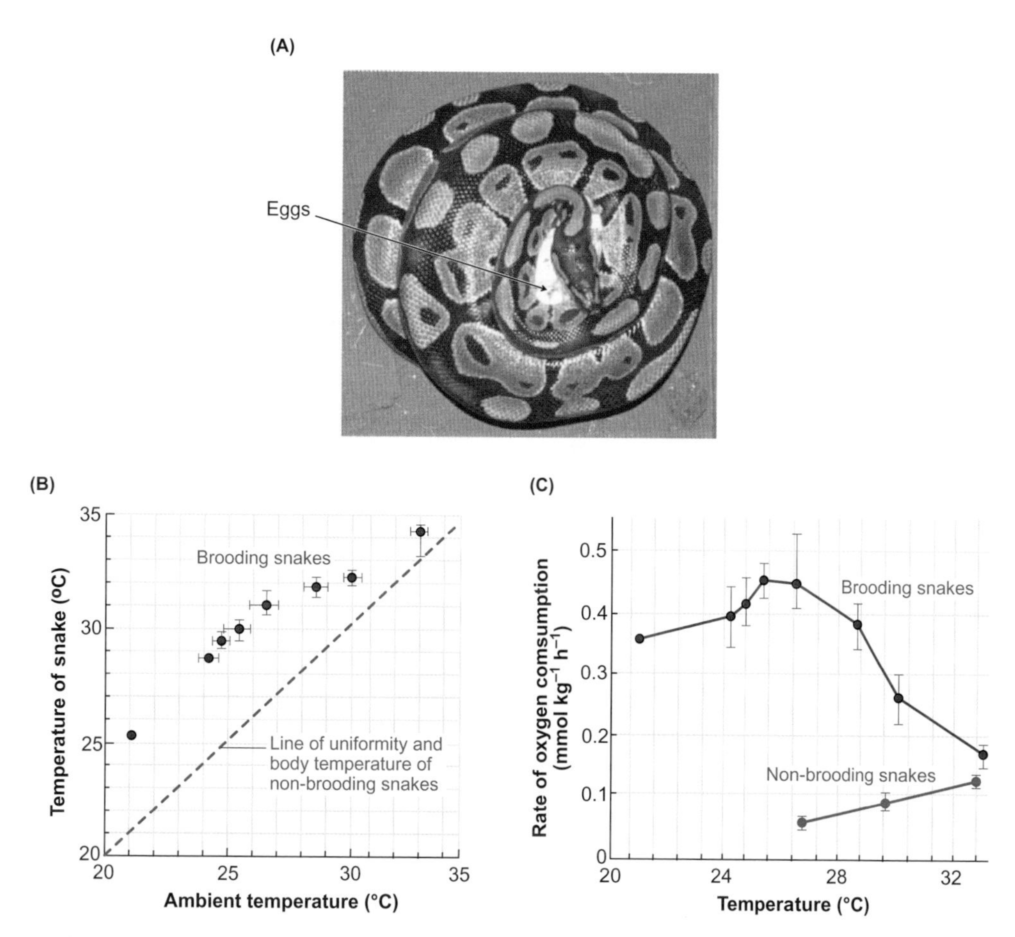

Figure 10.46 Brooding pythons

(A) Ball python (*Python regius*) wrapped around a clutch of eggs. This not only provides protection for the clutch, but also provides a stable thermal environment that is maintained by heat generated by the female shivering when it gets cold. (B) Body temperature of a brooding Indian python (*Python molorus*) relative to the ambient temperature. The brooding pythons maintain a higher body temperature than the environment. Non-brooding snakes have the same temperature as the environment. Horizontal lines, range of ambient temperature and vertical lines, range of temperature of snakes. (C) Rate of oxygen consumption in relation to ambient temperature for pythons that are brooding and the same animals when not brooding. When brooding, the rate of oxygen consumption is higher and increases as the ambient temperature becomes colder like a typical endotherm, but when the pythons are not brooding the relationship is the opposite and like that seen in a typical ectotherm. Vertical lines, range.

Reproduced from: A: Aubret F et al (2005) Why do female ball pythons (*Python regius*) coil so tightly around their eggs? Evol. Ecol. Res 7:743-758 B, C: Hutchinson VH et al (1966). Thermoregulation in a brooding female Indian python, *Python molurus bivittatus*. Science 151: 694-695.

wrapped around her clutch of eggs, which is typically located in a small burrow. The female python will stay with the eggs without feeding until they hatch, thus defending the clutch from predation.

However, her input of parental care extends beyond protection from predators. If a non-brooding female python is experimentally exposed to declining ambient temperature (T_a), her body temperature follows the change, as shown in Figure 10.46B. In contrast, if a brooding python is experimentally exposed to a similar decline in T_a, her body temperature does not fall as fast, and she can sustain differences between her body temperature and that of the environment of 10°C or more. This potentially gives a developing clutch of eggs a degree of thermal independence from environmental variations that would otherwise affect its rate of development.

This defence of clutch temperature requires an increased rate of heat production which the female generates by rhythmically contracting her skeletal muscles: she shivers. When shivering, the female's oxygen requirements can increase by a factor of about 10, as shown in Figure 10.46C and, since she was not feeding, this increase in metabolic rate must be fuelled by stored fat reserves. However, this was at ambient temperatures of around 26°C.

The use of shivering thermogenesis in brooding females under natural conditions depends on the latitude and, therefore ambient temperature, at which the species lives and the amount of heat the snakes can obtain from external sources during the day. In species that live at higher latitudes, such as Australian diamond pythons (*Morelia spilota*), body temperature of brooding females is significantly higher than that of non-brooding animals and females can lose over 15 per cent of their body mass, most likely as a result of the shivering thermogenesis they employ. In contrast, in ball pythons which, like most species are equatorial, the loss of body mass may be more modest (6 per cent) because there is less of a requirement for the snakes to activate endothermic heat production. The major benefit of continued brooding in these conditions may be to reduce water loss from the eggs[69].

Endothermy in dinosaurs?

The dinosaurs were an ancient group of reptiles that lived during the Triassic, Jurassic and Cretaceous periods. Although they shared the land and seas for this 150 million year period with several other groups of reptiles, and with birds and mammals during the second half of the period, they were the dominant vertebrate life form throughout almost the entire period. There has been controversy about the thermoregulatory capabilities of the dinosaurs ever since their first discovery, with an important factor influencing these capabilities being the size of the animals, as we discuss in Section 10.2.1.

The estimated body masses of dinosaurs vary widely, with one of the heaviest, estimated from good skeletal remains, being the herbivorous *Dreadnoughtus schrani,* first reported in 2014. *Dreadnoughtus schrani* is thought to have weighed around 40,000 kg (although the original estimate was 59,000 kg). Other estimates are based on fewer fossil remains and are more variable. Herbivores, such as *Argentinosaurus* spp., may have weighed up to 90,000 kg, while one of the largest carnivores, *Spinosaurus* spp., may have been just over 8000 kg. At the other extreme, the smallest known dinosaurs were about the size of common pigeons (*Columba livia*).

Even with the rate of heat generation achieved by most ectothermic animals, the largest dinosaurs ran the risk of overheating because of their relatively small surface area to volume ratio, whereas the smallest ones could have easily lost any metabolically generated heat because of their relatively large surface area to volume ratio.

During the first half of the 1900s, reptiles were generally seen as sluggish cold-blooded animals, which were unable to thermoregulate; dinosaurs were not thought to have been any different. However, opinions started to shift during the late 1960s and early 1970s, largely based on a number of correlations between metabolic rate and such factors as predator–prey ratios and fossilized trackways, which indicated stride length. Subsequent scrutiny has concluded that these correlations are not very sound. Nonetheless, the discussion continues and some of the evidence for and against endothermy in dinosaurs is outlined in Case Study 10.2.

Endothermy in insects

Insects that fly require high body temperatures to enable the high frequency contractions of their flight muscles. To initiate increases in the temperature of the thorax, where these muscles are located, many flying insects generate heat internally. There were once thought to be two internal heat-generating mechanisms. The first involves the metabolic cycle, which is shown in Figure 10.47A. The normal course of glycolysis involves the conversion of glucose to glucose-6 phosphate and then to fructose-6 phosphate. This latter substrate is converted to fructose-1,6-bisphosphate by the enzyme phosphofructokinase (PFK), utilizing ATP[70]. The reverse conversion of fructose 1,6-bisphosphate to

[69] Water loss from incubating eggs is discussed in Section 12.3.2.

[70] Intermediary metabolism is discussed in Section 2.4.

Case Study 10.2 Some evidence for and against endothermy in dinosaurs

A novel approach to the debate of endothermy versus ectothermy in dinosaurs has involved estimating the rate at which oxygen was consumed (represented by $\dot{M}O_2$) when bipedal dinosaurs of various body masses walk and run slowly. The methods involve studying the anatomy of the locomotor system and have been well validated with living animals, so the estimates should be reasonably accurate.

The estimates indicate that $\dot{M}O_2$ of dinosaurs larger than 20 kg, exceeded the maximum sustainable $\dot{M}O_2$ ($\dot{M}O_2$max) of present-day ectotherms of similar mass and fall within the range of that for present day birds and mammals. Of course, living ectotherms do not reach the body mass of the larger dinosaurs, so the data from living ectotherms were extrapolated to match the body masses of the larger dinosaurs, which is not statistically sound. Bearing this in mind, these extrapolated data suggest that dinosaurs functioned like present-day endotherms. If these conclusions are correct, then the dinosaurs with high $\dot{M}O_2$max would have required respiratory and circulatory systems similar to those present in birds and mammals[1] in order to provide the locomotor muscles with sufficient oxygen.

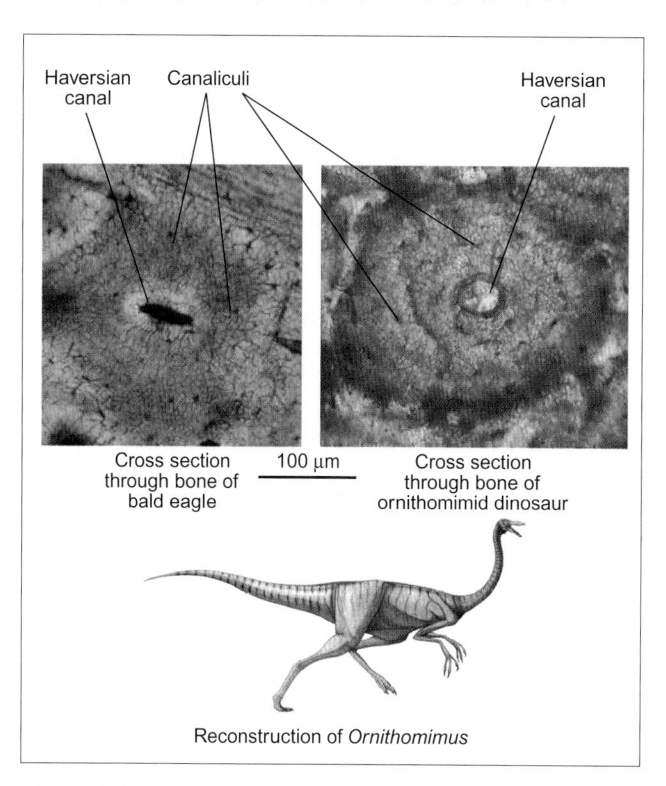

Figure A Fine structure of the bone of a modern bird and an ornithomimid dinosaur

The bone of modern birds, such as bald eagles (*Haliaeetus leucocephalus*) has well-developed, branching canaliculi arising from Haversian canals. These structures are also apparent in the bones of some dinosaurs, such as ornithomimids. One specimen of *Ornithomimus* has been estimated to be about 2 m tall, 3.8 m long and weigh 170 kg. Also, evidence from three other specimens indicates that they were covered in feathers.

Source: JM Rensberger: unpublished, with permission. Image: Universal Images Group North America LLC / Alamy Stock Photo.

Composition of bones in dinosaurs and modern endotherms

The conclusions reached by estimating $\dot{M}O_2$ are at least partly supported by studies of the structure of bones from dinosaurs and modern endotherms. The bones of most modern endotherms are **fibrolamellar bones**, which are ramified by small holes that carry blood vessels in them. This system of Haversian canals and canaliculi is associated with the high growth rate of bone, which is assumed only to be possible in animals with the high metabolic rates associated with endothermy. Haversian canals and canaliculi are far less well developed in most living ectotherms in which the metabolic rate of bone tissue is much less than that of endotherms.

It has been claimed that dinosaur bones are closer to those of modern endotherms, as shown in Figure A, implying a higher rate of blood flow to the tissue together with a high rate of metabolism. However, the differences in bone structure of endotherms and ectotherms are not always clear, and the extent to which dinosaurs conform to a supposed endothermic pattern is at best tenuous. For example, fibrolamellar bone is not found in many small, rapidly growing endotherms, but has been identified in some fossil amphibians, and in wild adult American alligators (*Alligator mississippiensis*). It has been suggested that the presence of fibrolamellar bone in dinosaurs may be more related to a high stable body temperature than to a high metabolic rate. A high stable body temperature may have occurred in many large dinosaurs which were homeothermic[2] because of their thermal inertia, but still possibly ectothermic.

It was thought for many years that the growth rings found in the bones of some dinosaurs are indicative of ectothermy, which gives rise to relatively rapid growth during the warm period of the year and slower or no growth during the colder months. This pattern of growth is, therefore, cyclical. In contrast, the bones of endotherms are thought to have a more steady growth rate because of their more stable, higher body temperature. However, a comprehensive study of the bones of modern ruminants from a wide variety of climate zones, has demonstrated that a cyclical pattern of growth does occur in living endotherms, thus undermining the conclusion that the presence of such a pattern in dinosaur bones indicates that these animals were ectotherms. They could well have been endotherms, or at least homeotherms.

A more recent study estimated growth rates of various species of dinosaurs and living vertebrates, from the annual growth rings in their bones. A metabolic scaling factor was used to relate growth rate to resting rate of oxygen consumption in these species. The study concluded that the metabolic rates of dinosaurs were intermediate between those of living endotherms and ectotherms and similar to those of regional endotherms (also called mesotherms in this study), such as leatherback turtles and active tuna and lamnid sharks.

The variability in the concentration of the oxygen isotope ^{18}O in bone allows us to assess how stable body temperature was during the growth of the bone. Relatively high variability in ^{18}O in the bones of an animal indicates variability in its body temperature, which in turn would indicate an ectothermic existence. Within the bones of juveniles of the large (up to 4000 kg) herbivore dinosaur, *Hypacrosaurus*, the variability in ^{18}O is relatively small—equivalent to only about 2°C variation in temperature. These data suggest that *Hypacrosaurus* was homeothermic, but not necessarily endothermic. However, data obtained using similar methodology, have demonstrated that at least some species of theropods[3] were endothermic homeotherms and were able to maintain a constant body temperature at any latitude.

Did dinosaurs have nasal turbinates like modern endotherms?

The main problem in answering the question of whether the dinosaurs were endothermic or ectothermic is that these traits are aspects of physiology, but physiology does not fossilize well. One possible indicator of endothermy is the structure of nasal turbinates, which we discuss in Figure 10.25. The respiratory turbinates of endotherms are elaborate structures involved in reducing heat loss in respiratory air, but they are absent in ectotherms, with only the olfactory turbinates being present.

An analysis of the sizes of the nasal cavities compared to body mass of present-day endotherms and ectotherms, as well as of fossils of early mammals and dinosaurs, is shown in Figure B. The three dinosaurs that have been examined, which includes *Hypacrosaurus*, fit onto the line for modern ectotherms, while early fossil mammals conform to the modern endotherm line. There is currently no evidence that dinosaurs had respiratory turbinates. This suggests that they had relatively low rates of ventilation, below the rates needed to support the high rates of metabolism that characterize endothermy. Convincing as this evidence is, so far only a few animals have been examined and these do not include the newly discovered feathered dinosaurs.

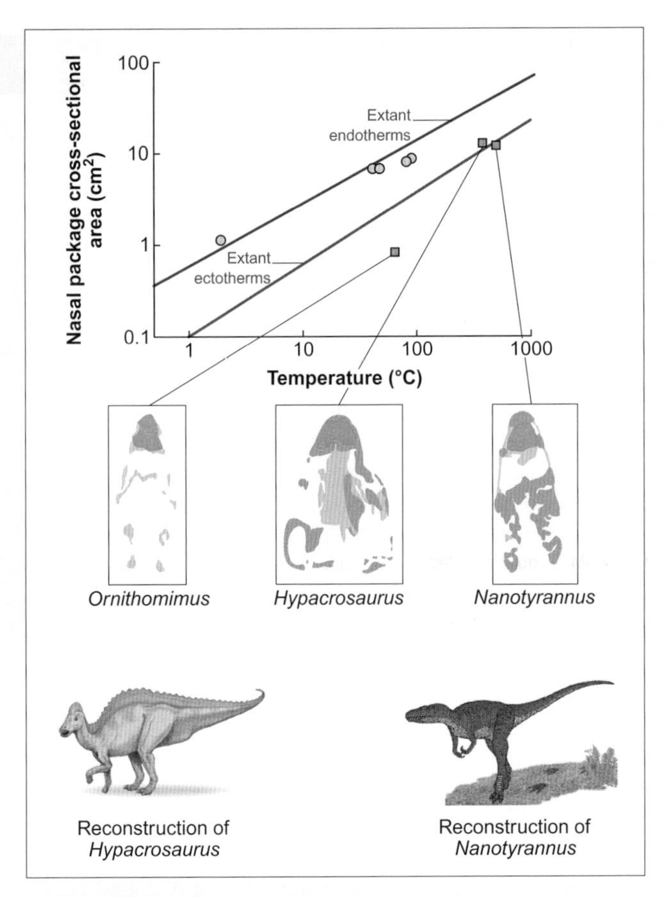

Figure B Sizes of the nasal passage in living endo- and ectotherms (solid lines)

Endotherms have larger nasal passages than ectotherms to accommodate the respiratory turbinates. Data for three dinosaur fossils (blue squares) and five early fossil mammals (red circles) are shown. The data for dinosaur nasal passages clearly lie on the ectotherm rather than the endotherm line. In the drawings of the samples at the bottom, the nasal passages are shown in orange.

Reproduced from: Hillenius WJ and Ruben JA (2004). The evolution of endothermy in terrestrial vertebrates: Who? When? Why? Physiological and Biochemical Zoology 77: 1019-42. Images: Debivort/Wikimedia Commons.

Did dinosaurs have feathers?

A consistent feature of endotherms is that their bodies have peripheral insulation (fur, feathers or fat), which serves to trap the heat they generate internally, while most species of ectotherms lack insulation because it would hinder their uptake of heat. As far as we can tell, large dinosaurs had no insulation on their bodies. However, large terrestrial endotherms also lack external insulation[4], whereas smaller mammals and birds (and pterosaurs) invariably do have insulative coverings.

Until the late 1990s, it was thought that small dinosaurs did not have external insulation. However, our views of small dinosaurs have been transformed by some startlingly well-preserved small theropod dinosaurs[4] from China, which appear to have feathers. The existence of feathers in these small dinosaurs strongly suggests they were generating heat internally and were endothermic thermoregulators. Were these the norm, or were they an unusual group that are unrepresentative of most other small dinosaurs? Unfortunately, skin and feathers are not well preserved, so at present we do not know how representative the fossils are of most other small dinosaurs. The consensus among most palaentologists is that all small dinosaurs, and even the young of larger dinosaurs, may have had feathers.

Concluding comments

The discovery of the feathered dinosaurs has completely changed our views of dinosaur thermoregulation, although the question of endothermy in dinosaurs remains a maze of conflicting evidence. Some scientists have suggested that it is not really sensible to ask whether the dinosaurs were endotherms or not. The outcome of modelling of thermal relationships of a large number of dinosaurs in the context of their habitat and the palaeoclimate suggests that many larger dinosaurs (>10,000 kg) would have had high (>30°C) and stable body temperatures, with less than 2°C variation over 24 h, independent of any capacity for endothermy. Indeed, the larger dinosaurs would probably have faced problems of overheating in tropical latitudes and would have needed to offload heat by immersing themselves in water and/or seeking shade. Calculations suggest that even large sauropods[5] could have dissipated sufficient heat by wallowing in water just like large present day herbivores. In contrast, smaller dinosaurs would have been unable to maintain high body temperatures particularly at higher latitudes.

Perhaps the answer to the question about endothermy in dinosaurs is that they were all in fact, mesotherms.

› Find out more

Tumarkin-Deratzian AR (2007). Fibrolamellar bone in wild adult *Alligator mississippiensis*. Journal of Herpetology 41: 341–345.

Barrick RE, Showers WJ (1995). Oxygen isotope variability in juvenile dinosaurs (*Hypacrosaurus*): evidence for thermoregulation. Paleobiology 21: 552–560.

Fricke HC, Rogers RR (2000). Multiple taxon-multiple locality approach to providing oxygen isotope evidence for warm-blooded theropod dinosaurs. Geology 28: 799–802.

Grady JM, Enquist BJ, Dettweiler-Robinson E, Wright NA, Smith FA (2014). Evidence for mesothermy in dinosaurs. Science 344: 1268–1272.

Hillenius WJ, Ruben JA (2004). The evolution of endothermy in terrestrial vertebrates: who? when, why? Physiological and Biochemical Zoology 77: 1019–1042.

Kohler M, Marin-Moratalla N, Jordana X, Aanes R (2012). Seasonal bone growth and physiology in endotherms shed light on dinosaur physiology. Nature 487.

Pontzer H, Allen V, Hutchinson JR (2009). Biomechanics of running indicates endothermy in bipedal dinosaurs. PLoS ONE 4: e7783.

Rensberger JM, Watabe M (2000). Fine structure of bone in dinosaurs, birds and mammals. Nature 406: 619–622.

Ruben J (1995). The evolution of endothermy in mammals and birds: from physiology to fossils. Annual Review of Physiology 57: 69–95.

Seebacher F (2003). Dinosaur body temperatures: the occurrence of endothermy and ectothermy. Paleobiology 29: 105–122.

[1] Respiratory and circulatory systems of birds and mammals are discussed in Sections 12.3.3 and 14.4.4.

[2] Homeothermy is defined in Section 8.5.

[3] Theropod dinosaurs include the largest terrestrial carnivores that have ever lived, and birds most likely evolved from small theropods.

[4] Refer back to Section 10.2.1 to remind yourself why large terrestrial endotherms lack external insulation.

[5] Sauropods were herbivorous dinosaurs which included the largest terrestrial animals that ever lived. A well-known example is *Diplodocus*.

fructose-6-phosphate by the enzyme fructose-1,6-bisphosphatase (FbPase) is normally inhibited by intracellular adenosine mono-phosphate (AMP), as shown in Figure 10.47A.

The muscles of European bumblebees (*Bombus* sp.) have activities of FbPase, which are approximately 30 times that in other tissues. This FbPase is reversibly insensitive to AMP, allowing a futile cycle[71] of conversion between the two substrates to proceed at low temperatures, as indicated in Figure 10.47A. It has been suggested that this process may be important as a mechanism for initiating the warming of the thoracic cavity prior to flight. Once the flight muscles start to warm up, the futile cycle is inhibited, probably by elevated calcium levels released from the sarcoplasmic reticulum that are important for initiation of muscle contraction[72] but which also inhibit FbPase activity. This allows normal glycolysis to proceed, providing substrates that feed into the tricarboxylic acid (TCA) cycle generating ATP. This ATP is utilized to support contraction of the flight muscles initiated by the calcium release.

Despite its elegance, this hypothesis has received a number of challenges. For example, the rate of heat production is insufficient to maintain the temperature of the flight muscles at about the required 27°C on a cold day. (It appears to be <7 per cent of that required). What is more, in 1991 the first direct evidence demonstrated that contractions of the flight muscles occur at all thoracic temperatures during pre-flight warm up in common eastern bumblebee (*Bombus impatiens*), not just in cold conditions. Also, high levels of FbPase activity have only been found in one species of North American bumblebee.

It seems, therefore, that the major (and possibly the only) heat-generating mechanism used prior to flight in insects involves contractions of the flight muscles. During flight, the muscles controlling the upstroke and downstroke of the wings are contracted alternately but, when warming the thorax, they contract simultaneously; the animals shiver. By doing this, the muscles generate substantial amounts of heat but do not cause movement of the wings. Figure 10.47B shows that the heating effect of these contractions can be sufficient to generate a 15–20°C elevation in thoracic temperature, although the temperature of the abdomen hardly changes. As a result, metabolic rate can increase by over 40 times the resting value[73]. In many species of flying insects, the thorax is covered in setae which provide thermal insulation. Overheating can be prevented by allowing exchange of haemolymph, and hence of heat, between the thorax and abdomen, which is not so well insulated.

[71] Other examples of futile cycles are discussed in Sections 10.2.5 and 10.3.1.

[72] Muscle contraction is discussed in Sections 18.1.2 and 18.1.3.

[73] Elevated metabolic rate during exercise is discussed further in Section 15.2.

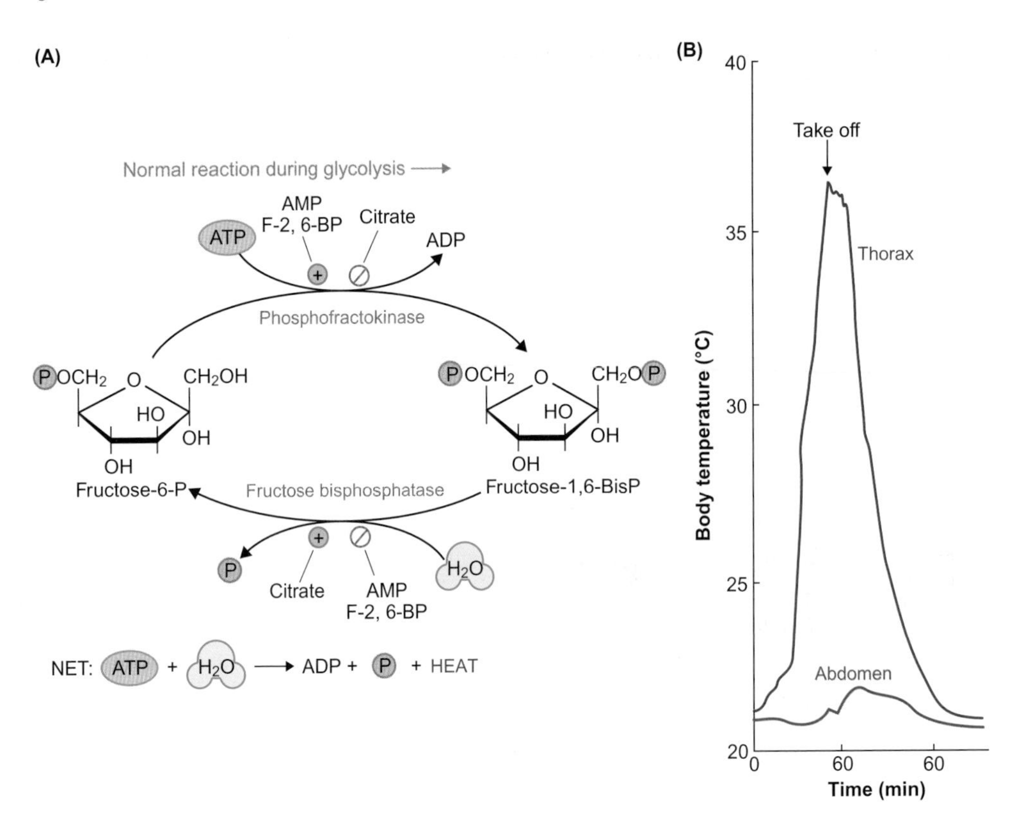

10

Figure 10.47 Heat generation in insects

(A) Futile metabolic cycle that was previously thought to generate heat in insects was based on conversion of fructose-6-phosphate to fructose-1,6-bisphosphate and back. The normal reaction during glycolysis, where AMP and fructose-2,6-bisphosphate act as activators and citrate acts as an inhibitor, is shown at the top. The reverse reaction at the bottom is normally inhibited by intracellular AMP and fructose-2,6-bisphosphate and is activated by citrate. The net reaction shows that the overall effect is the hydrolysis of ATP to produce ADP and heat. There is now little support for this 'futile' cycle being of major importance in generating heat prior to flight. (B) Temperatures in the thorax and abdomen of a dragonfly, the common green darner (*Anax junius*) during pre-flight warm up. The thorax heats rapidly but increases in the temperature of the abdomen are negligible until flight is initiated.

Sources: reproduced from: A: Atkinson DE (1977). Cellular Energy Metabolism and its Regulation. Academic Press Inc. New York; B: May ML (1976). Thermoregulation and adaptation to temperature in dragon flies. Ecological Monographs 46: 1-32.

Endothermy in insects is not only associated with flight. For example, male rain beetles (*Pleocoma* spp.) maintain their thoracic temperature (T_{th}) above environmental (ambient) temperature (T_a) when searching for females in winter, whether they search by flying or by walking, as shown in Figure 10.48A,B. For the male rain beetles, a high body temperature may be essential for the mating success and ability to avoid predators.

Most cicadas are ectotherms and use solar radiation to maintain their T_{th} above T_a; they call only during the day when the sun is out. However, a few species, such as *Platypleura hirtipennis*, generate heat almost entirely by endothermy, which enables them to call when the sun is not shining, as shown in Figure 10.48C. Heat generation may be the result of tetanic contraction[74] of muscles, as there is no wing movement, nor detectable vibrations. Facultative endothermy allows species such as *P. hirtipennis* to call during dawn and dusk and to call from shaded areas in the canopy, both of which can improve their chances of avoiding predators.

Like pythons, most species of social insects are able to regulate the temperature of their nest. One way they do this is by the workers clustering around the brood area when ambient temperature falls below a certain level (about 15°C in honey bees, *Apis mellifera*) and generate heat primarily by contracting their flight muscles. During the formation of a colony, a queen yellow-faced bumblebee (*B. vosnesenskii*) produces heat by contracting the flight muscles in her thorax and distributes it to the abdomen. When incubating, queen bumblebees are able to maintain their thorax at between 35 and 38°C and abdominal temperatures at between 31 and 36°C against an ambient temperature between 3 and 33°C.

[74] Tetanic contraction is a continuous contraction resulting from a number of relatively high frequency stimuli which do not give the muscle sufficient time to relax between stimuli, as we discuss in Section 18.2.2.

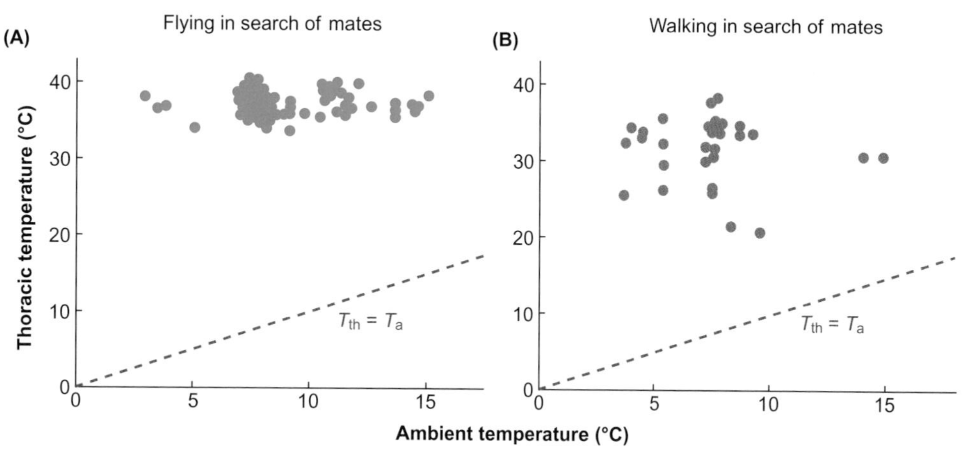

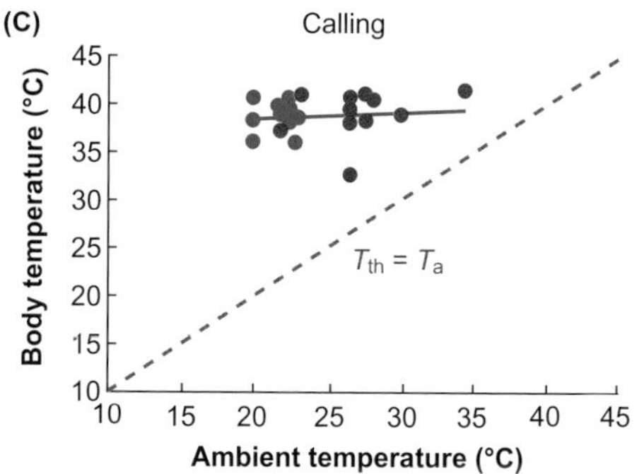

Figure 10.48 Temperature regulation in rain beetles and cicadas Thoracic and ambient temperatures (T_{th} and T_a, respectively) of male rain beetles (*Pleocoma* sp.): (A) during mating flights, (B) when searching for mates on the ground. (C) Body and ambient temperatures (T_b and T_a, respectively) in endothermic *Platypleura hirtipennis* when calling during the day (red circles) and when calling without access to solar radiation (blue circles). In all three cases, thoracic temperature is regulated at a constant value which is substantially above ambient temperature over a range of temperatures.
Reproduced from: (A, B) Morgan KR (1987). Temperature regulation, energy metabolism and mate searching in rain beetles (*Pleocoma* spp), winter-active, endothermic scarabs (Coleoptera). Journal of Experimental Biology 128: 107-122. (C) Sanborn AF et al (2004). Endothermy in African platypleurine cicadas: the influence of body size and habitat (Hemiptera: Cicadidae. Physiological and Biochemical Zoology 77: 816-823.

› *Review articles*

Bakker RT (1972). Anatomical and ecological evidence of endothermy in dinosaurs. Nature 238: 81–85.
Dickson KA, Graham JB (2004). Evolution and consequences of endothermy in fish. Physiology, Biochemistry and Zoology 77: 998–1018.
Heinrich B (1989). Beating the heat in obligate insect endotherms: the environmental problem and the organismal solutions. American Zoologist 29: 1157–1168.
Jones JC, Oldroyd BP (2007). Nest thermoregulation in social insects. Advances in Insect Physiology 33: 153–191.

Checklist of key concepts

- Endotherms compensate for heat loss to their environment in order to maintain their body temperature several degrees above ambient.
- Animals decrease their **thermal conductance** to minimize their rate of heat loss.
- The decrease in thermal conductance can be achieved by erecting hairs or feathers to increase the insulating layer of air, or by the operation of countercurrent heat exchangers in the extremities.
- **Basal metabolic rate** is the minimal rate of cellular metabolism (heat production) endotherms need to maintain their essential functions, including temperature regulation.
- Basal metabolic rate can be maintained over a range of environmental temperature by changing thermal conductance.
- The **thermoneutral zone** is the ambient temperature range within which an animal does not need to change its rate of heat production to regulate its temperature.
- Endotherms may regulate their body temperatures to different levels at different times of the day.
- Larger animals find it easier to keep warm than smaller ones because big objects cool down more slowly than small ones: they have a smaller **surface area to volume ratio**.
- Most ectotherms cannot maintain their body temperature above ambient by generating extra heat internally.

Regulating body temperature in hot conditions

- Animals may regulate their body temperatures in hot conditions by:
  - **increasing blood flow** to the periphery, which leads to increased heat loss as long-wave radiation, and by **conduction** and **convection**;
  - increasing forced convection (e.g. fanning);
  - **evaporation** from the body surfaces (e.g. sweating and panting); and
  - by storing heat.
- When generating excess heat, such as during exercise, animals can offload the excess heat via their **thermal windows**.

Regulating body temperature in cold conditions

- Once ambient temperature falls lower than the **lower critical temperature**, an animal must increase its rate of heat production to regulate its temperature effectively.
- Endotherms can increase their body temperatures by **shivering** or via **non-shivering thermogenesis**.
- Non-shivering thermogenesis is a characteristic of the **brown adipose tissue** found in placental mammals, which uses oxygen to produce heat rather than ATP from oxidation of fat molecules.
- Endotherms regulate their body temperature at a **set point** using a combination of peripheral and central **temperature sensors**:
  - **peripheral temperature sensors** include **cold receptors** and **warm receptors**;
  - **central temperature sensors** include thermosensitive neurons found in the pre-optic area of the hypothalamus in mammals.
- Larger animals have lost their external insulation to counter the low surface area to volume ratio their bodies exhibit.
- Some animals avoid heat stress by restricting physical activity to the coolest part of the day, or by seeking shade to reduce exposure to the sun.

Regulating body temperature through the seasons

- During winter, animals reduce their rate of heat loss by moulting from their thinner summer coat to a thicker winter one.
- During winter, larger animals tend to have thicker fur than smaller ones.
- Larger animals can also reduce their energy demand during winter by **reducing their body mass**.
- Loss of heat during expiration is dramatically reduced by **respiratory turbinates** in the nasal cavities of birds and mammals.
- Some animals, mainly birds, avoid cold stress by **migrating** to warmer climates in winter.
- If an animal cannot move to a warmer climate, it may store additional food reserves, either internally (as fat), or externally, in a food store.
- Species that hibernate during winter exist in a state of **torpor**: energy demands are reduced, body temperature drops and is regulated at a new set point, and the animals survive on their stored fat reserves.
- The torpid state of many hibernating animals is punctuated by periods of **arousal**, normally lasting for periods of about 10–20 hour, during which their body temperature returns to normal.
- We do not know exactly what the function of arousal is.

Regulating body temperature in water

- The rate of heat loss increases enormously when a mammal enters water because of its **high thermal conductivity** compared with that of air.
- Animals can overcome this increased rate of heat loss in water by:
  - trapping a layer of air within their fur;
  - replacing fur with a layer of insulating blubber;
  - exhibiting **regional heterothermy** in extremities.

Endothermy in fish, reptiles and insects

- Some species of ectotherms exhibit **regional endothermy**: they generate heat and retain it in specific regions of the body.
- Some fish species keep parts of their bodies warmer than their aquatic surroundings, either by conserving the heat generated by their locomotory muscles during exercise, or by non-shivering thermogenesis in modified muscles.
- Some species of reptiles and insects also use muscle activity to generate heat, either as a by-product of exercise or, in insects, by shivering thermogenesis

Study questions

*** Answers to these numerical questions are available online. Go to www.oup.com/uk/butler**

1. The figure below shows the relationship between rate of energy expenditure and ambient temperature in a typical resting mammal (red line).

   Answer the following questions: (Hint: Figure 10.2)

   a. What is happening at point A?
   b. Why does rate of energy expenditure remain stable when ambient temperature is higher than that at A?
   c. What is the region called where the energy expenditure is stable between points A and F?
   d. A colleague suggests to you that the body temperature of the animal in question is at either B or C. Which is correct and why is it correct?
   e. What problems do you think the animal will encounter if it remains at point D?
   f. What is the rate of energy expenditure at E called?

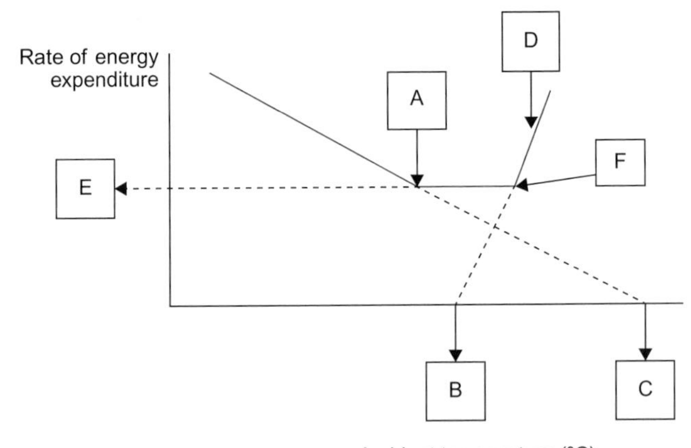

2.* The thermal conductance of a mammal (the gradient of the relationship between rate of energy expenditure and ambient temperature below the lower critical temperature) is measured at 1.54 μmol O_2 min^{-1} $°C^{-1}$. If the mammal has a body temperature of 38°C and a basal metabolic rate of 27.7 μmol O_2 min^{-1}, would it be in its thermoneutral zone at a temperature of 13°C? Show how you reached your conclusion. (Hint: Figure 10.2.)

3. Table 1 shows measurements of oxygen consumption at different ambient temperatures made on two different animals X and Y. Plot the data for both animals and answer the following questions, with an explanation of how you reached your answer.

   a. Which animal is an ectotherm and which is an endotherm?

   b. What do you estimate the lower critical temperature is for the endotherm? Show how you estimated this.

   c. What do you estimate the body temperature of the endotherm is? How did you obtain this value?

   d. Measurements are made on another species of endotherm that has a BMR of 0.38 μmol O_2 g^{-1} h^{-1}. Does this second endothermic species have a higher or lower BMR than the endotherm in the table?

   e. Is the endotherm for which data are given in Table 1 bigger or smaller than that the endotherm mentioned in question (d)?

   f.* Estimate the rates of oxygen consumption of both animals X and Y at 0°C. (Hint: Figure 10.2.)

Ambient temperature	Rate of oxygen consumption	Rate of oxygen consumption
	Animal X	Animal Y
°C	μmol $O_2 g^{-1} h^{-1}$	μmol $O_2 g^{-1} h^{-1}$
5	1.78	17.8
10	2.90	15.2
15	3.92	12.5
20	5.67	9.38
25	8.93	6.69
30	11.4	5.80
35	17.1	5.80
40	24.8	8.92
45	21.3	16.1

4. Discuss whether it is better to be larger or smaller in a cold environment. (Hint: Section 10.2.7, Reducing body mass during winter.)

5. Outline the responses that endotherms may use when faced with a short-term increase in ambient temperature that reaches temperatures above the upper critical temperature. (Hint: Section 10.2.2.)

6. Brown adipose tissue is the key source of heat production by non-shivering means in mammals. Describe the molecular basis by which it generates heat without producing ATP. (Hint: Section 10.2.5.)

7. Describe the adaptations of endothermic animals to living in persistently high ambient temperature environments. (Hint: Section 10.2.3.)

8. How do mammals and birds dissipate the heat that is generated during exercise? Compare the mechanisms used by a running mammal and a flying bird. (Hint: Section 10.2.4.)

9. Birds often avoid the cold temperature and arctic winter by migrating, while mammals seldom do this and instead retreat from the cold either with or without hibernation. Why do these groups adopt such different strategies? (Hint: Section 10.2.7, Migrating to warmer climates during winter.)

10. Hibernating animals periodically undergo arousal, heat themselves to euthermic body temperatures, stay there for a while and then go back into hibernation. We assume that this behaviour must have a function because it is the major drain on fat stores deposited prior to winter. Outline the different ideas that have been proposed for the function of these periodic arousals. (Hint: Section 10.2.7, Characteristics and possible function of arousals.)

11. Some endothermic mammals and birds have become secondarily aquatic. This presents them with some unique thermal challenges. Describe the ways that mammals have solved these problems. (Hint: Section 10.2.8.)

12. In the film *Jurassic Park* there is one scene where a *Velociraptor* breathes on a cold window and its breath condenses on the glass. Explain what that scene tells us about the film-maker's interpretation about whether dinosaurs were ectotherms or endotherms. Was the film-maker correct in this interpretation? Make a critical evaluation using all the available scientific evidence. (Hint: Case Study 10.2.)

13. Describe how some species of fish maintain regions of their body above ambient temperature (Hint: Section 10.3.1.)

Bibliography

Cossins AR, Bowler K (1987). Temperature Biology of Animals. London and New York: Chapman and Hall.

McNab BK (2002). The Physiological Ecology of Vertebrates. A View from Energetics. Ithaca and London: Comstock Publishing Associates.

Bar-headed geese (*Anser indicus*) breed in Mongolia and Tibet, but many of them spend their winter in India. This involves migrating over the Himalayas at altitudes of around 5500 m, where atmospheric pressure (and, therefore, the pressure of oxygen), is about half of its value at sea level.

http://www.barelyimaginedbeings.com/2009_07_01_archive.html

Part 4

Oxygen

In this part we explore how cells in the various tissues and organs of animals obtain oxygen and metabolic substrates, the basic materials they require to generate the energy needed to function. Energy is required by every active cell in an animal's body—by secretory and nerve cells every bit as much as it is by muscle cells. Most of this energy is liberated from the metabolism of two components of their food, carbohydrates and fat—very often in a way that requires oxygen, and generates carbon dioxide.

In many species of animals, the respiratory and cardiovascular systems work together to transport oxygen from the outside environment to the cells, and to transport carbon dioxide in the opposite direction The 'fuel' for metabolism—fats, carbohydrates and amino acids—must also be transported to the cells that need them.

Throughout **Chapters 11, 12, 13 and 14** we look at the mechanisms of gas exchange and transport around the body, and the storage and transport of metabolic substrates. We end the part by discussing in **Chapter 15** the coordinated responses of the respiratory and circulatory systems both to environmental changes—variations in temperature and oxygen availability, for example—and during such behaviours as exercise, feeding and torpor. We discuss the control and integration of the respiratory and cardiovascular systems in Chapter 22.

11

The respiratory gases, gas exchange and transport

Key characteristics and principles

In this chapter, we discuss the general characteristics of the respiratory gases and the principles underlying the processes involved in respiratory gas exchange. These characteristics and principles are fundamental to understanding how respiratory and circulatory systems function in different groups of animals, which we discuss in Chapters 12, 13 and 14.

We begin this chapter by discussing the origin and variation of oxygen in the atmosphere and the physical properties of the two respiratory gases, oxygen and carbon dioxide, in air and water. We then discuss movement of the gases between the cells and the environment which usually involves a combination of **passive diffusion** and the bulk movement of air or water and blood by **forced convection**[1].

11.1 The respiratory gases

The respiratory gases (oxygen and carbon dioxide) have varied considerably in their concentration in the atmosphere over evolutionary time. This variation, combined with different properties of oxygen and carbon dioxide in air and water, has had profound influences on the evolution of the respiratory systems of animals.

11.1.1 Origin and rise of atmospheric oxygen

As we cannot survive without oxygen for more than a few minutes, it is difficult to imagine the Earth with an atmosphere lacking oxygen. Yet this was initially the case. The Earth was formed approximately 4.5 billion years ago and composition of the Earth's atmosphere during its early history remains a subject of speculation. One possibility is that the Earth's early atmosphere consisted mainly of nitrogen, carbon dioxide and hydrogen, with some water vapour and trace amounts of other gases. This compares with the current (November 2020) composition of 20.95 per cent oxygen, 78.08 per cent nitrogen, 0.93 per cent argon and 0.0412 per cent carbon dioxide, as recorded at the Mauna Loa Obervatory, Hawaii. This means that the concentration of CO_2 is now approximately 30 per cent higher than the 1959 value of 0.0316 per cent.

Where did all that oxygen in the present-day atmosphere come from? Until about 2.5 billion years ago, the concentration of oxygen in the atmosphere was probably less than 0.02 per cent, as shown in Figure 11.1A. Figure 11.1A also shows that the concentration of atmospheric oxygen began to increase around 2.5 billion years ago[2], and this increase coincided with the presence of photosynthesizing bacteria, known as **cyanobacteria**. We are uncertain about what happened next, but from about 1.7 billion to about 800 million years ago, the concentration of oxygen remained at around 2 per cent. This stability resulted from there being a balance between the rate of oxygen production by the cyanobacteria and its rate of consumption by respiring organisms and reactions with the Earth's crust, such as the formation of iron oxide (rust).

During this time there was sufficient oxygen for the evolution of nucleated (**eukaryotic**) cells. Then, sometime between 1 billion and 650 million years ago, the level of atmospheric oxygen rose dramatically. Figure 11.1B shows that between 550 and 490 million years ago (during the Cambrian period), oxygen reached its present level—and this was when the multicellular organisms, the metazoans, evolved.

[1] Diffusion and convection are discussed in general terms in Sections 3.1.2 and 3.3.1, respectively.

[2] Recent evidence suggests the increase in atmospheric oxygen concentration may have begun about 2.7 billion years ago.

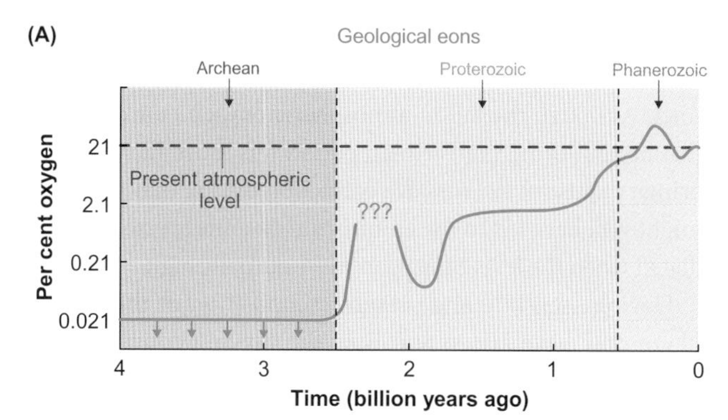

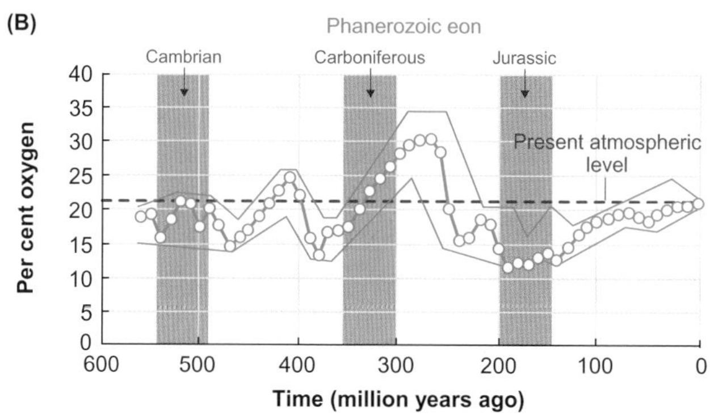

Figure 11.1 Proposed concentrations of atmospheric oxygen through geological time

(A) Question marks indicate period of particular uncertainty. (B) Best estimate of atmospheric oxygen concentration, ± margin of error, over the last 600 million years (the Phanerozoic eon). The absence of a margin of error below 12% oxygen at approximately 400 and between 200 and 150 million years ago is explained by the fact that 12% oxygen is the lowest level required to support forest fires and yet there is evidence for fires at these times in the form of charcoal.

A: Adapted from: Canfield DE (2005). The early history of atmospheric oxygen: homage to Robert M. Garrels. Annual Review of Earth and Planetary Sciences 33: 1-36; B: Berner RA (2006). GEOCARBSULF: A combined model for Phanerozoic atmospheric O_2 and CO_2. Geochimica et Cosmochimica Acta 70: 5653-5664.

Attempts to determine accurately the historical amount of oxygen in the atmosphere are based on estimates of the rates of oxygen production and of its utilization. Although there is much uncertainty, these estimates suggest that about 300 million years ago the concentration of oxygen peaked at about 30 per cent, as illustrated in Figure 11.1B. This high concentration of atmospheric oxygen is thought to have been the result of the evolution of the land plants, in particular woody plants (trees), and the associated widespread burial of organic carbon. This burial of carbon is reflected by the abundance of coal in deposits from the period (the Carboniferous period). Such a high concentration of oxygen would increase the chances of combustion, so the palaeontological evidence of more fire-resistant plants at that time is not surprising.

Associated with the dramatic increase in the concentration of oxygen about 300 million years ago, was an equally dramatic decrease in the concentration of carbon dioxide from approximately 0.7 per cent during the Cambrian period to around 0.03 per cent. However, atmospheric oxygen fell to about 12 per cent around 200 million years ago (early part of Jurassic period), as illustrated in Figure 11.1B, and carbon dioxide increased to about 0.1 per cent, maybe as a result of a drop in sea level and a general drying of the land masses. Atmospheric oxygen level then recovered, gradually increasing and then settling at its current level of around 21 per cent, with carbon dioxide falling back to around 0.03 per cent.

The large changes in the concentration of oxygen that occurred around 300 million years ago are thought to have had significant effects on a number of biological processes, including the rate of diffusion[3], which is a fundamental component of gas exchange. An atmospheric concentration of oxygen of 30 per cent is about 1.4 times greater than the current value of 21 per cent and, other things remaining equal, can cause a similar increase in the rate of diffusion[3]. Such high rates of diffusion may have led to the evolution of very large animals (**gigantism**), in particular insects, such as giant dragonflies (*Meganeura* sp.), the fossil of one of these is shown in Figure 11.2[4]. *Meganeura* lived about 300 million years ago and had a wing span of up to 70 cm and a thorax with a diameter of almost 3 cm. In comparison the largest modern dragonflies, such as species of *Anax* and *Aeshna*, have wingspans of about 10 cm.

What evidence do we have that high concentrations of oxygen may have led to gigantism? Studies on present-day insects indicate that the concentration of oxygen in the atmosphere limits their maximum size and hence suggest that the evolution of gigantic insects may have been made possible by high levels of atmospheric oxygen:

- Phylogenetically corrected data[5] from four species of darkling beetles (*Tribolium castaneum, Tenebrio*

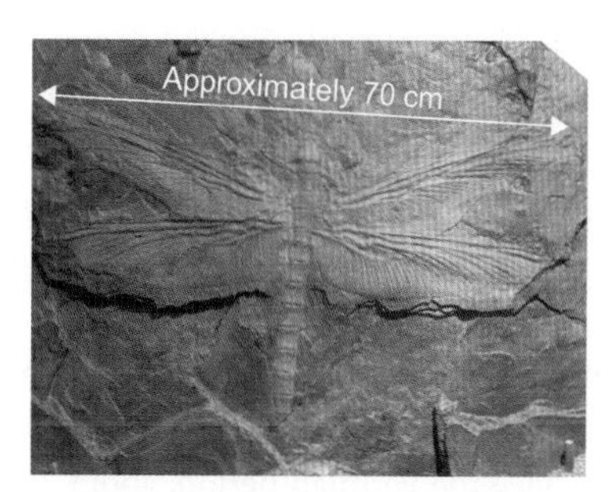

Figure 11.2 Fossil of giant dragonfly (*Meganeura monyi*)

These dragonflies had wingspans ranging from about 65 cm to over 70 cm.

Source: Alexandre Albore/Wikimedia Commons.

[3] Factors affecting rate of diffusion are discussed in Section 11.2.1.
[4] The supply of oxygen to the muscle fibres of present-day dragonflies is discussed in Section 12.4.1.
[5] We briefly explain the concept of phyologenetically corrected data analysis in Section 1.3.2.

molitor, Eleodes armata, Eleodes obscura) indicate that the respiratory system in the larger beetles[6] occupies a greater proportion of their body mass than it does in the smaller species. In other words, the respiratory system displaces other tissues, and this trend is greatest in the legs, which consist mainly of locomotor muscles. Figure 11.3A shows how the volume of the **tracheae**[7] in the body as a whole and the cross-sectional area of the trachea in in the legs, vary with body mass. It also shows how the size of the entrance from the body to the legs through which the tracheae have to pass varies with body mass. Insects have an external skeleton, which suggests that the space available for tracheae at the entrance from the body to the legs, may ultimately limit body size of the animals, as illustrated by the data in Figure 11.3B.

- Studies on other species of insects have shown that if they are raised over several generations in an atmosphere with a high concentration of oxygen, the size of the respiratory system is reduced. This suggests that in the natural conditions of high oxygen, which occurred around 300 million years ago, the increased supply of oxygen would have enabled insects to evolve larger bodies before the external skeleton became a constraint.

Data from fossils of other taxonomic groups also support the hypothesis that larger animals existed when the concentration of oxygen in the atmosphere was higher. For example, the near doubling of oxygen concentration in the atmosphere, from 12 to 21 per cent, over the last 200 million years or so, largely corresponds with three major aspects of the evolution of vertebrates: endothermic birds and mammals, the origin of placental mammals and an increase in body mass in mammals. This suggests that the increase in concentration of oxygen could have been an important factor in these evolutionary trends.

The decrease in atmospheric oxygen concentration from its peak at around 30 per cent to around 12 per cent, may have led to the extinction of many of the gigantic forms of different groups of animals. Most of the various species of insects that reached exceptionally large sizes during the Carboniferous period did not survive the Permian period, about 250 million years ago. Similarly, the diversification of the land vertebrates, which began about 400 million years ago, increased rapidly during the Carboniferous period when at least 11 of the 16 known phyla all appeared. By the end of the Permian, however, 75 per cent of the lineages of the land vertebrates were extinct. This was a far more extensive extinction than occurred around 65 million years ago, when the dinosaurs became extinct. Although the fall in atmospheric oxygen concentration that began during the Permian period was most likely not the primary cause of the massive extinction during that period, it might well have led to the elimination of some species that radiated when the level of oxygen was unusually high.

Having seen how the present composition of the atmosphere may have arisen and how the varying concentration of oxygen may have influenced the size of the animals that existed at the time, we now investigate the physical properties of the respiratory gases, oxygen and carbon dioxide in both water and air. These properties are important in the functioning of the gas-exchange systems in aquatic and terrestrial animals, which we discuss in detail in Chapter 12.

11.1.2 Physical properties of the respiratory gases in air and water

Aerobic (oxidative) metabolism involves the continuous utilization of oxygen as the final electron receptor of the electron transfer chain, and the continuous production of carbon dioxide and water[8]. Not all electrons reach the end of the chain, however. Some electrons leak directly on to oxygen during their transfer along the chain and generate free radicals[9]. Despite the potential danger of oxidative damage to the membranes, proteins and DNA of cells by the free radicals, the evolutionary advantage of the presence of a relatively high concentration of oxygen is that aerobic metabolism releases much more energy from its metabolic substrate than **anaerobic metabolism**. The full oxidation of 1 mole of glucose results in the maximum production of 27 or 29 mol ATP—far in excess of that produced during anaerobic metabolism, when only 2 mol ATP are produced from 1 mole glucose[10]. In the following sections we discuss the principles underlying the various mechanisms that have evolved in different organisms to provide the metabolizing cells with the oxygen they require and to remove the primary waste product of aerobic metabolism, carbon dioxide. The evolution of these mechanisms has been heavily influenced by the properties of oxygen and carbon dioxide in the media in which the animals live (air and water), so we must first consider these properties.

Volumes and pressures of gases

Unlike liquids and solids, gases are compressible; they take up less space when pressure is exerted upon them. The relationship between pressure, volume and temperature in

[6] The respiratory system of insects is discussed in Section 12.4.

[7] The tracheae are the air-filled tubes that form the respiratory system of insects.

[8] We discuss the electron transport chain in Section 2.4.2.

[9] A free radical has one unpaired electron in its valence shell. Having an unpaired electron makes a free radical unstable and extremely reactive. Free radicals are discussed in Section 2.4.3.

[10] Aerobic and anaerobic metabolism are discussed in Section 2.4.

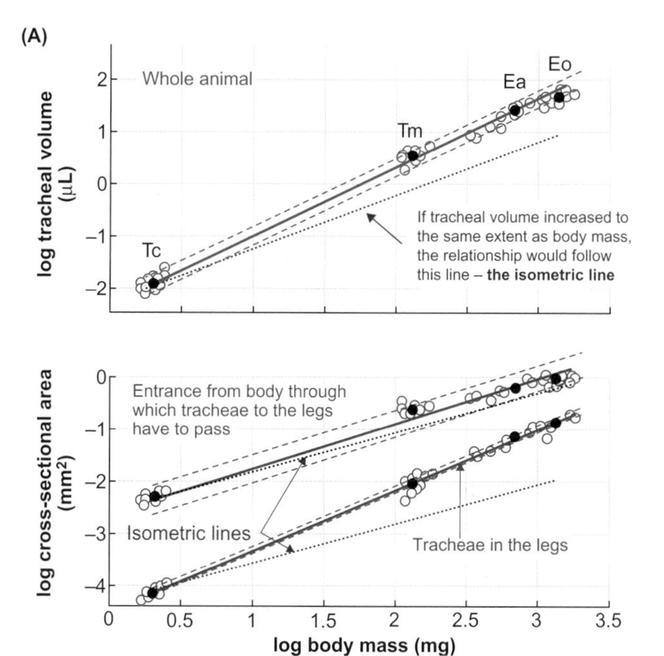

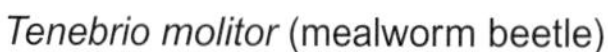

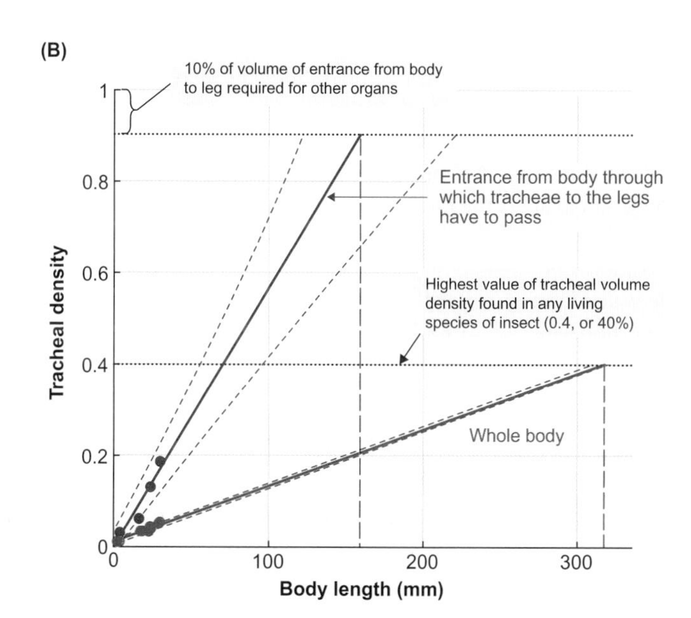

Figure 11.3 The tracheal system may limit the size of insects

(A) In the whole animal, the volume of the tracheal system increases to a greater extent than body mass (the relationship between tracheal volume and body mass has a greater slope than the isometric line) in four species of darkling beetles, *Tribolium castaneum*, Tc, *Tenebrio molitor*, Tm, *Eleodes armata*, Ea, and *Eleodes obscura*, Eo. Also, the relative area occupied by the trachea in the legs increases to a greater extent than the area of the entrance from the body through which they have to pass. (B) By extrapolating the tracheal volume density of the whole body beyond the values measured in the study to 0.4 (40%), which represents the highest value reported for any insect, darkling beetles should be able to reach a length of about 320 mm (solid blue line and dashed vertical line). This is far beyond the size of any known present-day darkling beetle, so suggests that the volume density of the tracheal system does not limit the size of the beetles. However, if it is assumed that at least 10% of the entrance from the body to the leg is required for other organ systems, such as nerves, tendons, i.e. that the tracheae cannot occupy more than 90% of the entrance, then no living beetle should be longer than about 160 mm (solid red line and dashed vertical line). The largest living beetle is, in fact, about 170 mm long, which suggests that constraints in the legs limit the size of the bodies of beetles. The thin red dashed lines are 95% confidence limits.

Reproduced from: Kaiser A et al (2007). Increase in tracheal investment with beetle size supports hypothesis of oxygen limitation on insect gigantism. Proceedings of the National Academy of Sciences of the United States of America 104: 13198-13203.

gases is discussed in Box 11.1. When the volume of a gas in a container is decreased its pressure increases proportionately, conversely, when the volume is increased, its pressure decreases. While liquids and solids change volume to a relatively small extent with changes in temperature, gases change volume to a much larger extent.

Physiologists often use volume (V) to denote the amount of respiratory gases that are exchanged by an animal. However, the effects of temperature and pressure on the volume of a gas could potentially lead to problems, particularly if two volumes of a respiratory gas obtained at different temperatures and/or different pressures are being compared. Fortunately, it is possible to use the information in Box 11.1, and particularly equation (i) in the box, to make such a comparison. For example, it is possible to compare the amount of oxygen consumed by fish at 20°C at sea level (say, 2 mL min^{-1} kg^{-1}) with that consumed by fish at 10°C in a high mountain stream at an altitude of approximately 4000 m (say, 1.5 mL min^{-1} kg^{-1}), where atmospheric pressure is 60 per cent that at sea level, as depicted in Figure 11.4.

There are two steps we need to take to make this comparison. The first step is to convert the volume of oxygen consumed under the two different conditions to the volume it would occupy under standard conditions of temperature (273 K[11] or 0°C) and atmospheric pressure, i.e. that at sea level (101.3 kPa or 760 mmHg) and when it is dry. The standard conditions are known as **standard temperature and pressure dry** or **STPD**, as discussed in Boxes 11.1 and 11.2. The second step is then to convert the volume of oxygen at STPD into the number of moles of oxygen (also known as the *amount* of oxygen). A mole is defined as the number of atoms in exactly 12 g of isotopically pure ^{12}C (^{12}C has an atomic mass of

[11] More accurately, 273.15 K.

Box 11.1 Properties of gases

The molecules of a gas, or liquid, are in continual random motion[1]. It is useful to think of the pressure (P) exerted by a gas, as it undergoes such motion, as the change in momentum (mass × velocity) when its molecules hit a solid object, such as the walls of a container. Figure A illustrates this process.

Pressure is related to the activity (kinetic energy) of the molecules of the gas. If the container shown in the figure is held at constant temperature (T, in kelvin, K) and the volume (V) of the container is reduced, the gas is correspondingly compressed. As a result, the number of molecules hitting the wall of the container per unit time increases, leading to a proportional increase in the pressure exerted by the gas.

Conversely if the volume of the container is increased, there is a proportional decrease in the pressure of the gas inside and fewer molecules will hit the wall of the container per unit time. In other words, at a constant temperature in a compressible container, the pressure of a gas is inversely related to its volume:

P is proportional to $\frac{1}{V}$; that is, PV is constant. This relationship is called **Boyle's Law.**

On the other hand, if the volume of the container is fixed, an increase in temperature will cause the molecules to move more quickly (their kinetic energy will increase). They will hit the wall of the container more frequently, again causing a proportional increase in pressure. A decrease in temperature has the opposite effect; it causes a decrease in pressure. This means that, at a constant volume, the pressure of a gas is directly related to its temperature:

P is proportional to T, that is $\frac{P}{T}$ is constant. This is known as **the Pressure or Amontons' Law.**

Finally, for a constant mass (number of molecules) of a gas that is not in a rigid container (i.e. its volume can change with a change in temperature), the increase in movement of the molecules resulting from an increase in temperature causes a proportional increase in volume of the gas. By contrast, a decrease in temperature causes a proportional decrease in the volume of the gas. Consequently, if it is allowed freely to expand or contract, the volume of a gas is directly related to its temperature:

V is proportional T, that is $\frac{V}{T}$ is constant, which is called **Charles' Law.**

Combining all three gas laws gives the combined gas law:

$\frac{PV}{T}$ is constant for any given system.

In other words, if you have one set of conditions, a pressure of P_1, a volume of V_1 and a temperature of T_1, and you change one of the variables (moving to another temperature, T_2 for example), at least one of the other variables must change as well so that:

$$\frac{P_1V_1}{T_1} = \frac{P_2V_2}{T_2} \qquad \text{Equation A}$$

Therefore, if P_1, T_1 and V_1 are those that exist during an experiment with an animal, and P_2 and T_2 are those at standard conditions, 101.3 kPa (= 760 mm Hg) and 273 K (= 0°C), respectively, the volume of gas measured during the experiment (V_1) can be corrected to the volume (V_2) it would occupy when dry (Box 12.2 gives an explanation of this) at standard conditions of pressure and temperature (STPD). An example of such a calculation is given in Section 11.1.2.

[1] The random motion of molecules is discussed in Section 3.1.2.

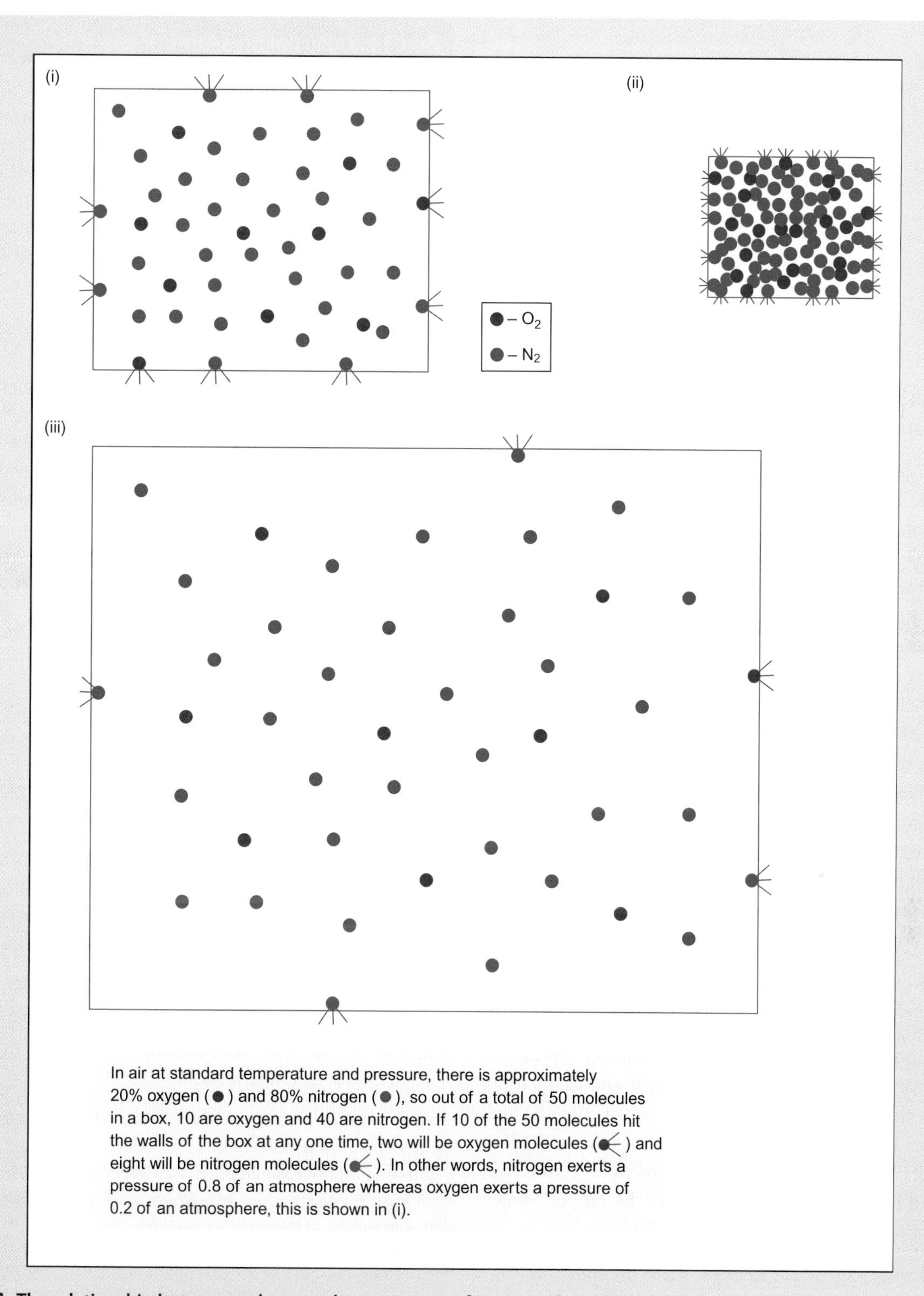

Figure A The relationship between volume and temperature of a gas on the pressure it exerts.

(i) Diagram of a square as representative of a box in which the gas is at standard temperature and pressure. (ii) If the volume of the box is halved there will be 20 hits in total (four oxygen and 16 nitrogen); total pressure of the air in the box doubles to two atmospheres.
(iii) If the volume of the box is doubled there will be five hits in total (one oxygen and four nitrogen); total pressure of the air in the box halves to 0.5 atmosphere. With a 10% increase in temperature (in K) of air in a rigid container which has the same volume as that in (i), the pressure also increases by 10%. In other words, there is a 10% increase in the number of hits compared to those in (i) – 2.2 by oxygen molecules and 8.8 by nitrogen molecules.
In a flexible container, a 10% increase in temperature (in K) leads to a 10% increase in volume of the container.

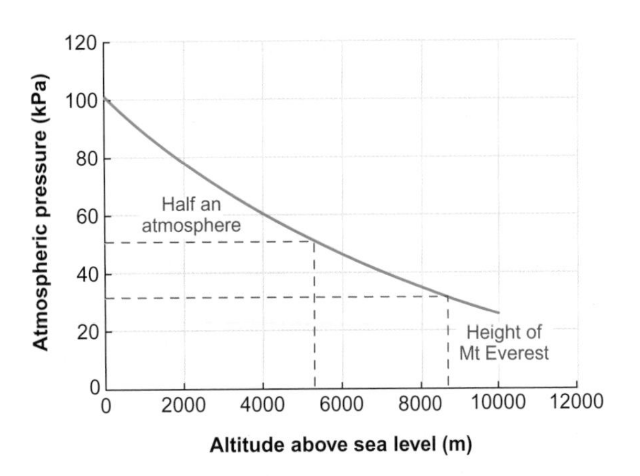

Figure 11.4 The relationship between altitude above sea level and atmospheric (barometric) pressure

12 atomic mass units or amu); this number has been experimentally determined to be 6.02×10^{23} and is known as **Avogadro's number**. So, a mole of atoms, or indeed, a mole of molecules, of any substance is 6.02×10^{23} (Avogradro's number) of atoms or molecules of that substance.

Fortunately, it is simple to convert between a volume of a gas and its amount in moles, as 1 mole of oxygen at STPD occupies 22.39 litres and 1 mole of carbon dioxide at STPD occupies 22.26 litres (although for simplicity, a value of 22.4 L is often used for both gases). Throughout this part of the book, we use the mole (abbreviated to mol, or *M*) as the unit for the amount of a gas. The abbreviation *M* is commonly used to signify the molar concentration (mol L^{-1}) of a substance, but it is often also used to signify the amount (mol) of a respiratory gas as, for example, in Equation 11.2.

Let us now work through the problem we set above:

For the fish at sea level, we will assume that barometric pressure is the standard value of 101.3 kPa, although it could be a value either side of this depending on weather conditions. Nonetheless, by making this assumption, we do not have to worry about barometric pressure in our calculation. We know that temperature of the water is 20°C (293 K) and, from Box 11.1, that a decrease in temperature from 293 K to 273 K will cause a decrease in the standard volume of oxygen consumed by the fish per unit time. So, the rate of oxygen consumption by the fish at STPD would be:

$$2 \text{ mL min}^{-1} \text{ kg}^{-1} \times 273 \text{ K}/293 \text{ K} = 1.86 \text{ mL min}^{-1} \text{ kg}^{-1}$$

We can now convert this volume to mmol min^{-1} kg^{-1} by dividing it by 22.4 mL (which is the volume occupied by a mmole of oxygen):

$$1.86 \text{ mL min}^{-1} \text{ kg}^{-1}/22.4 \text{ mL mmol}^{-1} = 0.083 \text{ mmol min}^{-1} \text{ kg}^{-1} \text{ or } 83 \text{ µmol min}^{-1} \text{ kg}^{-1}.$$

For the fish at an altitude of 4000 m, barometric pressure would be 0.61 times that at sea level, as shown in Table 11.1—that is $101.3 \times 0.61 = 61.8$ kPa. We know that water temperature is 10°C or 283 K. From equation (A) in Box 11.1, both an increase in atmospheric pressure from 61.8 kPa to 101.3 kPa (standard pressure at sea level) and a decrease in temperature from 283 K to 273 K (standard temperature) will cause a reduction in the volume of oxygen

Box 11.2 Partial pressures of gases

The air we breathe is a mixture of gases, and the pressure exerted by an individual gas in a mixture of gases is the pressure it would exert if it was the only gas in the volume occupied by the mixture. This describes **Dalton's Law of partial pressures.** In other words, the pressure that oxygen exerts in the atmosphere at standard pressure, the partial pressure of oxygen (PO_2), is simply its fraction in the atmosphere (0.2095) multiplied by standard atmospheric pressure (101.3 kPa). This works out to be 21.22 kPa.

In reality, it is necessary to deduct water vapour pressure (PH_2O) from atmospheric pressure before performing the above calculation and water vapour pressure in air can be calculated from the relative humidity (RH) or per cent saturation. The pressure exerted by water vapour when saturation is 100 per cent is known as the saturated water vapour pressure, and this varies with temperature as illustrated in the following table:

Saturated water vapour pressure at different temperatures

Temperature (°C)	Water vapour pressure (kPa)
0	0.61
5	0.87
10	1.23
15	1.71
20	**2.33**
25	3.17
30	4.24
35	5.62
37	6.28
40	7.37

So, in pure water at 20°C which is equilibrated with air at standard atmospheric pressure (101.3 kPa), water vapour pressure is 2.33 kPa, which means that the partial pressure of oxygen in the water is:

$$(101.3 \text{ kPa} - 2.33 \text{ kPa}) \times 0.2095 = 20.73 \text{ kPa}$$

Water vapour pressure should always be subtracted from atmospheric pressure before calculating PO_2 in the atmosphere or rate of oxygen consumption of an organism. So, in air at 20°C and standard atmospheric pressure and if the relative humidity of the air is 50 per cent:

$$PO_2 = (101.3 \text{ kPa} - (2.33/2 \text{ kPa})) \times 0.2095 = 21.11 \text{ kPa}$$

Table 11.1 **Partial pressure of oxygen at different altitudes**

Height above sea level (m)	Fraction of standard atmosphere	Partial pressure of oxygen (kPa)
0	1.00	21.2
1000	0.89	18.8
2000	0.78	16.7
3000	0.69	14.7
4000	0.61	12.9
5000	0.53	11.3
5500	0.50	10.6
6000	0.47	9.9
7000	0.40	8.6
8000	0.35	7.5
9000	0.30	6.4
10,000	0.26	5.5

Source: Dejours P (1981). Principles of Comparative Respiratory Physiology. 2nd edn. Amsterdam: Elsevier.

consumed by the fish per unit time. So at STPD, the rate of oxygen consumption by the fish would be:

$$1.5 \text{ mL min}^{-1} \text{ kg}^{-1} \times 61.8 \text{ kPa}/101.3 \text{ kPa} \times 273 \text{ K}/283 \text{ K} = 0.88 \text{ mL min}^{-1} \text{ kg}^{-1}$$

We now need to convert this value to mmol min^{-1} kg^{-1} as described for the fish at sea level:

$$0.88 \text{ mL min}^{-1} \text{ kg}^{-1}/22.4 \text{ mL mmol}^{-1} = 0.039 \text{ mmol min}^{-1} \text{ kg}^{-1}, \text{ or } 39 \text{ µmol min}^{-1} \text{ kg}^{-1}.$$

This calculation demonstrates that it appeared before correction that the fish at high altitude was consuming oxygen at 75 per cent the rate of the fish at sea level (1.5 mL min^{-1} kg^{-1}/2.0 mL min^{-1} kg^{-1}), when, in fact, it was only consuming oxygen at around 50 per cent of the rate of the fish at sea level (39 µmol min^{-1} kg^{-1}/83 µmol min^{-1} kg^{-1}).

The volume of a gas measured at any combination of temperature, pressure and relative humidity can be converted to the volume it would occupy if it were dry at standard temperature and pressure (STPD), as discussed in Boxes 11.1 and 11.2. It can then be converted to moles by dividing the volume at STPD by 22.4 . In effect then, by using moles we will always be stating the number of molecules of oxygen that have been consumed by an animal per unit time, or the number of molecules of carbon dioxide that have been released. This is important, because the molecules of oxygen that are consumed are those that oxidize the metabolic substrates during oxidative metabolism.

Although the proportion (percentage) of each gas in the atmosphere is constant, their **partial pressures** are not. Table 11.1 shows how the partial pressure of oxygen (PO_2) varies with changes in the height above sea level. As you will know from watching weather forecasts, atmospheric pressure can be relatively low or relatively high, but these changes are small compared to the changes in atmospheric pressure that occur with changes in altitude. For example, Figure 11.4 illustrates that atmospheric pressure at an altitude of 5500 m above sea level is half that at sea level (that is 50.65 kPa). So, although the proportion of oxygen in the atmosphere at 5500 m is the same as that at sea level (i.e. 20.95 per cent), PO_2 is only 50 per cent of that at sea level (21.2 kPa), i.e. approximately 10.6 kPa, or 380 mmHg. At this altitude, therefore, 1 mol of atmospheric gas would occupy approximately 44.8 L at standard temperature (22.4 L × 101.3 kPa/50.65 kPa), instead of the 22.4 L it would occupy at sea level. In other words, in a given volume of air at 5500 m, there are half as many molecules as in the same volume at sea level. This is why the air at high altitude is described as being 'thinner' than that at sea level[12].

Figure 11.4 shows that at the top of Mount Everest, which is 8848 m above sea level, the atmospheric pressure (and therefore partial pressure of oxygen) is approximately one third its value at sea level. Mammals, including humans, find it very difficult to survive at such altitudes without supplementary oxygen to increase PO_2. In contrast, birds seem to be better adapted to survive at high altitude. Indeed, some species, such as bar-headed geese (*Anser indicus*), undergo migratory fights at altitudes of 5500 m and above.

The cabins of commercial aircraft, which fly at altitudes over 9000 m above sea level, are pressurized to the pressure characteristic of an altitude of not more than 2440 m. In practice, most passenger aircraft are routinely pressurized to lower altitudes. This provides a tolerable PO_2 for healthy passengers and crew. However, passengers with lung diseases may need to be provided with supplementary oxygen.

Gases in water

The amount of a gas dissolved in a volume of water is directly related to the partial pressure of the gas above the water, as described by **Henry's Law**, and to the solubility of that gas in water. The solubility of a gas is described by its **solubility coefficient** (**α**), which is the increase in the amount of gas dissolved in a given volume of water, or any liquid, per unit increase in partial pressure of the gas. Solubility varies with temperature.

Different gases have different solubilities in water. For example, Figure 11.5A shows that carbon dioxide is much more soluble than oxygen in water, which is why it is used to produce the bubbles in fizzy drinks. Carbon dioxide is forced into a watery drink at high pressure and low temperature. It is released from solution and forms bubbles when the bottle or can is opened, and the pressure reduced. In contrast,

[12] We discuss the implications of low PO_2 for animals that live at high altitudes in Section 15.3.1.

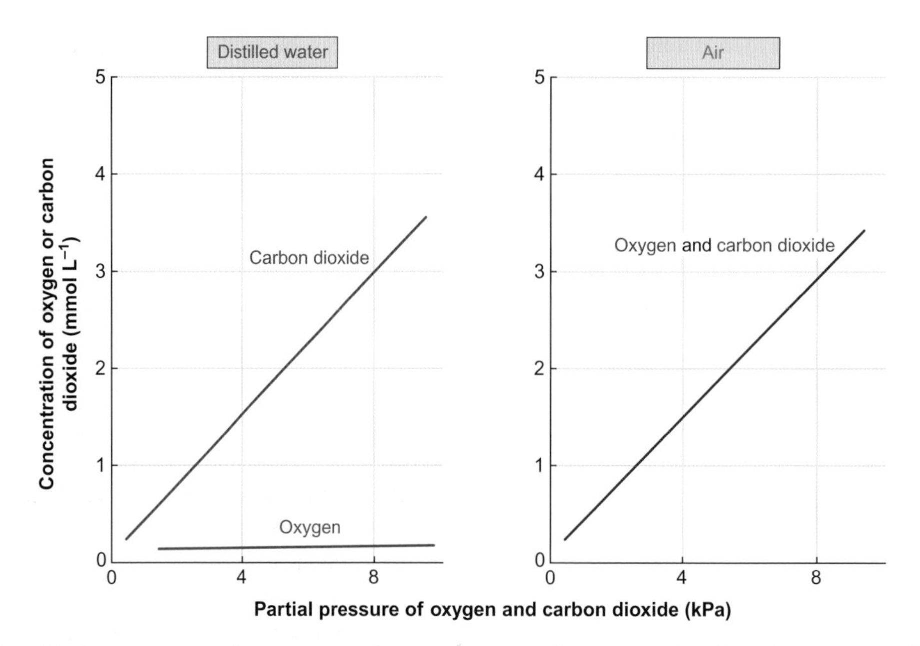

Figure 11.5 Relationship between partial pressures and concentrations of oxygen and carbon dioxide in distilled water and air at 15°C

The slopes of the lines are the capacitance coefficients. The capacitance coefficient of a gas in air is $1/RT$, where R is the universal gas constant (0.0083 L kPa K^{-1} mmol^{-1}) and T is absolute temperature (K). Note that when in water, the capacitance coefficient of oxygen is much less than that of carbon dioxide, whereas in air the capacitance coefficients are the same for the two gases. Adapted from: Dejours P (1981). Principles of Comparative Respiratory Physiology. 2nd edn. Amsterdam: Elsevier.

Figure 11.5B shows that oxygen and carbon dioxide behave similarly to each other in air: their concentrations increase by the same amount for a given increase in their partial pressures.

In some bodies of water and in the blood or haemolymph of many species of animals, the gas may be combined with other molecules or ions, as well as be in solution. For example, oxygen and carbon dioxide may be combined with a respiratory pigment, haemoglobin or haemocyanin, in the blood or haemolymph, respectively[13].

Carbon dioxide may form bicarbonates/carbonates in many natural bodies of water. So in order to include both the gas that is in solution and that which is in combination with other molecules or ions, we use the term **capacitance coefficient (β)** instead of solubility coefficient. The capacitance coefficient is the increase in the total amount of a gas in a given volume of liquid per unit increase in partial pressure of the gas. The use of the term 'solubility' is not strictly correct for gases in air but 'capacitance' is appropriate. Capacitance is also an appropriate term for the situation in which a gas is only in solution in a liquid. Values for β for O_2 and CO_2 in air and water are given in Table 11.2. Note that β for O_2 and CO_2 in air are the same, as also illustrated in Figure 11.6.

The capacitance coefficient of different gases is lower at higher temperatures. However, for CO_2 this effect is much smaller in air than it is in water, as shown in Figure 11.6.

[13] Respiratory blood pigments are discussed in Section 13.1.1.

Table 11.2 Capacitance coefficients (β) of oxygen and carbon dioxide ($\beta_W O_2$ and $\beta_W CO_2$, respectively) in distilled water and seawater, and in air (β_a) at various temperatures

Note that at 40°C, the capacitance of oxygen in air is 38 times that in distilled water, whereas at 0°C it is only 20 times greater. β in µmol L^{-1} kPa^{-1}.

	Distilled water			Seawater (salinity 35 practical salinity units)		Air
Temperature (°C)	$\beta_W O_2$	$\beta_W CO_2$	Ratio of $\beta_W CO_2$: $\beta_W O_2$	$\beta_W O_2$	$\beta_W CO_2$	$\beta_a O_2$ and CO_2
40	10.20	234.0	23.0	8.47	205.5	384.0
37	10.57	249.7	23.6	8.77	218.2	387.7
30	11.55	294.0	25.4	9.52	253.5	396.7
20	13.65	385.5	28.2	11.10	327.7	410.2
10	16.80	529.5	31.5	13.42	444.7	424.5
0	21.60	765.0	35.4	17.02	637.5	440.2

Source: Dejours, 1981.

Figure 11.6 also shows that in water the influence of temperature is greater for CO_2 than for O_2. This is why warm beer goes flatter faster (it loses its 'fizz') than cold beer. Gases are also less soluble in water containing solutes, e.g. in seawater. Take a look at Table 11.2 to see values for β of O_2 and CO_2 in fresh water and in seawater. At any given

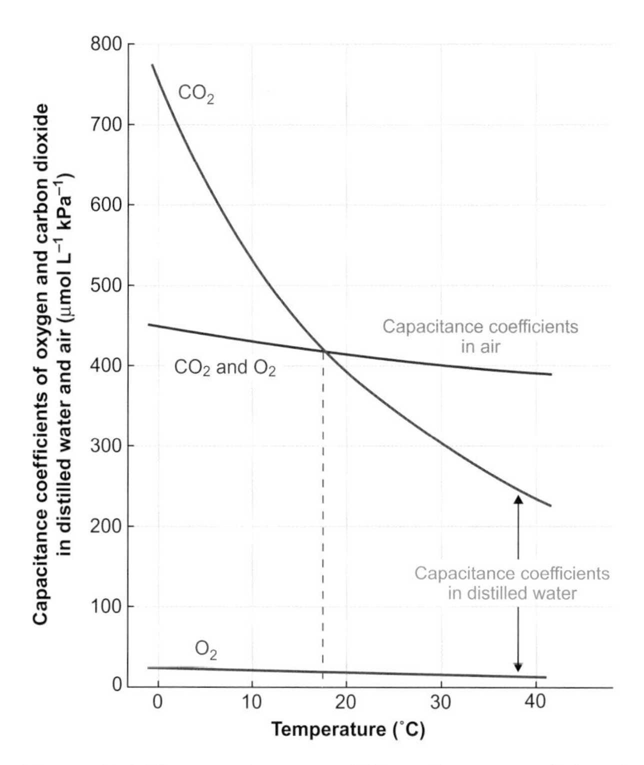

Figure 11.6 The capacitance coefficients for oxygen (O_2) and carbon dioxide (CO_2) in distilled water and air at different temperatures

The capacitance coefficient for gases in air is 1/RT (R is the universal gas constant, 8.31 L kPa K^{-1} mol^{-1}). Note that at 17.5°C, the capacitance coefficients of carbon dioxide in water and in air are identical (indicated by dashed, vertical red line).

Adapted from: Dejours P (1981). Principles of Comparative Respiratory Physiology. 2nd edn. Amsterdam: Elsevier.

temperature, both gases have a lower value of β in seawater than in fresh water.

You can demonstrate the effect of solutes on the capacitance coefficient of CO_2 in water by putting half a teaspoon of salt into a glass of fizzy drink. Try it and see what happens. The increased rate of bubble formation occurs because the salt reduces the capacitance coefficient of the dissolved CO_2 and drives it out of solution. Therefore, there is less of any gas in a given volume of seawater than in the same volume of fresh water at a particular temperature and partial pressure of the gas.

Figure 11.5 shows that there is a linear relationship between PO_2 and the concentration of oxygen in distilled water at a given temperature. The same applies to CO_2 in distilled water, but the situation is slightly more complex. When a gas mixture containing CO_2 is equilibrated with distilled water, the carbon dioxide combines with water to form carbonic acid (H_2CO_3), which dissociates into protons (H^+), bicarbonate ions (HCO_3^-) and carbonate ions (CO_3^{2-}), as illustrated in Equation 11.1:

$$H_2O + CO_2 \rightleftarrows H_2CO_3 \rightleftarrows H^+ + HCO_3^- \rightleftarrows 2H^+ + CO_3^{2-}$$

Equation 11.1

The situation is different, however, in natural waters, which contain varying amounts of bicarbonate and carbonate. When CO_2 is added to carbonated water, such as seawater, some bicarbonate is buffered by the CO_3^{2-}–HCO_3^- system that is present. Consequently, the change in PCO_2 is much less than it would be if CO_2 simply went into solution. This relationship is shown in Figure 11.7, which also illustrates the relationship between PCO_2 and the concentration of carbon dioxide (CCO_2) in distilled water or carbonate-free fresh water[14].

At a PCO_2 above 0.1–0.2 kPa, little carbonate is available to be converted to bicarbonate and practically all the CO_2 goes into solution. So, above a PCO_2 of about 0.1–0.2 kPa, the decrease in the amount of CO_2 in the water for a given decrease in its partial pressure is related to its physical solubility (α), as illustrated in Figure 11.7. However, below a PCO_2 of 0.1–0.2 kPa, the seawater line becomes curved because carbon dioxide is released from the carbonate/bicarbonate system. This results in a rapid decrease in the concentration of carbon dioxide in the water. Some natural bodies of fresh water are much more carbonated than seawater[15], so a small

[14] The influence on the increase in CO_2 in the atmosphere and acidification of the oceans is discussed in Case Study 1.1, Section 1.2.1.

[15] The presence of cations, such as Ca^{2+} in water, can lead to the formation of dissolved calcium bicarbonate, $Ca(HCO_3)_2$, which increases the total amount of CO_2 present.

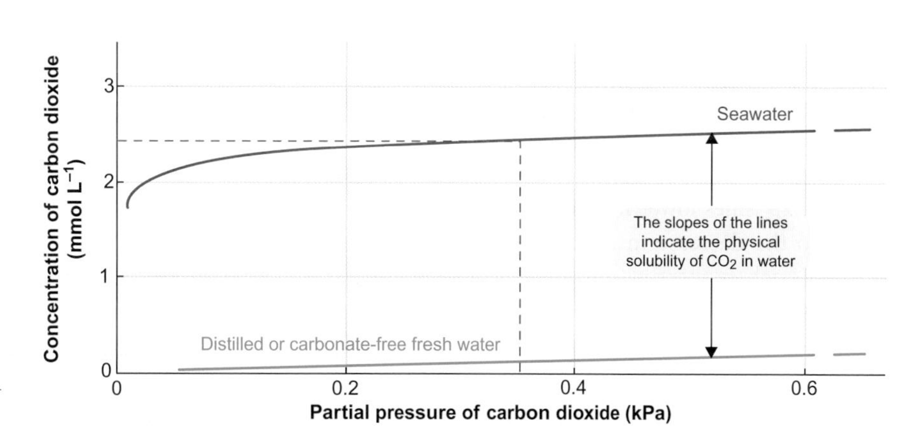

Figure 11.7 Relationships between concentrations and partial pressures of carbon dioxide in distilled or carbonate-free fresh water and in seawater at 13°C

Note the difference in the concentrations of CO_2 in seawater and distilled water at the same partial pressure of CO_2 (dashed red lines).

Adapted from: Dejours P (1981). Principles of Comparative Respiratory Physiology. 2nd edn. Amsterdam: Elsevier.

change in PCO_2 which covers the range 0 to 0.1–0.2 kPa, will cause an extraordinarily large change in CCO_2.

An example of the effect of carbonation of water on the relationship between PCO_2 and CCO_2 is found in small volumes of seawater, such as intertidal rock pools at night. The living organisms in these rock pools consume oxygen and release carbon dioxide in the water. They can reduce PO_2 from well above 20 kPa to as low as 0.2 kPa and yet PCO_2 increases from virtually 0 kPa to around only 0.35 kPa. At this partial pressure, the concentration of carbon dioxide in the seawater is approximately 2.4 mmol L^{-1} (see dashed lines in Figure 11.7). However, in carbonate-free fresh water, such a concentration of carbon dioxide would only be reached at a PCO_2 of approximately 5 kPa.

Gases in air versus gases in water

An important effect of the difference in the capacitances of oxygen and carbon dioxide in air and water is seen at the gas exchange surfaces of water- and air-breathing animals. The rate at which carbon dioxide is being produced by the cells in moles of carbon dioxide per unit time ($\dot{M}CO_2$, where the dot over the top of the 'M' signifies 'rate') divided by the rate at which oxygen is consumed by metabolizing cells of an animal ($\dot{M}O_2$), is known as the **respiratory quotient, RQ**[16]:

$$\text{Respiratory quotient, RQ} = \dot{M}CO_2 \,/\, \dot{M}O_2$$

An RQ of 1 indicates that only carbohydrates are being metabolized. RQ is more commonly about 0.85, which indicates that a combination of carbohydrates, fats and proteins are being metabolized. We assume that RQ is 1 in the following discussion.

When RQ equals 1, the amount of oxygen in the respiratory medium, air or water, is reduced by the same amount as the increase in the amount of carbon dioxide. As illustrated in Figure 11.6, the relationships between partial pressures and concentrations of oxygen and carbon dioxide in air are the same. So, with an RQ of 1, PO_2 and PCO_2 will change by the same amount. However, this is not the case in distilled or carbonate-free fresh water. For example, at 20°C the capacitance coefficient (β) of CO_2 in water is approximately 28 times higher than that of O_2, as shown in Figure 11.6 and Table 11.2.

The effect of this difference in β for CO_2 and O_2 in water is illustrated in Figure 11.8. For a water breather, a given increase in the amount of O_2 consumed and CO_2 released causes a 28 times lower increase in PCO_2 at the gills compared to the decrease in PO_2. This is not the case for air breathers, where there is an equivalent change in PCO_2 and PO_2 in the lungs.

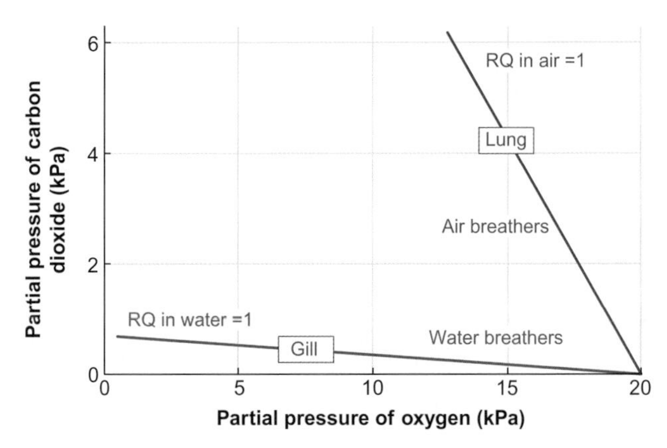

Figure 11.8 Relationship between the partial pressures of oxygen and carbon dioxide at the gills and lungs of water-breathing and air-breathing animals with a respiratory quotient (RQ) of 1

Adapted from: Rahn H (1966). Aquatic gas exchange: theory. Respiration Physiology. 1: 1-12.

This difference in the change in PCO_2 at the gas exchanger between water breathers and air breathers is also reflected in the PO_2 and PCO_2 in arterial blood of these animals. Although the values for PO_2 on the blood side of the gas exchanger in air breathers and water breathers tend to be similar (often within the range 10–15 kPa), the values of PCO_2 are much lower in those animals that only breathe water than in those that only breathe air. For example, PCO_2 in arterial blood of water-breathing molluscs, crustaceans and fish is less than 0.5 kPa, whereas in birds and mammals it is 4.5–5.5 kPa. This difference in PCO_2 in the blood between water and air breathers had an important impact on the acid-base balance of animals during their evolution from living in water to living on land[17].

As well as the respiratory gases having different physical properties in air and water, the two media themselves have different physical properties, such as differences in viscosity, density and thermal conductivity. Some of these differences, listed in Table 11.3, have influenced the evolution of animals from living in water to living in air. For example, water-breathing animals tend to have a uni-directional flow

Table 11.3 Differences in viscosity and density between air and distilled water at 20°C

	Air	Water	Ratio of air:water
Viscosity (mPa s)	0.018	1.002	1/55
Density (kg m^{-3})	1.2	998	1/832

Source: http://hyperphysics.phy-astr.gsu.edu; Kestin J et al (1978). Viscosity of liquid water in the range -8 °C to 150 °C. Journal of Physical and Chemical Reference Data 7: 941-948.

[16] We discuss RQ in more detail in Section 2.2.2.

[17] Acid–base balance in animals is discussed in Section 13.3.

of the respiratory current, whereas air-breathing animals tend to have a tidal flow. One possible advantage of this difference is that it would be energetically more costly to change the direction of flow of the more viscous and dense water than it is to change the direction of flow of air.

› *Review articles*

Canfield DE (2005). The early history of atmospheric oxygen: homage to Robert M. Garrels. Annual Review of Earth and Planetary Sciences 33: 1–36.
Graham JB, Dudley R, Aguilar NM, Gans C (1995). Implications of the late Palaeozoic oxygen pulse for physiology and evolution. Nature 375: 117–120.
Harrison JF, Kaiser A, VandenBrooks JM (2010). Atmospheric oxygen level and the evolution of insect body size. Proceedings of the Royal Society B 277: 1937–1946.

Table 11.4 **Values of Krogh's diffusion constant (*K*, nmol cm^{-1} s^{-1} kPa^{-1}) for oxygen (O_2) and carbon dioxide (CO_2) in air at 0°C, pure water and muscle at 20°C and chitin**

	Air	Water	Frog muscle	Chitin
Oxygen	78.4	0.00034	0.000105[1]	0.000010
Carbon dioxide	61.2	0.007	0.004	
Ratio of O_2:CO_2	1/0.78	1/20.6	1/38.1	

[1] Equivalent to 0.000638 μmol O_2 cm^{-1} min^{-1} atm^{-1} for Equation 11.3.
Source: Dejours P (1981). Principles of Comparative Respiratory Physiology. 2nd edn. Amsterdam: Elsevier.

11.2 Principles of gas exchange and transport

The major processes that are responsible for the exchange of the respiratory gases are diffusion and forced convection (bulk movement of air, water or blood). The factors that influence both of these processes were elucidated in the late 1800s by the German physiologist, Adolf Fick.

11.2.1 Diffusion of gases

The diffusion of a gas is the result of the random movement of its molecules, as discussed in Box 11.1, and, therefore, of the partial pressure (*P*) of the gas involved. An increase in partial pressure can be thought of as an increase in the number of collisions that molecules of a gas make with the walls of a container, as illustrated in Figure A in Box 11.1. If the walls of the container are permeable to the gas, then the increase in the number of collisions of the molecules with the wall will lead to an increase in the number of molecules passing through the wall. This process is called diffusion[18].

The data in Table 11.4 indicate that both oxygen and carbon dioxide diffuse far more rapidly through air than they do through water, and 2–3 times more rapidly through water than through tissues such as muscle. Also, carbon dioxide diffuses more rapidly than oxygen through water and muscle. This is because the rate of diffusion of a gas through a material is directly proportional to the capacitance coefficient of that gas in the material[19]. On the other hand, diffusion is faster for smaller molecules, being inversely proportional to approximately the square root of the molecular mass of the gas. However, oxygen (molecular mass 32) is only 30 per cent smaller than carbon dioxide (molecular mass 44), so this does not compensate for the much lower capacitance coefficient of oxygen in water and tissue. In contrast, the relatively smaller molecular mass of oxygen means that in air, it diffuses faster than carbon dioxide.

The concentration of a gas on either side of a permeable wall/barrier is $C = \beta \times P$, where β is the capacitance coefficient of the permeable wall and *P* is the partial pressure of the gas on either side of the barrier. The rate of diffusion across the permeable barrier is directly proportional to the difference in the concentration (*C*) of the gas on either side but is inversely proportional to the thickness of the barrier (*d*) and the proportionality constant is its **diffusion coefficient** (*D*). The diffusion coefficient is defined as the amount of a gas diffusing per unit time through a unit cross-sectional area and unit thickness of permeable barrier for a unit difference in concentration of the gas. Now, if the diffusion coefficient is multiplied by the barrier's capacitance coefficient, β, then the new constant $D \times \beta$ is called the **Krogh's diffusion constant** (*K*), named after the famous Danish physiologist, August Krogh. The rate of diffusion of the gas is thus proportional to the difference in partial pressure of the gas across the barrier and inversely proportional to the thickness of the barrier (*d*) and the proportionality constant is its Krogh's diffusion constant, *K*. The Krogh's diffusion constant for a gas (K_{gas}) represents the amount of the gas diffusing per unit time through a unit cross-sectional area and unit thickness of the gas-exchange surface for a unit difference in partial pressure of the gas across the exchange surface.

Thus, irrespective of the type of media that are present on either side of a gas-exchange surface, such as water and blood at the gills of aquatic animals or air and blood in air breathers with lungs at any given temperature, the rate at which a given amount of a gas such as oxygen, diffuses across a gas-exchange surface depends on four factors:

[18] Diffusion is discussed in general terms in Section 3.1.2.
[19] As shown in Table 11.2, at 20°C, carbon dioxide is approximately 28 times more soluble in water than oxygen.

- the difference in the average partial pressure[20] of the gas on either side of the exchange surface ($P_1O_2 - P_2O_2$);
- the area of the exchange surface (A);
- the Krogh's diffusion constant for oxygen (KO_2);
- the thickness (d) of the exchange surface.

The rate of diffusion will be faster if there is any one or more of the following:

- an increase in the difference in average partial pressure between the two sides of the exchange barrier;
- the permeability of the exchange barrier increases;
- the area of the exchange barrier increases.

Conversely, the rate of diffusion will be slower if the exchange barrier is thicker.

In terms of the rate of oxygen diffusion across an exchange surface in an animal (or rate of oxygen consumption) is concerned, the factors that are most likely to change in response to the oxygen demands of the animal are ($P_1O_2 - P_2O_2$)[21] and the effective area for gas exchange, (A)[22].

[20] Partial pressures in the respiratory medium (water or air) and in the blood change as they pass through the gas exchanger as illustrated in Figures A and B in Box 12.1, A in Experimental Panel 12.1, and Figures 12.1 and 12.29, so average values need to be calculated.

[21] Respiratory and circulatory responses to changes in oxygen demand and supply are discussed in Sections 15.2 and 15.3, respectively.

[22] The change in the effective surface area of a gas exchanger is discussed in Sections 12.2.2 and 14.4.1.

All of the factors affecting rate of diffusion of oxygen across a gas exchanger are represented diagrammatically in Figure 11.9A and the relationship between them is described by **Fick's first law of diffusion**[23]:

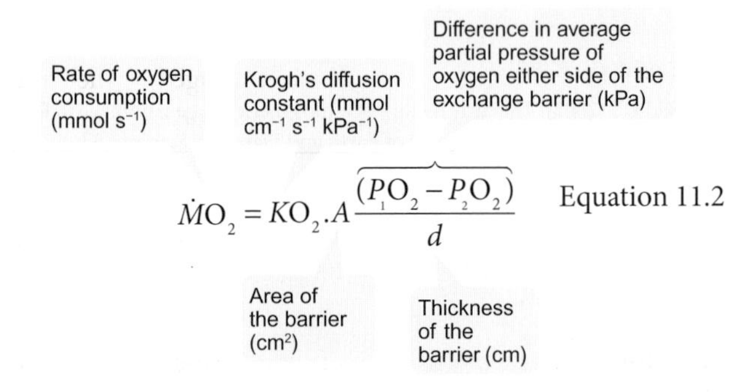

$$\dot{M}O_2 = KO_2.A\frac{(P_1O_2 - P_2O_2)}{d} \qquad \text{Equation 11.2}$$

If the barrier consists of a particular medium, then the Krogh's diffusion constant for that medium is the product between the diffusion coefficient of the gas passing through that medium multiplied by the medium's capacitance coefficient β. Values of KO_2 and KCO_2 in a number of different media are given in Table 11.4.

It is important to note that, as diffusion is dependent on the difference in partial pressure of oxygen (or carbon dioxide) across the gas exchanger, it is possible in some situations for oxygen to diffuse against a difference in **concentration** between the media. An example of this is exhibited by water

[23] Fick's law of diffusion and its similarity to Fourier's law of heat conduction are demonstrated in Sections 3.1.1 and 3.1.2.

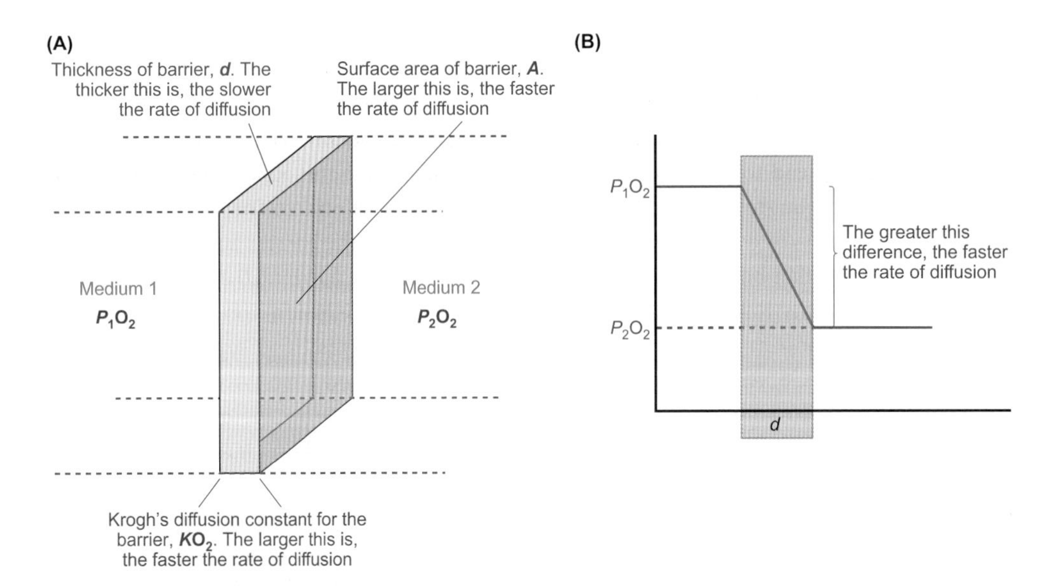

Figure 11.9 Diffusion of oxygen through a permeable barrier, such as a gas-exchange surface
(A) Diagram of factors affecting rate of diffusion of oxygen between two media, 1 and 2, through a permeable barrier of surface area A and thickness d. P is partial pressure. (B) Change in partial pressure of oxygen (PO_2) through a permeable barrier of thickness d from medium 1 to medium 2. The **difference** in partial pressure of oxygen across the barrier is ($P_1O_2 - P_2O_2$), whereas the **gradient** of the partial pressure of oxygen through the barrier is ($P_2O_2 - P_1O_2$)/d. Note, this is negative as PO_2 decreases from left to right. A: modified from Dejours P (1981). Principles of Comparative Respiratory Physiology. 2nd edn. Amsterdam: Elsevier.

and blood at the gills of fish, where the oxygen concentration of the blood is enhanced by the presence of the pigment haemoglobin. As a result, the venous blood returning from the body may have a **greater concentration** of oxygen within it than the surrounding water which ventilates the gills, but the water will have a **greater partial pressure** of oxygen. This is because the concentration of dissolved oxygen in the water is greater than the concentration of dissolved oxygen in the venous blood, where much of the oxygen is combined with haemoglobin. Only the dissolved oxygen determines the PO_2 in the blood. The greater PO_2 in the water means that oxygen will diffuse from the water into the venous blood.

An increase in the thickness of a gas-exchange barrier will slow down the rate of diffusion of a gas across it. This reduction in the rate of diffusion is because the resistance (R) to diffusion increases with the increase in thickness of the barrier. The opposite of resistance is conductance ($G = 1/R$) so, as conductance of a gas-exchange barrier increases, the rate of diffusion across it also increases.

Conductance of a gas-exchange surface can, therefore, be thought of as the ease with which O_2 or CO_2 diffuse across the skin or gills of aquatic animals, lungs of air-breathing animals, tracheoles of insects, or from the blood/haemolymph to the metabolizing cells, and is known as the **diffusion conductance** (G_{diff}). Diffusion conductance is used by respiratory physiologists to indicate the effectiveness of a gas-exchange system in an animal[24]. For example, the gills of many highly active fish, such as skipjack tuna (*Katsuwonus pelamis*), have a greater diffusion conductance than the gills of more sedentary fish, such as dogfish (*Scyliorhinus* spp.), as demonstrated by the data in Table 12.2.

Diffusion alone can only supply oxygen at the required rate for most organisms over very short distances. Krogh demonstrated this fact by using the following equation derived for the condition that the rate of oxygen diffusion into a spherical cell is greater than the rate of oxygen consumption by the cell[25]:

Partial pressure of oxygen in the surrounding medium (atmospheres – atm)

Rate of oxygen diffusion into a spherical cell – rate of oxygen consumption by the cell (μmol g^{-1} min^{-1})

Radius of the cell (cm)

$$PO_2 = \frac{\dot{M}O_2 r^2}{6KO_2} \qquad \text{Equation 11.3}$$

Krogh's diffusion constant (μmol O_2 cm^{-1} min^{-1} atm^{-1})

Table 11.4 gives KO_2 in the units μmol O_2 cm^{-1} min^{-1} atm^{-1}. Krogh found that diffusion alone would only supply the oxygen needs of an idealized spherical cell in water under certain conditions:

- the cell has a radius of less than 1 mm;
- its rate of oxygen consumption is no greater than 0.074 μmol O_2 g^{-1} min^{-1};
- the water is equilibrated with air (with a partial pressure of oxygen of 0.21 atm or 21 kPa), i.e. there is adequate movement of the water to maintain this partial pressure of oxygen next to the cell;
- the partial pressure of oxygen at the centre of the cell is zero.

If the movement of water outside the cell is inadequate, then a layer of oxygen-depleted water, a boundary layer[26], will quickly develop around the cell and the rate of diffusion of oxygen from the water will be reduced. So, even if diffusion is adequate for the transport of oxygen within the tissues of an animal, forced convection[27] must be involved to maintain the external PO_2 at a sufficiently high level to drive the diffusive movement of the gas.

Another problem with gas being exchanged across the outer surface of an animal is that the ratio between its surface area (for gas exchange) and its volume (which consumes the oxygen) gets smaller as the animal gets larger[28]. The surface area of a sphere is the smallest possible for a given volume, but Figure 11.10 illustrates that a long thin object with the same volume as a sphere has a larger surface area/volume ratio than the sphere. So a thread-like shape, as found among the nematodes and small annelids, and a flat shape, as seen in the platyhelminths (flatworms), is more beneficial for gas exchange across the general body surface than a spherical shape. There is, however, a limit to how long and thin or flat an animal can be, so many species have evolved specialized gas-exchange organs, gills or lungs, which provide a large surface area within a relatively small volume.

Despite the limitations of diffusion, a large number of small animals appear to rely on it for their gas exchange, e.g. planarians, small embryos and many eggs[29]. However, in many of these animals, there is streaming within the protoplasm inside the cell and this distributes the respiratory gases inside the animal. This internal convection current, together with a similar current maintaining a high external PO_2,

[24] Diffusion conductance is discussed further in Section 12.2.2.
[25] Note, Equation 11.3 is based on the assumption that 1 g = 1 cm^3.

[26] Boundary layer is discussed in Section 12.2.1.
[27] Forced convection is discussed in Sections 8.4.3 and 11.2.2.
[28] You can find further details on surface area to volume ratio in Section 10.2.1.
[29] We discuss diffusion across the outer surface of eggs in Section 12.3.2 and Figures 12.12 and 12.14.

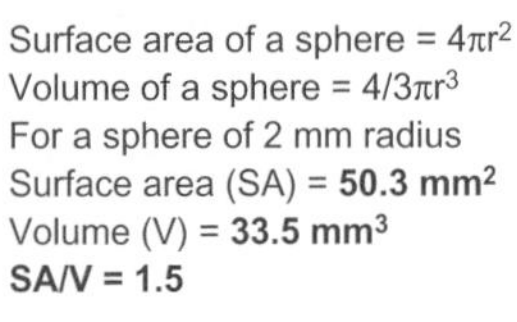

For a flat object of the same volume as a sphere of 2 mm radius and with a width of 2 mm and a thickness of 1 mm, and a length of 16.57 mm, the surface area = **104.5 mm²**

SA/V = 3.12

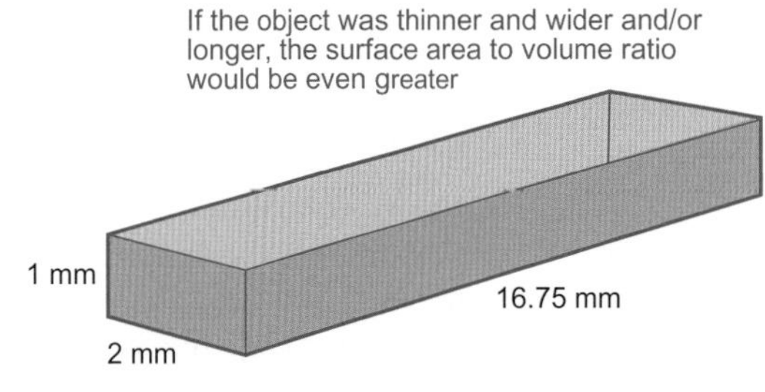

Figure 11.10 The surface area of a flat object is greater than that for a sphere, for a given volume

enhance gas transport by maintaining a large difference between internal and external PO_2 at the cell surface.

11.2.2 Forced convection

Forced convection[30] is the bulk movement of a fluid, such as water, air or blood, and usually involves the expenditure of energy by some form of pumping mechanism to generate the flow. Some organisms, such as oligochaete annelids (earthworms), lack a specialized respiratory system, while others, such as insects, do not use their circulatory system to any large extent in convective gas transport, as air-filled tubes called tracheae[31] transport the respiratory gases to and from the metabolizing cells. However, in many species of animals, forced convection is present on both sides of the gas-exchange surface and is provided in larger and/or more active animals by the respiratory and circulatory systems, as represented in Figure 11.11[32]. Figure 11.11 also shows that diffusion is important in the movement of the respiratory gases between the blood and the metabolizing cells, or from the tracheoles to the cells in insects, as well as across the gills or lungs.

We now investigate the principles of both external and internal convective transport of oxygen and carbon dioxide. The principles underlying the convective transport of the respiratory gases were first described by Adolf Fick when he devised a method to determine **cardiac output** (the volume of blood pumped from the heart per unit time, $\dot{V}_b$) in humans. Fick proposed that the rate at which oxygen is consumed as the blood passes around the body must be equal to the rate at which oxygen is taken up by the lungs from the atmosphere. In other words, the rate

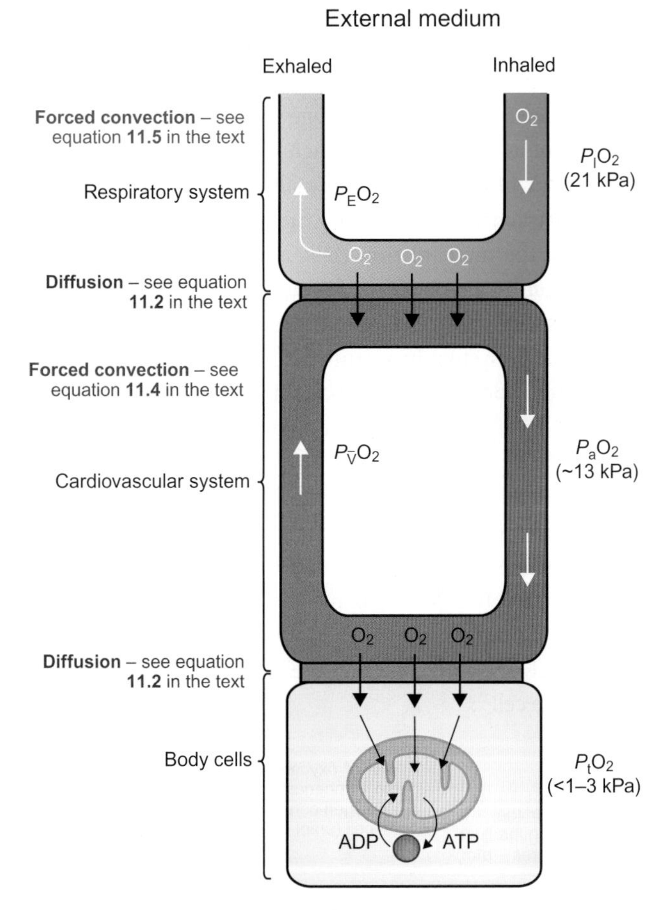

Figure 11.11 General model of the respiratory gas transport systems of animals which involve a circulatory system

The model is based on aquatic animals with gills (e.g. *Octopus*, crabs and fish), but the same principles apply to animals with lungs. Oxygen flows from the external medium (air or water) to the metabolizing cells and carbon dioxide flows in the opposite direction. Approximate partial pressures of oxygen are given for inhaled air or water (P_IO_2), arterial blood (P_aO_2) and tissue (P_tO_2). P_EO_2 is the partial pressure of oxygen in exhaled air or water and $P_{\bar{v}}O_2$ is the partial pressure of oxygen in mixed venous blood.

Adapted from: Taylor and Weibel (1981). Design of the mammalian respiratory system, 1. Problem and strategy. Respiration Physiology 44: 1-10.

[30] Forced and passive convection are discussed in Section 8.4.3.

[31] The respiratory system of insects is discussed in Section 12.4.

[32] The different types of arrangements of the respiratory and circulatory systems at the gas-exchange surfaces are discussed in Sections 12.2 and 12.3

of oxygen consumption ($\dot{M}O_2$) is equal to the amount of blood pumped around the body per unit time (cardiac output) multiplied by the amount of oxygen removed from the blood as it passes round the body, which is known as the **oxygen extraction** from the blood.

Oxygen extraction is the difference in concentration (C) of oxygen between the arterial (a) blood leaving the gas exchanger (C_aO_2), which is oxygen rich, and that about to enter the gas exchanger, which is a mixture of venous ($\bar{v}$) blood from all parts of the body ($C_{\bar{v}}O_2$), and is oxygen poor. The line over the top of the 'v' signifies that it is an average value for all the blood returning to the heart. This relationship between $\dot{M}O_2$, cardiac output and oxygen extraction is known as the **Fick principle of convection** in the cardiovascular system:

Rate of oxygen consumption (mmol min^{-1})
Cardiac output (L min^{-1})
Difference in oxygen concentration between inspired water or air and expired water or air (mmol L^{-1})

$$\dot{M}O_2 = \dot{V}_b \overbrace{(C_aO_2 - C_{\bar{v}}O_2)} \quad \text{Equation 11.4}$$

The relative importance of the different components of the cardiovascular system in delivering oxygen to the metabolizing cells in an animal at rest and during exercise is illustrated in Figure 11.12, which shows how cardiac output and oxygen extraction from the blood increase during exercise. Cardiac output ($\dot{V}_b$) is the product of heart rate (f_H) and **cardiac stroke volume** (the amount of blood ejected from the heart per beat, V_s).

The Fick principle can also be applied to the respiratory system:

Rate of oxygen consumption (mmol min^{-1})
Minute ventilation volume (L min^{-1})
Difference in oxygen concentration between inspired water or air and expired water or air (mmol L^{-1})

$$\dot{M}O_2 = \dot{V}_m \overbrace{(C_IO_2 - C_EO_2)} \quad \text{Equation 11.5}$$

Minute ventilation volume ($\dot{V}_m$) is the product of respiratory frequency (f_{resp}) and **respiratory tidal volume** (the amount of respiratory medium moved over the gills or in and out of the lungs per ventilation cycle, V_T). For an air-breathing animal, respiratory minute volume is written as $\dot{V}_a$, to signify the volume of air, whereas for a water breather, it is written as $\dot{V}_w$ to signify the volume of water.

Similar equations relate to the release of carbon dioxide at the gas-exchange surface, except that, in contrast to the situations for oxygen, the higher concentrations of carbon dioxide occur in the mixed venous blood and the expired water or air.

As we discuss in Section 11.1.2, the concentration of a gas in water, air or blood is dependent on its partial pressure (P) and its capacitance coefficient (β). However, the capacitance coefficient of oxygen in blood (β_bO_2) varies between different species of animals and can be over 20 times that in water in some species of mammals. Also, as shown back in Table 11.2, βO_2 is at least 20 times greater in air

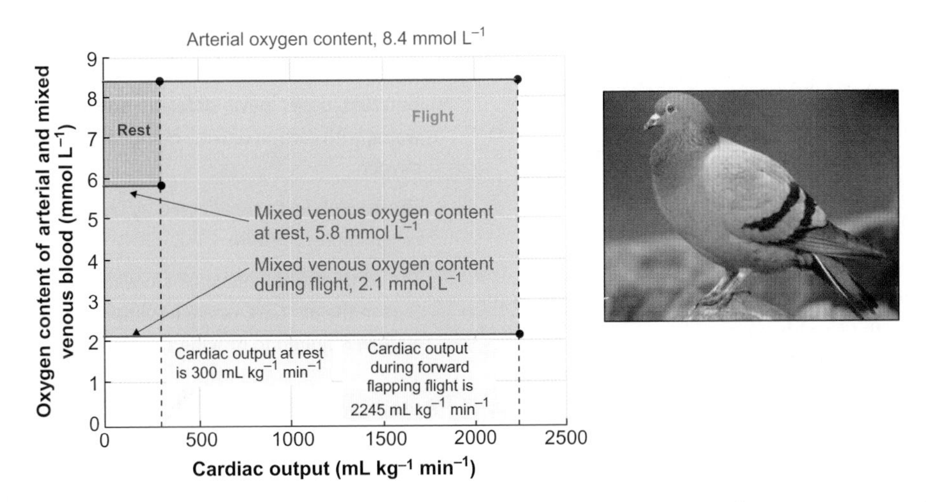

Figure 11.12 Contributions of different components of the cardiovascular system to rate of oxygen consumption in resting and flying common pigeons (*Columba livia*)

The values for cardiac output and oxygen content of arterial mixed venous blood in the resting birds support a rate of oxygen consumption of 0.8 mmol O_2 kg^{-1} min^{-1}, which is represented by the area of the pink quadrangle. The values for cardiac output and oxygen content of arterial mixed venous blood in flying birds support a rate of oxygen consumption of 13.8 mmol O_2 kg^{-1} min^{-1} which is 17 times that during rest and this is represented by the total area of the pink and green quadrangles. Source of data: Peters GW et al (2005). Cardiorespiratory adjustments of homing pigeons to steady wind tunnel flight. J. Exp. Biol. 208, 3109-3120. Image: © The Company of Biologists Limited.

(β_aO_2) than in water (β_wO_2). These differences mean that different cardiac outputs or ventilation volumes may be required for similar rates of oxygen consumption in different species of animals, depending on whether they breathe water or air[33].

The relative sequences of respiratory convection, respiratory diffusion, circulatory convection and tissue diffusion are illustrated in Figure 11.11. This figure also shows that PO_2 falls from 21 kPa in the environment (a value that is only true for water if it is equilibrated with air) to approximately 13 kPa in the arterial blood and to as low as <1–3 kPa in tissues, depending on the tissue, the closeness to a capillary and the level of demand for oxygen. It has been suggested that PO_2 in tissue needs to be so low to minimize the formation of reactive oxygen species (free radicals), which can damage membranes, proteins and DNA of cells[34]. In fact, PO_2 in tissue may still be higher than really necessary, as observations have demonstrated that oxidative metabolism within muscle cells can be maintained at a PO_2 of around 0.4 kPa.

› Review articles

Butler PJ, Metcalfe JD (1988). Cardiovascular and respiratory systems. In: Shuttleworth TJ (ed.) Physiology of Elasmobranch Fishes. pp 1–47. Berlin: Springer-Verlag.

Dejours P (1979). Oxygen demand and gas exchange. In: Wood SC, Lenfant C (eds.) Evolution of Respiratory Processes: A Comparative Approach. pp 1–49. New York: Marcel Dekker, Inc.

Wittenberg BA, Wittenberg JB (1989). Transport of oxygen in muscle. Annual Review of Physiology 51: 857–878.

[33] We investigate the relationship between $\dot{M}O_2$, $\dot{V}_b$ and $\dot{V}_m$ in water-breathing and air-breathing animals in Box 12.2.

[34] We discuss oxidative damage in Section 2.4.3.

Checklist of key concepts

- When it was originally formed, about 4.5 billion years ago, the atmosphere of the Earth contained no oxygen.
- Photosynthesizing bacteria, **cyanobacteria**, were most likely the cause of the initial increase in the concentration of oxygen, about 2.7 billion years ago.
- The subsequent changes in the concentration of oxygen in the atmosphere coincided with major changes in the evolution of organisms:
  - **gigantism**;
  - diversification of land vertebrates when oxygen levels reached 30 per cent;
  - extinctions as the level subsequently fell to about 12 per cent.
- **Aerobic metabolism** produces about 15 times more ATP per mole of glucose than **anaerobic metabolism**.

Gases are compressible and have different properties in air and water

- The relationship between pressure, volume and temperature of gases is described by the **combined gas law**.
- Volumes of gases measured at different temperatures and pressures should be converted to the volumes occupied by the gases under standard conditions: when **dry** and at **standard temperature (273 K) and pressure (101.3 kPa—STPD)**.
- It is possible to convert the volume of a gas at STPD to its molar equivalent (mol).
- Each gas in the atmosphere, including water vapour, exerts is own pressure, known as its **partial pressure**.
- The partial pressures of all the gases in the atmosphere contribute to **total atmospheric pressure**.
- The proportion of oxygen to nitrogen in the atmosphere is **constant** up to an altitude of approximately 100 km above the Earth's surface.
- The **molar concentrations** of oxygen and nitrogen change with altitude.
- Atmospheric pressure varies with different meteorological conditions, and decreases as altitude above sea level increases.
- The partial pressure of each gas in the atmosphere, including oxygen, mirrors the change in atmospheric pressure.
- Gases behave differently in water and air, with profound consequences for the evolution of animals.
- Different gases have different **solubilities (capacitances)** in water, with carbon dioxide being much more soluble in water than oxygen.
- These solubilities are affected by factors, such as temperature, and other solutes present.
- The concentration of oxygen in water equilibrated with air is lower than in the air itself, which has implications for the rate of water or air flow required to provide sufficient oxygen to the animal.

The transport of respiratory gases between an animal and its environment

- Metabolizing cells obtain oxygen from the environment and lose carbon dioxide to the environment via **passive diffusion** and **forced convection**.
- Forced convection includes:
  - the active pumping of water or air from the environment over the gas-exchange surfaces—**body surface, gills or lungs**—by a **ventilatory system**;

- in many species of animals, the pumping of blood from the gas-exchange surface to the metabolizing cells by a **circulatory system**.

- The rate of oxygen delivery and carbon dioxide removal depends on the rate of output from the pumps (**ventilation volume** or **cardiac output**) and the amount of oxygen extracted from the ventilatory medium or from the blood (the **oxygen extraction**).
- The low solubility of oxygen in water is overcome in animals by the presence of a **respiratory blood pigment**, mainly haemoglobin or haemocyanin, which enhances the **oxygen carrying capacity** of the blood.
- Some species, such as insects, possess a system of air-filled tubes called **tracheae**, which transport the respiratory gases directly to and from the cells.
- The passive diffusion of gases across the exchange surface is driven by a **difference in partial pressure** of the respiratory gases across the gills, the lungs, or at the metabolizing cells.
- Other factors which have a **positive effect** on the rate of diffusion across the gas-exchange surface are:
  - area of the surface;
  - diffusion characteristics of the surface as defined by **Krogh's diffusion constant**.
- The thickness of the surface has a **negative effect** on the rate of diffusion.
- More active animals tend to have gas exchangers that are thinner and have greater surface areas.

Study Questions

*** Answers to these numerical questions are available online. Go to www.oup.com/uk/butler**

1. What is a possible explanation for the evolution of gigantic insects approximately 300 million years ago? (Hint: Section 11.1.1.)
2.* The amount of oxygen consumed in 1 hour by a 400 g fish resting at 15°C was measured as 51 mL. Barometric pressure was 99 kPa. Convert this to the volume the oxygen would occupy at standard conditions of pressure and temperature. What would the amount of the gas be in mmol? (Hint: Section 11.1.2.)
3. What is the difference between solubility coefficient and capacitance coefficient of a respiratory gas in water or blood? (Hint: Section 11.1.2.)
4. Explain why, in general, only the smallest of animals are able to rely on diffusion alone to support their oxygen requirements, but how certain shapes have enabled some species to be larger yet still rely on diffusion. (Hint: Section 11.2.1.)
5. Using your knowledge of the equations that describe the factors affecting diffusion, explain the morphological differences you would expect to find in the lungs of two species of mammals, one that can run very fast and one that that is more sedentary. (Hint: Section 11.2.1.)
6.* A small spherical aquatic organism has an oxygen consumption of 0.05 μmol g^{-1} min^{-1}. The external partial pressure of oxygen is 21 kPa and that in the centre of the organism is 0 kPa. If Krogh's diffusion constant for oxygen (K_{O2}) for the organism is 0.000638 μmol O_2 cm^{-1} min^{-1} atm^{-1}, what is its diameter? (Hint: Section 11.2.1.)
7.* A 1.2 kg rainbow trout when swimming in water at 10°C had an oxygen consumption of 234 μmol min^{-1}, a concentration of oxygen in its arterial blood of 4.3 mmol L^{-1}, a concentration of oxygen in its mixed venous blood of 0.6 mmol L^{-1}, and a heart rate of 51 beats min^{-1}. Calculate its cardiac output in mL min^{-1} and cardiac stroke volume in mL. (Hint: Section 11.2.2.)
8.* The same fish as described in question 6 had a ventilation volume of 2.04 L min^{-1}, and the partial pressure of oxygen in the inspired water was 20.2 kPa. If the capacitance coefficient of water for oxygen at 10°C is 16.8 μmol O_2 L^{-1} kPa^{-1}, calculate the partial pressure of oxygen in the expired water. (Hint: Section 11.2.2.)

Bibliography

Dejours P (1981). Principles of Comparative Respiratory Physiology 2nd edn. Elsevier/Nort Holland, Amsterdam.

12

Respiratory systems

In Chapter 11, we discussed the physical characteristics of water and air and the basic principles involved in the exchange of respiratory gases between metabolizing cells and the environment. With the exception of some groups of arthropods, such as insects, which exchange the respiratory gases directly between the environment and the metabolizing cells, there are usually two systems involved in gaseous exchange: the respiratory and cardiovascular systems. We discuss in this chapter how the respiratory systems have evolved to cope with the different characteristics of the environments in which the animals live: water and air. We discuss circulatory systems in Chapter 14.

The major organs of gas exchange in the respiratory systems of aquatic animals are outgrowths (evaginations) of the body surface, the **gills**. In contrast, the gas-exchange organs of terrestrial animals are ingrowths (invaginations), the **lungs**. The body surface may also exchange gases, particularly in aquatic animals. The respiratory gases pass across respiratory surfaces by diffusion, so all of these organs need to have a sufficiently large surface area, a sufficiently thin diffusion distance and a sufficiently large difference in partial pressure of oxygen across the respiratory surface to enable the organism's maximum demand for oxygen to be met[1].

12.1 Types of gas exchanger

The evolution of gas-exchange organs accompanied the evolution of larger animals. Larger animals have a relatively smaller surface area than smaller animals. In other words, for animals of similar shape, the surface area/volume ratio decreases with increasing size[2]. The surface area can be increased with no change in volume by an animal being long and thin. In general, however, diffusion alone can only supply the oxygen requirements of very small animals[1]. Most larger animals have evolved specialized gas-exchange organs which have very large surface area to volume ratios because of their intricate shapes. A few species, however, rely to a greater or lesser extent on their body surface to supply their oxygen requirements and to remove carbon dioxide.

There are four basic types of gas-exchange surface:

- The simplest is the skin, because there is no specific pumping mechanism to maintain a regular flow of water or air over it. This is called an **infinite pool gas exchanger** and is the least effective[3] type.
- In the gills of most aquatic animals, the flow of water and blood are in opposite directions, so these are known as **countercurrent gas exchangers**. These are potentially the most effective of the four types of gas exchanger.
- In all air-breathing vertebrates, except birds, and in some species of invertebrates, air is pumped in and out of the sac-like lungs. These are known as **ventilated pool gas exchangers**.
- The respiratory system of birds consists not only of lungs, but also of a number of air sacs connected to the lungs. Air flows through the major tubes of the lungs (parabronchi) in the same direction into and out of the air sacs during both phases of ventilation. The parabronchi are at right angles to the corresponding blood vessels. The lungs of birds are examples of **cross-current gas exchangers**. The potential effectiveness of bird lungs lies between that of countercurrent gills and ventilated pool lungs.

In this chapter we consider the characteristics of the four different types of gas exchanger, before exploring the solution to the problem of gas transport that has evolved in some arthropods such as insects.

> *Review article*

Scheid P, Piiper J (1982). Models for a comparative functional analysis of gas exchange organs in vertebrates. Journal of Applied Physiology 53: 1321–1329.

[1] Section 11.2.1 discusses the factors affecting diffusion across the respiratory surfaces.

[2] Surface area/volume ratio is discussed in Section 10.2.1.

[3] Effectiveness of gas transfer at the gas exchange organ is discussed in Box 12.1.

12.2 Gas exchange in water

The vast majority of aquatic animals are **ectotherms**[4]. The body temperature of an ectotherm in water is essentially the same as that of its environment. The metabolic rate, and hence the oxygen requirements of ectotherms, depend on the intensity of their activity[5] and on water temperature. As we discuss in Section 11.1.2, the capacitance coefficients of oxygen and carbon dioxide in water decrease with increasing temperature and with increasing concentration of solutes. So, a given volume of warm water has less oxygen dissolved in it than a similar volume of colder water, and a given volume of seawater has less oxygen dissolved in it than a similar volume of fresh water at the same temperature.

In addition, in some bodies of water, there may not be complete equilibration of dissolved gases with the atmosphere, so the partial pressure of oxygen (PO_2) may be reduced. Consequently, the water may have a lower PO_2 than that of the atmosphere; it will be **hypoxic** (*hypo*, Greek—under, below). Conversely, photosynthesis by aquatic plants and algae may increase PO_2 to more than that in the atmosphere. The water is then said to be **hyperoxic** (*hyper*, Greek—over, above) or supersaturated. We now look at how the gas-exchange organs have evolved to meet the variations in oxygen supply and demand. We begin with the body surface.

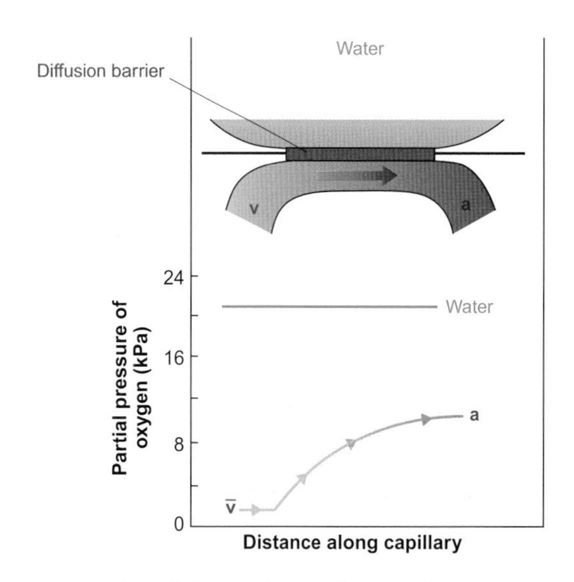

Figure 12.1 The infinite pool type of gas exchanger

Gases diffuse across skin perfused with blood. There is no ordered flow of the external medium over the gas exchange surface, so there is no inspired or expired water. As the mixed venous ($\bar{v}$) blood passes through the capillaries in the skin, the partial pressure of oxygen (PO_2) increases but, because of the large resistance to diffusion, partial pressure of oxygen in the arterial (a) blood does not reach that in the outside water. In other words, the rate of transport of oxygen is diffusion limited.

Adapted from Piiper J and Scheid P (1982). Models for a comparative functional analysis of gas exchange organs in vertebrates. Journal of Applied Physiology 53: 1321-1329

12.2.1 Gas exchange across the body surface: an infinite pool gas exchanger

Gas exchange across the body surface (called cutaneous gas exchange) is important for virtually all aquatic animals, except aquatic birds and mammals. As blood passes through the skin, the rate at which the blood pigment picks up oxygen by diffusion decreases. This is because PO_2 in the blood increases as it passes through the capillaries and the difference in PO_2 between the blood and the water decreases, as illustrated in Figure 12.1. As the difference in PO_2 between the blood and the water declines, so does the rate of diffusion[6].

Because the skin is exposed to the environment, it has to be tougher, and often thicker, than the exchange surfaces of a lung or a gill, which in most species are protected within a special cavity. Thicker tissues slow down the rate of diffusion, so the rate at which oxygen diffuses through the skin for a given difference in partial pressure is less than it would be in gills and lungs. The result of the lower rate of diffusion is that PO_2 in the arterial blood (P_aO_2) does not reach that in the surrounding environment during the time the blood passes through the capillaries in the skin, as shown in Figure 12.1. The rate of transport of oxygen across the skin is, therefore, said to be **diffusion limited**.

This limitation in diffusion is reduced and gas exchange is enhanced in some species by a reduction in the distance between the outside of the skin and the underlying blood vessels and an increase in the skin's surface area. Examples of such species are hellbenders (*Cryptobranchus alleganiensis*), which are large aquatic salamanders with no gills and poorly functioning lungs, and Lake Titicaca frogs (*Telamatobius culeus*), which live at an altitude of over 3800 m (where PO_2 is approximately 12.5 kPa[7]) and have never been seen to surface for air. In these species, the skin has a high density of blood capillaries close to its surface and its area is increased by a number of folds or flaps.

Even though animals that predominantly use their skin for gas exchange have no specific ventilatory system, movements of the animal and the natural flow of water over the skin increase the rate of gas exchange. When there is no flow of water across the skin, a layer of water with low PO_2 builds up immediately next to the skin. Such a layer of stagnant water adds to the resistance to diffusion of oxygen from the

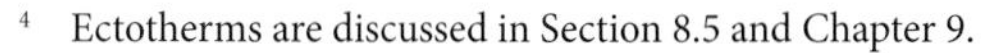

[4] Ectotherms are discussed in Section 8.5 and Chapter 9.

[5] Exercise in ectotherms is discussed in Section 15.2.2.

[6] Diffusion is discussed in Section 11.2.1.

[7] in the atmosphere at sea level is approximately 21 kPa.

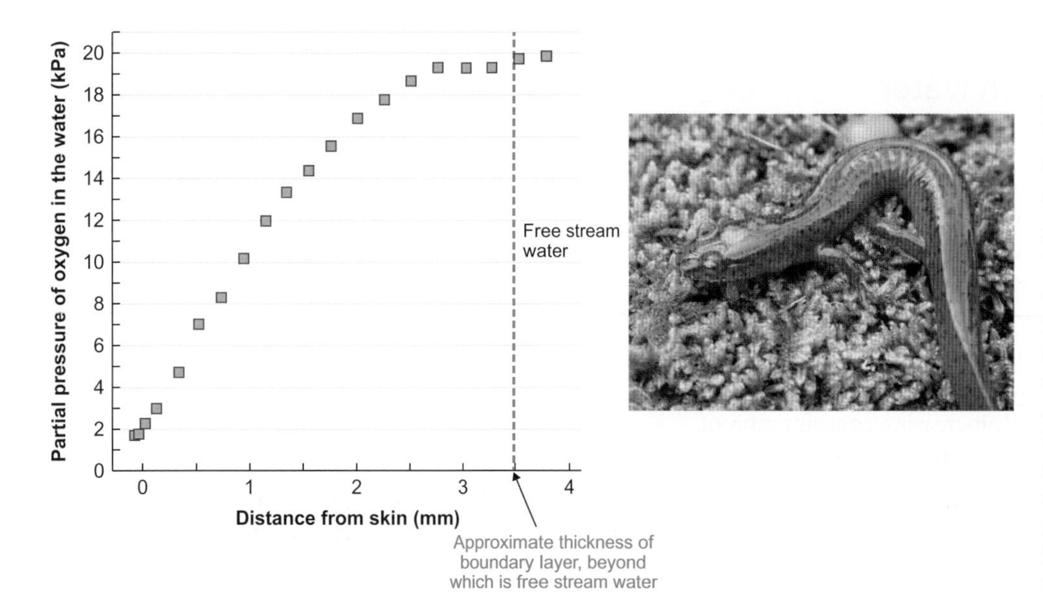

Figure 12.2 Profile of the partial pressure of oxygen in the boundary layer around the skin of a spring salamander (*Gyrinophilus porphyriticus*)

This amphibian relies entirely on its skin for gas exchange as it has no gills or lungs. Measurements were made in animals submerged in unstirred, normoxic water at 20°C.

Adapted from Feder ME and Booth DT (1992). Hypoxic boundary layers surrounding skin-breathing aquatic amphibians: occurrence, consequences and organismal responses. Journal of Experimental Biology 166: 237-251. Photo © John D. Willson

water to the blood and is known as a boundary layer[8]. The thickness of a boundary layer is defined as the distance from the respiratory surface to the point where the partial pressure of oxygen is 99 per cent of that in the free stream water, as illustrated in Figure 12.2.

Boundary layers have been studied by using very small oxygen electrodes to measure PO_2 in the water at increasing distances from the skin of a motionless lungless salamander (*Gyrinophilus porphyriticus*). Figure 12.2 shows that PO_2 is only 1–2 kPa at the interface between the surface of the skin and the water, compared with about 20 kPa toward the edge of the boundary layer and in the free stream water. The low value of PO_2 in the water next to the skin determines the maximum difference in partial pressure that drives diffusion of oxygen through the skin. However, the thickness of the boundary layer can be reduced and PO_2 next to the skin increased by movements of the animal if it is in still water or by movements of the water itself.

The effects of disrupting the boundary layer are shown in Figure 12.3. In motionless American bullfrogs (*Lithobates catesbeianus*, formerly, *Rana catesbeiana*), an increase in the velocity of water flow over the skin causes a decrease in the thickness of the boundary layer, as shown in Figure 12.3A, and an increase in PO_2 at the skin/water interface and in arterial blood, which is illustrated in Figure 12.3B. Increased rate of water flow also increases the rate of oxygen uptake ($\dot{M}O_2$)[9] across the skin of a motionless frog. However, when the frog moves, a boundary layer hindering diffusion does not develop, so there is no effect of water velocity over the skin on $\dot{M}O_2$, as shown in Figure 12.3C.

[8] Boundary layers are discussed in Section 8.4.3.

[9] $\dot{M}O_2$ is the rate of oxygen consumption in mol per unit time, as discussed in Section 11.1.2.

Movement of an animal that relies on its skin for the uptake of oxygen can be very important. Even at water flow rates that are much higher than those expected in ponds or lakes, the boundary layer still has a substantial effect on the rate of diffusion of oxygen. At the relatively high water flow rate of 5 cm s^{-1} across the skin of a motionless animal, 35 per cent of the total resistance to the diffusion of oxygen from the water can be due to the boundary layer. This rises to over 90 per cent at a water flow rate of 0.1 cm s^{-1}, although movements of the animal, even at a rate of once per minute, can prevent a boundary layer developing and so can maintain oxygen uptake in still water.

Relative importance of gas exchange across the skin

Oxygen uptake across the skin may not provide any more oxygen than the skin itself uses. This seems to be the case in many species of fish, although in some, such as channel catfish (*Ictalurus punctatus*) and flounder (*Platichthys flesus*), more oxygen diffuses across the skin than is consumed by it. Whether or not the oxygen is used only by the skin, it still contributes to the animal's overall oxygen requirement. Even in some species of air-breathing fish, amphibians and aquatic reptiles, the skin plays an important role in gas exchange between the animal and water, particularly at low temperatures, when demand for oxygen is reduced.

Some examples of the percentage of total rates of oxygen uptake and carbon dioxide elimination that occur across the skin of some species of amphibians and reptiles are given in Table 12.1. At higher temperatures, the relative importance of the skin for the uptake of oxygen tends to be reduced, although the skin may still be the more important route for the elimination of carbon dioxide. This is because carbon

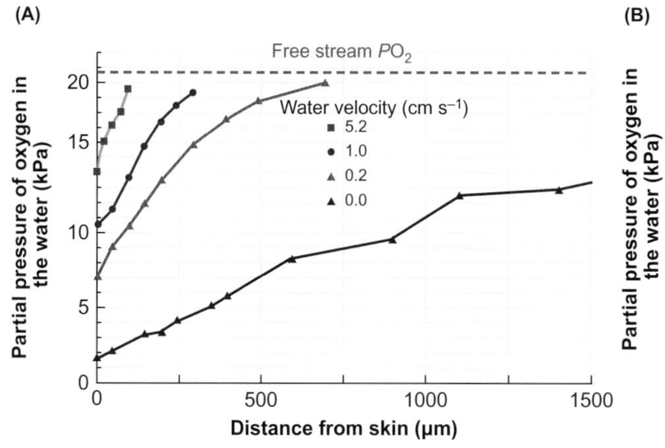

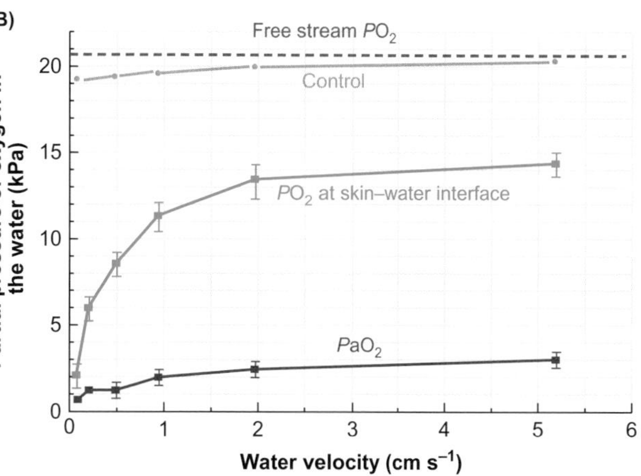

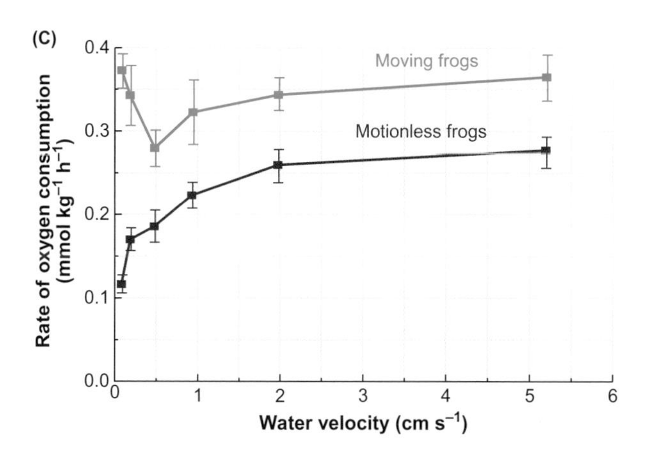

Figure 12.3 Effects of boundary layer on oxygen uptake of American bullfrogs (*Lithobates catesbeianus*, formerly *Rana catesbeiana*) submerged in water at 5°C

(A) With an increase in the velocity of water flowing over the skin, there is a reduction in the thickness of the boundary layer from well over 1.5 mm at no flow to approximately 100 μm at a flow of 5.2 cm s^{-1}. The values for no water flow (unstirred layer) were obtained after 1 hour of no water movement. The free stream is the water beyond the boundary layer. (B) There is an increase in the partial pressure of oxygen (PO_2) at the skin–water interface and in arterial blood (P_aO_2) with increasing water velocity. The line labelled 'Control' is for PO_2 measured next to a clean rubber stopper. (C) An increase in the rate of oxygen consumption occurs with increasing rate of water flow over the skin in motionless bullfrogs, but there is little or no effect on oxygen consumption in moving bullfrogs. Data in B and C are mean values ± standard error of the mean for seven bullfrogs.

Adapted from Pinder AW and Feder ME (1990). Effect of boundary layers on cutaneous gas exchange. Journal of Experimental Biology 143: 67-80.

Table 12.1 The relative importance of the skin for the exchange of oxygen and carbon dioxide in some species of amphibians and reptiles in water at different temperatures

Species	Temperature (°C)	% oxygen uptake	% of carbon dioxide released
Two-toed amphiuma (*Amphiuma means*)	5	92	100
	25	44	77
Hellbender (*Cryptobranchus alleganiensis*)	5	94	97
	25	92	98
Northern leopard frog (*Lithobates pipiens*)	5	53	76
	25	34	58
Red-eared slider (*Trachemys scripta elegans*)	20		10
Smooth soft-shelled turtle (*Apalone mutica*)	20		64
Yellow-bellied sea snake (*Hydrophis platura*)	30	12	
Elephant trunk snake (*Achrochordus javanicus*)	20–22	7	33

Sources: Guimond RW and Hutchison VH (1974). Resp Physiol 20: 147-159; Guimond RW and Hutchison VH (1973). Science 182: 1263-1265; Guimond RW and Hutchison VH (1968). Comp Biochem Physiol 27: 177-195; Jackson DC et al (1976). Comp Biochem Physiol 55A: 243-246; Graham JB (1974). Resp Physiol 21: 1-7; Standaert T and Johansen K (1974). J Comp Physiol 89: 313-320.

dioxide has a greater capacitance coefficient and rate of diffusion in water than oxygen[10].

It was thought for many years that gas exchange across the skin is a passive, uncontrolled process that is mainly limited by diffusion, rather than by the rate of perfusion of the skin by blood. If this was the case, the rate of oxygen uptake across the skin would be determined by the difference in PO_2 across it. This difference could be maximized by reducing the boundary layer, as we discuss earlier in this section, and by reducing PO_2 in the blood perfusing the skin. However, it is now apparent that another way of controlling the rate of oxygen consumption across the skin is by varying the rate of blood flowing through it, and there is evidence that this occurs in species of lungless salamanders that rely entirely on their skin for gas exchange.

Lungless salamanders are able to rapidly increase their rate of oxygen consumption at the onset of exercise. If cutaneous gas exchange was diffusion limited, the increase in the rate of oxygen consumption at the beginning of exercise would be relatively slow, as it would rely on a decrease in

[10] Diffusion rates of CO_2 and O_2 in water are given in Table 11.4.

the PO_2 of the blood perfusing the skin. However, the actual response of the salamanders at the onset of exercise suggests that they are able to control gas exchange across their skin. Data from other species of amphibians, such as American bullfrogs and northern leopard frogs (*Lithobates pipiens*), indicate how they may achieve this; they change the microcirculation of the skin.

The rate of blood flow through the microcirculation of the skin can be controlled by closing and opening different capillary beds and by changing the rate of blood flow through already open capillaries. These changes can influence the rate of gas exchange across the skin. For example, when a frog is at rest in aerated water, some capillary beds in the skin may not be perfused with blood. However, when the animal starts to exercise or during hypoxia, more capillary beds may be perfused. This increased perfusion increases the effective surface area of the skin for gas exchange. This means that, although the actual surface area of the skin cannot be varied, its functional surface area can.

Although gas exchange across the skin of animals that live in water can be significant, most animals rely on specific gas exchange organs such as gills, lungs or tracheae[11]. In species that undergo metamorphosis during development, such as fish and amphibians, there is a progression from gas exchange occurring largely across the skin in the larval stage to gas exchange across gills and/or lungs in adults. Such metamorphoses have been studied in detail in American bullfrogs and in some species of fish. The importance of the skin in gas exchange decreases and the lungs become more important as a bullfrog develops from a tadpole to an adult, as shown in Figure 12.4.

The situation is simpler in chinook salmon (*Oncorhynchus tshawytscha*). In larvae of the fish, uptake of oxygen across the skin quickly decreases from over 80 per cent of the total when they hatch to approximately 30 per cent by the time the fish weighs 1 g. So the gills play the dominant role in gas exchange even in very small fish, largely as a result of the unfavourable surface area/volume ratio of the body of the fish.

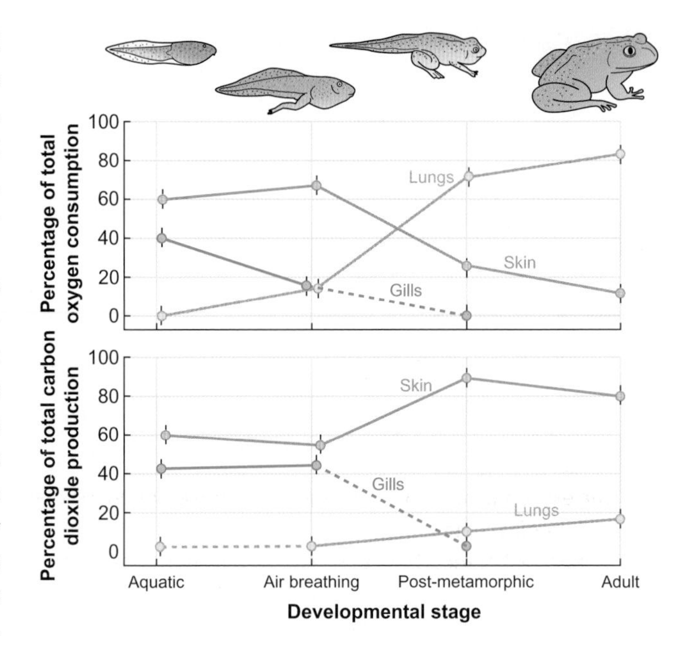

Figure 12.4 Relative importance of the skin, gills and lungs for gas exchange in developing American bullfrogs (*Lithobates catesbeianus*)

In tadpoles, the skin and the gills are responsible for all gas exchange. By the time the animal begins to breathe air, the importance of the lungs in oxygen uptake increases, while that of the gills decreases. The skin maintains its level of importance. Carbon dioxide continues to be lost mainly across the skin and gills. By the time the animals have metamorphosed into adults and the gills have disappeared, most of the oxygen uptake is across the lungs and the remainder is across the skin. Values are means ± standard error for 5 to 8 animals at each stage of development. Approximate developmental stages are shown.

Adapted from Burggren WW and West NH (1982). Changing respiratory importance of gills, lungs and skin during metamorphosis in the bullfrog Rana catesbeiana. Respiration Physiology 47: 151-164.

12.2.2 Gas exchange across gills: countercurrent gas exchangers

Gills are extensions of the body wall and may be relatively simple finger-like extensions into the water, as in many species of annelids, such as lugworms (*Arenicola* spp.), which live in U-shaped burrows[12] in the intertidal regions of the seashore. The surface area of gills in some species may be greatly increased by the presence of highly branched filaments or by a series of flattened plates, which are stacked upon each other. These gills are generally contained within a special chamber which has to be ventilated. Ventilation can be achieved by cilia on the gills themselves, by a special water-pumping apparatus, or by passive ventilation as the animal moves through the water. Different animal groups have evolved different structures that work in slightly different ways. To illustrate these differences, this section looks at the gills and their ventilation in different groups of animals.

Gills of molluscs, crustaceans and fish

In molluscs, decapod crustaceans and fish, the gills are protected within a special cavity. which allows the water to flow in one direction throughout the respiratory cycle[13]. Generally, the blood also flows in one direction through the gas exchange region of the gills of these animals. Water and blood may flow in the same direction (co-current), or in opposite directions (**countercurrent**). The advantages and limitations of a countercurrent system are discussed in Box 12.1.

[11] Gas exchange in aquatic insects is discussed in Section 12.4.2.

[12] A diagram of a lugworm in its burrow is shown in Figure A, Case Study 20.1.

[13] Notice the direction of water flow in Figures 12.5B and 12.6B.

Box 12.1 Countercurrent gas exchangers

In this box, we consider the ideal characteristics of gas exchangers and to what extent they are achieved in the countercurrent type of exchangers found in the gills of many species of aquatic animals.

The ideal conditions for gas exchange at any gas-exchange surface would be:

- no barrier to diffusion;
- for all the external medium ventilating the gas-exchange surface and all the blood perfusing it to be fully involved in the gas-exchange process;
- for there to be complete separation of the inhaled medium from the exhaled medium and of oxygen-rich blood leaving the exchanger from the oxygen-poor blood entering the exchanger from the body;
- a perfect balance between the rate at which oxygen is presented to the gas-exchange surface by the ventilatory system and the rate at which it is removed by the circulation. This means that for a water-breathing animal, the product of rate of water flow over the gills (minute ventilation volume) and the amount of oxygen per given volume of water (the capacitance coefficient[1] for oxygen in water) should equal the product of the rate of blood flow through the gills (cardiac output) and the amount of oxygen per given volume of blood (the capacitance coefficient for oxygen in blood):

Minute ventilation volume (L min^{-1})

Capacitance coefficient for oxygen in water (μmol L^{-1} kPa^{-1})

$$\dot{V}_w \times \beta_w O_2 = \dot{V}_b \times \beta_b O_2 \qquad \text{Equation A}$$

Cardiac output (L min^{-1})

Capacitance coefficient for oxygen in blood (μmol L^{-1} kPa^{-1})

$\dot{V}_w \times \beta_w O_2$ and $\dot{V}_b \times \beta_b O_2$ are known as the **convective conductances** of the ventilatory system and circulatory system, respectively, and are indicators of the ease with which these two systems transport oxygen.

Because of the presence of a respiratory pigment in blood or haemocyanin, the capacitance coefficient of blood/haemocyanin for oxygen ($\beta_b O_2$) is usually greater than that of water ($\beta_w O_2$). This means that for the convective conductances to be matched under ideal conditions, ventilation volume per unit time should be greater than cardiac output, and the difference should be in proportion to the difference in the oxygen capacitances of the blood and water. For example, with a capacitance coefficient of blood that is 10 times greater than that of water, the ventilation volume per unit time in a fish should, ideally, be 10 times greater than cardiac output.

Under ideal conditions in which water and blood flow in the same direction (i.e. flow is co-current) the partial pressure of oxygen in arterial blood (P_aO_2) and in the water leaving the gills (P_EO_2) would be identical, as shown Figure A(i).

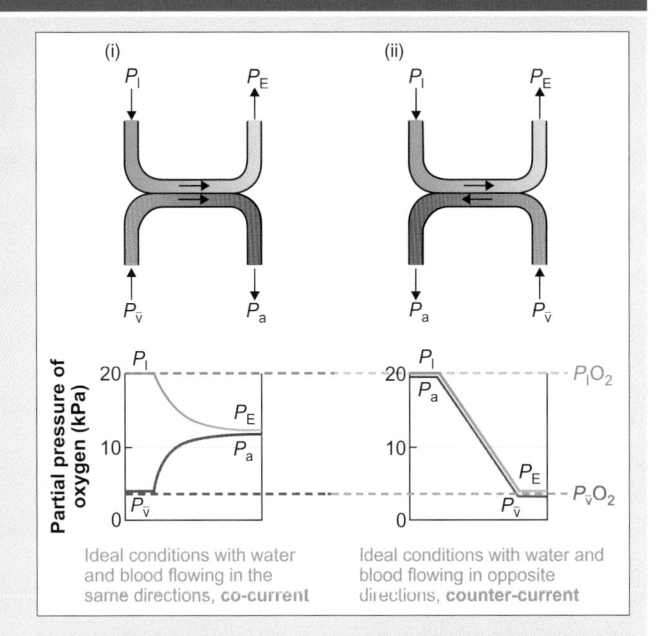

Figure A Gas exchange across gills perfused with blood under ideal conditions

The diagrams and graphs show the partial pressures of oxygen in inhaled and exhaled water (P_IO_2 and P_EO_2 respectively) and in arterial and mixed venous blood (P_aO_2 and P_vO_2, respectively) as they pass through the gills. (i) With the flows in the same direction, P_EO_2 and P_aO_2 are the same, whereas with the flows in opposite directions (ii), P_aO_2 is the same as P_IO_2 and $P_{\bar{v}}O_2$ is the same as P_EO_2. The countercurrent arrangement is, therefore, potentially much more effective at gas exchange than the co-current system.

Adapted from Butler PJ (1976). Gas exchange. In Environmental Physiology of Animals, eds, Bligh J et al. Blackwell Scientific Publications, Oxford.

In many gills, blood and water flow in opposite directions (countercurrent). In this situation, mixed venous blood that is low in oxygen (that is, it is returning from the body) is continuously 'meeting' incoming water from the environment, which is depicted in Figure A(ii). Under ideal conditions in a countercurrent arrangement, P_aO_2 would be the same as PO_2 in the inhalent water (P_IO_2), and P_EO_2 would be the same as PO_2 of the mixed venous blood arriving at the gas-exchange surface from the body ($P_{\bar{v}}O_2$). In other words, in an ideal co-current system, the maximum proportion of the oxygen that could be transferred is represented by 50 per cent of the difference between P_IO_2 and $P_{\bar{v}}O_2$, whereas in an ideal countercurrent system it is represented by 100 per cent of this difference, as indicated in Figure A. So, the countercurrent system is said to be 100 per cent effective. To obtain the actual amounts of oxygen transferred from the water, the partial pressures need to be multiplied by the capacitance coefficient of oxygen in water ($\beta_w O_2$) as shown in equation B.

The **effectiveness** of the removal of oxygen from the respiratory medium (water or air) by a gas exchanger of an animal is the ratio between actual rate of oxygen transfer of the gas by the medium (in effect, rate of oxygen uptake as indicated in

Equation 11.5) and the maximum possible rate of transfer. For an aquatic animal, this would be:

Difference between partial pressure in inspired and expired water (kPa) – the actual rate of oxygen transfer

$$\text{Effectiveness} = \frac{\dot{V}_w \times \beta_w O_2 (P_I - P_E)}{\dot{V}_w \times \beta_w O_2 (P_I - P_{\bar{v}})} \quad \text{Equation B}$$

Difference between partial pressure in inspired water and mixed venous blood (kPa) – the maximum possible rate of oxygen transfer

A similar equation can be used to calculate the effectiveness of the delivery of oxygen to the blood by a gas exchanger.

Real conditions for gas exchange

In reality, ideal conditions are never achieved: there will always be a barrier to diffusion, both in the water and in the tissue, as shown in Figure B(i), and the two convection conductances may not always be perfectly matched. In addition, the water and/or blood that leave the gas-exchange area, may include some that has not passed the exchange surface at all, as shown in Figure B(ii), and/or some that is too far away from the exchange surface to have fully participated in gas exchange as shown in Figure B(iii). These are known as **shunts** and effectively 'dilute' the oxygen-rich blood that has undergone gas exchange with oxygen-poor blood that has not undergone gas exchange. Similarly, the shunts enhance the level of oxygen in expired water. The diffusion barrier and shunts reduce the overall effectiveness of a countercurrent exchanger so that P_aO_2 is less than P_IO_2, as shown in Figure B.

As a rule of thumb, the presence of a functional countercurrent-exchange system is demonstrated if P_aO_2 is greater than P_EO_2. Figure A(i) shows that a co-current system could never achieve this level of oxygen exchange. In fact, it is not always achieved, even in those species in which the anatomical basis for a countercurrent-exchange system exists, as shown by the data in Table 12.2.

[1] Capacitance coefficient is discussed in Section 11.1.2.

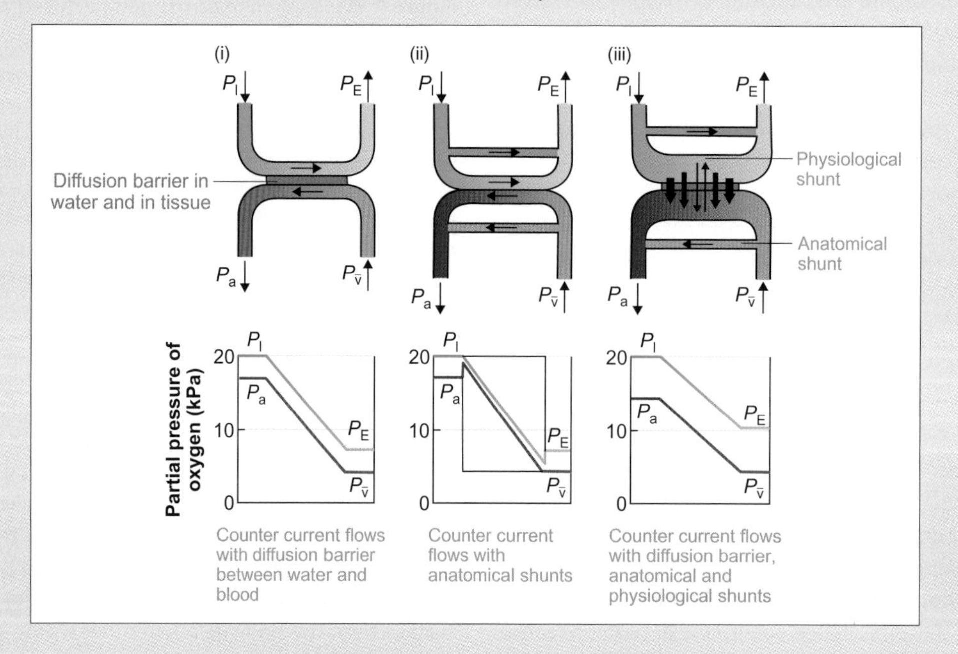

Figure B Gas exchange across gills perfused with blood under realistic conditions

The diagrams and graphs show the partial pressures of oxygen in inhaled and exhaled water (P_IO_2 and P_EO_2, respectively) and in arterial and mixed venous blood (P_aO_2 and $P_{\bar{v}}O_2$, respectively) as they pass through the gills. The presence of a **barrier to diffusion** between the water and blood (i) and **anatomical** (ii) and **physiological shunts** (iii) all reduce the effectiveness of gas exchange in real animals so that, even with countercurrent flow, P_aO_2 is less than P_IO_2. A physiological shunt results from the fact that the further the blood and water are away from the gas-exchange surface, the lower the degree of equilibration of oxygen, as indicated by the thickness of the arrows in (iii). Beyond a certain distance, no gas exchange occurs at all.

Adapted from Butler PJ (1976). Gas exchange. In Environmental Physiology of Animals, eds, Bligh J et al. Blackwell Scientific Publications, Oxford.

The gills (ctenidia) of most molluscs are located in a **mantle (gill) cavity**, and the water current that flows through this cavity is generated by cilia on the gills. Figure 12.5B shows this process in abalones (*Haliotis* spp.). Water flows in the opposite direction to the blood, so there is the anatomical basis for a countercurrent exchange system. However, data in Table 12.2 show that the partial pressure of oxygen in the arterial blood (P_aO_2) does not exceed that in the expired water (P_EO_2), as might be expected from an effective countercurrent exchange system, which we discuss in Box 12.1.

(A)

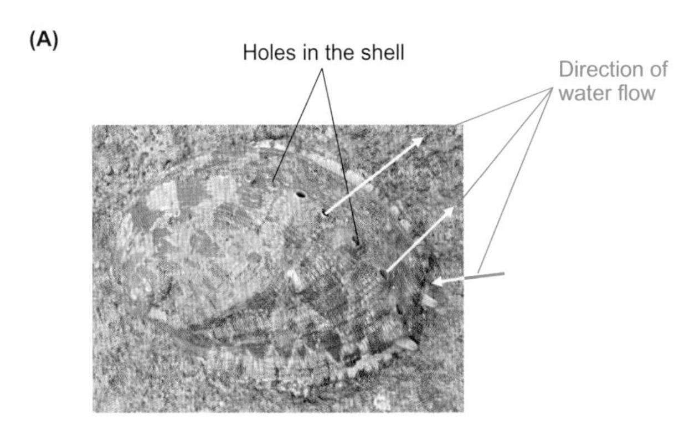

(B)

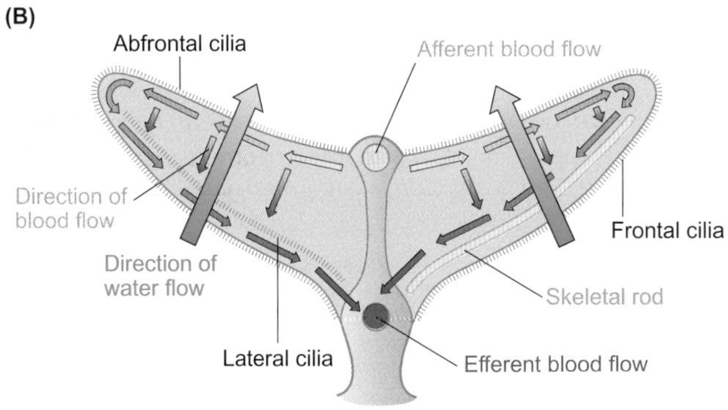

Figure 12.5 Gills of a gastropod mollusc, the abalone (*Haliotis tuberculata*)

(A) Photograph of the abalone showing the inward current of water at the front of the animal and its exit through the holes in the shell. (B) Lateral view of pair of lamellae from an abalone gill. Lateral cilia generate the water current that enters the gill chamber. Haemolymph flows from the afferent blood vessel to the efferent vessel in the opposite direction to the water (countercurrent). The frontal and abfrontal cilia are involved in the removal of particles carried in by the respiratory current.

Adapted from Yonge CM (1947). The pallial organs in the aspidobranch gastropoda and their evolution throughout the mollusca. Philosophical Transactions of the Royal Society 232B: 443-518; Image: photograph of abalone kindly supplied by Florence Gully/Estran22.

Most crabs have flattened leaf-like structures that are attached to the gill axis, as illustrated in Figure 12.6A. By contrast, the surface area of the gills in cephalopods and fish is greatly enhanced by folds or plates on both sides of each filament and at right angles to the long axis. This arrangement is shown for fish in Figure 12.6B. The folds or plates are called lamellae and are the sites at which the water and blood flow in opposite directions and where gas exchange occurs. In fish, water flows through a 'mesh' or 'sieve' formed by the lamellae, which is illustrated in Figure 12.6C.

Structure of the gills

The structure of the leaf-like gills of crabs is similar to that of the lamellae in the gills of fish, which are shown in Figure 12.7A. Each lamella of a fish gill consists of a double-layer of epithelial cells on each side of blood-filled channels. The space between the epithelial layers is divided into the blood-filled channels by **pillar cells**, with the projections or flanges of these cells forming the lining of each channel. Oxygen must pass through the double layer of epithelial cells, the basement membrane and the flange of the pillar cells as it diffuses from the water to the blood. The blood channels are so small, that the red blood cells (erythrocytes) are close to the flanges of the pillar cells. Nonetheless, oxygen still has to pass across the cell wall of the erythrocytes before it can combine with the respiratory pigment[14]. Figure 12.7B shows that the gills of crustaceans are covered with a thin (1–3 μm) layer of a cuticle containing chitin, which is only about 10 per cent as permeable to oxygen as muscle[15], but this does not seem to impair overall gas exchange.

The importance of diffusion conductance in the functioning of gas exchangers

The **diffusion conductance** (G_{diff}) of a gas exchanger for oxygen describes the ease with which oxygen diffuses across the exchanger and is the reciprocal of resistance to diffusion, R_{diff}. So $G_{diff} - 1/R_{diff}$. There arc two measures of G_{diff}, the **morphological diffusion conductance** and the **physiological diffusion conductance**, which can be obtained by dividing Equation 11.2 by the difference in average P_{O_2} either side of the gas exchanger ($P_1O_2 - P_2O_2$):

$$G_{diff} = \underbrace{\frac{\overbrace{\dot{M}O_2}^{\text{Rate of oxygen consumption (mmol } O_2 \text{ s}^{-1})}}{P_1O_2 - P_2O_2}}_{\text{Physiological diffusion conductance}} = \frac{KO_2A}{d} \qquad \text{Equation 12.1}$$

Physiological diffusion conductance (mmol O_2 s^{-1} kg^{-1} kPa^{-1})
Krogh's diffusion constant (mmol cm^{-1} s^{-1} kPa^{-1})
Rate of oxygen consumption (mmol O_2 s^{-1})
Morphological diffusion conductance (mmol O_2 s^{-1} kg^{-1} kPa^{-1})
Diffusion conductance (mmol O_2 s^{-1} kg^{-1} kPa^{-1})
Area of the gas-exchange surface (cm^2)
Difference in average partial pressure of oxygen either side of the exchange barrier (kPa)
Thickness of the gas-exchange surface (cm)

Equation 12.1 shows that estimates of diffusion conductance can be obtained in one of two ways:

- Physiological diffusion conductance is the rate that oxygen will diffuse across an exchange surface for a given difference in partial pressure across the exchange surface.
- Morphological diffusion conductance is directly related to the surface area of the gas-exchange surface and the Krogh's diffusion constant of the gas through that surface, and is inversely related to the overall thickness of the exchange surface.

[14] We discuss respiratory pigments in Section 13.1.

[15] Table 11.4 gives values of the diffusion constant (an indicator of permeability) of different tissues, including chitin.

Table 12.2 Morphometry of the gas exchange organs, and physiological characteristics of gas exchange in some resting, aquatic animals

Species	Gill area [mm^2 (g body mass)$^{-1}$]	Diffusion distance (μm)	Diffusion conductance (μmol O_2 min^{-1} kg^{-1} kPa^{-1})	Extraction of oxygen from the water (%)	Ventilation convection requirement [mL (μmol O_2)$^{-1}$]	Partial pressure of O_2 in arterial blood (kPa)	Partial pressure of O_2 in expired water (kPa)	Partial pressure of CO_2 in arterial blood (kPa)
Abalone *(Haliotis iris)*			1.3 (physiological)	40, 15°C		10.3	11.1	
Common octopus *(Octopus vulgaris)*	290	10	16.7 (morphological) 4.1 (physiological)	35, 23°C	16	10.4	12.1	0.29
Blue crab *(Callinectes sapidus)*	1367			53, 25°C	12	10	9.1	0.2
Lobster *(Homarus vulgaris)*				40, 15°C	14	6.5	10.6	0.26
Shore crab or European green crab *(Carcinus maenas)*	765	6	18 (morphological)	30–35, 10–18°C		10	12.5	0.25
Dogfish *(Scyliorhinus* spp.*)*	135–175	10	9 (morphological) 4 (physiological)	25–65, 15°C	7–21	13	7.5	0.3
Rainbow trout *(Oncorhynchus mykiss)*	240	6	17 (morphological) 3 (physiological)	30–35, 10°C	8.4	18	13.7	0.3
Atlantic horse mackerel *(Trachurus trachurus)*	1100	2.2						
Skipjack tuna *(Katsuwonus pelamis)*	1840	0.6	38 (physiological)	56		12	14.5	

Sources: Ragg and Taylor, 2006; Eno, 1994; Houlihan et al., 1982; 1986; Gray, 1957; Cameron, 1989; Butler et al., 1978; McMahon et al., 1978; Hughes, 1983; Taylor and Butler, 1978; Taylor et al., 1977; Taylor and Wheatly 1979; Hughes et al., 1986; Hughes, 1972; Butler and Metcalfe, 1988; Kiceniuk and Jones, 1977; Brauner and Randall, 1998; Hughes, 1966; Hughes, 1970; Muir and Hughes, 1969; Stevens, 1972. Full citations available online.

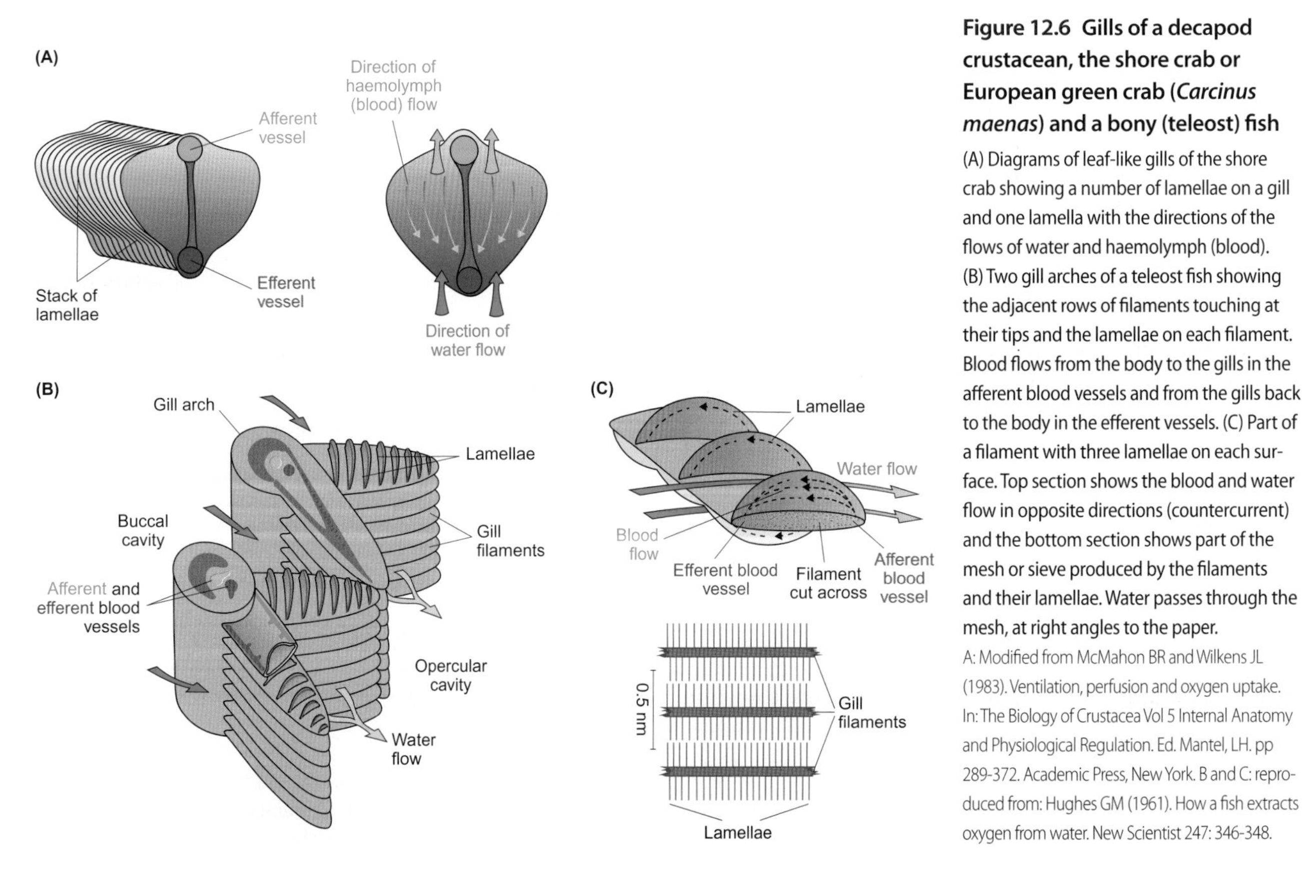

Figure 12.6 Gills of a decapod crustacean, the shore crab or European green crab (*Carcinus maenas*) and a bony (teleost) fish

(A) Diagrams of leaf-like gills of the shore crab showing a number of lamellae on a gill and one lamella with the directions of the flows of water and haemolymph (blood). (B) Two gill arches of a teleost fish showing the adjacent rows of filaments touching at their tips and the lamellae on each filament. Blood flows from the body to the gills in the afferent blood vessels and from the gills back to the body in the efferent vessels. (C) Part of a filament with three lamellae on each surface. Top section shows the blood and water flow in opposite directions (countercurrent) and the bottom section shows part of the mesh or sieve produced by the filaments and their lamellae. Water passes through the mesh, at right angles to the paper.

A: Modified from McMahon BR and Wilkens JL (1983). Ventilation, perfusion and oxygen uptake. In: The Biology of Crustacea Vol 5 Internal Anatomy and Physiological Regulation. Ed. Mantel, LH. pp 289-372. Academic Press, New York. B and C: reproduced from: Hughes GM (1961). How a fish extracts oxygen from water. New Scientist 247: 346-348.

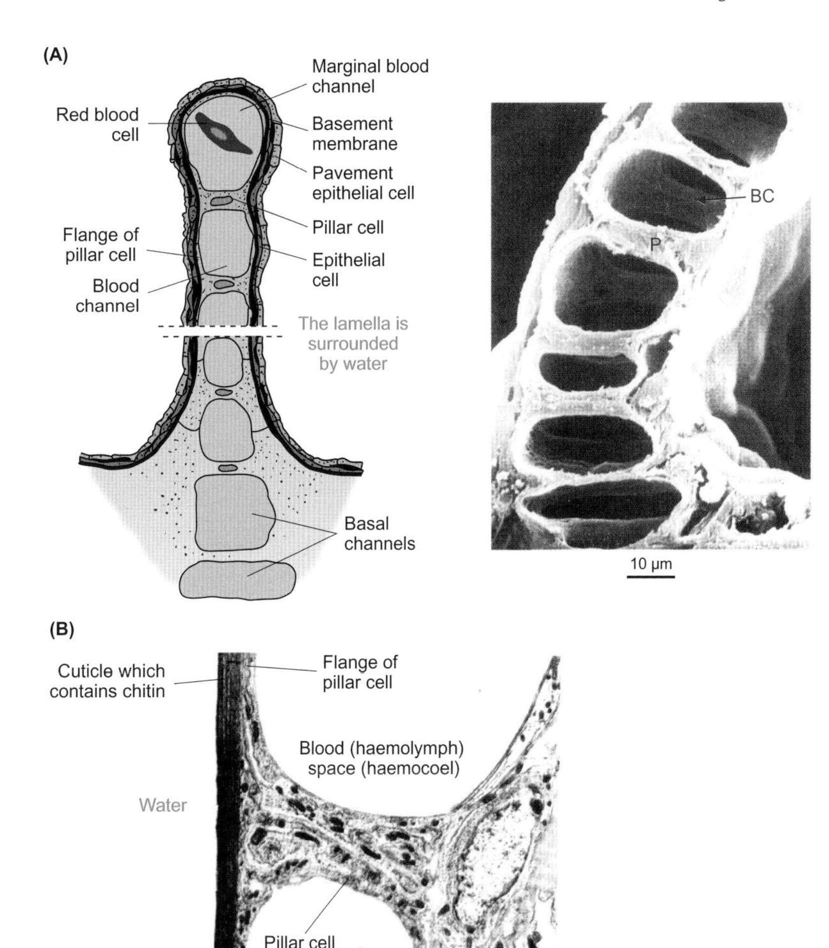

Figure 12.7 Fine structure of gills of fish and crabs

(A) Diagram of transverse section through a lamella of the gill of the lesser spotted dogfish or small spotted cat shark (*Scyliorhinus canicula*) showing the cell layers, blood channels including the larger basal and marginal channels, and scanning electron micrograph of a transverse section through a lamella. BC, blood channel; P, pillar cell. (B) Electron micrograph of a longitudinal section through a pillar cell of a lamella from a gill of the shore crab (*Carcinus maenas*). The chitinous cuticle is also visible.

Reproduced from: Butler PJ (1999). Respiratory system. In: Sharks, Skates and Rays. The Biology of Elasmobranch Fishes, ed Hamlett, WC. pp 174-197. The Johns Hopkins University Press, Baltimore. Taylor EW and Butler PJ (1978). Aquatic and aerial respiration in the shore crab, *Carcinus maenas* (L), acclimated to 15º C. Journal of Comparative Physiology 127: 315-323.

A simplification of the morphological diffusion conductance is the anatomical diffusion factor, which is used by some scientists[16] and is the area of the gas exchange surface divided by its average thickness A/d. In other words, the anatomical diffusion factor for oxygen does not include Krogh's diffusion constant for oxygen.

The morphological G_{diff} of any given gas-exchange system is probably close to the maximum for G_{diff} for that system because it considers the whole of the area of the gas exchanger. Table 12.2 shows that values for morphological G_{diff} are similar in rainbow trout, octopus and shore crab, and are between 2 to 6 times greater than the physiological estimates of G_{diff} in resting animals. These data suggest that physiological G_{diff} may not always give the maximum value of G_{diff}.

One possible explanation for the difference between physiological and morphological diffusion conductances is that not all the gill area is used for gas exchange while the animal is resting. In contrast, experiments with larger spotted dogfish or nursehound (*Scyliorhinus stellaris*) have demonstrated that the values of the two diffusion conductances are almost identical when the fish are exercising. This observation suggests that virtually all of the gas-exchange area is perfused and ventilated during exercise.

Many of the major factors that affect gas diffusion across the gills, such as surface area and thickness of water–blood barrier, are similar in cephalopod molluscs, decapod crustaceans and fish, and examples of these are given in Table 12.2. However, the gills of more active species, such as blue crabs (*Callinectes sapidus*), Atlantic horse mackerel (*Trachurus trachurus*) and skipjack tuna (*Katsuwonus pelamis*), have relatively larger surface areas and smaller diffusion distances, or thicknesses. Tuna, including the skipjack, are shown in Figure 12.8. What is more, surface area and diffusion distance of the gills of tunas may even approach those of the lungs of mammals of similar mass. One impact of these morphological differences is that the more active species, tuna and blue crab, can extract a higher percentage of oxygen from the water[17] than less active fish and crustaceans, as shown in Table 12.2.

[16] The use of anatomical diffusion factor is illustrated in Figure 12.8C.

[17] Percentage extraction of oxygen from respiratory water is the ratio of the amount of oxygen extracted from the water flowing over the gills to the total amount of oxygen in the inhalant water and is discussed in Box 12.2.

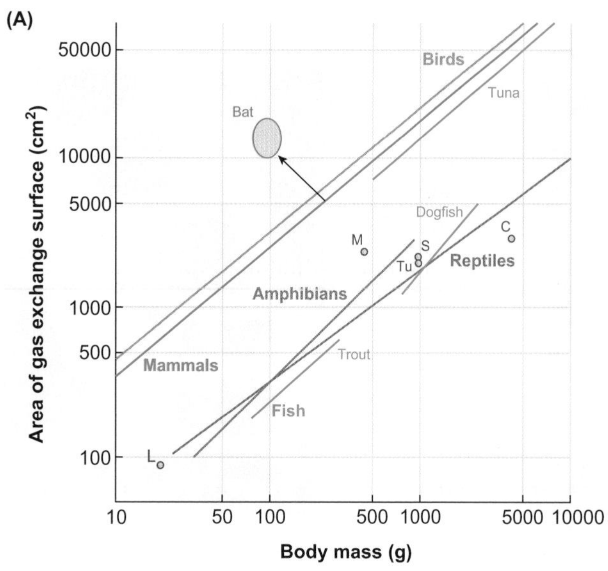

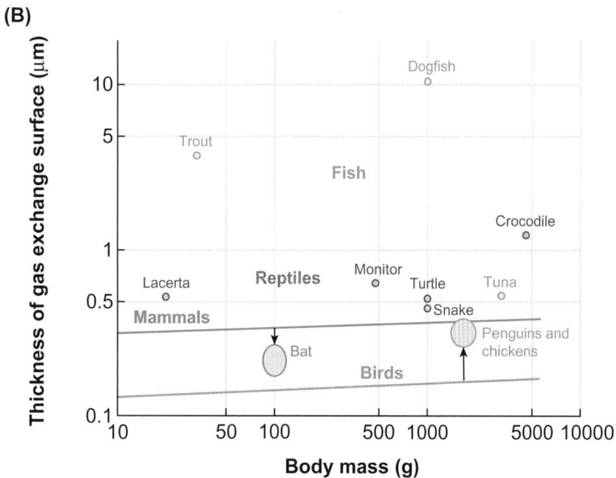

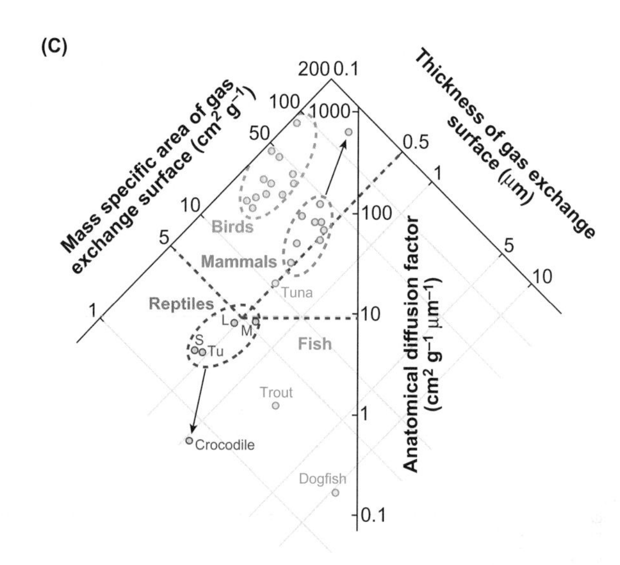

Figure 12.8 Anatomical diffusion properties of the gas exchangers in vertebrates

(A) Double logarithmic plot of the area of the gas-exchange surface as a function of body mass. This shows that the area is greater in birds and mammals than in most fish (with the notable exception of tuna), amphibians and reptiles of similar body mass. (B) Double logarithmic plot of the thickness of the gas-exchange surface as a function of body mass. The thinnest gas-exchange surfaces are in the lungs of birds and the thickest are in the gills of fish, except for tuna, which are similar to those in mammals. (C) Triple logarithmic plot of mass-specific surface area (area per unit of body mass) of the gas-exchange surface, thickness of the gas-exchange surface and anatomical diffusion factor. The anatomical diffusion factor is the mass-specific area of the gas-exchange surface divided by its thickness, as discussed in Section 11.2.1. Note that the anatomical diffusion factor is similar in bats to that in birds. Dashed red lines indicate values of the three variables for the lizard *Lacerta* (L).

Ventilation of gills

The use of cilia to generate the respiratory water current is fine for relatively inactive animals, such as the abalones we discussed earlier. However, cilia are not able to produce the levels of ventilation required by more active animals. The cephalopod molluscs (cuttlefish, squid, octopus), decapod crustaceans (crabs, lobsters and crayfish) and fish, all use muscular action to ventilate their gills, which enables larger ventilation volumes to be generated to meet the higher oxygen demands of these animals.

Although muscular activity is used to generate the respiratory current in decapod crustaceans, cephalopod molluscs and fish, the details of exactly how the water current is created in these groups differ. Figure 12.9 shows that water flow through the gill chamber of decapod crustaceans is generated by a modified head appendage known as the **scaphognathite** (which means boat jaw) or **bailer**, as shown in Figure 12.9A. The bailer oscillates within a narrow channel and reduces pressure in the gill chamber to below ambient, which is shown in Figure 12.9B. Water is drawn into the gill chamber around the base of the legs, between the gills and is forced out through the exhalent canal at the front of the animal, as shown in Figure 12.9A. In cephalopod molluscs, the respiratory water current is created by alternate contractions of the circular and radial muscles of the mantle (body wall).

In bony fish, water movement through the gills involves the buccal cavity (mouth) and opercular cavities, which are illustrated in Figure 12.10. Figure 12.11 demonstrates how these cavities function as two pumps during separate phases of the respiratory cycle. The **opercular suction pump** operates during phase 1 of the ventilatory cycle and involves the enlargement of the opercular cavities. This enlargement causes a reduction in pressure to below ambient, as shown in Figure 12.11B, which sucks water from the buccal cavity, through the gill mesh and into the opercular cavities. The **buccal force pump** operates during phase 3 of the ventilatory cycle. Figure 12.11B shows that as the mouth closes the pressure rises in the buccal cavity. This increase in pressure

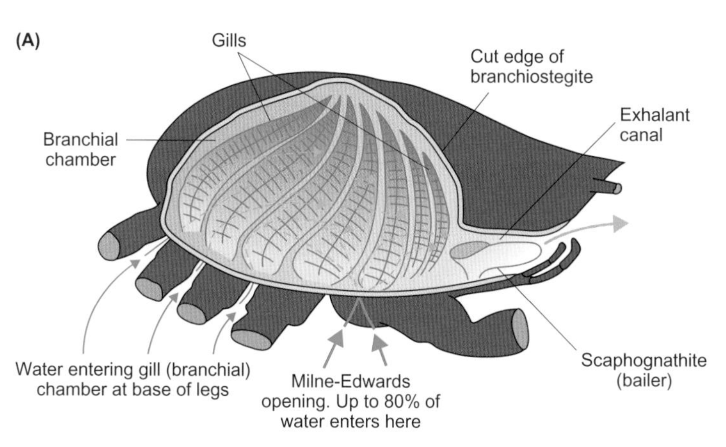

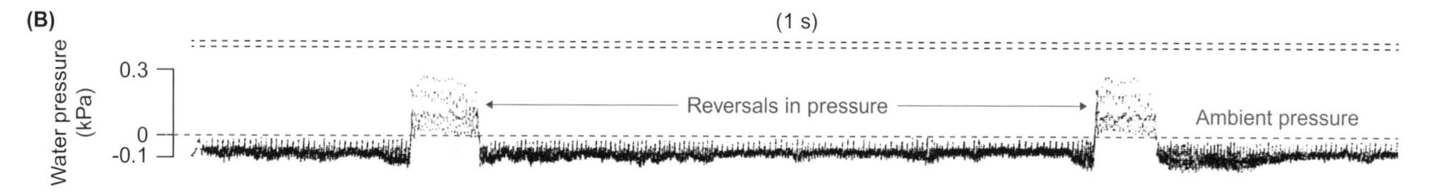

Figure 12.9 Traces of water pressure in a single gill chamber from a shore crab or European green crab in seawater

Sub-ambient pressure generated by the scaphognathite draws water in through the base of the legs. Periodic reversals in pressure cause water to flow in through the normally exhalent opening and out around the legs. These reversals may function in a similar way to a cough in as much as they may clear debris from the gills. The branchiostegites are extensions of the carapace which form one wall of each of the gill chambers.

Source: A: Taylor EW (1982). Control and co-ordination of ventilation and circulation in crustaceans: responses to hypoxia and exercise. J. Exp. Biol. 100: 289-319. B: Taylor EW et al (1973). The respiratory and cardiovascular changes associated with the emersion response of *Carcinus maenas* (L.) during environmental hypoxia at three different temperatures. Journal of Comparative Physiology 86: 95-115. Image: courtesy of D.Hazerli.

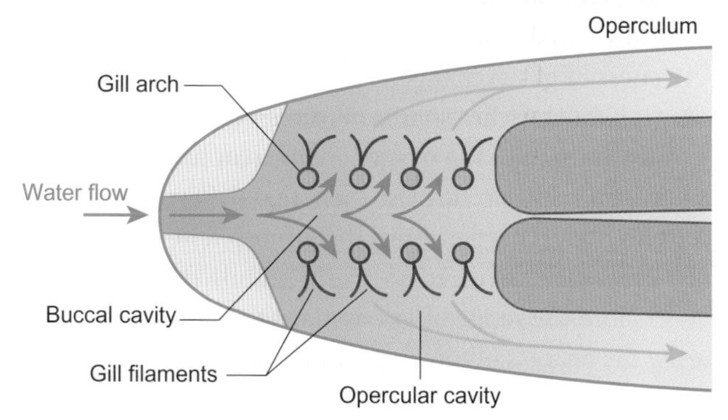

Figure 12.10 Respiratory system of bony (teleost) fish

Horizontal section through the pharyngeal region of a bony fish showing buccal and opercular cavities, the position of the four gill arches and the V-shaped arrangement of the gill filaments on the arches.

Image: http://www.canoethecaney.com/canoe-kayak-trips/trout-fishing.html.

forces water from the mouth, through the gill mesh and into the opercular cavities. Phases 2 and 4 are transition phases between the two pump phases. The whole of the ventilatory cycle in bony fish is powered by the activity of specific respiratory muscles.

By contrast, although cartilaginous fish use a similar system to ventilate their gills, their suction pump is passive. They have a series of gill slits and parabranchial cavities along each side instead of an operculum and opercular cavity. Soon after the mouth begins to close in dogfish (*Scyliorhinus* spp.), a sheet of superficial muscle contracts against the elastic skeleton of the gills and this assists in forcing the water out through the gill slits. When these superficial muscles relax, the parabranchial cavities expand passively as a result of the elasticity of the skeletal structure. This provides the suction required to draw the water across the gills.

The energetic cost of moving water across the gills is determined by the resistance offered by the gills, which is affected by the viscosity and density of the water. Both viscosity and density are much greater for water than they are for air[18], so the cost of ventilating gills is greater than that of ventilating lungs. Consequently, the amount of oxygen consumed by the respiratory muscles in fish is a significant

[18] Table 11.3 gives the viscosities and densities of water and air.

(A)

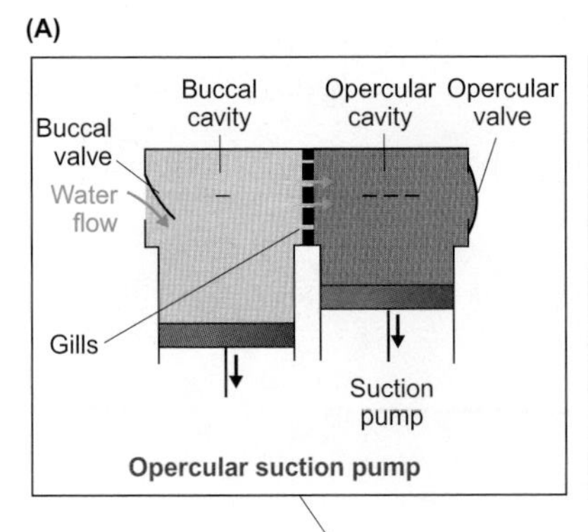

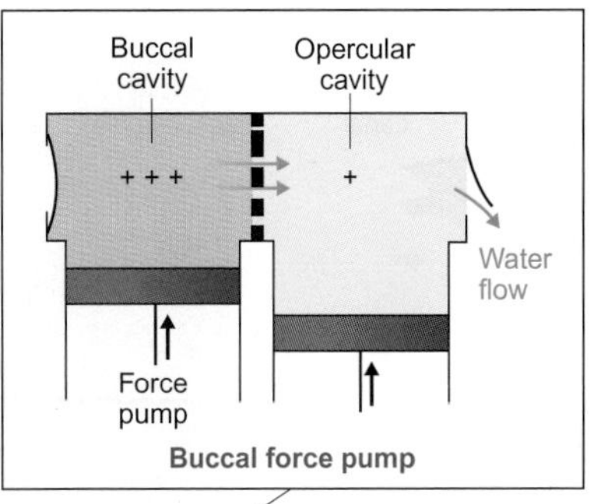

(B)

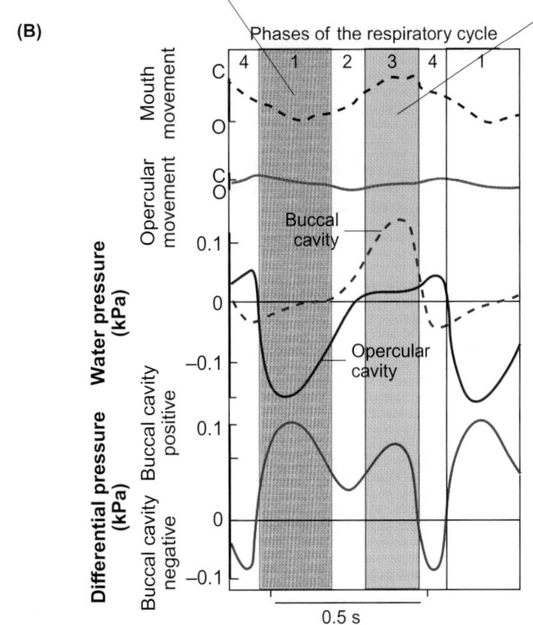

Figure 12.11 Ventilation in bony (teleost) fish

(A) Diagrammatic representation of the two pumps used in ventilating the gills of teleost fish. The expansion of the opercular cavity causes water to be sucked through the gills, while the reduction in volume of the buccal cavity (mouth) causes water to be forced through the gills. – sign indicates reduction in pressure and + sign indicates increase in pressure. (B) The movements of the mouth and opercula and the pressure changes in the buccal and opercular cavities of the rainbow trout (*Onchorhynchus mykiss*) during a ventilatory cycle (O – open, C – closed). The pressure difference between these two cavities shows that the pressure in the buccal cavity is greater than that in the opercular cavities for most of the ventilation cycle, thus causing water to flow between the gill lamellae for most of the cycle. For further details see the text.

Source: adapted from Hughes GM and Shelton G (1958). The mechanism of gill ventilation in three freshwater teleosts. Journal of Experimental Biology 35: 807-823.

proportion of their total oxygen consumption, being about 11 per cent in rainbow trout[19].

Active species, such as rainbow trout, Atlantic mackerel (*Scomber scombrus*), tunas and leopard shark (*Triakis semifasciata*) hold their mouths and opercula (or gill slits) open when swimming so that water passes in a continuous stream over the gills. This is called **ram ventilation** and does not involve the activity of any specific respiratory muscles. However, when swimming speed falls below a certain level (for example, 80–60 cm s^{-1} in the Atlantic mackerel), ventilation becomes active and similar to that shown in Figure 12.11 for rainbow trout. The rate of oxygen consumption falls by 10 per cent when rainbow trout change spontaneously from active to ram ventilation, at the same swimming speed. This reduction in the rate of oxygen consumption during ram ventilation is similar in magnitude to the estimated cost of active ventilation in resting rainbow trout mentioned above and, therefore, gives some validity to the estimated value. Tunas have largely lost their ability to pump sufficient water over their gills to satisfy their oxygen needs and so must swim continuously to ram ventilate. One advantage of this may be that it avoids a pulsatile flow of water over the gills that could reduce the effectiveness of gas exchange.

Air is a much richer source of oxygen than water; there is a greater amount of the oxygen in air than in the same volume of water. It is no surprise then, that the evolution of the ability to obtain oxygen from the atmosphere, and become fully terrestrial, occurred in many groups of animals. We now explore the range of gas exchange organs that evolved during the process of moving from water to land.

› *Review articles*

Butler PJ, Metcalfe JD (1988). Cardiovascular and respiratory systems. In: Shuttleworth TJ (ed.) Physiology of Elasmobranch Fishes. pp 1–47. Berlin: Springer-Verlag.

Feder ME, Burggren WW (1985). Cutaneous gas exchange in vertebrates: design, patterns, control and implications. Biological Reviews 60: 1–45.

[19] This value is compared with that in air-breathing vertebrates in section 12.3.

Piiper J (1990). Modeling of gas exchange in lungs, gills and skin. In: Boutilier RG (ed.) Advances in Comparative & Environmental Physiology, vol 6. pp 15–44. Berlin: Springer-Verlag.

Taylor HH, Taylor EW (1992). Gills and lungs: the exchange of gases and ions. In: Harrison FW, Humes A (eds.) Microscopic Anatomy of Invertebrates Vol 10: Decapod Crustacea. 203–293. New York: Wiley.

12.3 Gas exchange in air

While some groups of animals, such as the annelids and crustaceans, did not colonize land in a major way, in contrast, many species of molluscs, arachnids (spiders and scorpions), insects, millipedes, centipedes and vertebrates spend the whole of their life cycles on land and are, therefore, surrounded by air rather than water. The different physical properties of these two environments have important consequences for the process of ventilation:

- At any partial pressure of oxygen (PO_2), there is a greater concentration of oxygen in air than in water[20], which means for a given rate of oxygen consumption, air-breathing animals ventilate their gas exchangers at a lower rate than water breathers. However, the difference in concentration of oxygen in air and water does not fully explain the difference in respiratory minute volume between air and water breathers, as discussed in Box 12.2.
- Air is much less dense and viscous than water, so the energetic cost of ventilation for air breathers is less than that for water breathers.
- The capacitance coefficient of carbon dioxide in water is about 28 times greater than that of oxygen, whereas in air, the capacitance coefficients of carbon dioxide and air are the same[21]. As a result, and as shown in Tables 12.2, 12.3 and 12.5, the partial pressure of carbon dioxide (PCO_2) is lower in the blood of water-breathing animals than in that of air-breathers, and in animals that can breathe either air or water (which are known as **bimodal breathers**).
- Apart from living at or ascending to higher altitudes or living in poorly ventilated burrows, there is no significant variation of PO_2 in air[22]. Hence, low PO_2 in the environment (hypoxia) is uncommon among terrestrial animals, although internal hypoxia (**hypoxaemia**, hypoxia of the blood) resulting from long pauses in ventilation, or a malfunctioning gas-exchange surface, may occur.

[20] Figure 11.5 illustrates the difference in concentration of oxygen in air and water at similar.

[21] Capacitance coefficients of oxygen and carbon dioxide in air and water are discussed in Section 11.1.2.

[22] Hypoxia at high altitude and in burrows is discussed in Sections 15.3.1 and 15.3.2, respectively.

Box 12.2 Ventilation in water and in air

In this box, we look at why water-breathing and air-breathing animals exhibit different ventilation rates for a given rate of oxygen consumption (the **ventilation convection requirement**, which we discuss in Section 12.3.3).

The first reason is that the capacitance coefficient of oxygen in air (β_aO_2) is many times greater than that in fresh water (β_wO_2), and at 25°C, the difference is approximately 30-fold. It might be expected therefore, that for the same rate of oxygen consumption ($\dot{M}O_2$) at 25°C, the amount of the respiratory medium (air or water) passing over the gas exchange surface per minute (minute ventilation volume, $\dot{V}_m$), would be around 30 times greater in a water-breathing animal than in an air-breather.

The importance of the capacitance coefficients for oxygen in air and water can be illustrated by substituting the concentration of oxygen in Equation 11.5 with the partial pressure and capacitance coefficient for oxygen in respiratory medium (β_mO_2, i.e. β_aO_2 for air or β_wO_2 for water). This means that C_IO_2 is now represented by ($\beta_mO_2\,P_IO_2$) and C_EO_2 is represented by ($\beta_mO_2\,P_EO_2$):

Rate of oxygen consumption (mmol min⁻¹) — Minute ventilation volume (L min⁻¹) — Difference in oxygen concentration between inspired and expired the medium (air or water, mmol L⁻¹)

$$\dot{M}O_2 = \dot{V}_m \times \overbrace{\beta_mO_2\underbrace{(P_IO_2 - P_EO_2)}}\qquad \text{Equation A}$$

Capacitance coefficient for oxygen in the medium (air or water, mmol L⁻¹ kPa⁻¹) — Difference in partial pressure of oxygen between inspired and expired medium (kPa)

Equation A shows that $\dot{M}O_2$ is not only dependent on $\dot{V}_m$ and the capacitance coefficient for oxygen of the medium (β_m), but also on the difference in PO_2 between inspired and expired medium.

This relationship means that, for a given $\dot{M}O_2$ at 25°C, minute ventilation volume of a water breather ($\dot{V}_w$) would only be around 30 times greater than that of an air breather ($\dot{V}_a$) if the difference in $\dot{M}O_2$ between inspired and expired water or air [($P_IO_2 - P_EO_2$)] was the same in both cases. ($P_IO_2 - P_EO_2$) is directly related to percentage extraction of oxygen from water or air[1].

Data in Table 12.3 show that the percentage extraction of oxygen from the inspired medium is, in fact, far greater in bimodal freshwater crabs (*Austrothelphusa transversa*) at 25°C when they are breathing water (46 per cent) than when they are breathing air (12 per cent). This difference in percentage extraction of oxygen is reflected to a large extent in the values of the ventilation convection requirements in the two environments: 9.5 in water and 1.6 in air, a difference of only six times instead of the expected 30 times based on the values of capacitance coefficients of water and air for oxygen alone.

[1] Percentage extraction of oxygen from the respiratory medium is the ratio of the amount of oxygen extracted from the medium ventilating the gas exchange organ to the total amount of oxygen in the inspired medium: $\beta_mO_2(P_IO_2 - P_EO_2)/\beta_mO_2P_IO_2 = P_IO_2 - P_EO_2/P_IO_2$

Table 12.3 Morphometry of the gas-exchange organs and physiological characteristics of gas exchange in some decapod crustaceans and fish that are able to breathe water and air (bimodal breathers)

All values are from resting animals.

Species	Habitat	Area of gas exchange surface [mm^2 (g body mass)$^{-1}$]	Diffusion distance (μm)	Diffusion conductance (μmol O_2 min^{-1} kg^{-1} kPa^{-1})	Extraction of O_2 from water or air (%)	Ventilation convection requirement [mL (μmol O_2)$^{-1}$]	P_aO_2 (kPa)	P_aCO_2 (kPa)
Blue land crab (*Discoplax* spp.)	Burrows above high-tide mark		0.2–0.3 (lungs)	2.6 (physiological)	12 (air) 25°C	1.3 (air)	9.1 (air)	1.23 (air)
Coconut crab (*Birgus latro*)	Jungles of Pacific Islands	12 (gills)	0.5–1.2 (lungs)	1.4 (physiological)	5.2 (air) 28°C	2.4 (air)	10.4 (air)	0.8 (air)
Trinidad mountain crab (*Pseudothelphusa garmani*)	Mountainous rain forests		0.4 (lungs)				3.6 (water), 17 (air) 25°C	0.7 (water), 1.3 (air)
Freshwater crab (*Austrothelphusa transversa*)	Arid areas of Australia		5–8 (gills), 0.25–0.3 (lungs)		46 (water), 12 (air) 25°C	9.5 (water), 1.6 (air)	2.4 (water), 7.5 (air)	0.8 (water), 1.3 (air)
Climbing perch (*Anabas testudineus*)	Water/land	47 (gills), 29 (ABO)	10–15 (gills), 0.21 (ABO)	2.3 (gills, morphol) 76 (ABO, morphol)	71 (ABO)			
Bowfin (*Amia calva*)	Water/facultative air breather	195 (gills)	8 (gills)				4.8	0.7
Pirarucu (*Arapaima gigas*)	Water/obligate air breather						5.3	3.5
African lungfish or marbled lungfish (*Protopterus aethiopicus*)	Water/obligate air breather	1430 (lungs)	0.37 (lung)	64 (morphol)	82 (lungs), 27 (gills) 20°C		3.6	3.4
S. American lungish (*Lepidosiren paradoxa*)	Water/obligate air breather	850 (lungs)	0.86 (lung)	93 (morphol)	22 (gills), 20 °C		5.1	2.9

P_aO_2 is partial pressure of oxygen in arterial blood, P_aCO_2 is partial pressure of carbon dioxide in arterial blood, ABO is air breathing organ and morphol is morphological.

Sources: 1984; Burggren and McMahon, 1988; Farrelly and Greenaway, 1993; Wood and Randall, 1981; Cameron, 1981; Cameron and Mecklenburg, 1973; Innes and Taylor, 1986; Taylor and Greenaway, 1979; Greenaway et al., 1983a; 1983b; Hughes and Datta Munshi, 1979; Graham, 1997; Singh and Hughes, 1971; Daxboeck et al., 1981; Randall et al., 1981; Randall et al., 1978; Maina and Maloiy, 1985; Lenfant and Johansen, 1968, McMahon, 1970; Hughes and Weibel, 1976; Johansen and Lenfant, 1967; Johansen, 1970. Full citations available online.

- Air does not provide structural support for gills, and many species of terrestrial animals have evolved internal gas-exchange structures: lungs or tracheae. This internal location is also important in reducing rate of water loss across the respiratory surfaces.[23]

Although the evolution toward breathing air involved quite dramatic changes in the overall form of the gas exchanger, there are remarkable similarities between aquatic and aerial gas exchangers at the point where gaseous diffusion between the external and internal environments occurs. The organs for gas exchange in air breathers can be modified gills, regions of the body such as the lining of the gill chamber or mouth cavity in some fish, lungs, or tracheae in insects[24].

The body surface can only be of significance for gas exchange if the skin is thin enough to enable the gases to diffuse at a sufficiently high rate. This means that water will also diffuse from the body, making the surface of the skin moist, and that the skin will be prone to damage. The animal must, therefore, be protected from dessication and/or damage. Because oxygen diffuses more rapidly in air than in water, oxygen uptake across the skin of terrestrial animals is not constrained by the problems of boundary layers to anything like the same extent as it is in those animals that live in water[25].

We begin this section by discussing the limited role of the body surface in gas exchange in terrestrial animals and how adequate gas exchange is achieved across the outer surfaces of developing eggs of terrestrial vertebrates and insects, before going on to air-breathing organs, lungs and tracheae.

12.3.1 Gas exchange across the body surface

Terrestrial annelids, such as common earthworms (*Lumbricus terrestris*), which live in moist soil, have no special respiratory organs. Instead, they exchange gases across the general body surface. Such cutaneous gas exchange is also important in slugs and snails when they are active and emerged from their shells, even though most of them have a lung. There is, however, a group of tropical slugs that have no lung and which rely entirely on gas exchange across their skin.

In air, the skin of some fish and amphibians can also provide an effective gas-exchange surface, but the animal must be in a moist atmosphere to prevent excessive water loss across the skin. However, cutaneous gas exchange is unimportant in birds and mammals except in bats and smaller species of newborn marsupials. The exchange of oxygen and carbon dioxide across the folded wings of anaesthetized Wahlberg's epauletted fruit bats (*Epomophorus wahlbergi*) at 33°C is about 10 per cent of total gas exchange. In awake, active bats with their wings extended, it might be expected that gas exchange across the wings would be even greater.

The relative significance of the skin for gas exchange in newborn marsupials is dependent on the size of the newborn (which affects its surface area/volume ratio) and the stage of development of its lungs. For example, in tiny (approximately 17 mg) newborn Julia Creek dunnarts (*Sminthopsis douglasi*), there is a relatively high surface area/volume ratio and the lungs are poorly developed. Consequently, almost 100 per cent of gas exchange is across the skin immediately after birth. This falls to about 20 per cent when the animals are 3 weeks old (about 300 mg body mass). In contrast, the larger tammar wallabies (*Macropus eugenii*) are about 400 mg at birth, have a relatively smaller surface area/volume ratio and well-developed lungs. Cutaneous gas exchange is only about 30 per cent of the total in 1-day-old animals, and it drops to about 10 per cent by the time the wallabies are 4 days old.

12.3.2 Gas exchange across the eggs of birds, reptiles and insects

The problem faced by developing eggs of terrestrial animals is how to get oxygen in, and carbon dioxide out, without losing too much water, as eggs have no mechanism to replenish lost water. The eggs of birds and many reptiles are contained within a special shell that reduces the rate of water loss while still enabling gas exchange to occur. These are known as **cleidoic eggs** (Greek, *kleistos* = closed). Most is known about gas exchange across the shell of the eggs of birds, but the general principles established in the studies of birds' eggs also apply to gas exchange in the eggs of reptiles and insects, as they all depend on the process of diffusion.

The shell of a bird's egg consists mainly of calcium carbonate crystals, penetrated by pores through which gas exchange occurs. There are also inner and outer shell membranes, which consist of a meshwork of protein fibres, and the **chorioallantoic membrane** where gas exchange with the circulation occurs. As well as oxygen and carbon dioxide diffusing between the outside air and the blood in the circulatory system of the embryo, water vapour also diffuses out of the egg. The liquid water that is lost is replaced by air to form an air cell at the blunt end of the egg, as shown in Figure 12.12A. The rate of diffusion of the respiratory gases and water vapour is related to the number and size of the pores, the thickness of the shell and the thickness of the membranes between the shell and the capillary bed in the chorioallantoic membrane. These layers are shown in Figure 12.12B.

[23] Water loss across respiratory surfaces is discussed in Section 6.1.2.

[24] The tracheae in insects are discussed in Section 12.4.

[25] We discuss boundary layers in more detail in Section 12.2.1.

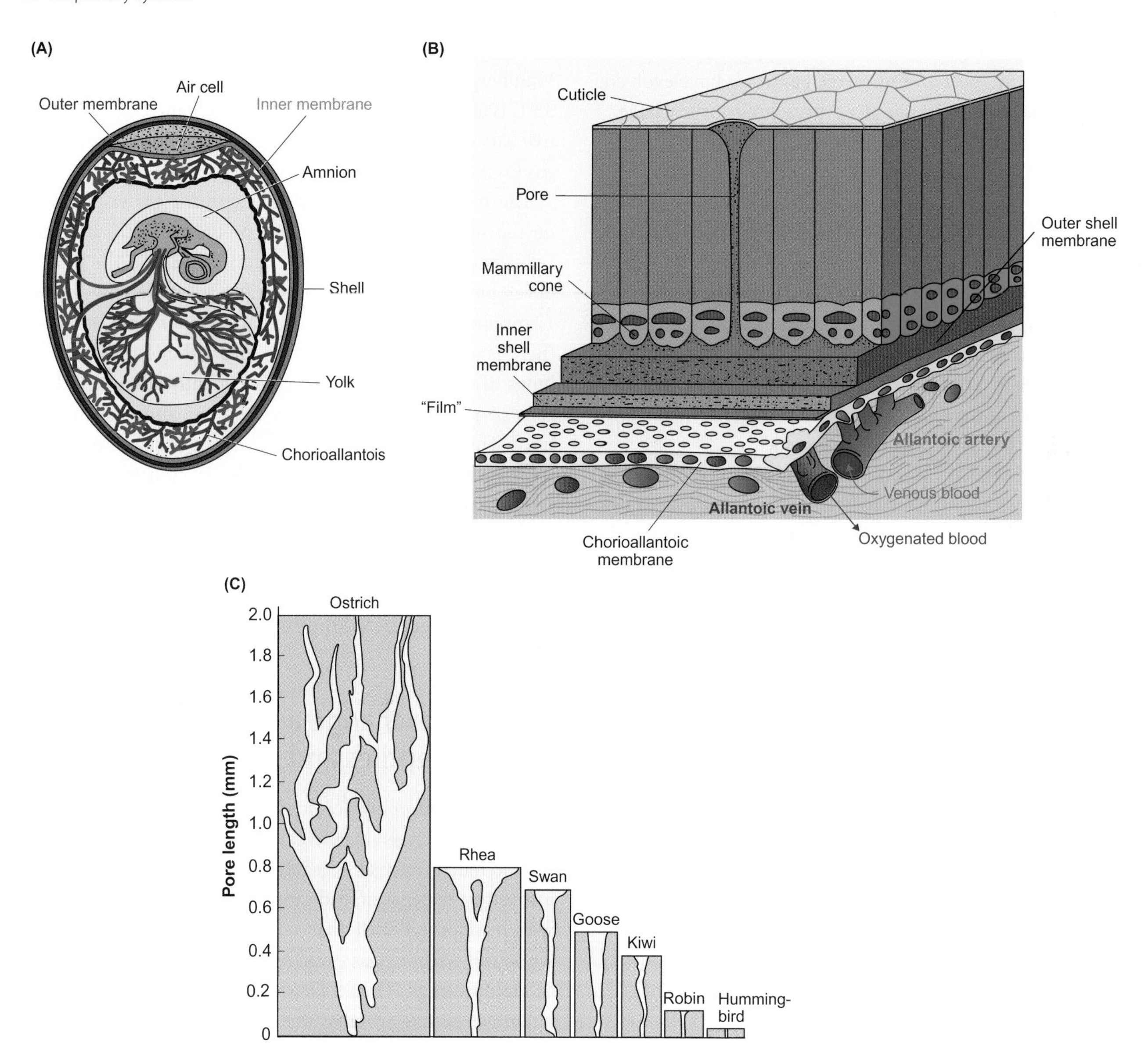

Figure 12.12 Gas exchange across the shells of bird eggs

(A) Embryo of a chicken during the second/third week of development showing the chorioallantois which extends from the embryo and covers the inner shell membrane with a network of capillaries. Venous blood (in blue) from the embryo is oxygenated in the chorioallantois and the oxygenated blood (in red) then flows back to the embryo. (B) Outer layers of the egg of a bird showing the shell with its pores, the inner and outer shell membranes and the chorioallantoic membrane where gas exchange with the circulation occurs. Note that oxygen-poor (venous) blood flows in the allantoic artery and oxygen-rich (oxygenated or arterial) blood flows in the allantoic vein. The chorioallantois is functionally equivalent to the placenta of mammals. (C) The size and shape of the pores in the shell of eggs vary considerably between species of birds, from 2 mm long in the ostrich to 0.4 mm long in hummingbirds.

Reproduced from Rahn H et al (1979). How bird eggs breathe. Scientific American 240: 46-55.

The shells of birds' eggs vary in thickness and there are also differences in the shape of the pores, as shown in Figure 12.12C. Despite this variability in the pores and the large variation in the egg mass and incubation time, the average diffusion conductance for oxygen of individual pores is similar in the eggs of most birds, at about 0.55 μmol O_2 day^{-1} kPa^{-1}. On average, surface area of eggs increases with mass (M), in the same way as rate of oxygen consumption ($M^{0.77}$), and this is reflected in the density of the pores (measured as pores mm^{-2}), which increases only slightly with increased egg mass. For example, a 1 g egg of an olive tree warbler (*Hippolais olivetorum*) has approximately 230 pores at a density of about 0.63 pores mm^{-2}, whereas a 150 g egg of a Canada goose (*Branta canadensis*) has over 21,000 pores at a density of about 1.06 pores mm^{-2}.

The study of the eggs of domestic chickens (which have been artificially selected) have demonstrated that partial pressusres of oxygen and carbon dioxide (PO_2 and PCO_2,

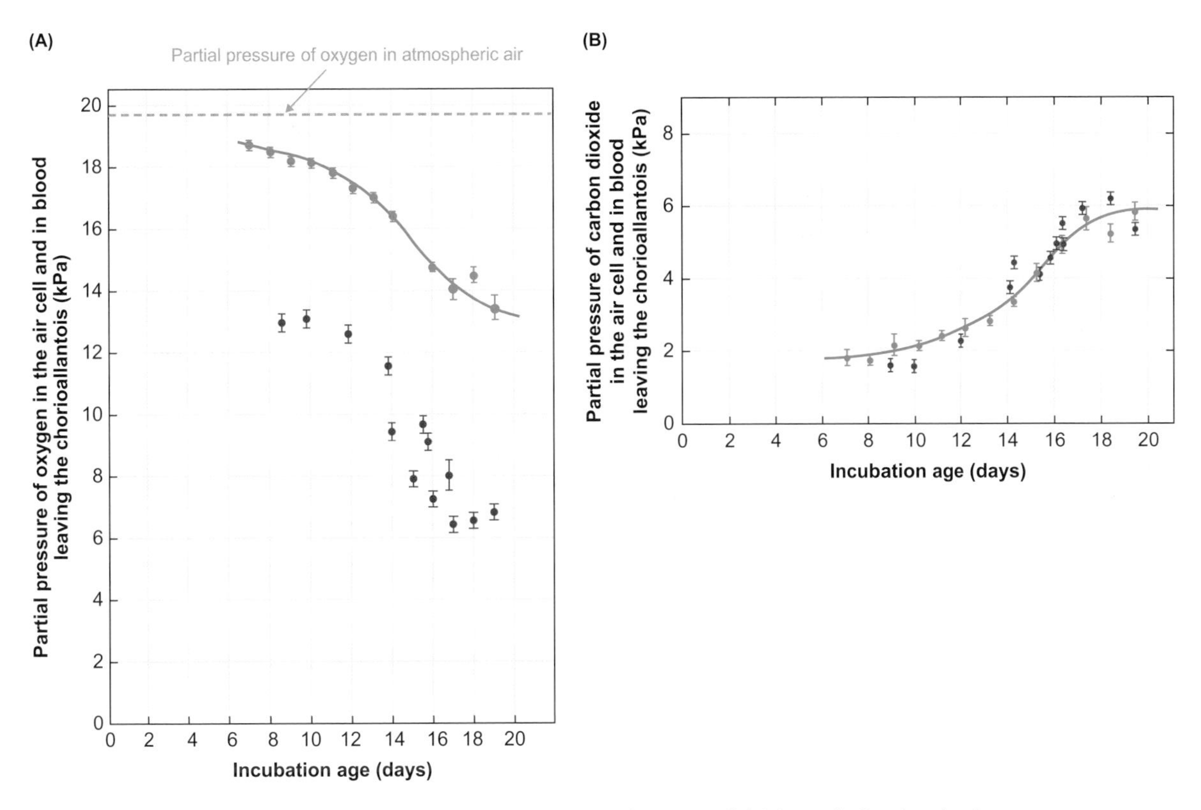

Figure 12.13 Changes in partial pressures of respiratory gases in the eggs of chickens during incubation

Partial pressures of oxygen (A) and carbon dioxide (B) in the air cell (●) and oxygenated blood leaving the chorioallantois in the allantoic vein (●) in the eggs during the second and third weeks of incubation. These data indicate that there is a large difference in the partial pressures of oxygen between the air cell and the arterialized blood of the chorioallantois but that a similar difference does not exist for carbon dioxide.

Reproduced from: Wangensteen OD (1972). Gas exchange by a bird's embryo. Respiration Physiology 14: 64-74.

respectively) in the air cell of an egg are not constant throughout incubation. Figure 12.13 shows that PO_2 decreases and PCO_2 increases as the embryo grows. The lowest PO_2, just before the embryo begins to make a hole in the shell (known as pipping), is 13 to 15 kPa and the highest value for PCO_2 is 4.4 to 5.8 kPa. These values are similar to those in the lungs of birds at hatch, so it has been suggested that the morphological diffusion conductance of the egg shell is set at laying to match the rate of oxygen consumption at hatching and to produce partial pressures of the respiratory gases in the air cell that stimulate pipping and ventilation at the time of hatching.

The overall exchange of oxygen from the air cell to the blood in the chorioallantois is not very effective. The result of this is shown in Figure 12.13A by the much lower PO_2 in the oxygen-rich blood leaving the chorioallantois in the allantoic vein compared to that in the air cell. The embryonic chicken becomes progressively more hypoxaemic after about day 10 of incubation. Under normoxic conditions, a shunt (similar to the shunts shown in Box 12.1) may be responsible for the major part (> 90 per cent) of the difference between PO_2 in the air cell and PO_2 in the oxygenated blood leaving the chorioallantois, with the remainder due to the 15 μm thick diffusion barrier. There is no such problem with the diffusion of carbon dioxide from the blood to the air cell because of the high rate of diffusion of carbon dioxide in blood. The partial pressure of carbon dioxide in the allantoic vein and air cell are, therefore, very similar, as shown in Figure 12.13B.

The shells of the eggs of reptiles are more variable in their structure than those of birds. At one extreme the rigid eggs of geckos have a lower diffusion conductance for oxygen than that of the shells of birds, while the flexible shells in many snakes and lizards have up to 100 times greater diffusion conductance than that for birds' eggs. Such a greater diffusion conductance in the eggs of some reptiles enables gas exchange to progress at appropriate rates under the conditions in which the eggs are laid. Unlike the nests of birds, those of reptiles are generally located in areas where humidity is high, such as soil, sand or decaying vegetation. This reduces the rate of water loss, but PO_2 in the environment surrounding the eggs is lower than that in the atmosphere. Such a reduction in PO_2 surrounding the eggs reduces the driving pressure for the diffusion of oxygen, but this is partly compensated by the greater diffusion conductance of the eggshell.

Unlike the situation in the eggs of birds and reptiles, there is little or no circulation within the eggs of insects, so diffusion is the only means of providing their oxygen

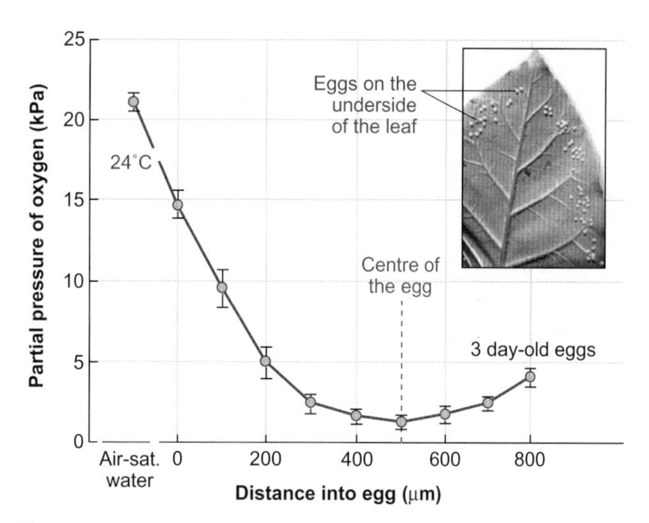

Figure 12.14 Oxygen profile in eggs of the tobacco hornworm or the Carolina sphinx moth (*Manduca sexta*)

Partial pressures of oxygen (PO_2) from air-saturated water and at increasing distances into 3-day-old eggs of the tobacco hornworm at 24°C. Note that PO_2 is less than 2.0 kPa at the centre of the eggs, which suggests that they are at the limits of diffusion for supplying the oxygen they require. See text for further details. Mean values ± SEM, n = 7.

Adapted from Woods HA and Hill RI (2004). Temperature-dependent oxygen limitation in insect eggs. Journal of Experimental Biology 207: 2267-2276.

requirements. Figure 12.14 shows that, in 3-day-old eggs of a tobacco hornworm or the Carolina sphinx moth (*Manduca sexta*) at 24°C and at a constant environmental PO_2 of 21 kPa, PO_2 at their centre is less than 2 kPa. This suggests that under these conditions, the eggs are probably at the limits of diffusion to supply their oxygen needs, despite their very high surface area/volume ratio. This is not too surprising as the eggs have a rate of oxygen consumption ($\dot{M}O_2$) of approximately 0.3 μmol O_2 min^{-1} g^{-1}, and a radius (r) of 0.5 mm, as shown in Figure 12.14, whereas Krogh calculated that diffusion alone could only account for an $\dot{M}O_2$ of no more than 0.074 μmol O_2 g^{-1} min^{-1} for a spherical cell of diameter less than 1 mm and with an external PO_2 of 0.21 atm (21 kPa)[26]. The discrepancy between the measured and calculated values may indicate that diffusion alone is not the only factor involved in the supply of oxygen to the eggs, and that some convection also occurs.

It is apparent from the previous section that the embryos of a number of species of animals (and maybe even, the majority of species) develop under conditions of low PO_2 and mammals are no exception. The partial pressure of oxygen in the umbilical vein is about 4 kPa. The arrangement of the gas exchange region in the human placenta is shown in Figure 20.3. What this figure does not show in any detail is the arrangement between the maternal and foetal blood flow. The arrangement of these two blood flows could indicate the theoretical effectiveness of the exchange region[27], and be one of the possible reasons for the relatively low PO_2 in the umbilical vein.

In fact, there does not seem to be a widely accepted model of oxygen transport across the human placenta. What we do know is that the minimum diffusion distance across the combined membranes in the placenta is about 3–4 μm compared to a diffusion distance of about 0.5 μm in mammalian lungs, as indicated in Table 12.5. This, together with a number of other factors, such as shunts, insufficient blood perfusion and diffusion limitations, could contribute to the low PO_2 in the umbilical vein. At the moment, however, the evidence is inconclusive.

[26] This calculation is discussed in Section 11.2.1.

[27] Effectiveness of gas exchangers is discussed in Box 12.1.

12.3.3 Gas exchange across air-breathing organs and lungs

With the general body surface being of limited use for gas exchange in air and because of their low surface area/volume ratio, air-breathing animals require specialized organs for gas exchange. We now investigate the gas exchangers present in air-breathing molluscs, crustaceans and vertebrates.

The relatively simple lungs of molluscs

Air breathing has evolved in two groups of molluscs. The more well-known are the pulmonate or lunged gastropods (true snails and slugs). The opening of the mantle cavity in pulmonates is reduced to a narrow hole called the **pneumostome**, which is shown in Figure 12.15A, B, and the roof of the mantle cavity is covered with well-vascularized ridges which form a lung. Some species, such as great pond snails (*Lymnaea stagnalis*) and ramshorn snails (*Planorbis corneus*), need to surface periodically to fill their lungs with air. They surface at an increasing frequency as temperature increases and/or when PO_2 decreases[28]. It has been calculated that, with the pneumostome open, diffusion alone easily supplies the oxygen needs of snails.

In pulmonate snails, gas exchange with the external environment is not always continuous. Milk snails (*Otala lacteal*) are edible snails introduced to the US from Europe. During unfavourably dry conditions the snails become dormant and the pneumostome may open only intermittently, as illustrated in Figure 12.15C. This means that, although the animal is continuously using oxygen and producing carbon dioxide, it only exchanges these gases with the environment on an irregular basis. The ratio of the rate of production of carbon dioxide ($\dot{M}CO_2$) and $\dot{M}O_2$, $\dot{M}CO_2$ / $\dot{M}O_2$, at the level of the

[28] Molluscs are ectotherms, so their $\dot{M}O_2$ increases as temperature rises, and the rate at which oxygen diffuses across the skin when they are under water decreases when there is a fall in PO_2. Also, the amount of oxygen dissolved in water at a given PO_2 is less in water at higher temperatures.

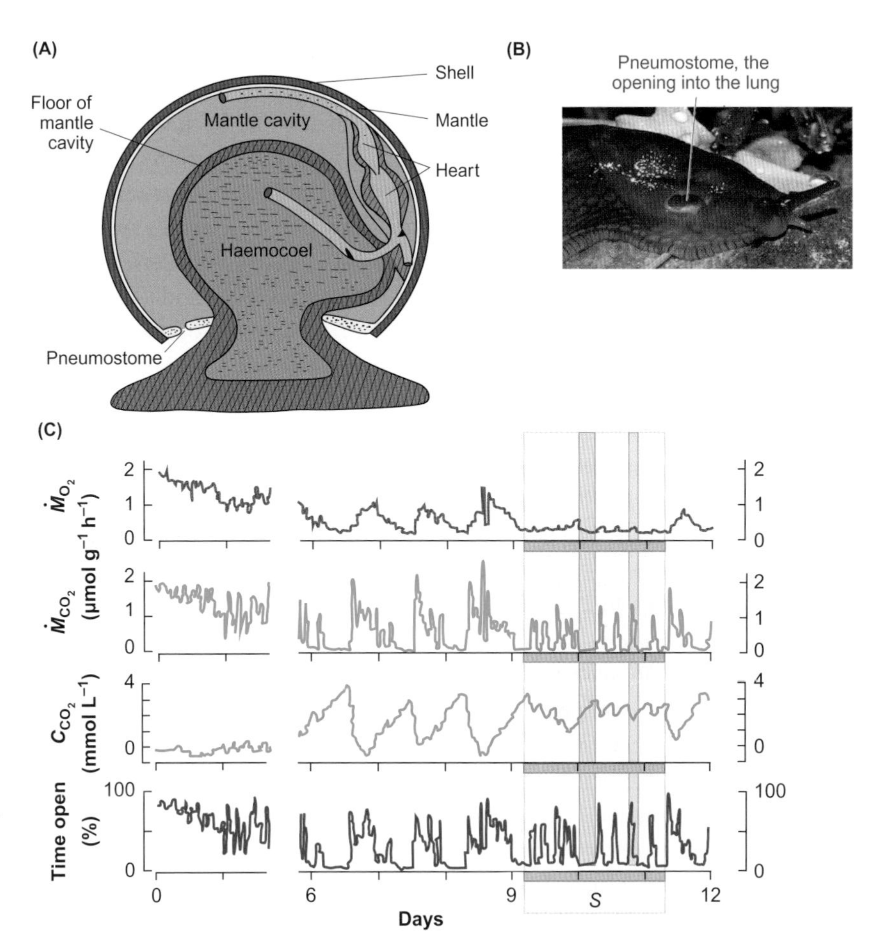

Figure 12.15 Gas-exchange system of terrestrial gastropod molluscs (snails and slugs)

(A) Diagrammatic cross-section of a common garden snail (*Cornu aspersa*) showing the mantle cavity (lung) and the opening into the lung, the pneumostome. (B) Terrestrial slug showing the open pneumostome. (C) Intermittent gas exchange in a dormant milk snail (*Otala lactea*). Traces of rates of oxygen consumption and carbon dioxide production ($\dot{M}O_2$ and $\dot{M}CO_2$, respectively), of whole-body concentration of CO_2 (CCO_2) and percentage of time the pneumostome was open during 30-min periods. These traces illustrate the intermittent nature of gas exchange in this species when it is dormant. During section S, there are periods when $\dot{M}CO_2$ is very low, but $\dot{M}O_2$ is relatively high, such as in the pink-shaded area, and other periods when there are relatively large increases in $\dot{M}CO_2$ which are not accompanied by similar increases in $\dot{M}O_2$, such as in the green-shaded area.

A: Adapted from Alexander R McN (1979). The Invertebrates. Cambridge University Press, Cambridge. C: adapted from Barnhart MC and McMahon BR (1987). Discontinuous carbon dioxide release and metabolic depression in dormant land snails. Journal of Experimental Biology 128: 123-138.

cells of an animal is called the respiratory quotient, RQ[29], and can only be determined in dormant snails over a period of several days.

During intermittent ventilation, the relationship between the rates of oxygen uptake from the air and carbon dioxide release does not necessarily reflect what is happening in the cells at the same time. This is particularly apparent during the period labelled *S* in Figure 12.15C. During the bursts, when the pneumostome is briefly open, $\dot{M}CO_2 / \dot{M}O_2$ is greater than 1, and may reach almost 5 (green-shaded area in Figure 12.15C). By contrast, between the bursts, when the pneumostome is mostly closed, the rate of oxygen uptake is greater than the rate of carbon dioxide release and $\dot{M}CO_2 / \dot{M}O_2$ is less than 1 (average minimum, 0.2, pink-shaded area). These ratios are known as the respiratory exchange ratio, RER[29].

The cutaneous lungs of crustaceans

Three groups of crustaceans have invaded the land. Although the isopods (pillbugs or woodlice) have been the most successful, the land crabs (decapods) have attracted the most attention from physiologists[30]. The third group is the amphipods (beach fleas, sand hoppers and leaf litter hoppers).

Some species of crabs can use their gills for air breathing, as well as for gas exchange in water. Crabs that live in the intertidal region, such as shore crabs or European green crab (*Carcinus maenas*), are often stranded in intertidal rock pools and may voluntarily move into air for brief periods in response to an increase in temperature and a decrease in PO_2 of the water in the pool. Their gills are relatively well spaced and covered in a thin layer of chitin-rich cuticle (see Figure 12.7B) which prevents them from collapse, and enables them to exchange gases with air. Measurements of pressure in the gill chambers and airflow around the exhalent opening demonstrate that the bailers, which create water flow over the gills when the animal is submerged (see Figure 12.9 for details), periodically move air through the gill chambers when the animal is out of water. Although the PO_2 of the arterial haemolymph (P_aO_2) is much lower after shore crabs have been in air for 3 h (2.5 kPa) than the 10 kPa recorded when they are in water, $\dot{M}O_2$ is actually 20 per cent higher.

[29] Respiratory quotient and respiratory exchange ratio are discussed in Section 2.2.2.

[30] Chapter 6 discusses water balance in terrestrial animals.

Some amphibious crabs carry a reservoir of water in their gill chambers; this water is aerated by the air that is pumped through the chambers. This enables the crabs to use the oxygen in air, as the gills exchange gases with the water. Although all land crabs, even the most terrestrial, retain gills, these gills are probably of more use for removal of carbon dioxide when water is available, than for oxygen uptake. PCO_2 in arterial blood (P_aCO_2) of amphibious crabs is almost as low as that in fully aquatic species (around 1 kPa).

The evolution of what has been called the cutaneous lung within the gill chamber improved the ability of land crabs to obtain oxygen from air. Cutaneous lungs may be little more than the highly vascularized surface of the greatly expanded gill covers, as found in blue land crabs (*Cardiosoma* spp.). Despite their relatively simple structure, the lungs of land crabs are very effective gas exchangers. The P_aO_2 of resting *Cardisoma* in air (9.1 kPa) is similar to that of aquatic decapods, as shown back in Table 12.2, and much higher than that of the shore crab in air (2.5 kPa). Air flow through the gill chambers of the common land crab is due to intermittent bursts of pressure generated by the bailers.

The most terrestrial crabs, such as coconut crabs (*Birgus latro*), Trinidad mountain crabs (*Pseudothelphusa garmani*) and freshwater crabs (*Austrothelphusa transversa*), have more complex lungs, reduced gills (compare the surface areas of the gills of *Birgus* – 12 mm^2 (g body mass)$^{-1}$, and *Carcinus* – 765 mm^2 g body mass^{-1} in Tables 12.3 and 12.2) and carry little or no water in their gill chambers. Diffusion distances across the lungs of the most terrestrial land crabs are considerably lower than those across the gills of both fully aquatic and air-breathing crabs, as shown in Figure 12.16. Some diffusion distances are comparable to those in the lungs of birds and mammals, which are also shown in Figure 12.16.

Trinidad mountain crabs are particularly well adapted for air-breathing. They live in air-filled burrows near freshwater streams in the mountainous rain forests of Trinidad and South America. Their lungs are confined to a thick area in the roof of the gill chamber and the lining of the gill cover is perforated by many holes which lead to air channels within each lung, as shown in Figure 12.17A,B. When mountain crabs are in water, they ventilate their reduced gills using the bailers, but these stop working when the animals are in air and a unidirectional air current is generated by movements of the walls of the gill chambers.

Figure 12.17C shows the pressure changes in the gill chamber when a crab is in water and when it is in air. A unidirectional air current in a respiratory system means that there is minimal deadspace[31]. The unusually high P_aO_2 (17 kPa), indicates an excessively high level of ventilation for the rate of oxygen consumption, which is known as **hyperventilation**. The primary function of the hyperventilion may be to maintain a relatively low P_aCO_2 (1.3kPa), as there is little or no water in the gill chamber within which CO_2 could dissolve.

An important aspect of breathing air compared with breathing water is that, for a given rate of oxygen consumption ($\dot{M}O_2$), air-breathing animals pump less air through their lungs ($\dot{V}_a$), than the volume of water aquatic animals pump over their gills ($\dot{V}_w$). The ratio between the rate of ventilation and $\dot{M}O_2$ is known as the ventilation-convection requirement. The values in Table 12.2 for the ventilation convection requirement of water-breathing crabs ($\dot{V}O_2 / \dot{M}O_2$) is about 6.5 times greater than those for air-breathing crabs ($\dot{V}_a / \dot{M}O_2$) given earlier in Table 12.3. The difference between these values results in part from the difference in the capacitance coefficients of oxygen in air and water[32]. Box 12.2 provides one explanation for the difference in the ventilation convection requirement between air-breathing and water-breathing animals being less than might be expected on the basis of the different capacitance coefficients of oxygen in air and water (approximately 30:1).

The most complex air-breathing organs have evolved in vertebrates, particularly in mammals and birds. However, many species of fish are also able to exchange gases with the atmosphere, and we look at these next.

A range of air-breathing organs in fish

The ability of fish to obtain oxygen from the air evolved in both marine and freshwater species about 415 million years ago. It has been estimated that over time, air breathing has independently evolved among fish at least 38 times and maybe as many as 67 times. There are currently approximately 25,000 known species of fish, but only about 2 per cent of them are able to exchange oxygen with the air.

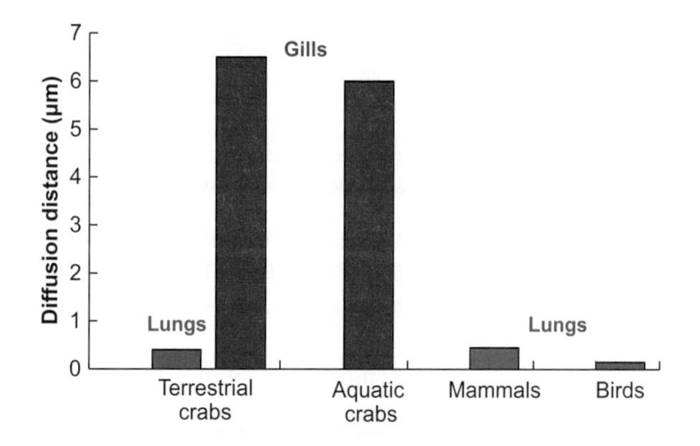

Figure 12.16 Diffusion distances across the gas-exchange organs of some species of terrestrial and aquatic crabs, and comparison with the lungs of mammals and birds
Average values from Tables 12.2, 12.3 and 12.5.

31 Deadspace is described in Section 12.2.3.
32 Values for the capacitance coefficients of oxygen in water and air are given in Table 11.2.

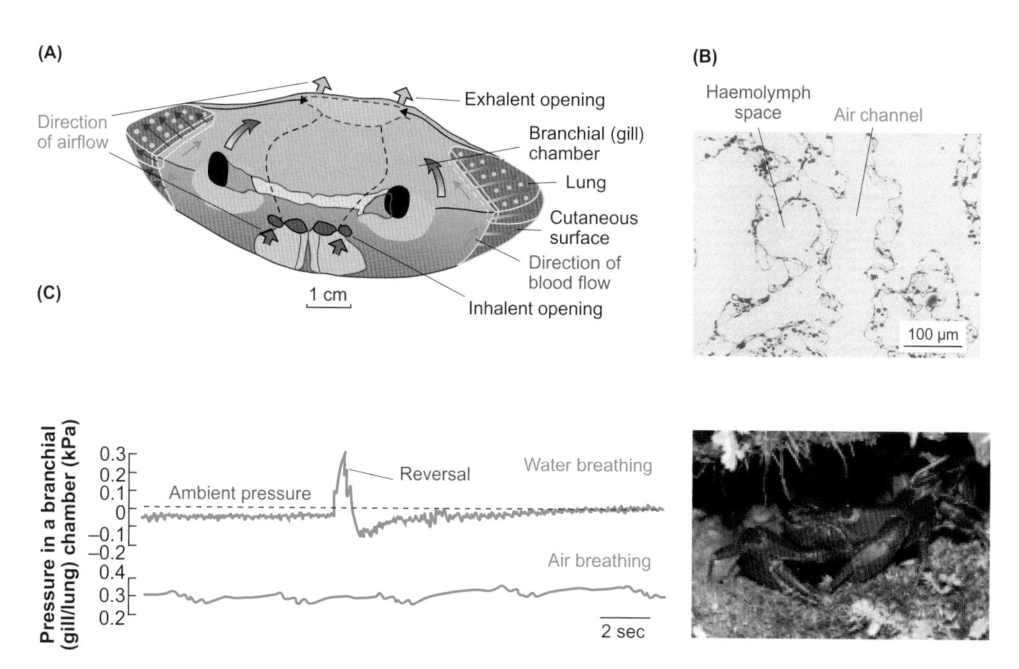

Figure 12.17 Gas exchange system of the Trinidad mountain crab (*Pseudothelphusa garmani*)

(A) Diagramatic section through the lung region of the branchial (gill/lung) chamber showing the directions of the flows of haemolymph (blood–blue arrows) and air (green arrows) through the gill/lung chamber when the crab is breathing air. Air flow is unidirectional. Note also the regular pores in the lungs which lead to the air channels. (B) Histological section through the lung showing the very thin tissue barrier between the haemolymph spaces and air channels. (C) The pressure changes in the branchial chamber while the crab is submerged under water are below atmospheric for most of the time and similar to those of the shore crab, as shown in Figure 12.6. There are even reversals of the beat of the scaphognathite (bailer – see Figure 12.6 for details). When the crab is in air, the pressure in the chamber is above that in the atmosphere and causes air to flow in one direction through the chamber.

A, C: Reproduced from: Innes AJ et al (1987). Air-breathing in the Trinidad mountain crab: a quantum leap in the evolution of the invertebrate lung? Comp. Biochemistry and Physiology 87A: 1-8. B: Taylor HH and Taylor EW (1992). Gills and lungs: the exchange of gases and ions. In: Microscopic Anatomy of Invertebrates vol 10: Decapod Crustacea. ed., Harrison FW and Humes A, pp. 203-293. Wiley, New York. Crab image © Tom Murray.

Air-breathing in fish is related to two factors: a reduction in the supply of oxygen (hypoxia) in the environmental water and an increase in demand for oxygen resulting from an increase in temperature and/or activity. Some species of fish breathe air even when they are in well-aerated water; these are known as **obligate air breathers**. Other species only breathe air when the water is particularly hypoxic, the temperature is high and/or they are particularly active. These are called **facultative air breathers**.

The evolutionary pressure for the development of an air-breathing organ (ABO) was most likely a limitation in the availability of oxygen from the surrounding water, so that the air-breathing organ was often the major source of oxygen. However, no present day air-breathing fish has lost its gills completely; in most species, the gills serve as a major route for the removal of carbon dioxide, much as in the terrestrial crustaceans we discussed in the previous section.

In modern air-breathing fish, three different regions of the body have evolved to form air-breathing organs:

- branchial (gill), opercular and pharyngeal cavities,
- the gastrointestinal (GI) tract,
- the air bladder.

The air bladder evolved into the lungs of the terrestrial vertebrates. We now consider each of these regions.

Some of the most effective air-breathing organs (ABO) in fish are found in climbing perch (*Anabas testudineus*) and are illustrated in Figure 12.18. Notice in this figure how the ABO consists both of the lining of the suprabranchial chambers and intricately layered bony structures known as the labyrinthine organs, which are attached to the dorsal regions of the gill arches. The ABO are highly effective gas-exchange organs with a high morphological diffusion conductance for oxygen, as indicated earlier in Table 12.3. The diffusion distance across them is less than that of other air-breathing organs in fish. In contrast, Tables 12.2 and 12.3 show that the diffusion distance across their gills is larger than that in other species of air-breathing fish, such as the bowfin, so the morphological diffusion conductance of the gills for oxygen is very low. Consequently, *Anabas* is probably an obligate air breather.

Parts of the gastrointestinal tract, especially the stomach or part of the small intestine, are modified in a few species of modern fish to allow gas exchange between the blood and rhythmically swallowed air. The air is then expelled either via the anus or mouth. Although the gastrointestinal tract does not make a major contribution to gas exchange in

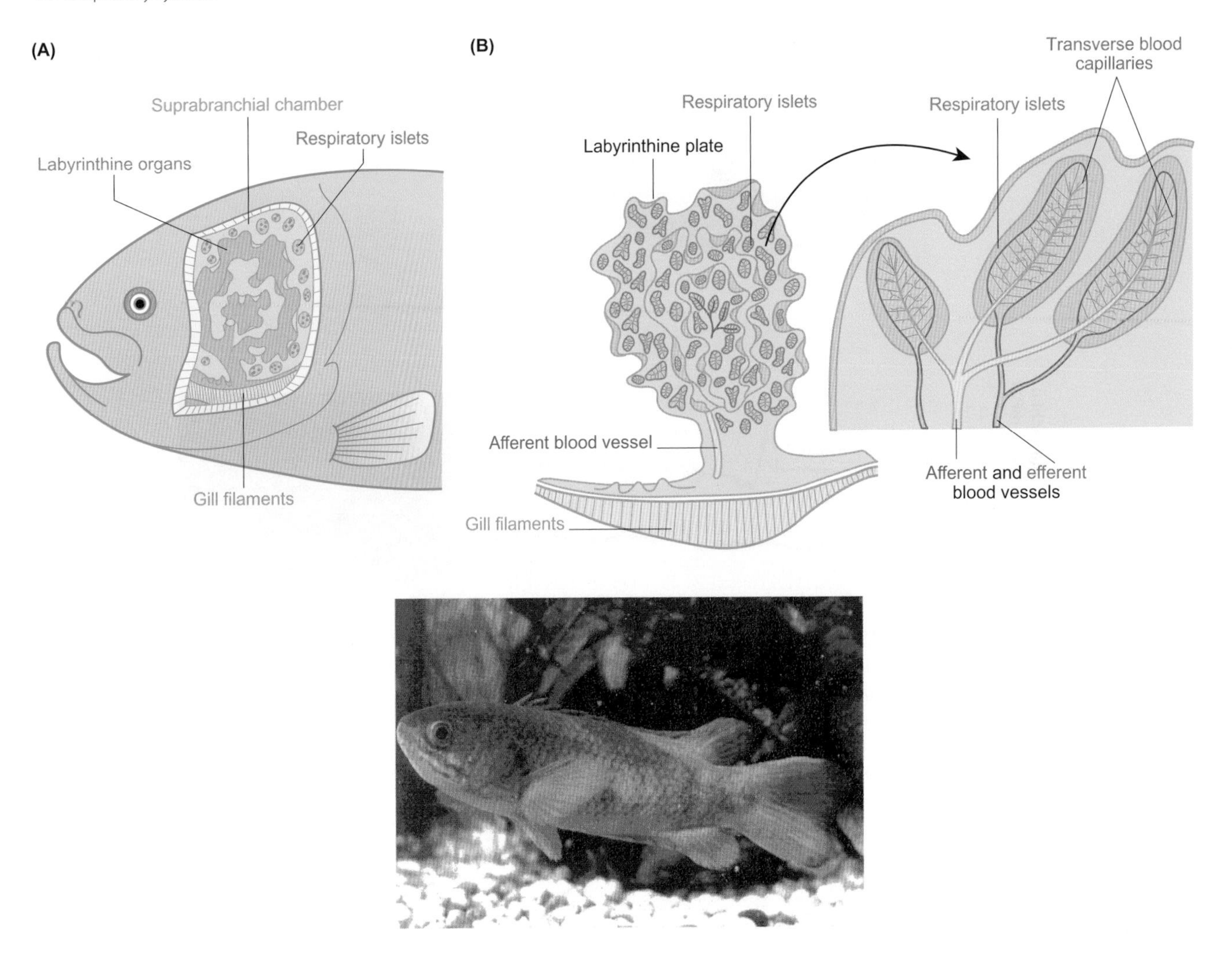

Figure 12.18 Air-breathing organs in climbing perch (*Anabas testudineus*)

(A) Location of the labyrinthine organs and respiratory islets in the lining of the suprabranchial chamber in the climbing perch. (B) Diagram of the entire labyrinthine plate on the first gill arch and close-up of a part of the surface of the plate showing the blood supply to (afferent vessels) and from (efferent vessels) three respiratory islets, which are where gas exchange occurs. Reproduced from Hughes GM and Datta Munshi JS (1973). Fine structure of the respiratory organs of the climbing perch, *Anabas testudineus* (Pisces: Anabatidae). Journal of Zoology 170: 201-225. Image: courtesy of Klaus de Leuw.

modern fish, the air bladder was derived from the anterior region of the gastrointestinal tract and gave rise to the lungs of the land vertebrates. We explore the lungs of land vertebrates later in this section.

Before exploring the structure and function of air bladders in modern fish, we need to consider the relationship between the main groups (clades) of bony fish: the Actinopterygii (also known as **ray-finned fish**) and the Sarcopterygii (known as **lobe-finned fish**)[33]. The Actinopterygii includes most of the bony fish, while the Sarcopterygii includes the lungfish and gave rise to the land vertebrates.

The original function of the air bladder in fish was as an accessory respiratory organ. It was derived from the posterior region of the pharynx known as the respiratory pharynx. The air bladder in most ray-finned fish arises from the dorsal surface of the pharynx whereas the lungs of the lobe-finned fish open ventrally from the pharynx. In more ancient ray-finned fish, such as bowfin (*Amia* spp.), long-nosed gar (*Lepisosteus* spp.) and bichir (*Polypterus* spp.), the air bladder still serves as a gas-exchange organ, but in most bony fish it serves primarily as a buoyancy, sound production or sound detection organ. In a few species of teleost fish (the largest group of ray-finned fish), such as S. American jeju or red wolf fish (*Erythrinus* spp.) and pirarucu (*Arapaima* spp.), the air bladder has secondarily reverted to its original use as a gas-exchange organ.

Most of the fish that use the air bladder for gas exchange use the mouth (buccal cavity) as a pump to force air into the bladder. This is similar to the way in which water is

[33] Section 1.4.3 explains the terms ray-finned fish and lobe-finned fish in more detail.

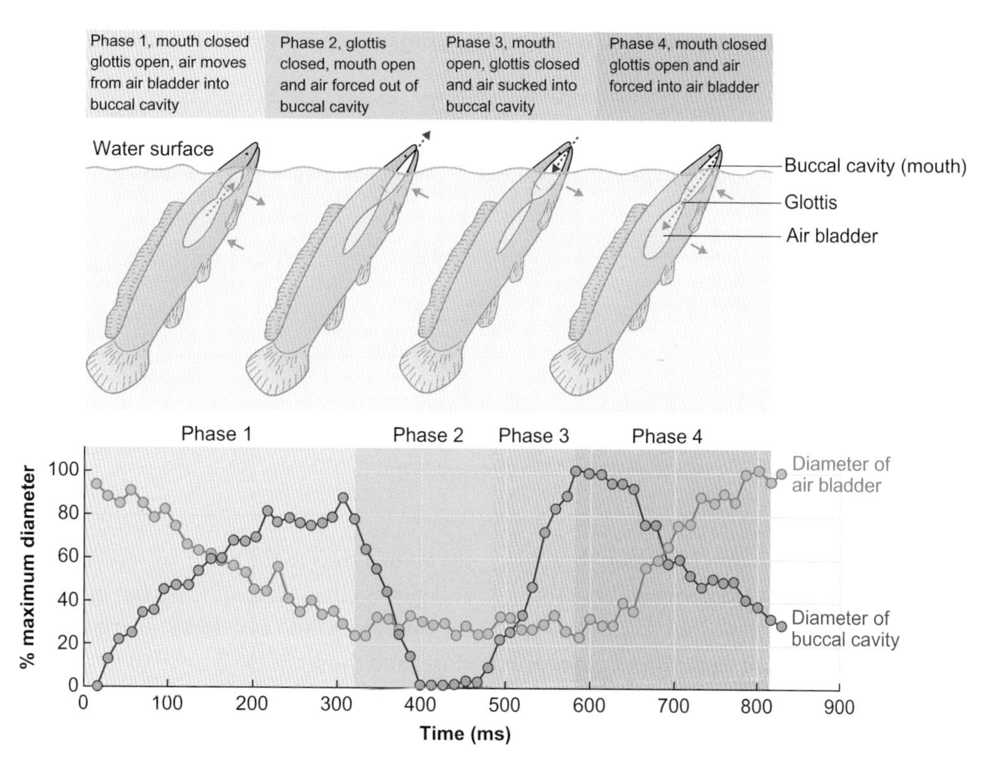

Figure 12.19 Ventilation cycles of air bladder in bowfin (*Amia calva*)

There are four phases to the ventilation cycle in this species. After the fish has reached the water surface, the buccal (mouth) cavity expands and compresses twice during each ventilation cycle, once to expel gas from the air bladder during phases 1 and 2 and again during phases 3 and 4 to force air into the air bladder. Dashed arrows in diagrams indicate direction of air flow and solid arrows indicate expansion or contraction of buccal cavity and air bladder.

Adapted from Brainerd EL (1994). The evolution of lung-gill bimodal breathing and the homology of vertebrate respiratory pumps. American Zoologist 34: 289-299.

forced between the gills in aquatic fish, as shown earlier in Figure 12.11.

The ventilatory cycle of the air bladder in bowfin has four phases, as shown in Figure 12.19. The buccal cavity expands and compresses twice during each ventilation cycle, once to expel gases from the lungs and again to force air into the lungs, so it is possible that the two streams of gas are maintained separately.

Some species, such as bichir, use a suction (**aspiration**) method to fill the air bladder. Exhalation in bichir results from contraction of muscles in the walls of the lungs, but the fish is encased in a stiff scale jacket which stores elastic energy as the volume of the lungs is reduced. Consequently, as soon as the muscular activity of exhalation ceases, the elastic recoil of the body causes the pressure in the lung to fall below atmospheric pressure, and air to be sucked into the lung through the open mouth and glottis. This suction mode of ventilation may be of significance in relation to the evolution of an aspiration (suction) system in the early land vertebrates, which also had a thick body covering, much like that in bichir.

Lungfish are among the lobe-fin group of fish and have retained and further evolved the respiratory function of the air bladder.

The lungs of lungfish and the pulmonary circulation

In all the air-breathing fish discussed so far, the circulation of blood is similar to the **single circulation** seen in water-breathing fish, whereby blood is pumped from the single ventricle of the heart, through the gills, and is distributed round the body in the arterial system before travelling back in the venous system to the single atrium of the heart[34]. With this type of circulation, the oxygen-rich blood from the air-breathing organ returns to the venous system where it mixes with oxygen-poor blood from the rest of the body before returning to the heart. However, lungfish have a **double circulation**: oxygen-rich blood from the lungs and oxygen-poor blood from the body are partially separated within the heart.

The blood vessels from the lungs, the pulmonary veins, go directly to the heart and the heart is partially divided into two. The right side of the heart receives oxygen-poor blood from the body and sends it to the lungs, and the left side of the heart receives oxygen-rich blood from the lungs and

[34] The circulatory system of water-breathing fish is discussed in Section 14.4.1.

Table 12.4 The relative importance of the gills/skin and air-breathing organ/lung in fish that can breathe water and air (bimodal breathers)

Species	Per cent O_2 uptake		Per cent CO_2 release		Respiratory exchange ratio (CO_2 release/O_2 uptake)	
	Gills/skin	Air-breathing organ/lung	Gills/skin	Air-breathing organ/lung	Gills/skin	Air-breathing organ/lung
Climbing perch *(Anabas testudineus)*	45	55 (25°C)			2.3	0.2
Bowfin *(Amia calva)*	100 60	0 (25°C) 40 (30°C)	100 92	0 8	1.6	0.2
Pirarucu *(Arapaima gigas)*	22	78	63	37 (29°C)	2.3	0.4
African lungfish or marbled lungfish *(Protopterus aethiopicus)*	8	92	68	32 (24°C)	5.4	0.27

Sources: Hughes and Datta Munshi, 1979; Randall et al., 1981; Randall et al., 1978; McMahon, 1970.

sends it to the body. This is similar to the circulation through the heart found in land vertebrates[35]. Consequently, the three species of lungfish, *Neoceratodus forsteri* from Australia, *Protopterus aethiopicus* from Africa and *Lepidosiren paradoxa* from South America, have attracted considerable attention.

Neoceratodus has only a single lung and well-developed gills with lamellae on all of the gill arches; it is a facultative air-breather. Virtually all of its gas exchange occurs in water, provided it is well aerated. In contrast, the two other species of lungfish have a pair of lungs and are obligate air breathers. Their poorly developed gills contribute little to oxygen uptake. For example, the gills of *Lepidosiren* contribute less than 0.002 per cent of total morphological diffusion conductance (gills plus skin and lungs) for oxygen, while the lungs contribute over 99 per cent. However, in *Protopterus*, most of the loss of carbon dioxide is via the gills and skin, as shown in Table 12.4.

The lungs of lungfish have regular subdivisions which increase their effectiveness as gas-exchange organs. For example, the internal surfaces of the lungs of *Protopterus* have septa, ridges and pillars that divide the air space into smaller compartments which open into a large central cavity, as shown in Figure 12.20. These subdivisions increase the area of the gas-exchange surface without increasing the overall volume of the lung, in much the same way as the lamellae increase the surface area of the gills of fish and which are shown back in Figure 12.6.

When submerged under water, *Protopterus* irrigate their gills by a buccal force pump mechanism, similar to that illustrated back in Figure 12.11 for fully aquatic fish, although the opercular suction phase is less important in *Protopterus*. The buccal force pump is also used to ventilate the lungs, but unlike that of bowfin, shown in Figure 12.19, the lung ventilatory cycle of lungfish consists of only two phases. When the animal rises to the water surface to breathe air, its mouth and glottis open. Stale air flows out of the lungs and the buccal cavity expands and sucks in fresh air. The mouth then closes and the air in the buccal cavity is forced into the lungs.

The fact that there are only two phases in the cycle means that it is possible for the stale gas leaving the lungs to mix with the fresh air in the buccal cavity. Such mixing would reduce the overall effectiveness of gas exchange. Mixing of fresh air with some exhaled stale and reduction in overall effectiveness may also occur in some species of amphibians.

The lungs of amphibians and reptiles

Among amphibians and reptiles, we see a greater dependency on air as their source of oxygen. This transition involves loss of gills and the development of more complex lung structures.

Structure of the respiratory systems of amphibians and reptiles

In adult amphibians, a short or non-existent trachea keeps the anatomical deadspace[36] to a minimum, which is probably related to the relatively small tidal volumes (2–3 mL kg^{-1}) in these animals. The paired lungs are well-vascularized sacs with internal compartments divided by septa to form a honeycomb-like structure that increases the surface area for gas exchange. Such compartments are called **faveoli** to distinguish them from the sac-like **alveoli** in the lungs of mammals. The average diameter of the faveoli varies between 0.8 to 2.3 mm, which is larger than the alveoli of mammalian lungs.[37] In reptiles there is an obvious trachea and the lungs have a greater variety in their internal structure than those of lungfish and amphibians, as shown in Figure 12.21.

[35] Figures 14.2, 14.40 and 14.42 show the circulation through the hearts of mammals, African lungfish, African clawed toad and turtles, respectively.

[36] The proportion of respiratory tidal volume (the volume of air inhaled with each breath) that does not reach the gas exchange area of the lungs and, therefore, is not involved in gas exchange is known as the anatomical deadspace. It is discussed in more detail later in this section and shown in Figure 12.26B.

[37] Examples of alveoli in mammalian lungs are shown in Figure 12.28.

12

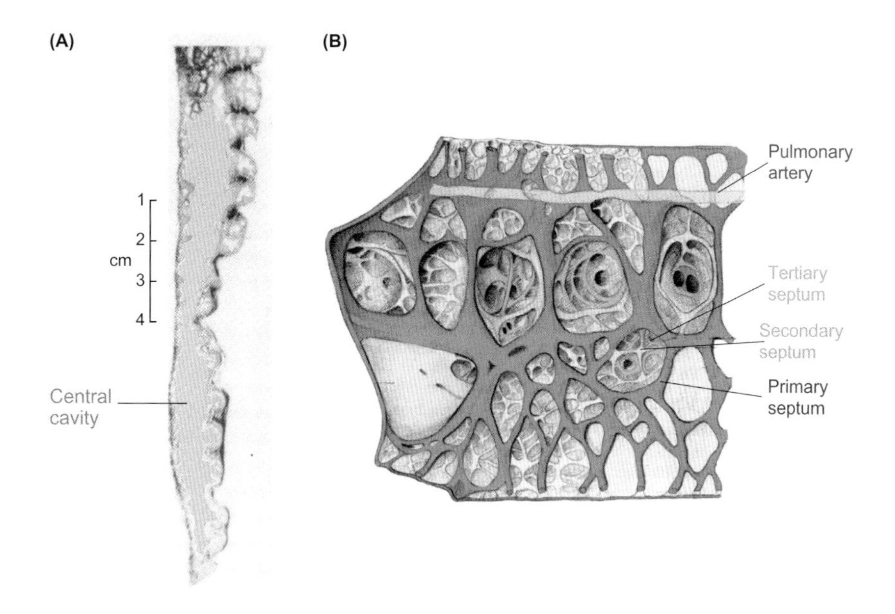

Figure 12.20 Structure of the lung in African lungfish (*Protopterus aethiopicus*)

(A) Longitudinal section through the posterior end of one lung showing coarse alveoli-like structures along the periphery and the large central cavity. (B) Internal structure of the lung showing the three levels of internal septa which divide the lung into alveoli-like pockets which are called faveoli. Some of the second and third levels of septa are shown in colour.

Reproduced from Jesse MJ et al (1967). Lung and gill ventilation of the African lung fish. Respiration Physiology 3: 267-287. reproduced from Johansen K (1970). Air breathing in fishes. In Fish Physiology, vol 4. Hoar WS and Randall DJ (Eds.) pp 361-411. Academic Press, New York.

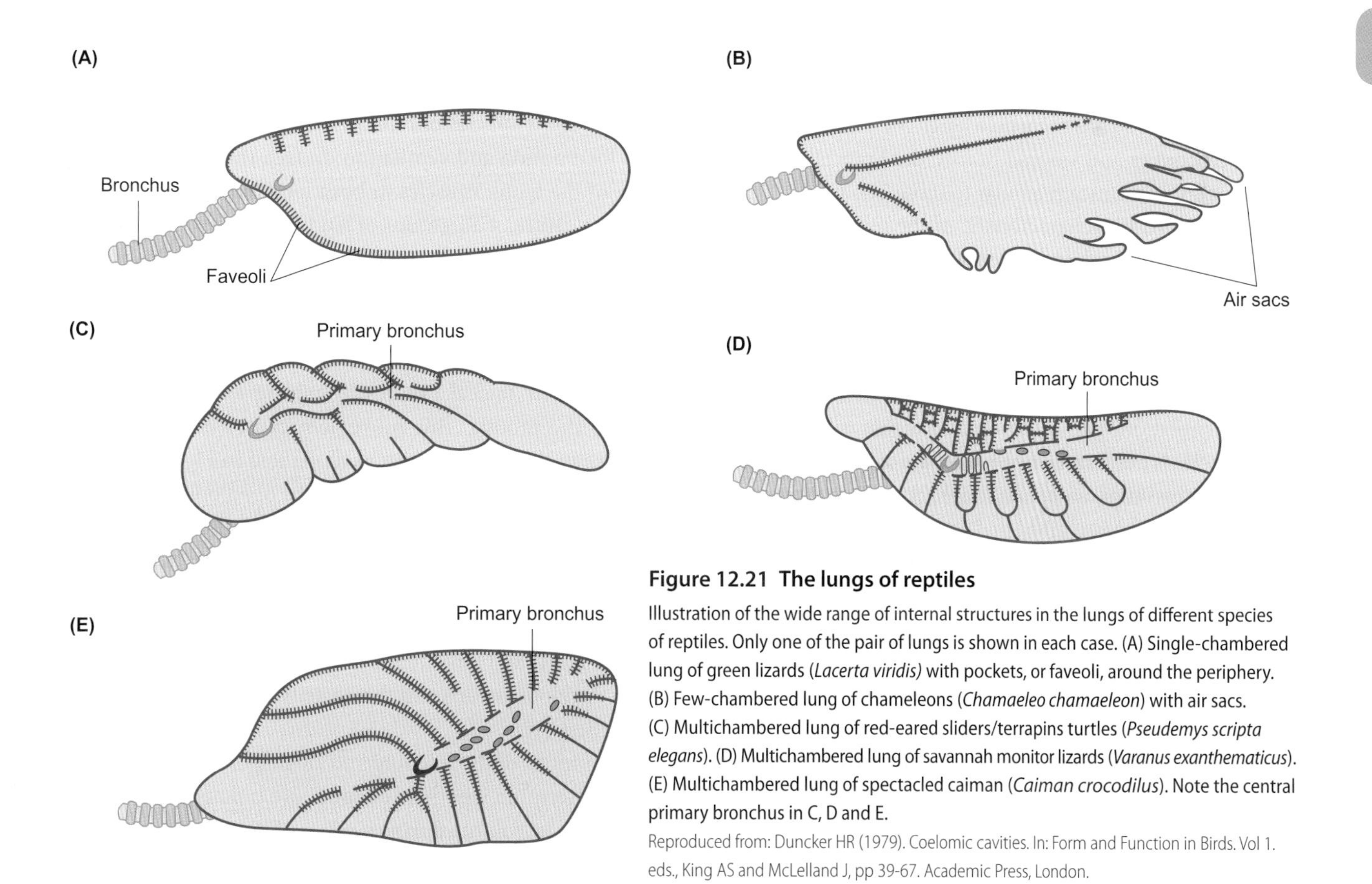

Figure 12.21 The lungs of reptiles

Illustration of the wide range of internal structures in the lungs of different species of reptiles. Only one of the pair of lungs is shown in each case. (A) Single-chambered lung of green lizards (*Lacerta viridis)* with pockets, or faveoli, around the periphery. (B) Few-chambered lung of chameleons (*Chamaeleo chamaeleon*) with air sacs. (C) Multichambered lung of red-eared sliders/terrapins turtles (*Pseudemys scripta elegans*). (D) Multichambered lung of savannah monitor lizards (*Varanus exanthematicus*). (E) Multichambered lung of spectacled caiman (*Caiman crocodilus*). Note the central primary bronchus in C, D and E.

Reproduced from: Duncker HR (1979). Coelomic cavities. In: Form and Function in Birds. Vol 1. eds., King AS and McLelland J, pp 39-67. Academic Press, London.

Table 12.5 Morphometry of lungs and physiological characteristics of gas exchange in the four Classes of terrestrial vertebrates. These data are from resting animals. Where there is a single value, it is a mean value for a number of species, unless otherwise stated.

Class	Volume [ml (kg body mass)$^{-1}$]	Surface area [cm^2 (g body mass)$^{-1}$]	Diffusion distance (μm)	Diffusion conductance (μmol or mmol O_2 min^{-1} kPa^{-1} kg^{-1})	Extraction of oxygen from air (%)	Ventilation convection requirement [mL (μmol O_2)$^{-1}$]	P_aO_2 (kPa)	P_aCO_2 (kPa)
Amphibians	60–90	2–8	1.0–2.3	9 μmol, physiological, bullfrog			variable	1.8
Reptiles	85–307	0.7–5.4	0.5–1.4	31 μmol, physiological, *Varanus* 0.85 mmol, morphological *Varanus & Lacerta* 0.11 mmol, morphological crocodile	5–15, lizards, 30°C 7–27, snakes, 30°C 18–25, turtles, 30°C 15, alligator, 30°C	0.24-0.56, turtles, 35°C 0.84, lizards, 37°C	variable	2.4
Birds	30 (160 including air sacs)	18–87 (excluding chicken)	0.1–0.23 (excluding chicken and penguins)	0.43 mmol, physiological, duck 1.4 mmol, morphological	20	0.51	12.5	4.4
Mammals (excluding bats)	46	65 etruscan shrew 48 horses	0.27 etruscan shrew 0.6 horses	0.24 mmol, physiological, canids 1.3 mmol, morphological	12–15	0.85	13.3	5.3

P_aO_2 is partial pressure of oxygen in arterial blood, P_aCO_2 is partial pressure of carbon dioxide in arterial blood.

Sources: Burggren, 1989; Glass et al., 1981a; 1981b; Perry, 1992; Ultsch, 1987; Perry, 1998; Glass and Wood, 1983; Lasiewski and Calder, 1971; Maina, 2002; Glass, 1989; Maina, 2000; Frappell et al., 2001; Powell, 2000; Gehr et al., 1981; Weibel et al., 1983; Weibel, 1984. Full citations available online.

The lungs of some groups of reptiles have a number of features that are similar to those seen in the lungs of birds, which we discuss later in this section. The lungs of the more active species, such as turtles, monitor lizards and crocodilians, are multichambered. In these species each bronchus runs the whole length of the lungs, connecting with all of the chambers. The bronchus also runs the whole length of the lungs in birds, and in crocodiles and varanid lizards, the branches from the bronchus are anatomically similar to those in birds[38].

Another bird-like characteristic of the respiratory system of reptiles is seen in chameleons and some species of snakes, in which the caudal (posterior) part of the lung is not involved in gas exchange and may form air sacs, as shown in Figure 12.21B. These non-respiratory regions or air sacs, store air that passes over the anterior gas-exchange surface during ventilation. This situation means that tidal volume in these species is not limited by the volume of the gas-exchange regions of the lungs, as it is in amphibians. The presence of such non-respiratory regions makes the lungs of reptiles relatively larger (85 to over 300 mL kg^{-1}) than those of amphibians, as shown by the data in Table 12.5.

Despite the differences in gross morphology, the data in Table 12.5 and earlier Figure 12.8 illustrate that the diffusion properties of the lungs of reptiles are not very different from those of amphibians and that the morphological diffusion conductance of the lungs is very similar between most reptiles, despite their different lung structures.

[38] Structure of the lungs of birds is illustrated in Figure 12.32.

Ventilation of the lungs: from force to suction

The lungs of amphibians and reptiles are ventilated differently. Amphibians use a buccal force pump, like fish, but reptiles have evolved a suction (aspiration) pump. The suction pump of reptiles is powered by muscles in the thorax and abdomen (trunk), but the trunk muscles are used for locomotion in fish, and some species of reptiles use their trunk muscles for both locomotion and ventilation. In fact, the involvement of some of the trunk muscles in both locomotion and ventilation is a feature of all groups of land vertebrates.

Ventilation of the lungs is intermittent in amphibians and reptiles, with periods of breath-holding following single or multiple inflations of the lungs. However, many species of frogs and toads (anurans) with access to air continuously ventilate their buccal cavities by raising and lowering the floor of the mouth. This movement causes fluctuations in pressure in the buccal cavity, as illustrated in Figure 12.22A, which move air in and out through the open nostrils. At this time, the glottis is closed and the pressure in the lung is above atmospheric pressure, which is shown in Figure 12.22B. The buccal cavity is filled with fresh air after a number of buccal ventilations and periodically much larger pressure changes in the buccal cavity inflate the lungs. Figure 12.22B indicates when the glottis opens and the nostrils close during the lung ventilation cycle.

The timings of the opening and closing of the nostrils and glottis determine whether the lungs are inflated, deflated or unchanged in volume during a series of lung ventilations. As we discuss for lungfish, the air forced into the lungs of these

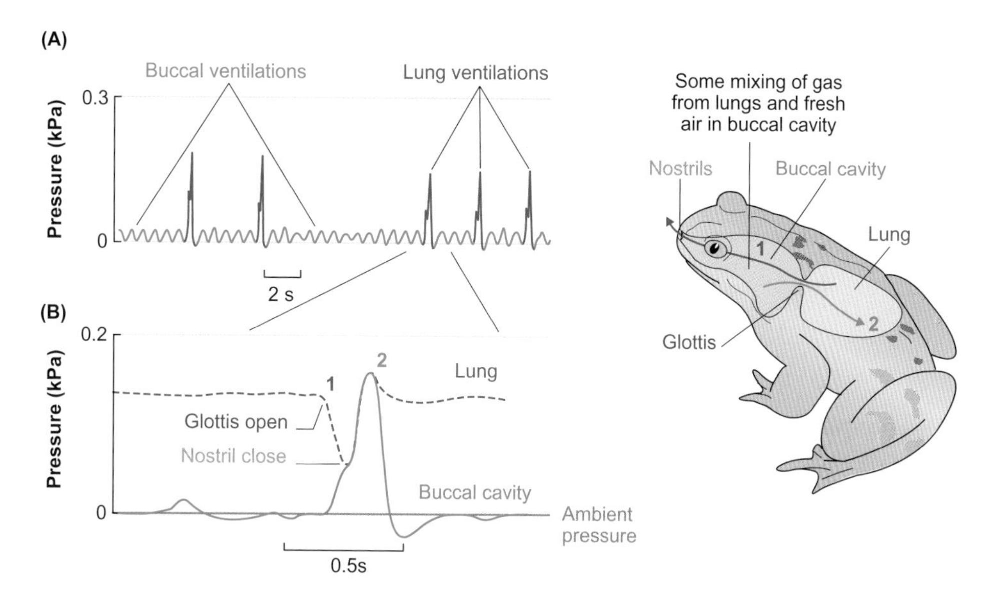

Figure 12.22 Ventilation of the buccal (mouth) cavity and lungs in frogs

(A) Pressure in the buccal cavity of a northern leopard frog (*Lithobates pipiens*, formerly *Rana pipiens*), showing continuous oscillations of the floor of the buccal cavity which move air in and out of the cavity, and five ventilations of the lungs. (B) Pressures in the lung and buccal cavity of a leopard frog during a single lung ventilation cycle. The lung ventilation cycle is also depicted in the diagram and shows that the first step in the cycle is when the glottis opens and air leaves the lungs, with some flowing out of the nostrils (1). The nostrils then close and the floor of the buccal cavity is raised to force air from the buccal cavity into the lungs (2).

A: Reproduced from Vitalis TZ and Shelton G (1990). Breathing in *Rana pipiens*: the mechanism of ventilation. Journal of Experimental Biology 154: 537-556. B: Reproduced from West NH and Jones DR (1975). Breathing movements in the frog *Rana pipiens*, 1. The mechanical events associated with lung and buccal ventilation. Canadian Journal of Zoology 53: 332-344. Diagram adapted from de Jongh HJ and Gans C (1969). On the mechanism of respiration in the bullfrog, *Rana catesbeiana*: a reassessment. Journal of Morphology 127: 259-289.

amphibians is a mixture of the 'stale' air from the lungs and fresh air from the buccal cavity.

The energetic cost of lung ventilation in northern leopard frogs (*Lithobates pipiens*) is approximately 5 per cent of total metabolic rate, compared with 10 per cent for active ventilation of the gills of fish[39]. This small difference in the cost of ventilation between amphibians and fish is despite the fact that air is much less dense and viscous than water[40], and suggests that the buccal force pump in amphibians is not very efficient.

Exhalation is passive in most frogs and toads due to the elastic recoil of the lungs and body. In contrast, many newts and salamanders (urodeles) actively expire by contraction of the *transverse abdominal* muscle. This muscle also plays an important role in lung ventilation of a number of species of reptiles, as we discuss later.

Reptiles use an aspiration (suction) mechanism to ventilate their lungs. However, different groups ventilate their lungs in different ways because of the variations in their body structures. The most straightforward ventilation system occurs in lizards and snakes. The muscles in the thorax move the ribs and change the pressure in the thoracic cavity, initially forcing gas out of the lung and then sucking it in. After the glottis has opened, gas is forced out of the lungs by the action of two deep muscles which move the ribs and reduce the volume of the thoracic cavity. One of the expiratory muscles is the **transverse abdominal**, the same muscle that is used to power expiration in some newts and salamanders. Inspiration then follows, resulting from contraction of the muscles between the ribs, the **internal and external intercostal muscles**, of the anterior four rib segments. The contractions of these muscles cause expansion of the thoracic cavity, drawing air into the lungs.

The impact of the use of what were exclusively locomotor muscles in fish for ventilation in reptiles is seen in some species of lizards, which cannot ventilate their lungs adequately while running. The reason for this is that most of the muscles that are used for the sideways flexing of the body during locomotion of lizards (including the respiratory muscles mentioned above), are also attached to the ribs. The consequence of the attachments of these muscles to the ribs is that the respiratory muscles on each side of the body contract alternately when the animal is running but need to contract simultaneously when it ventilates its lungs. Some species of lizards have evolved other methods of ventilation which overcome the conflict caused by the same muscles potentially being used for both ventilation and for locomotion. For example, some species of varanid lizards ventilate their lungs using a gular (throat) pump when running, and this compensates for the restriction on the use of the costal (rib) muscles.

[39] Energy cost of ventilation in fish is discussed in Section 12.2.2.

[40] Values for viscosities and densities of air and water are given in Table 11.3.

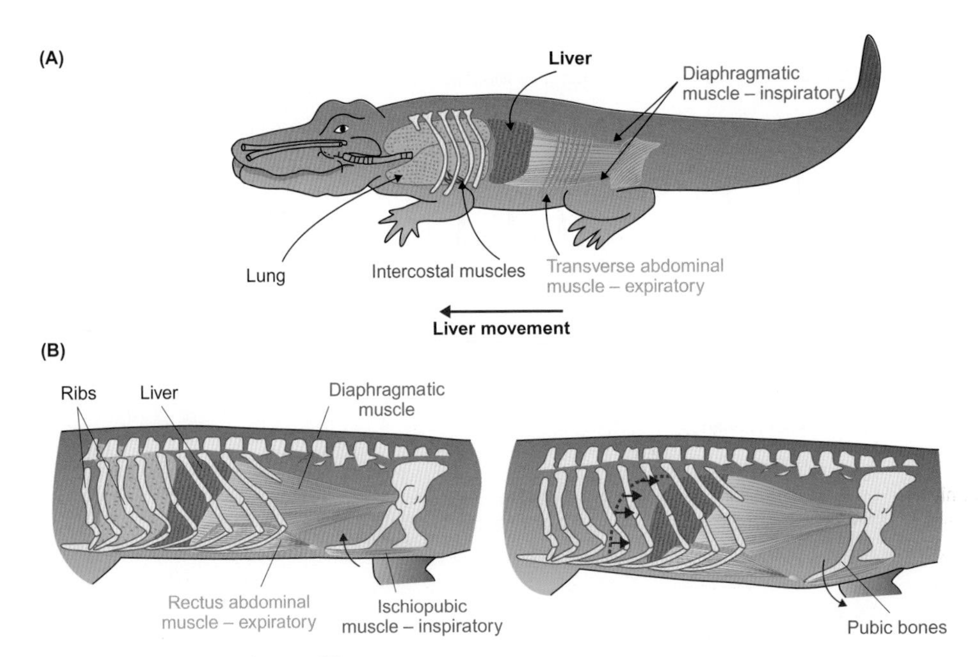

Figure 12.23 Ventilation of the lungs of crocodilians

(A) Diagram to illustrate the major muscles involved in lung ventilation in caiman (*Caiman crocodylus*). The diaphragmatic and transverse abdominal muscles function to move the liver backward and forward, respectively, so that it acts as a piston, sucking air into the lungs and then forcing it out. (B) Diagrams to illustrate the muscles involved in the movement of the pubic bones which may assist ventilation in crocodilians by making space for the backward movement of the viscera during inspiration. The rectus abdominal muscle pulls the pubic bones forward during expiration and the ischiopubic muscle pulls them backward during inspiration.

Reproduced from Gans C (1970). Strategy and sequences in the evolution of the external gas exchangers of ectothermal vertebrates. Forma et Functio 3: 61-104, and from Farmer CG and Carrier DR (2000). Pelvic aspiration in the American alligator (*Alligator mississippiensis*). Journal of Experimental Biology 203: 1679-1687.

Tortoises and turtles are unable to ventilate their lungs like lizards and snakes, as their ribs are fused to the dorsal carapace (shell) which, together with the ventral plastron[41] (breastplate), forms a rigid box-like exoskeleton enclosing the lungs. The only areas of soft moveable tissue are around the legs, neck and tail. The tortoise makes use of these areas during lung ventilation, when the limbs—and particularly the forelimbs—can often be seen moving in and out.

Lung ventilation in crocodiles is unusual in that it involves the use of the liver as a piston. This method of lung ventilation means that the respiratory muscles of crocodiles are independent of the locomotory muscles, unlike the situation in some lizards. Figure 12.23A shows the two main muscles that are used. Contraction of the **transverse abdominal** muscles reduces the volume of the abdominal cavity which pushes the liver forward, forcing gas out of the lungs. Contraction of the **diaphragmatic** muscle pulls the liver back, sucking air into the lungs. The **intercostal** muscles may assist in one or more of these phases, by adding some rigidity to the thoracic cavity as the liver moves back and forth. Dorsal and ventral movements of bones of the pelvic girdle may also contribute to ventilation by making space for the backward movement of the viscera during inspiration, as shown in Figure 12.23B. So, crocodiles can increase lung ventilation while running and may have been specialized for high levels of aerobic activity early in their evolution.

[41] Not to be confused with the layer of air trapped by fine hairs in some species of aquatic insects. We discuss plastrons and their role in gas exchange in aquatic insects in Section 12.4.2.

Some consequences of bimodal breathing in vertebrates

Most fully aquatic fish ventilate their gills continuously so that when in a steady state the PO_2 and PCO_2 in the blood leaving the gills are more or less constant. In contrast, resting, air-breathing fish, amphibians and reptiles ventilate their air-breathing organs (ABO) or lungs intermittently, with pauses of varying duration between bouts of ventilation. These pauses can lead to large variations in the partial pressures of the respiratory gases, both in the ABO or lungs, as illustrated in Figure 12.24A, and in the blood leaving the lungs in the pulmonary vein. At the end of a bout of ventilation, the air within the ABO or lung remains there until the next bout, which may be several minutes later.

The total amount of oxygen extracted from the air in the ABO or lungs increases with the duration of the pause in ventilation. Almost 30 per cent of the oxygen can be extracted from the air, as the data in Table 12.5 show, although some species of aquatic snakes and turtles have very long respiratory pauses and may extract up to 50 per cent of the oxygen in their lungs. In aquatic bimodal breathers, PCO_2 is typically much less variable, as shown in Figure 12.24A. African clawed frogs (*Xenopus laevis*) spend much of their time in water but visit the surface to breathe. Soon after a toad submerges, carbon dioxide is lost across the skin as fast

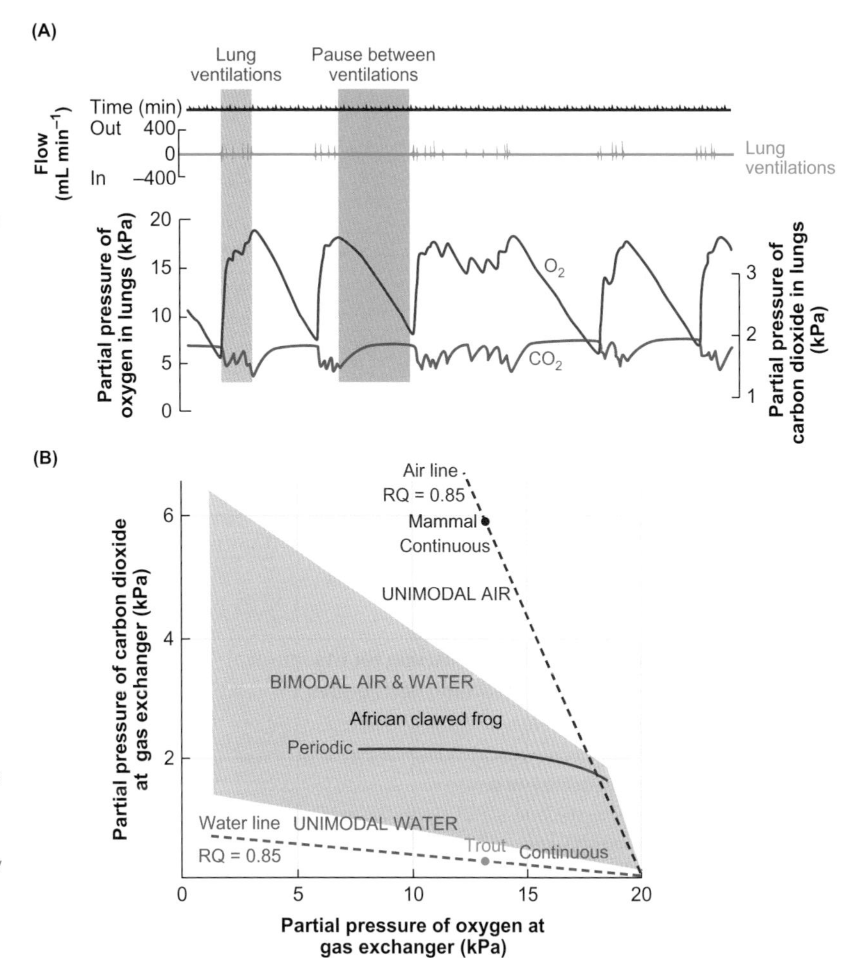

Figure 12.24 Bimodal breathing and gas exchange as illustrated by an African clawed frog (*Xenopus laevis*)

(A) Continuous recording of lung ventilation, as indicated by flow trace, and associated changes in partial pressures of oxygen and carbon dioxide (PO_2 and PCO_2, respectively) in the lungs of freely diving and surfacing *Xenopus* in water at 25°C and equilibrated with air. Note that although PO_2 falls continuously in between lung ventilations (red shading), PCO_2 reaches a plateau during the longer pauses as CO_2 diffuses into the water across the skin. (B) Different amounts of O_2 and CO_2 are exchanged by the lungs and gills/skin of bimodal breathers, as illustrated in Table 12.4, and they ventilate their lungs periodically. For these reasons, the relationship between PO_2 and PCO_2 in the lungs is different in the African clawed frog (solid line) from that in the expired water at the gills of exclusively water breathing animals (unimodal water) and in the lungs of the purely air breathing birds and mammals (unimodal air—dashed lines). The shaded area indicates the range of the relationships found in bimodal breathers. See Section 11.1.2 and Figure 11.8 for discussion of the two extreme conditions.

Reproduced from Boutilier RG and Shelton G (1986). Gas exchange, storage and transport in voluntarily diving *Xenopus laevis*. Journal of Experimental Biology 126: 133-155. adapted from Shelton G et al (1986). Control of breathing in ectothermic vertebrates. In: Handbook of Physiology – The Respiratory System II, section 3. The Control of Breathing. Cherniack NS and Widdicombe JG (Eds.) pp 857-909. American Physiological Society, Bethesda, MD.

as it is produced, so PCO_2 in the lungs remains constant. Figure 12.24B shows that the relationship between PO_2 and PCO_2 in the lungs of *Xenopus* and other bimodal breathers is part way between that exhibited in expired water at the gills of water-breathing, continuously ventilating animals such as fish and those in the lungs of air-breathing, continuously ventilating animals, such as mammals[42].

The overall respiratory quotient (RQ) for an animal will be approximately 0.85. In many species of bimodally breathing ectotherms, however, the relative amounts of oxygen and carbon dioxide exchanged at the gills/skin and lungs will give different respiratory exchange ratios (RER), as shown in Figure 12.25[43]. Generally, more carbon dioxide than oxygen is exchanged across the gills/skin because the capacitance coefficient of carbon dioxide in water is about 28 greater than that of oxygen, and so RER at these gas-exchange surfaces in bimodal breathers is above 0.85. The opposite occurs at the ABOs and lungs where the respiratory exchange ratio is less than 0.85. Some examples of these differences are given in Table 12.4.

[42] The extreme conditions between continually ventilating water breathers and continually ventilating air breathers is discussed in Section 11.1.2 and Figure 11.8.

[43] We discuss RQ and RER at the beginning of this section and in Section 2.2.2.

Even though the gills of fish that have more effective air-breathing organs are used primarily for the removal of carbon dioxide, their reduced diffusion conductance leads to an increase in the PCO_2 in their arterial blood compared to that in purely aquatic species, as indicated by the data back in Tables 12.2 and 12.3. This trend is even more apparent in amphibians, and tends to cause the pH of the blood to be lower than in completely aquatic species.

The internal structures of the lungs of reptiles vary in complexity, and the lungs themselves are ventilated by different mechanisms. Such variability is not seen within mammals and birds, although there are fundamental differences in lung structure, patterns of airflow and ventilation mechanisms between these two groups, which we now explore.

Mammalian lungs: ventilated pool gas exchangers

In mammals, the bronchus that enters each lung branches and divides like a tree, with branches that are progressively reduced in diameter, as shown in Figure 12.26A. In the lungs of humans, for example, we see 23 levels of branching with blind-ending sacs called **alveoli** occurring at increasing frequency beyond the terminal bronchioles at level 15, as illus-

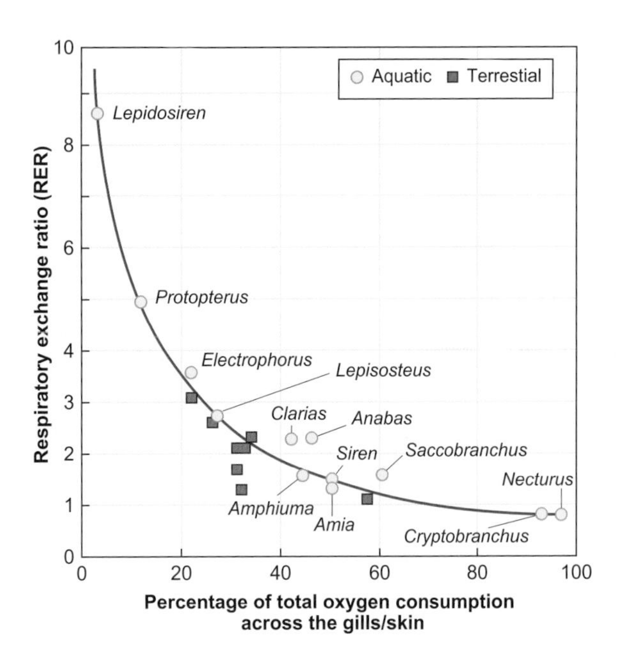

Figure 12.25 The importance of the gills/skin for the removal of carbon dioxide in bimodally breathing vertebrates

The respiratory exchange ratio (RER, $\dot{M}CO_2/\dot{M}O_2$) of the gills/skin system as a function of the percentage of total oxygen consumption that occurs across the gills/skin. The lower the percentage of oxygen consumption that occurs across the gills/skin, the greater the value of RER, i.e. the gills/skin remain important for the elimination of CO_2.

Reproduced from Rahn H and Howell BJ (1976). Bimodal gas exchange. In: Respiration of Amphibious Vertebrates. Hughes GM (Ed.), pp 271-285. Academic Press, London.

12

trated in Figure 12.26B. From level 19 onward, the ducts are completely lined with alveoli and terminate as alveolar sacs.

Gas exchange occurs exclusively within the alveoli. This means that the airways down to and including the terminal bronchioles are used purely for transporting (conducting) the respiratory gases and constitute the **anatomical dead-space**, as depicted in Figure 12.26B. In humans, the volume of the deadspace is approximately 150 mL.

The anatomical deadspace accommodates fresh air that does not reach the gas-exchange regions of the lungs during inspiration and expired air that has undergone gas exchange in the alveoli. This 'stale' air is then inhaled during the next inspiration, thus reducing the overall effectiveness of the lungs[44].

The respiratory zone is also shown in Figure 12.26B and extends from the end of the terminal bronchioles to the alveolar sacs and makes up the vast bulk of the lung. The enormous branching causes a dramatic increase in the total cross-sectional area of the airways at the transition between the conducting and respiratory zones and in the surface area for gas exchange, as shown in Figure 12.27A.

To give some idea of the exquisite design of the lungs, the total surface area of the alveoli in the lungs of a young, healthy human is approximately 140 m^2. This is about a third the area of a standard basketball court, yet it is contained in a volume of about 4 L. By comparison, a hollow sphere with a volume of 4 L has an internal surface area of only about 0.12 m^2.

Figure 12.27B gives an example of the distance over which the respiratory gases have to diffuse between the air in the alveoli and the red blood cells in the capillaries. For further comparison, the thickness of a page in this book is approximately 100 μm, whereas the average thickness of the exchange barrier in the alveoli of the mammalian lung is less than 1 μm, as the data in Table 12.5 illustrate.

[44] This reduction in effectiveness is illustrated in Figure 12.29A.

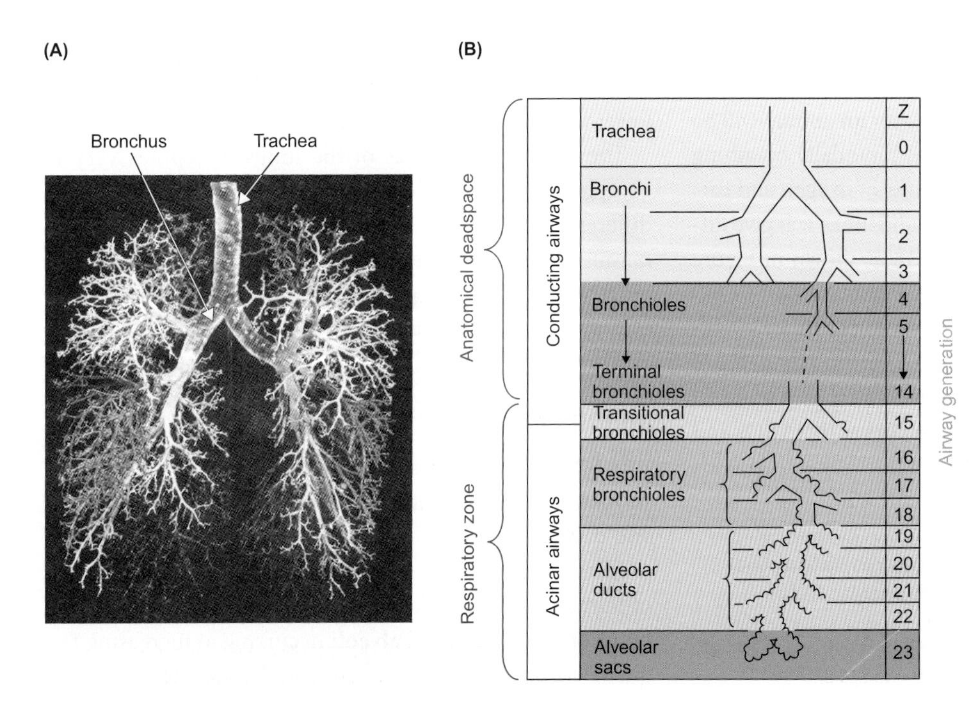

Figure 12.26 Lungs of mammals

(A) Cast of the airways of a pair of human lungs showing the network of branching bronchi and bronchioles. (B) The 23 generations (levels) of branching of the airways in the human lung from the trachea to the alveolar sacs. The regions where the alveolar sacs occur are known as the acinar airways and this is where gas exchange occurs. The acinar airways are not ventilated and constitute the respiratory zone

Reproduced from West JB (1974). Respiratory Physiology, the essentials. The Williams and Wilkins Company, Baltimore. Adapted from Weibel ER (2008). Modelling structure-function interdependence of pulmonary gas exchange. In Advances in Experimental Medicine and Biology Series, Vol 606: Integration in Respiratory Control, From Genes to Systems. Poulin MJ and Wilson RJA (Eds.), pp 193-200. Springer Publishers, New York.

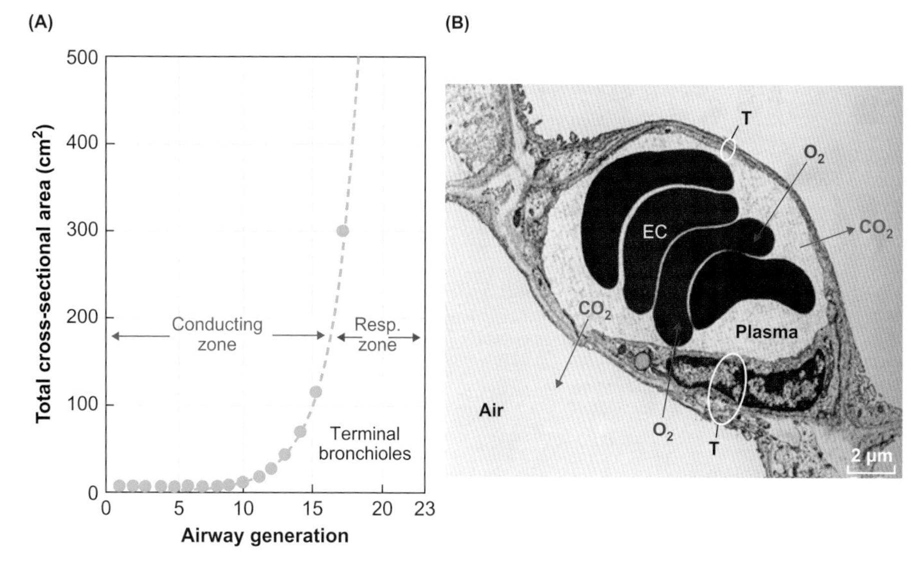

Figure 12.27 Characteristics of different regions of the lungs of mammals

(A) Plot of total cross-sectional area of the airways in a human lung at the different levels of branching showing the large increase in area that occurs at the transition between the conducting and respiratory zone (resp. zone). (B) Electron micrograph of an alveolar capillary in a human lung showing the thickness (T) of the barrier to diffusion between the blood cells (erythrocytes, EC) in the capillary and the air in the neighbouring alveolus.

Reproduced from West JB (1974). Respiratory Physiology, the essentials. The Williams and Wilkins Company, Baltimore. reproduced from Weibel ER (1984). The Pathway for Oxygen. Structure and Function in the mammalian respiratory system. Harvard University Press, Cambridge, Massachusetts.

Figure 12.8A and B, and Table 12.5 indicate that the surface area for gas exchange of mammals is related to body mas, with the mean diffusion distance being almost completely independent of body mass. Figure 12.8C also shows that the anatomical diffusion factor[45] of mammalian lungs is similar to that of the lungs of birds but substantially greater than that of the lungs of reptiles. This difference reflects the greater metabolic rate of mammals and birds, which are endotherms, compared to that of reptiles, which are ectotherms[46].

In general, the diameter of the alveoli increases gradually with increasing body mass. However, the alveoli of the lungs of the smallest mammals, Etruscan shrews (*Suncus etruscus*), shown in Figure 12.28 are so small, it has been suggested that the term 'alveolus' may not be appropriate and that the lung may be better described as a maze of interlacing blood and air capillaries. The lungs of Etruscan shrews seem to be at the lower limit of bioengineering.

The lungs of one group of mammals stand out from the rest: the bats. For animals of similar body mass, the surface area of the lungs of some species of bats, such as Wahlberg's epauletted fruit bats (*Epomorhorus wahlbergi*), is 138 cm^2 g body mass^{-1}. This is not only greater than that of other mammals but also than that of birds, as illustrated in Figure 12.8A. The thickness of the diffusion barrier for the lungs of bats is at the bottom end of the range for mammals, from 0.12 to 0.3 μm, while the anatomical diffusion factor is at the top of the range seen in birds (shown in Figure 12.8B,C). The morphological characteristics of the lungs of bats are thought to be related to the relatively high oxygen demands of flight[47].

[45] We discuss anatomical diffusion factor in Section 12.2.2. It is a simplified measure of morphometric diffusion conductance.

[46] We discuss metabolic rate of ectotherms and endotherms in Sections 2.2.3 and 2.2.4.

[47] We discuss the energy demands of flight in Section 18.2.3.

Ventilation of mammalian lungs involves the use of a muscular diaphragm

As for all of the air-breathing vertebrates we discuss in the previous sections, air is moved in and out of the sac-like lungs of mammals in a tidal fashion. This is called a ventilated pool gas-exchange system and is illustrated in Figure 12.29. In mammals, the thorax is divided from the abdomen by the muscular **diaphragm** (shown in Figures 12.30A and

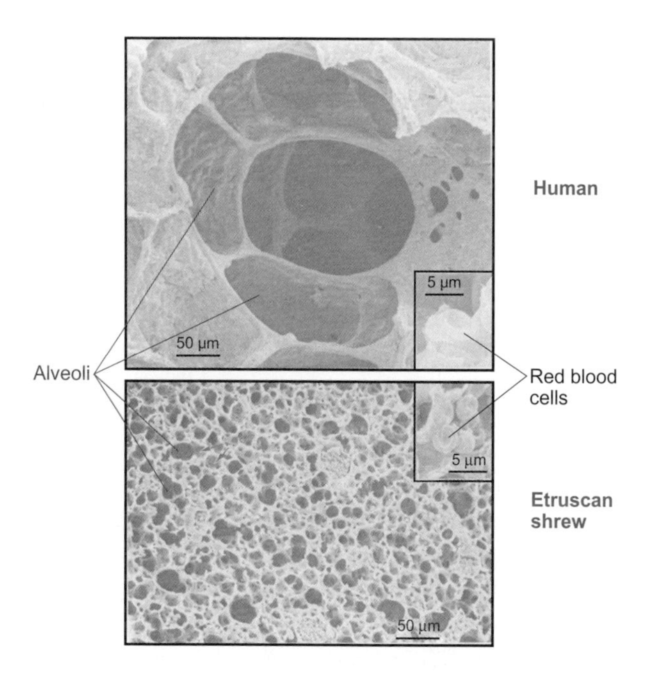

Figure 12.28 Gas-exchange surface (alveoli) of mammalian lungs

Scanning electron micrographs of the alveoli of the lung of a human and of an Etruscan shrew (*Suncus etruscus*) at the same magnification, with accompanying insets of some red blood cells at a greater magnification.

Reproduced from Weibel ER (1979). Oxygen demand and the size of respiratory structures in mammals. In: Evolution of Respiratory Processes A Comparative Approach. Wood SC and Lenfant C (Eds.), pp 289-346. Marcel Dekker Inc., New York.

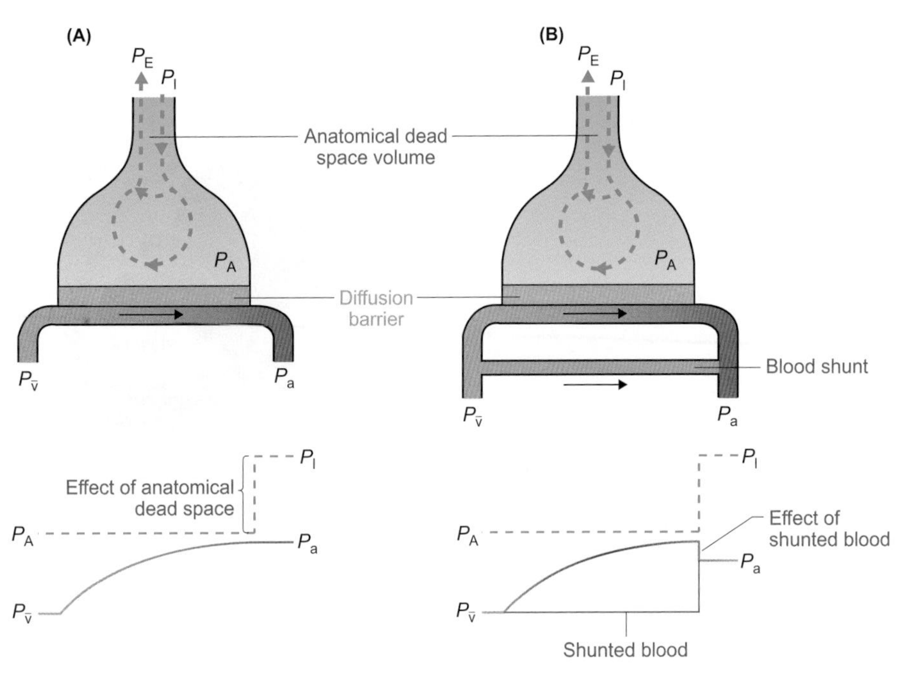

Figure 12.29 The ventilated pool type of gas exchanger

(A) Partial pressures of oxygen in: inspired (P_I) air expired (P_E) gas, gas in the alveoli (P_A), arterial (P_a) and mixed venous ($P_{\bar{v}}$) blood. Mainly because of the anatomical deadspace in the upper airways of the lungs of mammals, the partial pressure of oxygen in the alveoli of the lungs, P_AO_2, and in the arterialized blood leaving the lungs, P_aO_2, are substantially lower than that in the inspired air, P_IO_2. (B) Blood shunts cause a further reduction in P_aO_2, and hence in the overall effectiveness of the exchanger.

Adapted from Piiper J (1990). Modeling of gas exchange in lungs, gills and skin. In: Advances in Comparative & Environmental Physiology, vol 6. Boutilier RG (Ed.), pp 15-44. Springer-Verlag, Berlin; and Butler PJ (1976). Gas exchange. In, Environmental Physiology of Animals. Bligh J et al (Eds.), pp 164-195. Blackwell Scientific Publications, Oxford.

12.31A). The first muscles to become active during a respiratory cycle are those of the upper airway, which open the glottis. As the muscle fibres of the diaphragm contract, it flattens, and the thoracic cavity is enlarged, as shown in Figure 12.31A. The pressure in the lungs falls below that of the surrounding atmosphere and air is sucked into the lungs as a result. The intercostal muscles, shown in Figure 12.30A, are also involved, both in inspiration and expiration, although expiration during normal ventilation at rest, occurs passively because of the elastic recoil of the lungs. However, the expiratory muscles do become active during forced expiration and during exercise.

The idea that the external and internal intercostal muscles in mammals have completely separate functions (externals being active on inspiration and internals on expiration) is based primarily on the orientation of their fibres, as shown in Figure 12.30A, and stems from a study published in 1749 by Hamberger. However, we now know that the factors affecting contraction of the intercostal muscles are more complex. The activity of the **external intercostals** during inspiration is greatest in the dorsal portion of the most anterior areas of the rib cage, as indicated in Figure 12.30B and C, where the muscles have the greatest mechanical advantage[48] during inspiration. No inspiratory activity is ever present in the ventral portion of the external intercostals in the 6th and 8th rib spaces where the muscles have a mechanical advantage during expiration.

In contrast, the activity of the **internal intercostals** during expiration is greatest in the dorsal portions of the posterior segments, where the muscles have the greatest mechanical advantage during expiration, as shown in Figure 12.30B. No expiratory activity is present in the middle and ventral portion of the internal intercostal muscles in the most anterior segments where the muscles have a mechanical advantage during inspiration.

The intercostal muscles also have a locomotory function in some species of mammals.[49] In dogs, the majority of the intercostal musculature has a dominant locomotory function, probably to stabilize the rib cage and trunk. During resting and panting in dogs, the activity of the intercostals is like that described above. During trotting, however, ventilation and locomotion are normally coupled and the intercostals contribute to both ventilation and locomotion. On a few occasions, limb movement and ventilation are uncoupled during trotting. During such occasions, the part of both intercostal muscles between the bony part of the ribs (shown in Figure 12.30A) is locked to locomotion and drifts in and out of phase with ventilation. Activity of the rest of the internal intercostal muscles, between the sternum and the cartilaginous junctions (shown in Figure 12.30A), is always exclusively associated with ventilation.

The basic structure of the thorax and its muscles is the same in all mammals and the lungs are surrounded by two delicate and well-vascularized membranes, the **pleurae** (singular, pleura); one lines the chest wall and the other lines

[48] Mechanical advantage is the amplification of the force applied, as in a lever.

[49] We discuss the locomotory function of intercostal muscles in some species of lizards earlier in this section.

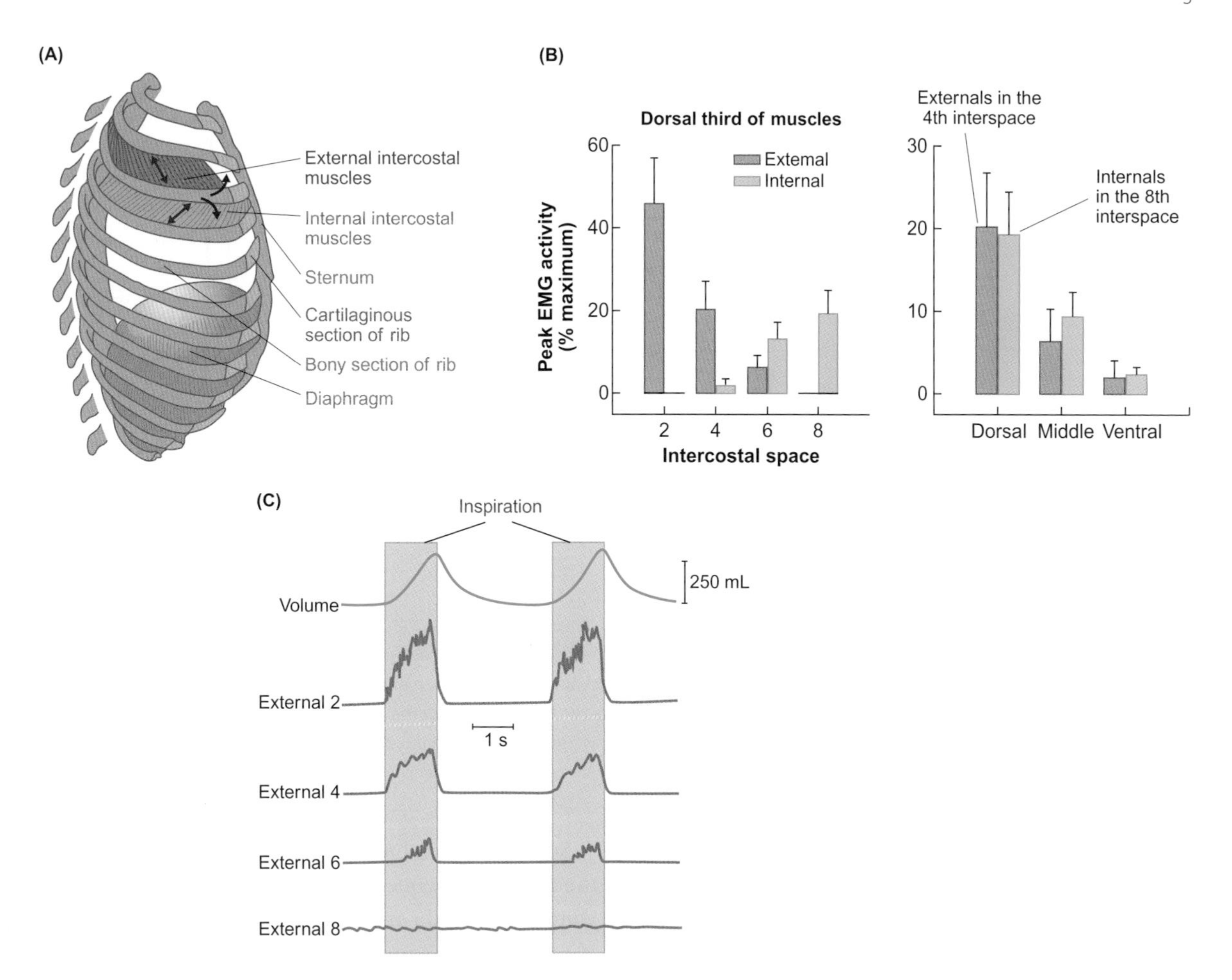

Figure 12.30 Ventilation of the lungs of mammals

(A) The orientation of the internal and external muscles between the ribs and the location of the diaphragm between the thorax and abdomen of a human. Double-ended arrows indicate the orientation of the muscle fibres and the single ended arrows indicate the supposed mechanical effect of these fibres—external intercostals cause inspiration, internal intercostals cause expiration. Note, in quadrapeds, such as dogs, the rib cage is horizontal and not vertical. (B) The electrical activity (EMG, electromyogram) of external intercostal muscles during inspiration (brown bars) and internal intercostal muscles during expiration (blue bars), is related to their antero-posterior (intercostal space) and dorso-ventral locations. For further details see the text. The EMG activity of each set of muscles is expressed as a percentage of the activity recorded during maximum stimulation of the external or intercostal nerve. Mean data ± SEM from eight dogs. (C) Traces of electrical activity of external intercostal muscles of dogs during two lung ventilation cycles, demonstrating that the more anterior ones (2 and 4) are active during inspiration, (up on volume trace) but that this activity is reduced or absent in the more posterior ones.

Sources: reproduced from: A: Weibel ER (1984). The Pathway for Oxygen. Structure and Function in the mammalian respiratory system. Harvard University Press, Cambridge, Massachusetts. B, C: Legrand A and De Troyer A (1999). Spatial distribution of external land internal intercostals activity in dogs. Journal of Physiology 518: 291-300.

the lungs. The two pleurae are separated by a narrow, fluid-filled space, the pleural space, as indicated in Figure 12.31A. This space gives the lungs some freedom of movement within the thorax.

The lungs are elastic structures, giving them a tendency to collapse to their resting volume. This collapse is opposed by the chest wall, which has its own tendency to spring out. The tendency for the chest wall to spring out causes the pressure in the pleural space to be below atmospheric pressure. If the pleurae are punctured and air enters the pleural cavity, a condition known as a pneumothorax, the lung will collapse and the chest will spring outward. However, the right and left pleural cavities are not connected, so damage to the pleurae on one side does not cause both lungs to collapse.

Uniquely in elephants (*Loxodonta* sp.), the pleurae are not delicate membranes but consist of dense connective tissue which contains few blood vessels. The cavity between the pleurae is filled with loose connective tissue to allow some sliding movement between them. It has been suggested that the structure of the pleurae in elephants is an adaptation to enable them to use their trunk to snorkel while walking or swimming under water.

Individuals who go snorkelling will know that you cannot buy a snorkel tube longer than about 30 cm. This is because the pressure of the water column increases by 1 atmosphere for every 10 m of depth, so the deeper you submerge, the greater the hydrostatic pressure. The pressure on the outside of your body is transmitted to all the

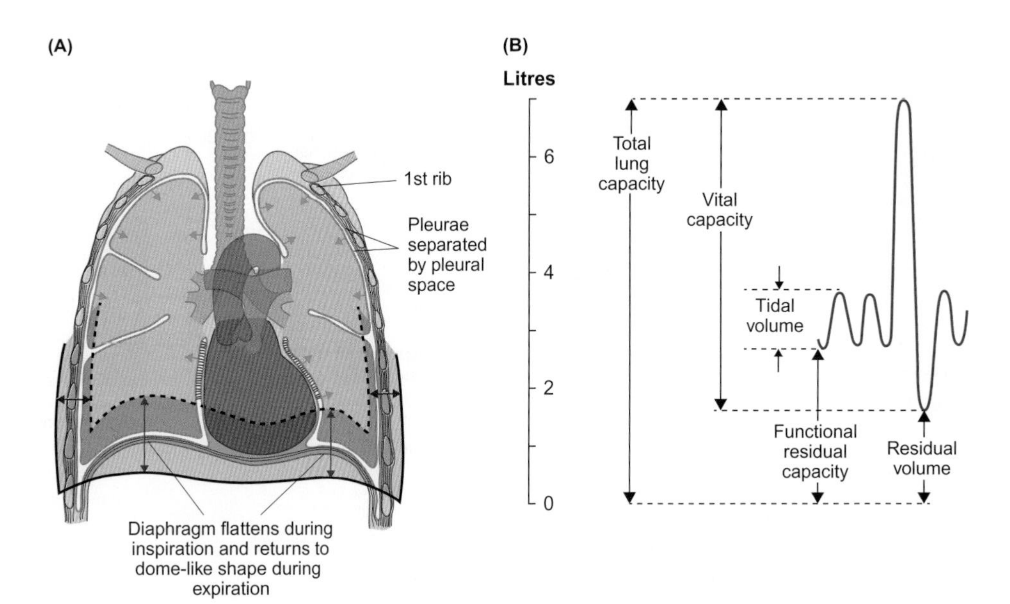

Figure 12.31 Volumes of lungs of human males

(A) Section through the thorax of a human showing the position of the pleural membranes and the movement of the diaphragm and rib cage during deep inspiration and expiration (double-ended arrows). The single-ended arrows indicate the retractile force of the lungs. (B) Various volumes of the lungs of a healthy, young, resting human male. (A): reproduced from Weibel ER (1984). The Pathway for Oxygen. Structure and Function in the mammalian respiratory system. Harvard University Press, Cambridge, Massachusetts. (B): reproduced from West JB (1974). Respiratory Physiology, the essentials. The Williams and Wilkins Company, Baltimore.

body tissue. As a result, blood pressure increases by the same amount, otherwise perfusion of the tissues would not be possible. However, the lungs are in contact with atmospheric air via the snorkel, which means that pressure inside the lungs is the same as that in the atmosphere, which is substantially less than that in the surrounding tissues and blood vessels. Therefore, if the snorkel tube was longer than about 30 cm, there would be an excessively large pressure difference between the blood inside the blood vessels supplying the pleurae and the inside of the lung which could lead to rupture of the vessels and/or exudation of fluid (oedema). The lungs of an elephant may be 200 cm below water (where the hydrostatic pressure is about 20 kPa), but capillary rupture or oedema within the pleurae are prevented by their unique structure.

The various volumes within the lungs during ventilation are well known for humans and approximate values are shown in Figure 12.31B. These values are for healthy males, 25 years old and 1.75 m tall. Values for females of similar age, but 1.62 m tall are approximately 75 per cent of those for males. The volume of air moving in and out of the lungs of a healthy, resting male adult (his **tidal volume**) is approximately 500 mL. The volume of the lungs following a normal resting expiration is approximately 3 L and is called the functional residual capacity. The maximum amount of air that can be expired after maximum inspiration is just over 5 L and is called the vital capacity. However, there is still some gas remaining in the lungs after such an expiration; this is called the residual volume and is just over 1.5 L. Thus, total lung capacity is 6.5 to 7 L.

So, when the subject is at rest, about 350 mL of air enters the respiratory zone of the lungs during each breath (tidal volume, 500 mL, minus anatomical deadspace, 150 mL). If respiratory frequency is 15 breaths min^{-1}, this gives an alveolar ventilation (i.e. the amount of inspired air available for gas exchange) of approximately 5.25 L min^{-1}. The energy cost of ventilating the lungs in resting humans is relatively low at 1 to 2 per cent of total metabolic rate, compared with 5 per cent in frogs and 10 per cent in water-breathing fish.

Animals with a long neck, like giraffes (*Giraffa camelopardalis*), also have a long trachea and may, therefore, have a proportionately larger anatomical deadspace. A larger tidal volume would therefore be required to maintain adequate ventilation of the respiratory zone. In reality, the longer length of the trachea in giraffes is partly compensated for by its diameter being narrower than expected for a similar-sized mammal. This partial compensation means that deadspace is still larger in giraffes than expected, but a larger than expected tidal volume means that the deadspace/tidal volume ratio is the same as in other mammals. However, the relatively long length and narrow diameter of the trachea means the resistance to air flow along the trachea is greater than in other similar-sized mammals[50].

Bird lungs: cross-current gas exchangers

The lungs are positioned in the dorsal region of the thorax and have a number of air sacs attached to them, which are shown in Figure 12.32A. The volume of the whole respira-

[50] Section 14.2.1 discusses fluid dynamics and explains how length and radius of a tube affect its resistance to the flow of a fluid through it.

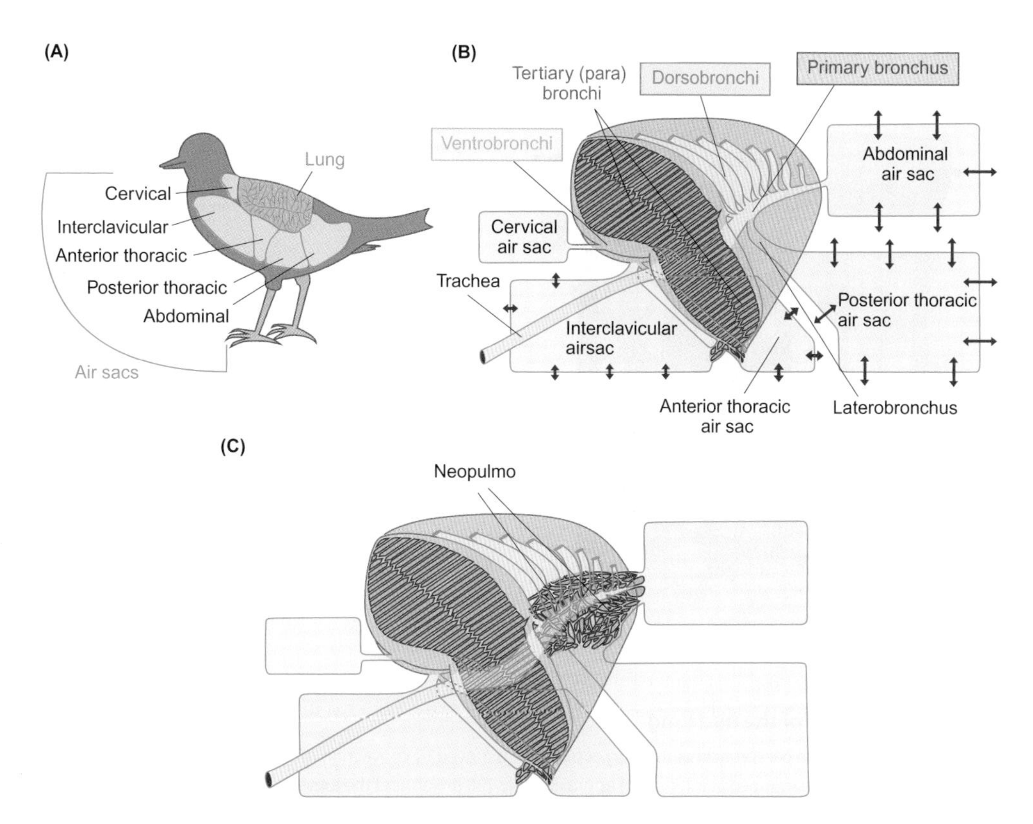

Figure 12.32 The respiratory system of birds

(A) Location of the lungs and associated air sacs in birds. (B) The basic arrangement of the pairs of air sacs and major airways. This arrangement has been called the palaeopulmo, which means old lung. The anterior (cranial) group of air sacs consists of the cervical, interclavicular (which are fused to form one sac) and anterior thoracic air sacs. The posterior (caudal) group of air sacs consists of the posterior thoracic air sacs and the abdominal air sacs. The lengths of arrows indicate the estimated expansion and compression of the air sacs during inspiration and expiration. (C) In most species of birds there is an additional network of parabronchi known as the neopulmo, which means new lung.

A: reproduced from Scheid P (1979). Mechanisms of gas exchange in bird lungs. Rev Physiol Biochem Pharmacol. 86: 137-186. B, C: adapted from Duncker H-R (1972). Structure of avian lungs. Respiration Physiology 14: 44-63.

tory system of birds, which includes the air sacs, is relatively much larger than that of mammals, as shown in Table 12.5. The length of the trachea is, on average, 2.7 times greater in birds than in similar-sized mammals. Unlike the situation in giraffes, however, the diameter is also 30 per cent greater so the resistance to air flow along the trachea is essentially the same in birds and mammals (other than giraffes).

Structure of bird lungs

The basic organization of a bird lung is shown in Figure 12.32B. The trachea divides into two bronchi which continue through each lung as the **primary bronchi**. As it enters a lung, each primary bronchus initially gives rise to four **secondary bronchi** (called **ventrobronchi**), and then gives rise to another 7–10 secondary bronchi (called **dorsobronchi**). Opposite the dorsobronchi, laterobronchi originate from the primary bronchus. The dorsal and ventral secondary bronchi in each lung are connected by the **tertiary bronchi** or **parabronchi**, which are long narrow tubes. Gas exchange occurs in the tissue surrounding these parabronchi. Figure 12.33 shows that a network of fine air capillaries radiates from the parabronchi and is interspersed with a similarly arranged network of blood capillaries.

The bronchi that pass through the lungs are joined to a set of associated **air sacs**, as illustrated and described in Figure 12.32B,C. The anterior (cranial) group of air sacs consists of the cervical and interclavicular sacs (the interclavicular sacs are fused to form one sac) and anterior thoracic air sacs all of which are attached to the ventrobronchi. The posterior (caudal) group of air sacs consists of the posterior thoracic air sacs which connect with the primary bronchi via the lateral bronchi and the abdominal air sacs, which connect directly with the primary bronchi. The attachments of some of the air sacs to the bronchi are important for determining the pattern of airflow through the lungs during ventilation, which we discuss later.

As well as the arrangement of parabronchi we have described, which is known as the palaeopulmo (meaning old lung), most species of birds have an additional network of parabronchi called the neopulmo (meaning new lung—as shown in Figure 12.32C). The majority of oxygen uptake occurs across the parabronchi of the palaeopulmo, with minimal proportions occurring in the parabronchi of the

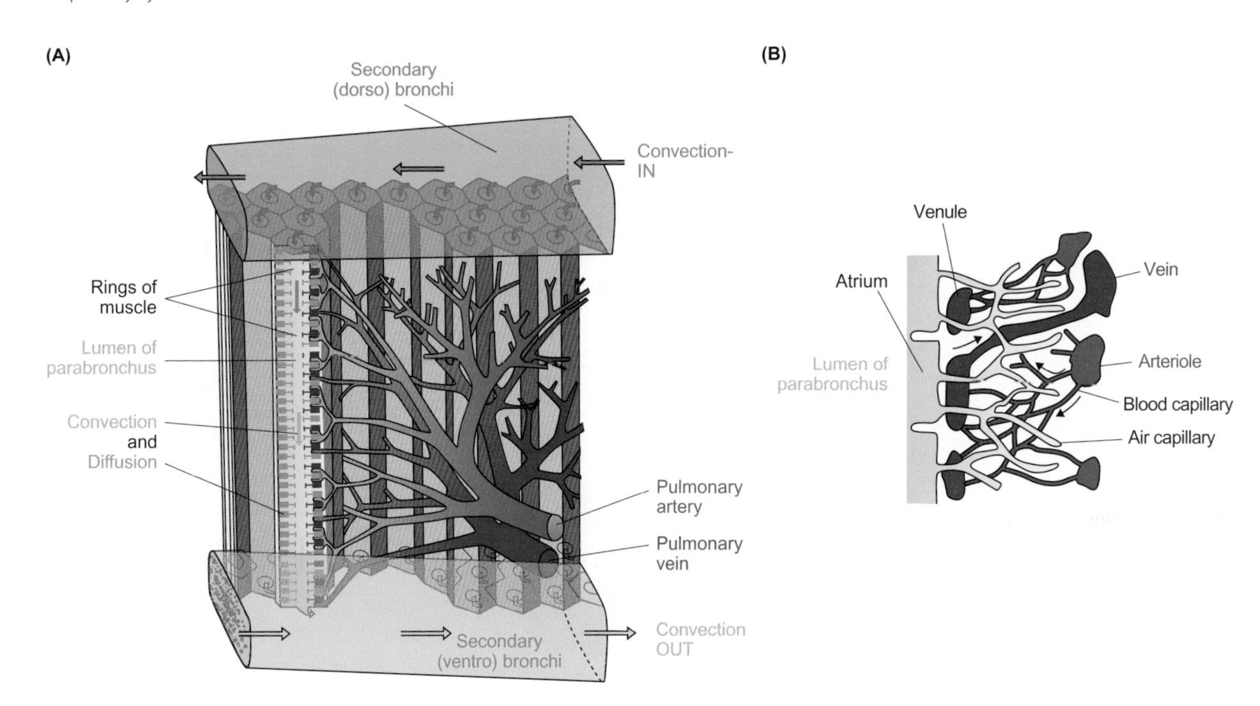

Figure 12.33 Gas-exchange region of the bird lung

(A) Arrangement and internal structure of the parabronchi of the lung of birds. Air is moved by convection from the dorsobronchi, through the parabronchi and on to the ventrobronchi. Rings of muscle are arranged at intervals of approximately 100 μm around the lumen of a parabronchus and between these rings are openings to depressions in the wall known as atria, which are separated from each other by sheets of tissue rich in elastic fibres. (B) Diffusion of oxygen occurs from the lumen of the parabronchi into the atria, from which a network of fine (3–10 μm diameter) air capillaries radiate, branch and reconnect (anastomose) with each other in all directions. They are interspersed with a similarly arranged network of blood capillaries and these networks form a 100–500 μm thick wall around the lumen of the parabronchi. This is where gas exchange occurs.

Reproduced from Duncker H-R (1974). Die Anordnung des Gefäßsystems in der Vogellunge. Verh. Anat. Ges. 68: 517-523. Scheid, P (1979). Mechanisms of gas exchange in bird lungs. Reviews of Physiology, Biochemistry and Pharmacology 86: 137-186.

neopulmo (about 10 per cent of the total) and across the air-sacs (less than 5 per cent of the total).

Ventilation of bird lungs

Ventilation of the lungs of birds involves movement of the sternum and rib cage, which is shown in Figure 12.34A. There is muscular activity during all phases of the cycle, but passive recoil of the thoracic cage aids both inspiration and expiration. The traditional view is that the intercostal muscles are active during both phases of ventilation, but it has more recently been demonstrated that the trunk muscles of birds are involved in locomotion, as well as in ventilation, as in some species of lizards and dogs. Figure 12.34C,D show that in standing, resting Canada geese (*Branta canadensis*), the external intercostal muscles show little activity and what activity there is does not follow a pattern that is linked to ventilation, but they are active and in phase with foot-fall when the bird is running. The major respiratory muscles are the **external oblique**, which is active during expiration and the **appendicocostal**, which is active during inspiration, but also during footfall when the bird is running, as shown in Figure 12.34C,D.

Each **appendicocostal** muscle is attached at one end to the **uncinate process** on a rib, as illustrated in Figure 12.34B. These structures are characteristic of the avian respiratory system and increase the mechanical advantage of the appendicocostal muscle by a factor of 2–4. It has also been found that the length of the uncinate processes is related to the main form of locomotion of the bird. The uncinate process is relatively short in non-flying species, such as ostriches (*Struthio camelus*), relatively long in diving birds, such as razorbills (*Alca torda*), and of intermediate length in others, such as common pigeons (*Columba livia*). This suggests that different ventilatory mechanics occur in birds adapted to different forms of locomotion.

The lungs of birds do not change volume during ventilation. Instead, air passes through the lungs to the air sacs, which act like bellows, drawing air through the lungs and then forcing gas out in the opposite direction. A question that intrigued physiologists for many years was: What pathway does the air take when passing through the palaeopulmo of the lungs during inspiration and expiration? It is now accepted that air flows along the primary bronchus during inspiration, with some flowing directly into the posterior group of air sacs while the rest travels through the parabronchi (where gas exchange occurs) and into the anterior group of air sacs, as shown in Figure 12.35A. During expiration, the air in the posterior air sacs also flows through the parabronchi, as shown in Figure 12.35B.

Although air flows in and out of the trachea in a tidal fashion, it flows through the parabronchi in the same direction during both phases of ventilation, so air flow is changed from being bidirectional in the trachea to being unidirec-

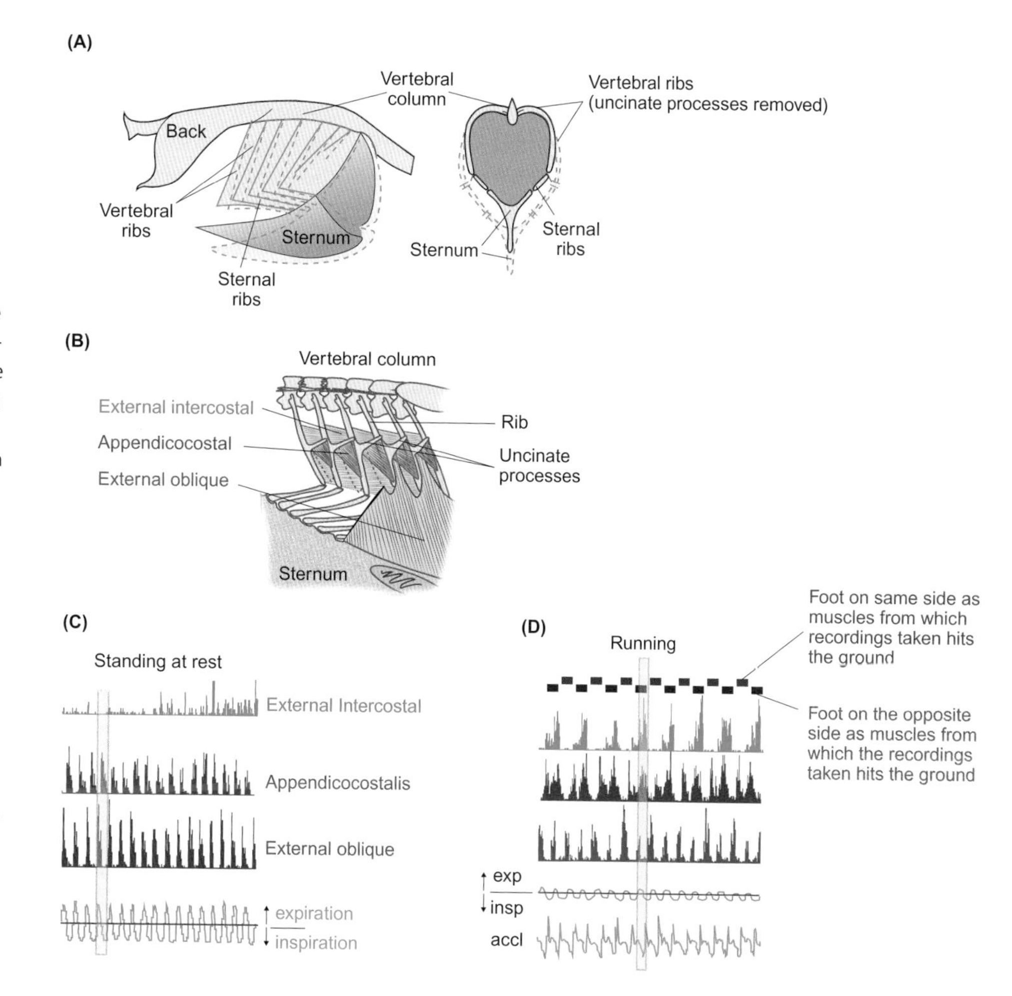

Figure 12.34 Ventilation of the bird lung

(A) Side and front views of the skeleton of a standing bird showing changes in the position of the thoracic skeleton during ventilation. Solid lines represent the position at the end of expiration and dotted lines that at the end of inspiration. (B) Part of rib cage showing the three muscles from which electrical recordings were taken from a Canada goose (*Branta canadensis*) while standing at rest (C) and while running (D). All recordings were made from muscles on the same side of the bird. Expiratory and inspiratory airflow are indicated in both (C) and (D). Acceleration (accl) is also indicated in (D). The external intercostal muscle, which is not active in phase with ventilation in the resting bird (shaded area in C), is active when the foot on the other side of the body hits the ground during running (shaded area in D).

Reproduced from Fedde MR (1976). Respiration. In: Avian Physiology, Sturkie PD (Ed.), pp 122-145. Springer-Verlag, New York. Adapted from Codd JR (2010). Uncinate processes in birds: morphology, physiology and function. Comparative Biochemistry and Physiology A 156: 303-308.

tional in the lung[51]. In birds, as in crocodiles and varanid lizards, there is no sign of anatomical valves that could control the direction of flow through the lungs. Instead, early studies concluded that unidirectional flow through the lungs of birds must be controlled by more subtle anatomical and aerodynamic factors. More recent studies on domestic geese

[51] The pattern of airflow through the parabronchi of the neopulmo is tidal.

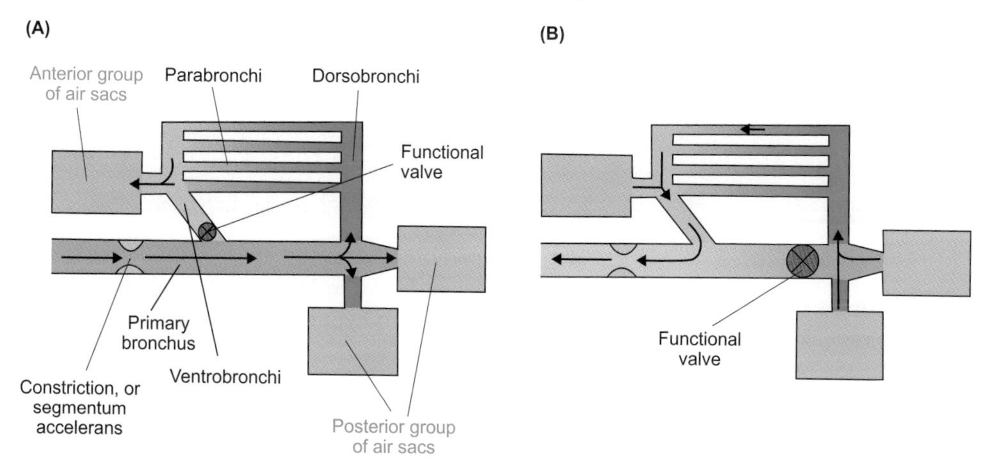

Figure 12.35 Air flow through the lungs of domestic geese (*Anser anser*)

(A) Direction of air flow through the lungs during inspiration, showing no flow along the ventrobronchi, which suggests the presence of a functional valve, indicated by the symbol ⊗. Also indicated is the location of a constriction, or segmentum accelerans, caused by a ring of smooth muscle at the entrance to the primary bronchus of the lung. This constriction causes the inspired air to accelerate and thus contributes to the inspiratory valve. (B) Direction of air flow through the lungs during expiration. There is no flow along the primary bronchus and the symbol ⊗ indicates the possible location of a functional valve. The presence of such a valve forces air through the parabronchi in the same direction as that during inspiration.

Adapted from Banzett RB et al (1987). Inspiratory aerodynamic valving in goose lungs depends on gas density and velocity. Respiration Physiology 70: 287-300 and Banzett RB et al (1991). Pressure profiles show features essential to aerodymanic valving in geese. Respiration Physiology 84: 295-309.

(*Anser anser*) have provided much of our current understanding of how unidirectional flow can be achieved in birds' lungs without the use of anatomical valves.

The most important aerodynamic factor is the velocity of the air. During inspiration, the greater its velocity, the more likely it is for air to flow past the ventrobronchi and continue along the primary bronchus as shown in Figure 12.35A. The velocity depends on several factors. For example, it is low when the bird is at rest and much higher during exercise. Velocity can also be controlled by varying the diameter of the primary bronchus before the entrances to the dorsobronchi; this variation is achieved by a ring of smooth muscle, a special region of the primary bronchus called the segmentum accelerans, which responds to changes in PCO_2. The decrease in PCO_2 during inspiration causes constriction of the region and an increase in the velocity of inspiratory airflow.

Thus, the segmentum accelerans could be an important feature for aerodynamic 'valving' during inspiration. However, at high levels of ventilation the constriction would increase the work of ventilation. Under such conditions, an increase in PCO_2 could cause the muscles of the segmentum to relax. It is not known whether or not there is such an increase in PCO_2 in the vicinity of the segmentum during the increased ventilation of exercise.

Another possible factor involved in the functioning of the inspiratory valve is thought to be the angle of insertion of the ventrobronchi into the primary bronchus, which is illustrated in Figure 12.35. This arrangement may encourage the air to bypass the ventrobronchi, as to enter them would involve the air making an obtuse[52] change in direction. The mechanism responsible for producing the valving that occurs during expiration is less clear.

[52] An obtuse angle is greater than 90° but less than 180°.

As we discuss earlier, the lungs of crocodiles and varanid lizards have similarities in their internal structure to those of birds. These similarities suggest the presence of aerodynamic valves in the lungs of crocodiles and varanid lizards. It is of great interest, therefore, that studies on normally breathing crocodiles have also shown that air flows in the same direction through their lungs during both phases of ventilation. There is both unidirectional and tidal flow in different regions of the lungs of varanid lizards, but the overall pattern is predominantly unidirectional. A similar pattern of airflow is also present in the relatively more simple lungs (that is, similar to those of chameleons, as shown back in Figure 12.21B) of green iguanas (*Iguana iguana*). This means that anatomical features characteristic of bird lungs are not necessary for unidirectional airflow.

This similarity between the airflow in the lungs of birds, crocodiles and iguanid lizards suggests that the character of unidirectional airflow through lungs may be ancestral for all diapsid reptiles[53], both living and extinct. However, this does not rule out the possibility of convergent evolution, and other species of air-breathing ectotherms need to be studied before any definitive conclusion can be reached.

Despite the fact that air passes through the lungs of birds in the same direction, the blood does not pass in the opposite direction to give a countercurrent arrangement for gas exchange. In fact, experimental evidence indicates that blood flows at right angles to gas flow in a **cross-current** arrangement, as discussed in Experimental Panel 12.1. The energy cost of breathing in birds is between 1 and 3 per cent of resting metabolic rate.

[53] Diapsid reptiles include those in which the ancestral condition was the presence of two posterior openings on either side of their skull. Traditionally, the living crocodilians and birds are members of one group, the Archosauria, and the lizards and snakes are members of another group, the Lepidosauria.

Experimental Panel 12.1 Experimental evidence for the existence of cross-current gas exchange in the lungs of birds

Background

When it was confirmed that the flow of air through the parabronchi of the lungs of birds is in the same direction during both phases of ventilation, the possibility existed that blood flow is in the opposite direction, thus providing a countercurrent exchange system. However, the anatomy of the lung does not support the presence of blood capillaries that are in contact with the parabronchi for the whole of their length. On the contrary, the blood capillaries are restricted to rather a small fraction of the total length of a parabronchus. Consequently, a cross-current arrangement was proposed whereby bulk blood flow is at right angles to bulk gas flow through the parabronchi. In the early 1970s, Peter Scheid and Johannes Piiper from the Max Planck Institute in Göttingen, Germany, tested this hypothesis in an elegant experiment on domestic ducks.

Experimental approach

Scheid and Piiper took advantage of the anatomy of the respiratory system of birds which allowed them experimentally to pump air through the parabronchi of an anaesthetized duck either entering from the trachea and exiting through the caudal air sacs, or in the opposite direction, from the caudal air sacs, through the parabronchi and out through the trachea. This reversal in flow was achieved by blocking the primary bronchus with an inflatable balloon. Figure A(i) illustrates how inflation of the balloon made it possible to prevent air flow through the primary bronchus, irrespective of which end of the respiratory system the air was introduced.

The partial pressures of oxygen and carbon dioxide were measured in the air entering (inspired) and leaving (expired) the respiratory system and in arterial and mixed venous blood. Blood samples were obtained through small plastic catheters placed into a major artery and vein. The catheter in the vein was advanced into the left ventricle of the heart so that it sampled a mixture of venous blood returning from the whole body of the bird.

Scheid and Piiper argued that if a countercurrent arrangement normally exists between blood flow and air flow through the parabronchi of the lungs, the reversal of the direction of air flow would create a co-current system and the relationship between the partial pressures of arterial and of expired respiratory gases would change, as shown in Figure A(i). In fact, these relationships did not change, as shown in Figure A(ii), and this provided physiological evidence for the presence of a cross-current exchange system.

Overall findings

Figures A(ii) and (iii) show that a cross-current system results in air always encountering venous blood from the body with the same PO_2 and PCO_2 in whichever direction it passes through the parabronchi. In ideal conditions, such a cross-current arrangement would lead to an overlap between the arterial and expired partial pressures of the respiratory gases, as shown in Figures A (ii) and (iii). Such a cross-current system means that the lungs of birds are potentially more effective than those of mammals, although, as indicated above, the unidirectional flow of air through the bird lung is not actually necessary to maintain the high level of effectiveness; the flow could be tidal in nature. The functional significance of unidirectional airflow may be to keep anatomical deadspace to a minimum by not repeatedly changing the direction of airflow through the lungs.

› Find out more

Duncker H-R (1972). Structure of avian lungs. Respiration Physiology 14: 44–63.

Scheid P, Piiper J (1972). Cross-current gas exchange in avian lungs: effects of reversed parabronchial airflow in ducks. Respiration Physiology 16: 304–312.

Scheid P (1979). Mechanisms of gas exchange In bird lungs. Reviews of Physiology, Biochemistry and Pharmacology 86: 137–186.

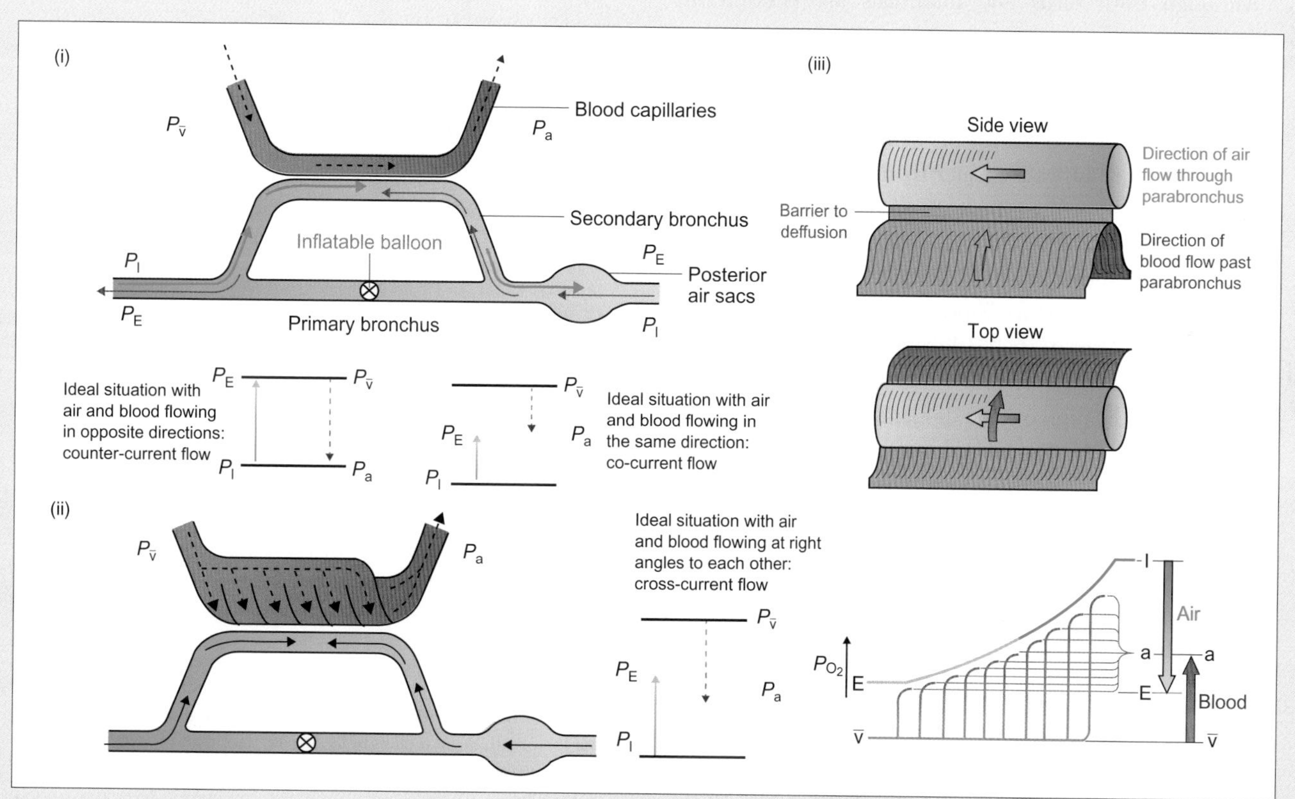

Figure A Demonstration of cross-current model of gas exchange in the bird lung

The effect of experimentally changing the direction of air flow through the lung, by blocking the primary bronchus (⊗), on the relationship between partial pressure (*P*) of oxygen in inspired (I) and expired (E) air and in arterial (a) and mixed venous ($\bar{v}$) blood in: (i) A parallel model of gas exchange. In this situation under ideal conditions, when blood and air flow in opposite directions (countercurrent), P_aO_2 tends to equal P_EO_2, whereas when they flow in the same direction (co-current) P_aO_2 tends to equal P_IO_2, as explained in Box 12.1. (ii) A cross-current model of gas exchange. Here, the flows of air and blood are at right angles to each other, and changing the direction of the flow of air has no effect on the relationship between P_aO_2, P_EO_2 and P_IO_2. It is this cross-current condition that exists in the lungs of birds. Under ideal conditions, P_aO_2 and P_EO_2 overlap. (iii). Diagrammatic representation of cross-current gas exchange model in the lungs of birds from the top and from the side, plus the profile of the partial pressure of oxygen in the inspired air and in the mixed venous blood as they pass through the lung and become expired air and arterial blood. This shows the origin of the theoretical overlap between P_EO_2 and P_aO_2.

Comparisons between the respiratory systems of birds and mammals

Birds and mammals show differences in their respiratory variables. For example, on average, tidal volume is approximately twice as large in birds as in mammals of similar mass. This difference may be related to the fact that the trachea is longer and has a larger volume, which is deadspace, in birds than in mammals. However, on average, mammals have a greater respiratory frequency which leads to a relatively higher minute ventilation volume ($\dot{V}_E$, 518 mL min^{-1} for a 1 kg mammal, compared to 385 mL min^{-1} for a 1 kg bird), despite a 20 per cent lower resting rate of oxygen consumption ($\dot{M}O_2$) for a given body mass in mammals. This means that birds ventilate their lungs less than mammals for a given rate of oxygen consumption, i.e. they have a lower ventilation convection requirement ($\dot{V}_E/\dot{M}O_2$), and extract a greater proportion of oxygen from the inhaled air than mammals, as shown in Table 12.5.

Although both birds and mammals are endotherms with high maximum rates of oxygen consumption, they achieve this with lungs of completely different structures. It has been suggested that the lungs of birds are superior to those of mammals. Indeed, the physiological diffusion conductance[54] of bird lungs is greater than that in mammalian (dog) lungs, as shown in Table 12.5. However, Table 12.5 also shows that PO_2 in the arterial blood is similar in both groups and the morphological characteristics of the lungs of the bats, are at least as impressive as those of birds, if not better, as demonstrated earlier in Figure 12.8. In other words, the basic structure of the mammalian lung is capable of meeting the high oxygen demands of flight equally as well as that of birds. The real advantage the lungs of birds may have over those of mammals is that they are better at extracting oxygen from the air, not only at sea level, but also at medium to high altitudes[55].

12.3.4 Surface tension in alveoli and surfactants in the lungs of mammals and other animals

Because of their extreme thinness and permeability to the respiratory gases, the walls of the gas-exchange areas of the lungs of air-breathing animals are also permeable to the fluid which lines their outer surface. This fluid gives rise to a fluid/air interface. The surface tension at this fluid/air interface tends to pull the walls of an alveolus toward its centre of curvature, or to collapse. This tendency for the sphere of fluid that lines the alveoli to collapse is directly proportional to its surface tension and inversely proportional to its radius. This means that the relatively small alveoli of the lungs of mammals have a particular tendency to collapse and a greater pressure is required to expand the lung than would be the case without the presence of the fluid/interface lining the alveoli.

This concept was demonstrated by an elegant experiment which demonstrated that lungs were easier to distend when filled with saline, which removes the air/fluid interface, than when filled with air. In other words, the lungs become more **compliant**[56] when they are filled with saline and there is a greater increase in volume for a given increase in pressure. The tendency for the alveoli to collapse is prevented in part by the presence of compounds called **surfactants**, which reduce the surface tension of the fluid/air interface. Case Study 12.1 discusses the role of these compounds in the lungs of air-breathing animals.

[54] Refer back to Section 12.2.2 and Equation 12.1 to remind yourself of what physiological diffusion conductance is.

[55] Birds and mammals at high altitude is discussed in Section 15.3.1.

[56] Compliance in blood vessels is discussed in Section 14.1.3.

› *Review articles*

Brainerd EL (1994). The evolution of lung-gill bimodal breathing and the homology of vertebrate respiratory pumps. American Zoologist 34: 289–299.

Brainerd EL, Owerkowicz T (2006). Functional morphology and evolution of aspiration breathing in tetrapods. Respiratory Physiology & Neurobiology 154: 73–88.

Burggren WW (1989). Lung structure and function in Amphibians. In: Wood SC (ed.) Comparative Pulmonary Physiology Current Concepts. pp 153–192. New York: Marcel Dekker, Inc.

Cieri RL, Farmer CG (2016). Unidirectional airflow in vertebrates: a review of structure, function and evolution. Journal of Comparative Physiology 186: 541–552.

Codd JR (2010). Uncinate processes in birds: morphology, physiology and function. Comparative Biochemistry and Physiology – Part A 156: 303–308.

Deeming DC (2006). Ultrastructural and functional morphology of eggshells supports the idea that dinosaur eggs were incubated buried in a substrate. Palaeontology 49: 171–185.

Farmer CG (2010). The provenance of alveolar and parabronchial lungs: insights from paleoecology and the discovery of cardiogenic, unidirectional airflow in the American alligator (*Alligator mississippiensis*). Physiological and Biochemical Zoology 83: 561–575.

Graham JB (1997). Air-Breathing Fishes–Chapter 3 Respiratory organs, pp 65–133 and Chapter 5 Aerial and aquatic gas exchange, pp 153–181. San Diego: Academic Press.

Innes AJ, Taylor EW (1986). The evolution of air-breathing in crustaceans: functional analysis of branchial, cutaneous and pulmonary gas exchange. Comparative Biochemistry and Physiology A 85: 621–637.

Maina JN (2000). What it takes to fly: the structural and functional respiratory refinements in birds and bats. Journal of Experimental Biology 203: 3045–3064.

Perry SF (1990). Recent advances and trends in the comparative morphometry of vertebrate gas exchange organs. In: Boutilier RG (ed.) Advances in Comparative and Environmental Physiology, Vol. 6: pp. 45–71. Berlin: Springer-Verlag.

Perry SF (1998). Lungs: comparative anatomy, functional morphology and evolution. In: Gans C, Gaunt AS (eds.) Biology of Reptilia. 1–92. Ithaca, NY: Society for the Study of Amphibians and Reptiles.

Perry SF, Sander M (2004). Reconstructing the evolution of the respiratory apparatus of tetrapods. Respiratory Physiology and Neurobiology 144: 125–139.

Case Study 12.1 The structure and function of surfactants in the respiratory systems of air-breathing animals

A respiratory surfactant is a complex substance consisting of phospholipids, neutral lipids (particularly cholesterol) and proteins. One of its important functions is to reduce the surface tension at the fluid/air interface in the gas exchange regions of lungs, and maybe in the tracheae of insects. The gas-exchange regions of the lungs of mammals are the alveoli.

The earliest demonstrations that a surfactant exists in mammalian lungs was published in 1955 by the English biophysicist Richard Pattle. Not only did he study surfactants in mammalian lungs, but also in those of birds, reptiles and amphibians. Recent leaders in the comparative studies of surfactants include Sandra Orgeig and Christopher Daniels at the University of South Australia, Adelaide.

The most abundant epithelial cells in the alveoli of mammalian lungs are the type I cells, but the phospholipids and proteins are produced in the endoplasmic reticulum and Golgi apparatus of the type II cells of the epithelium and stored in the lamellar bodies until exocytosis, as illustrated in Figure A(i). After release into the alveoli, the lamellar bodies swell and unravel into a structure called tubular myelin, which supplies the lipids for the surface film of the alveoli. There are four surfactant proteins. Two of these, B and C, are hydrophobic and are directly involved in the regulation of surface tension, while the other two, A and D, are involved in the defence mechanisms of the lung, as well as in stabilizing the surfactant.

The functional significance of surfactant protein B has been demonstrated by respiratory failure in adult mice after genetic disruption of its synthesis. Also, infant respiratory distress syndrome (IRDS) is the result of markedly reduced lung surfactant in premature human infants, which increases the tendency for the alveoli to collapse, making breathing more difficult.

The importance of surfactant is not restricted to the spherical alveoli of the mammalian lung; it is also important in narrow tubes within the respiratory systems of animals such as the terminal bronchioles in the lungs of mammals and possibly in the tracheae of insects. Despite the fact that the lungs of birds do not change volume during the ventilatory cycle, surfactant is present in the lining of the air capillaries and is assumed to be of importance in preventing their tendency to collapse.

During ventilation, surface tension in the alveoli changes with their changing surface area. This change in surface tension is the result of interactions between the various components of the film of surfactant. During expiration, compression of the film of surfactant leads to the 'squeezing out' of the unsaturated phospholipids and cholesterol so that the concentration of

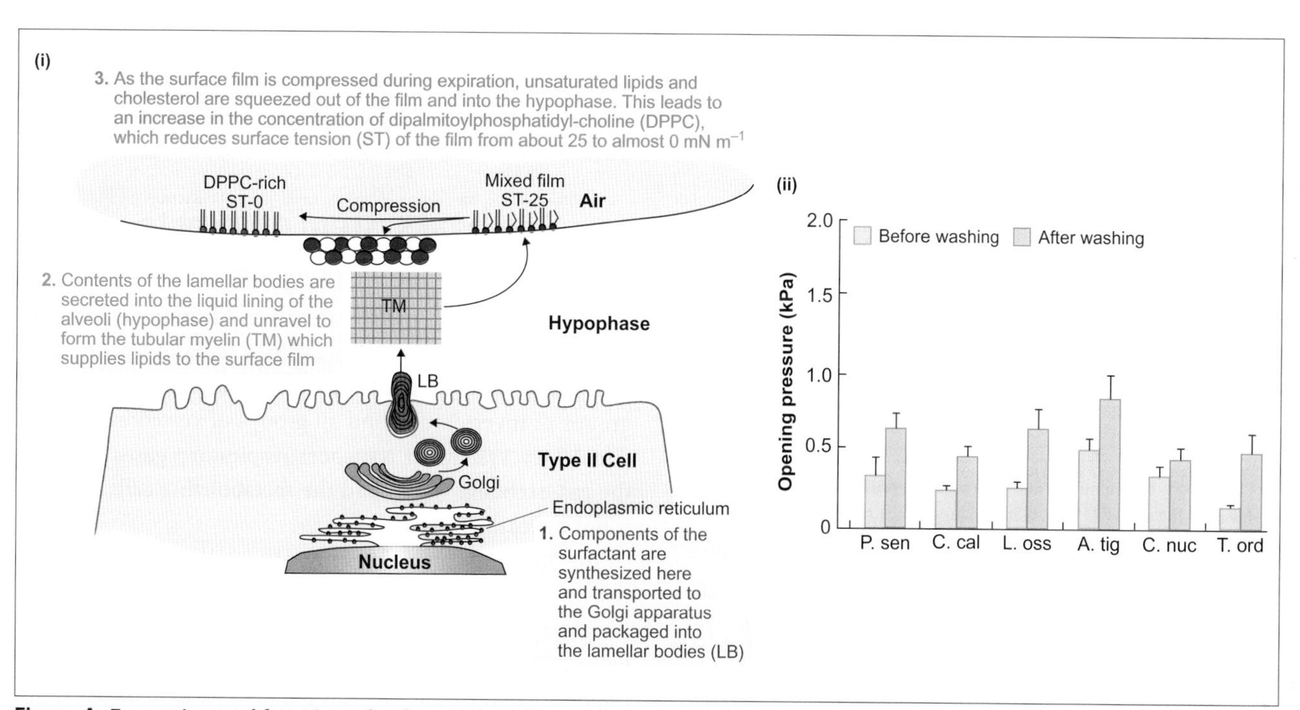

Figure A Formation and function of pulmonary surfactant in vertebrates

(i) Schematic showing the synthesis and functioning of the pulmonary surfactant. (ii) Histograms of the pressures required to open the lungs of some non-mammalian vertebrates before and after removal of the surfactant by washing. P. sen, Senegal bichir (*Polypterus senegalus*); C. cal, reedfish or snakefish (*Erpetoichthys calabaricus*); L. oss, longnose gar (*Lepisosteus osseus*); A. tig, tiger salamander (*Ambystoma tigrinum*); C. nuc, central netted dragon (*Ctenophorus nuchalis*); T. ord, northwestern garter snake (*Thamnophis ordinoides*).

Source: adapted from Daniels CB and Orgeig S (2003). Pulmonary surfactant: the key to evolution of air breathing. Neural Information Processing Systems, 18: 151-157.

dipalmitoylphosphatidyl-choline (DPPC) in the film increases. The molecules of DPPC can be compressed together as a result of their two fully saturated fatty acid chains, thereby excluding water molecules from the air–liquid interface and reducing surface tension. Below a temperature of 41°C, which is known as the transition temperature, DPPC exists in a solid-gel state that would not disperse over the surface of the alveoli. During inspiration, the reintroduction of the unsaturated fatty acids and cholesterol into the surface film lowers the transition temperature of the mixture, which is then able to exist in a fluid state at mammalian body temperature and spread out over the lining of the alveoli.

So, temperature has a strong influence on the structure and function of lung surfactants, which has implications for the air-breathing ectotherms and for endotherms that lower their body temperature and go into torpor[1]. In fact, the surfactant cholesterol increases during torpor in a number of species of mammals. Also, because of the effect of temperature on the functioning of the mixture of surfactant lipids, lungfish and amphibians have a lower ratio of DPPC relative to unsaturated phospholipids in the mixture (10–30 per cent) compared to that in mammals (40–50 per cent). This keeps the mixture of surfactant lipids in the liquid phase at the lower body temperatures of the ectotherms.

Although surfactant has been found in the lungs of all classes of vertebrates, as well as in the lung of the snails (*Helix* sp.), its surfactant properties are lowest in the lungs of air-breathing fish and greatest in those of mammals. This probably relates to the fact that the respiratory units (faveoli) of the lungs of air-breathing fish, amphibians and reptiles are up to 100 larger, and thus inherently more stable, than the alveoli of mammals. Nevertheless, surfactant in the lungs of air-breathing fish, amphibians and reptiles may serve an anti-adhesive function, assisting in the reopening of completely collapsed lungs, as indicated in Figure A(ii). Air-breathing fish, amphibians and reptiles tend to ventilate their lungs intermittently, so the lungs may collapse, bringing the epithelial surfaces into contact during the long periods between inflations. It has been suggested, therefore, that the original surface-active function of lung surfactants in vertebrates was anti-adhesive, which then evolved into the role it plays in the lungs of mammals.

› Find out more

Clements JA, Avery ME (1999). Lung surfactant and neonatal respiratory distress syndrome. American Journal of Respiratory and Critical Care Medicine 157: S59–S66.

Daniels CB, Orgeig S (2003). Pulmonary surfactant: the key to the evolution of air breathing. News in Physiological Sciences 18: 151–157.

McGregor LK, Daniels CB, Nicholas TE (1993). Lung ultrastructure and the surfactant-like system of the central netted dragon, *Ctenophorus nuchalis*. Copeia 2: 326–333.

Orgeig S, Bernhard W, Biswas SC, Daniels CB, Hall SB, Hetz SK, Lang CJ, Maina JN, Panda AK, Perez-Gill J, Possmeyer F, Veldhuizen RA, Yan W (2007). The anatomy, physics, and physiology of gas exchange surfaces: is there a universal function for pulmonary surfactant in animal respiratory structures? Integrative and Comparative Biology 47: 610–627.

Pattle RE (1955). Properties, function and origin of the alveolar lining layer. Nature 175: 1125–1126.

Pattle RE, Hopkinson DAW (1963). Lung lining in bird, reptile and amphibian. Nature 200: 894.

[1] Torpor in endotherms is discussed in section 10.2.7

12

Pinder AW (1997). Modelling gas exchange in embryos, larvae and foetuses. In: Burggren WW, Keller BB (eds.) Development of Cardiovascular systems. pp 240–258. Cambridge: Cambridge University Press.

Powell F (2000). Respiration In: Whittow GC (ed.) Sturkie's Avian Physiology, 5th Edn. San Diego: Academic Press.

Rahn H, Howell BJ (1976). Bimodal gas exchange. In: Hughes GM (ed.) Respiration of Amphibious Vertebrates pp 271–285. London: Academic Press.

Rahn H, Ar A, Paganelli CV (1979). How bird eggs breathe. Scientific American 240: 46–55.

Scheid P (1979). Mechanisms of gas exchange in bird lungs. Reviews of Physiology, Biochemistry and Pharmacology 86: 137–186.

Taylor HH, Taylor EW (1992). Gills and lungs: the exchange of gases and ions. In: Harrison FW, Humes A (eds.) Microscopic Anatomy of Invertebrates Vol 10: Decapod Crustacea. pp 203–293. New York: Wiley.

Wang T, Smits AW, Burggren WW (1998). Pulmonary function in reptiles. In: Gans C, Gaunt AS (eds.) Biology of Reptilia, Morphology G, Visceral Organs, pp. 297–374. Ithaca, NY: Society for the Study of Amphibians and Reptiles.

Weibel ER (1979). Oxygen demand and the size of respiratory structures in mammals. In: Wood SC, Lenfant C (eds.) Evolution of Respiratory Processes A Comparative Approach. pp 289–346. New York: Marcel Dekker Inc.

West JB (2002). Why doesn't the elephant have a pleural space? News in Physiological Sciences 17: 47–50.

West JB, Watson RR, Fu Z (2007). The human lung: did evolution get it wrong? European Respiratory Journal 29: 11–17.

12.4 Tracheal system of insects

Unlike the situation in crustaceans and fish, the respiratory system of insects most likely evolved on land, and became secondarily adapted to life in water in aquatic insects. Most groups of invertebrates and all groups of vertebrates have a circulatory system to transport respiratory gases between the gas-exchange organ and the metabolizing cells, but this is not the case for insects.

The major means of respiratory gas transfer in insects is by a series of branching, air-filled tubes called **tracheae** (singular: trachea) which penetrate directly to the metabolizing cells. Although the circulatory system is relatively unimportant in the transport of respiratory gases in most species of insects, oxygen-binding pigments, such as haemoglobin and haemocyanin are important for oxygen transport in the larvae of some species and in more ancestral orders of insects. The tracheal system is organized on a segmental basis with a maximum of 10 pairs of openings called **spiracles** present in

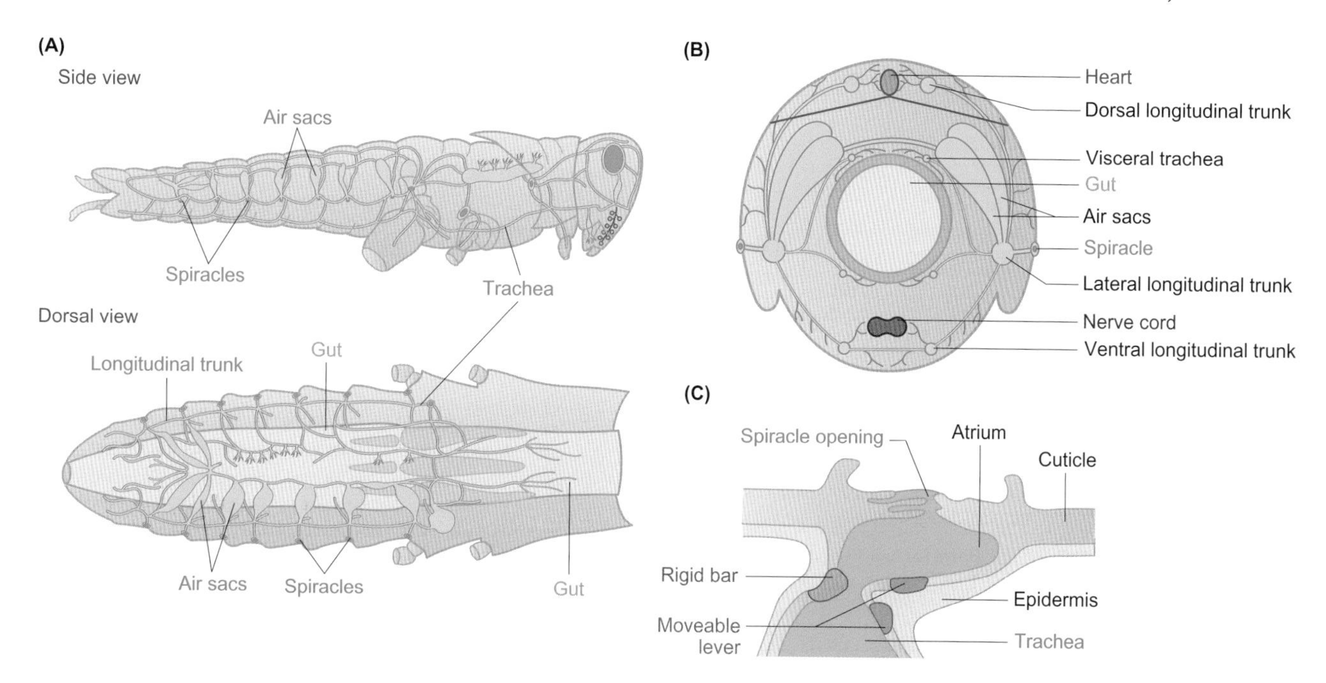

Figure 12.36 The tracheal and air sac system of insects

A Tracheal and air sac system of a Carolina locust, (*Dissosteira carolina*). **B** Cross-section of the abdomen of a locust to show the major tracheae, tracheal trunks and air sacs. **C** Section through an abdominal spiracle of a caterpillar showing parts of the closing mechanism – moveable lever and rigid bar.

(A) reproduced from Kerkut GA (1959). The Invertebrata. Cambridge University Press, Cambridge. (B) modified from and (C) reproduced from Chapman RF (1998). The Insects. Structure and Function. Cambridge University Press, Cambridge.

most terrestrial insects; two pairs are on the thorax and eight pairs are on the abdomen, as illustrated in Figure 12.36A. This basic design has been modified according to lifestyle of the animal. For example, the larvae of mosquitoes have spiracles at the end of a breathing tube and the animals hang at the surface of the water when replenishing their oxygen.

The spiracles generally lead into a cavity known as an atrium, shown in Figure 12.36C, and then into the tracheae. In most terrestrial insects the spiracles have a closing mechanism which may consist of either one or two moveable valves, or of the constriction and closing off of the atrium from the trachea. When the spiracles are open, water will be lost[57], so they usually remain open for the minimum amount of time required to meet the insect's oxygen requirements. Consequently, the level of activity and temperature of the animal are two important factors that influence the duration of spiracular opening.

The tracheae emerge from the atrium and are invaginations of the body wall. Tracheae are lined with cuticle which forms spiral thickenings, the **taenidia**, as shown in Figure 12.37A,B. These spiral thickenings prevent the tracheae collapsing. The main network of tracheae is shown in Figure 12.36. The tracheae may be expanded in places to form thin-walled **air sacs**, as shown in Figure 12.36A,B. Air sacs have poorly developed taenidia or no taenidia at all, so they collapse when exposed to pressure and play an important role in the ventilation of the tracheal system.

The tracheae divide into smaller branches, eventually reaching a diameter of approximately 2 μm. They then give rise to much finer tubes 0.6–0.8 μm in diameter, called **tracheoles**. Tracheoles are formed within single tracheolar cells, as shown in Figure 12.37A. In some, but not all species, the tracheoles may retain their lining of cuticle during moult. The tracheoles may appear to be intracellular, although they probably only form a deep fold in the plasma membrane of the cell, as shown in Figure 12.37C,D.

The distribution and density of tracheae and tracheoles reflect the oxygen demand of different tissues; flight muscle and the nervous system, which have the highest oxygen demands, tend to have the richest supply.

Morphological characteristics of the respiratory system have been determined for Indian stick insects (*Carausius morosus*). The surface area of the tracheoles for a given body mass is almost nine times greater than that for the largest tracheae (11 and 1.3 cm^2 g body $mass^{-1}$, respectively) and the diffusion distance in the tracheoles is approximately 14 per cent of that in the largest tracheae (0.2 and 1.4 μm, respectively). As a result, more than 70 per cent of the morphological diffusion conductance[58] for oxygen of the respireatory system resides in the tracheoles.

[57] Section 6.1.2 discusses evaporative water loss from respiratory surfaces.

[58] Morphological diffusion conductance is discussed in section 12.2.2 and Equation 12.1.

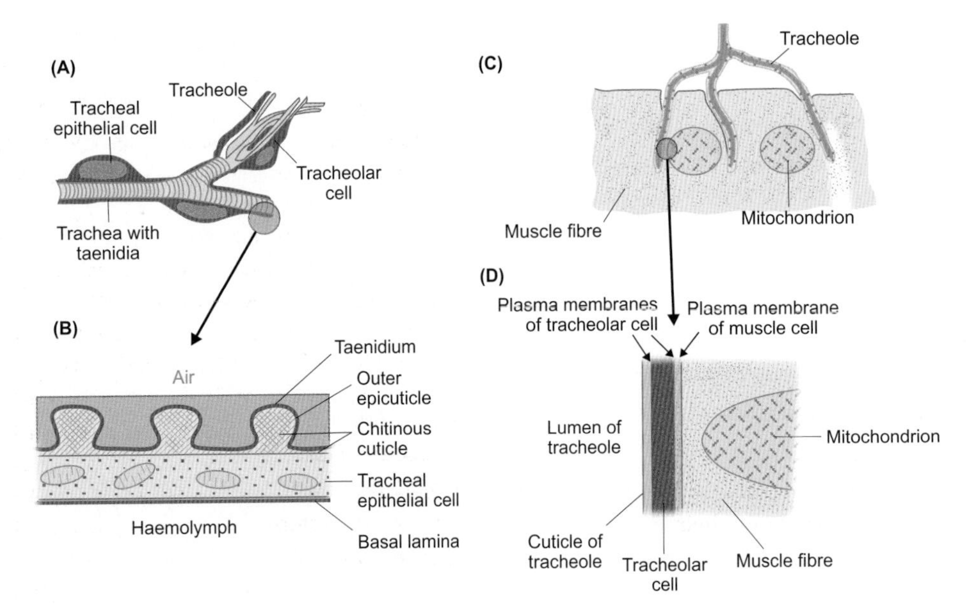

Figure 12.37 Structure of the tracheae of insects

(A) Trachea and tracheoles. (B) Longitudinal section of the wall of a trachea showing the taenidia. The lumen of the trachea (air) is one side and the haemolymph (blood) is the other. (C) Tracheoles indenting a muscle fibre. (D) Enlargement of the area indicated in (C).

Reproduced from Chapman RF (1998). The Insects. Structure and Function. Cambridge University Press, Cambridge.

The greater permeability of the ends of the tracheoles means that they normally contain variable amounts of liquid, both in terrestrial and aquatic insects, with the amount seeming to be related to oxygen demand and supply. The traditional view is that there is a balance between capillarity drawing more liquid along the tracheoles and the colloid osmotic pressure of the surrounding cytoplasm sucking the liquid out of the tracheoles. There is less fluid in the tubes when the PO_2 in the surrounding tissues falls (for example, during exercise), which means that the surface area of the tracheae filled with air is increased, although the mechanisms by which this could be achieved have yet to be demonstrated.

12.4.1 Respiratory gas transport in insects

Movement of gases through the tracheal system of terrestrial insects consists of a combination of active ventilation (forced convection) and passive diffusion. In southern hawker dragonflies (*Aeshna cyanea*) and other large insects, only the primary tracheae supplying the wing muscles need to be ventilated; diffusion is sufficient in the remaining parts of the tracheal system, even at the highest metabolic rates. For dragonflies, it has been calculated that a difference in PO_2 of about 13 kPa is required between the primary tracheae and the muscle mitochondria (a distance of about 1 mm) to ensure an adequate supply of oxygen during flight. As the primary tracheae are well ventilated, most of the drop in partial pressure of oxygen (11.5 kPa) occurs in the secondary tracheae.

In some tissues, such as the flight muscles of dragonflies, the tracheoles do not indent the muscle fibre membrane, but form an envelope or 'a layer of air' around the fibre. Calculations indicate that, in order to meet the metabolic demands of flight, the maximum diffusion distance from a tracheole to the mitochondria of the muscle can only be approximately 10 μm. In other words, the tracheoles should be less than 20 μm apart. This imposes a limit on the maximum diameter of the muscle fibres, and the muscle fibres of dragonflies are close to this limit[59]. However, in most insects, the tracheoles do indent the flight muscle, as shown in Figure 12.37C. Consequently, the tracheoles can be closer together and the muscle fibres can be much larger. In dipterans, such as European blowflies (*Calliphora vicina*), the tracheoles are only 2–3 μm apart; this ensures adequate diffusion of oxygen, even at the highest metabolic rates that have been measured in this species.

Three types of ventilation occur in terrestrial insects:

- passive, or suction, ventilation;
- active ventilation resulting from changes in the pressure in the air sacs during the heartbeat cycle, rhythmic compression of the trachea, and/or contraction of specific respiratory muscles;
- autoventilation resulting largely from the activity of flight muscles.

We now consider each of these in turn.

Passive ventilation and the discontinuous gas-exchange cycle

Resting insects display a large range of patterns of spiracular opening, from continuous through cyclic to discontinous. As at 2005, a **discontinuous gas-exchange cycle (DGC)** had

[59] The relationship between increased partial pressure of oxygen in the atmosphere 300 million years ago and the evolution of giant dragonflies is discussed in Section 11.1.1.

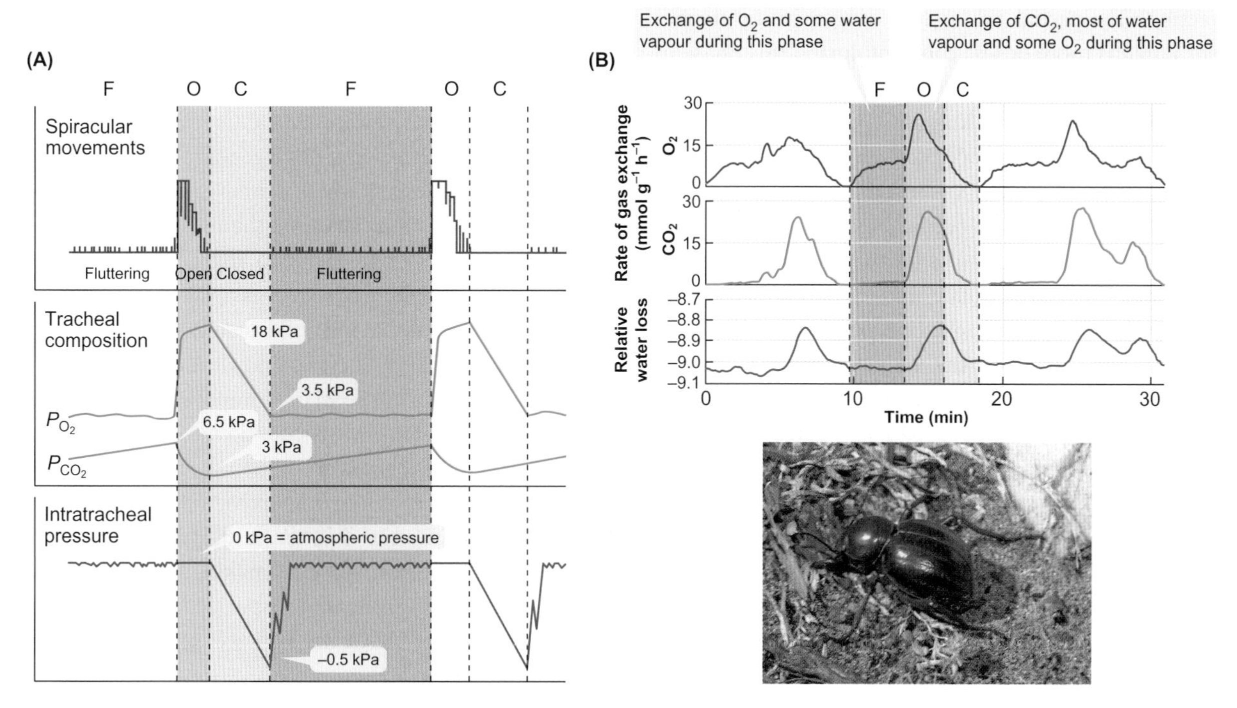

Figure 12.38 Discontinuous gas exchange in insects.

(A) The relationships between the activity of the spiracles, composition of oxygen and carbon dioxide and air pressure inside the tracheal system during two respiratory cycles in a pupa of cercropia moths (giant silk moths) (*Hyalophora cercropia*). Note that the partial pressure of oxygen (PO_2) inside the tracheal system remains more or less constant during the fluttering phase, i.e. the rate of oxygen movement inwards matches the rate at which it is being used by the metabolizing cells. On the other hand, the partial pressure of carbon dioxide (PCO_2) continues to rise, indicating that its rate of removal does not match its rate of production. F, fluttering phase; O, open phase; C, closed phase. (B) Rates of gas exchange and the relative rate of water loss during ventilation cycles in a tok-tok beetle (*Psammodes striatus*). Note that, although exchange of oxygen occurs during the fluttering phase, exchange of carbon dioxide and most water loss occurs during the open phase. Much exchange of oxygen also occurs during the open phase.

Adapted from Levy RI and Schneiderman HA (1966). Discontinuous respiration in insects – IV. Changes in intratracheal pressure during the respiratory cycle of silkworm pupae. Journal of Insect Physiology 12: 465-492. reproduced from Lighton JRB (1988). Simultaneous measurement of oxygen uptake and carbon dioxide emission during discontinuous ventilation in the tok-tok beetle, *Psammodes striatus*. Journal of Insect Physiology 34: 361-367. Image: © Dave Mangham.

been recorded in 59 species out of a total of 100 for which data were available. Thirty-five species showed a cyclic pattern and 31 showed a continuous pattern. Thus, the DGC is not common to all insects, but it is the most widely described and discussed of the patterns seen in resting insects.

The discontinuous pattern of gas exchange was first described in the pupae of some moths, and data from an early study are depicted in Figure 12.38A. The pattern also occurs in practically all adult ants and in some bees, cockroaches, beetles, bugs and orthopterans (locusts, crickets and grasshoppers). Studies on several species of insects have identified three distinct phases of the cycle, which are illustrated in Figure 12.38: the C phase, F phase and O phase.

The spiracles are held tightly closed most of the time; this is known as the C phase, or closed phase. During this phase oxygen is being consumed and carbon dioxide is being produced by the metabolizing cells. However, carbon dioxide is very soluble in the body fluids[60], where it forms mainly bicarbonate, so much less carbon dioxide enters the tracheal system than the amount of oxygen that is removed. As a result of this imbalance, the pressure in the tracheal system falls below atmospheric pressure, as illustrated in Figure 12.38A. Consequently, when the spiracles open, air is passively sucked in.

Changes in air pressure and in partial pressures of O_2 and CO_2 (PO_2 and PCO_2, respectively) in the tracheal system during such respiratory cycles have been measured in the pupa of cercropia moths (giant silk moths) (*Hyalophora cecropia*) and are shown in Figure 12.38A. When PO_2 reaches approximately 3.5 kPa and the air pressure in the tracheal system is approximately 0.5 kPa below atmospheric, the closer muscles of the spiracles periodically relax and contract, causing the spiracular valves to open and close partially and to flutter rapidly. This is the F phase, or flutter phase, and it enables small amounts of air to be sucked into the tracheal system and only relatively small amounts of water vapour and carbon dioxide to diffuse outward, against the inward flow of air, as shown in Figure 12.38B.

Although some inward diffusion of oxygen along the difference in partial pressure may occur during the F phase, most of the inward movement of oxygen results from forced

[60] More accurately, water has a high capacitance for carbon dioxide, as discussed in Section 11.1.2.

12

convection. The partial pressure of oxygen in the tracheal system remains more or less constant during this phase indicating that the amount of oxygen entering the system is approximately equal to the amount consumed by the cells. It is this phase that is generally considered to be important in keeping respiratory water loss to a minimum[57], while enabling exchange of oxygen to occur.

During the flutter phase of silkworm pupae, carbon dioxide continues to accumulate until a partial pressure of approximately 6.5 kPa is reached, at which point the closer muscles to some or all of the spiracles completely relax, enabling the spiracles to open fully. This is known as the O phase, or open phase. Figure 12.38B shows that during the O phase, carbon dioxide and water vapour escape while oxygen diffuses into the tracheal system. In the pupae of silkworms, approximately 90 per cent of carbon dioxide release occurs during the O phase. The spiracles close once PCO_2 in the tracheal system has fallen to about 3 kPa and PO_2 has reached 18–20 kPa, as illustrated in Figure 12.38A.

Diffusion during the open phase may be aided in some species by abdominal ventilation, as in tok-tok beetles (*Psammodes striatus*) and in some species of cockroach, such as *Nauphoeta cinerea*. In some species, the F phase is not always present.

The available evidence suggests that cyclic gas exchange is the ancestral form and that discontinuous gas exchange may have evolved at least five times in the insects. A cyclic or discontinuous pattern of gas exchange is reminiscent of the intermittent pattern of ventilation seen in other resting air-breathing ectotherms such as lungfish, amphibians and reptiles. A more continuous form of ventilation is present in most ectotherms, including insects, when metabolic rate increases.

What is the functional significance of the DGC? It was originally proposed that the DGC is an adaptation to reduce water loss (the hygric hypothesis), and some species of insects may well modify their discontinuous gas-exchange cycle in a manner that is consistent with water conservation[61]. However, it is now evident that the discontinuous gas-exchange cycle is not confined to insects that live in dry environments, neither is it found in all species that inhabit dry environments.

Another possibility is that the DGC in resting insects may serve to keep the partial pressure of oxygen as low as possible in the tracheoles, and therefore at the cells, and thus reduce the toxic effect of free-radicals produced by oxygen[62] (the oxidative damage hypothesis). However, this suggestion is not supported by studies on adult migratory locusts (*Locusta migratoria*), which indicate that when breathing a hyperoxic gas mixture, PO_2 in the trachea increases. Another proposal is that discontinuous gas exchange is an adaptation to life in burrows where partial pressure of oxygen is low (hypoxia) and that of carbon dioxide is high (hypercapnia—the chthonic hypothesis). However, both hypoxia and hypercapnia stimulate ventilation and actually oppose cyclic or discontinuous patterns of gas exchange.

The functional significance of DGCs in terms of these three hypotheses was tested in a phylogenetically based analysis[63] on 40 species of insects which exhibit discontinuous gas exchange. The study indicated that the most likely function of discontinuous gas exchange is to reduce respiratory water loss[64] while maintaining adequate gas exchange. Indeed, it has been demonstrated that in speckled cockroaches (*Nauphoeta cinerea*), DGCs extend survival time during water restriction. However, DGCs do not seem to be important in water conservation as such, as the cockroaches do not exhibit them when both water and food stress are at their greatest.

If a particular behaviour is adaptive[65], it must be passed on from one generation to another; it must be heritable, and it must provide some benefit, thus making it a target for natural selection[66]. Experiments on speckled cockroaches have demonstrated that DGCs are heritable and thus of possible adaptive significance.

To complicate matters further, it has been noted that all species of insects that display DGCs have either reduced or absent brain activity. This is certainly the case in the pupae in which DGC was first discovered; the electrical activity in the brain[67] of a larva ceases the day before it moults into a pupa. All the other ganglia in the insect's ventral nerve cord are active during pupal diapause.

Experiments on speckled cockroaches have demonstated that the discontinuous gas-exchange pattern resides in the thoracic and/or abdominal ganglia of the central nervous system[68] and this pattern can be transformed into a continuous pattern by input from the animal's brain. On the basis of this—and much other data—it has been proposed that discontinuous gas exchange is not necessarily an adaptation to fulfil a particular function, but is the result of reduced brain activity. Such reduced brain activity may occur, for example, during the inactive part of the daily cycle, a sleep-like state or diapause, any one of which may be part of an energy conserving strategy. This has been called the neural hypothesis.

Experimental support for both the hygric and neural hypotheses in some species has led to the suggestion that natural selection may act on the tendancy for some individuals to enter a sleep-like state rather than on DGCs as such. When in a sleep-like state, and exhibiting DGCs, it is highly likely that the insects will save water and if this occurs under conditions of water shortage, then the reduction in water loss could lead to positive selection for DGCs. There is now

[61] Section 6.1.2 and online Experimental Panel 6.1 examine the water saving in DGC of some insects.

[62] The production of free radicals by oxygen is discussed in Section 2.4.3.

[63] Phylogenetically based data analysis is discussed in Section 1.3.2.

[64] Respiratory water loss in insects is discussed in Section 6.1.2.

[65] Adaptation is discussed in Section 1.5.2.

[66] Natural selection is discussed in Section 1.5.3.

[67] The brain of insects is discussed in Section 16.1.1.

[68] Generation of the respiratory rhythm in insects is discussed in Section 22.1.1.

a need for future studies to concentrate on heritability and responses to selection in the wide range of orders of insects that display DGCs.

Active muscular ventilation in insects

Many insects actively ventilate their tracheal system by muscular activity. An important feature of the tracheal system for active ventilation is the presence of the air sacs, which are shown back in Figure 12.36A,B. Air sacs are found in many larger insects, particularly those that fly; they are most highly developed in the dipterans (flies) and hymenopterans (bees and wasps). There is considerable variation in the overall pattern of active ventilation in insects. Dragonflies, locusts and small flies actively ventilate more or less continuously, whereas others such as wasps, actively ventilate only during and immediately after flight, and some cockroaches actively ventilate intermittently.

Ventilatory movements commonly occur in the abdomen of the larger insects. In locusts, expiration and—to a lesser extent—inspiration are achieved by muscular activity. By contrast, in species, such as cockroaches, expiration is muscular and inspiration is purely the result of elastic recoil of the system. The basic expiratory movements are the result of the contraction of dorsoventral muscles, which causes either depression of the dorsal part of the abdomen (tergum), or a combination of depression of the tergum and raising of the ventral surface (sternum). During and after periods of high metabolic demand, e.g. flight, these basic movements may be supplemented by other movements such as the longitudinal telescoping of one segment within the adjacent segment.

In one genus of cockroach (*Periplaneta* spp.), airflow may be tidal, with air flowing in and out of the same spiracles, which remain open continuously during ventilation. However, in many insect species, including other species of cockroach, there is flow-through ventilation and activity of the spiracles is synchronized with ventilatory movements; some spiracles open during inspiration and others open during expiration. In the larger insects, air usually enters through the anterior spiracles and leaves via posterior ones. For example, in resting desert locusts (*Schistocerca gregaria*), air enters via spiracles 1, 2 and 4 and leaves via spiracle 10. When the animal is more active, spiracle 3 may also open during inspiration and spiracles 5–9 may open during expiration.

In addition to the contraction of specific respiratory muscles, the movement of haemolymph during the heart beat cycle may cause changes in pressure in air sacs in the head and thorax in some species, such as adults of the fly species *Calliphora* and *Drosophila*. These changes in pressure in the air sacs cause air flow within the tracheal system. The use of X-ray imaging has also revealed alternate compression and relaxation of the tracheae in the head and thorax of some species, such as ground beetles (*Platynus decentis*), although the cause and function of rhythmic tracheal compression are not yet understood.

Autoventilation and ram ventilation in insects

Ventilation of the tracheal system can be assisted as an insect moves through air, powered by the flight muscles. During the flight of locusts, oxygen consumption increases 24-fold, and abdominal ventilation is assisted by the activity of the flight muscles. Spiracles 2 and 3 remain open, thereby providing tidal flow of air through the tracheae, which is powered largely by the flight muscles. This is known as **autoventilation**. Autoventilation is also present in many smaller species, such as blowflies (*Calliphora vicina*), in which the air flow is unidirectional, entering through spiracle 1 during the downstroke of the wings and leaving via spiracle 2 during the upstroke.

Abdominal, unidirectional, pumping via spiracles 1 and 5–10 remains at an almost constant 0.5 mL g^{-1} min^{-1} during all levels of activity in locusts and is most likely responsible for supplying the central nervous system during flight. In addition, the abdomen produces tidal flow during horizontal flight at 28–29°C of about 2.5 mL g^{-1} min^{-1}, of which 1.2 mL g^{-1} min^{-1} goes to the thorax and the remaining 1.3 mL g^{-1} min^{-1} supplies other parts of the body. However, the total ventilation of the thorax during horizontal flight is 5.3 mL g^{-1} min^{-1}, with the additional 4.1 mL g^{-1} min^{-1} being provided by autoventilation.

In American cockroaches (*Periplaneta americana*) and some species of beetles with few or no air sacs, such as giant African longhorn beetles (*Petrognatha gigas*), abdominal pumping is weak or absent during flight, so autoventilation would appear to be of particular importance in these species. In *Petrognatha*, autoventilation is supplemented by **ram (or draught) ventilation**[69], whereby air is ducted through large tracheae as the insect moves through the air.

12.4.2 Gas exchange in aquatic insects: tracheal gills, air bubbles and plastrons

In many aquatic insects the tracheae open to the outside via spiracles, as shown earlier in Figure 12.36A. Aquatic insects with spiracles have the problem of preventing water entering the tracheal system, which is achieved either by the presence of protective, water-repellent hairs, or by the spiracles being withdrawn and covered by fleshy valves. In some aquatic larvae, such as the small larvae of midges, there is no spiracle and the tracheae divide to form a network beneath the body surface (Figure 12.39A) which enables gas exchange to occur across the surface of the body.

[69] Ram ventilation in some species of fish is discussed in Section 12.2.2.

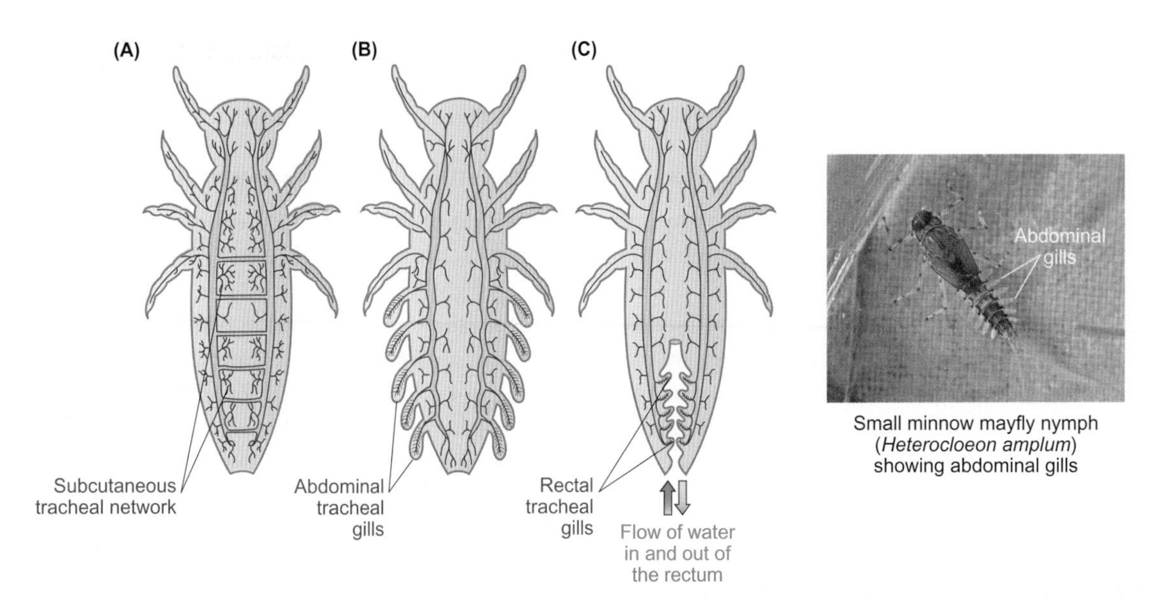

Figure 12.39 Schematic diagrams of different types of respiratory systems of aquatic insects

(A) Closed system, with no spiracles and with a network of fine tracheae beneath the general body surface thus enabling gas exchange to occur across the body surface. (B) Closed system with tracheal gills on the abdomen. (C) Closed system with tracheal gills in the rectum.

Adapted from Wigglesworth VB (1984). Insect Physiology. Chapman and Hall, London. Image © Robert G Henricks.

Tracheal gills of aquatic insects

Some species of aquatic insects which lack spiracles have specialized **tracheal gills** for gas exchange, as illustrated in Figure 12.39B and C. Tracheal gills are almost exclusively confined to the larvae of some species of insects. They are usually external and are often 'ventilated' when the insect makes movements through the water. Their location and numbers differ between insect groups. For example, Figure 12.39B shows some located on the abdominal segments, as found in the larvae of mayflies (Ephemeroptera). In larvae of dragonflies (Odonata), however, the gills are found in the rectum, as illustrated in Figure 12.39C.

Water is moved in and out of the rectal chamber of dragonfly larvae by active expulsion (expiration) caused by contraction of dorso-ventrally located muscles, and passive filling (inspiration) resulting from the elastic recoil of the exoskeleton. This tidal flow is unusual among aquatic organisms as it creates a relatively large volume of water, which does not reach the gas-exchange area of the gills, instead constituting what is known as a **dead space**. The presence of a deadspace reduces the overall effectiveness of gas exchange and is a feature of the respiratory systems of birds and mammals[70]. Although the majority of the diffusion barrier in the gills of the dragonfly is provided by chitin, the diffusion conductance of the gills is sufficiently large to support aerobic metabolism at all levels of activity of the larvae.

[70] Deadspace and gas exchange in birds and mammals is discussed in Section 12.3.3.

Some adult aquatic insects with open tracheal systems take a bubble of air underwater with them and are able to use the air bubble to exchange gases with the surrounding water. We now discuss how this is achieved.

Air bubbles and plastrons of aquatic insects

The air bubble that some species of aquatic insects take under water is in contact with the spiracles, but is in different locations in different species. For example, in great diving beetles (*Dytiscus marginalis*), the air bubble is held beneath the hard coverings over the wings and is renewed by the beetle approaching the water surface tail first. In contrast, water boatmen (*Notonecta glauca*) hold the bubble by long, water-repellent hairs on the ventral surface and under the wings.

We can calculate the maximum time an insect could stay submerged using the oxygen contained in a bubble of air if we know (a) the rate at which the oxygen is consumed by the animal and (b) the volume of the bubble. The amount of oxygen in the bubble can be calculated from its volume, assuming that air consists of 21 per cent oxygen. Such calculations indicate that many insects stay submerged for much longer than would seem possible.

The gas bubble that some aquatic insects carry with them under water is used not only as a store of oxygen, but also as a means of obtaining oxygen from the surrounding water. To understand how the bubble functions, we need to consider

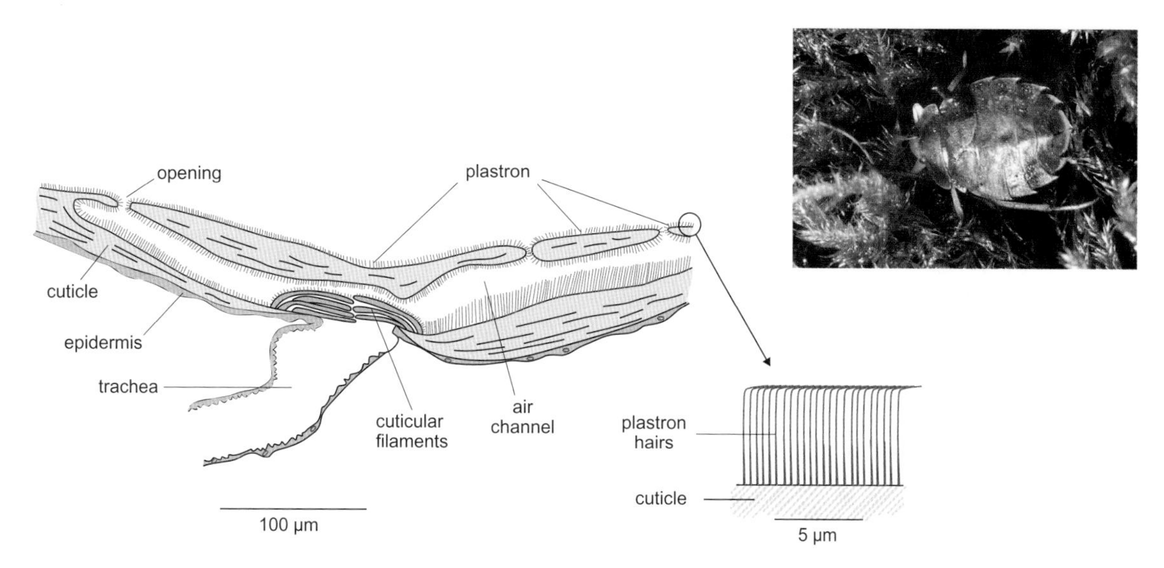

Figure 12.40 Plastron of river bugs (*Aphelocheirus aestivalis*)

Section through a spiracle showing the junction of a trachea with a system of channels in the cuticle connecting with the plastron, a portion of which is shown in detail. Note cuticular filaments guarding the entrance to the trachea.

Reproduced from: Chapman RF (1998). The Insects. Structure and Function. Cambridge University Press, Cambridge. Image: © H. Günther Germany / Rheinland-Pfalz, Ingelheim 1985.

the physical properties of the gases in the bubble. Carbon dioxide has a capacitance coefficient in water that is about 30 greater than oxygen, as shown in Table 11.2. As a result, the carbon dioxide produced by the insect enters the bubble but diffuses out into the water almost immediately. This means that the bubble consists essentially of around 21 per cent oxygen and 79 per cent nitrogen.

The oxygen that is used by the insect causes a decrease in the partial pressure of oxygen in the bubble and an increase in the partial pressure of nitrogen. These changes in partial pressure cause oxygen to diffuse from the surrounding water into the bubble and nitrogen to diffuse from the bubble into the water. The loss of nitrogen from the bubble means that it progressively shrinks to the point at which the animal has to visit the surface to collect another bubble of air. But, as there is $(0.79 / 0.21) = 3.8$ times as much nitrogen in the bubble as there is oxygen, the bubble should last the insect at least 3.8 times as long as it would if it just used the oxygen that was initially stored in the bubble. There is, however, another advantage for the insect; nitrogen diffuses out of the bubble 45 per cent more slowly than oxygen diffuses in. This means that the bubble actually lasts for about $(3.8 / 0.45) = 8.4$ times longer than if the bubble was used only as an oxygen store.

If the air bubble could be maintained at a constant volume, it would potentially function as a gas-exchange surface indefinitely, with oxygen diffusing into the bubble from the water at the same rate that it is used by the insect. This is exactly what happens in some species of insects such as the river bugs (*Aphelocheirus aestivalis*). In the adults of these species, a very dense pile of water-repellent hairs holds a thin layer of air, known as a **plastron**, on the outside of the body; the spiracles open into this layer of air.

The volume of the plastron is usually small, but it is held constant by the hairs, so it acts only as a gas-exchange surface and not as a store of oxygen. *Aphelocheirus* probably has the most effective plastron, which is shown in Figure 12.40. In order to maintain the integrity of the plastron, the insect must not venture too deeply into the water column, as the increased hydrostatic pressure will compress the layer of air and may disrupt it. Some practical applications of the plastron principle are outlined in Case Study 12.2.

› *Review articles*

Harrison JF (2009). Respiratory system. In: Resh VH, Cardé RT (eds.) Encyclopedia of Insects, 2nd Edn. pp 889–895. Amsterdam: Elsevier.

Harrison JF (2009). In: Resh VH and, Cardé RT (eds.) Tracheal system. In: Encyclopedia of Insects, 2nd Edn. pp 1011–1015. Amsterdam: Elsevier.

Harrison JF, Waters JS, Cease AJ, VandenBrooks JM, Callier V, Klok CJ, Shaffer K, Socha JJ (2013). How locusts breathe. Physiology 28: 18–27.

Matthews PGD, White CR (2011). Discontinuous gas exchange in insects: is it all in their heads? The American Naturalist 177: 130–134.

12

Case Study 12.2 Developing biomimetic technologies inspired by plastrons

Materials scientists and engineers have been inspired to mimic the fundamental design of the plastron to create surface materials for a variety of applications.

Plastrons for supplying fuel cells with oxygen

Investigators at Nottingham Trent University, UK, created a sol-gel foam material of a superhydrophobic substance (MTEOS, methyltriethoxysilane) that mimics the role of the chitin hair-like structures responsible for plastron formation in biological systems. When submerged in water, the superhydrophobic sol-gel foam material develops a silvery sheen which is caused by a film of air at the surface of the material, much the same as an insect's plastron. The researchers enclosed an oxygen sensor in a cavity surrounded by this material and found that the concentration of oxygen (per cent) in the cavity followed changes in the concentration of dissolved oxygen in the water, indicating that oxygen readily equilibrates between the air in the cavity and the surrounding water through the thin film of air on the surface of the cavity wall and through the foam itself.

Figure A shows the results of experiments that demonstrated the effectiveness of the MTEOS foam as a plastron-like structure. A zinc–oxygen fuel cell, that had an oxygen consumption rate of 246 $\mu L\ h^{-1}$ as it generated electricity by oxidizing zinc, was placed within the cavity of a MTEOS foam cell. The oxygen consumed by the fuel cell was partially replaced from the surrounding water until a dynamic equilibrium was reached. In contrast, when the fuel cell was placed in a sealed solid-walled chamber incapable of gas exchange, oxygen in the chamber progressively decreased to zero.

The foam material is not very rigid, so a carbon nanotube array can be used to mimic the plastron chitin 'hairs' to create a much stronger plastron. Such a structure has been demonstrated to maintain its integrity, and hence the enclosed air layer, down to a depth of 37.5 m. Such biomimetic plastron-like enclosures could theoretically allow small autonomous underwater vehicles or other underwater miniaturized machines to operate on the electricity generated by fuel cells without the need to carry stored oxygen supplies for those fuel cells. Although it is not yet possible to scale this design up to meet much higher levels of oxygen demand, it has been estimated that a MTEOS foam chamber with a surface area of 90 m^2 (equivalent to a sphere with a radius of 2.8 m) could provide enough oxygen for a human to survive underwater.

Plastrons for reducing drag on boats

Another possible application of mimicking the air-trapping design of insect plastrons would be to make the surface of boat or ship hulls more 'slippery' (that is, to reduce frictional drag) by reducing the 'no-slip' interaction of the water with the surface material of the boat's hull. Such a reduction in drag would enhance fuel economy.

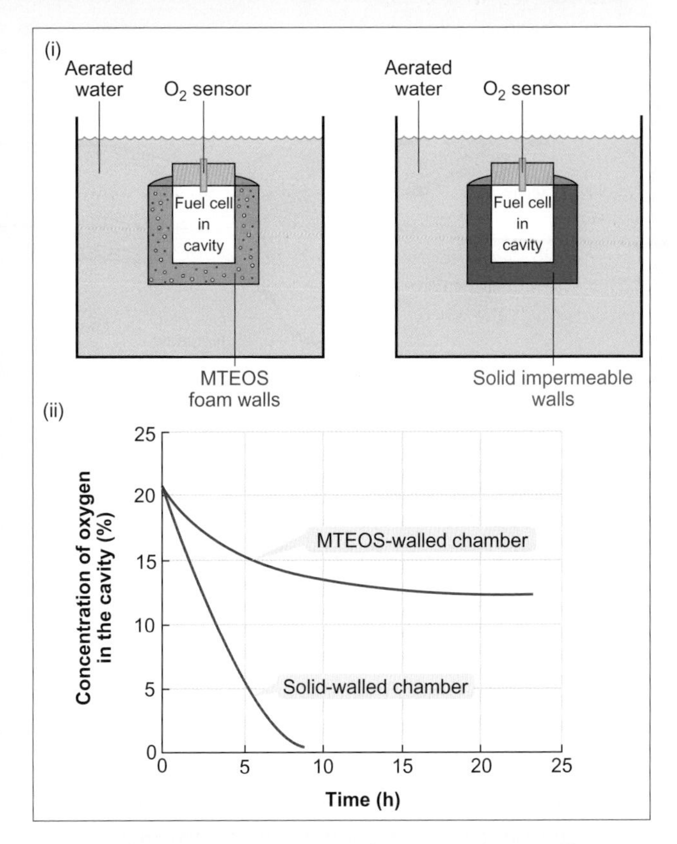

Figure A Effectiveness of MTEOS foam as a plastron-like structure

(i) The experimental set-up. A closed chamber (2.5 cm^3) built with either the MTEOS foam walls (4 mm thick) or solid impermeable walls, sits in well-aerated water and contains both an oxygen sensor and a fuel cell that consumed oxygen. (ii) Comparison of the oxygen levels recorded over time in the solid-walled chamber (blue curve) and in one enclosed by the superhydrophobic but gas-permeable foam (red curve) that allowed oxygen replenishment from the water.

Source: adapted from Shirtcliffe et al, 2006.

The concept of frictional drag stems from the basic idea that when a fluid moves over a solid surface the flow velocity at the surface is zero. The lack of flow at the surface forms a boundary layer or no-slip zone[1]. Beyond this surface zone, flow velocity gradually increases with increasing distance from the surface until it reaches a maximum.

The investigators at Nottingham Trent University performed a classic 'terminal velocity'[2] experiment to test the idea that a plastron-like surface would reduce drag on a solid ball dropped into water. The terminal velocity of a solid sphere dropped in water is determined by the drag imposed upon it by the surrounding water and is not normally affected by the chemistry of the surface material. These investigators used acrylic spheres. Some spheres had roughened surfaces and were sprayed with one of several hydrophobic or superhydrophobic coatings. By filming the spheres as they fell through the column of water, the investigators could measure their terminal velocities.

The spheres with a superhydrophobic coating developed a silvery sheen when submerged in water, indicating the presence of a thin film of air held at their surfaces. Those spheres that had the plastron-like thin film of air exhibited a 5–15 per cent increase in terminal velocity indicating a reduction in drag. The rougher the spheres were before treatments with the superhydrophobic compound, the thicker the air film after treatment, and the greater the reduction in drag. These data suggest that the addition of a plastron-like coating on the hulls of ships could reduce drag and therefore increase their fuel efficiency as they move through the water.

› Find out more

Lee C, Kim CJ (2011) Underwater restoration and retention of gases on superhydrophobic surfaces for drag reduction. Physical Review Letters 106: 014502.

McHale G, Shirtcliffe NJ, Evans CR, Newton MI (2009). Terminal velocity and drag reduction measurements on superhydrophobic spheres. Applied Physics Letters 94: 064104.

Shirtcliffe NJ, McHale G, Newton MI, Perry CC, Pyatt FB (2006). Plastron properties of a superhydrophobic surface. Applied Physics Letters 89: 104106.

[1] Boundary layer is discussed in Section 8.4.3 and shown in Figure 8.13A.

[2] The terminal velocity of a falling object is its velocity when the sum of the drag force acting on it and its buoyancy is equal to the downward force of gravity.

Checklist of key concepts

- Gas exchange occurs across the general **body surface, gills, lungs or tracheoles**.
- There are four basic types of gas exchanger:
  - the simplest and least effective is the skin, which is an **infinite pool** exchanger;
  - the most effective are the **countercurrent** exchangers of the gills of most aquatic animals, where the water and blood flow in opposite directions;
  - less effective than the gills of aquatic animals are **ventilated pool** gas exchangers, such as the lungs of vertebrates and some species of invertebrates, in which air is pumped in and out;
  - **cross-current** exchangers, such as the lungs of birds, have an effectiveness that is midway between that of gills and lungs; air flows through the lungs of birds at right angles to the flow of blood.
- **Diffusion conductance** (G_{diff}) of a gas exchanger for oxygen is the ease with which oxygen diffuses across the exchanger.
- G_{diff} is the reciprocal of **resistance** to diffusion, R_{diff}. So $G_{diff} = 1/R_{diff}$.
- **Morphological diffusion conductance** is directly related to the **surface area** of the exchanger and the **diffusion constant** of the gas through that surface and indirectly related to its **thickness**.
- **Physiological diffusion conductance** is the **rate of oxygen exchange** for a given **average difference in partial pressure of oxygen** across the gas-exchange surface.
- Morphological diffusion conductance is fixed for any animal, whereas physiological conductance can vary, depending on the effective area for gas exchange.
- More active animals tend to have greater morphological diffusion conductances.
- Gas exchange in water and in air are influenced by the physical properties of the two environments:
  - air is much less dense and viscous than water;
  - for a given partial pressure of oxygen, there is a greater concentration of oxygen in air than in water: the **capacitance coefficient of oxygen** in air is greater than that in water;
  - oxygen and carbon dioxide have similar capacitance coefficients in air;
  - carbon dioxide has a capacitance coefficient in water that is about **30 times greater** than that of oxygen.
- The **ventilation-convection requirement** of an animal breathing water is greater than that when it is breathing air.

Gas exchange across the body surface: an infinite pool gas exchanger

- The simplest and least effective gas exchanger is the skin, because there is no specific pumping mechanism to maintain a regular flow of water or air over it.
- Gas exchange across a body surface is reduced by the relative thickness of the skin and the build up of a **boundary layer** in the surrounding fluid, particularly for oxygen in water.
- Movements of the animal and/or of the surrounding fluid reduce the thickness of the boundary layer.
- Gas exchange across the body surface occurs in many water-breathing species, but the **skin is usually the more important route for the removal of carbon dioxide than for the uptake of oxygen**.
- The rate of gas exchange across the skin can be controlled to a certain extent by varying the rate of blood perfusion through it.
- Some species of **ectothermic** terrestrial animals exchange respiratory gases across their skin
- Among **endotherms**, only some very small newborn marsupials rely to any large extent on their skin for gas exchange.
- Gas exchange across some very small insect eggs may be entirely by diffusion.

- The embryos inside the larger eggs of reptiles and birds have a circulatory system which aids gas exchange by **forced convection**.
- Developing embryos, whether in eggs or in the uterus of a mammal, are **hypoxic**.

Gas exchange across gills: countercurrent gas exchangers

- Gills are delicate organs and are normally located and protected within a **gill chamber**.
- The surface area of gills is usually increased by the presence of folds or plates called **lamellae** across which gas exchange occurs.
- The ventilation current is generated by:
  - **cilia** in some species of molluscs,
  - a modified head appendage, **the scaphognathite**, in decapod crustaceans,
  - **muscular action forcing or sucking** water into the gill chamber in cephalopod molluscs and fish.
- Teleost (bony) fish have a **buccal force pump** and an **opercular suction pump**.
- More active fish, such as mackerel, tuna and sharks, use **ram ventilation** when swimming.
- In fully aquatic animals, partial pressure of carbon dioxide in the blood is very low as it readily diffuses out across the gills and/or skin.

Animals that undergo gas exchange in water and in air: bimodal breathers and the evolution of the vertebrate lungs

- Some species of animals that live in water but have access to air may use **lungs** in addition to their skin and/or gills for gas exchange.
- Animals that can breathe both water and air are known as **bimodal breathers**.
- Many species of pulmonate gastropod molluscs live on land, although they must live in moist conditions to avoid dessication.
- Land crabs retain their gills and tend to carry water in their gill chambers for the removal of carbon dioxide.
- The **cutaneous lungs** of land crabs vary from being a highly vascularized surface of the expanded gill chambers to the more complex structures of Trinidad mountain crabs.
- The **diffusion distances** across some cutaneous lungs are similar to those in the lungs of birds and mammals.
- Some present day fish have **air-breathing organs (ABO)** which may have evolved from:
  - branchial (gill), opercular and pharyngeal cavities,
  - the gastrointestinal (GI) tract,
  - the **air bladder**, which evolved into **lungs** of the terrestrial vertebrates.
- No present day air-breathing crab or fish has completely lost its gills, which are the major route for the removal of carbon dioxide.
- Some air-breathing fish are **obligate air breathers** whereas others are **facultative air breathers**.
- The original function of the air bladder of fish was as an accessory respiratory organ.
- In more ancient **ray-finned fish**, the air bladder still functions as a gas-exchange organ, but in most bony fish it functions as a buoyancy, sound production or sound detection organ.
- In most fish that use an air bladder for gas exchange, it is ventilated by way of a **four-phase buccal force pump**.
- Air-breathing ray-finned fish have a **single circulation** of blood, as in water-breathing fish.
- In lungfish and in other air-breathing vertebrates, there is what is known as a **double circulation**.
- Lungs of lungfish have many internal divisions, which increase the surface area for gas exchange.
- When in water, lungfish irrigate their gills using a buccal force pump, which is also used to ventilate the lungs.

From relatively simple to more complex lungs in amphibians and reptiles

- Amphibians and reptiles depend more on air as their source of oxygen.
- The lungs of amphibians and reptiles have internal compartments called **faveoli** divided by septa to form a honeycomb-like structure.
- The lungs of reptiles are more complex than those of lungfish and amphibians.
- The lungs of the more active species of reptiles have anatomical features that are similar to those in birds.
- Amphibians use a buccal force pump to ventilate their lungs, like lungfish, but reptiles have evolved a **suction (aspiration) pump**, which is powered by muscles in the thorax and abdomen.
- Both locomotion and ventilation in land vertebrates involves the use of some trunk muscles, a consequence being that some species of lizards are unable to ventilate their lungs adequately while running.
- Air-breathing ectotherms ventilate their gas-exchange organ **intermittently**, which causes large variations in the partial pressure of the respiratory gases in the blood.
- In aquatic amphibians and reptiles, carbon dioxide diffuses out across the skin.

Mammalian lungs: ventilated-pool gas exchangers

- The lungs of mammals are a **ventilated-pool** gas-exchange system in which air is moved in and out in a tidal fashion.
- In each lung, the bronchus divides into bronchioles and (eventually) terminal bronchioles, which are lined with blind-ending sacs called **alveoli**.
- Gas exchange only occurs across the alveoli.
- Airways down to and including the terminal bronchioles only conduct air: they constitute the **anatomical deadspace**.
- Mammalian lungs are ventilated in part by a muscular **diaphragm** which draws air into the lungs.
- The alveoli in the lungs of mammals have a tendency to collapse because of the surface tension at the fluid/air interface.

- The tendency of alveoli to collapse is prevented in part by the presence of compounds called **surfactants**.

Bird lungs: cross-current gas exchangers

- The **(primary) bronchus** of each lung produces a set of secondary bronchi (**ventrobronchi**) as it enters the lung followed by another set (**dorsobronchi**).
- The two sets of secondary bronchi in each lung are connected by the **tertiary bronchi** (also called **parabronchi**).
- These parabronchi constitute the **palaeopulmo** and are responsible for most of the gas exchange in the lungs.
- Most species of birds have another network of parabronchi, the **neopulmo**.
- The bronchi that pass through each lung are joined to two sets of associated **air sacs**—the anterior (cranial) set and the posterior (caudal) set which change volume during ventilation.
- Air flow through the lungs is **unidirectional**: air passes through the parabronchi of the palaeopulmo in the same direction during both inspiration and expiration.
- Bulk blood flow through the gas-exchange area of the lungs is at right angles to bulk air flow, giving rise to a **cross-current** gas exchanger.
- The physiological diffusion conductance of the lungs of birds is greater than that of mammalian lungs, with the exception of bats.

The tracheal system of insects

- The gas-exchange systems of insects, millipedes and centipedes do not rely to any great extent on a circulatory system to transport the respiratory gases between the environment and the metabolizing cells.
- The respiratory gases are transported through air-filled tubes called **tracheae**, which divide into smaller branches and eventually give rise to the **tracheoles**, where gas exchange occurs.
- The tracheae open to the surrounding air via pairs of segmentally arranged **spiracles** which can be actively opened or closed.
- Movement of gases through the tracheal system of terrestrial insects consists of a combination of **ventilation** and passive **diffusion**.
- Ventilation of the respiratory system in terrestrial insects can be:
  - **passive ventilation** (suction),
  - **active ventilation**, mainly by the contraction of specific respiratory muscles,
  - **autoventilation** and **ram ventilation** resulting largely from the activity of flight muscles.
- **Discontinuous gas-exchange cycles (DGC)** occur in many, but not all, of the species of insects that have been studied.
- DGCs consist of three phases:
  - **closed (C)**, when the spiracles are tightly closed most of the time,
  - **flutter (F)**, when the spiracles partially open and close and flutter rapidly,
  - **open (O)**, when the spiracles are open.
- DGCs are normally only apparent during the inactive part of the daily cycle or diapauses, when activity in the brain is reduced or absent altogether.
- Some species of insects became secondarily aquatic.
- Modifications to the tracheal system in aquatic species include:
  - a network of tracheae beneath the body surface,
  - protection for the spiracles, where they persist,
  - **tracheal gills**,
  - **air bubbles and plastrons**.

Study questions

*** Answers to these numerical questions are available online. Go to www.oup.com/uk/butler**

1. What is a boundary layer and why is it a hindrance to the exchange of oxygen across the skin of aquatic animals? (Hint: Section 12.2.1.)
2. What is the deadspace in the respiratory system of air-breathing vertebrates? What is its impact on the overall effectiveness of gas exchange? (Hint: Section 12.3.3.)
3. What is the ventilation convection requirement? Why is it higher in water-breathing animals than in air-breathers? (Hint: Section 12.3.3.)
4. What is the difference between a facultative air breather and an obligate air breather? (Hint: Section 12.3.3.)
5. Why do you think that air-breathing ectotherms ventilate their gas-exchange organs intermittently when at rest whereas endotherms (birds and mammals) and most water-breathing ectotherms ventilate their gas-exchange organs continuously? (Hint: Section 12.3.3.)
6. How are the eggs of birds able to exchange adequate amounts of respiratory gases without losing too much water? (Hint: Section 12.3.2.)
7. What is the difference between respiratory quotient and respiratory exchange ratio? (Hint: Section 12.3.3.)
8. What is the advantage of the suction (aspiration) pump used by reptiles, birds and mammals to ventilate their lungs over the buccal force pump used by lungfish and amphibians? (Hint: Section 12.3.3.)
9. Why in theory is the cross-current gas exchanger present in the lungs of birds more effective than the ventilated pool type seen in the lungs of mammals? (Hint: Section 12.3.3.)
10.* If tidal volume of a resting mammal is 100 mL, the anatomical deadspace is 30 mL and respiratory frequency is 25 breaths min^{-1}, what is alveolar ventilation, in litres min^{-1}? (Hint: Section 12.3.3.)

11. Surfactant is present in the lungs of most air-breathing animals, but why is it so important, particularly in mammals? (Hint: Section 12.3.4.)

12.* A water-breathing animal and an air-breathing animal have the same rate of oxygen consumption. In the water-breather, percentage extraction from the water ventilating the gills is 40 per cent and the capacitance coefficient for oxygen in water at the experimental temperature is 13.65 $\mu mol\ L^{-1}\ kPa^{-1}$. For the air breather, percentage extraction from the air ventilating the lungs is 12 per cent and the capacitance coefficient for oxygen in air at the same temperature is 410 $\mu mol\ L^{-1}\ kPa^{-1}$. How much greater is the ventilation volume in the water-breather compared to that in the air-breather? (Hint: Box 12.2.)

13.* If the initial volume of a bubble of atmospheric air trapped beneath the wings of an aquatic insect is 0.1 mL and the rate at which the insect consumes oxygen from the bubble is 0.6 $\mu mol\ min^{-1}$, how long would it be before the insect has to go to the water surface to replenish the air bubble? Assume that the insect lives just below the surface of the water and that the water is equilibrated with atmospheric air. (Hint: Section 12.4.2.)

Bibliography

Nilsson GE (2010). Respiratory Physiology of Vertebrates. Life with and without oxygen. Cambridge: Cambridge University Press.

Weibel ER (1984). The Pathway for Oxygen. Structure and Function of the Mammalian Respiratory System, Harvard University Press, Cambridge, Massachusetts.

12

13

Transport in respiratory systems and acid–base balance

Most of the animals we discussed in Chapter 12 rely on a cardiovascular system to transport respiratory gases between the gas exchanger and the metabolizing cells. This system also distributes metabolic substrates from the gastrointestinal tract and/or the body stores to the same cells.

In this chapter we discuss the nature of the fluid within the circulatory system that transports gases and metabolic substrates around the body[1]. The fluid circulated within blood vessels of vertebrates is normally referred to as blood, whereas that circulated within a haemocoel[2] is called **haemolymph**. When referring to the circulated fluid in general terms, we often call it blood for simplicity.

We begin by discussing the transport of the respiratory gases. We also discuss the transport of metabolic substrates in the blood, paying particular attention to lipids, which are not soluble in water. Finally, we discuss aspects of pH regulation in blood and cells.

13.1 Transport of oxygen and carbon dioxide

The fluid component of blood is essentially salty water. However, the solubility, or capacitance[3], of oxygen in water is low compared with that in air, it also decreases the warmer and saltier the water becomes. Animals have evolved special proteins with associated metal ions to which oxygen binds reversibly. These protein–metal complexes are coloured and are therefore called respiratory pigments.

A full understanding of the transport of the respiratory gases is only possible if we understand the characteristics of the respiratory pigments.

[1] We consider the structure and function of the circulatory systems themselves, as seen in different groups of animals, in Chapter 14.

[2] The different cavities in the bodies of animals are discussed in Section 4.1.

[3] Solubility and capacitance of oxygen in blood are discussed in Section 11.1.2.

13.1.1 Respiratory pigments

Respiratory pigments greatly enhance the capacitance of oxygen in the blood. There are two major types of respiratory pigment, depending on whether the metal ion is iron (Fe^{2+}) or copper (Cu^{2+}). The iron-based respiratory pigments are: **haemoglobins (Hb)**, erythrocruorin (sometimes called haemoglobin), chlorocruorin and haemerythrin and the copper-based pigment is **haemocyanin (Hcy)**. These pigments may be contained within cells or they may circulate freely in the blood. Relatively small molecules of haemoglobin are contained within red blood cells (**RBCs** or **erythrocytes**), whereas larger, giant, molecules of haemoglobin are not contained within cells. Erythrocytes are found in some species of annelids and molluscs and in all vertebrates, except some species of Antarctic fish which have a greater cardiac output per rate of oxygen consumption than those with haemoglobin (see later).

Haemerythrin is also contained within cells called pink blood cells, whereas chlorocruorin circulates freely and gives the blood a greenish colour. Both of these pigments are found in the blood of some species of annelids. Relatively large molecules of circulating Hb are found in some species of annelids, molluscs and arthropods. Haemocyanins are blue when oxygenated but colourless when deoxygenated. They are not contained within cells, and are found in the haemolymph of most species of molluscs and arthropods.

The significance of respiratory pigments in blood is illustrated in Figure 13.1. At the normal partial pressure of oxygen (PO_2) in arterial blood in a mammal (13.5 kPa), the concentration of oxygen in distilled water at 37°C would be approximately 0.14 mmol O_2 (L water)$^{-1}$ (0.32 vol per cent). However, the presence of a respiratory protein in blood can raise the oxygen concentration, at the same PO_2, to 9 mmol O_2 (L blood)$^{-1}$ (20 vol per cent), or more.

The maximum amount of oxygen that can be carried by a given volume of blood is called the **oxygen-carrying capacity** and depends on the concentration of pigment in the blood. This means that the rate at which oxygen can be delivered to

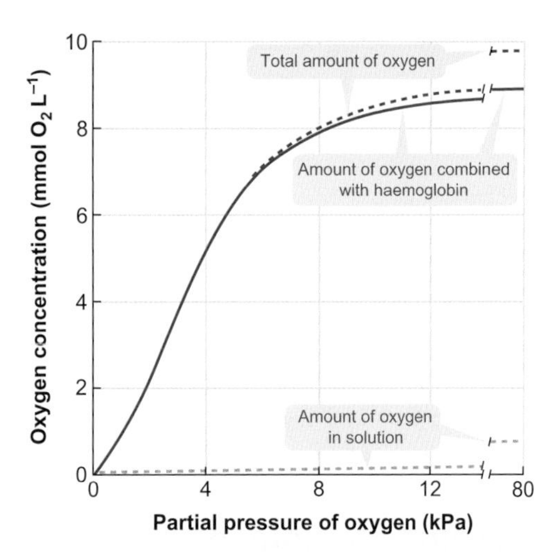

Figure 13.1 Enhancement of oxygen transport by the presence of a blood pigment

Oxygen equilibrium curve, plotting concentration of oxygen in the blood against its partial pressure. Most of the oxygen in the blood is bound to the respiratory blood pigment (in this case haemoglobin) and follows an S-shaped curve to full saturation. A relatively small amount of oxygen is in solution and increases linearly without limit as partial pressure of oxygen increases.

Reproduced from West JB (1974). Respiratory Physiology, the essentials. The Williams and Wilkins Company, Baltimore.

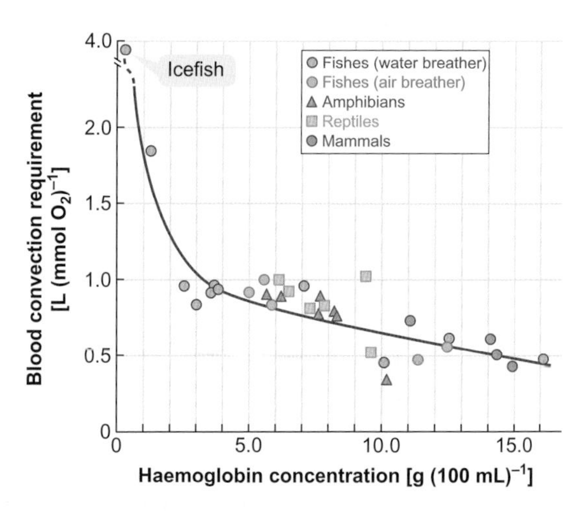

Figure 13.2 Relationship between the blood convection requirement and the concentration of haemoglobin in the blood and in vertebrates

As the concentration of the blood pigment (in this case haemoglobin) increases, the amount of blood flowing around the body (cardiac output) per given amount of oxygen consumed (known as the blood convection requirement) decreases.

Reproduced from Wood SC and Lenfant C (1979). Oxygen Transport and Oxygen Delivery. In: Evolution of Respiratory Processes, a comparative approach. Wood, SC and Lenfant C (Eds.), pp 193-223. Marcel Dekker, Inc, New York.

metabolizing cells depends on the rate at which blood flows around the body (cardiac output), the concentration of pigment in the blood and the extent to which the pigment is saturated with oxygen[4]. Therefore, at lower concentrations of blood pigment, the **blood-convection requirement** (cardiac output per given rate of oxygen consumption) is greater, as shown in Figure 13.2. Related to this, the haemoglobin-less icefish have hearts that are much larger per given body mass than those of icefish with haemoglobin and which result in a four- to five-fold greater relative cardiac output.

The concentration of red blood cells per unit volume of blood in vertebrates is known as the **haematocrit** and the fluid within which the red corpuscles and other cells are suspended is called the **plasma**[5]. Plasma contains proteins, ions, glucose, lipids, hormones, etc., and among the proteins are the clotting agents such as **fibrinogen**. Plasma without these clotting agents is called **serum**.

In addition to the respiratory pigments in the blood, vertebrates also have iron-linked respiratory proteins of the globin family in other tissues and organs. **Myoglobin (Mb)** is found predominantly in oxidative muscles, and also in the liver, gill and brain of some species such as common or European carp (*Cyprinus carpio*). Cytoglobin is found in many different organs and neuroglobin is found predominantly in nervous tissue. A common function of these globins may be to act as oxygen stores for use during periods of oxygen scarcity. For example, common carp tolerate periodic hypoxia in their natural environment, and in humans, neuroglobin may enhance survival of nerve cells during a stroke.

Before going on to the respiratory pigments found in the blood, we discuss myoglobin, the respiratory protein found in oxidative muscle fibres[6].

Myoglobin

Only one molecule of oxygen binds with a molecule of myoglobin. This is unlike the situation with vertebrate haemoglobin, where four molecules of oxygen bind with a single molecule of the pigment. Also, myoglobin has a very high **affinity** for oxygen (oxygen binds to it very easily), so it does not release its oxygen until the partial pressure of oxygen in muscle is extremely low—below 1 kPa. This means that, if it was a respiratory pigment in the blood, myoglobin would pick up oxygen very easily at the gas-exchange organ, but would not release it in the capillaries supplying metabolizing cells because of the partial pressures of oxygen that normally exist there. Nonetheless, myoglobin is thought to have a transport function inside the cells of hearts and oxidative muscle fibres of vertebrates.

Myoglobin combines with oxygen at the muscle capillaries and diffuses to the mitochondria where the partial pressure is low enough for the oxygen to be released. The

[4] Delivery of oxygen to the metabolizing cells by the blood is discussed in Section 11.2.2.

[5] Ion composition of human plasma is approximately: Na^+ 140 mmol L^{-1}, K^+ 4.2 mmol L^{-1}, Cl^- 100 mmol L^{-1} and Ca^{2+} 2.5 mmol L^{-1}.

[6] The different types of muscle fibres are discussed in Section 18.1.4.

myoglobin then diffuses in the opposite direction to collect more oxygen from the capillaries. This process is an example of facilitated diffusion. Although myoglobin diffuses at about 1/20th of the rate of free oxygen at 37°C, the concentration of Mb is about 30 times greater than that of oxygen in the heart or oxidative muscle cells of vertebrates at the same temperature, so the diffusion of myoglobin inside the cell is still beneficial in the delivery of oxygen to the mitochondria.

What is the evidence that facilitated diffusion of myoblobin is important in animals? Mice without myoglobin have been generated by gene-knockout technology and appear to function normally during exercise. This is the result of compensatory factors such as increased capillarity[7] and increased expression of the hypoxia-inducible transcription factor, HIF-1[8], which indicate that the level of oxygen in the muscles of these knockout mice is unusually low (hypoxia). These observations suggest that myoglobin is not essential for the delivery of oxygen to the mitochondria, providing other factors can compensate for its absence.

More natural evidence of such compensatory mechanisms is found in some species of Antarctic icefish, such as blackfin icefish (*Chaenocephalus aceratus*). These fish have no haemoglobin or myoglobin, but they do have greater mitochondrial density in their muscles and wider capillaries than close relatives, which possess respiratory pigments. They also have a larger blood convection requirement than other species of fish, as illustrated in Figure 13.2.

As well as its transport function, myoglobin also acts as a store of oxygen in animals that have no oxygen available to them from their environment, such as aquatic birds and mammals when diving under water[9].

Facilitated diffusion of oxygen and oxygen storage by Mb may be important during the contractile phase (systole) of the cardiac cycle in birds and mammals when flow in the coronary circulation to the heart muscle is temporarily reduced and the partial pressure of oxygen in the muscle falls. It has been calculated that Mb contributes about one third of the oxygen required by the heart during systole. This suggests, of course, that other sources of oxygen, such as haemoglobin in the capillaries, supply the other two thirds of the oxygen required during systole.

The use of knockout mice not only reveals that Mb is not essential for mice to be able to exercise normally, as we discuss above, but that its major function may be to protect the cells of the heart, and perhaps oxidative skeletal muscles, from the damaging effects of the reactive oxygen species (ROS) produced during the electron transport of oxidative metabolism[10]. ROS belong to a group of highly reactive, uncharged molecules known as **free-radicals**[11].

Deoxygenated Mb converts circulating nitrite (NO_2^-) into nitric oxide (NO). Nitric oxide has an inhibitory effect on the electron transport chain, which limits the production of ROS. Nitric oxide also reduces the workload of the heart by reducing the rate of oxidative metabolism. When oxygenated, Mb rapidly oxidizes NO to form nitrate (NO_3^-). A simplified diagram illustrating these events is given in Figure 13.3. The nitric oxide produced also causes dilation of blood vessels. Knockout mice without Mb cannot generate

[7] Capillarity of a muscle is the number of capillaries per unit area of muscle (capillary density) and the number of capillaries per muscle fibre (capillary to fibre ratio).

[8] HIF-1 is a transcription factor which accumulates in cells when they become hypoxic. Among the important physiological processes regulated by HIF-1 target genes are the production of red blood cells (erythropoiesis) and the growth of new blood vessels (angiogenesis). HIF-1 is discussed further in Section 22.2.3.

[9] Physiology of diving is discussed in Section 15.3.3.

[10] Oxidative (aerobic) metabolism is discussed in Sections 2.4.2 and 2.4.3.

[11] A free radical has an unpaired electron in its outer (valence) shell. Having an unpaired electron makes a free radical unstable and extremely reactive. Free radicals of oxygen are produced during oxidative phosphorylation, as discussed in Section 2.4.3, and can cause damage to DNA and cell membranes.

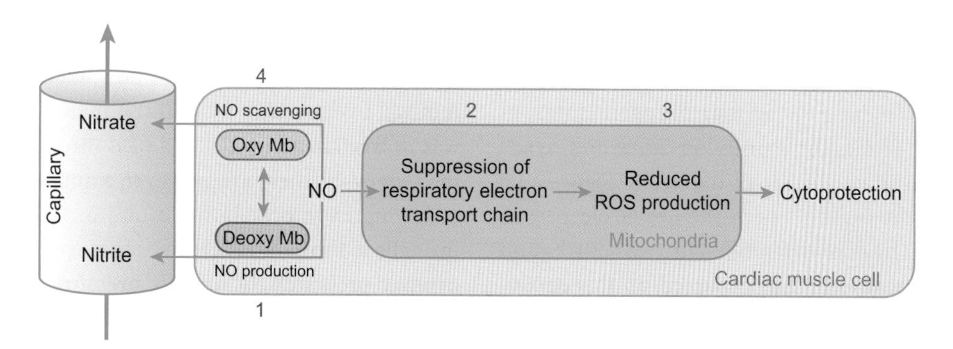

Figure 13.3 Role of myoglobin in the protective function of nitric oxide during restricted blood flow to heart muscle

Deoxygenated myoglobin (deoxy Mb) in a hypoxic muscle cell of the heart produces nitric oxide (NO) from nitrite in the circulation. The NO suppresses the electron-transport chain which results in reduced production of damaging reactive oxygen species (ROS) in mitochondria, protecting muscle cells from damage. The suppression of the electron-transport chain by NO also leads to a reduction in the rate of energy production and hence reduces the work of the heart. Excess NO is converted to nitrate by oxygenated myoglobin (oxy Mb).

Reproduced from Cossins A and Berenbrink M (2008). Myoglobin's new clothes. Nature 454: 416-417.

NO from NO_2^-, so knockout mice do not benefit from these protective mechanisms.

It has also been shown that nitrite is the source of NO that protects the heart of mice from the damaging effects of a reduction in perfusion of the coronary arteries (cardiac ischaemia, similar to a heart attack) followed by reperfusion. This is known as ischaemia/reperfusion (I/R) injury. Knockout mice without Mb do not show any recovery from I/R injury because they cannot convert nitrite to nitric oxide.

Structure of respiratory pigments

Most studies of respiratory pigments have been performed on mammalian haemoglobin. The structure of adult mammalian haemoglobin (HbA) was determined by X-ray analysis in 1960 by Max Perutz and colleagues at the Cavendish Laboratory, Cambridge, UK. Perutz was awarded the Nobel Prize for this work in 1962.

Mammalian haemoglobin consists of four polypeptide chains, and has a total molecular mass of around 65 kDa (each mol has a mass of 65,000 g). There are two α chains, each comprising 141 amino acid residues, with a total molecular mass of around 30.5 kDa, and two β chains, each with 146 residues and a total molecular mass of around 32 kDa. These four chains fit snuggly together in a tetrahedral arrangement, as shown in Figure 13.4A. Each individual polypeptide chain has essentially the same tertiary structure as myoglobin, which is shown in Figure 13.4B. The four chains of haemoglobin are held together mainly by non-covalent attractions such as ionic and hydrogen bonds. Note that a molecule of myoglobin (Mb) consists of just one polypeptide chain, which explains why each molecule of Mb only combines with one molecule of oxygen, as discussed above.

Each polypeptide chain has an iron-containing haem group (molecular mass about 2.5 kDa), whose structure is illustrated in Figure 13.4C. Each haem group is located in a characteristic V-shaped globin fold as shown in Figure 13.4A,B. This location of the haem group is necessary for the reaction between the ferrous ion (Fe^{2+}) and oxygen molecule to be reversible. The iron lies in the centre of the porphyrin ring where it is bonded to the four nitrogen atoms, as illustrated in Figure 13.4C. The ferrous ion can form two additional bonds, one on either side of the haem. Figure 13.5 shows that one of the bonds is with the imidazole ring of a histidine residue. In deoxyhaemoglobin (without bound oxygen), the other position is not occupied. In this deoxygenated state, the ferrous ion is slightly too large to fit into the hole within the porphyrin ring, so it resides approximately 0.04 nm outside the plane of the ring. However, when oxygen binds to the unoccupied site, the effective size of the ferrous ion reduces, enabling it to move into the plane of the porphyrin ring. The histidine residue at the fifth position moves with the ferrous ion, resulting in changes to the quaternary structure[12] of haemoglobin. These changes in structure have an important effect on the affinity of the haemoglobin for oxygen as it moves from the oxy- to the deoxy-state. As oxygen binds to the other haems of the haemoglobin, the affinity of the haemoglobin molecule for oxygen increases, as we discuss in more detail in Section 13.1.2.

Under normal conditions, some oxidation of the ferrous ions occurs, forming ferric ions (Fe^{3+}). Ferric haemoglobin is known as **methaemoglobin (metHb)** and is unable to combine with oxygen, so its presence reduces the overall oxygen-carrying capacity of the blood. The concentration of methaemoglobin depends on the balance between the rate of its formation and of its reduction back to ferrous haemoglobin by the enzyme methaemoglobin reductase. The concentration of metHb is normally low in the red cells of mammals (< 1 per cent), and less than 3 per cent in most other vertebrates, although it may be as high as 10 per cent in some species of bony fish. Elevated levels of nitrite in the environment can cause the concentration of metHb to increase, especially in fish living in fresh water which may contain large amounts of decaying organic matter or may receive nitrogenous effluents. For fish living in brackish water or sea water, the high concentration of chloride ions (Cl^-) inhibits the uptake of nitrite across the gills[13].

Extracellular versus intracellular respiratory pigments

The polypeptide chains of haemoglobins that occur in blood cells rarely have a molecular mass larger than 68 kDa. In contrast, the molecules of haemoglogin and haemocyanin that circulate freely in annelids, molluscs and arthropods are very large and bind to many more oxygen molecules than do the molecules of vertebrate haemoglobin. The basic subunit of the haemocyanin in arthropods has a molecular mass of 75 kDa and many of these subunits combine to form much larger extracellular molecules. For example, the molecular mass of the haemocyanin of cephalopod molluscs is around 4000 kDa. A possible explanation for this is that molecules in the plasma or haemolymph must be sufficiently large to avoid removal by excretory filtration. If mammalian haemoglobin was in free solution, it would pass through the glomeruli of the kidneys[14]. There are, however, a number of consequences of having large protein molecules in the circulating liquid. Proteins in the blood exert an osmotic pressure, called colloid osmotic pressure[15], or **oncotic pressure** which is a function

[12] The primary, secondary, tertiary and quaternary structures of proteins are discussed in Box 2.1.

[13] Case study 5.2 discusses further aspects of nitrite toxicity in freshwater fish.

[14] Kidney function is discussed in Sections 7.1 and 7.2.

[15] Colloid osmotic pressure is discussed in Section 14.1.5.

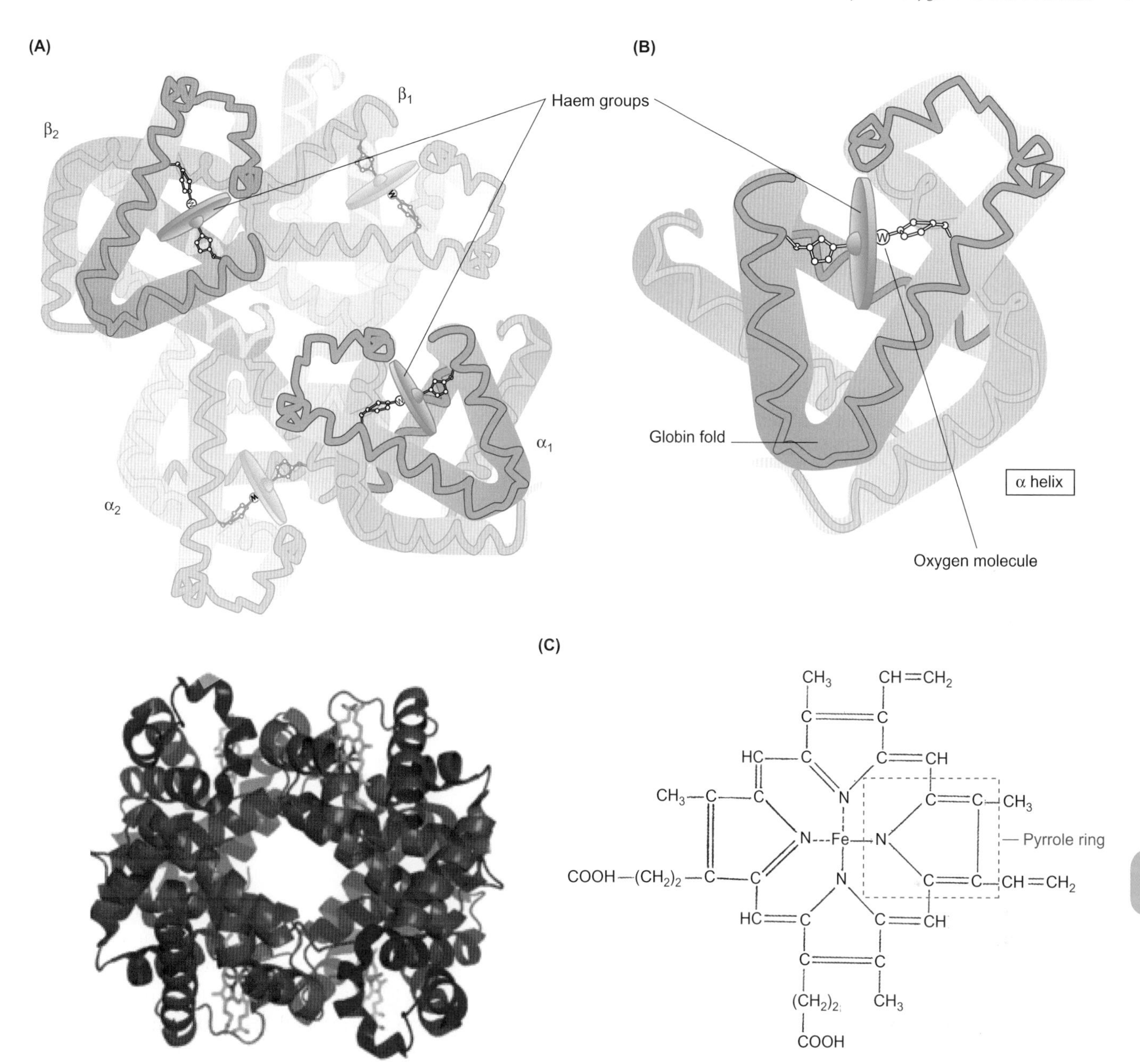

Figure 13.4 Structures of mammalian haemoglobin, myoglobin and haem

(A) Front view of the four chains of horse methaemoglobin (α_1, α_2, β_1, β_2), indicating the positions of the haem groups in the V-shaped globin folds of two of them. Horse methaemoglobin was the first to have its structure determined. It closely resembles oxyhaemoglobin. Molecular structure shown below, with α subunits in red, β subunits in blue and haem groups in green. (B) Front view of the single chain of myoglobin showing the location of the haem group in the V-shaped globin fold. (C) Haem consists of four pyrrole rings which form a porphyrin ring with a ferrous ion (Fe^{2+}) in the centre

A, B: adapted from Dickerson RE and Geis I (1983). Hemoglobin: Structure, Function, Evolution, and Pathology. The Benjamin/Cummings Publishing Company, Inc. Menlo Park, California. Model supplied by J Crowe. C: Reproduced from: Jones JD (1972). Comparative Physiology of Respiration. Edward Arnold, London.

of the concentration of the protein in mol L^{-1}. For a given protein mass per litre, the oncotic pressure will, therefore, depend on the protein molecular mass. The smaller the protein, the larger the colloid osmotic pressure it exerts. If haemoglobin in vertebrate blood with its relatively low molecular mass of 65 kDa was free in solution, (i.e. not within cells) and was not excreted by the kidneys, the colloid osmotic pressure of the plasma would be approximately three times greater than normal, i.e. approximately 10 kPa. Without a comparable increase in blood pressure, this osmotic pressure would suck water from the interstitial fluid into the blood capillaries. On the other hand, the high molecular mass of the extracellular haemocyanins of cephalopod molluscs minimizes the colloid osmotic pressure to as low as 0.4 kPa.

Figure 13.5 The binding of oxygen to the haem group

Before the binding of oxygen, the iron ion lies approximately 0.04 nm outside the plane of the porphyrin ring because in this form, the iron ion is too large to fit into the hole within the ring. However, the binding of oxygen rearranges the electrons within the iron ion so that the ion becomes smaller, allowing it to fit into the hole and move into the plane of the porphyrin ring.

When measured *in vitro* by conventional methods, suspensions of mammalian red blood cells display higher viscosities[16] than similar concentrations of haemoglobin in free solution. However, when the viscosity of a suspension of red blood cells is measured in a glass capillary tube, the viscosity of the suspension depends on the diameter of the tube below a diameter of about 0.3 mm, but below a tube diameter of 0.2 mm, the viscosity declines quite steeply. This is known as the Fåhraeus-Lindqvist phenomenon and is due to the ability of the red blood cells to congregate in the centre of the blood vessel and leave the plasma, which has a lower viscosity, closest to the walls of the vessel. The smallest blood vessels other than capillaries in the mammalian circulation have diameters that are less than this critical value (arterioles and venules about 0.025 mm)[17], which enables red cells to produce relatively low apparent viscosities when circulating through them.

Size of the red blood cells in vertebrates

Figure 13.6A shows that the red blood cells (RBCs) in vertebrates vary in size; there is also a positive relationship between the size of the RBCs and the diameter of the blood capillaries through which they flow. The capillaries in newts and salamanders are much larger than those in other vertebrates. It has been proposed that such relatively large capillaries enabled lungfish and early amphibians to have a low pressure circuit to their lungs that was supplied from a single-chambered ventricle in their heart[18]. The evolution of a double-chambered ventricle, with one side supplying the lungs and the other side supplying the rest of the body, was accompanied by smaller capillaries and higher pressures in the body (systemic) circuit.

The demand for oxygen is highest in endotherms, and birds and mammals have the smallest capillaries and RBCs. Figure 13.6B shows the possible functional significance of these small sizes. The smaller red cells of goats combine with

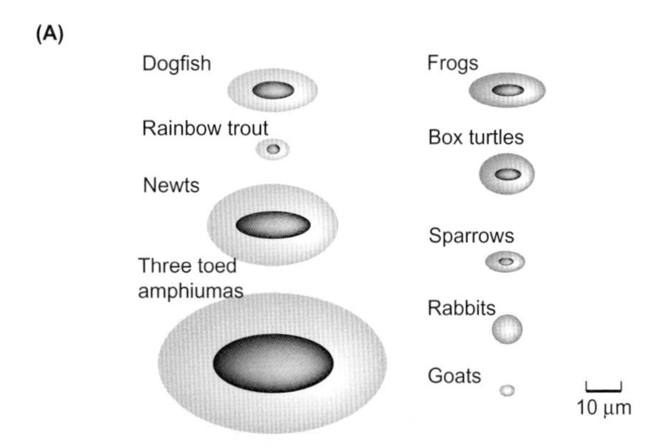

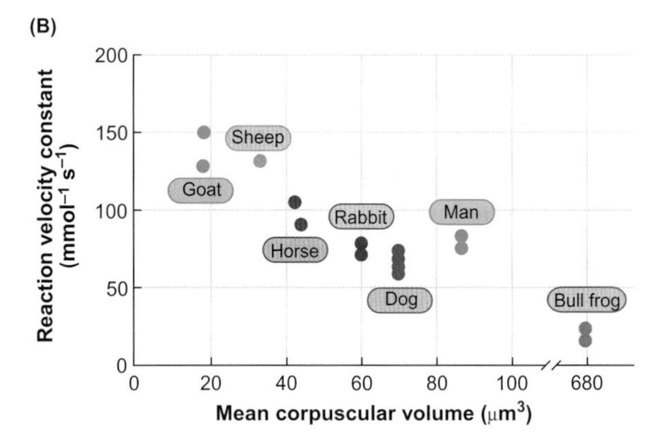

Figure 13.6 The size of red cells in the blood of vertebrates

(A) The relative dimensions of the red blood cells of some vertebrates. Note that all are ovoid in shape and have a nucleus, except those of mammals, which lack a nucleus and are normally round. *Scyliorhinus* dogfish; *Salmo* (now *Onchorhynchus*) rainbow trout; *Triturus* newts; *Amphiuma* three toed amphiumas, formerly known as congo eels (a salamander); *Rana* frogs; *Terrapene* box turtles; *Passer* sparrows; *Oryctolagus* rabbits and *Capra* goats. (B) The relationship between size of red blood cells and their rate of oxygenation indicating that the smaller ones are oxygenated faster than the larger ones.

Adapted from Nikinmaa M (1990). Oxygen Transport. In: Vertebrate Red Blood Cells. Zoophysiology, vol 28, Springer-Verlag, Berlin. Holland RB and Forster RE (1966). The effect of size of red cells on the kinetics of their oxygen uptake. Journal of General Physiology 49: 727-742.

[16] As blood is a non-Newtonian fluid, its viscosity is often referred to as 'apparent viscosity', as discussed in Section 14.2.3.

[17] Dimensions of blood vessels in mammals are discussed in Section 14.1.3 and illustrated in Figure 14.6.

[18] The circulatory systems of vertebrates are discussed in Section 14.4.

oxygen almost twice as quickly as the larger red cells of humans, and five times faster than the much larger red cells of bullfrogs.

The RBCs of mature mammals are unusual among vertebrates in that they have no nucleus, which gives them a biconcave shape[19]. They also contain no other organelles such as mitochondria, so are unable to produce ATP aerobically. However, they do contain a nucleus during the early stages of their development and the haemoglobin they contain when mature is effectively produced while the nucleus is still present during the first few days of development. After the nucleus has disappeared, any remaining RNA continues to produce haemoglobin until this supply of RNA has been exhausted.

Despite the fact that they have no nucleus or other organelles, the RBCs of mammals have a similar concentration of haemoglobin per unit volume of red cells (mean corpuscular haemoglobin concentration, MCHC) as those of birds—which is approximately 330 g (L red cells)$^{-1}$. The RBCs of birds, like those of other non-mammalian vertebrates, do contain a nucleus. Red blood cells of mammals are larger than the capillaries through which they flow, but are also flexible, which enables them to deform as they pass through the capillaries. This deformation facilitates diffusion of the respiratory gases between the haemoglobin in the RBCs and the tissues by minimizing the diffusion distance between the two.

[19] The biconcave shape of red mammalian red blood cells can be seen in Figure 13.14A.

The RBCs of vertebrates are produced in a range of locations within the body and have a lifespan that varies between species. The production and lifespan of red blood cells are discussed in Box 13.1.

All the cells in the blood are known as **haemocytes**; those that do not contain a respiratory pigment have other functions, which are similar in both invertebrates and vertebrates. These functions include coagulation of the blood, storage of metabolic substrates and immunological defence such as phagocytosis, aggregation, encapsulation and destruction (lysis) of foreign cells.

13.1.2 Factors affecting the binding of oxygen to respiratory pigments

Oxygen binds with the respiratory pigment (haemoglobin or haemocyanin) at the gas-exchange organ, where it diffuses[20] from the environment into the blood. The pigment then transports the oxygen in the circulation to the metabolizing cells. At the cells, oxygen is released from the pigment, diffuses from the blood into the cells, and plays its pivotal role in aerobic metabolism; it is the final electron acceptor of the electron transport chain[21].

[20] The factors affecting the diffusion of the respiratory gases are discussed in Section 11.2.1.

[21] Aerobic respiration and the electron transport chain are discussed in Sections 2.4.2 and 2.4.3.

Box 13.1 Production and lifespan of red blood cells

Red blood cells are produced by a process known as **erythropoiesis**. In the embryos of reptiles, birds and mammals, erythropoiesis occurs in the yolk sac and in the analogous structures of amphibians and fish. In foetal mammals, the liver and spleen may also be involved. In most adult teleosts, the kidney is the major site of erythropoiesis, whereas in amphibians it is the spleen. The bone marrow is also erythropoietic in some salamanders and *Rana* species. In adult reptiles, birds and mammals, the bone marrow is the major site of erythropoiesis, together with the thymus in birds.

The lifespan of red blood cells in the circulation varies from about 12 days in ducks to approximately 120 days in humans and dogs, but can be markedly longer in hibernating animals. For example, the lifespan of blood cells in marmots (*Marmota* spp.) is 36 days in active animals at room temperature but increases to 112 days in conditions that are right for hibernation. At the extreme, red cells of common box turtles (*Terrapene carolina*) at 17°C survive for more than 11 months.

In adult mammals, the kidneys (and to a lesser extent the liver), play an important role in the control of erythropoiesis by the production of the haematopoietic growth factor, **erythropoietin (Epo)**. Epo is essential for erythropoiesis in the bone marrow as demonstrated by the fact that knockout mice with disrupted genes for Epo or the Epo receptor (Epo-R) die at embryonic day 12–13 from anaemia. The most powerful stimulants for Epo production are hypoxia and anaemia, which lead to a reduction in the supply of oxygen. However, production of Epo by the kidney is not affected by reductions in renal blood flow until it is less than 10 per cent of its normal value. This makes the kidney an ideal organ to sense those changes in oxygen delivery that are independent of the rate of perfusion. Epo genes have also been reported in some species of teleost fish and in African clawed frogs (*Xenopus laevis*).

Strangely, perhaps, the carotid body oxygen sensors[1] of mammals are not involved in the regulation of Epo production. Rather it is regulated by local sensors in the kidneys and liver. Both the rate of oxygen delivery and the rate of local oxygen consumption by the proximal tubule cells of the kidney are also thought to be important in regulating Epo production. So, when the rate of local oxygen consumption is greater than the rate of oxygen delivery, there is a local hypoxia which causes an increase in the activity of hypoxia-inducible transcription factor (HIF[2]). Increased HIF then triggers expression of the Epo gene. A flow diagram illustrating this sequence of events is shown in Figure A.

Endurance athletes sometimes train at high altitude to benefit from increased production of red blood cells in response to the hypoxic environment. Such high-altitude training enhances the oxygen carrying capacity of their blood, which persists after their return to sea level. As a result, their athletic performance is enhanced. An easier way to achieve the same result is the illegal use of Epo to boost red cell production. The increase in the concentration of red blood cells (haematocrit) persists long after Epo is cleared from the system, so such misuse is not easy to detect. However, a high oxygen-carrying capacity does not always stem from the misuse of Epo: some individuals appear to have naturally high haematocrits. For example, a Finnish cross-country skier and Olympic champion had a haematocrit of 68 per cent (normal haematocrit is around 40 per cent) due to a mutation which increased the sensitivity of the Epo receptor on their erythropoietic cells.

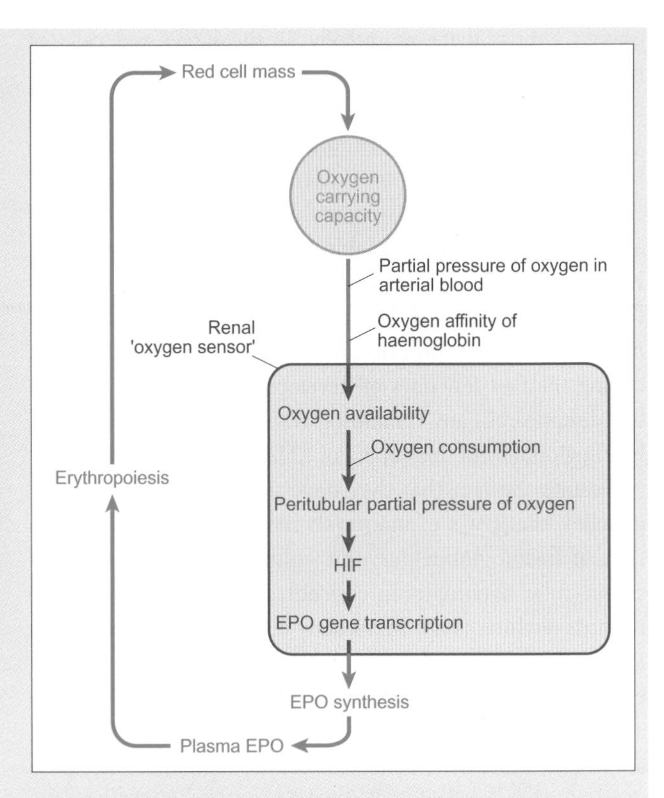

Figure A Oxygen sensing in the kidney controls the production of erythropoietin (EPO) in mammals

A reduction in the delivery of oxygen to the tissues is sensed by the kidneys leading to stimulation of EPO production. HIF, hypoxia-inducible transcription factor.

Adapted from: Wenger RH and Kurtz A (2011). Erythropoietin. Comprehensive Physiology 1: 1759-1794.

[1] Carotid body oxygen sensors are discussed in Sections 22.2.3 and 22.4.1.

[2] HIF-1 is a transcription factor which accumulates in cells when they become hypoxic. Among the important physiological processes regulated by HIF-1 target genes are the production of red blood cells (erythropoiesis) and the growth of new blood vessels (angiogenesis). HIF-1 is discussed further in Section 22.2.3.

The respiratory pigments are able to perform this task because of the ease with which oxygen binds to and dissociates from the pigments. The affinity of the pigment for oxygen varies:

- as the blood circulates around the body;
- in response to changes in the blood caused by different factors.

Changes in temperature, partial pressure of carbon dioxide, pH, organic phosphates and inorganic ions, such as chloride ions, all affect the binding characteristics of oxygen to respiratory pigments. Moreover, almost all the respiratory pigments are multi-unit proteins which gives them far greater functional flexibility than if they were single unit proteins such as myoglobin.

Co-operative binding in haemoglobin and oxygen carrying capacity

Experiments on vertebrate haemoglobin have shown that when deoxygenated, it is in the **T (tense) state** and has an affinity for oxygen that is lower than that of each of the four free subunits. This is because the ion pairs that cross-link the subunits stabilize the individual chains in such a way that they resist the binding of oxygen. As the partial pressure of oxygen progressively increases, an oxygen molecule eventually binds to a haem site. This initial binding leads to the change in configuration of the quaternary structure of the haemoglobin[22], as shown in Figure 13.7. This change increases the oxygen-binding affinity of the remaining, unoxygenated, haem sites. The haemoglobin is then in the **R (relaxed) state**.

In mammalian haemoglobin, the first oxygen molecule attaches itself very weakly to the haem, while the second and third oxygen molecules bind more strongly. The last oxygen molecule to bind has a two to three orders of magnitude greater affinity than the first one, as illustrated in Figure 13.8. This process is known as **co-operative binding**. Conversely, the unloading of oxygen from one haem facilitates the unloading of oxygen at the other haems.

The degree of co-operativity of a blood pigment can be quantified by plotting the logarithm of the ratio between oxyhaemoglobin (Y) and deoxyhaemoglobin ($1 - Y$), i.e. $\log(Y/1-Y)$, against the logarithm of the partial pressure

[22] A component of the change in configuration is shown in Figure 13.5.

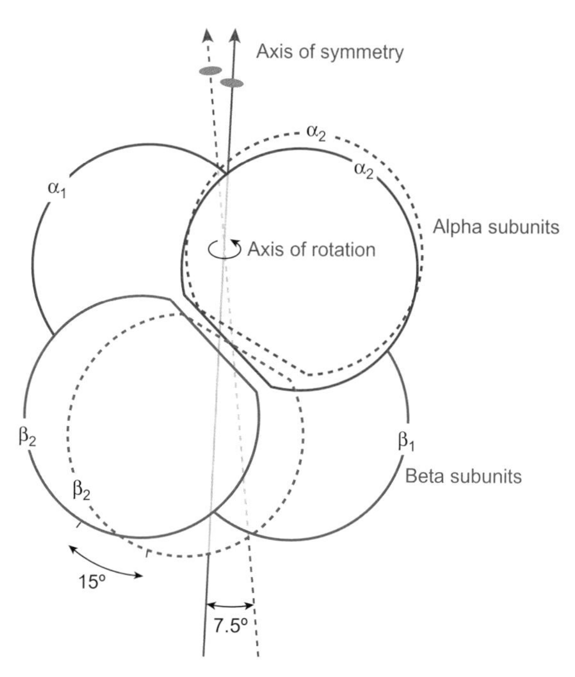

Figure 13.7 Effect of oxygenation on the quaternary structure of haemoglobin

During the transition from the deoxygenated (T-state) to the oxygenated (R-state) of haemoglobin there is a rotation of one pair of subunits relative to the other pair. One dimer is held fixed and the other turns by 15° around an off-centre axis and shifts slightly along it. The axis of symmetry is rotated by 7.5°.
Adapted from Perutz MF (1980). Stereochemical mechanism of oxygen transport by haemoglobin. Proceedings of the Royal Society B 208: 135-162.

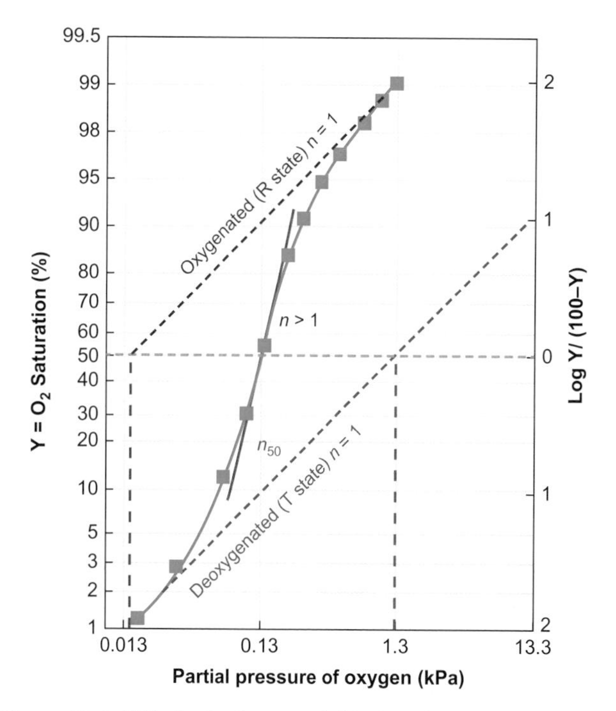

Figure 13.8 Hill plot for haemoglobin (erythrocruorin) of giant Australian earthworm (*Megascolides australis*)

This plot shows the low-affinity deoxygenated (T) and high-affinity oxygenated (R) states if there was no co-operative binding of oxygen to the pigment ($n = 1$) as interrupted lines. During the transition between these states there is co-operative binding of oxygen indicated by slope $n > 1$, which reaches a maximum value (n_{50}) around the P_{50} (50% saturation). The intersects of the dashed asymptotes of the upper and lower extremes with the 50% saturation line give a graphical illustration of oxygen affinities of the oxygenated and deoxygenated states (dashed red and blue lines, respectively). The oxygenated state has about a 100-fold greater affinity for oxygen than the deoxygenated state.
Adapted from Weber RE and Baldwin J (1985). Blood and erythrocruorin of the giant earthworm, *Megascolides australis*: respiratory characteristics and evidence for CO_2 facilitation of O_2 binding. Molecular Physiology 7: 93-106.

of oxygen (log PO_2). This plot is called a Hill plot, after AV Hill, who first produced it, and it approximates to a straight line, particularly in the mid region, as shown in Figure 13.8. The slope of the line at its midpoint (n_{50}) is called the Hill coefficient (n). Pigments to which oxygen molecules bind independently of each other and there is no co-operativity have a Hill coefficient of 1, as we see for myoglobin. By contrast, n is >1 for those pigments where there is co-operativity and the binding of oxygen at one haem makes it easier for oxygen to bind to the other haems of the same molecule. In other words, the Hill coefficient is an indicator of the degree of co-operativity of binding to a blood pigment. However, there is not necessarily a direct relationship between the number of functional units of a pigment and the value of n. For example, there are four functional units in human haemoglobin but its Hill coefficient is approximately 2.8. Further, the presence of multiple functional units does not necessarily mean that co-operativity occurs. Nevertheless, the Hill coefficient gives an indication of the smallest number of units that act co-operatively. Thus, a Hill coefficient of 2.8 indicates that there are at least three units that act co-operatively.

Co-operative oxygen binding gives rise to the characteristic S-shape of the **oxygen equilibrium curve (OEC)** of haemoglobin. An OEC plots the relationship between the partial pressure of oxygen and its concentration (or percentage saturation) in the blood, as shown in Figure 13.1. Figure 13.9 compares the OECs for myoglobin (a single unit or monomeric protein-metal complex) and haemoglobin (a four-unit or tetrameric complex). The oxygen equilibrium curve of myoglobin is hyperbolic in shape and reaches saturation at a much lower partial pressure of oxygen (PO_2) than the sigmoidal curve of the tetrameric haemoglobin.

The fact that myoglobin reaches saturation at a much lower PO_2 means that it has a considerably higher affinity for oxygen than haemoglobin. Affinity for oxygen is normally quantified by the PO_2 at which a respiratory pigment is 50 per cent saturated—that is, its $\boldsymbol{P_{50}}$. This means that the higher the $P_{50,}$ the lower the affinity, and *vice versa*. The P_{50} of myoglobin shown in Figure 13.9 is 0.3 kPa while that of haemoglobin is 3.7 kPa.

The use of percentage saturation to plot an oxygen equilibrium curve, as in Figure 13.9, does not give a complete picture of how the respiratory pigments transport oxygen around the body. It is much more informative to use concentration of oxygen in the blood instead. The maximum

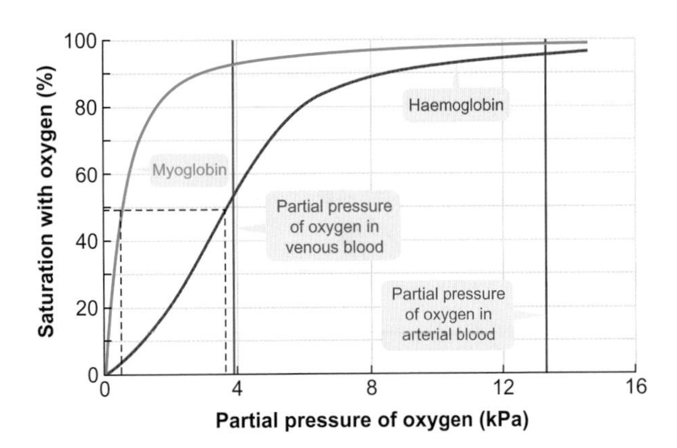

Figure 13.9 Oxygen-binding characteristics of the respiratory pigments myoglobin and haemoglobin

Oxygen equilibrium curves relating the percentage maximum saturation of the pigment with oxygen to the partial pressure of oxygen. The affinity (P_{50}) of the pigment for oxygen is indicated by the partial pressure at which it is 50% saturated (black dashed lines). Note the different shapes of the curves, with that for myoglobin being hyperbolic and that for haemoglobin being sigmoid.

Modified from Perutz MF (1978). Hemoglobin structure and respiratory transport. Scientific American 239: 68-86.

concentration of oxygen that is carried by the blood when the respiratory pigment is fully oxygenated, the oxygen carrying capacity, will depend on the type and concentration of the pigment in the blood. For example, a given amount of low molecular weight vertebrate haemoglobin combines with more oxygen than the same amount of high molecular weight haemocyanin from arthropods:

- For fully saturated vertebrate haemoglobin (Hb), 4 mol of oxygen combine with each mol of Hb, i.e. there are 4 mol of oxygen for each 65,000 g Hb, which gives 4 mol/65,000 g = 62 μmol O_2 (g Hb)$^{-1}$ or 1.39 mL O_2 (g Hb)$^{-1}$.
- On the other hand, for fully saturated haemocyanin (Hcy) of arthropods, 1 mol O_2 combines with each 75 kDa (75,000 g) subunit, resulting in only 13 μmol O_2 (g Hcy)$^{-1}$ or 0.3 mL O_2 (g Hcy)$^{-1}$.

Figure 13.10 gives examples of the ranges of maximum oxygen carrying capacity and affinity for oxygen for the respiratory pigments in the blood of different species of animals.

The concentration of haemoglobin in the blood of vertebrates varies considerably between species. Some species of fish living in the Antarctic (the icefish) have no haemglobin at all, whereas the Weddell seal (*Leptonychotes weddellii*), which also lives in the Antarctic, has a haemoglobin concentration of 260 g Hb (L blood)$^{-1}$. At 0°C, the blood of a haemoglobin-less icefish at a PO_2 of 13.5 kPa, has an oxygen-carrying capacity of approximately 0.26 mmol O_2 (L blood)$^{-1}$, whereas the blood of a Weddell seal at 37°C has an oxygen-carrying capacity of approximately 16 mmol O_2 (L blood)$^{-1}$—that is, over 60 times greater. In comparison, the oxygen-carrying capacity of the blood of humans is approximately 9 mmol O_2 (L blood)$^{-1}$.

Cephalopod molluscs such as octopus, and salmonid fish such as rainbow trout (*Oncorhynchus mykiss*), are both active groups of animals, yet the oxygen-carrying capacity of the blood of octopus [1.7 mmol O_2 (L blood)$^{-1}$] is much less than that of trout [4.6 mmol O_2 (L blood)$^{-1}$]. This difference means that, for a similar rate of oxygen consumption, octopus will need a greater cardiac output (as illustrated back in Figure 13.2) and/or greater extraction of oxygen from the circulating blood than trout.

Some species, such as water fleas (*Daphnia* spp.—branchiopod crustaceans), synthesize haemoglobin when exposed to hypoxic conditions. Depending on the severity of the hypoxia and ambient temperature[23], more of the same type of haemoglobin may be produced or a different type with a higher affinity for oxygen may be synthesized. For example, *Daphnia magna* are able to increase their haemoglobin

[23] As *Daphnia* is an ectotherm, its oxygen demand increases with increasing temperature.

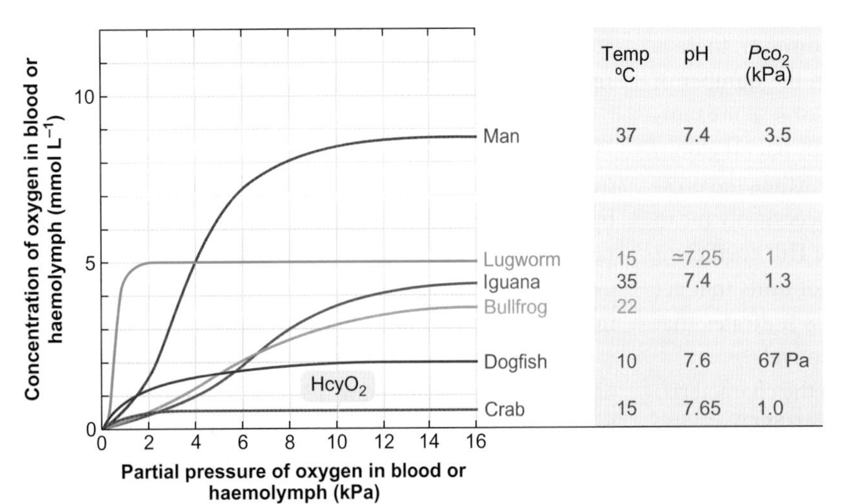

Figure 13.10 Oxygen equilibrium curves for the respiratory pigments of a number of species of animals

These curves indicate the ranges of affinities for oxygen, maximum oxygen-carrying capacities, and shapes of the curves (degree of sigmoidness). The curve for the lugworm (*Arenicola marina*) is sigmoid when plotted on the appropriate scale for partial pressure of oxygen, whereas that for the Pacific spiny dogfish (*Squalus suckleyi*), is hyperbolic, like that for myoglobin, as shown in Figure 13.9.

Modified from Dejours P (1981). Principles of Comparative Respiratory Physiology. Elsevier, Amsterdam.

by 15–20 times, with measurements of solutions of purified haemoglobin giving P_{50}s of 0.5 kPa and 0.2 kPa in normoxic and hypoxic animals, respectively.

The effect of temperature on binding of oxygen to respiratory pigments

When oxygen combines with a respiratory pigment, heat is produced (it is an exothermic reaction[24]). Exothermic reactions are favoured by the surroundings being at a lower temperature. Consequently, a decrease in temperature facilitates the loading of oxygen. Conversely, an increase in temperature favours endothermic reactions (ones in which heat is absorbed from the surroundings), and so tends to drive the reaction in the opposite direction, facilitating the unloading of oxygen from the pigment. In other words, an **increase** in temperature **decreases** the affinity of respiratory pigment for oxygen, or increases in P_{50}, as demonstrated in Figure 13.11. This is of significance for exercising animals in which the temperature is higher in the active muscles than at the respiratory surface. The unloading of oxygen is facilitated in the high-temperature, active muscles, but loading of oxygen is enhanced at the cooler respiratory surface[25]. The temperature coefficient is the ratio of the change in log P_{50} for a given change in temperature ($\Delta\log P_{50}/\Delta T$) and indicates the magnitude of the effect of temperature on the P_{50} of a blood pigment.

The sensitivity of a blood pigment to an increase in environmental temperature is different in some closely related species of ectotherms in a manner that reflects the thermal characteristics of their habitat. Look again at Figure 13.11 and note that the haemoglobin of intertidal lugworms (*Arenicola marina*), which may warm up in the shallow waters of low tide, is less sensitive to changes in temperature than that of subtidal lugworms (*Abarenicola* sp.) that are permanently submerged. The haemoglobin of some species of tuna fish is also relatively insensitive to changes in temperature, particularly at higher temperatures.

Southern bluefin tuna (*Thunnus maccoyii*) have large temperature differences between their body core and the peripheral tissues[26] so a normal sensitivity of the haemoglobin to temperature could lead to excessively high rates of unloading of oxygen in the warmer locomotor muscles. Studies on this species have demonstrated not only that there is no effect on P_{50} at higher temperatures (23–36°C), but that there is a reversed temperature effect at lower temperatures, with P_{50} increasing from 1.7 kPa at 23°C to 2.9 kPa at 10°C. The observed temperature sensitivities in the blood of southern bluefin tuna may, therefore, ensure adequate oxygen supply to the colder tissues, such as the heart and liver.

[24] Exothermic reactions are briefly discussed in Section 2.1.2.

[25] Experimental Panel 13.1 provides an illustration of loading of oxygen at the lungs and unloading at the muscles during exercise.

[26] Temperature regulation in tuna is discussed in Section 10.3.1.

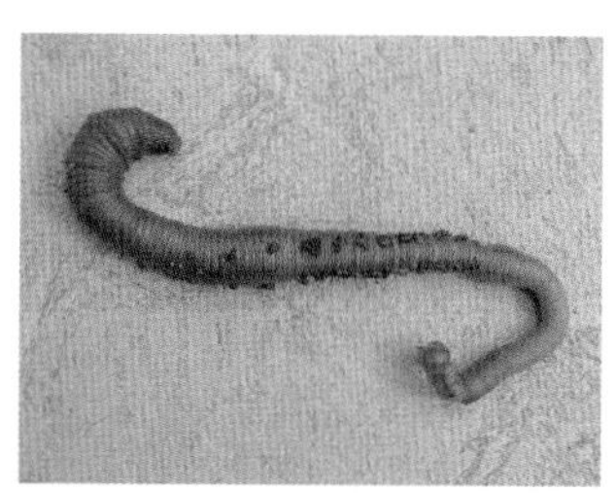

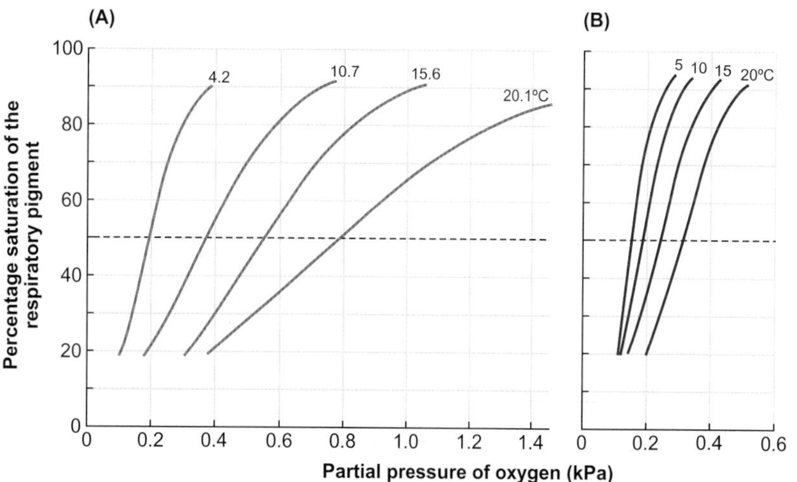

Figure 13.11 Effect of temperature on the oxygen binding properties of respiratory pigments

(A) Oxygen equilibrium curves of haemoglobin of the subtidal lugworm (*Abarenicola claparedii*) at different temperatures, compared to those of: (B), the intertidal lugworm (*Arenicola marina*) which is exposed to larger changes in temperature than *Abarenicola,* as the tide goes in and out.

Adapted from Weber RE (1978). Respiratory pigments. In: Physiology of Annelids. Mill PJ (ed.), pp 393-446. Academic Press, London. Photo: © David Fenwick.

Conversely, the haemoglobin of some cold-tolerant mammals, such as polar bears (*Ursus maritimus*) and reindeer (caribou) (*Rangifer tarandus*), have reduced temperature sensitivity compared to that of other species of mammals. This reduced sensitivity is thought to be an adaptation to ensure the unloading of oxygen at their cold extremities. A similar adaptation has been discovered in the genetically retrieved and resurrected haemoglobin from a preserved bone of a 43,000-year-old Siberian mammoth (*Mammuthus primigenius*). When compared with the Hb of closely related Asian elephants (*Elephas maximus*), two amino acid substitutions in the woolly mammoth (101 Glu → Gln and 12 Thr → Ala) correlate with the reduction in its temperature sensitivity. These substitutions are associated with increased binding of Cl^- and the organic phosphate, 2,3-bisphosphoglycerate (2,3-BPG)[27]. Both Cl^- and 2,3-BPG reduce the affinity of haemoglobin for oxygen and may have contributed to the ability of woolly mammoths to colonize higher, colder latitudes when they migrated from equatorial Africa 1–2 million years ago.

The effects of carbon dioxide and pH on the binding of oxygen to respiratory pigments

Carbon dioxide and pH can have two effects on the ease with which blood pigments combine with oxygen. In most species, changes in the partial pressure of carbon dioxide (PCO_2) and/or pH of the blood can affect the affinity (P_{50}) of the respiratory pigment for oxygen, and in some species of bony fish, pH and PCO_2 can also affect the oxygen-carrying capacity of the haemoglobin. We now look at these two influences in more detail.

The Bohr effect

The effect of carbon dioxide on P_{50} is known as the **Bohr effect**, after Christian Bohr, who, together with Karl Hasselbalch and August Krogh, first discovered the phenomenon in 1904. Carbon dioxide dissolved in water generates carbonic acid, which dissociates to release protons (H^+ ions)[28]; causing the pH of the blood to decrease. Protons and oxygen can be thought of as competitive antagonists; when one binds with haemoglobin, the other is released. The protons produced from the presence of carbon dioxide in the plasma combine with several amino acid residues on both the α and β chains of haemoglobin and alter their oxygen affinity. The major groups involved in this process are the uncharged α-imidazole side chains of histidine. The protons that combine with the haemoglobin are known as the Bohr protons. Sources of protons other than carbon dioxide, such as lactic acid, also contribute to this influence of pH on P_{50} of the haemoglobin.

So what effect do these bound protons actually have? Within physiological ranges of pH (normally between pH 9 and 6), the binding of protons to specific amino acid residues of the globin chains results in the stabilization of the deoxygenated (T) form of haemoglobin, which decreases its affinity for oxygen and releases oxygen. This is known as the normal or alkaline Bohr effect. A decline in pH below about 6 increases the oxygen affinity of the blood pigment and is known as the acid or reversed Bohr effect.

The alkaline Bohr effect causes the oxygen equilibrium curve to shift to the right as shown in Figure 13.12. The Bohr factor (φ) is the ratio of the change in the logarithm of P_{50} ($\Delta \log P_{50}$) to the associated change in pH (ΔpH) and indicates the extent to which a change in pH will affect the P_{50} of the blood pigment:

$$\varphi = \Delta \log P_{50} / \Delta \, pH. \qquad \text{Equation 13.1}$$

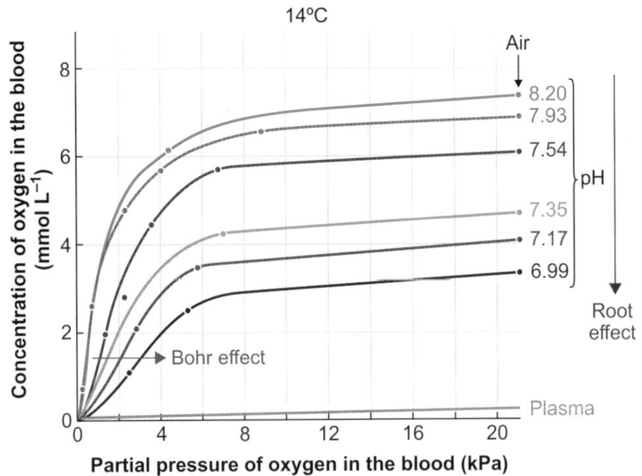

Figure 13.12 Effects of pH on the oxygen binding properties of respiratory pigments of some species of teleost fish

Oxygen equilibrium curves for the blood of European eels (*Anguilla anguilla*), at 14°C and a range of pH from 8.20 to 6.99. The line at the bottom is for plasma alone. Note how the increase in acidity not only decreases the affinity of the blood for oxygen (Bohr effect), it also reduces the maximum oxygen-carrying capacity of the blood (Root effect).

Source: Modified from Steen JB (1971). Comparative Physiology of Respiratory Mechanisms. Academic Press, London. Photo: © Ron Offermans & Bert Voorhorst.

[27] Formerly known as 2,3-diphosphoglycerate (2,3-DPG).

[28] $H_2O + CO_2 \leftrightarrows H_2CO_3 \leftrightarrows H^+ + HCO_3^-$. This reaction is discussed in Section 11.1.2.

13

As indicated above, the change in pH may be induced solely by a change in partial pressure of carbon dioxide or by a change in pH at a constant partial pressure of carbon dioxide. These are known as the CO_2 Bohr effect (φCO_2) and fixed acid Bohr effect (φ_{AH}), respectively. In human blood, the CO_2 Bohr effect is greater than the fixed acid Bohr effect. About 20 per cent of the Bohr effect is due to carbon dioxide reversibly combining with the α–NH_2 groups of the four N-terminal amino acids of haemoglobin to form carbamino compounds:

$$\alpha - NH_2 + CO_2 \leftrightarrow \alpha - NHCOO^- + H^+ \quad \text{Equation 13.2}$$

These compounds form a type of ionic interaction called salt bridges, which stabilize the T form.

The various respiratory pigments exhibit quite a large range of φ, from zero in the haemoglobin of Pacific spiny dogfish (*Squalus suckleyi*) to −1.7 in the haemocyanin of octopus. Such a range in values means that the fall in pH that usually occurs in the blood as it passes through the metabolizing cells will increase P_{50} (decrease affinity of pigment for oxygen) and facilitate the unloading of oxygen to differing degrees in different species. Conversely, as blood passes close to the surface of the gas exchanger, oxygen binds to the pigment, Bohr protons are released, which are then involved in the dehydration of bicarbonate producing carbon dioxide that is released. This process causes a decrease in P_{50} (and so increases the oxygen affinity of haemoglobin), thus facilitating the loading of oxygen to differing degrees in different species.

Experimental evidence for the combined effect of changes in temperature, pH, and partial pressure of carbon dioxide on oxygen uptake and delivery during the circulation of blood has been obtained from resting and exercising thoroughbred racehorses, among other species. This evidence is presented in Experimental Panel 13.1.

Experimental Panel 13.1 Construction of functional oxygen equilibrium curves in thoroughbred racehorses

Background

Much of what we know about the way blood pigments work is by studying them *in vitro*. Samples of blood (or haemolymph) are exposed to different conditions that are known to affect the binding properties of oxygen to the pigment. The objective of this study was to test the hypothesis that the effect of changes that occur in the blood perfusing the locomotor muscles during exercise in thoroughbred racehorses favour unloading of oxygen from the blood to the muscles. The changes studied were temperature, pH and partial pressure of carbon dioxide (PCO_2)

The first step was to produce an equation which describes the oxygen equilibrium curve (OEC) of the horses under resting (standard) conditions of temperature, pH and PCO_2. The equation was then used to produce functional OECs for resting and exercising horses.

Experimental approach

In order to construct the OEC, blood samples were taken from the jugular vein of thoroughbred racehorses (*Equus ferus caballus*). Heparin was added to prevent the blood from clotting and the blood then placed in the U-shaped cuvette of a spinning tonometer at a specific, controlled temperature. Gas-mixing pumps were used to produce a gas mixture of known concentrations of oxygen and carbon dioxide. Each gas mixture was humidified and passed through the cuvette, which was spun alternately in opposite directions for at least 15 min to ensure that the blood had equilibrated with the gas mixture. When the cuvette was spinning, the blood was thrown up its sides, so forming a thin film of relatively large surface area to facilitate diffusion of oxygen and carbon dioxide into the blood.

A single oxygen equilibrium curve was constructed by passing nitrogen through the cuvette to desaturate the blood completely; the blood was then slowly resaturated by increasing the oxygen concentration in the gas mixture. Samples were taken from the cuvette at points along the resaturation curve. For each curve, the temperature, pH, PCO_2 and haemoglobin (Hb) concentration were measured. Nine different concentrations of oxygen were used for each curve, up to a maximum of 25 per cent to ensure saturation of the haemoglobin. After equilibration of the blood with the gas mixture, a sample of blood was withdrawn from the cuvette into a syringe which had been flushed with the appropriate gas mixture.

Based on data from blood taken from resting horses, resting values for arterial blood from a thoroughbred racehorse were taken to be: 37°C, pH 7.4 and PCO_2 5.33 kPa. Also, the effects of temperature, pH and PCO_2 were examined by exposing blood in the cuvette to those values observed during exercise: 37–41°C, pH 7.01–7.4 and PCO_2 6.3–16.9 kPa. Three temperatures were used: 37, 39 and 41°C, which enabled the temperature coefficient, $\Delta \log P_{50}/\Delta T$, to be calculated.

The fixed acid Bohr coefficient was determined by adding 0.1 mol L^{-1} HCl to the blood until the required pH was achieved. The CO_2 Bohr effect was determined by increasing PCO_2. However, no attempt was made to maintain a constant pH, so the net CO_2 Bohr coefficient was obtained by subtracting the fixed acid coefficient from the determined CO_2 effect. Curves were also

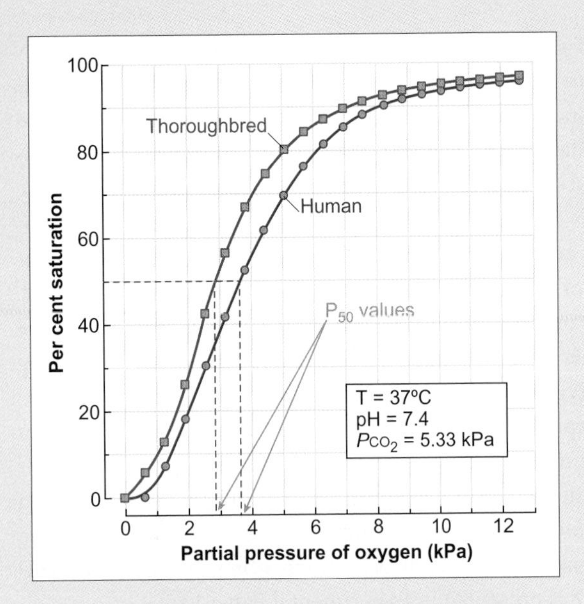

Figure A Oxygen equilibrium curve for thoroughbred racehorses under resting (standard) conditions of temperature (T), pH and PCO_2

The curve for humans is given for comparison.

Modified from: Smale K and Butler PJ (1994). Temperature and pH effects on the oxygen equilibrium curve of the thoroughbred horse. Respiration Physiology 97: 293-300.

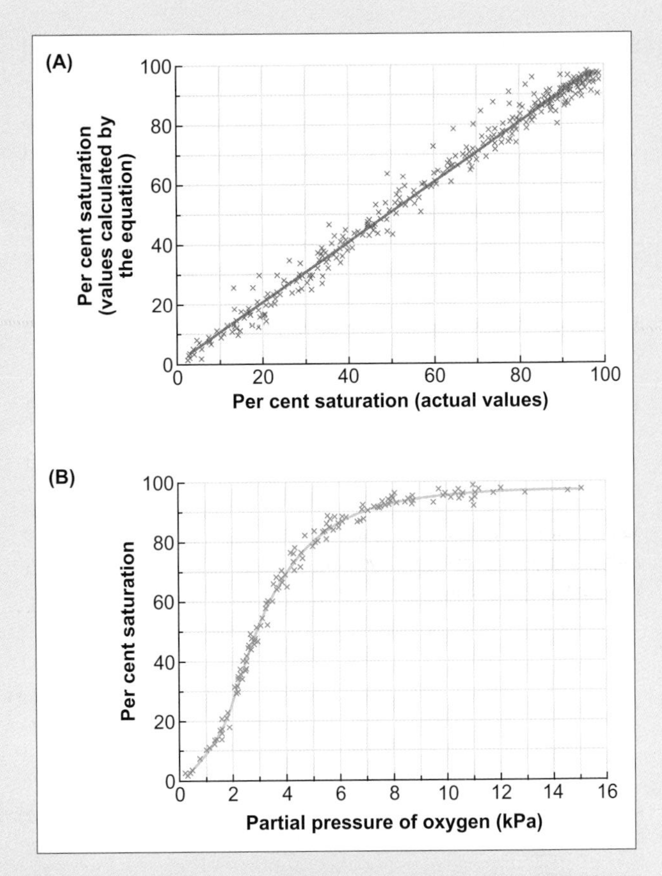

Figure B Testing the accuracy of an equation describing the oxygen equilibrium curve of thoroughbred racehorses

(i) The relationship between the measured values of % saturation of blood, as shown in Figure A, and those derived from the equation. The line is the line of equality. (ii) Crosses indicate measured values of partial pressure of oxygen and % saturation in arterial blood, as shown in Figure A, and the curve is derived from the equation.

Reproduced from: Smale K et al (1994). An algorithm to describe the oxygen equilibrium curve for the thoroughbred racehorse. Equine Veterinary Journal 26: 500-502.

constructed by adding lactic acid to the blood to mimic the *in vivo* conditions during exercise.

Experiment 1

The OEC for resting thoroughbred horses is shown in Figure A. A curve for humans is included for comparison. The P_{50} for the horse blood is 2.83 kPa, which is lower than that for human blood (approximately 3.5 kPa). The amount of oxygen combined with Hb when fully saturated is 62 μmol g^{-1} (1.39 mL g^{-1}).

An equation used to describe the OEC for human blood was modified by replacing the various coefficients for human blood with those obtained for the blood of resting thoroughbred horses, as described above. The accuracy of the equation for horses was confirmed by comparing:

- measured values of percentage saturation with values obtained from the equation, as shown in Figure B(i);
- the actual OEC with that derived from the equation, as illustrated Figure B(ii).

The data show that there is excellent agreement between actual and calculated values.

Experiment 2

Using data obtained from the blood of exercising thoroughbred horses (see Experiment 1) and the equation describing the OEC of these horses at rest, it was possible to construct OECs for arterial and mixed venous blood of the animals when they were running at a speed of 12 m s^{-1}. These curves are shown in Figure C. These curves are shifted to the right of that from horses at rest as a result of the increase in temperature and decrease in pH that occur in the locomotor muscles during exercise. The mean values for PO_2 in arterial and mixed venous blood (P_aCO_2 and $P_{\bar{v}}CO_2$ respectively) at rest and during exercise are indicated on the curves.

The combined effects of increased temperature and decreased pH during exercise reduce the P_{50} compared with that in resting horses. This comparison shows that, even at very low values of PO_2 in mixed venous blood of around 2 kPa, the unloading of oxygen at the active muscles is enhanced, as indicated by the dashed green line in Figure C. At any $P_{\bar{v}}CO_2$ above 2 kPa, the facilitation of oxygen unloading during exercise is even greater.

Overall findings

Although it is quite common to present OECs with the concentration of oxygen presented as per cent saturation (as in Figure C), such a presentation does not indicate the actual amount of oxygen present in a given volume of blood. As is illustrated by Figure 13.10, the maximum concentration of oxygen in blood can vary considerably between species. The actual concentration of oxygen in the blood can also vary during exercise in some species of vertebrates as a result of

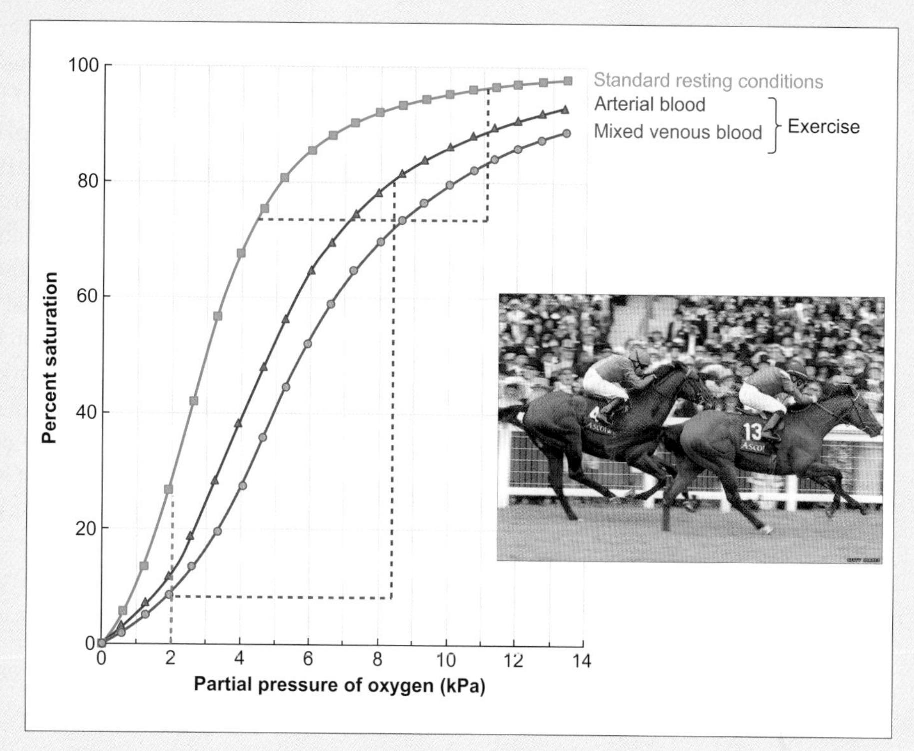

Figure C Oxygen equilibrium curves (OECs) constructed from the equation for the blood of thoroughbred horses in resting and exercising animals

Standard resting conditions for both arterial and mixed venous blood are taken as: 37°C, pH 7.4, PCO_2 5.33 kPa. During exercise, the conditions are 41°C, pH 7.13, PCO_2 7.9 kPa in arterial blood and 41°C, pH 7.01, PCO_2 16.9 kPa in mixed venous blood. Vertical lines indicate values for arterial blood (red) and horizontal lines indicate values for mixed venous blood (blue) in resting horses and in exercising horses. Note the lower PO_2 in arterial blood in exercising horses. ■, OEC for resting animals; ▲, OEC for arterial blood during exercise; ●, OEC for mixed venous blood during exercise. The green dashed vertical line indicates the enhanced unloading of oxygen from venous blood of exercising horses.

Adapted from Smale K et al (1994). An algorithm to describe the oxygen equilibrium curve for the thoroughbred racehorse. Equine Veterinary Journal 26: 500-502. Photo: Noah Salzman/Wikimedia Commons.

the release of red blood cells (RBCs) from the spleen. Thoroughbred horses are an extreme example of such a release of RBCs during exercise.

Haemoglobin concentration in the blood increases from 15 g $(100\ mL)^{-1}$ at rest to 22 g $(100\ mL)^{-1}$ when racehorses are running at 12 m s^{-1} and this increases the concentration of oxygen in arterial blood (C_aO_2) from 9.4 mmol L^{-1} at rest to a maximum of 12 mmol L^{-1} during exercise. The increase in C_aO_2 more than compensates for the decrease in partial pressure of oxygen in arterial blood (P_aO_2) during exercise, as shown in Figure C, which is a characteristic of thoroughbred racehorses. The increased concentration of oxygen in the blood and the enhanced unloading of oxygen during exercise combine to increase the amount of oxygen delivered to the muscles from what would have been 6.8 mmol $(L\ blood)^{-1}$ to what is, in fact, 10.6 mmol $(L\ blood)^{-1}$.

The results obtained lend support to the hypothesis that the effect of changes to oxygen transport to the muscles that occur in the blood during exercise in thoroughbred racehorses favour unloading of oxygen from the blood to the locomotor muscles.

› Find out more

Butler PJ, Woakes AJ, Smale K, Roberts CA, Hillidge CJ, Snow DH, Marlin DJ (1993). Respiratory and cardiovascular adjustments during exercise of increasing intensity and during recovery in thoroughbred racehorses. Journal of Experimental Biology 179: 159–180.

Kelman GR (1966). Digital computer subroutine for the conversion of oxygen tension into saturation. Journal of Applied Physiology 21: 1375–1376.

Smale K, Anderson LS, Butler PJ (1994). An algorithm to describe the oxygen equilibrium curve for the Thoroughbred racehorse. Equine Veterinary Journal 26: 500–502.

Smale K, Butler PJ (1994). Temperature and pH effects on the oxygen equilibrium curve of the thoroughbred racehorse. Respiration Physiology 97: 293–300.

The Root effect

Many bony (teleost) fish, including eels (*Anguilla* spp.) and rainbow trout (*Onchorhynchus mykiss*), possess multiple haemoglobins. Some of these haemoglobins have a positive net charge and are known as cathodic haemoglobins; by contrast, some have a negative net charge and are anodic haemoglobins. For the anodic haemoglobins of fish, an increase in the partial pressure of carbon dioxide and a fall in pH not only shift the oxygen equilibrium curve to the right (Bohr effect), but also reduce the oxygen-carrying capacity of the haemoglobin, as shown in Figure 13.12. This is known as the **Root effect**, after R.W. Root, who described it in a number of species of marine teleosts in 1931. The Root effect is widespread among the haemoglobins of fish, and has also been reported in the haemocyanins of cuttlefish (*Sepia* spp), and in octopus, although its function in these species is not known.

The strength of the Root effect in fish varies from species to species. For example, approximately 50 per cent of the haemoglobins in eels and 60 per cent in trout are anodic haemoglobins and exhibit a Root effect. The strength of the Root effect can be measured by the degree of deoxygenation of functional haemoglobin caused by acidification at a high partial pressure of oxygen, such as that of air (approximately 21 kPa).

At first sight, the Root effect is a strange phenomenon, for if there is a general reduction in pH of the blood (as, for example, during strenuous exercise when protons are produced) the oxygen-carrying capacity of the blood would be reduced, which would impair the delivery of oxygen to the active muscles. However, the release of the hormone adrenaline (epinephrine) during exercise enables some species of fish to maintain the pH inside their red blood cells above that of the surrounding plasma so that the Root effect does not come into effect[29].

The Root effect is important, however, in the retina and swimbladder of fish. Unlike in most other vertebrates, the retina of fish is not supplied with oxygen directly by blood vessels and the distance over which diffusion occurs can exceed 360 μm. This distance is six times larger than that in the retina of primates and, therefore, slows the rate of diffusion of oxygen[30]. In both the retina and the swimbladder of fish, the production of lactic acid and carbon dioxide cause a reduction in blood pH, which initiates the Root effect and an increase in the partial pressure of oxygen (PO_2). The presence of a countercurrent arrangement of capillaries[31] (known as a rete mirabile) enables the build-up of PO_2 to levels far in excess of those normally in arterial blood (12 to 18 kPa). The functional significance of the Root effect in the eye of fish is illustrated by the fact that the electrical activity in the retina drops markedly if the partial pressure of oxygen in the eye of a rainbow trout falls below approximately 33 kPa. The significance of the Root effect in the functioning of the swimbladder of bony fish as a buoyancy organ is discussed in Case Study 13.1.

[29] Intracellular pH of red blood cells of fish during strenuous exercise is discussed in Experimental Panel 15.2.

[30] Factors affecting the rate of diffusion of oxygen, including the distance over which diffusion occurs, are discussed in Section 11.2.1.

[31] The countercurrent arrangement is discussed in general terms in Section 3.3.1.

Case Study 13.1 Significance of the Root effect in the functioning of the swimbladder as a buoyancy organ in teleost fish

The density of many aquatic animals is greater than that of natural bodies of water, mainly because of their muscles and calcified skeletal systems. For example, fresh water has a density close to 1000 kg m^{-3} and sea water has an average density of 1025 kg m^{-3} (range 1020 to 1030 kg m^{-3}, depending on temperature and salinity). In contrast, teleost fish would have a density of between 1060 to 1090 kg m^{-3} if it were not for the presence of a gas-filled bladder, a **swimbladder**, in many species. In the absence of a swimbladder, teleost fish would need to expend energy to generate lift by continuously swimming in order not to sink in the water column. The density of the gas in the swimbladder is 700 to 800 times less than that of fresh water[1]; by increasing its volume with gas, a fish could decrease its density so that it becomes the same as that of the surrounding water. The fish would be **neutrally buoyant**.

What would the volume of the swimbladder need to be to make a fish neutrally buoyant?

If we assume a marine fish with a mass of **500 g**, has no swimbladder and a density of **1075 kg m^{-3}**, its volume would be: 0.5 kg/1075 kg m^{-3} = 0.000465 m^3 = **465 cm^3**. So, in order for the fish to have the same density as the surrounding sea water (**1025 kg m^{-3}**), it would need to have a volume of 0.5 kg/1025 kg m^{-3} = 0.000488 m^3 = **488 cm^3**. In other words (provided we ignore the density of the swimbladder), a swimbladder would need to have a volume of 488 cm^3 – 465 cm^3 = **23 cm^3** which is 4.7 per cent of the total volume of the fish. The volume of a swimbladder would need to be a greater proportion of the volume of a freshwater fish because of the lower density of fresh water.

Unfortunately, there are problems with using a gas to provide neutral buoyancy in a fish: the gas is compressible and the pressure in a column of water increases by 1 atmosphere (101.3 kPa) for every 10 m of descent[2]. This means that the pressure exerted on the gas in a swimbladder doubles to 2 atmospheres and the volume of the swimbladder halves when the fish moves from the surface to 10 m depth. The pressure doubles again to 4 atmospheres when the fish moves to 30 m depth, and the volume is now 25 per cent of what it was at the surface, and so on. In other words, the greatest effect of hydrostatic pressure on the volume of the bladder occurs in the upper layers, and many species of fish that occupy the upper 1000 m of the sea have a swimbladder. In order for the fish to maintain the volume of the swimbladder as it descends in the water column, it needs to add gas to the bladder. Conversely, when it ascends, the fish needs to remove gas from the bladder as it expands.

The general structure of swimbladders is shown in Figure A. Note how in one type, the physostome (Greek, *phys*—bladder, *stom*—open) type, the bladder is connected to the oesophagus[3] by the pneumatic duct, as shown in Figure A(i). In the other type, the physoclist (Greek, *clist*—closed) type, shown in Figure A(ii), the bladder is completely closed. Physostome bladders are the primitive type and can be filled at the surface by the fish

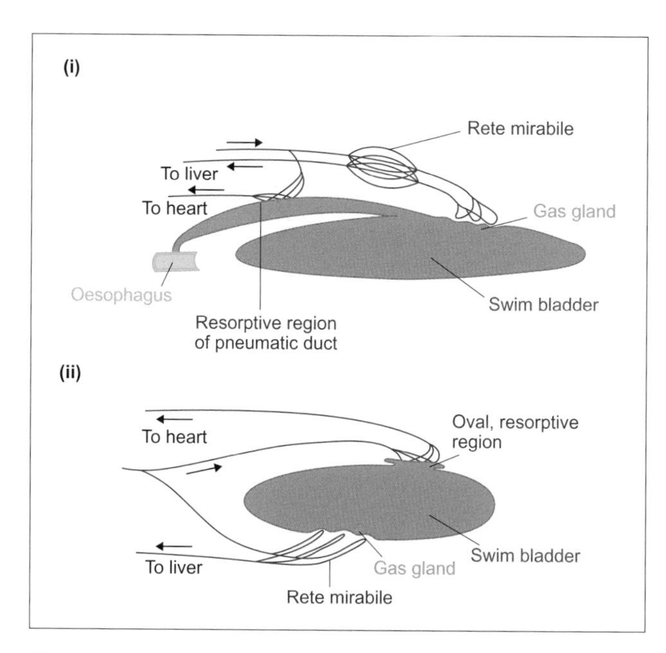

Figure A The two main types of swimbladders in teleost fish

(i) Physostome swimbladder as found in European eels (*Anguilla anguilla*). These swimbladders are connected to the oesophagus by a pneumatic duct. (ii) Physoclist swimbladder as found in perch (*Perca fluviatilis*). In these swimbladders, the duct is absent and the bladder is completely closed.

Adapted from Denton EJ (1961). The buoyancy of fish and cephalopods. Progress in Biophysics and Biophysical Chemistry 11: 177-234.

swallowing air or emptied via the pneumatic duct. However, in physoclist bladders the gas has to diffuse into the bladder from the blood that supplies it. This is achieved by the rete mirabile, which forms a countercurrent multiplier system[4] and a gas gland that produces lactic acid and CO_2. Both of these structures are shown in Figure A.

The composition of gas in the swimbladder varies extensively between species, being mainly nitrogen in salmonids, but in most fish living at greater depth, oxygen makes up the largest proportion. Freshly deposited gas may contain up to 25 per cent CO_2, but this is soon replaced by oxygen. For fish at a depth of 2000 m, the partial pressure of oxygen (PO_2) in the swimbladder may be almost 180 atmospheres (atm)[5] and that of nitrogen may be around 20 atm.

The gas gland uses glucose to produce lactic acid via anaerobic glycolysis[6] and CO_2 via another anaerobic pathway which operates in parallel to glycolysis called the pentose-phosphate shunt[7]. Oxidative metabolism in the gas gland is minimal, with only about 1 per cent of the glucose being completely oxidized. The remaining 99 per cent is used for anaerobic metabolism to produce lactic acid and CO_2, which have three important effects:

- The presence of increased concentrations of lactic acid in the blood as it passes through the gas gland reduces the solubility of all gases in the plasma of the blood, with the result of increasing their partial pressure[8]. This is known as the **salting out effect**.
- The lactic acid also reduces the pH of the blood, thereby reducing the affinity of the haemoglobin for oxygen by the Bohr effect. In many species of teleost fish, it also reduces the carrying capacity of the haemoglobin for oxygen by the Root effect. The oxygen released from the haemoglobin causes an increase in PO_2 in the blood
- The presence of the protons from lactic acid cause in increase in PCO_2 in the blood, as shown in Equation 11.1, and the CO_2 produced by the pentose-phosphate shunt in the gas gland increases PCO_2 in the gland. As a result of this increase in PCO_2 in the gas gland, CO_2 diffuses into the swimbladder and into the blood.

Much of the work on how swimbladders function has been performed on European eels (*Anguilla anguilla*). This is somewhat ironic, for, as shown in Figure A(i), the eel has a physotome bladder, although it does also possess a rete mirabile and a gas gland. The countercurrent multiplier system consists of about 30,000 to 50,000 capillaries which supply the gas gland. These capillaries run parallel to and are intermingled with about 20,000 to 40,000 venous capillaries which eventually send blood to the liver. The capillaries in the rete can be several mm long (capillaries in muscles are about 0.5 mm long) and the diffusion distance between the capillaries ranges from between 1–2 μm. Fish that live at the greatest depth have the longest capillaries which can be up to 25 mm long.

The partial pressures of gases in the arterial blood supplying the rete and in the surrounding tissues are going to be no higher than those in the atmosphere (about 0.2 atm for O_2, about 0.8 atm for N_2 and negligible for CO_2). The three processes outlined above, cause an increase in the partial pressure of all gases in the blood as it passes through the gas gland, as shown in Figure B for oxygen. If the partial pressures of gases in the blood within the gas gland are less than those in the swimbladder, gases diffuse out of the bladder, and the gases in the venous (efferent) capillaries will be higher than those in the arterial (afferent) capillaries. These differences in partial pressure mean that the gases diffuse from the venous to the arterial capillaries.

However, more importantly as far as oxygen is concerned, it appears that the diffusion of lactic acid and CO_2 from the venous to the arterial capillaries cause an increase in PO_2 in the arterial capillaries by the Bohr and Root effects, as shown in Figure B(ii) and (iii). These processes produce the **multiplying effect** and continue until the partial pressure of any gas exceeds that in the swimbladder, when the gas diffuses from the blood into the bladder, shown in Figure B(iii). Eventually, the diffusion process stops when the pressure reached in the swimbladder is sufficient to counteract the surrounding hydrostatic pressure and the fish is neutrally buoyant.

Despite such high pressures of gases in the swimbladder, very little gas diffuses out into the surrounding tissue. Any gas diffusing back into the blood as it passes through the gas gland also diffuses from the venous capillaries to the arterial capillaries of the rete. In addition, the wall of the swimbladder is about 100 times less permeable to O_2, N_2 and CO_2 than connective tissue in some deeper-dwelling species of fish as the result of the presence of a layer containing crystals of guanine.

Fish that undergo diurnal vertical migrations need to deposit and remove gas rather rapidly in order to maintain a constant

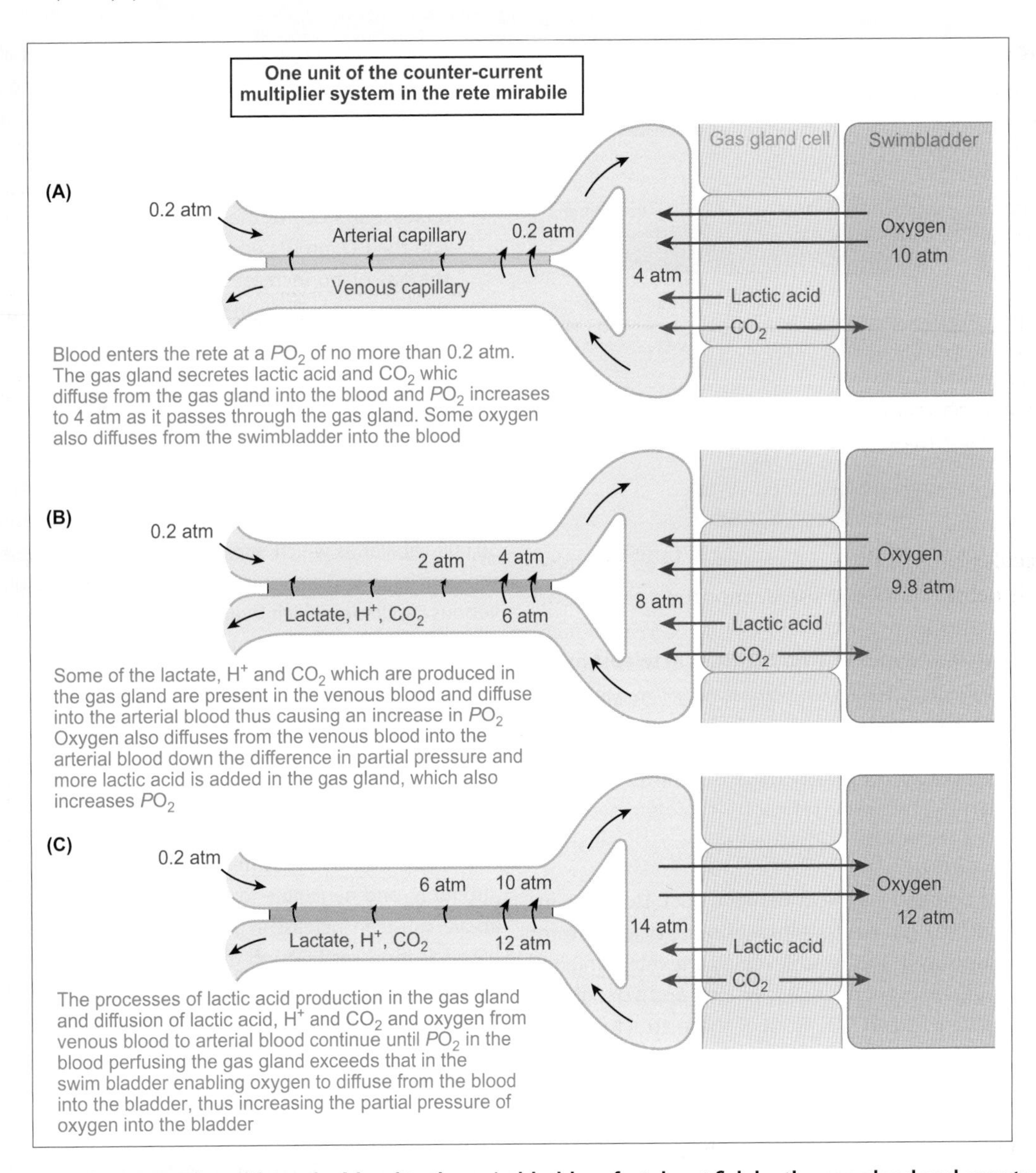

Figure B How oxygen is transferred from the blood to the swimbladder of a teleost fish by the gas gland and countercurrent arrangement of the rete mirabile

Imagine the purely hypothetical situation where oxygen is the only gas in the swimbladder of a fish that is at a depth of 90 m and is neutrally buoyant. The gas gland is not producing lactic acid and is not being perfused with blood. The pressure in the bladder is 10 atm. The fish then begins to descent to 110 m, so has to increase the pressure in the swimbladder to 12 atm to maintain its neutral buoyancy. (i) The gas gland begins to be perfused with blood and to secrete lactic acid and CO_2 (blue lines) into the blood. Because of the large pressure difference between the oxygen in the swimbladder and in the blood, oxygen diffuses from the bladder into the blood (red line). These processes lead to the partial pressure of oxygen (PO_2) in the blood increasing from 0.2 atm as it enters the rete to 4 atm as it passes through the gas gland. The lactic acid dissociates into lactate and H^+. (ii) Some oxygen, lactate and H^+ and CO_2 diffuse back from the venous to the arterial capillaries. The lactate and H^+ cause an increase in PO_2 by way of the salting out effect and the Root and Bohr effects, respectively. Back diffusion of CO_2 is also important in increasing PO_2 in the arterial blood. There is a further increase in PO_2 to 8 atm as the blood passes through the gas gland, but this is insufficient to cause diffusion of oxygen from the blood to the bladder. Oxygen diffuses from the venous capillary to the arterial capillary down the difference in partial pressure. (iii) This process is repeated until PO_2 in the blood exceeds that in the swimbladder, leading to diffusion of oxygen into the bladder until the required pressure of 12 atm is reached.

Based on Pelster B and Scheid P (1992). Countercurrent concentration and gas secretion in the fish swimbladder. Physiological Zoology 65: 1-16.

volume of their swimbladder. The rate of gas deposition is affected by the rate at which the gas gland produces protons and CO_2 and the rate at which it is perfused with blood. An increase in these rates leads to an increase in rate of deposition. Removal of gas from the bladder occurs at the resorbing part of the pneumatic duct in physostomes or at the oval organ in physoclists and the rate of resorbtion is controlled by the rate of perfusion. Both of the resorptive areas are separated from the main bladder by muscles, which relax during resorption. Both of areas are well vascularized.

› Find out more

Denton EJ, Liddicoat JD, Taylor DW (1972). The permeability to gases of the swimbladder of the conger eel (*Conger conger*). Journal of the Marine Biological Association of the United Kingdom 52: 727–746.

Pelster B, Scheid P (1992). The influence of gas gland metabolism and blood flow on gas deposition into the swim bladder of the European eel, *Anguilla anguilla*. Journal of Experimental Biology 173: 205–216.

Pelster B, Scheid P (1992). Countercurrent concentration and gas secretion in the fish swim bladder. Physiological Zoology 65: 1–16.

Pelster B, Scheid P (1993). Glucose metabolism of the swimbladder tissue of the European eel, *Anguilla anguilla*. Journal of Experimental Biology 185: 169–178.

Scheid P, Pelster B (1995) Gas exchange in the fish swimbladder. In: Heisler N (ed) Advances in Comparative and Environmental Physiology Vol 21, pp 41–59, Berlin: Springer-Verlag.

1 The densities of air and water are given in Table 11.3.
2 The relationship between volume and pressure of a gas is discussed in Section 11.1.2.
3 The origin of the swimbladder and its relationship to air-breathing organs in fish is discussed in Section 12.3.3.
4 Countercurrent multipliers are discussed in general terms in Sections 3.3.1 and with respect to the mammalian kidneys in Section 7.2.3.
5 1 atmosphere equals 101.3 kPa at STPD.
6 We discuss anaerobic glycolysis in Section 2.4.1.
7 The pentose-phosphate shunt consists of an oxidative phase when CO_2 and NADPH are produced and a non-oxidative phase when 5-carbon sugars (pentoses) are produced.
8 The effect of increasing the solute concentration in water on the solubility of gases is discussed in Section 11.1.2.

The molecular basis of the Root effect has been the subject of much debate. One suggested important feature of those haemoglobins showing a Root effect is the presence of serine instead of cysteine at position 93 on the β chains. The serine strengthens the salt bridge formed by the terminal residue of the β chains (histidine 146, His-146 β) in the T (deoxygenated) state, thus reducing their tendency to bind oxygen. However, this cannot be the only factor, as the haemoglobin of the African clawed frog (*Xenopus laevis*) also possesses the serine/cysteine substitution, but does not display a Root effect.

A study of the haemoglobin of tuna has revealed novel mechanisms that appear to account for the Root effect, at least in this species. These include:

- movement of a particular residue on one α chain;
- a salt bridge between two β residues;
- proton binding between residues on the α_1 and β_2 chains.

There was no indication in this study that a salt bridge is formed between His-146 β and any other group, so it is possible that the Root effect results from different mechanisms in different species. This possibility has also been suggested by a comprehensive phylogenetic analysis. This analysis also demonstrates that changes in haemoglobin function may in some cases be the result of a varying number of amino acid substitutions, rather than substitution of just a few amino acids in key locations.

The same phylogenetic analysis indicates that a pronounced Root effect of 20 per cent to 40 per cent evolved well before the evolution of the retia mirabilia (plural of rete mirabile) in the eyes or the swimbladder, and that the retia in the eyes evolved some 100 million years before the evolution of those in the swimbladder, as shown by the cladogram[32] in Figure 13.13. In other words, these data suggest that the functional significance of the Root effect may initially have been the secretion of oxygen in the eyes.

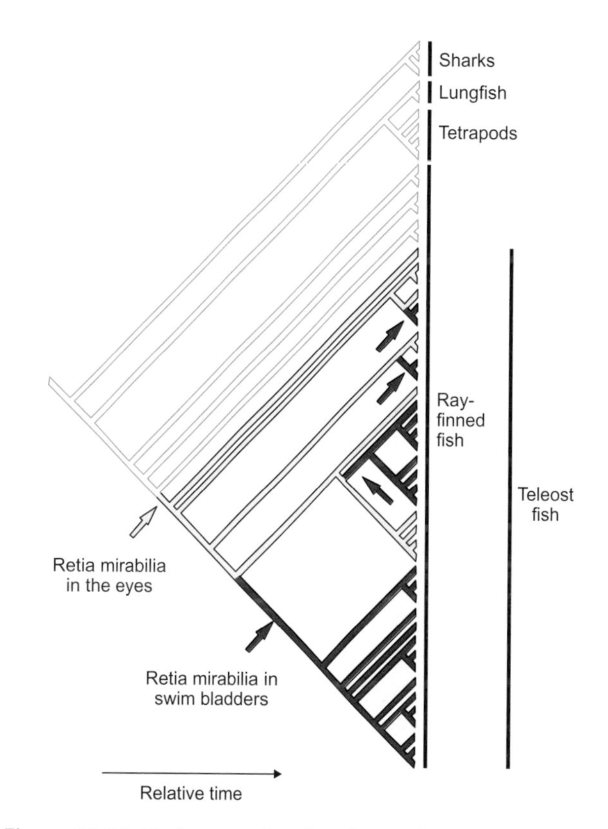

Figure 13.13 Cladogram showing the evolution of retia mirabilia in the eyes and swimbladders of ray-finned fish

Yellow lines and red lines indicate the presence of retia in the eyes and swimbladder, respectively. Open lines indicate the absence of retia. Retia in the eyes of fish evolved once about 250 million years ago (indicated by the yellow arrow), whereas they evolved in the swimbladder independently in four major teleost groups (indicated by the red arrows), with the earliest appearing about 130 million to 140 million years ago. Ray-finned fish are discussed in Section 1.4.2.
Modified from Berenbrink M et al (2005). Evolution of oxygen secretion in fishes and the emergence of a complex physiological system. Science 307: 1752–1757. Berenbrink M (2007). Historical reconstruction of evolving physiological complexity: O_2 secretion in the eye and swimbladder of fishes. Journal of Experimental Biology 209: 1641–1652.

32 Cladograms are discussed in Section 1.3.3.

13

The influences of organic phosphates and other organic ions on oxygen binding

Organic phosphates also affect the P_{50} of haemoglobin. Organic phosphates are produced in the red blood cells of vertebrates, with the exception of hagfish, crocodiles and ruminants, and cause a rightward shift in the oxygen equilibrium curve and a decrease in the affinity of the haemoglobin for oxygen.

The major organic phosphate in the red blood cells of mammals is 2,3-bisphosphoglycerate (2,3-BPG, formerly known as 2,3-diphosphoglycerate, 2,3-DPG), which is present at approximately equimolar concentration to haemoglobin. Other organic phosphates, such as adenosine triphosphate (ATP), guanosine triphosphate (GTP) and inositol pentakisphosphate (IP5, formerly known as inositol pentaphosphate, IPP), have been found to have a similar influence on P_{50} of haemoglobin of other vertebrates.

If the respiratory pigment is contained within a cell, then the conditions within the cell may be different from those in the surrounding blood. Consequently, the pH or partial pressure of carbon dioxide in the plasma does not necessarily indicate the conditions the pigment experiences within the blood cell.

In the absence of any modulating influences, the haemoglobin of most vertebrates has a high affinity for oxygen, and it is the modulators, such as carbon dioxide and organic phosphates, which reduce the affinity. The affinity of a solution of mammalian haemoglobin alone, i.e. in the absence of all other influences (often called stripped haemoglobin), is much higher than when carbon dioxide and the organic phosphate 2,3-BPG are present in the concentrations found in whole blood, as shown in Figure 13.14. Therefore, factors within the red blood cells influence the way in which haemoglobin functions in the whole animal by lowering its affinity for oxygen, thereby enabling it to release oxygen at relatively high partial pressures at the metabolizing tissues.

For most intracellular haemoglobins, increases in the concentrations of organic phosphates reduce the affinity of the pigment for oxygen by favouring the deoxygenated (T) state. This is because the organic phosphates form ionic bonds with specific amino acid residues of each β chain when haemoglobin is in its deoxygenated form. Upon oxygenation (R state), the formation of ionic bonds by the organic phosphates is prevented.

In some species of crustaceans, the presence of L-lactate and urate, both end products of anaerobic metabolism, like lactate in vertebrates, cause an increase in the affinity of extracellular haemocyanin for oxygen.

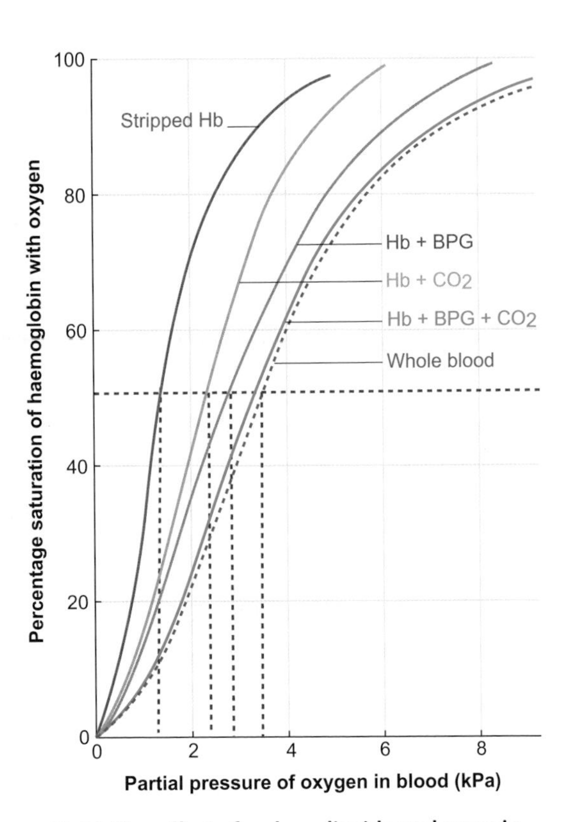

Figure 13.14 The effect of carbon dioxide and organic phosphates on the oxygen equilibrium curve of human haemoglobin (Hb)

Oxygen equilibrium curves of a solution of human haemoglobin, i.e. in the absence of any modulating influences (stripped haemoglobin), and in the presence of carbon dioxide and/or the organic phosphate, bisphosphoglycerate (BPG), all of which reduce the affinity of the haemoglobin for oxygen (i.e. they increase its P_{50} as indicated by the vertical red lines). Note how the addition of BPG and CO_2 to a solution of stripped haemoglobin produces a P_{50} which is almost identical to that of whole blood.

Adapted from Ingermann RL (1997). Vertebrate hemoglobins. In: Handbook of Physiology Section 13; Comparative Physiology Vol 1, Dantzler WE (Ed.), pp 357-408. Oxford University Press, New York.

The effects of amino acid substitutions on haemoglobin function

We discuss in the previous section how the substitution of a few amino acids by others can alter the characteristics of haemoglobin. There is one substitution in particular that is detrimental to those humans who possess it. Some humans have a genetic mutation that leads to the replacement of glutamic acid by valine at position 6 on the β chains. This change causes haemoglobin molecules to clump together, particularly at relatively low partial pressures of oxygen, i.e. on the venous side of the circulation. This clumping distorts the red cells into sickle shapes as illustrated in Figure 13.15A, leading to a decrease in their elasticity. These distorted, relatively rigid cells are unable to flow through the narrow capillaries, which impairs the delivery of oxygen

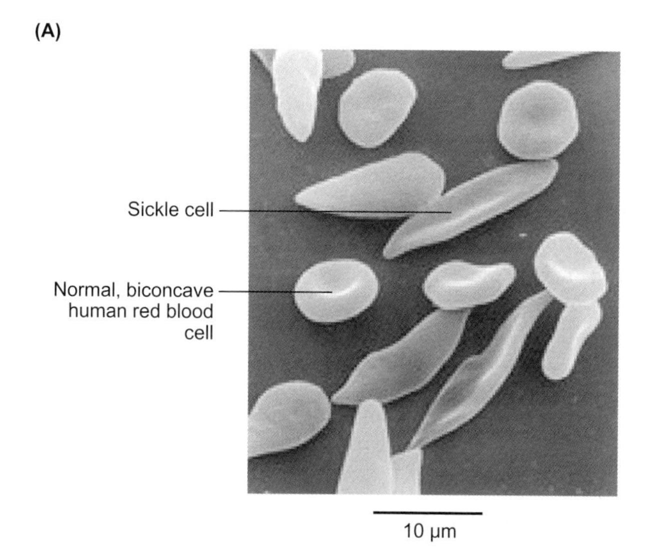

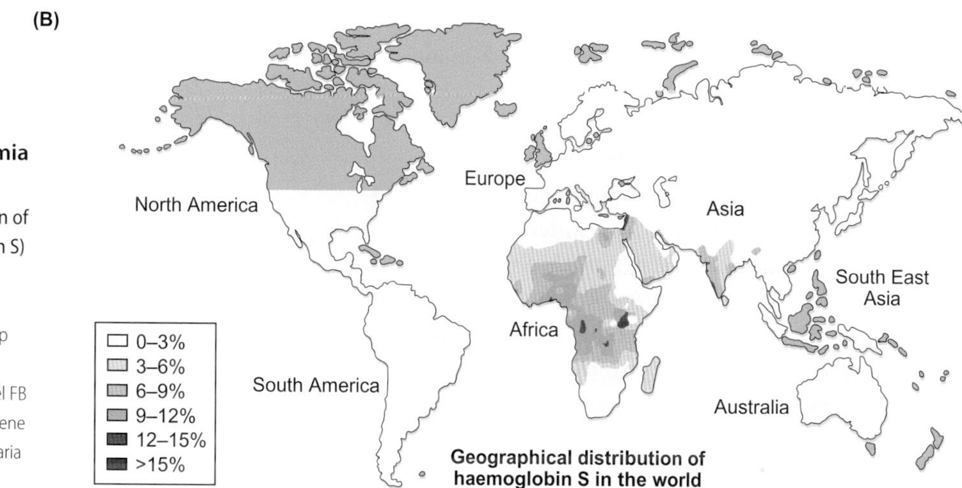

Figure 13.15 Sickle cell anaemia

(A) shows sickle cells among normal human red blood cells. (B) Distribution of sickle cell haemoglobin (haemoglobin S) around the globe.

Sources: (A) National Institute of Diabetes and Digestive and Kidney Diseases: (B) Map redrawn with permission from Macmillan Publishers Ltd: Nature Communication. Piel FB et al. Global distribution of the sickle cell gene and geographical confirmation of the malaria hypothesis Vol. 1, Article 104, © 2010.

13

to the tissues and can cause severe pain, organ failure and death. The lifetime of the affected cells is also reduced—in the order of 10–20 days compared with about 120 days for normal red cells—causing anaemia. The disease is known as sickle cell anaemia.

On the positive side, however, those people with sickle cell haemoglobin have a greater resistance to malaria, although the exact basis of this resistance in uncertain. In homozygotes, sickle cell anaemia usually leads to damage of internal organs, resulting in a premature death. By contrast, heterozygotes that possess both normal and sickle cell haemoglobin suffer only mild debilitation, but they still have greater resistance to malaria than individuals with completely normal haemoglobin. For this reason, the gene responsible for sickle cell haemoglobin persists, particularly in Africa, India and the Middle East, as depicted in Figure 13.15B.

Well-documented beneficial amino acid substitutions occur in species of geese that inhabit high altitudes. Bar-headed geese (*Anser indicus*) cross the Himalayas when they migrate from the Tibetan lakes and Mongolia, where they breed, to India, where they spend winter. Andean geese (*Chloephaga melanoptera*) live in the high Andes all year (as their name suggests). Both species have haemoglobins with a higher affinity for oxygen (lower P_{50}–4 kPa and 4.5 kPa, respectively) than that for greylag geese (*Anser anser*—P_{50}, 5.3 kPa), which live in the Indian plains the whole year.

The reasons for these differences in P_{50} are:

- Key amino acids are substituted by other amino acids: proline is replaced by alanine at position 119 in the α-chains of bar-headed geese, and leucine is replaced by serine at position 55 of the β -chains of Andean geese. These substitutions result in the loss of a hydrogen bond between the $\alpha_1\beta_1$ interface, which normally stabilizes the T-structure (the deoxy form of haemoglobin), and produce small increases in intrinsic oxygen affinity (i.e. that of stripped haemoglobin).
- These small intrinsic increases in oxygen affinity are amplified by the presence of organic phosphates.

Amino acid substitutions are also involved in the high oxygen affinity of the haemoglobins of high altitude mammals such as llama (*Lama glama*) and vicuña (*Vicugna vicugna*). The significance of the amino acid substitutions has been elegantly demonstrated by introducing such substitutions into human haemoglobin synthesized in *E. coli*—using a technique called site-specific mutagensis. The intrinsic oxygen affinities of the reconstituted (effectively, stripped) haemoglobins are greater (P_{50}, 0.44 kPa) than that of normal, stripped human haemoglobin (P_{50}, 0.77 kPa). However, there is no such adaptation in the oxygen affinity of the haemoglobin of snow leopards (*Panthera uncia*), which can live at altitudes between 3500 and 5000 m, where atmospheric PO_2 is between approximately 65 per cent and 55 per cent of that at sea level. The lack of any adaptation in the haemoglobin is even more surprising as it has a relatively low affinity for oxygen[33] compared with that of other mammals. It is currently unclear exactly how snow leopards manage to survive at such high altitudes.

Embryonic/foetal haemoglobins

Embryos and foetuses of vertebrates develop in a low PO_2 environment. They also tend to have different haemoglobins from those in adults, which influence the transfer of oxygen to the developing embryo or foetus. For example, the haemoglobin in the embryos of dimaondback terrapin (*Malaclemys terrapin*), has a P_{50} which is about half of that in the adult, favouring the transfer of oxygen across the egg shells and shell membranes to the embryo. Embryos that develop inside their mothers, such as placental mammals, obtain their oxygen from the mother's circulation in the placenta, which means that oxygen is passed from the mother's haemoglobin to that of the foetus. This process is facilitated by foetal haemoglobin having a higher affinity for oxygen than maternal haemoglobin.

Adult haemoglobin of mammals has two α-chains and two β-chains. In contrast, human foetal haemoglobin has the two α-chains but has two γ-chains in place of the two β-chains. There is no difference in the intrinsic affinities of adult and foetal haemoglobins for oxygen (when there is a solution of haemoglobin alone). However, in the presence of the organic phosphate 2,3-bisphosphoglycerate (2,3 BPG), the affinity for oxygen of foetal haemoglobin decreases less than that of maternal haemoglobin. This means that foetal haemoglobin has a greater affinity for oxygen than haemoglobin of the mother. After birth, there is a gradual transition from foetal to adult haemoglobin and the affinity of the newborn's haemoglobin for oxygen decreases. By the age of 4 to 6 months, all of the foetal haemoglobin has disappeared.

[33] A relatively low-affinity haemoglobin is common among all members of the cat-like group of carnivores.

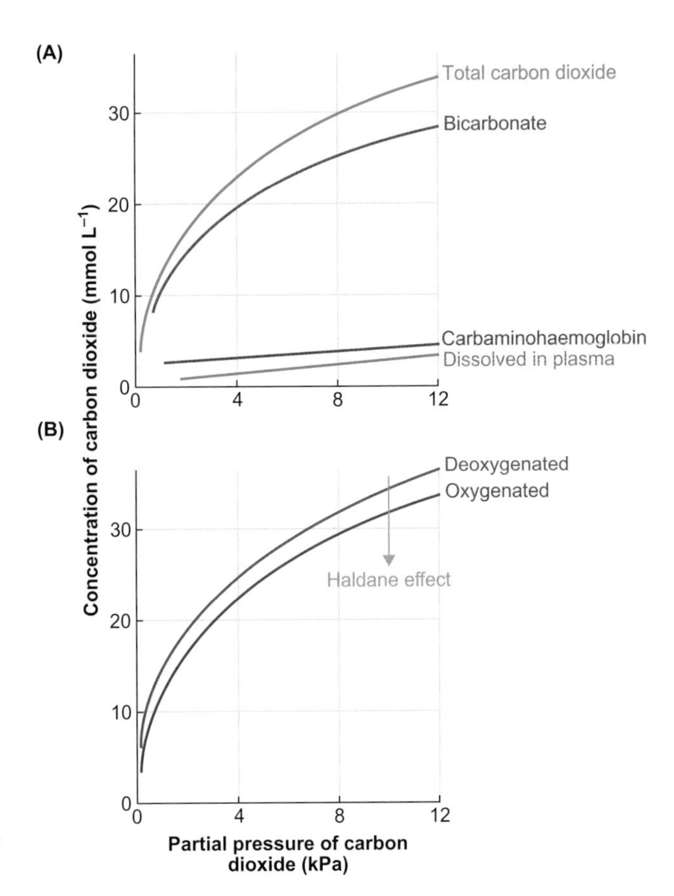

Figure 13.16 Carbon dioxide transport in the blood

(A) The amount of carbon dioxide in human blood at 38°C which is dissolved in the plasma, in the form of carbaminohaemoglobin and as bicarbonate. Note that by far the most CO_2 is in the form of bicarbonate. (B) Haldane effect in human blood at 37°C.

Modified from Dejours P (1981). Principles of Comparative Respiratory Physiology. Elsevier, Amsterdam.

13.1.3 Transport of carbon dioxide

The other major gas to be transported in blood is the primary end product of aerobic metabolism, carbon dioxide[34]. Carbon dioxide is transported by three mechanisms, as illustrated in Figure 13.16A:

- CO_2 is highly soluble in water and forms carbonic acid, H_2CO_3[35], which dissociates into protons (H^+) and bicarbonate ions (HCO_3^-), so CO_2 is carried mostly as bicarbonate ions in the plasma of vertebrates.
- Carbon dioxide is also transported as carbamino compounds by association with amino acid residues on respiratory pigments, as shown in Equation 13.2.
- A small amount is carried in solution, as illustrated in Figure 13.16A.

[34] Carbon dioxide production during metabolism is discussed in Section 2.2.

[35] The formation of H_2CO_3 when CO_2 dissolves in water is shown in footnote number 25.

Carbamino compounds account for approximately 10 per cent of total carbon dioxide transported to the lungs from the metabolizing tissues in humans, but are probably of little significance in amphibians and fish, where the N-terminal amino acids of the α-chains are acetylated and therefore not available for combining with carbon dioxide.

The protons formed from carbon dioxide dissolving in water combine with deoxygenated haemoglobin, so the greater the concentration of haemoglobin in the blood, the more carbon dioxide can be taken up by the plasma. In other words, the forward reaction is favoured. Consequently, the carbon dioxide carrying capacity of blood is dependent on the concentration (buffering capacity[36]) of the pigment. When the buffering capacity is high, large changes in the content of carbon dioxide may result from relatively small changes in the partial pressure of carbon dioxide.

Another important factor affecting maximum *in vivo* carbon dioxide carrying capacity of a pigment is the range of partial pressures of carbon dioxide (PCO_2) in the blood. In the blood of birds and mammals, PCO_2 is over the range of 4.5 and 5.5 kPa, whereas it is invariably below 1 kPa in water-breathing fish and crustaceans[37]. This is another consequence of the relatively high capacitance coefficient of carbon dioxide in water compared with that of oxygen[38].

The Haldane effect

Deoxygenated haemoglobin has a greater carbon dioxide carrying capacity than oxygenated haemoglobin, as shown in Figure 13.16B. This phenomenon was first described by J. S. Haldane and his colleagues in 1914 and is, therefore, known as the **Haldane effect**. The Bohr protons[39] are released from the haemoglobin upon its oxygenation as it moves through the gas exchanger. This causes the release of carbon dioxide from the blood: the leftward shift of the reaction $H_2O + CO_2 \leftrightarrows H_2CO_3 \leftrightarrows H^+ + HCO_3^-$ is favoured. Conversely, when the Bohr protons are incorporated into the structure of haemoglobin upon deoxygenation as it moves through the metabolizing tissues, carbon dioxide is taken up by the blood. Also, if carbamino compounds are formed, they incorporate carbon dioxide with deoxygenated haemoglobin, as shown in Equation 13.2; by contrast, carbon dioxide is released from oxygenated haemoglobin.

The magnitude of the Haldane effect can be quantified by determining the moles of protons released per mole of oxygen which binds to the haemoglobin. This is known as the Haldane coefficient.

[36] Buffering capacity is discussed in Section 13.3.2.

[37] Partial pressures of carbon dioxide in the blood of some aquatic and air-breathing animals are given in Tables 12.2 and 12.5.

[38] The effects of the different capacitance coefficients of oxygen and carbon dioxide in water are discussed in Section 11.1.2.

[39] Bohr protons are discussed in the section on the Bohr effect.

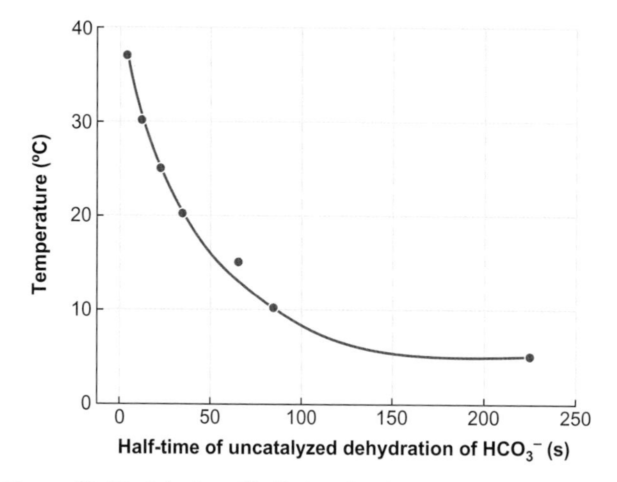

Figure 13.17 Calculated half-time for the production of CO_2 from HCO_3^- at different temperatures

These data are for the uncatalysed reaction, that is, in the absence of the enzyme carbonic anhydrase.

Adapted from Randall DJ et al (1981). The Evolution of Air Breathing Vertebrates. Cambridge University Press, Cambridge.

As the Haldane effect is dependent on the presence of the blood pigment, the difference in maximum carbon dioxide carrying capacity between deoxygenated and oxygenated pigments depends on the concentration of the pigment. In an analogous way to the Bohr effect, the Haldane effect is important in the unloading of carbon dioxide at the lungs, as the blood becomes oxygenated, and in the uptake of carbon dioxide at the tissues, as the blood becomes deoxygenated.

The role of carbonic anhydrase in the transport of carbon dioxide

The enzyme **carbonic anhydrase (CA)** plays an important role in the transport of CO_2 in the blood as it increases the rate of the reaction between carbon dioxide and water. Figure 13.17 shows that the reaction times of the formation of carbon dioxide from carbonic acid (and of the reverse reaction) in the absence of carbonic anhydrase are relatively slow, especially at low temperature, with half times (the time required for the reaction to be 50 per cent complete) between around 10 and 250 s, However, in the presence of carbonic anhydrase, the rate of these reactions is usually increased by a factor of at least 100.

Carbonic anhydrase has been identified in the gas-exchange organs, the blood and the metabolizing tissues of a number of species. In crustaceans, carbonic anhydrase occurs in the gills of aquatic species, in the lung tissue of terrestrial species and, to a lesser degree, in the muscles and heart. In most vertebrates, a relatively high concentration of carbonic anhydrase is found within the red blood cells, with a smaller and more variable amount in the inner endothelial wall of the capillaries of lungs and tissues.

Figure 13.18A shows that, when present in the inner endothelial wall of capillaries, carbonic anhydrase is available to the plasma. However, it is not present on the inner, plasma-facing side of the gills teleost fish, although it is present on the outer, water-facing surface of the gill lamellae, in red blood cells and in muscles. So, unlike the situation in terrestrial vertebrates, there is no carbonic anhydrase activity available to the plasma of teleost fish as it flows through the gas-exchange organ. In those tissues that do contain carbonic anhydrase, the movement of carbon dioxide is enhanced, or facilitated, by the simultaneous diffusion of bicarbonate ions. Hence, this process is known as **facilitated diffusion**. Carbonic anhydrase rapidly catalyses the interconversion of carbon dioxide and bicarbonate ions, both of which can diffuse from high to low concentration.

To maintain electrical balance, the passage of bicarbonate ions (HCO_3^-) between the red blood cells and the plasma is accompanied by the opposite movement of chloride ions (Cl^-) via a membrane-bound anion exchanger (also known as band 3 protein) shown in Figure 13.18A. This exchange is known as the chloride shift. Figure 13.18A also shows the linkage between the exchange of carbon dioxide and oxygen (Bohr and Haldane effects) by the proton binding and release reactions of haemoglobin and carbon dioxide.

The above description demonstrates that the transfer of carbon dioxide between metabolizing tissue and the red blood cells, and between the red blood cells and water at the gills or air in the lungs of vertebrates, is more complicated than simple diffusion. The limitation of the chemical reactions, such as the chloride shift, may be more important than the simple diffusivity of the gas through the different tissues

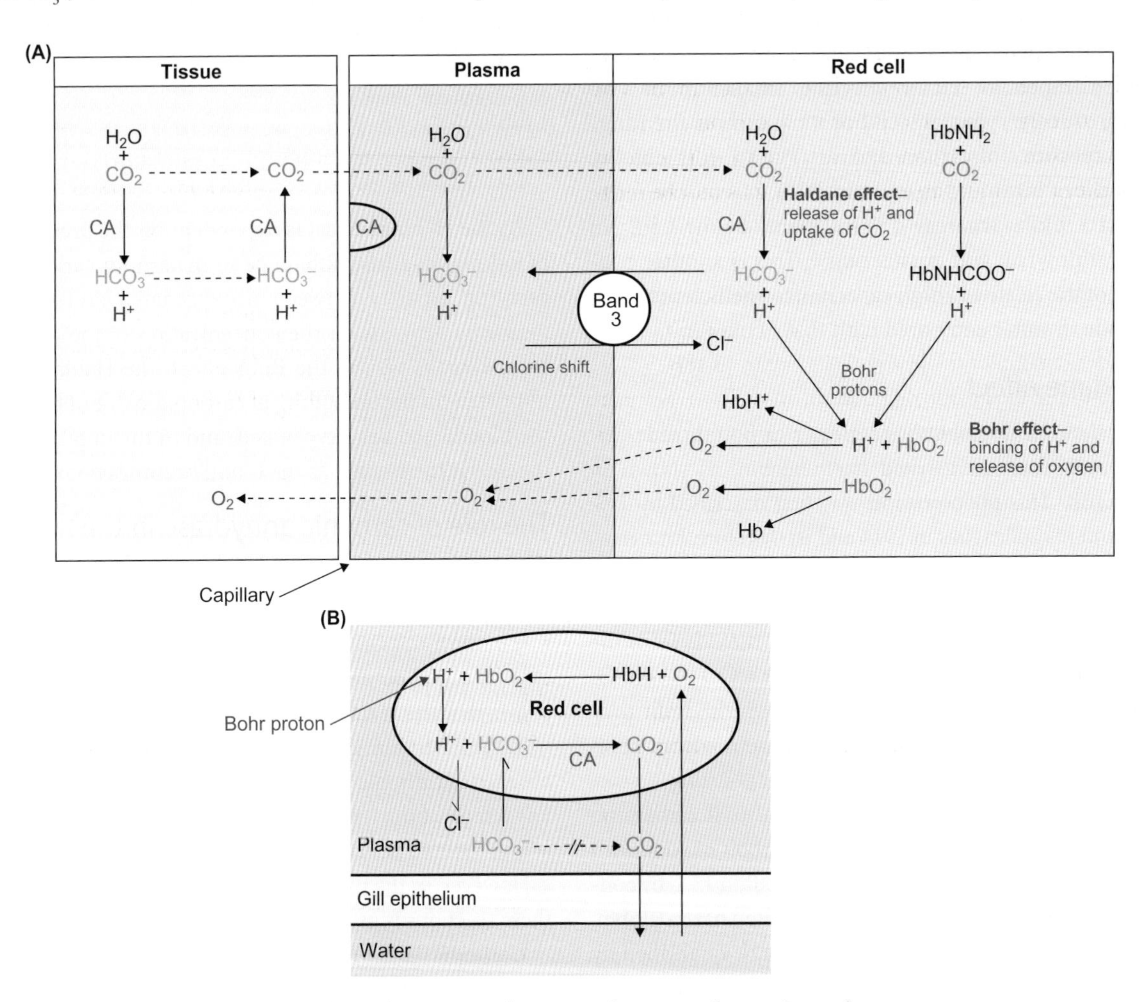

Figure 13.18 Reactions of carbon dioxide and oxygen at the gas exchange surfaces of vertebrates

(A) Reactions at the tissues: release of O_2 and uptake of CO_2. Carbonic anhydrase (CA) is either unbound in the tissue or red cells or membrane-bound, as in the endothelium of the capillaries, thus making the enzyme available to the plasma. At the surfaces of the external gas exchangers, the reactions occur in the opposite direction (uptake of O_2 and release of CO_2), except that in teleost fish, where there is no carbonic anhydrase available to the plasma (see B). The exchange of bicarbonate (HCO_3^-) for chloride (Cl^-) across the membrane of the red cells is effected via the anion exchange (band 3) protein. Diffusion is depicted by dashed arrows and chemical reactions by straight arrows. The linkage between the exchange of oxygen and carbon dioxide (Bohr and Haldane effects) is illustrated by the proton (H^+) binding and release reactions of haemoglobin (Hb) and carbon dioxide. (B) Gas exchange across the gills of fresh water teleost fish. Because of the absence of carbonic anhydrase in the plasma, the dehydration of bicarbonate is negligible in the plasma as it flows through the gills. This means that the bicarbonate diffuses into the red blood cells.

Adapted from Swenson ER (1990). Kinetics of oxygen and carbon dioxide exchange. In: Boutlier RG (ed.) Vertebrate Gas Exchange. Advances in Comparative and Environmental Physiology Vol 6. Pp 163-210. Springer-Verlag, Berlin. Randall DJ and Brauner C (1998). Interactions between ion and gas transfer in freshwater teleost fish. Comparative Biochemistry and Physiology 119A: 3-8.

and fluids. So, despite the fact that carbon dioxide diffuses approximately 20 times faster than oxygen through water and tissues, its equilibration between the alveoli and capillaries of the lungs of mammals is slower, in comparison to that of oxygen, than the relative rates of diffusion through tissues suggest. Thus, in resting humans, there is equilibration between blood and alveolar CO_2, which becomes more limited during moderate exercise, and more severely so at higher levels of exercise.

In smaller mammals, the time it takes for blood to pass through the capillaries of the lung, which is known as the **capillary transit time**[40], is much shorter than in larger mammals (0.6 s in resting mouse, 4.6 s in resting cow). Such short transit times through the lungs of smaller mammals may make it more likely for the equilibration of CO_2 to be limited. This may be compensated for by a higher concentration of carbonic anhydrase in the red blood cells of smaller animals by facilitating the conversion of bicarbonate to carbon dioxide.

The significance of the absence of carbonic anhydrase in the plasma of teleost fish as it passes through the gills may be related to the fact that the haemoglobins of some fish demonstrate the Root effect, which reduces the oxygen-carrying capacity of the blood as pH falls. The lack of carbonic anhydrase may minimize changes in pH in the red cells as they pass through the gills. The mechanism for this is shown in Figure 13.18B.

Oxygenation of the haemoglobin in the red cells as they pass through the gills of fish results in the release of Bohr protons (as we discuss above). These protons are involved in the dehydration of bicarbonate, which produces carbon dioxide. This carbon dioxide diffuses from the red cells, through the plasma and across the gill epithelium to the water, as illustrated in Figure 13.18B. As the dehydration of bicarbonate in the plasma is very slow in the absence of carbonic anhydrase, bicarbonate enters the red cell from the plasma and is available to react with the Bohr protons, a process also shown in Figure 13.18B. Carbonic anhydrase is present within the red cells, so the formation of CO_2 from Bohr protons and bicarbonate is fast and the CO_2 eventually diffuses out into the water. The exchanges of oxygen and carbon dioxide are linked such that the Bohr protons produced by oxygenation of the haemoglobin are involved in the production of carbon dioxide from bicarbonate. Consequently, changes in pH in the red cells as they pass through the gills are minimized.

During the acidic stress caused by exhaustive exercise, the release of the hormone adrenaline (epinephrine) enables some species of fish to maintain the pH inside their red blood cells above that of the surrounding plasma so that the Root effect does not come into effect[41]. This works in favour of oxygen loading of the haemoglobin at the gills. However, such a mechanism would work against the unloading of oxygen in muscles. Recent studies have found that the carbonic anhydrase in the capillaries of muscles effectively bypasses the adrenaline-stimulated process so that pH of the red cells is reduced. This reduction in pH induces the Root effect, thus enhancing the unloading of oxygen to the tissues. It has been suggested, therefore, that the Root effect may have evolved in teleost fish to enhance the general delivery of oxygen during stress, rather than specifically to provide oxygen to the eyes or swimbladder, as we discussed earlier in this section.

In addition to transporting oxygen and carbon dioxide, it has been demonstrated that the haemoglobin of humans transports nitric oxide (NO) from the lungs to the tissues. We discuss how this is achieved, and its functional significance in controlling the supply of oxygen to tissues in Case Study 13.2 (see Online Resources).

[40] Capillary transit time is discussed in more detail in Section 15.2.3.

[41] The effect of exhaustive exercise on the pH of red blood cells of teleost fish is discussed in Section 15.2.4.

› *Review articles*

Bridges CR, Morris S (1989). Respiratory pigments: interactions between oxygen and carbon dioxide transport. Canadian Journal of Zoology 67: 2971–2985.

Burmester T, Hankeln T (2004). Neuroglobin: a respiratory protein of the nervous system. News in Physiological Sciences 19: 110–113.

Campbell KL, Roberts JEE, Watson LN, Stetefeld J, Sloan AM, Signore AV, Howatt JW, Tame JRH, Rohland N, Shen T-J, Austin JJ, Hofreiter M, Ho C, Weber RE, Cooper A (2010). Substitutions in woolly mammoth haemoglobin confer biochemical properties adaptive for cold tolerance. Nature Genetics 42: 536–540.

Cossins A, Berenbrink M (2008). Myoglobin's new clothes. Nature 454: 416–417.

Ingermann RL (2011). Vertebrate hemoglobins. Comprehensive Physiology, Supplement 30: Handbook of Physiology Comparative Physiology pp 357–408. First published in print 1997.

Mairbäurl H, Weber RR (2012). Oxygen transport by haemoglobin. Comprehensive Physiology 2: 1463–1489.

Pelster B (2001). The generation of hyperbaric oxygen tensions in fish. News in Physiological Sciences 16: 287–291.

Randall DJ, Brauner C (1998). Interactions between ion and gas transfer in freshwater teleost fish. Comparative Biochemistry and Physiology 119A: 3–8.

Riggs AF, Gorr TA (2006). A globin in every cell? Proceedings of the National Academy of Sciences of the United States of America 103: 2469–2470.

Sidell BD, O'Brien KM (2006). When bad things happen to good fish: the loss of haemoglobin and myoglobin expression in Antarctic icefishes. Journal of Experimental Biology 209: 1791–1802.

Snyder GK, Sheafor BA (1999). Red blood cells: centrepiece in the evolution of the vertebrate circulatory system. American Zoologist 39: 189–198.

Terwilliger NB (1998). Functional adaptations of oxygen-transport proteins. Journal of Experimental Biology 201: 1085–1098.

Weber RE (2007). High-altitude adaptations in vertebrate hemoglobins. Respiratory Physiology & Neurobiology 158: 132–142.

Weber RE, Fago A (2004). Functional adaptation and its molecular basis in vertebrate hemoglobins, neuroglobins and cytoglobins. Respiratory Physiology & Neurobiology 144: 141–159.

Weber RE, Campbell KL (2010). Temperature dependence of haemoglobin-oxygen affinity in heterothermic vertebrates: mechanisms and biological significance. Acta Physiologica 202: 549–562.

Wenger RH, Kurtz A (2011). Erythropoietin. Comprehensive Physiology 1: 1759–1794.

Wittenberg JB, Wittenberg BA (2003). Myogobin function reassessed. Journal of Experimental Biology 206: 2011–2020.

Wood SC, Lenfant C (1979). Oxygen transport and oxygen delivery. In: Wood SC, Lenfant C (eds.) Evolution of Respiratory Processes, a Comparative Approach. pp 193–223. New York: Marcel Dekker, Inc.

13.2 Transport and storage of metabolic substrates

As well as transporting the respiratory gases to and from the metabolizing cells, the blood also transports the metabolic substrates from the gastrointestinal tract or from body stores to the cells, as illustrated in Figure 13.19.

The two most important metabolic substrates for generating ATP are monosaccharides (glucose, fructose and galactose) and fatty acids[42]. Amino acids may also be a source of energy during exercise but are mainly required for the synthesis of proteins, which include enzymes, transporter proteins and muscles. The transport of these substrates from the gastrointestinal tract or body stores to the metabolizing cells is not always simple, but may involve transporter proteins in cell membranes and, in the case of lipids, transport proteins in the blood. Water-soluble molecules, such as monosaccharides and amino acids, may be absorbed from the gut to a limited extent by passive diffusion, but most are absorbed by carrier-mediated transport systems located in the membranes of epithelial cells. This is the principal means of uptake at low concentrations in the lumen of the gut, at least in vertebrates. We now discuss how metabolic substrates are transported in the blood from the intestine and into the metabolizing cells, looking firstly at carbohydrates and proteins and then at lipids.

[42] Metabolic substrates are discussed in Section 2.3.

13.2.1 Transport of carbohydrates and proteins

The sodium-independent glucose transporter, GLUT2, is responsible for the movement of glucose, fructose and galactose from the cells of the intestine into the interstitial fluid, as discussed in Section 2.3.3. These monosaccharides then enter the blood, presumably by similar mechanisms, and other glucose transporters are involved in the movement of glucose from the blood into the metabolizing cells. GLUT1 is widely distributed in the tissues of the body,

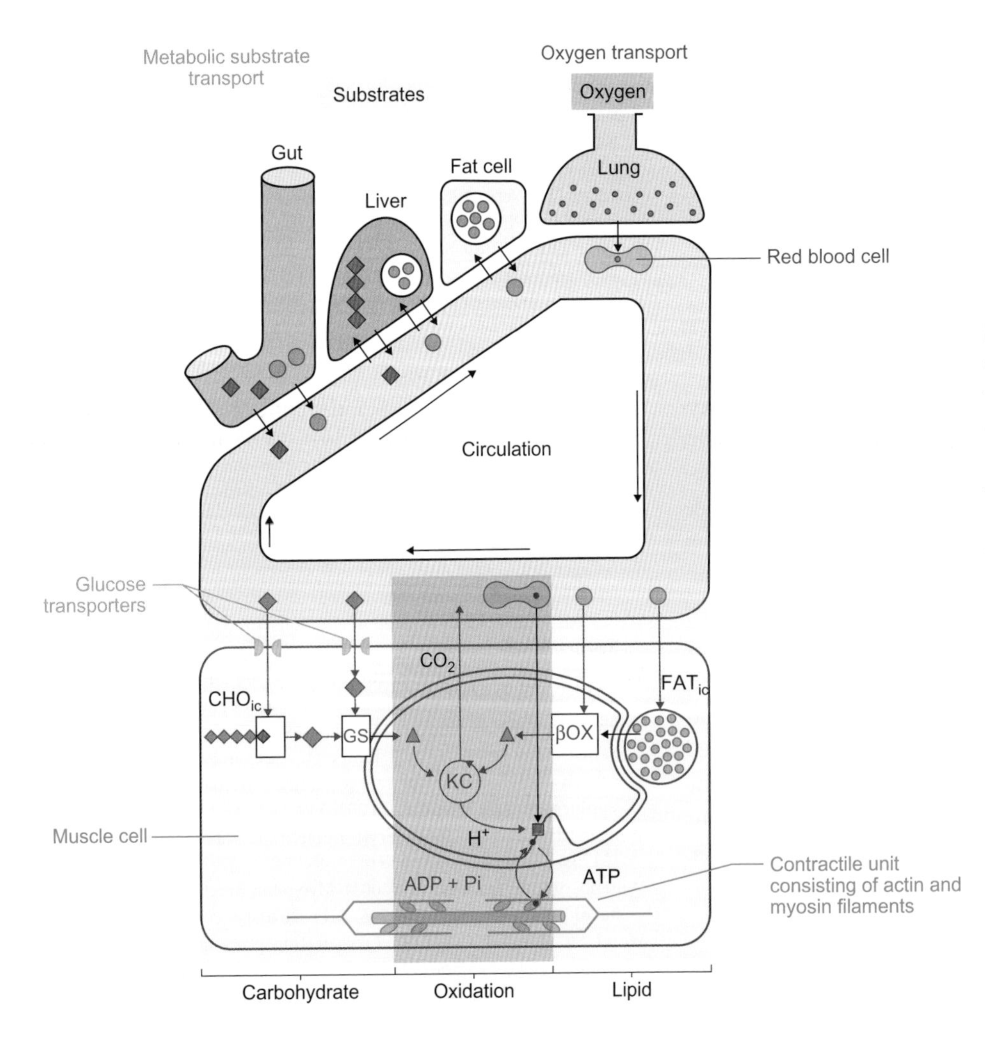

Figure 13.19 Role of the circulatory system in transporting the respiratory gases and metabolic substrates around the body

The diagram is based on the situation in mammals. As well as transporting the respiratory gases, oxygen and carbon dioxide, to and from the metabolizing cells, the circulatory system also transports carbohydrates (◆, CHO) and fatty acids (●, FAT), both directly from the gut and from stores in the liver and fat cells. The transported carbohydrates and fatty acids go either directly into the glycolysis (GS) or β-oxidation (βOX) pathways, respectively, and then to the Krebs cycle (KC) or to replenish the intracellular stores (CHOic, and FATic). Adapted from Hoppeler H and Weibel ER (1998). Limits for oxygen and substrate transport in mammals. Journal of Experimental Biology 201: 1051-1064.

13

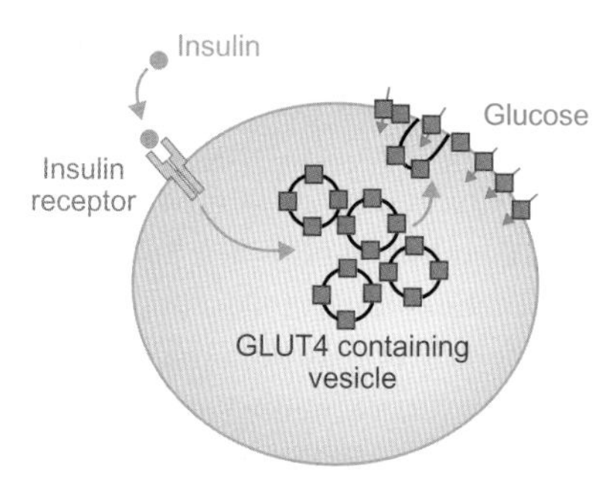

Figure 13.20 The transport of glucose from the intestine and into muscle cells

In muscle and fat cells (adipocytes), glucose uptake is stimulated by insulin and involves the translocation of vesicles containing GLUT4 from an intracellular site to the plasma membrane, with which they fuse.

Reproduced from Uldry M and Thorens B (2004). The SLC2 family of facilitated hexose and polyol transporters. Pflügers Archiv: European Journal of Physiology 447: 480-489.

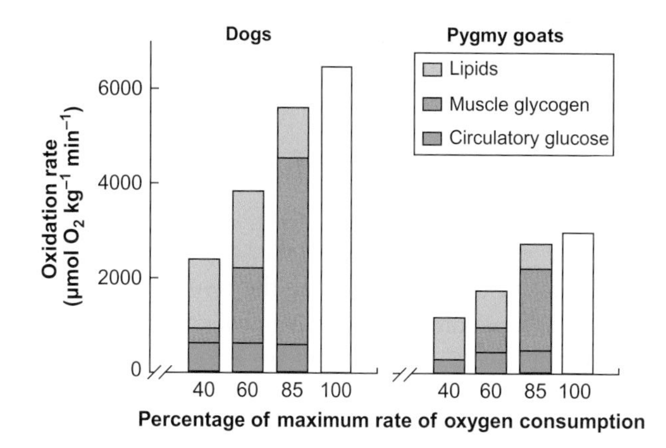

Figure 13.21 Substrate oxidation rates at different intensities of exercise in two species of mammals with different aerobic capacities

The large difference in the maximum rates of oxygen consumption of these two species is mainly the result of the larger rate at which glycogen is supplied from stores within the locomotor muscles themselves. Open bars represent maximum rate of oxygen consumption.

Adapted from Weber J-M et al (1996). Design of the oxygen and substrate pathways III Partitioning energy provision from carbohydrates. Journal of Experimental Biology 199: 1659-1666.

GLUT3 is preferentially found in the brain and GLUT4 is located in fat cells (adipocytes) and skeletal muscles, as shown in Figure 13.20. Glucose uptake in skeletal muscle and fat is stimulated by insulin[43] and involves the translocation of vesicles containing GLUT4 from an intracellular site to the plasma membrane, and subsequent fusion with that membrane.

Despite these complex transport systems, the circulation is responsible for supplying a relatively small proportion of the total carbohydrate required for oxidative metabolism in most species of mammals, with the majority of it coming from glycogen stored in the active muscles, as shown in Figure 13.21. Only a relatively small proportion of the total carbohydrate oxidized during maximum levels of exercise is supplied by the circulation in mammals: about 13 per cent and 23 per cent of the carbohydrates oxidized at exercise levels approaching the mass-specific maximum rate of oxygen consumption in dogs and in relatively sedentary pygmy goats (*Capra hircus hircus*), respectively. The muscle stores are replaced when the animals are not exercising. Muscle fatigue ensues when the glycogen stores become depleted.

Figure 13.21 also illustrates that the maximum rate of glucose transport by the circulation is similar in active dogs to that in pygmy goats despite the 2.2-fold difference in their mass-specific maximum rate of oxygen consumption. In other words, as far as metabolic substrate is concerned, the difference in the rate of aerobic production of ATP in these two species is largely the result of the greater capacity to supply carbohydrate from intramuscular stores of glycogen and not the result of the greater transport of glucose by the circulatory system.

Nectar-feeding bats and hummingbirds are notable exceptions to this general scheme. Anna's and rufous hummingbirds (*Calypte anna* and *Selasphorus rufus*, respectively) use recently ingested sugars (which are transported by their circulatory systems) to support about 95 per cent of their metabolism during hovering. For nectivorous bats, such as Pallas's long tongued bats (*Glossophaga soricina*), the value is around 80 per cent[44]. These data indicate that these animals do not rely on endogenous carbohydrates to anything like the same extent as most mammals.

The most important carbohydrate for relatively short flights in insects is trehalose (a disaccharide), which is present in the haemolymph at much higher concentrations (around 0.06 mmol L^{-1}) than the concentration of glucose in the blood of mammals (0.005 mmol L^{-1}). It can, therefore, be delivered to the flight muscle at a very high rate. Trehalose is synthesized in the fat body from monosaccharides, such as glucose and fructose, or is produced from glycogen.

The transport of amino acids into the interstitial fluid in mammals is discussed in Section 2.3.3 (Figure 2.16). Both monosaccharides and amino acids are soluble in water and are transported in the blood without modification, but this is not the case for lipids.

13.2.2 Transport of lipids

The transport of lipids is complicated by the fact that they are insoluble in water (they are **hydrophobic**). In order to cope

[43] Insulin is discussed in Sections 19.1.4 and 19.3.2.

[44] High capacity absorption of glucose by nectar-feeding hummingbirds and bats is discussed in Section 2.3.3.

13

with this problem, a number of mechanisms have evolved along the transport chain from the intestine or fat storage sites (adipocytes, liver) to the metabolizing cells.

Transport of lipids in mammals

In mammals, ingested lipids are incorporated into relatively large (100–500 nm) lipoproteins called **chylomicrons** in the intestinal cells. Instead of entering the bloodstream immediately, chylomicrons enter the lymphatic system[45] before entering the blood[46]. They bypass the liver before entering the main circulation. In birds, however, the intestinal lymphatic system is poorly developed and smaller lipoproteins called portomicrons, with a mean diameter about 150 nm, are formed which are able to enter the hepatic portal system via the capillaries of the villi.

The **triacylglycerol** in the chylomicrons is hydrolysed in the blood by the enzyme lipoprotein lipase, which is found on the outer surface of the endothelial cells of capillaries in muscles and adipose tissue and on the outer surface of the hepatocytes. The fatty acids released by this hydrolysis bind to the protein albumin in most species of vertebrates for transport in the blood. In some species of fish and birds, lipoproteins may also be involved in this process. The remainder of the chylomicrons, which now consist mainly of **cholesterol** and phospholipids, are known as chylomicron remnants. These are taken up by hepatocytes where the cholesterol is partly converted into bile acids; these bile acids are excreted in the bile together with some of the unaltered cholesterol.

[45] The lymphatic system is discussed in Section 14.1.5.

[46] Absorption of lipids by the gut is discussed in Section 2.3.3.

The fatty acids in the blood may be used immediately as a source of energy for contracting muscles or they may be stored predominantly in fat cells or **adipocytes**, or in the liver. In adipocytes, fatty acids are transformed into triacylglycerols, which are stored as anhydrous droplets and may occupy more than 95 per cent of the volume of the adipocyte. Triacylglycerols synthesized in the liver may be combined with cholesterol and phospholipids to form a number of different types of lipoproteins which are subsequently released into the circulation for uptake by tissues such as muscle.

There is some debate as to how fatty acids cross the cell membrane, enter the cytosol (cytoplasm) of the cell and are then transferred to the mitochondria. Until the early 1980s it was thought that the uptake of long-chain fatty acids into cells was by passive diffusion. However, evidence now suggests that there are specific **fatty acid transport proteins** both in the cell membrane and in the cytosol of cells. It is possible, therefore, that there is both passive and protein-mediated transport of fatty acids into cells and into their organelles, such as the mitochondria.

Figure 13.22 depicts how the fatty acids may cross the plasma membrane. Once dissociated from the albumin in the plasma, the fatty acids may cross the plasma membrane by:

- passive diffusion and/or membrane bound fatty acid binding proteins ($FABP_{pm}$) and fatty acid translocase (FAT);
- fatty acid transport protein (FATP), which is also located in the plasma membrane.

Whichever is the case, once in the cytosol, the fatty acids are converted to acyl-CoA esters, either directly or via a cytosolic fatty acid binding protein called heart-type fatty acid

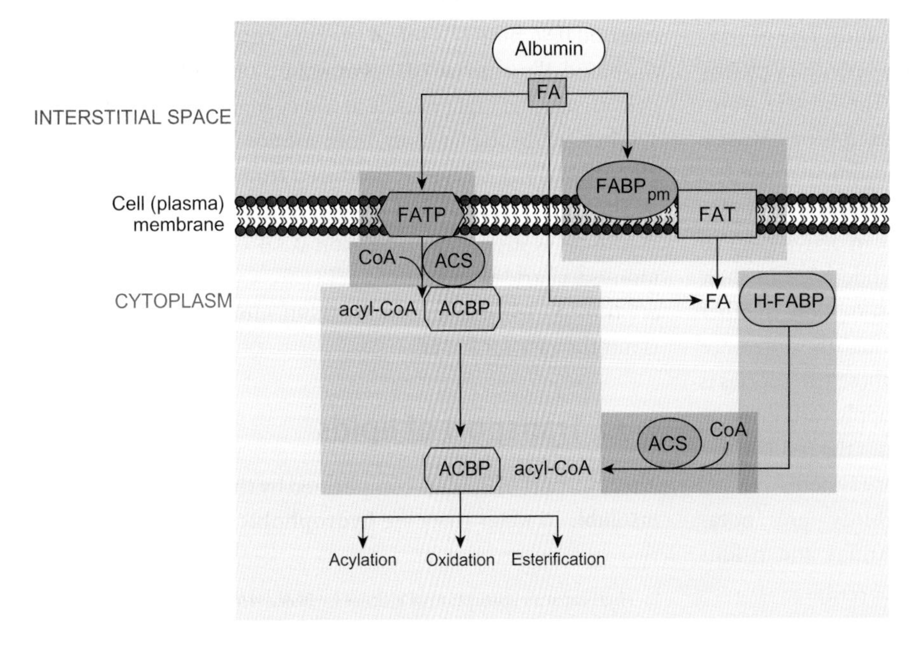

Figure 13.22 Transport of long-chain fatty acids (FA) into muscle cells

Long-chain fatty acids are transported in the plasma bound to albumin and, following their dissociation from the albumin, their movement across a cell membrane into the cytoplasm is either: (i) by passive diffusion, or by membrane-bound fatty acid binding proteins ($FABP_{pm}$) and fatty acid translocase (FAT), or by a combination of the two (red-shaded box); (ii) by fatty acid transport protein (FATP), which is also located in the plasma membrane (blue-shaded box). Once inside the cytoplasm, the fatty acids are converted to acyl-CoA esters by acyl-CoA synthetase (ACS, brown-shaded boxes), either via binding to a binding protein called heart-type fatty acid binding protein (H-FABP, green-shaded box), or directly. Finally, they bind to acyl-CoA binding protein (ACBP, orange-shaded box).

Adapted from Glatz JFC et al (2003). Cytoplasmic fatty acid-binding protein facilitates fatty acid utilization by skeletal muscle.Acta Physiologica Scandinavica 178: 367-371.

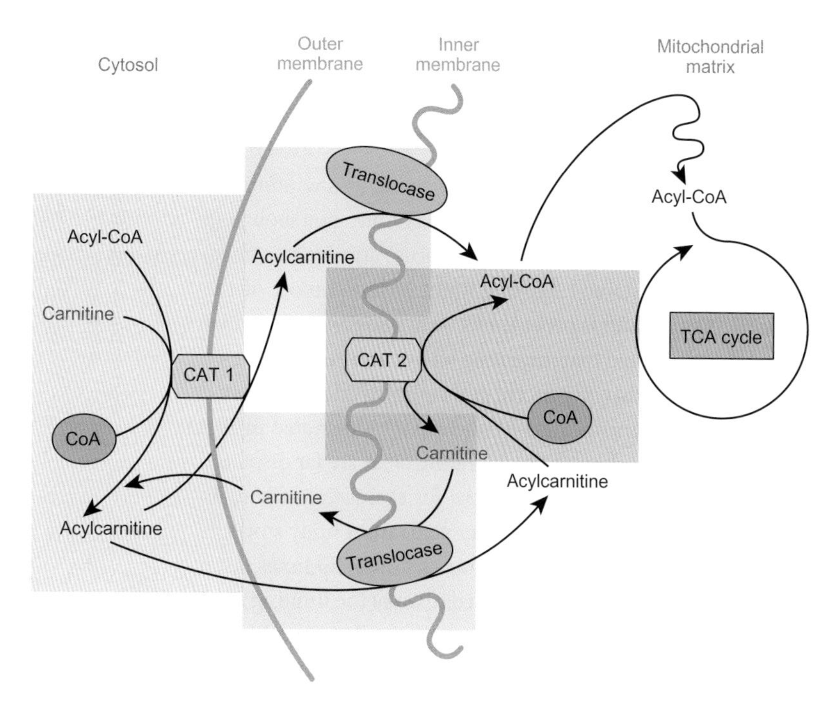

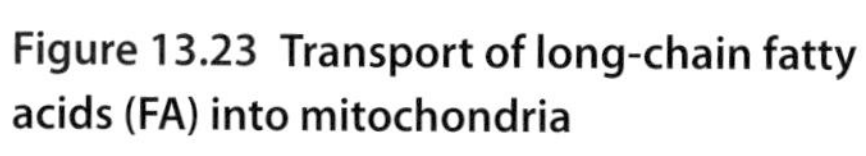

Figure 13.23 Transport of long-chain fatty acids (FA) into mitochondria

Acyl-CoA esters are unable to cross the inner membrane of the mitochondria so they are firstly converted to their acylcarnitine esters by carnitine acyl transferase 1 (CAT 1, red-shaded box), which is present on the outer membrane. Acylcarnitine crosses the inner membrane via the translocase system (green-shaded boxes) and, once inside the mitochondrion, acyl-CoA and free carnitine are re-formed by CAT 2 (brown-shaded box), which is located on the inner membrane. TCA cycle, tricarboxylic acid cycle.

Adapted from Kiens B (2006). Skeletal muscle lipid metabolism in exercise and insulin resistance. Physiological Reviews 86: 205-243.

binding protein (H-FABP)[47]. The acyl-CoA esters then bind to acyl-CoA binding protein (ACBP) and are translocated to mitochondria within the cell.

Recent evidence suggests that the membrane-bound transport proteins play the major role in fatty acid transport across the plasma membrane in mammals. Fatty acid binding protein within the cytoplasm acts as a 'sink' for fatty acids, thereby increasing their solubility in the cytosol, their rate of removal from the inner surface of the plasma membrane, and their rate of intracellular diffusion.

There is strong evidence that fatty acid uptake in the muscles of rainbow trout (*Oncorhynchus mykiss*) is via a carrier-mediated process, but the exact nature of the proteins involved is not yet known.

Cytosolic fatty acid binding protein is found in the flight muscles of desert locusts (*Schistocerca gregaria*) and at particularly high concentrations in the flight muscles of migrating birds, both of which rely almost exclusively on fatty acids as metabolic substrates when flying.

There is still one more step to go before the fatty acids can enter the mitochondria for β-oxidation, as acyl-CoA esters cannot pass the inner membrane of the mitochondria directly. The processes involved are shown in Figure 13.23.

- Firstly, the acyl-CoA esters are converted to their acylcarnitine[48] esters by the enzyme carnitine acyl transferase (CAT-1)[49], which is present on the outer mitochondrial membrane.
- The acylcarnitine then crosses the inner mitochondrial membrane by the acylcarnitine/carnitine translocase system.
- Once inside the mitochondria, carnitine acyl transferase-2 (CAT-2), which is located on the inner membrane of mitochondria, regenerates acyl-CoA and free carnitine.

This whole transport process could be the main rate-limiting step in the use of fatty acids by active muscles, with either carnitine or carnitine acyl transferase-1 being the important factors.

Figure 13.21A shows that fatty acids provide approximately 80 per cent of the energy for locomotion in mammals at relatively low levels of activity, that is, at less than 40 per cent of maximum oxygen consumption ($\dot{M}O_{2max}$). However, as a result of the transport limitations for fatty acids, they provide a decreasing amount of the energy required above 40 per cent $\dot{M}O_{2max}$. Indeed, at close to maximum rate of oxygen consumption, fatty acids supply only 10 to 20 per cent of the energy being used. Carbohydrates, therefore, provide an increasing amount of energy at increasing levels of exercise in mammals.

Even at low levels of exercise, 75–85 per cent of the fatty acids in mammals are obtained from the fat droplets within the muscle cells themselves and not via the circulation, which is unable to provide metabolic substrates at the rate

[47] Nine different cytosolic FABPs have been identified and different types are found in different tissues. The H-type is found in heart and skeletal muscle.

[48] Carnitine is synthesized in the liver and is normally particularly abundant in tissues which are able to oxidize long-chain fatty acids.

[49] Carnitine acyl transferases are also known as carnitine palmitoyl transferases.

they are being used. The circulation does, however, replenish the intracellular stores at the end of the period of exercise.

Transport of lipids in flying birds and insects

The minimum metabolic rate of birds flying in a wind tunnel is approximately twice as great as the maximum metabolic rate ($\dot{M}O_2$) of similar sized mammals and birds running on a treadmill[50]. During sustained flapping flight of birds, most of this huge energy demand is provided by fatty acids from stored fat. It has been estimated that migrating birds exercise at 70–90 per cent $\dot{M}O_{2max}$ and that 90 per cent of their energy demand comes from fatty acids. The inevitable conclusion is illustrated in Figure 13.24A, which shows how migrating birds are far displaced from the typical fuel selection curve for exercising mammals. Migrating birds may transport fatty acids from body stores to the mitochondria of the working muscles up to 20 times faster than running mammals. In mammals, two of the limiting steps to the transport of fatty acids occur in the circulation and across the plasma membrane of the muscle cell (sarcolemma), so we might expect migratory birds to show differences from mammals in these locations. This is indeed the case.

Mammals transport non-esterified fatty acids in the blood by binding them to albumin. However, it would not be feasible for migrating birds to increase the concentration of plasma albumin drastically as this would lead to a large increase in plasma colloid osmotic pressure[51]. So, instead of transporting lipids in the blood solely by binding them to albumin, migratory birds may also use lipoproteins.

For smaller migrants such as a number of species of passerines, it has been suggested that fatty acids are taken up by the liver, where they are transformed to triglycerides and very low density lipoproteins (VLDLs), as shown in Figure 13.24B. The VLDLs are then released back into the blood and transported to the muscles where they are converted back to fatty acids by lipoprotein lipase present in the endothelium of the capillaries. The fatty acids are then taken up by the muscles.

This mechanism enables the concentration of plasma protein in migrating passerine birds to be maintained at a physiological level. In addition, since the proportion of triglycerides in VLDL is 60–70 per cent (compared to only 3 per cent in albumin) there is a dramatic increase in the rate of fatty acid transport by the circulatory system of these birds. However, this suggested mechanism has not been completely supported by other studies and further investigations are required in order to determine what role, if any, VLDLs play in the transport of fatty acids in the blood of migrating birds.

The movement of fatty acids across the cell membrane of muscle fibres in birds appears to be similar to that in mammals. Fatty acid binding protein bound to the plasma membrane has recently been identified in the flight muscles of migratory white-throated sparrows (*Zonotrichia albicollis*), where it has been found to double in abundance during migration. In addition, the concentration of fatty acid binding protein in the cytoplasm of muscle cells in a number of species of small migratory birds such as western sandpipers (*Calidris mauri*) is several times greater than that in any mammalian muscles and almost doubles during migration. Although this protein is not directly involved in the transfer of fatty acids across cell membranes, as shown back in

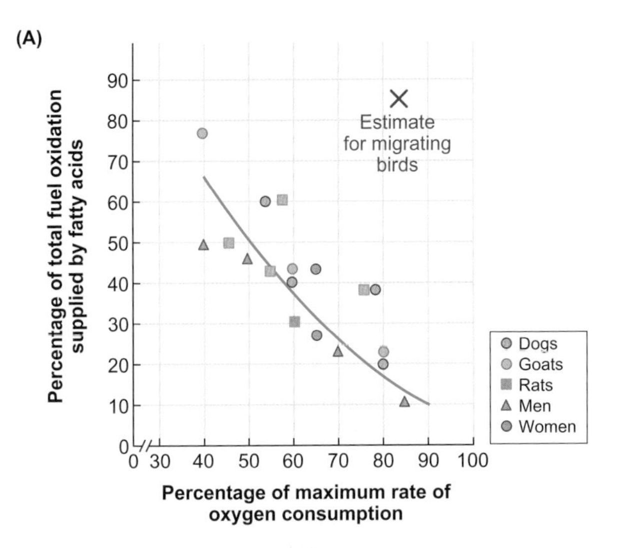

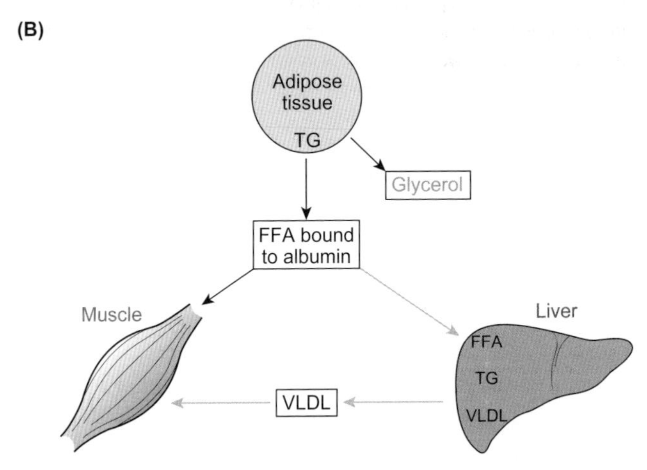

Figure 13.24 Supply of metabolic substrate to the muscles of mammals and birds during exercise

(A) Energy requirements during increasing exercise intensity (indicated by percentage of maximum rate of oxygen consumption) for a number of running mammals. **X** indicates an estimated value for flying birds.
(B) Simplified model of fatty acid transport from adipose tissues to the muscles of birds. The double-lined arrows show the pathway that may be accelerated during flight. FFA, free fatty acids, TG, triglycerides, VLDL, very low-density lipoproteins.

Adapted from: (A) McClelland GB (2004). Fat to the fire: the regulation of lipid oxidation with exercise and environmental stress. Comparative Biochemistry and Physiology B 139: 443-460. (B) Jenni-Eiernann S and Jenni L (1992). High plasma triglyceride levels in small birds during migratory flight: a new pathway for fuel supply during endurance locomotion at very high mass-specific metabolic rates? Physiological Zoology 65: 112-123.

[50] Energy cost of exercise in birds is discussed in Sections 15.2.3 and 18.2.3.

[51] Colloid osmotic pressure is discussed in Section 14.1.5.

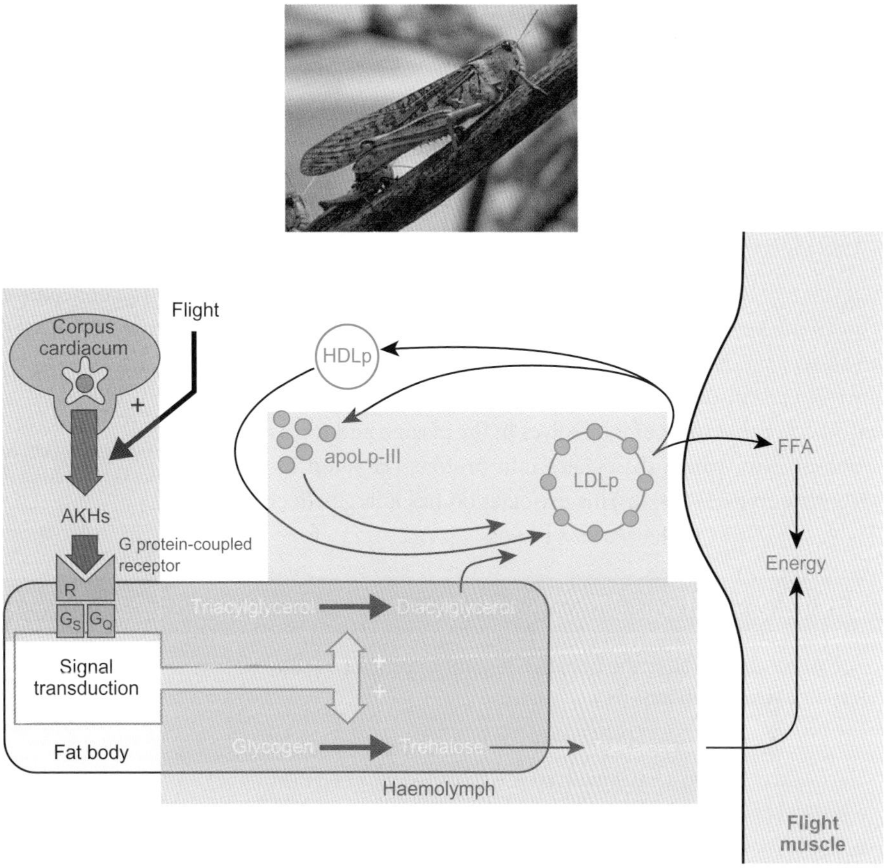

Figure 13.25 Control of metabolic substrate release from the fat body of migratory locusts (*Locusta migratoria*)

During flight, adipokinetic hormones (AKHs) are released from the corpus cardiacum and bind to G protein-coupled receptors in the fat body (red-shaded box). This leads to the release of the carbohydrate, trehalose, and the lipid diacylglycerol (blue-shaded box). Trehalose is transported freely in the haemolymph, whereas diacylglycerol is transported by a lipophorin. Diacylglycerol combines with high-density lipophorin (HDLp) to form low-density lipophorin (LDLp), which is stabilized by apolipoprotein III (apoLp-III) (green-shaded box). The LDLp transports diacylglycerol to the working muscle where apoLp-III dissociates from the particle thus regenerating HDLp, which is re-used. Free fatty acids derived from the diacylglycerol are oxidized for the production of ATP.

Adapted from Van der Horst DJ (2003). Insect adipokinetic hormones: release and integration of flight energy metabolism. Comparative Biochemistry and Physiology B 136: 217-226. Photo © Jonathan Hornung.

Figure 13.22, it may serve as an intracellular sink that is important for maintaining a high flux rate of fatty acids across the cell membrane by transporting fatty acids away from the cell membrane and toward the mitochondria. In addition, carnitine acyl transferase-1 doubles in activity during the migratory season in white-throated sparrows.

Although the circulatory system of insects is not involved in the transport of the respiratory gases, it does transport hormones and metabolic substrates, among other things. Metabolism of flight muscles of insects can be based on a number of different fuels which vary between different species and duration of a particular flight. Several species that fly for long distances mobilize both carbohydrate and lipid reserves as fuels from the fat body, which combines many of the functions and properties of the liver and adipose tissue in vertebrates.

One species that has been well studied and is an accepted model for short- and long-distance flight in insects is the migratory locust (*Locusta migratoria*). At the onset of flight, **adipokinetic hormones** are released from a neuroendocrine gland, the corpus cardiacum[52]. These hormones bind to G protein-coupled receptors[53] in the fat body, which leads to the release of lipids and the carbohydrate trehalose, via a number of co-ordinated processes, which are shown in Figure 13.25. These substrates are transported by the haemolymph to the flight muscles.

During prolonged flight, fuel utilization in insects is gradually shifted from carbohydrates to lipids. Triacylglycerol is converted to diacylglycerol, which is the principal fuel for long-distance flight, and it is transported in the haemolymph by lipophorin. Lipophorin can alternate between a relatively lipid-poor form, high-density lipophorin, and a relatively lipid-rich form, low-density lipophorin, as shown in Figure 13.25. At the flight muscles, fatty acids derived from the diacylglycerol are oxidized. The lipophorin is now in its high-density form and is returned to the fat body for another round of lipid uptake and transport.

[52] The endocrine system of insects is discussed in Section 19.5.2.

[53] G-protein receptors are discussed in Section 16.3.4.

› Review articles

Bröer S (2002). Adaptation of plasma membrane amino acid transport mechanisms to physiological demands. Pflügers Archiv: European Journal of Physiology 444: 457–466.

Glatz JFC, Luiken JJFP, Bonen A (2010). Membrane fatty acid transporters as regulators of lipid metabolism: implications for metabolic disease. Physiological Reviews 90: 367–417.

Guglielmo CG (2010). Move that fatty acid: fuel selection and transport in migratory birds and bats. Integrative and Comparative Biology 50: 336–345.

Kiens B (2006). Skeletal muscle lipid metabolism in exercise and insulin resistance. Physiological Reviews 86: 205–243.

Schwenk RW, Holloway GP, Luiken JJFP, Bonen A, Glatz JFC (2010). Fatty acid transport across the cell membrane: regulation by fatty acid transporters. Prostaglandins, Leukotrienes and Essential Fatty Acids 82: 149–154.

Suarez RK, Herrera LG, Welch KC (2011). The sugar oxidation cascade: aerial refuelling in hummingbirds and nectar bats. Journal of Experimental Biology 214: 172–178.

Van der Horst DJ (2003). Insect adipokinetic hormones: release and integration of flight energy metabolism. Comparative Biochemistry and Physiology B 136: 217–226.

13.3 Acid–base balance

One of the products of aerobic metabolism of carbohydrates and fatty acids is carbon dioxide. When discussing the transport of carbon dioxide around the body in Section 13.1.3, we discover that most of it dissolves in the plasma and forms carbonic acid, which dissociates into protons (H^+) and bicarbonate ions (HCO_3^-). This dissociation has a large effect on the acid–base balance of both the extra- and intracellular fluids of an animal.

Many biochemical processes, such as the catalysis of metabolic reactions by enzymes[54] and the transport of oxygen by blood pigments[55], depend on the concentration of H^+, which we represent as $[H^+]$. More accurately, these processes depend on the concentration of the hydrated form of a proton, the hydronium ion, $[H_3O^+]$, but we refer to $[H^+]$ for simplicity. The concentration of protons in solution is normally represented by the term **pH**. We discuss the basics of pH in Box 13.2.

[54] Some functions of enzymes is discussed in Sections 2.1 and 2.3.3.

[55] We discuss the transport of oxygen by blood pigments in Section 13.1.2.

Box 13.2 The basic principles of pH

The **pH** of a solution is the negative logarithm of the concentration of hydrogen ions $[H^+]$:

$$pH = -\log[H^+]$$

The reason for using a logarithmic scale is that the range of $[H^+]$ in aqueous solutions is huge. At pH = 1, the proton concentration is 0.1 mol L^{-1}, whereas at pH = 13, it is 10^{-13} (or 0.0000000000001) mol L^{-1}. Figure A illustrates that the use of a logarithmic scale means a change of one pH unit equates to a 10-fold change in $[H^+]$.

It is possible to calculate $[H^+]$ from pH and vice versa by using the relationship $pH = -\log[H^+]$. For example, if the pH of normal arterial blood of a mammal at 37°C is 7.41, what is its hydrogen ion concentration, $[H^+]$?

The negative logarithm of $[H^+]$ is 7.41, which means that the logarithm is −7.41. The antilog of this is 3.9×10^{-8}, which means that $[H^+]$ at a pH of 7.41 is 3.9×10^{-8} mol L^{-1}.

Two important points to note are that:

1. It is not correct to calculate the arithmetical mean of a number of values of pH; they should first be converted to the corresponding values for $[H^+]$. The mean of these values can be calculated and reconverted to pH.
2. It is not correct to present percentage changes in pH, as percentages are only appropriate for a linear scale and not for a logarithmic one like the pH scale.

The concentration of hydrogen ions in a solution determines its acidity. Solutions containing high $[H^+]$ have a low pH and are acidic, while those with a low $[H^+]$ have a high pH and are alkaline. An **acid** in water dissociates to produce a proton, which can be accepted by another substance (a base). On the other hand, a **base** is a substance that can accept a proton from another substance:

$$HA \rightleftharpoons H^+ + A^-$$

Acid Proton Base

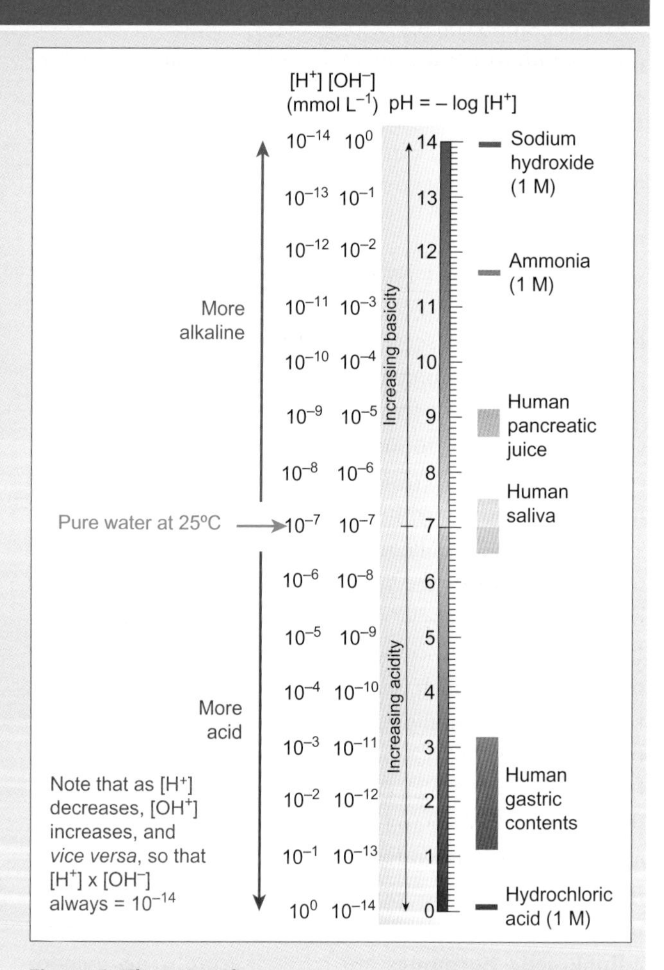

Figure A The pH scale

This figure shows the hydrogen ion concentration $[H^+]$, corresponding hydroxyl ion concentration $[OH^-]$ and pH for various fluids, including some from humans, at 25 °C. Note that pure water at 25 °C has a neutral pH of 7

Modified from: Horton RH et al (1993). Principles of Biochemistry. Prentice Hall, Englewood Cliffs NJ.

This reaction scheme illustrates how an acid, HA, becomes a base, A^-, when it gives up a proton. The pair of species, which differ only in the presence or absence of a proton, are called a **conjugate acid–base pair**:

$$HA \rightleftharpoons H^+ + A^-$$

conjugate acid pair

The ratio between the product of the concentration of A^-, $[A^-]$, and of H^+, $[H^+]$, and the concentration of HA, [HA] is known as the acid dissociation constant, K_a:

$$K_a = \frac{[A^-][H^+]}{[HA]} \qquad \text{Equation A}$$

The acid dissociation constant therefore represents, the degree of dissociation of an acid in water—that is, its strength. The value of K_a is large for strong acids, such as hydrochloric acid, which dissociate completely, and small for weak acids, such as amino acids, which do not dissociate completely.

It is more usual to convert K_a to pK_a, which is $-\log_{10} K_a$. Notice how this expression mirrors that of pH ($-\log_{10} [H^+]$). By taking the logarithm of both sides of the above equation, we obtain:

$$\log K_a = \log[H^+] + \log\frac{[A^-]}{[HA]}$$

This can be rearranged to give:

$$-\log[H^+] = -\log K_a + \log\frac{[A^-]}{[HA]}$$

By substituting pK_a for $-\log K_a$, we can write:

$$pH = pK_a + \log\frac{[A^-]}{[HA]} \qquad \text{Equation B}$$

Equation B is known as the **Henderson–Hasselbalch equation**.

The pK_a is the pH at which an acid is half dissociated or ionized. This means that when the pH of a solution is below the pK_a of a compound in that solution, the undissociated form of the compound predominates. In contrast, when pH is above the pK_a, the dissociated form of the compound predominates.

Different acids have different pK_as, depending on their tendency to dissociate—that is, their strength. The lower the pK_a, the stronger the acid, and *vice versa*.

13.3.1 The effect of temperature on pH

When pure water dissociates to form ions, H^+ and OH^- ions are produced in equal quantities—thus $[H^+] = [OH^-]$. This means that both $[H^+]$ and $[OH^-] = 10^{-7}$ mol L^{-1}, such that the pH of pure water = $-\log(10^{-7}) = 7$, as illustrated in Figure A, Box 13.2. The pH at neutrality is also known as the **pN**. However, the ratio of $[H^+]$ to $[OH^-]$ varies with temperature, so the pN of water is only 7 at a temperature of 25°C, as indicated in Figure 13.25a. Below 25°C, the pN of water is above 7, whereas at temperatures above 25°C it is below 7.

Relative alkalinity

Ionic strength also influences pN. Figure 13.26A shows that the pN of an ionic solution, such as sea water, is different from that of pure water at a given temperature. This means that pH alone is not an indicator of neutrality, acidity or alkalinity (also known as basicity). Instead, one way to assess the acid–base status of an aqueous solution is to determine $[OH^-]/[H^+]$ ratio: a neutral solution has a ratio of 1, an alkaline solution has a ratio of >1 and an acidic solution has a ratio of <1. This is known as the **relative alkalinity**, which is also indicated by value of pH – pN. For a neutral solution, pH – pN = 0, for an alkaline solution pH – pN >0, and for an acid solution pH – pN <0.

For example, water at 25°C has a pH = 7.0. Since the pN = 7.0 under these conditions, it follows that pH – pN = 0 and therefore, the solution is neutral. However, water of pH = 7.0 at 30°C is an alkaline solution because the pN value for 30°C is 6.92 (Figure 13.26A) and therefore, pH – pN = 0.08, i.e. >0. In contrast, if water had a pH = 7.0 at 10°C, then the solution would be considered acidic because the pN value at 10°C is 7.27 and therefore, pH – pN = –0.27, i.e. <0. Generally, if the pH of a solution changes with temperature, but the value pH – pN remains constant as the temperature changes, it means that the solution maintains a constant relative alkalinity or acidity.

Most species of birds and mammals maintain a more or less constant body temperature, so the relationship of pH with temperature is mainly of particular significance in ectotherms. However, it is also important in endotherms which hibernate or live in cold environments where their extremities may cool down. Figure 13.26b shows that the pH of mammalian blood at 37°C is relatively alkaline, at 7.4, compared to that of pure water (pH 6.8) at the same temperature.

Many ectotherms maintain almost the same relative alkalinity in their blood and many of their tissues, irrespective of their body temperature; there is a change in pH of approximately 0.017 pH unit for every °C change in temperature. Figure 13.26B shows many ectothermic animals at 10°C have a blood pH of around 8.0, whereas at 30°C, blood pH is around 7.6.

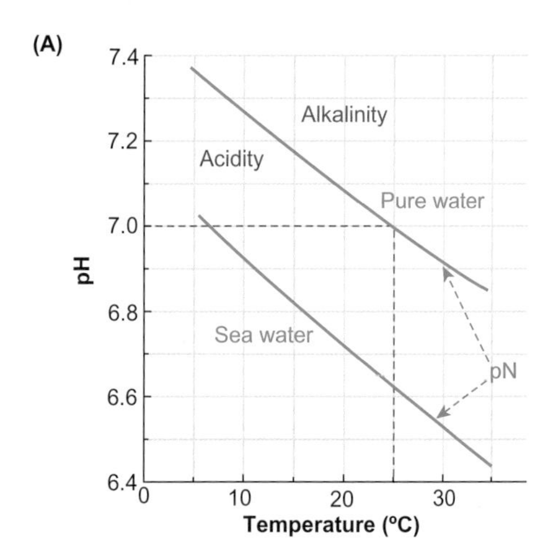

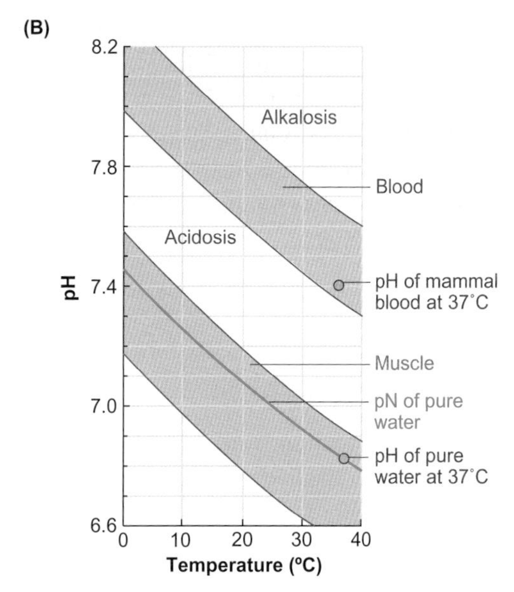

Figure 13.26 Temperature and pH

(A) The relationship between neutral pH (pN) and temperature in pure water and sea water (35‰). Note that the pH of pure water is 7.0 only at 25°C. (B) Range of pH in blood and striated muscle cells (intracellular pH) of ectotherms at different temperatures. pH of the blood of mammals at 37°C is also shown. Note that both of these ranges are parallel to the pH of pure water (pN, dark line), with the range of values for intracellular pH of muscle at 37°C being close to that for pure water (6.8). However, the pH of blood of mammals at 37°C is 7.4.

Adapted from Truchot JP (1987). Comparative Aspects of Extracellular Acid–Base Balance. Springer-Verlag, Berlin. Rahn H et al (1975). Hydrogen ion regulation, temperature, and evolution. American Review of Respiratory Disease 112: 165-172.

Figure 13.26B also illustrates that the pH inside the cells of tissues is usually about 0.7 units[56] lower than that for blood and similar to that of the neutral pH (pN) of water. However, the physiological significance of there being a constant relative alkalinity is not clear. So what is the significance of the change in pH of the blood and tissues of ectotherms at different temperatures? The answer may be that it maintains a constant net charge on protein molecules, which would ensure that enzymes function optimally.

Net protein charge

The net charge (Z) on proteins is the algebraic sum of cationic (n^+) and anionic (n^-) groups per molecule. The extent of ionization of the side chains of most amino acids does not change within the physiological range of pH (6–8) because their p*K* values are outside of this pH range. However, the p*K* value of one important group does fall within this range that is—the imidazole groups of histidine—histidine imidazole. The effect of temperature on the p*K* of imidazole (pK_{Im}) is similar to the effect of temperature on pH in the blood and cells of ectotherms: approximately −0.017 pH units $°C^{-1}$. This implies a constant fractional dissociation (α) of the imidazole groups (the ratio of unprotonated imidazole [Im] to total imidazole [Im] + [HIm^+], also known as α-imidazole), as well as a constant relative alkalinity, which we discuss above.

The fractional dissociation of imidazole groups provides a measure of the changing contribution of histidine imidazole groups to the net protein charge of each and every protein species. It has been proposed, therefore, that acid–base status in ectotherms is regulated to maintain a constant fractional dissociation of imidazole groups as temperature changes. This has been called the **alphastat hypothesis** of acid–base regulation, with there being direct evidence from some species to support this hypothesis.

Nuclear magnetic resonance (NMR) spectroscopy has been used to demonstrate that the pH inside the cells of a tail muscle of a species of eastern newt (*Notophthalmus viridescens*) decreases by 0.3 pH unit as temperature increases by 20°C (−0.015 pH units $°C^{-1}$), whereas the fractional dissociation of imidiazole remains constant. These data are shown in Figure 13.27. However, this does not appear to be the case for all ectothermic vertebrates: the relationship between intracellular pH and temperature can be non-linear in many tissues and there can be large differences in this relationship between different tissues.

In many species of reptiles, such as varanid lizards, the change in pH with temperature of arterial blood and of the intracellular compartments of the heart and locomotor muscles is less than predicted by the alphastat hypothesis. Also, there is no constant fractional dissociation of imidazole in these species, which suggests that enzyme function in these species is not dependent on a constant fractional dissociation of imidazole.

In some species of insects, such as two-striped grasshoppers (*Melanoplus bivittatus*), the pH of the haemolymph is constant between 10 and 25°C but decreases by 0.017 pH units $°C^{-1}$ at temperatures above 25°C. This pattern maintains protein charge at a constant level when the grasshoppers are active in the field at temperatures above 25°C but not when they are quiescent below this temperature.

[56] More recent intracellular pH measurements in mammalian muscle fibres indicate that the difference is closer to 0.3 pH units.

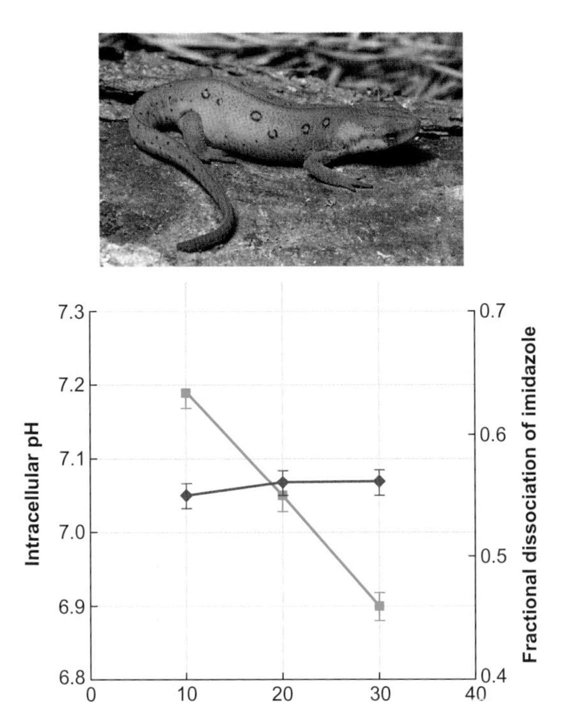

Figure 13.27 Effect of temperature on the fractional dissociation of imidazole

Intracellular pH (■) and the fractional dissociation of imidazole (◆) as determined by nuclear magnetic resonance spectroscopy in white skeletal muscle of eastern newts (*Notophthalmus viridescens*) at different temperatures.

Adapted from Hitzig BM et al (1994). ^{1}H-NMR measurement of fractional dissociation of imidazole in intact animals. American Journal of Physiology 266: R1008-R1015. Image: © Patrick Coin.

The way in which the pH of the blood changes in response to variations in temperature differs between water-breathing and air-breathing ectotherms. In many species of reptiles, such as red-eared sliders/terrapins (*Trachemys scripta*), there is a decrease in the ratio of respiratory minute ventilation volume to the rate of oxygen uptake (the ventilation convection requirement)[57] as body temperature increases (i.e. the animals hypoventilate). As such, the partial pressure of carbon dioxide increases and the pH of the blood of the animal falls as temperature increases, as predicted by the alphastat hypothesis.

In contrast, ventilation in savannah monitor lizards (*Varanus exanthematicus*) matches the increase in rate of oxygen consumption as temperature increases. Consequently, PCO_2 in arterial blood does not increase sufficiently to maintain a constant fractional dissociation of imidazole. The concentration of bicarbonate ions decreases with increasing temperature, with the net result that pH remains almost unchanged, as discussed above. This suggests that pH regulation in these highly active reptiles is linked to oxygen demand rather than to the maintenance of a constant fractional dissociation of imidazole, as predicted by the alphastat hypothesis.

[57] The ventilation convection requirement is discussed in Section 12.3.3.

Whatever the functional significance of the change in pH with temperature may be in ectotherms, ventilation is involved in this process in amphibians and reptiles (with the exception of varanid lizards) to a greater or lesser extent. However, ventilation is not involved in controlling pH in water-breathing animals.

The relatively high capacitance coefficient of carbon dioxide in water compared to that of oxygen leads to the partial pressure of carbon dioxide in the blood of water-breathers being low compared to that in air-breathers[58]. It follows that there is little scope for varying the partial pressure of carbon dioxide in the blood of most aquatic animals by hyper- or hypoventilation. Instead, acid–base regulation in water-breathing animals is achieved by the regulation of the concentration of plasma bicarbonate ions, or more accurately, the difference in the concentration of strong ions[59]. This can involve the active transport of strong ions (mainly, Na^+, Cl^-) across the gills[60]. In both Dungeness crabs (*Cancer magister* or *Metacarcinus magister*) and rainbow trout, a decrease in pH of the haemolymph/blood with an increase in temperature is largely the result of a decrease in the concentration of bicarbonate, with PCO_2 remaining constant.

13.3.2 Buffering of pH and the pH-bicarbonate diagram

An important relationship exists between the concentrations of H^+ and OH^- ions in aqueous solutions at 25°C: $[H^+] \times [OH^-]$ always remains at 10^{-14}, as illustrated in Box 13.2[61]. As a consequence of this relationship, if the concentration of protons, $[H^+]$, increases when an acid is dissolved in water, the concentration of hydroxyl ions, $[OH^-]$, will decrease so that $[H^+][OH^-]$ remains at 10^{-14}. Taking this further, as the strength of an acid increases (such that it dissociates more readily to yield protons), we see a higher concentration of protons and a relatively lower concentration of hydroxyl ions. The opposite occurs when a basic (alkaline) substance is added. However, the pH of the blood, and other body fluids, is maintained within normal ranges by the **buffer systems** that are present in the body. This buffering is essential. As the functioning of many proteins, including enzymes and blood pigments, depend on the pH of their environment, it is important that pH does not change dramatically in the body fluids in response to the addition of acid or base.

The pH of a system that is well buffered changes very little, within a certain range, when a relatively large amount of an

[58] Partial pressure of carbon dioxide in the blood of water- and air-breathing vertebrates is discussed in Section 11.1.2.

[59] Strong ion difference is discussed in Section 13.3.2.

[60] Ion transport across the gills of fish is discussed in Chapter 5.

[61] However, as discussed in Section 13.3.1, this is not the situation at other temperatures: at 10°C the product is 0.293×10^{-14} and at 30°C, the product is 1.47×10^{-14}. More generally $[H^+] \times [OH^-] = 10^{-2pN}$

13

acid or a base is added. A buffer must, therefore, consist of an acid, which donates protons (H^+) to neutralize any base that is added, and a base (A^-) that can bind protons, to neutralize any acid that is added. The ability of a solution to resist changes in pH is called its **buffer capacity**. As weak acids do not dissociate completely, solutions of weak acids contain a large reservoir of HA which can neutralize any added base. Consequently, these solutions act as buffers.

Buffering of pH

The buffer capacity of a solution can be determined by adding small amounts of an acid or base to a solution and measuring the change in pH (a process called titration). The result of titrating acetic acid with NaOH is shown in Figure 13.28A. Over a certain range of added NaOH, there is a relatively small change in pH. This is the range over which acetic acid has its greatest buffer capacity. If pH is plotted against the number of equivalents of strong base added during the titration, the pH at which the acid has been titrated with 0.5 of an equivalent of base can be identified, as shown in Figure 13.28B. This pH represents the point at which the concentration of undissociated acetic acid, $[CH_3COOH]$, equals the concentration of the acetate ion, $[CH_3COO^-]$. This is the pK_a for acetic acid, which has a value of 4.8 at 25°C. In other words, an acid's pK_a value is the pH at which the acid is 50 per cent dissociated, as discussed in Box 13.2.

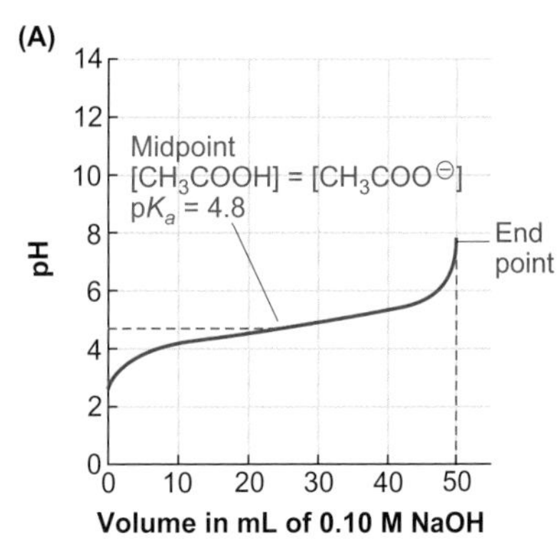

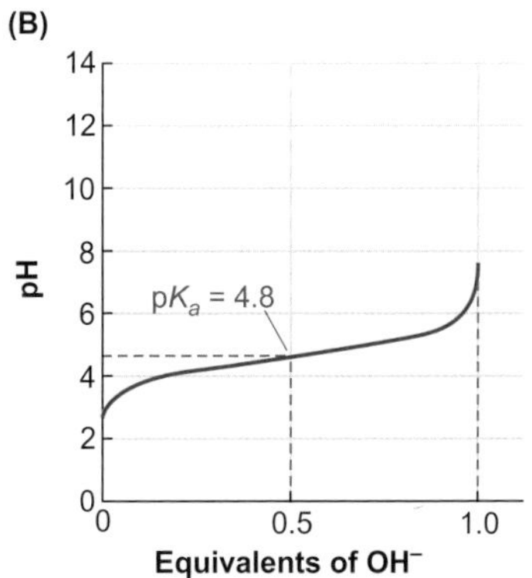

Figure 13.28 Buffering of pH in solutions of weak acids

Titration of 50 mL of 0.1 M acetic acid (CH_3COOH) with 0.1 M aqueous sodium hydroxide (NaOH) at 25°C. (A) There is a point of minimum slope, known as the midpoint of titration, when 25 mL of NaOH have been added. At this point, the concentration of CH_3COOH $[CH_3COOH]$ is equal to the concentration of CH_3COO^- $[CH_3COO^-]$ and pH = pKa. At the end point, all of the NaOH has been added and all molecules of CH_3COOH have been converted to acetate (CH_3COO^-). (B) Here, the same data are expressed in terms of the number of equivalents of strong base, OH^-, added rather than a volume of a solution of specific concentration.

Adapted from Horton RH et al (1993). Principles of Biochemistry. Prentice Hall, Englewood Cliffs, NJ.

The most important inorganic buffer systems in body fluids are the bicarbonates and phosphates; organic buffers include amino acids, peptides and proteins. The effectiveness of a physiological buffering system is illustrated by the fact that the addition of 1.0 mL of 10 M HCl to 1 L of water, initially at pH 7, causes the pH to decrease to 2 (that is, $[H^+]$ is 10^{-2} M). By contrast, if 1.0 mL of 10 M HCl is added to 1 L of human blood plasma at pH 7.4, the pH only falls to pH 7.2.

When CO_2 dissolves in water to form the CO_2/HCO_3 conjugate base pair[62], the Henderson–Hasselbalch equation (shown in Box 13.2) can be written as:

$$pH = pK_a + \log\frac{[HCO_3^-]}{[CO_2]}$$

We can calculate the concentration of carbon dioxide in solution in the blood, $[CO_2]$ from the partial pressure of carbon dioxide, PCO_2, and the capacitance coefficient, β, of carbon dioxide in the blood, βCO_2[63], so that:

$$pH = pK' + \log\frac{[HCO_3^-]}{\beta CO_2 \cdot PCO_2} \qquad \text{Equation 13.3}$$

pK' is known as the apparent pK as it is the result of the equilibrium constant of the combined reactions of carbon dioxide with water, the solubility of carbon dioxide, and temperature. Therefore, different values of pK' have to be used under different conditions.

The Henderson–Hasselbach equation is of fundamental significance to the study of acid–base balance in animals. For example, it is possible to calculate PCO_2 and $[HCO_3^-]$ of human arterial blood from the following data:

- the pH of the blood at 37°C is 7.44;
- pK' is 6.1;
- The total concentration of carbon dioxide in the plasma of the blood, ($[CO_2] + [HCO_3^-]$), is 26.7 mmol L^{-1};
- βCO_2 for the plasma is 0.23 mmol L^{-1} kPa^{-1}.

The concentration of bicarbonate ions, $[HCO_3^-]$, in the plasma is the difference between the total concentration of carbon dioxide and that dissolved in the plasma = 26.7 – ($\beta CO_2 \cdot PCO_2$) mmol L^{-1}.

[62] This relationship is shown in Box 13.2.

[63] The capacitance coefficient of carbon dioxide is discussed in Section 11.1.2.

Substituting the above values into equation 13.3, we get:

$$7.44 = 6.1 + \log\left(\frac{26.7 - 0.23PCO_2}{0.23PCO_2}\right)$$

This becomes: $1.34 = \log\left(\frac{26.7 - 0.23PCO_2}{0.2PCO_2}\right)$

If we take the antilog of both sides: $21.88 = \left(\frac{26.7 - 0.23PCO_2}{0.23PCO_2}\right)$

Therefore, $21.88\ (0.23PCO_2) = 26.7 - 0.23PCO_2$

$$5.03\ PCO_2 + 0.23\ PCO_2 = 26.7$$

$$5.26\ PCO_2 = 26.7$$

$$PCO_2 = 26.7 / 5.26 = \mathbf{5.1\ kPa}$$

The amount of carbon dioxide dissolved in the plasma is 5.1 kPa × 0.23 mmol L^{-1} kPa^{-1} = 1.2 mmol L^{-1} so that $[HCO_3^-]$ is 26.7–1.2 = **25.5 mmol L^{-1}**.

The pH–bicarbonate diagram

Equation 13.3 demonstrates that changes in pH will affect the ratio of $[HCO_3^-]$ to PCO_2, and *vice versa*. The complex relationships contained within the Henderson–Hasselbalch equation can be plotted on a **pH-bicarbonate (pH – $[HCO_3^-]$) diagram**[64]. Figure 13.29 gives two examples of a pH–bicarbonate diagram, one for an air-breather and one for a water-breather. Note that the values on the coordinates may vary between species. In the two examples given, this difference is largely related to the greater capacitance coefficient of carbon dioxide in water. As a result, there is a lower PCO_2 in the blood of water-breathing animals compared with that in air-breathing species[65].

Figure 13.29A shows that there is a unique value of PCO_2 at any point on the graph where the values pH and $[HCO_3^-]$ meet. Many other phenomena can be represented on the pH–$[HCO_3^-]$ diagram. For example, the **buffer line** shows the relationship between pH and $[HCO_3^-]$ as PCO_2 changes. As PCO_2 decreases, pH increases and $[HCO_3^-]$ falls. This relationship is (almost) linear and the slope of the buffer line is known as the buffer value, β (but not to be confused with the capacitance coefficient of fluids for gases, as used in Equation 13.3). The unit for the buffer value is the slyke [mmol L^{-1} (pH unit)$^{-1}$], named after D. van Slyke, the US chemist who first described it.

$$\text{The buffer value, } \beta = \frac{\Delta[HCO_3^-]}{\Delta pH} \qquad \text{Equation 13.4}$$

The position and slope of buffer lines on a pH – $[HCO_3^-]$ diagram vary considerably between species, and even between individuals within a particular species; they are also temperature-dependent.

The relationship between PCO_2, $[HCO_3^-]$ and pH can be influenced by changes in ventilation that do not match changes in metabolic rate and by internal changes in the concentration of protons. If an animal under- (hypo-) ventilates with respect to its oxygen requirements, carbon dioxide

[64] As pH is the independent variable, it should correctly be called the bicarbonate–pH ($[HCO_3^-]$–pH) diagram.

[65] Partial pressure of carbon dioxide in the blood of water- and air-breathing vertebrates is discussed in Section 11.1.2.

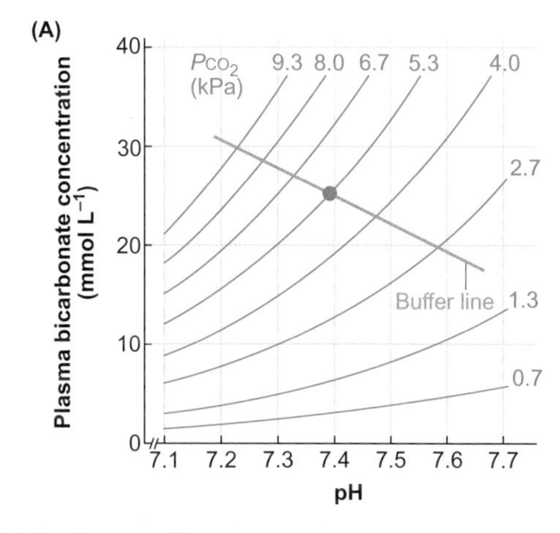

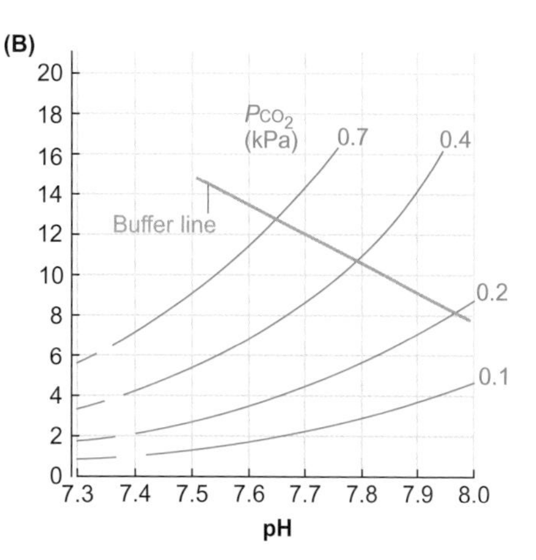

Figure 13.29 pH–bicarbonate diagrams

(A) Human blood at 37°C. ● indicates the normal values of pH, PCO_2 and HCO_3^- concentration. The lines of constant partial pressure of carbon dioxide are known as PCO_2 isopleths or CO_2 isobars. (B) Haemolymph of the rock-pool prawn (*Palaemon elegans*) at 15°C. Note different values of bicarbonate, pH and PCO_2 compared with those for human blood in (A). The buffer lines in (A) and (B) show the relationship between bicarbonate concentration and pH as PCO_2 changes. Note, these diagrams should more accurately be called bicarbonate–pH diagrams, as pH is the independent variable.

Adapted from Dejours P (1981). Principles of Comparative Respiratory Physiology, Elsevier/North Holland, Amsterdam. Bridges CR and Morris S (1989). Respiratory pigments: interactions between oxygen and carbon dioxide transport. Canadian Journal of Zoology 67: 2971-2985.

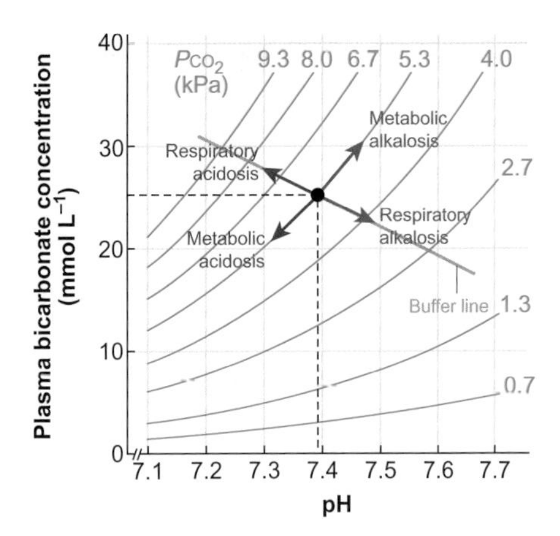

Figure 13.30 Respiratory and metabolic changes in pH

Human blood at 37°C ● indicates the normal values of pH, PCO_2 and HCO_3^- concentration. During respiratory alkalosis or acidosis caused by over- or under-ventilation (hyperventilation or hypoventilation, respectively), the associated changes in pH and HCO_3^- concentration are described by a buffer line. However, the changes in pH and HCO_3^- concentration during metabolic acidosis or alkalosis occur at constant PCO_2, assuming there is no compensatory change in ventilation.

Modified from Dejours P (1981). Principles of Comparative Respiratory Physiology, Elsevier/North Holland, Amsterdam.

will accumulate in the body (hypercapnia) and pH will fall below its normal level. This is known as a **respiratory acidosis**. On the other hand, if an animal over- (hyper-) ventilates with respect to its oxygen requirements, as often occurs during excessive panting, carbon dioxide will be removed from the body faster than it is being produced. Consequently, the partial pressure of carbon dioxide will fall (hypocapnia) and pH will increase above its normal level, a condition called **respiratory alkalosis**. These changes can be represented on a pH–bicarbonate diagram as shown in Figure 13.30. However, the situation in the field may be more complex than observations in the laboratory might otherwise indicate, particularly in ectotherms. Case Study 13.3 considers crayfish (*Austropotamobius pallipes*) as an example.

If an animal undergoes strenuous exercise involving anaerobic metabolic pathways, the production of protons in the muscles will increase, and intramuscular and blood pH will fall. This is known as **metabolic acidosis**, with the changes occuring at a constant PCO_2, as illustrated in Figure 13.30. However, a respiratory alkalosis caused by hyperventilation can partly compensate for the metabolic acidosis. Conversely, vomiting the contents of the stomach, which are acid, will lead to a **metabolic alkalosis**. Metabolic alkalosis can also occur in some species of reptiles which feed intermittently and secrete a large amount of acid into the stomach, although respiratory acidosis can compensate for this alkalosis to a greater or lesser extent[66]. In addition to the respiratory adjustments we have discussed, the gills and kidneys can also be involved in compensating for these acid–base disturbances by controlling the rate of excretion of bicarbonate ions and protons. By constructing a pH–bicarbonate diagram, it is possible to identify what type of acid–base disturbance exists, whether or not it is compensated and, if it is, by what mechanisms.

When a cell, such as a muscle cell, becomes acidic as a result of the production of protons during anaerobic metabolism, protons leave the cell at a rate that is proportional to the decrease in pH of the cell. Some of this efflux is related, at least during the initial stages of recovery, to the co-transport of lactate and protons, and some as the result of sodium/proton antiport and bicarbonate/chloride exchange. Acidic and basic equivalents in the plasma can be excreted via the kidney in terrestrial vertebrates, the gut of some vertebrates and the gills and skin of aquatic animals[67].

The pH–bicarbonate diagrams present the traditional view of acid–base regulation, but another explanation is based on the fact that body fluids are electrically neutral, which means that the sum of the charges carried by cations (positively charged ions) must equal the sum of the charges carried by anions (negatively charged ions). The cations are the inorganic ions, with the major one being sodium, and the minor ones being potassium, calcium and magnesium. The anions include chloride (Cl^-), and sulfate (SO_4^-) plus bicarbonate, proteins (Pr) and phosphate buffers (HPO_4^{2-} + $H_2PO_4^-$) (the latter three are collectively known as the buffer base), and lactate (La^-) when present. Thus:

$$Na^+ + K^+ + 2\,Ca^+ + 2\,Mg^{2+}$$
$$= Cl^- + 2\,SO_4^{2-} + \underbrace{HCO_3^- + Pr^- + (2\,H-PO_4^{2-}) + H_2PO_4^-}_{\text{Buffer base}} + La^-$$

In order to maintain electroneutrality, a reduction in the concentration of plasma bicarbonate is associated with an equivalent increase in the concentration of chloride ions and/or an equivalent decrease in the concentration of sodium ions. The opposite occurs for an increase in the concentration of bicarbonate. Those ions that do not form part of the buffer base are known as **strong ions**, with the difference between the sum of the strong cations ($Na^+ + K^+ + Ca^{2+} + Mg^{2+}$) and the sum of the strong anions ($Cl^- + SO_4^{2-}$) being known as the **strong ion difference (SID)**.

As a change in pH is usually accompanied by a change in the buffer base, there must also be a change in strong ion difference. In fact, it has been convincingly argued that variables such as [HCO_3^-], [OH^-] and [H^+], or pH, are dependent

[66] Alkaline tide in intermittent feeders is discussed in Section 15.2.1.

[67] Ionic transport processes in aquatic animals are discussed in Chapter 5 and kidney function is discussed in Sections 7.1 and 7.2.

Case Study 13.3 Annual changes in acid–base balance in crayfish (*Austropotamobius pallipes*) in the field

The study of wild white-clawed or Atlantic stream crayfish (*Austropotamobius pallipes*) explored in this case study illustrates the importance of obtaining a combination of data from animals in their natural habitat with measurements of physicochemical variables in the laboratory throughout the season in order to understand complex phenomena, such as acid–base regulation in the wild, especially in aquatic ectotherms.

The crayfish studied live in a large shallow pool fed by a stream which undergoes wide changes in environmental variables. The crayfish undergo a seasonal moult[1]. As the temperature increases from February to June, Figure A illustrates that the pH of the arterial haemolymph decreases, much as predicted by the alphastat hypothesis. However, from June to July/August, the pH of the haemolymph actually increases, despite the continued increase in environmental temperature. Then, as temperature falls from July/August to September, the pH of the haemolymph increases (as predicted by the alphastat hyothesis) along a line that runs parallel to the decrease between February and June. There is then a fall in pH as temperature continues to decrease. From October to the following January, pH of the haemolymph remains almost constant, despite a continued decline in temperature by 9°C.

According to the alphastat hypothesis, this latter decrease in temperature should result in an increase in pH of about 0.14 pH units. How can the more unexpected aspects of these data be explained?

In July/August, the crayfish are in various stages of post-moult and the elevated pH of the haemolymph may have resulted from the increased permeability of the cuticle following moult and/or the conversion of metabolically produced carbon dioxide to bicarbonate ions for calcification of the new exoskeleton. In addition, the pH of the water rises in the summer, which may induce an alkalosis (increase in pH) in the haemolymph because of reduced transfer of ammonia across the gills[2], as has been found for rainbow trout in alkaline environments.

Between October and December/January, the concentration of bicarbonate ions in the haemolymph increases at a constant PCO_2 of 0.4 kPa (that is, along the 0.4 kPa PCO_2 isopleth) to position *A*, as shown in Figure B. However, the increase in the concentration of bicarbonate ions is compensated for by

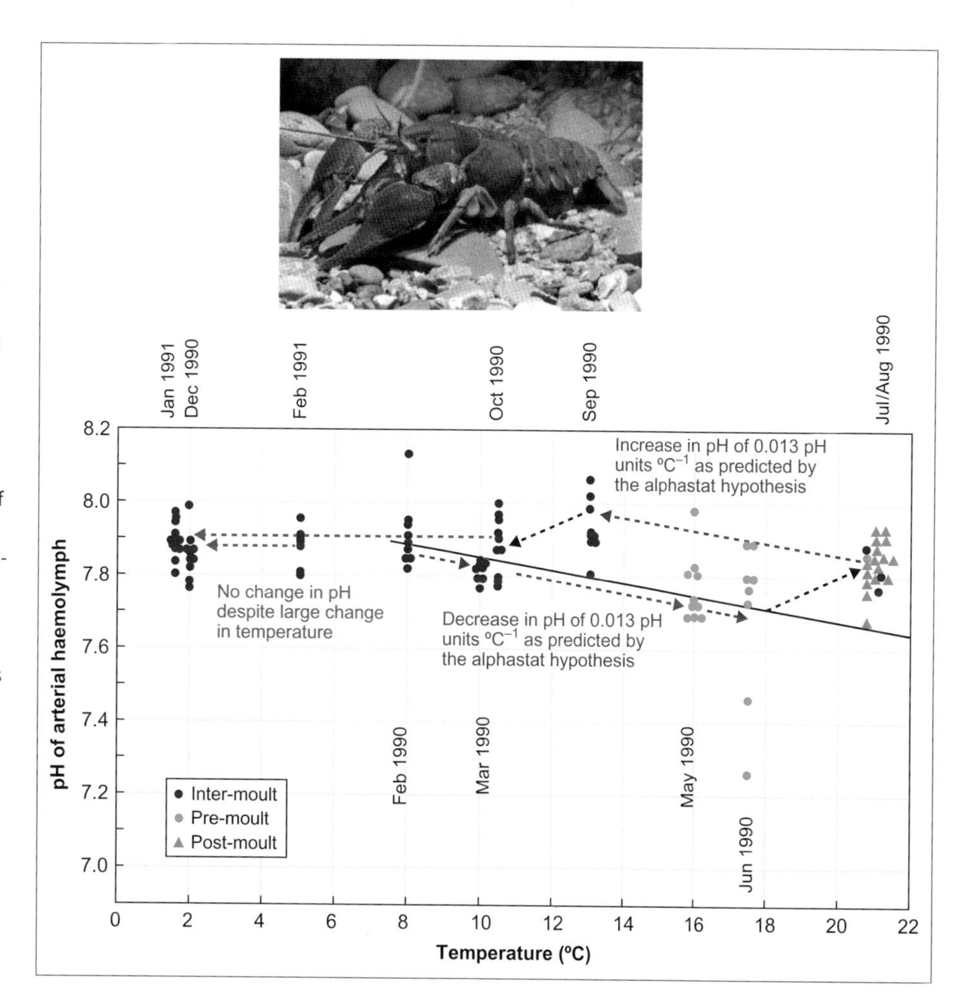

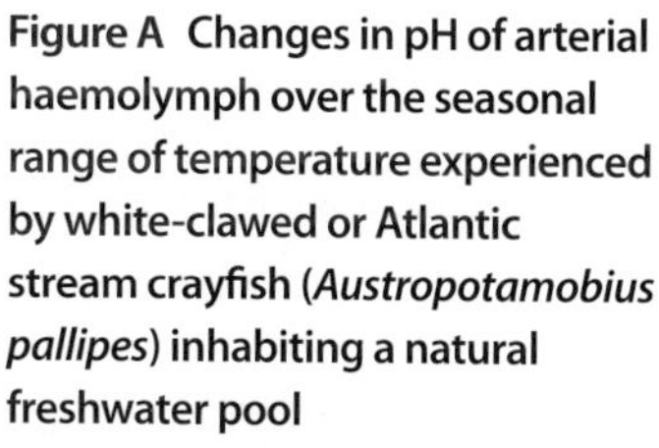

Figure A Changes in pH of arterial haemolymph over the seasonal range of temperature experienced by white-clawed or Atlantic stream crayfish (*Austropotamobius pallipes*) inhabiting a natural freshwater pool

Dashed lines and the arrows represent the direction of the changes in pH throughout the year. Between February and June, pH of the haemolymph decreased much as predicted by the alphastat hypothesis. The continuous line represents the regression line for this relationship. Conversely, between July/August and September, pH increased much as predicted. However, at other times of the year, pH did not respond to changes in temperature as predicted, and the possible reasons for this are discussed in the text and illustrated in Figure B.

Adapted from: Whiteley NM and Taylor EW (1993). The effects of seasonal variations in temperature on extracellular acid-base status in a wild population of the crayfish, *Austropotamobius pallipes*. Journal of Experimental Biology 181: 295-311.

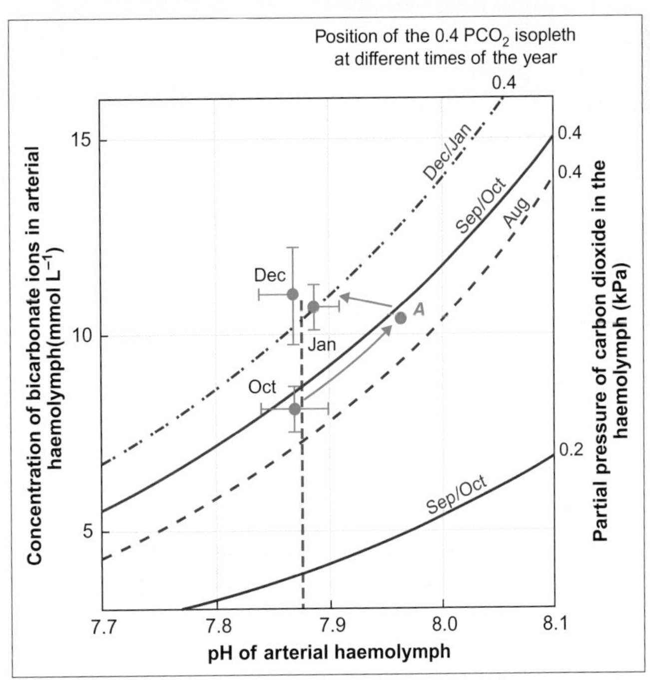

Figure B pH–bicarbonate diagram illustrating the acid–base variables throughout the year in the arterial haemolymph of white-clawed or Atlantic stream crayfish (*Austropotamobius pallipes*) inhabiting a freshwater pool

The arrows along the Sep/Oct PCO_2 isopleth and across to the Dec/Jan PCO_2 isopleth indicate a possible pathway for adjustments in haemolymph acid–base status between October and December/January. As indicated in Figure A, the temperature of the water varied between 21°C in July/August and 1°C in the following January during which time the animals progressed from post-moult to inter-moult. Changes in the solubility of carbon dioxide in the water and in the measured p*K′* (p*K′* is explained in the text) lead to a shift in the position of the 0.4 kPa PCO_2 isopleth for the haemolymph. As a result of this, pH remained almost unchanged between October and December/January (dashed blue line), despite a 9°C fall in temperature. Mean values ± SEM, n = 7–10.

Modified from: Whiteley NM and Taylor EW (1993). The effects of seasonal variations in temperature on extracellular acid-base status in a wild population of the crayfish, *Austropotamobius pallipes*. Journal of Experimental Biology 181: 295-311.

temperature-related, seasonal changes in the capacitance coefficient of carbon dioxide in the water[3] and in the measured value of p*K′*[4]. The change in these physicochemical variables between October and December/January results in a shift in the 0.4 kPa isopleth for PCO_2, as shown in Figure B. This means that the increase in the concentration of bicarbonate between October and December/January is along a single PCO_2 isopleth –but the change in isopleth position means that pH is maintained at an almost constant value of 7.87, as illustrated in Figure B.

Whiteley NM, Taylor EW (1993). The effects of seasonal variations in temperature on extracellular acid-base status in a wild population of the crayfish, *Austropotamobius pallipes*. Journal of Experimental Biology 181: 295–311.

Wilkie MP, Wood CM (1991). Nitrogenous waste excretion, acid-base regulation and ionoregulation in rainbow trout (*Oncorhynchus mykiss*) exposed to extremely alkaline water. Physiological Zoology 64: 1069–1086.

13

› Find out more

Cameron JN and Wood CM (1985). Apparent H^+ excretion and CO_2 dynamics accompanying carapace mineralization in the blue crab (*Callinectes sapidus*) following moulting. Journal of Experimental Biology 114: 181–196.

Mangum CP, McMahon BR, deFur PL and Wheatly MG (1985). Gas exchange, acid-base balance and the oxygen supply to the tissues during moult of the blue crab, *Callinectes sapidus*. Journal of Crustacean Biology 5: 188–206.

1 Moult in arthropods is discussed in Section 19.5.3.

2 The transport of ammonia across gills of aquatic animals is discussed in Section 7.4.1.

3 The effect of temperature on the capacitance of CO_2 in water is discussed in Section 11.1.2.

4 p*K′* is known as the apparent p*K*; it is the result of the equilibrium constant of the combined reactions of carbon dioxide with water, the solubility of carbon dioxide and temperature. Therefore, different values of p*K′* have to be used under different conditions.

on variables such as the partial pressure of carbon dioxide (PCO_2), strong ion difference and total weak acid, which is usually protein. In other words, it is only changes in these independent variables that can affect [H^+] or pH. There is experimental evidence to support this view, such as that from rainbow trout following exhaustive exercise, where it has been demonstrated that the movements of metabolically produced H^+ are related to the movements of strong ions, mainly across the gills.

› *Review articles*

Burton RF (2002). Temperature and acid-base balance in ectothermic vertebrates: the imidazole alphastat hypothesis and beyond. Journal of Experimental Biology 205: 3587–3600.

Fencl V, Leith DE (1993). Stewart's quantitative acid-base chemistry: applications in biology and medicine. Respiration Physiology 91: 1–16.

Harrison JF (2001). Insect acid-base physiology. Annual Review of Entomology 46: 221–250.

Rahn H, Howell BJ (1978). The OH^-/H^+ concept of acid-base balance: historical development. Respiration Physiology 33: 91–97.

Reeves RB (1977). The interaction of body temperature and acid-base balance in ectothermic vertebrates. Annual Review of Physiology 39: 559–586.

Checklist of key concepts

- Oxygen is not very soluble in the fluid component of the blood of animals, as this fluid is essentially salty water.
- The evolution of **respiratory blood pigments** to which oxygen can reversibly bind, enabled the amount of oxygen carried in a given volume of blood to increase dramatically.
- Two major respiratory pigments are found in the blood: the copper-based **haemocyanin** and the iron-based **haemoglobin**.
- There are relatively small and relatively large (giant) molecules of haemoglobin.
  - The **smaller** molecules are contained within the **red blood cells (erythrocytes)**.
  - The **giant** molecules of haemoglobin and haemocyanin are not contained within cells.
- The maximum concentration of oxygen in the blood when the blood pigment is fully saturated is the **oxygen-carrying capacity**.
- Oxygen-carrying capacity depends on the type of pigment and on its concentration in the blood.

Many different factors affect the binding of oxygen to blood pigments

- The relationship between the concentration of oxygen in the blood and partial pressure is known as the **oxygen equilibrium curve (OEC)** and is generally **sigmoidal** in shape.
- Myoglobin is a pigment found in muscles and has an OEC that is **hyperbolic** in shape.
- The sigmoid shape of haemoglobins is the result of **cooperativity** between the haem groups, that is, the binding of oxygen at one haem makes it easier for oxygen to bind to the other haems of the same molecule.
- The **Hill coefficient** indicates the degree of cooperativity between the haem groups within a particular haemoglobin molecule
- The partial pressure of oxygen at which a respiratory pigment is 50 per cent saturated is known as the $\boldsymbol{P_{50}}$.
- The lower the P_{50}, the greater the affinity of the pigment for oxygen.
- Myoglobin has a very low P_{50} (very high affinity for oxygen).
- Increases in temperature of blood and partial pressure of CO_2 (PCO_2), and/or decrease in pH *increase* the P_{50} (decrease the affinity) of the pigment.
- The effects of PCO_2 and pH on the P_{50} are known as the **Bohr effect**.
- The Bohr effect favours the loading of oxygen onto the pigment at the gas-exchange surface and the unloading of oxygen in the tissues.
- The elevated temperature in active muscles relative to that at the gas-exchange surface during exercise contributes to the unloading of oxygen in the muscles.
- In some species of fish, a decrease in pH causes a reduction in the oxygen carrying capacity of the haemoglobin, known as the **Root effect**.
- The Root effect is important in the functioning of the swimbladder and in supplying oxygen to the retina.

Transport of carbon dioxide, pH and acid–base balance

- Carbon dioxide is carried mostly as bicarbonate ions (HCO_3^-) in the plasma of vertebrates, some as **carbamino compounds**, after it combines with amino acid residues on respiratory pigments, and a small amount is in solution.
- The state of oxygenation of haemoglobin determines its CO_2 carrying capacity: the deoxygenated pigment has a greater CO_2 carrying capacity than the oxygenated form. This is known as the **Haldane effect**.
- As a result of the Haldane effect, oxygenation of blood as it passes through the gas exchanger favours the release of CO_2, and its deoxygenation favours the uptake of CO_2 as it passes through the tissues.
- The enzyme **carbonic anhydrase** increases the rate of reaction between CO_2 and water to produce carbonic acid, which dissociates into H^+ and HCO_3^-.
- This dissociation has a large effect on the **acid–base balance** of an animal.
- The concentration of H^+, $[H^+]$, in a solution is normally represented by the term pH, which is equal to $-\log [H^+]$.
- An acid (HA) in water dissociates to produce a proton (H^+) and a base (A^-).
- The pH of blood changes with temperature, which can affect the functioning of proteins, such as enzymes, in ectotherms.
- At a constant temperature, **buffering** of the blood and other body fluids ensures pH is maintained within a certain range when a relatively large amount of acid or base is added.
- The most important inorganic buffers in body fluids are bicarbonates and phosphates; organic buffers include amino acids, peptides and proteins.
- The relationship between pH, the concentration of CO_2, $[CO_2]$ and the concentration of bicarbonate ions, $[HCO_3^-]$, is represented by the **Henderson–Hasselbach equation**.

Transport of metabolic substrates in the blood to the metabolizing cells

- Glucose and amino acids are transported from the intestine into the blood and from the blood into the metabolizing cells by **sodium-independent glucose transporters** or by **antiporters**, respectively.
- In most species of vertebrates, fatty acids bind to **albumin**.
- The fatty acids may be stored in the liver or in **adipocytes** or used immediately as a source of energy for active muscles.
- Both passive and **protein-mediated transport** of fatty acids into cells can occur.

- Fatty acids are transported through the inner membrane of mitochondria (in the form of acyl-CoA esters) by a **translocase system**.
- The relative slowness of the fatty acid transport system in mammals means that it only supplies the majority of the energy required for exercise below about 40 per cent maximum oxygen consumption. Above this level, carbohydrates provide an increasing proportion of the energy.
- During migration and other long duration flights of birds and insects, fatty acids provide most of the energy, so transport of lipids to the working muscles is faster in these groups than in mammals.

Study questions

*** Answers to these numerical questions are available online. Go to www.oup.com/uk/butler**

1. Why is it necessary for the blood of animals to contain a respiratory pigment? (Hint: Section 13.1.1.)
2. Draw an oxygen equilibrium curve for a respiratory blood pigment and for myoglobin and explain why they are different in shape. (Hint: Section 13.1.2.)
3. What is the P_{50} of a respiratory pigment? (Hint: Section 13.1.2.)
4. Explain the difference between the Bohr effect and the Root effect. (Hint: Section 13.1.2.)
5. What is the functional significance of the Root effect? (Hints: Section 13.1.2, Case Study 13.1.)
6. Explain how the Bohr and Haldane effects influence the transport of oxygen and carbon dioxide between the environment and the metabolizing cells. (Hint: Section 13.1.3.)
7. How is carbon dioxide transported in the blood of animals? (Hint: Section 13.1.3.)
8. What are the roles of nitric oxide in regulating the demand and supply of oxygen to cardiac and skeletal muscles? (Hints: Section 13.1.3, Case Study 13.2.)
9. Lipids are insoluble in water, so how are they transported in the blood of animals? (Hint: 13.2.2.)
10. What are the major sources of energy during sustained aerobic exercise in mammals and birds? Explain the significance of any differences that you describe between the two groups. (Hints: Sections 13.2.1, 13.2.2.)
11. What is an acid and what is the difference between weak and strong acids? (Hint: Box 13.2.)
12.* If the pH of the blood of a trout at °C is 7.95, what is the hydrogen ion concentration in the blood? (Hint: Box 13.2.)
13. What are pN and pK_a? (Hints: Section 13.3.1, Box 13.2.)
14.* If pH in the haemolymph of a crayfish at 16°C is 7.77, pK' is 6.08, the total amount of carbon dioxide in the haemolymph is 7.2 mmol L^{-1}, and βCO_2 for the haemolymph is 0.4 mmol L^{-1} kPa^{-1}, calculate PCO_2 and concentration of bicarbonate ions, $[HCO_3^-]$, of the haemolymph. (Hint: Section 13.3.2.)
15. What effect does temperature have on pH of plasma in ectotherms and what is the explanation for this effect. (Hint: Section 13.3.1.)
16. Explain the difference between respiratory acidosis and metabolic acidosis. (Hint: Section 13.3.2.)
17. What is strong ion difference and what is its relationship to pH? (Hint: Section 13.3.2.)

Bibliography

Cameron JN (1989). The Respiratory Physiology of Animals. Oxford University Press. New York, Oxford. pp. 353

Davenport HW (1974). The ABC of Acid-Base Chemistry, 6th Ed. The University of Chicago Press. Chicago and London.

Dejours P (1981). Principles of Comparative Respiratory Physiology, Elsevier/North Holland, Amsterdam.

Horton RH, Moran LA, Ochs RS, Rawn JD, Scrimgeour KG (1993). Principles of Biochemistry. Neil Patterson Publishers Prentice Hall, Englewood Cliffs, NJ.

Maughan R, Gleeson M, Greenhaff PL (1997). Biochemistry of Exercise and Training. Oxford University Press. Oxford. pp. 234.

Nikinmaa M (1990). Vertebrate Red Blood Cells. Springer-Verlag, Berlin.

14

Cardiovascular systems

An overview

In Chapter 13, we discussed the characteristics of the blood or haemolymph that is pumped around the body within the cardiovascular system. In this chapter we take a close look at the physiology of the cardiovascular system itself: the heart and the blood vessels.

The circulatory system of animals is central to the way they respond to changes in the environment, as evidenced by examples we describe throughout the book. For example, we discuss the role of the circulatory system in the thermoregulatory responses of ectotherms and endotherms to changes in environmental temperature in Section 9.2.3 and Chapter 10, respectively. We consider environmental influences on respiratory gas transport in Chapter 13; the influence of exercise and a decrease in oxygen supply on circulatory systems in Sections 15.2.2 and 15.3; and nervous control of the cardiovascular system, including the sensing of environmental oxygen, in Chapter 22.

When we refer to the circulating fluid in general terms during this chapter, we call it blood for simplicity (as we do in Chapter 13). Before looking at the circulatory systems of specific groups of animals, we discuss the basic characteristics of the circulation.

14.1 General characteristics of circulatory systems

Traditionally, circulatory systems have been divided into '**closed systems**' and '**open systems**'. However, as a result of more recent studies, the distinction between open and closed systems is not clear-cut, as we discover later in this section.

14.1.1 Open and closed circulatory systems

The movement of fluids inside cells (the streaming of protoplasm) and/or inside of organisms—that is, the existence of some form of circulatory system—plays an important part in the transfer of respiratory gases between the environment and the metabolizing cells[1]. In its simplest form, the movement of fluids within the body is seen in the coelom (the secondary body cavity)[2] of some annelid worms, such as some polychaetes. In these annelids, the coelomic fluid is mixed by cilia that line the coelom and by contractions of the body wall muscles. However, there is no ordered direction to the flow, so there is no circulation as such.

A clearly defined circulation lies at the other extreme. Flow of a specialized respiratory fluid (blood) is generated by a single pulsatile pump (the heart) and circulates around the body within closed tubes (arteries, arterioles and capillaries) which are lined with endothelium or endothelium-like cells. Blood returns to the heart and respiratory exchange surface by way of larger veins, which can function with thinner walls than the arteries because of the lower pressure within them. Such a **closed circulatory system** is found in vertebrates and in cephalopod molluscs. The generalized vertebrate system is shown in Figure 14.1A.

In between the two extreme cases—of random mixing in the coelom and the completely closed circulatory systems—we find a number of systems that are often described as **open circulatory systems**. An open circulatory system is diagrammatically represented in Figure 14.1B.

Open systems have certain common features:

- The propulsive force is generated by one or more contractile hearts.
- The blood enters a branching arrangement of arteries and finer vessels (which are not lined by endothelium or endothelium-like cells) from which it emerges into a system of sinuses or smaller spaces (lacunae) collectively known as the **haemocoel**. The haemocoel is the remnant of the primary body cavity (the blastocoel)[2].
- The coelom is small and may remain as a cavity surrounding the heart (pericardial cavity), e.g. in molluscs.
- The sinuses have no cell lining. Consequently, the haemolymph (blood lymph—functionally equivalent to the blood in closed circulatory systems) bathes the tissues directly.

[1] Processes involved in the transfer of respiratory gases between the environment and the cells are discussed in Section 11.2.

[2] Body cavities are discussed in Section 4.1.

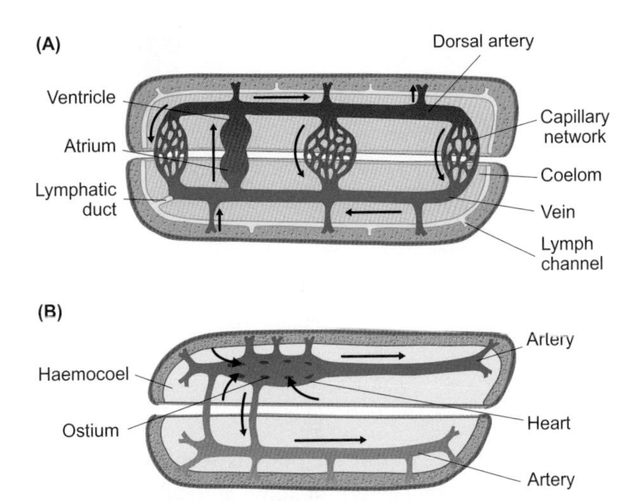

Figure 14.1 Circulatory systems in animals

(A) Closed circulatory system, of a generalized vertebrate, with coelom and lymphatic system which returns fluid (lymph) that exudes from the blood in the capillary beds to the circulation via the lymphatic ducts. (B) Open circulatory system, as in some species of crustaceans and molluscs, (except the cephalopods) with part of the circulation in the haemocoel. The heart shown is similar to that in crustaceans (the haemolymph enters the heart by way of small openings – ostia) rather than to that in molluscs.

Adapted from Prosser CL and Brown FA (1961). Comparative Animal Physiology, 2nd edn. Philadelphia: W.B. Saunders Company.

- Pressures generated within open systems are generally lower and blood volume is generally greater than those in closed systems, as shown later in Tables 14.2 and 14.3.

We now discuss the general characteristics and function of the heart and blood vessels.

14.1.2 The pumping cycle of the heart

Hearts of different shapes and structures are seen throughout the animal kingdom. In all cases, however, the heart (or hearts) generates the pressure, *P* (force per unit area), that propels the blood around the body. In many species of molluscs, decapod crustaceans and vertebrates, the propulsive part of the heart is one muscular chamber, the **ventricle**. In contrast, the hearts of birds and mammals consist of two atria, which receive blood from the body or lungs, and two ventricles, which pump blood around the body or lungs, as shown in Figure 14.2. We know most about the heart of birds and mammals, so for the purposes of this discussion, we concentrate on the left side of the heart of mammals, which receives blood from the lungs and pumps it around the body. By contrast, the right side of the heart receives blood from the body and pumps it round the lungs at a pressure lower than that in the circulation around the body.

The general principles are the same in the hearts from other groups of animals, which are discussed in more detail in Sections 14.3 and 14.4.

Mechanical events of the heart

Hearts consist of muscles which contract and relax in a rhythmic fashion. Equally as important to the pumping function of the heart as its contractility is the presence of valves between the chambers of the heart and at the base of the arteries leaving the ventricles. These valves are shown in Figure 14.2; they prevent backflow of the blood, thus ensuring its movement in one direction During the contraction phase of the ventricle (called **systole**), blood is pumped out of the heart so that its volume decreases; during the relaxation phase (**diastole**), blood returns to the heart and its volume increases. These phases are shown in Figure 14.3A.

At the onset of contraction of the ventricle, the valves between the **atrium** and ventricle (the atrioventricular, AV, or bicuspid valves) close and the pressure in the ventricle increases. However, its volume does not change until pressure in the ventricle reaches that in the blood vessel leaving the heart (the aorta). At this stage, the valves at the base of the aorta open. Thus, the first phase of systole is known as the phase of isovolumetric (sometimes called isovolumic) contraction.

Once the aortic valves have opened, there is a phase of rapid ejection of blood from the ventricle which is followed by a phase of reduced ejection. The transition between these two phases is characterized by the peaks in systolic pressure in the ventricle and the aorta and in blood flow along the aorta, as shown in Figure 14.3A. During the period of reduced ejection, the flow of blood from the aorta to the periphery exceeds that leaving the ventricle so that the pressure in the aorta begins to fall. Note that there is a progressive increase in left atrial pressure throughout systole as a result of the blood returning from the lungs.

At the end of the ejection phases, a volume of blood, approximately equal to that ejected during systole (the **cardiac stroke volume**), remains in the ventricle. This remaining volume of blood is known as the residual or **end systolic volume**.

The fall in pressure in the ventricle as the rate of ejection decreases, eventually leads to the closure of the aortic valves, which causes the notch in the aortic pressure trace, shown in Figure 14.3A. The aortic pressure continues to fall rather slowly until the aortic valves open again during the next ventricular contraction. From the time that the aortic valves close, pressure in the ventricle falls rapidly towards zero, but there is no change in ventricular volume. This first phase of diastole is known, therefore, as isovolumetric relaxation. The pressure in the aorta and in all arteries remains well above zero during diastole. This is the result of the resistance to flow produced by the peripheral blood vessels, primarily the small arteries and arterioles.

When ventricular pressure has fallen sufficiently, the AV valves open and the ventricle fills relatively rapidly with

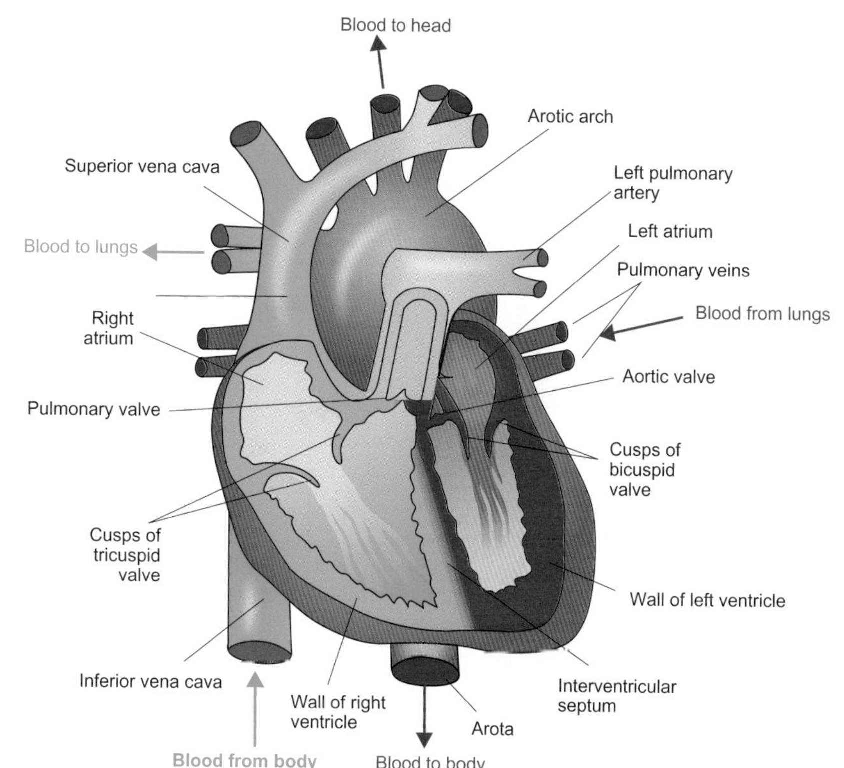

Figure 14.2 Structure of the mammalian heart

Section through the heart of a human showing the valves between the atria and the ventricles (tricuspid and bicuspid valves), between the left ventricle and the dorsal aorta (aortic valve) and between the right ventricle and the pulmonary arteries (pulmonary valves).
Adapted from Fig 21.1, Pocock G, Richards CD, Richards D (2013). Human Physiology, 4th edn., Oxford University Press.

blood from the atrium. There then follows a phase during which the ventricle is filled more slowly. Contraction of the atrium is not essential for the ventricle of mammals to fill as blood is sucked into the ventricle as it relaxes. However, the importance of the atrium increases at higher heart rates. The return of blood from the venous system into the right atrium is the result both of the pressure remaining in the veins from the previous contraction of the left ventricle and also a transient reduction in pressure in the atrium as the ventricle contracts.

The relationship between pressure and volume changes of the ventricle throughout the cardiac cycle can be plotted, as shown in Figure 14.3B. The area enclosed by the **pressure-volume loop** represents the amount of work performed by the ventricle.

In the heart muscle of vertebrates, as in their skeletal muscle, there is a relationship between the length of a muscle fibre and the force it generates during contraction[3]. According to the sliding filament theory of muscle contraction, the force generated during contraction is related to the amount of overlap of the thick (myosin) and thin (actin) filaments within a sarcomere and the number of cross-bridge attachments that are made. However, it is also now known that stretching of the muscle fibres[4] of the heart, during diastolic filling for example, causes an increase in the sensitivity to calcium ions (Ca^{2+}) of the contractile machinery, which leads to an increase in the force of contraction[5].

Up to a certain point of stretch, the force of contraction increases as the length of the muscle fibres increase. This is known as the **Frank–Starling relationship** after the physiologists Otto Frank and Ernest Starling, who first developed the idea at the beginning of the 20th century. Figure 14.4 represents this relationship by plotting the peak systolic pressure, which represents the force generated by the contracting muscle fibres during systole, against the volume of blood in the ventricle at the end of diastole (end diastolic volume), which influences the length of the muscle fibres in the ventricle. Figure 14.4 also illustrates the relationship between end diastolic pressure and end diastolic volume in the ventricle; as end diastolic volume increases, so does end diastolic pressure.

In a normal human heart, the peak force of contraction is attained at a filling pressure of approximately 1.5 kPa (11 mmHg). Usually, end diastolic pressure in the ventricle of a normal human heart is 0–1 kPa (0–7.5 mmHg), indicating that the normal human heart operates over the ascending portion of the Frank–Starling curve, which is illustrated in

[3] The contraction of muscles is discussed in Section 18.1.3.

[4] Types of muscle fibres are discussed in Section 18.1.4.

[5] Ionic events associated with contraction of the heart are discussed in Section 18.1.5: stretching of the cardiac myocytes increases the affinity for Ca^{2+} of troponin C, which is responsible for binding calcium to activate muscle contraction.

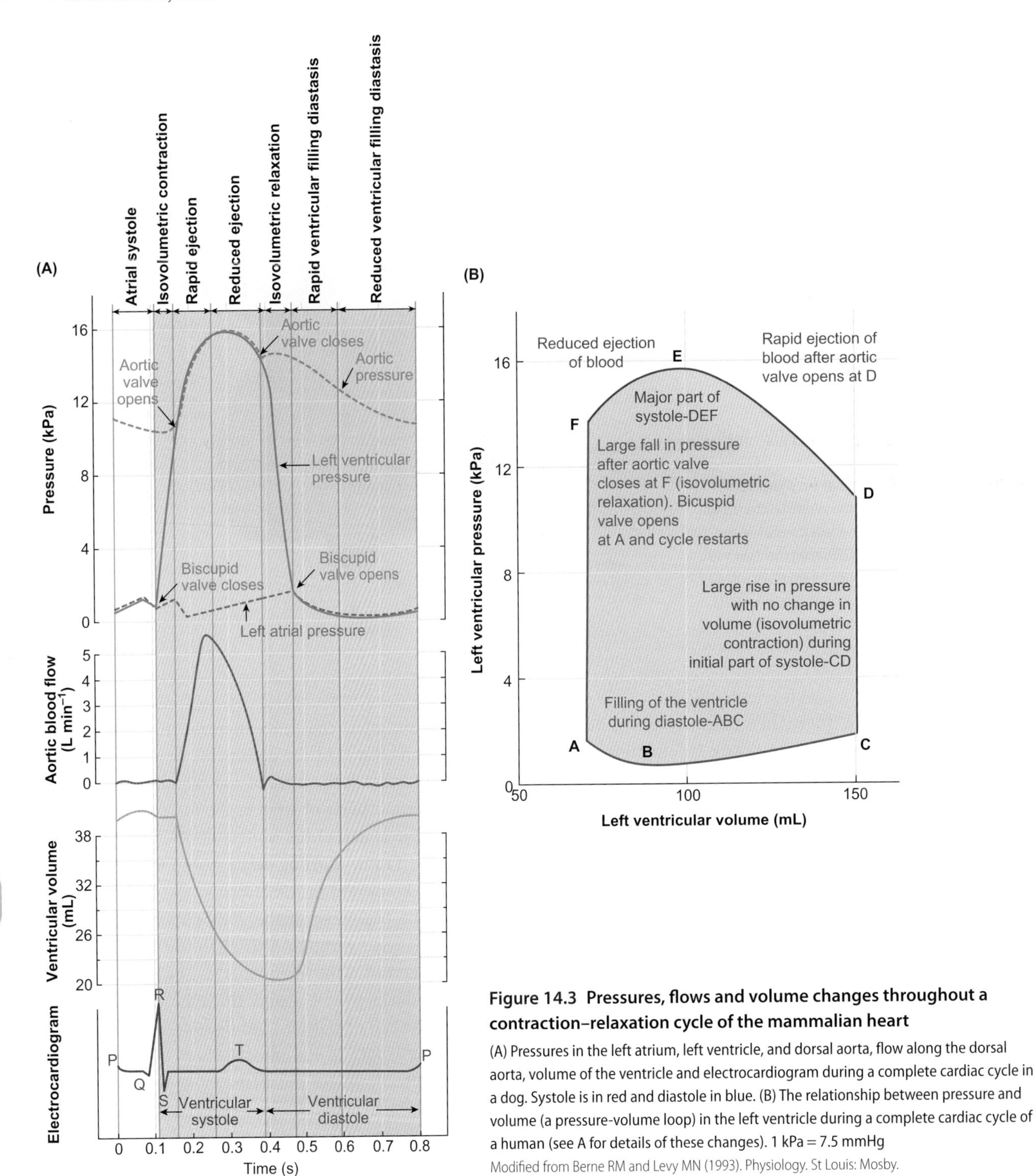

Figure 14.3 Pressures, flows and volume changes throughout a contraction–relaxation cycle of the mammalian heart

(A) Pressures in the left atrium, left ventricle, and dorsal aorta, flow along the dorsal aorta, volume of the ventricle and electrocardiogram during a complete cardiac cycle in a dog. Systole is in red and diastole in blue. (B) The relationship between pressure and volume (a pressure-volume loop) in the left ventricle during a complete cardiac cycle of a human (see A for details of these changes). 1 kPa = 7.5 mmHg

Modified from Berne RM and Levy MN (1993). Physiology. St Louis: Mosby.

Figure 14.4. Over the normal range, as end diastolic pressure (called pre-load) increases, and the wall of the ventricle is increasingly stretched, the ventricle performs more work and cardiac stroke volume increases. Peak systolic pressure in the aorta (called after-load) indicates the pressure that the ventricle must generate in order for it to eject blood into the circulation. The greater the after-load, the greater the proportion of the work performed by the heart goes into generating this pressure rather than into ejecting blood. Consequently, as after-load increases, cardiac stroke volume decreases.

14.1.3 Structure of blood vessels in vertebrates

Blood vessels are not rigid structures; in fact, they can be quite elastic—particularly the larger arteries. The walls of

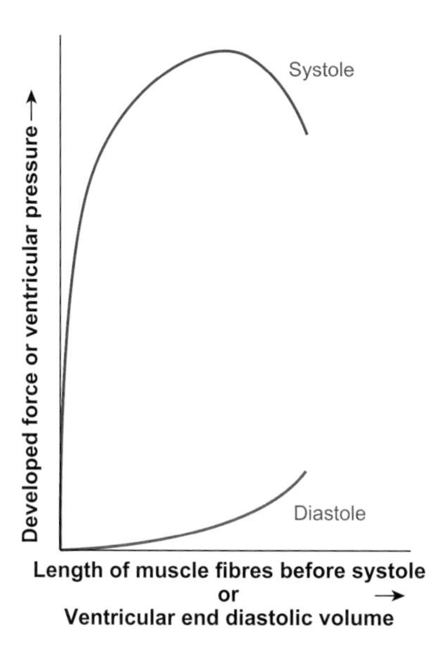

Figure 14.4 The Frank–Starling relationship of the heart

The red curve shows the relationship between the force developed by the muscle fibres (which can be represented by the peak systolic pressure) in the ventricle of a dog and the length of muscle fibres of the heart following diastolic filling and before systole (this can be represented by the volume of ventricle at the end of diastole). This is known as the Frank–Starling curve. Also, illustrated by the blue curve, is the relationship between end diastolic pressure and end diastolic volume in the ventricle.

Reproduced from Berne RM and Levy MN (1993). Physiology. St Louis: Mosby.

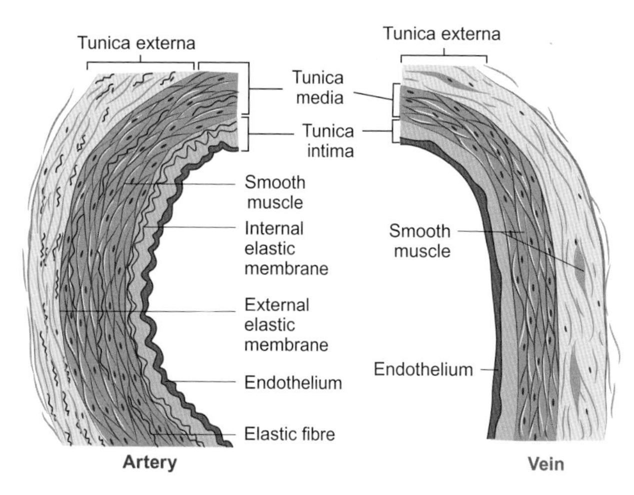

Figure 14.5 Structure of the major blood vessels in mammals

Cross-sections of the walls of part of a typical artery and part of a typical vein showing the three layers. Note the predominance of elastic fibres in the artery and the presence of smooth muscle cells in both. Both types of vessels are lined by a single layer of endothelial cells. These are not drawn to the same scale.

Reproduced from Martini FH et al (2012). Fundamentals of Anatomy & Physiology, 9th edn. San Francisco: Benjamin Cummings.

the blood vessels of vertebrates are composed of three layers, which are shown in Figure 14.5.

- The innermost layer is the tunica intima (sometimes called *tunica interna*—Latin for inner coat). It consists mainly of a layer of endothelial cells which line the vessel and an underlying layer of connective tissue containing a variable number of elastic fibres[6] which provide the blood vessel with elasticity. In arteries, the outer margin of the tunica intima contains a thick layer of elastic fibres (the internal elastic membrane).
- The middle layer is the tunica media. Although some smooth muscle contributes to the tunica media in large arteries, elastic fibres predominate in the larger conducting arteries. The outer layer of the tunica media also contains a layer of elastic fibres (the external elastic membrane). In arterioles and large veins, on the other hand, the tunica media consists mainly of smooth muscle. The smooth muscle fibres are arranged around the circumference of the vessel.
- The outermost layer is the tunica externa and consists mainly of fibrous connective tissue. In arteries, this layer contains collagen fibres and elastic fibres, whereas in veins there are smooth muscle cells, as well as elastic and collagen fibres.

The relative proportions of the different layers in a range of blood vessels are shown in Figure 14.6.

The smooth muscle in the walls of the peripheral blood vessels enables the diameter of the vessels to be reduced when it contracts (the process of vasoconstriction, which increases the resistance to blood flow) or to increase when it relaxes (called **vasodilation**, which decreases the resistance to flow). Because the pressure inside veins is relatively low[7], these vessels have valves which prevent the backflow of blood towards the **capillaries**. This is of particular significance when the veins are compressed as a result of the contraction of skeletal muscles; the valves ensure that blood flows toward the heart and not back toward the capillaries.

The relatively large amount of elastic fibres in the larger arteries close to the heart gives them **elasticity**; they can be stretched and will return to their original form. The change in volume, ΔV, for a given change in pressure, ΔP, is known as **compliance**[8] (C) and the degree of compliance is determined by the proportion of elastic fibres the arteries contain. By contrast, collagen is relatively inextensible and acts as a stiff reinforcing component in the walls of blood vessels.

[6] Elastic fibres consist of a central core of elastin surrounded by microfibrils rich in glycoproteins called fibrillins.

[7] The pressures in different parts of the circulation of mammals are shown in Figure 14.21B.

[8] Compliance in lungs is discussed in Section 12.3.4.

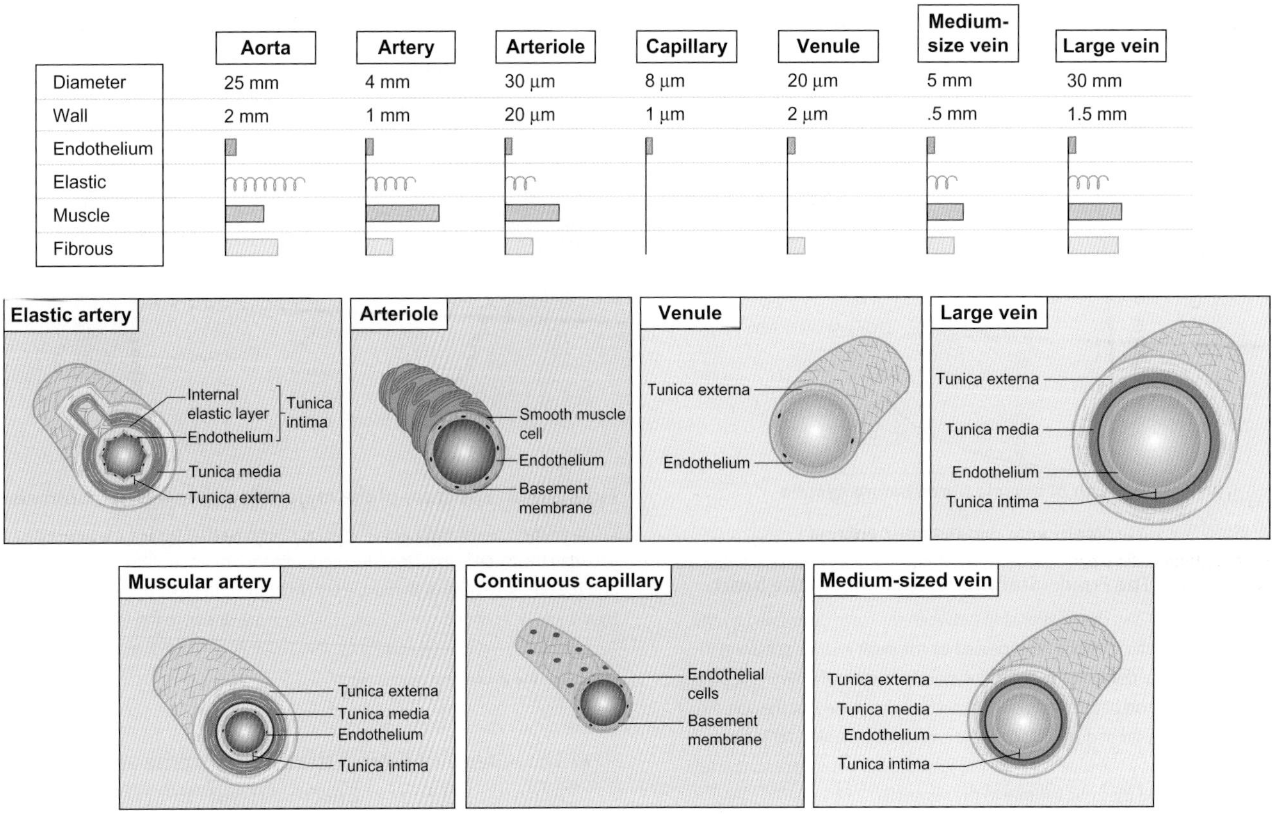

	Aorta	Artery	Arteriole	Capillary	Venule	Medium-size vein	Large vein
Diameter	25 mm	4 mm	30 μm	8 μm	20 μm	5 mm	30 mm
Wall	2 mm	1 mm	20 μm	1 μm	2 μm	.5 mm	1.5 mm
Endothelium							
Elastic							
Muscle							
Fibrous							

Figure 14.6 The dimensions and relative amounts of elastic and muscle fibres and fibrous tissues in vessels of different sections of the circulatory system of humans

Also shown are diagrams of cross-sectional views of the walls of the major types of blood vessels to illustrate the relative sizes of their different layers. Note, the vessels are not drawn to scale.

Reproduced from Rushmer RF (1976). Cardiovascular Dynamics 4th edn, W.B. Saunders & Company. and Martini F et al (2012) Fundamentals of Anatomy & Physiology, 9th edn., San Francisco: Pearson/Benjamin Cummings.

Both the compliance and elasticity of the major arteries play an important part in maintaining an energy-efficient relationship between the pressure generated by the heart and the flow of blood around the body.

14.1.4 Pulsatile blood flow

Flow through the relatively large elastic arteries is highly pulsatile; it reaches a peak during systole when the heart contracts and then declines to zero during diastole as it relaxes, as shown in Figures 14.3A and 14.7A. However, in the more peripheral, more muscular arteries, flow does not reach zero. Instead, Figure 14.7A shows how there is a large flow during diastole with superimposed pulses created by the surge in flow during systole. By the time the blood reaches the capillary beds, flow is quite steady. Pressure pulses change in a similar manner, with pressure reaching zero in the ventricle, at least in birds and mammals, but not in the arteries, as illustrated for the dorsal aorta in Figure 14.3A.

This smoothing of the flow of blood towards the periphery is only possible when the large elastic arteries are highly compliant (stretchy) and can expand during the surge of flow during systole, as shown diagrammatically in Figure 14.7B. By stretching, they store some of the energy generated by the heart during its contraction. As a result, not all of the blood that leaves the heart during systole flows into the resistance vessels of the peripheral circulation. Rather, some is stored in the expanded elastic arteries. As these expanded arteries return to their resting state during diastole, they continue to propel blood into the peripheral circulation. This smoothing of the flow of blood minimizes the work load of the heart, as it is energetically more costly to maintain a pulsatile flow, with alternating acceleration and deceleration of the mass of blood, than to maintain a more constant flow.

Unfortunately, the elasticity of the major arteries decreases with age, at least in humans and a number of species of herbivores. Consequently, the ability of the arteries to act as elastic reservoirs also decreases. Figure 14.8A depicts the extreme condition—of completely rigid arteries. Here, the whole of the blood ejected from the heart during systole would flow through the capillary beds and there would be no flow through them during diastole. In

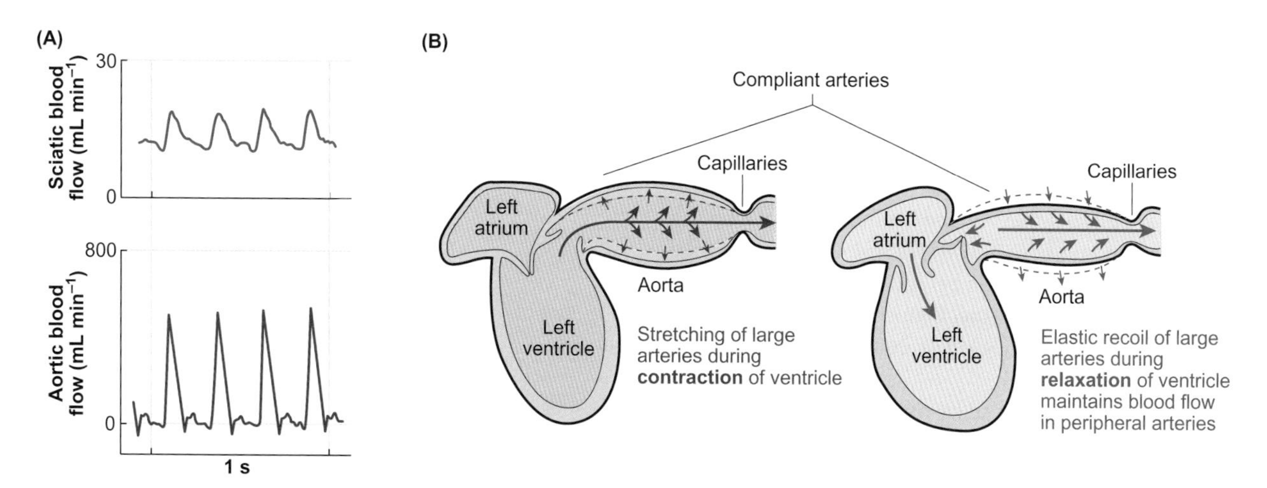

Figure 14.7 Pulsatility of flows and pressures in the arterial system of vertebrates, as illustrated by birds and mammals

(A) Blood flows in the dorsal aorta and sciatic (leg) artery of a chicken. (B) The compliance (stretchiness) of the arteries, particularly the larger ones, creates the more continuous flow pattern in the peripheral regions of the circulation, particularly in the capillaries.

Reproduced from Butler, 1967 (unpublished data); Berne RM and Levy MN (2001). Cardiovascular Physiology 8th edn, St Louis: Mosby.

other words, flow would be highly pulsatile throughout the whole of the circulation. In reality, there is still some compliance in the arteries of older people, but it is much lower than that in younger individuals. This decrease in compliance is indicated by the relationship between the increase in volume of an isolated large artery (e.g. the aorta) in response to increasing pressure—a relationship known as the pressure–volume relationship. It takes a greater increase in pressure to produce a given increase in volume in less compliant arteries, indicating an increase in the workload of the heart under such conditions.

Figure 14.8B shows that for a group of young people (20–24 years), the volume–pressure relationship is sigmoid, with the maximum increase in volume occurring over the normal pressure range of 10–18.7 kPa (75–140 mmHg). By contrast, a more hyperbolic curve exists for the aortae of

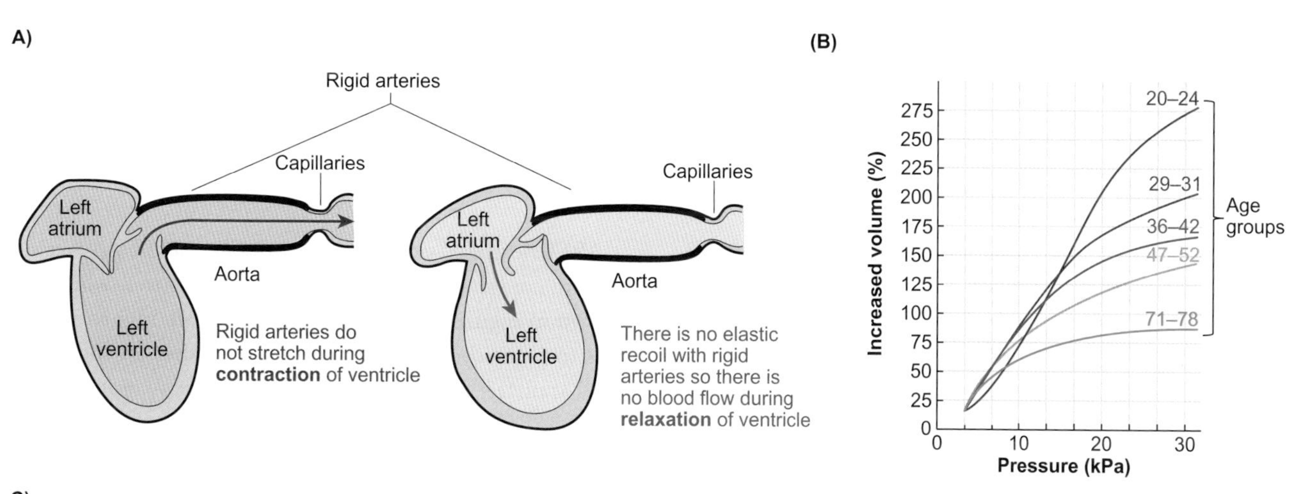

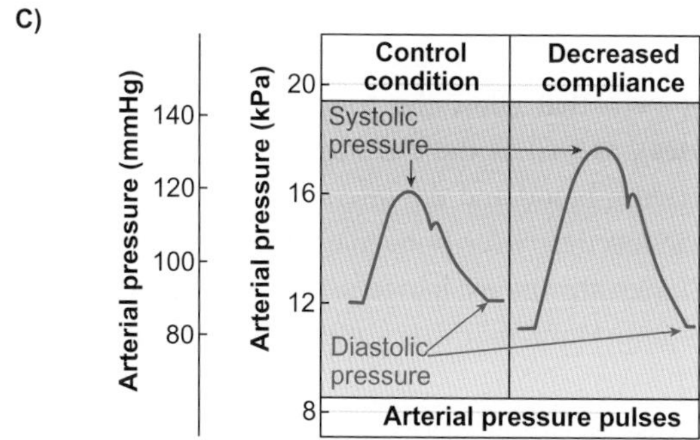

Figure 14.8 Effect of loss of compliance in the major arteries on pressure and flow in the circulatory system of humans

(A) If the major arteries lose their compliance completely and are rigid, all of the blood ejected from the heart during systole enters the peripheral circulation.
(B) Volume–pressure curves for the dorsal aorta obtained at autopsy from humans of different age groups. (C) A decrease in compliance of the major arteries causes a relatively large increase in systolic pressure and a slight decrease in diastolic pressure.

Reproduced from Berne RM and Levy MN (2001). Cardiovascular Physiology 8th edn. St Louis: Mosby. Berne RM and Levy MN (Eds) (1998) Physiology, 4th edn. St Louis: Mosby. Berne RM and Levy MN (2000). Principles of Physiology, 3rd edn. St Louis: Mosby.

Box 14.1 Blood pressure measurements in humans

Those of you who have had your blood pressure measured by your medical practitioner will have noticed that members of the medical profession use the millimetre of mercury (mmHg) as the unit of measurement, even though the Pascal (Pa, a force of one newton per square metre, Nm^{-2}) is the international unit for pressure. Converting between mmHg and Pa, or more realistically kPa, is quite simple: 1 kPa = 7.5 mmHg.

So, a blood pressure reading of 120/80 means that the peak pressure during systole—the **systolic pressure (SP)**—in the vessel in which the pressure is measured is 120 mmHg and the minimum during diastole—**diastolic pressure (DP)**—is 80 mm Hg. The difference between systolic and diastolic pressures (40 mm Hg) is known as **pulse pressure (PP)**. These pressures are equivalent to:

- SP = 120 mmHg/7.5 mmHg = 16 kPa,
- DP = 80 mmHg/7.5 mmHg = 10.7 kPa
- PP = 40 mmHg/7.5 mmHg = 5.3 kPa.

Mean blood pressure is calculated as DP + 1/3 PP. So, in this example, mean blood pressure is 93.3 mmHg (12.4 kPa).

older people for whom the increase in volume for a given increase in pressure is progressively less. The stiffening of the large arteries (atherosclerosis) in most older people, but also unusually in some younger people, characteristically increases systolic pressure and slightly decreases diastolic pressure, as shown in Figure 14.8C. Blood pressure measurements in humans are discussed in Box 14.1.

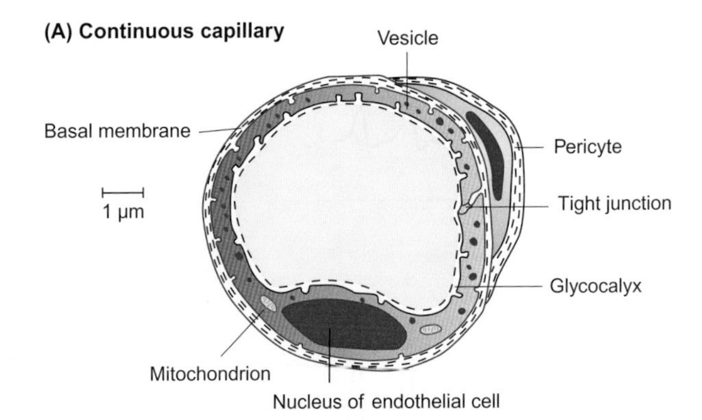

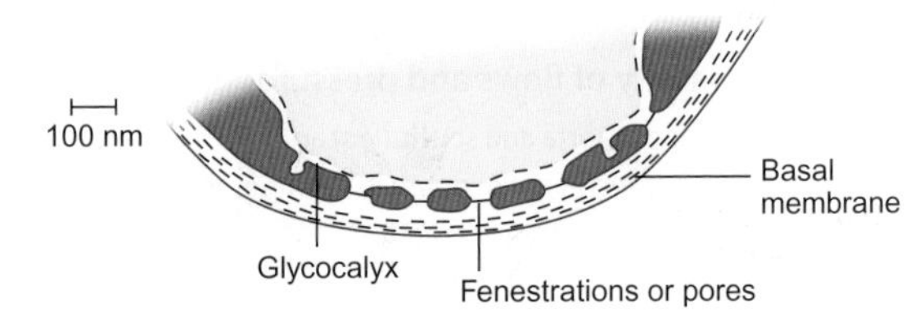

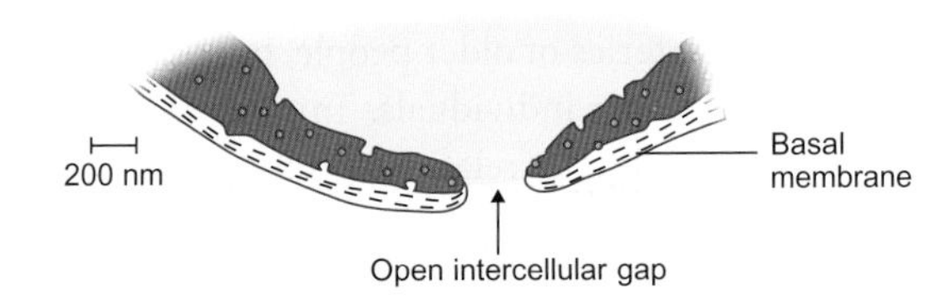

Figure 14.9 Capillaries in the circulatory system of mammals

(A) A continuous capillary showing the tight junctions between the evenly thick endothelial cells. (B) A fenestrated capillary showing the pores or 'windows' that penetrate the endothelial lining. (C) Discontinuous capillary showing wider gap between adjacent endothelial cells.

Reproduced from: Pocock G et al (2013). Human Physiology, 4th edn. Oxford: Oxford University Press.

14.1.5 Capillaries: the exchange vessels

The capillaries of the closed circulatory systems of vertebrates consist only of a tube of endothelial cells inside a delicate basement membrane, as shown in Figure 14.9. These are the exchange vessels of closed circulatory systems. It is across these delicate vessels that the exchange of gases, water, ions, metabolic substrates, metabolites and hormones take place between the organs/tissues and the circulating blood.

Figure 14.9A shows that the continuous capillaries have a complete endothelial lining and, as a result, are the least permeable of the capillaries. Continuous capillaries occur in skin, lung, muscle, fat and connective tissue. They permit the diffusion of gases, water, small solutes and lipid-soluble materials into the surrounding interstitial fluid. In the brain they are the 'supertight' capillaries, which do not allow the passage of any small solutes such as amino acids[9].

The fenestrated capillaries are of intermediate permeability and contain round pores 50 nm in diameter, which are shown in Figure 14.9B. These capillaries are highly permeable to water, solutes and small peptides. They are located in fluid-transferring tissues such as the kidneys[10], gut mucosa, exocrine and endocrine glands.

The discontinuous capillaries are the most permeable. They are similar to fenestrated capillaries, but are flattened and have wide, leaky gaps between adjacent endothelial cells (shown in Figure 14.9C) that allow the transfer of red blood cells (**erythrocytes**), as well as the exchange of water and plasma proteins between blood and the interstitial fluid. Discontinuous capillaries occur in liver, bone marrow and spleen.

As capillaries are the exchange vessels of a closed circulation, it is important that every metabolizing cell is close

[9] The blood–brain barrier is discussed in Box 16.1.

[10] Filtration in the mammalian kidney is discussed in Section 7.1.

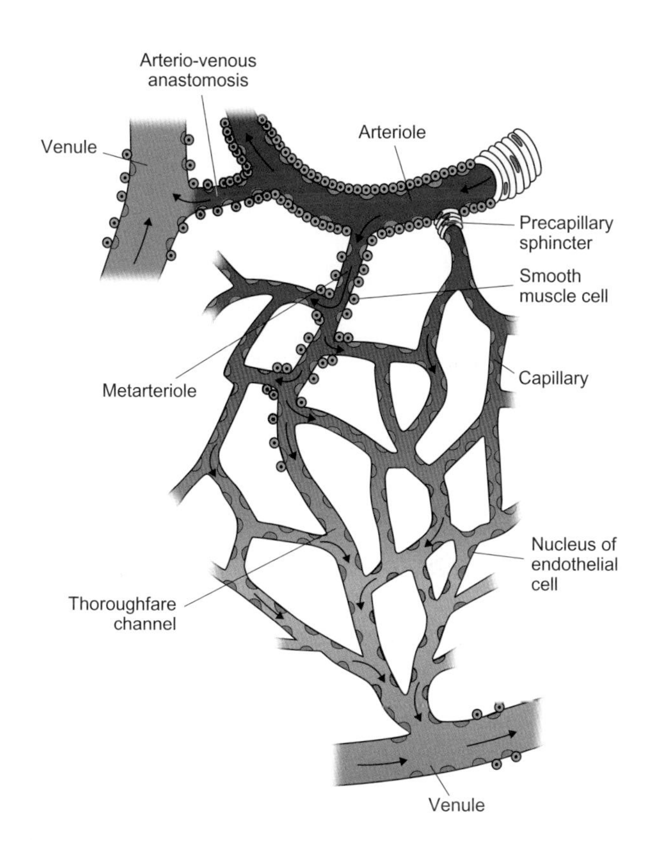

Figure 14.10 A capillary bed indicating a thoroughfare vessel and an arterio-venous anastomosis.

Precapillary sphincter muscles and/or smooth muscle around the length of a vessel control the flow of blood through parts of the capillary bed.

Adapted from: Hill RW et al (2016) Animal Physiology, 4th edn. Sinauer Associates: Sunderland, MA.

to a capillary so that it can obtain the oxygen, metabolic substrates and hormones it requires and can dispose of carbon dioxide and metabolic waste products. In mammals, most capillaries are approximately 8 μm in diameter and about 1 mm in length, with the average distance between them being approximately 40 μm. Most cells lie within 1–3 cell widths of a capillary. Thus, a complex bed of capillaries occurs between the arterioles and the venules, with each arteriole giving rise to many capillaries. This arrangement is shown diagrammatically in Figure 14.10.

At the entrance to each capillary is a band of smooth muscle known as a precapillary sphincter, the function of which is to control the flow of blood into the particular capillary. There are a number of direct connections between arterioles and venules. The initial segment of these connections is called the metarteriole and its walls contain smooth muscle. The remainder of each connection, which is like a typical capillary, is the thoroughfare channel. In addition, there are arterio-venous anastomoses, which also provide direct pathways from the arterioles to the venules. The flow of blood through these anastomoses is regulated by smooth muscle.

Gas exchange across capillaries

The factors that influence the exchange of oxygen and carbon dioxide across the capillaries in organs and tissues are much the same as those that influence their exchange across the respiratory organs—gills, lungs and tracheae—in particular those related to diffusion, as defined by Fick's principle[11]. These factors are:

- the difference in partial pressure of the particular gas between the capillary and the tissue it supplies[12];
- the area of capillary available for such exchange;
- the distance over which the exchange occurs;
- Krogh's diffusion constant for the gas through the tissue[11];

At the respiratory organs, the blood in the capillaries is loaded with oxygen; in the peripheral capillaries, this oxygen passes by diffusion to the metabolizing tissues.

Gas exchange across capillaries in organs/tissues

Ultrastructural studies of mammalian skeletal muscle have given us a reasonable idea of the relationship between the capillaries, which supply the oxygen, and the mitochondria in the cells, which consume it. Cross-sections of muscle fibres indicate that they are irregularly shaped and have a number of capillaries located predominantly around their exterior, as illustrated in Figure 14.11A. Figure 14.11B shows that capillaries tend to run parallel to the length of the fibres. So, we can assume that oxygen will diffuse out radially from the capillaries as indicated by the circles in Figure 14.11A. As the capillaries are relatively long, they will supply a cylinder of tissue, which is shown diagrammatically in Figure 14.11C.

The radius of such a cylinder will depend on the number of capillaries per unit cross-sectional area of tissue, called the **capillary density**. In the leg muscles of humans, capillary density is approximately 400 mm^{-2}, the average radius of each cylinder of tissue supplied by a capillary is 28 μm and the length of a capillary from arteriole to venule is approximately 0.5–1.0 mm. This gives some idea of the volume of the cylinder of tissue that is supplied by each length of capillary in the skeletal muscle of humans. As August Krogh was the first to discuss the supply of oxygen from capillaries in such terms, the cylindrical unit is commonly known as a **Krogh cylinder**.

The oxygen demand of each cylindrical unit of tissue within a cell depends on the volume of mitochondria within each cylinder. Some muscle fibres are highly oxidative

[11] The Fick principle of diffusion and Krogh's diffusion constant in respiratory systems are discussed in Section 11.2.1.

[12] Partial pressures of gases are discussed in Section 11.1.2.

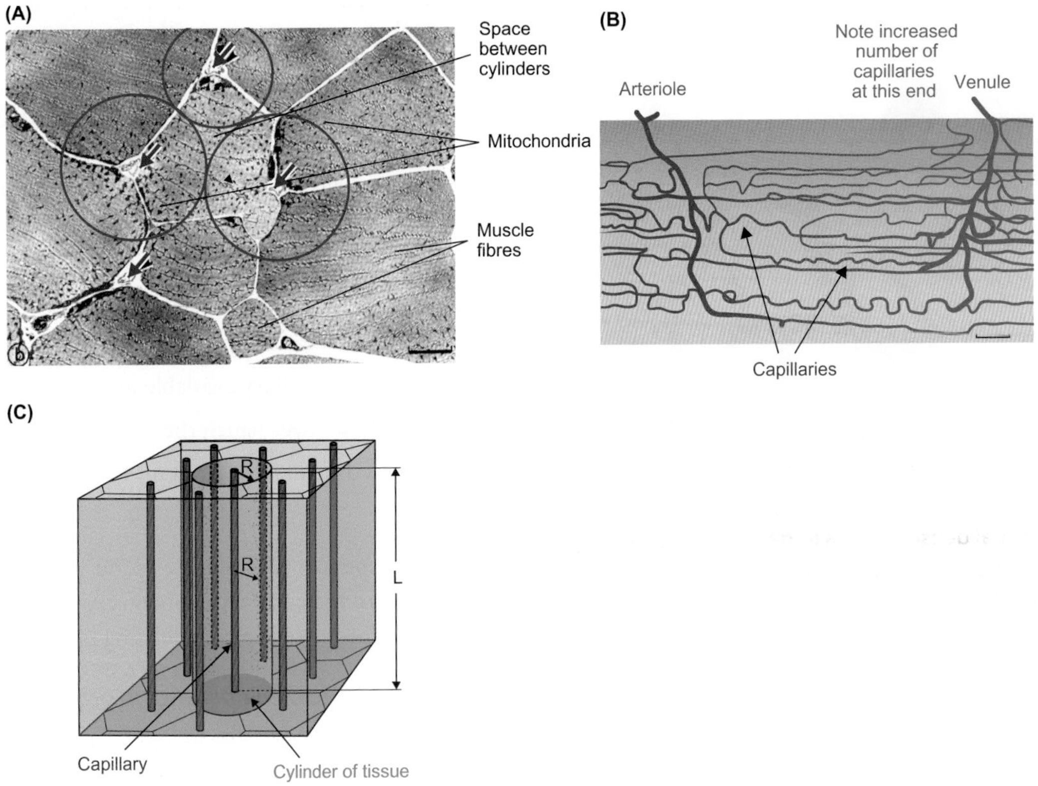

(A)

(B)

(C)

Figure 14.11 Arrangement of blood capillaries in skeletal muscle of mammals

(A) Cross-section of muscle fibres in a leg muscle of a small gazelle showing the location of the capillaries, indicated by the purple arrows, and mitochondria. The red circles indicate the areas over which oxygen will need to diffuse from three capillaries in order to supply the local mitochondria. Scale bar, 10 μm. (B) Capillaries in skeletal muscle of a mammal from the arteriole to the venule. Scale bar, 100 μm. (C) Idealized model to show the cylinder of tissue (Krogh cylinder) of radius, R, and length, L, which is supplied by oxygen from the capillary in its centre. In this model, the capillaries are parallel and evenly spaced, but this is not always the case, as shown in B.

Adapted from Weibel ER (1984). The Pathway for Oxygen. Cambridge, MA: Harvard University Press.

and hence have a relatively large density of mitochondria. For example, there are 0.34 cm^3 of mitochondria per cm^3 of muscle in the flight muscle of hummingbirds, *Selasphorus rufus*. By contrast, other muscle fibres are less oxidative, with relatively low mitochondrial densities (0.06 $cm^3\ cm^{-3}$ in sartorius muscle of gray tree frogs, *Hyla versicolor*)[13]. As might be expected, there is a positive relationship between the density of capillaries in a muscle and the density of mitochondria, as shown in Figure 14.12. These data were obtained from various types of skeletal muscle from 13 species of African mammals ranging in mass from 0.4 to 250 kg, plus some data from brown rats (*Rattus norvegicus*) and shrews. On average, each μm^3 of capillary blood supplies about 2.6 μm^3 of mitochondria.

As blood passes through the capillary, oxygen is consumed by the organ/tissue. In humans, the partial pressure of oxygen falls from approximately 12.5 kPa at the arteriolar end of a capillary to about 5.5 kPa at the venular end. As a result, the difference in partial pressure of oxygen between the blood in the capillaries and in the mitochondria of the surrounding cell decreases as the blood moves along the capillary. This means that the partial pressure of oxygen in the mitochondria close to the venular end of the capillary will not only be lower than that at the arteriolar end, but will be even lower still in mitochondria in the surrounding tissue, some distance from the capillary.

If it is assumed that the mitochondria are evenly distributed throughout a cell, the partial pressure of oxygen (PO_2) will fall in a parabolic fashion with increasing distance from the capillary. Figure 14.13 shows that if the radius of the cylinder is sufficiently large, PO_2 will eventually reach zero at the periphery. In addition, PO_2 of the blood in the capillary will be lower toward the venular end as a result of the oxygen that has been consumed as the blood passes along the capillary. Figure 14.13 also shows that this consumption of oxygen leads to the profile of PO_2 from the capillary into the tissue at the venular end being different from that at the arteriolar end. Thus, a PO_2 of zero will be reached closer to the capillary at the venular end.

There is, therefore, a maximum distance that oxygen can diffuse, beyond which the tissue becomes devoid of oxygen

[13] Types of muscle fibres are discussed in Section 18.1.4.

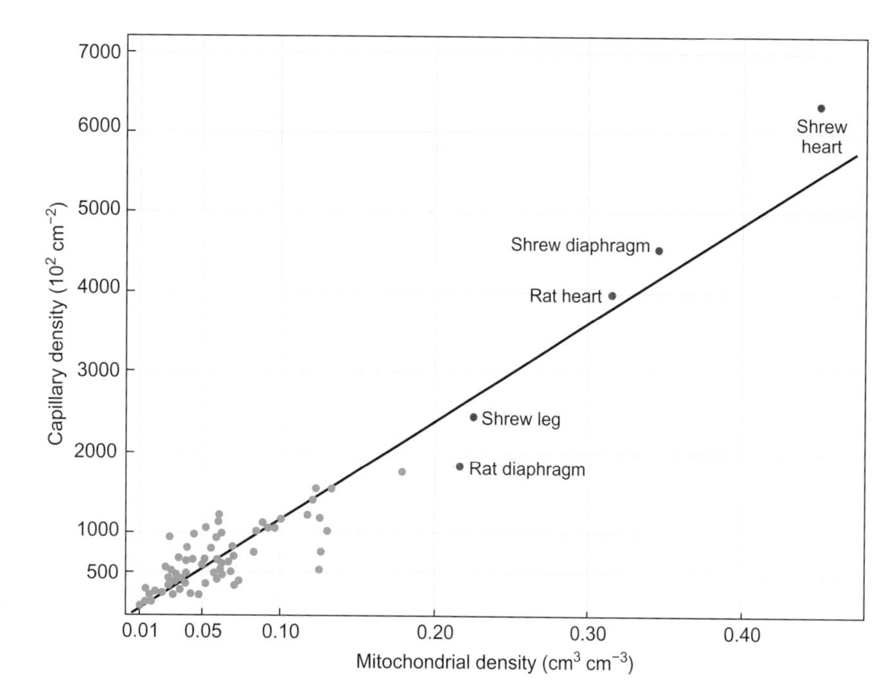

Figure 14.12 Relationship between capillary density (number of capillaries per square cm of muscle cross-section) and mitochondrial density from a large range of mammals and types of muscles

Adapted from Weibel ER (1984). The Pathway for Oxygen. Cambridge, MA: Harvard University Press.

(it would be **anoxic**). It is calculated that the maximum diffusion distance before the partial pressure of oxygen reaches zero, at maximum rate of oxygen consumption and for values of PO_2 of approximately 5 kPa at the venular end of capillaries, is 25–50 μm. The average radius of a Krogh cylinder in the leg muscle of humans (28 μm) is well within this range, so there may be no anoxic region in muscle tissue, except perhaps at very high rates of oxygen consumption, and then only at the venular end. Nonetheless, the lower PO_2 at the venular end raises two questions: Is tissue at the venular end supplied less well with oxygen? If so, does it, function less well than that at the arteriolar end?

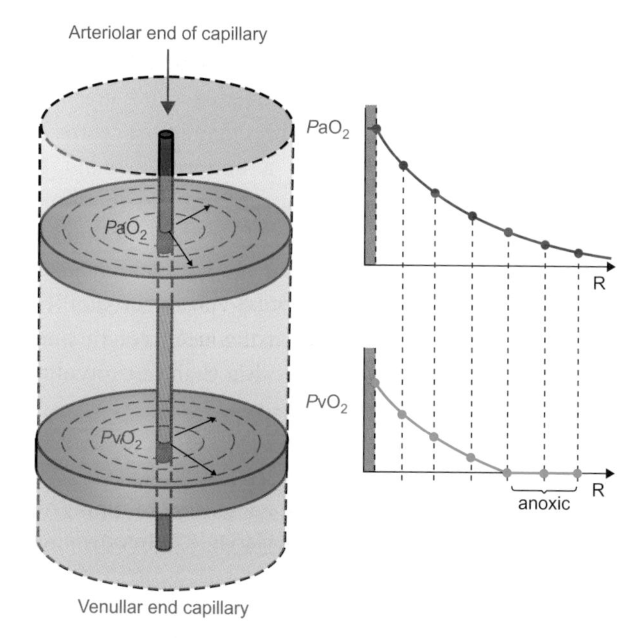

Figure 14.13 Profiles of partial pressure of oxygen along a Krogh cylinder

As the partial pressure of oxygen in the blood (PO_2) is higher at the arteriolar end of the capillary (P_aO_2) than at the venular end (P_vO_2), oxygen diffuses to a greater distance into the cylinder at the arteriolar end. Thus, at a given distance from the capillary, the partial pressure of oxygen is lower at the venular end, where PO_2 may be zero (anoxic) at the most distant locations. R is the radial distance.

Adapted from Weibel ER (1984). The Pathway for Oxygen. Cambridge, MA: Harvard University Press.

Both physiological and morphological factors may reduce the diffusional limitations to oxygen transfer from the capillaries to the centre of their surrounding cylinders of tissue. The physiological factors are the **Bohr shift** and **facilitated diffusion** by way of myoglobin[14].

The diffusion of oxygen out of the blood in the capillaries to the mitochondria is accompanied by the inward diffusion of carbon dioxide, which leads to a reduction in pH as the blood passes from the arteriolar to the venular end. This causes a rightward shift of the oxygen equilibrium curve (the Bohr shift), leading to a higher PO_2 in the blood at a given concentration (or per cent saturation) of oxygen than it would be the case at a higher pH, as shown in Figure 14.14A.

The oxidative muscle fibres contain more myoglobin than the glycolytic fibres, and the facilitated diffusion of oxygen by myoglobin may play an important role in maintaining an adequate supply of oxygen to the regions of the cell farthest from the capillary at its venular end.

In addition to these physiological factors, morphological characteristics may contribute to the maintenance of an

[14] The Bohr effect and facilitated diffusion by myoglobin are discussed in Sections 13.1.2 and 13.1.1, respectively.

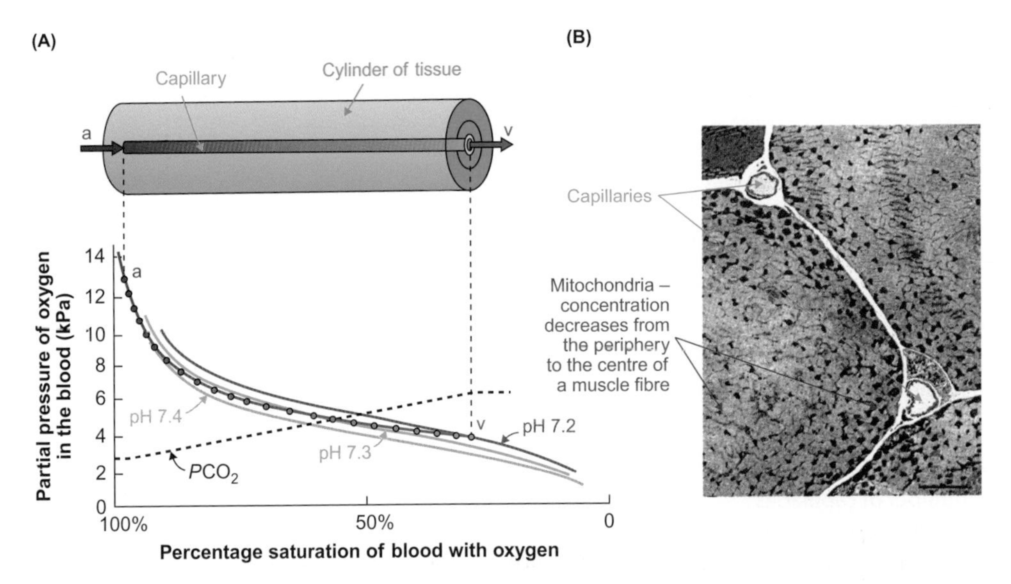

Figure 14.14 Two factors affecting the profile of partial pressure of oxygen along a Krogh cylinder

(A) Uptake of carbon dioxide by the blood as it passes along the capillary increases partial pressure of carbon dioxide (PCO_2) causing a reduction in pH of the blood (from 7.4 at the arteriolar end, a, to 7.2 at the venular end, v). The reduction in pH causes a rightward shift in the oxygen equilibrium curve (Bohr shift). This shift means that at a given percentage saturation of the blood with oxygen, the partial pressure of oxygen is higher at the venular end (v) than it otherwise would be. Dots indicate values as pH changes from 7.4 to 7.2 along a capillary. (B) Micrograph showing decreasing concentration of mitochondria from the periphery to the centre of a muscle cell. This means that the demand for oxygen is not constant throughout the Krogh cylinder from the capillary. Scale bar 5 μm.

Adapted from Weibel ER (1984). The Pathway for Oxygen. Cambridge, MA: Harvard University Press.

adequate supply of oxygen to the mitochondria throughout the cell. One of these characteristics is an increase in the number of capillaries toward their venular end, which is demonstrated back in Figure 14.11B. This means that the distance into the cell supplied by one capillary would decrease, thus ensuring a more homogeneous oxygenation of the muscle fibres along their length. Nonetheless, PO_2 would still be higher close to the capillary than deeper in the cell. Thus, although mitochondria can function at a PO_2 of 0.13–0.25 kPa, there would be less oxygen available at the centre of a muscle fibre than at its periphery. However, there is some evidence that the density of mitochondria in muscle cells is not uniform throughout the cell, but decreases from the periphery to the centre, as shown in Figure 14.14B. Such an arrangement would suggest that the demand for oxygen is greater in the peripheral regions of muscle fibres than in the centre.

The Krogh cylinder assumes a uniform demand for oxygen around a capillary and a uniform distribution of capillaries. In fact, the distribution of capillaries in muscles is heterogeneous and they are not always surrounded by muscle fibres with the same metabolic characteristics. For example, slow oxidative and slow oxidative/glycolytic fibres will have a greater demand for oxygen than fast glycolytic fibres, as illustrated in Figure 14.15A.

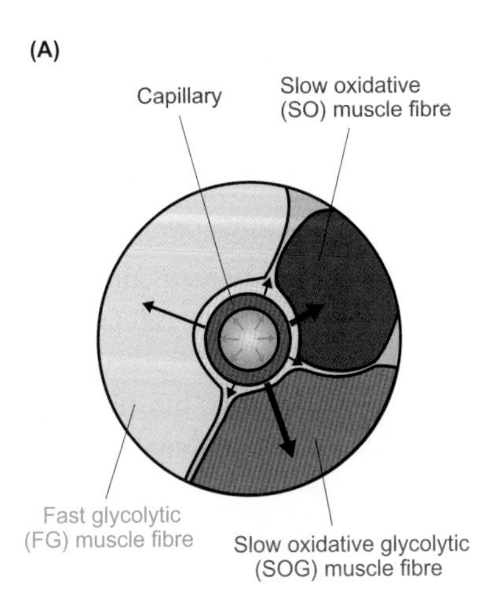

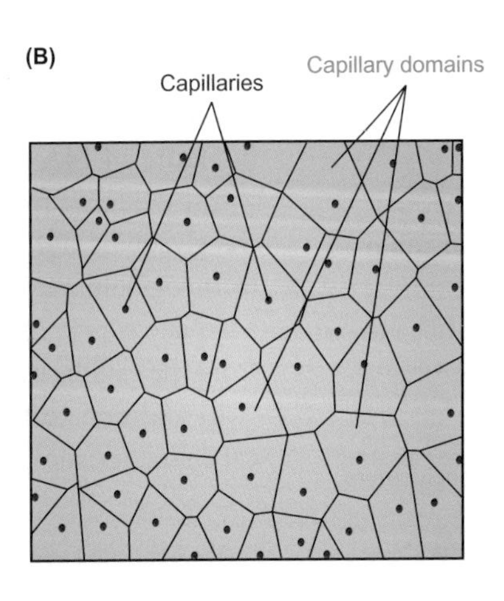

Figure 14.15 Characteristics of supply of oxygen from capillaries in heterogeneous muscles

(A) Capillary surrounded by three different types of muscle fibres with different demands for oxygen. Length of arrows indicates distance over which oxygen diffuses and their thickness indicates the amount of oxygen that diffuses. (B) Diagram showing uneven distribution of capillaries. Each polygon signifies the area of tissue potentially supplied by the capillary enclosed by the polygon. These areas are known as capillary domains.

Reproduced from Egginton S and Gaffney E (2010). Tissue capillary supply – it's quality that counts! Experimental Physiology 95: 971-979.

A shortcoming of the Krogh cylinder model is that there are spaces between adjacent cylinders, as shown in Figure 14.11A. However, a recent modification of the Krogh cylinder model describes a space-filling equivalent of the Krogh cylinder. This new model is illustrated in Figure 14.15B; it considers capillaries as the sources of oxygen and identifies the area of tissue closer to one capillary than another. This produces a pattern of polygons, known as capillary domains, as shown in Figure 14.15B. This recent model casts doubts on some of the conclusions reached based on more conventional approaches, but more studies are needed to determine whether or not the doubts are justified. All following discussions are based on data obtained from the traditionally-used methods.

Some muscle fibres are predominantly glycolytic in nature; the capillary density in such fibres is not always related to rate of oxygen demand. For example, in the glycolytic gracilis muscles of cats (superficial muscles found in the thigh), capillary density is proportional to the rate of lactate clearance.

Capillary density and angiogenesis in skeletal muscles

Table 14.1 demonstrates that different species exhibit a wide range of capillary densities in their muscles, even in the most oxidative of their muscle fibres. Within oxidative muscle fibres, the density of capillaries is matched to the oxygen demands of those muscles. The data in Table 14.1 indicate that capillary density in oxidative (red) muscle fibres of brown trout (*Salmo trutta*) is comparable to that in the limb muscles of mammals, but skipjack tuna (*Katsuwonus pelamis*) have a density that is almost twice that of trout. In contrast, capillary density is relatively low in the red muscle of haemoglobinless black ice fish (*Chaenocephalus aceratus*).

Most of the locomotory muscle fibres in amphibians and reptiles have a relatively low capacity for oxidative metabolism, hence the density of blood capillaries in these muscles is also relatively low. It is of note that the density of the capillaries is much lower in an oxidative muscle of the body wall that is involved in ventilation and jetting in cuttlefish (*Sepia* spp.) than in the leg muscles of brown rats (*Rattus norvegicus*). This suggests that the aerobic capacity of cuttlefish is also low.

High capillary densities are not only found in locomotor muscles. The calling muscles of some frogs, such as the external oblique muscle in gray tree frogs (*Hyla versicolor*), have higher capillary densities than those in the locomotor muscles from the thigh of the same species. This difference is related to the high oxidative capacity of the muscles and to the ability of tree frogs to maintain rates of oxygen consumption while calling (around 2500 calls h^{-1} for 2–4 h) that are almost twice those obtained during short bouts of vigorous exercise lasting just a few minutes. In desert iguanas (*Dipsosaurus dorsalis*), capillary density is 8.5 times greater in the red, oxidative fibres of one of its hindlimb locomotor muscles than in the white fibres of the same muscle, and virtually the same as that in the calling muscles of frogs. However, as shown in Table 14.1, they are both about half that in the soleus muscle from the calf of a rat.

Table 14.1 Density of capillaries in muscles of cephalopod molluscs and vertebrates

The capillary density is greater in the more aerobic muscle fibres of a given species and in the locomotor muscles of more active species (in red).

	Capillary density (caps mm^{-2})	
	'Red', aerobic muscle fibres	'White', glycolytic muscle fibres
Sepia, cuttlefish	160 (body wall muscle)	
Chaenocephalus, haemoglobinless black icefish	550	
Salmo, brown trout	1800	230
Katsuwonus, skipjack tuna	3400	
Hyla, gray tree frog	700 external oblique muscle (trunk muscle)	230 sartorius muscle (thigh muscle)
Dipsosaurus, desert iguana	630 iliofibularis muscle (hindlimb muscle)	75 iliofibularis muscle
Columba, homing (common) pigeon	6180 flight muscle	
Selasphorus, hummingbird	8775 flight muscle	
Rattus, brown rat	1300 soleus muscle (calf muscle)	
Eptesicus, brown bat	6395 flight muscle 2865 hindlimb muscle	

References: Abbott and Bundgaard, 1987; Fitch et al., 1984; Day and Butler, 2005; Mathieu-Costello et al., 1996; Marsh and Taigen, 1987; Gleeson et al., 1984; Mathieu-Costello, 1991; Mathieu-Costello et al., 1992a; Mathieu-Costello et al., 1992b. Full references available online.

The highest capillary densities in locomotor muscles among the vertebrates are found in the flight muscles of birds and bats, as demonstrated in Table 14.1. These are almost 5–7 times as great as that in the soleus muscles of brown rats, with that in hummingbirds being the highest of all. The high level in hummingbirds relates to the fact that during hovering flight, their flight muscles have the highest maximum rate of oxygen uptake per unit mass of any vertebrate muscle (90 μM g^{-1} min^{-1} compared to an estimated maximum of 39 μM g^{-1} min^{-1} in the flight muscles of tufted ducks, *Aythya fuligula*).

The close relationship between capillary density in an oxidative muscle and its oxidative metabolic rate may be the result of local hypoxia (unusually low PO_2) being an important stimulus for growth of the vascular system. Any fall in the supply of oxygen in relation to its demand, causes local hypoxia which stimulates capillary growth. In early development, cells are oxygenated by diffusion, but when the

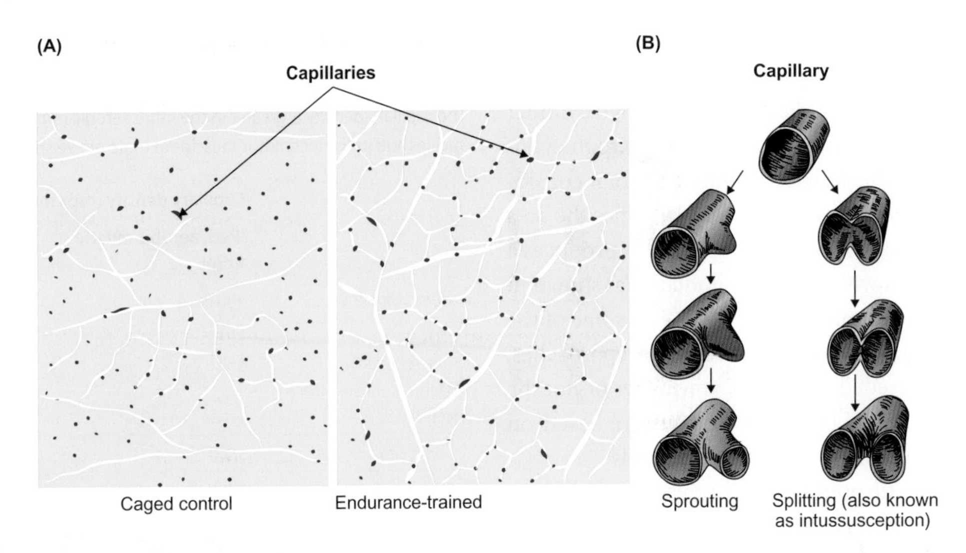

Figure 14.16 Angiogenesis in skeletal muscles of mammals

(A) Increased capillarity (capillaries are the dots) as a result of angiogenesis in the superficial gastrocnemius muscle of a rat following endurance-type exercise training. The two photomicrographs are at the same magnification. (B) Angiogenesis can occur by endothelial cells of a capillary branching out (sprouting) into the surrounding tissue to form a tube which eventually joins with another capillary or venule, or by a single capillary splitting into two capillaries (splitting).

Reproduced from Prior B et al (2004). What makes vessels grow with exercise training? Journal of Applied Physiology 97: 1119-1128.

embryo grows beyond the limit of diffusion, hypoxia stimulates growth of blood vessels by way of hypoxia-inducible transcription factors (HIFs). The formation of the blood vessels in the embryo is known as vasculogenesis, which is the differentiation of precursor cells (angioblasts) into endothelial cells. The endothelial cells then form a network of small capillaries. The growth and remodelling of this early network of endothelial-lined blood vessels into a more complex network is known as **angiogenesis**.

Angiogenesis in adults is caused by a number of factors, including the menstrual cycle and pregnancy in placental mammals, wound healing and endurance exercise, as shown in Figure 14.16A. Angiogenesis is also a fundamental step in the transition of a benign tumour into a malignant one and in diabetic retinopathy. All of these processes involve a local shortage of oxygen, either as a result of an increase in demand for oxygen or a local reduction in supply, such as capillary damage during wounding. We concentrate here on angiogenesis associated with exercise.

Angiogenesis can occur in one of two ways:

- by a capillary splitting into two capillaries, (also known as intussusception), or
- by a new capillary branching out (sprouting), from an existing capillary.

Both of these processes are illustrated in Figure 14.16B. Two of the most important factors associated with angiogenesis accompanying exercise are nitric oxide and vascular endothelial growth factor (VEGF). Production of nitric oxide can be critical to VEGF signalling, but exactly how exercise induces VEGF and causes angiogenesis is unclear. We explore some of the factors which may mediate exercise-induced angiogenesis in Experimental Panel 14.1, see Online Resources.

Capillaries in gas-exchange organs

Capillaries in gas-exchange organs form dense mesh-like networks composed of very short segments, with their length barely greater than their diameter. The meshwork is so dense that it has been referred to as a sheet of blood. So, the pattern of blood flow through the alveolar walls of the lungs of mammals has been called sheet flow, rather than flow through a system of interconnected tubes. This arrangement is shown in Figure 14.17A.

The walls of each alveolus in the lungs of adult mammals contains a sheet of blood bounded by two flat membranes of endothelium held and supported by 'struts' or 'columns' which are the diameter of one capillary apart. When blood flows through the sheet, it is free to move along a tortuous route between the columns. This arrangement explains why blood flow is not interrupted when some parts of the sheet become flattened at high lung inflation pressure. In addition, the sheet is continuous through the walls of many alveoli and most likely throughout an entire unit of gas exchange (an acinus).

This general pattern is found in both gill, and lungs, as illustrated in Figure 14.17B, and of virtually all species except birds. This pattern provides the greatest possible area for gas exchange between the blood and the external environment. In the lungs of birds, however, blood capillaries inter-digitate

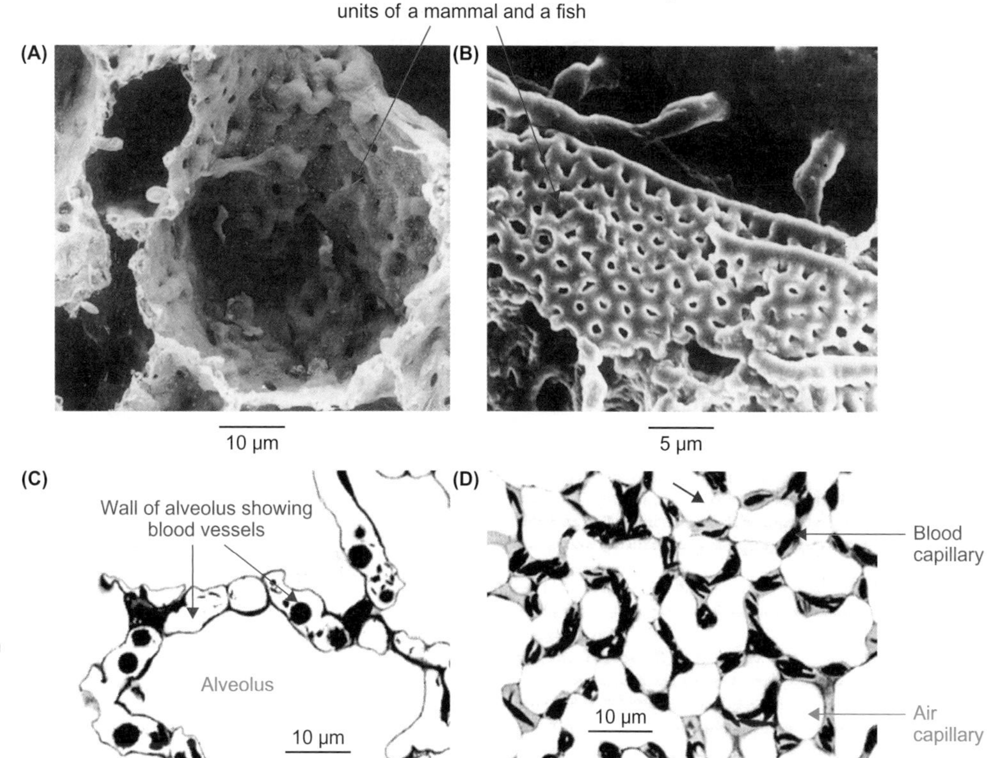

Figure 14.17 Microvasculature of gas exchangers

(A) Cast of the blood vessels of an alveolus of the lung of a bushbaby (*Galago* spp.). (B) Cast of the blood vessels of a gill filament from the Lake Magadi tilapia (*Alcolapia grahami*). (C) Sections through the blood vessels in the alveolar wall of a rabbit lung (left) and within the parabronchi of the lung of a chicken (right). Note the dense packing of the air capillaries around the blood capillaries and the thinner walls of the blood capillaries in the lungs of birds. Purple arrow indicates strand of tissue supporting the blood capillaries.

A, B: Reproduced from Maina JN (2000) Is the sheet-flow design a 'frozen core' (a Bauplan) of the gas exchangers? Comparative functional morphology of the respiratory microvascular systems: illustration of the geometry and rationalization of the fractal properties. Comp Biochem Physiol A 126: 491-515. C, D: West JB et al (2006). The honeycomb-like structure of the bird lung allows a uniquely thin blood-gas barrier. Respiratory Physiology & Neurobiology 152: 115-118.

extensively with the air capillaries[15], as shown in Figure 14.17C. The arrangement of the capillary blood in the lungs of birds creates a three-dimensional volume of blood rather than a two-dimensional sheet of blood. In fact, this dense packing of relatively rigid air capillaries around the blood capillaries may provide the blood capillaries with the necessary mechanical strength for their walls to be so thin and uniform compared with those of mammals[16].

The PO_2 in the blood changes asymtotically during its passage through the capillaries of an alveolus in the mammalian lung. The shape of this profile results mainly from the decrease in the difference of partial pressure of oxygen in the capillary and the lumen of the alveolus as the blood passes through the capillary system. When an animal is at rest, the blood equilibrates with oxygen in the alveolus after less than half of its passage through the capillaries. By contrast, during exercise, blood flows through the capillaries much faster and equilibration occurs only toward the end of the capillaries. The time taken for the blood to pass through the capillaries is known as the **pulmonary capillary transit time**[17]; for humans this is approximately 2.8 s at rest and 0.6 s during exercise. In some smaller animals, such as mice, the capillary transit time during exercise (approximately 0.12 s) may not be sufficient to allow complete saturation of the haemoglobin.

Fluid exchange across capillaries and movement through the lymphatic system

Capillaries are permeable to water, ions, and small solutes as well as to gases. As with so many other substances we have discussed, the movement of water across capillaries depends on the difference in pressure (kPa) across the walls of the capillaries, and the permeability of the capillary walls to water.

Fluid exchange across capillaries

There are two sources of pressure as far as water movement is concerned:

- the pressure generated by the heart, that is, **blood pressure** (a hydrostatic pressure), which tends to force water out of the plasma into the interstitial space;
- the osmotic pressure created by colloidal macromolecules—proteins. This is known as **colloid osmotic pressure** (COP) or **oncotic pressure**. COP is greater in the blood plasma than in the interstitial fluid because of the relative concentrations of proteins[18], so it tends to draw water out of the interstitial fluid into the blood.

[15] Air capillaries in the lungs of birds are discussed in more detail in Section 12.3.3.

[16] The structure of the lungs of birds and mammals are discussed in Section 12.3.3.

[17] Capillary transit time is discussed in Section 15.2.3.

[18] Similar principles apply in the kidney and determine filtration in the glomeruli, which we discuss in Section 7.1.2.

14

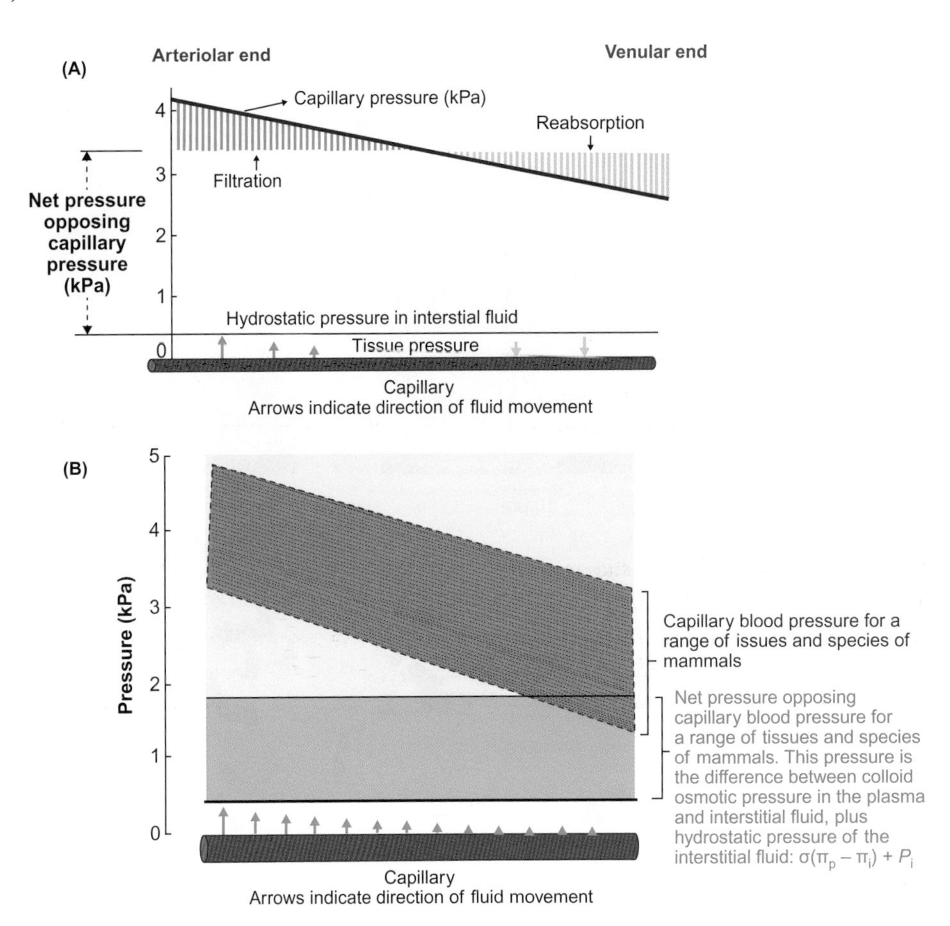

Figure 14.18 Fluid exchange across the capillaries of mammals

(A) The traditional model in which there are similar volumes of fluid reabsorbed at the venular end of a capillary as filtrated at the arteriolar end. (B) More recent model in which blood pressure at the venular end of capillaries is rarely less than colloid osmotic and tissue hydrostatic pressure. This model incorporates more recently measured values of interstitial colloid osmotic and hydrostatic pressures and the fact that the capillary does not retain (reflect) all of the solute in the plasma, i.e. some passes into the interstitial fluid in most tissues. σ, reflection coefficient, π_p, colloid osmotic pressure in plasma, π_i, colloid osmotic pressure in interstitial fluid, P_i, hydrostatic pressure in interstitial fluid.

A: Reproduced from Rushmer RF (1976). Cardiovascular Dynamics 4th edn. Philadelphia: W.B. Saunders Company. B: adapted from: Levick JR (1995). Changing perspectives on microvascular fluid exchange. In: Cardiovascular Regulation. Jordan D and Marshall J (eds.), pp 127-152. Portland Press.

Figure 14.18 shows that blood pressure is substantially higher at the arteriolar end of capillaries than at the venular end. The traditional view, shown in Figure 14.18A, is that blood pressure is higher than tissue hydrostatic pressure and plasma COP (net pressure in Figure 14.18A) at the arteriolar end of a capillary and lower than tissue hydrostatic pressure and COP at the venular end. The amount of fluid that is filtered at the arteriolar end is more or less balanced by the amount that is absorbed at the venular end. Any fluid remaining in the interstitial fluid, along with proteins, such as albumin, that have escaped into the interstitial space, enters the **lymphatic system** and returns to the central circulatory system.

The rate of movement of fluid (J_v) across capillaries is dependent on the relationship between:

- hydrostatic pressures (P) in the capillaries (P_c) and interstitial fluid (P_i)
- COP (π) in the plasma (π_p) and interstitial fluid (π_i).

Their importance in controlling the movement of fluid across the capillaries is described as follows:

Rate of movement of fluid across capillaries of unit area (mL min^{-1} mm^{-2})

Filtration coefficient (mL min^{-1} kPa^{-1} mm^{-2})

$$J_v / A = K_f[(P_c - P_i) - (\pi_p - \pi_i)] \quad \text{Equation 14.1}$$

Difference in hydrostatic pressure between capillaries and interstitial fluid (kPa)

Difference in colloid osmotic pressure between the capillaries and interstitial fluid (kPa)

where A is the area of capillaries across which filtration occurs (mm^2), K_f is the permeability or, more correctly, the filtration coefficient (mL min^{-1} kPa^{-1} mm^{-2}) of the capillaries. When the value of J_v is positive, fluid is filtered from the capillaries into the interstitial fluid; when it is negative, fluid is absorbed from the interstitial fluid into the capillaries.

The COP of plasma is, on average, generally relatively low in most invertebrates (<1 kPa), and slightly higher in vertebrates (<1.4 kPa), excluding mammals. In mammals it is around 3.0 kPa (2.6 kPa in cats, 3.3 kPa in humans). The traditional model illustrated in Figure 14.18A assumes that interstitial COP is essentially zero and includes a slightly positive (i.e. above ambient) interstitial hydrostatic pressure. However, accurate measurements of these variables and recognition that effective osmotic pressure of plasma proteins at the walls of capillaries is slightly less than expected, has led to a different model, shown in Figure 14.18B, in which almost no absorption occurs at all.

Pressures measured in human skin indicate that interstitial hydrostatic pressure is actually slightly sub-ambient (negative, −0.3 kPa), but interstitial COP is approximately 2.1 kPa. Both of these situations favour fluid filtration. The reason for the presence of a COP in the interstitial fluid is because most continuous and all fenestrated capillaries allow some small proteins to pass through them and into the interstitial space, as discussed above.

The supertight capillaries of the brain do not allow the passage of any plasma proteins and are said to have a reflection coefficient (σ) of 1[19]—that is, they 'reflect back' all of the plasma proteins, allowing none to pass through into the interstitial space. However, only about 80–95 per cent of proteins are held back by most continuous and all fenestrated capillaries, so they have reflection coefficients of 0.8–0.95. This means that the plasma proteins only exert 80–95 per cent of their COP when determining the filtration of fluid across the capillaries. This again favours fluid filtration. So, it is necessary to modify equation 14.1 by incorporating the reflection coefficient:

Reflection coefficient

$$J_v / A = K_f[(P_c - P_i) - \sigma(\pi_c - \pi_i)] \qquad \text{Equation 14.2}$$

All of these modifications mean that, for most capillary beds, there is a progressive decrease in filtration along the vessels, reaching virtually zero at the venular end. Having said that, absorption of interstitial fluid by capillaries does occur in some tissues specialized for fluid uptake such as intestinal mucosa, parts of the kidney in birds and mammals[20] and the lymph nodes. Also, a reduction in capillary hydrostatic pressure, either as a result of constriction of arterioles or blood loss (haemorrhage), can lead to the transient absorption of interstitial fluids.

The model depicted in Figure 14.18B has been further modified to incorporate the idea that the glycocalyx[21] (which is shown in Figure 14.9A and B) is the semi-permeable membrane of the endothelium and that colloid osmotic pressure beneath the glycocalyx, π_g, should be used in equation 14.2 in place of π_i. This modification leads to the net filtration force being smaller than that indicated in Figure 14.18B. Any fluid that does filter into the interstitial space, along with proteins, such as albumin, that escape from the plasma, enters the lymphatic system.

The secondary circulation of fish and the lymphatic systems of other vertebrates

Teleost fish do not have a lymphatic system as such, but they are unique among the vertebrates in possessing a secondary circulation in parallel with the primary circulation[22]. The secondary circulation is formed from the walls of certain vessels of the primary system by way of many small (7–15 μm diameter) arterioles, which join to form progressively larger secondary arteries. Secondary capillaries have been found in many locations, such as the gills, skin and lining of the mouth. Their endothelium is thin-walled and reminiscent of the lymphatic vessels in mammals, which we discuss below.

The secondary veins drain into large cutaneous veins and from there into the primary venous system. The central sinus and associated blood vessels (interlamellar system) in the gill filaments[23] are probably part of the secondary circulation. Blood flow through the secondary venous circulation is assisted by one or more caudal pump (sometimes called a caudal heart), which consists of venous sinuses that are rhythmically compressed by the muscles of the tail during swimming. The physiological function of the secondary circulation is uncertain, but the lack of a clearly identified lymphatic system in fish has led to the idea that the secondary system is analogous to, or the precursor of, the lymphatic system in other vertebrates.

Lungfish do not possess a secondary circulation as found in teleosts. Instead they have a distinct lymphatic system, i.e. there is no direct connection with the blood (primary) circulation; instead, the blood transports any surplus interstitial fluid and protein back to the venous side of the circulation. Lymphatic capillaries are distributed throughout the body, except the central nervous system. They form small units which are interconnected and the lymph is pumped into adjacent blood vessels by a number of muscular micropumps.

[19] Glomerular capillaries in the kidneys of most vertebrates have a reflection coefficient close to 1.

[20] Reabsorption of water in the kidneys of birds and mammals is discussed in Section 7.2.3.

[21] The glycocalyx is found on the apical surface of vascular endothelial cells which line the lumen of capillaries. It affects the filtration of fluid from capillaries into the interstitial space, among other things.

[22] The secondary circulation in fish is shown diagrammatically in Figure 14.30A.

[23] Circulation through the gills of fish is discussed in Section 14.4.1.

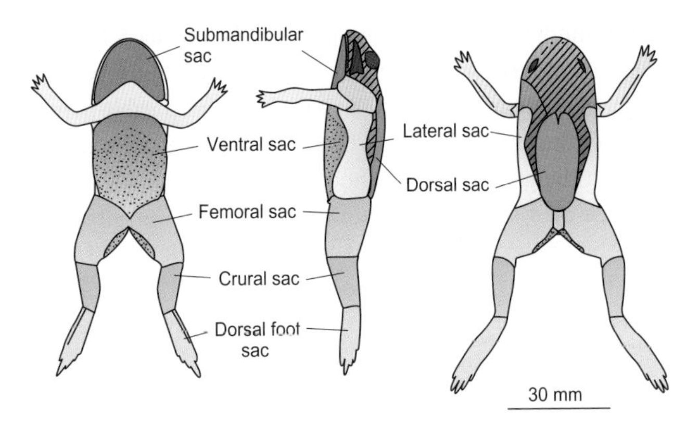

Figure 14.19 Lymphatic system of amphibians

Location of major lymph sacs in terrestrial amphibians, Mallee spadefoot toads (*Neobatrachus pictus*).

Reproduced from: Toews DP and Wentzell LA (1995). The role of the lymphatic system for water balance and acid-base regulation in the Amphibia. In: Adv Comp Env Physiol. 21. Heisler N (ed.), 201-214. Heidelberg: Springer-Verlag.

Figure 14.20 The lymphatic system of mammals

Network of blood capillaries and lymph capillaries. Green arrows show movement of fluid from the blood vessels and, via the interstitial tissue, to the lymph vessels.

Reproduced from Martini FH et al. (2012). Fundamentals of Anatomy & Physiology, 9th edn. San Francisco: Pearson/Benjamin Cummings.

The situation is somewhat modified in anuran amphibians (frogs and toads). Lymph is collected in the lymph capillaries which join together to form larger vessels and empty into lymph sacs located beneath the skin, which are shown in Figure 14.19. There are also four distinct and pulsatile lymph hearts. Although these hearts have a role in returning the lymph from the sacs to the blood circulation, the contraction of skeletal muscle is thought to be an important factor in ensuring that lymph returns to the lymph hearts in the first place, even in resting animals. Such a mechanism also seems to occur in reptiles. In contrast, urodele amphibians (newts and salamanders) lack lymph sacs and probably do not rely on contraction of skeletal muscles to return lymph to the lymph hearts.

The lymphatic system of mammals consists of a widespread network of blind ending, highly permeable capillaries, shown in Figure 14.20, and a number of fine filaments which anchor the lymph vessels to the surrounding connective tissue. The contraction of neighbouring skeletal muscles causes the filaments to pull on lymph vessels, thus opening spaces between the endothelial cells, allowing proteins, other large peptides, and fluid to enter the lymphatic system. The lymph capillaries drain into larger vessels, the largest of which is the thoracic duct. Lymph drains from the thoracic duct into the low-pressure veins close to the heart. The lymphatic system not only returns fluid and protein to the central circulation, but also carries fats absorbed from the gastrointestinal tract.

As well as the activity of the skeletal muscles, the contraction of the lymphatic vessels themselves, together with an extensive network of one-way valves, creates the flow of the lymph from the periphery to the central circulation. The lymph is filtered at the lymph nodes where foreign particles, such as bacteria, are removed, making the lymphatic system an important part of the body's defences against infection.

› *Review articles*

Burggren WW, Reiber CL (2007). Evolution of cardiovascular systems and their endothelial linings. In Baird WC (ed.) Endothelial Biomedicine pp 29-39. Cambridge: Cambridge University Press.

Egginton S (2009). Activity induced angiogenesis. Pflügers Archiv—European Journal of Physiology 457: 963–977.

Egginton S, Gaffney E (2010). Tissue capillary supply—it's quality not quantity that counts! Experimental Physiology 95: 971–979.

Levick JR, Michel CC (2010). Microvascular fluid exchange and the revised Starling principle. Cardiovascular Research 87: 198–210.

Olson KR (1996). Secondary circulation in fish: anatomical organization and physiological significance. Journal of Experimental Zoology 275: 172–185.

Risau W (1997). Mechanisms of angiogenesis. Nature 386: 671–674.

Roddie IC (1990). Lymph transport mechanisms in peripheral lymphatics. News in Physiological Sciences 5: 85–89.

Shadwick RE (1999). Mechanical design in arteries. Journal of Experimental Biology 202: 3305–3313.

Suarez RK (1992). Hummingbird flight: sustaining the highest mass-specific metabolic rates among vertebrates. Experientia 48: 565–570.

West JB, Watson RR, Fu Z (2006). The honeycomb-like structure of the bird lung allows a uniquely thin blood-gas barrier. Respiratory Physiology & Neurobiology 152: 115–118.

14.2 Fluid dynamics

The force generated by a heart or hearts when they contract must be sufficient to overcome the resistance to the movement of the blood through the blood vessels and to return it to the heart. In closed circulatory systems, blood is contained within blood vessels throughout the whole of its circulation, although this is not the case in more open systems.

14.2.1 Pressure, resistance and flow

The rate at which a volume of a fluid ($\dot{V}$) flows through a tube is directly related to the pressure difference along the length of the tube ($P_1 - P_2$) and is inversely related to the resistance to that flow (R):

$$\dot{V} = \frac{(P_1 - P_2)}{R} \qquad \text{Equation 14.3}$$

As the pressure difference increases and/or the resistance decreases, there is an increase in the rate of flow of the fluid. Conversely, fluid flow rate decreases with a reduction in the pressure difference and/or an increase in resistance.

The resistance to flow is determined by the viscosity[24] of the fluid and the length and radius of the tube through which the fluid flows. For example, more force (a greater pressure difference) is required to propel honey through a tube of given length and radius than is required to propel water through a tube of similar dimensions. Also, more force is required to propel either honey or water through a longer and/or narrower tube.

The amount of an ideal (Newtonian) fluid (that is, one that does not change its viscosity at any given temperature) that will flow in a steady fashion through a long, straight, rigid tube per unit of time is directly proportional to:

- the difference in pressure at either end of the tube (ΔP);
- the radius (r) of the tube raised to the fourth power (r^4).

In contrast, the rate of fluid flow is inversely proportional to:

- the viscosity (η) of the fluid;
- the length (l) of the tube.

However, the relationships are not simple ones, as they also depend on viscosity multiplied by 8 and r^4 multiplied by π, as indicated by Poiseuille's equation, which brings all these factors together (as we will see later).

[24] Viscosity is discussed in Section 14.2.3.

It is not entirely appropriate to apply Poiseuille's equation to the circulatory system, as blood is not a Newtonian fluid, blood vessels are not straight and rigid, and blood flow is not steady (at least on the arterial side). Nonetheless, Poiseuille's equation has been used to describe the factors affecting the rate of blood flow (cardiac output, $\dot{V}_b$) around the body of an animal:

Difference in pressure between arterial and venous systems (kPa)

Radius of blood vessel (mm) raised to the fourth power

Cardiac output (mL min⁻¹)

$$\dot{V}_b = \frac{(P_a - P_v)\pi r^4}{8\eta l} \qquad \text{Equation 14.4}$$

Viscosity of blood (mPa s)

Length of blood vessel (mm)

Note that equations 14.3 and 14.4 are similar, except that $1/R$ has been replaced by $\pi r^4/8\eta l$. This means that $R = 8\eta l/\pi r^4$ indicating that by far the greatest influence on resistance is the radius of the blood vessels, as it is raised to the power of four. So, all other things remaining equal, a doubling of the radius of a blood vessel will give rise to a 16-fold (2^4) decrease in resistance and a 16-fold increase in blood flow through the vessel. Conversely, halving the radius of a blood vessel will increase resistance 16-fold and reduce flow through the vessel to one sixteenth of its original value.

A reduction in the radius of a vessel is of clear significance in coronary disease when the blood vessels supplying the muscles of the heart become narrow as the result of the build up of fatty deposits thereby dramatically reducing the supply of blood and oxygen to the heart.

Dilation of peripheral blood vessels in one region of the body of a vertebrate (e.g. the gastrointestinal tract after a meal) favours perfusion of that region and will often be accompanied by constriction of blood vessels in another region, such as the locomotory muscles. The opposite occurs during exercise[25]. The resistance to blood flow through the peripheral blood vessels of vertebrates is varied almost exclusively by changes in the radius of the vessels, with viscosity playing a role in only some extreme circumstances, e.g. an increase in the concentration of haemoglobin in the blood as a result of severe dehydration, or perhaps as a consequence of 'blood doping'—the artificial increase in the number of red blood cells and hence in haemoglobin concentration[26].

The total resistance offered by a system of tubes, such as the vessels of the vascular system, depends on whether

[25] Differential distribution of blood to peripheral tissues during feeding, exercise and diving is discussed in Sections 15.2.1, 15.2.3 and 15.3.3, respectively.

[26] Two of the methods used in blood doping are discussed in Box 13.1.

(A)

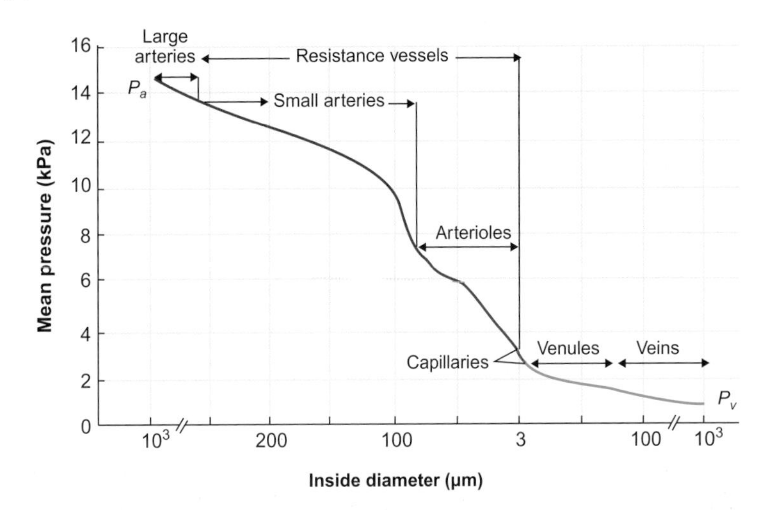

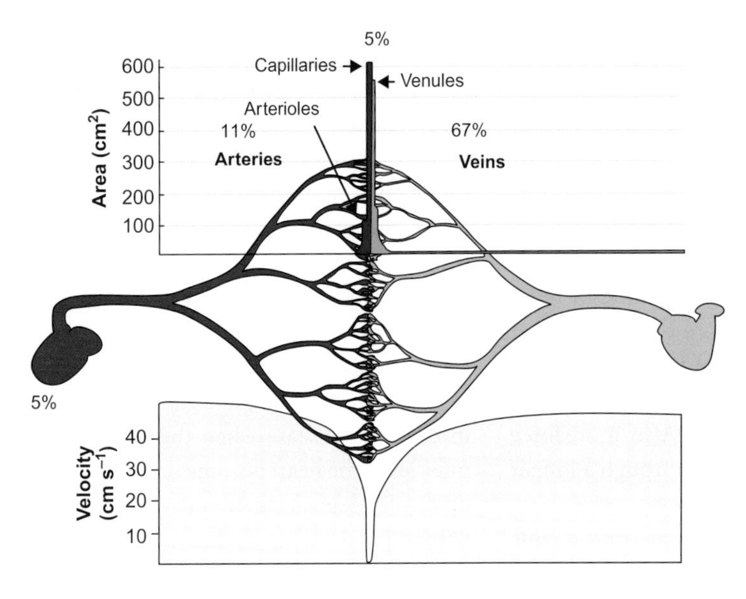

Figure 14.21 Pressure and flow throughout the circulatory system of mammals

(A) Mean pressures in different regions of the vascular system in the cheek pouch of a hamster. (B) Relationship between total cross-sectional area and velocity of blood flow in different regions of the circulatory system of a 13 kg dog. The percentages of total blood volume in each section are given. The remainder of the blood is in the pulmonary circuit. Note that about two thirds of the volume of blood is contained within the systemic venous system, with only 11% in the systemic arterial system. The increase in total cross-sectional area leads to a reduction in the velocity of blood flow through the capillaries compared with that through the arteries and veins. At the capillaries of mammals, velocity of the blood is about 0.07 cm s^{-1} compared to about 50 cm s^{-1} in the major arteries and veins. *Pa*, mean arterial pressure, *Pv*, venous pressure. 1 kPa = 7.5 mmHg

A: Reproduced from: Ruch TC and Patton HD (1965) Physiology and Biophysics. Philadelphia: W.B. Saunders B: Berne RM and Levy MN (Eds) (1998). Physiology, 4th edn. St Louis: Mosby.

they are in series (that is arranged one after the other) or in parallel. Figure 14.21A demonstrates that the arteries, arterioles and capillaries are in series. As a result, the total resistance (R_{tot}) of tubes in series is the sum of their individual resistances:

$$R_{tot} = R_1 + R_2 + R_3$$

However, Figure 14.21A also shows that all of the arteries, all of the arterioles and all of the capillaries lie in parallel with each other; the total resistance of the vessels in parallel is calculated differently from when they are in series. For example, if three tubes are in parallel, the reciprocal of their total resistance is the sum of the reciprocals of the individual resistances:

$$\frac{1}{R_{tot}} = \frac{1}{R_1} + \frac{1}{R_2} + \frac{1}{R_3} \qquad \text{Equation 14.5}$$

So, if all of the resistances are equal, $1/R_{tot} = 3/R_1$, and, therefore, $R_{tot} = R_1/3$. This means that for resistances in parallel, the total resistance is less than that of any individual component. This applies whether the individual resistances are the same or whether they are different; total resistance of blood vessels organized in parallel is always less than the lowest individual resistance. As a result, blood will tend to flow more easily through blood vessels that are in parallel with each other than through those in series.

As blood flows through the different sections of a circulatory system, the pressure becomes progressively lower. Figure 14.21B shows that the largest drop in pressure is seen across the small arteries and arterioles in vertebrates because they produce the greatest resistance to flow in the circulation, even though their total cross-sectional area is greater than that in the larger arteries. In contrast, the pressure drop across the capillary beds is much lower than that across the small arteries and arterioles, even though capillaries are the narrowest of the blood vessels.

We can explain these apparently contradictory pressures and resistances on the basis of:

- the influence of the fourth power of the radius of a blood vessel on its resistance to flow, as demonstrated in equation 14.4;
- all the small arteries, all the arterioles and all of the capillaries run parallel to one another such that the total resistance of each section of the circulation is less than that of any one of the blood vessels within that section;
- the total cross-sectional area of the capillaries is much greater than that of the arterioles, as shown in Figure 14.21.

When the number of blood vessels in parallel drops below a certain level, their radius has a greater influence on their total resistance; above a certain number however, their parallel arrangement has the greater influence, as explained in Box 14.2, which is available online. Thus, when a large number of small blood vessels (capillaries) occur in parallel, their total resistance to blood flow is lower than that in fewer or single larger vessels.

The validity of equation 14.4 when applied to circulatory systems has been tested for the pulsatile nature of pressure and flow in birds and mammals. For the larger arteries of birds and mammals, some of the assumptions made in equation 14.3 are not valid, for example, pressures and flows are highly pulsatile rather than steady and the blood vessels are distensible rather than rigid. For smaller arteries, capillaries and the smaller veins, however, it has been found that equation 14.4 can be used.

14.2.2 The influence of gravity on the circulatory system

So far, we have considered the flow of blood in horizontal tubes, when there is no change in the gravitational potential energy from one end of the tube to another. However, there are animals where some parts of the body are substantially higher above the ground than others, in which case we need to consider the effect of gravity on the circulatory system. The potential energy of 1 L blood due to the Earth's gravity (the gravitational potential energy) is given by the relationship $\rho \times g \times h$ (where ρ is the density of the blood in kg L^{-1}, g is the gravitational constant, 9.81 m s^{-2} and h is the height (in m) above the heart). As the height of blood above the heart of terrestrial animals increases, gravitational potential energy also increases, but blood pressure decreases. In contrast, as the height of the blood below the heart decreases, gravitational potential energy decreases and blood pressure increases as a result of the standing column of blood (hydrostatic pressure). For example, when lying down, blood pressure in humans is around 13 kPa (100 mmHg) at the heart and at the head and feet. However, when standing, blood pressure in the head is about 6.5 kPa, while that in the feet is about 24 kPa.

Some species of birds and mammals, such as ostriches and giraffes, have long necks and legs, and a vertical stance. In these species, gravity has a large influence on the pressures within the circulation. This is not a problem for aquatic animals as the densities of blood and water are very similar, so the pressure difference between the top and bottom of a vertical column of blood is cancelled out by the surrounding column of water.

So, when giraffes (*Giraffa camelopardalis*) are standing upright, their hearts would need to generate additional pressure to overcome the effect of gravity on the column of blood from the heart to the head. However, it has been suggested that these animals have an effective siphon operating to overcome this problem. The basic concept of the siphon effect is that gravity has opposite effects on the column of blood in the arteries and the column of blood in the veins, thereby 'siphoning' the blood back to the heart. Consequently, the heart would not need to generate any additional pressure to circulate blood through the head of a giraffe standing in an upright position compared with when the head was held out horizontally. We discuss the arguments for and against a functioning siphon in giraffes in Case Study 14.1.

The greatest challenge to the circulatory system of a giraffe is when it lowers its head to drink, as shown in Figure 14.22.

Figure 14.22 The change in the height of the head relative to the heart in an upright giraffe (*Giraffa camelopardalis*) and in one bending down to drink

As in other mammals, the heart lies anteriorly and ventrally in the thorax.

Image: Chloe Rutter, with permission, © Albie Venter/ Shutterstock.

Case Study 14.1 The case of giraffes (*Giraffa camelopardalis*): siphon or no siphon?

Some studies have demonstrated that mechanical models can produce a siphon effect which could function in giraffes. The data from one of these experiments are shown in Figure A. However, despite a rigorous defence of the hypothesis by some authors, most experiments in which pressures have been measured in the blood vessels of upright giraffes do not support the existence of a siphon, while other studies using mechanical models have demonstrated that a siphon effect only occurs if all the tubes are rigid, but not if the descending tube is collapsible, as is the case for the jugular veins in giraffes.

If the heart of giraffes in an upright position had to generate sufficient pressure to overcome the hydrostatic pressure of the column of blood between it and the head, pressure in the carotid arteries close to the heart would be considerably higher than that close to the head. The difference in pressure would be directly related to the height of the head above the heart. Figure B illustrates that approximately a 10 kPa difference in carotid artery pressure exists over the 1 m distance along the neck of an upright giraffe. This pressure difference along the carotid artery is compatible with the heart having to generate sufficient pressure to overcome the hydrostatic pressure of the column of blood between it and the head, as indicated in Figure B. Blood pressure at the level of the brain in upright giraffes is similar to that in other mammals.

Compared with mammals with shorter necks, therefore, the heart of a giraffe has to generate a high blood pressure. However, unlike the situation in most other species of mammals, the existence of relatively high blood pressure, at and below the level of the heart, is a natural condition in giraffes. So do they show any adaptations associated with this?

Early reports stated that the heart of giraffes is proportionally larger than that of other mammals (2.5 per cent body mass vs 0.55 per cent body mass), presumably in order to generate the high pressures. However, more recent studies have demonstrated that this is not the case. In fact, the high blood pressure of giraffes is not associated with a heart that is relatively larger than that of other mammals, as shown in Figure C(i). For example, a giraffe weighing 800 kg has a heart mass of approximately 0.5 per cent body mass.

However, the giraffe heart does have relatively thick left ventricular and interventricular walls. Indeed, their thickness is related to the length of the neck in growing animals, as shown in Figure C(ii). It is these thick walls that generate the high arterial pressure required to overcome the hydrostatic pressure in the column of blood above the heart in upright giraffes. A similar increase in thickness occurs in the walls of the arterioles in order for them to withstand the high pressures.

On the venous side of the circulation, pressure is positive and slightly higher near the head (2 kPa)[1] than near the heart (1 kPa). These pressures are difficult to explain, as the difference is the reverse of what would be expected if the blood in the jugular vein was a supported column of fluid in which the pressure increases toward the bottom.

An explanation could be that pressure in the jugular vein is influenced by pressures in the external tissues, such as the inflexible skin, and by the viscosity of the blood (which is four times more viscous than water). A sophisticated mechanical model has enabled this hypothesis to be tested and has supported this explanation: applying an external 'tissue' pressure of around 0.5 kPa produced pressures in the descending limb

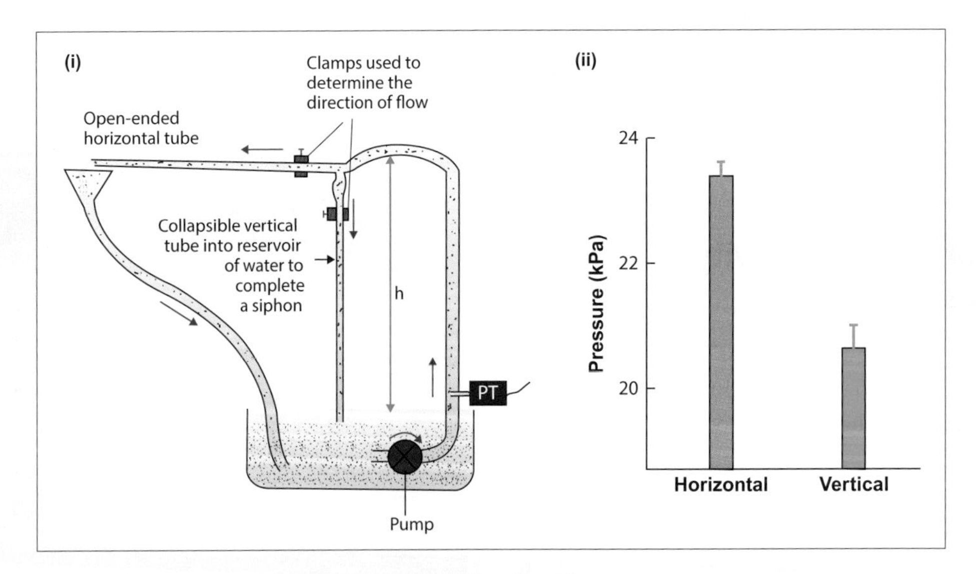

Figure A Demonstration of the siphon mechanism in a circulating liquid

(i) Diagram of mechanical model used to demonstrate the advantage of a siphon mechanism compared with a horizontal tube of equal length. PT, pressure transducer, which is an instrument for converting pressure into an electrical output. (ii) When water is pumped through the horizontal tube only, the pump has to overcome the resistance of water flowing in the tubes and gravity in order to raise the water to height, h. When water is pumped through the vertical tube only, the pump only has to overcome the resistance of water in the tubes as the effect of gravity is reduced by the siphon. As such, the pressure required to do this is also reduced. Mean values + SEM (standard error of the mean), n = 5.

Modified from Hicks JW and Badeer (1989). Siphon mechanism in collapsible tubes: application to circulation of giraffe. American Journal of Physiology 256: R567-R571.

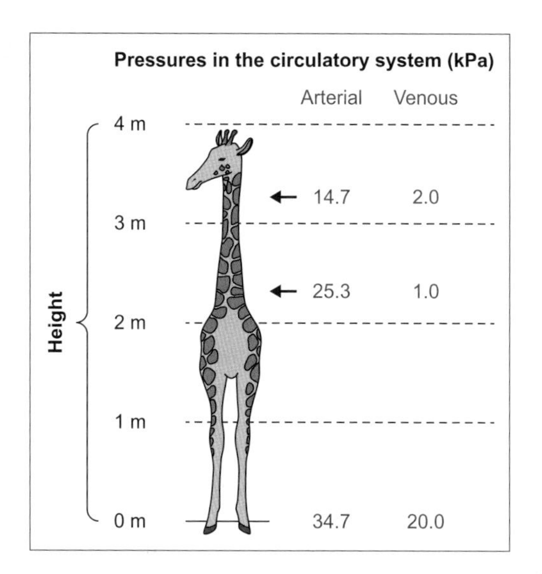

Figure B Pressures in the arterial and venous system of an upright giraffe (*Giraffa camelopardalis*)

Note that the difference in the arterial blood pressures at two points along the neck (approximately 10 kPa) is equal to the hydrostatic pressure created by the vertical distance between the points (approximately 1 m). This conclusion is based on the following: the density of water (1000 kg m^{-3}) is similar to that of blood (1060 kg m^{-3}), and a 1 m column of water produces a hydrostatic pressure of 9.8 kPa. Note, 1 kPa = 7.5 mmHg

Modified from Hargens AR et al (1987). Gravitational haemodynamics and oedema prevention in the giraffe. Nature 329: 59-60.

of the model (equivalent to the venous jugular vein) similar to those recorded in upright giraffes.

Figure B shows that pressures at the level of the feet of standing giraffes are very high (arterial pressure 34.7 kPa, venous pressure 20.0 kPa). The pooling of blood in the veins and accumulation of fluid (oedema) in the tissues is prevented by giraffes having a tight skin on their legs, one-way valves in their veins, a prominent lymphatic system and capillaries with a low permeability (high reflection coefficient[2]) to plasma proteins, so that the colloid osmotic pressure does not increase in the surrounding tissues.

› Find out more

Hicks JW, Badeer HS (1992). Gravity and the circulation: 'open' vs 'closed' systems. American Journal of Physiology 262: R725–R732.

Mitchell G, Maloney SK, Mitchell D, Keegan DJ (2006). The origin of mean arterial and jugular venous blood pressures in giraffes. Journal of Experimental Biology 209: 2515–2524.

Mitchell G, Skinner JD (1993). How giraffes adapt to their extraordinary shape. Transactions of the Royal Society of South Africa. 48: 207–218.

Seymour RS (2000). Model analogues in the study of cephalic circulation. Comparative Biochemistry and Physiology Part A 125: 517–524.

[1] 1 kPa = 7.5 mmHg

[2] Reflection coefficient is discussed in section 14.1.5 and equation 14.2

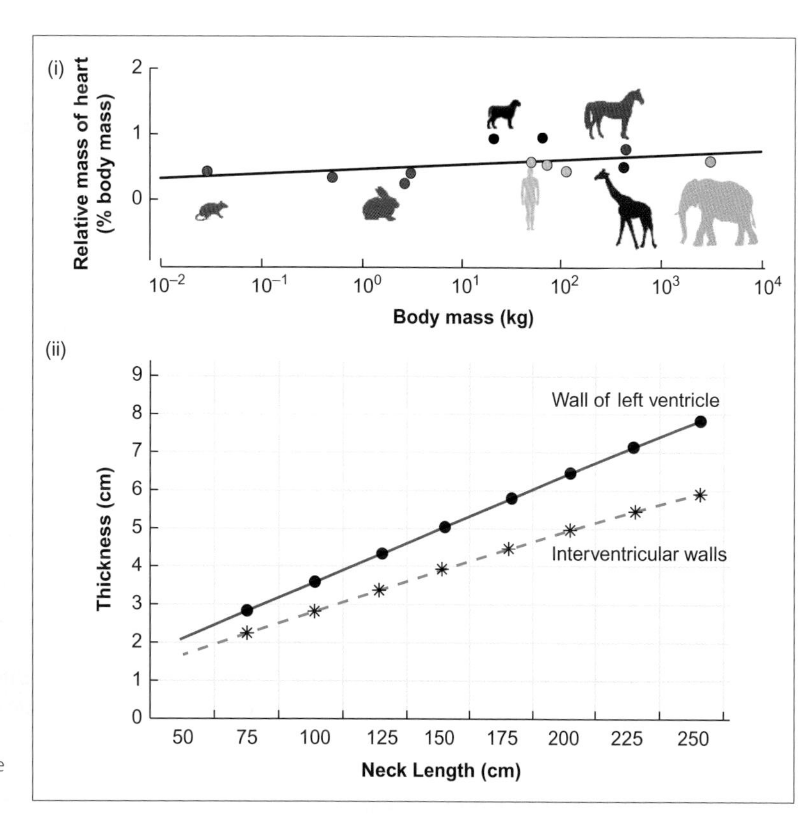

Figure C Characteristics of the heart of giraffes (*Giraffa camelopardalis*)

(i) Relative mass of heart of giraffes is similar to that of other mammals. (ii) The thickness of the walls of the heart increases with an increase in the length of the neck as giraffes grow.

Reproduced from Østergaard KH et al (2013). Left ventricular morphology of the giraffe heart examined by stereological methods. Anat Rec. 296: 611-621. Mitchell G and Skinner JD (2009). An allometric analysis of the giraffe cardiovascular system. Comparative Biochemistry and Physiology A 154: 523-529.

The brain is moved from 2–3 m above the heart to around 2 m below within a few seconds. How is the brain protected from the pressure change caused by such a large variation in height with respect to the heart? Studies on anaesthetized giraffes have shown that the jugular vein expands during lowering of the head and accommodates an increased volume of blood. It has been suggested that this accumulation of blood influences the mean arterial blood pressure which falls from 25.7 kPa when the animal is upright to 17.5 kPa when the head is lowered, thereby contributing to the protection of the capillaries of the brain when the head is lowered.

14.2.3 The effect of viscosity on blood flow

Sir Isaac Newton described the **viscosity** of a fluid as a degree of slipperiness as the different parts of the fluid pass one another. This led to the concept of the movement of a fluid being that of infinitesimally thin layers, or laminae, of fluid sliding over one another. When a fluid flows through a tube, a very thin layer next to the wall of the tube known as the **boundary layer**[27] does not move, but the next layer of fluid moves over the boundary layer, and so on, with each successive layer moving at greater velocity. This results in a parabolic flow profile, which is illustrated in Figure 14.23A. The maximum flow velocity occurs in the centre and is twice the average velocity of the fluid flow through the tube. Viscosity is, therefore, a measure of the resistance to sliding between the different layers of fluid. Hence, an increase in viscosity of a fluid requires a greater pressure difference to maintain the same flow rate of the fluid through a tube (literally, to force the less 'slippery' fluid forward).

The viscosity of blood plasma depends on the concentration of proteins within it, so is a particularly important consideration in those animals whose respiratory pigments (which are proteins) are in solution. The presence of plasma protein means that even the plasma of mammals has a viscosity that is slightly greater than that of water (around 1.4 Pa s vs 1.0 Pa s), as a result of the presence of plasma proteins. Newton's concept of viscosity is accurate for a homogeneous fluid where all the particles are the size of small molecules[28] (a **Newtonian fluid**), but is inappropriate for vertebrate blood which contains blood cells (erythrocytes, leucocytes and thrombocytes), as these will determine the thickness of the laminae.

The blood of vertebrates is, therefore, a non-Newtonian fluid. Consequently, the term **apparent viscosity** is often used instead of viscosity when discussing blood. The blood of birds and mammals behaves as if it is three to four times more viscous than water. However, the apparent viscosity of blood is related in an exponential fashion to the concentration of blood cells present (the haematocrit). So, a 50 per cent increase in haematocrit causes a 70 per cent increase

[27] Boundary layers are also discussed in Sections 8.4.3 and 12.2.1.

[28] Most molecules are no more than a few nm in diameter, whereas a human red blood cell has a diameter of approximately 8 μm.

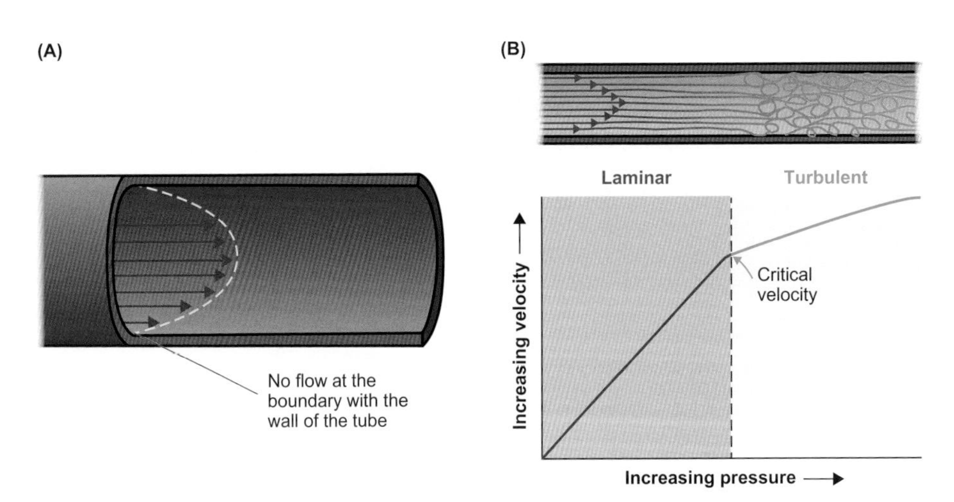

Figure 14.23 Effect of velocity on flow profiles of a fluid through a straight, rigid tube

(A) Parabolic profile of laminar flow of a fluid in a tube. There is no flow at the boundary with the wall of the tube, but the flow rate progressively increases away from the periphery, reaching its highest value in the centre. All the flow is parallel to the axis of the tube. (B) There is a linear relationship between pressure and velocity of a fluid through a tube, up to a certain velocity (the critical velocity) while flow is laminar. The relationship above this velocity becomes non-linear as flow becomes turbulent. Under this condition, the fluid moves in many directions and a greater increase in pressure is required to generate a given increase in velocity through the vessel than during laminar flow.

A: Reproduced from: Berne RM and Levy MN (2000). Principles of Physiology, 3rd edn. St Louis: Mosby. B: Fiegl EO (1974). Physics of the cardiovascular system. In: Ruch TC and Patton HD (eds) Physiology and Biophysics 20th edn. II Circulation, Respiration and Fluid Balance, pp 10-22. Philadelphia: W. B. Saunders Company.

in apparent viscosity of human blood, but a doubling of haematocrit causes almost a four-fold increase in apparent viscosity.

Other non-Newtonian features of the blood of vertebrates are the dependence of its apparent viscosity on both its flow rate (particularly at low flow rates) and the diameter of the tube (blood vessel) through which it flows.

At low flow rates, red blood cells in mammals tend to aggregate, forming large particles which increase the blood's apparent viscosity. Conversely, as flow rate increases, these aggregates break up, thereby decreasing the apparent viscosity.

Another flow-related factor affecting viscosity in the blood of mammals is the fact that red cells are deformable. At low-to-medium blood flow rates, mammalian red cells have a biconcave shape, but deform in the direction of flow and become flattened and more ellipsoid at higher flow rates. This change in shape reduces the apparent viscosity of the blood.

The ability of the red cells to deform is important in determining their flow through small capillaries. Mammalian red cells (erythrocytes) are approximately 8 μm in diameter, yet their deformation means they can pass through capillaries with a diameter of only 3 μm.

Erythrocytes tend to accumulate in the centre of the blood vessel where the flow rate is highest, which is shown in Figure 14.23A. This means that the walls of the blood vessel are relatively free of red cells and any blood flowing from this area into small branches of the vessel has a low concentration of erythrocytes. This is known as plasma skimming. The accumulation of red cells in the centre of a blood vessel means that the viscosity of the blood is lower at the margins, with the region of low viscosity being relatively larger in smaller blood vessels. This could help to explain why the apparent viscosity of mammalian blood in tubes below a diameter of 0.3 mm decreases with decreasing diameter[29].

14.2.4 Laminar and turbulent flow in blood vessels

Flow of an ordered, streamlined nature through a tube, as indicated in Figure 14.23A, is known as laminar flow. However, when the velocity of a fluid flowing in a tube exceeds a certain value, the flow is no longer laminar, but becomes turbulent. When flow is turbulent, fluid moves at different angles to the main direction of flow, as illustrated in Figure 14.23B.

A combination of four factors determines whether the flow of fluid in a tube is laminar or turbulent. These factors give rise to a dimensionless quantity known as the Reynolds number, *Re*, after Osborne Reynolds:

Density of the fluid (kg m^{-3}) — Diameter of the tube (m) — Average velocity of fluid through the tube (m s^{-1})

Reynolds number

$$Re = \frac{\rho D \bar{u}}{\eta} \qquad \text{Equation 14.6}$$

Viscosity of the fluid (Pa s)

In general, flow through a tube is usually laminar when the Reynolds number is below about 2000; above about 3000, however, turbulence normally occurs. The velocity at which this transition occurs is called the **critical velocity**. A variety of flow patterns may develop within the transition zone (Reynolds number between 2000 and 3000). The smaller a blood vessel and the slower the blood flow through it, the greater the chance of the flow being laminar, as indicated by equation 14.6. In fact, flow is usually laminar in most blood vessels.

Turbulent flow usually causes vibrations in the fluid, some of which may be heard as murmurs. For example during severe anaemia, the number of erythrocytes, and hence the apparent viscosity of the blood, are low. Consequently, in order to supply sufficient oxygen to the tissues, the rate of blood flow from the heart (cardiac output) is unusually high. Both of these factors—the low apparent viscosity and the high rate of blood flow—would lead to turbulent flow and murmurs in and around the heart.

Although the circulatory systems of animals are similar in their basic design, there are considerable differences between the various groups, largely depending on whether they have an open or closed system. In the next section, we discuss how some of the different designs meet the demand of transporting substances around the body.

> *Review articles*

Smith FM, West NH, Jones DR (2000). The cardiovascular system. In Whittow GC (ed.) Sturkie's Avian Physiology. San Diego: Academic Press, pp. 141–231.

14.3 Circulatory systems of invertebrates

The circulatory systems of invertebrates vary in their complexity. In annelids, the coelom expands and the vascular space is greatly reduced, being contained within distinct tubes or vessels. For example, oligochaete annelids, such as giant Australian earthworms (*Megascolides australis*) have well-developed circulatory systems with blood contained largely within blood vessels. These blood vessels are not lined

[29] The relationship between the diameter of a small blood vessel and the apparent viscosity of the blood flowing through it is discussed section 13.1.1

with endothelium, so this system is technically classified as being open. In contrast, part of the open circulatory systems of crustaceans and molluscs (except the cephalopods) are in the haemocoel, as shown back in Figure 14.1B. However, the circulatory systems of some species of molluscs and decapod crustaceans such as crabs and lobsters, have been described as being incompletely closed rather than open[30].

In molluscs, such as land snails (*Helix* spp.) and decapod crustacean, some organs and tissues, such as nerve ganglia, feature very fine blood vessels akin to capillaries. The fine capillary-like vessels in *Helix* are lined with an endothelium. In cephalopod molluscs, the circulatory system is essentially closed. The blood vessels are of mesodermal origin and are lined with endothelial-like cells, although this lining may be incomplete in the smallest vessels (sometimes called capillaries).

Whatever the origin of the blood vessels or tubes, one of the advantages of circulatory systems in which the blood is contained within vessels or tubes throughout most or all of its circulation is that a relatively small volume of blood is circulated. The variability in blood volume within a number of species of molluscs and crustaceans is demonstrated by the data in Table 14.2. The major reason why blood volume is so high in some molluscs, such as the lamellibranch *Anodonta* sp. and the gastropods *Haliotis* spp. and *Busycon* spp., is that it serves a hydraulic role during extension of the foot, as well as a transport role.

In molluscs (with the exception of cephalopods) and all arthropods, the heart pumps oxygen-rich haemolymph out through the arteries. The haemolymph makes its way to sinuses where it comes into contact with the organs of the body. From the sinuses the now oxygen-poor haemolymph passes through the gas-exchange organs before returning to the heart.

[30] Crustacean circulatory systems are discussed in Section 14.3.3.

Table 14.2 Characteristics of the cardiovascular systems of resting molluscs and arthropods

Blood volume is lower and pressure is higher in *Octopus* than in the other species as it has a closed circulatory system. Blood (haemolymph) volume is greater in mussels and whelks because they use their circulatory system as a hydrostatic skeleton during locomotion.

Species	Blood pressure (kPa) systolic/diastolic	Blood volume (% body mass)
Anodonta (duck mussel)	0.5/0.1 (ventricle)	55 (body mass excluding shell)
Haliotis (abalone) or *Busycotypus* (channelled whelk)	0.9/0.3 (aorta)	57 (body mass excluding shell)
Homarus (American lobster)	2.0/0.2 (heart)	17
Octopus (common octopus)	4.1/2.5 (aorta)	4.9

References: Brand, 1972; Jones, 1983; Bourne and Redmond, 1977a; Bourne and Redmond, 1977b; Jones, 1988; Taylor and Ragg, 2005; McMahon et al., 1997; McMahon and Wilkens, 1983; Wells et al., 1987; O'Dor and Wells, 1984. Full references available online.

14.3.1 Circulation in non-cephalopod molluscs

Many aspects of the circulatory system of non-cephalopod molluscs are quite variable. Bivalve molluscs, such as duck mussels (*Anodonta anatina*), have an open-type circulatory system with a relatively low pressure generated by the ventricle of the heart, as shown in Table 14.2. The organization of the circulatory system in *Anodonta* is shown in Figure 14.24.

The two atria and the ventricle are surrounded by a pericardium, shown in Figure 14.25A. The pressure in the atria and pericardial cavity drops as the ventricle contracts, which has been taken to indicate that haemolymph is sucked into the atria from the veins as the volume of the

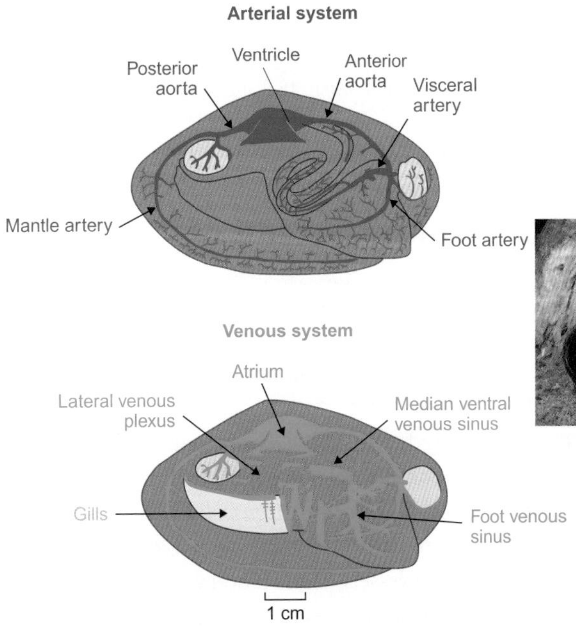

Figure 14.24 Circulatory system in lamellibranch (bivalve) molluscs as exemplified by duck mussels (*Anodonta anatina*)

Haemolymph is pumped from the ventricle to the anterior and posterior aorta. The anterior aorta supplies, among other areas, the intestine, foot and mantle lobes. Each branch divides many times to form a number of small vessels which open into haemocoelic spaces between muscle fibres. The posterior aorta supplies blood to the posterior abductor muscles and the wall of the mantle cavity. Oxygen-depleted blood collects into a number of sinuses from where it enters the gills. After passing through the gills where it becomes oxygenated, the haemolymph is returned to the atria via the lateral plexus.

Adapted from Brand AR (1972) The mechanism of blood circulation in *Anodonta anatina* (L) (Bivalvia, Unionidae). Journal of Experimental Biology 56: 361-379. Image: © Biopix: N Sloth.

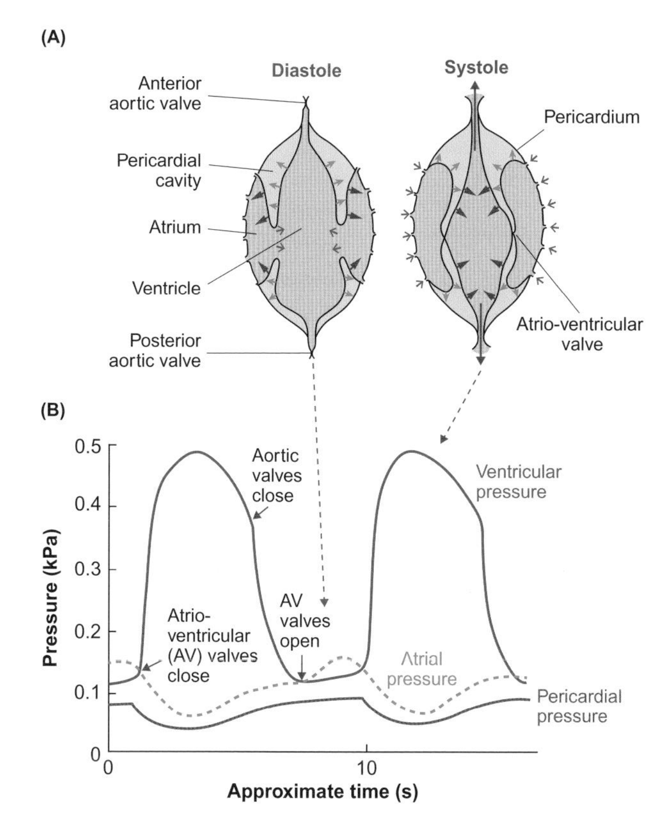

Figure 14.25 Cardiac function in duck mussels (*Anodonta anatina*)

(A) Longitudinal horizontal sections through the heart during relaxation (diastole) and contraction (systole) of the ventricle. Red arrows represent the contracting chamber and green arrows represent the relaxing chamber. Blue arrows represent the flow of haemolymph. During contraction of the ventricle, haemolymph is sucked into the atria, via the lateral venous plexus, if the pericardial cavity maintains a more or less constant volume. (B) Pressure changes in the atrium, ventricle and pericardial cavity during two beats of the heart. The shape of the pressure profile in the ventricle is very similar to that in vertebrates, as shown in Figure 14.2B, with inflections indicating when valves, which prevent backflow, open and close.

A: Modified from Brand AR (1972). The mechanism of blood circulation in *Anodonta anatina* (L) (Bivalvia, Unionidae). Journal of Experimental Biology 56: 361-379. B: Jones HD (1983). The circulatory systems of the gastropods and bivalves. In: The Mollusca. Vol 5, Physiology, Pt 2. Saleuddin ASM and Wilbur KM (eds), pp189-238. Paris: Academic Press.

ventricle decreases. This sucking requires the volume of the pericardium to remain more or less constant throughout the cardiac cycle, as suggested in Figure 14.25A. This process is known as the **constant volume mechanism** and is also found in other groups of molluscs, in some arthropods and in vertebrates. The crucial requirement for this mechanism to work is for the pressure in the atrium to fall below that in the veins during contraction of the ventricle. This difference is not obvious from the information in Figure 14.25B, but there is evidence that such a pressure difference exists in channelled whelks (*Busycotypus canaliculatus*). The use of magnetic resonance imaging with Mediterranean mussels (*Mytilus galloprovincialis*) has largely supported the constant volume hypothesis in these animals.

The majority of bivalve molluscs have no respiratory pigment in their haemolymph. Consequently, the haemolymph plays only a minor role in the transport of respiratory gases around the body when these molluscs are in well-oxygenated water. The haemolymph is important, however, during burrowing when it acts as a **hydrostatic skeleton** during the extension and retraction of the foot caused by the alternate contraction and relaxation of longitudinal (retractor) and transverse (extensor) muscles. This locomotor activity affects the pressures in the circulatory system. The systolic pressure in the ventricle increases during burrowing in order to maintain the supply of haemolymph to the foot for most of the digging cycle. However, the general anatomy of the circulatory system effectively isolates the heart from the pressure changes that occur in the haemocoel of the foot during burrowing.

The ventricle is more muscular in gastropods than in bivalves. As such, the systolic pressure generated by the heart of channelled whelks is greater than that in duck mussels, as shown in Table 14.2, although their pressure profiles are very similar. Most water-breathing gastropods have a single gill and atrium, whereas some, such as abalone, have two gills and atria.

The muscle fibres of the ventricle of whelks are arranged in a complex network of branching and anastomosing bundles (trabeculae) and contain myoglobin. No blood vessel supplies the muscle of the ventricle (coronary vessels), so it is assumed that oxygen diffuses from the haemolymph in the ventricle into the cells of the ventricular muscle. This process would be aided by the **spongy** nature of the inner wall of the ventricle and presence of the myoglobin. A similar situation is seen in the hearts of cephalopods, some fish and amphibians[31].

14.3.2 The circulatory system of cephalopods

The circulatory systems of the highly active cephalopod molluscs have many features in common with those of vertebrates. In particular, the circulatory system is entirely enclosed within blood vessels, the relative volume of the haemolymph is much lower and the pressure generated by the heart is higher in cephalopods than that in other species of molluscs and crustaceans, as shown in Table 14.2.

In most species of cephalopods, two gills drain into the corresponding atria of the heart and then into a single ventricle, as illustrated by Figure 14.26. The ventricle of the heart of octopuses has a relatively thick muscular wall (**myocardium**) which needs to be provided with an adequate supply of oxygen. The lumen of the ventricle is partly divided by a septum and the inner surface of the ventricular wall is highly trabeculated, while the outer section is more compact. Thus, oxygen-rich blood within the lumen of the ventricle permeates spaces between the

[31] Blood supply to the heart of vertebrates is discussed in Section 14.4.5.

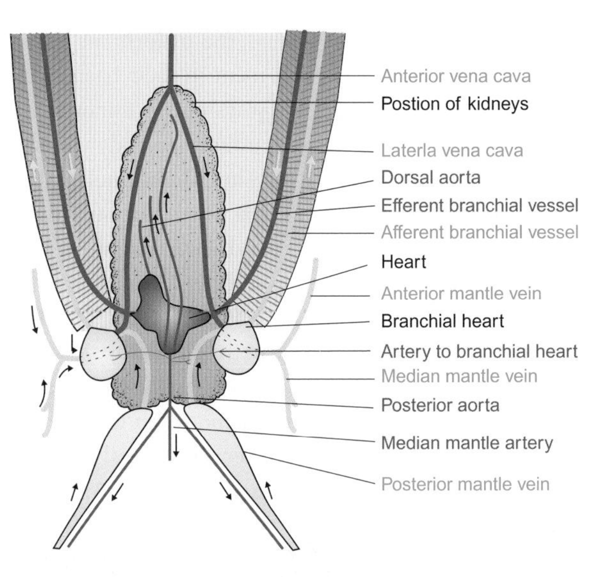

Figure 14.26 Circulatory system of cephalopod molluscs
Blood vessels around the main heart and branchial hearts of squid (*Loligo* sp.). Arrows indicate direction of blood flow. Only two of the main arteries are shown, the dorsal and posterior aortae. The one not shown is the gonadal artery.
Reproduced from Bullough (1960) Practical Invertebrate Anatomy 2nd edn London: Macmillan & Co. Photo © Hans Hillewaert.

trabeculae and some enters a rich network of capillaries which arise from the trabecular spaces. These capillaries supply the outer, more compact layer of the ventricular wall with oxygen-rich blood and form a number of superficial veins on the outer surface of the ventricle. They converge into two coronary veins which join the lateral vena cavae. Unlike the situation in mammals, blood flows through the coronary veins during ventricular contraction in octopus.

The ventricle gives rise to three main arteries, the largest of which is the dorsal (or cephalic) aorta, as shown in Figure 14.26. The other main arteries are the posterior aorta and the gonadal artery. The wall of the dorsal aorta consists of layers of circular and longitudinal muscle and the lumen is lined with a dense layer of connective tissue, which contains elastic fibres with mechanical properties similar to those of the elastic fibres in the blood vessels of vertebrates. The initially pulsatile flow of haemolymph is smoothed in the larger elastic arteries and becomes more continuous as it reaches the periphery. Even though these arteries subdivide and eventually form small, thin-walled vessels that are called capillaries, the equivalence of these to the capillaries of vertebrates is still a matter of debate, so they are sometimes referred to as microvessels.

There are relatively large gaps between the endothelial-like cells of the microvessels. However, molecules of respiratory pigment (haemocyanin), which have a diameter of approximately 35 nm, are retained within the microvessels by a very compact covering of cells, the pericytes, outside the basement membrane. Despite the small size of the haemocyanin molecule, microvessels of less than 2 μm diameter are rare in cuttlefish, (*Sepia* spp.), and their mean diameter is approximately 4.5 μm—similar to that of capillaries in vertebrates.

Having perfused the microvessels in the tissues of the body, the haemolymph enters the relatively large sinuses of the venous system at a low pressure, but then has to perfuse the gills before it returns to the main heart. Ultrasound studies have demonstrated that the anterior and lateral venae cavae contract peristaltically toward the heart, which suggests that they aid the return of the venous blood to the branchial hearts. Also, a valve ensures that blood flows toward the branchial hearts, despite the fact that the anterior and lateral vena cavae do not contract in synchrony.

Branchial hearts are situated between the sinuses (veins) returning haemolymph from the body and the afferent branchial vessels supplying the gills. These accessory hearts are shown in Figure 14.26. The pressures generated by the branchial hearts are about 20 per cent of that generated by the ventricle and boost the pressure produced by the veins. They contract in synchrony with the ventricle and play an important role in propelling the haemolymph through the gills.

14.3.3 The circulatory system of crustaceans

The circulatory system of insects and crustaceans are very similar in their general anatomy. However, the circulatory system of insects does not play a major role in the transport of respiratory gases[32], so we concentrate here on the crustaceans. Crustaceans exhibit diverse cardiovascular systems. For example, the dorsally placed heart can vary in length and position even within the same group of crustaceans,

[32] Transport of respiratory gases in insects is discussed in Section 12.3.5.

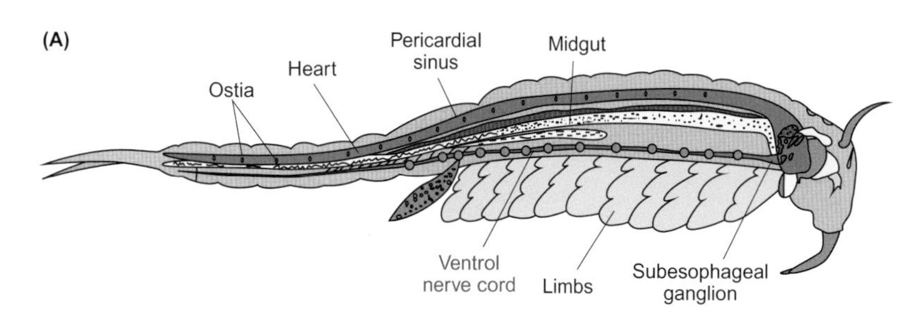

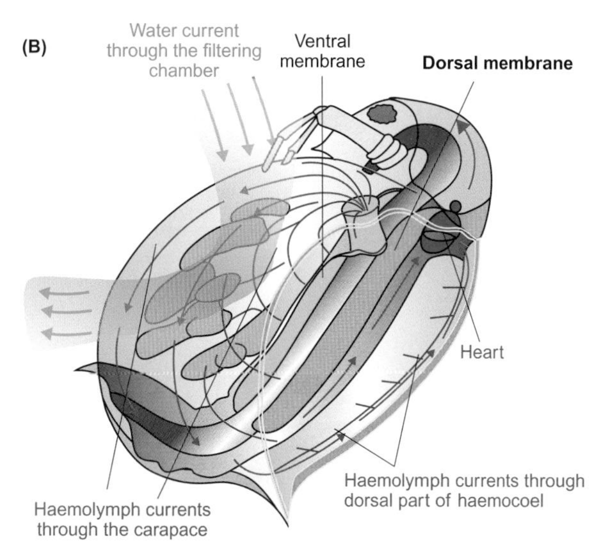

Figure 14.27 Circulatory systems of non-decapod crustaceans

(A) Elongated heart and lack of arteries in brine shrimps (*Artemia* sp.). (B) Diagram showing the small round heart and lack of arteries in water fleas (*Daphnia magna*), which are related to brine shrimps. The diagram also shows the main currents of haemolymph flow through the haemocoel obtained from spectral imaging of haemoglobin oxygen saturation. Haemolymph from the first four pairs of limbs takes curved, radial paths through the carapace haemocoelic space where oxygen is transferred from the water currents inside the filtering chamber. Note, under certain conditions, some species of *Daphnia* can synthesize haemoglobin, as discussed in Section 13.1.2.

A: Reproduced from McMahon BR et al (1997). Invertebrate circulatory systems. In: Dantzler WH (ed.) Handbook of Physiology, Section 13: Comparative Physiology, vol II pp 931-1008. New York: Oxford University Press. B: Reproduced from Pirow et al. (1999) The sites of respiratory gas exchange in the plantonic crustacean *Daphnia magna*: an in vivo study employing blood haemoglobin as an internal oxygen probe. Journal of Experimental Biology 202: 3089-3099.

as shown in Figure 14.27. Studies on water fleas (*Daphnia magna*) indicate that haemolymph can flow in an ordered fashion around the body, even in the absence of blood vessels, as shown in Figure 14.27B. There is no atrium to receive haemolymph returning from the body and the functional role of the atria is taken by the cavity surrounding the heart, the pericardial sinus.

Haemolymph returning from the body (a process known as **venous return**) enters the pericardial cavity through a series of channels and then enters the heart through pairs of segmentally arranged valved openings, called ostia, which lie along its length, as shown in Figure 14.27A, which is of a brine shrimp, *Artemia* spp. The filling of the heart is achieved primarily by the suspensory ligaments, which are shown in Figure 14.28A. The suspensory ligaments are stretched when the heart contracts and restore the heart to its original volume by their elastic recoil, when the heart relaxes. Pressure is often lower in the pericardial cavity than in the venous sinuses during contraction of the heart, thereby 'sucking' the blood into the pericardial space[33].

After entering the heart, the haemolymph leaves via arteries and eventually enters the sinuses, which are collectively known as the haemocoel. The extent of the division of the arteries into smaller vessels varies between different groups; the smallest vessels of some species are comparable to the size of capillaries in vertebrates, especially in nervous tissue. The muscles of each thoracic limb (pereiopod) of decapod crustaceans are perfused by many fine 'capillary-like' vessels, which are shown in Figure 14.28B. The smallest vessels in the nervous tissue of decapod crustaceans are less than 10 μm in diameter, which is well within the size range of capillaries in vertebrates.

Having perfused the various parts of the body, the oxygen-depleted (venous) haemolymph collects in sinuses, which in decapods are thought to be discrete channels rather than arbitrary spaces between organs. All of the sinuses on one side of the body eventually drain into a single sinus (the infrabranchial sinus) as shown in Figure 14.29. This sinus runs along the posterior edge of the gills and supplies haemolymph to the afferent branchial veins. The haemolymph passes through the gills, where it undergoes gas exchange. Oxygen-rich haemolymph from the gills then flows along the efferent branchial veins and into three pairs of branchiocardiac veins which empty into the pericardial sinus.

In most decapod crustaceans, the pericardial sinus also receives haemolymph that has drained from the inner wall

[33] This is similar to the constant volume mechanism of bivalve and gastropod molluscs.

14

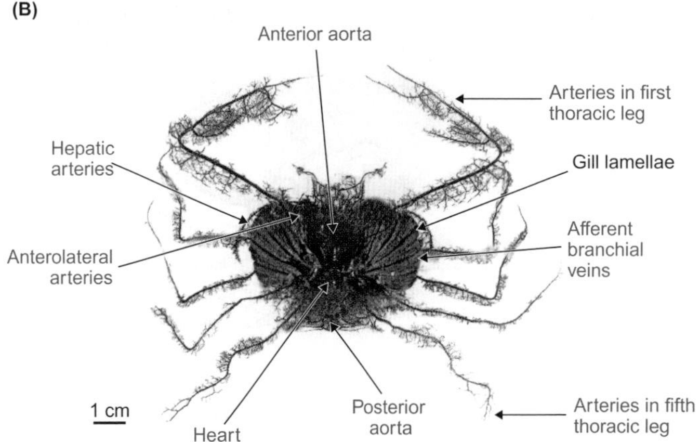

Figure 14.28 Circulatory system of decapod crustaceans

(A) Diagram showing the compact heart and major arteries of a decapod crustacean such as a crab. (B) Dorsal view of resin cast of the extensive arterial system of a blue crab (*Callinectes sapidus*).

A: Reproduced from McMahon BR et al (1997). Invertebrate circulatory systems. In: Dantzler, W.H. (ed). Handbook of Physiology, Section 13: Comparative Physiology, vol II pp 931-1008. New York: Oxford University Press. B: Reproduced from McGraw and Reiber (2002). Cardiovascular system of the blue crab *Callinectes sapidus*. Journal of Morphology 251: 1-21.

14

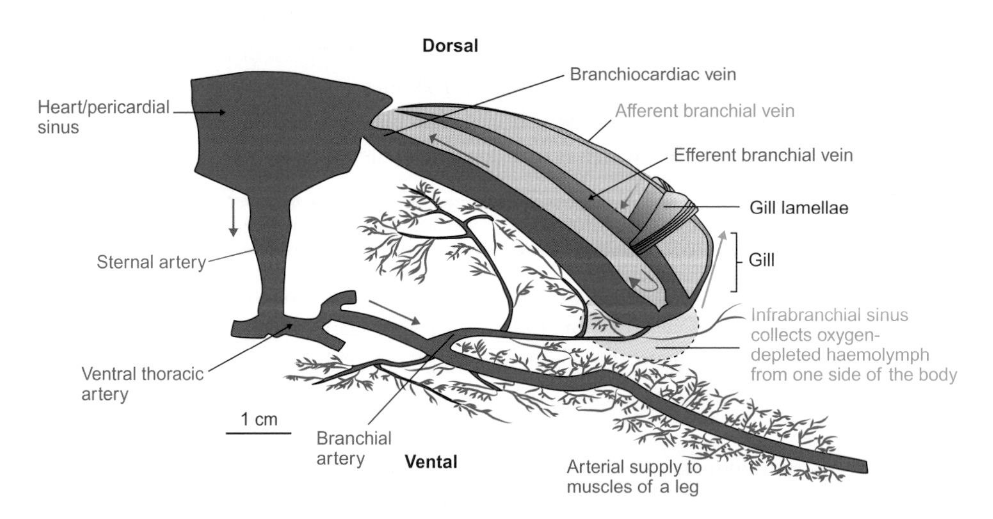

Figure 14.29 Diagram of gill vasculature of blue crabs (*Callinectes sapidus*)

Arrows indicate direction of haemolymph flow.

Reproduced from: McGraw and Reiber (2002). Cardiovascular system of the blue crab *Callinectes sapidus*. Journal of Morphology 251: 1-21.

of the gill covers. In air-breathing crabs the linings of the gill covers are often expanded and act as lungs. Oxygen-depleted haemolymph from the rest of the body perfuses the gas exchange area of the lungs and oxygen-rich haemolymph returns to the pericardial sinus.

The arteries of decapods contain some elastic fibres and are sufficiently compliant and elastic to serve a pulse-smoothing function in the circulation. All arteries of decapod crustaceans, except the dorsal abdominal artery, are devoid of muscle layers and yet it has been demonstrated that the flow of haemolymph through a number of them can be independently varied during exercise, when exposed to hypoxic water and following feeding[34]. The mechanisms responsible for this distribution of the haemolymph are not completely understood. What we do know is that the vessels leaving the heart in decapod and isopod crustaceans have semi-lunar valves containing innervated muscle. These valves could, therefore, be under nervous control and provide a mechanism whereby haemolymph flows from the heart to particular regions of the body.

› *Review articles*

McMahon BR (2001). Control of cardiovascular function and its evolution in Crustacea. Journal of Experimental Biology 204: 923–932.

McMahon BR, Wilkens JL, Smith PJS (1997). Invertebrate circulatory systems. In: Dantzler WH Ed. Handbook of Physiology, Section 13: Comparative Physiology, Vol II pp 931–1008. New York: Oxford University Press.

Reiber CL, McGaw IJ (2009). A review of the 'open' and 'closed' circulatory systems: new terminology for complex invertebrate circulatory systems in light of current findings. International Journal of Zoology Article ID 301284, 8 pages

Shadwick RE (1999). Mechanical design in arteries. Journal of Experimental Biology 202: 3305–3313.

Taylor HH, Taylor EW (1992). Gills and lungs: the exchange of gases and ions. In: Harrison FW, Humes AG (eds.) Microscopic Anatomy of Invertebrates Vol 10, pp. 203–293. New York: Wiley.

Wells MJ, Smith PJS (1987). The performance of the octopus circulatory system: a triumph of engineering over design. Experientia 43: 487–499.

14.4 Circulatory systems of vertebrates

The circulatory system of vertebrates is much less variable in its basic design than that of invertebrates and is considered to be completely closed because all blood vessels are lined with endothelial cells. The blood itself is contained within discrete transport vessels (arteries/arterioles, venules/veins) and exchange vessels (capillaries)[35]. The heart is located ventrally (anteriorly in upright vertebrates such as humans) and the major artery or arteries direct the blood toward the head, before they bend toward the tail/posterior region.

Completely aquatic fish (teleosts and elasmobranchs) have a heart with a single atrium which receives blood from the body and a single ventricle which pumps blood around the body, as shown in Figure 14.30A. This arrangement is sometimes called a **single circulation**. However, in all vertebrates, oxygen-depleted venous blood is not pumped through the gas-exchange organs before entering the heart, as is the case in crustaceans and molluscs. In fish, the blood leaving the heart is pumped through the gills before supplying the body.

In air-breathing vertebrates (lungfish, amphibians, reptiles, birds and mammals), the heart has two atria, as shown in Figure 14.30B, C, D. The right atrium receives oxygen-depleted blood from the body and the left one receives oxygen-rich blood from the lungs. This arrangement is sometimes called a **double circulation**. However, the ventricle is only completely divided into two chambers in crocodilians, birds and mammals, as indicated in Figure 14.30D.

14.4.1 The circulation in water-breathing fish

In most species of fish, blood returns from all parts of the body via the anterior and posterior cardinal veins and from the liver via the hepatic vein, as indicated in Figure 14.30A. Venous return collects in the sinus venosus before entering the heart. From the ventricle of the heart, the blood enters the ventral aorta from where vessels travel along the paired gill arches[36].

In fish, and at some stage during the development of all other vertebrates, the gill arches are derived from six pairs of pharyngeal pouches. Each pouch contains an aortic arch (blood vessel). Figure 14.31 shows the theoretical ancestral arrangement, with the aortic arches numbered according to convention with Roman numerals. In contrast, Arabic numerals (1, 2, 3, etc.) are used for the gill (branchial) arches, as illustrated in Figure 14.32.

In bony fish (teleosts) and elasmobranchs, the heart consists of four chambers in series: the sinus venosus, the atrium, the highly muscular ventricle and either a muscular conus arteriosus in cartilaginous fish (elasmobranchs) or an elastic bulbus arteriosus in bony fish, as shown in Figure 14.33. The bulbi of bony fish are more compliant than the aortae of mammals and frogs because they have a

[34] The response of the cardiovascular system of crustaceans to feeding is discussed in Section 15.2.1.

[35] The characteristics of blood vessels in vertebrates are shown in Figure 14.6.

[36] The gill arches are structures that support the gills and contain the gill blood vessels.

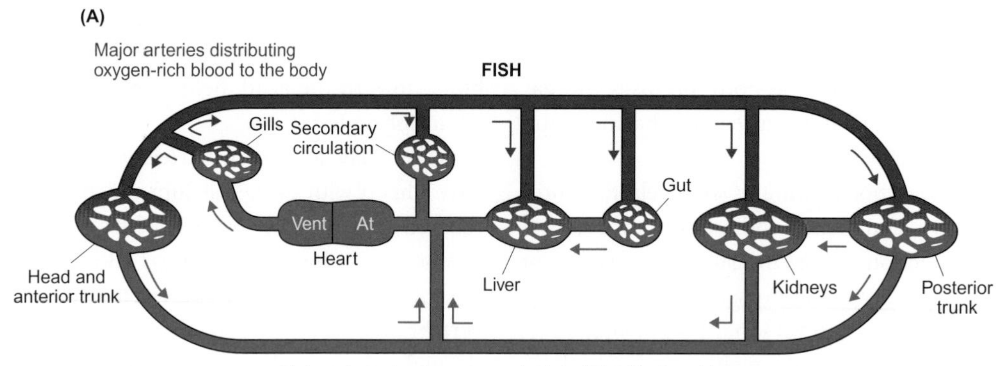

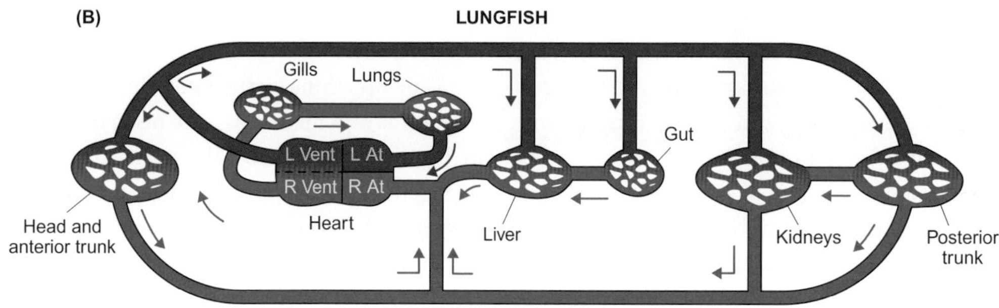

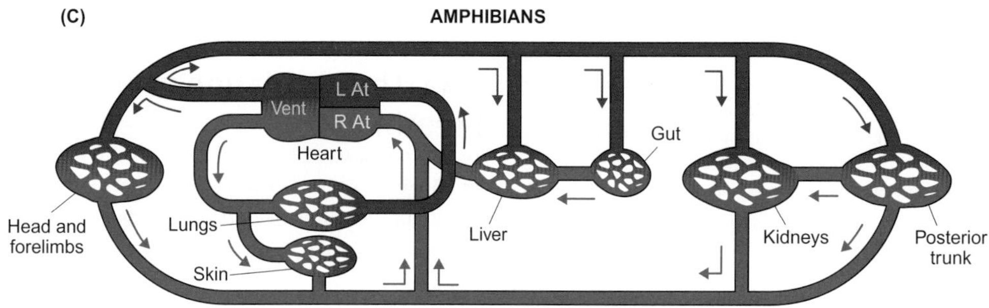

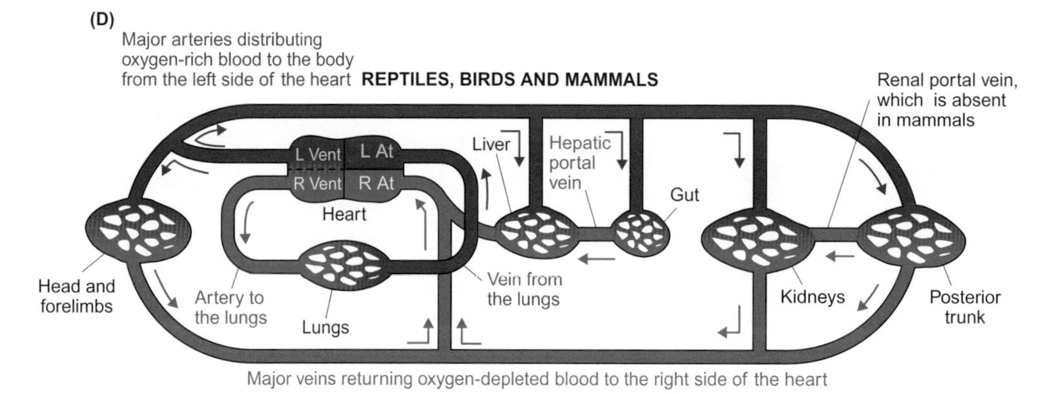

Figure 14.30 General arrangement of circulatory systems of water-breathing vertebrates (fish) and air-breathing vertebrates (lungfish, amphibians, reptiles, birds and mammals)

(A) In completely aquatic fish, the heart consists of two major chambers, the atrium (At) and the ventricle (Vent), and there is a secondary circulation in parallel to the primary circulation. (B) In lungfish, the heart is divided into two atria and an incompletely divided ventricle and there are gills as well as lungs. (C) The hearts of amphibians have two atria and one undivided ventricle. Gas exchange occurs across the skin as well as across the lung, but oxygen-rich blood from the skin returns to the venous system rather than to the left atrium, which receives blood from the lungs. (D) In reptiles, birds and mammals, the heart is divided into two atria, and in crocodilian reptiles, birds and mammals, there is complete division of the ventricle, whereas in non-crocodilian reptiles the ventricle is partly divided.

Modified from Schmidt-Nielsen K (1975). Animal Physiology: Adaptation and Environment. Cambridge: Cambridge University Press.

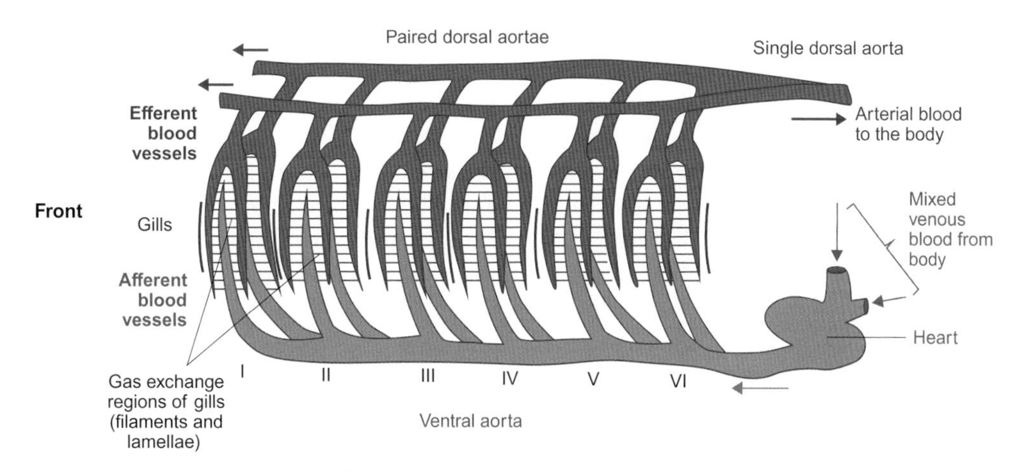

Figure 14.31 Arrangement of the major blood vessels in the theoretical ancestor of the jawed vertebrates

The six unspecialized aortic arches are conventionally numbered with Roman numerals.

Adapted from Romer AS and Parsons TS (1977). The Vertebrate Body. 5th edn. Philadelphia: W.B.Saunders Company.

much larger ratio of elastin to collagen (14:1 in trout, 1.5:1 in mammals and 0.4:1 in frogs). All compartments of the heart, except the bulbus arteriosus, are contractile; as such, the conus arteriosus is thought to be part of the heart itself, whereas the bulbus arteriosus is considered to be a part of the ventral aorta.

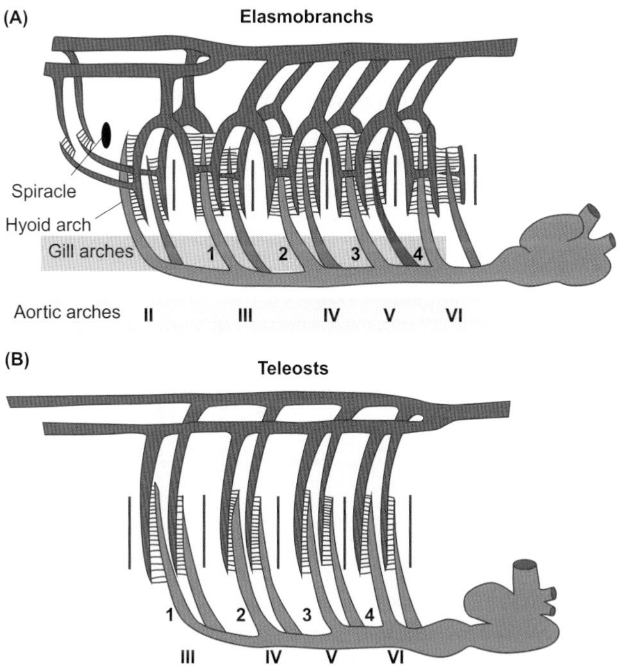

Figure 14.32 Arrangement of the main blood vessels from the hearts of fish

The aortic arches are numbered with Roman numerals and the gill arches with Arabic numerals. Blood vessels as labelled in Figure 14.31. (A) Elasmobranchs (carilagenous fish). The remains of the slit associated with the first aortic arch is known as the spiracle and the second aortic arch (hyoid arch) has one set of gill filaments only and is known as a **hemibranch**, which means half gill. The first complete gill arch corresponds to the third aortic arch. (B) Teleosts (bony fish). Aortic arches I and II have disappeared, which means that the first gill arch corresponds to aortic arch III.

Reproduced from Romer AS and Parsons TS (1977). The Vertebrate Body. 5th edn. Philadelphia: W.B. Saunders Company.

Structure and function of the heart in fish

Our understanding of the mechanisms involved in filling the atrium and ventricle in the hearts of fish has changed over the last few years. It was once thought that filling of the atrium in cartilaginous fish was similar to that proposed in some species of molluscs and arthropods as a result of the heart being in a rigid pericardium: contraction of the ventricle causes a reduction in pressure below ambient in the pericardial cavity which leads to blood being sucked into the atrium from the veins. This mechanism is known as *vis-a-fronte* (force from the front) by vertebrate physiologists, as opposed to *vis-a-tergo* (force from behind). *Vis-a-tergo* refers to the influence of the pressure remaining in the venous system, which causes blood to enter the atrium. However, it is now clear that pressure in the pericardial cavity of fish does not have to be below ambient for suction to occur.

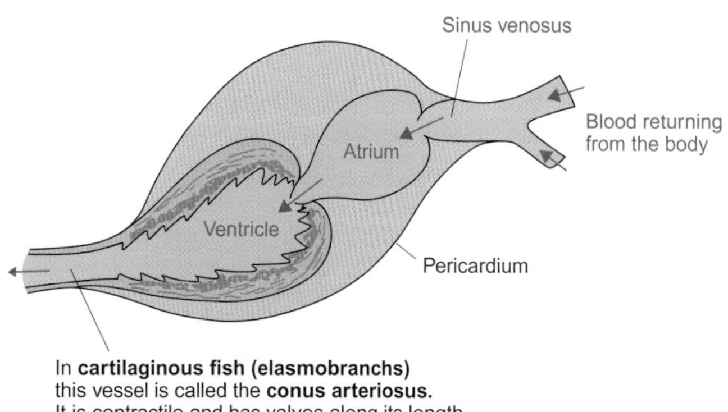

Figure 14.33 Structure of the heart of fish

The heart of fish consists of four chambers. The fourth chamber is muscular in elasmobranchs (cartilaginous fish), whereas in the teleosts (bony fish) it is elastic.

It was also once thought that contraction of the atrium was the main (if not the only) means of filling the ventricle in fish. However, it now seems that there is an early filling phase during relaxation of the ventricle, which causes blood to be sucked from the atrium, and that the filling occurring as a result of subsequent contraction of the atrium may be of less significance. Data that support these ideas are shown in Figure 14.34.

The ventricle of fish is capable of pumping out 80–95 per cent of its entire volume so that the volume of blood in the ventricle (**end systolic volume**) by the end of systole is low. However, a number of species of fish are able to increase ventricular filling during diastole so that cardiac stroke volume can be increased by way of the Frank–Starling relationship.

During contraction of the ventricle, the pressure measured in the ventral aorta is greater than that in the dorsal aorta, as shown by the data in Table 14.3. The lower pressure in the dorsal aorta is caused by the blood leaving the heart in the ventral aorta and passing through the relatively high resistance channels of the gill lamellae, where gas exchange occurs, before being distributed to the rest of the body in the dorsal aorta[37].

[37] Equation 14.4 shows that if blood flow and blood pressure one side of a resistance (in this case ventral aortic flow and pressure) do not change, an increase in total resistance (provided by the gills) leads to a reduction in blood pressure on the other side of the resistance (dorsal aortic pressure).

Blood flow through the gills of fish

For gas exchange in the gills to be effective, the flow of water over them and blood flow through them should be continuous. Continuous blood flow may be achieved by the stretching and subsequent elastic recoil of the ventral aorta during the ejection of blood from the ventricle. The extent of this can be investigated by measuring flow along the ventral aorta. Such measurements illustrate major differences between the flow patterns of cartilaginous and bony fish. Figure 14.35A shows that flow through the ventral aorta of elasmobranchs occurs almost exclusively during cardiac contraction (systole); by contrast, there is zero flow during relaxation (diastole). In teleosts, however, there is normally a continual—albeit pulsatile—flow along the ventral aorta throughout the cardiac cycle. Flow never reaches zero, as shown in Figure 14.35B.

These differences in flow along the ventral aorta are most likely the result of the different characteristics of the conus arteriosus in elasmobranchs and the bulbus arteriosus in teleosts. The conus contracts during ventricular systole, which means that its walls are stiff; by contrast, the non-contractile bulbus is extremely stretchy (compliant). In fact, when the relationship between pressure and volume (pressure–volume loops) are determined in isolated lengths of the bulbus and ventral aorta from teleost fish, the aortae have J-shaped curves whereas the bulbi have r-shaped curves, as shown in Figure 14.36A.

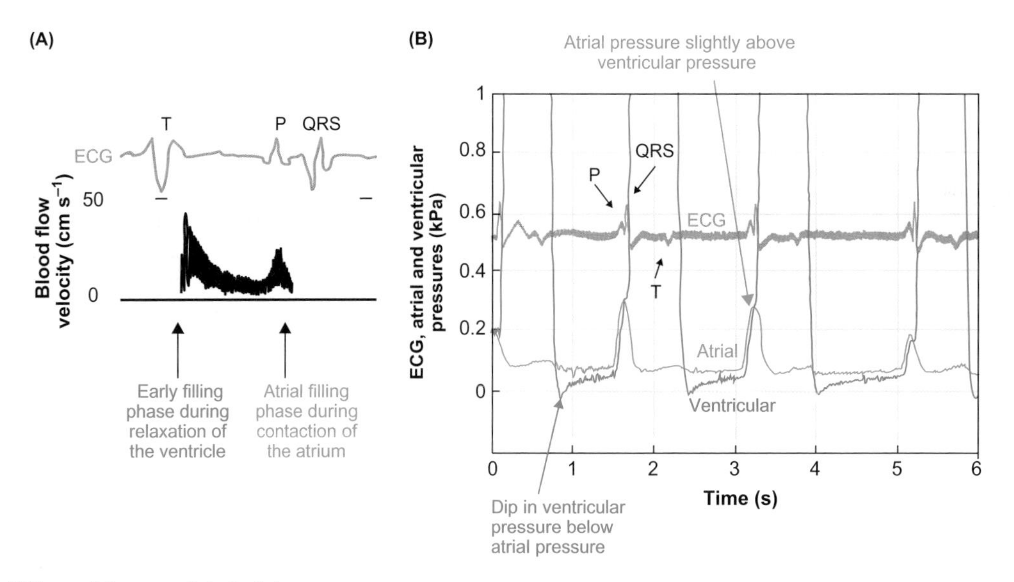

Figure 14.34 Filling of the ventricle in fish

(A) Electrocardiogram (ECG) (top) and velocity of blood flow at the entrance of the ventricle from the atrium of the heart of sea bass (*Paralabrax* sp.). The velocity of blood flow was determined by Doppler echocardiography and demonstrates that there is an early surge of blood (early filling phase) during diastole as the ventricle relaxes, indicated by the T wave of the ECG and a later, smaller surge, as the atrium contracts, indicated by the P wave of the ECG. (B) These traces illustrate the pressure changes that cause the surges in flow. At the end of the relaxation phase of the ventricle, there is a dip in pressure below that in the atrium and during contraction of the atrium, its pressure is slightly above that in the ventricle.

Reproduced from Lai NC et al (1998). Echocardiograpic and hemodynamic determinations of the ventricular filling pattern in some teleost fishes. Physiological Zoology 71: 157-167.

Table 14.3 Characteristics of the cardiovascular systems of resting vertebrates

Blood pressure is lower in the dorsal than in the ventral aorta of fish. In amphibians and most reptiles, pressure in the systemic and systemic circuits are similar, whereas in varanid lizards, crocodilians, birds and mammals, pressure is higher in the systemic circuit but substantially lower in the pulmonary circulation.

Species	Blood pressure (kPa) systolic/diastolic or mean		Blood volume (% body mass)
Oncorhynchus (rainbow trout)	6.2/4.6 (ventral aorta)	4.6/3.9 (dorsal aorta)	3.5
Katsuwonas (skipjack tuna)	11.6 (ventral aorta)	5.4 (dorsal aorta)	
Xenopus (African clawed toad) when not ventilating lungs	5.0/3.0 (systemic artery)	5.0/3.3 (pulmonary artery)	11.2
Trachemys (red-eared slider) when not ventilating lungs	3.4/2.5 (right aorta)	3.4/1.3 (common pulmonary artery)	6.9
Varanus (monitor lizard) when not ventilating lungs	11.2/7.8 (right aorta)	5.7/3.0 (pulmonary artery)	
Alligator or *Crocodylus* when not ventilating lungs	5.2 (aortae)	2.1 (pulmonary)	
Birds	18.2 (aorta)	2.3 (pulmonary)	7.1 (non-aquatic)
Mammals	13.8 (aorta)	2.3 (pulmonary)	6.8 (non-aquatic)

References: Kiceniuk and Jones, 1977; Olsen et al., 2003; Bushnell and Brill, 1992; Shelton, 1970; Hillman, 1976; Shelton and Burggren, 1976; Smits and Zozubrowski, 1985; Burggren and Johansen, 1882; Gleeson et al., 1980; Jones and Shelton, 1993; Bishop, 1997; Faraci et al., 1984; Grubb, 1983; Bond and Gilbert, 1958; Birchard and Tenney, 1991. Full citations available online.

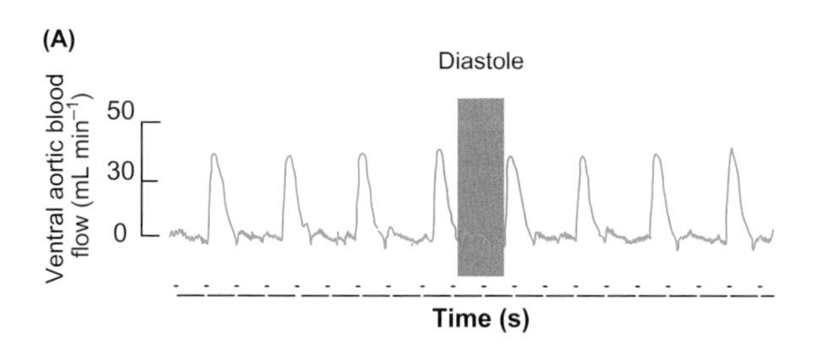

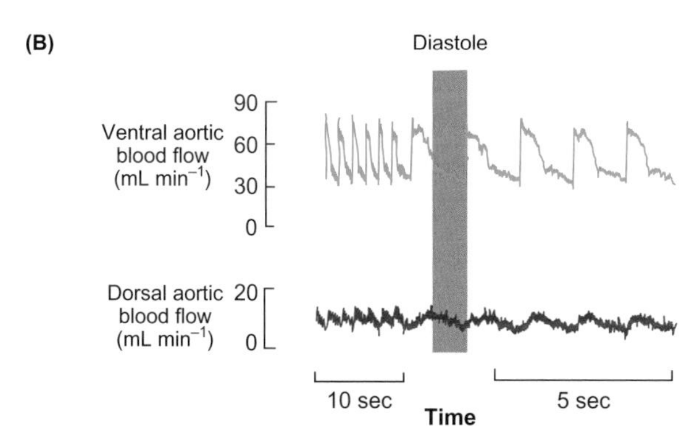

Figure 14.35 Blood flow through the ventral aortae of elasmobranch and teleost fish

(A) Blood flow through the ventral aorta in a dogfish (*Scyliorhinus canicula*). Note that there is no forward flow during the relaxation phase (diastole) of the cardiac cycle, in fact there is a brief reversal of flow at the beginning of this phase. (B) Blood flow through the ventral (top) and dorsal aortae of a cod (*Gadus morhua*). Although flow along the ventral aorta is highly pulsatile, it never reaches zero. The pulsatility is much reduced by the time the blood reaches the dorsal aorta because of the large compliance in the ventral aorta and gill capillaries.

Reproduced from Butler PJ and Taylor EW (1975). The effect of progressive hypoxia on respiration in the dogfish (*Scyliorhinus canicula*) at different seasonal temperatures. Journal of Experimental Biology 63: 117-130; Jones DR, Langille BL, Randall DR, and Shelton G (1974) Blood flow in the dorsal and ventral aortas of the cod, *Gadus morhua*. American Journal of Physiology 226: 90-95.

Initially upon filling, the isolated bulbus is very stiff and requires a large increase in pressure to cause a relatively small change in volume, but then it becomes extremely compliant. It is so compliant in fact, that a more or less constant pressure is reached (a plateau) despite further increases in volume. Figure 14.36A shows that the plateau pressure corresponds to the mean pressure in the ventral aorta of live fish, in this case yellowfin tuna (*Thunnus albacores*) and Atlantic blue marlin (*Makaira nigricans*). This relationship has also been reported for a range of other species, as shown in Figure 14.36B.

Following the initial expulsion of blood upon contraction of the ventricle, the bulbus expands with very little change in pressure and the blood which accumulates is then released during relaxation of the ventricle, again with very little change in pressure. As a result, this relatively small section of the arterial system in teleost fish smooths blood flow and pressure to a similar extent as all the major arteries in mammals[38]. After having travelled through the gill lamellae, the pulsatility of the blood is less in the dorsal aorta than in the ventral aorta, as shown in Figure 14.35B. This suggests that the pulsatility of blood flow through the gills is also reduced.

Circulation through the gills of teleost fish is quite complex. Figure 14.37 shows that the major route is from the ventral aorta → afferent arch artery → afferent filament artery → afferent lamellar arterioles → lamellae → efferent lamellar arterioles → efferent filament artery → efferent arch artery → dorsal aorta. In addition, oxygenated blood from the efferent filament arteries is distributed in two different ways:

- via connections with the central sinus or interlamellar system;

[38] Pulsatility in the major arteries of mammals is discussed in Section 14.1.4.

14

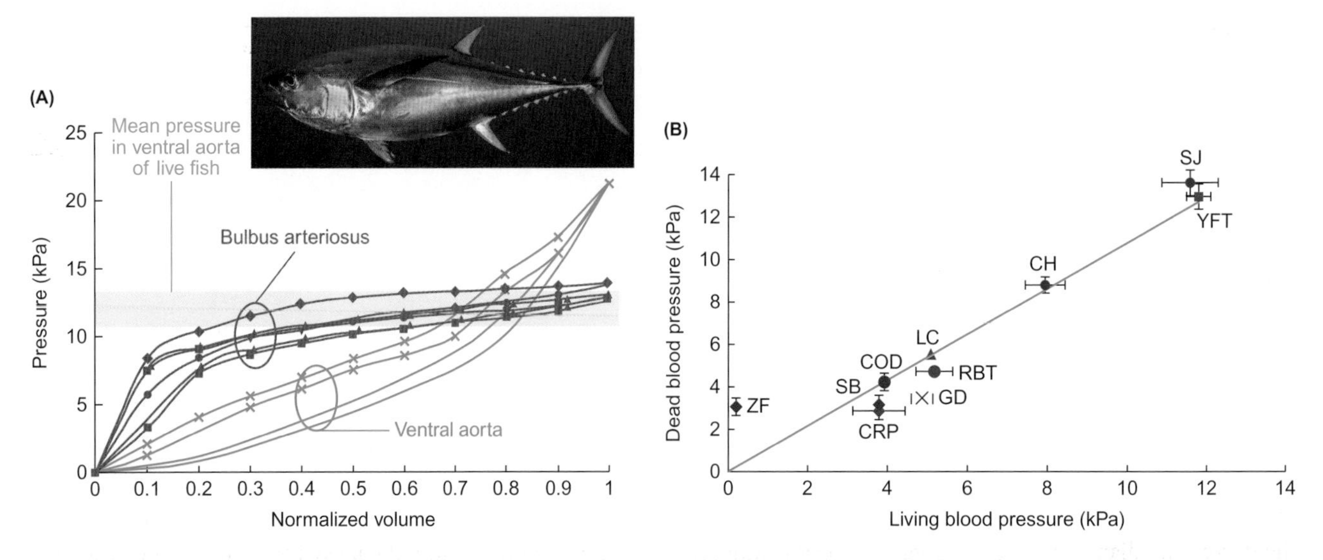

Figure 14.36 Pressure–volume relationships in the ventral aorta and bulbus arteriosus of teleost fish

(A) Pressure–volume loops for the ventral aorta and bulbus arteriosus from yellowfin tuna (*Thunnus albacares*) and Atlantic blue marlin, (*Makaira nigricans*). Note that the loops for the bulbus arteriosus flatten out after a relatively steep increase in pressure and that the plateau pressure is very similar to the range of physiological pressure measured in the ventral aorta of live fish of each species, which is indicated by the blue bar on the y-axis. (B) The similarity between plateau pressure in the isolated bulbus arteriosus and mean physiological pressure in the ventral aorta has been demonstrated in 10 species of fish, which means that inflation of the bulbus in recently dead fish can be used to determine ventral aortic blood pressure in the living animal of the same species. ZF, zebra fish (*Danio reiro*); SB, stickleback (*Gasterosteus aculateus*); CRP, carp (*Cyprinus carpio*); COD, cod (*Gadus morhua*); GD, giant danio (*Davario. aequipinnatus*); LC, lingcod (*Ophiodon elongatus*); RBT, rainbow trout (*Oncorhynchus mykiss*); CH, coho salmon (*Oncorhynchus kisutch*); SJ, skipjack tuna (*Katsuwonus pelamis*); YFT, yellowfin tuna (*Thunnus albacares*).

A: Modified from Braun MH et al (2003). Form and function of the bulbus arteriosus in yellowfin tuna (*Thunnus albacares*), bigeye tuna (*Thunnus obesus*) and blue marlin (*Makaira nigricans*). Journal of Experimental Biology 206: 3311-3326. B: Modified from Jones DR et al (2005). Necrophysiological determination of blood pressure in fishes. Naturwissenschaften 92: 582-585. Image © Shane Gross/ Shutterstock.

- via a nutritive pathway which supplies oxygen and nutrients to the tissues of the gill arch and filaments, and is also connected to the interlamellar system.

These two routes are called arteriovenous anastomoses (efferent, abbreviated to A-V_{eff}) and are present in most species of fish. The presence of A-V_{eff} means that some of the arterial blood is returned via the central venous sinus back to the heart instead of being transported around the body. These routes may be of significance in species with a completely or predominantly spongy myocardium[39]. In at least

[39] Composition of the heart muscle (myocardium) in fish is discussed in Section 14.4.5.

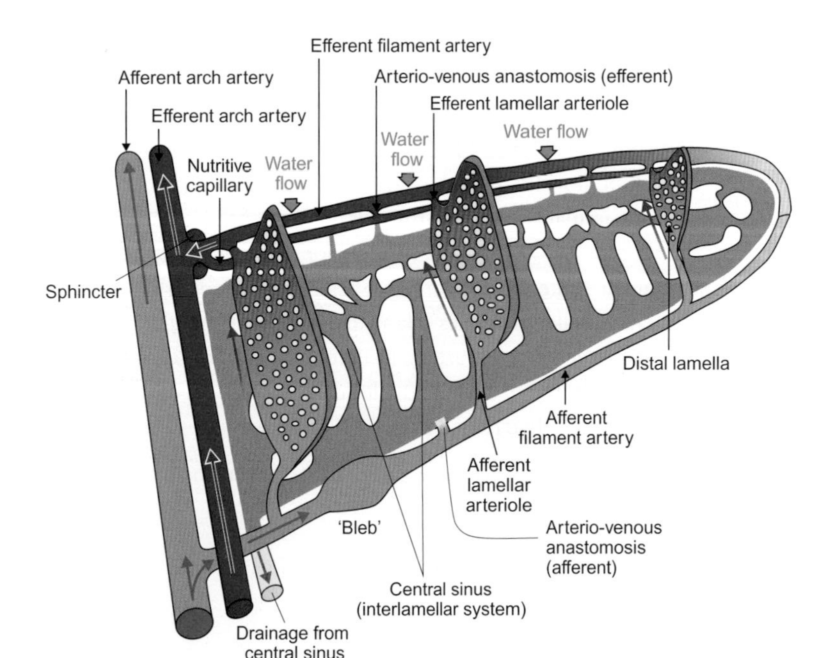

Figure 14.37 Circulation through the gills of teleost fish

Blood circulatory system in a gill filament and lamellae of a typical teleost fish. Blue–red arrows indicate flow of blood through the lamellae. Water flows in the opposite direction.

Adapted from Randall DJ et al (1982). Gas transfer and acid/base regulation in salmonids. Comparative Biochemistry and Physiology 73B: 93-103.

three species of fish, trout, eel (*Anguilla* sp.) and catfish (*Ictalurus* spp.), there are also arteriovenous anastomoses between the *afferent* filament artery and the central venous sinus (A-V_{aff}), as shown in Figure 14.37, although the significance of these is unknown. An additional feature which is, no doubt important in determining the pattern of the flow of blood through the gills, is the presence of a sphincter muscle at the base of each efferent filament artery.

Blood flow between or within the lamellae is not uniform. In lingcod (*Ophiodon elongatus*), almost 90 per cent of the area of the gill lamellae is perfused with blood, but in rainbow trout resting in well-aerated water, only about 60 per cent of the lamellae are perfused, as shown in Figure 14.38A. The perfused lamellae tend to be in the basal two-thirds of each filament. This distribution is most likely the result of the decreasing diameter of the filamental arteries along the length of the filament. The proportion of perfused lamellae increases when the fish are exposed to hypoxic water (partial pressure of oxygen, 4 kPa) and, it is assumed, during exercise, when overall pressure and pulse pressure increase in the ventral aorta. In other words, a reduction in oxygen supply or an increase in oxygen demand causes the recruitment of additional lamellae (**interlamellar recruitment**) and, therefore, an increase in the surface area for gas exchange.

There is also evidence that channels within a lamella are recruited (**intralamellar recruitment**). In resting lingcod, blood tends to flow preferentially through the larger diameter basal and marginal channels, which are shown in Figure 14.38B. The basal channel is buried in the epithelium of the filament, so blood flowing through it undergoes little or no gas exchange. As perfusion pressure and flow increase during exercise, for example, the more central channels expand and blood redistributes towards them. The expansion of more central channels takes the blood closer to the surface of the lamellae, reducing the diffusion distance between the blood and water. This process is illustrated in Figure 14.38B.

14.4.2 The circulatory system of air-breathing fish

Some species of teleost fish have evolved modifications to various parts of their bodies for use as air breathing organs[40]. In most air-breathing fish, the oxygen-rich blood from the air-breathing organ returns to the venous system, where it is mixed with oxygen-depleted blood returning from the body, and then goes to the heart. This type of circulation reduces the effectiveness of the air-breathing organ supplying oxygen to the body.

However, some species have evolved an air bladder[40] which they use as a lung—including bichirs (*Polypterus* sp.) and bowfins (*Amia* sp.), for example—and in these species paired arteries (pulmonary arteries) arise from aortic arch VI and supply oxygen-depleted blood to the air-breathing organ. Paired veins (pulmonary veins) transport oxygen-rich blood either directly to, or into the general vicinity of, the heart. Whether

[40] Air breathing in fish and the air bladder in fish are discussed in Section 12.3.3.

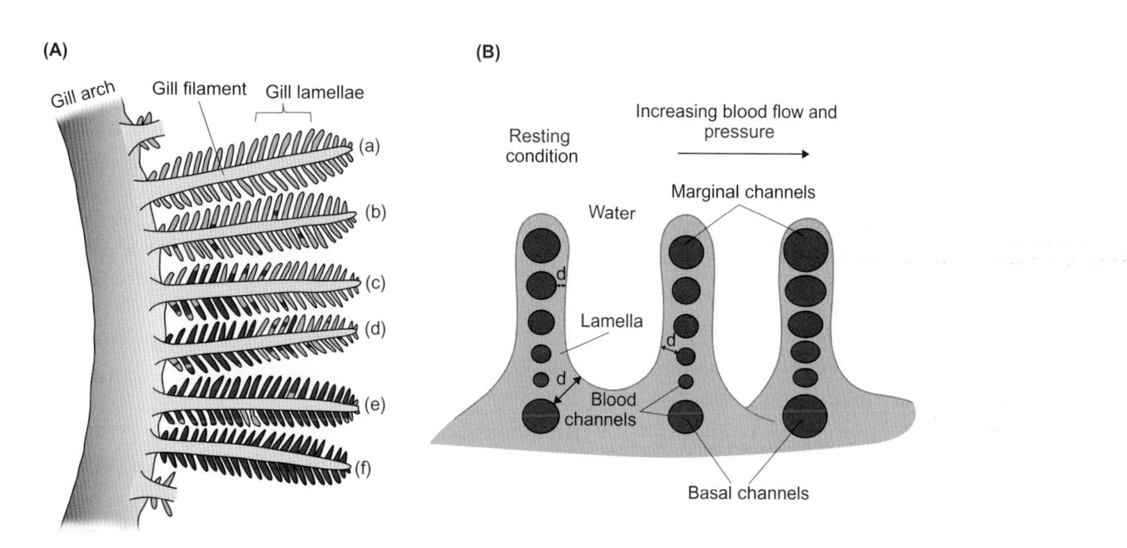

Figure 14.38 Blood flow through the filaments and lamellae of teleost fish

(A) Part of a gill arch of a rainbow trout (*Oncorhychus mykiss*) in well aerated water at 5°C showing the typical distribution patterns of blood cells (dark regions) in the lamellae of the gill filaments. Patterns (a) and (f) are less frequent than the others. (B) Blood channels in the lamellae of a lingcod (*Ophiodon elongatus*) illustrating the likely resting condition on the left and the probable changes that occur when flow and pressure are elevated on the right. Not only does more blood flow through the more central channels, but the diffusion distances (d) are shorter in the lamellae depicted on the right.

A: Reproduced from Booth JH (1978). The distribution of blood flow in the gills of fish: application of a new technique to rainbow trout (*Salmo gairdneri*) Journal of Experimental Biology 73: 119-129. B: Reproduced from Randall D and Daxboeck C (1984). Oxygen and carbon dioxide transfer across fish gills. In: Fish Physiology, vol 10. Pt A, Hoar WS and Randall DJ (eds.), pp 263-314.

or not the two bloodstreams remain separated as they pass through the hearts of these species of fish is unclear, although anatomical and physiological evidence suggests that separation occurs in some air-breathing teleosts, such as northern snakeheads, (*Channa argus*). Snakeheads have two branches to their ventral aorta, with the concentration of oxygen in the blood in the posterior branch, which sends blood to all parts of the body, always being greater than that in the anterior branch, which sends blood to the air-breathing organ.

Those teleost fish with both an air-breathing organ and functional gills tend to breathe air more frequently when there is increased demand for oxygen—for example, if water temperature increases, and/or if there is a reduction in the partial pressure of oxygen in the water (i.e. it becomes hypoxic). This means that if any blood that has been oxygenated by the air-breathing organ passes through functional gills, some oxygen is likely to be lost to the hypoxic water. This is exactly what happens in bowfins, in which the gills are not reduced and there is no shunt bypassing them. In order to prevent (or at least to reduce) this problem, many air-breathing fish, such as electric eels (*Electrophorus electricus*), have reduced gills or even direct connections between the ventral and dorsal aorta in some (if not all) gill arches.

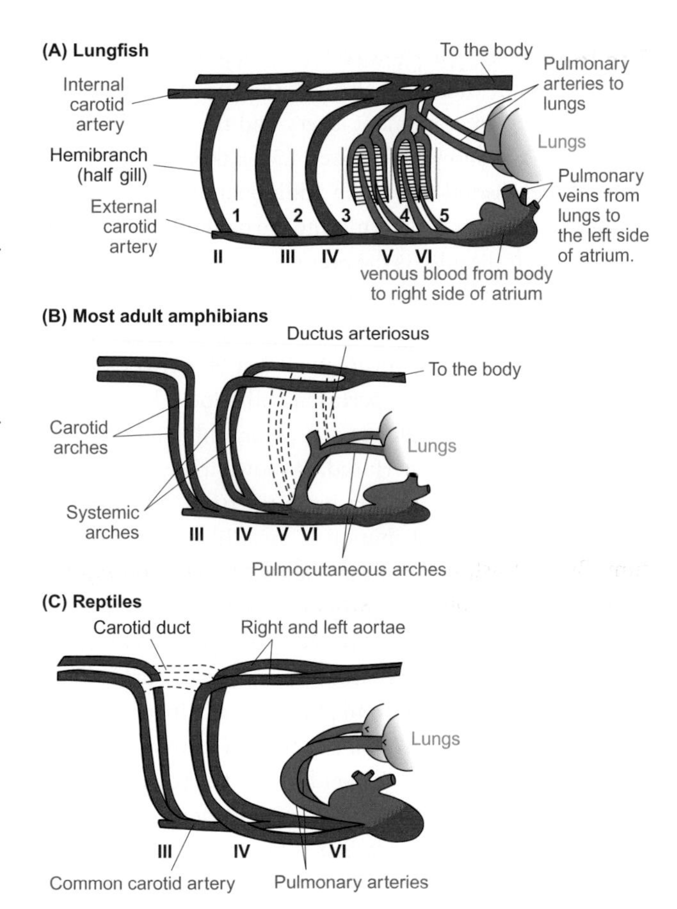

Figure 14.39 Aortic arches in lungfish, amphibians and reptiles

(A) In lungfish (*Protopterus* sp. and *Lepidosiren* sp.) gill arches 2 and 3 (aortic arches III and IV) have no gas-exchange areas (filaments and lamellae) and make direct connections between the ventral and dorsal aortae. Each aortic arch VI gives rise to a pulmonary artery which carries oxygen-depleted blood to the lungs. A pair of pulmonary veins carries oxygen-rich blood from the lungs to the heart. (B) Most adult amphibians have no gills. In anurans only three pairs of the aortic arches remain, the IIIrd become the carotid arches which supply the head, the IVth become the systemic arches which supply the remainder of the body and the VIth become the pulmocutaneous arches. In urodeles, aortic arch V and the connection between the pulmocutaneous and systemic arches, the ductus arteriosus, remain. (C) In reptiles, the situation is similar to that in amphibians, except that in lizards the connection between the aortae and internal carotids, the carotid duct, remains and there is no cutaneous supply from aortic arch VI.

Adapted from Romer AS and Parsons TS (1977). The Vertebrate Body. 5th edn. Philadelphia: W.B. Saunders Company.

14.4.3 The incompletely divided hearts of lungfish, amphibians and non-crocodilian reptiles

The evolution of lungs and an increasingly greater reliance upon air as the medium for gas exchange, brought about clear changes in the structure of the hearts and arrangement of the blood vessels coming from those hearts in ectothermic, air-breathing vertebrates (lungfish, amphibians and reptiles). As we discuss in Section 14.4.2 for bichirs and bowfins, the VIth aortic arches supply oxygen-depleted (venous) blood to each lung in the pulmonary arteries, while a pair of pulmonary veins return oxygen-rich (arterial) blood directly to the heart, i.e. separately from the return of oxygen-depleted blood from the rest of the body. This arrangement is shown in Figure 14.39.

This nomenclature may sound somewhat confusing given that all other arteries in the body contain oxygen-rich blood and all other veins contain oxygen-depleted blood. However, this terminology is consistent with the anatomical definition that arteries carry blood **away** from the heart and veins carry blood **toward** the heart.

The presence of the two separate groups of veins entering the hearts of these animals (those from the body and those from the lungs) means that the two bloodstreams—one oxygen-depleted and one oxygen-rich—could remain completely separate during their passage through the hearts. However, none of these groups of animals has a completely divided heart (with the exception of crocodiles and alligators among the reptiles), so there is the opportunity for at least some mixing of the two bloodstreams as they pass through it. Such incomplete division of the heart also enables blood to flow around the pulmonary and systemic (body) circuits at unequal rates. Indeed, changes in the resistance to blood flow in the pulmonary circulation is the major influence on the distribution of blood from the heart.

The low-pressure circulation of lungfish and amphibians

Figure 14.39A shows that gill arches 2 and 3 of the lungfish, *Protopterus* and *Lepidosiren,* do not possess any filaments and their aortic arches pass between the ventral and dorsal aortae without interruption. There is, therefore, no gas exchange between water and blood in these anterior gill arches. Even the lamellae on the last two pairs of gill arches, supplied by aortic arches V and VI, have a reduced surface area and increased diffusion barrier compared with gills in other fish, and their major role in gas exchange is the removal of carbon dioxide[41].

In lungfish, the atrium and ventricle, the latter of which is shown in Figure 14.40A, are both partly divided by a septum. By contrast, most species of amphibians and reptiles, possess two anatomically separate atria with the left one receiving blood from the lung and the right one receiving blood from the body. However, there is no anatomical division in the ventricle of anuran amphibians (frogs and toads), although the trabeculae of the inner wall form deep, blind-ended pockets, as shown in Figure 14.40B. These pockets collect and hold blood during the diastolic filling of the ventricle and are thought to be important for maintaining some separation of blood from the left and right atria.

In addition to the anatomical features of the heart itself, anatomical structures in the vessels leaving the ventricle are important in maintaining separation between the two blood streams. In lungfish, the vessel is the non-contractile bulbus cordis; in amphibians, the vessel is the contractile conus arteriosus. On the inner wall of the bulbus cordis of lungfish, two folds become fused at the end furthest from the ventricle (distal end) to form a complete division into two channels, as shown in Figure 14.40A. Similarly, a spiral valve occurs along the inner wall of the conus arteriosus in amphibians, which is illustrated in Figure 14.40B. Three pairs of blood vessels emerge from the conus of amphibians, as indicated in Figures 14.39B and 14.40B. These are:

1. The carotid arches;
2. The systemic arches;
3. The pulmo-cutaneous arches (in the anurans).

As their name suggests, the pulmo-cutaneous arches not only supply blood to the lungs, but also to the skin.

Both lungfish and amphibians ventilate their lungs intermittently[42], and the perfusion of the lungs with blood matches their ventilation in both groups. Figure 14.41A shows how most of the oxygen-rich blood from the left side of the heart in African lungfish (*Protopterus aethiopicus*) is directed to

[41] Removal of carbon dioxide in lungfish is discussed in Section 12.3.3.

[42] Ventilation in lungfish and amphibians is discussed in Section 12.3.3.

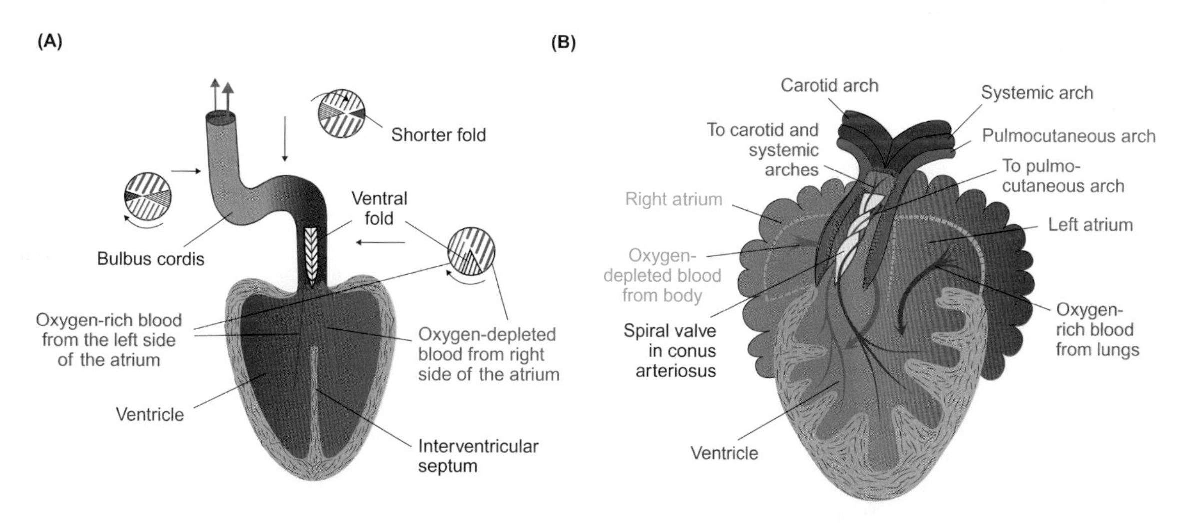

Figure 14.40 Structures and pathways of blood through the hearts of lungfish and amphibians

(A) The ventricle of the heart of African lungfish (*Protopterus aethiopicus*) is partly divided by the interventricular septum which maintains partial separation between the oxygen-rich and oxygen-depleted blood from the left and right atria, respectively, following ventilation of the lungs. This separation is maintained in the bulbus cordis by the two folds. Cross-sections of the bulbus cordis along its length show the separation of the two blood streams. The straight arrows indicate the level at which cross-sections were taken and the curved arrows indicate the direction of rotation of the bulbus along its length. (B) There is no anatomical division in the ventricle of the heart of frogs and toads as illustrated by African clawed frogs (*Xenopus, laevis*), but the wall of this chamber is very spongy and blood from the right and left atria may remain partially separate in the ventricle as they are 'absorbed' into the ventricular wall. Note, right and left are depicted on different sides of these diagrams, depending on whether they are viewed from above or from below.

A: Adapted from Johansen K and Burggren W (1980). Cardiovascular function in the lower vertebrates, In: Hearts and Heart-like Organs, vol 1, Bourne G (ed.), pp 61-117, New York: Academic Press. B: Adapted from Shelton G (1976). Gas exchange, pulmonary blood supply, and the partially divided amphibian heart. In: Perspectives in Experimental Biology, Vol 1. Spencer Davies P (ed.), pp 247-259. Oxford: Pergamon Press.

the anterior gill arches following lung ventilation. Partial separation achieved in the ventricle is maintained by two folds in the bulbus cordis: a more prominent ventral fold and a shorter fold, which does not extend as far back.

More anteriorly, the two folds fuse to form two complete channels in the bulbus; a ventral channel and a dorsal channel. The bulbus bends first to the left and then to the right and the folds rotate by approximately 270° as they extend forward. This rotation is illustrated by the three cross-sections of the bulbus taken at the locations indicated by the straight arrows in Figure 14.40A. Oxygen-rich blood is conveyed in the ventral channel of the bulbus of lungfish and goes on to the dorsal aorta via the anterior gill arches to be distributed around the body. The dorsal channel of the bulbus directs oxygen-deficient blood from the right side of the ventricle to the posterior gill arches and to the lungs.

Just after a breath, the amount of blood perfusing the lungs of African lungfish is relatively high and may be as much as 70 per cent of cardiac output. In other words, there is an overall shunting of blood from the left side of the heart to the right side, known as a **left to right (L → R) shunt**. The proportion of oxygen-rich blood perfusing the filament-free anterior arches to be distributed around the body, gradually declines from almost 100 per cent immediately following lung ventilation to almost 60 per cent by the time of the next ventilation, when it again increases, as shown in Figure 14.41A.

The amount of blood perfusing the lungs during the period between lung ventilations also decreases. This could be as low as 20 per cent of cardiac output, which means there is an overall shunting of blood from the right to the left side of the heart, a **right to left (R → L) shunt**. The reduction in lung perfusion occurs because vessels somewhere in the pulmonary circulation (possibly the pulmonary artery itself) constrict as oxygen is depleted; they dilate again when the lungs are inflated.

The lungs of amphibians are also variably perfused depending on whether or not they are being ventilated, and there is evidence from experiments on African clawed frogs (*Xenopus laevis*) that the two bloodstreams through the heart are reasonably well (but not perfectly) separated during periods of ventilation. For example, during lung ventilation, the partial pressure of oxygen is much higher in the blood supplying the body (systemic circulation) than in that perfusing the lungs (pulmocutaneous circuit) with values of 11 kPa and 8 kPa, respectively, which does suggest separation of the two bloodstreams.

This separation is not only achieved by the spongy wall of the single ventricle, but also by the spiral valve within the outflow tract (conus arteriosus) from the heart. The spiral valve maintains the separation achieved in the ventricle. When the two bloodstreams enter the conus arteriosus, the spiral valve channels oxygen-rich blood predominantly to the carotid and pulmocutaneous arches and oxygen-poor blood predominantly to the lungs. This scheme applies especially during and after ventilation of the lungs when their blood vessels are dilated. During the breath-hold phase, there is a right to left shunt as perfusion of the lungs declines, as indicated in Figure 14.41B.

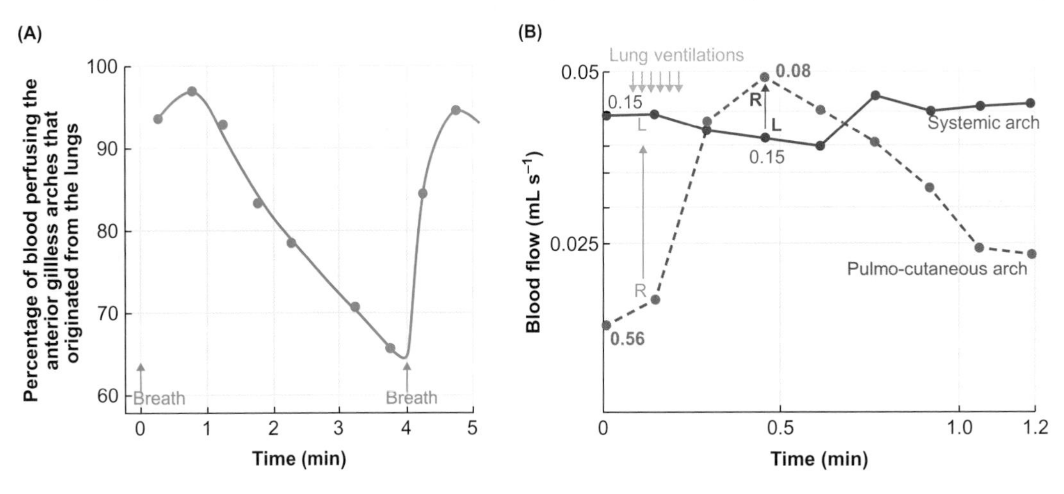

Figure 14.41 Effect of lung ventilation on distribution of blood from the heart of African lungfish (*Protopterus aethiopicus*) and African clawed toads (*Xenopus laevis*)

(A) The proportion of oxygen-rich blood (i.e. that from the lungs) perfusing the first, filament-free, arches is almost 100% following lung ventilation, but this declines to almost 60% following a breath-hold of 4 min duration in an African lungfish. (B) Blood flow through the lungs of *Xenopus* is low during the periods of breath-hold but increases substantially following ventilations of the lung. There is little change in flow along the systemic arch. These changes in blood flow are achieved by changes in resistance to flow, shown by numbers next to the lines on graph (units are kPa mL^{-1} min^{-1} kg^{-1}). Toward the end of a period of breath-hold, when resistance in the lung circulation is high, most of the blood leaving the heart circulates round the body, i.e. there is a large overall right to left (R ⟶ L) shunt. On the other hand, following a period of lung ventilation, resistance to flow of blood through the lungs is reduced and slightly more of the blood leaving the heart is from the lungs, i.e. there is a small overall left to right (L ⟶ R) shunt.

A: Reproduced from Johansen K et al(1968). Cardiovascular dynamics in the lungfishes. Zeitschrift für vergleichende Physiologie 59: 157-186. B: Adapted from Adapted from Shelton G (1970). The effect of lung ventilation on blood flow to the lungs and body of the amphibian, *Xenopus laevis*. Respiration Physiology 9: 183-196.

As in lungfish, inflation of the lungs of amphibians is associated with dilation of the pulmonary circulation so that more blood perfuses the lungs. For a short period slightly more blood flows along the pulmo-cutaneous arches than along the systemic arches; there is a slight, overall left to right shunt, as shown in Figure 14.41B.

The vasodilation in the pulmonary circulation during lung inflation is accompanied by a relatively large drop in pressure in the pulmo-cutaneous arches. However, note from Table 14.3 that the pressures in the two circuits between ventilations are very similar and relatively low, with systolic pressure similar to that in most teleost fish.

There is very little change in the flow of blood through the systemic arches which supply blood to the body throughout these cycles of ventilation and breath-hold, as shown in Figure 14.41B. Consequently, total cardiac output is greater during lung inflation. This greater output is achieved, at least in part, by an increase in heart rate.

The emergence of a high-pressure systemic circulation in reptiles

The heart and circulatory systems of reptiles have much in common with those of amphibians, as shown back in Figure 14.39. However, reptiles exhibit two basic patterns of cardiac structure with the most common being found in the non-crocodilian reptiles—the snakes, lizards (except the varanids) and the turtles and tortoises.

The circulation of non-crocodilian reptiles

The structure of the heart in non-crocodilian reptiles is illustrated by the heart of freshwater turtles (*Pseudemys scripta*) in Figure 14.42. The large, single ventricle is incompletely divided into three chambers by muscular ridges that protrude from the inner ventricular wall. The passage of blood through the ventricle is shown in Figure 14.42. The right atrium, which receives oxygen-depleted blood from the body, opens into one of these chambers, the cavum venosum. The left atrium, which receives oxygen-rich blood from the lungs, opens into another chamber, the cavum arteriosum. Oxygen-depleted blood moves from the cavum venosum into the third chamber, the cavum pulmonale, from where the blood enters the pulmonary arteries. Oxygen-rich blood moves from the cavum arteriosum into the cavum venosum and from there into the major arteries supplying the body (the right and left aortae and the brachiocephalic artery). Such complicated flow patterns suggest that some mixing of the two bloodstreams must occur, and this is confirmed by measurements of blood flows and partial pressures of oxygen in the right aorta and the left pulmonary artery.

As in amphibians, separation of the oxygen-rich and oxygen-depleted blood streams is greatest following lung inflation and there is an overall L → R shunt. In contrast, there is an overall R → L shunt during extended breath holding. During ventilation of the lungs, almost 90 per cent of the blood flowing along the pulmonary arteries is from the right atrium (oxygen-depleted) and almost 70 per cent of the blood flowing along the right aorta is from the left atrium (oxygen-rich). However, variations in the perfusion of the lungs of turtles are not always associated with lung ventilation. It seems that large overall R → L shunts only occur when the animals are resting. When the demand for oxygen increases during exercise or an increase in temperature, or when the supply of oxygen is reduced during hypoxia, the R → L shunt is reduced. For example, during exercise in aquatic turtles, perfusion of the lungs is maintained even during periods of breath-hold[43].

The direction and magnitude of the cardiac shunts in most non-crocodilian reptiles is determined by the resistance of the pulmonary and systemic circulations.

[43] Exercise in reptiles is discussed in Section 15.2.2.

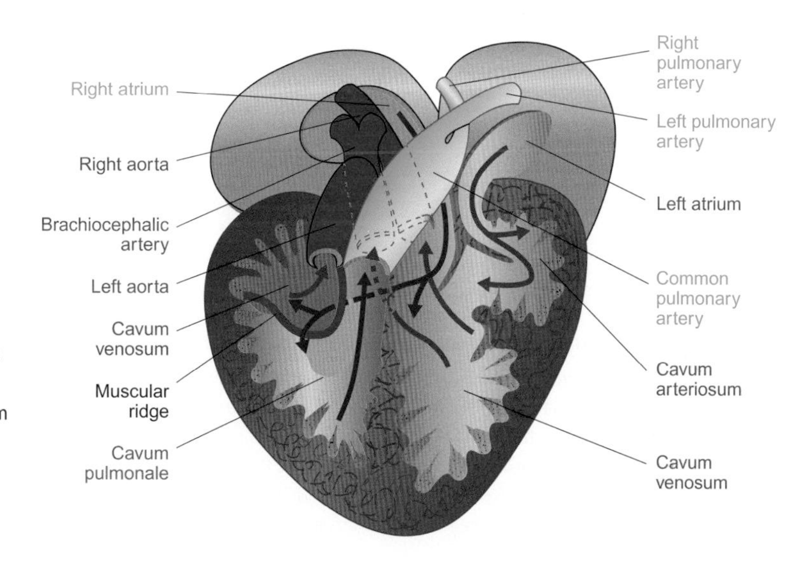

Figure 14.42 Anatomy in the different chambers of the ventricles of the hearts of turtles

The single ventricle of non-crocodilian reptiles, in this case red-eared sliders (*Trachemys scripa*), is divided into three chambers by muscular ridges. From the location of these chambers and the blood vessels leaving them, it appears that oxygen-deficient blood enters the cavum venosum and leaves predominantly from the cavum pulmonale. whereas oxygen-rich blood enters the cavum arteriosum and leaves largely from the cavum venosum.

Reproduced from Shelton G and Burggren W (1976). Cardiovascular dynamics of the Chelonia during apnoea and lung ventilation. Journal of Experimental Biology 64: 323-343.

In turtles, peak systolic blood pressures are the same in both the cavum pulmonale and cavum venosum and in the right aorta and common pulmonary artery, as shown in Figure 14.43A. Such a similarity indicates anatomical continuity throughout the ventricle during the whole of the cardiac cycle. However, this is not the case in the varanid lizards and in some species of snakes.

Varanids are the most active of the lizards[41] and have slight modifications in the anatomy of their ventricle which relate to their active lifestyle. Compared to those in the hearts of other non-crocodilian reptiles, the muscular ridge between the cavum pulmonale and cavum venosum is more prominent in varanids to the extent that it presses against the opposite side of the ventricle during contraction (systole), thereby anatomically separating these two chambers. Also, the cavum venosum is greatly reduced in size, compared with that in other non-crocodilians, making the cavum arteriosum the dominant chamber on the systemic side of the circulation. Indeed, the cavum arteriosum is larger and more muscular than that in other non-crocodilians.

The anatomical separation between the two chambers during systole enables the cavum arteriosum to produce a higher blood pressure on the systemic side of their circulation than the cavum pulmonale produces on the pulmonary side, as shown in Figure 14.43B and Table 14.3. Despite the anatomical separation during systole, some shunting of blood still occurs during diastole, but the adaptive significance of the shunts is uncertain. It has been suggested that the cardiac shunts in reptiles (and, it is assumed, in amphibians) exist not because they have been selected for, but because they have not been selected against.

The high systemic blood pressure in varanids is thought to be related to their active lifestyle, enabling blood to circulate round the body more quickly and to supply oxygen to the metabolizing organs and tissues at a higher rate. It would also enable processes, such as kidney filtration, to proceed more quickly[44]. Keeping pressure in the pulmonary circuit relatively low reduces the likelihood of the passage of fluid across the lung epithelium (pulmonary oedema) and into the gas-exchange regions (the **faveoli**[45]), which would impair gas exchange.

The ventricle of pythons, such as Burmese pythons (*Python bivittatus*), is also anatomically divided during its contraction phase, unlike that in other species of snakes that have been studied. While pythons are not particularly active, the anatomical division of their ventricles may be related to their high rate of oxygen consumption during digestion and/or to their use of shivering thermogenesis during incubation of their eggs[46].

Effect of gravity on the circulation of climbing snakes

For terrestrial animals that have a relatively long body and adopt a vertical position, such as tree-climbing snakes, gravity tends to cause blood to pool in the lower part of the body[47]. Some of the compensatory mechanisms that terrestrial climbing snakes possess are:

- an ability to generate higher blood pressures when resting in a horizontal position;
- the ability to increase blood pressure even further when the head is tilted upward, as illustrated in Figure 14.44.

[44] Ultrafiltration in vertebrate kidneys is discussed in Section 7.1.2.
[45] Faveoli are equivalent to the alveoli in the lungs of mammals and are discussed in Section 12.3.3.
[46] Thermogenesis during incubation in *Python molurus* is discussed in Section 10.3.2.
[47] The effect of gravity on the circulation is discussed in Section 14.2.2.

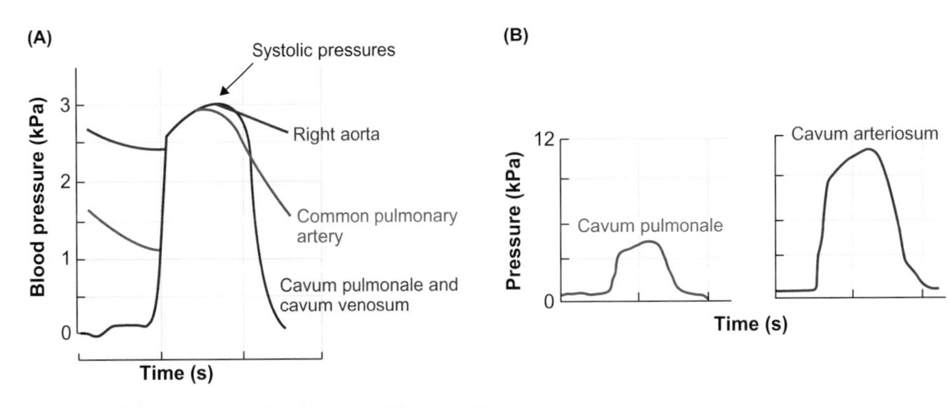

Figure 14.43 Pressures in the ventricles of non-crocodilian reptiles

(A) In red-eared sliders (*Trachemys scripa*) the pressure pulses in the cava pulmonale and venosum are similar, which indicates that are in contact during the whole of the cardiac cycle in turtles. (B) In varanid lizards, such as savannah monitor lizards (*Varanus exanthematicus*), peak systolic pressure in the cavum arteriosum is considerably greater than that in the cavum pulmonale. This is achieved by the muscular ridge, which is larger in these species, being pressed against the opposite side of the ventricle during systole, thus separating the two chambers.

A: Reproduced from Shelton G and Burggren W (1976). Cardiovascular dynamics of the Chelonia during apnoea and lung ventilation. Journal of Experimental Biology 64: 323-343.
B: Reproduced from Burggren W and Johansen K (1982). Ventricular haemodynamics in the monitor lizard, *Varanus exanthematicus*: pulmonary and systemic pressure separation. Journal of Experimental Biology 96: 343-354.

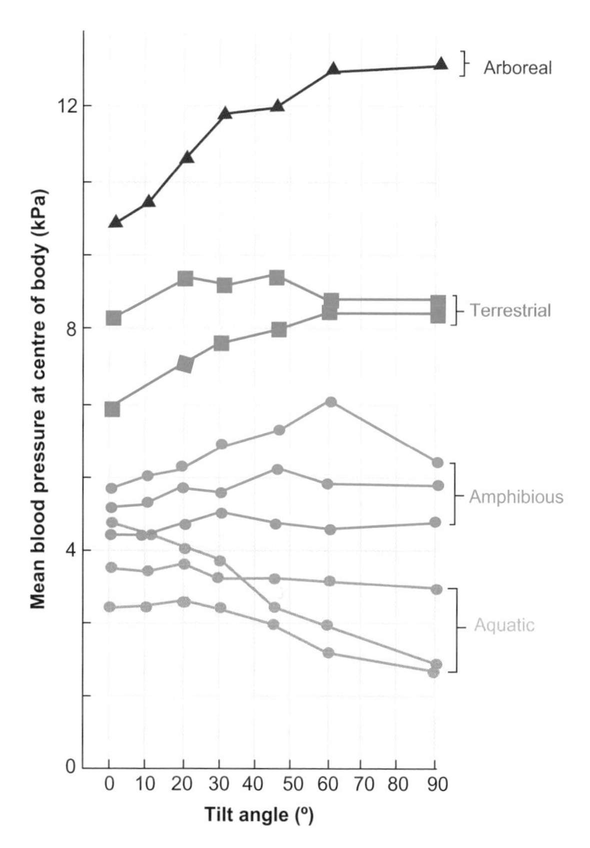

Figure 14.44 Adaptations of climbing snakes to counteract tendency for blood to pool in the tail when they are in a vertical position

Tree-climbing (arboreal) snakes can generate greater blood pressures when at rest than other species of snakes and are able to increase this pressure even further when the head is tilted upward.

Reproduced from Lillywhite HB (1996). Gravity, blood circulation and the adaptation of form and function in lower vertebrates. Journal of Experimental Zoology 275: 217-225.

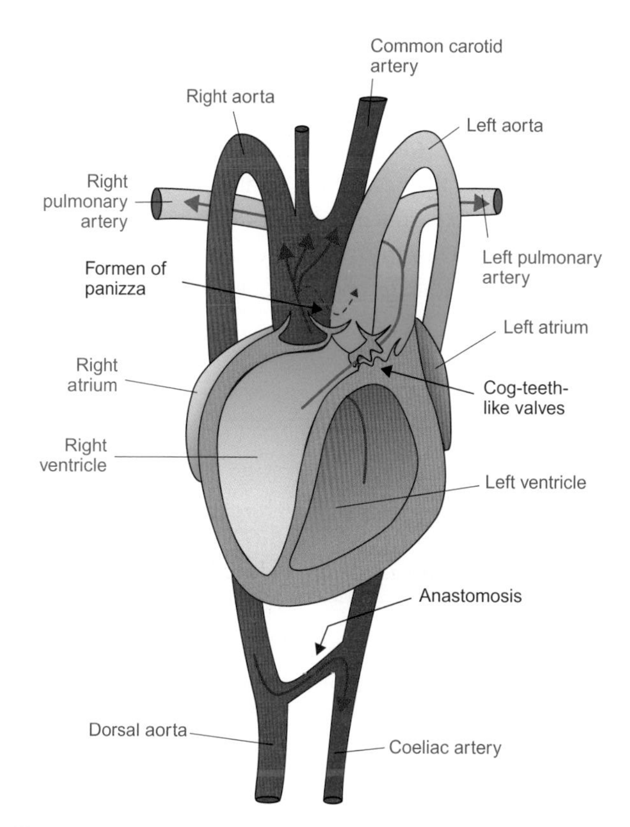

Figure 14.45 Anatomy, pressures and flows in the ventricles and major blood vessels of the crocodilian heart

The ventricle is completely divided but the right aorta leaves the left ventricle and gives rise to the right (dorsal) aorta, whereas the left aorta leaves the right ventricle and becomes the coeliac artery. There is a connection between the two aortae at their bases by way of the foramen of Panizza and by an anastomosis between the dorsal aorta (right aorta) and the coeliac artery (left aorta). Cog-teeth-like valves occur at the base of the outflow tract of the pulmonary arteries.

Reproduced from Axelsson M et al (1996). Dynamic anatomical study of cardiac shunting in crocodiles using high-resolution angioscopy. Journal of Experimental Biology 199: 359-365.

Blood pressure is increased in order to counteract the greater **hydrostatic pressure** in the column of blood as the head is raised above the heart enabling adequate perfusion of the cerebral blood vessels.

It has also been claimed that climbing snakes tend to be more slender than non-climbing members of the same family in order to reduce or even prevent the pooling of blood in the lower part of the body. This is because a smaller circumference provides a greater resistance to distension than a larger circumference. A similar effect is also achieved in these snakes by their less compliant (less stretchy) skin and underlying structures compared with those in non-climbing snakes.

The circulation of crocodilian reptiles

The heart of crocodilians is completely divided into left and right sides as shown in Figure 14.45. As in birds and mammals, the left ventricle has a thicker wall than the right ventricle, although the wall of the right ventricle is also relatively thick. The crocodilian heart has other unique features:

- the origin of the left aorta from the right ventricle; the right aorta is the main artery supplying oxygen-rich blood to the body and is supplied by the left ventricle;
- the presence of a direct communication, the foramen of Panizza, between the left and right aortae at their bases;
- the presence of nodules of connective tissue, cog teeth, at the base of the pulmonary outflow tract;
- an anastomosis connecting the two aortae just posterior to the heart; the left aorta (coeliac artery) supplies blood to the viscera.

Figure 14.45 demonstrates that there is no opportunity for the mixing of blood within the heart itself. It is theoretically possible, however, for oxygen-depleted blood to enter the left aorta directly from the right ventricle (R → L shunt), provided pressure in the right ventricle exceeds that in the left aorta, but not for any of the blood in the left

14

ventricle to enter the pulmonary circulation (L → R shunt). It is also potentially possible for blood to pass between the right and left aortae, via the foramen of Panizza, as indicated in Figure 14.45. However, this foramen is usually covered by a flap valve at the base of the right aorta during systole. We discuss in Experimental Panel 14.2, experiments that have sought to determine whether or not a R → L exists in the heart of crocodilians and, if it does, what its functional significance is.

14.4.4 The completely divided hearts of birds and mammals

The hearts of birds and mammals are completely separated into right and left sides, with two atria and two ventricles. Hence the rate of blood flow around the lung and body circuits must be well matched. In both birds and mammals, the left ventricle is cone-shaped and extends to the apex of the heart and its wall is 2–3 times thicker than that

Experimental Panel 14.2 Characterizing the right-to-left shunt in the hearts of crocodilians and its significance

Background

Despite the fact that the heart of crocodilians is divided into left and right sides, it is possible for blood to pass from the right ventricle into the left aorta (R → L shunt) if pressure in the right ventricle exceeds that in the left aorta. The objective of these experiments was to determine to what extent such a shunt exists and, if it does, what its significance is.

Experimental approach

Three series of experiments were performed.

Experiment 1

Eight alligators (*Alligator mississippiensis*) were anaesthetized and their hearts and arterial arches exposed. Catheters were inserted into the ventricles and the major arteries. The catheters were connected to devices which recorded the blood pressures. Blood flow was measured in the right and left aorta by placing flow sensors around the vessels. The animals remained anaesthetized throughout the experiment.

The pressures measured in the ventricles and major arteries are shown in Figure A. These data demonstrate that the pressure in the left ventricle and the right aorta are similar to those in the hearts of birds and mammals[1]. In contrast, the profile of the pressure generated in the right ventricle has a unique shape but does not reach the pressures in the right or left aortae, as shown in Figure B(i). This means that there is normally no flow of blood from the right ventricle to the left aorta. Part way through right ventricular systole, closure of the cog-teeth valve at the base of the pulmonary outflow increases the resistance, which leads to the secondary increase in right ventricular pressure and the end of ejection of blood into the pulmonary arteries.

As diastole begins, the valve at the base of the right aorta closes and moves away from the foramen, enabling blood to

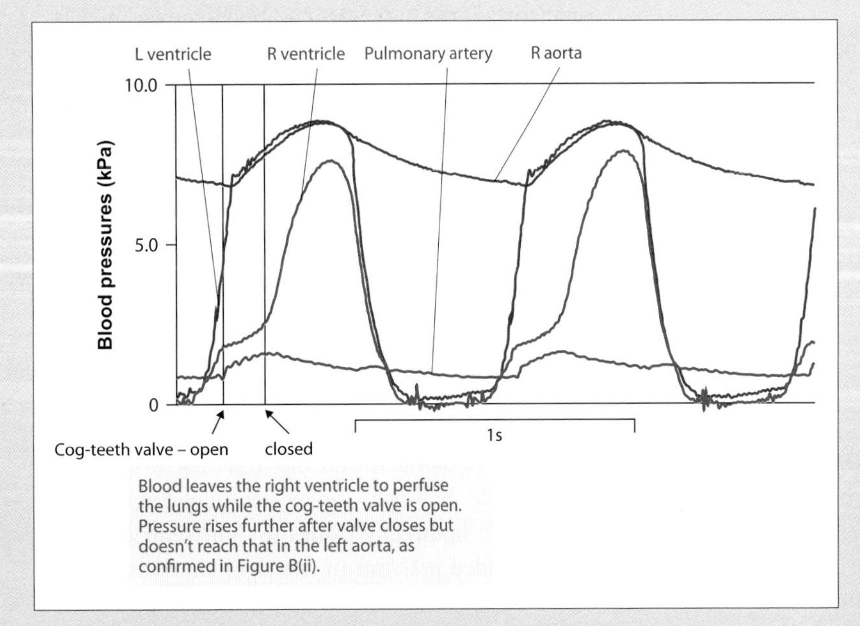

Figure A Pressures in the heart of anaesthetized alligators

Pressure profiles in the left ventricle and the right aorta are, as would be expected, similar to those in other reptiles, birds and mammals. However, pressure in the right ventricle is quite unique, being two-phased and greater than that in the pulmonary artery.

Adapted from: Shelton G and Jones DR (1991). The physiology of the alligator heart: the cardiac cycle. Journal of Experimental Biology 158: 539-564.

14

flow from the right to the left aorta, as shown in Figure 14.45 and Figure B(i). This flow represents only a small proportion (around 4 per cent) of the output of the left ventricle. When the valve at the base of the right aorta opens during systole and covers the foramen of Panizza, there is a brief reversal of flow along the left aorta, but not back into the right ventricle as the valve at the base of the left aorta remains closed.

The timing of the pressure changes in the left aorta is similar to that in the right aorta, but the mean systolic pressure in the left aorta is slightly lower than that in the right aorta (9.6 kPa vs 10.2 kPa, respectively).

Experiment 2

In another group of alligators, the catheters and leads from the flow sensors were taken through the abdominal body wall and the incision made to expose the heart and arterial arches was closed with sutures and made watertight with tissue cement. The animals were allowed to recover from the anaesthesia and placed in tanks with 5–10 cm water.

When the unanaesthetized animals were disturbed, the pressure and flow profiles in the ventricles and major blood vessels were similar to those shown in Figures A and B(i). However, Figure B(ii) shows that, when calm and completely undisturbed, unanaesthetized animals develop a R → L cardiac shunt (pulmonary bypass) as a result of the decrease in their aortic blood pressures. This enables pressure in the right ventricle to exceed that in the left aorta, allowing the ventricular–aortic valves to open. The resulting cardiac R → L shunt is relatively small (on average, 14 per cent of output from the right ventricle), although shunting occurs for at least 85 per cent of the time in completely undisturbed crocodilians.

Unfortunately, we do not know the relationship between the cardiac R → L shunt and lung ventilation. For example, we do not know whether or not this relationship still holds true when the animals are submerged under water for any length of time.

It has been suggested that a R → L cardiac shunt aids digestion, in particular by delivering CO_2-rich venous blood to the

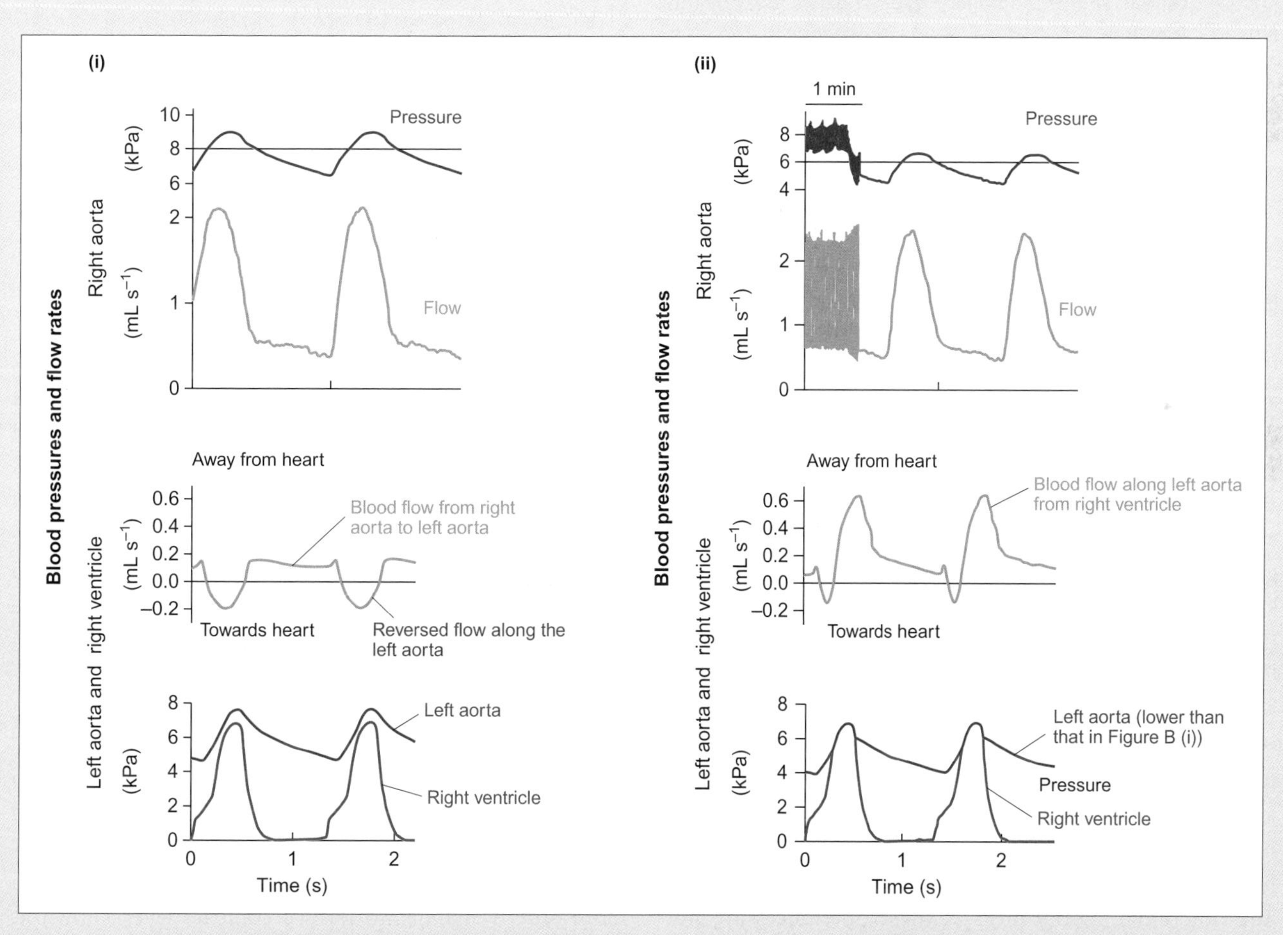

Figure B Pressures and flows in right ventricle and major blood vessels of anaesthetized alligators and development of right to left shunt in the heart of undisturbed, unanaethetized alligators

(i) In anaesthetized alligators, the pressure in the right ventricle of crocodilians never exceeds that in the left ventricle (and, therefore, in the left aorta). Thus, the only route for blood to enter this vessel is from the right aorta during diastole when the valves at the base of the aortae close and move away from the foramen of Panizza. There is reverse flow during systole. This is the situation the majority of the time in anaesthetized animals and in disturbed unanaesthetized animals. (ii) When unanaesthetized animals are completely calm and resting, aortic pressure falls (compare with that shown in B(i)), so that the pressure in the right ventricle exceeds that in the left aorta and blood flows from the right ventricle into the left aorta during systole. In other words, there is an overall right to left shunt.

Adapted from: Shelton G and Jones DR (1991). The physiology of the alligator heart: the cardiac cycle. Journal of Experimental Biology 158: 539-564. Jones DR (1995). Crocodilian cardiac dynamics: a half–hearted attempt. Physiological and Biochemical Zoology 68: 9-15.

gut where the CO_2 is used in the production of gastric acid. This process may be important after a particularly large meal. Indeed, when the R → L cardiac shunt is surgically eliminated, the rate of digestion of bone is reduced in this group of animals compared with that in a control group.

Experiment 3

This series of experiments was designed to investigate whether or not the R → L cardiac shunt in crocodilians confers any selective advantage to them. The R → L shunt was removed by suturing the left aorta upstream and downstream of the foramen of Panizza in hatchling alligators. This procedure had no effect on the rate of growth compared with age-matched control groups comprising either animals that were rested or that underwent regular exhaustive exercise. It was calculated that the alligators in which the R → L shunt was eliminated would have reached sexual maturity during the same breeding season as those in the control group.

Overall findings

A R → L cardiac shunt (pulmonary bypass) exists most of the time in completely undisturbed, unanaethetized alligators as a result of the decrease in their aortic blood pressures enabling right ventricular pressure to exceed that in the left aorta. However, it has not been possible as yet, to demonstrate any selective advantage of this R → L shunt in crocodilians.

› Find out more

Axelsson M, Franklin CE, Löfman CO, Nilsson S, Grigg GC (1996). Dynamic anatomical study of cardiac shunting in crocodiles using high-resolution angioscopy. Journal of Experimental Biology 199, 359–365.

Eme J, Gwalthney J, Owerkowicz T, Blank JM, Hicks JW (2010). Turning crocodilian hearts into bird hearts: growth rates are similar for alligators with and without right-to-left cardiac shunt. Journal of Experimental Biology 213: 2673–2680.

Farmer CG, Uriona TJ, Olsen DB, Steenblick M, Sanders K (2008). The right to left shunt of crocodilians serves digestion. Physiological and Biochemical Zoology 81: 125–137.

Jones DR (1995). Crocodilian cardiac dynamics: a half-hearted attempt. Physiological Zoology 68: 9–15.

Jones DR, Shelton G (1993). The physiology of the alligator heart: left aortic flow patterns and right-to-left shunts. Journal of Experimental Biology 176: 247–269.

Shelton G, Jones DR (1991). The physiology of the alligator heart: the cardiac cycle. Journal of Experimental Biology 158: 539–564.

[1] Pressures measured in the heart of mammals are shown in Figure 14.3.

of the right ventricle, thus enabling it to generate a higher pressure.

Despite the many similarities between them, the structure and function of the cardiovascular systems of birds and mammals do show some differences. In both groups, only one of the systemic arches remains: the right one in birds and the left one in mammals, as indicated in Figure 14.46. Birds tend to have larger hearts than mammals of a similar body mass and related to this birds also have a greater resting cardiac stroke volume (averages of 1.7 mL kg^{-1} in birds and 0.7 mL kg^{-1} in mammals). However, birds have a lower average resting heart rate (180 beats min^{-1} for a 1 kg bird, vs 235 beats min^{-1} for a 1 kg mammal) but this does not fully compensate for the higher cardiac stroke volume in birds. Consequently, resting cardiac output is greater in birds (290 mL min^{-1} kg^{-1} vs 165 mL min^{-1} kg^{-1}). Systemic blood pressure is also higher in birds, as indicated in Table 14.3. In contrast, blood pressure in the pulmonary arteries is similar in both birds and mammals. The functional significances of these differences are not obvious, although the higher cardiac outputs and systemic arterial pressures could be related to the higher rates of metabolism in flying birds compared to those in running mammals of a similar size[48].

14.4.5 Blood and oxygen supply to the heart

In order to beat continuously, the heart requires a continuous supply of oxygen. In some species of vertebrates this supply is obtained from the blood in the lumen of the chamber of the heart. In other species, including birds and mammals, special blood vessels, **coronary vessels**, transport arterial blood (and hence oxygen) to the muscles of the heart.

The supply of blood to the muscular wall (myocardium) is quite variable among fish. Basically, the muscle cells in the walls of the ventricle, the cardiac myocytes, are arranged in one of two ways, giving either a spongy or compact structure. The spongy arrangement consists of interconnecting thin bundles, or bars, of muscle (trabeculae) which form a complex meshwork[49] known as the spongiosa, as shown in Figure 14.47. **Spongy myocardium** has no coronary blood vessels supplying it. The spongiosa may have an outer covering of dense bundles of muscle cells arranged in a much more orderly manner to form a compact layer known as the **compacta, or compact myocardium**. This layer does receive arterial (coronary) blood, as shown in Figure 14.47. The coronary supply comes from vessels supplying the head (cranial) and anterior (pectoral) fins.

[48] Exercise in birds and mammals is discussed in Section 15.2.3.

[49] This meshwork is similar to the situation we discuss in Sections 14.3.1 and 14.3.2 for the hearts of some species of mollusc.

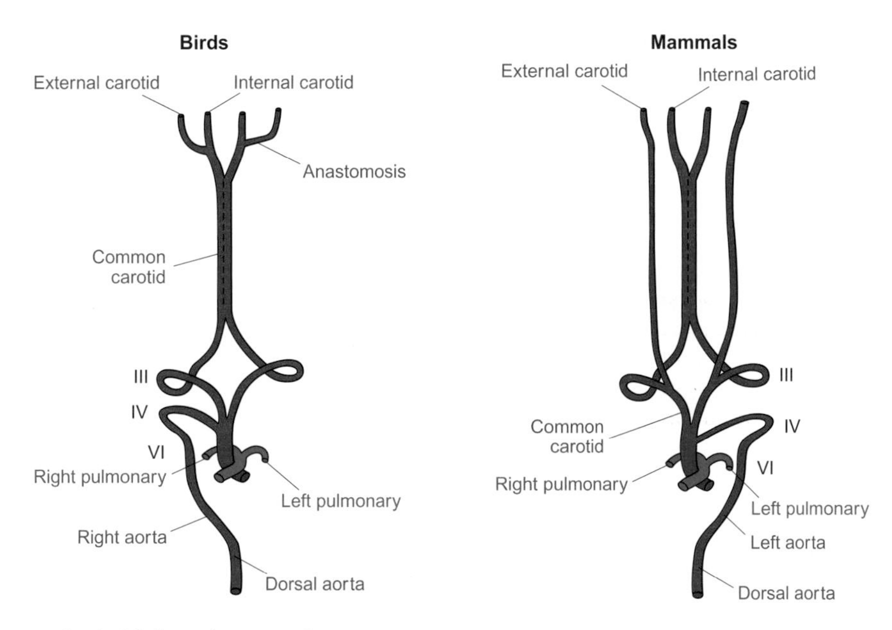

Figure 14.46 Major arteries in birds and mammals

The three remaining aortic arches in birds and mammals are: III, IV and VI. Only one of the IVth (systemic) arches remains in birds and mammals; it is the right one in birds and the left one in mammals. The original division of each common carotid artery into the internal and external carotids is at the angle of the jaw, which in most species of mammals is close to the heart, and this persists in mammals. In birds, however, most of which have a relatively long neck, the external carotid artery disappears from the point of the initial division to a point higher in the neck, near the angle of the jaw. An anastomosis joins the internal carotid with the remainder of the external carotid.

Adapted from West NH et al (1981). Cardiovascular system. In: Form and Function in Birds vol II, eds King, A. and McLelland, pp 235-339. New York: Academic Press.

The hearts of fish can be divided into four types based on the proportion of compacta in their walls, as shown in Figure 14.47. The majority of species of fish—for example, plaice (*Pleuronectes* spp.) and icefish (those with or without haemoglobin)—have only a spongy myocardium (type I). This means that the supply of oxygen and nutrients to the myocytes of the ventricle of these fish is from the venous blood returning to the heart from the body. (Note, the spongy arrangement increases the area of contact of the ventricle wall with the blood.) Most teleosts with a compacta have type II hearts; most elasmobranchs have type III hearts. The more active sharks and tunas have type IV hearts in which more than 30 per cent of the wall of the ventricle is compacta; at the extreme this rises to almost 74 per cent in bigeye tuna (*Thunnus obesus*).

The coronary circulation is not well developed in lungfish and amphibians. In South American lungfish (*Lepidosiren* spp.), the coronary arteries originate from the afferent (oxygen-depleted) side of aortic arch IV. Although there are coronary vessels in urodele and anuran amphibians, they do not penetrate the myocardium and there is no compacta in the ventricle. There are, however, compacta and a coronary supply to the conus arteriosus.

Coronary blood vessels are present in the hearts of reptiles, but are restricted mainly to the ventricle and are most likely confined to the compacta, although this is not certain. Perhaps not surprisingly, the extent of the coronary vessels seems to be related to the level of activity, and hence to the level of oxygen demand, of the heart muscle. The compact layer is thicker in varanid lizards and snakes than in non-varanid lizards and turtles. In addition, the atria of varanids and snakes have a compact layer and well-developed coronary supplies. There seems to be little information on the extent of the coronary circulation in the ventricle of crocodiles, but it would be expected to be relatively large in view of the pressures that are generated on both the right and the left sides of the heart, as indicated in Experimental Panel 14.2.

In the mammalian heart, the right and left coronary arteries originate from the base of the aorta, as shown in Figure 14.48. These arteries principally supply the right atrium and ventricle and left atrium and ventricle, respectively, although there is some overlap. After passage through the capillary beds, most of the venous coronary blood returns to the right atrium via the coronary sinus. Because of the squeezing effect of the contracting heart muscle on the coronary vessels, particularly in the left ventricle of birds and mammals during systole, blood flow in the large left coronary artery is briefly reversed during early systole. Maximum flow along the left coronary artery occurs during early diastole, when the ventricles have relaxed and compression of the coronary vessels is virtually absent.

The capillaries in the heart are not as regular as those in skeletal muscle so are not so easy to study. An exception are the papillary muscles. These muscles are attached to the

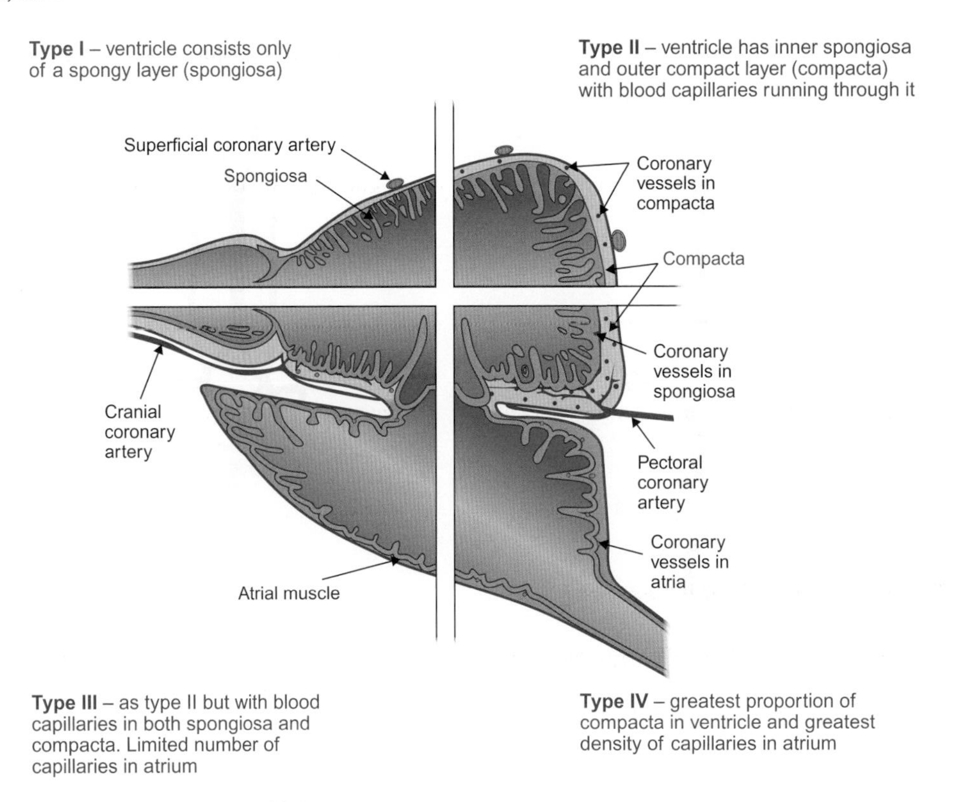

Figure 14.47 Oxygen supply to the heart of fish

Cross-section through the heart of a hypothetical fish. The heart is divided into four sections, each representing a different type of heart based on the extent of the coronary blood supply to both the ventricle and the atrium.

Reproduced from Farrell AP and Jones DR (1992) The heart, In: Fish Physiology, vol 12 Pt A, Hoar WS et al (eds.), 1-88. New York: Academic Press.

wall of each ventricle and to the tricuspid or bicuspid valves, which are shown back in Figure 14.2. As an illustration of the high oxidative capacity of heart muscles, capillary density is greater in the papillary muscles than in a calf muscle of both flying and non-flying mammals. For example, in greater mouse-eared bats (*Myotis myotis*), capillary density is 6748 mm^{-2} in the papillary muscle compared with 2216 mm^{-2} in a calf muscle. The respective values in golden hamsters (*Mesocricetus auratus*) are 5445 mm^{-2} and 1472 mm^{-2}. Compare these values with those for skeletal muscles given earlier in Table 14.1.

› *Review articles*

Burggren W, Farrell A, Lillywhite H (1997). Vertebrate cardiovascular systems. In: Dantzler WH Ed. Handbook of Physiology, Section 13: Comparative Physiology, Vol I pp 215–308. New York: Oxford University Press.

Bushnell PG, Jones DR, Farrell AP (1992). The arterial system. In: Hoar WS, Randall DJ, Farrell AP (eds.) Fish Physiology, Vol XII, Part A The Arterial System. pp 89–139. San Diego: Academic Press.

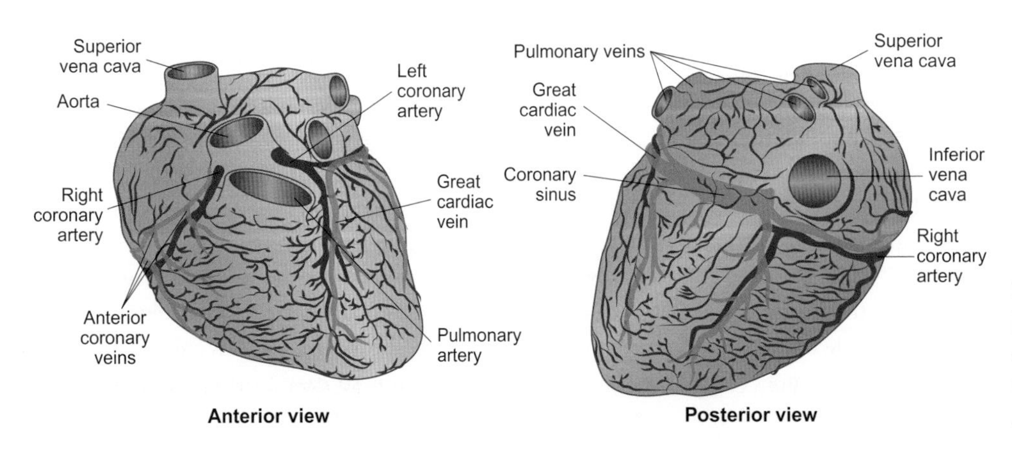

Figure 14.48 Coronary blood vessels of the heart of mammals

Reproduced from Berne RM and Levy MN (1993). Physiology. St Louis: Mosby.

Grubb BR (1983). Allometric relations of cardiovascular function in birds. American Journal of Physiology 245: H567–H572.

Hicks JW (2002). The physiological and evolutionary significance of cardiovascular shunting patterns in reptiles. News in Physiological Sciences 17: 241–245.

Olson KR (2002). Vascular anatomy of the fish gill. Journal of Experimental Zoology 293: 214–231.

Randall DJ (1994). Cardio-respiratory modelling in fishes and the consequences of the evolution of airbreathing. Cardioscience 5: 167–171.

Shadwick RE (1999). Mechanical design in arteries. Journal of Experimental Biology 202: 3305–3313.

Wang T, Krosniunas EH, Hicks JW (1997). The role of cardiac shunts in the regulation of arterial blood gases. American Zoologist 37: 12–22.

Checklist of key concepts

- Most invertebrates have an **open circulatory system**, although decapod crustaceans and some species of non-cephalopod molluscs have an **incompletely closed** system.
- Fine, **capillary-like** blood vessels have been found in some species of terrestrial molluscs and decapod crustaceans.
- Vertebrates have a **closed circulatory system**: all blood vessels are lined with endothelium.
- Cephalopod molluscs also have a closed system, and the blood vessels are lined with endothelium-like cells.
- The **ventricle** of the heart contracts rhythmically to generate sufficient pressure to overcome the resistance to the flow of blood around the body.
- Water-breathing invertebrates and fish have a **single circulation**.
- In water-breathing invertebrates, the ventricle propels the blood around the body and then through the gills for gas exchange.
- In fish, the ventricle sends the blood through the gills and then round the body. Vertebrates with lungs possess a **double circulation**.
  - Oxygen-poor blood returning from the body enters the **right atrium** and then the **right side of the ventricle**, which may or may not be anatomically divided into right and left sides.
  - The oxygen-poor blood in the right side of the ventricle is sent to the lung for gas exchange.
  - Oxygen-rich blood returns to the **left atrium** from where it enters the **left side of the ventricle**.
  - The oxygen-rich blood is then sent around the body.

How the heart works

- Contraction of the ventricle is known as **systole** and the peak pressure generated is called the **systolic pressure**.
- The **residual** or **end systolic volume** is the volume of blood that remains in the ventricle at the end of systole.
- Relaxation of the heart is known as **diastole** and the lowest pressure reached is called the **diastolic pressure**.
- A series of valves ensures that the blood passes through the heart in only one direction.
- Filling of the ventricle after the blood has circulated round the body or lungs causes the muscle fibres in the wall of the ventricle to stretch.
- The **Frank–Starling relationship** states that the force of contraction increases with stretch up to a certain point.
- The greater the **end diastolic volume**, the greater the force of contraction during the subsequent systole.
- The volume of blood ejected by the ventricle during systole is the **cardiac stroke volume**.

Structure and function of blood vessels

- In vertebrates, blood vessels consist of three layers:
  - **tunica intima** (sometimes called **tunica interna**),
  - **tunica media**,
  - **tunica externa**,
- The **larger arteries** close to the heart of cephalopod molluscs, decapod crustaceans and vertebrates are quite **compliant** and smooth out the pulsatility of the blood flow and pressure.
- Smooth muscle in the walls of **smaller arteries** and **arterioles** can contract and reduce the diameter of the vessels, a process known as **vasoconstriction**.
- Relaxation of the smooth muscle causes the vessels to dilate, known as **vasodilation**.
- Pressure inside the **veins** is relatively low and these vessels have relatively thin walls and valves to prevent backflow.
- There are three different types of **capillaries** :
  - **continuous**,
  - **fenestrated**,
  - **discontinuous**.
- Gas exchange across capillaries is governed by **Fick's law of diffusion** and is often represented in the form of a **Krogh cylinder**.
- **Capillary density** tends to be greatest in the highly oxidative locomotor muscles of vertebrates, such as the flight muscles of hummingbirds.
- New capillaries are formed by a process called **angiogenesis**.
- The short capillaries of gas-exchange organs make the network look like a sheet of blood, giving rise to a pattern of blood flow known as **sheet flow**.
- Fluid movement across capillaries is determined by the factors first identified by **Ernest Starling**, plus the **reflection coefficient**.
- Any fluid that filters into the interstitial space enters the **lymphatic system** and returns to the general circulation.
- Fish have a **secondary circulation**, although its relationship to the lymphatic system of other groups of vertebrates is uncertain.

Fluid dynamics

- The rate of flow of a fluid through a long straight rigid tube is described by **Poiseuille's equation**.
- The major factor affecting the resistance to blood flow is the **radius** of a blood vessel.
- The total resistance to the flow of blood through the circulation depends on whether the blood vessels are in **series** or in **parallel**.
- In tall terrestrial animals, such as giraffes, **gravity** also has a large effect on the pressure the heart has to generate in order to circulate the blood.
- Blood is a **non-Newtonian** fluid and so the term **apparent viscosity** is often used instead of viscosity.
- Flow of fluid through a tube can be **laminar** or **turbulent** and the transition from one to the other depends on the factors incorporated in the **Reynolds number**.

Circulatory systems

- The hearts of molluscs with two gills have two atria and a ventricle surrounded by a pericardium.
- The hearts of those molluscs with one gill or no gill have one atrium.
- In molluscs, the haemolymph enters one or two **atria** before returning to the ventricle.
- In arthropods there is no atrium and the haemolymph enters the **pericardial sinus** and then the ventricle, via valved **ostia**.
- The **closed circulatory systems** of cephalopod molluscs and vertebrates generally have a smaller blood volume and generate higher pressures than those of other species of molluscs or crustaceans.
- Blood flow through different blood vessels can be independently varied in decapod crustaceans.
- Oxygen supply to the muscle of ventricles may be from the blood permeating the **spongy layer** of their inner walls and/or by arterial blood vessels (**coronary vessels**) supplying a more **compact layer** of ventricular walls.
- More active species have a greater proportion of compact layer.
- The hearts of lungfish, amphibians and non-crocodilian reptiles have **two atria** for receiving oxygen-rich or oxygen-poor blood from the lungs or body, and an **incompletely divided ventricle**.
- Incomplete anatomical division of the heart provides some opportunity for **right to left** or **left to right shunts** to occur within the heart, and for some mixing of the two blood streams.
- Folds in the **bulbus cordis** of lungfish and the **spiral valve** in the **conus arteriosus** of amphibians help to maintain some separation of the two bloodstreams following ventilation of the lungs.
- Peak systolic pressures are similar in the systemic circulation and pulmonary circulation in lungfish, amphibians and most species of non-crocodilian reptiles.
- Crocodilians have complete anatomical division of the heart, although there is the opportunity for a right to left shunt as the left aorta arises from the right ventricle.
- Crocodilians are able to generate a higher systolic pressure in the systemic circulation than that in the pulmonary circulation.
- Varanid lizards and pythons are also able to generate high systemic pressures as their ventricles become anatomically divided during systolic contraction.
- Birds and mammals have completely divided circulations with no opportunity for any shunts.

Study questions

*** Answers to these numerical questions are available online. Go to www.oup.com/uk/butler**

1. Are the terms 'open' and 'closed' circulatory systems useful concepts? (Hint: Sections 14.1.1, 14.3.)
2. What are meant by systolic and diastolic blood pressures and how and why do they vary throughout the circulatory system? (Hint: Section 14.1.2, 14.1.4, Box 14.1.)
3. Discuss how the muscles of hearts from different groups of animals obtain sufficient oxygen from the blood. (Hint: Section 14.3, 14.4.5.)
4. What are the major factors that determine the rate of flow of a liquid along a straight, rigid tube of uniform diameter? To what extent do these factors apply to the cardiovascular system of animals? (Hint: Section 14.2.1.)

5.* (a) If cardiac output of resting rainbow trout is 22 mL min^{-1} and mean difference in pressure between the dorsal aorta and the venous system is 4 kPa, what is the average resistance to blood flow in the systemic circulation? (b) At maximum sustainable exercise, cardiac output increases to 50 mL min^{-1} and mean difference in pressure across the systemic vascular bed increases to 4.75 kPa. What happens in the systemic circulation during exercise? (Hint: Section 14.2.1.)

6.* If the resistance in each of six straight, rigid tubes is 5, 2.5, 2, 3.3, 5 and 2.5 kPa mL^{-1} min^{-1}, what is the total resistance (a) if the tubes are arranged in series and (b) if they are arranged in parallel? What is the relevance of these answers to the circulatory system? (Hint: Section 14.2.1.)

7. What is Reynold's number and how does it relate to laminar and turbulent flow patterns of a liquid through a tube? (Hint: Section 14.2.4.)
8. What are the major features of the structures of arteries, capillaries and veins? Discuss the functional significance of these differences. (Hint: Section 14.1.3.)

9. What is a Krogh cylinder and how accurately does it explain the transfer of oxygen from the blood to metabolizing cells? (Hint: Section 14.1.5.)

10.* At the venular end of a capillary, the hydrostatic and colloid osmotic pressures (COP) are:

    Capillary hydrostatic pressure, 1.0 kPa, interstitial hydrostatic pressure, −0.3 kPa.

    Capillary COP, 3.3 kPa, interstitial COP, 2.1 kPa and reflection coefficient is 0.85.

    In which direction does fluid flow? Give the evidence for your conclusion. (Hint: Section 14.1.5.)

11. Discuss what is meant by a constant volume mechanism in the hearts of some species of molluscs and crustaceans. To what extent does a similar mechanism, known as vis-a-fronte, exist in the hearts of fish? (Hint: Sections 14.3.1, 14.3.3, 14.4.1.)

12. What are the functional differences between the conus arteriosus and the bulbus arteriosus in the hearts of fish? (Hint: Section 14.4.1.)

13. Discuss the functional significance of cardiac shunts in the amphibians and reptiles. (Hint: Section 14.4.3, Experimental panel 14.2.)

14. Outline the major differences between the cardiovascular systems of birds and mammals and suggest possible reasons for these differences. (Hint: Section 14.4.4.)

Bibliography

Berne RM, Levy MN, Eds. (1998). Physiology 4th Ed. Mosby, St Louis.

Duncan G (1990). Physics in the Life Sciences. 2nd Ed. Blackwell Scientific Publications, Oxford.

Weibel ER (1984). Pathway for Oxygen. Harvard University Press, Cambridge, MA.

15 Environmental and behavioural influences on the cardiorespiratory system

In this chapter, we explore in more detail the respiratory and circulatory systems introduced in Chapters 12, 13 and 14, and find out how they respond to different environmental and behavioural challenges. For example, we investigate the integrated responses of these two systems when demand for oxygen increases during feeding and digestion and during exercise; when there is a reduction in the availability of oxygen during environmental hypoxia (including living in burrows and diving by birds and mammals) and when the demand for oxygen is reduced during torpor.

Where appropriate, we also discuss any associated changes in the acid–base status[1] of an animal when responding to these environmental and behavioural challenges.

15.1 Responding to a change in oxygen demand and supply

How do the respiratory and cardiovascular systems respond when an organism experiences a change in its demand for oxygen or a decrease in its supply? In order to answer this question, we need to remember the **Fick principle of convection**[2]: the rate of oxygen consumption is equal to the amount of respiratory medium or blood pumped per minute (**respiratory minute volume** or cardiac output, respectively), multiplied by the **amount of oxygen extracted** by the gas-exchange system from the respiratory medium (air or water) or by the metabolizing cells from the blood. The ventilatory system and the heart control the amount of respiratory medium and blood they pump per minute by varying the **rate** at which they pump (**respiratory frequency** or **heart rate**) and the **volume** they pump during each cycle (respiratory tidal volume or cardiac stroke volume).

These relationships can be described by the following two equations—the first relates to the respiratory system and the second relates to the cardiovascular system:

Rate of oxygen consumption (mmol min^{-1})
Respiratory frequency (breaths min^{-1})
Tidal volume (L)
Concentration of oxygen extracted from air or water passing over the respiratory surface (mmol L^{-1})

$$\dot{M}O_2 = \underbrace{(f_{resp} \times V_T)}_{\text{Respiratory minute volume (L min}^{-1})} \times \underbrace{(C_IO_2 - C_EO_2)}_{\text{Oxygen concentration in inspired air or water minus oxygen concentration in expired air or water (mmol L}^{-1})} \qquad \text{Equation 15.1}$$

Rate of oxygen consumption (mmol min^{-1})
Heart rate (beats min^{-1})
Cardiac stroke volume (L)
Concentration of oxygen extracted from blood (or haemolymph) as it circulates round the body (mmol L^{-1})

$$\dot{M}O_2 = \underbrace{(f_H \times V_S)}_{\text{Cardiac output (L min}^{-1})} \times \underbrace{(C_aO_2 - C_{\bar{v}}O_2)}_{\text{Oxygen concentration in arterial blood minus oxygen concentration in mixed venous blood (mmol L}^{-1})} \qquad \text{Equation 15.2}$$

Any change in the demand for oxygen (increase or decrease) or a decrease in its supply can be met by a change in one or more of the factors on the right-hand side of equations 15.1 and 15.2, but the relative importance of these factors varies in response to different environmental conditions or behavioural demands for oxygen. We begin by looking at the effect of feeding and digestion on the various components of equations 15.1 and 15.2.

[1] The factors affecting acid-base balance are discussed in Section 13.3.

[2] The Fick principle of convection is discussed in Section 11.2.2.

15.2 Responding to an increase in demand for oxygen

A number of factors can increase the demand for oxygen. In ectotherms, a rise in environmental temperature can increase oxygen demand, and this topic is discussed in Box 8.2[3]. In this section, however, we discuss the increase in oxygen demand associated with feeding and digestion and with exercise.

15.2.1 Feeding and digestion

The increase in the rate of oxygen consumption ($\dot{M}O_2$) following a meal is known as **specific dynamic action (SDA)**[4] or **heat increment of feeding (HIF)**. In this book we use the term specific dynamic action. Specific dynamic action is present in every species that has been studied so far; it is the result of a number of processes that occur before, during and after absorption of the ingested food. Processes that occur before absorption include enzyme secretion, breakdown of proteins and remodelling of the intestine. Absorption involves the transport of nutrients across the outer layers of the intestine into the bloodstream[5] and postabsorptive processes include protein synthesis and amino acid deamination[6].

In essence, the responses of the cardiovascular systems of most species of animals to the increase in oxygen requirements following feeding are similar, and those for blue crabs (*Callinectes sapidus*) are shown in Figure 15.1. The increase in $\dot{M}O_2$ following a meal is shown in Figure 15.1A. There is a steep increase in $\dot{M}O_2$ for the first 2–4 hours following feeding to about twice the resting value; this elevated level is maintained for at least another 20 hours. There are increases in blood flow to parts of the body involved in the processes of feeding and digestion, as shown in Figure 15.1B. Heart rate, cardiac output and blood flow through the sternal and hepatic arteries increase immediately food is detected and reach their maximum values within 30 min to 1 hour after feeding.

The sternal artery supplies the claws (chelae) and muscles of the mouthparts (mandibles), which are involved in the handling and tearing of the food. However, the sternal artery also supplies the hepatopancreas (functionally equivalent to the liver in vertebrates) and muscles surrounding the gut, which explains why the flow of blood through this vessel remained elevated throughout the monitoring period. Blood flow through the hepatic artery, which also supplies the hepatopancreas and parts of the gastrointestinal tract, takes a little longer to reach its maximum and remains elevated throughout the period of digestion (approximately 8–10 hours).

These increases in blood flow may be the result of an increase in cardiac output by itself or in combination with reduced blood flow to other parts of the body. In blue crabs, the increase in blood flow to the mouthparts, hepatopancreas and gastrointestinal tract appears to be met entirely by an increase in cardiac output, as there is no associated reduction in blood flow through vessels supplying the rest of the body, such as the anterior aorta. Cardiac stroke volume changes little, which means that the increase in cardiac output is almost exclusively the result of the increase in heart rate.

The response of the cardiovascular system of blue crabs is remarkably similar to those of many species of vertebrates, despite the fact that vertebrates have a closed circulatory system, whereas crustaceans have an incompletely closed system[7]. For example, in unfed fish, 16–40 per cent of cardiac output perfuses the gastrointestinal tract, whereas after feeding blood flow to the gut increases by 40–100 per cent.

In most species of fish, as in blue crabs, the increase in blood flow to the gut following feeding is mainly the result of a similar increase in cardiac output rather than by the redistribution of blood away from other vascular beds. This is similar to the situation in dogs and humans, but unlike that in resting baboons, where the increase in flow to the gut is largely at the expense of a reduction in flow to the limbs.

The magnitude of the response of the cardiovascular system to feeding is related to the size and composition of the meal and involves both mechanical and chemical stimuli. For example, in humans, the increase in blood flow through the superior mesenteric artery[8] is greater for larger, higher-calorie meals. Also, humans only exhibit a significant reduction in blood flow to the periphery (as represented by the calf muscles) with the largest and highest calorie (3 MJ) meal. Even then, the reduction is only during the first 15 min following feeding. In contrast, blood flow to the intestine following the largest, highest calorie meal remains elevated for at least 120 min.

The initial increase in blood flow to the oesophagus and stomach of mammals is mediated by the sympathetic nervous system. In the remainder of the gastrointestinal (GI) tract of mammals, the mass of partly digested food that passes from the stomach to the duodenum (called **chyme**, Greek, *khymos*—juice) causes dilation of the local blood vessels and an increase in blood flow. The influence of the chyme on blood flow as it passes from the stomach to the ileum is shown in Figure 15.2A.

[3] Chapter 9 discusses the responses of ectotherms to changes in environmental temperature.
[4] Specific dynamic action is discussed in Section 2.3.4.
[5] Section 2.3.3 discusses digestion and absorption of foodstuffs.
[6] Deamination is discussed in Section 7.4.1.

[7] Open, closed and incompletely closed circulatory systems are discussed in Sections 14.1.1, 14.3 and 14.4.
[8] The superior mesenteric artery supplies the small intestine and part of the large intestine.

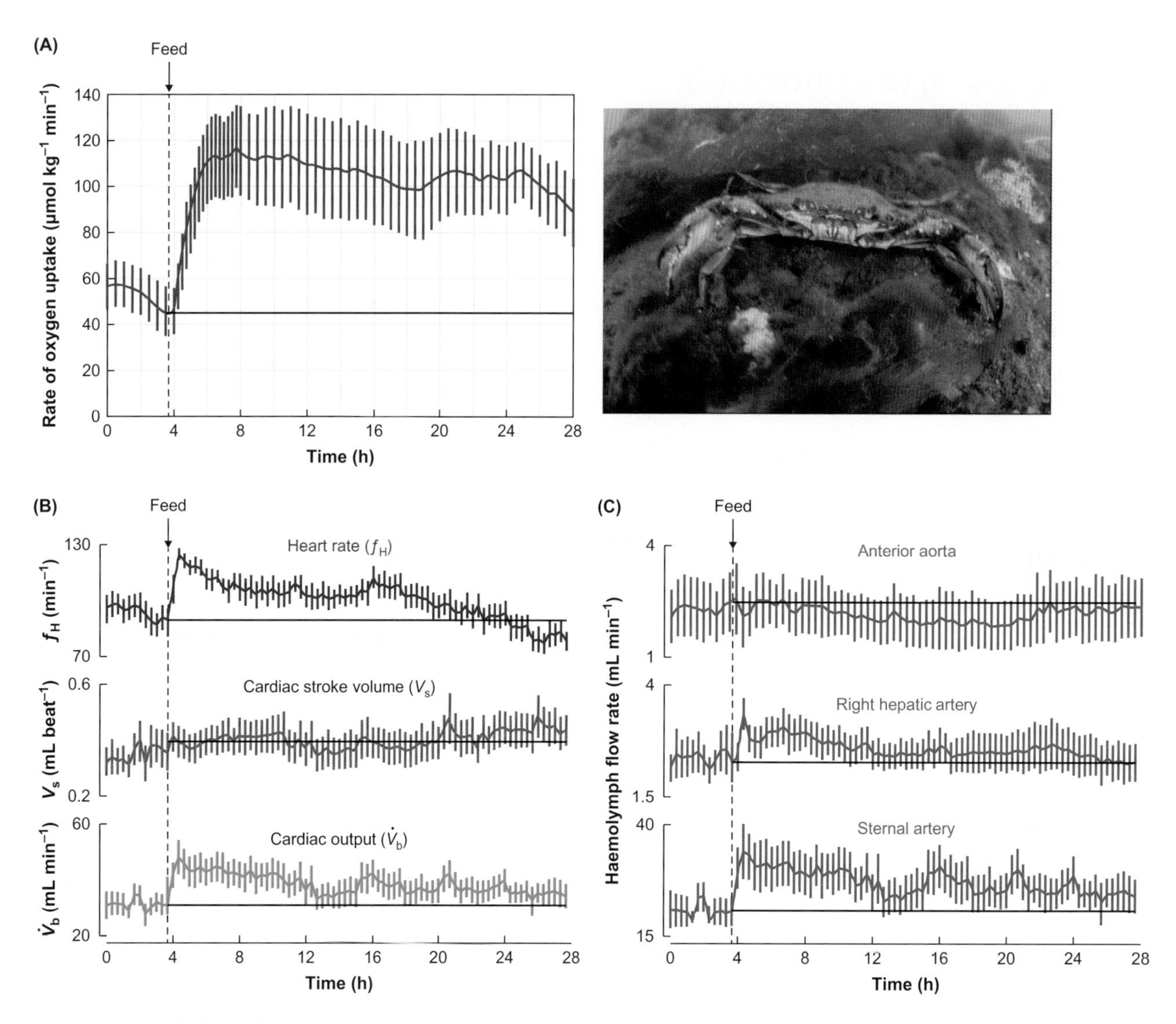

Figure 15.1 Metabolic and associated cardiovascular responses to feeding in blue crabs (*Callinectes sapidus*)

(A) Rate of oxygen consumption; (B) heart function; (C) flow rate of haemolymph through selected blood vessels. For details, see text. Mean values ± SEM, n = 10. The horizontal black lines indicate the values of the variables before feeding.

Adapted from: McGaw IJ and Reiber CL (2000). Integrated physiological responses to feeding in the blue crab *Callinectes sapidus*. Journal of Experimental Biology 203: 359-368. Photo: Ethan Daniels/Shutterstock.com.

Is there any particular constituent of the chyme that is responsible for the increase in blood flow? In the intestine of dogs, the single most potent mediator of increased blood flow when administered alone is glucose, followed by various long-chain fatty acids. Proteins and the products of protein digestion are the least potent components of chyme in terms of causing vasodilation in the intestine. One suggested explanation for dilation of the blood vessels supplying the gut wall as food passes through is that glucose initiates a mechanism that causes the production of the powerful vasodilators, adenosine and nitric oxide[9], as illustrated in Figure 15.2B.

Some of the largest increases in metabolic rate following feeding occur in carnivorous reptiles, particularly if there has been a long interval since the last meal. In extreme situations in which a snake such as a Burmese python (*Python molurus*) has been fasting at least a month and then eats a large meal (equivalent to its own body mass), maximum $\dot{M}O_2$ ($\dot{M}O_2$ max) may increase to over 40 times the resting value and remain elevated for as long as 14 days[10].

Such a large increase in $\dot{M}O_2$ is far in excess of that during maximal sustainable exercise, which we discuss in Section 15.2.2. Under such circumstances, the ventricles of the heart increase in the mass (hypertrophy, Greek—*hyper*—above

[9] The role of nitric oxide in the control of local blood flow is discussed in Case Study 13.2.

[10] The increase in metabolic rate following feeding is discussed in Section 2.3.4.

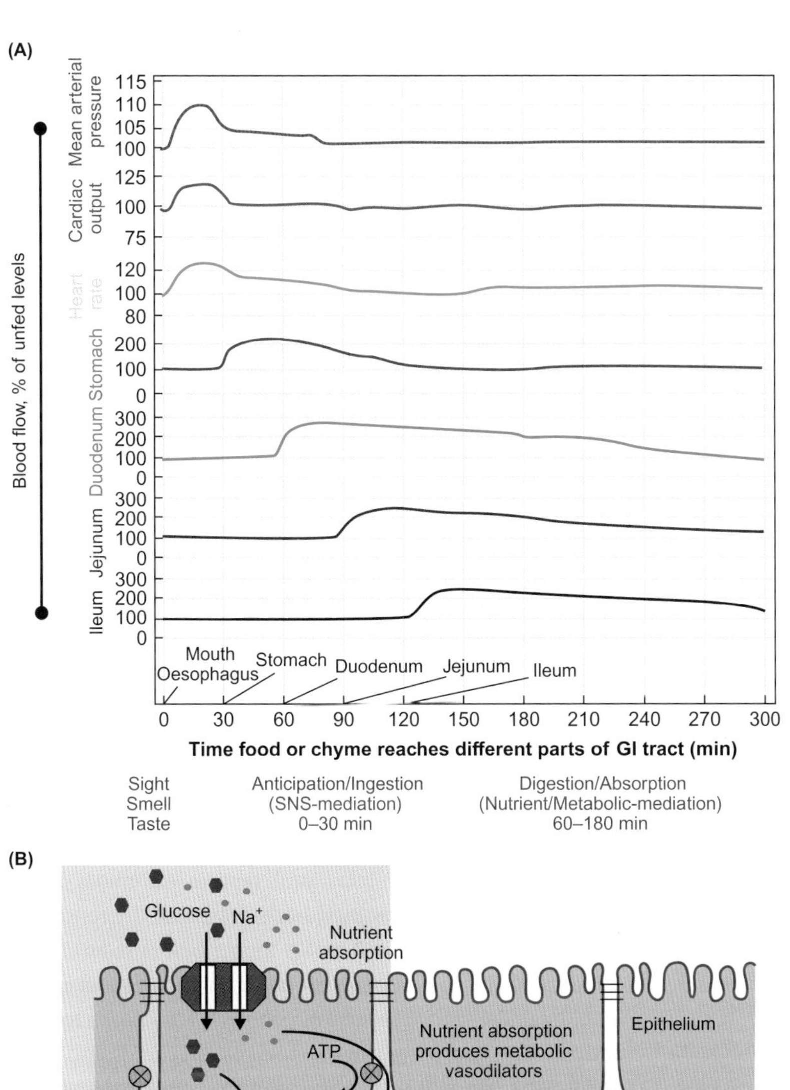

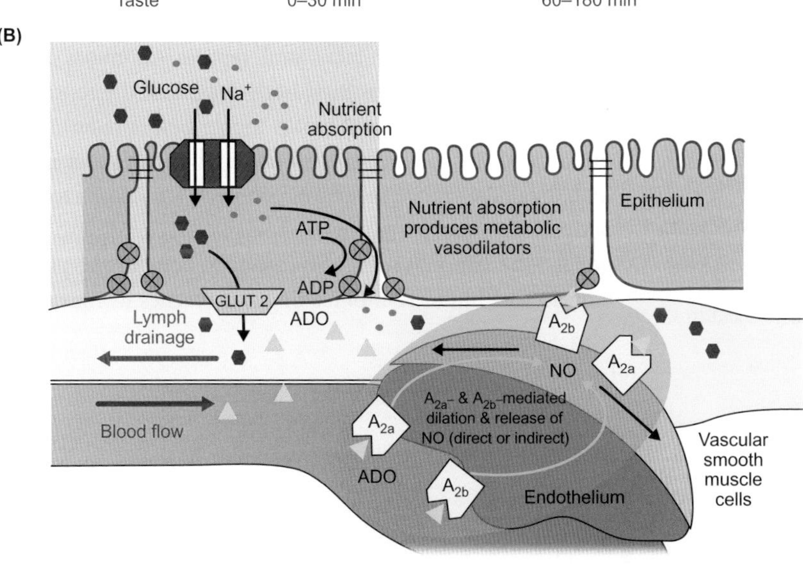

Figure 15.2 Regulation of blood flow in the gastro-intestinal (GI) tract of mammals following feeding

(A) Summary of the cardiovascular responses to feeding. Initially, sight, smell, taste and ingestion cause changes in cardiac output and blood pressure, which are mediated by the sympathetic nervous system (SNS). The passage of the chyme through the gastro-intestinal (GI) tract then causes sequential increases in blood flow to the different regions of the GI tract. Onset of these changes is indicated by the labels along the x-axis. (B) A proposed mechanism causing the release of vasodilators during the absorption of nutrients. The Na^+ (●) linked absorption of glucose (⬢) increases the use of ATP, thereby producing adenosine (ADO, ▲ blue shading), which not only causes vasodilatation via adenosine A_{2a} and A_{2b} receptors, but also causes the production and release of the potent vasodilator, nitric oxide (NO, green shading).

Adapted from Matheson PJ et al (2000). Regulation of intestinal blood flow. Journal of Surgical Research 93: 182-196.

normal, *trophe*—nourishment) within two days of feeding. This increase in ventricular mass has a beneficial impact on oxygen transport by enabling cardiac stroke volume to increase. The increase in the expression of mRNA for heavy-chain cardiac myosin after feeding, shown in Figure 15.3, indicates that the increase in mass of the ventricles results from an increase in the expression of contractile proteins.

However, such an increase in heart mass following feeding in pythons does not always occur. One possible explanation for this inconsistent response is that it depends on there being an imbalance between the increased demand for oxygen following feeding and the ability of the cardiovascular system to deliver the additional oxygen. If, for any reason (such as a low concentration of oxygen in arterial blood, for example) the cardiovascular system is unable to meet the increased demand, there is hypertrophy of the heart.

The effect of feeding on blood flow to the gut means that feeding may affect how the cardiovascular system responds to situations which require increased blood flow to other regions of the body. For example, the varanid lizard, Rosenberg's monitor lizard (*V. rosenbergi*), takes 1.7 times longer to heat up from 19°C to 35°C immediately after feeding than when it has been fasted for

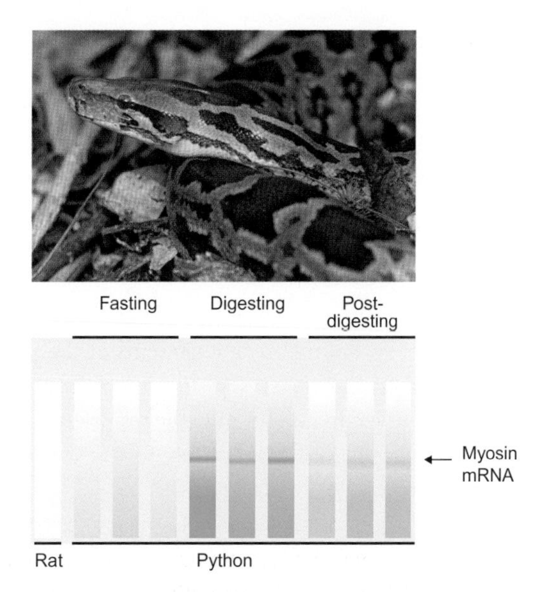

Figure 15.3 Effect of feeding on myosin mRNA in heart ventricles of Burmese pythons (*Python bivittatus*)

Northern blot of myosin mRNA from fasting, digesting and post-digesting pythons. Each lane of the northern blot represents RNA from an individual python, hybridized to a probe which is specific to the isoform of the ventricular myosin heavy-chain in pythons. The lane on the extreme left contains ventricular RNA from rats showing that the probe does not cross-react.

Adapted from Andersen JB et al (2005). Postprandial cardiac hypertrophy in pythons. Nature 434: 37-38. Photo © Bernard Castelein/ Naturepl.com.

a few days, suggesting that blood flow to the skin is lower after feeding and that digestion has priority over thermoregulation[11].

Acid–base balance during and after feeding

Feeding leads to the secretion of acid into the lumen of the stomach of vertebrates. Hydrochloric acid is secreted into the stomach by special cells in the stomach wall—the parietal cells. Figure 15.4A shows that the secretion of acid by the parietal cells leads to an increased concentration of bicarbonate ion (HCO_3^-) in the cells and also shows how HCO_3^- is transferred to the plasma in exchange for Cl^-. The transfer of HCO_3^- to the plasma is known as the **alkaline tide**, but does not necessarily result in an increase in pH. Secretions from the pancreas into the intestine are alkaline and tend to restore pH of the chyme to a more neutral value as it leaves the stomach and enters the small intestine.

Thus, the blood pH in animals that feed frequently changes relatively little. In these species there is almost continuous passage of food from the stomach to the intestine and more or less simultaneous secretion of acid in the stomach and base from the pancreas into the intestine. In animals that feed infrequently, however, large alkaline tides occur as a result of the delay between secretion of acid in the stomach and base from the pancreas into the intestine. Meal size is also important in generating an alkaline tide. Animals which eat frequently tend to eat smaller meals than those that eat infrequently, so pH of the plasma of those species that eat frequently, relatively small meals rarely increases significantly.

Compared to most mammals (with the exception of large, wild carnivores), amphibians and reptiles eat much larger meals as a proportion of their body mass and the food normally stays in the stomach much longer before entering the intestine; these animals eat less frequently than most mammals. In all the species that eat larger meals (compared to their body mass)—including frogs, toads, snakes, lizards and crocodiles—the concentration of bicarbonate ions in the plasma increases within 6–12 h of feeding and reaches its maximum

[11] Thermoregulation in reptiles is discussed in Sections 9.1, 9.2 and 10.3.

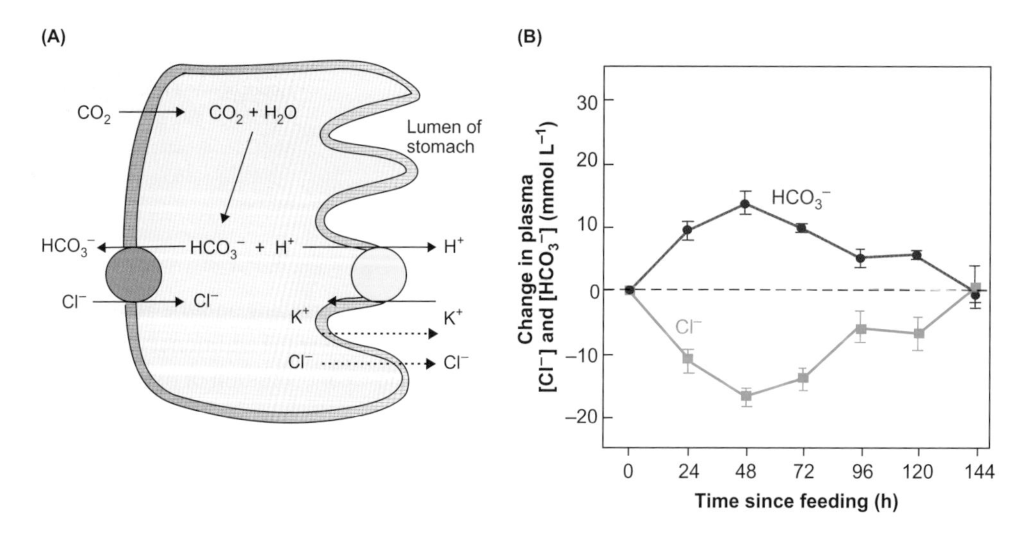

Figure 15.4 Acid secretion into the stomach of vertebrates and the alkaline tide after feeding in anurans

(A) Diagram of a parietal cell from the stomach showing the secretion of protons (H^+) following a meal. The secretion of protons leads to a rise in bicarbonate ions (HCO_3^-) which are transferred to the plasma. (B) Changes in concentrations of chloride ions, [Cl^-], and bicarbonate ions [HCO_3^-] in the plasma following a meal in American bullfrogs (*Lithobates catesbeianus*). Mean values ± SEM, n = 8.

Adapted from Wang T et al (2005). Effects of digestion on the respiratory and cardiovascular physiology of amphibians and reptiles. In: Physiological and Ecological Adaptations to Feeding in Vertebrates. Starck JM and Wang T (eds.), pp 279-304. Enfield (NH), USA: Science Publishers, Inc.

18–36 h later, as shown in Figure 15.4B. These species also experience an increase in PCO_2 in the blood following feeding, which is the result of under-ventilation (also known as hypoventilation), i.e. the increase in ventilation is not sufficient to match the increases in $\dot{M}O_2$ and rate of CO_2 production.

The increase in PCO_2, and accompanying decrease in pH (respiratory acidosis), compensates to a greater or lesser extent for the metabolic alkalosis caused by the alkaline tide and thus appears to be an active regulator of plasma pH. Evidence for this has been obtained by using a proton pump inhibitor to prevent the secretion of acid in the stomach. These experiments demonstrated that, following administration of a proton pump inhibitor, cane toads (*Rhinella marina*) and boa constrictors (*Boa constrictor*) not only show a reduction in the post-feeding rise in HCO_3^-, but also no increase in PCO_2.

An alkaline tide does occur in elasmobranch and teleost fish following feeding, but this is not compensated for by an increase in PCO_2 in the arterial blood (P_aCO_2). This is because of the much higher capacitance coefficient of CO_2 relative to O_2 in water[12]. In order to obtain sufficient oxygen, the gills of water-breathing animals are greatly over ventilated with respect to the removal of CO_2. Hence, changes in ventilation volume have little effect on P_aCO_2 in water-breathers[13].

The hypoventilation following feeding in amphibians and reptiles leads to a reduction in the PO_2 in the lungs, which might be expected to impair the transfer of oxygen from the lungs to the blood. However, PO_2 in the blood leaving the lungs actually increases following a meal, even though PCO_2 also increases. One possible explanation for this increase in PO_2 leaving the lungs is that as the demand for oxygen increases, ventilation of the lungs increases partially to meet this demand. There is also an accompanying reduction of the right- to left- shunt of blood as it flows through the incompletely divided heart of these species[14], so that less oxygen-poor blood enters the body circulation.

15.2.2 Sustained exercise in ectotherms

Another activity that increases the demand for oxygen is exercise. Again, this increase in demand is met by changes in the components on the right-hand side of equations 15.1 and 15.2. There are basically two types of exercise in animals, sustained (or endurance) and burst (or exhaustive). As these names suggest, sustained exercise can be maintained for relatively long periods and is fuelled mainly by aerobic metabolism. By contrast, burst exercise can only be maintained for relatively short periods and is fuelled mainly by anaerobic metabolism[15].

During sustained exercise, such as migration of some species of birds, the increase in oxygen demand by the oxidative muscle fibres[16] and the removal of the metabolic end product, carbon dioxide, are met by adjustments of the ventilatory and cardiovascular systems so that the animal is in a steady state. However, in insects, only the respiratory system is involved in the exchange of the respiratory gases between the environment and the locomotor muscles.

Metabolism during flight of insects is aerobic. In small insects such as Hawaiian fruit flies (*Drosophila mimica*), in which the increased $\dot{M}O_2$ is met almost entirely by diffusion, the opening of the spiracles is controlled to meet changing energy requirements, as shown in Figure 15.5A. In larger insects, the increased oxygen demands are met by ventilation of the tracheae[17]. The effectiveness of the ventilatory responses during flight is demonstrated by experiments on potato hawkmoths

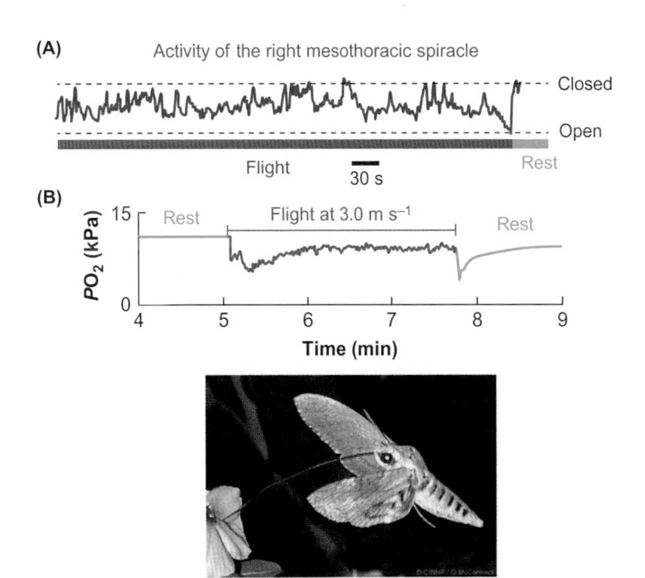

Figure 15.5 Respiratory function during flight of insects

(A) Closing and opening behaviour of the right mesothoracic spiracle in an Hawaiian fruit fly (*Drosophila mimica*) during flight and when at rest. (B) Partial pressure of oxygen (PO_2) in the flight muscle of a sweet potato hawkmoth (*Agrius convolvuli*) while at rest and when flying with a wind stimulus of 3 m s^{-1}.

Adapted from A: Lehmann F-O (2001). Matching spiracle opening to metabolic need during flight in *Drosophila*. Science 294: 1926-1929. B: Komai Y (1998). Augmented respiration in a flying insect. Journal of Experimental Biology 201: 2359-2366. Photo: © CINHP/G. McCormack.

[12] Capacitance coefficients of oxygen and carbon dioxide in water are discussed in Section 11.1.2.

[13] The effect of ventilation volume on PCO_2 of blood in water breathers is also discussed in Section 13.3.1.

[14] Blood flow through the hearts of amphibians and reptiles is discussed in Section 14.4.3. The incompletely divided heart of amphibians and most reptiles enables some of the oxygen-poor blood returning from the body to pass from the right side of the ventricle to the left side and to circulate around the body again. This is known as a right to left shunt.

[15] Aerobic and anaerobic metabolism are discussed in Section 2.4.

[16] Muscle fibre types are discussed in Section 18.2.3.

[17] Ventilation in insects is discussed in Section 12.4.1.

(*Agrius convolvuli*). Some data from these experiments are shown in Figure 15.5B. After an initial dip, PO_2 in the flight muscle is maintained close to that recorded when the insect is at rest.

In most species of animals, the ventilatory system meets the increased demand for oxygen during exercise by a similar, or even greater, proportional increase in minute ventilation volume, as illustrated by the data in Table 15.1. This means that there is no change (or only a slight decrease) in the amount of oxygen extracted from the water or air that ventilates the gas exchange surface. The net effect is that PO_2 in arterial blood (P_aO_2) is maintained at or close to the value seen in the resting animal, as shown in Tables 15.1 and 15.3.

In contrast to the respiratory system, Table 15.2 shows that heart rate and cardiac output rarely increase to the same extent as the increase in $\dot{M}O_2$ during sustained exercise, which means that the amount of oxygen extracted from the blood as it passes through the active, aerobic muscle fibres increases. What is more, the contributions of heart rate and cardiac stroke volume to the increase in cardiac output vary between species. The increase in heart rate during exercise is often referred to as **exercise tachycardia**.

Sustained exercise in crustaceans

Blue crabs (*Callinectes sapidus*) are excellent swimmers, and in this species both heart rate and cardiac stroke volume contribute almost equally to the increase in cardiac output during swimming, with only a slight increase in the concentration of oxygen extracted from the haemolymph. In resting crabs, there is a substantial amount of oxygen in the mixed venous haemolymph (0.25 mmol L^{-1}); this is known as the **venous oxygen reserve**. Venous oxygen reserve can be depleted to a greater or lesser extent during exercise. During moderate swimming activity, oxygen in venous blood of blue crabs is reduced to 0.12 mmol L^{-1}, thus still maintaining a venous oxygen reserve which could be used at higher levels of exercise. However, pH of arterial haemolymph decreases substantially during sustained exercise. This decrease in pH is related to a 14-fold increase in lactate concentration, suggesting that anaerobic metabolism[18] makes a substantial contribution to generation of ATP, even during moderate levels of sustained exercise in this species.

[18] Lactic acid is a product of anaerobic metabolism, which is discussed in Section 2.4.1.

Table 15.1 Ventilatory responses to endurance exercise in ectotherms

The numbers in parentheses give the factorial increase in that variable during exercise compared to the resting value and indicate to what extent the different components of the ventilatory system (shown in equation 15.1) contribute to the increase in rate of oxygen consumption.

Species	Mass (kg)	Activity	Temperature (°C)	Rate of O_2 consumption (mmol min^{-1} kg^{-1})	Respiratory frequency (min^{-1})	Ventilation volume (mL min^{-1} kg^{-1})	O_2 extracted from water or air (mmol L^{-1})	PO_2 in arterial blood (kPa)
Callinectes sapidus (blue crab)	0.14	Rest	20	0.05	94	490	0.1	10.4
		Swimming		0.13 (× **2.6**)	312 (× **3.3**)	1400 (× **2.9**)	0.08 (× **0.8**)	9.1
Oncorhyncus mykiss (rainbow trout)	0.9–1.5	Rest	9.5	0.026		211	0.1	18.3
		Swimming at 0.25–0.5 m s^{-1} (80–90 per cent max)		0.194 (× **7.5**)		1700 (× **8.1**)	0.1 (× **1**)	17.1
Katsuwonus pelamis (skipjack tuna)	1–2	Swimming at approx. 1 m s^{-1} (routine)	25	0.51		3800	0.12	9.3
Alligator mississippiensis (American alligator)	1.34	Rest	30	0.024	0.92	19.3	1.77	
		Walking at 0.44 m s^{-1}		0.41 (× **17**)	13.6 (× **14.8**)	1200 (× **62**)	0.56 (× **0.32**)	
Python bivittatus (Burmese python)	1.4	**Fasting**						
		Rest	30	0.034	1.8	27.8		
		Crawling at 0.11 m s^{-1}		0.34 (× **10**)	14.7 (× **8.2**)	660 (× **23.7**)		
		Digesting (72 h after feeding)						
		Rest		0.20 (× **5.9**)	6.0 (× **3.3**)	118 (× **4.2**)		
		Crawling at 0.11 m s^{-1}		0.40 (× **11.8**)	18.0 (× **10**)	604 (× **21.7**)		

PO_2, partial pressure of oxygen.
References Brill and Bushnell, 2001; Booth et al, 1982; Farmer and Carrier, 2000; Kiceniuk and Jones, 1977; Secor et al, 2000. Full citations available online.

Table 15.2 Circulatory responses to endurance exercise in ectotherms

The numbers in parentheses give the factorial increase in that variable during exercise compared to the resting value and indicate to what extent the different components of the circulatory system (shown in equation 15.2) contribute to the increase in rate of oxygen consumption.

Species	Mass (kg)	Activity	Temperature (°C)	Rate of O_2 consumption (mmol min^{-1} kg^{-1})	Heart rate (beats min^{-1})	Cardiac output (mL min^{-1} kg^{-1})	O_2 extracted from blood (mmol L^{-1})
Callinectes sapidus (blue crab)	0.14	Rest	20	0.05	89	151	0.33
		Swimming		0.13 (× **2.6**)	143 (× **1.6**)	345 (× **2.3**)	0.37 (× **1.1**)
Oncorhyncus mykiss (rainbow trout)	0.9–1.5	Rest	9.5	0.026	38	18	1.5
		Swimming at 0.25–0.5 m s^{-1} (80–90 per cent max)		0.194 (× **7.5**)	51 (× **1.35**)	43 (× **2.4**)	3.3 (× **2.2**)
Katsuwonus pelamis (skipjack tuna)	1–2	Swimming at approx 1 m s^{-1} (routine)	25	0.51	79	132	2.2
Iguana iguana (iguana)	0.71	Rest	35	0.14	48	202	0.8
		Running at 0.14 m s^{-1}		0.62 (× **4.4**)	100 (× **2.1**)	343 (× **1.7**)	1.7 (× **2.1**)
Varanus exanthematicus (savannah monitor lizard)	1.02	Rest	35	0.14	45	112	1.16
		Running at 0.33 m s^{-1}		0.94 (× **6.7**)	110 (× **2.4**)	342 (× **3.0**)	2.72 (× **2.3**)
Python bivittatus (Burmese python)	1.4	**Fasting** Rest	30	0.034	24.7	18.9	1.8
		Crawling at 0.11 m s^{-1}		0.34 (× **10**)	56.9 (× **2.3**)	54.9 (× **2.9**)	5.8 (× **3.22**)
		Digesting (72 h after feeding) Rest		0.20 (× **5.9**)	55.8 (× **2.26**)	58.9 (× **3.1**)	3.12 (× **1.73**)
		Crawling at 0.11 m s^{-1}		0.40 (× **11.8**)	66.6 (× **2.7**)	62 (× **3.28**)	6.7 (× **3.72**)

References: Brill and Bushnell, 2001; Booth et al, 1982; Gleeson et al, 1980; Kiceniuk and Jones, 1977; Secor et al, 2000. Full citations available online.

In arthropods, such as Dungeness crabs (*Cancer magister* or *Metacarcinus magister*), the muscles of the walking legs and bailers (the structures which generate the ventilatory current of water) are supplied by the sternal artery. Flow of haemolymph through the sternal artery increases when a fasting Dungeness crab walks. At the same time, flow through the hepatic artery, which supplies the hepatopancreas and other parts of the gastrointestinal system, decreases and may virtually cease. Thus, in a fasting crab, blood is shunted away from the gastrointestinal tract to the muscles of the legs during sustained exercise. This is not the case in crabs that have recently fed, as we discuss later in this section.

In general, land crabs have similar responses to moderate exercise as their aquatic relatives, with their $\dot{M}O_2$max during exercise being 3–6 times the resting value. However, in many of the species of land crabs, such as blackback land crabs (*Gecarcinus lateralis*), $\dot{M}O_2$ does not reach a steady-state during exercise periods of 20 min or more and may not peak until after the period of exercise has finished, as shown in Figure 15.6A. Nonetheless, the energy demand of the locomotor muscles of the crabs increases as soon as they begin to run. So, as the data in Figure 15.6A indicate, the cardiorespiratory system of many species of land crabs is not able to provide oxygen at a sufficiently rapid rate to satisfy the initial needs of the locomotory muscles during sustainable levels of exercise.

In contrast, the fastest species of land crabs (painted ghost crabs, *Ocypode gaudichaudii*) have a sustainable $\dot{M}O_2$ which is about five times their resting rate, as shown in Figure 15.6B. Figure 15.6B also shows that the rate of increase of $\dot{M}O_2$ at the beginning of exercise is relatively rapid in ghost crabs compared to that in blackback land crabs. These two features indicate that the performance of the respiratory and incompletely closed circulatory systems of some species of crustaceans is comparable to that of birds and mammals with their completely closed circulatory systems[19].

It is possible to compare the aerobic abilities of different species and of the same species under different environmental

[19] Incompletely closed, closed and open circulatory systems are discussed in Sections 14.3 and 14.1.1.

15

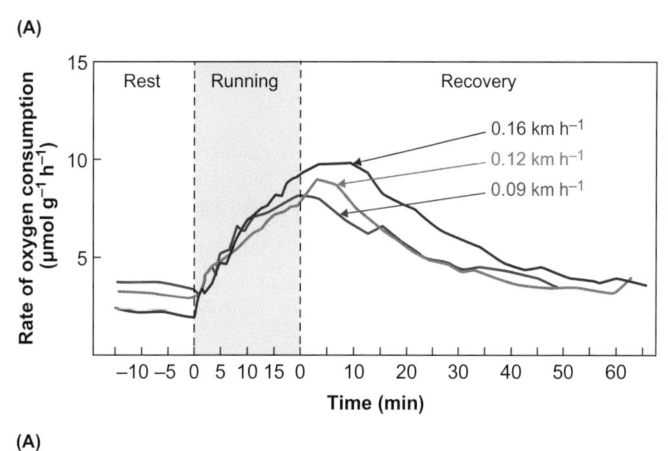

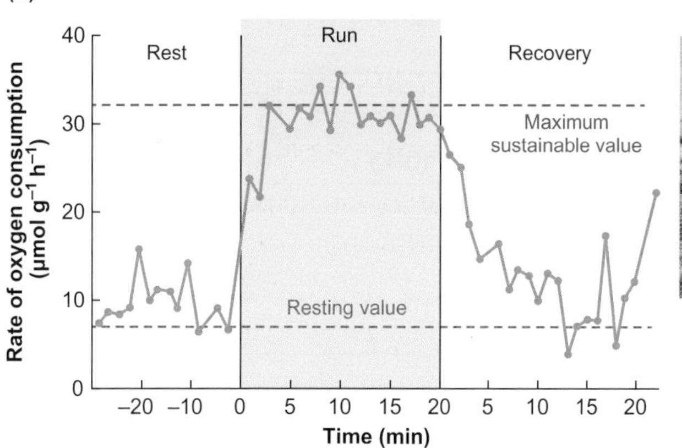

Figure 15.6 Rate of oxygen consumption of land crabs at rest and while running for 20 min

(A) *Gecarcinus lateralis*, (B) *Ocypode gaudichaudii*. Note that in (B), maximum sustainable rate of oxygen consumption is 32 μmol g^{-1} h^{-1} and resting rate of oxygen consumption is 7 μmol g^{-1} h^{-1}. The data from *Gecarcinus* are the mean values from five crabs running at three running speeds (0. 09, 0.12 and 0.16 km h^{-1}) at 25°C while those from *Ocypode* are from one crab running at 0.19 km h^{-1} at 22°C.

Adapted from: (A) Herreid CF (1981). Energetics of pedestrian arthropods. In: Herreid CF and Fourtner CR (eds) Locomotion and Energetics in Arthropods. pp 491-526. New York and London: Plenum Press. (B) Full RJ and Herreid CF (1983). Aerobic response to exercise in the fastest land crab. American Journal of Physiology 244: R530-R536. Top photo: Bhny/Wikimedia Commons; Bottom photo: © Hans Hillewaert en.wikipedia.org/wiki/File:Ocypode_quadrata_(Cahuita).jpg.

conditions, such as temperature for ectotherms, by calculating aerobic scope or factorial aerobic scope. **Aerobic scope** is the difference between maximum sustainable $\dot{M}O_2$ and resting $\dot{M}O_2$ which, for the data in Figure 15.6B, is 32 − 7 = 25 μmol O_2 g^{-1} h^{-1}. **Factorial aerobic scope** is the ratio between maximum sustainable $\dot{M}O_2$ and resting $\dot{M}O_2$, which is 32/7 = 4.6 for the data in Figure 15.6B.

Sustained exercise in teleost fish

The most studied species of fish is rainbow trout (*Oncorhynchus mykiss*). In this species, increases in cardiac output and in the amount of oxygen extracted from the blood contribute equally to the increase in $\dot{M}O_2$ during sustained exercise. Cardiac stroke volume is the major contributor to the increase in cardiac output, as indicated in Table 15.2.

The ability of the gills of trout to transfer oxygen (as indicated by the diffusion conductance, G_{diff}[20]) increases by 5–6 times during exercise. This suggests that there is an increase in the effective surface area of the gills and/or a reduction in the diffusion distance between blood and water. These changes could be achieved by increasing the number of gill lamellae that are perfused with blood (known as interlamellar recruitment) and/or by increasing the area of each gill lamella that is perfused (known as intralamellar recruitment)[21]. Increasing the effective surface area of the gills in swimming fish can, however, increase the passive flux of water and ions across the gills[22].

Tuna swim continuously[23], so it is not possible to obtain resting data from these species. The oxygen content of their arterial blood is, on average, over twice that of rainbow trout (15.3 g dL^{-1} vs 6.4 g dL^{-1})[24] and is approaching that of some mammals. Therefore, greater extraction of oxygen from the blood can contribute to the higher relative $\dot{M}O_2$ in tuna, shown in Table 15.2. The heart rate of free-swimming southern bluefin tuna (*Thunnus maccoyii*) is not exceptionally high compared with that in rainbow trout (if compared at comparable body temperatures), but they do have a relatively larger cardiac

[20] Diffusion conductance is the ease with which respiratory gasses diffuse across the gas exchanger and is discussed in Section 12.2.2.

[21] Inter- and intralamellar recruitment in the gills of fish are discussed in Section 14.4.1.

[22] Passive ion and water fluxes across fish gills are discussed in Sections 5.1.3, 5.2.

[23] Tuna and lamnid sharks have many adaptations for their active lifestyle, such as relatively large surface area and small diffusion distance of the gills, ram ventilation (discussed in Section 12.2.2) and regional endothermy (discussed in Section 10.3.1).

[24] Concentration of oxygen is represented as g dL^{-1} and mol L^{-1}. One mol O_2 has a molecular mass of 32 g.

stroke volume than trout. In fact, the hearts of tuna most likely function close to their maximum stroke volume all the time so that, unlike many other species of teleosts, any change in cardiac output may be solely the result of a change in heart rate.

The higher performance of the heart of skipjack tuna (*Katsuwonus pelamis*) compared to that of rainbow trout is related to the fact that the compact layer of the ventricle, which receives a coronary blood supply of arterialized blood, makes up around 65 per cent of the ventricle compared to about 35 per cent in the trout. The remainder of the ventricle is composed of the spongy layer, which receives its oxygen from the venous blood returning from the body[25]. Also, the mass of the heart relative to body mass is about four times larger in skipjack tuna than in rainbow trout, which means that coronary blood flow per kg of fish is over seven times greater in the tuna. It seems that the performance of skipjack tuna is dependent on the full functioning of their coronary blood supply, whereas rainbow trout swim reasonably well after the coronary supply is surgically removed. Swimming performance of trout is limited by a threshold amount of oxygen in mixed venous blood returning to the heart, which progressively declines as swimming speed increases. The oxygen in the mixed venous blood supplies the spongy layer of the ventricle.

The distribution of blood flow during exercise in unfed trout is similar to that in unfed blue crabs. Flow to the aerobic, red, locomotor muscle fibres increases by an order of magnitude or more in rainbow trout, depending on swimming speed. On the other hand, blood flow to the gastrointestinal (GI) tract decreases quite substantially.

As well as increasing the flow of blood to the active organs and tissues, the oxygen-carrying capacity of the blood could also increase during exercise. The spleen acts as a reservoir of red blood cells in vertebrates and it may contract during exercise in response to direct adrenergic nervous stimulation and/or an increase in circulating catecholamines. However, it has been suggested that an optimum concentration of red blood cells exists in the blood of vertebrates. The concentration of red blood cells is often presented as the percentage of red cells in a given volume of blood (called the **haematocrit**—Hct) and the optimum Hct is a balance between an adequate concentration of oxygen in the blood at low haematocrits and a high viscosity in the blood at high haematocrits. This suggestion has been tested in rainbow trout.

Figure 15.7 shows what happens to a range of variables when haematocrit (Hct) is experimentally varied in the blood of trout. There are modest increases in maximum sustainable swimming speed and in $\dot{M}O_2$ max with an increase in Hct above the lower end of the normal range. Although $\dot{M}O_2$ max peaks at an Hct of about 42 per cent, which is well above the normal range for trout, swimming performance and maximum diffusion conductance continue to increase up to the maximum Hct tested. These data suggest that an increase in viscosity of the blood is not a limiting factor to swimming performance in trout. However, if Hct falls below the normal range for any reason, swimming performance declines relatively steeply. This indicates that there is a lower critical level for Hct and hence, for the oxygen carrying capacity of the blood.

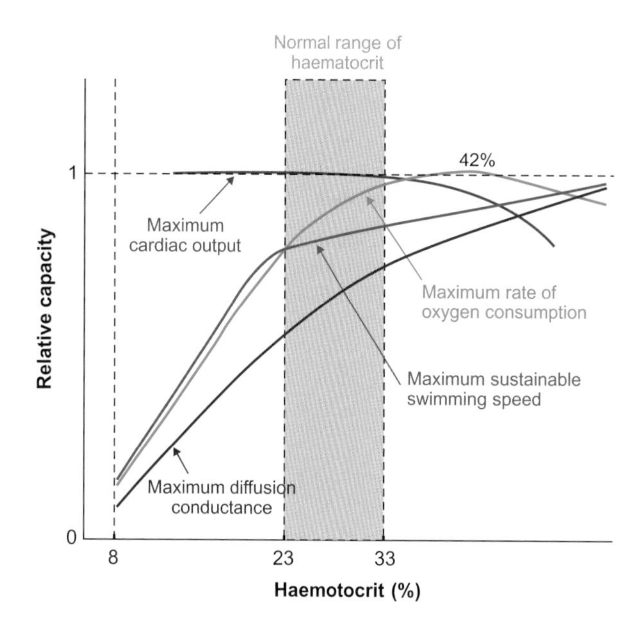

Figure 15.7 Influence of haematocrit (Hct) on maximum sustainable swimming performance and maximum rate of oxygen consumption in rainbow trout (*Oncorhynchus mykiss*)

Also included are maximum cardiac output and maximum diffusion conductance. Although maximum swimming performance continues to increase when Hct is increased above the normal range, maximum rate of oxygen consumption reaches a peak at a Hct of 42%. There is a sharp fall in maximum swimming speed when Hct falls below about 20%.

Adapted from Gallaugher P et al (1995). Hematocrit in oxygen transport and swimming in rainbow trout (*Oncorhynchus mykiss*). Respiration Physiology 102: 279-292.

Sustained exercise in terrestrial ectotherms

In general, the exercise abilities of terrestrial ectotherms are similar to those in water-breathing animals. Also, with a few exceptions that include marine turtles and varanid lizards, a much greater proportion of locomotor behaviour of amphibians and reptiles is fuelled by anaerobic metabolism (burst exercise)[26] compared with that in many species of fish, birds and mammals. Nonetheless, some species, such as leatherback sea turtles (*Dermochelys coriacea*), can migrate over distances of up to 15,000 km, which suggests the involvement of mainly aerobic metabolism.

[25] Compact and spongy layers of fish hearts are discussed in Section 14.4.5.

[26] We discuss burst exercise in Section 15.2.4.

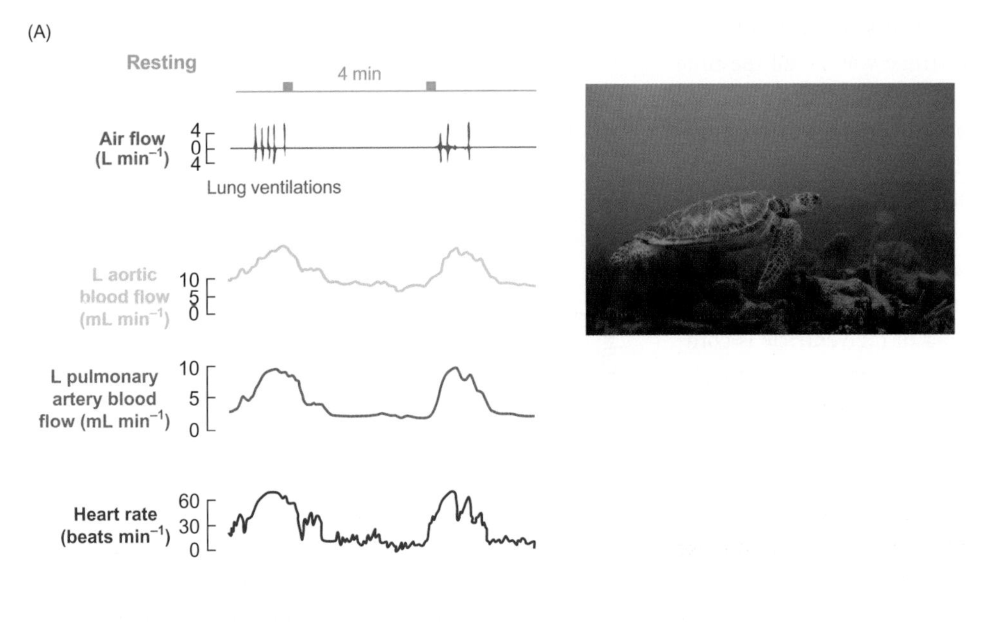

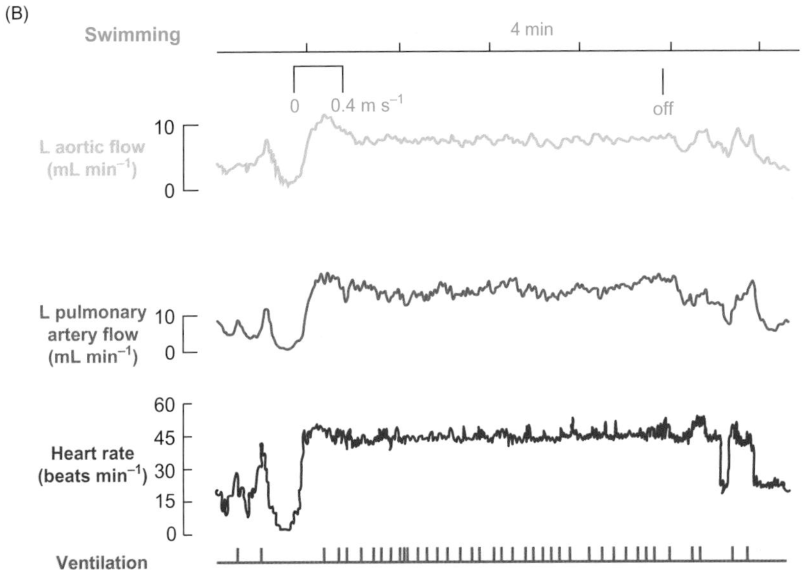

Figure 15.8 Distribution of blood flow in resting and swimming green sea turtles (*Chelonia mydas*)

(A) In resting animals, heart rate and blood flows along the aortic and pulmonary arteries are relatively low when the animals are not ventilating their lungs, but increase during periods of ventilation. (B) When the animals are swimming (0.4 m s^{-1}), heart rate and blood flow along both arteries remain high and lung ventilation is more continuous. Transition between rest (0 m s^{-1}) and swimming at 0.4 m s^{-1} is indicated. Off, indicates when the water flow rate returns to 0 m s^{-1}.

Adapted from West NH et al (1992). Pulmonary blood flow at rest and during swimming in the green turtle, *Chelonia mydas*. Physiological Zoology 65: 287-310. Photo: © Pete Oxford/naturepl.com.

Amphibians and most species of reptiles have incompletely divided hearts, and blood flow through the major arteries increases during periods of lung ventilation and falls during periods when the animals are not ventilating their lungs[27]. These changes in blood flow are illustrated for resting green turtles (*Chelonia mydas*) in Figure 15.8A. It is important that oxygen is continually presented to the locomotory muscles when the animal is swimming, so when the turtles are swimming, blood flow along the major arteries is constant and not obviously related to ventilation, which is shown in Figure 15.8B.

[27] The undivided hearts of amphibians and reptiles and the shunting of blood between the two sides of the ventricle are discussed in Section 14.4.3.

Monitor lizards (*Varanus* spp.) have a greater ability to perform aerobic exercise than most other species of lizards, but how is this achieved? The data in Table 15.2 indicate that monitor lizards have a higher sustained running speed and a higher $\dot{M}O_2$ at that speed than iguanas. These differences in aerobic performance between monitor lizards and iguanas seem to be related to a greater oxygen concentration in the arterial blood in monitor lizards (3.35 mmol O_2 L^{-1} vs 2.48 mmol O_2 L^{-1}), thus enabling a higher extraction of oxygen by the tissues. In addition, cardiac stroke volume contributes to a larger proportional increase in cardiac output during exercise in monitor lizards than it does in iguanas.

In many species of lizards, the increase in $\dot{M}O_2$ during sustained exercise is matched by a similar increase in ventilation.

However, in savannah monitor lizards (*V. exanthematicus*), Burmese pythons (*Python bivittatus*) and American alligators (*Alligator mississippiensis*) ventilation increases by a greater proportion than $\dot{M}O_2$. Experiments on monitor lizards indicate that this may be because the animal operates near the functional limits of the lung and the hyperventilation during exercise facilitates oxygen loading in the lung. However, this does not seem to be the case in pythons in which ventilation responds differently to exercise and feeding.

In pythons, similar increases in $\dot{M}O_2$ are associated with hyperventilation during exercise and hypoventilation following feeding, as we discuss in Section 15.2.1. These data do not suggest gas exchange at the lungs is limited during either situation. Hence, the functional significance of the hyperventilation in this species during exercise is not known, but it may serve to maintain pH of the blood close to the resting levels in both monitor lizards and alligators.

Hyperventilation during sustained exercise in monitor lizards and alligators causes more CO_2 to be removed than is being produced, leading to a decrease in PCO_2. This decrease in turn causes a potential increase in arterial pH[28] (respiratory alkalosis). At speeds that cause relatively modest increases in the concentration of lactic acid in monitor lizards, the hyperventilation and potential alkalosis, maintains pH close to its resting value.

The effect of temperature on exercise in ectotherms

Resting metabolic rate is greater at higher temperatures than at lower temperatures in ectotherms, as illustrated in Figure 15.9B, and sustained exercise becomes more difficult for these animals under such conditions. Figure 15.9 shows that when wild sockeye salmon (*Oncorhynchus nerka*), are swimming at a sub-maximal constant speed, heart rate, cardiac output and $\dot{M}O_2$ reach plateaux at temperatures above 19–21°C. Resting $\dot{M}O_2$ continues to increase with increasing environmental temperature, so the fish are unable to increase $\dot{M}O_2$ during exercise sufficiently above the resting level to supply adequate oxygen to the locomotor muscles. The fish fatigue at temperatures between 21°C and 24°C.

All species of Pacific salmon and most Atlantic salmon (*Salmo salar*) die after their first spawning migration, so they only get one chance to produce offspring. In the Fraser River, Canada, there seems to be a relationship between the temperature at which different populations of sockeye salmon exhibit their maximum aerobic scope and the historical temperatures that the populations experience during their upstream migrations. Some populations make relatively long (up to 1100 km) migrations when the river temperature is around its seasonal maximum, whereas other populations migrate over shorter distances (as low as 100 km) later in the year when the temperature is lower. Those fish that migrate in summer have a greater aerobic scope, a relatively larger ventricle to their heart, a better coronary supply to the ventricle and can swim faster than those that migrate during autumn. These factors are also discussed in Experimental Panel 15.1.

The peak summer temperature in the Fraser River has increased by almost 2°C over the last 60 years, which could be pushing the salmon close to their limits. For example, for almost 10 years, a population of salmon that had historically migrated later in the year had, for some currently unknown reason, been entering the river earlier than usual. In 2004, a year in which river temperature was unusually high, the majority of more than 120,000 fish of this population did not reach their spawning ground. There is strong evidence that the higher than usual temperature was the major reason for this failure. If these fish had continued to migrate later in the year, they might well have survived the unusually high temperatures of 2004, but their new behaviour and the abnormally high temperatures of that year illustrate what could happen to populations of wild salmon if river temperatures continue to rise at the rate they have done over the last 60 years.

These data are useful from a conservation and management perspective. For example, it is not sensible to replace lost salmon from upriver locations with individuals from coastal populations, as we know they are relatively poor swimmers and cannot cope with higher river temperatures than they usually experience. As more physiological data become available, both for different populations of the same species and for different species of Pacific salmon, the better equipped fisheries' managers will become to sustain the fish populations in the face of a changing climate.

Figure 15.9 demonstrates that salmonid fish from temperate latitudes are able to swim aerobically over a reasonably wide range of temperatures. This temperature range seems to relate to the environmental temperatures to which the fish are normally exposed. However, fish that live in Polar seas experience a much narrower range of temperatures. One example of such a species is the Antarctic fish, the bald rockcod, or bald notothen (*Pagothenia borchgrevinki*).

Pagothenia borchgrevinki lives in temperatures close to the freezing point of seawater (−1.9°C), and which vary annually by less than 1°C. It would be expected, therefore, that this species would not be able to maintain the same aerobic scope and maximum sustainable swimming speed at temperatures much above 0°C. This is certainly the case if fish are taken from an acclimation temperature of −1°C and tested at a range of temperatures up to 8°C. Above a temperature of 2°C, maximum sustainable swimming speed decreases.

[28] The relationship between PCO_2 and pH is discussed Section 13.3.2.

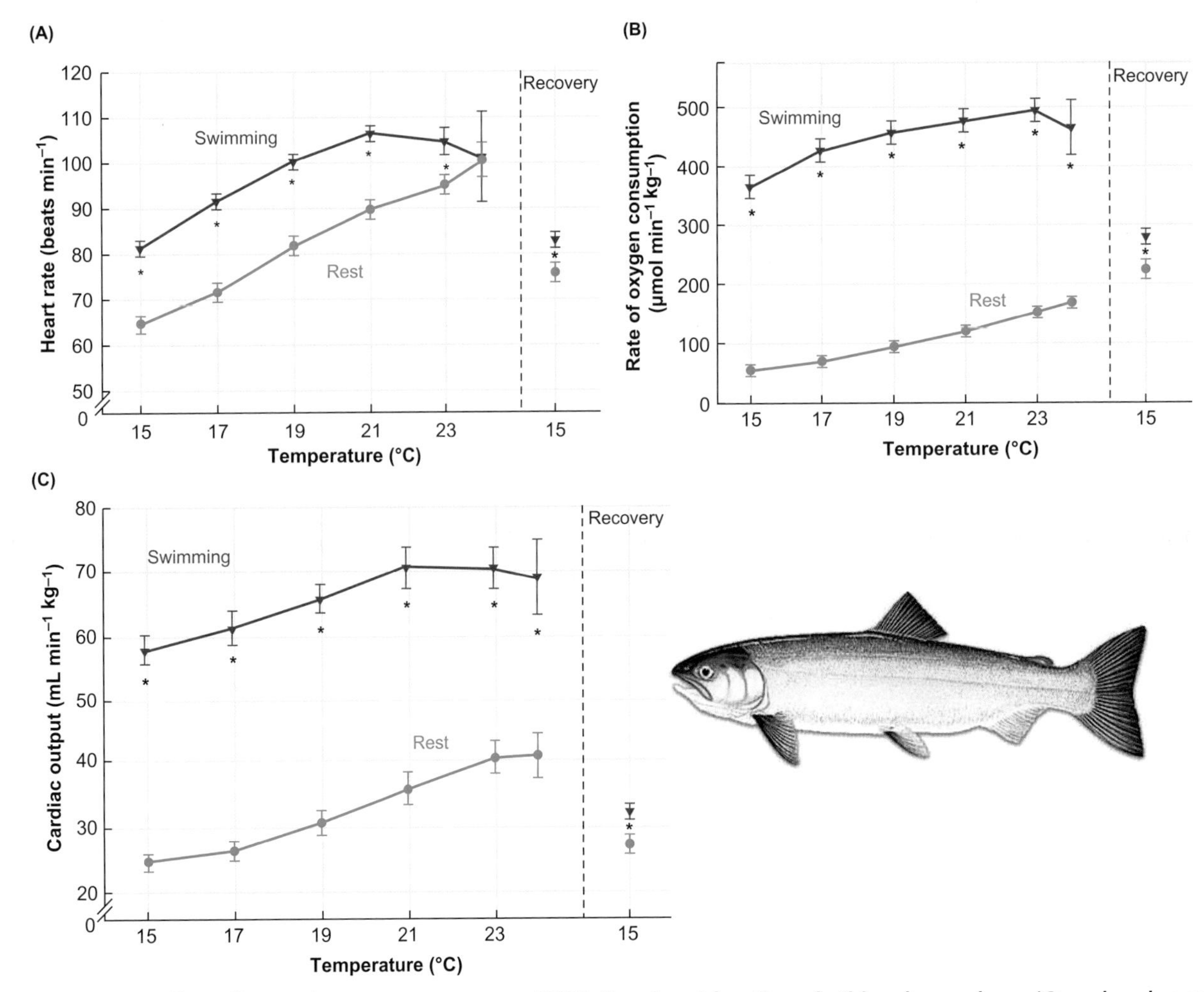

Figure 15.9 The effect of increasing water temperature (2°C h^{-1}) on heart function of wild sockeye salmon (*Oncorhynchus nerka*) while at rest and while swimming.

(A) Heart rate; (B) rate of oxygen consumption, (C) cardiac output. The fish swam at 0.85 m s^{-1} (≈ 1.35 body lengths s^{-1}), which is approximately 75% of their maximum sustainable swimming speed. At the end of the period of temperature increase, water speed was reduced to 0.2 m s^{-1} and water temperature to 15°C. The fish were left for an hour before the 'recovery' values were taken. Values are mean ± SEM of 18 resting fish and 15 swimming fish. Asterisks indicate significant differences between resting and swimming fish at a given temperature.

Adapted from Steinhausen MF et al (2008). The effect of acute temperature increases on the cardiorespiratory performance of resting and swimming sockeye salmon (*Oncorhynchus nerka*). Journal of Experimental Biology 211: 3915-3926.

Experimental Panel 15.1 The role of the heart in determining the exercise ability of salmonid fish

Background

There is no evidence that training of captive fish affects the relative size of their heart. So, the objective of this series of experiments was to test the hypothesis that the natural variation in the form and functioning of the heart plays a crucial role in determining the exercise ability of individual animals, using salmonid fish as the test animals.

Experimental approach

Juvenile, hatchery-raised rainbow trout (*Oncorhynchus mykiss*) were divided into two groups on the basis of their swimming ability: good and poor swimmers. Their swimming ability was determined in batches of 100 fish by varying the flow of water around a circular tank between 0.2 and 1.2 m s^{-1}. The first 10 fish to fatigue were termed poor swimmers. The water current was maintained and the last 10 to fatigue were designated as the good swimmers. Nine months later, six fish from each group had a cuff which measured blood flow surgically placed around their ventral aorta in order to measure cardiac stroke volume, heart rate and cardiac output. They were allowed to recover for 2 days after which they were placed in a tube through which the flow rate of water could be controlled and from which water samples could be taken in order to measure rate of oxygen consumption ($\dot{M}O_2$)[1].

[1] Details can be found in Section 2.2.3 and Box 2.2

The swimming performance of the fish was tested by determining their critical swimming speed (U_{crit}). The fish were initially swum at a relatively low water speed with the speed being increased by 10 cm s^{-1} every 30 min until they fatigued. Critical swimming speed is the last speed at which the fish could swim for the full 30 min plus the proportion of 30 min completed at the highest speed. So, if the last speed at which a fish swam for 30 min was 80 cm s^{-1}, and it swam for 20 min at 90 cm s^{-1}, its U_{crit} was 86.7 cm s^{-1}.

Overall findings

Some of the results from these tests are shown in Figure A. The good swimmers could not only swim faster, but they had a greater maximum $\dot{M}O_2$ ($\dot{M}O_2$ max), and a greater cardiac stroke volume, although maximum heart rate was no different between the two groups. The anatomy of the heart was also different between good and poor swimmers, with the length to width ratio of the ventricle being significantly larger in the better swimmers (1.01 vs 0.88).

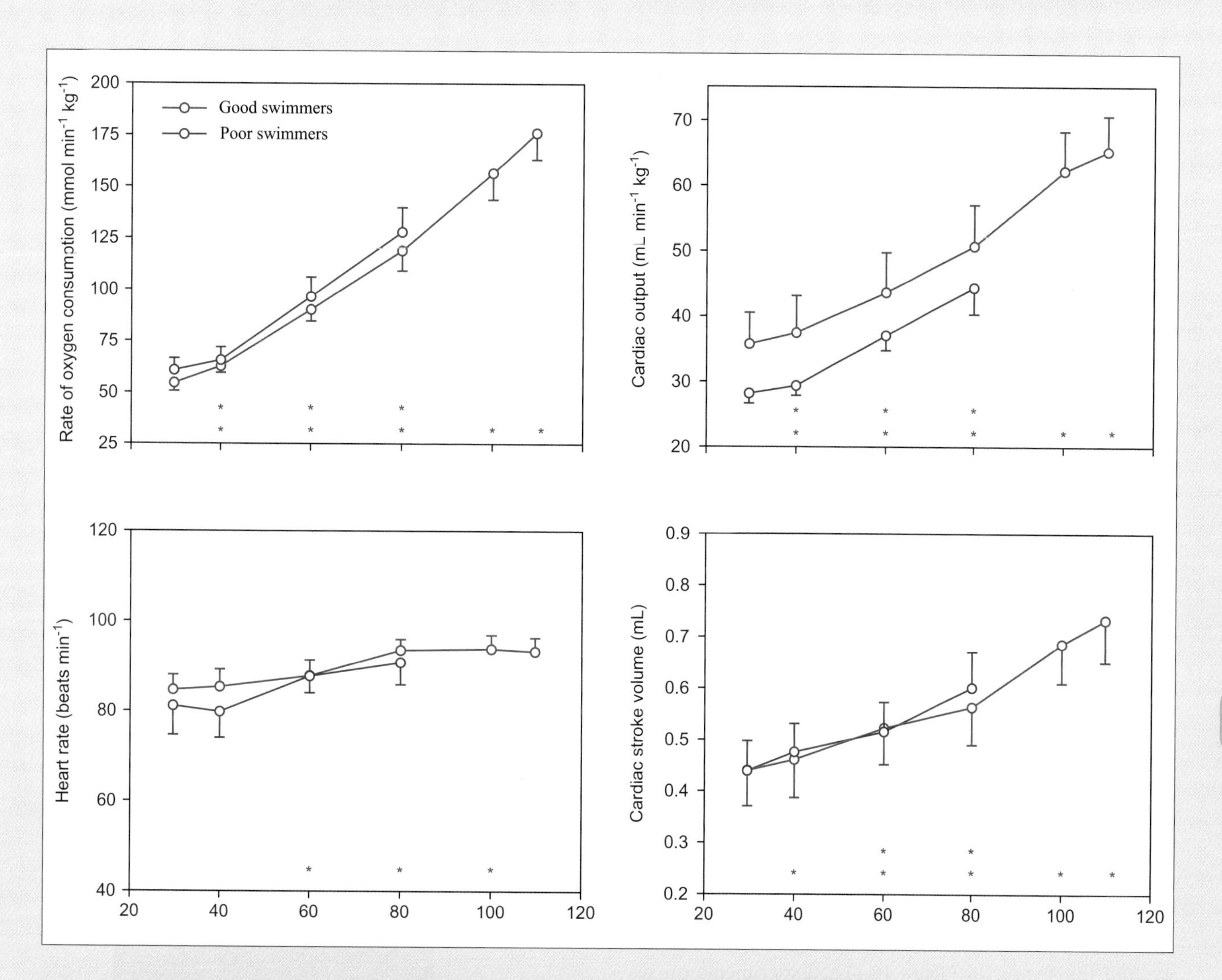

Figure A Comparison of swimming performance, rate of oxygen consumption and cardiac performance between two groups of rainbow trout (*Oncorhynchus mykiss*) that had been identified as good or poor swimmers

Red lines and symbols indicate poor swimmers, blue lines and symbols indicate good swimmers. Values are mean ± SEM of six fish in each group. Asterisks indicate a significant change in the variable relative to that at a swimming speed of 30 m s^{-1} within the two groups.

Reproduced from: Claireaux G et al (2005). Linking swimming performance, cardiac pumping ability and cardiac anatomy in rainbow trout. Journal of Experimental Biology 208: 1775-1784.

These results indicate that the relative size and performance of the heart are linked to $\dot{M}O_2$ max and swimming performance in individual trout. These data formed the basis of field studies on wild populations of sockeye salmon (*Oncorhynchus nerka*).

Subsequent studies on wild fish

Different populations of sockeye salmon that spawn in the Fraser River, British Columbia, Canada, migrate different distances (100 to 1100 km) and undergo different gains in elevation (10 to 1200 m) during their once in a lifetime migration up river to their spawning grounds. Those populations with the most challenging migrations have greater aerobic scope (maximum $\dot{M}O_2$ – resting $\dot{M}O_2$), larger hearts and a better coronary supply to their hearts. This study suggests that natural variability in cardiac function is an important factor in determining the physical performance of wild fish and most likely of other vertebrates.

› Find out more

Claireaux G, McKenzie DJ, Genge AG, Chatelier A, Aubin J, Farrell AP (2005). Linking swimming performance, cardiac pumping ability and cardiac anatomy in rainbow trout. Journal of Experimental Biology 208: 1775–1784.

Eliason EJ, Clark TD, Hague MJ, Hanson LM, Gallagher ZS, Jeffries KM, Gale MK, Patterson DA, Hinch SG, Farrell AP (2011). Differences in thermal tolerance among sockeye salmon populations. Science 332: 109–112.

However cardiac function in this species of fish can vary, potentially enabling it to maintain its aerobic scope and to support aerobic swimming in response to a modest warming of the Southern Ocean. If different groups of fish are acclimated to −1°C and to 4°C, those acclimated to −1°C have their maximum factorial increase in cardiac output at their acclimation temperature, whereas the factorial increase of cardiac output and maximum sustainable swimming speed of those acclimated to 4°C were maintained up to 8°C.

Reptiles show similar responses to changes in temperature as fish. For example, Figure 15.10 shows that aerobic scope in Rosenberg's monitor lizard or health monitor (a varanid lizard, *Varanus rosenbergi*) is much reduced at temperatures below their thermal optimum of around 35°C. However, the factorial aerobic scope is constant at around 7 (this is not shown in Figure 15.10). Related to a constant factorial aerobic scope is the fact that the relative contributions of heart rate, cardiac output and extraction of oxygen from the blood to the increase in $\dot{M}O_2$ are similar at different temperatures.

The effect of feeding on the response of the cardiovascular system to exercise in ectotherms

In Section 15.2.1 we discuss how blood flow to the GI tract increases after feeding. So what happens to the perfusion of the GI tract in recently fed animals during exercise? For fish, we find the answer in Figure 15.11A. Experiments on sea bass (*Dicentrarchus labrax*) have shown that blood flow to the GI tract is higher in fed fish during swimming, but that it decreases as swimming speed (indicated by rate of oxygen consumption, $\dot{M}O_2$) increases. Despite this reduction in perfusion of the gut, and hence in its supply of oxygen, Figure 15.11B shows that the difference in $\dot{M}O_2$ between fed and unfed fish (known as the specific dynamic action, SDA[29]) does not change as swimming speed increases. This suggests that SDA is the result of the metabolism of other tissues and processes in addition to those of the GI tract.

In some species of reptiles, the increase in $\dot{M}O_2$ following feeding can be in excess of that seen during sustainable

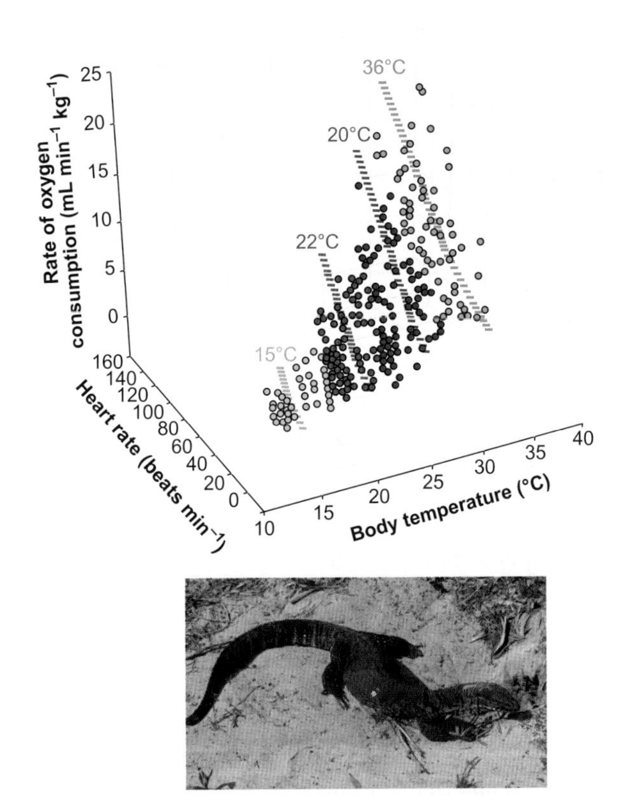

Figure 15.10 Effect of temperature on metabolic and cardiac responses to exercise in Rosenberg's monitor lizard, or health monitor (*Varanus rosenbergi*)

Relationship between body temperature, heart rate and rate of oxygen consumption in six monitor lizards at different levels of exercise. The dashed lines indicate the relationships between heart rate and rate of oxygen consumption during exercise at four different temperatures.

Reproduced from Clark TD et al (2006). Factors influencing the prediction of metabolic rate in a reptile. Functional Ecology 20: 105-113. Own photo.

[29] Specific dynamic action is discussed in Section 2.3.4.

15

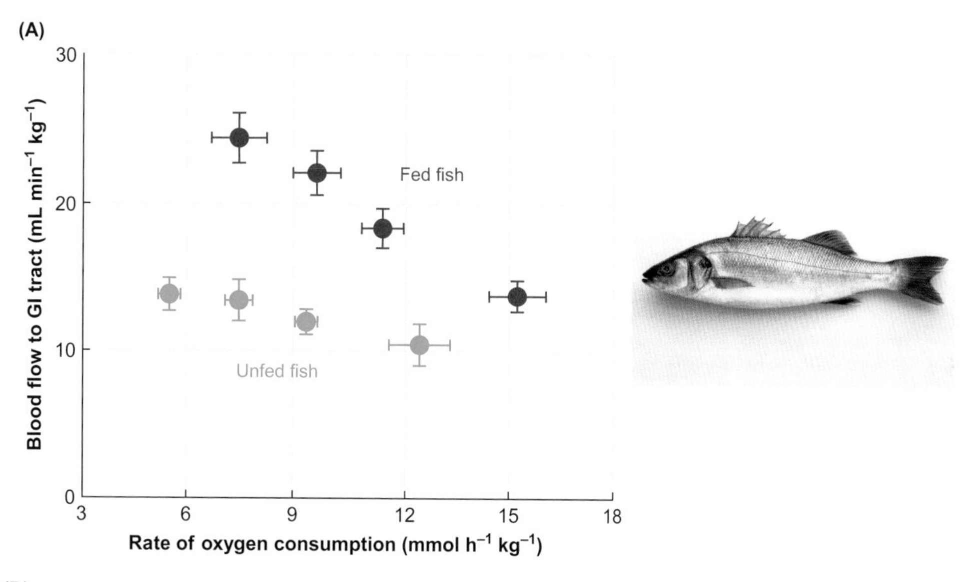

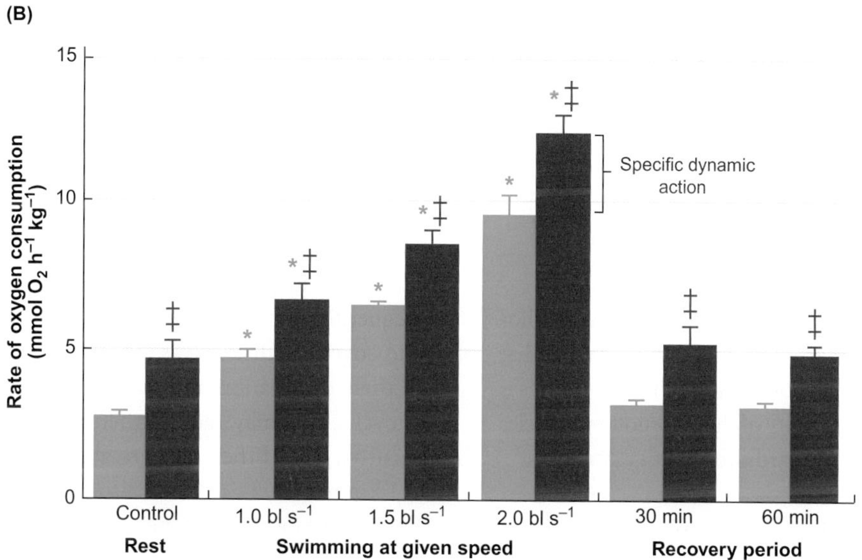

Figure 15.11 Effect of feeding on blood flow to the gastrointestintal (GI) tract and on overall rate of oxygen consumption in sea bass (*Dicentrachus labrax*) swimming at different speeds

(A) Blood flow to GI tract in fed and unfed fish. Swimming speed is indicated by rate of oxygen consumption. (B) Rate of oxygen consumption in unfed (blue bars) and fed (filled red bars) fish at rest and swimming at different speeds. In both (A) and (B), values are mean ± SEM for six fish and in (B) asterisks indicate significant differences between successive swimming speeds and double crosses indicate significant differences between fed and unfed fish. Note, bl s^{-1} = body lengths per second and is explained in Experimental Panel 15.1.

Adapted from Altimiras J et al (2008). Gastrointestinal blood flow and postprandial metabolism in swimming sea bass *Dicentrarchus labrax*. Physiological and Biochemical Zoology 81: 663-672. Image: Steve Cavalier/Alamy Stock Photo.

aerobic exercise. So, what happens when one of these species of reptiles exercises after it has fed? Is one or the other process curtailed or is the cardiorespiratory system able to provide the necessary oxygen for each to proceed at the rate it would in the absence of the other? Figure 15.12 shows that savannah monitor lizards exhibit a higher $\dot{M}O_2$ during exercise after feeding than during exercise after fasting. Also, the increase in $\dot{M}O_2$ following feeding in monitor lizards is not as great as that during exercise. However, $\dot{M}O_2$ during exercise in fed monitor lizards is equal to the sum of the values obtained following feeding when at rest and during exercise alone; there is an additive effect.

In contrast, the data in Table 15.2 show that $\dot{M}O_2$ in Burmese pythons during exercise following feeding is about 75 per cent of that expected from simply adding the values seen during exercise and feeding alone. Nonetheless, these experiments on monitor lizards and pythons demonstrate that maximum $\dot{M}O_2$ during sustained

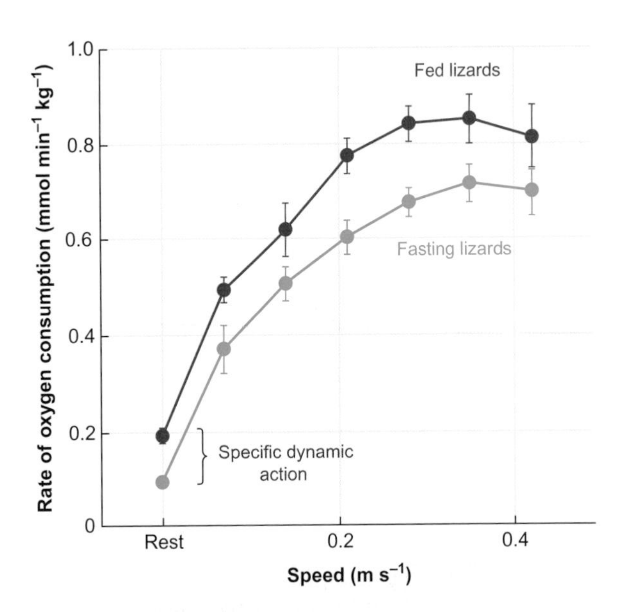

Figure 15.12 Rates of oxygen consumption at rest and during exercise in fasting and fed savannah monitor lizards (*Varanus exanthematicus*)

Values are means ± SEM from eight animals.

Adapted from Wang T et al (2005). Effects of digestion on the respiratory and cardiovascular physiology of amphibians and reptiles. In: Physiological and Ecological Adaptations to Feeding in Vertebrates. Starck JM and Wang T (eds.), pp 279-304. Enfield (NH), USA: Science Publishers, Inc.

exercise is not necessarily the maximum that the cardiorespiratory system can support. This suggests, therefore, that maximum $\dot{M}O_2$ during sustainable exercise in these species is determined by the aerobic capacity of the locomotory muscles rather than by the limits of the cardiorespiratory system.

Exercise in wild animals and migration

Many of the experiments performed to study the effects of exercise on animals are conducted on those that have been held in captivity for varying durations, sometimes for all of their lives. Although this has the clear advantage that the animals are used to being handled and are readily available, there is potentially a major disadvantage. Most species in the wild have to be prepared to become highly active, either to escape a predator, chase some prey or move from one location to another, sometimes for long distances and durations during migrations. In contrast, those held in captivity are mostly relatively inactive. As such, recently caught wild fish have greater aerobic capacity and maximal sustainable swimming speeds than fish held in captivity for long periods. In many species of non-salmonid fish, endurance training leads to a proportional increase in the mass of aerobic muscle and to an increase in its capillarity[30], although similar changes have not been reported for salmonid fish. These changes in relative mass of the oxidative muscle and in capillarity in response to training are examples of phenotypic flexibility[31].

Strangely, there is no evidence that training influences the relative size of the heart in captive fish, which only increases in direct proportion to any increase in body mass. There is, however, an increase in the activity of the aerobic enzyme, citrate synthase which is an important component of the tricarboxylic acid cycle (Kreb's cycle)[32]. The characteristics of the heart may, however, be important in determining the swimming ability of different individuals of a population, as we discuss in Experimental Panel 15.1.

While the physiological responses of wild animals to exercise are often different from those obtained from animals that have been held in captivity for long periods of time, wild animals that migrate are often in a different physiological state during their migratory period than when in their non-migratory phase. An example is terrestrial gecarcinid crabs, such as Christmas Island red crabs (*Gecarcoidea natalis*), which migrate from their inland locations to the coast to mate and spawn. Migration is stimulated by the arrival of the monsoon rains, and may occur over distances of more than 4–5 km. The breeding activities are synchronized with the lunar cycle and occur at the end of migration. Consequently, if the rains are later than usual, the crabs may have to cover the distance as rapidly as possible in order to reach their destination and still be in synchrony with the lunar cycle. This may require them to travel for up to five consecutive days. (The crabs are able to travel distances of up to 1 km per day.)

Data from laboratory experiments on red crabs caught in the wild one week before their annual migration are shown in Figure 15.13. These data demonstrate that, like other species of crustaceans (such as blue crabs, which we discuss earlier), red crabs accumulate large amounts of lactic acid in their muscles and haemolymph during sustained exercise and become acidotic (pH is reduced). What is more, they have an aerobic scope of only just over two. In contrast, the concentration of lactate[33] in migrating crabs in the field remains low, around 1–2 mmol L^{-1}, suggesting that any reduction in pH that is present during the migratory period is the result of an increase in PCO_2 in the haemolymph.

[30] Capillarity of a muscle is the number of capillaries per unit area of muscle (capillary density) and the number of capillaries per muscle fibre (capillary to fibre ratio).

[31] Phenotypic flexibility is the ability of an organism reversibly to modify its phenotype. This often occurs in response to the environment, but it can also occur during or before long term changes in the level of activity, for example, before and after migration.

[32] The enzyme citrate synthase catalyses the reaction between oxaloacetate and acetyl-CoA to produce citric acid (citrate) in the tricarboxylic acid cycle, which is discussed in Section 2.4.2.

[33] Lactic acid dissociates into protons (H^+) and lactate ions (La^-).

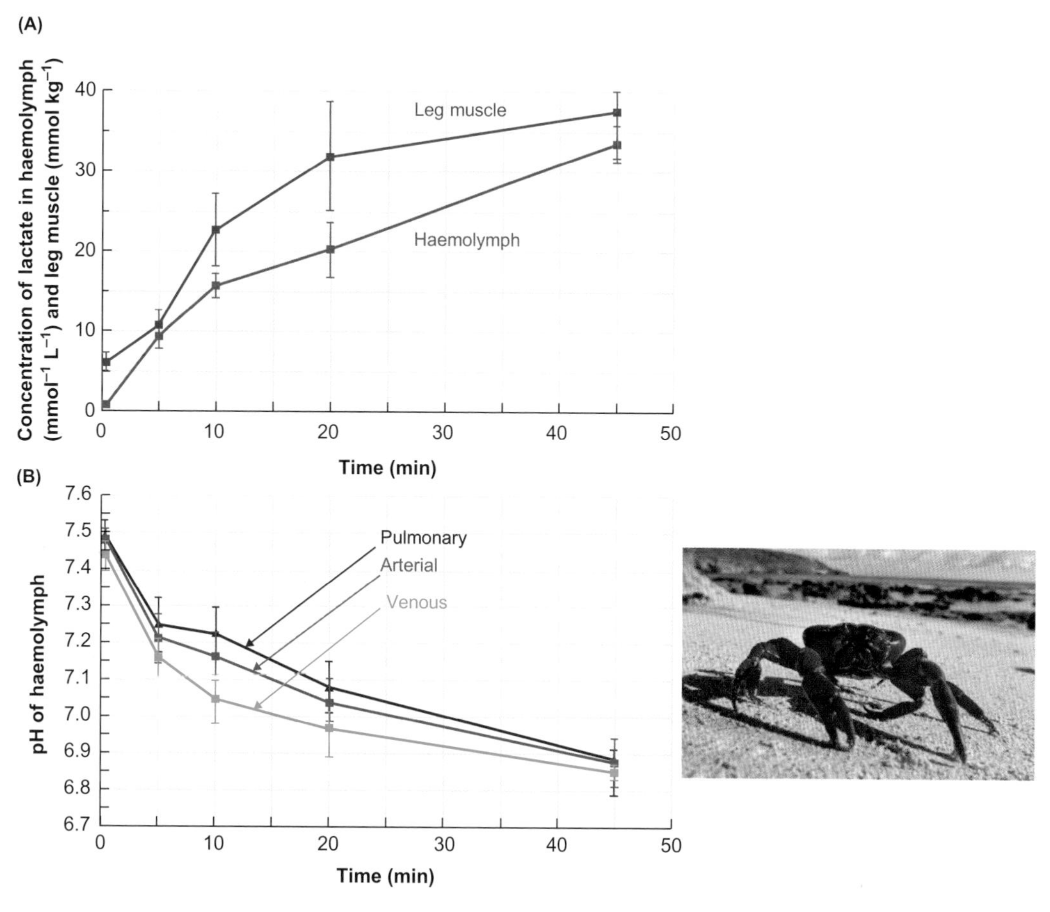

Figure 15.13 Changes in the concentration of lactate ([La⁻]) in the haemolymph and leg muscle, and pH of the haemolymph, of the Christmas Island red crab (*Gecarcoidea natalis*) during 45 min of walking at their chosen speed

(A) [La⁻] in haemolymph and leg muscle. (B) pH in haemolymph from the pulmonary vein, heart (arterial blood) and venous sinus. Crabs were caught from the wild 1 week before their annual migration. Mean values ± SEM from six animals.

Reproduced from A: Adamczewska AM and Morris S (1994). Exercise in the terrestrial Christmas crab *Gecarcoidea natalis*. II Energetics of locomotion. J. Exp. Biol. 188, 257-274. B: Adamczewska AM and Morris S (1994) Exercise in the terrestrial Christmas crab *Gecarcoidea natalis*. I Blood gas transport. Journal of Experimental Biology 188: 235-256. Image: © Kirsty Faulkner.

Thus, there is no reliance on anaerobic metabolism during migration in this species and the data suggest that the physiological state of migrating red crabs is different from that of non-migratory crabs kept in the laboratory for a short period.

15.2.3 Sustained exercise in birds and mammals

The data in Table 15.3 show that birds and mammals have much greater maximum rates of oxygen consumption (aerobic capacities) than ectotherms, so they need respiratory and cardiovascular systems which are able to meet these higher demands. Unlike most mammals, however, birds have two different locomotor systems, their legs for walking/running/swimming and their wings for flying[34]. In those migratory species of birds that can fly for relatively long periods, maximum rate of oxygen consumption ($\dot{M}O_2$ max) and heart rate when flying are greater than those when running, as shown in Figure 15.14A for barnacle geese (*Branta leucopsis*). This is not only because of the larger mass of the flight muscles compared to that of the leg muscles, but also because of the relatively higher activity of aerobic enzymes, such as citrate synthase, in the flight muscles of many species of birds, including barnacle geese.

The response of the respiratory system to exercise may differ not only between the two modes of locomotion, but also among species. In most species of birds, the increase in ventilation volume during flapping flight matches the increase in $\dot{M}O_2$ so that the amount of oxygen extracted from the respired air does not change, as shown in Table 15.3. In contrast, some species such as common pigeons (*Columba livia*) hyperventilate during flight, and oxygen extraction from the respired air decreases.

As far as the cardiovascular system is concerned, the data in Table 15.3 show that a higher heart rate and greater

[34] Bats can use all four limbs for both moving on land and flying, but most species are very awkward on land and take to flight as soon as possible.

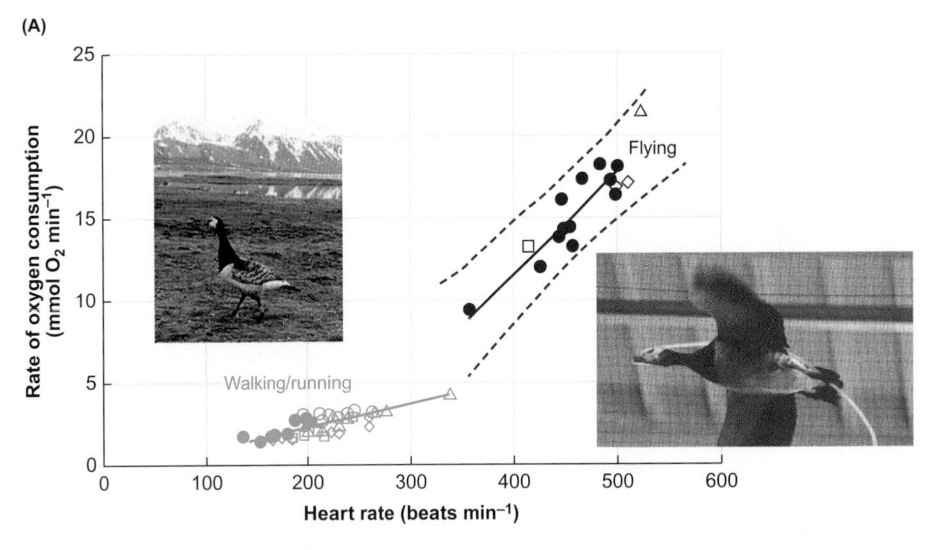

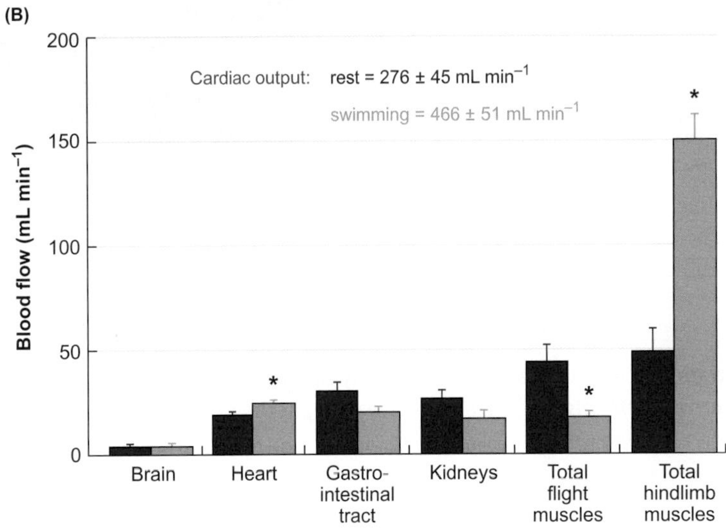

Figure 15.14 Metabolic and cardiovascular responses of birds to exercise

(A) Relationships between heart rate and rate of oxygen consumption in barnacle geese (*Branta leucopsis*) when flying or walking. Filled circles indicate data from a single bird which performed 12 flights and the open symbols indicate data from three other birds during flight and from four other birds when walking/running. The regression lines are for all of the walking data but only for the single flying bird. The dashed lines show the 95% prediction interval for the flying data from the single bird. (B) Blood flow to some of the major vascular beds of six tufted ducks (*Aythya fuligula*) at rest (dark blue bars) and while swimming at an average speed of 0.7 m s^{-1} (orange bars). Values are mean ± SEM and asterisks indicate significant differences between resting and swimming values.

Reproduced from A: Ward S et al (2002). Heart rate and the rate of oxygen consumption of flying and walking barnacle geese (*Branta leucopsis*) and bar-headed geese (*Anser indicus*). JJournal of Experimental Biology 205: 3347-3356. B: Butler PJ et al (1988). Regional distribution of blood flow during swimming in the tufted duck (*Aythya fuligula*). Journal of Experimental Biology 135: 461-472.

extraction of oxygen from the circulating blood make the major contributions to the additional oxygen that is required during flight compared to that required during running/swimming. But how is cardiac output distributed during exercise? Figure 15.14B shows that swimming tufted ducks (*Aythya fuligula*) exhibit increases in blood flow to the heart and leg muscles, but no significant change in blood flow to the brain, gastrointestinal tract and kidneys. There is, however, a decrease in flow to the flight muscles. Similar data have been obtained from exercising dogs with a decrease in blood flow to the non-active muscles. On the basis of the data from swimming ducks, it is assumed that there are similar changes during flapping flight, with blood being directed away from the leg muscles to the flight muscles.

Training of captive birds and migration of wild birds

Training of captive tufted ducks leads to increases in $\dot{M}O_2$ max and in maximum swimming speed, although maximum

Table 15.3 Ventilatory and circulatory responses to endurance exercise in birds and mammals

The numbers in parentheses give the factorial increase in that variable during exercise compared to the resting value and indicate to what extent the different components of the respiratory and circulatory systems (shown in equations 15.1 and 15.2) contribute to the increase in rate of oxygen consumption.

Species	Mass (kg)	Activity	Temperature (°C)	Rate of O_2 consumption (mmol min^{-1} kg^{-1})	Respiratory frequency (min^{-1})	Ventilation volume (mL min^{-1} kg^{-1})	O_2 extracted from air (mmol L^{-1})	Heart rate (beats min^{-1})	Cardiac output (mL min^{-1} kg^{-1})	O_2 extracted from blood (mmol L^{-1})	PO_2 in arterial blood (kPa)
Aythya fuligula (tufted duck)	0.6	Rest	18 (water)	0.9	13.5	445	2.6	149	341		11.0
		Swimming at 0.8 m s^{-1}		2.43 (× **2.7**)	36.0 (× **2.67**)	1520 (× **3.4**)	2.03 (× **0.78**)	257 (× **1.7**)	595 (× **1.7**)	(× **1.6**)	11.0
Columba livia (common pigeon)	0.34	Rest	11 (air)	0.8				110	302	2.63	14.8
		Flying at 18.4 m s^{-1}		13.8 (× **17.2**)				663 (× **6.0**)	2244 (× **7.43**)	6.16 (× **2.34**)	14.1
Corvus ossifragus (fish crow)	0.28	Rest	20	1.38	27.3	0.81 L min^{-1} kg^{-1}	1.78				
		Flying at 10 m s^{-1}		11.8 (× **8.6**)	120 (× **4.4**)	6.51 L min^{-1} kg^{-1} (× **8.0**)	1.78				
Capra hircus (goat)	30	Rest		0.37				86	186	1.96	14.0
		Running at 3.7 m s^{-1} on the flat		2.54 (× **6.9**)				268 (× **3.1**)	556 (× **3.0**)	4.60 (× **2.35**)	16.4
Canis lupus familiaris (domestic dog)	13.6	Rest	18	0.39	13	297	1.66	83	143	2.77	12.5
		Running at 2.2 m s^{-1} and incline of 6°		2.1 (× **5.4**)	46.2 (× **3.55**)	1563 (× **5.26**)	1.68	188 (× **2.26**)	378 (× **2.64**)	5.52 (× **1.99**)	13.4

PO_2, partial pressure of oxygen

References: Bernstein, 1976; Bevan and Butler, 1992; Bouverot et al., 1981; Peters et al. 2005; Taylor et al., 1987; Woakes and Butler, 1986. Full citations available online.

15

heart rate does not change. This means that trained ducks at $\dot{M}O_2$ max must have a greater cardiac stroke volume and/or level of extraction of oxygen from the blood perfusing the locomotor muscles than untrained birds. The major change in the cardiovascular system of trained ducks is an increase in the capillary to fibre ratio in the locomotor muscles. There is also an increase in citrate synthase activity in these muscles.

So, how close can we get by training captive animals to the level of fitness present in wild animals?

It has only been possible to begin to answer this question with the advent of miniature electronic storage devices that monitor and store physiological data such as heart rate from wild animals. Figure 15.14A illustrates that there is a linear increase in $\dot{M}O_2$ associated with an increase in heart rate during exercise. If this relationship is rigorously calibrated, heart rate can be used to give an estimate of actual $\dot{M}O_2$ of wild animals in the field[35]. However, there is currently no evidence that training of captive birds causes changes in locomotor muscles and in the cardiovascular system that match those present in wild birds, particularly those that exist just before migration.

Normal everyday activity of wild animals may be sufficient to maintain an appropriate level of fitness for survival. Figure 15.15A indicates that there are characteristic increases in heart rate when birds fly. On the basis of such changes in heart rate, there is no evidence that wild barnacle geese 'train' at their breeding grounds in the high Arctic by increasing their time spent flying before they set off on their autumn migration to southern Scotland.

Studies of wild and captive barnacle geese have demonstrated that, for a given $\dot{M}O_2$, heart rate is substantially lower in wild birds compared to that in captive birds. For example, heart rate is 50 beats min^{-1} in wild birds resting at night and 75 beats min^{-1} in captive birds, whereas there is no difference in $\dot{M}O_2$ between the two groups. This lower heart rate in wild geese for the same $\dot{M}O_2$ is possible because of their 30 per cent larger heart (and hence cardiac stroke volume) compared to that in captive geese, despite the captive geese having been trained by regularly running them on a treadmill.

The relative mass of the pectoral (flight) muscles in post-hatch wild barnacle geese is no different from that of captive geese. There is also no difference in the mass-specific activity of citrate synthase (CS) in the flight muscles between the two groups up to the time they become flighted at about 7 weeks of age. However, beyond this time, there is a greater activity of CS (and, it is assumed, a greater capillary density) in the flight muscles of the wild birds. It seems, therefore, that the locomotor muscles of wild animals have a greater aerobic capacity than those that have been held in captivity, and associated adjustments in the cardiovascular system of wild geese enable a greater amount of oxygen to be delivered to their flight muscles. These differences may be related entirely to the greater level of activity in the wild birds.

Despite their increased level of physical fitness before migration, birds may use behavioural mechanisms, such as flying in V-formation or using up-currents of wind, to keep the burden on their cardiorespiratory system as low as possible during migration.

Data from satellite transmitters attached to the backs of wild barnacle geese migrating from their Arctic breeding grounds are shown in Figure 15.15B. These data indicate that the birds fly along the Norwegian coast for a large part of their migration. Figure 15.15A shows that during this phase, heart rate during flight falls progressively, indicating a fall in $\dot{M}O_2$. This decline in heart rate during migration could at least in part be the result of the fact that the birds lose mass as they use the stored fat during migration. However, heart rate then increases as the birds leave the coast of Norway and fly across the North Sea, as also shown in Figure 15.15A.

A possible explanation for this is that the birds obtain some uplift from rising air deflected upward by the Norwegian coast, thereby reducing the cost of flight and hence, heart rate. When the geese fly across the North Sea, however, this assistance no longer exists and heart rate increases. Evidence to support this possibility has been obtained from barnacle geese flying in a wind tunnel. If a goose moves progressively closer to the trainer in the wind-tunnel, there are associated reductions in both heart rate and $\dot{M}O_2$, presumably as the bird benefits from updrafts created by the trainer.

Matching oxygen supply to demand during sustained exercise

The transport of oxygen from the environment to the metabolizing cells relies entirely on the cardiorespiratory systems. A question that has often been asked about this pathway from the environment to the mitochondria in the cells, particularly in mammals, is: Is there a step at any stage which could set the limit for maximum rate of oxygen transfer—a limiting step? Alternatively, is each step matched to the maximum demand, i.e. without imposing a limit, and without having excess capacity, as suggested by the concept of **symmorphosis**[36]?

[35] The use of heart rate to estimate rate of oxygen consumption in wild animals is discussed in Section 2.2.6.

[36] The term symmorphosis was coined from two Greek words meaning 'balanced measures'. It is used to describe the concept that each step along the pathway transporting oxygen from the environment to the mitochondria of active tissues (usually muscles) is matched in its capacity to meet the maximum demands of those tissues.

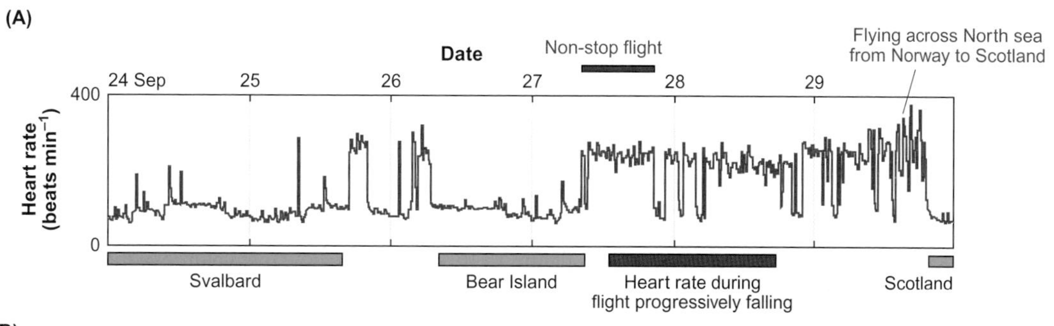

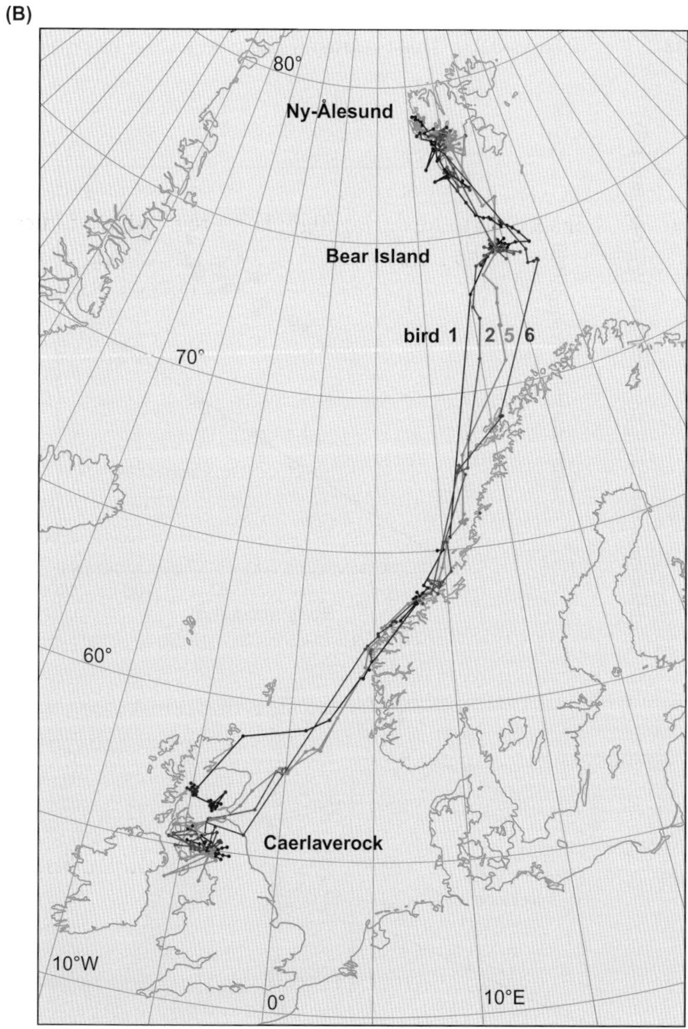

Figure 15.15 Behaviour and physiology of barnacle geese (*Branta leucopsis*) during their autumn migration

(A) Heart rate of a barnacle goose during its migration from Svalbard to Scotland. High heart rate indicates when the bird was flying. (B) Routes taken by four barnacle geese migrating from their breeding grounds at Ny-Ålesund, Svalbard, to their wintering grounds at Caerlaverock, Scotland.

Reproduced from Butler PJ et al (1998). Behaviour and physiology of Svalbard barnacle geese *Branta leucopsis* during their autumn migration. Journal of Avian Biology 29: 536-545.

We consider two approaches that have been used to answer this question by Ewald Weibel at Berne University and his colleagues:

- comparing exercise-induced maximum rate of oxygen consumption ($\dot{M}O_2$ max or $\dot{V}O_2$ max) of a range of species of mammals of different body masses, with various structural components of the cardiorespiratory system;
- comparing animals of similar body mass but with different aerobic capacities, such as Arctic foxes (*Vulpes lagopus*) and agoutis (*Cuniculus paca*), domestic dogs (*Canis lupus familiaris*) and goats (*Capra hircus*)[37], ponies (*Equus ferus caballus*) and domestic calves (*Bos taurus*), horses and steers.

Although the following discussion is focused on mammals, the general principles apply to all species with respiratory and circulatory systems, and even to the respiratory system of insects.

The allometric relationship between $\dot{V}O_2$ max and body mass for a number of species of mammals has an exponent of

[37] Data for dogs and goats are given in Table 15.3.

0.86[38], as shown in Figure 15.16A, and the maximum demand for oxygen is determined by the activity of the oxidative enzymes in the mitochondria of the active organs. It is not surprising, therefore, that Figure 15.16A also shows that the mitochondrial volume in the locomotor muscles has a similar slope to its relationship with body mass (0.87) as $\dot{V}O_2$ max. What is more, the corresponding values for the four more athletic species used in the analysis are systematically higher for both variables.

Let us now look at how the circulatory system meets the increased demand for oxygen during exercise. At first sight, Figure 15.16B appears to show that volume of the capillaries supplying blood (and therefore oxygen) close to the mitochondria has a similar allometric relationship as $\dot{V}O_2$ max to body mass (0.86 vs 0.89). This is certainly the case for the more sedentary

[38] Allometry and exponents are discussed in Section 2.2.4 and Box 2.4. An exponent of 1 means that the variable changes in direct proportion to body mass (body mass raised to the power of 1), for example cardiac stroke volume in Figure 15.16C. An exponent <1 means that the proportional change in the variable is less than that for body mass (body mass raised to a power <1), for example maximum rate of oxygen consumption in Figure 15.16. An exponent >1 indicates that the proportional change in the variable is greater than that for body mass (body mass raised to a power >1), such as diffusion conductance in Figure 15.18.

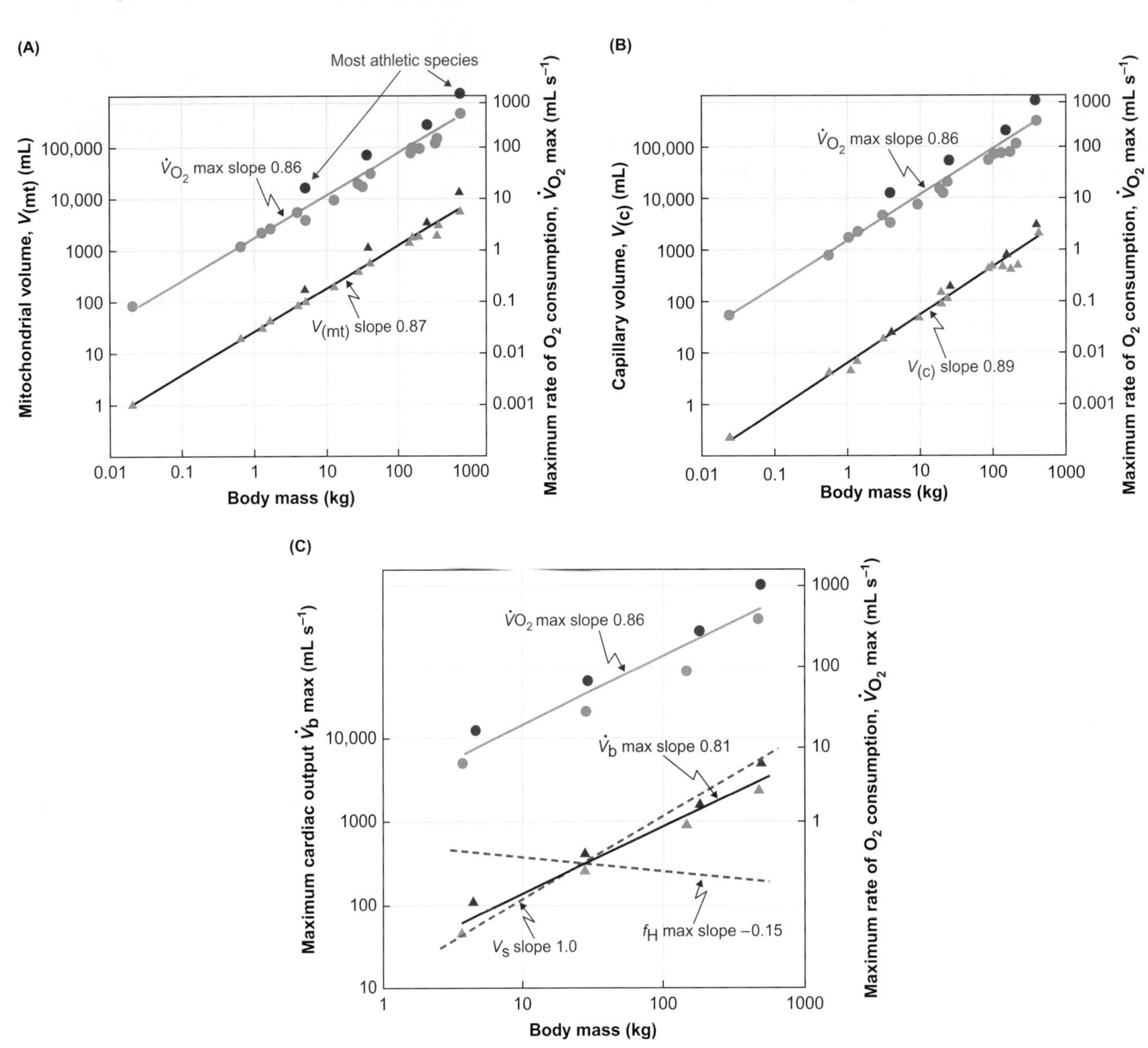

Figure 15.16 Allometric plots of maximum rate of oxygen consumption ($\dot{V}O_2$max) and factors affecting the demand for oxygen and its supply by the cardiovascular system of mammals

(A) It is the mitochondria in cells that consume oxygen and their volume, $V_{(mt)}$, has the same relationship with body mass as $\dot{V}O_2$ max (●, ▲, more sedentary species; ●, ▲, more athletic species). (B) However, in more active species capillary volume, V(c), does not match their proportionately greater $\dot{V}O_2$ max. (C) Maximum cardiac output, $\dot{V}O_2$ max, also has the same relationship with body mass as $\dot{V}_b$ max. However, maximum heart rate ($f_H$max) is lower in larger species and is not affected by their degree of athleticism, whereas cardiac stroke volume (V_s) increases in direct proportion to body mass in the sedentary species, which means that, compared with $\dot{V}O_2$ max, it is relatively greater in the larger species. The relatively greater V_s in larger species compensates for their lower maximum heart rate. 22.4 mL O_2 = 1 mmol O_2 Adapted from Weibel ER (2000). Symmorphosis On Form and Function in Shaping Life. Cambridge, MA: Harvard University Press.

species that were studied. However, closer inspection of Figure 15.16B indicates that capillary volume of the more athletic species does not match their proportionately greater $\dot{V}O_2$ max. The capillary volume is only 1.7 times greater in the more athletic dogs and horses compared with that in the more sedentary goats and steers, but the $\dot{V}O_2$ max is about 2.5 times greater. How, then, is sufficient oxygen delivered to the mitochondria? To answer this, we have to look beyond the structural features of the cardiovascular system, as rate of oxygen delivery depends not only on the volume of arterial blood that can be delivered per unit time, but also on the concentration of oxygen in that blood, as shown in equation 15.2.

The concentration of oxygen in the arterial blood is related to the percentage concentration of red cells in the blood (the haematocrit). Haematocrit (Hct) tends to increase during exercise in some species of vertebrates, as a result of the release of red blood cells from the spleen. In addition, at $\dot{V}O_2$ max, Hct is approximately 50 per cent higher in dogs and ponies, than in goats and calves. Thus, it is the product of capillary volume and Hct, the **capillary red cell volume**, that has a similar allometric relationship with body mass as $\dot{V}O_2$ max.

Maximum perfusion of the capillaries with blood depends on maximum cardiac output, which also has a similar allometric relationship with body mass as $\dot{V}O_2$ max, as shown in Figure 15.16C. However, Figure 15.16C also shows that the components of cardiac output, namely heart rate and cardiac stroke volume, do not have the same allometric relationships with body mass as $\dot{V}O_2$ max. Maximum heart rate is lower in larger species, and has an allometric slope of −0.15, whereas maximum cardiac stroke volume in sedentary species is directly (isometrically) related to body mass: it has a slope of 1.

Maximum cardiac output is higher in the more athletic species like dogs and ponies, as shown in Figure 15.16C. This is due entirely to cardiac stroke volume (which is directly related to heart size) being approximately 50 per cent greater than that of more sedentary species. Maximum heart rate is similar in the two groups.

In summary, when we compare a range of species of different body mass, capillary volume of sedentary species and maximum cardiac output are well matched to $\dot{V}O_2$ max. However, cardiac stroke volume and, therefore, heart mass increase isometrically with increasing body mass, so this increase occurs to a greater extent than $\dot{V}O_2$ max. This apparent discrepancy is compensated for by the fact that maximum heart rate decreases with increasing body mass.

The increase in cardiac output that accompanies exercise, means that the rate at which blood flows through the gas exchanger, be it gills or lungs, is greater during exercise than it is at rest. This means that a bolus of blood spends less time in gas exchange regions of the gills or lungs during exercise than it does when the animal is at rest. Consequently, it has less time to equilibrate with the water between the lamellae of the gills or air in the exchange regions of the lungs. The time a bolus of blood spends in the gas-exchange regions of capillaries of the gills or lungs is known as the **capillary transit time**. Figure 15.17 shows a diagrammatic representation of the alveoli and associated capillary system in a mammalian lung (a single unit of alveolus and capillary). The shorter capillary transit time during exercise requires a greater proportion of the capillary length for equilibration of oxygen in the blood with that in the air, or water in aquatic animals.

The next question is: How well is the mammalian respiratory system matched to $\dot{V}O_2$ max? Using a similar approach to that just described for the cardiovascular system, the main conclusions are:

- Diffusion conductance (G_{diff}) across the lung[39] (sometimes called lung diffusing capacity, D_L) and $\dot{V}O_2$ max have different relationships with body mass, with G_{diff} increasing more steeply than $\dot{V}O_2$ max and having

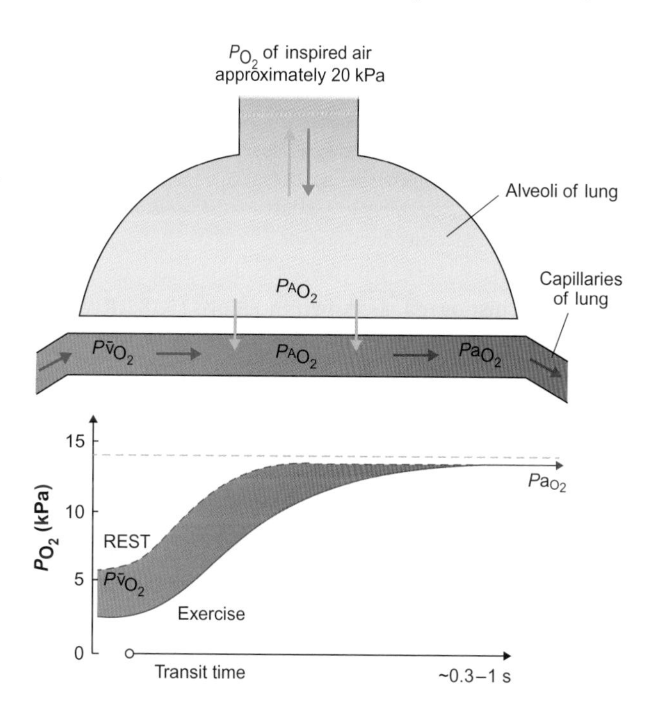

Figure 15.17 Change in the partial pressure of oxygen (PO_2) in the blood passing through the capillaries of the lung of resting and exercising mammals

Capillary transit time when animal at rest is approximately 1 s, but when the animal is exercising, it is only about 0.3 s. Note that the oxygen in the blood equilibrates earlier in its transit through a capillary in resting animals. P_AO_2 is PO_2 in alveolar gas, $P_{\bar{v}O_2}$ is PO_2 in mixed venous blood and P_aO_2 is PO_2 in arterial blood.
Modified from Weibel (2000) Symmorphosis On Form and Function in Shaping Life. Cambridge, MA: Harvard University Press; and Longworth KE et al (1989). High rates of O_2 consumption in exercising foxes: large PO_2 difference drives diffusion across the lung. Respiration Physiology 77: 263-276.

[39] Diffusion conductance (G_{diff}) is discussed in Section 12.2.2, Equation 12.1, and is the rate at which oxygen diffuses across the barrier between the alveoli of the lungs (A) to the blood in the pulmonary capillaries (c) per unit difference in PO_2 between these two locations ($P_AO_2 - P_{\bar{c}}O_2$, or PO_2, where $P_{\bar{c}}O_2$ is mean PO_2 in capillary blood). As the rate at which oxygen diffuses across the lungs is equivalent to the rate of oxygen consumption ($\dot{V}O_2$), $G_{diff} = \dot{V}O_2/\Delta PO_2$.

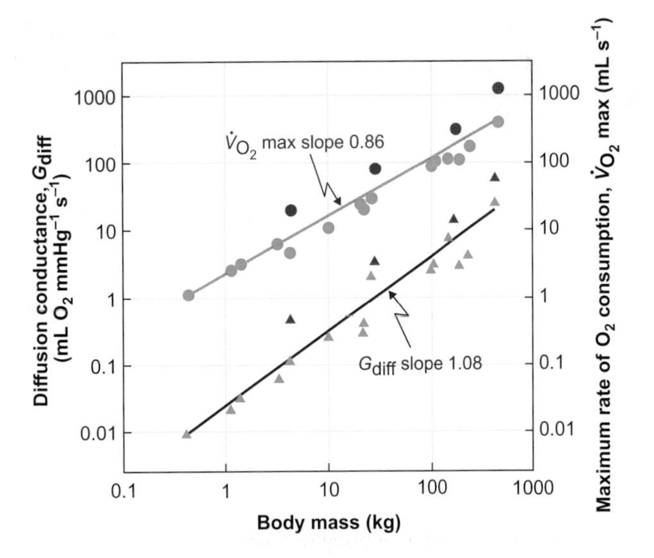

Figure 15.18 Allometric plot of maximum rate of oxygen consumption and diffusion conductance

Diffusion conductance (G_{diff}) has almost a direct relationship with body mass (it has a slope of approximately 1), whereas rate of oxygen consumption ($\dot{V}O_2$ max) has an allometric relationship (a slope of 0.86). ●, ▲, more sedentary species; ●, ▲, more athletic species). 22.4 mL O_2 = 1 mmol O_2.

Adapted from Weibel ER (2000). Symmorphosis On Form and Function in Shaping Life. Cambridge, MA: Harvard University Press.

a slope just over 1, as shown in Figure 15.18. (Recall that $\dot{V}O_2$ max has a slope of approximately 0.86.) This means that a 30 g mouse has a mass-specific G_{diff} that is about three times that of a 500 kg cow, but has a mass-specific $\dot{V}O_2$ max that is about eight times that of the cow. So, if G_{diff} is relatively low in smaller animals, we might expect the difference in PO_2 between the air in the alveoli and the blood in the pulmonary capillaries (ΔPO_2) to be greater in order to meet the greater mass-specific $\dot{V}O_2$ max.

- The general principle of there being a greater ΔPO_2 with greater mass-specific $\dot{V}O_2$ max is supported by data from a range of mammals with different mass-specific $\dot{V}O_2$ max.

The fact that G_{diff} scales almost isometrically with body mass (that is, it has a slope of just over 1) and yet $\dot{V}O_2$ max scales allometrically with a slope of approximately 0.86, means that G_{diff} for larger species of mammals may be greater than that required to meet $\dot{V}O_2$ max; in other words, there may be excess capacity for diffusion. Having excess G_{diff} appears to contradict the concept of symmorphosis. This certainly seems to be the case for more sedentary species such as goats and humans, but not for the more athletic species such as dogs, with the notable exception of pronghorn antelopes (*Antilocapra americana*), which are the most athletic of all mammals that have been studied.

Pronghorn antelopes can run at a speed of 60 km h^{-1} (16.7 m s^{-1}) for up to 40 min and have a $\dot{V}O_2$ max twice that of a similar-sized dog. However, unlike athletic dogs, pronghorn antelopes have lungs with excess diffusion conductance, at least under the conditions that exist at sea level. The natural habitat of the pronghorn antelope is in the high plains of the Rocky Mountains at altitudes between 2000 and 3000 m above sea level where partial pressure of inspired oxygen is approximately 70 per cent of that at sea level. It is possible, therefore, that pronghorn antelopes have an apparent excess G_{diff} in their lungs in order to cope with heavy exercise under the hypoxic conditions of high altitude. Similarly, burrowing Middle East blind mole rats (*Spalax ehrenbergi*) have about a 50 per cent excess lung diffusing capacity which is most likely related to the hypoxic conditions they experience in their burrows[40].

Although the heart and capillaries in the skeletal muscle can grow in response to training[41], it is almost impossible for adult mammals to increase the surface area of the alveoli in their lungs, so excess G_{diff} in the lungs of some species enables them to adjust other components of the cardiorespiratory system to meet an increase in demand and/or a reduction in oxygen supply. For example, goats do not increase their $\dot{V}O_2$ max when breathing 25 per cent oxygen, but can maintain their $\dot{V}O_2$ max when breathing a gas mixture containing only 15 per cent oxygen. They achieve this by increasing their cardiac output by 20 per cent above that at $\dot{V}O_2$ max when breathing normal air. In this species, $\dot{V}O_2$ max is set by the oxidative capacity of the mitochondria and not by the cardiorespiratory system, which appears to have excess capacity.

Artificial selection can distort the natural balance between the demand for oxygen and the means of provision of that demand. No animals exemplify this more than thoroughbred racehorses, which have been selectively bred for several thousand years for their racing performances. As a result, there seem to be limits within the oxygen transport system in these animals. This has been demonstrated by running the animals in 25 per cent oxygen, which leads to a 11 per cent increase in $\dot{V}O_2$ max, indicating that the mitochondria are not setting $\dot{V}O_2$ max when the horses are exercising in normal air. In Case Study 15.1, we discuss the effect of artificial selection on the oxygen transport system of thoroughbred racehorses.

Oxygen deficit and excess post-exercise oxygen consumption (EPOC)

Despite immediate increases in ventilation volume and cardiac output at the onset of exercise in most animals, there is invariably a lag before the rate of oxygen consumption ($\dot{M}O_2$) increases. The difference between the rate at which energy is required by an animal and the rate at which it is provided by the uptake of oxygen before attainment of the steady-state rate, is known as the **oxygen deficit**. This is shown for humans inFigure 15.19A.

[40] We discuss mole rats further in Section 15.3.2.

[41] The growth of capillaries (angiogenesis) in response to exercise is discussed in Section 14.1.5 and Experimental Panel 14.1.

Case Study 15.1 Artificial selection and the thoroughbred racehorse

All modern thoroughbred horses (*Equus ferus caballus*) can trace their ancestry to a small number of stallions imported to England from the Middle East at the turn of the 17th century. The three most important were named after their English owners: the Darley Arabian, the Godolphin Arabian and the Byerley Turk. There were a larger number of foundation mares consisting predominantly of British and Irish horses with fewer from the Middle East and western Asia. The stallions were mated with the mares to produce a breed that could run at high speed over extended distances while carrying a rider. Such horses were ideal for the rapidly developing sport of the aristocracy, horse racing. The best stallions were mated with the best mares, based on their performance on the race track. During the 18th and 19th centuries, thoroughbreds spread throughout the world.

Thoroughbreds were selected mainly for racing and can reach speeds of around 65 km h^{-1} (18 m s^{-1}). For their body mass, thoroughbred horses have an extraordinarily high maximum rate of oxygen consumption ($\dot{M}O_2$ max), as shown in Figure A. Note in this figure that thoroughbred horses and greyhounds (who have also been selectively bred for speed) have much higher values of $\dot{M}O_2$ max than non-athletic horses and dogs of similar body mass.

The exceptional speed of thoroughbred horses relates to their relatively high skeletal muscle mass. Between 52–55 per cent of body mass is skeletal muscle in thoroughbreds compared with about 30–40 per cent in non-athletic horses and most other species of mammals. However, mitochondrial and capillary densities in the muscles of thoroughbreds are not different from those in the muscles of non-athletic species such as calves and goats. It is, therefore, the total mass of skeletal muscle that creates the demand for such a high $\dot{M}O_2$ max. During the process of their selective breeding, the cardiovascular system of thoroughbreds has been able to meet the high demand for oxygen by the skeletal muscles, whereas the constraints placed on the respiratory system has led to severe hypoxaemia during exercise.

Heart mass is the main determinant of cardiac output and hence $\dot{M}O_2$ max. Heart mass of thoroughbreds is more than 1 per cent of body mass, and can be as high as 2 per cent in exceptional individuals, compared to about 0.6 per cent of body mass in human athletes. The net result is that cardiac output during maximum exercise is around 10-fold

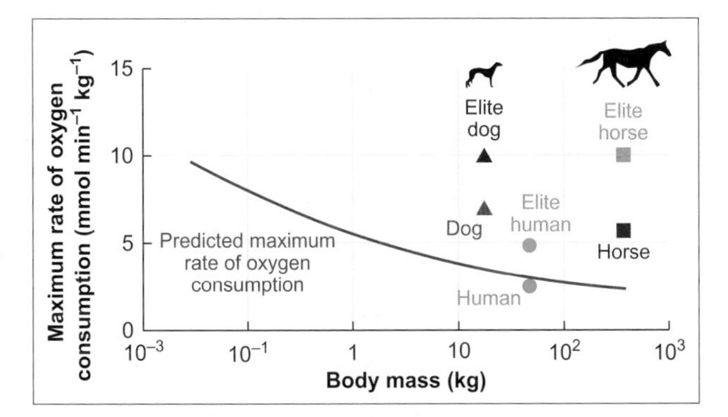

Figure A Mass-specific maximum rate of oxygen consumption in a range of mammals of different body masses

Solid line represents predicted values for mammals. Non-athletic dogs and horses have higher maximum rates of oxygen consumption than predicted, but elite dogs and horses, which have been selectively bred for racing, have higher maximum rates of oxygen consumption than non-athletic dogs and horses of similar body mass.

Modified from Poole DC and Erickson HH (2011). Highly athletic terrestrial mammals: horses and dogs, Comprehensive Physiology 1: 1-37.

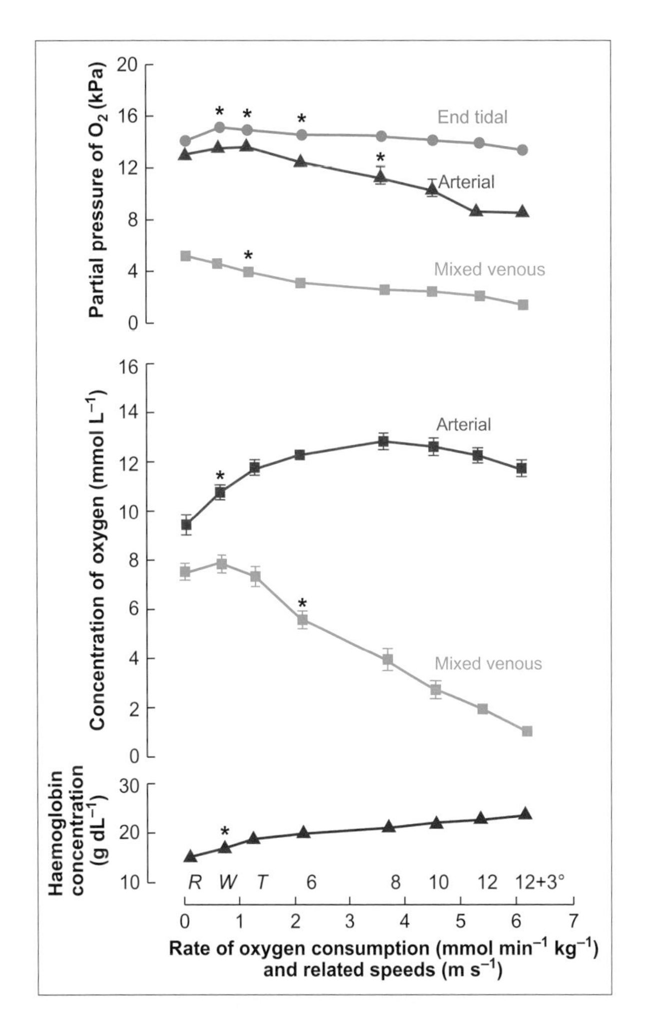

Figure B Blood gases and haemoglobin concentration in thoroughbred racehorses at increasing levels of exercise on a treadmill

Partial pressure of oxygen (PO_2) in the lung (end tidal) is maintained at all levels of exercise, but PO_2 of arterial blood falls from a speed of about 8 m s^{-1}. However, because haemoglobin concentration increases as a result of red blood cells being released by the spleen, the concentration of oxygen in arterial blood actually increases as soon as the horse begins to walk (W) and remains above the resting (R) value at all levels of exercise. * Indicates when a variable first becomes significantly different from the value at rest. In the case of end tidal PO_2, the asterisks indicate those values that are significantly different from that at rest. T, trot; 12+3°, speed of 12 m s^{-1} with the treadmill at an incline of 3°. Values are mean ± SEM from seven horses.

Modified from Butler PJ et al (1993). Respiratory and cardiovascular adjustments during exercise of increasing intensity and during recovery in thoroughbred racehorses. Journal of Experimental Biology. 179: 159-180.

greater than that at rest. When at rest, about 15 per cent of cardiac output perfuses the skeletal muscles, whereas during maximum exercise more than 80 per cent of the greatly increased cardiac output perfuses the exercising muscles. Despite these cardiovascular adaptations, a characteristic feature of racehorses when exercising maximally is severe hypoxaemia, with PO_2 in arterial blood falling from 13 kPa at rest to approximately 9 kPa.

The hypoxaemia is illustrated in Figure B; it suggests that there may be diffusion limitation[1] in the lungs. Under-ventilation (hypoventilation) of the lungs at maximum exercise may contribute to the hypoxaemia, although some studies do not support this. Data from one such study are shown in Figure C. An increase in $\dot{M}O_2$ max of 25.7 times resting is matched by an increase in ventilation of 26.6 times resting, which suggests there is no overall hypoventilation.

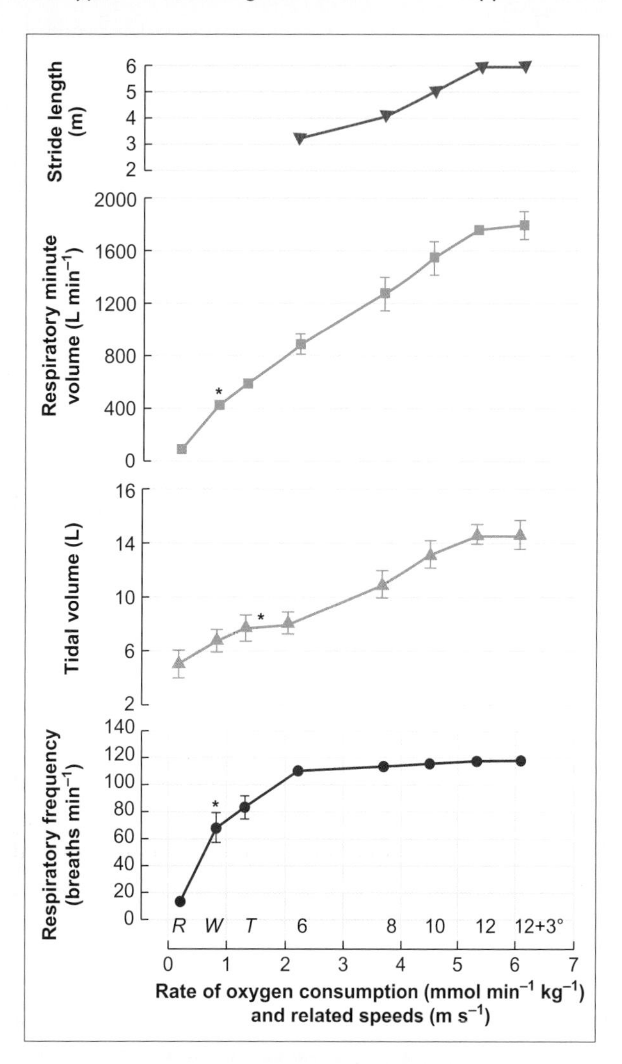

Figure C Response of the respiratory system of thoroughbred racehorses to increasing exercise intensity

Once the horses begin to canter at a speed of about 6 m s^{-1}, there is a 1:1 relationship between respiratory frequency and limb frequency and neither variable increases further as the horses gallop faster. This means that further increases in respiratory minute volume and speed are the result of increases in respiratory tidal volume and stride length, respectively. * Indicates when a variable first becomes significantly different from the value at rest (R). W, walk; T, trot; 12+3°, speed of 12 m s^{-1} with the treadmill at an incline of 3°. Mean values ± SEM from seven horses.

Reproduced from: Butler PJ et al (1993). Respiratory and cardiovascular adjustments during exercise of increasing intensity and during recovery in thoroughbred racehorses. Journal of Experimental Biology 179: 159-180.

Whatever the cause of the hypoxaemia, it is more than compensated for by the release of red blood cells (RBCs) from the spleen. This release of RBCs causes an exceptionally large increase in haematocrit (Hct), from 40 to over 60 per cent, and thus in the haemoglobin concentration of the blood, as indicated in Figure B. Such an increase in red blood cells and haemoglobin concentration is known as **polycythaemia**.

Polycythaemia leads to the concentration of oxygen in arterial blood being higher than the resting value at all levels of exercise. The difference between the concentration of oxygen in arterial and mixed venous blood increases from 1.9 mmol L^{-1} at rest to 10.6 mmol L^{-1} when the horses are running at a speed of 12 m s^{-1} on a 3° incline (12 m s^{-1} +3°). The concentration of oxygen in

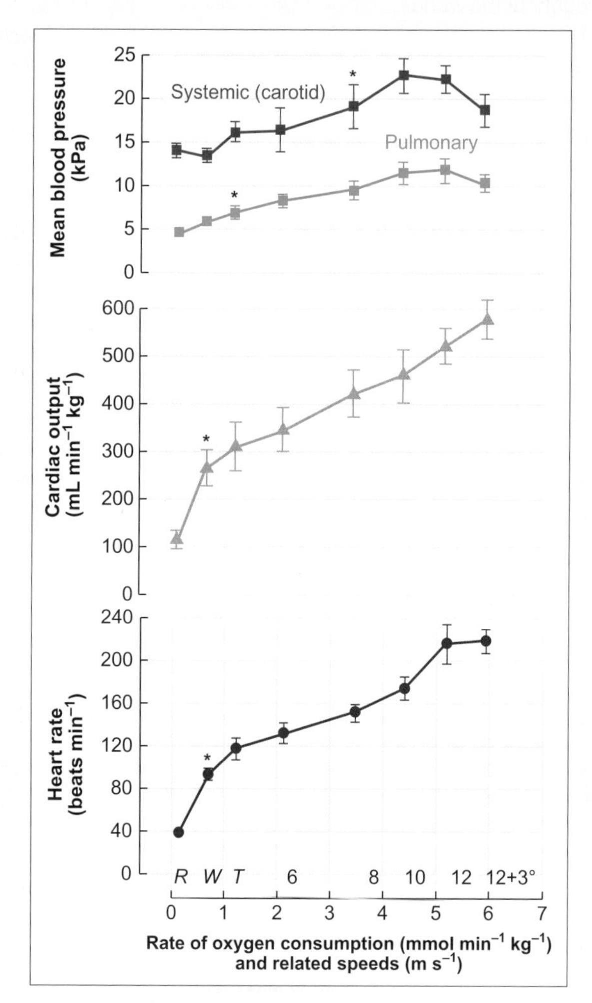

Figure D Cardiovascular changes in thoroughbred racehorses at increasing levels of exercise on a treadmill

A large increase in cardiac output is accompanied by increases in both systemic and pulmonary blood pressures. * Indicates when a variable first becomes significantly different from the value at rest (R). W, walk; T, trot; 12+3°, speed of 12 m s^{-1} with the treadmill at an incline of 3°. Values are mean ± SEM from seven horses.

Modified from Butler PJ et al (1993). Respiratory and cardiovascular adjustments during exercise of increasing intensity and during recovery in thoroughbred racehorses. Journal of Experimental Biology 179: 159-180.

arterial blood at 12 m s^{-1} +3° is 14 per cent greater than that in resting animals. Despite the large increase in Hct, the high shear rate[2] during exercise may mean that the apparent viscosity of the blood is little higher than that at rest, when haematocrit is much lower.

The large increase in cardiac output at high levels of exercise, as illustrated in Figure D, leads to high systemic and pulmonary blood pressures. The pressures in the capillaries of the lungs are often so large that they cause the vessels to rupture and for blood to appear in the lungs. This bleeding is known as exercise-induced pulmonary haemorrhage. Pulmonary haemorrhage is rare in other species of mammals when exercising maximally. It would appear that the artificial selection process has resulted in the oxygen demands of the locomotor muscles exceeding the ability of the respiratory system to meet those demands. Consequently, the oxidative capacity of the skeletal muscle strains the cardiorespiratory system to its limits, despite the plasticity of the circulatory system.

› Find out more

Bower MA, Campana MG, Whitten M, Edwards CJ, Jones H, Barrett E, Cassidy R, Nisbet RER, Hill EW, Howe CJ, Binns M (2011). The cosmopolitan maternal heritage of the Thoroughbred racehorse breed shows a significant contribution from British and Irish native mares. Biological Letters 7: 316–320.

Butler PJ, Woakes AJ, Smale K, Roberts CA, Hillidge CJ, Snow DH, Marlin DJ (1993). Respiratory and cardiovascular adjustments during exercise of increasing intensity and during recovery in thoroughbred racehorses. Journal of Experimental Biology 179: 159–180.

Fedde MR, Wood SC (1993). Rheological characteristics of horse blood: significance during exercise. Respiration Physiology 94: 323–335.

Poole DC, Erickson HH (2011). Highly athletic terrestrial mammals: horses and dogs. Comprehensive Physiology 1: 1–37.

[1] Diffusion limitation is discussed in Section 12.2.1.

[2] In non-Newtonian fluids like vertebrate blood, viscosity decreases as shear rate increases.

During this period, energy is obtained from ATP and creatine phosphate[42] in the muscle cells and, if PO_2 falls to sufficiently low levels, oxygen is released from myoglobin in the muscles. At the end of the exercise period, $\dot{M}O_2$ remains elevated for a period. This excess oxygen, above the pre-exercise resting level, was originally known as the oxygen debt, but has more recently become known as the **excess post-exercise oxygen consumption** or **EPOC**.

If the exercise was sustainable and relatively mild, $\dot{M}O_2$ returns to the pre-exercise resting level within a few minutes of the end of exercise, as shown for humans in Figure 15.19B. On the other hand, if the exercise level was not sustainable and relatively difficult, Figure 15.19C shows that $\dot{M}O_2$ may remain elevated for several hours after exercise stops. Figure 15.19C also indicates that exhaustive exercise is followed by a relatively fast component to the $\dot{M}O$ recovery curve and a much slower component. We discuss these two components in more detail in the next section.

Both the oxygen deficit and the EPOC increase with increasing intensity of exercise. EPOC also increases with duration of exercise, particularly when the intensity is above 30 per cent of maximum rate of oxygen consumption ($\dot{M}O_2$ max).

15.2.4 Exhaustive or burst exercise

Burst exercise, such as sprinting, relies to a large extent on glycogen stored in the glycolytic muscle fibres[43] and results in the acidic metabolic end-products of anaerobic metabolism, such as lactic acid in vertebrates and many species of invertebrates[44], accumulating in the active muscles and circulating blood. Eventually the animal becomes exhausted (fatigued) and the level of exercise can no longer be maintained. The accumulation of acidic metabolites during exhaustive exercise may have a large effect on the acid–base status[45] of the animals, from which they may take several hours to recover.

Endothermic and ectothermic vertebrates exhibit specific differences in their use of and recovery from exhaustive exercise. These differences include a greater reliance on anaerobic metabolism in amphibians and reptiles than in mammals, although this does depend on the duration of the exercise, as shown in Figure 15.20A. For example, during a 10-second sprint, a human athlete obtains over 90 per cent of his or her energy from anaerobic sources—glycolysis, ATP and phosphocreatine (PCr). This decreases to around 50 per cent during a 2-minute run and to less than 20 per cent for exercise longer than 4.5-minute duration. Also, anaerobic sources account for only about 10 per cent of energy during maximal activity for 5 minutes in the mountain vole (*Microtus montanus*) and the Merriam kangaroo rat (*Dipodomys merriami*). These examples are in contrast to the situation in some species of lizard, such as desert iguanas *Dipsosaurus dorsalis*, shown in Figure 15.20B. In some species of lizards, the anaerobic contribution to 2 minutes of burst exercise at 35°C is almost 70 per cent, and this increases with decreasing temperature.

[42] Creatine phosphate is discussed in more detail in Section 18.5.2.

[43] Muscle fibre types are discussed in Section 18.2.3.

[44] In some species of active molluscs, such as squid (*Lolliguncula brevis*), the end products of anaerobic metabolism include octopine and succinate.

[45] Acid–base balance of animals is discussed in Section 13.3.

15

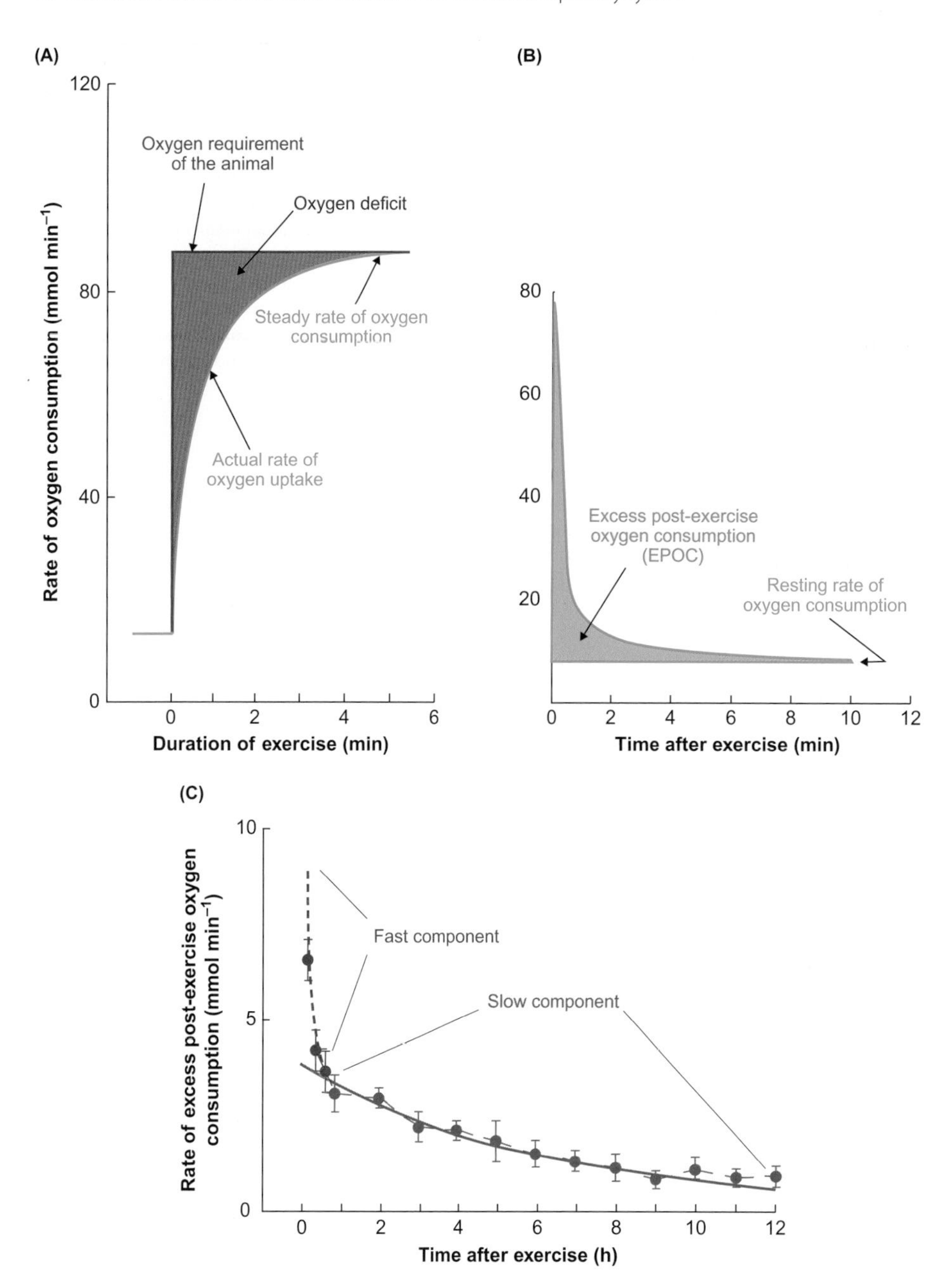

Figure 15.19 Oxygen deficit at the beginning of exercise and excess post-exercise oxygen consumption (EPOC) during the recovery period in humans

(A) Rate of oxygen consumption before and during sustainable submaximal exercise. This shows how the oxygen deficit is calculated. (B) Rate of oxygen consumption after a period of sustainable submaximal exercise showing how the excess post-exercise oxygen consumption, EPOC, is determined.
(C) Rate of decline of excess post-exercise oxygen consumption following exhaustive, submaximal exercise. Mean values ± SEM from 12 subjects.

Adapted from A, B: Gaesser GA and Brooks GA (1984). Metabolic bases of excess post-exercise oxygen consumption: a review. Medicine and Science in Sports and Exercise 16: 29-43. reproduced from C: Bahr R (1992). Excess postexercise oxygen consumption – magnitude, mechanisms and practical implications. Acta Physiologica Scandinavica 144, Suppl. 605: 1-70.

Exhaustive exercise in birds and mammals

In active mammals, such as domestic dogs (*Canis lupus familiaris*) and juvenile lions (*Panthera leo*), exercise levels below 50 per cent of maximum sustainable $\dot{M}O_2$ ($\dot{M}O_2$ max) are purely aerobic with no accumulation of anaerobic end products such as lactic acid. Under these conditions, the rate at which lactic acid is produced is matched by the rate it is removed. However, at an exercise intensity of about 75 per cent $\dot{M}O_2$ max, there is an initial increase in the concentration of blood lactate which returns toward the resting level after a few minutes, as shown in Figure 15.21. During the initial few minutes, the rate of lactic acid production is greater than that of its removal, but the rate of removal then exceeds the rate of production, thus leading to a reduction in its concentration. At about 92 per cent $\dot{M}O_2$ max, the concentration of blood lactate is maintained at a new stable level after the initial increase as the rate of lactate production is balanced by its rate of removal. Under these conditions, ATP production is supplemented by the anaerobic metabolism of glycogen. This leads to the accumulation of lactic acid and the reduction in pH, i.e. the animal experiences a **metabolic acidosis**[46].

At extremely high, exhaustive, levels of exercise, when $\dot{M}O_2$ is well above $\dot{M}O_2$ max, blood lactate increases continually throughout the exercise period until the animal becomes exhausted. We should point out that the occurrence of muscle fatigue at high levels of blood lactate and intracellular H^+ does not necessarily mean that lactate and/or H^+ are the *cause* of the muscle fatigue. In fact, it has been suggested that plasma

[46] Metabolic acidosis is discussed in Section 13.3.2.

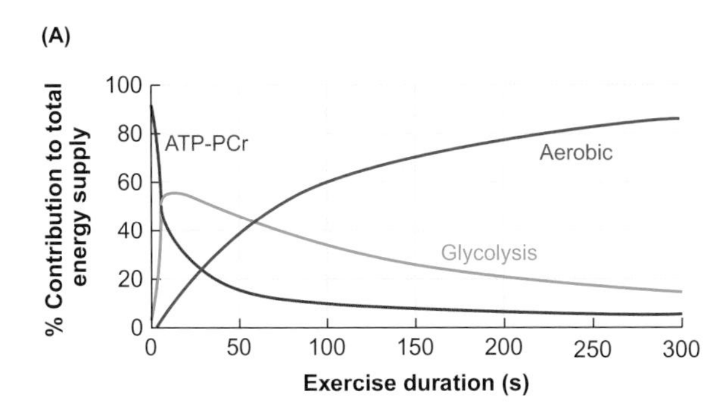

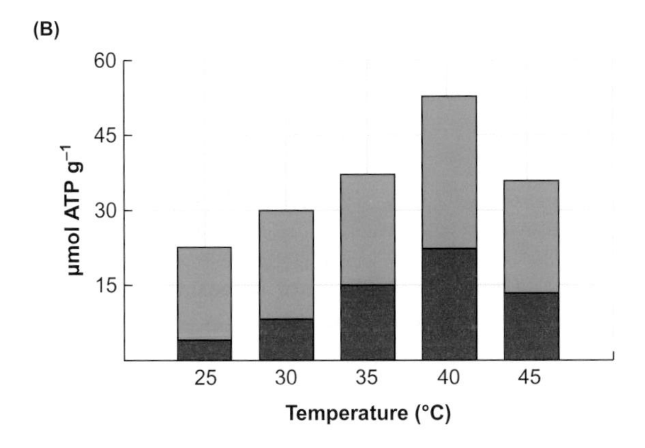

Figure 15.20 Sources of energy during exercise

(A) Proportion of energy supplied by aerobic or anaerobic (ATP-PCr plus glycolysis) sources during different durations of maximal exercise in humans. PCr, phosphocreatine. (B) Contributions of aerobic (red section of bars) and anaerobic (blue sections of bars) to energy production during 2 min of burst activity at different acclimation temperatures by lizards (*Dipsosaurus dorsalis*). Note the stronger effect of temperature on the aerobic component compared with that on anaerobic metabolism.

Reproduced from A: Gastin PB (2001). Energy system interaction and relative contribution during maximal exercise. Sports Medicine 31: 725-741. B: Bennett AF (1994). Exercise performance in reptiles. In: Jones JH (ed) Advances in Veterinary Science and Comparative Medicine Vol 33B, Comparative Vertebrate Exercise Physiology: Phyletic Adaptations. pp 113-138. Academic Press, San Diego.

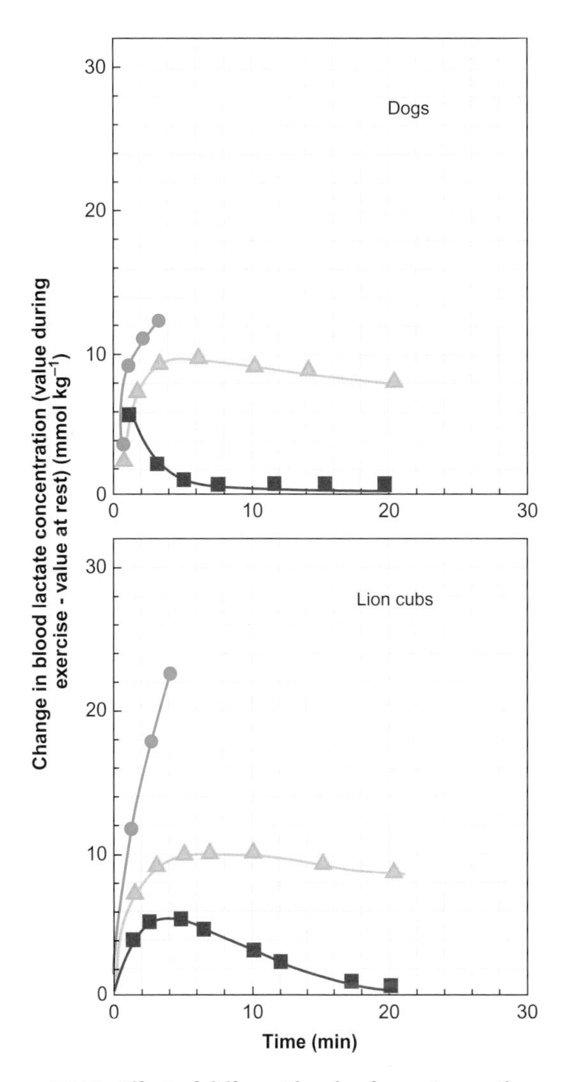

Figure 15.21 Effect of different levels of exercise on the change in blood lactate concentration in dogs and lions

Change in blood lactate concentration in two domestic dogs, *Canis lupus familiaris* (mean body mass 21 kg) and three lion cubs, *Panthera leo* (mean body mass 30 kg) exercising at approximately 75% maximum sustainable rate of oxygen consumption, $\dot{M}O_2$ max, (■), 92% $\dot{M}O_2$ max (▲) and well above $\dot{M}O_2$ max (●).

Modified from Seeherman HJ et al (1981). Design of the mammalian respiratory system. II Measuring maximum aerobic capacity. Respiration Physiology 44: 11-23.

acidosis may impair exercise performance by reducing the output from the central nervous system to the muscles. However, most studies have been on isolated muscles or muscle fibres and indicate that changes within the muscles themselves are the major limiting factors. Such local factors include a decrease in the release of Ca^{2+} from the sarcoplasmic reticulum and the effects of reactive oxygen species[47].

When high levels of power output are required, either to capture mobile prey, escape a predator or during take-off by some species of birds, the glycolytic fibres in the locomotor muscles are utilized and the rapid production of ATP is achieved by anaerobic metabolism. The flight muscles of some species of birds such as the Galliformes, which includes red jungle fowl (*Gallus gallus*) pheasants (*Phasianus* spp.) and wild turkeys (*Meleagris gallopavo*), have a preponderance of glycolytic fibres in their flight muscles. These species of birds are restricted to anaerobic flights of relatively short duration. In contrast, the flight muscles of other species, such as pigeons, ducks and geese, which can fly non-stop for many hours, have a preponderance of oxidative fibres. The colour of the flight muscles is pale pink in chickens, which are thought to have originated from the red jungle fowl, and dark red in ducks and pigeons. Why do you think there is this difference in colour? Look at Section 18.1.4 for clues.

The transition between sustained and exhaustive exercise during progressively higher work rates has been studied most in humans. One indicator of this transition is known as the **anaerobic** or **lactate threshold**. This is the point at which the release of lactate from the glycolytic muscle fibres causes the concentration of lactate in the blood to increase above its resting level, as shown in Figure 15.22. At the same

[47] Muscle fatigue is discussed in Section 18.2.3 and Experimental Study 18.2.

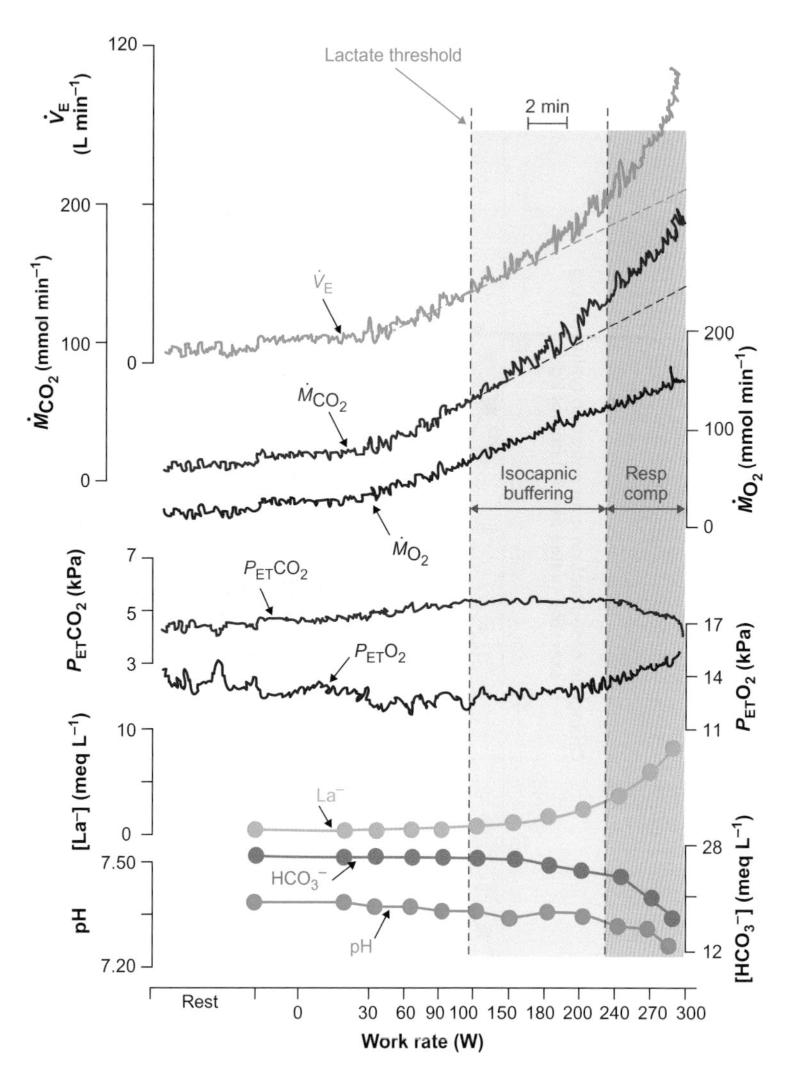

Figure 15.22 Effects of increasing work rates on ventilation, end tidal gases and acid–base balance in humans

$\dot{V}_E$, minute ventilation volume; $\dot{M}O_2$ and $\dot{M}CO_2$, rates of O_2 consumption and CO_2 production, respectively; $P_{ET}O_2$ and $P_{ET}CO_2$, partial pressures of end tidal O_2 and CO_2, respectively, which are similar to partial pressures of arterial O_2 and CO_2; [La$^-$] and [HCO_3^-], concentrations of lactate and bicarbonate respectively; pH, pH in arterial blood; Resp Comp, respiratory compensation.

Adapted from Wasserman K and Casaburi R (1991). Acid-base regulation during exercise in humans. In: Whipp BJ and Wasserman K (eds) Exercise, Pulmonary Physiology and Pathophysiology, pp 405-448. New York: Marcel Dekker, Inc.

time, ventilation volume begins to increase by a greater proportion than the increase in $\dot{M}O_2$.

The increase in the concentration of lactate ions (La$^-$) from lactic acid is accompanied by an increase in the concentration of protons (H^+), which leads to a decrease in pH. The increase in H^+ concentration is partly buffered by the bicarbonate (HCO_3^-) stores in the body which leads to the release of CO_2 from these stores[48] and, the refore, to a reduction in the concentration of bicarbonate in the blood. It has traditionally been thought that the production of CO_2 and lactic acid by the metabolizing muscles are the main determinants of changes in H^+ concentration during and after exercise above the anaerobic threshold. However, other factors, such as efflux of K^+ from the muscles and an increase in the concentration of Cl^- in the muscles, may contribute to a reduction in the strong ion difference[49] (SID) and, therefore, to a fall in intracellular pH.

Figure 15.22 shows that at work rates just above the lactate threshold, the increase in ventilation volume matches the increase in release of CO_2 both from the metabolizing muscles and from the bicarbonate stores, and so partial pressure of CO_2 in the expired breath ($P_{ET}CO_2$), and in arterial blood (P_aCO_2), remain more or less constant, but the respiratory exchange ratio (RER)[50] increases. This period has been called the zone of isocapnic[51] buffering and is highlighted in Figure 15.22. However, as the intensity of exercise increases further, ventilation volume increases by a greater proportion than the release of CO_2, so that $P_{ET}CO_2$ (and P_aCO_2) falls and $P_{ET}CO_2$ (and P_aCO_2) increases. This is a period of respiratory compensation and is also shown in Figure 15.22.

The fall in P_aCO_2 begins partly to compensate for the continuing decrease in pH; in other words, there is a respiratory alkalosis[52] accompanying the metabolic acidosis. Nonetheless, blood pH can be below 6.9 and blood lactate concentration can be around 25 mmol L^{-1} at the end of exhaustive exercise.

[48] Equation 11.1 shows the details of this reaction.

[49] Strong ion difference and its influence on pH is discussed in Section 13.3.2.

[50] Respiratory exchange ratio is the ratio of the rate of release of carbon dioxide to the rate of uptake of oxygen at the gas-exchange surface and is discussed in Sections 2.2.2 and 12.3.3.

[51] Isocapnic means equal carbon dioxide. *Isos* is Greek for equal and *kapnos* is Greek for smoke, which is composed mainly of carbon dioxide.

[52] Respiratory alkalosis is discussed in Section 13.3.2.

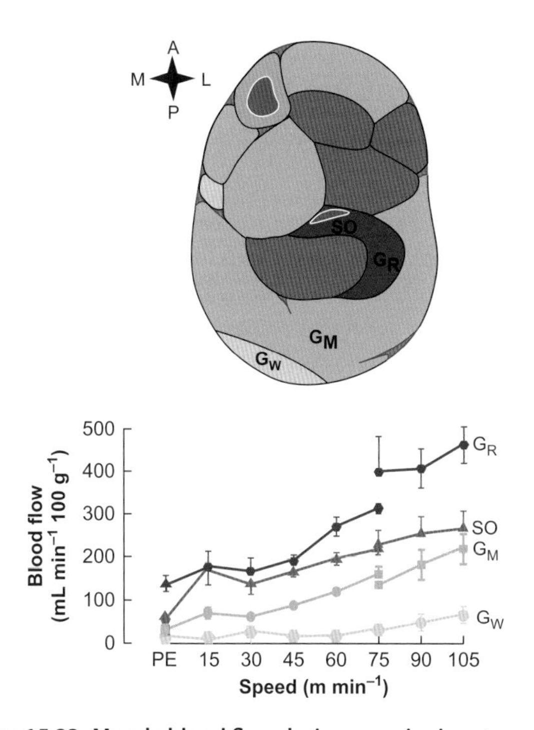

Figure 15.23 Muscle blood flow during exercise in rats

Cross-section of the leg of a rat showing the proportion of slow oxidative plus fast oxidative glycolytic fibres in the different muscles. These are the most oxidative (aerobic) fibres. The more red the shading the greater the proportion of slow oxidative plus fast oxidative glycolytic fibres. A, anterior; P, posterior; M, medial; L, lateral. The labelled muscles are: SO, soleus muscle; G_R, red gastrocnemius; G_M, middle gastrocnemius and G_W, white gastrocnemius. The GW fibres are the most glycolytic (anaerobic) fibres. The graph shows blood flow to these muscles after running for 1 min at different speeds. PE, pre-exercise values.

Sources: Armstrong RB and Laughlin MH (1985). Metabolic indicators of fibre recruitment in mammalian muscles during locomotion. Journal of Experimental Biology 115: 201-213; Armstrong RB and Laughlin MH (1985). Rat muscle blood flows during high speed locomotion. Journal of Applied Physiology 59: 1322-1328.

Most of the increase in cardiac output that occurs during exercise of all intensities in vertebrates supplies the oxidative fibres in locomotor muscles with flow to the glycolytic fibres increasing very little. Figure 15.23 illustrates the changes in blood flow to different muscle fibres of rats during increasing intensity of exercise. There is a further increase in blood flow to the muscles after exhaustive exercise has stopped, and this increase is partly the result of massive dilation of the blood vessels in the muscles, which leads to a reduction in arterial blood pressure known as **post-exercise hypotension**.

Recovery from exhaustive exercise in ectotherms

Qualitatively, the responses to exhaustive exercise are similar in ectotherms to those we have seen in mammals, although there are specific differences in the rate at which the accumulated lactate is removed. Figure 15.24 shows that rats and humans require little longer than 60 min for blood lactate to return to resting levels following exhaustive exercise, whereas in ectotherms, it can take several hours for lactate levels to return to normal.

In many species of ectotherms, blood pH returns to its resting value faster than the concentration of La^-, although there are clear differences between more active fish such as skipjack tuna and rainbow trout on the one hand, and more sluggish species such as starry flounder (*Platichthys stellatus*) on the other. The more active species have high aerobic scopes, and the change in the concentration of La^- following exhaustive exercise greatly exceeds that of H^+. The tuna is an extreme example, with H^+ concentration back to its resting level within 50 min despite the concentration of La^- continuing to rise. On the other hand, in the relatively sluggish species with their low aerobic scopes, change in concentration of H^+ greatly exceeds that of La^-, with La^- concentration barely increasing.

It has been suggested that the adaptive significance of these differences is the importance for the more active

Figure 15.24 Change in the concentration of lactate in the blood or haemolymph following exhaustive exercise

Percentage change following the end of exercise (time 0 min) in a range of animals. rat; humans; Galápagos marine iguana (*Amblyrhynchus cristatus*); desert iguana (*Dipsosaurus dorsalis*); Pacific tree frog (*Pseudacris regilla*); the California slender (lungless) salamander (*Batrachoseps attenuatus*); rainbow trout (*Oncorhynchus mykiss*); blue crab (*Callinectes sapidus*).

Modified from Gleeson TT (1991). Patterns of metabolic recovery from exercise in amphibians and reptiles. Journal of Experimental Biology 160: 187-207.

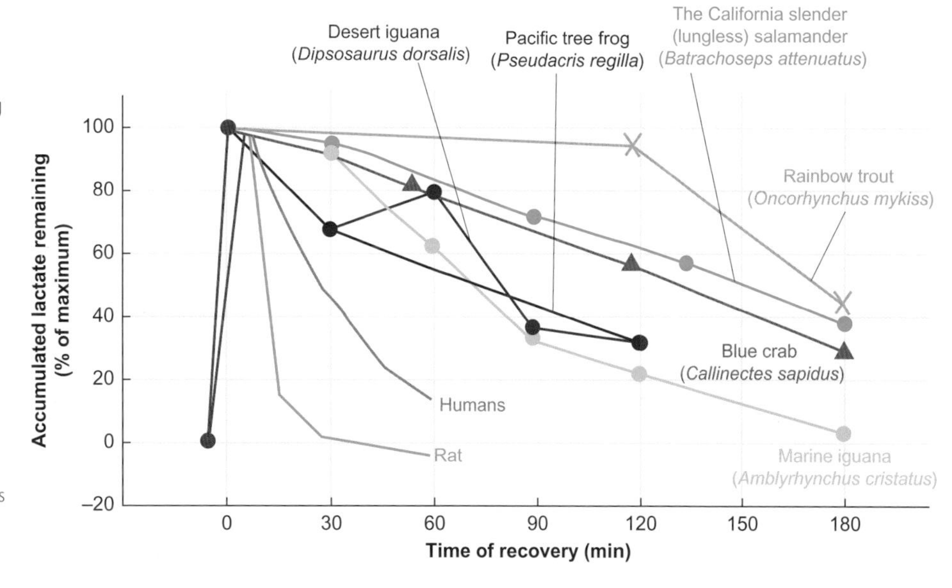

species to minimize acidification of the blood by the high acid load in the muscles, which would reduce affinity of Hb for O_2[53]. It is essential that pelagic fish such as tuna, are able to maintain their ability to swim. The lactate in the blood may be oxidized by the heart and red muscles to provide energy, as occurs in mammals.

In contrast, it may be more important for the more sluggish species to minimize the release of lactate from the muscle so that the majority can be resynthesized to glycogen within the muscle itself. A return of intracellular pH within the muscle to a level that is suitable for glycogenesis may be more important than minimizing acidification of the blood in these species, in which oxygen transport to oxidative muscle would be less important.

The physiological basis for the differences between the two types of fish is discussed in Experimental Panel 15.2, available online.

› Review articles

Allen DG, Lamb GD, Westerblad H (2008). Skeletal muscle fatigue: cellular mechanisms. Physiological Reviews 88: 287–332.

Butler PJ (2016). The physiological basis of bird flight. Philosophical Transactions of the Royal Society 371: 20150384.

Butler PJ, Bishop CM, Woakes AJ (2003). Chasing a wild goose: posthatch growth of locomotor muscles and behavioural physiology of migration of an arctic goose. In Berthold P, Gwinner E, Sonnenschein E (eds.) Avian Migration. pp 527–542. Berlin and Heidelberg: Springer-Verlag.

Cooke SJ, Hinch SG, Donaldson MR, Clark TD, Eliason, EJ, Crossin GT, Raby GD, Jeffries KM, Lapointe M, Miller K, Patterson DA, Farrell AP (2012). Conservation physiology in practice: how physiological knowledge has improved our ability to sustainably manage Pacific salmon during up-river migration. Philosophical Transactions of the Royal Society B 367: 1757–1769.

Farrell AP, Hinch SG, Cooke SJ, Patterson DA, Crossin GT, Lapointe M, Mathes MT (2008). Pacific salmon in hot water: applying aerobic scope models and biotelemetry to predict the success of spawning migrations. Physiological and Biochemical Zoology 81: 697–708.

Gastin PB (2001). Energy system interaction and relative contribution during maximal exercise. Sports Medicine 31: 725–741.

Matheson PJ, Wilson MA, Garrison RN (2000). Regulation of intestinal blood flow. Journal of Surgical Research 93: 182–196.

McCue MD (2006). Specific dynamic action: a century of investigation. Comparative Biochemical and Physiology A 144: 381–394.

Piersma T, Drent J (2003). Phenotypic flexibility and the evolution of organismal design. TREE 18: 228–233.

Wang T, Andersen JB, Hicks JW (2005). Effects of digestion on the respiratory and cardiovascular physiology of amphibians and reptiles. In: Physiological and Ecological Adaptations to Feeding in Vertebrates. Eds. Starck JM and Wang T. pp 279–304. Enfield, NH: Science Publishers, Inc.

Wasserman K (1987). Determinants and detection of anaerobic threshold and consequences of exercise above it. Circulation 76 (suppl VI): VI-29–VI-39.

Wasserman K, Casaburi R (1991). Acid-base regulation during exercise in humans. In Whipp BJ, Wasserman K (eds.) Exercise, Pulmonary Physiology and Pathophysiology. pp 405–448. New York: Marcel Dekker, Inc.

Weibel ER (1999). Understanding the limitations of O_2 supply through comparative physiology. Respiration Physiology 118: 85–93.

[53] The effect of pH on the oxygen affinity of haemoglobin is discussed in Section 13.1.2.

15.3 Responding to a decrease in oxygen supply

Many species of animals live in an environment in which the partial pressure of oxygen (PO_2) is less than that in air at sea level (which we'll refer to as 'normal' or 'normoxia'), or display a particular behaviour pattern that periodically exposes them to a PO_2 that is lower than normal.

Many species of animals experience a low PO_2 during development, for example in the uterus of mammals, or inside an egg[54]. A more dramatic situation occurred between about 150 million years and 200 million years ago when atmospheric PO_2 fell to about 12 kPa, compared to almost 21 kPa at present. Thus, a PO_2 below the current normal level (known as hypoxia) is not unusual both in terms of evolutionary time and during development of many species of animals. In this section, we discuss hypoxia, with or without an associated increase in the partial pressure of CO_2.

If the partial pressure of oxygen (PO_2) in the environment is lower than normal, the animal is exposed to environmental hypoxia. In some circumstances, PO_2 may reach zero, or very close to zero, in which case the animal experiences environmental **anoxia**.

Anoxia is tolerated by many species of invertebrates and some ectothermic vertebrates. Various species of marine molluscs and two groups of vertebrates, crucian carp (*Carassius carassius*) and the freshwater turtles, red-eared sliders (*Trachemys scripta elegans*) and western painted turtles (*Chrysemys picta belli*), have been particularly well studied. Under anoxic conditions, all of these species rely on ATP being provided by anaerobic metabolism using glycogen as the metabolic substrate and survive longer at lower temperatures.

As there is no oxygen to deliver to the metabolising tissues, the role of the respiratory and cardiovascular systems under these conditions is, at most, limited. Crucian carp produce ethanol as the end product of anaerobic metabolism. They maintain gill ventilation and cardiac output during anoxia, most likely in order to remove the ethanol across the gills.

For animals that live in a confined space such as a burrow, where there is limited exchange with the outside air, PO_2 is lower than that outside and the partial pressure of carbon dioxide (PCO_2) is higher. Consequently, the animals will be exposed to external hypoxia and **hypercapnia**. In contrast, when an aquatic bird or mammal dives under water, its body becomes a closed system and the amount of oxygen in its blood will progressively decline while the amount of carbon dioxide progressively accumulates; it will experience internal hypoxia and hypercapnia.

[54] PO_2 inside the eggs of birds and insects and in the foetus of mammals is discussed in Sections 12.3.2 and 12.3.3, respectively.

Internal hypoxia and hypercapnia can also occur whenever there is a mismatch between the demand of oxygen and its delivery to the metabolizing tissues, such as during extreme forms of exercise, high environmental temperature in ectotherms, anaemia and failure of the respiratory and/or cardiovascular systems in some way.

Exposure to environmental hypoxia means that the concentration of oxygen in inspired air or water (C_IO_2) is lower than when the animals are exposed to normoxia. So, in order to maintain a constant rate of oxygen consumption ($\dot{M}O_2$), animals exposed to environmental hypoxia must change the other components of the respiratory system shown on the right-hand side of equation 15.1. If these changes within the respiratory system are not sufficient to maintain the oxygen concentration in arterial blood (CaO_2), or for aquatic birds and mammals when diving, then the components of the cardiovascular system shown in equation 15.2 will need to respond appropriately in order to maintain $\dot{M}O_2$ max. It would also be possible for an animal when faced with hypoxia to reduce its oxygen demand. We begin by looking at how the respiratory and cardiovascular systems respond to environmental hypoxia.

15.3.1 Environmental hypoxia

Factors that affect the level of oxygen in the environment are different in air and water. Partial pressure of oxygen in the air is dependent mainly on atmospheric pressure which changes with altitude; the greater the altitude above sea level, the lower the PO_2[55]. So, air-breathing animals living at high altitude experience environmental hypoxia, whereas those living at or near sea level rarely, if ever, do.

In contrast, oxygen in water is obtained from the air above it and/or from the photosynthesis of aquatic plants. It is removed from water by aerobic respiration of organisms, by the decay of organic matter and by degradation of organic pollutants. If the rate at which oxygen enters a body of water is less than the rate at which it is used, the water will become hypoxic. Hypoxia can occur naturally in stagnant bodies of water if there is an increase in the level of nutrients, called eutrophication, by the upwelling of deeper water.

Eutrophication stimulates the growth of certain species of phytoplankton and may cause an algal bloom. Although these algae are net producers of oxygen during the day when they are photosynthesizing, they can no longer photosynthesize at night and become net consumers, at which point they can rapidly deplete the oxygen in the water column. Also, when algal blooms die, oxygen is further depleted as bacteria decompose the dead cells. There are also human causes of eutrophication and algal blooms, such as leaching of chemical nutrients (mainly nitrogen and phosphorus used in fertilisers) from the soil which can enter the water system.

Naturally occurring hypoxia can be quite common in bodies of fresh water, such as swamps and backwaters which circulate poorly and contain heavy loads of organic matter, and in slow-moving parts of rivers, particularly if there is a high density of water plants. Natural hypoxia is also a feature of some parts of the marine environment, such as deep basins, upwelling areas of eastern boundary currents and fjords. Indeed, the development of hypoxia in relatively shallow coastal regions has increased over the last two centuries as a result of human activity.

The level of oxygen in water is often measured by field biologists in units of mg O_2 (L water)$^{-1}$ and is known as the dissolved oxygen (DO). As a result, all water quality criteria (which set out the conditions needed for particular species of animals) state the amount of oxygen in natural water bodies in mg O_2 L^{-1}. The relationships between different units for quantifying DO and PO_2 at two different temperatures in fresh water and seawater are given in Figure 15.25. We continue to use mol L^{-1} for concentration and kPa for PO_2 as these are the units used by physiologists when discussing diffusive gas exchange and transport of oxygen between the environment and the metabolizing cells[56].

Temperature regulation during hypoxia

The effect of hypoxia on an organism depends to a large extent on the rate at which oxygen is being consumed, so body temperature is an important factor. Many species of animals use a variety of ways to maintain a more or less constant body temperature, at least during some parts of the day[57]. Ectotherms such as reptiles use mainly behavioural means to make use of heat from the sun, whereas endotherms use some form of internal thermogenesis. Either way, the nervous system is involved in the control of the animal's behaviour and/or thermogenesis in order to maintain it at its preferred internal temperature.

Body temperature (T_b) can be reduced by behavioural or physiological means. If Q_{10} (discussed in Box 8.2) is about 2.5, both ectotherms and endotherms could save at least around 11 per cent on their resting $\dot{M}O_2$ per °C drop in body temperature. For endotherms, there would be additional savings from the reduction in thermogenesis itself. Also, the

[55] The relationship between PO_2 in air and atmospheric pressure is discussed in Section 11.1.2.

[56] Diffusion of respiratory gases and their transport within animals are discussed in Section 11.2.

[57] Chapters 9 and 10 discuss temperature regulation in ectotherms and endotherms, respectively.

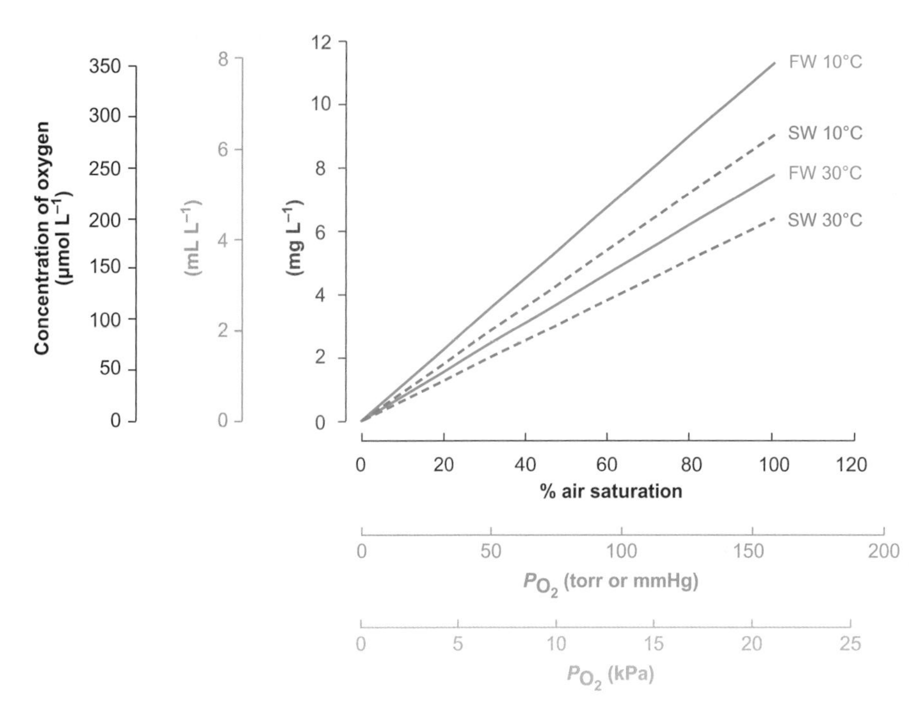

Figure 15.25 Relationships between different units for quantifying the relationship between partial pressure of oxygen (PO_2, % air saturation) and its concentration in fresh water (FW) and seawater (SW) at 10 and 30°C.

Adapted from Diaz RJ and Breitburg DL (2009). The hypoxic environment. In: Richards JG et al (eds) Fish Physiology, vol 27 Hypoxia, pp 1-23. Elsevier.

reduction in T_b will cause a leftward shift of the oxygen equilibrium curve (decrease in P_{50}, increase in affinity for oxygen) of the blood pigment[58] and, therefore, will facilitate the uptake of oxygen at the gas exchange surface.

When most species of animals are placed in a thermal gradient and exposed to progressively more severe hypoxia, they move to a lower ambient temperature (T_{amb}) as PO_2 falls below a well-defined threshold, as demonstrated in Figure 15.26. In mammals, the major mechanisms for reducing body temperature are physiological rather than behavioural. However, Figure 15.26 shows that there is a behavioural component in mice; they also move into cooler air when PO_2 falls below a certain value.

Some species of animals have different genotypes which influence their behavioural and/or physiological responses to changes in the environment. For example, Atlantic cod (*Gadus morhua*) can have two different homozygous types of haemoglobin (Hb): cod with the HbI-1 type of haemoglobin occur predominantly in warmer waters and have a preferred temperature of about 15°C. In contrast, those with the HbI-2 type are found more frequently in more northerly, colder parts of the Atlantic and have a preferred temperature of 8°C. When exposed to hypoxia, the HbI-1 cod reduce their preferred temperature from around 15°C to 10°C, whereas the preferred temperature of the HbI-2 cod remains at 8°C. Thus, HbI-1 cod could be at an advantage if there is any increase in temperature of the Atlantic resulting from climate change; they could merely move north. However, as they prefer a lower temperature when exposed to hypoxia, HbI-1 cod could be vulnerable if they encounter a combination of hypoxia and higher temperature in the Atlantic, as occurs in some coastal areas and the Baltic Sea.

For endotherms, the reduction in T_b during hypoxia is similar to the regulated hypothermia (anapyrexia) that occurs during daily torpor and hibernation[59]. The controlled nature of this response is illustrated by Figure 15.27A; although T_b falls during a progressive decline in ambient temperature (T_{amb}) in rats exposed to hypoxia, it is still maintained above T_{amb}. Figure 15.27B shows the saving in metabolic rate (rate of oxygen consumption, $\dot{M}O_2$) associated with this regulated hypothermia. Rate of oxygen consumption is lower at a given T_{amb} below 25 °C for rats exposed to hypoxia than for those in normoxia.

In some species of endotherms, the lower critical temperature[60] (T_{lc}) is lower when the animals are in hypoxia than when they are breathing normal air. For example, golden-mantled ground squirrels (*Spermophilus lateralis*), spend at least part of their lives in hypoxic burrows (as their common name suggests). When breathing air, their T_{lc} is about 15°C, whereas when breathing 7 per cent O_2, T_{lc} is as low as 6°C.

Environmental hypoxia is not a common experience for most terrestrial animals, which live at or near sea level, so we concentrate first on hypoxia in water.

15

[58] The effect of temperature on the oxygen equilibrium curve is discussed in Section 13.1.2.

[59] Daily torpor and hibernation are discussed in Section 10.2.7

[60] Lower critical temperature is discussed in Section 10.1.

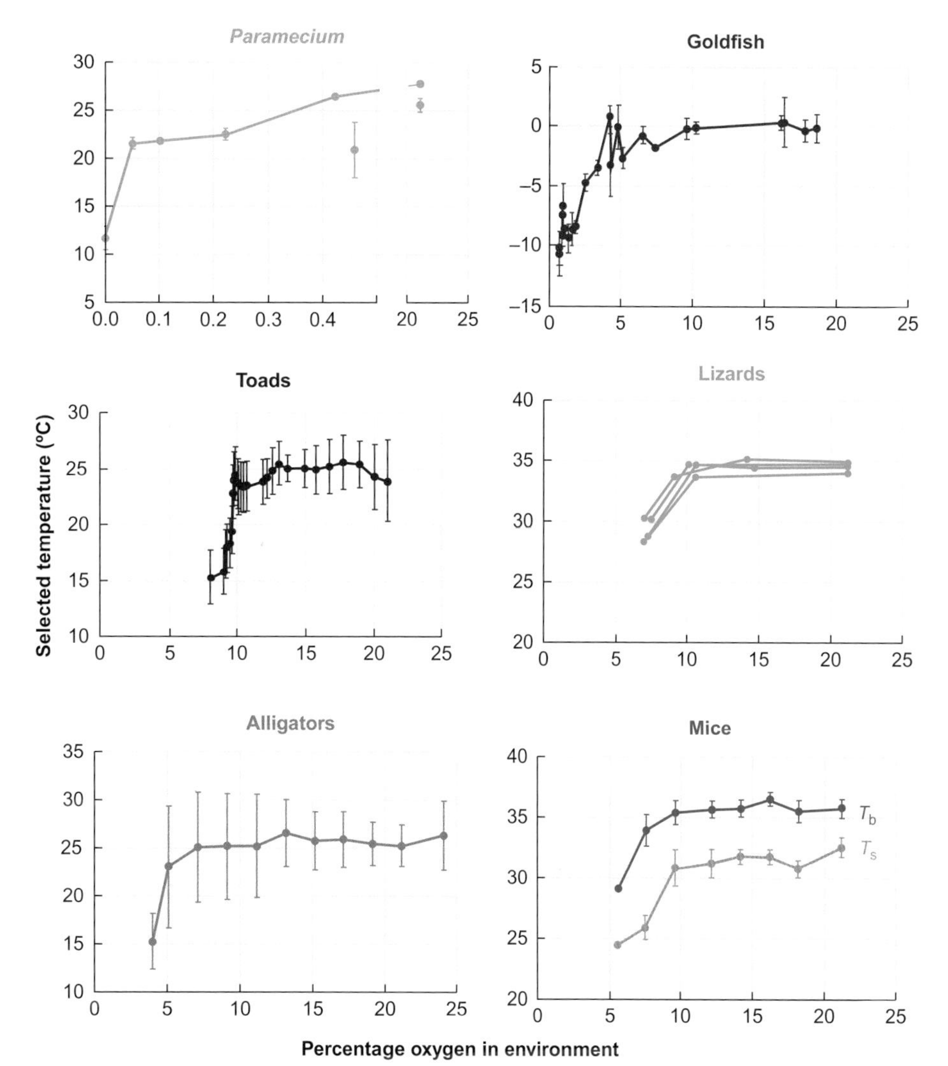

Figure 15.26 Effect of ambient level of oxygen on temperature selection of various species of animals during exposure to hypoxia in a temperature gradient

Data for mice show both selected environmental temperature (T_s) and associated body temperature (T_b). Note that mice, like the other species, move into colder air as percentage oxygen falls below a threshold value. Note, *Paramecium* is now considered to be a protist and not an animal.

Reproduced from Wood SC (1995). Interrelationships between hypoxia and thermoregulation in vertebrates. In: Heisler N (ed). Advances in Comparative and Environmental Physiology, vol 22. pp 209-231. Berlin and Heidelberg: Springer-Verlag.

Hypoxia in water

In almost all water-breathing animals, exposure to environmental hypoxia causes a reflex increase in ventilation (hyperventilation). At any given level of hypoxia, this hyperventilation is greater at higher temperatures. In facultative air-breathing aquatic animals[61] such as the catfish (*Clarias batrachus*), the increase in ventilation during hypoxia includes an increase in the frequency at which the animal visits the surface to ventilate its air-breathing organ.

Just as some ectotherms are adapted to live in cold conditions while others are adapted for living in warmer environments, so some have evolved to live in hypoxic, and often warm, environments while others need well aerated, often cool, water to survive. Classic examples of these are the hypoxia-tolerant carp and, in contrast, the hypoxia-intolerant salmonid fish and tuna. During progressively developing hypoxia, carp (*Cyprinus carpio*) maintain the percentage of oxygen extracted from inspired water (per cent Ext_w)[62], whereas it decreases by over 50 per cent in rainbow trout (*Oncorhynchus mykiss*). What is more, per cent Ext_w during normoxia is higher in carp, at 80 per cent, than in rainbow trout, where it is 35–55 per cent.

Any large increase in ventilation in response to hypoxia could make a significant contribution to the oxygen requirements of the animal. For example, in carp, an almost six-fold increase in ventilation volume is associated with a 30 per cent increase in $\dot{M}O_2$.

Some closely related species show subtle differences in their response to hypoxia. European plaice (*Pleuronectes platessa*) are usually found in relatively deep, well-oxygenated water, whereas flounder (*Platichthys flesus*) are common in shallower coastal waters which can become hypoxic. When in water at 15°C, European flounder will die at a PO_2 of about 2.5 kPa, whereas plaice will die at a PO_2 of 5 kPa. When exposed to similar levels of hypoxia, flounder have

[61] Facultative air breathers are discussed in Section 12.3.3.

[62] Percentage extraction of oxygen from water is discussed in Box 12.2.

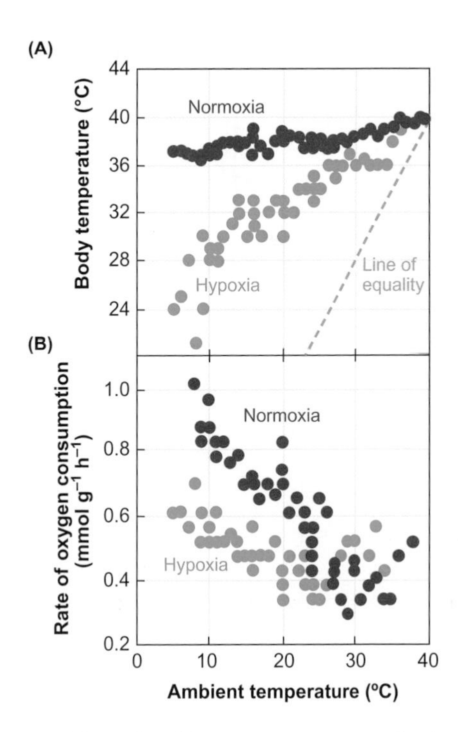

Figure 15.27 Effect of ambient temperature on body temperature (A) and rate of oxygen consumption (B) of rats when breathing air (normoxia) and 10% oxygen (hypoxia).

Reproduced from Wood SC (1995). Interrelationships between hypoxia and thermoregulation in vertebrates. In: Heisler N (ed). Advances in Comparative and Environmental Physiology, vol 22. pp 209-231. Berlin and Heidelberg: Springer-Verlag.

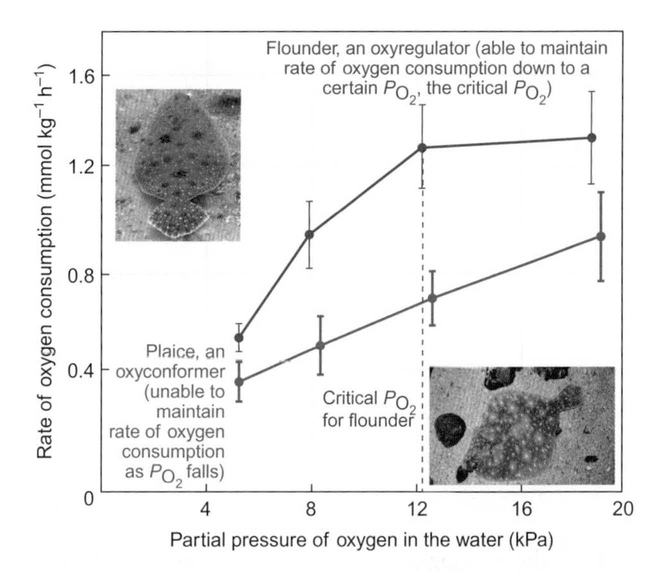

Figure 15.28 Effect of progressive hypoxia on rate of oxygen uptake in European plaice (*Pleuronectes platessa*) and European flounder (*Platichthys flesus*) at 10°C

Note that flounder are able to maintain rate of oxygen consumption ($\dot{M}O_2$) down to an environmental partial pressure of oxygen (PO_2) of 12 kPa. They are known as oxyregulators. In contrast, $\dot{M}O_2$ of plaice falls continuously with progressive hypoxia. They are oxyconformers. Mean values ± SD for 13 plaice and seven flounder.

Modified from Steffensen JF et al (1982). Gill ventilation and O_2 extraction during graded hypoxia in two ecologically distinct species of flatfish, the founder (*Platichthys flesus*) and plaice (*Pleuronectes platessa*). Environmental Biology of Fishes 7: 157-163. Images: © Florida Division of Historical Resources; ©John Mason/ardea.com..

a larger increase in ventilation and a higher per cent Ext_w for a given ventilation volume than plaice. These responses contribute to flounder at 10°C being able to maintain their $\dot{M}O_2$ at a constant value down to an inspired PO_2 (P_{IO_2}) between 8 and 12 kPa, below which there is a progressive decline in $\dot{M}O_2$. Flounders are, therefore, **oxyregulators**, as shown in Figure 15.28. The P_{IO_2} at which flounder are no longer able to maintain their $\dot{M}O_2$ is known as the **critical** PO_2, or $\boldsymbol{P_{crit}}$.

In contrast, plaice are not able to maintain $\dot{M}O_2$ as P_{IO_2} declines. This species is an **oxyconformer**. However, an ectothermic species may be an oxyregulator at a relative low environmental temperature, but an oxyconformer at a higher temperature, and so P_{crit} may also be temperature-dependent. crucian carp (*Carassius carassius*) have a particularly low P_{crit} of less than 1 kPa at 17°C and this is an adaptation to their natural habitat of lakes which become covered in ice and severely hypoxic (eventually anoxic) in winter.

A detailed study of different species of sculpins (fish of the Family Cottidae) has examined the factors related to hypoxic tolerance in those species living in the hypoxic intertidal zones compared with those living in well-aerated subtidal zones or in fresh water.

The three major factors that account for 75 per cent of the tolerance to hypoxia of the intertidal species:

- a lower routine $\dot{M}O_2$[63] and, therefore, lower metabolic demands than species that are less tolerant to hypoxia;
- a greater mass-specific gill surface area;
- higher affinity of haemoglobin for oxygen (lower P_{50}), which enhances the uptake of oxygen at the gills.

The last two of these factors can be influenced by exposure to hypoxia itself. For example, the gill area of crucian carp increases in response to environmental hypoxia. When crucian carp, are in well aerated water, their gills have been described as looking like 'sausages' in as much as they have no obvious lamellae. However, when they are kept in hypoxic water, lamellae protrude from the filaments. During normoxia, the lamellae are embedded in a mass of cells (called the interlamellar cell mass) which fills the space between the lamellae. These cells regress when the animals are in hypoxic water.

The affinity of haemoglobin for oxygen can be increased (P_{50} decreased) by three basic mechanisms[64]. These are:

- a reduction in PCO_2 and associated increase in pH in arterial blood which may result from the hyperventilation caused by hypoxia;

[63] Routine $\dot{M}O_2$ in this study was similar to standard metabolic rate, as discussed in Section 2.2.3.

[64] These three mechanisms are discussed in Section 13.1.2.

- a decrease in temperature resulting from behavioural hypothermia;
- a decrease in the concentration of organic phosphates, such as guanosine triphosphate (GTP) in red cells of fish. (Note that a similar effect in the haemolymph of hypoxia-tolerant crustaceans is the result of increases in the concentrations of urate and/or lactate.)

The effect of hypoxia on the concentration of organic phosphates in the red cells of fish was first described for the European eel (*Anguilla anguilla*). Eels migrate between fresh water and seawater and are exposed to different environmental conditions, including low PO_2. When eels are exposed to hypoxia in the laboratory for 1–2 weeks, the P_{50} of their haemoglobin decreases from 2.2 kPa to 1.4 kPa. As such, the affinity of the haemoglobin for oxygen increases. This decrease in P_{50} is related to a reduction in the concentrations of the organic phosphates, such as ATP and GTP in the red cells of the blood. GTP is the more important of these two, as its concentration is almost four times that of ATP, and a greater proportional decrease in its concentration during hypoxia means it has a greater influence on P_{50} than ATP.

In addition, the concentration of haemoglobin, [Hb], in eels increases by almost 50 per cent during exposure to hypoxia, giving rise to a similar increase in the oxygen-carrying capacity of the blood. In general, relatively short-term increases in [Hb] result from the release of red blood cells from the spleen. The two responses of decreased P_{50} and increased [Hb] mean that not only is oxygen transfer at the gills enhanced during exposure to hypoxia, but that the arterial blood carries more oxygen to the tissues at any given PO_2.

We might think that aquatic ectotherms would increase cardiac output (the amount of blood leaving the heart per minute) to ensure an adequate supply of oxygen to the tissues during exposure to hypoxia. However, in most species, heart rate actually decreases in response to hypoxia. A decrease in heart rate is a **bradycardia**. The bradycardia in response to hypoxia may be offset by an increase in cardiac stroke volume resulting in maintained or even slightly increased cardiac output. What then is the functional significance of the bradycardia which, at first sight, seems to be maladaptive? Despite a number of studies attempting to answer this question, we do not, as yet, have a general explanation for the significance of the hypoxia-induced bradycardia in aquatic ectotherms.

Hypoxia in air

Hypoxia in air occurs at high altitude, but the increased diffusion coefficient (diffusivity) of oxygen[65] and low temperatures of high altitude (approximately −6 °C (km increase in altitude)$^{-1}$) plus the relatively low metabolic rate of ectotherms, probably mean that hypoxia is not a major problem for most air-breathing ectotherms. For example, Lake Titicaca frogs (*Telametobius coleus*)[66] live at an altitude of over 3800 m. They have poorly developed lungs, but are able to meet most of their oxygen requirements from the lake water, partly as a result of their haemoglobin having a high oxygen affinity and oxygen-carrying capacity compared to amphibians living at sea level. Also, insects respond to hypoxia by the growth of more tracheoles in metabolizing tissues and by increasing the diameter of their primary tracheae. In contrast, those endotherms which spend all or part of their lives at high altitude have to maintain a relatively high rate of oxygen delivery to the tissues in the face of low environmental temperature and PO_2 in order to maintain their high body temperature and metabolic rate.

Birds at high altitude

Perhaps the most well-known endotherms to spend at least part of their lives at high altitude are bar-headed geese (*Anser indicus*) which breed on the high plateaus of central Asia. Between 25 and 50 per cent of the population migrates twice a year over the Himalayas to the lowlands of India, where they spend the winter. Their renowned status is based on auditory and visual observations of their flying over the highest peaks of the Himalayas at altitudes in excess of 8000 m, where PO_2 of inspired air (P_{IO_2}) is less than 35 kPa, or around a third that at sea level (at standard atmospheric pressure). However, Figure 15.29 shows data from satellite transmitters attached to migrating geese and indicates that they mostly fly no higher than 6000 m. Also, they fly along valleys rather than over the highest peaks of the Himalayas. A possible explanation for the earlier observations of geese at altitudes over 8000 m is based on the fact that satellite-tracked birds ascended more rapidly than normal with no increase in heart rate on a few rare occasions. This suggests the birds were obtaining some uplift from favourable air movements, which could well assist them to altitudes over 8000 m.

Nonetheless, even to fly at an altitude of 5000–6000 m is still a great achievement: in addition to the low P_{IO_2}, the lower air density means that the power required to generate lift is at least 30 per cent greater than that at sea level. For comparison, the maximum rate of oxygen consumption ($\dot{M}O_2$ max) a human can achieve at an altitude of 5500 m is only about 60 per cent that at sea level. Needless to say, the respiratory and cardiovascular systems of bar-headed geese are adapted to cope with the challenges these birds face at high altitude.

[65] The diffusion coefficient of oxygen at high altitude is discussed in Case Study 15.2.

[66] Lake Titicaca frogs are also discussed in Section 12.2.1.

15

(A)

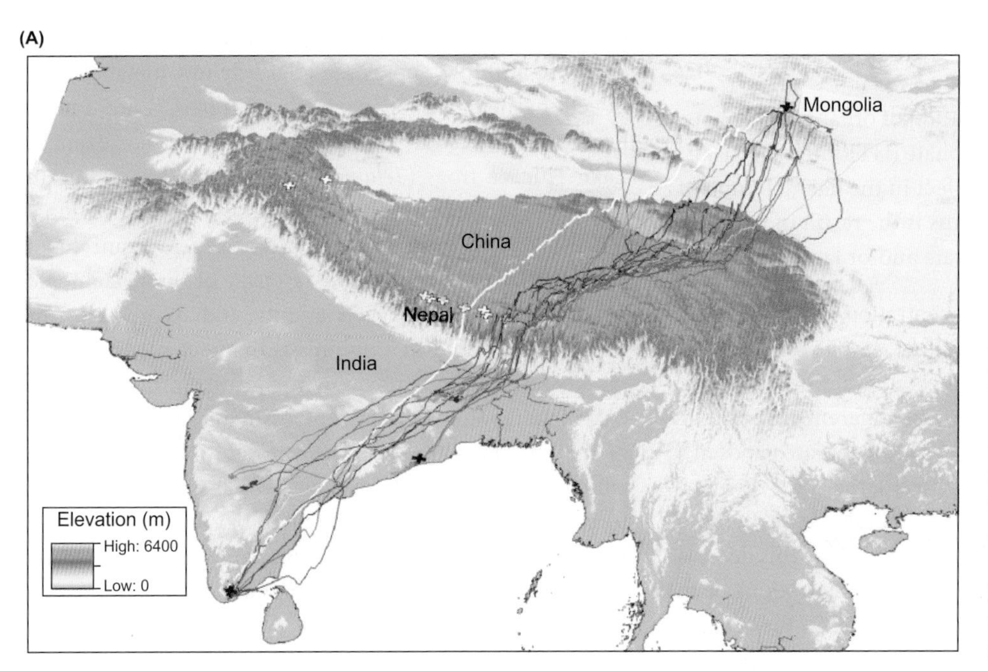

(B)

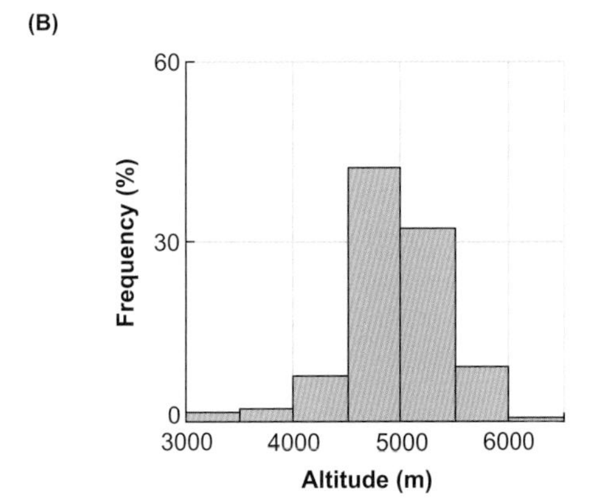

Figure 15.29 Biannual migration of bar-headed geese (*Anser indicus*) between India and China or Mongolia

(A) Three-dimensional map showing release sites (black crosses) of geese in India (two sites) and Mongolia (one site). Coloured lines represent 16 different birds and coloured background indicates elevation – see key next to map. Solid white line shows the shortest route, the great circle route, and the white crosses show locations of the world's highest mountains, those over 8000 m in altitude. (B) Frequency distribution of altitudes at which the geese fly. 95% of flights are below altitudes of 5800 m and the majority are less than 600 m above the ground (not shown).

Reproduced from: Hawkes LA et al (2013). The paradox of extreme high altitude migration in bar-headed geese *Anser indicus*. Proceedings of the Royal Society B 280: 20122114. Image: N.Batbayar, with permission.

15

In comparison with geese that spend their lives at low altitude, such as greylag geese (*Anser anser*) and pink-footed geese (*A. brachyrhynchus*), bar-headed geese have a number of adaptations for living and flying at high altitude.

- Although ventilation volume increases (hyperventilation) in response to hypoxia by similar proportions in bar-headed geese and lowland geese, tidal volume is substantially greater and respiratory frequency correspondingly lower in bar-headed geese. The greater tidal volume reduces the relative contribution of deadspace[67] to total ventilation, resulting in a more effective ventilation of the gas exchange surfaces of the lungs. The functional significance of this is shown in Figure 15.30, where PO_2 in arterial blood (P_aCO_2) is highest in bar-headed geese at all levels of hypoxia.
- Another aspect of the hyperventilation is the difference between P_{IO_2} and P_aCO_2. When breathing air at sea level, the difference is about 8 kPa, whereas the breathing of 7 per cent oxygen, which produces a P_{IO_2} equivalent to that at the top of Mt Everest, reduces the difference to 2.1 kPa in the lowland geese but to only 1.6 kPa in bar-headed geese. This means that, when at high altitude, bar-headed geese have a higher P_aO_2, and hence higher concentration of oxygen in arterial blood (C_aO_2) than lowland geese.
- Haemoglobin of bar-headed geese has a higher affinity for oxygen[68] than that of lowland geese. This is also the case for mammals that live at high altitude such as vicuñas (*Vicugna vicugna*) compared with lowland dwelling species, although mammals, in general, have Hb with a higher affinity for oxygen than that of birds.

[67] The relationship between tidal volume and deadspace is discussed in Section 12.3.3.

[68] The affinity for oxygen of haemoglobin in bar-headed geese is discussed in Section 13.1.2.

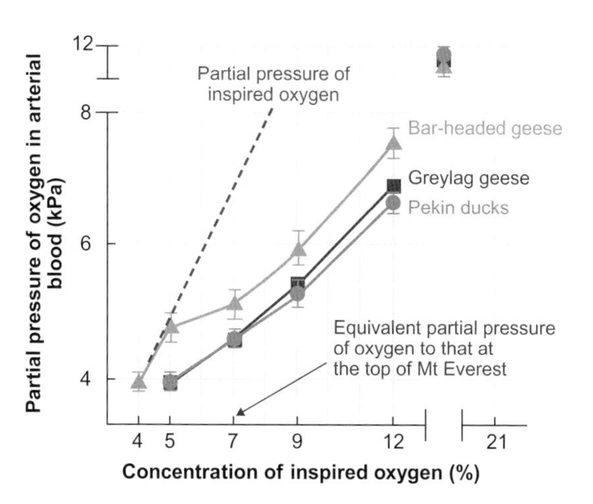

Figure 15.30 The effect of hypoxia on partial pressure of arterial blood in different species of water fowl

Partial pressure of oxygen in arterial blood (P_aO_2) is greater in bar-headed geese (*Anser indicus*) at all levels of hypoxia than in greylag geese (*Anser anser*) and Pekin ducks (*Anas platyrhynchos domesticus*). P_aO_2 in bar-headed geese is close to that in the inspired air when concentration of oxygen is 5% and less. Data points are mean values ± SEM for eight bar-headed geese, six greylag geese and 10 Pekin ducks.

Modified from Scott GR and Milsom WK (2007). Control of breathing and adaptation to high altitude in bar-headed geese. American Journal of Physiology 293: R379-R391.

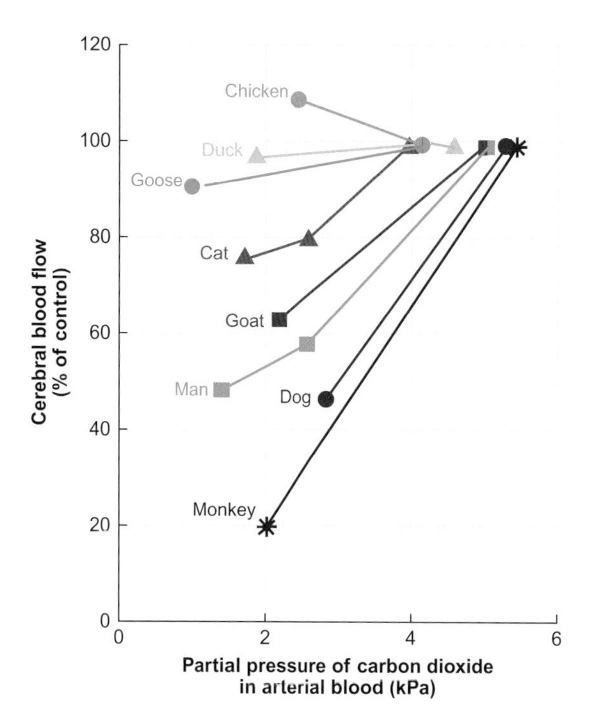

Figure 15.31 The effect of a reduction in partial pressure of CO_2 in arterial blood on cerebral blood flow in birds, including the bar-headed goose (Goose, *Anser indicus*) and mammals Reproduced from Faraci FM and Fedde MR (1986). Regional circulatory responses to hypocapnia and hypercapnia in bar-headed geese. American Journal of Physiology 19: R499-R504.

- Hyperventilation during hypoxia leads to reduced P_aCO_2 (hypocapnia) and increased pH, which in turn will further increase the affinity of haemoglobin for oxygen. Unlike lowland species, bar-headed geese do not exhibit an increase in the concentration of haemoglobin during exposure to hypoxia. However, the increased affinity of their haemoglobin for oxygen means that the concentration of oxygen in arterial blood is greater at the same P_aO_2 in bar-headed geese than in lowland birds.
- The ratio of blood capillaries to muscle fibres in the flight muscles is about 25 per cent greater in bar-headed geese than in lowland species, enabling a greater extraction of oxygen from the blood. However, overall capillary density is similar in bar-headed geese to that in barnacle geese, which also migrate long distances between the high Arctic and southern Scotland.

A potential disadvantage of the hyperventilation during exposure to hypoxia relates to the resulting hypocapnia. Hypocapnia causes a dramatic reduction in cerebral blood flow and supply of oxygen to the brain in mammals, but Figure 15.31 shows that there is no such effect in birds. This lack of effect of hypocapnia on cerebral blood vessels in birds may be one explanation as to why birds in general are more tolerant of high-altitude hypoxia than mammals. Assuming that bar-headed geese migrate at altitudes of around 6000 m under their own power without the aid of any assistance from the wind, the adaptations we have highlighted enable the respiratory and cardiovascular systems to cope with the associated high demand of the flight muscles for oxygen.

Birds are not the only species that can fly at high altitudes. Bumble bees (*Bombus* spp.) have been collected at altitudes above 5600 m, where they actively forage among the widely dispersed flowers, and are regularly found above 4000 m. Studies on *B. impetuosus* indicate that aerodynamic limitations resulting from the low air density impair flight performance at such altitudes. This limitation is largely overcome by the insects increasing the amplitude of their wing beats. The importance of hypoxia has yet to be determined.

Mammals at high altitude

A number of species of mammals also live at high altitudes and many of the adaptations they have are similar to those already discussed for birds, although their demands for oxygen are not as great as those of birds actively migrating at or above 5500 m.

Detailed studies of different subspecies of deer mice (*Peromyscus maniculatus*) living at a range of altitudes across western North America have demonstrated that those with the lowest P_{50} of their haemoglobin—

the highest affinity for oxygen—live at the highest altitudes. The genetic basis for these differences is related to two closely linked gene duplicates that encode the α-chains of adult haemoglobin[69]. The alleles of the two genes most commonly occur in the following combinations: a^0c^0/a^0c^0, a^0c^0/a^1c^1 and a^1c^1/a^1c^1. There is a clear rank order of P_{50} values and the three genotypes: mice with the a^0c^0/a^0c^0 genotype have haemoglobin (Hb) with the lowest P_{50} (highest affinity for oxygen); those with the a^1c^1/a^1c^1 genotype have Hb with the highest P_{50}, while those with the a^0c^0/a^1c^1 genotype have an intermediate P_{50}.

Also related to these genotypes, maximum rate of oxygen consumption is highest in the a^0c^0/a^0c^0 mice when tested at an altitude of 3800 m, whereas it is highest for the a^1c^1/a^1c^1 mice when tested close to sea level. There are clear advantages in the ability to exercise fairly vigorously at high altitude without incurring an oxygen debt and to be able to maintain body temperature by thermogenesis when exposed to the cold environmental temperatures of high altitude.

Humans living at high altitude

Based on archaeological data, the Tibetan Plateau has been populated by humans for about 25,000 years, whereas the Andean Altiplano has only been populated for about 11,000 years. Not only are there differences in the physiological adaptations to high altitude between humans in these two locations, but also between humans and other mammals.

High-altitude native (HAN) mammals such as llamas (*Llama glama*) have a similar ventilatory response to hypoxia as low-altitude natives (LAN) that have lived for several generations at sea level, as shown in Figure 15.32A. In contrast, as Figure 15.32B shows, high-altitude native humans have a reduced or blunted ventilatory response to hypoxia compared to that of low altitude individuals.

More detailed studies on humans have shown that the situation is not quite as simple as just described. There are indications that Tibetan HAN are better adapted to living at high altitude than Andean HAN. Studies on both populations at an altitude of about 4000 m have shown that, when resting, Tibetan HAN ventilate about 50 per cent more (at 15 L min^{-1}) than Andean HAN (at 10.5 L min^{-1}). What is more, haemoglobin (Hb) concentrations in Tibetan and Andean men are 15.6 g dL^{-1} and 19.2 g dL^{-1}, respectively, whereas they are 14.2 g dL^{-1} and 17.8 g dL^{-1} respectively for Tibetan and Andean women. The values for Tibetans are similar to those of LAN; by contrast, Hb concentrations in Andeans are similar to those of LAN recently exposed to hypoxia. Thus, Tibetan HAN are similar to bar-headed geese when exposed to hypoxia, in not having increased haemoglobin concentration, whereas Andean HAN are similar to lowland birds exposed to hypoxia; they have increased haemoglobin concentration.

The eggs of species of birds that live at high altitude also have adaptations to cope with the low environmental PO_2; we discuss these adaptations in Case Study 15.2.

[69] The structure of haemoglobin is discussed in Section 13.1.1

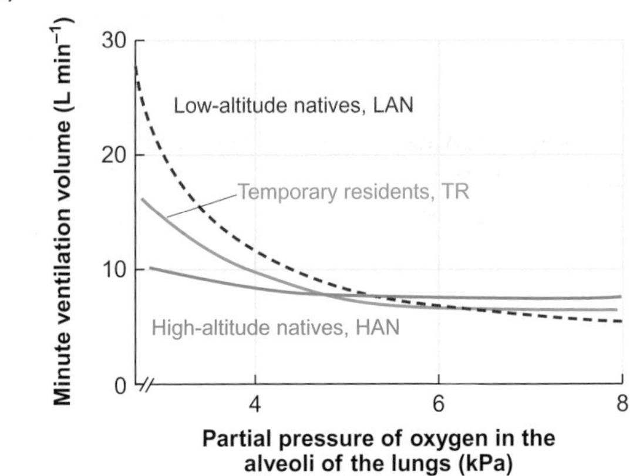

Figure 15.32 Effect of hypoxia on ventilation in individuals that are low altitude natives, LAN, high altitude natives, HAN, and temporary residents, TR, of various species of mammals, including man

(A) Relative increases in ventilation in humans, goats and llamas to a decrease in partial pressure of oxygen in the alveoli of the lungs (P_AO_2) from 13 to 6 kPa. (B) Changes in total ventilation volume in response to a reduction in P_AO_2 in humans.
Modified from Lenfant C (1973). High altitude adaptation in mammals. American Zoologist 13: 447-456.

15.3.2 Life in burrows

Burrows represent another environment in which PO_2 is below the normoxic level, and where there is also an accompanying increase in PCO_2. Both aquatic and terrestrial animals can live in burrows. Among burrowing aquatic animals, most studies have been performed on some species of polychaete annelids, such as

Case Study 15.2 Gas exchange across birds' eggs at high altitude

Gas exchange across the egg shell occurs by diffusion, so eggs that are incubated at relatively high altitudes, where partial pressure of oxygen (PO_2) is reduced, must cope with a reduced pressure difference between the surrounding air and the metabolizing cells of the embryo compared with eggs that are incubated at or close to sea level. This effect is partly offset by the increase in the diffusion coefficient (diffusivity) with increased altitude[1], although this effect would also increase the rate of water and CO_2 loss from the egg.

Up to an altitude of about 3000 m, the number of pores in the eggshell, and hence its total pore area is reduced so that the conductance to water, and therefore to O_2, is reduced to match the increase in diffusivity. This reduction in pore number means that the rate of water loss does not increase above that at sea level. In other words, conservation of water takes precedence over the supply of oxygen up to an altitude of about 3000 m. This means that the egg is more hypoxic than it would otherwise be if conductance of the eggshell was the same as that in lowland birds. However, the conductance of the eggs of birds living above 3000 m is not less than that of lowland birds. Consequently, oxygen delivery is not compromised, although water loss is greater but apparently tolerable. Different species that live above 3000 m have adopted at least two different strategies to deal with the low environmental PO_2.

These two strategies are illustrated in Figures A–C, which show data from studies on the eggs of Peruvian coots (*Fulica americana*), which breed at both sea level and at high altitude (over 4000 m), in the Andes, and of Puna teal (*Anas versicolor puna*), which breed only at high altitude. One strategy is used by the eggs of coots laid at over 4000 m, which have a lower rate of oxygen consumption ($\dot{M}O_2$) than that of eggs laid at sea level, as shown in Figure A. Conductance of oxygen from the surrounding air to the blood in the chorioallantois of coot eggs remains constant as the embryo develops, as depicted in Figure B, so there is very little change in the partial pressure of oxygen in the air cell, which is illustrated in Figure C.

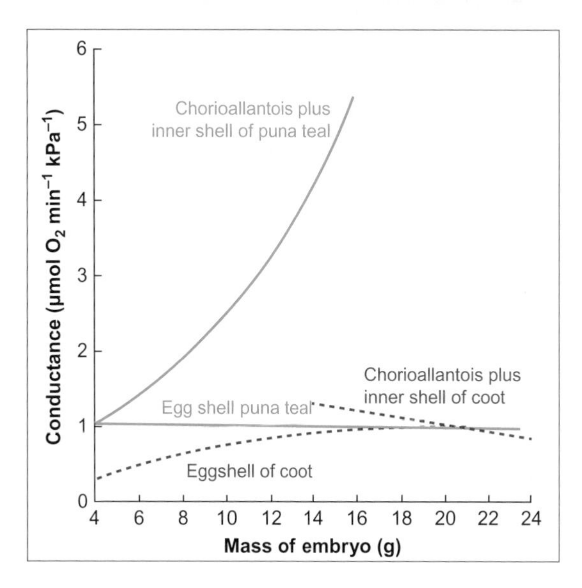

Figure B Changes in conductance for oxygen of the eggshell and chorioallantois during development of Puna teal and coot eggs at high altitude

Modified from Carey C et al (1989). Variation in eggshell characteristics and gas exchange of montane and lowland coot eggs. Journal of Comparative Physiology 159: 398-400; Carey C et al (1994). Gas exchange and blood gases of Puna teal (*Anas versicolor puna*) embryos in the Peruvian Andes. Journal of Comparative Physiology 163: 649-656.

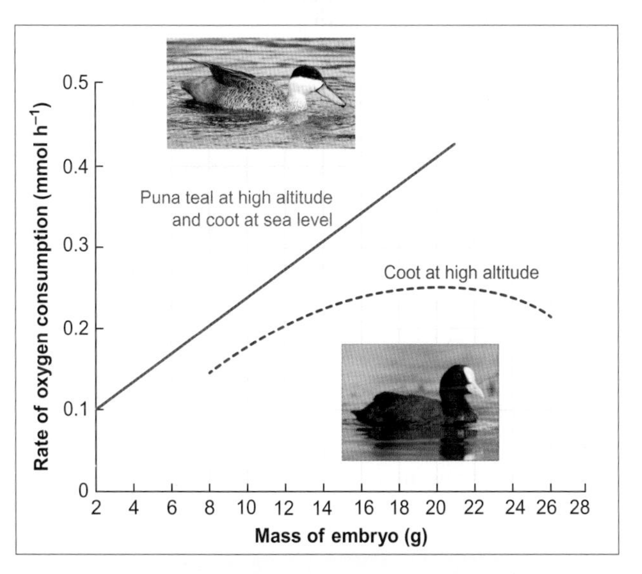

Figure A Rate of oxygen consumption (metabolic rate) of the eggs of Puna teal (*Anas versicolor puna*) at high altitude and of coot (*Fulicula americana*) at both sea level and high altitude

Modified from Léon-Velarde, F et al (1997). Physiological strategies of oxygen transport in high altitude bird embryos. Comparative Biochemistry and Physiology 118A: 31-37. Images: flickr.com; animal.discovery.com.

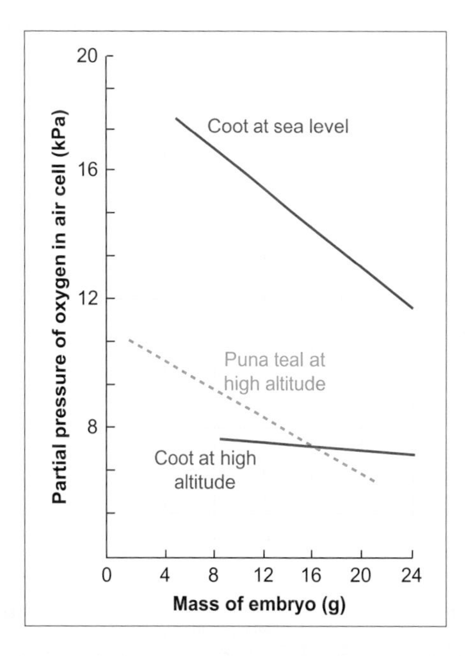

Figure C Partial pressures of oxygen in the air cell of eggs during development of Puna teal at high altitude and coots at both sea level and high altitude

Redrawn from Carey C et al (1989). Variation in eggshell characteristics and gas exchange of montane and lowland coot eggs. Journal of Comparative Physiology 159: 398-400; Carey C et al (1994). Gas exchange and blood gases of Puna teal (*Anas versicolor puna*) embryos in the Peruvian Andes. Journal of Comparative Physiology 163: 649-656.

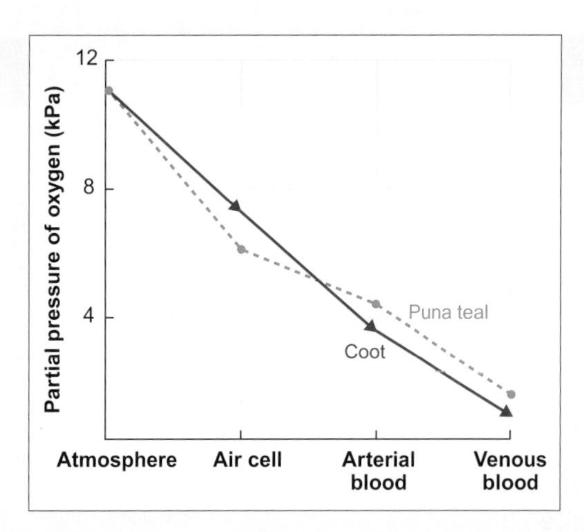

Figure D Partial pressures of oxygen in the environment, air cell of the eggs and in arterial and venous blood of near-term embryos of Puna teal and coots at high altitude

Modified from Léon-Velarde, F et al (1997). Physiological strategies of oxygen transport in high altitude bird embryos. Comparative Biochemistry and Physiology 118A: 31-37.

Eggs of teals laid at high altitude use an alternative strategy. Figure A shows that their $\dot{M}O_2$ increases continuously during development. This increase in $\dot{M}O_2$ is possible because the combined diffusion conductance of the inner membrane and the chorioallantois also increases throughout development, as shown in Figure B. As a result, and as demonstrated in Figure C, PO_2 in the air cell falls considerably, as it does in coot eggs at sea level. So, despite using these two different ways to cope with high altitude, the net result is that the PO_2 in the arterial or venous blood of the near-term embryos of teal and coot are similar, as shown in Figure D.

› Find out more

Carey C, Dunin-Borkowski O, Leon-Velarde F, Espinoza F, Monge C (1994). Gas exchange and blood gases of Puna teal (*Anas versicolor puna*) embryos in the Peruvian Andes. Journal of Comparative Physiology B 163: 649–656.

Carey C, Leon-Velarde F, Dunin-Borkowski O, Bucher TL, de la Torre G, Espinoza D, Monge C (1989). Variation in eggshell characteristics and gas exchange of montane and lowland coot eggs. Journal of Comparative Physiology B 159: 398–400.

Léon-Velarde F, Monge C, Carey C (1997). Physiological strategies of oxygen transport in high altitude bird embryos. Comparative Biochemistry and Physiology 118A: 31–37.

Monge C, Léon-Velarde F, Carey C (1997). Oxygen transport system of high-altitude bird embryos. News in Physiological Sciences 12: 121–125.

Rahn H, Carey C, Balmas K, Bhatia B, Paganelli C. (1977). Reduction in pore area of the avian eggshell as an adaptation to altitude. Proceedings of the National Academy of Sciences U S A 74: 3095–3098.

[1] Diffusion coefficient, *D* is defined as the quantity of a substance diffusing per unit time through a unit cross-sectional area and unit thickness of permeable barrier for a unit difference in concentration of the substance. It is discussed further in Section 11.2.1.

lugworms (*Arenicola marina*)[70], decapod crustaceans and fish. Among terrestrial animals, burrowing owls (*Athene cunicularia*) and mole rats (members of the family Bathyergidae) have been relatively well studied.

As a result of the intermittent irrigation of their burrows by aquatic animals, PO_2 in the burrow can vary between 8 and 17 kPa (compared to 20 kPa in the water above the burrows) and PCO_2 can increase to over 0.2 kPa (compared to 0.05 kPa, above the burrows). There is no evidence that the respiratory responses of burrowing decapod crustaceans and fish to hypoxia are substantially different from those of non-burrowing species. However, the respiratory pigments of burrowing crustaceans have higher affinities for oxygen (lower P_{50}s), higher oxygen carrying capacity and moderately high Bohr effects[71], all of which facilitate oxygen uptake and delivery during hypoxia. High-affinity respiratory pigment is also a feature of freshwater species of fish that experience hypoxia on a regular basis.

PCO_2 is much higher in terrestrial burrows and can reach 7–10 kPa in the burrows of the naked mole rat (*Heterocephalus glaber*). The lower PCO_2 in aquatic burrows than in terrestrial ones is because of the high capacitance coefficient of CO_2 in water[72]. Burrowing birds and mammals have adapted to a higher PCO_2 by having a reduced sensitivity to CO_2 compared with non-burrowing species. A contributing factor to this reduced ventilatory response to elevated CO_2 in burrowing owls is the lower sensitivity of the CO_2 receptors in their lungs (the intrapulmonary chemoreceptors[73]).

PO_2 in the burrows of terrestrial animals is generally within the range of that found in aquatic burrows, at between 7 to 14 kPa, but can reach as low as 6 kPa in the burrows of some subterranean mammals, and the mass huddling behaviour of some species, such as the naked mole rat, can exacerbate the level of hypoxia. Naked mole rats are able to tolerate an environmental PO_2 of 5 kPa for 5 h with no obvious adverse effects, compared with house mice (*Mus musculus*), which die in less than 15 min at a PO_2 of 5 kPa.

A reduced ventilatory response to hypoxia is not common in burrowing mammals, and is absent in burrowing owls. The oxygen affinity of the haemoglobin is similar in

[70] A diagram of a lugworm in its burrow is shown in Figure A, Case Study 20.1.

[71] Oxygen affinity, carrying capacity and the Bohr effect of blood pigments are discussed in Section 13.1.2.

[72] Capacitance coefficients of CO_2 in air and water are discussed in Section 11.1.2.

[73] Intrapulmonary chemoreceptors of birds are discussed in Experimental Panel 22.1.

15

burrowing owls and non-burrowing birds, with a P_{50} around 5.7 kPa. In contrast, the haemoglobin of burrowing mammals does have a higher oxygen affinity than that of non-burrowing species.

A number of other characteristics are also present in burrowing mammals, which enable them to survive under the hypoxic conditions found in their burrows. These characterisitics have been studied most extensively in Middle East blind mole rats (*Spalax ehrenbergi*).

Burrowing mammals in general have lower basal metabolic rates (BMR) than non-burrowing species. Despite this, however, maximum $\dot{M}O_2$ during hypoxia (11 per cent O_2) is higher in mole rats compared with that in white rats. This is because of the following adaptations to the oxygen transport system of mole rats:

- a 43 per cent greater lung diffusion conductance;
- haemoglobin concentration in the upper range for terrestrial mammals;
- a 31 per cent greater capillary density in skeletal muscles;
- three-fold greater concentration of myoglobin in skeletal muscles to facilitate diffusion of oxygen within the muscles and maybe, to supply oxygen when delivery by the circulatory system is temporarily insufficient to meet the demand of the muscles.

Although burrowing birds and mammals are exposed to varying degrees of hypoxia and hypercapnia, they do have continual access to the surrounding air and are therefore able continually to ventilate their lungs. The same is not the case for air-breathing vertebrates that dive under water, either to forage or to travel from one location to another. We discuss the adaptations to diving in air-breathing vertebrates next.

15.3.3 Diving birds and mammals

Intermittent, or irregular, ventilation of the lungs is common in amphibians and reptiles and occurs when the animals are in air, as well as when they are submerged under water. Responses of the cardiovascular system to irregular ventilation in amphibians and reptiles are discussed in Chapter 14, so in this section we concentrate on those aquatic birds and mammals which routinely undergo irregular ventilation when diving.

Oxygen stores and the effects of hydrostatic pressure in diving birds and mammals

When submerged under water, diving birds and mammals are unable to ventilate their lungs, so must rely entirely on the oxygen carried in various body stores: the respiratory system, the blood (associated with haemoglobin) and the muscles (associated with myoglobin). To maximize these stores, it might be expected that the volume of the respiratory system, blood volume, and the concentrations of haemoglobin (Hb) and myoglobin (Mb) would be greater in diving animals than in non-divers—and this is indeed the case.

Total oxygen stores in diving birds and mammals

Table 15.4 shows that total oxygen stores are larger in diving than in non-diving birds and mammals and that they tend to be greater for animals that dive the deepest and the longest. In some diving animals, such as Weddell seals (*Leptonychotes weddellii*), the spleen releases red blood cells during dives so that haematocrit and Hb concentration rise substantially (by about 50 per cent) above the non-diving values.

In more general terms, the change in size of the oxygen stores is in direct proportion to body mass ($M_b^{1.0}$), whereas basal metabolic rate (BMR) scales[74] somewhere around $M_b^{0.67}$. These different scaling factors suggest that that dive duration could be longer in larger animals. Indeed, an extensive phylogenetic analysis of diving birds and mammals in 2006 indicates that both mean and maximum dive durations scale with $M_b^{0.33}$. Figure 15.33 illustrates these different relationships. It has been found that a multiple regression with body mass and maximum concentration of myoglobin in skeletal muscle as factors, explains over 80 per cent of the variation in maximum dive duration across the whole range of diving ability in mammals.

The relationship between myoglobin concentration and diving ability has been used to construct an evolutionary tree of diving behaviour of living and extinct species of terrestrial and aquatic mammals, as illustrated in Figure 15.34. For example, *Pakicetus*, from the early Eocene, was about the size of a wolf and is thought to have been amphibious. It had an estimated maximum dive duration of about 1.5 min. In contrast, *Basilosaurus* from the late Eocene, weighed approximately 6500 kg and had an estimated maximum dive duration of about 17.5 min. This duration falls within the range of some living dolphins but is far below that of the longest-diving whales. For example, Cuvier's beaked whales, *Ziphius cavirostris*, have a median dive duration of 67.4 min, with a maximum of 137.5 min. The estimated diving capacity of an early seal, *Enaliarctos*, is at the lower end of the range of living eared seals (fur seals and sea lions).

The store of oxygen in the lungs could be problematic for deeper diving birds and mammals. One problem with a relatively large amount of air in the respiratory system is that it makes the animal more buoyant, although descent through

[74] The scaling of basal metabolic rate against body mass is discussed in Section 2.2.4.

Table 15.4 Dive behaviour and usable oxygen stores in diving and non-diving birds and mammals

Species	Common (and max) duration (min)	Common (and max) depth (m)	Diving respiratory volume (mL kg^{-1})	Blood volume (mL kg^{-1})	Haemoglobin concentration (g dL^{-1})	Muscle mass (per cent body mass)	Myoglobin concentration (g 100 g^{-1})	Total body oxygen stores (mL O_2 kg^{-1}, mmol O_2 kg^{-1})
Mallard duck *Anas platyrhynchos*			112	91	17.1	25	0.55	29, 1.3
Tufted duck *Aythya fuligula*	0.37 (0.77)	2 (3)	180	114	18.4	25	0.9	42, 1.9
Adélie penguin *Pygoscelis adéliae*	1.7 (3)	26 (98)	200	87	16	40	3.0	63, 2.8
King penguin *Aptenodytes patagonicus*	3–5 (8.7)	70–160 (304)	125	83	18	33	4.3	55, 2.45
Emperor penguin *A. forsteri*	4–5 (21.8)	20–40 (564)	117	100	18	25	6.4	68, 3.0
Human			78	77	14.8	35	0.47	24, 1.1
Bottlenose dolphin *Tursiops truncatus*	1 (8)	20 (390)	40	71	14	30	3.3	29, 1.3
California sea lion *Zalophus californianus*	2 (10)	62 (274)	48	120	18	37	5.4	55, 2.45
Weddell seal *Leptonychotes weddellii*	10–12 (82)	150–400 (726)	27	210	26	38	5.4	89, 4.0
Sperm whale *Physeter macrocephalus*	40–60 (138)	400–900 (2250)	28	200	22	34	5.4	74, 3.3
Northern elephant seal *Mirounga angustirostris*	23 (119)	437 (1581)	27	216	26	28	6.5	88, 3.9

References: Janssen et al., 2000; Keijer and Butler, 1982; Moore et al., 1999; Ponganis, 2011; Ponganis et al., 2011; Wienecke et al., 2007. Full citations available online.

the water column compresses the air so buoyancy is reduced. For every 10 m descent in the water column, hydrostatic pressure increases by approximately 1 atmosphere (101.3 kPa).

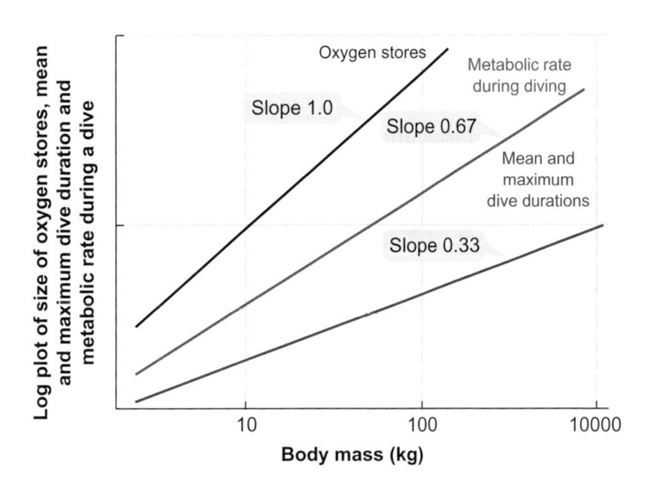

Figure 15.33 Relative relationships between size of oxygen stores, mean and maximum dive durations and metabolic rate during diving with body mass of diving birds and mammals

Oxygen stores scale directly with body mass (slope of 1.0), whereas mean and maximum dive durations scale with body mass to the power of 0.33. These suggest that maximum metabolic rate during diving scales with body mass to the power of 0.67.

So, going from the water surface, where the pressure is 1 atmosphere, to a depth of 10 m doubles the hydrostatic pressure surrounding the animal. Providing the respiratory system is sufficiently compliant[75], the doubling of pressure will halve the volume of the respiratory system. Going to a depth of 30 m doubles hydrostatic pressure again (to four times that at the surface) and volume of the respiratory system is then reduced to a quarter of its surface value, and so on[76].

The changes in lung volume cause inverse changes in the partial pressure of the gases in the lungs. For example, if lung volume is reduced to a quarter, partial pressures of gases in the lungs, which includes nitrogen, increase by a factor of four. Blood flowing through the gas exchange regions of the lungs equilibrates with the gases in the lungs, so any changes in partial pressures of lung gases will also occur in blood gases, including the partial pressure of nitrogen (PN_2). An increase in PN_2 in the blood could potentially have deleterious effects on a diving bird or mammal. Why is this?

Nitrogen gas is not involved in any metabolic process, so it accumulates in the blood and tissues. This accumulation

[75] Compliance of the respiratory system is discussed in Section 12.3.4.

[76] The relationship between pressure and volume of a gas is discussed in Section 11.1.2.

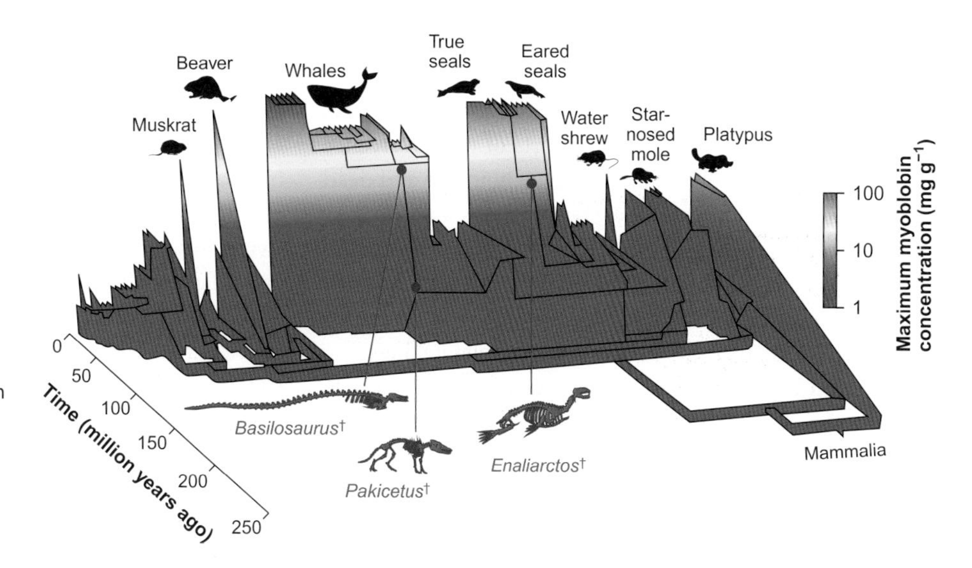

Figure 15.34 Evolutionary reconstruction of maximum myoglobin concentrations in terrestrial and aquatic mammals

All living elite mammalian divers (upper silhouettes) have high concentrations of myoglobin. The diving capacity of extinct species representing the stages during the transition from land to water (†) were deduced from estimates of their maximum myoglobin concentrations.

Modified from Mirceta S et al (2013). Evolution of mammalian diving capacity traced by myoglobin net surface charge. Science, 340, Issue 6138, 1234192.

will be greater for longer and deeper dives. The increase in PN_2 in the blood during a dive may cause unconsciousness, but if ascent from a dive is too rapid, nitrogen will not leave the blood and tissues via the lungs. Instead, it will form bubbles in the tissues. Formation of nitrogen bubbles in tissues is known as **decompression sickness**, or **the bends**. Decompression sickness is not likely to be a problem for relatively shallow divers, and deeper divers are thought to have evolved mechanisms to prevent its development. In humans, the bends can be very painful and, at worse, lethal.

One of the solutions evolved by deeper divers is for the volume of the respiratory system to be as small as possible. In birds, about 90 per cent of the volume of the respiratory system consists of the air sacs where there is no substantial gas exchange. However, as we discuss later, there is evidence that the gas in the air sacs can be moved across the gas-exchange surfaces of the lungs during dives and thus undergo gas exchange. Such gas exchange would not only enhance PO_2 in the blood, but also PN_2.

The reduction in the proportion of oxygen stored in the respiratory system is more than offset by increased stores in the blood and muscles as illustrated in Table 15.4 and Figure 15.35. Note the extent to which blood volume, haemoglobin concentration and myoglobin concentration are elevated in the deeper- and longer-diving species and, therefore, how they lead to increases in the proportional contribution of the blood and muscles to the total stores of oxygen.

In addition to the extreme compressibility (compliance) of the lungs, the lower airways may be reinforced. For example, in sea lions and cetaceans, the lower airways are reinforced with cartilage and seals have connective tissue and smooth muscle in their lower airways. These reinforcements prevent the lower airways from collapsing as the animal dives deeper. It has been proposed that, as the lungs collapse, air in the alveoli is forced into upper airways where gas exchange with the blood is drastically reduced. Measurements of PN_2 in the blood during natural dives and during experimental exposure to high hydrostatic pressure support this proposal, although do not exclude the possibility of other factors being involved.

Calculations based on these data indicate that lung collapse occurs at depths of 20 m in northern elephant seals (*Mirounga angustirostris*) and at 50 m in Weddell seals.

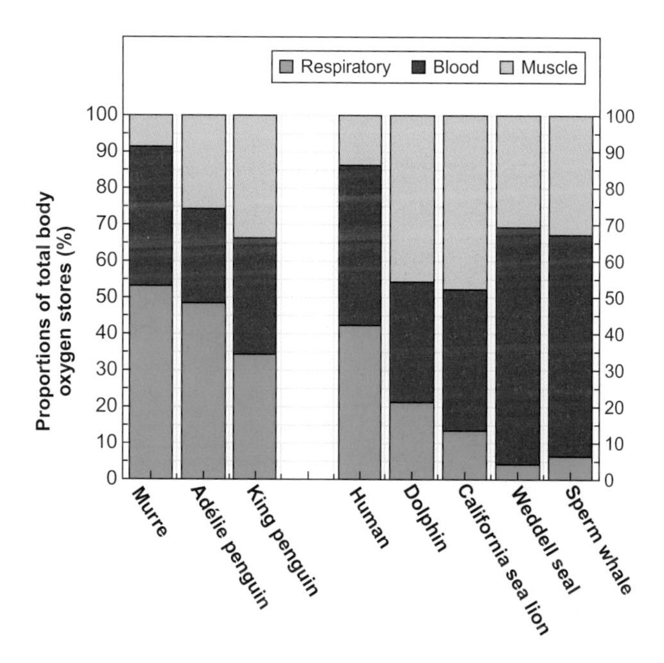

Figure 15.35 Distribution of body oxygen stores between respiratory system, blood and muscles in diving birds and mammals

Thick-billed murre, or Brünnich's guillemot (*Uria lomvia*); Adélie penguin (*Pygoscelis adéliae*); king penguin (*Aptenodytes patagonicus*); bottlenose dolphin (*Tursiops truncatus*); California sea lion (*Zalophus californianus*); Weddell seal (*Leptonychotes weddellii*); sperm whale (*Physeter macrocephalus*).

Modified from Ponganis PJ et al (2011). In pursuit of Irving and Scholander: a review of oxygen store management in seals and penguins. Journal of Experimental Biology 214: 3325-3339.

However, there is more recent evidence that the trachea is compressed in some species and that alveolar collapse may not occur until depths of over 100 m have been reached. There have been relatively recent reports of the occurrence of gas bubbles in the blood and tissues of both live and dead stranded marine mammals. These observations indicate that, particularly in deep diving mammals, the presence of gas bubbles (composed of over 70 per cent nitrogen) may be more prevalent than previously thought, but that as yet unknown adaptations normally enable the animals to tolerate them.

Adélie and gentoo penguins (*Pygoscelis adéliae*, and *P. papua*, respectively) carry a large volume of gas in their lungs and air sacs in relation to their body mass—some seven times greater than that in Weddell and elephant seals. It is suggested, however, that the dives of gentoo penguins are sufficiently shallow and of sufficiently short duration to provide adequate protection against dangerously high levels of PN_2 in the blood, although this may not be the full explanation.

King and emperor penguins (*Aptenodytes patagonicus* and *A. forsteri*, respectively) dive for longer durations and to greater depths than gentoo penguins, with more recent studies on king penguins showing that PN_2 in their blood does not increase above the minimum PN_2 of 330 kPa necessary for bubble formation upon decompression in non-diving species of mammals, even to a depth of 136 m. An important factor in king penguins seems to be the relatively small volume of their respiratory system, as shown in Figure 15.35.

The diving response and management of the oxygen stores

As well as possessing relatively large oxygen stores, diving birds and mammals must ensure that the stores are depleted at an optimum rate in order to sustain their underwater activity, which is often associated with foraging. During the early days of the studies of diving physiology, in the 1930s and 1940s, physiologists forcibly submersed their study species. Under such conditions, all birds and mammals show a similar response, the **diving response**, which is illustrated in Figure 15.36. The most obvious response is a gradual reduction in heart rate, the so-called **diving bradycardia**, which is shown in Figure 15.36A. The bradycardia results in a reduction in cardiac output. Functionally, the most significant component of the diving response is a decrease in blood flow to all parts of the body except to the brain, which is illustrated in Figure 15.36B.

The reduction in peripheral blood flow seen during the diving response results from constriction of the blood vessels in most parts of the body. Blood pressure does not increase, as cardiac output is reduced. By reducing the flow of blood (and, hence, the supply of oxygen) to most of the body, the animals conserve oxygen for the part of their body that is most sensitive to lack of oxygen, the brain. Notice in Figure 15.36B that blood flow decreases to the heart, as well as to other organs, such as skeletal muscle. However, the fall

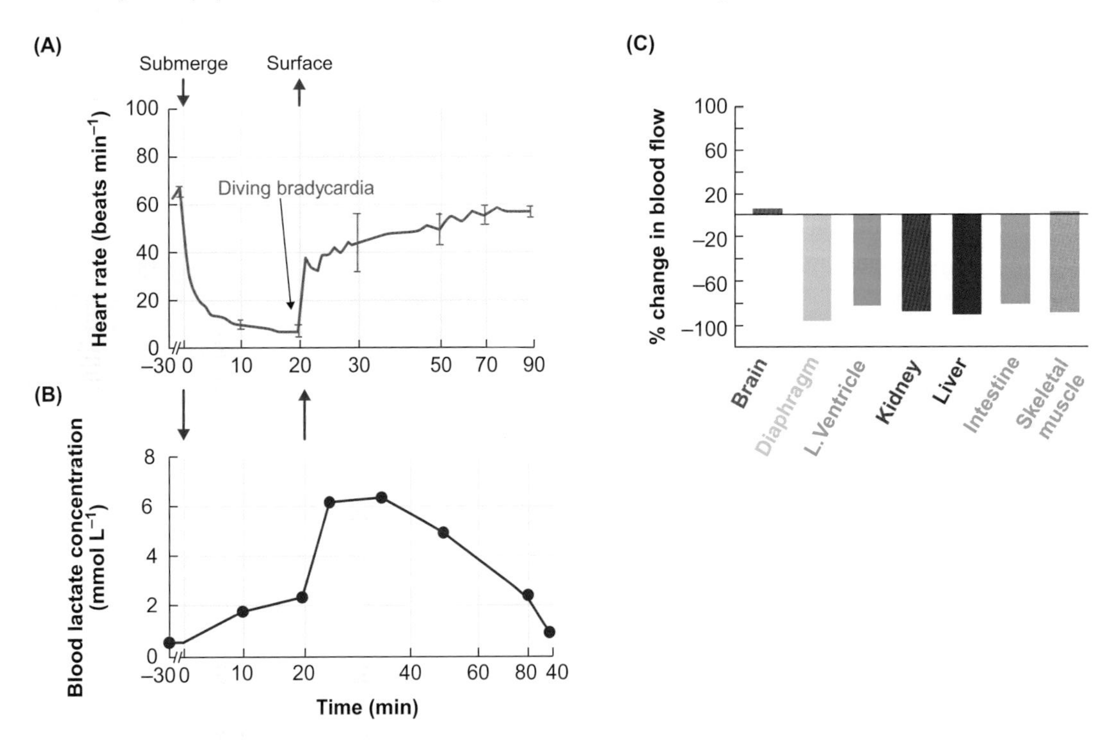

Figure 15.36 The diving response to involuntary submersion of Weddell seals (*Leptonychotes weddellii*)

(A) Mean (± SD) heart rate during and after submersion of four seals. (B) Percentage change in blood flow to various organs after 8–12 min submersion of up to six seals. (C) Concentration of lactate in arterial blood during and after submersion of one seal.

Reproduced from Butler PJ and Jones DR (1997). Physiology of diving birds and mammals. Physiological Reviews 77: 837-899.

15

in perfusion of the heart is in proportion to the reduction in its work load resulting from the bradycardia.

Although enforced submersion severely reduces the supply of oxygen to most tissues and organs, they continue to require a supply of ATP. This is provided to a much greater extent than normal by phosphocreatine and by anaerobic glycolysis, with the production of lactic acid. Note in Figure 15.36C that there is a relatively small rise in the concentration of blood lactate during the submersion period when there is peripheral vasoconstriction, but that it increases much more dramatically when the animal surfaces and dilation of the peripheral blood vessels occurs. In the example shown in Figure 15.36C, it took 2 h (six times longer than the duration of the submersion period) for the concentration of blood lactate to return to its presubmersion value. These data indicate that lactate is accumulating in many tissues and organs during enforced submersion and that it is flushed out upon surfacing when the tissues and organs are reperfused.

This view of how diving animals survive under water was largely unchallenged until the 1970s and early 1980s, although some investigators recognized in the early 1940s two facts about naturally diving birds and mammals:

- During many, if not most dives, diving birds or mammals are actively diving to depth to forage and pursue prey, which means that their locomotor muscles need a continual supply of oxygen.
- The majority of their dives are of relatively short duration and the stores of oxygen would be sufficient for these dives to be fuelled completely by aerobic metabolism without an accumulation of lactate.

This prediction was confirmed in 1980. The investigators transported Weddell seals or emperor penguins long distances from natural cracks in the ice and bored an artificial hole in the thick ice. They attached time depth recorders to the seals and removed blood samples before releasing the instrumented seals into the water at the hole. The animals had to return to the same hole to ventilate their lungs, when the recorders could be removed and more blood samples taken. Some of the results are shown in Figure 15.37.

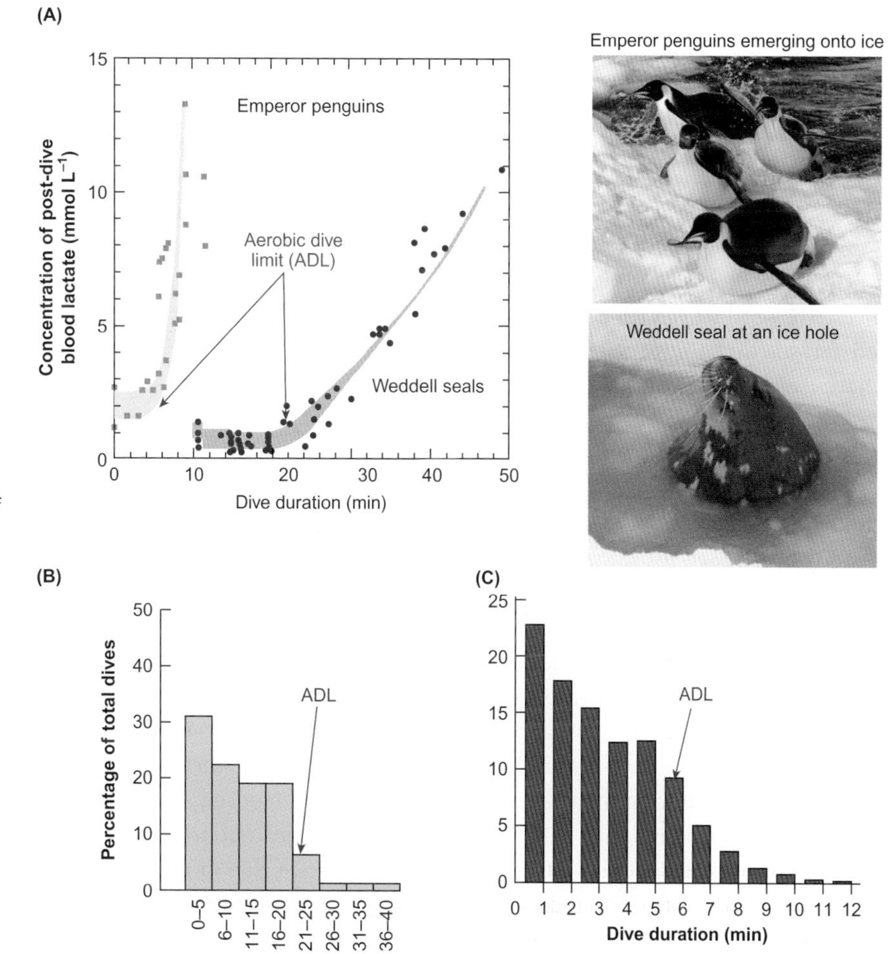

Figure 15.37 Aerobic dive limit and frequency distribution of dive durations in emperor penguins (*Aptenodytes forsteri*) and Weddell seals (*Leptonychotes weddellii*)

(A) Determination of dive duration, after which there is an increase in the concentration of blood lactate—the aerobic dive duration. It is about 5.5 min in the penguins and around 20 min in large male seals. (B) Frequency distribution of dive durations of Weddell seals. (C) Frequency distribution of dive durations of emperor penguins.

Reproduced from A: Kooyman GL and Ponganis PJ (1998). The physiological basis of diving to depth: birds and mammals. Annual Review of Physiology 60: 19-32. B: Kooyman GL et al (1980). Aerobic and anaerobic metabolism during voluntary diving in Weddell seals: evidence of preferred pathways from blood chemistry and behavior. Journal of Comparative Physiology B 138, 335-346. C: Kooyman GL and Kooyman TG (1995). Diving behavior of emperor penguins nurturing chicks at Coulman Island, Antarctica. The Condor 97: 536-549. Photos: PJ Ponganis, with permission and RW Davis, with permission.

Figure 15.37A shows that peak concentration of blood lactate in large (450 kg) adult male seals does not increase above the resting value during recovery from dives of up to 20 min duration. This duration is called the **aerobic dive limit (ADL)**, and can be compared to the anaerobic or lactate threshold seen in animals exercising at increasing intensity, which we showed previously in Figure 15.22.

Figure 15.37B shows that over 90 per cent of the dives of free-ranging Weddell seals are of relatively short duration and below the aerobic dive limit. However, the ADL is reached at shorter dive durations in smaller, immature Weddell seals. Figures 15.37A and C show that the aerobic dive limit of emperor penguins is about 5.5 min and around 80 per cent of dives at sea are of shorter duration than this. In dives that are of longer duration than the ADL, blood lactate concentration increases after the dive. The longer the dive, the greater the lactate concentration, and the longer it takes for blood lactate to return to the pre-dive value: after a 60 min dive, Weddell seals are exhausted and it takes up to 2 h for their blood lactate to recover, twice as long as the duration of the dive. Recall that the time it takes to clear lactate following 20 min of forced submersion is also 2 h, as shown in Figure 15.36, indicating that forced submersion is not representative of natural diving.

The data in Table 15.4 illustrate that the maximum dive durations of Weddell seals and emperor penguins are around four times longer than their ADL, suggesting that they have not exhausted their oxygen stores at this time. Sufficient oxygen must be available to supply the brain and locomotor muscles. However, some authors have calculated ADL by dividing the usable oxygen stores of an animal with some measure of the rate of oxygen usage ($\dot{M}O_2$) by the animal during dives. The product of such a calculation is called the **calculated aerobic dive limit (cADL)**, but it tells us when animals have *completely* exhausted their supply of oxygen, which they do not do. Hence, cADL is of limited scientific value and the only method to determine the true aerobic dive limit of an animal is to measure the post-dive lactate concentration in its blood.

In one study of Weddell seals, the animals dived for 10–12 h around mid-day. They performed a number of dives in relatively quick succession. Such a sequence of dives is known as a dive bout. Each dive bout is followed by a period when the animals do not dive. The average dive duration during these dive bouts was 11.5 min and the surface interval between each dive lasted 2–4 min. More impressive are female elephant seals during the 10-week period at sea between nursing their pups and their return to the rookery to moult. The females perform dives more or less continuously during this 10-week period, and only rarely is this behaviour interrupted. For most dives (99.5 per cent), the surface interval remains relatively constant at around 2 min, even after the longest dives of around 50 min. As yet, we do not know exactly how these seals recover so quickly after such long dives. The remaining 0.5 per cent of dives are followed by extended surface intervals which have a mean duration of 52 min.

The majority of natural dives performed by emperor penguins and Weddell seals are aerobic, and their oxygen requirements during these dives are provided by their oxygen stores, which are replaced during the relatively short period of time the animals are at the surface. The pattern of mainly aerobic dives is seen in almost all aquatic birds and mammals that have been studied. But how do birds and mammals diving in their natural environment manage their stores of oxygen effectively while foraging and travelling under water?

Management of the oxygen stores during diving

One important factor in optimizing the use of stored oxygen is to keep $\dot{M}O_2$ as low as possible. Deep and long-duration divers are very well streamlined, which is very important in reducing drag as the animals move through the water[77]. Some species such as Weddell and elephant seals and bottlenose dolphins make use of decreases in buoyancy as depth increases and cease using propulsive strokes below the depths at which they become negatively buoyant; they then glide passively down. A number of species of penguins can control the volume of air in their relatively large respiratory system to match the depth to which they will dive. However, none of these tactics can reduce metabolic rate to below its normal resting value. It has been suggested that metabolic rate may fall below the resting level during the dives of the female elephant seals we discussed earlier in this section. In fact, a reduction in temperature in some part of the body during bouts of diving could reduce metabolic rate, especially in some species of birds such as king penguins.

In some species, the fact that a large proportion of the stored oxygen is located in the muscles is a potential problem, as Mb has a much higher affinity for oxygen (low $P_{50} \approx 0.3$–0.4 kPa) than Hb ($P_{50} \approx 3.6$ kPa in Weddell seals). Unless PO_2 in the blood flowing to the muscles is sufficiently low, the oxygen stored in the muscles would only be usable by the muscles themselves. However, the oxygen stores in the muscles are not always sufficient to support muscle oxygen consumption, so it is not clear how diving birds and mammals are able to use all of their stored oxygen effectively.

A mathematical model suggests that the dive response of reduced cardiac output (bradycardia) and selective vasoconstriction is important in ensuring that diving mammals are able to use oxygen from both their Mb and Hb stores. The challenge is for Hb-bound O_2 and Mb-bound O_2 to be

[77] Locomotion through water is discussed in Section 18.6.2.

depleted at the same time. To achieve this, cardiac output (heart rate) and blood supply to the muscles must be carefully matched to the rate at which the muscles are using oxygen. The predicted outcome is that as transport of oxygen by blood to the muscles progressively decreases during a dive, the divers become increasingly more reliant on the muscles' own myoglobin stores of O_2.

The model predicts that if the level of physical exertion (and hence $\dot{M}O_2$) increases during a dive, the required decreases in cardiac output and heart rate are less. These predictions are shown in Figure 15.38. The implications of the predictions are that heart rate would be lower and aerobic dive limit (ADL) longer during relatively inactive dives, when $\dot{M}O_2$ is low, compared to those during more energetic dives, when $\dot{M}O_2$ is high. This emphasizes that the ADL is not a fixed value for any naturally diving species.

The aerobic dive limit depends on the extent to which the oxygen stores have been replenished between dives and the rate at which they are consumed during the subsequent dive. In other words, there seems to be a balance between the dive response (bradycardia, decreased cardiac output) and the exercise response (tachycardia, increased cardiac output) which varies with the intensity of exercise during diving.

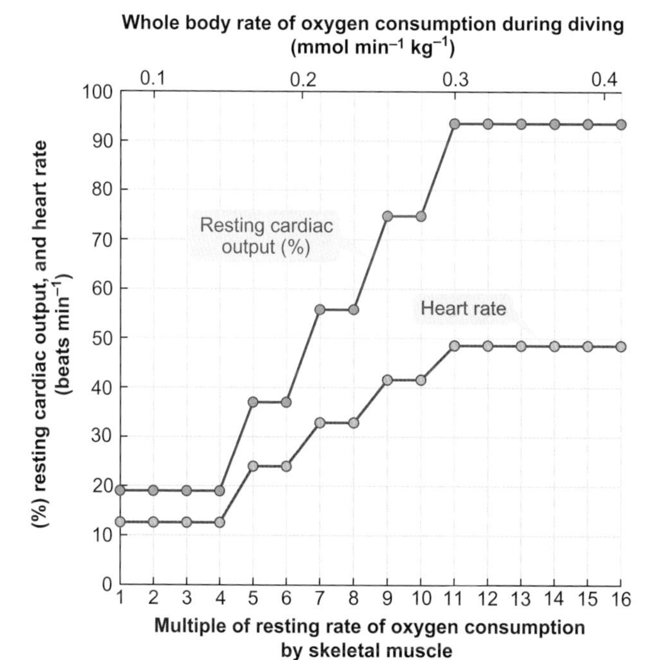

Figure 15.38 Predicted optimum cardiac output and heart rate as functions of rate of oxygen consumption by a Weddell seal and by its skeletal muscles during a dive

Optimum values enable the maximum aerobic dive limit (ADL) to be achieved. At higher rates of oxygen consumption during a dive, heart rate and cardiac output must be higher in order to maximize ADL.

Reproduced from Davis RW and Kanatous SB (1999). Convective oxygen transport and tissue oxygen consumption in Weddell seals during aerobic dives. Journal of Experimental Biology 202: 1091-1113.

There is some evidence for this model from naturally diving birds and mammals. However, a number of studies do not report a correlation between heart rate during a dive and level of exercise.

Recent advances in technology have enabled us to investigate how effectively free-ranging aquatic birds and mammals manage their oxygen stores during natural dives. For example, there is evidence that the relatively large volume of oxygen present in the respiratory systems of most species of birds can be accessed, possibly by the movements of the limbs when swimming underwater causing air to move between air sacs and through the gas-exchange regions of the lungs. Figure 15.39 illustrates that even in emperor penguins, where the relative size of the oxygen store of the respiratory system is lower than that in other species of birds, some oxygen is removed from the respiratory system during dives. These impressive and unique data were obtained by using implantable miniature oxygen sensors attached to data recorders and releasing the birds into isolated, artificial holes in the Antarctic ice. This is a far cry from the early studies on restrained animals in the laboratory.

Note in Figure 15.39 that the initial increase in PO_2 in the air sacs as they are compressed during descent is followed by a decrease as oxygen is consumed and as hydrostatic pressure declines during ascent. This profile of an initial increase in PO_2 followed by a decrease, also occurs in the arterial and venous blood as a result of the change in hydrostatic pressure during descent and surfacing.

The haemoglobin (Hb) in the blood of penguins has a higher affinity for oxygen than that of other birds. As a result, during some dives by emperor penguins, percentage saturation of arterial blood with oxygen can remain close to 100 per cent for almost an entire dive. However, percentage saturation of the Hb in arterial blood does decline during ascent as PO_2 decreases. The situation is different in elephant seals where saturation of Hb in arterial blood is normally between 70 and 30 per cent during the majority of a dive. This difference may be related to the lung oxygen store being relatively much smaller in seals than that in penguins, as shown by the data in Table 15.4.

There is much variability in PO_2 and saturation of venous blood before, during and after dives of emperor penguins compared with that of arterial blood. High values before some dives indicate hyperventilation, which may serve to reduce blood PCO_2 as well as to raise the level of venous PO_2. Initial or peak saturation may be as high as 90 per cent or more in venous blood, whereas saturations at the end of dives range from 75 to 3 per cent. This range of oxygen saturation of venous blood suggests that blood flow to the locomotor muscles during dives is variable to the extent that at one extreme, the whole of the venous oxygen could have been extracted by the muscles.

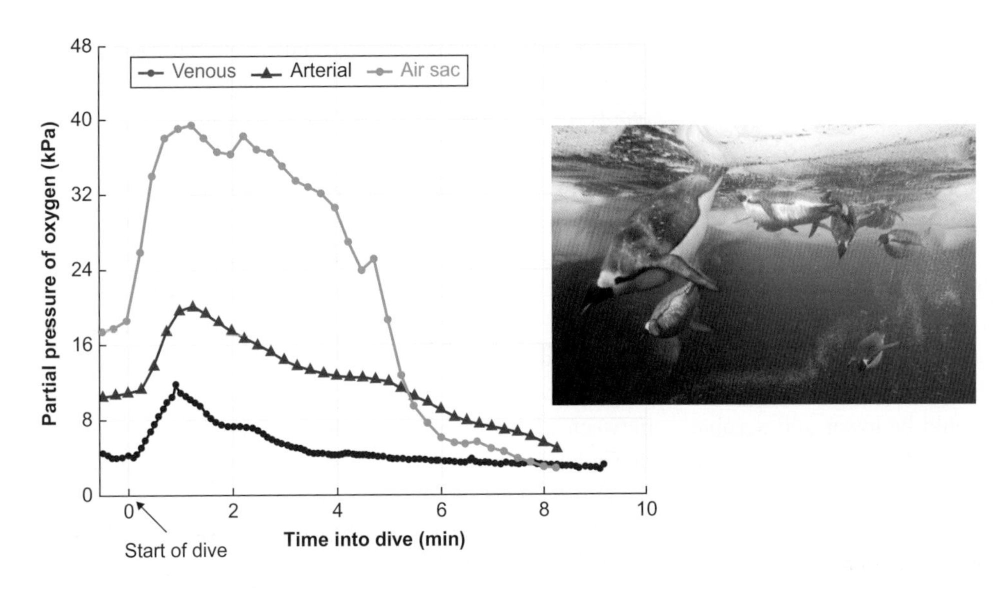

Figure 15.39 Depletion of oxygen stores in lungs and blood of emperor penguins (*Aptenodytes forsteri*) during natural dives

Partial pressure of oxygen (PO_2) in air sac, arterial and venous blood during different dives of 8–9 min duration. The initial increase in PO_2 is the result of the increase in hydrostatic pressure as the bird dives deeper. PO_2 in the air sac then decreases as oxygen is transferred to the blood and eventually used by the animal.

Reproduced from Ponganis PJ et al (2011). In pursuit of Irving and Scholander: a review of oxygen store management in seals and penguins. Journal of Experimental Biology 214: 3325-3339. Photo: PJ Ponganis, with permission.

The variability in the rate of depletion of oxygen from the venous blood is reflected in the rate of depletion of the oxygen combined to the myoglobin (Mb) in the muscles. Figure 15.40A shows that saturation of the Mb can decline continuously at a more or less steady rate during a dive, signifying that there is little or no blood supply to the muscles and the oxygen bound to the myoglobin is being used. On the other hand, Mb saturation can decline at a more irregular rate during submersion, as shown in Figure 15.40B, indicating that blood flow to the muscles is variable, with some oxygen in the blood being utilized.

The average rate at which all the body oxygen stores are depleted in emperor penguins diving for a duration of about 10 min is 0.3 mmol O_2 kg^{-1} min^{-1} (6.8 mL O_2 kg^{-1} min^{-1}), which is similar to their resting rate of oxygen consumption. The rate of oxygen depletion in unperfused muscle, as depicted in Figure 15.40A, is 0.55 mmol O_2 (kg muscle)$^{-1}$ min^{-1} (12.4 mL O_2 (kg muscle)$^{-1}$$min^{-1}$). However, it must be emphasized that the rate of oxygen depletion (and, therefore, metabolic rate) during diving is highly variable, and depends mainly on the level of activity during a particular dive. When oxygen supply to the muscles is unable to match demand, such as at the beginning of exercise or toward the end of a long dive, animals rely more on anaerobic sources of ATP.

Measurements of arterial PO_2 during routine dives of 10 min or more duration in northern elephant seals, *Mirounga angustirostris*, demonstrate that these animals routinely tolerate extreme hypoxaemia during such dives. As already indicated, the brain and central nervous system of mammals are extremely sensitive to low levels of oxygen, so how are some species of marine mammals able to survive under such circumstances? Studies on hooded seals, *Cystophora cristata*,

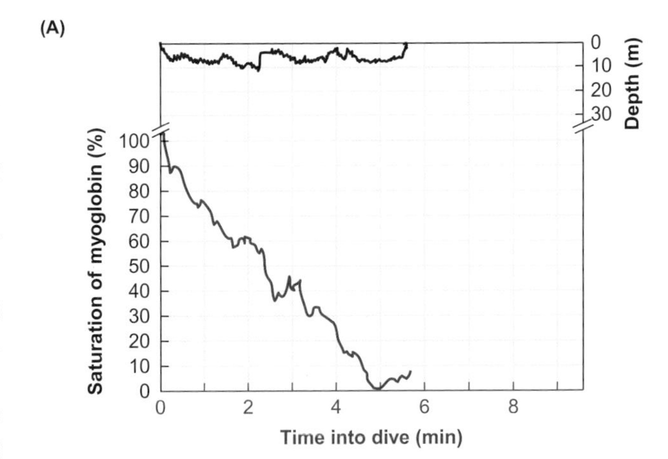

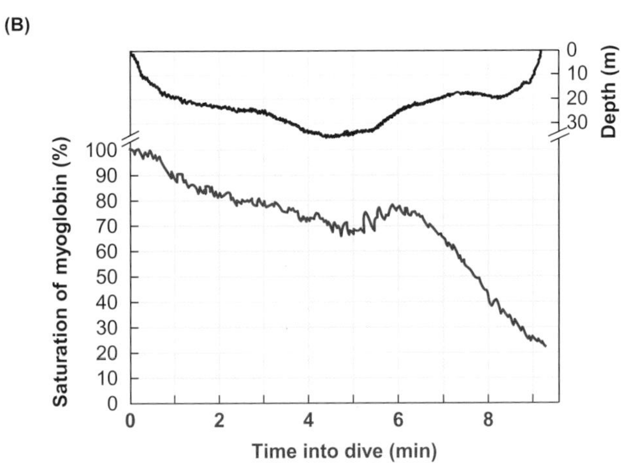

Figure 15.40 Depletion of oxygen stores in the myoglobin of emperor penguins (*Aptenodytes forsteri*) during natural dives

(A) During this relatively shallow dive, saturation of myoglobin (Mb) decreased relatively rapidly, indicating that the muscles were not being perfused with blood to any great extent, if at all. (B) During this deeper dive, Mb saturation initially declined relatively slowly, indicating perfusion of the muscle.

Reproduced from Ponganis PJ et al (2011). In pursuit of Irving and Scholander: a review of oxygen store management in seals and penguins. Journal of Experimental Biology 214: 3325-3339.

have shown that their brains may have a lower rate of aerobic metabolism and a greater expression of stress-related genes during hypoxia than those of terrestrial mammals.

Anaerobic metabolism during diving

In an unperfused, resting muscle of an emperor penguin, the anaerobic stores for ATP production are over eight times larger than the aerobic store. An important anaerobic source of ATP in muscles is phosphocreatine (PCr, as illustrated earlier in Figure 15.20), which can donate its phosphorus to ADP leading to the production of ATP. Phosphocreatine concentration in the muscles of emperor penguins is around 27 mmol (kg muscle)$^{-1}$ which, if completely depleted, can produce 27 mmol ATP (kg muscle)$^{-1}$. The glycogen store in the muscles can produce up to 163 mmol ATP (kg muscle)$^{-1}$, although it is unlikely that the complete store of glycogen would be consumed because of the acid–base implications. These values compare with the 23 mmol ATP (kg muscle)$^{-1}$ that could be produced aerobically from the oxygen stored in muscle [3.84 mmol (kg muscle)$^{-1}$].

Based on a rate of ATP usage of 3.32 mmol ATP (kg muscle)$^{-1}$ min^{-1} [= 0.55 mmol O_2 (kg muscle)$^{-1}$ min^{-1}], the anaerobic stores in the muscles (PCr and glycogen) could provide energy for dives that are well in excess of the longest recorded dive (22 min) for emperor penguins. However, dives of such extreme duration and which cause such depletion of the anaerobic stores would cause an extremely large accumulation of lactate [110 mmol lactate (kg muscle)$^{-1}$] and protons, cause large disturbances to acid–base balance and require long periods at the surface for complete re-synthesis of the glycogen.

› *Review articles*

Beall CM (2007). Two routes to functional adaptation: Tibetan and Andean high-altitude natives. Proceedings of the National Academy of Sciences of the USA 104: 8655–8660.

Boggs DF, Kilgore DL, Birchard GF (1984). Respiratory physiology of burrowing mammals and birds. Comparative Biochemistry and Physiology 77A: 1–7.

Butler PJ (2010). High fliers: the physiology of bar-headed geese. Comparative Biochemistry and Physiology A 156: 325–329.

Butler PJ, Jones DR (1997). Physiology of diving birds and mammals. Physiological Reviews 77: 837–899.

Butler PJ, Bishop CM, Woakes AJ (2003). Chasing a wild goose: posthatch growth of locomotor muscles and behavioural physiology of migration of an Arctic goose. In Berthold P, Gwinner E, Sonnenschein E (eds.) Avian Migration. pp 527–542. Berlin: Springer-Verlag.

Davis RW, Polasek L, Watson R, Fuson A, Williams TM, Kanatous SB. (2004). The diving paradox: new insights into the role of the dive response in air-breathing vertebrates. Comparative Biochemistry and Physiology Part A 138: 263–268.

Diaz RJ, Breitburg DL (2009). The hypoxic environment. In: Richards JG, Farrell AP, Brauner CJ (eds.) Fish Physiology, Vol 27 Hypoxia, pp 1–23. Amsterdam: Elsevier.

Dillon ME, Frazier MR, Dudley R (2006). Into thin air: physiology and evolution of alpine insects. Integrative and Comparative Biology 46: 49–61.

Farrell AP (2007). Tribute to P.L. Lutz: a message from the heart—why hypoxic bradycardia in fishes? Journal of Experimental Biology 210: 1715–1725.

Hooker SK, Fahlman A, Moore MJ, Aguilar de Soto N, Bernaldo de Quiros, Y, Brubakk AO, Costa DP, Costidis AM, Dennison S, et al. (2012). Deadly diving? Physiological and behavioural management of decompression stress in diving mammals. Proceedings of the Royal Society of London B 279: 1041–1050.

McMahon BR (2001). Respiratory and circulatory compensation to hypoxia in crustaceans. Respiration Physiology 128: 349–364.

Nilsson GE (2007). Gill remodelling in fish—a new fashion or an ancient secret? Journal of Experimental Biology 210: 2403–2409.

Perry SF, Gilmour KM (2010). Oxygen uptake and transport in water breathers. In: Nilsson GE (ed.). Respiratory Physiology of Vertebrates: Life with and without Oxygen. pp 49–94. Cambridge: Cambridge University Press.

Ponganis PJ, Meir JU, Williams C (2011). In pursuit of Irving and Scholander: a review of oxygen store management in seals and penguins. Journal of Experimental Biology 214: 3325–3339.

Ponganis PJ (2011). Diving mammals. Comprehensive Physiology 1: 517–535.

Rabalais NN, Diaz RJ, Levin LA, Turner RE, Gilbert D, Zhang J (2010). Dynamics and distribution of natural and human-caused hypoxia. Biogeosciences 7: 585–619.

Storz JF (2007). Hemoglobin function and physiological adaptation to hypoxia in high-altitude mammals. Journal of Mammalogy 88: 24–31.

Taylor AC, Atkinson RJA (1991). Respiratory adaptations of aquatic decapod crustaceans and fish to a burrowing mode of life. In: Woakes AJ, Grieshaber MK, Bridges CR (eds.). Physiological Strategies for Gas Exchange and Metabolism. pp 211–234. Cambridge University Press.

Wood SC (1995). Interrelationships between hypoxia and thermoregulation in vertebrates. In: Heisler N (ed.) Advances in Comparative and Environmental Physiology, Vol 22 pp 209–231. Berlin Heidelberg: Springer-Verlag.

15.4 Responding to a decrease in oxygen demand—hibernation

Many species of ectotherms, such as amphibians and reptiles, enter what is sometimes called torpor or hibernation[78] in as much as they become less active, or even inactive, and have a reduced metabolic rate during cold winter months. In most species of ectotherms, hibernation is mainly a passive response to the reduction in environmental temperature. By contrast, in endotherms (i.e. mammals) it is a controlled reduction in metabolic rate and body temperature.

The main function of daily and seasonal torpor (hibernation) in endotherms is to reduce the rate of energy expenditure, to enable the animal to survive a period of food shortage[79]. Thus, the role of the respiratory and cardiovascular systems during this state is to ensure that the brain and the rest of the body are supplied with the oxygen they require at the lowest possible energetic cost. During arousal from hibernation, metabolic rate increases markedly as the body quickly warms up. Equations 15.1 and 15.2 illustrate the components of the respiratory and cardiovascular systems which can change to match the different metabolic demands associated with the different phases of hibernation.

[78] Hibernation in ectotherms is discussed in Section 9.2.2.

[79] The metabolic aspects of torpor are discussed in Section 10.2.7.

15.4.1 Entry into hibernation

Upon entry into hibernation, a rapid decline in ventilation volume occurs before the fall in metabolic rate; the animals hypoventilate, leading to an increase in the partial pressure of CO_2 in arterial blood (P_aCO_2) and respiratory acidosis (decrease in pH). It has been suggested that this acidosis is involved in the initial suppression of metabolic rate, although there is some evidence against the idea. For example, breathing a gas mixture containing 6 per cent CO_2 by some species of mammalian hibernators does not facilitate entrance into hibernation.

Two different ventilation patterns occur during hibernation. In some species, such as Alpine marmots (*Marmota marmota*), ventilation slows and breaths become further apart, with the period of apnoea usually lasting 1–6 minutes, but sometimes up to 14 minutes. In other species, such as European hedgehogs (*Erinaceus europaeus*), periodic bursts of ventilation (called **episodic ventilation**) are separated by longer periods of apnoea lasting from some minutes to over an hour. The longest apnoeic period reported is 150 minutes for hedgehog. Figure 15.41 shows that ventilation in golden-mantled ground squirrels (*Callospermophilus lateralis*) can change between the two patterns, being episodic or slow and regular.

As body temperature (T_b) falls during the entrance to hibernation, neither tidal volume nor duration of each lung ventilation cycle change. Overall ventilation volume is modified by altering the duration between each ventilation in those species with a regular ventilation pattern or by altering the duration of the period of apnoea in episodic ventilators.

Heart rate may drop by over 50 per cent upon entry into hibernation before there is a fall in T_b. The decreases in heart rate and cardiac output upon entry to hibernation are accompanied by a significant decline in arterial blood pressure, shown in Figure 15.42A and B. Despite the fall in arterial blood pressure, blood flow to critical organs such as brain, kidneys, liver and heart is maintained. Although the hearts of non-hibernating mammals become arrhythmic and cease to function between 10 and 15°C, hearts of hibernators function down to 0°C, or less.

This difference in behaviour at low temperatures has been explained by studies on isolated myocytes from the left ventricles of hibernating and non-hibernating woodchucks (*Marmota monax*). The myocytes from the hearts of hibernators manage to prevent the build-up of Ca^{2+} (Ca^{2+} overload), which is a major cause of ventricular arrhythmias and fibrillation in the hearts of non-hibernators at low temperature. Significantly greater uptake of Ca^{2+} by the sarcoplasmic reticulum and a shorter duration of the action potential resulting from a reduction in Ca^{2+} channels contribute to maintaining normal function in the hearts of hibernators[80].

The initial fall in heart rate during entry into hibernation results from an increase in vagal (parasympathetic) activity. The heart may become arrhythmic but eventually settles into a slow uniform rate, as shown in Figure 15.42B and C. As body temperature falls, the activity of the vagus nerve is progressively withdrawn and there may also be a decrease

[80] Section 18.3.1 discusses action potentials and excitation–contraction coupling in cardiac muscle fibres.

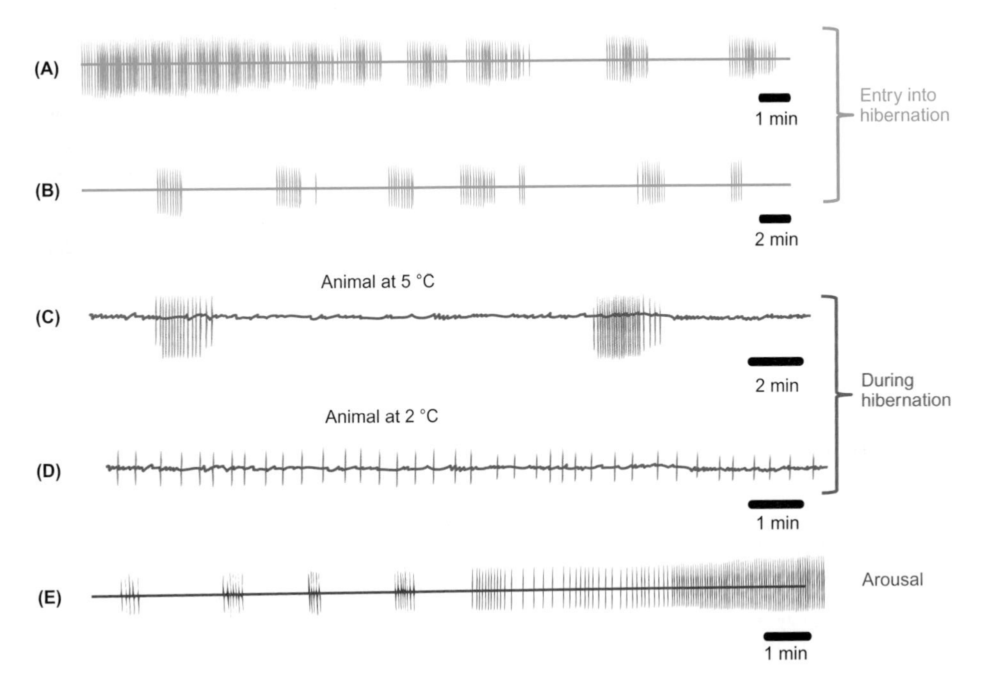

Figure 15.41 Ventilation of hibernating ground squirrels while entering and during hibernation and upon arousal

Ventilation patterns of golden-mantled ground squirrel (*Spermophilus lateralis*) while entering hibernation, (A) and (B); during hibernation, (C) and (D); and upon arousal, (E). Note the two different patterns of ventilation during hibernation—episodic when animal at 5°C and regular when it is at 2°C

Modified from: Milsom WK and Jackson DC (2011). Hibernation and gas exchange. Comprehensive Physiology 1: 397-420.

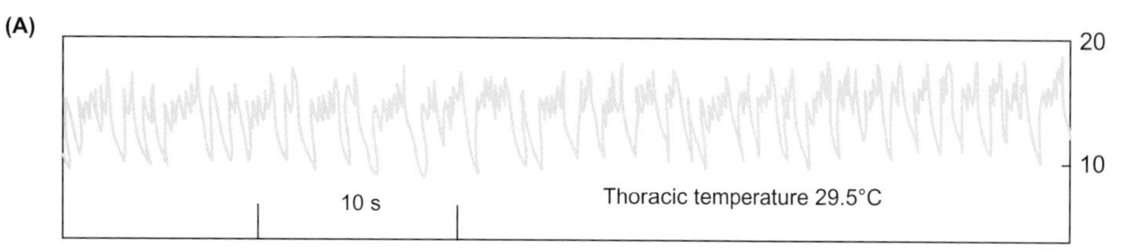

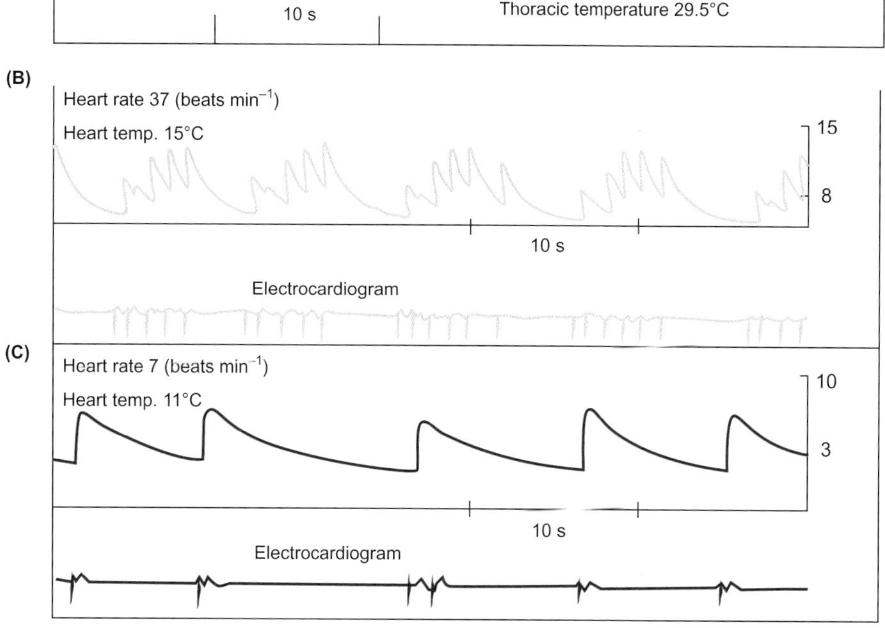

Figure 15.42 Heart rate, blood pressure and body temperature during various phases of hibernation in thirteen-lined ground squirrels (*Ictidomys tridecemlineatus*)

(A) Early phase of entrance into hibernation showing high (value not given) heart rate and skipped beats (fall in blood pressure). (B) Mid phase of entrance into hibernation showing irregular (arrhythmic) heart rate, as indicated by the electrocardiogram. (C) Late phase of entrance into hibernation showing slow, more uniform heart rate.

Reproduced from Milsom WK and Jackson DC (2011). Hibernation and gas exchange. Comprehensive Physiology 1: 397-420.

in sympathetic activity to the heart. The relative influences of the parasympathetic and sympathetic nervous systems on the heart during hibernation, if any, are unclear.

15.4.2 During hibernation

In some species of hibernators such as Columbian ground squirrels (*Urocitellus columbianus*), hypoxia does not stimulate ventilation during hibernation and the animals do not begin to awake, even if inspired O_2 is reduced to extremely low levels of 1–3 per cent. In contrast, the related golden-mantled ground squirrel does respond to hypoxia by increasing ventilation volume, but the threshold partial pressure of oxygen (PO_2) at which it responds is lower than that in non-hibernating individuals. The response remains after denervation of the carotid body chemoreceptors, which suggests that the aortic body chemoreceptors[81] may play a dominant role in this response.

In all species studied so far, hypercapnia strongly stimulates ventilation in hibernating mammals, and if CO_2 in inspired air is raised to 5–7 per cent, ventilation becomes continuous and animals often begin to awake. As with the response to hypoxia, the hyperventilation in response to hypercapnia is not affected by denervation of the carotid bodies.

The fact that ventilation is intermittent in some species during hibernation means that blood gases and pH can vary quite considerably. For example, during ventilatory pauses lasting 50 to 70 minutes in European hedgehogs, PO_2 in arterial blood (P_aO_2) may fall from around 15 kPa to as low as 1.5 kPa and pH may fall by 0.04 pH units. However, such a large reduction in P_aO_2 during hibernation is not as bad as it seems.

The oxygen affinity of haemoglobin increases dramatically during hibernation; the equilibrium curve is shifted to the left with P_{50} falling from around 2.41 kPa in summer animals to 0.77 kPa in hibernating golden-mantled ground squirrels, as shown in Figure 15.43. Thus, despite the low P_aO_2 that can occur, venous blood is 60–70 per cent saturated in hibernating hedgehogs, which is higher than that in non-hibernating hedgehogs. However, during arousal, the increase in ventilation may not match the rise in oxygen demand as the animal re-warms. In Arctic ground squirrels, P_aO_2 may fall to as low as 0.9 kPa and can be below 4.7 kPa for as long as 4 h during arousal.

During hibernation, blood pH is similar to or slightly above that in the non-hibernating state. For example, in European or common hamsters (*Cricetus cricetus*), plasma pH is 0.17 pH units higher during hibernation than when the animals are not hibernating. However, because of the large fall in body

[81] The influences of carotid and aortic bodies on the respiratory system are discussed in Section 22.2.3.

15

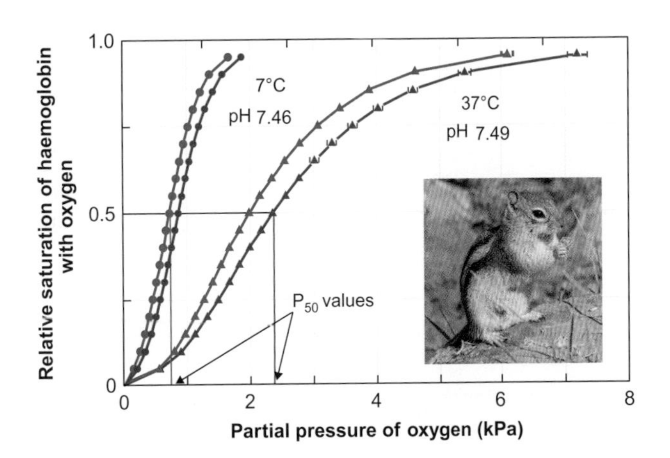

Figure 15.43 Effect of temperature on oxygen equilibrium curves (OECs) for winter hibernating (●, ▲) and summer active (●, ▲) golden-mantled ground squirrels (*Callospermophilus lateralis*)

At 7°C, the curves for both sets of animals are left shifted, there is an increase in the affinity of the haemoglobin for oxygen (decrease in PO_2 at which haemoglobin is half-saturated, P_{50}). There is also a direct effect of hibernation on P_{50}: at both temperatures there is a significant leftward shift of the OEC for the blood from the winter hibernators. The combined effect produces a reduction in P_{50} from 2.41 kPa in blood from summer active squirrels at 37°C to 0.77 kPa in blood from hibernating animals at 7°C.

Modified from Maginniss LA and Milsom WK (1994). Effects of hibernation on blood oxygen transport in the golden-mantled ground squirrel. Respiration Physiology 95: 195-208. Photo: gailhampshire/WikiMedia Commons.

temperature from 37 to 8°C when this species hibernates, the pH would be expected to increase by approximately 0.016 pH unit °C^{-1}, i.e. by 0.46 pH units based on the alphastat model of acid–base regulation[82] to maintain a constant net charge on the plasma proteins. Since 0.17 pH units is less than 0.46 pH units, the hibernating hamster actually exhibits relative respiratory acidosis., As a result of this, the net charge on the proteins would be expected to be less negative, as would happen if the pH becomes more acidic. It has been suggested that this relative acidosis may contribute to the metabolic depression during entry into hibernation.

However, the relative acidosis that occurs in blood plasma, and to a similar extent in brain tissue, does not occur in all tissues. The pH increases substantially more than 0.17 units in liver and heart muscle, and therefore, the relative acidosis is slight. Intracellular pH in skeletal muscle is intermediate between these two extremes.

15.4.3 Arousal from hibernation

The initial rapid increase in ventilation at the onset of arousal, shown in Figure 15.40E, causes a large amount of CO_2 to be removed and for plasma pH to increase. European hedgehogs hyperventilate and their respiratory exchange ratio is much greater than 1: plasma pH increases by as much as 0.17 pH units during the first 15–30 min of arousal thereby removing the inhibitory effect of the relative acidosis on metabolic rate, especially from thermogenic tissue such as brown adipose tissue[83].

The rapid increase in heart rate, and cardiac output, at the start of arousal is the result of an increase in sympathetic activity to the heart. Heart rate peaks together with ventilation and metabolic rate during mid to late arousal when shivering and non-shivering thermogenesis[80] are at their maximum. Studies on little brown bats (*Myotis lucifugus*) have demonstrated that the increased rate of blood flow during arousal is targeted to the brown adipose tissue and skeletal muscle to supply them with the oxygen required for thermogenesis. Heat generated in these tissues is distributed around the body, in particular to the brain. The rate of blood flow to organs such as kidneys, liver, stomach and intestine actually decreases during arousal, compared with that during hibernation.

› *Review articles*

Milsom WK, Jackson DC (2011). Hibernation and gas exchange. Comprehensive Physiology 1: 397–420.

[82] The alphastat model of acid–base regulation is discussed in Section 13.3.1.

[83] Brown adipose tissue, shivering and non-shivering thermogenesis are discussed in Section 10.2.5.

Checklist of key concepts

- Changes in the demand for and supply of oxygen are met by varying components of the respiratory and circulatory systems contained within the **Fick principle of convection**.

Responses to an increase in the demand for oxygen

- Feeding and digestion lead to an increase in the rate of oxygen consumption ($\dot{M}O_2$) known as the **specific dynamic action (SDA)** or **heat increment of feeding (HIF)**.
- The rate of blood flow increases to parts of the body involved in the processes of feeding and digestion mainly because of an increase in cardiac output.
- The size and composition of the meal determines the magnitude of the increase in blood flow.
- In mammals, blood flow increases sequentially as the **chyme** leaves the stomach and passes along the gastrointestinal tract.

- The constituent of chyme that increases blood flow most potently is **glucose**.
- Some carnivorous reptiles, such as pythons, may fast for several weeks; upon feeding, SDA may increase by up to 40 times the resting value and there may be an increase in heart ventricular mass.
- Secretion of acid by the parietal cells in the stomach of vertebrates tends to lead to the **alkaline tide**.
- In intermittent feeders, the potential to increase pH in the blood is normally offset by hypoventilation causing a **respiratory acidosis**.
- A respiratory acidosis is not possible in fish because of the higher capacitance coefficient of CO_2 in water compared to that in air.
- Exercise also causes an increase in $\dot{M}O_2$.
- The response of the respiratory system during sustained exercise is normally to increase **minute ventilation volume** by a similar proportion with little or no increase in the **amount of oxygen extracted from the respiratory medium**.
- The proportional increase in **cardiac output** in response to sustained exercise rarely matches the increase in $\dot{M}O_2$, so the **amount of oxygen extracted from the blood** increases by making use of the **venous reserve**.
- **Anaerobic metabolism** makes a substantial contribution to sustained levels of exercise in decapod crustaceans.
- Those wild Pacific salmon with the most demanding migratory journeys up river have a greater aerobic scope, a relatively larger ventricle to their heart, a better coronary supply to the ventricle and can swim faster than those with less demanding journeys.
- Training can increase aerobic scope and maximum exercise speeds in captive animals, but these do not reach the levels found in their wild counterparts.
- Wild animals that migrate are physically fitter before migration than during their non-migratory period.
- Blood flow increases to the locomotory muscles in exercising birds and mammals but decreases to non-locomotory muscles.
- **Artificial selection** can distort the natural balance between the demand for oxygen and the means of provision of that demand, as exemplified by thoroughbred racehorses.
- At the onset of exercise there is an **oxygen deficit**.
- **Excess post-exercise oxygen consumption** (**EPOC**) follows a period of aerobic exercise.
- Amphibians and reptiles rely more on anaerobic metabolism during exercise than birds and mammals.
- **Burst** or **exhaustive exercise** results in the production of acidic metabolic end-products such as lactic acid.
- At the transition between sustainable (aerobic) exercise and exhaustive exercise, the **anaerobic** or **lactate threshold**, lactic acid begins to accumulate in the blood.
- A potential **metabolic acidosis** above the lactate threshold is counteracted by a **respiratory alkalosis** in humans.

Responses to a decrease in oxygen supply

- Naturally occurring **hypoxia** can occur in bodies of fresh water and in some parts of the marine environment, but rarely occurs in air, except with an increase in altitude.
- Both ectotherms and endotherms can reduce their rate of oxygen consumption ($\dot{M}O_2$) during severe hypoxia by seeking colder surroundings.
- Most species of animals **hyperventilate** when exposed to hypoxia.
- **Oxyregulators** are able to maintain $\dot{M}O_2$ down to a certain environmental partial pressure of oxygen (PO_2) during exposure to progressive hypoxia.
- **Oxyconformers** are unable to maintain their $\dot{M}O_2$ as PO_2 falls.
- The environmental PO_2 below which an oxyconforming ectotherm is unable to maintain its $\dot{M}O_2$ is the **critical** PO_2 or $\boldsymbol{P_{crit}}$.
- Hypoxia tolerant fish have a relatively low P_{crit} and have lower routine $\dot{M}O_2$, greater gill surface area and higher affinity haemoglobin for oxygen than those that are less tolerant of hypoxia.
- Birds that migrate at high altitudes have higher affinity haemoglobin and a greater ratio of blood capillaries to muscle fibres than lowland species.
- Unlike mammals, birds are generally tolerant of the alkalosis caused by hyperventilation when exposed to hypoxia.
- High-altitude native humans have a **blunted ventilatory response** to hypoxia compared to that of low altitude individuals.
- Living in burrows involves not only hypoxia but also **hypercapnia**.
- The partial pressure of CO_2 (PCO_2) in terrestrial burrows is much higher than that in the surrounding environment, unlike the situation in aquatic burrows.
- The respiratory response of terrestrial burrowing birds and mammals to hypercapnia is less than that of non-burrowers.
- The blood pigments of burrowing crustaceans, fish and mammals have a higher affinity for oxygen than those of non-burrrowing species.
- Burrowing mammals have a greater lung diffusion conductance, and a greater capillary density and myoglobin concentration in skeletal muscles, than non-burrowers.
- Diving birds and mammals also experience hypoxia and hypercapnia when they are under water, but they are unable to exchange gases with the surrounding water so have to take sufficient oxygen with them.
- Oxygen is stored in the lungs, the haemoglobin of the blood and the myoglobin of the muscles of diving birds and mammals.
- The progressive compression of air in the lungs as an animal descends in the water column reduces **buoyancy** of the animal, but increases the partial pressure of the gases, including nitrogen, in the lungs and blood.
- Increased partial pressure of N_2 may cause **decompression sickness**, or **the bends** as the animal surfaces.
- In deeper and longer diving birds and mammals the relative size of the lungs is less and the relative amount of oxygen stored in the blood and muscles is greater than in shallower and shorter divers.

- In mammals, the diving response of **bradycardia** and **selective reduction in perfusion of peripheral tissues** varies according to the level of activity during a dive and to its duration, ensuring maximum use of the various oxygen stores.
- As such, the **aerobic dive limit (ADL)** also varies in duration.

Responses to a decrease in oxygen demand—hibernation

- Hibernation by mammals is a controlled process involving a reduction in metabolic rate (MR) and body temperature (T_b) so that the animals can survive the cold winter months.
- As an animal enters hibernation, ventilation volume decreases before the decline in MR so that the animal becomes acidotic, which may aid in the suppression of MR.
- During hibernation, ventilation may be **episodic** or slow and regular.
- The affinity of haemoglobin for oxygen increases substantially during hibernation.
- Heart rate may drop by over 50 per cent before there is a fall in body temperature.
- **Cardiac arrhythmia** is prevented during hibernation by the absence of an accumulation of calcium ions in the **myocytes**.
- Hibernating animals experience a **relative acidosis**, although this is modest in the liver and heart.
- A rapid increase in ventilation upon arousal removes a large amount of CO_2, enabling plasma pH to increase and thereby reducing the relative acidosis.
- Peaks in heart rate, minute ventilation volume and metabolic rate during mid to late arousal coincide with maximum **shivering** and **non-shivering thermogenesis**.
- The increased rate of blood flow during arousal is targeted to the brown adipose tissue and skeletal muscle to supply them with the oxygen required for thermogenesis.

Study questions

1.* A pigeon, *Columba livia*, flying in a wind tunnel at a speed of 18 m s^{-1}, increases its rate of oxygen consumption by 15 times above its resting value. Heart rate increases six times and cardiac output increases eight times. What proportional increases occur in (a) cardiac stroke volume and (b) in the amount of oxygen extracted from the blood circulating round the body? (Hint: Equation 15.2 and Table 15.3.)

2. What is chyme and what role does it play in the circulatory response to feeding? (Hint: Figure 15.2 and Section 15.2.1.)

3. Describe the cause of the alkaline tide after feeding and discuss why its magnitude varies between different species of animals. (Hint: Section 15.2.1.)

4. What is venous oxygen reserve, and what role does it play during endurance exercise? (Hint: Section 15.2.2, Table 15.2.)

5. What is critical swimming speed and how has it been used to test the role of the heart in determining the exercise ability of fish? (Hint: Experimental Panel 15.1.)

6. Describe the major effects of exercise training in birds and the differences between captive and wild birds (Hint: Section 15.2.3.)

7. What is meant by the term symmorphosis and what evidence is there that it exists in the respiratory and cardiovascular systems of mammals? (Hint: Section 15.2.3, Figures 15.16 and 15.18.)

8. Describe what occurs in exercising endotherms at the lactate threshold (Hint: Figure 15.26 and Section 15.2.4.)

9. How do animals reduce their oxygen requirements when exposed to environmental hypoxia? (Hint: Section 15.3.1, Figures 15.26 and 15.27.)

10. What adaptations do bar-headed geese (*Anser indicus*) possess to enable them to migrate at altitudes of around 5500 m? (Hint: Section 15.3.1, Figures 15.30, 15.31.)

11. Discuss the strategies used by birds that breed at high altitudes to enable sufficient gas exchange to occur across their eggs. (Hint: Case Study 15.2.)

12. Discuss the adaptation of birds and mammals to life in burrows. (Hint: Section 15.3.2.)

13.* A 50 kg California sea lion, *Zalophus californianus*, has a lung volume of 50 mL kg^{-1} when at the surface. It has a completely compliant chest and lungs and it dives to a depth of 70 m. What will the absolute volume of its lungs be at this depth? (Hint: Section 15.3.3.)

14. Briefly describe how diving birds and mammals are thought to manage their oxygen stores when submerged. (Hint: Section 15.3.3 and Figures 15.37 and 15.39.)

15. What is the explanation of the relative acidosis during hibernation and what influence might it have on the entry to and maintenance of hibernation? (Hint: Section 15.4.)

Bibliography

Kooyman GL (1989). Diverse Divers. Physiology and Behavior. Springer-Verlag, Berlin. pp. 200.

Weibel ER (2000). Symmorphosis. On Form and Function in Shaping Life. Harvard University Press, Cambridge, Massachusetts. pp. 263.

A cheetah (*Acinonyx jubatus*) chasing after prey at top speed.
Source: © Jonathan C Photography/ Shutterstock.

Part 5

Coordination and integration

In the final part of this book, we explore the physiological systems that synchronize, coordinate and integrate an animal's bodily functions to give the animal unity of action in its environment. Such coordination and integration is essential for each individual animal's survival, the survival of populations, and the evolution of its species.

In **Chapter 16**, we focus on the nervous system, one of the two main communication systems in animals that coordinate and integrate cellular, tissue and organ activities in different parts of the body. We learn how electrical and chemical signals are initiated and integrated in neurons, how memory is formed, and how rhythmic activity is generated in neural circuits.

Chapter 17 explores mechanisms that link the animal to the outside world. We learn how animals sense objects and other facets of their environments in a way that is essential for their survival and reproduction capacity.

In **Chapter 18**, we consider muscles as biological motors that convert chemical energy to mechanical energy for powering all forms of animal movement: from pumping blood around the body to changing location. Such movement often involves the exquisite coordination of many different muscles, as illustrated in the photograph here, where a cheetah is chasing at top speed after prey.

In **Chapter 19**, we explore how the hormonal system—the second major communication system in animals—uses chemical signals to regulate and coordinate body functions in close cooperation with the nervous system.

In **Chapter 20**, we examine reproduction, the overall process by which offspring come into existence. Without an ability to reproduce, species are doomed to extinction as no organism can live indefinitely.

Maintenance of water, sodium and calcium balance is critical to animal life as we discuss in Parts 1 and 2. In **Chapter 21** we learn about the coordination of the regulatory processes that permit animals to control their body fluid composition with respect to sodium and calcium concentrations under different environmental conditions.

Finally, in **Chapter 22** we look at the control systems responsible for the integration of the activity of the cardiovascular system with that of the respiratory system to deliver the necessary amount of oxygen for aerobic metabolism to the tissues and to remove the carbon dioxide produced.

16

Neurons, nerves and nervous systems

An overview

The **nervous system** is the main communication system that links an animal to the outside world; it also generates an animal's perception of the world and, jointly with the endocrine system[1], links different parts of the body and determines an animal's behaviour. Large and mobile animals have well-developed nervous systems, which may have played a major role in their evolution.

The nervous system monitors and coordinates the functions and behaviour of the body through a series of positive and negative feedback loops[2]. In humans, a part of the nervous system is also the seat of the mind, consciousness and thought, processes about which we understand only little; as such, the study of neurons, nerves and the nervous systems (called neurophysiology) continues to be a major area of study in animal physiology.

In this chapter, we learn:

- the general organizational plan of the nervous system in invertebrates and vertebrates;
- how electrical and chemical signals are initiated in neurons and other excitable cells[3];
- how the transmission and processing of information takes place within and between neurons;
- how animals are able to respond to such information.

Electrical signals initiated in neurons[4] are very fast, dependable and energy-efficient, and can travel over long distances within an animal's body. It is, therefore, no surprise that electrical signals play a major role in conveying information from the external and internal environments to coordination areas in the nervous system. Here, the signals are integrated (that is, 'interpreted') before being relayed to effector cells, such as muscle[5], secretory and endocrine cells, thus enabling animals to respond effectively to the input signals.

1 We discuss the endocrine (hormonal) system in Chapter 19.

2 We discuss feedback loops in Section 3.3.3.

3 Excitable cells respond promptly to stimuli with a change in their membrane properties, which, in turn, initiate an electrical or chemical signal: we say cells become excited. Typical examples of excitable cells are neurons, sensory cells and muscle cells.

4 Electrical signals in neurons and excitable cells are not caused by a flow of electrons, but are usually generated by changes in the flow of ions (electrically charged atoms or molecules) as we discuss in Sections 16.2 and 16.3.

16.1 Nervous systems in animals

The nervous system in animals comprises the totality of nerve cells (neurons) and specialized cells called **glial cells** that provide structural and metabolic support to neurons. Figure 16.1 illustrates the four anatomical parts of a neuron, which have different properties:

- The **soma** (or cell body) includes all the cellular organelles and is responsible for the metabolic maintenance of the entire cell.
- The **dendrites** provide a large area of contact with other neurons from which they receive signals.
- The **axon** (nerve fibre) carries rapid signals of nerve impulses, often over long distances to other neurons or to effector cells.
- The **axon terminals** (nerve terminals), which allow the transmission of signals from one neuron to another neuron or to the responding cells.

Neurons always function in combination with other neurons to which they connect through **synapses**[6]. As illustrated in Figure 16.1, not all neurons are the same: they have different morphologies—ranging from simple to very complex—depending on the species, location in the body and their function. Neurons are structurally classified based on the number of processes that are attached to the cell body: unipolar neurons (Figure 16.1A) have one process attached to the soma; bipolar neurons (Figure 16.1B) have two processes attached to the soma; and multipolar neurons have multiple processes attached (Figure 16.1C–D).

5 We discuss the different types of muscle in Chapter 18.

6 We discuss how synapses function in Section 16.3.

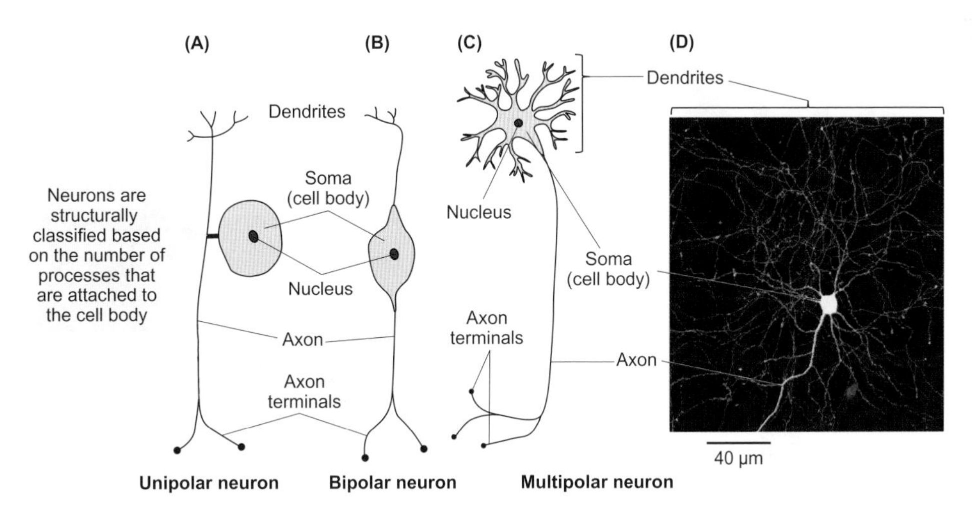

Figure 16.1 Anatomical regions of a neuron and neuron classification

(A) A unipolar neuron; sensory neurons in animals are of this type. (B) A bipolar neuron; neurons connecting to other neurons in the central nervous system are often bipolar. (C) A multipolar neuron; motor neurons and most neurons in the brain are multipolar. (D) Micrograph of a multipolar neuron expressing a fluorescent protein showing the complexity of the neuronal processes.

Source: D: © Professor Paul De Koninck, Laval University.

The evolutionary origin of the nervous system can be traced back to the most basic animal phylum[7], Ctenophora, whose members evolved diffuse neural systems consisting of cells organized in 'nerve nets' with limited means of processing information to make complex specific responses. Interestingly, members of the next two phyla on the phylogenetic tree[8]—Placozoa and Porifera (sponges)—do not have a nervous system per se, but have genes that encode sets of proteins expressed in neurons of animals with well-defined nervous systems belonging to the clade Bilateria (animals with bilateral symmetry, like us). Based on current available evidence it appears that the neural system in ctenophores evolved independently from nervous systems of other animals—an example of **convergent evolution**. However, we cannot fully exclude the alternative possibility that a nervous system existed in the common ancestor to Animalia and was lost in certain lineages including Placozoa and Porifera.

Cnidaria—the sister clade of Bilateria—includes animals, such as jellyfish, sea anemones, corals, sea pens, box jellies and hydra, which have a simple nervous system organized as a nerve net shown in Figure 16.2. Electrical signals in the form of action potentials[9] propagate in all directions within the nerve net of cnidarians, which can only process information in a limited way. Bilateria have more advanced nervous systems than Cnetophora and Cnidaria; these systems generally contain aggregations of interconnected neurons and neuronal components called **ganglia** (the plural of ganglion). Note that if neurons are close together, the processing of information is faster because the delays associated with communication between neurons are minimized. This type of structural organization, called **centralization**, allows for increased power of processing information.

Furthermore, the concentration of the major sensory organs[10] in the anterior part of the body in Bilateria led to the evolution of large clusters of neurons in the head region for processing and integrating the incoming information about the environment. These ganglia gradually acquired dominance over other clusters of neurons in the body to coordinate the behaviour of the animal. This arrangement, called **cephalization**, facilitates rapid and efficient processing of incoming information and increases the chance of animal's survival in the environment.

The part of the nervous system in Bilateria that acts as the processing centre for the nervous system as a whole, is called the central nervous system (CNS). The other part of the nervous system, which is located outside the CNS and connects the CNS with various parts of the body to enable the animal to respond specifically to various stimuli, is called the peripheral nervous system. Communication between the CNS and the various sensory and effector structures takes place via **nerves**, normally made of thousands of axons (nerve fibres) bunched together. (Nerves and neurons are distinct entities, which should not be confused.)

[7] We discuss the phylogenetic origin of the main animal phyla in Sections 1.4.1 and 1.4.2.

[8] Figure 1.15 displays the phylogenetic relationships between the main animal phyla.

[9] We discuss how action potentials are generated and propagate in Sections 16.2.5 and 16.2.7, respectively.

[10] We discuss sensory systems in Chapter 17.

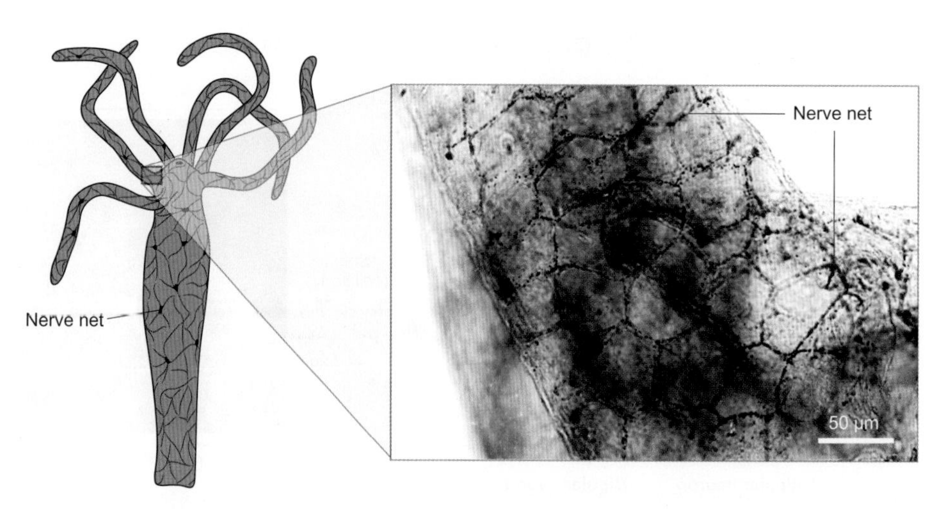

Figure 16.2 The nerve net in hydra (Cnidaria, Hydrozoa, *Hydra oligactis*)

The micrograph on the right shows the nerve net stained in black at the base of a tentacle in a specimen of *Hydra oligactis*

Source: Micrograph: Pierobon P (2012). Coordinated modulation of cellular signalling through ligand-gated ion channels in *Hydra vulgaris* (Cnidaria, Hydrozoa), International Journal of Developmental Biology 56: 551–565.

Afferent or **sensory neurons** carry inputs from sensory structures to the CNS and **efferent neurons** or **motoneurons** (motor neurons) carry signals away from the CNS to muscles and other responding organs.

The peripheral nervous system is functionally subdivided into the **sensory-somatic**, or voluntary, **nervous system** and the **autonomic** or **visceral nervous system**. The sensory-somatic nervous system connects the CNS with sensory structures[11] associated with the external environment in which the animal is situated, and with effector organs, such as the musculo-skeletal system, surface epithelia and their derivatives to regulate the animal's activities in the context of the external environment in which it finds itself. In turn, the autonomic (visceral) nervous system connects the CNS with sensory structures and internal organs (viscera) associated with the internal environment to control body homeostasis as we discuss in Chapter 3.

There is now molecular evidence that somatic motoneurons in animals displaying bilateral symmetry—including ascidians, nematodes, annelids, molluscs, insects and vertebrates—are all homologous and that the distinction between the somatic and autonomic/visceral divisions of the peripheral nervous systems pre-dates the evolutionary appearance of Bilateria.

The interaction between neurons and glial cells can be rather complex. Some glial cells, like the **Schwann cells** and the **oligodendrocytes** in vertebrates, produce the electrically insulating sheath of myelin around the axons as we discuss in Section 16.2.7. This sheath plays an important role in reducing the energetic cost and speeding up the velocity of signal transmission. Similar glial cells in invertebrates, such as the peripheral and neuropil glial cells in insects, also produce insulating sheath-like structures around axons.

Other types of glial cells in vertebrates, shown diagrammatically in Figure 16.3, are the astrocytes, which take their name from their star-like shape, and the **microglia**. The astrocytes are the most abundant type of glial cells and provide metabolic support to neurons by acting as cellular conduits to supply oxygen and nutrients to synapses on target neurons[12]. Astrocytes are in close contact with other astrocytes, vascular cells and neurons, and can send separate chemical signals to each cell in their network. The microglia perform mainly immune-like functions such as the removal of dead neurons and other type of glial cells.

In insects, the **cortex glial cells** fulfil a similar role to the astrocytes in vertebrates by ensheathing synapses between neurons. However, there are no microglia-like glial cells in insects. Instead, the immune function in insects is performed by cortex and neuropil glial cells.

16.1.1 Nervous systems of invertebrates

The invertebrate ganglia are encapsulated clusters of interconnecting neurons and glial cells, which receive inputs from sensory structures and transmit signals to effector structures, such as muscle, cells to produce a response. In animals displaying segmentation (in which the body is made of a series of repetitive segments) each body segment

[11] We discuss sensory systems in Chapter 17 and the musculo-skeletal system in Chapter 18.

[12] Some astrocytes also act as neural stem cells in both the normal and regenerating mammalian brain.

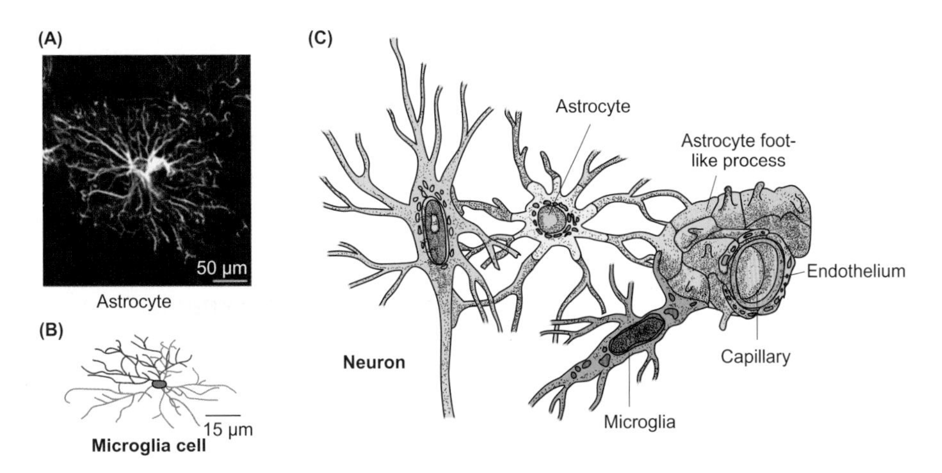

Figure 16.3 Morphological appearance of astrocytes and microglia and their relationship with neurons and blood capillary vessels in the CNS of vertebrates

(A) An astrocyte from the human cortex labelled with a fluorescent dye. (B) Microglia cell from marmoset (*Callithrix penicillata*) cortex with multiple processes shown in different colours. (C) Astrocytes act as a cellular conduit for the delivery of oxygen and energy substrates to neurons in the CNS; astrocytes make contact with neurons and enwrap blood capillaries in the CNS with their processes. Microglia cells are also in contact with astrocytes in the resting state. When activated, microglia migrate between other cells in the CNS and play the major role in the active immune defence of the CNS.

Source. A. Kimelberg HK, Nedergaard M (2010). Functions of astrocytes and their potential as therapeutic targets. Neurotherapeutics 7: 338-353; B: Diniz DG et al. (2016). Hierarchical Cluster Analysis of Three-Dimensional Reconstructions of Unbiased Sampled Microglia Shows not Continuous Morphological Changes from Stage 1 to 2 after Multiple Dengue Infections in *Callithrix penicillata*, Frontiers in Neuroanatomy 10: 23.

contains one or two ganglia that respond to sensory inputs from that segment and control it. In order to coordinate the segments, ganglia in each segment are connected in a chain-like manner to ganglia in other segments by ventrally placed nerve cords, which act like the dorsally located spinal cord in vertebrates.

Some invertebrates including flatworms, some annelids, molluscs and arthropods have two parallel net cords linked by transverse connections called **commissurae**, forming a structure resembling a rope ladder as shown in Figures 16.4, 16.5 and 16.6. Longitudinal connections between ganglia are called **connectives**. The anterior ganglia in invertebrates normally receive sensory information and control an animal's behaviour. These ganglia sit above the gut and are called cerebral ganglia, or sometimes simply, the brain, as shown for annelids in Figure 16.4.

Nervous system organization in molluscs

Figure 16.5A illustrates the general organization of the CNS in molluscs. The CNS comprises the buccal, the cerebral, the pleural, the pedal, the parietal and the visceral ganglia, which serve different parts of the body as indicated in the legend of

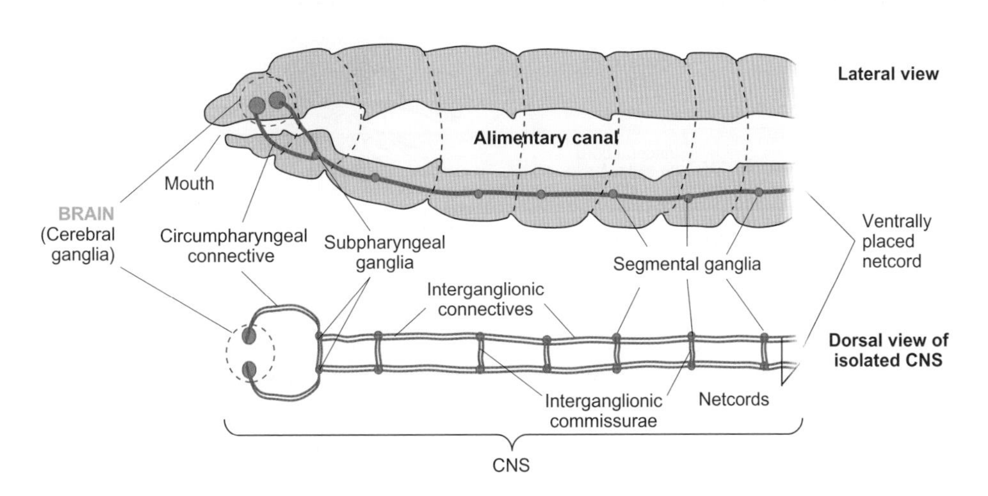

Figure 16.4 Diagrammatic representation of the earthworm CNS (Annelida, *Lumbricus* spp.)

Notice that the connections between ganglia are called commissurae if the ganglia are in the same segment and connectives if the ganglia are in different segments.

16

Figure 16.5A. The parietal and visceral ganglia are often fused into one single ganglion known as the abdominal ganglion.

The brain in molluscs generally represents the entire set of ganglia that encircle the oesophagus (cerebral, pleural and pedal ganglia) as shown in Figure 16.5A. Note that, based on this definition, bivalve molluscs do not have a 'brain' because they lack pleural ganglia.

The sensory-somatic nervous system in molluscs is represented by the pedal (somatic) cord, which connects cerebral ganglia with the pedal ganglia, as shown in Figure 16.5. The pedal cord controls the movements of the foot. In turn, the visceral cord shown also in Figure 16.5 interconnects the visceral ganglia with the parietal, pleural and the buccal ganglia and controls the feeding apparatus, the mantle, the respiratory apparatus and the internal organs.

Cephalopod molluscs, such as the cuttlefish, squid and octopus, have a well-developed, differentiated brain. Indeed, the octopus has the largest and most complex brain of any invertebrate. As shown in Figure 16.5B, the octopus brain has very large optic lobes, which support sophisticated eyesight. The vertical lobe, the inferior frontal lobe and the basal lobes play an important role in the storage and processing of visual information. The median superior frontal lobe mediates sensory information received from the arms before transmitting it to the vertical lobe, which has a key role in memory formation and learning. The peduncle lobes are paired visual-motor centres that are associated with the execution of tactile and visual discrimination. The octopus is likely the most intelligent invertebrate, displaying an ability to learn and use retained knowledge. For example, at least one octopus species, the giant Pacific octopus, (*Enteroctopus dofleini*) recognizes individual humans.

Nervous system organization in arthropods

Arthropods generally have the 'ladder-type' CNS with paired ganglia in each segment connected to each other and to ganglia in other segments via paired ventral cords running through all segments. Their brains encircle the oesophagus and result from the fusion of ganglia in the segments that form their heads.

Insects

In insects, three pairs of ganglia in the head fuse together to form the supraoesophageal ganglion, which acts as the brain. Figure 16.6A shows the three regions of the supraoesophageal ganglion: the **protocerebrum**, which innervates the eyes and processes visual information, the **deutocerebrum**, which processes sensory inputs from the antennae, and the **tritocerebrum**, which innervates the upper lip (called labrum) of an insect's mouth. The

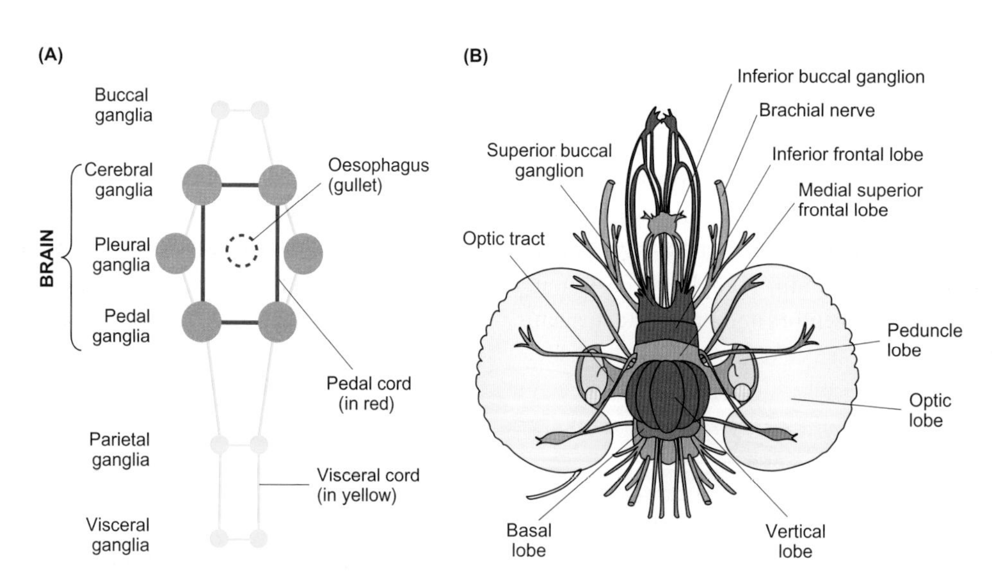

Figure 16.5 The plan of organization of the CNS in molluscs

(A) The general plan of organization of the CNS in molluscs. The buccal ganglia serve the feeding apparatus and the anterior part of the digestive system; the cerebral ganglia receive sensory information from the sensory organs in the head and coordinate the overall behaviour of the animal; the pleural ganglia innervate the mantle— the dorsal body wall, which covers the visceral mass and usually protrudes beyond the visceral mass in the form of flaps; the pedal ganglia innervate the foot and control its movement; the parietal ganglia innervate the respiratory organ (gills or siphon) and visceral ganglia control the intestine, the main digestive gland, the heart, the excretory organ and part of the genital apparatus. The connections shown in red between the cerebral ganglia and the pedal ganglia form the pedal cord, while the connections shown in yellow between the buccal, cerebral, pleural, parietal and visceral ganglia form the visceral cord. (B) The CNS in octopus (*Octopus vulgaris*). The vertical lobe receives sensory information from arms and eyes and plays a key role in memory formation. The peduncle lobes are associated with the execution of tactile and visual discrimination.

Source: B: Young JZ (1971). The anatomy of the nervous system of *Octopus vulgaris*. Clarendon Press, Oxford.

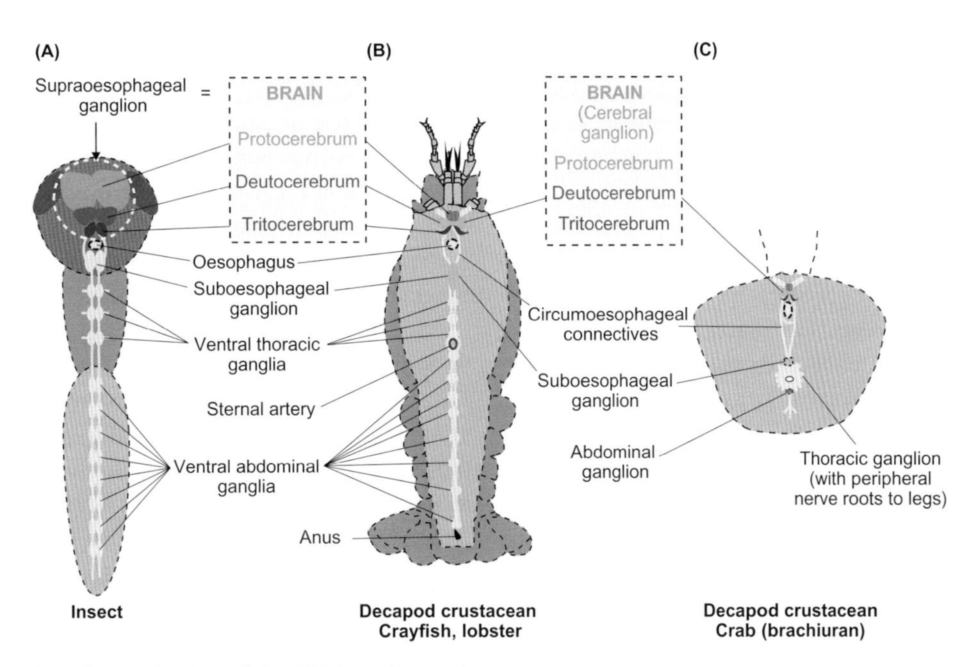

Figure 16.6 The plan of organization of the CNS in arthropods

(A) General plan of organization of the CNS in insects (subphylum Hexapoda (meaning 'six-footed')). (B) Plan of organization of CNS in a group of crustacea which includes crayfish and lobsters (Subphylum Crustacea, class Malacostraca, order Decapoda, meaning 10-footed). (C) Plan of organization of CNS in crabs which belong to a group of decapod crustaceans called brachiurans (in Greek brachy = short, oura = tail) because their abdomen is reduced in size and tucked under the thorax. Notice also that in crabs the suboesophageal ganglion, the thoracic ganglia and the abdominal ganglion are fused to form a large thoracic ganglion mass from which peripheral nerve roots emerge.

tritocerebrum also integrates sensory information from the proto- and deutocerebrum and connects to the stomatogastric nervous system, which controls the function of the internal organs.

The suboesophageal ganglion innervates the feeding apparatus, part of the pharynx and the salivary glands. The three pairs of thoracic ganglia receive inputs from sensory receptors on the thorax and innervate the legs and the wings of the insect. The abdominal ganglia receive inputs from abdominal sensory receptors and innervate the genitalia, the anus and the abdominal muscles. The thoracic and abdominal segmental ganglia control most of the insect's general behaviour, such as locomotion and mating, as evidenced by the fact that a headless insect continues to display a 'normal' type of activities with respect to locomotion and mating behaviour until it dies of dehydration or starvation.

The sensory-somatic nervous system in insects consists of the sensory neurons, which receive visual, mechanical, chemical and thermal information about the external environment from the sense organs located on the surface of the insect, and the motoneuron axons that branch out from the CNS ganglia shown in Figure 16.6 to the muscles that control the movement of the legs, wings, mouth parts, antennae, body segments and other body appendages.

The autonomic/visceral system in insects has three components: the **stomatogastric nervous system**, the **ventral visceral nervous system** and the **caudal nervous system**, which function almost independently of the CNS. The stomatogastric nervous system comprises a small set of peripheral ganglia including the frontal ganglion and the hypocerebral ganglion, which together function as central pattern generators[13] capable of producing rhythmic motor patterns independent of sensory input to elicit foregut movements. Branches from the stomatogastric nervous system also innervate the salivary glands and two important endocrine organs: the **corpora allata** and the **corpora cardiaca**, which release hormones that regulate moulting and metamorphosis as we discuss in Section 19.5.3. The ventral visceral system innervates the midgut, the spiracles of the tracheal system and the heart, and the caudal visceral system innervates the hindgut and the reproductive organs.

Crustaceans

The organization of the nervous system in crustaceans is similar to that described for insects. The brain in crustaceans is generally called the cerebral ganglion, and consists of proto-, deuto- and tritocerebrum (Figure 16.6B, C) like

[13] We discuss central pattern generation in Sections 16.3.7 and 22.1.1.

in insects. The general plan of the CNS in crayfish (Figure 16.6B) is very similar to that described in insects. In crabs (Figure 16.6C), however, the abdomen is concealed under the thorax, the thoracic ganglia are fused together into a single thoracic ganglion, and this thoracic ganglion is further fused with the suboesaophageal ganglion and the abdominal ganglion to form the thoracic ganglion mass from which all peripheral nerve roots emerge.

Crustaceans, like insects, also have a sensory-somatic nervous system and an autonomic/visceral system. As we discuss in Section 16.3.7, some of the most easily accessible (and therefore most intensely studied) neural circuits are located in the stomatogastric ganglion of the stomatogastric nervous system of decapod crustaceans, which controls the movements of the foregut and oesophagus.

The nervous system of invertebrates as a valuable resource to neuroscience

The CNS of invertebrates is generally responsible for an elaborate but invariant and repetitive type of behaviour (stereotyped behaviour), which may involve a relatively small number of neurons that are accessible for experimentation. When such preparations are identified, great progress can be made in understanding major aspects of brain function and development and how nerve cells respond to and integrate signals to elicit coordinated behaviour.

For example, the nematode worm, *Caenorhabditis elegans*, is a 1 mm-long soil roundworm that is completely transparent and has a very simple and stereotyped nervous system, as illustrated in Figure 16.7. The hermaphrodite morph of *C. elegans* contains exactly 302 neurons, which occupy the same body location and make identical connections with one another in all individuals. This means that the nervous system in *C. elegans* is completely specified by its genome and does not display any plasticity. (Compare this with the many billions of neurons in the CNS of vertebrates which may differ in their precise location and number and type of synapses with other neurons!)

This lack of plasticity has allowed the three-dimensional mapping of each individual neuron within the body of the

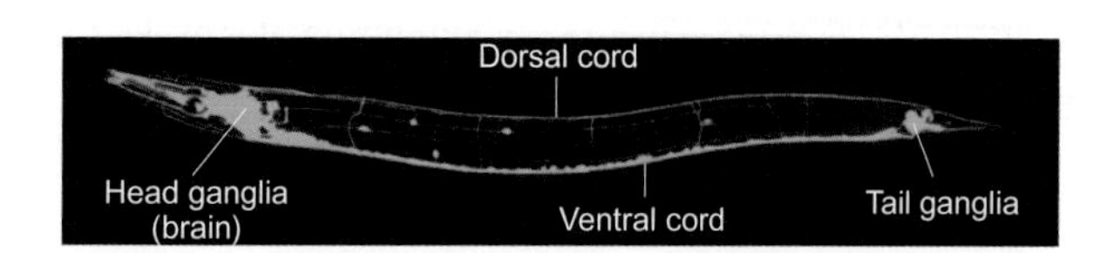

Figure 16.7 Microscope image of the nervous system of *Caenorhabditis elegans*

Caenorhabditis elegans is a soil-roundworm (nematode) of about 1 mm length with a nervous system containing precisely 302 neurons. The nervous system glows green due to labelling with a green jellyfish protein.

Source: © Professor Harald Hutter, Simon Fraser University, Canada.

animal with its synapses to other neurons using electron microscopical techniques. A laser microbeam can then be used to kill one specific neuron at a time to observe whether a particular behavioural response changes. If the behaviour is affected after the neuron is destroyed, then that neuron is deduced to participate in the neural circuit responsible for the respective behaviour. The whole neural circuit and the function of each neuron in it can then be identified.

Caenorhabditis elegans was chosen as the animal model by Sydney Brenner, the winner of the 2002 Nobel Prize for Physiology or Medicine, to understand how neural signals are processed in the living animal to produce coordinated movement in response to touch. Brenner entitled his Nobel lecture 'Nature's gift to science' as a homage to this modest creature, in which he pointed out that choosing the right organism was as important as posing the right questions to answer important problems in science. *Caenorhabditis elegans* is now also widely used as a model organism in studies of the mechanisms by which genes control development.

Another example of an invertebrate CNS preparation that has greatly advanced our knowledge of the molecular and cellular basis of learning and memory formation is the abdominal ganglion of the giant sea slug (*Aplysia* spp.), shown in Figure 16.8. In *Alpysia* spp. the visceral and parietal ganglia present in other molluscs (Figure 16.5) are fused together into a large abdominal ganglion. The ganglion mediates a family of behaviours that can be altered by learning and contains some of the largest neurons in the animal kingdom, with cell bodies up to 1 mm in diameter. After removal of the sheath surrounding the ganglion, readily identified neurons can be easily accessed for various types of experimentation while eliciting and observing specific behaviours.

The large size and fixed location of readily identified neurons in the abdominal ganglion allows for simultaneous measurements of electrophysiological activity from several neurons connected by synapses, the injection of labelled compounds, antibodies and genetic constructs in specific neurons, and the collection from identified neurons of sufficient material for biochemical and molecular biology analyses such as extraction of mRNA for making a cDNA library.

Eric Kandel and colleagues found out that the simple process of gill retraction upon stimulation of the siphon in *Aplysia californica* can be modified by learning. They identified the neurons that are involved in this behaviour, located the critical neurons that had been modified by learning and stored memory, and analysed changes that occurred in these neurons at a cellular and molecular level. Based on this research we now know that short-term memory is linked to changes in the functional properties of existing synapses, while long-term memory is associated with protein synthesis and formation of new synaptic connections, as we discuss in Section 16.3.6. For his impressive body of work on the

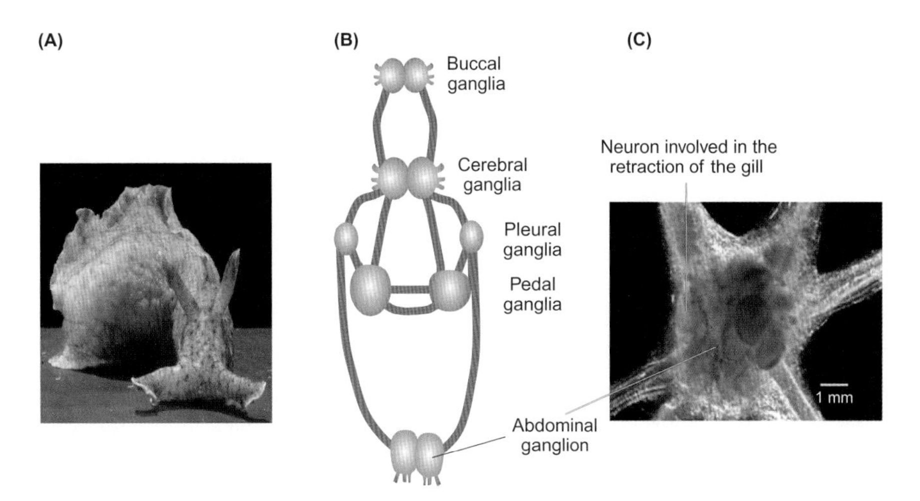

Figure 16.8 The CNS of the sea slug *Aplysia californica*

(A) Photograph of an *Aplysia californica* specimen, which on average is about 20 cm long and weighs 1 kg or more. (B) Diagrammatic representation of the *Aplysia* CNS. Note that the abdominal ganglia in *Aplysia* are fused together and appear as one single mass of neurons, called the abdominal ganglion. (C) Photomicrograph of the de-sheathed abdominal ganglion (dorsal view) illustrating individual identifiable neurons that can be easily recorded intracellularly. The neuron involved in the retraction of the gill is located on the left side of the ganglion.

Sources: A: Genevieve Anderson; C: Courtesy Professor Mike Schmale, University of Miami.

abdominal ganglion of *Aplysia californica*, Eric Kandel was awarded the 2000 Nobel Prize in Physiology or Medicine.

Fruit flies (*Drosophila* spp.) have been the object of many studies on the role of genes in brain development because of the large number of existing techniques for studying their genetics. Despite the large evolutionary distance between insects and mammals, many aspects of neurogenetics discovered in *Drosophila* appear to be relevant to humans. For example, the first biological clock genes were identified in *Drosophila* and a search for analogous genes in mammals has discovered that these genes play a similar role in mammalian biological clocks.[14]

Many invertebrates as diverse as squids, crayfish, earthworms, cockroaches and flies evolved giant axons (>100 μm diameter) in their peripheral nervous system to increase the speed of propagation of signals[15] from the CNS to effector organs, such as muscle, or from sensory cells to CNS to facilitate rapid predator-avoidance responses. The sheer size of these giant axons was critical for performing the key studies that helped understand the physiological basis of electrical signalling in neurons as we discuss in Section 16.2.1. In squids (*Loligo* spp.), giant axons can be as large as 1 mm in diameter, propagating electrical signals at a speed of 100 km h^{-1}. Experimental Panel 16.1 explores the key role played by the squid giant axon in understanding the physical and chemical bases of the membrane potential.

16.1.2 CNS of vertebrates

The organization of the CNS is common to all vertebrates. It consists of the brain, located in the cranial cavity[16], and the spinal cord, which is located dorsally, above the gut, in the vertebral canal (as we discuss later in this Section). Parts of the CNS consist of a high density of neuron cell bodies, dendrites and unmyelinated axons that are darker in colour and are called **grey matter**, while parts consist of myelinated fibres that are white in colour and are called **white matter**. The sheath of myelin around axons in the CNS is produced by oligodendrocytes.

The CNS in vertebrates is covered over its entire surface by an impermeable membrane (not to be confused with a cellular membrane); this structure is called **meninx** if the membrane structure is made of one single layer or **meninges** (plural of meninx) if the structure is multilayered.

In fish, the CNS is covered by a single-layer structure called the primitive meninx. In amphibians, reptiles and birds, the CNS is wrapped by meninges made of two layers: a tougher, thicker outside layer called dura mater and a thin inner layer called secondary meninx. In mammals, the meninges consist of three layers: the dura mater on the outside, the arachnoid in the middle and the pia mater on the inside. The space between CNS tissue and the meninges is filled with a clear, colourless fluid called **cerebrospinal fluid** or **CSF**.

[14] We discuss biological clocks in Section 17.2.2.

[15] We learn in Section 16.2.7 how electrical signals propagate faster in axons of larger diameter.

[16] Figure 3.27 in Chapter 3 shows the brain morphology in different groups of vertebrates.

Experimental Panel 16.1 Separation of electrical currents carried by Na^+ and K^+ across axonal membranes

Background to the study

The voltage-clamp technique[1] permits the study of electrical current flowing through ion channels in the membrane when the membrane potential is set and 'clamped' at a particular level. In order to understand how ion channels open and close when the membrane potential changes, it was critical to separate the total electrical current flow across membrane into components carried by the sodium and the potassium ions.

Experimental approach

Alan Hodgkin and Andrew Huxley took advantage of the accessibility and extraordinary size of the giant axons in the peripheral nerves of the veined squid (*Loligo forbesi*)[2], which allowed them to insert wire electrodes for voltage-clamping without damaging the surface membrane. The wire electrodes were wrapped around a glass capillary inserted in the centre of the axon, as illustrated in Figure A. In order to separate the ionic currents carried by the sodium and potassium ions across the membrane, Hodgkin and Huxley also took advantage of the fact that there is no net current carried by the sodium ions when the membrane potential is clamped at a value corresponding to the equilibrium potential[3] for sodium.

Results

Trace (i) in Figure B shows the time course of the total current carried by the sodium and potassium ions ($I_{Na} + I_K$) across the membrane of a squid axon when the membrane was suddenly depolarized and voltage-clamped from the resting membrane potential (V_M) to a value close to 0 mV. Notice that the current is first negative, indicating an inward flow of positive charges, after which it changes direction and becomes positive, indicating an outward flow of positive charges (as we discuss in Box 16.5).

The current component carried by the sodium ions was eliminated when the membrane was voltage-clamped at 0 mV by reducing the sodium concentration in the external solution to a value close to that in the axon such that the equilibrium potential for Na^+ was 0 mV. Since at 0 mV sodium was now at equilibrium across the membrane, all the current shown by trace (ii) in Figure B was carried by potassium ions. The positive sign of the current indicates the outward flow of potassium

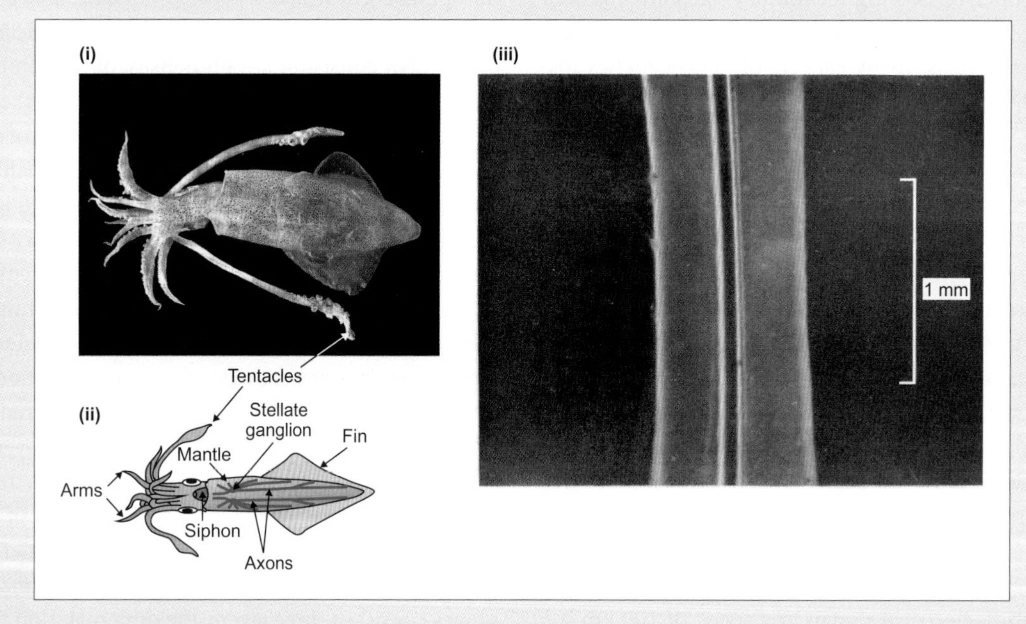

Figure A Squids (*Loligo* spp.) have giant axons that were used in key experiments that helped formulate and test the ionic theory of membrane excitability

(i) The adult males of the common squid *L. vulgaris* reach a length of 700 mm or more. (ii) Diagram showing the location of the giant axons, which were initially confused with blood vessels. (iii) An isolated squid axon with a 100 µm glass capillary inserted in preparation for experimentation.

Sources: (i) © Hans Hillewaert.jpg; (iii) Hodgkin AL (1958). The Croonian Lecture: Ionic movements and electrical activity in giant nerve fibres. Proceedings of the Royal Society of London Series B 148: 1–37.

ions when the membrane was depolarized from the resting level (V_M) to 0 mV.

The potassium current (I_K) is expressed as $I_K = g_K$ (0 mV – V_K), as we discuss in Box 16.5. The potassium conductance, g_K, is proportional to the number of open potassium channels, V_K is the equilibrium potential for potassium, and 0 mV is the 'clamped' membrane potential for the duration of the measurement. The equilibrium potential for potassium is negative and does not change for the duration of the experiment; therefore, the potassium current is directly proportional to g_K, which in turn is proportional to the number of open potassium channels. Thus, the time course of the potassium current directly follows the time course of open potassium channels. Thus, from trace (ii) in Figure B, we can conclude that, upon membrane depolarization to zero mV, potassium channels open over a period of several milliseconds and stay open for the duration of depolarization.

The sodium current when the membrane was voltage-clamped at 0 mV was then obtained by subtracting the potassium current (trace B(ii)) from the total current (trace B(i)) with the result shown in Figure B(iii). The sodium current ($I_{Na} = g_{Na}$ (0 mV–V_{Na})) tracks the time course of the sodium conductance g_{Na} because (0 mV–V_{Na}) stays constant during the voltage-clamp experiment. In turn, g_{Na} is proportional to the number of open sodium channels. Therefore, trace B(iii) tracks the time course of the open sodium channels and shows that, upon depolarization from the resting membrane potential to 0 mV, there is a very rapid rise (within a millisecond or so) in the number of sodium channels that open and that most sodium channels close soon thereafter. The negative values of the sodium current indicate that sodium ions enter the axon when the membrane is depolarized from its resting level to 0 mV.

Separation of the total current across the membrane into components carried by sodium and potassium ions has also been achieved using specific channel blockers for the voltage-dependent sodium and potassium channels. For example, we can resolve the component carried by potassium ions by blocking the sodium channels in the presence of tetrodotoxin[4]. The sodium component can then be obtained by subtracting the potassium component from the total current in the absence of tetrodotoxin. Alternatively, we can resolve the current component carried by the sodium ions by blocking the voltage-dependent potassium channels with tetraethylammonium, a positively charged quaternary ammonium ion. In this case the potassium component is obtained by subtracting the sodium component from the total current in the absence of tetraethylammonium.

Conclusions reached

The experiments on the giant axon of the squid by Hodgkin and Huxley were used to determine the kinetic properties of

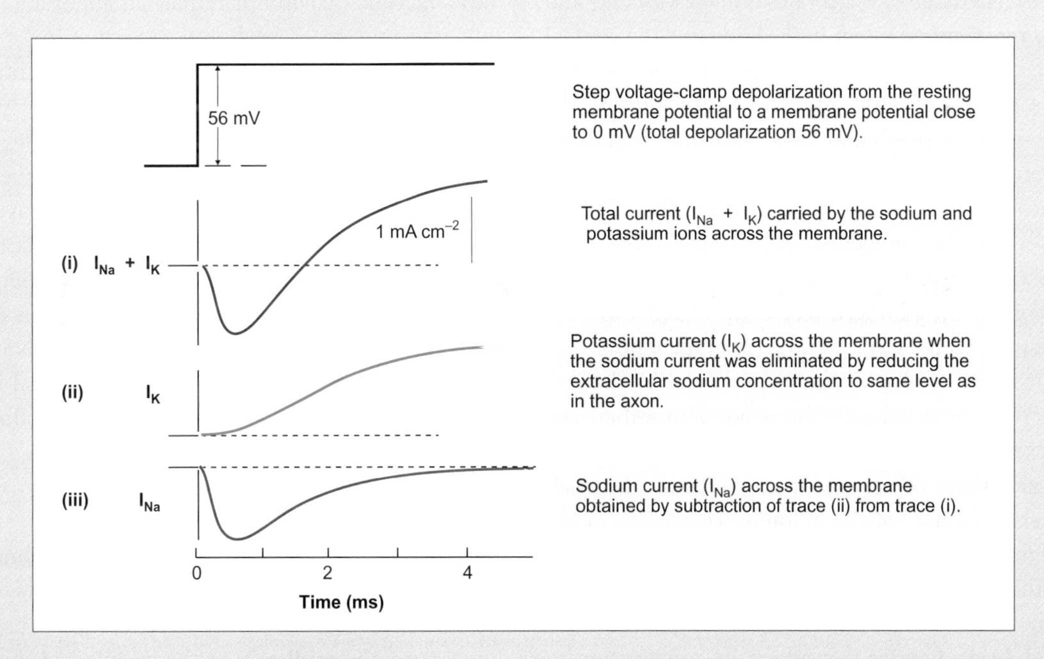

Figure B Separation of sodium (I_{Na}) and potassium (I_K) currents across the membrane of a giant axon of *Loligo* during a step depolarization

Outward and inward currents are shown respectively, as deflections of the current trace above and below the dashed line (representing the zero current).

Source: Adapted from Hodgkin AL, Huxley AF (1952). Currents carried by sodium and potassium ions through the membrane of the giant axon of *Loligo*. Journal of Physiology 116: 449-472.

the opening and closing of the sodium and potassium channels and to develop the ionic theory of membrane excitability to the point at which it could unify complex experimental data. For their seminal contributions, Alan Hodgkin and Andrew Huxley received the 1963 Nobel Prize in Physiology or Medicine.

The choice of preparation on which Hodgkin and Huxley worked was critical for elucidating the basis of membrane excitability, as highlighted by Alan Hodgkin in his 1972 address as president of the Royal Society of London when he referred candidly to a remark that 'It's the squid that really ought to be given the Nobel Prize!' The giant axon could also be gently squeezed to remove the axoplasm (whose composition could be chemically analysed) with little contamination of the extracellular fluid, and then could be refilled with solutions of known ionic composition. This allowed for the most rigorous testing of the ionic theory of membrane excitability by comparing experimental results with theoretical predictions.

› Find out more

Hodgkin AL (1958). The Croonian Lecture: Ionic movements and electrical activity in giant nerve fibres. Proceedings of the Royal Society of London. Series B, 148: 1–37.

Hodgkin AL (1973). Presidential address. Proceedings of the Royal Society of London. Series B, 183: 1–19.

Hodgkin AL, Huxley AF (1952). Currents carried by sodium and potassium ions through the membrane of the giant axon of *Loligo*. Journal of Physiology 116: 449–472.

Young JZ (1936). The giant nerve fibers and epistellar body of cephalopods. Quarterly Journal of Microscopy Sciences 78: 367–386.

[1] We discuss the voltage-clamp technique in Box 16.4.

[2] The British anatomist J.Z. Young reported in 1936 that the giant axons in fully grown specimens of the veined squid can reach up to 1000 μm in diameter.

[3] We introduce the concept of equilibrium potential for a specific ion in Section 3.1.3.

[4] We discuss tetrodotoxin in Section 16.2.4.

The multilayered meninges occurring in terrestrial vertebrates allows CSF to flow between the meningeal layers and offers protection to the CNS tissue from injury caused by jolts sustained particularly during terrestrial modes of locomotion.

The CNS tissue of adult vertebrates develops from the embryonic neural tube; the CNS is therefore hollow in the centre for its entire length. In contrast, the CNS tissue in invertebrates is solid in appearance. The hollow space in the centre of the CNS tissue of vertebrates is filled with CSF and is called the **ventricular system** in the brain and the **central canal** in the spinal cord. The CSF acts as a transport vehicle within the CNS; it is largely produced by cells lining the ventricular system (**ependymal cells**) and then flows into the space between the nervous tissue and the meninx in fish (or meningeal layers in terrestrial vertebrates) and is subsequently absorbed in the blood (venous system)[17].

Arguably the most important function of the CSF is to provide a consistent, pathogen-free extracellular environment for neurons in the CNS. Neurons are much more sensitive to infections by viruses and bacteria and to changes in the external environment compared with other cell types. For example, as we discuss in Sections 16.2 and 16.3, the electrical activity of neurons is very sensitive to small changes in extracellular potassium ion concentration, or the presence of neurotransmitters at synapses. It is therefore not surprising that the composition of the CSF is very tightly regulated by various membrane interfaces that form a so-called **blood–brain barrier** that allows its composition to differ from that of blood plasma in several important ways. We discuss how the composition of the CSF is regulated by the blood–brain barrier in Box 16.1 (see Online Resources).

[17] The CSF in humans has a volume of about 100–140 mL and it turns over about 3.5 times in one day.

For example, mammalian CSF has a potassium concentration around 3 mmol L^{-1}, which is about 67 per cent of that in plasma, and varies very little even when the plasma potassium concentration changes. The CSF calcium ion concentration is also kept constant, at around 1–1.5 mmol L^{-1}, which is half the concentration in the plasma. Conversely, the CSF magnesium ion concentration is about 50 per cent greater than that in the plasma. Both calcium and magnesium ion concentrations in the extracellular environment play an important role in synaptic transmission, as we discuss in Section 16.3.

The CSF sodium and chloride ion concentrations are about the same as in the plasma, but, unlike plasma and similar to interstitial fluid, the CSF is almost devoid of proteins. The CSF glucose concentration is effectively buffered at a value that is about 60–80 per cent of the average value in plasma and changes little even when the plasma glucose concentration decreases or increases several fold.

Astrocytes vastly outnumber the neurons in the CNS of vertebrates and cover the outer surface of CNS capillaries with their processes to form the so-called glial barrier between the blood and the CSF as we discuss in Box 16.1 (see Online Resources).

Divisions of the vertebrate brain

The general plan of organization of the vertebrate brain is illustrated in Figure 16.9. The three major parts of the brain are the **hindbrain**, at the interface between the brain and the spinal cord, the **midbrain** in the middle and the **forebrain** at the front.

The hindbrain consists of three regions: the **medulla** (also called medulla oblongata), the **pons** and the **cerebellum**. The functions of the hindbrain are to maintain the constant state of the internal environment (homeostasis), posture and balance and coordinate movement.

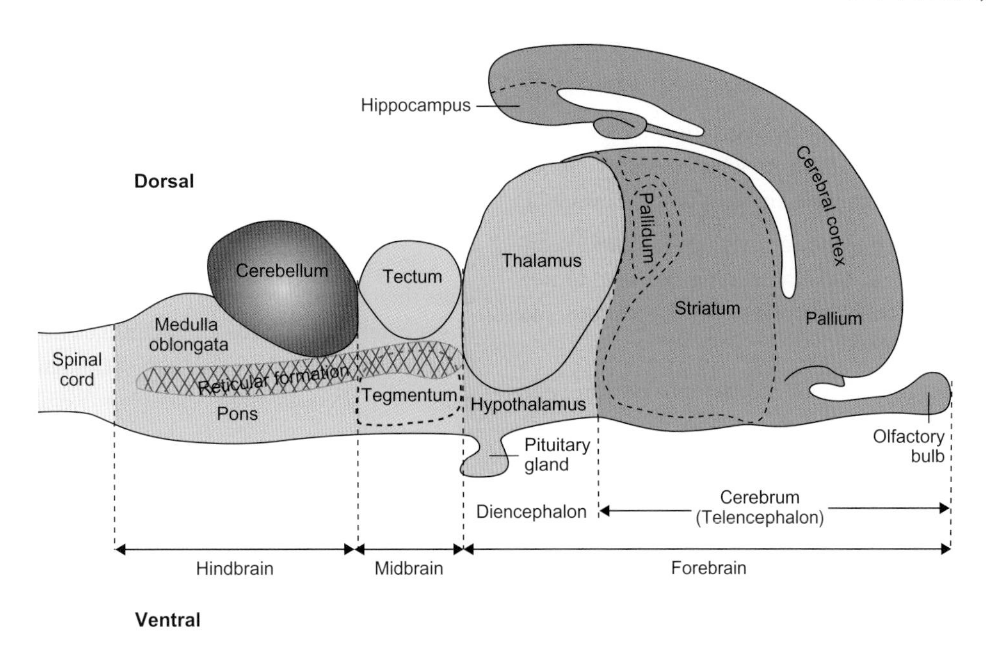

Figure 16.9 General layout of anatomical structures in the vertebrate brain

The midbrain together with the pons and the medulla form the brainstem. In mammals, four basal nuclei—the caudate nucleus, the putamen, the accumbens and the amygdala—are located in the striatum. The pallidum contains the nucleus globus pallidus; one other basal nucleus, the subthalamic nucleus, is located in the diencephalon, and another, called substantia nigra, is located in the midbrain tegmentum.

The midbrain contains the centres for the reception and integration of visual and acoustic sensory information and includes a sensory part called **tectum** and a motor region[18] called **tegmentum**. The midbrain together with the pons and the medulla of the hindbrain are called the **brainstem**.

The forebrain in all vertebrates consists of the **diencephalon** and the **cerebrum** (or **telencephalon**). As shown in Figure 16.9, the diencephalon contains the **thalamus**, **hypothalamus** and the **pituitary gland** (the posterior part). The hypothalamus links the nervous system to the endocrine system via the pituitary gland, as we discuss in Chapter 19, and controls numerous other vital functions, such as appetite, thirst, energy balance, body temperature, circadian rhythm and various activities of the autonomic nervous system, which is involved in complex homeostatic mechanisms[19].

The cerebrum is divided into two hemispheres (**cerebral hemispheres**) and each hemisphere in mammals is anatomically subdivided into four lobes: the **frontal lobe** (at the front), the **parietal lobe** (positioned behind the frontal lobe), the **temporal lobe** (located below the parietal lobe and below and behind the frontal lobe) and the **occipital lobe** (at the back).

Developmentally, the cerebrum is organized into three main distinct domains: pallium, striatum and pallidum. These domains are homologous in fish, amphibians, reptile, birds and mammals. According to this (relatively new) terminology, the cerebral cortex, the olfactory bulb and the hippocampus are components of the pallium (with the hippocampus being referred to as the medial pallium), as shown in Figure 16.9. The olfactory sensory information projects directly into the cerebrum via the olfactory bulb in all vertebrates.

Not shown in Figure 16.9 is a group of relatively compact clusters of neuron cell bodies called the basal ganglia (nuclei[20]) located at the base of the cerebral hemispheres[21]. The basal nuclei act as a cohesive functional unit, connecting to each other, to the cerebral cortex and to the thalamus. Together, these structures are involved with instinctive behaviour, movement and motor control of the limbs. Disruption of their function leads to the involuntary, purposeless motion of the limbs.

The **reticular formation** refers to a region in the brainstem shown diagrammatically in Figure 16.9 that consists of a mesh of neurons and their fibres distributed through the medulla, pons and midbrain. The reticular formation

16

[18] Motor regions contain neurons that generally carry signals to muscles or glands.

[19] We discuss various aspects of hypothalamic function in Chapters 10, 17, 19, 22.

[20] In vertebrates, clusters of neuron cell bodies are preferentially referred to as nuclei in the CNS and as ganglia in the peripheral nervous system.

[21] Other nuclei in the brain of vertebrates involved in the generation of the respiratory rhythm are discussed in Section 22.1.2.

contains a large number of neural circuits[22], which allows it to perform many diverse functions. The reticular formation:

- plays an essential role in the regulation of the sleep-wake cycle because it controls which sensory signals reach the cerebrum for the animal to become aware of;
- filters out background sensory signals that are meaningless or repetitive, while remaining sensitive to new signals to be passed to the cortex via the thalamus, like the sound of an alarm;
- helps maintain muscle tone, balance and posture, particularly during body movements;
- relays visual and acoustic signals to the cerebellum so that it can integrate these signals with other sensory signals to maintain the positional equilibrium of the animal;
- coordinates autonomic functions like breathing and heartbeat[23], because it contains the cardiac and vasomotor centres of the medulla oblongata;
- helps modulate the sensation of pain by either allowing pain signals from the lower body to reach the cerebral cortex or by preventing the transmission of pain signals to the cortex.

Figure 3.37 in Chapter 3 shows the location and relative size of the main anatomical structures in the brains of different vertebrate groups. In fishes, the hindbrain is the dominant part of the brain. It controls the respiratory and cardiovascular systems and movement. The optic lobes in the midbrain process visual information and, in bony fish, form a large part of the brain. Olfactory lobes in the fish forebrain are also well developed and process odour signals. The cerebral cortex in fish is relatively small and is mainly concerned with processing olfactory information.

During the evolution of vertebrates from fishes to mammals, the hindbrain (except for the cerebellum, which controls posture, balance and coordinates movement) becomes less and less prominent, while the midbrain and the forebrain increase considerably in size. In amphibians, the midbrain is the major integrative region of the brain. The diencephalon becomes increasingly important in reptiles, while the cerebrum is particularly important in birds and mammals, processing information related to learning, voluntary movement and interpretation of sensation. The basal ganglia (nuclei) are large in birds and mammals, and mammals have the most developed cerebral cortex, where most neural processing takes place.

A new brain feature that appeared during the evolution of placental mammals is the **corpus callosum**, which connects the two hemispheres of the forebrain and facilitates communication between them. Generally, the bundles of nerve fibres that connect different parts of the CNS are called **nerve tracts**.

As shown in Figure 16.10, birds and mammals evolved, on average, a 10-fold greater brain mass relative to body mass than fish and reptiles, particularly because of the much larger cerebrum in birds and mammals than in fish and reptiles. For example, the cerebrum of humans weighs up to 80 per cent of the total brain mass, but is only a small fraction of the total mass in fish. Studies suggest that species differences, and differences in lifestyle—such as dietary habits[24] and social interactions between individuals—affect brain and brain component size independently of body mass and evolutionary history. Such differences explain, for example, why the brain of the ostrich is several fold smaller and that of humans is several fold greater than the projected brain size of an 'average' bird or mammal of same mass.

The spinal cord

The other component of the CNS in vertebrates is the **spinal cord**, which contains nerve tracts carrying signals to and from the brain and acts as the site for simple reflexes as we discuss in Section 18.5.3. As shown in Figure 16.11, the core of the spinal cord consists of grey matter with dorsal and ventral

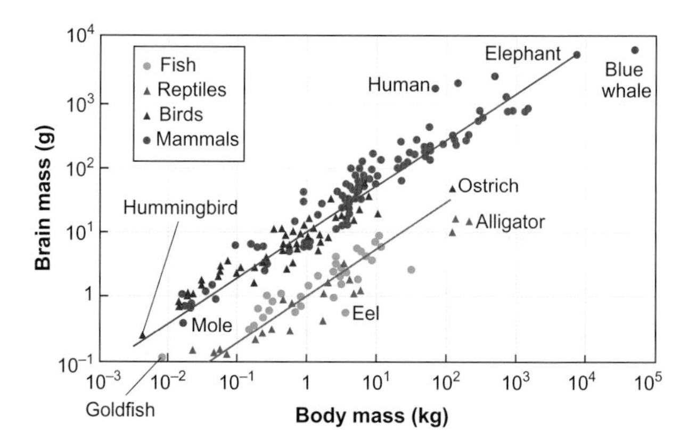

Figure 16.10 Scaling of brain mass with body mass in fish, reptiles, birds and mammals

The data points plotted on double logarithmic axes are fitted by two straight lines derived from the allometric equation $y = ax^b$ (discussed in Box 2.4), where y = *Brain mass*, x = *Body mass* and b = *scaling factor*. One line fits the data points for birds and mammals and the other the data points for fish and reptiles. The two lines run parallel to each other (both have same scaling factor $b = 0.67$) and are separated by one $\log_{10}$ unit, indicating that the brain weighs, on average, 10 more in mammals and birds than in fish and reptiles of same body mass.

Source: modified from Jerison HJ (1973). The evolution of the brain and intelligence. New York: Academic Press.

[22] Neural circuits are functional entities of interconnected neurons that regulate their own activity using feedback loops.

[23] We discuss the involvement of the reticular formation in the generation of the respiratory rhythm in vertebrates in Section 22.1.2.

[24] After controlling for body size and phylogeny, the primate brain size is primarily predicted by diet.

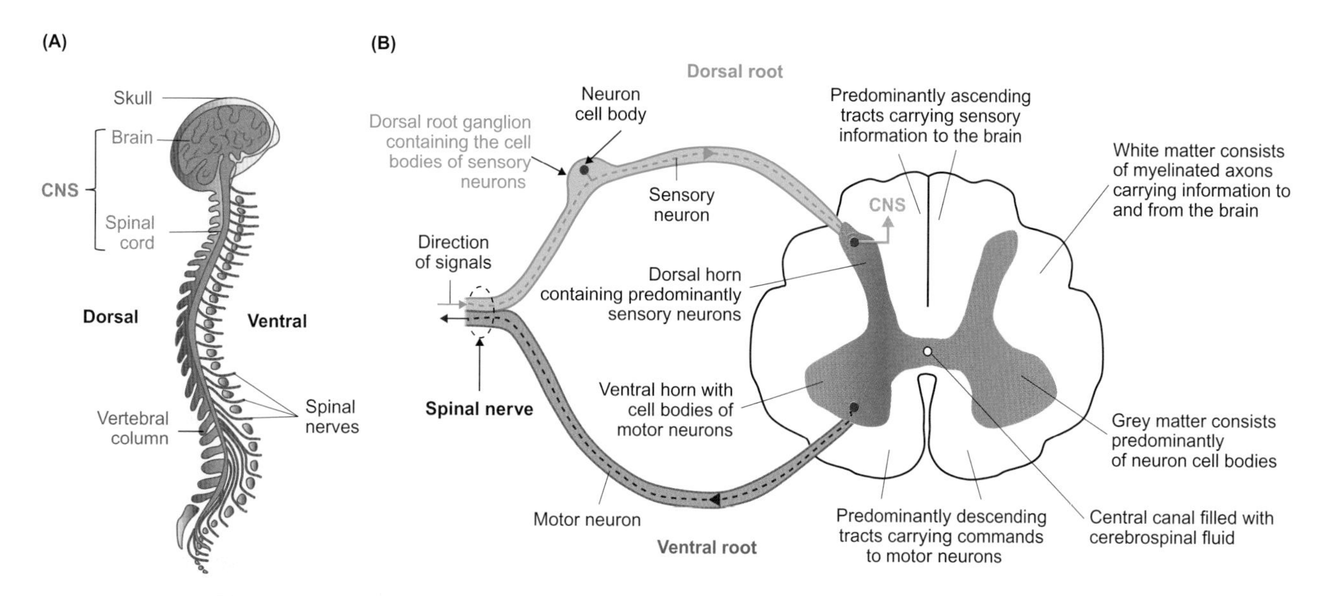

Figure 16.11 Schematic representation of the spinal cord location and transverse section through the spinal cord showing a spinal nerve with its roots in vertebrates

(A) Side-view longitudinal section through the human CNS, showing the location of the spinal cord in the vertebral canal. (B) Transverse section through the spinal cord showing a spinal nerve with its dorsal and ventral roots. Located on the dorsal root, just outside the spinal cord, is the dorsal root ganglion which contains the cell bodies of sensory neurons. Sensory information is transmitted further to different regions of the CNS via neurons located in the dorsal horn. The cell bodies of the motor neurons are located within the grey matter of the ventral horns of the spinal cord.

extensions called **dorsal** and **ventral horns**, respectively. The dorsal horns contain predominantly sensory neurons that relay information received from (peripheral) sensory neurons to different regions of the CNS. The bodies of these peripheral sensory neurons are located in the dorsal root ganglia just outside the spinal cord. The ventral horns contain the cell bodies of motor neurons that send nerve impulses to effector cells via axons leaving the spinal cord through the ventral roots. The white matter region of the spinal cord contains nerve tracts of myelinated nerve fibres linking different levels of the spinal cord with each other and with the brain.

16.1.3 Peripheral nervous system in vertebrates

The vertebrate peripheral nervous system is divided into sensory-somatic and visceral/autonomic components. The sensory-somatic nervous system regulates activities that are under voluntary control, carrying information from sensory structures to the CNS and signals from the CNS to somatic tissues such as skeletal muscle and surface epithelia. The vertebrate autonomic system monitors the internal environment of an animal (e.g. blood pressure, core and skin temperature, and oxygen and carbon dioxide partial pressures), and controls organ function. The function of the autonomic nervous system is largely involuntary.

Morphologically, the peripheral nervous system in vertebrates consists of the **cranial nerves** (described in Box 16.2), which emerge in pairs (bilaterally) from the brain, and the **spinal nerves**, which also emerge in pairs from the spinal cord. Unlike the CNS, the peripheral nervous system is not protected by bone, meninges or the blood–brain barrier, leaving it exposed to toxins and mechanical injuries.

Figure 16.12 shows the location on the brain surface where the cranial nerves emerge in the shark (*Scymnus* spp.) and human (*Homo sapiens*). The cranial nerves are numbered in order from the front of the brain using Roman numerals, except for the first pair, called the terminal nerve or nerve 0. The terminal nerve was first identified in sharks, but after many years of controversy, its presence has been established in all other vertebrate groups. We now believe that it transmits, to the brain, signals carried by sexual pheromones[25]. Reptiles, birds and mammals have—in addition to the terminal nerve—12 other pairs of cranial nerves numbered from I to XII. There are no cranial nerves XI and XII in fish and amphibians. The roots of the cranial nerves are generally enclosed in the braincase except for the terminal (0) nerve, the olfactory nerve (I) and optic nerve (II) (see Box 16.2). Notice in Figure 16.12 how cranial nerves 0 to X in sharks and humans emerge from the brain in similar locations and innervate similar structures. Most cranial nerves (listed in Box 16.2) have mixed sensory and somatic components.

Spinal nerves have dorsal and ventral roots as shown in Figure 16.11B and are named according to the region of

[25] Pheromones are chemicals secreted in the environment by one animal which impact on the behaviour of other individuals.

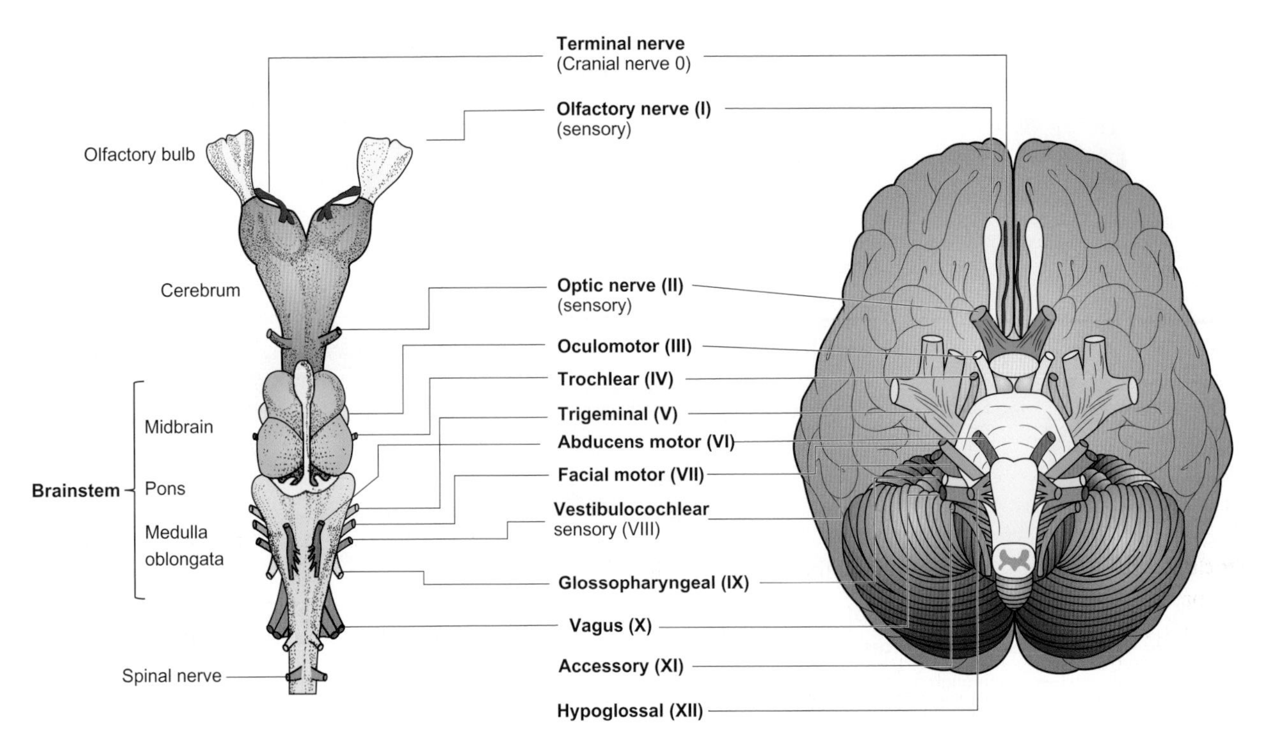

Figure 16.12 Ventral view of cranial nerves (CNs) emerging from the shark (left) and human (right) brains

CNs 0, I, II and VIII are sensory nerves. CN 0 and CN I originate in the telencephalon and send signals to the brain associated with sexual pheromones and with the sense of smell, respectively. CN II originates in retinal ganglion cells in the eyes and conveys visual signals to the brain and CN VIII originates in the pons and sends signals to the brain associated with the sound, rotation and gravity, the latter signals being essential for balance and movement. CNs III, IV, VI, XI and XII are mainly motor nerves. CNs III, IV and VI predominantly innervate muscles associated with the movement of the eyes. CNs III and CN IV originate in the midbrain and CN VI originates in the pons. CN XI has cranial and spinal roots and controls muscles in the neck and upper back. CN XII originates in the medulla and controls most movement of the tongue. CNs V, VII, IX and X are mixed nerves. CNs V and VII originate in the pons and CNs IX and X originate in the medulla. CN V receives sensation from the face and innervates the mastication muscles. CNs VII and IX receive signals associated with the sense of taste and innervate salivary glands and muscle responsible for facial expression (CN VII). CN X (the vagus nerve) provides parasympathetic fibres to essentially all thoracic and abdominal internal organs (viscera). It also receives signals related to the sense of taste from the epiglottis and innervates most laryngeal and pharyngeal muscles which control the act of swallowing and vocalization.

Source: Right-hand image: Wikimedia Commons/Patrick J Lynch.

the spinal cord from which they originate: cervical, thoracic, lumbar and sacral (see also Figure 16.13). Located in the dorsal roots, just outside the spinal cord, are the dorsal root ganglia, which contain the cell bodies of sensory neurons. In mammals and birds, the dorsal roots of the spinal cord carry sensory information (both somatic and visceral) and the ventral roots carry almost exclusively motor information (both somatic and visceral). In fish, amphibians and reptiles, the dorsal root can also carry visceral motor fibres.

The nerve fibres in the vertebrate peripheral nervous system are categorized into three major groups (A, B and C) based on their diameter, action potential conduction velocity and time course of excitability recovery during the relative refractory period[26].

- Group A fibres are myelinated, have large diameters (up to 22 μm) and high conduction velocities[27]. These fibres are primarily involved in carrying afferent signals from sensory structures to the CNS and efferent signals from the CNS to motor structures[28].
- Group B fibres are also myelinated but have a small diameter (<3 μm) and therefore have a lower conduction velocity compared with Group A fibres. Preganglionic fibres of the autonomic nervous system (Figure 16.13) generally belong to this group.
- Group C fibres are unmyelinated and, like the Group B fibres, have a small diameter (0.1–1.5 μm). Consequently, Group C fibres have the slowest conduction

[26] We discuss action potential refractory periods and action potential conduction velocity in Sections 16.2.5 and 16.2.7.

[27] The action potential conduction velocity is greater in larger and better insulated axons as we discuss in Section 16.2.7.

[28] Group A fibres are further divided into four subgroups of decreasing fibre diameter (Aα to Aδ).

Box 16.2 Cranial nerves of vertebrates

Terminal (0) emerges from the telencephalon and innervates blood vessels in the olfactory epithelium. It was first identified in sharks and consists mainly of a thin bundle of nonmyelinated sensory fibres. It has been suggested that one of its functions is to transmit sexual signals carried by pheromones to the brain.

Olfactory (I) also emerges from the telencephalon. As its name suggests, this nerve carries sensory signals form the olfactory epithelium.

Optic (II) has its roots in the ganglion cells of the retina. By convention this nerve is called the 'optic nerve' although its correct name should be the optic **tract** if the ganglion cells were considered to be part of the brain.

Oculomotor (III) is a motor nerve that innervates four of the six eye muscles. It originates in the midbrain.

Trochlear (IV) is a **motor** nerve to one of the eye muscles. It originates in the midbrain.

Trigeminal (V) is a mixed nerve which consists of three main branches, two of which innervate the upper (maxillary) and lower (mandibular) jaws. It originates in the pons.

Abducens (VI) sends motor fibres to one of the eye muscles. It originates in the pons.

Facial (VII) includes sensory fibres from the head, including the taste buds, but also sends motor fibres which affect facial expression. It originates in the pons.

Vestibulocochlear (or auditory) (VIII) carries sensory fibres from the inner ear, which is concerned with balance and hearing. It originates in the medulla.

Glossopharyngeal (IX) is a mixed nerve containing sensory fibres from the taste buds and the adjacent lining of the pharynx. There are also motor fibres to the salivary glands originating in the medulla.

Vagus (X) means 'wandering' in Latin, which reflects the fact that it is not confined to the head region. It is formed from the combination of nerves across several head segments. It is both sensory and motor. It provides parasympathetic fibres to essentially all thoracic and abdominal internal organs (the **viscera**). It also receives signals related to the sense of taste from the epiglottis and innervates most laryngeal and pharyngeal muscles.

Spinal accessory (XI) is an accessory motor nerve to the vagus. It originates from the medulla and supplies the pharynx, larynx and perhaps the heart.

Hypoglossal (XII) is a motor nerve that supplies the muscles of the tongue. It originates from the medulla.

The ventral view of the shark and human brains in Figure 16.12 shows the locations on the brain from where the cranial nerves emerge.

velocity. Importantly, C fibres react to a variety of stimuli that are thermal, mechanical, or chemical in nature and are involved in thermoreception and nociception, as we discuss in Chapter 17. A network of C fibres innervates receptors in the lungs of mammals, as we discuss in Section 22.2.1. Postganglionic fibres in the autonomic nervous system also generally belong to the C group.

The vertebrate autonomic nervous system

The autonomic nervous system in vertebrates comprises the **sympathetic system**, the **parasympathetic system** and the **enteric system**. The sympathetic and the parasympathetic systems act antagonistically to control the activity of visceral organs. The sympathetic division mobilizes body systems during activity, while the parasympathetic division is responsible for maintaining normal body functions in the animal at rest. The enteric system comprises the neurons that are intrinsically embedded in the lining of the digestive tract.

The sympathetic and the parasympathetic divisions of nervous systems

The sympathetic division of the autonomic nervous system in mammals (shown in Figure 16.13) consists of:

- preganglionic nerve fibres originating from the thoracic and lumbar regions of the spinal cord;
- the pair of sympathetic chain ganglia located just ventrally and laterally to the spinal cord;
- the collateral ganglia, which together with the chain ganglia contain the cell bodies of the sympathetic postganglionic neurons;
- the postganglionic nerve fibres extending from the sympathetic ganglia to their target tissue.

The pair of sympathetic chain ganglia extend for the length of the vertebral column from the upper neck down to the coccyx (tailbone). The preganglionic fibres enter the chain ganglia at the same level at which they emerge from the ventral roots of the spinal nerves. Once in the chain ganglia, they can make synapses with postsynaptic neurons at the level of entry, they can run up or down from the level of entry and make synapses with 15–20 postsynaptic neurons at different locations in the chain ganglia, or move through the chain of ganglia to make synapses in the collateral ganglia. The last ganglion in the cervical region of each sympathetic chain of ganglia is merged with the first ganglion in the thoracic region of the same chain, forming the so-called 'stellate ganglion'.

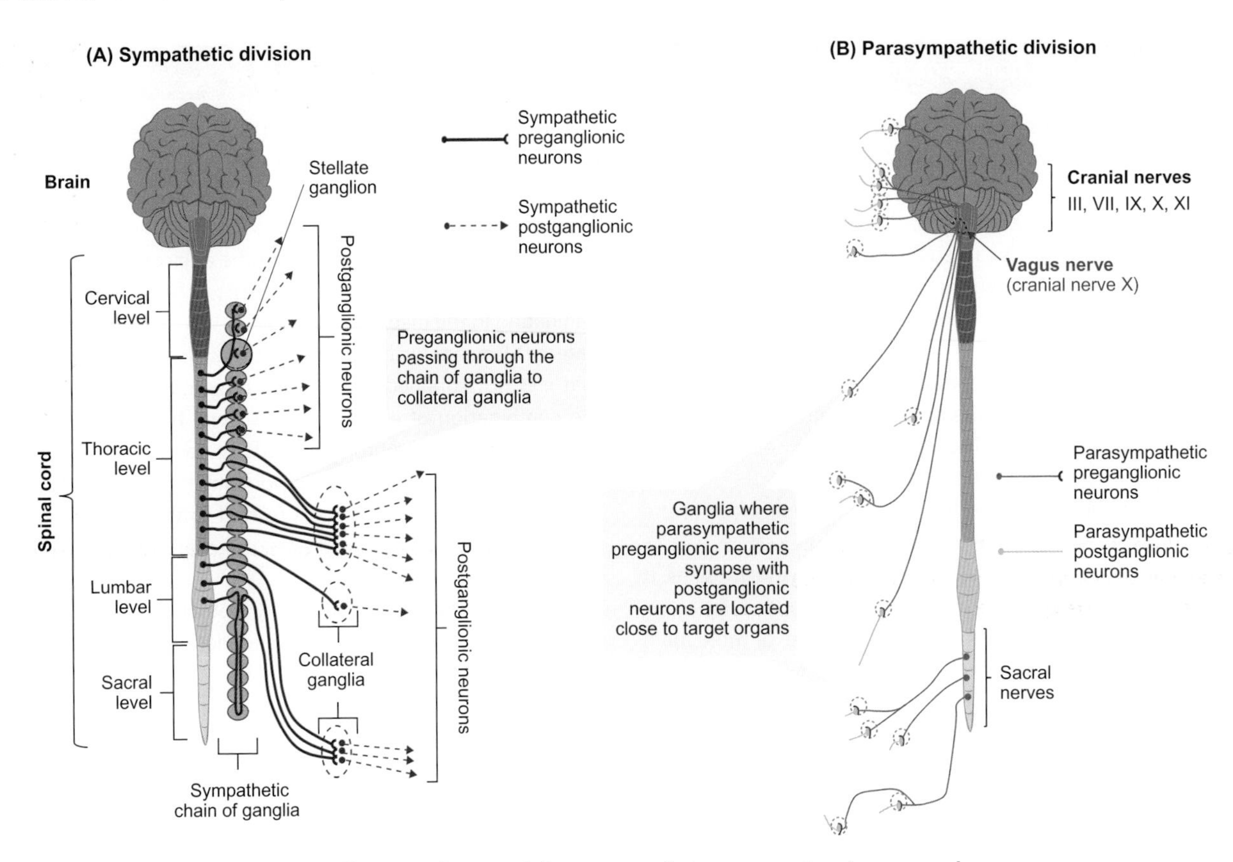

Figure 16.13 Organization of the efferent pathways of the autonomic nervous system in mammals

(A) Sympathetic preganglionic neurons have the cell body in the ventral horn of grey matter at the thoracic or lumbar levels of the spinal cord. Their axons enter the chains of ganglia on either side of the spinal cord, where some synapse with postganglionic neurons, while others pass through to collateral ganglia, where they synapse with postganglionic neurons. (B) The parasympathetic division originates from the brainstem as part of cranial nerves III, VII, IX, X, XI and from the sacral region of the spinal cord. Preganglionic neurons synapse with postganglionic neurons in ganglia close to the target organs.

By contrast, the parasympathetic division in mammals originates from the brainstem as part of the cranial nerves III, VII, IX and X (vagus nerve), and XI, and from the sacral region of the spinal cord, as illustrated in Figure 16.13. Like the sympathetic division, the parasympathetic division also consists of preganglionic and postganglionic nerve fibres, but the ganglia in this case are located close to the target organ, as shown diagrammatically in Figures 16.13 and 16.14.

Paired sympathetic chains of ganglia are also present in teleost fish, reptiles and birds. In elasmobranchs the segmental sympathetic ganglia are not fully connected longitudinally and the sympathetic chains may be reduced in limbless caecilian amphibians (Order Gymnophiona) and lungfish (Subclass Dipnoi). Birds and reptiles have parasympathetic pathways in cranial nerves III, VII, IX and X, while jawed fish (gnathostome fish) have parasympathetic pathways only in the oculomotor (III) and vagus (X) nerves. In vertebrates other than mammals, it is difficult to distinguish a sacral parasympathetic part of the autonomic nervous system from the sympathetic part.

Activation of the sympathetic system has a variety of effects (for an expanded list see Figure 3.24):

- increases the heart rate and blood pressure,
- dilates coronary and skeletal muscle blood vessels,
- causes dilation of the bronchi[29],
- dilates the pupil,
- stimulates the adrenal gland,
- mobilizes glucose from glycogen stores in the liver,
- causes release of extra blood cells from the spleen,
- inhibits the activity of the digestive tract.

It is commonly said that activation of the sympathetic system induces the **defence-alerting response**[30] by preparing vertebrate species to either fight the danger or flee. Indeed, the sympathetic nervous system becomes most active during times of stress[31], when the animal uses more energy

[29] We discuss the involvement of the autonomic nervous system in the regulation of the respiratory and cardiovascular systems of vertebrates in Sections 22.4.1 and 22.4.3.

[30] We discuss the defence-alerting response in Section 22.5.2.

[31] We discuss the link between activation of adrenal gland and stress in Chapter 19.

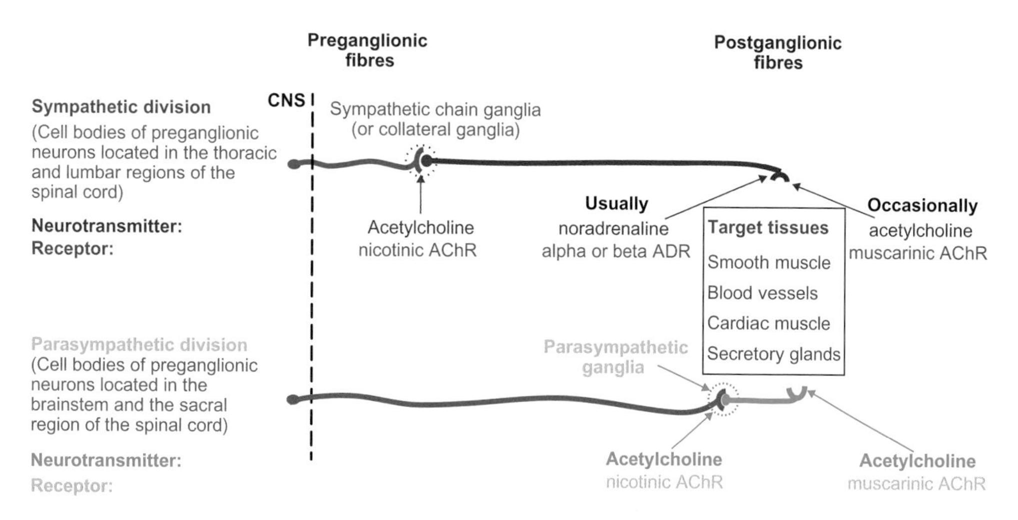

Figure 16.14 Types of synapses in the efferent pathways of the sympathetic and parasympathetic nervous systems in mammals. We discuss synapses and neurotransmitters in Section 16.3.
AChR = acetylcholine receptor, ADR = adrenergic receptor (adrenoceptor).

to react to stressing stimuli. In contrast, activation of the parasympathetic system returns the body to a restful state, which conserves and replenishes the energy stores by slowing the heart rate, reducing the blood pressure, causing constriction of the pupils, coronary arteries and bronchi, promoting glycogen formation in the liver and enhancing digestion.

Most control of visceral organs is based on simple reflex actions, which at their simplest involve four neurons that anatomically form a so-called autonomic reflex arc: one visceral sensory neuron, two visceral motor neurons and one interneuron[32]. The sensory neuron has its cell body located in the dorsal root ganglion of the spinal cord and synapses with an interneuron located in the dorsal horn of the spinal cord (Figure 16.11). The interneuron connects to a preganglionic motor neuron located in the ventral horn cord and to other nuclei in the CNS. The preganglionic motor neuron makes an excitatory synapse[33] with the postganglionic motor neuron either in the sympathetic chains of ganglia located on each side of the spinal cord as shown in Figure 16.13, or in a collateral ganglion, or in the wall of the visceral organ. The postganglionic neuron innervates the effector visceral organ.

As shown in Figure 16.14, the neurotransmitter[34] involved in the excitatory synapses between the two motor neurons is **acetylcholine** irrespective of whether the motor neurons involved are part of the sympathetic or the parasympathetic circuits. However, the neurotransmitter released at the *end* of the circuit is usually **noradrenaline** (also called **norepinephrine**) or **adrenaline** (also called **epinephrine**) in the sympathetic system and acetylcholine in the parasympathetic system. Nerve fibres that release acetylcholine at their synapses are called cholinergic fibres and nerve fibres that release noradrenaline or adrenaline are called adrenergic fibres.

Types of receptors present at sympathetic and parasympathetic synapses

The complex specific effects elicited by the activation of the sympathetic/parasympathetic divisions of the autonomic nervous system are associated not only with the type of neurotransmitter molecules secreted, but also with the type of receptors to which the neurotransmitter molecules bind at synapses.

Acetylcholine released at synapses between preganglionic and postganglionic neurons in the autonomic nervous system binds and opens ligand-gated channels on the postganglionic neurons. These acetylcholine ligand-gated channels are **ionotropic receptors** because they form an ion pore when acetylcholine binds to them. When open, the acetylcholine ionotropic receptors are not selective for sodium and potassium ions and cause depolarization and activation of the postganglionic neurons, as we discuss in Section 16.3.3. Since this type of channel is also activated by nicotine, these channels are called nicotinic acetylcholine receptors or nAChRs.

Furthermore, most of the parasympathetic postganglionic neurons release acetylcholine at their synapse with target cells, such as smooth and cardiac muscle cells, and secretory and endocrine cells. However, in this case, acetylcholine

[32] Autonomic reflex arcs in the respiratory and cardiovascular systems are discussed in Section 22.5.

[33] We discuss how excitatory synapses function in Sections 16.3.2 and 16.3.3.

[34] We discuss neurotransmitter molecules in Section 16.3.4.

binds to a different type of receptors called **metabotropic receptors**[35]. Metaboreceptors are coupled to GTP-binding proteins (G-proteins)[36] and are indirectly linked with ion channels via second messengers such as cyclic AMP (cAMP). When activated, the metaboreceptors mediate changes in target cell excitability. This type of acetylcholine receptors is sensitive to muscarine (a toxin produced by certain mushrooms) but not to nicotine and, therefore, such receptors are called muscarinic acetylcholine receptors, or mAChRs. There are many subtypes of mAChRs, enabling great flexibility in the actions of the parasympathetic nervous system on the target tissues.

[35] We discuss in Section 16.3.4 how metabotropic receptors work.

[36] We discuss the signalling mechanism involving G-proteins in Chapter 19 (Figure 19.4).

In contrast to parasympathetic postganglionic neurons, most sympathetic postganglionic neurons release noradrenaline at their synapse, as shown in Figure 16.14. Noradrenaline binds and activates receptors, called **adrenoceptors** located on target cells of the sympathetic nervous system. The adrenoceptors are metabotropic receptors, coupled to G-proteins and are broadly classified as alpha- and beta-adrenoceptors as shown in Table 16.1. The alpha-adrenoceptors are subdivided into two groups: alpha-1 and alpha-2; the beta-adrenoceptors are subdivided into three groups: beta-1, beta-2 and beta-3. The adrenoceptor subtype present in the target tissue of the sympathetic circuit determines the type of response that sympathetic stimulation elicits as indicated in Table 16.1.

Some sympathetic postganglionic neurons that project to sweat glands or blood vessels release acetylcholine, which

Table 16.1 Adrenoceptor types in the sympathetic nervous system of vertebrates

Receptor type	Subtype	Location	Receptor activation[1] effects
Alpha (α)	α-1: α-1A, α-1B, α-1D	*Smooth muscle of blood vessels supplying:* the skin, mucosae (plural of mucosa), salivary glands, gastrointestinal tract, kidney, brain	*Vasoconstriction decreases perfusion of respective organs* α-1A and α-1D subtypes are involved in the regulation of blood pressure
		Smooth muscle of: ureter, urethral sphincter, uterus, eye iris	*Contraction of respective tissue/organ;* urine retention, pupil dilation
	α-2: α-2A, α-2B, α-2C, α-2D only in vertebrate groups other than mammals and crocodiles	*Presynaptic terminals of adrenergic axons*	*Inhibition of noradrenaline (norepinephrine) release at synapses*
		Smooth muscle of internal anal sphincter	*Contraction*
		Pancreas	*Inhibition of insulin[2] release*
		Adipose (fatty) tissue	*Inhibition of lipolysis*
Beta (β)	β-1	*Heart*	*Cardiac output increases because: Heart rate increases, Strength of contraction increases Impulse conduction velocity over the heart increases*
		Kidney	*Release of renin[3] from modified smooth muscle cells (juxtaglomerular cells)*
		Stomach	*Release of ghrelin[4] from specialized cells in the stomach wall*
	β-2	*Smooth muscle in airways (bronchi)*	*Bronchodilation increases ventilation*
		Uterus, urinary bladder	*Relaxation*
		Blood vessels in the lung, skeletal muscles, liver	*Vasodilation: increases perfusion of target organs needed during the defence-alerting response*
		Pancreas	*Release of insulin*
		Gastrointestinal tract smooth muscle	*Decreased motility: delays digestion during the defence-alerting response*
	β-3	*White adipose (fatty) tissue* *Brown adipose tissue*	*Lipolysis* (breakdown of fat) *Lipolysis* and *thermogenesis[5]* (heat generation)
		Gallbladder	*Relaxation*
		Urinary bladder	*Relaxation*

[1] We discuss the different modes of action of various types of receptors in smooth muscles in Section 18.4.

[2] We discuss insulin in Section 19.1.4.

[3] We discuss the physiological role of renin in Section 21.1.1.

[4] We discuss ghrelin in Section 2.3.5.

[5] We discuss the role of brown adipose tissue in thermoregulation in Section 10.2.5.

acts on mAChRs in the target cells to stimulate secretion from sweat glands and promote vasodilation. There are also instances in which the postsynaptic neuron is neither cholinergic nor adrenergic, because it releases other neurotransmitter molecules, as we discuss in Section 16.3.4.

The enteric division of the nervous system in vertebrates

The enteric autonomic system in vertebrates consists of a mesh-like network of neurons embedded in the lining of the gastrointestinal tract. The neurons are collected into two types of ganglia: myenteric ganglia, located between the longitudinal and the circular layers of muscle surrounding the alimentary canal, and submucosal ganglia, close to the mucosa, which contains the secretory glands.

The enteric system is made of:

- sensory neurons that report to the enteric ganglia on the mechanical and chemical conditions in the gastrointestinal tract,
- interneurons that connect the ganglia, and
- motor neurons that control the movements of the gastrointestinal tract through the longitudinal and circular layers of smooth muscle, and the secretion of enzymes into the lumen of the alimentary canal.

As a whole, the enteric system is made of more neurons than the spinal cord. It also contains support cells that are similar to the astrocytes in the brain and a diffusion barrier around the capillaries surrounding ganglia, similar to the blood–brain barrier in the brain of vertebrates as we discuss in Box 16.1, available online.

Normally, the enteric nervous system communicates with the CNS via the sympathetic and parasympathetic nervous systems through the sympathetic chain and vagus nerve, respectively. However, the enteric nervous system is capable of carrying on many of its functions after connections with the CNS are severed, acting as a 'second brain' in the vertebrate body. In addition to acetylcholine and noradrenaline, more than 30 other neurotransmitters[37] are released at synapses of the enteric system.

› *Review articles*

Avian Brain Nomenclature Consortium (2005). Avian brains and a new understanding of vertebrate brain evolution Nature Reviews Neuroscience, 6: 151–159.

Bertucci P, Arendt D (2013). Somatic and visceral nervous systems—an ancient duality. BMC Biology 11: 54.

Chotard C, Salecker I (2004). Neurons and glia: team players in axon guidance. Trends in Neurosciences 27: 655–661.

Fields DR (2007). Sex and the secret nerve. Scientific American Mind February/March: 21–27.

Figley CR, Stroman PW (2011) The role(s) of astrocytes and astrocyte activity in neurometabolism, neurovascular coupling, and the production of functional neuroimaging signals. European Journal of Neuroscience 33: 577–588.

Freeman MR, Doherty J (2006) Glial cell biology in *Drosophila* and vertebrates. Trends in Neurosciences 29: 82–90.

Hochner B, Shomrat T, Fiorito G (2006) The Octopus: A model for a comparative analysis of the evolution of learning and memory mechanisms. Biological Bulletin 210: 308–317.

Nilsson S (2011) Comparative anatomy of the autonomic nervous system. Autonomic Neuroscience 165: 3–9.

Di Terlizzi R, Platt S (2006) The function, composition and analysis of cerebrospinal fluid in companion animals: Part I—Function and composition. The Veterinary Journal 172: 422–431.

[37] We discuss neurotransmitter molecules in Section 16.3.4.

16.2 The ionic basis of electrical activity in neurons

One of the greatest achievements in neurophysiology was to reveal the nature of electrical signals in neurons and other excitable cells. This achievement required new methods to measure the electrical potential difference across the plasma membrane of nerve cells under different conditions. As we discuss in Chapter 3, the difference between electrical potential in the cytosol (V_{in}) and that in the external environment (V_{out}), i.e. $V_M = V_{in} - V_{out}$, is called the membrane potential (V_M).

The membrane potential is generally measured using small glass pipettes called micropipettes or glass microelectrodes, as shown in Figure 16.15. Measurements of changes in the membrane potential when nerve fibres were exposed to solutions of different ionic composition gave insight into the origin of the membrane potential in nerve and other excitable cells.

16.2.1 The resting membrane potential

The **resting membrane potential**, as the term suggests, refers to the membrane potential of cells when the membrane potential is steady and the cells are 'at rest'. In neurons, the resting membrane potential under physiological conditions ranges from approximately −60 to −70 mV in both invertebrates and vertebrates.

Alan Hodgkin and Bernard Katz, both Nobel laureates working at the Laboratory of Marine Biological Association in Plymouth, UK, made careful observations of changes in the resting membrane potential of the giant axon of squid in response to changes in the ionic composition of the external environment. First, they measured the concentrations of the potassium, sodium and chloride ions in the cytosol of the squid giant axon (called axoplasm) ($[K_{in}] = 345$; $[Na_{in}] = 72$ and $[Cl_{in}] = 61$ mmol L^{-1}). Then, they showed that changes in the resting membrane potential could be quantitatively predicted by the equation that bears their name:

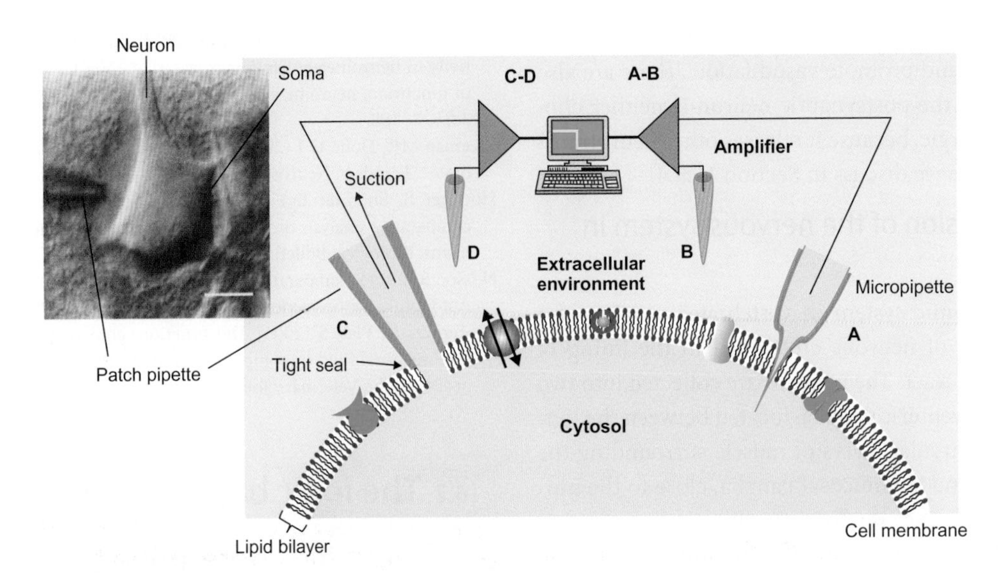

Figure 16.15 Methods for measuring the membrane potential V_M (=V_{in} – V_{out}) using glass microelectrodes
Classical method: A glass micropipette, with a very sharp tip (<1 μm), (A) filled with 3 M KCl (which dissociates in solution to form K^+ and Cl^- ions) is inserted through the membrane into the cytosol, while another electrode (B) is in electrical contact with the extracellular environment. The membrane potential $V_M = V_{in} - V_{out}$ = A–B is measured with a special type of amplifier and is recorded on a computer. The downward deflection on the computer screen indicates the moment when pipette A penetrates the membrane. **Patch electrode method** (whole-cell recording): A patch pipette, which has a larger (1–2 μm), polished tip (C) is placed on the cell surface and under gentle suction makes a very tight seal with the membrane. The membrane patch under the tip is then ruptured under increased suction and the membrane potential difference is measured between the patch pipette (electrode C) now in contact with the cytosol and electrode D in contact with the external environment. The blue shapes in the membrane indicate various membrane proteins (from left to right): receptor protein, ion pump, carrier protein, membrane enzyme and ion channel. **Inset:** Image of a patch micropipette approaching the surface of a rat brain neuron (scale bar 10 μm)
Source: Inset: Stuart GJ et al (1993). Patch-clamp recordings from the soma and dendrites of neurons in brain slices using infrared video microscopy. Pflügers Archiv 423: 511-518.

the Goldman–Hodgkin–Katz (GHK) equation (which we discuss in Chapter 3 (Equation 3.9)) when the membrane permeability ratios for potassium, sodium and chloride ions are P_K:P_{Na}:P_{Cl} = 1:0.04:0.45.

The reliability of the GHK equation to predict the resting membrane potential in neurons was also tested in experiments in which the axoplasm was replaced with artificial solutions. In these experiments, the axon was cut and a small glass cannula was attached to its distal end. The axon was then placed on a rubber pad and the axoplasm was extruded with the help of a small rubber roller, just like squeezing the paste out of a toothpaste tube. Finally, the axon was perfused through the cannula with solutions containing different concentrations of potassium and other monovalent ions and the axon was returned to a chamber with seawater. The membrane potential changed in accord with predictions based on the GHK equation for the respective conditions.

The GHK equation describes the electrical potential difference that develops across a barrier separating two solutions with different monovalent ion compositions when no net electrical current flows through the barrier—that is, when the electrical current carried by a particular ion is electrically compensated for by currents carried by the other ions. Importantly, according to the GHK equation, the electrical potential difference that establishes across the barrier depends only on the concentrations of the permeable ions in the two compartments separated by the barrier and the relative (not absolute) permeability of the barrier to these ions.

The demonstration that the GHK equation describes the membrane potential at rest when the ion composition changes on either side of the membrane is of fundamental importance because it shows that the resting membrane potential of the squid axon (and other cell types) is a direct consequence of:

- the different ionic composition of the aqueous media on either side of the plasma membrane,
- the specific permeability ratios of the plasma membrane to the sodium, potassium and chloride ions.

Radioactive isotopes of potassium (K^{42}) and sodium (Na^{24}) were used to measure the sodium and potassium ion fluxes across the membrane of giant axons from cephalopod molluscs (cuttlefish (*Sepia officinalis*) and squid). In freshly dissected axons at rest, there was an efflux of potassium ions of 290 nmol m^{-2} s^{-1} and a comparable influx of sodium ions. These fluxes are very small. For example, a potassium efflux of 290 nmol m^{-2} s^{-1} would decrease the axoplasmic potassium concentration of a giant axon approximated by a cylinder of radius R (0.3 mm) and length Lg at rest by only about 1 mmol L^{-1} from 345 mmol L^{-1} to 344 mmol L^{-1} after 10 minutes:

The amount of potassium loss (290 nmol m^{-2} s^{-1}) after 10 minutes (600 s) across the surface area of the axon ($2\Pi RLg$) divided by the volume of the axon ($\Pi R^2 Lg$)

$$= 290\,\text{nmol m}^{-2}\,\text{s}^{-1} \times 600\,\text{s} \times 2\Pi RLg / (\Pi R^2 Lg)$$

$$= 348\,\mu\text{mol m}^{-2}\,R^{-1} = 348\,\mu\text{mol} / (0.3\,\text{mm m}^2)$$

$$= 348\,\mu\text{mol} / 0.3\,\text{L} = 1.16\,\text{mmol L}^{-1}$$

The small influx of sodium ions and efflux of potassium ions at rest are efficiently compensated for through the action of the ATP-driven Na^+/K^+ pump, thus maintaining the uneven distribution of potassium and sodium ions across the membrane on a long-term basis.

How is it possible that the potassium efflux and the sodium influx have essentially the same magnitude, even though the membrane at rest is some 25-fold more permeable to potassium than to sodium ions (P_K:P_{Na} = 1 : 0.04)? The answer is that the potassium and the sodium fluxes across membranes depend not only on the ion concentrations across the membrane and the permeability of the membrane for the respective ions, but also on the action of electrical forces acting on the ions flowing across the barrier, as we discuss in Section 3.1.3.

For example, the positively charged potassium ions on the cytosolic side of the membrane are inhibited from crossing the membrane by attractive electrostatic forces acting on them; these electrostatic forces stem from the axoplasm being more negatively charged than the external environment. In contrast, the positively charged sodium ions on the extracellular side of the membrane are pulled across the membrane by the negative membrane potential. The overall effect is that when the membrane potential is around −60 mV, the potassium efflux is considerably smaller and the sodium influx is considerably greater than in the absence of a membrane potential such that they become similar in magnitude.

More complicated expressions than the GHK equation have been derived when not only monovalent ions, but also divalent ions, such as Ca^{2+}, are considered. The contribution of calcium ions to the resting membrane potential is negligible, but Ca^{2+} ions can exert a physiologically critical influence on the membrane potential when the Ca^{2+} permeability rises due to the opening of channels in the membrane that are permeable to Ca^{2+} as it happens during the activation of cardiac and some smooth and skeletal muscle cells.

16.2.2 How does the membrane permeability ratio for sodium and potassium ions determine the membrane potential?

Box 16.3 shows how the membrane permeability ratio for sodium and potassium ions (P_{Na}:P_K) determines the electrical potential difference across membranes of excitable cells. More specifically, Figure 16.16 illustrates how the membrane potential (V_M) changes in the squid axon when the permeability ratio for the sodium and potassium ion (P_{Na}:P_K) changes in two situations. In one case (curve 1), the ratio between the permeability for the chloride and potassium ion stays the same as it was at rest (P_{Cl}:P_K = 0.45). In the other case (curve 2), it was assumed that the ratio P_{Cl}:$P_K \approx 0$, which happens when the potassium permeability rises and becomes very much greater than the chloride permeability, or when the membrane is not permeable to chloride. The two curves 1 and 2 intersect at a point close to the resting membrane potential; in the squid axon this is −60 mV.

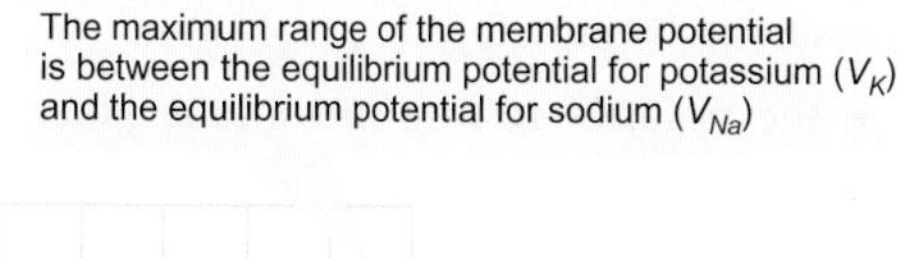

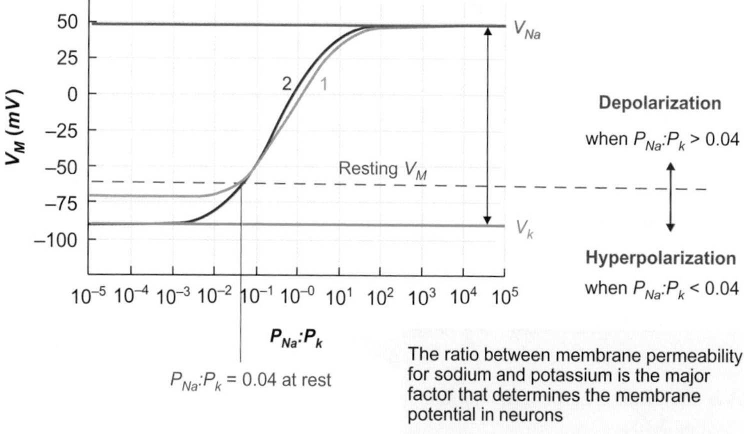

Figure 16.16 Dependence of the membrane potential V_M on the ratio between membrane permeability for sodium and potassium ions (P_{na}:P_K) in the giant axon of the squid (*Loligo* spp.)

The ratio between membrane permeability for chloride and potassium ions (P_{Cl}:P_K) was 0.45 for curve 1 and 0 for curve 2. When the membrane potential is less negative (more positive) than at rest, it is said that the membrane is depolarized. When the membrane potential is more negative than at rest, then the membrane is said to be hyperpolarized.

When the ratio $P_{Na}:P_K$ reaches high values, the membrane potential approaches the equilibrium potential for sodium (V_{Na})[38] in both situations. For the squid axon:

$$\begin{aligned} V_{Na} &= 59 \text{ mV} \log_{10}([\text{Na}_{out}]/[\text{Na}_{in}]) \\ &= 59 \text{ mV} \log_{10}(465 \text{ mmol L}^{-1}/72 \text{ mmol L}^{-1}) \\ &= +48 \text{ mV}. \end{aligned}$$

At the opposite end of the scale, when $P_{Na}:P_K$ is very small, the membrane potential approaches the equilibrium potential for potassium, V_K for curve 2 (when $P_{Cl}:P_K = 0$):

$$\begin{aligned} V_K &= 59 \text{ mV} \log_{10}([\text{K}_{out}]/[\text{K}_{in}]) \\ &= 59 \text{ mV} \log_{10}(10 \text{ mmol L}^{-1}/345 \text{ mmol L}^{-1}) \\ &= -91 \text{ mV for curve 2.} \end{aligned}$$

However, when $P_{Na}:P_K$ is very small and the ratio between membrane permeability for chloride and potassium stays constant ($P_{Cl}:P_K$ =0.45) (curve 1), then according to the GHK equation:

$$\begin{aligned} V_M &= 58 \text{ mV} \log_{10} \frac{P_K[\text{K}_{out}] + P_{Cl}[\text{Cl}_{in}]}{P_K[\text{K}_{in}] + P_{Cl}[\text{Cl}_{out}]} \\ &= 58 \text{ mV} \log_{10} \frac{\left(1\times 10\frac{\text{mmol}}{\text{L}}\right) + \left(0.45\times 61\frac{\text{mmol}}{\text{L}}\right)}{\left(1\times 345\frac{\text{mmol}}{\text{L}}\right) + \left(0.45\times 587\frac{\text{mmol}}{\text{L}}\right)} \\ &= 59 \text{ mV} \log_{10} (37.45/609) = -71.5 \text{ mV}. \end{aligned}$$

Therefore, for curve 1, the membrane potential approaches −71.5 mV when $P_{Na}:P_K$ approaches zero.

Thus, we can conclude that:

- the membrane potential in neurons is intrinsically linked to the membrane permeability ratio for sodium and potassium ions ($P_{Na}:P_K$), and
- the negative resting membrane potential of neurons (and other cells) takes the value that it does because cellular membranes at rest are more permeable to potassium than to sodium ions and the potassium ion concentration in the cytosol is much greater than that in the extracellular environment.

As shown in Figure 16.16, the membrane potential, V_M, is more positive than the resting value when the ratio $P_{Na}:P_K$ is greater than 0.04 (the resting value in this case); by contrast, V_M is more negative than the resting value when the $P_{Na}:P_K$ ratio is smaller than 0.04. By convention, a membrane potential that is less negative (i.e. more positive) than at rest is called **depolarization** because it implies a *reduction* in the amount of charge separation[39] across the membrane compared with the situation at rest. A more negative membrane potential than at rest is called **hyperpolarization** because it implies an *increase* in charge separation across the membrane.

In more general terms, depolarization refers to a change in the membrane potential that makes it less negative, while hyperpolarization refers to a change in the membrane potential that makes it more negative. For example, when the membrane potential changes from −60 mV to −40 mV, we say that the membrane depolarized by 20 mV (−40 − (−60) = +20 mV), while a change from −60 mV to +40 mV equals a depolarization of 100mV (+40 − (−60) = +100 mV). Conversely, when the membrane potential changes from −60 mV to −70 mV, then we say that the membrane hyperpolarized by 10 mV (−70 − (−60) = −10 mV).

16.2.3 Voltage-gated ion channels in neurons

We discuss in Section 3.2.3 how ion channels that open (activate) or close in response to changes in the membrane potential are called voltage-gated ion channels. Molecular studies have shown that voltage-gated channels are part of a superfamily of related proteins[40] that share a number of structural and functional features. The voltage-gated ionic channels are essential for the function of neurons and other excitable cells.

Some voltage-gated channels, called **voltage-activated Na^+ channels**, have a high **selectivity**[41] for sodium ions and are central to the generation of action potentials in excitable cells. Other voltage-gated channels with high selectivity for potassium ions, called **voltage-activated K^+ channels**, are essential for determining and restoring the resting membrane potential in excitable cells. There are also **voltage-gated Ca^{2+} channels** with high selectivity for calcium ions that transmit signals between excitable cells. In the heart of vertebrates and some invertebrate muscle, voltage-gated Ca^{2+} channels are also involved in the generation of action potentials.

38 We discuss the concept of equilibrium potential for specific ions in Section 3.1.3.

39 We discuss in Chapter 3.2.3 that the amount of electrical charges separated by plasma membranes is directly proportional to the membrane potential.

40 We discuss the general properties and structure of proteins in Box 2.1.

41 As we discuss in Chapter 4, the selectivity of channels for particular ions depends on a combination of factors, including the pore size of the channel, positive or negative electrical charges on protein residues at the opening of the channel and inside the channel, and ion binding sites in the channel.

Box 16.3 The origin of the membrane potential in neurons

Figure A(i) shows a diagram approximating a cellular membrane consisting of the lipid bilayer and two types of ion channels[1]: K^+-specific channels that are only permeable to potassium ions when open and Na^+-specific channels that are only permeable to sodium ions when open. The cytosol-like compartment on the inner side of the membrane has a high concentration of K^+, a low concentration of Na^+ and a concentration of organic anions $[A^{n-}]$ that carry *n* negative charges per particle, which counterbalances the concentration of positive charges carried by Na^+ and K^+ ions. The extracellular-like compartment on the outer side of the membrane has a high concentration of Na^+, a low concentration of K^+ and a concentration of chloride ions $[Cl^-]$ that counterbalances the concentration of positive charges carried by the potassium and sodium ions.

No electrical potential difference develops across the membrane if all ion channels stay closed

Since ions cannot pass through the lipid bilayer, it means that the membrane is impermeable at all times to Cl^- and A^{n-} ions. When all channels are closed, as shown in Figure A(i), the membrane is also impermeable to K^+ and Na^+ ions and there is no net electrical charge difference across the membrane. Consequently, there is no electrical potential difference across the membrane, i.e. the membrane potential is zero ($V_M = V_{in} - V_{out} = 0$).

Opening K^+ channels causes the membrane to charge more negatively on the cytosol-like side, where the K^+ concentration is higher

Imagine now that some K^+ channels open, while all Na^+ channels remain closed (Figure A(ii)). Then, some K^+ ions move down their concentration difference from the cytosol-like compartment to the extracellular-like compartment. This leads to an accumulation of positive charges (carried by the K^+ ions that moved across the membrane) on the 'extracellular' side of the membrane and negative charges (carried by the excess anions left behind because they cannot cross the membrane) in the cytosolic compartment. Accumulation of these positive and negative charges establishes a negative membrane potential ($V_M = V_{in} - V_{out} < 0$), which starts to oppose the diffusion of

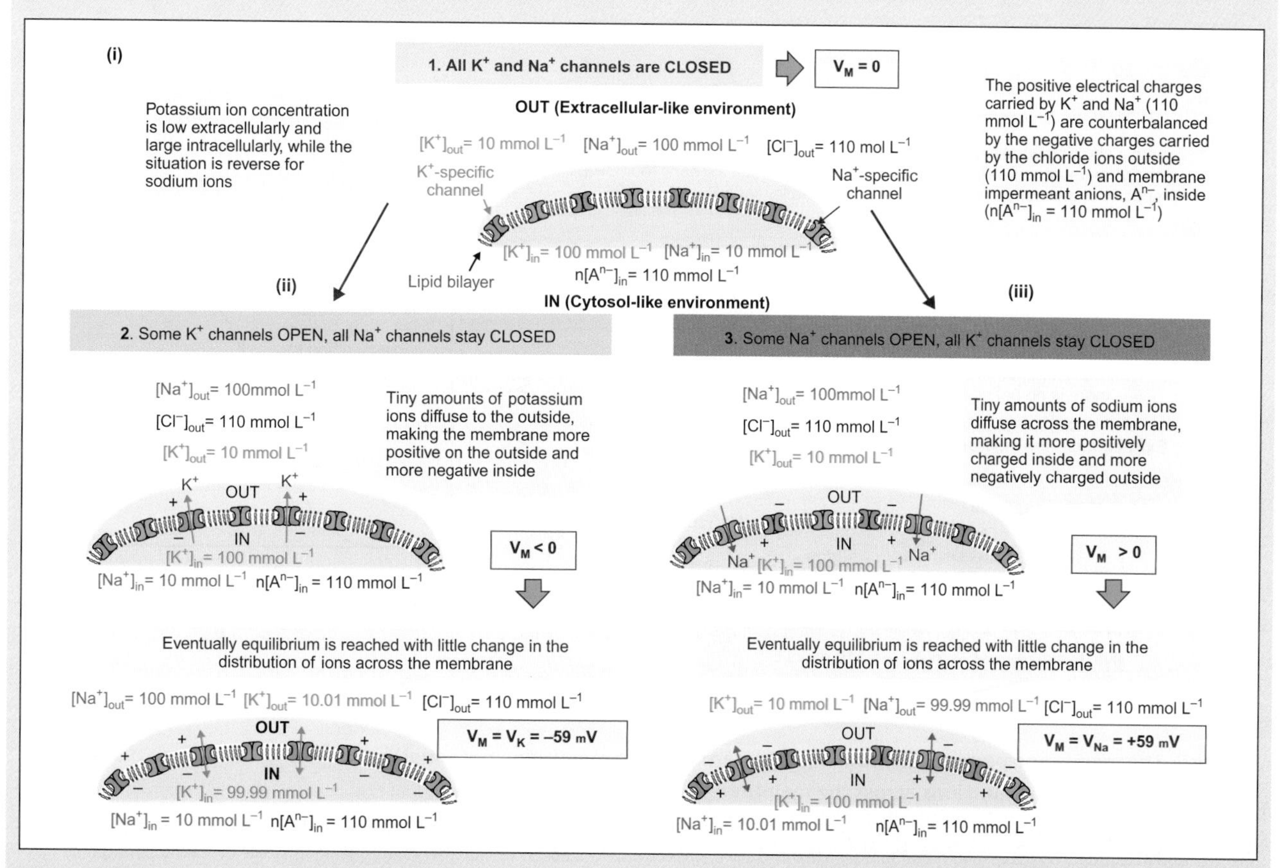

Figure A The selective opening/closure of potassium ion-specific and sodium ion-specific channels in a cellular membrane causes the change of a cell's membrane potential ($V_M = V_{in} - V_{out}$) between the equilibrium potential for potassium ion (V_K) and the equilibrium potential for sodium ion (V_{Na}) with little change in the ion distribution across the membrane

The equilibrium potentials V_K and V_{Na} in mV are given by the Nernst Equation (3.5) in Chapter 3: $V_K = 59 \log([K^+]_{out}/[K^+]_{in})$; $V_{Na} = 59 \log([Na^+]_{out}/[Na^+]_{in})$.

16

more potassium ions down their concentration difference: the accumulating positive charges on the extracellular side of the membrane increasingly repel K^+ ions and push them back into the cytosol-like compartment.

The net flow of potassium ions across the membrane eventually stops when the membrane potential reaches the equilibrium potential for potassium given by the Nernst equation: $V_K = 59\ mV\ log_{10}([K_{out}]/[K_{in}])$, which we discuss in Section 3.1.3. Since the ratio between the potassium concentrations $[K_{out}]/[K_{in}] = 0.1$, it follows that $V_K = 59\ mV\ log_{10}(0.1) = -59\ mV$. Then, the system is in electrical and chemical equilibrium (electrochemical equilibrium) with respect to K^+.

Opening of Na^+ channels causes the membrane to charge more positively on the cytosol-like side, where the Na^+ concentration is lower

Now imagine the reverse situation that some Na^+-channels are open and all K^+ channels are closed (Figure A(iii)). Then, the positively charged Na^+ ions start moving down their concentration difference from the 'extracellular' compartment to the 'cytosolic' compartment. This leads to an accumulation of positive charges carried by the Na^+ ions on the 'cytosolic' side of the membrane and build up of an equivalent amount of negative charges on the 'extracellular' side of the membrane: the anions that were initially balancing the charges associated with the Na^+ ions that passed through the membrane cannot cross the membrane. Hence, a positive membrane potential now develops ($V_M > 0$), producing an electrical potential opposing the diffusion of the positive Na^+ ions down their concentration difference. Again, when the two opposing forces balance each other, there is no net movement of Na^+ across the membrane at the equilibrium potential for sodium, $V_{Na} = 59\ mV\ log_{10}([Na_{out}]/[Na_{in}]) = 59\ mV\ log_{10}(10) = 59\ mV$.

Note that as we calculate at the end of Section 3.2.3, only a tiny amount of the total ions in a cell are needed to produce a membrane potential difference of 50 mV across its surface membrane. Therefore, the membrane potential of a neuron can change on hundreds of occasions from negative potential values (when the membrane is permeable to potassium ions) to positive values (when the membrane is permeable to sodium ions) with little change in the actual distribution of the potassium and sodium ions across the membrane.

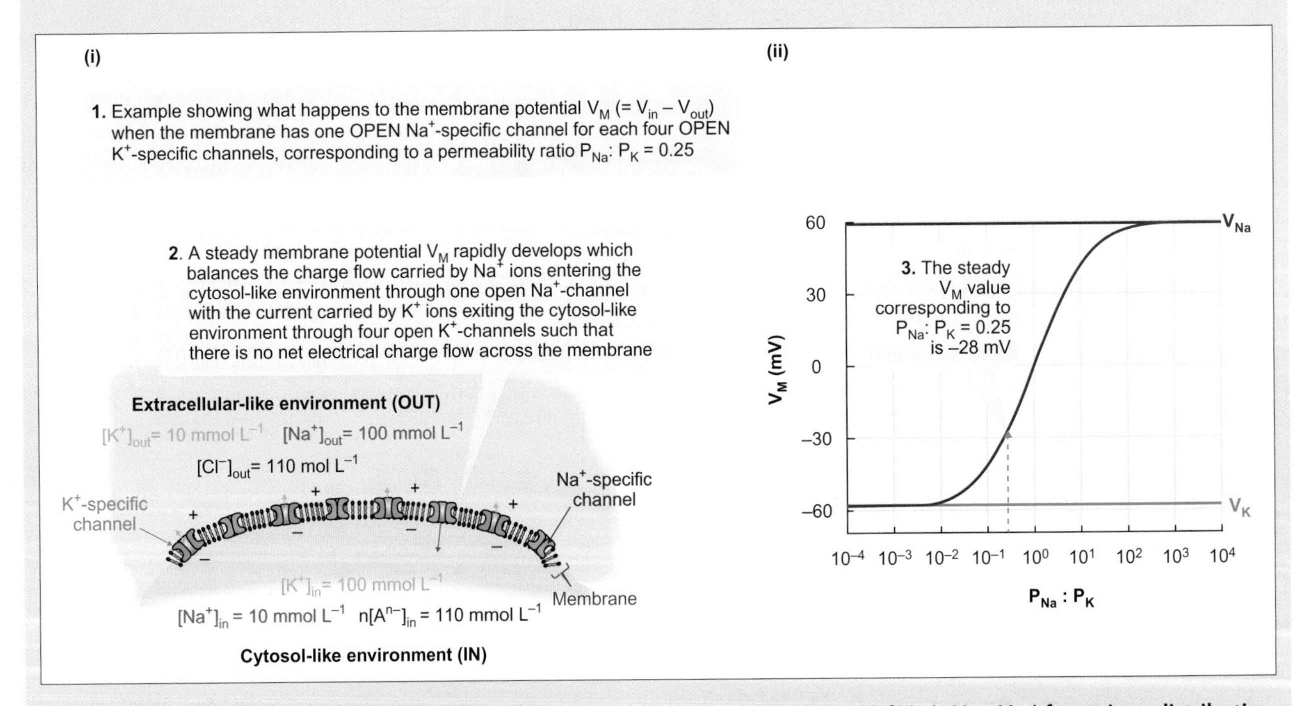

Figure B The membrane permeability ratio P_{Na}:P_K determines the membrane potential V_M (=$V_{in} - V_{out}$) for a given distribution of the sodium and potassium ions across the membrane

(i) The membrane permeability ratio P_{Na}:P_K depends on the number of open channels through which sodium and potassium ions can pass through. Generally, the Na^+- and K^+-currents across cellular membranes are tiny and the distribution of ions across cellular membranes remains stable over time through the action of the sodium/potassium pump. (ii) The dependence of the membrane potential (V_M) on the permeability ratio P_{Na}:P_K. When Na^+-channels suddenly open (or K^+-channels suddenly close) more sodium ions enter than potassium ions leave the cytosol causing a shift in V_M toward a more positive value; this corresponds to a higher P_{Na}:P_K value (membrane depolarization). Conversely, when K^+-channels suddenly open (or Na^+-channels suddenly close), more potassium ions leave than sodium ions enter the cytosol causing a shift in V_M toward a more negative value; this corresponds to a lower P_{Na}:P_K value (membrane hyperpolarization).

The electrical potential difference across cellular membranes stabilizes at a level that is intermediate between the equilibrium potential for K^+ and the equilibrium potential for Na^+

Cellular membranes are at all times permeable to certain degrees to both Na^+ and K^+. Therefore, we encounter the question: what happens to the membrane potential in such situations? When the membrane is permeable to both potassium and sodium ions, the membrane potential, V_M, stabilizes when the net movement of sodium ions across the membrane is compensated for by the net movement of potassium ions in the opposite direction such that there is no net (electric) current across the membrane as illustrated in Figure B(i). The value of the membrane potential V_M when there is no current across the membrane is given by the GHK equation (equation 3.9), which can be written in a slightly different form, where P_{Na}/P_K is the ratio between the membrane permeability for sodium and potassium ions:[2]

$$V_M = 59\ \text{mV} \log_{10} [(P_K[K_{out}]+P_{Na}[Na]_{out}) / (P_K[K_{in}]+P_{Na}[Na]_{in})]$$
$$= 59\ \text{mV} \log_{10} [([K_{out}]+(P_{Na}/P_K)[Na]_{out})/([K_{in}]+(P_{Na}/P_K)[Na]_{in})]$$

Equation A

In Figure B(i), four potassium-specific channels are open for each open sodium-specific channel. Assuming that potassium ions pass as easily through open K^+-channels as sodium ions pass through open Na^+-channels, then the membrane permeability for potassium (P_K) would be four-fold greater than the permeability for sodium (P_{Na}), i.e. $P_{Na}/P_K = 1/4 = 0.25$. From Equation A we calculate the value of the membrane potential for this scenario as

$$V_M = 59\ \text{mV} \log_{10} [((10+0.25)\times 100) / ((100+0.25)\times 10)]$$
$$= 59\ \text{mV} \log_{10} (35/102.5)$$
$$= 59\ \text{mV} (-0.47) = -28\ \text{mV}.$$

The relationship described by the GHK equation applies when V_M is steady—that is, when there is no net electric current across the membrane. This condition is satisfied not only in resting neurons, or when the membrane potential changes slowly, but also at the peak of an action potential and when the membrane is maximally hyperpolarized during the after-potential.

We also need to remind ourselves that the amount of sodium and potassium ions crossing the plasma membrane of excitable cells in general and neurons, in particular, is extremely small in absolute terms and that these small fluxes are efficiently compensated for by the action of the Na^+/K^+ pumps such that the uneven distribution of potassium and sodium ions across the cellular membranes (and therefore, V_M) is maintained on a long term basis.

Figure B(ii) illustrates how the membrane potential can attain any value between the equilibrium potentials for potassium (V_K) and sodium ions (V_{Na}) by altering the permeability ratio ($P_{Na}:P_K$) without changing in any significant way the actual ion concentrations across the membrane.

[1] We discuss properties of cellular membranes and the concepts of membrane potentials, ion channels and ion distribution across cellular membranes in Sections 3.2.3–3.2.5.

Genomic analysis combined with heterologous expression of cloned channels[42] and patch-clamp studies of their properties have shown voltage-gated channels to be hugely diverse, with individual neurons often having several types of potassium, sodium and calcium voltage-gated channels. For example, neurons in the vertebrate brain express at least two different types of sodium voltage-gated channels, six to eight types of potassium voltage-gated channels, and four to five types of calcium voltage-gated channels.

The voltage-gated channels have electrically charged transmembrane segments that act as voltage-sensing elements, which cause the individual channels to open or close abruptly when the membrane potential changes. At a whole-cell level, the overall response is less abrupt because of random differences in opening, closing and reopening of individual channels.

The change in the physiological state (open or closed) of voltage-sensitive ionic channels alters the membrane permeability for the ions that flow through the respective channels; changes in relative membrane permeability for ions, in turn, alter the membrane potential as described by the GHK equation and illustrated in Figure 16.16. Therefore, in order to study how the membrane potential controls the opening and closing of the voltage-gated channels, it is critical to ensure that the membrane potential is kept constant and is rapidly changed (within a millisecond or less) from one constant level to other predetermined constant levels. This is achieved by the **voltage-clamp technique**[43], which uses an electronic negative feedback loop to hold the membrane potential at predetermined, constant levels. The electrical current flowing across the membrane to hold the membrane potential constant quantitatively describes the net flux of electrical charges carried by ions moving through channels across the membrane as we discuss in Box 16.4.

[42] Heterologous expression involves the expression of channel proteins in cells that do not normally express them. For example, oocytes from clawed African toads (*Xenopus* spp.) are often used as host cells to express channel proteins by injecting complementary RNA or DNA into their cytoplasm or nucleus, respectively.

[43] The voltage-clamp technique was invented by the American physiologists George Marmont and Kenneth S. Cole in 1947; it was extensively improved and used by Alan Hodgkin and Andrew Huxley soon thereafter, and it is now widely used as a basic tool in electrophysiology.

In Box 16.5 we show that ionic currents flowing through an ion-specific channel are directly proportional to the difference between the membrane potential, V_M, and the equilibrium potential for that respective ion, V_{ion}. This difference, $V_M - V_{ion}$, acts as the driving force[44] for moving ions through the channel. The proportionality coefficient, g_{ion}, is called the membrane conductance for that ion and is proportional to the number of open channels through which the respective ion can pass.

[44] Notice that the driving force is zero when the membrane potential is equal to the equilibrium potential for that ion.

Notice that when the membrane potential is held by voltage-clamping at a constant level, V_h, the difference between the membrane potential and the equilibrium potential for that ion, $V_h - V_{ion}$, is constant. Then, the electrical current flowing through the membrane is proportional to the membrane conductance for the respective ion (and, in turn, to the number of open channels through which the ion can pass). If the membrane potential is allowed to change, the measurements taken do not allow us to deduce how channels open and close. Thus, changes in the membrane current directly reflect changes in the opening and closing of the specific channels only when the voltage-clamp technique is used.

Box 16.4 The voltage-clamp technique

The voltage-clamp technique permits measurement of transmembrane currents, such as those carried by ions moving through channels in the membrane, when the membrane potential, V_M, is held at a specific level or is suddenly changed from one level to another.

At the core of the voltage-clamp technique is an electronic negative feedback loop that clamps the membrane potential V_M to a specified command value V_{clamp} as illustrated in Figure A using the giant axon of the squid as an example. If V_M differs from V_{clamp}, then an electrical current, I, of appropriate direction and magnitude is almost instantaneously generated to keep V_M as close as possible to the command value, V_{clamp}.

Let us imagine that the sudden change of the membrane potential to a command value V_{clamp} causes some ion channels in the axon's membrane to open or close. The opening or the closing of channels in the membrane cause a change in the net flux of electrical charges carried by ions across the membrane. Since the total current across the membrane must be zero for the membrane potential to stay constant at the chosen command value V_{clamp}, the change in the net electrical current carried by ions across the membrane is matched (with opposite sign) by the current delivered by the voltage-clamp system. Thus, the current I, delivered by the system to maintain the membrane potential steady at the V_{clamp} value, quantitatively describes the ***net*** flux of electrical charges carried by ions across the axonal membrane and can be used directly to study the voltage-dependent behaviour of specific ion channels in the membrane as we discuss in Experimental Panel 16.1 and Box 16.5.

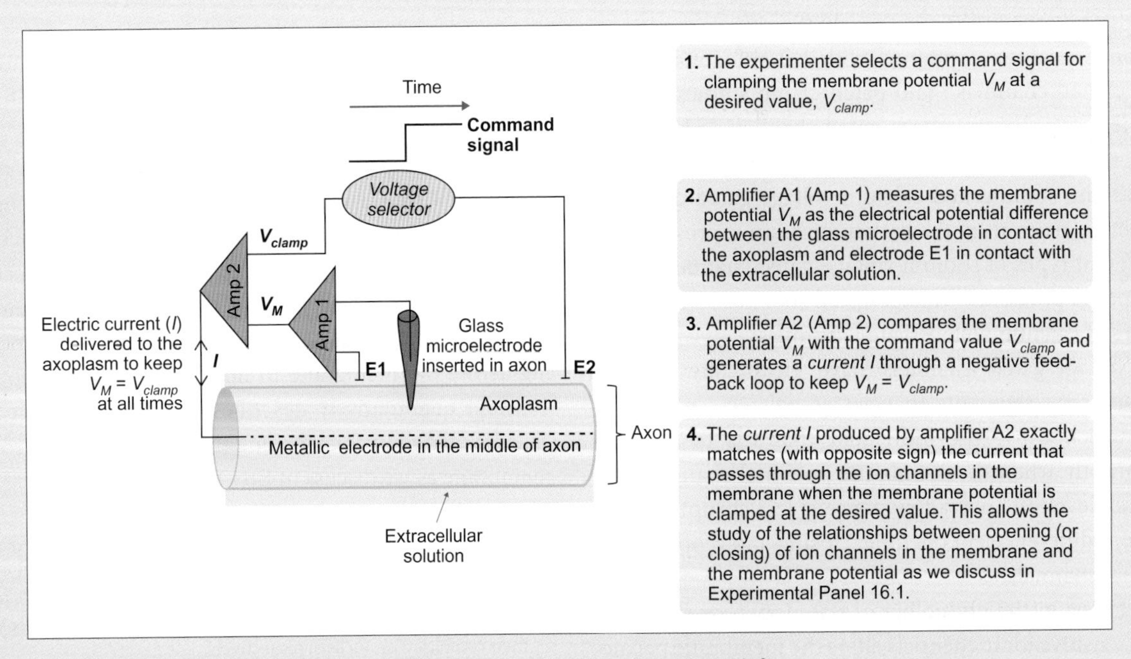

Figure A Simplified schematic diagram for explaining the voltage clamp principle

The voltage-clamp technique can also be applied in combination with the patch-clamp technique to record currents passing through single channels in a patch of membrane attached to the tip of the glass pipette.

Box 16.5 Membrane currents

As shown in Figure A, an inward (electrical) current is defined by convention as the movement of positive charges into the cell and is displayed as a downward (negative) deflection. Conversely, an outward (electrical) current is defined as the movement of positive charges out of the cell and is displayed as an upward (positive) deflection. Therefore, cations moving into the cell carry an inward current and cations moving out of a cell carry an outward current. In the case of anions, which are negatively charged, their movement out of a cell is electrically equivalent to an inward movement of positive charges (that is, an inward current) and their movement into the cell is electrically equivalent to the outward movement of positive charges (an outward current).

The direction of the electrical current carried by the sodium or potassium ions is inward, i.e. from the 'extracellular' to the 'cytosolic' side of the membrane, when the membrane potential is smaller (i.e. more negative or less positive) than the equilibrium potential for sodium or potassium ions, i.e. when $V_M < V_{Na}$ and $V_M < V_K$. Conversely, the direction of the current carried by the sodium or potassium ions is outward, i.e. from the 'cytosolic' to the 'extracellular' side of the membrane, when the membrane potential is greater than the equilibrium potential for sodium or potassium, i.e. when $V_M > V_{Na}$ and $V_M > V_K$. Thus, we can quantitatively describe the sodium- (I_{Na}) and potassium currents (I_K) using the following expressions:

$$I_{Na} = g_{Na}(V_M - V_{Na}) \quad \text{Equation A}$$

$$I_K = g_K(V_M - V_K) \quad \text{Equation B}$$

where the proportionality coefficients g_{Na} and g_K are the conductances (measured in siemens, S) of the membrane for the sodium and potassium currents[1], respectively. Note that the term 'ion conductance' is different from the term 'ion permeability', although high ion permeability generally implies high ion conductance and *vice versa*.

Equations A and B show that the equilibrium potentials for potassium (V_K) and sodium (V_{Na}) determine not only the lower and the upper range for V_M, as we show in Figure 16.16, but also the direction of the currents carried by the sodium and potassium ions (inward currents for Na^+ and K^+ when $V_M < V_{Na}$ and $V_M < V_K$, respectively, and outward currents for Na^+ and K^+ when $V_M > V_{Na}$ and $V_M > V_K$, respectively).

Qualitatively, for the membrane potential to change there must be a net movement of electrical charges carried by ions across the membrane. However, quantitatively, the number of ions that need to cross the membrane to produce a change in the membrane potential is very small indeed as we discuss in Box 16.3. Therefore, for most situations, the distribution of ions across the membrane does not change when the membrane potential changes.

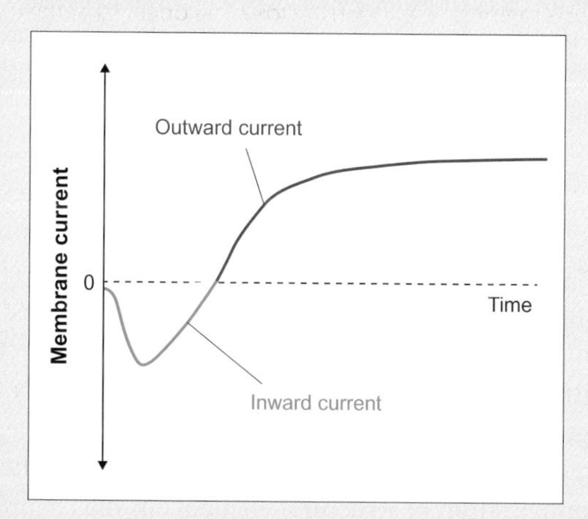

Figure A Inward and outward membrane currents

Conventionally, an inward current has a negative value and refers to either entry of cations into the cell or exit of anions from the cell. Conversely, an outward current has a positive value and refers to either exit of cations from the cell or entry of anions into the cell.

[1] We explain the meaning of the term conductance in Box 3.1.

In order to study the opening and closing properties of a particular type of channel with the voltage-clamp technique, it is necessary to eliminate electrical currents passing through all other types of channels. This is achieved by:

(i) using specific blocking agents for other channels,

(ii) removing the types of ions in which we are not interested from the extracellular environment and 'cytosolic' compartments,

(iii) reducing the driving force to zero for those ions we are not studying.

The voltage-clamp technique can be applied in conjunction with a patch electrode consisting of a glass micropipette with a fire-polished tip of about 1 µm in diameter attached by gentle suction to a patch of membrane to record current flow through single ion channels. The membrane patch can be left attached to the cell or detached from the rest of the membrane and the properties of the individual channels in that patch of membrane can be investigated. The principle of this technique, called the 'patch-clamp' technique,[45] is shown in Figure 16.17[46].

[45] The patch-clamp technique was developed by two German scientists: Erwin Neher and Bert Sakmann for which they received the 1991 Nobel Prize in Physiology or Medicine.

[46] Look at Figures 16.15 and 16.17 and notice the difference between the patch electrode method, where there is no membrane patch attached to the tip of the pipette, and the patch-clamp method where a patch of the membrane is firmly attached to the tip of the pipette.

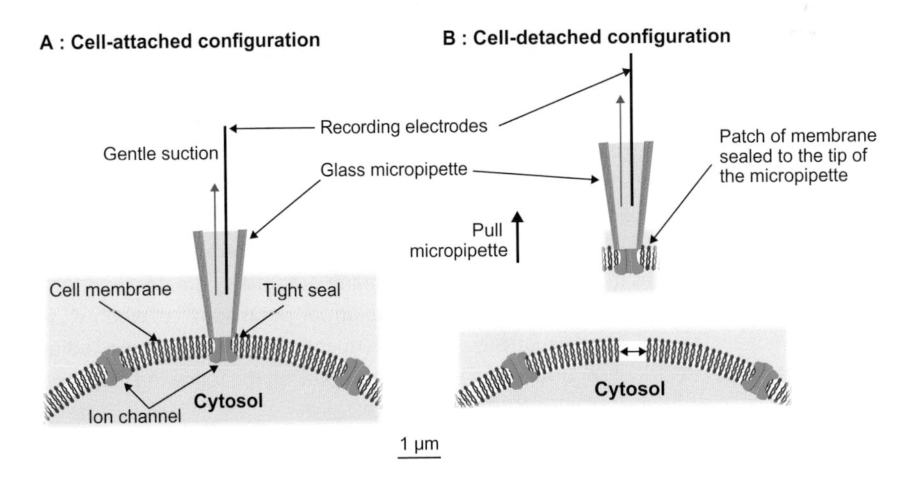

Figure 16.17 The principle of the 'patch-clamp' technique

(A) A glass micropipette with a fire-polished tip of about 1 μm in diameter is brought in contact with the surface of a cell. Gentle suction is then applied to produce a tight mechanical and electrical seal (in the order of 10^9 ohm, known as a gigaseal) between pipette and membrane. The patch of membrane attached to the tip of the pipette is left attached to the cell. (B) The membrane patch is detached from the cell by pulling away the micropipette and the membrane re-seals as indicated by the double-headed arrow. Properties of channels in the membrane patch under the tip of the pipette are studied with voltage-clamp techniques in either cell-attached (A) or cell-detached (B) configurations. Notice that the membrane patch under the tip of the pipette is left intact in both these configurations, while it is ruptured in the whole-cell recording configuration shown in Figure 16.15.

The patch-clamp technique can provide direct answers to basic questions regarding the function of specific ion channels such as: How fast do they open and close? How long do they stay open? How much current (i.e. how many ions) can single channels carry under specific conditions? Therefore, the patch-clamp technique in its different variations is widely used to identify and characterize the function of different channel types. Figure 16.18 shows an example of the first published recordings from one potassium channel in the squid giant axon.

Let us now discuss the main features of three major types of voltage-gated ion channels: the voltage-gated Na^+, K^+ and Ca^{2+} channels.

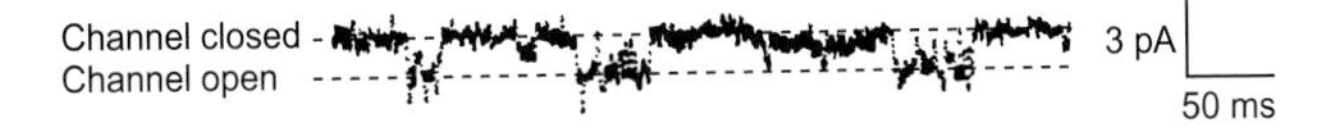

Figure 16.18 First published recordings of single K^+ channel activity from the squid giant axon using the patch-clamp technique

In this experiment the patch pipette was applied from the inside of the axon to permit a good electrical seal around the tip of the patch pipette. Potassium concentration was low in the pipette and high outside the axon to maximize the electrical current entering the pipette (downward deflection) carried by the potassium ions when the potassium channel opened at a holding membrane potential of −25 mV.

Source: Conti F and Neher E (1980). Single channel recordings of K currents in squid axons. Nature 285: 140–143.

Voltage-gated Na^+ channels

Structural studies have shown that the voltage-gated ion channels in all animals have a conserved architecture: a central pore surrounded by four **voltage sensors**. The voltage sensors sense the change in the electrical potential difference across the membrane and respond by altering the configuration of structural elements in the channel to open or close the pore to the passage of ions through the channel.

All neurons contain voltage-gated Na^+-channels. As is diagrammatically shown in Figure 16.19, each Na^+-channel consists of one large α-subunit comprising one long polypeptide chain and two smaller β-subunits, each consisting of a shorter polypeptide chain. The α-subunit has four similar (homologous) domains called 'repeats' I to IV and forms the channel pore. Each repeating unit (domain) has a positively-charged segment (S4 in Figure 16.19B), which acts as a sensor to the change in the electrical potential difference across the membrane. The β-subunits are accessories: they are not directly responsible for the opening or closing the channel, but can alter how long the channels stay open or how fast the channels close and open, for example.

Sodium channels have two types of gates: an **activation gate** associated with the S4 segments, and an **inactivation gate** associated with the intracellular loop between domains III and IV in the α-subunit shown in Figure 16.19. Note that the voltage-gated sodium channel is closed when either (or both) the activation or the inactivation gates

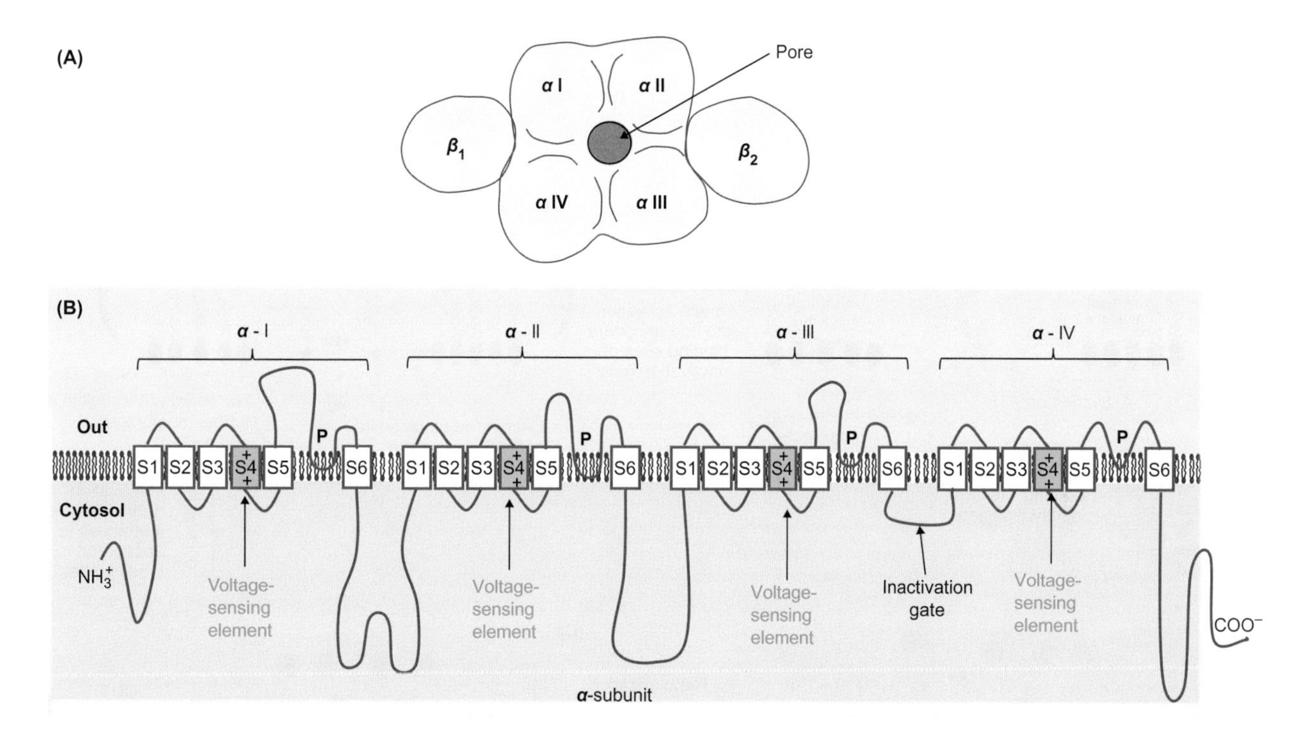

Figure 16.19 Schematic representation of the structure of the voltage-gated Na⁺ channel

(A) General representation of the subunit structure of the Na^+-channel consisting of one α-subunit and two β-subunits. (B) The primary structure of the α-subunit consisting of four domains (I–IV). Each domain contains six transmembrane segments (S1–S6) and pore-forming loops (P). The positively charged S4 segments act as transmembrane voltage sensing elements for the activation gate and the loop between domains III and IV is associated with the inactivation gate.

are closed. Figure 16.20 shows a schematic diagram of a sodium channel in cross-section with the two positively charged residues representing two of the four voltage sensors, and the inactivation loop shown as a round particle tethered to the rest of the channel protein. The inactivation gate closes the channel when it attaches to exposed binding site(s) on the channel.

When the membrane potential is more negative than about −60 mV, the sodium channels close because the electrical forces between the positive charges on the voltage sensors and the potential difference across the membrane keep the voltage sensors in a position in which the activation gate is closed. This situation is shown diagrammatically in Figure 16.20A. In contrast, the inactivation gate is open when the membrane potential is more negative than about −60 mV because its binding site is covered.

When the membrane rapidly depolarizes by more than 20 mV (membrane potential becomes less negative, i.e. more positive) the electrical forces that keep the activation gate closed become weaker, or even change direction if the membrane potential becomes positive, such that the voltage sensors now move outward and rotate, as illustrated in Figure 16.20B. This causes the opening of the activation gate and opening of the channel. The binding site of the inactivation gate also becomes uncovered when the activation gate opens causing the inactivation gate to close soon after the membrane becomes depolarized, as shown in Figure 16.20C; consequently, the Na^+ channel opens only transiently when the membrane is rapidly depolarized and then closes and stays closed if the membrane stays depolarized. In this state, the channel is said to be inactivated because it cannot be open by further depolarization of the membrane.

Thus, sodium channels can only open briefly after rapid depolarization when the activation gates open and before the inactivation gates close. Note, however, that if the depolarization of the membrane is relatively slow, the inactivation gates close as the activation gates open and, therefore, the channels stay closed and become inactivated.

In order to allow sodium channels to open again by depolarization, the inactivation gates must move back to their open position in a process called **repriming** as shown in Figure 16.20(A,D). The repriming of the sodium channels occurs when the membrane potential returns to, or becomes more negative than, the resting level (−60mV). However, when this happens (i.e. the membrane repolarizes) the activation gates rapidly close before the inactivation gates open. Thus, the sodium channels stay closed while the repriming of the sodium channels takes place.

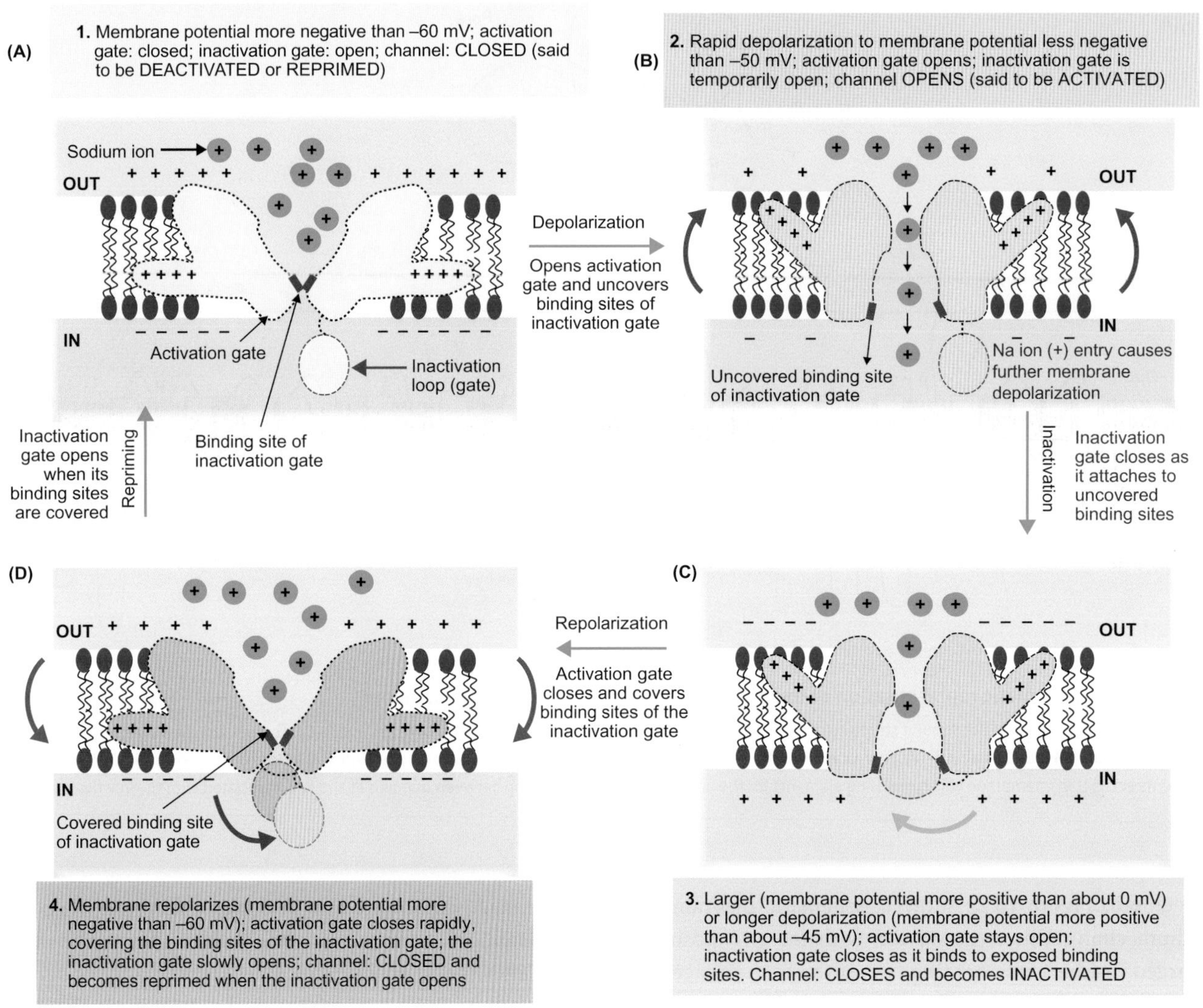

Figure 16.20 Functional states of the voltage-activated sodium channel

The channel is open only shortly after the rapid depolarization of the membrane. Note that the values given for the membrane potential are only indicative; at steady membrane potential < –60mV, the majority of the sodium channels are reprimed and at a steady membrane potential > –40 mV, the majority of the channels are inactivated.

The voltage-gated sodium channel is specifically blocked by tetrodotoxin (TTX), a toxin from puffer fish of the genus *Takifugu*[47] and other fish belonging to the Order Tetraodontiformes. TTX causes prey paralysis by blocking the generation of action potentials in nerve and muscle cells. TTX played a central role in electrophysiological studies by allowing functional separation of voltage-gated sodium channels from other types of ion channels in excitable cells as described in Experimental Panel 16.1.

Voltage-gated K⁺ channels

Most voltage-gated potassium channels have the same overall structure of the conductance pore as that of the α-subunit of the sodium channels. However, instead of being formed from one polypeptide chain with four repeating domains, the conductance pore of the voltage-gated K^+ channels consist of four separate polypeptide chains called α-subunits. The four α-subunits assemble into a structure called a tetramer, as shown in Figure 16.21. The positively charged S4 segments in the four α-subunits of the tetramer act as the voltage sensors of the channel, as in voltage-gated sodium channels. The voltage-gated potassium channels are very diverse. For example in humans there are 40 different types of α-subunits grouped in 12 different classes.

In addition to the conductance pore, many voltage-gated potassium channels have auxiliary β-subunits associated with the α-subunits. The β-subunits influence the kinetics of the conductance pore, such as how fast the pore closes and opens or how long the pore stays closed or open.

[47] This fish is considered a delicacy in Japan, but eating it can be lethal due to its high TTX content.

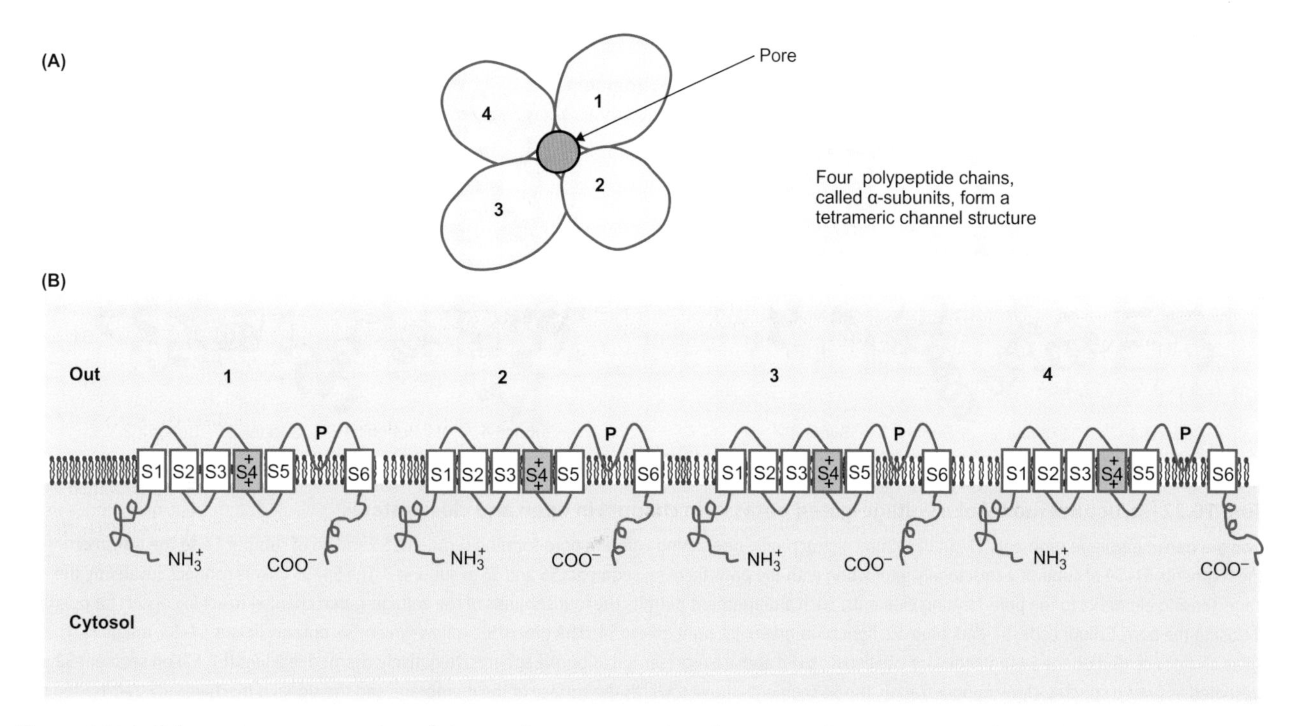

Figure 16.21 Schematic representation of the conductance pore in voltage-gated potassium (K^+) channels

(A) The K^+ channel consists of four polypeptide chains (1–4), called α-subunits that assemble in a structure homologous to that of the conductance pore of the sodium channel. Accessory subunits, called β-subunits, are usually associated with the α-subunits. The β-subunits can alter the functional properties of the conductance pore. (B) Each α-subunit consists of six transmembrane segments (S1–S6) and pore-forming loops (P). The positively charged S4 segments act as transmembrane voltage sensing elements for the activation of the gate.

Source: Catterall WA (2010). Ion channel voltage sensors: structure, function, and pathophysiology. Neuron. 67, 915–928.

Figure 16.22 shows a structural three-dimensional model of the open and closed configurations of the voltage-gated potassium channel. The four S4 segments together with their respective S1, S2 and S3 segments form the voltage-sensing element module, which connects to the pore-forming module via the S4–S5 linkers. Importantly, the voltage-sensing element (S1–S4) in each of the four subunits interacts structurally with the pore-forming elements of an adjacent subunit (S5, S6). This arrangement allows a concerted movement of the four subunits to open the pore when the membrane becomes depolarized and to close the pore when the membrane repolarizes or hyperpolarizes.

Some voltage-gated potassium channels lack the positively charged S4 transmembrane segments, but are sensitive when the membrane potential becomes more positive than the equilibrium potential for potassium ($V_M - V_K > 0$). Under these conditions, the channels are blocked by intracellular magnesium ions (Mg^{2+}) and endogenous polyamines[48], such as spermine, which act as a plug in the channel pore, and prevent potassium ions from flowing out when the membrane is depolarized. This type of K^+ channel occurs in some neurons, but plays a particularly important role in heart and skeletal muscles by stabilizing the membrane potential close to the equilibrium potential for potassium, and preventing potassium loss from cells when they become depolarized.

[48] Polyamines are organic molecules with two or more primary amino groups ($-NH_2$).

Voltage-gated Ca^{2+} channels

The overall structure of voltage-gated Ca^{2+}-channels is analogous to that of the sodium channels. As with sodium channels, the pore of voltage-gated Ca^{2+} channels is formed from a single polypeptide chain (α_1-subunit) with four repeating domains, each containing six transmembrane sequences and one pore-forming loop. Also, similarly to the sodium channel, membrane potential sensitivity is conferred by the positively charged S4 transmembrane segments in each domain of the α_1-subunit. In addition to the α_1-subunit there are several other subunits (α_2, β, γ and δ) which confer specific properties to Ca^{2+} channels.

Ca^{2+} channels occur in all excitable cells and their properties can be very diverse. Ten different genes in mammals encode different forms (isoforms) of α_1-subunit alone. As we discuss in Box 16.3, the concentration of potassium and sodium in the cytosol changes only very slightly when the membrane potential changes following the opening of the sodium or potassium channels for short periods of time. However, this is not the case regarding the cytosolic calcium ion concentration. In the resting cell, the calcium ion concentration is extremely low at around 0.0001 mmol L^{-1}—many orders of magnitude smaller than the potassium, sodium and chloride ion concentrations. For this reason, the cytosolic calcium ion concentration is likely to increase many fold when Ca^{2+} channels briefly open. Major cellular

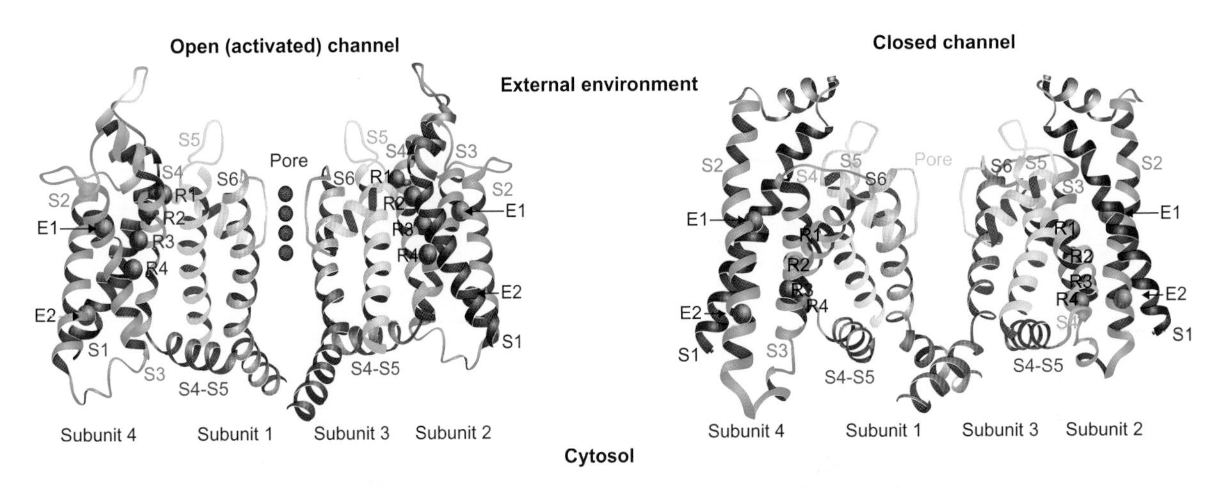

Figure 16.22 Structural model of a voltage-gated potassium channel in open and closed states

Shown are transmembrane segments S1–S4 of subunit 4 structurally interacting with the pore-forming segments S5 and S6 of subunit 1 and the transmembrane segments S1–S4 of subunit 2 structurally interacting with the pore-forming segments S5 and S6 of subunit 3. The S4–S5 linkers connect covalently the voltage-sensing elements to the pore-forming elements. Such arrangement permits the four subunits of the voltage-gated channel to act in concert for opening and closing the pore. Colour code: S1, dark blue; S2, light blue-green; S3, light green; S4, dark green; S5, yellow-green; S6, orange; linkers S4–S5, magenta. The positive charges carried on the S4 segments are labelled R1 to R4 and are represented as purple spheres. Negatively charged residues (E1, E2) on segment S2 are represented as brown spheres. Upon depolarization, the S4 segments move towards the surface of the membrane and the signal is mechanically transmitted to the pore-forming elements opening the channel. All transmembrane segments are in the form of α-helices as we discuss in Box 2.1.

Source: Catterall WA (2010). Ion channel voltage sensors: structure, function, and pathophysiology. Neuron 67: 915–928.

processes, including synaptic transmission, which we discuss in Section 16.3, and cardiac and smooth muscle contraction[49] depend on this increase in cytosolic calcium ion concentration following the opening of voltage-gated Ca^{2+} channels in plasma membrane.

16.2.4 Physiological basis of action potentials in nerve fibres

The fast and reliable transmission of information in the body of animals over long distances is achieved by electrical signals in the form of action potentials. Action potentials are all-or-none electrical responses triggered in the membrane of excitable cells by local depolarization of the membrane above a certain level, called the **threshold**. The threshold for triggering an action potential in neurons is a rapid local membrane depolarization by about 10–15 mV from its normal resting membrane potential level. For example, if the resting membrane potential rapidly changes from about −65 mV to about −50 mV in a nerve fibre (axon), then the threshold is reached and an action potential is triggered.

Figure 16.23 shows recordings of two action potentials from the giant axon of squid. One recording was made from an intact giant axon in the squid's mantle at 8.5°C and the other from the isolated contra-lateral axon from the same squid at 12.5°C. As indicated in Figure 16.23C, once the threshold is reached, the action potential is characterized by a rapid depolarization of the membrane, called the **upstroke**, to levels at which the polarity of the membrane reverses: the membrane becomes more positive on the cytosolic side than on the extracellular side so that the membrane potential becomes positive.

The action potential reaches its peak within milliseconds (0.001s) and the amount by which the membrane potential shoots over the zero line is called the **overshoot**[50]. Notice that the amplitude of an action potential is measured from the resting membrane potential to the peak of the action potential. The amplitude of the action potential in a nerve fibre is about 90–120 mV.

After the membrane potential reaches its peak, the membrane repolarizes rapidly and soon becomes even more negative inside than at rest (i.e. it hyperpolarizes) before the membrane potential returns more slowly to its resting level—a period called the **afterpotential**.

The British physiologists Alan Hodgkin and Andrew Huxley unravelled the underlying sequence of events responsible for the generation of an action potential. Their seminal work, for which they received the Nobel Prize for Physiology and Medicine in 1963, was performed on the giant axon of the

[49] We discuss mechanisms of intracellular Ca^{2+} rise in cardiac and smooth muscles in Sections 18.3 and 18.4, respectively.

[50] The overshoot reflects an increase in (opposite) polarity compared with that of the resting membrane potential and is described as depolarization because it is more positive than the resting membrane potential as we discuss in Figure 16.16.

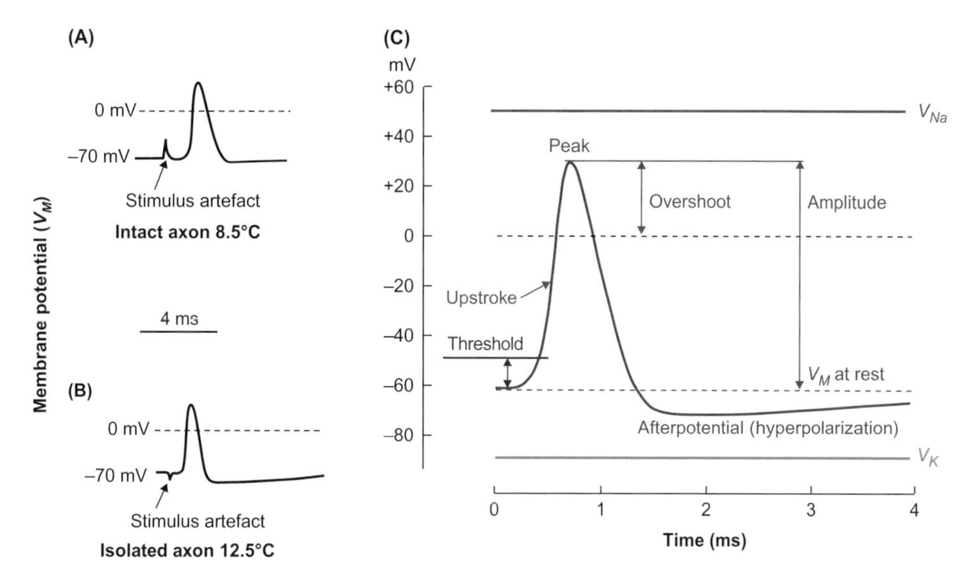

Figure 16.23 Changes in the membrane potential (V_M) during an action potential in the squid axon

(A) Action potential recorded with glass microelectrodes in an intact giant axon in the body of a squid (*Loligo forbesi*). (B) Action potential recorded in the isolated axon from the other side of the mantle of the same squid. The axon was placed in seawater. The action potentials were triggered by external stimulation which produces an artefact on the trace recordings. Notice that there are only minor differences between the two types of recordings if we take into account the difference in temperature. (C) Diagrammatic representation of an action potential with its main characteristics.

Source: traces (A) and (B) are reproduced from Hodgkin AL (1958). The Croonian Lecture: Ion movements and electrical activity in giant nerve fibres. Proceedings of the Royal Society B 148: 1–37.

squid using the voltage-clamp technique, but the same basic mechanisms were found to be involved in all excitable cells regardless of the species. The underlying mechanism for the generation of an action potential involves a large transient change in the relative permeability of the membrane for sodium and potassium ions (P_{Na}:P_K) due to the presence in the membrane of voltage-gated sodium and potassium channels, which open and close as shown in Figure 16.24.

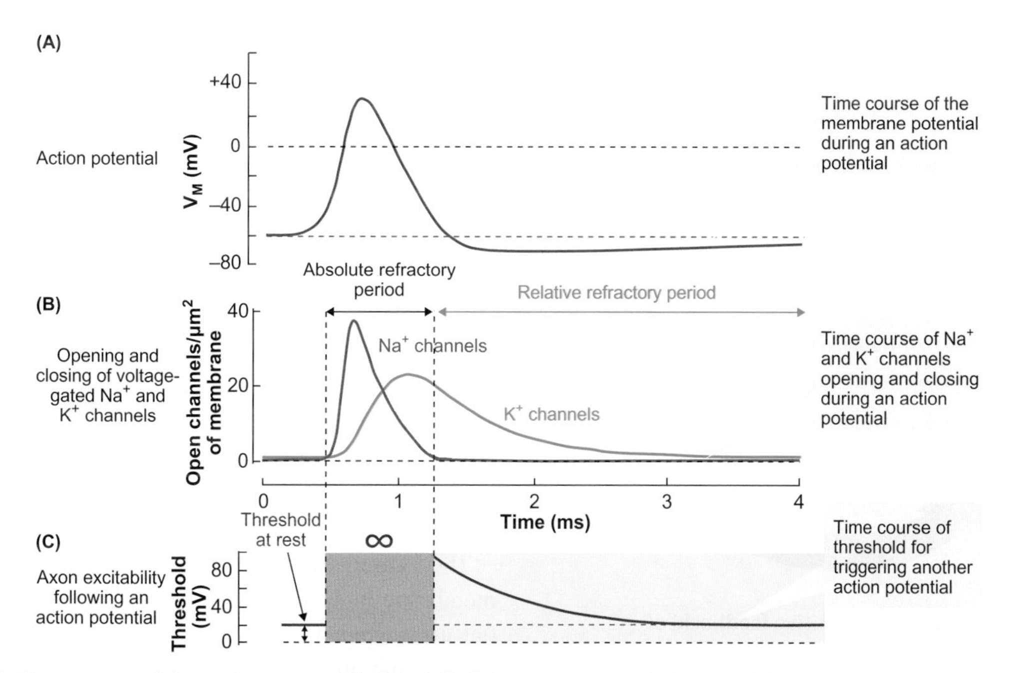

Figure 16.24 Time course of the action potential, the underlying opening and closing of the voltage-gated Na^+ and K^+ channels and changes in membrane excitability following initiation of an action potential

The absolute refractory period occurs over the period when the Na^+ channels open and then close (inactivate), while the relative refractory period is associated with the period when the Na^+ channels reprime (while remaining closed) and more K^+ channels are open than at rest.

Source: A, B: Hodgkin AL, Huxley AF (1952). A quantitative description of membrane current and its application to conduction and excitation in nerve. Journal of Physiology 117: 500–544.

If we look back at Figure 16.16, we are reminded that the membrane potential V_M can only change between the equilibrium potential for potassium (which in the squid axon is −91 mV) and the equilibrium potential for sodium (which in the squid axon is +48 mV). The actual change depends on the change in relative membrane permeability for sodium and potassium (P_{Na}:P_K). It is important to know that the membrane permeability ratio P_{Na}:P_K in neurons (and other excitable cells) is around 0.04 at rest because some potassium channels that are not sensitive to the membrane potential are open.

Prior to the generation of an action potential, the membrane becomes locally depolarized (less negative than at rest); when the threshold condition is reached, a regenerative, positive feedback cycle develops as illustrated in Figure 16.25. The local depolarization of 10–20 mV to the threshold of the membrane causes the rapid opening of the activation gate of some voltage-gated sodium channels, causing them to open as shown in Figures 16.20 and 16.24. This increases the number of open sodium channels, thereby increasing the permeability of the membrane for sodium. The increase in the sodium permeability leads, in turn, to the entry of a small number of positive charges carried by the sodium ions into the cell, causing further depolarization of the membrane. The rapidly developing depolarization causes more sodium channels to open and further increases the sodium permeability relative to potassium permeability, as shown in Figure 16.24.

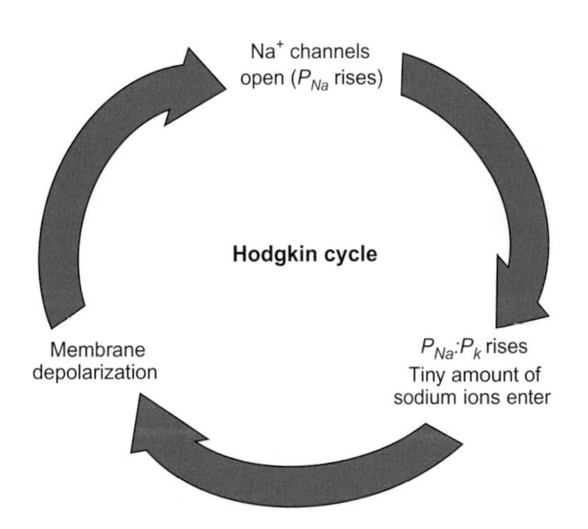

Figure 16.25 Reinforcement (positive feedback) of membrane depolarization and increase in sodium (Na^+) permeability during an action potential

Initial depolarization of the membrane causes the opening of some Na^+ channels, which increases the P_{Na}: P_K ratio, causing further depolarization, opening of more Na^+ channels and further increase in P_{Na}: P_K ratio until all Na^+ channels open.

This cycle repeats itself until sodium permeability rises rapidly, within 1 ms, so that P_{Na}:P_K increases from a resting value of about 0.04 to ≈ 30. As shown in Figure 16.16, when the P_{Na}:P_K ratio is about 30, the membrane potential approaches the equilibrium potential for sodium (V_{Na} ≈ +48 mV) and explains why the peak of the action potential is close to V_{Na}. It must be re-emphasized that only a tiny number of sodium ions need to enter the cell, down their electrochemical potential difference, to change the membrane potential rapidly from its resting level (−61 mV) to a value close to that of V_{Na}. When the membrane potential has this positive value, the inactivation gates on the sodium channels shut rapidly, closing the sodium channels and inactivating them, as illustrated in Figures 16.20B and 16.24. At the same time, as the membrane becomes depolarized and changes polarity, a large fraction of voltage-gated potassium channels become activated, albeit more slowly than the voltage-gated sodium channels. This is also shown in Figure 16.24. Therefore, after the peak of the action potential there are many more open potassium channels than in the membrane at rest.

The rapid inactivation of the sodium channels, combined with the increased activation of the potassium channels, causes an abrupt reduction in the ratio between sodium and potassium permeabilities (P_{Na}:P_K) to values that are even lower than at rest (<0.04). The increase in potassium permeability is associated with a tiny number of positive charges carried by potassium ions leaving the cell, which causes the return of the membrane potential towards the resting potential (i.e. repolarization of the membrane). When the ratio P_{Na}:P_K becomes even smaller than the value at rest, i.e. <0.04, the membrane potential becomes slightly more negative than the value at rest (that is, the membrane becomes hyperpolarized) according to the curves shown back in Figure 16.16; this produces the afterpotential.

When the membrane potential returns towards the resting level, the activation gates on the potassium channels also close and the voltage-gated potassium channels shut. The transient hyperpolarization during the afterpotential in nerve fibres speeds up the deactivation (closure) of the voltage-gated K^+-channels and the repriming of the Na^+-channels as described earlier in Figure 16.20D.

The amounts of sodium and potassium that cross the membrane in opposite directions with each action potential are extremely small: measurements with radioactive isotopes of potassium (^{42}K) and sodium (^{24}Na) indicate that the Na^+ and K^+ fluxes amount to only about 4×10^{-12} mol per 1 cm^2 of membrane in squid axons. The return of these small amounts of Na^+ and K^+ to their original compartments requires the use of about 2×10^{-12} mol of ATP by the Na^+/K^+ pump, as the pump transports two potassium ions per molecule of hydrolysed ATP[51]. From this value, it is possible to estimate that the aerobic degradation of 1 g of glucose would provide sufficient energy to compensate for the movement of sodium and potassium across the membrane following the generation of 0.8 billion (8×10^8) action potentials over a total length of 100 m of 30 μm diameter of unmyelinated axons. This estimation is based on the following information:

(i) about 28 mol of ATP are regenerated from the aerobic degradation of 1 mol of glucose[52],

(ii) 1 cm^2 of membrane can cover a 30 μm diameter axon for about 1 m in length, and

(iii) the membrane properties of the 30 μm diameter axon are similar to those of the squid axon.

This calculation indicates that the energetic cost associated with the maintenance of the asymmetric distribution of sodium and potassium ions across the membrane associated with the generation and propagation of action potentials is very small.

16.2.5 Refractory period

Soon after an action potential is triggered, it is not possible to trigger another action potential regardless of how strong the depolarization stimulus is, because the sodium channels are inactivated and closed. This period is called the **absolute refractory period** and determines the shortest possible time interval between two consecutive action potentials as shown in Figure 16.24.

Looking back at Figure 16.20A shows how the inactivation gates of the Na^+ channels need to be shifted to the open position (that is, the sodium channels need to reprime) to permit the Na^+ channels to open again upon depolarization to generate another action potential. Na^+ channels can reprime only after the membrane potential returns to levels that are as negative, or more negative, than at rest. Hence, as time passes, more and more sodium channels reprime and can open again upon depolarization. Stronger stimulation (depolarization) than at rest is necessary, however, to ensure that a sufficient number of Na^+ channels activate to start the regenerative cycle. This period, which follows the absolute refractory period, is called the **relative refractory period**. When the number of voltage-gated Na^+ channels in the membrane reprime to the same level as at rest and the voltage-gated potassium channels deactivate to the same level as at rest, the threshold returns to the initial level.

As we discuss in Section 17.2, the change in the threshold level for triggering an action potential during the relative refractory period plays an essential role in coding the intensity of a certain stimulus in terms of frequency of action potentials.

16.2.6 Post-inhibitory rebound excitation

Post-inhibitory rebound excitation refers to the intrinsic ability of a variety of neurons in the CNS of both vertebrates and invertebrates to respond with one action potential or a train of action potentials at the termination of a stimulus that causes membrane hyperpolarization (compared with the 'resting' membrane potential), as illustrated in Figure 16.26.

Post-inhibitory rebound excitation occurs in neurons in which the number of sodium channels that can open upon depolarization is considerably greater when the membrane is hyperpolarized than when the membrane is 'at rest'. The larger number of sodium channels available for activation increases the excitability of the membrane when it is hyperpolarized, such that when the membrane potential returns to its normal resting level at the termination of hyperpolarization, the condition for threshold is exceeded, and an action potential is triggered. Hyperpolarization of the membrane occurs, for example, when inhibitory synapses[53] on these neurons become activated, or when a small, persistent depolarizing (inward)[54] current in the 'resting' neuron becomes inactivated. Alternatively, termination of hyperpolarization can activate non-specific cationic channels, which cause rapid entry of positive charges such as Na^+ (i.e. inward current) and the membrane depolarizes above the threshold for triggering an action potential.

Post-inhibitory rebound excitation is pivotal for producing rhythmic activity of neurons in **neural circuits**[55] without sensory feedback. Such neural circuits occur in the CNS of both vertebrates and invertebrates and can produce rhythmic patterned outputs without needing sensory

[51] Note that the Na^+/K^+ pump works relatively slowly compared with the duration of an action potential and that it does not transport back the potassium and sodium ions immediately after each action potential.

[52] We discuss in Section 2.4.3 that the yield of ATP regeneration during the aerobic oxidation of glucose is about 28 molecules of ATP per mol of glucose.

[53] We discuss inhibitory synapses in Section 16.3.3.

[54] An inward current refers to a current carried by cations (positive charges) entering the neuron, or anions (negative charges) exiting the neuron as we discuss in Box 16.5.

[55] Neural circuits refer to a functional entity of interconnected neurons that is able to regulate its own activity using a feedback loop.

16

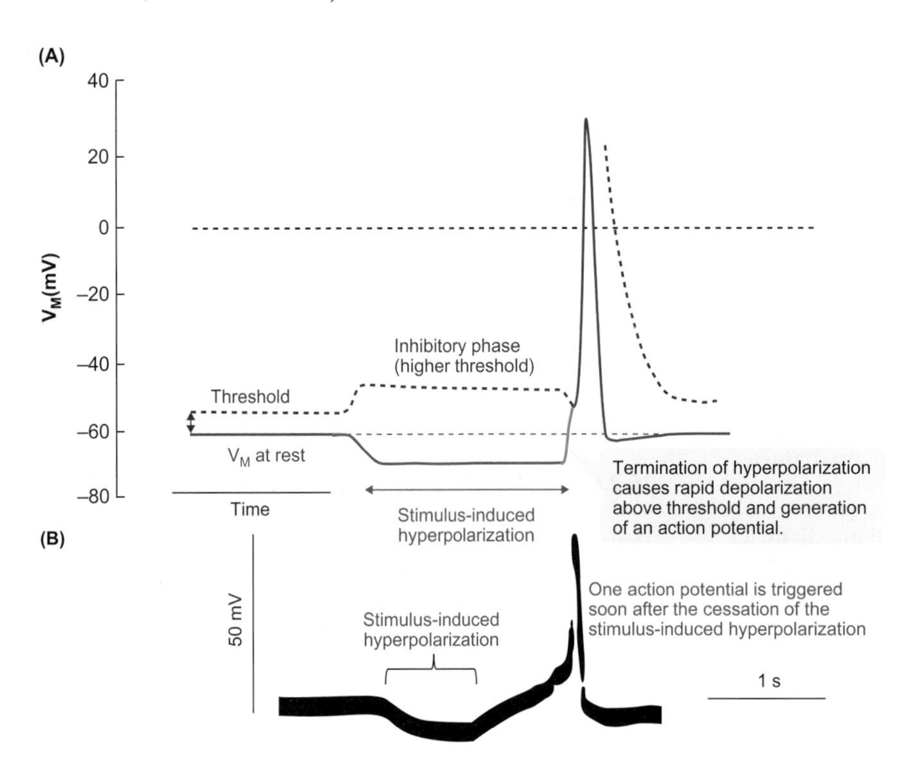

Figure 16.26 Post-inhibitory rebound excitation

(A) Diagrammatic representation of the underlying processes during post-inhibitory rebound excitation in neurons. (B) Original intracellular recording in a neuron from a marine snail (*Clione limacine*) that uses its foot as wings for swimming. The membrane was hyperpolarized by the application of a hyperpolarizing pulse on another neuron and an action potential was triggered soon after the termination of the hyperpolarization pulse.

Source: B: Satterlie RA (1985). Reciprocal inhibition and postinhibitory rebound produce reverberation in a locomotor pattern generator. Science 229: 402-404.

feedback[56]. These neural circuits are called **central pattern generators**[57].

16.2.7 Propagation of action potentials: the need for speed

A rapid and powerful escape response to danger decreases the predation risk and increases the chances of an animal's survival in the environment. Just as important, a rapid response for catching prey is critical for an animal's ability to survive. If the velocity at which the action potential is conducted increases, the speed at which the nervous system can process and respond to the information it receives also increases. The fastest response times are achieved by having the simplest neuronal elements and fastest conduction velocity of action potentials.

The mechanism for the propagation of an action potential along the giant axon of the squid is illustrated in Figure 16.27. In principle, the action potential triggered in the membrane at a specific site acts as a stimulus for triggering an action potential in the adjacent region, where the membrane is at rest.

When the action potential reaches its peak at a specific site on the surface of the axon, the membrane reverses its polarity and becomes more positive inside than outside (compared with its polarity when at rest). The difference in electrical potential along the axon between the site where an action potential is generated and adjacent sites where the membrane is at rest produces local currents, which are carried by ions along the membrane through the axoplasm and the extracellular environment. These local currents, shown in Figure 16.27, cause depolarization of the membrane ahead of the site at which the action potential is generated in the first place. As a result of this depolarization, the threshold condition is reached along the membrane for a certain distance from where the action potential occurs, triggering an action potential over this adjacent membrane region.

The cycle repeats itself, with the new region at which the action potential develops acting as a depolarizing stimulus for an adjacent region where the membrane is at rest. This ensures that the action potential moves along areas of the membrane that are at rest and explains why the magnitude of the action potential does not decrease as it propagates away from its initiation site.

The large amplitude of the action potential in nerve fibres provides a safety measure to ensure that the threshold condition is reached as far as possible ahead of the place where the action potential occurs. Note that an action potential cannot propagate backward in the direction from which it comes because the membrane immediately behind the region where an action potential occurred is in a refractory state.

[56] We discuss central pattern generators in Section 16.3.7.

[57] We discuss, in Section 22.1.1, that post-inhibitory rebound excitation is central for the generation of the respiratory rhythm in invertebrates.

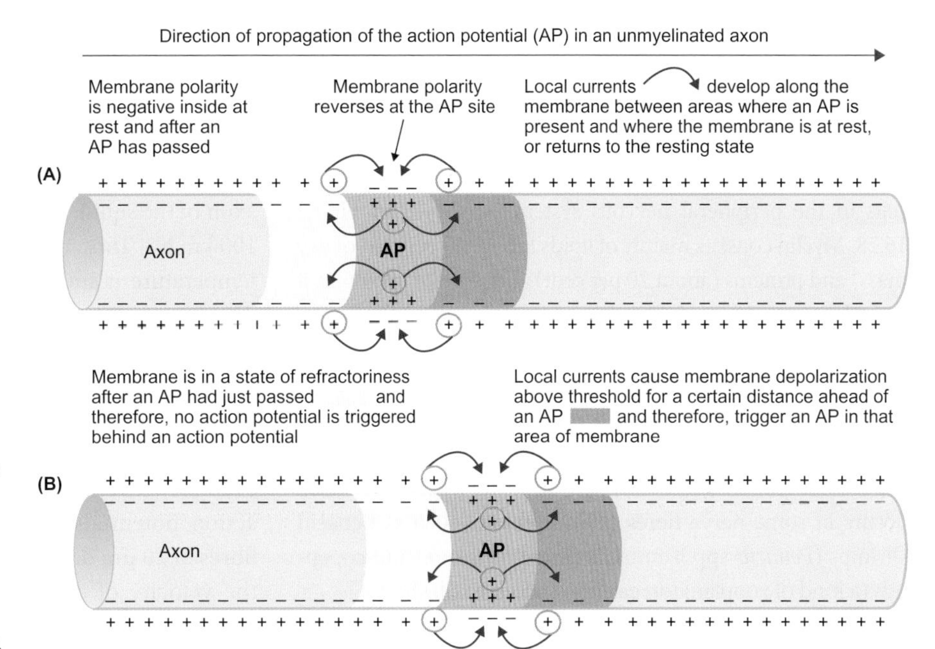

Figure 16.27 Mechanism of propagation of an action potential (AP) along an unmyelinated axon

(A) The region of the membrane where an action potential occurs is polarized differently (red-brown areas) from regions, where the membrane is at rest (green regions) and where the membrane is refractory (yellow region). Local currents carried by ions along the axon (indicated by arrows) cause membrane depolarization above threshold for a certain distance ahead of an AP (purple areas), which triggers an AP further along the membrane (B).

Axon size and insulation determine the speed of action potential propagation

The speed of action potential propagation depends on how far local currents can reach along the axon to depolarize the membrane above threshold.

Larger axons propagate action potentials faster

When the cross-sectional area of the axon is greater, local currents reach further to depolarize the membrane to threshold because the increased cross-sectional area decreases the internal electrical resistance of the axon. This means that, once an action potential is generated, it will trigger an action potential over a longer distance ahead of it, which translates into the action potential being propagated more quickly. Experimental evidence suggests that the conduction velocity increases proportionally with the diameter of the axon.

The development of large diameter axons is thought to have occurred as a result of selection pressure for fast escape behaviour in animals from several phyla including Mollusca, Arthropoda and Annelida. For example, the giant axons of squid can propagate action potentials at a speed of up to 100 km h^{-1}, which is critical for the squid's jet-propelled escape response. The axons run for most of the animal's body length from a compacted cluster of neurons called the stellate ganglion, to the muscles of the mantle that they innervate. When the muscles contract, water is rapidly ejected from the siphon cavity and the animal is propelled backward. This allows the squid to make brief—but very fast—movements through water and escape predators.

Insects have also evolved giant axons that participate in escape behaviour. For example, flies like the housefly (*Musca domestica*) and the fruit fly (*Drosophila melanogaster*) take off in tens of milliseconds in response to a visual stimulus such as the approach of a predator (or something that mimics it—a fly swatter, for example). The speed of this reflex response is because of the presence of giant interneurons, which collect incoming information from the eyes and fire action potentials directly to neurons that innervate jumping muscles in the legs and the flight motor apparatus.

Some annelids, such as the common earthworm (*Lumbricus terestris*), also display rapid escape responses that are mediated by giant nerve fibres. The escape response in the earthworm manifests itself as a rapid withdrawal of the part that is stimulated and offers some degree of protection against the many animal species that prey on them.

Fibre insulation increases the propagation velocity of action potentials

The electrical insulation of the membrane along most of the length of a nerve fibre, with small gaps at regular intervals where the membrane is in contact with the extracellular environment can greatly increase the conduction velocity of action potentials. For the same fibre diameter, such an arrangement vastly increases the distance along the nerve fibre over which the threshold condition is reached ahead of an action potential. This translates to an action potential being propagated much more quickly. Such an arrangement evolved typically in the nerve fibres of all vertebrates with the exception of cyclostomes[58] as an evolutionary adaptation for increasing the velocity of signal transmission in larger organisms.

The majority of axons in vertebrates are surrounded and electrically insulated by **myelin**, a sheath of multilayered membranes produced by oligodendrocytes in the CNS and Schwann cells in the peripheral nervous system, as shown in Figure 16.28. Myelin consists mainly of lipids (about 80 per cent of dry mass) and proteins (about 20 per cent) and is white in colour; it provides electrical insulation around the axons except for regular 1–10 μm wide gaps along the axon's length, which occur at about 1–2 mm intervals called the **nodes of Ranvier**[59]. Action potentials in myelinated fibres can only be generated at the level of the nodes of Ranvier, where the membrane of the axon is exposed to the extracellular environment. Myelination also occurs in some nerve fibres of invertebrates such as penaeid shrimps (*Penaeus* spp.), annelids (*Lumbricus* spp.) and copepods (a kind of zooplankton eaten by baleen whales).

An action potential generated at one node of Ranvier can act as stimulus to initiate an action potential 2–3 nodes ahead, where the membrane is at rest. The principle of this mechanism of action potential propagation is illustrated in Figure 16.28. Since action potentials occur only at the level of the nodes of Ranvier, observing the propagation of an action potential along a myelinated fibre gives the impression that the action potential jumps from one node to another[60]. Therefore, this type of action potential propagation is called **saltatory[61] conduction**.

[58] The oldest surviving class of vertebrates, which includes lampreys and hagfish, as we discuss in Section 1.4.2.

[59] Louis-Antoine Ranvier was a 19th century French physician and histologist, who discovered the regularly spaced interruptions in the myelin sheath, named after him.

[60] Note that the longitudinal electrical currents that cause depolarization above threshold at Ranvier nodes ahead of an action potential do not reflect individual ions jumping from one node to the next.

[61] From the Latin *saltare* which means to hop, leap or jump.

How much faster is the conduction velocity of action potentials in myelinated compared to unmyelinated fibres? As an example, myelinated fibres of only 20 μm diameter in mammals can conduct action potentials at 400 km h^{-1} compared with unmyelinated fibres 30 times wider like the giant axon of the squid, which conducts action potentials at about 100 km h^{-1}. Thus, myelination combined with a higher body temperature is much more efficient at increasing the conduction velocity of action potentials than an increase in fibre diameter. In evolutionary terms, myelination precluded the need for vertebrates to evolve 'giant' nerve fibres.

The conduction velocity of action potentials in myelinated fibres increases linearly with the fibre diameter. For example, myelinated fibres of 10 μm diameter conduct action potentials at about half the speed of myelinated fibres of 20 μm diameter. In general, myelination increases the velocity of action potential propagation by a factor of about 10 or more compared to unmyelinated fibres of same diameter.

Another great advantage of myelination is that it reduces the membrane area in contact with the extracellular environment. Since action potentials occur only at the nodes of Ranvier, the ion fluxes associated with the propagation of one action potential per unit length of axon are reduced. This reduction decreases the amount of ATP that is necessary to maintain homeostasis with respect to the ion composition of the axoplasm for a given axon length. Therefore, myelination considerably reduces the energetic cost for signal transmission along the nerve fibres in vertebrates.

There is generally a positive correlation between age-dependent changes in brain myelination and speed for performing specific tasks. For example, the reaction time to different stimuli is several fold slower in very young mammals when the myelination process of the brain is not complete, compared with young adults, when the myelination process is

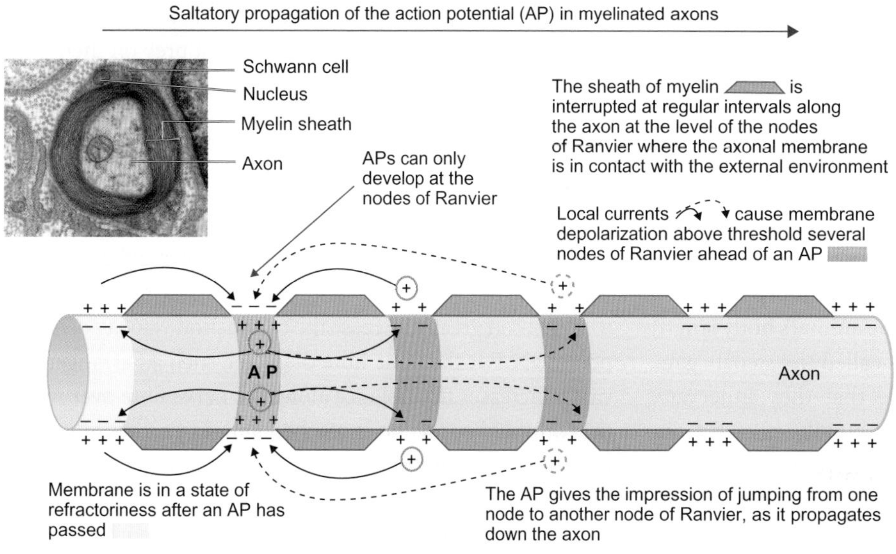

Figure 16.28 Mechanism of propagation of an action potential (AP) along a myelinated axon (saltatory conduction)

Local currents along the axon (indicated by arrows) cause membrane depolarization above threshold for several nodes of Ranvier ahead of an AP (red-brown area). APs triggered at these nodes further along the membrane give the impression of the AP jumping from one node of Ranvier to another node, as it propagates down the axon. Image: Roadnottaken/Wikimedia Commons

Transmission electron micrograph of a myelinated axon, generated at the Electron Microscopy Facility at Trinity College, Hartford, CT (https://en.wikipedia.org/wiki/Myelin#/media/File:Myelinated_neuron.jpg).

complete. Moreover, after myelination reaches a peak (around mid-thirties in humans), the level of myelination begins to decline as the animal ages (after about 40 years of age in humans). This decline in myelination is paralleled by the reduction in the speed at which specific motor and cognitive tasks are performed. This is well-known in elite sports, where most sports(wo)men lose competitiveness after their mid-thirties.

16.2.8 Intracellular and extracellular recordings of action potentials

Figures 16.27 and 16.28 show that the propagation of action potentials along nerve fibres is associated with a wave of negative electrical charge (or negative potential) spreading down the fibres. The change in electrical potential on the surface of nerve fibres associated with the propagation of action potentials can be readily recorded by placing two metallic electrodes at some distance from each other along a nerve fibre (or a bundle of nerve fibres) and measuring the potential difference between the electrodes. When an action potential arrives at one electrode, the electrical potential around that electrode becomes slightly more negative than before and the small potential difference between electrodes (less than 1 mV) can be recorded. Such measurements are called extracellular recordings of action potentials to distinguish them from the intracellular type of recordings discussed so far in this chapter.

Figure 16.29 shows intracellular and extracellular recordings of the action potential in the cell body of the same rat neuron. Notice that the peak of the extracellular signal (20 µV) was 1500 times smaller than that of the intracellular recording (30 mV). In addition, the time course of the action potential is different if recorded intracellularly or extracellularly. This is because the extracellular recordings depend on experimental conditions in addition to changes in membrane permeability to specific ions. The size of intracellularly recorded action potentials is smaller in the cell body of the neuron (about 30 mV in Figure 16.29) than in the axon (about 120 mV in Figure 16.23) because of the lower density of voltage-gated sodium channels in the cell body.

Extracellular recordings of action potentials are technically less demanding than intracellular recordings and are often used as a tool to give direct information about the time when action potentials are generated and about the frequency of action potentials[62]. Extracellular recordings are also useful for measuring the conduction velocity of action potentials. In this case, action potentials are measured at two points along a nerve and the velocity of propagation is calculated by dividing the distance over which the action potential travelled by the time difference between the two extracellularly recorded action potentials.

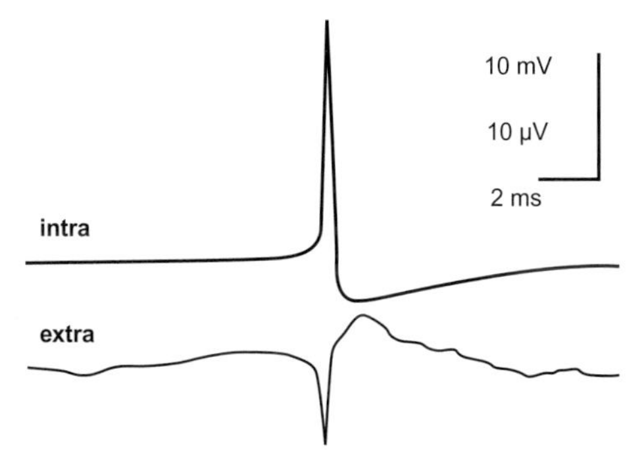

Figure 16.29 Simultaneous intracellular (intra) and extracellular (extra) recordings of an action potential in the cell body of same rat brain neuron

The 10 mV calibration bar refers to the intracellular trace recording and the 10 µV calibration bar refers to the extracellular trace recording.

Source: Henze DA et al (2000). Intracellular features predicted by extracellular recordings in the hippocampus in vivo. Journal of Neurophysiology 84: 390-400.

16.2.9 Non-spiking neurons

The vast majority of neurons propagate signals via action potentials. However, some neurons, called non-spiking neurons, do not generate signals in the form of action potentials because they have a relatively low density of voltage-gated sodium channels in their membrane. These neurons form connections between other neurons particularly in the CNS of vertebrates and invertebrates and transmit signals over short distances via graded changes in their membrane potential. Non-spiking neurons are integrated with spiking neurons in the same neural circuits and often function in a modulatory role, helping to establish a certain type of behaviour, as we discuss in Section 22.2.2.

16.2.10 Pacemaker potentials

Some neurons and other type of excitable cells spontaneously generate oscillations of their membrane potential at frequencies that are sensitive to specific factors, such as the release of neurotransmitters from nerve terminals, the release of hormones or even partial pressure of carbon dioxide (CO_2). These cells, called **pacemaker cells**, do not have a resting membrane potential as such, because they never rest. The spontaneous oscillation in the membrane potential of pacemaker cells is called **pacemaker potential**, and is the result of the activation of specific voltage-gated ion channels, which display certain voltage- and time-dependent characteristics. The pacemaker potential can manifest itself

[62] Figures 16.34, 16.43 and 22.12B, for example, contain extracellular recordings of action potentials.

in regular triggering of action potentials (spiking) or in non-spiking oscillations in the membrane potential.

As an example, Figure 16.30 displays intracellular recordings of spontaneous spiking activity in two neurons from the sea hare (*Aplysia californica*), a marine snail. Following an action potential, the membrane repolarizes and then gradually depolarizes until it reaches the threshold for triggering another action potential; the cycle is then repeated. The frequency with which action potentials are triggered is determined by the rate of membrane depolarization to the threshold at which an action potential is triggered. In turn, the rate of membrane depolarization is mainly determined by the rate of increase in the membrane permeability ratio $P_{Na}:P_K$[63].

Neuronal spiking and non-spiking pacemaker cells occur in both vertebrates and invertebrates and play important roles in determining certain rhythms in the body of animals (Section 16.3.7)[64].

› Review articles

Armstrong CM, Hille B (1998). Voltage-gated ion channels and electrical excitability. Neuron 20: 371–380.
Bean BP (2007). The action potential in mammalian central neurons. Nature Reviews Neuroscience 8: 451–465.
Bezanilla F (2006). The action potential: from voltage-gated conductances to molecular structures. Biological Research 39: 425–435.
Catterall WA (2010). Ion channel voltage sensors: structure, function, and pathophysiology. Neuron 67: 915–928.
Hodgkin AL (1958). The Croonian lecture: Ionic movements and electrical activity in giant nerve fibres. Proceedings of the Royal Society of London. Series B 148: 1–37.
Keynes RD (1983). The Croonian Lecture. Voltage-gated ion channels in the nerve membrane. Proceedings of the Royal Society of London, Series B 220: 1–30.
Xu K, Terakawa S (1999). Fenestration nodes and the wide submyelinic space form the basis for the unusually fast impulse conduction of shrimp myelinated axons. Journal of Experimental Biology 202: 1979–1989.

[63] Note that a decrease in the potassium permeability causes an increase in the permeability ratio $P_{Na}:P_K$.
[64] We also discuss pacemakers involved with the generation of the cardiac rhythms in Sections 18.3.2 and 22.3.1.

16.3 How neurons communicate with one another

The transmission of signals between neurons takes place at specialized junctions called **synapses**, where one neuron sends information about its own activity to the receptive surfaces of other neurons. There are two general types of neuro-neuronal synapses: electrical and chemical synapses.

16.3.1 Electrical synapses

Electrical synapses are very tight structures in which the membranes of the two cells are kept in very close proximity: the gap across the synapse is about the same as the thickness of the plasma membranes (about 5–6 nm). Figure 16.31 shows how electrical synapses feature channels that pass through both membranes; these are called **gap junction** channels. Gap junction channels consist of two nonselective channels (called hemi-channels) joined together, one in

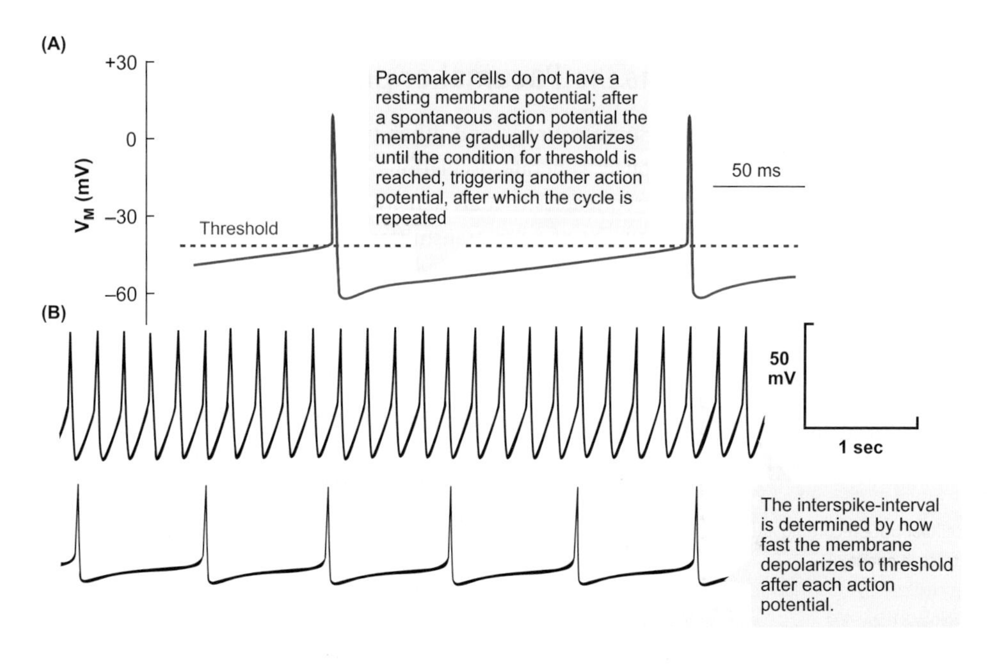

Figure 16.30 Pacemaker potentials in neuronal pacemaker cells

(A) Diagrammatic representation of spontaneous activity in pacemaker cells. (B) Intracellular recordings of spontaneous activity in two pacemaker neurons from the visceral ganglion of the sea snail (*Aplysia californica*). The interspike interval is determined by how fast the membrane depolarizes to threshold after each action potential.

Source: B from Junge D, Moore GP (1966). Interspike-interval fluctuations in *Aplysia* pacemaker neurons. Biophysical Journal 6: 411-434.

each membrane. The hemi-channels are called connexons in vertebrates and innexons in invertebrates. Each connexon or innexon is made of six subunits called connexins or innexins, respectively, forming a pore-like structure when open, as shown in Figure 16.31.

Gap junction channels permit ions, such as sodium, potassium and calcium, and molecules smaller than about 1000 Dalton (D) to pass through. Consequently, the channels allow ionic currents to flow unimpeded from one cell to another. However, gap junction channels can open and close according to the environmental conditions they experience. For example, gap junction channels close when exposed to Ca^{2+} concentrations in the millimolar range, or if the pH becomes too acidic. Such situations arise, for example, when one of the cells becomes metabolically impaired or is damaged, allowing the pH to decrease or ionized calcium to rise. In such instances, the gap junction channels close to protect the other cell from becoming damaged as well.

The principle of synaptic transmission in electrical synapses is essentially the same as for action potential propagation along an unmyelinated axon. The local current carried by ions can flow from one cell into the other through the gap junction channels causing depolarization of the membrane in the second cell. Upon reaching the synapse, an action potential in one cell depolarizes the membrane of the second cell above threshold, triggering an action potential in the other cell.

Electrical signals at electrical synapses, including action potentials, can generally propagate in either direction[65] and the transmission does not involve any delay. Thus, the depolarization or hyperpolarization of one neuron causes depolarization or hyperpolarization, respectively, of the other neuron connected by an electrical synapse. Because of their speed of transmission, neuro-neuronal electrical synapses occur in situations when quick responses are needed, such as escape mechanisms in most animals, or synchronized actions involving large numbers of excitable cells.

For example, the escape response in flies and the tail-flip escape response in goldfish (*Carassius auratus*) involve electrical synapses. The individual polyps in coral colonies, such as those of reef-building species (Class Scleractinia)[66], are also connected by a nerve net, which is based on electrical synapses, and which coordinates the activities of the entire colony. Electrical synapses are also involved in the defence mechanism of the sea hare (*Aplysia* spp.). When disturbed, the sea hare releases a defensive screen of ink to obscure the vision of predators. The neurons that stimulate the ink gland are interconnected by electrical synapses to ensure that they fire together. This synchronized action is critical for this defensive mechanism to be as efficient as possible. Electrical synapses also occur in other tissues, such as the vertebrate heart and smooth muscle, as we discuss in Sections 18.3 and 18.4.

16.3.2 Chemical synapses

Chemical synapses are specialized communication sites between two excitable cells, where signal transmission occurs via chemical transmitter molecules. A diagram and an electron micrograph of a chemical synapse between two neurons is shown in Figure 16.32, which illustrates the existence

[65] There are exceptions in which the signal can pass more easily in one direction than in the other direction. These electrical synapses are called rectifying or unidirectional electrical synapses. Rectifying gap junctions occur in both vertebrates and invertebrates and are thought to result from the joining of two hemi-channels made of different combinations of connexin/innexin isoforms.

[66] Reef-building corals are one of the simplest but most important animals on Earth as they are responsible for building the most diverse marine communities on the planet.

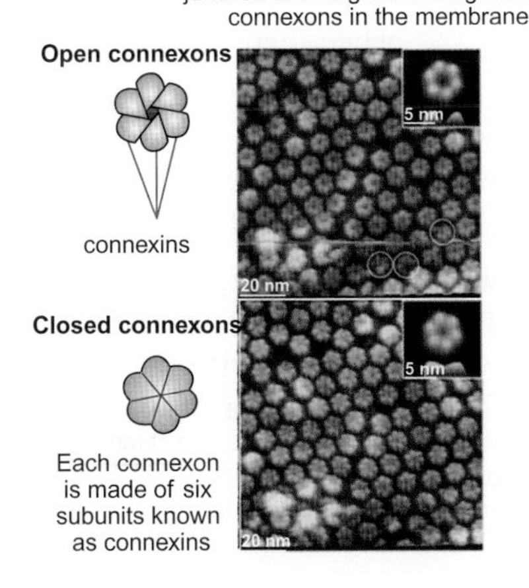

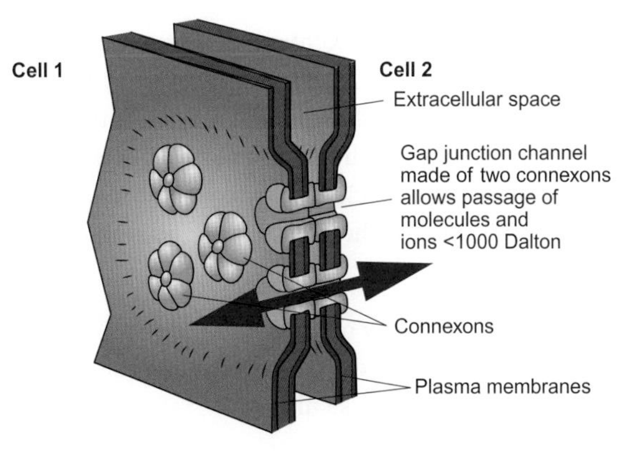

Figure 16.31 Morphology of an electrical synapse

Gap junction channels allow bidirectional passage of electrical signals between excitable cells. The atomic force microscopy of rat liver connexons shows the hexameric arrangement of connexons in the membrane and that the pores in the centre of the connexons close when the connexons are exposed to extracellular levels of ionized calcium (mmol L^{-1} range).

Source: left panel: Sosinsky GE, Nicholson BJ (2005). Structural organization of gap junction channels. Biochimica et Biophysica Acta 1711: 99-125.

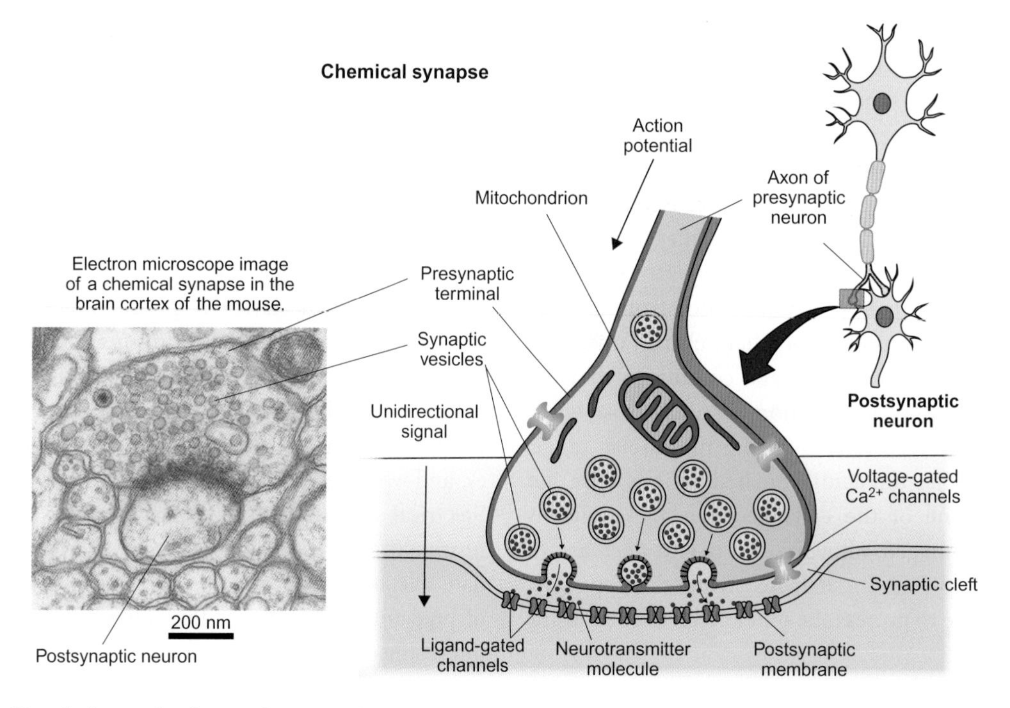

Figure 16.32 Morphology of a chemical synapse between two neurons

There are major differences between the structure and the properties of the presynaptic membrane compared with those of the postsynaptic membrane. The electrical signal reaching the presynaptic membrane is converted into a chemical signal that is transmitted to the postsynaptic membrane, where it is converted back into an electrical signal. This arrangement enables unidirectional passage of signals at the level of chemical synapses from the presynaptic to the postsynaptic neuron.

Source: Electronmicrograph from Bundesen L (2004). Biography of Pietro de Camilli. Proceedings of the National Academy of Sciences USA 101: 8259–8261.

of major structural differences between the synaptic terminals of the two neurons. These differences confer asymmetric properties to the synapse such that the signals can only travel in one direction (generally from the presynaptic to the postsynaptic membrane). Activation of the chemical synapse generally produces a change in the electrical properties of the postsynaptic membrane due to changes in its permeability to monovalent ions (Na^+, K^+, Cl^-). These changes can have an excitatory or an inhibitory effect on the postsynaptic neuron.

Chemical transmission involves transmitter release at the presynaptic membrane

A change in electrical potential at the presynaptic terminal is converted into a secretory process through which a chemical signal is released. As shown in Figure 16.32, the presynaptic terminal contains all the subcellular elements necessary for the synthesis and calcium-ion dependent release of transmitter molecules in the synaptic cleft in response to depolarization:

(i) a large density of mitochondria to produce the vast amounts of ATP required for the synthesis of transmitter molecules;

(ii) a high density of vesicles filled with transmitter molecules[67]; and

(iii) voltage-activated calcium channels through which calcium ions enter the cytosol in the presynaptic neuron when the action potential reaches the synapse.

The membranes of the transmitter-filled vesicles in the presynaptic terminal contain two proteins: synaptotagmin and synaptobrevin. As illustrated in Figure 16.33, these proteins play an important role in the mechanism of transmitter release. Synaptotagmin acts as a calcium ion-sensitive switch by preventing vesicles from fusing with the presynaptic membrane at low intracellular calcium ion concentrations when there is no calcium bound. Synaptobrevin is essential for facilitating fusion of vesicles with the presynaptic membrane.

Two other membrane proteins located in the presynaptic membrane, syntaxin and snap25, are critical for mediating vesicle fusion with the presynaptic membrane. Together, synaptobrevin, syntaxin and snap25 are called **snare proteins**.

The arrival of an action potential at the presynaptic membrane triggers the series of events shown in Figure 16.33B,

[67] We discuss different types of transmitter molecules in Section 16.3.4.

which results in the calcium ion concentration ([Ca^{2+}]) in the presynaptic terminal rising briefly, and the discharge of the vesicles' content into the synaptic cleft, which separates the presynaptic membrane from the postsynaptic membrane. The process of cellular secretion in which substances contained in intracellular vesicles are discharged from the cell by fusion of vesicular membranes with the outer cell membrane is called **exocytosis**[68].

[68] Exocytosis is a common process in neurosecretion, for example of neurohormones, which we discuss in Chapter 19.1.1.

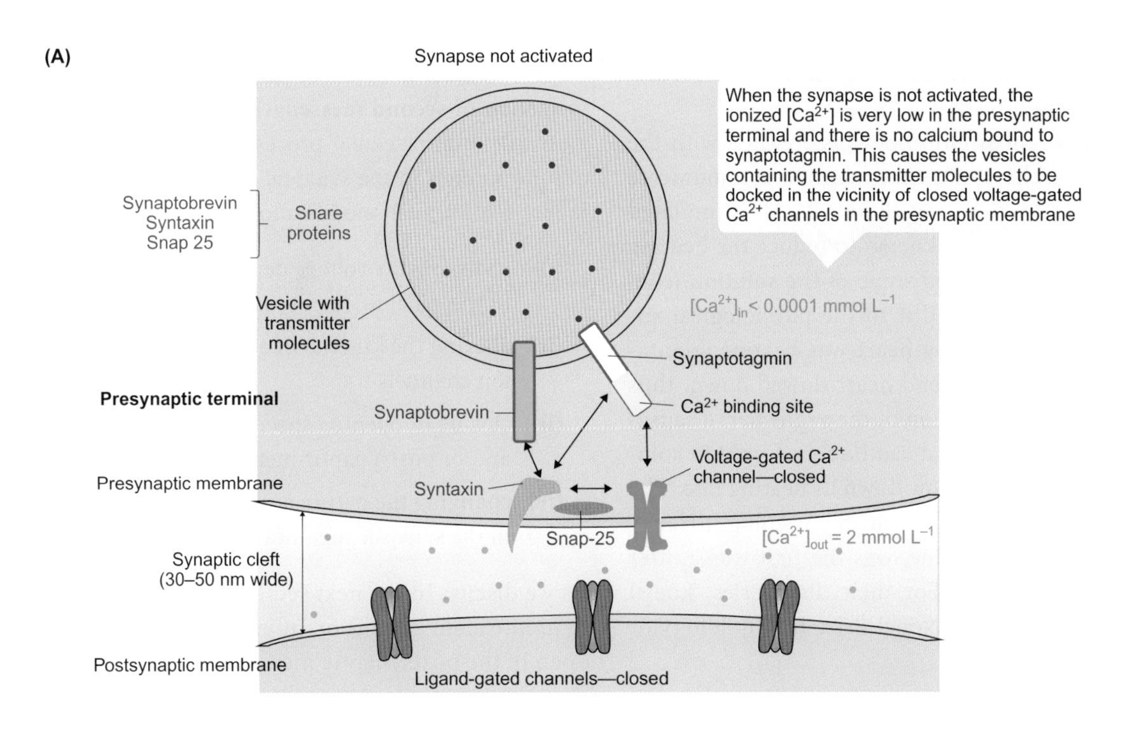

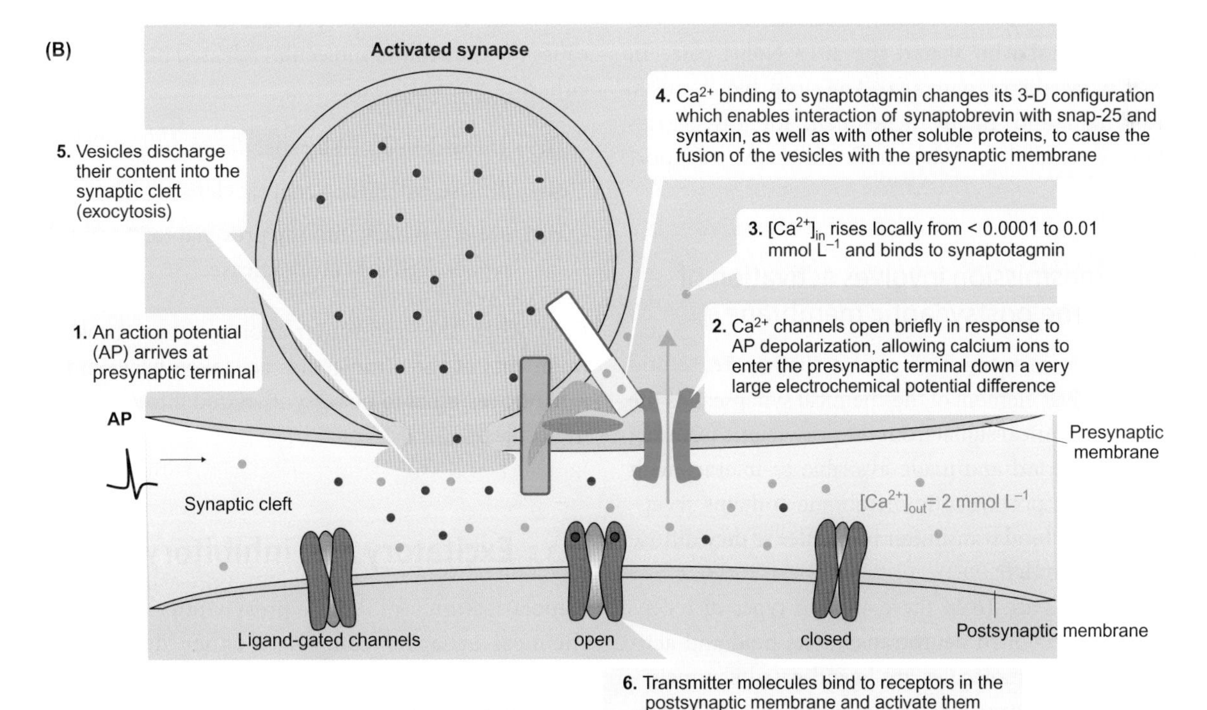

Figure 16.33 Mechanism of transmitter release at the presynaptic membrane of a chemical synapse

(A) Structural elements in a chemical synapse that is not activated. (B) Changes in the presynaptic terminal of a chemical synapse when is activated by arrival of an action potential.

After an action potential, the membrane potential at the presynaptic terminal returns to its resting level, causing the closure of the voltage-gated Ca^{2+} channels; $[Ca^{2+}]$ then returns to its resting level and calcium ions dissociate from synaptotagmin. The inhibitory action exerted by synaptotagmin on the interaction between synaptobrevin and the presynaptic membrane is reinstated and further fusion of vesicles with the presynaptic membrane is prevented. Consequently, there is no further discharge of transmitter molecules into the synaptic cleft.

The German scientist Otto Loewi is credited with the demonstration in the 1920s that neurons communicate with one another by releasing chemicals. He stimulated the vagus nerve, which was known to reduce the beating rate of a frog heart, collected some of the solution from around the heart, and added it to the physiological solution in which another frog heart was beating spontaneously. The rate of the second heart slowed down, thus demonstrating that stimulation of the vagus nerve causes the release of a substance that can be transferred in solution to the second heart to slow down its beating rate. The substance, later isolated by the British physiologist Henry Dale and named acetylcholine, was the first transmitter molecule to be identified. For their discoveries, Loewi and Doyle shared the 1936 Nobel Prize in Physiology or Medicine.

The machinery mediating neurotransmitter release at the presynaptic membrane was discovered more recently, much to the credit of the German-American neuroscientist Thomas Südhof who shared the 2013 Nobel prize in Physiology or Medicine with two American biochemists, James Rothman and Randy Schekman, for 'discoveries of machinery regulating vesicle traffic, a major transport system in our cells'.

16

Chemical transmission involves activation of receptors in the postsynaptic membrane

The postsynaptic membrane shown in Figures 16.32 and 16.33 is the receiver element of the chemical synapse: it is the site at which a chemical signal from the presynaptic terminal is received, converted and made available to interact with other signals. The postsynaptic membrane contains receptor molecules that bind transmitter molecules as they diffuse across the synaptic cleft.

As shown in Figure 16.34 there are two types of receptors at synapses to which neurotransmitters bind and activate. Some receptor molecules are ligand-gated channels[69], or **ionotropic receptors**, which open when specific neurotransmitters bind to them as described in Figures 16.33 and 16.34A. Ionotropic receptors are similar in structure to the voltage-gated channels[70] and have, in general, broad ionic selectivity to either monovalent cations (Na^+, K^+) or monovalent anions (Cl^-)[71]. The other type of receptors, called **metabotropic receptors**, are not ion channels but generally are coupled to GTP-binding proteins (G-proteins)[72] and can be located not only in postsynaptic, but also in presynaptic membranes. Upon binding the transmitter, the metabotropic receptors initiate a sequence of events leading to the formation of **second messengers**, which, in turn, promote a cascade of intracellular processes that alter the properties of ion channels in the synaptic membranes as described in Figure 16.34B. Such modifications include:

(i) changing the voltage dependence of ion channels in the synaptic membranes
(ii) altering the kinetics of activation/inactivation of ion channels
(iii) closing or opening specific ion channels in the pre- and/or postsynaptic membrane
(iv) changing the nature and distribution of receptors in the synaptic membranes.

As we discuss in the next section (16.3.3), the selectivity to monovalent cations or anions of the ion channels that open in the postsynaptic membrane determines whether the respective synapse is excitatory or inhibitory, respectively.

Overall, chemical synapses are only briefly activated because the transmitter molecules released in the synaptic cleft either:

(i) are broken down by specific enzymes,
(ii) diffuse out of the synaptic cleft, or
(iii) are taken back into the presynaptic terminal to be repackaged for future release.

Even when the transmitter molecules are broken down, fragments of these molecules are taken back into the presynaptic terminal to be re-synthesized into new transmitter molecules.

16.3.3 Excitatory and inhibitory synapses

Depolarization of the presynaptic membrane in chemical synapses can cause either depolarization or

[69] We discuss ligand-gated channels in Section 3.2.

[70] We discuss the structure of voltage-gated channels in Section 16.2.4.

[71] We discuss in Chapter 4 how selectivity of channels in general is achieved.

[72] We discuss the signalling mechanism involving G-proteins in Figure 19.4.

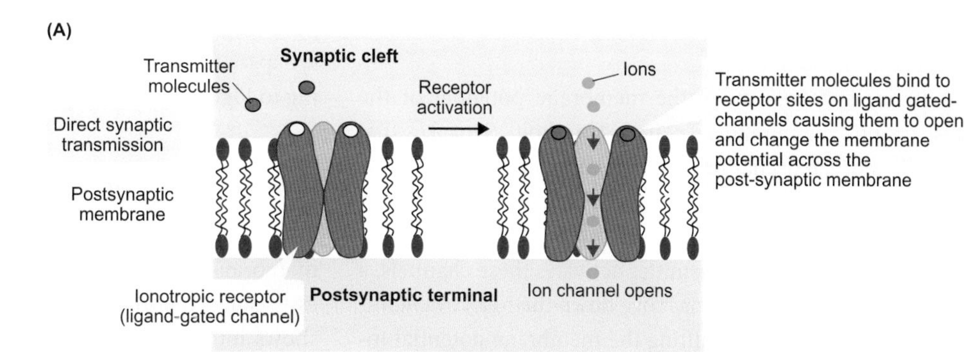

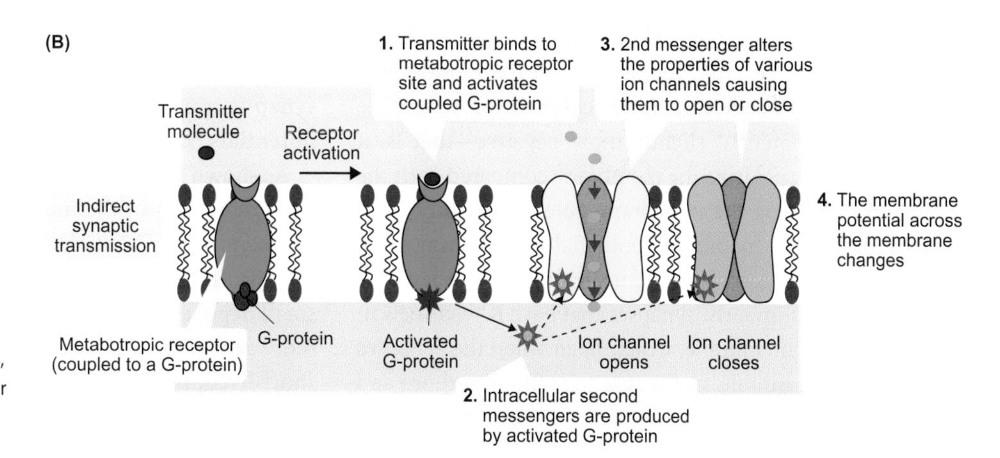

Figure 16.34 Diagrammatic representation of ionotropic (A) and metabotropic (B) receptor activation

Direct transmission occurs when the transmitter molecules bind to and open ligand-gated ion channels (ionotropic receptors) in the postsynaptic membrane. In indirect transmission, the transmitters bind to a receptor molecule (metabotropic receptor) in the postsynaptic or presynaptic membrane that is not an ion channel but is coupled to a G-protein, which becomes activated. The activation of the receptor affects the production of second messengers, which in turn influence the opening or closing of specific ion channels in the membrane.

hyperpolarization of the postsynaptic membrane, which moves the membrane potential of the postsynaptic neuron closer or further from the threshold for triggering an action potential. This is unlike electrical synapses where the electrical signal crossing the synapse is always of the same sign such that depolarization or hyperpolarization of the presynaptic membrane invariably causes depolarization or hyperpolarization, respectively, of the postsynaptic membrane. Thus, activated chemical synapses can be either excitatory, whereby they facilitate the generation of action potentials in the postsynaptic neuron, or inhibitory, whereby they oppose the generation of action potentials in the postsynaptic neuron.

Excitatory synapses

When an excitatory synapse is activated, the postsynaptic membrane depolarizes and the membrane potential moves closer to the threshold. Chemical synapses that have ligand-gated channels in the postsynaptic membrane that are equally permeable to monovalent cations (i.e. to sodium and potassium ions) are **excitatory**. This is because if the permeability of the postsynaptic membrane for the sodium and potassium ions increases by the same amount, a, invariably the ratio of sodium and potassium ion permeability after activation of the synapse (i.e. $(P_{Na}+a):(P_K+a)$) is greater than the permeability ratio before the synapse was activated ($P_{Na}:P_K$)[73]. According to previous discussion relating to Figure 16.16 an increase in the ratio of sodium and potassium ions permeability causes depolarization.

When activated, an excitatory neuro-neuronal synapse causes only a very small depolarization in the postsynaptic membrane, normally in the order of several mV or less. However, the small depolarization lasts longer than the action potential that activated the synapse. In order to trigger an action potential in the postsynaptic neuron, many excitatory synapses need to be activated at about the same time. For example, assume that one excitatory synapse causes a depolarization of, say, 1.5 mV in the region of the membrane of the postsynaptic neuron where action potentials are generated, and the threshold for triggering an action potential is 15 mV. Then, at least 10 excitatory synapses need to be activated at about the same time to trigger an action potential in the postsynaptic neuron.

[73] For example if $P_{Na}:P_K = 0.04:1$ and $a = 0.1$, then $(P_{Na}+a):(P_K+a) = 0.14:1.04 = 0.135$ which is greater than 0.04.

16

Inhibitory synapses

Inhibitory synapses move the membrane potential of the postsynaptic membrane away from threshold. Synapses that have ligand-gated channels in the postsynaptic membrane that, when open, are permeable to monovalent anions—of which the chloride ion (Cl^-) is the most abundant—are **inhibitory**. When the transmitter activates these channels, a tiny number of chloride ions move down their electrochemical potential difference, shifting the membrane potential towards the equilibrium potentials for chloride, which is about −71 mV in mammalian neurons. Since at rest the membrane potential is about −61 mV, the opening of the ligand-gated channels that are permeable to the chloride ion will make the membrane potential slightly more negative—that is, it will slightly hyperpolarize the membrane compared with the resting level. Making the membrane potential slightly more negative moves the membrane potential further away from threshold, which means it is now more difficult to trigger an action potential. Similarly, synapses that have K^+ channels in the postsynaptic membrane, which open when the synapses are activated, are inhibitory, because the ratio of sodium and potassium ion permeability decreases after activation of the synapse and this is associated with hyperpolarization according to Figure 16.16.

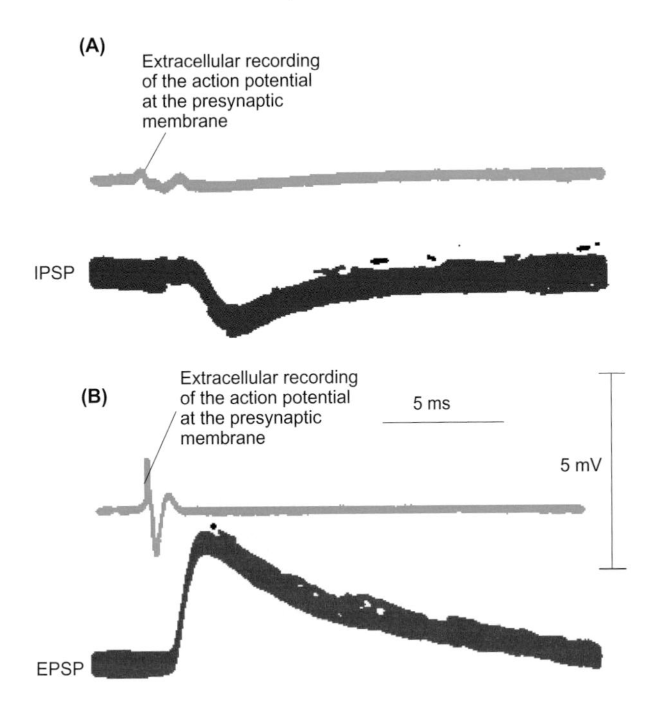

Figure 16.35 Inhibitory (IPSPs) and excitatory postsynaptic potentials (EPSPs)

(A) Intracellular recording (thicker traces) of direct inhibitory postsynaptic potentials. (B) Intracellular recording (thicker traces) of excitatory postsynaptic potentials in cat motoneurons at normal body temperature. The thinner traces are extracellular recordings of action potentials in the presynaptic neurons. Notice that the postsynaptic potential follows the action potential in the presynaptic membrane with a delay of several milliseconds.

Source: A, B from Coombs JS et al (1955). The specific ion conductances and ionic movements across the motoneuronal membrane that produce the inhibitory postsynaptic potential. Journal of Physiology 130: 326-373.

Postsynaptic potentials

The change in the membrane potential of the postsynaptic membrane in response to the activation of chemical synapses is called the **postsynaptic potential (PSP)**. Figure 16.35 shows intracellular recordings of PSPs when an inhibitory synapse and an excitatory synapse were activated on a cat motoneuron[74]. The PSPs associated with activation of inhibitory or excitatory synapses are called **inhibitory postsynaptic potentials (IPSPs)** or **excitatory postsynaptic potentials (EPSPs)**, respectively.

As shown in Figure 16.35A, the IPSP is generally a downward deflection of the membrane potential, indicating that the membrane potential becomes more negative than at rest. In contrast, Figure 16.35B shows that an EPSP is generally a small depolarization, which makes the postsynaptic neuron more sensitive for triggering an action potential. Although short-lived, the IPSPs and EPSPs last generally longer (10–20 ms) than an action potential and are slightly delayed in relation to the action potential that has activated the synapse. This is important for understanding how information is integrated at the level of chemical synapses, which we discuss in Section 16.3.5.

Presynaptic inhibition and facilitation

Axons of neurons, in the CNS, synapse not only with the dendrites of other neurons, but also with presynaptic terminals of other axons. The activation of the axo-axonic synapse on presynaptic terminals can either increase or decrease the amount of transmitter released from the presynaptic terminal. **Presynaptic facilitation** occurs when the amount of transmitter released from the presynaptic terminal *increases*; by contrast, **presynaptic inhibition** occurs when the amount of transmitter released *decreases* in response to the activation of the axo-axonic synapse[75].

16.3.4 Synaptic transmitter molecules

Neurophysiologists have established a set of experimentally verifiable criteria, which are used to decide whether a

[74] Motoneurons generally innervate muscle, secretory and endocrine cells.

[75] Presynaptic inhibition occurs when alpha-2-adrenoceptors on presynaptic terminals of adrenergic neurons in the autonomic nervous system of vertebrates become activated, as indicated in Table 16.1.

particular chemical can be classified as a synaptic transmitter. The criteria can be summarized as follows:

(i) Precursors and enzyme systems for the synthesis of the respective chemical are present in the presynaptic neuron.

(ii) The respective chemical is present in the presynaptic terminal.

(iii) Activation of the presynaptic neuron causes the release of the respective chemical in sufficient amounts to affect the postsynaptic neuron.

(iv) There are receptors in the postsynaptic membrane to which the respective chemical can bind with high affinity.

(v) Direct application of the respective chemical onto the postsynaptic membrane mimics the action of the endogenously released neurotransmitter when the synapse is activated.

(vi) There is a specific (biochemical) mechanism present at the level of the synapse (synaptic cleft) for removal/inactivation/breakdown of the respective chemical.

While it is often difficult to show experimentally that all these criteria are fulfilled at one particular synapse, it is generally agreed that each neuron makes use of the same combination of chemical messengers at all of its synapses. Therefore, one can study several synapses on the same neuron to show that all six criteria are satisfied.

There are several classes of neurotransmitter molecules.

Amines[76]

Figure 16.36 shows the chemical structure of acetylcholine, which was the first identified transmitter molecule as we discuss in Section 16.3.2. Acetylcholine is released in excitatory synapses between neurons in the autonomic nervous system[77] and in all excitatory synapses between neurons and skeletal muscle fibres of vertebrates. Acetylcholine is also present in inhibitory synapses between neurons and cardiac or smooth muscle cells in vertebrates. In invertebrates, acetylcholine is a major neurotransmitter at excitatory synapses in the CNS. Indeed, many current pesticides have active ingredients that interfere with signal transmission at synapses where acetylcholine is released, as we discuss in Case Study 16.1. Acetylcholine receptors are of two types: ionotropic and metabotropic, as we discuss in Section 16.1.3. The ionotropic acetylcholine receptors are sensitive to nicotine, and therefore are called **nicotinic acetylcholine receptors** or **nAChRs**, while the metabotropic receptors are sensitive to

[76] Amines are organic compounds that contain a positively charged nitrogen at physiological pH.

[77] The autonomic nervous system is part of the peripheral nervous system as we discuss in Section 16.1.3.

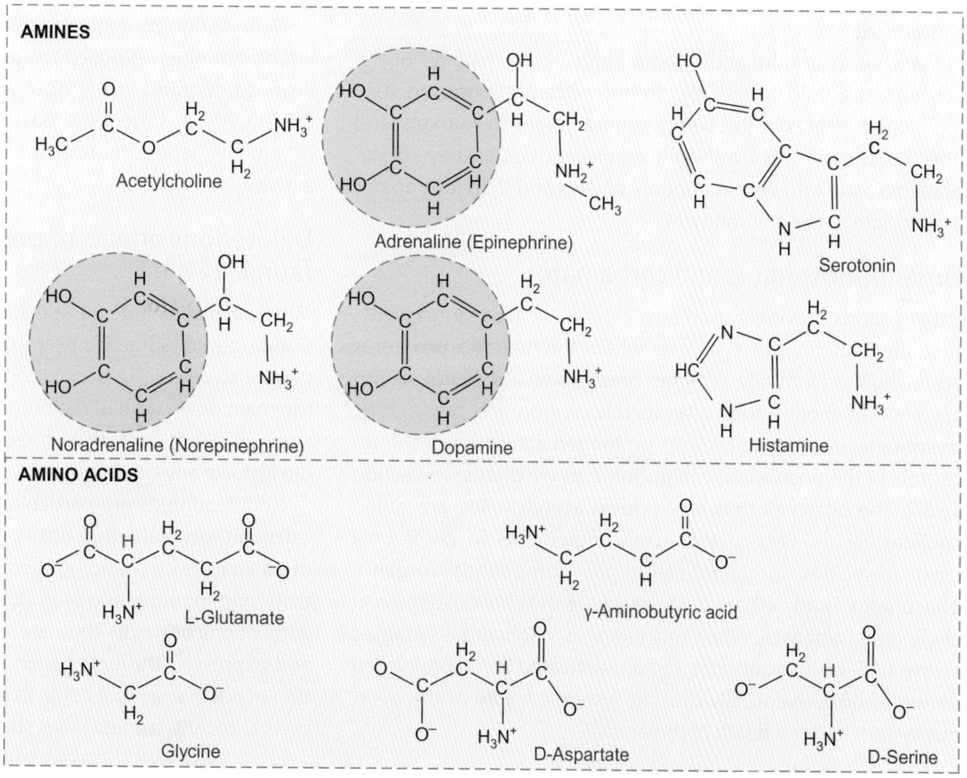

Figure 16.36 Chemical structure of common neurotransmitters classified as amines and amino acids

Amines that contain the catechol group (shown enclosed in the dashed circles) are often called catecholamines. According to the convention described in Box 2.1 (Figure E) for L/D amino-acid isomers, the hydrogens shown in red, point upward towards the viewer.

Case Study 16.1 Impact of pesticides on the nervous system

Pesticides are chemical or biological agents meant for deterring, debilitating or killing target pests, which spread disease, destroy property (including crops) or cause nuisance. Target pests include individual species from most animal taxa such as nematodes, molluscs, arthropods, fish, amphibians, reptiles, birds and mammals.

Most commonly used pesticides are directed particularly against insects and other invertebrates, and contain active ingredients that interfere with signal transmission in the nervous system.

Organochlorines and pyrethroids

Organochlorines[1] and pyrethroids[2] act on the voltage-gated sodium channels in neurons by slowing the rate of closing of the activation gates, as we discuss in Section 16.2.4. This slows the inactivation of the sodium channels upon depolarization and results in intense repetitive activity in neurons, which causes rapid failure of the nervous system and death of the animal.

The best-known organochlorine pesticide is DDT (dichlorodiphenyl trichloroethane), which was widely used as an insecticide during the Second World War until the mid-1970s, when it became banned for agricultural use in many developed countries due to its chronic toxic effects on humans and other fauna and its very long persistence in the environment for up to several decades. It continues, however, to be used for combating malaria and other insect-borne diseases. The potent insecticide properties of DTT were discovered by the Swiss scientist Paul Müller, who was awarded in 1948 the Nobel prize in Physiology or Medicine.

Pyrethroids are the active ingredients in the majority of the current household insecticides. In the concentrations used, they rapidly kill insects while being generally safe to humans and pets. The pyrethroids are short-lived insecticides, as they are degraded by sunlight within a couple of days and therefore do not accumulate in the environment.

Organophosphates and carbamates

Other groups of widely used pesticides have active ingredients that affect cholinergic synapses, where the neurotransmitter is acetylcholine. Normally, enzymes break down acetylcholine into acetate and choline soon after its release from the presynaptic membrane, in order to prevent prolonged activation of the receptors in the postsynaptic membrane, as we discuss in Section 16.3.2. The enzymes that break down acetylcholine are called cholinesterases. Two of the active ingredients in commonly used pesticides, organophosphorous compounds (organophosphates[3]) and carbamates[4], inactivate the cholinesterases at cholinergic synapses. When this happens, cholinergic synapses in the CNS of invertebrates remain activated for a prolonged period of time, eventually causing systemic failure of the nervous system and the death of the animal.

The inhibitory action of carbamates on the activity of cholinesterases is reversible, meaning that the animal can gradually recover from sublethal doses without further detrimental effects. However, cholinesterases are irreversibly inactivated by organophosphates. Therefore, the effects of organophosphates are more profound and longer lasting. Carbamates and organophosphates remain in the environment for days to months after their application and are subsequently degraded into harmless compounds. The degradation of organophosphates and carbamates in the environment is faster at more alkaline pH and higher temperatures.

Neonicotinoids

A different group of pesticides, developed only relatively recently and directed particularly against insect and nematode pests, contain compounds that mimic nicotine in its ability to activate ionotropic acetylcholine receptors (nAChR). Recall from Sections 16.1.3 and 16.3.4 that these receptors are used in autonomic synapses between pre- and postganglionic neurons[5]. The neonicotinoids bind to insect nAChRs with particularly high affinity, and can thus activate the receptor for a prolonged period of time. The nAchRs spontaneously inactivate when activated for more than a few seconds.

Although the mechanism of action of neonicotinoids at cholinergic synapses is different from that of carbamates and organophosphates, the overall outcome is similar: cholinergic synapses in the CNS of invertebrates stay activated for prolonged periods of time, eventually causing failure of the CNS as a whole and subsequently death.

Neonicotinoids degrade over a period of months in the environment when exposed to sunlight, but they can take several years to degrade in the absence of sunlight and microorganism activity. It is therefore possible that nicotinoids accumulate in aquifers, where there is no sunlight and no microorganism activity.

Deleterious effects of pesticides on non-target fauna

Most pesticides are not species specific and therefore impact non-targeted fauna. For example, insecticides used to kill insect pests also affect the population of beneficial insects, which are the main pollinators of major food crops. If non-targeted insects come in direct contact with recently sprayed insecticides, they die just like any other insect.

Neonicotinoids are particularly dangerous to pollinators because, at low, sublethal doses, they affect the nervous system of pollinators by changing their feeding behaviour, foraging activity and memory and learning ability. When seeds are coated with neonicotinoids, they are taken up by the growing plant and expressed through pollen and nectar, causing contamination of pollinators that come in contact with the flowers. Indeed, several recent studies have shown that low concentrations of neonicotinoids interfere with the navigation system in foraging honey bees (*Apis mellifera*), increasing the rate at which they fail to return to the hive after foraging. In turn, the increased number of foraging bees that do not return to their hives decreases the hive population, and this increases the hive death rate. A marked decline in honey bee populations in the Northern Hemisphere since 2006 has been linked to neonicotinoid usage. Moreover, other studies have shown that the decline in the state of health of bumblebees (*Bombus* spp.), which are wild pollinators, is also linked to neonicotinoid usage. Based on this research, the European Commission imposed a number of use restrictions in May 2013 on neonicotinoid insecticides.

Birds are another group of non-target fauna that is particularly affected by pesticides. They are very mobile and difficult to exclude from pesticide treated areas; have a high rate of ventilation, which causes efficient inhalation of pesticides applied as vapour or fine droplets; and ingest pesticides through their food (including pesticides applied in the form of granules) and absorb pesticide through their skin by preening and grooming. Exposure of birds to sublethal doses of carbamates, organophosphates and neonicotinoids has been shown to impact behaviours involved with foraging, mating, nest-building activity, parenting and predator avoidance. For example, the bird's ability to sing is impaired, which decreases its chances of successfully attracting a mate or establishing a territory. Impact on parenting reduces the bird's caring ability for its young, causing the nestlings to starve to death.

Often it is difficult to judge the deleterious effects of pesticides on non-target wildlife because, when an animal dies in the wild, no obvious trace of the dead animal remains after several hours. For example, studies have shown that more than 90 per cent of bird carcasses are scavenged and removed from the area within 24 hours. Absence of corpses can lead to the false assumption that particular pesticides are harmless to the wildlife. However, pesticide-killed insects become a readily available food source for birds, and birds that die from pesticide poisoning will be likely eaten by a mammalian scavenger or a bird of prey. The high accessibility of birds to pesticides combined with their vulnerability to pesticides makes them a sensitive indicator of the quality of the environment. In March 2013, the American Bird Conservancy called for a ban on neonicotinoid use for seed treatments because of their toxicity to birds, aquatic invertebrates, and other wildlife.

› Find out more

Costa LG, Giordano G, Guizzetti M, Vitalone A (2008). Neurotoxicity of pesticides: a brief review. Frontiers in Bioscience 13: 1240–1249.

Desneux N, Decourtye A, Delpuech J-M (2007). The sublethal effects of pesticides on beneficial arthropods. Annual Review of Entomology. 52: 81–106.

Fukuto TR (1990). Mechanism of action of organophosphorus and carbamate insecticides. Environmental Health Perspectives. 87: 245–254.

Mickaël H, Béguin M, Requier F, Rollin O, Odoux J-F, Aupinel P, Aptel J, Tchamitchian S, Decourtye A (2012). A common pesticide decreases foraging success and survival in honey bees. Science 336: 348–350.

Osborne JL (2012). Ecology: Bumblebees and pesticides. Nature 491: 43–45.

Whitehorn PR, O'Connor S, Wackers FL, Goulson D (2012). Neonicotinoid pesticide reduces bumble bee colony growth and queen production. Science 336: 351–352.

[1] Organochlorines, also known as organochlorides, chlorocarbons, chlorinated hydrocarbons or chloroalkanes are organic compounds containing at least one covalently bonded atom of chlorine.

[2] Pyrethroids are organic compounds similar to the natural pyrethrins found in the chrysanthemum (pyrethrum) flowers.

[3] Organophosphates are organic compounds containing one or more phosphate ester groups.

[4] Carbamates are organic compounds derived from carbamic acid (NH_2COOH).

[5] Note that neonicotinoids activate only the nAChRs, and do not activate the muscarinic type of acetylcholine receptors (mAChRs).

muscarine and are called **muscarinic acetylcholine receptors** or **mAChRs**. The nAChRs are involved in excitatory synapses, while the five subtypes of mAChRs (M_1 to M_5) are coupled to G-proteins and modulate neuron excitability through signalling pathways that involve either a rise in intracellular Ca^{2+} or a decrease in cAMP concentration.

Noradrenaline[78] (also called norepinephrine) and adrenaline[79] (also called epinephrine) are two other major amine neurotransmitters and hormones (see Figure 16.36 for their chemical structure) that operate in both excitatory and inhibitory neuro-neuronal synapses of the CNS and autonomic nervous system[80] in vertebrates. Noradrenaline is also released in excitatory neuromuscular (cardiac and smooth muscle) synapses. Noradernaline and adrenaline receptors are called adrenoceptors.

As we elaborate in Section 16.1.3 and Table 16.1, there are different types of adrenoceptors (α_1, α_2, β_1, β_2, β_3) which play specific roles in signal transmission in the autonomic nervous system of vertebrates. Adrenoceptors are also located on neurons in the CNS. All adrenoceptors are metabotropic receptors, coupled to G-proteins and mediate responses that result either in a rise in intracellular Ca^{2+}-concentration (α_1 adrenoceptors), or changes in cAMP concentration (α_2 and β adrenoceptors). In the CNS, α_1-adrenoceptors are located in postsynaptic membranes and mediate excitatory responses; α_2-adrenoceptors are located on both pre- and postsynaptic membranes and mediate inhibitory responses by reducing the intracellular cAMP concentration, which in turn causes membrane hyperpolarization. Hyperpolarization of the

[78] The terms noradrenaline (or noradrenalin) and norepinephrine are interchangeable, with noradrenaline being the common name in most parts of the world except the US, where norepinephrine is preferred.

[79] The terms adrenaline (or adrenalin) and epinephrine are also interchangeable.

[80] We discuss the autonomic nervous system in Sections 3.3 and 16.1.3.

presynaptic membrane inhibits the release of noradrenaline at synapses[81], while hyperpolarization of the postsynaptic membrane increases the threshold for action potential generation in the postsynaptic neurons.

Chemicals that bind and activate α_2-adrenoceptors (called α_2-adrenoceptor **agonists**) have a strong sedative effect on the animal and also act as muscle relaxants. For example, in vertebrates the agonists bind to α_2-adrenoceptors in the reticular formation (shown earlier in Figure 16.9) and other parts of the CNS and block transmission of pain and anxiety signals to the brain, while also keeping the muscles in the animal relaxed. α_2-adrenoceptor agonists are therefore widely used in animal conservation management and veterinary anaesthesia. The anaesthetic action of the agonists is rapidly reversed by α_2-adrenoceptor **antagonists**, which bind to α_2-adrenoceptors and displace the agonists from the receptor, without activating it. The β-type adrenoceptors are coupled to G-proteins that cause subsequent rise in intracellular cAMP when activated.

Other major neurotransmitters belonging to the amine group are **serotonin**, **dopamine** and **histamine** (Figure 16.36), which function in neuro-neuronal synapses in the CNS of vertebrates and invertebrates. Serotonin is mainly implicated in inhibitory synapses in vertebrates, but is also present in excitatory synapses of some invertebrates, while dopamine and histamine are involved in both excitatory and inhibitory synapses.

Noradrenaline (norepinephrine), adrenaline (epinephrine) and dopamine contain the catechol group (benzene with two hydroxyl side groups) in their molecule as shown in Figure 16.36 and are therefore often grouped together under the name of **catecholamines**.

Amino acids

The most important amino acid neurotransmitters are **L-glutamate**, **γ-aminobutyric acid** (GABA), **glycine**, **aspartate** and **serine** (Figure 16.36). L-Glutamate is by far the most prevalent transmitter in the CNS of vertebrates and invertebrates, being used by more than 90 per cent of the excitatory synapses in the human brain. L-Glutamate is also the transmitter in excitatory neuro-muscular synapses in arthropods. The next most prevalent amino acid transmitter is GABA, which is used at inhibitory synapses in the CNS of vertebrates and invertebrates and in inhibitory neuromuscular synapses in arthropods.

The L-glutamate receptors can be either ionotropic or metabotropic. The ionotropic L-glutamate receptors are of three subtypes, named by the compound that specifically activates them: **AMPA** ((2-amino-3-(3-hydroxy-5-methyl-isoxazol-4-yl) propanoic acid)) **receptors**, **kainate receptors**, activated by a drug isolated from the red alga *Digenea simplex* called kainate and **NMDA** (*N*-methyl-D-aspartate) **receptors**. The AMPA and the kainate receptors, when activated are permeable to both sodium and potassium ions causing depolarization. The AMPA receptors are the most abundant L-glutamate receptors in the vertebrate CNS where they mediate the majority of the excitatory synapses. Kainate receptors are present not only on the postsynaptic membrane of excitatory synapses, but also on presynaptic membranes of inhibitory synapses, where they regulate the release of the neurotransmitter. Generally, the function of the kainate receptors is less well-defined than that of the AMPA receptors.

The NMDA receptors are a special type of ionotropic receptors that play a major role in synapse plasticity and memory formation. When activated, the NMDA receptor is nonselective to calcium, sodium and potassium ions; it causes a relatively large rise in the intracellular calcium ion concentration, but only minor changes in the intracellular potassium and sodium concentrations. As shown in Figure 16.37B, the NMDA receptor is both a ligand-gated and voltage-dependent ion channel. When the membrane is normally polarized, extracellular magnesium ions are attracted into the channel by the more negative potential on the cytosolic than the extracellular side and bind at a site deep in the channel, blocking it. Membrane depolarization removes the block as potassium ions start flowing through the channel in the opposite direction and displace the magnesium ions from their binding sites. Structurally, the NMDA receptor is made of four subunits. Two of these subunits have binding sites for glutamate and two have binding sites for glycine or D-serine[82]. Diverse glutamate-binding and glycine-binding isoforms (subunits) exist which, when combined, form receptors with specific functional properties that are expressed at specific times during development and at specific locations in the brain of animals.

Physiological activation of the NMDA receptor requires both simultaneous binding with high affinity of glutamate and glycine/D-serine, and membrane depolarization to remove the Mg^{2+} block. When the NMDA receptors are activated, calcium enters across the postsynaptic membrane and raises the intracellular ionic concentration of calcium (Ca^{2+}), which acts as a second messenger in various signalling pathways, causing changes at the respective synapse. For example, the postsynaptic terminal contains more than 1500 different proteins that are organized in dynamic protein–protein interaction complexes, which can change composition in response to the level of activation of the postsynaptic terminal.

[81] We discuss, in Section 16.3.3, how membrane hyperpolarization reduces the release of neurotransmitters at synapse.

[82] D-serine has a different configuration than the L-serine, which occurs in proteins, as we discuss in Box 2.1.

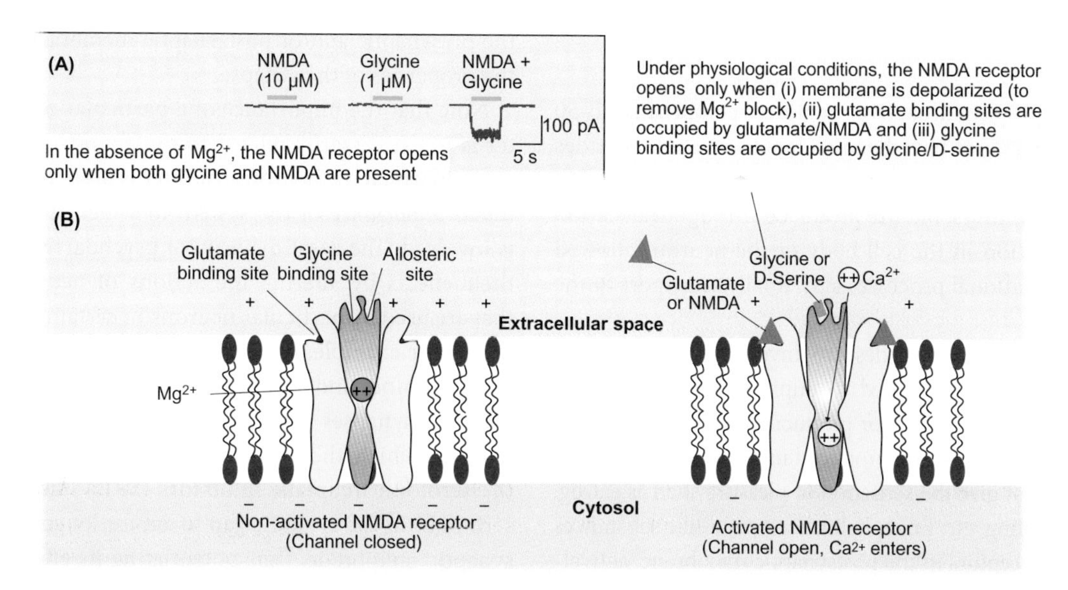

Figure 16.37 Conditions for the activation of the NMDA receptor

Original whole-cell patch-clamp current recordings from NMDA receptors in a cultured mouse brain neuron. The holding potential was –50 mV and there was no magnesium in the extracellular solution to prevent channel blocking by magnesium at –50 mV. The channel opened (indicated by a downward deflection of the current trace) only when both glycine and NMDA were present. Schematic diagram of the NMDA receptor which is made of four subunits (for clarity only three subunits are shown). Two subunits bind glutamate and two subunits bind glycine or D-serine. Under physiological conditions, one magnesium ion blocks the channel when the membrane is fully polarized. In order to become activated and allow calcium ions to enter the cytosol, both glutamate and glycine need to be bound to their respective sites and the membrane needs to be depolarized to facilitate the removal of magnesium ion which blocks the channel. The NMDA receptors are subject to regulation by an unusually large number of agents. This is indicated by the presence of allosteric sites to which these agents can bind and alter the channel's properties.

Source: A: Johnson JW, Ascher P (1987). Glycine potentiates the NMDA response in cultured mouse brain neurons. Nature 325: 529–531.

Such changes at chemical synapses form the basis for memory formation as we discuss in Section 16.3.6 and in Experimental panel 16.2 (see Online Resources).

Several metabotropic subtypes of L-glutamate receptors exist, which are present at synapses located in different parts of the brain. These receptors are linked to different signalling pathways and can be involved in specific aspects of memory formation.

Purines

The neurotransmitters belonging to the purines class include adenosine, **ATP**, **GTP** and their derivatives and are involved in synapses of the central, peripheral and autonomic nervous systems of various organisms, including humans.

Nitric oxide

Nitric oxide (NO) does not satisfy all the strict criteria for synaptic transmitters mentioned above, but is an important messenger molecule that affects synaptic transmission. NO is a small, uncharged, lipid soluble molecule that is produced in a variety of neurons in the CNS and peripheral nervous system. In all animal species investigated except *C. elegans*, NO is released in response to a rise in intracellular Ca^{2+} concentration. When produced, it diffuses into nearby neurons, even those that are not connected by a synapse. Since the half-time of NO is relatively short (in the order of few seconds), its action is limited to a restricted volume around the site of its production.

The receptor molecule for NO is not a postsynaptic membrane receptor but a cytoplasmic enzyme (guanylyl cyclase) which produces cyclic GMP (cGMP). cGMP acts as a second messenger to produce downstream effects which include the up- and down-regulation of Ca^{2+} channels (which in turn increases or decreases neurotransmitter release in presynaptic terminals, respectively), and the alteration of the membrane potential of the postsynaptic membrane by upregulating potassium channels in membranes. The release of NO from postsynaptic terminals can act on the presynaptic membranes, reducing the amount of synaptic transmitter released at the synapse. Thus, unlike classical synaptic transmitters, the signal carried by nitric oxide can move in opposite directions (retrograde transmission) as compared to the classical directionality of signal transmission at chemical synapse from pre- to postsynaptic membrane.

Nitric oxide plays an important role in various parts of the gastrointestinal tract, causing predominantly relaxation of the gastrointestinal smooth muscle.

Peptides

More than 100 neuroactive peptides comprising 2–50 amino acid residues have been so far discovered. Peptides are not synthesized in the nerve terminals like the other types of transmitters, but are produced through translation and transcription in the cell body of the neuron followed by post-translational processes and, finally, transport to the synapse.

Some neuroactive peptides are involved in neural and behavioural processes related to euphoric states and pain relief. Peptides responsible for producing euphoric states are called **enkephalins** and are produced in the brain in response to activities that give the sensation of pleasure such as eating, jogging, gambling, etc. Enkephalins are opiate-like substances that bind to receptors in the postsynaptic membrane, activating the production of second messengers[83], which then change the excitability of the postsynaptic membrane.

Peptides involved in pain relief are called **endorphins**[84]. Interestingly, these natural painkillers act on the presynaptic membrane and inhibit transmitter release, thereby preventing the signals from reaching the pain perceiving areas in the brain.

We now know that two or more neurotransmitter molecules are released at many synapses: generally one of the transmitters binds to ionotropic receptors (ligand-gated ion channels) in the postsynaptic membrane, and one or more neurotransmitters (generally peptides) bind to metabotropic receptors. Peptides act on various metabotropic receptors on the presynaptic and/or postsynaptic membrane and change the properties of the synapse.

[83] We discuss some second messenger systems in Section 19.1.3.

[84] Release of β-endorphin and met-enkephalin during stress are discussed in Section 19.3.2.

Note that the importance of a particular neurotransmitter is not necessarily related to the abundance of the synapses in which the neurotransmitter is used, but to the *functional importance* of the system in which that transmitter is involved. The great majority of psychoactive drugs exert their effects by altering the actions of neurotransmitters that are used by particular neurons in certain regions of the brain. For example, some addictive drugs, such as heroin, amphetamines and cocaine, appear to exert their action by affecting synapses involved in the brain's reward mechanism in which the transmitter is dopamine. In contrast, the serotonin reuptake inhibitors (SRIs) like Prozac leave serotonin in the synaptic gap to act for longer on the postsynaptic membrane, thus potentiating its effect. Serotonin is involved in synapses associated with memory, emotions, mood, wakefulness and sleep.

16.3.5 Synaptic integration

It is important to emphasize that, on average, one neuron in the mammalian brain forms thousands of chemical synapses with other neurons, as shown in Figure 16.38. For example, 1 μL (= 1 mm^3) of cortex in humans contains more than one billion chemical synapses, suggesting that a healthy adult brain may hold 1–2 quadrillion (10^{15}) synapses, which is thousands of times more than the catalogued number of stars in the whole universe. The presence of chemical synapses, which, when activated, usually produce small local changes in the electrical properties of the postsynaptic membrane, allows integration of information at the level of a neuron.

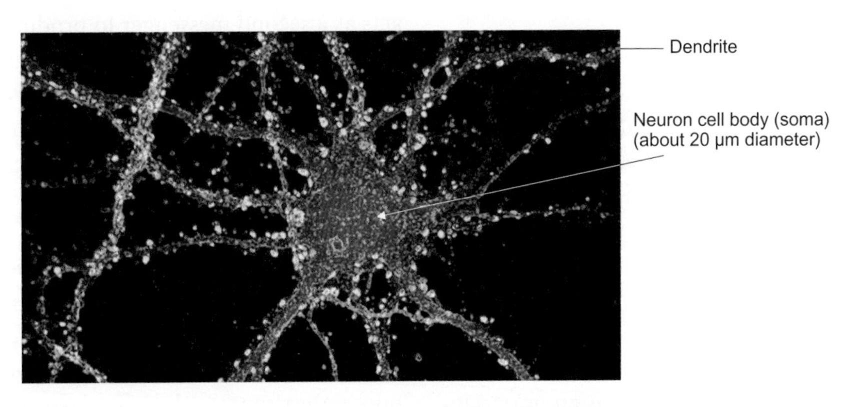

Figure 16.38 Neuron from a brain region of the mouse involved in memory formation (hippocampus)

The neuron, labelled in blue, has many dendrites and each dendrite can have thousands of tiny membranous protrusions called dendritic spines. Each spine has generally a bulbous head connected to the dendrite by a narrow neck and typically receives input at a synapse from a single axon. The synapses on the dendritic spines are labelled in yellow and red.

Source: Fereira JS et al (2015). GluN2B-Containing NMDA Receptors Regulate AMPA Receptor Traffic through Anchoring of the Synaptic Proteasome. Journal of Neuroscience 35: 8462–8479.

Figure 16.39A shows how an EPSP summates with an IPSP. In the example given, the resulting EPSP is smaller than the original. The summation process illustrated in Figure 16.39A is linear, meaning that the resulting effect is a simple subtraction of the IPSP from the EPSP. However, often the resulting effect could be nonlinear, meaning that the outcome could differ considerably from the simple subtraction of IPSPs from EPSPs.

Neural integration is the overall process by which one neuron responds to all inputs from other neurons reaching it within a given time interval. The process of integration takes place at the **spike-initiating zone** of a neuron, also called the **axon initial segment** or **axon hillock**. The axon hillock is located immediately after the cell body of the neuron where the axon emerges from the cell body, as shown in Figure 16.40. Integration at the neuron level is made possible by the phenomena of **temporal** and **spatial summation**.

Temporal summation refers to the situation in which two or more electrical signals that individually are below threshold for generating an action potential in the postsynaptic neuron may trigger an action potential if they arrive in quick succession at the spike-initiating zone, as shown in Figure 16.39B. Spatial summation refers to the situation in which signals generated at different locations on a postsynaptic neuron summate at the spike-initiating zone, where an action potential may be triggered when the summated response is above threshold, as shown in Figure 16.39C. Synapses closer to the spike-initiating zone generally produce greater PSPs at the spike-initiating zone than synapses that are further away.

Thus, the possibility that an action potential is generated in a neuron depends on the location and intensity of activation of the whole set of synapses (excitatory and inhibitory) on that neuron at a particular moment. The phenomenon can be likened to a decision-making process when a course of action is taken after all pros and cons are weighed against each other. This is illustrated in Figure 16.40. Importantly, once an action potential is generated at the spike-initiating zone, it propagates not only forward along the axon, but also backward to the cell body of the neuron and to the dendrites modifying the efficacy of synapses on the postsynaptic neuron. Thus, it can be argued that each chemical synapse in the brain operates like

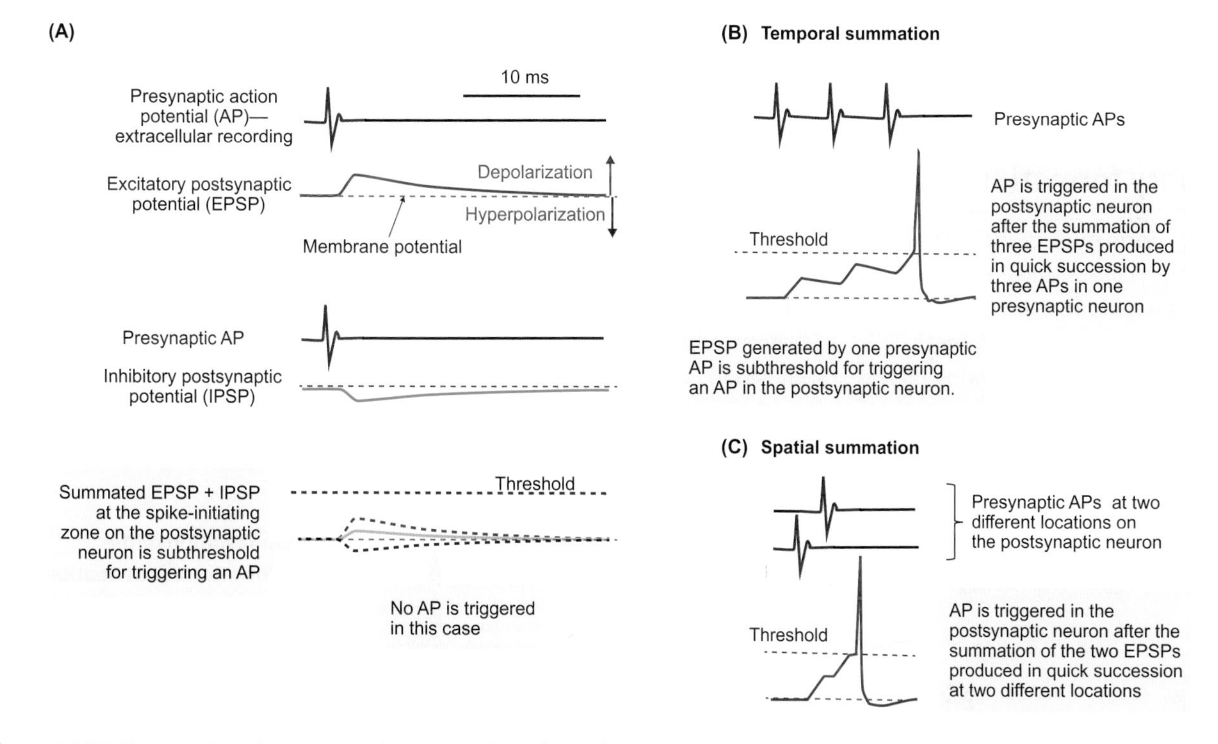

Figure 16.39 Summation of postsynaptic potentials at the spike-initiating zone

(A) An excitatory (small depolarization) and an inhibitory (small hyperpolarization) postsynaptic potential reach the spike initiating zone of the postsynaptic neuron at the same time. The summated response (continuous green trace) is below threshold and no action potential is triggered in the postsynaptic neuron. (B) Temporal summation of subthreshold EPSPs after three action potentials arrive in quick succession at the presynaptic membrane and trigger one action potential in the postsynaptic neuron. (C) Spatial summation of EPSPs generated at two locations on postsynaptic neuron trigger an action potential when the summated response is above threshold. Synapses closer to the spike initiating zone generally produce greater postsynaptic potentials at the spike initiating zone than synapses that are further away.

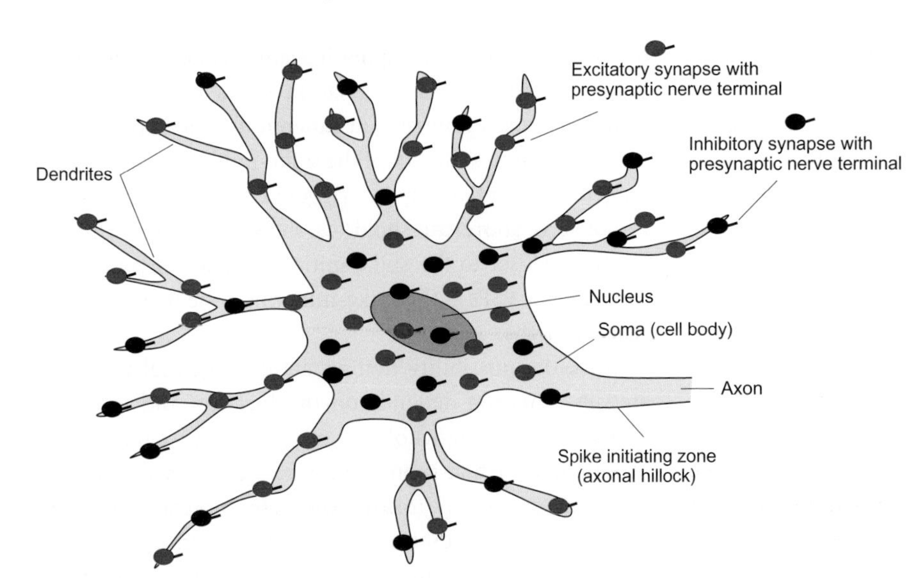

Figure 16.40 Schematic representation of the principle responsible for signal integration in neurons

The possibility that an action potential is generated in one neuron depends on the location and intensity of activation of the whole set of synapses (excitatory (red) and inhibitory (black)) on that neuron at a particular moment. The process of integration takes place at the spike-initiating zone of a neuron and results in either an action potential being triggered or not triggered at a particular time.

a nanocomputer in which the amount of neurotransmitter released and its effect on the postsynaptic membrane depends on that particular synapse's past and current experience. A whole field in computer science called artificial neural networks was inspired by how neurons process information. As we discuss in Case Study 16.2, available online, artificial neural networks have wide-ranging applications in the real world.

16.3.6 Long-term-potentiation and memory formation

Memory formation is attributed to changes at the level of chemical synapses, and is thought to involve a process called **long-term potentiation** (LTP). LTP is a long-lasting enhancement in **synaptic strength** at an excitatory synapse between two neurons following increased stimulation of the synapse. LTP was initially discovered in neurons from rabbit **hippocampus**[85], which is a central site for memory formation in vertebrates. It now appears that LTP occurs in neurons from many parts of the CNS in both vertebrates and invertebrates. An example of LTP is illustrated in Figure 16.41. Here, the amplitude of the EPSP to one stimulus increased after each bout of intense stimulation and this increase persisted for several hours.

The underlying mechanism of LTP in the CNS can be understood by looking at the properties of the NMDA receptor illustrated back in Figure 16.37B. The NMDA receptor only opens when three conditions are simultaneously satisfied:

[85] The hippocampus is a brain structure in the form of a horseshoe located in the medial region of the **temporal lobe**, as we discuss in Section 16.1.2. Each vertebrate brain contains two hippocampi, one in each temporal lobe.

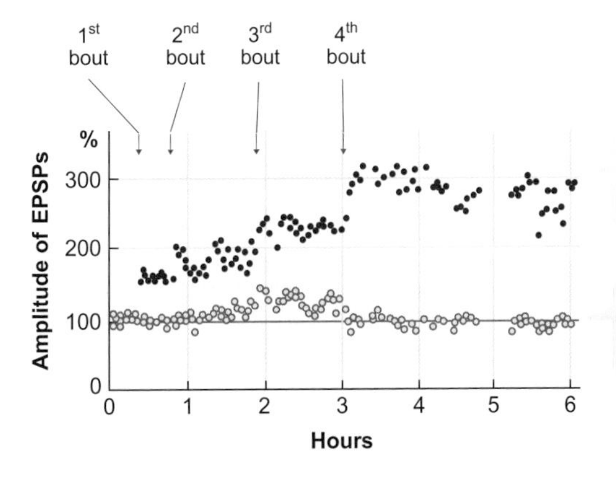

Figure 16.41 Manifestation of the phenomenon of long-term potentiation (LTP)

Demonstration of the LTP phenomenon at the level of excitatory synapses between hippocampal neurons of the rabbit following application of four bouts of stimulations (each bout consisted of 150 stimuli at 15/second) at the times indicated by arrows. The amplitude is expressed as a percentage of the amplitude before the conditioning stimulation. Notice that the amplitude of the EPSPs rises and remains elevated for hours after each bout of stimulation.

Source: Bliss TV, Lømø T (1973). Long-lasting potentiation of synaptic transmission in the dentate area of the anaesthetized rabbit following stimulation of the perforant path. Journal of Physiology 232: 331-356.

(i) the membrane is depolarized sufficiently to remove the Mg^{2+} block,

(ii) glutamate is bound to the glutamate binding sites,

(iii) glycine or D-serine are bound to the glycine binding sites.

Thus, when a glutamate excitatory synapse shown in Figure 16.42 containing AMPA and NMDA receptors in the postsynaptic membrane is activated by one action potential, glutamate and glycine are released from the presynaptic vesicles; glutamate binds to some AMPA ionotropic receptors in the postsynaptic membrane, which open and produce a small depolarization of the postsynaptic membrane. The glutamate and glycine also briefly bind to NMDA receptors in the postsynaptic membrane but the NMDA receptors remain closed because the depolarization is too short and too small to remove the Mg^{2+} block.

If the synapse is, however, intensely stimulated for a short time by a high frequency train of action potentials, more transmitter is released in the synaptic cleft, the depolarization of the postsynaptic membrane is greater and more prolonged, and the chance of opening some NMDA receptors increases. When NMDA receptors open, Ca^{2+} enters the postsynaptic terminal. Depending on how much the level of ionized calcium rises in the postsynaptic terminal, various Ca^{2+}-dependent signalling pathways are activated. For example, Ca^{2+}-dependent phosphorylation[86] of the AMPA glutamate-gated channels in the synapse increases their activity. Ca^{2+}-dependent phosphorylation also facilitates translocation of AMPA receptors from pools adjacent to the postsynaptic membrane and their insertion into the postsynaptic membrane as shown in Figure 16.42 (lower right panel). These changes strengthen the synapse by producing a greater EPSP when the synapse is later stimulated by one action potential (lower left panel in Figure 16.42).

After several bouts of stimulation Ca^{2+}-induced changes may activate transcription factors, gene expression and formation of new synapses between the two neurons, strengthening the connection between them. Indeed, structural changes in the **dendritic spines**[87] and associated synapses (shown earlier in Figure 16.38) were observed on the same time scale with LTP formation in the same neuron.

[86] Covalent binding of a small phosphate group to another molecule, which changes that molecule's activity.

[87] Tiny membranous protrusions on the dendrites of neurons that receive input at a synapse from one single axon as shown in Figure 16.38.

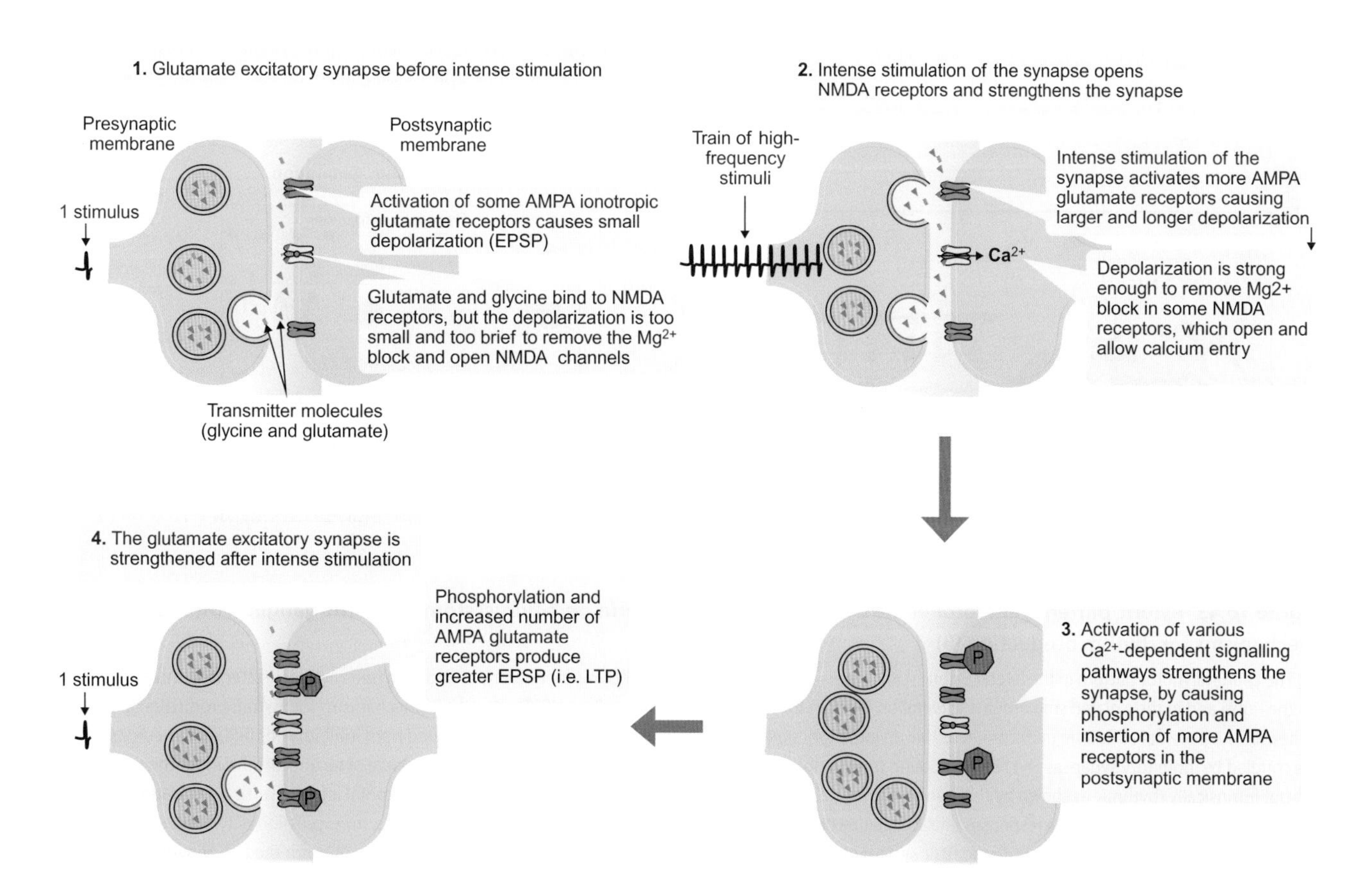

Figure 16.42 The mechanism of NMDA-dependent strengthening of excitatory synapses causing long-term potentiation (LTP)

Not all LTPs depend on the NMDA receptor. Some of LTPs depend on metabotropic L-glutamate receptors, or even other receptors in the synaptic membranes. The variety of signalling pathways that contribute to LTP, and the wide distribution of these various pathways in the brain, are reasons that the type of LTP exhibited between neurons depends in part upon the anatomic location in which LTP is observed.

A process that is opposite to LTP, called **long-term depression** or LTD, can also affect synapses and be involved in memory formation. At least some forms of LTD occur when ionized calcium concentrations in the postsynaptic terminal are lower than those causing LTP and are associated with the decrease in the level of phosphorylation of the NMDA receptors.

NMDA receptor-dependent LTP appears to work in specific types of learning, such as the encoding of space or the avoidance of unpleasant experiences, as shown in Experimental Panel 16.2, available online. LTD has also been implicated in some types of learning such as the encoding of specific features within space (the position of a tree in a landscape, for example). Sleep is thought to play an important role in long-term memory formation by facilitating synapse remodelling. In this process, information stored in the wakeful state in specific parts of the brain, such as the hippocampus, is processed and transferred to other parts of the brain.

16.3.7 Central pattern generators

Central pattern generators are neuronal circuits that, when activated, produce **rhythmic motor patterns**, such as those involved in walking, breathing, flying, swimming, chewing, swallowing and coordinated gut movements, in the absence of sensory stimulation or inputs carrying specific timing information from other levels of the nervous system.

The most convincing evidence that a certain region of the nervous system is capable of generating rhythmic motor patterns is to remove it from the animal and place it in a dish filled with physiological solution, such that all sensory pathways and communication with the external environment and other parts of the nervous system are severed. Many such isolated neuronal preparations from both invertebrates and vertebrates have been shown to generate motor patterns that would drive muscle movement[88], if the muscles were present.

An excellent preparation for studying rhythm generation in neural circuits is the stomatogastric ganglion[89] in European lobster (*Homarus gammarus*). Lobsters swallow their meal whole and chewing occurs in their stomach. The pylorus is part of the lobster stomach, and consists of a system of valves and sieve plates that move in a coordinated fashion to sort out food particles for further digestion. Figure 16.43A displays extracellular recordings of action

[88] We discuss in Section 18.5.3 neural control of muscle activity.
[89] We discuss the stomatogastric ganglion of invertebrates in Section 16.1.3.

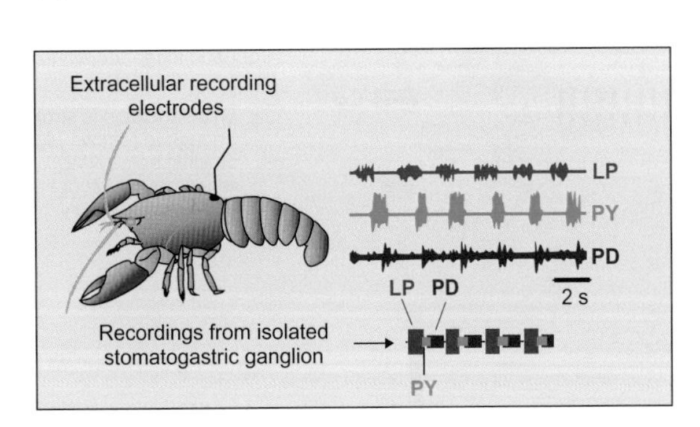

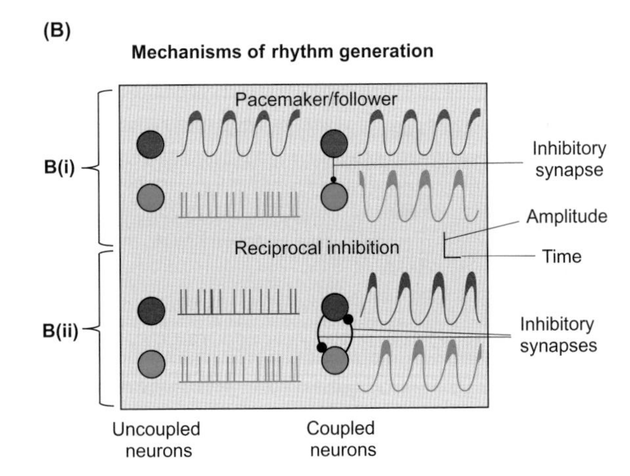

Figure 16.43 Pyloric pattern generator in the European lobster (*Homarus gammarus*) stomatogastric ganglion (A) and the basic mechanisms for rhythm production (B)

(A) The triphasic motor pattern in the lateral pyloric (LP), pyloric (PY) and pyloric dilator (PD) motor neurons recorded in free-moving lobster (top three recordings on the right) and in the isolated stomatogastric ganglion (bottom trace). Extracellular recordings. (B) The two basic mechanisms for rhythm production involve either. (B(i)) A pacemaker neuron (red) connected via an inhibitory synapse to a neuron (green) that fires on rebound from inhibition, or (B(ii)) two neurons that are coupled by inhibitory synapses and fire on rebound from inhibition. These neurons fire alternatively as a consequence of reciprocal inhibition. The neurons are not intrinsically rhythmic as shown by their random firing pattern when they are isolated from each other. The horizontal and the vertical calibration lines indicate time and amplitude of the responses (intracellular recordings). (Recordings from: Clemens S et al. (1998). Long-term expression of two interacting motor pattern-generating networks in the stomatogastric system of freely behaving lobster. Journal of Neurophysiology 79: 1396-1408 for the free-moving lobster; and Marder and Bucher (2001) for the isolated stomatogastric ganglion, see Source below).

Source: Marder E, Bucher D (2001). Central pattern generators and the control of rhythmic movements. Current Biology 11: R986–R996.

potential activity in pyloric muscles of free moving lobsters. No movement occurs in lobster's stomach unless motor neurons in the stomatogastric ganglion stimulate the muscles to contract.

The pyloric muscles are innervated by three motor nerves from the stomatogastric ganglion: the lateral pyloric nerve (LP), the pyloric nerve (PY) and pyloric dilator nerve (PD). Figure 16.43A shows that the firing of the pyloric motor neurons follows a highly regular, triphasic rhythm pattern. First, the LP motor neurons fire followed in strict order by firing from the PY and then from the PD motor neurons. The cycle is then repeated after a brief pause. The trace at the bottom of Figure 16.43A shows the firing pattern of the same motor neurons after the stomatogastric ganglion was dissected out of the lobster's body and placed in a dish filled with physiological solution. Notice that the same strict pattern of firing of the pyloric motor neurons is evident in the isolated preparation as in the living animal, indicating that the pyloric motor neurons are activated by a central pattern generator located in the stomatogastric ganglion.

Two basic mechanisms for rhythm generation are shown in Figure 16.43B(i) and 16.43B(ii). Some circuits have pacemaker neurons (which we discuss in Section 16.2.10 and Figure 16.30) connected via inhibitory synapses to neurons that fire on rebound from inhibition (which we discuss in Section 16.2.6 and Figure 16.26). If the two types of neurons are uncoupled, the pacemaker neuron fires at its intrinsic rate, while the other neuron fires randomly as shown in the right panel in Figure 16.43B(i). When the two neurons are coupled, however, they fire antagonistically: when the pacemaker fires, the other neuron is inhibited and when the pacemaker is not firing, the other neuron starts to fire on rebound from inhibition. The rhythm in this situation is determined by the intrinsic firing rate of the pacemaker neuron, which acts as a core oscillator. The pyloric rhythm of the crustacean stomatogastric ganglion shown in Figure 16.43A and the vertebrate respiratory rhythms[90] are pacemaker driven. A characteristic of the pacemaker-driven rhythms is that they are continuously active in the animal.

The other basic mechanism for rhythm production involves reciprocal coupling by inhibitory synapses of two neurons that fire on rebound from inhibition as shown in Figure 16.43B(ii). When the two neurons are uncoupled, they fire randomly. However, when they are reciprocally coupled by inhibitory synapses, the firing of one neuron in response to an external stimulus inhibits the firing of the other neuron—but this second neuron fires immediately on rebound from inhibition after the initial neuron stops firing. This pattern of antagonistic firing continues until at least one neuron is inhibited by an outside stimulus. This pattern of firing is therefore temporary, and occurs in activities such as walking, swimming, swallowing or chewing.

An important characteristic in the operation of robust central pattern generators is the phenomenon of burst firing, where silent periods follow periods of rapid spiking as shown in Figure 16.43A. Most neurons exhibit burst firing when they are driven from outside. However, some neurons have an endogenous burst firing mechanism, which involves calcium channels. When some of these channels open, a calcium ion current enters the cell, driven by the large electrochemical potential difference across the membrane, and the membrane depolarizes. If the depolarization is above threshold, repetitive action potential firing occurs for as long as the calcium channels stay open.

The rising calcium ion concentration in the cytoplasm, however, triggers a second-messenger cascade within the cell, which causes the closure of the calcium channels. As the calcium channels close, the membrane potential becomes more negative, below the threshold for triggering action potentials and the period of rapid spiking stops; the neuron becomes quiescent. After the calcium channels close, the ionized calcium concentration in the cell decreases and the inhibition exerted on the calcium channels by the second messenger cascade is removed, allowing some of these channels to open. The membrane potential depolarizes again above threshold and repetitive action potential firing resumes, thus generating a repetitive bursting pattern.

[90] We discuss respiratory and cardiac rhythms in vertebrates in Section 22.1.2 and 22.5.

› *Review articles*

Bennett MV, Zukin RS (2004). Electrical coupling and neuronal synchronization in the mammalian brain. Neuron 41: 495–511.

Briggman KL, Kristan WB (2008). Multifunctional pattern-generating circuits. Annual Reviews in Neuroscience 31: 271–294.

Friesen WO (1994). Reciprocal inhibition: A mechanism underlying oscillatory animal movements. Neuroscience and Behavioral Reviews 18: 547–553.

Marder E, Bucher D (2001). Central pattern generators and the control of rhythmic movements. Current Biology 11: R986–R996.

Snyder SH (1984). Drug and neurotransmitter receptors in the brain. Science 224: 22–31.

Sosinsky GE, Nicholson BJ (2005). Structural organization of gap junction channels. Biochimica et Biophysica Acta 1711: 99–125.

Stevens CF (1987). Molecular neurobiology: Channel families in the brain. Nature 328: 198–199.

Südhof TC (2004). The synaptic vesicle cycle. Annual Review of Neuroscience 27: 509–547.

Walmsley B, Alvarez FJ, Fyffe REW (1998). Diversity of structure and function at mammalian central synapses. Trends in Neuroscience 21: 81–88.

Bibliography

Cole KS (1968). Membranes, Ions and Impulses. Berkeley, CA. University of California Press.

Hille B (2001). Ion channels of excitable membranes. 3rd Edition Sinauer Associates, Inc.

Kandel ER, Schwartz JH, Jessell TM, Siegelbaum SA, Hudspeth AJ, Eds. (2012). Principles of Neural Science. 5th Edition, McGraw-Hill, New York.

Katz B (1966). Nerve, Muscle and Synapse. New York: McGraw-Hill, p.193.

Nicholls JG, Martin AR, Fuchs PA, Brown DA, Diamond ME, Weisblat DA (2012). From Neuron to Brain, 5th Edition, Sinauer Associates, Sunderland, Massachusetts, USA.

Purves D, Augustine GJ, Fitzpatrick D, Hall WC, LaMantia A-S, Mooney RD, Platt ML, White LE Eds. (2017). Neuroscience. 6th Edition. Oxford University Press.

Robertson D, Biaggioni I, Burnstock G, Low PA, Paton FR Eds. (2011). Primer on the Autonomic Nervous System. 3rd Edition Academic Press, London.

Young J Z (1971). The anatomy of the nervous system of *Octopus vulgaris*. Clarendon Press, Oxford.

Checklist of key concepts

Nervous systems in animals

- All animals except Placozoa and Porifera (sponges) have a well-defined **nervous system**, which is the main means of communication between the animal and the outside world.
- The nervous systems consist of **neurons** and **glial cells** that provide structural and metabolic support to neurons.
- Animals with bilateral symmetry have a **central** and a **peripheral nervous system**.
- The **central nervous system (CNS)** in **invertebrates** is organized in **ganglia** close to sensory structures.
- The CNS in **vertebrates** consists of the **brain** and the **spinal cord**.
- The presence of a functional **blood–brain barrier** provides a stable, pathogen-free extracellular environment for CNS neurons to function.
- The vertebrate brain consists of three major parts:
  - the **hindbrain** (comprising the medulla, the pons and the cerebellum),
  - the **midbrain** (comprising a sensory and a motor region called tectum and tegmentum, respectively),
  - the **forebrain** (comprising the diencephalon and the cerebrum).
- The diencephalon contains the thalamus, hypothalamus and the pituitary gland.
- The **cerebrum** is divided into two **hemispheres** and is organized into three domains:
  - **pallium**,
  - **striatum**,
  - **pallidum**.
- The pallium comprises the **cerebral cortex**, the **hippocampus** and the **olfactory bulb**.
- The **midbrain** together with **pons** and **medulla** are referred to as the **brainstem**.
- The brainstem contains a region called **reticular formation**, which consists of neural circuits involved in diverse functions.
- The **peripheral nervous system** in general comprises the **sensory-somatic system** and the **autonomic/visceral system**.
- The sensory-somatic system controls the animal's activities in the surrounding environment by carrying information from sensory structures to the CNS and signals from the CNS to somatic tissue.
- The autonomic/visceral system controls body homeostasis by connecting the CNS with sensory structures and internal organs.
- The autonomic nervous system in vertebrates comprises the **parasympathetic division**, the **sympathetic division** and the **enteric system**.
- The sympathetic and the parasympathetic divisions act **antagonistically** to control the activity of visceral organs:
  - the parasympathetic division maintains normal body functions in the animal at rest,
  - the sympathetic division mobilizes body system during activity.
- The enteric system regulates the activity of the **digestive tract**.

The ionic basis of electrical activity in neurons

- The **resting membrane potential** in neurons (and other cells) is determined by the unequal distribution of potassium, sodium and chloride ions across the plasma membrane and generally by the relative permeability of the membrane for K^+, Na^+ and Cl^-.
- The **Goldman–Hodgkin–Katz equation** quantitatively predicts the value of the membrane potential when there is no net flow of electrical charges across the membrane.
- The range of the membrane potential in neurons is between the equilibrium potential for potassium ion (V_K around −70 to −95 mV) and the equilibrium potential for sodium ion (V_{Na} around +40 to +55 mV).
- Membrane potential is determined to a large extent by the **membrane permeability ratio** for sodium and potassium ions ($P_{Na}:P_K$).
- At rest, the membranes are more permeable to K^+ than to Na^+ ($P_{Na}:P_K < 0.05$) and, therefore, the membrane potential is close to V_K.

- $P_{Na}:P_K$ depends to a large extent on whether the voltage-gated Na^+- and K^+-channels are **open** or **closed**, which depends, in turn, on the membrane potential.
- **Action potentials** are stereotyped responses of the membrane potential in neurons triggered by membrane depolarization above threshold.
- The time course of the action potential is determined by the opening and closing of voltage-dependent Na^+- and K^+-channels, which alter $P_{Na}:P_K$.
- Neurons communicate over **longer distances** through signals encoded by action potentials.
- Action potentials propagate along axons faster if axons are **electrically insulated** by myelin and/or have a **larger diameter**.
- Myelination also reduces the energetic cost for signal transmission along axons.
- **Pacemaker cells** generate spontaneous oscillations in their membrane potential.
- These oscillations can be in the form of action potentials (spiking) or non-spiking oscillations.

How neurons communicate with each other

- Signal transmission between neurons takes place at **synapses**, which are of two types:
  - electrical synapses
  - chemical synapses.
- Neurons at electrical synapses are connected by **gap junctions**, which allow electrical signals to pass rapidly across the synapse in both directions.
- At chemical synapses, signals are transmitted more slowly in only one direction from a presynaptic neuron to a postsynaptic neuron via the release of **transmitter molecules**.
- Different types of transmitter and receptor molecules occur at different chemical synapses.
- Individual neurons in the CNS are connected to other neurons by thousands of **excitatory** and **inhibitory** chemical synapses.
  - **Excitatory synapses** move the membrane potential of the postsynaptic neuron closer to threshold.
  - **Inhibitory synapses** move the membrane potential further from the threshold for the generation of action potentials in the postsynaptic neuron.
- Neurons continuously integrate the inputs received from all their synapses by either not triggering or triggering an action potential.
- Neural integration afforded by the presence of chemical synapses is the foundation on which all signal processing takes place in the nervous systems.
- Memory formation and learning is attributed to changes that occur at chemical synapses in the CNS.
- Increased stimulation of an excitatory synapse leads to its strengthening by a process referred to as **long-term-potentiation**, which plays a key role in memory formation.
- Circuits made of neurons connected by chemical synapses can generate **rhythmic patterns** that play a central role in the coordination of complex body movements, such as walking, breathing, flying, swimming, chewing, swallowing and coordinated gut movements.

Study questions

***Answers to these quantitative questions are available online. Go to www.oup.com/uk/butler**

1. Give two examples of how studies of the nervous systems in invertebrates have advanced the basic knowledge in neuroscience. (Hints: Sections 16.1.1, 16.2.1 and 16.2.4.)
2. Depict the general plan of organization of the vertebrate brain. (Hint: Section 16.1.2.)
3. Describe the general plan of organization of the sympathetic and parasympathetic divisions of the peripheral nervous system in vertebrates and highlight the major functional differences between the sympathetic and parasympathetic divisions. (Hint: Section 16.1.3.)
4. Explain what cranial nerves are and briefly describe their function. (Hint: Section 16.1.3 and Box 16.2.)
5. Discuss the importance of the blood–brain barrier for the function of the central nervous system of vertebrates. (Hint: Section 16.1.2, Box 16.1 available online.)

6.* What happens to the resting membrane potential of a neuron when (i) the membrane permeability for sodium increases by a factor of two and the permeability for potassium ion increases by a factor of three, (ii) the membrane permeability for both Na^+ and K^+ both decrease by a factor of two and (iii) the membrane permeability for Na^+ rises and the permeability for K^+ does not change. (Hint: Section 16.2.2.)

7.* Use four of the expressions mentioned in parentheses (it depolarizes, it hyperpolarizes, it becomes less negative, it becomes more positive, it becomes more negative, it becomes less positive, it does not change, it repolarizes, it increases, it decreases) to describe what happens to the membrane potential of a neuron when the extracellular concentration of the potassium ion (K^+) rises from 2.5 to 4 mmol L^{-1}. (Hint: Section 16.2.2.)

8. Describe the patch-clamp technique and explain how it can be used to study properties of ion channels in neurons. (Hint: Section 16.2.3.)

9.* Explain what will happen to the voltage-gated sodium channels when the membrane potential changes from −60 mV to +40 mV and how this is relevant to the generation of an action potential. (Hint: Sections 16.2.3 and 16.2.4.)

10. Describe the mechanism responsible for action potential propagation in myelinated axons and explain why action potentials propagate faster in myelinated than in unmyelinated axons of same diameter. (Hint: Section 16.2.7.)
11. What are the major differences between intracellular and extracellular recordings of action potentials. (Hint: Section 16.2.8.)
12. Outline the mechanisms of signal transmission at chemical synapses and describe the difference between excitatory and inhibitory chemical synapses. (Hint: Sections 16.3.2 and 16.3.3.)
13. Explain the difference between ionotropic and metabotropic receptors at chemical synapses. (Hint: Sections 16.3.2 and 16.3.4.)
14. What is the role of the NMDA receptors in the mechanism of long-term-potentiation (LTP) and discuss how LTP is relevant for memory formation. (Hint: Sections 16.3.4 and 16.3.6.)
15. Describe the principle behind the phenomenon of post-inhibitory rebound excitation and explain how this phenomenon is relevant to the generation of rhythmic patterns in neural circuits. (Hint: Sections 16.2.6 and 16.3.7.)

17

How animals sense their environments

Animals have evolved amazing senses that facilitate their survival and reproduction. For example, some moths and butterflies detect potential mates at a distance of several kilometres; birds and sharks sense the Earth's magnetic field and use it for orientation and navigation; birds of prey locate small rodents from a height of several kilometres; polar bears smell a seal from 1–2 kilometres away and even humans can distinguish up to one trillion different odours[1]!

Specific physical and chemical factors in the external and internal environment act as stimuli (singular stimulus) on sensory structures; these stimuli elicit signals that animals use to interpret their environment. In this chapter we discuss first the general principles of how information contained in stimuli is converted to neural signals. We then focus on the specific mechanisms responsible for sensing and processing particular types of stimuli.

17.1 Principles of sensory processing

The world as animals experience it is defined by their senses, which receive and interpret only a small fraction of the potential sensory information available in their environment. Thus, the same physical environment is perceived differently by different animals, with each having evolved different sensory priorities and specific abilities that facilitate their capacity to survive and reproduce in that particular environment with varying degrees of success. For conservation purposes it is important to understand the range of sensory information that is actually available to specific groups of animals and how specific physiological processes are affected by various changes in the environment.

Stimuli in physiology refer to detectable factors in the external or internal environment that are associated with particular forms of energy. Animals have specialized cells called **sensory receptors**; these receptors contain structures that are sensitive to specific forms of stimulus energy.

The ability of an organism to respond to stimuli at low levels of energy is called **sensitivity**. **Stimulus modality (or sensory modality)** refers to what the animal perceives after a stimulus. Basic stimulus modalities include light, sound, smell, taste, temperature and pressure.

The most common and useful classification of sensory receptors is based on the type of stimulus energy that activates them:

- **photoreceptors** are sensitive to electromagnetic radiation in the form of light;
- **chemoreceptors** are sensitive to the presence of specific molecules and ions;
- **mechanoreceptors** are sensitive to mechanical stimuli, such as stretch and pressure;
- **thermoreceptors** are sensitive to heat or changes in temperature;
- **electroreceptors** are sensitive to electrical fields;
- **magnetoreceptors** are sensitive to Earth's magnetic field;
- **nociceptors** are sensitive to stimuli that are capable of causing damage to the animal.

Sensory receptors are also classified depending on their location: **interoceptors** monitor variations within the body of the animal—such as the composition of gases in the blood, or the stretch levels of various tissues, while **exteroceptors** monitor variations in the external environment. The ability of animals to interpret the surrounding environment by processing information that is acquired through exteroceptors is called **sensory perception**.

In all instances, a neural signal in the form of action potentials[2] is produced in response to a stimulus acting on

[1] We discuss later in this chapter the physiological bases for these amazing senses.

[2] We discuss action potentials in Section 16.2.

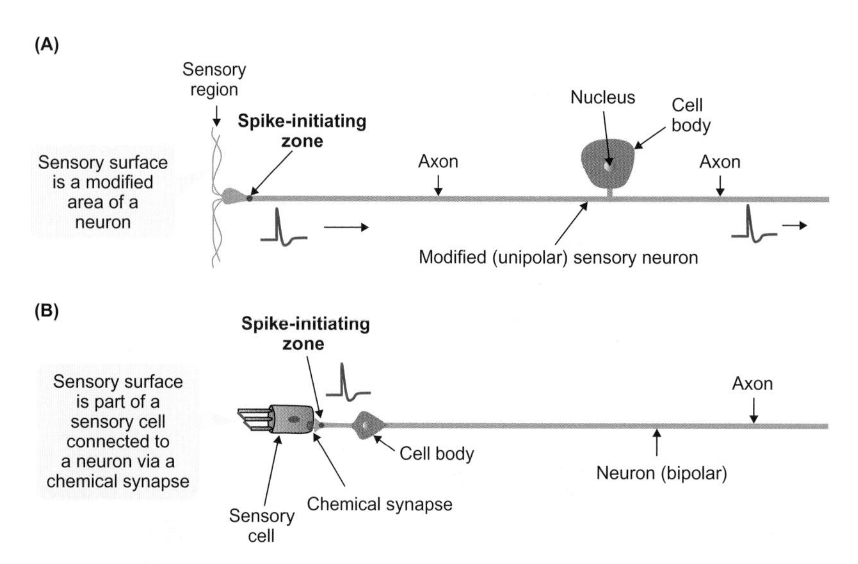

Figure 17.1 Sensory receptors are of two types

(A) Modified neurons with sensory surfaces that respond to stimulus modalities such as temperature, pressure, stretch, smell and taste. (B) Sensory cells connected to neurons via chemical synapses that respond to stimulus modalities such as sound and light. The 'hair cell' shown diagrammatically here is located in the auditory apparatus of mammals as we discuss in Section 17.4.1. Auditory nerve fibres are somewhat unique in that action potentials pass through the cell body. Photoreceptors in vertebrate eyes make synapses with non-spiking neurons, which, in turn, make synapses with other neurons in which propagating action potentials are generated. Sensory surfaces are separated from the sites where propagating action potentials are generated (spike-initiating zone on neurons). Note that some sensory neurons are unipolar, other are bipolar, while many others are multipolar, as we discuss in Section 16.1.

a sensory receptor. Figure 17.1 shows that sensory receptors are either modified neurons or sensory cells connected to neurons via chemical synapses. Thermoreceptors, most mechanoreceptors, and some chemoreceptors belong to the first group, while most photoreceptors and electroreceptors, and some chemoreceptors and mechanoreceptors belong to the second group.

The conversion of stimulus energy into an action potential is called **sensory transduction**. The neural signals from sensory receptors propagate in the nerve nets of Ctenophora and Cnidaria; in Bilateria[3] the neural signals are transmitted to the central nervous system (CNS) for processing and interpretation via distinct **afferent neural pathways**. When signals reach the CNS, they either elicit a reflex response and/or are progressively integrated at several levels in the CNS enabling animals to interpret and respond to the world around them.

Sensory receptors are often grouped together in larger multicellular structures called **sensory organs**, which are specialized for acquiring a particular type of stimulus at low levels of energy while filtering out other types of stimuli. Sensory organs, and parts of the nervous system that participate in processing the information acquired through the respective organs, form what is commonly called a **sensory system**.

[3] We discuss the nervous system in animals in Section 16.1.

17.1.1. Sensory receptors act as selective amplifiers for specific stimuli

At weak-to-moderate levels of stimulus intensity, each sensory receptor is sensitive to only one type of stimulus (Figure 17.2A). However, if the intensity of other types of stimuli is very high, then a sensory receptor may also respond to these different stimuli (Figure 17.2B). For example, a mechanical blow applied to the eye can activate photoreceptors in the eye, despite these receptors primarily detecting light rather than pressure. Whatever the nature of the incoming stimulus—whether a mechanical blow or any other type of stimulus—the photoreceptors always give a visual signal, like seeing stars in the eyes.

Sensory receptors also act as amplifiers for weak stimuli. For example, the binding of one pheromone[4] molecule to a chemoreceptor could trigger a train of action potentials that modifies the behaviour of the animal. As shown in Box 17.1, available online, amplification factors in the order of 10^4–10^6 are achieved when the binding of one pheromone molecule to a chemoreceptor, or the absorption of one photon[5] by a photoreceptor, triggers one action potential in the sensory neuron.

[4] We discuss pheromones in Section 17.3.1.

[5] A photon is a particle of light or other electromagnetic radiation. The photon is not electrically charged and has no static mass but carries energy that is inversely proportional to its wavelength.

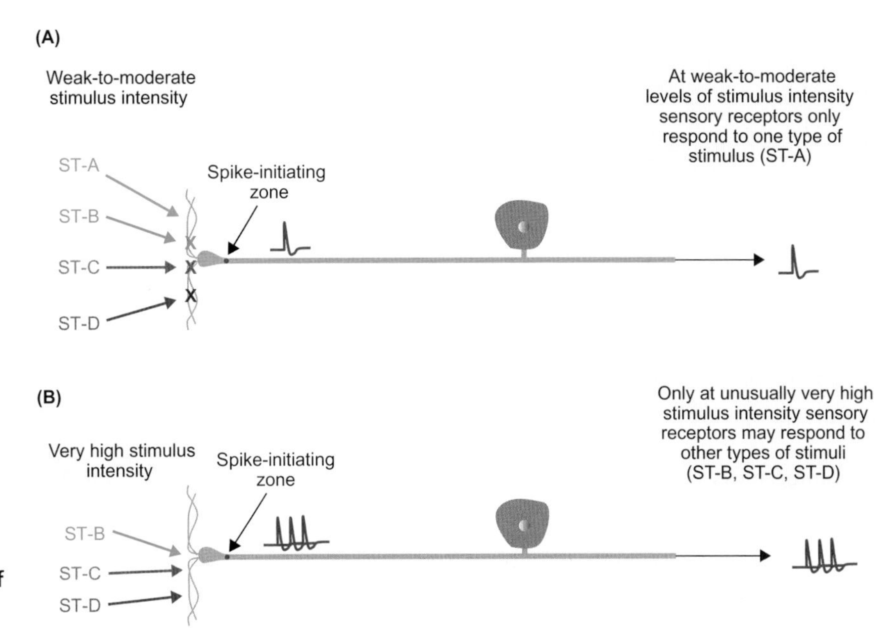

Figure 17.2 Sensory receptors are generally sensitive to only one type of stimulus at weak-to-moderate levels of intensity (energy) (ST-A)

17.1.2 Major steps between stimulus input and sensory output

Sensory reception—the process of transduction in which stimuli are converted into action potentials—involves a series of intermediary steps.

First, a specific type of stimulus acts either directly or indirectly on a sensory receptor to produce a structural change[6] in a specific receptor protein on the sensory surface, which affects the opening of particular ion channels[7] in the membrane.

The receptor protein can be an **ionotropic** or **metabotropic receptor**. An ionotropic receptor is itself an ion channel or a protein that *directly* affects the opening of the ion channel when a stimulus acts on it[8]. The receptor proteins in some chemoreceptors, and in most mechanoreceptors and thermoreceptors, are ionotropic receptors. Other receptor proteins located on sensory surfaces of photoreceptors and many chemoreceptors affect *indirectly* the opening of ion channels through G-proteins[9] and second messengers; these are metabotropic receptors[10].

Second, the direct or indirect opening of particular ion channels changes the membrane permeability to specific ions, which produces a change in the membrane potential[11] of the sensory receptor. This change in the membrane potential of a sensory receptor in response to a stimulus is called the **receptor potential**.

Central to the generation of receptor potentials in many sensory receptors are the so-called **transient receptor potential** (TRP) **ion channels**. The TRP ion channels[12] are an ancient group of relatively non-selective cation channels (permeable to cations, including Na^+, K^+ and Ca^{2+}) that cause membrane depolarization when activated. This membrane depolarization is based on the same mechanism we describe for excitatory chemical synapses in Section 16.3.3. The TRP ion channels were initially discovered in a mutant of the fruit fly (*Drosophila melanogaster*), which showed only a transient response to continuous illumination. The TRP ion channels were later found in all animal species and are involved in the generation of the receptor potential in many types of sensory cells. These include chemoreceptors, mechanoreceptors and

6 Note that 'structural change' includes mechanical distortion.
7 We discuss ion channels in Section 3.2.3 and 4.2.2.
8 We discuss ionotropic receptors in Section 16.3.2.
9 We discuss the role of G-proteins in cellular signalling in more detail in Chapter 19.
10 We discuss metabotropic receptors in Section 16.3.2.

11 We discuss, in Section 16.2, how changes in membrane permeability to ions alters the membrane potential.
12 The TRP channels belong to the large superfamily of cation channels with six transmembrane-spanning helices forming a transmembrane pore loop, as shown in Figure 16.21. The TRP channels normally require four individual subunits to form a functional pore. In humans, there are at least 31 channel subunit genes that encode seven families of TRP channels (TRPC(1–7), TRPV (1–6), TRPA(1), TRPM(1–8), TRPP (1–5), TRPML (1–3) and TRPN), which are involved in various cellular functions. The TRP channels are classified based on their amino acid sequence homology rather than by their ionic selectivity or their ligand function, as is the case with the voltage- and ligand-gated channels.

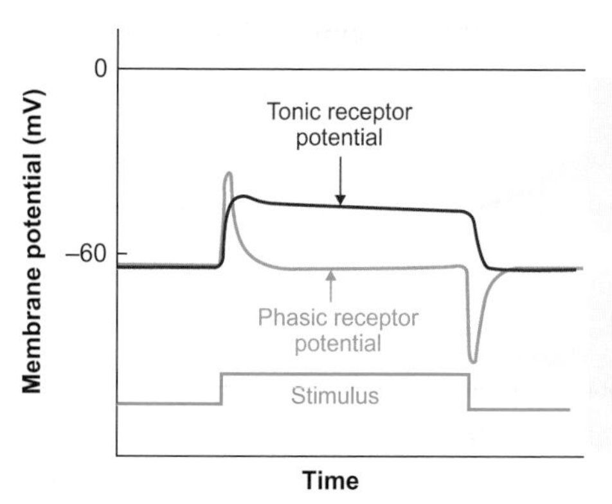

Figure 17.3 Schematic representation of tonic and phasic receptor potentials induced by a stimulus

thermoreceptors[13] in all animal groups, and photoreceptors in insects and other invertebrates.

Third, the information contained in the receptor potentials is converted into action potentials as we will discuss.

Receptor potentials

The receptor potential can be either a depolarization (whereby the membrane potential becomes *less* negative) or a hyperpolarization (whereby it becomes *more* negative), depending on the relative change in membrane permeability to specific ions induced by the stimulus[14]. A depolarizing receptor potential is common for most sensory cells, but hyperpolarizing receptor potentials occur in some sensory cells, such as the vertebrate photoreceptor cells that we discuss in Section 17.2.1.

[13] Section 9.2.2 and Figure 9.12 discuss the role of TRP ionic channels in behavioural thermoregulation of reptiles.

[14] As we discuss in Section 16.3.3.

Broadly speaking, receptor potentials are classified into two categories, as illustrated in Figure 17.3:

- **tonic receptor potentials** are maintained for the duration of the stimulus;
- **phasic receptor potentials** are short-lived and occur when there is a change in stimulus intensity.

Prolonged stimulation in general causes the receptor potential to decrease in amplitude, in a process known as **sensory adaptation**[15]. In effect, phasic receptors adapt very quickly to the stimulus, while tonic receptors adapt more slowly.

Receptor potentials are **graded potentials**, varying in magnitude with stimulus intensity (Figure 17.4). The

[15] 'Sensory adaptation' should not be confused with 'evolutionary adaptation', as we discuss in Section 1.5.

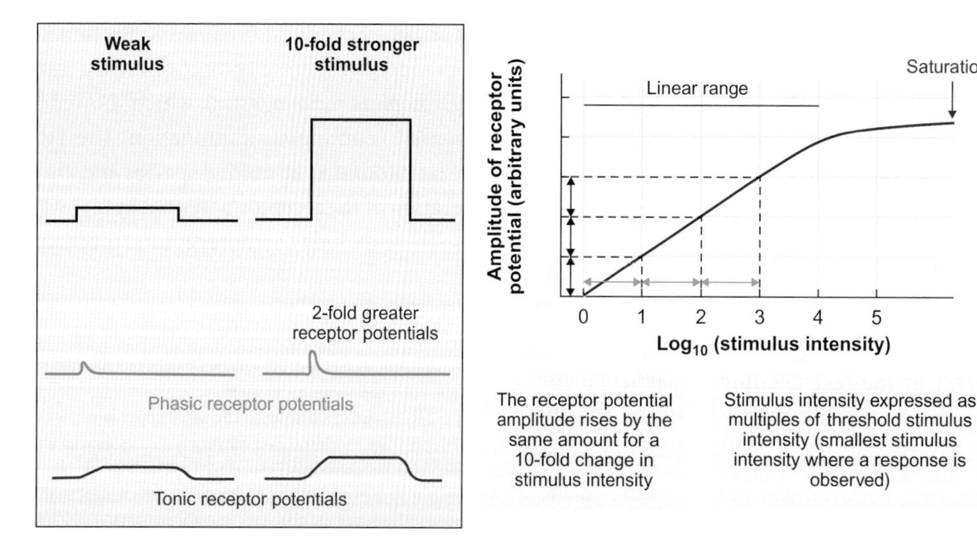

Figure 17.4 Logarithmic relationship between amplitude of the receptor potential and stimulus intensity

The linear logarithmic relationship holds for both phasic and tonic receptor potentials before reaching saturation. The amplitude of the receptor potential, measured in mV, is sensitive to the exact location of the recording electrodes with respect to the sensory surface and, therefore, all recordings are made with electrodes in the same precise location while the stimulus intensity is changed.

amplitude of the receptor potential rises linearly with the logarithm of the stimulus intensity, but only up to a certain value; beyond that value it remains constant due to a process called *saturation*. Saturation occurs because:

- The changes in permeability for particular ions initiated by the stimulus are limited by the total number of available channels in the membrane, so when the stimulus increases beyond a certain level, there is no further change in membrane permeability.
- The membrane potential cannot exceed the range delineated by the equilibrium potentials of the ions to which the membrane is or becomes permeable, so there is no further change in the membrane potential beyond a certain level of permeability changes as we discuss in Section 16.2.2.

Conversion of receptor potentials into action potentials

If the sensory receptor is a modified area of a neuron, then the appropriate stimulus acting on the sensory surface generates a depolarizing receptor potential. The depolarization decreases in amplitude as it spreads from the sensory surface to the spike initiation zone of the neuron (Figure 17.2), where action potentials are generated.

An action potential is triggered when the receptor potential at the spike initiating zone exceeds the threshold level in the resting neuron (Figure 17.5). Once the action potential is triggered, no further action potential can be triggered during the absolute refractory period, regardless of how intense the stimulus is. The threshold then decreases quasi-exponentially during the **relative refractory period** toward its initial (resting) level[16]. When a stimulus is strong and the amplitude of the receptor potential is high, the threshold condition for triggering another action potential is reached earlier than when the stimulus is weak and the amplitude of the receptor potential is low. Therefore, action potentials are triggered closer together—that is, at higher frequencies—when the intensity of the stimulus is high than when it is low.

The duration of the absolute refractory period determines how much close together action potentials can be generated.

Other receptors, such as the 'hair cells' in the auditory apparatus of mammals and photoreceptors, form chemical synapses[17] with neurons as shown in Figure 17.1B. In these sensory cells the stimulus generates either depolarizing or hyperpolarizing receptor potentials, which spread to the chemical synapse. A stronger stimulus causes a higher amplitude receptor potential as compared with a weaker stimulus; this stronger stimulus alters the rate of transmitter release at the synapse and results in a higher firing rate of action potentials in associated neurons.

We generally observe a linear relationship between the amplitude of receptor potentials and frequency of action potential generation, as shown in Figure 17.6A, where the recordings were made from a frog muscle mechanoreceptor[18] using extracellular electrodes[19]. The maximum frequency at which action potentials can be generated is limited by the

[16] We discuss in Section 16.2.5 the mechanisms responsible for the absolute and relative refractory periods.

[17] We discuss the mechanism of signal transmission at chemical synapses in Section 16.3.2.

[18] We discuss how muscle spindles work in Section 17.4.4.

[19] We discuss, in Section 16.2.8, and show, in Figure 16.29, that changes in the membrane potential measured with extracellular electrodes are much smaller in absolute magnitude compared with intracellular recordings.

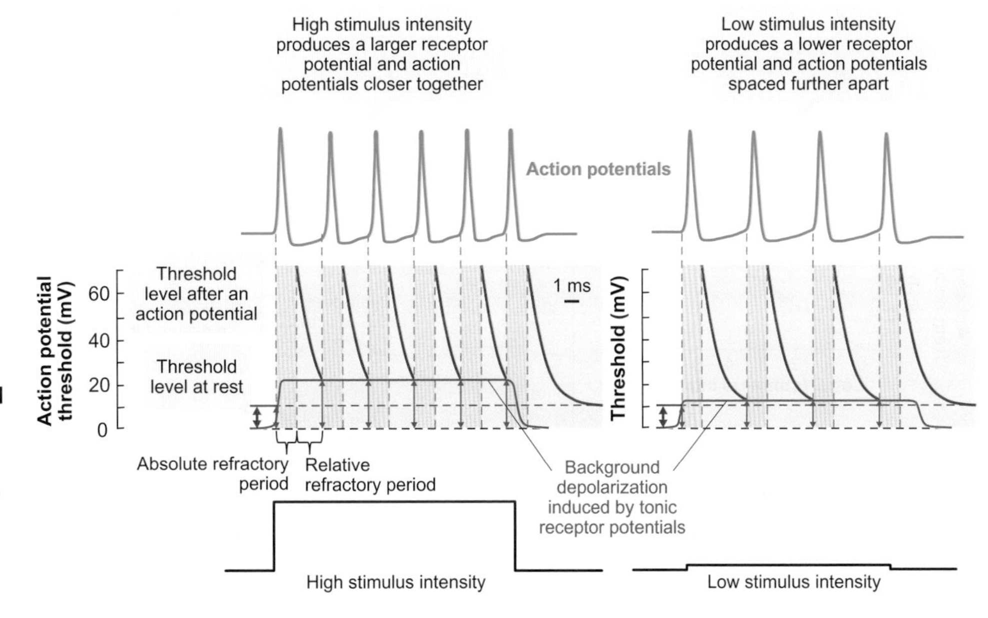

Figure 17.5 Diagrammatic representation of the general principle for the conversion of stimulus intensity into action potentials at the spike initiating zone on modified sensory neurons

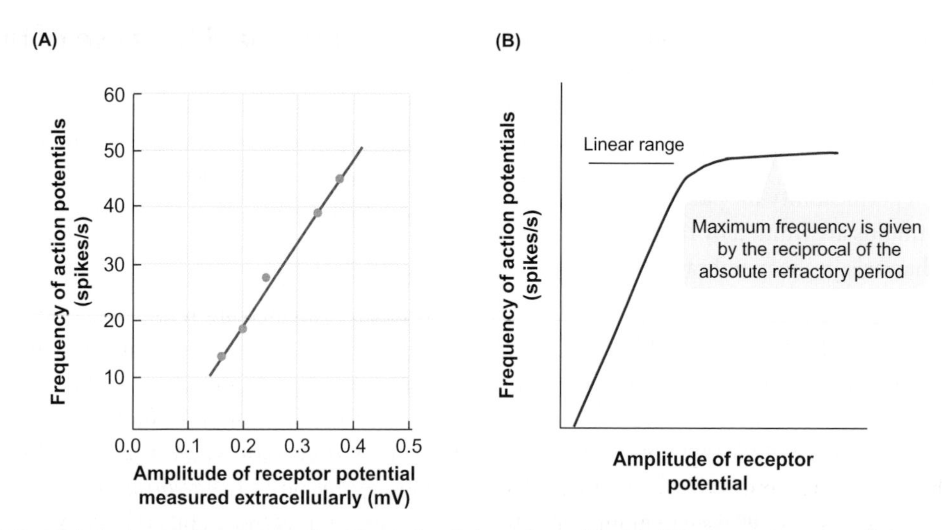

Figure 17.6 Relationship between frequency of action potentials and amplitude of the receptor potential

(A) Measurements of action potential frequency in a mechanoreceptor (muscle spindle of the frog) as a function of the receptor potential amplitude (measured extracellularly). (B) Generally, the frequency of action potentials rises linearly with the amplitude of the receptor potential up to a certain point and then the curve flattens out. Note that the maximum frequency of action potentials that can be triggered in a neuron is limited by the duration of the absolute refractory period. For example, if the absolute refractory period is 1 ms, then the maximum frequency of action potentials that can be carried by that axon is $(1\ \text{ms})^{-1} = 1000$ spikes s^{-1}.

Source: A: Ottoson D, Shepherd GM (1970). Steps in Impulse Generation in the isolated Muscle Spindle. Acta Physiologica Scandinavica 79: 423-430.

reciprocal value of the absolute refractory period. Therefore, beyond a certain magnitude of the receptor potential, the frequency at which action potentials are generated cannot increase any further; this is illustrated in Figure 17.6B.

By combining the logarithmic relationship in Figure 17.4 with the relationship in Figure 17.6B, we obtain the relationship in Figure 17.7A. This relationship shows that the frequency of action potentials is linearly dependent on the logarithm of the stimulus intensity up to a certain level; beyond this level, however, it displays saturation, meaning that the frequency can increase no more.

The linear relationship between frequency of action potentials and the logarithm of stimulus intensity allows animals to sense an enormous range of stimulus intensities. For example, the human ear can perceive a 1,000,000,000 range of stimulus intensities without significant distortion. Similarly, the vertebrate eye can register differences in light intensities greater than 1,000,000, which is the difference between full sunlight and full moonlight intensities.

Sensory transduction encodes the key characteristics of a stimulus

Action potentials carried in sensory neurons from specific sensory receptors contain information about stimulus modality, location, strength and its temporal characteristics.

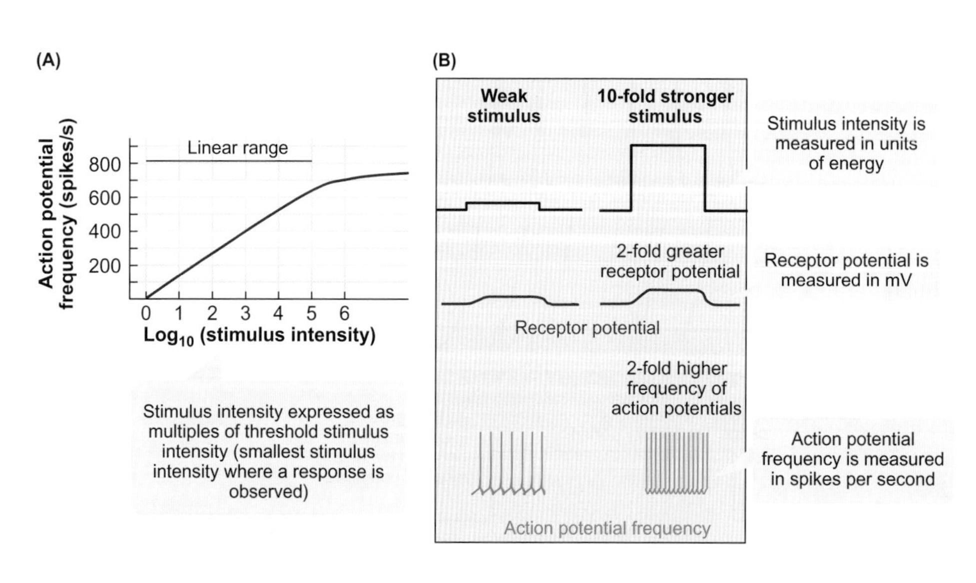

Figure 17.7 Relation between frequency of action potentials in sensory neurons and stimulus intensity

(A) The logarithmic relation between action potential frequency and stimulus intensity has major implications for understanding how sensory information is coded and processed.
(B) Graphic representation of frequency modulation of action potentials in response to two stimuli of different intensities.

The stimulus *modality* (light, sound, temperature, smell, taste, etc.) is encoded by the particular type of sensory receptors that respond to that stimulus at weak to moderate level of stimulus intensity (Figure 17.2A). These sensory receptors are connected to specific parts of the CNS and allow the animal to perceive a particular type of sensation or **sensory modality**.

Stimulus *strength* is encoded by the frequency of action potentials, as shown in Figure 17.7B. This relationship between stimulus strength and action potential frequency occurs in all animals and permits them to sense a very wide range of stimuli intensities.

The *location* of the stimulus is encoded by the anatomical position of the sensory receptors that respond to that stimulus. Sensory receptors also encode temporal stimulus characteristics such as the onset, offset and duration of stimulation.

› Review articles

Clapham DE (2003). TRP channels as cellular sensors. Nature 426: 517–524.
Frings S (2009). Primary processes in sensory cells: current advances. Journal of Comparative Physiology A 195: 1–19.

17.2 Photoreception

We now consider in more detail the ways animals sense specific types of stimuli, beginning with photoreception.

Photoreception refers to the ability of animals to sense light. **Visible light** refers to the part of the electromagnetic spectrum to which our eyes are sensitive. As shown in Figure 17.8, the spectrum of visible light encompasses electromagnetic radiation of wavelengths between about 400 nm for violet light and 700 nm for red light[20]. The visible spectrum of electromagnetic radiation propagates through the atmosphere with little loss of intensity; it is also the region of the spectrum that is least absorbed by water compared with other regions of the electromagnetic spectrum. This fact strongly supports the idea that animals evolved in water.

Photoreception is made possible by the presence of light-sensitive proteins called **photopigments** that change configuration when they absorb photons of specific wavelength (energy). Some types of photopigments are involved in vision as we discuss in the next section, while others are involved in the synchronization of biological clocks as we discuss in Section 17.2.2.

[20] Electromagnetic radiation of 400 and 700 nm wavelength has frequencies of 750 and 429 THz, respectively (1 THz = 1 teraherz = 10^{12} s^{-1}). Frequency = speed of light (3×10^8 m s^{-1})/wavelength.

17.2.1 Vision: making sense of light from animal's surroundings

Vision is the ability of animals to interpret their surrounding environment by processing information contained in the electromagnetic radiation spectrum, which they acquire through their photoreceptive organs.

Generally, animals have photoreceptive organs that are sensitive to the radiation spectrum between about 400 to 700 nm. However, some animals are also sensitive to electromagnetic radiation outside this range. For example, most insect pollinators also detect ultraviolet light in the range 315 to 400 nm (UVA in Figure 17.8). As such, their visual perception of the environment depends on how the environment appears in both UV *and* visible light. Most diurnal birds can also see in the UV range down a wavelength of 300 nm. Indeed, some birds have sex-dependent plumage markings that can be seen *only* in UV light. Figure 17.9A shows images of a myna bird (*Acridotheres* spp.) as seen through human eyes and how another bird would see it.

Intriguingly, recent genetic studies have shown that humans and mice also express a UV-sensitive photopigment called neuropsin in their nervous system. Neuropsin absorbs photons with a wavelength around 380 nm, but the physiological function of neuropsin remains unknown. The lens in the human eye absorbs radiation below a wavelength of 400 nm. The absence of the lens in some humans (condition called *aphakia*) permits them to detect (some) UVA light.

Some animals can also detect infrared (IR) radiation. For example, piranhas[21] (Figure 17.9B) and dwarf cichlids (*Pelvicachromis taeniatus*), have photoreceptors in their eyes that are sensitive to photons in the near IR range, giving them infrared vision. These fishes live in murky waters that are more transparent to longer wavelength infrared radiation than shorter wavelength visible light[22]; they use the infrared vision for hunting their prey. Interestingly, goldfish (*Carassius auratus*), a domesticated carp, has one of the broadest visual spectra among animals, encompassing electromagnetic radiation from 350 to 750 nm, which includes ultraviolet, visible and infrared radiation. The broad visual spectrum of carp (*Cyprinus carpio*), in general, enhances their fitness, helping them out-compete native fish when introduced to new habitats.

Infrared 'vision' also evolved in pit snakes (pythons, boas and pit vipers), which use infrared sensors that detect heat (thermoreceptors[23]) rather than photoreceptors that detect photons.

[21] Piranhas are omnivorous South American freshwater fish which attack mammals, including humans and cattle.

[22] As shown in Figure 17.9, a 10 m column of water is not transparent to IR light. However, over short distances (in the order of 10–20 cm), sufficient light can be transmitted through water in the infrared band for the fish to see.

[23] We discuss thermoreceptors in Section 17.5.

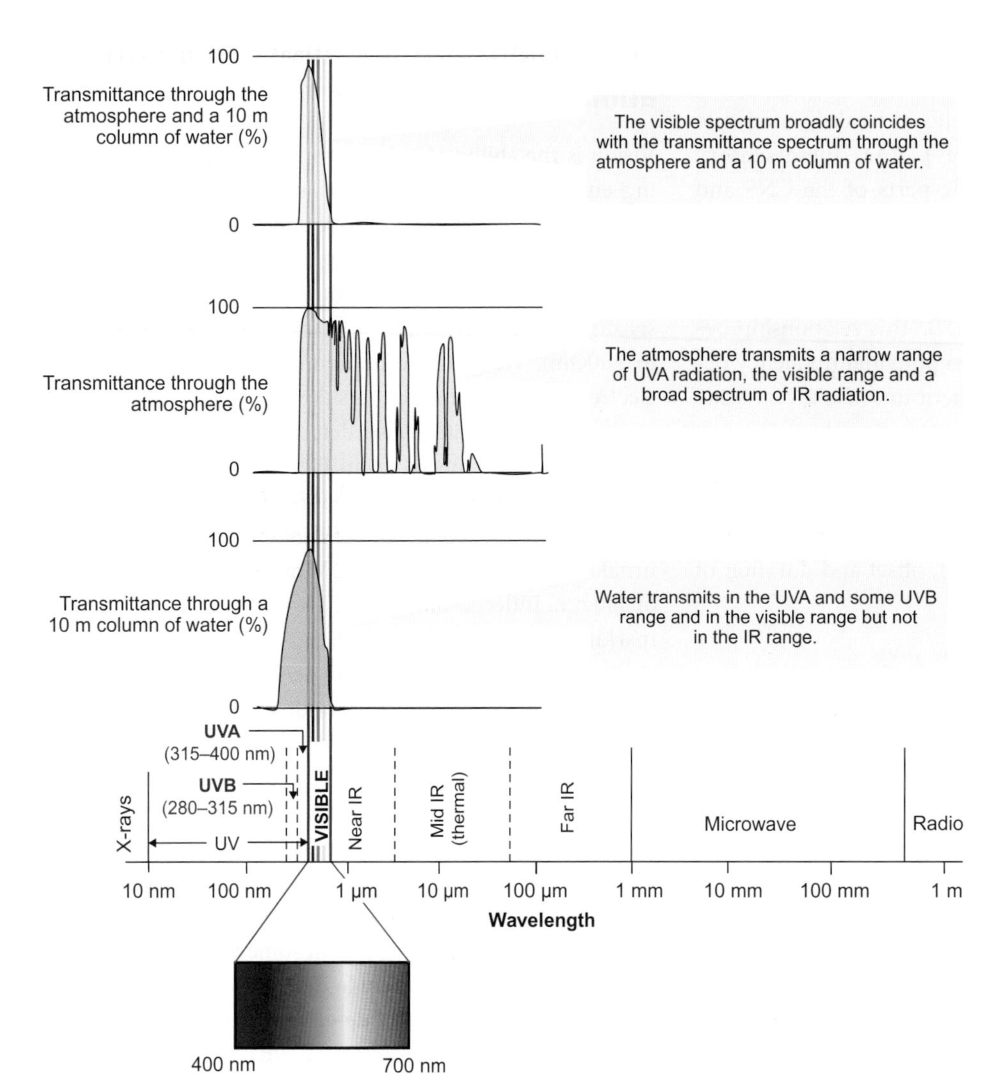

Figure 17.8 Divisions of the electromagnetic radiation spectrum from X-rays to radiowaves and transmittance of electromagnetic radiation of 0.1 to 100 μm wavelength through atmosphere and water

Transmittance refers to % of radiation of certain wavelength transmitted through a medium. Only electromagnetic radiation close to the visible range propagates with relatively little loss through both the atmosphere and water. UV, ultraviolet radiation; IR, infrared radiation.

The thermoreceptors are located in small indentations known as facial pits or pit organs around their mouths, nostrils and eyes (Figure 17.9C). The pit organs contain a vascularized membrane innervated by terminal endings of nerve fibres, which have temperature-sensitive cation channels of the type we discuss in Section 17.5. Infrared radiation, from objects warmer than the environment, raises the temperature, locally, of the pit membrane and activates the temperature sensitive cation channels, causing depolarization of the nerve terminals and a rise in the firing rate (see Section 17.5). Both anatomical and functional evidence suggests that the facial pits in snakes are functionally integrated with the eyes to produce a generalized multispectral visual sense[24].

Blood-sucking insects, such as mosquitoes, have infrared-sensitive receptors located in their antennae, which allow them to determine the direction towards the source of heat with such accuracy that they land on the skin of warm-blooded animals in close proximity to blood vessels carrying blood that is warmer than the skin[25].

The signal transduction cascade in the visual sensory systems

The visual sensory systems in animals comprise two parts: the photoreceptive organs that acquire visual information from the surrounding environment, and parts of the nervous system that process this information. The most common photoreceptive organ in animals is the eye. Eyes vary in complexity from a group of photoreceptive cells to complex structures that can form sharp images focused onto photoreceptor cells. Photoreceptors absorb the photons that fall on them and convert the information into a receptor potential.

The photopigment molecules in visual systems consist of a **chromophore** (a coloured compound), which is usually

[24] Pit organs are also used in selecting their thermal environment for behavioural thermoregulation, as we discuss in Section 9.2.1.

[25] Note that mosquitoes also use olfactory receptors for carbon dioxide and scent to locate prey.

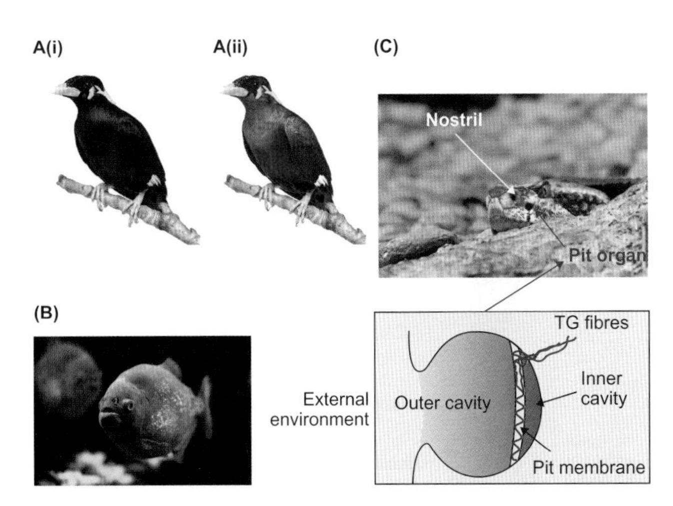

Figure 17.9 Examples of animals that are sensitive to UV (A) and infrared (B, C) radiation

(A) Most diurnal birds can detect feather markings that can only be seen in UV light. A(i) Image of a myna bird (*Acridotheres* spp.) seen through human eyes and A(ii) image of the same bird as it would appear to another bird. (B) Piranhas have infrared vision based on photochemical transduction, which allows them to detect prey in murky waters. (C) Pit-snakes, such as pit vipers (Crotalinae family), have pit organs (indicated by the red arrow) which are sensitive to infrared radiation through a different mechanism which involves thermoreceptors rather than photoreceptors. The pit membrane is highly vascularized and is innervated by terminal endings of nerve fibres belonging to the trigeminal (TG) cranial nerve. Infrared radiation from sources warmer than the environment is detected by thermoreceptors in the nerve endings causing an increase in the firing rate of the nerve terminals as we discuss in the text.

Source: A(i) and A(ii) John Courtney-Smith; B: Luc Viatour/Wikimedia Commons; C: © Mark Mannetti.

a derivative of Vitamin A[26]. The chromophore is covalently bound to a protein called **opsin**, as shown in Figure 17.10. The chromophore exists in two molecular conformations: **11-*cis*-retinal** and **all-*trans*-retinal**[27]. 11-*cis*-retinal rapidly changes conformation to all-*trans*-retinal upon the absorption of a photon of appropriate wavelength (step 1 in Figure 17.10). This process, called **photoisomerization**, initiates a series of conformational changes in opsin. One of the intermediates (called metaopsins) activates a specific G-protein[28], which starts a downstream signal transduction cascade (step 2). This mechanism produces a high level of signal amplification since:

- one photon captured by one photopigment molecule can activate many G-protein molecules until the intermediate breaks down, and
- each activated G-protein-dependent enzyme produces many second messenger molecules until the enzyme becomes uncoupled from the activating G-protein subunit.

This cascade converts light intensity into a receptor potential (step 3), which in turn alters the firing rate of action potentials in associated sensory neurons (step 4).

Photoreceptor types

There are two major types of photosensitive cells in animals:

- **ciliary photoreceptors**, found in all vertebrates and some invertebrates, and
- **rhabdomeric photoreceptors**, found in all arthropods and many other invertebrate groups.

The two types of photoreceptors differ in their structure and use distinct transduction signalling mechanisms as

[26] Vitamin A is an alcohol, also known as all-*trans*-retinol or retinol. A slightly different form of vitamin A, termed vitamin A_2 exists in freshwater fishes and in freshwater stages of amphibians.

[27] The all-*trans*-retinal molecule is the aldehyde (–CHO) of vitamin A. The aldehyde of vitamin A_2 is termed retinal$_2$. We consider in our discussion only the derivatives of vitamin A.

[28] We discuss signalling via G-proteins in Chapter 19.

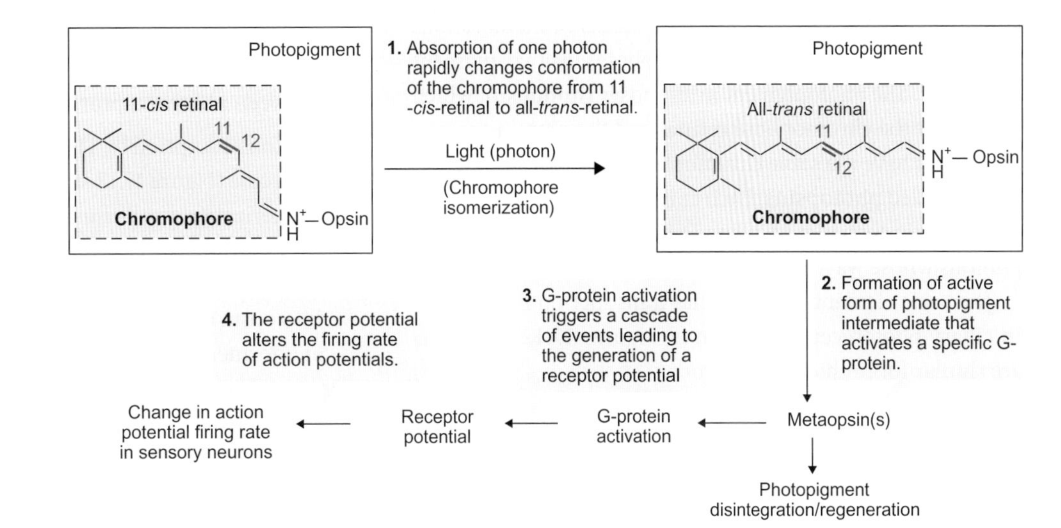

Figure 17.10 The general principle of signal transduction in photoreceptors

Animal opsins are seven transmembrane proteins that are coupled to G-proteins.

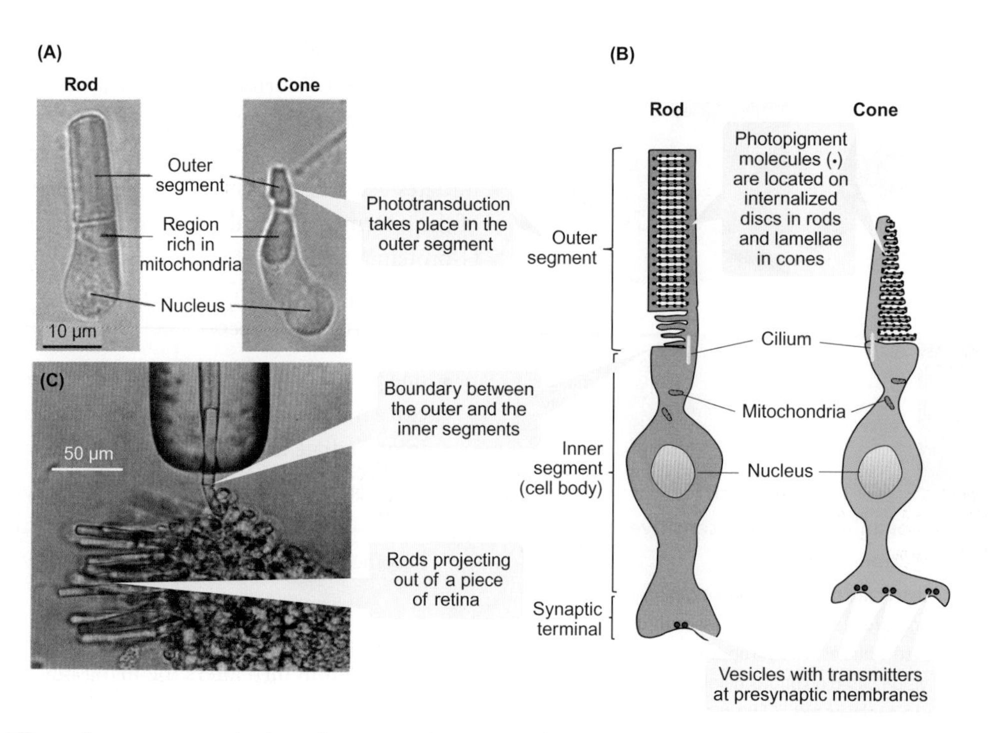

Figure 17.11 Ciliary photoreceptors (rods and cones) in the retina of vertebrates

(A) Photomicrographs of intact rod and cone photoreceptors isolated from the retina of tiger salamander (*Ambystoma tigrinum*). (B) Diagram of one rod and one cone showing the distribution of photopigment molecules (rhodopsin in rods and cone visual pigments in cones) on stacks of membranous structures in the outer segments of the cells (freely floating discs in rods and lamellae in cones). The outer segments are connected to the cell body (inner segment) by one cilium. (C) Suction glass electrode with outer segment of one rod in position for making current measurements across the inner segment membrane.

Source: A: Yiannis Koutalos, Medical University of South Carolina; C: Baylor DA, Lamb TD, Yau K-W (1979). The membrane current of single rod outer segments. Journal of Physiology 288: 589-611.

described below. While all vertebrates and arthropods have only one type of photoreceptor—ciliary and rhabdomeric photoreceptors, respectively—the two types of photoreceptors coexist in several bilaterian groups, including platyhelminthes (flat worms), polychaetes (bristleworms), tunicates (sea squirts) and molluscs.

Ciliary receptors (Figure 17.11) have a narrow cilium connecting the **outer segment**, which contains the light sensitive structures, to the main body of the cell (**inner segment**). The outer segments in vertebrates are in the form of either **rods** or **cones** (Figure 17.11B); these contain stacks of membranous structures with a high density of photosensitive pigment molecules. The photopigment in all rods is **rhodopsin**, which consists of the chromophore **retinal** and the protein **scotopsin**. Cones have the same chromophore, retinal, but different opsins, called **photopsins**. Each cone expresses only one type of photopsin such that each cone photopigment type absorbs photons whose wavelengths fall into narrow bands.

The basic elements of the signal transduction mechanism in ciliary photoreceptors shown in Figures 17.12 and 17.13 are similar for both cones and rods:

- The plasma membrane of the outer segments contains non-selective cationic channels that open when cyclic-GMP (cGMP) binds to them (Figure 17.12A). The opening of these ligand-gated cGMP channels causes membrane depolarization by the same mechanism, as described for excitatory chemical synapses in Section 16.3.3.
- When ciliary photoreceptors are in the dark (Figure 17.13A), the intracellular concentration of cGMP is relatively high (about several μmol L^{-1}); at this concentration some cGMP-gated channels are open and the membrane is depolarized (less negative inside, about −40 mV compared with about −60 to −70 mV in neurons). The depolarization causes continuous release of transmitter molecules (glutamate) at synapses between photoreceptor cells and sensory neurons.
- Photon absorption by 11-*cis*-retinal in the photopigment molecule (Figure 17.12 step 1) induces chromophore isomerization to all-*trans*-retinal and initiates a series of conformational changes in opsin, producing a series of intermediates before the photopigment dissociates into opsin and all-*trans*-retinal. One of the intermediates formed during this process, metaopsin II, (step 2) activates a G-protein called **transducin** (step 3); this, in turn, activates an enzyme (phosphodiesterase) (step 4) that hydrolyses cGMP to GMP, causing a light-intensity dependent decrease in cGMP concentration (step 5).

17

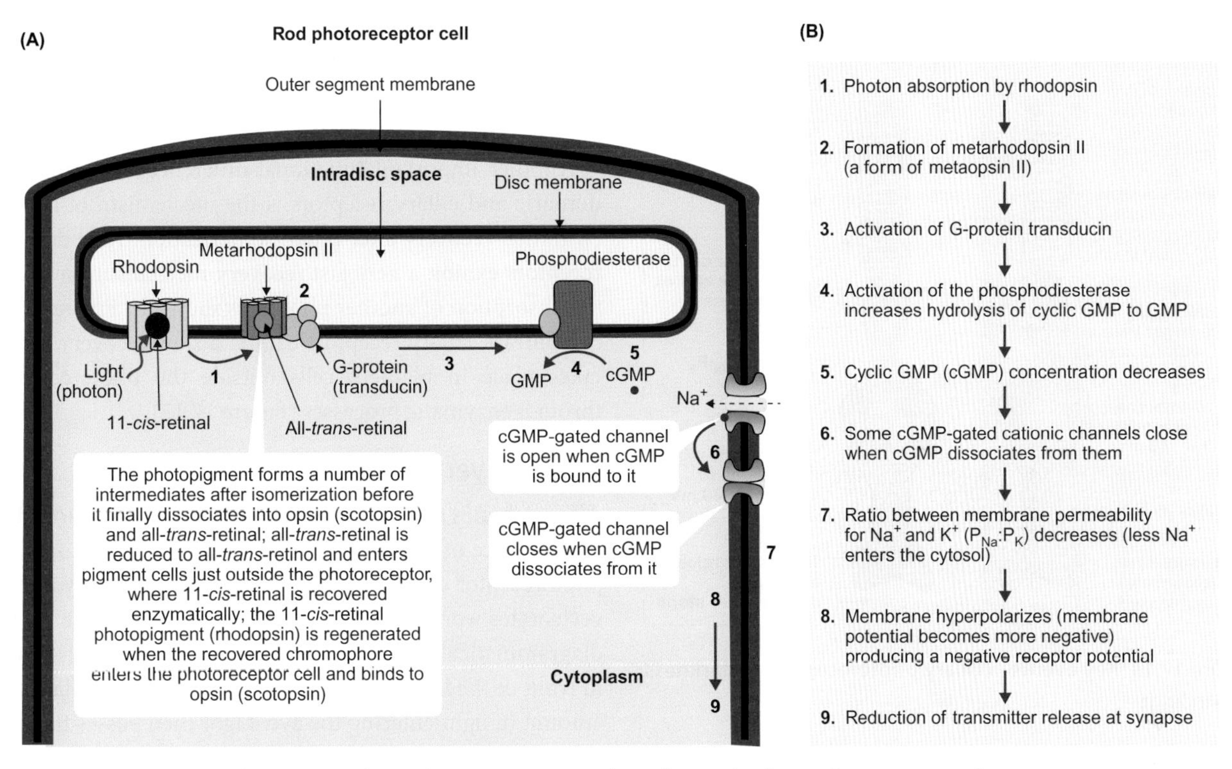

Figure 17.12 Sequence of events in ciliary photoreceptors (rods and cones) using rods as an example

(A) Diagram of key structures involved in phototransduction in ciliary receptors. The membrane of the outer segment contains photopigments (rhodopsin; metarhodopsin II is the intermediary form that activates the G-protein), G-protein (transducin), an enzyme (phosphodiesterase) that hydrolyses cGMP to GMP and non-selective cGMP-gated cationic channels, which cause membrane depolarization when open and hyperpolarization when closed. Na^+ is the main cation which passes through the cGMP-gated cationic channels when open. (B) Sequence of events in ciliary photoreceptor phototransduction.

- The number of cGMP-gated channels that are open in the membrane decreases (step 6) as the light intensity increases, reducing the overall membrane permeability ratio for sodium and potassium ions (P_{Na}:P_K) (step 7) and making the membrane more negative inside (hyperpolarization) (Figure 17.12, step 8, and Figure 17.13B). Thus, the receptor potential in ciliary receptors is a graded hyperpolarization, depending on the light intensity as shown in Figure 17.14A.
- The more negative membrane potential at the presynaptic membrane reduces the amount of glutamate released at the synapse with secondary neurons (Figures 17.12 (step 9) and 17.13B). The reduction of glutamate release with light causes depolarization of some secondary neurons and hyperpolarization of others due to different types of glutamate receptors in the postsynaptic membrane[29]. In turn, these secondary neurons make synapses with other neurons, which fire action potentials at frequencies that are proportional to the receptor potential of ciliary photoreceptors.

[29] We discuss the different types of glutamate receptors in Section 16.3.4.

- Photopigment recovery in ciliary photoreceptors is a slow process. The isomerization of 11-*cis*-retinal to all-*trans*-retinal in ciliary photoreceptors ultimately leads to its disintegration into all-*trans*-retinal chromophore and opsin. The all-*trans*-retinal by itself is a toxic compound, which is reduced in the photoreceptor cells to all-*trans*-retinol (vitamin A) and is then transported to pigment cells just outside the rods and cones (Figure 17.16B), where the 11-*cis*-retinal is enzymatically recovered. The 11-*cis*-retinal photopigment is regenerated when the recovered 11-*cis*-retinal enters the ciliary receptors and binds to opsin.

Although the mechanism of phototransduction is the same in both cones and rods, in principle, there are some important functional differences between these photoreceptor cells. For example, rods are about 100 times more sensitive to a single photon than cones, but respond more slowly. The much higher sensitivity to photons of rhodopsin than cone pigments is due to the higher activation efficiency of transducin by rhodopsin than by cone pigments, which results in a greater amplification factor of the signal in rods than in cones.

Rhabdomeric photoreceptors shown in Figure 17.15A have a highly folded apical surface, which forms microvillar

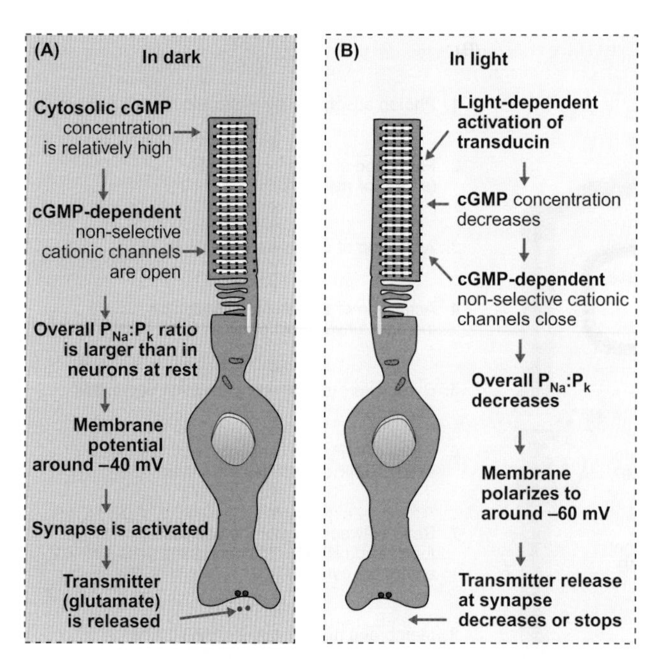

Figure 17.13 Key events in ciliary photoreceptors in darkness and in light

projections that contain the photopigment molecules. The transduction mechanism, shown diagrammatically in Figure 17.15B and C, involves the activation of phospholipase C (PLC) to produce diacylglycerol (DAG) and inositol trisphosphate (IP_3). DAG is further hydrolysed to polyunsaturated fatty acids (PUFA), which open TRP non-selective cation channels, causing depolarization of the membrane.

The breakdown of phospholipids in the lipid bilayer under the action of PLC may also contribute to the opening of the TRP channels. Thus, the rhabdomeric receptors respond to light with a depolarizing receptor potential as shown in Figure 17.14B. The level of depolarization modulates the frequency of action potentials generated, which is ultimately perceived by an animal as light of a certain intensity.

Another important functional difference between rhabdomeric and ciliary receptors is the mechanism of the 11-*cis* retinal recovery after photoisomerization. Rhabdomeric receptors exhibit a very fast recovery mechanism involving the absorption of another photon as shown in Figure 17.15; by contrast, ciliary receptors exhibit a slow recovery, which takes place in pigment cells outside the photoreceptors through enzymatic processes (Figure 17.12).

Vision of vertebrates

All vertebrates have **vesicular camera eyes** characterized by the presence of a lens which focuses the image onto ciliary photoreceptor cells in the retina. In principle, the eyes of all vertebrates work in a similar way to the human eye shown diagrammatically in Figure 17.16A. The **iris**, which is placed in front of the lens and gives eyes their colour, regulates the amount of light entering the eye. Light then passes through the lens, which changes curvature under the action of ciliary muscles, to focus the image onto the back of the eye, which is covered by retina.

In vertebrates, the retina is **inverted**, meaning that the photoreceptor cells are not located on the side facing the incoming light, but at the base of the retina after light passes through several layers of transparent neurons (Figure 17.16B). This arrangement is favourable for supplying the highly active photoreceptor cells with nutrients and oxygen from a network of capillaries situated immediately outside the photoreceptor cells.

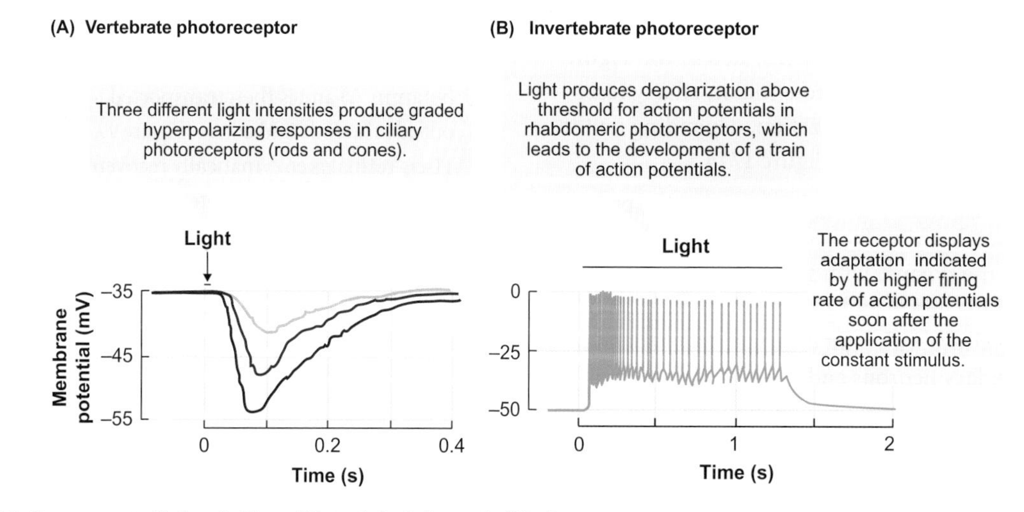

Figure 17.14 Response to light of ciliary (A) and rhabdomeric (B) photoreceptors

(A) Intracellular membrane potential recordings from a red-eared slider (terrapin, *Trachemys scripta elegans*) cone photoreceptor in response to a short pulse of light of three different intensities. (B) Intracellular membrane potential recording from a rhabdomeric photoreceptor of a horseshoe crab (*Limulus* spp.) exposed to a constant light intensity stimulus.

Sources: A: Baylor DA et al (1971). Receptive fields of cones in the retina of the turtle. Journal of Physiology 214: 265-294. B: Fuortes MGF (1959). Initiation of impulses in visual cells of *Limulus*. Journal of Physiology 148: 14-28.

17

Figure 17.15 Sequence of events in rhabdomeric photoreceptor cells of arthropods and other invertebrates

(A) General structure of fruit fly (*Drosophila* spp.) photoreceptors. (B) Key elements in a microvillus involved in phototransduction. (C) Sequence of events in rhabdomeric phototransduction. Note that the term rhodopsin is used as a generic name to describe all visual pigments in rhabdomeric photoreceptors and that invertebrate rhodopsins differ in structure from the vertebrate rhodopsin.

Source: A: Hardie RC, Raghu P (2001). Visual transduction in *Drosophila*. Nature 413: 186–193.

(A)

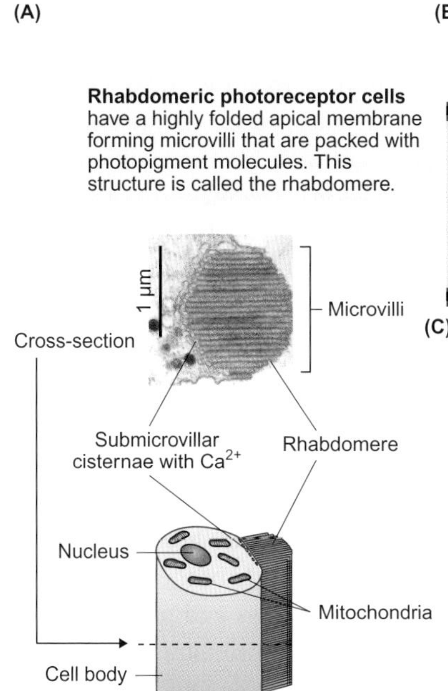

(B)

11-*cis*-retinal is recovered through absorption of another photon
Phospholipase C
Light
Microvillus
1
2
3
4
IP_3 + DAG
5
Inactive rhodopsin with 11-*cis*-retinal
G-protein (G_q)
PUFA
Active rhodopsin with all-*trans*-retinal
Non-selective cationic TRP channels
Na^+
6
8
7

(C)

1. Photon absorption by inactive rhodopsin with 11-*cis*-retinal causes formation of stable active rhodopsin with all-*trans*-retinal
2. Activation of G-protein (G_q)
3. Activation of phospholipase C (PLC)
4. Production of Inositol trisphosphate (IP_3) and diacylglycerol (DAG)
5. DAG lipase produces polyunsaturated fatty acids (PUFAs) from DAG
6. PUFAs activate non-selective cationic TRP channels
7. Ratio between membrane permeability for Na^+ and K^+ (P_{Na}:P_K) increases causing depolarization
8. Membrane depolarization above threshold at spike-initiating zone triggers trains of action potentials

The retina is a multilayered structure of neurons containing ciliary photoreceptors (rods and cones) which make chemical synapses with secondary neurons (**bipolar neurons** and **horizontal cells**), as shown in Figure 17.16B. In turn, the bipolar neurons make synapses with two other types of neurons in the retina: the **amacrine** and **ganglion cells**, also shown in Figure 17.16B. Signal transmission is generally from photoreceptors through to ganglion cells via bipolar neurons. The horizontal and amacrine cells provide lateral connections between photoreceptors, bipolar neurons and ganglion cells, suggesting that significant signal processing occurs within the retina before signals are transmitted to the brain via the optic nerve. Action potentials are only generated in the axons of ganglion cells.

The axons of the ganglion cells join together to form the optic nerve, which exits the eye at a point known as the **optic disc**, where there are no photoreceptors. This means that the optic disc produces a blind spot, as no signals can be transmitted to the brain from the part of the image that forms in the optic disc area.

Figure 17.16B also shows the layer of pigmented cells containing melanin at the base of the retina where 11-*cis* retinal recovery takes place. This pigmented epithelium, together with the **choroid**[30], helps improve vision by minimizing uncontrolled light reflection within the eye. In many crepuscular

[30] The choroid in humans and most primates contains the darkly coloured pigment melanin.

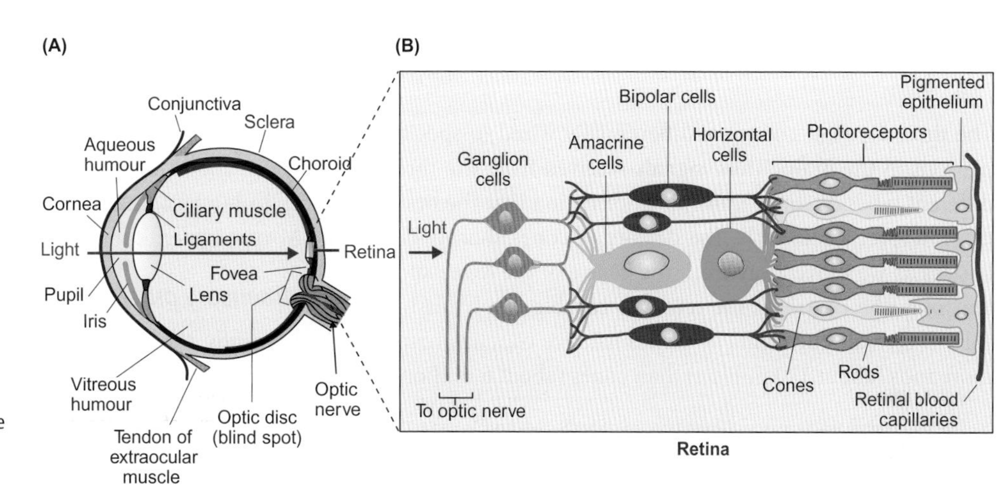

Figure 17.16 Structure of the mammalian vesicular eye

(A) General structure. (B) The multilayered cellular structure of the 'inverted retina' on an expanded scale, showing that the photoreceptors (i) are placed behind several layers of neurons and (ii) have their photosensitive structure facing away from the direction of light.

and nocturnal vertebrates, however, a section of the choroid is devoid of melanin and contains a layer of a highly reflective tissue called the **tapetum lucidum** (from the Latin meaning bright tapestry) that enhances the sensitivity of the eyes to see in the dark. This structure acts as a concave mirror, reflecting back onto the retina in a controlled fashion all of the light that was not absorbed by the photoreceptors when the light first fell on them. The tapetum lucidum is responsible for eyes shining in the dark as is demonstrated so strikingly by cats' and alligators' eyes.

Birds have the largest eyes relative to body mass among vertebrates; the cross-sectional area of their optic nerve is greater than that of the spinal cord, emphasizing the pre-eminent importance of vision in birds. Ostriches (*Struthio camelus*) in particular have the largest eyes of any land vertebrate, which evolved to support vision at great distances in open habitats. This is an important evolutionary adaptation used by ostrich to locate predators.

A small central area of the retina called **fovea centralis** (from Latin meaning 'central pit'), or simply fovea, shown in Figure 17.16A, contains the highest density of cones packed closely together, which allows for maximum sharpness (acuity) of vision. In humans, the fovea (which is clinically called the **macula**) has an area of about 1 mm^2 and contains three types of cones, each type having a different photopsin. The different photopsins confer different spectral properties to the three types of cones, forming the basis for the **trichromatic colour vision** of humans.

Colour vision of vertebrates

The three types of cones in the human fovea (macula) are called S-cones, M-cones and L-cones:

- **S-cones** are sensitive to shorter wavelengths in the visual spectrum with peaks in the blue range (around 420–440 nm)
- **M-cones** are sensitive to medium wavelengths with peaks in the green range (around 535–555 nm)
- **L-cones** are sensitive to longer wavelengths with peaks approaching the red range (around 565–580 nm).

The three types of cones are differentially wired to specific neurons in the brain. Thus, signals received from the three different types of cones transmit information that the brain can interpret as colour[31]. The density of cones sharply decreases as we move away from the centre of the fovea such that only the centre of the retina is capable of processing colour. Overall, the retina of the human eye contains many more rods (about 120 million) than cones (about 6 million).

Most vertebrates perceive colour in a different way from humans. Teleost fish, reptiles, birds and some amphibians have at least four distinct cone populations, each with different spectral characteristics, such that they can discern light of at least four different wavelengths (**tetrachromatic vision**). The presence of a cone population that is sensitive to UV radiation explains why the visual spectrum of most diurnal birds, some fish and reptiles includes UV light. The complexity of colours and shades in tropical fish and some birds indicate that these animals can detect subtle shades of colour, which are likely used for signalling purposes.

Most mammals and some amphibians, however, express only two types of photopsins in their cones; they therefore have **dichromatic vision**. Such vision also occurs in people who have red–green colour blindness because they lack either the M- or the L-cones and hence cannot discriminate reds, yellows and greens from one another. Among mammals, the primates closely related to humans—known as Catarrhini, or Old World monkeys and apes[32]—and some marsupials like fat-tailed dunnarts (*Sminthopsis carassicaudata*) have trichromatic colour vision.

All marine mammals studied so far, including whales, dolphins, porpoises, seals, sea lions and walruses, lack S-opsin and have only one type of functional cone (L-cones). In bright light these animals therefore have **monochromatic vision**. By contrast, in dim light, when both the rods and the cones are active, there is the possibility that some interaction between these two types of photoreceptors takes place to allow some form of colour discrimination. At low levels of illumination when only the rods are excited, humans too have monochromatic vision.

Phylogenetic studies have shown that the visual pigments in vertebrates evolved from four types of cone opsins—LWS (long-wave sensitive), SWS1 (short-wave sensitive 1), SWS2 (short-wave sensitive 2) and Rh2 (rhodopsin-like)—which were expressed in the earliest vertebrates as shown in Figure 17.17. These four types of opsins diverged into four cone opsin families by gene duplication. The vertebrate rod opsin (rhodopsin, Rh1) family diverged last from Rh2. The Rh2 opsin is extinct in all mammals and the SWS2 opsin is only present in monotremes (e.g. platypus). This knowledge helps explain why teleost fish, reptiles, birds and some amphibians have at least tetrachromatic vision, while most mammals have dichromatic vision. In humans, the LWS opsin diverged into the L- (red) and M-type (green) cone opsins, enabling trichromatic vison.

Visual sensory processing

Until recently, the 'inverted retina' of the vertebrate eye was considered incompatible with the evolution of the high spatial resolution vision of vertebrates. However, it now appears that

[31] The spectral characteristics of each type of cone (S-, M- and L-cones) may differ between individuals. This is responsible, at least in part, for the different way in which individuals perceive a particular colour.

[32] Old World monkeys and apes are natives of Africa and Asia.

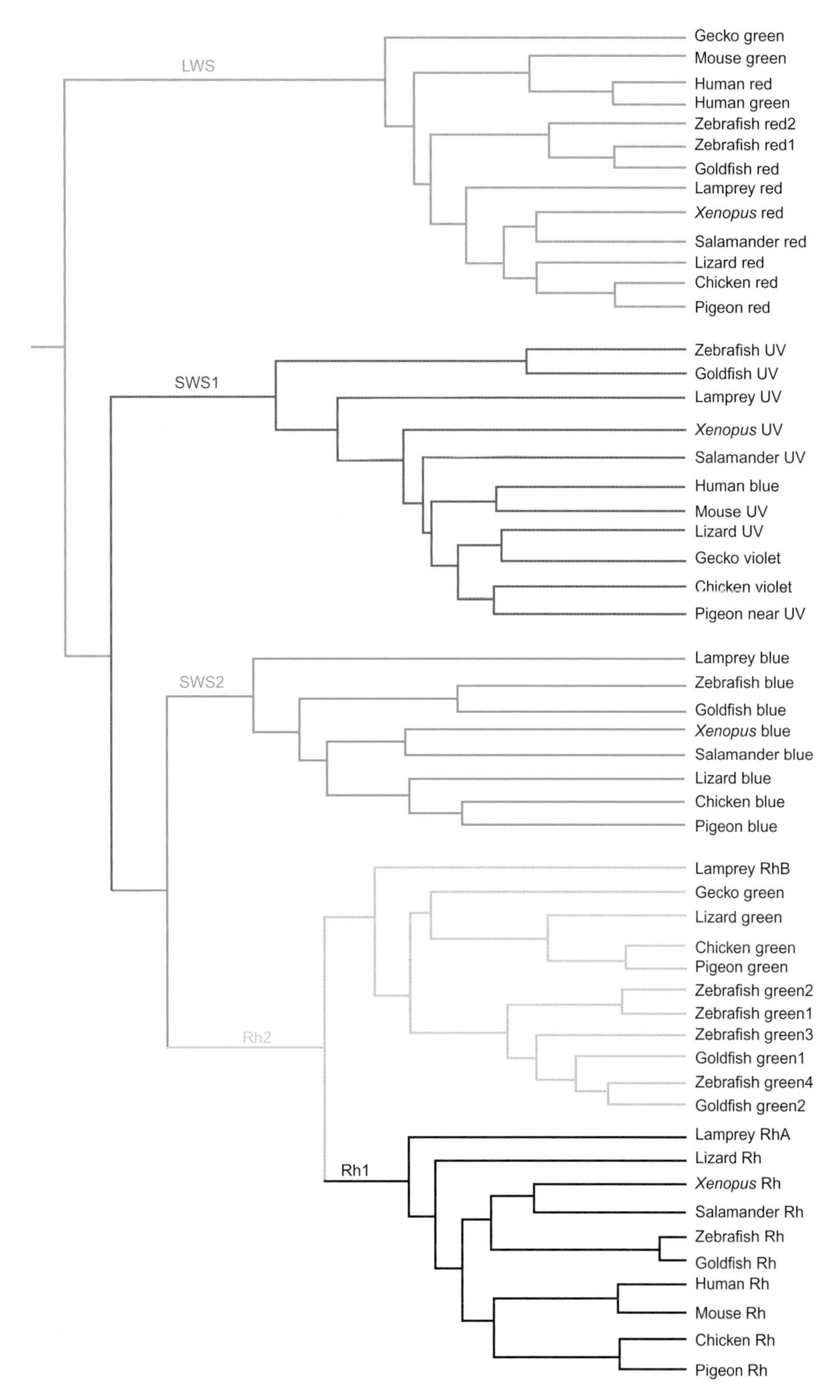

Figure 17.17 Phylogeny of the four families of cone opsins (LWS, SWS1, SWS2 and R2 (rhodopsin-like)) and one family of rod opsin (Rh1) in vertebrates

Notice that zebrafish have eight cone opsins; this allows them to detect subtle shades of colour. Also note that this phylogenetic tree is only based on differences in the visual opsin genes and therefore, does not necessarily reflect the true phylogenetic relationship of organisms.

Source: Shichida Y, Matsuyama T (2009). Evolution of opsins and phototransduction. Philosophical Transactions of the Royal Society 364: 2881–2895.

17

the occurrence of an 'inverted retina' in earliest vertebrates enabled the evolution of a transparent network of neurons in the protected space between the lens and the retina, where visual information could be processed close to the photoreceptive surfaces such as to increase the spatial resolution of the eye without requiring extra space and addition of extra weight. Thus, the posterior location of ciliary receptors with respect to the surface of the retina in vertebrates was most likely an important driving (rather than restricting) factor in the evolution of the high resolution, image-forming vertebrate eyes.

After retinal processing, the visual signals travel from the eyes to the brain via the optic nerve. The optic nerve fibres originate in the retinal ganglion cells and, in mammals, end in the thalamus[33] as shown in Figure 17.18. In primates and other mammals with frontally located eyes—including cats, bats and some marsupials—the left side of the retina in each eye receives images from the right part of the visual field and *vice versa*. Furthermore, the left side of the retina in each eye projects to the left lateral geniculate nucleus in the thalamus and the right side of the retina in each eye projects to the right lateral geniculate nucleus, as illustrated in Figure 17.18. From the geniculate nuclei, the visual signals travel to a region of the brain known as the **primary visual cortex**, which lies in the occipital lobes.

Vertebrates with laterally placed eyes and forelimbs, such as most amphibians and birds, and some mammals (e.g. rabbits and mice), also process visual information from the left visual field in the right cerebral hemisphere and from the right visual field in the left hemisphere. The left eye in these vertebrates covers the left visual field and the right eye the right visual field. Most nerve fibres from the retina in one eye cross over at the optic chiasm to the other hemisphere. This pattern of processing can be regarded as an evolutionary adaptation for optimizing eye–forelimb (hand) coordination: the motor and somato-sensory areas associated with limb movements in vertebrates are largely located in the cerebral hemisphere on the opposite side of the body to the limb involved. Consequently, this arrangement reduces the time lag for the integration of visual, motor and somato-sensory information and the number of connections between the cerebral hemispheres. Note that the cerebral hemispheres in limbless vertebrates, such as snakes and fish, which have no need for eye–forelimb coordination, receive visual information in various proportions, from both the left and right visual fields.

[33] We discuss the divisions of the vertebrate brain in Section 16.1.2.

Vertebrates with lateral eyes have a forward-facing horizontal field of view approaching 360°. This compares with a horizontal field of view of only about 180° in vertebrates with frontally positioned eyes. However, the binocular field of vision, which permits depth perception, is much greater in vertebrates with frontal eyes (about 120°) than in vertebrates with lateral eyes (about 0–20°).

The pathways of the visual signals through the brain from the primary visual cortex are complex and not fully understood. It is known, however, that signals received by the brain via the optic nerve are processed in a parallel fashion in different parts of the brain. Although we do not know exactly how this happens, it has been established that different parts of the brain process different aspects of the image—for example, resolving colour, determining shape and detecting movement. In Case Study 17.1 we discuss how it is possible to detect which specific parts of the brain are involved in processing various types of visual information.

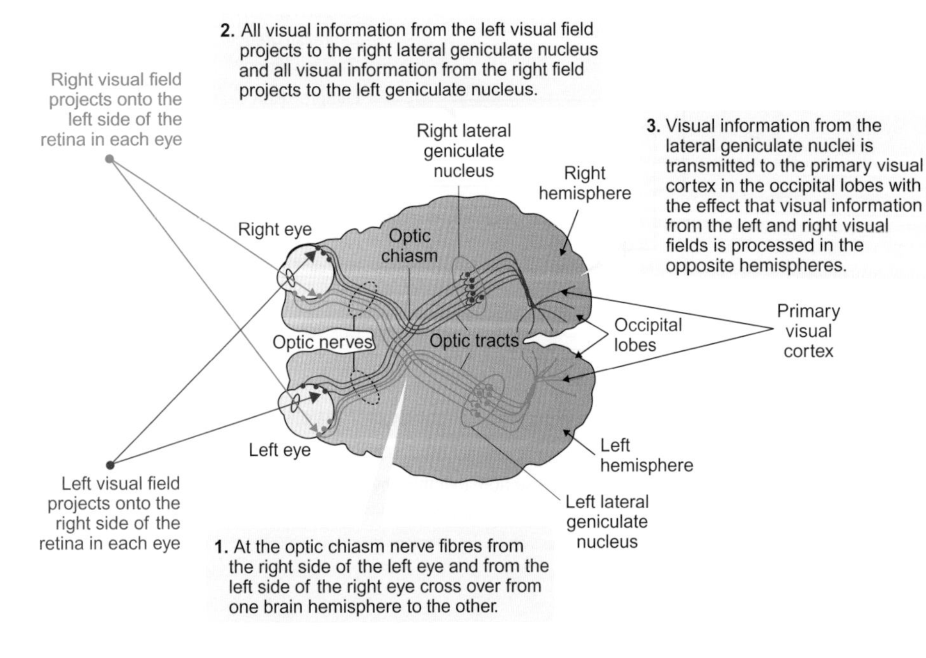

Figure 17.18 Visual pathways in the primate brain

Visual information is transmitted from the retinal ganglion cells via their axons, which form the optic nerves, to the lateral geniculate nuclei in the thalamus. The two large bundles of axons connecting the eyes with the brain are called optic nerves between the eyes and the optic chiasm and optic tracts beyond the chiasm.

Case Study 17.1 Functional magnetic resonance imaging: an important tool for measuring activity within the vertebrate brain

Functional magnetic resonance imaging, or fMRI, is a neuroimaging procedure that uses magnetic resonance imaging (MRI) technology to map neural activity within the vertebrate brain by imaging the change in blood flow associated with energy use by the brain cells.

MRI technology uses a strong magnetic field and radio waves to generate detailed 2- and 3D images of tissues within the body. A powerful magnet is used first to align the hydrogen nuclei in the water molecules in the body with the magnetic field. Radio waves are then pulsed, causing the nuclei to change alignment with respect to the magnetic field. When the hydrogen nuclei return to their previous alignment after each pulse, they generate weak radio signals with distinct characteristics dependent on their surroundings, which provide a means of discrimination between various structures in the body.

Glucose is the major source of energy in neurons[1]. The brain cannot store large amounts of glucose, so the delivery of glucose by the blood circulation is finely tuned to the local needs of neurons. This means that the rate of glucose utilization rises soon after specific regions of the brain increase their level of activity (synaptic transmission, action potential generation) and the flow of oxygenated blood carrying glucose to these regions also increases. In this process, well-oxygenated blood, containing a high concentration of oxyhaemoglobin, replaces blood that is partly depleted of oxygen and which therefore has a higher concentration of deoxyhaemoglobin. The net effect is a decrease in deoxyhaemoglobin in the blood vessels located in the vicinity of the regions of the brain in which an increase in neuron activity takes place.

fMRI technology can detect the decrease in deoxyhaemoglobin in the blood vessels associated with brain areas of increased activity because deoxyhaemoglobin and oxyhaemoglobin have different magnetic properties. The signal produced can measure increased neural activity in a volume of ~1 mm^3 with a time resolution of ~1 s and is mapped using a colour scale as shown in Figure A where yellow refers to areas of highest neural activity and light blue of areas of lowest neural activity.

As shown in Figure A, fMRI was used to demonstrate that the visual system in humans is not strictly organized in a modular fashion, with completely distinct areas for faces and other categories of visual stimuli, as it was previously believed. Instead, the fMRI studies have evidenced that the regions of the visual system activated by different visual stimuli overlap extensively.

The fMRI technique has become an important tool for investigating brain function not only in humans, but also in other vertebrates, as it presents several major advantages compared with other neurophysiological procedures:

- it is safe for the subject (does not involve ionizing radiation and it is not invasive);
- it has a good spatial and temporal resolution for all parts of the brain;
- it is relatively simple to use, as activation maps are produced by computer analyses at the press of a button.

Indeed, task-based fMRI has provided new insight in all areas of brain research by highlighting the specific areas in the brain in which various types of signals are processed.

Nevertheless, it is important to know that fMRI is prone to artefacts because it is an indirect measurement of neural activity

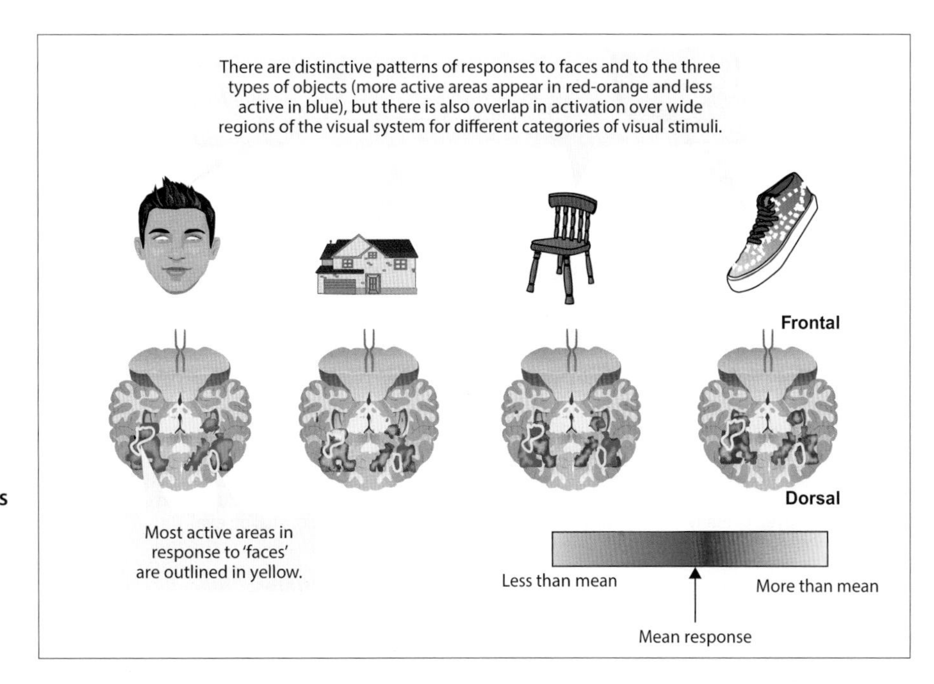

Figure A Functional MRI activation maps of a section through the ventral temporal cortex of an individual presented with four types of visual stimuli

Source: Haxby JV, Gobbini MI (2012). Distributed Neural Systems for Face Perception, in Oxford Handbook of Face Perception, Eds Rhodes G et al. Oxford: Oxford University Press.

17

and the signal is extracted from the background noise using complex computer analyses.

› Find out more

Haxby JV, Gobbini MI, Furey ML, Ishai A, Schouten JL, Pietrini P (2001). Distributed and overlapping representations of faces and objects in ventral temporal cortex. Science 293: 2425–2430.

Voss HU, Tabelow K, Polzehl J, Tchernichovski O, Maul KK, Salgado-Commissariat D, Ballon D, Helekar SA (2007). Functional MRI of the zebra finch brain during song stimulation suggests a lateralized response topography. Proceedings of the National Academy of Science U S A 104: 10667–10672.

[1] We discuss the mechanisms of energy production in cells in Chapter 2.

After all of the basic visual information is assembled, it is again processed by the brain, probably by comparing the gathered information with a database of reference information which allows the brain to interpret what was actually observed.

Invertebrate visual systems

General patterns of eye organization in invertebrates are shown in Figure 17.19. Cephalopods (squids, octopuses, cuttlefish) have vesicular eyes with spherical lenses (Figure 17.19A), which are similar to those seen in fish, while spiders have vesicular eyes with **cornea** optics (Figure 17.19B).

There are several important differences between the morphology and physiology of the vertebrate eye and the vesicular invertebrate eye. For example, the vesicular invertebrate eyes:

- have rhabdomeric photoreceptors instead of the ciliary photoreceptors (rods and cones) of vertebrates[34];
- have the photoreceptors located on the side of the retina facing the incoming light rather than at the back of the retina as in vertebrates (compare Figures 17.19A and 17.16);
- lack secondary neurons and ganglion cells in the retina meaning there is little visual signal processing in the retina of invertebrates.

Nevertheless, the vesicular eyes of cephalopods support sophisticated eyesight, comparable for example with that found in sharks, but less advanced than that of birds and mammals. The octopus brain has very large optic lobes and can process

[34] We consider the different types of photoreceptors in Figures 17.12 and 17.15.

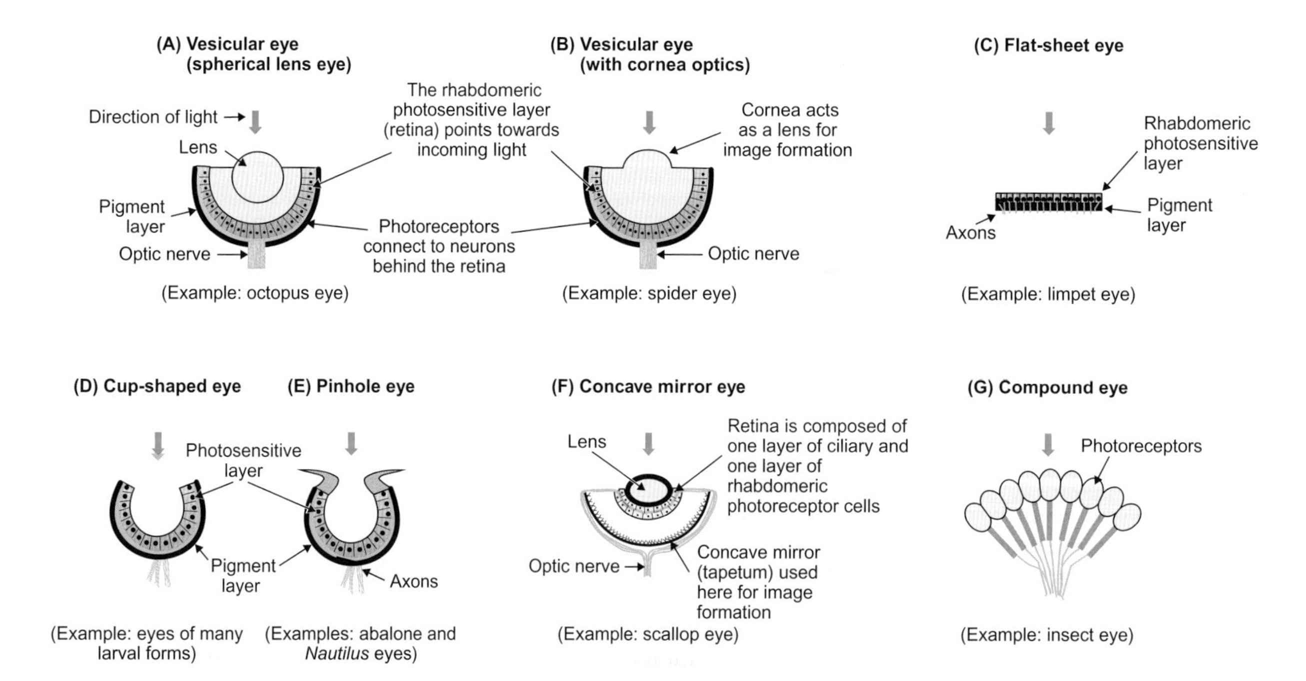

Figure 17.19 Several patterns of eye organization in invertebrates

Unlike the vertebrate eyes which have an 'inverted retina', the photosensitive surfaces in the invertebrate eyes face the incoming light and the photoreceptors make synapses with neurons behind the retina.

complex visual information. For example, it has been shown that common octopuses (*Octopus vulgaris*) have the ability to distinguish size, shape, orientation (vertical or horizontal) and brightness of objects. Another octopus species (*Enteroctopus dofleini*) can even recognize individual humans.

Cuttlefish (*Sepia officinalis*) also express photoreceptor proteins throughout the skin, raising the possibility that cuttlefish can perceive light with their whole body, a feature that may be related to their ability to camouflage themselves. Many species of cephalopods have the ability to match rapidly the specific colour and pattern of their background, as shown in Figure 17.20. This camouflaging ability gives cephalopods considerable protection from predators but does not appear to be related to their eyesight, which is, with few exceptions, monochromatic.

Flat-sheet, **cup-shaped** and **pinhole eyes** shown diagrammatically in Figure 17.19C, D and E, respectively, consist of a layer of rhabdomeric photoreceptor cells lined by a layer of pigmented epithelium. As such, the flat-sheet, cup-shaped and pinhole eyes are simpler structures than the vesicular eyes shown Figure 17.19A, B.

The flat-sheet and **cup-shaped** eyes are common in the larval form of many animal groups and only provide some sense of direction of the light being detected. The pinhole eyes (Figure 17.19E) consist of a folded layer of photoreceptor cells (lined by a pigmented epithelium) with a narrow opening. Light passes through the narrow opening and projects an upside down image on the layer of photoreceptor cells with colour and perspective preserved. However, there is a trade-off between image intensity, which improves as the amount of light that passes through the pinhole increases, and image sharpness, which deteriorates as the pinhole size is increased. Pinhole eyes occur in some gastropods, such as abalone (*Haliotis* spp.), and in some cephalopods such as nautilus (*Nautilus* spp.). The pinhole and vesicular eyes are camera-type image-forming eyes.

Some invertebrates like scallops (*Pecten* spp.) have developed concave eyes (Figure 17.19F) consisting of a reflective tapetum behind the retina, which enhances the sensitivity of the eyes; this layer acts as a concave mirror and focuses the light that passed through it back onto the retina. Scallops have up to 200 single eyes along the edge of the mantle.

Most spiders have four pairs of eyes located on the top-front area of their cephalothorax. The main pair of eyes consists of vesicular eyes with cornea type optics (Figure 17.19B) capable of forming clear images, while the other smaller pairs of secondary eyes have a fast response for detecting low levels of light. Similar secondary pairs of eyes are found in many arthropods, including insects.

Convex eyes, in which photoreceptor cells radiate outward forming a convex photoreceptive surface (Figure 17.19G), occur in arthropods and many annelids and molluscs.

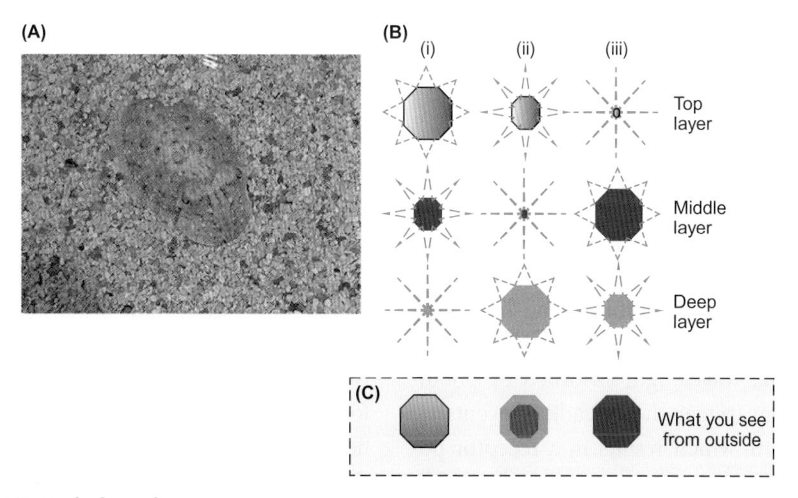

Figure 17.20 Camouflage in cephalopods

(A) Camouflage displayed by an unidentified infant cuttlefish (*Sepia* spp.). Colour change in cephalopods is made possible by the presence in their skin of chromatophores, a type of cell containing sacs filled with pigment granules of specific colours that can change shape. (B) Schematic diagram of three layers of chromatophores of different colours in the skin of cephalopods. Each chromatophore is surrounded by small radial muscles under the control of the central nervous system (CNS) that can change the area of the pigment-filled sac by hundreds of times when the radial muscles contract and the chromatophore is stretched. The three layers of chromatophores act independently. (C) Surface colour corresponding to the three different combinations of chromatophores shown in columns B(i), B(ii) and B(iii). Directly underneath the chromatophores is a layer of cells of a species-specific colouration, and below this layer is a reflecting layer of white cells. Under the direct control of the CNS, the chromatophores change their size and other optical properties until the skin matches the colour and the pattern of the local environment. Source: A: Wikimedia Commons/Raul654.

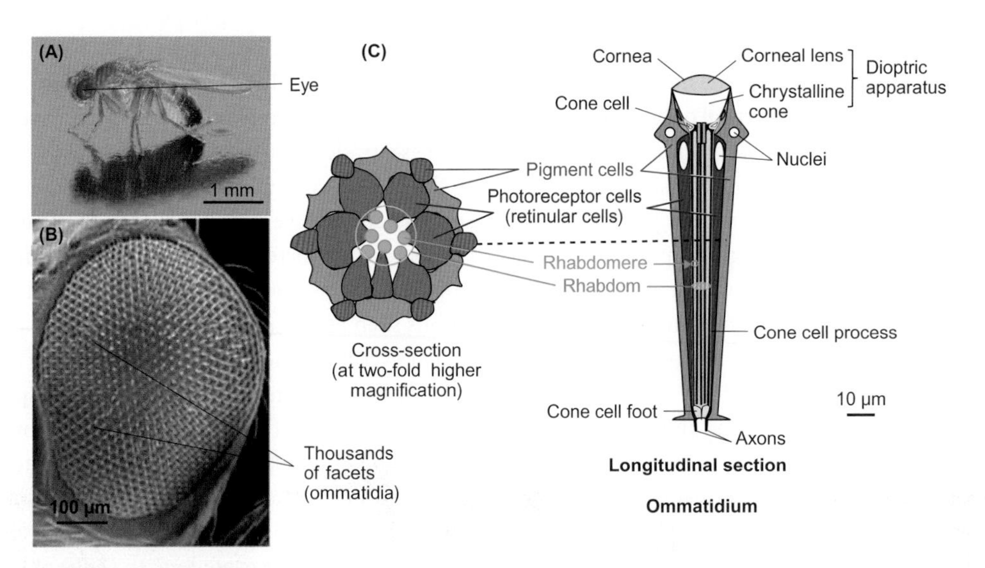

Figure 17.21 Compound eyes of insects

(A) Adult fruit flies (*Drosophila melanogaster*) have two red convex eyes, each made of several thousand hexagonal facets. (B) Scanning electron micrograph showing the facets at higher magnification. Each facet is the upper part (corneal lens) of a core unit called an ommatidium (plural = ommatidia). (C) Diagram of an ommatidium in cross-section and longitudinal section showing that under the corneal lens is the crystalline cone, made by the cone cells that have projections for the entire length of the ommatidium to the cone cell foot. The photoreceptor cells have rhabdomeres facing the core of the ommatidium. Collectively, the rhabdomeres in each ommatidium form the rhabdom. Photons that enter the ommatidium through the dioptric apparatus are captured by the photopigments in the rhabdom causing depolarization of the respective photoreceptor cells, which triggers trains of action potentials in their axons. The pigment cells surrounding the photoreceptor cells capture all photons that do not enter through the dioptric apparatus.

Source: A: André Karwath/Wikimedia at Commons B: de Iongh RU et al (2006). WNT/Frizzled signaling in eye development and disease. Frontiers in Bioscience 11: 2442–2464

Compound eyes

The most complex convex eyes, known as **compound eyes**, are found in arthropods. Figure 17.21B shows that compound eyes normally consist of many individual units called **ommatidia**, which are arranged to form a dome-like structure. Light enters each ommatidium through the **dioptric apparatus**, which consists of a corneal lens and a crystalline cone situated above a ring made out of 7–9 elongated, closely packed rhabdomeric photoreceptor cells called **retinular cells**. The photoreceptor cells are arranged with their photopigment-containing microvilli pointing inward to form a structure called the **rhabdom**.

Photons of a particular wavelength absorbed by photopigments in the rhabdom trigger the cascade of events we examined in Figure 17.15, which results in a receptor potential and generation of trains of action potentials in each retinular cell as shown back in Figure 17.14B. The animal perceives the set of signals from each ommatidium as light of certain intensity and wavelength, depending on the spectral characteristics of the photopigments present in the retinular cells that have absorbed the respective photons.

The photoreceptor cells of many diurnal insects, such as fruit flies (*Drosophila melanogaster*), are surrounded by pigment cells to absorb any light that does not enter the ommatidium (Figure 17.21C). This absorption of light ensures that each ommatidium functions independently of its neighbours. Therefore, the resolving power of the compound eyes increases in direct proportion with the number of ommatidia present, mirroring the way that the quality of a digital image depends on the total number of pixels in the image. Nevertheless, even when the number of ommatidia is very high, as in emperor dragonflies (*Anax imperator*) (Figure 17.22A), which has about 25,000 ommatidia arranged in a compact hexagonal pattern, the resolving power of the compound eye is several orders of magnitude lower than that of a vesicular eye of similar size. This is because the compact eyes capture images from a much wider field compared with globular eyes, which focus on narrow regions of the visual field. Compact eyes are, however, much better at rapidly capturing images from many directions at the same time due to their wider visual field.

Nocturnal insects, such as moths, have also evolved highly reflective structures in their eyes, similar to the tapetum lucidum in vertebrates and some invertebrates, to enhance the sensitivity of their eyes for night vision. Moreover, moths evolved structures on the corneal lens that reduce the amount of light reflected back to the environment. These eyes have inspired improvements in the efficiency of solar cells, which we discuss in Case Study 17.2.

Figure 17.22 Images of invertebrates with remarkable vision systems

(A) The adult emperor dragonfly (*Anax imperator*) is about 80 mm long and has the largest number of ommatidia (about 25,000) in each eye. (B) The mantis shrimps, which grow to about 100 mm or more, have the most complex colour vision system in the animal kingdom; their eyes can move independently of each other and contain up to 16 types of photoreceptors located on four rows of specialised ommatidia in the mid-band region of their eyes. Source: (A) Quartl/Wikimedia Commons; Duncan Iskandar/Wikimedia Commons (B) Professor Roy L. Caldwell, UC Berkeley.

Case Study 17.2 Insect eyes inspire engineers to improve the efficiency of solar cells

Being nocturnal, moths need to direct as much light as possible through the multiple ommatidia of their compound eyes[1] to the photoreceptors within. As many of us know from experiencing dirty spectacles, dust particles collected on the eyes of moths would reduce the amount of light entering the ommatidia. Furthermore, light that enters moths' ommatidia, but is not absorbed by photoreceptors, is normally reflected back towards the photoreceptor cells by a structure similar to the tapetum lucidum in vertebrates[2]. Reducing the light being reflected from the moths' eyes back to the environment would keep more light within the ommatidia to be used for visual purposes and would also help moths to be less visible to potential predators as they fly around at night. So how do moths deal with these issues?

The answer is remarkably simple. Some insects have arrays of protuberances, 50–300 nm in diameter and 25–70 nm tall, called **corneal nipples** or **ommatidial gratings** on the surface of the dioptric apparatus, as shown in Figure A. The presence of such corneal nipples has been shown to confer both anti-reflective and anti-adhesive properties to transparent surfaces, due to their decreased real contact area for potentially contaminating particles.

The characteristics of reduced light reflection, redirected reflection back to the sensor, and reduced dust particle adhesion all enhance light collection by a moth's eyes and are exactly the kinds of improvements needed in solar collectors, including photovoltaic arrays (solar cells).

Figure B shows that investigators in Germany have demonstrated, in the laboratory, a marked increase in the quantum efficiency (defined as the ratio of the power absorbed by the solar cell to the total power of incident light) of silicon thin film solar cells over a broad wavelength band when they applied nanostructures of similar shape and periodicity onto the silicon film

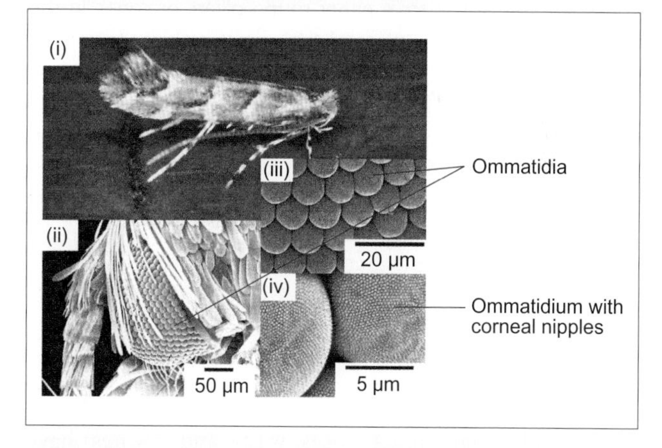

Figure A The compound eye of chestnut leafminer moths (*Cameraria ohridella*)

(i) Image of adult moth (about 3 mm length); (ii) Scanning electron microscope image of the moth compound eye; (iii) higher magnification of the eye showing ommatidia in rows; (iv) still greater magnification of ommatidia revealing corneal nipples over their entire surface.

Source: Dewan R et al (2012). Studying nanostructured nipple arrays of moth eye facets helps to design better thin film solar cells. Bioinspiration and Biomimimetics 7: 016003.

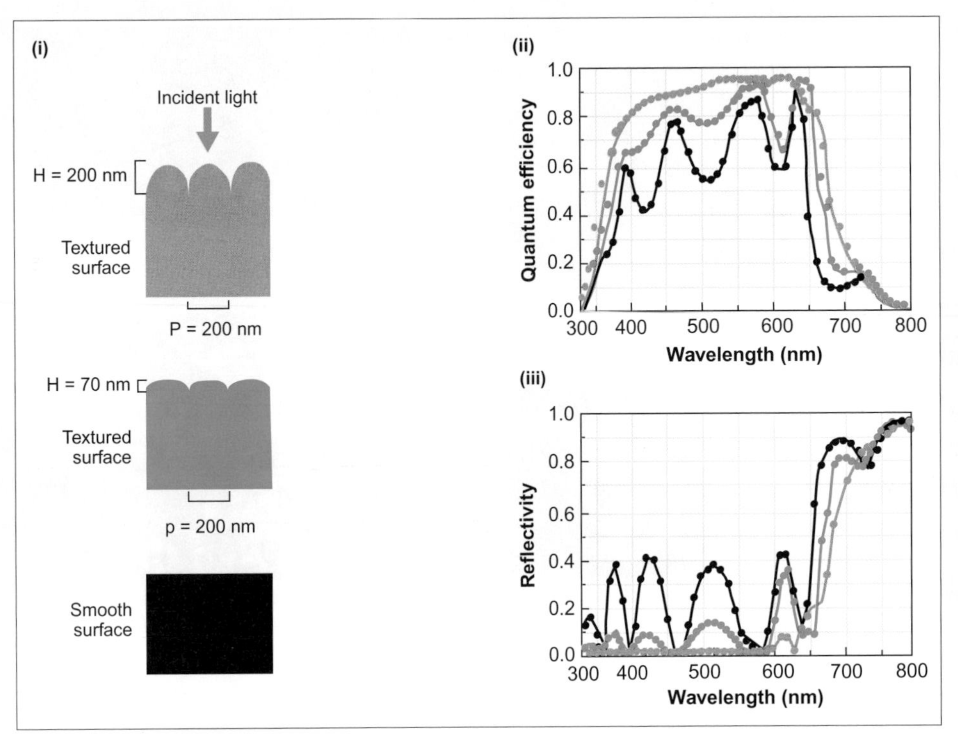

Figure B Characteristics of silicon solar cells with smooth and textured surfaces in all layers of material

(i) Range of experimental surfaces. The textured solar cells have moth-eye-like parabola shaped nipple arrays. The period (P) of the parabolas is 200 nm and the height (H) is either 70 nm or 200 nm. (ii) Data on comparative efficiency, i.e. the ratio of power absorbed by the solar cell to the total incident power of incident light, of the various surfaces, indicated by their respective colour. (iii) Data on comparative reflectivity (the fraction of incident radiation reflected by a surface) of textured and smooth solar cells.

Source: modified from Dewan et al (2012).

as in the moth ommatidia. The anti-reflective properties of such textured materials compared with smooth surfaces were found to be even more impressive, as shown in Figure B(iii). These authors also showed that the parabolic shape of moth corneal nipples was better than either round pillars or cones in the man-made versions, because of their greater ability to reduce reflectivity.

The same concept, based on properties of corneal nipples on moth's ommatidia, has also been applied to reduce reflected light from all kinds of screens, from computers to cell phones, and will have further application in production of optical sensors and light-emitting diodes.

› Find out more

Bernhard CG, Miller WH, Moller AR (1965). The insect corneal nipple array: A biological broad band impedance transformer that acts as an antireflection coating. Acta Physiologica Scandinavica 63 (Supplementum 243): 1–79.

Dewan R, Fischer SV, Meyer-Rochow VB, Özdemir Y, Hamraz S, Knipp D (2012). Studying nanostructured nipple arrays of moth eye facets helps to design better thin film solar cells. Bioinspiration and Biomimimetics 7: 016003.

Peisker H, Gorb SN (2010). Always on the bright side of life: anti-adhesive properties of insect ommatidia gratings. Journal of Experimental Biology 213: 3457–3462.

[1] We discuss insect compound eyes in Section 17.2.1 and Figure 17.21.

[2] We discuss the tapetum lucidum in Section 17.2.1.

17

Colour vision is common in animals with compound eyes. For example, some insects (bees, wasps and sawflies) have trichromatic colour vision which is sensitive to UV light but insensitive to red light (wavelength above 600nm). Butterflies also generally have trichromatic colour vision, but some swallowtail butterflies (*Papilio* spp.) possess six types of photoreceptors and may therefore have more complex colour vision, which enables them to discern the colour pattern on flowers and the wings of conspecifics.

Arguably the most complex colour vision system in animals occurs in mantis shrimps, a group of predatory marine crustaceans. Their compound eyes are mounted on mobile stalks, as shown in Figure 17.22B, and can be moved independently of each other. The eyes carry up to 16 different types of photoreceptors. Twelve of these types cover the entire spectral range from UV to red light and four other types are sensitive to polarized light[35]. This colour vision system is used by mantis shrimps to hunt for prey species and recognize different types of coral, mating partners, as well as predators, but we do not yet understand why such a complex system is needed.

Animal eyes evolved from one common precursor

The comparison of eyes based on anatomy and photoreceptor cell types (ciliary or rhabdomeric) led to the view

[35] Polarized light is a beam of light in which most photons vibrate in one single plane.

that eyes evolved independently multiple times in different groups of animals. However, genetic and molecular comparisons indicate that animal eyes evolved from one common, simple precursor called the **proto-eye**, which consisted of an opsin–Vitamin A-based precursor photoreceptor cell and associated pigment cell to detect the direction of light. This proto-eye is suggested to have evolved in pre-bilaterian animals, where the opsin-based precursor photoreceptor cell would have diversified into two sister opsin-based photoreceptor cell types: ciliary and rhabdomeric. Additional cell types, such as lens cells and support cells, were added later during eye evolution in Bilateria, reaching maximum diversity and complexity in the vertebrate and cephalopod camera eyes, and in the arthropod convex eyes.

17.2.2 The role of photoreception in biological clocks

Photoreception is central not only for animal vision, but also for non-vision processes, such as the functioning of timing mechanisms termed **biological clocks**. Biological clocks are inbuilt, endogenous rhythms that allow organisms to co-ordinate their physiological functions and behaviour with their environments by anticipating regular changes in the day–night cycle or food abundance associated with different seasons.

In order to be useful, biological clocks, just like ordinary clocks, need to be adjustable to the local conditions. The adjustment of the biological clocks is called **entrainment** and is done using external cues called **zeitgebers** (meaning 'time givers' in German), of which day time and day length are the most important. The period of the intrinsic rhythm observed under constant conditions is called the **free-running period**.

Biological clocks, and the photoreceptor proteins capable of sensing light that synchronize biological clocks to the day-time in the external environment, evolved early in unicellular organisms. The replication of nucleic acids is sensitive to damage from UV light. As such, natural selection would have favoured organisms that could anticipate the times of low UV intensity and could replicate their DNA during these times to minimize the risk of DNA damage. It is therefore not surprising that biological clocks occur in many forms of life including cyanobacteria, fungi, plants and animals.

Circadian clocks

Endogenous rhythms that have a free-running period of about 24 h are called **circadian clocks** or circadian rhythms (in Latin 'circa' means 'about' and 'dies' means 'day'). Circadian clocks have great value to an animal in relation to the external environment because they allow the animal to prepare for the regular environmental changes associated with the day–night cycle. There are clear circadian patterns driven by circadian clocks including the sleep–wake cycle,

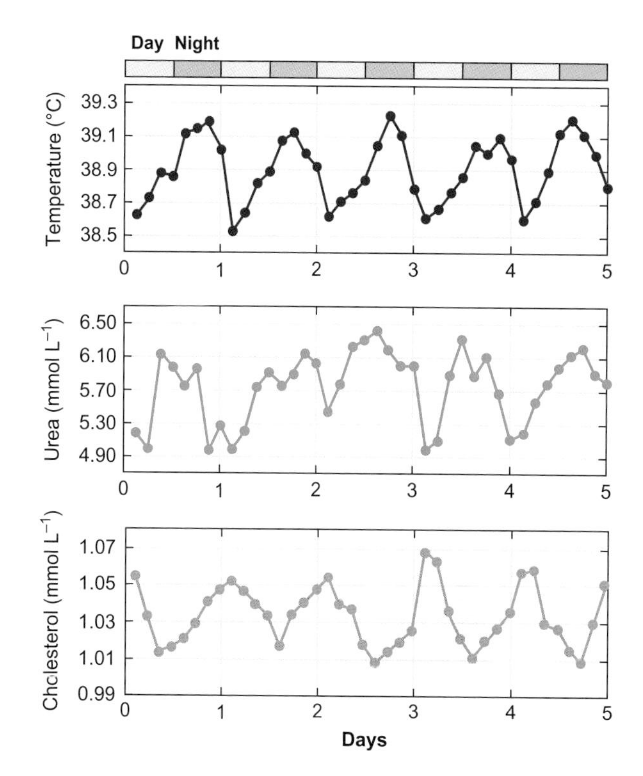

Figure 17.23 Circadian oscillations of three physiological parameters in a female goat (*Capra hircus*)

The horizontal bars at the top indicate the timing of the light–dark cycle. The rectal temperature, plasma urea concentration and plasma cholesterol concentration were measured simultaneously. Notice that the rhythms of rectal temperature and urea concentration have similar phases (peaking in the middle of the night), but the rhythm of cholesterol concentration has the opposite phase (peaking in the middle of the day).

Source: Dr Roberto Refinetti, Boise State University.

changes in core body temperature, brain wave activity, cell regeneration, hormone production[36] and feeding behaviour. Figure 17.23 shows diurnal patterns of body core temperature, plasma urea concentration and plasma cholesterol concentration measured simultaneously in a female goat (*Capra hircus*).

The presence of a master circadian clock in our body can be promptly revealed by travelling across several time zones within a short period of time. Then, we normally experience the condition of 'jet lag' characterized by one or more of fatigue, sleep disturbance, lack of concentration, problems with digestion, and increased irritability and headaches, until our circadian clock becomes synchronized with local time.

In multicellular organisms, different cells synchronize their circadian clocks by communicating with each other, resulting in periodic electrical signals. These electrical signals may interface with endocrine glands in the brain to cause rhythmic release of hormones, which in turn, synchronize the clocks of peripheral cells and organs.

[36] Section 19.2 discusses circadian rhythms in hormone production and release.

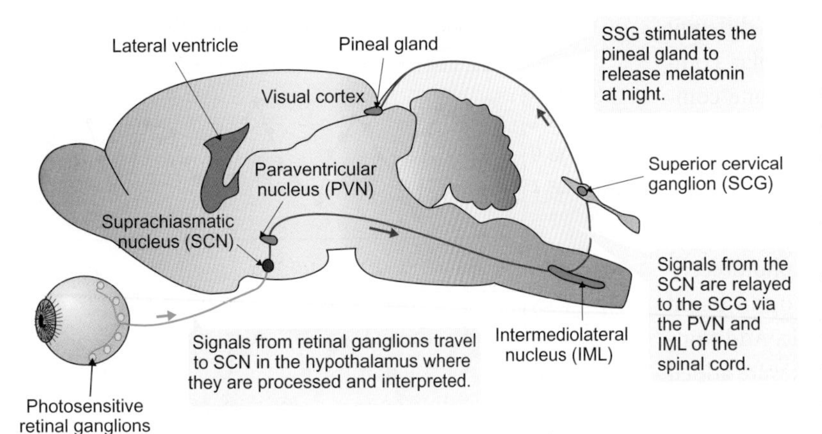

Figure 17.24 The neural pathway that drives circadian rhythms in mammals using the diagram of the mouse brain as an example

Signals from the photosensitive retinal ganglion cells entrain the master circadian clock in the superchiasmatic nuclei. Signals from the master clock are then passed to the pineal gland to release melatonin at night. The presence of melatonin in the blood carries information about the night length and synchronizes all circadian rhythms in the body to the rhythm of the SCN master clock.

Source: Berson DM (2003). Strange vision: ganglion cells as circadian photoreceptors. Trends in Neurosciences 26: 314–320.

The master circadian clock in mammals is located in a pair of neuron groups in the hypothalamus above the **optic chiasm**, the **suprachiasmatic nuclei (SCN)**. These neurons maintain their own rhythm when isolated and kept in cell culture in the absence of external cues. Animals in which the SCN are not functional, lack a regular sleep–wake rhythm.

The SCN receive information about daylight through a group of specialized ganglion cells in the retina of the eye, as shown in Figure 17.24. Unlike the great majority of ganglion cells in the retina, which are connected to the typical photoreceptors (cones and rods), these specialized ganglion cells contain the photopigment **melanopsin**. These ganglion cells make synapses with neurons in the SCN and send signals that entrain the master clock to local conditions. Signals from the master clock are then passed to the **pineal gland**[37], which releases **melatonin** at night. The presence of melatonin in the blood carries information about the night length and synchronizes all circadian rhythms in the body to the rhythm of the SCN master clock. The signaling cascade associated with melanopsin is similar to that shown back in Figure 17.15 for invertebrate photoreceptors.

Since circadian systems are sensitive to the night length, they play a key role in the measurement and interpretation of photoperiod, which is the most predictive environmental cue for the change in seasons at higher latitudes. This information is critical for preparing the animals for the upcoming change in seasons, and determines the timing of seasonal behaviour such as migration, hibernation and reproduction[38].

Even mammals that lack image-forming eyes, such as subterranean Middle East blind mole rats (*Spalax* spp.), maintain a circadian rhythm. In blind mole rats, the melanopsin-containing ganglion cells in their retina detect light when the animals periodically come to the surface. This behaviour ensures the entrainment of their master clock to local conditions.

Birds possess photoreceptors located deep within the hypothalamus that sense light directly and are implicated in synchronizing their circadian master clock to local conditions. The photoreceptors contain an opsin-based photo pigment called the **vertebrate ancient opsin,** which is also found in jawless (Agnatha) and teleost fish. A different type of opsin, called encephalopsin, which may also be involved in entraining the circadian clock of mammals was located in the brain of mice and humans.

Cryptochromes are a highly conserved group of photoreceptor proteins that directly modulate light input into the circadian clocks of various animal groups. The cryptochromes belong not to the opsin family of proteins, but to a different group of evolutionarily older and highly conserved photoreceptor proteins that evolved from light-activated bacterial enzymes involved in the repair of UV-induced DNA damage. Cryptochromes are sensitive to blue light; in *Drosophila* and other insects, the cryptochromes act as the blue-light photoreceptors for the entrainment of the circadian clock to local conditions. In mammals, however, the cryptochromes act as light-independent regulators of circadian clocks[39].

Some animals, such as the monarch butterflies (*Danaus plexippus*), have both a *Drosophila*-like version and a mammal-like version of cryptochromes, suggesting that cryptochromes both directly modulated light input into the ancestral circadian clock and regulated its function.

Circadian systems are also used for navigation. For example, monarch butterflies use a circadian clock in their antennae to determine not only the time of migration to their overwintering grounds in central Mexico, but also to sense direction of movement to their destination over

[37] Section 19.2.3 discusses the functions of the pineal gland in vertebrates.

[38] Migration, hibernation and seasonal breeding are discussed in Sections 10.2.5, 10.2.8 and 20.2.1, respectively.

[39] Cryptochromes are also involved in magnetoreception, as we discuss in Section 17.8.

thousands of kilometres using a sun compass orientation system[40].

The molecular workings of circadian clocks

The first genetic component of a circadian clock was the '*period*' gene called *Per*, which was discovered in fruit flies (*Drosophila melanogaster*)[41]. *Per* genes have been discovered in other species (including mammals) by screening their genome[42] for DNA sequences that were homologous with the *Drosophila Per* gene, thus demonstrating the conserved nature of the molecular circadian clock through evolution.

The molecular circadian clocks consist of several components that interact in a negative feedback loop to generate periodic fluctuations in the concentration of certain molecules, which are interpreted as a specific time of the day by the cell. The simplest molecular circadian clock, discovered in cyanobacteria *Synechococcus elongatus,* consists of only three proteins, called KaiA, KaiB and KaiC, and depends on the rhythmic KaiC phosphorylation. The clock can sustain a 22-h rhythm in the presence of ATP.

In animals, the circadian clock consists of a network of transcriptional–translational[43] feedback loops in which proteins participate in the negative feedback loop that controls their production. The **core clock** consists of a group of four genes/proteins shown in Figure 17.25. Two of the proteins act as transcriptional activators (**activator proteins** or **positive regulators**). The activator proteins form a heterodimer that binds to a specific enhancer sequence on genes that share that sequence and increase production in the cytoplasm of two **repressor proteins** (or **negative regulators**), as well as other proteins, whose genes share that sequence.

The repressor proteins accumulate with delay in the cytoplasm, form a complex, and then enter the nucleus and inhibit the action of the transcriptional activators, thus repressing the transcription of their own genes as well as the transcription of other genes that are activated by the positive regulators. The repressor proteins are gradually proteolysed, relieving the inhibition on the activator proteins. This enables the cycle to restart, thus leading to oscillations in the levels of mRNAs, of repressor proteins, and of the products of the other genes that are activated by the positive regulators as illustrated in Figure 17.25.

The two most intensely studied models of circadian clocks are the mammalian and the insect circadian clocks. In mammals, the activator clock proteins and their genes are CLOCK/*Clock*[44] (Circadian-Locomotor-Output-Cycles-Kaput) and BMAL1/*Bmal1* (Brain and Muscle ARNT-Like-Protein-1). CLOCK and BMAL1 proteins form heterodimers, which act

[40] Case Panel 9.2 discusses the spectacular migration of the monarch butterflies.

[41] Three American scientists: Jeffrey Hall, Michael Rosbash and Michael Young were awarded the 2017 Nobel prize in Physiology or Medicine for their work related to the identification of the *Per* gene in fruit flies.

[42] We discuss genes and genomes in Section 1.3.

[43] We discuss the flow of genetic information from genes to proteins in Section 1.3.1.

[44] By convention gene products appear in capital letters, while the name of a gene is italicized and only starts with a capital letter.

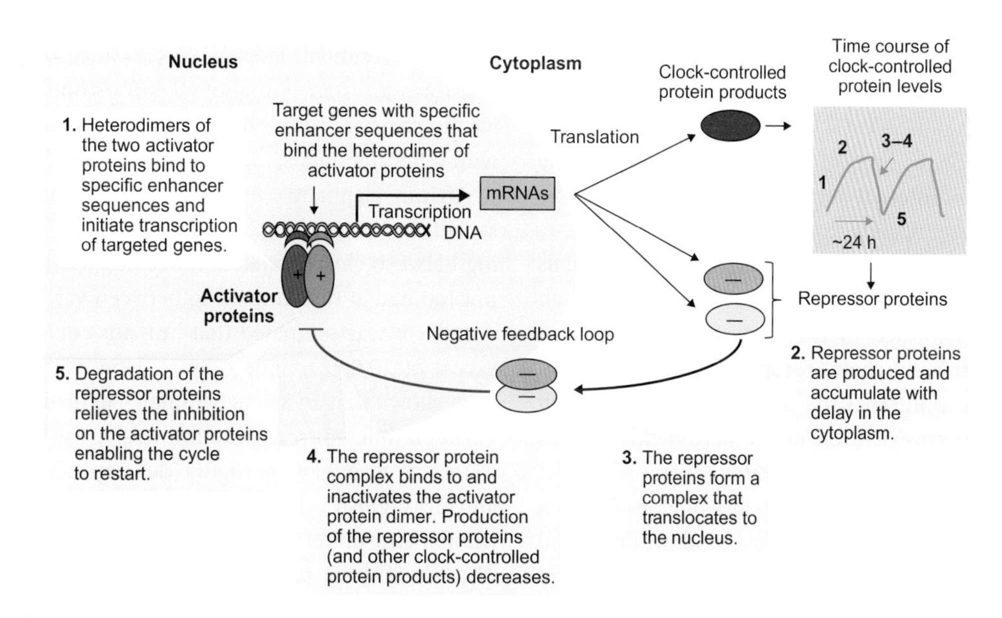

Figure 17.25 The general molecular principle of the core circadian clock operating in animal cells

The two activator proteins together with the two repressor proteins and their genes form the essential components of the core circadian clock. The oscillation in the mRNA and protein levels has approximately a period of 24 h. Importantly, the heterodimer of the activator proteins regulates the transcription of other genes that share the enhancer sequences of the genes coding for the repressor proteins. In turn, the concentrations of proteins controlled by these genes oscillate and exert their influence on a wide array of physiological functions external to the oscillatory mechanism of the circadian clock.

as a transcription factor by enhancing the transcription of E-box[45] containing genes. The negative regulator proteins and their genes in mammals are the PER1,2,3/*Per1,2,3* (Periodic) and the CRY1,2/*Cry1,2* (Cryptochrome)[46]. The CRY1,2 proteins in mammals act as light-independent negative regulators. The *Per1* and *Per2* genes are necessary for the daily resetting of the circadian clock to normal light cues, while mutations or deletions in the Clock gene cause obesity and alterations in glucose and lipid metabolism in mice and sleep disorders in humans.

› Review articles

Arendt D (2003). Evolution of eyes and photoreceptor cell types. International Journal of Developmental Biology 47: 563–571.
Berson DM (2003). Strange vision: ganglion cells as circadian photoreceptors. Trends in Neurosciences, 26: 314–320.
Hardie RC, Franze K (2012). Photomechanical responses in *Drosophila* photoreceptors. Science 338: 260–263.
Hardie RC, Raghu P (2001). Visual transduction in *Drosophila*. Nature 413: 186–193.
Kojima D, Mori S, Torii M, Wada A, Morishita R, Fukada Y (2011). UV-sensitive photoreceptor protein OPN5 in humans and mice. PLoS ONE 6(10): e26388.
Lamb TD, Collin SP, Pugh EN Jr (2007). Evolution of the vertebrate eye: opsins, photoreceptors, retina and eye cup. Nature Reviews in Neurosciences 8: 960–976.
Land MF (2005). The optical structures of animal eyes. Current Biology 15: R319–R323.
Larsson M (2013). The optic chiasm: a turning point in the evolution of eye/hand coordination. Frontiers in Zoology 10: 41.

17.3 Chemoreception

Chemoreception refers to the ability of animals to respond to the presence of chemical stimuli in their external and internal environments. Most chemoreceptors respond to stimuli from outside the body of the animal. However, some chemoreceptors are internal and monitor variations in the concentration of chemicals within the blood.

Central chemoreception refers to the ability of animals to detect changes in internal chemical parameters through sensory structures located in the central nervous system (CNS), while **peripheral chemoreception** refers to the detection of changes in internal chemical parameters through structures that are an extension of the peripheral nervous system. For example, peripheral respiratory chemoreception detects the variation in the partial pressure of gases (oxygen and carbon dioxide) and pH within the blood, while central respiratory chemoreception detects the oxygen and carbon dioxide partial pressure in the cerebrospinal fluid[47] surrounding the brain tissues. As we discuss in Sections 22.2.3 and 22.4.1, inputs from central and peripheral respiratory chemoreceptors are received by the vasomotor centre of the medulla, which modulates breathing, airway resistance and blood pressure in air-breathing vertebrates to maintain the levels of oxygen, carbon dioxide, and pH in the blood within narrow limits.

The hypothalamic osmoreceptors discussed in Section 21.1.4 are another great example of central chemoreceptors that play a key role in maintaining body volume homeostasis. Together, peripheral and central chemoreception are central to body homeostasis[48].

In this section, we focus on the most commonly recognized chemosensory systems that are responsible for the senses of smell (**olfaction**) and taste (**gustation**). The combined sensory experience of olfaction and gustation is called **flavour**.

17.3.1 Olfaction: making sense of molecular compounds in the surrounding medium

Olfaction, the sense of smell, is critical for the survival and reproductive success of animals because olfactory signals play key roles in finding food, stimulating feeding[49], avoiding predators, detecting warning signals like smoke from an approaching fire, locating mates for reproduction, identifying offspring and bonding. For example, polar bears (*Ursus maritimus*) have a most acute sense of smell (the ability to detect specific odorants at low concentrations) among terrestrial mammals, which is essential for locating food in the arctic environment. Polar bears can smell the presence of a seal from a distance of about 2 km or under one metre of ice. In terms of acuity, humans rank rather poorly compared with other mammals, including dogs, cats and even mice.

African elephants (*Loxodonta africana*) have the most sensitive olfactory sense among mammals for discriminating between closely related molecules. We humans are also not too bad at discriminating between various odours. For decades it was estimated that humans could distinguish in excess of 10,000 unique odours. However, recently, it has been shown that an average human can distinguish between 1 trillion different scents: an impressive feat!

Insects also have a highly developed ability to detect and distinguish odours. For example, many lepidopterans (moths and butterflies) can detect potential mates from a distance of about 10 km.

[45] An E-box (Enhancer box) is a DNA sequence found in promoter regions of some genes of eukaryotes.

[46] PER1,2,3 and CRY1,2 proteins are produced by three and respectively two genes: *Per1,2,3* and *Cry1,2*.

[47] We discuss cerebrospinal fluid in Section 16.1.2 and Box 16.1 (available online).

[48] We discuss chemoreceptors for oxygen and carbon dioxide in Sections 22.3.1, 22.3.3 and 22.4.1.

[49] It has been estimated that the sense of smell contributes about 80 per cent of the food flavour perceived by humans.

Olfactory system of vertebrates

The sense of smell is mediated by olfactory neurons located in the olfactory epithelium. Each olfactory neuron expresses only one type of **odorant receptor**, as shown in Figure 17.26; receptors for a particular odorant are randomly distributed throughout the olfactory epithelium.

In fish, the olfactory epithelium is located in a pair of small pouches behind the nostrils, through which water flows. The pouches are not connected to the mouth and act as a simple olfactory organ. Olfaction is a major sense for fishes; it is used for kin recognition, predator avoidance, location of spawning grounds and finding food. For example, salmon use their sense of smell as a navigation aid when travelling up the river to find the spawning ground on the tributary where they were born. Sharks also have a very acute sense of smell. The lemon shark (*Negaprion brevirostris*) can detect tuna oil at a concentration of about 50 drops in an average sized swimming pool (1 part per 25 million) or one drop of fish blood in an Olympic-sized swimming pool (1 part per 10 billion). Decomposing shark material has also been found to act as a powerful shark repellent.

The olfactory epithelium in mammals and other air-breathing vertebrates is located in the upper part of the nasal cavity and is covered by a protective layer of mucus (about 0.05 mm thick) which is completely replenished in a matter of minutes by mucus secreting cells. Hydrophilic odorants inhaled through the nose dissolve in the mucus and then bind to odorant receptors located on the sensory cilia of the olfactory receptor neurons (Figure 17.26). Hydrophobic odorants bind first to odorant-binding proteins, which act as vehicles for transporting these molecules through the mucus layer to the surface of the sensory cilia.

Signals from the olfactory epithelium of vertebrates are transmitted to the olfactory bulb via the axons of the olfactory receptor neurons, as illustrated in Figure 17.27. The axons project through a thin portion of bone to the olfactory bulb forming the olfactory nerve. The olfactory nerve is the first and the shortest of the cranial nerves in vertebrates.

Odorant specificity, the ability to distinguish between different odours, is achieved by two main mechanisms:

- The axons of all receptor cells expressing one specific odorant receptor converge to one globular anatomical structure consisting of nerve endings called the glomerulus[50], as illustrated in Figure 17.27. The mechanism of how this anatomical convergence of axons is achieved, given that the olfactory receptor neurons are continuously replaced throughout the life of the animal, remains a mystery. Furthermore, all axons in one glomerulus synapse to one single neuron, called the **mitral cell**, thus ensuring that neuronal activity in a given mitral cell reflects the stimulation of one specific type of olfactory receptor.
- One specific odorant activates a precise combination of different types of odorant receptors, which translates to an exact combination of mitral cells in the olfactory bulb becoming activated. Thus, the combination of signals from activated mitral cells directly reflects the combination of the type of olfactory receptors that have been activated by the odorant.

[50] The glomerulus made of nerve endings in the olfactory bulb should not be confused with the kidney glomerulus, which is made of blood capillaries and plays a central role in kidney function, as we discuss in Chapter 7.

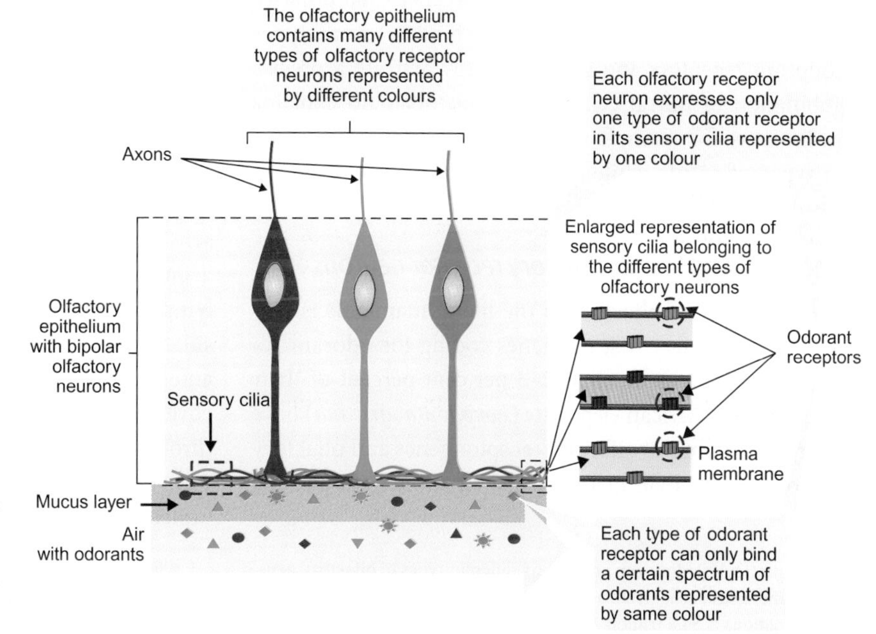

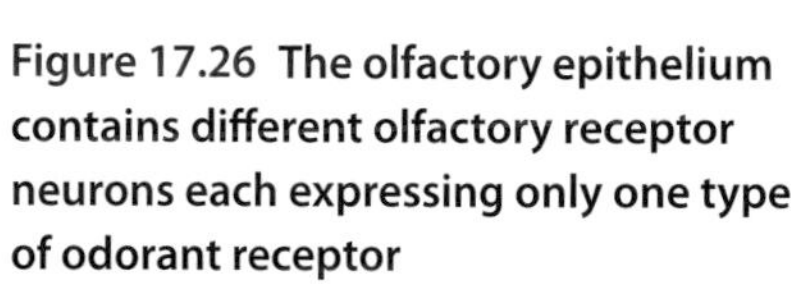

Figure 17.26 The olfactory epithelium contains different olfactory receptor neurons each expressing only one type of odorant receptor

All odorant receptor proteins are closely related to one another; they all have seven transmembrane segments and differences in their ability to bind different odorants is achieved by subtle differences in their amino acid composition. The odorant receptors are located on the cilia of the olfactory receptor neurons that are located on the surface of olfactory epithelium.

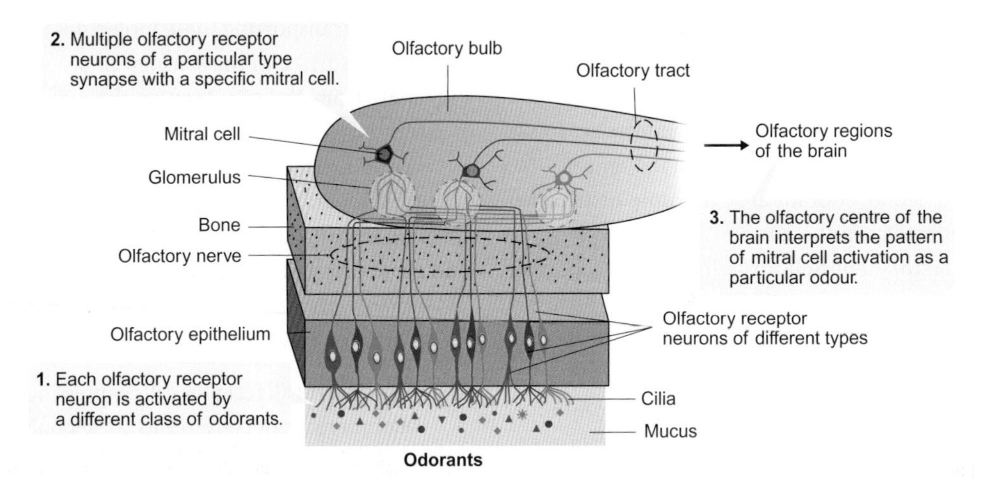

Figure 17.27 Diagram of signalling pathways in the olfactory system of air-breathing vertebrates

Odorants bind to receptors expressed on a specific combination of olfactory receptor neuron types and trigger action potentials; signals from randomly distributed receptor neurons of the same type converge to the same glomerular structure—called a glomerulus—in the olfactory bulb. Each glomerulus consists of nerve endings of olfactory neurons of the same type, which synapse with the dendrites of one mitral cell neuron; mitral neurons forward the signals to higher regions of the bran via their long axons which form the olfactory tract; the pattern of mitral cell activation is interpreted in the brain as a particular odour.

For example, assume that an animal has 200 different types of olfactory receptor neurons. If odorants could only bind to one type of odorant receptor, then the animal would potentially have the ability to distinguish between a maximum of 200 types of odours, resulting in one sensation for each type of stimulated olfactory receptor neuron. However, if odorants could bind to two different type of receptors and activate them, then potentially there would be combinations of 200 receptors taken in groups of two, which is $200 \times 199/2 = 19{,}900$ distinct combinations in the pattern of olfactory stimulation. Such binding properties would give an animal an ability to distinguish between 19,900 putative odours. Thus, the ability of odorants to bind to several types of odorant receptors and simultaneously activate a precise combination of mitral cells in the olfactory bulb explains how humans, for example, can distinguish more than 1 trillion types of scents while having only about 400 different types of odorant receptors[51].

Signal transduction in olfactory receptor neurons

The importance of olfaction in the life of mammals is emphasized by the fact that the genes coding for odorant receptors comprise as much as 2–3 per cent percent of their entire genome. African elephants (*Loxodonta africana*) have the largest number of olfactory receptor genes and olfactory receptors of any mammal investigated so far (about 2000 compared with 1200 in mice, 800 in dogs and 400 in humans), which would potentially allow them the highest level of scent discrimination.

The mechanism of signal transduction in a vertebrate olfactory receptor neuron is summarized in Figure 17.28. Each odorant receptor is coupled to a G-protein[52] which becomes activated when an odorant binds to the receptor. The activation of the G-protein starts a cascade of signalling events triggering trains of propagating action potentials in the axon of the respective olfactory neuron. Since one odorant can activate more than one type of olfactory receptor neurons, each odour is defined by the precise combination of mitral cells in the olfactory bulb that become activated.

Olfactory system of insects

Among invertebrates, the best understood olfactory system is that of insects. The olfactory receptor neurons are located in sensory hairs called **olfactory sensilla** (singular sensillum) as shown in Figure 17.29. Sensilla are filled with a mucus-like liquid packed with proteins called sensillum lymph, which is secreted by the support cells in the sensilla. The majority of olfactory sensilla are situated on the antennae and the rest on maxillary palps[53]. The sensilla are covered by a thin layer of cuticle containing many pores through which odorants can enter. Each sensillum contains 2–4 receptor neurons, which extend their dendrites into the lumen of the sensillum.

[51] If one particular odour can activate six different types of olfactory neurons (and mitral cells) from a total of 400, then the number of possible combinations is 54.8 trillion.

[52] This signalling system is outlined in Figure 19.4.

[53] Maxillary palps are sensory appendages of the maxillae.

(A) Cilium of an olfactory receptor neuron
Odorant
Odorant receptor
G-protein coupled to the receptor
Plasma membrane
1 2
Adenylyl cyclase
3
4
ATP
cAMP
Cytosol
5
Na^+
6
7
cAMP non-selective ligand-gated cationic channel
(B)
1 Odorant binds to odorant receptors
2 Activation of coupled G-protein (G_{olf})
3 Activation of adenylyl cyclase
4 ATP → Cyclic AMP (cAMP)
5 Activation of cAMP-dependent non-selective cationic channels in the plasma membrane
6 Ratio between membrane permeability for Na^+ and K^+ (P_{Na}/P_K) increases, entry of Na^+, membrane depolarization
7 Depolarization of membrane above threshold at spike initiating zone triggers trains of action potentials in the respective olfactory receptor neuron for as long as the depolarization persists

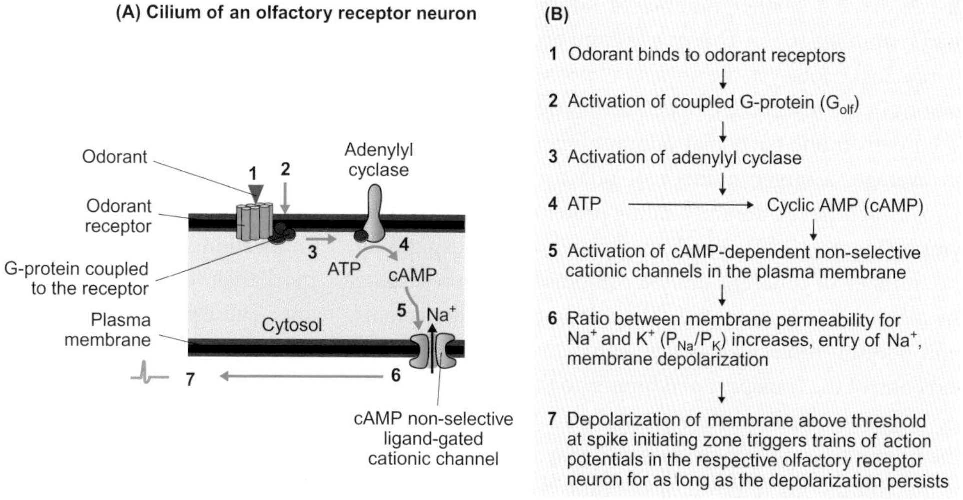

Figure 17.28 Signal transduction in olfactory receptor neurons of vertebrates

(A) Diagram of key elements in a cilium of an olfactory receptor neuron involved in olfactory signal transduction. (B) Sequence of events in olfactory transduction.

There are striking similarities between the organization and physiological properties of peripheral olfactory circuits in insects and mammals:

- Olfactory receptor neurons in insects generally express only one type of odorant receptor.
- The axons of olfactory receptor neurons that express the same type of odorant receptor converge onto the same glomerulus in the antennal lobe of the insect brain (which is the insect equivalent of the olfactory bulb in mammals).
- One odorant binds to several types of odorant receptors and activates a precise combination of glomeruli in the antennal lobe, which defines the neuronal representation of odour in the insect's brain.

A(i)
A(ii)
Antenna
A(iii) Sensillum
Cuticle
Pores
25 µm
1 µm
B(i)
Lymph
Dendrites
Sensillum
Pores
Cuticle
Receptor neurons
Support cells
Axons
B(ii)
Dendritic membrane
Odorant binding protein
Heteromeric odorant receptor
Cuticle
Pore
1
2
Non-selective cation channel opens when the odorant binds to the receptor
3
Na^+
4
Odorant
6
Enzymatic degradation of odorant
Dendritic cytosol
Sensillum lymph
AIR
5
B(iii)
1 Odorant enters sensillum and binds to binding protein
2 Odorant transported to heteromeric receptor
3 Odorant binds to receptor and opens non-selective cation channel
4 P_{Na}/P_K increases, Na^+ enters the cytosol and membrane depolarizes
5 Depolarization at spike initiating zone reaches threshold and triggers trains of action potentials
6 Odorant molecule disintegrates

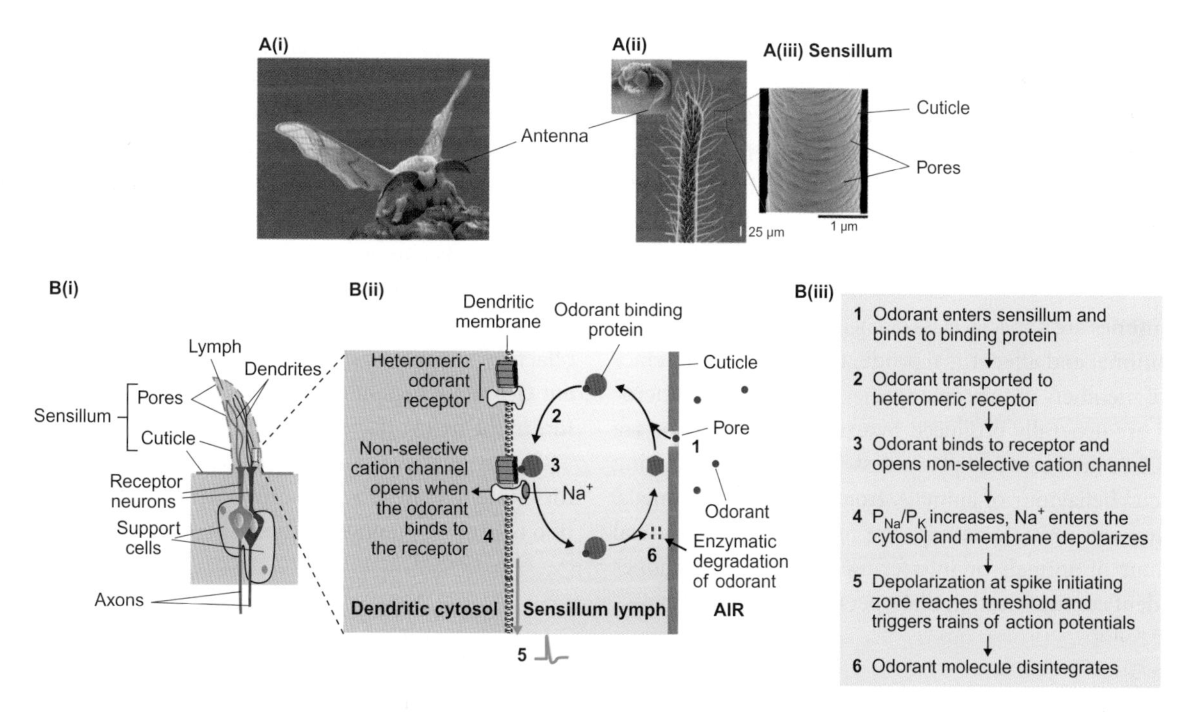

Figure 17.29 Signal transduction in olfactory receptor neurons of an insect's olfactory sensillum

(A(i)) Frontal view of a domestic silk moth's head (*Bombyx mori*) with the antennae. Adults are about 75 mm long and have a wing span of about 50 mm. (A(ii)) Scanning electron micrograph of a branch of the antennae with numerous olfactory sensilla. (A(iii)) Micrograph of a single olfactory sensillum showing the location of the pores. (B(i)) Diagram of an olfactory sensillum with two receptor neurons and two supporting cells. (B(ii)) Key elements on the dendrite of an olfactory receptor neuron. (B(iii)) Sequence of events in olfactory transduction in insects. Source: A(i), A(ii), A(iii): Images provided by Prof. Ryohei Kanzaki, Tokyo University.

17

Insects have a smaller number of odorant receptor types than mammals (62 in *Drosophila* compared with about 400 in humans). However, the number of possible unique combinations in the insect olfactory code may be even greater than in humans if one particular odorant binds to and activates, on average, a larger number of olfactory receptor types. Also, insects have odorant-binding proteins in the sensillum lymph (Figure 17.29), which may play a role in expanding the number of different unique combinations representing the olfactory code of insects. The odorant-binding proteins bind odorant molecules as they enter the sensillum lymph and control the transport of odorants to the receptors by directing or restricting delivery to specific type of receptors. The odorant-binding molecules also play an important role in the continuous removal of odorant molecules from the sensillum lymph by enzymatic degradation.

Despite fundamental similarities between the organization of peripheral olfactory circuits in insects and mammals, the structural and functional properties of their odorant receptors exhibit important differences. The odorant receptors in insects are more complex than in mammals, consisting of two different components (heteromeric odorant receptors). One component is the usual mammal-like, seven-transmembrane-segment receptor molecule that binds the odorant; the other component forms a nonselective cation channel. This channel opens rapidly when the odorant binds to the seven-transmembrane-segment component, causing depolarization and generation of action potentials in the receptor neuron when the depolarization is above threshold (Figure 17.29).

The ion channel receptors of insect olfactory neurons transduce the signal more rapidly than their G-protein-linked[54] mammalian counterparts, allowing insects to respond more quickly to olfactory cues than mammals.

Pheromones and their perception

Pheromones are a special class of chemicals that are secreted by an animal and affect the reproductive and social interactions of members of the same species (conspecifics). Pheromones are generally produced by exocrine glands[55], or are excreted in the urine and play an essential role in regulating the social behaviour of animals, from insects to mammals. Pollution can severely damage the reproductive and social behaviour of animals by interfering with the pheromone-dependent communication that operates between them.

Some pheromones, called **alarm pheromones**, are released when animals are attacked by a predator and trigger either aggression in the other members of the community—as seen when ants, bees and termites are stimulated to defend their nest—or flight to escape predation, as in aphids and mice. Other pheromones are used to mark the territory of individual animals, as in dogs, cats and social seabirds.

Social insects, such as ants, produce **trail pheromones** to mark their foraging paths, which are used as a guide by other ants. These pheromones are very volatile and need to be continuously renewed by ants as they return with food to the nest. Larvae of sea lampreys (*Petromyzon marinus*) produce a pheromone which indicates the location of habitats suitable for spawning. This pheromone attracts maturing adults from the open sea over hundreds of kilometres into streams where they grow and develop.

Aggregation pheromones are used by non-social insects to bring together, at a certain location, a large number of individuals for defence against predators or to overcome host resistance by mass attack. For example, some species of bark beetle (Coleoptera, Scolytidae) can overcome active defences of healthy trees by means of synchronous attack by many individuals brought together by the release of aggregation pheromones.

Sex pheromones are associated with mating behaviours or dominance. Generally, sex pheromones produced by females indicate their availability for mating. However, males also produce sex pheromones, which the females use for choosing their mates.

In insects, pheromones are detected by the sensilla located on the antennae. In vertebrates, some pheromones are detected by the main olfactory system. However, most reptiles, amphibians and non-primate mammals also have a well-developed specialized organ called the **vomeronasal organ**[56] at the base of the nasal septum between nose and mouth, which detects pheromones. The vomeronasal organ is part of the accessory olfactory system[57], which provides a chemoreception pathway that is independent from that of the main olfactory system. The vomeronasal receptors activate different G-protein dependent pathways from the odorant receptors in the main olfactory system, leading to the activation of TRP non-selective cationic channels. As in the main olfactory system, these channels cause depolarization and generation of action potentials in the respective receptor neuron.

[54] In insects there is also the possibility that the odorant binding to the receptor initiates a slower, G-protein- and cAMP-dependent pathway that causes a slower and more prolonged depolarization of the olfactory receptor neurons.

[55] Exocrine glands produce and secrete substances onto epithelial surfaces by way of a duct (e.g. sweat glands).

[56] The vomeronasal organ is absent in birds, and underdeveloped in adult catarrhine monkeys (monkeys that have nostrils facing downwards) and apes, including humans.

[57] The accessory olfactory system consists of the vomeronasal receptor neurons, the accessory olfactory bulb, situated on the dorsal-posterior region of the main olfactory bulb, and projections from the mitral cells in the accessory olfactory bulb to the amygdala and the hypothalamus.

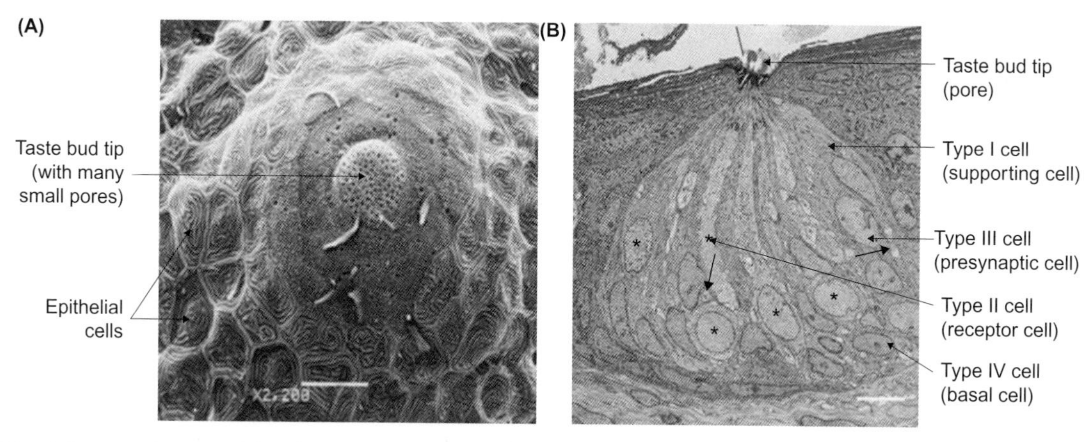

Figure 17.30 Micrographs of taste buds from a fish and a mammal

(A) Scanning electron micrograph of a taste bud from the buccal cavity of the African sharptooth catfish fish (*Clarias gariepinus*). (B) Electron micrograph of a longitudinal section through a taste bud from a rabbit (*Oryctolagus cuniculus*) showing the four types of cells in a mammalian taste bud. Type II receptor cells are marked by asterisks and neurons by short arrows. Calibration bars: 4 μ for A and 5 μm for B.

Source: A: Gamal AM et al (2012). Morphological adaptation of the buccal cavity in relation to feeding habits of the omnivorous fish *Clarias gariepinus*: A scanning electron microscopic study. The Journal of Basic & Applied Zoology. 65: 191–198. B: Royer SM, Kinnamon JC (1991). HVEM serial-section analysis of rabbit folate taste buds. I. Type III cells and their synapses. Journal of Comparative Neurology 306: 49–72.

17.3.2 Gustation: distinguishing between nourishing and toxic compounds in the environment

Taste (or 'gustation') refers to the sensory ability of organisms to distinguish between various forms of nutrients while avoiding toxic and indigestible substances. Taste stimuli, called **tastants**, are compounds that act on taste receptor cells.

All animals experience various types of taste qualities associated with identifying the nutritional values of foods and avoiding toxic substances. Humans exhibit five conventional basic taste qualities that are associated with the nutritional or physiological requirements of the individual: sweet, salty, sour, umami[58] and bitter.

Sweet-tasting food is associated with the presence of carbohydrates, which are the major source of energy in animals. Salty food indicates the presence of salts, particularly sodium chloride, which is essential for maintaining salt and water balance. Sour-tasting food indicates the presence of acids, and umami-tasting food indicates the presence of amino acids, from which proteins are made. The bitter taste of foods relates to the presence of potentially toxic substances, many of which are produced by plants to protect themselves from herbivores. Accumulating evidence also suggests that mammals experience two additional taste qualities: one for fatty acids and another for calcium.

[58] Umami is a taste quality for L-glutamate containing foods such as meat, fish, mushrooms and cheese. It originates from a Japanese word meaning 'deliciousness' associated with the taste of monosodium glutamate (MSG).

Gustatory receptor cells of vertebrates

Among vertebrates, the gustatory (taste) receptor cells are found in structures called **taste buds**. Taste buds comprise clusters of up to 100 **neuroepithelial** cells that form compact 'islands' embedded in the surrounding epithelium as shown in Figure 17.30. Taste buds are normally distributed across the oral cavity, on the tongue, the palate and the epiglottis and serve similar functions regardless of their location.

Fish and sharks also have taste buds located on the surface of their body. For example, channel catfish (*Ictalurus punctatus*) of North America have taste buds distributed over their entire body, including the fins, earning them the nickname the 'swimming tongue'. Taste buds are particularly concentrated on the catfish's four pairs of 'whiskers' around the mouth, called barbells. Catfish also have many olfactory receptors in their pouches behind the nostrils, which give them an exceptional ability to find food with relative ease in the dark or muddy waters at the bottom of lakes and rivers. Sharks and teleost fish have been seen brushing their bodies across potential prey, possibly testing whether it is actually food.

Taste buds contain four types of cells as shown in Figure 17.30B:

- Type I cells (or supporting cells) are the most abundant type of cells in taste buds. These cells generally support the function of other cells, for example by degrading or absorbing ATP, which is the major signalling molecule between the gustatory cell types in taste buds. Type I cells also regulate the extracellular K^+ concentration. A subset of type I cells may also respond to salty stimuli.

- Type II cells (or receptor cells) respond individually to either sweet, umami/amino acids or bitter stimuli. These cells do not respond to sour or salty stimuli.
- Type III cells (also called presynaptic cells) have neuron-like properties and form identifiable chemical synapses with nerve terminals. They respond directly to sour stimuli and carbonated solutions.
- Type IV cells (or basal cells) are taste stem/progenitor cells that generate the other types of taste cells.

Mechanisms of taste transduction in vertebrates

The complete mechanism for encoding gustatory information in vertebrates has yet to be fully elucidated. Nevertheless, we know many aspects of the different processes involved in the transduction of gustatory signals in vertebrates, particularly mammals; these are summarized in Figures 17.31 to 17.34.

The general signalling pathways from the three types of taste bud cells that respond to gustatory signals are shown in Figure 17.31. Compounds with a salty taste act on a subset of type I cells that release ATP when stimulated, which is sensed by ATP receptors on the membrane of type III cells. There are several classes of ATP receptors. Some act as ligand-gated non-selective cationic channels and cause rapid depolarization of cellular membranes, while others are coupled to G-proteins and respond more slowly. The binding of ATP to ATP receptors on type III cells causes depolarization of the membrane, which triggers action potentials and activation of chemical synapses[59] with sensory taste neurons.

[59] We discuss excitatory and inhibitory chemical synapses in Section 16.3.3.

Any given type II cell is 'tuned' to either sweet, bitter or umami taste by expressing specific types of **taste receptors** that recognize a specific type of tastant. When stimulated, type II cells also release ATP, which acts as a signal transmitter on endings of sensory neurons in their vicinity, or bind to ATP receptors on type III cells, as shown in Figure 17.31.

Type III cells respond specifically to sour taste compounds and carbonated liquids triggering action potentials in sensory neurons to which they are connected by chemical synapses. Type III cells also affect the sensitivity of type II cells to tastants as shown in Figure 17.31 and integrate responses received from type I and type II cells. Gustatory sensory information is then transmitted through sensory nerves to the rostral (more forward) solitary nucleus in the brainstem via three separate cranial nerves: the facial (VII), the glossopharyngeal (IX) and the vagus (X) nerves[60]. From the solitary nucleus the gustatory information is transmitted via the thalamus to the primary gustatory cortex, which is the brain structure responsible for the perception of taste.

What are the specific mechanisms of signal transduction in the taste receptor cells?

Type I cells have potassium channels over their entire surface and express **epithelial sodium channels** only on their tips. Normally, the type I cells are polarized (negatively charged inside due to a low permeability ratio for sodium and potassium ions, P_{Na}/P_K). However, when the sodium ion concentration in food is high (greater than about 150 mM), sodium ions enter the cytosol via epithelial Na^+ channels and cause

[60] We discuss cranial nerves in Box 16.2.

[61] The level of depolarization is determined by the Goldman–Hodgkin–Katz equation (equation 3.11 in Chapter 3).

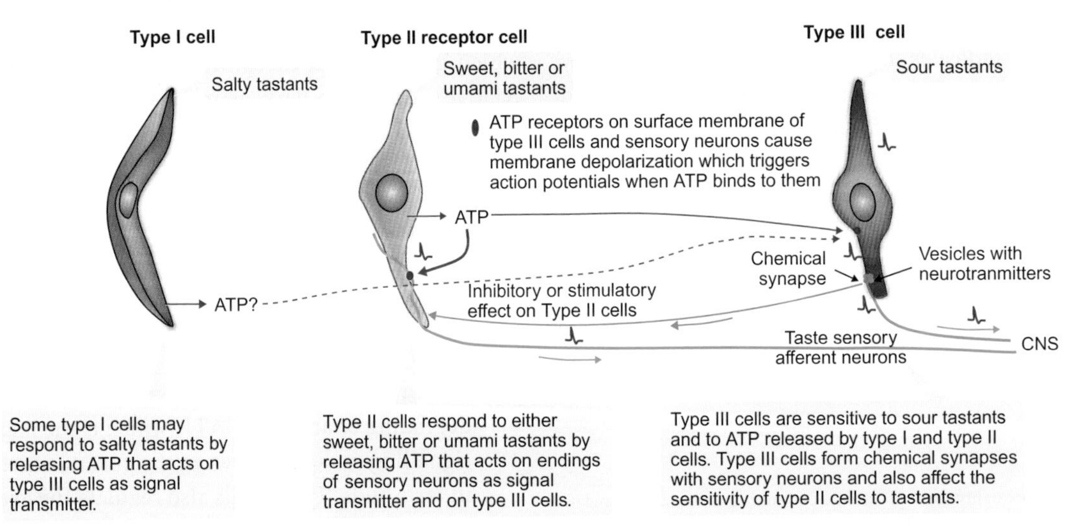

Figure 17.31 The three cell types involved in taste transduction in mammalian taste buds and the general pathways of signal transmission to the central nervous system (CNS)

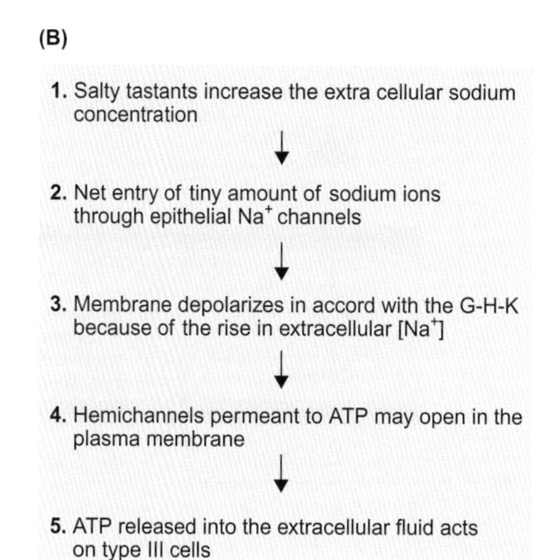

Figure 17.32 Signal transduction in a subset of type I taste bud cells of vertebrates

(A) Diagram of key elements in type I cells involved in the transduction of salty tastants. (B) Sequence of events in type I taste bud cells.

membrane depolarization[61]. The depolarization activates the opening of **hemichannels**[62] through which ATP is released, as illustrated in Figure 17.32. The signal carried by the ATP released from type I cells is then transmitted to the type III cells as discussed.

[62] Hemichannels are large transmembrane channels connecting the intracellular and extracellular space allowing passage of ions and small molecules such as ATP across membranes. Gap junctions discussed in Section 16.3.1 and shown in Figure 16.31 have two hemichannels, one in each membrane, joined together.

Type II cells are electrically polarized, excitable cells and contain G-protein-coupled taste receptors with seven-transmembrane-segment domains as shown in Figure 17.33A. Individual receptor cells respond to sweet, umami or bitter taste compounds. The taste receptors for bitter tastants are particularly diverse, encoded by one large family of G-protein-coupled taste receptor genes and may have evolved to prevent ingestion of a wide range of potentially toxic substances.

As shown in Figure 17.33B, the principal pathway for the activation of the type II receptor cells involves

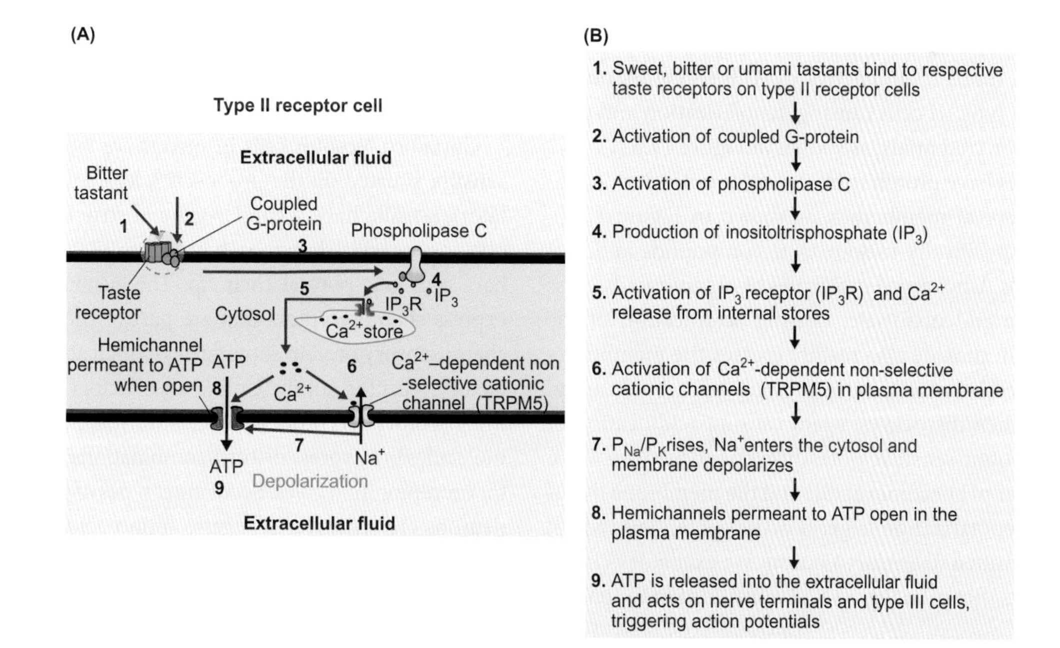

Figure 17.33 Principal signal transduction pathway in type II receptor cells of vertebrates

(A) Diagram of key elements in type II receptor cells involved in the transduction of bitter tastants. The taste receptors for bitter tastants consist of seven-transmembrane-segment G-protein-coupled molecules encoded by one large family of G-protein-coupled taste receptor genes. Multiple forms of bitter taste receptors are expressed in one bitter-taste receptor cell. Taste receptors for sweet and umami tastants consist of heterodimers of seven-transmembrane-segment G-protein-coupled molecules encoded by a different small family of G-protein-coupled taste receptor genes. (B) Sequence of events in type II receptor cells.

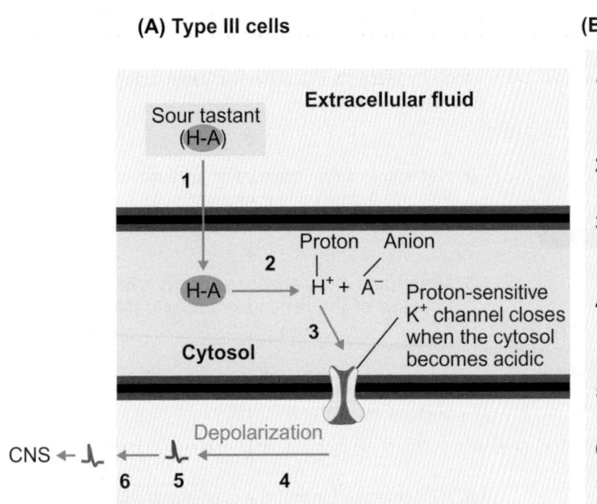

(B)

1. Sour tastant (H-A) diffuses through the membrane in the uncharged form
↓
2. Tastant dissociates in the cytosol releasing protons
↓
3. Proton-sensitive potassium channels close when the cytosol becomes acidic
↓
4. P_{Na}/P_K rises and membrane depolarizes in accord with the G-H-K because P_{Na}/P_K rises
↓
5. Action potentials are generated in type III cell
↓
6. Chemical synapse with sensory neuron becomes activated and action potentials generated in the sensory neuron propagate to CNS

Figure 17.34 Signal transduction in type III cells of vertebrates in response to acidic (sour) tastants

(A) Diagram of key elements in type III cells involved in the transduction of acidic tastants. (B) Sequence of events in the transduction of acidic tastants.

G-protein-dependent activation of phospholipase C (PLC), which stimulates production of inositol trisphosphate (IP_3)[63]. IP_3 opens Ca^{2+} channels on the endoplasmic reticulum (called IP_3 receptors) causing a rise in intracellular Ca^{2+} concentration. In turn, the rise in intracellular Ca^{2+} opens TRP channels[64] (specifically TRPM5 expressed in taste receptors), which causes depolarization of the membrane. The combined effect of elevated ionic calcium concentration and membrane depolarization opens hemichannels in the membrane through which ATP is released.

Similar to type I cells, type II receptor cells do not form chemical or electrical synapses with secondary neurons. However, the ATP released through the hemichannels directly activates taste neurons located in the proximity of type II cells; it also binds to and activates ATP receptors on the membranes of type III cells causing depolarization and generation of action potentials, as shown in Figure 17.31.

Type III cells have proton (pH)-sensitive potassium channels on their apical membranes, as shown in Figure 17.34, and respond specifically to sour taste compounds and carbonated liquids. Uncharged molecules of organic acids cross the membrane and dissociate, causing acidification of the cytosol. Carbon dioxide also readily crosses the membranes and produces carbonic acid in the presence of carbonic anhydrase, which then dissociates, again causing acidification of the cytosol. Proton-sensitive potassium channels close when the intracellular pH becomes acidic and the membrane depolarizes. This depolarization triggers action potentials, which are then transmitted to sensory neurons via excitatory chemical synapses as shown in Figure 17.31. In addition to sensing sour tastants, type III cells also integrate responses received from type I and type II cells as discussed earlier.

We are yet to fully understand how the specificity of the different taste qualities, associated with the activation of particular taste receptor types, is preserved when gustatory signals are transmitted via sensory taste neurons to the CNS.

Taste transduction in insects

Among invertebrates, taste transduction is most intensely studied in insects. We know, for example, that the specificity of the different taste qualities is maintained when the signals are transmitted from gustatory receptor neurons to different ganglia in the insect's brain. However, little is known about the molecular mechanisms by which sugars, amino acids, carbon dioxide and bitter-testing compounds cause membrane depolarization to generate action potentials in the receptor neurons.

Gustatory receptor cells in insects are located in **gustatory sensilla** situated on the proboscis[65], legs and wings. The gustatory sensilla have a similar general structure to that of the olfactory sensilla shown back in Figure 17.29, except that they have only one pore at their tip. The gustatory receptor cells express chemoreceptors that are part of the wider multigene chemosensory protein superfamily, which also includes the olfactory receptor family. Unlike an olfactory receptor neuron, the gustatory receptors in insects do not express co-receptor molecules but express distinct combinations of different gustatory receptor molecules. So, distinct types of gustatory receptor neurons are sensitive to sugars, amino acids, carbon dioxide and bitter-testing compounds, and express distinct subsets of gustatory receptor genes. The receptor neurons expressing a particular subset of gustatory receptors connect to same specific ganglia in the insect's brain, preserving the specific activation of particular ganglia in the brain by different types of tastant.

[63] The concentration of cAMP is also altered in type II receptor cells through the action of a G-protein called gustducin, which is similar in structure to transducin found in ciliary photoreceptors that we discuss in Section 17.2.1.

[64] We discuss TRP channels in Section 17.1.2.

[65] A proboscis is an elongated sucking tube generally found in siphoning insects like moths and butterflies.

17

› Review articles

Buck L, Axel R (1991). A novel multigene family may encode odorant receptors: a molecular basis for odor recognition. Cell 65: 175–187.
Chaudhari N, Roper SD (2010). The cell biology of taste. Journal of Cell Biology 190: 285–296.
Corey EA, Bobkov Y, Ukhanov K, Ache BW (2013). Ionotropic crustacean olfactory receptors. PLoS ONE 8(4): e60551.
Galizia CG, Rössler W (2010). Parallel olfactory systems in insects: anatomy and function. Annual Review of Entomology 55: 399–420.
Jacquin-Joly E, Merlin C (2004). Insect olfactory receptors: contributions of molecular biology to chemical ecology. Journal of Chemical Ecology 30: 2359–2397.
Kaupp, UB (2010). Olfactory signalling in vertebrates and insects: differences and commonalities. Nature Reviews Neuroscience 11: 188–200.
Pellegrino M, Nakagawa TJ (2009). Smelling the difference: controversial ideas in insect olfaction. Journal of Experimental Biology 212: 1973–1979.
Sánchez-Garcia A, Vieira FG, Rozas J (2009). Molecular evolution of the major chemosensory gene families in insects. Heredity 103: 208–216.

17.4 Mechanoreception

Mechanoreception refers to the ability of an organism to detect mechanical stimuli, which are responsible for a variety of sensory functions. These include:

- sound detection,
- balance maintenance (orientation relative to gravity),
- detection of water currents,
- identification of shapes and textures,
- detection of the position of objects by touch or echolocation,
- detection of relative position of different parts of the body,
- detection of blood pressure in the circulatory system, and
- detection of deflection or stretch in structures of the respiratory system.

Mechanical stimuli are converted into a depolarizing receptor potential by mechanoreceptors acting on stretch-sensitive ion channels in the cell membrane. These channels can be activated either directly in response to the force applied to the cell membrane, or indirectly if a tether attached to the channel is stretched, as is the case with mechanoreceptors in the vertebrate ear[66]. The stimulus information contained in the receptor potential is then converted into action potentials whereupon the sensory information is transmitted via afferent neurons to the CNS where it is processed.

[66] Figures 4.20 and 17.36 consider such stretch activated mechanosensors.

17.4.1 Hearing: making sense of vibrations in the surrounding environment

Hearing is the ability of animals to perceive sound by detecting vibrations in their surrounding environment through a specialized organ—for example, the ear of vertebrates. The sense of hearing is very important for communication between animals and for the detection of sounds associated with specific conditions in their environment such as presence of prey, approaching predators, or changing weather patterns.

The auditory sensory system consists of the auditory organ, which contains the mechanoreceptors involved in acquiring the sound information from the surrounding environment, and parts of the nervous system that process this information.

The auditory system of vertebrates

The most intensely studied auditory sensory system in vertebrates is that of terrestrial mammals, which is shown diagrammatically in Figure 17.35. Sound waves are funnelled by the visible part of the ear (**pinna**) into the ear canal, where they induce vibrations in the eardrum (**tympanum**)[67]. The vibrations in the tympanum are efficiently transmitted via three **middle ear ossicles** (malleus, incus and stapes—Latin for hammer, anvil and stirrup, respectively) to the **oval window**, which is in contact with fluids of the **inner ear**. This arrangement acts as a hydraulic press in which the force acting on the relatively large area of the tympanum is concentrated onto the much smaller area of the stapes in contact with the oval window, resulting in a pressure increase[68] on the fluid behind the oval window.

The mechanoelectrical transduction of the sound waves into action potentials takes place in the **auditory apparatus**. In mammals, the auditory apparatus resides in a structure resembling a snail's shell called the **cochlea** (Greek for 'spiral'), which is divided into three longitudinal compartments, as shown in Figure 17.35. The central compartment, called the cochlear duct, is separated from the other two compartments (the vestibular and tympanic ducts) by the **Reissner's membrane** and the **basilar membrane** on which the **organ of Corti** is located. The organ of Corti runs along the entire length of the basilar membrane and contains the acoustic sensory cells, called **hair cells**, which act as mechanoreceptors.

The cochlear duct is filled with **endolymph**, a fluid containing a high K^+ concentration (about 150 mmol L^{-1}, similar to that of the cytosol) and 20 μmol L^{-1} ionized calcium.

[67] In aquatic mammals such as whales and dolphins the sound waves travel through water, enter the lower jaw and are then transmitted to the inner ear.

[68] The middle ear ossicles increase the pressure exerted on the oval window by about 27-fold in humans.

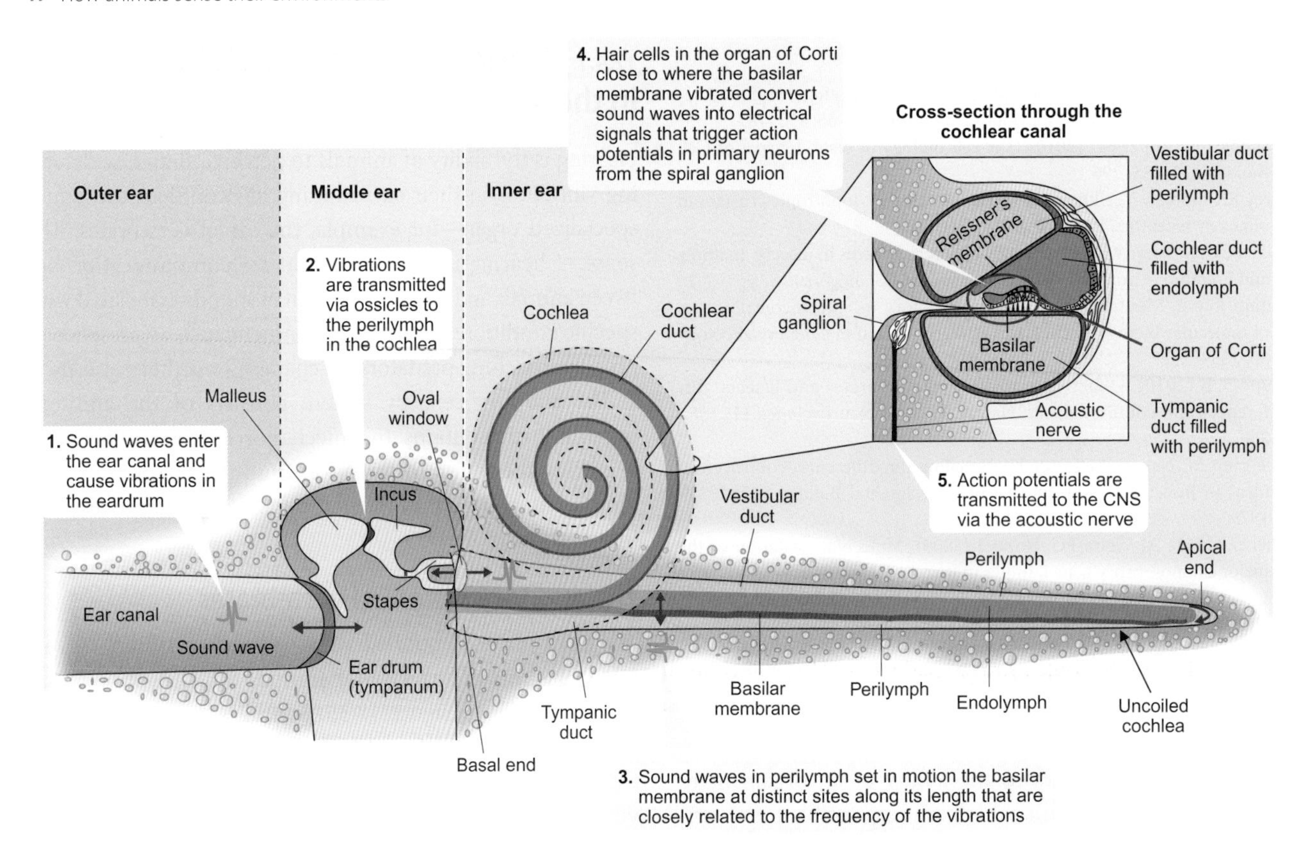

Figure 17.35 The sound conduction pathway and signal transduction in the mammalian ear

Source: main panel: modified from Fettiplace R, Hackney CM (2006). The sensory and motor roles of auditory hair cells. Nature Reviews Neuroscience 7: 19–29. Inset: Avan P et al (2013). Auditory distortions: origins and functions. Physiological Reviews 93: 1563–1619.

In contrast, the other two compartments, which communicate with each other at the apical end of the cochlea, are filled with **perilymph** that is typical of extracellular fluids in having a high Na^+ concentration (about 150 mmol L^{-1}).

Sound waves transmitted to the perilymph in the vestibular duct of the inner ear set in motion-specific regions of the basilar membrane that vibrate at particular frequencies. Higher sound frequencies cause displacement of the basilar membrane closer to its base (as it enters the cochlea) and lower frequencies closer to its tip. Auditory hair cells become activated when the basilar membrane is driven by the sound wave to move upward towards the vestibular duct. Responses from a particular hair cell are thus intrinsically linked to a characteristic frequency of the sound wave, which causes vibrations in a particular segment of the basilar membrane.

Hair cells make synapses with bipolar primary afferent auditory neurons, whose cell bodies are located in the spiral ganglion as shown in Figure 17.35. Action potentials generated in these afferent neurons carry auditory sensory information to the medulla of the brainstem via the auditory nerve (cranial nerve VIII)[69]. This ensures that neuronal activity from a particular hair cell reflects a characteristic frequency of the sound when auditory signals are processed. From the medulla, second-order nerve fibres project to the midbrain where they synapse with third-order neurons projecting to the thalamus[70]. The ascending pathway continues from the thalamus to the primary auditory cortex located in the temporal lobes in mammals, where most processing of auditory information takes place.

There are differences between the auditory system in terrestrial mammals and other vertebrates. The pinna, for example, is absent in non-mammalian vertebrates, and amphibians and some reptiles do not have an ear canal, in which case the tympanum is at the body surface. Vibrations of the tympanum in birds, reptiles and amphibians are transmitted by movements of a single bone, the **columella** (Latin for 'little column') through the middle ear cavity to the oval window and the fluids of the inner ear. In fish, which lack external and middle ears, sounds from the external environment are transmitted to the inner ear via various routes including the jawbone and the swim bladder, which can act as resonators to enhance sound detection. In snakes,

[69] We discuss cranial nerves in Box 16.2.

[70] The thalamus is made of two symmetrical bulb-shaped masses located within the vertebrate brain between the midbrain and the cerebral cortex as we discuss in Section 16.1.2.

17

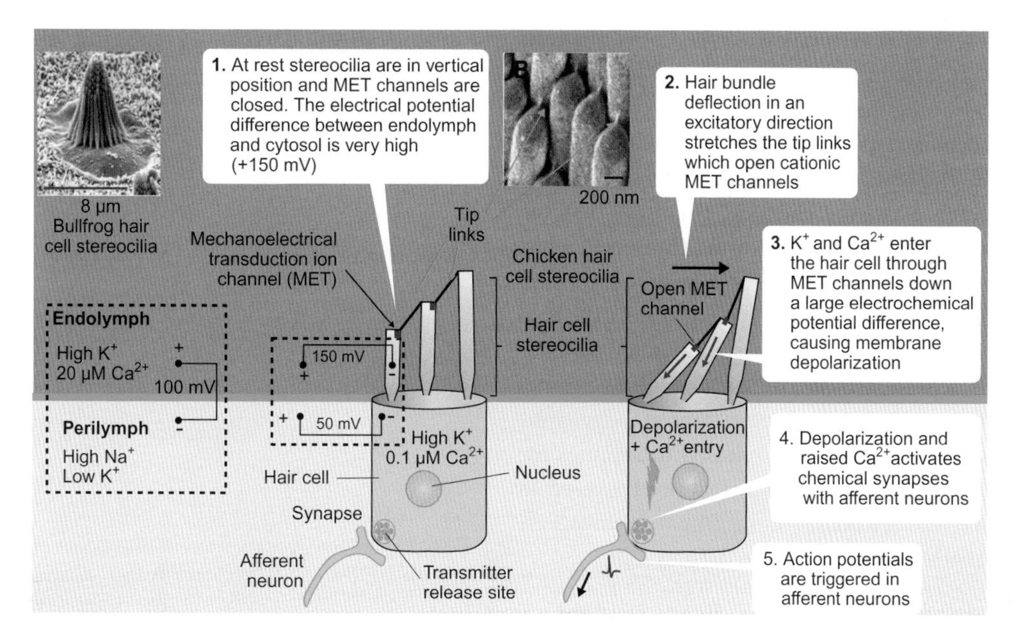

Figure 17.36 Mechanoelectrical transduction (MET) mechanism in hair cells

Hair cells are part of epithelial sheets that separate the endolymph from perilymph.

Source: bullfrog (*Rana catesbeiana*) stereocilia: Dr David P Corey, Harvard University. Chicken (*Galus galus domesticus*) hair cell stereocilia: Kachar B et al (2000). High-resolution structure of hair-cell tip links. Proceedings of the National Academy of Science. 97: 13336-13341.

the columella connects the jaw apparatus to the inner ear, allowing them to sense vibrations that travel through the ground via their jawbone. Snakes lack tympana (plural of tympanum), but can also sense relatively loud airborne sounds through sound-induced vibrations of the skull that are relayed to the inner ear.

Birds, reptiles and some amphibians have the auditory apparatus in a blind-ended tubular structure called the **basilar papilla**, which is homologous to the organ of Corti in mammals. Amphibians that lack basilar papillae have sensory cells for hearing, located in a compartment called **saccule** of the **vestibular organ**, which in vertebrates is used to maintain an animal's balance[71]. Bony fish use sensory cells located in a different vestibular compartment, called the **utricle**, and on their surface (lateral line, described in Section 17.4.2) to transduce sound waves into electrical signals.

Mechanism of signal transduction in hair cells

The sensory receptors in the auditory apparatus of all vertebrates, the hair cells, are part of epithelial sheets that separate the endolymph from the perilymph. As shown in Figure 17.36, many microvilli-like structures called **stereocilia** (numbering 20 to hundreds) are located at the apical end of hair cells, where they are in contact with the endolymph, while the basal end of the hair cells is in contact with the perilymph.

[71] We discuss the vestibular organ in vertebrates in Section 17.4.2.

A large electrical potential difference of about +100 mV exists between the endolymph and the perilymph, which translates into an even greater electrical potential difference of about +150 mV between the endolymph and the cytosol of the hair cells, which have a resting membrane potential of around −50mV relative to perilymph. These membrane potential differences are shown in Figure 17.36. The very large potential difference between endolymph and the hair cell cytosol is important for the rapid mechanoelectrical transduction function of the hair cells.

Each stereocilium behaves as a rigid rod that can only bend at its point of insertion into the hair cell. As shown in Figure 17.36, each stereocilium is connected by protein filaments called **tip links** to its taller neighbour such that stereocilia move together as a unified bundle of rigid rods. Mechanotransduction occurs when stereocilia are deflected by mechanical stimulation. When the taller stereocilia bend away from their shorter neighbours, the links are stretched and pull open **mechanoelectrical transduction (MET) ion channels**. Movement in the opposite direction releases the tension in the tip links and the channels close. Therefore, more MET channels open when the movement of the shorter stereocilia is toward the longest stereocilia of the hair bundle and more MET channels close when stereocilia move in the opposite direction.

The MET channels are non-selective cation channels that are permeable to Na^+, K^+ and Ca^{2+}, but the molecular structure of these channels is unknown. When MET channels open, K^+ and Ca^{2+} ions, in particular, enter the hair cell from the endolymph causing the rapid depolarization[72] of the hair cell. This rapid depolarization opens Ca^{2+}-channels at the base of the hair cells. The rapid rise in cytosolic calcium ion concentration activates excitatory chemical synapses[73] between one hair cell and up to 20 primary neurons. Conversely, closing MET channels causes rapid hyperpolarization of the hair cells and deactivation of the chemical synapses. Thus, the level of depolarization of the hair cells (receptor potential) depends on the angle of deflection of the stereocilia and determines the amount of transmitter (glutamate) released at the synapse. In turn, the amount of transmitter released determines the spacing between the action potentials that are triggered in primary neurons. The cell bodies of the primary neurons are organized in ganglia in close vicinity to the hair cells.

Coding of information in the auditory system

The two major sound characteristics encoded in the auditory systems are the sound frequency range and the sound intensity.

Frequency range of sound perception

The frequency of the sound waves is encoded by the precise location of the hair cells that become activated along the basilar membrane. Therefore, responses from a particular hair cell are intrinsically linked to a characteristic sound frequency. This ensures that neuronal activity from that particular hair cell reflects a characteristic frequency of the sound when auditory signals are processed within the hindbrain and the auditory cortex.

The frequency range of vibrations that each species can perceive as sound is closely related to the length of their basilar membrane. Generally, the frequency range along the basilar membrane is logarithmic, meaning that an equivalent distance along the basilar membrane corresponds to a certain factor by which the frequency that can be perceived increases. For example, the basilar membrane in humans is about 30 mm long and the 'normal' frequency hearing range is between 20 Hz and 20 kHz. This means that the frequency changes by a factor of 10 for every 10 mm travelled along the basilar membrane (20 kHz at the base of the membrane; 2 kHz at 10 mm from the base; 200 Hz at 20 mm from the base; and 20 Hz at the tip of the basilar membrane).

Relative to the 'normal' frequency range in humans (20–20,000 Hz), low-frequency sound ('**infrasound**') refers to sound vibrations between 1 and 20 Hz, while high-frequency sound ('**ultrasound**') refers to vibrations >20 kHz. The hearing frequency range varies between different vertebrates, as illustrated in Table 17.1. The frequency range also changes with the age of the animal. Humans, for example, lose their ability to hear sounds in the high-frequency range with increasing age (above ~8 years), but 9-year-old green turtles respond to higher sound frequencies (3 kHz) than 5-year-old (1.9 kHz) and 2-year-old (1.6 kHz) turtles.

Table 17.1 Hearing frequency range in a selected group of vertebrates

Animal	Frequency range (Hz)
Baleen whales	10–31,000
Elephants	15–10,000
Ferrets	15–45,000
Goldfish	20–3000
Cows	20–35,000
Humans	20–20,000
Atlantic salmon	40–350
Tree frogs	50–4000
Cats	55–75,000
Dogs	60–44,000
Green turtles	60–1600
Snakes	80–160
Chickens	125–2000
Sheep	125–42,500
Bottlenose dolphins	150–150,000
Owls	200–12,000
Canaries	250–8000
Mice	900–80,000

Several animals use infrasound for long-distance communication, because low-frequency sound travels more effectively over long distances with less distortion than high frequency sound. For example, females in oestrus from matriarchal breeding herds of Asian elephants (*Elephas maximus*) produce very loud vocalizations below 20 Hz that travel through the air for up to 5 km or through the ground as seismic signals for tens to hundreds of kilometres to attract bulls for mating. Baleen whales use infrasound calls at frequencies as low as 10 Hz that propagate through water at a speed of 1.5 km s^{-1} (which is about 4.3 times faster than in air) to communicate over distances of hundreds to thousands kilometres. Giraffes, rhinoceros, hippopotamuses and alligators also use infrasound to communicate with other members of their species (conspecifics).

At the other end of the frequency scale, many animals detect ultrasound; these include whales, bats, rodents, cats, dogs and sheep, as well as insects like grasshoppers, locusts and moths. Ultrasound is used for communication, navigation, foraging and capture of prey, as well as for avoiding capture.

[72] Only a tiny amount of cations need to enter the cytosol of the hair cell to cause depolarization by 10–20 mV, as we discuss in Section 16.2. When MET channels close, the hair cell rapidly repolarizes as an equivalent amount of potassium ions leave the hair cell to the perilymph through potassium channels located at the base of the hair cell.

[73] We discuss synaptic transmission in Section 16.3.

Figure 17.37 The organ of Corti is the peripheral auditory organ of mammals where the mechanoelectrical transduction takes place

The organ of Corti consists of one longitudinal row of inner hair cells involved in sound detection and three rows of hair cells that change their length in response to stimulation by efferent neurons and alter the sensitivity of the sound-detecting hair cells.

Source: modified from Fettiplace R, Hackney CM (2006). The sensory and motor roles of auditory hair cells. Nature Reviews Neuroscience 7: 19–29

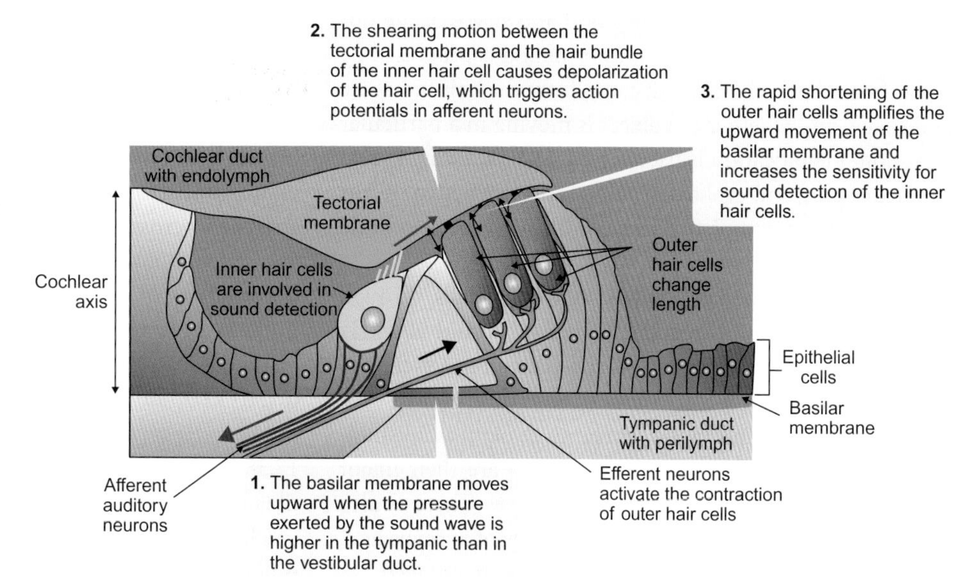

Coding of sound intensity

The sound intensity at a particular frequency is coded by the deflection level of stereocilia, which in turn determines the level of depolarization of the respective auditory hair cell. The depolarization translates to action potentials in the afferent neurons being triggered closer together when the sound intensity is greater and being spaced further apart at lower sound intensities as we discuss in Section 17.1.2.

The hair cells in the organ of Corti involved in sound transduction are called **inner hair cells** due to their location on the inner side of the basilar membrane, close to the cochlear axis, as illustrated in Figure 17.37. Covering the hair cells is a non-cellular, gel-like structure called the **tectorial membrane**. Movement of the basilar membrane by the sound wave induces a shearing movement between the hair cells and the tectorial membrane, causing a sound-intensity-dependent deflection of stereocilia, graded depolarization of the hair cell and generation of action potentials in the afferent auditory neurons whose frequency depends on the sound intensity.

In addition to the inner hair cells that are located in one row on the internal side of the basilar membrane, the organ of Corti also contains **outer hair cells** that are located in three rows on the outer side of the basilar membrane as shown in Figure 17.37. The outer hair cells contain on their surface the protein **prestin** (after 'presto' in music, which indicates a quick tempo). Prestin changes shape upon membrane depolarization, causing rapid cell shortening upon depolarization and lengthening upon hyperpolarization. The outer hair cells connect directly to the tectorial membrane through their tallest stereocilia and play a key role in the process of **cochlear amplification**, which refers to the ability of mammals to amplify very faint sounds by employing a positive feedback loop that enhances the amplitude of the basilar membrane vibration.

In the process of cochlear amplification, the outer hair cells are stimulated by efferent (motor) neurons from the medulla, causing depolarization and shortening, as shown in Figure 17.37. Since the outer hair cells are attached to the tectorial membrane their active shortening permits the basilar membrane to move closer to the tectorial membrane. This movement increases the amplitude of the basilar membrane vibration, which enhances the sensitivity of the sound detecting inner hair cells. Thus, these outer hair cells increase the sensitivity of hearing when they shorten.

Sound localization and echolocation

Sound localization refers to the ability of animals to identify their orientation and distance from the origin of a sound. Terrestrial vertebrates, in general, can horizontally locate the direction of sound sources with good accuracy by turning their heads until there is no time difference between the sound waves reaching the two ears. Vertical localization can be achieved by having asymmetrical earflaps: one collecting sound from above for one ear and one from below for the other ear. For example, barn owls (*Tyto alba*) have asymmetrical ears on their face and can locate sound sources like mice with exceptional accuracy.

Echolocation or **biosonar** refers to the ability of some mammals like the microchiropteran bats and odontocetes (including sperm whales, killer whales, dolphins and porpoises) to locate and identify objects by listening to echoes of brief calls made by the animal itself. The echoes returning to the two ears arrive at slightly different times and intensity levels, depending on the position and shape of the object generating the echoes. This information is processed by the animal to perceive their distance from the object, its size and shape, and whether the object is moving in a particular direction. Echolocation is used for navigation, foraging and capture of prey, and evolved in animals that are active and/or forage in almost complete darkness, where vision is of little or no use.

Microchiropteran bats emit echolocation calls through their open mouth at a frequency range of 11 and 200 kHz depending on both the particular environment in which they forage and the food source, which can be as diverse as insects, frogs, nectar, fruit and blood. Foraging at night is advantageous to these bats because insects, which are their major food source, are more abundant after sunset. There is also less competition for food from other groups of animals and, importantly, the dangers posed by predators are reduced.

The echolocation calls of bats are produced in the larynx and have durations between 0.2 and 100 ms. When searching for prey, bats in flight produce 10–20 calls per second at no apparent additional energetic cost. The repetition rate and the duration of their calls increases once they get closer to their prey so that they can benefit from faster updates about their target.

Anatomically, the ear of bats is similar to the ear of other mammals. How then can bats hear vibrations in the ultrasound range when other mammals cannot? The answer is that bats using echolocation have a section of the basilar membrane in the cochlea that is considerably thicker and adapted to resonate at the frequency of the echo produced by the respective bat's call. Also, the primary auditory neurons in this section of the basilar membrane respond most strongly to the frequency range of the echoes of the bat's call.

Odontocetes, such as killer whales, have a similar cochlear specialization as the bats. They live underwater where vision is limited but where acoustic waves propagate over vast distances. Toothed whales emit a focused beam of high frequency clicks in the direction that their head is pointing by passing air from the bony nares through a morphological structure called **phonic lips**. Incoming echoes are received through fatty structures around the jaw and the ears and are transmitted to the inner ears. The odontocetes have the highest density of ganglion cells per unit length of the basilar membrane (about 2500 ganglion cells/mm)[74] in the section of the basilar membrane that resonates to the echoes of their clicks.

Much simpler forms of echolocation systems compared with those present in bats and odontocetes have been more recently discovered in some cave dweller birds like the Palawan swiftlet (*Aerodramus palawanensis*), which can navigate in complete darkness in caves, and in some shrew species like the common Eurasian shrew (*Sorex araneus*).

Auditory system in arthropods

Insects have **tympanal organs** that are used for hearing. The organ consists of a thin cuticular membrane (tympanum) stretched across a frame over an air-filled cavity and an associated internal mechanoreceptor organ called the **chordotontal organ**, which senses the sound-induced vibrations in the tympanum. Depending on the group of insects, tympanal organs respond to frequencies ranging from 100 Hz to 240 kHz, are generally paired, and occur on any part of the body (legs, abdomen, thorax, base of the wing). Figure 17.38A shows tympanal organs located at the base of the tibia on the front legs of a meadow katydid[75].

Chordotonal organs are made of basic units called **scolopidia** (singular scolopidium, shown in Figure 17.38B). Each scolopidium consists of one or more ciliary bipolar sensory neurons, a **scolopale cell** and a **scolopale cap cell**. Scolopidia convert mechanical vibrations into action potentials (Figure 17.38B) that propagate to ganglia where information is combined and processed into a resultant behaviour. In the tympanal organ, each scolopidium responds to a particular frequency of the sound wave. A collection of individually tuned scolopidia on top of the trachea behind the tympanum can discriminate vibration frequencies in a manner similar to the ear of mammals.

Chordotonal organs also occur in crustaceans and are used in insects and crustaceans to detect not only vibrations in the tympanum but also movements of the substratum on which the animal stands, or the relative position of different parts of the exoskeleton—for example, the position of the antennae relative to the body.

17.4.2 Balance and spatial orientation

The ability of organisms to maintain balance and orient themselves in space is essential for their survival. Without maintaining balance and/or spatial orientation, organisms cannot move in a coordinated way to forage for food, avoid predation or find mates for reproduction.

Vertebrates use a vestibular organ for balance and spatial orientation

The **vestibular organ** in vertebrates is located in the inner ear labyrinth and plays a key role in maintaining balance by sensing the position and motion of the head. As illustrated in Figure 17.39, the vestibular organ of mammals generally consists of three **semicircular canals**, which are broadly oriented

[74] The density of ganglion cells on the basilar membrane in humans is about 1000 ganglion cells/mm.

[75] Katydids are closely related to crickets.

17

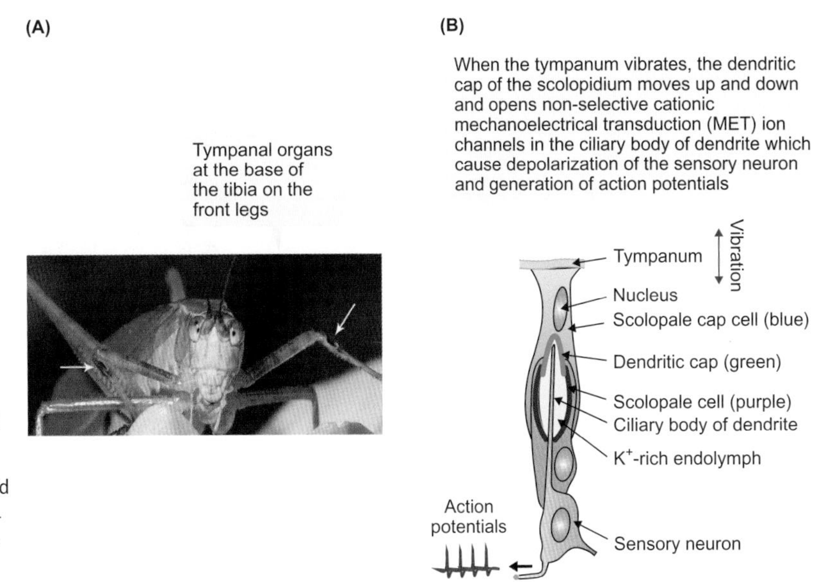

Figure 17.38 Tympanal organs in insects and how they work

(A) External view of tympanal organs in a meadow katydid indicated by the white arrows. (B) Structure of a scolopidium, the basic unit of chordotonal organs responsible for the transduction of sound waves into action potentials.
Source: A: http://songsofinsects.com/biology-of-insect-song.

in the three planes of space, and two chambers: the utricle (or utriculus) and the saccule (or sacculus). The vestibular organ is filled with K^+-rich endolymph and is surrounded by Na^+-rich perilymph. The organ communicates with the cochlear duct of the auditory apparatus and functions by activating hair cells contained in two types of sensory structures shown in Figure 17.40: the **cristae**, located in enlargements at the base of the semicircular canals (termed **ampullae**), and the **maculae,** present in the saccule and utricle.

The cristae consist of a collection of hair cells, supporting cells and afferent nerve fibres, which form a crest covered by a gelatinous cap called **cupula** in which the stereocilia of the hair cells are embedded. The cupula extends across the ampulla and is attached to the opposite wall such that

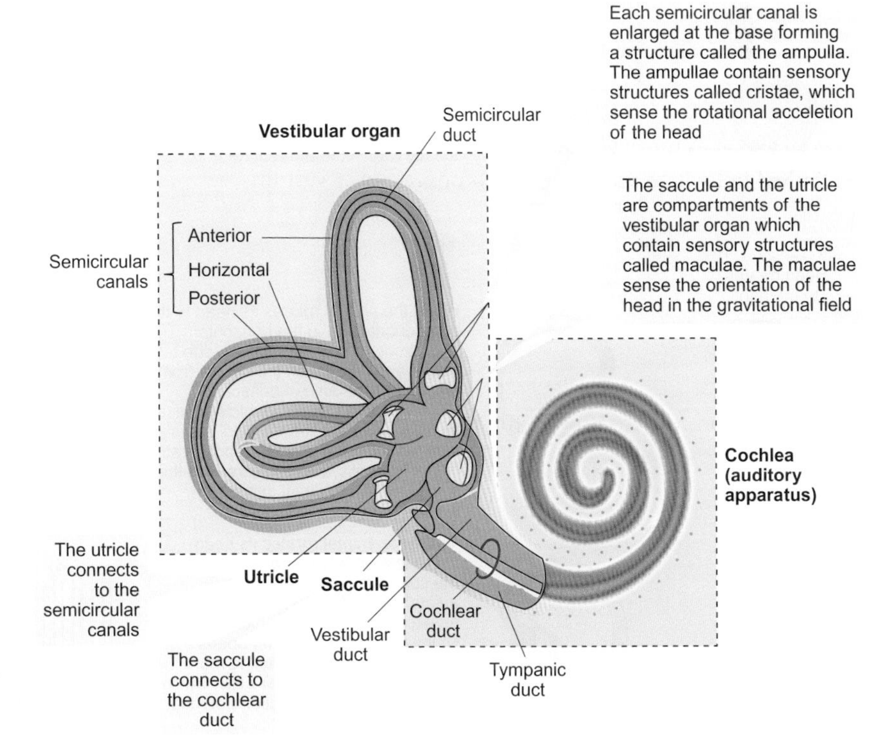

Figure 17.39 The inner ear labyrinth of mammals

The inner ear labyrinth of mammals consists of the vestibular organ, which contributes to the maintenance of an animal's balance, and the auditory apparatus. The vestibular organ comprises the semicircular canals, the utricle and the saccule.
Source: modified from Martini F (2012). Human anatomy, 8th edition. Pearson Education.

17

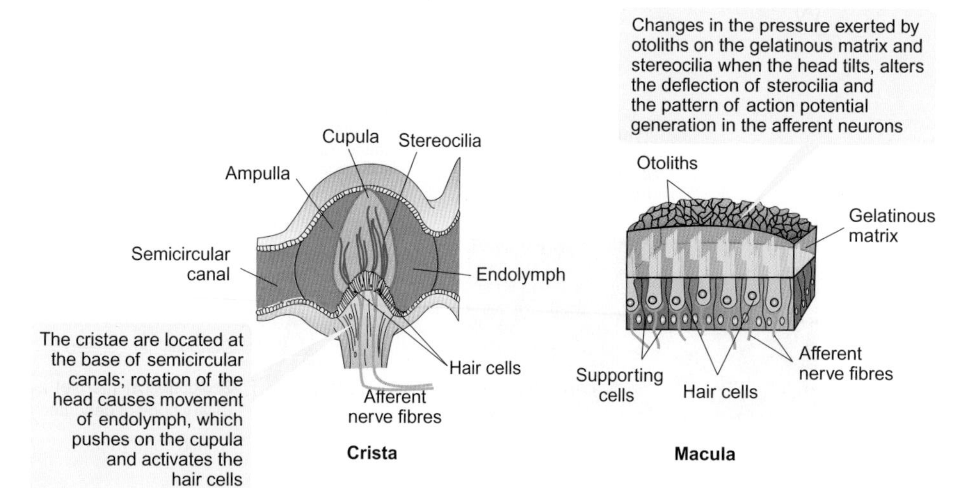

Figure 17.40 Sensory structures in the vestibular organs of vertebrates

rotation of the head from one direction to another causes endolymph in the semicircular canals to push on the cupula and deflect sterocilia in the opposite direction as shown in Figure 17.41. The direction and intensity of stereocilia deflection determines the level of depolarization of the hair cells, which, in turn, determines the pattern of action potential generation in the sensory neuron.

The sensory receptors within the utricle and saccule are called maculae. The maculae also consist of a collection of hair cells, supporting cells, nerve cells and a gelatinous matrix, but also contain **otoliths**, which are tiny dense particles of calcium carbonate embedded on the surface of the cupula as shown in Figure 17.40B[76]. The otoliths have a higher density than the endolymph and therefore exert a small pressure on the gelatinous matrix and on stereocilia of the underlying hair cells.

When the position of the head changes, the maculae tilt under a different angle in the gravitational field, changing

[76] Maculae are also called **otolith receptors** due to the presence of otoliths.

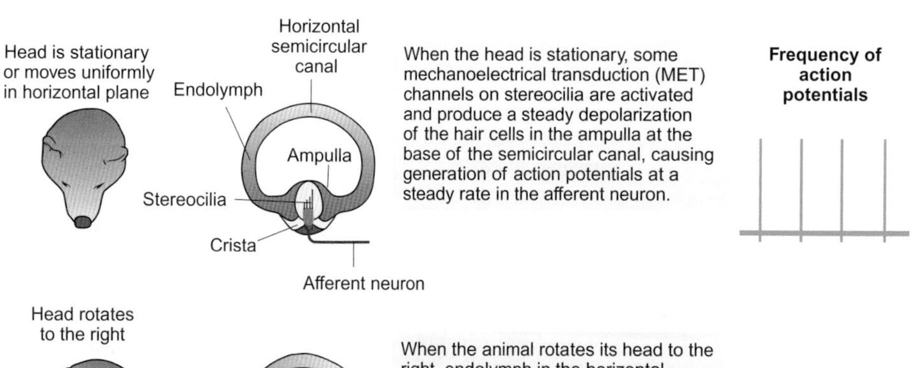

Head rotates to the right

When the animal rotates its head to the right, endolymph in the horizontal semicircular canal lags behind due to inertia and pushes on the cupula from the right of the animal deflecting stereocilia in a direction which releases tip links and closes more MET channels thus reducing the level of depolarization of the hair cells, which decreases the rate of action potential generation.

Head rotates to the left

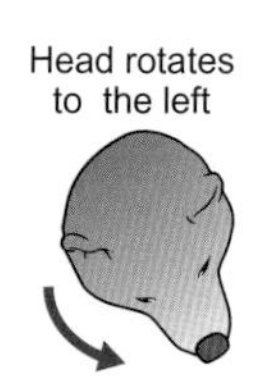

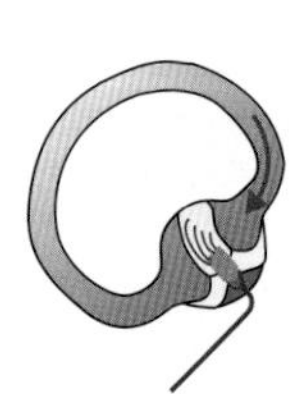

When the animal rotates its head to the left, endolymph pushes on the cupula from the left of the animal and deflects stereocilia in an excitatory direction which stretches tip links on stereocilia and opens more MET channels causing more depolarization of the hair cells, which increases the rate of action potential generation.

Figure 17.41 Transduction of rotational acceleration in semicircular canals

The hair cells in the ampullae of the semicircular canals have similar properties to the sound detecting hair cells in the organ of Corti: MET channels at the tip of stereocilia open when the deflection stretches the tip links, and close when the deflection, in opposite direction, releases the tip links.

the force exerted by otoliths on the stereocilia and the amount by which stereocilia are deflected. The change in the deflection of the stereocilia alters the level of polarization of the hair cell, as we discuss in Figures 17.36 and 17.41; this, in turn, alters the pattern of action potential generation in sensory neurons. The maculae are responsive not only to changes in orientation within the gravitational field, but also to changes in linear acceleration, since movement with acceleration in one direction causes the otholiths to deflect stereocilia in the opposite direction due to inertia.

Thus, the utricle and the saccule respond to changes in head orientation in the gravitational field, as well as to linear acceleration, while the semicircular canals respond to angular acceleration associated with the rotation of the head. Vestibular signals travel to the brain of vertebrates via sensory nerve fibres of the cranial nerve VIII, keeping the CNS informed about the orientation of the animal in the gravitational field and whether the animal is at rest or in motion. The vestibular signals also mediate automatic behaviours that maintain stability of gaze, body posture and balance.

Lateral line systems in fish and aquatic amphibians are multifunctional sensory systems

Fish and aquatic amphibians have a **lateral line** system present in their skin which provides information about the direction of water movement, and the presence of objects that cause specific disturbances in water. The lateral line is one of the main sensory systems in fish and is used to maintain balance in water, navigate around obstacles, locate prey, detect sound waves, identify predators and find social partners.

The lateral line system consists of strings of lateral line organs called **neuromasts**, which are located along lateral line canals on the head and extend along the entire body. In many species of fish, the canals are covered by skin that has pores at regular intervals through which water enters and flows over the neuromasts, as shown in Figure 17.42. Neuromasts consist of hair cells with their cilia embedded in a wedge-shaped cupula, which projects into the centre of the lateral line canal. The most sensitive axis of the neuromast is parallel to the canal axis. Neuromasts respond directly to minute changes in the water flow through the canal by changing the pattern of action potentials generated in the afferent neurons associated with the hair cells.

Aquatic invertebrates use statocysts for balance and spatial orientation

Some aquatic invertebrates, including bivalve molluscs, cnidarians, echinoderms, crustaceans and cephalopods, have balance sensory organs called **statocysts** that function in a similar manner to the vestibular apparatus in vertebrates. The statocysts consist of a mineralized mass called

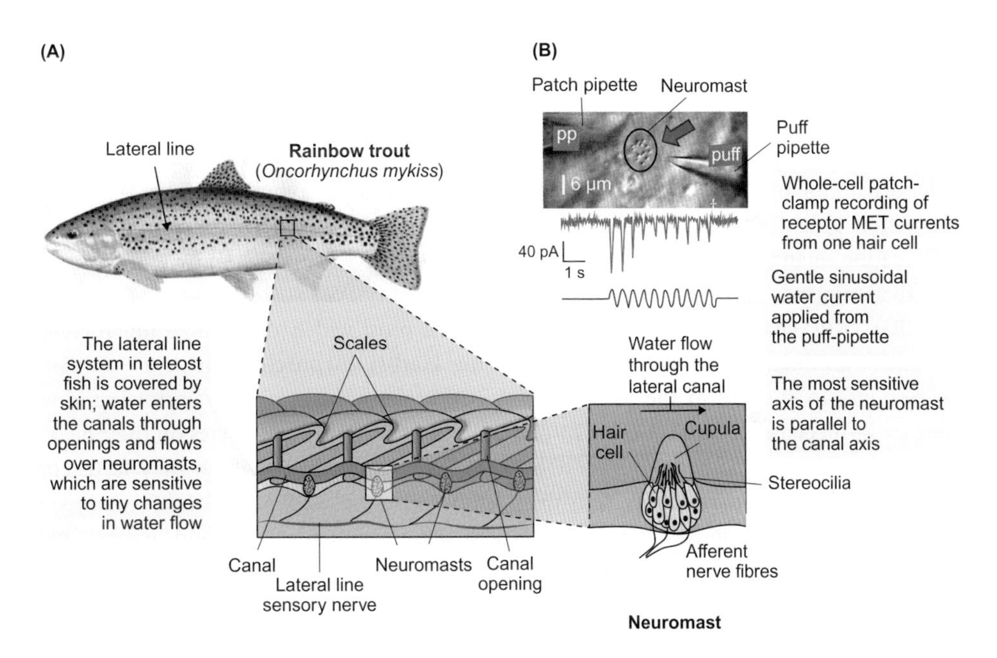

Figure 17.42 Lateral line sensory system in fish

(A) General appearance and structural organization of the lateral line in teleost fish. (B) Microphotograph of a neuromast in a young zebrafish (*Danio rerio*) before scale formation and whole-cell patch recordings (see Section 16.2) of depolarizing mechanoelectrical transduction (MET) currents from one hair cell in response to the application of gentle sinusoidal water currents over the cupula from a puff-pipette (puff). The location of the patch-pipette (pp) is also indicated on the microphotograph. The decrease in amplitude of the MET currents during the period of sinusoidal stimulation resembles sensory adaptation.

Source: A: US Fish and Wildlife Service; B: Ricci AJ et al (2013). Patch-Clamp Recordings from Lateral Line Neuromast Hair Cells of the Living Zebrafish. Journal of Neuroscience 33: 3131–3134.

the **statolith** surrounded by an epithelial layer containing sensory hair cells, forming a sac-like structure. When animals accelerate, the statoliths' inertia causes them to touch and deflect stereocilia of the sensory hairs lining the sac. The deflection of stereocilia alters the level of polarization of the hair cells, which changes the action potential signalling pattern to the CNS. This change in pattern provides feedback to the animal on the change in orientation to help them maintain their balance. Removal of statocysts interferes with the ability of animals to maintain balance: cephalopods that have had their statocysts removed lose their ability to swim and walk in a coordinated fashion.

17.4.3 Touch: making sense of shape, form and texture of objects in the environment

Touch is one of the five classical senses in humans (the others being vision/sight, hearing, smell, and taste)[77] and refers to the ability to perceive direct contact of the body with relatively solid objects in the external environment. Touch is essential for a myriad of behaviours in animals, from avoiding bodily harm arising from contact with external objects, to seeking social interactions conducive to mating and rearing the young. The sense of touch is mediated by different types of mechanoreceptors on the surface of animals, which respond to local deformations of the body surface in response to direct mechanical contact.

The touch mechanoreceptors have stretch-activated channels that are either non-selective cationic channels or are selective to sodium ions and protons. Activation of these channels when the body comes into contact with an external object causes local depolarization of the nerve terminals and initiates propagated action potentials. Although the exact type of non-selective cation channel present in each type of mechanoreceptor that responds to touch is not always known, we do know that TRP[78] channels play a significant role in mediating signal transduction in touch mechanoreceptors, either in response to changes in the membrane lipid curvature or due to the mechanical application of force.

Often, a certain part of an animal's body has a high density of mechanoreceptors, making it particularly sensitive to touch—for example, the arms of octopuses, the hands of primates, the snouts of pigs and rodents, or the antennae of insects and other arthropods. Different types of touch receptors are sensitive to different features of mechanical stimulation due to specific properties of accessory structures associated with the nerve terminals.

[77] The Greek philosopher Aristotle (384BC to 322BC) is credited with the original classification of senses in humans.

[78] We discuss TRP channels in Section 17.1.2.

The cell bodies of the sensory neurons that respond to touch in vertebrates are located in dorsal root ganglions. Signals from touch receptors travel to the spinal cord and, from there, use distinct pathways according to the functional category and location on the body of the mechanoreceptors involved. The signals eventually reach the cerebral cortex where they are interpreted as sensations.

Touch receptors

Figure 17.43 shows the different types of mechanoreceptors present in the skin of mammals. The largest touch receptor, called the **Pacinian corpuscle**, is located in the innermost layer of the skin and responds to sudden changes in pressure both when pressure is applied and when the pressure is removed. The Pacinian corpuscle is therefore a phasic receptor and is well suited to detect mechanical vibrations of up to several hundred Hz. The corpuscles are common in the fingertips of primates and around the footpads and claws of terrestrial mammals. **Meissner's corpuscle** is another type of phasic touch receptor in mammals that confers exquisite mechanical sensitivity to lower-frequency vibrations up to 50 Hz.

The skin of mammals also contains two types of tonic touch receptors, the **Ruffini's corpuscles** and the **Merkel's discs**, which remain stimulated for as long as pressure is applied to the skin. Such receptors encode information about how hard the foot is pressing onto the ground or how firm the hand is grasping an object.

Hairs on the skin of mammals also act as tactile organs, as hair contact with external objects stimulates hair follicle receptors and activates sensory neurons. Moreover, whiskers found on the snouts of rodents, cats, dogs, seals, sea lions and walruses are typically thicker than normal hairs and are very sensitive to touch and vibrations as the whisker follicles are heavily innervated by hundreds of primary afferent neurons.

The most widely distributed touch receptors in birds are the **Herbst corpuscles**, which are similar to the Pacinian corpuscles of mammals. Waterfowls and wading birds have numerous Herbst corpuscles on their beaks, which they use to locate animals on which they feed. For example, sandpipers like the red knots (*Calidris canatus*) have been shown to sense prey under wet sand or soil. Other birds, such as woodpeckers, have a high density of Herbst corpuscles on their tongues, which are used to sense the presence of moving invertebrates (their food) under tree bark.

Stretch-sensitive mechanoreceptors attached to the skin, or the cuticle of invertebrates, also function as touch receptors. For example, chordotonal organs of insects made of scolopidia are not only used for hearing but are used as very sensitive touch receptors when they are attached to bristle-like structures on an insect's surface.

17

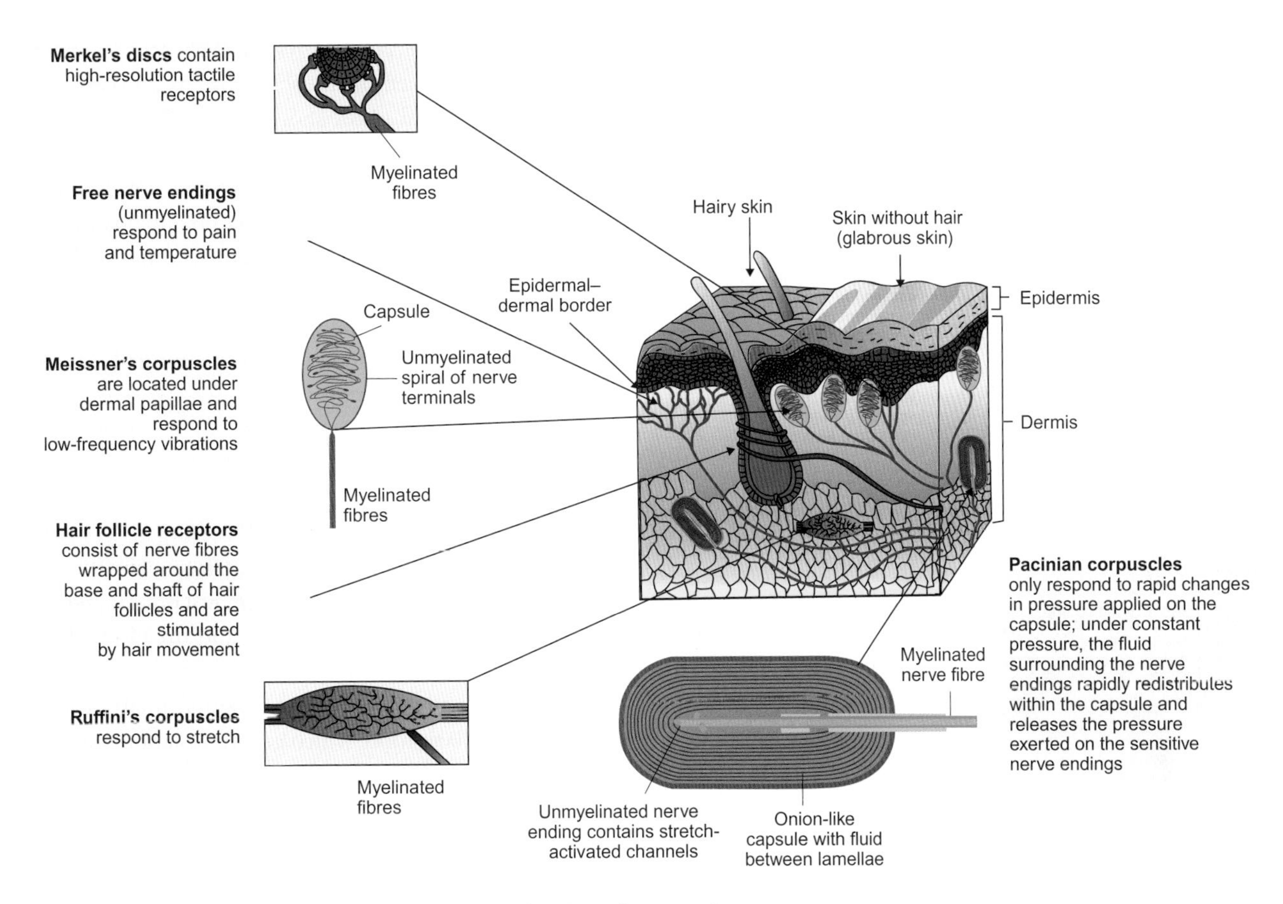

Figure 17.43 Touch sensory receptor structures in the skin of mammals

17.4.4 Internal mechanoreception

Internal mechanoreception is the sense that allows animals to detect their internal mechanical environment, including the orientation and displacement of different parts of their body, strength of their muscle contraction, and pressure in their circulatory system.

Proprioception

Proprioception (from Latin '*proprius*' meaning 'one's own') refers to the ability of animals to 'feel' themselves by sensing the relative position of their different body parts and the strength of contraction of the muscles used to maintain posture when the animal is stationary or when it moves. Proprioception is made possible by the presence of mechanosensitive receptors located in muscles[79], tendons, joint structures and body surfaces that provide information about muscle length, muscle force, speed of muscle shortening and joint angles.

Signals from proprioceptors travel to the CNS, where the proprioceptive information is integrated with information from mechanoreceptors involved in balance and spatial orientation, to produce an overall sense of body orientation in the environment. Proprioception in humans also has a conscious component that makes us aware of the location and position of different parts of our body even when our eyes are closed.

There is strong evidence that the mechano-sensitive channel in the proprioceptive organs of several groups of animals, including nematodes (*Caenorhabditis elegans*), flies (*Drosophila melanogaster*), zebra fish (*Danio rerio*) and African clawed toads (*Xenopus laevis*) is a member of the TRP family of channels[80], the TRPN channel. In mammals, however, no mechano-sensitive channel in proprioceptor organs has, as yet, been discovered.

Pacinian and Ruffini's corpuscles, located around joints and in the joint-supporting ligaments of mammals, function as proprioceptors and encode information related to joint angle. Vertebrates also have specialized proprioreceptors called **muscle spindles** (due to their shape), which provide information about changes in muscle length, and **Golgi tendon organs**, which provide information about the amount of force produced by a particular muscle.

[79] We discuss muscles, tendons and joints in Chapter 18.

[80] We discuss TRP channels in Section 17.1.2.

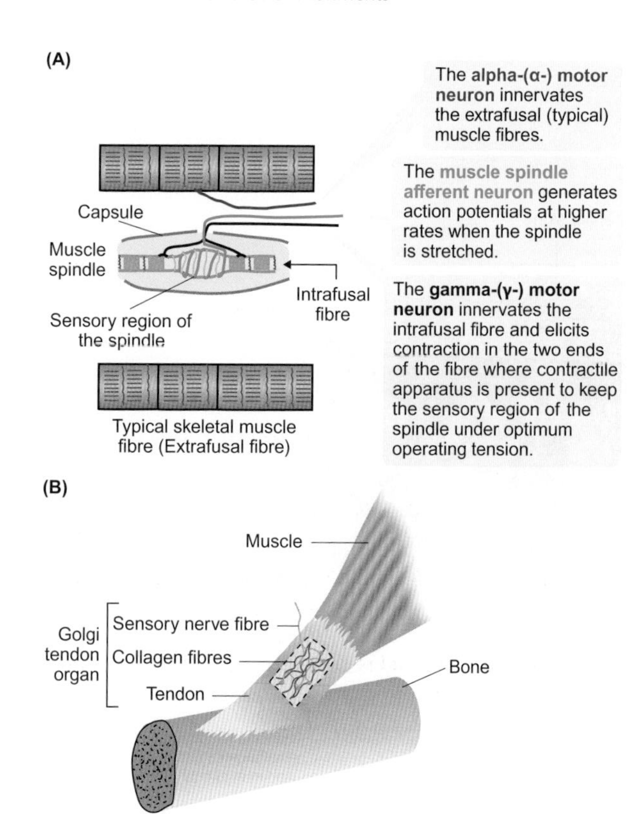

Figure 17.44 Mechanosensitive organs involved in internal mechanoreception

(A) Muscle spindles are embedded in skeletal muscles between typical muscle fibres and detect absolute changes in length and the rate of change in muscle length. (B) Golgi tendon organs consist of sensory neurons intertwined with collagen fibres and provide information about the level of force generated in that particular muscle.

Muscle spindles

Muscle spindles are stretch receptors located throughout skeletal muscles of vertebrates in parallel with the muscle axis. They consist of small bundles of fibres called intrafusal[81] fibres, enclosed in a capsule, as shown in Figure 17.44A. Unlike typical muscle fibres (called extrafusal fibres), where the contractile apparatus is continuous from one end of the fibre to the other, the intrafusal fibres are devoid of contractile machinery in their central region. This area is filled with nuclei and is wrapped by dendrites of primary sensory (afferent) neurons of relatively large diameter[82].

When the muscle lengthens, the muscle spindles become stretched; this in turn stretches the dendrites in the middle of the fibre. This causes stretch-sensitive, non-selective cation channels in the sensory dendrites to open, and a depolarizing receptor potential to develop. As a result, the rate of action potential generation in the primary sensory neuron rises. Conversely, when the muscle shortens, the central region of the intrafusal fibres shortens as well, reducing the stress on the stretch-sensitive channels in the endings of the primary neurons. This makes the membrane potential more negative and reduces the rate of action potential generation in the primary sensory neuron. Thus, the endings of the primary sensory neurons in muscle spindles detect absolute changes in muscle length and the rate of change in muscle length, both of which contribute to the animal's sense of limb position and movement.

[81] The Latin word '*fusus*' means spindle, so 'intrafusal' means inside the spindle.

[82] We discuss nerve fibre classification in Section 16.1.3.

Golgi tendon organs

The Golgi tendon organs (Figure 17.43B) are located in tendons that attach muscles to bone. These organs consist of sensory dendrites interwoven with collagen fibrils in the tendon. When the muscle produces force, it pulls on the tendon, straightening the collagen fibrils and stretching the associated nerve terminals. This stretching activates mechanosensitive ion channels in the nerve terminals, causing membrane depolarization, which generates action potentials in sensory fibres at rates that increase with the increase in force produced by the muscle. Therefore, the Golgi tendon organs provide information about the amount of force produced by the muscle.

The proprioceptive information in vertebrates is integrated with information from mechanoreceptors involved in balance and spatial orientation in the cerebellum, the part of the brain responsible for maintenance of balance.

Proprioceptors in arthropods

Insects and crustaceans have well developed proprioreceptor organs to convey information to the animal about joint movements and the position of various body parts. Crustacean stretch receptors have cell bodies lying in isolation close to a bundle of muscle fibres into which the dendrites are inserted and have axons projecting to a segmental ganglion in the CNS, from which the receptors also receive inhibitory innervation. This morphological arrangement makes it relatively easy to gain the access needed to take accurate physiological recordings from single sensory cells using microelectrodes, which has contributed greatly to our current understanding of how primary sensory receptors work.

Two types of stretch receptors in crustaceans are involved in proprioception:

- Rapidly adapting receptors are phasic receptors that respond well only at the beginning of the stretch, after which their response rapidly wanes.
- Slowly adapting receptors are tonic receptors that maintain their response during prolonged stretch.

Like the muscle spindles in mammals, the dendrites of the stretch receptors in arthropods contain stretch-sensitive channels that become activated when the muscle fibres in which the dendrites are embedded are stretched.

Chordotonal organs made of **scolopidia** (described earlier in Figure 17.38B) also function as proprioceptors in insects and crustaceans when they span the two arms of a joint and report the joint angle. For example, **Johnston's organ** in insects consists of several hundred scolopidia located in the second segment of the antennae; it detects fine motions in the uppermost segment of the antenna, called the flagellum.

Baroreception

Baroreception refers to the ability of animals to sense pressure; we describe in Section 22.4.2 how all vertebrates have mechanosensory neurons, called **baroreceptors**, which sense the blood pressure at strategic locations in the circulatory system. The baroreceptors have mechanosensitive cationic channels that open and cause membrane depolarization (and, hence, an increased rate of action potential generation) when the blood vessels in which the baroreceptors are located become stretched in response to increased blood pressure.

The change in the firing rate of action potentials by the baroreceptors is recognized by neurons in the medulla, which, in turn, influence the cardiac outputs and vascular resistance to blood flow in order to return the mean arterial blood pressure to its 'normal' level. The baroreceptors are thus part of a fast negative-feedback system, called the **baroreflex**, which helps regulate the blood pressure on a short-term basis.

Mechanoreceptors also play an important role in the control of the respiratory system as we discuss in Section 22.2.1.

› *Review articles*

Avan P, Büki B, Petit C (2013). Auditory distortions: origins and functions. Physiological Reviews 93: 1563–1619.
Fettiplace R, Hackney CM (2006). The sensory and motor roles of auditory hair cells. Nature Reviews Neuroscience 7: 19–29.
Gillespie PG, Walker RG (2001). Molecular basis of mechanosensory transduction. Nature 413: 194–202.
Lumpkin EA, Marshall KL, Nelson AM (2010). The cell biology of touch. Journal of Cell Biology 191: 237–248.

17.5 Thermoreception

Thermoreception refers to the ability of animals to detect temperature on their body surface and in tissues deeper inside their bodies; it plays a major role in protecting animals from damaging temperatures and maintaining thermal homeostasis in mammals and birds[83].

The sensation of temperature arises largely from the activity of narrow nerve fibres (nonmyelinated group C fibres and myelinated group Aδ fibres)[84] which lack any accessory structures and end freely in tissues. The free endings of these fibres form dense networks with multiple branches and become activated when specific transient receptor potential (TRP) channels[85], called **thermoTRPs,**[86] open in their membranes. The opening of these channels causes membrane depolarization and the initiation of trains of action potentials at the nerve terminals, which travel to specific regions in the brain.

Different subpopulations of thermoreceptive neurons express different thermoTRP channels, which have different temperature thresholds for activation and are responsible for producing sensations that range from noxious cold to burning hot in humans. Mammals express at least six different thermoTRP channels in sensory neurons:

- the TRPA1 channel is activated when the temperature drops below 17°C and is responsible for triggering the sensation of noxious cold;
- the TRPM8 channel is activated below 25°C;
- TRPV4 is activated above 25°C and reaches a peak around 40°C, after which its activity decreases;
- TRPV3 is activated when the temperature rises above 33°C;
- TRPV1 is activated when the temperature reaches 42°C and is responsible for the burning hot sensation;
- finally, TRPV2 is activated when the temperature reaches 52°C.

Interestingly, thermoTRP channels are also receptors for various chemicals found in the environment. For example, **capsaicin**, the active ingredient in chilli peppers, also binds to the TRPV1 channel in the thermoreceptive neurons in the mouth causing the burning hot sensation we experience when eating chilli peppers. Menthol, on the other hand, binds to the TRPM8 channels and induces the sensation of cold. Chemical pollution in the environment, which contains molecules that can bind to peripheral thermoTRP channels, can seriously compromise the ability of animals to avoid harmful temperatures because it activates an animal's TRP channels, interfering with their thermoreceptive function.

All animals investigated, including nematodes, flies, fish, anurans, reptiles and birds, express homologues of at least

[83] We discuss temperature and endothermy in Chapter 10.

[84] We discuss nerve fibre classification in the peripheral nervous system in Section 16.1.3.

[85] We discuss TRP channels in Section 17.1.2.

[86] We also discuss thermoTRPs in Section 10.2.5.

some of the thermoTRP channels reported in mammals. However, the various homologues may have either a different temperature activation threshold than in mammals or may not be sensitive to temperature at all.

For example, the diamondback rattle snake (*Crotalus atrox*) TRPA1 has a threshold for activation of around 28°C. This contrasts with mammalian TRPA1 channels, which activate at temperatures <17°C, and with fish TRPA1-like channels, which are not activated by heat. These observations show the broad potential of TRP channels in general to act as thermosensors for a specific temperature range or as receptors for various chemicals in the environment.

Pit snakes detect infrared signals through thermoreceptor organs located in facial pits as shown back in Figure 17.9C. The thin, highly vascularized and densely innervated membrane, divides the pit into a larger section facing the outside and a smaller section facing the inside. Thermal infrared radiation (wavelength range 5–1000 μm) from a source that is warmer than the environment, such as a small rodent or a bird, enters the pit and warms up a small area of the membrane in which exceptionally temperature sensitive TRPA1 channels are present in the nerve terminals. A tiny rise in temperature relative to the immediate environment (in the order of only 0.001°C) activates heat-sensitive TRPA1 cationic channels, causing local depolarization in the nerve terminals, which, in turn, increases the firing rate of the respective neurons. Signals from the pit organ are then conveyed to the optic tectum of the brain where they are integrated with visual signals. The heavy vasculature around the thermoreceptors ensures that after stimulation, the receptors are rapidly cooled down to their neutral temperature by blood flowing through the blood vessels[87].

The presence of pit organs on either side of the head gives pit snakes the ability to perceive depth and locate the source of infrared radiation; this ability is of great value for hunting prey at night and finding suitable places for thermoregulation.

> *Review article*

Bandell M, Macpherson LJ, Patapoutian A (2007). From chills to chillis: mechanisms for thermosensation and chemesthesis via thermoTRPs. Current Opinion in Neurobiology 17: 490–497.

17.6 Nociception

Nociception (from Latin '*nocere*', meaning hurt/harm) refers to the ability of animals to perceive noxious or damaging stimuli. It occurs when intense mechanical, thermal or chemical stimuli increase the excitability of specific primary neurons that function as nociceptors, signalling to the CNS the presence of adverse stimuli. The ability of organisms to detect noxious or damaging stimuli and to respond effectively is an important trait for their survival. It is therefore not surprising that nociception is not restricted to vertebrates but is present in most animal groups, including medicinal leeches (*Hirudo medicinalis*) and sea slugs (*Aplysia californica*), which have become model systems for studying it.

Similar to thermoreceptors, the nociceptors consist of free afferent nerve terminals of small-calibre primary fibres (non-myelinated group C fibres and myelinated group Aδ fibres). In order to initiate electrical signals at the nerve terminal, a nociceptive stimulus needs to produce depolarization of the neurons above the threshold to trigger action potentials.

Some molecular mechanisms involved in the detection of noxious stimuli have been identified in mammals. For example, painful heat (hotter than 43°C) causes nonselective cation channels (TRPV1) to open in group C nociceptive fibre endings, causing membrane depolarization and the generation of action potentials. If the localized temperature rises above 52°C, the TRPV2 cation channels discussed in the previous section also become activated. Acids may also act directly on acid-sensitive channels found in nociceptive neurons and cause depolarization. Mechanical stimuli leading to skin damage can also directly activate nociceptive fibres.

Nociceptors also respond to chemical activators from damaged cells. For example, ATP released from damaged cells activates ATP-gated ion channels in nociceptive endings, causing depolarization. **Bradykinin**, a nine-amino-acid peptide formed from the action of cytoplasmic proteases released from damaged cells on serum proteins, is a particularly potent activator of nociceptive fibres. The action of bradykinin is mediated by metabotropic[88] receptors and also increases the sensitivity of the nociceptive fibres to other stimuli.

Nociception can lead to complex unconscious physiological and behavioural responses, as well as to learning avoidance reactions, but nociceptive reactions cannot be equated to pain. The ability of organisms to experience pain requires more than the activation of nociceptors because pain always encompasses a conscious response and an emotional experience (suffering). In humans, signals from nociceptors are processed into a sensation of pain in the cerebral cortex. For organisms other than humans it is not possible to know directly if they experience pain, but we try to infer that this is the case from physiological and behavioural reactions to noxious stimuli that are similar to those observed in humans, and from fMRI[89] studies in which animals are subjected to noxious stimuli.

Based on such reactions, it is mostly agreed that vertebrates other than fish and some invertebrates, such as cephalopod

[87] We discuss temperature and ectothermy in Chapter 9.

[88] We refer to metabotropic receptors in Section 17.1.2.

[89] We discuss fMRI in Box 17.2.

molluscs and decapod crustaceans, experience pain when subjected to noxious or damaging stimuli. The question of whether fish experience pain remains a controversial issue, however.

› Review articles

Rose JD, Arlinghaus R, Cooke SJ, Diggles BK, Sawynok W, Stevens ED, Wynne CDL (2012). Can fish really feel pain? Fish and Fisheries 15: 97–133.

Sneddon LU (2004). Evolution of nociception in vertebrates: comparative analysis of lower vertebrates. Brain Research Reviews 46: 123–130.

Smith ES, Lewin GR (2009). Nociceptors: a phylogenetic view. Journal of Comparative Physiology A 195: 1089–1106.

17.7 Electroreception

Electroreception refers to the ability of some species, both invertebrate and vertebrate, to sense electric fields[90] in their environment to augment or replace their other senses when foraging, for prey detection, predator avoidance and communication.

Passive electroreception occurs when animals sense electric fields produced by various sources in their habitat. In contrast, **active electroreception** occurs when animals generate an electric field and then sense distortions of this electric field by objects in the environment that have different electrical properties.

[90] The term 'electric field' is defined in Box 3.1.

17.7.1 Passive electroreception

Passive electroreception has been observed in both aquatic and terrestrial animals belonging to groups as diverse as insects, fish, amphibians and mammals. It is used in foraging and prey and predator detection.

Flying insects, such as bumblebees (*Bombus terrestris*), become positively charged (compared to the ground) when flying through the air, while flowers are generally negatively charged, compared to the ground, due to an atmospheric electrical field. This electrical charge difference between flying insect and flower generates an electric field, which changes depending on the pollination status of the flower and whether the flower was recently visited by a pollinator. Such a visit causes some negative charges on the flower to be neutralized by positive charges flowing from the surface of the visiting pollinator.

From evidence presented in Experimental Panel 17.1, bees sense the presence of electric fields in their environment via mechanoreceptors in their antennae (Johnston's organs[91]) and can assess whether the flower has been recently visited by another bee. When bees encounter an electric field, an electric force[92] develops, acting on the flagellum (the uppermost segment of the antenna), which is sensed by mechanoreceptors in the Johnston's organ.

[91] We discuss Johnston's organ in Section 17.4.4.

[92] The electrical force is the product of the electric charge (on the flagellum) and the intensity of the electric field.

Experimental Panel 17.1 Floral electric fields act as sensory cues for insect pollinators

Background to the study

Floral cues, such as colour, shape, patterns, petal texture and fragrances, play a central role in the complex interactions between plants and pollinators by enhancing the foraging efficiency of pollinators, which, in turn, makes pollination possible. The transfer of pollen between flowers and insect pollinators and *vice versa* is facilitated by the presence of electric fields between flowers and insect pollinators. The question that arises is: Can insect pollinators perceive floral electric fields and use them as foraging cues?

Experimental approach

A group of researchers at the University of Bristol, UK used bumblebees (*Bombus terrestris*) and artificial flowers (E-flowers) with controlled electric fields to test the hypothesis that bumblebees have the ability to discriminate between E-flowers which differed only with respect to their electric fields.

Bumblebees are relatively large insect pollinators which can be followed visually with ease and also permit reliable measurements of the electric charges carried on them. The flying bumblebee is generally positively charged, carrying, on average, 32 pC[1] of positive charges.

The E-flowers consisted of 35 mm diameter × 1.5 mm thick steel base discs with a smaller, thin purple epoxy disc on top. The steel base discs are electric conductors and can be kept at different electric potentials to produce different electric fields around them. A +30 V difference between the E-flower and its environment was considered to be biologically relevant for a flower standing 0.3 m high in a typical fair-weather atmospheric electric field[2] of −100 Vm^{-1}. The electric field profile around the E-flower kept at +30 V is shown in Figure A. When the E-flower is kept at 0 V, there is no electric potential difference between the E-flower and the surrounding air.

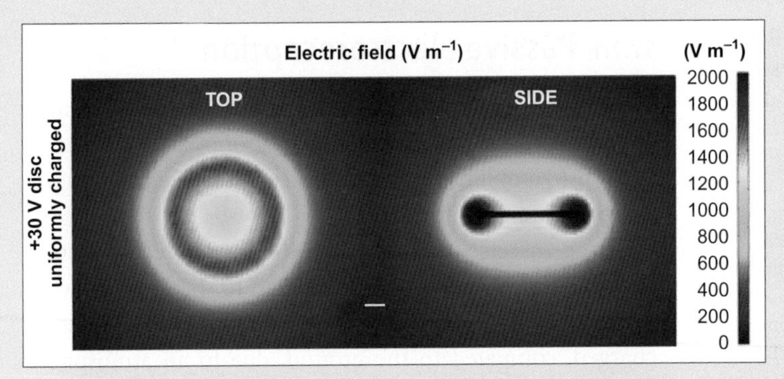

Figure A Colour-based representation of the electric field around an E-flower kept at +30 V viewed from the top and from the side

The horizontal calibration bar represents 1 cm.

Source: adapted from Clarke D et al (2013). Detection and Learning of Floral Electric Fields by Bumblebees. Science 340: 66-69.

Results

In the main test, half of the otherwise identical E-flowers had a small droplet of sucrose solution placed in the middle, on the epoxy disc, and were kept at +30 V bias potential, which produced the electric field shown in Figure A. The other half had a droplet of an aversive solution[3] instead and were kept at 0 V. Bumblebees were then allowed to make 50 visits to the E-flowers.

The accuracy with which the bees detected the rewarding E-flowers with the sucrose solution increased steadily and reached 81 ± 3 per cent in the last 10 visits to the E-flowers, as shown in Figure B(i). The result was highly statistically significant when compared to random choice ($P < 0.0001$), indicating that the presence of the electric field on the E-flowers acted as a cue for bumblebees to distinguish between rewarding and aversive E-flowers.

As a control experiment, the electric potential on the rewarding E-flowers was then reduced to 0 V, like that of the aversive E-flowers and the performance of the same group of bumblebees was recorded. The level of accuracy with which the bees detected the rewarding E-flowers dropped suddenly to 54 ± 4 per cent, which was not statistically different from random choice ($P = 0.35$). This result confirms that the presence of the electric field around the E-flowers kept at +30 V acted as a cue for distinguishing between rewarding and aversive E-flowers.

The bumblebees could not distinguish between rewarding and non-rewarding E-flowers when the rewarding E-flowers were kept at +10 V instead of +30 V, suggesting there is a threshold below which bumblebees cannot detect electrical fields. Since positively charged bumblebees landing on negatively charged natural flowers reduce the electrical potential of the flowers within a matter of seconds for a period of time lasting between minutes to hours (depending on the atmospheric conditions), the reduced electric field may not act as a strong cue for other bumblebees to visit the respective flowers, which would have been depleted of nutrients.

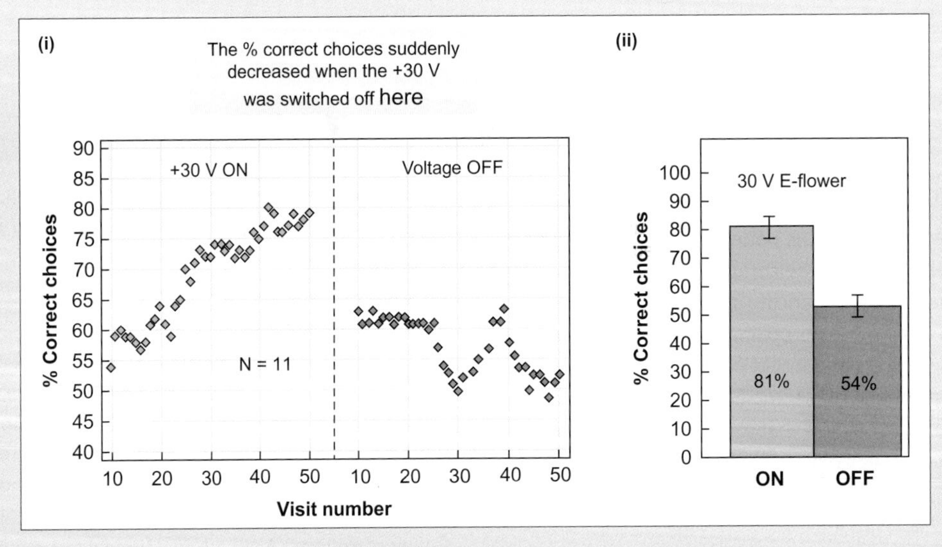

Figure B Learning curve of foraging bumblebees trained to E-flowers

(i) Half of the E-flowers had a small droplet of sucrose solution on them (rewarding E-flowers) and were kept at +30 V potential and the other half had a droplet of aversive (bitter) solution on them (aversive E-flowers) and were kept at 0 V. After 50 visits, the electric field of the rewarding E-flowers was switched off such that now all E-flowers were at 0 V. (ii) The % correct choices refers to the percent of total visits to rewarding E-flowers. The bar graph shows the % correct choices (± SEM) for visits 41 to 50 when the 30 V was on for the rewarding E-flowers and after the voltage for the rewarding E-flowers was brought down to 0 V.

Source: Clarke D et al (2013). Detection and Learning of Floral Electric Fields by Bumblebees. Science 340: 66-69.

In a different test, in which the geometry of the electric field was changed (flowers with bull's eye pattern with the centre ring held at −10 V and the outer ring held at +20 V vs flowers that had both rings held at +20 V), the bumblebees were also able to discriminate between the two different electric field geometries.

Moreover, the bumblebees required significantly fewer visits to E-flowers that differed in both colour and electric field to reach 80 per cent accuracy compared with E-flowers that differed only in colour. This result shows that the combination of flower colour and electric field acts as a significantly stronger cue than flower colour alone.

Conclusions

Taken together, the results from the above experiments show that floral electric fields act as sensory cues for bumblebees, which can distinguish between fully electrically charged and uncharged or less electrically charged flowers and also between charged flowers that have different geometries of electric field. Since visiting pollinators affect the floral electric fields, such information can be used by other potential pollinators to asses whether the flower has been recently visited by another pollinator and is therefore likely to have a reduced concentration of nutrients.

› Find out more

Clarke D, Whitney H, Sutton G, Robert D (2013). Detection and learning of floral electric fields by bumblebees. Science 340: 66–69.

Rycroft MJ, Israelsson S, Price C (2000). The global atmospheric electric circuit, solar activity and climate change. Journal of Atmospheric and Solar-Terrestrial Physics 62: 1563–1576.

Vaknin Y, Gan-Mor S, Bechar A, Ronen B, Eisikowitch D (2000). The role of electrostatic forces in pollination. Plant Systematics and Evolution 222: 133–142.

[1] The coulomb (C) is the SI unit for electrical charge as we discuss in Box 3.1. 1 pC = 10^{-12} C.

[2] The electrical potential of the atmosphere around the flower at 0.3 m from the ground is −30 V (−100 V m^{-1} × 0.3 m = −30 V). Since the flower is connected electrically to the ground through the stem and the roots of the plant, it follows that the potential difference between the flower and its surroundings is +30 V (0 V − (−30V) = +30V).

[3] The aversive solution contained quinine hemisulfate, which is bitter to humans.

The vast majority of fish and amphibians do not possess the sense of electroreception. However, some fish species—including lampreys, lungfish, elasmobranchs (sharks, stingrays, skates), sturgeons, paddlefish and catfish—and some amphibians—including the giant salamander (*Andrias davidianus*) and the axolotl (*Ambystoma mexicanum*)—possess electroreceptive organs derived from mechanoreceptors in the lateral line[93] (see Figure 17.42) to detect electrical fields produced by potential prey and predators.

[93] We discuss lateral line mechanoreceptors in Section 17.4.2.

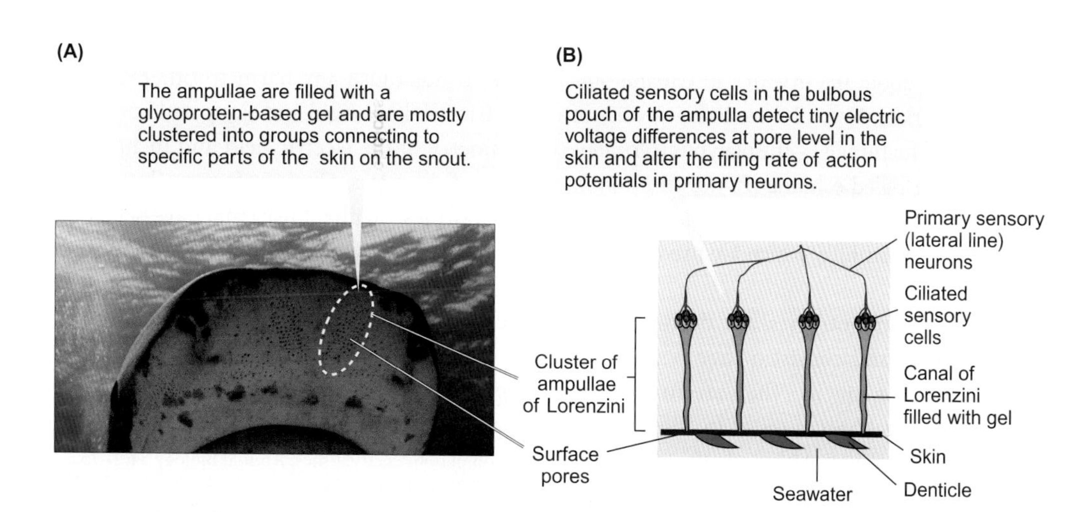

Figure 17.45 The electroreceptive organs of sharks, called the ampullae of Lorenzini, are located on the shark's snout

(A) Image of the snout of the tiger shark (*Galeocerdo cuvier*) viewed from below. (B) Structural elements of the ampullae of Lorenzini.

Source: A: Wikimedia Commons/ Albert Kok.

All animals emit electric signals as a consequence of the activity of their nerves and muscles. These signals are transmitted much more effectively through water than through air because water, in general, and seawater in particular, is a much better electrical conductor than air due to the presence within it of electrically charged ions.

Sharks, for example, have the most sensitive passive electroreceptive system among animals, allowing them to detect electric fields with threshold sensitivities as low as 0.5 $\mu V\ m^{-1}$. The electroreceptive organs in sharks, called the **ampullae of Lorenzini**, are located on their snouts (Figure 17.45A). The ampullae detect electric voltage differences between the pore openings at skin level and the pouch of the ampullae, where ciliated sensory cells are located (Figure 17.45B). A negative pore stimulus causes activation of Ca^{2+} channels in the ciliated cells, but we are yet to elucidate the exact molecular mechanism for this happening. The ensuing depolarization and increased Ca^{2+} at synapses between ciliated cells and primary sensory neurons enhance transmitter release at the synapse which increases the action potential frequency in primary neurons; conversely, a positive pore stimulus decreases activity in sensory neurons.

The pollution of oceans with man-made objects that can produce local electric fields, such as discarded electrical batteries, discarded metallic objects and improperly insulated cables[94], can be problematic: these electric fields can attract sharks and cause their injury or death when they attempt to attack or ingest them.

Passive electroreception also occurs in some mammals. For example, at least one species of cetacean, the Guiana dolphin (*Sotalia guianensis*), has developed passive electroreception. The electroreceptor organs in Guiana dolphins evolved from mechanosensory structures originally associated with the mammalian whiskers and are primarily used for prey detection. The sensitivity for detecting electric fields is 1000 times lower in Guiana dolphins than in sharks. (The threshold for electrical field detection is about 0.5 $mV\ m^{-1}$ in Guiana dolphins vs 0.5 $\mu V\ m^{-1}$ in sharks.) Also, all three living species of monotremes (the egg lying mammals)—the platypus (*Ornithorhyncus anatinus*), the long-billed echidna (*Zaglossus brujnii*), and the short-billed echidna (*Tachyglossus aculeatus*)—have electroreceptors located on their bills, which are used for prey detection.

The electroreceptors in monotremes are derived from cutaneous secretory glands and consist of free nerve endings embedded in a mass of mucus. They are innervated by trigeminal sensory nerve fibres, similar to the electroreceptors in the Guiana dolphins. Although the detection threshold for electric fields is four times higher in platypus (about 2 $mV\ m^{-1}$) than in Guiana dolphin, the electroreceptive ability of the platypus is well suited to the detection of bioelectric fields emitted by their prey.

17.7.2 Active electroreception

Active electroreception occurs in some slow-moving or sedentary fish that are active at night or in areas of low visibility. These fish, called electric fish, have electric organs that generate a specific electric field. Their electroreceptors sense distortions in this specific field by objects in the environment that have electric properties.

The **electric organs**, which evolved independently from skeletal muscle fibres at least six times in six unrelated orders belonging to 13 families (an example of parallel evolution), consist of flat, disc-like cells called **electrocytes, electroplaques** or **electroplaxes**; like muscle fibres,[95] these cells can be stimulated by neurons to produce action potential-like pulses of 0.15 V amplitude. Connecting electrocytes electrically in a series of hundreds to thousands can generate very large electrical potential differences when the electrical organ is neurally discharged. Indeed, some electric fish can produce very large electric shocks, which can stun or even kill prey, or act as a defensive weapon. These fish, called strongly electric fish, belong to five families and include electric eels (*Electroporus electricus*), which can produce pulses of up to 650 V in fresh water, and torpedo or electric rays (*Torpedo* spp.)[96], which produce electric discharges with an intensity up to 220 V and 30 A in seawater—enough to kill a person.

Weakly electric fish generate weaker electric fields that are used for locating hidden prey, for navigating at night, in caves or in turbid waters, and for communicating with members of their own species. They can produce either pulses of electricity or continuous discharges that can be modulated and do not cause any harm to other animals. The weakly electric fish have rigid and elongated bodies to ensure lateral symmetry of the electric field and include members of two teleost fish orders: Gymnotiformes (knifefishes) and Osteoglossiformes (elephant fish) that have been intensely studied. The electric eel, which is actually a knifefish, in addition to producing strong electric fields can also produce weak electric fields that are used in navigation and communication.

› Review articles

Collin SP, Whitehead D (2004). The functional roles of passive electroreception in non-electric fishes. Animal Biology 54: 1–25.

[94] Recent information suggests that even insulated submarine cables are detectable by sharks. There are therefore proposals to increase the insulation thickness.

[95] We discuss the electrical properties of skeletal muscle fibres in vertebrates in Chapter 18.

[96] Torpedo, the naval weapon, was named after this genus of rays.

Fields RD (2007). The shark's electric sense. Scientific American 297: 74–81.
Pettigrew JD (1999). Electroreception in monotremes. Journal of Experimental Biology. 202: 1447–1454.
Zakon HH, Zwickl DJ, Lu Y, Hillis DM (2008). Molecular evolution of communication signals in electric fish. Journal of Experimental Biology 211: 1814–1818.

17.8 Magnetoreception

Magnetoreception (or magnetoception) refers to the ability of some animals, both invertebrates and vertebrates, to detect magnetic fields. Indeed, fruit flies, lobsters, bees, sharks, rays, turtles, fish, birds and some mammals can detect the Earth's magnetic field, which has an intensity of about 50 microtesla (μT), and use this sense to orientate and navigate.

As yet, no specific magnetic sensory receptor has been identified in any species, which makes it difficult to understand how the transduction of magnetic fields takes place and how such information is processed. Our difficulty in locating a specific magnetic sensory receptor in animals may be due to the fact that we as humans lack this sense of magnetoreception. Nevertheless, three potential mechanisms have been proposed:

- An induction mechanism may operate in marine animals that have very sensitive electroreceptors. According to Faraday's law of induction, an electric potential difference is generated when an electric conductor (in this case, the body of the animal) moves through a magnetic field (in this case, the Earth's magnetic field). Thus, sharks and rays could, in principle, sense the magnetic field through their electroreceptors, but this possibility is not easy to demonstrate.
- The second putative mechanism for magnetoreception involves the photon-induced formation of paired radicals[97], which interact with the ambient magnetic field to produce a magnetic field-sensitive chemical signal.

  Genetic studies on fruit flies (*Drosophila melanogaster*) trained to respond to an artificial magnetic field have shown that flies lose their magnetosensitivity when their only type of cryptochrome[98] is not functional (i.e. when the cryptochrome gene is knocked out or altered, or flies are exposed to filtered light above 420 nm that lacks the range (350 to 400 nm) to which the cryptochrome is sensitive).

  When exposed to light in the range 350 to 400 nm, cryptochromes are activated and form paired radicals which can exist in two spin states (parallel and antiparallel). The distribution of the paired radicals between the two states is sensitive to the magnetic field. Assuming that the light-sensitivity of photoreceptive cells is differentially affected by the distribution of the paired radicals between the two states, this scenario can provide the basis for fruit flies being able to sense magnetic fields in a light- (and cryptochrome-) dependent fashion. Magnetoreception in birds, salamanders and marine turtles is also sensitive to blue light illumination of the eyes, suggesting that a cryptochrome-dependent mechanism also operates in birds and other vertebrates.
- The third potential mechanism for magnetoreception may involve the presence of **magnetite**, a specific form of iron oxide (Fe_3O_4) crystals whose magnetic properties are related to their size. European robins (*Erithacus rubecula*) have magnetite in their upper beak and, as we discuss in Experimental Panel 17.2, experiments have clearly shown that their sense of magnetoreception used in migration is affected by man-made electromagnetic radiation associated with radiotransmission. Other birds capable of magnetoreception, include domestic hens (*Gallus gallus domesticus*) and homing pigeons (*Columba livia domestica*). Homing pigeons are known to use the sun for navigation whenever possible. However, on overcast days[99], they use their sense of magnetoreception for navigation. Magnetoreception in pigeons is selectively impaired by anaesthetic application on the olfactory mucosa or by sectioning the trigeminal nerve, but not by sectioning the olfactory nerve. These and other observations on migratory robins suggest that magnetoreception may stem from magnetite located in the beak of the birds, which acts via the trigeminal nerve, but the exact mechanism is not known.

There is now broad agreement that birds must have both a light- (and cryptochrome-) dependent, and a magnetite-based, sense of magnetoreception, which allow them to perceive direction with respect to the Earth's magnetic field and/or small local fluctuations in the Earth's magnetic field, both of which they use in orientation and navigation.

The sense of magnetoreception has also been demonstrated in some species of mammals, including rodents (Zambian mole rats, *Cryptomys* spp.), bats (the big brown bat, *Eptesicus fuscus* and the greater mouse-eared bat, *Myotis*

[97] Radicals are molecules with a single, unpaired electron that may be found in one of two spin states: ↑ or ↓. The unpaired electrons on a radical pair can have either antiparallel (↑↓) or parallel (↑↑) spins. The chemical properties of a radical pair can be different when the spin of the unpaired electrons is antiparallel from when the spin is parallel.

[98] We discuss cryptochromes in Section 17.2.2.

[99] There is also the suggestion that on overcast days, homing pigeons may also use infrasonic (soundwaves below the threshold of detection by human ear, <20 Hz) map cues for orientation.

17

Experimental Panel 17.2 Radiowaves interfere with biomagnetic compass in migratory birds

Background to the study

It has been known for many years that night-migratory song birds use the geomagnetic field to orient spontaneously in their migratory direction in spring and autumn[1]. However, when a group of researchers from the University of Oldenburg in Germany tried to replicate this basic experiment, which has been replicated many times in other locations, they failed. Preliminary data suggested that the failure to reproduce the experiment may have been caused by electromagnetic noise in the environment, which interfered with the magnetosensory system in birds despite the level of the electromagnetic noise being well below the safety guidelines for human exposure to non-ionizing radiation adopted by the World Health Organization.

Mindful of the potentially serious ramifications of a clear-cut result implying that man-made electromagnetic background noise can disrupt a sensory system in vertebrates, this group of researchers embarked on a rigorous study of the effect of background electromagnetic noise on the ability of European robins (*Erithacus rubecula*) to orient in the geomagnetic field. Their hypothesis was that man-made electromagnetic background noise can disrupt the magnetosensory system of European robins.

Experimental approach

The researchers used bird-orientation cages placed inside wooden huts whose walls and ceiling were lined with electrically connected aluminium plates. When grounded, the aluminium plates acted as a Faraday cage, reducing the electromagnetic noise in the huts by about two orders of magnitude, but leaving the geomagnetic field and other static magnetic fields intact. No other cues were available to the birds when placed in the orientation cages. The huts could be grounded randomly without the knowledge of the experimenters working inside. The experiments were conducted after sunset in spring, when the European robins display a northerly migratory direction.

The orientation cages were made of aluminium in the form of upside-down funnels (35 cm diameter at the base, walls inclined at 45°, 15 cm high with a hole of 5 cm at the top) and were lined with scratch-sensitive paper on which the birds left marks as they moved during 1 hour of testing.

The overlap line of the lining paper was randomly oriented by an experimenter to coincide with one of the four compass points (N, S, E or W). Two independent evaluators, who did not know the direction of orientation of the overlap line and whether the huts in which the experiments were performed were grounded or not, determined the mean direction of movement for each bird with respect to the overlap line from the distribution of the scratches on the paper.

The codes that identified (i) which hut was grounded and which was not and (ii) to which compass point the overlap line of the scratching paper was oriented were only revealed after the data were already analysed by the evaluators.

This type of experiment is said to be a **double-blind experiment**, because the evaluators of the birds' direction of orientation did not know whether the hut in which the bird was kept was grounded or not, and to which cardinal point the overlap line was oriented. Double-blind experiments considerably enhance the rigour of a study because they eliminate subjective, unrecognized biases of individual experimenters.

Results

Figure A(i) shows the effect of simply connecting and disconnecting the grounding of the aluminium screens inside the wooden huts, where the double-blind experiments were conducted. When the aluminium screens were connected to the ground, the birds oriented themselves in the northerly direction (blue circle), but when the grounding was disconnected, the same birds lost the ability to orient themselves in the geomagnetic field.

Figure A(ii) shows the magnetic component of the frequency-dependent electromagnetic noise in the huts when the grounding was connected (blue trace) and when the grounding was disconnected (red trace). The intensity of the magnetic field measured in nT (10^{-9} tesla) at different frequencies is plotted on a logarithmic scale on the vertical axis. Grounding of the aluminium screens attenuated the magnetic component of the electromagnetic noise by about 100-fold for frequencies up to 3 megaHz (MHz), and by gradually less from 3 to 5 MHz. The peak magnetic field intensity of the electromagnetic noise was close to 30 nT for frequencies around 0.01–0.05 MHz (10–50 kHz) and the total magnetic field intensity for the frequency range 0.01–5MHz did not exceed 1100 nT. These intensity values are much weaker than the geomagnetic field intensity of about 50,000 nT (50 μT).

The researchers showed that birds placed in grounded huts lost their ability to sense the geomagnetic field when a broadband electromagnetic noise was introduced inside the grounded hut. This experiment indicates that the electromagnetic noise *per se* was directly responsible for birds' loss of ability to orient in the geomagnetic field.

Further experiments indicated that the loss of birds' ability to use their magnetic compass was not due to a specific frequency as one would expect if the mechanism of magnetoreception was based on the photon-induced formation of paired radicals hypothesis, but to broad bands in the frequency range between 2 kHz and 5 MHz. Electromagnetic noise in this frequency range originates mainly from AM radio signals and from various electronic equipment and cannot be produced by either powerlines, which have frequencies between 16.7 and 60 Hz, or by mobile phone signals, which operate with GHz (10^9 Hz) frequency.

When experiments were performed in a rural location, where the level of electromagnetic noise was considerably less than in town, the birds could use their magnetic compass orientation without the need for screening the ambient noise.

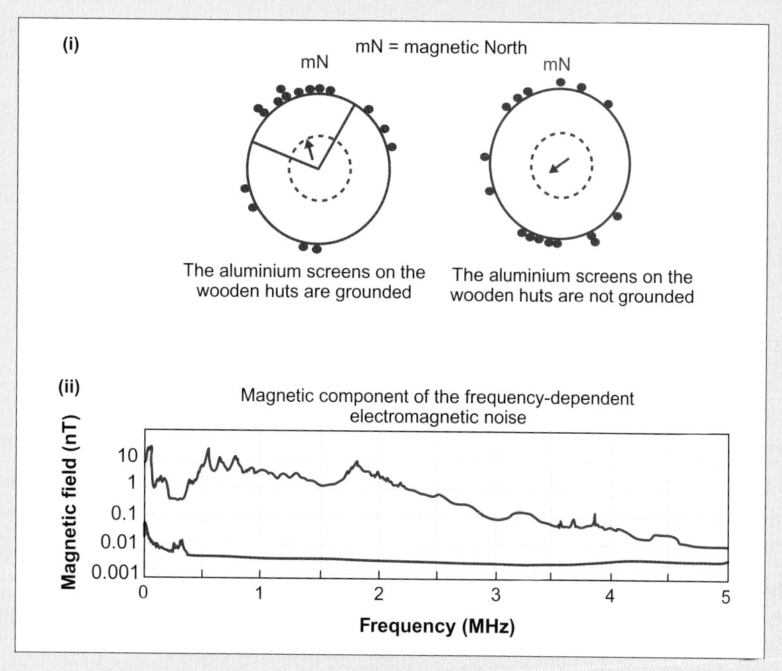

Figure A European robins maintain compass orientation in the northerly direction of migration only when the man-made electromagnetic noise is attenuated by grounding

(i) Each dot on the circles indicates the mean orientation for one individual bird for the respective condition. The arrows in the middle of the circles show group mean direction of movement. The radius of the small dashed inner circles indicates the minimum length of the group mean arrow for statistical significance ($p < 0.05$). The same group of 16 birds were tested for the two sets of conditions. Statistical significance was reached ($p = 0.04$) only when the screens were grounded and the direction of movement was 341° +/−40°. The blue arrow is flanked by the 95% confidence interval limits. (ii) The magnetic component of the frequency-dependent electromagnetic noise (red trace) is greatly attenuated when the aluminium screens are grounded (blue trace).

Results from Figure 2 of Engels S et al (2014). Anthropogenic electromagnetic noise disrupts magnetic compass orientation in a migratory bird. Nature 509: 353-356.

Conclusions reached

Taken together the results from this study show that:

- Man-made electromagnetic noise in the medium band of AM radio signals (2 kHz and 5 MHz) interferes with the magnetic compass orientation in European robins.
- The intensity of the electromagnetic noise which causes disorientation in robins is at least 300 times lower than the reference levels for general public exposure to time-varying magnetic fields adopted by the World Health Organization in the relevant frequency band.
- The mechanism of magnetoreception in European robins is unlikely to be based on photon-induced formation of paired radicals from cryptochromes, which suggests that the biomagnetic compass in migratory birds may be based on magnetite crystals as we discuss in Section 17.8.

› Find out more

Engels S, Schneider N-L, Lefeldt N, Hein CM, Zapka M, Michalik A, Elbers D, Kittel A, Hore PJ, Mouritsen H (2014). Anthropogenic electromagnetic noise disrupts magnetic compass orientation in a migratory bird. Nature 509: 353–356.

Wiltschko W, Wilitschko R (1972). Magnetic compass of European robins. Science 176: 62–64.

[1] We discuss in Section 17.8, two potential mechanisms for sensing magnetic fields in birds, although no specific magnetic sensory receptor has been identified, as yet, in any animal.

myotis) and deer (roe deer, *Capreolus capreolus*, and red deer, *Cervus elaphus*), but not humans. In mole rats, the sense of magnetoreception appears to be magnetite-based, as a strong magnetic pulse, which can remagnetize magnetite particles, causes a persistent change in the geomagnetic orientation of their nests.

› *Review articles*

Rodgers CT, Hore PJ (2009). Chemical magnetoreception in birds: the radical pair mechanism. Proceedings of the National Academy of Sciences of the United States of America 106: 353–360.

Wiltschko W, Wiltschko R (2005). Magnetic orientation and magnetoreception in birds and other animals. Journal of Comparative Physiology A 191: 675–693.

Bibliography

Fain GL (2003). Sensory Transduction. Sinauer, Sunderland, MA.

Checklist of key concepts

Principles of sensory processing

- The world as perceived by animals is defined by their senses, which receive and interpret only a small fraction of the potential sensory information available in their environment.
- Each individual group of animals has evolved different sensory priorities and specific abilities to ensure their survival in a particular environment.
- Sensory information in the environment is captured by specific **sensory receptors**, which act as selective amplifiers of specific types of stimulus energy.
- Sensory receptors are either **modified neurons** or are closely associated with **sensory neurons**.
- Information carried by specific stimuli is converted into electrical signals in the form of **receptor potentials** and then into action potentials by a process called **sensory transduction**.
- The frequency of action potentials generated in sensory neurons is linearly related to the **logarithm of stimulus intensity**, which allows animals to sense a very large range of stimulus intensities in their environment.
- Each type of stimulus is processed by distinct parts of the central nervous system before information acquired by different types of sensory receptors is integrated into a unified view of the world perceived by the animal.

Photoreception

- **Photoreception** refers to the ability of animals to interpret their surroundings by acquiring and processing information contained in the **electromagnetic radiation spectrum**.
- Photoreception is made possible by photoreceptor cells containing **photopigments** that change configuration when they absorb photons of specific wavelength (energy) in the range 380 to 700 nm.
- The photopigments in the visual system of animals consist of a **chromophore** bound to a protein called **opsin**.
- When a photopigment absorbs a photon, opsin is activated and triggers a cascade of events via a G-protein, leading to a change in the membrane potential of the photoreceptor cell.
- The photoreceptor cells in animals are of two types: **ciliary** and **rhabdomeric**.
  - ciliary photoreceptors are **depolarized** in the dark and become **hyperpolarized** in the presence of light;
  - rhabdomeric receptors are **polarized** in the dark and become **depolarized** in the presence of light.
- Vertebrates have only ciliary receptors, arthropods have only rhabdomeric receptors and several groups of invertebrates have both types of receptors.
- Ciliary receptors are of two kinds: **rods** and **cones**.
  - Rods contain the photopigment **rhodopsin**, consisting of the chromophore **retinal** and the protein scotopsin.
  - The photopigments in cones also have retinal but have different opsins called **photopsins**.
- Rods are 100-fold more sensitive to light but respond more slowly than cones to changes in light intensity.
- Colour vision in vertebrates is based on the presence of at least two classes of cones with different spectral characteristics.
  - Humans, primates closely related to humans and some marsupials have three classes of cones in the retina: they have **trichromatic** vision.
  - Fish, reptiles, birds and some amphibians have four distinct classes of cones and have **tetrachromatic** vision.
  - Most terrestrial mammals have only two classes of cones (**dichromatic** vision)
  - Marine mammals have only one type of cones (**monochromatic** vision).
- The most common photoreceptive organ in animals is the **eye**, which varies in complexity between different animal groups.
- All vertebrates have **vesicular eyes** characterized by a lens focusing the image onto the retina, where significant processing of visual signals takes place.
- **Compound** eyes in arthropods are complex convex eyes consisting of many units, called **ommatidia**, organized in a dome-like structure.
- The signals from each ommatidium are perceived by animals as light of certain wavelength and intensity, allowing insects and other arthropods to have colour vision.
- Photoreception plays a key role in synchronizing biological clocks to the local time in the environment.

Chemoreception

- Chemoreception refers to the ability of animals to interpret their external and internal environments by acquiring and processing information associated with the presence of different **chemicals** in the environment.
- The **transduction** mechanism in chemoreceptors is rather diverse: some chemoreceptors are coupled to G-proteins while others are not.

- All animals have evolved extraordinarily sensitive **olfactory** systems that can distinguish between tens and hundreds of thousands of different **odours**.
- **Olfactory receptors** in terrestrial vertebrates are located in the upper part of the **nasal cavity**.
- In insects and other invertebrates, olfactory receptors are located in **sensilla**.
- **Pheromones** are a special class of chemicals secreted by animals to communicate to members of their own species.
- **Taste** chemoreceptors play a key role in guiding animals to distinguish between different forms of nutrients and to avoid toxic and indigestible substances.
- The taste receptor cells in vertebrates are located in **taste buds** and are sensitive to either salt, sweet, umami, acid and bitter tastants.
- The taste receptors of insects are located in **sensilla**.

Mechanoreception

- Diverse sensory organs are involved in sound detection, balance maintenance, identification of shapes and texture, detection of position of objects outside the body, detection of relative position of different body parts and detection of blood pressure.
- These organs operate on the basis of mechanical stimuli acting directly or indirectly on **stretch-sensitive ion channels** located on receptor cells, which convert the respective mechanical stimuli into depolarizing receptor potentials and action potentials.
- Hair cells with mechanoelectrical (MET) ion channels on their stereocilia are involved in:
  - sound transduction in the vertebrate ear,
  - **balance** and **spatial orientation** in the gravitational field in both vertebrates and aquatic invertebrates,
  - **multifunctional lateral line systems** in fish.
- The **auditory system** in arthropods comprises **chordotonal organs** containing ciliary mechano-sensitive neurons that respond to vibrations of membranes stretched over air-filled cavities.
- The sense of **touch** refers to the ability of animals to perceive direct contact of the body with solid objects in the external environment.
- Touch in mammals is mediated by **phasic receptors** (Pacinian and Meissner's corpuscles) and **tonic receptors** (Ruffini's corpuscles and Merkel's discs) as well as **hair follicle receptors** containing stretch-sensitive cationic channels.
- The most common touch receptors in birds are the **Herbst receptors** and the **Merkel's cells**.
- In insects, touch is mediated by **chordotonal organs** attached to bristle-like structures on the surface of their body and in the antennae (**Johnston's organ**).
- The sense of **proprioception** enables animals to sense their position with respect to the environment and to maintain posture and balance.
- **Baroreceptors** are stretch-sensitive receptors which detect blood pressure in the circulatory system and are part of a fast negative-feedback system that regulates blood pressure on a short-time basis in vertebrates.

Thermoreception

- Thermoreception refers to the ability of animals to detect **temperature** on their body surface and in body tissues.
- Thermoreception plays a key role in protecting animals from damaging temperatures and maintaining thermal homeostasis in mammals and birds.
- **Thermoreceptors** are thermoreceptive nerve terminals containing non-selective cation TRP channels called **thermoTRPs**, which have different temperature thresholds for opening in different animals.
- Activated thermoTRPs cause membrane depolarization and generation of action potentials.
- The thermoTRPs in humans are responsible for producing sensations from noxious cold to burning.

Nociception

- Nociception refers to the ability of all animals to perceive **noxious** or **damaging stimuli**, helping them to avoid further damage.
- Nociceptors are nerve terminals containing non-selective cation channels that open in response to damaging mechanical, thermal or chemical stimuli, or to compounds produced following cell damage.
- Behavioural reactions in response to noxious stimuli in animals can be **unconscious** or **conscious**.
- Animals may experience pain when such reactions are conscious because pain always encompasses a conscious response and an emotional experience (suffering).

Electroreception

- Electroreception refers to the ability of some animals to sense **electric fields** in their environment.
- These electric fields may be:
  - generated by external sources (**passive electroreception**)
  - generated by them and distorted by the environment (**active electroreception**).
- The most sensitive electroreceptive system among animals is found in sharks, which have an electroreceptive organ on their snouts called the **ampullae of Lorenzini**.
- A significant number of both vertebrates and invertebrates—including insects, fish and mammals—can sense electric fields.

Magnetoreception

- Magnetoreception refers to the ability of some animals to detect the **geomagnetic field** and use it for **orientation** and **navigation**.
- No specific magnetic sensory receptors have been identified in any animal, but three potential mechanisms for sensing the geomagnetic field have been proposed:
  - some animals, like sharks and rays, may sense the geomagnetic field through their electroreceptive organs by detecting an electrical potential difference induced as they move through the geomagnetic field;

- blue-light-induced activation of cryptochromes forms paired radicals in two states, with the distribution between the two states being sensitive to the geomagnetic field; and
- a mechanism mediated by the presence of magnetic crystals of iron oxide, called magnetite, found in the beak of some migratory birds.

Study questions

*** Answer to this numerical question is available online. Go to www.oup.com/uk/butler**

1.* Calculate the overall amplification factor of a rhabdomeric photoreceptor which triggers one action potential with an associated energy of 5×10^{-11} J in response to the absorption of one photon of blue light (λ =420 nm). (Hint: The energy of one photon = hc/λ, where h is Planck's constant (6.626×10^{-34} J s) and c is the speed of light (2.998×10^{8} m s^{-1}); see also Section 17.1.1.)
2. Draw a graph showing the relationship between action potential frequency in a sensory neuron and stimulus intensity. Explain the physiological basis of this relationship and highlight its physiological significance for sensory processing. (Hint: Section 17.1.2.)
3. Compare and contrast the signal transduction cascade in the rod photoreceptor cells of vertebrates to that in the rhabdomeric photoreceptor cells of arthropods. Where is rhodopsin located in rods and in the rhabdomeric photoreceptors? How is the receptor potential generated in rods and in rhabdomeric photoreceptors? (Hint: Section 17.2.1.)
4. Define the term 'circadian clock' and indicate how the master circadian clock in mammals synchronizes all circadian rhythms in the body. Also, outline the molecular basis of the core circadian clock in animal cells. (Hint: Section 7.2.2.)
5. Describe the signal transduction pathway in olfactory receptor neurons of terrestrial vertebrates and explain how odorant specificity is achieved. (Hint: Section 17.3.1.)
6. Outline the major functions of type I, type II and type III taste bud cells and the signal transduction pathways in these cells. (Hint: Section 7.3.2.)
7. Explain the mechanism of mechanoelectrical transduction (MET) in hair cells. Describe two types of sensory receptors in which MET channels are involved. (Hint: Sections 17.4.1, 17.4.2.)
8. Animals have the general ability to sense their body orientation in the environment. Taking a mammal as an example, indicate three types of sensory receptors that are important for this capability. (Hint: Section 17.4.4.)
9. What is nociception? Discuss whether nociception is related to pain. (Hint: Section 17.6.)
10. Give three examples of animals that sense the geomagnetic field and describe the three potential mechanisms for magnetoreception. (Hint: Section 17.8.)

18

Muscles and animal movement

Most of the movement that occurs within an animal—and the movement of the whole organism itself—is driven by the contraction of muscles, biological motors that convert chemical energy into mechanical energy. Muscle contraction is central to an animal's survival: it underpins feeding and the processing of food, maintaining posture and balance, breathing, the pumping of blood around the body, locomotion and many aspects of animal behaviour.

In the first sections of this chapter, we learn about the diversity of muscles found in animals, the general principles of how muscles work, the molecular mechanism involved in converting chemical energy into mechanical energy and the regulation of muscle contraction. In the last sections of the chapter, we then discuss how animals use muscles to ensure their survival in the environment.

18.1 Muscle form and function

Muscles vary widely in terms of size, morphology, microscopic structure, strength of contraction, velocity of contraction, endurance and energy efficiency. All muscles are made of relatively long and narrow excitable cells called **muscle fibres**, which contract in response to excitation of their surface membrane, which is called the **sarcolemma**[1]. Muscle contraction is activated by a rise in the concentration of cytosolic **calcium ion** (Ca^{2+}); the sequence of events that begins with muscle fibre excitation and ends with muscle fibre contraction is called **excitation–contraction coupling**. Muscles can only act by pulling (never by pushing) on various structural components of an animal's body such as the skeleton or other tissues (like the eyes and muscular tongues of vertebrates), or by exerting pressure on a body cavity that is surrounded by muscle, like the heart or gut.

18.1.1 General muscle types

The broadest type of muscle classification, which applies to all animal species, is based on the general appearance of the constituent muscle fibres when viewed under a light microscope. Muscles made of fibres that display a regular banded pattern (Figure 18.1A, B, D) are called **striated muscles**. Other muscles, made of fibres that are uniform in their appearance, such as that shown in Figure 18.1C, are called **smooth (or non-striated) muscles**.

In vertebrates, the striated muscles are characterized by cross-striations relative to the fibre axis (also called transversal striations), as shown in Figure 18.1A, B; striated muscles are further subdivided into **skeletal muscles** and **cardiac (or heart) muscles**. Skeletal muscles in vertebrates are connected to the skeleton and are under an animal's voluntary control. As such, skeletal muscles are also known as **voluntary muscles**. In contrast, cardiac muscles are **involuntary muscles** because their pattern of contraction cannot be deliberately controlled. The smooth muscles in vertebrates are also involuntary muscles and generally occur as sheets of muscle fibres that wrap around tubular structures (such as blood vessels or the gastrointestinal tract), around the pupil in the eye, and in the wall of various internal organs. Therefore, smooth muscles in vertebrates are also known as **visceral muscles**. Table 18.1 and Figure 18.1A-C show some basic structural characteristics of the three different muscle types found in vertebrates.

The muscles of invertebrates lack the clear distinction that exists in vertebrates between muscle appearance (striated or smooth) and muscle type (skeletal, cardiac and visceral). For example, arthropods lack smooth muscles: the visceral muscles, as well as those associated with the exoskeleton, are cross-striated. In other invertebrate species, such as bivalve molluscs, muscles attached to the shells can be either cross-striated or non-striated.

Many soft-bodied invertebrates, such as nematodes, annelids, cephalopod molluscs and tunicates[2], have obliquely striated muscles, where the striations run obliquely rather than transversally to the fibre axis, as shown in Figure 18.1D. The obliquely striated muscles have similar properties to cross-striated invertebrate muscle. Smooth muscles with diverse functions occur in cnidarians, ctenophores, annelids, molluscs and echinoderms.

[1] Cellular structures in muscle often start with the prefix 'sarco-' derived from the Greek 'sark-', meaning 'flesh'.

[2] We discuss animal diversity in Section 1.4.

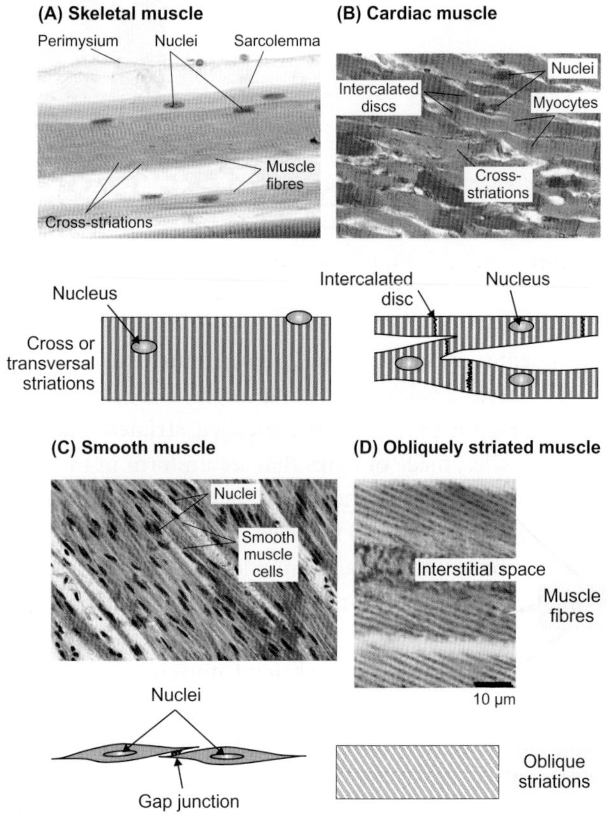

Figure 18.1 General appearance of fibres from different types of muscle viewed under a light microscope at similar magnification

(A) Bundle of human skeletal muscle fibres displaying cross-striations and multiple nuclei located under their surface membrane (sarcolemma); the bundle of fibres is covered by a layer of connective tissue called perimysium. (B) Human cardiac muscle showing fibre branching, nuclei, intercalated discs (or disks) that provide mechanical and electrical connections between individual fibres (see Figure 18.22) and weaker cross-striations than skeletal muscle at this magnification. (C) Human smooth muscle showing the spindle-shaped form of the smooth muscle cells that are much smaller in size than skeletal and cardiac fibres, lack striations, contain a single central nucleus and are often electrically coupled via gap junctions to other cells. (D) Longitudinal section through a nematode (*Ascaris lumbricoides*) muscle containing obliquely striated muscle fibres. Note that the calibration bar (10 μm) shown in panel (D) also applies to panels (A), (B) and (C).
Source: A, B, C: © Christopher D. Richards; D: Rosenbluth J (1965). Ultrastructural organization of obliquely striated muscle fibers in *Ascaris Lumbricoides*. Journal of Cell Biology 25: 495-515.

18.1.2 The muscle contractile apparatus consists of myosin- and actin-containing filaments that face each other

The **contractile apparatus** in all muscles is composed of two kinds of myofilaments[3]:

- a thicker myofilament type (diameter >12 nm) containing **myosin** molecules[4], which act as molecular motors and convert the energy released from the hydrolysis of MgATP[5] into mechanical work;
- a thinner myofilament type (about 6 nm diameter), composed primarily of **actin** molecules with which the myosin molecules interact.

All animal groups except sponges and placozoans have differentiated muscle cells. Genetic studies have nevertheless found that even sponges and placozoans have genes for the two major muscle proteins: myosin (class II) and actin. This indicates that an ancestral form of an actin–myosin-based contractility is present in all animals. In sponges, this type of contractility may help to regulate the circulation of water through the many pores in the animals' body. Moreover, we now know that acto-myosin (class II) contractility underlies not only muscle contraction, but also basic cellular processes, which include cell division and cell migration.

[3] Prefix 'myo-' denotes a relationship to muscle in general.
[4] Specifically, these are class II myosin molecules. Other classes of myosin molecules are involved in specific cellular functions, such as transport of vesicles and cytoplasmic streaming.
[5] ATP is the common currency for energy transfer and utilization in animal cells, as we discuss in Section 2.4.2; myosin-based molecular motors use MgATP (which is a complex between ATP and magnesium ion, Mg^{2+}), instead of simply ATP, for providing the necessary energy for contraction.

Table 18.1 **General characteristics of the major muscle types in vertebrates**

	Striated		Smooth
	Skeletal	Cardiac	
General appearance	Regular transversal striations	Regular transversal striations	Uniform appearance
Muscle cells	Elongated, multinucleated cells, organized in bundles and connected by tendons to bones	Shorter, single or binucleated, branched cells connected mechanically and electrically to other cells	Fusiform cells, connected mechanically and electrically to other cells
Control	Voluntary (somato-sensory innervation)	Involuntary (autonomic nervous system and hormonal control)	Involuntary (autonomic nervous system and hormonal control)

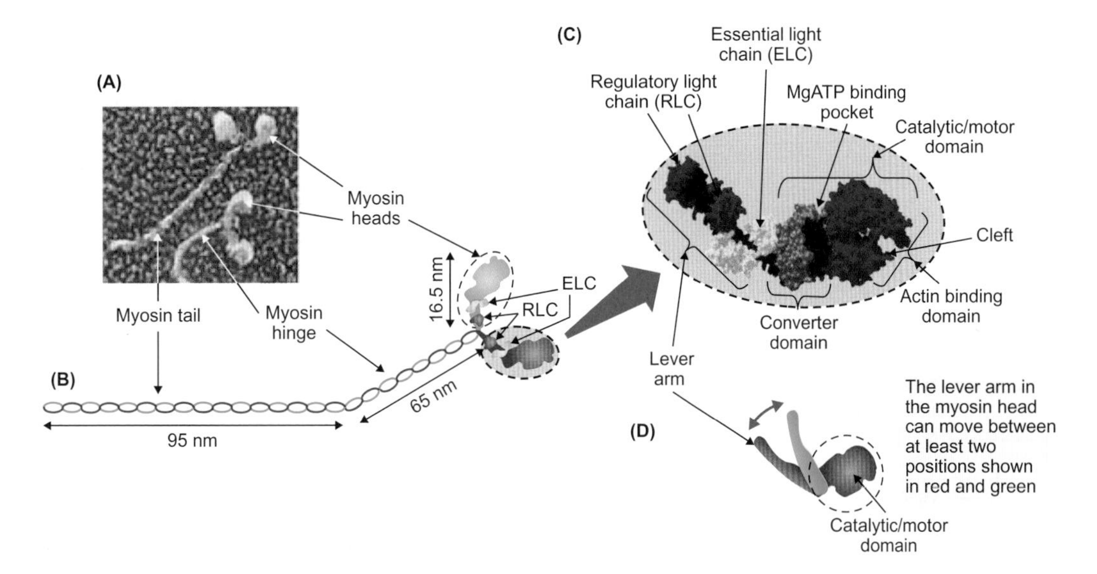

Figure 18.2 Structure of muscle-type (class II) myosin molecules

(A) High-resolution electron microscope image of single myosin molecules. (B) Diagrammatic representation of the myosin molecule consisting of six polypeptide chains: two heavy chains displayed in red and blue and four light chains: two essential light chains (ELCs) and two regulatory light chains (RLCs). The two heavy chains each have one globular head and long helical structures which associate to form the hinge and the tail domains of the myosin molecule. The four light chains are attached in pairs at the base of the globular heads of the heavy chains, such that each myosin head consists of one heavy chain globular head, one ELC and one RLC. (C) Space-filled representation of all atoms in the myosin head showing the lever arm, the nucleotide-binding pocket, the cleft in the actin binding domain and the converter domain. Notice that the two light chains attached to the myosin head (ELC and RLC) are part of the lever arm. (D) Diagrammatic representation of the myosin head with the lever arm that can change position with respect to the catalytic/motor domain.

Sources: A: Schliwa M, Woehlke G (2003). Molecular motors. Nature 422: 759-765. C: Rayment I et al (1993). Three-dimensional structure of myosin subfragment-1: a molecular motor. *Science* 261: 50–58.

General structure of myosin filaments

The myosin molecules in striated muscle fibres spontaneously aggregate to form filaments. Each **myosin molecule** (Figure 18.2A, B) is a very large MgATP-dependent motor protein (about 540 kD mass) consisting of six polypeptide chains[6]: two heavy chains, two regulatory light chains (which play a role in the regulation of myosin function) and two essential light chains (which are necessary for force production).

At one end, the myosin molecule has two globular heads called **myosin heads**, each consisting of the globular ending of one heavy chain together with one essential and one regulatory light chain (Figure 18.2B and 18.2C). Beyond the globular heads, the two heavy chains are intertwined for most of their length to form a **hinge region** and a **long tail**, as illustrated in Figure 18.2B. The general three-dimensional molecular structure of the myosin head (Figure 18.2C) is highly conserved among animals in terms of structural motifs: each myosin head has a **motor domain** (or **catalytic domain**, where MgATP is hydrolysed) and a **lever arm** that can rotate with respect to the motor domain, as shown in Figure 18.2C and D.

[6] A polypeptide chain is a complex made of different peptides that are not covalently linked to each other. We discuss protein structure in Box 2.1.

The tails of different myosin molecules strongly interact with each other to form the backbone of the **myosin filament**. In striated muscles of both vertebrates and invertebrates, the myosin filaments contain several hundred myosin molecules whose heads are oriented in opposite directions in the two halves of the myosin-filament (Figure 18.3A). Since the heads of the myosin molecules point away from the filament's centre, it means that the myosin filaments in striated muscles are **bipolar**. Notice that there are no myosin heads in the middle of bipolar myosin filaments. This region is called the **bare zone**.

The backbone of myosin filaments in invertebrate muscles also contain **paramyosin**, a protein that plays a structural role in determining the length, diameter and overall rigidity of the myosin-filament.

The organization of the myosin-filaments is different in vertebrate smooth muscle, as illustrated in Figure 18.3B: the myosin molecules are oriented at a low angle to the filament axis and form flat sheets with the heads on each side oriented in the one direction along the filament. On the other side of the filament, all the myosin heads are oriented in the opposite direction. This organization forms what we call

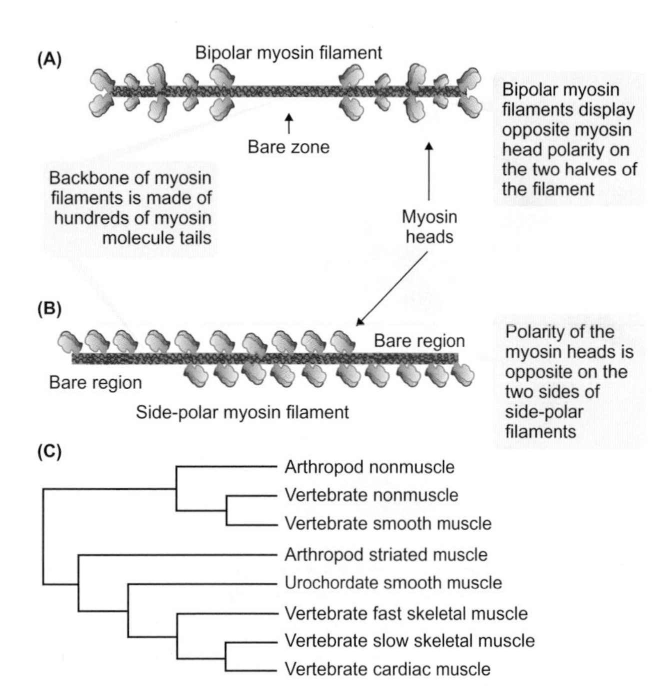

Figure 18.3 Organization of myosin molecules into myosin filaments and evolutionary relationship of muscle tissues

(A) Bipolar myosin filaments with bare zone in the middle occur in striated muscles. (B) Side-polar myosin filament with bare regions at the ends occur in vertebrate smooth muscles. (C) Phylogenetic tree of muscle tissues based on protein genes expressed in muscle, including myosin heavy chains, myosin regulatory light chains, myosin essential light chains and actin. This phylogenetic tree shows that (i) arthropod striated muscle and vertebrate skeletal and cardiac muscles share a common ancestor, (ii) urochordate smooth muscle shares an ancestor with vertebrate skeletal and cardiac muscles, (iii) urochordate smooth muscle evolved independently of vertebrate smooth muscle and (iv) vertebrate smooth muscle evolved independently of other muscles. Note that the length of the branch lengths is not proportional to evolutionary time.

Source: C Oota S, Saitou N (1999). Phylogenetic relationship of muscle tissues deduced from superimposition of gene trees. Molecular Biology and Evolution 16: 856–867.

side-polar filaments and is important for force development in smooth muscle, as we discuss later. Side-polar myosin filaments have no central bare zone, but have bare zones on both ends of the filament.

The different organization of the myosin filaments in smooth vertebrate and striated muscles is consistent with the phylogeny of the muscle tissue (Figure 18.3C), indicating that smooth vertebrate muscle evolved independently of the other muscle types.

General structure of actin filaments

Actin filaments (also called filamentous actin, or F-actin) are formed from the polymerization of globular actin (G-actin) molecules in the presence of ATP. G-actin is a highly conserved asymmetric globular protein (Figure 18.4A) among eukaryotes with a molecular mass of 42 kD. The actin filaments have a double-stranded helical structure, with two chains of G-actin monomers wrapped around each other at regular intervals as shown in Figure 18.4C. The asymmetric structure of the actin molecule confers **polarity** to the entire actin filament, which has a plus (+) and a minus (–) end (Figure 18.4C).

The actin filaments in both vertebrates and invertebrates also contain additional proteins that are part of the **calcium regulatory system**. This system keeps the muscle fibres relaxed when the Ca^{2+}-concentration is low and makes the contractile apparatus sensitive to Ca^{2+}, as we discuss later.

The phylogenetic tree of muscle tissues shown in Figure 18.3C is based on various protein genes expressed in muscle, including myosin heavy chains, myosin regulatory chains, myosin essential light chains and actin. It shows that cross-striated muscles in arthropods and vertebrates, as well as the smooth muscles in urochordates share a common ancestor, while the vertebrate smooth muscle evolved independently of the other muscles.

18.1.3 The banded pattern of striated muscle fibres reveals the length and location of the myosin and actin filaments in the fibre

As shown in Figure 18.5, skeletal muscles in vertebrates are connected via **tendons** to the skeleton. Tendons are a type of connective tissue made primarily of **collagen** fibres that confer tensile strength[7] and elasticity and act as shock absorbers particularly in terrestrial vertebrates, allowing them to jump and run. Skeletal muscles are covered by a thick layer of connective tissue called **epimysium** and are organized in bundles of muscle fibres aligned in parallel arrays called **fascicles**. The fascicles are surrounded by a layer of connective tissue called **perimysium**. A very thin layer of connective tissue called **endomysium** surrounds each muscle fibre in the fascicle, together with its associated blood capillaries and the nerve tissue.

At the level of a single skeletal muscle fibre, the myosin and the actin filaments are highly organized, giving the fibre its characteristic banded pattern. Figure 18.6 illustrates the organization of a cross-striated vertebrate skeletal muscle fibre, where thin (about 1–2 μm diameter) quasi-cylindrical structures called **myofibrils**[8] run in parallel from one end of the muscle fibre to the other.

[7] Tensile strength refers to the stress that tendons can withstand without breaking while being stretched.

[8] Myofibrils and myofibres are different terms which should not be confused. 'Myofibre' is synonymous with muscle cell or muscle fibre containing all cellular components, while the myofibril is only a small component of the myofibre, containing a small fraction of the entire contractile apparatus of the muscle fibre.

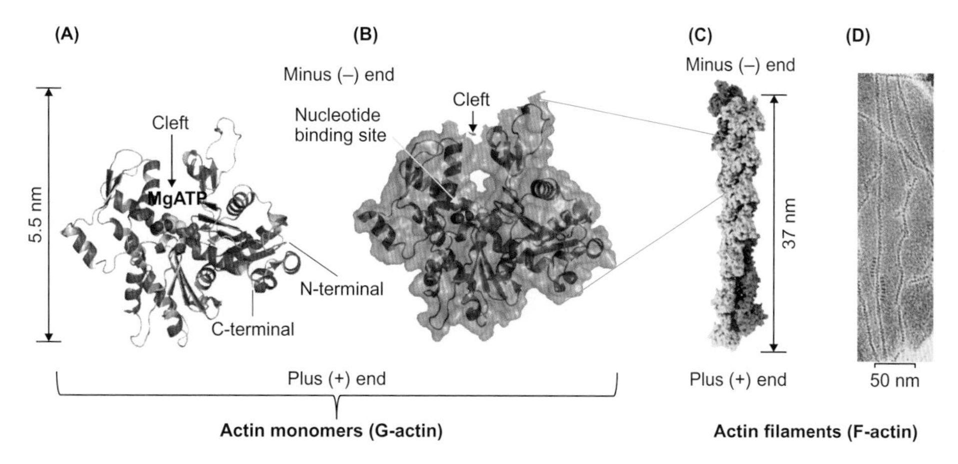

Figure 18.4 Structure of the globular actin (G-actin) and the actin filament (F-actin)

(A) Three-dimensional representation (ribbon model) of the G-actin molecule showing a large cleft containing the nucleotide (MgATP) binding site and the structural asymmetry of the molecule. (B) The ribbon model of the G-actin molecule shown in A with surface plot overlaid, showing that MgATP makes extensive contacts with the two lobes of the protein. (C) Binding of MgATP to the nucleotide-binding site promotes the polymerization of actin with the hydrolysis of MgATP to MgADP, removal of phosphate and formation of actin filaments (F-actin). The actin filaments are double-stranded helical structures made of two chains of G-actin crossing each other every 37 nm. The asymmetric structure of the G-actin molecules confers asymmetric properties to the entire actin filament, which has a plus (+) and a minus (–) end. (D) Actin filaments visualized by electron microscopy techniques. Note the regularly spaced lines which permit counting the number of G-actins per turn of the double stranded helix of the F-actin.

Sources: A, B: J Crowe from the Protein Data Bank file: 1ATN; C: US National Library of Medicine (Actin); D: Hanson J, Lowy J (1963). The Structure of F-Actin and of Actin Filaments Isolated from Muscle. Journal of Molecular Biology 6: 46–60.

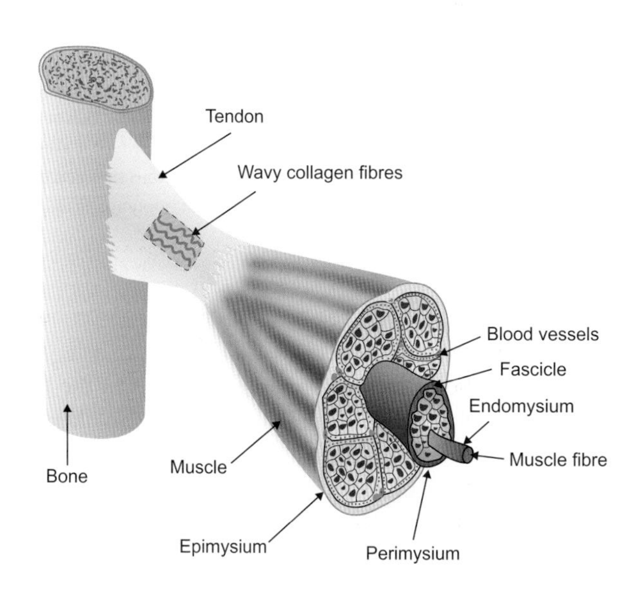

Figure 18.5 General organization of skeletal muscle in vertebrates

Muscles are connected to the skeleton via tendons. The collagen fibres in the tendon are wavy when there is little force exerted by the muscle. When the collagen fibres are straightened, the tendon increases its stiffness and transmits efficiently the force exerted by the muscle to the bone. Each muscle fibre is covered by endomysium; bundles of parallel fibres are covered by perimysium to form muscle fascicles, and fascicles together with the nerve and blood supplies to the muscle fibres are covered by epimysium.

The longitudinal banded pattern displayed by the myofibril in Figure 18.6B is due to the regular arrangement of the thicker myosin filaments and the thinner actin filaments. The very regular arrangement of the thicker myosin filaments in the cross-section of the myofibril (Figure 18.6D) causes those regions where myosin filaments are located to appear as darker bands, called anisotropic[9] bands, or **A-bands.**

The lighter bands, called isotropic bands or **I-bands**, occur where only the thinner actin filaments are present in myofibrils. The darker line in the middle of the I-band indicates the location of the **Z-disc**[10] (also called the **Z-line** due to its zigzag shape when visualized in an electron microscope) to which the myofilaments are anchored. Z-discs are made of a protein called **actinin.**

The structural and functional unit of the contractile apparatus in striated muscle is the **sarcomere**, which runs between two adjacent Z-lines as shown in Figures 18.6C and E. Each myofibril can have hundreds to thousands of sarcomeres along its length.

[9] 'Anisotropic'—as opposed to 'isotropic'—means that its optical properties are not uniform in all directions.

[10] From the German 'Zwischenscheiben' meaning 'in between discs'; note that many terms related to muscle structure are derived from German words because prior to the1940s, most scientific publications on muscle were published in German.

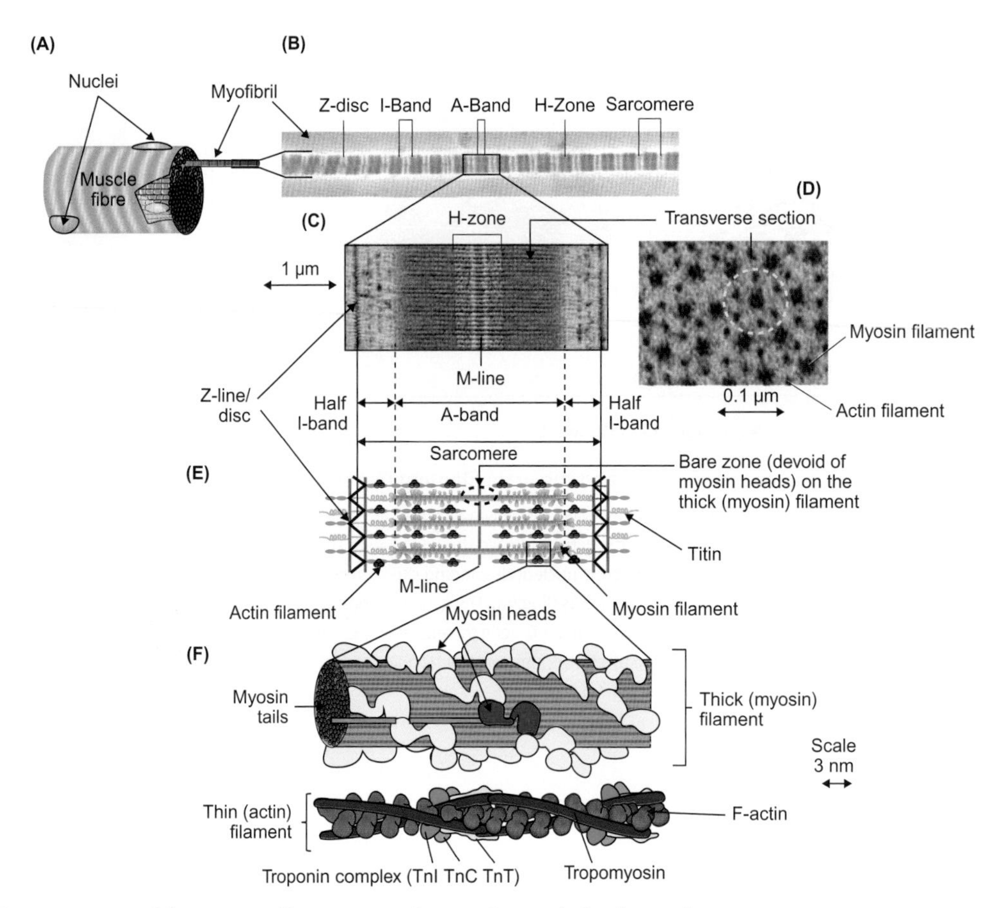

Figure 18.6 Fine structure of the contractile apparatus in vertebrate skeletal muscles

The contractile apparatus is organized in myofibrils (A, B) which consist of regular double hexagonal arrays of thick and thin filaments (C, D). The regular organization of the filaments confers the banded pattern to myofibrils and to the entire muscle fibre. The thick filaments are located in the middle of the sarcomere (E) and are connected to the Z-lines by titin (E). The thick filaments are made mainly of myosin (F). The thin filaments made predominantly of F-actin (F) attach at the plus (+) ends to the Z-lines, thus projecting in opposite directions from each Z-line (E). The thin filaments also contain the Ca^{2+} regulatory system, which makes the contractile apparatus sensitive to calcium ions. The Ca^{2+} regulatory system consists of the troponin (Tn) complex (TnI, TnC, TnT) and filamentous tropomyosin. Tropomyosin stabilizes and stiffens the actin filament and prevents other proteins from accessing the filament; Tn controls the positioning of tropomyosin along the grooves of the actin filament.

Sources: (B): Leonard TR, Herzog W (2010). Regulation of muscle force in the absence of actin-myosin-based cross-bridge interaction. American Journal of Physiology 299: C1398-C1401. D: Millman BM (1988). The filament lattice of striated muscle. Physiological Reviews 78: 359-391.

The striation pattern and the length and location of the myofilaments in the sarcomeres of cross-striated muscle fibres are closely related. The somewhat lighter zone in the middle of an A-band, called the **H-zone**[11], corresponds to the region where only myosin filaments are present in cross-section. The darker zones on both sides of the H-zone band correspond to the regions in which the myosin and actin filaments overlap.

The myosin filaments are anchored to the Z-lines by an elastic protein called **titin**[12], which keeps the thick filaments in the centre of the sarcomere. In the middle of the H-zone is the **M-line**[13], which corresponds to a structure that keeps the myosin filaments together.

[11] From the German 'heller', meaning 'lighter'.

[12] Skeletal muscle titin is the largest protein in the body of most animals, reaching a molecular mass of up to 3.9×10^6 D.

[13] M- from 'Mittelscheibe' in German, meaning 'middle disc'.

The actin filaments connect to the Z-line at their plus (+) end. They therefore have opposite polarities in the two half-sarcomeres such that the polarity of each actin filament matches the polarity of the myosin filament in each half-sarcomere. The actin filaments are located between Z-lines and the edge of the H-zone closest to them.

The Ca^{2+} regulatory system associated with the thin actin filaments in vertebrate striated muscles comprises one **troponin complex** and one rod-like protein, **tropomyosin**, for every seven actin monomers, as shown in Figure 18.6F. The troponin complex consists of three subunits: **troponin C**, the Ca^{2+} binding subunit, **troponin T**, the tropomyosin binding subunit, and **troponin I**, the inhibitory troponin subunit.

Cytoskeletal structures keep the Z-lines in adjacent myofibrils aligned to each other such that the regular banded pattern of myofibrils—which contains information on the

18

position and length of the myosin and actin filaments in the sarcomere, as well as the amount of overlap between filaments—extends to the entire muscle fibre.

18.1.4 Interactions between myosin and actin generate force in muscle

Figure 18.7 summarizes the current view regarding the molecular basis for force and movement generation in muscles. Optimal conditions for force production occur when the polarity of the actin filament matches the polarity of the myosin heads on the myosin filament (Figure 18.8A).

Each myosin head has a **lever arm** and a **motor domain** (Figure 18.2C, D). The motor domain contains:

- the nucleotide-binding pocket, where MgATP binds and is then hydrolysed;
- the converter domain, where small structural changes in the nucleotide-binding pocket are amplified and translated into rotations of the lever arm;
- the actin-binding domain, which has a cleft that can open and close, altering the properties of the nucleotide-binding pocket and the orientation of the myosin head with respect to the actin molecule(s) to which it is bound.

The many α-helices in a myosin molecule's head and hinge regions confer a degree of elasticity to the myosin molecule, which is represented in Figure 18.7 by a spring connecting the myosin head to the myosin filament. This elasticity is central to the mechanism of force generation that underpins muscle contraction.

Myosin heads interact with actin filaments to form cross-bridges that produce force

Force is produced from cyclical interactions of myosin heads on a myosin filament with an actin filament in the presence of MgATP as summarized in Figure 18.7. In *Step 1*, the MgATP molecule already bound to the myosin head that caused its detachment from the actin filament in the previous step of the cycle (*Step 5*) is rapidly hydrolysed to MgADP and inorganic phosphate (P_i). The products of reaction (MgADP and P_i) remain attached to the myosin head and inhibit further

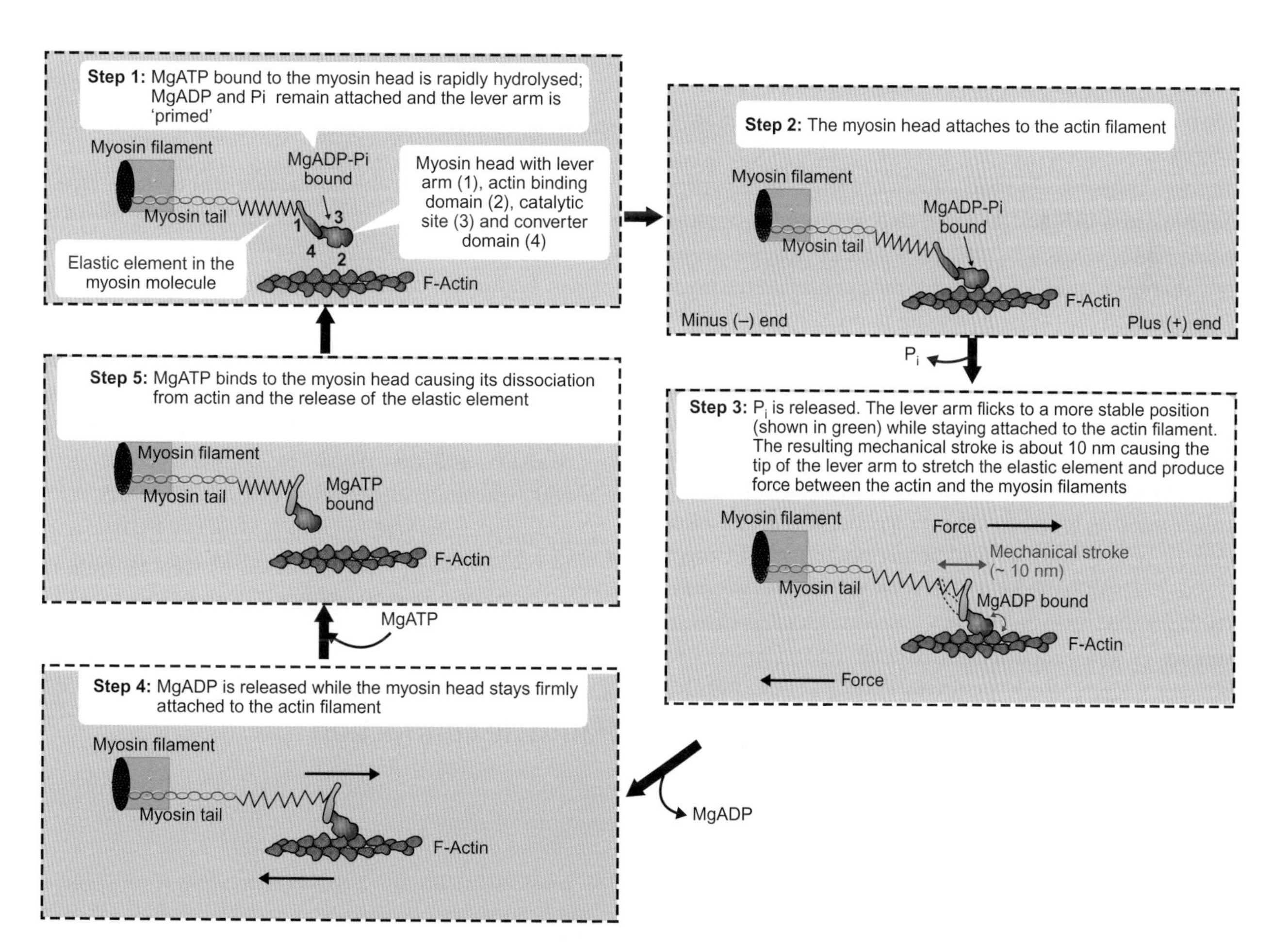

Figure 18.7 The chemo-mechanical cycle for force generation in muscle

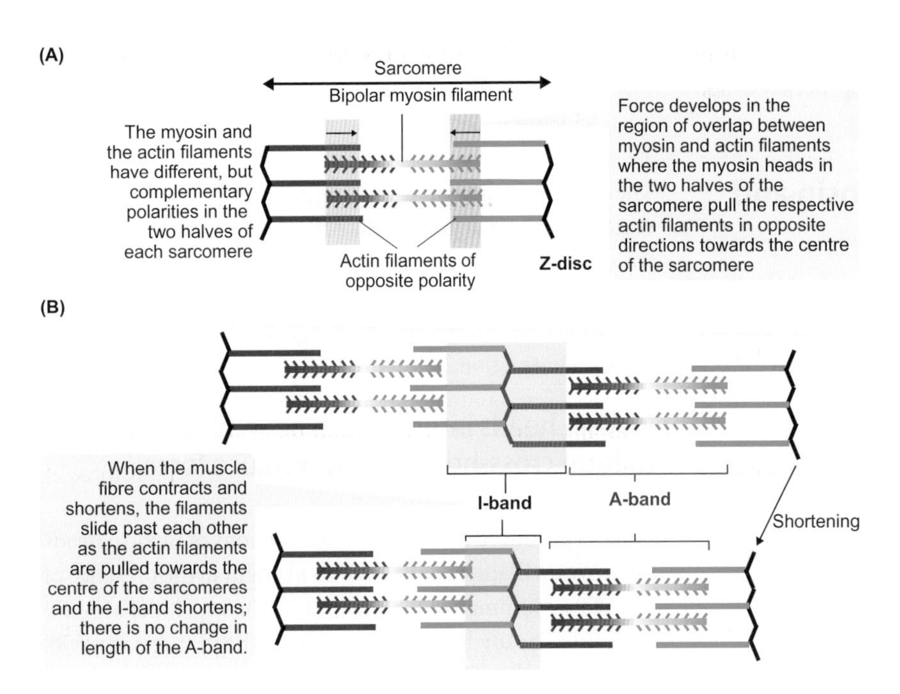

Figure 18.8 The myosin and actin filaments slide past each other when the striated muscle contracts and shortens

(A) Force produced by myosin heads interacting with actin filaments of complementary polarity in the region of overlap between the myosin and actin filament causes shortening of the sarcomeres without changing the length of the filaments. (B) Principle of the sliding filament theory of muscle contraction.

binding and hydrolysis of MgATP. The energy released from the hydrolysis of MgATP is not dissipated, but is stored by virtue of subtle configuration changes in the structure of the myosin head. These changes cause the lever arm on the myosin head to adopt its pre-power stroke (or primed) position (shown in red) and the affinity of the myosin head for actin to increase.

In *Step 2*, the myosin head attaches to the actin filament to form a **cross-bridge**. The large cleft in the actin-binding domain on the myosin head (Figure 18.2C) closes, causing subtle changes to the nucleotide-binding pocket on the myosin head, which results in the release of phosphate.

In *Step 3*, the release of phosphate is coordinated with the rotation of the lever arm, which performs a mechanical stroke of about 10 nm. This movement stretches the elastic element of the myosin head pulling on the actin filament and produces force between the filaments, without the actin filament needing to move. On average, the amount of force produced by one myosin head acting on an actin filament of same polarity is about 10^{-11}N (10 pN). Thus, we need 100,000,000,000 myosin heads located on parallel myosin filaments to pull randomly on parallel actin filaments to produce 1N force, which is close to a 100 g weight. Relative movement of the filaments ensues only when the force produced by cross-bridges[14] is greater than the force exerted on the filaments in the opposite direction, called **load**[15].

In *Step 4*, MgADP is released from the myosin head, which is very strongly attached to the actin filament. A new molecule of MgATP can now bind to the nucleotide-binding site on the myosin head, which causes the myosin head to detach rapidly from the actin filament (*Step 5*) and the elastic element in the myosin molecule to be released.

This cycle, called the **chemo-mechanical cycle**, is repeated through Steps 1–5 for as long as the myosin head can interact with actin[16] and MgATP is present.

From Figure 18.7 we can draw the following important conclusions:

- MgATP is rapidly hydrolysed by myosin, but the products of reaction (MgADP and P_i) remain attached to the myosin head and inhibit further hydrolysis of MgATP until the myosin head interacts with actin, which facilitates the removal of P_i and MgADP;
- the mechanical cycle does not coincide with the chemical cycle: MgATP is hydrolysed in Step 1, while the power stroke occurs in Step 3.

18.1.5 The sliding filament theory of muscle contraction

Independent electromicroscopic and light microscopic observations of single striated muscle fibres by two groups of

[14] Force produced by the muscle is often referred to as 'tension' in muscle physiology.

[15] We discuss load further in Section 18.5.1.

[16] As we discuss later in this Section (18.1.6), the ability of the myosin heads to interact with actin and produce force is regulated by the concentration of calcium ions.

researchers[17] in the early 1950s showed that the length of the myofilaments does not essentially change when cross-striated muscle fibres contract and shorten or when the muscle fibres are stretched. These scientists further interpreted their results to indicate that:

- the two sets of filaments slide past each other when the muscle fibre length changes, and
- force in the contracting fibres is produced by cross-bridges that form and work independently of each other in the region of overlap between the myosin filament and the actin filament.

The mechanism became known as the **sliding filament theory of muscle contraction** and was shown to apply universally to all muscle types. The underlying principles of this theory are illustrated in Figure 18.8B. Note that the sliding filament concept of muscle contraction does not in itself indicate *how* force is generated at cross-bridge level.

The strongest support for the sliding filament theory of muscle contraction came from experiments described in Experimental Panel 18.1. During these experiments, single muscle fibres were stimulated to produce maximal force at

[17] Hugh Huxley and Jean Hanson were the members of one group and Andrew Huxley and Rolf Niedergerke were the members of the other group. Hugh Huxley and Andrew Huxley were not related to one another.

Experimental Panel 18.1 Support for the sliding filament theory of muscle contraction

Background to the study

The sliding filament theory of muscle contraction was based on electron-microscopic and microscopic observations that the length of the thin (actin) and thick (myosin) filaments does not change when skeletal muscle contracts and shortens, or when the muscle is stretched[1]. When the theory was formulated in 1954 by both Hugh Huxley and Jean Hanson, and Andrew Huxley and Rolf Niedergerke, the suggestion was also made that force may be produced in the region of overlap between the two types of filaments. The strongest support for the sliding filament theory of muscle contraction was provided by careful measurements of the capacity of single frog skeletal muscle fibres to produce force at various sarcomere lengths where the amount of overlap between filaments was different.

Experimental approach

Advances in dissection, isolation, mounting and electrical stimulation of single intact muscle fibres from frog leg muscles together with the development of an ingenious feedback mechanism that kept the length of sarcomeres constant when the fibre was stimulated to produce maximal force permitted accurate measurements of maximal force production at different sarcomere lengths. The general experimental set-up is shown in Figure A.

Results

The relationship between the force produced by single intact frog muscle fibres when maximally stimulated at specific sarcomere lengths together with the relative position of the myofilaments within the sarcomere at critical sarcomere lengths are shown in Figure B. The relationship, known as the **force-length diagram**, displays an **ascending limb** between 1.3 and 2.05 μm sarcomere length, a distinct **plateau** between 2.05 and 2.25 μm, and a **descending limb** between 2.25 μm to 3.65 μm. In frog muscles, myosin (thick) filaments are 1.6 μm long with a bare region of 0.2 μm in the middle, devoid of myosin heads; actin (thin) filaments projecting from either side of the Z-line are 1.0 μm long and the thickness of the Z-line is 0.05 μm.

As shown in Figure B, frog muscle fibres cannot produce force when stimulated at sarcomere lengths greater than 3.65 μm.

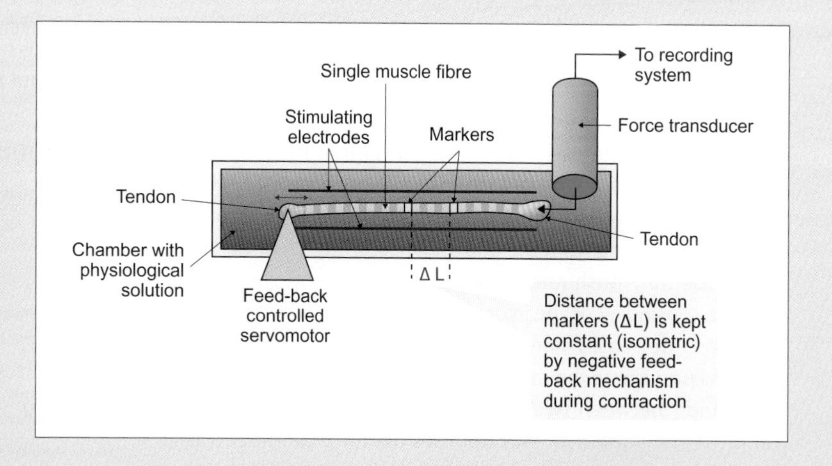

Figure A Experimental arrangement for making force measurements from single frog muscle fibres at chosen sarcomere lengths

The muscle fibre is electrically stimulated to contract at chosen length values of sarcomeres in a small central segment of the fibre between markers placed on the fibre surface. The distance between markers is monitored by optical means and is kept constant during contraction by a feedback controlled servomotor. Note that when the ends of the frog muscle fibre are fixed, sarcomeres in the middle of the fibre shorten more during contraction than sarcomeres at the ends of the fibre such that it is not possible to accurately ascribe the force response to a precise sarcomere length.

Source: adapted from Gordon AM et al (1966). Tension development in highly stretched vertebrate muscle fibres. Journal of Physiology 184: 143-169.

18

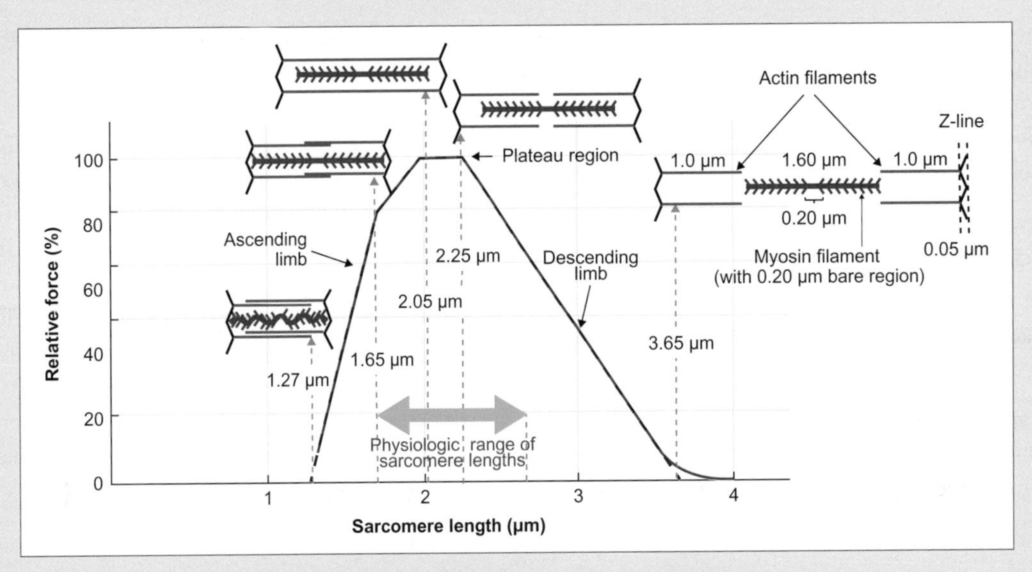

Figure B The maximal force—sarcomere length relationship for striated frog muscle fibres

There is an almost perfect overlap between the red curve representing the experimental data for maximal force production in frog muscle fibres at different sarcomere lengths and the predicted (blue) curve based on relative number of cross-bridges in the region of overlap between myosin and actin filaments at sarcomere lengths >2.0 µm. The predicted force curve at sarcomere lengths below 2.0 µm was drawn to indicate a shallow linear decrease in force between 2.05 and 1.65 µm due to the double overlap of actin filaments which causes some disruption in the arrangement of the actin filaments around the myosin filaments, and a sharper linear decrease in force below 1.65 µm to zero force at 1.27 µm due to the compression of the myosin filaments between the Z-lines. The physiologic sarcomere range corresponds to fibre lengths between about 80% and 130% of resting fibre length in the frog. Source: Experimental curve taken from Gordon et al (1966).

Notice that at sarcomere lengths >3.65 µm there is no overlap between the myosin and actin filaments to support the formation of force-producing cross-bridges between the two types of filaments.

The decrease in sarcomere length from 3.65 to 2.25 µm causes a linear increase in the force produced. This coincides with the linear increase in the number of force-producing cross-bridges that can be made in the region of overlap between the myosin and actin filaments.

Maximum force is produced at sarcomere lengths between 2.05 µm to 2.25 µm, where a plateau is reached. This range coincides with sarcomere lengths where the maximum number of force-producing cross-bridges can be made between the actin and myosin filaments. Note that over this range of sarcomere lengths, actin filaments overlap the bare zone on myosin filaments, where no myosin heads are present, and therefore no additional force-producing cross-bridges can be formed.

At sarcomere lengths shorter than 2.05 µm, the capacity to produce force decreases. Below 2.05 µm the actin filaments from the two halves of the sarcomere start to overlap, which causes some disruption in the arrangement of the actin filaments around the myosin filaments. We also know that myosin heads develop force in the direction of the plus end of the actin filaments, where the filaments attach to the Z-line. Therefore, when actin filaments overlap, myosin heads can also attach to actin filaments projecting from the opposite half of the sarcomere, in which case they push rather than pull on the respective actin filament, causing a decline in overall force.

The force decreases more steeply at sarcomere lengths shorter than 1.65 µm, where the myosin filaments become compressed, like springs, between two Z-lines. The force developed in the compressed myosin filaments opposes the force developed by cross-bridges, which explains why the force generated during maximal muscle contraction declines more steeply at sarcomere lengths shorter than 1.65 µm, until eventually no force can be generated by muscle fibres at sarcomere lengths around 1.3 µm.

Conclusions

- The very close agreement between measured and predicted maximal force responses at different sarcomere lengths based on known myofilament dimensions shown in Figure B provides the strongest support for the sliding filament theory of muscle contraction[2].
- Frog skeletal muscles cannot develop force at sarcomere lengths greater than 3.65 µm or shorter than 1.3 µm. Considering that the sarcomere length in resting muscle fibres of the frog is around 2.1 µm, it shows that the frog muscle can shorten to no more than about 60 per cent of its resting length (1.3 µm/2.1 µm = 0.62 = 62 per cent). More generally, the physiologic range of sarcomere lengths in frog muscle fibres is between about 1.7 and 2.7 µm which corresponds to about 80–130 per cent of the resting length.

› Find out more

Gordon AM, Huxley AF, Julian FJ (1966). The variation in isometric tension with sarcomere length in vertebrate muscle fibres. Journal of Physiology 184: 170–192.

18

1 More recent studies have shown that there is a small change in filament lengths (about 0.3 per cent) during maximal contractions of skeletal muscle, which amounts to less than 0.006 µm at a sarcomere length of 2 µm. This filament length change is therefore negligibly small for the overall force-sarcomere length relationship shown in Figure B.

2 Many studies on muscle preparations from different animals largely confirm these findings.

various sarcomere lengths. Results from such studies were quantitatively explained by predictions based on the number of cross-bridges that can form at the level of one sarcomere in the region of overlap between the myosin filament and the actin filaments of complementary polarity.

Thicker fibres produce more force than thinner fibres of the same type, irrespective of their length

Consider the force generated by one myofibril consisting of one sarcomere as shown in Figure 18.8A. The force produced between the ends of the myofibril is made of two components, which pull the actin filaments and the Z-lines in the two (end) half-sarcomeres toward the centre of the myofibril. If the myofibril is made of several sarcomeres in series, equal opposing forces pull each Z-line between adjacent sarcomeres to the left and to the right; as such, these forces cancel each other out. The only forces that remain are those exerted by the end two half-sarcomeres in the myofibril. Therefore, the net force exerted between the ends of a myofibril is, in principle, the same regardless of whether the myofibril is one or many sarcomeres long.

Now, consider two identical myofibrils next to each other. The net force produced by the two myofibrils will be twice the force produced by each myofibril. More generally, the total amount of force produced by a striated muscle fibre is proportional to the number of myofibrils in the cross-section of the fibre, implying that a thicker fibre produces more force than a thinner fibre under otherwise identical conditions.

Thus, an important conclusion from the sliding filament theory of muscle contraction is that the force output of a muscle fibre or a whole muscle is proportional to the number of myofibrils in its cross-sectional area and is independent of the muscle's length. Therefore, for comparative purposes, the force-generating capacity of a muscle is generally expressed as force per muscle cross-sectional area and is called **specific force**[18].

18.1.6 Muscle fibres are sensitive to Ca^{2+} concentration

The Ca^{2+} regulatory system in muscle, which makes the contractile apparatus sensitive to Ca^{2+}, is associated either with the actin filament, the myosin filament, or with both filaments.

In skeletal and cardiac muscles of vertebrates, the Ca^{2+} regulatory system is associated with the actin filament (Figures 18.6F and 18.9) and comprises a pair of filamentous

[18] Specific force has the same dimensions as stress ($N\ m^{-2}$) and the two terms are often used interchangeably in muscle research.

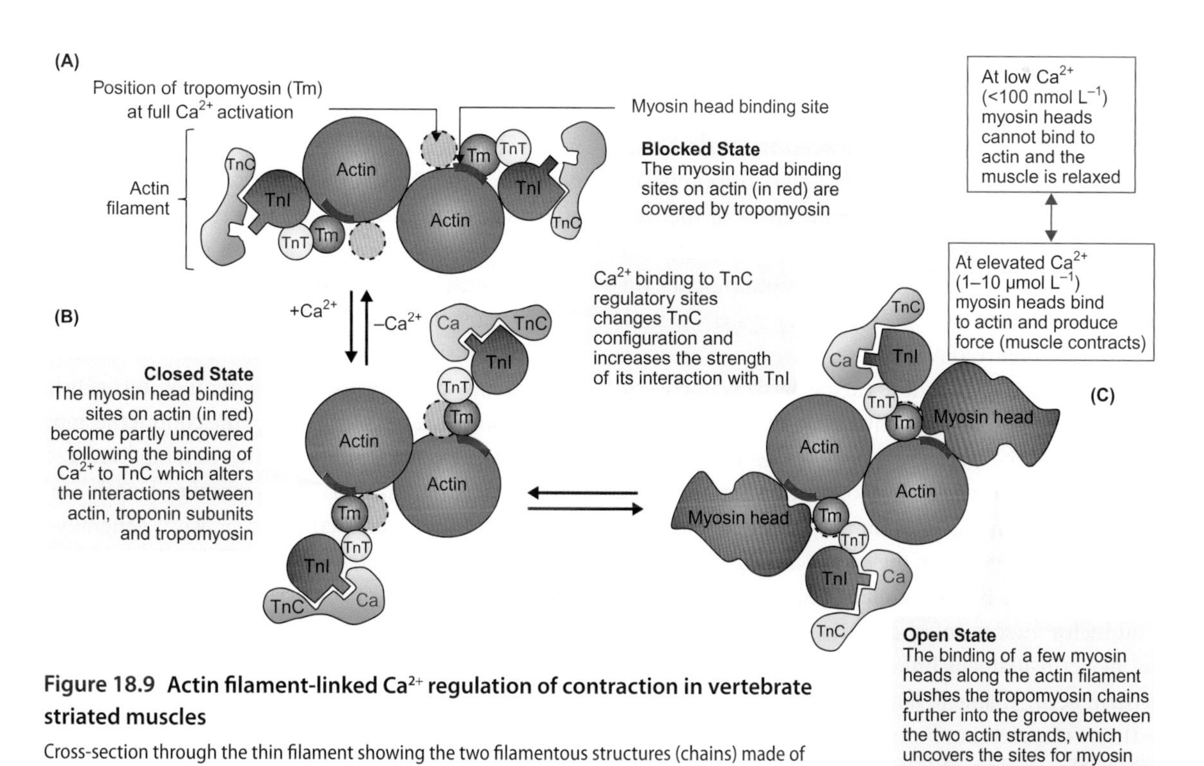

Figure 18.9 Actin filament-linked Ca^{2+} regulation of contraction in vertebrate striated muscles

Cross-section through the thin filament showing the two filamentous structures (chains) made of tropomyosin (Tm) molecules attached end to end along the actin filament and the troponin complex consisting of three subunits for each tropomyosin molecule (troponin C (TnC), the Ca^{2+}-binding subunit; troponin T (TnT), the tropomyosin binding subunit and troponin I (TnI), the inhibitory troponin subunit).

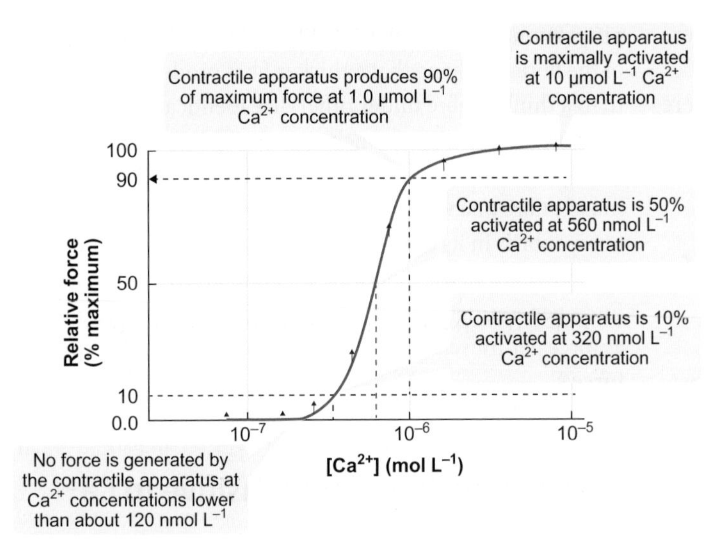

Figure 18.10 Force Ca^{2+} activation curve for a rat skeletal muscle fibre

A rise in Ca^{2+} concentration by a factor of 3 from 320 to 1000 nmol L^{-1} increases the level of activation of the contractile apparatus from 10% to 90%.

Source: Stephenson DG, Williams DA (1981). Journal of Physiology 317: 281–302.

structures (chains) made of tropomyosin molecules attached end to end along the entire length of the actin filament together with a troponin complex consisting of three subunits (troponin C, troponin T and troponin I) for each tropomyosin molecule.

When the Ca^{2+} concentration is low at myofibrils, such that no Ca^{2+} is bound to troponin C, the tropomyosin chains cover the sites on the actin-filament where the myosin heads bind (Figure 18.9A): the Ca^{2+} regulatory system prevents myosin heads interacting with actin and the system is in the *blocked state*. (The myosin heads are at *Step 1* of Figure 18.7.) Therefore, no active force is generated in the myofibrils, and the muscle is relaxed.

When Ca^{2+} concentration rises, Ca^{2+} binds to troponin C and increases its affinity for troponin I, which, in turn, decreases the strength of interaction between troponin I and actin. The binding between troponin T and tropomyosin is also reduced, allowing a rotational movement of the tropomyosin chains on the surface of the actin filament. This movement partially uncovers the myosin binding sites on the actin monomers and the Ca^{2+} regulatory system is now in the *closed state* (Figure 18.9B).

The binding of a few myosin heads to the partially uncoverd myosin sites along the actin filaments drives further the rotational movement of the tropomyosin chains deeper into the groove between the two actin strands of the thin filament, uncovering all myosin binding sites on the actin filament. The Ca^{2+} regulatory system is now in the *open state* (Figure 18.9C).

Striated muscles in bivalve molluscs such as the cross-striated adductor muscle of the scallop (*Pecten* spp.) do not have an actin-filament-linked Ca^{2+} regulatory system because their actin filament lacks the troponin complex. Instead, these muscles have myosin-filament-linked Ca^{2+} regulation. In the absence of Ca^{2+}, the MgATPase activity of the myosin heads (*Step 1* in Figure 18.7) is inhibited. This inhibition is removed when Ca^{2+} binds to the regulatory light chain on the myosin head. Echinoderms and brachiopods also have myosin-linked Ca^{2+} regulation[19].

Striated muscles in arthropods have *double* Ca^{2+} regulation, i.e. they exhibit both actin- and myosin-filament-linked Ca^{2+} regulation. Therefore, in arthropod muscles, both systems must be activated by Ca^{2+} to allow the myosin heads to interact with actin and produce force.

The activation of the contractile apparatus by Ca^{2+} is studied in myofibrillar preparations[20] incubated in MgATP containing media with buffered Ca^{2+} concentrations. Figure 18.10 shows the relationship between force and Ca^{2+} concentration in rat skeletal muscle; this is an example of a **force Ca^{2+} activation curve**. In this example, no force develops at Ca^{2+} concentrations less than 120 nmol L^{-1} and maximum force is produced at 10 μmol L^{-1}. Sensitivity to Ca^{2+} generally refers to the Ca^{2+} concentration corresponding to 50 per cent maximum force. Later in this chapter we discuss how the concentration of Ca^{2+} ions changes in the different types of muscle fibres in response to excitation.

Ca^{2+} activation curves similar to that in Figure 18.10 were obtained for striated muscle fibres of many animals, including amphibians, crustaceans, insects, reptiles, birds, fish and bivalves. Note that the sensitivity of the contractile apparatus to Ca^{2+} changes when conditions such as temperature, cytosolic Mg^{2+} concentration, pH and sarcomere length are altered. For example, an increased cytosolic Mg^{2+} generally requires a higher Ca^{2+} concentration to produce the same amount of force in both vertebrate and crustacean skeletal muscle.

[19] Vertebrate smooth muscles also have a myosin-linked type of Ca^{2+} regulation but, as we discuss in Section 18.1.8, the mechanism of action is different from that in molluscs.

[20] Myofibrillar preparations consist of bundles of myofibrils that are structurally and functionally intact and are made by rendering the surface membrane permeable to extracellular solutions by various chemical or mechanical means.

18

> *Review articles*

Batters C, Veigel C, Homsher E, Sellers JR (2014). To understand muscle you must take it apart. Frontiers in Physiology 5:90.
Gordon AM, Homsher E, Regnier M (2000). Regulation of contraction in striated muscle. Physiological Reviews 80: 853–924.
Geeves MA, Holmes KC (2005). The molecular mechanism of muscle contraction. Advances in Protein Chemistry 71: 161–193.
Reconditi M, Linari M, Lucii L, Stewart A, Sun Y-B, Boesecke P, Narayanan T, Fischetti RF, Irving T, Piazzesi G, Irving M, Lombardi V (2004). The myosin motor generates a smaller and slower working stroke at higher load. Nature 428: 578–581.
Schliwa M, Woehlke G (2003). Molecular motors. Nature 422: 759–765.
Spudich JA (2012). One path to understanding energy transduction in biological systems. Nature Medicine 18: 1478–1482.

18.2 Voluntary muscle fibres: trusted followers of the nervous system

Skeletal muscles of vertebrates and voluntary muscles of invertebrates are under the direct control of the somatic nervous system and are responsible for maintaining posture and generating all movement of the animal in its environment. In vertebrates, skeletal muscles represent between 40 and 70 per cent of total body mass. They consist of muscle fibres that are approximately cylindrical in shape, 5 to 120 μm in diameter, and which vary in length between less than one mm to tens of centimetres, depending on the size of the animal and location in the body. All voluntary muscle fibres are multinucleated. Some voluntary muscle fibres of invertebrates, such as the giant muscle fibres of the barnacle *Balanus nubilus* and other crustaceans, are up to 2 mm in diameter. These giant fibres enabled pioneering investigations into the role of calcium ions in the regulation of muscle contraction.

18.2.1 Neural control of muscle fibres is exerted at neuromuscular junctions

Neuromuscular junctions are chemical synapses between motor neurons and voluntary muscle fibres. The synapses become activated by action potentials[21] in the motor neurons.

The sarcolemma contains various types of ion channels and Na^+/K^+-pumps, and acts as a barrier for large transmembrane concentration differences of Na^+, K^+ and Ca^{2+} in a similar way to plasma membrane in neurons[22]. At rest, the sarcolemma has a higher permeability to K^+ relative to Na^+ in skeletal muscle than in neurons, which translates into a more negative resting membrane potential[23], closer to the equilibrium potential for K^+ (V_K). The membrane potential of the sarcolemma changes in response to stimulation of neuromuscular junctions by action potentials in the motor neurons.

The neuromuscular junction of vertebrate skeletal muscle is an excitatory ionotropic chemical synapse

Figures 18.11A and B illustrate typical neuromuscular junctions[24] in vertebrates. The motor neurons make multiple synaptic contacts with the sarcolemma over an area known as the **motor end-plate**. When action potentials reach the nerve terminals, acetylcholine (ACh) is released from vesicles in the nerve terminals and binds to the ACh receptors[25] found at high density on the crests of the extremely folded postsynaptic membrane, which is part of the sarcolemma.

The ACh receptors are ionotropic receptors that function as ACh-gated channels. When open, the ACh-gated channels are equally permeable to Na^+ and K^+, causing depolarization of the sarcolemma, as we discuss in Section 16.3.3. The high density of ACh channels at the postsynaptic membrane ensures that one action potential arriving at the neuromuscular junction causes a much greater depolarization of the sarcolemma (in the order of 10–50 mV) than it would in excitatory synapses between neurons, where the depolarization is about 1–2 mV or less.

Twitch fibres develop propagated action potentials in response to neural stimulation

The majority of skeletal muscle fibres of vertebrates have only one neuromuscular junction located in the middle of the fibre[26]. In these fibres, an action potential is generated in the sarcolemma for each neuronal action potential reaching the neuromuscular junction. The action potential spreads over the entire fibre and the fibre responds with a set brief

[21] We discuss action potentials and chemical synapses in Sections 16.2.4 and 16.3.2.
[22] In Section 16.2, we discuss the ionic basis of electrical activity in neurons.
[23] We discuss the basis for membrane potential generation across cellular membranes in Section 16.2.
[24] We discuss how chemical synapses work in Section 16.3.3.
[25] The ACh receptors in neuromuscular junctions are of the nicotinic type, which we discuss in Section 16.3.4 Curare (the common name of a group of alkaloids in plants from South America) binds tightly to the nicotinic ACh receptors in skeletal muscle and causes muscle paralysis as it prevents ACh binding and opening of the ACh channels. Curare and its derivatives are used in conjunction with general anaesthetics during surgical interventions on animals and humans to prevent uncontrolled body movements during surgical procedures.
[26] Twitch muscle fibres in some fish, like the zebrafish (*Danio rerio*), have two or more neuromuscular junctions and can be innervated by different motor neurons.

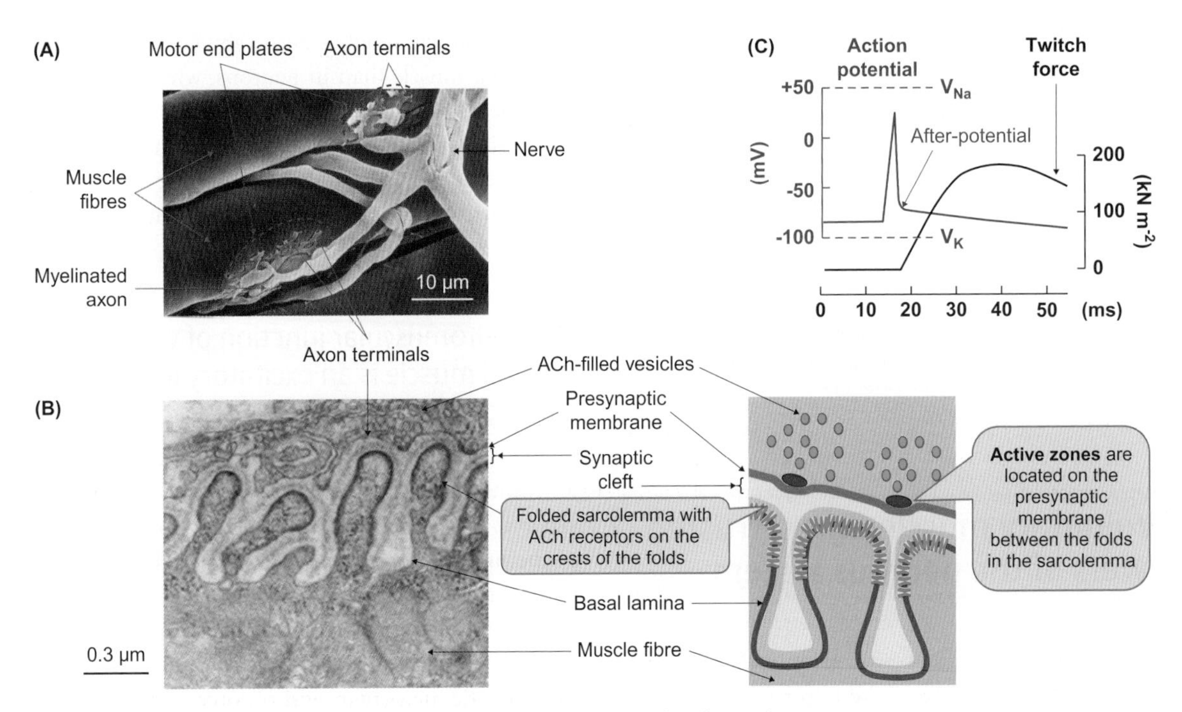

Figure 18.11 Structure of the neuromuscular junction in skeletal muscle of vertebrates

(A) Scanning electron micrograph of frog neuromuscular junctions showing the termination of axons on skeletal muscle fibres. The axon terminals make contact with the muscle fibre over a relatively large area called the motor end plate. (B) Higher magnification electron micrograph across a neuromuscular junction and diagram showing acetylcholine (ACh)-containing vesicles in the axon terminals and the folds in the sarcolemma with ACh receptors located on the crests of the folds. ACh is released at the level of active zones. The basal lamina in the sarcolemmal folds contains the enzyme acetylcholine esterase, which breaks down the ACh molecules. (C) Action potential and twitch force response (in kN m^{-2}) in a single twitch skeletal muscle fibre of the European common frog (*Rana temporaria*) at 18°C. The upstroke of the action potential in skeletal muscle fibres is caused by a rapid increase in the number of open voltage-gated Na^+ channels which displaces the membrane potential towards the equilibrium potential for Na^+ (V_{Na}). Repolarization results from inactivation of the Na^+ channels and delayed opening of K^+ channels. Immediately after the spike of the action potential the membrane potential is slightly less negative (depolarized) compared with the resting potential, i.e. muscle fibres have a depolarizing after potential.

Sources: A: Getty Images/Don W Fawcett; B: National Institute of Mental Health, US; C: Hodgkin A L, Horowicz P (1957). The differential action of hypertonic solutions on the twitch and action potential of a muscle fibre. Journal of Physiology 136: 17–18P.

contraction, called a **twitch**. This type of fibre is a **twitch fibre**.

Figure 18.11C shows the time course of the action potential followed by the twitch force response in a frog muscle fibre. The nature of action potentials in the twitch skeletal muscle fibres of vertebrates is essentially the same as in neurons (Section 16.2.4): voltage-gated Na^+ channels rapidly open causing rapid depolarization of the membrane, and then close (inactivate). The delayed opening of K^+ channels causes repolarization of the membrane. The duration of the action potential spike in twitch muscle fibres is about 2–5 ms, depending on body temperature and animal species. Twitch fibres respond with a set contraction in response to an action potential.

Tonic fibres respond with graded depolarizations to neural stimulation

Vertebrates also have **tonic muscle fibres**, which produce graded force responses depending on the stimulus intensity. Tonic fibres occur usually in specialized muscles that are responsible for fine and/or complex movements. Such muscles include extraocular muscles, which move the eyeballs; wing muscles in gliding birds for fine control of the wing positions; muscles of the middle ear of mammals; and laryngeal muscles, which apply varying force to the vocal chords to produce specific sounds, and are therefore responsible for vocalization.

Tonic muscle fibres have multiple neuromuscular junctions from the same neuron over their length, as shown in Figure 18.12B. The Na^+ channels in tonic fibres respond more slowly to depolarization than those in twitch fibres and therefore do not generate action potentials. Instead, they sustain graded depolarizations in response to neural stimulation: a higher frequency causes a greater level of depolarization of the sarcolemma, more Ca^{2+} release and a greater contractile response. This graded response enables more complex mechanical responses in tonic fibres than in twitch fibres, which respond with a set contraction.

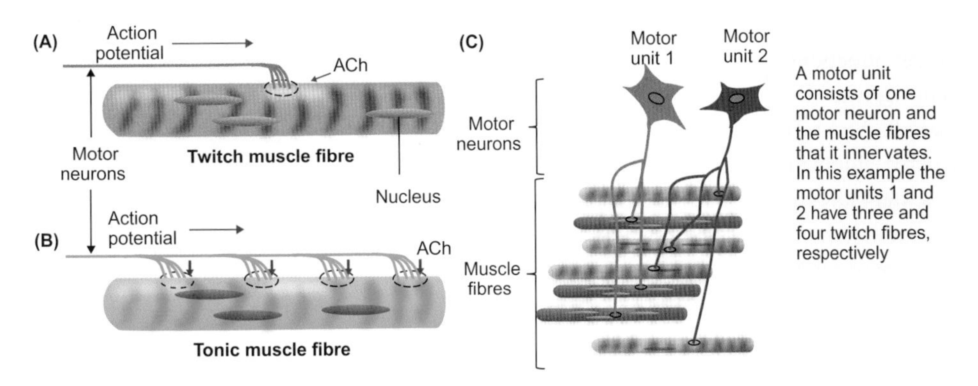

Figure 18.12 Motor neuron innervation of vertebrate skeletal muscle fibres

Each vertebrate skeletal muscle fibre is generally innervated by only one motor neuron which makes either one excitatory synapse like in twitch fibres (A) or multiple excitatory synapses like in tonic fibres (B). The transmitter in all vertebrate neuromuscular junctions is acetylcholine (ACh). Each muscle in vertebrate skeletal muscles has several motor units which contract independently of each other (C).

The strength of contraction at the level of whole muscle in vertebrates is regulated by the number of fibres that are simultaneously stimulated

Among vertebrates (except fish[27]), each skeletal muscle fibre is innervated by only one motor neuron, as shown in Figure 18.12A and 18.12B, but one motor neuron normally innervates many other muscle fibres. The motor neuron and the muscle fibres that it innervates form a **motor unit**, as shown in Figure 18.12C. The muscle fibres in a motor unit are not located next to each other but are scattered throughout a particular muscle.

All muscle fibres in a motor unit contract simultaneously when the motor neuron is activated. Thigh muscles (quadriceps) that are used for power movements in terrestrial mammals have motor units consisting of hundreds of muscle fibres. In contrast, muscles involved in fine movements, like the finger muscles in humans, have motor units consisting of only a few muscle fibres. The primary mechanism for producing a particular level of force in a vertebrate muscle is usually determined by the size and number of motor units that are activated.

[27] Fish muscle fibres can be innervated by two or more motor neurons.

The neuromuscular junctions of invertebrate voluntary muscle fibres

The voluntary muscle fibres of invertebrates do not normally generate action potentials. Most studies of skeletal muscle excitation among invertebrates have been conducted on arthropods (mainly crustaceans and insects). Unlike vertebrates, in which each skeletal muscle fibre is generally innervated by one single motor neuron, skeletal fibres of arthropods are innervated by several different motor neurons, which make multiple neuromuscular synapses along the fibre, as shown in Figure 18.13. One muscle fibre can be supplied by two different excitatory neurons, one fast and one slow, as well as one inhibitory neuron. All these neurons make multiple neuromuscular synapses along the fibre.

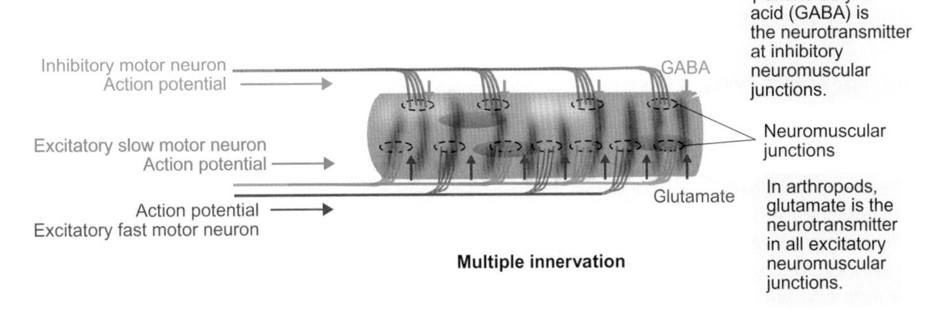

Figure 18.13 Motor neuron innervation of arthropod muscle fibres

The arthropod skeletal muscle fibres are supplied by three (or more) different nerve fibres: one for fast contractions, another for slow contractions and one for inhibition of contraction.

All excitatory neurons in arthropods release glutamate[28] at the neuromuscular junctions, which causes depolarization of the sarcolemma, while the inhibitory neurons release γ-amino butyric acid (GABA). The release of GABA increases the membrane permeability for chloride ions (Cl^-), which, in turn, reduces the depolarizing effect of glutamate on the sarcolemma. Thus, the action potential activity in the different axons that innervate one muscle fibre at a particular time determines the level of depolarization of the sarcolemma, which, in turn, determines the level of intracellular Ca^{2+} rise and the size of the mechanical response in the muscle fibre. Accordingly, the strength of the muscle response to excitation in arthropods is regulated at the single muscle fibre level.

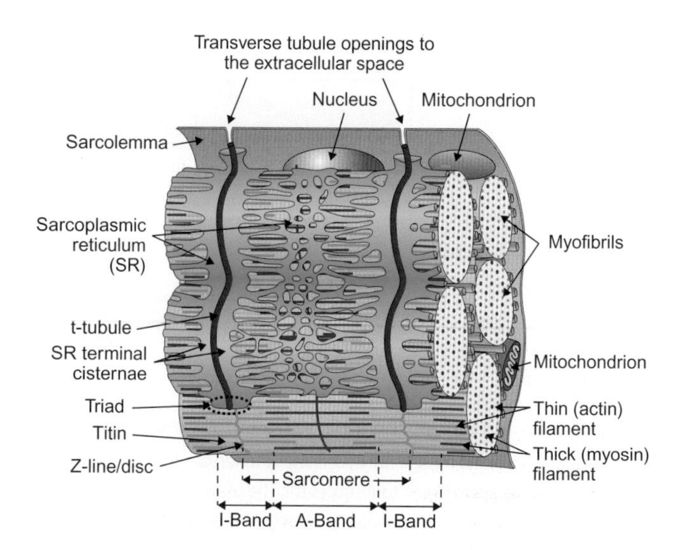

Figure 18.14 Spatial relationship of membranous structures (sarcolemma, t-tubules and sarcoplasmic reticulum) with respect to the myofibrils and the contractile apparatus in frog skeletal muscle fibres

In frog skeletal muscle the t-tubules run at the level of the Z-lines. Skeletal muscles fibres of mammals, reptiles and crustacea, however, have two sets of t-tubules per sarcomere, running at the junction between the A- and I-bands in the sarcomere.

Source: modified from Hill RW, Wyse GA, Anderson M, (2016). Animal Physiology, 4th Edition, Sinauer Associates.

18.2.2 Excitation–contraction coupling in voluntary muscle fibres

Excitation from the sarcolemma—in the form of membrane depolarization—spreads rapidly to intracellular Ca^{2+} stores surrounding each myofibril, via a network of tubules called **t-tubules** that run transversely to the fibre axis, as shown in Figure 18.14 for frog skeletal muscle. The intracellular Ca^{2+} store in muscle is the **sarcoplasmic reticulum**[29], which contains enlarged compartments called **terminal cisternae** (singular cisterna). Terminal cisternae contain a protein called calsequestrin, which binds Ca^{2+} reversibly with low affinity[30]. The t-tubules open to the extracellular space at specific, regular points along the sarcolemma and form specialized junctions with the terminal cisternae called dyads or triads depending on whether the t-tubule forms junctions with one or two neighbouring cisternae, respectively. Triads occur typically in vertebrate skeletal muscles as shown in Figures 18.14 and 18.15A, B while dyads are prevalent in muscle fibres of invertebrates.

Transmission of excitation from t-tubules to the sarcoplasmic reticulum involves two types of coupled Ca^{2+} channels

The t-tubule membrane at triads (or dyads) contains one type of voltage-gated Ca^{2+} channel[31] called **DHP receptors** because they bind with high-affinity compounds known as dihydropyridines. DHP receptors have positively charged transmembrane segments, which make them sensitive to changes in the electrical potential difference across the t-tubule membranes and enable them to function as voltage sensors. DHP receptors are organized in groups of four, called tetrads, as shown in Figure 18.15C for vertebrate skeletal muscle. Note that the DHP receptors in twitch vertebrate fibres do not function as Ca^{2+} channels because the action potential is too short for them to open and actually permit the passage of Ca^{2+}.

Each tetrad on the t-tubule faces a different type of Ca^{2+} channel in the membrane of the terminal cisterna of the sarcoplasmic reticulum as shown in Figure 18.15B, C for vertebrate skeletal muscle. This type of Ca^{2+} channel is a very large protein[32] called either the **Ca^{2+} release channel** or the **ryanodine receptor** (**RyR**), because it binds the alkaloid ryanodine[33] with very high affinity. The tetrads and the RyRs are generally organized in long arrays that face each other (Figure 18.15C).

18

[28] We discuss transmitter molecules at chemical synapses in Section 16.3.4.

[29] The endoplasmic reticulum in muscle fibres is called the sarcoplasmic reticulum.

[30] The presence of calsequestrin ensures that the sarcoplasmic reticulum can store the necessary amount of calcium needed for contraction in a relatively small volume; a much greater volume would be needed to store the same amount of calcium without affecting the osmotic balance, if calcium were not bound.

[31] We discuss in Section 16.2.3 the general structure of Ca^{2+} channels.

[32] The RyR has a molecular mass of 2,240 kD and consists of four identical subunits.

[33] Alkaloids are alkali-like (basic) nitrogen-containing organic substances that occur mainly in plants.

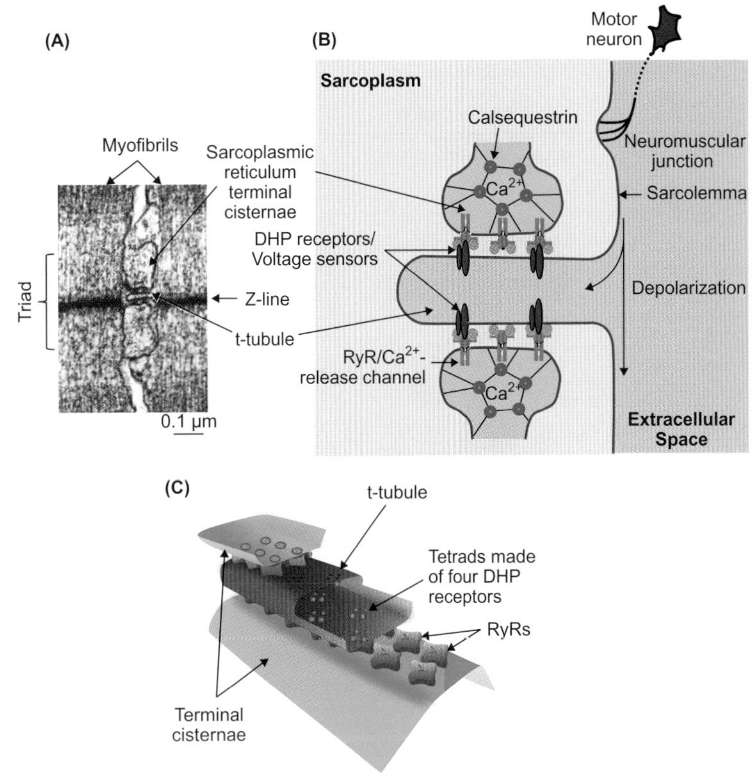

Figure 18.15 Structure of the triad in vertebrate skeletal muscle fibres

(A) Electron micrograph of a triad in the twitch muscle fibre of the frog. (B) Diagrammatic representation of the triad showing the location of the DHP receptors in the t-tubule membrane and RyRs in the terminal cisternae membrane. The terminal cisternae contain a Ca^{2+}-binding protein calsequestrin that keeps Ca^{2+} concentration buffered at around 1 mmol L^{-1}. (C) The DHP receptors in the t-tubule are organized in groups of four (tetrads) and function as voltage sensors responding to membrane depolarization. Each tetrad in the t-tubule membrane faces one Ca^{2+}-release channel (RyR) in the sarcoplasmic reticulum membrane. Both the tetrads and the RyRs are organized in long arrays facing each other such that every second RyR faces one tetrad.

Source: A: Franzini-Armstrong C (1970) Studies of the triad. I. Structures of the junction in frog twitch fibers. Journal of Cell Biology 47: 488–499. C: Modified after Felder E, Franzini-Armstrong C (2002). Type 3 ryanodine receptors of skeletal muscle are segregated in a parajunctional position. Proceedings of the National Academy of Sciences USA 99: 1695–1700.

The Ca^{2+} release channels have three important sites that modulate their function, as illustrated in Figure 18.16:

- an ATP-binding site, to which ATP binds to stimulate channel opening;
- a Ca^{2+} activation site that needs to have Ca^{2+} bound to it for the channel to open fully;
- a Ca^{2+}/Mg^{2+} inhibition site which needs to be free of Ca^{2+} or Mg^{2+} for the channel to open.

Under resting conditions, Mg^{2+} powerfully inhibits the sarcoplasmic reticulum Ca^{2+} release channels in twitch skeletal muscle fibres of vertebrates as shown in Figure 18.17A. (Mg^{2+} is bound to both the Ca^{2+} activation site and the Ca^{2+}/Mg^{2+} inhibition site.) Therefore, although primed to open (cocked) by ATP, the Ca^{2+} release channels are closed under resting conditions and only open when the inhibitory action exerted by Mg^{2+} is removed or bypassed.

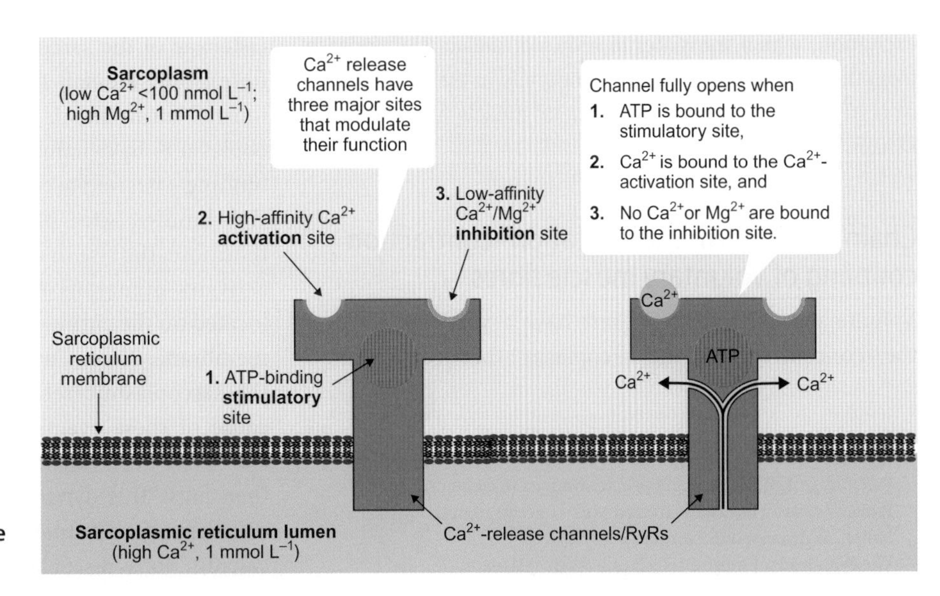

Figure 18.16 Modulatory sites on sarcoplasmic reticulum Ca^{2+} release channels (RyRs)

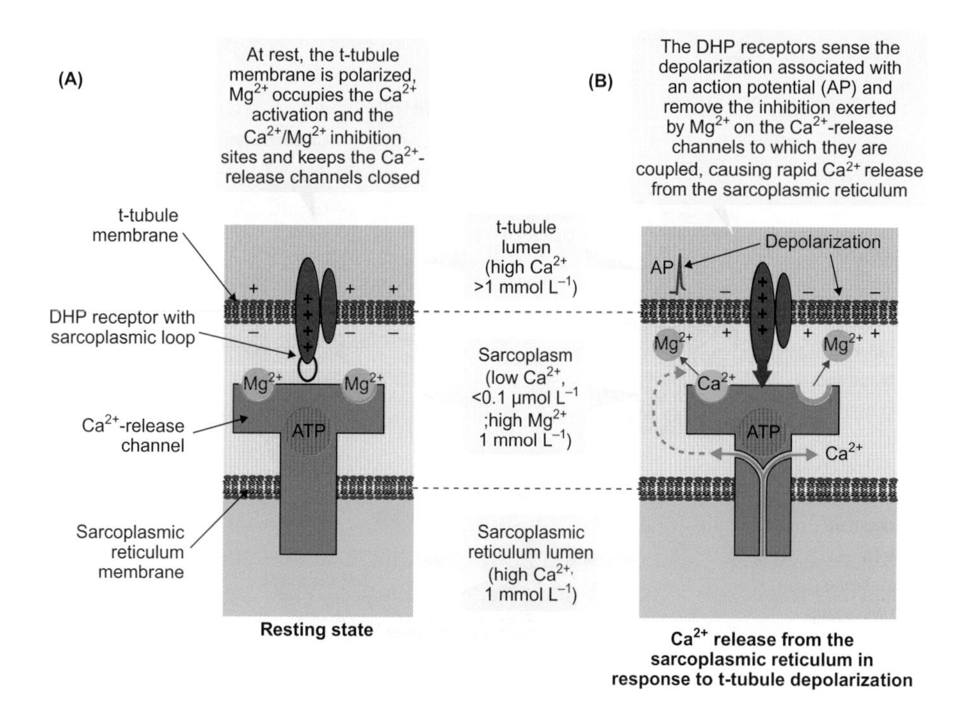

Figure 18.17 Signal transmission from DHP receptors to Ca^{2+} release channels in twitch vertebrate muscle fibres

A sarcoplasmic loop in the DHP receptor complex is critical for signal transmission to the Ca^{2+}-release channels. Note that the polarity of the t-tubule membrane changes from (A) to (B), as the membrane becomes excited following the arrival of an action potential. The DHP receptors have positively charged transmembrane segments, which make them sensitive to changes in the electrical potential difference across the t-tubule membranes and enable them to function as voltage-sensors.

Activation of the DHP receptors by t-tubule membrane depolarization in twitch fibres is thought to remove the magnesium inhibition of the Ca^{2+}-release channels to which they are mechanically coupled. This activation process may involve displacement of Mg^{2+} from the Ca^{2+} activation and the Ca^{2+}/Mg^{2+} inhibition sites. The lifting of inhibition causes the channels to open partially until some of released Ca^{2+} binds to the Ca^{2+}-activation site after which they open fully as shown in Figure 18.17B. After the t-tubule membrane repolarizes, the DHP receptors deactivate, the inhibitory action of Mg^{2+} on the Ca^{2+}-release channels is reinstated, and the Ca^{2+}-release channels close.

Note that the inhibitory action of Mg^{2+} is not so readily removed at higher cytosolic Mg^{2+} concentrations following excitation, which translates to reduced rates of Ca^{2+} release from the sarcoplasmic reticulum not only in the twitch fibres of vertebrates, but also in the skeletal muscle fibres of crustaceans[34].

18

Chain of events in the excitation–contraction coupling of voluntary muscle fibres

Voluntary muscle fibres at rest have a very low ionized Ca^{2+}-concentration in the sarcoplasm[35] (between about 30 and 100 nmol L^{-1}) due to the action of the ATP-dependent plasma membrane Ca^{2+} pump, which is located throughout the sarcolemma and the t-tubule membrane. At this low Ca^{2+} concentration, the Ca^{2+} regulatory system associated with the contractile apparatus (Figures 18.9 and 18.10) prevents myosin heads from interacting with the actin filament and muscle fibres are relaxed. As summarized in Figure 18.18, depolarization of the sarcolemma activates the DHP receptors, which, in turn, cause activation of Ca^{2+} release channels in the sarcoplasmic reticulum, a rise in intracellular Ca^{2+} concentration, the activation of the Ca^{2+} regulatory system, and, finally, activation of the contractile apparatus.

Of the total amount of Ca^{2+} that is released from the cisternae of the sarcoplasmic reticulum (≤ 1 mmol per L fibre), only a small fraction (<10 per cent), stays in the ionized form. This is because the ionized Ca^{2+} in the sarcoplasm is in dynamic equilibrium with numerous Ca^{2+}-binding sites exposed to the sarcoplasm, including the Ca^{2+}-binding sites on the Ca^{2+}-regulatory system. Furthermore, the rise in sarcoplasmic Ca^{2+} concentration also activates numerous ATP-dependent **Ca^{2+} pump** proteins located on the entire surface of the sarcoplasmic reticulum membrane. These Ca^{2+} pumps bring back to the lumen of the sarcoplasmic reticulum Ca^{2+} that was released. Because of these events, after each excitation, the ionized Ca^{2+} concentration in the sarcoplasm rapidly rises and then falls. This dynamic change in Ca^{2+} concentration is called the **Ca^{2+} transient**. The muscle fibre relaxes when calcium ions dissociate from the binding sites on the

[34] The depressing effect of Mg^{2+} on Ca^{2+}-release and force Ca^{2+}-sensitivity (Section 18.1.6) can explain the link observed in crustaceans between reduced activity level and elevated Mg^{2+} concentration in the haemolymph, as discussed in Section 5.1.1.

[35] The cytoplasm of muscle cells is called sarcoplasm.

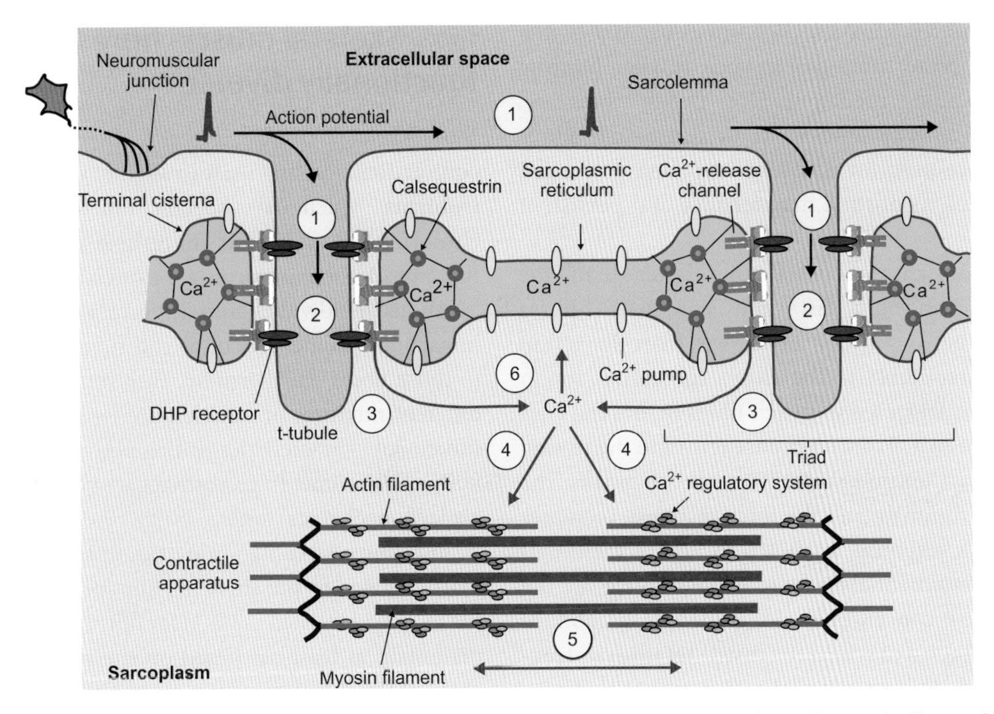

Figure 18.18 Overview of events in the excitation–contraction coupling and relaxation of twitch muscle fibres of vertebrates

1. Propagation of the action potential along sarcolemma and into the t-tubules. 2. Activation of the DHP receptors and signal transduction to sarcoplasmic reticulum Ca^{2+}-release channels. 3. Ca^{2+} release from the sarcoplasmic reticulum cisternae, where Ca^{2+} is loosely bound to calsequestrin, increases the concentration of ionized Ca^{2+} in the sarcoplasm from about 0.1 to 1–10 μmol L^{-1}. 4. Diffusion and binding of Ca^{2+} to sites on the Ca^{2+} regulatory system and removal of inhibition of cross-bridge formation (as discussed in Figure 18.9). 5. Force production by the contractile apparatus (as discussed in Figure 18.7). 6. Reuptake of Ca^{2+} into the sarcoplasmic reticulum by the Ca^{2+}-ATPase/pump, decline of sarcoplasmic Ca^{2+}, dissociation of Ca^{2+} from Ca^{2+}-binding sites on the Ca^{2+}-regulatory system and reinstatement of the inhibition of cross-bridge formation exerted by the Ca^{2+}-regulatory system.

Ca^{2+}-regulatory system as Ca^{2+} returns toward the resting level (Figure 18.19).

As we discussed in Section 18.2.1, tonic/slow muscle fibres in vertebrates and voluntary muscle fibres in invertebrates do not have action potentials but sustain graded depolarizations in response to neural stimulation. The number of DHP receptors that become activated, the ensuing Ca^{2+} transients, and the contractile responses are also graded, ultimately depending on the level of membrane depolarization.

Ca^{2+} transients and tetanic contractions in twitch fibres

Physiological measurements of how fast-twitch muscle fibres contract led to the subdivision of vertebrate skeletal twitch fibres into **fast-twitch** and **slow-twitch fibres**. As shown in Figure 18.19, the peak of the Ca^{2+} transient in fast-twitch fibres is about twice that in slow-twitch fibres, and the twitch force reaches its peak considerably earlier in fast-twitch than in slow-twitch fibres. Fast-twitch fibres also produce more power than slow-twitch fibres, as we

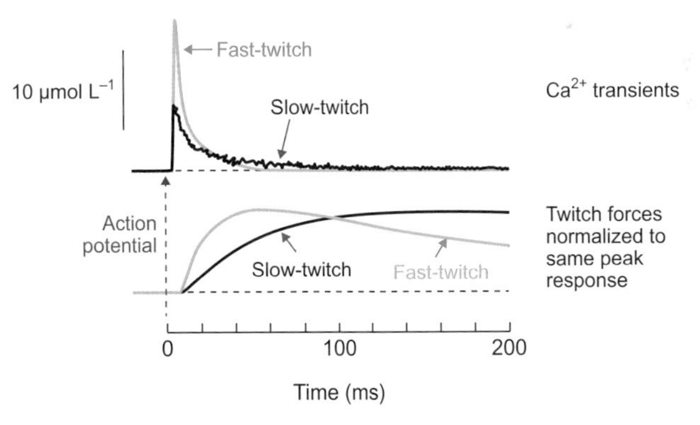

Figure 18.19 Ca^{2+} transients and twitch force responses in fast- and slow-twitch skeletal muscle fibres of the rat

The action potential (not shown) was initiated at time 0 (red arrow). There is a very short lag time (few milliseconds) between the initiation of the action potential and the peak Ca^{2+} transients and a much longer lag time (tens to hundred milliseconds) between peak force and peak Ca^{2+} transient. There is a clear distinction between fast- and slow-twitch fibres based on how fast force reaches the peak and relaxes during a twitch. The dotted lines indicate the resting levels for Ca^{2+} in the sarcoplasm and force. The temperature in this experiment was kept at 16°C for technical reasons, to prevent deterioration of the isolated muscle fibres used in experiments. Responses are about 4–6-fold faster at 37°C.
Source: Baylor SM, Hollingworth S (2003). Sarcoplasmic reticulum calcium release compared in slow-twitch and fast-twitch fibres of mouse muscle. Journal of Physiology 551: 125–138.

discuss in Section 18.5.2. As such, animals use fast-twitch fibres for rapid, powerful responses such as when hunting for prey or escaping predation. Slow-twitch fibres are, however, more energy efficient than fast-twitch fibres and therefore are used for activities involving prolonged contractions such as posture maintenance and persistent movements.

The Ca^{2+} transient is considerably faster than the twitch response in both fast- and slow-twitch fibres (Figure 18.19), but is substantially slower than the action potential (Figure 18.11C). Therefore, successive action potentials can be triggered in twitch fibres before the Ca^{2+} transient and the twitch force response return to the resting levels between action potentials. When this happens, the Ca^{2+} transients summate and the force response increases above the twitch response, as shown in Figure 18.20A. If the rate of action potentials is rapid enough to elevate the sarcoplasmic Ca^{2+} concentration long enough to produce maximal force, then a maximal (or fused) **tetanic contraction** is produced. At lower rates of action potential stimulation, incomplete (or unfused) tetanic contractions occur, as shown in Figure 18.20B, because sarcoplasmic Ca^{2+} decreases sufficiently between action potentials to cause partial relaxation.

Twitch skeletal muscle fibres normally operate in the body at near maximum tetanic force. The maximum force output of skeletal muscles of vertebrates and invertebrates is in the range 100–400 kN m^{-2} (see, for example, Figure 18.20B).

Striated muscles in animals broadly operate between about 80 to 130 per cent of their resting length because the capacity of the contractile apparatus to produce force outside this range is greatly reduced as illustrated by Experimental Panel 18.1 Figure B for frog skeletal muscle.

18.2.3 Skeletal muscle fibres are functionally diverse

Most skeletal muscle proteins involved in excitation–contraction coupling exist in slightly different forms called **isoforms**, which have specific functional characteristics. (Isoforms possess similar amino acid sequences and overall functions.) These isoforms are expressed in particular combinations in individual muscle fibres, with each combination imparting specific structural and functional characteristics to those fibres. Furthermore, each muscle often consists of different mixtures of muscle fibre types, each with specific functional characteristics, which permit each muscle in the body of an animal to perform a particular set of highly specialized mechanical tasks.

Since the contractile properties of a muscle fibre are primarily related to the characteristics of its molecular motor, the most widely used method for fibre type classification—in mammals, at least—is based on the type of myosin heavy chain (MHC) isoform expressed in that fibre, as shown in Figure 18.21. Trunk and limb muscles of adult mammals express four MHC isoforms: three MHC isoforms are expressed in fast-twitch fibres (MHC IIa, MHC IId/X and MHC IIb) and one (MHC I) is expressed in slow-twitch fibres. Fibres that express only one MHC isoform are called pure fibres; Table 18.2 shows the main characteristics of the pure fibre types: type I, type IIA, type IID/X and type IIB. Type I fibres are slow-twitch fibres, while type IIA, IID/X and IIB fibres are fast-twitch fibres.

Muscle fibres that express different combinations of MHC isoforms are called hybrid fibres and display intermediate characteristics to those associated with pure fibres. The presence of hybrid fibres enables muscles to fine tune their

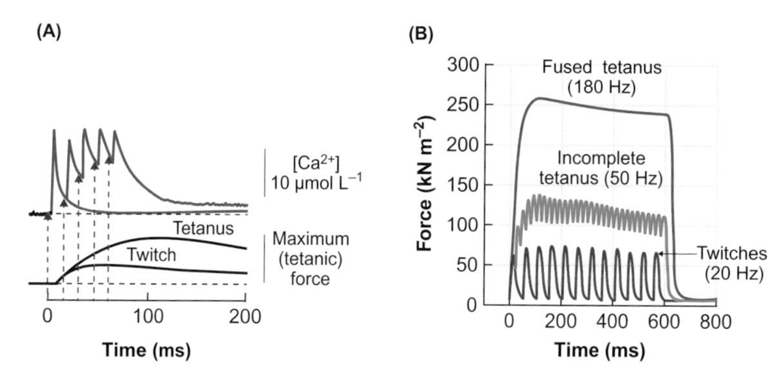

Figure 18.20 Twitch and tetanic force responses in mammalian muscle fibres

(A) Summation of Ca^{2+} transients elicited by action potentials generated every 15 ms in a fast-twitch muscle fibre from the rat at 16°C produces a fused tetanus that is twice as large in magnitude than the twitch response. The red arrows indicate the time when action potentials were generated in the fibre. (B) Repetitive stimulation of a muscle strip from the mouse diaphragm at 37°C over 600 ms produces only twitch responses when stimulated every 50 ms (20 Hz stimulation), an incomplete (unfused) tetanus when stimulated every 20 ms (50 Hz stimulation) and a fused tetanus when stimulated every 5.5 ms (180 Hz stimulation). The fused tetanic force response is about 250 kN m^{-2} and is 3.5-fold greater than the twitch force.

Sources: A: Baylor SM, Hollingworth S (2003). Sarcoplasmic reticulum calcium release compared in slow-twitch and fast-twitch fibres of mouse muscle. Journal of Physiology 551: 125–138. B: Murray JD et al. (2012). The force-temperature relationship in healthy and dystrophic mouse diaphragm; implications for translational study design. Frontiers in Physiology 3: 422.

18

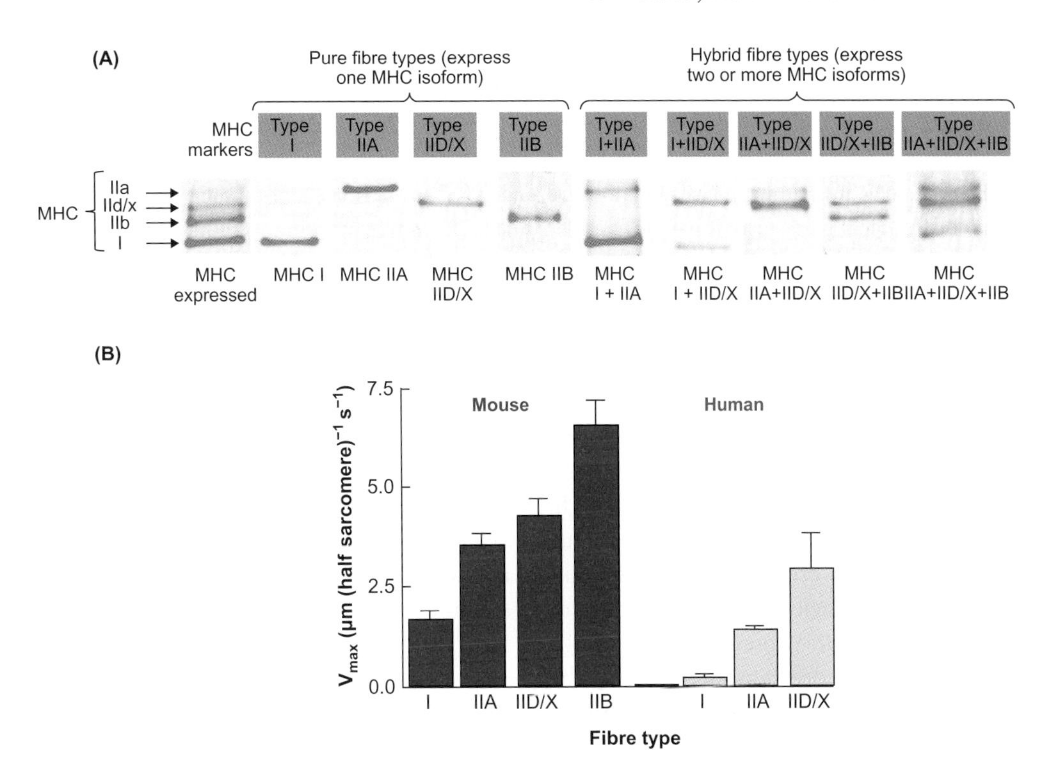

Figure 18.21 Muscle fibre type classification in mammals based on the myosin heavy chain (MHC) isoform(s) expressed in that fibre and relationships between maximum shortening velocities and fibre types in mice and humans

(A) Identification of myosin isoforms expressed in single skeletal muscle fibres of the rat. Muscle proteins are first solubilized using a detergent (sodium dodecyl sulfate, SDS) and then are separated on polyacrylamide gels by electrophoresis (PAGE). The different MHC isoforms run with different mobility on the gel. In rats, MHC I has the highest mobility and is followed in order by MHC IIb, MHC IId/x and MHC IIa. (B) Comparison of maximum shortening velocities (expressed in µm per half sarcomere per second, as discussed in Section 18.5.2) of different fibre types from humans and mice measured under the same experimental conditions. Notice that mouse fibres of given type shorten faster than human fibres of same type and that humans do not have type IIB fibres.

Sources: A: Bortolotto SK et al (2000). MHC isoform composition and Ca^{2+}– or Sr^{2+}-activation properties of rat skeletal muscle fibers. American Journal of Physiology, Cell Physiology 279: C1564–C1577. B: Schiaffino S, Reggiani C (2011). Fiber Types in Mammalian Skeletal Muscles. Physiological Reviews 91: 1447–1531.

Table 18.2 Fibre type characteristics of limb and trunk skeletal muscle fibres in mammals

Fibre type#	Type I	Type IIA	Type IID/X	Type IIB*
Myosin heavy chain (MHC) expressed	MHC I	MHC IIA	MHC IID/X	MHC IIB
Twitch time course	Slow	Moderately fast	Fast	Very fast
Twitch/tetanic force ratio	Small	Medium	Large	Large
Power produced	Low	Medium	High	Very high
Speed of shortening	Slow	Moderately fast	Fast	Very fast
Fatigue resistance	High	Moderately high	Low	Very low
Myofibrillar ATPase	Low	Moderately high	High	Very high
Force cost (ratio between ATPase and force produced)	Low	Moderately high	High	Very high
Endurance	Hours	<30 min	< Few min	<1 min
Oxidative capacity	High	High	Low	Very low
Glycogen content	Low	High	High	High
Mitochondrial density	High	Medium to high	Medium to low	Low
Capillary density	High	Medium to high	Medium to low	Low
Metabolic activity	Aerobic	Aerobic	Short term aerobic	Mainly anaerobic

Hybrid fibres express two or more MHC isoforms and have intermediate characteristics to those associated with fibres that express only one of the respective MHC isoforms.

* Humans and large mammals lack type IIB muscle fibres.

function to a wider range of functional characteristics such as velocity of shortening, power level and resistance to fatigue. **Muscle fatigue** refers to the reduced ability of muscle to produce force after a period of intense muscle activity[36].

Other methods of skeletal muscle fibre classification are based on various histological and functional differences noted in Table 18.2. These include oxidative capacity, mitochondrial content, glycogen content, pH-dependent myofibrillar ATPase activity, blood supply, speed of contraction and fatigue resistance.

Limb muscles of adult amphibians, like mammals, also express four MHC isoforms: three different MHC isoforms are expressed in twitch fibres and one is expressed in tonic fibres. The twitch fibres display different, but relatively high, shortening velocities and power, but generally are not fatigue resistant, while the tonic fibres are slow, fatigue resistant and have a higher oxidative capacity. A large proportion of amphibian muscle fibres are hybrid fibres: they express two or more MHC isoforms, with expression varying along the fibre length. This variation causes sarcomeres in the muscle fibre to contract nonuniformly along the fibre length.

Adult skeletal muscles in birds express at least six MHC isoforms, two in slow-tonic fibre types and four (or more) in fast-twitch fibre types. Since there is uncertainty regarding the number of fast MHC isoforms, particularly in the flight muscles of adult birds, the most widely used method of avian muscle fibre classification is based on contraction properties, oxidative capabilities and colouration associated with blood supply:

- SO (slow-tonic oxidative) coloured red;
- FO (fast-twitch oxidative), highly oxidative for pectoral muscles of hummingbirds and finches, coloured red;
- FOG (fast-twitch, oxidative glycolytic) often coloured pink;
- FG (fast-twitch glycolytic), white.

SO, FO and FOG fibres to some extent are generally fatigue resistant and use aerobic metabolism[37] (lipid-based, oxidative), while FG fibres store more glycogen, have a higher capacity for anaerobic metabolism and fatigue more easily.

Skeletal muscle in fish is generally classified as red or white:

- The red muscle is slow, aerobic, richly supplied by blood and is fatigue resistant; it contains SO fibres and is generally used for steady swimming.
- The white muscle is fast, high-power, fatiguable and anaerobic, and is used for short powerful bursts of movement; it is made up of FG fibres.

Some fish such as the scup (*Stenotomus chrysops*) also have intermediate 'pink' muscle, which contains FOG[38] fibres. Intriguingly, genetic studies have shown that the common carp (*Cyprinus carpio*) has genes for 29 fast skeletal MHCs, which are expressed differentially depending on various environmental conditions such as water temperature[39]. The very large number of MHC isoforms that can be expressed in the skeletal muscle of fish allows them to acclimatize[40] to various environmental conditions by expressing MHCs that function optimally under the prevailing conditions[41].

Three different muscle fibre types recognized in crustaceans are the fast, the slow-twitch (S_1) and the slow-tonic (S_2) fibres. These fibres express three different MHC isoforms: MHC fast, MHC S_1 and MHC S_2. Crustacean muscle fibres also often express combinations of MHC isoforms, forming a continuum between pure slow-twitch (S_1) and pure slow-tonic (S_2) fibres within one muscle.

18.2.4 Muscle fibre fatigue is a multifactorial process that protects muscle from damage

As we discuss in Experimental Panel 18.2, muscle fatigue is not due to the accumulation of lactate that is associated with intense exercise[42]. Instead, it is the result of a multitude of contraction-dependent processes that affect specific steps in excitation–contraction coupling to prevent ATP depletion below levels that can cause irreversible cell damage.

Broadly speaking, there are two types of muscle fatigue due to processes that occur within the muscle cells, depending on the intensity and duration of stimulation:

- high frequency fatigue, and
- metabolic fatigue.

The high frequency stimulation of muscles by nerve fibres necessary to maintain maximum tetanic force leads to rapid build-up of K^+ in the restricted space of the t-system, due to K^+ movements associated with the generation of action potentials in the muscle fibre. As the extracellular K^+ concentration in the t-system rises, the sarcolemma depolarizes,

[36] We discuss muscle fatigue later in this Section, in Section 15.2.4 and in Experimental Panel 18.2.

[37] We discuss aerobic and anaerobic metabolism in Section 2.4.

[38] Similar terminology is in common use for non-mammalian muscle classification as we discuss in Sections 14.1.5 (Figure 14.15) and 15.2.4 (Figure 15.23).

[39] We discuss the effect of temperature on fish muscle performance in Chapter 9.

[40] We discuss acclimatization in Section 1.5.2.

[41] We discuss some other characteristics of vertebrate skeletal muscle fibres related particularly to their capillary supply in Section 14.1.4 and Table 14.1.

[42] Lactate can accumulate under conditions of oxygen depletion as we discuss in Section 2.4.1.

Experimental Panel 18.2 Does lactic acid contribute to muscle fatigue?

Background to the study

During vigorous exercise, if insufficient oxygen is available to degrade glucose to CO_2 and H_2O, then anaerobic respiration[1] causes lactic acid to accumulate in skeletal muscles. In the sarcoplasm, lactic acid dissociates into lactate and protons (H^+) causing the sarcoplasm to become more acidic, a process known as acidosis (pH<7.0). For most of the last century, acidosis and lactate accumulation were commonly thought to be the major causes of muscle fatigue[2].

Experimental approach

In order to test whether lactate and acidosis affect the contractility of skeletal muscle, a fully functional muscle preparation was used, which enabled:

(i) electrical stimulation to trigger propagated action potentials in the t-system and activate the contractile apparatus by the normal sequence of events in excitation–contraction coupling;

(ii) unrestricted access to the sarcoplasmic environment to rapidly add and remove lactate and change the pH.

The mechanically skinned fast-twitch rat muscle fibre is just such a preparation. In this preparation the transverse tubules pinch off and the whole t-system seals when the surface membrane (sarcolemma) is peeled off by microdissection. The Na^+/K^+ pump establishes the normal Na^+ and K^+ concentration differences across the tubular membranes, when the preparation is placed in an artificial 'internal solution' that mimics the internal sarcoplasmic environment with respect to the main ions, including K^+, ATP, Mg^{2+}, creatine phosphate, Ca^{2+} and pH. As a result, the t-system becomes normally polarized and responds to electrical stimulation. This preparation can be activated at different steps in excitation–contraction coupling, when lactate in the sarcoplasmic environment is present or absent, or when the sarcoplasmic pH is altered.

Results

Figures A(i) and A(ii) show that the twitch and the tetanic responses at 50 Hz stimulation are effectively the same in the presence and absence of 30 mmol L^{-1} lactate in the sarcoplasmic environment. These results indicate that the presence of the lactate ion in the sarcoplasm of fast-twitch muscle fibres does not interfere with the excitation–contraction coupling.

When the t-system is depolarized, as is often the case with working fast-twitch muscle, the excitability of the t-system decreases because some sodium channels are inactivated by prolonged depolarization and the amplitude of the action potentials is consequently reduced. This is shown in Figure A(iii), where a mechanically skinned fibre was first tetanically stimulated when the t-system was fully polarized at –79 mV and again when the t-system was depolarized at –66mV. The response at –66 mV was only about 20 per cent of the response at –79 mV.

The pH of the 'internal solution' was then decreased from 7.1 to 6.6, with the membrane potential of the t-system maintained at –66 mV. The tetanic force response more than doubled, indicating that acidosis increased the excitability of the fibre when the t-system was depolarized. Thus, acidosis had a beneficial effect on the force output, by increasing the excitability of the muscle fibre when the t-system is depolarized, as is often the case with fatigued muscle.

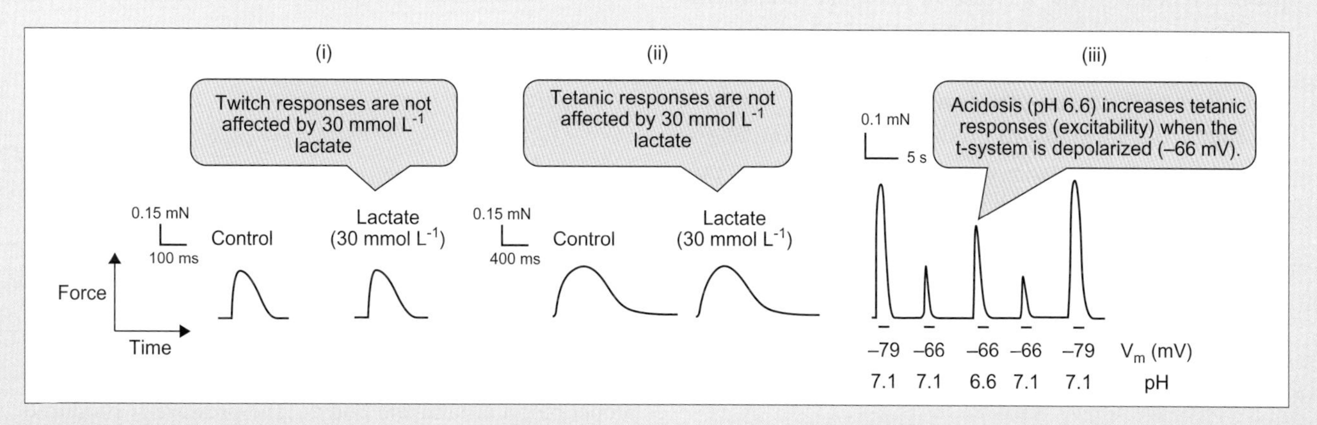

Figure A Lactate and acidosis do not cause muscle fatigue

(i) The presence of 30 mmol L^{-1} lactate in the sarcoplasmic environment does not alter the twitch response in mechanically skinned fast-twitch muscle fibres from the rat. (ii) The tetanic response at 50 Hz stimulation for 400 ms also is not affected by the presence of 30 mmol L^{-1} lactate in the sarcoplasmic environment. (iii) Sequence of tetanic responses at 25 Hz stimulation for 1.25 s when the t-system membrane was depolarized from –79 to –66 mV by reducing the K^+ concentration difference across the t-system membrane. The depressed tetanic response under depolarized conditions increased markedly when the pH in the sarcoplasmic environment was decreased from 7.1 to 6.6.

Source: (i) and (ii) from Posterino GS et al (2001). L(+)-lactate does not affect twitch and tetanic responses in mechanically skinned mammalian muscle fibres. Pflügers Archiv - European Journal of Physiology 442; 197–203. (iii) from Pedersen TH et al (2004). Intracellular acidosis enhances the excitability of working muscle. Science 305: 1144-1147.

Conclusions

Taken together, the results from these experiments show that a build-up of lactic acid in the sarcoplasmic environment does not have a detrimental effect on the force developed during contraction. Instead, acidosis associated with lactic acid production could be beneficial for preserving muscle fibre excitability.

› Find out more

Lamb GD, Stephenson DG (2006). Point: Counterpoint: Lactic acid accumulation is an advantage/disadvantage during muscle activity. Journal of Applied Physiology 100: 1410–1412.

Posterino GS, Dutka TL, Lamb GD (2001). L(+)-lactate does not affect twitch and titanic responses in mechanically skinned mammalian muscle fibres. Pflügers Archiv 442: 197–203.

Pedersen TH, Nielsen OB, Lamb GD, Stephenson DG (2004). Intracellular acidosis enhances the excitability of working muscle. Science 305: 1144–1147.

[1] We discuss anaerobic glycolysis in Section 2.4.1.

[2] We discuss factors contributing to muscle fatigue in Section 18.1.4.

because the concentration difference across the membrane is reduced[43]. This depolarization causes Na^+ channels to become inactivated and membrane excitability to be reduced[44]. In turn, the reduced excitability reduces the amount of Ca^{2+} released from the sarcoplasmic reticulum and consequently the level of activation of the contractile apparatus to produce force. Muscles recover from high-frequency fatigue shortly after the cessation of stimulation, as potassium ions are pumped back into the sarcoplasm by the Na^+/K^+-ATPase located in the t-system membrane.

Metabolic fatigue arises from direct and indirect effects induced by the accumulation in the sarcoplasm of metabolites such as inorganic phosphate (P_i), ADP, Mg^{2+}, protons and reactive oxygen species, or a decrease in ATP concentration and glycogen. Accumulation of some of these metabolites (P_i, Mg^{2+}, ADP) and depletion of ATP and glycogen reduce the amount of Ca^{2+} released from the sarcoplasmic reticulum. The increase in phosphate and proton concentrations reduces the sensitivity to Ca^{2+} of the contractile apparatus and its ability to develop force. Production of reactive oxygen species affects both the ability of the contractile apparatus to produce force and the release of Ca^{2+} from the sarcoplasmic reticulum. Moreover, prolonged elevation of intracellular Ca^{2+} concentration above the resting level causes the uncoupling of signal transmission from the voltage sensors in the t-tubules to the Ca^{2+}-release channels (RyRs) in the sarcoplasmic reticulum (Step 2 in Figure 18.18). Recovery from metabolic fatigue requires a prolonged period of hours to days.

[43] We discuss in Section 16.2 the basis for membrane potential generation across cellular membranes.

[44] We discuss in Section 16.2 that membrane depolarization causes inactivation of sodium channels, which are essential for maintaining membrane excitability with respect to the generation of action potentials.

Plasma acidosis associated with the production of lactate during anaerobic glycolysis may also affect the function of the central nervous system in the whole animal and reduce neural signalling to the muscles. This type of muscle fatigue is called central muscle fatigue. However, considerable controversy surrounds the importance of central muscle fatigue: some researchers believe that it may play only a limited role in the overall muscle fatigue experienced by animals during strenuous activity.

› *Review articles*

Allen DG, Lamb GD, Westerblad H (2008). Skeletal muscle fatigue: cellular mechanisms. Physiological Reviews 88: 287–332.

Pette D, Staron RS (2000). Myosin isoforms, muscle fiber types and transitions. Microscopy Research and Technique 50: 500–509.

Schiaffino S, Reggiani C (2011). Fiber types in mammalian skeletal muscles. Physiological Reviews 91: 1447–1531.

Stephenson DG, Lamb GD, Stephenson GMM (1998). Events of the excitation-contraction coupling in fast- and slow-twitch mammalian muscle fibres relevant to muscle fatigue. Acta Physiologica Scandinavica 162: 229–245.

Stephenson GMM (2001). Hybrid skeletal muscle fibres: A rare or common phenomenon? Proceedings of the Australian Physiological and Pharmacological Society 32: 69–87.

18.3 Cardiac myocytes: muscle cells that never pause to rest

The heart in all animals generates the pressure required to propel blood around the body[45]. The pressure is produced by the contraction of heart muscle cells, called **cardiac myocytes**, located in the muscular wall of the heart. All vertebrates and many invertebrates, including arthropods,

[45] We discuss the anatomy and physiology of the heart of animals in Sections 14.1.2 and 14.4.

gastropods and cephalopod molluscs, have striated cardiac myocytes that share major structural and functional characteristics with voluntary striated muscle fibres discussed in Section 18.2. These characteristics include:

- regular arrangement of actin and myosin filaments in myofibrils and sarcomeres;
- presence of a Ca^{2+}-regulatory system which keeps the myocytes relaxed at low Ca^{2+} concentration;
- sarcoplasmic reticulum acting as a Ca^{2+} store surrounding each myofibril;
- presence of dyadic/triadic junctions between the sarcoplasmic reticulum and the plasma membrane;
- excitable surface membrane, which responds to depolarization by facilitating the Ca^{2+} rise in the sarcoplasm, which causes the activation of the contractile apparatus.

By contrast, the heart cells of tunicates (sea squirts and ascidians) and bivalve molluscs such as clams have 'smooth cardiac' cells. In this section we will focus on the cardiac muscle of vertebrates, which has been systematically investigated.

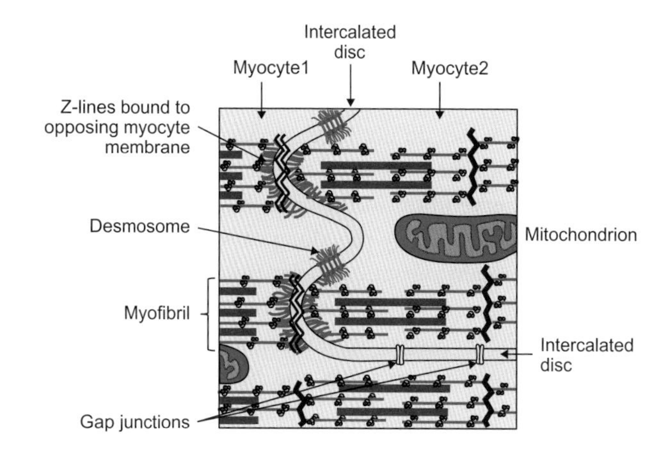

Figure 18.22 Diagrammatic representation of structures present in the intercalated discs connecting cardiac myocytes
The mechanical contacts which connect the cytoskeletons of neighbouring myocytes at the level of desmosomes and the connections at Z-lines between myofibrils in adjacent myocytes are important for transmitting force between cardiac cells, while the gap junctions facilitate electrical connection between myocytes for the propagation of action potentials. Figure 22.23 shows an electron micrograph of a mammalian myocyte.

18.3.1 Structural and functional characteristics of vertebrate cardiac myocytes

Despite many similarities between cardiac myocytes and skeletal muscle fibres, there are also important differences in their organization, structure and function. Thus, unlike skeletal muscle fibres, the cardiac myocytes in vertebrates:

- are relatively short (about 100 μm length) and narrow (diameter 2–10 μm) depending on the species and location in the heart;
- contain one or two centrally located nuclei;
- have a lower density of myofibrils, which branch at the ends to form complex connections with multiple neighbouring myocytes, as illustrated in Figure 18.1B[46];
- are mechanically joined end to end by **intercalated discs** (Figure 18.22), which anchor myofibrils from neighbouring cells at the level of the Z-lines across the narrow gap between the surface membranes of the two cells;
- have **desmosomes,** which are tight mechanical connections between the cytoskeleton of one cell with the cytoskeleton of a neighbouring cell in the region between myofibrils (Figure 18.22); and
- are electrically connected by gap junctions[47] between membranes of adjacent cells, which act as electrical connections (synapses) between cells (Figure 18.22).

These mechanical and electrical contacts between cardiac myocytes allow the rapid transmission of electrical signals and force from one myocyte to another within the major heart compartments (atria and ventricles). The contraction of the ventricle is called **systole** and the period of time over which the ventricle relaxes is called **diastole**[48].

In addition to atrial and ventricular myocytes, the vertebrate heart also contains a specialized system of modified myocytes, which has two functions:

1. To generate pacemaker potentials[49] that manifest themselves in regular triggering of action potentials in the absence of neural stimulation.
2. To conduct the stimulus for the heart beat from atria to the ventricles[50], which are otherwise electrically insulated from each other.

Therefore, unlike skeletal muscle, which physiologically contracts only in response to neural stimulation, the isolated heart of vertebrates can contract rhythmically in the absence

[46] Figure 22.23 in Chapter 22 shows an electron micrograph of a mammalian myocyte.

[47] We discuss gap junctions in Section 16.3.1.

[48] We discuss the pumping cycle of the heart in more detail in Section 14.1.2.

[49] We discuss pacemaker potentials in general in Section 16.2.11 and myogenic pacemakers in Section 22.3.1.

[50] We discuss how this specialized system of modified myocytes works in Section 22.3.1.

of neural inputs after it is removed from the animal's body[51]. The rate of action potential firing by pacemaker cells is, however, modulated by neural stimulation.

Cardiac myocytes do not have specific postsynaptic specializations with neurons like skeletal muscle fibres do. Instead, the nerve fibres in the heart have many swellings called **varicosities** as they run along cardiac myocytes. The varicosities contain synaptic vesicles from which transmitter molecules are released when action potentials propagate down the neurons. The transmitter molecules diffuse to neighbouring myocytes and bind to metabotropic receptors[52] in the sarcolemma coupled to G-proteins and initiate responses that change the rate of depolarization between action potentials in pacemaker cells and alter the strength of contraction of ventricular cardiac myocytes.

Cardiac muscle fibres of non-flying vertebrate species are also more richly supplied with blood than skeletal muscles and contain a much higher density of mitochondria to support their high-energy demand—the result of continual cycles of contraction and relaxation. Flight muscles in some bird species, however, have an even higher blood capillary density than their heart[53].

The action potential is much longer in cardiac myocytes than in skeletal muscle fibres

The sarcolemma of cardiac myocytes contains (L-type[54]) Ca^{2+} channels[55] that open rapidly upon depolarization from a resting level around −70 to −80 mV and stay open for as long the depolarization lasts. The presence of these Ca^{2+} channels, in addition to voltage-sensitive Na^+ channels and a diversity of K^+ channels in the plasma membrane of cardiac myocytes, causes the generation of action potentials that persist for hundreds of ms compared with that in skeletal muscle, which lasts for only several ms.

Figure 18.23 shows the action potential in a rabbit ventricular myocyte and the underlying activities of the ion channels responsible for the five phases (0–4) that are characteristic of the action potential in vertebrate cardiac myocytes. The magnitude of the early repolarization phase (Phase 1) varies between species and is generally greater in atrial than in ventricular myocytes. The long plateau (Phase 2) can be relatively flat, show a gradual decline, or be shaped like a dome.

Arrival of an action potential opens the L-type Ca^{2+} channels, which leads to the rise in sarcoplasmic Ca^{2+} concentration by a calcium-induced calcium release mechanism (Figure 18.24).

Keeping the membrane potential around zero mV during the plateau phase keeps the Na^+ channels inactivated for a relatively long time, increasing the refractory period[56] of cardiac myocytes to such extent that the heart cannot be tetanized[57] under physiological conditions because the myocytes relax before the action potential terminates (Figure 18.23). The long refractory period in myocytes is also fundamental to the organized conduction of excitation across the heart[58].

51 Isolated myocytes in culture can also display spontaneous activity.
52 We discuss metabotropic receptors in Section 16.3.4.
53 We discuss this issue in Section 14.4.5 and Table 14.1.
54 L- from 'long lasting'.
55 These cardiac L-type Ca^{2+} channels also act as voltage sensors/DHP receptors but are unlike the DHP receptors in skeletal muscle fibres of adult vertebrates, which do not function as Ca^{2+} channels, as we discuss in Section 18.2.2.

56 We discuss the basis of the refractory period for action potential generation in Section 16.2.5.
57 This means that Ca^{2+} transients and force responses cannot summate in cardiac myocytes, unlike twitch muscle fibres.
58 We discuss the spread of excitation over the mammalian heart in Section 22.3.1.

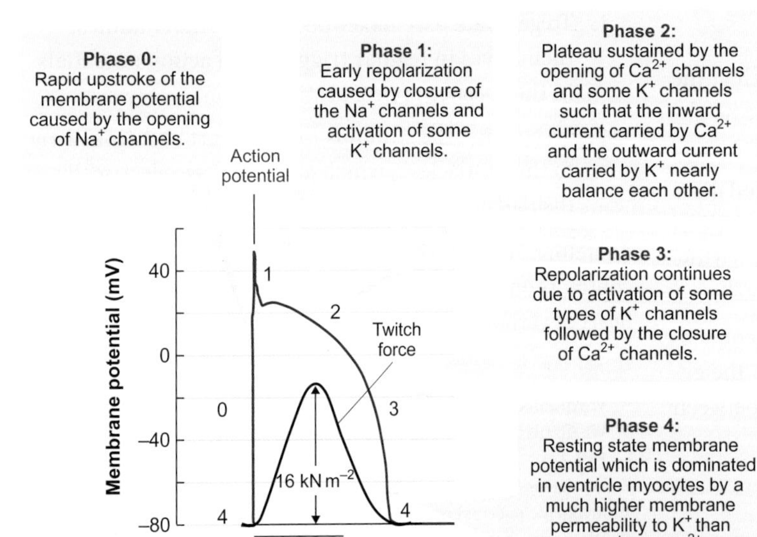

Figure 18.23 The five characteristic phases of an intracellularly recorded action potential from vertebrate cardiac myocytes

Intracellular recording of an action potential from a rabbit ventricular myocyte and overlaid, the time course of the twitch force response in rabbit ventricular muscle at 35–37.5°C.

Sources: Action potential recording trace from Go A et al (2005). Negative inotropic effect of nifedipine in the immature rabbit heart is due to shortening of the action potential. Pedriatic Research 57: 399–403; twitch response trace from Endoh M, Blinks JR (1988). Actions of sympathomimetic amines on the Ca^{2+} transients and contractions of rabbit myocardium: Reciprocal changes in myofibrillar responsiveness to Ca^{2+} mediated through α- and β-adrenoceptors. Circulation Research 62: 247–265.

Excitation–contraction coupling in cardiac myocytes

All cardiac myocytes in vertebrates have a sarcoplasmic reticulum compartment, but its volume (and therefore the pool of Ca^{2+} available for release) varies considerably among vertebrates. For example, the sarcoplasmic reticulum occupies up to 10 per cent cell volume in mammalian myocytes, but only 0.5–1.0 per cent cell volume in cardiac myocytes of birds and most ectothermic vertebrates. Mammalian cardiac myocytes also have a t-system, like in the skeletal muscle fibres, but cardiac myocytes in fish, amphibians, reptiles and birds lack a t-system and have dyads between sarcolemma and the sarcoplasmic reticulum cisternae, where signal transduction takes place.

The chain of events in the excitation–contraction coupling of mammalian cardiac myocytes is similar to that summarized in Figure 18.18 (steps 1–5) for the twitch vertebrate muscle fibres:

- propagation of action potential into the t-system;
- signal transduction at dyads/triads causing the rise in sarcoplasmic Ca^{2+} concentration;
- binding of Ca^{2+} to the troponin C-subunit of the actin-filament linked Ca^{2+} regulatory system and removal of inhibition exerted on cross-bridge formation; and
- force production at the level of the contractile apparatus.

The major functional difference in the process of excitation–contraction coupling between skeletal and cardiac muscle fibres in vertebrates is that, upon excitation, *cardiac myocytes require Ca^{2+} entry from the extracellular environment to initiate further Ca^{2+} release from the sarcoplasmic reticulum and promote contraction*, while skeletal muscle fibres contract, when stimulated, in the absence of Ca^{2+} entry from outside.

Figure 18.24 illustrates the mechanism of signal transduction at dyads/triads in cardiac myocytes. At rest, the Ca^{2+}-release channels/RyRs[59] in the sarcoplasmic reticulum are closed because the Ca^{2+} activation sites on the Ca^{2+}-release channels are occupied by Mg^{2+}. During the plateau phase of the cardiac action potential, the L-type Ca^{2+} channels in the sarcolemma and tubular membrane stay open. Ca^{2+} enters

[59] The Ca^{2+}/Mg^{2+}-inhibition site on cardiac Ca^{2+}-release channels RyRs have a low affinity for Mg^{2+} or Ca^{2+}; therefore, under physiological conditions this site is unoccupied. For the channels to open fully, Ca^{2+} needs to bind to the Ca^{2+}-activation site and ATP to the stimulatory site on the RyR as we discuss in Figure 18.17.

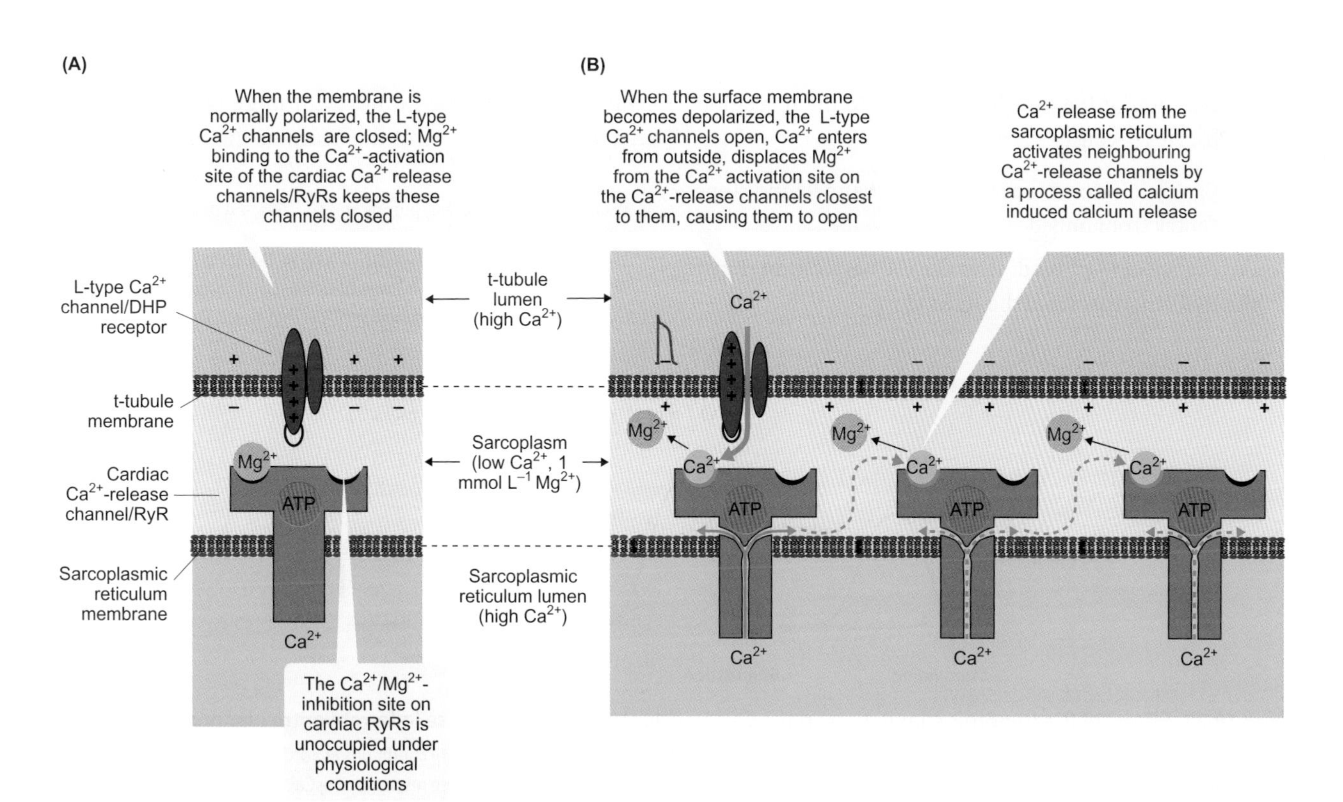

Figure 18.24 Mechanism of sarcoplasmic Ca^{2+} rise in mammalian cardiac myocytes upon membrane depolarization

(A) At normal resting level of polarization, the L-type Ca^{2+} channels and the Ca^{2+} release channels/ RyRs are closed. (B) Arrival of an action potential opens the L-type Ca^{2+}-channels which leads to the rise in sarcoplasmic Ca^{2+} concentration by a calcium-induced calcium release mechanism.

across the plasma membrane and displaces Mg^{2+} from the Ca^{2+}-activation sites of nearby Ca^{2+}-release channels/RyRs, causing them to open; Ca^{2+} released from the sarcoplasmic reticulum further amplifies displacement of Mg^{2+} from the Ca^{2+}-activation sites of neighbouring RyRs by Ca^{2+}, opening them and causing more Ca^{2+} release from the sarcoplasmic reticulum. This process is called **Ca^{2+}-induced Ca^{2+} release.**

The relative contribution of Ca^{2+} released from the sarcoplasmic reticulum and Ca^{2+} entry across the plasma membrane to the calcium transient of cardiac myocytes varies greatly among vertebrates; the sarcoplasmic reticulum is the major contributor in mammals, but plays a lesser role in birds and most ectothermic vertebrates, where the sarcoplasmic reticulum Ca^{2+} content is reduced. The Ca^{2+} transient in cardiac myocytes:

- generally peaks below 1 μM (i.e. around 10-fold smaller than in fast-twitch muscle fibres), which induces only a small force response (Figure 18.25A), and
- is commonly shorter than the action potential: compare the duration of the Ca^{2+} transient in Figure 18.25A with that of the action potential in Figure 18.23.

These characteristics ensure that there cannot be summation of Ca^{2+} transients to cause tetanization and that force fully relaxes between action potentials, which is critical for the heart to work as an efficient pump.

Repolarization of the sarcolemma leads to the relaxation of the myocytes because:

- repolarization closes the L-type Ca^{2+} channels and stops Ca^{2+} entry from outside;
- the chance of Mg^{2+} binding to the Ca^{2+}-activation site of the RyRs increases and the RyRs close;
- calcium that entered the myocytes and that was released from the sarcoplasmic reticulum is pumped into the sarcoplasmic reticulum and extruded across the plasma membrane by a Ca^{2+}/Na^{+} exchange system[60];
- Ca^{2+} concentration in the sarcoplasm decreases;
- Ca^{2+} dissociates from troponin C, and inhibition of cross-bridge cycling is reinstated by the Ca^{2+}-regulatory system; and
- the cardiac myocytes relax.

18.3.2 Control of cardiac myocyte function

Unlike the twitch skeletal muscle fibres, which function at near-full tetanic force capacity in the body, cardiac myocytes commonly operate at relatively low, submaximal force levels. This permits force in cardiac myocytes to be readily increased or decreased by changing the magnitude of the Ca^{2+} transient, the sensitivity of the contractile apparatus to Ca^{2+} concentration, or both. The major factors that modulate the cardiac function in vertebrates include the stretch level of the myocytes, the level of activity in the autonomic nervous system, and the presence of hormones.

Myocyte stretching increases contractile force

The vertebrate heart responds with a stronger, more powerful contraction when it fills with more blood. This property, known as the law of the heart, or the Frank–Starling[61]

[60] We discuss about Na^{+}/K^{+} pumps and Ca^{2+}/Na^{+} exchangers in Section 3.2.4.

[61] The law of the heart was first discovered by the Italian physiologist Dario Maestrini in 1914 and was then validated and popularized by the German and English physiologists Otto Frank and Ernest Starling.

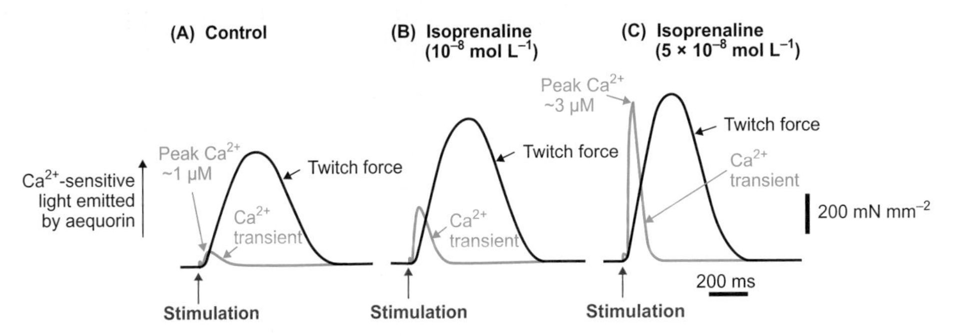

Figure 18.25 Effects of β-adrenergic stimulation on the Ca^{2+} transient and force response in ferret ventricular muscle

(A) Ca^{2+}-transient and twitch force in ferret ventricular muscle in response to electrical stimulation under control conditions at 30°C. (B, C) Effects of 10 and 50 nmol L^{-1} isoprenaline (a specific agonist for β-adrenergic receptors), respectively, on the Ca^{2+} transient and force response. The Ca^{2+} transient was measured with the Ca^{2+}-sensitive luminescent protein aequorin, which emits more light in the presence of higher Ca^{2+} concentrations. Notice that isoprenaline causes a marked increase of the Ca^{2+} transient and the twitch response in ventricular muscle.

Source: Okazaki O et al (1990). Modulation of Ca^{2+} transients and contractile properties by β-adrenoceptor stimulation in ferret ventricular muscles. Journal of Physiology 423: 221–240.

18

relationship, has an important autoregulatory function as it increases the stroke volume (amount of blood pumped by the heart in one beat) when the amount of blood returned to the heart increases[62]. This autoregulatory process results from a stretch-induced rise in Ca^{2+} affinity of troponin C, which increases the sensitivity to Ca^{2+} of the contractile apparatus, such that more force is produced for a given Ca^{2+} transient.

Regulation by the autonomic nervous system

The function of the heart in vertebrates is mostly controlled via reflex responses to sensory inputs from baroreceptors and chemoreceptors[63] in the cardiovascular system[64]. The neurotransmitter released at the end of the circuit is usually noradrenaline[65] in the sympathetic system[66] and acetylcholine in the parasympathetic system.

Stimulation of the sympathetic nervous system increases the heart rate and blood pressure by activating both α- and β-adrenergic receptors[67] in the heart, which are coupled to G-proteins[68]. The major effects on the vertebrate heart in general, and mammalian heart in particular, are caused by the activation of β-adrenergic receptors, which stimulate production of cyclic AMP (cAMP) by adenylate cyclase. In turn, cAMP activates cAMP-dependent protein kinase[69], which phosphorylates key proteins in the cardiac myocytes.

As shown in Figure 18.25, the net effects of β-adrenergic activation are:

- an increased force output;
- a large increase in the peak and faster decline of the Ca^{2+} transient;
- a faster relaxation.

The action potential also tends to become shorter following activation of β-adrenergic receptors and the heart rate increases by affecting the activity of the pacemaker cells[70]. The overall effect of β-adrenergic stimulation is a large increase in cardiac output.

Stimulation of the parasympathetic nervous system releases acetylcholine, which activates muscarinic receptors[71] on pacemaker cells. This activation leads to a decrease in the rate of depolarization between action potentials, and therefore causes the heart rate to decrease.

Myocyte activity is also affected by hormones in the blood

Adrenaline and noradrenaline released from adrenal glands under stressful situations increase the heart rate and the force of contraction by acting primarily on β-adrenergic receptors on pacemaker cells and ventricular myocytes, helping the body respond to such situations.

› *Review articles*

Bers DM (2008). Calcium cycling and signaling in cardiac myocytes. Annual Review of Physiology 70: 23–49.
Moss R L, Razumova M, Fitzsimons DP (2004). Myosin crossbridge activation of cardiac thin filaments: implications for myocardial function in health and disease. Circulation Research 94: 1290–1300.
Yoshida S (2001). Simple techniques suitable for student use to record action potentials from the frog heart. Advances in Physiology Education 25: 176–186.

18.4 Smooth muscle fibres: the 'invisible' achievers

Smooth muscle is specialized for slow and sustained contractions and is generally found in the walls of hollow visceral organs in vertebrates, such as:

- the digestive tract, where it propels and mixes the food as it passes through the gut, by the process of peristalsis[72];
- the respiratory tract of terrestrial animals, where it controls the flow of air to the lungs;
- blood vessels, where it serves to control the blood pressure and distribution of blood to various organs;
- the urinary tract, where it propels urine via the ureter;
- ducts of various secretory glands, where it moves forward various secretions;
- the uterus of mammals, where it contracts during labour to give birth to the foetus.

The iris of the eye is also made of smooth muscle, where it serves to control the amount of light that enters the eye. The thermal insulation power of the fur in mammals and plumage in birds is also controlled by tiny smooth muscles at the base of hairs and feathers, which alter the amount of still air trapped around the surface of the animal by varying the level

[62] We discuss this point in more detail in Section 14.1.2.
[63] We discuss chemoreceptors and baroreceptors in general terms in Sections 17.3.3 and 17.4.5, respectively.
[64] We discuss control of the cardiovascular system in Sections 22.4 and 22.5.
[65] The names noradrenaline and norepinephrine are interchangeable. The chemical structure of noradrenaline is shown in Figure 16.37.
[66] We discuss the autonomic nervous system in Section 16.1.3.
[67] We discuss different types of adrenergic receptors in Section 16.1.3.
[68] We discuss G-proteins in Section 19.1.4.
[69] This process is illustrated in Figure 19.6.
[70] We discuss myogenic pacemakers in Section 22.3.1.
[71] We discuss muscarinic receptors in Section 16.1.3.
[72] Peristalsis refers to a series of wave-like contractions of circular and longitudinal muscles surrounding a tube-like structure like the gut that moves its content along it, or acts on a hydrostatic skeleton to facilitate animal movement as we discuss in Section 18.5.1

of erection or compression of the hairs and feathers with respect to the skin surface.

The smooth muscle cells are thin (2–10 μM diameter), spindle shaped, 20–500 μm long cells that can be connected to other cells by gap junctions and have central nuclei as shown back in Figure 18.1C. Generally, the smooth muscles are categorized as **single-unit** or **multi-unit**, depending on whether the muscle contracts and relaxes as one unit, or as multiple individual units, respectively.

Single-unit muscles are innervated by only a few neurons and consist of mechanically and electrically connected smooth muscle cells organized in large sheets. The excitation spreads through gap junctions between cells like in cardiac muscle and the muscles often contract when stretched. Smooth muscles in most blood vessels, the gut and the uterus (in late pregnancy in mammals) are of single-unit type.

Multi-unit smooth muscles consist of muscle cells that are individually innervated, and are often mingled with connective tissue fibres. Multi-unit muscles produce more graded responses than single-unit muscles, permitting fine control of specific tasks; the iris of the eye and the musculature of large arteries contain multi-unit smooth muscles.

18.4.1 The different myofilament organization in smooth muscle permits it to function over broader length changes than striated muscle

As shown in Figure 18.26, the actin filaments in smooth muscle cells anchor to so-called **dense bodies** in the cytosol (made of the same protein as the Z-line in striated muscle) and to the **extracellular matrix**. Smooth muscle actin filaments are organized in fibrils that are not aligned to each other, like in striated muscle, but instead criss-cross the cellular space in all directions.

Smooth muscle myosin filaments are organized as flat sheets between actin filaments, and display side polarization, with myosin heads oriented in opposite directions on the two sides of the myosin filament, as shown in Figure 18.3B. The polarity of the myosin heads matches the polarity of the actin filaments in their vicinity. This arrangement is functionally important because it provides optimal conditions for force development for the entire length of actin–myosin filament overlap. This is unlike the situation with striated muscle where force declines when actin filaments from one half of the sarcomere start to overlap myosin heads of opposite polarity at shorter sarcomere lengths, as we discuss in Experimental Panel 18.1.

Also importantly, the random orientation of the actin–myosin fibrils in the smooth muscle (which gives the uniform appearance to smooth muscle cells) enables smooth muscle fibres to generate force even when extremely stretched. This is because even at such extreme lengths there will be overlap between myosin heads and actin filaments to produce force in some actin–myosin fibrils, as shown in Figure 18.26C. The strict parallel organization of myofibrils in striated muscles prevents force generation when the fibres are stretched beyond about 180 per cent of their resting length where there is no overlap between actin and myosin filaments.

The ability of the smooth muscle to function over a very wide range of lengths is critically important to allow blood vessels to change diameter to regulate blood flow, or

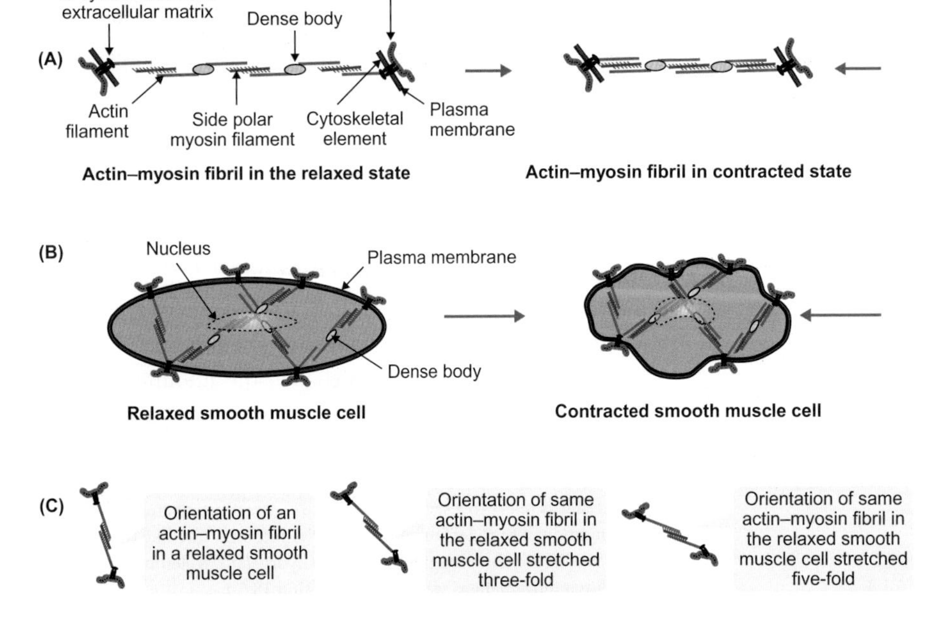

Figure 18.26 Diagrammatic representation of the arrangement of the contractile apparatus in smooth muscle cells

(A) The contractile apparatus in smooth muscle cells is organised in fibrils consisting of actin filaments attached to dense bodies in the cytosol and to the extracellular matrix via cytoskeletal elements and adhesion proteins; side-polar myosin filaments (refer to Figure 18.3B) are located between actin filaments. (B) The fibrils criss-cross through the cellular space. Myosin filaments, whose number rises when the smooth muscle contracts, pull on the actin filaments of complementary polarity causing shortening of the actin–myosin fibrils which changes the shape of the smooth muscle cell. (C) Change in the orientation of an actin–myosin fibril when a relaxed smooth muscle cell is stretched three- and five-fold.

the uterus to stretch to accommodate a growing foetus, or pythons and humans alike to eat large meals that stretch the musculature of the gastrointestinal tract.

18.4.2 Excitation–contraction coupling in smooth muscle occurs by two mechanisms

Similar to cardiac cells, smooth muscle cells:

- do not have specific postsynaptic specializations with the neurons that innervate them;
- are regulated by the autonomic nervous system and hormones;
- contain a variety of Ca^{2+} channels, K^+ channels, non-specific cationic channels and G-protein-coupled (metabotropic)[73] receptors in the sarcolemma;
- have a negative membrane potential at rest (between −40 to −70 mV), and
- contract in response to a rise in intracellular Ca^{2+} concentration.

[73] We discuss ionotropic and metabotropic receptors in Section 16.3.4.

The rise in intracellular Ca^{2+} concentration occurs by two mechanisms in smooth muscle fibres, as summarized in Figure 18.27. In one mechanism, called **electromechanical coupling**, neurotransmitters released from swellings (varicosities) on neurons that innervate smooth muscles, as well as hormones transported in the blood, can cause depolarization of the sarcolemma. This depolarization is the result of neurotransmitters either binding to and opening non-selective cationic ligand-gated ion channels, or by binding to receptors in the sarcolemma linked to signalling pathways that close some of the K^+ channels, which reduces the overall K^+ permeability of the sarcolemma.

If the depolarization is strong enough, action potentials are triggered in smooth muscle cells. The action potentials in smooth muscle cells involve the much slower voltage-gated Ca^{2+} channels instead of the much faster voltage-gated Na^+ channels; action potentials are therefore very slow when compared with those in neurons and skeletal muscle fibres. The entry of Ca^{2+} in smooth muscle fibre through the open voltage-gated Ca^{2+} channels cause the rise in intracellular Ca^{2+} and activation of the contractile apparatus.

Contraction of the majority of smooth muscle cells in single-unit smooth muscle, in which action potentials

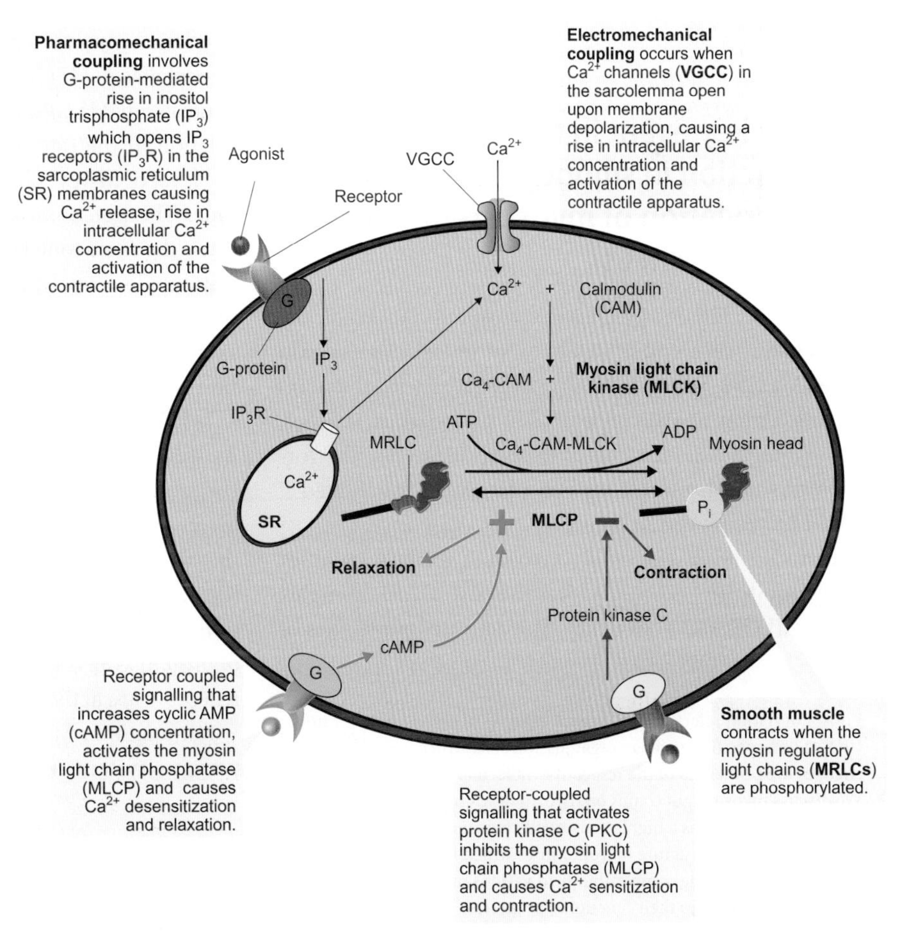

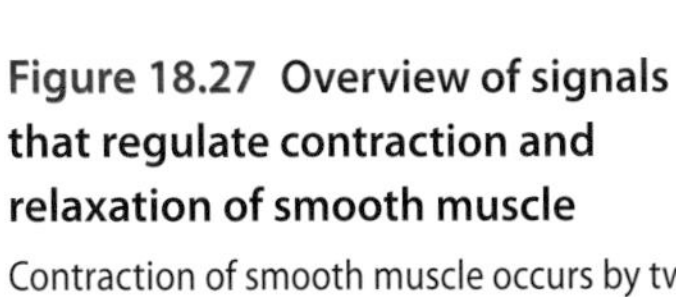

Figure 18.27 Overview of signals that regulate contraction and relaxation of smooth muscle

Contraction of smooth muscle occurs by two different mechanisms: pharmacomechanical coupling (which occurs without a change in the membrane potential) and electromechanical coupling, both of which cause a rise in sarcoplasmic Ca^{2+} concentration. The rise in Ca^{2+} activates the myosin light chain kinase (MLCK) with the formation of a complex calcium–calmodulin–myosin-light-chain-kinase (Ca_4-CAM-MLCK). This complex phosphorylates the regulatory light chain on the myosin molecules (MRLC). Phosphorylation of the MRLC activates the contractile apparatus and dephosphorylation of the MRLC by myosin light chain phosphatase (MLCP) causes relaxation. Receptor coupled signalling alters the activity of the MLCP which causes either sensitization or desensitization of the smooth muscle.

propagate through gap junctions from cell to cell, occurs by electromechanical coupling. Some smooth muscle cells, like in the gut musculature of vertebrates, have a spontaneous cycle of depolarization and act as pacemaker cells similar to the pacemaker cells in the vertebrate heart. Action potentials are triggered in these pacemaker cells when the level of depolarization reaches the threshold. The action potential then spreads to all smooth muscle cells that are electrically connected by gap junctions, synchronizing their activity.

The other mechanism responsible for the rise in intracellular Ca^{2+} concentration and activation of contraction is called **pharmacomechanical coupling**. This mechanism operates in response to neurotransmitters from the autonomic nervous system or hormones binding to metabotropic receptors associated with the production of **inositol trisphosphate (IP_3)**. IP_3 binds to IP_3 receptors in the membrane of the sarcoplasmic reticulum opening them causing Ca^{2+} release and the rise in sarcoplasmic Ca^{2+} concentration.

The rise in Ca^{2+} concentration by either mechanism initiates the phosphorylation[74] of the **myosin regulatory light chain** (**MRLC**)[75] on the myosin molecules followed by contraction. Relaxation occurs when the MRLCs become dephosphorylated. Generally, smooth muscles contract and relax much more slowly than striated muscles; therefore, smooth muscles can maintain force for longer durations. The level of contraction of the smooth muscle is also called **tone**.

18.4.3 The key event in the activation of contraction in smooth muscle is the phosphorylation of myosin regulatory light chains (MRLCs)

As shown in Figure 18.27, the phosphorylation of the MRLCs takes place under the action of **myosin light chain kinase** (**MLCK**), which is loosely associated with the actin filaments. Unlike myosin molecules in striated muscle fibres, which aggregate spontaneously to form filaments, the formation of smooth muscle myosin filaments is facilitated if the MRLCs are phosphorylated.

When Ca^{2+} levels rise in the cytosol, it forms a complex with **calmodulin**[76] (Ca_4-CAM). The Ca_4-CAM complex binds to MLCK and activates it. The activated molecules of MLCK dissociate from the actin filaments, bind to myosin molecules in the vicinity of the actin filaments and phosphorylate the regulatory light chains. The phosphorylation of the MRLCs has two main outcomes:

- It facilitates formation of myosin filaments in the vicinity of actin filaments because myosin molecules with phosphorylated MLCs aggregate more readily than myosin molecules with dephosphorylated MLCs to form filaments.
- It allows myosin heads to interact with actin and produce force, as we discussed in Figure 18.7, because the myosin heads of smooth muscle myosin molecules do not have a high affinity for actin unless their regulatory light chains are phosphorylated.

When smooth muscle cells contract, their general shape changes markedly due to the contraction of the randomly distributed actin–myosin fibrils, as shown in Figure 18.26B. The muscle relaxes when the MRLCs become dephosphorylated under the action of another enzyme, called the **myosin light chain phosphatase (MLCP)**.

Regulation of smooth muscle tone in vertebrates

The most important mechanism of smooth muscle tone regulation involves a change in the level of myosin light chain phosphatase (MLCP) activity, as shown in Figure 18.27. A decrease in MLCP activity reduces the rate of dephosphorylation. If the rate of phosphorylation is maintained, then the level of MRLC phosphorylation rises, as does the muscle tone. The outcome is equivalent to Ca^{2+} sensitization, because the same tone is now produced at a lower Ca^{2+} concentration. Conversely, an increase in MLCP activity decreases the phosphorylation level of MRLCs and causes Ca^{2+} desensitization because, in this instance, a higher Ca^{2+} concentration is needed to produce the same tone.

MLCP activity is regulated by receptor-coupled signalling. As shown in Figure 18.27, inhibition of MLCP activity and therefore Ca^{2+} sensitization occurs, for example, through the increase in protein kinase C (PKC) activity associated with the activation of α_1-adrenergic receptors[77]. Conversely, an increase of MLCP activity, and therefore Ca^{2+} desensitization (and even relaxation), follows an increase in the activity of protein kinase A (PKA), which occurs when β-adrenergic receptors are activated[78].

Stimulation of the sympathetic division of the autonomic nervous system in vertebrates, which releases noradrena-

[74] Phosphorylation refers to the covalent binding of a phosphate group (PO_4^{3-}) to a specific site on an organic molecule, which generally changes the properties of the respective molecule.

[75] We discuss myosin light chains on myosin heads in Section 18.1.2.

[76] Calmodulin (CAM) is a highly conserved cytosolic Ca^{2+}-binding protein, similar in structure to troponin C, which forms a complex with four calcium ions. The complex binds to various molecules (including enzymes) and changes their functional properties.

[77] We discuss different types of adrenergic receptors in Section 16.1.3.

[78] This process is illustrated in Figure 19.6.

line at varicosities in smooth muscles, prepares the body for activity such as in the **defence-alerting response**[79]. Smooth muscles, which have predominantly β_2-adrenergic receptors relax, while smooth muscles that have predominatly α_1-adrenergic receptors contract. Smooth muscles that relax include the smooth muscle lining the bronchi of the airway[80] and the smooth muscle in arteries that deliver blood to skeletal muscles. Conversely, smooth muscles in arteries that supply blood to the kidneys, skin, gastrointestinal tract and brain contract and induce vaso-constriction. The overall effect is an increased delivery of oxygen-rich blood to skeletal muscles in the body and reduced blood flow to other organs that are not essential in the defence-alerting response[81].

Smooth muscle actin filaments contain tropomyosin in the same proportion to G-actin as in striated muscle (i.e. 1:7), but they do not contain troponin. Two other proteins associated with the actin filament—calponin and caldesmon—exert some inhibitory action on the interaction between myosin heads and actin filaments at low Ca^{2+} concentration when these proteins are not phosphorylated. Therefore, phosphorylation of caldesmon and calponin, which occurs at higher Ca^{2+} concentration, also increases the muscle tone.

Muscle tone can also be maintained with little energy expenditure when the smooth muscle enters the state of '**latch**'. In this state, the myosin heads stay attached to the actin filaments for a long time and maintain the muscle tone without further hydrolysis of MgATP. The 'latch' state is believed to occur when the RMLCs become dephosphorylated while the myosin heads are still attached to the actin filaments. Then, the myosin heads bind very slowly MgATP to induce their detachment from the actin filaments (Figure 18.7, step 4).

18.4.4 The molluscan catch muscle maintains force highly efficiently

Smooth muscles of bivalve molluscs in general display the unusual property of '**catch**', where force is maintained with little consumption of metabolic energy for many hours to keep the shells tightly shut. The diameter of smooth muscle cells in the molluscan catch muscle is very small (about 2 μm), which is about half that in vertebrate smooth muscle. However, the molluscan catch muscle can produce very large specific forces (up to 1500 kN m^{-2}) compared with vertebrate smooth muscle (up to 400 kN m^{-2}).

The most intensely studied catch muscle is the anterior byssus retractor muscle of the blue mussel (*Mytilus edulis*), which consists entirely of smooth muscle fibres and is relatively easy to isolate. The byssus is a collection of sticky threads used by mussels to attach to a variety of substrates; the anterior byssus retractor muscle controls the position of the mussel on the substrate and the tightness of the attachment to the substrate through the byssus.

Essential for the development of the 'catch' state is the presence of the protein **twitchin**, which is a myosin-filament-binding protein similar to titin in vertebrate striated muscle. Twitchin is phosphorylated by the cAMP-dependent protein kinase (PKA) and is dephosphorylated by a Ca^{2+}-dependent phosphatase. The affinity of twitchin for the actin filament is low when twitchin is phosphorylated and increases when twitchin is dephosphorylated. In the resting muscle twitchin is phosphorylated.

Release of acetylcholine from nerve terminals causes membrane depolarization and opening of Ca^{2+} channels through which Ca^{2+} enters the sarcoplasm. Molluscan myosin is activated by direct binding of Ca^{2+} to the myosin heads—as we describe for molluscan striated muscle in Section 18.1.6—and force develops.

The rise in Ca^{2+} concentration also activates the Ca^{2+}-dependent phosphatase, which causes the dephosphorylation of twitchin. Twitchin can now also bind to the actin filament, producing a network of *passive linkages between the actin and the myosin filaments*. These linkages maintain force even after the myosin heads stop cycling when the Ca^{2+} concentration returns to the resting level. The muscle is now in a state of **catch**, in which passive connections between filaments maintain the force exerted on the ends of the muscle without consumption of metabolic energy. This situation is similar to zipping up a bag full of clothes: force is needed to push the clothes into the bag and zip it up, but once the zip is pulled, no extra force is necessary to keep the bag closed.

The muscle relaxes following release of serotonin from nerve terminals. Serotonin increases the production of cAMP and activation of protein kinase A, which phosphorylates twitchin. The affinity of twitchin for actin decreases and the linkages between myofilaments are undone.

[79] Sections 19.3 and 22.5.3 cover the defence-alerting response in more detail.

[80] The effect of allergens that induce asthma (constriction of the airways causing difficulty in breathing) can be counteracted by the use of 'puffers' that deliver agonists of β_2-adrenergic receptors on the smooth muscle lining the airways, inducing relaxation.

[81] We discuss the sympathetic tone in the blood vessels of vertebrates in Section 22.4.3.

› Review articles

Berridge MJ (2008). Smooth muscle cell calcium activation mechanisms. Journal of Physiology 586: 5047–5061.

Puetz S, Lubomirov LT, Pfitzer G (2009). Regulation of smooth muscle contraction by small GTPases. Physiology 24: 342–356.

Somlyo AP, Somlyo AV (2003). Ca^{2+}-sensitivity of smooth muscle and non-muscle myosin II: modulated by G proteins, kinases, and myosin phosphatase. Physiological Reviews 83: 1325–1358.

18.5 The muscular system is the engine that provides the power for an animal's movements and behaviour

The force produced by contracting muscles described in Sections 18.1–18.4 powers all types of animal movement. In this and the following section we focus on the performance of voluntary muscles, which use up to 75 per cent of total body energy[82], and examine how the muscular system facilitates the coordinated body movements used in locomotion.

The work output of voluntary muscles depends on various factors including:

- the intrinsic properties of the constituent muscle fibre types,
- fibre organization within muscles,
- precise pattern of neural stimulation of the individual fibres,
- sarcomere length,
- level of activation of individual fibres,
- the elastic properties of the connective tissues that link the fibres to other fibres and the skeleton, and, very importantly,
- the load against which the muscles contract.

[82] We discuss energy requirements and utilization in animals in Section 2.2.

Since muscles can only pull, they must act in a coordinated fashion with other muscles by pulling on various body structures to deliver a particular type of movement or achieve stability of **joints**. Joints function as part of pulley and lever systems in animals with stiff external or internal skeletons, which are involved in locomotion as shown in Figure 18.28. The presence of positional sensing receptors (proprioceptors[83]) at each joint conveys information to the central nervous system (CNS)[84] about the three-dimensional position of limbs with respect to the body. Neural circuits in the CNS produce rhythmic motor patterns when activated[85] and the muscular system responds to the particular pattern of stimulation dictated by the nervous system.

18.5.1 Muscles and skeletons work together to produce animal movement

Arthropods evolved an external skeleton (**exoskeleton**) with relatively simple joints shown in Figure 18.28A, which can be rotated by muscles only in one plane. By contrast, vertebrates evolved internal skeletons (**endoskeletons**) with more versatile joints that can be rotated by muscles in different planes as shown in Figure 18.28B. The exoskeleton provides a very efficient protective layer between an animal's body and the external environment, but needs to be replaced regularly by moulting the old exoskeleton[86]. The exoskeleton increases

[83] We discuss proprioception in Section 17.4.4.

[84] We discuss the CNS in Section 16.1.

[85] We discuss central pattern generators in Section 16.5.4.

[86] We discuss moulting in insects in Section 19.5.3.

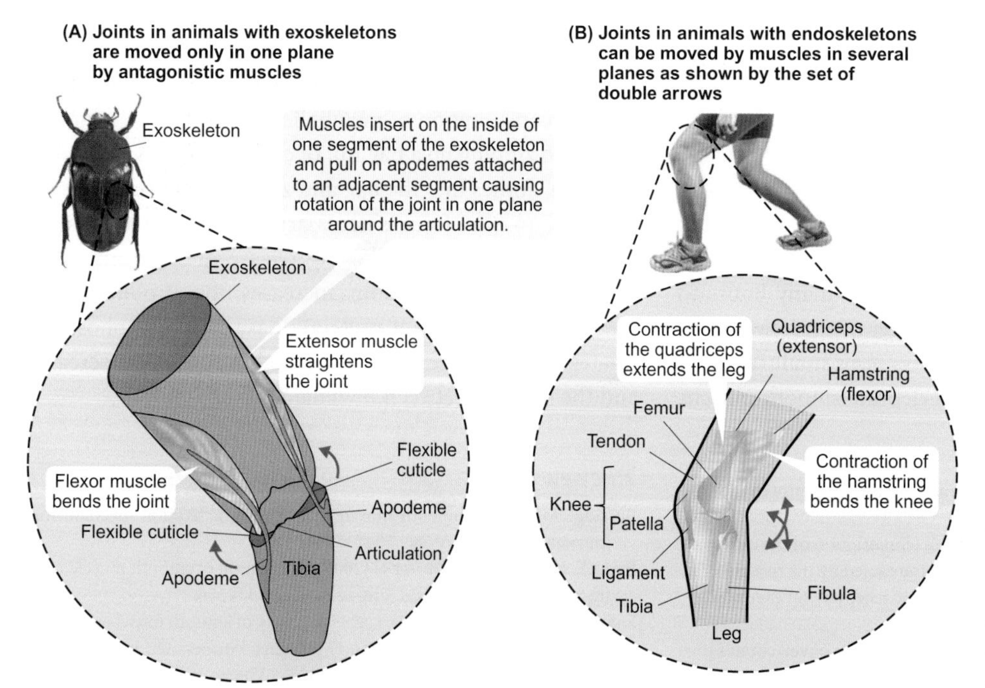

Figure 18.28 Joints function as pulley systems driven by antagonist muscles

(A) Diagram of a joint between the femur and the tibia in the leg of a beetle (*Ischiopsopha jamesi*) that can grow to 30 mm length. Muscle fibres are attached directly to the exoskeleton (cuticle) at one end and via apodemes, which connect to the other segment at the other end. (B) Diagram of the knee in humans. Muscles connect to bones via tendons and bones connect to other bones by ligaments.

in thickness disproportionately as animals get bigger, becoming increasingly very heavy which eventually can limit the size to which animals can physically grow. Animals are also most vulnerable to predation during the moulting period until the new exoskeleton hardens. In contrast, the endoskeletons can be continuously remodelled[87], increasing in size and strength as animals grow, providing the necessary support to the animal's body as it changes. In addition, endoskeletons can support much larger bodies than exoskeletons of same weight.

The major muscle in fish is the trunk muscle, which is used in swimming and can amount to up to 70 per cent of body mass. The trunk muscle is organized in a longitudinal series of complex segmental structures called **myomeres** (or **myotomes**) that are separated by collagen sheets called **myoseptae**, on which the muscle fibres from myotomes attach. The myoseptae connect to the backbone and to the skin. The segmented arrangement of the axial musculature shown in Figure 18.29 permits the freedom to delay muscle excitation along the trunk of the fish to produce body and tail fin oscillations for efficient swimming. In other vertebrates, skeletal muscles connect to the skeleton via tendons as we discussed earlier (Figure 18.5). The slow, red muscle fibres in fish run in a band along the midline on each side, immediately under the surface of the skin (Figure 18.29A); the white, fast-twitch fibres are located deeper in the sideways W-shaped myomeres shown in Figure 18.29B.

Skeletal muscles of invertebrates do not connect to tendons but connect directly to the cuticle or to **apodemes**, as shown in Figure 18.28A. Apodemes are ingrowths of the exoskeleton and, similar to tendons, can stretch to store elastic energy.

Muscles surrounding cavities filled with fluid exert hydrostatic pressure on the (incompressible) fluid in the cavity, which can act as **hydrostatic skeletons**. For many soft-body invertebrates, like worms and insect larvae, the body cavity (coelom) is surrounded by one layer of longitudinal muscle with fibres running parallel to the body axis, and another layer of circular muscle with fibres running around the body circumference. Since animals have essentially a constant volume, consisting of relatively incompressible fluids and tissues, the decrease in one dimension of the body caused by muscle contraction raises the hydrostatic pressure in the soft-body animal and results in an increase in another dimension.

A wide range of movements is produced by animals with hydrostatic skeletons depending on the direction of force generated by the muscles and the organization of connective tissue in the body. For example, earthworms, snails and slugs[88] use their hydrostatic skeleton to crawl by coordinated contraction and relaxation of longitudinal and circular muscles, which permit the generation of a series of wave-like muscle contractions called **peristaltic movements.** Figure 18.30 illustrates how earthworms use their hydrostatic skeleton for locomotion.

[87] Bone remodelling is a process that occurs throughout an animal's life and involves removal of mature bone tissue (bone resorption) and new bone formation (ossification); bone remodelling is under hormonal control, as we discuss in Section 21.2.1.

[88] We discuss the hydrostatic skeleton in the foot of molluscs in Section 14.3.1.

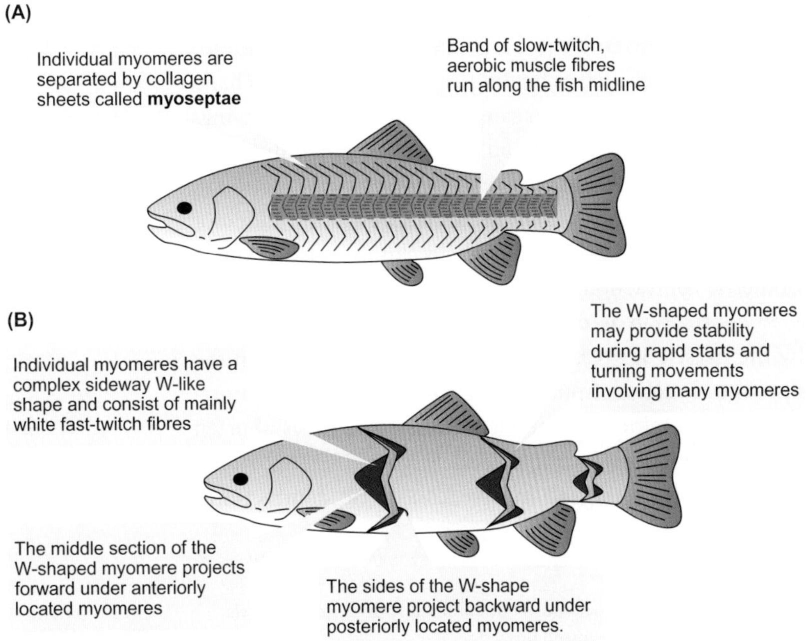

Figure 18.29 Diagram of trunk muscle of the brown trout (*Salmo trutta*)

(A) Myomeres as they appear under the skin with one myomere per each vertebra and with a longitudinal band of red-coloured, slow-twitch, aerobic muscle fibres. (B) Individual myomeres have a typical sideways W-shape and consist mainly of white, fast-twitch, anaerobic muscle fibres.

Source: A, B Coughlin DJ. (2002). Aerobic muscle function during steady swimming in fish. Fish and Fisheries 3: 63–78.

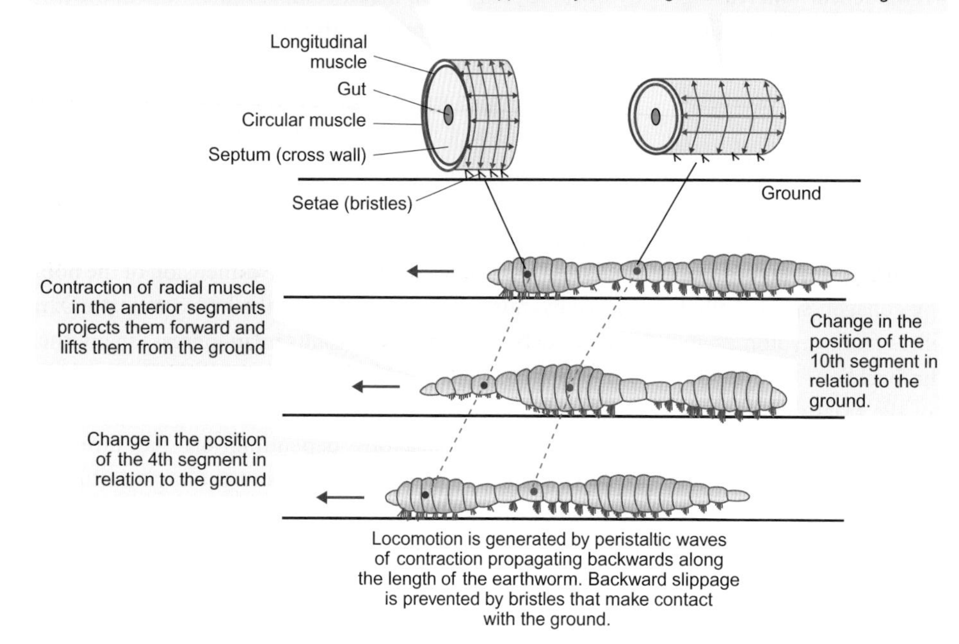

Figure 18.30 Diagram showing how circular and longitudinal muscles work together with the hydrostatic skeleton to produce movement in the common earthworm (*Lumbricus terrestris*)

Types of muscle contraction

Examples of the different types of muscle contraction are shown diagrammatically in Figure 18.31. If the muscle contracts without changing length (or average sarcomere length), the muscle is said to undergo an **isometric contraction**[89] (Greek 'isos' meaning equal and 'metria' meaning measurement). Activities such as holding an object still in one's hand (Figure 18.31A), pushing against a wall, keeping the wings of gliding birds extended and generally maintaining posture while standing still involve the isometric contraction of muscles that stabilize joints. The contraction of the adductor muscle of bivalves to keep their shells closed is also a type of isometric contraction.

A contraction is **concentric** when the force produced by the muscle is greater than the load (force acting in the opposite direction from that in which the muscle shortens) and therefore the muscle and its sarcomeres shorten when contracting. Lifting a load (Figure 18.31B), riding a bike or swimming all involve concentric contractions.

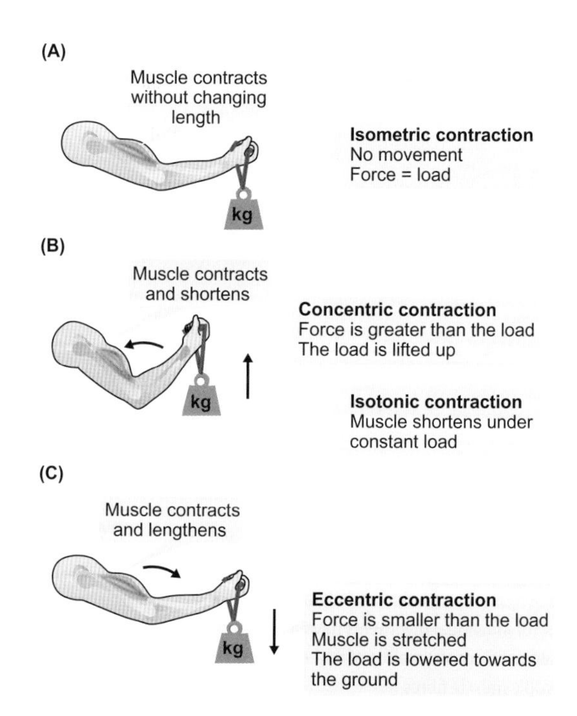

Figure 18.31 Several types of muscle contraction

[89] Isometric contraction should not be confused with an isometric relationship, which we discuss in Box 2.3.

18

If the load stays constant when the muscle shortens, like in Figure 18.31B, then the contraction is also isotonic (Greek 'isos' meaning equal and 'tonos' meaning tension).

Eccentric or **lengthening contractions** occur when muscles (and sarcomeres) lengthen while contracting because the force produced by the respective muscles (sarcomeres) is less than the applied load. For example, muscles perform eccentric contractions when a load is lowered toward the ground (Figure 18.31C), or when walking downhill, in the latter case because contracting calf muscles in the leg lengthen under the weight of the body with each step that is taken downhill.

18.5.2 Muscle mechanical performance and energetics

The mechanical performance of muscle is best described in terms of its shortening velocity (speed), the power[90] it generates, and its efficiency[91] in the same way as the mechanical performance and efficiency of a car are judged in terms of speed, engine power and fuel comsumption per distance travelled.

Longer muscles shorten faster than shorter muscles of the same type

The shortening velocity of striated muscle is characterized by how fast actin- and myosin filaments slide past each other in one half of the sarcomere. Thus, if one half sarcomere shortens at, say, 1 μm s^{-1}, then a 1000 sarcomere long muscle (2000 half-sarcomeres) shortens at a velocity of 2 mm s^{-1} (2000 × 1 μm s^{-1} = 2000 μm s^{-1} = 2 mm s^{-1}) and a muscle of 20,000 half sarcomeres, which is 10 times longer, shortens at 20 mm s^{-1}. Therefore, to compare the intrinsic velocities of shortening of different muscles, the shortening velocity is expressed either in μm per half sarcomere per s, or in muscle lengths per s, as shown in Figures 18.21B and 18.32, respectively.

The maximum velocity of shortening depends on how fast the myosin heads cycle

The *maximum* velocity of shortening (v_{max}) of a muscle is implicitly linked to how fast the myosin heads cycle and to the ATPase activity of the myosin isoform(s) expressed in that muscle. Generally, muscles that display a higher myofibrillar ATPase activity shorten faster than muscles that display a lower myofibrillar activity. For example, in individual mammals, v_{max} and ATPase activity are highest in type IIB fast-twitch fibres[92] and lowest in type I slow-twitch fibres (type IIB> type IIX > type IIA > type I).

Fibres of a given myosin heavy chain (MHC) type from smaller animals shorten faster than fibres of the same type from larger animals belonging to the same group, as shown in Figure 18.21B for mouse vs human. Moreover, type I slow-twitch fibres in mice shorten even faster than type IIA fast-twitch fibres in humans. This is in line with smaller animals having generally higher mass-specific metabolic rates[93] than larger animals of the same group.

The shortening velocity depends on the load carried by the muscle

Figure 18.32 illustrates the **force-velocity curve** for a fast-twitch frog muscle. The velocity of shortening is zero

[90] Power is work done (or converted) per unit time and is measured in watts (1 W = 1 joule s^{-1}; Appendix 1).

[91] Thermodynamic efficiency of the muscle is defined as the fraction of free energy available from the hydrolysis of ATP that is converted into mechanical work.

[92] General properties of different fibre types in mammals are summarized in Table 18.2.

[93] We discuss the relationship between metabolic rate and body mass in Section 2.2.4.

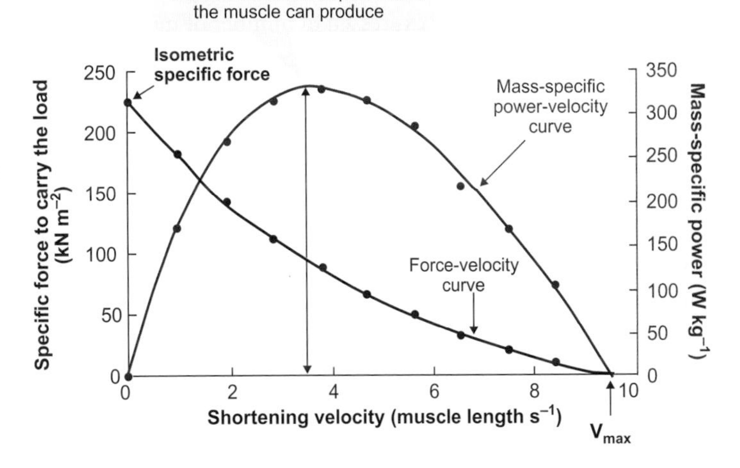

Figure 18.32 Force-velocity and power-velocity curves for the semimembranous muscle from the thigh of the leopard frog (*Rana pipiens*) at 25°C

The force-velocity curve is a hyperbola and the power-velocity curve is obtained from the force-velocity curve by multiplying the velocity of shortening (at a particular load) by the load. The mean jump velocity measured in the living animal provides maximal power for this fast-twitch muscle as indicated by the up-down arrow.

Source: Lutz GJ, Rome LC (1994). Built for jumping: The design of the frog muscular system. Science 263: 370–372.

when the load equals the isometric force and rises toward its maximum value (v_{max}) as the force needed to carry the load decreases toward zero.

The shape of the force-velocity (F-v) curve for all striated muscles investigated is described by an expression known as the **Hill equation**, which is characterized by three parameters: F_{max} (=isometric force), v_{max} (=maximum velocity of shortening) and b (a constant with dimension of velocity that is specific for a particular muscle):

b = constant with dimension of velocity

F_{max} = isometric force

v = velocity of shortening

$$v(F + bF_{max}v_{max}^{-1}) = b(F_{max} - F)$$

F = force produced to carry the load

v_{max} = maximum velocity of shortening

Equation 18.1

The curvature of the force-velocity plot is given by the value of b/v_{max}.

The power produced by a muscle is proportional to its volume (or mass)

The power of a muscle refers to the amount of mechanical work (in joules) done by the muscle per second and is measured in watts (1W = 1J s^{-1} = 1N m s^{-1}). The power (in W) is calculated by multiplying the force produced to carry the load (F, measured in N) by the shortening velocity (v measured in m s^{-1}):

$$\text{Power} = F \times v \quad \text{(W)}\ \text{(N)}\ (\text{m s}^{-1}) \qquad \text{Equation 18.2}$$

The amount of force produced (in N) by the muscle equals the specific force (in N m^{-2}) × cross-sectional area (CSA, in m^2). The muscle shortening velocity (in m s^{-1}) equals the nominal shortening velocity expressed in number of muscle lengths per s (s^{-1}) × muscle length in m. Since CSA (m^2) × muscle length (m) represents the muscle volume in m^3 = 1000 L, it follows that:

Power (W) = 1000 × Specific force to carry the load (N m^{-2}) × Nominal shortening velocity (s^{-1}) × muscle volume (L)

Considering now that 1 kN = 1000 N and that: muscle volume = muscle mass ÷ muscle density, where the muscle density is about 1.06 kg L^{-1}, it follows that:

$$\text{Power (W)} = (1.06\ \text{kg L}^{-1})^{-1} \times \text{Specific force to carry the load (kN m}^{-2}) \times \text{Nominal shortening velocity (s}^{-1}) \times \text{muscle mass (kg)} \qquad \text{Equation 18.3}$$

The term: '(1.06 kg L^{-1})$^{-1}$ × Specific force to carry the load (kN m^{-2}) × Nominal shortening velocity (s^{-1})' is called mass-specific power (in W kg^{-1}) or **power density**; it is a measure of the intrinsic power-generating capacity of a muscle type under given conditions. In order to obtain the actual power generated by a muscle (in W), we multiply the mass-specific power by the muscle mass.

The power generated by the muscle depends on its shortening velocity

Let us look at the force-velocity curve in Figure 18.32 for the semimembranosus muscle[94] of the frog and calculate the mass-specific power corresponding to different shortening velocities. For example, the muscle shortens at a velocity of two muscle lengths s^{-1}, when it develops a specific force of 140 kN m^{-2} to carry the load. The mass-specific power of the muscle under these conditions, according to equation 18.3, is (1.06 kg L^{-1})$^{-1}$ × 140 kN m^{-2} × 2 s^{-1} = 263 W kg^{-1}. In fact, all data points on the force-velocity curve can be converted into power points by multiplying the factor 1/1.06 by the specific force produced by the muscle when carrying a particular load and by the corresponding velocity of shortening on the force-velocity curve.

When the load is zero, the muscle shortens at maximum velocity, v_{max}, but the power produced by the muscle is zero because the specific force produced is zero: $0 \times v_{max} = 0$. Similarly, when the load is equal to the isometric force (F_{max}), the muscle does not shorten and the power is zero, because the shortening velocity is zero ($v = 0$): $F_{max} \times 0 = 0$. The mass-specific power-velocity curve in Figure 18.32 reaches a maximum when the velocity is about 35 per cent of v_{max}.

The striated muscles in most species deliver their highest power at velocities around 30 per cent v_{max}. Generally, the peak mass-specific power produced by slow-twitch muscles in vertebrates is less than one third of the peak mass-specific power produced by the fastest fast-twitch muscles in the same animal. In mammals[95], muscle fibres deliver mass-specific power in descending order: type IIB, type IIX, type IIA, type I. Therefore, muscles involved in power movements generally consist of fast-twitch muscle fibre types.

The mean jump velocity of the frog corresponds to the maximal power of the semimembranosus muscle (Figure 18.32), showing that frogs can recruit muscles to develop maximal power (about 350 W kg^{-1}) during jumping. The pectoralis muscle in birds, a fast-twitch muscle that powers the downstroke movement of wings, has a mass-specific power up to 400 W kg^{-1}, similar to that of the semimembranosus frog

[94] The semimembranosus muscle is a thigh muscle, which together with the semitendinosus and the biceps femoris muscles make up the hamstring.

[95] Table 18.2 describes the main characteristics of the different fibre types in mammals.

muscle. In contrast, the peak mass-specific muscle power is much smaller in mammals (~70 W kg^{-1} in the fast-twitch leg muscles of the mouse), fish (91 W kg^{-1} in the white (fast) myotome muscle of the dogfish (*Scyliorhinus canicula)* and 60 W kg^{-1} in the red (slow) muscle of the carp *(Cyprinus carpio)*) and marine invertebrates (18 W kg^{-1} in the mantle muscle of the squid *(Alloteuthis subulata)*).

Muscle efficiency for converting chemical energy into mechanical work is similar to that of a car engine

Muscle efficiency refers to the fraction of free energy available from the hydrolysis of ATP that is converted into mechanical work. The remainder is dissipated as heat. Generally, slower contracting muscles are more efficient than faster muscles at producing force and mechanical work because the myosin heads in slower fibres stay attached longer to the actin-filament producing force for the same amount of MgATP consumed. Yet, faster muscles produce greater power than slower muscles and are crucial to an animal's survival in escape responses, when hunting for prey, or indeed for flying.

The maximum efficiency of the contractile apparatus in mammalian slow-twitch (type I) and fast-twitch (type IIB) fibres is around 50 per cent and 30 per cent, respectively. In all muscle types, the maximum efficiency occurs at velocities of shortening in the range 20–35 per cent v_{max} and the efficiency decreases steadily as the muscle shortens more quickly because muscles produce force less economically with faster cycling cross-bridges under low load conditions.

In muscle, ATP is not only used for mechanical power generation, but also by other ATP-dependent processes. For example, during muscle contraction, myosin is responsible for 60–70 per cent of the total ATP consumption on average, the remaining 30–40 per cent being associated primarily with ion transport across membranes. Thus, only about 20 per cent of the total free energy liberated from the hydrolysis of ATP is converted into mechanical work in fast-twitch muscle (30 per cent efficiency of the 60–70 per cent total ATP free energy that is used by myosin). The slow-twitch muscle is more efficient, with about 30 per cent of the total free energy from the hydrolysis of ATP converted into mechanical work (50 per cent efficiency of the 60–70 per cent total ATP free energy that is used by myosin). Overall, skeletal muscle efficiency is comparable to that of modern petrol car engines if the fuel is ATP and petrol, respectively.

The ATP pool in muscle must be continuously replenished to support contraction

The ATP concentration in the cytosol[96] of vertebrate skeletal muscle varies between about 5 and 10 mmol L^{-1}, with the highest concentrations measured in the pectoral muscle of birds such as the domestic fowl (*Gallus gallus domesticus*), pheasants (*Phasianus colchicas*), hummingbirds and pigeons, even though these species have completely different patterns of flight. (Domestic fowl do not fly or fly over short distances; pheasants fly intermittently; hummingbirds fly to stay stationary in the air and pigeons fly over long distances.)

Invertebrate muscles exhibit a much wider range of ATP concentrations. For example, the foot muscle of the edible snail (*Helix pomatia*) contains <1 mmol L^{-1} ATP, while the extremely aerobic insect flight muscles of honeybees (*Apis melliphera*), blowflies (*Calliphora vicinia*) and garden fruit chafers (*Pachnoda sinuata*) contain up to 10 mmol L^{-1} ATP. The abdominal muscles in decapod crustaceans such as lobsters (*Homarus vulgaris*) contain 6–8 mmol L^{-1} ATP, which is similar to that measured in fast- and slow-twitch mammalian muscles.

The rate of ATP consumption is very low (<0.01 mmol L^{-1} s^{-1}) when muscles are at rest, but can increase rapidly by several hundred fold upon activation. The total rate of ATP consumption during sustained flight is about 30 mmol L^{-1} s^{-1} in insect flight muscle, 10 mmol L^{-1} s^{-1} in hummingbirds' flight muscle, and 7 and 1.5 mmol L^{-1} s^{-1}, respectively, in fast- and slow-twitch muscle fibres in humans during high-intensity exercise. At these rates, the entire ATP pools would be depleted in a matter of seconds or less in all maximally activated voluntary muscle fibres of both vertebrates and invertebrates, unless ATP was rapidly regenerated.

Voluntary muscle fibres in all animals contain a sizeable pool of high-energy phosphates (1.5 to 9 times greater than the ATP pool) called **phosphagens**, which act as rapid buffers to maintain a high ATP concentration and low ADP concentration. At least six different phosphagen molecules have been identified, of which **creatine phosphate** and **arginine phosphate** are the most important: creatine phosphate[97] is present in all vertebrate and some invertebrate muscles, while arginine phosphate is the most widely distributed form of phosphagen in invertebrate muscles. Annelids have a different phosphagen called **phospholombricine.**

As shown in Figure 18.33 for creatine phosphate and arginine phosphate, the phosphagens rapidly donate their phosphate to ADP to re-synthesize ATP in the presence of phosphokinases, when the ADP concentration rises in the sarcoplasm and the equilibrium is displaced to the right. Conversely, the phosphate is transferred from ATP to resynthesize the phosphagen when the muscles are at rest and the equilibrium shifts to the left.

The phosphagen pool is sufficient to prolong the time muscles can function at maximum intensity from several

[96] Expressed in mmol L^{-1} water.

[97] We discuss in Section 15.2.3 how creatine phosphate is used in muscles at the beginning of exercise.

18

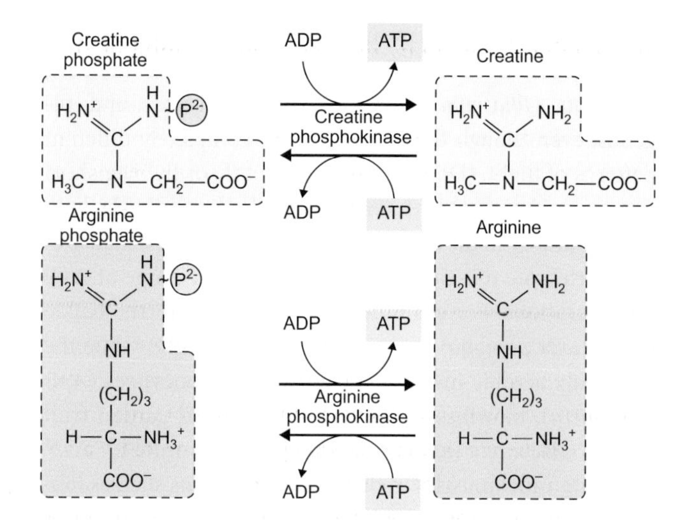

Figure 18.33 Phosphagens are molecules that include creatine phosphate in all vertebrates (and some invertebrates) and arginine phosphate in most invertebrates

Phosphagens can transfer phosphate groups (shown as P^{2-}) to ADP to regenerate ATP in muscle and can be regenerated with the help of ATP. Equilibrium under physiological conditions is achieved when the ratio between ATP and ADP concentrations, [ATP]/[ADP] = 100 × [creatine phosphate]/[creatine] for the creatine phoshokinase reaction and [ATP]/[ADP] = 15 × [arginine phosphate]/[arginine], for the arginine phosphokinase reaction. The equilibria shift to the right when the ADP concentration rises and [ATP]/[ADP] < 100 × [creatine phosphate]/[creatine] for the creatine phoshokinase reaction and [ATP]/[ADP] < 15 × [arginine phosphate]/[arginine] for the arginine phosphokinase reaction.

seconds to up to tens of seconds, but not for much longer. For extended periods, ATP is regenerated from metabolic reactions associated with the degradation of glycogen, glucose and fatty acids[98]. Glycogen and fatty acids are stored in muscle fibres in the form of granules and lipid droplets, respectively. The amount of glycogen and fatty acids stored within the muscle fibres can be increased by exercise. Glucose and fatty acids are delivered from the gut or from their stores in the body to the muscle in the blood[99].

Most of the energy needs of muscles at rest or at low to moderate levels of activity are provided by the oxidation of fatty acids[100] in mitochondria. Slow muscles in general are economical at generating force and low power with low to moderate ATP consumption and can be maintained at a relatively high level of activity for several hours using predominantly fatty acid oxidation. Consequently, slow muscle fibres are generally used for lower power, prolonged or repetitive activity, such as joint stabilization and posture maintenance in vertebrates. During the long duration exercise in migrating birds, lipids are also the primary source of energy[101].

Fast muscles contract more quickly, generate more power and are involved in high-energy consumption activities such as jumping, running, flapping wings, flicking the tail or expelling water from a syphon like in cephalopod molluscs. Fast muscles use primarily glycogen and glucose to cover their high-energy requirements for producing short, powerful contractions.

18.5.3 Neural control of muscle type and activity

Slow muscles—those generally involved in activities that place a prolonged demand on them such as posture maintenance—are innervated by slowly firing motor neurons for prolonged periods. For example, in rats, individual slow-twitch (type I) muscle fibres[102] are stimulated at about 20 Hz for 5–8 h in a 24-h period, on average.

In contrast, fast muscle fibres that are used for bursts of rapid power movements are stimulated by fast firing motor neurons, on average, for only a very short time. For example, type IIB fast-twitch fibres in rats are stimulated at about 70–90 Hz for only 30 to 180 s in a 24-h period. Type IIA fast-twitch fibres in the rat, which are closest to type I fibres and are involved in sustained exercise, are stimulated at about 50–80 Hz, for only 20 to 70 mins in a 24-h period.

Thus, based on results obtained in rats, it appears that individual muscle fibres of different types in non-flying mammals are at rest for most of the time (an average of 66–80 per cent for type I, 94.8–98.5 per cent for type IIA and 99.8–99.9 per cent for type IIB type). This ensures that an adequate ATP pool can be maintained in each fibre, enabling high muscle power capacity to be produced at a moment's notice.

Flight muscles in flies, bats and birds are of the fast type to provide the high power needed for aerial locomotion; they generally use aerobic metabolism for sustained ATP production.

Skeletal muscle is a plastic tissue

Skeletal muscle is capable of marked changes in structure and physiological characteristics when the demand placed on it changes. This property, generally known as **muscle**

[98] We discuss in Section 2.4 metabolic pathways for ATP production from foodstuff.

[99] We discuss digestion and absorption of foodstuff in Section 2.3.3.

[100] We discuss ATP production from oxidation of fatty acids in Section 2.4.3.

[101] We discuss in Section 13.2 the proportion of lipids and carbohydrates utilization during exercise in birds and mammals.

[102] The major characteristics of mammalian muscle fibre types are shown in Table 18.2.

plasticity, is caused by changes in fibre size and fibre-type composition in response to changes in the requirements imposed on the muscle. Physical training, for example, increases the muscle mass, while muscle disuse causes it to rapidly decline.

We now know that the neural pattern of stimulation largely determines the type of muscle protein isoforms expressed in that fibre and consequently determines the contractile properties of the fibre concerned. Indeed, a change in the stimulation pattern of slow-twitch fibres from prolonged, low-frequency stimulation, to brief, high frequency stimulation induces slow-to-fast fibre transformation. Conversely, fast-to-slow fibre transformation occurs by applying prolonged, low-frequency stimulation resembling the firing pattern of slow motor neurons. Endurance training favours fast-to-slow fibre transformation while power training favours slow-to-fast fibre transformation.

An evolutionary change in the way wings are used in penguins compared with other birds (penguins use wings for swimming instead of flying as we discuss in Section 18.6.2) have placed different demands on penguins' wing muscles, and have led to a change in the fibre-type composition of specific muscles. For example, the anterior latissimus dorsi (ALD) muscle is normally used by birds to stabilize the folded wings onto the body when perching and roosting. In such a situation, this muscle consists almost entirely of slow-type fibres. In penguins, however, whose rigid wings do not fold up, the ALD muscle participates in wing-propelled locomotion and is composed predominantly of fast-type muscle fibres.

Motor coordination is based on reflex behaviour

Motor coordination is achieved through rapid, involuntary and stereotyped reactions of skeletal muscles to sensory stimulation. The anatomical basis for reflex behaviour involved in motor coordination is the **somatic reflex arc**, where sensory receptors and skeletal muscle are linked through the central nervous system (CNS). Sensory input pathways and the final common pathway of motor output are, in principle, the same from the most primitive invertebrate to the most complex vertebrate. The simplest somatic reflex arc involves only one single somatic sensory neuron[103] and one single motor neuron, which are connected via an excitatory chemical synapse within the CNS.

Figure 18.34 shows the knee-jerk reflex as an example of reflex behaviour and **reciprocal inhibition** involving skeletal muscles in vertebrates. Muscle reciprocal inhibition is a coordinated process of contraction and relaxation of antagonist muscles such that when muscles on one side of a joint contract, antagonist muscles on the other side of the joint relax. Such coordination allows the body to work effectively. Thus, the knee-jerk reflex is possible because the hamstring muscles are kept relaxed when their antagonists, the quadriceps muscles, contract to cause extension of the leg. In contrast, lifting the leg from the ground when walking involves the contraction of the hamstring muscles, which bend the knee to cause flexion

[103] We discuss the mechanisms of sensory transduction in Section 17.1.2.

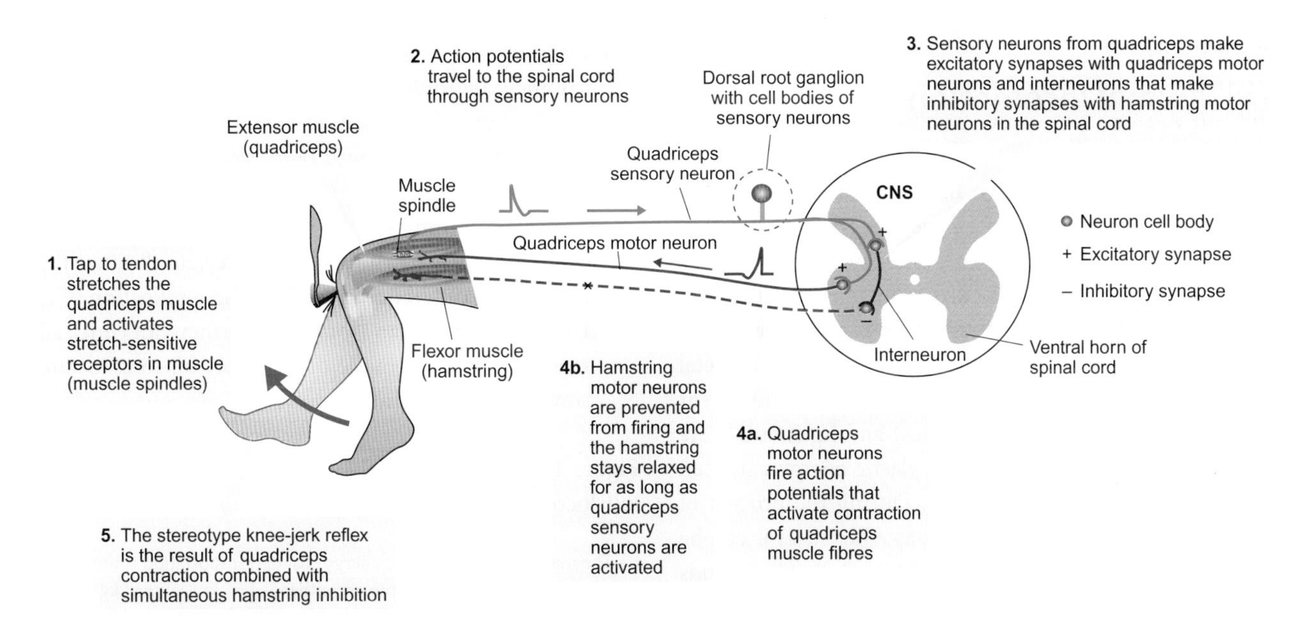

Figure 18.34 The knee-jerk reflex as example of reflex behaviour and reciprocal inhibition involving muscles

Flexor muscles bend the knee joint (the leg) and extensor muscles straighten the knee joint (the leg). In vertebrates the cell bodies of sensory neurons are located in the dorsal root ganglia and the sensory neurons synapse the motor neurons in the ventral horn of the spinal cord.

of the leg, and the simultaneous relaxation of the quadriceps muscles.

Reciprocal inhibition is a specific example of the more general principle of reciprocal innervations, which implies that any process that has an excitatory or an inhibitory influence on the motor neurons of a specific muscle also inhibits, or respectively stimulates, the motor neurons of its antagonist muscle. For example, the withdrawal reflex—when we touch a hot object with the hand—involves the involuntary action of the body to remove the hand from the object by contracting the biceps while relaxing the triceps muscle.

The presence of an interneuron between the sensory neuron and the motor neuron permits information about the reflex action to be transmitted to other levels in the CNS. Moreover, the presence of interneurons permits somatic reflexes to involve sensory receptors and muscle effectors on opposite sides of the spinal cord or at different levels up or down the cord. For example, lifting one leg off the ground causes reflex contractions of many muscles in the body on both sides of the spinal cord to maintain balance.

› Review articles

Biewener AA (2016). Locomotion as an emergent property of muscle contractile dynamics. The Journal of Experimental Biology 219: 285–294.

Rall JA (1985). Energetic aspects of skeletal muscle contraction: implications of fiber types. Exercise and Sport Sciences Reviews 13: 33–74.

Smith NP, Barclay CJ, Loiselle DS (2005). The efficiency of muscle contraction. Progress in Biophysics and Molecular Biology 88: 1–58.

Rome LC (2006). Design and function of superfast muscles: New insights into the physiology of skeletal muscle. Annual Reviews of Physiology 68: 193–221.

18.6 Animal locomotion

Animals generally move from one location to another in search of food and shelter, to escape predators, find mates, or extend their habitat. As such, animals employ diverse self-propelling mechanisms largely driven by muscular contraction. Muscles also play the key role in connecting the different elements of the skeleton and stabilizing joints. Skeletal muscles can also act as brakes and together with tendons (or tendon like-structures like apodemes) and ligaments store elastic energy when stretched. This elastic energy can then be used by animals in locomotion. During the process of evolution, each species has become as efficient as it can be in its environment for its size and shape such that it expends the least amount of energy to move a certain distance.

In this section, we explore how muscles help to mediate locomotion in a range of environments and how animals, particularly those that migrate, move as efficiently as possible.

18.6.1 Energy cost of active locomotion

An estimate of the energy cost of locomotion (often called **energetic cost of transport**, COT) is based on the amount of energy consumed (in joules) to transport one unit of body mass (g) over one unit of distance (km) when animals move at particular speed (km h^{-1}). We obtain the energy cost for a particular mode of locomotion (in J g^{-1} km^{-1})[104] from the value of metabolic rate associated with that particular mode of locomotion (in J g^{-1} h^{-1}) divided by the animal speed (in km h^{-1}):

$$\text{Energy cost (J g}^{-1}\text{km}^{-1}) = \frac{\text{Metabolic rate (J g}^{-1}\text{ h}^{-1})}{\text{speed (km h}^{-1})}$$

Equation 18.4

Figure 18.35 shows the relationships between minimum net energy cost and body mass for swimming, flying and running animals, from which we can draw several important conclusions:

- COT decreases as the mass of the animals increases, regardless of the mode of locomotion. The data points for each type of locomotion fit regression lines derived from the allometric equation[105] with a scaling factor around –0.3 for all three modes of locomotion, indicating that COT halves when the body mass increases by a factor of 10.
- Swimming is more economical than flying mainly because flight requires the generation of considerably more upward lift than swimming and this comes at a substantial extra cost; on average, animals of the same size consume about 2.7 times more energy flying over a certain distance than swimming. On the other hand, animals can fly much faster than they can swim.
- Running is less economical than flying over a given distance. In addition to overcoming gravity, each step involves the generation of a horizontal braking (decelerating) force followed by an accelerating force and uses considerable energy simply to maintain balance; on average, animals of the same size consume about 7.5 times more energy running over a certain distance than flying. This difference in COT is also a reflection of the fact that animals can fly faster than they can run, making flight an ideal mode of transport for long distance travel, such as migration.

Figure 18.35 also shows the measured energy cost of pre-kill locomotion in hunting pumas (*Puma concolor*) using

[104] The numerical value for the energy cost of transportation is the same when the energy cost is expressed either as J g^{-1} km^{-1} or as J kg^{-1} m^{-1} = N m kg^{-1} m^{-1} = N kg^{-1} = kg m s^{-2} kg^{-1} = m s^{-2}. Some researchers prefer to present the energy cost of transportation as a dimensionless quantity by dividing it by the gravitational acceleration (9.8 m s^{-2}).

[105] We discuss allometric equations: $y = a\,x^b$ in Box 2.3, where a = proportionality coefficient and b = scaling factor.

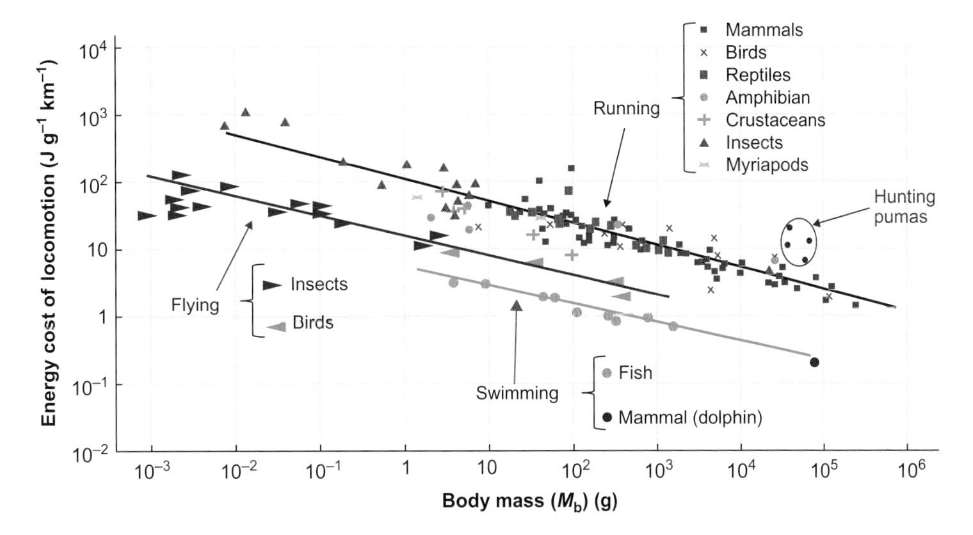

Figure 18.35 Minimum energy cost of locomotion (COT) for running, flying and swimming animals as a function of body mass (M_b) and the measured energy cost of hunting pumas (*Puma concolor*)

The data points are plotted on double logarithmic scales and were fitted by regression lines log (Energy cost) = log a + b log (M_b), where a is the energy cost at M_b = 1 g, and b is the scaling factor in the scaling equation Energy cost (J g^{-1} km^{-1}) = a (M_b)^b = 120 (M_h)$^{-0.32}$ for running; = 16 (M_b)$^{-0.30}$ for flying and 6 (M_b)$^{-0.32}$ for swimming. (Myriapods are arthropods related to insects, which have numerous paired jointed legs such as centipedes and millipedes.)

Sources: Data points: Alexander R McN (2005) Journal of Experimental Biology 208: 1645–1652; Schmidt-Nielsen K (1972). Science 177: 222–227; Williams TM et al (2014). Science 346: 81–85.

accelerometers[106]. The pre-kill hunting energy cost is surprisingly high (almost four-fold greater than the minimum COT for running pumas) but is in line with measurements for other mammal carnivores, including foxes, wild dogs and tigers. Recent measurements in the field show that pumas have evolutionarily adapted to limit the high energy cost of ensnaring prey by evolving a stalk-and-pounce tactic whereby the animal matches the force used for the pounce when killing the prey, with the mass of the prey. The high hunting energetic costs of mammal carnivores need to be considered in efforts to conserve these species as their habitats decline.

18.6.2 Locomotion in aquatic environments

Active forms of aquatic locomotion that involve muscular contraction include:

- swimming through water like fish, marine mammals and penguins,
- swimming in water while floating, like many aquatic birds,
- jet propulsion like jellyfish and cephalopods,
- walking on water without breaking the water surface tension like water striders, or
- walking on the bottom of aquatic environments (benthic locomotion) like crabs and echinoderms.

When swimming through water, animals encounter the force of gravity and water resistance in the form of **drag force**, which must be overcome if they are to propel themselves in a particular direction. Gravity can be overcome by increasing buoyancy, which is achieved by various mechanisms that require relatively little energy expenditure. For example, elasmobranchs and marine mammals have large deposits of fat or oil (which are less dense than water) in their body to increase buoyancy. Another energy efficient mechanism used by fish is to alter the gas pressure in the swim bladder to maintain position in the water column[107].

Many aquatic species generate upward lift by using their body and/or appendages as **hydrofoils**[108]. The principle of lift generation by hydrofoils is illustrated in Figure 18.36 using a hydrofoil in the shape of a dolphin's flipper[109]. The term **'lift'** has an exact meaning when discussing hydrodfoils; it refers to the force generated on the hydrofoil that is perpendicular on the direction of the water flow over the hydrofoil, as shown in Figure 18.36. However, the lift does not necessarily act in an upward vertical direction: for example, it can be anteriorly directed (with respect to the animal), as shown in Figure 18.37.

The magnitude of the drag is a complex function of shape, size and speed and depends critically on whether the

[106] We discuss the use of accelerometers for measuring energy expenditure in active animals in the field in Box 2.4.

[107] We discuss swim bladder function in fish in Case Study 13.1.

[108] A hydrofoil is a streamlined surface that generates lift when water flows over it; the hydrofoil has a rounded leading edge, a sharp trailing edge and curved upper and lower surfaces as shown in Figure 18.36.

[109] Dolphins use the flippers (pectoral fins) for directional control and the flukes (dorsal fin) for propulsion.

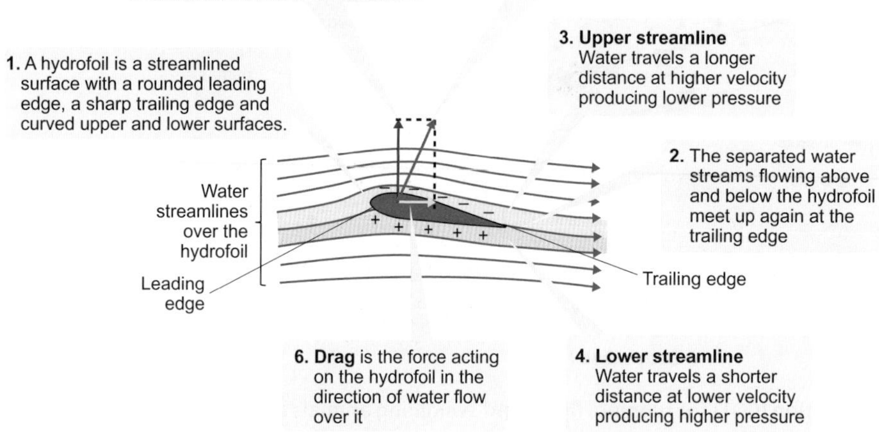

Figure 18.36 The general principle of lift generation in a hydrofoil that has the profile of a dolphin's flipper

The drag is in the direction of the water flow over the hydrofoil and the lift is perpendicular to the direction of water flow pointing towards the upper surface of the hydrofoil. The angle of the hydrofoil with respect to the direction of water flow is called the angle of attack.

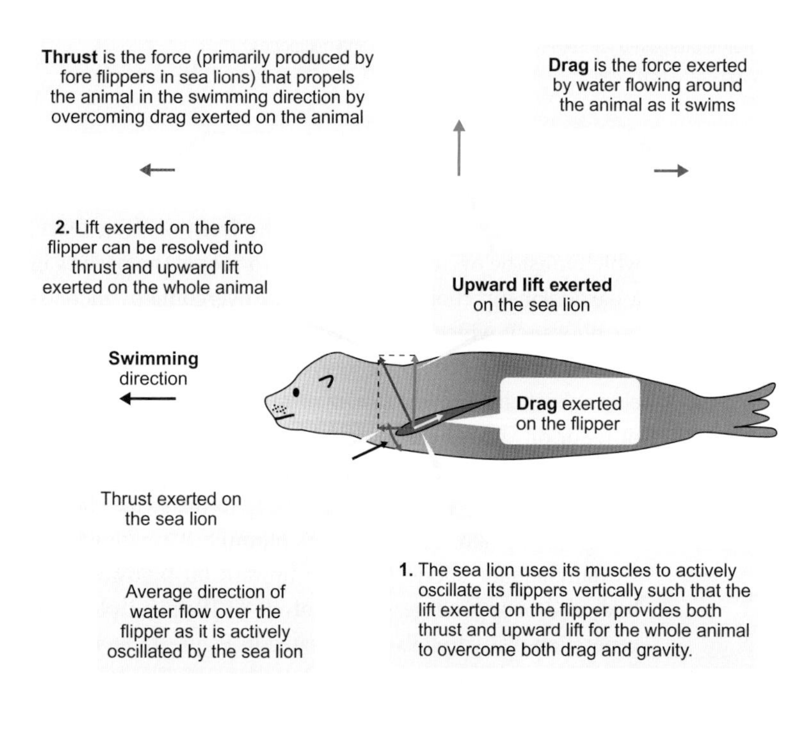

Figure 18.37 Principle of thrust generation in lift-based swimmers

The animal uses its hydrofoil-shaped appendages to generate a water flow over the hydrofoils such as to produce anteriorly directed lift that resolves into a vertical component (upward lift) to counteract gravity and a thrust component to counteract drag exerted on the whole animal.

object is streamlined. The shape of fish and marine animals evolved to become streamlined with a rounded leading edge and a slowly tapering tail, as shown in Figure 18.38A. This shape minimizes drag regardless of the type of movements animals make for generating the thrust needed for propulsion.

Mechanisms for thrust generation in aquatic locomotion

Most fish routinely swim by lateral movements of their bodies and/or tails, pushing water backward. This type of locomotion is called **body and/or caudal fin (BCF) swimming**. Other fish use their paired pectoral and pelvic fins for their routine mode of locomotion, called **median and/or paired fin (MPF) swimming**. A large proportion of typical BCF swimmers employ MPF swimming at lower speeds since MPF swimming offers greater manoeuvrability.

Depending on the movement characteristics of the propulsive structure, BCF and MPF swimming are further classified as **oscillatory swimming** if the propulsive surface swivels on its base without displaying a wave formation, or **undulatory swimming**, if a wave forms along the propulsive surface. BCF undulatory swimming is the most common form of locomotion in fish, whereby sequential activation of muscle fibres in paired myomeres (Figure 18.29) on either side of the body work

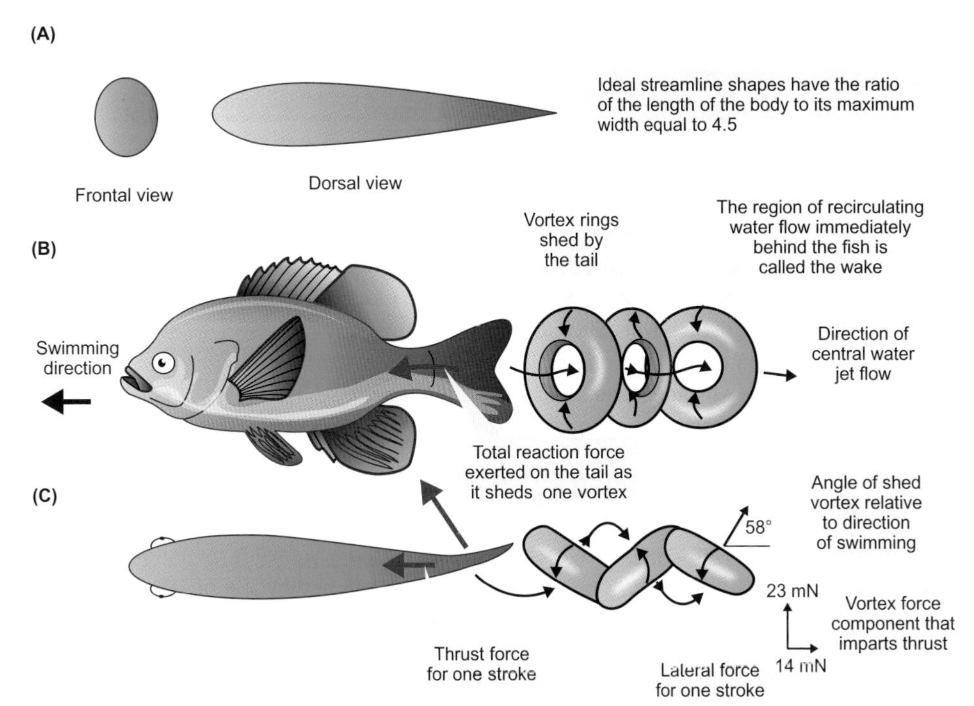

Figure 18.38 Swimming fishes shed vortex rings from their moving tail to generate thrust

(A) Streamline shape of fishes. (B) Side view of bluegill sunfish (*Lepomis macrochirus*) swimming with caudal fin (BCF swimming), which generates a chain of linked vortex rings in the wake. Each vortex has a central high-velocity jet flow shown by the black curved arrows. (C) Dorsal view showing the mean jet angle measured relative to the path of motion and the lateral and thrust forces determined per vortex ring shed by the tail. Note that the lateral forces exerted on the fish by vortices shed from the tail cancel out in a complete tail stroke.

Source: B, C: Lauder GV, Drucker EG (2002). Forces, Fishes, and Fluids: Hydrodynamic Mechanisms of Aquatic Locomotion. News in Physiological Sciences 17: 235–240.

antagonistically[110] to produce a rearward propagating undulating wave along the body that generates thrust.

The force generated by propulsive structures in swimming fish is assessed from the pattern of vortices[111] shed by that structure. Figures 18.38B and C show how thrust is generated by BCF-swimming bluegill sunfish (*Lepomis macrochirus*), which sheds water vortex rings from its oscillating tail. The precise pattern of vortices in the wake of the swimming fish depends on the species and the swimming speed.

Lift-powered swimming is produced by the active movement of hydrofoil-shaped appendices, which shed vortices and generate anteriorly directed lift. Many marine species, including marine mammals, sharks and tunas, have evolved hydrofoil-shaped appendages, which are used to produce thrust to overcome drag[112].

The principle of thrust generation from the active vertical oscillations of horizontal, hydrofoil-shaped structures is illustrated in Figure 18.37 using the sea lion flipper as an example. The oscillations force water to flow over the hydrofoil to produce an anteriorly directed lift with respect to the whole animal. The lift resolves into a thrust component, which counteracts drag exerted on the whole animal, and an upward lift, which counteracts gravity. Horizontal hydrofoils that oscillate vertically for generating thrust also occur in sea turtles and penguins, which use their wings as oscillating hydrofoils. Moreover, whales and dolphins use their powerful horizontal hydrofoil-shaped tails to produce both thrust and positive or negative lift by oscillating their tails vertically.

Lateral movements of vertical hydrofoil-shaped tails also generate anteriorily oriented lift, which resolves into a thrust component exerted on the animal, and a lateral component, similar to the diagram shown in Figure 18.38C. The most energetically efficient mode of aquatic locomotion is lift-powered swmimming with vertical hydrofoil-shaped tails hinged to rigid bodies. This type of BCF swimming is called **thunniform swimming** after the name of one of five genera (*Thunnus*) of fish that are collectively known as tunas. Thunniform swimming involves very little bending of the body and thrust is generated by the undulatory movement of the hydrofoil-shaped vertical tail. The vortices shed by the tail produce an almost linear water current flow in the wake, which generates the thrust that propels the fish forward. The Pacific bluefin tuna (*Thunnus orientalis*) can reach swimming speeds of up 100 km h^{-1} when chasing prey.

The amazing ability of tunas to accelerate through water to such high speeds is because they are continuous swimmers and their body retains heat; this allows their muscles to operate at significantly higher temperatures than that dictated by the ambient environment[113]. In tunas, the red muscle, which powers its movements during migration and steady movement, is located deeper within the body where the temperature is up to 10 °C higher than in the ocean. In Case Study 18.1, we consider the energetic advantages of thunniform swimming. Tunas can bring their fins close to the body to reduce drag still further when chasing prey.

[110] This is another example of reciprocal inhibition as noted above.

[111] A vortex (plural vortices) is a quickly rotating mass of fluid (water or air) like a whirlpool into which fluid (or other objects) may be drawn, as illustrated in Figure 18.38B. Vortices in the wake of swimming fish are visualized by using laser light reflected by small shiny particles added to the water in which the fish swim.

[112] This is an example of convergent evolution, as we discuss in Section 1.5.1.

[113] We discuss regional endothermy in ectotherms in Section 10.3.1

Case Study 18.1 Energy conservation and conversion technologies inspired by aquatic vertebrate locomotion

As animals move through water they have to overcome drag and generate lift, as we discuss in Section 18.6.2. Natural selection has acted on the shape of aquatic animals to minimize drag and maximize lift. Studies of animal hydrodynamics can therefore teach us a great deal about optimal designs for reducing drag, and enhancing lift and manoeuverability, as well as conserving energy and converting mechanical energy into other forms of energy. In this panel, we explore three examples of biomimetic design inspired by the locomotion of aquatic vertebrates.

Penguin and sea turtle-inspired propulsive mechanism

The propulsive mechanism of penguins and sea turtles—rigid bodies propelled by oscillating flippers—has inspired potential propulsion mechanisms for boats and autonomous underwater vehicles (AUVs). The propellers currently used on boats operate with efficiencies of less than 50 per cent to a maximum of 70 per cent. However, engineers at the Massachusetts Institute of Technology (MIT) have developed a prototype propeller with an efficiency of 87–90 per cent. Called 'Proteus, the penguin boat', the propeller features flipper-like propulsors at the rear whose oscillating hydrofoils produce less turbulence than conventional rotating propellers. The concept of oscillating propulsors has yet to be applied to full-sized boats, but has been used in a variety of AUVs.

One MIT graduate student studied the swimming mechanisms of a green sea turtle at the New England Aquarium (Figure A(i)) and designed the 'Finnegan' AUV (Figure A(ii)), which has four independent oscillating propulsors designed to mimic the shape and propulsory movement pathway of turtle flippers.

AUVs driven by oscillating hydrofoils are more energy efficient and far more manoeuverable than conventional torpedo-like propeller-driven devices, making them perfectly suited to working in relatively unsteady conditions and obstacle-ridden underwater environments, such as around coral reefs. Oscillating hydrofoil propulsion AUVs are now used in a variety of underwater exploration, research, mapping and monitoring tasks.

Power generation inspired by thunniform swimming

Large fast-swimming fish, such as tuna, mackerel and sharks, use the 'thunniform' mode of swimming, in which their large crescent-shaped tail fins generate powerful thrust by oscillating from side to side, while the rest of their streamlined body remains quite rigid, as we discuss in Section 18.6.2. The design principle of tail fin propulsion has been applied in electricity generation, as shown in Figure B.

Instead of an animal moving its fin against the water for propulsion, a water current oscillates the tail-fin-like paddle to generate electricity. The shark tail-like oscillating hydrofoil shown in Figure B can be placed in current speeds of 2.5 ms^{-1} or greater. Assemblages of many such paddles to harness tidal energy are being developed to produce 250 kW to 1000 kW of electrical power. The oscillating hydrofoil is purported to be of lower construction and maintenance cost than other tidal energy systems, less susceptible to severe weather damage, and is of course more environmentally benign than energy generation that relies on fossil fuels.

Tubercle technology inspired by leading edge tubercles on humpback whale flippers

Humpback whales use the technique of 'bubble netting' for feeding, in which the whale swims in tight upward spirals around a school of small fish or krill, releasing bubbles that effectively

Figure A Propulsion mechanisms with oscillating flippers

(i) Swimming green sea turtle (*Chelonia mydas*); (ii) 'Finnegan' the autonomous underwater vehicle swimming in a pool. The Finnegan is 2 m long, 0.6 wide and 0.5 m in height; it weighs about 200 kg out of water. The four independent flapping foils can propel the AUV at a speed of up to 1.2 ms^{-1} under water, keep it stationary and give the AUV a high level of manoeuverability with low turning radii and high turning speeds.

Source: (i) Alejandro Fallabrino, (ii) MIT/Woods Hole.

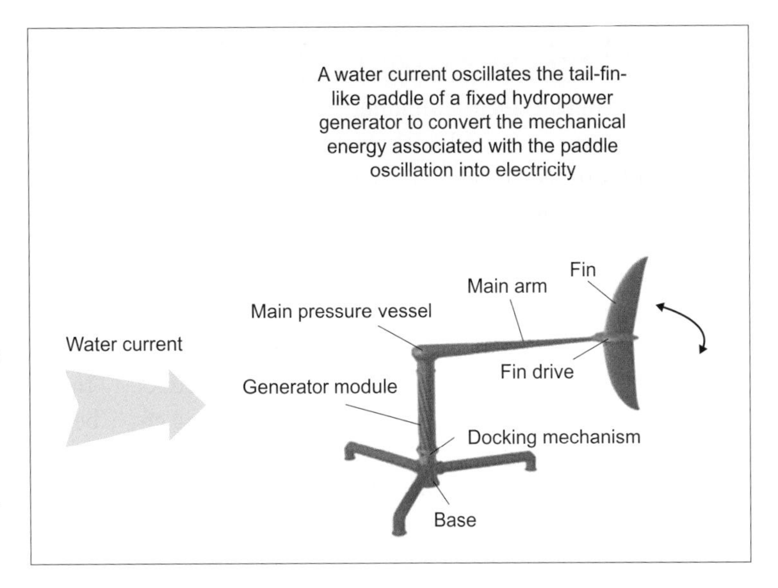

Figure B The BioStream power generator based on the hydrofoil-shaped tail of a shark

Water flow over the vertical (double) hydrofoil-shaped paddle generates lateral lift as we discuss in Section 18.6.2. The lateral lift moves the paddle sideways in one direction. As the angle between the paddle and the direction of the water flow changes, the direction of the lift also changes, causing the paddle to swing to the opposite side whereupon the lift again changes direction, causing the paddle to oscillate from side to side in the water current.

Source: Kloos G et al (2009). The bioSTREAM tidal current energy converter. Proceedings of the 8th European Wave and Tidal Energy Conference, Uppsala, Sweden, 2009.

concentrate and corral the prey. The whale then lunges with its mouth open to scoop up dinner.

It appears that the remarkable maneuverability of such a large marine animal to swim in tight upward spirals is related to the presence of bumps, called tubercles, on the leading edge of their flippers, as shown in Figure C(i). The bumpy shape of tubercles could be considered a surprise given the usual need for airfoils or hydrofoils to be smooth to reduce drag. However, tubercles on models of whale flippers were experimentally found to channel the water flow, enhance lift and reduce drag, thus contributing to the whale's ability to avoid stalling during the tight upward spiralling turns of bubble netting. Indeed, research has demonstrated that the tubercles provide an 8 per cent increase in lift and 40 per cent increase in angle of attack without stall, as well as providing a 30 per cent reduction in drag.

'Tubercle technology' has been applied to the leading edge of large wind turbines and small fans, as shown in Figure C(ii) and C(iii), improving their energy efficiency, reducing the noise and strain on the equipment, and significantly increasing the range of wind speeds over which power can be generated, even in gusty conditions.

› Find out more

Fish FE, Lauder GV (2006). Passive and active flow control by swimming fishes and mammals. Annual Review of Fluid Mechanics 38: 193–224.

Fish FE, Weber PW, Murray MM, Howle LE (2011). The tubercles on humpback whales' flippers: Application of Bio-Inspired Technology. Integrative and Comparative Biology 51: 203–213.

Kloos G, Gonzalez CA, Finnigan TD (2009). The bioSTREAM tidal current energy converter. Proceedings of the 8th European Wave and Tidal Energy Conference, Uppsala, Sweden, 2009 http://www.homepages.ed.ac.uk/shs/Wave%20Energy/EWTEC%202009/EWTEC%202009%20%28D%29/papers/235.pdf

Stanway J (2008). The turtle and the robot. Oceanus Magazine 47: 22–24.

Triantafyllou MF, Techet AH, Hover FS (2004). Review of experimental work in biomimetic foils. IEEE Journal of Oceanic Engineering 29: 585–594.

Wiley D, Ware C, Bocconcelli A, Cholewiak D, Friedlaender A, Thompson M, Weinrich M (2011). Underwater components of humpback whale bubble-net feeding behaviour. Behaviour 148: 575–602.

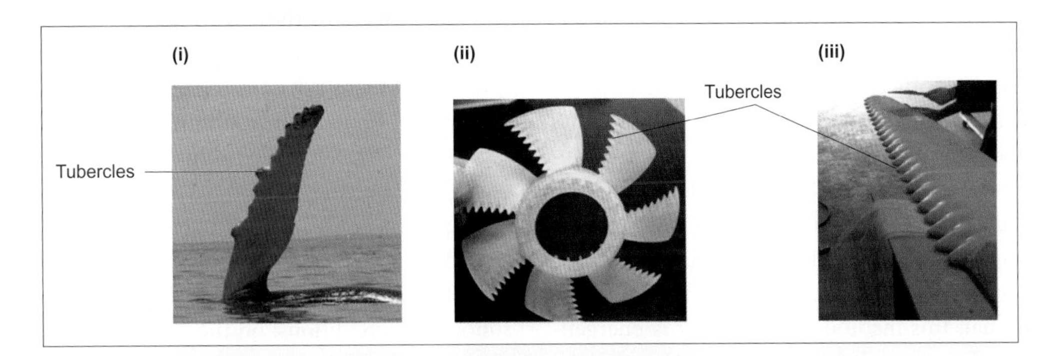

Figure C Examples of 'tubercle technology' inspired by the presence of tubercles on whale flippers

(i) Humpback whale (*Megaptera novaeangliae*) flipper with leading edge tubercles. (ii) Whale power prototype fan for cooling computers. (iii) Prototype wind turbine blade with 'tubercle technology'.

Source: (i) Fish FE and Lauder GV (2006). Passive and active flow control by swimming fishes and mammals. Annual Review of Fluid Mechanics 38: 193-224. (ii) and (iii) from Whalepower corporation; http://www.whalepowercorporation.com

Sharks belonging to the Lamnidae family, such as the great white and salmon sharks, are also thunniform swimmers and maintain their swimming muscles at higher temperatures than in the ambient environment in a manner similar to tunas.

Drag-based swimming is achieved by paddling and is used by terrestrial and semiaquatic mammals, most aquatic birds and aquatic insects. Drag-based swimming involves cyclic motions of limbs or other appendages during which water is pushed back in a power stroke after which the limbs return to their original position in a recovery stroke. Reducing drag on the recovery stroke is critical for improving efficiency.

A common adaptation in drag-based aquatic bird swimmers such as sea gulls, pelicans, ducks and geese is the presence of webbed feet. The webbing spreads out during the power stroke, increasing thrust, and is pulled together during the recovery stroke, reducing drag. Many aquatic beetles (Coleoptera) and bugs (Hemiptera) also use their middle and/or hind legs as oars for swimming or diving. These legs are usually flattened or have hairs on them to improve the efficiency of the power stroke. During the recovery stroke, the legs twist slightly and the hairs bend onto the leg, reducing drag and improving performance.

Lift-powered swimming is more energetically efficient than either undulatory- or drag-based swimming. Thus, thrust power is generally >80 per cent of total mechanical power produced in lift-based swimming compared with 50–80 per cent in undulatory swimming and only about 30 per cent in paddle-based swimming. Lift-based swimmers like whales, some turtles, sharks and tunas can cover long distances in the order of thousands of kilometres migrating between feeding and breeding grounds. Lift-based swimmers are generally also very good divers, diving to depths of up to several hundred meters[114] and can maintain fast pursuit of prey for feeding. Drag-based swimmers like aquatic birds, on the other hand, live at the interface between water, air and land, are buoyant and spend little energy just floating when they feed.

Jet propulsion is another method of aquatic locomotion, which is used by different groups of invertebrates such as jellyfish, cephalopod molluscs, some bivalve molluscs, tunicates and some larvae of aquatic insects. Jet propulsion involves squirting out water from a muscular cavity, which propels the animal in the opposite direction. Since filling up the cavity with water increases the body mass and the drag of the animal, this method of swimming is energetically a rather inefficient mode of locomotion. However, jet propulsion is a very efficient escape mechanism, as employed by cephalopod molluscs.

[114] We discuss diving in mammals and birds in Section 15.3.3.

A few aquatic insects, such as water striders, can skate on the water surface. This is made possible by the presence of hydrophobic hairs on the tips of their feet, which prevent the legs from breaking the surface tension of the water. Benthic dwellers such as crabs and echinoderms, on the other hand, walk on the substrate bottom of aquatic environments.

Behavioural strategies for energy conservation in aquatic locomotion

The optimal energetic cost for achieving a certain speed or behavioural response in fish depends on the different pattern of recruitment of the red and white muscle fibres. At the steady pace of swimming used during foraging and migration, only the slow, low-energy-consuming red muscle fibres contract in fish. The red fibres operate aerobically using lipid-based fuels at their optimum length for force production (Figure 18.39A), and at velocities of 20–40 per cent of their maximal velocity of shortening, where power is optimal. For example, Figure 18.39C shows that the brown trout (*Salmo trutta*) uses only red fibres (muscle) at a speed of 0.5 × body length s^{-1} in clean water at pH 7.0 and that it recruits also fast, high-energy-consuming white fibres at a speed of 1.0 × body length s^{-1}, when higher power is needed to overcome the markedly increased drag[115]. This strategy helps conserve energy by recruiting high-energy-consuming white muscle only when increased power is needed.

Also, when a fish is startled, a powerful escape reflex response is triggered. This response consists of two large sequential contractions. In the first contraction, all fibres in the myomeres on one side of the fish—both white and red—contract, while the fibres in the myomeres on the other side relax, causing the fish's body to bend in the form of letter C. (It is therefore known as a C-start.) This is followed by the powerful contraction of the tail in the opposite direction of the first contraction, which suddenly and unpredictably changes the direction of swimming and helps the fish avoid sudden dangers in the environment. This reflex response, in which all trunk muscle fibres are recruited for maximal power, also helps fish conserve energy because trying to escape predation by simply swimming at full power would be less energy efficient than making one sudden, unpredictable change of direction.

As shown in Figure 18.39B, in the trout escape response, the white fibres operate at their optimum length for force, and at peak power. The red fibres, however, operate under suboptimal conditions: on the ascending limb of the force-sarcomere length curve, and at suboptimal power. Interestingly, if the water in which brown trout swim is made acidic (pH 4.0), which is often the case with polluted waters, then

[115] The drag force is proportional to speed cubed.

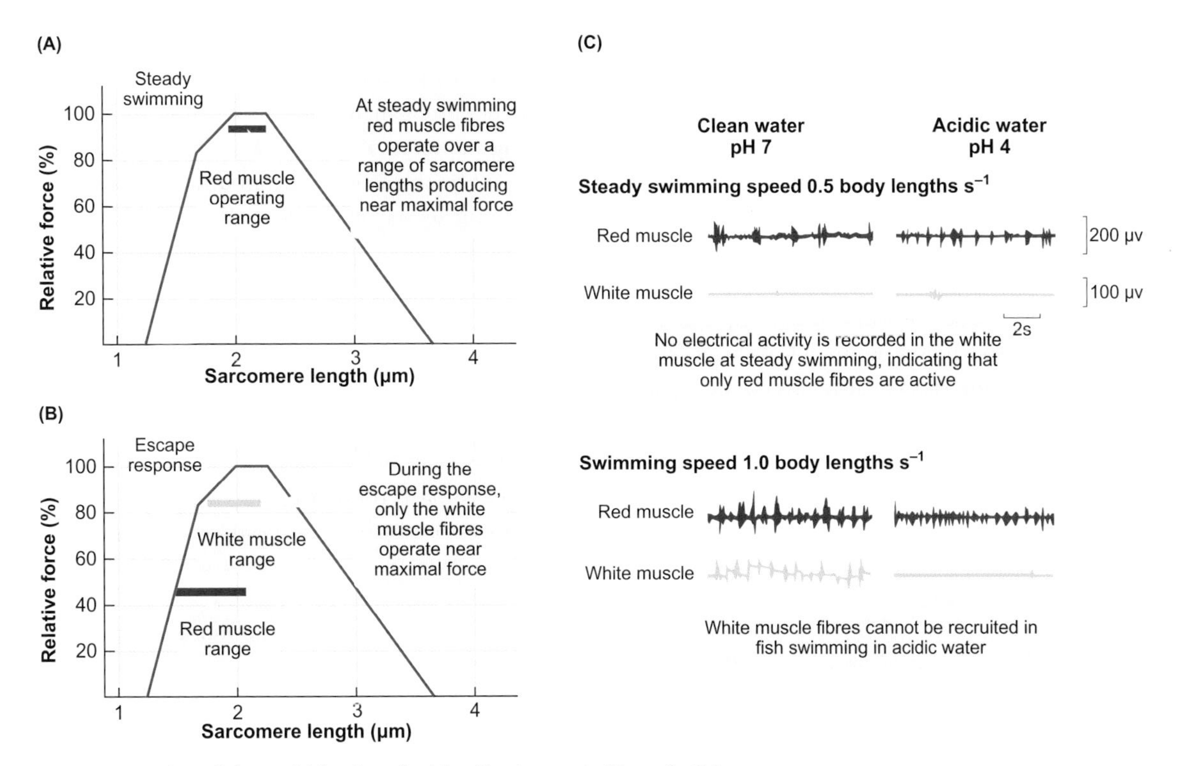

Figure 18.39 Properties of the red (slow) and white (fast) muscle fibres in fish

(A) The red muscle fibres in the carp (*Cyprinus carpio*) myomeres operate at near optimum sarcomere length for force production during steady swimming. (B) The white muscle fibres in the carp myomeres function near the optimum range for force production during the escape response. This is not the case for red fibres. (C) Electrical activity in red and white muscle of the trout (*Salmo trutta*) swimming at two speeds in water at 5°C in winter at neutral pH 7 and sublethal acidic pH 4. At a speed of 1 body length s^{-1}, there is no activity in the white muscle of the fish exposed to pH 4 and this fish would not swim at a higher speed, whereas the fish in water at pH 7 swam at over 2 body lengths s^{-1}.

Sources: A, B: Rome LC, Sosnicki AA (1991). Myofilament overlap in swimming carp II. Sarcomere length changes during swimming. American Journal of Physiology (Cell Physiology) 260: C289–C296. C: Day N, Butler PJ (1996). Environmental acidity and white muscle recruitment during swimming in the brown trout (*Salmo trutta*). Journal of Experimental Biology 199: 1947–1959.

the fish cannot swim much faster than 1.0 × body length s^{-1}, because white fibres cannot be recruited. Separate studies have suggested that the acidic environment causes chronic depolarization of the white muscle fibres, which are unable to respond to neural stimulation. Thus, polluted waters of acidic pH would impair the escape response, making the trout more vulnerable to predation.

A significant reduction in energetic cost of transportation (COT) has been reported for **fish schooling**, where large numbers of fish of the same species synchronize their swimming, moving together at the same speed, in the same direction, in a highly regular formation pattern. An important benefit of fish schooling is a significant reduction in COT by up to 20 per cent for individuals that take advantage of vortices produced by the fish swimming in front of them. Interestingly, it now appears that even individual fish swimming at the front of the school benefit from an approximate 10 per cent reduction in COT, possibly because the fish swimming behind exert some pressure on water ahead of them. This is similar to the pressure exerted by moving boats and ships, which is used by dolphins for bow riding.

About 50 per cent of all species of fish school at some stage in their life. Highly migratory fish such as tunas and herrings (*Clupea* spp.) maintain a very precise arrangement when schooling, which allows them to maintain a constant cruising speed during migration.

Fish and other aquatic animals such as marine mammals[116] can reduce their COT by alternating short bursts of active swimming with long passive coasts or glides over relatively long distances. During the coast phase, the body is kept rigid to minimize drag and the muscular effort is minimal. Many fish use this behavioural strategy, which can reduce COT by up to 50 per cent or even more, particularly in larger fish that can accelerate to a higher speed during the burst phase. Fish that have negative buoyancy, like mackerel tuna (*Euthynnus affinis*), use one variant of this type of behaviour. These fish swim actively toward the water surface and then glide downward, conserving energy. Mackerel tuna can save about 20 per cent of the energy needed to cover the same horizontal distance using this strategy.

[116] We discuss this strategy in Section 15.2.4.

18

Swordfish (*Xiphias gladius*), reach speeds in excess of 100 km h^{-1}. Recent studies identified an oil gland near the base of their sword-like upper beak (rostrum) which secretes a mixture of highly hydrophobic compounds that distribute over the front part of their head and forms an oil layer. The oil layer reduces friction drag and increases the swordfish's swimming efficiency, making them possibly the fastest swimmers on Earth.

Flying fish, penguins and the fastest swimming mammals, such as dolphins and porpoises, jump out of the water and move through the air for various distances. Under specific circumstances, this mode of locomotion can be energetically more efficient than moving through water for the same horizontal distance because the drag is negligibly small when moving through air relative to water. 'Porpoising' refers to the mode of locomotion often employed by porpoises and dolphins, which consists of sequential numbers of leaps above the water surface. Flying fish (64 species in the family Exocoetidae) can glide for 50 m or more through the air using rapid movements of the tail to exit the water and then stretching lateral fins to provide some lift while gliding through the air. This adaptation is also a very effective way to escape predators.

Many aquatic animals also use passive transportation as a supplementary or sole mode of locomotion to conserve energy. Some turtles such as the leatherback sea turtle[117] (*Dermochelys coriacea*) and the green sea turtle[118] (*Chelonia mydas*) migrate for up to 10,000 km between spawning, feeding and nursery grounds as they normally feed in colder waters and lay eggs on tropical and subtropical beaches. Even though they are good lift-based swimmers, they also use ocean currents to conserve energy when travelling such large distances.

Other animals such as the Portuguese man o'war (*Physalia physalis*), a cnidarian, lives at the surface of the sea and moves by a combination of sea currents, tides and winds. It lacks any mechanism for propulsion but controls the level of inflation of its gas-filled bladder, which acts as a sail to increase or decrease the influence of winds on its movements. When the bladder is rapidly deflated, the animal sinks—a mechanism used to avoid predation. Barnacles also employ passive locomotion by attaching to the hull of boats and other floating objects and being transported with them. Some fish, such as remoras or suckerfish (*Echeneididae*), can swim freely in water but more often attach onto other aquatic animals such as sharks, whales, turtles and manta rays, and are transported by the host.

[117] Leatherback sea turtles are the largest living turtles, weighing between 250 and 700 kg as adults.

[118] The green sea turtle takes its name from the green colour of its fat; adults weigh between 70 and 400 kg.

18.6.3 Aerial locomotion

The major problem for animals travelling through air is overcoming the force of gravity. Animals using aerial locomotion must be able to leave the ground and generate sufficient upward lift to stay airborne[119]. The use of wings to power flight occurs in only three extant groups of animals: insects, birds and bats (mammals). Similar to lift-based swimming, powered flight is based on the use of wings in the shape of **aerofoils** to provide both uplift and thrust to overcome gravity and drag, respectively. Aerofoils have similar shape to hydrofoils and function on the same principles as outlined in Figures 18.36 and 18.37. Since air is less dense than water, air must move over the wing's surface at much faster speeds than water over hydrofoils to generate the necessary uplift.

Insects were the first group of animals to evolve wings and flight ability about 400 million years ago (Ma) as plants grew taller. Early in their evolution, insects would walk to the top of a plant and then jump off. Eventually they evolved stiff lateral extensions on the thoracic segments, which helped them to direct their aerial gliding descent from the taller plants. These lateral extensions became articulated and mobile appendages later in their evolution; in due course—helped by natural selection—they became wings. The wings of insects are reinforced by riblike structures called veins and are shaped in the form of aerofoils, to provide lift and thrust when moving through the air.

These wings were powered by two pairs of antagonistic skeletal muscles connected directly to their bases; the wings themselves were hinged in such a way that contraction of one pair of muscles made the wings move up and contraction of the other pair moved the wings down in a uniform and regular fashion. Insects with flapping wings could then exploit the growing diversity of taller plants.

This mode of powering wings by attaching muscles directly to them is called a **direct flight mechanism** and still occurs in two living insect orders: Ephemenoptera (mayflies) and Odonata (dragonflies and damselflies). In these insects the flight muscles contract and then relax in synchrony with each nerve impulse, at 5 to 200 times per second (Hz). They are therefore called **synchronous flight muscles**. The motor neurons that innervate the upstroke and the downstroke muscles are activated in antiphase by central pattern generators within the nervous system[120].

All other living species of flying insects use an **indirect flight mechanism**, which evolved later and enabled insects to flap their wings at rates of up to 1000 strokes per second—rates much greater than those at which muscles can contract

[119] Jumping is not considered a form of aerial locomotion because animals do not create lift to stay airborne when jumping.

[120] We discuss central pattern generators in Section 16.5.4.

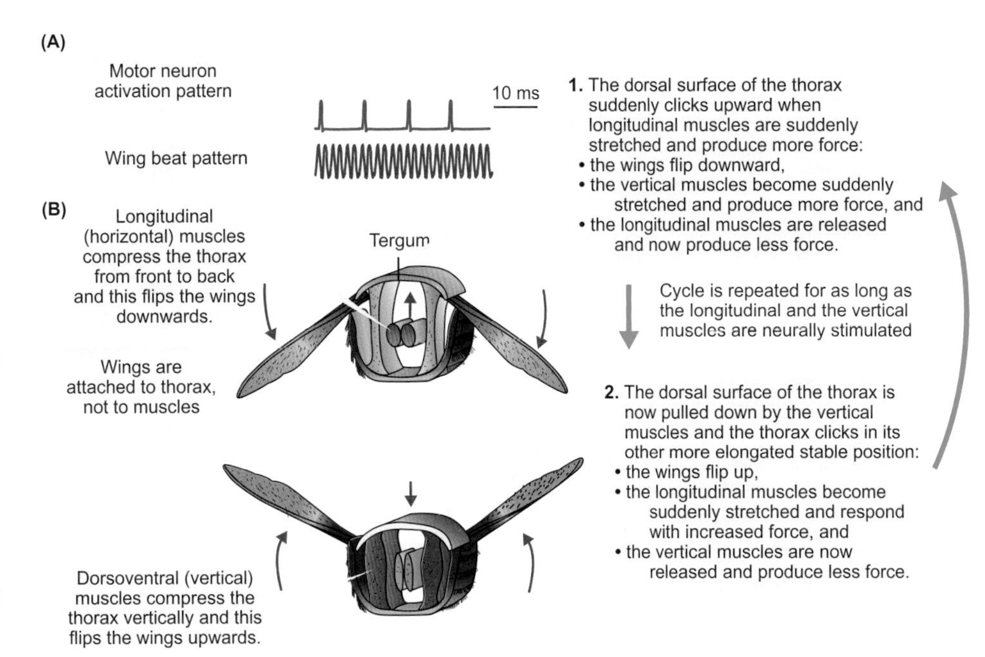

Figure 18.40 Mechanism of indirect flight muscle activation in insects

(A) The frequency of nerve stimulation is much slower than the wing beat frequency. (B) The elastic thorax together with the two sets of muscles act as a mechanical oscillator moving the wings up and down at the resonant frequency of the mechanical oscillator for as long as the muscles are neurally stimulated. Note that force produced by a stimulated muscle rises when the muscle is slightly, but rapidly stretched and drops when the muscle is only slightly, but rapidly released.

and relax in response to nerve stimulation. As shown in Figure 18.40, the flight muscles in this case are not directly connected to the wings, but attach to an elastic thorax that flicks between two configurations at considerably higher rates than the rate of nerve stimulation. Therefore, these muscles are called **asynchronous (or fibillar) muscles**. Tiny muscles in the thorax can control the tilt, stiffness and the flapping frequency of the wings.

Most insects fly by rapidly flipping their wings through two basic strokes. During the downstroke, the dorsal surface of the wing faces up and the leading edge of the aerofoil (wing) is pushed downward and forward, generating lift. The wing then flips, with the leading edge now pointing backward and the wing is pushed upward and backward during the upstroke, again generating lift. The wing then flips over to its initial position (with the leading edge pointing forward) such that another downstroke can occur. This movement of the wing generates a rotational lift like a propeller. Furthermore, after reversing direction, the wing collides with vortices shed in the previous stroke and this can produce additional lift. Thrust is generated by a faster upstroke than downstroke and hovering occurs when the two strokes are equally fast. The mechanical efficiency (the ratio of mechanical power output to metabolic power input) of the insect flight muscle is, however, relatively low: generally around 10 per cent.

Birds evolved around 150 Ma in the late Jurassic, probably from a group of small dinosaurs called **theropod dinosaurs**. Bird adaptations for flight include the presence of a large keel to support a powerful flight musculature for flapping the aerofoil-shaped wings, a streamlined body to reduce drag, a hollow skeleton and a lightweight beak to reduce weight. However, the unidirectional pulmonary system that evolved in birds may not confer any particular advantage (as previously thought) compared with the mammalian lung with respect to oxygen extraction from the environment[121].

The downstroke flap of the wing in birds forces air to flow fast over the aerofoil-shaped surface of the wing (shown in Figure 18.41B) usually producing an **anteriorly directed lift** illustrated back in Figure 18.37. This lift can be resolved into an upward lift component, which counteracts gravity, and a thrust component, which overcomes drag and/or increases the speed of the animal as it moves through the air. The upstroke movement of the wing can also provide upward lift by changing the angle of the wing as it moves through the air. For example, hovering hummingbirds adopt a vertical position so the wings tend to move in a more anterior/posterior direction, with the wing becoming partially inverted during the upstroke, resembling the movement of the insect wing. The downstroke and upstroke flaps of the wing in hovering hummingbirds produce about 75 per cent and 25 per cent, respectively, of the upward lift necessary to support the bird's weight.

All birds vigorously flap their wings at take-off to gain height and at certain times during flight to produce lift and increase thrust for higher speed. As shown in Figure 18.41A, the two main wing muscles in birds are the **pectoralis** (70–80 per cent of flight musculature), which shortens and powers the downstroke flap of the wing to produce thrust and positive uplift, and the **supracoracoideus** (10–30 per cent of flight musculature), which shortens to raise the wing[122]. Both muscles generally consist of fast-twitch muscle fibres and can

[121] We discuss gas exchange in birds in Section 12.3.3.

[122] The coracoideus muscle is relatively large in birds that produce thrust also during the upstroke of the wing such as hummingbirds and penguins.

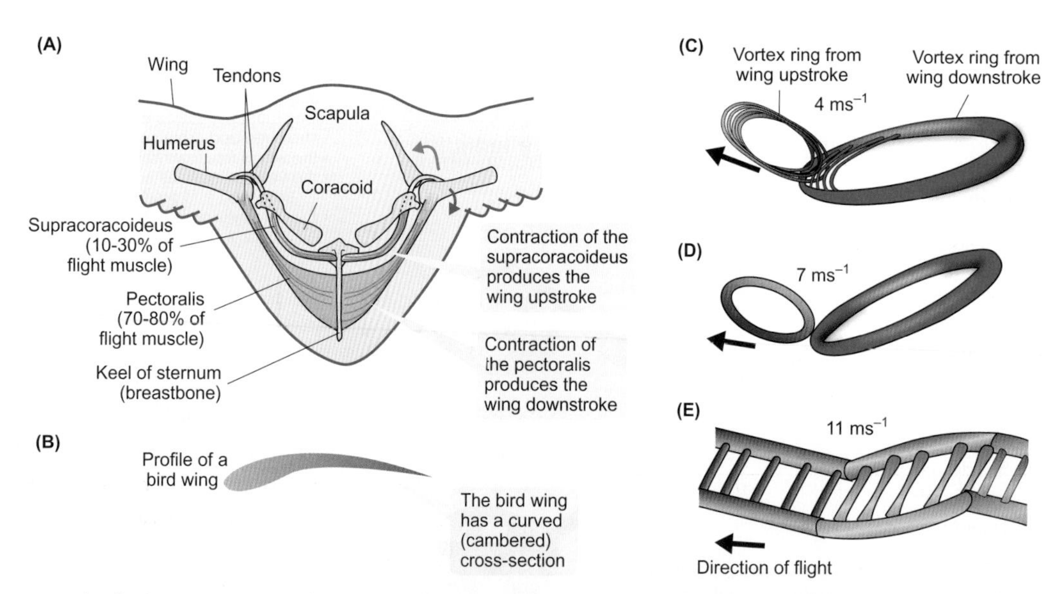

Figure 18.41 The flight apparatus and patterns of vortex wakes shed from the wings of flying birds at different speeds

(A) General plan of the flight apparatus in birds seen from the front. (B) The aerofoil shape approximating a bird wing. (C–E) Pattern of vortex wakes from a thrush nightingale *Luscinia luscinia* at (C) slow (4 m s^{-1}), (D) medium (7 m s^{-1}) and (E) fast (11 m s^{-1}) flight speeds. (C–E) indicates the vortex rings from the wing upstroke coloured violet and the vortex rings from the wing downstroke coloured blue. The bird produces upward lift both during the downstroke and the upstroke at each speed. At lower flight speeds (C, D) the wings shed closed-loop, discrete vortices associated with the wing upstroke and downstroke. As the speed increases, a constant-circulation, continuous vortex develops with primary structures oriented mainly in the direction of the air stream. The wakes are scaled to occupy approximately the same length on the page. Without scaling the length of the wake pattern at high speed is almost three times that at low speed.

Source: (C), (D), (E): Spedding GR, Rosén M, Hedenström A (2003). Journal of Experimental Biology 206: 2313–2344.

deliver a high mass-specific power output of up to 400 W kg^{-1}. As the pectoralis muscle is generally several fold larger than the supracoracoideus, the overall power output by the pectoralis is much greater than that of supracoracoideus.

Birds adapted for rapid take-off followed by short flights to escape predation, such as the junglefowl and the domestic chicken, have mainly fast-twitch glycolytic (FG) muscle fibres in their flight muscles, which function anaerobically and fatigue quickly. In contrast, birds such as ducks, geese, swans and pigeons, which can sustain prolonged flights over relatively long distances, have predominantly fatigue-resistant fast-twitch oxidative (FOG) fibres interspersed with FG fibres in their flight musculature. In order to produce the high power output needed for flight, both the pectoralis and the supracoracoideus muscles are stretched by 20–30 per cent beyond their resting length before they start contracting, and rapidly shorten with velocities between 4–10 muscle lengths s^{-1} to about 10 per cent below their resting length when they become relaxed. The corresponding change in sarcomere length during the power production phase is between about 2.5 and 1.8 μm. Over this sarcomere range, muscles develop close to maximal (above 85 per cent) force and power. (The force-sarcomere length diagram shown back in Figure 18.39B for white fibres in fish also applies for bird flight muscle.) The mechanical efficiency of the flight muscle in birds is between 13 and 23 per cent, depending of the species and flight speed. Generally, the mechanical efficiency is greater in larger birds and reaches the highest values during migratory climbing flights.

Figure 18.41C, D, E shows the pattern of vortices shed from the wings of a thrush nightingale (*Luscinia luscinia)* at different flight speeds. Notice that at lower speeds, the vortex rings associated with the wing downstroke are larger than the vortex rings associated with the wing upstroke indicating the difference in power between the two strokes. At higher flight speed, where the mechanical efficiency is reduced, the vortex structure is more complex, with cross-stream vortices throughout the wingbeat cycle.

The power required for flight typically varies with the flight speed according to the U-shaped curve shown in Figure 18.42. Greater power is needed at zero speed, when birds are hovering; lower power is required at intermediate flight speeds; and greater power is needed at higher speeds. Birds can regulate the flight power as needed by altering the fraction of motor units that are recruited in the flight musculature, by varying the shortening distance of the muscle fibres or by modifying the timing and duration of muscle fibre activation.

The speed where the power required for flight is lowest is called the **minimum power speed** or V_{mp}. Since the amount of energy used for a certain duration of flight is lowest at this speed, birds would be expected to fly at V_{mp} when foraging. The maximum distance covered for a given amount of energy is achieved at a slightly higher speed, called **maximum**

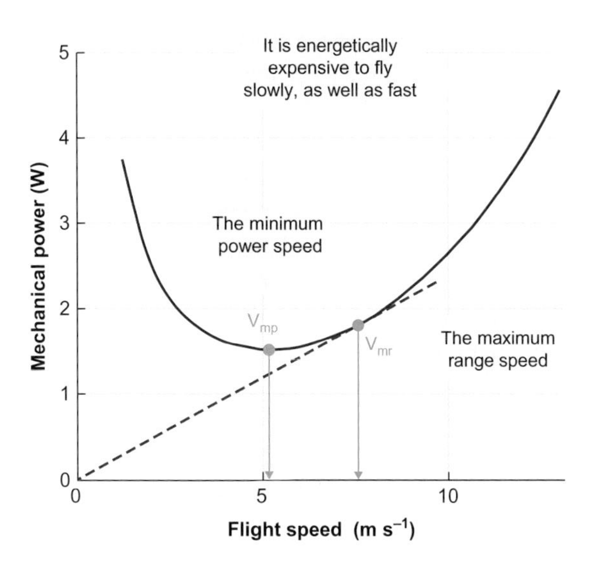

Figure 18.42 U-shaped curve of total mechanical power output as a function of flight speed in birds

Estimated mechanical power output required for flight in a European kestrel (*Falco tinnunculus).* The minimum power speed (V_{mp}) corresponds to the minimum of the *Power* curve. The maximum range speed (V_{mr}) is the most energetically efficient flight speed to cover a certain distance. The total energy expended for covering a certain distance is the product between *Power* and duration of flight (*Distance* × *(speed of flight)*$^{-1}$). Therefore, V_{mr} corresponds to the minimum value of the *Power* × *(speed of flight)*$^{-1}$ curve. In practice, V_{mr} is obtained by drawing the tangent to the *Power* curve from the origin of the graph as indicated by the dotted line.

Source: Rayner JMV (1999). Estimating power curve of flying vertebrates. Journal of Experimental Biology 202: 3449–3461.

range speed or V_{mr} (Figure 18.42). V_{mr} is also called **migration speed** because birds are expected to select flying at this speed when travelling long distances during migration.

Bats are the third group of extant animals that evolved powered flight around 60 Ma. The bat wing is made of a frame of finger bones covered by a stretched double membrane. Like the wings of birds and insects, the bats' wings approximate in cross-section an aerofoil, which provides the necessary lift to keep the animal airborne. The smallest bats have mainly fast-twitch fibres in their pectoral muscles, but larger bats also have fatigue resistant slow-twitch fibres. Most bats cannot take off from the ground like insects and birds do, because they cannot develop the significantly greater power needed to fly at low speed. By hanging upside down from various structures and then dropping with the wings stretched for a short distance, they gain enough speed for flying.

Behavioural strategies for energy conservation in aerial locomotion

Powered flight is energetically demanding; on average, the minimum rate of oxygen consumption during forward flapping flight in birds is 2.2 times that of mammals of similar body mass running at their maximum sustainable speed. Birds have evolved various behavioural strategies, which enable them to minimize the energy cost of flying. One strategy, called **intermittent flight**, is used by most birds and involves the interruption of wing flapping. This strategy is similar to the intermittent generation of thrust in fish.

The intermittent flight is of two main types: **flap-gliding** and **flap-bounding**. In flap-gliding, the wings are extended when the flapping of wings is interrupted and the bird glides. During glides, the flight muscles contract isometrically using little power compared with continuous flapping. This makes the glides energetically economical for covering a certain distance particularly at slow flight. In flap-bounding flight, the wings are folded close to the body and the flight muscles are relaxed during the period of wing flapping interruption. Flap-bounding is efficient at minimizing energy cost at fast speeds. Small birds (<30 g) generally use only bounds, while large birds (>300 g) birds use only glides during intermittent flight. Many intermediate species use glides at slow speeds and switch to bounds at faster speeds.

Large flying birds also resort to passive locomotion to maintain and even gain height without flapping their wings whenever possible; this passive locomotion is called **soaring**. Soaring flight is a special type of glide in which the bird glides in raising air currents called **thermals**. Thermals occur because of uneven heating of the ground by the sun. Since the air is rising, the soaring bird can 'ride' the rising air to increase altitude.

Soaring birds like albatrosses (e.g. *Diomedea* spp.) use rising air currents to maintain or increase altitude. Albatrosses also evolved a locking system in the shoulder to enable their wings to stay extended with little use of energy. This locking system involves a rigid limiting tendon that acts much like the device on an umbrella that keeps it opened. Deeper layers of pectoralis muscle of soaring birds also contains slow-tonic muscle fibres that help control wing position, such as tilt. Albatrosses can remain airborne for weeks and cover distances of several thousand kilometres in one flight.

Migrating birds also conserve energy by climbing to altitudes where the temperature and prevailing winds are beneficial. They often fly in V-formation, which permits individuals to save energy by taking advantage of the air vortices produced by the wings of the bird flying in front of them. This strategy is similar to schooling in fish.

Prevailing winds also play an important role in the migration and dispersal of insects. Dispersal of insects is readily attained simply by flying upward into moving air masses, which takes them for distances of up to several thousand kilometres at an altitude up to 10 km above the Earth's surface.

18.6.4 Locomotion on land

Terrestrial animals are subjected to gravity and require well-developed muscular and skeletal systems for structural support. Most terrestrial animals with exo- and endoskeletons, including arthropods and most terrestrial vertebrates, use limbs for locomotion.

Limbs consist of a series of jointed components connected to muscles that enable animals to extend, flex and partially rotate their limbs with respect to the body, as we discuss in Section 18.5.1. The pattern of limb movement during locomotion over a solid substrate is called **gait**. Gaits are controlled by central pattern generators[123] and are classified based on the mechanics of movement. Types of gaits used by terrestrial vertebrates include walking, trotting and galloping, where trotting and galloping are forms of running.

The metabolic cost of terrestrial locomotion in legged animals largely reflects the cost of force production by muscles involved in rotating and stabilizing the joints. One important factor in determining the energy cost of terrestrial locomotion in legged animals is the posture of the animal in general and leg profile in particular. For example, smaller mammals move with more flexed limbs than do larger mammals. Keeping the legs flexed at all times requires increased muscle activity to support the body of the animal than keeping legs straight and stiff. Therefore, the energetic demand for muscular activity to maintain posture is relatively greater in smaller than in larger mammals.

Walking is characterized by 'vaulting' movements of the body over largely stiff limbs such that the centre of gravity is alternately raised and lowered. This movement resembles an 'inverted pendulum'. In a swinging pendulum, a continuous transformation takes place between gravitational potential energy and kinetic energy. At its highest point in the swing, the pendulum has the highest gravitational potential energy but no kinetic energy while at its lowest point in the swing the pendulum has the highest kinetic energy and the lowest gravitational potential energy. Vertebrates of all sizes have the ability to use some of the mechanical energy associated with the **pendular movement** of the centre of gravity during locomotion and reduce the muscular energy needed for the particular mode of locomotion. The capacity to utilize some energy from the pendular mechanics depends on the size of the animal and the mode of locomotion, being highest in walking and least in trots and gallops, as shown in Figure 18.43.

Most modes of terrestrial locomotion, including walking, trotting and galloping, involve spring-mass mechanics, allowing with each step spring-like elements of the musculoskeletal system to stretch and passively store energy when the leg hits the ground. This elastic energy is then returned when the leg leaves the ground in the next step. We consider in Case Study 18.2 an application of the elastic energy storage-release principle. The capacity of animals to reduce muscle work and metabolic energy by conserving energy from spring-mass mechanics is greatest in larger vertebrates when using faster gaits such as trotting and galloping as shown in Figure 18.43.

[123] We discuss central pattern generators in Section 16,5.4.

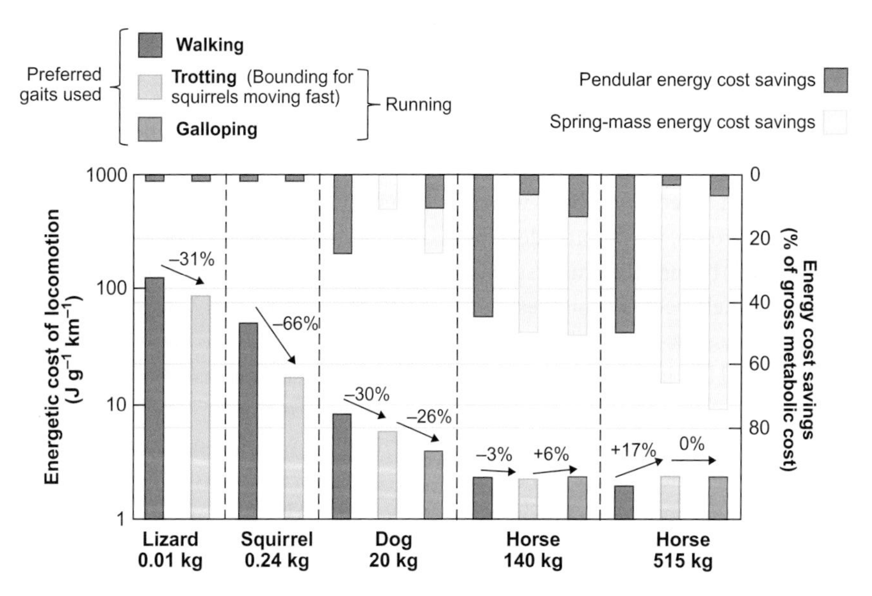

Figure 18.43 Energetic cost of locomotion and energy cost savings across gait changes in a size range of animals

The energetic cost of locomotion (left axis log scale) was measured at the preferred speed within each gait. Percentages over the black arrows indicate the change in cost incurred in changing gaits (magnitudes of differences are obscured by the log scale on the cost axis). Note that faster gaits are cheaper for lizards, squirrels and dogs, but not for horses, where running (trots and gallops) becomes more expensive than walking.

Hanging bars indicate the realized cost savings resulting from pendular and limb-spring savings (right axis, linear scale). Note that realized pendular and spring savings are insignificant to the cost of locomotion in small animals but become increasingly more important in larger animals.

Animal species used: lizards: walk *Cnemidophorus tigris*; trot, *Callisaurus draconoides;* ground squirrel *Spermophilus saturates*; dog *Canis lupus familiaris* and horse *Equus caballus*. Source: Reilly SM et al (2007). Posture, gait and the ecological relevance of locomotor costs and energy-saving mechanisms in tetrapods. Zoology 110: 271–289.

Case Study 18.2 The carbon fibre prosthetic foot: applying the energy storage–release principle of elastic energy used in animal locomotion

The locomotor systems of many species of terrestrial animals involve cyclic storage and release of elastic energy, as we discuss in Section 18.6.4. Among the most dramatic and familiar examples of the use of an elastic energy storage—release cycle in animal locomotion is the elastic energy storage in the tendons of the gastrocnemius muscles in the hind legs of large macropods, such as kangaroos and wallabies, as illustrated in Figure A(i).

When the animal's feet hit the ground, energy is stored in the stretched tendons. A large part of this energy is released during the next hop, enabling the wallaby (*Macropus eugenii*) to increase its hopping speed above 3 m s^{-1} with little increase in energetic cost, as shown in Figure A(ii). The recovery of metabolic power by the tendon elastic power increases linearly with the speed of locomotion and reaches 25 per cent at a hopping speed of 7 m s^{-1}. Importantly, there is no lactate accumulation as the hopping speed increases from 3.9 m s^{-1} to 7.9 m s^{-1}. This indicates that no significant anaerobic glycolysis occurs at up to 7.9 m s^{-1} and, therefore, the energetic cost during hopping was accurately reflected by the rate of oxygen consumption.

It has been estimated that at speeds measured in the field, the energy cost of locomotion of large macropods is less than one-third the cost of running for quadrupeds of equivalent body mass. This adaptation enables macropods to forage over large areas in semi-arid environments.

Tendons in the lower limbs and feet of quadruped mammals like horses and cheetahs, bipedal humans and bipedal running birds like turkeys, also play a similar elastic energy storage role during running (although to a less dramatic extent). These principles have inspired the development of the prosthetic foot and lower limb that achieves high energy return along with shock absorption. 'The Cheetah' is one example of a prosthetic foot that is specialized for sprinting (so-named because the cheetah is the master sprinter), and has been used by many runners in the Paralympic Games.

Tendons of the kangaroo have an elastic energy storage efficiency of about 90 per cent while the carbon graphite Cheetah foot has an efficiency of about 82 per cent. Aside from the material properties of the carbon-fibre construction, its dynamic shape allows the foot to flex more than other models, offering more spring (that is, energy release from recoil). Its longer and flatter toe also enhances push off, while a specially designed pad provides good traction.

› Find out more

Baudinette RV, Snyder GK, Frappell PB (1992). Energetic cost of locomotion in the tammar wallaby. American Journal of Physiology, (Regulatory Integrative and Comparative Physiology) 262: R771–R778.

Biewener AA, Baudinette RV (1995). In vivo muscle force and elastic energy storage during steady-speed hopping of tammar wallabies, (*Macropus eugenii*). Journal of Experimental Biology 198: 1829–1841.

Roberts TJ, Azizi E (2011). Flexible mechanisms: the diverse roles of biological springs in vertebrate movement. Journal of Experimental Biology 214: 353–361.

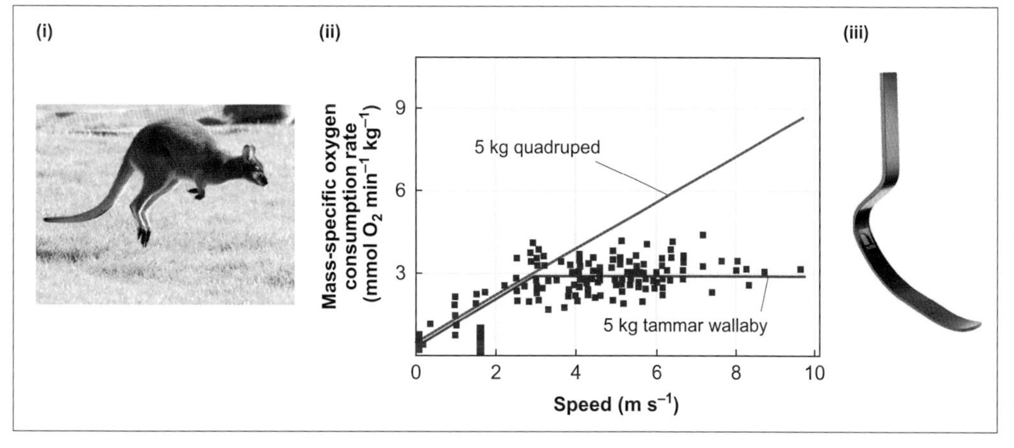

Figure A The elastic energy storage–release mechanism used in macropod locomotion inspired the development of elastic prosthetic feet

(i) Hopping tammar wallaby (*Macropus eugenii*); (ii) relationship between mass-specific oxygen consumption rate and locomotion speed in a 5 kg hopping wallaby compared to a typical 5 kg quadruped runner; (iii) the 'Cheetah' carbon-fibre prosthetic foot.

Sources: (i) © Dennis Jacobsen/ Shutterstock; (ii) Baudinette RVet al (1992). Energetic cost of locomotion in the tammar wallaby. American Journal of Physiology (Regulatory Integrative and Comparative Physiology) 262: R771-R778. (iii) Ossur Prosthetic Solutions.

Both vertebrates and invertebrates use jumping as a defensive mechanism to escape predation or for capturing prey. Only a few vertebrate species use jumping as a primary mode of locomotion: these include frogs, hares, rabbits and various macropods such as kangaroos. Kangaroos store elastic energy in contracting muscles and tendons every time they land on their hind legs after a jump and release it in the following jump as we discuss in Case Study 18.2.

Some insects that also use jumping as a form of locomotion, which includes grasshoppers, locusts and fleas, employ an elastic energy storage-release mechanism for propulsion instead of direct muscular contraction. First, energy produced by a slowly contracting muscle is gradually stored in an elastic structure made of a composite material of hard cuticle and an elastic protein called **resilin**. Then, the stored energy is suddenly released within a millisecond to extend the hind legs, which causes the insect to jump.

Not all terrestrial vertebrates have or use legs for locomotion. Snakes, for example, do not have legs but can be very mobile on land, using muscles and scales on their body to perform four types of movement:

- Slithering or lateral undulation occurs when snakes alternately contract and relax muscles along each side of their body to produce rearward travelling horizontal waves. At the same time snake use muscles attached to the ventral scales to push their body forward against resistance points on the ground.
- Sidewinding is a variation of the slithering motion that snakes use to move over loose or sandy soil, or slippery surfaces. During this form of locomotion, snakes successively throw their head laterally and maintain their body in an S-shape form that has only two points of contact with the ground. Muscles at the points of contact with the ground push laterally onto the ground, flinging the snakes sideways.
- A concertina movement is used by snakes to move along vertical surfaces: the front part of the body is extended along the vertical surface and finds a place to grip onto with the ventral scales; the middle part of the body is then bunched up into tight curves; and the lower part of the body is pulled up. The motions are then repeated, with the front part of the body being pushed further forward to find a new place to grip with the scales.
- Caterpillar or rectilinear locomotion occurs when snakes stalk prey: the movement is slow, straight and noiseless and involves contraction of the body into small vertical curves. Ventral scales in contact with the ground push against the ground, moving the the upper part of the curve forward and create a rippling effect similar to that observed in moving caterpillars.

Behavioural strategies for energy conservation in terrestrial locomotion

Generally, animals realize a decrease in the energetic cost of locomotion (COT) by changing gaits and selecting the most appropriate gait for the type of terrain and speed required. Each gait has an optimum speed, at which the COT is lowest. COT increases at slower or faster speeds. Gait transitions occur near the speed where the COT at the respective gait (like fast walking) becomes greater than the COT at the next gait (like a slow trot).

The decrease in COT in small animals such as the lizard and squirrel when they change gaits from walking to running (trotting and bounding, respectively) is considerable: 31 per cent in lizards and 66 per cent in squirrels as shown in Figure 18.43. The decrease is primarily due to the decrease in the number of strides they make per unit distance travelled. The energy savings from pendular and spring-mass mechanics is negligible in these small vertebrates.

Dogs save about 25 per cent of gross metabolic costs from pendular mechanics when walking and decrease their COT by 30 per cent and 26 per cent when change gait from walking to trotting and from trotting to galloping, respectively.

Horses, which are much heavier than dogs, save about 45 per cent of gross metabolic costs from pendular mechanics when walking. However, their faster gaits (trotting and galloping) are actually slightly more energetically expensive than walking, even though horses save around 60–70 per cent of the gross metabolic costs in their faster gaits primarily from spring-mass mechanics as shown in Figure 18.43.

Penguins employ a different behavioural strategy for energy conservation in terrestrial locomotion. Penguins have short legs and waddle with a side-to-side swaying motion as they walk on hard surfaces as opposed to up and down movement of the gravity centre. Swaying helps penguins conserve momentum between steps and use less energy for locomotion. Some penguins also use tobogganing in snow, which is more energy efficient than waddling. When tobogganing, penguins drop down on their chest and use their flippers and toes to push their body forward, moving faster than humans can under these conditions.

› *Review articles*

Alexander R McN (2005). Models and the scaling of energy costs for locomotion. Journal of Experimental Biology 208: 1645–1652.

Biewener AA (2011). Muscle function in avian flight: achieving power and control. Philosophical Transactions of the Royal Society, B 366: 1496–1506.

Bishop CM, Butler PJ (2015). Flight, Chapter 39 in Sturkie's Avian Physiology, Scanes CG Ed, 6th edition, pp 919–974 Elsevier, Amsterdam.

Butler PJ (1991). Exercise in birds. Journal of Experimental Biology 160: 233–262.

Coughlin DJ (2002). Aerobic muscle function during steady swimming in fish. Fish and Fisheries 3: 63–78.

Dickinson MH, Farley CT, Full RHJ, Koel MAR, Kram R, Lehman S (2000). How animals move: an integrative view. Science 288: 100–106.

Lauder GV, Drucker EG (2002). Forces, fishes, and fluids: hydrodynamic mechanisms of aquatic locomotion. News in Physiological Sciences 17: 235–240.

Fish FE (1996). Transitions from drag-based to lift-based propulsion in mammalian swimming. American Zoologist 36: 628–641.

Reilly SM, McElroy EJ, Biknevicius AR (2007). Posture, gait and the ecological relevance of locomotor costs and energy-saving mechanisms in tetrapods. Zoology 110: 271–289.

Shadwick RE (2005). How tunas and lamnid sharks swim: an evolutionary convergence. American Scientist 93: 524–531.

Schmidt-Nielsen K (1972). Locomotion: Energy cost of swimming, flying and running. Science 177: 222–228.

Tobalske BW (2007). Biomechanics of bird flight. Journal of Experimental Biology 210: 3135–3146.

Checklist of key concepts

Muscle form and function

- Animal movement and behaviour in the environment is powered by the **pulling force** produced by **contracting muscles**.
- The contractile apparatus in all muscles consists of **myofibrils** made up of **myosin-containing** and **actin-containing** filaments.
- The longitudinal striation pattern of striated muscles reveals the location of the two types of filaments.
- Smooth muscles do not display a regular striation pattern because myofibrils are not uniformly oriented within the fibres.
- All muscles generate force through a **chemo-mechanical cycle** in which the globular heads of the myosin molecules hydrolyse MgATP and then pull on the actin-filament, producing force between filaments.
- The two filament types slide past each other only when the force produced is greater than the load.
- Force generation is regulated by the **concentration of Ca^{2+}** in sarcoplasm: the chemo-mechanical cycle is stopped at low Ca^{2+} (<100 nmol L^{-1}) and proceeds when Ca^{2+} concentration rises.
- Different Ca^{2+} regulatory systems exist in different muscle types and in different animal species.

Voluntary muscle fibres

- Contraction of **voluntary muscles** is under the **direct control** of the nervous system.
- Voluntary muscles are functionally diverse.
- **Excitation–contraction coupling** refers to the sequence of events that starts with excitation at the sarcolemma, which causes a rise of Ca^{2+} in the sarcoplasm, and finishes with force production by the contractile apparatus.
- Myosin exists in different forms called **isoforms**, which determine the maximum speed of muscle shortening.
- In vertebrate skeletal and cardiac muscles the Ca^{2+}-regulatory system is associated with the actin filament.
- **Muscle fatigue** is a multifactorial process that protects muscle from damage.

Cardiac myocytes

- The heart in vertebrates contracts rhythmically in the absence of neural inputs, but neural inputs modulate the **strength** and the **rate** at which the muscle contracts.
- The action potential is very **prolonged** in cardiac myocytes.
- **Contractile force** in cardiac muscle is sensitive to stretch and is regulated by the autonomus nervous sytem and hormones.

Smooth muscle fibres

- **Smooth muscle** in general **contracts slowly** in response to a broad range of stimuli.
- Activation of smooth muscle involves phosphorylation of myosin regulatory light chains.
- Smooth muscle maintains force highly efficiently.

The muscular system is the engine that provides the power for an animal's movements and behaviour

- The mechanical power generated by muscles is proportional to the muscle volume (mass) and depends on how fast muscles shorten when carrying loads.
- Fast muscles can produce considerably more power than slow muscles, but consume more energy to produce the same amount of force.
- **Slow** muscles are used in **lower power, prolonged and/or repetitive activity**, while **fast** muscles are essential for **short and powerful movements**:
  - in escape responses,
  - when chasing prey,
  - for generating upward lift at take-off in airborne species.
- The coordination of muscle activity in an animal's body is based to a large extent on spontaneous reflex behaviour involving **sensory receptors** that synapse in the central nervous system with interneurons, and **motor neurons**, which ultimately control muscle activity.
- Muscles are **plastic tissues**, which can adapt to changing demands placed on them by changed conditions in the environment, though rapid changes in environmental conditions can disrupt muscle functioning.

Animal locomotion

- The **energetic cost of transport (COT)** of unit mass decreases as the size of the animal increases and is lowest for swimming followed by flying and running.
- Fish generally use either their body and/or caudal fin (BCF) or their paired pectoral and pelvic fins (MPF) for their routine mode of locomotion.
- **Lift-powered swimming** is produced by the active movement of hydrofoil-shaped appendices, which shed vortices and generate **anteriorly directed lift**.
- **Lift-based swimming** is used by diverse species including fast swimming fish like tunas, sharks, sea turtles, penguins and marine mammals.
- **Drag-based swimming** is achieved by **paddling** and is used by terrestrial and semiaquatic mammals, most aquatic birds and aquatic insects.
- **Powered flight** occurs in insects, birds and bats, which use **aerofoil-shaped wings** to provide both **uplift** and **thrust**.
- The wings of flying insects are either powered directly by **synchronous flight muscles** or indirectly by **asynchronous muscles**.
- The two main wing muscles in birds consist of **fast-twitch fibres** to provide the high power output necessary for flight:
  - the **pectoralis** powers the downstroke flap of the wing to produce thrust and positive uplift,
  - the **supracoracoideus** raises the wing which can also produce some thrust and positive uplift.
- The power required for flight typically varies with the flight speed according to a **U-shaped curve** indicating that it is energetically expensive to fly slowly as well as fast.
- Most terrestrial animals with exo- and endoskeletons use **limbs** for locomotion.
- Limbs consist of a series of jointed components connected to muscles that enable animals to extend, flex and partially rotate their limbs with respect to the body.
- Types of **gaits** used by terrestrial vertebrates include **walking, trotting** and **galloping**, where trotting and galloping are forms of running.
- Terrestrial locomotion generally involves up and down movement of the centre of gravity (pendular motion) and spring-mass mechanics.
- The energy cost of active transportation can be significantly **reduced** by:
  - the intermittent generation of **thrust** in aquatic species (burst swimming) and birds (flap-gliding or flap-bounding),
  - by swimming or flying in **formation patterns** (schooling in fish, flying in V-formation in birds), which take advantage of vortices produced by individuals swimming or flying in front.
- Terrestrial animals **decrease** their energetic cost of transportation by
  - selecting the most appropriate gait for the type of terrain and speed required,
  - optimizing the energy associated with pendular movement of the centre of gravity and the elastic energy storage-release principle.

Study questions

* Answers to these numerical questions are available online. Go to www.oup.com/uk/butler

1. Explain how force is generated in muscle from the interactions between actin and myosin filaments. (Hint: Section 18.1.4.)
2. Animal carcasses become stiff soon after animals die. Explain why this happens. (Hint: Figure 18.7.)
3. Highlight the major events involved in the excitation–contraction coupling of twitch vertebrate skeletal muscle and explain the difference between a twitch and a tetanic contraction. (Hint: Section 18.2.2.)
4. What are the main differences between the action potentials in skeletal and cardiac muscle fibres of vertebrates and how are these differences important for their function? Explain your answer. (Hint: Sections 18.2.1 and 18.3.1.)
5. Two skeletal muscles, A and B, contain 1000 and 200 twitch muscle fibres, respectively. Muscle A has 10 motor units, each consisting of one neuron and 100 muscle fibres. Muscle B has 20 motor units, each consisting of 10 muscle fibres and one neuron. Which muscle can produce a wider range of forces? Explain your answer. (Hint: Section 18.2.1.)
6. Highlight the differences in how signals are transmitted between motor neurons and skeletal muscle fibres of vertebrates and arthropods. (Hint: Figures 18.12 and 18.13.)
7. Magnesium ions bind to ATP to form the MgATP complex, which is hydrolysed by the myosin heads to produce force. Do magnesium ions play another major role in the function of skeletal muscle fibres? Explain how. (Hint: Figure 18.16.)
8. Two smooth muscles that exhibit the same level of tone (contraction) are exposed to noradrenaline. One smooth muscle contains α_1-adrenergic receptors, while the other contains β-adrenergic receptors in the sarcolemma. Explain what will happen to the muscle tone in the two muscles. (Hint: Section 18.4.3 and Figure 18.27.)

9.* One 10-cm-long skeletal muscle with a cross-section area of 1 cm^2 produces a force of 20 N and 2 W maximum power. Estimate the force and power that similar muscles of following dimensions produce: (i) 20 cm long and 1 cm^2 in cross-section, (ii) 20 cm long and 0.5 cm^2 in cross-section, (iii) 5 cm long and 2 cm^2

18

in cross-section and (iv) 10 cm long and 0.5 cm^2 in cross-section. Explain which of these muscles would be better suited for power generation and which for joint stabilization. (Hint: Sections 18.1.5 and 18.5.2.)

10*. The energy cost for flying is described by the following allometric equation: Energy cost (in $J\ g^{-1}\ km^{-1}$) = 16 $(M_b)^{-0.30}$, where M_b is the body mass in g. A migratory bird of an average mass of 1 kg used 200 g of its fat reserves purely for flying during a non-stop flight. Calculate the distance travelled by the bird assuming that the mass of the bird did not change during flight. (Hint: energy content of 1 g fat is 35.5 $kJ\ g^{-1}$, Table 2.1.)

11*. The energy content of two slices of toast, consisting mainly of carbohydrates, is about 590 kJ. Considering that about 10 per cent of the energy content is used for processes associated with the ingestion, digestion, absorption and distribution of the carbohydrates within the body (Section 2.3.4), calculate how many kilometres an average person weighing 70 kg needs to run in order to 'burn' the energy gained from the consumption of the two slices of toast. Assume that the relationship between energy cost for running and body mass (M_b expressed in g) is described by the following allometric equation: Energy cost (in $J\ g^{-1}\ km^{-1}$) = 120 $(M_b)^{-0.32}$.

12. Compare and contrast the mechanisms of excitation and regulation of force production between skeletal, cardiac and smooth muscle of vertebrates.

13. Explain the principle behind lift-power swimming. (Hint: Section 18.6.2.)

14. Highlight two similar behavioural strategies for energy conservation in aquatic and aerial locomotion. (Hint: Sections 18.6.2 and 18.6.3.)

15. Provide three supporting arguments to explain the large difference in the energy cost of locomotion between a terrestrial and a marine mammal of similar mass. (Hint: Sections 18.6.1, 18.6.2 and 18.6.4.)

Bibliography

Bers DM (2001). Excitation-Contraction Coupling and Cardiac Contractile Force. Kluwer Academic Publishers. 2nd Edition, Dordrecht.

Engel AG, Franzini-Armstrong C (1994). Myology. 2nd Edn, Vol I and II. McGraw-Hill.

McNeill AR (2003). Principles of Animal Locomotion. Princeton University Press, Princeton, NJ.

Rüegg JC (2011). Calcium in Muscle Contraction: Cellular and Molecular Physiology. 2011. 2nd Edition, Springer.

Woledge RC, Curtin NA, Homsher E (1985). Energetic Aspects of Muscle Contraction. London: Academic Press.

19 Hormones

Hormones produced by **endocrine glands**, and related cell secretions, are chemical signals that affect the function of other cells. Hormones are involved in regulating the internal environment of animals—their ion and water balance, appetite and metabolism, development, growth and reproduction; they also prepare animals for fighting, flight and other activities.

In this chapter, we learn about the hormonal systems (also called endocrine systems) that have evolved in animals and examine some of the systems that secrete the hormones responsible for regulating the physiological processes discussed in earlier chapters. We also learn about the hormone systems that regulate processes such as reproduction, and ion and water balance, which we go on to explore in later chapters. We explore how stress influences hormone function and how hormone studies give insights into the well-being of captive animals and wildlife.

We begin by examining the dynamics of hormonal processes—how a range of different hormones are synthesized and transported in the blood, how they interact with receptors, and how hormonal systems are themselves regulated.

19.1 Dynamics of hormonal processes

Figure 19.1 shows the four sub-categories of chemical communication that operate in animals, and which are recognizable by the type of cell that produces the chemical and the distance over which the chemicals exert effects. These categories are:

- **Neuroendocrine secretion** of **neurohormones** synthesized and secreted by neurosecretory cells. Neurosecretory cells contain hormone-filled vesicles, distributed throughout the cell body, the axon and the enlarged terminal endings. (They are not just restricted to a synapse as in nerves.) The nerve endings often terminate on blood vessels and form **neurohaemal organs** with stores of hormone that are released in response to neural stimulation. Fusion of the vesicles with the cell membrane results in release of the neurohormone, which diffuses away in extracellular fluid and enters any nearby capillaries. Thus, neurohormone secretions act on *many* cells in tissues at some distance from their release. This contrasts with action of neurotransmitter molecules released from axons, at synapses, which act on receptors on a *single* postsynaptic cell[1].
- **Endocrine secretions** from endocrine glands. A single endocrine gland may secrete many hormones. Hormones released from endocrine cells enter extracellular fluid from where they subsequently enter nearby capillaries. By being circulated in the blood[2], hormones initiate physiological responses in many tissues that are distant from the endocrine gland. Sometimes hormones act on another endocrine tissue, which stimulates synthesis and release of another hormone. An example is thyroid stimulating hormone (TSH), which maintains the structure and activity of the thyroid gland and its production of thyroid hormones[3]. The stimulating hormone and the combination of tissues on which is acts form an **endocrine axis**.
- **Paracrine secretions (paracrines)**, which act more locally than classical hormones, usually on nearby cells within the same organ.
- **Autocrine secretions (autocoids)** that have rapid and usually brief actions on the cells that secrete them. The release of larger amounts may result in actions at greater distances.

Grouping chemical communication into these four categories can be useful when exploring hormone actions, but these categories are not absolute. In many cases, the same chemical substance may function as a classical hormone after its release into the blood while also functioning locally as a paracrine.

Hormones, paracrines and autocoids differ in their biochemical structures, with four main types being recognized: peptide and glycopeptides, steroids, amines and eicosanoids. We outline each of these in more detail in Box 19.1.

[1] We discuss neurotransmitter release by axons in Section 16.3.2.
[2] Invertebrates with more open circulations, which we discuss in Sections 14.1.1 and 14.3, distribute hormones via the haemolymph.
[3] Section 19.4 explores the thyroid axis.

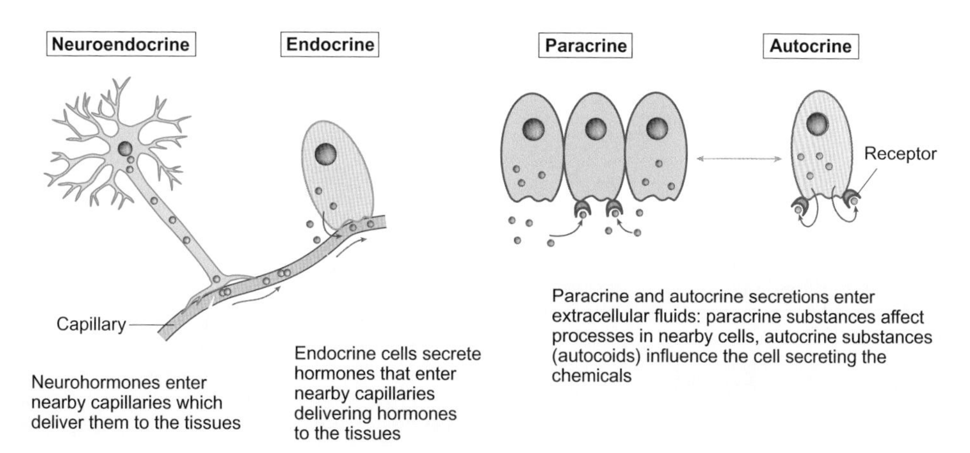

Figure 19.1 Chemical communication by neuroendocrine secretion of neurohormones, endocrine secretion of hormones from the cells of endocrine glands, paracrine secretion and autocrine secretion

19.1.1 Chemical differences determine the synthesis, storage and secretion of hormones

The chemical differences between hormones result in differences in the structure of endocrine tissues. Figure 19.2 compares the structural features of an endocrine cell producing a peptide hormone with those of a cell producing a steroid hormone.

Synthesis of peptide hormones

Peptide hormones are synthesized on the ribosomes, as large precursor proteins (so-called pre-pro-hormones) and are secreted into the lumen of the endoplasmic reticulum. Therefore, cells producing peptide hormones have a

Box 19.1 Major types of endocrine and paracrine substances

The major types of endocrine/paracrine substances, recognized by their chemical structures as shown in Figure A, are:

1. Peptide and glycopeptide hormones (and proteins), which make up the largest group of hormones, and consist of a pre-determined sequence of amino acids in a chain of variable length. At one extreme, there may be small chains of as few as three amino acids, as in thyrotropin-releasing hormone (TRH)[1], while longer chains of amino acids make up proteins. However, many protein hormones are referred to as peptides, so we will use this term to describe any hormones made up of amino acids. Glycopeptide hormones, such as **follicle stimulating hormone** (FSH), **luteinizing hormone (LH)** and **thyroid stimulating hormone** (TSH) consist of amino acids with carbohydrate groups attached to the side chains of the amino acid residues[2].
2. Steroid hormones derived from sterols, such as cholesterol, which gives them the common core structure shown in Figure A. Perhaps the most well-known steroid hormones among vertebrates are **testosterone** and **oestrogen**. In Section 19.3, we learn about steroid hormones produced by the adrenal glands and gonads of the vertebrates and in Section 19.5.3 we examine hormonal control of **moulting** and **metamorphosis** of insects, which depends on ecdysteroids.
3. Amine hormones, with one or more $-NH_2$ groups in their chemical structures, which are derived from amino acids that change structurally during hormone synthesis. Examples that we explore in this chapter include **melatonin** (secreted by the **pineal gland**), which is a derivative of tryptophan, and the hormone derivatives of tyrosine: **thyroid hormones**, and **adrenaline** and **noradrenaline**. (Adrenaline and noradrenaline are also known as epinephrine and norepinephrine and are collectively called **catecholamines**.)
4. Eicosanoids are synthesized from fatty acids such as **arachidonic acid**[3]. This group includes **prostaglandins**, **leukotrienes** and **thromboxanes**, which are substances that have a wide range of paracrine and autocrine actions, such as vasodilation or vasoconstriction of vascular smooth muscle, stimulation of inflammation and activation of immune responses. These substances are synthesized by many cell types rather than one or more endocrine glands.

[1] TRH is discussed in Sections 19.2.2 and 19.4.
[2] FSH, LH and TSH are discussed in Section 19.2.2.
[3] Box 9.2 gives details on fatty acids and the structure of arachidonic acid.

(i)

Cortisol

(ii)

$H\text{-}Cys^1\text{-}Tyr^2\text{-}Phe^3\text{-}Gln^4\text{-}Asn^5\text{-}Cys^6\text{-}Pro^7\text{-}Arg^8\text{-}Gly^9\text{-}NH_2$

Arginine vasopressin

(iii)

Thyroxine (T_4)

(iv)

Prostaglandin E_2

Figure A Examples of the four main types of hormone structure

(i) Steroid hormones, such as cortisol and testosterone, have a characteristic hydrocarbon ring structure because of their synthesis from cholesterol. All steroids have a 17-carbon core structure in 3 × 6 carbon rings (labelled A, B, C), one 5-carbon ring (labelled D), and a small number of carbon atoms in side chains. The steroid hormone shown (cortisol) is a 21-carbon steroid. (ii) Peptide hormones, such as arginine vasopressin, consist of a chain of amino acids. (iii) The amine group of hormones (such as thyroxine) contain an $-NH_2$ group as they are synthesized from amino acids. (iv) Prostaglandins, such as prostaglandin E_2 are synthesized from fatty acids, and hence show similarities to these in their core structure.

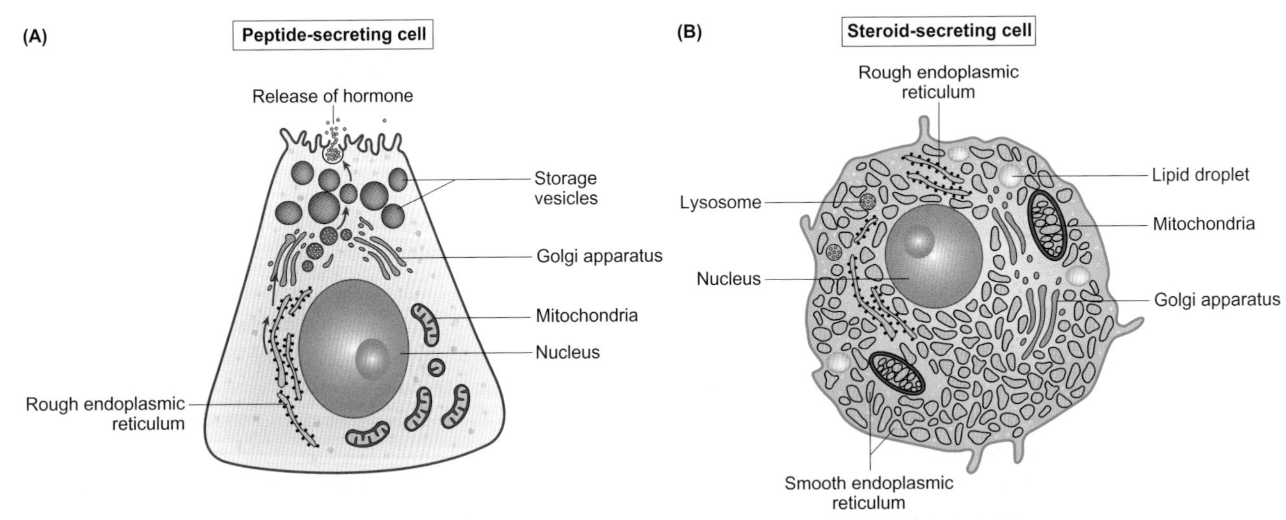

Features	Peptide-secreting cells	Steroid secreting cells
Synthetic machinery	Prominent *rough* endoplasmic reticulum and Golgi	Extensive *smooth*-surfaced endoplasmic reticulum and Golgi
Precursors	Amino acids	Cholesterol stored in lipid droplets
Storage of hormone	Storage vesicles of peptide hormone	Not stored in cells
Secretion mechanism	Exocytosis	Via membrane transporter; some diffusion across plasma membrane
Transport in blood	Mostly dissolved in plasma	Mostly bound to carrier proteins

Figure 19.2 Structural features of cells synthesizing peptide and steroid hormones

(A) Cell synthesizing and secreting a peptide hormone. Red arrows indicate progression of hormone through rough endoplasmic reticulum into Golgi forming storage vesicles of hormone for later release by exocytosis. (B) Cells synthesizing steroid hormone do not store hormone products. Hormone is synthesized on demand.

(A) and (B): adapted from Goldsworthy GJ et al (1981). Endocrinology. Glasgow and London: Blackie.

well-developed rough endoplasmic reticulum and Golgi apparatus, as shown in Figure 19.2A.

The pre-pro-hormone migrates through the endoplasmic reticulum to the Golgi apparatus, where packaging creates secretory vesicles, each surrounded by a membrane. During migration through the endoplasmic reticulum, removal of fragments of the amino acid chain first forms the pro-hormone and then the active hormone. Consequently, secretory vesicles may contain more than one hormone together with inactive peptide fragments.

The secretory vesicles vary in shape, size and electron density (as seen in the transmission electron microscope), allowing us to characterize particular endocrine/neuroendocrine cells and their activity. Some cells contain vesicles that are highly electron dense (called dense-cored vesicles), while cells that secrete peptide hormones generally contain paler storage vesicles. The store of vesicles may be small when tissues secrete relatively small amounts of hormone continuously, or much greater if larger quantities are secreted intermittently.

These hormone-containing vesicles are stored in the cytoplasm until the cell receives the signal to trigger hormone release. The membrane of the vesicle has proton pumps[4] that maintain a low pH inside the vesicles, so that the peptide aggregates and remains stable during storage. A rise in intracellular concentration of calcium ions is a prerequisite for the activation of the secretory process that results in the release of the contents of the pre-formed vesicles. When this happens, some vesicles fuse with the plasma membrane and release their contents in a similar way to neurosecretion by neurons[5].

Synthesis of steroid hormones

The cells secreting steroid hormones are strikingly different from those secreting peptide hormones. Notice in Figure 19.2B that a large amount of smooth endoplasmic reticulum dominates the structural appearance of steroid-secreting cells. In addition, the prominent mitochondria, with tubulo-vesicular cristae, have an unusual open appearance, much like the smooth endoplasmic reticulum. Together, the mitochondria and smooth endoplasmic reticulum are responsible for the synthesis of steroid hormones.

Cells producing steroid hormones do not store the steroids they synthesize, so do not contain vesicles of hormone. Rather, they contain droplets of lipid containing precursor molecules for the synthesis of the steroid hormone, as shown in Figure 19.2B. **Cholesterol** is a precursor for all steroid hormones. However, many animals, such as insects, cannot synthesize cholesterol and so must obtain it in their diet. Vertebrates can synthesize small amounts of cholesterol in their steroid-secreting cells and store it as cholesterol-esters, but most cholesterol is synthesized in the liver. Low-density lipoproteins in the blood are the major carriers of cholesterol to endocrine tissues in which steroid-producing cells take up and store cholesterol-esters in lipid droplets.

Various circulating hormones stimulate the chain of reactions that result in the hydrolysis of the cholesterol-esters to generate the free cholesterol that is the starting point for **steroidogenesis** (literally the generation of steroids). This deceptively simple term hides the multitude of enzymatic steps that add chemical side groups to (or remove them from) the basic ring structure of cholesterol.

Figure 19.3 illustrates the main pathways of steroidogenesis in mammals. Steroidogenesis initially involves the cleavage of cholesterol by enzymes in the inner mitochondrial membrane to form pregnenolone. After this rate-limiting step, subsequent enzymatic processes in the smooth endoplasmic reticulum and further steps in the mitochondria lead to the formation of an array of steroids and many intermediate products.

Among vertebrates, steroidogenesis occurs in three specialized organs: the ovary, testis, and the adrenal gland[6]. Production of particular steroids can only take place in the specific organs that have the correct complement of enzymes to produce particular steroid molecules, so the hormones generated are specific to the particular organ or tissue and, in some cases, are species dependent.

Steroid hormones have high lipid solubility, so for many years the release of steroid hormones from the synthesizing cell was thought to rely entirely on diffusion through the lipid bilayer of the plasma membrane. However, in the 1990s it became clear that **solute carriers** in the basal membrane of the cells facilitate the release of steroid hormone whenever a steroid-synthesizing cell is stimulated, at least in some tissues such as the adrenal gland of mammals.

19.1.2 Hydrophilic and hydrophobic hormones are transported in different ways

The way that hormones are distributed in the blood and extracellular fluids is highly dependent upon their chemical composition. Peptide hormones and catecholamines are highly soluble in water (they are hydrophilic), and hence they are transported readily in solution.

All hormones bind non-specifically to plasma albumin to some extent. However, the importance of such binding for hydrophilic molecules is unclear. In contrast, hydrophobic hormones, which have low water solubility (such as steroid

[4] Section 4.2.2 provides further detail on vesicular proton pumps.
[5] Section 16.3.2 discusses neurosecretion by chemical synapses in more detail.
[6] Steroids are synthesized by the adrenal cortex of the adrenal gland or its functional equivalent, which we discuss in Section 19.3.1.

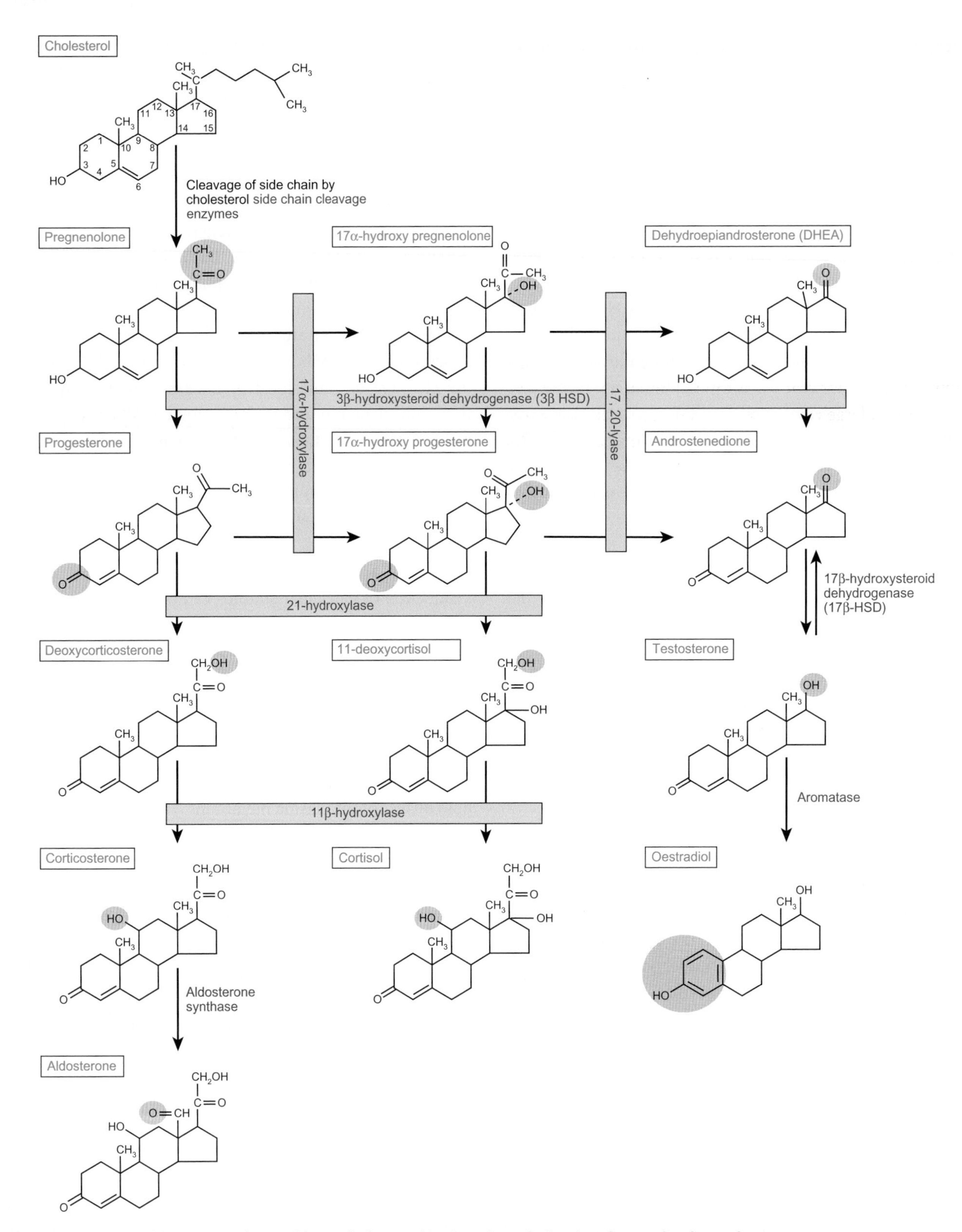

Figure 19.3 Steroid hormones derived from cholesterol in the adrenal gland and gonads of vertebrates

Cholesterol is the precursor for all types of steroid hormones. Steroidogenesis commences by side chain cleavage of cholesterol in the inner mitochondria to form pregnenolone. Pregnenolone and its products (labelled in green) are modified by different enzymes (labelled in red) to form different steroids in different tissues. Not all of the reactions, enzymes and intermediates are shown in the schematic diagram. The enzymes shown are named according to their biochemical actions. Regions of chemical modification of the products are indicated by the pink overlaid circles.

Based on scheme in Hanukoglu I (1992). Steroidogenic enzymes: structure, function and role in regulation of steroid hormone biosynthesis. Journal of Steroid Biochemistry Molecular Biology 43: 779–804.

hormones and thyroid hormones), bind to specific carrier proteins. For instance, a specific globulin protein binds thyroid hormones in mammals, birds and some ectothermic animals such as turtles.

Only the small proportion of a circulating hydrophobic hormone that is unbound is free to diffuse across the capillary walls and through extracellular fluids, where it can interact with receptors on or in the target cells that initiate physiological effects. The free concentrations of steroid hormones and thyroid hormones in solution in the blood plasma are typically 10^{-9} to 10^{-12} mol L^{-1}, which is sufficient to exert physiological effects.

Bound and free hormone are in dynamic equilibrium. Consequently, hydrophobic hormones bound to carrier proteins act as a large reserve of stored hormone in the blood. This reserve is important for the maintenance of physiological concentrations of free hormone in the plasma: once some of the free hormone binds to hormone receptors, or is metabolized, its concentration in the plasma declines; the increasing difference between bound and free concentrations of the hormone results in detachment of some of the bound hormone from its carrier.

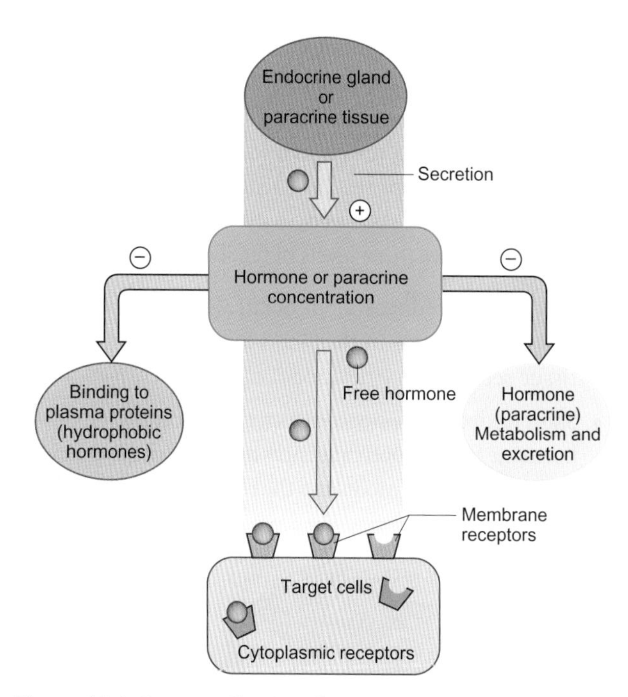

Figure 19.4 Factors affecting the concentration of hormone in extracellular fluids (blood and interstitial fluids) and their physiological actions

+ = positive effect on free hormone concentration; – = negative effect on free hormone concentration.

19.1.3 Circulating concentrations of hormones are regulated by balancing their secretion and removal

Figure 19.4 is a schematic diagram of the key factors that determine the concentration of hormones and paracrines in the extracellular fluids—something that has a major effect on physiological actions. The concentration of free (unbound) hormone in the blood results from the balance between secretion, which increases the circulating concentration, and removal—either by metabolism or by binding to receptor molecules and carrier proteins.

Enzymes in the blood degrade molecules of unbound hormones, and their products are excreted. Small peptide hormones and catecholamines are rapidly degraded, so these hormones have relatively short half-lives[7]: small peptides comprising just a few amino acids have half-lives of seconds to a few minutes at most. Larger peptides (proteins) such as **growth hormone** have half-lives of up to 30 minutes but even these are much less persistent than steroid hormones.

Steroids bind to carrier proteins. When bound to proteins they are not lost in the urine[8] and are more resistant to enzyme degradation. Steroids persist for hours and thyroid hormones persist for days in the blood.

[7] The half-life of a hormone is the time taken to remove half of the hormone molecules present in the blood.

[8] The large molecular size and molecular charge impedes filtration by the kidney, as we discuss in Section 7.1.1.

All hormones are eventually inactivated by metabolism (by the liver, for example, or in the target cells). Consequently, the rate of hormone metabolism and excretion, in faeces or urine, influence their concentrations in the blood, but usually has less impact on circulating concentrations than secretion rates. However, the excretion of hormones and their metabolites in urine and faeces provides an ideal way of assessing the hormonal status of animals without the need to collect blood samples. This approach is the basis of the pregnancy test in humans. Similar approaches are also used to assess the breeding condition of captive animals by analysis of faecal or urine samples. It is also a strategy for understanding stress in animals, as we learn in Section 19.3.

Secretion rates of endocrine systems are controlled by feedback loops. Diseases characterized by over-production (hypersecretion) or under-production (hyposecretion) of a hormone may occur because of the failure of such regulatory systems. For hormones carried in the circulation by carrier proteins, only the smaller, free component of hormone is physiologically active and can exert feedback effects, not the total hormone concentration.

Most endocrine feedback loops inhibit further hormone release, that is, they are negative feedback loops, along the lines of a set point system[9]. Negative feedback operates when the free hormone concentrations rise *above* a desirable set

[9] Section 3.3.3 explains the principles of feedback systems.

point, while reduced inhibition (less feedback) occurs when free concentrations of the hormone fall *below* the set point. Negative feedback tends to stabilize the concentrations of circulating hormones and results in a steady state. More rarely, **positive feedback** loops result in rising concentrations and culminate in a climax, such as ovulation in vertebrates[10].

Figure 19.5 shows three levels of feedback loop:

- Ultra-short feedback loops occur when the concentration of a hormone in the blood directly influences the tissue secreting the hormone.
- Short feedback loops occur when the concentration of hormone in the blood influences a higher part of an endocrine cascade that modulates the release of the hormone.
- Long feedback loops occur when an outcome from the actions of a hormone influence steps further back up a cascade of hormones or higher centres such as parts of the central nervous system.

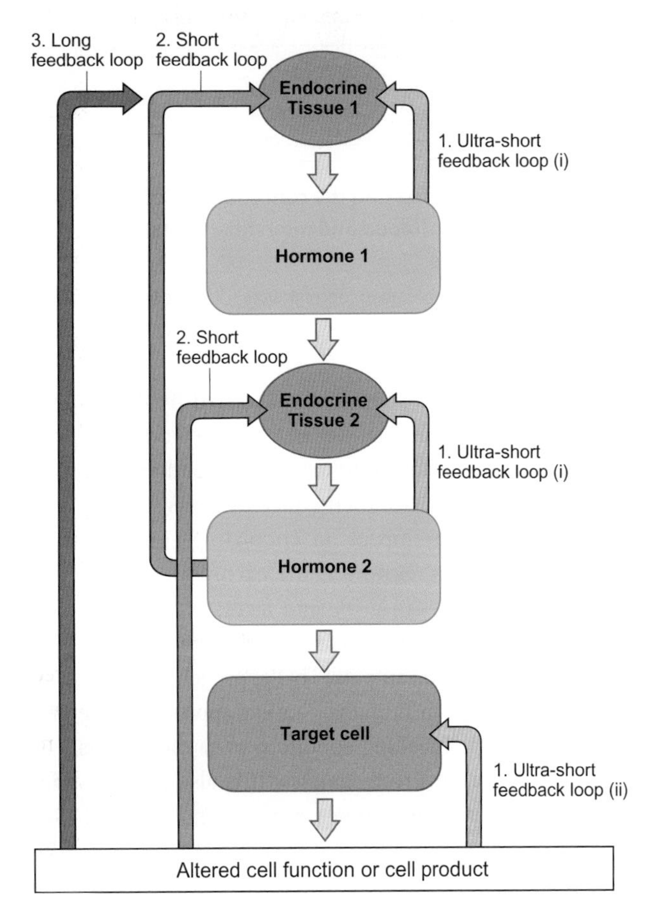

Figure 19.5 Feedback loops in endocrine cascades

Three levels of feedback loops are shown, defined by the number of steps across which they exert feedback.

1 Ultra-short feedback (pale orange arrows) involves either (i) direct feedback by a hormone on the tissue producing the hormone, or (ii) direct feedback by a cell product or the result of altered cell function on the activated cell.

2 Short feedback loops (dark orange arrows) involve feedback two steps back of an altered cell product or the concentration of free hormone in the blood on a higher part of an endocrine cascade.

3 Long feedback loops (brown arrow) occur when altered cell function and/or cell products influence tissues three or more steps back up the endocrine cascade.

Ultimately, circulating concentrations of hormone are mainly influenced by the secretion of active substances, which often occurs in pulses. Pulsatile secretion of neurohormones, such as those of the vertebrate hypothalamus, often results in pulsatile secretion of other hormones in an endocrine axis[11]. Generally, pulses of secretion overlay a low level of basal secretion and determine the average concentrations of a particular hormone in the blood via changes in one or more of the following variables:

- the amplitude of each pulse of secretion, which determines the rise in hormone concentration from minimum to peak levels during the pulse;
- the frequency of secretory pulses;
- the accumulative area under the curve of changing hormone concentration over time.

The changing amplitude and/or frequency of pulses of hormonal secretion often result in a circadian rhythm (an approximately daily pattern) of fluctuating hormone concentrations. Such hormonal rhythms are entrained by environmental factors, often light–dark cycles (photoperiod), acting on the central nervous system[12].

19.1.4 Hormones affect specific tissues by binding to specific receptors in target cells and activating specific signalling molecules

All hormones exert their actions through interactions with specific receptors. Frequently a hormone acts on more than one type of receptor and hence produces multiple effects[13]. With the multiplicity of cellular receptors available in any one cell and multiple hormone actions, the interactions of hormones determine the actual functional changes. Antagonistic hormones oppose each other's effects, while two hormones acting toward a common endpoint that have greater effect than the sum of their individual actions are termed

[10] We discuss the positive feedback regulation controlling ovulation in Section 20.4.1.

[11] We discuss the pulsatile release of particular hormones in Sections 19.2.1 and 19.3.1.

[12] For vertebrates, melatonin released from the pineal gland entrains the rhythm as we discuss in Section 19. 2.3. We explore biological clocks and circadian rhythms in greater detail in Section 17.2.2.

[13] For an example, look at Box 21.1, which discusses the effects of arginine vasopressin acting on V_2 receptors in kidney tubules and V_1 receptors in vasculature.

synergistic. For instance, two insect hormones that we explore in Section 19.5.2—juvenile hormone and ecdysone—have a synergistic effect on moulting. A further interaction involves **permissive hormones**, when the presence of one hormone primes a tissue so that another hormone can be effective.

The extent of a physiological response resulting from binding of a hormone to receptors depends on the number of occupied receptors, which, in turn, depends on:

- the concentration of free hormone molecules available to bind to receptors, and the binding affinity of these receptors;
- the density of the hormone's target receptors in particular cells. A decrease in receptor density inevitably decreases the responsiveness of a target tissue, while increasing receptor density increases tissue responsiveness. As such, the adjustment of the density of a receptor is an important factor in regulating tissue responsiveness. When hormone concentrations increase for prolonged periods, the number of available receptors usually decreases.

But how does the binding of a hormone to its receptors initiate specific physiological actions? The process of hormone detection by a specific receptor, and transformation of the message into a cellular response, is called **signal transduction**. Signal transduction occurs by different mechanisms depending on the chemical messenger and the type of receptor[14].

Hydrophobic hormones such as steroids and thyroid hormones usually enter cells through the plasma membrane and bind to receptors inside the cells, although interaction with additional receptors in the plasma membrane may also occur. Water-soluble hormones and paracrines cannot easily gain entry to cells because of their poor solubility in the lipid bilayer of the plasma membrane, and usually bind with receptors in the plasma membrane of target cells at locations that face the extracellular environment. The binding of molecules of hormone (the *first* messenger) triggers intracellular events that often (but not always) involve **second messengers**.

Second messenger systems are important in amplification of the hormone signal. The circulating concentrations of peptide hormones are low—often as little as 10^{-12} mol L^{-1}, or around 1 picogram (10^{-12} g) per mL of plasma—but a single molecule of hormone binding to a specific receptor results in the formation of many molecules of the second messenger. Then, each molecule of second messenger goes on to generate many molecules of cell proteins in a cascade of events that are ultimately responsible for physiological effects.

[14] Signal transduction also occurs in response to neurotransmitters acting on ligand-gated ion channels in postsynaptic membranes, as we discuss in Section 16.3.

Most water-soluble hormones act on G-protein-coupled receptors in the plasma membrane of target cells

Figures 19.6 and 19.7 are schematic diagrams of the workings of the most common second messenger systems for the receptors of water-soluble hormones. In each case, the transmembrane receptor is coupled to a **G-protein**—so named because of their high affinity for guanine nucleotides: guanosine diphosphate (GDP) and guanosine triphosphate (GTP). Binding of the hormone (the first messenger) to the extracellular side of a specific receptor activates the **G-protein-coupled receptor** (GPCR) on the cytosolic face of the membrane.

In fact, several types of G-proteins bind to different receptors. Hence, GPCRs are responsible for sensing light by photoreceptors, for olfaction and taste[15], most responses to neurotransmitters, and the majority of hormonal responses. Because of the prevalence of GPCRs, about half of all existing therapeutic drugs act as ligands for GPCRs, and yet we have only recently begun to understand how the binding of an agonist or antagonist influences the ligand–receptor–G-protein complex.

Recent advances in understanding GPCRs are in large part due to the groundbreaking discoveries of two US scientists, Robert Lefkowitz and Brian Kobilka, who received the 2012 Nobel Prize in Chemistry for research that revealed some of the inner workings of the GPCRs. They found that the family of GPCRs look alike and function in the same basic way.

All GPCRs pass across the cell membrane seven times, so they are also called seven-transmembrane receptors. The G-protein with which these receptors are associated consists of three subunits: Gα, Gβ, and Gγ, as shown in Figure 19.8. These subunits easily separate from the receptor protein, so working out the receptor activation process has not been easy. Tricky stabilization of the entire three-component complex (agonist, receptor and the whole G-protein) was necessary so that it could be crystallized and examined by X-ray crystallography to create structural models of the events occurring during signal transduction. Figure 19.8A shows a molecular model of the three-dimensional structure of an active complex of agonist bound to one of the receptors for adrenaline and noradrenaline (the **β_2 adrenergic receptor** or **adrenoceptor**)[16] and the associated G-protein subunits.

When the G-protein is inactive, GDP is bound to the Gα subunit. When hormone binding activates the receptor, however, GTP replaces GDP, and the Gα subunit dissociates from Gβ and Gγ. Looking again at Figures 19.6 and 19.7 we find that interaction of the Gα subunit alone with the relevant membrane-bound enzyme leads to formation of second messengers.

[15] Section 17.3.1 outlines the process of sensory conduction in olfaction.

[16] Sub-types of adrenergic receptors (adrenoceptors) occur, and are outlined in Table 16.1. Adrenaline and noradrenaline are also called epinephrine and norepinephrine, respectively.

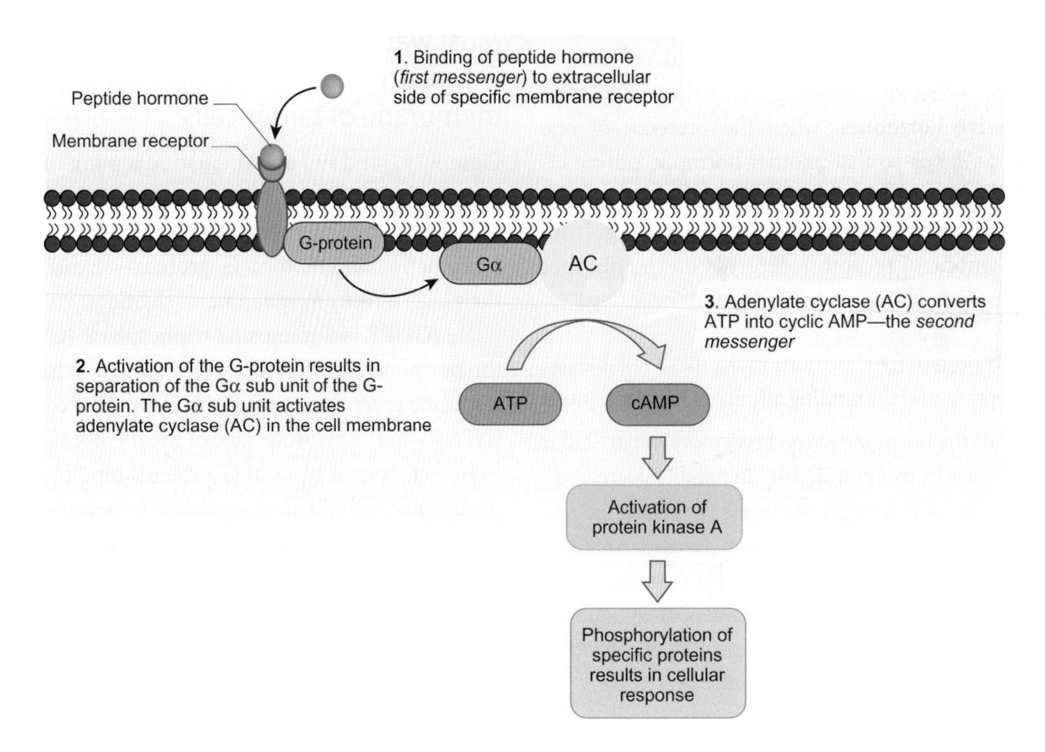

Figure 19.6 Cyclic AMP second messenger system of G-protein coupled hormone receptor

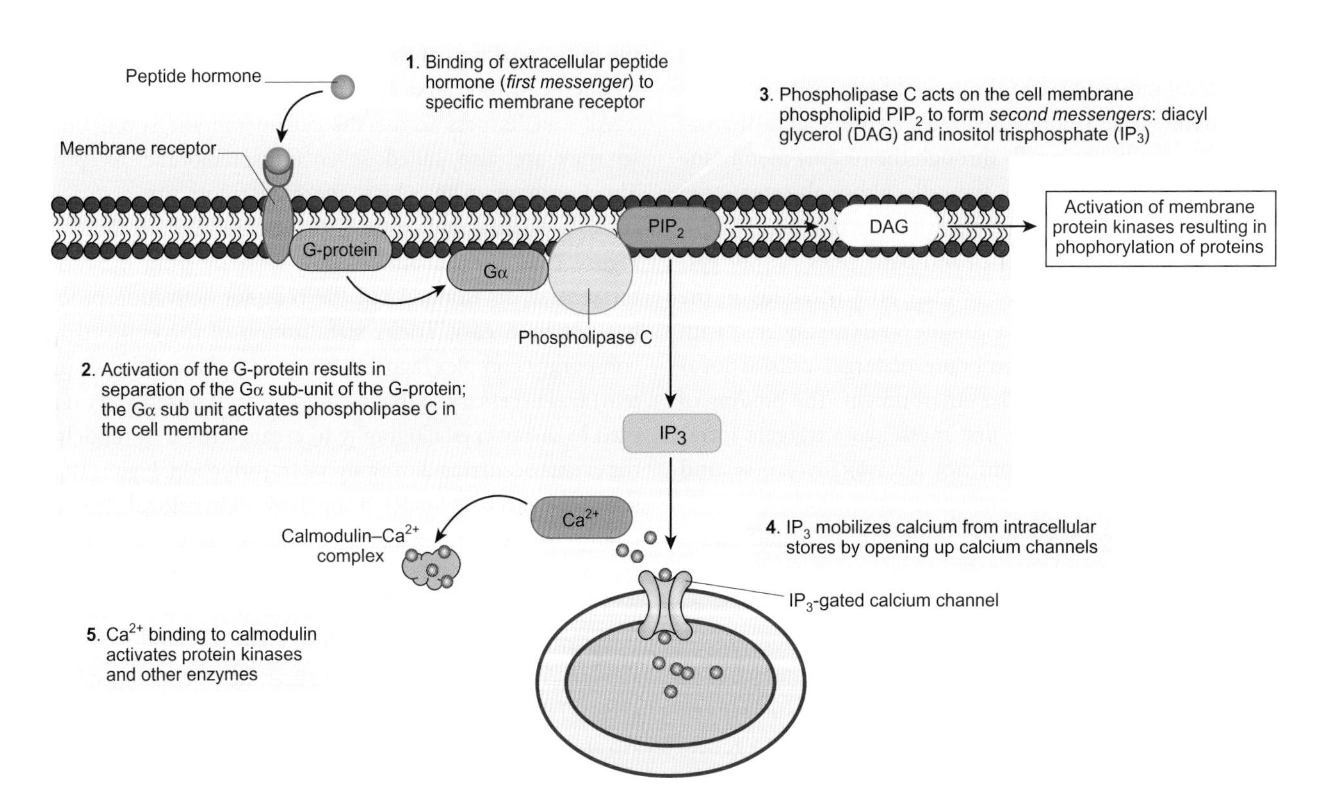

Figure 19.7 Steps in G-protein coupled hormone receptor activation of the second messengers: inositol trisphosphate (IP_3), diacyl glycerol (DAG) and Ca^{2+}

DAG as a second messenger activates membrane kinases and results in protein phosphorylation. The IP_3 signalling mechanism mobilizes intracellular calcium (Ca^{2+}) stores. The increase in the cytoplasmic concentration of Ca^{2+} results in binding to a cellular protein, calmodulin, which activates protein kinases and other enzymes that affect cell function.

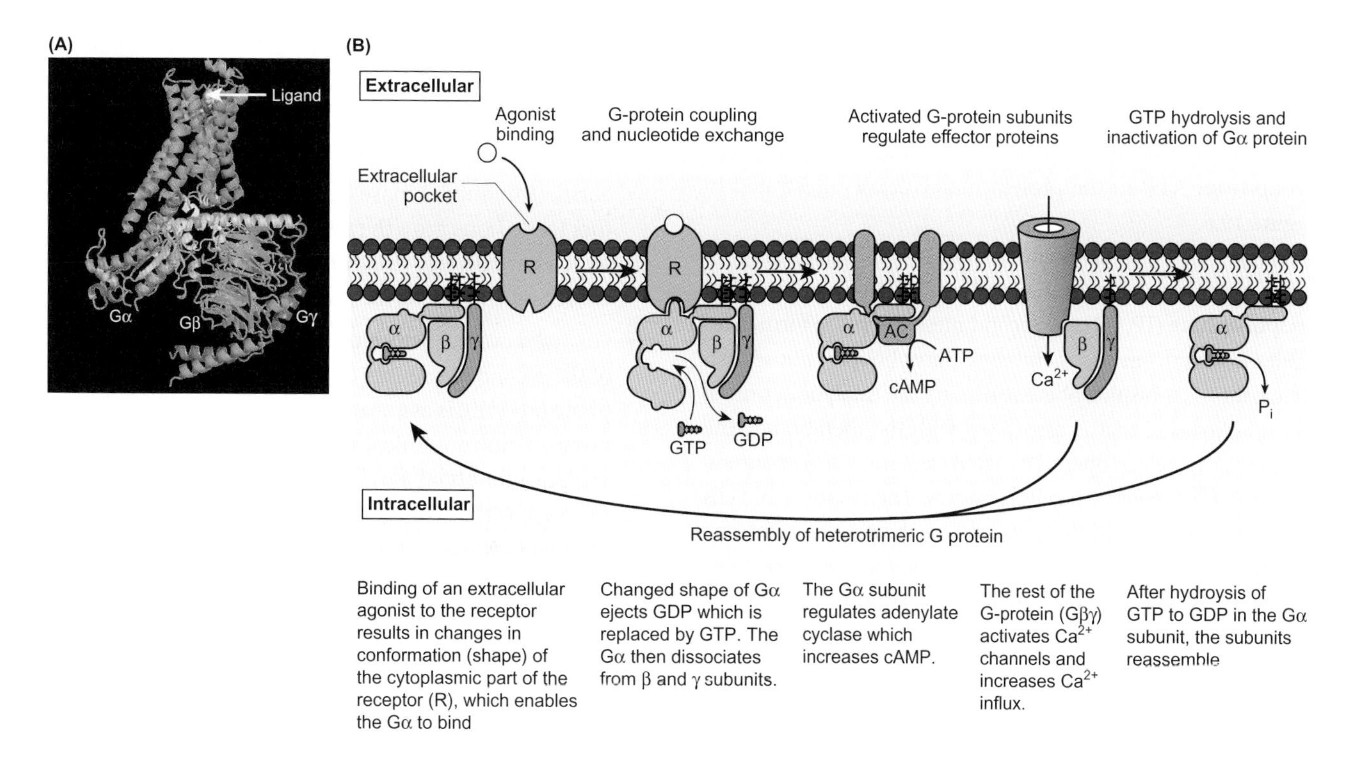

Figure 19.8 G-protein coupled receptor complex and its activation

(A) Structural model of β_2 adrenoceptor (adrenergic receptor, AR) showing seven transmembrane loops (in green) passing across the membrane. The ligand (yellow) is in the binding pocket. The activated complex is linked to the three subunits of the G-protein: Gα (orange), Gβ (turquoise), Gγ (blue). (B) G-protein cycle for the β_2 AR-stimulated G-protein complex.

Sources: A: Buchen L (2011). Cell signalling caught in the act. Receptor imaged in embrace with its G protein. News in Focus. Nature 475: 273–274; B: Rasmussen SGF et al (2011). Crystal structure of the β_2 adrenergic receptor-Gs protein complex. Nature 477: 549–557.

What is happening at the molecular level during this process? It turns out that once the hormone binds to the receptor, the change in shape of the receptor allows activation of the G-protein via the Gα subunit. Contortion forces guanosine diphosphate (GDP) out of the α-subunit, with GTP taking its place. This change results in dissociation of the combined β-γ subunit and the α-subunit from the receptor complex. The breaking apart of the activated G-protein allows the component subunits to activate separate signalling mechanisms. The Gα subunit moves freely in the membrane until it meets and activates adenylate cyclase, or phospholipase C, as illustrated in Figures 19.6 and 19.7, respectively.

Figure 19.8B shows how these subunits recycle. After GTP hydrolyses to GDP on the Gα subunit, the G-protein subunits reassemble and recycle; once a molecule of hormone attaches to the receptor, it can activate hundreds of G-proteins.

With so many different GPCRs, an obvious question is: how are specific responses achieved? The specificity of a receptor for a particular hormone depends on whether or not it binds to the extracellular pocket, as illustrated in Figure 19.8B. Inside the cell, interaction of GPCR signalling units with specific signalling proteins such as enzymes and ion channels can determine the signal specificity of particular tissues. Even in a single cell type, the same G-protein can elicit different cellular responses depending upon which signalling proteins are present in the membrane domains, and the second messengers that are activated.

Adrenaline and many of the peptide hormones use **cyclic adenosine monophosphate (cAMP)** as the second messenger. Activation of the receptor by binding of hormone increases cAMP by the action of **adenylate cyclase** (or more accurately adenylyl cyclase) on ATP, as shown in Figure 19.6. A number of processes involve cAMP signalling; these include the control of water reabsorption in the kidneys (by arginine vasopressin)[17], and stimulation of the heart by adrenaline[18]. In the same way, signalling via cAMP occurs when follicle stimulating hormone stimulates the ovaries[19] or thyroid stimulating hormone stimulates the thyroid gland[20].

As so many hormones act via cAMP, we might ask: how are such diverse functions achieved? The answer lies in the occurrence of specific cellular responses to the second messenger system in particular target cells. In this way, a common second messenger system can achieve specific responses

17 We learn about the action of arginine vasopressin on the kidney in Section 21.1.4.

18 Section 22.4.3 discusses the actions of adrenaline on the heart.

19 We discuss the actions of follicle stimulating hormone in Section 20.4.

20 Section 19.4 discusses the actions of thyroid stimulating hormone.

according to the type of cell involved and the proteins it produces. Second messenger systems most commonly involve the activation of intracellular **protein kinases**, which phosphorylate particular cell proteins (i.e. transfer a phosphate group from ATP to the protein molecule), as illustrated in Figure 19.6.

Figure 19.7 illustrates some hormone receptors working via other second messengers: **diacyl glycerol (DAG)** and **inositol trisphosphate** (IP_3). These second messengers are formed simultaneously from the membrane phospholipid, phosphatidylinositol 4, 5-bisphosphate (PIP_2), by the catalytic action of phospholipase C that is associated with the membrane. The IP_3 in the cytoplasm opens IP_3-gated calcium (Ca^{2+}) channels, releasing Ca^{2+} from intracellular stores. The response of cells to hormones acting in this way, such as the peptide hormone **angiotensin II** that causes vasoconstriction, are detectable by measuring the cytoplasmic concentration of Ca^{2+} or tissue activity of IP_3. Calcium ions may themselves act as further second messengers—for instance, activating the cytoplasmic protein **calmodulin** and thereby activating cellular enzymes.

Some cell surface receptors for hormones are enzyme-linked receptors

Some water-soluble hormones, such as **insulin**, which regulates blood glucose, and various growth factors, such as **insulin-like growth factor 1** (IGF-1), act on enzyme-linked receptors. These transmembrane proteins have either intrinsic enzyme activity or associate with enzymes in the cytoplasm (rather than associating with G-proteins as is the case with G-protein coupled receptors (GPCRs)). Instead of the seven transmembrane domains of GPCRs, each monomer of the enzyme-linked receptors has a single transmembrane domain. These monomers usually form dimers (pairs) during their activation.

Enzyme-linked receptors exhibit three main types of enzyme activity:

- Tyrosine kinase activity accounts for most enzyme-linked receptors, including many of the receptors for growth factors and insulin receptors. Hormone binding to two nearby receptor sites results in formation of receptor dimers (or tetramers). A key factor in all tyrosine kinase receptors is their autophosphorylation, which Figure 19.9 illustrates. Receptor binding activates the enzyme site on the cytosolic side of the transmembrane receptor, which catalyses the transfer of phosphate groups from ATP to the amino acid tyrosine on the substrate protein. This autophosphorylation results in the formation of binding sites for a whole series of intracellular signalling proteins. Once these proteins bind, subsequent phosphorylation or dephosphorylation initiates complex cellular responses. Different receptor protein kinases bind different combinations of signalling proteins and hence achieve different cellular responses.
- Serine-threonine kinase activity occurs in the receptors for various growth factors and tissue cytokines, which have diverse functions in cell signalling by autocrine and paracrine mechanisms.
- Guanylyl cyclase, which catalyses cytosolic production of the second messenger **cyclic guanosine monophosphate (cGMP)**, is associated with some receptors. This group of enzyme-linked receptors includes some of the receptors for **atrial natriuretic peptide** (ANP). ANP is secreted by the heart and acts on vasculature, the kidney and osmoregulatory organs[21]. The cGMP acting as a second messenger for such receptors activates specific proteins in a similar way to the activation of intracellular proteins by the GPCRs that act via cAMP.

Lipophilic hormones pass through the plasma membrane and promote synthesis of new proteins

Steroid hormones (for example, oestrogens, androgens, cortisol and corticosterone shown in Figure 19.3, ecdysone in insects), thyroid hormones, and active forms of vitamin D[22] are all lipophilic hormones that act by a similar mechanism. Figure 19.10 illustrates this process.

The initial step involves passive entry of the hormone into cells via the lipid bilayer of the plasma membrane. Some hormones, such as thyroid hormones, also enter the cell via transporter proteins. Once inside the cells, interaction with intracellular receptors occurs. These receptors are called **nuclear receptors** because of the direct interaction of the hormone–receptor complex with DNA in the cell nucleus. Most intracellular receptors for lipophilic hormones are in the cytosol and enter the nucleus after binding to hormone, but cellular receptors for some lipophilic hormones, such as the thyroid hormone receptor, occur in the nucleus.

All nuclear receptors have a similar structural organization. The carboxy-terminal part of these receptors encompasses a region called the ligand-binding domain, which binds with the hormone (ligand), and ensures hormone recognition. The binding of a hormone to its nuclear receptor acts like a molecular switch that activates the hormone–receptor complex. In an inactive state, the nuclear receptors are bound to an inhibitory protein complex that prevents binding to DNA. However, hormone binding causes the inhibitory complex

[21] We discuss the actions of atrial natriuretic peptide on salt and water balance in Section 21.1.2.

[22] Section 21.2.1 discusses the role of vitamin D in control of calcium balance.

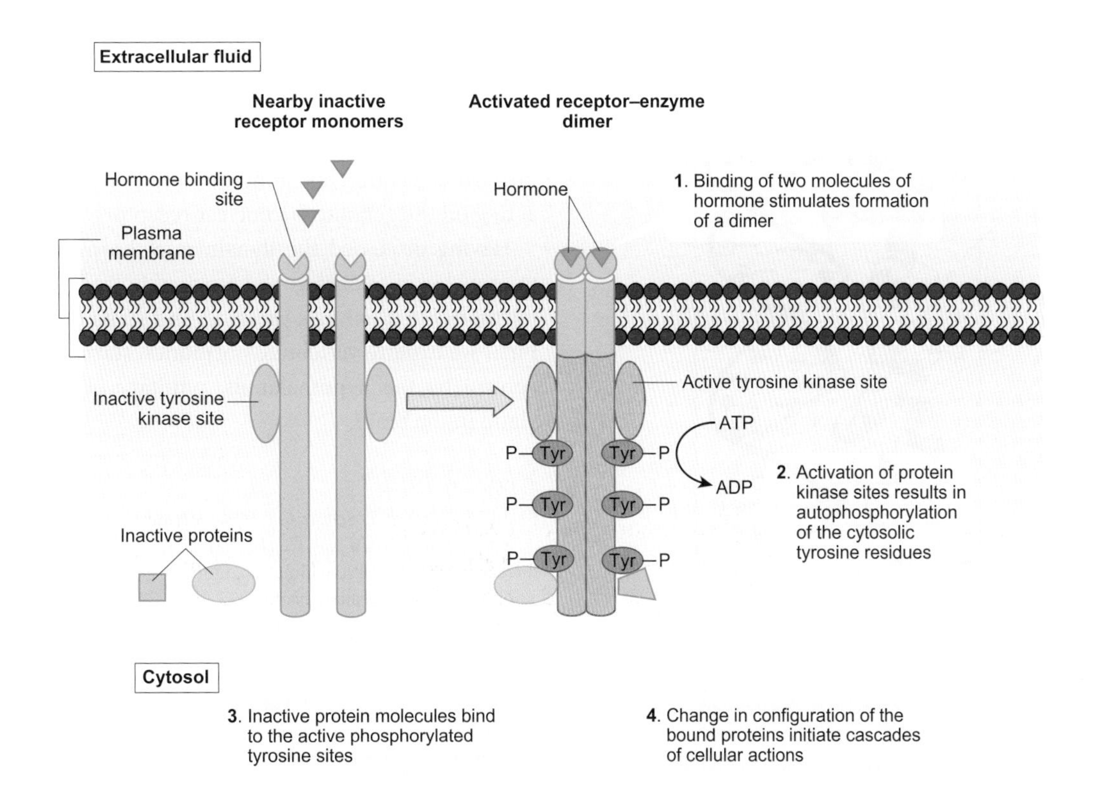

Figure 19.9 Simplified mechanism of functioning of receptor–enzyme molecules

The enzymatic site (tyrosine kinase) is activated when a hormone molecule binds to the extracellular side of the receptor causing their association to form cross-linked dimers, which phosphorylate each other at tyrosine (Tyr) sites. This cross-phosphorylation creates binding sites for a multiplicity of cellular proteins, which bind other cell proteins and trigger cascades of cell signalling pathways that result in specific cellular responses.

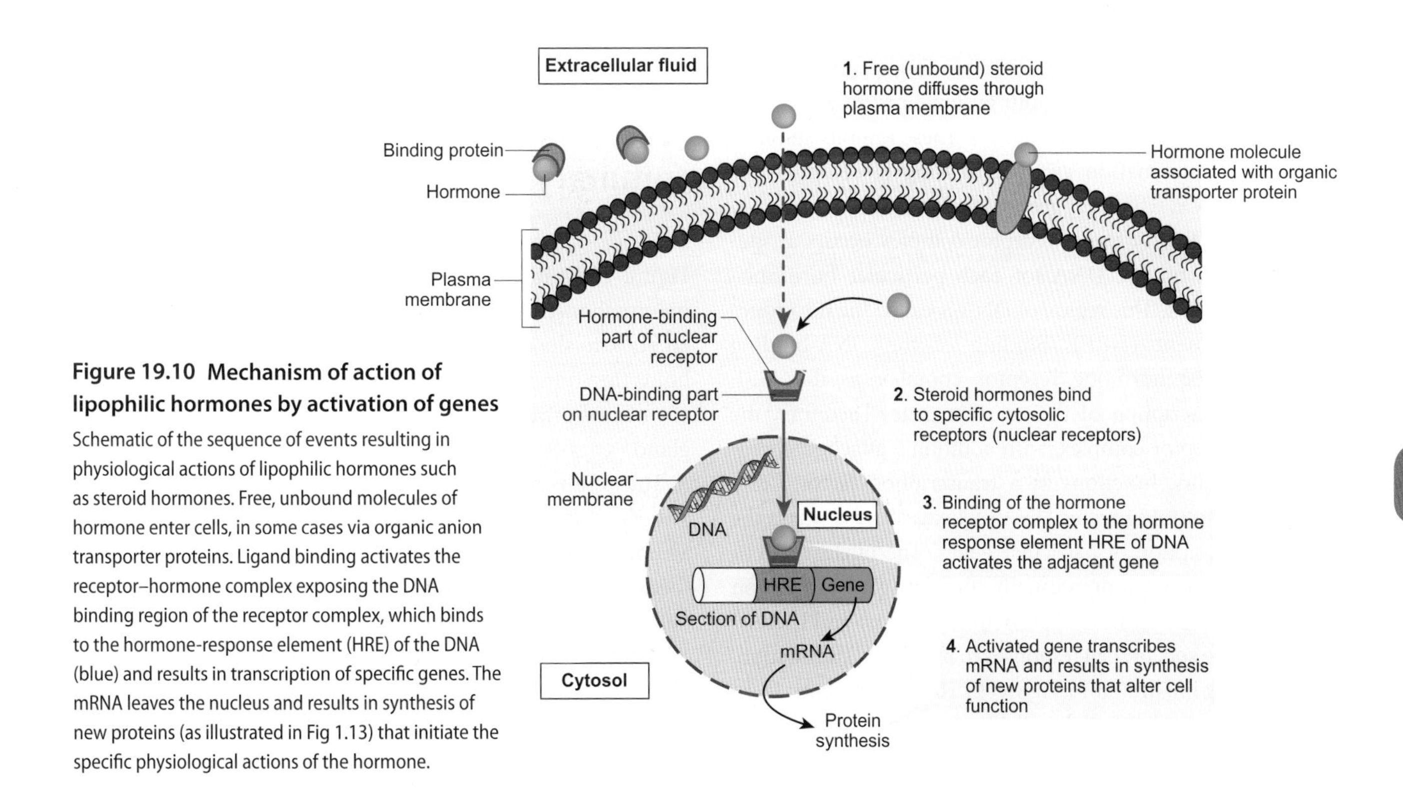

Figure 19.10 Mechanism of action of lipophilic hormones by activation of genes

Schematic of the sequence of events resulting in physiological actions of lipophilic hormones such as steroid hormones. Free, unbound molecules of hormone enter cells, in some cases via organic anion transporter proteins. Ligand binding activates the receptor–hormone complex exposing the DNA binding region of the receptor complex, which binds to the hormone-response element (HRE) of the DNA (blue) and results in transcription of specific genes. The mRNA leaves the nucleus and results in synthesis of new proteins (as illustrated in Fig 1.13) that initiate the specific physiological actions of the hormone.

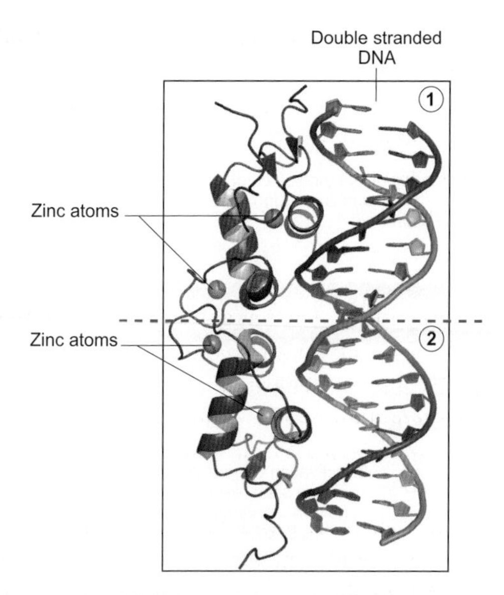

Figure 19.11 Molecular configuration of DNA-binding domain of a steroid hormone nuclear receptor

Steroid hormone receptors (shown in purple) bind to DNA as dimers of two adjacent molecules, labelled 1 and 2 either side of the horizontal blue dashed line in the diagram. Each monomer contains two zinc atoms (shown as orange spheres; two labelled in each adjacent receptor molecule). Each zinc atom is each held in place by four conserved cysteine residues, and forms two 'zinc fingers' per molecule that stabilize the binding of the receptor to the major groove of double-stranded DNA. The diagram shows a glucocorticoid receptor; we discuss glucocorticoids in Section 19.3.

Image derived from Protein Data Bank entry 1GLU.

to be released, exposing the central part of the hormone–receptor complex that contains a DNA-binding domain.

The DNA-binding domain of all of the nuclear receptors contains two highly conserved 'zinc fingers' that sets these receptors apart from other DNA-binding proteins. Each zinc atom is held in place by four cysteine residues (the 'fingers' that give the domain its name). These regions stabilize molecular conformation and the binding of receptors, as dimers, to DNA, which Figure 19.11 illustrates.

Binding of the hormone–receptor complex occurs at specific locations on the DNA for each particular hormone–receptor complex. This region of DNA is called the hormone-response element (HRE). Binding of the DNA-binding domain of the hormone receptor complex to the HRE regulates transcription of the adjacent gene. Therefore, the hormone–receptor complex, with additional attached coregulatory proteins, functions as a transcription factor, i.e. a protein that binds to specific DNA sequences and controls the rate of transcription of particular genes. The nuclear receptors of lipophilic hormones usually stimulate gene expression (and, hence, the synthesis of particular proteins) and thereby determine the specific cell responses to hormonal stimulation.

The events illustrated in Figure 19.10 typically take 15 to 30 minutes to occur, but the effects are relatively persistent compared to the faster but less persistent responses when peptide hormones activate membrane receptors. However, recent studies indicate that membrane receptors for lipophilic hormones can initiate faster non-genomic actions on cell enzymes or ion fluxes in a similar way to the signal transduction of peptide hormones, in addition to the hormones acting on the genome via nuclear receptors.

Having discussed the dynamics of hormonal processes, we now begin to explore the range of hormones and neurohormones that animals produce, and some of their actions. First, we explore vertebrate hormones, starting with central control by the hypothalamus, pituitary gland and pineal gland in the brain.

› *Review articles*

Beato M, Herrlich P, Schütz G (1995). Steroid hormone receptors: many actors in search of a plot. Cell 83: 851–857.

Buchen L (2011). Cell signalling caught in the act. Receptor imaged in embrace with its G protein. Nature 475: 273–274.

Evans RM (1988). The steroid and thyroid hormone receptor superfamily. Science 240: 889–895.

Hanukoglu I (1992). Steroidogenic enzymes: structure, function and role in regulation of steroid hormone biosynthesis. Journal of Steroid Biochemistry Molecular Biology 43: 779–804.

Kaltenbach JC (1988). Endocrine aspects of homeostasis. American Zoologist 28: 761–773.

Mangelsdorf DJ, Thummel C, Beato M, Herrlich P, Schütz G, Umesono K, Blumberg B, Kastner P, Mark M, Chambon P, Evan RM (1995). The nuclear receptor superfamily: The second decade. Cell 83: 835–839.

Rosenbaum DM, Rasmussen SGF, Kobilka BK (2009). The structure and function of G-protein-coupled receptors. Nature 459: 356–363.

Simons SS (2008). What goes on behind closed doors: physiological vs. pharmacological steroid hormone actions. Bioessays 30: 744–756.

Ullrich A, Schlessinger J (1990). Signal transduction by receptors with tyrosine kinase activity. Cell 61: 203–212.

19.2 Central control processes of vertebrates—the hypothalamus, pituitary gland and pineal gland

Figure 19.12 illustrates the anatomical arrangement of the mammalian hypothalamus[23] connected by the infundibular stalk to the pituitary gland lying just beneath. This stalk holds the connecting nerve fibres and blood vessels that convey neurohormones from the hypothalamus to the pituitary gland.

The pituitary gland consists of two distinct lobes: the **anterior pituitary** and the **posterior pituitary**. This basic arrangement occurs throughout the vertebrates, although there are marked morphological variations at a more detailed level. For example, a simpler, looser arrangement of pituitary regions occurs in the jawless fish (hagfish and

[23] The schematic in Figure 16.9 shows the location of the hypothalamus (and pituitary gland) in the forebrain of the vertebrate brain.

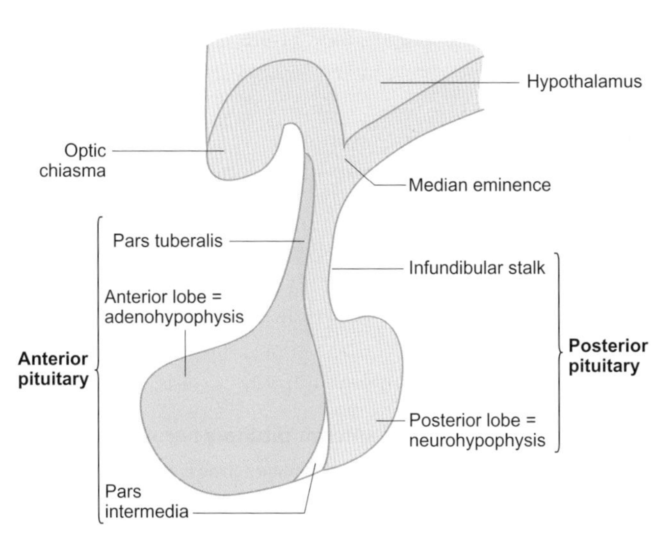

Figure 19.12 Diagram of the mammalian pituitary gland showing the anatomical relationships between the anterior and posterior pituitary and the hypothalamus

The anterior pituitary consists of the non-neural lobe (adenohypophysis), pars tuberalis, and the pars intermedia (some species). The posterior pituitary consists of the stalk and the neural lobe (neurohypophysis).

lampreys). A small intermediate lobe occurs in many vertebrate groups—jawless fish, amphibians, reptiles and most mammals—but is rudimentary in humans.

The pronounced division of the pituitary into two lobes reflects its embryonic origin as a down-growth of neural tissue from the brain (which forms the posterior lobe) and an up-growth called Rathke's pouch, from the roof of the stomodeum (the beginnings of the oral cavity). Once these two growths have come together to form a pituitary gland the connection to the stomodeum is lost, but the gland remains attached to the brain by the stalk.

The posterior pituitary lobe consists primarily of nerve terminals that arise from specific neurons in the hypothalamus, as shown in Figure 19.13, together with some supporting cells. This composition explains the alternative name of **neurohypophysis**: neuro = neural structure; hypophysis refers to the pituitary. The neurohypophysis does not synthesize hormones. Instead, it acts as a storage depot for peptide hormones that are synthesized by specific groups of neurons in the hypothalamus and conveyed to the neurohypophysis, before later release into the circulation.

In stark contrast, the anterior pituitary consists of non-neural glandular tissue. Its alternative name of **adenohypophysis** literally means the glandular part of the hypophysis (pituitary). The anterior pituitary synthesizes and secretes hormones, and usually connects to the hypothalamus by a portal system of blood vessels running from the median eminence to the pituitary, through the stalk, as shown in Figure 19.14. However, teleost fish lack this portal system of connecting vessels.

19.2.1 The hypothalamus and neurohypophysis function together to secrete neurohypophysial peptides

Two different types of neurohormones (neurohypophysial peptides) occur in their neurohypophysis of most vertebrate species, which reflects the evolution of two lineages that are traceable from bony fish to mammals. The occurrence of similar peptides among invertebrate animals, such as molluscs and

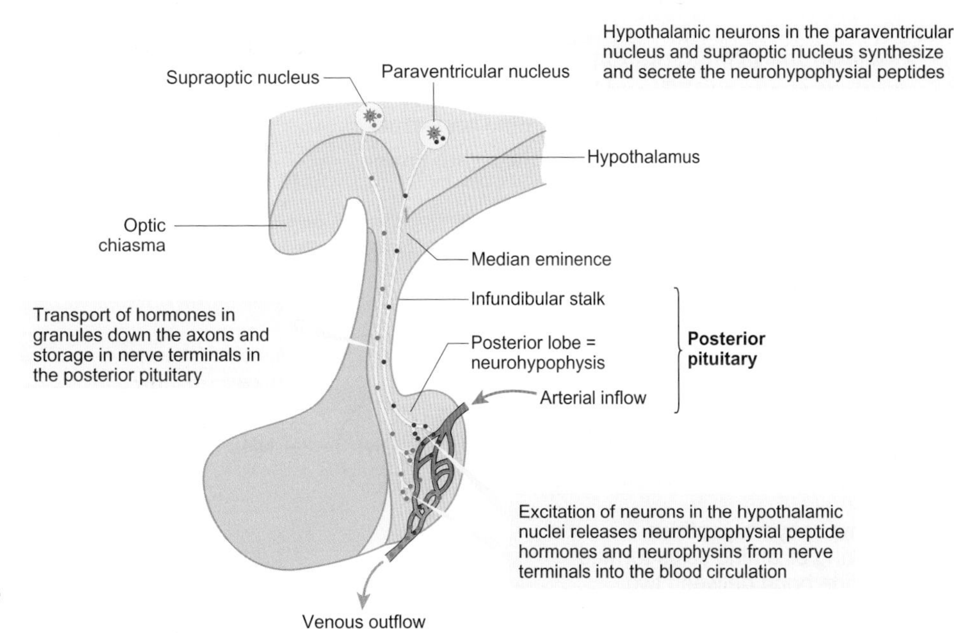

Figure 19.13 Hypothalamic–pituitary regulation of neurohypophysial peptide synthesis, storage and release in birds and mammals

19

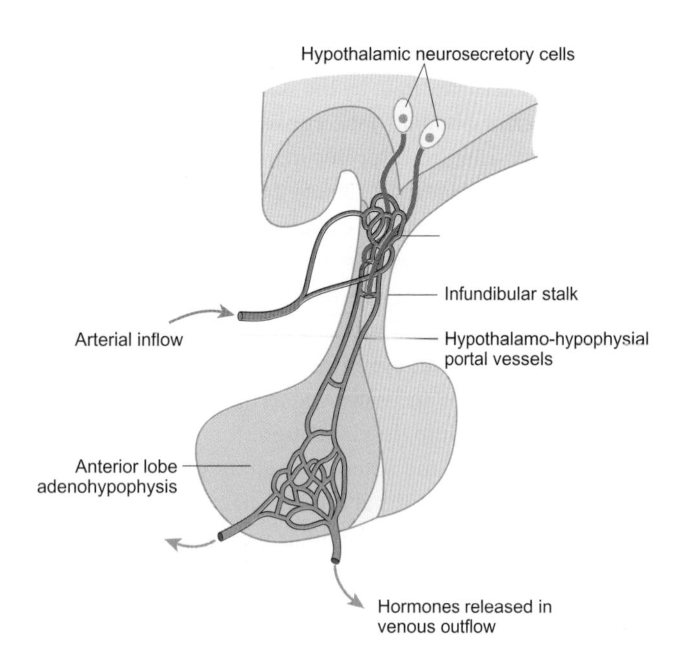

Figure 19.14 Secretion of anterior pituitary hormones

Secretion of hormones from the anterior pituitary gland is controlled by the hypothalamic neurohormones which are released by neurosecretory cells in the hypothalamus and transported along axons to the median eminence. These regulatory hormones are transported in the hypothalamo-hypophysial portal system (except in teleost fishes that lack the portal system) to the anterior pituitary.

annelids, suggests an earlier origin of these peptides. For example, in the marine polychaete worm *Platynereis dumerilii* these peptides have been identified in neurosecretory cells in the brain where they have been proposed to convey sensory cues.

All neurohypophysial peptides consist of nine amino acid residues (they are nonapeptides) and have a similar ring structure formed by the disulfide bond between two cysteine residues, and a side chain of three amino acids, as shown in Figure 19.15. Table 19.1 shows the amino acid structures of the main neurohypophysial peptides identified in particular vertebrate groups. From these structures, we can see that six of the nine amino acids are highly conserved, while significant differences occur in the other three amino acids. These differences allow us to distinguish two major peptide groups:

- Vasotocin-like peptides: arginine vasopressin (AVP), arginine vasotocin (AVT) and lysine vasopressin (LVP), named partly based on their amino acid structures, which are shown in Table 19.1; the eighth amino acid residue is an arginine or lysine. Lysine vasopressin only occurs in a limited number of mammalian species: pigs and related placental mammals and some of the marsupial mammals. Some marsupials produce a different vasotocin-like peptide, phenylpressin, which has a similar structure to AVP but with phenylalanine as the second amino acid residue.
- **Oxytocin-like peptides**. Oxytocin-like peptides are more varied than the vasotocin-like peptides, particularly among cartilaginous fish. Table 19.1 shows the variations that occur particularly at the fourth and eighth amino acid residues. Originally, all placental mammals were thought to possess oxytocin, but recent studies identified a novel oxytocin-like peptide, Pro^8-oxytocin, in South American common squirrel monkeys (*Samiri sciureus*) and DNA coding for Pro^8-oxytocin in other South American primates (tufted capuchin *Cebus apella*, Nancy Ma's night monkeys *Aotus nancymaae* and common marmosets *Callithrix jacchus*).

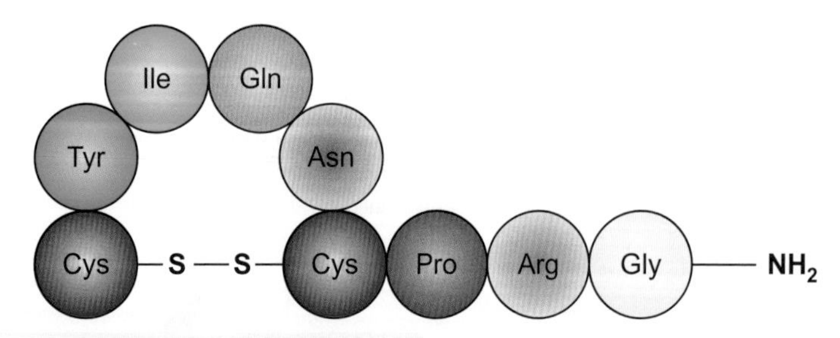

Figure 19.15 Diagram of structure of neurohypophysial peptides

The example shown is arginine vasotocin which occurs in all non-mammalian vertebrates.

Table 19.1 **Neurohypophysial hormones of vertebrates**

All neurohypophysial hormones are peptides consisting of a ring of amino acid residues resulting from a disulfide bridge between cysteine residues (1 and 6; indicated by the red bracket), and a tail of three amino acid residues. The amino acid residues in blue font are conserved in all animal groups. Variable amino acid residues occur at positions 3, 4 and 8.

Peptide	Animal group	Amino acid residue								
Vasotocin-like (basic peptides)		1	2	3	4	5	6	7	8	9
Arginine vasopressin (AVP)	Most mammals	Cys	Tyr	Phe	Gln	Asn	Cys	Pro	Arg	Gly(NH_2)
Lysine vasopressin (LVP)	Pigs, peccary, hippopotamus; several marsupials e.g. eastern grey kangaroo (*Macropus giganteus*)	Cys	Tyr	Phe	Gln	Asn	Cys	Pro	Lys	Gly(NH_2)
Arginine vasotocin (AVT)	All non-mammalian vertebrates (birds, reptiles, amphibians, fish, including lampreys and hagfish)	Cys	Tyr	Ile	Gln	Asn	Cys	Pro	Arg	Gly(NH_2)
Oxytocin-like (neutral peptides)										
Oxytocin	Placental mammals Cartilaginous spotted ratfish (*Hydrolagus colliei*)	Cys	Tyr	Ile	Gln	Asn	Cys	Pro	Leu	Gly (NH_2)
Pro[8]-Oxytocin	Some placental mammals: e.g. common squirrel monkey (*Saimiri sciureus*) and northern treeshrew (*Tupaia belangeri*)	Cys	Tyr	Ile	Gln	Asn	Cys	Pro	Pro	Gly (NH_2)
Isotocin	Bony fish	Cys	Tyr	Ile	Ser	Asn	Cys	Pro	Ile	Gly(NH_2)
Mesotocin	Birds, amphibians, lungfish, marsupials	Cys	Tyr	Ile	Gln	Asn	Cys	Pro	Ile	Gly (NH_2)
Glumitocin	Cartilaginous fish—rays	Cys	Tyr	Ile	Ser	Asn	Cys	Pro	Gln	Gly (NH_2)
Valitocin	Cartilaginous fish—spiny dogfish (*Squalus acanthias*)	Cys	Tyr	Ile	Gln	Asn	Cys	Pro	Val	Gly (NH_2)
Aspargtocin	Cartilaginous fish—spiny dogfish (*Squalus acanthias*)	Cys	Tyr	Ile	Asn	Asn	Cys	Pro	Leu	Gly (NH_2)
Asvatocin	Cartilaginous fish—spotted dogfish (*Scyliorhinus canicula*)	Cys	Tyr	Ile	Asn	Asn	Cys	Pro	Val	Gly (NH_2)
Phasvatocin	Cartilaginous fish—spotted dogfish (*Scyliorhinus canicula*)	Cys	Tyr	Phe	Asn	Asn	Cys	Pro	Val	Gly (NH_2)

Sources of data: Archer T (1996). Molecular evolution of fish neurohypophysial hormones: Neutral and selective evolutionary mechanisms. General and Comparative Endocrinology 102: 157–172; Lee AG et al (2011). A novel form of oxytocin in New World monkeys. Biology Letters 7: 584–587; Wallis M (2012). Molecular evolution of the neurohypophysial hormone precursors in mammals: Comparative genomics reveals novel mammalian oxytocin and vasopressin analogues. General and Comparative Endocrinology 179: 313–318.

Arginine vasotocin occurs in all non-mammalian species of vertebrates so is likely to be the ancestral molecule that gave rise to arginine vasopressin and lysine vasopressin. The replacement of AVT by AVP in mammals, results from a single amino acid substitution at the third residue (isoleucine replaced by phenylalanine), which arises from a change in just one base pair in the gene encoding AVT. Similarly, the change of arginine in AVP to lysine in LVP only requires a single base pair change in the codon for the eighth amino acid. From these examples, we can see how gene duplication and subsequent single base pair changes may account for evolution of the neurohypophysial peptides.

The evolution of two lineages of neurohypophysial peptides—the vasopressin line and the oxytocin line—are thought to be associated with the evolution of their different actions. Arginine vasopressin has a pronounced vasoconstrictor action, increasing blood pressure, and acts as an **antidiuretic hormone** (ADH)[24]. Oxytocin stimulates the smooth muscle contraction linked to milk ejection in lactating mammals.

[24] We discuss the antidiuretic actions of AVP in mammals and AVT in non-mammalian vertebrates in more detail in Section 21.1.4.

As AVT acts on both vasopressin and oxytocin receptors, retention of this hormone would be problematic for lactating mammals. However, modification of AVT to AVP separates these actions and increases the antidiuretic activity.

Hypothalamic neurons control the posterior pituitary

Figure 19.13 shows the location of two distinct clusters of neurosecretory neurons (nuclei) in the hypothalamus, which exert control on the posterior pituitary in mammals. These are the paraventricular nucleus, close to the ventricle of the brain, and the supraoptic nucleus, closer to the optic chiasma. Mammals synthesize AVP mainly in the supraoptic nucleus and oxytocin mainly in the paraventricular nucleus, but there is overlap between the two areas and a capacity of the large magnocellular neurons in both nuclei to produce both peptides. These neurosecretory nuclei also occur in the hypothalamus of birds and reptiles. In other vertebrate groups, the neurosecretory neurons linking the pituitary and hypothalamus are in a preoptic nucleus.

The nuclei of the neurosecretory neurons have two linked functions:

19

- synthesis of prepro-neuropeptide hormones and packaging of the hormones into large dense-cored vesicles, which pass along the axon of the neurosecretory cell into the posterior pituitary where they remain until release;
- control of the release of the neurohypophysial hormones. Invasion of the neuronal terminal by an action potential causes depolarization of the membrane and triggers **exocytosis** of a small quantity of peptide, which enters the blood circulation.

Each prepro peptide consists of the active peptide attached to a specific carrier protein called a **neurophysin**, which appears to be important for correct protein folding, and in some cases other peptides of unknown physiological roles. Proteolytic cleavage of the peptide hormones from its neurophysin occurs during axonal transport, although the hormone remains loosely associated with the neurophysin during storage in the neural lobe of the posterior pituitary. Neural stimulation results in release of all components of the prepro peptides into the circulation, as indicated in Figure 19.13.

Figure 19.16 gives a better idea of the very large store of neurohypophysial peptide hormones in the posterior pituitary. In rats, 9000 to 18,000 magnocellular neurons each give rise to a single axon that stores hormone vesicles in the nerve endings and swellings within the neural lobe. Based on the number of vesicles per swelling or nerve ending, there are between 1×10^{15} to 2×10^{15} molecules of hormone stored in 1.3 to 2.6×10^{10} vesicles[25].

The hypothalamic neurons that secrete neurohypophysial peptides receive neural input from many other neurons that influence the rate of hormone secretion. For cells synthesizing oxytocin, suckling by a newborn mammal acts as an important signal for synthesis and release.

For neurons synthesizing AVP or AVT, blood volume and plasma osmolality are two major factors that determine peptide release[26]. When unstimulated, most of the neurosecretory neurons show a slow irregular firing pattern. In response to blood loss, progressive dehydration, or injection of hypertonic saline, however, the irregular firing ceases and is replaced by increased firing in a phasic pattern, which features bursts lasting for tens of seconds to 1 minute or more, separated by silences of similar duration. An increase in the firing rate of AVP or oxytocin neurons enhances the amount of hormone secretion per spike.

[25] From data for rats in Figure 19.16, the number of vesicles per axon = (436 swellings × 2240 vesicles) + (1840 nerve endings × 258 vesicles) = 1451360. For 9000 axons the total number of vesicles = 1,451,360 × 9000 = 1.3×10^{10} vesicles per neurohypophysis. In this case, the number of molecules of hormone per neurohypophysis = $1.3 \times 10^{10} \times 85000 = 1.1 \times 10^{15}$.

[26] Figure 21.20 explores the factors regulating release of AVP/AVT.

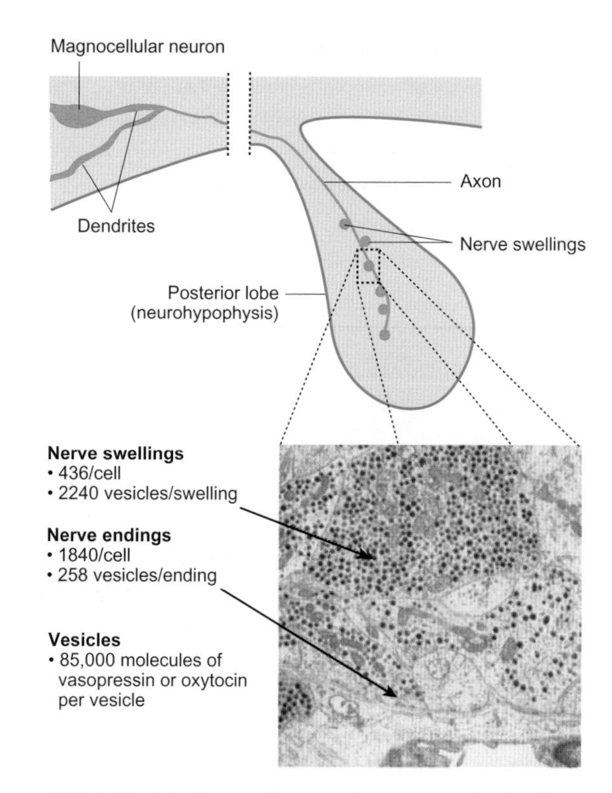

Figure 19.16 Vesicles in swellings and nerve terminals of magnocellular neurons of the rat

Diagram illustrates the route of an axon from a single magnocellular neuron. Numerous nerve swellings contain large numbers of vesicles of neuropeptide hormone. The large numbers of nerve endings add about half as much again to the stores of vesicles.

Source: Leng G, Ludwig M (2008). Neurotransmitters and peptides: whispered secrets and public announcements. Journal of Physiology 586: 5625–5632.

19.2.2 Hormones secreted by the anterior pituitary

The hormones secreted by the anterior pituitary are synthesized within the gland by specific groups of cells that form two major groups: (i) hormones acting directly on non-endocrine tissues, and (ii) hormones that stimulate growth and proliferation of another endocrine tissue and its release of further hormones, forming an endocrine axis as illustrated by Figure 19.17. These are 'trophic' hormones (from the Greek *trophikos* meaning nourishment)[27]. In the absence of an appropriate trophic hormone, the unstimulated endocrine tissue decreases in size and becomes relatively inactive.

Figure 19.18 illustrates the trophic hormones secreted by the anterior pituitary, and their actions, and the anterior pituitary hormones that act directly on non-endocrine tissues. The trophic hormones are:

[27] The use of the term tropic hormones (dropping the 'h') is widespread, but tropic has a slightly different origin and means 'turning' (for example, as in plant tropisms). The hormones in the anterior pituitary are best considered trophic hormones.

- **Thyroid stimulating hormone (TSH)** also called thyrotrophin, a glycoprotein produced by cells called thyrotrophs. TSH stimulates proliferation of endocrine cells in the thyroid gland and their synthesis and release of thyroid hormones[28].
- **Adrenocorticotrophic hormone (ACTH)**, produced by cells called corticotrophs. ACTH stimulates the growth and proliferation of cells in the outer layer of the adrenal gland—the **adrenal cortex**, or their functional homologue, and their synthesis and release of the steroid hormones, cortisol and/or corticosterone[29].
- **Growth hormone** (GH) produced by cells in the pituitary called somatotrophs; GH has metabolic actions on many tissues and stimulates release from the liver of other growth-promoting hormones.
- **Gonadotrophins—follicle stimulating hormone (FSH)** and **luteinizing hormone** (**LH**), which stimulate the growth of the gonads and production of gametes. These hormones operate differently in males and females. In females, FSH and LH stimulate development of ova (eggs) and production and release of **oestrogens**. In males, FSH and LH are necessary for production of sperm and secretion of **androgens** (usually **testosterone**) by the testes[30].

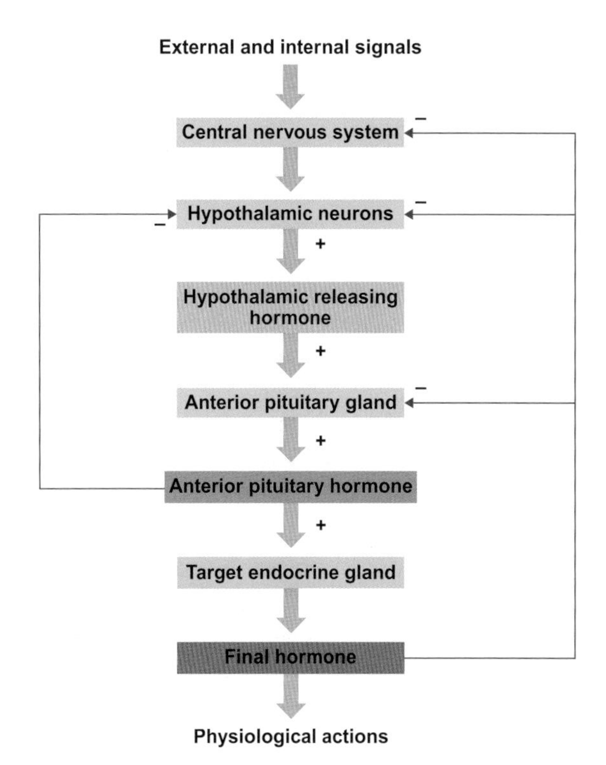

Figure 19.17 Schematic illustration of a typical endocrine axis involving the hypothalamus, anterior pituitary and an endocrine organ

External and internal signals impinge on the central nervous system and hence on hypothalamic neurons, the anterior pituitary gland and the final endocrine gland. Tissues are shaded blue. Hormones secreted at each stage in the endocrine axis are shaded in purple, with an increasing intensity to indicate increasing amounts of hormone secreted in the consecutive stages in the cascade; the initial signal is thus amplified. Blood concentrations of anterior pituitary hormones and final hormone regulate the hormone axis via both long and short feedback loops. + indicates a positive effect, – indicates a negative effect.

The anterior pituitary also synthesizes and secretes **prolactin**, which has a number of actions that vary according to the vertebrate group. All known prolactins consist of 197 to 199 amino acids, except those of fish, which are slightly smaller because they lack a section of 12 amino acids at the N-terminal end of the molecule.

Early in the 20th century, the observation of changes in the histology of the prolactin cells in the anterior pituitary gland gave the first clues as to the action of prolactin in pregnant mammals. Stimulation by prolactin of the development of the mammary glands during pregnancy and milk production was subsequently identified and remains perhaps the best-known action of this hormone. However, it soon became clear that prolactin has many other actions.

Prolactin occurs in all vertebrates, not just mammals. In the pigeon, prolactin stimulates secretion of 'milk' by the crop sac. In fish and amphibians, prolactin has actions that influence calcium balance[31]. Receptors for prolactin occur in a wide variety of tissues, including many regions of the brain, skin, sweat glands, bone, the gills of fish and larval amphibians, lung, reproductive system, gastrointestinal tract, crop sac of birds, liver, lymphoid tissue and **lymphocytes**. With this diversity of receptor locations, the identification of more than 300 separate actions of prolactin is not that surprising.

Recent studies have shown that prolactin is also produced at many sites outside the pituitary gland—for example, in the brain, thymus, spleen, sweat glands, in circulating lymphocytes and lymphoid cells of bone marrow. Consequently, prolactin occurs in cerebrospinal fluid, tears, milk and sweat, as well as in the blood. It therefore seems that prolactin acts both as a classical hormone, after release from the pituitary gland, and in an autocrine or paracrine way[32]. Such local actions include modulation of the immune

[28] We learn more about the thyroid axis and actions of thyroid hormones in Section 19.4.

[29] Section 19.3 examines the involvement of cortisol and corticosterone in stress.

[30] We examine further the role of FSH and LH in regulation of gonadal function in Chapter 20.

[31] Section 21.2.2 discusses the influence of prolactin on calcium balance among vertebrates.

[32] Figure 19.1 outlines the distinction of classical endocrine, paracrine and autocrine communication.

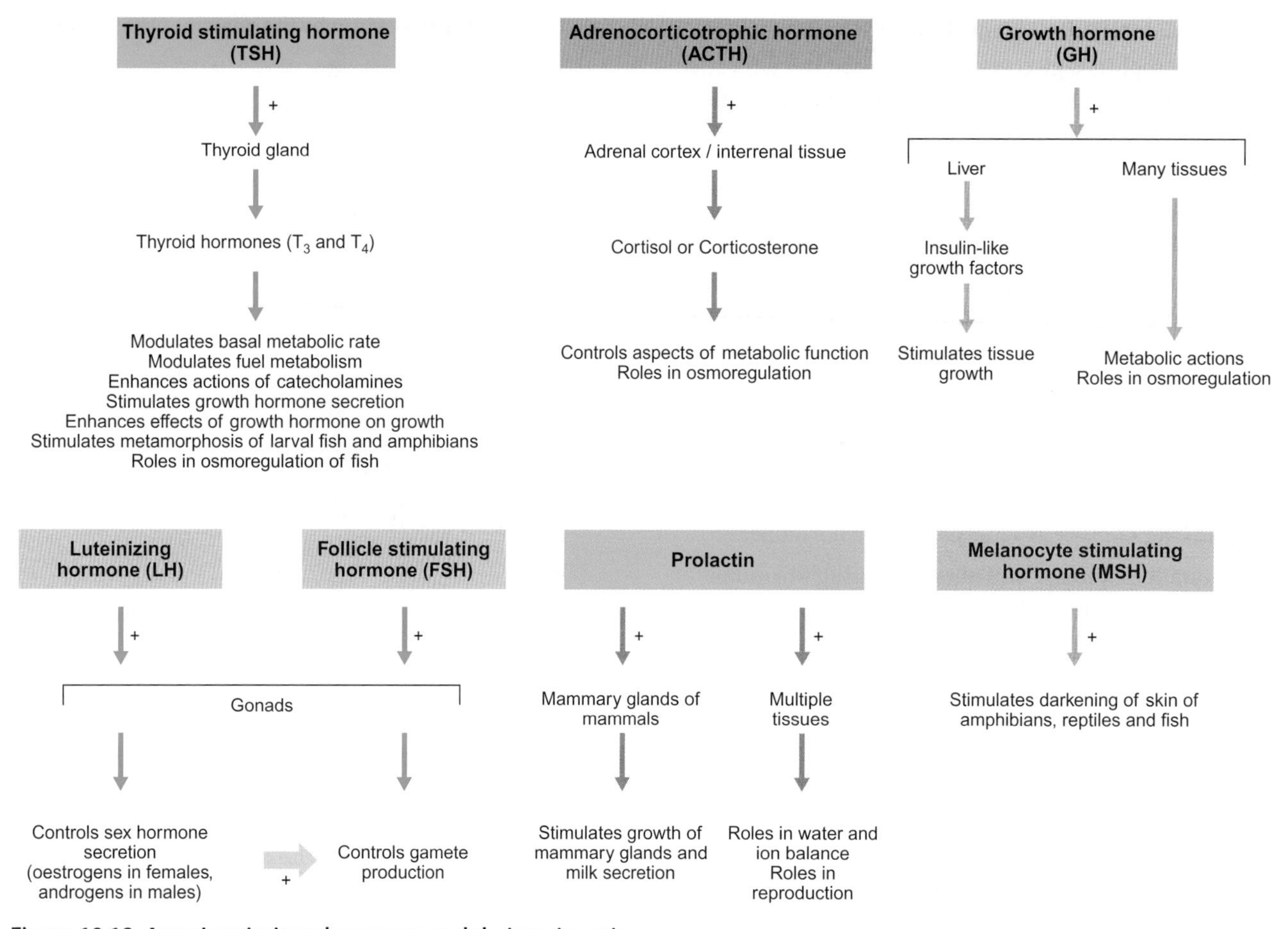

Figure 19.18 Anterior pituitary hormones and their main actions

Six groups of pituitary cells produce seven hormones. Actions of these hormone are shown including stimulatory actions (+) on other endocrine tissue in the thyroid gland, adrenal cortex or interrenal tissue, gonads or liver.

system, acting as a growth factor (for example, by stimulating development of blood vessels—**angiogenesis**), and acting on brain function. For example, prolactin appears to be involved in the regulation of the parental behaviours of fishes, birds and mammals.

Hypothalamic neurohormones control the synthesis and release of hormones by the anterior pituitary

In most vertebrate groups, a **hypothalamo-hypophysial portal system** carries venous blood from a capillary bed in the median eminence, at the lower boundary of the hypothalamus, to another capillary bed in the anterior pituitary gland. Figure 19.14 gives a simplified idea of this arrangement. The portal system rapidly conveys neurohormones from the hypothalamus to the anterior pituitary, without dilution by the general circulation, to control the synthesis and release of specific hormones by the anterior pituitary. Venous vessels that drain the pituitary then carry hormones from the anterior pituitary into the general circulation, which in many cases act on further endocrine tissues to release the final hormones.

A hypophysial portal system does not exist in teleost fish; the hypothalamic neurons terminate close to or make direct connections with the endocrine cells of the anterior pituitary lobe. In lampreys and the elasmobranchs (sharks and rays), which are the living representatives of the oldest fish lineages, neither a hypophysial portal system nor nervous connections occur between the hypothalamus and anterior pituitary cells, and regulation of the pituitary gland may be more diffuse but nevertheless involves control by hypothalamic neurohormones.

Figure 19.17 illustrates the typical endocrine cascade resulting from hypothalamic control of the release of an anterior pituitary trophic hormone that ultimately controls a further endocrine organ. Such hormonal cascades amplify the initial signal from the central nervous system to the final hormone. Typically, nanogram amounts of releasing hormone result in release of μg amounts of the relevant anterior pituitary hormone, which stimulates synthesis

Table 19.2 Hypothalamic neurohormones controlling the secretion of hormones synthesized by the anterior pituitary gland

Hypothalamic neurohormone		Main actions
Abbreviated name	Full name	
Stimulatory hormones		
TRH	Thyrotrophin releasing hormone = Thyroid stimulating hormone releasing hormone	Stimulates release of thyroid stimulating hormone (TSH) Stimulates prolactin release Stimulates release of growth hormone (GH)
GnRH	Gonadotrophin releasing hormone	Stimulates release of follicle stimulating hormone (FSH) and luteinizing hormone (LH)
CRH	Corticotrophin releasing hormone	Stimulates release of adrenocorticotrophic hormone (ACTH) Stimulates release of TSH in larval amphibians
GHRH	Growth hormone (GH) releasing hormone	Stimulates release of growth hormone (GH)
Inhibitory hormones		
Dopamine	Prolactin inhibiting hormone	Inhibits TRH-stimulated prolactin release
Somatostatin		Inhibits TRH-stimulated release of growth hormone
GnIH	Gonadotrophin inhibiting hormone	Inhibits release of follicle stimulating hormone (FSH) and luteinizing hormone (LH)
MIH	Melanocyte inhibiting hormone MSH release-inhibiting hormone	Inhibits release of melanocyte stimulating hormone (MSH); potentiates effects of melatonin

and release of μg to mg amounts of final hormone into circulation. Such processes are regulated by feedback loops controlling both the functioning of neurosecretory neurons in the hypothalamus, and the release of anterior pituitary hormones. The hypothalamus does not have the blood–brain barrier[33] that impedes the passage of large molecules such as peptide hormones into the cerebrospinal fluid and interstitial tissue of the brain.

Although it was originally thought that individual hypothalamic neurohormones regulate individual hormones secreted by the anterior pituitary on a one-to-one basis, it is now clear that the control processes are more complex, which Table 19.2 illustrates. This table shows the main hypothalamic releasing hormones and inhibiting hormones that control secretion of hormones by the anterior pituitary gland.

In some cases, a particular releasing hormone stimulates the release of more than one hormone from the anterior pituitary gland. For example, **thyrotrophin releasing hormone** (TRH), so named because of its stimulatory action on the thyrotrophs of the anterior pituitary cells, which results in release of thyroid stimulating hormone (TSH), also stimulates growth hormone release, and prolactin release in mammals, birds, reptiles and amphibians.

The secretion of some of the anterior pituitary hormones is stimulated by more than one hypothalamic neurohormone. For example, both TRH and **corticotrophin releasing hormone** (CRH) regulate the release of TSH in larval amphibians, although CRH is usually emphasized as the principal stimulant of the corticotrophs, which release adrenocorticotrophic hormone (ACTH) as part of the hypothalamus–pituitary–adrenal axis[34]. In other cases, both stimulatory and inhibitory neurohormones from the hypothalamus regulate the secretion of an anterior pituitary hormone. A good example is prolactin; while TRH stimulates release, dopamine inhibits release, as indicated in Table 19.2. Similarly, the secretion of the gonadotrophins (FSH and LH) is regulated by **gonadotrophin releasing hormone** (GnRH) and **gonadotrophin inhibiting hormone** (GnIH)[35].

We have seen that the endocrine secretions from both the posterior and anterior pituitary gland are under the control of neurohormones. The posterior pituitary is a storage depot for peptide neurohormones. In the anterior pituitary, neurohormones control the on-site production of hormones and their release. In both cases, the central nervous system impinges on the neurons in the hypothalamus and may exert overriding control. The hypothalamus–pituitary axis

[33] For further information on the blood–brain barrier look at Online Box 16.1.

[34] Section 19.3.1 examines the hypothalamus–pituitary–adrenal axis.

[35] We examine the hypothalamic–pituitary–gonadal axis and other controls of gonadal function in Sections 20.3.4 and 20.4.

is influenced by higher brain functioning and the effects of environmental conditions, emotional state and stress, which we explore in Section 9.3.

Underlying these external influences, brain neural biological clocks[36] set a daily rhythm of hypothalamic function that results in daily cycles in the circulating concentrations of pituitary hormones and the production of hormones by endocrine glands under the control of the pituitary hormones. Biological clocks work in conjunction with the pineal gland and its production of **melatonin**.

19.2.3 The pineal gland synthesizes and releases melatonin

The pinealocytes of the pineal gland of vertebrates synthesize and release melatonin from tryptophan after its uptake from the blood. Figure 19.19 illustrates the steps in melatonin biosynthesis, which involves four enzymes; the activity of serotonin N-acetyltransferase is the rate-limiting step in formation of an intermediate that generates melatonin from serotonin. Melatonin synthesized by the pinealocytes diffuses into the circulating blood.

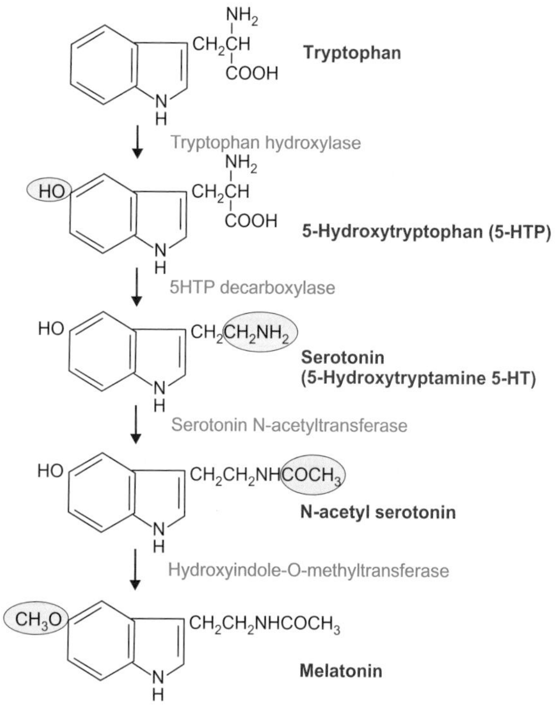

Tryptophan

Tryptophan hydroxylase

5-Hydroxytryptophan (5-HTP)

5HTP decarboxylase

Serotonin (5-Hydroxytryptamine 5-HT)

Serotonin N-acetyltransferase

N-acetyl serotonin

Hydroxyindole-O-methyltransferase

Melatonin

Figure 19.19 Steps in synthesis of serotonin from tryptophan Substrates and products are indicated in red; the four enzymes involved are shown in green. Encircled chemical groups indicate those added or the region of the molecule where a chemical group has been removed. Tryptophan uptake by the pinealocytes initially leads to its conversion into 5-hydroxytryptophan. This product is converted into serotonin (5-hydroxytryptamine, 5-HT) which is acetylated by serotonin N-acetyltransferase; this is the rate-limiting step in forming melatonin. N-acetyl serotonin is converted to melatonin by addition of a methyl group (methylation).

Normally, hormones are named after the first identification of their action. The first identified action of melatonin was the lightening of the skin of frogs, some reptiles and lampreys (but this does not occur in humans). As little as 10^{-12} g mL^{-1} of melatonin lightens the skin of frogs by stimulating the aggregation of **melanin** granules in the melanocytes, which antagonizes the darkening actions of the pituitary hormone, melanocyte stimulating hormone.

We now know that melatonin is not restricted to vertebrates with a pineal gland, but also occurs in various invertebrates and even bacteria. Actions of melatonin in these organisms are poorly studied, but recent evidence that melatonin release occurs at night and controls the patterns of daily activity of a marine polychaete *Platynereis dumerilii* support the idea that a fundamental and ancestral action of melatonin is to synchronize the timing of natural circadian rhythms, a **chronobiotic** action.

In all vertebrates, whether they are diurnally active, nocturnal, or crepuscular (i.e. active at dusk and dawn), light is the dominant environmental signal that regulates the biosynthesis of melatonin. Light suppresses the activity of serotonin N-acetyltransferase and hence inhibits melatonin synthesis and release. Instead, melatonin is synthesized and released during the dark phase of natural or imposed light cycles when the activity of serotonin N-acetyltransferase increases. Hence, melatonin is often called 'the hormone of darkness'. However, there are differences among vertebrates as to how light signals are conveyed to the pinealocytes that synthesize melatonin.

The pineal body of fish, amphibians and reptiles (but not snakes) consists mainly of photoreceptive pinealocytes that function as a 'third eye' by responding directly to light perceived through the skull. The pineal gland of birds also has photosensory ability but bird pineal glands vary from a hollow structure much like the pineal gland of fish and amphibians to a more solid endocrine gland.

The pineal gland of adult mammals has lost an ability to respond directly to light. Instead, photosensitive cells in the retina of the eyes convey the signal of light to the **suprachiasmatic nuclei** (SCN) in the anterior hypothalamus. The SCN acts via inhibitory and stimulatory controls on other nuclei in the brain that convey signals to a spinal cord ganglion that in turn regulates the rhythmic synthesis and release of melatonin from the pinealocytes. This neural pathway is illustrated back in Figure 17.24.

An important component of the release of melatonin during darkness is that release varies in proportion to the *duration* of the period of darkness, which signals photoperiod (hours of light per day). In temperate zones, seasonal changes

[36] See Section 17.2.2 for further discussion of biological clocks.

in photoperiod result in a relatively long daily peak concentration of melatonin in winter, when day length is short compared to summer, when nights are short. Hence, the duration of the peak of melatonin conveys information about the time year.

In mammals, seasonal changes in photoperiod that result in an annual cycle of changing duration of melatonin secretion acts as a time signal for control of seasonal events, such as reproduction. For instance, increasing duration of the peak of secretion of melatonin triggers reproduction in short-day breeders such as goats and deer.

Recent studies indicate that melatonin influences the energy metabolism of mammals. Melatonin supplementation of the diet or drinking water of rats reduces body weight and abdominal fat, even without a reduction in food intake or any increase in physical activity. Melatonin acts on two types of G-protein coupled receptors[37] in brown and white adipose tissue, as illustrated in Figure 19.20.

An additional component of melatonin activity is its inhibitory action on β-cells of the pancreas that secrete **insulin**, which results in lower concentrations of insulin during darkness. The lowering of insulin reduces negative feedback on the pinealocytes that produce melatonin, which increases melatonin secretion in darkness. The outcome is the stimulation of heat production from brown adipose tissue, and energy expenditure and lipolysis in white adipose tissue. Humans often interfere with these natural controls by using artificial lighting, which reduces the duration of melatonin secretion. The resulting decrease in metabolism may be one factor leading to the higher levels of obesity among some human populations.

Having examined the main central endocrine glands of vertebrates (hypothalamus, pituitary gland and pineal), we next turn attention to peripheral endocrine glands. First, we explore the adrenal gland (and its functional homologues) and its role in stress responses, before moving to the **thyroid gland**.

[37] G-protein coupled receptors are discussed in Section 19.1.4.

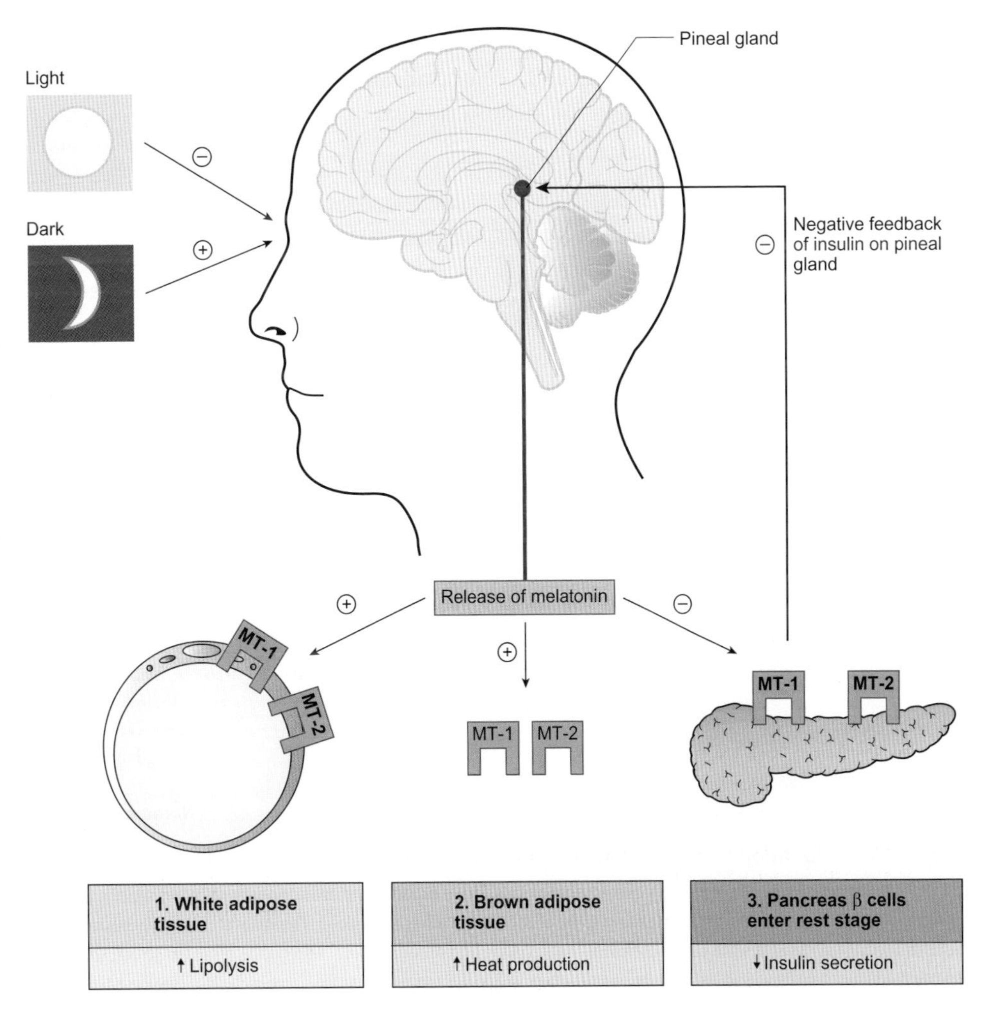

Figure 19.20 Interactions between pineal gland, melatonin secretion, adipose tissue and pancreatic β cells

Melatonin secretion from the pineal gland (red circle) during darkness, acts on two types of melatonin receptors (MT-1, MT-2) in three main tissues: **1** White adipose tissue, where melatonin increases lipolysis. **2** Brown adipose tissue, where melatonin stimulates an increase in heat production (Section 10.2.5 discusses brown adipose tissue in thermogenesis of endotherms). **3** Pancreas β cells, where melatonin inhibits insulin secretion, so these cells enter a rest stage during darkness. Insulin exerts a negative feedback (−) via insulin receptors in the pineal gland that influence melatonin release. The reduced insulin concentrations during darkness reduces this feedback, and therefore acts as a stimulant of melatonin secretion during darkness. + = stimulatory effects; − = inhibitory effects

Source: adapted from Cizza G et al (2011). Chronic sleep deprivation and seasonality: Implications for the obesity epidemic. Journal of Endocrinological Investigation. 34: 793–800.

› Review articles

Archer R (1996). Molecular evolution of fish neurohypophysial hormones: Neutral and selective evolutionary mechanisms. General and Comparative Endocrinology 102: 157–172.

Cerdá-Reverter JM, Canosa LF (2009). Neuroendocrine circuits of the fish brain. In Fish Physiology. Volume 28. Fish Neuroendocrinology. pp 4–75. Eds Bernier NJ, Kraak, G Van Der, Farrell, AP, Brauner CJ. Academic Press.

Cizza G, Requena M, Galli, G, de Jonge L (2011). Chronic sleep deprivation and seasonality: Implications for the obesity epidemic. Journal of Endocrinological Investigation 34: 793–800.

Ekström P, Meissl H (2003). Evolution of photosensory pineal organs in new light: the fate of neuroendocrine photoreceptors. Philosophical Transactions of the Royal Society of London B 358: 1679–1700.

Leng G, Ludwig M (2008). Neurotransmitters and peptides: whispered secrets and public announcements. Journal of Physiology 586: 5625–5632.

Walker JJ, Terry JR, Lightman SL (2010). Origin of ultradian pulsatility in the hypothalamic–pituitary–adrenal axis. Proceedings of the Royal Society of London B 277: 1627–1633.

19.3 The vertebrate adrenal gland and stress

The adrenal glands of mammals and birds are a discrete pair of endocrine glands that lie close to the kidneys but have their own blood supply. In mammals, the adrenal gland consists of a distinct outer **adrenal cortex** and a central **adrenal medulla**, as shown in Figure 19.21. The medulla consists of modified post-ganglionic neurons that form an extension of the sympathetic nervous system[38]. These cells are also called chromaffin cells because of their affinity for chromium salts. In histological sections, chromium salts stain the cells brown because chromium polymerizes with the catecholamines (adrenaline, noradrenaline and dopamine) that are synthesized and stored in granules in the **chromaffin cells**.

In mammals, adrenaline is usually the dominant catecholamine in the adrenal medulla, but this is not universal and variable proportions of adrenaline and noradrenaline occur in different species. Pre-ganglionic sympathetic nerve fibres innervate the chromaffin cells of the adrenal medulla such that activation of the sympathetic nervous system results in exocytosis of the catecholamines from chromaffin tissue, driving an increase in the concentration of circulating catecholamines.

The distinct core of chromaffin cells surrounded by an outer cortex does not occur in non-mammalian vertebrates. Instead, the homologues of the cortical cells and chromaffin cells are intermingled. Figure 19.22 shows schematic drawings of the arrangement of chromaffin tissue and the functional homologue of the adrenal cortex in relation to the kidneys and gonads of different vertebrate groups. In teleost fish, the homologue of the adrenal cortex forms collections of glandular cells embedded in the anterior part of the kidney and is therefore called interrenal tissue.

[38] Figure 3.24 illustrates the arrangement of sympathetic innervation in mammals.

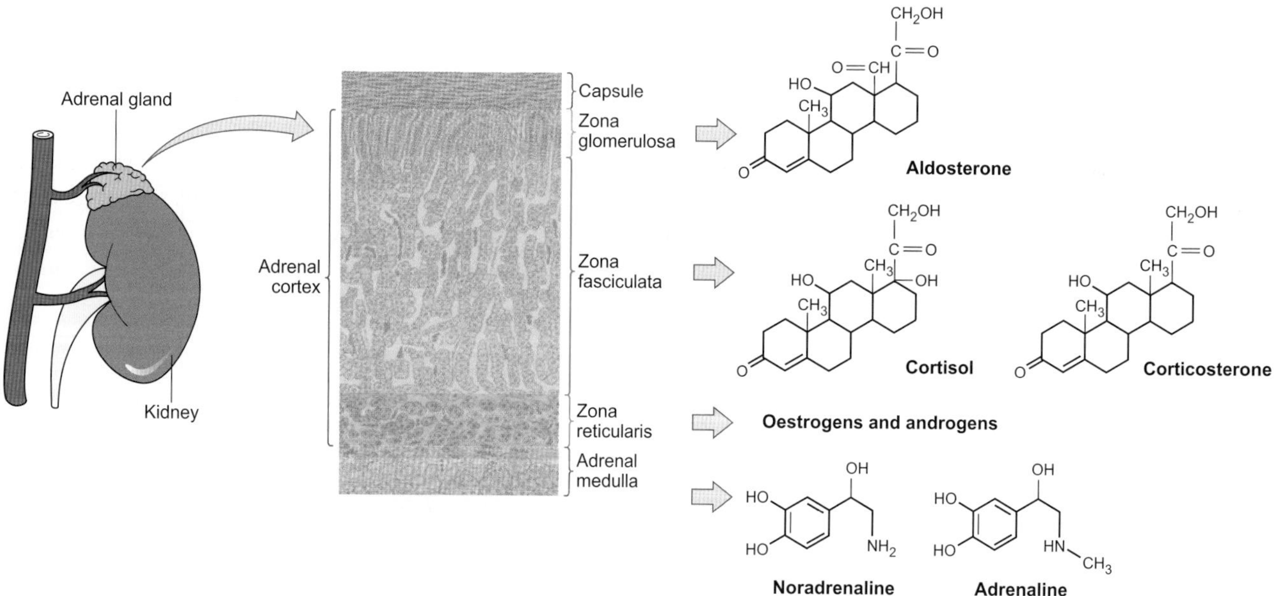

Figure 19.21 Zonation of mammalian adrenal gland and hormones secreted

(A) Diagram of one of the pair of kidneys and adrenal glands. These organs have an independent blood supply. (B) Illustration of the zonation of the mammalian adrenal gland showing the adrenal cortex and some of the central adrenal medulla. The main steroids synthesized by each zone of the cortex are shown. The adrenal medulla synthesizes the catecholamines—adrenaline and noradrenaline. Adrenaline differs in structure from noradrenaline by an extra methyl group.

Source: B: adapted from Bentley PJ (1976). Comparative Vertebrate Endocrinology. Cambridge: Cambridge University Press.

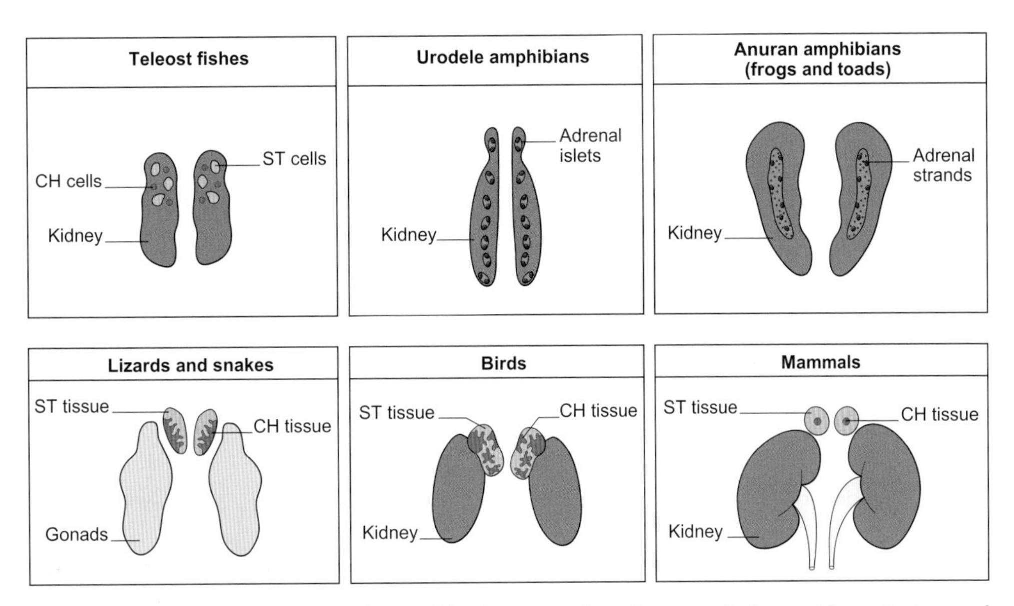

Figure 19.22 Schematic drawings showing the relationships between the adrenocortical steroidogenic tissue, chromaffin tissue and kidney or gonads of vertebrates.

Steroidogenic tissue/cells (ST) is shown in orange, chromaffin tissue (CH) is shown in red. In mammals, the central medulla is distinct from an outer cortex of steroidogenic tissue but this arrangement does not occur in other vertebrates. In teleost fish, clusters of chromaffin cells are only loosely associated with the steroidogenic tissue within the kidney. In most urodele amphibians, the steroidogenic tissue forms small bodies, partially embedded in the ventral surface of the kidney, with groups of associated chromaffin cells. Frogs and toads have a continuous strand of adrenocortical steroidogenic tissue containing clusters of chromaffin tissue on each kidney. In lizards and snakes the adrenal glands are distinct bodies separated from the kidney in a close association with the gonads. The adrenal gland of birds may be a single structure or two organs near the kidneys; the steroidogenic tissue and chromaffin tissue are intermingled.

Source: Perry SF, Capaldo A (2011). The autonomic nervous system and chromaffin tissue: Neuroendocrine regulation of catecholamine secretion in non-mammalian vertebrates. Autonomic Neuroscience: Basic and Clinical 165: 54–66.

19.3.1 Steroid-secreting cells in adrenal cortex and interrenal tissue

The outer layer of the mammalian adrenal gland—the adrenal cortex—generally consists of three layers (zones), as shown in Figure 19.21. An outer zona glomerulosa consists of balls of cells, a middle zona fasciculata consists of columns of cells, and a zona reticularis forms an inner network of cells. The main secreted steroid hormones are segregated within these zones, with three types of hormone being recognizable, based on their primary action:

- Sex steroids—the zona reticularis secretes small amounts of oestrogens and androgens, which are similar or identical to those produced by the gonads[39].
- **Mineralocorticoids**—the main action of mineralocorticoids is on salt (mineral) balance. The zona glomerulosa synthesizes the mineralocorticoid, aldosterone[40].
- **Glucocorticoids**—the main role of glucocorticoids is to increase the availability of glucose and other metabolic fuels either as part of normal regulatory mechanisms or in response to stress. The zona fasciculata synthesizes and secretes the main glucocorticoids, **cortisol** and/or **corticosterone**.

[39] We discuss sex steroids in some detail in Chapter 20.

[40] Figure 21.3 in Section 21.1.1 discusses the control of aldosterone release from the zona glomerulosa, and its actions on salt balance.

The relative proportion of the glucocorticoids varies between species—from 100 per cent cortisol in teleost fish to 100 per cent corticosterone in birds. Most mammalian species secrete mainly cortisol, but some, such as rodents (e.g. rats and mice) and lagomorphs (rabbits), solely or primarily secrete corticosterone. The distinction between mineralocorticoids and glucocorticoids is useful, but not exact, particularly among non-mammalian vertebrates. For example, cortisol is important in determining the mineral balance of fish and their glucose regulation.

The hypothalamus–pituitary–adrenal (HPA) axis regulates glucocorticoid secretion

A similar hormonal axis regulates the secretion of cortisol/corticosterone in all vertebrates. The HPA axis (or hypothalamus–pituitary–interrenal (HPI) axis of fish) begins with the synthesis and release of the neuropeptide hormone, corticotrophin releasing hormone (CRH), by hypothalamic nuclei. The central nervous system and various transmitter substances (including catecholamines) released by higher brain

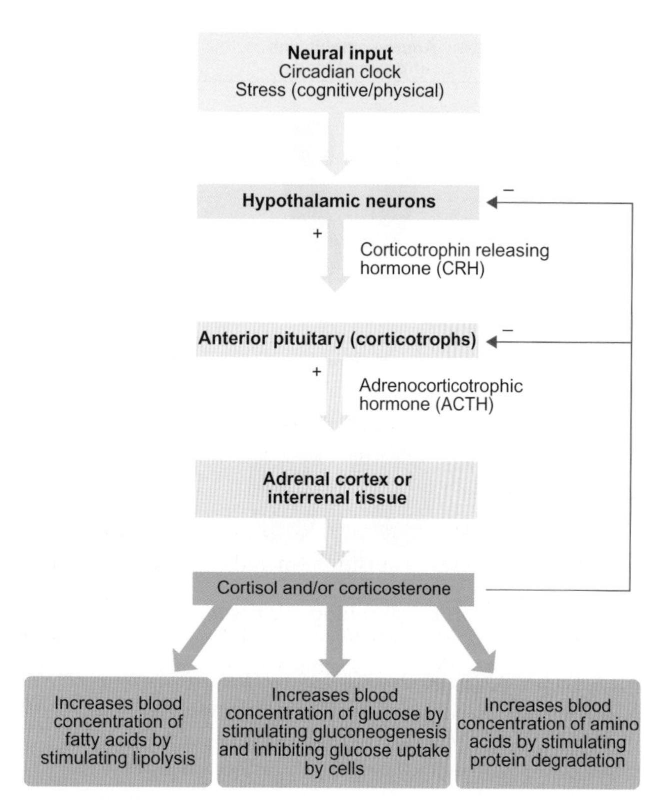

Figure 19.23 The hypothalmic–pituitary–adrenal/interrenal axis

Secretion of cortisol and/or corticosterone by the adrenocortical tissue of the adrenal cortex or interrenal tissue is stimulated by adrenocorticotrophic hormone (ACTH) which is secreted by the anterior pituitary. The corticotrophs in the anterior pituitary are stimulated by corticotrophin releasing hormone (CRH) released from hypothalamic neurons. Neural input to the hypothalamic neurons also influences CRH release. Cortisol/corticosterone exert negative feedback on the pituitary secretion of ACTH and to a lesser extent on the hypothalamic neurons that secrete CRH. The metabolic actions of cortisol and corticosterone are indicated. + signifies stimulation; – indicates negative feedback effects.

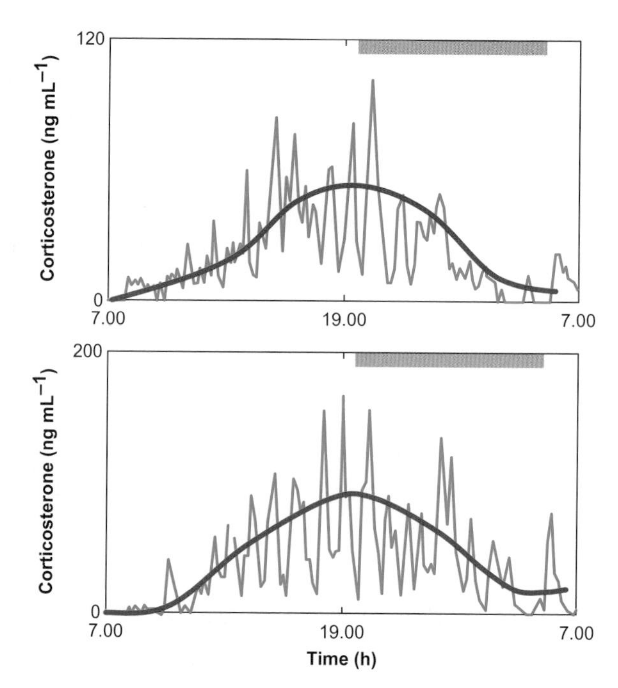

Figure 19.24 Circadian rhythm of corticosterone in blood of rats and underlying ultradian rhythm

Graphs show concentrations of blood corticosterone in blood samples collected every 10 minutes using an automated system over a 24 h period in two individual male rats. Blue bars indicate the dark phase from 19.15 h to 05.15 h. Green line shows ultradian rhythm. A circadian rhythm is indicated by the red curve showing approximate mean values superimposed on the measured fluctuations.

Source: adapted from Walker JJ et al (2010). Origin of ultradian pulsatility in the hypothalamic–pituitary–adrenal axis. Proceedings of the Royal Society of London B 277: 1627–1633.

centres impinge on these nuclei. Consequently, stress activates CRH release and the subsequent hormonal cascade, as indicated in Figure 19.23.

In addition to any effects of stress, the neuronal biological clock controls the activity of the hypothalamic neurons. For diurnal species active in daylight, large pulses of CRH are released in the early morning followed by smaller pulses throughout the day, reaching the lowest level in the evening, before rising again overnight. Pulses of CRH stimulate synthesis and release of pulses of adrenocorticotrophic hormone (ACTH) by the corticotrophs of the anterior pituitary gland. The rise in ACTH then stimulates pulses of secretion of cortisol and/or corticosterone.

Figure 19.24 shows the daily rhythm of corticosterone concentrations in the blood of two rats. In this case, maximum concentrations occur in the early evening when the rats start to become active. An ultradian rhythm of peaks and troughs (a rhythm that repeats in less than one day) occurs throughout the circadian cycle. Some researchers argue that pulsatile release of CRH lies behind the ultradian rhythm of cortisol/corticosterone in the blood. Whatever its cause, large fluctuations in the concentration of corticosteroids in the blood occur, particularly when the average concentrations are at their highest. Such fluctuations are likely to prevent the de-sensitization of target cells that could occur if there were persistently high circulating concentrations of glucocorticoids.

An elaborate set of checks and balances controls the concentration of circulating glucocorticoids. Self-regulation of the HPA/HPI axis by multiple types of negative feedback processes turns off secretion of glucocorticoids. Slow feedback mechanisms involving binding of free (unbound) cortisol or corticosterone to specific nuclear receptors that affect gene transcription take more than 30 minutes and possibly several hours to achieve. Slow feedback effects operating mainly at the level of the pituitary reduce the sensitivity of the corticotrophs in the anterior pituitary to CRH, which reduces further secretion of ACTH and hence further release of cortisol/corticosterone.

In addition to feedback effects on the pituitary, there is evidence of a long feedback loop acting on the hypothalamus, so Figure 19.23 includes both levels of feedback. Recent studies indicate that feedback on the hypothalamus includes fast feedback mechanisms, operating within 10 to 15 minutes, via non-genomic signalling pathways that are independent of mRNA and protein synthesis. We look at some of these findings in Figure 19.25. These data show a 30 to 40 per cent reduction in the frequency of excitatory postsynaptic potentials of hypothalamic neurons secreting CRH within 3 to 5 min of the application of corticosterone or a synthetic steroid (dexamethasone).

Other hormones and neurohormones interact with regulators of the HPA axis. Some of the magnocellular neurons of the hypothalamus that release arginine vasopressin (AVP) have terminals in the median eminence. AVP released here enters the anterior pituitary and increases the effect of CRH on ACTH release. A similar synergistic effect occurs in the interrenal tissue of fish; their homologue of AVP is AVT, as indicated in Table 19.1. AVT enhances the action of ACTH on cortisol release from interrenal tissue.

In unstressed animals, a basal level of glucocorticoid hormones in the plasma, usually 1–10 ng mL^{-1}, is essential for basic glucose and salt regulation, although basal levels of cortisol/corticosterone vary by species. Within a species, many factors influence the basal concentration of glucocorticoids, including age, life history stage, social status and the genetics of the individuals.

Low to moderate levels of glucocorticoids maintain a balance between energy acquisition (including appropriate feeding behaviour), deposition of energy stores, and mobilization of glucose and fatty acids from energy stores. Seasonal or even daily increases in the energy requirement can result in an in crease in the circulating levels of glucocorticoids above basal levels, which is illustrated in the schematic diagram in Figure 19.26. However, when an unexpected challenge occurs, such as an exposure to a noxious pollutant, energy demands may increase to such an extent that a state of stress occurs.

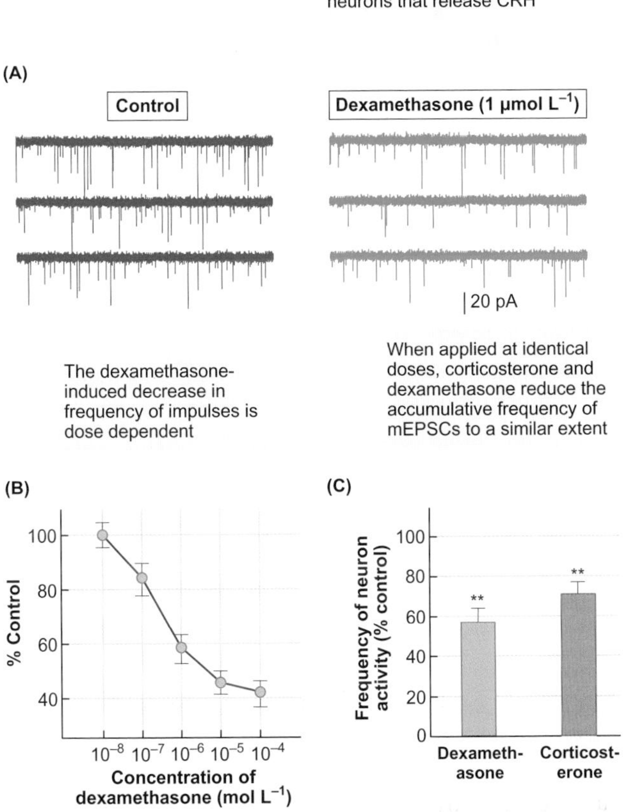

Figure 19.25 Fast feedback inhibition by glucocorticoids of activity of hypothalamic neurons secreting CRH in rats

Neurons in the paraventricular nucleus (parvocellular neurons) that secrete CRH in mammals were investigated *in vitro* in slices from 80 rats by making whole-cell patch-clamp recordings (Figure 16.16 illustrates methods for studying neuron activity). A 10-minute baseline was established before treatments. Effects were monitored after 7-minute applications. (A) Sample recordings in the control period and during treatment with the synthetic glucocorticoid dexamethasone. (B) Dose–response relationships for dexamethasone on frequency of impulses. (C) Comparison of effects of 1 μmol L^{-1} corticosterone (seven cells) and 1 μmol L^{-1} dexamethasone (18 cells) on accumulative mean frequency of mEPSCs compared to untreated controls. In both cases there was a significant decrease in mEPSCs compared to controls (***P*<0.01).

Source: adapted from Shi Di et al (2003). Nongenomic glucocorticoid inhibition via endocannabinoid release in the hypothalamus: a fast feedback mechanism. The Journal of Neuroscience 23: 4850–4857.

19.3.2 Coordination of acute stress responses

In the 1930s, the recognition of a common set of responses to stress in all vertebrates led Hans Selye to put forward the concept of a General Adaptation Syndrome (GAS) that

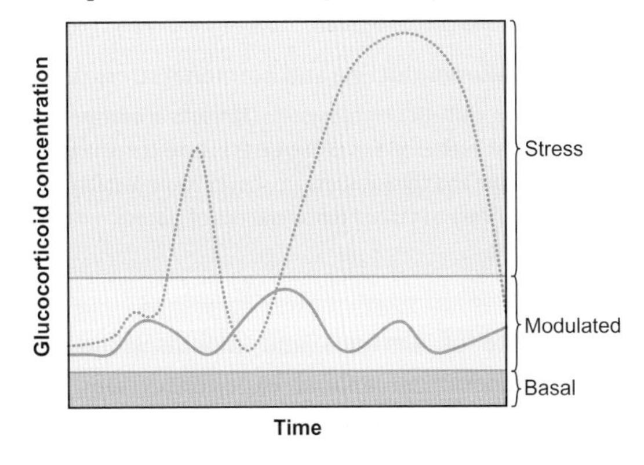

Figure 19.26 Variations in the plasma concentration of glucocorticoid hormones

A basal level of glucocorticoid is necessary for basic regulatory functions. Modulated levels reflect increases and decreases resulting from daily, tidal, seasonal and life-history cycles. Acute noxious, physical or psychological stimuli increase glucocorticoid concentrations into the stress range. Each incursion into the stress range indicates an emergency response.

Source: Busch DS, Hayward LS (2009). Stress in a conservation context: A discussion of glucocorticoid actions and how levels change with conservation-relevant variables. Biological Conservation 142: 2844–2853.

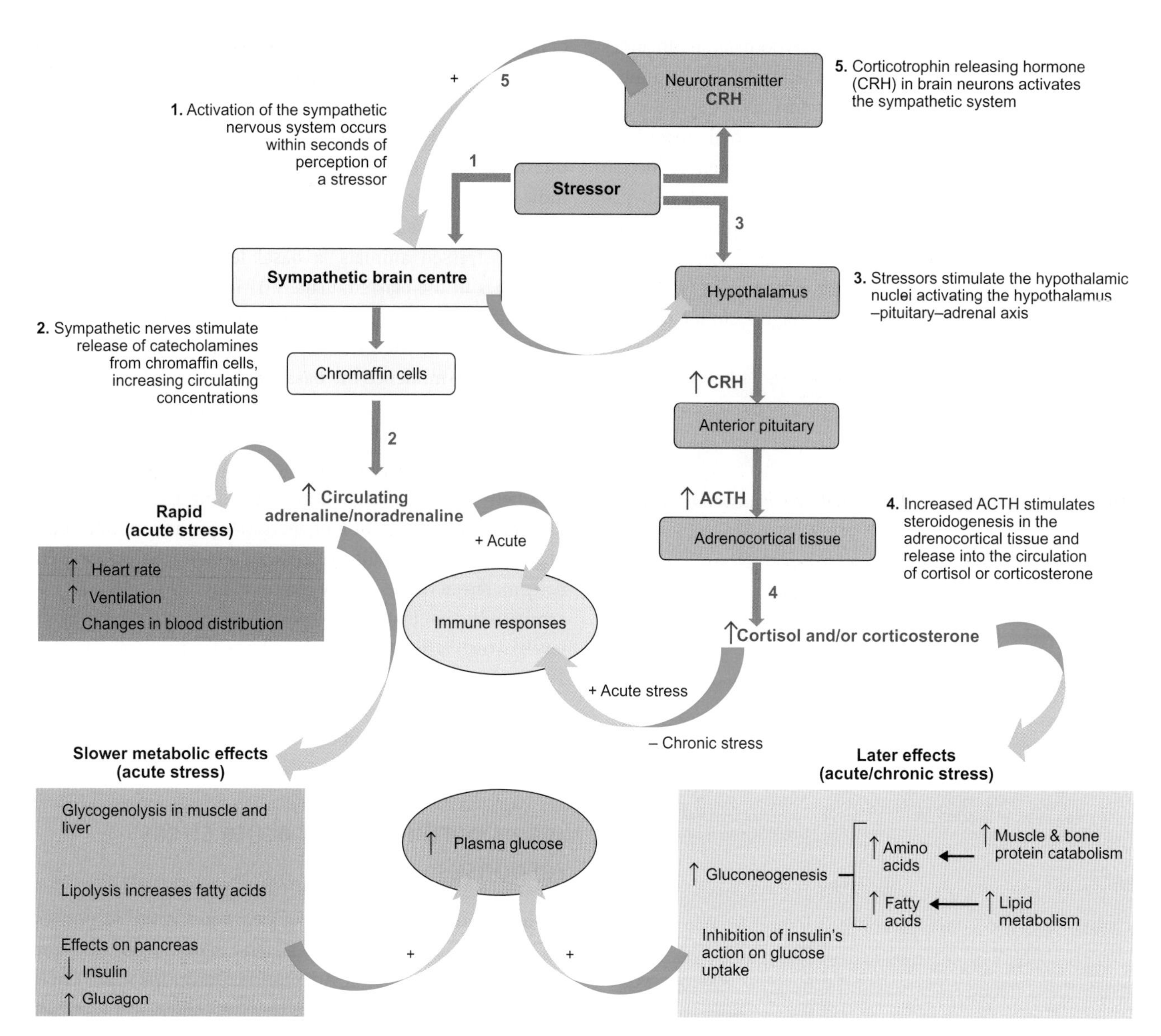

Figure 19.27 Neuroendocrine and hormonal stress response of vertebrates

Activation of the sympathetic nervous system (1) results in increased circulating concentrations of adrenaline/noradrenaline (2). Activation of the hypothalamus–pituitary–adrenal (or interrenal) axis (3) increases plasma concentrations of adrenocorticotrophic hormone (ACTH) and cortisol/corticosterone (4). Rapid and slower effects of acute and chronic stress are shown. Acute metabolic responses to stress resulting from the combined actions of catecholamines and glucocorticoids result in an increase in blood concentrations of glucose. + = positive (stimulatory) effects; – = negative (inhibitory) effects; ↑ represents an increase in a particular parameter.

initiates a state of alertness and mobilises biochemical resources whenever an animal experiences any stressful stimuli (stressors). The core elements of the GAS are endocrine responses that involve the adrenal gland. Figure 19.27 illustrates the primary endocrine responses to stress, and the secondary effects.

Within seconds of the perception of a stressor, activation of the sympathetic nervous system occurs. Sympathetic nerves innervate the chromaffin tissue; release of acetylcholine from the nerve endings, which stimulates exocytosis of granules stored in the chromaffin cells, results in an increase in the circulating concentration of catecholamines. Such responses are evident in all vertebrate groups. For example, handling of birds elevates their circulating catecholamines and aggressive social interactions raise the concentration of plasma catecholamines in reptiles. In teleost fish, stressors such as handling and physical disturbance increase their plasma concentrations of catecholamines.

The second part of the generalized stress response involves the hypothalamus–pituitary–adrenal/interrenal axis. Stressors stimulate the hypothalamic nuclei, activating secretion of corticotrophin releasing hormone (CRH) and hence stimulate the corticotrophs of the anterior pituitary gland. Their release of ACTH increases its circulating concentration within a few seconds of the perception of stress. The increased ACTH stimulates steroidogenesis by the adrenal gland (or its functional homologue) followed by release into the circulation of cortisol or corticosterone. These hormones are not stored in the adrenal cortex or the functional equivalents, but glucocorticoid concentrations typically increase by 4 to 10-fold within a few minutes of the perception of stress. Peak concentrations in the blood usually occur after about an hour.

We have considered the release of catecholamines and activation of the HPA axis as two separate events. In reality, however, the two systems interact and stimulate one another. Much of their interaction involves higher regions of the brain. CRH occurs as a neurotransmitter in neurons in many areas of the brain, and CRH neurons and noradrenergic neurons within brain innervate and stimulate each other in a positive feedback mechanism, such that activation of one leads to activation of the other. Both noradrenaline and CRH act as neurotransmitters and modulators in areas of the brain (amygdala and hippocampus) that are linked to formation of memories of emotional events.

In the natural world, the ubiquitous endocrine responses to acute stressors are widely considered to allow animals to cope with short-term stressors such as pursuit by a predator, brief exposure to a noxious chemical, or stressful social interactions. A high concentration of circulating catecholamines resulting from acute stress generally reinforces the effects of the sympathetic nervous system on cardiovascular and respiratory function, resulting in increases in heart and breathing rate, increases in blood flow to some organs and tissues (such as the heart), while reducing blood flow to others (such as skin and gastrointestinal tract)[41]. While such acute responses might be beneficial in the wild they may be less of an advantage when an animal is in captivity and unable to escape from the stressor, or when—like some humans—they adopt a sedentary life style.

Figure 19.27 summarizes the acute metabolic responses to stress that result from the combined actions of catecholamines and glucocorticoids, which trigger an increase in blood glucose concentrations. An increase in circulating glucose can maintain high rates of metabolism or may fuel an escape from the stressor. The actions of catecholamines and glucocorticoids that increase plasma glucose include:

- glucocorticoid stimulation of the catabolism of muscle protein and lipid stores, which increases the circulating concentrations of amino acids, and glycerol and fatty acids, which provides the substrates for **gluconeogenesis** (glucose formation from non-carbohydrate sources) primarily by the liver;
- glucocorticoid stimulation of liver gluconeogenesis;
- catecholamine stimulation of **glycogenolysis** (the breakdown of glycogen stores in muscles and liver to release glucose that enters the pathways of **glycolysis**[42]) and lipolysis of fat stores;
- catecholamine effects on the islets of Langerhans of the pancreas, stimulating release of the hyperglycaemic hormone, **glucagon**, and inhibiting release of the hypoglycaemic hormone, insulin;
- glucocorticoid inhibition of the action of insulin on the uptake of glucose by cells, which helps to maintain high glucose concentrations in the blood.

In acute stress, both catecholamines and corticosteroids (glucocorticoids) stimulate inflammatory responses and activate the immune system. Such effects could be beneficial to a wounded or physically damaged animal. There may also be decreased pain perception because production of ACTH in the pituitary can generate other peptide fragments, including the analgesics β-endorphin and met-enkephalin. Looking at Figure 19.28, note that synthesis of ACTH involves its cleavage from the much larger peptide pro-opiomelanocortin (POMC), which occurs in several tissues including the hypothalamus and the anterior and intermediate lobes of the pituitary. The products of POMC cleavage are determined by tissue-specific processing mediated by enzymes present in the particular tissue; these products include endorphins and met-enkephalin.

Selye's original idea of the 'General Adaptation Sndrome' implied that stress responses have the potential to be adaptive. To be adaptive in an evolutionary sense implies improved long-term survival and reproductive fitness. It is widely considered that short-lived (acute) stressor leads to 'good' endocrine responses and secondary effects, for free-living animals at least. For glucocorticoids, it is thought that a good response is a fast increase from a low baseline concentration, coupled with a rapid induction of negative feedback regulation, which avoids persistent high levels of glucocorticoids. In acute stress, high glucocorticoid responses seem to channel energy into coping with a short-term threat to survival rather than making long-term investments in functions such as courtship, reproduction, growth and/or immune defences.

Once a stressor ends—when a stressful social interaction stops, for example—the sympathetic neural response will rapidly decline. Inhibitory feedback of the circulating

[41] Section 22.4.3 provides a detailed discussion of the action of catecholamines on vertebrate cardiovascular function.

[42] Figure 2.19 illustrates the set of enzyme-catalysed reactions in glycolysis.

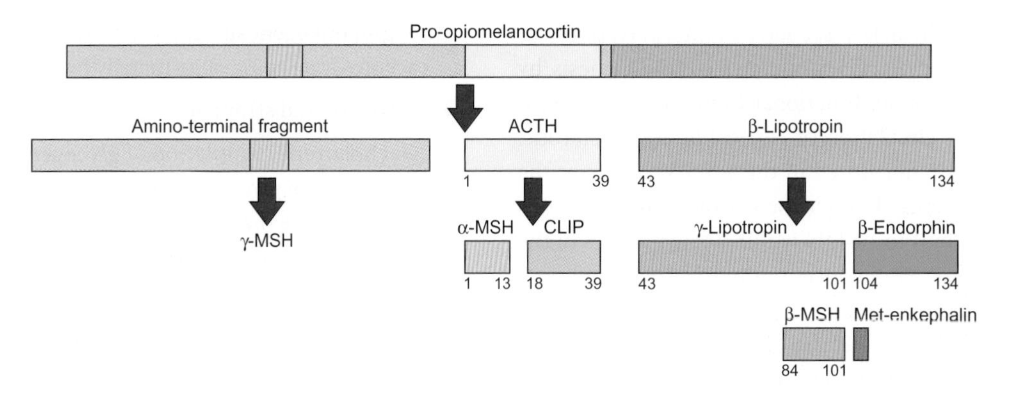

Figure 19.28 Simplified schematic of steps in processing of pro-opiomelanocortin (POMC) in the pituitary gland to form active peptide fragments

In the adenohypophysis, cleavage of the 241 amino-acid-residue POMC generates adrenocorticotrophic hormone (ACTH) which consists of 39 amino acids. Further cleavage of the first 13 amino acids forms α-MSH (melanocyte stimulating hormone) which darkens the skin, and CLIP (corticotropin-like intermediate peptide) whose activity is unclear. The β-lipotropin fragment, also formed from POMC, is important in generating the analgesics β-endorphin and met-enkephalin. There are species differences and not all cleavages always occur. Pituitary corticotrophs produce mainly ACTH and β-endorphin.

Data from Humphreys MH (2004). γ-MSH, sodium metabolism, and salt-sensitive hypertension. American Journal of Physiology Regulatory Integrative and Comparative Physiology 286: R417–R430.

corticosteroids on pituitary corticotrophs and hypothalamic CRH cells will reduce the circulating concentration of corticosteroids that ultimately return to normal concentrations.

19.3.3 Adverse effects of chronic stress

Animals confined in an unnatural environment, for example fish caged for aquaculture, or chickens in intensive agriculture, may suffer the damaging effects of chronic stress due to the persistent activation of the HPA/HPI axis and elevated circulating concentrations of glucocorticoids. Whether in captivity or in the wild, social interactions such as holding a low rank in a dominance hierarchy can also be persistently stressful, while retaining a high rank, with the continual threat of displacement can be equally stressful for some species. Similarly, animals exposed to persistent pollutants or repeated episodes of pollution can suffer chronic stress.

It is generally assumed that chronic activation of the HPA/HPI axis in contrast to acute responses has adverse effects on the biological fitness of animals (their survival and lifetime reproductive success), although this is poorly studied in wild animal populations. In Experimental Panel 19.1, we explore one of the few studies that examined the relationship between the plasma corticosterone, body condition and the survival of wild populations of a species, in this case Galápagos marine iguanas (*Amblyrhynchus cristatus*). These studies show a clear negative relationship between plasma corticosterone and survival during severe food shortage.

Investigation of the long-term stress (and health) of animals, including wild animals, has been considerably advanced by the development of non-invasive approaches that avoid the need for blood sampling. Such methods generally rely on the measurement of the concentration of glucocorticoid metabolites excreted by animals in their faeces or urine, and in some cases extracted from the water in which animals, such as fish, are swimming.

The transit time of faeces passing through the gut means that measurement of faecal glucocorticoid metabolites provides a picture of events some time earlier. Herbivores have relatively fast gut transit times, although these still extend many hours, while transit times for carnivores and omnivores are highly variable and may extend to several days. Thus, the measurement of glucocorticoid metabolites in faeces provides an integrated measure of stress levels typically over several hours or days.

During chronic stress, the effects of CRH and/or glucocorticoids on other endocrine systems can adversely affect growth and reproductive function. For example, the effects of chronic stress on gonad development can result from the inhibitory actions of CRH, *β*-endorphin and glucocorticoids on the secretion of gonadotrophin releasing hormone (GnRH) by hypothalamic neurons, resulting in reduced release of follicle stimulating hormone (FSH) and luteinizing hormone (LH) from the anterior pituitary.

Effects on growth can result from the inhibitory effects of glucocorticoids on growth hormone release from the pituitary gland, or because persistent protein catabolism (stimulated by glucocorticoids) results in muscle wasting. In mammals, elevated concentrations of circulating glucocorticoids also inhibit secretion of thyroid stimulating hormone (TSH) by the pituitary, thereby interfering with thyroid function.

Persistently high concentrations of catecholamines and glucocorticoids are immunosuppressive and anti-inflammatory. As a result, chronically stressed animals are more vulnerable to pathogens. Glucocorticoids inhibit proliferation of lymphocytes and increase the rate of **apoptosis** (cell death)

Experimental Panel 19.1 Impact of stress responses on condition and survival of Galápagos marine iguanas

Background

Ever since Hans Selye put forward the concept of a 'General Adaptation Syndrome' (GAS) to stressors, there has been an implicit assumption that *acute* stress responses are adaptive and improve fitness and survival. Longer-term *chronic* stress responses are often considered more deleterious and are hypothesized to have negative effects on fitness or survival. Galápagos marine iguanas (*Amblyrhynchus cristatus*) are an ideal species to examine the effects of chronic stress on the health and survival of the population on different islands.

In El Niño years, warm ocean water temperatures develop off the Pacific coast of South America and result in a failure of the nutrient-rich upwelling necessary for the growth of the algae on which marine iguanas feed. The different island populations of iguanas are affected to different extents during the El Niño, with populations on some islands declining by up to 90 per cent in a severe El Niño year. This raises the questions: do corticosterone concentrations in Galápagos marine iguanas indicate the health and body condition of the population on an island, and their likely survival in El Niño years?

Methods

To answer the question, marine iguanas were captured from six of the Galápagos Islands at the end of a severe El Niño period (1997–1998) in which water temperature of up to 32°C had persisted for almost 18 months (normally water temperatures are 18–23°C). Marine iguanas were also collected from the same islands one year later (1999), when algae had returned and the surviving animals had recovered.

Within 1–2 minutes of capture, a blood sample was collected from individual iguanas to measure baseline concentrations of corticosterone, as a measure of the chronic conditions.

Each iguana was then held in an individual bag for 30 minutes to investigate its response to the standardized acute stress of confinement. Use of a standardized stressor allows comparisons to be made between the populations of iguanas on different islands. Further blood samples were collected after 15 minutes and 30 minutes. At the end of the experiment, all iguanas were weighed and measured to calculate their body condition index from the formula: (body mass/snout to vent length^3) $\times 10^6$.

Experimental results and interpretation

This study allowed several things to be examined:

(i) population differences in baseline corticosterone as a measure of chronic conditions;

(ii) whether individual differences in baseline corticosterone are related to body condition;

(iii) whether population differences in stress responsiveness (to a standardized acute stressor) affect survival.

(i) Population differences in baseline concentrations of corticosterone in response to chronic conditions

The populations of marine iguanas on three islands (Fernandina, Seymour and Santa Fe) were found to have higher baseline plasma concentrations of corticosterone during the El Niño than in the following year when there was a good food supply. Figure A shows

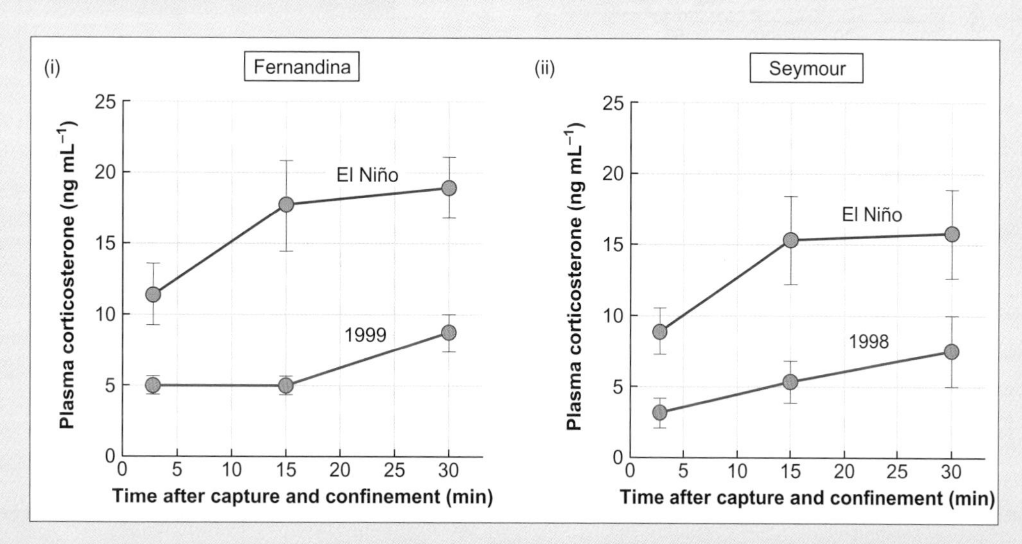

Figure A Corticosterone responses to the stress of capture, handling and confinement in marine iguanas captured on two Galápagos Islands

Iguanas were captured on Fernandina (i) and Seymour (ii) during an El Niño year (1998; red symbols and lines) and the following year (1999) when algae were plentiful (blue symbols and lines). Initial plasma concentrations of corticosterone, 2.5 min after capture, are taken as baseline values. Two subsequent values were obtained 15 min and 30 min later. Each point represents the mean ± standard error (11 iguanas from Fernandina and 8 iguanas from Seymour).

Source: Romero LM, Wikelski M (2001). Corticosterone levels predict survival probabilities of Galápagos marine iguanas during El Niño events. Proceedings of the National Academy of Sciences of the United States of America 98: 7366–7370.

the results for marine iguanas on Seymore and Fernandina. These results suggest that food shortage in the El Niño is sufficiently severe on some islands for the *population* of iguanas to activate the hypothalamic–pituitary–adrenal (HPA) axis, which could help to mobilize their energy reserves.

(ii) Individual differences in baseline concentrations of corticosterone in relation to body condition index

At an *individual* level, it is apparent that the baseline corticosterone of individual iguanas within a population is influenced by their body condition index. Figure B(i) shows the results for two islands, which show that some iguanas maintain good condition and have a relatively low plasma corticosterone, even in the famine conditions of an El Niño year.

The baseline corticosterone of iguanas is generally only elevated if body condition index of the individual iguana falls below a critical threshold of 35. At this threshold value, any further reduction in condition is associated with a steep rise in the plasma corticosterone concentration. The severe activation of the HPA axis could mobilize further energy reserves, but at this point an iguana is likely to be in a poor condition and its survival is threatened. Generally, iguanas die if their body condition index falls below 25. In an El Niño year, many iguanas can be in extremely poor condition, close to the minimum for survival; for example, they may show the severe muscle wastage exhibited by the iguana in the lower panel of Figure B(ii).

(iii) Do population differences in stress responsiveness (to a standardized acute stressor) affect survival?

Figure A shows that the standardized confinement stress resulted in the expected increase in corticosterone. All six populations of marine iguanas showed a stress response to confinement, with those from five of the six islands exhibiting stress-induced increases in plasma concentrations of corticosterone at both 15 min and 30 min that were significantly higher in the El Niño year than when there was a good food supply.

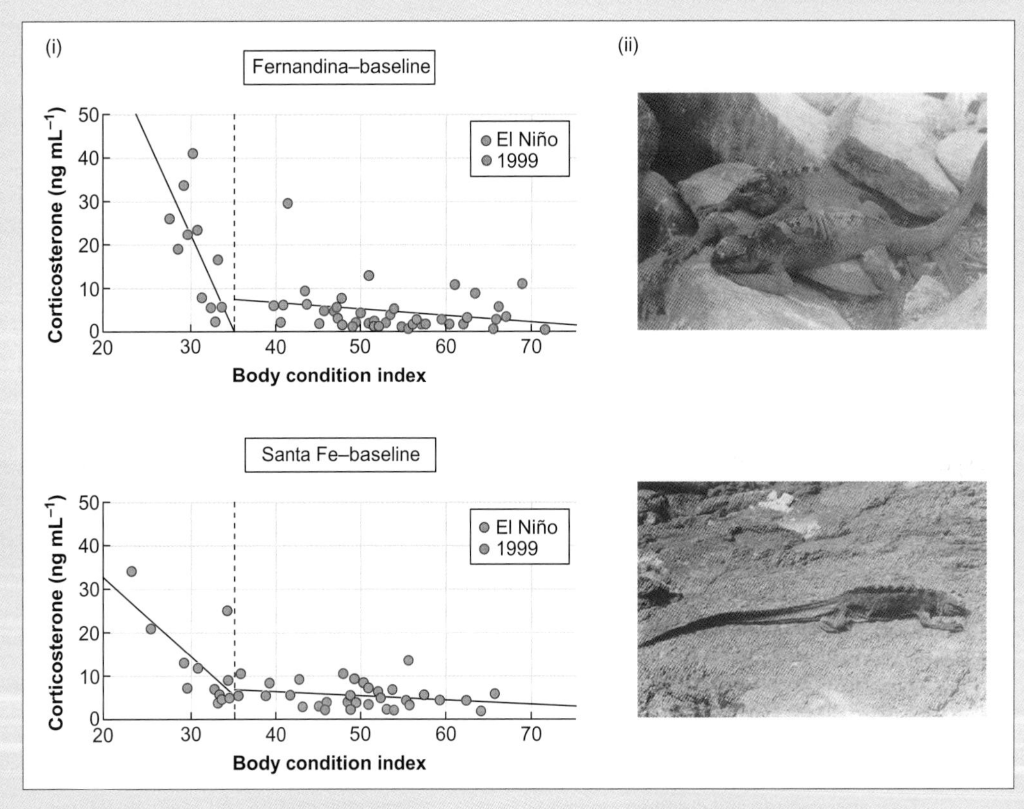

Figure B Relationship between baseline plasma corticosterone concentrations and body condition of iguanas on Fernandina and Santa Fe

Data for El Niño year (1998) are shown by red symbols. Data for the following year (1999) when algae were plentiful are shown by blue symbols. (i) At body conditions of more than 35, corticosterone is relatively low in all individuals and not significantly related to body condition. The vertical dashed line indicates the threshold point, at a condition index of about 35. Below this threshold, plasma corticosterone concentrations increase dramatically and there is a significant negative correlation between plasma corticosterone and body condition. Lines show regression lines fitted to data above and below the critical threshold. (ii) Examples of iguanas with condition indices of near 60 (upper panel when algae are plentiful) and about 30 (lower panel in El Niño year).

Source: Romero LM, Wikelski M (2001). Corticosterone levels predict survival probabilities of Galápagos marine iguanas during El Niño events. Proceedings of the National Academy of Sciences of the United States of America 98: 7366–7370.

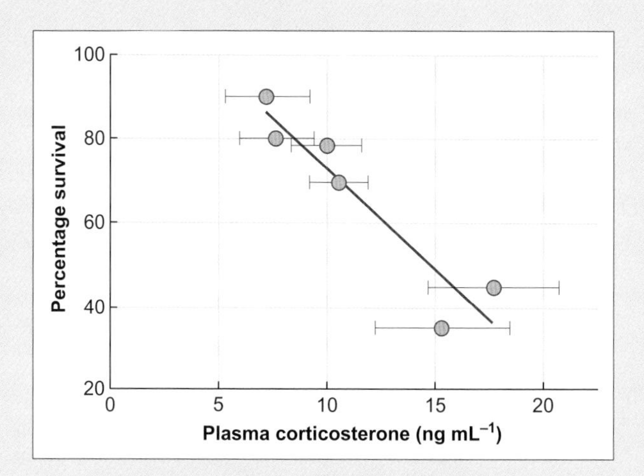

Figure C Relationship between plasma corticosterone concentrations of acutely stressed Galápagos marine iguanas from each island and survival rates on the respective island during the El Niño period

Percentage survival was based on counts of iguanas on each island. Corticosterone was measured in blood samples collected 15 min after capture, handling and restraint. The line shows the regression line for the relationship between survival and circulating corticosterone. Data show means ± standard errors for corticosterone measurements.

Source: adapted from Romero LM, Wikelski M (2001). Corticosterone levels predict survival probabilities of Galápagos marine iguanas during El Niño events. Proceedings of the National Academy of Sciences of the United States of America 98: 7366–7370.

The elevation in plasma corticosterone concentrations of the different populations of marine iguanas in responses to the standardized stressor was found to be a good predictor of their relative survival during the severe food shortage resulting from El Niño. Figure C shows the data for the six islands. The highest corticosterone concentrations induced by the acute stress occurred in the Seymour and Fernandina populations, and were correlated with the heavy mortalities (poor survival) of iguanas on these islands. By contrast, iguanas on Santa Cruz and Isabela were barely affected.

› Find out more

Busch DS, Hayward LS (2009). Stress in a conservation context: A discussion of glucocorticoid actions and how levels change with conservation-relevant variables. Biological Conservation 142: 2844–2853.

Romero LM, Wikelski M (2001). Corticosterone levels predict survival probabilities of Galápagos marine iguanas during El Niño events. Proceedings of the National Academy of Sciences of the United States of America 98: 7366–7370.

of lymphocytes, so the number of circulating lymphocytes decreases. Similarly, the number of **thrombocytes** (platelets), eosinophils and monocytes[43] decreases during chronic stress.

The inhibition of immune and inflammatory responses by glucocorticoids during prolonged stress often results from the reduction in the expression of various cytokines that promote immune and/or inflammatory responses. The term 'cytokine' is used to describe a diverse range of proteins and peptides that are synthesized on demand and act at low concentrations to regulate cell function. Cytokines may enter circulation and have actions at a distance from their origin, aside from actions on immune responses, so there is no distinct boundary between cytokines and hormones.

Glucocorticoids decrease the transcription of the gene for **tumour necrosis factor** (TNF) that was first observed to cause necrosis of tumour cells (hence its name), but whose primary role is in the regulation of various immune cells: activated **macrophages**[44] produce TNF as part of the immune response to bacteria. Glucocorticoids also decrease the expression of pro-inflammatory **interleukins** such as IL-1 and IL-6. Interleukins were first named to reflect their role in communication (hence 'inter-') and their production by, and action on, leucocytes ('-leukins'). For example, activated macrophages produce interleukin 1 (IL-1) in response to infection; IL-1 stimulates inflammation and activates lymphocyte proliferation. Interleukin-6 (IL-6) stimulates the final maturation of B-lymphocytes, the cells producing immunoglobulins as part of the immune response. However, many cell types other than immune cells are now known to produce interleukins.

Further inhibition of the immune and inflammatory responses by glucocorticoids results from the reduction of circulating monocytes and inhibition of the production of **leukotrienes** and prostaglandins that are often produced alongside leukotrienes as part of an inflammatory response. Leukotrienes (produced by leucocytes and other immune cells, such as macrophages) promote migration of neutrophils to areas of tissue damage and stimulate production of inflammatory cytokines by immune cells.

19.3.4 Stress responsiveness

In recent years, it has become increasingly apparent that the response of the vertebrate hypothalamic–pituitary–adrenal/interrenal axes to stress is not fixed, but instead shows phenotypic flexibility[45]. Chronic stress can result in feedback

[43] Monocytes are a type of white blood cell (**leukocyte**) involved in the innate (non-acquired) immune system of all vertebrates; monocytes take up microbes by phagocytosis.

[44] Macrophages are large phagocytic cells, one of the white blood cells, which occur in tissues at sites of infection.

[45] Phenotypic flexibility (plasticity) generally refers to the ability of an animal of one genotype to result in more than one phenotype under different environmental conditions, as we discuss in Section 1.5.1.

mechanisms that alter the baseline concentrations of glucocorticoids and influence responses to new stressors, although such responses vary between species.

Within the populations of many species of fish, birds and mammals, including humans, there is a high level of variability in the response to a standardized stressor (such as the confinement test used in the studies in Experimental Panel 19.1). Such variability in the magnitude of HPA/HPI responses indicates some are low responders and some are high responders. These differences are thought to represent alternative coping strategies that may each be beneficial depending on the particular environmental conditions and the personality characteristics of an individual. Among birds, for example, at high population densities the low responders with a proactive personality will benefit provided there is food available.

Beyond the responses of individual animals, an emerging new concept is that a transgenerational change in responses to stress can occur, with maternal stress affecting the HPA axis of the offspring, and even subsequent generations. In mammals, exposure to stress during pregnancy increases the maternal plasma concentrations of ACTH, cortisol and catecholamines. The placenta normally forms a barrier to the majority of maternal hormones; however, changes in placental function during stress appear to allow more glucocorticoids to pass to the foetus and may explain changes in the activity of the hypothalamus–pituitary–adrenal (HPA) axis of the offspring.

Generally, there is an increase in the activity of the HPA axis of the offspring of stressed mothers, although responses vary and appear to depend on the type of experimental manipulation, its timing and duration. For example, in one study of guinea pigs, maternal stress in late gestation elevated the HPA activity of the offspring in unstressed conditions (baseline cortisol concentrations), but not their responses to stress, while the same maternal stress experienced closer to the end of gestation resulted in offspring with normal baseline cortisol concentrations, but an increased HPA responsiveness to stressful events.

Furthermore, some studies suggest that maternal stress/anxiety during pregnancy can lead to transmission of stress sensitivity by non-genomic (epigenetic)[46] mechanisms that affect not just the F_1 generation but also the F_2 generation or later generations. For example, studies of guinea pigs with food restriction during pregnancy showed increased basal cortisol and altered HPA responsiveness to stress of F_1 and F_2 generations. It is uncertain whether similar effects apply in humans; if they do, adverse environmental influences on one generation could adversely influence the physiological responses and disease risks of later generations. However, we need much more information to assess long-term consequences and the mechanisms responsible for changes in stress responsiveness.

[46] We discuss epigenetic mechanisms in Section 1.5.1.

› *Review articles*

Busch DS, Hayward LS (2009). Stress in a conservation context: A discussion of glucocorticoid actions and how levels change with conservation-relevant variables. Biological Conservation 142: 2844–2853.

Breuner CW, Patterson SH, Hahn TP (2008). In search of relationships between the acute adrenocortical response and fitness. General and Comparative Endocrinology 157: 288–295.

Landys MM, Ramenofsky M, Wingfield JC (2006). Actions of glucocorticoids at a seasonal baseline as compared to stress-related levels in the regulation of periodic life processes. General and Comparative Endocrinology 148: 132–149.

Matthews SG, Phillips DIW (2010). Minireview: Transgenerational inheritance of the stress response: A new frontier in stress research. Endocrinology 151: 7–13.

O'Connor TM, O'Halloran DJ, Shanahan F (2000). The stress response and the hypothalamic-pituitary-adrenal axis: from molecule to melancholia. Quarterly Journal of Medicine 93: 323–333.

Perry SF, Capaldo A (2011). The autonomic nervous system and chromaffin tissue: Neuroendocrine regulation of catecholamine secretion in non-mammalian vertebrates. Autonomic Neuroscience: Basic and Clinical 165: 54–66.

19.4 The vertebrate thyroid gland

The thyroid gland is structurally similar in all vertebrates: it consists of many follicles, each with an outer layer of follicular epithelial cells surrounding a centre of colloid that contains the store of thyroid hormones, as shown in Figure 19.29. This arrangement is unique among endocrine glands and may be an adaptation to allow storage of a large quantity of hormone containing the trace element, iodine.

In mammals, the thyroid gland consists of a right and a left lobe, joined by an isthmus lying over the trachea, just below the larynx (in the neck). In non-mammalian vertebrates, the thyroid is more diffuse but with the same basic follicular structure that is fundamental to the functioning of the gland and its production and storage of thyroid hormones. In mammals, further endocrine cells called **C-cells** occur between the thyroid follicles, The C-cells secrete **calcitonin**, which has a role in calcium balance[47].

The hypothalamus–pituitary–thyroid axis, part of which is shown in Figure 19.18, ultimately regulates the circulating concentrations of thyroid hormones. Thyrotrophin releasing hormone (TRH), a tripeptide released by hypothalamic neurons, stimulates the thyrotrophs of the anterior

[47] We discuss the actions of calcitonin on calcium balance in Section 21.2.4, alongside the involvement of other endocrine secretions that influence calcium balance.

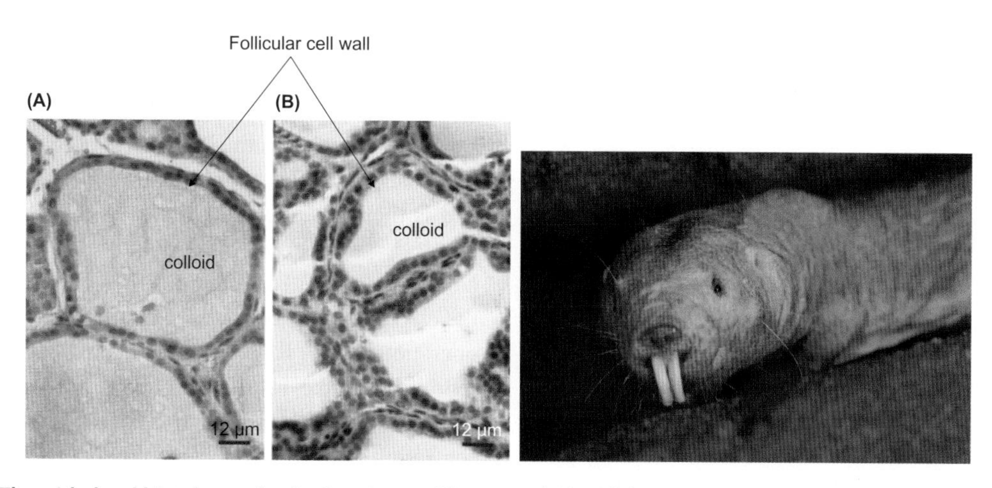

Figure 19.29 Thyroid gland histology of naked mole rats (*Heterocephalus glaber*)

Photomicrographs of histological sections of thyroid glands of naked mole rat in two temperature conditions: (A) In thermoneutral conditions (30°C), transverse sections of the spherical follicles show relatively low tissue activity. The follicles are filled with colloid stores of thyroid hormone surrounded by a flat epithelial cell wall. (B) When acclimated to relatively cold conditions (25°C), the epithelial cells increase in height and colloid stores are reduced, which indicates increased thyroid gland activity stimulated by thyroid stimulating hormone (TSH).

Source: Buffenstein R et al (2001). Cold-induced changes in thyroid function in a poikilothermic mammal, the naked mole-rat. American Journal of Physiology Integrative Comparative Physiology 280: R149–155. Image: Frans Lanting Studio/Alamy Stock Photo.

pituitary, which secrete **thyroid stimulating hormone** (TSH). Figure 19.30 illustrates an example of TRH stimulation of the pituitary–thyroid axis in lizards (Carolina anoles, *Anolis carolinensis*).

TSH binds to receptors on the basal membrane of the follicular cells, stimulating all aspects of the production of thyroid hormones. Thyroid hormones usually exert negative feedback on the hypothalamus–pituitary–thyroid axis mainly by changing the sensitivity of the thyrotrophs to TRH. If the concentration of thyroid hormones is low, a reduction in negative feedback results in higher levels of TSH, which stimulates the follicular cells as a mechanism to increase hormone levels. Figure 19.29B shows an example of the resulting hypertrophy (increased cell size) and hyperplasia (increased cell number) that occurs under persistent thyroid stimulation (in this case cold acclimation of naked mole rats, *Heterocephalus glaber*). Epithelial cell height is a good index of the level of activation of thyroid follicles. Note in the micrographs in Figure 19.29 that at 30°C the follicular epithelium is flat, indicating a low level of activity. Naked mole rats have limited insulation and their relatively low metabolic rate (in comparison with other mammals) probably reflects their lower levels of thyroid hormones—an order less than is typical of mammals[48]. Nevertheless, naked mole rats show typical mammalian responses to cold exposure; a decline in ambient temperature of just 5°C is sufficient to increase the height of the follicular epithelial cells by more than 50 per cent.

A marked exception to the regulation of the thyroid axis by TRH occurs in larval amphibians in which corticotrophin (CRH) from the hypothalamus stimulates the thyrotrophs of the anterior pituitary in addition to their more usual stimulation of pituitary corticotrophs that secrete ACTH. Hence, the control of the thyroid function of amphibians alters during metamorphosis.

[48] We discuss the actions of thyroid hormones including effects on metabolic rate, in Section 19.4.2.

19.4.1 Synthesis and secretion of thyroid hormones

Figure 19.31 illustrates the steps involved in the production of thyroid hormones, which are synthesized as part of the matrix of a large glycoprotein, **thyroglobulin** (molecular weight ~660,000 Daltons), and which ultimately accumulate as colloid in the lumen of the thyroid follicles, as shown in Figure 19.29. Thyroglobulin is synthesized and packaged into membrane-bound vesicles by the endoplasmic reticulum and Golgi apparatus of follicular cells and is secreted by exocytosis at the apical side of the cells.

The active hormones are thyroxine (tetraiodothronine) (T_4) and triiodothyronine (T_3). The naming of these hormones tells us how many iodine atoms they contain: the prefix tetra- and subscript 4 tells us that there are four iodine atoms in a molecule of thyroxine, while the prefix tri- and subscript 3 tells us there are three iodine atoms in triiodothyronine. Figure 19.32 shows how the addition of an appropriate number of iodine atoms to tyrosine that is incorporated into thyroglobulin forms both T_4 and T_3.

The thyroid follicular cells take up the two ingredients needed to synthesize thyroid hormones—tyrosine and iodine from the blood. Organisms can synthesize tyrosine

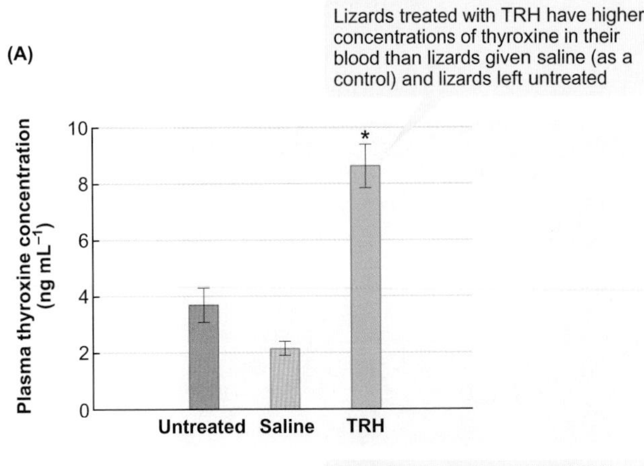

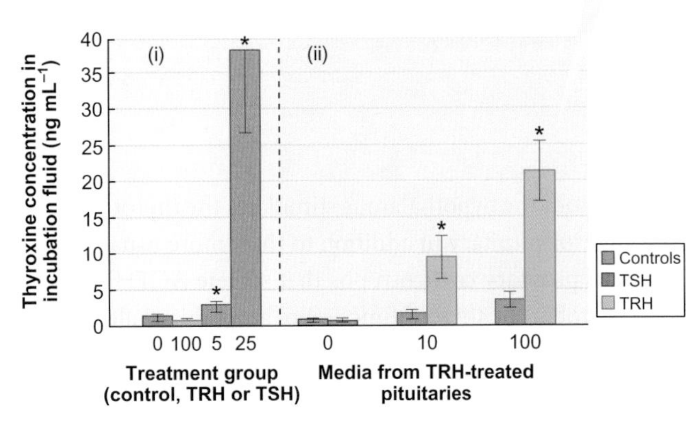

Figure 19.30 The hypothalamic–pituitary–thyroid axis of the lizards, Carolina anoles (*Anolis carolinensis*)

(A) Effects of *in vivo* infusion of thyrotrophin releasing hormone (TRH at 5 μg h^{-1} for 10 h) on the plasma concentration of thyroxine (T_4) compared to controls: untreated lizards and lizards infused with saline. (B) Regulation of thyroxine release from thyroid glands *in vitro*: (i) thyroid stimulating hormone (TSH; 5 or 25 ng) stimulates thyroxine release from incubated thyroid glands compared to controls incubated in media only. TRH (100 ng mL^{-1}) has no direct on thyroid glands, but acts via TSH as shown by the results in B (ii). (ii)Thyroxine release by thyroid glands incubated in fluid media in which pituitaries had been previously incubated with TRH at 0, 10 or 100 ng mL^{-1} compared to controls incubated in media only. Data are means and standard errors for 10 lizards in (A); six incubations per group in (B); asterisks indicate significant differences: $P<0.001$ in (A); $P<0.05$ in (B).

Source: Licht P, Denver RJ (1988). Effects of TRH on hormone release from pituitaries of the lizard, *Anolis carolinensis*. General and Comparative Endocrinology 70: 355–362.

in sufficient quantities so there is no dietary need for this precursor. In contrast, external sources of iodine (I_2), iodate (IO_3^-) or ionic iodide (I^-) are required. Iodine or iodate in the diet is chemically reduced to iodide before absorption from the gastrointestinal tract. Fish have the capacity to take up I^- from water across their gills as well as from the diet, and fish living in seawater absorb I^- and iodate from the seawater they drink[49].

The thyroid gland has the ability to take up iodide across the basal membrane of the cells from the interstitial fluids, against a large concentration difference. Iodide uptake is via a **symport carrier** in the basolateral border of the follicular cells that co-transports sodium ions (Na^+) and I^- into the cell and is therefore often referred to as NIS: N = sodium, I = iodide, S = symporter. The simultaneous uptake of Na^+ (and co-transport of I^-) in the ratio of 2:1 (Na^+:I^-) is driven by the electrochemical difference for Na^+ that results from the operation of **Na^+, K^+-ATPase** pumps[50].

Iodide uptake results in higher concentrations in the thyroid gland than in the blood. In humans, the usual concentration ratio (thyroid: blood) is about 20:1 and increases to 100:1 or more if the thyroid gland is stimulated by TSH, which stimulates the Na^+/I^- symporters.

Iodide shuttles across the follicular cells to their apical membranes, where transport into the lumen of the follicle occurs. In mammalian thyroid follicles, the apical transporter is generally considered to be a sodium independent chloride/iodide transporter named pendrin, although not all results support this idea.

Iodide needs to be oxidized by membrane-bound thyroid peroxidase (which uses hydrogen peroxide as the oxidizing agent) on the luminal membrane; it then attaches to the tyrosine residues of the thyroglobulin in the colloid initially to form monoiodotyrosine (MIT) and di-iodotyrosine (DIT), as shown in Figure 19.32. Thyroid stimulating hormone increases thyroid peroxidase activity, which contributes to the increased production of thyroid hormones.

Note in Figure 19.32 that coupling of MIT and DIT in the colloid forms T_3, while coupling of two DITs forms T_4, which is the major hormone stored in association with the thyroglobulin until required. A supply of T_4 capable of lasting several months is stored at any one time; for instance, about 2 months' supply of T_4 is stored in the thyroid gland of humans.

Because stored thyroid hormones are attached to thyroglobulin, they need to be taken back into the cells by endocytosis, and then released from the thyroglobulin to enter circulation. When stimulated to do so by TSH, thyroglobulin is taken up into the cells and lysosomes coalesce with the absorption droplet. The lysosomal membranes become permeable and proteolytic digestion releases non-active compounds and active hormones. The inactive compounds—the iodotyrosines (MIT and DIT)—are then rapidly deiodinated and the iodide is reused, as illustrated in Figure 19.31.

Thyroid hormones have high lipid solubility, so readily pass through the basal membrane of the cells and into the blood. Thyroxine (T_4) is the main hormone secreted by the thyroid gland of all vertebrates, but T_4 is deiodinated by

[49] Section 5.1.3 discusses the need to drink seawater for fluid balance in marine teleosts.

[50] Section 3.2.4 outlines these principles and the main categories of ion transport proteins. NIS proteins also occur in the gastrointestinal tract where they are responsible for iodide absorption from the diet.

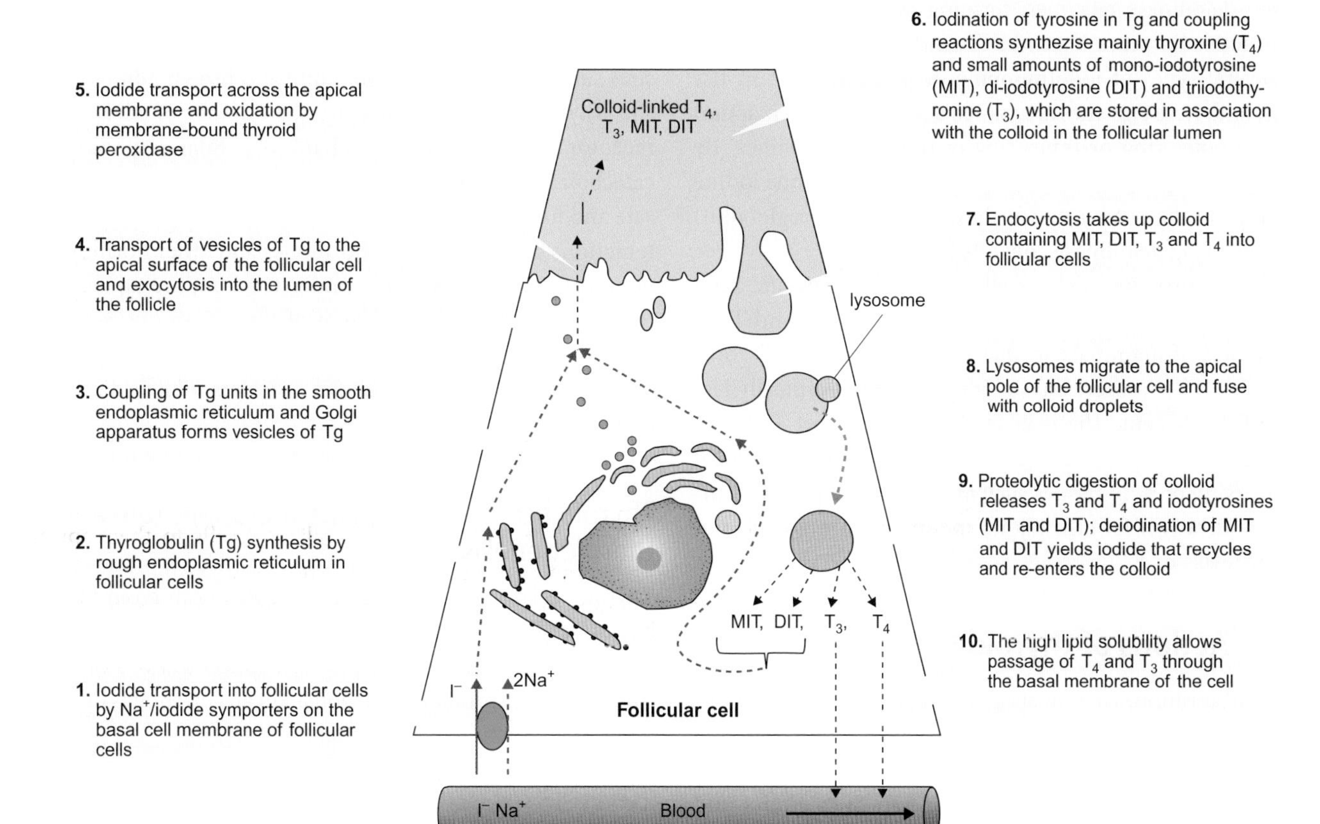

Figure 19.31 Schematic diagram showing stages in synthesis, storage and secretion of thyroid hormones

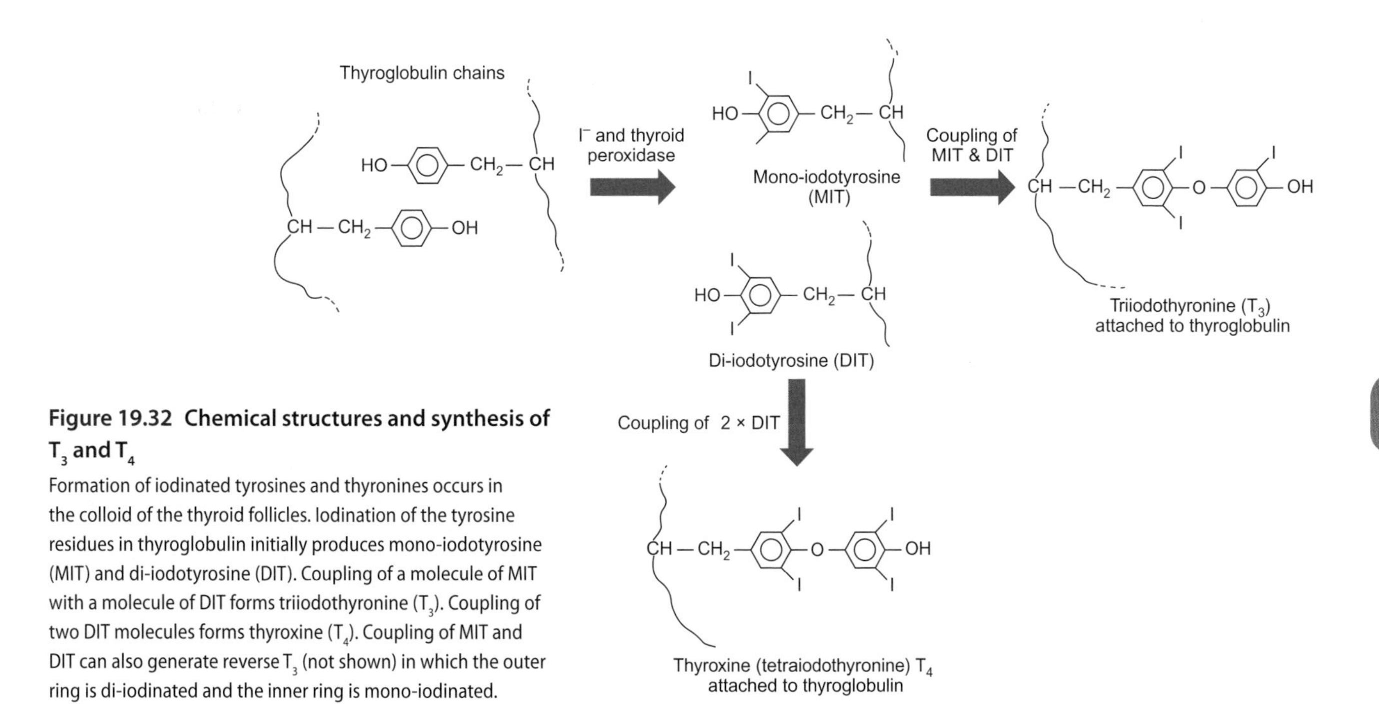

Figure 19.32 Chemical structures and synthesis of T_3 and T_4

Formation of iodinated tyrosines and thyronines occurs in the colloid of the thyroid follicles. Iodination of the tyrosine residues in thyroglobulin initially produces mono-iodotyrosine (MIT) and di-iodotyrosine (DIT). Coupling of a molecule of MIT with a molecule of DIT forms triiodothyronine (T_3). Coupling of two DIT molecules forms thyroxine (T_4). Coupling of MIT and DIT can also generate reverse T_3 (not shown) in which the outer ring is di-iodinated and the inner ring is mono-iodinated.

tissue deiodinase enzymes, forming the main physiologically active hormone T_3, and DITs and MIT.

Several types of iodothyronine deiodinases exist in tissues, and these enzymes are named according to their action on the outer ring or inner ring of the iodothyronines. By examining Figure 19.32, we find how removal of one iodine atom from the outer ring (furthest from the thyroglobulin) by outer-ring deiodinase converts T_4 to T_3.

Plasma proteins such as albumins and a specific high affinity T_4-binding globulin bind most of the T_4 and T_3 in the blood of mammals so that usually only a small component is free to exert physiological actions—typically less than 0.1 per cent. The liver and kidney convert T_4 to T_3, which then circulates to other tissues, but additional local conversion of T_4 in tissues such as the brain, anterior pituitary and brown fat also allows independent regulation of tissue levels of T_3.

19.4.2 Actions of thyroid hormones

Thyroid hormones act on specific nuclear receptors belonging to the superfamily of steroid hormone receptors with the characteristics outlined earlier in Figure 19.10. Three forms of receptors for thyroid hormones (TRs): TRα, TRβ-1 and TRβ-2 occur to different extents in different tissues. Cardiac and skeletal muscle contain high levels of TRα, but this receptor type is widely distributed; TRβ-1 is predominantly in the brain, kidney and liver; TRβ-2 is mainly restricted to the hypothalamus and pituitary.

Looking back at Figure 19.10 notice how hormone binding to nuclear receptors activates gene expression. However, this diagram oversimplifies the events leading to the synthesis of particular proteins in different cell types in response to thyroid hormones. A complex array of coactivator proteins (that activate gene expression) and corepressor proteins (that inhibit the expression of genes) of thyroid hormone receptors have been identified and are presumed to allow the particular actions of thyroid hormones in specific tissues, but these complexities are not understood.

The main actions of thyroid hormones, which are summarized in Figure 19.18, influence normal growth and development, and metabolism and thermogenesis. Thyroid hormones have direct actions and indirect actions (permissive actions) that involve the stimulation of other hormones. For example, thyroid hormones stimulate the secretion of growth hormone (GH), but growth is affected by the combined effect of thyroid hormones, growth hormone and insulin growth factor (produced by the liver).

Thyroid hormones also act synergistically with other hormones and the sympathetic nervous system. A good example is the synergistic effect of thyroid hormones and catecholamines (adrenaline and noradrenaline) in tissues such as liver, heart, and white and brown adipose tissue. Thyroid hormones increase the number of adrenergic receptors, which results in enhanced cellular responses to catecholamines, for example resulting in an increase in the rate and force of contraction of the heart[51]. The complex interactions between thyroid hormones and other hormonal/neural systems are not fully understood but are an area of immense interest; manipulation of these pathways is considered to have the potential importance for development of drug treatments for metabolic disorders such as diabetes and obesity.

Thyroid hormones regulate metabolic rate and thermogenesis

Thyroid hormones are important regulators of basal metabolic rate (BMR) of mammals (endotherms) and the resting metabolic rate of ectotherms[52]. High concentrations of thyroid hormones promote metabolism in all vertebrates by two types of interacting mechanism:

- peripheral regulation of the rate of cellular metabolism of the peripheral tissues, particularly muscle and liver;
- central regulation by the actions of thyroid hormones on the hypothalamus of the brain.

In peripheral tissues, thyroid hormones influence carbohydrate, protein and lipid metabolism. For instance, when high concentrations of thyroid hormones are present in the blood, conversion of glycogen stores in liver and muscle to glucose occurs, and glucose acts as a fuel for an increased metabolic rate. However, the overall effect of actions of thyroid hormones on fuel metabolism has a relatively small effect on the BMR of mammals. Stimulation of Na^+, K^+-ATPase activity in most cells by thyroid hormones has a greater effect in stimulating metabolic rate. A third process is the peripheral stimulation of metabolic rate via actions of thyroid hormones on the skeletal muscle in which thyroid hormones stimulate Ca^{2+}-ATPase in the sarcoplasmic/endoplasmic reticulum (SERCA)[53], which increases the production of heat during ATP hydrolysis.

[51] The actions of catecholamines on the cardiovascular system are discussed in Section 22.4.3.

[52] Basal metabolic rate of mammals/birds (the minimum metabolic rate at rest), and resting metabolic rate of ectotherms, in which metabolic rates and rates of heat production are too low to affect body temperature, and high rates of heat loss occur, are discussed in Section 2.2.3.

[53] The use of ATP by Ca^{2+}-ATPase in the sarcoplasmic reticulum (SERCA) of the skeletal muscle cells (myocytes) transfers Ca^{2+} from the cell cytosol into the lumen of the sarcoplasmic reticulum, during relaxation of the myocytes, as we discuss in Section 18.2.2.

Mammals have evolved an ability to maintain core body temperatures above environmental temperatures during exposure to cold environments. This response is partly the result of the actions of thyroid hormones in heat generation, which is known as **thyroid thermogenesis**. Thyroid thermogenesis was initially thought to result from T_3 activation of a general increase in the metabolic rate of most body cells, activation of Na^+, K^+-ATPase activity in many cell types and activation Ca^{2+}-ATPase (SERCA) in skeletal muscle cells. However, recent studies emphasize the influence of thyroid hormones on non-shivering thermogenesis by brown adipose tissue[54], especially during acclimation to cold conditions[55].

Iodothyronine deiodinase in tissues rapidly converts circulating T_4 into intracellular T_3, which exerts physiological actions. In brown adipose tissue, type 2 iodothyronine deiodinase (D2) is essential in generating T_3 locally, which stimulates thermogenesis. The importance of this local action in temperature regulation becomes evident in gene knockout mice lacking the gene for D2.These mice become hypothermic as they must largely rely on ineffective shivering for body temperature regulation.

The generation of heat by brown adipose tissue involves the synergistic actions of the T_3 formed in the adipocytes of brown adipose tissue and noradrenaline released from the sympathetic nerve fibres that innervate the brown adipose tissue. Noradrenaline and T_3 acting separately generate a two-fold increase in the **uncoupling protein** (UCP-1) present in the inner mitochondrial membranes in brown adipose tissue and which is critical for heat generation[56]. In contrast, the synergistic effects of noradrenaline and T_3 result in a 20-fold increase in UCP-1 in brown adipose tissue.

While T_3 has direct effects on adrenergic-stimulated thermogenesis in brown adipose tissue, recent studies suggest the additional central regulation on the brain. Specifically, rats given injections of T_3 into the hypothalamus increase their sympathetic output to brown adipose tissue, which results in increased thermogenesis and metabolic fuel expenditure. Central regulation of the sympathetic nervous system by thyroid hormone has recently been localized to previously unrecognized neurons, in the anterior hypothalamus.

[54] The high density of mitochondria in brown adipose tissue, which is illustrated in Figure 10.16, is crucial to the generation of heat by this tissue.

[55] The typical mammalian responses to cold exposure of non-shivering thermogenesis does not occur during **hibernation** when thyroid hormones in the blood decrease and basal metabolic rate declines, as we discuss in Section 10.2.7.

[56] We explore the mechanism by which UCP1 generates heat by decoupling the proton gradient in the mitochondria of brown adipose tissue in Section 10.2.5.

While thyroid thermogenesis is well described in cold-acclimated mammals, it is unlikely to be a significant influence on the body temperatures of ectotherms. Nevertheless, in an evolutionary context, thermogenic actions of thyroid hormones in ectotherms have been argued to be the forerunners to the involvement of thyroid hormones in the evolution of endothermy.

In cold-acclimated zebrafish (*Danio rerio*), thyroid hormones have been shown to maintain heart function and muscle function and result in enhanced swimming performance. These actions include an increase in SERCA activity in skeletal and cardiac muscle cells, which also occurs in mammals. Thyroid hormones also appear to act centrally in zebrafish in a similar way to mammals, by increasing the sympathetic outflow to the heart during cold acclimation. As a result, thyroid hormones increase the heart rate of zebrafish. Overall, these actions are one factor resulting in an improvement in the swimming performance of zebrafish during cold acclimation.

Thyroid hormones are essential for normal developmental processes. In birds and mammals, thyroid hormones increase during moulting and the formation of new feathers/hair. In migrating species of salmon, an increase in thyroid hormones (alongside several other hormones) in juvenile fish is a pre-requisite for transformation of parr into smolts (smoltification). The elevated concentrations of thyroid hormones initiate body silvering and result in changes in gill function so that the salmon are ready to osmoregulate in seawater after their migration downstream and into the sea[57].

Developing embryos do not have an endogenous supply of thyroid hormones until the thyroid follicles form, and the cellular machinery required for the synthesis of thyroglobulin and iodinated thyronines develops. Yolk laid down in the eggs of oviparous vertebrates[58] contains lipid soluble hormones (which include thyroid hormones) that provide the developing larva/embryo with thyroid hormones before it can make its own. In mammals, we might ask whether there is transfer of thyroid hormones from the mother to the embryo. Case Study 19.1 examines this question and reveals that maternal provision of T_4 in the first few weeks of embryonic development is important for neural development. Thyroid hormones influence neuronal maturation, synapse formation and myelination. Low maternal levels of thyroid hormones can therefore have damaging effects on neural development of the offspring.

[57] We explore some of the events associated with osmoregulation during upstream and downstream migration in teleost fish in Section 5.3.3.

[58] Section 20.1.2 discusses yolk provision in eggs.

Case Study 19.1 Maternal provision of thyroid hormones for developing mammalian embryos

Thyroid hormones are required for normal development and organ maturation and for normal brain development of mammals, including humans. Therefore, maternal hypothyroidism[1] (low levels of free thyroid hormone) may have disastrous consequences for the developing foetus. Such risks are prevalent in iodine-deficient parts of the world, but are not restricted to these areas, since other factors such as inadequate function of thyroid peroxidase can also result in hypothyroidism.

Work on rats has revealed that foetuses from mothers made hypothyroid by chemical treatments (such as methimazole), in which the production of thyroid hormones is inhibited, have impaired neuronal development, particularly in the cerebellum and cortex[2]. Hypothyroidism in rats results in decreased whole-brain weight and densely packed neurons in the cerebellum with less branching of dendrites, and a decrease in the density of neurons in the cerebral cortex. However, the retarded neuronal development of these foetuses can be reversed by thyroid hormone replacement, which suggests that normal neuronal development relies on the transfer of maternal thyroid hormones across the placenta.

Timing of the thyroid development in mammalian foetuses

There is a critical period in embryonic development before the thyroid gland develops in the foetus. In rats, development of the spinal cord and brain starts at day 10–11, but the foetal thyroid of rats only starts to produce its own hormones at day 16 of gestation.

In humans, as in rats, neural development starts before the functional foetal thyroid starts to produce thyroid hormones. Neural development of the neural tube starts from just over 2 weeks of gestation, and at weeks 3–4 the spinal cord, brain pons and medulla begin to develop; at around week 6 development of the cortex of the brain commences. While the thyroid gland begins to develop in week 3 of gestation, hypothalamus–pituitary control of thyroid tissue does not commence until weeks 12–14. Increased thyroid hormone production due to increases in follicular cell uptake of iodide commences in weeks 18–20 and there is subsequent maturation of the hypothalamus–pituitary–thyroid axis. These patterns of development raise the question: does the human embryo and developing foetus rely on maternal thyroid hormones for proper neuronal development, particularly during the first trimester (week 1–12) of gestation?

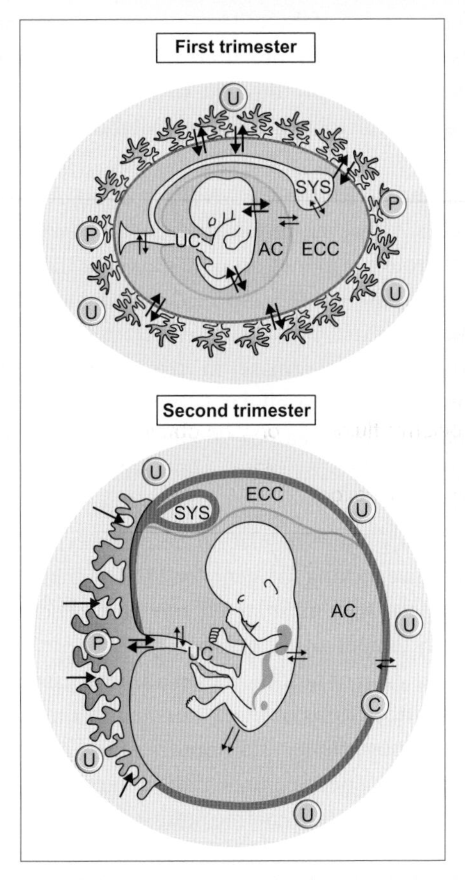

Figure A Schematic representation of the maternal–foetal unit during the first two trimesters of pregnancy

For most of the first trimester the human foetus, held in the uterus (U), is surrounded by two distinct fluid cavities: an inner amniotic cavity (AC) containing the foetus and an outer, exocoelomic cavity (ECC) or primary yolk sac, which contains the secondary yolk sac (SYS), connected to the foetal digestive tract and circulation via the umbilical cord (UC). In the second trimester, in which the foetus develops its own thyroid gland, the placenta (P) develops. The secondary yolk sac degenerates. The amniotic cavity progressively increases in volume and the exocoelomic cavity is progressively obliterated. Arrows show main routes of exchange in each trimester. In the first trimester the main route for molecular exchange between the foetus and mother is via the exocoelomic cavity. In the second trimester, the main exchange is via the placenta.

Source: Calvo RM et al (2002). Fetal tissues are exposed to biologically relevant free thyroxine concentrations during early phases of development. The Journal of Clinical Endocrinology & Metabolism 87: 1768–1777.

Route for maternal provision of thyroid hormones

During early development, the human foetus is surrounded by two distinct fluid-filled cavities, which are separated by a thin membrane. These cavities are an inner amniotic cavity containing the foetus, and an outer exocoelomic cavity, as shown in Figure A. Looking at the routes for molecular exchanges shown in this figure, if thyroid hormones are conveyed from the mother to the foetus, then we might expect to find them in the amniotic and coelomic fluid before the foetus makes its own thyroid hormones. Fluid in the coelomic fluid of the foetus is an ultrafiltrate of the mother's serum, which means that molecules of T_4 or T_3 bound to plasma proteins will not enter the coelomic fluid: only free (unbound) hormone could enter the foetal fluids. Does this allow biologically relevant concentrations of thyroid hormones to pass to the foetus? To answer this question, samples of coelomic and amniotic fluid from foetuses in the early stages of pregnancy should ideally be analysed.

Although such procedures would be unethical in a continuing pregnancy, samples of amniotic fluid and coelomic fluid have been obtained, by consent, before elective termination of pregnancies.

Measurable concentrations of thyroid hormone occur in foetal fluids before the foetal thyroid develops

Highly-sensitive radioimmunoassays allow measurement of the concentration of thyroid hormones in the coelomic and amniotic fluids of human foetuses. Figure B summarizes results for coelomic and amniotic fluids taken from foetuses at 5 to 17 weeks gestation, compared to the serum of the foetus and their mothers. Coelomic fluids were found to contain measurable concentrations of total T_4 within 5 to 6 weeks, gestation, before control of an embryonic thyroid develops, so some T_4 must be passed from the mother to the foetus. Coelomic fluids can only be obtained up until 12 weeks. Figure A shows that, thereafter, the volume of the coelomic cavity diminishes as the foetus grows, the placenta develops and the volume of the amniotic cavity increases.

Figure B(i) shows that in the first 12 weeks gestation, the concentration of total T_4 in the coelomic fluids is at least 100 times less than that in the maternal serum. We might wonder whether the low concentration in the foetus is biologically relevant. Is there enough present to exert a physiological effect? In answering this question we need to remember that it is the *free* hormone that is biologically active. Measuring total concentrations can be misleading when dealing with physiological actions, especially if there is a change in the concentration of binding proteins. Measuring free T_4 in foetal fluids was therefore important and some unexpected results emerged.

In the first trimester (up to week 12 gestation), even before the foetus has its own capacity to make thyroid hormones, the concentration of free T_4 in foetal fluids is similar to that in maternal blood, as shown in Figure B(ii), despite the bulk of maternal thyroid hormones being unable to pass across the membrane barriers between the foetus and mother. About 0.5 per cent of the total T_4 in the coelomic fluid is free T_4. It appears that in the first trimester, maternal free T_4 passes into the excoelomic cavity and subsequently enters the amniotic fluid from where it enters the developing foetus. This has important implications: the human foetus is potentially vulnerable to the effects of low maternal concentrations of free T_4.

Low concentrations of maternal thyroid hormones have adverse effects on foetal development

Low maternal concentrations of free T_4 will reduce transfer to the foetus, reduce the availability of free T_4 for embryonic and foetal tissues, and could result in adverse effects on development of the human foetus. Such conditions may occur during maternal hypothyroidism, or if there is severe iodide deficiency. Less severe maternal hypothyroxinaemia (defined as free T_4 below the lowest tenth percentile with normal TSH) can occur even in areas in which there is sufficient iodine intake in the general population.

The effects of untreated maternal hypothroxinaemia at 12-weeks gestation, on subsequent infant development, has been investigated by comparison of infants born to mothers (matched for parity) with free T_4 concentrations between the 50th and 90th percentile concentrations. Recognized scales allow assessment of motor and cognitive development. Comparison of the mental and motor function of 63 individual cases and 62 control individuals at 1 year old, and 57 cases and 58 controls at 2 years old, showed small but persistent delays in mental and motor development of infants born to hypothroxinaemic mothers. These findings implicate thyroid hormones conveyed from the mother

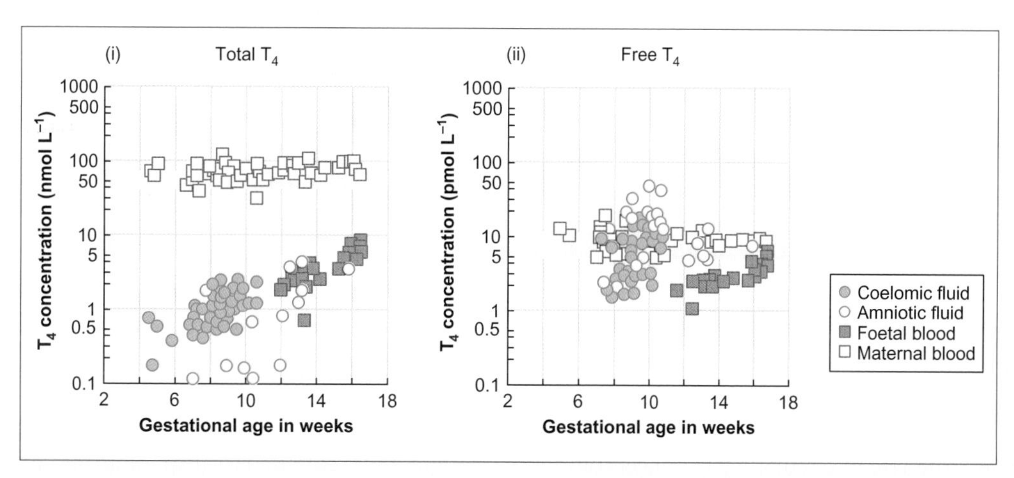

Figure B Concentrations of total T_4 and free T_4 in coelomic fluid, amniotic fluid, foetal blood and maternal blood over the first 17 weeks of gestation

Note the logarithmic scale for T_4 concentrations in both graphs. Values for total T_4 are given in nmol L^{-1}, while concentrations of free T_4 are much lower and given in pmol L^{-1} (1000 times difference). Total T_4 is much higher in maternal blood than foetal blood or foetal fluids, although total T_4 is detected in the foetal fluids by 5-6 weeks gestation and increases through early gestation. Note in panel (ii) that free T_4 is similar in coelomic and amniotic fluids to that in maternal blood. Shaded area = 95% confidence interval for normal (euthyroid) adult population.

Source: Calvo RM et al (2002). Fetal tissues are exposed to biologically relevant free thyroxine concentrations during early phases of development. The Journal of Clinical Endocrinology & Metabolism 87: 1768–1777.

to the foetus in the normal brain development, and illustrate the importance of assessing thyroid status during pregnancy.

› Find out more

Calvo RM, Jauniaux E, Gulbis B, Asunción M, Gervy C, Contempré B, Morreale de Escobar G (2002). Fetal tissues are exposed to biologically relevant free thyroxine concentrations during early phases of development. The Journal of Clinical Endocrinology & Metabolism 87: 1768–1777.

Howdeshell KL (2002). A model of the development of the brain as a construct of the thyroid system. Environmental Health Perspectives 110 Suppl 3: 337–348.

Pop VJ, Brouwers EP, Vader HL, Vulsma T, van Baar AL, de Vijlder JJ (2003). Maternal hypothyroxinaemia during early pregnancy and subsequent child development: a 3-year follow-up study. Clinical Endocrinology 59: 282–288.

Pathak A, Sinha RA, Mohan V, Mitra K, Godbole MM (2011). Maternal thyroid hormone before the onset of fetal thyroid function regulates reelin and downstream signaling cascade affecting neocortical neuronal migration. Cerebral Cortex 21: 11–21.

[1] Hypothyroidism and hyperthyroidism are discussed in Section 19.4.3.

[2] The areas of the vertebrate brain are outlined in Section 16.1.2.

Amphibian metamorphosis requires activation of the thyroid axis

Among amphibians, the most striking stage in development is the metamorphosis from an aquatic larva to an adult, which involves many morphological and physiological events. There is extensive reorganization of the brain and changes in the eye and visual pigments, loss of gills and development of lungs, changes in the liver, gut and kidneys linked to the shift to a carnivorous diet, and a shift from excretion of ammonia to predominantly urea[59].

The morphological development of frogs and toads is particularly well described for African clawed frogs (*Xenopus* spp.) in which 66 distinct developmental stages are recognized. Figure 19.33A illustrates this scheme. Starting at fertilization (stage 1), a free-swimming larva or tadpole has formed by stage 46. The next period—between stages 47 and 52—is premetamorphosis when larval growth occurs in the absence of thyroid hormones being synthesized. Prometamorphosis begins at stage 53, when functioning of the thyroid gland commences, as shown in Figure 19.33B. Stage 58 marks the end of prometamorphosis when the tadpole enters the metamorphic climax that culminates in formation of a small froglet.

Metamorphic climax is dependent on the elevation of circulating concentrations of thyroid hormones. Notice in Figure 19.33B that the pronounced rise in plasma T_3 in *Xenopus* tadpoles precedes the increase in the metamorphic index. This index measures hind limb length relative to tail length. As hind limbs increase in length, the tail length decreases. The rise in circulating T_3 concentrations continues until metamorphic climax, when it abruptly declines to low levels.

Two types of thyroid receptors (TRα and TRβ) occur in all vertebrates. Biochemical studies in the American bullfrog (*Lithobates catesbeianus*) have shown that the amount of TRβ mRNA and receptor protein increases during metamorphosis, by 50–100 times during the metamorphic climax and is induced by T_3. This increase in TRβ is thought to be a key element in driving the transformation of larval tissues into adult tissues.

Typically, among vertebrates, T_3 exerts a negative feedback effect on the hypothalamus and pituitary axis much as we see in Figure 19.23 for the hypothalamus–pituitary–adrenal axis. At first, it was thought that the peak of T_3 in metamorphic climax could reflect a lack of negative feedback effects in tadpoles, but experiments in bullfrogs show that negative feedback does occur at the level of the pituitary in pre- and prometamorphic tadpoles (as in juveniles and adults).

Amphibians show a striking degree of variability in the duration of the larval period both within and between species. Genetics set the upper and lower limits to the length of the larval period, so species living in temporary ponds usually have a shorter larval phase, but environmental factors such as temperature, crowding and food availability also exert strong influences, resulting in phenotypic plasticity. Environmental factors influence the endocrine systems that control the balance between the period of larval growth and the shift into metamorphosis. While a longer period of growth could maximise size at metamorphosis, an earlier metamorphosis could be advantageous if food is short due to overcrowding, or if predators are present in the water body.

Figure 19.34 summarizes the endocrine interactions thought to occur. In larval amphibians, a single neurohormonal stimulus, corticotropin releasing hormone (CRH), activates pituitary corticotrophs and thyrotrophs, so in stressful conditions both the thyroid gland and the interrenal tissue are stimulated, increasing circulating concentrations of both thyroid hormones and corticosterone. This means that stressful environmental conditions such as overcrowding or food shortage that result in CRH release

[59] We discuss the patterns of nitrogenous excretion of amphibians in Section 7.4.2 and look at some examples in Figure 7.47.

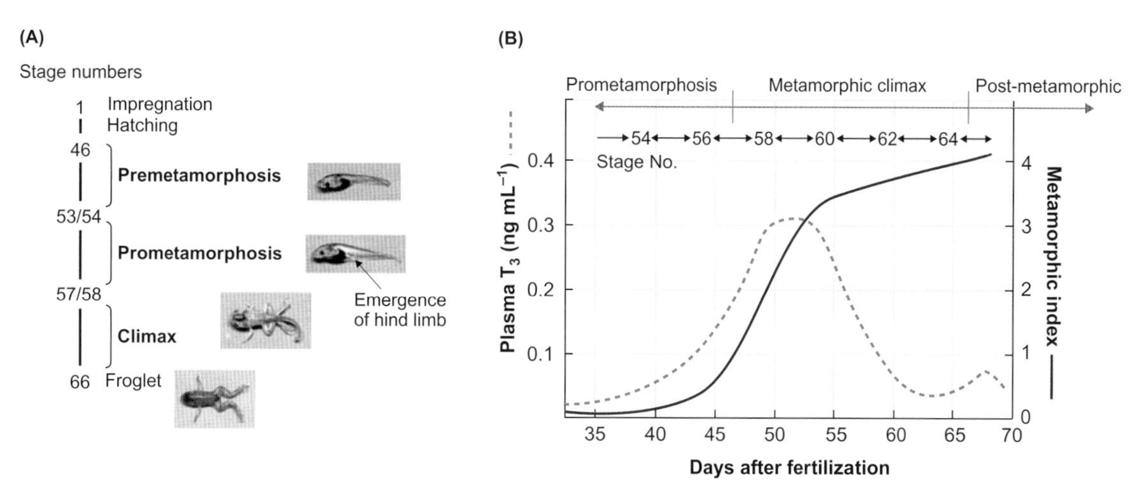

Figure 19.33 Thyroid hormones and amphibian metamorphosis, as illustrated by *Xenopus*

(A) Stages in amphibian development (as described by Nieuekoop and Faber). Metamorphosis is divided into three components: premetamorphosis which lasts until the appearance of the hind limbs, prometamorphosis which is the phase between appearance of the hind limbs and the later appearance of the fore limbs at the beginning of the metamorphic climax in which resorption of the tail and gills occurs and lungs develop. (B) Changes in plasma T_3 concentration (dashed line) during metamorphosis. The rise in plasma T_3 is followed by an increase in the metamorphic index (hind limb length: tail length). In other words, hind limbs increase in length and the tail decreases in length following the increase in T_3 concentration.

Sources: A: Miyata K, Ose K (2012). Thyroid hormone-disrupting effects and the amphibian metamorphosis assay. Journal of Toxicology and Pathology 25: 1–9. B: Tata JR (1993). Gene expression during metamorphosis: an ideal model for post-embryonic development. BioEssays 15: 239-248.

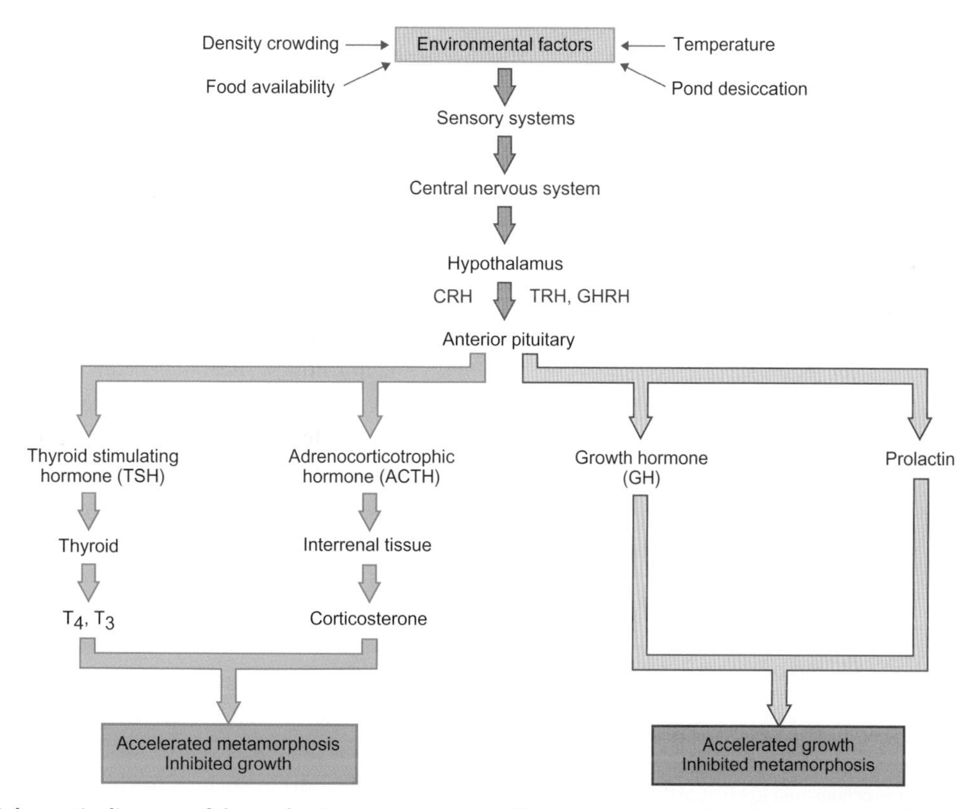

Figure 19.34 Schematic diagram of the endocrine systems controlling tadpole growth and metamorphosis

Environmental factors affect the functioning of the central nervous system in different ways depending on the stage of development. In prometamorphosis, stressors (such as overcrowding or food shortage) with a negative effect on survival stimulate the release of corticotrophin releasing hormone (CRH) from the hypothalamus, which stimulates release of ACTH from the anterior pituitary and increases circulating corticosterone. In amphibians, CRH stimulates the release of TSH, which activates thyroidal output of T_4 that is converted in tissues to T_3. These hormones accelerate metamorphosis and inhibit further growth of the tadpoles. However, if conditions are beneficial for further growth of larvae in premetamorphosis, then hypothalamic release of TRH and GHRH stimulate release of growth hormone and prolactin from the anterior pituitary, which accelerates growth and inhibits metamorphosis.

Source: adapted from Denver RJ (1997). Proximate mechanisms of phenotypic plasticity in amphibian metamorphosis. American Zoologist 37:172-184.

will stimulate development and both accelerate metamorphosis and inhibit growth (via the actions of corticosterone).

On the other hand, if conditions are beneficial for further growth of larvae in premetamorphosis, the hypothalamic release of TRH and growth hormone releasing hormone stimulates release of growth hormone and prolactin from the anterior pituitary, which accelerates growth and inhibits metamorphosis. For example, prolactin reduces the rate of tail reabsorption as shown by the data in Figure 19.35.

19.4.3 Thyroid dysfunction

Abnormalities in thyroid function are among the most common endocrine disorders in mammals. Normal activity is called euthyroidism ('eu' is from the Greek for well or good), while under-activity of the thyroid gland results in hypothyroidism, and an overactive thyroid results in hyperthyroidism. A look back at Figures 19.18 and 19.31, which show the endocrine cascade and the production of thyroid hormones, provides the key to understanding how hyperthyroidism or hypothyroidism might occur. Hypothyroidism in general can result from a deficiency in TRH or TSH (or both) and therefore a reduced stimulation of the thyroid gland or may reflect a deficiency in the thyroid tissue itself despite adequate TRH and TSH activation.

In endotherms (mammals and birds), the most obvious symptoms of hypothyroidism are an overall reduction in metabolic rate (and heat production), weight gain, reduced heart rate, slow reflexes and poor memory, and fur or feathers that are in poor condition. Hypothyroidism is treatable by oral administration of synthetic forms of T_4 or T_3 such as levothyroxine or liothyronine.

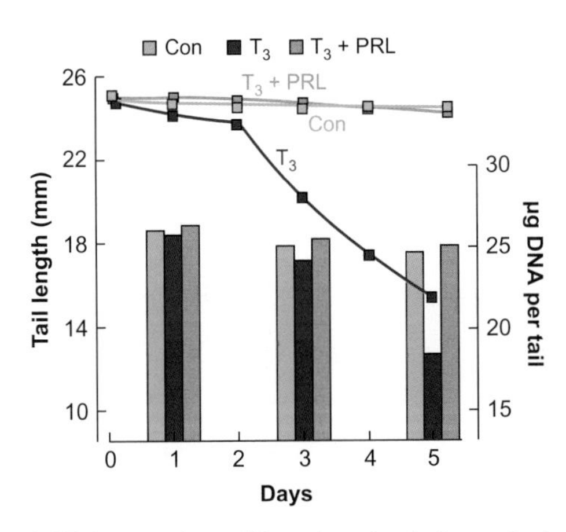

Figure 19.35 Interaction of T_3 and prolactin in control of tail resorption of *Xenopus* tadpoles

Tail buds from *Xenopus* were removed at stage 54/55 when animals are in prometamorphosis, as shown in Figure 19.33A. The tails buds were cultured *in vitro* for 8 days in the absence of hormones (control), in the presence of T_3 at 2×10^{-9} mol L^{-1}, and in the presence of the same concentration of T_3 with prolactin (PRL) at 0.2 units mL^{-1}. Tail lengths over the next 5 days, given by symbols and lines, show that T_3 alone results in tail resorption (purple symbols and line), while prolactin inhibits this process (orange symbols and line). The bars show μg DNA per tail after 1, 3 and 5 days. During the resorption process, T_3 treatment results in cell death which reduces tail DNA after 5 days (purple bars), and again prolactin inhibits this process.

Source: Tata JR (1993). Gene expression during metamorphosis: an ideal model for post-embryonic development. BioEssays 15: 239-248.

An inadequate supply of dietary iodine or poor functioning of the symporter transporting iodide into the follicular cells may limit the synthesis of thyroid hormones and hence reduce the negative feedback on the hypothalamus–pituitary axis. With reduced negative feedback, an increased production of TSH occurs, increasing the stimulation of the thyroid gland, which continues unchecked. If the deficiency persists, then continued stimulation of the thyroid results in enlargement of the thyroid gland but the release of thyroid hormones remains low. An enlarged thyroid is often visible as a swelling of the neck of hypothyroid humans, and the swelling is called a non-toxic **goitre** (non-toxic because levels of thyroid hormones in circulation remain low). Goitres are common among humans in parts of the world with iodine-poor soils, but consumption of iodized salt can provide the necessary iodine for synthesis of thyroid hormones.

Hyperthyroidism (or thyrotoxicosis) occurs when excessive and persistent stimulation of the thyroid gland results in growth of the thyroid gland to form a toxic goitre. In clinical hyperthyroidism, there are high levels of thyroid hormones in circulation. The most common form of hyperthyroidism in humans is Graves' disease (named after the Irish doctor, Robert Graves, who described a case of thyrotoxicosis in a patient in 1835). Graves' disease results from production of antibodies by an individual against his/her own TSH receptors on the thyroid follicles. A hyperactive thyroid gland typically has follicles made up of larger, taller cells containing large droplets of colloid, indicating their increased secretion of thyroid hormones. The follicles contain smaller amounts of colloid because of the increased release of thyroid hormones.

An excess of thyroid hormones in the blood results in an elevated metabolic rate and elevated heat production. Heart rate is typically elevated, and an overactive nervous system leads to hyperactivity, irritability and anxiety. Food intake typically increases to support the higher metabolic rate, but body weight nevertheless typically declines, with reduced carbohydrate and fat stores. A reduction in skeletal muscle protein may leads to muscular weakness.

Hyperthyroidism can be treated by inhibition of thyroid peroxidase on the apical border of the follicular cells, using pharmaceutical drugs such as methimazole, or by inhibiting iodide uptake on the basolateral border of the cells. Thiocyanate ions (SCN^-) compete with iodide for uptake and hence reduce the ability of thyroid follicles to produce the iodinated hormones. Perchlorate ions (ClO_4^-) have a much higher affinity than iodide for the symporter and if present in drinking water could inhibit production of thyroid function. For aquatic animals, if the water contains perchlorate it can disrupt normal thyroid functioning, as we learn in Case Study 19.2.

Case Study 19.2 Thyroid disruption: inhibition of amphibian metamorphosis by perchlorate

Many chemicals are thought to act as potential disruptors of thyroid function by acting as agents that interfere with the normal steps in synthesis of T_4 shown in Figure 19.31. These concerns have led to regulatory requirements involving the use of amphibian models to screen chemicals for their potential disruption of the thyroid axis. Potential thyroid disruption by the perchlorate anion (ClO_4^-) is of particular concern.

In laboratory experiments, the perchlorate anion competes with iodide for uptake into follicular cells by the sodium-iodide symporter, which is a prerequisite step for synthesizing thyroid hormones. Although most experiments deliberately using perchlorate to manipulate the uptake of iodide use very high concentrations of perchlorate (250–1000 mg L^{-1}), its mode of action raises questions as to whether lower levels of perchlorate in the environment could interfere with normal thyroid function.

Perchlorate is a strong oxidizing agent that is used mainly in the form of ammonium perchlorate as a military propellant (in jet and rocket fuel) and, domestically, in car air bags or fireworks. Ammonium perchlorate is soluble in water and persists in the environment for decades. Environmental contamination of ground and surface waters has occurred in numerous locations, so there are wide areas of low level contamination of surface waters in the µg range (8 µg L^{-1} or more) with hot spots of higher contamination (up to about 30 mg L^{-1}), for instance near military ammunition plants. Thus, there are plausible routes of exposure for animals. Coupled with the known mechanism of toxicity, ecological concerns for wildlife have consequently arisen. Adverse effects have been reported in mammals, birds and fish species. Equally, as perchlorate is measurable in drinking water in some locations in the USA, possible impacts on human health are also debated.

Ecological concerns about perchlorate and its potential disruption of thyroid function are especially relevant to amphibians that need to increase thyroid hormone synthesis to undergo metamorphosis (as shown in Figure 19.33): interference with metamorphosis could result in reduced populations.

African clawed frogs (*Xenopus* spp.) are often employed as a test species to investigate the impacts of pollutants on amphibian metamorphosis. Figure A shows some of the results from one of the first studies exploring the impacts of ammonium perchlorate in the concentration range seen in the natural environment. In these studies, fore limb emergence was significantly inhibited by ammonium perchlorate at 5 µg L^{-1} or higher. Tail resorption was significantly inhibited by ammonium perchlorate at 18 µg L^{-1} or above. Hind limb growth after exposure to ammonium perchlorate was similarly inhibited by concentrations of 18 µg L^{-1} or above. At higher concentrations, morphological changes are completely blocked. Thus, perchlorate exposure leads to oversized larvae that do not proceed through metamorphosis effectively.

Inhibition of the synthesis and secretion of T_4 by perchlorate can be predicted to reduce negative feedback on the hypothalamic–pituitary axis (and, hence, increase circulating concentrations of thyroid stimulating hormone (TSH)) as a mechanism to attempt to restore T_4 levels. Figure B shows the results from some studies that examined thyroid histology after exposure to sodium perchlorate and shows that perchlorate exposure

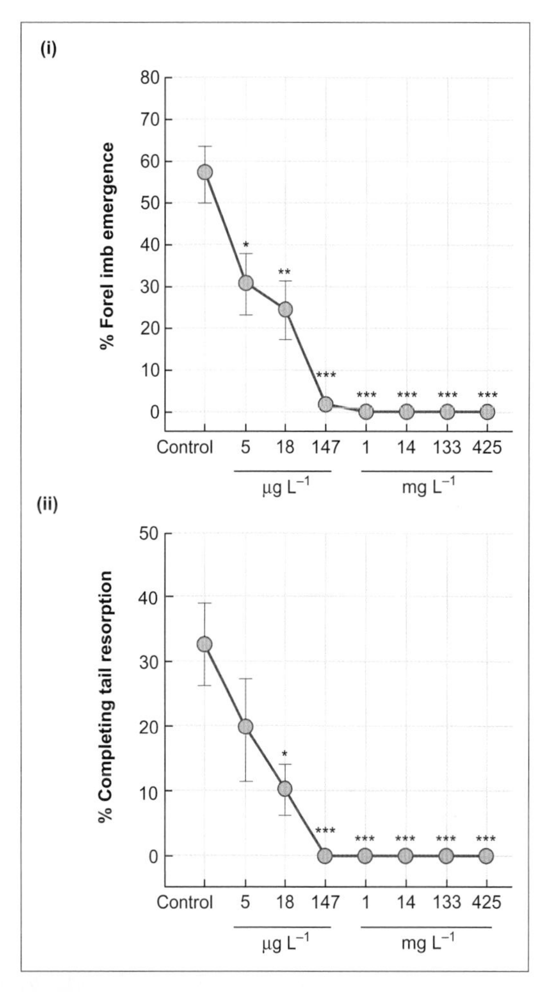

Figure A Effects of ammonium perchlorate on metamorphosis of *Xenopus*

These experiments involved continuous exposure of *Xenopus* eggs to water containing no perchlorate (control) and water containing environmentally realistic concentrations (5 µg L^{-1}–14 mg L^{-1}) and two higher concentrations of perchlorate. Exposures lasted for 70 days from the point of fertilization. The graphs show the effects of exposures to ammonium perchlorate on: (i) per cent success in fore limb emergence and (ii) per cent completing tail resorption at the completion of the metamorphic climax. In the control larvae (not exposed to perchlorate) on average almost 60 per cent of the larvae showed fore limb emergence within the experimental period and more than 30 per cent completed tail resorption. Exposure to environmentally realistic concentrations of perchlorate reduced the percentage of animals in which the fore limbs emerged during the 70-day period and reduced the percentage of animals showing complete tail resorption. Metamorphosis was totally inhibited at perchlorate concentrations of 147 µg L^{-1} or higher.
Data are means ± standard errors; * $P<0.05$, ** $P<0.01$, *** $P<0.001$.

Source: Goleman WL et al (2002). Environmentally relevant concentrations of ammonium perchlorate inhibit development and metamorphosis in *Xenopus laevis*. Environmental Toxicology and Chemistry 21: 424–430.

19

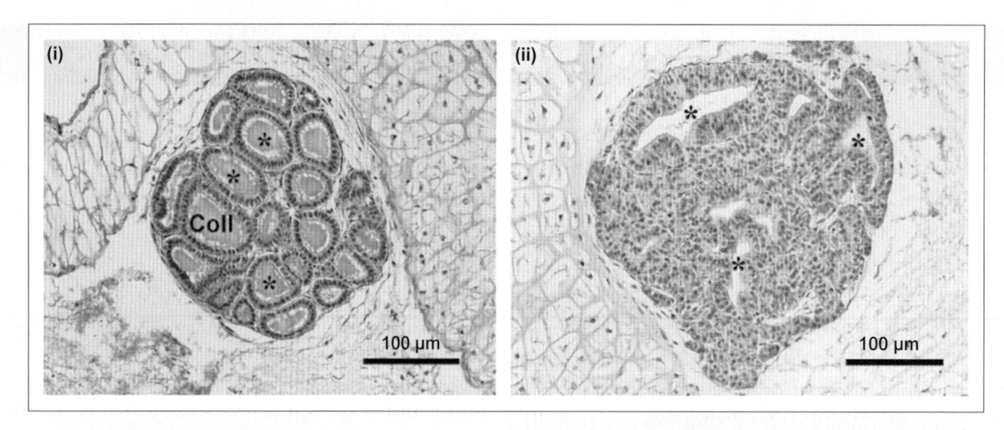

Figure B Representative histological sections of thyroid glands from *Xenopus laevis* tadpoles showing impacts of exposure to perchlorate for 38 days

(i) Thyroid from a representative control tadpole reared in the absence of perchlorate. Note the cuboidal epithelium and central colloid (Coll) in the lumen of each follicle (asterisks indicate examples). (ii) Thyroid from tadpole exposed to 93 µg L^{-1} perchlorate showing follicle cell hypertrophy (increased cell size) resulting in a columnar epithelium and hyperplasia (increased cell number) resulting in many tightly packed cells, and a relatively small collapsed lumen. Where a lumen is visible there is a marked depletion or absence of colloid (indicated by asterisks).

Source: Hu F et al (2006). The colloidal thyroxine (T_4) ring as a novel biomarker of perchlorate exposure in the African clawed frog *Xenopus laevis*. Toxicological Sciences 93: 268–277.

does result in feedback stimulation. Persistent stimulation by TSH causes follicular cell hypertrophy (increased cell size) and hyperplasia (an increase in cell number), and depletion of colloid, as predicted.

Histological studies have suggested that the assessment of changes in colloid provides the most sensitive and relatively easily quantifiable means of investigating thyroid disruption. The T_4 associated with colloid can be measured by immunocytochemistry, as shown in Figure C. In control *Xenopus* tadpoles the distinct colloidal T_4 ring close to the epithelium is thought to result from T_4 accumulation. Exposure to 8 µg L^{-1} sodium perchlorate results in a much fainter ring in the thyroid follicles of tadpoles and a ring of low intensity in young postmetamorphic frogs; presumably reduced synthesis of T_4 reduces the accumulation in the colloidal ring. From these and other studies it appears that perchlorate is a thyroid disruptor chemical that poses a significant threat to normal development of natural amphibian populations in some locations.

Field studies of northern cricket frogs (*Acris crepitans*) living at sites contaminated by perchlorate show a significant positive correlation between follicle cell height and mean perchlorate concentrations measured over a 2-year period. Whether such effects translate into physiological effects and have deleterious impacts on the fitness of wild adult frogs is not certain. Feedback

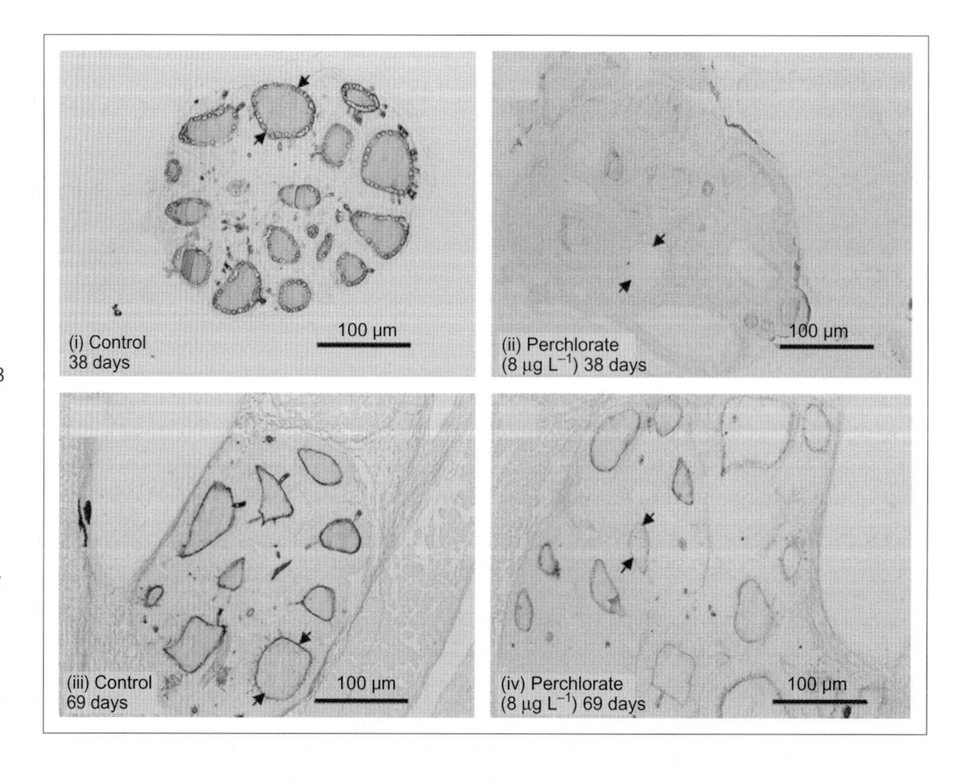

Figure C Effects of perchlorate on colloidal T_4 ring in thyroid glands of *Xenopus*

Photomicrographs show immunostained sections of thyroid glands collected after 38 days: (i) control, (ii) perchlorate-exposed, and in young post-metamorphic froglets after 69 days of exposure: (iii) control, (iv) perchlorate-exposed. Concentration of perchlorate = 8 µg L^{-1}. Arrows indicate the position of the colloidal ring of T_4 that is distinct in control animals but much fainter after exposure to perchlorate.

Source: Hu F et al (2006). The colloidal thyroxine (T_4) ring as a novel biomarker of perchlorate exposure in the African clawed frog *Xenopus laevis*. Toxicological Sciences 93: 268–277.

mechanisms may initially be sufficient to overcome the inhibitory actions of perchlorate and could allow normal development to proceed, even in tadpoles. However, perchlorate may reach critical concentrations when metamorphic development is impeded which seems likely to have adverse effects of wild amphibian populations.

› Find out more

Crane HM, Pickford DB, Hutchinson TH, Brown JA (2005). Effects of ammonium perchlorate on thyroid function in developing fathead minnows *Pimephales promelas*. Environmental Health Perspectives 113: 396–401.

Goleman WL, Urquidi LJ, Anderson TA, Smith EE, Kendall RJ, Carr JA (2002). Environmentally relevant concentrations of ammonium perchlorate inhibit development and metamorphosis in *Xenopus laevis*. Environmental Toxicology and Chemistry 21: 424-430.

Hornung MW, Degitz SJ, Korte LM, Olson JM, Kosian PA, Linnum AL, Tiege JE (2010). Inhibition of thyroid hormone release from cultured amphibian thyroid glands by methimazole, 6-propylthiouracil, and perchlorate. Toxicological Sciences 118: 42–51.

Hu F, Sharma B, Mukhi S, Patiño R, Carr JA (2006). The colloidal thyroxine (T_4) ring as a novel biomarker of perchlorate exposure in the African clawed frog *Xenopus laevis*. Toxicological Sciences 93: 268–277.

Miyata K, Ose K (2012). Thyroid hormone-disrupting effects and the amphibian metamorphosis assay. Journal of Toxicology and Pathology 25: 1–9.

Tiege JE, Holcombe GW, Flynn KM, Kosian PA, Korte JJ, Anderson LE, Wolf, DC, Degitz SJ (2005). Metamorphic inhibition of *Xenopus laevis* by sodium perchlorate: effects on development and thyroid histology. Environmental Toxicology and Chemistry 24: 926–933.

Theodorakis CW, Rinchard J, Carr JA, Park J-W, McDaniel L, Liu F, Wages M (2006). Thyroid endocrine disruption in stonerollers and cricket frogs from perchlorate-contaminated streams in east-central Texas. Ecotoxicology 15: 31–50.

› *Review articles*

Boelaert K, Franklyn JA (2005). Thyroid hormone in health and disease. Journal of Endocrinology 187: 1–15.

Cannon B, Nedergaard J (2010). Thyroid hormones: igniting brown fat via the brain. Nature Medicine 16: 965–967.

Denver RJ (1997). Proximate mechanisms of phenotypic plasticity in amphibian metamorphosis. American Zoologist 37: 172–184.

Denver RJ (1998). Review. The molecular basis of thyroid-dependent central nervous system remodelling during amphibian metamorphosis. Comparative Biochemistry and Physiology Part C 119: 219–228.

Eales JG, Brown SB (1993). Measurement and regulation of thyroidal status in teleost fish. Reviews in Fish Biology and Fisheries: 3: 299–347.

Little AG, Seebacher F (2014). The evolution of endothermy is explained by thyroid hormone-mediated responses to cold in early vertebrates. Journal of Experimental Biology 217: 1642–1648.

Muller R, Liu Y-Y, Brent GA (2014). Thyroid hormone regulation of metabolism. Physiological Reviews 94: 355–382.

19.5 Invertebrate hormones

Invertebrates, in common with vertebrates, have two kinds of tissues that secrete hormones:

- Endocrine organs that secrete classical hormones. From an evolutionary perspective, distinct endocrine glands first appeared in nemertean worms that have paired cerebral organs close to and linked to their brain. The endocrine glands of invertebrates are fewer in number and are generally simpler than those that occur among the vertebrates.
- Groups of neurosecretory neurons with large cell bodies in the ganglia of the nervous systems, often in the brain, but also more widely distributed throughout the body; neurohormones pass along the axons of neurosecretory neurons into terminals where release occurs, often in neurohaemal organs.

Neurosecretory neurons in leeches, oligochaete worms (such as earthworms) and polychaete worms are thought to be involved in the regulation of growth, osmoregulation and reproduction[60]. However, the neuroendocrine system and endocrine organs of arthropods are by far the best studied, so we focus on these in this section.

The arthropod brain contains a wide variety of neurosecretory cells. Many are scattered and the location of the end of their projections is uncertain. However, some occur in conspicuous clusters with axons projecting into specific neurohaemal organs that release neuropeptide hormones into the blood. Arthropods also possess a small number of endocrine organs secreting hormones directly into the haemolymph.

19.5.1 Crustacean hormones

Among crustaceans, we know most about the endocrine systems of decapods (crabs, prawns and lobsters), largely because they have prominent eyestalks that are easy to remove. Why should removal of the eyestalks matter? The answer lies in Figure 19.36, which is a schematic diagram showing the location of the main neurosecretory cells and endocrine glands of decapod crustaceans. Each eyestalk contains an **X-organ** connected to a **sinus gland**. In crustaceans without stalked eyes, such as water fleas (*Daphnia*), the X-organs and sinus glands are inside the head.

The neurosecretory cells of the paired X-organs synthesize neuropeptides that pass via axons to terminals in the paired sinus glands, which act as storage–release centres (neurohaemal organs). Neuropeptides stored in the sinus glands are released into the haemolymph.

[60] Case Study 20.1 looks at a scheme for endocrine regulation of egg and sperm maturation and synchronous spawning of polychaete lugworms (*Arenicola marinus*).

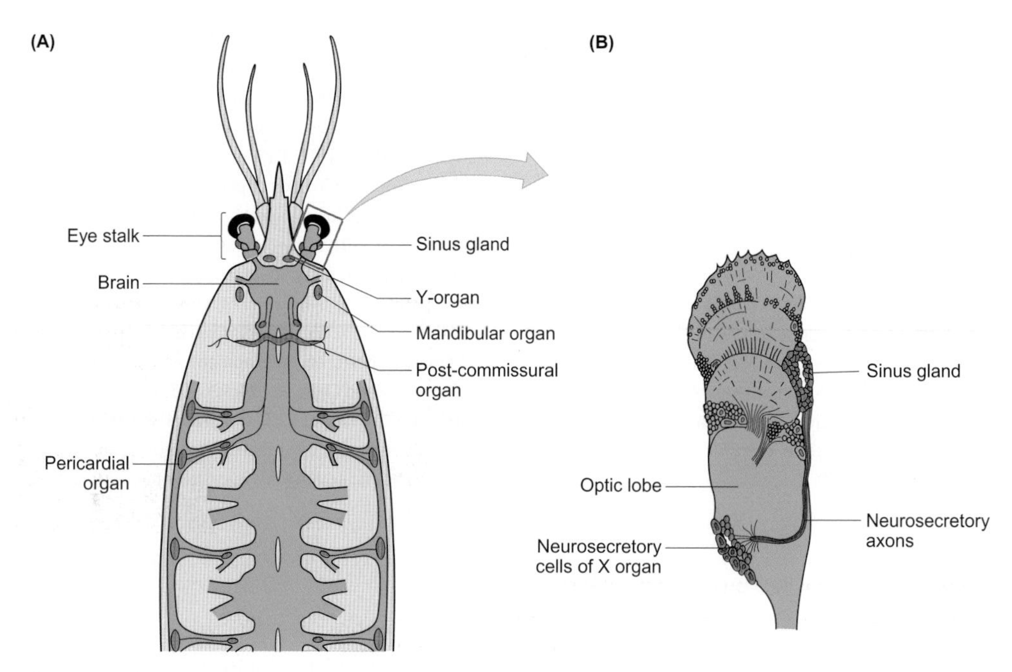

Figure 19.36 Schematic of crustacean brain, neuroendocrine and endocrine system

(A) Head end of a decapod crustacean showing brain and anterior part of central nervous system (green) and neurosecretory cells within the central nervous system and endocrine glands (red). The paired Y-organs lie in front of the brain. Mandibular organs occur behind the Y-organs. Separate clusters of neurosecretory cells occur in the paired 'X-organs' in the optic lobe in the eyestalk, the paired post-commissural organs (behind the brain) and pericardial organs (more posteriorly). The eye stalk (boxed in grey) is expanded in (B). (B) Eyestalk showing location of one of the X-organs. The neurosecretory cells of the X-organs are linked by nerve axons to paired sinus glands.

Source: adapted from Hartenstein V (2006). The neuroendocrine system of invertebrates: a developmental and evolutionary perspective. Journal of Endocrinology 190: 555–570.

The X-organ–sinus gland axis releases a family of neuropeptides

The first action of the X-organ–sinus gland neurohormones to be identified at the end of the 1920s was control of the colour of the exoskeleton mediated by a hormone that concentrates black pigment in the chromatophores. Injecting prawns or shrimps with either an extract of eyestalks after grinding them in physiological saline, or an extract of the peptide hormones contained in the eyestalks, can demonstrate this action. The sinus gland represents only about 1 per cent of the volume of the eyestalk, but contains 80 per cent of the chromatophore-concentrating activity of the eyestalk. However, control of the chromatophores is not entirely dependent on the sinus gland; we learn later that the post-commissural organs (shown in Figure 19.36A) also influence pigmentation.

Colour control is not the only function of the X-organ–sinus gland axis; a whole family of neuropeptides are released by this neuroendocrine axis. One of these neuropeptides is gonad-inhibiting hormone (GIH), which inhibits reproduction, and in female crustaceans this hormone is given the alternative name of **vitellogenesis inhibiting hormone**, which indicates the hormone's inhibition of egg development[61]. This action explains why early studies in which the eyestalks of prawns were removed induced rapid maturation of the ovaries (having lifted the inhibition of development by GIH), and why eyestalk ablation is now widely used in aquaculture of shrimps/prawns to induce gonadal maturation and egg production.

The sinus gland of some species also secretes a **diuretic hormone** (which increases fluid excretion)[62], crustacean **hyperglycaemic hormone** (CHH) (which increases the concentration of glucose in the haemolymph) and **moult-inhibiting hormone** (MIH). The hormones GIH, CHH and MIH form a unique family of peptide hormones in crustaceans, all consisting of 71–78 amino acid residues.

Secretion of MIH by the X-organ–sinus gland axis has an important role in the inhibition of moulting. Crustaceans have an exoskeleton, which eventually becomes too small for them as they grow. Consequently, they must moult before further growth is possible. Prior to moulting their fluid volume is increased to such an extent that the old cuticle splits.

Shore crabs (*Carcinus maenas*) increase fluid volume by about 80 per cent during their moult, which triggers the reduction of MIH secretion that is necessary for moulting to occur: when the exoskeleton is stretched, MIH secretion is reduced. The practice of eyestalk removal to induce rapid maturation of the ovaries for egg collection removes MIH

[61] We discuss vitellogenesis and incorporation of vitellogenins (yolk proteins) into eggs in Section 20.1.2.

[62] Section 21.1.4 discusses arthropod diuretic hormones.

19

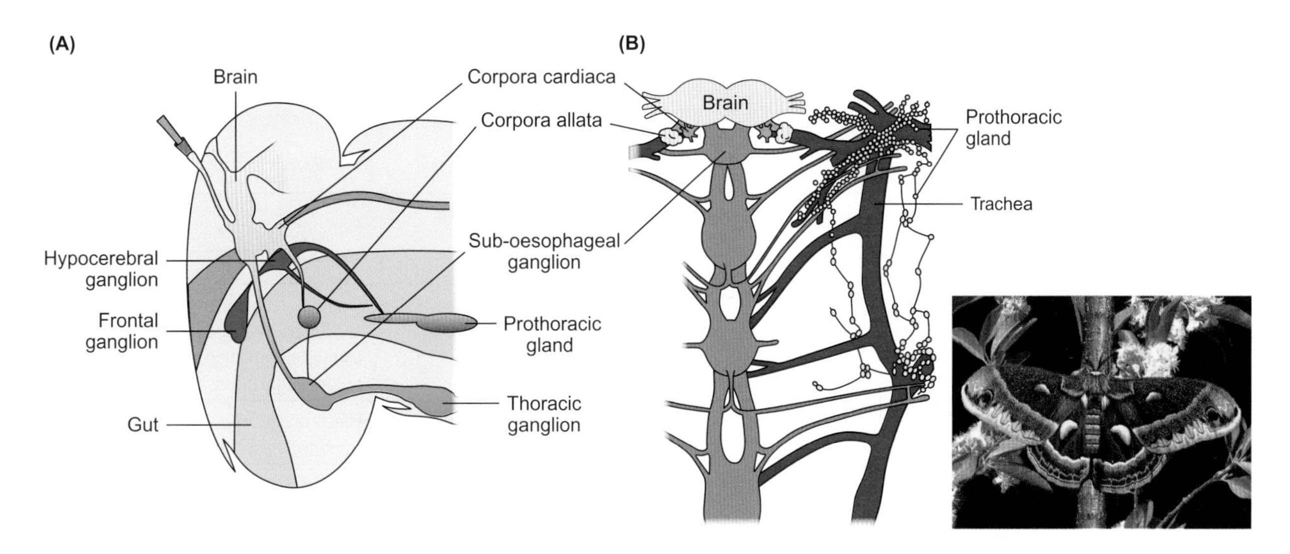

Figure 19.37 Endocrine and central nervous systems of insects

Diagrams show anterior parts of the central nervous system including the brain and some of the ganglia, and the endocrine system (corpora cardiaca, corpora allata and prothoracic glands) of (A) locust and (B) giant silkmoth (*Hyalophora cecropia*). The prothoracic glands of lepidopterans such as the silkmoth are loosely connected strands of large secretory cells.

Sources: A: adapted from Ayali A et al (2002). The locust frontal ganglion: a central pattern generator network controlling foregut rhythmic motor patterns. Journal of Experimental Biology 205: 2825–2832. B: Nijhout HF (1994). Insect Hormones. Princeton University Press. Image: © Bill Oehlke.

and results in precocious moulting. A new exoskeleton is in place beneath the hard shed exoskeleton but is soft, so crustaceans are particularly vulnerable to predation at this stage.

Aside from the X-organ–sinus gland axis, crustaceans have further neurosecretory and endocrine organs, shown in Figure 19.36.

Other endocrine tissues and hormones in crustaceans

The **Y-organ**, at the base of each eyestalk in the head is named to distinguish it from the X-organ. The Y-organ of crustaceans secretes a steroid hormone, **ecdysone**, which stimulates the process of moulting (culminating in ecdysis, when the animal sheds its exoskeleton).

Post-commissural organs comprise the terminals of neurosecretory axons and form a neurohaemal organ rich in peptide neurohormones that stimulates dispersion of black pigment in the chromatophores, resulting in darkening of the exoskeleton.

Pericardial organs are another pair of neurohaemal organs in crustaceans that vary in precise position but always occur around the heart. Experimental injections of extracts of the pericardial organs have been shown to increase the rate and the strength of the heart beat in European lobsters (*Homarus gammarus*), brown crabs (*Cancer pagurus*) and mantis shrimps (*Squilla mantis*). Peptides isolated from the pericardial organs stimulate heart rate.

Crustaceans also possess several non-neural endocrine glands. Male crabs and lobsters have androgenic glands that control development of their gonads (testes) and their **secondary sexual characteristics**[63]. Mandibular organs located close to the brain, as shown in Figure 19.36A, secrete methyl farnesoate, the precursor of juvenile hormone, which regulates development and metamorphosis of crustaceans.

19.5.2 Insect hormones

Insect endocrinology is one of the oldest areas of insect physiology as hormones regulate almost all aspects of the life of insects. Just some of the roles of insect hormones include:

- stimulation of growth, moulting and metamorphosis
- regulation of water balance and metabolism
- control of development of different social castes and social dominance
- control of reproductive processes and diapause
- initiation of colour change.

With so many actions, inevitably we discuss insect hormones many times in this book[64].

Despite the huge number and variation of insect species, their endocrine and neuroendocrine systems are similar in overall design. Figure 19.37 shows two examples of the

[63] Secondary sexual characteristics are recognizable features of males and females that may be important in the functional aspects of reproduction and behaviour but are not directly involved in the formation or delivery of gametes.

[64] Section 19.5.3 examines the role of hormones in moulting and metamorphosis; Section 9.2.1 discusses insect diapause and its endocrine control; Section 21.1.4 discusses the role of insect hormones in control of fluid balance.

Table 19.3 **Main hormones of insects**

Hormone (structural type)	Secretion site	Structural group	Target tissues	Action
Prothoracicotrophic hormone	Anterior brain: lateral neurosecretory neurons	Protein	Prothoracic glands	Stimulates release of α-ecdysone
Ecdysteroid hormones	Prothoracic glands Gonads in adults	Steroid	Juvenile epidermis Adult—fat body, ovary/testes	Promotes moulting, digestion of old cuticle and secretion of new cuticle
Juvenile hormone	Corpora allata	Fatty acid derivative	Juvenile epidermis Adult ovary	Promotes retention of juvenile structures and inhibits metamorphosis in juveniles Stimulates formation of yolk proteins and sex accessory glands in adults
Adipokinetic hormones	Corpora cardiaca	Peptide	Fat body Flight muscle	Effects on lipid metabolism and carbohydrate metabolism
Eclosion hormone	Brain	Peptide	Inka cells on the tracheae	Stimulates release of ecdysis triggering hormone from inka cells on tracheae
Ecdysis triggering hormone; pre-ecdysis triggering hormone	Inka cells of tracheae	Peptides	Central nervous system	Coordinates movements resulting in ecdysis
Bursicon	Brain and nerve cord thoracic and abdominal ganglia	Peptide	Epidermis	Plasticization of cuticle prior to ecdysis Promotes tanning of new cuticle
Diuretic hormones	Diverse sources: abdominal and/or other body ganglia; neurohaemal sites; brain	Peptides	Malpighian tubules	Increases volume of urine excretion

arrangement of these tissues, in locusts and giant silkmoths (*Hyalophora cercopia*). Table 19.3 lists the main hormones and neurohormones produced by the neurosecretory cells and glandular tissues of insects, and their main actions.

The brain is the most important neuroendocrine organ of insects. Indeed, large neurosecretory neurons are visible in intact insect brains, partly because of their size and partly because the neurosecretory granules in the cell bodies of these neurons scatter light. The image in Figure 19.38 shows some of these neurons in the brain of tobacco hornworms (*Manduca sexta*). These neurons synthesize and secrete a neuropeptide with a molecular mass of about 28 kDa that was the first neurohormone to be discovered in the early 1920s, in the caterpillars of gypsy moths (*Lymantria dispar*). Experiments on these moths led to the conclusion that the moths need a 'brain hormone' in order to progress through normal development. When it became clear that the brain hormone stimulates the prothoracic glands the brain hormone was named **prothoracicotrophic hormone (PTTH)**.

Figure 19.38 shows the entire system that produces and releases PTTH, visualized by exposing whole brains to PTTH-specific antibodies that were pre-tagged with a fluorescent marker, so that the locality of binding is visible by fluorescence microscopy. Two large neurosecretory cells on each lateral side of the anterior brain (**protocerebrum**) synthesize and secrete PTTH; these are the lateral neurosecretory cells (L-NSC type III). Their axons run through two areas called the **corpora cardiaca** (singular = corpus cardiacum, so named because in many species they are close to the heart). The axons end in a branching area of dendrites in the two **corpora allata**, in the neck region behind the brain.

The corpora allata act as **neurohaemal organs** in lepidopterans (butterflies and moths), releasing PTTH into the haemolymph in a similar way to the posterior pituitary of vertebrates. In other insects, the corpora cardiaca act as the neurohaemal organs. In all cases, however, the brain, associated glands and neurohaemal organs form an integrated system for the synthesis and release of hormones. The PTTH–corpora cardiaca/allata axis is crucial in controlling the moulting of insects.

In addition to acting as neurohaemal organs, in some insects the corpora cardiaca contain neurosecretory cells that synthesize **adipokinetic hormone** (AKH). As the name suggests, AKHs have important actions on fat mobilization. The 8–10 amino acid AKHs differ in structure in different insects and may have different effects in larvae than adults. For example, AKH controls carbohydrate metabolism in the larvae of tobacco hornworm, increasing blood **trehalose**[65], but stimulates lipid mobilisation in adults. In insects, lipids are stored in the **fat body** as **triacylglycerols**. AKH receptors on the fat body stimulate the cAMP second messenger

[65] Box 2.1 outlines the structure of carbohydrates, including trehalose, and the structures of fats, including triacylglycerols (or triglycerides).

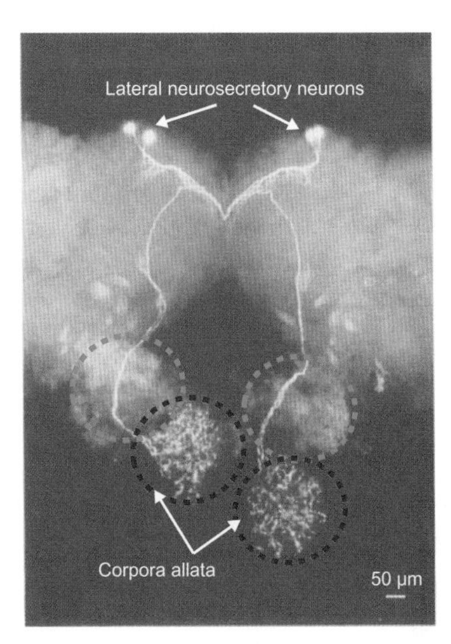

Figure 19.38 Entire neuroendocrine axis for secretion of prothoracicotrophic hormone (PTTH) in tobacco hornworm (*Manduca sexta*)

This specimen is a mount of a whole brain from a 1-day-old pupa of tobacco hornworm. The bright white areas are the result of binding of fluorescently labelled antibodies, raised to PTTH, to the granules of hormone contained in the two large neurosecretory neurons in the lateral part of each side of the anterior part of the brain (the protocerebrum) their axons and the dendritic terminals. These neurosecretory cells are one of several types of neurosecretory cells in the brain and referred to as lateral neurosecretory cells type III (LNSC III). The axons of each pair running towards the medial axis of the brain have descending dendritic processes. Axons leave the brain, passing over the corpora cardiaca (green circles) without any terminals before ending in a dendritic network in the neurohaemal organs—the corpora allata (red circles).
Source: O'Brien MA et al (1988). A monoclonal antibody to the insect prothoracicotropic hormone. The Journal of Neuroscience 8: 3247-3257.

system, resulting in mobilization of diacylglycerol (glycerol plus two fatty acids), which enters the haemolymph and binds to carrier proteins (lipophorins).

The secretion of AKH is sensitive to the metabolic demands of an insect. When available, carbohydrates are used, but if the concentration of trehalose declines, secretion of AKH increases, which mobilizes lipids from the fat body. Such events are particularly important for fuelling insect flight. Migratory locusts are a good species to study when exploring such events. Carbohydrates and some lipids fuel the first 30 minutes of their migratory flight; thereafter, however, lipids are the sole fuel that sustains the flight until the locusts become exhausted. Release of AKH stimulates lipid mobilization soon after the onset of flight. Severing nervous connections between the brain and the corpora cardiaca blocks the increase in lipids during flight, which suggests that brain neurons control AKH secretion.

In addition to acting as neurohaemal organs, in some species the corpora allata contain glandular tissue that secretes **juvenile hormone** (a fatty acid derivative), which is important in retaining juvenile features. After metamorphosis, juvenile hormone takes on a new action of stimulating the gonadal development of adult insects.

19.5.3 Insect moulting and metamorphosis: the role of hormones

The hormones that we explore in Section 19.5.2 play important roles in regulating the processes of moulting and metamorphosis. Before we discuss their interactions, we should examine the various patterns of metamorphosis among insects and the process of moulting.

Most insects are **holometabolous**—they undergo the process of complete metamorphosis that is familiar among butterflies and moths, beetles, bees and wasps, and flies. The stages in this development are summarized in Figure 19.39. The adults lay eggs that hatch into larvae (caterpillars or grubs), which are the feeding stage; these larvae develop into pupae, which metamorphose into adults.

Figure 19.39A shows a second pattern of *partial* metamorphosis, which occurs in **hemimetabolous** insects such as bugs, cockroaches and grasshoppers. These insects start life as pronymphs, which grow and moult through a series of nymphs (or instars) becoming larger but similar-looking insects, before finally forming a winged adult[66].

Insects, like all arthropods, have an external cuticle[67], which confines them as they grow. Consequently, they need to shed their cuticle to become larger, by moulting. Moulting is an imprecise term that can refer to the entire process of manufacturing a new cuticle, digesting the old one for reuse of its components, and shedding the old one, or just the final step of casting off the old cuticle. The term moulting cycle is preferable when discussing the complete repertoire of processes from the first endocrine change to finally shedding the old cuticle. This final step is called ecdysis in juvenile insects or **eclosion** when an adult emerges.

Ecdysis and eclosion involve elaborate stereotyped behaviours. Most insects begin by swallowing air or water, which increases internal pressures and results in splitting of the old cuticle along its lines of weakness. In a series of peristaltic movements and abdominal rotations, the old cuticle slips off and the next developmental stage emerges. The new cuticle is initially soft and very extensible. Intake of air into the respiratory system forces air into the space between the old and new cuticle, and puffs up the insect, stretching the cuticle. The cuticle then hardens in a process called tanning, which leaves space for growth.

[66] Additionally, a few groups of ametabolous insects such as silverfish do not undergo metamorphosis: the young look like mini-adults.

[67] Figure 6.17 illustrates the structural layers in insect cuticle.

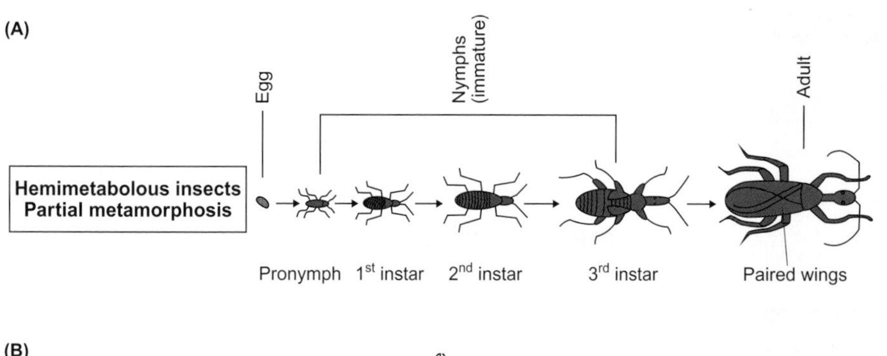

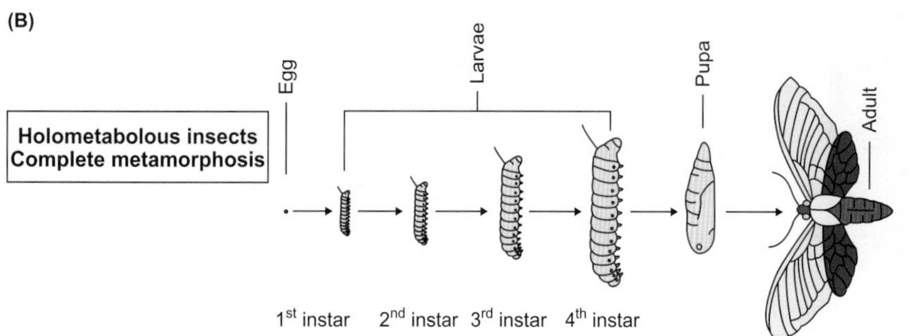

Figure 19.39 Hemimetabolous and holometabolous developmental patterns

(A) Hemimetabolous insects go through a partial metamorphosis, starting as a small pronymph and progressing through a series of wingless instar nymphs, of increasing size at each moult, with gradual morphological changes, before emerging as a winged adult after the final moult cycle. (B) Holometabolous insects undergo a more profound complete metamorphosis, at the pupal stage; prior to this eggs hatch into larvae that feed, grow and moult at each instar.

Source: adapted from Kroening M (2007). Core Manual. Master Gardener. Insects. University of Missouri Extension. University of Missouri-Columbia.

What determines the start of a moulting cycle?

The triggers for the moulting cycle are best understood in hemipterans (true bugs), largely because of the work of Vincent Wigglesworth, a British physiologist who studied moulting of South American kissing bugs (*Rhodnius prolixus*). Kissing bugs feed exclusively on the blood of reptiles, birds and mammals (including humans—hence their common name, which reflects the way it feeds on human faces).

In his now classical studies on kissing bugs, Wigglesworth showed that each large blood meal initiates a moulting cycle. In contrast, a small meal or a series of small meals does not trigger moulting. Wigglesworth showed that meals need to be about 100 mL to result in moulting of fifth stage nymphs and concluded that stretching of the abdomen triggers the start of the moulting cycle.

The first stage in the moulting cycle, apolysis, involves separation of the existing cuticle from the underlying epidermis, which occurs within two or three days of a large blood meal in kissing bugs. The subsequent secretion and deposition of a new cuticle follows some 2–4 weeks after feeding (in fourth- and fifth-stage nymphs).

Stretch-induction of the moulting cycle is not restricted to blood-feeding bugs like *Rhodnius* that take large meals. Bugs feeding on plant sap generally begin their moulting cycle when they reach a critical mass that creates sufficient tension on abdominal stretch receptors. Hence, injection of saline into nymphs stimulates moulting of some species. It seems that as insects grow (get heavier) their stretched cuticle often acts as the stimulus for moulting. So what is the downstream effect of this stimulus?

Stretching of the cuticle stimulates secretion of prothoracicotrophic hormone (PTTH) by the neurosecretory cells in the brain. Hence, control over the precise timing of the onset of moulting resides in the brain. Figure 19.40 shows the cascade of endocrine events that is stimulated by the release of PTTH from the neurohaemal organs.

Every moulting cycle starts with PTTH secretion. Release of PTTH stimulates a pair of endocrine glands located at the back of the head: the prothoracic glands. These glands show enormous morphological diversity among different types of insect. For example, prothoracic glands form loose strands of large secretory cells in butterflies and moths (Lepidoptera), as shown back in Figure 19.37B, but are more compact in true bugs. Despite differences in gross morphology, their cellular features (e.g. large amounts of smooth endoplasmic reticulum) indicate a common function in secreting steroid hormones[68].

The prothoracic glands secrete ecdysteroids that result in ecdysis or eclosion

In response to stimulation by PTTH, the prothoracic glands secrete the steroid, α-**ecdysone**, which was the first classical hormone extracted from an insect endocrine gland, in 1954; about 500 kg of pupae of silkmoths (*Bombyx mori*) was needed for the extraction. In a whole insect, α-ecdysone is not the active hormone: it is transformed into **20-hydroxyecdysone** (initially called β-ecdysone) in peripheral tissues, particularly by cells of the fat body, which

[68] Figure 19.2A shows the ultrastructural features of steroid-secreting cells.

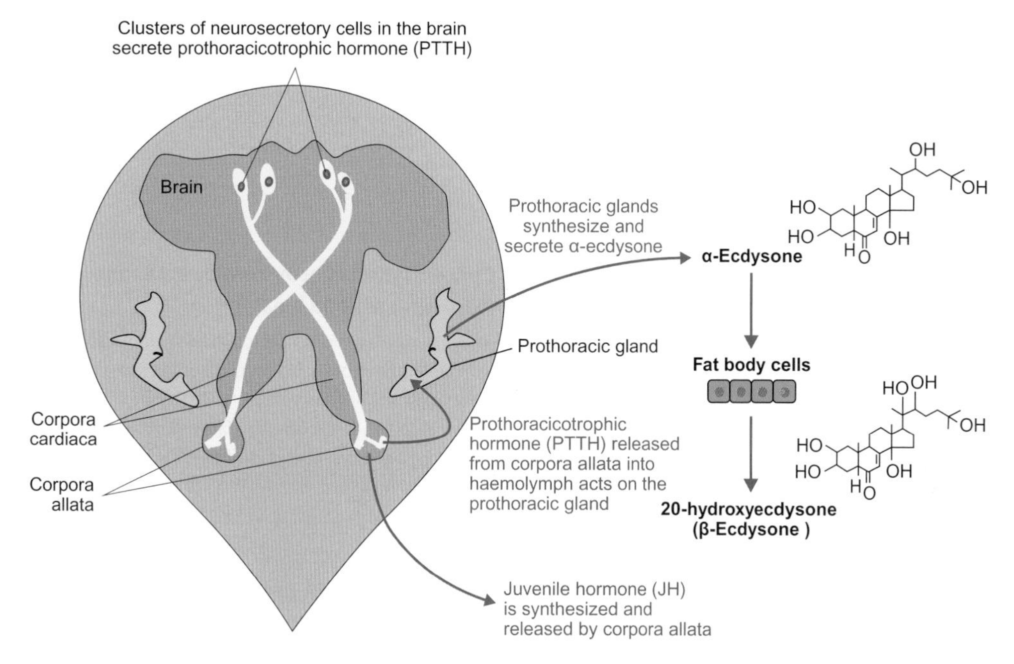

Figure 19.40 Some of the neuroendocrine clusters in the insect brain and endocrine glands involved in moulting and metamorphosis of insects

Lateral neurosecretory cells in the brain secrete PTTH, which passes along axons to a neurohaemal area of the corpora allata (in lepidopterans—butterflies and moths) or corpora cardiaca. PTTH released into the haemolymph from the neurohaemal organs stimulates the secretion of the steroid hormone, α-ecdysone (structure shown) from the prothoracic glands. Fat body cells convert α-ecdysone into 20-hydroxyecdysone (β-ecdysone—structure shown). The corpora allata synthesizes and secretes juvenile hormone.

in insects performs many of the functions of the vertebrate liver. The 20-hydroxyecdysone in the haemolymph then stimulates moulting.

Figure 19.40 shows the chemical structure of both α-ecdysone and 20-hydroxyecdysone, which are collectively called ecdysteroids. More than 60 ecdysteroids have been identified in insects and other arthropods, and more than 100 phytoecdysteroids have been isolated from plants, some identical to those in insects. Some of the phytoecdysteroids occur only in plants but are nevertheless biologically active in insects and may have evolved as protection against herbivorous insects.

Once ecdysteroid concentrations increase beyond a certain point, the moulting cycle continues without the need for continual PTTH secretion. Early studies investigated the point at which the moulting cycle continues independently of brain hormones, by removing the brain, decapitating the animal, or by placing a ligature around the neck to prevent blood movement through the body. At some point, animals still undergo a moult, because events are already set in motion by 20-hydroxyecdysone. One way to assess the point at which moulting becomes independent of brain neurohormones is to monitor the 'head critical period', which is the number of days when half of the animals still undertake a normal moult in the absence of brain hormones.

Figure 19.41 shows the five nymph stages of kissing bugs. First-stage nymphs require 3 days of PTTH secretion after a meal, while fifth-stage nymphs require PTTH secretion for about 7 days; thereafter, blood concentrations of ecdysteroids increase independently of PTTH secretion, as shown in Figure 19.41.

In contrast to the situation in kissing bugs, in which several days of sustained secretion of PTTH are required, just a few hours of PTTH secretion is sufficient to trigger ecdysteroid secretion and moulting of some insect species. Figure 19.42 shows a short-lived peak in PTTH occurs in tobacco hornworms (*Manduca sexta*); when secretion stops, PTTH in the haemolymph declines, but ecdysteroid concentrations rise *after* the drop in PTTH levels and generally remain elevated for one to five days.

Ecdysteroid secretion causes a sequence of dramatic changes in behaviour, such as the cessation of feeding and, in the case of holometabolous species that undergo complete metamorphosis, the seeking out of a site suitable for pupation. In Lepidoptera and **Diptera** this is called the 'wandering phase'. The final escape from the old cuticle involves the actions of a further peptide hormone, **eclosion hormone**, released under the control of the insect's biological clock.

Eclosion hormone stimulates the **Inka cells**, which are located in the tracheal system (the respiratory system of insects). These cells synthesize and release ecdysis triggering

(A)

1 2 3 4 5 Adult

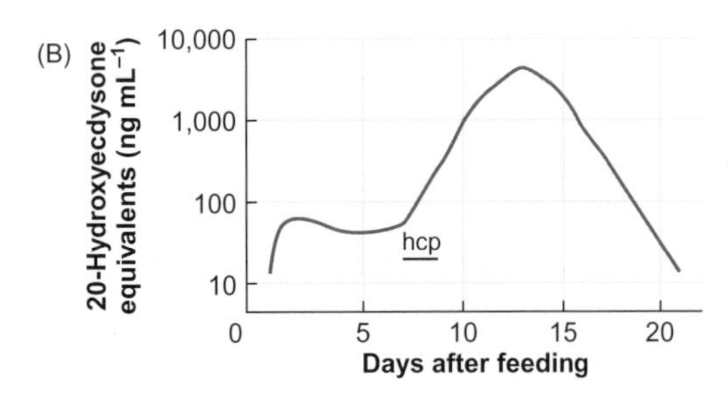

Figure 19.41 Concentration of ecdysteroids in the blood of kissing bugs (*Rhodnius prolixus*) after a blood meal

(A) Photograph shows gradual development through five nymph stages (instars) and winged adult in the partial metamorphosis of kissing bugs. (B) Graph shows changes in ecdysteroids (equivalent to concentration of 20-hydroxyecdysone) of the fifth (final) nymph. The initial small peak in ecdysteroid concentration is followed by a larger increase soon after the head critical period (hcp) when moulting becomes independent of PTTH secretion by brain neurons. Ecdysis occurred on day 20 after feeding.

Source: Nijhout HF (1994). Insect Hormones. Princeton University Press. image © Thierry Heger Wikimedia.

hormones (ETHs), usually of two forms: **pre-ecdysis triggering hormone** (PETH) and **ecdysis triggering hormone** (ETH). Figure 19.43 shows some examples; Figure 19.43C shows an example of the large segmentally arranged paired Inka cells that occur in some holometabolous insect groups (Lepidoptera, Diptera, and some **Coleoptera** and **Hymenoptera**). In other holometabolous insects, including most beetles and bees, and all hemimetabolous insects examined so far, small Inka cells scattered throughout the tracheal system secrete ETHs.

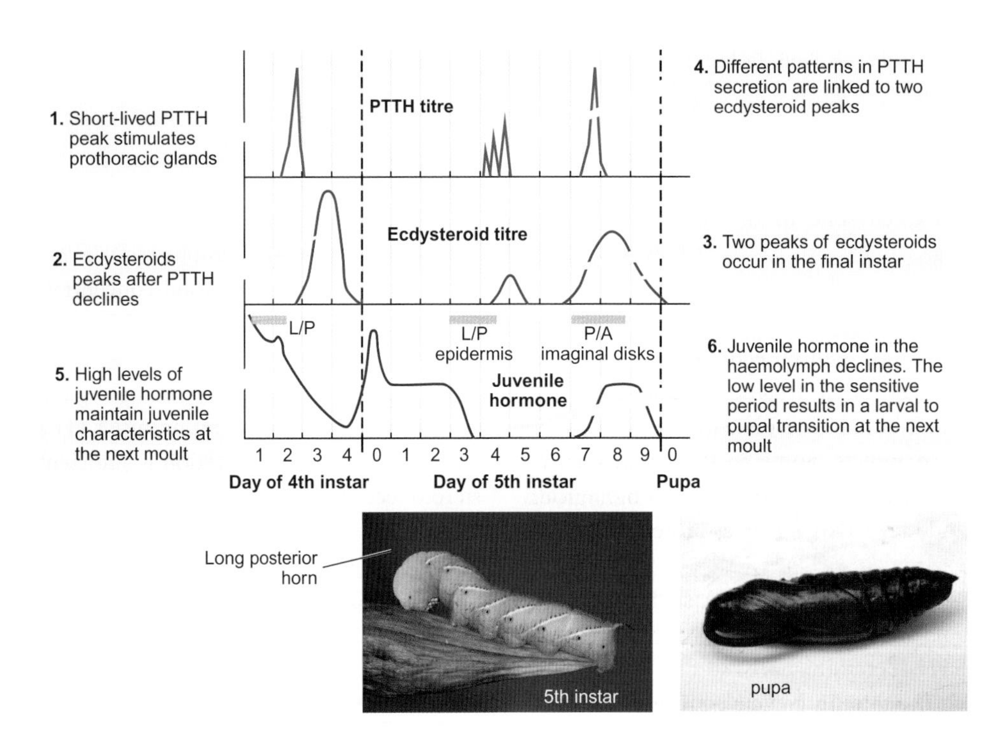

Figure 19.42 Relative concentrations of prothoracicotrophic hormone (PTTH), ecdysteroids and juvenile hormone in the haemolymph during the last two larval stages (instars) of tobacco hornworms (*Manduca sexta*)

The 4th and 5th instars have long posterior horns, hence the name, hornworm. The graphs showing hormonal changes are derived from many sources to show the pattern of changes relative to each other as the hornworm proceeds through the final two instars and forms a pupa. The graphs do not show actual hormone concentrations because of different methodologies. Specific times when particular cells are sensitive to juvenile hormone are shown in grey and relate to potential larva to pupal transformation (L/P) at the next moult and P/A = pupa to adult transformation at the next moult. In the 4th instar, high concentrations of juvenile hormone result in a larval moult to the next larval instar. In the 5th instar there are two periods of juvenile hormone sensitivity. The first influences epidermal cells; the relatively low amount of juvenile hormone at this time results in transformation of larval epidermis into pupal epidermis. The second sensitive period influences the imaginal discs that determine wing development; the relatively high levels of juvenile hormone in the 5th instar prevent differentiation of this tissue.

Source: adapted from Nijhout HF (1994). Insect Hormones. Princeton University Press. 5th instar image: Daniel Schwen/Wikimedia Commons; pupa image: Smithsonian Institution/Flickr.

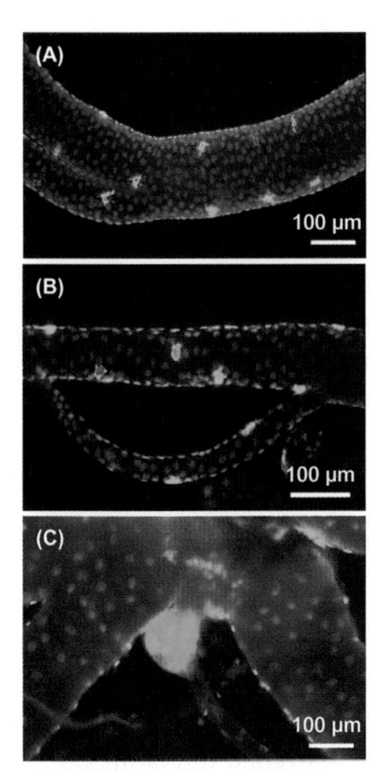

Figure 19.43 Inka cells in insect tracheae

Inka cells are scattered on the tracheae of most species. The Inka cells contain ecdysis triggering hormones (ETHs) and can be visualised by immunocytochemistry, using an antibody to pre-ecdysis triggering hormone (PETH). The cells containing PETH are green. Blue stain shows nuclei of the trachea. The examples show Inka cells on the surface of the trachea of nymphs ready for ecdysis, having developed a new cuticle in: (A) cockroach (*Periplaneta*), (B) locust (*Locusta*), and (C) one of the paired large oval Inka cells (with many nuclei) of a beetle (*Leptinotarsa*). These Inka cells are attached to the tracheal bush near each spiracle.

Source: Roller L et al (2010). The ecdysis triggering hormone signaling in arthropods. Peptides 31: 429–441.

Experimental studies demonstrate that ecdysis is induced after injection of tracheal extracts containing PETH and ETH; without these peptides, ecdysis fails to occur. The Inka cells release their full content of PETH and ETH during each larval, pupal or adult ecdysis. These hormones act on the central nervous system, triggering a pre-programmed, co-ordinated pattern of nervous and muscular activity that ultimately sheds the cuticle. A neat way of examining the importance of ETHs has been through a genetic approach using *Drosophila* mutants with the *eth* gene deleted. Such mutants die at the time of the first ecdysis, having failed to perform the ecdysis behavioural sequences or to inflate their new respiratory system.

Bursicon has actions on the mechanical properties of the cuticle

In addition to stimulation of the release of ETHs, eclosion hormone also stimulates the release of another brain neuropeptide hormone, **bursicon**. Plasticization of the new cuticle by bursicon just before the ecdysis of moths, locusts and blowflies softens the new cuticle and increases its extensibility. Plasticization of the cuticle of the wings of adult Lepidoptera is important in allowing them to expand after emergence from the pupa.

When an insect sheds its cuticle, the underlying new cuticle is still soft and vulnerable to injury. To survive, the new cuticle must be 'tanned', which involves hardening and darkening (melanization). Bursicon stimulates this tanning process. Hence, experimental prevention of the release of bursicon from brain neurons and its entry into the haemolymph, by placing a ligature around the neck, prevents tanning, as shown in Figure 19.44. In such experiments, injection of bursicon restores the tanning process.

Bursicon consists of two components (bursicon α and bursicon β) that form heterodimers, which act on G-protein-coupled receptors in the integument. Figure 19.44 shows that heterodimers are required for stimulation of the tanning process. Monodimers made up of two similar bursicon components (two α-bursicons or two β-bursicons) do not stimulate tanning and their biological function is uncertain. However, a clue about their possible function has arisen from genomic investigations of fruit flies (*Drosophila*).

After neck ligation to block passage of brain neuropeptides to the body, fruit flies treated with heterodimer bursicon show changes in the expression of 87 genes, most of which act during tanning and wing maturation. A few genes (seven) affected by bursicon treatment are immune response genes, which suggests a novel action for bursicon. Subsequent

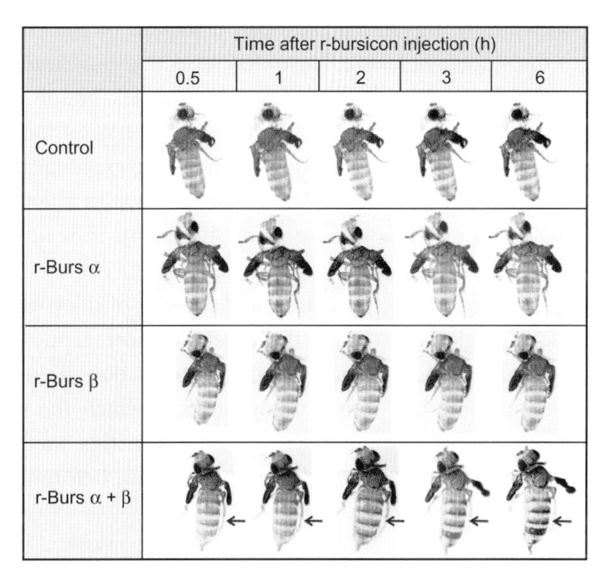

Figure 19.44 Tanning of cuticle of *Drosophila* by bursicon

The necks were ligated in newly emerged flies to prevent the passage of brain neuropeptides into the haemolymph. These flies were injected with purified recombinant bursicon (r-Burs) of three types in equal doses (60 ng in 0.5 μL): r-Burs α+β (heterodimer), r-Burs α (homodimer) or r-Burs β (homodimer). The control group of flies were injected with a blank cDNA plasmid. Tanning (darkening due to melanization and hardening) occurred only in the flies given the heterodimer (arrows) and commenced within 30 minutes, progressing over the 6-hour study.

Source: An S et al (2012). Insect neuropeptide bursicon homodimers induce innate immune and stress genes during molting by activating the NF-κB transcription factor relish. PLoS One 7: e34510.

treatment of fruit flies with the two homodimers of bursicon stimulates the production of peptides with antimicrobial activity. Insects are vulnerable to potential attack and infection in the moulting period, and antimicrobial activity could be beneficial in preventing infections.

Juvenile hormone is necessary to retain juvenile characteristics after ecdysis

The profound changes of a complete metamorphosis raise the question: what determines the type of moult? That is, whether the moult results in another nymph or larvae, or whether the insect moults into a pupa before developing into an adult. This question intrigued insect endocrinologists for many years. Elegant experiments on species such as silkworms and tobacco hornworms provided answers to the question, and show the critical role of juvenile hormone (JH, synthesized by the corpora allata) in determining the outcome of a moulting cycle.

A decrease in the concentration of JH (as well as an increase in ecdysteroids) is necessary for metamorphic changes to occur. Hence, experimental removal of the corpora allata (a procedure called an allectomy), which removes the source of juvenile hormone, results in a pupa after the next moult. Allectomy at an early stage in development results in smaller pupae and tiny adults.

An alternative experiment is to give an insect extra JH—for instance, by implanting extra corpora allata, or applying JH to the surface of larvae. The outcome of this experiment is the development of larger larvae after the next moult. If performed on late instars the result is an oversized larva that then forms an oversized pupa. Commercial exploitation of this approach by spraying late-instar larvae with chemical analogues of JH can create giant silkworms: instead of moulting into pupae, extra-large larvae emerge. When these larvae do pupate, they spin extra-large cocoons of silk (from their salivary glands) to encase the larger than normal pupae.

In the haemolymph, juvenile hormone is bound to a protein (juvenile hormone-binding protein) which the fat body synthesizes. Juvenile hormone bound in this way is usually protected from degradation by enzymes. Before the final moult, however, the fat bodies increase production of JH esterase that degrades JH, even when bound to the binding protein. As a result, the amount of JH decreases, while the amount of 20-hydroxyecdysone increases.

There are specific times when particular cells are sensitive to JH; if JH is present at those times, the current state of development is maintained. If JH is absent, however, development proceeds to a more mature stage. Figure 19.42 shows these periods in the last two instars of tobacco hornworms, and their influence on the outcome of the moults. Thus, ecdysteroids initiate the moult, while JH determines its nature. Juvenile hormone is normally present in larvae, so they progress through a series of instars until they reach a suitable size for metamorphosis. At this point, PTTH release governed by internal factors such as size, and external factors such as photoperiod and temperature, drives the increase in ecdysone needed for moulting to take place.

The prothoracic glands are lost after metamorphosis of a holometabolous juvenile insect into an adult, or during the first few days as an adult in hemimetabolous species. Although ecdysteroids still occur in the haemolymph of adults, the gonads become the primary source. In adults, ecdysones produced by the ovaries under the control of a brain neuropeptide stimulates the production of **vitellogenins** by the fat body, and the deposition of vitellogenin in the developing eggs[69].

› *Review articles*

Fingerman M (1997). Crustacean endocrinology: a retrospective, prospective and introspective analysis. Physiological Zoology 70: 257–269.

Hartenstein V (2006). The neuroendocrine system of invertebrates: a developmental and evolutionary perspective. Journal of Endocrinology 190: 555–570.

Lafont (2000). Endocrinology of invertebrates. Ecotoxicology 9: 41–57.

Truman JW, Riddiford LM (2002). Endocrine insights into the evolution in metamorphosis of insects. Annual Review of Entomology 47:467–500.

Truman JW (2006). Steroid hormone secretion in insects comes of age. Proceedings of the National Academy of Sciences of the United States of America 103: 8909–8910.

[69] Section 20.1.2 discusses vitellogenesis as a means of nutrient provision for eggs.

19

Checklist of key concepts

Dynamics of hormonal processes

- **Steroidogenesis** requires the correct complement of enzymes to produce particular steroid hormones from cholesterol.
- Binding of hydrophobic hormones to **carrier proteins** provides large hormonal reserves that are in dynamic equilibrium with physiologically active free hormone.
- Circulating concentrations of hormones are determined by **secretion rates**, under the control of feedback loops, balanced by binding to receptors or carrier proteins, metabolism and excretion.
- **Pulsatile secretion** of **neurohormones** often results in pulsatile secretion of further hormones in an **endocrine axis**.

- **Excretion** of a hormone or its metabolites in urine and faeces can provide an ideal non-invasive way of assessing the **hormonal status** of animals.
- Hormone signals are amplified by binding to specific receptors in **target cells** and **signal transduction**.
- **Water-soluble hormones** (e.g. peptides) often bind with cell-surface receptors coupled to **G-proteins**, but some act via **enzyme-linked receptors**.
- Upon binding, the hormone (**first messenger**) rapidly initiates a cascade of events involving specific **second messengers** that results in specific cellular responses.
- **Lipid-soluble hormones** (e.g. steroids) usually act more slowly by binding to **nuclear receptors** that interact with the specific **hormone response element** of DNA, activating the adjacent gene, which results in synthesis of new proteins.

Vertebrate hypothalamus–pituitary axis and the pineal gland

- Hypothalamic neurons and higher brain centres control the synthesis, storage and release of **vasotocin-like** and **oxytocin-like** peptides from the **neurohypophysis** (posterior pituitary).
- Releasing hormones and inhibitory factors produced by hypothalamic neurons regulate the synthesis and release of **prolactin** and five **trophic** hormones acting on other endocrine tissues.
- Hypothalamic functions exhibit natural rhythms set by **biological clocks**, which result in the daily cycles in secretion of many hormones.
- Secretion of **melatonin** during darkness has a **chronobiotic** action in a wide range of organisms.
- The **duration** of the period of darkness controls the period of melatonin secretion, which signals photoperiod and seasonal changes in photoperiod.
- Light detection by the **pineal gland** (third eye) of most vertebrate groups (but not mammals) inhibits its melatonin secretion.
- In mammals, light detected by the **eyes** signals to the main biological clock to regulate melatonin secretion.

The vertebrate adrenal gland and stress

- Adrenal steroid hormones are commonly divided into:
  - **mineralocorticoids**, acting on salt (mineral) balance,
  - **glucocorticoids**, which control carbohydrate metabolism.
- Low to moderate levels of glucocorticoids usually maintain a balance between energy acquisition, deposition of energy stores, and their mobilization.
- A common set of responses to stress results from:
  - stimulation of the brain,
  - rapid activation of the sympathetic nervous system,
  - slower activation of the HPA/HPI axis.
- These responses trigger multiple secondary effects.
- **Short-term (acute) stressors** stimulate a rise in corticosteroids from a low baseline followed by inhibition through feedback mechanisms.
- **Chronic stress** often results in persistently high levels of corticosteroids.
- These high levels can have adverse effects that include inhibition of thyroid function, inhibition of growth and reproduction, immunosuppression and anti-inflammation, and decreased stress responses to new stressors.
- **Maternal stress** during pregnancy may result in transgenerational changes in the stress response of offspring that may include non-genomic mechanisms.

The vertebrate thyroid gland

- All aspects of thyroid function are stimulated by **thyroid stimulating hormone** (TSH).
- The main hormone secreted by thyroid follicles is **thyroxine (T_4)**.
- The main physiologically active hormone is **triiodothyronine (T_3)** formed by tissue deiodination of T_4.
- Thyroid hormones are required for normal **growth** and **development** and regulate the rate of metabolism and thermogenesis.
- High concentrations of thyroid hormones promote **metabolic rate** by peripheral and central mechanisms.
- Normal development requires thyroid hormones:
  - **yolk** in eggs can provide the necessary supply for some animal types;
  - **maternal thyroid hormones** conveyed to the foetus are essential for normal brain and neural development in mammals.
- Increasing amounts of T_3 stimulate amphibian metamorphosis, but environmental factors and other hormones also influence the timing of metamorphosis.
- Abnormalities in thyroid function (e.g. hypothyroidism or hyperthyroidism) are common endocrine disorders, with multiple causes.
- Many chemicals may act as potential disruptors of the synthesis of thyroid hormones.

Invertebrate hormones

- Invertebrate physiological processes controlled by hormones include:
  - pigment aggregation or dispersion,
  - fluid balance,
  - reproduction and growth,
  - moulting and metamorphosis.
- When the exoskeleton of arthropods becomes stretched by growth, it must be shed and replaced with a new one if growth is to continue.
- Insects secrete **prothoracicotrophic hormone (PTTH)** from brain neurons in response to cuticular stretching, which initiates a **moulting cycle**.
- Moulting of insects results from the cascade of events stimulated by PTTH acting on prothoracic glands, which release **α-ecdysone**; its conversion to **20-hydroxyecdysone** triggers the moult (**ecdysis**).

- **Complete metamorphosis** involves **pupation**: ecdysteroid secretion causes behavioural changes such as the cessation of feeding and seeking a site suitable for pupation.
- While ecdysteroids initiate a moult, **juvenile hormone** (JH) determines its nature: when JH is present the larvae moult into larger instars, until they reach a suitable size for metamorphosis when circulating JH declines.
- Escape from the old cuticle requires the stimulation of **Inka cells** in the tracheal system.
- Inka cells release **pre-ecdysis triggering hormone** and **ecdysis triggering hormone** that trigger coordinated nervous and muscular activity to shed the cuticle.
- When an insect sheds its old cuticle, **bursicon** acts to harden the soft cuticle.

Study questions

1. Based on chemical structures, distinguish four types of hormones giving examples of each type, and explain their differences. How are the structures of steroid hormones described and compared? (Hints: Box 19.1, Figure 19.3, Figure 19.15, Section 19.1.1.)
2. Two students are examining electron micrographs of sections taken from endocrine glands and trying to decide on the type of hormone that the tissue secretes. What features should they look for to distinguish an endocrine tissue that produces a peptide hormone from an endocrine tissue synthesizing a steroid hormone? Explain why the effects of steroid hormones take longer to occur that those of peptide hormones. (Hints: Section 19.1; Figure 19.2; Figure 19.10.)
3. Outline the feedback mechanisms that regulate endocrine systems, giving specific examples. (Hints: Section 19.1.3; examples appear in Sections 19.2, 19.3 and 19.4; Chapter 20 contains additional information on hormone systems controlling reproductive processes.)
4. How are hormones that act via G-protein coupled receptors able to initiate many different physiological responses? (Hint: Section 19.1.4.)
5. In what ways does the operation of the hypothalamic–posterior pituitary axis differ from the hypothalamic–anterior pituitary axis? (Hints: 19.2.1 and 19.2.2.)
6. What are the secondary effects of the primary endocrine responses to acute stress? (Hints: Section 19.3.1 and 19.3.2.)
7. What are the potential adverse consequences of chronic stress? What advice could you provide to a person in charge of the husbandry of animals of conservation importance held in captivity in relation to assessing their welfare? (Hints: Section 19.3.3 and Experimental Panel 19.1.)
8. Explain how thyroid stimulating hormone (TSH) influences the synthesis and secretion of thyroid hormones. What experimental approaches enable assessment of thyroid function? (Hint: Section 19.4; Case Studies 19.1 and 19.2.)
9. How can exposure to specific chemical pollutants influence the processes involved in amphibian metamorphosis? (Hints: Case Study 19.2 and Section 19.4.)
10. Discuss some of the experimental evidence indicating that maternal provision of thyroid hormones is important for foetal development. (Hints: Section 19.4; Case Study 19.1.)
11. Explain the role of the X-organ–sinus gland complex and relevant neuropeptides in crustaceans. (Hint: Section 19.5.1.)
12. What does the term 'neurosecretory–neurohaemal system' mean? Give examples of such hormonal systems of vertebrate and invertebrates. (Hints: Sections 19.2.1 and 19.5.)
13. Outline the role of specified hormones and neurohormones in the moulting cycle of insects. (Hint 19.5.3.)

Bibliography

Nijhout HF (1994). Insect Hormones. Princeton: Princeton University Press.

Norman AW, Henry HL (2014). Hormones. 3rd Edition. Academic Press.

Norris D, Carr J (2013). Vertebrate Endocrinology. 5th Edition. Academic Press.

20

Reproduction

Many of the previous chapters examine how animals survive—for instance, how they obtain oxygen and how they maintain salt and water balance. These processes allow animals to reach maturity, at which point they can breed to maintain or expand populations. While reproduction is not essential for an animal's personal survival, it is an absolute requirement for passing on genes. In this chapter, we explore the process of sexual reproduction through which gametes produced by males and females fuse to produce new offspring. We look at how environmental conditions affect sexual reproduction and can even change the sex of some species.

Animals vary tremendously in their sexual reproductive patterns. We explore the differences between species that reproduce every year, and those breeding only once in their lifetime, and between species that provide their eggs with little nutrient material and those that protect their developing embryos and provide them with a steady supply of nutrients.

Toward the end of the chapter, we learn how some animals can reproduce asexually (without fusion of gametes) to expand their population rapidly and how some animals switch to asexual reproduction when mating becomes problematic.

20.1 Characteristics of sexual reproduction

Sexual reproduction involves the meeting of two specialized cells called gametes, one from each parent, normally of the same species[1]. Male gametes are **spermatozoa** (sperm) and female gametes are **ova** (eggs or oocytes). Fertilization—the point at which a sperm successfully penetrates the egg—can occur either within the female (when it is referred to as internal fertilization) or in the surrounding environment (in which case, fertilization is external).

20.1.1 Internal and external fertilization

Aquatic animals usually release their gametes into the water, in a process known as spawning. Consequently, external fertilization occurs in most frogs and toads, most bony fish and most marine invertebrates. However, some aquatic animals such as many species of sharks, skates and rays use internal fertilization. Figure 20.1A shows the leathery egg case that some sharks, skates and rays use to protect the fertilized egg and in which their young develop.

External fertilization seems to be a risky strategy given the volume of surrounding water. However, **synchronized spawning**, in which males and females spawn at the same time, increases the likelihood of sperm and egg encountering each other, and so improves the chance of fertilization. Environmental factors such as temperature, **photoperiod**, lunar periodicity and tidal cycles are important in triggering the maturation and synchronized spawning of marine intertidal species, but the exact mechanisms are often not clear. In animals at the top of the shore, only large spring tides (1–3 days after each new moon and full moon) stimulate spawning. Marine polychaetes are amongst the best-studied species. These include lugworms (*Arenicola marina*); in Case Study 20.1 we examine the synchronized mass spawning (epidemic spawning) of lugworms.

Internal fertilization is obligatory for animals that retain the fertilized egg within the female's body. For animals that enclose their eggs in a protective case that sperm could not penetrate, the addition of protective layers or a shell must occur after fertilization of the egg.

All birds, most reptiles and the terrestrial insects lay self-contained eggs (called **cleidoic eggs**, from the Greek meaning closed). The yolk and albumen in cleidoic eggs fulfil the developing embryo's energy needs, and the waste produced during development is stored within the egg[2]. Among mammals, egg laying only occurs in the few

[1] Although hybridization between different animal species is rare it is not uncommon in some species of amphibians, for example water frogs of the *Pelophylax* genus. The fertile hybrids are mostly females and are called kleptons (from the greek *kleptein* for steal) as they cannot reproduce without 'stealing' sperm from one of the parental species. During formation of gametes, the hybrid frogs discard the genome of one of their parental species, so the gametes are exclusively of the genome of the other parental species.

[2] Look at Figure 12.12 in Section 12.3.2 for a description of the membranes and compartments within bird eggs.

(A) (B)

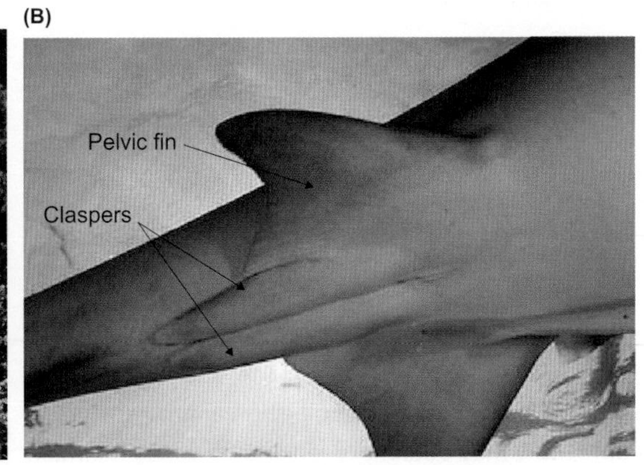

Figure 20.1 Internal fertilization occurs in sharks, skates and rays

(A) Mermaid's purse of lesser spotted dogfish, also called small spotted cat shark (*Scyliorhinus canicula*). After internal fertilization the embryo develops in a leathery case. Tendrils at each corner of the egg case attach the egg case to algae. (B) Ventral view of a male shark. Male sharks have modified pelvic fins with two claspers. These rod-like structures are inserted into the female to channel the passage of a spermatophore into the oviduct.

Sources: Dogfish egg case and embryo: Nature Picture Library/Alamy Stock Photo. Male Lemon Shark: ©Jonathan Bird/Photolibrary/Getty Images.

Case Study 20.1 Epidemic spawning of lugworm (*Arenicola marina*)

The lugworm is a common polychaete which lives in a burrow in muddy sand in the intertidal regions of northern Atlantic shores, as shown in Figure A. Their use of synchronized spawning—and additional behavioural differences between males and females—mean that high rates of fertilization are achieved despite it being external. Synchronized spawning that involves a single species in a locality is called epidemic spawning.

Female lugworms pass their eggs into their coelomic cavity and then out of the body cavity via their nephridia (their equivalent of a kidney)[1]. The eggs are retained in the horizontal part of the burrow, which is an important step in increasing the chance of fertilization, provided the females can gain access to sperm—and they have an interesting way of doing so.

Spawning in males involves intermittent muscular contractions of the body wall that result in bursts of sperm being ejaculated in small streams, from the tail shaft of the burrow, onto the surface of sediment where they coalesce in dense puddles. Figure B shows the irregular surface of the sediment that results from the feeding activity of a population of lugworms and their passage of sediment out of the burrow. These irregularities hold the milky puddles of inactive sperm when the burrows and casts are exposed by the ebb tide. These sperm puddles disperse

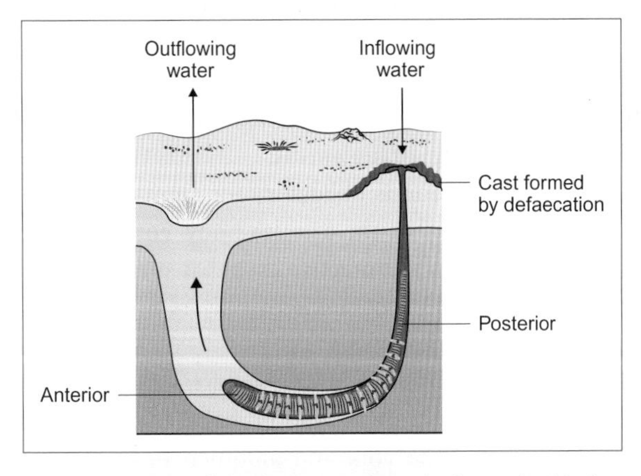

Figure A Diagram of the lugworm (*Arenicola marina*) in its burrow

Arrows indicate direction of water flow pumped through the burrow, periodically when submerged, and used for respiration. Sand moves in the opposite direction: ingestion of sand on the left is followed by intermittent defaecation of casts on the right.

Source: adapted from Ruppert EE, Barnes RD (1994). Invertebrate Zoology 6th Edn. Saunders College Publishing.

Figure B Photograph of an estuarine mudflat showing the irregular surface of sandy mud created by lugworm casts in the low intertidal region

The irregularities create an ideal surface for coalescence of puddles of sperm.

Image © Phil Smith, Aquatonics Ltd.

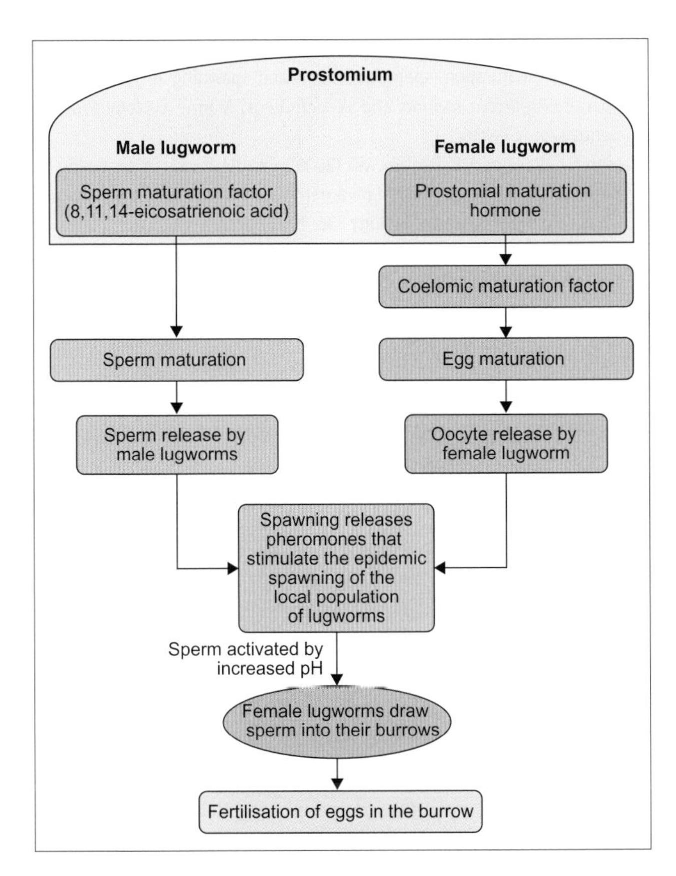

Figure C Interacting factors regulating gamete maturation, spawning and fertilization

The anterior-most segment of the lugworm (in front of the mouth: the prostomium) contains the hormones that are the first step in triggering maturation prior to spawning. In male lugworms (blue boxes), the active substance has been identified as a 20-carbon fatty acid: 8,11,14-eicosatrienoic acid (for an explanation of the naming of fatty acids based on their chemical structures see Box 9.2). In female lugworms (purple boxes) a two-step hormonal control process occurs. In the breeding season, prostomial maturation hormone (PMH) increases in the 2–3 weeks prior to spawning. PMH triggers production of a coelomic maturation factor that stimulates egg (oocyte) maturation and spawning. Epidemic spawning of the population of lugworms results from the cascade effect of the release of pheromones during spawning.

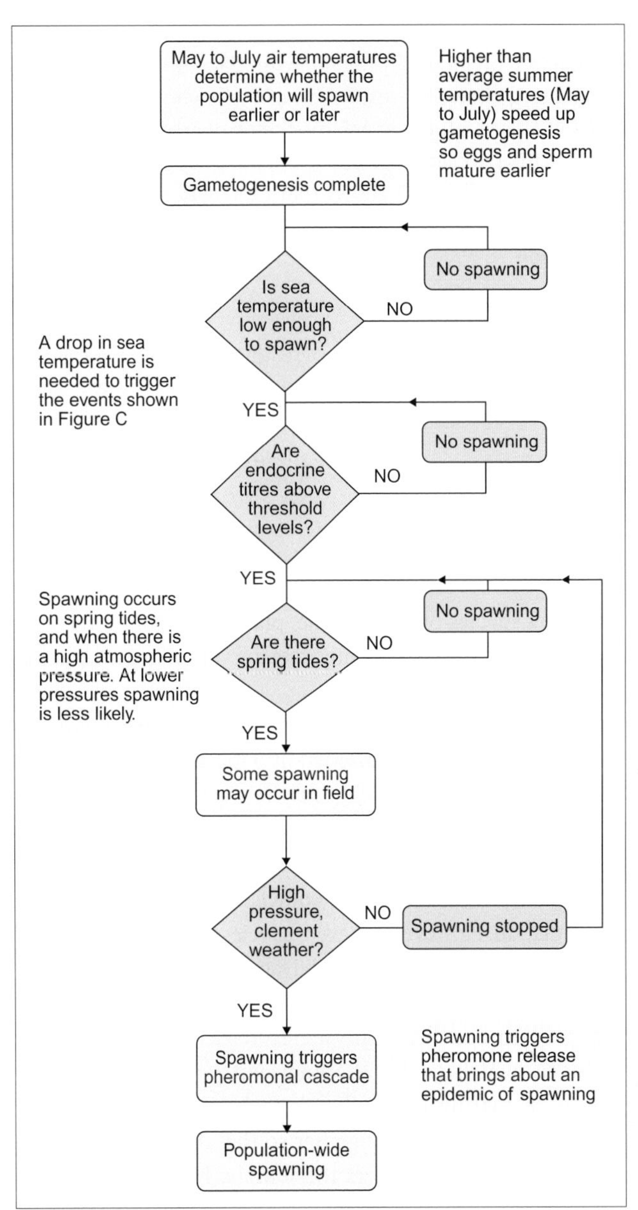

Figure D Flow diagram of environmental factors and control mechanisms that determine when an epidemic of spawning by lugworms occurs

Source: Watson GJ et al (2000). Can synchronous spawning be predicted from environmental parameters? A case study of the lugworm, *Arenicola marina*. Marine Biology 136: 1003-1017.

across the muddy sediment, giving female lugworm the opportunity to 'collect' active sperm for fertilization (which they must do before the sperm are too diluted by seawater).

The sperm do not begin swimming until the puddles of sperm are diluted by some seawater at a higher pH, usually on the incoming tide. The pumping into the burrow of a female lugworm of active sperm is more persistent than the pumping to draw in water for respiration (which only lasts a few minutes); females pump in activated sperm for up to an hour, and sometimes change the direction of pumping from that shown in Figure A. The pump also generates 30–150 times more suction than normal burrow ventilation.

It is unlikely that all oocytes are fertilized by a single exposure to sperm. Experimental studies show that the eggs are viable for several days, which allows fertilization to be spread over several days of spawning. This means that a batch of eggs released into the burrow is likely to be fertilized by several males, which increases the genetic diversity of the offspring.

The larvae undergo early development in the burrow of the female (where fertilization occurs). Later, however, they move to the surface and are transported by the tide to a new patch of sandy mud where they settle and develop in mucous tubes until they are carried to new areas of sediment where they then burrow.

So what regulates the synchronous epidemic spawning of lugworms, in just a few days in late autumn? Figure C summarizes the hormonal events that were compiled as a result of more than a decade of investigations by Matthew Bentley and co-workers at the University of St Andrews, UK. This is the first annelid for which the chemical control systems regulating spawning have been identified.

The collection of 13 years of data on spawning for a Scottish population of lugworm, along with data on environmental parameters, allowed investigation of the environmental factors that influence maturation and the timing of epidemic spawning, which in this population is restricted to a 4–5 day period in October to November. Figure D shows that in lugworms a hierachy of environmental signals influence the epidemic spawning of the population.

› Find out more

Hardege JD, Bentley MG (1997). Spawning synchrony in *Arenicola marina*: evidence for sex pheromonal control. Proceedings of Royal Society London B 264: 1041–1047.

Watson GJ, Cadman PS, Paterson LA, Bentley MG, Aukland MF (1998). Control of oocyte maturation, sperm activation and spawning in two lugworm species: *Arenicola marina* and *A. defodiens*. Marine Ecology Progress Series 175: 167–176.

Watson GJ, Williams ME, Bentley MG (2000). Can synchronous spawning be predicted from environmental parameters? A case study of the lugworm, *Arenicola marina*. Marine Biology 136: 1003–1017.

Williams ME, Bentley MG (2002). Fertilization success in marine invertebrates: the influence of gamete age. Biological Bulletin 202: 34–42.

[1] Section 7.3 outlines the excretory functioning of the nephridia of invertebrates.

surviving species of monotreme: platypus (*Ornithorhynchus anatinus)* and echidnas (spiny anteaters) that live in Australia and New Guinea[3].

We are familiar with the process of insemination in mammals in which the male penis inserted into the female reproductive tract ejects sperm-rich fluid into the tract. A similar process occurs in some amphibians belonging to the group known as caecilians. These worm-like or snake-like amphibians without legs use a protrusible organ known as the phallodeum for insemination. However, not every species that employs internal fertilization has a penis or an equivalent structure.

In almost all birds and reptiles, elasmobranch fish and lobe-finned fish[4], the males release sperm through a **cloaca**, which is also the final exit point for urine and faeces. In mating birds, the cloacal opening of the male and female are pressed together so that the released semen (containing sperm) enters the female. However, males of about 3 per cent of bird species, waterfowl in particular, have developed a pronounced phallus that protrudes prior to copulation.

In some species of salamanders (a group of amphibians (*Ambystoma* spp.)) the male drops a packet of sperm (a **spermatophore**) after elaborate courtship behaviours, which the female picks up using her cloacal lips. She then stores the spermatophore internally for use in fertilization.

The process of transferring sperm into the female's reproductive tract in spermatophores also occurs in some sharks and rays. Figure 20.1B shows the pronounced rod-like clasper on the modified pelvic fin that male sharks, skates and rays insert into the female's oviduct (the duct along which eggs pass from the ovary to the cloaca). Once in place, seawater washes a spermatophore into the female along a groove in the claspers. The female can then use the sperm for internal fertilization for a prolonged period.

Many of the terrestrial arthropods, such as hermit crabs, mites, pseudoscorpions and scorpions, also transfer sperm in spermatophores. In scorpions, the transfer involves four behavioural stages: initiation, followed by a dance known as the 'promenade à deux', which leads to sperm transfer, and then separation. Male scorpions usually initiate courtship by approaching a female and grasping her pedipalps, and then lead the female round and round in the courtship dance that can last from 10 minutes to many hours. Once a suitable surface is located, the male scorpion touches his genital aperture to the ground and extrudes a stalked spermatophore containing a sperm mass, which is protected from desiccation in the stalk. Scorpion spermatophores differ in appearance but always have a sticky basal plate by which they attach to the ground. The male scorpion next carefully pulls the female into position over the spermatophore.

The exact means by which the spermatophore latches perfectly into the genital pore of the female scorpion and the sperm break away differs between species. In some species, the female lowers herself onto the large leaf-like end of a lamelliform spermatophore. This triggers structural changes in the spermatophore that ensure the safe transfer of the sperm mass through the female's genital pore. In buthid scorpions (the majority of scorpions venomous to humans), the flagelliform spermatophores have a long extension from their tip (the 'flagella') which remains connected to the male at his genital pore during mating. The connection is only broken after the female is inseminated, when the male rapidly pulls away, perhaps to avoid the possibility of cannibalism.

[3] Figure 1.22 in Section 1.4.2 examines the evolution and types of mammals.

[4] The evolutionary relationships between these vertebrate groups is illustrated in Figure 1.21.

20.1.2 Number of eggs, energy and protein provision

The number of eggs produced by different animals varies enormously. Those using external fertilization tend to release high numbers of eggs. For example, reproductive female lugworms (*Arenicola marina*)[5] hold more than 300,000 oocytes. However, only a few will be fertilized when spawned into the burrow.

Most marine teleosts release half a million to several million eggs each time they spawn; if males release sperm at the same time, the number of fertilized eggs can be high. However, the tiny amount of yolk per egg provides only a low level of energy (lipid) and protein, which the developing embryo soon uses up. The embryos must therefore hatch and find their own food within a few days, as indicated by the graph in Figure 20.2A. Most will not survive, but the huge initial numbers mean that some do. At the other end of the scale, many mammals and birds produce few offspring at a time but invest much more energy in each offspring. For example, female emperor pigeons (*Aptenodytes forsteri*) and king penguins (*Aptenodytes patagonicus*) lay a single large egg[6].

[5] Case Study 20.1 discusses the epidemic spawning of lugworms.

[6] In Case Study 2.2, we examine the high level of parental investment by male emperor penguins, which lose about 40 per cent of their body mass, as a result of mainly lipid metabolism during their starvation when they are incubating the egg.

In the 1960s and 1970s, the theory of r/K selection was proposed as an explanation for the evolution of different life history (reproductive) strategies. The r-selected species are characterized by the production of large numbers of eggs (or offspring) with little nutritional investment; the emphasis is on population growth (represented by *r*). The K-selected species invest heavily in a small number of eggs (or offspring) with a larger amount of nutrients. K-selected species live at high densities, close to the maximum carrying capacity (*K*) of the environment.

The basic ideas behind the original r/K selection theory have since been developed into more complex 'demographic models' of life history strategies that incorporate both density-dependent factors, such as resource availability, and the many density-independent factors such as the age structure of the population, age at maturity, environmental fluctuations and mortality.

Eggs contain different amounts of yolk

Yolk consists of the nutrients (lipids and proteins) and the enzymes that are necessary for development of fertilized eggs; its distribution differs among different species:

- Isolecithal eggs with evenly distributed yolk occur in mammals.
- Centrolecithal eggs in which yolk is concentrated in the centre of the egg occur in insects.

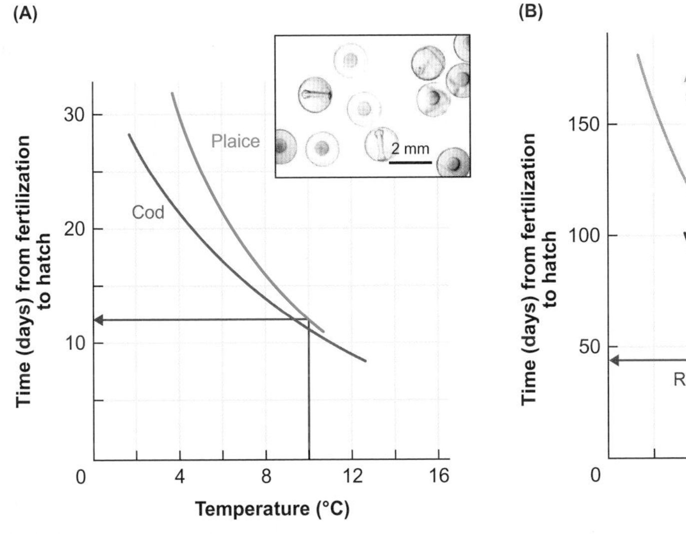

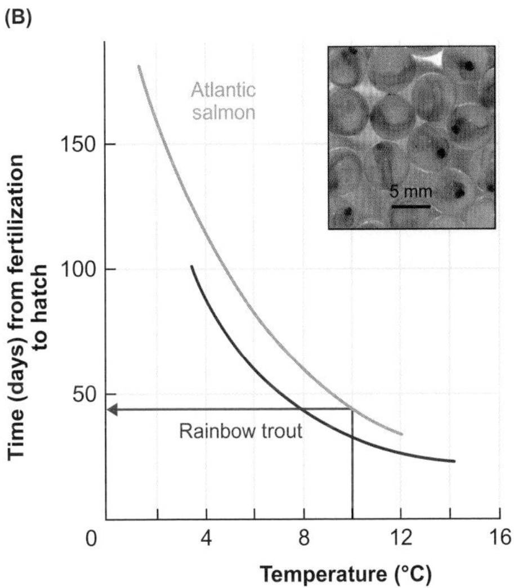

Figure 20.2 Development of eggs of fish

Time from fertilization to hatching of (A) marine species: Atlantic cod (*Gadus morhua*) and European plaice (*Pleuronectes platessa*); eggs shown are cod eggs, and (B) two salmonid species breeding in rivers: Atlantic salmon (*Salmo salar*) and rainbow trout (*Oncorhynchus mykiss*). Note different scales on y-axis when comparing data. Reading off times for development until hatching, at temperatures of 10°C (red lines and arrows) shows that embryos in the larger eggs of salmon and trout, of > 5 mm diameter, develop for about four times longer than those of cod or plaice that produce smaller eggs.

Sources: adapted from Jobling M (1995). Environmental Biology of Fishes. London: Chapman & Hall. Images: A: © Crown Copyright; B: E Peter Steenstra/USFWS

Table 20.1 Egg numbers and mass, and percentage of endogenous protein and lipid reserves of laying birds required during egg laying

Capital breeders (such as Adélie penguin and common eider) draw almost exclusively on endogenous reserves for nutrient provision (protein and lipid) for the clutch of eggs and their own needs during the laying period, while income breeders (such as kestrels, zebra finches and blue tits) seek exogenous supplies of lipid and protein; endogenous provision by income breeders is usually less than 25 per cent and may be zero.

Species	Body mass of laying female (g)	Number of eggs in clutch	Mean mass per egg (g)	Protein	Lipid
				Per cent endogenous reserves used	Per cent endogenous reserves used
Adélie penguin (*Pygoscelis adeliae*)	3400	2	59.5	96	99
Common eider (*Somateria mollissima*)	2163	4	26	100	100
Mallard (*Anas platyrhynchos*)	1300	8–11	6.5	4–5	6–45
Common kestrel (*Falco tinnunculus*)	300	5	4.2	0	0
Zebra finch (*Taeniopygia guttata*)	15	4	0.25	26	10
Eurasian blue tit (*Cyanistes caeruleus*)	11	11	0.11	0	0

Source of data: Meijer T, Drent R (1999). Re-examination of the capital and income dichotomy in breeding birds. Ibis 141: 399-414.

- Teleolecithal eggs in which yolk is concentrated at one side, and the embryo develops at the other side occur in amphibians.

Eggs also fall into four categories based on the *amount* of yolk they contain:

- alecithal eggs with no yolk, which occur in placental mammals;
- microlecithal eggs containing small amounts of yolk, which occur in marsupials and echinoderms;
- mesolecithal eggs containing intermediate amounts of yolk occur in most fish and amphibians and enable embryos to hatch at a later stage of development;
- macrolecithal eggs with a large amount of yolk, which occur in insects, some cartilaginous and bony fish, reptiles and birds, enable the young to hatch at a later stage of development than those from mesolecithal eggs.

The inclusion of larger amounts of yolk in an egg inevitably results in a larger egg and allows a longer period of development before hatching. As an example, Figure 20.2 shows the contrasting size of the eggs of two fish species breeding in fresh water and two marine species. The poorer yolk provision in the smaller eggs of the marine species leads to earlier hatching than in the salmon and trout species with greater yolk provision. The graph also shows how water temperature affects the rate of development, with larvae hatching more rapidly at higher temperatures, whatever the species.

After hatching, the amount of yolk remaining in the yolk sac and the rate at which this is used (which depends on the environmental temperature) determines the maximum time before feeding becomes essential. The yolk sac of salmon and trout species contains sufficient yolk for the larvae to survive up to 300 days at 2°C while cod and plaice must find food within about 30 days at the same temperature because of the poorer yolk provision.

Birds, turtles and lizards produce a relatively small number of macrolecithal eggs after internal fertilization. For example, green sea turtles (*Chelonia mydas*) produce clutches of 100 to 200 eggs. Sharks and rays take this a step further and usually produce less than 20 macrolecithal eggs per occasion.

Table 20.1 gives data for the nutrient provision of the eggs of several species of bird. In some species, such as Adélie penguins (*Pygoscelis adeliae*), the clutch of eggs comprises as few as one or two eggs, while other species, such as mallards (*Anas platyrhynchos*) or Eurasian blue tits (*Cyanistes caeruleus*) produce clutches of 10 or more eggs. Not surprisingly, birds laying large eggs that contain more yolk incubate their eggs for longer and the chicks hatch at a later stage of development than those from smaller eggs.

Bird species that hatch as well-developed mobile chicks with eyes open, covered with down, and able to thermoregulate[7], are called **precocial** (from the same Latin route as precocious). At the other end of the spectrum, species that hatch at a much earlier stage in development are known as **altricial** (derived from the Latin for nourish). The chicks of altricial species hatch with their eyes closed; they are incapable of moving around to find food and have no down present so they are unable to thermoregulate.

[7] Chapter 10 discusses the thermoregulation of birds.

Different bird species fall along a continuum between the two extremes of precocial and altricial. The most precocial species are the Australian megapodes (literally 'with large feet'), the brush turkeys and scrub fowl, which are completely independent of parental care—but this is a rarity. Adélie penguin chicks are precocial, but their development is about midway between the brush turkeys and the altricial passerines (a bird grouping that includes more than half of the bird species, including house sparrows, *Passer domesticus*).

Precocial chicks generally hatch from larger eggs with a higher energy content, which may increase their chances of survival compared to altricial chicks. However, the outcome is not so straightforward. While small eggs with less yolk hatch at an earlier stage in development, and the chicks are more vulnerable to predation, both parents can contribute to parental care and food provision for the dependent offspring.

Lipid and protein provision within an egg has cost implications for the female animal producing the eggs. Food intake must exceed the animal's own needs or the animal must use its own reserves when laying down yolk in the eggs. Some large species of birds breeding in cold areas accumulate sufficient reserves for these reserves to be drawn upon during egg formation and laying. These are **capital breeders**, so named because they use existing energy stores (i.e. they draw on their 'capital'). Table 20.1 includes two good examples of capital breeders: Adélie penguins and common eiders (*Somateria mollissima*).

The Adélie penguin stops feeding almost two weeks before laying starts. Producing a clutch of two eggs uses almost all of the female's endogenous lipid (96 per cent) and protein (99 per cent) reserves, including her muscle tissue. Similarly, female eider ducks are capital breeders, which increase their body mass by about 20 per cent above winter levels while feeding in the high Arctic near the nesting islands, before laying eggs. They then fast completely during laying and incubation of the eggs, losing about 46 per cent of their pre-laying body mass. During egg laying, the female eider duck loses about 113 g of protein and 262 g of lipid from endogenous reserves, transferring half of the protein and about 25 per cent of the lipid to the eggs.

In contrast to the use of endogenous reserves, small birds tend to be **income breeders**, reliant on continual food intake to supply protein/lipid to the eggs. Table 20.1 shows some examples: blue tits, zebra finches (*Taeniopygia guttata*) and common kestrels (*Falco tinnunculus*).

Deposition of yolk in eggs occurs by vitellogenesis

The process of depositing nutrients into the oocytes that will develop into yolky eggs is similar in all animals. This process—called **vitellogenesis**—involves the deposition of lipids and proteins, mainly **vitellogenin**, in the developing eggs. The term vitellogenin originated in the 1960s to describe specific proteins in the haemolymph (blood) of insects but is now used widely to describe the precursor proteins incorporated into the eggs of fish, amphibians, reptiles, birds and most invertebrates.

Synthesis of vitellogenins by invertebrates can occur in different tissues: the intestine of nematodes and sea urchins, the ovary of molluscs, the ovary and fat body of insects. By contrast, all vertebrates synthesize vitellogenins in the liver. In laying hens, vitellogenin accounts for 10–20 per cent of the protein synthesized by the liver. Vitellogenins are secreted into the body fluids where they form complexes with calcium. The blood circulation delivers vitellogenin to the ovary where the vitellogenin binds to receptors on the plasma membrane of the developing gametes that will form the ova. The vitellogenin–receptor complex is then internalized within the cytoplasm of the egg: the ooplasm.

The formation of vitellogenin is normally associated with female animals and their yolk provision in eggs; the gene for vitellogenin production also occurs in male animals but is normally silent. Experimental Panel 20.1 discusses some circumstances in which the vitellogenin gene is activated and considers the use of vitellogenins as biomarkers for exposure to **environmental oestrogens**—chemical pollutants that mimic the natural oestrogens (female sex steroids)[8] which normally regulate vitellogenesis.

We have seen how animals can make provisions for their developing offspring by including nutrients in their eggs. Taking this a step further, the retention of the fertilized egg and internal development of the young has evolved in many animal groups, as we learn in the next section.

20.1.3 Oviparity, ovoviviparity and viviparity

Animals that release eggs in which there has been little or no embryonic development are known as **oviparous** (meaning egg-laying from the Latin for egg = *ovum* and *parere* = to bring forth or produce). This definition clearly applies to animals that release ova (and sperm) for external fertilization. The term oviparous also applies to the many invertebrates (crabs, insects, scorpions and spiders), most reptiles and all birds, in which internal fertilization is followed soon after the eggs emerge from the ovary.

Some animals retain their fertilized eggs and give the developing young protection, although the nutrients that

[8] Figure 20.11 examines the structure of the natural oestrogens.

Experimental Panel 20.1 Vitellogenins and related proteins as biomarkers for exposure to environmental oestrogens in wildlife

Background

Endocrine disrupting chemicals (EDCs) entering the environment are a potential threat to wildlife. These chemicals include natural oestrogens such as 17β-oestradiol and oestrone (which is also formed by degradation of 17β-**oestradiol**), synthetic oestrogens such as 17α-ethinyl-oestradiol and progesterone from the contraceptive pill and medical treatments, and chemicals such as organochlorine pesticides, alkylphenolic compounds (produced by degradation of detergents), bisphenol A (used in manufacturing polycarbonates) and phalates (used as plasticizers). These environmental oestrogens mostly originate from industrial effluents, agricultural run-off, livestock wastes, and effluents from sewage and water treatment works. Once in an aquatic environment, these lipophilic compounds are liable to bioaccumulate in aquatic organisms to levels at which they may interfere with the hormonal processes controlling sexual differentiation and reproduction.

Predicting the oestrogenic activity of chemicals based on their structures proved to be highly unreliable, so alternative approaches were necessary. The synthesis of vitellogenins and related proteins that are incorporated into eggs is regulated by oestrogens acting on oestrogen receptors[1], so the concentration of vitellogenin in the blood could act as a biomarker for exposure to environmental oestrogens. Vitellogenin not taken up by eggs accumulates in the blood.

In this Panel we explore two studies of vitellogenins and related proteins: (1) in a fish species—the common roach (*Rutilus rutilus*); and (2) in an invertebrate—the zebra mussel (*Dreissena polymorpha*). In each case, animals were exposed to environmental oestrogens in the effluent from sewage treatment works.

1. Roach study

Adult roach were exposed as mixed-sex groups to effluent from a sewage treatment works, which had been diluted to simulate the normal dilution of effluent in rivers. Two groups of control roach were also studied: (i) roach held in river water collected upstream of the treatment works, and (ii) roach held in tap water. At various time points (day 0 (pre treatment), and after 1, 2 and 4 months) groups of 10 fish were removed from each tank. A blood sample was collected from each fish for measurement of vitellogenin.

Results and discussion

Figure A shows results for the vitellogenin concentrations in the blood plasma of male roach after various periods of exposure to diluted effluent from the water treatment works and in control male roach. Male fish usually lack significant levels of vitellogenin in their plasma; they have the gene for vitellogenin, but it remains silent in the absence of oestrogens. Figure A shows there is a low concentration of vitellogenin present even in the control males, which indicates some activation of the vitellogenin gene. More importantly, a dramatic increase in plasma concentrations of vitellogenin occurred in the male fish living in the diluted effluent. (Notice that the y-axis of the graph in Figure A has a logarithmic scale, so the differences are larger than they might at first appear.)

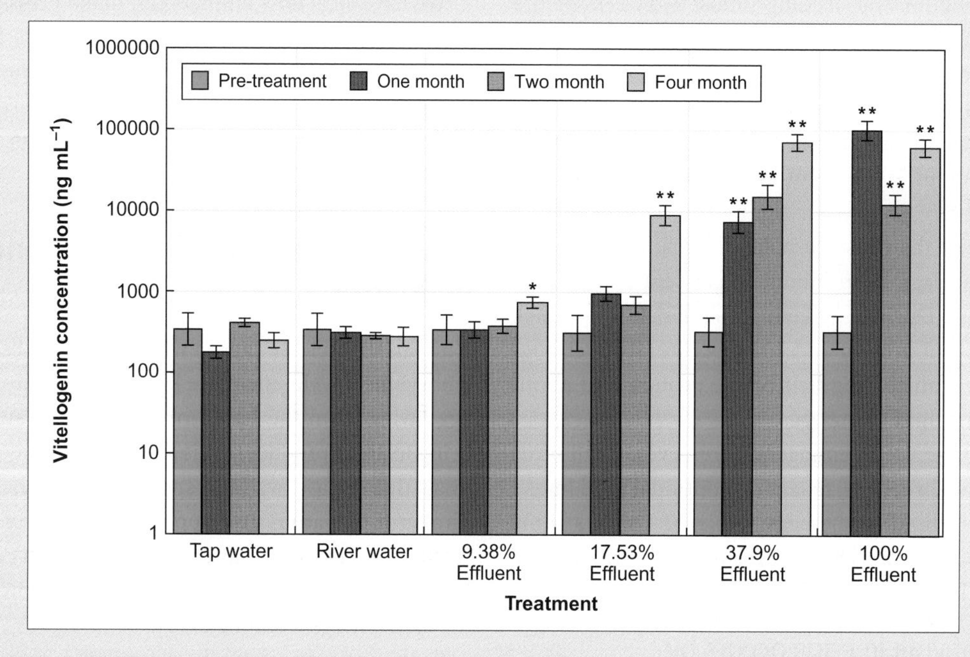

Figure A Plasma vitellogenin concentration (ng mL^{-1}) in adult male common roach (*Rutilus rutilus*) exposed to various concentrations of treated sewage effluent to simulate the normal dilution of final sewage effluent in rivers

Data are means ± standard error; 10 fish sampled at each time point in each exposure concentration. Asterisks denote levels of significance compared to controls: ** $P < 0.01$, * $P < 0.05$. Source: Rodgers-Gray TP et al (2000). Long-term temporal changes in the estrogenic composition of treated sewage effluent and its biological effects on fish. Environmental Science & Technology 34: 1521-1528.

At about 10–20 per cent effluent the male roach had significantly increased concentrations of plasma vitellogenin within the 4-month exposure period. At higher concentrations of effluent, plasma vitellogenin increased more rapidly and to higher levels. The vitellogenin concentrations of the male roach in this study reached similar levels to those that occur in female roach (data not shown). This scale of response (by male fish) to the oestrogenic cocktail in the effluent indicates that vitellogenin can be a very useful biomarker for exposure of fish to environmental oestrogens.

2. Study of zebra mussels

In comparison to the large number of studies in which vitellogenin acts as a biomarker for endocrine disruption in fish, fewer studies have examined invertebrates, and these have mostly been of bivalve molluscs. One such study exposed zebra mussels for more than three months to effluent from a sewage and waste water treatment works. The effluent had been subjected to tertiary (advance) treatment before discharge. These studies were timed to coincide with the period of seasonal **gametogenesis** of zebra mussels (December to mid-March), so that any interference with gametogenesis could also be examined.

At the end of the exposure period, tissue sections of some mussels were stained to examine the gonad and make comparisons with stained sections from the gonads of control mussels (unexposed to the effluent). The flesh from whole mussels was homogenized to assay the vitellogenin-like protein content. Specific antibodies are used to monitor these proteins in fish but do not cross-react against invertebrate vitellogenins. All vitellogenins are phospholipoproteins so an assay that measures alkali-labile protein-bound phosphate (ALP) can be used to monitor vitellogenin-like proteins of invertebrates.

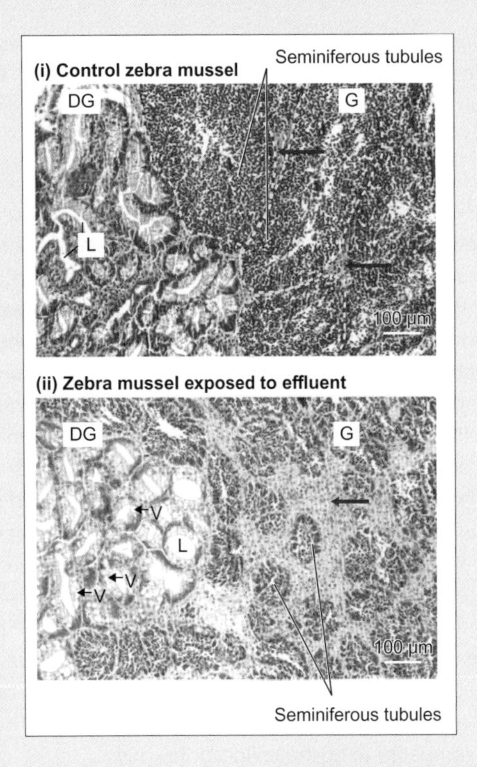

Figure B Photomicrographs showing the histology of digestive gland (DG) and gonad (G) of male zebra mussels

(i) Control mussel, in which only a thin border of interstitial tissue (red arrows) occurs between the seminiferous tubules containing the spermatocytes. Three seminiferous tubules are outlined by a dashed yellow line. The digestive gland (DG), adjacent to the testes, consists of tubules with narrow lumens (L).
(ii) Male zebra mussel after exposure to effluent from the sewage and waste water treatment works for 112 days, showing a large amount of undifferentiated interstitial tissue (red arrow) between poorly developed seminiferous tubules. The lumen (L) of the digestive gland tubules is enlarged and most cells show vacuolation (V).

Source: Quinn B et al (2004). The endocrine disrupting effect of municipal effluent on the zebra mussel (*Dreissena polymorpha*). Aquatic Toxicology 66: 279–292.

Results and discussion

Figure B shows the tissue sections of male zebra mussels after exposure to the effluent compared to the gonad of control mussels. In control mussels, the gonadal tissue was packed with seminiferous tubules containing spermatocytes[2]. In contrast, the gonad of mussels exposed to the effluent, shown in Figure B(ii), was relatively undifferentiated; the large amount of interstitial tissue between the small seminiferous tubules indicates an inhibition of the development of the testes. Adverse effects on the digestive gland, which is important in providing energy supplies during gametogenesis, were also seen. In contrast to the male mussels, gonadal development of female mussels was unaffected (histology not shown).

Figure C shows that exposure of the zebra mussels to the effluent significantly increased ALP, with similar increases in both male and female mussels. Figure C suggests similar amounts of vitellogenin-like proteins per gram of total protein in both male and female control mussels. This finding suggests there is a high background level in males, resulting from prior exposure to oestogenic compounds. The mussels were collected from a site where there are many existing discharges from sewage treatment works, so vitellogenin-like proteins were likely to already be activated. This demonstrates some of the pitfalls of using animals from wild

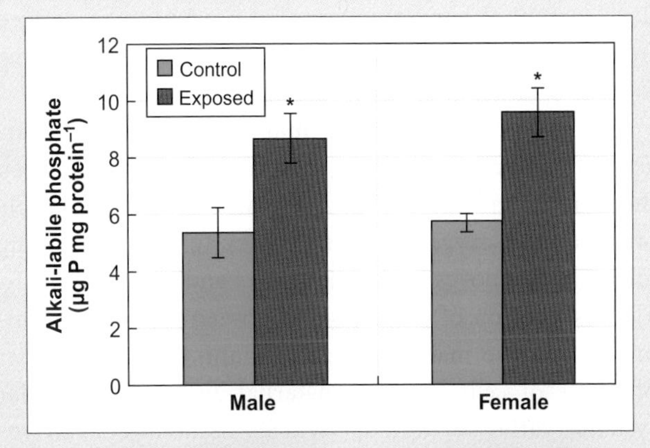

Figure C Alkali-labile phosphate of homogenates of body tissues of male and female zebra mussels exposed to effluent from a sewage treatment works for 112 days compared to control (unexposed) mussels

Data are normalised against total protein content of the mussel homogenates. Bars show means and standard errors; *indicates a significant difference ($P < 0.05$) between control mussels and mussels exposed to effluent containing endocrine disruptor chemicals.

Source: Quinn B et al (2004). The endocrine disrupting effect of municipal effluent on the zebra mussel (*Dreissena polymorpha*). Aquatic Toxicology 66: 279–292.

populations to explore the impact of experimental exposure to contaminants. Nevertheless, the experiments showed further activation of vitellogenins and an associated disruption of testes development in the mussels exposed to the effluent.

Wider issues and implications

These and other studies indicate that vitellogenin and related proteins are useful biomarkers for exposure to oestogenic compounds. However, an additional functional role of vitellogenins in immune defences has been suggested, which may complicate interpretations. Hence, use of vitellogenin as the sole biomarker of oestrogenic activity should be avoided. Instead, a combination of biomarkers will provide a more robust indication of the impacts of oestrogenic contaminants.

Evidence of the inhibition of testes development and sperm production in males by histological studies acts as an important functional biomarker of oestrogenic effects. Such effects on sexual differentiation can be predicted to have potential impacts at a population level—for example, influencing sex ratios. There is evidence of such effects in wild mussels (*Elliptio complanata*) in the Saint Lawrence River, in Canada, where the sex ratio of mussels downstream of a major urban effluent is skewed toward females compared to upstream locations.

› Find out more

Gagné F, Bouchard B, André C, Farcy E, Fournier M (2011). Evidence of feminization in wild *Elliptio complanata* mussels in the receiving waters downstream of a municipal effluent outfall. Comparative Biochemistry and Physiology, Part C 153: 99–106.

Matozzo V, Gagné F, Marin MG, Ricciardi F, Blaise C (2008). Vitellogenin as a biomarker of exposure to estrogenic compounds in aquatic invertebrates: A review. Environment International 34: 531–545.

Porte C, Janer G, Lorusso LC, Ortiz-Zarragoitia M, Cajaraville MP, Fossi MC, Canesi L (2006). Endocrine disruptors in marine organisms: Approaches and perspectives. Comparative Biochemistry and Physiology, Part C 143: 303–315.

Quinn B, Gagné F, Costello M, McKenzie C, Wilson J, Mothersill C (2004). The endocrine disrupting effect of municipal effluent on the zebra mussel (*Dreissena polymorpha*). Aquatic Toxicology 66: 279–292.

Rodgers-Gray TP, Jobling S, Morris S, Kelly C, Kirby S, Janbakhsh A, Harries JE, Waldock MJ, Sumpter JP, Tyler CR (2000). Long-term temporal changes in the estrogenic composition of treated sewage effluent and its biological effects on fish. Environmental Science & Technology 34: 1521–1528.

[1] Oestrogen receptors are a type of nuclear receptor, which we discuss in Section 19.1.4

[2] Box 20.1 examines the stages in spermatogenesis.

they need are contained with the eggs. This mode of reproduction is called **ovoviviparity**. The term ovoviviparity distinguishes the reproductive mode of retaining eggs for protected development until the birth of live young from true **viviparity** in which the mother provides the daily nutritional requirements of the embryo/foetus throughout development. (Viviparous means live bearing and is derived from the Latin *vivus* = living or alive and *parere* = to bring forth or produce).

The **eutherian (placental) mammals** develop a characteristic **placenta**, which forms a connection between the mother and the foetus, and which provides a supply of nutrients, oxygen and hormones to the developing foetus and removes metabolic wastes and carbon dioxide. The distribution of contact sites between the foetal membranes and the maternal endometrium, and the number of layers of cells between maternal and foetal vascular systems, varies between species. Six tissue layers can potentially separate maternal and foetal blood, all of which occur in the diffuse placentae of horses and pigs. These are called epithelialchorial placentae as they consist of maternal epithelial and foetal chorial layers. In humans and other primates, invasion of the maternal epithelium and underlying connective tissues layers forms a haemochorial placenta in which a single endothelial cell layer separates foetal and maternal blood as a placental disc. Figure 20.3 illustrates this arrangement.

The embryos of marsupial mammals do not have a true placenta, but some species are thought to receive some nutrition via a yolk-sac type of placenta that makes contact between the embryonic yolk sac and the mother's uterine wall during the short period that the embryo remains in the uterus[9]. Something akin to placental viviparity also occurs in representatives of many other animal groups, but is never a characteristic feature of all species in the group. For example, a type of placental viviparity occurs in some species of snakes, lizards (the skinks), salamanders and frogs, and some sharks. In these cases, developing embryos receive nutrients and oxygen across a highly vascularized region of the reproductive tract in which finger-like protrusions (microvilli) from the extra-embryonic layers form a placenta-like structure.

[9] Figure 20.34 gives further information on the yolk-sac placenta of marsupial embryos.

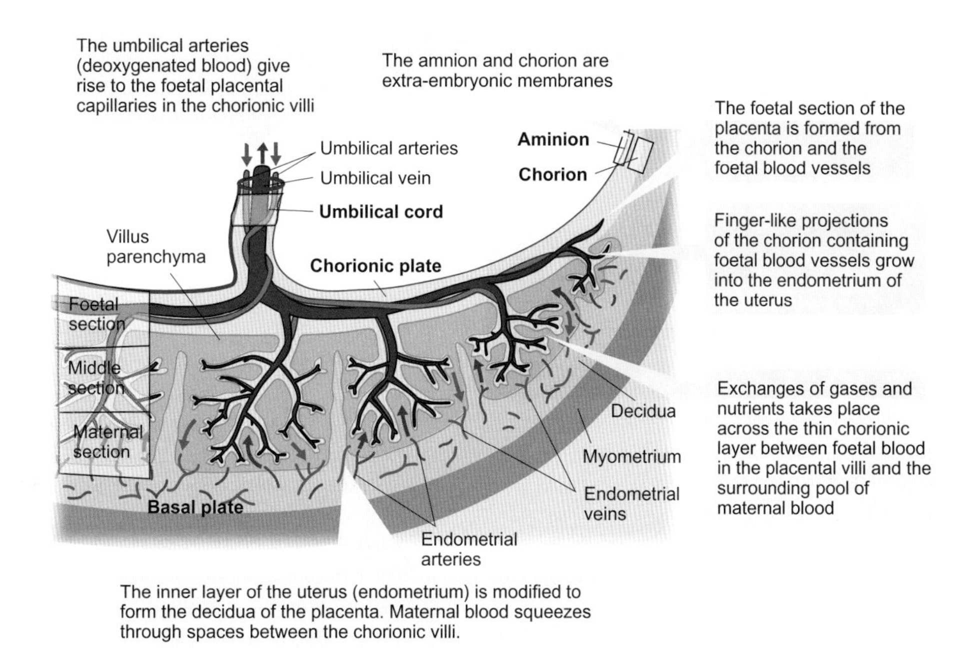

Figure 20.3 Placenta of human at 44 weeks' gestation

The section across the placenta shows basal and chorionic plates, and division of the placenta into foetal, middle and maternal sections.

Source: Sood R et al (2006). Gene expression patterns in human placenta. Proceedings of the National Academy of Sciences of the United States of America 103: 5478-5483.

20.1.4 When do animals reproduce—and how often?

Viviparous animals have the capacity to reproduce more than once in their lifetime. Oviparous animals may also produce eggs many times in their life. However, an alternative scenario of once-in-a-lifetime breeding occurs in some species. These animals are **semelparous** (from the Latin for once = *semel)* and their reproductive strategy is semelparity. A pronounced feature of most semelparous animals is the extent to which their body becomes dedicated to reproduction, so much so that the animal's own tissues are often metabolized and used to provide the nutrients for gamete synthesis.

Semelparity occurs in many groups of invertebrates. Among insects, many species of butterflies and mayflies are semelparous. Most of a mayfly's life in fresh water is as an immature nymph. After up to a year in water, the nymph metamorphoses into a flying adult that emerges from the water. It lives for at most a few days and cannot feed because it lacks effective mouthparts. During the short period as an adult, mating occurs and the females lay eggs.

Among marine polychaetes, many nereid and syllid worms develop into a reproductive stage called an epitoke in which the entire worm becomes packed with gametes. Prior to sexual maturation, a juvenile hormone secreted by cerebral ganglia (brain) inhibits maturation, but maturation of gametes occurs when secretion of the juvenile hormone decreases. Once the final stages of **gametogenesis** are underway, the process is irreversible: feeding ceases, the gut breaks down, and the worm abandons repair of its body tissues and becomes an epitoke.

The sole purpose of the epitoke is to spawn. The epitoke, packed with gametes, moves from its usual benthic location and floats toward the surface of the water, attracted by light. After bursting to release gametes for external fertilization, the epitoke often breaks up. Consequently, reproduction is semelparous, although in some cases, the worm survives and returns to a benthic lifestyle before reproducing again, as illustrated in Figure 20.4.

In some syllid polychaetes, a series of epitokes called **stolons** develop at the posterior end of the worm, as shown by the striking image in Figure 20.4A. Each stolon has a head, but no mouth or pharynx. Gametes form in the maturing stolons, which separate and live only a brief life swarming at the sea surface; their sole purpose is to act as a locomotive vessel for the gametes. However, the front end of the worm may enter a new reproductive cycle, so the worm is **iteroparous**—it breeds more than once during its lifetime, rather than semelparous.

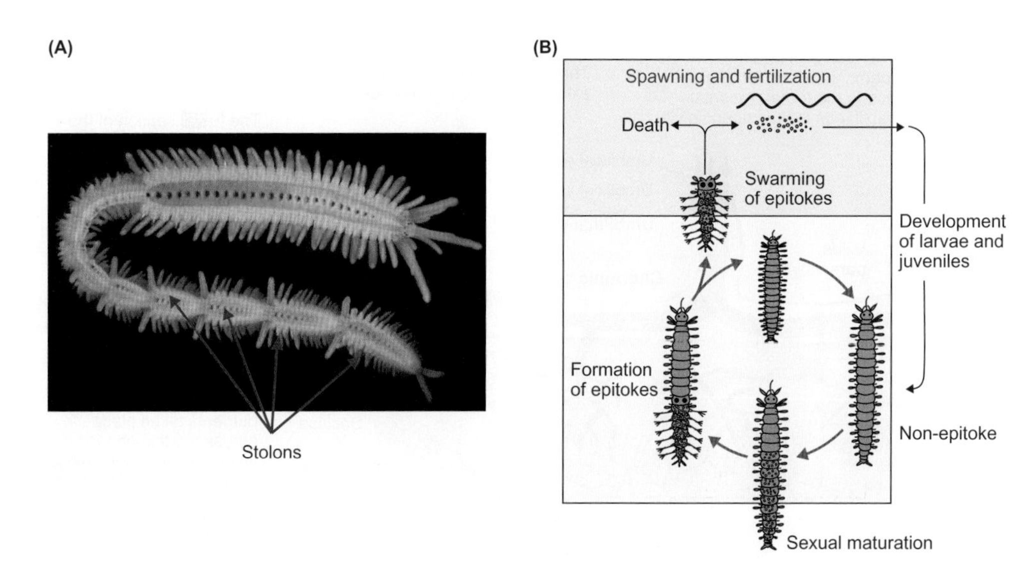

Figure 20.4 Epitoky in the polychaete *Myrianida pachycera*

(A) Stolons form at the posterior end of the worm; up to 29 stolons can occur in a single individual. (B) Schematic diagram of the stages in forming epitokes and spawning. Blue-shaded region represents the pelagic reproductive stages; yellow-shaded region represents stages in the benthic environment. Once the epitokes detach they live briefly in the water column before breaking up, releasing the gametes and dying; the remaining front end of the worm can enter a new reproductive cycle.

Sources: (A): Dr Greg Rouse, Scripps Institution of Oceanography; (B): Adapted from Franke H-D (1999). Reproduction of the Syllidae (Annelida: Polychaeta). Hydrobiologia. 402: 39–55.

In comparison to invertebrate animals, semelparity is infrequent among the vertebrates, but does occur in some species of fish. All species of Pacific salmon (chinook, coho, sockeye, pink and chum) are semelparous and die after breeding in rivers. Prior to breeding, the eggs develop synchronously in the ovary, which changes from a thin empty strand in the immature fish, to an ovary packed with gametes that occupies much of the body cavity by the time of spawning.

Among mammals, semelparity is rare, and seems to be restricted to a few short-lived small marsupials in which all males die after several days of almost continuous mating. This unusual reproductive behaviour may be explained by the intense competition of males for females in a short breeding season and the competition for fertilization between sperm of different males (**sperm competition**) that occurs within the female reproductive tract after multiple matings. The intense period of mating is sufficient to cause high levels of stress in the males that culminates in breakdown of the immune system and death[10].

Rather than breeding in a one-off event, most animals breed more than once during their lifetime: they are iteroparous. Some, such as humans, can reproduce at any time of year. For animals living in tropical or sub-tropical zones, however, fluctuations in rainfall often act as the cue to trigger reproduction. In temperate zones, factors such as food availability and its nutritional quality, the changing period of light and dark and ambient temperatures, act as cues for seasonal breeding.

[10] Section 19.3.3 discusses the adverse effects of stress on immune function.

Changes in photoperiod act as a signal for seasonal breeding

At relatively high latitudes, the changing number of daylight hours (the **photoperiod**) results in a predictable photoperiodic cycle throughout the year and is therefore an accurate indicator for the time of year. Consequently, many animals rely on the changing photoperiod as a signal that triggers seasonal development of gametes and reproduction. Photoperiod influences occur in mammals, birds, reptiles, many fish, insects, terrestrial slugs and some intertidal marine invertebrates.

Such effects were demonstrated in birds by the classical studies of William Rowan, a Canadian zoologist who in the 1920s systematically studied gonad development of the dark-eyed junco (*Junco hyemalis*). His studies showed that the birds' gonads could be stimulated to develop by changes in the light cycle. Exposure to long day lengths (at a time of year when the birds would normally experience short days) resulted in the male juncos beginning to sing and their gonads increasing in size.

In seasonal breeders, the transition of males from non-breeding to breeding is usually associated with enlargement of the testes and production of sperm. As an example, Figure 20.5 shows some data for Japanese quail (*Coturnix japonica*) in which a change to a long day length (20 h light: 4 h

20

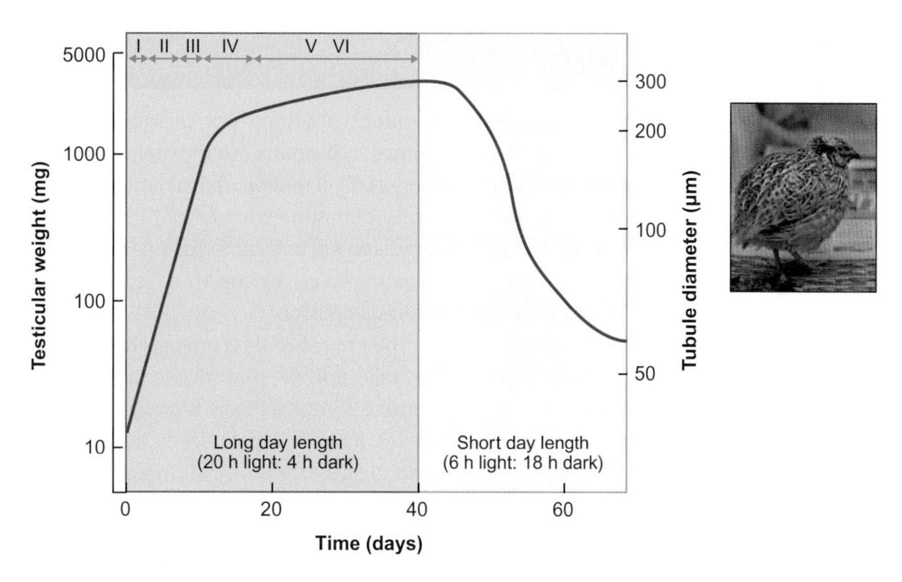

Figure 20.5 Photoperiod stimulation of growth of the testes in Japanese quails (*Coturnix japonica*)

Long day length (green-shaded region between day 0 and day 40) of 20 h light: 4 h dark stimulates testicular development. An increase in tubule diameter in the testes occurs as cells progress through stages I to VI. Stage I = spermatogonia only, II = dividing spermatogonia some spermatocytes, III = many spermatocytes, IV = spermatocytes and spermatids, V/VI = spermatids and mature sperm. The yellow-shaded region represents a period of short day length (6 h light: 18 h dark) during which the testes regress.

Source: adapted from Bentley PJ (1976). Comparative Vertebrate Endocrinology. Cambridge: Cambridge University Press. Quail image: Ingrid Taylar/Flickr.

dark) stimulates testes development. The diameter and length of the coiled **seminiferous tubules** that make up the testes increase[11] so the testes increase in weight. Gametogenesis by the germ cells results in a progression through the various stages that we examine in Box 20.1. When the quail are returned to a short day length (6 h light: 18 h dark) the testes regress. A similar seasonal growth of the ovaries occurs in seasonally breeding birds.

Long-day breeding also occurs in species of mammals such as horses, donkeys, cats, ferrets and racoons. In contrast, short-day breeding occurs in goats, deer and sheep. However, after several months of pregnancy (the **gestation period**), the offspring are born in the spring. In keeping with this timing, the annual increase in the size of the testes of male sheep occurs when day length decreases. The peak maturation of spermatozoa in the northern hemisphere occurs in October–December. After breeding, the testes regress and production of spermatozoa ceases. With the next phase of declining day lengths, testicular **recrudescence** (revival and re-growth of the testes) occurs.

Photoperiod should not be thought of as the direct driving force behind gonad development but instead as an external (exogenous) signal or cue, or more precisely a zeitgeber (from the German for time giver or synchronizer) that entrains internal (endogenous) **circannual rhythms**. The changing photoperiod alters the duration of the secretion of a hormone, **melatonin**[12], which is secreted by the pineal gland during the hours of darkness. Hence, in temperate zones, a relatively long daily peak concentration of melatonin occurs in winter, when day lengths are short, compared to summer when nights are short. As such, the duration of the peak of melatonin conveys information about the time year and acts as a time signal for seasonal reproduction. For instance, when the peak of melatonin secretion is of a long duration, the hormonal control processes (discussed later in this chapter) that stimulate gonad development are triggered in short-day breeders such as goats and deer.

The effect of photoperiod on gonad development allows its artificial manipulation in agriculture or aquaculture. Many fish are seasonal breeders with fry usually emerging soon after the water temperature and food availability are at their most favourable. Early studies of salmonid fish species showed that compressing the normal annual cycle into a 6-month photoperiod cycle leads to earlier spawning, by 3 months; by contrast, stretching the 12-month cycle over 18 months delays spawning by 4 months. These changes in spawning time suggest that photoperiod is the zeitgeber for the endogenous annual rhythm of gonad development and spawning. If this idea is correct, then perturbation of the endogenous rhythm by photoperiod shifts will have different effects depending on the phase of the normal rhythm that is perturbed.

[11] Section 20.3.2 discusses the functioning of seminiferous tubules in more detail.

[12] Section 19.2.3 discusses light detection and its effects on secretion of melatonin by the pineal gland; Section 17.2.2 gives information on the communication of the master biological clock with the pineal gland.

Box 20.1 Gametogenesis: formation of gametes by spermatogenesis in males and oogenesis in females

Figure A shows the steps in gametogenesis in males and females. Most animals are diploid, which means that their chromosomes occur in homologous pairs. Figure A shows two homologous pairs of chromosomes with one of the pair from the female (maternal chromosome shown in red) and one from the male (paternal chromosome shown in green).

The initial step (1) is mitotic division of cells[1]. To simplify the diagram we show the chromosomes as unreplicated, although the chromosomes are replicated before cell division starts. From the perspective of gametogenesis for reproduction the important point is that mitotic proliferation produces a large number of the diploid spermatogonia in the testes and large numbers of diploid oogonia in the ovary.

The next step (2) is **meiosis** I, which forms the haploid gametes, each with one chromosome from the original pair (1*n*). Random distribution of maternal and paternal chromosomes between daughter cells in meiosis I results in many genetic variations. For example, the 23 pairs of chromosomes in humans (46 in total) create 2^{23} possible combinations, i.e. >8 million combinations. The real number of variations is far more because of the exchange of genetic material between each homologous pair of chromosomes (known as crossing-over) before they separate at the end of the first meiotic division[2].

The next step (3) is meiosis II, which generates spermatids in males and the mature ova (plural of ovum) in females. The completion of meiosis II occurs just after fertilization in females. If the first polar body has not already degenerated, it also divides. However, all the polar bodies soon degenerate and only the mature fertilized ovum remains. During fertilization, the offspring acquires one of each pair of chromosomes from the ova and one of each pair of chromosomes from the sperm.

[1] For a description of the stages of **mitosis** see Box 3.2.

[2] The phases in meiosis are outlined in Box 3.2 which also explains how these result in genetic variability.

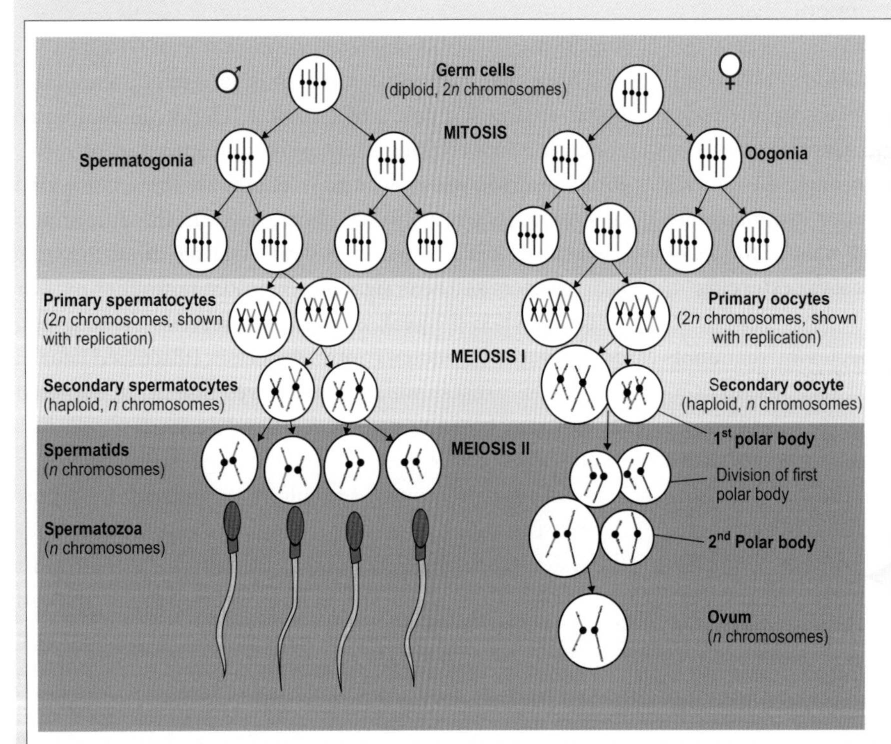

1. **Mitotic proliferation** produces large numbers of spermatogonia in males and oogonia in females. Two chromosome pairs are shown in each diploid (2*n*) cell; in each pair one is the maternal chromosome (red) and one the paternal chromosome (green).
2. **Meiosis I** forms haploid gametes (1*n* chromosomes). Note the random patches of red and green within the chromosomes in the secondary oocytes and spermatocytes due to exchange of chromosome material between each pair. In females a secondary oocyte and the first polar body are formed.
3. **Meiosis II** generates many spermatids in males that go on to form haploid spermatozoa. A second polar body is extruded when meiosis II is completed, just after fertilization. An ovum is haploid (1*n*) until fusion with genetic material from the spermatozoa occurs.

Figure A Steps in gametogenesis in males and females

Figure 20.6 shows the resulting phase-response curve of spawning by female rainbow trout (*Oncorhynchus mykiss*) exposed to 2 months of continuous light (perceived as long day lengths) during rearing in a seasonally changing day length. Long days occurring earlier than they should for the natural changes in photoperiod are perceived by the central nervous system as an indication that the internal clock is running slowly. The result is a corrective forward adjustment, i.e. an advance phase-shift causing earlier spawning. On the other hand, when the long day lengths occur later

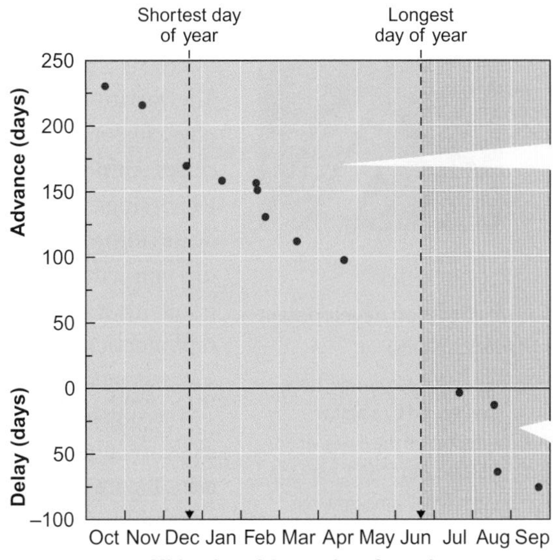

Figure 20.6 Phase-response relationship for spawning of female rainbow trout (*Oncorhynchus mykiss*)

The trout in these experiments came from a northern latitude domesticated stock and naturally spawned, under ambient day length, in November–January. Exposure to 2 months of continuous light (perceived as long day lengths) at different times of the year (data points are placed at mid-point of the 2-month period) results in an advance in spawning by up to 232 days when applied before the longest day in the year (green panel), or a delay in spawning of up to 80 days when applied after the longest day (pink panel). The further away from the period of natural long days that the two months of continuous light occurs, the greater the advancing or delaying of spawning time.

Source: adapted from Randall CF et al (1998). Photoperiod-induced phase-shifts of the endogenous clock controlling reproduction in the rainbow trout: a circannual phase-response curve. Journal of Reproduction and Fertility 113: 399-405.

than they would under a natural photoperiodic cycle, the endogenous clock is perceived to be running faster than usual; backward adjustments or a delay phase-shift result in later spawning than usual.

So far, we have looked at the effects of day length in isolation from other variables, but this oversimplifies natural environments. Photoperiod may act as the dominant cue for reproduction, but other factors, such as temperature, also have an impact, and in some cases dominate over the effects of photoperiod.

Temperature effects on seasonal breeders

Many animals (most invertebrates, fish, amphibians and reptiles) do not regulate their body temperature, except by behavioural means. These animals are ectotherms[13]; for such animals, the ambient temperature of the environment can have significant effects on physiological processes, including reproduction.

As an example, Figure 20.7 examines the frequency of egg laying in California sea hares (*Aplysia californica*) held at two different temperatures and in two photoperiods. Notice that the animals held in warmer water (20°C) lay eggs more frequently than animals held in colder water (15°C), regardless of the photoperiod. We can conclude that warm water is necessary for optimal egg laying by the sea slugs. However, temperature and photoperiod have interactive effects on egg laying. Even in the warmer sea water, more than double the frequency of egg laying occurs when there are fewer daylight hours than in the long day photoperiod.

In reptiles, temperature is a more important cue for gonad development than photoperiod. Notice in Figure 20.8 the pronounced effect of environmental temperature on the development of the ovaries of lizards (*Xantusia vigilis*).

[13] We discuss ectotherms in detail in Chapter 9.

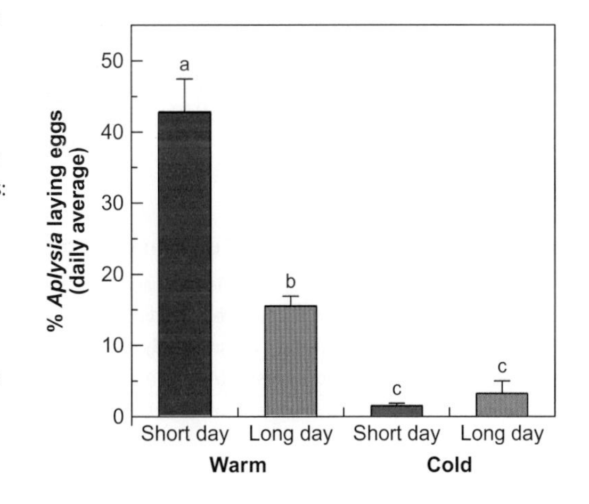

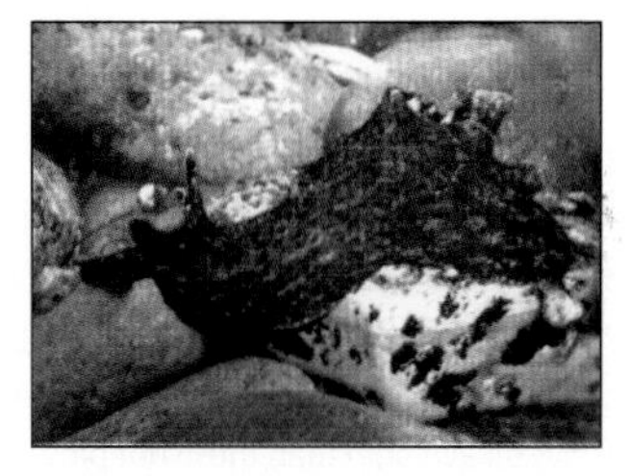

Figure 20.7 Percentage of California sea hares (*Aplysia californica*) laying eggs in warm seawater (20°C) and colder seawater (15°C) at two photoperiods

Sea slugs were captured in autumn (fall) and kept at two temperatures under two different photoperiods: short day (8 h light: 16 h dark) and long day (16 h light: 8 h dark).

Data are means ± standard errors for four groups in each environmental condition. Different letters indicate significantly different values ($P<0.05$). Source: Wayne NL, Block GD (1992). Effects of photoperiod and temperature on egg-laying behavior in a marine mollusk, *Aplysia californica*. Biological Bulletin 182: 8-14. Image: © Peter Bryant, University of California, Irvine.

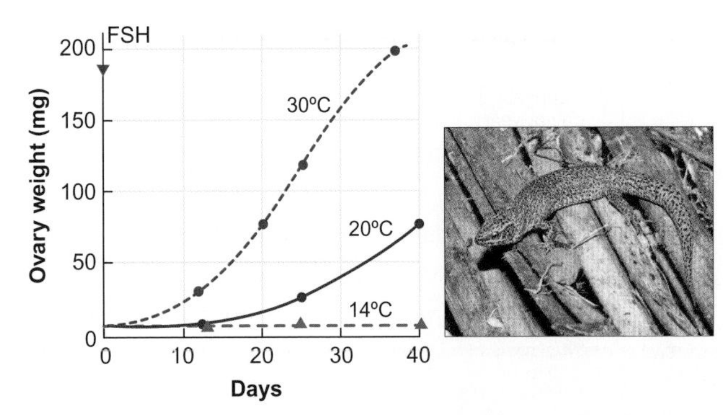

Figure 20.8 Environmental temperature effects on development of the ovaries in desert night lizards (*Xantusia vigilis*)

Hormonally stimulated development of the ovary, resulting from injections of follicle stimulating hormone (FSH) at time 0, is enhanced by a higher environmental temperature. At 14°C, FSH has little effect on the ovary. At an intermediate temperature (20°C) ovarian growth occurs, but at 30°C over the 40-day period of study the impact on ovarian growth is 3–4 times greater.

Data are means from six individuals per group. Source: Licht P (1972). Environmental physiology of reptilian breeding cycles: role of temperature. General and Comparative Endocrinology Supplement 3: 477-488.; image of desert night lizard: Bill Boulton/Flickr.

At 14°C, there is little stimulation of gonad development after injection of follicle stimulating hormone (FSH)[14], whereas ovarian growth occurs at 20°C, and at 30°C it is three to four times greater than at 20°C.

For ectotherms living in temperate zones, the spring rise in ambient temperature is likely to stimulate gonad maturation. Birds and mammals, however, are endotherms—they actively regulate their own body temperature[15]—so direct effects of environmental temperature on gonad development and reproduction are generally less important.

Nutritional effects on seasonal breeders

The variations in food intake or food quality that affect the physical condition of an animal can affect reproductive processes within the same season, or exert carry-over effects, in which single events or the accumulation of many small events occurring in one season influence an individual's performance in a non-lethal way in the following season. Carry-over effects occur most clearly in birds, in which a poor diet can delay reproduction, while a good diet can enhance reproductive success (measured as the number of offspring produced).

The supplementation of food for garden birds by humans is likely to enhance their reproductive success by carry-over effects. Figure 20.9A shows that supplementary feeding results in a slightly earlier laying of eggs by Eurasian blue tits. These birds are income breeders: they rely on exogenous resources for the supply of protein/lipid to the eggs. However, there is likely to be a threshold in the condition that a female bird needs to reach before it can start laying eggs. Overwinter feeding presumably enables the birds to reach this threshold slightly earlier.

There is no difference in clutch or brood sizes when comparing blue tits with access to extra food with unfed blue tits, and no apparent correlation between the date on which the tits lay eggs and the fledging success. However, the data in Figure 20.9B show a higher fledging success at fed sites compared to unfed sites. The improved fledging success is likely to reflect the improved condition of the parents, which enabled them to provide better care for their chicks.

Carry-over effects also occur in mammals in which they affect the timing of seasonal breeding. For example,

[14] Follicle stimulating hormone (FSH) is important in hormonal regulation of gonadal development, as we discuss in Section 20.4.1.

[15] Chapter 10 discusses endothermy in detail.

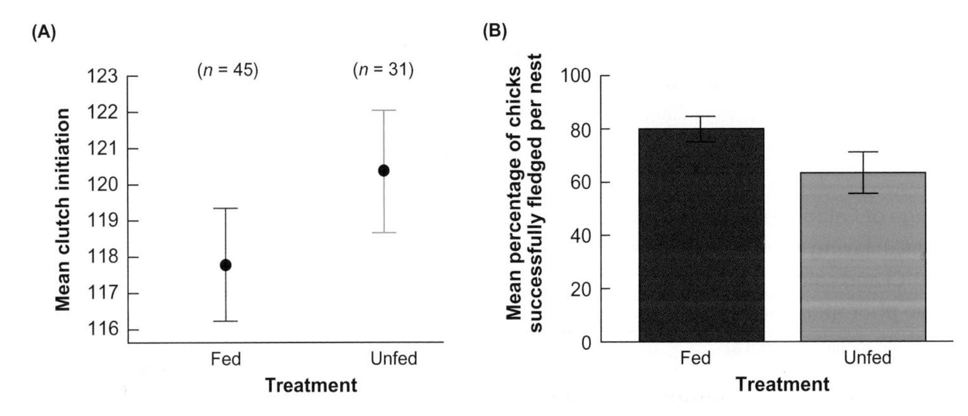

Figure 20.9 Carry-over effects of supplemental winter feeding of Eurasian blue tits (*Cyanistes caeruleus*)

The effect of supplemental winter feeding on (A) the start of laying eggs (days after 1st Jan), and (B) percentage of chicks successfully fledged is examined by comparison of data for the 'fed' group of blue tits to data for tits not given supplemental feeding ('unfed'). (A) Fed birds started laying significantly earlier (2.5 days on average) than those not provided with extra food. (B) Fed birds fledged a significantly higher percentage of chicks than unfed birds.

Data are means ± 95% confidence intervals (sample sizes in brackets above bars) in (A); mean ± standard errors in (B).

Source: Robb GN et al (2008). Winter feeding of birds increases productivity in the subsequent breeding season. Biology Letters 4: 220-223.

in North American elk (*Cervus canadensis*) a threshold of 8–10 per cent body fat appears to be necessary by the end of the non-breeding season for there to be a high probability of pregnancy. Individuals with poor nutrition also become pregnant later than those that have better quality nutrition, suggesting it takes longer to reach the threshold for reproduction when feeding on a lower quality diet.

We have explored the timing of sexual reproduction and different reproductive modes. To understand the underlying processes, we next explore how an animals' sex is determined and what regulates the development of the gonads and gametes.

› *Review articles*

Bromage N, Porter M, Randall C (2001). The environmental regulation of maturation in farmed finfish with special reference to the role of photoperiod and melatonin. Aquaculture 197: 63–98.
Costa DP, Sinervo B (2004). Field physiology: physiological insights from animals in nature. Annual Review of Physiology 66: 209–238.
Drent RH, Daan S (1980). The prudent parent: energetic adjustments in avian breeding. Ardea 68: 225–252.
Harrison XA, Blount JD, Inger R, Norris DR, Bearhop S (2011). Carry-over effects as drivers of fitness differences in animals. Journal of Animal Ecology 80: 4–18.
Meijer T, Drent R (1999). Re-examination of the capital and income dichotomy in breeding birds. Ibis 141: 399–414.
Twan W-H, Hwang J-S, Lee Y-H, Wu H-F, Tung Y-H, Chang C-F (2006). Hormones and reproduction in scleractinian corals. Comparative Biochemistry and Physiology Part A 144: 247–253.
Yoshimura T (2006). Molecular mechanisms of the photoperiodic response of gonads in birds and mammals. Comparative Biochemistry and Physiology Part A 144: 345–350.

20.2 Sex determination and sexual differentiation

Sexual differentiation involves a cascade of events that result in development of the gonads (testes in males and ovaries in females) and gender-specific features of the reproductive tract, and external genitalia. These are the primary sex characteristics of a species: their *primary* function is gamete production and fertilization. The primary sex characteristics are distinct from the secondary sex characteristics that are not directly involved in gamete formation or delivery, but are nevertheless recognizable features of males and females and may be important in functional aspects of reproduction and behaviour.

Sex determination involves **sex chromosomes** in many species, although the processes are often still unclear, and it is widely believed that environmental sex determination was the ancestral condition. The clearest examples of existing environmental sex determination are the temperature-dependent mechanisms of reptiles, so we examine these first.

20.2.1 Temperature-dependent sex determination

Madeleine Charnier working in the Faculté des Sciences in Dakar in the 1960s discovered that that the incubation temperature of the eggs of an African lizard (common agamas, *Agama agama*) affects the sex of their offspring. Her findings led to a multitude of investigations of temperature effects on sex determination in reptiles. These studies have shown that the temperature at which eggs are incubated determines the sex in many species of reptile, including many turtles (all sea turtles), all crocodiles, the tuatara (*Sphenodon punctatus*) and several lizard species. These species have no known sex chromosomes.

Studies have identified a thermosensitive period of development in many reptilian species during which the incubation temperature for the eggs can alter gonadal development. For turtles, this transition can occur up to about half way through development of the embryo and sometimes up to the last third of development. For crocodiles and the tuatara, the period is shorter, at about a third of the incubation period.

The change in gonadal development occurs over a surprisingly narrow temperature range of just 1 to 2°C, as shown by the examples in Figure 20.10 where we see two distinct patterns:

- Pattern I animals have a single temperature transition zone, with one sex dominating at one side of the zone, and the other sex dominating at the other side. Species with a single transition zone are of two sub-types: sub-groups IA and IB; in sub-group IA lower temperatures result in *male* dominance while in sub-group IB below the transition zone there is *female* dominance. Figure 20.10C shows the temperature effects on sex determination of red-eared sliders/terrapins (*Trachemys scripta elegans*), which show Pattern IA: cool nests produce more males while warm nests produce more females; there is a mixture of sexes at intermediate temperatures. This pattern also occurs in some lizards.
- Pattern II animals have two temperature transition zones, with males dominating at intermediate temperatures and females dominating at *both* extremes, as shown by American alligators (*Alligator mississippiensis*) and alligator snapping turtles (*Macrochelys temminckii*) in Figures 20.10A and 20.10B respectively.

For the most part, studies of temperature-dependent sex determination have been in the controlled environment of a laboratory. In the wild, however, temperature-dependent effects are likely to result in different outcomes depending on the location of a nest of eggs, which has implications for

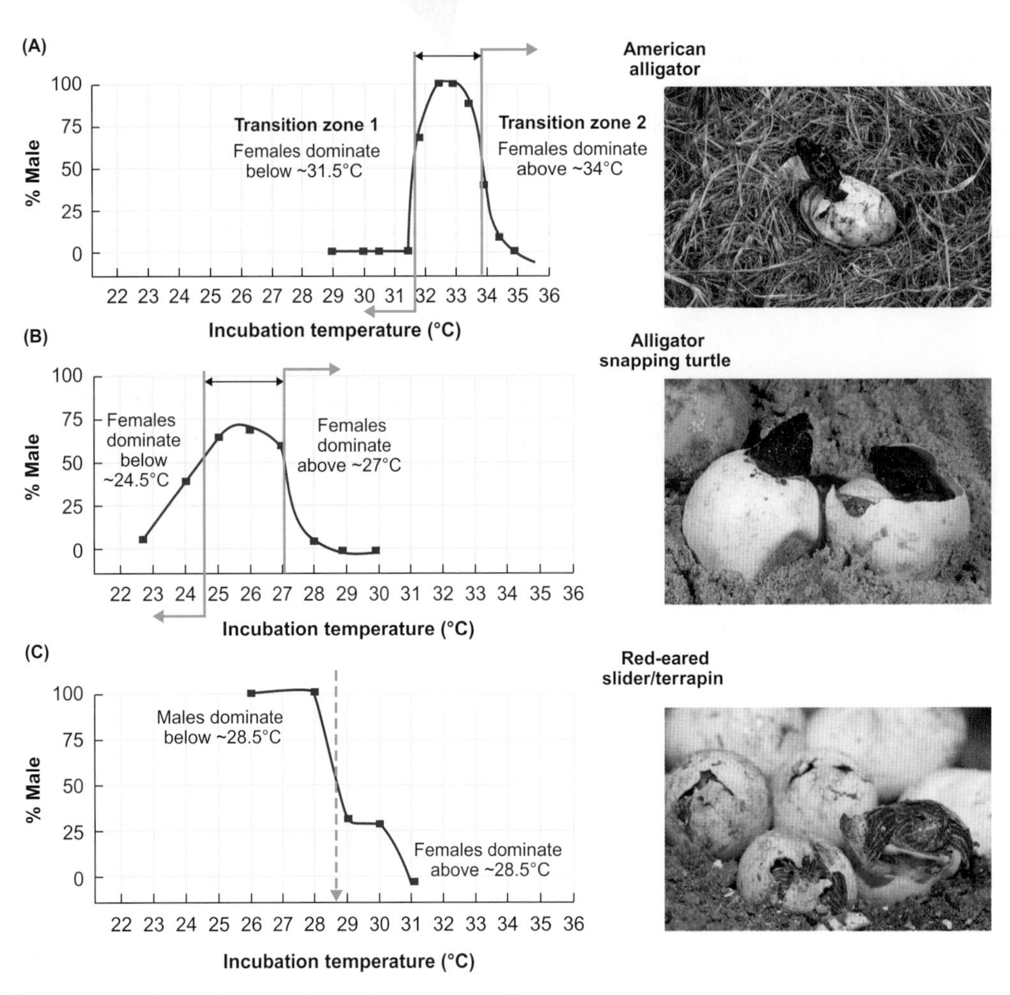

Figure 20.10 Temperature-dependent sex determination in reptiles

The three graphs show the effects of the incubation temperature of the eggs on the percentage of male offspring for (A) American alligators (*Alligator mississippiensis*), (B) alligator snapping turtles (*Macrochelys temminckii*) and (C) red-eared slider/terrapin (*Trachemys scripta elegans*). Photographs show each of the species hatching from their eggs. Transition zones, when the change in temperature has a significant effect on the sex ratio (above or below 50% males) are indicated by blue lines and arrows. Notice that American alligators (A) and alligator snapping turtles (B) show two transition zones (Pattern II)—females dominate at both low and high incubation temperatures. Temperatures at which there are more than 50% males are indicated by the temperature zone between the blue lines and arrows. The red-eared slider (C) has a single transition zone (Pattern I) – in this case pattern IA as males dominate at lower incubation temperatures.

Source: adapted from Crain DS, Guillette LJ (1998). Reptiles as models of contamination-induced endocrine disruption. Animal Reproduction Science 53: 77-86. Images: (A): Robert H. Potts/Science Source; (B): © Leszczynski, Zigmund / Animals Animals – All rights reserved; (C): © Kerstin Hinze/ Naturepl.com.

conservation. For instance, the movement of a nest of turtle eggs to a safer location, perhaps away from predators, will potentially affect the sex ratio of the offspring. A further effect is that thermal fluctuations occur in the nest.

So how does temperature determine the sex of reptilian embryos? Although environmental sex determination is not genetically driven by sex chromosomes, it *is* due to the expression of genes that determine the production of hormones. The differentiation of an ovary in reptiles will not occur unless **oestrogens** are present. Oestrogens are a group of steroid hormones[16] named because of their action on the oestrous cycle that we explore in Section 20.4. Figure 20.11 shows the structure of **oestradiol**, which is the principal natural oestrogen of vertebrates. The importance of oestrogens in initiating development of the ovaries becomes clear after exogenous administration. Ovaries develop after

[16] We discuss the common structure of functionally distinct groups of steroid hormones in Box 19.1 in Section 19.1.1; their formation during the processes of steroidogenesis is illustrated in Figure 19.3.

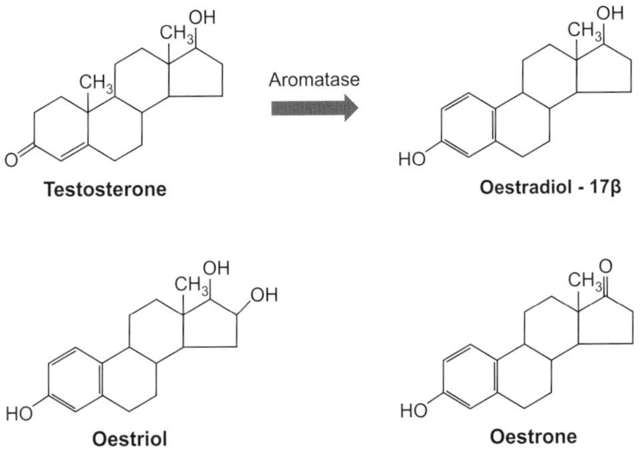

Testosterone — Aromatase → Oestradiol - 17β

Oestriol   Oestrone

Figure 20.11 Structure of female and male sex hormones

Notice the similarities in the chemical structure of these hormones, which are all derivatives of cholesterol. Testosterone is a masculinizing hormone (an androgen). Testosterone is converted by the enzyme aromatase to oestradiol-17β, which is the principal feminizing hormone (an oestrogen) and is generally referred to simply as oestradiol. Oestriol is abundant during pregnancy as a result of placental synthesis. Oestrone is the dominant female sex hormone of echinoderms and molluscs, but less common among vertebrates, including humans (except when postmenopausal).

injection of embryos with oestrogens, even in embryos incubated at male-producing temperatures. Environmental oestrogens (chemicals that mimic oestrogens and interact with oestrogen receptors) may also influence sex determination; we discuss such actions on wildlife in Experimental Panel 20.1. By contrast, testes develop in the absence of oestrogens.

The enzyme **aromatase** converts testosterone, which is the main **androgen** (male sex hormone), and whose structure is shown in Figure 20.11, to oestradiol in most vertebrates[17]. This conversion determines the relative amounts of testosterone and oestradiol present. The data in Figure 20.12 show how the increase in aromatase activity in female turtle embryos starts during the temperature sensitive period for sex determination. An increase in aromatase will convert testosterone to oestradiol and initiate the development of an ovary. Aromatase activity, in the turtles, peaks at the time of female sex differentiation. This raises the question: how are males (with testes) produced? What regulates the levels of aromatase?

Experiments on olive ridley sea turtles (*Lepidochelys olivacea*) suggest *SOX9*, a sex-determining gene that occurs in all vertebrates, plays an essential role in sex determination of turtles. These experiments involved separating the gonads and incubating them at male- and female-producing temperatures alongside the investigation of the development of gonadal tissue and the expression of the *SOX9* gene in the gonadal tissue. Figure 20.13 summarizes how the cascade of events could work. At female-producing temperatures, conversion of testosterone to oestradiol by aromatase inhibits the expression of the *SOX9* gene. In contrast, at male-producing temperatures, the increased expression of the *SOX9* gene in gonadal tissue of the embryos inhibits aromatase, which increases testosterone concentrations such that male gonads can develop.

Although the temperature-dependent system for sex determination is believed to have been the ancestral condition in reptiles, some species have a pair of sex chromosomes that can result in genetic sex determination. Laboratory studies

[17] In fish, the main active androgen is 11-keto testosterone. Various oestrogens and androgens have also been identified in molluscs (bivalves, gastropods and cephalopods) and in arthropods.

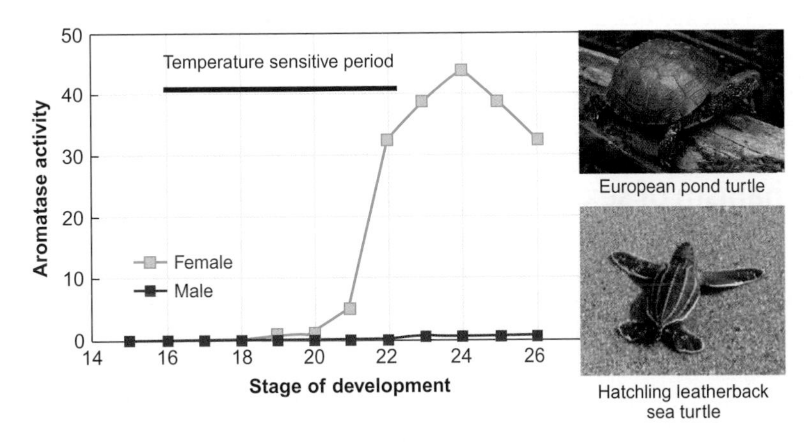

Figure 20.12 Aromatase activity in the gonads of male and female turtles

Aromatase activity data (in arbitrary units) are combined for two turtle species: European pond turtles (*Emys orbicularis*) and leatherback sea turtles (*Dermochelys coriacea*). Aromatase activity is first detectable (in female embryos only) during the period of gonadal differentiation (stage 19-20). Aromatase activity in the gonads peaks in female embryos at the time of sex differentiation (stage 24) before hatching (stage 26).

Source: Lance VA (1997). Sex determination in reptiles: an update. American Zoologist 37: 504-513. Images: European pond turtle: © Alberto Nardi; Leatherback sea turtle: © Mathew Godfrey.

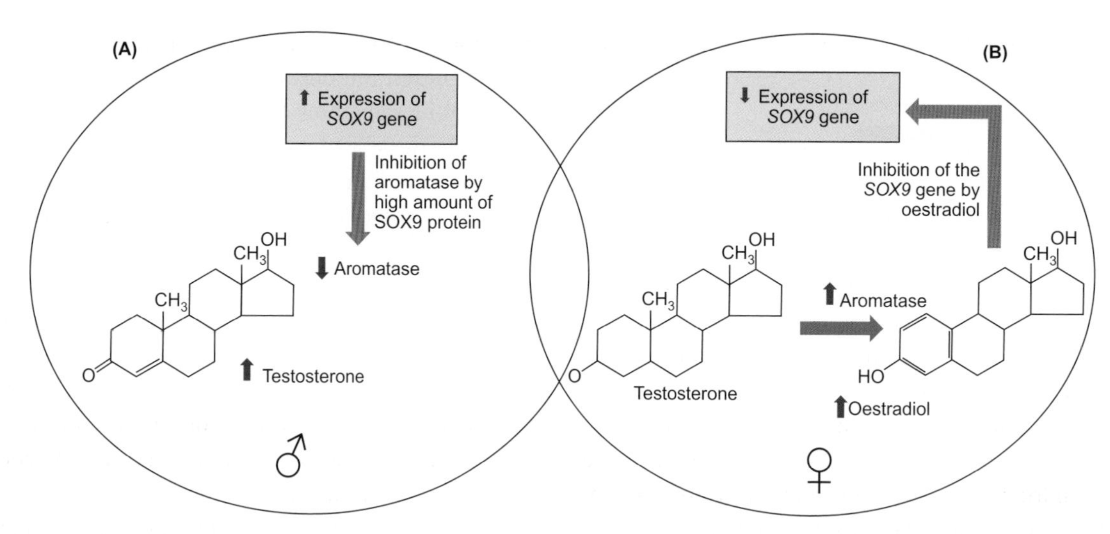

Figure 20.13 Model of the cascade of events for regulating temperature-dependent sex determination in turtle gonadal tissue by the SOX9 protein and aromatase

↑ Indicates an increase; ↓ indicates a decrease. The SOX9 protein inhibits aromatase activity. Aromatase converts testosterone to oestradiol. (A) At male-producing temperatures, an increase in the expression of the *SOX9* gene inhibits aromatase activity so that conversion of testosterone to oestradiol is inhibited and male gonads differentiate in the embryos. (B) At female-producing temperatures, an increase in aromatase activity increases oestradiol formation and ovarian tissue develops in the embryos. Oestradiol inhibition of the expression of the *SOX9* gene acts in a positive feedback way on the development of ovarian tissue rather than testes.

of some lizard species that have sex chromosomes (central bearded dragons (*Pogona vitticeps*) and eastern three-lined skinks (*Bassiana duperreyi*), both found in Australia) show that incubation temperature can override their sex chromosomes in determining gonadal development (the gonadal sex) and determine the phenotypic sex (based on all external morphology, physiology and behaviour). Cool incubation conditions (16°C) that mimic the lower temperatures at high altitude produce offspring with a male phenotype, despite a female genotype.

Sex chromosomes involved in genetic sex determination systems occur in some fish, some reptiles, and all birds and mammals. In the next section, we explore the most common of these.

20.2.2 Genetic mechanism of sex determination

The most common system of sex chromosomes among animals is the XY/XX sex chromosome system. This system may be familiar to you as it occurs in humans, but it is also very common among many other species, including most mammals, some amphibians (frogs), some fish and some invertebrates, such as fruit flies (*Drosophila* spp).

In the XY/XX system, genetic males have both X and Y sex chromosomes and so are the heterogametic sex, while genetic females have two X chromosomes and are the homogametic sex. During formation of gametes (gametogenesis), as outlined in Box 20.1, the chromosome pairs separate so each gamete cell has only one of the sex chromosomes. Since females have two X chromosomes, all the ova have the X chromosome. Males are XY, so half of the sperm have an X chromosome and half a Y chromosome. Hence, the fertilizing sperm determines the genetic sex of the embryo: fertilization of an ovum produces a zygote that is either XX (genetic female) or XY (genetic male).

Some placental mammals (including humans) and marsupial species with the XY/XX sex chromosome system have a gene in the Y chromosome that is responsible for maleness. The expression of this **s**ex-determining **r**egion of the **Y** chromosome (abbreviated to *SRY* gene), results in development of the male gonad. The *SRY* gene evolved only in the mammalian line, although related genes occur in other animal groups, such as birds, which have a different sex chromosome system in which the females are the heterogametic sex.

In birds, some fish, some reptiles (all snakes and some species of turtles and lizards) and some amphibians (newts and salamanders), there is a ZW sex chromosome system (named to avoid confusion with the XY system). In the ZW system, males are homogametic (ZZ), while females are heterogametic (ZW). As a result, the sex chromosome carried by the ovum (Z or W) determine the genetic sex of the offspring.

Insects are extremely diverse, so it is perhaps not surprising that they have particularly diverse mechanisms of sex determination. The majority of insects have sex chromosomes, like other animal groups, but with a wide variety of sex chromosome systems. Butterflies and moths (lepidopterans) have the ZW system outlined above. Some grasshoppers have only a single sex chromosome and use an X0/XX sex determination system in which females are homogametic (XX) and males, with only one sex chromosome, are X0; such males are still considered heterogametic because they produce two different kinds of gametes—one with a sex chromosome, and one without.

The genetic sex of offspring obviously determines the genetically driven development of the gonads. However, other factors are important in the differentiation of the gonads, so phenotypic sex may differ from the genetic sex. Sexual differentiation involves morphological and physiological development of gonadal tissues, and differentiation of the neural and endocrine tissues that can result in the different hormonal and behavioural patterns in males and females.

20.2.3 Gonadal sex differentiation

Once fertilization occurs in animals that rely on genetic sex determination (and carry sex chromosomes) the genetic machinery is immediately in place to enable differentiation of the gonads and reproductive tract, but embryos retain the potential for their gonads to develop into either testes or ovaries, whatever their genetic sex. Vertebrate embryos have two systems of ducts: Wolffian ducts that can develop into the male reproductive tract and Müllerian ducts that can develop into the female reproductive tract[18].

In placental mammals, and marsupials, a delay occurs before the sex-regulating region of the Y chromosome (*SRY* gene) triggers the events that begin to masculinize the gonads. In humans, there is a delay of about 6 weeks of gestation before differentiation of the testes begins. A female foetus has no Y chromosome (so no *SRY* gene). Consequently, the gonads start to differentiate into ovaries. In humans, this occurs at 9 weeks of gestation.

Like mammals, birds have an undifferentiated gonad for several days of embryonic development (4 days in the chicken egg), after which time an ovary or a testis starts to develop, but the trigger for gonad differentiation in birds is uncertain. Oestrogens are required earlier in sexual differentiation in birds than mammals, and may be necessary for the formation of the ovary, as among reptiles.

In mammals, once the *SRY* gene product (the testis-determining factor) of genetic males stimulates the undifferentiated gonads to develop into testes, the testes begin secreting testosterone. In all vertebrates, testosterone or its active metabolites stimulate the continued development of the testes. The developing testes of mammals and birds secrete the peptide hormone anti-Müllerian hormone (AMH), which causes degeneration of the Müllerian duct that would otherwise develop into the female type of reproductive tract.

The release of testosterone (the main androgen in mammals) from the developing testes results in differentiation of the characteristic features of the male reproductive tract and any external genitalia characteristic for the species. In humans, the testes start to develop in the first 3 months of pregnancy; by week 12–14 of gestation, the foetus has external genitalia.

In the absence of androgens and AMH, the gonadal tissue will develop into an ovary. Oestrogens trigger development of the Müllerian duct to form an oviduct and specialized regions such as the uterus.

The genetic sex and phenotypic sex of animals are usually identical, but differences can arise—for example, if the animal lacks **androgen receptors** so that the masculinizing effects of testosterone cannot occur in genetic males, or if there are high concentrations of sex steroids during early development that change the phenotypic sex of the animal. The mammalian adrenal cortex[19] secretes a weak androgen (dehydroepiandrosterone, DHEA) but in amounts that are usually insufficient to cause masculinization of a genetic female. However, excessive secretion of DHEA during early development *can* result in masculinization.

The opposite change of feminization can occur in genetically male animal—for example, in chicken embryos in which the egg yolk is injected with a synthetic oestrogen such as ethinyl oestradiol (used in the contraceptive pill). Laboratory studies suggest that oestrogenic contaminants can feminise male bird embryos in which gonad development is incomplete. Similarly, exposure to environmental oestrogens can potentially feminize animals that are genetic males[20].

20.2.4 Some animals produce both male and female gametes

So far, we have considered the sex of animals as male or female, but many invertebrate species and some vertebrates are **hermaphrodite**—they produce both male and female gametes. Two forms of hermaphroditism exist:

- simultaneous hermaphroditism, in which an ovotestis produces male and female gametes simultaneously;
- sequential hermaphroditism, in which the gonads initially form one type of gamete, but then change to the other type.

[18] Wolffian ducts and Müllerian ducts are the names given to the ducts because of their description by the early German physiologists Caspar Friedrich Wolff and Johannes Peter Müller.

[19] Section 19.3.1 discusses the steroid hormone secretions of the adrenal cortex.

[20] Experimental Panel 20.1 examines some studies of environmental oestrogens.

Simultaneous hermaphrodites usually avoid self-fertilization

Simultaneous hermaphrodites could, in theory, self-fertilize, but usually avoid doing so unless mates are in short supply. Instead, they usually exchange gametes, often at the culmination of an elaborate and prolonged mating process. Earthworm species (oligochaete worms) are good examples of hermaphrodites that cross-fertilize after copulation that involves two worms lying in opposite directions, as shown in Figure 20.14.

Earthworms have paired gonads in their anterior segments, in front of the clitellum. Testes occur on the septum dividing the 9th and 10th segments and the septum between segments 10 and 11 of the earthworm. Maturation of the sperm is completed in the three pairs of seminal vesicles in segments 9, 11 and 12. The ovaries occur in segment 13 and the eggs mature in the ovisac on the septum between segments 13 and 14.

During copulation, sperm enter the testes sac from where they pass along a ciliated sperm duct, leaving the earthworm via the male gonopore on its ventral surface. These sperm enter the other worm via the opening to two pairs of spermatotheca, where the sperm are stored. The fertilization of one worm's eggs with the other worm's sperm takes place later.

Notice in Figure 20.14 that each worm has a prominent clitellum of several segments forming a girdle of almost continuous epidermis. The clitellum has three layers of cells: (i) mucous cells at the surface, (ii) a layer of cocoon-secreting glandular cells and (iii) a deep layer of albumen-secreting cells. When laying eggs for fertilization, the secretory layers of the clitellum secrete a sheath. As the earthworm backs out of the sheath, it deposits some of its eggs together with some sperm from the other individual into the sheath. It eventually slips away from the sheath, which closes to form a cocoon that encloses and protects the fertilized eggs during their development.

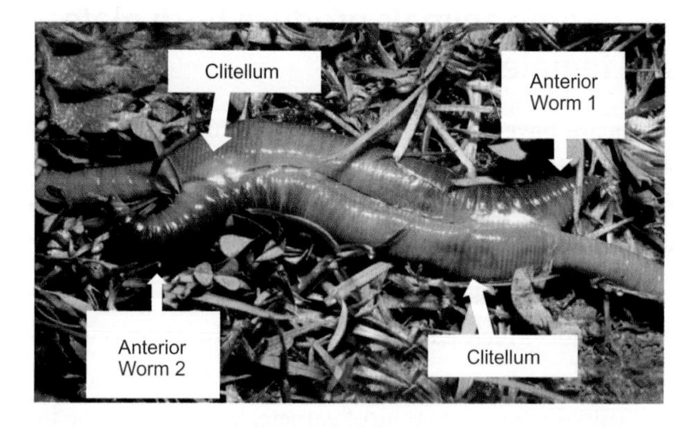

Figure 20.14 Mating in common earthworms (*Lumbricus terrestris*)

Mating of two earthworms which are lying in opposite directions involves indirect sperm transfer. Note white material at the point of direct contact between the two earthworms.

Image: © Nature Picture Library

Changing phenotypic sex of sequential hermaphrodites

There are two types of sequential hermaphroditism, in which male and female gametes are produced one after the other. **Protandry** refers to male-first sequential hermaphroditism, while the female-first version is **protogyny**.

The common slipper shell (*Crepidula fornicata*) is a striking example of a protandrous mollusc. Slipper shells are common in the intertidal and sub-tidal regions along the east coast of North America and are an invasive species in Europe, where they are known as common slipper limpets. The slipper shells (slipper limpets) have an unusual lifestyle, piling on top of one another in a chain, as shown in Figure 20.15A.

Slipper shells (slipper limpets) do not move around once the larvae settle on a hard substrate. Their close proximity in a chain is ideal for reproduction by internal fertilization. Insemination requires the male to insert a short penis (shown in Figure 20.15B) through the female's gonopore, so mating individuals must be near to one another to employ internal fertilization. Genetic studies confirm that >80 per cent of the fertilized eggs in the brooding female (usually the largest female at the bottom of the pile) have been fertilized by the male above her. The female has a seminal receptacle in which spermatozoa can be stored for the rest of her life and retains the fertilized eggs in a uterus for their initial development.

Larvae released into the plankton later settle on hard substrates. If a larva settles on a flat surface, it develops a flat base fixed to the surface. Young slipper shells are always male. Left on its own, without the influence of a female, an isolated male soon becomes a female, although it cannot reproduce until another individual settles on top of it. If a larva settles on another slipper shell, it grows round the convex surface of the shell beneath, and so a stack begins to form. Ultimately, chains of up to 12 individuals can occur.

Figure 20.15C shows an orderly progression of the phenotypic sex of the individuals within a chain. The oldest animals at the bottom of the chain are female, but would have originally been males like the younger (smaller) animals at the top of the chain, which have settled most recently.

Protandry involves gradual loss of the penis, and regression of the male gonad. After a period of transition (in an intersex state), a female gonad and the associated reproductive tract develops. Stem cells in the gonad give rise to spermatogonia when the gonad is male, or oogonia when the slipper shell becomes female. Usually, only the bottom-most males

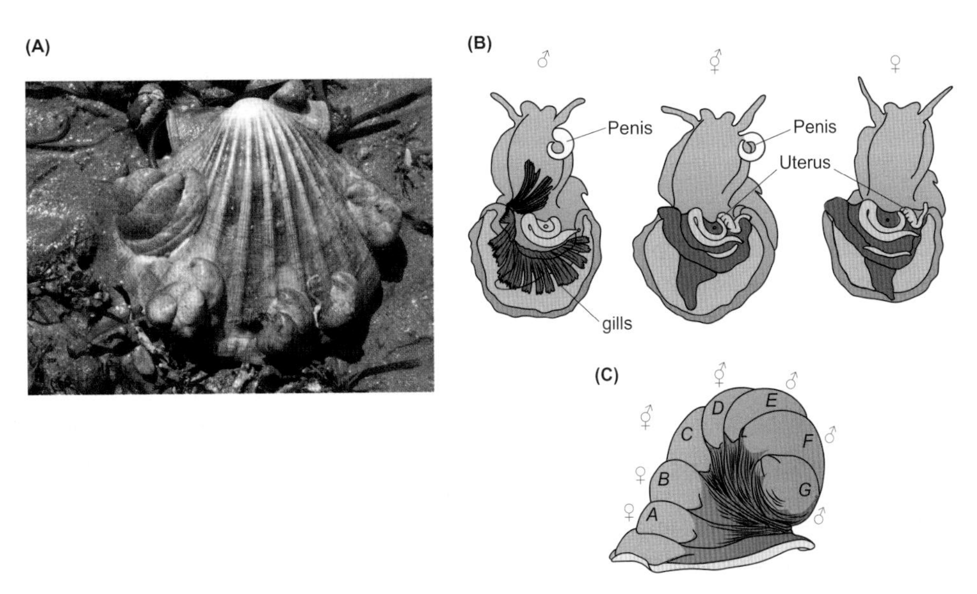

Figure 20.15 Protandry (male first) in common slipper shells (slipper impets) *Crepidula fornicata*

(A) Chains of slipper limpets on a scallop shell in the UK; the chains vary from two to four individuals. Note the changing size from the oldest and largest individual at the base of the chain to smaller ones above. (B) Diagrams of a male (♂) slipper limpet , an intersex individual (in process of becoming a female), and a female (♀) slipper limpet, seen from below, to show the penis in the male and intersex limpets and the uterus of the female and intersex limpets. Gills are shown only in the male. (C) Phenotypic sex of a chain of slipper shells/limpets (the series of letters A–G indicates the order of settling on one another). Those at the top of the pile (E, F, G) are younger males; beneath the males there are two intersex individuals (C & D) that are changing from male to female. Once these become female they will be fertilized by the males above them. The oldest animals (A & B) are female, although initially they were male. The bottom female (A) started the chain when the larva settled onto an oyster shell (at the bottom of the drawing).

Sources: (A): © Dominic Flint, Plymouth, UK. (B) & (C): drawings from one of the earliest studies, Orton JH (1909). On the occurrence of protandric hermaphroditism in the mollusc *Crepidula fornicata*. Proceedings of the Royal Society of London Series B, Biological Sciences 81: 468-484.

change sex, which results in a number of males on top of one to several females. So how are such changes in the gonad regulated?

Experimental studies suggest that during the male phase the cerebral ganglia of slipper shells produce a masculinizing factor, which is released into the haemolymph (blood), and stimulates development of a male gonad and penis, while secretion of a feminizing factor occurs in the female phase. Male gonads cultured *in vitro* continue the process of **spermatogenesis**[21] (sperm production) as long as they are cultured along with extracts of cerebral ganglia from an animal in the male phase. In contrast, resting gonads (after regression of the male gonad) become ovaries and start **oogenesis** (generation of ova) when cultured with haemolymph taken from slipper shells that are in the female phase.

Some of the best examples of protogyny—female before male—occur among wrasse species of fish, such as the California sheephead (*Semicossyphus pulcher*), a species currently listed as 'vulnerable' by the International Union for the Conservation of Nature and Natural Resources (IUCN). All California sheephead start as females but change to males when they are large enough to hold and defend a spawning territory, in which females congregate for courtship prior to spawning.

[21] Box 20.1 gives information on the stages in spermatogenesis and oogenesis.

This pattern of sex change has important implications for the conservation of sheepheads, which are exploited by sport fishing that generally removes larger (likely male) fish. Investigations of the changing demographics of sheephead populations in southern California have shown that removal of large male sheepheads results in earlier sex change by the females, before they reach their maximum reproductive capacity, which reduces egg release. Some overexploited populations that consist of smaller animals change sex at a body length of 23 cm, while unexploited populations of sheepheads change sex at 48 cm. One way to protect sheephead populations is to increase the size limits for fishing, so as to retain larger males and allow females to reach their peak reproductive output.

In the next sections, we look in more detail at the reproductive systems of males and females and their hormonal regulation. Our focus here is on vertebrates, where much of the research has been concentrated.

› *Review articles*

Brungström B, Axelsson J, Halldin K (2003). Effects of endocrine modulators on sex differentiation in birds. Ecotoxicology 12: 287–295.

Lance VA (1997). Sex determination in reptiles: an update. American Zoologist 37: 504–513.

Nakamura M (2010). The mechanisms of sex determination in vertebrates—are sex steroids the key factor? Journal of Experimental Zoology Part A. Ecological Genetics and Physiology 313A: 381–398.

Wilhelm D, Palmer S, Koopman P (2007). Sex determination and gonadal development in mammals. Physiological Reviews 87: 1–28.

20.3 Vertebrate male reproductive systems

The primary reproductive organs of male vertebrates are paired testes, which are responsible for production of spermatozoa (sperm) that are expelled via the reproductive tract. The reproductive tract encompasses the ducts through which the sperm must pass, and accessory sex glands that add secretions into the genital tract.

20.3.1 Internal or external testes

Fish, reptiles and birds have internal testes, making this the ancestral condition for mammals. However, most present-day land mammals have testes suspended in the scrotum, outside the abdominal cavity, at least during periods of spermatogenesis. In seasonal breeders, the testes descend into the scrotum during the reproductive period, when spermatogenesis occurs.

Some mammals such as monotremes, armadillos, sloths, elephants, rhinoceroses and marine mammals, such as whales and dolphins, still have internal testes. So why did external testes evolve? One idea is that the evolution of a higher core temperature necessitated the externalization of the testes to allow continued enzyme function and spermatogenesis. In the scrotum, tissue temperature is several degrees below that in the abdomen. Those species with internal testes tend to have a lower core temperature. However, this does not seem to be a satisfactory explanation on its own, because birds have internal testes but have a higher core temperature than mammals[22].

Textbooks often suggest that birds can use evaporative cooling in their air sacs[23] to cool the nearby internal testes. However, temperature measurements indicate no difference between the temperature of the lower air sacs (close to the testes) and core body temperature. We can conclude that cooling of the testes of birds by association with an air sac is unlikely. Instead, it seems that the testes of birds can function at their core body temperature: long-term simultaneous measurements of body temperature and the temperature of the testes of chickens shows that the testes operate effectively and spermatogenesis continues at a temperature equivalent to their core body temperature (40–41°C).

The testes of sea mammals such as dolphins, whales and seals are internal, but are at a lower temperature than core body temperature because of the operation of a countercurrent arrangement of arteries and veins. Heat exchange between the artery supplying the testes and cooled blood from the body surface enables cetaceans (whales, dolphins and porpoises) to regulate the temperature of blood flowing to the testes. Figure 20.16 illustrates the countercurrent arrangement of cooled blood flowing from the dorsal fin and flukes and warmed arterial blood running in the opposite direction. Measurements of temperatures close to the testes show a cooling effect, which is illustrated by the data for a prepubescent male in Figure 20.16D. The cooling effect reaches up to 1.3°C in sexually mature males, which reduces the temperature of the testes to the optimum for spermatogenesis.

20.3.2 Testes structure

The internal structure of the testes of vertebrate species reveals similarities that reflect the common features of spermatogenesis that occur in all male vertebrates. Looking at Box 20.1, we see that spermatogonia, spermatocytes, spermatids and spermatozoa (sperm) are the key stages in spermatogenesis.

In the elongated testes of elasmobranch fish (sharks, skates and rays), the testes are composed of spherical spermatocysts. Figure 20.17 shows the ring of **Sertoli cells** that line each spermatocyst and support the development of the spermatogonia (the germ cells). Sertoli cells occur in all vertebrate testes.

The spermatocysts of elasmobranch testes develop in synchrony across the testes until the mature spermatocysts rupture at the edge of the testes and release spermatozoa into the ducts leaving the testes. The ducts coalesce to form the main duct (the vas deferens). Each vas deferens enters a seminal vesicle, where sperm are stored before leaving via the cloaca.

In reptiles, birds and mammals the testes have evolved into a more compact rounded pair of organs packed with the seminiferous tubules that produce the sperm. In species that breed throughout the year, the testes remain similar in size and continually produce sperm. For example, testes of male humans produce around 50 to 230 million sperm each day (depending on the age of an individual after maturation at puberty), so fertilization can take place at any time after sexual maturation. However, in seasonal breeders the testes develop annually, regress, and redevelop the following year. Looking back at Figure 20.5, note the increase in the testicular weight of the Japanese quail, stimulated by long day lengths; this increase in weight reflects the annual cycle of increasing and then decreasing diameter and length of the seminiferous tubules.

The basic structure of avian and mammalian testes is similar. Figure 20.18A shows a transverse section of the seminiferous tubules in the bird testis, with cells at various stages of development. However, spermatogenesis in birds is about four times faster than in mammals, probably because of the poorer survival of spermatozoa in the sperm ducts of birds and possibly also due to the mating pattern of birds, with numerous copulations each day during the period of egg laying.

Looking more closely into the mammalian testis, Figure 20.18B shows how the spermatogonia pass through

[22] Section 10.2 examines body temperature regulation in endothermic birds and mammals.

[23] Figure 12.32 shows the arrangement of the air sacs of birds.

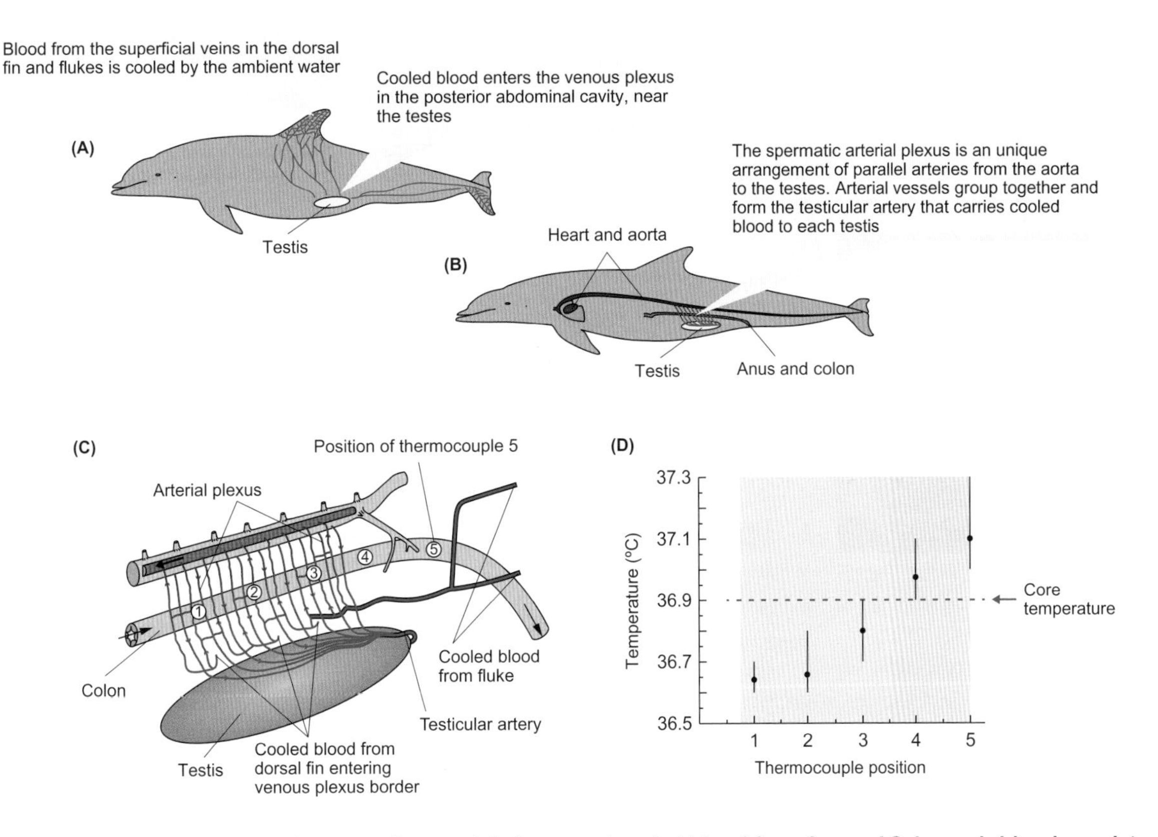

Figure 20.16 Countercurrent system between the arterial plexus and cooled blood from fins and fluke cools blood supplying the testes of the common bottlenose dolphin (*Tursiops truncatus*)

(A) and (B) show overall vascular arrangement for cooling the testes in dolphins. (C) shows one side of the countercurrent heat exchanger. Colouring of the vessels indicates relative temperatures: red warm grading through purple to blue for cooled blood. The arrowheads indicate the direction of blood flow in arterial blood and the opposite direction of blood flow in the venous plexus. Temperatures (means and ranges) monitored using thermocouples at positions 1 to 5 (marked in C) along the colon of a prepubescent male are shown in the graph in (D). The dotted line on the graph shows core body temperature of the dolphins. At positions 1 to 3 (blue area on graph) in the location of the testes temperature is lowered by the cooling effect of the countercurrent heat exchanger, while more posterior locations (4 & 5) are outside the heat exchanger for this dolphin.

Source: adapted from Rommel SA et al (1994). Temperature regulation of the testes of the bottlenose dolphin (*Tursiops truncatus*): evidence from colonic temperatures. Journal of Comparative Physiology B 164: 130-134. Pabst DA et al (1995). Thermoregulation of the intra-abdominal testes of the bottlenose dolphin (*Tursiops truncatus*) during exercise. Journal of Experimental Biology 198: 221-226.

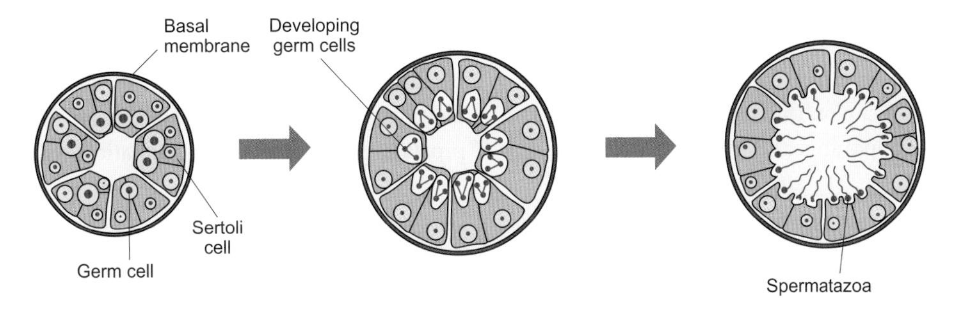

Figure 20.17 Developing spermatocysts in elasmobranch testes

Enclosed spherical spermatocysts consist of clones of germ cells associated with Sertoli cells. The developing spermatocysts pass across the testes, burst and release the spermatozoa into a network of ducts that coalesce to form the vas deferens that drains each testis.

Source: Jobling M (1995). Environmental Biology of Fishes. London: Chapman & Hall.

a number of stages of cell division while passing between a pair of Sertoli cells, and ultimately produce spermatozoa. Tight junctions[24] between the Sertoli cells create a barrier that prevents nutrient transfer from the extracellular fluid directly to the developing spermatocytes, spermatids and spermatozoa. Instead, the Sertoli cells provide the essential nutrients for the dividing cells, which pass through the tight junction as they move toward the lumen. The Sertoli cells are essential

[24] Section 4.2.1 gives information on the functioning of tight junctions.

20

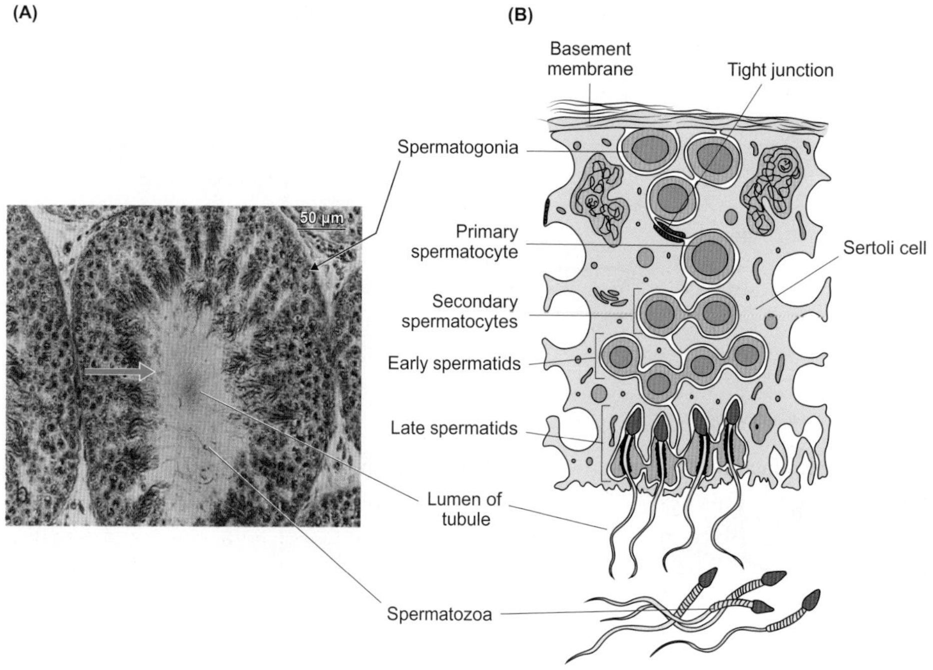

Figure 20.18 Stages in spermatogenesis in seminiferous tubules

(A) Transverse section through seminiferous tubule of chicken testis. Spermatogenesis proceeds in the direction shown by the green arrow. (B) Diagram of spermatogenesis in mammalian seminiferous tubules showing the stages in formation of spermatozoa.

Source: A: Beaupré CE et al (1997). Determination of testis temperature rhythms and effects of constant light on testicular function in the domestic fowl (*Gallus domesticus*). Biology of Reproduction 56: 1570-1575. B: Spence AP and Mason EB (1992). Human Anatomy & Physiology. 4th Edition. West-Publishing Co.

for all steps in spermatogenesis from the provision of the correct environment for the spermatogonia through to the final step of **spermiogenesis**, which involves remodelling of spermatids into motile spermatozoa.

20.3.3 Spermiogenesis

The transformation of spermatids into spermatozoa involves reducing the amount of cytoplasm, condensing the chromosome material into a head, and adding a motile tail. Fluids secreted by the Sertoli cells flush the spermatozoa from the tubules. Figure 20.19 shows the characteristic features of spermatozoa:

- a head, mainly occupied by genetic material in the nucleus;
- the **acrosome**—a cap on the head that develops from the Golgi apparatus and in many species is important in penetrating the ovum during fertilization. The spermatozoa of teleost fish do not have an acrosome; they enter the egg by a channel called the micropyle;
- a midpiece containing many mitochondria;
- a tail powered by mitochondria in the midpiece and actions of the microtubules, which provides motility.

In most animals, the spermatozoa are ready to fertilize an ovum immediately after release by the male, but mammalian spermatozoa need to undergo further changes before they have the capacity to fertilize the egg. These processes, collectively called **capacitation**, normally occur in the female's reproductive tract (but can take place *in vitro*). One reason for the delay is that a tripeptide called fertilization-promoting peptide (FPP), which is secreted into the seminal fluid by the prostate gland, impedes capacitation when it is present in

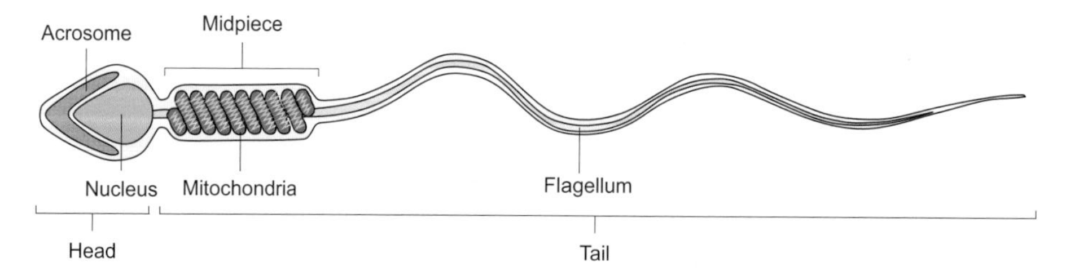

Figure 20.19 Human sperm

The head contains the genetic material in the nucleus. A cap on the head (acrosome) contains enzymes such as hyaluronidase, and acrosin (a proteinase) used to penetrate the outer coating of the ovum. The tail is powered by mitochondria in the midpiece which provide motility. The overall length of the human sperm is about 60 μm.

Source: Wolpert L, Tickle C (2011). Principles of Development, 4th edition. Oxford: Oxford University Press.

high concentrations prior to ejaculation. Once the seminal fluid enters the female's reproductive tract, dilution of FPP occurs, which allows capacitation to proceed.

Capacitation involves an increase in the fluidity of the acrosomal membrane, which results in a higher permeability of the membrane to calcium ions (Ca^{2+}). An influx of Ca^{2+} increases **sperm motility**, allowing them to swim rapidly up the oviducts. Capacitation also increases the ability of sperm to bind to the cell membrane of the ovum.

20.3.4 Hormones control male reproductive function

Figure 20.20 is a schematic diagram of the main factors controlling reproductive function of male vertebrates. This figure includes the interactions between hormones secreted in the hypothalamus of the brain, the pituitary gland and the testes, and negative feedback mechanisms.

In most vertebrate groups, the brain has several areas of neurons that secrete **gonadotrophin releasing hormone** (GnRH). Some of these neurons are involved in sexual behaviour, while others regulate pituitary function. Those in the preoptic area of the hypothalamus have nerve endings on a portal blood system (the **hypothalamo-hypophysial portal system**) that supplies the pituitary gland[25].

Bursts of GnRH release from the hypothalamus stimulate specific cells (gonadotrophs) in the anterior pituitary—that secrete **gonadotrophins**, which enter the blood circulation and stimulate gonadal development and functioning in both sexes. The control of the pituitary gonadotrophs was considered to be exclusively via GnRH until 2000 when a hypothalamic neuropeptide that inhibits pituitary release of gonadotrophins, named **gonadotrophin inhibiting hormone** (GnIH), was discovered in the Japanese quail (*Coturnix japonica*). GnIH has since been identified in the brain of all species of birds and mammals investigated, and some fish. The inhibitory actions of GnIH in birds and mammals influence reproductive function of both males and females, including effects on behaviour in some cases. These actions appear to result from dual inhibition of the hypothalamic

[25] Section 19.2.2 discusses the hypothalamic control of the anterior pituitary and alternative systems of direct diffuse control in animal groups lacking a hypophysial portal system (teleost and elasmobranch fish).

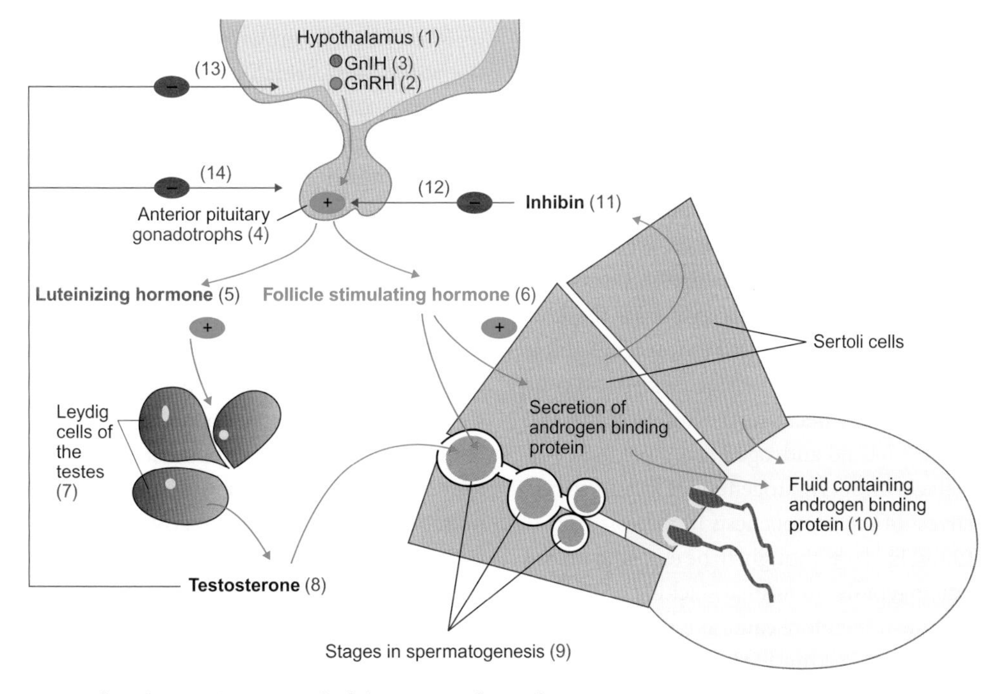

Figure 20.20 Hormonal and paracrine control of the testes of vertebrates

Green lines and arrows show stimulatory effects; red lines and arrows show negative feedback effects. The hypothalamus of the brain (1) secretes gonadotrophin releasing hormone (GnRH) (2), which stimulates the anterior pituitary gonadotrophs, and gonadotrophin inhibiting hormone (GnIH) (3), which inhibits the secretion and release of gonadotrophins from the anterior pituitary. The gonadotrophs (4) synthesize and release two hormones: luteinizing hormone (LH) (5) and follicle stimulating hormone (FSH) (6) which enter the blood circulation. In the testes, LH binds to receptors on the Leydig cells (7) which stimulate the synthesis and secretion of testosterone (8). Testosterone stimulates spermatogenesis (9) of the testes, and binds to androgen binding protein (10) secreted by the Sertoli cells of the seminiferous tubules, driven by FSH stimulation. As a result, testosterone becomes concentrated in the luminal fluid. Optimal spermatogenesis is by the synergistic actions of LH and FSH. Negative feedback regulates LH and FSH concentrations in the blood: secretion of inhibin by Sertoli cells (11), which is stimulated by FSH exerts negative feedback on the pituitary gonadotrophs (12); testosterone in the circulation inhibits hypothalamic release of GnRH (13) and exerts negative feedback on pituitary gonadotrophs (14).

neurons that secrete GnRH, and inhibition of the pituitary gonadotrophs[26] (although not all studies of mammals support a direct action of GnIH on the pituitary).

Two types of gonadotrophins: gonadotrophin I and gonadotrophin II have been identified; these hormones are named follicle stimulating hormone (FSH) and luteinizing hormone (LH) respectively (based on their actions in female mammals). The gonadotrophins are structurally similar throughout the vertebrates, but not identical.

The complete control of spermatogenesis by the gonadotrophins, and other substances released in the testes, varies among species and during their development. Although not understood fully, it is apparent that a subtle cooperation between FSH, LH, testosterone, various growth factors and **androgen binding protein** is necessary to maintain spermatogenesis and the production of large numbers of viable sperm in reproductive adult males. Figure 20.20 summarizes the main interacting effects and the negative feedback effects controlling the hypothalamus–anterior pituitary axis.

LH binds to surface receptors on interstitial cells called **Leydig cells** that lie between the seminiferous tubules and triggers the secretion of testosterone and various growth factors. The release of testosterone from the testes into circulating blood exerts negative feedback on the pituitary and hypothalamus, which results in low, stable concentrations of testosterone in the blood.

In the testes, testosterone binds to androgen receptors in Sertoli cells of the seminiferous tubules. Testosterone stimulates the proliferation of spermatogonia, which maintains a population of spermatocytes. More importantly, testosterone is essential for the completion of meiosis, which forms the spermatids necessary for sperm production (as shown in Box 20.1). This requirement for testosterone for the normal fertility of mammals is demonstrated in transgenic mice that lack androgen receptors in the Sertoli cells. These gene knockout mice are infertile because the testosterone (an androgen) they secrete has no androgen receptors to act on in order to exert effects on spermatogenesis; a lack of sperm results from the arrest of spermatogenesis in meiosis II.

Until the late 1990s, FSH was thought to be essential for the initiation of spermatogenesis, so animals lacking FSH were expected to be infertile. It therefore came as a surprise to discover that transgenic mice lacking FSH or its receptor were fertile, although with reduced numbers of spermatogonia, spermatocytes and spermatids. The transgenic mice studies show that FSH increases the number of spermatogonia and their entry into meiosis. The exact mechanisms of actions are uncertain but it is clear that maximal sperm numbers require the synergistic actions of FSH and testosterone.

FSH acts on the functioning of Sertoli cells in a number of important ways. FSH stimulates the Sertoli cell secretion of **inhibin** into the blood circulation, which exerts negative feedback on the anterior pituitary, as shown in Figure 20.20, and maintains low levels of FSH secretion. FSH also stimulates the secretion of androgen binding proteins by mammalian Sertoli cells, most of which passes into the fluid filling the lumen of the seminiferous tubules, as illustrated in Figure 20.20. The binding of testosterone to androgen binding protein is thought to partially explain the high concentrations of testosterone around the gametes (up to 50 times higher than in the blood), which is important for sperm viability.

20.3.5 Testosterone results in secondary sex characteristics and influences behaviour

In some species of vertebrates, such as humans, the large increase in the concentration of testosterone in the blood at puberty drives the development of secondary sexual characteristics, like hair growth in specific areas and thickening of the vocal cords, which lowers the tone of the voice. In contrast, in seasonal breeders, secondary sexual characteristics undergo cycles of development and regression in parallel with hormonal cycles. A striking example is the development of new antlers each year in males of most deer species.

Antlers in male deer grow from permanent pedicles that first develop during puberty when the testicles grow and testosterone concentrations in the blood increase. From this, we might conclude that testosterone has a role in the development of the pedicels that give rise to antlers. Experimental evidence supports this idea. First, castration of male deer (i.e. removal of the testes) inhibits development of the pedicles. Second, testosterone replacement (for example, by intravenous injection) in males castrated before puberty restores the development of pedicles. Third, testosterone administration to female red deer (*Cervus elaphus*) or white-tailed deer (*Odocoileus virginianus*) induces development of antlers, while normally females of these species very rarely produce antlers.

Among birds, testosterone is involved in territoriality and stimulates heightened aggression by some species. Among songbirds such as great tits (*Parus major*), Atlantic canaries (*Serinus canaria*) and zebra finches (*Taeniopygia guttata*), a component of territorial defence and attracting a mate involves singing. In these birds, testosterone stimulates development of the nuclei in the brain that control song production and hence stimulates singing activity. Castration of male songbirds decreases the rate of song production and the frequency and

[26] The additional presence of GnIH and receptors for GnIH in the gonads and *in vitro* studies suggest GnIH also functions as a paracrine/autocrine regulator of steroid synthesis and gamete maturation in the gonads of both sexes of birds and mammals. Environmental factors such as nest availability in birds affect GnIH expression in the gonads and in some species appears to be a significant factor in reproductive success.

intensity of aggressive interactions. On the other hand, injection of castrated males with testosterone restores singing activity and restores aggressive interactions, while testosterone injection of normal males heightens aggression.

So do wild birds show an increase in the blood concentrations of testosterone during periods of heightened territoriality? Field studies of some species have shown a significant increase in blood testosterone concentration about 10 minutes after introducing an intruder male. However, the extent of the aggressive behaviour by birds varies between species and according to their social context. It seems that minor intrusions do not generally trigger an increase in the blood concentrations of testosterone and aggressive interactions, while persistent challenges of monogamous species often do.

In bird species with high levels of aggression among males and competition for territories, the plasma concentrations of testosterone are usually at their highest when a male is first establishing his territory and facing many challenges from competing males. Figure 20.21 shows that the plasma concentrations of luteinizing hormone and testosterone in house sparrows increase during each of the brooding periods and decrease when aggression reduces and parental care begins. However, western gulls (*Larus occidentalis*) that pair bond for life show reduced aggression if there is no shortage of nest sites, and plasma testosterone shows only minor fluctuations.

Just as singing increases seasonally in male songbirds, vocalization of male frogs and toads calling for a mate is also seasonal. Calling behaviour requires testosterone, and is abolished by castration or treatment of frogs and toads with anti-androgen drugs that prevent the actions of testosterone. Figure 20.22 shows blood concentrations of testosterone are around 10-fold higher in southern toads (*Anaxyrus* (formerly *Bufo*) *terrestris*) that are calling (by inflation of their vocal sacs) during their seasonal reproduction than the non-calling toads. Similar relationships occur in several species of frogs and toads, but not all, and other developmental hormones such as thyroid hormones[27] are also necessary to masculinize the larynx. However, in African clawed frogs (*Xenopus laevis*), castration or anti-androgen treatment prevents the development of the cartilage and muscle fibres in the larynx that are required to make mating calls using the laryngeal vocal sacs.

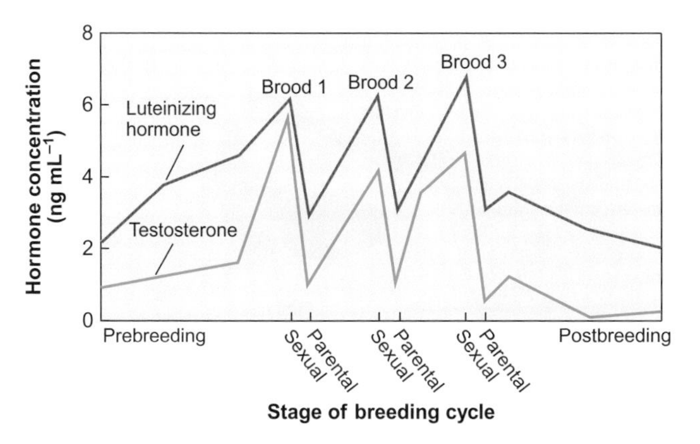

Figure 20.21 Plasma concentration of luteinizing hormone and testosterone in male house sparrows (*Passer domesticus*) throughout the breeding cycle

The sparrows have several broods during their seasonal breeding. During periods of sexual activity and aggression with other males, the concentrations of luteinizing hormone and testosterone increase, and decrease when parental care commences.

Source: Wingfield JC et al (1987). Testosterone and aggression in birds. American Scientist 75: 602-608.

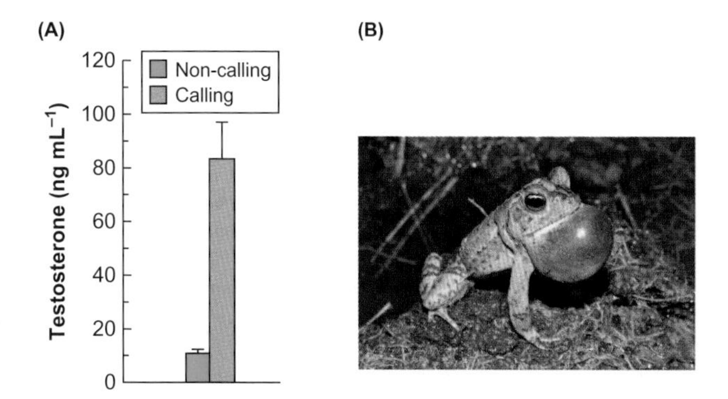

Figure 20.22 Testosterone is elevated in calling male toads

(A) Concentration of testosterone in the blood plasma of calling and non-calling southern toads (*Anaxyrus terrestris*). Values are means ± standard errors of means for 13 calling and 36 non-calling males captured in June and July. (B) Southern toad calling by inflation of vocal sacs, which is an energy demanding process.

Sources: A: Hopkins WA et al (1997). Increased circulating levels of testosterone and corticosterone in southern toads, *Bufo terrestris*, exposed to coal combustion waste. General and Comparative Endocrinology 108: 237-246; B: Daniel Parker/Flickr.

Androgens do not appear to be the ultimate controller of the calling behaviour of male frogs and toads. Areas of the brain of frogs and toads that are concerned with vocalizations contain neurons that synthesize **arginine vasotocin (AVT)**[28] and other neurons with AVT receptors. These findings suggest that brain AVT is involved in controlling the calling behaviour of frogs and toads, and an increase in brain AVT receptors provides a mechanism to maintain the sexual dimorphism of seasonal calling behaviour. In a similar way, AVT in birds, and arginine vasopressin (AVP) the functional analogue of AVT in mammals, have effects on seasonal patterns of social behaviour.

› *Review articles*

Ball GF, Riters LV, Balthazart J (2002). Neuroendocrinology of song behavior and avian brain plasticity: multiple sites of action of sex steroid hormones. Frontiers in Neuroendocrinology 23: 137–178.

Gribbins KM (2011). Reptilian spermatogenesis. A histological and ultrastructural perspective. Spermatogenesis 1: 250–269.

[27] Sections 19.4 discuss thyroid hormones, which are important in many developmental processes, including the metamorphosis of amphibians.

[28] Table 19.1 shows the structure of arginine vasotocin (AVT) and arginine vasopressin (AVP) and Section 19.2.1 discusses the hypothalamus–posterior pituitary axis that controls circulating concentrations of AVT.

Moore FL, Boyd SK, Kelley DB (2005). Historical perspective: Hormonal regulation of behaviors in amphibians. Hormones and Behavior 48: 373–383.

O'Shaughnessy PJ (2014). Hormonal control of germ cell development and spermatogenesis. Seminars in Cell and Developmental Biology 29: 55–65.

Tsutsui K, Ubuka T, Bentley GE, Kriegsfeld LJ (2012). Gonadotropin-inhibitory hormone (GnIH): discovery, progress and prospect. General and Comparative Endocrinology 177: 305–314.

Wingfield JC, Ball GF, Dufty AM Jr, Hegner RE, Ramenofsky M (1987). Testosterone and aggression in birds. American Scientist 75: 602–608.

20.4 Female reproductive systems of vertebrates

Most vertebrate species have paired ovaries and two oviducts. Ova released from the ovary enter an oviduct or fallopian tube through a funnel-like opening lying close to the ovary, as illustrated in Figure 20.23. Although bilateral symmetry is a general feature of embryos, one ovary and oviduct regress during the development of some members of some animal groups. For example, only one ovary and a single functional oviduct occur in the adults of some fish species, a few lizards, some mammals, in particular many species of bats, and nearly all bird species.

There is a long-held belief that retention of only a single ovary and oviduct was an important component in the evolution of flight (and weight reduction) by birds. This idea is supported by the recent discovery in China of fossils of very early birds with a single mature oviduct, which indicates that loss of one side of the reproductive system took place early in the evolution of birds. In contrast, the flightless kiwis (*Apteryx* spp.) retain both ovaries, and if more than one egg is laid in a season then eggs are ovulated alternatively from each ovary, but this does not apply to all flightless birds.

A layer of smooth muscle around the oviduct or fallopian tube propels the ovulated ova along the duct. Often the duct is merely a relatively unspecialized conduit. In some animal groups, however, specialized segments of the oviduct, or beyond, add outer layers and a protective outer casing to the eggs. This process occurs in some elasmobranch fish and all birds, reptiles and terrestrial insects that lay cleidoic eggs[29]. For example, the mermaid's purse surrounding the fertilized egg of the lesser spotted dogfish (*Scyliorhinus canicula*), shown in Figure 20.1A, is added when the egg passes from the short upper oviduct into a specialized region known as the oviducal gland. The gland has three distinct areas: an anterior albumen-secreting zone, a narrow mucous-secreting zone and a large shell-secreting zone. Similar staging posts occur in birds. Notice in Figure 20.23A a region called the magnum in the oviduct of the hen, where addition of albumin occurs over several hours. Toward the end of the oviduct, the addition of a calcium-rich shell occurs in the shell gland. In the hen, the egg spends 16–18 h in the shell gland and its passage through the entire oviduct takes just over 24 h, so eggs are laid progressively later each day.

In some viviparous animals, such as some sharks, the oviduct has a specialized region that holds the developing

[29] Cleidoic eggs are closed self-contained eggs, as we discuss in Section 20.1.1.

(A)

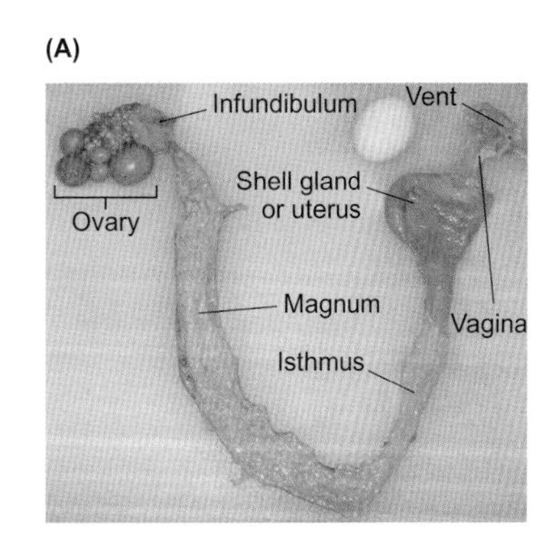

(B)

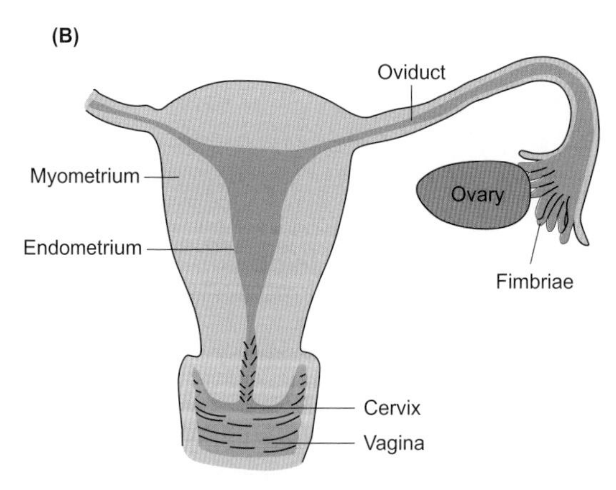

Figure 20.23 Specialization of the oviduct and reproductive tract

(A) Ovary and oviduct of the hen. Ova from the ovary pass through an oviduct that can reach 600 mm in length. There are five recognizable regions: (i) infundibulum that collects the ova, (ii) magnum—the longest and most conspicuous part where egg albumen is added, (iii) isthmus—a narrower region where the shell membranes are added, (iv) shell gland (or uterus). The egg remains in the shell gland for 16–18 h while the calcium-rich shell and outer pigments are added, (v) vagina through which the egg is passed. Finally, the egg is passed through the vent (cloaca) that connects with the gastrointestinal tract. (B) Female reproductive system in human. Fimbriae collect ova that pass along the oviduct into the uterus which has a thick muscular wall (the myometrium) and a well-vascularized inner lining (endometrium).

Source: A: © Extension.org.

embryo in the female's reproductive tract. In a similar way, the lower parts of the oviducts in some mammals, such as humans, fuse to form a uterus as shown in Figure 20.23B. Looking at this diagram, notice the thick muscular wall of the uterus, called the **myometrium** and the inner lining or **endometrium**. Before we examine the processes that result in implantation of a fertilized embryo in the uterus, we explore the developmental processes that produce the ova.

20.4.1 Development of ova by folliculogenesis leads to ovulation

In mammals, the mitotic expansion of the number of **oogonia** begins early in the life of the foetus. The primordial germ cells (also called oogonial stem cells or germ line stem cells) migrate to the developing ovary and proliferate by mitosis. For example, in humans, mitotic divisions in the first half of gestation result in anything from tens of thousands to millions of oogonia. Before birth (or soon afterward) oogonia begin the next stage of oogenesis, meiosis I (outlined in Box 20.1). The primary oocytes then enter a state of rest (in prophase). At birth, the mammalian ovary contains many primary oocytes in small **primordial follicles**. The ovary of a newborn human contains about 1 million primordial follicles.

Since the early 1950s, one of the cornerstones of reproductive biology has been the idea that gradual depletion of a fixed population of primordial follicles in the ovary of a newborn mammal occurs over its reproductive lifespan[30], in some cases lasting several decades: up to ~50 years in humans or 60 years in elephants. However, this dogma is now challenged by work on mice which suggests that primordial follicles in the ovaries of mice develop into oocytes about 10 times faster than expected. At this rate, their use would outstrip supply within the reproductive lifespan if the oocytes are derived only from the pool of primordial follicles available at birth.

A few oogonial stem cells were found to occur in the mouse ovary, and experiments suggest that these rare stem cells produce oocytes spontaneously in culture or when isolated and transplanted into young mice. However, isolation of oogonial stem cells from the ovaries of a wider range of mammals, including humans, remained elusive until 2012, when Jonathan Tilly's team at the Harvard Medical School (USA) identified oogonial stem cells in adult humans (of reproductive age), and again in mouse ovaries.

To make this discovery, they used a highly selective technique that differentiates between the few oogonial stem cells and oocytes and other cells in dispersed cells from the ovary. First, the oogonial stem cells were labelled with a fluorescently labelled antibody to a cell surface antigen that does not occur in later stage oogonia or oocytes. (The antigen is internalized in the cytoplasm of oocytes and can be detected with a different antibody.) Next, the stem cells were isolated using fluorescence-activated cell sorting, which separates the fluorescent cells from non-labelled cells as they pass through a detector, in single file.

The finding of the oogonial stem cells in the ovaries of human females is a major breakthrough. It offers the prospect of new approaches in human infertility treatment, if ovarian stem cells can be coaxed into developing into ova *in vitro*.

Figures 20.24 illustrates the stages in **folliculogenesis**, which lead to **ovulation**. Each of the primordial follicles contains a primary oocyte, surrounded by a single layer of granulosa cells. The primary oocyte secretes a thin layer of glycoproteins around the oocyte to form a transparent layer called the zona pellucida. This layer occurs in the ova of most species, not just among vertebrates.

Hormonal control of folliculogenesis

Figure 20.25 outlines the hormonal processes that regulate folliculogenesis and ovulation in mammals. Early in folliculogenesis, pulses of gonadotrophin releasing hormone (GnRH) stimulate the release of follicle stimulating hormone (FSH) from the pituitary gland. Spikes of neural activity in neurons of the preoptic area of the hypothalamus stimulate GnRH secretion[31]. The feedback response to oestrogens also influences the activity of these neurons. In seasonal breeders, environmental signals (such as the photoperiod) are a dominant influence on the activity of the preoptic nuclei secreting GnRH, which are active during the period leading up to ovulation and breeding. Outside the breeding season, secretion of GnRH decreases.

FSH stimulates the ovary, resulting in recruitment of several primordial follicles, which start to develop. The binding of FSH to receptors on the granulosa cells stimulates the division of the granulosa cells and oocyte growth so that some of the primary follicles form larger secondary follicles. Note in Figure 20.24 that the follicles grow multiple layers of granulosa cells, like layers of an onion.

Connective tissue at the outer margin of the follicle differentiates to form layers of **theca cells** outside the granulosa cells of the growing follicle. The theca cells have receptors for luteinizing hormone (LH). In response to LH, the theca cells synthesize androgens. Aromatase enzymes are synthesized by granulosa cells in response to FSH stimulation and convert androgens to oestrogens. As a result, the concentration of oestrogens in the follicle increases and stimulates further

[30] Proposed by Solly Zuckerman in 1951.

[31] Section 19.2.2 discusses the hypothalamic control of the anterior pituitary in more detail.

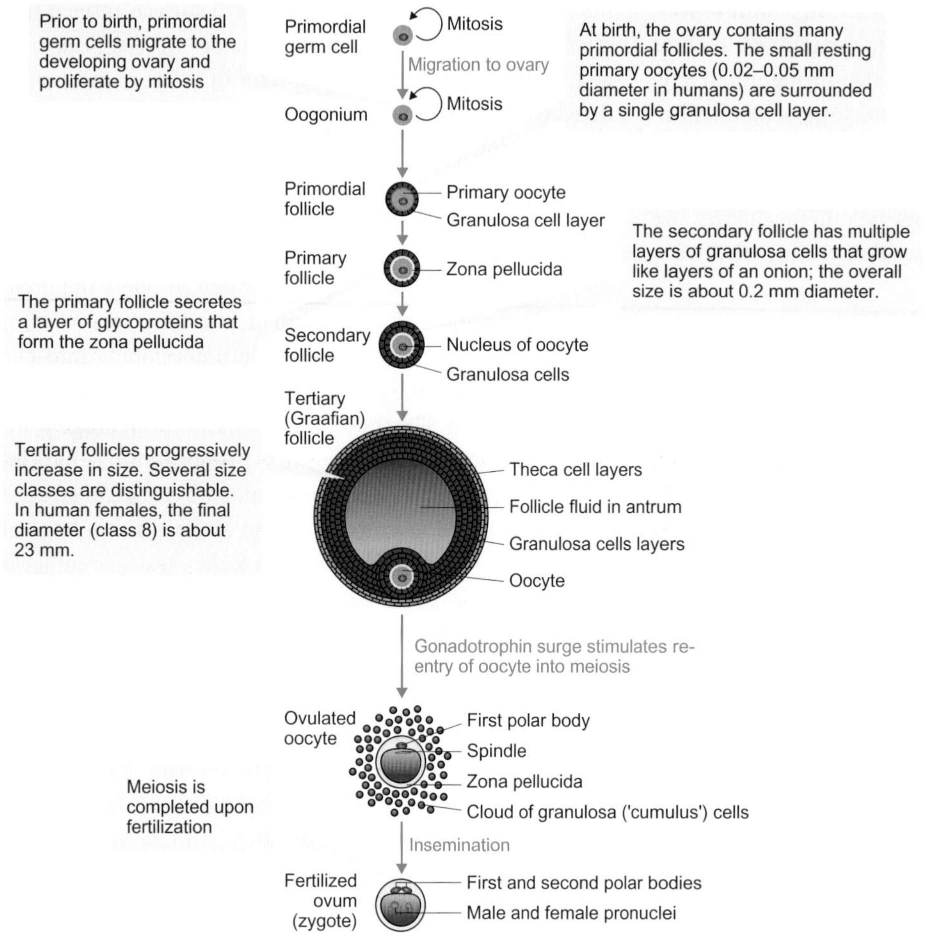

Prior to birth, primordial germ cells migrate to the developing ovary and proliferate by mitosis
Primordial germ cell
Mitosis
Migration to ovary
Mitosis
Oogonium
At birth, the ovary contains many primordial follicles. The small resting primary oocytes (0.02–0.05 mm diameter in humans) are surrounded by a single granulosa cell layer.
Primordial follicle
Primary oocyte
Granulosa cell layer
Primary follicle
Zona pellucida
The secondary follicle has multiple layers of granulosa cells that grow like layers of an onion; the overall size is about 0.2 mm diameter.
The primary follicle secretes a layer of glycoproteins that form the zona pellucida
Secondary follicle
Nucleus of oocyte
Granulosa cells
Tertiary (Graafian) follicle
Tertiary follicles progressively increase in size. Several size classes are distinguishable. In human females, the final diameter (class 8) is about 23 mm.
Theca cell layers
Follicle fluid in antrum
Granulosa cells layers
Oocyte
Gonadotrophin surge stimulates re-entry of oocyte into meiosis
Ovulated oocyte
First polar body
Spindle
Zona pellucida
Cloud of granulosa ('cumulus') cells
Meiosis is completed upon fertilization
Insemination
Fertilized ovum (zygote)
First and second polar bodies
Male and female pronuclei

Figure 20.24 Schematic diagram of the stages in follicle maturation forming mature oocytes in the mammalian ovary

Source: adapted from Gosden R, Lee B (2010). Portrait of an oocyte: our obscure origin. The Journal of Clinical Investigation 120: 973-983.

Hypothalamus
GnRH
Anterior pituitary
Follicle stimulating hormone (FSH)
Luteinizing hormone (LH)
Ovary
mitosis
Fluid secretion
Androgen synthesis by theca cells
Primordial follicle
Primary follicle
Secondary follicle
Androgen
Increase in oestrogens in follicular fluid
Aromatase activity
Oestrogens
Inhibin
Low blood oestrogens

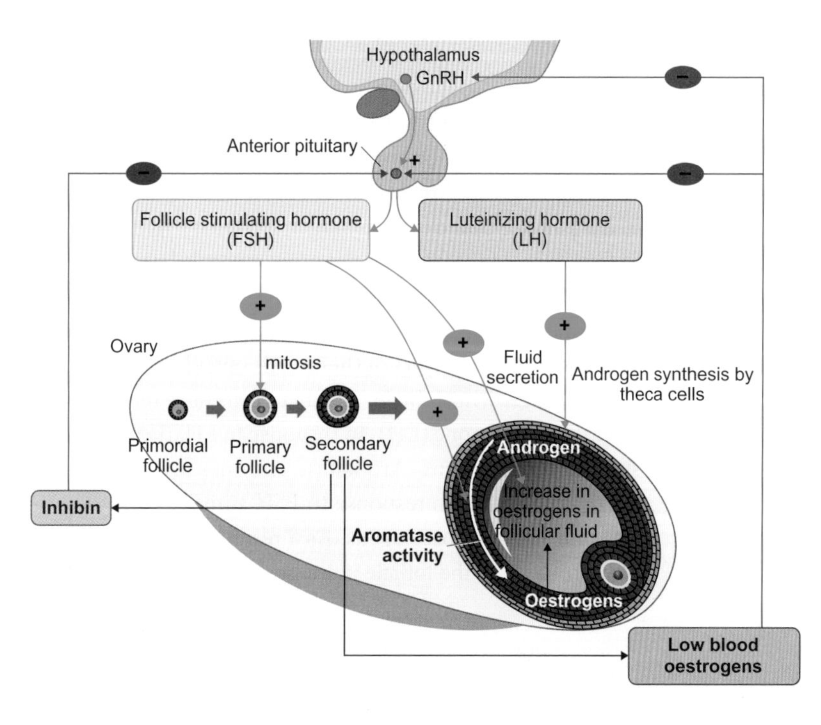

Figure 20.25 Folliculogenesis in female mammals

Stimulatory or positive effects are shown by + and green arrows; feedback inhibition or negative effects are shown by – and red arrows. Follicle stimulating hormone (FSH) and luteinizing hormone (LH) are regulated by the gonadotrophin releasing hormone (GnRH) secreted by hypothalamic neurons. FSH stimulates granulosa cells of follicles causing: 1. Cell division (mitosis), increasing the size of the follicles. 2. Increased aromatase activity. 3. Secretory activity and formation of follicular fluid. LH stimulates androgen synthesis by the outer theca cells of the growing follicle. Conversion of androgens to oestrogens (by aromatase) in the granulosa cells increases the concentration of oestrogens in the follicular fluid of developing follicles. FSH and LH concentrations remain low because of the negative feedback effects on the hypothalamus and pituitary of: 1. Relatively low blood concentration of oestrogens. 2. Inhibin secretion by granulosa cells of the developing follicles.

proliferation of granulosa cells. This cellular proliferation increases the size of the follicle.

Together, FSH and LH stimulate further ovarian synthesis of oestrogens, although circulating concentrations are relatively low at this stage and exert a *negative* feedback effect on the pituitary and hypothalamus, which Figure 20.25 illustrates. The negative feedback effect of oestrogens on the hypothalamic–pituitary axis throughout folliculogenesis is thought to involve an increase in gonadotrophin inhibiting hormone (GnIH). The developing follicles also secrete inhibin, which enters the blood circulation and inhibits FSH secretion by the pituitary, as in males, so FSH concentrations in the blood remain low.

At around this time, for those species in which a single ovum is released (monovular species, such as cattle, deer, horses and humans), the growth of a dominant follicle deviates from the rest of the developing follicles. The non-dominant follicles ultimately degenerate in a process called atresia. The deviation in growth of a dominant follicle is driven by an array of substances that the follicles secrete and which act within the ovary. This growth at least partially reflects a higher sensitivity of the dominant follicle than other follicles to FSH, probably due to a greater number of FSH receptors.

FSH increases the secretory activity of granulosa cells, forming the follicular fluid that collects in a cavity known as the antrum. Fluid collection pushes the oocyte to one side, as shown in Figures 20.24 and 20.25. The dominant follicle develops into the mature follicle, or **Graafian follicle**[32], which bulges from the surface of the ovary. Mature follicles have an average diameter of 23 mm in humans and up to 30 mm in horses.

The final step in folliculogenesis is the development of LH receptors on the surface of the granulosa cells of the mature Graafian follicle, a process stimulated by oestradiol. The Graffian follicle(s) secrete large amounts of oestradiol, which explains the dramatic increase in blood concentrations seen after day 13 of the hormonal (oestrous) cycle of sheep, which Figure 20.26 illustrates. Similar principles apply in the hormonal cycles of all mammals that ovulate spontaneously after a surge of LH.

[32] Named after the 17th century Dutch anatomist, Regnier de Graaf, who described their development in rabbits.

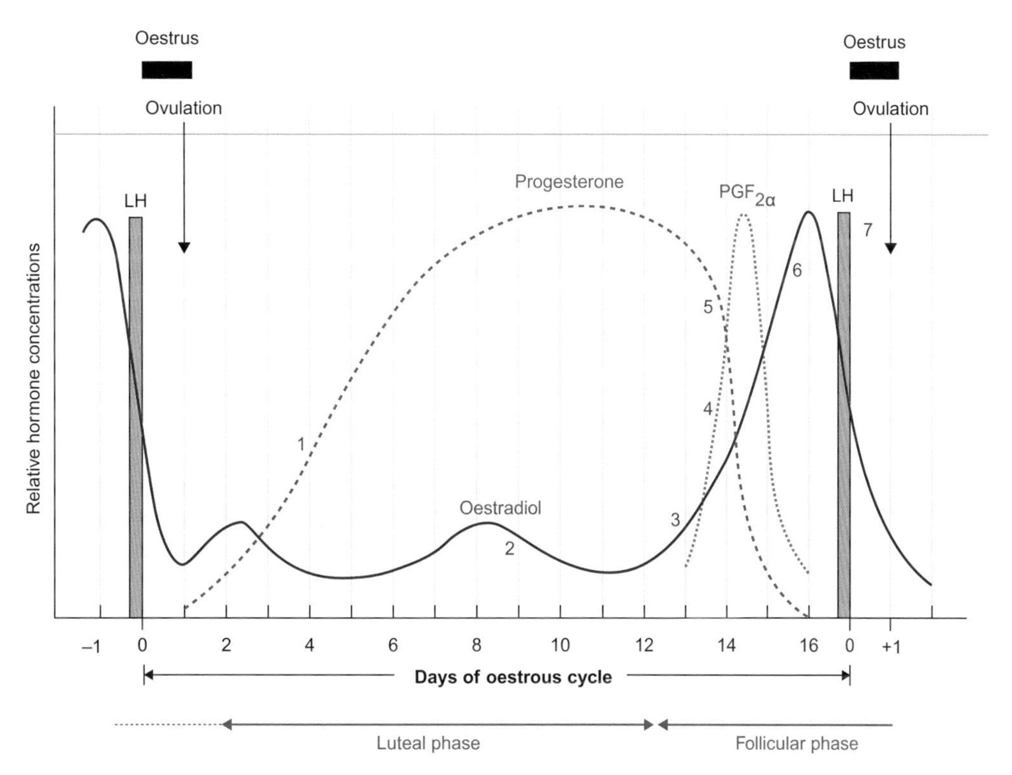

Figure 20.26 Sequence of hormonal secretions in the oestrous cycle of the sheep

Curves show blood concentrations of luteinizing hormone (LH), progesterone and oestradiol, and the surge of prostaglandin $PGF_{2\alpha}$ in the endometrium of the uterus. Starting at day two, in the luteal phase (after ovulation), the blood concentrations of progesterone increase (1) due to secretion from the corpus luteum that is formed from the remnants of the ovulated follicle. Blood concentrations of oestradiol are low in the luteal phase (2), but increase from about day 12 (3). Secretion of $PGF_{2\alpha}$ by the endometrium of the uterus (4) is stimulated by progesterone, and programmes the destruction of the corpus luteum in sheep (but not all mammalian species). When the corpus luteum breaks down, the secretion of progesterone declines so blood concentrations of progesterone decrease (5). The continuing rapid rise in oestradiol (6) stimulates a surge of LH release (7) that results in ovulation (arrow).

Graph adapted from Goldsworthy GJ et al (1981). Endocrinology. Glasgow and London: Blackie.

Oestradiol in circulation has a number of important effects in females:

- increased sexually receptivity, by actions on the central nervous system; during the state of **oestrus**, the receptive female allows mating, usually just prior to ovulation;
- mucus secretion from the wall of the vagina and thickening of the wall, in preparation for mating;
- swelling of the terminal part of the oviduct (infundibulum) in mammals, so that it clasps the ovary and guides the ova into the duct;
- preparation of the uterus lining (endometrium) for the implantation of fertilized zygote(s).

The increasing levels of blood oestrogens resulting from maturation of one or more follicles have important feedback effects on the hypothalamus–pituitary axis that lead to ovulation as we discover next.

A surge of luteinizing hormone stimulates ovulation

As a follicle matures, it secretes higher amounts of oestrogens (mainly oestradiol) leading to an increase in their blood concentrations through *positive* feedback effects, as illustrated on Figure 20.27. The high concentrations of oestradiol in the blood exert a positive feedback on the hypothalamus, which increases GnRH, possibly due to reduced GnIH[33]. Oestradiol also increases the sensitivity of the pituitary gland to GnRH.

In all mammals that show periodicity in their ovulation, only the mature follicles secrete sufficient oestradiol for the positive feedback to trigger the surge of LH release that initiates ovulation. Notice in Figure 20.26 that the concentration of LH in the blood peaks after the rising concentration of oestradiol, which reflects the positive feedback effect on the anterior pituitary.

The growth of a pre-ovulatory follicle brings it close to the epithelial surface of the ovary where the single layer of cells is intimately involved in the cellular events that result in ovulation. These cells have receptors for LH that increase in number prior to ovulation, under the influence of oestradiol. Surges of LH stimulate the epithelial cells of the ovary to release substances that cause tissue degradation, beginning at the surface of the ovary and progressing toward the outer surface of the mature follicle. The epithelial cells release substances that result in a local increase of a protease enzyme which activates collagenases and leads to disruption and death of the theca cells at the apex of the pre-ovulatory follicle. The action of collagenases also releases tumour necrosis factor-α (TNF-α) from its anchor to the thecal cells, and thereby promotes proteolysis and cell death.

Progressive thinning of the ovarian wall and weakening of the follicle wall at its apex result in release of the secondary oocyte[34] (or ovum) from one or more follicles. The ovulated ovum (egg), surrounded by the zona pellucida, carries with it an outer layer of about 5000 granulosa cells known as cumulus cells. Notice in Figure 20.27 that these cells form an outer layer, called the corona radiata.

In contrast to the **spontaneous ovulation** considered so far, some mammalian species only ovulate in response to mating. Such **induced ovulation** occurs in rabbits and hares, ferrets, mink, rats and camelids (alpacas, llamas and camels). The hormonal control of induced ovulation mirrors

[33] Look back to Section 20.3.4 for discussion of gonadotrophin inhibiting hormone (GnIH) and its effect on gonadotrophin releasing hormone (GnRH) and gonadal function, which apply to both sexes.

[34] Look at Box 20.1 for an explanation of the stages in oogenesis.

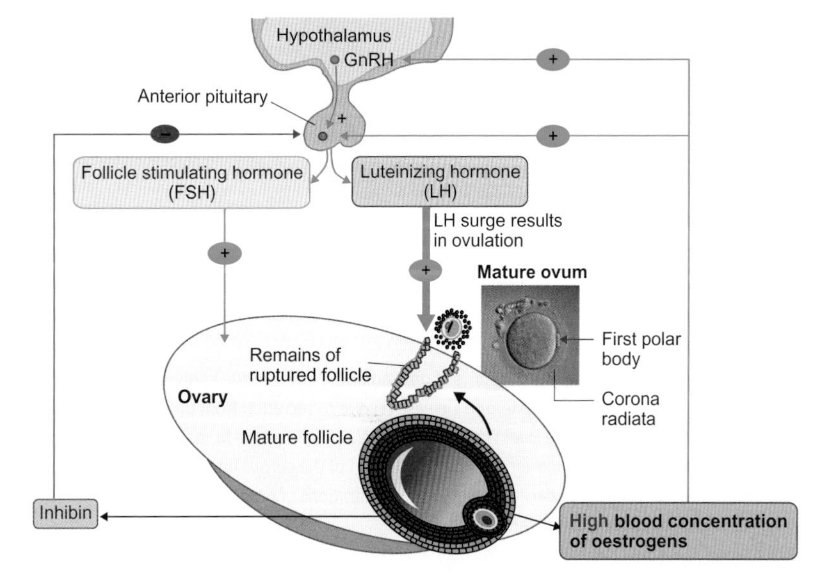

Figure 20.27 Reproductive hormones and feedback regulation leading to a surge in luteinizing hormone and ovulation in mammals

Stimulatory or positive effects are shown by + and green arrows; feedback inhibition or negative effects are shown by − and red arrows. The mature follicle (under the influence of FSH and LH) secretes: 1. Inhibin which exerts a negative feedback on pituitary gonadotroph release of FSH, and 2. High amounts of oestrogens, mainly oestradiol. High blood concentrations of oestrogens exert *positive* feedback on the hypothalamus and pituitary gonadotroph secretion of LH, leading to the surge of LH release that stimulates ovulation of a mature ovum (or several ova) surrounded by a layer of granulosa cells. In the photograph, notice the retained polar body resulting from meiosis (indicated by red arrow).

that for spontaneous ovulation: a surge of LH is the ultimate cause of induced ovulation, although the trigger for the LH surge is mainly the mechanical events during copulation that initiate a neuroendocrine (hormonal) reflex. The input is via nerves and the output is GnRH release from the hypothalamus, which leads to a surge of LH. An ovulation-inducing factor in seminal fluid also elicits LH release and ovulation.

20.4.2 The luteal phase

After ovulation, most species enter a **luteal phase** in which the remains of each ruptured follicle reorganize to form a **corpus luteum** (from the Latin for body (corpus) and yellow (luteum)). Each corpus luteum contains lipid droplets of cholesterol ester, the precursor for steroid hormone synthesis[35], and β-carotene, which explains its yellowish colour. In some species such as the horse, extra corpora lutea (the plural of corpus luteum) are formed from follicles that did not fully mature.

Under stimulation by luteinizing hormone and various growth factors, the corpus luteum grows in size and begins to secrete large amounts of **progesterone**, together with a smaller amount of oestradiol. Notice in Figure 20.26 that the concentration of progesterone in the blood of sheep begins to increase soon after ovulation and is at maximal concentrations between about 7 and 14 days after ovulation.

Figure 20.28 illustrates the interactions between hormones in the luteal phase. The increase in circulating progesterone inhibits the hypothalamus–pituitary axis, which reduces the release of gonadotrophins and prevents the next wave of folliculogenesis. Progesterone stimulates the final development of the uterine lining ready for implantation, after fertilization of the ovum (or ova).

[35] Figure 19.3 summarizes the synthesis of steroid hormones.

If an ovum is not fertilized, the corpus luteum ultimately regresses in the process of luteolysis, which terminates the luteal phase and returns the ovary to the next **follicular phase**. When the cells of a corpus luteum degenerate, a small scar of tissue called the corpus albicans persists in the ovary for several weeks.

Figure 20.29 shows some different patterns of hormone cycles in female mammals. Female primates have a long follicular phase, while domestic farm animals (e.g. sheep and pigs) have a shorter follicular phase. In rodents, the corpora lutea (from several ruptured follicles) do not form unless mating occurs; in unmated animals, the lack of progesterone from the corpora lutea avoids the inhibition of pituitary gonadotrophins, so ovulation occurs repeatedly, every 4 or so days, until they find a mate. Once mated, the secretory corpora lutea form, and persist during the pregnancy (or a **pseudopregnancy** if the ova are unfertilized). In a similar way, animals in which mating induces ovulation, such as rabbits, cats or ferrets, retain the corpora lutea even if fertilization does not occur after mating and a pseudopregnancy occurs before luteolysis of the corpora lutea.

Canine species such as dogs and wolves have two ovulatory cycles each year, and fully functional corpora lutea form after ovulation. In unmated animals (or after an infertile mating) the corpora lutea persist as long as a normal pregnancy. During pseudopregnancy, females exhibit behavioural changes, as if they were pregnant.

In many species, luteolysis of the corpus luteum (or several corpora lutea) is induced by the release of the prostaglandin $F_{2\alpha}$ ($PGF_{2\alpha}$)[36] from the endometrium of the uterus and its

[36] Prostaglandins are synthesized from fatty acids and generally exert local (paracrine) effects on tissues, as we discuss in Box 19.1.

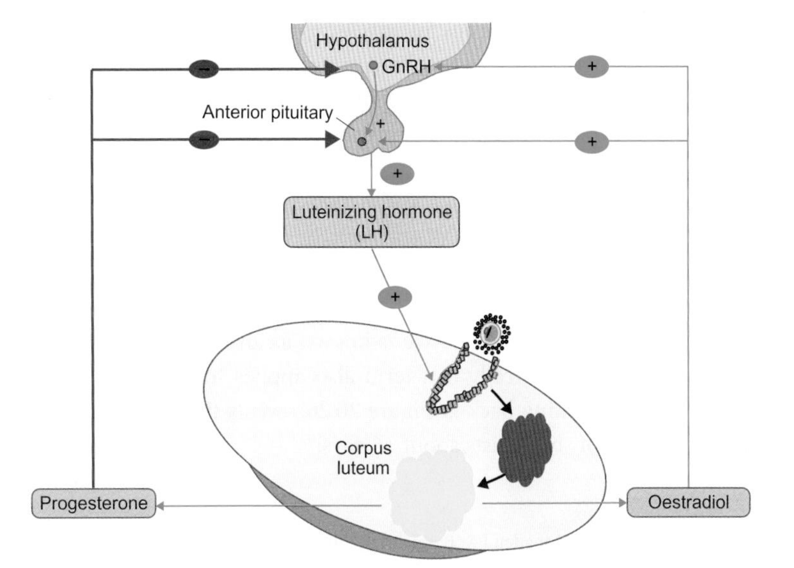

Figure 20.28 Reproductive hormones in the luteal phase, after ovulation in mammals

Stimulatory or positive effects are shown by + and green arrows; feedback inhibition or negative effects are shown by – and red arrows. The ruptured follicle reorganizes to form a corpus luteum, which synthesizes large amounts of progesterone and some oestradiol. The increase in circulating progesterone exerts the dominant effect and inhibits the hypothalamus–pituitary axis, preventing the secretion of LH and the next wave of folliculogenesis.

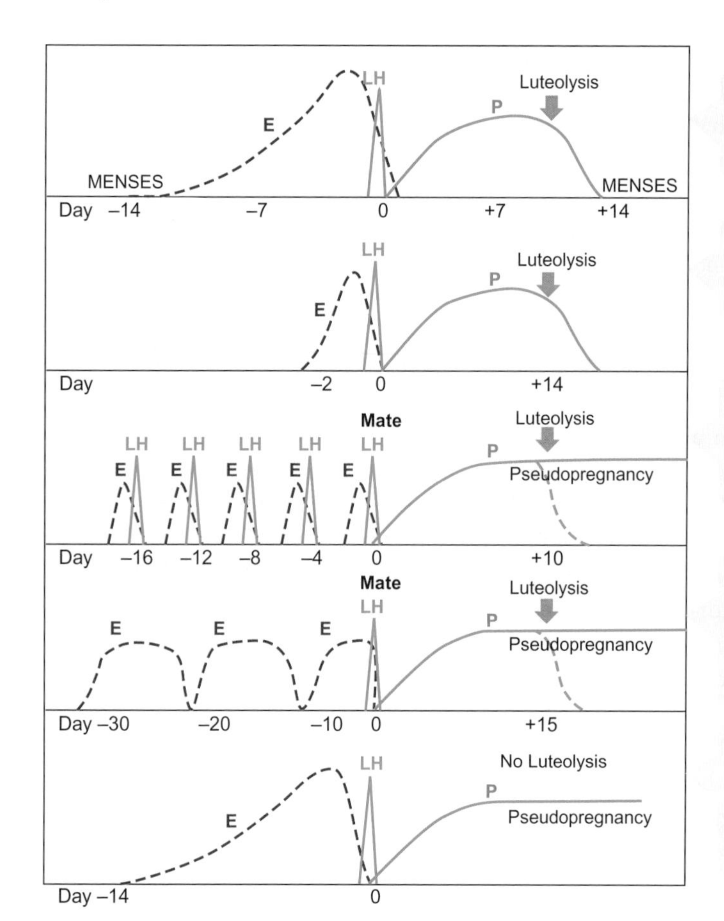

PRIMATES (e.g. humans, rhesus monkeys)
Long period of folliculogenesis, followed by luteal phase of 14 days. Menstruation (menses) occurs between the end of luteolysis and the start of folliculogenesis.

DOMESTIC FARM ANIMALS (e.g. sheep and pigs)
Short period of folliculogenesis. Longer luteal phase.

RODENTS (e.g. rats, mice, hamsters)
Ovulation occurs every few days, until mating. Functional corpora lutea form only after mating and are retained during pregnancy. If the ova are not fertilized, the retention of corpora lutea gives false endocrine signals of pregnancy, known as a pseudopregnancy.

INDUCED OVULATORS (e.g. rabbits and cats)
Mating stimulates ovulation. The corpora lutea remain intact even if fertilization does not occur. The long pseudopregnancy after an infertile mating may be as long as normal pregnancy, or be terminated by luteolytic mechanisms.

CANINE SPECIES (e.g. dogs and wolves)
Usually two ovulatory cycles per year. After ovulation a fully functional corpus luteum (or several corpora lutea) form. In unmated animals, or after an infertile mating, the corpus luteum persists in a pseudo-pregnancy similar in length to a normal pregnancy.

Figure 20.29 Patterns of luteolysis controlling the lifespan of the corpus luteum during cycling or in pseudopregnancy of different mammalian species

The changes in plasma concentrations of oestradiol (also known as estradiol) are shown by the red dashed lines labelled E; progesterone (P) by the green lines, and luteinizing hormone (LH) by the blue lines, for five types of cycles. The point at which luteolysis of the corpus luteum occurs and progesterone release decreases is indicated by the orange arrow for each example. Pseudopregnancies in induced ovulators and rodents are usually terminated by luteolytic mechanisms (dashed green lines). Time scales differ on the five panels.

Source: McCracken JA et al (1999). Luteolysis: a neuroendocrine-mediated event. Physiological Reviews 79: 263-324.

transfer to the ovary. This process occurs in guinea pigs, rabbits, and rats and many domestic farm animals (sheep, goats, cows, pigs, and horses), as illustrated in Figure 20.26 for sheep.

The luteolysis of the corpus luteum (or corpora lutea) determines the periodicity of the ovarian cycle in mammals. Declining progesterone leads to regression of the lining of the uterus. In primates (such as humans or gorillas), shedding of the uterine lining leads to menstruation (menses), so the ovulatory cycle is known as the **menstrual cycle**. The time between the beginning of menstruation and the next ovulation varies from eight to 21 days in primates, averaging about 14.5 days. In non-primate species, where there is no menstruation, the hormone cycle is known as an **oestrous cycle**, although technically this term also applies to menstrual cycles. Looking back at Figure 20.26 reveals that the oestrous cycle of sheep lasts about 16 days.

› *Review articles*

Ginther OJ, Beg MA, Donadeu FX, Bergfelt DR (2003). Mechanism of follicle deviation in monovular farm species. Animal Reproduction Science 78: 239–257.

Gosden R, Lee B (2010). Portrait of an oocyte: our obscure origin. The Journal of Clinical Investigation 120: 973–983.

McCracken JA, Custer EE, Lamsa JC (1999). Luteolysis: a neuroendocrine-mediated event. Physiological Reviews 79: 263–324.

Murdoch WJ, McDonnel AC (2002). Roles of the ovarian surface epithelium in ovulation and carcinogenesis. Reproduction 123: 743–750.

Telfer EE, Albertini DF (2012). The quest for human ovarian stem cells. Nature Medicine 18: 353–354.

20.5 Fertilization and subsequent events

In this section, we look at how sperm cross the outer layers of the egg and ultimately enter it—all of which must happen if fertilization is to occur. The first step toward fertilization of mammalian ova is for sperm to penetrate the corona radiata by digesting the cumulus cells forming this layer of cells around the oocyte, using enzymes released from the acrosome. Once the sperm reach the zona pellucida or an equivalent layer of jelly, an **acrosomal reaction** begins.

20.5.1 The acrosomal reaction

The acrosomal reaction was first described among invertebrate species such as sea urchins. While anatomical details differ between species, the same general principles apply to most animal groups. Figure 20.30 illustrates the major events in the acrosomal reaction of sea urchins and starfish:

- Once in contact with the jelly coat of the egg, the acrosome of the sperm releases protease enzymes that digest the jelly where the sperm head meets the egg.
- Actin, which is initially in globular form, polymerizes to form filaments in an acrosomal process that extends from the sperm head toward the vitelline layer of the egg.
- The plasma membrane of the acrosomal process binds to species-specific sites on the outer layer of the egg vitelline membrane, allowing the sperm to penetrate and fuse with the plasma membrane of the egg so that the cytoplasm of the sperm and egg are continuous.

20.5.2 Prevention of polyspermy

Most animal groups avoid the fertilization of the egg by more than one sperm (polyspermy), which is lethal for most species, by preventing the entry of further sperm. Some groups, such as sharks, are unusual in that more than one sperm normally enters the oocyte, although only one sperm nucleus fuses with the oocyte nucleus.

The attachment of the sperm plasma membrane to the oocyte plasma membrane induces several events that act as fast and slow blocks of polyspermy:

- A fast block, within 1–2 seconds of the first sperm attaching to an oocyte, results from a rapid inflow of sodium ions across the plasma membrane. The resulting change in membrane polarity creates an internal positive charge rather than the usual internal negative charge and stops further binding of sperm.
- Calcium release from intracellular stores, starting at the point of sperm entry, initiates a slower absolute block

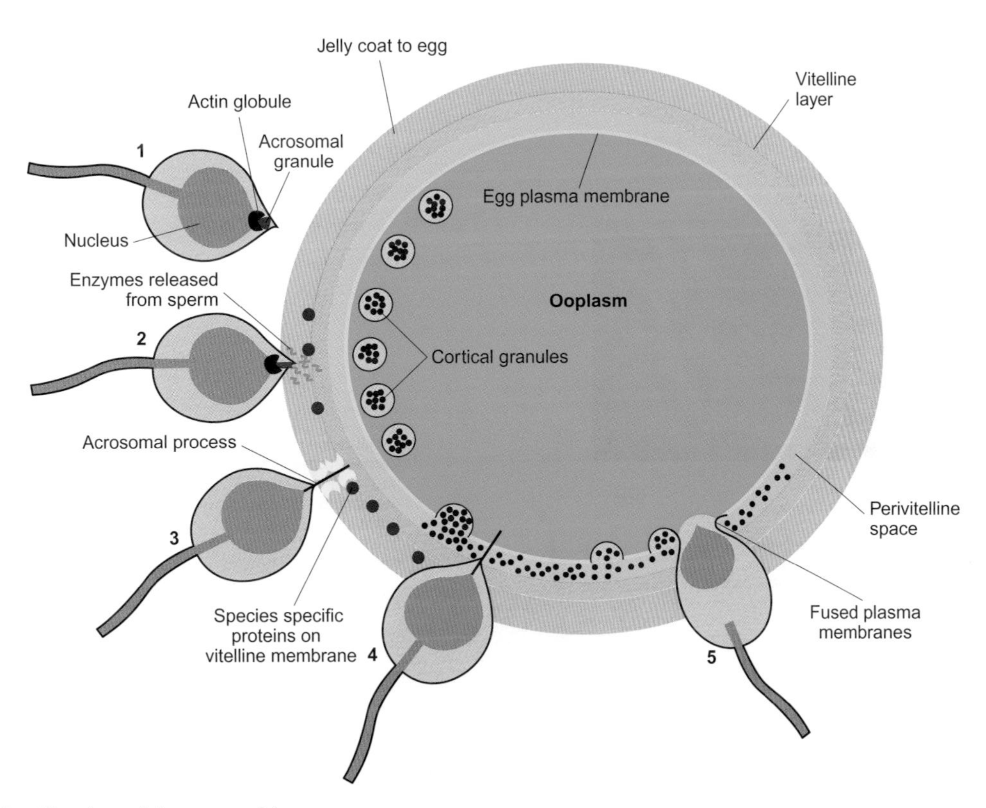

Figure 20.30 Fertilization of the sea urchin egg

The release of enzymes (sperm 2) and formation of an acrosomal process (sperm 3) is followed by binding of sites on the plasma membrane of the acrosomal process to species-specific binding sites on the vitelline layer of the egg (sperm 3), which enables sperm to penetrate the egg plasma membrane (sperm 4). Fusion of the plasma membrane of the egg and sperm (sperm 5) allows nuclear incorporation from the sperm into the egg. A cortical reaction by the egg releases cortical granules into the perivitelline space and causes a build-up of fluid in the perivitelline space, which prevents polyspermy (fertilization by multiple sperm).

of polyspermy by stimulating the fusion of cortical granules with the plasma membrane of the egg and the release of their contents into the perivitelline space, as illustrated in Figure 20.30. The cortical granules are membrane-bound vesicles derived from the Golgi apparatus that accumulate during oogenesis.

- Protease enzymes released from the cortical granules break connections between the vitelline membrane and the egg plasma membrane and block sperm binding, which avoids polyspermy. The molecular basis of these events remains poorly understood, but recent work has identified a protease, ovastacin, as a key component of the cortical granules of mice. Ovastacin cleaves one of the three glycoproteins in the zona pellucida of mice eggs, which prevents further sperm binding to the zona pellucida and ensures fertilization by just one sperm.
- Release of the contents of the cortical granules increases the osmotic concentration of the fluid between the plasma membrane and the vitelline layer, which draws in water by osmosis. The extra fluid lifts the vitelline layer away from the cytoplasm of the egg. Within about one minute of the first sperm attachment to the oocyte, the entire egg envelope separates from the plasma membrane of the oocyte.
- Hardening of the vitelline membrane forms a fertilization membrane and prevents the penetration of additional sperm.

The next stage is the formation of a fertilization cone that surrounds the nuclear material in the sperm head and draws the nuclear material into the egg. Penetration of the sperm into the ooplasm triggers the completion of the second division in meiosis. Fusion of nuclear material from the two gametes occurs soon afterward.

In teleost fish, most sperm do not have an acrosome. Instead, they gain access to the oocyte plasma membrane by passing through a funnel-shaped channel called the **micropyle**. Figure 20.31A shows an outer view of the micropyle in a fish egg, while Figure 20.31B shows a diagrammatic section through a micropyle. Notice that the micropyle has an outer wider section leading into an inner narrower channel. At its narrowest point the channel is only about as wide as the head of a single sperm of most species, at 1.5–2 μm wide, so just one sperm can pass through, which drastically reduces the chance of polyspermy.

Further events that occur after a sperm has made its way through the micropyle are also important in impeding polyspermy of teleosts. Figure 20.31C shows the formation of a fertilization cone that blocks the inner channel and a fertilization plug that fills the outer vestibule. These blockages result from the release of a burst of perivitelline fluid through the narrow channel. The fluid contains the contents of the cortical granules and interacts with superfluous sperm causing their immobilization.

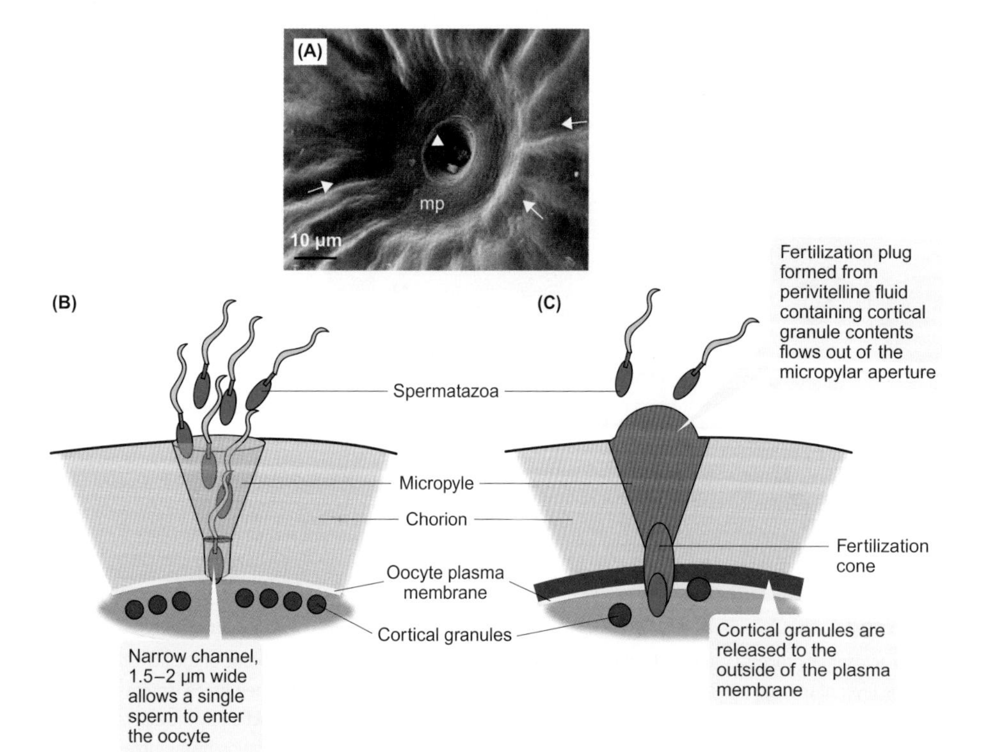

Figure 20.31 Micropyle of teleost fish oocytes and prevention of polyspermy

(A) Scanning electron micrograph of the surface of an ovulated oocyte from the marbled swamp eel (*Synbranchus marmoratus*) showing furrows (arrows) running towards the micropyle pit (mp) which opens into a central narrow channel (arrowhead). (B) Diagram of micropyle showing restricted route for passage through the thick egg envelope (conventionally called the chorion) to reach the oocyte plasma membrane. (C) Initiation of cortical granule release and formation of a fertilization plug.

Source of image (A): Ravaglia MA and Maggese MC (2002). Oogenesis in the swamp eel *Synbranchus marmoratus* (Bloch, 1795) (Teleostei; synbranchidae). Ovarian anatomy, stages of oocyte development and micropyle structure. Biocell 26: 325-337.

20.5.3 Luteal rescue and the placenta

Fertilization of the oocyte(s) of placental mammals results in 'rescue' of the corpus luteum (or corpora lutea), which does not degenerate, but instead has a critical role in maintaining the condition of the uterus. In some mammalian species, such as horses, humans and sheep, the need for one or more functional corpora lutea declines over time and the placenta gradually takes over, but in other mammals such as pigs, cattle and dogs the secretions of the corpora lutea are required throughout pregnancy.

The rescued corpus luteum increases in size at a rapid rate, due to hypertrophy (increased cell size) and cell division. In cows, for example, the corpus luteum is just over 0.5 g in mass three days after ovulation but increases to about 5 g over the next 11 days. The rescued corpus luteum secretes large amounts of progesterone together with small/minimal amounts of oestradiol. Secretion of progesterone maintains the endothelium of the uterus, which otherwise would be shed.

In primates and horses, rescue of the corpus luteum results from the actions of **chorionic gonadotropin** (CG)[37], which is a hormone secreted by the chorion—the outer membrane of the implanting **blastocyst** which forms by division of the fertilized ovum (**zygote**). The number of cells is doubled at each division. When there are 16 to 32 cells, the zygote is referred to as a morula because it resembles a mulberry (Latin for mulberry = *morus*). In mammals, the continued divisions form a blastocyst within a few days of fertilization. The photomicrograph in Figure 20.32 shows the fluid-filled space (blastocoel) in a human blastocyst and the beginnings of an embryo that are first recognizable as an inner cell mass.

Once in the uterus, the blastocyst breaks free of the outer zona pellucida—they 'hatch'. The blastocysts of most placental mammals begin to implant in the endometrial lining of the uterus within a few days of hatching (about three days in humans, for example).

[37] Chorionic gonadotropin is more commonly used than chorionic gonadotrophin because it is not strictly a trophic (nourishing) hormone.

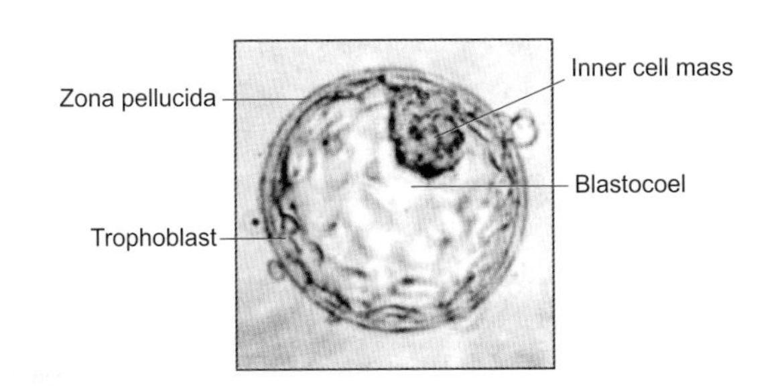

Figure 20.32 Early blastocyst

Source: © J Conaghan/Wikimedia Commons.

Some of the cells of the outer layers of the implanting blastocyst form the placenta[38]. From about the time that the blastocyst starts to implant in the uterine wall, the chorion (which forms a large part of the placenta) begins to secrete chorionic gonadotropin (CG) so blood concentrations of CG increase. Note the rising GC in pregnant humans illustrated in the graph in Figure 20.33. Initially, the blood serum concentrations of CG double about every 1.5 days, and then increase more slowly to reach a peak after 8 to 10 weeks of gestation (i.e. 8 to 10 weeks since the last menstrual period). The rise in blood CG results in excretion of CG in the urine of pregnant females soon after the blastocyst implants and forms the basis of the many available pregnancy tests. Then, blood concentrations of CG slowly decline to about 20 per cent of peak concentrations and remain at about this level throughout the rest of the pregnancy.

Once fully developed, the placenta supplies the developing foetus with oxygen and nutrients and removes nitrogenous waste by diffusion across the placenta. The placenta is also an important endocrine organ, which secretes a variety of hormones and takes over from the corpus luteum in maintaining the pregnancy and inhibiting ovulation in many (but not all) species. Placental secretion of progesterone and oestrogens suppresses the secretion of FSH and LH from the pituitary, which inhibits ovulation. The placental secretion

[38] Figure 20.3 illustrates the organization of the human placenta.

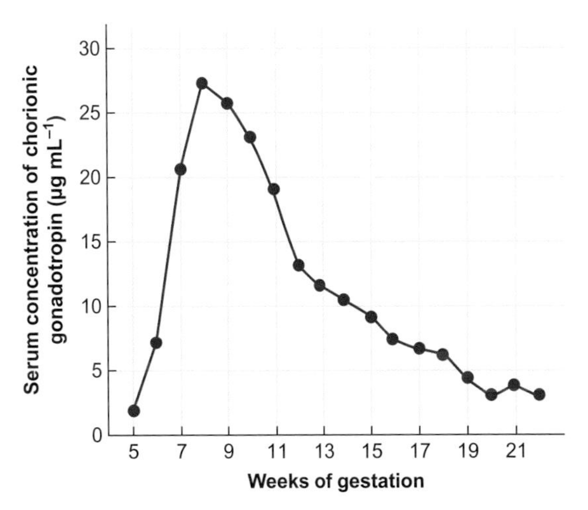

Figure 20.33 Blood concentration of chorionic gonadotropin (CG) in female humans during pregnancy

Graph shows concentrations of CG in maternal blood serum measured between 5 and 22 weeks of gestation (weeks after the last menstrual period).

Source: Ozturk M et al (1987). Physiological studies of human chorionic gonadotropin (hCG), αhCG, and βhCG as measured by specific monoclonal immunoradiometric assays. Endocrinology 120: 549-558.

of progesterone and oestrogens is also important for maintaining the endothelial lining of the uterus and stimulating uterine growth. Progesterone also suppresses contractions of the smooth muscle of the uterus during pregnancy, which avoids ejection of the developing embryo.

20.5.4 Delays in reproductive processes

The timing of spawning or mating, resulting in fertilization, and followed by embryonic development, usually sets in motion the timing of the emergence of new individuals. However, some species exert extra control over the timing of the stages of reproduction.

Sperm storage decouples copulation and insemination from fertilization

Storage of viable sperm in a female's reproductive tract allows fertilization of her ova over several days or weeks. For instance, sperm storage in birds allows a single insemination to produce fertilized eggs over several days. In some seasonally breeding animals, sperm storage after mating enables longer delays between ovulation and fertilization, which enables births to be timed to coincide with good food availability.

Many species of temperate insectivorous bats provide good examples of the use of sperm storage to delay annual births to the optimal times for food availability, after their winter hibernation. The seasonal spermatogenesis of male bats occurs in summer and reaches a peak in early autumn (fall), when mating occurs. Mating also extends into the period when bats are hibernating. If fertilization occurred then, the offspring would be born in winter when there is poor food availability. Instead, sperm are stored in the upper vagina, uterus or oviduct, and ovulation and fertilization are delayed until the following spring (after the winter hibernation). The extremely long viability of the stored sperm (within the female bats) is not well understood, but is thought to involve the establishment of a hyperosmotic environment around them that dehydrates the sperm, and then later removal of these conditions.

The delay in fertilization means that the young bats are born in summer, when there is a plentiful supply of insect food items. This timing is important both in providing the lactating females with sufficient energy and in later enabling the weaned bats to develop to a critical mass before hibernating.

Delayed embryonic development

A temporary arrest in the development of embryos (called **embryonic diapause**) occurs in some members of many animal groups, including nematodes, insects, mites, fish, lizards, birds, placental mammals and marsupial mammals. In these animals, seasonal factors such as photoperiod, food availability, temperature and rainfall affect the timing of the embryonic diapause such that reproduction can be synchronized to environmental events.

Among aquatic animals, water shortage is a common cause of embryonic diapause. For example, short-lived killifish (*Nothobranchius* spp.) spawn on flood plains that appear annually in East Africa; the fertilized eggs sink into the mud and, when the water dries up, these eggs spend the next 8 months in an arrested state of development. Development does continue but very slowly, and embryos hatch when the water returns.

In a similar way, bird embryos typically enter a short diapause after fertilization, in this case induced by the low temperature, which allows the female to build up a clutch of several fertilized eggs before incubating them all simultaneously. Incubation of the clutch of eggs as one batch results in the chicks hatching within a few hours of each other. In comparison, incubation of eggs from day one would result in hatching over several days, which would cause mayhem.

Many marsupials (more than 30 species, most of which are kangaroos and wallabies) synchronize births and mating to an annual cycle, by including a long period of embryonic diapause in their developmental cycle. Figure 20.34 illustrates the synchronization of the annual reproductive cycle of tammar wallabies (*Macropus eugenii*).

Births of tammar wallabies occur shortly after the summer solstice and are soon followed by oestrus and mating. However, embryonic development is held up at the blastocyst stage. A prolonged embryonic diapause extends throughout the entire time that the mother is suckling the joey produced by the previous years' mating. Reactivation of the new blastocyst occurs only when the young joey ceases suckling and leaves the pouch permanently, at about 10 months old.

Female tammar wallabies have their first oestrous cycle soon after leaving the pouch; as males are fertile all year round, fertilization is highly likely, but again the blastocyst enters diapause. Reactivation of blastocysts (in both mother and daughter) occurs soon after the summer solstice, which times the emergence of their foetus to about one month after the summer solstice.

So what controls the timing of an embryonic diapause? The induction of embryonic diapause in mammals is by one of two mechanisms, but both of these apply in tammar wallabies, depending on the time of year, as Figure 20.35 illustrates. The dominant mechanism, mediated by the effects of lactation, is called **lactational quiescence**. In females with a joey in their pouch, the suckling of the joey stimulates prolactin release from the anterior pituitary[39], which results

[39] We discuss the hormone prolactin in Section 19.2.2.

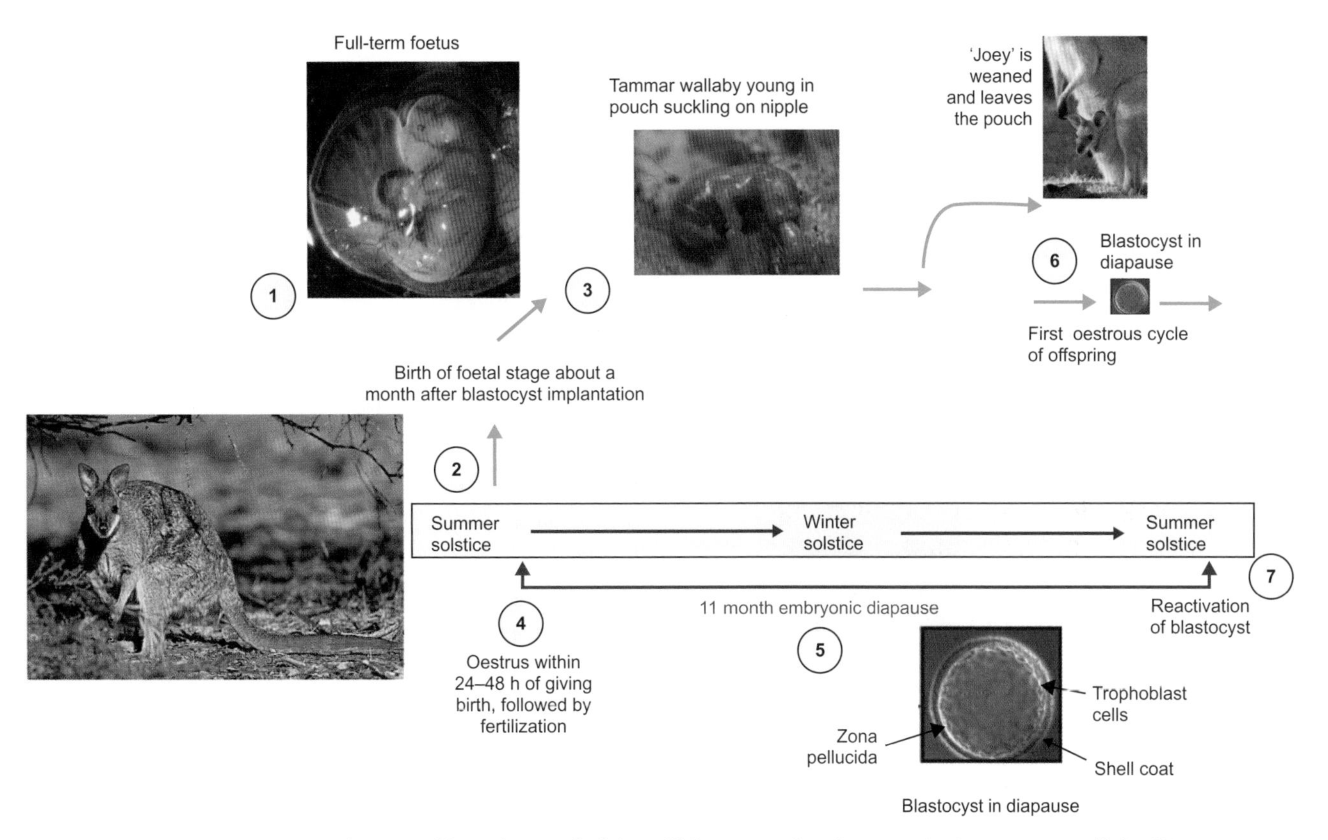

Figure 20.34 Stages in timing of seasonal breeding with delayed blastocyst development in the tammar wallaby (*Macropus eugenii*)

1. The full-term foetus is 16 mm in crown to rump length and 7.2 mm in head length. A yolk-sac placenta provides nutrition to the foetus. The allantois is covered by the yolk sac membranes. Prominent vitelline vessels make contact with the uterine endometrium and obtain nutrition from the mother. 2. The wallabies give birth in late January, soon after the summer solstice. 3. The poorly developed newborn crawls up to a nipple located in a pouch (a marsupium). 4. After giving birth, the females rapidly enter the state of oestrus, ovulate within 24 to 48 hours and mate a few hours later. 5. If fertilization occurs, the embryo develops to a blastocyst of 80–100 cells within about 7 days and then enters an embryonic diapause—a prolonged phase in which the blastocyst does not develop: that lasts up to 11 months; tammar wallabies are therefore pregnant almost all of the year, except for the day they give birth. The resting blastocyst is about 0.25 mm diameter and has a zona pellucida and an outer shell coat, which is added in the reproductive tract. Divisions form a single layer of trophoblast cells, but the resting blastocyst has no inner cell mass. 6. Female joeys are weaned in summer and enter their first oestrus soon afterward; mating results in a blastocyst that enters a short diapause. 7. The blastocysts of the mother and daughter are reactivated soon after the summer solstice.

Sources: Adult (female) tammar wallaby, Southern Australia © Dave Watts /Naturepl.com; Full-term foetus: Menzies BR et al (2011). Placental expression of pituitary hormones is an ancestral feature of therian mammals. EvoDevo 2: 16. Suckling foetus in pouch © Marilyn Renfree/Australasian Science. Blastocyst in diapause: Renfree MB and Shaw G (2000). Diapause. Annual Review of Physiology 62: 353–75. Joey in pouch © D & S Tollerton/age fotostock

in milk secretion (lactation) by the mammary glands (as in all lactating mammals). In tammar wallabies, prolactin inhibits the development of the corpus luteum and its synthesis of progesterone; progesterone is essential for continued development of the blastocyst and preparation of the uterus for implantation, so in its absence the blastocyst remains in a resting state.

If the joey is lost (or removed experimentally) during the first half of the year and suckling ceases, the corpus luteum develops; its secretion of progesterone reactivates the resting blastocyst. The outcome is the replacement of the lost suckling joey within less than a month. However, if the joey is lost after the winter solstice (the shortest day in June in the southern hemisphere), reactivation of the blastocyst is not immediate. This tells us that something other than suckling maintains the embryonic diapause in the second half of the year. The explanation is a **seasonal quiescence** between the winter and summer solstices[40] controlled by the changing photoperiod (day length). In the first pregnancy, seasonal quiescence is the only control mechanism inhibiting the development of the blastocyst.

The seasonal quiescence of the blastocysts of tammar wallabies is induced by the increasing day length, which shortens the period of peak secretion of melatonin by the pineal gland[41],

[40] December 22 is the summer solstice in the southern hemisphere.

[41] Section 19.2.3 discusses the influence of photoperiod on the pineal gland and melatonin secretion.

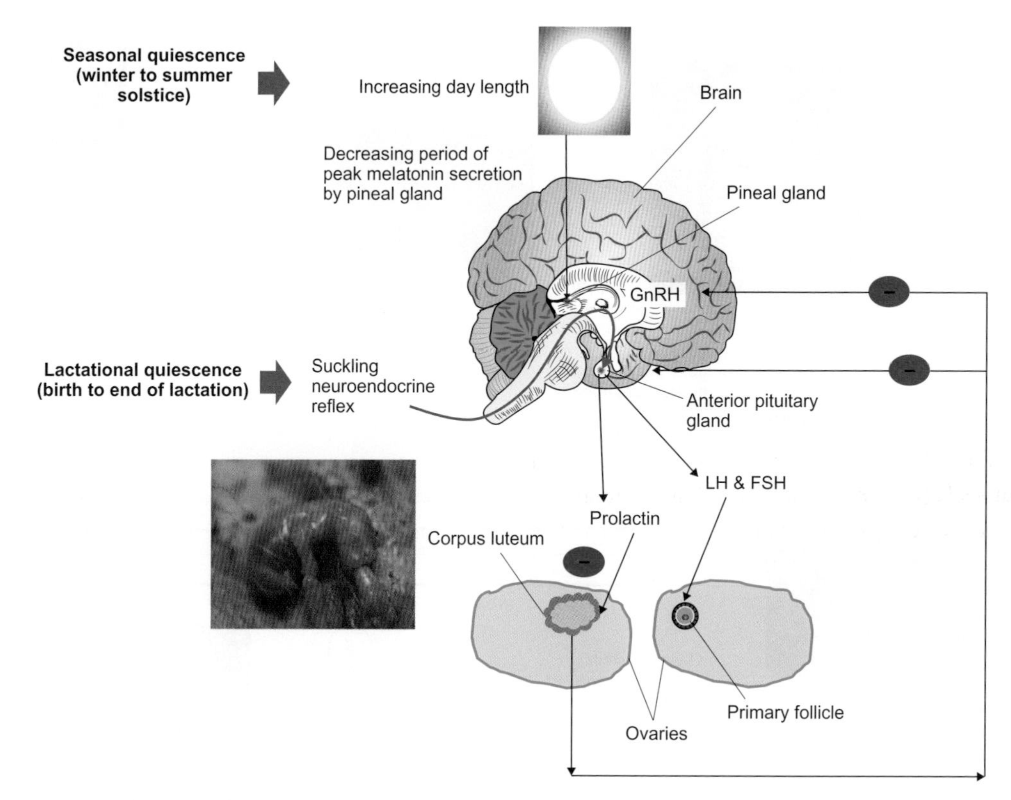

Figure 20.35 Schematic diagram of the control of lactational and seasonal phases of quiescence (diapause) of the blastocyst in tammar wallabies (*Macropus eugenii*).

Lactational quiescence exerts the dominant control and occurs between the birth of the young wallaby (end January) and the end of lactation, by which diapause is maintained by the suckling of the foetus, which stimulates prolactin release from the pituitary gland. Prolactin inhibits growth of the corpus luteum in the ovary and hence inhibits progesterone secretion. In wallabies, the resting corpus luteum inhibits secretion of gonadotrophin releasing hormone (GnRH) from the hypothalamus, and hence reduces the secretion of follicle stimulating hormone (FSH) and luteinizing hormone (LH) so that follicle growth (in both ovaries), and oestrus, are inhibited for most of the year. *Seasonal quiescence* occurs between the winter solstice (shortest day in June in the southern hemisphere) and the summer solstice (longest day in December). In this control mechanism, the increasing daylight hours reduce the period of peak concentrations of melatonin secretion by the pineal gland, which triggers a pulse of prolactin each morning, until the longest day.

Source: adapted from Renfree MB, Shaw G (2000). Diapause. Annual Review of Physiology 62: 353–75. Image: © Marilyn Renfree/Australasian Science.

during darkness. The outcome is a reduction of the inhibitory effect of melatonin on prolactin secretion by the pituitary gland, i.e. prolactin secretion increases. The pulses of prolactin secretion at dawn each morning have a critical role in maintaining the seasonal quiescence of the blastocysts of tammar wallabies[42], until the days start to shorten and nights lengthen, after the summer solstice. At this point, the increasing secretion of melatonin inhibits prolactin release, which allows the corpus luteum to develop and secrete progesterone that stimulates the reactivation of the blastocyst.

[42] This hormonal control scheme does not apply in all mammals. In American mink (*Neovison vison*) the seasonally induced diapause of blastocysts results from high melatonin secretion during short photoperiods which results in **low** concentrations of prolactin; reactivation of the blastocysts occurs when the photoperiod increases, circulating prolactin rises and results in increasing secretion of progesterone by the ovaries.

The outcome is that births occur each summer when food is abundant and provides the mother with the energy needed for lactation.

Many species of placental mammals (at least 94 species) control the time at which births occur after mating by a preimplantation embryonic diapause. During the embryonic diapause, cell division may stop altogether, as in mice and rats, or the embryo may develop very slowly with minimal cell divisions, as in bears, seals and European roe deer (*Capreolus capreolus*).

An embryonic diapause is a common occurrence among those rodents (such as rats and mice) that mate soon after giving birth (parturition), due to their post-parturition oestrus. Without the delayed implantation of the blastocyst, rats and mice would end up simultaneously feeding their newborn (by lactation) and providing nutrition for developing foetuses. The embryonic diapause, induced by lactation, avoids this double demand. An additional benefit is that if anything happens to the litter that are suckling, embryonic development can resume. In this way, rats and mice maximize the total number of litters within the breeding season.

Marine mammals such as seals that give birth on shore and mate soon afterward synchronize their annual reproductive cycle by delayed implantation of the blastocyst. Female seals are unusual in having a uterus with two horns: in one they carry a foetus, while the other is prepared for the next fertilization. After giving birth to a single pup, female seals enter oestrus within a few days and mate with males that gather when the females are ashore. However, the blastocyst does not implant in the uterine wall as long as the female is lactating, which can last several months in some species such as northern fur seals (*Callorhinus ursinus*). Once implanted, a placental phase of development continues through to the following year when seals gather again for giving birth and mating.

In some bat species, such as California leaf-nosed bats (*Macrotus californicus*), delays in blastocyst development occur *after* implantation. The delay in embryonic development lasts 4–5 months (October to March) and is unaffected by food availability and ambient temperature. The exact control mechanisms are unclear but morphological studies suggest there is poor development of the corpus luteum. This poor development is likely to account for reduced blood concentrations of oestrogens and progesterone that prevent the usual development of the uterine lining accompanying implantation.

Embryonic diapause among mammals shows no evidence of a phylogenetic distribution, which has led to the suggestion that it is an ancestral mechanism, but the geographic distribution of mammals using embryonic diapause suggests an alternative explanation. Embryonic diapause is rare among mammals living in the tropics, where environmental conditions are relatively stable. It seems that embryonic diapause evolved when mammals moved to temperate environments, with variable environmental conditions, where births need to be timed to occur in spring or summer[43]; the selective pressures could then have favoured evolution of embryonic diapause in some temperate species.

› *Review articles*

Mead RA (1993). Embryonic diapause in vertebrates. The Journal of Experimental Zoology 266: 629–641.

Primakoff P, Myles DG (2002). Penetration, adhesion and fusion in mammalian sperm-egg interaction. Science 296: 2183–2185.

Renfree MB, Shaw G (2000). Diapause. Annual Review of Physiology 62: 353–75.

Renfree MB, Fenelon JC (2017). The enigma of embryonic diapause. Development 144: 3199–3210.

[43] Look back to Section 20.1.4 for discussion of seasonal cycles of reproduction.

20.6 Asexual reproduction

Some animals can reproduce by asexual reproduction, without fusion of male and female gametes. This strategy allows rapid expansion of populations, but the lack of genetic change in the offspring could limit the scope of the population to cope with environmental changes, in which case the population expansion becomes undesirable as competition increases. A genetically diverse population will increase the chance of at least some members of the population being viable, but a population of genetically identical individuals risks no members being able to survive[44]. These factors are likely explanations of the alternation between asexual reproduction and sexual reproduction among many of the invertebrate species that use asexual reproduction.

Cnidarians (sea anemones, corals, jellyfish and hydroids) and tunicates (sea squirts)[45] can reproduce asexually by fragmentation or budding, and *Trichoplax*[46] generally reproduces asexually by binary fission (a division into two organisms). In fragmentation, a new, mature individual can grow from a fragment of the parent. Fragmentation occurs in many different types of invertebrate animal—for example, in some annelid worms, sea stars and flatworms—but these animals can also employ sexual reproduction.

In hydroids, asexual reproduction by budding involves cell division by mitosis, which forms a bud of new cells attached to the cylindrical polyp. The bud breaks away to form a new organism with an identical genetic make-up. Sea anemones can also split horizontally or at the base to form new individuals. In colonial animals such as corals and colonial sea squirts, incomplete separation after budding adds new individuals to the colony.

In some invertebrate species, including some species of ants, bees, aphids (sap-sucking insects) and water fleas (*Daphnia* spp.), *unfertilized eggs* can develop naturally into adults without any input from males, by a process called **persistent (obligate) parthenogenesis** (meaning 'virgin origin'). These animals produce egg cells by mitosis (rather than the usual process of meiosis shown in Box 20.1) so the diploid offspring are an exact full clone of their mother.

Among aphids, although persistent parthenogenesis occurs in a few species (~3 per cent) for most aphid species the

[44] We discuss the importance of genetic variation within populations of species when environmental conditions change, and as the driving force for evolution in Sections 1.3.2 and 1.5.

[45] Tunicates are one of the groups of chordates, which we discuss in Section 1.4.2.

[46] Figure 1.17C illustrates the simple flattened body form of *Trichoplax*, which is one of the three basal phyla lacking tissues, as we discuss in Section 1.4.1.

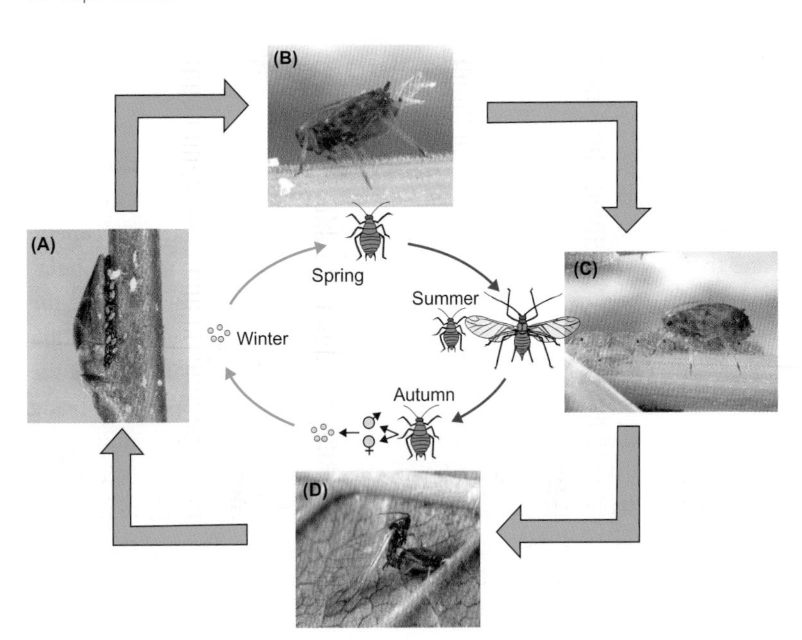

Figure 20.36 Cyclical parthenogenesis of aphids

Most aphid species typically employ cyclical parthenogenesis. The overwintering eggs (A) hatch in spring into female aphids that reproduce by parthenogenesis, giving birth to live young (B) of several generations within a few weeks. Parthenogenesis continues over the summer months building up clones of young (C). During summer, wingless and winged forms occur; winged forms can migrate to new plants. Sexual reproduction occurs in autumn; in (D) mating is between a winged male and wingless female. Mating produces overwintering eggs; in (A) the dark eggs have been fertilized.

Sources: Simon J-C et al (2002). Ecology and evolution of sex in aphids. Trends in Ecology and Evolution 17: 34-39.

typical reproductive pattern is **cyclical parthenogenesis**, in which there is an alternation between sexual and asexual reproduction. Figure 20.36 illustrates this reproductive cycle. In the spring and summer, parthenogenesis generally occurs, and is associated with viviparity. Within the body of a female, her developing daughters are already developing their own daughters. Hence, aphid populations can expand dramatically over a short timescale during parthenogenesis. However, in autumn the decreasing temperature and day length induces sexual reproduction and production of eggs that enter an overwinter resting phase (diapause)[47]. These eggs are cold (frost) resistant, whereas adult aphids are not, so sexual reproduction gives a specific physiological and ecological advantage. An additional advantage of cyclical parthenogenesis could be the increase in genotypic diversity and its evolutionary consequences. Studies indicate that the genetic diversity of lines of aphids reproducing by cyclical parthenogenesis may be as high as in fully sexual species.

20.6.1 Facultative parthenogenesis

In contrast to the common occurrence of long periods of persistent parthenogenesis in aphids and other invertebrates, a surprising number of animals that normally reproduce sexually can employ **facultative parthenogenesis** if there is a complete absence or low frequency of males for prolonged periods.

[47] We learn about diapause of overwintering eggs and adult insects in Section 9.2.1.

There are reports of facultative parthenogenesis among several species of crayfish and for all of the major vertebrate lineages, except mammals. The first report emerged in 2007 when a captive bonnethead shark (*Sphyrna tiburo*) gave birth to a single offspring after isolation from males for 3 years. Then, a captive blacktip shark (*Carcharhinus limbatus*) that had been isolated from males for 9 years was found to be nurturing a well-developed embryo. Both the bonnethead and the blacktip sharks are viviparous species (they give birth to live young).

Parthenogenesis can also occur in shark species that release egg cases. A captive whitespotted bamboo shark (*Chiloscyllium plagiosum*) reared from an egg case that had no contact with male sharks produced two parthenogens (offspring formed by parthenogenesis). Later, a zebra shark (*Stegostoma fasciatum*) held in an aquarium (without any male presence) was added to the list of shark species employing parthenogenesis. The zebra shark produced 15 pups (as live births) over four consecutive years, despite having had no contact with males. DNA analysis of the offspring and mother confirmed parthenogenesis for several of the zebra shark pups.

From these examples, it seems highly likely that facultative parthenogenesis is a common occurrence among isolated sharks. We know nothing of the behavioural, hormonal or neurological induction of facultative parthenogenesis. However, these findings mean that sharks in the wild could continue to breed even if populations get very low. This could sound like good news except for the inherent risk of reduced genetic fitness because of the inevitable reduction in genetic variability.

In solitary species that are aggressive to the opposite sex except during the periods of sexual reproduction, it is common practice in zoos to separate males and females. For example, Komodo dragons (*Varanus komodoensis*) are usually housed individually. Despite long-term isolation, two sexually mature female Komodo dragons held in zoos in England have produced healthy offspring; one produced 11 viable eggs the other four, from which healthy hatchlings emerged, as shown in Figure 20.37. Use of some of the eggs laid by these Komodo dragons for genetic fingerprinting showed that the genotype of the offspring matched the mother exactly. We can conclude that facultative parthenogenesis of Komodo dragons is likely to occur when males are absent and might occur in the wild: there are probably less than 3,000 Komodo dragon remaining, on 5 Indonesian islands, so finding males may often prove difficult.

The triggers of facultative parthenogenesis in species that normally use sexual reproduction are uncertain. Equally, the mechanism by which vertebrate species produce diploid parthenogens, when meiosis in oogenesis generates haploid eggs (as shown in Box 20.1) is not certain. Chromosome duplication could occur, but other explanations are possible. For instance, fusion of the ovum with the extruded polar body or a failure to extrude the polar body could give rise to $2n$ eggs. A similar event occurs in fish eggs when they are given an environmental shock (such as a rapid change in temperature or pressure). In this case, the application of an environmental shock during the critical period after fertilization, before extrusion of the second polar body, produces diploid ($2n$) eggs. Of course, this is not parthenogenesis. Fertilization of diploid eggs created by the aquaculture industry (by $1n$ sperm) creates *sterile*, triploid ($3n$) fish.

Figure 20.37 Hatchling Komodo dragon (*Varanus komodoensis*) an Indonesian lizard produced by parthenogenesis

The female Komodo dragon that laid the egg (one of four viable eggs in a clutch of 22 eggs) was captive bred in a zoo.

Source: Watts PC et al (2006). Parthenogenesis in Komodo dragons. Nature 444: 1021-1022.

› *Review articles*

Lampert KP (2008). Facultative parthenogenesis in vertebrates: Reproductive error or chance? Sexual Development 2: 290–301.

Neaves WB, Baumann P (2011). Unisexual reproduction among vertebrates. Trends in Genetics 27: 81–88.

Simon J-C, Rispe C, Sunnucks P (2002). Ecology and evolution of sex in aphids. Trends in Ecology and Evolution 17: 34–39.

Checklist of key concepts

Characteristics of sexual reproduction

- **Fertilization** occurs either within a female or in the surrounding environment, depending on the species.
- Environmental factors are important triggers of maturation and **synchronized spawning** of marine intertidal species.
- **Vitellogenesis** involves yolk deposition in eggs and has energetic implications for the females producing them but increases the extent of development before offspring must obtain exogenous food.
- Production of vitellogenin by **males** acts as a biomarker for exposure to **oestrogenic contaminants** in the environment.
- **Oviparous** animals release eggs when there has been little or no embryonic development.
- **Ovoviviparous** animals give the developing young protection.
- **Viviparous** animals provide the nutritional requirements of an embryo/foetus throughout development.
- The **placenta** of **eutherian (placental) mammals** forms a connection between the foetus and mother for provision of nutrients, oxygen and some hormones, and waste removal.
- Most animals reproduce **more than once** during their lifetime (= **iteroparity**).
- Some animals have a **once-in-a-lifetime** breeding event (= **semelparity**).
- Changing photoperiod often acts as a cue for **seasonal breeding**.
- Variations in food intake and its quality can affect reproductive success.

Sex determination and sexual differentiation

- **Genetic sex determination** occurs in most animal species, but not all.

- The most common system of genetic sex determination is the **XY/XX sex chromosome system**: genetic males are heterogametic (XY) while genetic females are homogametic (XX).
- Several other systems of genetic sex determination do occur.
- **Sexual differentiation**, the development of gonadal, neural and endocrine tissues, determines an animal's **phenotypic sex**.
- Embryonic gonads retain the potential to differentiate into either testes or ovaries for a short period, so genetic and phenotypic sex of animals can differ.
- Environmental factors including temperature regimes or chemical exposures in early development may influence whether testes or ovaries (or mixed gonads) develop.
- Most reptiles do not have sex chromosomes; their sex is determined during a **thermosensitive period** of embryonic development.
- **Hermaphroditic** animals are capable of producing both **male and female** gametes, either simultaneously (but usually avoid self-fertilization) or sequentially over a period of time:
  - a change from male to female is called **protandry**
  - a change from female to male is called **protogyny**.

Vertebrate male reproductive systems

- The stages of **spermatogenesis** common to all male vertebrates result in characteristic structural features in the seminiferous tubules of the testes.
- The testes of **iteroparous** seasonal breeders undergo an **annual cycle** of development and regression.
- Spermatozoa are usually ready to fertilize an ovum immediately after release, but mammalian spermatozoa must undergo **capacitation**, which increases **sperm motility** and their capacity to bind to an ovum.
- The hypothalamic–pituitary axis controls the release of **follicle stimulating hormone** (FSH) and **luteinizing hormone** (LH) which exert dual control of the **Sertoli cells** of the seminiferous tubules and the **Leydig cells**.
- Large increases in **testosterone secretion** at puberty drive development of **secondary sexual characteristics**; such changes occur annually in seasonal breeders.

Female reproductive systems of vertebrates

- At birth, the **ovary** contains many **primary oocytes** for use throughout reproductive life, but oogonial stem cells may add to the pool of oocytes.
- **FSH** stimulates follicular growth by cell division of the granulosa cells of primary follicles and secretion of follicular fluid.
- **LH** stimulates secretion of androgens by maturing follicles; **aromatase** enzymes convert the androgens to oestrogens.
- Mature **Graffian follicle**(s) secrete large amounts of **oestradiol**.
- Oestradiol enhances sexual receptivity and preparedness of the reproductive tract for mating and implantation of fertilized zygotes.
- Oestradiol also results in concentration-related feedback effects on the pituitary and hypothalamus: *negative* feedback early in **folliculogenesis**, *positive* feedback when oestradiol concentrations are higher.
- A surge of LH release stimulates **ovulation** in mammals with reproductive cycles.
- A neuroendocrine (hormonal) reflex resulting from the mechanical events during copulation cause the LH surge and ovulation of **induced ovulators**.
- A **corpus luteum** (or several corpora lutea) developed from the ruptured follicle(s) secrete **progesterone**.
- Progesterone stimulates development of the uterine lining and inhibits the next phase of folliculogenesis.

Fertilization and subsequent events

- The **acrosomal reaction** involves the release of protease enzymes that digest the outermost layer of the egg, an extension of the acrosomal process from the sperm head, and interaction of this process with species-specific sites on the egg.
- Fertilization by multiple sperm (**polyspermy**) is prevented by various mechanisms.
- Fertilization forms a **zygote**, which divides to form a **blastocyst**.
- Blastocysts of most placental mammals implant in the lining of the uterus within a few days of fertilization.
- Rescue of the corpus luteum of placental mammals and its secretion of progesterone is necessary to maintain uterine condition either until the placenta takes over, or (in some species) throughout pregnancy.
- The placenta secretes progesterone and oestrogens that inhibit ovulation and act on the uterus to maintain the pregnancy.
- The reproductive cycle of some seasonally breeding mammals is synchronized to the calendar year. Births occur at optimal times because of one of:
  - delayed fertilization,
  - delayed implantation of the blastocyst,
  - delayed blastocyst development.

Asexual reproduction

- Some animal species can reproduce asexually (as well as sexually) by fragmentation, budding, or by **parthenogenesis**.
- **Asexual reproduction** allows rapid expansion of populations, but without genetic change, which limits the population's scope to cope with environmental changes.
- In **persistent parthenogenesis** of some arthropod species, the *unfertilized eggs* produced by mitosis develop into diploid adults that are **exact clones** of their mother.
- **Facultative parthenogenesis** appears opportunistically in some vertebrate species when males are absent or rare for prolonged periods.

Study questions

* Answers to these numerical questions are available online. Go to www.oup.com/uk/butler

1. What is synchronized spawning? How do environmental factors influence this process? Discuss the control processes of a selected species that achieves synchronized spawning. (Hints: Section 20.1.1 and Case Study 20.1.)
2. Discuss the influences on the gonadal and phenotypic sex of an animal. Explain how environmental variables affect two named species in different ways. (Hints: Section 20.2; Experimental Panel 20.1.)
3. Explain why and how feeding of garden birds could influence their reproductive success. Which species are most likely to be affected and why? (Hints: Sections 20.1.2 and 20.1.4.)
4. Explain the difference between a long-day breeder and a short-day breeder and give an example of each type. How is seasonal breeding controlled? (Hint: Section 20.1.4.)
5. What do we mean by the term heterogametic sex? Describe the process of genetic sex determination known as the XY/XX system. How does this compare with the ZW system? (Hint: Section 20.2.2.)
6. What do we mean by hermaphrodite? Outline and discuss the difference between simultaneous and sequential hermaphrodites, giving examples of each. (Hint: Section 20.2.4.)
7. Explain the terms spermatogenesis, spermiogenesis and capacitation. What happens in each process? (Hints: Section 20.3 and Box 20.1.)
8.* What is the approximate age of the primary oocytes in follicles ovulated by a female gorilla when she reaches 20 years of age? What process might reduce this value? (Hints: Section 20.4.1 and Box 20.1.)
9. Explain the steps in folliculogenesis and ovulation in mammals. What are the similarities and differences between spontaneous ovulation and induced ovulation? What might be the advantage of an animal using induced ovulation? (Hint: Section 20.4.1.)
10. Outline the action of three named *peptide* hormones on the reproductive processes of particular animals or animal groups. (Hints: Sections 20.3.4, 20.4.1, 20.5.4; Sections 19.2.3 and 19.5.1 also contain relevant information.)
11. How could blood or urine samples be analysed to tell whether a captive primate is pregnant? (Hint: Section 20.5.3.)
12. Discuss three examples of species that delay one of the stages between insemination and giving birth. How is the delay achieved in each case? What are the potential advantages of the delay in each of your selected examples? (Hint: Section 20.5.4.)
13.* The period of development after implantation of the blastocyst of a fur seal until birth is 250 days. Oestrus occurs within 2 days of giving birth, followed by mating 5 days later. How long an embryonic diapause is required to synchronize reproductive events? (Hint: Section 20.5.4.)
14. What do we mean by asexual reproduction? Discuss, giving examples, how some animals can reproduce asexually. (Hint: Section 20.6.)
15. Three female sharks were brought into captivity and held together in an aquarium. Eighteen months later, two of the animals gave birth to live young. In another zoo, an isolated shark of the same species gave birth two months after captivity. What are the possible explanations of the births in each aquarium? How would you go about testing your ideas without killing the sharks? (Hints: Sections 20.6, 20.5.4, 20.1.1.)

Bibliography

Bentley PJ (1998). Comparative Vertebrate Endocrinology. 3rd Edition. Cambridge: Cambridge University Press.

Jobling M (1995). Environmental Biology of Fishes, Chapter 9; Reproducton. London: Chapman and Hall.

Norman AW, Henry HL (2014). Hormones. Academic Press.

Taylor EW (Ed.) (1996). Toxicology of Aquatic Pollution. Physiological, Molecular and Cellular Approaches. pp 283. Cambridge University Press.

21

Control of sodium, water and calcium balance

Hormones control many of the processes involved in sodium and water balance that we discuss in Part 2 of this book, such as the transport processes of fish gills, and the kidneys of mammals. Neural control also influences some organs, such as the salt glands of birds and reptiles that regulate salt balance in relation to salt influxes from the environment. These regulatory processes ultimately determine how animals adapt to their environment or how they acclimatize to new conditions and may also affect the ability of animals to cope with temporary disturbances—during exposure to pollutants, for example. In the first part of this chapter, we learn about these controlling processes and some of the changes during evolution.

The second part of the chapter examines the control of calcium balance. The most obvious biological function of calcium is in forming the internal skeleton of vertebrates, or the external body skeleton of many invertebrates, yet it has many other functions. Maintenance of calcium balance is crucial in determining membrane permeabilities, cell excitability and signalling[1], and in muscle contraction[2]. Calcium is also essential in many aspects of reproduction, including embryonic development, the production of shells in egg-laying animals, and vitellogenesis, in which yolk proteins (vitellogenins)[3] bound to calcium are deposited into eggs and provide embryos with the reserve of nutrients and calcium needed for their development. Hormonal regulation is central to controlling calcium balance, as we learn in this chapter.

21.1 Control of sodium and water balance among vertebrates

Most sodium chloride in the body is in the extracellular fluids (ECF) and is effectively barred from most of the cells[4]. Cell membranes are generally water permeable, and hence Na^+ and Cl^-, as the major ions in ECF, affect the distribution of fluid between ECF and the cytosol. These fluid movements are important in determining total ECF volume, and hence the volume of blood plasma (a component of ECF) circulating through the tissues, which influences arterial blood pressure[5]. An intake of sodium chloride tends to raise ECF volume and hence blood pressure, while low ECF concentrations of sodium chloride tend to reduce ECF volume.

Vertebrate animals have evolved hormonal control mechanisms that tightly control salt and water balance in response to disturbances such as dehydration or an increase in internal salt content. Figure 21.1 illustrates the two phases of response and some of the resulting changes in plasma hormone concentrations:

- Rapid (acute) responses enhance particular ion and/or water transport processes. These processes involve hormones that work on a timeframe of minutes to hours on existing cells and by activating existing transport mechanisms, as illustrated in Figure 21.2. Acute phase hormones may also activate the production of slower acting hormones.
- Slower responses take several hours or even days to change transport capacity and effects are generally longer lasting. Slow and longer-acting hormones stimulate the synthesis of new transport proteins and ion channels, or initiate the proliferation of new ion transporting cells, and tissue reorganization. These responses lead to changes in transport capacity that may last for several days and are often a feature of acclimatization[6].

1 We consider calcium's involvement in cell signalling in Section 19.1.4.
2 Chapter 18 examines muscle function.
3 Fish, amphibians, reptiles, birds and most invertebrates incorporate vitellogenins into their eggs, as we discuss in Section 20.1.2.

4 Figure 3.21 outlines the distribution of ions between extracellular fluids and cells.
5 Chapter 14 discusses blood pressure.
6 Acclimatization occurs in the wild, while acclimation occurs in the laboratory, as we discuss in Section 1.5.2. We discuss acclimatization to salinity in Section 5.3.

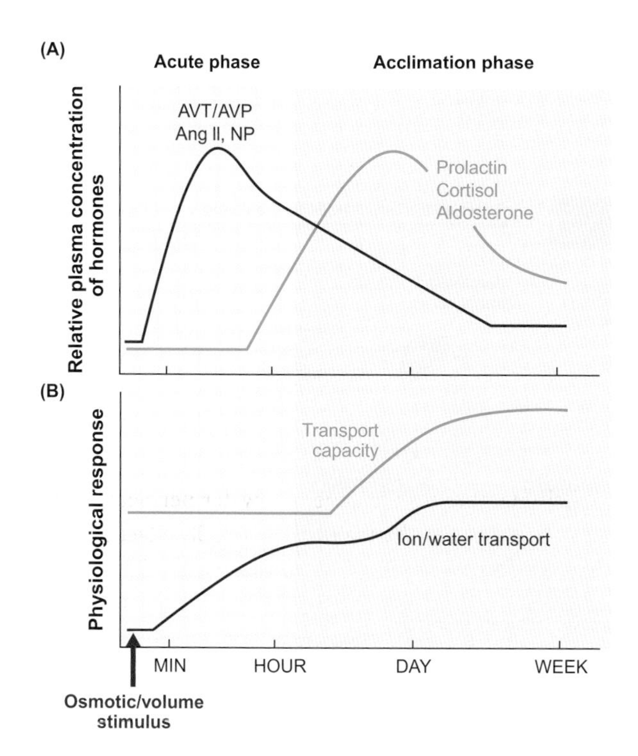

Figure 21.1 Schematic diagrams of hormonal control of ion and water balance in acute (rapid) phase, and slower longer-lasting (acclimation)

These diagrams are based on laboratory studies of responses to an osmotic/volume stimulus such as alteration in internal osmotic pressure caused by dehydration or exposure to seawater. In the acute phase (shaded in blue), an osmotic or volume stimulus (arrow) activates release of rapid-acting hormones, within seconds to hours (dark blue line), which rapidly activates the ion and/or water transport mechanisms shown in (B). Hormones released in the acute phase include arginine vasotocin (AVT) or arginine vasopressin (AVP) (depending on the species), angiotensin II (Ang II) and natriuretic peptides (NP). Osmotic stimuli and rapid-acting hormones increase longer-term acting hormones in the acclimation phase (cream panel) indicated by the pale blue line, and result in increased transport capacity by cell proliferation and reorganization.

Source: adapted from McCormick SD, Bradshaw D (2006). Hormonal control of salt and water balance in vertebrates. General and Comparative Endocrinology 147: 3–8.

Hormones regulate salt balance by controlling salt intake, salt conservation and salt excretion, and regulate water balance by regulating water intake and water conservation. We begin by examining the hormonal control of salt conservation of the terrestrial vertebrates.

21.1.1 Control of salt conservation and salt intake

Terrestrial vertebrates often have low levels of salt intake in their diet, so maximizing the reabsorption of salt by the kidney tubules[7] is usually important to minimize losses and maintain the high levels of sodium chloride in the extracellular fluid. **Aldosterone** is the key hormone reducing the excretion of sodium chloride in terrestrial tetrapods (amphibians, reptiles, birds and mammals).

Aldosterone stimulates salt reabsorption by the kidney tubules

Aldosterone is a steroid hormone synthesized by the adrenal cortex of mammals (specifically the zona glomerulosa), and the adrenocortical tissue in the adrenal glands of birds, reptiles and amphibians[8]. Figure 21.3 illustrates the control mechanisms that regulate the release of aldosterone in tetrapods in relation to their salt balance. The main control

[7] Figures 7.13, 7.16 and 7.17 give information on sodium chloride handling by vertebrate kidneys.

[8] Section 19.3.1, Figures 19.21 and 19.22 discuss secretion of aldosterone in different vertebrate groups in more detail. Teleost fish do not secrete aldosterone but secrete an alternative steroid (cortisol) that has actions on ion-regulatory processes.

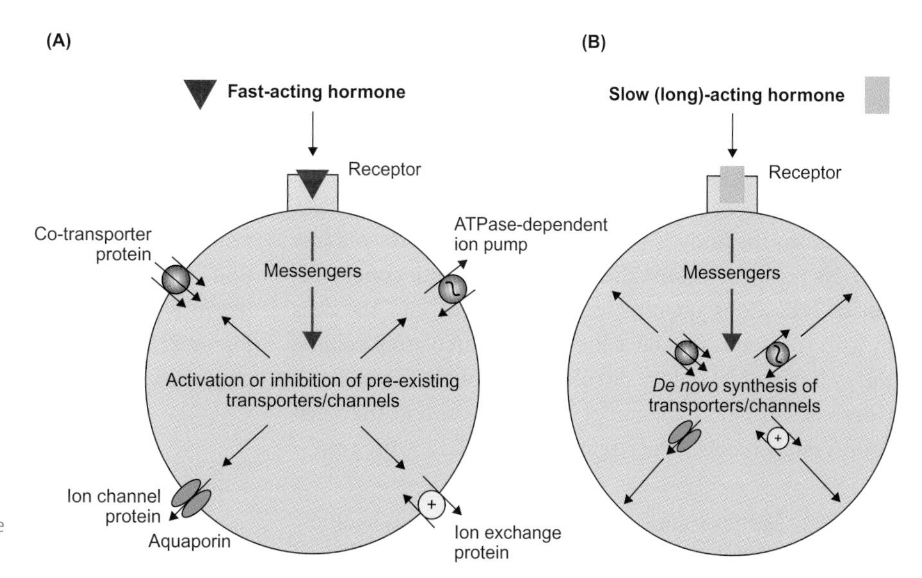

Figure 21.2 Cellular responses initiated by receptor binding of fast-acting hormones and slower (longer)-acting hormones

(A) Fast-acting hormones acting by cell messengers (discussed in Section 19.1.4) activate or inhibit existing transport mechanisms in existing ion/water transporting cells.
(B) Slower longer-acting hormones initiate synthesis of new transporter proteins and proliferation of the cells involved in transport processes.

Source: adapted from Takei Y, Hirose S (2002). The natriuretic peptide system in eels: a key endocrine system for euryhalinity? American Journal of Physiology - Regulatory, Integrative and Comparative Physiology 282: R940–R951.

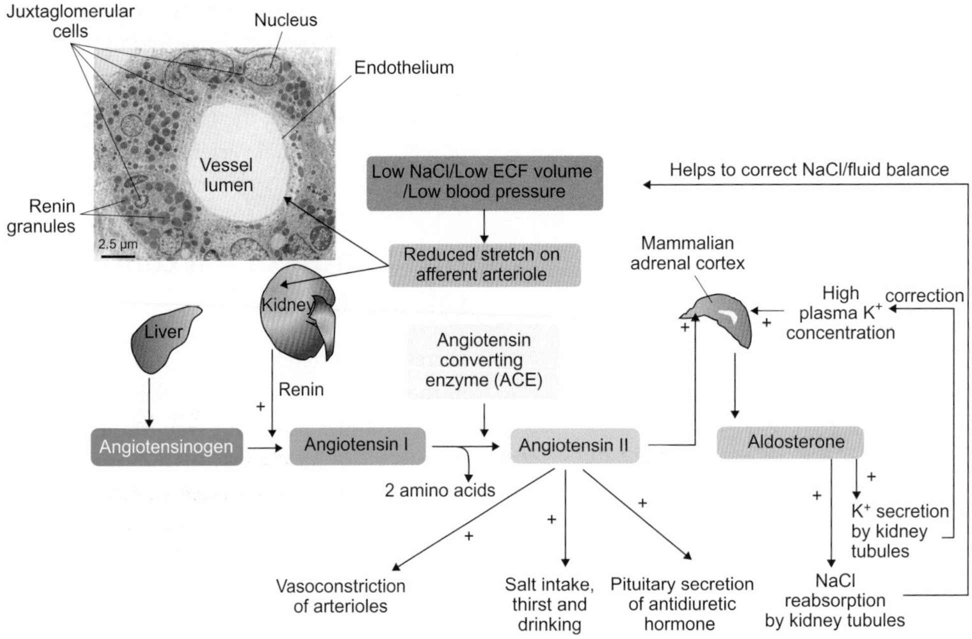

Figure 21.3 Renin–angiotensin–aldosterone system and main physiological actions

The liver synthesizes and secretes angiotensinogen which acts as the substrate for formation of angiotensin I. Renin is released from granules in secretory vesicles of juxtaglomerular cells in the wall of an afferent arteriole supplying a glomerulus. The electron micrograph shows renin granules in the mouse kidney. Renin release increases in response to reduced plasma NaCl, low extracellular fluid (ECF) volume or low arterial pressure, and is a major rate-limiting step in the formation of angiotensin I, which is converted by angiotensin converting enzyme to angiotensin II. The main physiological actions of angiotensin II are indicated; positive symbol (+) indicates stimulatory physiological effects. Angiotensin II stimulates (i) salt intake, thirst and drinking, (ii) vasoconstriction of arterioles, (iii) synthesis and release of antidiuretic hormones that conserve water in the kidney, and (iv) synthesis and secretion of steroids by the adrenal cortex. In tetrapods secretion of aldosterone has dual actions on the kidney, by actions that control NaCl and fluid balance, and actions that influence potassium balance.

Source: Micrograph from Jensen BL et al (1997). Giant renin secretory granules in beige mouse afferent arterioles. Cell Tissue Research 288: 399–406.

system is the **renin–angiotensin system**, whose principle components are shown in Figure 21.3.

The first component of the renin–angiotensin system is renin, which is an enzyme that occurs in the kidney. Renin is synthesized by modified smooth muscle cells in the wall of the afferent arterioles that supply the glomeruli of the kidney[9], or in other renal vessels. Figure 21.3 shows some renin granules in the granular cells that are visible in the electron microscope; the position of the granular cells close to the glomerulus explains their alternative name of **juxtaglomerular cells** (juxta = close to).

As illustrated in Figure 21.3, when body sodium content decreases, renin release increases. However, low total body sodium as such cannot act as a stimulus for renin release. Most Na^+ in the body is in the extracellular fluids, so a low body Na^+ content means there is a low plasma concentration of Na^+. Consequently, water movement into the cells (by osmosis) tends to reduce the effective circulating volume (the volume of blood in circulation through the tissues). A decrease in blood volume or blood pressure in the renal blood vessels reduces the stretch of the vessels with granular cells containing renin, and triggers release of the renin. Such events are the main stimulus for renin release in all vertebrates. However, additional factors also influence renin release. Nerve fibre endings of the sympathetic division of the autonomic nervous system[10] innervate the granular cells of many species and are activated by an acute reduction in blood volume/pressure, which stimulates the release of renin[11].

Chronic (long-term) hypotension (low blood pressure) increases the number of granules in the granular cells and extra granular cells may differentiate, such that more renin is released. On the other hand, a chronic increase in blood pressure (hypertension) reduces renin release.

Renin acts on the substrate, angiotensinogen, which is synthesized by the liver and always present in the circulation. Cleavage of angiotensinogen by renin forms angiotensin I (a decapeptide = 10 amino acids), as illustrated in Figure 21.3. The release of renin from the granular cells is the major rate-limiting step in producing angiotensin I and then

[9] Figures 7.5 and 7.6 show the vascular organization of the mammalian kidney and avian kidney, respectively.

[10] Section 3.3.3 outlines the sympathetic nervous system.

[11] The macula densa of the distal nephron, illustrated in Figure 7.10, is also involved in regulating renin release, as a means of controlling individual nephron function, which we discuss in Section 7.1.3.

angiotensin II, which is formed by cleavage of angiotensin I to remove two amino acids in a process catalysed by angiotensin converting enzyme (ACE). High levels of ACE occur in well-vascularized tissues such as the lungs and gills, in which ACE anchors to the vascular endothelium, so formation of angiotensin II occurs in the circulation of all vertebrates.

Circulating angiotensin II has several physiological actions in vertebrates, the main ones of which are summarized in Figure 21.3. These actions include:

- vasoconstriction of arterioles;
- stimulation of thirst and drinking, which we discuss in Section 21.1.3;
- synthesis and release of antidiuretic hormones that conserve water by actions on the kidney, as we learn in Section 21.1.4;
- stimulation of the synthesis and secretion of aldosterone, which forms the renin–angiotensin–aldosterone system in tetrapods, and/or other steroid hormones (e.g. corticosterone or cortisol) by the adrenocortical tissue. Aldosterone first appeared in the evolution of lobe-finned fishes (living representatives are the coelacanths and lungfish).

When and how did the renin–angiotensin–(aldosterone) system evolve? Physiological and morphological studies have explored this question over many years, but data from whole-genome sequence projects now allow us to search gene sequences for evidence of homologues of the genes coding for the proteins that make up a functional renin–angiotensin–(aldosterone) system. From the phylogenetic analysis in Figure 21.4, it is apparent that a step-wise emergence gave

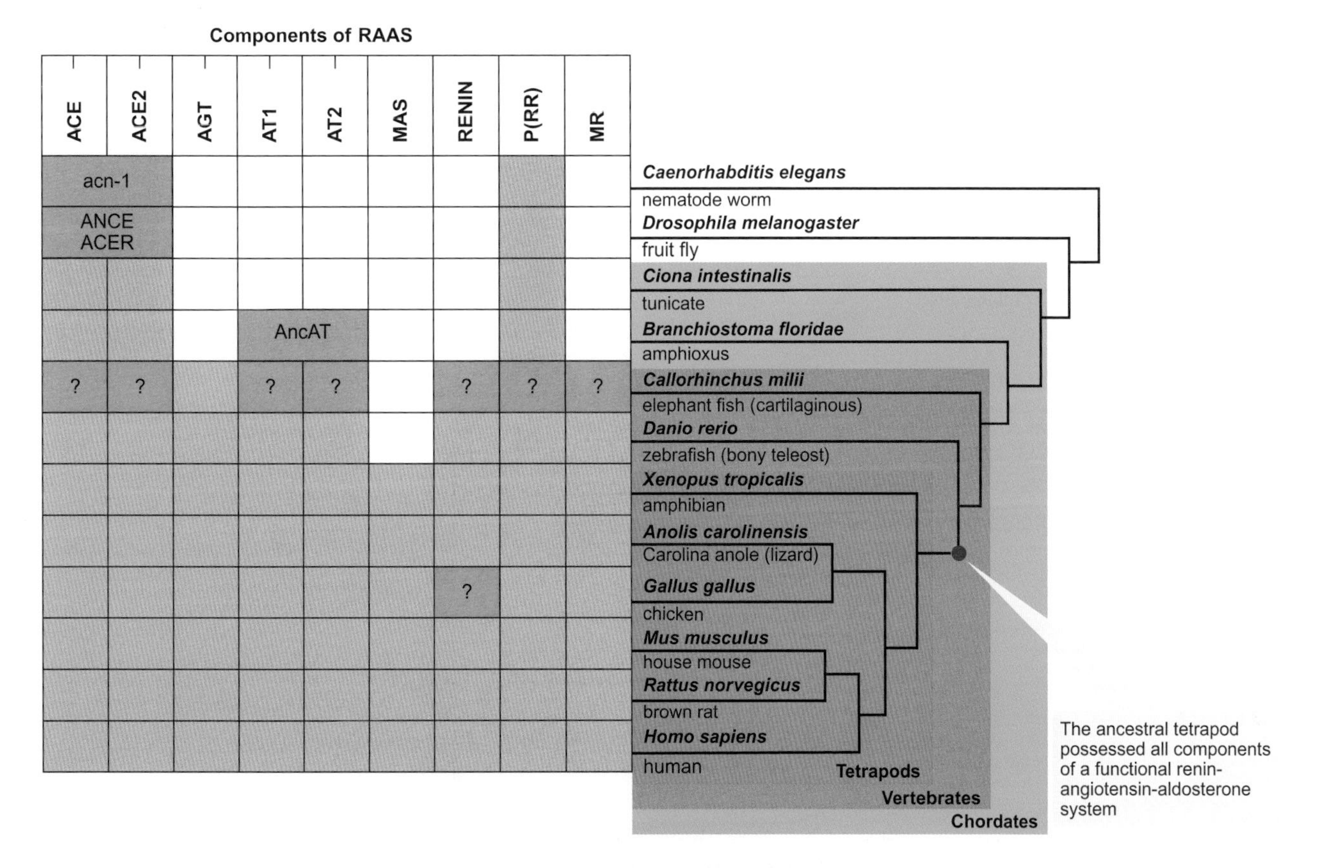

Figure 21.4 The evolution of the renin–angiotensin–aldosterone system (RAAS)

The grid shows the presence (blue), absence (white) or uncertainties (?) for each component of the RAAS in 12 species in which the entire genome has been sequenced. The components of the RAAS, which are itemized at the top of the grid are: angiotensin converting enzyme, (ACE), an alternative form of ACE (ACE2), angiotensinogen (AGT), two angiotensin receptor types (AT1, AT2), MAS (a receptor for a metabolite of angiotensin II, angiotensin 1–7), renin, a receptor for the precursor to renin–prorenin (P(RR)), and the mineralocorticoid receptor (MR) to which aldosterone binds. The animal species are listed to the right of the panel. These species include four invertebrates, two of which are chordates (within the green box), eight vertebrates (enclosed by brown box) and five species are vertebrate tetrapods (enclosed by red box). The analysis showed the tunicate has the two ACEs and the prorenin receptor, but many components are missing. Amphioxus has an additional component (AncAT), which is an ancestral angiotensin receptor. Bony fish (teleosts) have a mammalian-like system, except for the absence of the Mas receptor and lack of aldosterone (but angiotensin stimulates the release of cortisol that binds to the mineralocorticoid receptors). The tetrapods have the complete system, although the renin gene has not yet been detected in chicken. Phylogenetic analysis of the gene sequences of the RAAS shown to the right indicates a step-wise emergence of the system and that the ancestral tetrapod had all components of a functional RAAS.

Source: adapted from Fournier D et al (2012). Emergence and evolution of the renin-angiotensin-aldosterone system. Journal of Molecular Medicine 90: 495–508.

rise to this system and a complete system existed in the ancestor to all present-day tetrapods.

Notice in Figure 21.4 that a few of the component proteins in the renin–angiotensin system occur even among invertebrates, the nematode worm (*Caenorhabditis elegans*) and fruit flies (*Drosophila melanogaster*), although the ancestral functions of these proteins are uncertain. The emergence of further genes, including homologues of angiotensin converting enzyme, occurred with the evolution of non-vertebrate chordates (tunicates and amphioxus—a cephalochordate)[12], but these animals lack the full complement of genes necessary for a functional renin–angiotensin system. At the divergence of bony fish, as represented by zebrafish (*Danio rerio*), a functional renin–angiotensin system was present, except for the *MAS* gene, which encodes the Mas receptor that binds a metabolite of angiotensin II, and which evolved in the tetrapods.

In mammals, aldosterone acts on the late distal nephron: the second half of distal convoluted tubule, the connecting tubule and the collecting duct[13], all of which are often grouped together as the aldosterone-sensitive distal nephron. Figure 21.5 shows a simplified model of the main cellular steps involved in Na^+ conservation by the kidney under the stimulation of aldosterone. The rate-limiting step in this process is the luminal reabsorption of Na^+ via Na^+ channels. Notice that aldosterone enters the basolateral border of the cells and binds to specific receptors. The aldosterone–receptor complex binds to the hormone response element of DNA[14], which stimulates a genomic action by transcription of specific genes and translation of specific mRNAs. Recent studies have shown that aldosterone can also induce the repression of many genes, but the importance of these changes in cellular responses is not certain.

Aldosterone's stimulation of sodium reabsorption by the kidney occurs in two main phases. After a delay of 20 to 60 minutes, an initial regulatory phase occurs over a period of

[12] We discuss animal diversity including the tunicates and cephalochordates in Section 1.4.2.

[13] Figure 21.30 shows the basic layout of the mammalian nephron; Figure 7.5 gives more detail.

[14] We examine this mode of action by nuclear receptors more closely in Section 19.1.4 (Figures 19.10 and 19.11).

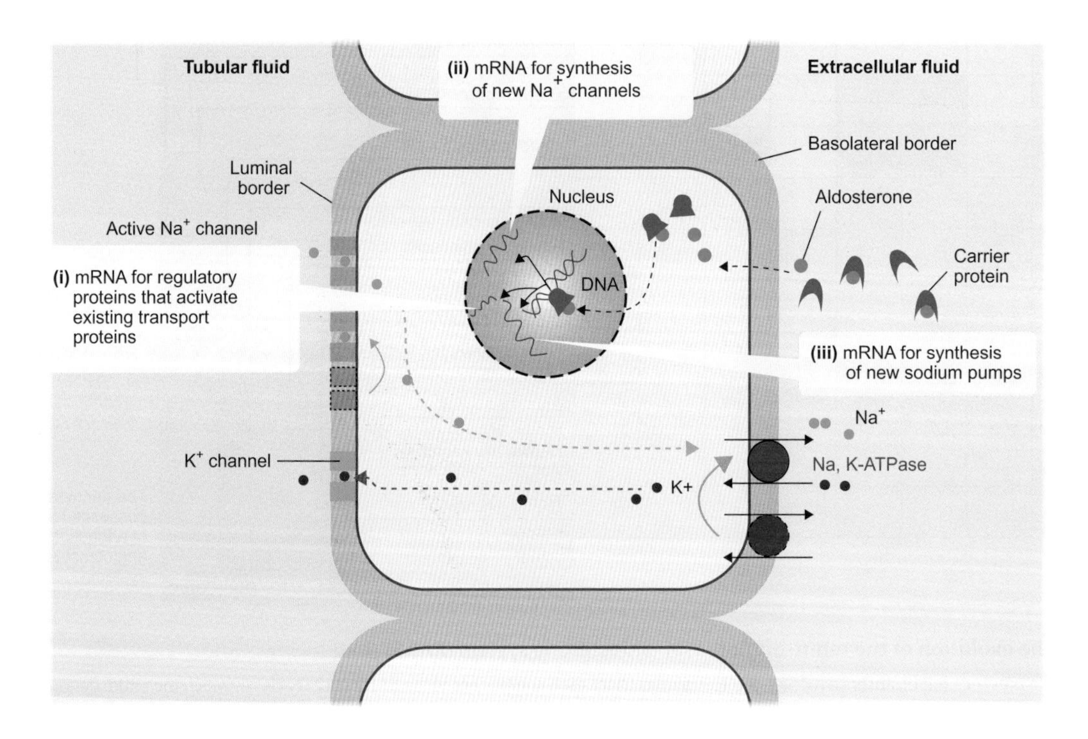

Figure 21.5 Key steps in aldosterone stimulation of sodium reabsorption and potassium secretion in the aldosterone-sensitive distal nephron of the mammalian kidney

Circulating aldosterone is mostly bound to molecules of carrier proteins. Free aldosterone enters the basolateral border of the epithelial cells. Within the cells, aldosterone interacts with nuclear receptors in the cytosol that bind to DNA and stimulate transcription of specific genes. Aldosterone stimulates the regulatory proteins that activate existing Na^+ channels in the luminal border and Na^+, K^+-ATPase (sodium pumps) in the basolateral border of the epithelium of the distal nephron (orange arrows). Aldosterone also stimulates synthesis of new Na^+ channels and sodium pumps. Thus, aldosterone stimulates uptake of Na^+ from the tubular fluid and increased passage across the basolateral border, in exchange with K^+. Aldosterone stimulates K^+ secretion by the enhanced Na^+, K^+-ATPase activity and K^+ exits via an increased number of open K^+ channels on the luminal border of the cells.

several hours. In this early regulatory phase, activation of existing ion channels in the luminal border of the epithelial cells increase Na^+ uptake into the cells, and activation of pre-existing Na^+, K^+-ATPase (sodium pumps) in the basolateral border of the cells increase Na^+ exit from the cells. Activation of additional proteins also occurs in some tissues but their involvement in the cellular responses is uncertain. Figure 21.6 shows a significant increase in Na^+ transport across collecting duct cells of mice after 90 min of treatment with aldosterone. The continuing rise in Na^+ transport over several hours possibly indicates the beginnings of the later regulatory phase, which lasts up to 24 hours. During the late regulatory phase, aldosterone stimulates the synthesis of new Na^+ channels and sodium pumps.

Aldosterone stimulates potassium excretion by the kidney

In addition to the effects of aldosterone on Na^+ reabsorption, aldosterone increases potassium secretion as the schematic in Figure 21.5 illustrates. In the aldosterone-sensitive part of the distal nephron:

- stimulation of basolateral Na^+, K^+-ATPase by aldosterone increases cellular K^+ concentrations and thus favours K^+ secretion;
- aldosterone stimulates an increase in the number of open K^+ channels in the luminal membrane of the tubular epithelium[15];
- aldosterone's stimulation of Na^+ absorption from the luminal fluids (discussed in the previous sub-section) increases the electronegativity of the lumen, which also favours K^+ secretion.

[15] Section 7.2.4 discusses the types of K^+ channels in the mammalian nephron that drive K^+ secretion.

Indeed, aldosterone is the main hormone responsible for controlling K^+ balance in mammals, as indicated by the rise in circulating aldosterone when dietary intake of K^+ increases plasma (extracellular fluid) concentrations of K^+.

An increase in the blood plasma concentration of K^+ in mammals, to 5 per cent above normal, (hyperkalaemia) is sufficient to act as a direct stimulant of the secretion of aldosterone by the zona glomerulosa of the adrenal gland, and the aldosterone secretion correlates with plasma K^+ concentrations. As such, plasma K^+ and aldosterone are two components of a feedback loop that regulates the plasma concentrations of both K^+ and Na^+.

The actions of aldosterone on the kidney examined in Figure 21.3 and Figure 21.5 suggest that elevated aldosterone simultaneously stimulates an increase in Na^+ absorption and K^+ secretion, by the nephron, which poses a possible conflict between maintaining sodium and volume balance and maintaining potassium balance. For example, volume depletion activates the renin–angiotensin system and results in an increase in circulating aldosterone, which stimulates sodium absorption from the kidney, and helps to restore extracellular fluid volume. As aldosterone also stimulates potassium secretion by the kidney, we might wonder whether a low ECF volume also results in excessive K^+ excretion and hence causes K^+ depletion. This does not occur, rather during volume depletion, aldosterone stimulates salt retention by the mammalian kidney *without* increasing K^+ excretion; K^+ secretion is normally independent of volume status. Also, the increase in aldosterone in hyperkalaemia stimulates K^+ excretion, but *without* increasing salt retention.

The apparent separation of effects on K^+ excretion and salt retention is known as the **aldosterone paradox** and is partly explained by the concomitant effects of changes in the delivery of Na^+ and water to the distal nephron on K^+ secretion. During volume depletion, the stimulation of K^+ secretion by the effects of an increase in aldosterone on the

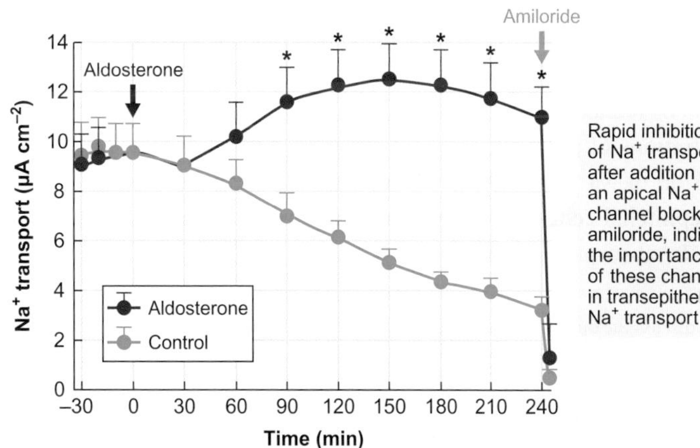

Figure 21.6 Aldosterone stimulation of Na^+ transport across monolayers of cultured cells from the collecting ducts (distal nephron) of mice kidneys

Na^+ transport is significantly increased within 90 min of adding aldosterone (at time 0 min) to fluid on the basolateral side of the cells, compared to Na^+ transport in control preparations without aldosterone.

Data are means + standard errors for eight preparations. The asterisks indicate significant effects ($P<0.05$ relative to pre-treatment values).

Source: Shane MA et al (2006). Hormonal regulation of the epithelial Na^+ channel: From amphibians to mammals. General and Comparative Endocrinology 147: 85–92.

distal nephron are largely offset by the inhibitory effects on K^+ secretion of the reduced delivery of Na^+ and fluid to the distal nephron[16] via two effects:

- Volume depletion results in an increased reabsorption of salt and water in the proximal tubule (partly through the action of angiotensin II). This results in a higher concentration of K^+ in the fluid reaching the distal nephron, which impedes K^+ secretion.
- The reduced Na^+ delivery to the distal nephron inhibits the more distal reabsorption of Na^+, which reduces the electronegativity of the tubular lumen, and impedes passage of K^+ via the luminal ion channels.

Physiological concentrations of aldosterone often have relatively minor effects on K^+ excretion, and dietary intake of large quantities of K^+ often results in an increase in K^+ excretion *before* significant elevation of the plasma concentration of aldosterone. Thus, aldosterone-independent factors are also implicated in K^+ secretion by the renal tubules. Perhaps the strongest evidence is that mice with deletion of the gene for aldosterone synthase—the enzyme that is necessary for synthesis of aldosterone[17]—maintain high levels of renal excretion of K^+ if they are given a high K^+ diet. There is evidence for an intestinal sensor for dietary K^+ (and Na^+ and phosphate) in mammals, and such a sensor may signal to the kidney causing alterations in ion excretion, but the signalling mechanisms are unknown.

Salt-conserving actions of aldosterone on non-renal tissues

In addition to the salt-conserving actions of aldosterone on the kidney, aldosterone stimulates salt conservation and/or salt uptake in a variety of other tissues in tetrapods:

- Aldosterone stimulates Na^+ reabsorption from the fluid secreted by the sweat glands of mammals.
- Aldosterone stimulates Na^+ and Cl^- absorption from fluid in the lower colon and lower parts of the gastrointestinal tract (and K^+ secretion) in a similar way to the salt absorption from the distal nephron. For example, in hens fed a low-salt diet, an increase in plasma concentrations of aldosterone results in an increased expression of Na^+ channels in the luminal membrane of the epithelial cells of the lower intestine, which enhances Na^+ absorption from the diet.
- In birds, salt absorption from the mixture of urine and faeces also occurs in the coprodeum, which forms part of the **cloaca**, and after refluxing of the urine into the colon[18]. An increase in circulating concentrations of aldosterone of hens acclimated to a low-salt diet is associated with more than a doubling of the luminal surface area of the epithelial cells of the coprodeum, due to the increased density and length of the apical microvilli, which enhances salt absorption and reduces salt loss in the urine–faecal mix.
- Among amphibians, the stimulation of salt-conservation by the urinary bladder by aldosterone is generally more important than the stimulation of Na^+ reabsorption from the kidney tubules. Aldosterone stimulates Na^+, K^+-ATPase activity (sodium pumps) in the basolateral membrane of the bladder epithelium, which increases in the retrieval of Na^+ and Cl^- from the urine before its excretion.
- Aldosterone stimulates Na^+ uptake via the skin of amphibians when they immerse in water. The uptake mechanism, via apical epithelial Na^+ channels, bears a striking resemblance to the transport processes in the mammalian distal nephron. The uptake of Na^+ across the skin of amphibians allows Cl^- uptake and draws in water, by osmosis, which is particularly important to compensate for dehydration during periods on land[19].

The influence of aldosterone on salt intake

Selection of food items is a significant factor in determining the amount of salt consumed by an animal in its diet. Salt depletion (when aldosterone is elevated) stimulates consumption of dietary items with a higher salt content—at concentrations normally avoided. The functioning of salty taste receptors[20] plays an important part in such salt intake. The presence of epithelial Na^+ channels in the apical membranes of these receptors is thought to enable them to function as Na^+ sensors.

Aldosterone regulates the number of Na^+ channels in the apical membranes of salty taste receptors, which will determine their salt sensitivity. Within a few hours of an increase in circulating concentrations of aldosterone, extra Na^+ channels insert in the apical membranes of the salty taste receptor cells, which increases their salt sensitivity and hence reduces salt intake. This seems a paradoxical response

[16] We discuss the opposite effects of an increase in fluid volume via effects on Na^+ and fluid delivery that enhance K^+ secretion in the distal nephron in Section 7.2.4.

[17] Figure 19.3 illustrates the pathways for steroid secretion, including aldosterone.

[18] Figure 6.33 illustrates the bird cloaca (proctodeum, urodeum and coprodeum), its communication with the colon and the channelling and processing of urine in the coprodeum and colon.

[19] We discuss the hormonal regulation of water uptake across the skin of amphibians in Section 21.1.4.

[20] Section 17.3.2 discusses taste cells in some detail.

given that aldosterone is elevated when salt concentrations in ECF are low. However, angiotensin II (which is elevated prior to aldosterone, as shown by the scheme in Figure 21.3) reduces the Na^+ channels in the apical membrane of salty taste cells.

How the opposite effects of angiotensin II and aldosterone operate together is not certain, but the current idea is that the initial effects of salt deficiency and increasing circulating concentrations of angiotensin II reduce the sensitivity of the salty taste receptors, which results in a rapid increase in salt intake. The actions of aldosterone subsequently increase the responsiveness of these cells, perhaps preventing excessive salt intake.

21.1.2 Excreting excess salt

Excretion of excess salt is necessary to maintain salt balance in animals that face high rates of salt influx, such as marine teleost fish, in which salt influx occurs via the gills and gastrointestinal tract[21], and in animals such as marine birds and humans that often consume an excess of salt in their diet. One way to eliminate some of these salts is by excretion in the urine. Another way is to excrete salts via salt glands, or for marine fish, via their gills.

Natriuretic peptides stimulate renal salt excretion

Adolf de Bold and his team in Canada in 1981 were the first to discover the vertebrate heart has hormonal activity that stimulates salt excretion when they identified modified cardiac myocytes in the atrial chambers of the heart that synthesize and store peptide hormones. Early attempts to isolate the active substance(s) initially obtained peptides of various lengths because various amounts of degradation occurred during purification. Eventually, a complete peptide was obtained from rat atria, and was named **atrial natriuretic peptide (ANP)** to describe its origin and its natriuretic action once released into circulation. A **natriuresis** is an increase in salt excretion (from the Latin *natrium* for sodium; and the Greek for urination).

Over the next 20 years or so, a whole family of structurally related peptides were identified among the vertebrates, and work on these peptides and understanding their evolution continues today. In addition to ANP in the mammalian heart, two related peptides occur (predominantly in the brain): B-type natriuretic peptide (BNP), functioning usually as a paracrine/autocrine factor, and C-type natriuretic peptide (CNP), which is the only natriuretic peptide in agnathans. An additional cardiac peptide, **ventricular natriuretic peptide (VNP)**, occurs in the ventricles of some types of fish (teleosts, and chondrosteans such as sturgeons), along with four different sub-types of CNP identified among fish. Therefore, we can see how fish show a particularly high level of variation in natriuretic peptides. All natriuretic peptides contain a ring of 17 amino acids formed by a disulfide bond between two cysteine residues, and there are conserved amino acids in the ring, as shown in Figure 21.7.

Given the wide distribution of ANP among vertebrates, the reptilian heart was for many years assumed to contain ANP. Indeed, use of mammalian ANP antibodies suggested some ANP-like immunoreactivity in atrial myocytes of two lizard species (*Anolis carolinensis* and *Lacerta viridis*) and a snake (*Python reticulatus*). However, these studies show the dangers of using antibodies raised against the peptides of a different species (heterologous antibodies): the risk of false-positive signals is high. Molecular studies to clone genes for natriuretic peptides in particular species, which enable generation of species-specific antibodies, are necessary to be certain that particular species produce particular peptides. Such studies have not found ANP in reptiles, except chelonians (tortoises, for example, eastern long-necked tortoise, *Chelodina longicollis*). From an evolutionary perspective, it is interesting that ANP is also absent in the heart of the few species of birds investigated (chicken and pigeon). It seems that the ANP gene was lost in non-chelonian reptiles and birds after the evolutionary branch of tortoises/turtles from the line that gave rise to other reptiles (crocodilians, lizards and snakes) and birds. These reptiles and birds instead release BNP from their cardiac myocytes.

Functional role and actions of natriuretic peptides

The main stimulus for the release of natriuretic peptides from the heart is stretching of the cardiac myocytes, by a high blood volume, or an increase in venous return to the heart. This is the starting point in Figure 21.8. The fundamental link between the stretch of the atrial myocytes and ANP release raised interest in the impact of hypervolaemia (high blood volume), hypertension and congestive heart failure on the circulating levels of the ANP in humans. Indeed, clinical measurements of ANP have proved to be a useful marker of heart failure because an increase in venous volume resulting from congestive heart failure results in atrial stretch, which stimulates secretion of ANP from the heart. In such circumstances, the release of ANP has potentially beneficial actions, and explains why some countries (e.g. Japan) use ANP as a therapeutic treatment for congestive heart failure.

[21] We examine salt influxes and effluxes of marine teleosts in Section 5.1.3.

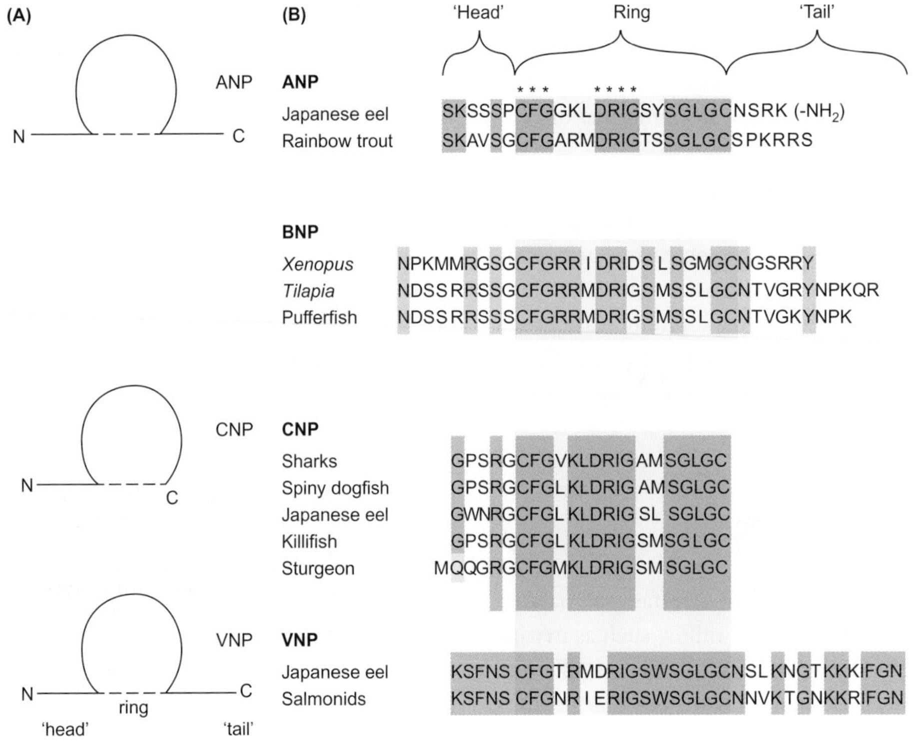

(A)

ANP
N C

CNP
N C

VNP
N C
ring
'head' 'tail'

(B)

'Head' Ring 'Tail'

ANP
Japanese eel SKSSSPCFGGKLDRIGSYSGLGCNSRK (-NH2)
Rainbow trout SKAVSGCFGARMDRIGTSSGLGCSPKRRS

BNP
Xenopus NPKMMRGSGCFGRRIDRIDSLSGMGCNGSRRY
Tilapia NDSSRRSSGCFGRRMDRIGSMSSLGCNTVGRYNPKQR
Pufferfish NDSSRRSSSCFGRRMDRIGSMSSLGCNTVGKYNPK

CNP
Sharks GPSRGCFGVKLDRIGAMSGLGC
Spiny dogfish GPSRGCFGLKLDRIGAMSGLGC
Japanese eel GWNRGCFGLKLDRIGSLSGLGC
Killifish GPSRGCFGLKLDRIGSMSGLGC
Sturgeon MQQGRGCFGMKLDRIGSMSGLGC

VNP
Japanese eel KSFNSCFGTRMDRIGSWSGLGCNSLKNGTKKKIFGN
Salmonids KSFNSCFGNRIERIGSWSGLGCNNVKTGNKKRIFGN

Figure 21.7 Structure of natriuretic peptides

All natriuretic peptides contain a 17 amino acid ring formed by a disulfide bond between two cysteine residues. (A) Schematic diagrams of the ringed structure, which has head and tail components extending either side of the ring, except in C-type natriuretic peptides (CNP), which lacks the tail; C = carboxy terminus, N = amino terminus. (B) Examples of the amino acid sequences, as single letter codes for atrial natriuretic peptide (ANP), B-type natriuretic peptides (BNP), CNP, and ventricular natriuretic peptide (VNP). Single letter codes for amino acids are explained in Box 2.1. CNP 'shark' sequences are for Japanese dogfish, European lesser spotted dogfish and porbeagle shark. VNP sequences for salmonid species are for rainbow trout and chum salmon. Grey shading indicates identical amino acid residues within the specific type of natriuretic peptide, for the sequences illustrated. Asterisks indicate amino acid residues that are identical in ANPs, most BNPs, CNPs and most occur in VNPs. Note that CNPs are smaller peptides without a tail section, while VNP has an extended tail. The amino acids forming the ring between two cysteines (C) are overlaid by yellow boxes.

Sources: Loretz CA, Pollina C (2000). Natriuretic peptides in fish physiology. Comparative Biochemistry and Physiology Part A 125: 169–187. Kawakoshi A et al (2004). Four natriuretic peptides (ANP, BNP, VNP and CNP) coexist in the sturgeon: identification of BNP in fish lineage. Journal of Molecular Endocrinology 32: 547–555.

The beneficial effects of ANP are due to its rapid vasodilatory action, including the dilation of coronary vessels.

While stretch of the myocytes usually acts as the main control of ANP release, an increase in blood Na^+ or osmolality may also trigger it. Large changes in blood volume can trigger the release of natriuretic peptides from the teleost heart. However, elevated plasma osmolality[22] and sodium concentrations have a much greater effect. The graphs in Figure 21.9 show the results of some experiments on freshwater-acclimated Japanese eels (*Anguilla japonica*) in which intravascular administration of hypertonic NaCl stimulates a rapid increase in plasma concentrations of ANP and VNP by up to 20–50 times the initial level, while intravascular administration of isotonic saline has no effect on circulating natriuretic peptides. The increased plasma osmolality seen in these eels is within the range seen in eels exposed to seawater. A sensitivity of isolated cardiac myocytes to Na^+ concentrations and osmolality is also apparent in mammals. However, in whole animals, an increase in blood (extracellular fluid) osmolality ultimately drives an increase in blood volume as water moves from the cells by osmosis.

We know most about the physiological role of natriuretic peptides in mammals, and specifically the actions of ANP, although even here the effects are not completely understood. ANP released by the heart is thought to act as a protective mechanism against volume expansion and its impact on blood pressure mainly through three interlinked effects that are summarized in Figure 21.8:

- rapid **vasodilation** of systemic arterioles that are pre-constricted;
- stimulation of a renal diuresis: an increase in the urine volume produced per unit time;
- stimulation of a renal natriuresis: an increase in the rate of sodium chloride excretion.

[22] Osmolality is a measure of the total osmotic concentration as explained in Section 4.1.1.

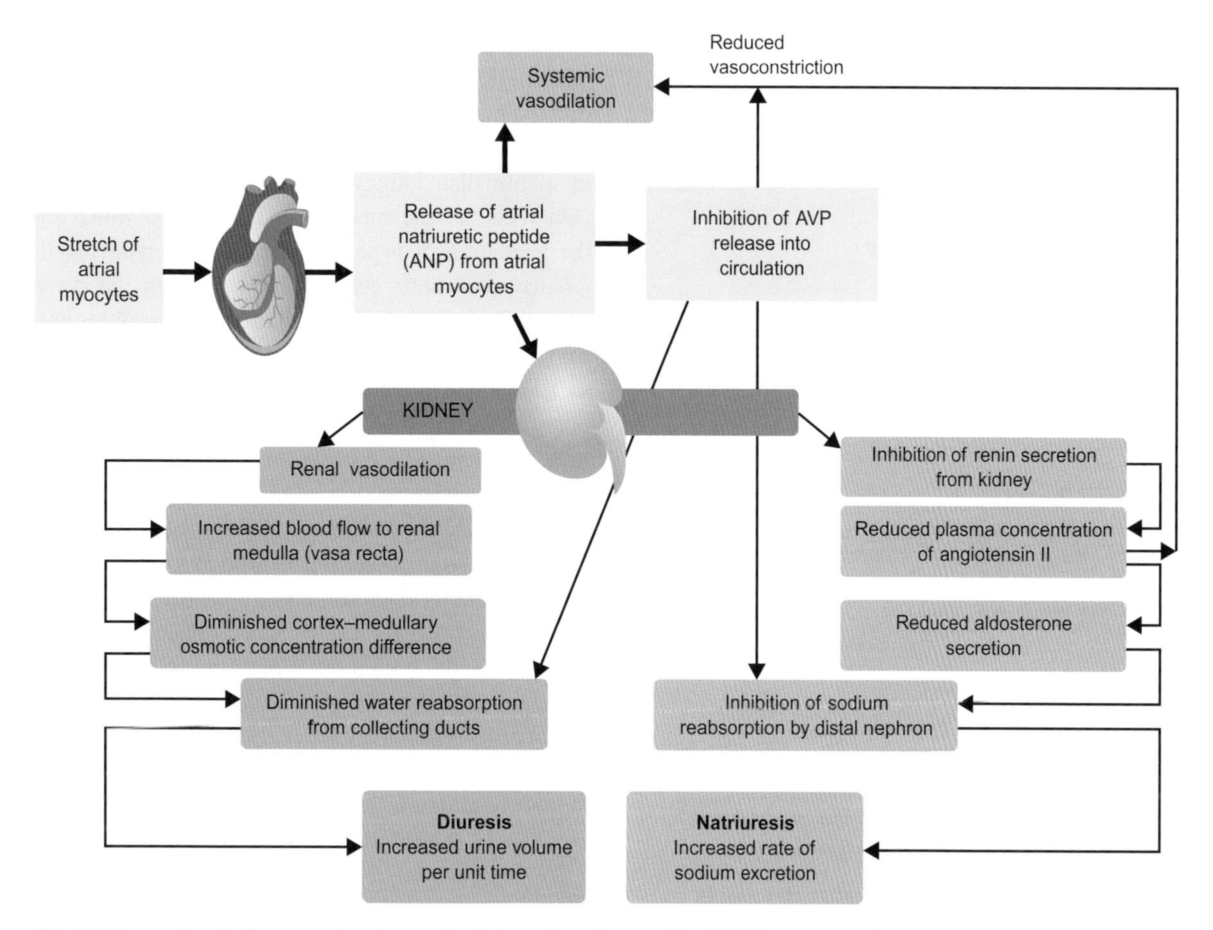

Figure 21.8 Schematic of release and actions of atrial natriuretic peptide (ANP) in mammals

Stretch of the atrial myocytes stimulates the release of ANP that results in systemic vasodilation and has renal actions that result in natriuresis and diuresis. The inhibition of the release of arginine vasopressin (AVP) and inhibition of the renin–angiotensin system reduces their vasoconstrictor actions, which enhances the vasodilatory effects of ANP on pre-constricted vessels. In the kidney, increased blood flow through the vasa recta diminishes the ability to reabsorb water and inhibits the formation of hyperosmotic urine. The decline in AVP (= antidiuretic hormone) resulting from the actions of ANP adds to its diuretic and natriuretic effects. The natriuresis induced by ANP results from inhibition of sodium reabsorption from the renal tubule due to direct action and by indirect effects by inhibition of AVP and the renin–angiotensin–aldosterone system.

These three types of action result from the direct actions of ANP on specific receptors in the systemic vasculature, the renal vasculature and the epithelium of kidney tubules, coupled with the indirect effects of other hormone systems under the influence of ANP[23].

The rapid vasodilatory action of ANP on systemic vasculature is enhanced by ANP's inhibition of vasoconstrictor hormone/neural systems:

- inhibition of the sympathetic nervous system;
- inhibition of the renin–angiotensin system (outlined in Figure 21.3), by inhibiting renin release and hence reduced generation of angiotensin II, which reduces the vasoconstriction of arterioles induced by circulating angiotensin II;
- inhibition of the secretion from the posterior pituitary gland of arginine vasopressin (AVP) which is a vasoconstrictor hormone[24].

The natriuretic action of ANP, and its accompanying diuretic action, is a somewhat slower event than its rapid vasodilatory actions. The excretion of salt and water stimulated by ANP and its indirect actions via other hormone systems results in longer-term depressive effects on ECF volume (and blood pressure[25]).

At first, the diuretic action of ANP in mammals was thought to result from the combination of an increase in

[23] ANP also inhibits drinking; we discuss the hormonal regulation of drinking in Section 21.1.3

[24] Arginine vasopressin is also called antidiuretic hormone (ADH) due to its renal actions, which we learn about in Section 21.1.4.

[25] Section 22.4.4 also discusses the long-term regulation of blood pressure.

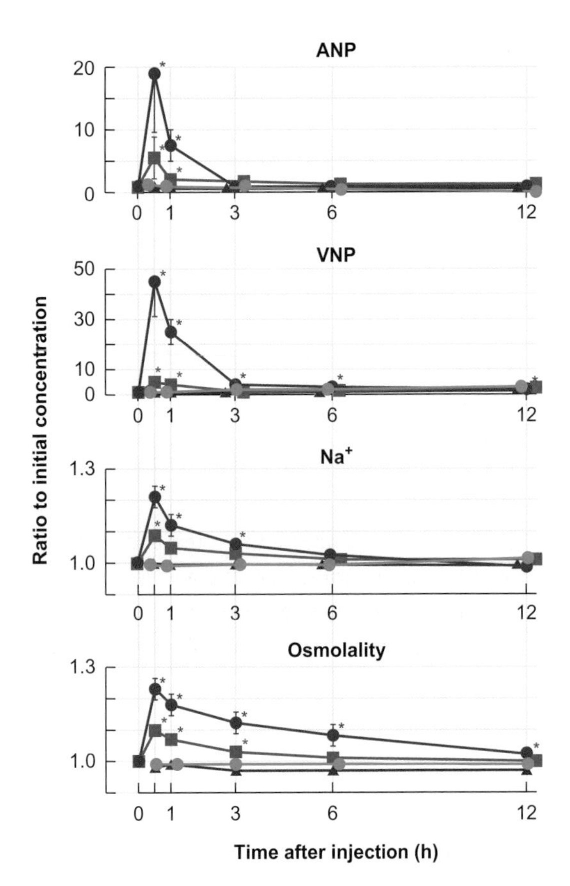

Figure 21.9 Effect of salt loading on relative changes in plasma concentrations of natriuretic peptides, plasma Na^+ concentrations and plasma osmolality of Japanese eels (*Anguilla japonica*) acclimated to fresh water

Graphs show time courses of changes in plasma concentrations of atrial natriuretic peptide (ANP), ventricular natriuretic peptide (VNP), Na^+ concentrations and plasma osmolality after injection of 2.5 mL of NaCl per kg body mass at various concentrations of NaCl: hypertonic NaCl (● 1.7 mol L^{-1} and ■ 0.85 mol L^{-1}), isotonic NaCl (● 0.15 mol L^{-1}), and hypotonic NaCl (▲ 0.017 mol L^{-1}). Hypertonic NaCl results in a transitory increase in plasma concentrations of ANP and VNP following a very similar pattern and time course to the changes in plasma osmolality and sodium concentrations. Data points are means ± standard errors of means; asterisks show significant changes, $P < 0.05$ compared with eels injected with isotonic saline (0.15 mol L^{-1} NaCl).

Source: Kaiya H, Takei Y (1996). Osmotic and volaemic regulation of atrial and ventricular natriuretic peptide secretion in conscious eels. Journal of Endocrinology 149: 441–447.

glomerular filtration rate (GFR)[26], i.e. a **glomerular diuresis**, and a decrease in water reabsorption by the renal tubules, i.e. a **tubular diuresis**. However, ANP's actions on mammalian GFRs are controversial. Generally, concentrations of ANP within the physiological range have relatively little effect on the GFR of mammalian kidneys. Hence, in mammalian kidneys a decrease in tubular water reabsorption, particularly by the collecting duct, is the main driving force for the ensuing diuresis. In non-mammalian vertebrates, however, in which kidney GFR is much more labile because of the variable number of filtering nephrons[27], natriuretic peptides typically do increase GFR, possibly due to an increase in renal blood flow.

As indicated in Figure 21.8 (brown-shaded panels), ANP in mammalian kidneys increases blood flow through the vasa recta, in the medulla of the kidney, which impedes the formation of hyperosmotic urine by reducing the difference in osmotic concentration between the cortex and medulla[28]. ANP also inhibits the actions of arginine vasopressin (AVP, also known as antidiuretic hormone, ADH), such as the increase in the osmotic water permeability of the medullary collecting duct action induced by AVP. Inhibition of such effects impedes water absorption by the collecting ducts.

The natriuretic action of ANP (blue-shaded panels in Figure 21.8) in the mammalian kidney partly results from direct inhibition of sodium reabsorption from the inner medullary collecting duct and partly the effects of inhibition of other salt-conserving mechanisms. A major effect of ANP in mammals is the inhibition of renin release from the kidney. In Section 21.1.1 and Figure 21.3, we learn how the renin–angiotensin system stimulates release of aldosterone. Therefore, inhibition of renin release by ANP ultimately reduces the salt-retaining actions of aldosterone and increases NaCl excretion by the kidney. This action is included in Figure 21.8.

By comparison to mammals, the actions of natriuretic peptides in non-mammalian groups of vertebrates are much less certain, in spite of considerable research effort, but probably because of the sheer variation in natriuretic peptides and species variations. The striking diversity of natriuretic peptides that occur in fish (seven types) and the multiplicity of receptors for natriuretic peptides (four types) largely remains a mystery. One idea is that the greater variation of natriuretic peptides among fish species reflects their roles in animals that have evolved profoundly different osmoregulatory strategies. Osmoconformation occurs in the jawless hagfish, slight **hyperosmoregulation** (with a higher plasma osmolality than the external medium) occurs among sharks and other cartilaginous species, while strong hyperosmoregulation occurs among freshwater teleosts[29]. Euryhaline teleosts shift from hyperosmoregulation to hypo-osmoregulation when they migrate from rivers or lakes into seawater. Exactly how natriuretic peptides map onto the control of the complexity of events in fish osmoregulation remains unclear[30].

Among amphibians like mammals, ANP is a diuretic and natriuretic hormone, but these effects result primarily from

[26] We discuss glomerular filtration in Section 7.1.

[27] The highly variable glomerular filtration rates (GFRs) are the main factor driving changes in urine excretion rates in non-mammalian vertebrates, as we discuss in Section 7.1.3.

[28] Production of hyperosmotic urine relies on the osmotic concentration difference between the cortex and medulla as we discuss in Section 7.2.3.

[29] We discuss these different patterns of osmoregulation in Chapter 5.

[30] Actions of natriuretic peptides on drinking and intestinal absorption of Na^+, which we discuss in the Section 21.1.3, are in keeping with the concept that the 'natriuretic' peptides are important for volume and salt balance.

an increase in glomerular filtration rates, rather than the more dominant tubular effects of ANP in mammals. These differences reflect fundamental differences in the way kidneys of different vertebrate groups function. Nevertheless, there are some similarities. Vasodilatory actions of the cardiac 'natriuretic' peptides appear to be a common feature among vertebrates. Also, recall that ANP in mammals inhibits the antidiuretic actions of arginine vasopressin (AVP), which we discuss above; in amphibian, a similar inhibition of the antidiuretic action of arginine vasotocin (their homologue of AVP) reduces water absorption by the bladder[31].

In some of the animals with salt-secreting glands (marine reptiles, cartilaginous fish and some birds), natriuretic peptides stimulate salt secretion by these glands, which we examine in the next section.

Control of salt excretion by salt glands

Marine birds and reptiles with an excess of salt in their diet maintain salt balance by excreting a hyperosmotic salt solution from a salt gland[32]. A similar salt-excreting gland, the rectal gland, occurs in cartilaginous fish and excretes excess salts acquired in their diet, by drinking seawater, or by influx through their gills. The transport mechanisms employed by salt and rectal glands are functionally similar, with the rate-limiting step being chloride secretion across the luminal membrane of the epithelial cells making up the secretory tubules. However, the control mechanisms determining the rates of salt secretion differ.

Control of the elasmobranch rectal gland

Nerve fibres occur throughout the rectal gland and surround the blood supply to the gland[33]. Some of these nerves release vasoactive intestinal peptide (VIP), a 28-amino-acid peptide that belongs to a family of brain–gut neuropeptides. VIP influences blood flow and water and ion secretion in many tissues, including the gut and various exocrine glands[34]. Hence, we might anticipate that VIP influences rectal gland function. Figure 21.10A illustrates such an

[31] Section 21.1.4 examines the action of arginine vasotocin on the amphibian urinary bladder.

[32] We discuss the functioning of salt glands of birds and reptiles and the elasmobranch rectal gland in Section 5.1.5. Figure 5.14 illustrates the transport processes for salt secretion by salt glands.

[33] Section 22.4.3 discusses neural control of blood vessels.

[34] Exocrine glands secrete fluid/substances via ducts onto epithelia and external surfaces. Sweat glands and salivary glands are examples of exocrine glands.

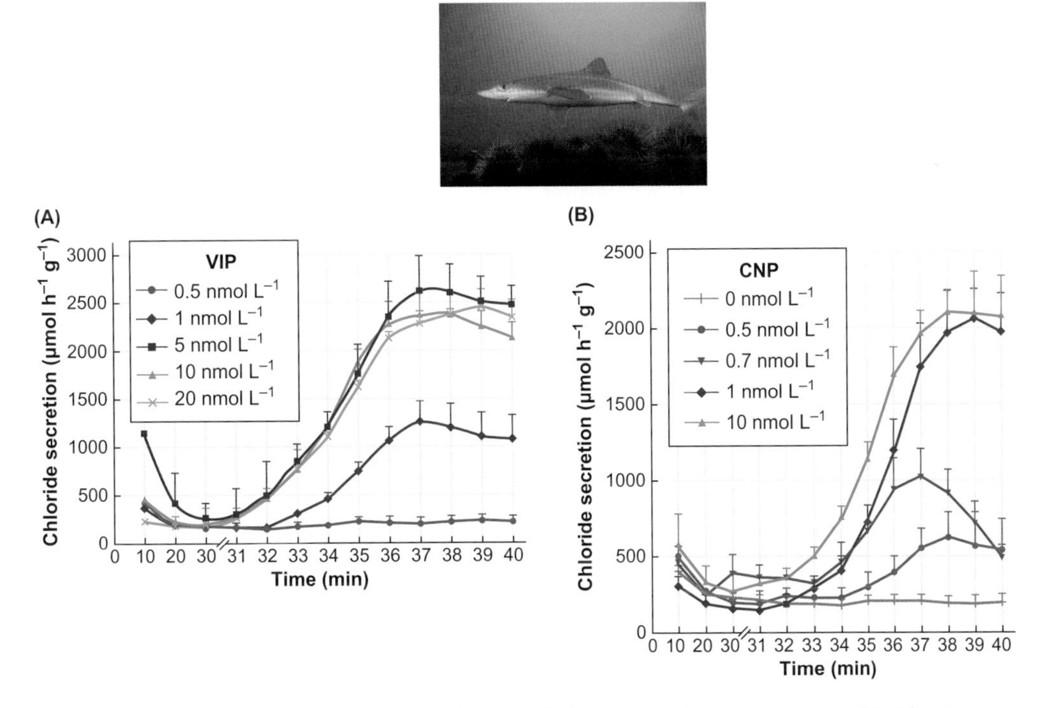

Figure 21.10 Stimulation of chloride secretion by the rectal gland of spiny dogfish (*Squalus acanthias*) by C-type natriuretic peptide (CNP) and vasointestinal peptide (VIP)

The rectal glands were perfused *in vitro* with solutions containing various concentrations of VIP or CNP. (A) VIP at 1 nmol L^{-1} or above stimulates chloride secretion; maximal stimulation occurs at 5 nmol L^{-1} VIP, and at higher concentrations no further increase in chloride secretion occurs. (B) CNP at 0.5 nmol L^{-1} or greater stimulates chloride secretion from perfused rectal glands within about 35 min. Maximal secretion occurs in the presence of 1 nmol L^{-1} CNP.

Data are means + standard errors of 4–5 preparations at each dose.

Sources: A: Bewley MS et al (2006). Shark rectal gland vasoactive intestinal peptide receptor: cloning, functional expression, and regulation of CFTR chloride channels. American Journal of Physiology - Regulatory, Integrative and Comparative Physiology 291: R1157–R1164. B: Aller SG et al (1999). Cloning, characterization, and functional expression of a CNP receptor regulating CFTR in the shark rectal gland. American Journal of Physiology (Cell Physiology 45) 276: C442–C449. Image of spiny dogfish: © Doug Perrine/Nature Picture Library.

action in the spiny dogfish (*Squalus acanthias*). However, despite the effects of VIP, and other neuropeptides, the dominant control of rectal gland secretion is thought to be hormonal in response to volume expansion, or as a response to a salt load.

Two peptide hormones, scyliorhinin II and C-type natriuretic peptide (CNP), stimulate rectal gland secretion. Scyliorhinin has been isolated from the gut wall so it is possible that food containing high amounts of salts stimulates the release of scyliorhinin II from the gut wall into circulation, which then stimulates rectal gland secretion. Figure 21.10B shows the potent stimulation of rectal gland secretion by CNP in spiny dogfish. Direct effects of CNP on salt secretion occur, as shown by additional studies on cultured monolayers of rectal gland cells. In such studies, indirect effects via increased blood flow are not possible, but chloride secretion increases by more than five times that observed in the absence of CNP.

Control of salt secretion by reptilian and avian salt glands

Nerve fibres of both the sympathetic division and parasympathetic division of the autonomic nervous system[35] densely innervate the salt glands of birds. These nerve fibres innervate the blood vessels supplying the salt glands and their secretory epithelia. Early studies demonstrated that acetylcholine, secreted by the parasympathetic nerves, stimulates salt secretion by avian salt glands. Stimulation of osmoreceptive areas of the brain that lie outside the blood–brain barrier[36] and are therefore exposed to extracellular fluid are thought to activate the parasympathetic nerve fibres and result in dilation of the blood vessels, which increases salt secretion. In the longer term, proliferation of extra salt secretory tissue occurs.

Adrenergic nerves that form part of the sympathetic nervous system cause vasoconstriction of blood vessels supplying the salt glands, which inhibits salt secretion. Hence, activation of the sympathetic nervous system during stress[37] may inhibit salt secretion. In addition, nerve fibres containing VIP occur in close association with blood vessels supplying the salt glands and near the secretory tubules on which VIP acts. As an example, Figure 21.11A shows the rapid stimulation of chloride secretion by the lingual salt glands (salt glands on the tongue) of crocodiles by VIP treatment, in this case administered intra-arterially.

Although neural control appears to dominate the control of salt secretion by the salt glands of marine birds, and possibly reptiles, circulating hormones also influence secretion. The B-type natriuretic peptide (BNP) of birds and reptiles, secreted by modified cardiac myocytes, is a potent stimulant of salt gland secretion. Figure 21.11B shows that BNP treatment results in a large increase in salt excretion by the salt glands of estuarine crocodiles (*Crocodylus porosus*). The increasing salt secretion occurs independently of any change in blood flow to the gland, which indicates direct actions of BNP on the secretory processes and suggests that maximal salt secretion does not require maximal blood flow.

[35] Section 3.3.3 outlines the parasympathetic and sympathetic divisions of the autonomic nervous system.

[36] See Online Box 16.1 for further information on the blood–brain barrier.

[37] We discuss the adrenergic responses to stress in Section 19.3.2.

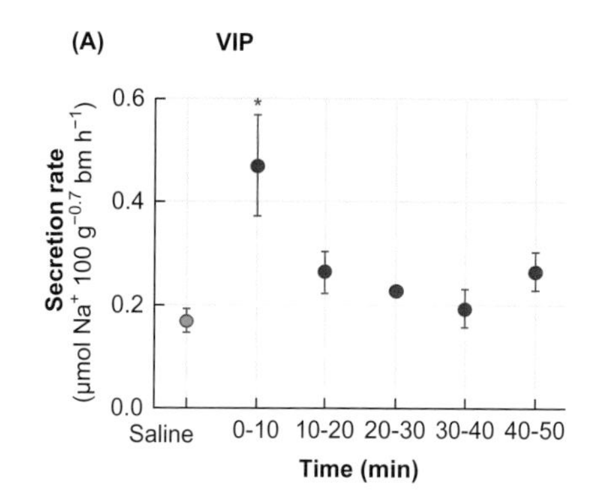

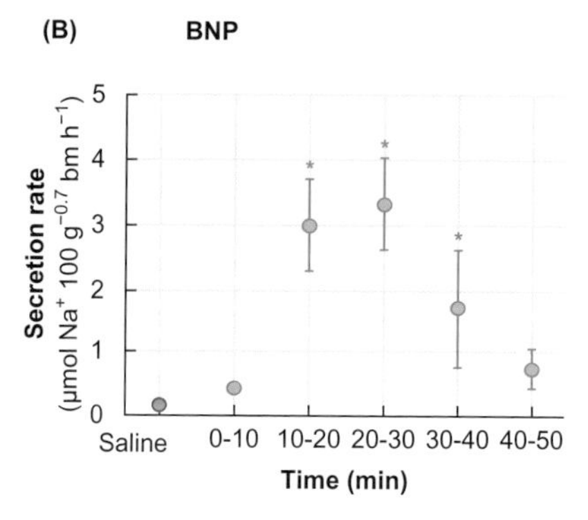

Figure 21.11 Vasointestinal peptide (VIP) and B-type natriuretic peptide (BNP) stimulate salt secretion from the lingual salt glands of estuarine crocodiles (*Crocodylus porosus*)

Administration of VIP (A) or BNP (B) intra-arterially rapidly increases the rate of Na^+ secretion by the lingual salt gland of the crocodiles. The measurements were obtained by absorbing the fluid secreted from the salt glands with filter paper and measuring Na^+ after placing the filters in a standard volume of ultrapure water. VIP administered at 100 pmol kg^{-1} body mass (bm); BNP administered at 1.58 pmol kg^{-1} body mass. Data points are mean ± standard errors; asterisks indicate significant effects ($P < 0.05$) for data during treatment compared with pre-treatment values during saline infusion of seven crocodiles.

Source: Cramp RL et al (2010). Hormone-dependent dissociation of blood flow and secretion rate in the lingual salt glands of the estuarine crocodile, *Crocodylus porosus*. Journal of Comparative Physiology B 180: 825-834.

21.1.3 Control of drinking and intestinal fluid absorption

In this sub-section, we explore the control of drinking behaviour, which involves integration of neural and hormonal control processes. First, we examine the perception of thirst by the brain, which initiates the drive to seek water and drink.

Thirst stimulates drinking by tetrapods

Small increases in plasma osmolality (1–2 per cent in mammals) stimulate osmosensitive neurons that lie outside the **blood–brain barrier** in the forebrain of tetrapods and are important for triggering drinking responses. The first insights into these control mechanisms arose in the late 1970s when destruction of the anterioventral wall of the third ventricle of goats and rats (the AV3V region, which is included in Figure 21.12) was shown to block the drinking responses to osmotic stimuli and in response to angiotensin II.

Further lesion studies and electrophysiological recordings subsequently identified two areas of the lamina terminalis, in the anterior wall of the third ventricle, as the most likely areas of osmoreceptive neurons. Figure 21.12 shows the location in the rat brain of these areas: the subfornical organ (SFO) and the organum vasculosum of the lamina terminalis (OVLT). Between these two **circumventricular organs (CVOs)** (but within the blood–brain barrier) lies the median preoptic nucleus (MnPO), which is thought to integrate neural input from both CVOs and additional input from neurons in another CVO the area postrema of the hindbrain. Thus, it is apparent that the lamina terminalis is sensitive to high plasma osmolality, which stimulates thirst and drinking.

The cells in the CVOs detect an increase in plasma osmolality, by the resultant intracellular dehydration, but it remains unclear how information from the CVOs generate the perception of thirst. While neural connections are uncertain, there *is* evidence of activation of the cerebral cortex during thirst. Figure 21.13 shows some of the results from studies of humans, using positron emission tomography to gather information on brain activity. These brain images show activation of the cortex during thirst induced by intravenous infusion of hypertonic saline, and diminished cortex activity soon after drinking.

Drinking for osmoregulation of teleost fish

Teleosts, living in seawater, drink continuously to obtain the water that compensates for the continual osmotic water loss across their body surfaces (mainly via their gills)[38]. Drinking in marine teleosts is a reflex swallowing response to the high

[38] Section 5.1.3 examines how drinking maintains the water balance of marine teleosts.

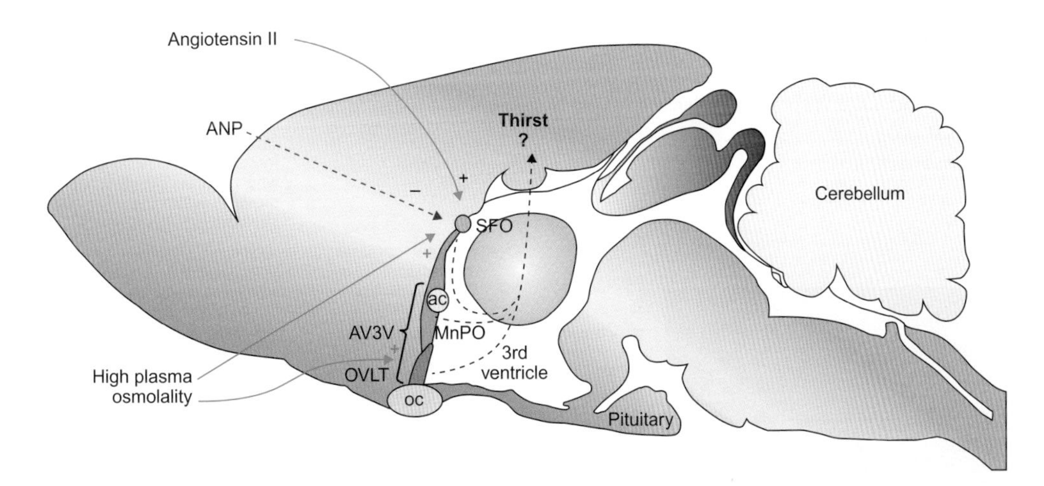

Figure 21.12 Diagram of a midline section of the rat brain showing circulating factors that act on areas of the lamina terminalis to influence thirst

The two parts of the lamina terminalis shown in blue lack the normal blood–brain barrier: the subfornical organ (SFO) and the OVLT (organum vasculosum of the lamina terminalis). These areas are known as circumventricular organs (CVOs) because of their proximity to the 3rd ventricle. The rest of the lamina terminalis, the median preoptic nucleus (MnPO), is shown as an orange area. The anterioventral wall of the third ventricle (AV3V region), which was destroyed in early studies of thirst control, is encompassed by the bracket. Input to the MnPO from the hindbrain conveys sensory input from baroreceptors in the systemic circulation, but these inputs are usually less important in controlling drinking by tetrapods than osmosensitive cells in the CVOs. High plasma osmolality stimulates the OVLT and SFO. Hormonal input to the SFO via circulating angiotensin II, stimulates thirst and increases drinking. The dashed red line and negative symbol indicates the inhibitory input of circulating atrial natriuretic peptide (ANP) on the SFO. The question mark indicates that the efferent pathways from the lamina terminalis that mediate thirst are uncertain. ac = anterior commissure; oc = optic chiasm.

Source: McKinley MJ, Johnson AK (2004). The physiological regulation of thirst and fluid Intake. News in Physiological Sciences 19: 1-6.

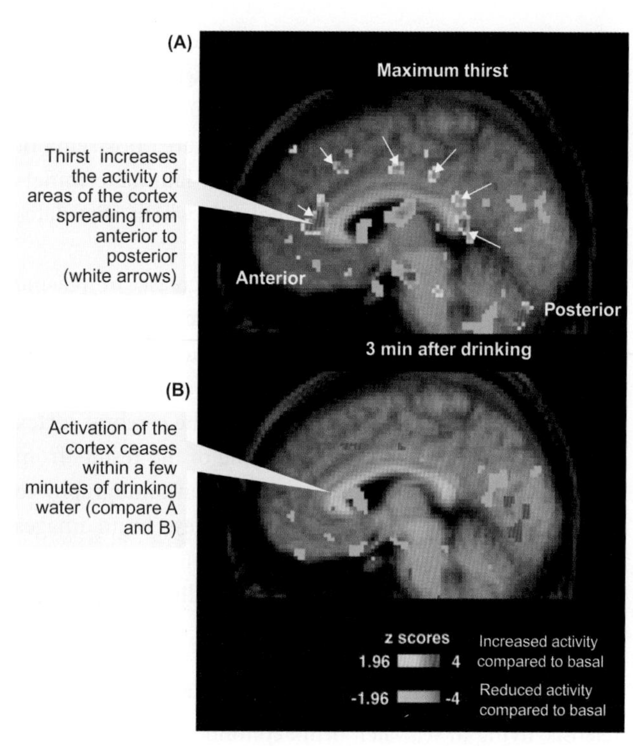

Figure 21.13 Brain images of humans made thirsty by intravenous infusion of hypertonic saline (A) and after drinking water (B)

The images shown were obtained by positron emission tomography which demonstrates increased or decreased neuron activity resulting from changes in blood flow to particular areas of the brain. The images are for the left side of the brain a few mm lateral to the midline. Brain activity is represented by the colour scoring scheme beneath the images. Areas with increased neural activity compared to basal have scores of >1.96 (yellow–orange/red); areas where there is reduced activity have scores of <–1.96 (green–blue). The images shown were derived by overlaying the images from 10 male subjects with 10 individual scans from each subject to produce a combined image.

Source: McKinley MJ, Johnson AK (2004). The physiological regulation of thirst and fluid Intake. News in Physiological Sciences 19: 1-6.

concentration of chloride ions in seawater that enters their buccal cavity during ventilation of the gills for gas exchange. However, in addition, the area postrema, which is a small area of the hindbrain, is important in stimulating drinking. The area postrema receives afferent neural input from volume receptors in the venous system. The lack of a blood–brain barrier in the area postrema means that it is likely to function as the CVO of marine teleosts and control their drinking rates in response to circulating hormones (rather than the CVOs of tetrapods shown in Figure 21.12, for which there is no evidence in fish).

The most persuasive evidence that the area postrema detects systemic factors in teleost fish comes from studies of Japanese eels (*Anguilla japonica*). These studies employed a neat way of investigating the control of drinking: a catheter was placed in the oesophagus to carry away any water that the eel drinks, which prevents water reaching the intestine where absorption would normally occur. In such eels (with an intact area postrema) the rate of drinking increases to higher levels than normal, as shown in 21.14. Notice in Figure 21.14 that after destruction of the area postrema there is little increase in the drinking rates of the eels when the oesophageal catheter is operational, despite the dehydration that would arise from draining away the seawater intake via the catheter. These results provide strong evidence that the area postrema of the brain is essential for producing an appropriate drinking response to dehydration in teleost fish. Such responses may reflect stimulation of the area postrema by osmotic signals and/or hormonal changes that arise from changes in extracellular fluid volume and composition.

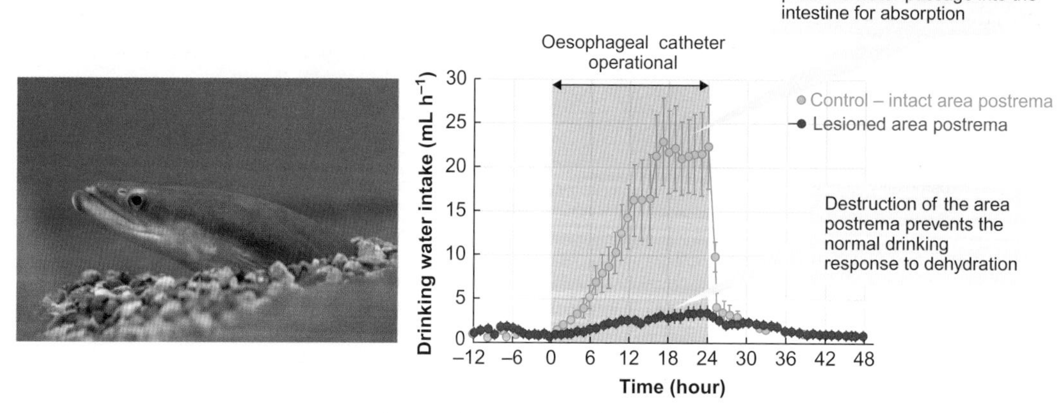

Figure 21.14 An intact area postrema in the hindbrain of seawater-acclimated Japanese eels (*Anguilla japonica*) is necessary to induce drinking responses to dehydration

The shaded area between 0 and 24 hours is the period when a catheter in the oesophagus is used to drain away any intake of drinking water, which prevents the seawater entering the intestine, where water would be absorbed. The resulting dehydration stimulates increased drinking in control eels, with a functional area postrema, by 30-fold. Lesioning of the area postrema blocks the normal response to dehydration.

Data are means ± standard errors for nine eels. Source: Nobata S, Takei Y (2011). The area postrema in hindbrain is a central player for regulation of drinking behavior in Japanese eels. American Journal of Physiology - Regulatory, Integrative and and Comparative Physiology 300: R1569–R1577. Image of eel: opencage/Wikimedia Commons

Hormonal control of drinking and intestinal salt absorption

Angiotensin II is a potent stimulant of drinking (a **dipsogenic** hormone). Dipsogenic actions of angiotensin II have been shown in many mammals (rats, goats, dogs, sheep), birds (ducks and pigeon), reptiles and many species of fish. Atrial natriuretic peptide (ANP), on the other hand, is an **antidipsogenic** hormone, i.e. it inhibits drinking.

To influence drinking, circulating hormones must first gain access to the brain via the areas that lie outside the blood–brain barrier (such as the circumventricular organs (CVOs) shown in Figure 21.12) to initiate signalling to the brain. The importance of the CVOs in responding to hormonal signals has been demonstrated by experiments in rats, in which injection of angiotensin II into the subfornical organ (SFO) stimulates drinking in a dose-dependent way, while destruction of the SFO blocks the dipsogenic action of angiotensin II.

Angiotensin II also occurs as a neurotransmitter in the mammalian brain in neurons that project into the SFO. In addition, a complete renin–angiotensin system occurs in the mammalian brain (that is, renin, angiotensinogen and converting enzyme all occur) and can generate angiotensin II locally, in a similar way to the circulating system illustrated back in Figure 21.3. Hence, angiotensin II in brain tissue could exert local (paracrine) control of drinking. However, the blockage of angiotensin receptors does not prevent drinking in response to osmotic stimuli despite the multiple ways in which angiotensin can be involved in the stimulation of drinking. This finding emphasizes the occurrence of multiple interactive back-up systems that control drinking.

In teleost fish, the area postrema in the hindbrain is sensitive to hormonal controls that appear to fine-tune the rate of spontaneous drinking. Release of ANP into the circulation occurs when there is a high blood (extracellular fluid) volume and inhibits drinking. The area postrema is implicated in detection of circulating ANP and the resulting inhibition of drinking (antidipsogenesis): lesioning of the area postrema almost abolishes the normal inhibition of drinking by ANP. Note in Figure 21.15 that circulating ANP reduces the water intake of eels to about 36 per cent of pre-treatment rates of drinking when the area postrema is intact, but not after its destruction.

Further evidence for an antidipsogenic action of ANP in teleosts has emerged from immunization studies, using specific antibodies to ANP. Such treatment increases drinking rates, presumably because of the blockage of the antidipsogenic actions of ANP. Ventricular natriuretic peptide (VNP) secreted by the heart ventricles[39] is equally as effective as ANP (and more effective than C-type natriuretic peptide, CNP) in inhibiting drinking by eels.

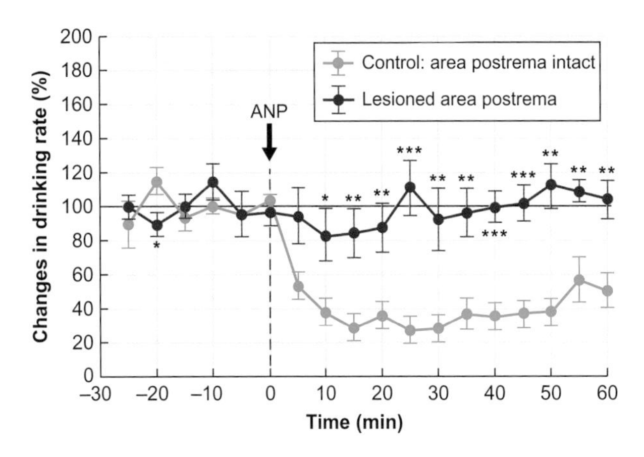

Figure 21.15 Atrial natriuretic peptide (ANP) effects on drinking rates of normal seawater-acclimated Japanese eels (*Anguilla japonica*) and after destruction of the area postrema

Drinking rates for each 5-minute period are expressed as percentage change relative to the average rates before injection of ANP (indicated by arrow). The horizontal line shows 100% values where there is no change; less than 100% represents a decrease in drinking rates; more than 100% represents an increase in drinking rates. In control eels, with an intact area postrema, ANP rapidly decreases the drinking rate (it is antidipsogenic). Destruction of the area postrema blocks the antidipsogenic response to ANP.

Data are means ± standard errors at each time point for nine eels in each group; asterisks indicate time points where there is a significant difference between the eels with a functional area postrema and eels in which this area has been destroyed by cautery (*$P<0.05$, **$P<0.01$, ***$P<0.001$).

Source: Nobata S, Takei Y (2011). The area postrema in hindbrain is a central player for regulation of drinking behavior in Japanese eels. American Journal of Physiology - Regulatory, Integrative and Comparative Physiology 300: R1569–R1577.

After drinking, the water balance of marine teleosts relies on what happens to the water in the intestine, where 70–85 per cent of the water is absorbed, by processes linked to the absorption of sodium chloride[40]. There is some evidence that ANP reduces the intestinal absorption of Na^+ in marine teleosts, as illustrated in Figure 21.16. The diagrams make a quantitative assessment of sodium balance for seawater-acclimated eels and show the influence of ANP. Although the antidipsogenic action of ANP is important for reducing Na^+ intake of the eels (to about a third), from the data in Figure 21.16, ANP's action in reducing Na^+ absorption by

[39] Section 21.1.2 discusses the many different natriuretic peptides produced by fish; Figure 21.7 illustrates the common features in their structures and some of their differences.

[40] Figure 5.4 gives an overview of the ion and water transport processes in the intestine of marine teleosts.

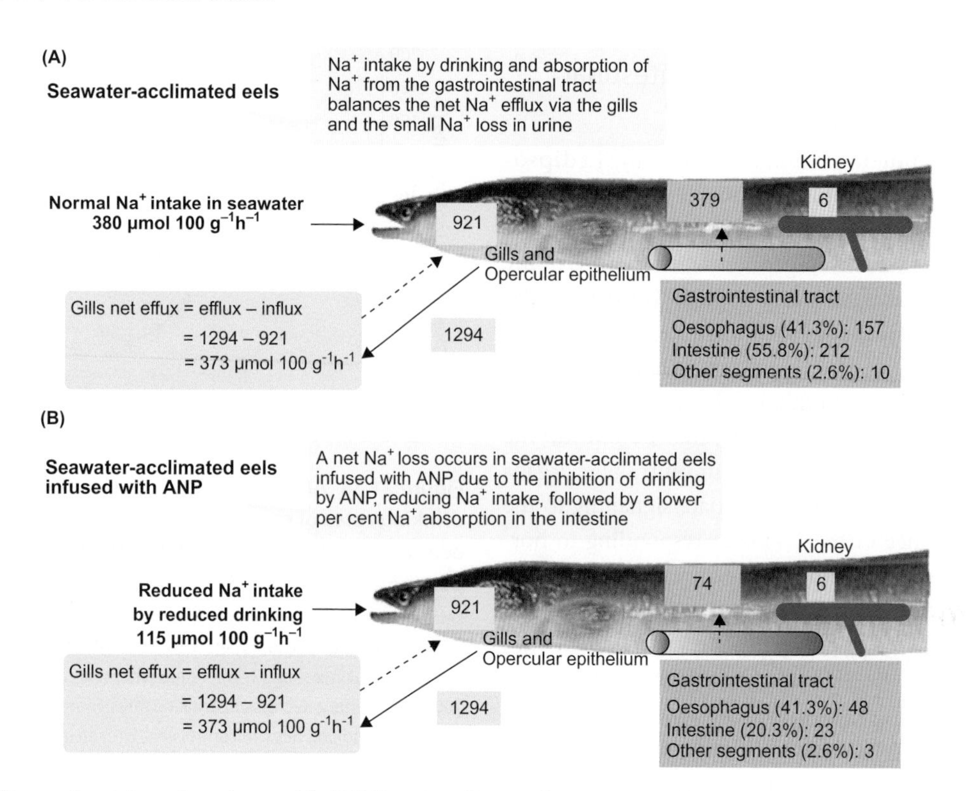

Figure 21.16 Effects of atrial natriuretic peptide (ANP) on Na^+ fluxes of seawater-acclimated Japanese eels (*Anguilla japonica*)

Normal seawater-acclimated eels (A) and seawater-acclimated eels infused with ANP (B) at 5 pmol kg^{-1} min^{-1} for one hour are illustrated. The main effect of ANP infusion is an inhibition of drinking rates, which reduces Na^+ intake, and subsequent reduction in Na^+ absorption from the gastrointestinal tract. Numbers given in boxes are Na^+ fluxes (µmol 100 g^{-1} body mass h^{-1}) at each location for each group of eels; green boxes show fluxes via the gills (dotted line = passive influx, solid line = active efflux); purple boxes show net Na^+ absorption from the gastrointestinal tract, and the percentage of the Na^+ influx via the oesophagus and intestine. ANP in these studies had no effect on the relatively small amount of Na^+ excretion by the kidney (data in pink boxes) or the much higher Na^+ fluxes via the gills (green boxes).

Source: adapted from Tsukada T, Takei Y (2006). Integrative approach to osmoregulatory action of atrial natriuretic peptide in seawater eels. General and Comparative Endocrinology 147: 31–38.

the intestine is more striking. Identification of ANP secreting neurons in the intestinal tissue of teleost fish suggests that, in addition to the effects of circulating ANP on intestinal absorption of Na^+, paracrine actions on intestinal function may also occur.

In contrast to the antidipsogenic effects of ANP and VNP in seawater-acclimated eels, injection of angiotensin II stimulates drinking, either after injection into the fourth ventricle of the brain near to the area postrema, or after injection of angiotensin II into the circulation. In eels, circulating angiotensin II causes a transient increase in water intake (for several minutes). However, as secondary effects emerge, a longer-lasting inhibition of drinking occurs. Lesioning of the area postrema blocks the transient dipsogenic effect of angiotensin II, which indicates the importance of this area for initiation of dipsogenic actions. However, it has no effect on the secondary antidipsogenic responses. It is apparent that control of drinking involves a multiplicity of systems whose interactions are not fully understood.

Increased drinking occurs in response to hypovolaemia (extracellular dehydration)

A decrease in plasma volume can occur clinically during haemorrhage, severe diarrhoea or vomiting. This extracellular dehydration is distinguishable from the intracellular dehydration resulting from an increase in plasma NaCl, which draws water from the cells by osmosis. Loss of extracellular fluid may be sufficient to activate the sympathetic nervous system, the renin–angiotensin system, or the release of arginine vasopressin (AVP) from the pituitary gland.

In mammals, an 8–10 per cent decrease in plasma volume results in thirst and stimulates an increase in drinking. Note that this is a less sensitive response than the response to an increase in plasma osmolality examined earlier. Hypovolaemia reduces the inhibitory input on drinking from stretch receptors[41] in low-pressure areas of the vascular systems (the

[41] The stretch (mechano) receptors in low pressure areas of the cardiovascular system seem to function in a similar way to baroreceptors in high pressure parts of the cardiovascular system, which are discussed in Section 22.4.2.

large veins and the heart wall) which function as volume receptors. An inhibitory action of these vascular receptors can be demonstrated in experiments in which a small balloon is inflated in a low pressure-sensitive area (e.g. at the junction of the right atrium and vena cava): inflation of the balloon distends the area, mimicking the effects of an excessive blood volume, and inhibits drinking.

Volume receptors convey afferent input to the brain via the IXth and Xth cranial nerves and terminate mostly in the **nucleus of the solitary tract** of the mammalian hindbrain[42]. From here, an inhibitory circuit and input from the nearby area postrema of the hindbrain communicate with the forebrain organs of the lamina terminalis shown in Figure 21.12, which control thirst and drinking.

21.1.4 Control of urine volume by diuretic and antidiuretic hormones

A high rate of urine production tends to reduce the volume of extracellular fluids, while a reduction in the volume of urine excreted helps to conserve fluids when animals are dehydrated. Both vertebrates and invertebrates use hormones to regulate the volume of urine they excrete. Hormonal control of fluid balance in insects is an area of particular interest, especially because a better understanding could ultimately allow targeted interference in the control processes and generation of more environmentally friendly alternatives to chemical pesticides—for instance, using gene knockout technology. We therefore start this section by examining the control of extracellular fluid volume of arthropods, and insects in particular.

Control of arthropod extracellular fluid volumes by diuretic and antidiuretic peptides

Various neurosecretory cells of arthropods secrete diuretic hormones. Neurosecretory cells in the X-organ of crustaceans secrete a diuretic hormone, which is stored and released from the sinus gland[43] into the haemolymph. Insects secrete diuretic hormones from various ganglia, such as the terminal abdominal ganglia of American cockroaches (*Periplaneta americana*), neurosecretory cells in the brain of various lepidopterans, and the mesothoracic ganglionic mass (a fusion of thoracic and abdominal neural ganglia) in kissing bugs (*Rhodnius prolixus*).

Blood-feeding insects such as mosquitoes, tsetse flies (*Glossina* spp.) and kissing bugs that rapidly consume large volumes of blood compared to the volume of their body fluids need to excrete large volumes of urine to restore the volume of their extracellular fluids. This is most obvious in kissing bugs, which ingest blood meals of up to 10 times their body volume in 20 min or less. The enormous blood meal is taken into the anterior midgut (crop) where blood cells and proteins are stored for digestion, while water and salts enter their haemolymph, expanding the body fluids of the bugs. Kissing bugs need to get into small crevices to avoid predation, so the excess fluid must be excreted as rapidly as possible for the bugs to survive.

Among insects, the day-to-day regulation of excretion involves the combined functioning of Malpighian tubules[44] that generate urine, and the hindgut, which receives the urine and creates the mixed excreta. The urine production of kissing bugs increases up to 1000 times after feeding, so in just a couple of hours about half the ingested volume of blood is excreted.

Such an impressive diuresis[45] encouraged physiologists to investigate its control. The control processes that result in fluid secretion by the Malpighian tubules, and excretion of the excess fluid, begin within 2–3 minutes of feeding, often while feeding itself is still ongoing. This has implications for the person providing the meal because kissing bugs in northern Central and South America are one of the principal vectors for the parasitic protozoan *Trypanosoma cruzi* that causes Chagas' disease. The disease affects 18 million people in the Americas and results in about 50,000 deaths each year. Transmission of the protozoan occurs via the mixture of urine and faeces of the bug, which allows the parasites to enter the wound left by the feeding bug.

Studies of individual isolated Malpighian tubules placed in physiological saline with the open end immersed in mineral oil allow collection of secreted fluid as a droplet under the oil. Tubules from unfed kissing bugs do not secrete much fluid, if any. However, secretion starts almost immediately after a drop of haemolymph from a bug that has fed recently is added to the saline around the tubule. This elegant but simple experiment indicates the presence of one or more diuretic hormones in the haemolymph of recently fed kissing bugs.

The addition of an extract of the mesothoracic ganglionic mass of kissing bugs to the saline also stimulates fluid secretion almost immediately, which tells us that this ganglionic mass contains one or more diuretic neurohormones. Two such hormones occur: serotonin (5-hydroxytryptamine)

[42] Box 16.2 and Section 16.1.3 give details of cranial nerves. Section 22.5 discusses the role of the nucleus of the solitary tract in the integration of respiratory and cardiovascular systems of mammals.

[43] Figure 19.36 shows the location of the sinus gland in the eyestalk of decapod crustaceans.

[44] Section 7.3.4 discusses the functioning of Malpighian tubules—the insect's equivalence to our kidneys.

[45] Diuresis refers to an increase in the discharge of urine.

and *Rhodnius* protein diuretic hormone (RhoprDH, which is similar in structure to vertebrate corticotrophin releasing hormone[46]).

Serotonin and the post-feeding diuresis of kissing bugs

Serotonin coordinates many of the feeding-related events in kissing bugs by its physiological effects on many tissues, including the salivary glands, digestive tract, cardiac muscle and Malpighian tubules. Serotonin acts both as a neurotransmitter, delivered to some tissues in their nerve supply, and as a neurohormone, released into the haemolymph. Malpighian tubules have no innervation, but serotonin in the haemolymph triggers a diuretic response.

Abdominal distension after a large blood meal stimulates stretch receptors in muscles in the abdominal wall and triggers serotonin release from neurohaemal tissues. Figure 21.17 shows the increasing concentrations of serotonin in the haemolymph of feeding bugs. Notice that the concentration of serotonin increases from about 7 nmol L^{-1} in unfed bugs to a peak of above 100 nmol L^{-1} within 5 min of gorging on blood.

In unfed bugs, the Malpighian tubules secrete small quantities of fluid containing mainly K^+ and Cl^-. The blood on which they feed contains mainly Na^+ and Cl^- and is hypo-osmotic to their haemolymph as shown in Figure 21.18. The release of serotonin during feeding stimulates secretion of these ions into the upper part of the tubule, and extra influx of water occurs via osmosis. An increased expression of water channels (aquaporins[47]) in the Malpighian tubules occurs after feeding, which increases the osmotic water influx. Note in Figure 21.18 that the upper Malpighian tubules of fed bugs secrete fluid that is almost isosmotic to their haemolymph and contains high concentrations of Na^+, Cl^- and K^+.

The speed of the processing of ions and water by kissing bugs after feeding is impressive. Secretion of fluid in the upper tubule generates fluid flowing through the lower tubule at 150 nL min^{-1}. Serotonin stimulates reabsorption of K^+ and Cl^- in the lower part of the Malpighian tubules, and decreases the already low water permeability, such that within about 30 seconds the osmotic concentration of the fluid is reduced from 360 to 240 mOsm kg^{-1} – a decline of 4 mOsm sec^{-1}. Notice in Figure 21.18 that K^+ concentrations decrease from 70 mmol L^{-1} in fluid generated by the upper half of the tubule to around 4 mmol L^{-1} at the point of excretion. The excreted fluid is hyposmotic to haemolymph but contains high levels of NaCl; the Na^+:K^+ concentration ratio of urine reaches 30–40 in fed bugs compared to 1.8–2.5 in unfed bugs.

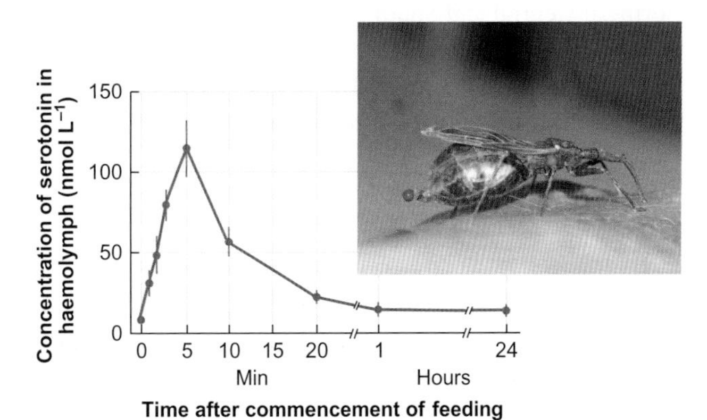

Figure 21.17 Haemolymph concentration of serotonin in kissing bugs (*Rhodnius prolixus*) after feeding

Graph shows that the serotonin concentration of unfed fifth-instar kissing bugs is very low (at time 0). Serotonin in the haemolymph rises rapidly after feeding to repletion (in this case on rabbit blood), peaking within 5 min, which stimulates a diuresis. The image of *Rhodnius* feeding on a human shows the distinct swelling of the abdomen because of the volume of blood consumed. A drop of urine has been excreted and another has just been expelled from the hind gut.
Data are means ± standard error of 5–13 measurements. Source: Lange AB et al (1989). Changes in haemolymph serotonin levels associated with feeding in the blood-sucking bug, *Rhodnius prolixus*. Journal of Insect Physiology 35: 393–399; Image: © Marcia Franco.

Other hormones controlling urine production of arthropods

Some bugs that feed on plant saps are unresponsive to serotonin but respond to other diuretic peptides that are structurally similar to the vertebrate corticotrophin releasing hormone (CRH). At least 26 species of insects and two mite species are known to possess such peptides in their brain, abdominal neurosecretory cells, or in the corpora cardiaca (neurohaemal organs close to the brain[48]). In kissing bugs, serotonin and a CRH-like peptide (RhodprDH) each stimulate fluid secretion by the Malpighian tubules, but their combined effect is four times greater than their individual effects, which presumably maximizes fluid excretion after engorgement with a blood meal.

A further group of arthropod diuretic hormones that are structurally related to (but functionally quite different from) the vertebrate hormone calcitonin[49] occurs in at least 15 species of insects, some crustaceans (such as American lobsters *Homarus americanus*) and some mites.

[46] Section 19.2.2 and Table 19.2 provide information on corticotrophin releasing hormone (CRH) in the vertebrate hypothalamus.

[47] We discuss water channels in Section 4.2.3.

[48] Figure 19.37 and 19.38 show the location of the corpora cardiaca.

[49] We discuss the actions of calcitonin among vertebrates in Section 21.2.4.

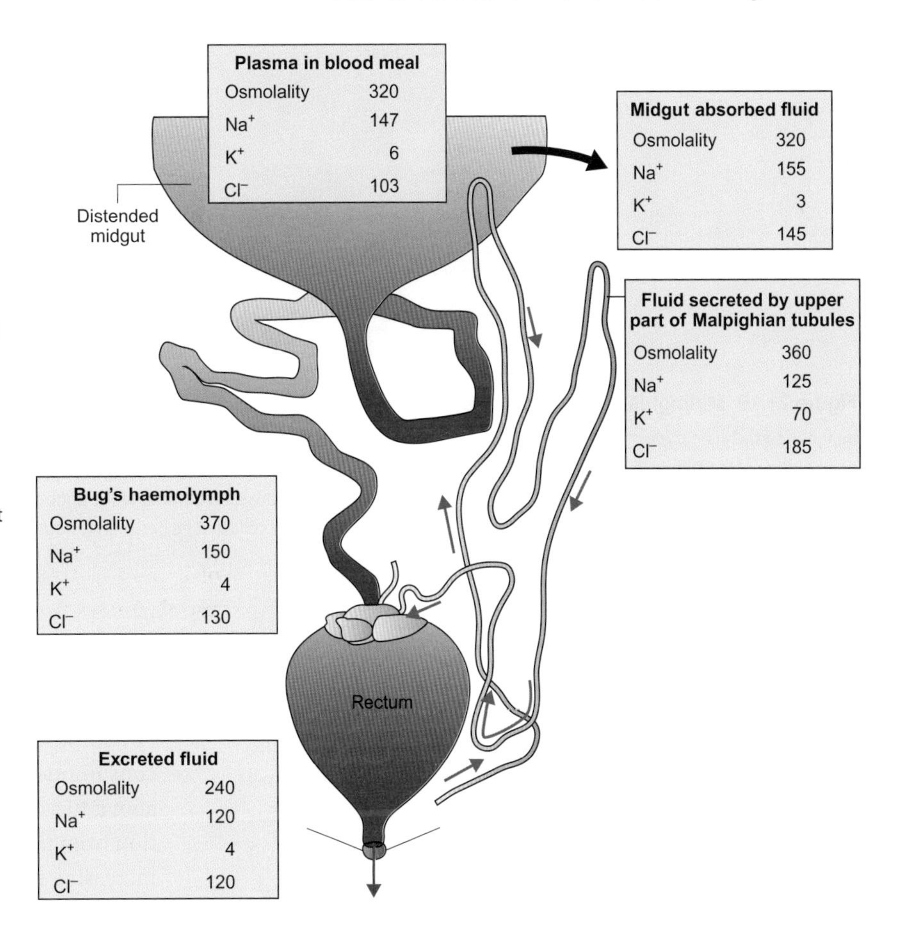

Figure 21.18 Arrangement of midgut, rectum and Malpighian tubules (only one shown) of kissing bugs (*Rhodnius prolixus*) showing sodium, chloride and potassium concentrations (mmol L^{-1}) and osmolality (mOsm kg^{-1}) of various fluids after feeding

The fluid absorbed from the midgut is isosmotic to the blood plasma consumed, but has a lower osmolality than the bug's haemolymph. The fluid secreted by the upper part of the Malpighian tubules is almost isosmotic to the bug's haemolymph and contains high concentrations of Cl^-, Na^+ and K^+. Subsequent reabsorption of K^+ and Cl^- in the second part of the Malpighian tubules results in excreted fluid that has a lower osmolality than the bug's haemolymph, and lower concentrations of K^+ and Cl^- than in the secreted fluid in the first part of the tubules. Arrows show direction of flow through the long Malpighian tubule.

Source: adapted from Maddrell SHP, Phillips JE (1975). Secretion of hypo-osmotic fluid by the lower Malpighian tubules of *Rhodnius prolixus*. Journal of Experimental Biology 62: 671–683.

All calcitonin-like diuretics identified so far consist of 31 amino acids, which explains their alternative name: **diuretic hormone**[31]. While these peptides increase fluid secretion by the Malpighian tubules of some insects, other actions such as mixing the haemolymph and stimulating hindgut contractility are thought to help maintain a diuresis.

The most recently identified diuretic hormones among insects are **leucokinins**, which are a diverse group of peptide hormones identified initially as stimulants of hindgut motility. Leucokinins stimulate chloride flux into Malpighian tubules, which ultimately drives fluid secretion, although the exact mechanisms appear to differ between species[50]. In *Drosophila* (fruit flies), which have two distinct types of cells in their Malpighian tubules (principal cells and smaller stellate cells) only stellate cells have leucokinin receptors, as shown in Figure 21.19. Leucokinin stimulation of fluid secretion by Malpighian tubules of *Drosophila* occurs in direct proportion to the number of stellate cells and appears to result from an increase in chloride channels in these cells. However, the presence of stellate cells is not a prerequisite for the diuretic action of leucokinins, and the relative importance of paracellular and transcellular components of the chloride flux is still controversial.

Antidiuretic peptides appear to co-ordinate the termination of diuretic responses of insects. In several species, neurosecretory cells in the abdomen secrete antidiuretic cardioactive peptides (named for their first discovered actions on heart function). In kissing bugs, the inhibition of the effect of serotonin on Malpighian tubules is thought to be important in reducing urine excretion to a more persistent but less pronounced diuresis that occurs sometime after feeding, during the digestion of the K^+-rich blood cells.

[50] Figure 7.37 provides a model of the secretory processes in the Malpighian tubules of *Drosophila*.

Male *Drosophila melanogaster*

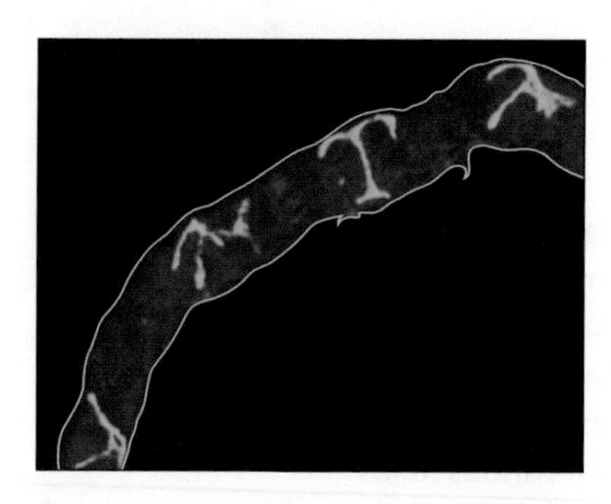

Figure 21.19 Malpighian tubule of *Drosophila* showing localization of a leucokinin (named drosokinin) to receptors in stellate cells

Image shows whole tubule of approximately 35 μm diameter examined in a fluorescence microscope. The blue lines have been added to indicate the approximate outline of the tubule. The bright-green cells are stellate cells showing binding of drosokinin to receptors on the basolateral borders. The tubules were incubated in a solution containing an antibody to drosokinin (raised in rabbits) and binding of the antibody was revealed by addition of an anti-rabbit second antibody labelled with fluorescein. No staining is seen in the large principal cells that lie between the stellate cells.

Source: Dow JAT (2012). The versatile stellate cell–more than just a space filler. Journal of Insect Physiology 58: 467-472. Image of male *Drosophila melanogaster* © Peter J. Bryant.

Vertebrate antidiuretic hormones

Arginine vasopressin (AVP) from the neural lobe of the pituitary gland of mammals and its homologues (lysine vasopressin (LVP) in pigs and some other vertebrates, and arginine vasotocin (AVT) function as the main antidiuretic hormones of vertebrates[51]. These hormones act on more than one type of receptor, as outlined in Box 21.1. The fact that V_1 receptors stimulate vasoconstriction of vascular smooth muscle, which may increase blood pressure, explains the name arginine vasopressin (or simply vasopressin) for the peptide that occurs in mammals, although arginine vasotocin has similar vasoconstrictor action in many non-mammalian vertebrates.

Figure 21.20 summarizes the two main factors controlling the release of AVP and which influence LVP or AVT release in some species:

- **Blood (extracellular fluid) volume**. A high blood volume increases the stretch in low pressure areas of the vascular system (venous system and atrial chambers of the heart) inhibiting AVP release from the posterior pituitary. On the other hand, low blood volume reduces the stretch of the volume receptors, reducing their inhibitory effects on hormone release.
- **Osmotic concentration of extracellular fluids**. Control of AVP release by the osmotic sensing systems is usually a more sensitive system than responses to volume. In mammals, increases of 1–2 per cent in plasma osmolality result in up to five-fold increase in plasma concentrations of AVP, but a 5–10 per cent decrease in blood volume is required for a similar increase in AVP. Osmosensitive cells occur in the lamina terminalis of the brain (shown back in Figure 21.12), close to the neurosecretory neurons producing AVP. If dehydration or excess salt consumption result in an increase in plasma osmolality above the osmotic threshold or set point, AVP secretion from the posterior pituitary increases and results in increasing circulating concentrations of AVP.

In normal circumstances, the two control systems (volume receptors and osmoreceptors) operate without conflict. In mammals, a high intake of water increases blood volume (inhibiting AVP release) while simultaneously reducing plasma osmolality (again inhibiting AVP release). On the other hand, dehydration raises plasma osmolality and reduces blood volume. Figure 21.20 illustrates the resultant cascade of events.

In mammals, both a high extracellular fluid osmolality and low extracellular fluid volume result in an increase in the reabsorption of water from the kidney and stimulation of thirst[52] that ultimately restore normal extracellular fluid osmolality. However, situations can arise when output from the two control systems regulating AVP release could be antagonistic.

For example, in a haemorrhage, there is no change in plasma osmolality, but the release of AVP that could occur in response to blood loss could stimulate an increase in water reabsorption by the kidney (which we explore in some detail later in this section). The resulting reduction in plasma osmolality could then switch off further AVP release. However, a reduced sensitivity of the osmosensitive neurons, as part of an 'emergency response' to blood loss in mammals, avoids the potential conflict.

[51] Table 19.1 shows the structures of these peptides (each of nine amino acids) and related peptides. Figure 19.13 illustrates the hypothalamic-posterior axis which regulates the storage and release of AVT/AVP/LVP.

[52] Section 21.1.3 discusses thirst and drinking in some detail.

21

Box 21.1 Receptors for arginine vasopressin (AVP) and arginine vasotocin (AVT)

Arginine vasopressin and arginine vasotocin act on membrane bound G-protein coupled receptors[1] in various tissues, which include vascular smooth muscle and epithelial cells in various organs. The different effects of AVT and AVP result from the location of the receptors in different cell types, the different receptor types, and their signalling mechanisms. Three main subtypes are recognized: V_1, V_2 and V_3 receptors.

V_1 receptors (previously called V_{1A}) occur at a high density on vascular smooth muscle and initiate vasoconstriction with a possible increase in blood pressure. They act via the inositol phosphate cascade[2], resulting in an increase in intracellular calcium. In the mammalian kidney, V_1 receptors on the vasa recta initiate vasoconstriction that restricts blood flow to the inner medulla during production of hyperosmotic urine. Figure A shows an example of V_1 receptors in the afferent and efferent glomerular arterioles of European flounders (*Platichthys flesus*) where they influence glomerular function.

V_2 receptors occur at high density in the distal tubule and collecting duct of the kidney where they are responsible for increasing the osmotic water permeability of the epithelial cells. These receptors signal via cAMP and adenylate cyclase, which triggers an increase in aquaporin-2 water channels.

V_3 receptors occur in the anterior pituitary gland. The V_3 receptors were initially considered a subtype of the V_1 receptor (called V_{1B}), but their characteristics in relation to interaction with drugs, their cell signalling, and sequence identities indicate that V_3 receptors are a distinct sub-category.

Table A summarizes the three subtypes, their signalling mechanisms and the main tissues where they act. Antagonists that block specific V_1 or V_2 receptors are sometimes used to treat clinical conditions such as congestive heart failure. These pharmaceutical tools are useful in investigating the receptor types in osmoregulatory tissues of non-mammalian species. Such studies and molecular investigations have identified V_1-like and V_2-like receptors that respond to AVT in non-mammalian vertebrates.

Recent analysis of genomic sequence data for various vertebrate species *in silico*[3] suggests that the two rounds of genome duplication in early vertebrate evolution resulted in the evolution of a family of vasopressin/vasotocin receptors that have been retained. The teleost fish have the largest repertoire with at least six types or subtypes. Interactions of specific receptor subtypes with specific ligands in particular tissues are likely to elicit a diverse range of physiological and behavioural effects.

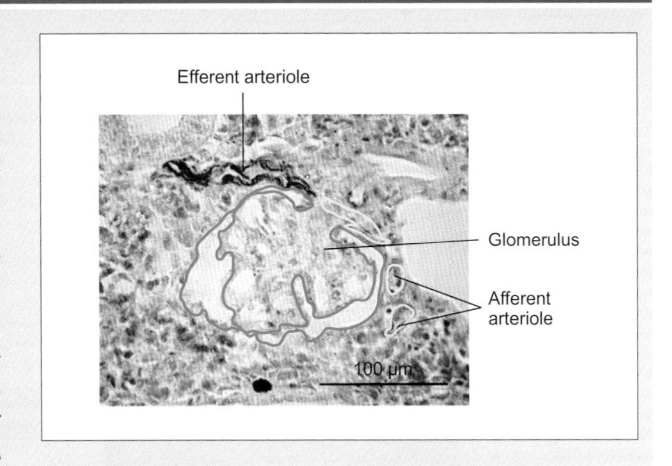

Figure A Localization of V_1-type receptors for arginine vasotocin (AVT) in afferent and efferent glomerular arterioles in the kidney of European flounders (*Platichthys flesus*)

Immunocytochemistry using antibodies to flounder V_1-type receptors (VTR) reveals V_1 receptors in the kidney. To help interpret the image, outlines of the glomerulus in its Bowman's capsule (green outline) and the afferent arterioles (yellow outline) have been added over the micrograph image. A high level of VTR immunostaining (red) occurs in the wall of the efferent arteriole and there is some weaker staining of the afferent arteriole.

Source: Balment RJ et al (2006). Arginine vasotocin a key hormone in fish physiology and behaviour: A review with insights from mammalian models. General and Comparative Endocrinology 147: 9–16.

1 Section 19.1.4 gives information on G-protein coupled receptors.
2 The inositol phosphate cascade is illustrated in Figure 19.7.
3 *in silico* is Latin for 'in silicon'. It implies the use of computer-based analysis of gene sequences, as opposed to *in vivo* or *in vitro* techniques.

Table A Summary of V_1 and V_2 characteristics and distribution

Receptor type	Signalling mechanism	Main tissues	Actions
V_1	G-protein coupling to phospholipase C, inositol 1,4,5-trisphosphate (IP_3) and diacyl glycerol (DAG) pathway, leading to an increase in intracellular calcium and increase in protein kinase C	Kidney vasculature, systemic vasculature	Vasoconstriction, glomerular filtration
V_2	G-protein coupling to adenylate cyclase, cAMP and protein kinase A signalling	Distal tubule and collecting duct of kidney	Increase in water permeability
V_3	G-protein-coupled, inositol trisphosphate pathway, intracellular Ca^{2+} and activation of protein kinase C	Anterior pituitary gland	Release of hormones: adrenocorticotrophic hormone (ACTH) and prolactin

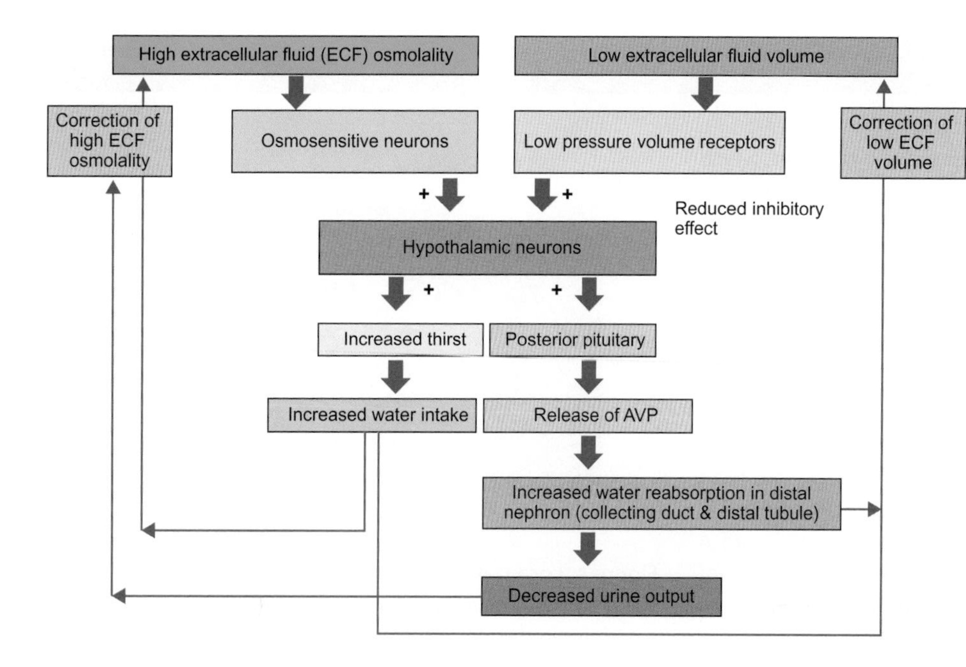

Figure 21.20 Control systems regulating the release of arginine vasopressin (AVP)

Osmoreceptors and volume receptors control the release of AVP (or LVP or AVT in some species), which stimulates (+) thirst and hence water intake by drinking, and changes in kidney function resulting in a reduction in urine output. These responses result in negative feedback effects.

Among non-mammalian vertebrates, AVT secretion by the pituitary increases during dehydration (as in mammals). In general, however, the influence of plasma osmolality appears much more important than volume responses. In birds, cerebral osmoreceptors seem to be the prime controllers of AVT release from the pituitary gland as demonstrated by injections of salt solutions into the third cerebral ventricle; hypertonic saline stimulates AVT release, while hypotonic saline inhibits release of AVT. Injection of solutions with a low NaCl concentration but made isosmotic to the cerebrospinal fluid by addition of other solutes also inhibits AVT release. This experiment illustrates the importance of Na^+ as a signal regulating AVT release.

In a similar way, release of AVT from the pituitary gland of amphibians and the few species of reptiles studied so far occurs in response to hyperosmotic stimuli and the associated cellular dehydration. Hence, plasma concentrations of AVT in cane toads (*Rhinella marina)* and Australian lizards (*Pogona minor*) correlate with their plasma osmolality during various osmotic challenges or dehydration.

Among teleost fish, the influence of plasma osmolality seems to be the only control mechanism that determines AVT release, with vascular volume having little effect. Certainly, in European flounders (*Platichthys flesus*), even large changes in vascular volume do not influence circulating concentrations of AVT, whereas increasing plasma osmolality has a significant impact. Figure 21.21 illustrates the positive relationship between circulating concentrations of AVT and plasma osmolality, whether the flounders are in fresh water or seawater. Plasma AVT concentrations increase in association with a rise in plasma osmolality after transfer from fresh water to seawater and *vice versa* or after intraperitoneal injection of hypertonic saline to mimic the salt-loading that accompanies the acute phase after transfer to seawater.

Antidiuretic actions of AVT in fish, amphibians and reptiles

The role of AVT in water conservation arose early in the evolution of the vertebrates. Among fish and amphibians, the dominant antidiuretic action of AVT results from a reduction in the rate of glomerular filtration by which their production of urine starts[53]. Effects of AVT on the subsequent water reabsorption by the renal tubule are generally absent or are at most relatively minor in fish, amphibians and reptiles.

The glomerular antidiuretic action of AVT is likely to result from stimulation of V_1 receptors in the renal vasculature. The V_1 receptors in the afferent and efferent glomerular arterioles of European flounders (*Platichthys flesus*), shown in Box 21.1, Figure A, provide the means to constrict these vessels, resulting in changes to glomerular haemodynamics. Such actions may explain the major effect of physiological concentrations of AVT is reducing the proportion of the glomeruli that are filtering. This action of AVT, which is shown for rainbow trout (*Oncorhynchus mykiss*) in Figure 21.22 is just one example of the hormonal control of the **glomerular intermittency** that allows nephrons to be switched on and off in the kidneys of fish, amphibians and reptiles[54]. Many hormones work in this way in non-mammalian kidneys, but

[53] Most fish and all amphibians produce the primary urine by glomerular filtration, as we discuss in Section 7.1.3.

[54] Section 7.1.3 discusses glomerular intermittency in more detail.

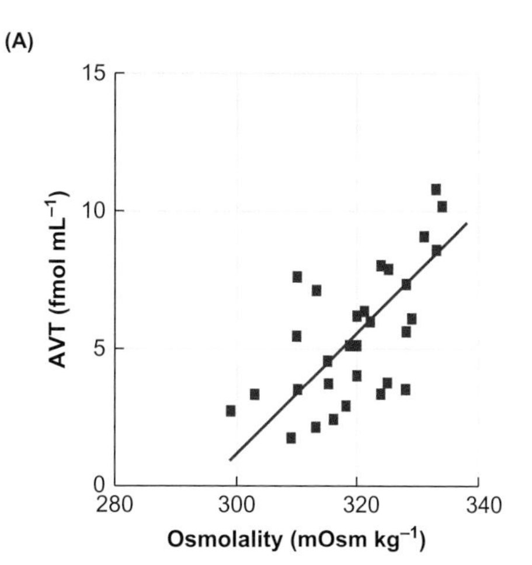

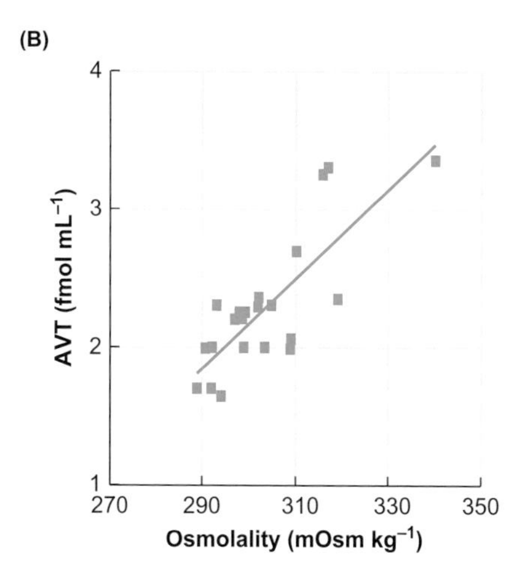

Figure 21.21 Relationship between plasma concentrations of arginine vasotocin (AVT) and plasma osmolality of European flounders (*Platichthys flesus*) held in seawater (A) or fresh water (B)

Data points are for individual flounders. In both groups of flounders plasma AVT concentrations are correlated with plasma osmolality.

Source: Balment RJ et al (2006). Arginine vasotocin a key hormone in fish physiology and behaviour: A review with insights from mammalian models. General and Comparative Endocrinology 147: 9–16.

the detailed mechanisms behind glomerular intermittency are unclear.

In addition to the actions of AVT on the glomerulus, AVT increases the density of **aquaporins** (AQP) in the kidney collecting ducts of several anuran species (e.g. Japanese tree frogs *Hyla japonica*), allowing increased water reabsorption. We examine this hormonal regulation in Case Study 21.1, along with the more pronounced actions of AVT on the specialized osmoregulatory organs and tissues of anurans. These are: (i) the large urinary bladder from where water is reabsorbed during dehydration, before excretion of the urine (ii) the ventral pelvic patch of skin that takes up water by a process also known as **cutaneous** drinking[55].

[55] We discuss cutaneous uptake of water by amphibians in relation to their water balance in Section 6.2.3.

In reptiles, the antidiuretic actions of AVT generally result from:

- decreased GFR, presumably through interaction with AVT V_1-type receptors in kidney vasculature;
- increased water permeability of the nephron, presumably by interaction of AVT with V_2-type receptors, such as those identified in the collecting duct of the agamid lizards *Ctenophorus ornatus*.

The extent of these two responses during dehydration varies between reptilian species, but there is no evidence that these variations relate to differences in water availability in their habitats. Nevertheless, a low GFR may be the first step in reducing urinary output, augmented by an increase in tubular reabsorption of water under more stringent conditions.

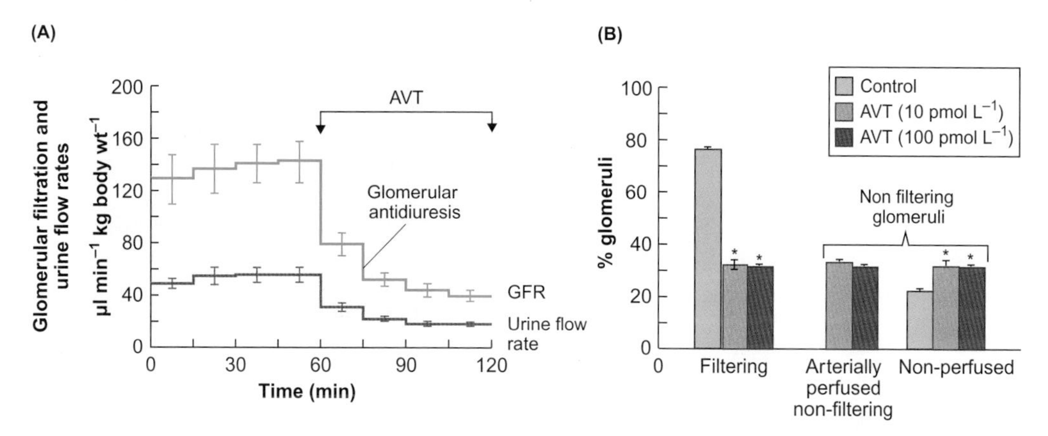

Figure 21.22 Effects of arginine vasotocin (AVT) on kidney function in trout (*Oncorhynchus mykiss*)

(A) Kidneys were perfused *in situ* at a constant pressure head and studied prior to and during addition of a physiologically relevant concentration of AVT (10 pmol L^{-1}). The decrease in glomerular filtration rate (GFR) induced by AVT results in a decrease in urine flow rate. This is known as a glomerular antidiuresis.
(B) Proportion of glomeruli in perfused kidneys that are (i) filtering, (ii) arterially perfused but non-filtering and (iii) non-perfused, in control kidneys and during AVT treatment. AVT was added to the perfusate at 10 pmol L^{-1} (light purple) or 100 pmol L^{-1} (dark purple). The results show that AVT reduces the percentage of glomeruli that are filtering and results in glomeruli that do not filter despite their perfusion with arterial blood (presumably at flow rates and/or pressure too low for filtration to occur).

Data are means ± standard errors for six trout in each group; *$P < 0.001$ compared to controls. Source: Amer S, Brown JA (1995). Glomerular actions of arginine vasotocin in the in situ perfused trout kidney. American Journal of Physiology (Regulatory, Integrative and Comparative Physiology 38): 269: R775–780.

Case Study 21.1 AVT-mediated control of aquaporins for water conservation and water uptake in frogs and toads

Frogs and toads are adapted to live in habitats with very different water availabilities—freshwater ponds, semi-aquatic habitats, trees (arboreal species) and even arid deserts. Their ability to cope with dehydrating conditions relies on AVT stimulating the expression of specific aquaporin (AQP) proteins in key osmoregulatory tissues: the skin, bladder and kidney. Three types of AQPs occur, named according to their main tissue location: 'kidney-type' AQP-2, 'ventral skin-type' AQP-a2 and 'bladder-type' AQP-a2. The 'a' in the names of two of the subtypes indicates these AQPs are specific to anurans and phylogenetically separated from those of all other animal groups.

Kidney aquaporins

Figure A illustrates the changing expression of kidney-type AQP-2 in Japanese tree frogs (*Hyla japonica*) in response to AVT. In hydrating conditions (unstimulated) non-functional water channels reside in vesicles distributed throughout the cytoplasm of the principal cells of the collecting duct. During dehydration, an increased concentration of AVT results in translocation of AQP-2 from the cytoplasmic pool and insertion in the apical membrane where it facilitates water absorption from the tubular fluid. Another AQP that is not subject to hormonal regulation (AQP-3) occurs in the basolateral membranes, where it allows water transport from the cells to the interstitial fluids.

Bladder aquaporins

Terrestrial and tree-living (arboreal) toads and frogs have large urinary bladders, capable of holding up to 50 per cent of the standard body mass (body mass with an empty bladder). In contrast, aquatic species, such as African clawed frogs (*Xenopus laevis*), have small bladders (holding 1 per cent of standard body mass) and inevitably a much poorer capacity to retrieve water from the bladder. Such a poor capacity is not a problem for animals dealing with osmotic water inflow from the freshwater environment across their skin.

For terrestrial species, increasing concentrations of circulating AVT occur during dehydration and increase the osmotic water permeability of the bladder by increasing bladder-type AQP-a2, as illustrated in Figure B. The large volume of the bladder of terrestrial species enables retrieval of water from the urine, which can potentially double their survival time when water availability is restricted. Stimulation of Na^+ transport

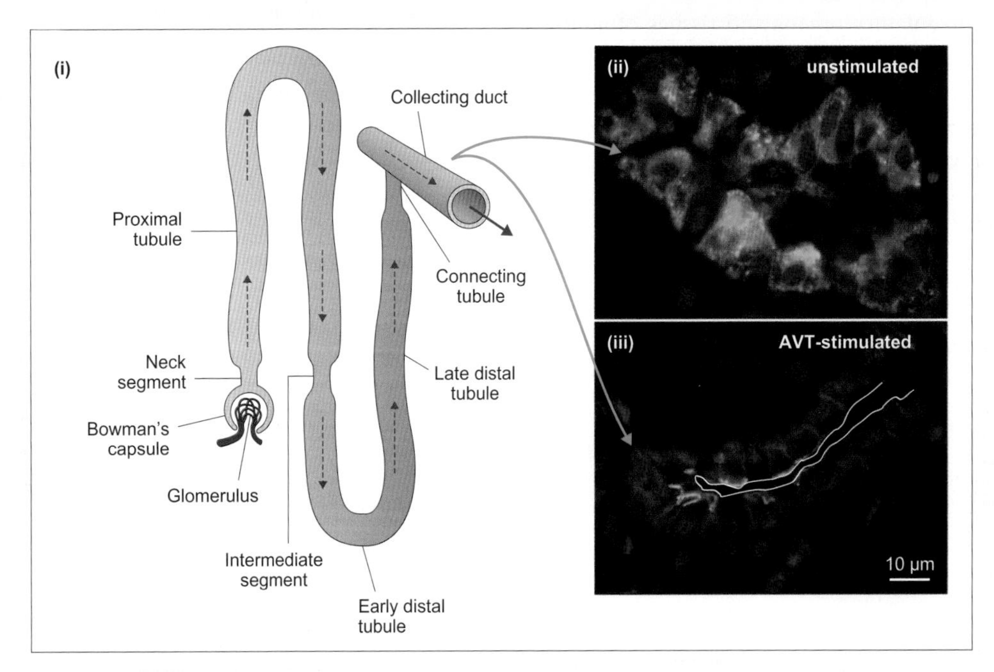

Figure A Arginine vasotocin (AVT) stimulates translocation of aquaporin-2-type water channels to the apical membrane of the principal cells of the collecting duct of Japanese tree frogs (*Hyla japonica*)

(i) Schematic illustration of an anuran nephron connected to a collecting duct. Red dashed arrows depict the flow of tubular fluid along the nephron. (ii) and (iii) are micrographs of cross-sections of collecting ducts showing immunofluorescence after binding of antibodies to two types of aquaporins in the principal cells when: (ii) unstimulated and (iii) AVT-stimulated. Green fluorescence indicates location of AQP-2 (AQP-h2K, i.e. a kidney, K type of AQP-2 in *Hyla*, h); red fluorescence shows the location of aquaporin 3 (AQP-h3); nuclei are stained blue. The white line in (iii) delineates the approximate position of the tubular lumen. In response to AVT, the AQP-2 (green fluorescence) distributed throughout the cell (shown in (ii)) is translocated to the luminal membrane of the principal cells. Type 3 AQP (red) remains in the basolateral membranes whether or not AVT is present.

Source: adapted from Suzuki M, Tanaka S (2009). Molecular and cellular regulation of water homeostasis in anuran amphibians by aquaporins. Comparative Biochemistry and Physiology A Molecular & Integrative Physiology 153: 231–241.

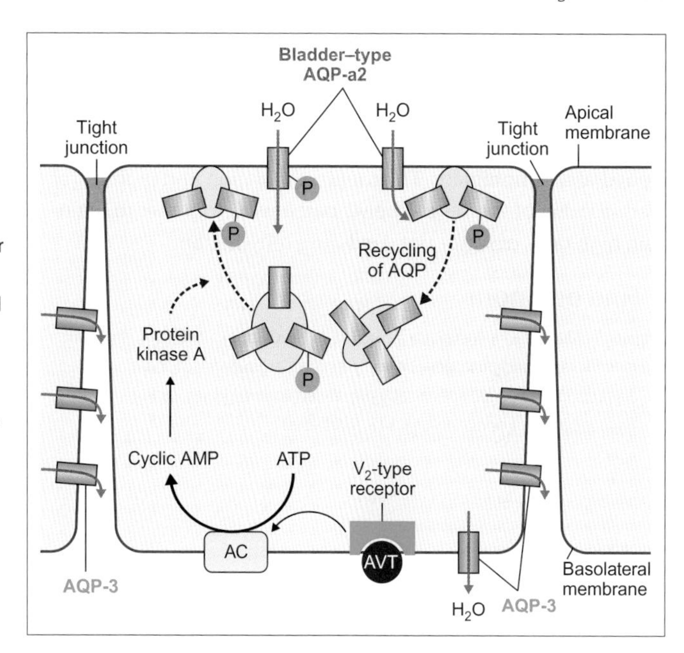

Figure B Model for AVT actions on aquaporins (AQPs) in urinary bladder of terrestrial anurans

The main route for water absorption (grey arrows) is transcellular since water impermeable tight junctions occur between the cells in the epithelial cell layer of the urinary bladder. AVT binds to V_2-type receptors in the basolateral membrane of these cells and triggers intracellular signalling processes: activation of adenylate cyclase (AC), formation of cyclic AMP and protein kinase A, which results in phosphorylation (P) of the urinary bladder-type AQP-a2. As a result of these cellular processes, vesicles holding bladder-type AQP-a2 are translocated to, and are inserted in, the apical membrane. These aquaporins allow water passage into the cells. Water leaves the cells across the basolateral membranes, via AQP-3 that is not hormonally-regulated, from where water enters the capillaries (not shown).

Source: adapted from Suzuki M, Tanaka S (2009). Molecular and cellular regulation of water homeostasis in anuran amphibians by aquaporins. Comparative Biochemistry and Physiology A Molecular & Integrative Physiology 153: 231–241.

across the bladder wall by AVT (and linked chloride transport) enhances water reabsorption via AQPs (by osmosis) which gradually reduces the volume of stored urine.

Aquaporins in ventral pelvic patch of skin

The type of aquaporin in the ventral pelvic patch of skin—AQP-a2—is located in the 'first reacting cell layer' of the epithelial cells of the pelvic patch, which forms an adjustable barrier between the cells inside of the body and the outer layers of dead cells forming the stratum corneum[1]. During dehydration, AVT stimulates water-seeking behaviour and postural movements that bring the ventral pelvic patch into contact with damp surfaces.

The fascinating discovery of both 'skin-type' AQP-a2 and 'bladder-type' AQP-a2 in the epithelial cells of the ventral pelvic patch of skin of Japanese common toads (*Bufo japonicus*) and arboreal species, such as Japanese tree frogs (*Hyla japonica*), shows that terrestrial frogs and toads are specially adapted to allow water uptake when they find water. AVT stimulates translocation of both types of AQP-a2 and their insertion into the apical membrane of the granular epithelial layer of skin (the first reacting layer).The basic process is much as shown for the bladder in Figure B, except for the incorporation of both skin and bladder AQP-a2 in the presence of high plasma concentrations of AVT, which probably explains the high sensitivity of their skin to AVT.

The data in Table 1 show that AVT increases the osmotic water permeability of the skin of terrestrial and amphibious frogs and toads by up to five times. Even though the ventral pelvic patch only represents about 10 per cent of the skin area, it is responsible for 70 per cent or more of the water uptake. Bladder-type AQP-a2 does not occur in the skin of semi-terrestrial or aquatic species of frogs and toads with poor responses to AVT. The aquatic African clawed frogs are at the extreme end of the spectrum: they have virtually no skin response to AVT because of the absence of significant amounts of any AQP in their skin. Thus, the evolution of particular AQPs in the osmoregulatory tissues of adult anurans and their responsiveness to AVT are an important component of adaptation to different ecological environments.

Water uptake could reduce the difference in osmolality across the skin that drives the process, but increased blood flow to the ventral pelvic patch during rehydration appears to avoid

Table A Effects of arginine vasotocin (AVT) on osmotic water permeability of ventral skin of anuran amphibians

Permeability data represent water transport per cm^2 of skin with an osmotic difference across the skin of 200 to 220 mOsm kg^{-1}.

Species	Habitat	Skin permeability (µL H_2O cm^{-2} h^{-1})	
		Without AVT	**with AVT**
Cane toad (*Rhinella marina*)	Terrestrial	28	85
European toad (*Bufo bufo*)	Terrestrial	19	44
American bullfrog (*Lithobates catesbeianus*)	Amphibious	13	61
African clawed frog (*Xenopus laevis*)	Aquatic	8	9

Source of data: Bentley PJ (1971). Endocrines and Osmoregulation. A Comparative Account of the Regulation of Water and Salt in Vertebrates. Springer Verlag.

this limitation. In toads, angiotensin II more than doubles blood flow to the ventral pelvic patch, while reducing blood flow to other areas of skin prone to evaporative water loss. However, other factors are also likely to regulate blood flow to the skin of dehydrated amphibians. Thus, rapid circulation and high water permeability of the ventral pelvic patch work together to maintain high rates of water uptake during rehydration.

› Find out more

Ogushi Y, Mochida H, Nakakura T, Suzuki M, and Tanaka S (2007). Immunocytochemical and phylogenetic analyses of an arginine vasotocin-dependent aquaporin, AQP-h2K, specifically expressed in the kidney of the tree frog, *Hyla japonica*. Endocrinology 148: 5891–5901.

Ogushi Y, Akabane G, Hasegawa T, Mochida H, Matsuda M, Suzuki M, Tanaka S (2010). Water adaptation strategy in anuran amphibians: molecular diversity of aquaporin. Endocrinology 151: 165–173.

Suzuki M, Tanaka S (2009). Molecular and cellular regulation of water homeostasis in anuran amphibians by aquaporins. Comparative Biochemistry and Physiology, Part A. Molecular & Integrative Physiology 153: 231–241.

Suzuki M, Tanaka S (2010). Molecular diversity of vasotocin-dependent aquaporins closely associated with water adaptation strategy in anuran amphibians. Journal of Neuroendocrinology 22: 407–412.

1 Section 6.1.3 and Figure 6.22 give further details on skin structure in amphibians.

An increase in water permeability of the nephron increases the chance of tubular fluid reaching the osmolality of the surrounding interstitial fluid and plasma, but it is not possible for reptiles, amphibians or fish (except lampreys) to produce urine with a higher osmotic concentration than the plasma (hyperosmotic urine). Production of hyperosmotic urine is, however, typical of mammals and birds with the aid of AVP or AVT.

Antidiuretic actions of AVT and AVP in birds and mammals

The production of hyperosmotic urine by birds and mammals relies on the structure–function relationships of the nephron and, in particular, the presence of a countercurrent loop and the parallel arrangement of collecting ducts[56]. In mammals, the main effect of AVP is an increase in the water permeability of the distal parts of the kidney nephrons (late distal segment, collecting tubule and collecting ducts[57]), which allows an increase in the reabsorption of water from the tubular fluid by osmosis, generating urine with a greater osmotic concentration.

Figure 21.23 illustrates the cellular actions of AVP on the water permeability of the distal nephron. Within minutes of AVP binding to V_2 receptors in the basolateral membranes of the tubular epithelium, intracellular trafficking of vesicles containing stored aquaporins (AQP-2)[58] results in their insertion in the luminal membrane, which increases the water permeability of the distal nephron. This response may have evolved in early terrestrial vertebrates, given that similar responses to AVT occur in amphibians, reptiles and birds. Water that passes into these cells from the tubular fluid, via AQP-2, diffuses through the cytosol and exits via AQP-3 and AQP-4 types of water channels in the basolateral membrane of the cells.

The AQP-2 inserted in the luminal membrane is rapidly recycled and not reinserted unless high concentrations of AVP persist, so short-term rapid control is possible. If dehydration continues for hours or days then persistently high concentrations of AVP occur, which result in increased expression of the gene for AQP-2. Further increases in the number of water channels in the luminal membrane of the principal cells of the collecting duct then occur, but the exact mechanisms responsible for long-term regulation of AQP-2 remain unclear. When vasopressin stimulation ends, endocytosis of water channels restores low basal water permeability.

AVP also increases the permeability to urea of the collecting duct in the inner medulla of mammalian kidneys, by about 100-fold in most species, by rapid increases in the membrane accumulation of urea transporters (UT-A1 and UT-A3)[59]. The late distal tubule and cortical (early) collecting duct have relatively low permeability to urea but reabsorb variable amounts of water. During dehydration, when AVP concentrations are high, more water is reabsorbed so the concentration of urea in the tubular fluid increases as it passes through the distal convoluted tubule and early parts of the collecting duct system. Once the tubular fluid reaches the inner medullary collecting duct, in the presence of AVP, an increase in the permeability to urea occurs.

Figure 21.23 illustrates the cellular actions of AVP on urea (and water) permeability of the inner medullary collecting ducts. The effects on urea transporters occur

21

56 Section 7.2.3 discusses the anatomical arrangements and transport processes in the loop of Henle, collecting ducts, and vasculature which are necessary for production of hyperosmotic urine by mammals and birds.

57 See Figure 21.30 for a diagram of the mammalian nephron.

58 We discuss aquaporins in a broad sense in Section 4.2.3.

59 We briefly examine the main types of urea transporters in Section 4.2.2.

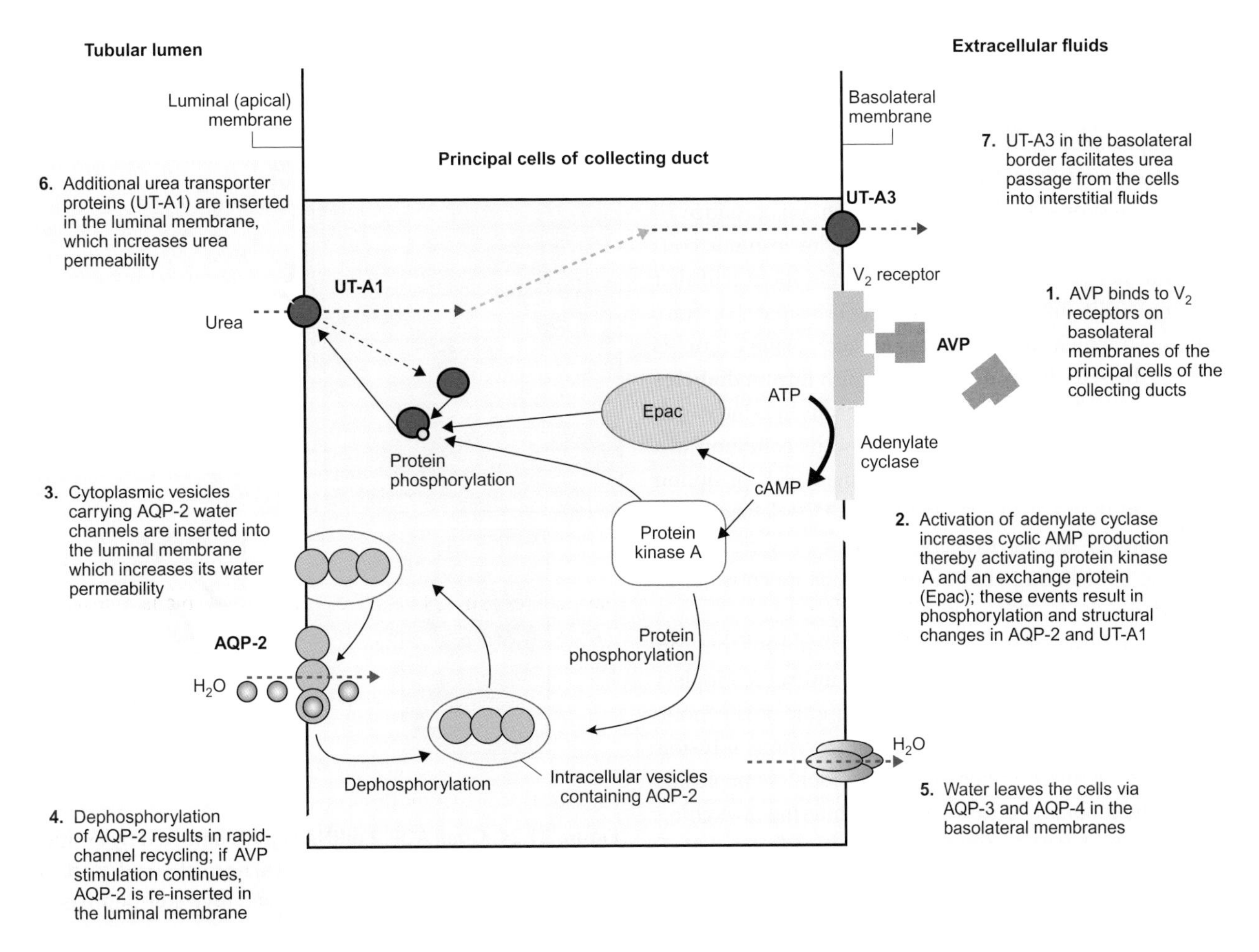

Figure 21.23 Schematic diagram of the actions of arginine vasopressin (AVP) on aquaporin-2 (AQP-2) and urea transporters (UT-A1) in principal cells of inner medullary collecting ducts of mammalian kidneys

via rapid activation of two cAMP-dependent signalling pathways: the exchange protein Epac and protein kinase A, both of which increase the phosphorylation of UT-A1 and its increased insertion in the luminal membrane of the tubular epithelial cells, which facilitates the passage of urea from the tubular fluid into the cells. During dehydration, urea passage from the cells occurs via UT-A3 in the basolateral membranes, which is also enhanced by AVP. From the interstitial fluid, urea enters a countercurrent exchange system of capillaries[60], which leads to urea recycling within the kidney.

The addition of urea to the interstitial osmolality along with the NaCl that results from the transport processes in the loop of Henle enables water reabsorption from the collecting duct, via AVP-stimulated AQP-2. The importance of urea transporters in the production of concentrated urine by some animal species becomes very apparent in 'gene knockout' mice lacking UT-A1 and UT-A3: the kidneys of these mice have a poor ability to generate concentrated urine and a very poor ability to increase urine concentration in response to AVP treatment. Urine osmolality is about 800 mOsm kg^{-1} in the gene knockout mice, whether or not they receive AVP treatment, whereas in wild-type mice the average osmolality of the urine increases from about 3000 mOsm kg^{-1} to an average of about 4500 mOsm kg^{-1} during AVP treatment.

AVP has further actions on the kidneys of some mammalian species that help to mediate antidiuretic responses. In some species, AVP stimulation of sodium chloride reabsorption from the tubular fluid in the collecting ducts contributes to the osmotic concentration of the inner medullary interstitium. AVP also reduces blood flow through the countercurrent system of blood vessels—the **vasa recta**—that runs alongside the loop of Henle and parallel to the collecting duct. The slower blood flow increases the time available for the exchange of water and ions with interstitial fluids, which helps to maintain the maximum difference in osmotic

[60] Figures 7.20 and 7.21 illustrate the processing of urea in the medulla of the kidney of dehydrated mammals.

21

concentrations between cortex and medulla[61] that is necessary for production of maximum hyperosmotic urine for a particular species.

The importance of AVP becomes strikingly clear when genetic defects result in AVP deficiency or a lack of the normal kidney response to AVP. AVP-deficient rats produce 10 times more urine than normal rats, with a low osmotic concentration (about 150 mOsm kg^{-1}, which is about half that of plasma). This dilute urine is tasteless (insipid); the rare condition in which the production of such urine occurs is known as **diabetes insipidus** to distinguish it from **diabetes mellitus**, in which excretion of glucose results in urine which has a sweet smell (and taste, which is how the condition was diagnosed historically; diabetes is from the Greek for siphon and refers to the excessive volume). The AVP-deficient rats need to compensate for the large amount of water loss in their urine by drinking about four times as much water as normal rats.

Kidneys of birds contain some looped nephrons that can produce urine hyperosmotic to plasma, and some loopless nephrons that can only generate hypo-osmotic or isosmotic urine[62]. In at least some bird species, intravenous infusion of AVT reduces the number of filtering loopless nephrons while all the looped nephrons continue filtering. Switching off loopless nephrons is reminiscent of the effects of AVT in fish and amphibians, shown in Figure 21.22. However, in birds the advantage lies in redistribution toward looped nephrons that can produce hyperosmotic urine. An additional possibility is that AVT may act preferentially on water conservation by the looped nephrons.

An AQP-2 type water channel similar to that in mammalian collecting ducts occurs in the collecting ducts of the kidneys of Japanese quails (*Coturnix japonica*). As in mammals, the AQP-2 protein of Japanese quails appears to be located in sub-apical vesicles of the epithelial cells when inactive. After dehydration of quails (which increases the plasma concentrations of AVT) or after injection of quails with AVT, AQP-2 mRNA increases and AQP-2 protein increases in apical areas of the cells.

Figure 21.24 shows the more pronounced expression of quail AQP-2 mRNA during water deprivation. However, notice in the micrographs that quail AQP-2 mRNA is higher in medullary collecting ducts than cortical collecting ducts in both hydrated and water-deprived quails. While the collecting ducts are likely to be the main site for water conservation, an increase in AQP-2 in both cortical and medullary nephrons suggests that both types of nephron may be involved in water conservation. One hypothesis is that an increase in water absorption by the cortical collecting ducts, stimulated by AVT, could reduce the flow of tubular fluid into the medullary collecting ducts. A reduced flow rate would increase the contact time between tubular urine and the epithelial cells of the medullary collecting ducts, thereby enhancing water reabsorption. Of course, these ideas need testing by further experiments and in other avian species.

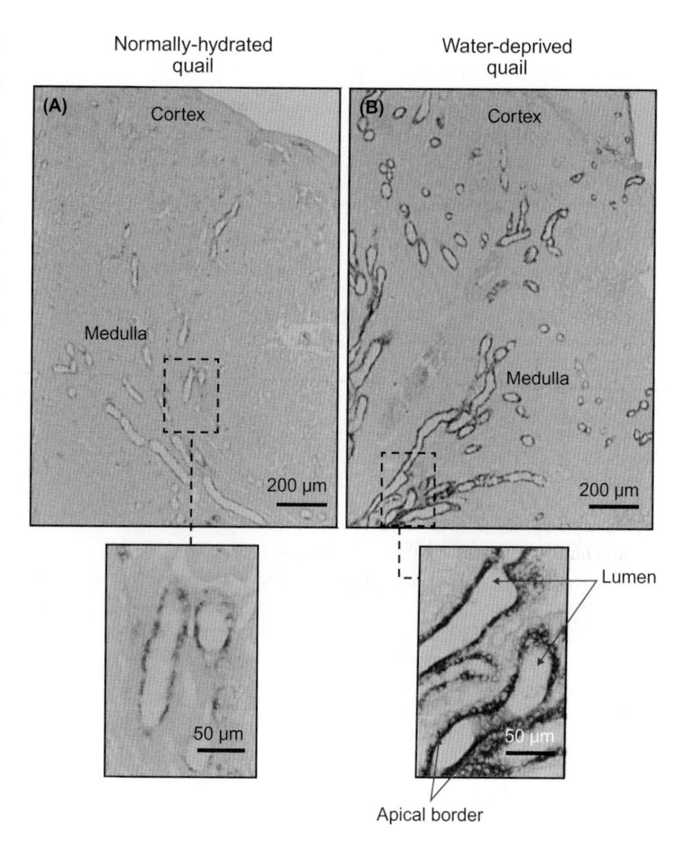

Figure 21.24 Quail AQP-2 mRNA signals detected by in situ hybridization in kidney slices from (A) normally-hydrated Japanese quails (*Coturnix japonica*) and (B) water-deprived quails

The larger images are orientated with the kidney cortex at the top and the medulla towards the bottom of the micrographs. A small area of the medulla is magnified for each image. The intensity of staining indicates the level of AQP-2 mRNA. In normally-hydrated quail a modest level of AQP-2 mRNA occurs in some epithelial cells of the collecting ducts, particularly in the medulla. Water-deprivation increases AQP-2 mRNA (shown by increased staining intensity), and results in a wider distribution of AQP-2 mRNA in both the medullary and cortical collecting ducts and their branches.

Source: Lau KK et al (2009). Control of aquaporin 2 expression in collecting ducts of quail kidneys. General and Comparative Endocrinology 160: 288–294.

› *Review articles*

Balment RJ, Lu W, Weybourne E, Warne JM (2006). Arginine vasotocin a key hormone in fish physiology and behaviour: A review with insights from mammalian models. General and Comparative Endocrinology 147: 9–16.

Beyenbach KW (2003). Transport mechanisms of diuresis in Malpighian tubules of insects. Journal of Experimental Biology 206: 3845–3856.

[61] The schematic in Figure 7.23A examines the operation of a countercurrent exchange system in the vasa recta.

[62] Figure 7.6 shows the structural arrangement of bird kidneys.

Dow JAT (2012). The versatile stellate cell—More than just a space-filler. Journal of Insect Physiology 58: 467–472.
Fournier D, Luft FC, Bader M, Ganten D, Andrade-Navarro MA (2012). Emergence and evolution of the renin-angiotensin-aldosterone system. Journal of Molecular Medicine 90: 495–508.
Hildebrandt J-P (2001). Coping with excess salt: adaptive functions of extrarenal osmoregulatory organs in vertebrates. Zoology 104: 209–210.
Holmgren S, Olsson C (2011). Autonomic control of glands and secretion: A comparative view. Autonomic Neuroscience: Basic and Clinical 165: 102–112.
Laverty G, Skadhauge E (2008). Adaptive strategies for post-renal handling of urine in birds. Comparative Biochemistry and Physiology Part A 149: 246–254.
Loretz CA, Pollina C (2000). Natriuretic peptides in fish physiology. Comparative Biochemistry and Physiology Part A 125: 169–187.
McCormick SD, Bradshaw D (2006). Hormonal control of salt and water balance in vertebrates. General and Comparative Endocrinology 147: 3–8.
McKinley MJ, Johnson AK (2004). The physiological regulation of thirst and fluid intake. Physiological Sciences 19: 1–6.
Nishimura H, Yang Y (2013). Aquaporins in avian kidneys: function and perspectives. American Journal of Physiology – Regulatory Integrative and Comparative Physiology 305: R1201–R1214.
Orchard I (2006). Serotonin: A coordinator of feeding-related physiological events in the blood-gorging bug, *Rhodnius prolixus*. Comparative Biochemistry and Physiology, Part A 144: 316–324.
Palmer BF (2015). Regulation of potassium homeostasis. Clinical Journal of American Society of Nephrology 10: 1050–1060.
Penton D, Czogalla J, Loffing J (2015). Dietary potassium and the renal control of salt balance and blood pressure. Pflügers Archiv – European Journal of Physiology 467: 513–530.
Rossier BC, Baker ME, Studer RA (2015). Epithelial sodium transport and its control by aldosterone: the story of our internal environment revisited. Physiological Reviews 95: 297–340.
Sands JM, Blount MA, Klein JD (2011). Regulation of renal urea transport by vasopressin. Transactions of the American Clinical and Climatological Association 122: 82–92.
Schweda F (2015). Salt feedback on the renin-angiotensin-aldosterone system. Pflügers Archiv – European Journal of Physiology 467: 565–576.
Tsukada T, Takei Y (2006). Integrative approach to osmoregulatory action of atrial natriuretic peptide in seawater eels. General and Comparative Endocrinology 147: 31–38.

21.2 Calcium balance

The 'total' calcium concentration of extracellular fluid in vertebrates is usually between 2 to 3 mmol L^{-1}, with the actual value depending on the species. Of this total calcium, however, approximately half is bound to circulating plasma proteins. Only the 'free' ionized calcium (Ca^{2+}) is biologically active and regulated at 1 to 1.5 mmol L^{-1}. Disturbed Ca^{2+} balance can lead to disturbed neural, neuromuscular and cardiovascular function[63], or skeletal damage.

The integration of the functioning of several tissues and organs determine the addition of Ca^{2+} to the extracellular fluids or its removal. The main influences on **calcium homeostasis** are:

- the exchange of Ca^{2+} between bony structures and extracellular fluids. Bone provides mechanical support in a protective skeleton and support for attachment of muscles, bony structures for impact absorption in some animals (such as the antlers of deer), and outer protection as in the scales of fish and shell of turtles[64]. Approximately 99 per cent of the total body calcium of bony vertebrates occurs in bony structures that consist mainly of crystals of the mineral hydroxyapatite $Ca_{10}(PO_4)_6(OH)_2$ deposited in a matrix of mainly **collagen** (protein) fibres. Hence, bony structures are an important reservoir for Ca^{2+} and inorganic phosphate anions, P_i[65];
- the rate of Ca^{2+} uptake from environmental sources—the diet, drinking water, and (for aquatic animals) from the surrounding water. Fish living in seawater, where Ca^{2+} concentrations are approximately 10 mmol L^{-1}, employ hormonally regulated processes to limit Ca^{2+} influx and excrete the excess. Fish living in rivers or ponds containing Ca^{2+} at 0.1 to 2.5 mmol L^{-1} take up Ca^{2+} through their gills by hormonally regulated processes, while amphibians take up Ca^{2+} via their skin[66];
- the rate of Ca^{2+} excretion in the urine.

Despite the enormous variation in the availability of calcium in the environment for different vertebrate groups, hormones control the free Ca^{2+} concentrations in their extracellular fluids in all cases. **Hypercalcaemic hormones** increase Ca^{2+} concentrations in response to hypocalcaemia (low concentrations of free Ca^{2+} in the blood or extracellular fluids; -aemia refers to blood), while **hypocalcaemic hormones** reduce Ca^{2+} levels in the blood in response to hypercalcaemia (elevated concentrations of free Ca^{2+} in the blood). In the next sections, we first explore the most important hypercalcaemic hormones in tetrapods (amphibians, reptiles, birds and mammals): **parathyroid hormone (PTH)** and **vitamin D**.

[63] Low concentrations of Ca^{2+} in the blood plasma cause neuromuscular excitability, because of an increase in the membrane permeability to Na^+; Section 16.2 discuss the effects on changing membrane permeabilities on nerve function; Chapter 18 discusses muscle function.

[64] Case Study 10.2 discusses the growth of bones, including those of dinosaurs.

[65] Inorganic phosphate is usually designated as P_i because it occurs as various ions depending on pH, as we discuss in Section 3.2.5.

[66] Section 5.2.2 discusses the osmoregulation of aquatic vertebrates in fresh water.

21.2.1 Parathyroid hormone and vitamin D hypercalcaemic actions

The chief cells of the **parathyroid glands** or **parathyroid tissue** are responsible for the synthesis and secretion of the parathyroid hormone (PTH). Although the precise composition of the 84 amino acids in PTH varies between species, the high level of conservation of the first 34 amino acids at the N-terminal end of the molecule indicates conservation of the active site in the molecule.

As their name suggests, the parathyroid glands lie close to the thyroid gland, at least in mammals, although in some species (e.g. humans) the parathyroid tissue is embedded within the thyroid gland; in birds, reptiles and amphibians, parathyroid glands/tissue lies away from the thyroid.

Fish have no parathyroid glands and for many years PTH was thought to have evolved with the emergence of tetrapods and their need to maintain Ca^{2+} balance on land, in the absence of the unlimited Ca^{2+} supply available to their aquatic ancestors. The discovery of PTH-like peptides and their receptors in fish overturned this idea and PTHs are now recognized to be a part of a family of related peptides, including PTH-related peptides (PTHrp), which appear to occur in all vertebrates, and PTH-L, which has been lost from the genome of placental mammals. Although not fully understood, it has become apparent that production of PTHrp and/or PTH-L occurs in a range of tissues and is associated with wide-ranging paracrine[67] (local) actions in all vertebrates. Although parathyroid glands do not occur in fish, the pituitary gland appears to be a major source of PTHrp, and circulating concentrations suggest that these peptides are involved in hormonal control of Ca^{2+} balance of fish in which they may exert hypercalcaemic effects, as in other vertebrates.

The release of PTH and its actions are best understood in mammals, in which **calcium-sensing receptors (CaSRs)**[68] in the plasma membrane of the chief cells detect the circulating concentrations of Ca^{2+}. Detection of high concentrations of Ca^{2+} inhibits the release of PTH, and hence inhibits its Ca^{2+}-retaining actions. By contrast, detection of a low concentration of Ca^{2+} by the chief cells removes the inhibition of PTH release, resulting in an increase in circulating concentrations of PTH, as shown in Figure 21.25.

Figure 21.26 illustrates the cascade of events that follows the release of PTH, and which restores the concentration of Ca^{2+} in the extracellular fluid. Note in the diagram that direct actions of PTH on kidney and bone raise Ca^{2+}

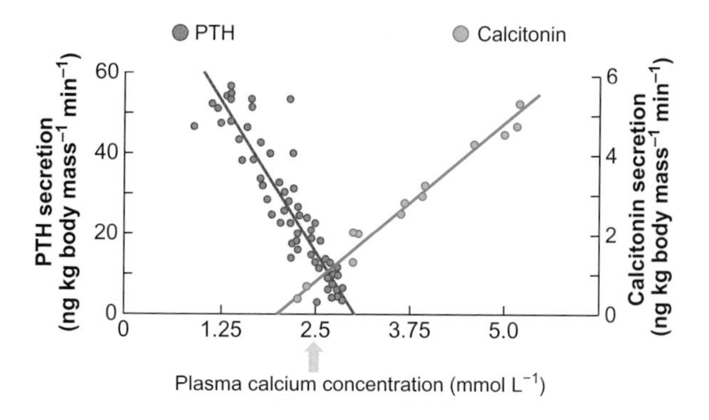

Figure 21.25 Effect of plasma calcium (total) on secretion of parathyroid hormone (PTH) and calcitonin by mammals

The graph shows PTH and calcitonin secretion occur at low rates when the plasma concentrations of total calcium are normal (blue arrow; approximately 2.5 mmol L^{-1}). A decline in plasma calcium results in an increase in secretion of PTH, while secretion of calcitonin decreases. At high plasma concentrations of calcium, calcitonin secretion increases and PTH secretion decreases. Although the data for PTH are for cow while data for calcitonin are for sheep, all mammals show similar patterns of response.

Source: Copp DH (1969). Calcitonin and parathyroid hormone. Annual Review of Pharmacology 9: 327–344.

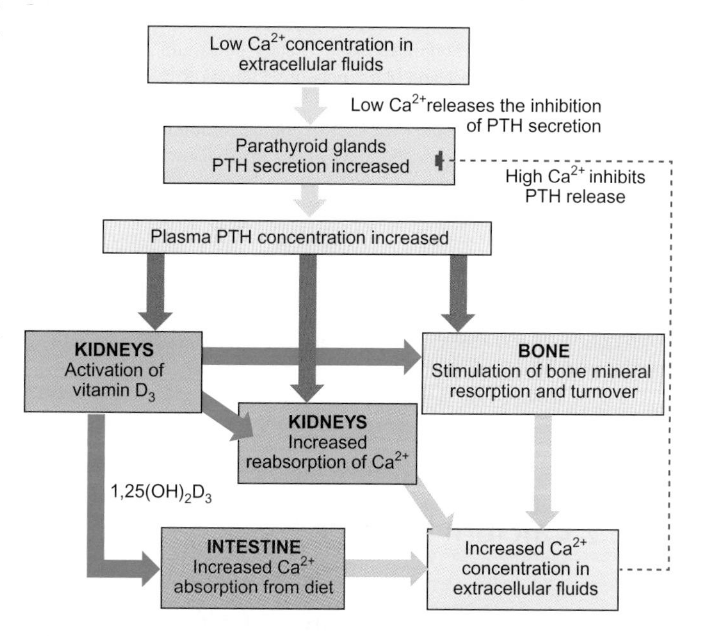

Figure 21.26 Effects of low extracellular fluid (plasma) concentrations of Ca^{2+} on PTH secretion and activation of vitamin D_3 resulting in physiological changes in kidneys, intestine and bone in mammals

Secretion of PTH is regulated by Ca^{2+}-sensing receptors in the chief cells of the parathyroid gland. PTH secretion is inhibited by high Ca^{2+} concentrations. Low Ca^{2+} concentrations release this inhibition, which results in higher rates of PTH secretion from the parathyroid glands. PTH stimulates production of active vitamin D_3 ($1,25(OH)_2D_3$) by the kidneys. Actions of $1,25(OH)_2D_3$ are shown by green arrows: increased Ca^{2+} absorption from the diet and drinking water, in the intestine, helps to restore the extracellular fluid concentrations of Ca^{2+}. Additional actions of PTH and $1,25(OH)_2D_3$ increase Ca^{2+} reabsorption from the kidney and stimulate bone turnover and resorption. Together these responses restore normal Ca^{2+} concentrations of extracellular fluids.

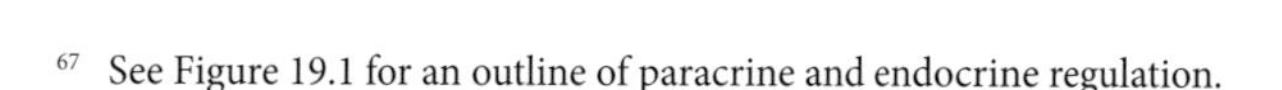

[67] See Figure 19.1 for an outline of paracrine and endocrine regulation.

[68] In Experimental Panel 21.1, we examine the role of CaSRs of fish in regulating their Ca^{2+} balance.

concentrations, but activation of vitamin D by PTH also occurs. In reality, Ca^{2+} balance depends on the integrated actions of active vitamin D_3 and PTH on Ca^{2+} transport by the intestine, kidney and bone. Before looking at these actions in more detail, we explore the activation of vitamin D.

The kidney activates vitamin D

Vitamin D is actually a group of related sterols (chemically similar to steroid hormones[69]). Land vertebrates often obtain sufficient vitamin D by synthesizing it in their skin, from 7-dehydrocholesterol when they are exposed to UV-B (with a wavelength of 295–300 nm) in sunlight. In birds and some mammals, feathers/fur block UV irradiation from reaching the skin, but intake of the vitamin D generated from skin secretions occurs during preening or grooming.

Vitamin D_3 (also known as cholecalciferol) is the most important D vitamin in controlling Ca^{2+} balance after conversion into its biologically active form by the steps illustrated in Figure 21.27:

- In the liver, incorporation of a hydroxyl group (OH) at carbon number 25 forms $25(OH)D_3$ (25-hydroxycholecalciferol), which enters the circulation.
- A second step, in the kidney, is crucial in forming the biologically active hormone: a second hydroxyl group is incorporated into some of the $25(OH)D_3$ to form $1,25(OH)_2D_3$ (1,25-dihydroxycholecalciferol), also known as calcitriol. This step requires the enzyme 1α-hydroxylase, which occurs in mitochondria in the epithelial cells of the proximal segment of the nephrons[70]. Hence, the 1α-hydroxylase activity of the kidney is a critical factor in determining the formation of $1,25(OH)_2D_3$ and its physiological effects. Both ionic and endocrine control factors regulate 1α-hydroxylase activity in the kidney, as shown in Figure 21.27. When extracellular concentrations of Ca^{2+} are low, the resulting increase in circulating PTH stimulates formation of $1,25(OH)_2D_3$. In addition, plasma concentrations of inorganic phosphate (P_i) directly influence 1α-hydroxylase activity; reduced P_i concentrations stimulate the synthesis of $1,25(OH)_2D_3$.

In circulation, most $25(OH)D_3$ and $1,25(OH)_2D_3$ are bound to plasma proteins; only the free component of $1,25(OH)_2D_3$ is physiologically active. PTH and $1,25(OH)_2D_3$ work together to determine the calcium balance of tetrapods through their combined actions on intestine, kidney and bone. These actions are summarized in Figure 21.26.

[69] We discuss the structure and synthesis of steroid hormones in Section 19.1.1.

[70] The first part of the nephron; we examine the organization of kidney nephrons in Section 7.2.

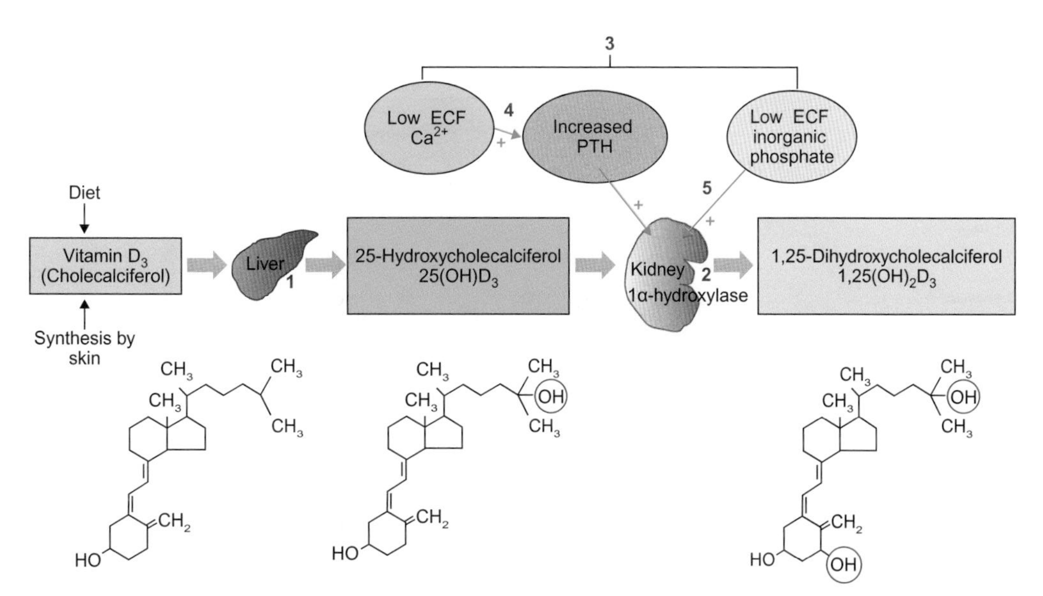

Figure 21.27 Activation of Vitamin D_3

(1) Vitamin D_3 is hydroxylated in the liver to form 25-hydroxycholecalciferol ($25(OH)D_3$). (2) A second hydroxylation step in the kidney by 1α-hydroxylase is the critical factor in determining the production of the physiologically active $1,25(OH)_2D_3$. The positions of hydroxylation are circled in red in the illustrations of the chemical structures. (3) Both ionic and endocrine control factors regulate kidney 1α-hydroxylase. Low extracellular fluid (ECF) concentrations of Ca^{2+} increase circulating concentrations of parathyroid hormone (PTH), (4) which *stimulates* 1α-hydroxylase by the kidney and thus increases formation of $1,25(OH)_2D_3$. (5) ECF concentrations of inorganic phosphate (P_i) also influence 1α-hydroxylase activity; a *decrease* in P_i concentrations *stimulates* synthesis of $1,25(OH)_2D_3$. Green arrows and + symbols indicate stimulatory effects.

21

$1,25(OH)_2D_3$ stimulates intestinal absorption of Ca^{2+}

In most tetrapods, absorption of Ca^{2+} from the intestine is a crucial component of calcium balance because it allows Ca^{2+} acquisition from the diet and drinking water. For terrestrial vertebrates this is the only source by which the body can replenish stores of Ca^{2+} in the bone.

After $1,25(OH)_2D_3$ passes across the basolateral border of the intestinal epithelial cells—the enterocytes, found mainly in the first part of the small intestine (the duodenum)—the hormone interacts with nuclear vitamin D receptors, which activates calcium absorption. Figure 21.28 illustrates the three main stages in transcellular absorption of Ca^{2+} that are activated by $1,25(OH)_2D_3$:

- Passage of Ca^{2+} into the enterocytes occurs via open **calcium channels**[71] in the luminal membrane of the enterocytes, which determines the rate at which Ca^{2+} enters the cells (Stage 2 in Figure 21.28). The number of open Ca^{2+} channels determines the rate of Ca^{2+} influx. The channels remain open as long as intracellular concentrations of free Ca^{2+} remain low (~0.1 mM), but close if intracellular Ca^{2+} concentrations increase.
- Binding of Ca^{2+} to Ca^{2+}-binding proteins called **calbindins**[72] (Stage 3 in Figure 21.28) keeps cellular concentrations of free Ca^{2+} low and results in carriage of Ca^{2+} across the cells. There is a linear relationship between the maximal Ca^{2+} passage across the intestine and the calbindin content of intestinal tissue, as shown in Figure 21.29.
- The third stage in transcellular calcium absorption is Ca^{2+} efflux via the basolateral border of the enterocytes (Stage 4 in Figure 21.28). Efflux occurs via active Ca pumps (Ca^{2+}-ATPase) and/or Na^+/Ca^{2+} exchange proteins.

The clearest evidence for $1,25(OH)_2D_3$ stimulation of transcellular Ca^{2+} absorption by the intestine comes from studies

[71] Section 4.2.2 outlines the concept of transport via ion channels. In the mammalian duodenum the specific Ca^{2+} channels are TRPV6 channels, which are a type of transient receptor potential (TRP) ion channel. Many types of TRPs are relatively non-selective ion channels responsible for the senses, as we learn in Chapter 17, Section 17.3.1 (olfaction), 17.4.3 (touch), and Sections 9.1.2 and 17.5 (thermoreception).

[72] Calbindins in different animal groups and among different tissues vary in molecular mass. In the intestinal epithelium of mammals, calbindin D has a molecular mass of 9 kDa and so is designated as calbindin-D_{9k}, whereas the calbindin found in the small intestine of birds has a larger molecule mass and is designated as calbindin-D_{28k}.

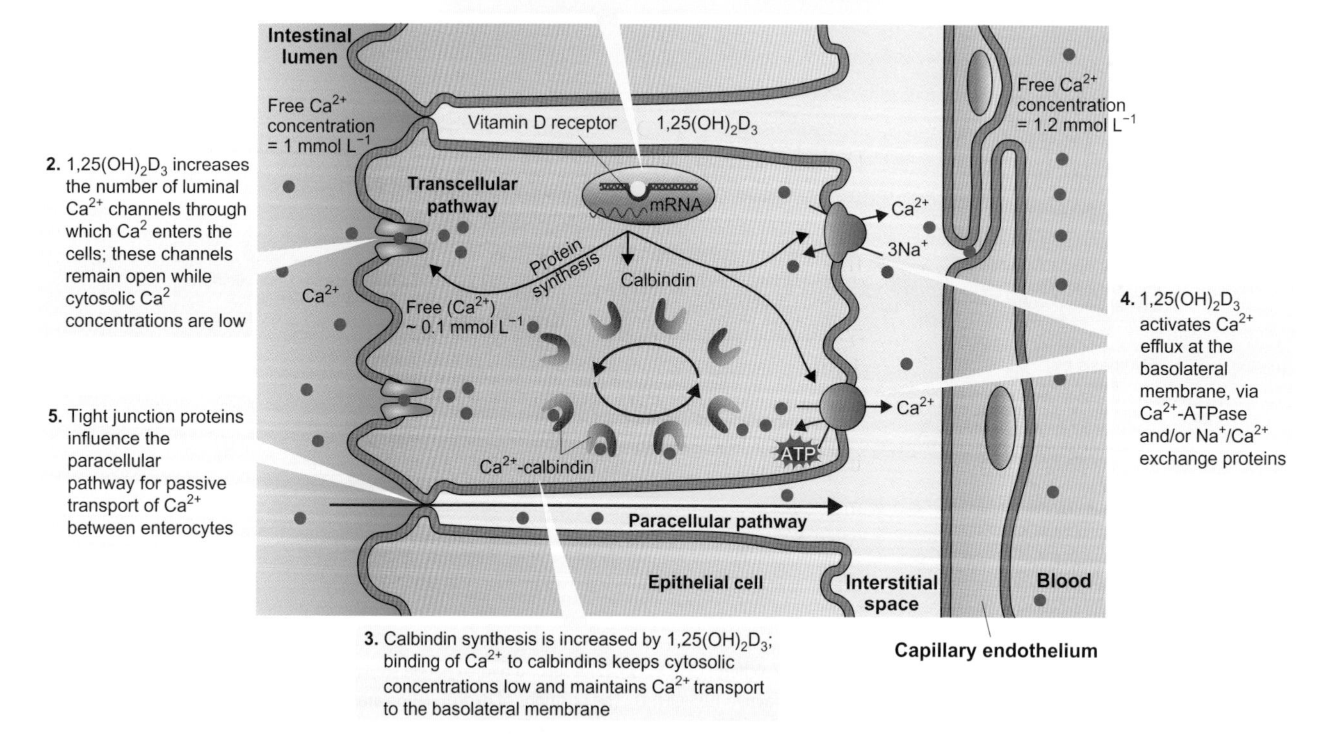

Figure 21.28 Model of intestinal Ca^{2+} absorption by transcellular and paracellular routes and stimulation by $1,25(OH)_2D_3$

Blue-filled dots indicate Ca^{2+} ions. 1 to 4 show steps in transcellular Ca^{2+} absorption. 5 indicates the paracellular route.

Source: adapted from Hoenderop JGJ et al (2005). Calcium absorption across epithelia. Physiological Reviews 85: 373–422.

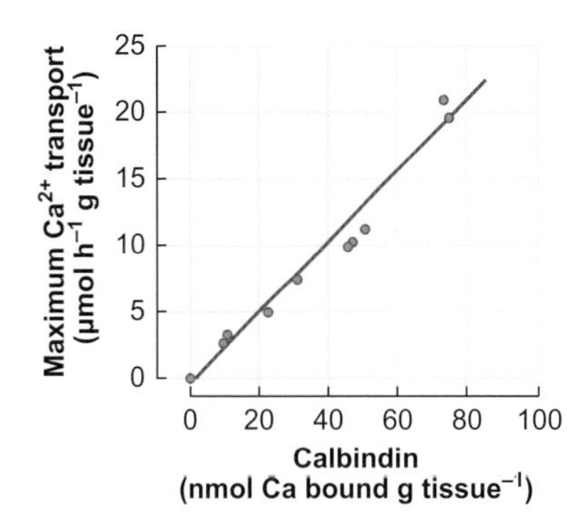

Figure 21.29 Linear relationship between maximum Ca^{2+} transport across the intestine and its calbindin-D_{9K} content

Source: Bronner F (2003). Mechanisms and functional aspects of intestinal calcium absorption. Journal of Experimental Zoology 300A: 47–52.

of mice with specific gene deletions (gene knockout mice). Mice lacking the 1α-hydroxylase gene are virtually incapable of producing $1,25(OH)_2D_3$ in their kidneys. Their low plasma Ca^{2+} concentrations correlate with reduced expression of TRPV6, calbindin and Ca^{2+}-ATPase in the intestine. Treatment of these mice with $1,25(OH)_2D_3$ restores the normal expression of Ca^{2+} transporting proteins and normal plasma concentrations of Ca^{2+}.

The gene deletion studies indicate that $1,25(OH)_2D_3$ activates all three of the main steps in transcellular Ca^{2+} absorption: steps 2 to 4 in Figure 21.28. An increase in Ca^{2+} channels by $1,25(OH)_2D_3$ will enhance Ca^{2+} entry into the cells on the luminal border. Increased expression of calbindin-D_{9K} maintains low intracellular concentrations of Ca^{2+} and enhances Ca^{2+} transport to the basal border. Increased Ca^{2+}-ATPase enhances Ca^{2+} efflux into the extracellular fluid. Hence, $1,25(OH)_2D_3$ increases the absorption of Ca^{2+} from the intestinal lumen to the extracellular fluids.

Looking again at Figure 21.28, notice that the passage of Ca^{2+} across the intestinal epithelium may also occur via a paracellular route, between the enterocytes. This route allows passive Ca^{2+} absorption along the entire intestine. The extent of passive absorption depends on three factors:

- the time that fluid spends in the intestine;
- tight junction functioning at the luminal border of the enterocytes. Calcium absorption via the paracellular route has been assumed to be beyond the influence of hormones. However, recent results suggest that $1,25(OH)_2D_3$ induces two claudins[73], which are membrane proteins that are major components of the tight junctions (claudin 2 and claudin 12), and which mediate the formation of Ca^{2+} permeable pores;
- the electrochemical potential difference[74] for Ca^{2+} across the intestinal epithelium. This factor considers the combined effect of electrical potential difference and the difference in Ca^{2+} concentrations of intestinal and extracellular fluids.

The dietary content of Ca^{2+} is the most significant factor to affect paracellular absorption of Ca^{2+} and determines the difference in Ca^{2+} concentrations between the gut lumen and extracellular fluids. Hence, dietary Ca^{2+} is a major factor in determining calcium balance. A high or normal dietary content of Ca^{2+} results in absorption of large amounts of Ca^{2+} by the paracellular route. In such conditions, inhibition of the vitamin D system reduces the number of Ca^{2+} channels in the enterocytes, which avoids flooding the enterocytes with Ca^{2+}, and low calbindin D_{9K} restricts Ca^{2+} transport across the cells, which reduces transcellular absorption of Ca^{2+}. However, when the dietary intake of Ca^{2+} is low and extracellular fluid concentrations of Ca^{2+} fall below normal, activation of the vitamin D system increases the circulating concentrations of $1,25(OH)_2D_3$, which stimulates transcellular absorption of Ca^{2+} by the intestinal enterocytes. In such circumstances, absorption of most of the Ca^{2+} in the diet occurs by the combination of paracellular and transcellular absorption.

Parathyroid hormone and $1,25(OH)_2D_3$ stimulate kidney reabsorption of Ca^{2+}

In tetrapods, the kidney is important in determining calcium balance. Of the total plasma concentration of calcium, about 60 per cent is filterable by the kidney glomeruli[75], while 40 per cent is associated with the non-filterable plasma proteins. Most of the filtered Ca^{2+} is reabsorbed in the kidney tubules.

Figure 21.30A illustrates the percent reabsorption of filtered Ca^{2+} from various parts of the mammalian nephron, starting in the proximal convoluted tubule, where 60–70 per cent of the filtered Ca^{2+} is reabsorbed, almost all of it passively (along with Na^+ and water)[76] via paracellular pathways between the tubular cells. A relatively small amount of the filtered Ca^{2+} is actively reabsorbed across the cells of the proximal segment and this process is stimulated by parathyroid hormone (PTH).

Further passive reabsorption of Ca^{2+} occurs in the thick ascending limb, as shown in Figure 21.30, where tight junction proteins (claudins) have been shown to influence the Ca^{2+} permeability of the paracellular pathways. Parathyroid

[73] We discuss claudins and tight junctions in Section 4.2.1.

[74] Section 3.1.3 explains the concept of electrochemical potential.

[75] Section 7.1 discusses glomerular filtration.

[76] We discuss NaCl and water reabsorption from the proximal tubule in section 7.2.1.

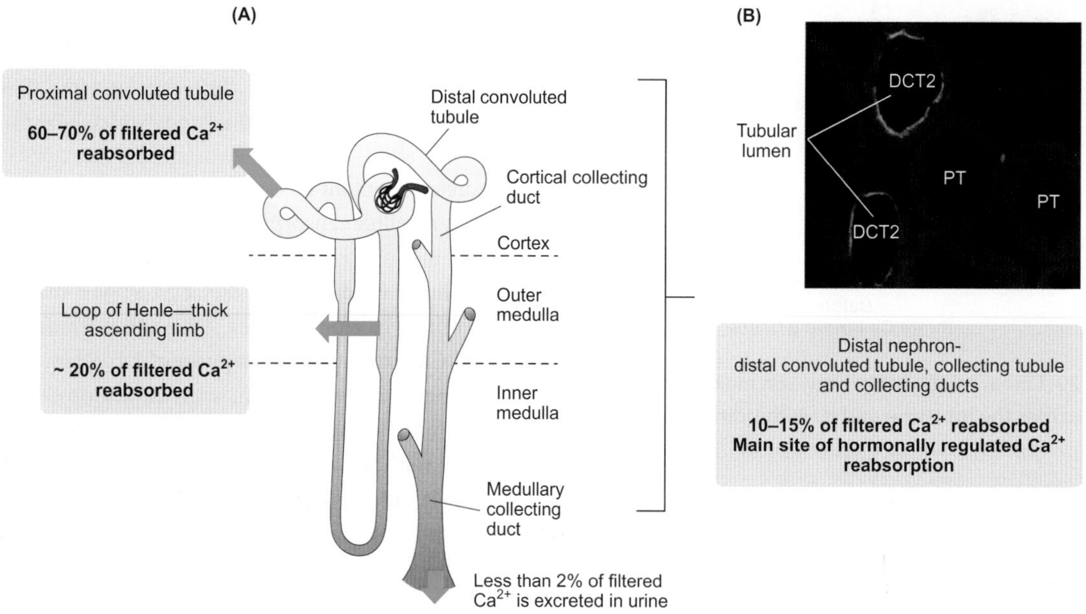

(A)

Proximal convoluted tubule

60–70% of filtered Ca^{2+} reabsorbed

Distal convoluted tubule

Cortical collecting duct

Cortex

Outer medulla

Loop of Henle—thick ascending limb

~ 20% of filtered Ca^{2+} reabsorbed

Inner medulla

Medullary collecting duct

Less than 2% of filtered Ca^{2+} is excreted in urine

(B)

Tubular lumen

DCT2

PT

PT

DCT2

Distal nephron- distal convoluted tubule, collecting tubule and collecting ducts

10–15% of filtered Ca^{2+} reabsorbed Main site of hormonally regulated Ca^{2+} reabsorption

Figure 21.30 Kidney reabsorption of filtered Ca^{2+} by mammalian nephrons

(A) Schematic drawing of mammalian nephron showing percentage of filtered Ca^{2+} reabsorbed by various parts of the nephron. (B) Immunohistochemical studies of mouse kidney show highest abundance (green staining) of Ca^{2+} channels (TRPV5) along the luminal membrane of the second part of the distal convoluted tubule (DCT2). The proximal tubules (PT) are difficult to see as they show no significant expression of Ca^{2+} channels.

Source: B: Hoenderop JGJ et al (2005). Calcium absorption across epithelia. Physiological Reviews 85: 373–422.

hormone stimulates active cellular absorption of Ca^{2+} by the thick ascending limb.

Once the tubular fluid enters the distal part of the mammalian nephron (distal convoluted tubule, collecting tubule and collecting duct), the tight junctions between the epithelial cells are relatively impermeable to Ca^{2+}. Consequently, Ca^{2+} reabsorption by the distal nephron occurs exclusively across the cells, against the electrochemical potential difference, and involves active processes.

Calcium reabsorption from the distal nephron has similarities to the transcellular absorption of Ca^{2+} by the intestine, illustrated in Figure 21.28, and usually leaves only a small proportion of the filtered Ca^{2+} (<2 per cent) in the urine. The initial step is entry into the epithelial cells via Ca^{2+} channels in the luminal membrane. Two types of Ca^{2+} channel occur in the luminal membranes of the distal nephron of mammals: TRPV6 channels are found along the entire distal nephron, while TRPV5 channels occur mainly in the second half of the distal convoluted tubule, as shown in Figure 21.30B. Studies of gene knockout mice in which the *TRPV5* gene is deleted suggest that TRPV5 channels are essential for Ca^{2+} balance, as *TRPV5* knockout mice suffer serious Ca^{2+} imbalance, resulting from excessive Ca^{2+} excretion. The Ca^{2+} concentration in the urine of these knockout mice reaches 20 mmol L^{-1} compared to 6 mmol L^{-1} in the wild-type littermates with intact *TRPV5* genes.

Parathyroid hormone is often considered the most important hormonal regulator of active reabsorption of Ca^{2+} by the distal nephron, but experimental findings indicate that $1,25(OH)_2D_3$ also increases the expression of Ca^{2+} transporter proteins in the tubular epithelium of the distal nephron and that both hormones play a part in stimulation of Ca^{2+} absorption. These experimental findings include:

- Parathyroidectomy (removal of the parathyroid glands) in rats reduces luminal Ca^{2+} channels (TRPV5), calcium binding proteins (calbindin-D_{28K}) and basolateral Na^+/Ca^{2+} exchange proteins in the distal nephron, and hence reduces plasma Ca^{2+} concentrations.
- Treatment of parathyroidectomized rats with PTH restores the level of Ca^{2+} transporter proteins and their plasma Ca^{2+} concentrations.
- Administration of $1,25(OH)_2D_3$ to rats depleted of vitamin D_3 and which have low plasma concentrations of Ca^{2+} increases luminal TRPV5 in the distal nephron and restores their plasma Ca^{2+} concentrations.
- The low plasma Ca^{2+} concentrations of gene knockout mice lacking the gene for 1α-hydroxylase, which are unable to produce adequate $1,25(OH)_2D_3$, correlates with low expression of TRPV5 and calbindin-D_{28K}. Treatment of these mice with $1,25(OH)_2D_3$ restores the expression of the Ca^{2+} transporter proteins.

21

Controversy surrounds the hormonal regulation of Ca^{2+} efflux across the basolateral membranes of the epithelial cells of the distal nephron. Some evidence suggests that Na^{+}/Ca^{2+} exchange proteins may be the main hormonally regulated mechanism for Ca^{2+} extrusion into the extracellular fluids. However, work on isolated kidney cells indicates that $1,25(OH)_2D_3$ also stimulates Ca^{2+}-ATPase activity in the basolateral membranes. Parathyroid hormone consistently affects Ca^{2+}-ATPase; for example, return of PTH after parathyroidectomy in rats stimulates the Ca^{2+}-ATPase, which could allow more Ca^{2+} transfer via this pump into the extracellular fluid.

Bone and the actions of PTH and $1,25(OH)_2D_3$

Continual turnover of bone results from two concurrent processes: (i) bone deposition by **osteoblasts** and (ii) bone resorption by **osteoclasts**. These processes preserve bone architecture and its integrity over many cycles of bone repair and replacement, while removing areas of micro-damage. During bone growth, the balance shifts toward bone deposition, which exceeds the rate of resorption; consequently, the dimensions of bones alter. This form of bone remodelling is at its greatest during early life, under the influence of growth hormone[77].

Mechanical loading on bones is a major influence on their remodelling as increased bone deposition occurs in areas of high stress, while resorption tends to occur in areas of lower stress. A lack of mechanical stress on the skeleton tips the balance toward bone resorption and bone mass decreases. This process increases the porosity of mature bones, resulting in osteoporosis (porous bones characterized by a decrease in bone mass and bone density) in which a progressive decrease in the mineral structure of the skeletal bones results in fragility and susceptibility to fractures.

An interesting case of temporary osteoporosis occurs in human astronauts that experience zero gravity in orbit, which reduces the mechanical stress on the skeleton. The rate of bone loss exceeds the rate of bone building such that a decrease in bone density exceeding 1 per cent per month results in spaceflight-induced osteoporosis. Over four decades of spaceflight, aerobic exercise regimes provided no significant benefit in preventing bone loss during spaceflight. However, recent studies have shown beneficial impacts and stable bone density of crews undertaking high resistance exercise during 4–6-month periods at the International Space Station.

Figure 21.31 illustrates the action of PTH and $1,25(OH)_2D_3$ on mammalian bone. This diagram simplifies the complex process, which is not fully understood. The combined influences of PTH and $1,25(OH)_2D_3$ acting on different receptors in bone-forming osteoblasts and bone-resorbing osteoclasts results in an increase in bone turnover. In the longer-term, proliferation of osteoblasts is impaired and new mature osteoclasts from precursor cells in the blood stream collect on the bone surfaces, shifting the balance between bone deposition and bone dissolution toward bone loss.

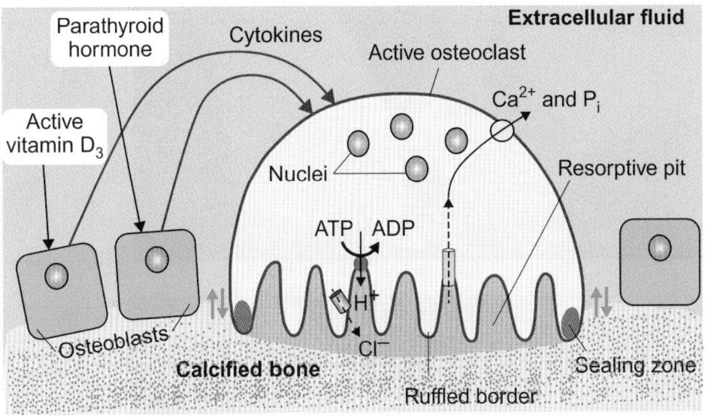

Figure 21.31 Parathyroid hormone and Vitamin D_3 control of bone turnover: exchange of Ca^{2+} between extracellular fluids and proteins at bone surfaces

Constant exchange of minerals occurs at the interface of bone mineral and extracellular fluids (indicated by small blue arrows in both directions). During bone resorption, the activated, dome-shaped multinuclear osteoclasts create a sealing zone that separates the extracellular fluid from resorptive pits. Secretion of enzymes and the operation of proton pumps results in dissolution of calcified bone. Bone minerals (Ca^{2+} and inorganic phosphate, P_i) are transported across the ruffled border via membrane channels such as Ca^{2+} channels into the cell cytoplasm of the osteoclasts and then into the extracellular fluid. Active vitamin D_3 ($1,25(OH)_2D_3$) and parathyroid hormone stimulate osteoblasts (by specific intracellular and membrane receptors, respectively). Their production of cytokines stimulates osteoclasts. Actions on both the bone-forming osteoblasts and the osteoclasts result in an increase in bone turnover.

Figure 21.31 includes a diagrammatic representation of an activated osteoclast in the process of mobilizing bone minerals: Ca^{2+} and inorganic phosphate (P_i). Notice the resorptive pit beneath the ruffled border of the osteoclast attached to the bone surface. Secretion of protease enzymes into a sealing zone between the ruffled border of the cell and the bone surface digests the bone matrix, while acid secretion dissolves the bone mineral, which releases Ca^{2+} and P_i that enter extracellular fluids. Bearing in mind that low extracellular fluid concentrations of Ca^{2+} act as a trigger for the release of PTH, and hence elevate circulating concentrations of $1,25(OH)_2D_3$, dissolution of bone helps to restore extracellular fluid concentrations of Ca^{2+}, as illustrated in Figure 21.26.

Birds that lay eggs with a calcified shell require large amounts of Ca (and P_i) during egg laying to the extent that plasma concentrations of Ca^{2+} may decrease as secretion of

[77] Section 19.2.2 gives further information on growth hormone.

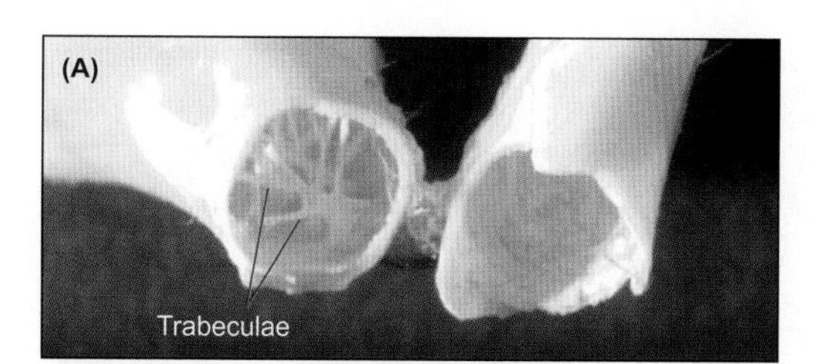

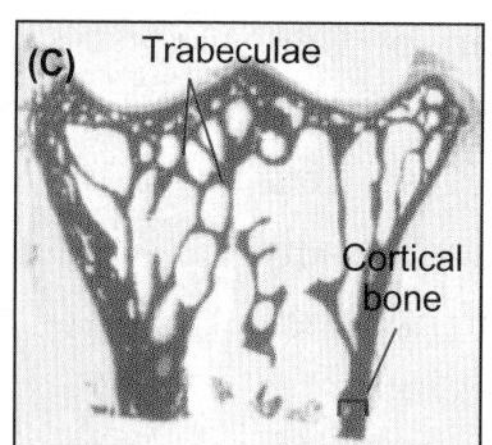

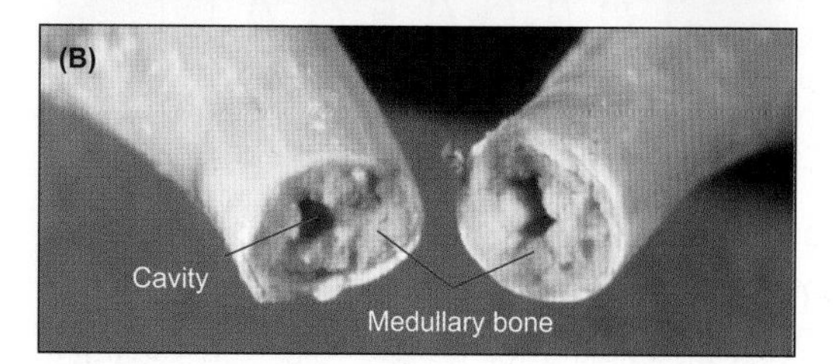

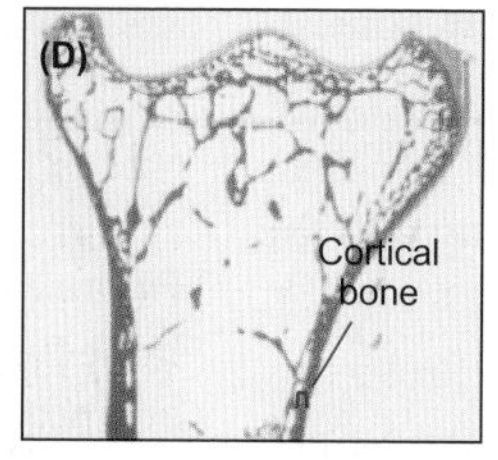

Figure 21.32 Bird bones mobilize calcium during egg laying

Photographs are of humerus (wing) (A & B) and femur (limb) bone sections (C & D) of hens. (A) and (C) are from young hens, (B) is from a mature hen and (D) is from a hen with osteoporosis. The long bones of birds are of pneumatic bone (hollow, with a central cavity). Note the cross-connecting trabeculae in (A) and (C) that strengthen structural (trabecular) bone. After maturation, the trabeculae are gradually replaced by medullary bone as shown in (B). In (D) osteoporosis is evident by the fewer and finer trabeculae, and the thinner layer of cortical bone in limb bones compared to this layer in (C).

Source: Whitehead CC (2004). Overview of bone biology in the egg-laying hen. Poultry Science 83: 193–199.

eggshells proceeds. The hypocalcaemia stimulates secretion of PTH, which stimulates resorption of Ca^{2+} from bone.

The long bones, especially those of the legs of mature female birds (and crocodilians), are unusual in having a cavity lined by a unique type of non-structural medullary bone, which Figure 21.32 illustrates. Formation of medullary bone occurs in mature female birds in response to the surge of plasma oestrogen, which precedes the production of Ca^{2+}-rich shells in the shell gland[78]. Medullary bone is poorer in strength than structural bone, but has a higher mineral content that can be released 10–15 times faster than from structural bone.

PTH acts on resorption by osteoclasts of both medullary and structural bone, although the medullary bone of egg-laying birds is most sensitive to PTH. If the diet is Ca^{2+}-deficient, the circulating concentrations of PTH are likely to increase further and stimulate significant resorption of structural bone, such that a temporary increase in bone porosity (temporary osteoporosis) occurs. Recent investigations indicate that the degree of osteoporosis is determined by the length of time that birds are in a continuous reproductive state, rather than the number of eggs they lay. Temporary osteoporosis is therefore a more likely problem in domesticated birds, such as chickens, which lay eggs over a long period. Typically, a progressive loss of structural bone occurs over the period of egg laying by chickens, and once egg laying ceases, the osteoblasts resume structural development of the bone. The temporary osteoporosis of some birds during egg laying, when concentrations of oestrogen are *elevated*, contrasts with the osteoporosis that may occur in postmenopausal humans, in which *declining* oestrogen inhibits bone formation and increases bone resorption.

Another case of temporary osteoporosis occurs in male deer in which annual development of antlers[79] creates a high demand for mineral precursors (Ca^{2+} and P_i). The mineral demand toward the end of antler growth often exceeds availability in the diet; partial resorption of the internal skeleton liberates the necessary minerals, and later reversal of the process occurs.

Vitamin D deficiency

Deficiency in vitamin D can arise in humans if inadequate provision in the diet and poor synthesis of vitamin D in the skin occur. The main consequence is poorer absorption of Ca^{2+} from the intestine. A resulting low plasma concentration of Ca^{2+} could stimulate PTH release (as indicated earlier in Figure 21.26) to maintain plasma Ca^{2+} at the expense of bone mineral density and with an increase in bone turnover. Bone-softening diseases, such as rickets in children and osteoporosis in adults, may result. Although it is almost a century since the role of vitamin D in preventing rickets was discovered, the exact mechanisms of action of its active metabolite ($1,25(OH)_2D_3$) on bone are still being explored. It is also argued that the skeletal actions of $1,25(OH)_2D_3$ may be redundant if normal mineral balance (Ca^{2+} and phosphate) can be maintained by their dietary intake.

Although rickets almost disappeared as a problem in much of the USA and Europe in the 1930s, largely through dietary supplementations, it has recently re-emerged as a major public health concern. Clinical assessment of vitamin D status is normally made by measuring blood serum concentrations of $25(OH)D_3$ as this is the main circulating form of vitamin D and has a long half-life (2–3 weeks). Consequently, serum concentration of $25(OH)D_3$ gives a good index of the vitamin D stores. The blood serum concentrations of $25(OH)D_3$ required for optimal health are, however, a subject of debate. A blood serum concentration of <20 ng mL^{-1} (50 nmol L^{-1}) of $25(OH)D_3$ in humans has been considered vitamin D deficient, based on the risk of rickets in children, or risk of defective bone mineralization causing the bone softening known as osteomalacia in adults. However,

[78] Figure 20.23A and Section 20.4 gives more details on the bird reproductive system and egg cycle.

[79] Section 20.3.5 discusses male secondary sex characteristics, including the development of antlers in male deer.

higher concentrations of 21–29 ng mL^{-1} may still be vitamin D insufficient so 30 ng mL^{-1} may be a better threshold for potential vitamin D sufficiency.

The most likely causes of vitamin D deficiency in humans become apparent when we remember that in humans about 90 per cent of vitamin D_3 is synthesized in the skin when it exposed to UV-B in sunlight. Any process that reduces the number of UV-B photons entering the epidermis reduces vitamin D_3 and hence reduces the synthesis of 25(OH)D_3 by the liver (following the processes illustrated back in Figure 21.27). Lifestyle factors such as use of concealing clothing and long periods of working indoors reduce 25(OH)D_3 synthesis and may lead to vitamin D deficiency. Protective sunscreens reduce UV penetration, and therefore reduce vitamin D_3 synthesis in the skin, by up to 99 per cent.

The skin pigment melanin absorbs UV-B photons and can reduce vitamin D_3 synthesis by up to 90 per cent in individuals with high levels of skin pigmentation. Consequently, African Americans could be at risk of vitamin D deficiency; indeed, higher levels of vitamin D deficiency have been reported in black mothers and their newborn than in white mothers and their newborn. This is a particular problem for dark-skinned migrants that move to higher latitudes.

Latitudinal, seasonal and daily alteration in the zenith angle of the sun[80] have a major influence on UV solar radiation and hence vitamin D_3 synthesis by the skin. Above and below latitudes of approximately 37° the seasonal reduction in radiation means that vitamin D_3 synthesis in the skin is very low or absent in winter, even in white-skinned individuals. On this basis, for six months of the year (October to April), all of Scandinavia, most of western Europe and about 50 per cent of Northern Canada are above the latitude where there is sufficient UV radiation for adequate vitamin D_3 synthesis. In such locations, 25(OH)D_3 is highly likely to fall to low levels, unless there is high dietary intake of vitamin D (for example, in a high-fish diet), or dietary supplementation with vitamin D. As an example, a winter and spring survey in the UK found ~50 per cent of the adult population had low levels of 25(OH)D_3 (<40 nmol L^{-1}) and ~16 per cent were severely deficient (<25 nmol L^{-1}).

The exact risks to particular populations depends on the conversion of 25(OH)D_3 to 1,25(OH)$_2D_3$, which has been found to be enhanced in black African American subjects compared to whites. Hence, few signs of vitamin D deficiency are actually apparent and bone density is higher among the black population. Genetic studies have identified a relationship with African ancestry as an important determinant of serum concentrations of 25(OH)D_3. Similar genetic factors also seem to apply among northern native Inuits that have normal serum Ca^{2+} and 1,25(OH)$_2D_3$ despite low serum 25(OH)D_3.

A few mammalian species are naturally vitamin D deficient because of the composition of their diet, coupled with a limited opportunity to synthesize vitamin D because of their way of life. For example, Egyptian fruit bats (*Rousettus aegyptiacus*) leave their roosts in caves to forage at night, and their diet of pulp and seeds does not contain measurable vitamin D, so both endogenous and dietary sources of vitamin D are very limited. Wild populations of these bats have undetectable concentrations of 25(OH)D_3, and very low amounts of 1,25(OH)$_2D_3$ compared to other mammals. Nevertheless, experimental studies suggest that wild Egyptian fruit bats have some small sources of vitamin D: in captivity (when these natural sources are likely to be reduced) their blood concentrations of 1,25(OH)$_2D_3$ decline to half the level that occurs in wild bats. Alternatively, their dusk or dawn forays for food may enable some vitamin D synthesis in the skin.

Despite low circulating concentrations of 1,25(OH)$_2D_3$, nocturnal Egyptian bats show no evidence of disturbed mineral balance or pathological problems associated with vitamin D deficiency. Similar findings apply to naked mole-rats (*Heterocephalus glaber*) that live underground and rarely see sunlight (if ever). Absorption of Ca^{2+} from the gastrointestinal tract, via the paracellular route (indicated in Figure 21.28), seems to be sufficient in both mole rats and fruit bats to maintain calcium balance, regardless of vitamin D status.

21.2.2 Hypercalcaemic actions of prolactin

The pituitary hormone prolactin[81] increases plasma Ca^{2+} concentrations of all vertebrate groups, although its actions in milk production by mammals and stimulation of parental behaviours are perhaps better-known actions. In fact, our understanding of the effects of prolactin in regulating epithelial Ca^{2+} transport of organs such as the intestine and kidney other than in fish and mammals is poor.

In species of teleost fish that can acclimate to a range of salinities, their transfer from seawater to fresh water generally results in the release of prolactin from the pituitary gland, but disturbed Ca^{2+} balance alone does not stimulate prolactin secretion. Instead, prolactin-producing cells are sensitive to the osmolality or sodium chloride concentrations of extracellular fluids. After the transfer of teleost fish to fresh water, the decreasing osmolality of extracellular fluids may open stretch-activated ion channels in prolactin-producing cells and allow Ca^{2+} entry to enhance release of prolactin. Thus, among fish, prolactin is most important in fish living in fresh water.

[80] The zenith angle of the sun is the angle measured directly overhead to the centre of the sun. The resulting global patterns in temperature together with global patterns in rainfall result in the global terrestrial biomes illustrated in Figure 1.9.

[81] Section 19.2.2 discusses prolactin secretion by the anterior pituitary and control of the secretion by hypothalamic neurohormones is outlined in Table 19.2.

Prolactin's effects on the gills of fish are important in regulating osmotic balance, by reducing the water and Na^+ permeability of the gills. Prolactin also has effects on Ca^{2+} balance of fish, via its hypercalcaemic action on the gills. These effects take several days to reach a maximum, which supports the idea that prolactin generally controls slow acclimatory responses to fresh water rather than rapid responses. Prolactin reduces the Ca^{2+} efflux across the epithelium of the gills and stimulates Ca^{2+} influx from the surrounding fresh water by increasing the activity of Ca^{2+}-ATPase pumps in the branchial epithelial cells.

Among amphibians and reptiles, there is evidence that prolactin has a hypercalcaemic effect, but the underlying mechanisms are unclear. Hypercalcaemia could result from enhanced intestinal absorption of Ca^{2+} or release of Ca^{2+} from bone and/or the endolymphatic sacs surrounding the brain, which are a unique store of calcium carbonate in amphibians. In anurans (frogs and toads), the sac extends down the vertebral column forming paravertebral lime sacs. Calcium carbonate stored here can be mobilized when required for bone repair or for use in metamorphosis or growth.

In birds, the hypercalcaemic activity of prolactin is likely to be largely indirect: prolactin stimulates 1α-hydroxylase activity in the kidney, which increases the production of $1{,}25(OH)_2D_3$, as shown in Figure 21.27. Early experiments showed a doubling of plasma $1{,}25(OH)_2D_3$ in chickens given daily injection of prolactin for several days, an action that stimulates calcium absorption by the intestine via mechanisms similar to those in Figure 21.28.

As in other vertebrates, prolactin induces hypercalcaemia in mammals as based on evidence from rodents, which have prolactin receptors in:

- renal tubular epithelium, where prolactin stimulates reabsorption of Ca^{2+}, which helps to minimize Ca^{2+} loss in the urine;
- intestinal epithelial cells, where prolactin triggers increased intestinal absorption of Ca^{2+}.

Prolactin stimulates a rapid increase in the absorption of Ca^{2+} from the duodenum (first part of the small intestine) of rats, which seems to result from luminal uptake via voltage-activated Ca^{2+} channels. Longer-term stimulation of Ca^{2+} absorption by the duodenum results from increased expression of TRPV5/6 Ca^{2+} channels, calbindin-D_{9k} and Ca^{2+}-ATPase. Figure 21.28 includes all of these components. An increase in these Ca^{2+} transporter proteins by prolactin allows a sustained increase in Ca^{2+} absorption from the diet, independently of, but entirely in cooperation with, the actions of $1{,}25(OH)_2D_3$.

Such actions are especially important in pregnant and lactating mammals whose plasma prolactin concentrations increase to 100–300 ng mL^{-1} from concentrations of 7–10 ng mL^{-1} in non-pregnant mammals. During suckling, surges of prolactin lead to even higher prolactin concentrations, reaching as much as 800 ng mL^{-1}. If the dietary supply of Ca^{2+} is inadequate for milk production, prolactin may also stimulate bone resorption of Ca^{2+} by osteoclasts.

In contrast to the situation in land vertebrates, aquatic animals live surrounded by Ca^{2+} in water. Marine teleosts rely on hypocalcaemic hormones to restrict Ca^{2+} influx and maintain a concentration of Ca^{2+} in their extracellular fluid (of 2–3 mmol L^{-1}) that is lower than the external concentrations of Ca^{2+} in seawater (at about 10 mmol L^{-1}).

21.2.3 Stanniocalcin and its hypocalcaemic actions

Stanniocalcin (so named because of its production by the **corpuscles of Stannius** of bony fish), and calcitonin are important hypocalcaemic hormones in teleosts. Figure 21.33 shows the location of the corpuscles of Stannius, which lie close to the ventral surface of the kidneys, and are made up of follicles of secretory tissue. Some species, such as eels, have discrete corpuscles of Stannius, as shown by Figure 21.33 that can be removed relatively easily. Their removal in European eels (*Anguilla anguilla*) (a process called a stanniectomy) results in an increase in plasma Ca^{2+} concentrations, which can reach up to 8 mmol L^{-1}. Re-implantation of the glands reverses the hypercalcaemia, which provides strong evidence that the corpuscles contain a hypocalcaemic hormone.

With a role in calcium balance in mind, another experimental approach is the injection of fish with calcium chloride

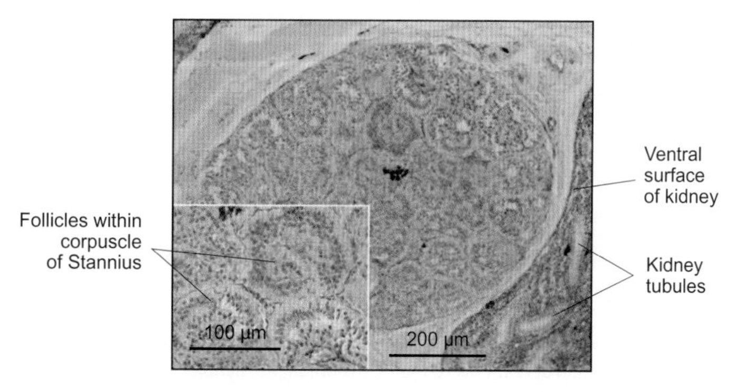

Figure 21.33 Corpuscle of Stannius of the Japanese eel (*Anguilla japonica*)

Tissue shown is from an eel adapted to seawater. Inset shows some corpuscle of Stannius follicles at higher magnification. The brown staining distributed over the glandular cells within the follicles is the result of an enzyme reaction with antibody bound to teleost calcium-sensing receptors (CaSR).

Source: Loretz CA et al (2009). Extracellular calcium-sensing receptor distribution in osmoregulatory and endocrine tissues of the tilapia. General and Comparative Endocrinology 161: 216–228.

to raise Ca^{2+} concentrations above normal levels. A resulting increase in the concentration of circulating stanniocalcin supports the idea that the corpuscles of Stannius respond to changes in the circulating concentration of Ca^{2+}. Such changes could occur naturally when there are environmental variations in water concentrations of Ca^{2+}.

To mimic natural variations, corpuscles of Stannius have been incubated *in vitro* with different concentrations of Ca^{2+} in the incubating medium. A relatively high Ca^{2+} concentration (2.5 mmol L^{-1}) increases the release of stored stanniocalcin and stimulates the synthesis of mRNA encoding stanniocalcin. If high Ca^{2+} concentrations stimulate release of stanniocalcin, we might anticipate that a low Ca^{2+} concentration could reduce the release of stanniocalcin.

One way to investigate such effects is to use calcium-binding substances to reduce the plasma concentrations of free, ionized Ca^{2+}. Figure 21.34 shows the results of such experiments in European flounders. Alongside the reduced plasma concentrations of Ca^{2+}, plasma concentrations of stanniocalcin decline. These results suggest that the low Ca^{2+} concentrations inhibit the secretion of stanniocalcin and that any existing stanniocalcin gradually degrades. Experimental Panel 21.1 explores some evidence for calcium-sensing receptors (CaSRs) in the glandular cells of the corpuscles of Stannius; their detection of Ca^{2+} concentrations in the extracellular fluid provides a mechanism to regulate stanniocalcin release.

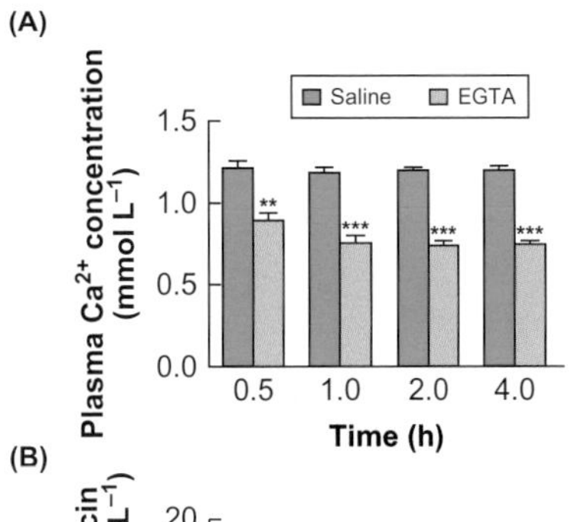

EGTA reduces plasma Ca^{2+} concentrations of flounders compared to those of time-matched control flounders given an injection of saline

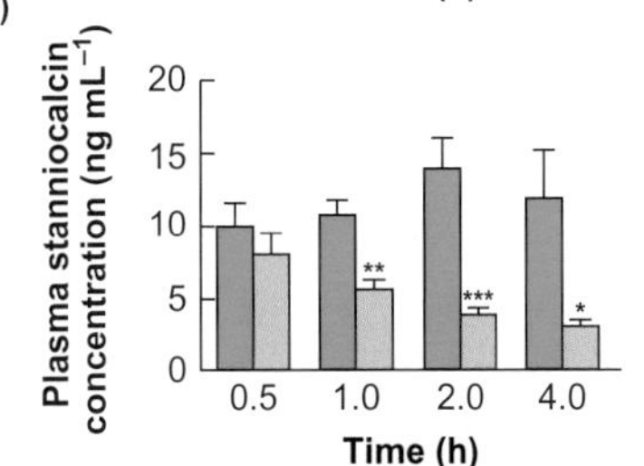

EGTA treatment results in a progressive reduction in stanniocalcin in flounders compared to time-matched control flounders given an injection of saline

Figure 21.34 The effects of lowering plasma Ca^{2+} with a calcium-binding agent on stanniocalcin of flounders (*Platichthys flesus*) living in fresh water

Flounders were given either a saline injection (as a control) or an injection of the calcium-binding agent EGTA (ethylene glycol tetra-acetic acid) at time 0 h. Blood samples were collected 0.5, 1, 2 and 4 hours later to examine the effects on plasma concentrations of Ca^{2+} and stanniocalcin.

Data are mean values + standard errors (n = 6–7 for each group); asterisks indicate significant differences compared to time-matched controls: $^{*}P < 0.05$; $^{**}P < 0.01$; $^{***}P < 0.001$. Source: Greenwood MP et al (2009). The corpuscles of Stannius, calcium-sensing receptor, and stanniocalcin: responses to calcimimetics and physiological challenges. Endocrinology 150: 3002–3010.

Experimental Panel 21.1 Calcium-sensing receptors controlling the release of stanniocalcin from the corpuscles of Stannius of teleost fish

Background and hypothesis

In mammals, calcium-sensing receptors (CaSRs) play a major role in regulating the secretion of the hormones involved in calcium regulation, but we know far less about the calcium-sensing receptors of other vertebrates. The Ca^{2+} concentrations of the extracellular fluids of species of teleost fish are influenced by external Ca^{2+} concentrations so such changes could act on CaSRs and trigger regulatory responses. In this panel, we explore the hypothesis that CaSRs regulate the release of stanniocalcin from the corpuscles of Stannius.

Experimental approach

Calcimimetic drugs are small organic compounds that are known to selectively activate mammalian CaSRs. The use of these drugs as an experimental tool provides a powerful means of studying the activation of potential CaSRs and their physiological effects. In such studies it is useful to compare the responses to injections of the calcimimetic drug with any effects following the injection of its enantiomer as a control. (Enantiomers are molecules whose structures are non-superimposable mirror images of each other—in this case, the control was the non-superimposable mirror image of the active drug, and should be inactive.) Two series of experimental studies have undertaken such investigations in rainbow trout and European flounders.

Experimental results and interpretation

Figure A shows the effects of a calcimimetic (R-467) on plasma concentrations of stanniocalcin in rainbow trout. Stimulation of the CaSRs with R-467 resulted in a rapid increase in stanniocalcin concentrations, as indicated by Figure A(i). Maximal stimulation was observed within 30 min of an injection of the calcimimetic and the effect persisted for at least 4 hours. Figure A(ii) shows that the increase in plasma concentrations of stanniocalcin followed a dose-related relationship at the lower end of the dose range for R-467, and plateaued at higher doses. At the maximal response, stanniocalcin increased by nine-fold.

Figure B shows that the effects of the calcimimetic drug are stereospecific in trout: the enantiomer was ineffective.

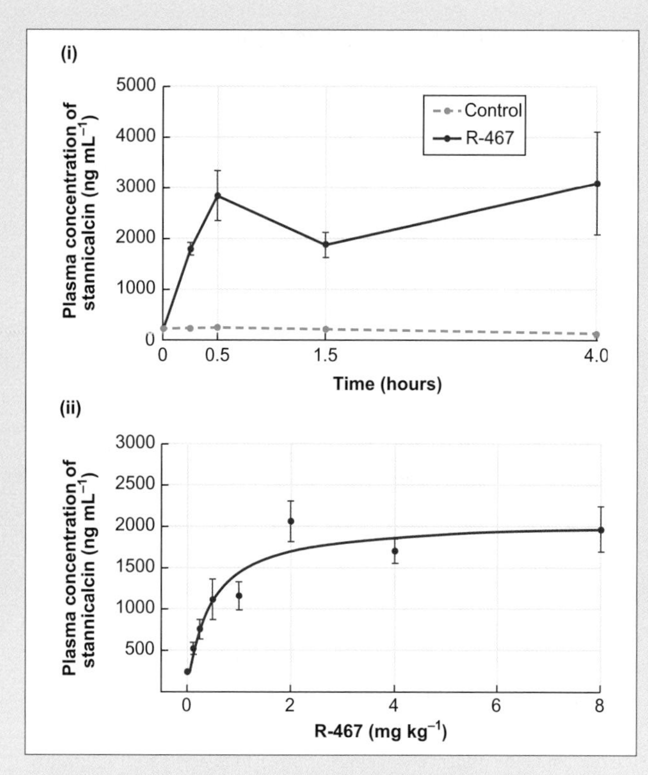

Figure A Effects of an intraperitoneal injection of a calcimimetic drug (R-467) on plasma concentrations of stanniocalcin in rainbow trout (*Oncorhynchus mykiss*)

(i) Time-related stimulation of stanniocalcin release by R-467. There is a significant increase in stanniocalcin at all time points compared to control (saline-injected) trout. (ii) Dose-related effect of R-467 one hour after injection. All doses significantly increased plasma stanniocalcin concentrations compared to control fish injected with saline for which data are not shown.

Data shown are means ± standard errors for (i) 10 fish per group; (ii) 15 fish per group. Source: Radman DP et al (2002). Evidence for calcium-sensing receptor mediated stanniocalcin secretion in fish. Molecular and Cellular Endocrinology 186: 111–119.

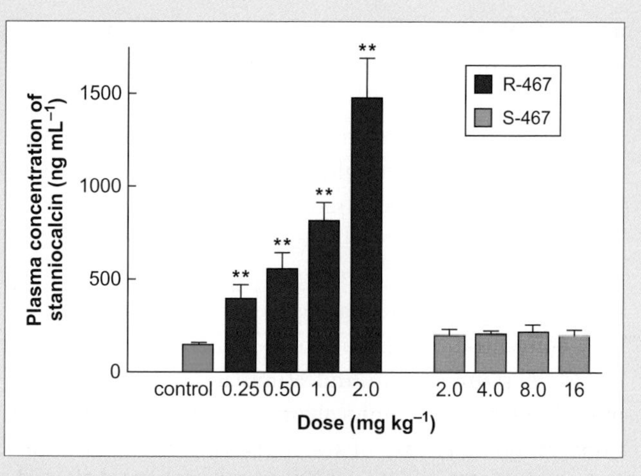

Figure B Stereospecific stimulatory effects of the calcimimetric drug (R-467) on stanniocalcin secretion in rainbow trout

R-467 stimulates secretion at all doses tested whereas the enantiomer (S-467) has no effect.

Data are means + standard errors for 15 fish per group. Significant effects were detected (**$P<0.01$) in all R-467 treated groups compared to controls. Source: Radman DP et al (2002). Evidence for calcium-sensing receptor mediated stanniocalcin secretion in fish. Molecular and Cellular Endocrinology 186: 111–119.

Figure C (i) shows similar findings in European flounders injected with a different calcimimetic drug compared to its enantiomer. In flounders, plasma concentrations of Ca^{2+} were also measured. Figure C(ii) shows a significant decrease in plasma concentrations of Ca^{2+} occurred after intra-arterial injection of the calcimimetric, which is likely to have resulted from the release of stanniocalcin as the enantiomer had no effect. Figure C(iii) shows an increase in urine concen-

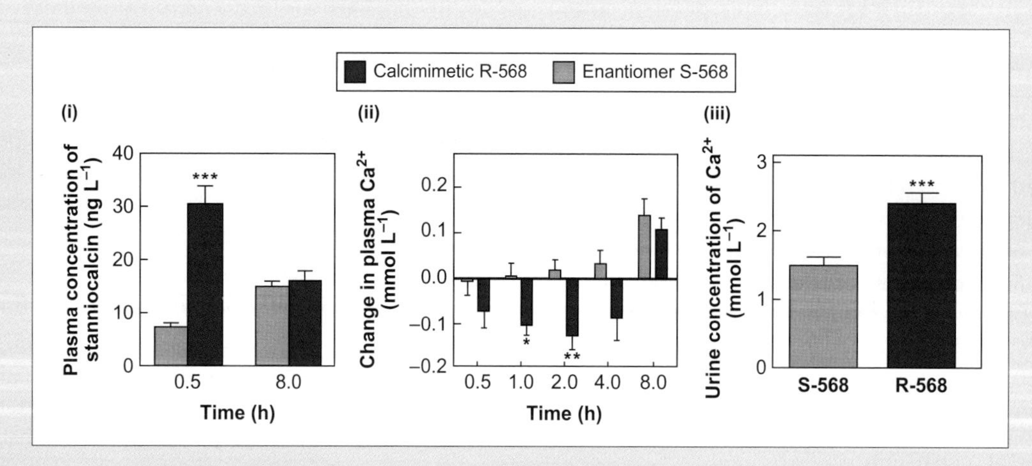

Figure C Effects of the calcimimetic drug (R-568) in flounders (*Platichthys flesus*)

(i) A large increase in plasma concentrations of stanniocalcin (purple columns) occurred within 0.5 h of the intra-arterial injection of the calcimimetric R-568 by comparison to control flounders injected with the enantiomer S-568 (orange columns). This effect was no longer apparent 8 h after the injection, when plasma concentrations of Ca^{2+} were similar in the two groups of flounders (panel ii). (ii) Plasma Ca^{2+} concentrations based on blood samples taken at 5 time points after injection of the calcimimetric or the enantiomer. The horizontal line indicates values where there is no net effect. After injection of the calcimimetic, plasma Ca^{2+} concentrations decreased progressively over the following 2 h relative to Ca^{2+} concentrations immediately before injection of the drug and then recovers. (iii) Urine sampled after injection of the calcimimetric contained a higher concentration of Ca^{2+} in flounders treated with the calcimimetric. Although the urine was collected 8 h after the injection, it reflects transport processes in the kidney several hours previously, presumably when stanniocalcin concentrations were elevated.

All values are means + standard errors for 8 fish per group; significant effects are indicated by asterisks: *$P< 0.05$; **$P< 0.01$; ***$P< 0.001$ comparing time-matched groups.

Source: Greenwood MP et al (2009). The corpuscles of Stannius, calcium-sensing receptor, and stanniocalcin: responses to calcimimetics and physiological challenges. Endocrinology 150: 3002–3010.

tration of Ca^{2+} after treatment of flounders with the calcimimetic which suggests at least a part of the induced hypocalcaemia by stimulation of CaSRs and after stanniocalcin release is due to increased kidney excretion of Ca^{2+}.

Implications of findings

The results give insights into how natural variations in Ca^{2+} concentrations externally, which influence extracellular Ca^{2+} concentrations, can affect circulating stanniocalcin. They provide a strong argument that CaSRs in the corpuscles of Stannius[1] regulate secretion of stanniocalcin into the circulation according to the variability in extracellular fluid concentrations of Ca^{2+}. For example, when fish move from fresh water into coastal waters, and extracellular fluid concentrations of Ca^{2+} increase; the rapid release of large amounts of stanniocalcin that is triggered is likely to initiate regulatory hypocalcaemic responses.

› Find out more:

Greenwood MP, Flik, G, Wagner GF and Balment RJ (2009). The corpuscles of Stannius, calcium-sensing receptor, and stanniocalcin: responses to calcimimetics and physiological challenges. Endocrinology 150: 3002–3010.

Loretz CA (2008). Extracellular calcium-sensing receptors in fishes. Comparative Biochemistry and Physiology Part A. Molecular & Integrative Physiology 149: 225–245.

Radman DP, McCudden C, James K, Nemeth EM, Wagner GF (2002). Evidence for calcium-sensing receptor mediated stanniocalcin secretion in fish. Molecular and Cellular Endocrinology 186: 111–119.

[1] Figure 21.33 shows immunohistochemical evidence of such CaSR in the corpuscles of Stannius of the Japanese eel (*Anguilla japonica*).

The major effect of stanniocalcin in teleost fish is the rapid inhibition of Ca^{2+} uptake from the water, across the gills. Figure 21.35 illustrates the series of steps thought to be involved in the passage of Ca^{2+} across the gills. Stanniocalcin inhibits the first (rate-limiting) step: the passage of Ca^{2+} across the apical membranes, via calcium channels. The action of stanniocalcin in reducing Ca^{2+} uptake across the gills is particularly important for fish living in seawater, in which there is a need to limit the apical entry of Ca^{2+}, and probably explains the higher levels of stanniocalcin of fish living in seawater.

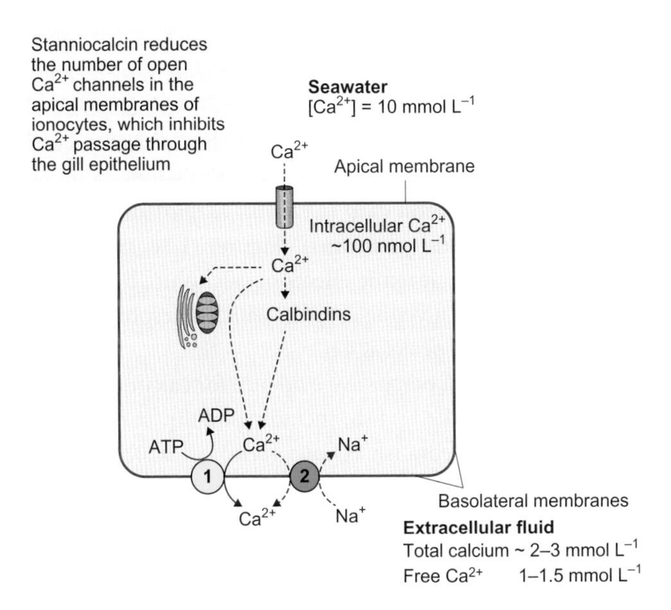

Figure 21.35 Model of action of stanniocalcin on gill ionocytes of marine teleost fish and its inhibition of Ca^{2+} transport across the gill epithelium

The rate-limiting step in Ca^{2+} passage through the gill epithelium is via open Ca^{2+} channels in the apical membrane of the ionocytes, which occurs due to the electrochemical difference across the membrane. Stanniocalcin inhibition of calcium channels reduces this influx, but any Ca^{2+} that does enter the cells is sequestered by cell organelles such as mitochondria or binds to calcium-binding proteins (calbindins) and is transported through basolateral Ca^{2+} transport mechanisms: (1) Ca^{2+}-ATPase and (2) Ca^{2+}, Na^+ exchange proteins. Stanniocalcin has no effect on these basolateral transport mechanisms or other transport processes that could indirectly influence these transporters. Solid lines show active transport; dashed lines show passive transport.

Source: adapted from Guerreiro PM, Fuentes J (2007). Control of calcium balance in fish. Chapter 15 in: Fish Osmoregulation, pp 427-495 (Eds. Baldisserotto B et al.) Science Publishers, Enfield, USA.

After stanniectomy, eels develop a more pronounced hypercalcaemia when they are in seawater than when they are in fresh water. In keeping with the hypocalcaemic actions of stanniocalcin, experiments on Japanese eels (*Anguilla japonica*), rainbow trout, and Atlantic cod (*Gadus morhua*) indicate that stanniocalcin also reduces the absorption of Ca^{2+} across the luminal membrane of the epithelial cells of the intestine.

The actions of stanniocalcin on the kidney are more controversial, although some studies suggest stanniocalcin inhibits Ca^{2+} reabsorption by the kidney tubules and hence increases Ca^{2+} excretion. Further studies of the effects of stanniocalcin on isolated kidney tubules of winter flounders (*Pseudopleuronectes americanus*) suggest that stanniocalcin stimulates reabsorption of phosphate ions, which could lead to binding of phosphate to Ca^{2+} and increased deposition of calcium phosphate in bone and scales.

The discovery of stanniocalcin in fish and the identification of stanniocalcin genes was a landmark that has led to the discovery of stanniocalcins in birds and mammals (including humans), none of which have corpuscles of Stannius. It was also discovered that vertebrates have two forms of stanniocalcins, both large glycoproteins (comprising 250 amino acids); these peptides have been named stanniocalcin 1 and 2 (STC 1 and ST2).

The principal function of STC1 is to regulate calcium balance. The function of STC2 is less certain, but the occurrence of stanniocalcins in many tissues suggests they have predominantly local **paracrine/autocrine**[82] actions, includ-

[82] Section 19.1 and Figure 19.1 outline the distinguishing features of paracrine and endocrine actions.

ing effects on tissue growth and an involvement in **angiogenesis**[83] (formation of new blood vessels) in tumours.

Most recently, the phylogenetic investigation of stanniocalcins and their genes have shown that these proteins have deep evolutionary roots among the **eukaryotes**[84]. Stanniocalcin homologs occur in most taxonomic groups of animals, including sponges, cnidarians, nematodes, annelids, molluscs, a cephalochordate (amphioxus: *Branchiostoma floridae*), the tunicate *Ciona intestinalis*, and vertebrates[85]. Only the echinoderms and arthropods seem to lack genes for stanniocalcin homologues. Hence, stanniocalcin homologues are predicted to have appeared in evolutionary terms as early as single celled eukaryotes: even single-celled ciliates, amoeboids and flagellates, and fungi have homologues of stanniocalcin genes, although their possible actions in these species are unclear.

While stanniocalcin generally appears to be the most important hypocalcaemic hormone of teleost fish, another hormone, calcitonin, also has hypocalcaemic actions and is the main hypocalcaemic hormone in some vertebrate groups, as we now go on to discuss.

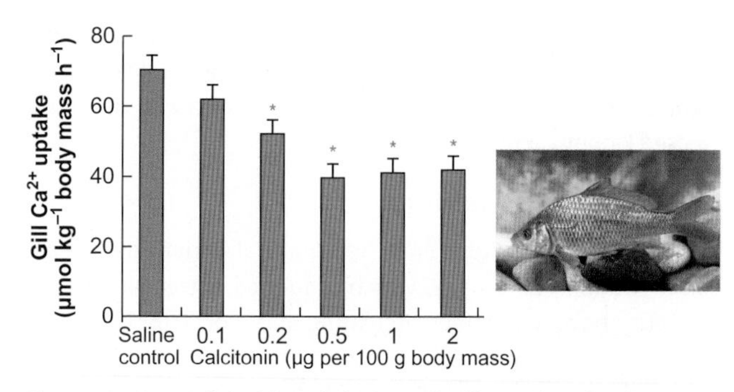

Figure 21.36 Calcitonin inhibition of gill uptake of Ca^{2+} in juvenile common carp (*Cyprinus carpio*)

Blood samples were collected 4 h after injection of various doses of calcitonin (the time when maximum effects are observed). A dose of 0.2 μg calcitonin per 100 g body mass is sufficient to significantly reduce Ca^{2+} uptake compared to the Ca^{2+} uptake of saline-injected control carp.

Data are mean + standard error for five fish in each group. Significant effects are indicated by asterisks: *$P < 0.05$ by comparison to saline-injected control fish. Source of data: Mukherjee D et al (2004). Inhibition of whole body Ca^{2+} uptake in fresh water teleosts, *Channa punctatus* and *Cyprinus carpio* in response to salmon calcitonin. Journal of Experimental Zoology 301A: 882–890. Image: © Michigan Sea Grant.

21.2.4 The role of calcitonin in calcium balance

Calcitonin comprises 32 amino acids. Although the amino acid sequence of calcitonin differs amongst species from different vertebrate groups, the N-terminal end of the molecule responsible for its biological activity is highly conserved. Calcitonin is one of a larger family of calcitonin-like peptides that occur in non-bony animal groups including invertebrate groups[86], the cartilaginous elasmobranchs, and jawless cyclostomes (hagfish and lampreys). Actions on Ca^{2+} balance occur in hagfish and elasmobranchs in which calcitonin-like peptides stimulate excretion of Ca^{2+} into the bile, by the liver.

Among non-mammalian vertebrates, calcitonin is primarily synthesized by the ultimobranchial bodies in the neck area. On an evolutionary timescale, these glands first appeared in fish in which the glands lie close to the gills ('branchial' refers to gills). The ultimobranchial bodies as such do not occur in mammals, but have become incorporated into other structures—usually the thyroid, as the C (clear) cells[87].

Experimental studies have found the effect of calcitonin on Ca^{2+} balance of bony fish to be inconsistent; its clearest effect is generally seen in young animals in which there is a faster turnover of Ca^{2+} linked to growth. Calcitonin consistently inhibits whole body Ca^{2+} uptake of young rainbow trout, juvenile common carp (*Cyprinus carpio*) and spotted snakeheads (*Channa punctatus*). Figure 21.36 shows the dose-related hypocalcaemic effects of calcitonin, by inhibiting Ca^{2+} uptake of juvenile carp, in which a response starts after about 1 h and peaks at 4 h.

Some of the most persuasive evidence for calcitonin regulation of Ca^{2+} uptake by fish, including adults, comes from investigations of zebrafish (*Danio rerio*), in which whole-genome sequencing has enabled us to identify genes for calcitonin, calcitonin receptors and Ca^{2+} transporters, and has facilitated the investigation of the expression of these genes. Figure 21.37 shows some results of studies of zebrafish held in low and high Ca^{2+} environments. Notice the gene expression for calcitonin in the ultimobranchial body is higher in zebrafish held in a high Ca^{2+} environment than for zebrafish held in a low Ca^{2+} environment. Accompanying the rise in calcitonin gene expression, the expression of the gene for calcitonin receptors in gill tissue is also elevated. At the same time, there is a decrease in expression of the gene for epithelial Ca^{2+} channels in the gill, which presumably reduces apical influx of Ca^{2+} at the gills.

[83] We discuss angiogenesis in relation to sustained exercise in Section 14.1.5 and Experimental Panel 14.1 in Online Resources.

[84] Eukaryotes are organisms in which the cells contain a nucleus and other membrane-enclosed organelles, and include the Protozoa and unicellular algae and fungi, as well as all multicellular animals, as we discuss in Section 1.3.3. Figure 1.15 illustrates the phylogenetic tree of the main animal groups.

[85] Figure 1.21 illustrates the evolutionary relationships of non-vertebrate chordates and vertebrate groups.

[86] Calcitonin-like peptides have actions on volume regulation among invertebrates, which we discuss in Section 21.1.4.

[87] C-cells are easily distinguishable from the mass of follicular cells in the thyroid gland that produce thyroid hormones, which we discuss in Section 19.4.

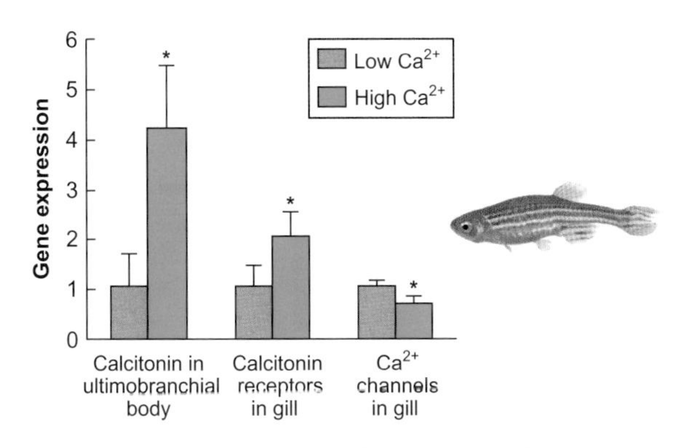

Figure 21.37 Gene expression of calcitonin, calcitonin receptors and Ca^{2+} channels in adult zebrafish acclimated to high and low calcium conditions

Zebrafish were held in water containing high concentrations of Ca^{2+} (2 mmol L^{-1}) or low concentrations of Ca^{2+} (0.02 mmol L^{-1}) for 14 days. The gene expression (mRNA) of calcitonin in the ultimobranchial body and calcitonin receptors in the gill tissue was significantly higher in fish exposed to the higher Ca^{2+} concentration than in fish exposed to the lower concentration of Ca^{2+}. Gene expression of epithelial Ca^{2+} channels in gill tissue significantly declined in fish exposed to high Ca^{2+}. Levels of mRNA expression are expressed relative to the mRNA of actin (a control gene).

Values are the means + standard errors for six fish. Asterisks indicate statistically significant differences ($P < 0.05$). Source: Lafont A-G et al (2011). Involvement of calcitonin and its receptor in the control of calcium-regulating genes and calcium homeostasis in zebrafish (*Danio rerio*). Journal of Bone and Mineral Research 26: 1072-1083. Image of zebrafish: © David Dohnal/Shutterstock.

Molecular studies of zebrafish also suggest calcitonin may have indirect actions on Ca^{2+} balance, by stimulating the release of stanniocalcin from the corpuscles of Stannius. Specifically, zebrafish larvae that have been genetically manipulated to increase their expression of calcitonin also show an increase in stanniocalcin mRNA. Overall, the studies of zebrafish provide good evidence that calcitonin is involved in Ca^{2+} regulation through direct or indirect (via stanniocalcin) inhibition of apical Ca^{2+} channels, which reduces the apical uptake of Ca^{2+} via ion-transporting cells in the gills.

Among adult amphibians, calcitonin stimulates Ca^{2+} uptake across their skin when they enter water[88], which might initially seem surprising, if we consider calcitonin to be a hypocalcaemic hormone. However, calcitonin may stimulate the subsequent transfer of calcium into storage (as calcium carbonate) in the endolymphatic sac surrounding their brain and the paravertebral lime sacs.

In mammals, the physiological actions of calcitonin are less clear than in non-mammalian vertebrates. Early experiments in the early 1960s, soon after the discovery of calcitonin, investigated the secretion of calcitonin by perfusing mammalian thyroid glands (including the C-cells) with blood containing high concentrations of calcium. Looking back at Figure 21.25, we find some of the results of these early studies, which show that high plasma concentrations of calcium increase calcitonin secretion.

Based on the results of such studies, the physiological action of calcitonin was for many years considered hypocalcaemic. In human patients, administration of calcitonin *does* reduce plasma Ca^{2+} in patients that are hypercalcaemic. However, a lack of hypocalcaemia in humans that secrete excessive amounts of calcitonin due to carcinoma of the thyroid gland makes the role of calcitonin in controlling the normal plasma concentrations of Ca^{2+} questionable.

Studies of the location of calcitonin receptors have helped us to understand the control of calcitonin release and physiological actions. The C-cells in the mammalian thyroid gland have Ca^{2+} sensing receptors that regulate calcitonin secretion, and stimulate calcitonin secretion in response to high blood Ca^{2+}. However, calcitonin receptors occur in many tissues, such as the central nervous system and heart, so actions other than in calcium regulation are likely.

Calcitonin receptors occur in the tubular epithelial cells in various parts of the nephrons of the mammalian kidney (thick ascending limb and early part of distal convoluted tubule, shown back in Figure 21.30), which are indicative of likely actions on the kidney. Calcitonin promotes renal excretion of calcium and phosphate, presumably by decreasing tubular reabsorption, but the mechanism of action is unclear.

Calcitonin receptors also occur on bone. Early studies showed that *high concentrations* of calcitonin inhibit the activity of osteoclasts and their proliferation. Within minutes of treatment with calcitonin, the osteoclasts lose their ruffled border and retract as bone resorption diminishes. These effects appear to be most pronounced in young animals in which bone growth is at its highest.

In mammals, bone remodelling diminishes with age following the declining sensitivity of osteoclasts to calcitonin. Figure 21.38 shows that within a few weeks of birth, calcitonin has a much reduced hypocalcaemic effect in young rats. As such, it seems that the action of calcitonin in regulating bone tissue may become vestigial in adult mammals. Nevertheless, administration of high doses of calcitonin is an approved treatment for osteoporosis and other diseases involving accelerated bone turnover. Interestingly, these treatments in humans use salmon calcitonin because it is 40–50 times more potent than human calcitonin.

Calcitonin-induced suppression of osteoclast activity also occurs in scales and bone of some fish species in which a prolonged elevation of calcitonin stimulates mineralization of the bone matrix. Dietary intake of calcium also has a significant influence on the mineralization of bone and scales of fish. In goldfish, for example, calcitonin increases Ca^{2+} deposition in the bone and scales of starved fish, but not in fed fish with ample dietary Ca^{2+}. Fish can mobilize the

[88] Active vitamin D_3 (1,25$(OH)_2D_3$) has also been shown to stimulate Ca^{2+} uptake across amphibian skin.

Figure 21.38 Effect of age on relative hypocalcaemic response of rats to calcitonin

Rats were given an injection of calcitonin (0.5 mg/100 g body mass) and hypocalcaemic responses were monitored as an integrative measure of the effects on bone. The diminishing response to calcitonin with age is reflected in the decreasing rate of skeletal remodelling.

Source: adapted from Copp DH (1969). Calcitonin and parathyroid hormone. Annual Review of Pharmacology 9: 327–344.

calcium stores of bone and scales during reproduction when increased synthesis of **vitellogenins**[89] occurs; vitellogenins bind to circulating Ca^{2+} before deposition as an energy and calcium store in the eggs.

Calcitonin in pregnancy and lactation

Circulating concentrations of calcitonin are elevated during the pregnancy of placental mammals so a hypothesis arose that the long-term reduction in bone resorption induced by calcitonin is important in protecting pregnant and/or lactating mammals from resorption of their bones. During pregnancy, maternal provision of Ca^{2+} and phosphate is necessary for growth of the developing foetus. Increased absorption of Ca^{2+} from the diet occurs in the intestine, under the stimulation of prolactin[90].

Initial investigations of the hypothesis that calcitonin protects the skeleton of pregnant mammals involved the removal of the C-cells (and associated thyroid glands, though the thyroid hormones were replaced). These studies were inconclusive, with some showing decreased bone mineral content after a full cycle of pregnancy and lactation compared to sham-operated animals, but others showing no effect. An added complication, not appreciated at the time, is that tissues other than the C-cells (such as the mammary glands and placenta) also secrete calcitonin. Secretion of calcitonin by these tissues is usually limited, but removal of C-cells actually stimulates *increased* secretion by them.

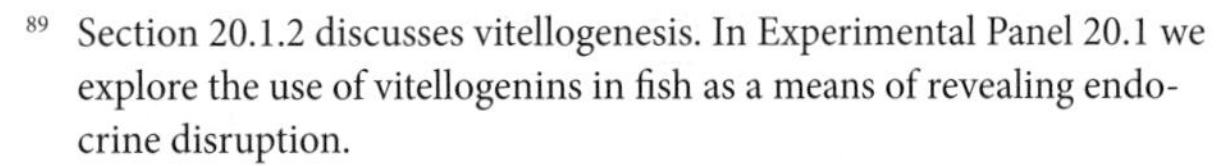

[89] Section 20.1.2 discusses vitellogenesis. In Experimental Panel 20.1 we explore the use of vitellogenins in fish as a means of revealing endocrine disruption.

[90] Section 21.2.2 examines the hypercalcaemic actions of prolactin including actions on the intestine.

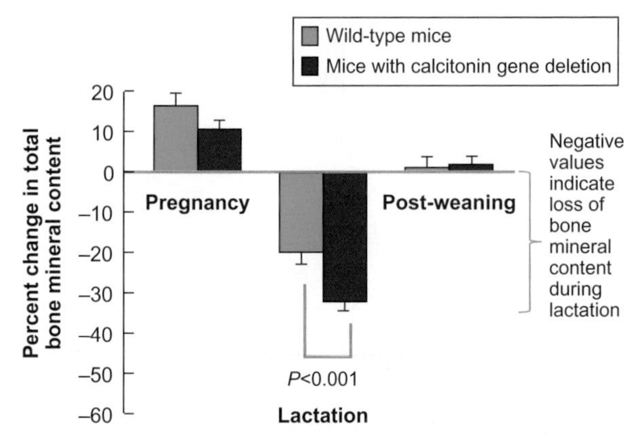

Figure 21.39 Calcitonin reduces the loss of bone mineral content during lactation

Wild-type mice (with calcitonin gene) maintain bone mineral content during pregnancy but lose about 20% of bone mineral content during lactation. The protective effects of calcitonin become apparent when the effects of deletion of the calcitonin gene are examined. Gene-deletion results in significantly greater loss of bone mineral density during lactation than in the wild-type mice, although these gene-deleted mice recover fully after weaning their young.

All values are means ± standard errors. Source: Woodrow JP et al (2006). Calcitonin plays a critical role in regulating skeletal mineral metabolism during lactation. Endocrinology 147: 4010–4021.

New genetic approaches offer a better way to investigate the effects of an absence of calcitonin, by producing animals unable to secrete it. As shown by the data in Figure 21.39, the deletion of the gene for calcitonin in mice does not significantly affect the mineral content of their bones during pregnancy. However, the loss of minerals from bones during lactation is greater in mice lacking calcitonin than in the 'wild-type' mice that are able to secrete it. The effect is specific to the loss of calcitonin (rather than the calcitonin gene-related peptide, which can be synthesized by alternate processing of the calcitonin gene) as indicated by the restoration of normal mineral balance by treatment with calcitonin. From these experiments, we can conclude that calcitonin is important in protecting the maternal skeleton against excessive resorption and fragility during lactation. Looking at Figure 21.39 again, note that the skeleton has a remarkable ability to recover rapidly after weaning, even when calcitonin is missing. This recovery occurs, despite the loss of about 30 per cent of total bone mineral content during lactation with more than 50 per cent loss of mineral content in the spine during the lactation of mice lacking calcitonin.

21.2.5 Calcium in invertebrates

Many invertebrates live in seawater, where large amounts of Ca^{2+} are readily available, and external concentrations of Ca^{2+} (approximately 10 mmol L^{-1}) are similar to the Ca^{2+} concentration of the extracellular fluids suggesting these animals are in Ca^{2+} balance with the external environment[91].

[91] We examine the ionic composition of seawater and the extracellular fluid composition of marine invertebrates in Section 5.1.1 & Table 5.1.

However, many marine invertebrates such as the crustaceans and molluscs need to use calcium which is incorporated into an organic matrix to produce their calcareous exoskeletons or shells in the process of **biomineralization**. Some marine invertebrates, such as squids and cuttlefish, have internal calcareous plates or other internal skeletal structures.

Invertebrates living in fresh water or in terrestrial habitats have much less exogenous Ca^{2+} available to them than marine animals, but some, such as freshwater crustaceans, still rely on a protective calcified exoskeleton. The production of these calcareous structures involves the incorporation of calcium carbonates into an organic matrix as the animal grows. Such processes are likely to involve hormonal control mechanisms, although we know far less about these than for vertebrates.

Calcium and shell formation by molluscs

The control of the growth of molluscs and their calcium balance has been a topic of research since the 1970s when various neurosecretory cells and ganglia were identified as involved in growth regulation and shell regeneration. Some of the earliest studies of pond snails revealed the importance of clusters of neuroendocrine cells (about 200 cells in total) for controlling growth. Figure 21.40 illustrates the location of these cells, as clusters of light green cells (named because of their colour after staining with alcian blue/yellow) in the cerebral ganglia[92], and as clusters of canopy cells, which are also thought to function as light green cells.

Removal or destruction of the light green cells brings shell growth to a halt. These effects are likely to be due to the absence of peptide hormones produced by these cells–**molluscan insulin-related peptides** (MIPs)–so named because of their structural similarity to vertebrate insulin. MIPs have a multiplicity of roles in regulating growth and metabolic processes, so their effects on shell growth may include indirect as well as direct actions.

Growth of molluscs involves growth of their shells, which largely consist of crystals of calcium carbonate ($CaCO_3$) making up more than 95 per cent of the shell, and which are deposited in an organic matrix mainly of proteins, chitin, lipids and polysaccharides. The eventual form of the mineralized crystals of calcium carbonate in molluscan shells is usually aragonite, but sometimes calcite. Recent studies show that the calcium-binding properties of matrix proteins control the particular type of $CaCO_3$ crystal formed and the growth of the crystals.

Shell growth and biomineralization occurs by enlargement at the edge of the shell where minute layers are added to adjacent layers, and by thickening of its inner surface. These processes result from the secretions of the mantle[93],

[92] Section 16.1.1 discusses invertebrate nervous systems.

[93] The mantle forms the mantle cavity, which functions as a chamber for respiration in most molluscs, as we discuss in Section 12.2.2.

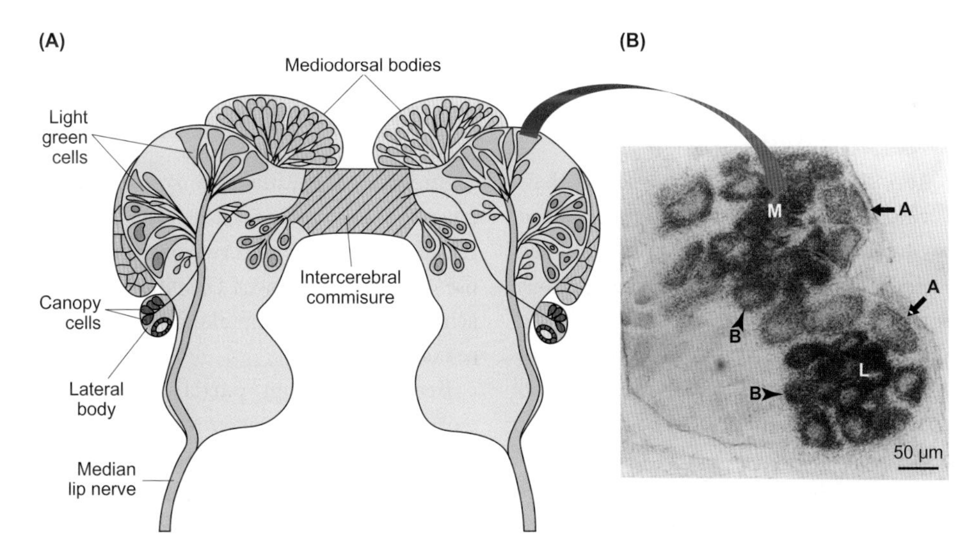

Figure 21.40 Light green cells and canopy cells of great pond snails (*Lymnaea stagnalis*)

(A) Diagrammatic illustration of the cerebral ganglia of the brain (blue) showing the light green cell system of two pairs of clusters of light green cells (green) and a pair of clusters of canopy cells (brown) in the lateral bodies. Neurosecretion from the light green cells passes into a neurohaemal storage area in the periphery of the median lip nerve. Other areas of neuroendocrine cells (shown in purple) occur in various clusters including the two large mediodorsal bodies. (B) Expression of molluscan insulin-related peptides (MIPs) in light green cells investigated by *in situ* hybridization. Cells in the right mediodorsal (M) and latero-dorsal (L) clusters of light green cells express *MIP* genes, which is seen as black dots in the images shown. Two types of light green cell are distinguishable: type-A cells occur mainly in the periphery of the clusters and express four *MIP* genes; image shows *MIP-III* gene expression: type-B cells toward the centre strongly express the *MIP-III* gene.

Sources: A: adapted from Hartenstein V (2006). The neuroendocrine system of invertebrates: a developmental and evolutionary perspective. Journal of Endocrinology 190: 555–570. B: Meester I et al (1992). Differential expression of four genes encoding molluscan insulin-related peptides in the central nervous system of the pond snail *Lymnaea stagnalis*. Cell and Tissue Research 269: 183-188.

which is a layer of tissue covering the visceral mass, and which extends to or beyond the edge of the shell. The mantle secretes both the organic matrix and the inorganic ions, Ca^{2+} and bicarbonate (HCO_3^-) needed for shell formation into a fluid space between the existing shell and mantle.

The intrinsic importance of Ca^{2+} in shell formation and growth of molluscs suggests that hormones similar to those involved in regulating the calcium balance and biomineralization of vertebrate bones may be involved in controlling shell growth of molluscs. Recent work suggests that peptides related to calcitonin and calcitonin gene-related peptides (CGRP) of vertebrates[94] control the deposition of the organic matrix and shell formation in molluscs. Evidence for this assertion includes the following:

- In European abalones (green ormers, *Haliotis tuberculata*) CGRP-like activity increases three-fold in the post-larval stage when formation of the adult shell commences.
- Receptors related to the vertebrate calcitonin/CGRP receptors have been identified in the mantle edge and gills of Pacific oysters (*Crassostrea gigas*).
- Experiments using the human CGRP resulted in an almost six-fold increase in the activity of the enzyme carbonic anhydrase of isolated mantle cells of European abalones. Carbonic anhydrase has long been considered to enhance shell formation by hydration of carbon dioxide to form bicarbonate ions and increase calcium carbonate precipitation[95].

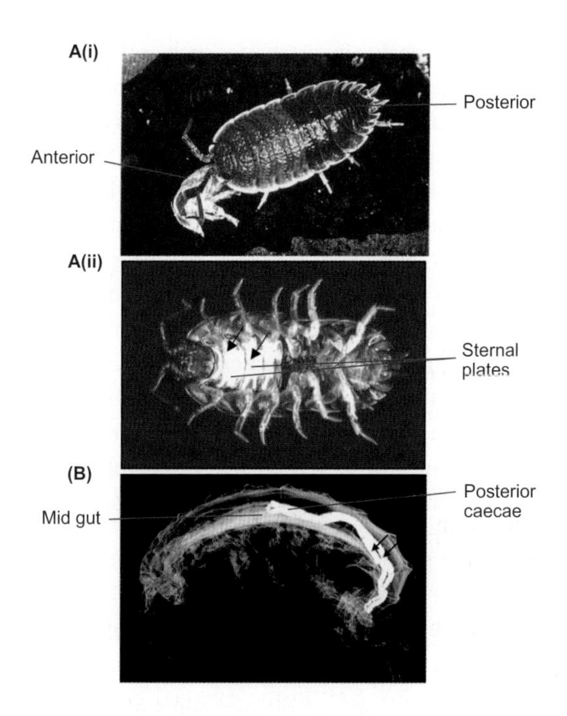

Figure 21.41 Calcium storage organs and moulting in crustaceans

(A) Common rough woodlouse (*Porcellio scaber*) showing (i) two halves at different stages in the unusual moulting pattern. This isopod moults the posterior part of its cuticle first. In the specimen in A(i) the new exoskeleton in the posterior half has hardened; the anterior half of cuticle has just been shed and is being eaten. A(ii) shows a ventral view of the anterior calcium storage areas (arrows) beneath the sternal plates. (B) X-ray photograph of the terrestrial amphipod *Cryptorchestia garbinii* previously known as *Orchestia cavimana*, just after ecdysis showing the pair of posterior caecae, that develop during the pre-moult phase, and which are filled with calcareous concretions (arrows).
Source: Luquet G, Marin F (2004). Biomineralisations in crustaceans: storage strategies. Comptes Rendus Palevol 3: 515-534.

Calcium and the exoskeleton of crustaceans

Crustaceans need one or more sources of Ca^{2+} to calcify their rigid exoskeleton, which is composed of three mineralized layers (the epi-, exo- and endo-cuticle) above a non-calcified layer. During biomineralization, precipitation of amorphous calcium and formation of calcite (a stable form of calcium carbonate) occurs in a protein matrix. Periodic shedding of the exoskeleton during ecdysis (moulting) is necessary to allow growth[96], so supplies of calcium need to be recycled or new supplies accessed.

In general, food including the rejected old cuticle (the exuviae), which is sometimes eaten (as shown by the isopod in Figure 21.41A(i)), makes only a minor contribution to calcification. For marine crustaceans, Ca^{2+} is highly available in the seawater, but many crustaceans live in fresh water, where Ca^{2+} availability is lower. Despite this, freshwater crustaceans achieve similar Ca^{2+} uptake rates to those of marine crustaceans.

Before moulting, partial digestion of the cuticular chitin and proteins occurs and dissolution of the cuticle results in a rise in the haemolymph concentration of Ca^{2+}. An increase in Ca^{2+} concentrations in the haemolymph generally increases the efflux of Ca^{2+} via the gills of crustaceans living in aquatic habitats, with relatively little being stored for use in calcification of the new cuticle. In contrast, most terrestrial crustaceans (and some aquatic species) store calcium as temporary stores of amorphous calcium carbonate that can be rapidly mobilized and reused to calcify the new cuticle.

The location of Ca^{2+} stores differ according to taxonomic groupings:

[94] In vertebrates, the alternate splicing of the calcitonin gene produces the calcitonin gene-related peptide, which is mainly present in peripheral and central nerves. We outline alternative splicing in Figure 1.13.

[95] We discuss the role of carbonic anhydrase in the reversible conversion of carbon dioxide and water to bicarbonate and protons in Section 13.1.3.

[96] Section 19.5.1 discusses hormonal control of moulting by crustaceans.

- Terrestrial isopods, such as woodlice, deposit calcium carbonate in an organic matrix beneath their anterior body plates (sternites), as shown in Figure 21.41A(ii). Before ecdysis, they resorb these calcium deposits and use the Ca^{2+} to calcify the new posterior cuticle; then ecdysis of the anterior half of the body occurs.
- Terrestrial amphipod crustaceans store precipitated calcium salts within an organic matrix in the lumen of a large pair of posterior caecae, as shown in Figure 21.41B. Mobilization of these calcareous concretions, in less than 48 h, provides up to 60 per cent of the Ca^{2+} needed to calcify a new cuticle.
- Decapods, particularly land crabs, store calcium carbonate in calcified discs, known as gastroliths, located in the stomach wall, or as calcified granules in the haemolymph or in the **hepatopancreas**. As moulting approaches, the gastroliths enlarge to reach their maximal size. During a moult, digestion of the gastroliths or mobilization of Ca^{2+} in the calcified granules releases Ca^{2+} that enters the haemolymph, which enables mineralization of the new cuticle.

Whatever the origin of the calcium used for calcification of the exoskeleton, the transfer of Ca^{2+} between the storage organs, haemolymph, and cuticle needs to be closely co-ordinated with moulting. Hence, **ecdysteroid** hormones that control moulting can be predicted to influence calcium dynamics.

A key hormone in stimulating moulting is ecdysone, while moult-inhibiting hormone produced by the X-organ/sinus gland complex in the eyestalk inhibits the release of ecdysone[97]. Hence, one way to study the actions of ecdysone is to remove the eyestalks to trigger a surge of ecdysone release. Eyestalk removal in decapod crustaceans results in the formation of the gastroliths, a reduction in haemolymph Ca^{2+} concentrations, and the acceleration of moulting. The identification of ecdysteroid binding sites in the epithelium of the gastroliths suggests direct actions of ecdysteroids in the elaboration and/or resorption of these calcium stores.

In amphipod crustaceans, ecdysteroid concentrations increase during pre-moult concomitant with the increase in calcium storage in the posterior caecae. Studies of the amphipod *Cryptorchestia garbinii* have shown increased expression of a gene encoding a Ca^{2+}-binding protein in the posterior caecae known as **orchestin** that appears to be involved in controlling the formation of the calcareous concretions. In addition, injection of *Cryptorchestia* with **20-hydroxyecdysone** (the active moulting hormone), which induces an early ecdysis, results in increased expression of the gene for orchestin. However, cycloheximide (a protein synthesis inhibitor) inhibits the effect of orchestin, which suggests that while ecdysteroids are involved in cuticle calcification and calcium storage, their actions are probably indirect.

In addition to hormones controlling ecdysis, several pieces of evidence have begun to suggest that calcitonin-like peptides (akin to vertebrate calcitonin) are involved in controlling the calcium balance of crustaceans:

- In common prawns (*Palaemon serratus*), haemolymph concentrations of a calcitonin-like peptide are inversely proportional to Ca^{2+} concentrations; in other words, high concentrations of calcitonin-like peptide occur when Ca^{2+} concentrations in the haemolymph are relatively low, but these studies do not identify the functional links.
- In the amphipod *Cryptorchestia garbinii*, antibodies to calcitonin-like peptide bind strongly to its central nervous system, suggesting neurosecretion of calcitonin-like peptide. Injection with salmon calcitonin results in hypocalcaemia. Calcitonin may regulate Ca^{2+} translocation into the posterior caecae since antibodies to calcitonin-like peptide also bind to these calcium storage organs.

› *Review articles*

Blaine J, Chonchol M, and Levi M (2015). Renal control of calcium, phosphate, and magnesium homeostasis. Clinical Journal of American Society of Nephrology 10: 1257–1272.

Bronner F (2009). Recent developments in intestinal calcium absorption. Nutrition Reviews 67: 109–113.

Davey RA, Findlay DM (2013). Calcitonin: physiology or fantasy? Journal of Bone and Mineral Research 28: 973–979.

Frost P (2012). Vitamin D deficiency among northern Native Peoples: a real or apparent problem? International Journal of Circumpolar Health 71: 10.3402/IJCH.v71i0.18001.

Guerreiro PM, Renfro JL, Power DM, Canario AVM (2007). The parathyroid hormone family of peptides: structure, tissue distribution, regulation, and potential functional roles in calcium and phosphate balance in fish. American Journal of Physiology–Regulatory, Integrative and Comparative Physiology 292: R679–R696.

Hoenderop JGJ, Nilius B, Bindels RJM (2005). Calcium absorption across epithelia. Physiological Reviews 85: 373–422.

Li Z, Kong K, Qi W (2006). Osteoclast and its roles in calcium metabolism and bone development and remodelling. Biochemical and Biophysical Research Communications 343: 345-350.

Talmage RV, Mobley HT (2008). Calcium homeostasis: Reassessment of the actions of parathyroid hormone. General and Comparative Endocrinology 156: 1–8.

Wagner GF, Dimattia GE (2006). The stanniocalcin family of proteins. Journal of Experimental Zoology 305A: 769–780.

Wongdee K, Charoenphandhu N (2013). Regulation of epithelial calcium transport by prolactin: From fish to mammals. General and Comparative Endocrinology 181: 235–240.

[97] Figure 19.36 illustrates the X-organ/sinus gland complex of crustaceans, and Figure 19.40 shows the structure of ecdysone and 20-hydroxyecdysone (the active moulting hormone) that also occur in insects.

Checklist of key concepts

Salt conservation, salt intake and potassium secretion

- The **renin–angiotensin system** is the main regulator of aldosterone secretion.
- **Aldosterone** is the main salt-conserving hormone of tetrapods.
- Aldosterone stimulates salt absorption from the distal nephron and fluids in various non-renal tissues:
  - amphibian urinary bladder
  - coprodeum and colon of birds
  - sweat glands of mammals.
- Aldosterone stimulates K^+ secretion by the distal nephron.
- The **aldosterone paradox** describes the ability to stimulate renal salt-retention during volume/salt depletion *without* stimulating K^+ secretion, and the ability during hyperkalaemia for aldosterone to stimulate K^+ excretion *without* causing Na^+ reabsorption.
- The sensitivity of salty taste receptors affected by angiotensin and aldosterone may influence dietary salt intake.

Excreting excess salt

- **Natriuretic peptides** are a protective mechanism against volume expansion.
- Natriuretic peptides are released from cardiac myocytes in response to elevated extracellular fluid volume (ECF) and/or high plasma osmolality/Na^+ concentrations.
- Atrial natriuretic peptide (ANP) stimulates:
  - rapid vasodilation of systemic arterioles
  - an increase in urine volume (**renal diuresis**)
  - increased salt excretion (**renal natriuresis**).
- ANP stimulates a **glomerular diuresis** in non-mammalian vertebrates, while a **tubular diuresis** dominates in mammals.
- The natriuretic action of ANP, in mammals, results from direct and indirect actions:
  - direct inhibition of sodium reabsorption from the collecting duct
  - via inhibition of the renin–angiotensin–aldosterone system.
- Salt excretion by elasmobranch rectal glands is primarily controlled by hormones such as C-type natriuretic peptide and scyliorhinin II.
- Regulation of salt secretion by the salt glands of marine birds and reptiles is mainly via the actions of neuropeptides released from nerve fibre endings that innervate the secretory epithelia and blood vessels of the glands.

Drinking behaviour and intestinal fluid absorption

- Circulating hormones gain access to the brain via areas outside the blood–brain barrier and influence the intake of drinking water:
  - Angiotensin II is a potent **dipsogenic** hormone: it **stimulates** drinking.
  - Atrial natriuretic peptide is an **antidipsogenic** hormone: it **inhibits** drinking.
  - Ventricular natriuretic peptide (VNP) of teleost fish is as effective an antidipsogenic peptide as ANP.
- Small increases in **plasma osmolality** detected by osmosensitive brain neurons outside the blood–brain barrier stimulate ingestion of water by tetrapods.
- Relatively large decreases in **blood volume** increase drinking.
- A reflex swallowing response to the high chloride concentration of seawater stimulates drinking by marine teleosts.
- Marine teleosts absorb 70–85 per cent of drinking water in their intestine.
- If ANP increases, Na^+ and water absorption by the intestine decrease in some species.

Control of urine volume—diuretic and antidiuretic hormones

- A diverse range of diuretic and antidiuretic hormones/neurohormones control the functioning of insect **Malpighian tubules**.
- Release of antidiuretic peptides by vertebrate animals (mainly **arginine vasopressin** (AVP) or **arginine vasotocin** (AVT)) is controlled by:
  - ECF osmotic or Na^+ concentrations detected by osmosensitive areas of the brain;
  - blood volume.
- In fish and amphibians, AVT reduces **glomerular filtration rates** by reducing the number of filtering glomeruli.
- In birds, AVT reduces the number of filtering loopless nephrons, but looped nephrons continue filtering.
- In mammals, AVP has a number of actions on the nephron that result in an antidiuresis.
- Water conservation by terrestrial and arboreal frogs and toads involves actions of AVT on the kidney and bladder.
- Rehydration of frogs and toads involves AVT stimulation of water uptake across the **ventral pelvic patch** of skin.

Calcium balance

- Free calcium ion (Ca^{2+}) concentrations of ECF in terrestrial vertebrates are determined by the integrated function of the **intestine** (Ca^{2+} absorption), **kidney** (Ca^{2+} reabsorption or excretion) and Ca^{2+} exchanges between ECF and **bone**.
- The bone is an important reservoir for Ca^{2+} and inorganic phosphate anions (P_i) that terrestrial vertebrates can only replenish by Ca^{2+} absorption from the diet and by drinking water.

- Intestinal absorption of Ca^{2+} occurs by:
  - passive transport between the enterocytes (paracellular pathways), and
  - hormonal regulation primarily of transcellular transport of Ca^{2+}.
- **Hypercalcaemic hormones** increase ECF Ca^{2+} concentrations in response to hypocalaemia.
- **Hypocalcaemic hormones** reduce ECF Ca^{2+} concentrations in response to hypercalcaemia.
- Detection of a low plasma concentration of Ca^{2+} results in a release of **parathyroid hormone (PTH)** from the parathyroid tissue of vertebrates and may result in net resorption of bone.
- Ca^{2+} balance depends on the integrated actions of active vitamin D_3 and PTH on Ca^{2+} transport by the intestine, kidney and bone.
- **Temporary osteoporosis** in female birds due to mineral mobilization from bones may occur, as a result of maintenance of Ca^{2+} balance in ECF during prolonged periods of egg laying.
- **Vitamin D deficiency** may reduce absorption of Ca^{2+} from the intestine, but the possibility of skeletal damage may be overcome by sufficient dietary intake of minerals.
- **Prolactin** in mammals stimulates Ca^{2+} reabsorption by the kidney and intestinal absorption of Ca^{2+}.
- Hormonal regulation of Ca^{2+} influx/efflux via the surface epithelia of aquatic animals has a major effect on calcium balance.
- **Stanniocalcin** is an important hypocalcaemic hormone secreted by the corpuscles of Stannius of teleosts.
- In marine teleosts, stanniocalcins restrict Ca^{2+} influx into gill epithelial cells.
- Paracrine/autocrine actions of stanniocalcins are probable in many tissues of animals lacking the corpuscles of Stannius.
- Hypocalcaemic actions of **calcitonin** include:
  - inhibition of Ca^{2+} uptake via fish gills
  - promotion of renal excretion of calcium (and phosphate)
  - suppression of bone osteoclasts
  - stimulation of mineralization of the scales and bone in fish.
- Calcitonin has a pronounced action on the bone of young mammals when growth is fastest.
- Production of the **exoskeleton** by crustaceans, and **shell growth** by molluscs, are likely to involve hormonal control mechanisms such as by calcitonin gene-related peptides.
- Dissolution of the calcified exoskeleton of crustaceans before moulting retrieves Ca^{2+} for use in calcifying a new exoskeleton.
- Transfer of Ca^{2+} between storage organs, haemolymph and the exoskeleton of crustaceans is likely to involve control by **ecdysteroid** hormones and be closely coordinated with moulting.

Study questions

*** Answers to these numerical questions are available online. Go to www.oup.com/uk/butler**

1. Which hormones influence water uptake across the skin of dehydrated amphibians and what are the effects of these hormones? (Hint: Section 21.1.4 and Case Study 21.1.)
2. Outline three experimental approaches by which you could investigate the control of salt secretion by rectal glands. (Hint: Section 21.1.2; Section 5.1.5 also examines rectal gland functioning.)
3. How are the drinking rates of vertebrates controlled? Give two examples of environments in which these processes are stimulated in named animals. (Hint: Section 21.1.3.)
4. What type of hormone is aldosterone? Outline its mechanism of action on the mammalian kidney. (Hint : Section 21.1.1; see also Sections 19.1.5, 19.3, and Figures 19.3 and 19.10.)
5. Discuss the actions of specified hormones that influence the fluid excretion of named arthropods. (Hint: Section 21.1.4.)
6. Explain how the actions of arginine vasotocin (AVT) in fish and amphibians influence salt and water balance. How do these actions compare to those of AVT in birds and arginine vasopressin (AVP) in mammals? (Hint: Section 21.1.4; see also Figure 7.20.)

7.* Imagine a 6 kg mammal containing 3.5 litres of water. Suppose it rapidly drinks 50 mL (cm^3) of water (about the volume in a glass of water), and that the entire volume is absorbed from the gut and rapidly distributed throughout the entire body water, without excretion, at least at first. If the animal was in fluid balance before drinking, and the blood plasma had an osmolality of 300 mOsm kg^{-1}:

   (i) by what percentage would the osmotic concentration of extracellular fluids decrease?

   (ii) how much in mOsm kg^{-1} would plasma osmolality decrease after consuming the water?

   (iii) is the water intake sufficient to influence the circulating concentration of arginine vasopressin?

   (Hint: Section 21.1.4.)

8. What is the main stimulus for an increase in the plasma concentration of parathyroid hormone (PTH)? Outline the direct and indirect effects of PTH that influence calcium balance. (Hint: Section 21.2.1.)
9. Why is vitamin D important? How can geographical factors influence vitamin D levels? (Hint: Section 21.2.1.)
10. Consider a euryhaline teleost fish, such as a flounder, as it moves from fresh water through an estuary and into the sea. What physiological processes maintain the calcium balance of teleost fish during such migrations? (Hints: Section 21.2 (subsections 1 to 4); Experimental panel 21.1.)

11. In which circumstances or environments do calcitonin or calcitonin gene related peptides influence ionic or water balance of (a) mammals (b) teleost fish (c) invertebrates? (Hints: Sections 21.2.4 and 21.2.5).

12.* The filterable calcium in the plasma of mice is 60 per cent of a total plasma calcium concentration of 2.5 mmol L^{-1}; the proximal tubule and loop of Henle reabsorb 90 per cent of the filtered calcium and the distal nephron reabsorbs a further 9 per cent of filtered calcium. The glomerular filtration rate (GFR) of mice is typically around 0.2 mL min^{-1}.

(i) How much Ca^{2+} do mice filter per hour? (calculate your answer in μmol h^{-1})

(ii) How much Ca^{2+} per hour is reabsorbed by the distal nephron? How is reabsorption in this part of the renal tubules controlled?

(iii) Most of the reabsorption of Ca^{2+} occurs in the proximal tubule. What is the dominant route for Ca^{2+} reabsorption by the proximal tubule?

(Hints: Section 21.2.1; to answer question 12 also needs an understanding of GFR, which Section 7.1.3 and Box 7.2 explain.)

Bibliography

Baldisserotto B, Mancera JM, Kapoor BG (Editors) (2007). Fish Osmoregulation. Science Publishers, Enfield USA.

Hazon N, Flik G (Editors.) (2002). Osmoregulation and Drinking in Vertebrates. Experimental Biology Reviews. Taylor & Francis Group.

Holick MF (Editor) (2010). Vitamin D. Physiology, Molecular Biology and Clinical Applications. 2nd Edition. Springer.

Norris DO, Carr JA. (2013). Vertebrate Endocrinology. 5th Edition. Academic Press.

Scientific Advisory Committee on Nutrition (2016). Vitamin D and Health https://www.gov.uk/government/groups/scientific-advisory-committee-on-nutrition

Yang B, Sands JM (Editors) (2014). Urea Transporters. Subcellular Biochemistry 73. Springer Science, Dordrecht.

22

Integration of the respiratory and circulatory systems

Rhythm generation and control

In many groups of animals, the transport of oxygen from the environment to their metabolizing cells and of carbon dioxide in the reverse direction depends on the continual rhythmic (oscillatory) activities of respiratory and circulatory pumps[1]. What is more, in order to provide the required amount of oxygen, the rate at which these pumps oscillate and the amount of the respiratory medium (air or water) or blood that they pump with each cycle (known as the stroke volume), must be tightly controlled. In many circulatory systems, the diameters of peripheral blood vessels are also under tight control to ensure those parts of the body that consume the most oxygen are preferentially perfused with blood. The control systems consist of sensory and motor elements.

The sensory side detects:

- mechanical aspects of the activity of the pumps,
- the levels of the respiratory gases in the blood, and maybe also in the environment,
- the pressure in the circulatory system.

This information is transmitted to the central nervous system.

The motor side sends the resulting outputs from the central nervous system to the rhythm generators and/or muscles of the pumps and blood vessels.

In this chapter, we discuss the means by which the rhythmic activities of the two pumps are generated and controlled by the nervous and endocrine systems and how these activities are integrated within the central nervous system[2] so that the outputs of both pumps and the distribution of blood can be matched to the requirements of the metabolizing cells in different regions of the body.

22.1 Generation of the respiratory rhythm

The respiratory muscles are normal skeletal muscles and so require input from nerves to stimulate their contraction. This means that the respiratory rhythm is generated within some part of the nervous system. There are two basic ways in which such a rhythm can be produced:

(i) by specialized neurons known as **oscillator** or **pacemaker cells**[3] which depolarize and repolarize spontaneously, and which create activity in other cells;

(ii) by at least two interconnected antagonistic networks of neurons which alternately generate bursts of activity, thereby forming a neuronal oscillating network or central rhythm (or pattern) generator[4].

Most aquatic animals, and most terrestrial endotherms (that is, birds and mammals) irrigate or ventilate their gas-exchange surfaces continuously, but most air-breathing ectotherms, such as many species of insects, lung fish, amphibians and reptiles, ventilate intermittently[5]. Consequently, the extent to which the generation of the respiratory rhythm of those organisms that ventilate continuously—particularly vertebrates—is different from that in those animals that ventilate intermittently has been the subject of debate. There is evidence in some species of fish that there may be a separate area of the brain that is responsible for the generation of intermittent ventilation.

Input from peripheral receptors may be important in regulating the volume of each ventilation, the frequency

1 Details of the respiratory and circulatory systems are discussed in Chapters 12 and 14.

2 Integration within the nervous system is discussed in Section 16.3.5.

3 Pacemaker cells are discussed in Section 16.2.10.

4 Central rhythm (pattern) generators are discussed in Section 16.3.7.

5 Patterns of ventilation are discussed Sections 12.2.2, 12.3.3 and 12.4.1.

of ventilation and the duration of the pauses (if present) between ventilatory activity. In fact, it could be argued that all vertebrates possess an underlying centrally generated respiratory *rhythm* that is influenced to a greater or lesser extent by peripheral receptors to produce a respiratory *pattern*. The pattern is based on the requirement for oxygen by the metabolizing cells and the availability of oxygen in the surrounding environment. The terms 'central (respiratory) rhythm generator' and 'central (respiratory) pattern generator' are generally, but not always, used in this context. The use of these terms in the following discussion of specific animal groups reflects their use in the relevant literature.

22.1.1 Generation of respiratory rhythm in invertebrates

In invertebrates, very few, specifically identifiable cells are involved in generating the respiratory rhythm. By contrast, vertebrates employ an unquantifiable number of unidentifiable cells. We now discuss the functional organization of the respiratory rhythm generators in the central nervous systems of decapod crustaceans, insects and molluscs[6].

Respiratory rhythm generation in decapod crustaceans

The respiratory water current in decapod crustaceans is generated by the sinusoidal movement of a pair of organs called bailers[7]. Each bailer is controlled by a small number of neurons that form a respiratory rhythm generator on each side of the central nervous system. This arrangement enables the bailers to work independently of each other.

The upstroke (levation) and downstroke (depression) of the bailers are generated by two sets of muscles known as levators and depressors. Each set is divided into two groups called D1 and D2 (depressors) and L1 and L2 (levators). Each muscle is innervated by two or three excitatory motor nerves; alternate bursts of activity in these nerves lead to the undulations of the bailers. The resulting movements are monitored and probably regulated by various mechanical receptors, most notably a receptor known as the **oval organ** which is within the bailer itself and responds to stretch. So, motor nerve fibres from the central nervous system (CNS) innervate the muscles of each bailer and sensory fibres return from a bailer to the CNS. The questions are, therefore:

- Where within the CNS is the rhythm generator that actives the respiratory muscles located?
- How is the rhythm generated?

Lobsters and crayfish have a long, cylindrical thorax and abdomen; the head and thorax of crabs are flattened and extend to the side, and the abdomen is reduced in length and tucked under the thorax. This shortening of the body of crabs is also apparent in the nervous system in as much as the ganglia in the thorax and abdomen are fused together to form the thoracic ganglion, as shown in Figure 22.1A. This mass of ganglia includes a ganglion known as the sub-oesophageal ganglion, which is a separate ganglion in lobsters and crayfish[6].

Neurons responsible for generating the respiratory rhythm are located in the sub-oesophageal ganglion of lobsters (*Homarus* spp.) and in the analogous region of the thoracic ganglion of shore crabs or European green crabs (*Carcinus maenas*). There are eight oscillator neurons on each side of the sub-oesophageal ganglion of shore crabs, which are shown diagrammatically for one side in Figure 22.1B. These neurons alternately depolarize and hyperpolarize[8]. Figure 22.1C shows that these oscillations in membrane potential relate to activity in the levator or depressor motor nerves.

The oscillator neurons do not produce action potentials[9], i.e. they are **non-spiking**. The eight neurons on each side of the sub-oesophageal ganglion form the **central pattern generator** (CPG) for the motor nerves to the bailer on the same side of the animal. The interconnections between the neurons of a central pattern generator are unknown, but there is evidence that they are affected by other regions of the nervous system. For example, the overall frequency of the respiratory rhythm is influenced by at least three different frequency-modulating neurons within the sub-oesophageal ganglion, which are shown diagrammatically in Figure 22.1b. As with the oscillator neurons, they do not generate action potentials but undergo changes in membrane potential. There is also a neuron, the reverse switch neuron, which depolarizes when the rhythm switches into reverse; this neuron remains depolarized during the whole of the period of reverse ventilation[7].

Interaction between respiratory rhythm generation and flight in insects

The control of ventilation and flight are closely linked in some species of insects. Two species of locusts (*Schistocerca gregaria* and *Locusta migratoria*) have been the subjects

[6] The basic structure of the nervous system arthropods (for example, decapods crustaceans and insects) and molluscs is discussed in Section 16.1.1.

[7] Ventilation in decapod crustaceans is discussed in Section 12.2.2.

[8] Hyperpolarization and depolarization are discussed in Section 16.2.2.

[9] Action potentials and non-spiking neurons are discussed in Sections 16.2.4, 16.2.8 (APs) and 16.2.9 (N-SNs).

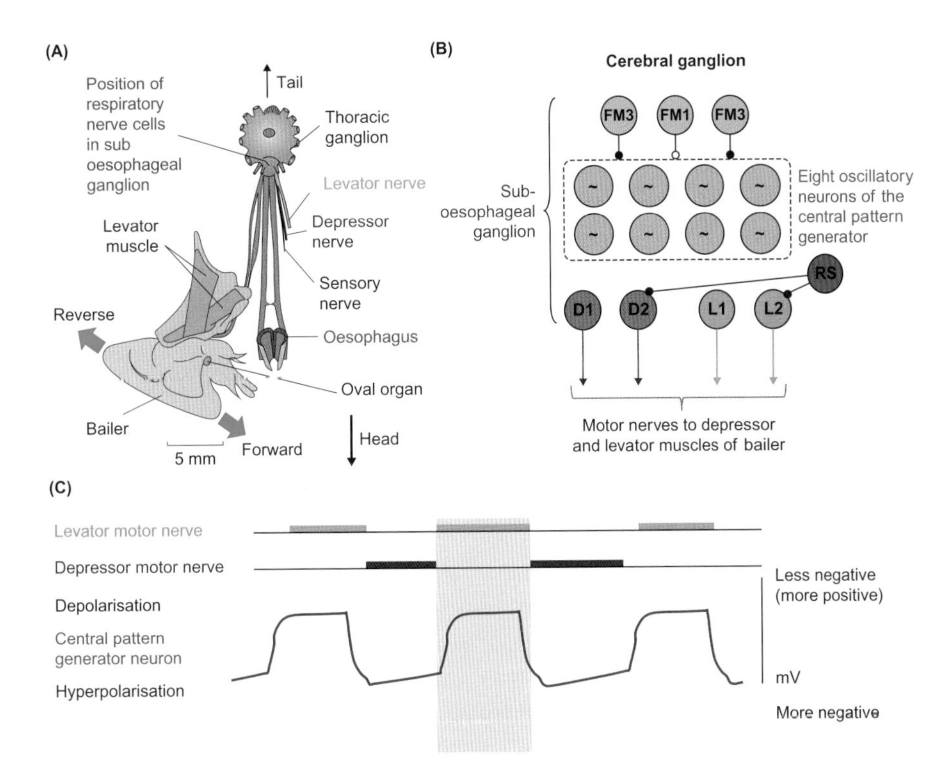

Figure 22.1 Generation of respiratory rhythm in decapod crustaceans

(A) The ventral nervous system of shore crabs or European green crabs (*Carcinus maenas*) related to the ventilatory system. (B) The main respiratory neurons on each side of the sub-oesophageal ganglion of the shore crab and some possible interactions between them. FM, frequency-modulating neurons; D and L, motor neurons innervating depressor and levator muscles, respectively; RS, reverse switch neuron. Open and closed circles represent excitatory and inhibitory synapses, respectively. (C) Diagram showing the relationships between activity in levator and depressor neurons and intracellular recording from a neuron in the central pattern generator (CPG). Note that the neuron in the CPG does not generate action potentials (spiking activity). However, during the depolarization phase of the oscillation in its membrane potential, there is activity in one set of motor nerves on the same side of the animal (e.g. those to the levator muscles—indicated by pink shading), and a cessation of activity in the other set, in this case those to the depressor muscles. The opposite occurs during the hyperpolarization phase.

Sources: (A): modified from Bush, BMH et al (1987). Neuronal control of gill ventilation in decapod Crustacea. pp 80-112 in: The Neurobiology of the Cardiorespiratory System. Taylor EW (ed.). Manchester: Manchester University Press.

of a number of studies on the generation of the respiratory rhythm in insects.

The organization of those neurons involved in the generation and coordination of the respiratory rhythm in locusts is quite complex, and we still don't know how the rhythm itself is generated—whether it is as the result of pacemaker neurons or oscillating neural networks. Figure 22.2 indicates the location and function of the neurons involved in the generation and coordination of the respiratory rhythm[10].

In resting locusts, air is moved in one direction through the network of tracheae, mainly as a result of the pumping movements of the abdomen[11]. When a locust is not flying, the main part of the respiratory cycle is expiration, during which muscles in the abdomen contract and force air out of the open spiracles 5 to 10, as shown in Figure 22.2B, while the first four spiracles are closed. Inspiration is largely passive as a result of the elastic recoil of the abdomen, although a few inspiratory muscles may assist this process. During inspiration, the first four spiracles are open and the last ones are closed.

All spiracles, except those of the second thoracic segment, have both opener and closer muscles; the second thoracic spiracles only have closer muscles. The cell bodies of the motor nerves innervating the abdominal muscles are in the metathoracic or abdominal ganglia and those to the muscles of the spiracles are within the ganglion of the appropriate segment. Note that the third thoracic ganglion, the metathoracic ganglion, includes the third thoracic ganglion and the first three abdominal ganglia all fused together, as indicated in Figure 22.2A.

The neurons in the isolated fused metathoracic ganglion display typical ventilatory pattern, suggesting that the respiratory rhythm generator is located in this ganglion. However, none of the neurons in the metathoracic ganglion itself (that is, the third thoracic ganglion) is thought to be part of the respiratory rhythm generator. There are, however, at least nine neurons in other ganglia which are thought to be part

[10] The basic structure of the insect central nervous system is discussed in Section 16.1.1.

[11] Ventilation in insects is discussed in Section 12.4.1.

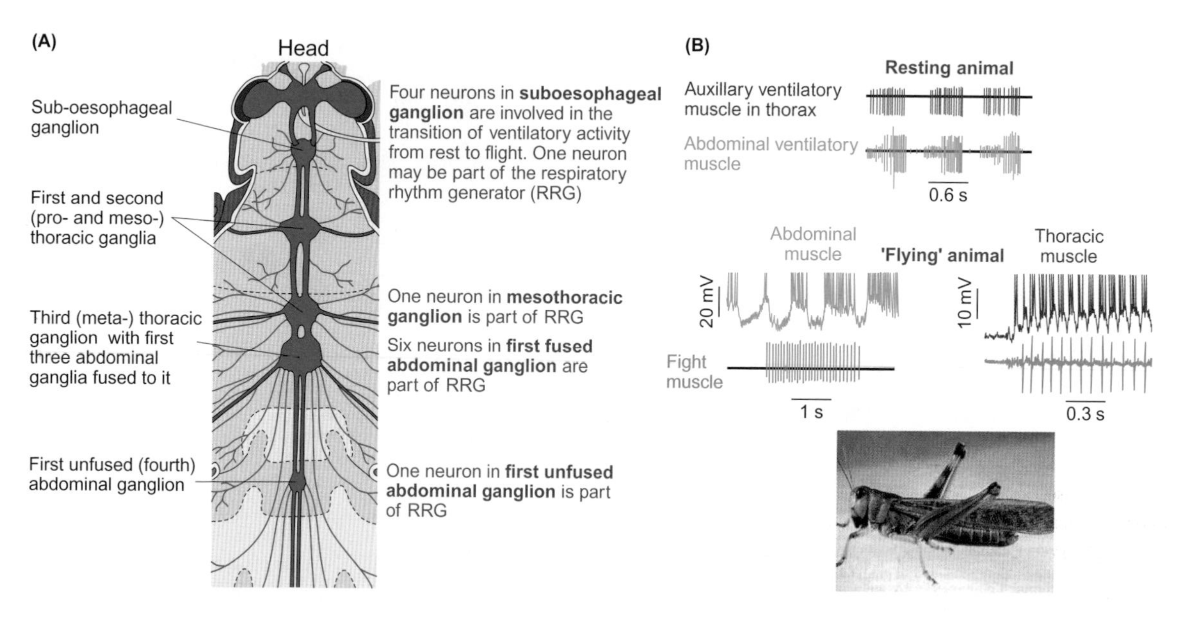

Figure 22.2 Generation of respiratory rhythm in an insect

(A) Diagram of the part of the ventral nervous system of grasshoppers (*Dissosteira* sp., which is similar to that in locusts) containing neurons involved in the generation of the respiratory rhythm, with notes on the location of respiratory neurons. (B) Extracellular recordings from ventilatory muscles in the thorax and abdomen of a resting migratory locust (*Locusta migratoria*) showing them contracting in synchrony during the expiratory phase of ventilation and intracellular recordings from motor neurons innervating the same thoracic and abdominal ventilatory muscles during flight. This time, the activity of the thoracic respiratory neurons is in phase with the activity of a flight muscle, whereas the activity of the abdominal respiratory neuron is not.

Sources: (A): modified from Snodgrass RE (1935). Principles of Insect Morphology. New York: McGraw-Hill. (B) Ramirez J-M (1998). Reconfiguration of the respiratory network at the onset of locust flight. Journal of Neurophysiology. 80, 3137-3147. Image: http://www.daff.qld.gov.au.

of the respiratory rhythm generator, all but one of which are active during expiration. Six of these are in the first fused abdominal ganglion (which is itself fused to the metathoracic ganglion), so it seems likely that the primary respiratory rhythm generator of locusts is located in the first fused abdominal ganglion.

A number of changes to the pattern of ventilation take place at the onset of flight. These changes are shown in Figure 22.2b. At the onset of flight, expiratory activity to the abdominal pumping muscles is initially inhibited and inspiratory activity is increased. These changes lead to an increase in respiratory frequency at the beginning of flight, but not to the same frequency as the beating wings. However, nervous activity to the thoracic muscles *is* at the same frequency as wing beating. This reconfiguration of the respiratory pattern of the thorax at the onset of flight is highly significant because the movements of the thorax and neck provide air, and hence oxygen, to the flight muscles. The flow of air is tidal through spiracles in the thorax, which remain open.

The neurons responsible for the reconfiguration of the ventilatory pattern during flight are located in the suboesophageal ganglion, four of which may be part of the respiratory rhythm generator. These neurons are involved in the transition of the ventilatory activity of the thorax from being in phase with that of the abdomen in the resting locust to being in phase with the wing muscles during flight. They also play a part in the initiation of flight.

A three-cell respiratory rhythm generator and memory storage in some molluscs

Great pond snails (*Lymnaea stagnalis*) are aquaic pulmonate gastropod molluscs and offer a unique opportunity to study how a rhythm can be generated within very few neurons. These snails periodically come to the water surface to ventilate their lung; Figure 22.3A shows a snail at the water surface with its pneumostome open[12]. The frequency of visits to the surface depends on the temperature and partial pressure of oxygen of the water: if temperature increases and the demand for oxygen also increases, and/or the water becomes hypoxic, the frequency of visits increases. The overall ventilatory behaviour is driven by as few as three relatively large (50–200 μm diameter) neurons located in the central ring of ganglia. The three neurons are shown in Figure 22.3B. These neurons are:

- the input 3 interneuron (IP3I), which is located in the right parietal ganglion,

[12] The respiratory system and gas exchange in air-breathing molluscs is discussed in Section 12.3.3.

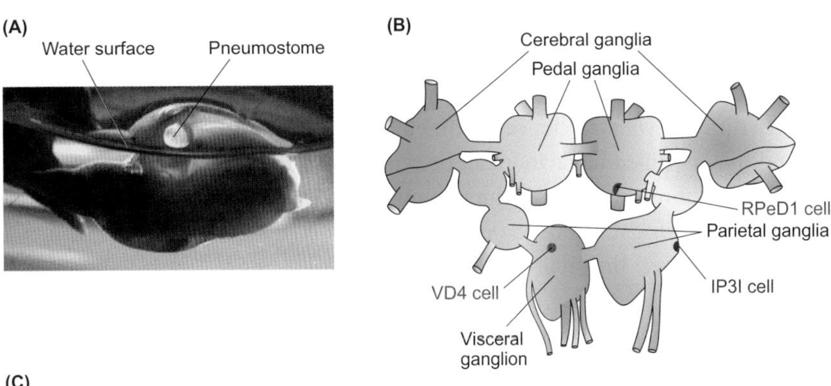

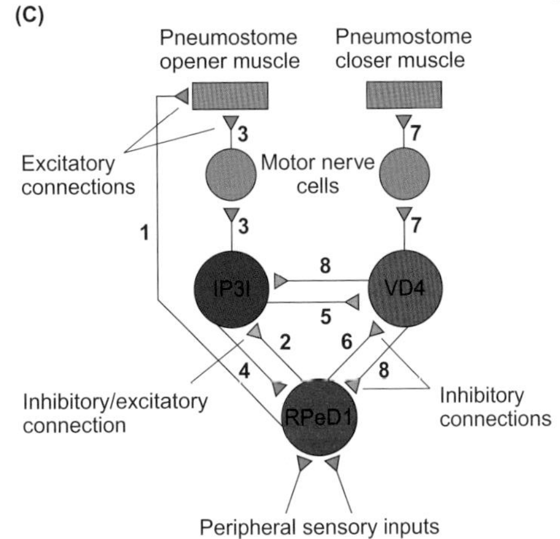

Figure 22.3 Generation of the respiratory rhythm in great pond snails (*Lymnaea stagnalis*)

(A) Adult snail at water surface with its pneumostome open. (B) The ring of ganglia in the snail (dorsal view), three of which contain the neurons that are involved in the generation of the ventilatory rhythm: the neuron in the right pedal ganglion, RePD1, the one in the right parietal, IP3I, and the one in visceral ganglion, VD4. (C) Schematic diagram showing the various interconnections between the three neurons of the central pattern generator, the outputs to motor neurons and inputs from peripheral sense organs. Green and orange triangles denote excitatory and inhibitory synapses, respectively. Light green triangle denotes a synapse that can be either inhibitory or excitatory.

Sources: (A): reproduced from Lukowiak K et al (2006). Modulation of aerial respiratory behaviour in a pond snail. Respiratory Physiology & Neurobiology, 154: 61-72. (B): modified from Syed NI et al (1990). In vitro reconstruction of the respiratory central pattern generator of the mollusk *Lymnaea*. Science. 250: 282-285. (C): Haque Z et al (2006). An identified central pattern-generating neuron co-ordinates sensory-motor components of respiratory behaviour in *Lymnaea*. European Journal of Neuroscience. 23: 94-104.

- the visceral dorsal 4 interneuron (VD4) in the visceral ganglion, and
- the giant dopamine cell in the right pedal ganglion (RPeD1).

Muscles in the mantle cavity contract to force 'stale' air out of the lung before relaxing, which draws fresh air into the lung. This ventilatory cycle occurs a number of times before the pneumostome closes and there may be several such bouts of ventilation before the snail eventually submerges.

Figure 22.3C shows that RPeD1 receives excitatory input from peripheral sense organs[13]. The sequence of events associated with the opening and closing of the pneumostome which follow the stimulation of RPeD1 are:

- Direct stimulation of the opener muscles to the pneumostome (1 in Figure 22.3C) initially inhibits IP3I (2 in Figure 22.3C).
- IP3I is then excited by a process known as post-inhibitory rebound excitation[14], and also stimulates the pneumostome opener muscles via motor nerve cells (3 in Figure 22.3C).
- Once activated, IP3I, in turn, excites RPeD1 (4 in Figure 22.3C) and inhibits VD4 (5 in Figure 22.3C).
- When active, RPeD1 also inhibits VD4 (6 in Figure 22.3C).
- By way of post-inhibitory rebound, VD4 eventually recovers from the inhibitory effect of IP3I and RPeD1 and sends a burst of activity to the motor nerve cells which innervate the pneumostome closer muscles (7 in Figure 22.3C).
- At the same time, cell VD4 also inhibits the activity of RPeD1 and IP3I (8 in Figure 22.3C).

The alternate, mutual inhibition of cells is known as reciprocal inhibition[15].

There is no evidence that either of the respiratory cells IP3I and VD4 is endogenously active, i.e. they are not pacemaker cells. However, they can be considered to be the oscillator of the circuit because they are mutually inhibitory and control the antagonistic motor functions of inspiration and expiration. It seems that the release of dopamine from RPeD1 is necessary for the generation of the rhythm from this network. In other words, the rhythm is a property of the

[13] We discuss the sense organs in the respiratory system of pond snails in Sections 22.2.2 and 22.2.3.

[14] Post-inhibitory rebound excitation is discussed in Section 16.2.6.

[15] Reciprocal inhibition is discussed in Section 16.3.7 and 18.5.3.

network as a whole, as none of the cells is individually able to produce a rhythmic output.

As in other groups of animals, generation of the respiratory rhythm in *Lymnaea* can be modified by external stimuli and this modified behaviour can be remembered by the three-cell rhythm generator. If the pneumostome is weakly touched every time it begins to open, the snail immediately closes it. However, the snail does not retract into its shell and remains at the surface of the water. Because *Lymnaea* are able to exchange respiratory gases across their skin, as well as across the walls of their lungs, they are not adversely affected by this procedure.

Naïve snails make 8–14 attempts to open their pneumostome during a 30–45 min training session, although Figure 22.4A shows they make fewer attempts during subsequent training sessions; they have learned not to open their pneumostome as frequently as they would do normally. If there is a gap of one hour between training sessions, the snails 'remember' what they have learned one day later. This has been called long-term memory (LTM). However, if the gap between training sessions is only 30 min, the snails remember for 2–3 hours, but not for a whole day, as shown in Figure 22.4B. This is called intermediate-term memory (ITM).

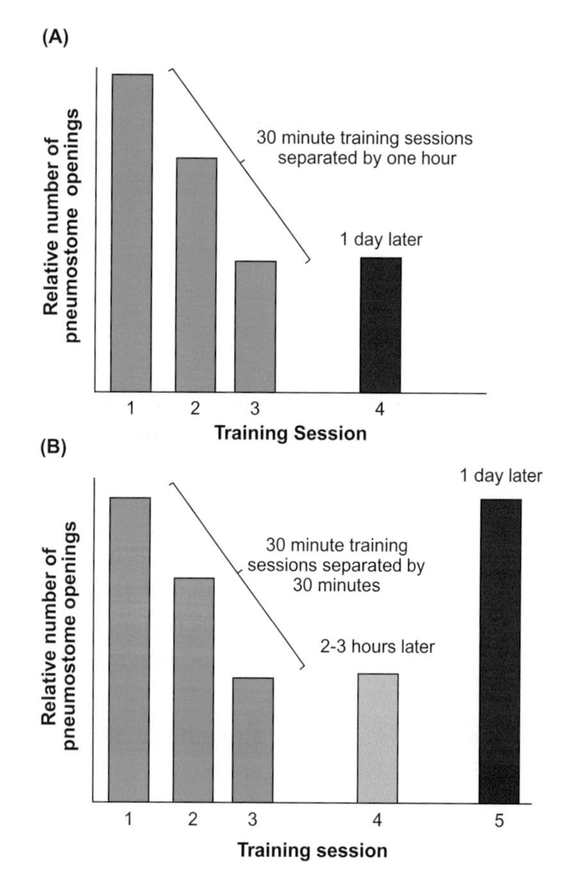

Figure 22.4 Learning and memory in the respiratory rhythm generator of a great pond snail (*Lymnaea stagnalis*)

The graphs show the decline in the number of times a snail visits the water surface and opens its pneumostome when the pneumostome is touched every time the snail begins to open it. These training periods last for 30 min but have rest periods of different duration between them. (A) Three training periods separated by 1 h produce learning, and the reduced response is remembered a day later. This has been called long-term learning. (B) Three training periods separated by only 30 min also produce learning, but the reduced response is only remembered 2–3 h later, not a day later. This has been called intermediate-term memory.

Source: modified from: Taylor BE and Lukowiak K (2000). The respiratory central pattern generator of *Lymnaea*: a model, measured and malleable. Respiration Physiology 122: 197-207.

Our understanding of memory formation and recall has been greatly enhanced by studies on molluscs, such as California sea hares, *Aplysia californica*[16]. Experiments investigating the cellular and molecular bases for memory in the respiratory rhythm generator of *Lymnaea* have demonstrated that the formation of memory resides in one neuron, LPeD1. Similar to the situation in *Aplysia*, ITM can form in the neurite (axon or dendrite) of the neuron and depends on protein synthesis from transcription factors that already exist. In contrast, LTM is formed in the soma (cell body) of LPeD1 and depends on the synthesis of proteins from new transcription factors produced in the soma. The soma is not, however, required for memory recall.

22.1.2 Respiratory rhythm generation in vertebrates

The nervous system of vertebrates is different from that of invertebrates[17] in respect to the numbers of neurons involved in generating the respiratory rhythm. Only a handful of cells are involved in the invertebrates, but millions are involved in the vertebrates.

Most of the research into the generation of the respiratory rhythm in vertebrates has taken place on mammals, but controversy still exists over the relative importance of spontaneously active pacemaker cells and networks of nerves in respiratory rhythm generation, despite the large amount of literature on the topic. The spinal cord of vertebrates is segmental and segmentally arranged ganglia, called sympathetic ganglia[18], lie outside the cord. The ganglia make up the **sympathetic nervous system**. In addition, pairs of **cranial nerves** (numbered 0 to XII), with a segmental origin from the brain, innervate various regions of the head and body[19].

[16] Memory formation in sea slugs is briefly discussed in Section 16.1.1.
[17] Nervous systems are discussed in Section 16.1.
[18] Sympathetic ganglia are discussed in Section 16.1.3.
[19] Cranial nerves are discussed in Section 16.1.2 and Box 16.2.

Fish and amphibians: rhythm generators for ventilating gills and lungs

The basic respiratory pump[20] found in fish is, with some small modifications, present in all living amphibians, although some urodeles (newts and salamander) may use parts of the body wall innervated by spinal nerves during expiration.

The motor and sensory innervations of the gills of fish by cranial nerves VII, IX and X are shown in Figure 22.5. In bony fish, the muscles of the respiratory pump which creates the current of water over the gills are innervated by cranial nerves V (trigeminal nerve) and VII (facial nerve), while cranial nerves IX (glossopharyngeal nerve) and X (vagus nerve) innervate muscles that stabilize and maintain the functional arrangement of the gill arches. In cartilaginous fish, nerves IX and X innervate the muscles of the gill slits. The majority of the respiratory muscles of amphibians are innervated by cranial nerves V and X, as in fish, plus cranial nerve XII—the hypoglossal. In reptiles, birds and mammals, the respiratory muscles are from the thorax and abdomen and are innervated by spinal nerves. These groups of animals use an aspiration (suction) pump to inflate their lungs.

The differing location of the respiratory muscles in different groups of vertebrates means that different areas of the brain are involved in generation of the respiratory rhythm in these groups. Case Study 22.1 discusses the development of the hind region of the vertebrate brain, from where the cranial nerves emerge and where generation of the respiratory rhythm occurs. It suggests that the respiratory rhythm generators in vertebrates develop as paired, segmentally arranged rhythm generators, and that there is a reduction in the number of paired rhythm generators to one or possibly two in adult amphibians, reptiles, birds and mammals.

Respiratory rhythm generation in fish

The recording of electrical activity from the brains of anaesthetized adult bony fish has shown that neurons which are active during some part of the ventilatory cycle are located in two longitudinal strips that run the length of the medulla oblongata. The location of these strips is shown in Figure 22.6A. Each strip consists of the motor nuclei[21] of the V, VII, IX and X cranial nerves, with the reticular formation[22] (reticulum—small net or mesh) alongside these nuclei. Neurons with a respiratory rhythm have been found in the reticular formation of fish; these cells are involved in the generation of the respiratory rhythm.

The respiratory nerve cells are not clustered into groups according to the relationship between their activity and a particular phase of the ventilatory cycle. Indeed, destruction of respiratory nerve cells at any particular site does not abolish the respiratory rhythm. So, we can conclude that the organization of rhythm generating nerve cells is rather diffuse. There is also some support for the presence of separate respiratory oscillators in both cartilaginous and bony fish. For example, Figure 22.6B shows that a burst of electrical activity in cranial nerve V of a lesser spotted dogfish or small spotted cat shark (*Scyliorhinus canicula*) is followed by a burst in cranial nerve IX and then by simultaneous bursts in the three respiratory (branchial) branches of cranial nerve X. However, these delays may merely be the result of conduction delays caused by motor neurons with axons of different length.

[20] Respiratory pumps of vertebrates are discussed in Sections 12.2.2 and 12.3.3.

[21] Clusters of neurons in the brain are known as nuclei and are discussed in Section 16.1.2.

[22] The reticular formation is discussed in Section 16.1.2.

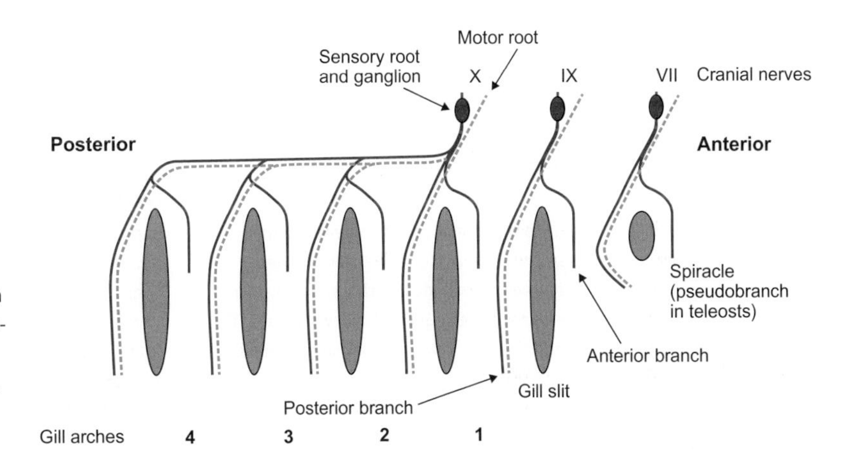

Figure 22.5 The nerves supplying the gills of fish, showing the facial (VII), glossopharyngeal (IX) and vagus (X) cranial nerves

Before reaching the gills or spiracle, if present, each nerve divides into an anterior and a posterior branch which straddle the gill slits. The anterior branches contain only sensory nerve fibres, whereas the posterior branches contain both sensory and motor nerve fibres. The result of this is that each gill arch is innervated by motor and sensory fibres in the posterior branch of cranial nerve X and the sensory fibres in the anterior branch of the subsequent branch of cranial nerve X.
Source: modified from: Sundin L and Nilsson S (2002). Branchial innervation, Journal of Experimental Zoology. 293: 232-248.

Case Study 22.1 Segmentation in the hindbrain of vertebrate embryos and the respiratory rhythm generators

By studying the early development of vertebrates, scientists have been able to gain some insight into the organization of the respiratory rhythm generators in this group of animals. At an early stage in the development of the brain of all vertebrates, there are seven or eight segmentally arranged bulges in the hindbrain which are known as rhombomeres, and are illustrated in Figure A. The rhombomeres eventually give rise to the regions of the brain known as the pons in adult birds and mammals, where the motor nuclei of cranial nerves V and VII are located, and the medulla oblongata in adults of all groups of vertebrates, where the motor nuclei of cranial nerves IX and X reside. These regions generate the respiratory rhythm in adults.

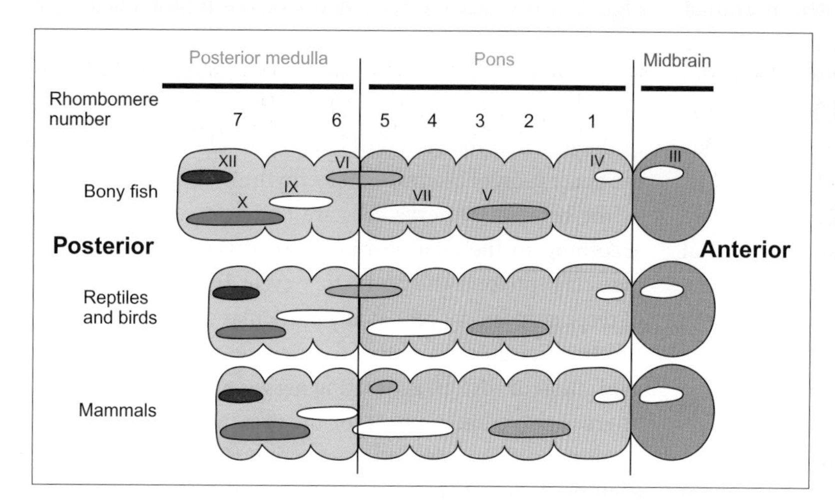

Figure A Schematic diagrams showing the relationships between the segments (rhombomeres) of the hindbrain and the motor nuclei of the cranial nerves (Roman numerals) in early development of a number of groups of vertebrates

Some of the motor nuclei are colour-coded.

Source: modified from Milsom WK et al (2004). Pontine influences on respiratory control in ectothermic and heterothermic vertebrates. Respiratory Physiology & Neurobiology. 143: 263-280.

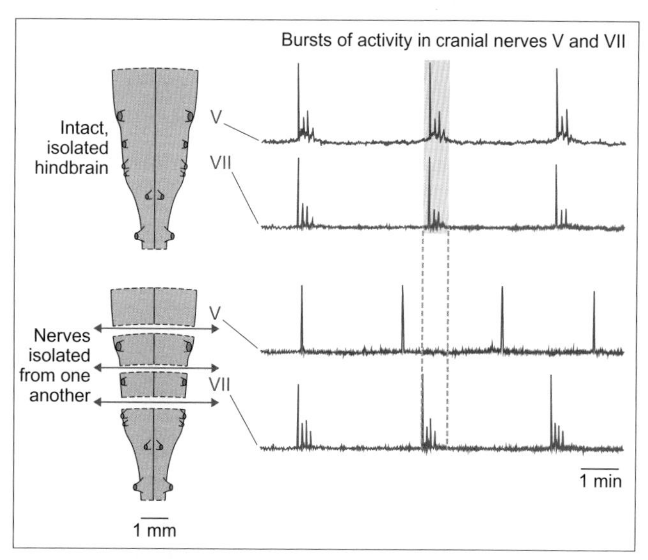

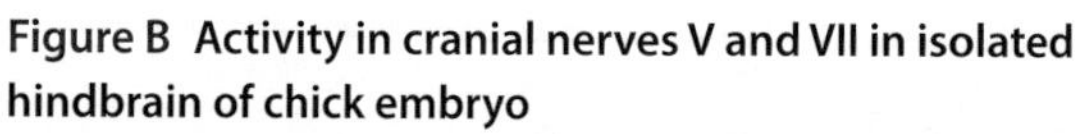

Figure B Activity in cranial nerves V and VII in isolated hindbrain of chick embryo

When the preparation is intact, rhythmic activity in both nerves is at the same frequency. When the two segments from which the nerves emerge are isolated from each other by transverse sections (blue lines), the rhythms continue, with that of nerve V being the faster, but that of nerve VII being the same as that in the 'intact' preparation. These data suggest that the caudal oscillators dominate in the intact preparation.

Source: modified from Champaganat J and Fortin G (1997). Primordial respiratory-like rhythm generation in the vertebrate embryo. Trends in Neurosciences, 20: 119-124.

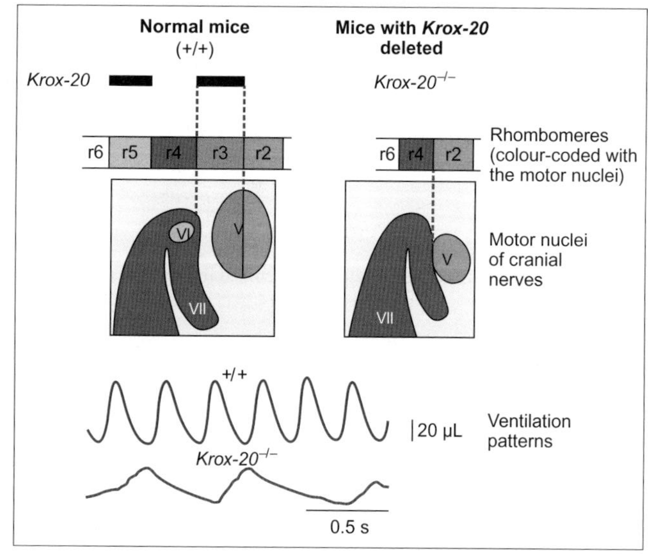

Figure C Effects of deletion of the *Krox-20* gene on the rhombomeres, motor nuclei and ventilation in newborn mice

In mutant mice with *Krox-20* deleted (*Krox-20*$^{-/-}$), rhombomeres 5 (green) and 3 (violet) are eliminated, together with the motor nucleus of cranial nerve VI (which is not involved in respiratory rhythm generation) and part of that for cranial nerve V. The motor nucleus of cranial nerve VII is not affected. These changes affect the ventilation pattern which is much slower in mice with *Krox-20* deleted (*Krox-20*$^{-/-}$) mouse than in a wild-type (+/+) littermate.

Source: modified from Champaganat J and Fortin G (1997). Primordial respiratory-like rhythm generation in the vertebrate embryo. Trends in Neurosciences, 20: 119-124.

This segmentation is not permanent and occurs during the third and fourth weeks of pregnancy in humans and between days 8–12 of development in mice. Although the arrangement of the rhombomeres is similar among the different groups of vertebrates, the location finally occupied by some of the motor nuclei of the cranial nerves varies. For example, the motor nucleus of cranial nerve V occupies rhombomeres 2 and 3 in all groups shown in Figure A, except in mammals, where it occupies rhombomeres 1, 2 and 3.

Toward the end of the period of segmentation in the chick embryo, rhythmic activity is present in cranial nerves V, VII, IX, X and XII. Each pair of rhombomeres contains its own rhythm generator; when isolated, those at the front end have a faster rhythm. However, those slower generators from the hind (caudal) end dominate when the hindbrain is intact. Figure B illustrates these characteristics for cranial nerves V and VII.

How could we find out if the rhythm generators in the embryo are related to the respiratory rhythm generators in adults? One powerful approach is to use knock-out (mutant) mice, which have specific genes deleted. *Krox-20*, *Hoxa-1* and *Kreisler* are three genes involved in the process of segmentation in the hindbrain. Figure C shows the effects of deletion of one of these genes, *Krox-20*, on specific rhombomeres and on the respiratory rhythm in newborn mice. In mice with *Krox-20* deleted, rhombomeres 3 and 5 are eliminated and the respiratory rhythm is very slow and shallow. During the first day after birth, about 70 per cent of the mice die. By contrast, mice with *Kreisler* deleted survive and are fertile. All mice with *Hoxa-1* deleted die.

Although this evidence is not conclusive, it does suggest that the respiratory rhythm generators in vertebrates develop as paired, segmentally arranged rhythm generators associated with cranial nerves V, VII, IX, X and XII. There is a reduction in the number of paired rhythm generators to one or possibly two in adult vertebrates with an anterior to posterior shift in the location of the dominant pair.

› Find out more

Borday C, Wrobel L, Fortin G, Champagnat J, Thaëron-Antôno C, Thoby-Brisson M (2004). Developmental gene control of brainstem function: views from the embryo. Progress in Biophysics & Molecular Biology 84: 89–106.

Champagnat J, Fortin G (1997). Primordial respiratory-like rhythm generation in the vertebrate embryo. Trends in Neurosciences 20: 119–124.

Fortin G, Kato F, Lumsden A, Champagnat J (1995). Rhythm generation in the segmented hindbrain of chick embryos. Journal of Physiology 486: 735-744.

Milsom WK, Chatburn J, Zimmer MB (2004). Pontine influences on respiratory control in ectothermic and heterothermic vertebrates. Respiratory Physiology and Neurobiology 143: 263–280.

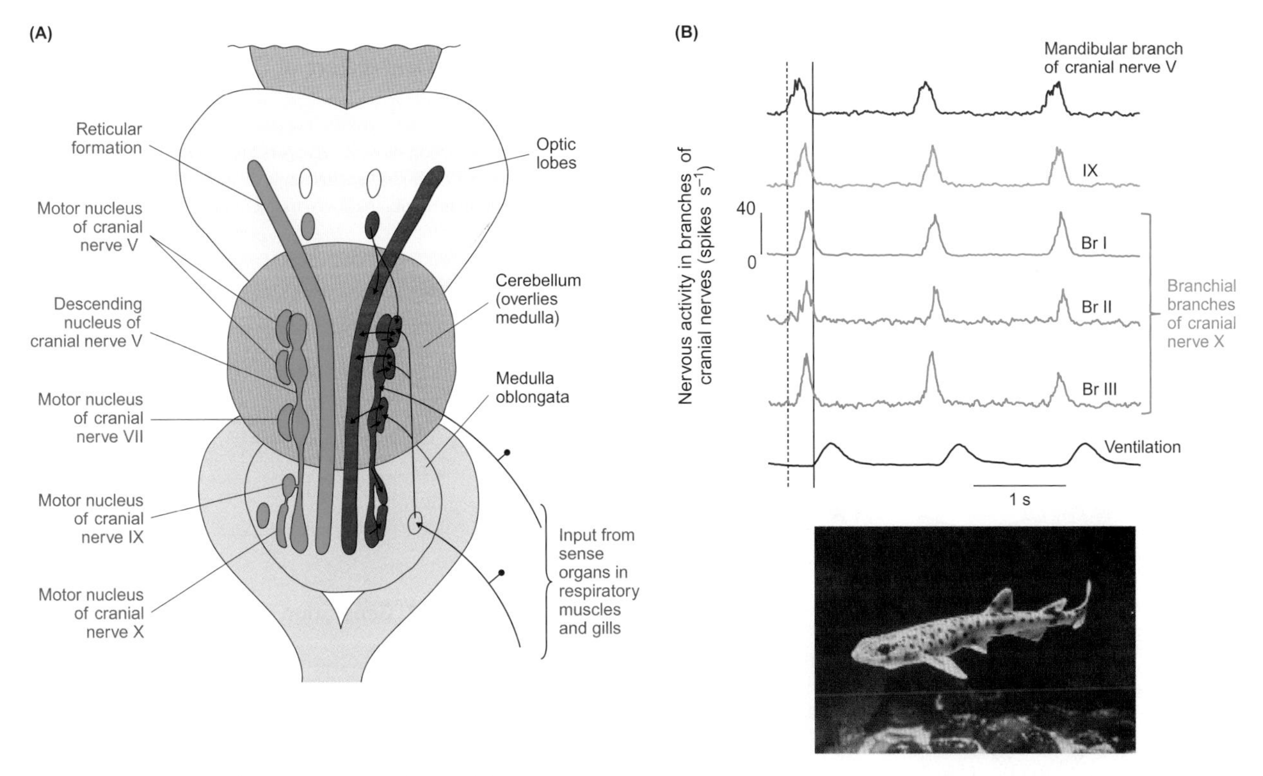

Figure 22.6 Generation of respiratory rhythm in fish

(A) Dorsal view of hindbrain of a bony fish to illustrate: on the left-hand side, the location of neurons which have a respiratory rhythm; on the right-hand side, the connections between these various areas. (B) The relationship between activity in branches of the motor nerves of cranial nerves V, IX and X (branchial, Br, branches I, II and III) and ventilatory movements in a lesser spotted dogfish (*Scyliorhinus canicula*). The dashed line indicates the start of activity in the branch of cranial nerve V, which begins before that in any of the other nerves. Activity in nerve IX begins next while activity in the branches of nerve X occur after that and simultaneously with each other. The bottom line indicates the contraction of the first gill septum (ventilation).

Sources: (A): modified from Butler PJ and Metcalfe JD (1983). Control of respiration and circulation. In: Control Processes in Fish Physiology. Rankin JC et al (eds.). London: Croom Helm. (B): Modified from Barrett DJ and Taylor EW (1985). Spontaneous efferent activity in branches of the vagus nerve controlling heart rate and ventilation in the dogfish. Journal of Experimental Biology, 117: 433-448. Image: (©) Uwe Waller

Respiratory rhythm generation in amphibians

Studies on the isolated brainstem of post-metamorphic tadpoles of American bullfrogs (*Lithobates catesbeianus*) indicate that there are two respiratory oscillators in adult frogs, one responsible for generating buccal ventilation and one for lung ventilation. Buccal ventilation in adult frogs is thought to be the remnant of gill ventilation in tadpoles, which appears to be homologous to gill ventilation in fish. Therefore, the oscillator responsible for generating buccal ventilation in frogs may have evolved from the circuit(s) controlling gill ventilation in fish.

The motor outputs to the respiratory muscles involved in buccal ventilation of adult frogs are provided by the mandibular branch of cranial nerve V and a minor branch from cranial nerve XII, the hypoglossal nerve, which causes the raising and lowering of the floor of the buccal cavity, respectively. The same nerves are also active during lung ventilation. In addition, however, the main branch of cranial nerve XII innervates muscles which assist in the raising of the floor of the buccal cavity so that air can be forced into the lung. Laryngeal branches from cranial nerve X supply muscles which open and close the glottis.

Figure 22.7A indicates that two separate oscillators in each half of the hindbrain are responsible for these activities. The one for lung ventilation is located just anterior to the roots of cranial nerves IX, X and XI, while that for buccal ventilation is posterior to these nerves. By cutting the connections between these two regions, it has been demonstrated that each region is independently able to produce respiratory-related bursts. This may represent the remains of the segmental organization of coupled rhythmic oscillators seen in embryos and shown in Figure B of Case Study 22.1.

A current model of how buccal and lung oscillators in amphibians function and interact is presented in Figure 22.7B. The buccal oscillator is continuously active and inhibits the lung oscillator until there is sufficient stimulation to the lung oscillator to initiate its activity. The source of the stimulation to the lung oscillator is unknown, but could be from the midbrain in an area equivalent to the pons in mammals, or from oxygen and carbon dioxide receptors[23]. Once active, the lung oscillator stimulates the buccal oscillator, thereby increasing its burst frequency. By the process of post-inhibitory rebound excitation[24], the buccal oscillator also causes an increase in bursting frequency of the lung oscillator. This excitatory interaction between the two oscillators maintains a relatively high frequency of ventilation for the duration of the lung cycle[25].

Lung ventilation eventually stops, through some as-yet-unknown process, and the endogenous frequency of buccal ventilation returns. The buccal oscillator seems to rely on inhibitory interactions between the nerve cells (reciprocal inhibition), whereas the lung oscillator does not. This could mean that the buccal oscillator is based on a network of interacting nerve cells, whereas the lung oscillator relies on the activity of pacemaker cells, as in neonatal (newborn) mammals. The important feature of the system in amphibians is

[23] Oxygen and carbon dioxide receptors are discussed in Section 22.2.3.
[24] Post-inhibitory rebound excitation is discussed in Section 16.2.7.
[25] Ventilation in amphibians is discussed in Section 12.3.3.

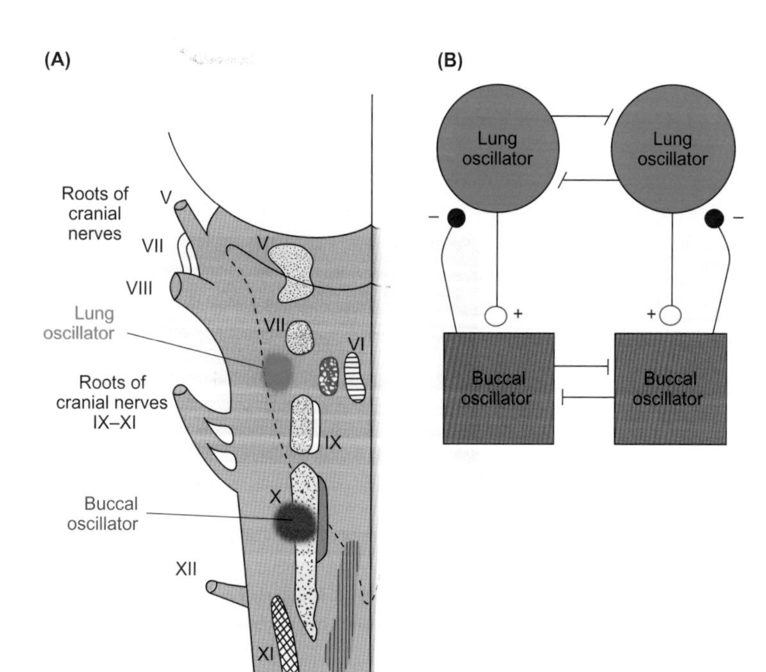

Figure 22.7 Generation of respiratory rhythm in amphibians

(A) Dorsal view of half of hindbrain of a frog showing the location of the motor nuclei and nerve roots of various cranial nerves, indicated by Roman numerals. Also shown are the locations of the buccal oscillator in solid red and the lung oscillator in green. (B) Model for bimodal ventilation in frogs involving the buccal and lung oscillators. + indicates excitatory synapses and – indicates inhibitory synapses

Sources: (A): reproduced from Vasilakos K et al (2004). Ancient gill and lung oscillators may generate the respiratory rhythm of frogs and rats. Journal of Neurobiology 62: 369-385. (B) modified from Wilson RJA et al (2002). Evidence that ventilatory rhythmogenesis in the frog involves two distinct neuronal oscillators. Journal of Physiology, 540: 557-570.

that the episodic pattern of lung ventilation is an intrinsic property of the central respiratory control system and does not depend on feedback from oxygen and carbon dioxide sensors. This does not mean, however, that input from peripheral receptors does not influence various aspects of lung ventilation in amphibians and other vertebrates.

A question that is often asked is: What is the relationship between the intermittent lung ventilation (**episodic ventilation**) of amphibians and reptiles and the continual ventilation of the gills or lungs of many species of fish, and in birds and mammals? It seems that the episodic pattern of lung ventilation in amphibians and reptiles depends on the integrity of the pons/midbrain. Isolation of this region from the medulla oblongata leads to a regular pattern of ventilation that is slower than the frequency during episodes of ventilation. These results suggest that the region in the midbrain has alternating positive and negative effects on the rhythm generators in the medulla, producing a tight clustering of faster ventilations separated by periods of **apnoea**[26]. The episodic pattern seems to be produced by 'skipping' breaths from the basic, regular rhythm. Any increase in stimulation of the respiratory system of amphibians and reptiles, such as by changes in inspired oxygen and/or carbon dioxide, an increase in temperature, or exercise, can cause the transition from episodic to more continuous ventilation.

Episodic ventilation is not confined to ectotherms; marine mammals resting on land and some species of hibernating mammals also exhibit such patterns[27]. It has been suggested that episodic ventilation minimizes the mechanical work of ventilation. For animals that do not need to ventilate continuously to meet their metabolic demands, such as resting air-breathing ectotherms at relatively low temperatures and hibernating mammals, it may be mechanically more efficient to intersperse ventilation occurring at an optimal volume and rate with pauses, rather than to ventilate continuously but sub-optimally.

The two oscillators of the mammalian respiratory rhythm generator

Although there have been very few studies on the origin of the respiratory rhythm in reptiles and birds, there is a wealth of information on mammals. Many techniques, including recording from neurons in the brainstem of anaesthetized adult animals, have been used to establish that the neurons which supply the respiratory muscles in mammals (either directly or indirectly via synapses) are located within two bilateral groupings known as the dorsal and ventral respiratory groups.

The neurons of the dorsal respiratory group are in the vicinity of the **nucleus of the solitary tract** (**NTS**, originally known and sometimes still referred to as the nucleus of the tractus solitarius)[28], while those of the ventral respiratory group are in the general area of the **nucleus ambiguus** (**NA**)[29] and the surrounding reticular formation. These two groups of respiratory cells are shown in Figure 22.8A. Motor outputs to the **phrenic nerves**, which innervate the diaphragm, arise mainly from the inspiratory neurons in the dorsal respiratory group. They leave the spinal cord via cervical (neck) spinal nerves C4 to C8 in cats (C3 to C5 in humans). Nerves to the intercostal muscles arise in the ventral respiratory group and leave the spinal cord in the thoracic (chest) region. Nerves which innervate regions of the upper airway, such as the glottis, are also located in the ventral respiratory group, but travel via cranial nerves for example, in the pharyngeal and laryngeal branches of cranial nerve X.

Only the ventral respiratory group is essential for generation of the respiratory rhythm. Neurons which are related to particular phases of the respiratory cycle are located closely together, as indicated in Figure 22.8A and B. The firing patterns of some of the neurons in adult cats are shown in Figure 22.8C. The inspiratory (I) and expiratory (E) neurons have an increasing frequency of firing (often called 'augmenting'); this pattern of activity (which is also apparent in the activity of the phrenic nerve to the diaphragm), together with other information, has led to the idea that these types of cells have excitatory synapses with one another and are, therefore, self re-exciting. Neurons that are active after the abrupt decline in the activity of the I neurons also exist; these are known as post inspiratory (post-I) neurons.

Self re-excitation leads to more and more cells becoming active; as the activity in individual cells begins to decrease, they are re-excited by others in the group, so maintaining overall inspiratory activity. Inspiration is eventually terminated by input from stretch receptors in the lungs[30] and neurons located in the pons, which is shown in Figure 22.8A. This termination process is known as an inspiratory-off switch.

Also, active inspiratory neurons inhibit the expiratory cells and vice versa (which is known as reciprocal inhibition, as we discuss for insects in Section 22.1.1). When this inhibition is removed (when the inspiratory neurons become inactive), active expiration can occur via the E neurons, if required. In resting mammals, however, expiration is often passive, although the post-I neurons are active during the

[26] Apnoea is derived from Greek and means absence of breathing.

[27] Episodic ventilation in marine mammals and hibernating mammals are discussed in Sections 15.3.3 and 15.4.2.

[28] NTS is, strictly speaking, the abbreviation for the nucleus of the tractus solitarius, but is also frequently used for the anglicised version: nucleus of the solitary tract.

[29] The locations of these two nuclei are shown in Figure 22.32.

[30] Sense organs in the lungs are considered in Section 22.2.1.

22

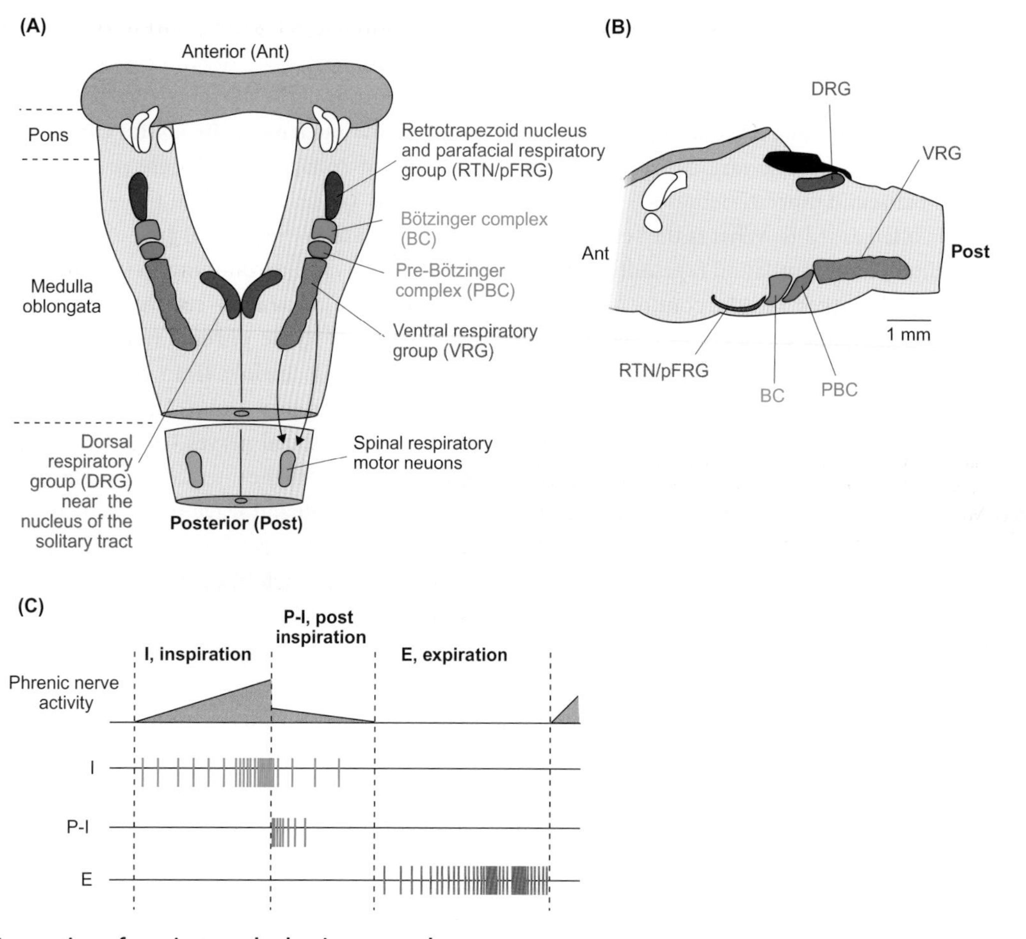

Figure 22.8 Generation of respiratory rhythm in mammals

(A) Dorsal and (B) vertical longitudinal section through the middle of the hindbrain of a rodent to illustrate the locations of the dorsal and ventral respiratory groups of neurons and the regions where predominantly inspiratory (I) and expiratory (E) neurons are found. The locations of the two oscillators (retrotrapezoid nucleus/parafacial respiratory group and pre-Bötzinger complex) that are thought to be involved in the generation of the respiratory rhythm are also indicated. (C) The activity of some of the respiratory neurons in cats that have been identified in relation to inspiration (I), post inspiration (post-I) and expiration (E) as indicated by the activity of the phrenic nerve to the diaphragm.

Source: modified from: Spyer KM (2009). To breathe or not to breathe? That is the question. Experimental Physiology 94: 1-10. Hilaire G and Pásaro R (2003). Genesis and control of the respiratory rhythm in adult mammals. Proceedings of Neural Information Processing Systems, 18: 23-28.

transition between inspiration and expiration—the post-inspiratory period. These neurons send axons to the laryngeal muscles which control airflow from the lungs to ensure a smooth transition between inspiration and expiration.

Reciprocal inhibition is a key component of this oscillating network and was initially thought to be the basis of respiratory rhythm generation in mammals. However, subsequent studies have indicated that the situation is slightly more complex. The blocking of inhibitory transmission with drugs in isolated brainstem–spinal cord preparations of newborn rats does not abolish the respiratory rhythm. Furthermore, some neurons in a region that is known as the **pre-Bötzinger complex**[31], shown in Figures 22.8A and B, continue to produce rhythmic activity when all synaptic transmission has been blocked with drugs—although it is of a gasping nature. In other words, the basic rhythm is not the result of reciprocal inhibition or self re-excitation within a network of cells. It has been postulated that these neurons act as inspiratory pacemakers and that they can impose the rhythm of their bursts on other inspiratory neurons and thereby drive the inspiratory rhythm. Despite this, the pacemaker neurons are not essential for rhythm generation and may not even display pacemaker activity in mature mammals.

It has also been proposed that there are two distinct but coupled rhythm generators in the medulla of mammals, the pre-Bötzinger complex and the ventrally located retrotrapezoid nucleus/parafacial respiratory group (RTN/pFRG), which are shown in Figures 22.8A and B. Both of these structures may be homologous to the two rhythm generators in amphibians, with the RTN/pFRG oscillator being equivalent to the buccal oscillator. Current evidence

[31] The Bötzinger complex consists of neurons active during expiration and is named after a German white wine. Although the pre-Bötzinger complex is situated behind the Bötzinger complex, it was called pre-Bötzinger to emphasize its proposed importance in the generation of the respiratory rhythm.

indicates that the pre-Bötzinger complex is the dominant rhythm generator in adult mammals. It also appears that the RTN/pFRG oscillator in mammals is more related to expiration than inspiration, such that only the pre-Bötzinger complex may be active a resting mammal, when there is typically passive expiration.

› Review articles

Bell HJ, Syed NI (2012). Control of breathing in invertebrate model systems. Comprehensive Physiology 2: 1745-1766.
Burrows M (1996). The Neurobiology of an Insect Brain. Ch 12, Breathing. pp 563–596. Oxford: Oxford University Press.
Feldman JL, Del Negro CA (2006). Looking for inspiration: new perspectives on respiratory rhythm. Nature Reviews Neuroscience 7: 232–242.
Feldman JL, Del Negro CA, Gray PA (2013). Understanding the rhythm of breathing: so near, yet so far. Annual Review of Physiology 75: 423–452.
Fong AY, Zimmer MB, Milsom WK (2009). The conditional nature of the 'central rhythm generator' and the production of episodic breathing. Respiration Physiology & Neurobiology 168: 179–187.
Hilaire G, Pásaro R (2003). Genesis and control of the respiratory rhythm in adult mammals. News in Physiological Sciences 18: 23–28.
Hooper SL, DiCaprio RA (2004). Crustacean motor pattern generator networks. Neurosignals 13: 50–69.
Lukowiak K, Martens K, Orr M, Parvez K, Rosenegger D, Sanga S (2006). Modulation of aerial respiratory behaviour in a pond snail. Respiration Physiology & Neurobiology 154: 61–72.
Milsom WK (1991). Intermittent breathing in vertebrates. Annual Review of Physiology 53: 87–105.
Spyer KM (2009). To breathe or not to breathe? That is the question. Experimental Physiology 94: 1–10.
Wilson RJA, Vasilakos K, Remmers JE (2006). Phylogeny of vertebrate respiratory rhythm generators: the oscillator homology hypothesis. Respiration Physiology & Neurobiology 154: 47–60.

22.2 Control of the respiratory system

Generation of the respiratory rhythm is only part of the story of how the respiratory system and cardiovascular systems work in concert to meet the metabolic demands of the cells and organs of the body; information on how the systems are performing is also required. There are two major sources of information which influence the respiratory system:

- mechanical sense organs, or **mechanoreceptors**, which are found in the respiratory system itself: in the gas-exchange organs, e.g. gills or lungs, and/or in the muscles associated with these organs;
- chemical sense organs, or **chemoreceptors**[32], which are found on the external surface of aquatic animals, where they monitor the composition of the surrounding water, and/or within the animal, where they monitor the composition of the air in the respiratory system of some air-breathing vertebrates and the blood or haemolymph. These peripherally located chemoreceptors are called the peripheral chemoreceptors. In vertebrates, some chemoreceptors located within the brainstem monitor the level of carbon dioxide and pH in the cerebrospinal fluid; they are known as the central chemoreceptors.

We know most about the sensory output from the various components of the respiratory system in the vertebrates, so most of the following discussion is based on this group of animals.

22.2.1 Mechanoreceptors and chemoreceptors in the respiratory system of vertebrates

Water-breathing vertebrates possess gills while air-breathers have lungs, so the sense organs associated with the gas exchangers are not exactly the same in the two groupings, although there are many similarities between them.

Mechanoreceptors in the respiratory system of water-breathing vertebrates

Receptors respond to the deflection of the gill filaments, gill arches and gill rakers (projections on the buccal (mouth) side of the gill arches). Studies on carp (*Cyprinus carpio*) and channel catfish (*Ictalurus punctatus*) have identified two types of receptor by the pattern of activity in their axons, as shown in Figure 22.9.

In one type of receptor, the activity coincides with either the expansion of the buccal and opercular cavities (inspiration) or with their contraction (expiration), as shown in Figure 22.9A(i). Sustained stimulation of this type of receptor (with a continuous jet of water for example) does not cause a continuous output; activity in the sensory nerves lasts only a few milliseconds, as shown in Figure 22.9A(ii). These rapidly adapting sensory receptors are known as **phasic receptors** and are found on the gill filaments and gill rakers.

The other type of receptor, which is found on the gill arches, exhibits continuous activity, but it increases during either inspiration or expiration, as shown in Figure 22.9B(i). These slowly adapting sensory receptors are therefore known as **tonic receptors**. The output from these receptors is almost linearly related to the position of the gill arch, as shown in Figure 22.9B(ii). The input

[32] Sense organs, including chemoreceptors and mechanoreptors, are discussed in, Sections 17.3 and 17.4.

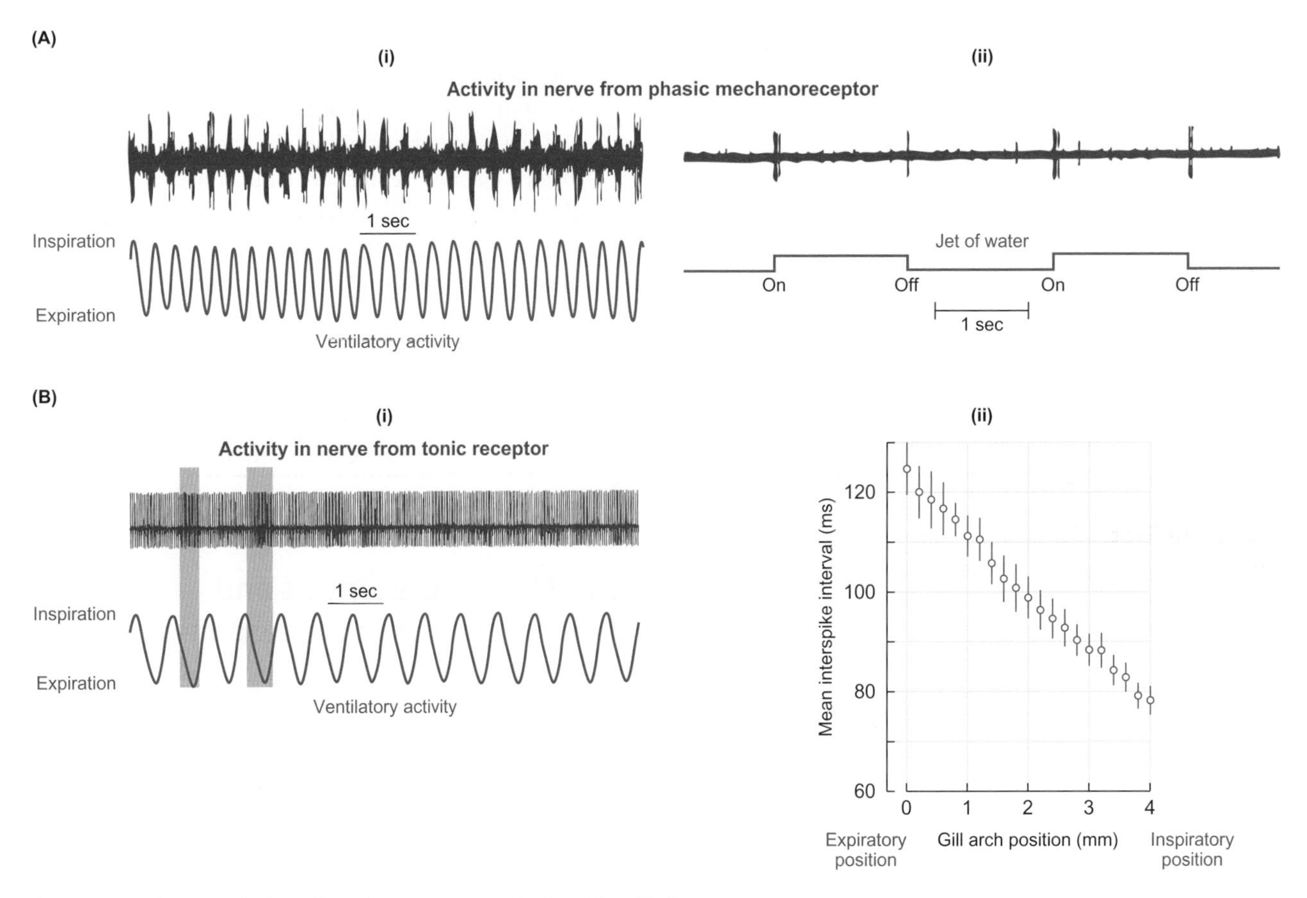

Figure 22.9 Characteristics of mechanoreceptors in the gills of fish

(A) Activity from: (i) a phasic mechanoreceptor in the gill of an anaesthetised channel catfish (*Ictalurus punctatus*) during spontaneous ventilation, and (ii) from a similar receptor in the gill of an anaesthetized, artificially ventilated carp (*Cyprinus carpio*) in response to two periods of sustained stimulation with a jet of water. (B) Activity from: (i) a tonic receptor in the gill of an anaesthetized catfish during spontaneous ventilation (note the increased activity during expiration—purple shading indicates two such instances), and (ii) from a similar receptor in the gill of a carp during movement of the gill arch from its expiratory (contracted) position to its inspiratory (expanded) position. Note that the y-axis is the time between each action potential (spike): as this time gets shorter, the frequency increases.

Sources: (A): modified from Burleston ML et al (2001). Branchial mechanoreceptor activity during spontaneous ventilation in channel catfish. Comparative Biochemistry and Physiology A, 128: 129-136. Modified from De Graaf PJF et al (1987). Mechanoreceptor activity in the gills of carp. I. Gill filament and gill raker mechanoreceptors. Respiration Physiology 69: 173-182. (B): modified from Burleston ML et al (2001). Branchial mechanoreceptor activity during spontaneous ventilation in channel catfish. Comparative Biochemistry and Physiology, 128: 129-136. De Graaf PJF and Ballintijn CM (1987). Mechanoreceptor activity in the gills of the carp. II. Gill arch proprioceptors. Respiration Physiology 69: 183-194.

to the brain from these receptors is via cranial nerves IX and X.

Although these mechanoreceptors are stimulated by the ventilatory movements of the gill filaments and gill arches, data from channel catfish indicate that rhythmic feedback from the phasic mechanoreceptors is not involved in the breath-by-breath control of the normal ventilatory cycle in resting fish. These receptors may, however, be important in limiting inspiratory activity when the system is stimulated during environmental hypoxia (a fall in partial pressure of oxygen in the water) and during exercise. It has been suggested that the stimulation of the tonic mechanoreceptors by an increasing flow of water over the gills as the fish swims faster may trigger the transition from active ventilation to ram ventilation in some species[33].

[33] Ram ventilation in fish is discussed in Section 12.2.2.

Air-breathing in fish can be achieved via modifications of the gill, opercular and pharyngeal cavities, or by the evolution of an air bladder or lung. It seems likely that the types of mechanoreceptors already described for water breathers would be sufficient when using the gill, opercular or pharyngeal cavities to provide mechanoreceptive feedback for the control of air breathing, although this has not been demonstrated. However, the use of an air-bladder or lungs requires the presence of specific sense organs associated with the air-breathing organ.

Receptors in the respiratory system of air-breathing vertebrates

The air bladders or lungs of all air-breathing vertebrates contain sense organs that respond to the state of inflation of the lungs. These receptors may be mechanoreceptors which respond only to the degree of stretch of the lungs and are

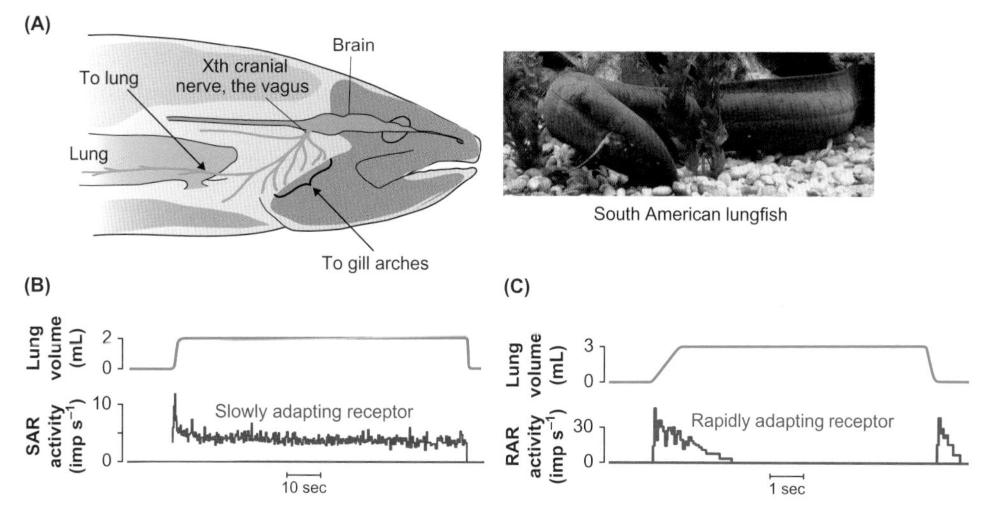

Figure 22.10 Types of mechanoreceptors in lungs of air-breathing vertebrates

(A) Innervation of the lungs of the air-breathing fish, bowfin (*Amia calva*) by a branch of the Xth cranial nerve, the vagus. (B) Response of a single slowly adapting receptor (SAR) in an isolated lung of the South American lungfish (*Lepidosiren paradoxa*) in response to rapid and sustained inflation. This shows the initial dynamic burst of activity followed by the slower, sustained tonic discharge which falls to zero upon deflation. (C) Response of a single rapidly adapting receptor (RAR) in a lung of the South American lungfish in response to inflation and deflation. The initial bursts of activity upon inflation and deflation decline relatively rapidly to zero. Note different time scales between B and C.

Sources: (A): modified from Milsom WK and Jones DR (1985). Characteristics of mechanoreceptors in the air-breathing organ of the holostean fish, *Amia calva*. Journal of Experimental Biology 117: 389-399. (B) & (C): DeLaney RG et al (1983). Pulmonary mechanoreceptors in the dipnoi lungfish, *Protopterus and Lepidosiren*. American Journal of Physiology, 244: R418-R428. Image: © Dr Solomon R David.

found in all air-breathing vertebrates except birds. A proportion of the mechanoreceptors in the lungs of some air-breathing fish such as gar (*Lepisosteus oculatus*), lungfish, amphibians and some species of reptiles are, to a greater or lesser extent, inhibited by increasing concentrations of carbon dioxide, while birds have receptors which are only sensitive to changes in the level of carbon dioxide. All of these groups of receptors are innervated by a branch of the Xth cranial nerve, the vagus, which is shown in Figure 22.10A.

Mechanoreceptors in the lungs of air-breathing vertebrates

There are two types of mechanoreceptors in the lungs of vertebrates, slowly adapting and rapidly adapting receptors. Although the details of the response may vary from species to species, the slowly adapting receptors (SARs) essentially display two types of nervous activity upon inflation, as shown in Figure 22.10B; an initial, rate-sensitive, dynamic burst of activity upon inflation of the lung is followed by a slower, sustained tonic discharge. The magnitude of the tonic discharge is dependent on the volume of lung inflation; the greater the volume of inflation, the greater the frequency of discharge.

In the lungs of all groups of air-breathing vertebrates that have been studied (except bowfin, *Amia calva*, which only have slowly adapting stretch receptors), there are also rate-sensitive receptors that adapt rapidly to changes in volume of the lung. These rapidly adapting receptors (RARs) are briefly active upon inflation of the lung and again upon deflation, as shown in Figure 22.10C.

There are basically three types of mechanoreceptors in the lungs of mammals, the main ones being slowly adapting receptors. In addition, there are rapidly adapting receptors which have also been called irritant receptors, or cough receptors for those located in the larynx, trachea and larger bronchi, and the so-called C-fibre receptors[34]. The role of the C-fibre receptors is controversial, but the body of evidence suggests that they cause the temporary cessation of ventilation and perhaps even the reflex inhibition of coughing.

In terms of the control of normal ventilatory activity in mammals, the slowly adapting receptors are responsible for the Breuer–Hering reflex, which was first described by Joseph Breuer and Ewald Hering in 1868. The receptors are located in the smooth muscle of the larger, extrapulmonary airways—those airways outside the lung itself. Between 40 and 85 per cent of slowly adapting receptors are found in the extrapulmonary airways in cats, dogs and rabbits and their stimulation during lung inflation inhibits inspiratory activity and prolongs the duration of expiration. This is the **Breuer–Hering inflation reflex** which, in humans at least, is probably of more functional significance when ventilation is increased above the resting level (during exercise, for example). It has been suggested that the lung stretch receptors monitor changes in the mechanical conditions of the lungs to optimize the pattern of ventilation in terms of mechanical work.

The inflation reflex, which operates via the slowly adapting stretch receptors in the lungs, has been

[34] A, B and C nerve fibres are discussed in Section 16.1.3.

demonstrated in all species of air-breathing fish, amphibians and reptiles that have been studied. These animals ventilate their lungs intermittently and it is likely that the inflation reflex helps to determine the volume of the lungs following a burst of ventilation. This is particularly important in air-breathing fish and amphibians which use a buccal pump to inflate their lungs in steps; the inflation reflex may determine the number of breaths in an episode of ventilation. The duration of the pause which follows such an episode of ventilation may also be influenced by the stretch receptors as they slowly adapt and their inhibitory effect reduces.

Breuer and Hering also found that abrupt deflation of the lungs stimulates inspiration, the **Breuer–Hering deflation reflex**. This reflex may be mediated by a reduction in activity of the slowly adapting receptors and/or by stimulation of other receptors such as the rapidly adapting receptors. It may be particularly important in newborn infants to prevent their lungs from deflating below their functional residual volume as a consequence of the inward recoil of their lungs being greater than the outward recoil of their chest wall (unlike the situation in adults)[35].

Although the primary stimulus to the stretch receptors in mammalian lungs is mechanical, it has been shown that they can be inhibited by increases in the concentration of carbon dioxide in the airways. However, the main range of sensitivity is below a partial pressure of carbon dioxide of 4 kPa in the lungs (compared with the normal level of approximately 5.5 kPa). It may be more accurate, therefore, to say that the receptors are stimulated by an unusually low partial pressure of carbon dioxide (hypocapnia), as may occur during the increased ventilation at high altitude, for example.

Notable by their absence from the above discussion on mechanoreceptors in the lungs are the birds. This absence stems from a very significant difference between the respiratory systems of birds and mammals. Unlike the lungs of all other air-breathing vertebrates, those of birds do not change volume to any notable extent during ventilation, which means that the presence of mechanoreceptors in their lungs would be of little use[36].

Chemoreceptors in the lungs of lungfish, amphibians, reptiles and birds

The lungs of birds possess receptors that respond only to changes in the level of carbon dioxide. Their activity increases as the level of CO_2 falls during inspiration such that they have a similar function to the slowly adapting mechanoreceptors in the lungs of other air-breathing vertebrates. The carbon dioxide-sensitive receptors in the lungs are often called **intrapulmonary chemoreceptors (IPCs)**. Experimental Panel 22.1 discusses how the characteristics of the pulmonary receptors in the lungs of birds were determined.

The CO_2-sensitive mechanoreceptors in the lungs of gar, lungfish, amphibians and some species of reptiles respond both to being stretched and to the associated decrease in carbon dioxide by increasing their activity, as shown in Figure 22.11. The opposite occurs during expiration as the lungs deflate and the level of carbon dioxide increases. However, the roles of these carbon dioxide-sensitive mechanoreceptors in the control of ventilation are uncertain.

Carbon dioxide-sensitive mechanoreceptors may influence lung volume and the duration of the pause between periods of ventilation. That said, exposure to relatively high levels of carbon dioxide in the environment is unlikely, except in some burrow-dwelling species.[37] However, when such exposure does occur, a reduction in output from both the CO_2-sensitive mechanoreceptors and of the CO_2-sensitive receptors would lead to an increase in tidal volume and increased removal of carbon dioxide.

22.2.2 Mechanoreceptors in the respiratory system of invertebrates

Although there is very little information about mechanoreceptive input to the respiratory system of invertebrates, these receptors do appear to play an important role in some species. For example, in air-breathing pond snails, one of the sensory inputs to the right pedal dorsal 1 (RPeD1) neuron of the respiratory rhythm generator[38] is from mechanoreceptors around the pneumostome. Under natural conditions these mechanoreceptors are stimulated as the animal surfaces and the pneumostome breaks the water/air interface. This stimulation, causes the pneumostome to open, as illustrated in Figure 22.12A; this opening, together with other sensory input (such as chemoreceptors, discussed in the next section), initiates the generation of the respiratory rhythm. Unfortunately, the mechanosensory neurons have not yet been identified.

The only known mechanoreceptor within the respiratory system of crustaceans is a single stretch receptor, the **oval organ**, found in bailers of decapod crustaceans, as shown back in Figure 22.1A. However, its function is uncertain.

[35] Characteristics and mechanics of lungs are discussed in Section 12.3.3.
[36] The respiratory system of birds is discussed in Section 12.3.3.
[37] Burrowing animals are discussed in Section 15.3.2.
[38] Look at Figure 22.3 to remind yourself of the organization of the respiratory rhythm generator in pond snails.

Experimental Study 22.1 Characteristics of the pulmonary receptors in birds

Background

It is known that birds respond to changes in the concentration of carbon dioxide in their lungs during the respiratory cycle. The following experiments were designed to answer the overall question: What stimuli influence the receptors in the lungs of birds?

Experimental approach

Two series of experiments were performed to answer this general question using neurophysiological methods in Experiment 1 and pharmacological techniques in Experiment 2.

Experiment 1

This experiment was designed to determine if CO_2-sensitive mechanoreceptors are present in the lungs of birds by recording the activity in nerves from either deflated or inflated lungs in response to changes in the concentration of CO_2.

Anaesthetized Muscovy (*Cairina moschata*) were ventilated continuously in one direction (unidirectional ventilation), with warm air flowing in through their trachea and out through a caudal air sac[1]. The percentage of carbon dioxide in the air stream that ventilated the lung was varied between 0 and 18 per cent while oxygen was maintained at between 21 and 25 per cent. The pressure in the lungs (intrapulmonary pressure) was varied between 0.15 and 1 kPa by altering the resistance at the outflow from the air sac.

Fine strands of nerve were dissected free from the right vagus nerve in the neck and single-unit sensory impulses were recorded from small filaments draped over two thin platinum electrodes. The signal was amplified and recorded. A nerve strand containing a unit sensitive to CO_2 was identified by suddenly switching the concentration of CO_2 to 0 per cent in the air flowing through the lungs. The sudden decrease in CO_2 concentration in the lungs causes a rapid increase in frequency of discharge of the receptor.

Figure A demonstrates that the CO_2-sensitive receptors are spontaneously active at low levels of CO_2. The effect of CO_2 concentration on discharge frequency is shown in Figure A(i). The high discharge frequency in the absence of CO_2 is reduced as the concentration of CO_2 increased. This pattern of response is unusual among chemoreceptors, such as mammalian carotid

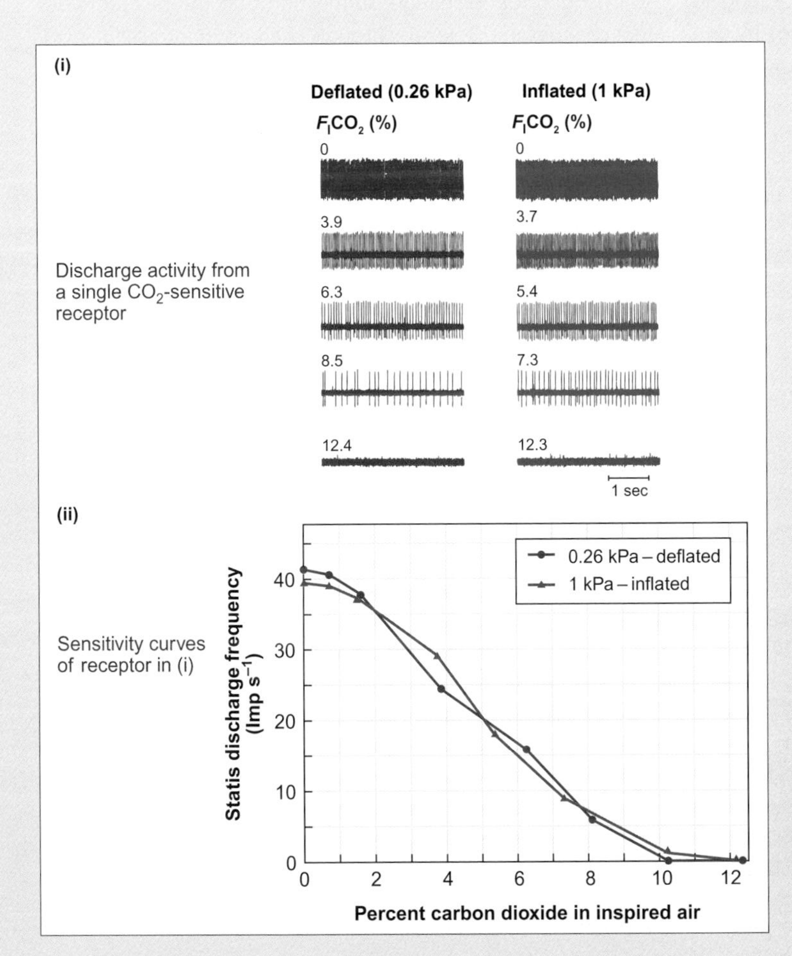

Figure A Discharge characterisitics of carbon dioxide receptors in the lungs of ducks

(i) Discharge from a receptor in the lung of an anaesthetized Muscovy duck (*Cairina moschata*) in response to increases in concentration of carbon dioxide at two different lung inflation pressures. (ii) Sensitivity curves of the data from the receptor showing that the responses of the receptors are almost identical in inflated and deflated lungs.

Source: modified from Fedde MR et al (1974). Intrapulmonary CO_2 receptors in the duck: I. stimulus specificity. Respiration Physiology 22: 99-114.

bodies, which normally increase their discharge frequency with increasing stimulus.

The response of the CO_2-sensitive receptors was not affected by intrapulmonary pressure, and thus by the degree of stretch of the lungs, as illustrated in Figure A(ii). Rapidly changing intrapulmonary pressure had no effect on discharge rate at any concentration of CO_2 tested. In other studies, it has been found that the intrapulmonary CO_2-sensitive receptors do not respond to changes in the concentration of oxygen in the lungs. In other words, these receptors respond only to the concentration of CO_2 and not to stretch of the lungs, nor to hypoxia or hyperoxia.

Experiment 2

A question that is asked about all receptors that are sensitive to changes in carbon dioxide is: Do the receptors actually respond to changes in CO_2 or to changes in pH? A second set of experiments was designed to answer this question.

These experiments involved the intravenous injection of drugs that block the action of carbonic anhydrase[2] (acetazolamide), the exchange of hydrogen ions (amiloride), or the exchange of bicarbonate ions across the cell membrane (DIDS[3]).

Figure B shows the average response of four receptors. Blocking carbonic anhydrase activity causes an elevated discharge rate of the receptors, as if they are experiencing very low CO_2, even when CO_2 in the ventilating gas is high. These data suggest that the activity of CO_2-sensitive receptors is determined by one of the products of the hydration of CO_2: H^+ or HCO_3^-, and not by molecular CO_2.

Administration of DIDS, which inhibits HCO_3^-/Cl^- exchange, causes a small increase in discharge rate of the receptors at all levels of CO_2, probably as a result of the accumulation of HCO_3^- in the receptor cells. In contrast, administration of amiloride, which inhibits Na^+/H^+ exchange, strongly depresses discharge rate at all levels of CO_2, as shown in Figure C. The response of the CO_2 receptor cells to the administration of amiloride is probably due to the accumulation of H^+ in the cells causing an acidosis.

Overall findings

These results indicate that the receptors in the lungs of birds respond only to the concentration of CO_2 and not to stretch of the lungs, nor to hypoxia or hyperoxia. As the concentration of CO_2 decreases during inspiration, activity of the CO_2-sensitive receptors in the lungs increases, while it decreases during expiration as the concentration of CO_2 rises. So, the responses of the carbon dioxide receptors to inspiration and expiration are similar to those of the slowly adapting mechanoreceptors in the lungs of other air-breathing vertebrates. The receptors in the lungs of birds also serve a similar function to those in other air-breathers, as increased activity in the intrapulmonary chemoreceptors inhibits inspiratory activity.

The response to CO_2 by the CO_2-sensitive receptors in the lungs of birds depends strongly on Na^+/H^+ exchange and less so on HCO_3^-/Cl^- exchange, suggesting that the receptors detect changes in CO_2 levels through changes in intracellular pH. Figure D presents a scheme of how this transduction mechanism may work.

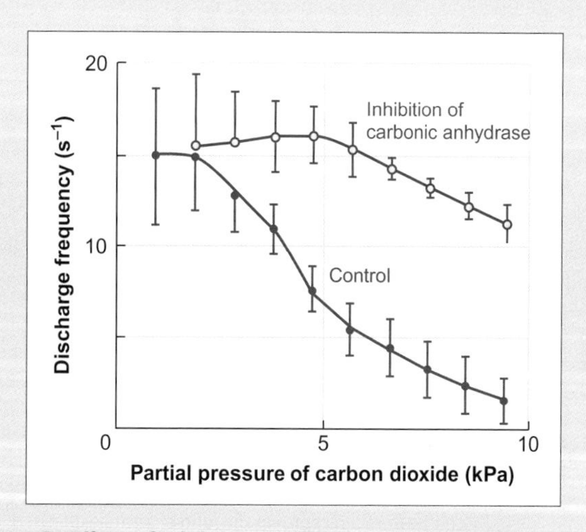

Figure B Effect of inhibiting carbonic anhydrase with acetazolamide on response of CO_2-sensitive receptors in lungs of ducks

Average (± SEM) sensitivity of four receptors to increasing partial pressure of CO_2 before (control) and after inhibition of carbonic anydrase with acetazolamide.

Source: modified from Scheid P et al (1978). Ventilatory response to CO_2 in birds, II. Contribution by intrapulmonary CO_2 receptors. Respiration Physiology 35: 361-372.

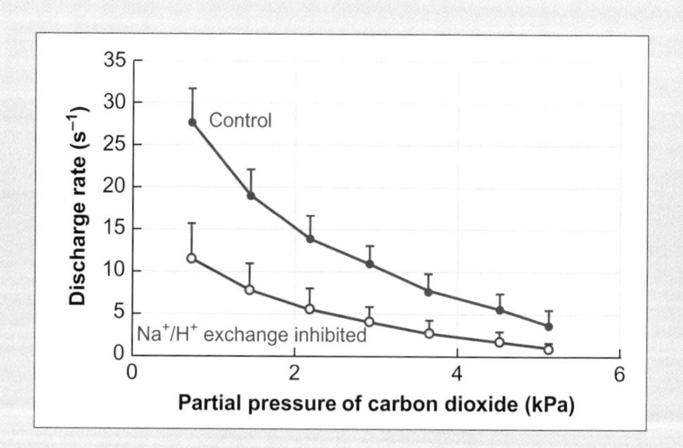

Figure C Effect of blocking Na^+/H^+ exchange with amiloride on the response of CO_2-sensitive receptors in the lungs of ducks to increasing partial pressure of CO_2

Mean sensitivity, + SEM, of receptors from eight animals before (control) and after inhibition of Na^+/H^+ exchange with amiloride.

Source: modified from Hempleman SC et al (2003). CO_2 transduction in avian intrapulmonary chemoreceptors is critically dependent on transmembrand Na^+/H^+ exchange. American Journal of Physiology, 284: R1551-R1559.

Figure D Schematic diagram of proposed model of how receptors in the lungs of birds sense carbon dioxide

The model proposes that intracellular hydrogen ion concentration (intracellular pH, pHi) is the stimulus for changing the activity of the receptors. pHi is determined by a dynamic balance between the rate of CO_2 hydration/dehydration catalysed by carbonic anhydrase (CA, pink-shaded box), and rates of acid–base extrusion, with the main factor being Na^+/H^+ exchange (blue-shaded box), and intracellular buffering (green-shaded box).

Source: modified from Hemplema SC and Posner RG (2004). CO_2 transduction mechanisms in avian intrapulmonary chemoreceptors: experiments and models. Respiratory Physiology & Neurobiology, 144: 203-214.

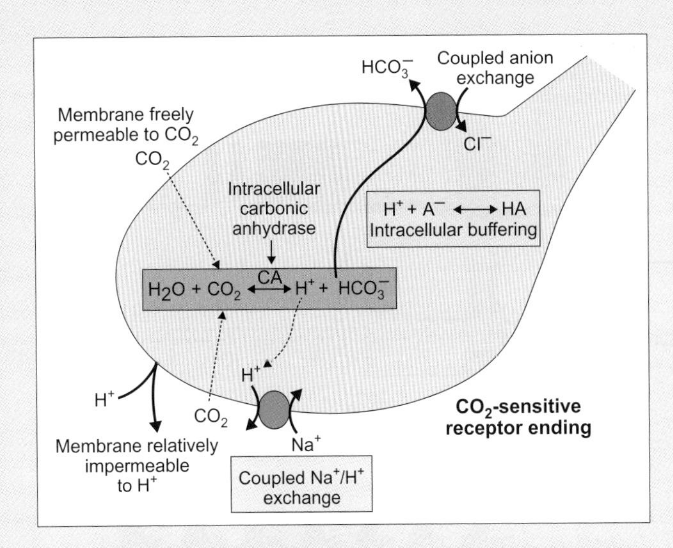

› Find out more

Fedde MR, Gatz RN, Slama H, Scheid P (1974). Intrapulmonary CO_2 receptors in the duck: I. Stimulus specificity. Respiration Physiology 22: 99–114.

Hempleman SC, Adamson TP, Begay RS, Solomon IC (2003). CO_2 transduction in avian intrapulmonary chemoreceptors is critically dependent on transmembrane Na^+/H^+ exchange. American Journal of Physiology 284: R1551–R1559.

Hempleman SC, Posner RG (2004). CO_2 transduction mechanisms in avian intrapulmonary chemoreceptors: experiments and models. Respiration Physiology & Neurobiology 144: 203–214.

Scheid P, Gratz RK, Powell FL, Fedde MR (1978). Ventilatory response to CO_2 in birds. II. Contribution by intrapulmonary CO_2 receptors. Respiration Physiology 35: 361–372.

[1] The anatomy of the respiratory system of birds is discussed in Section 12.3.3.

[2] Carbonic anhydrase catalyses the hydration of carbon dioxide or the dehydration of carbonic acid: $H_2O + CO_2 \leftrightarrows H_2CO_3 \leftrightarrows H^+ + HCO_3^-$, and is discussed in section 13.1.3.

[3] DIDS is 4,4′-Diisothiocyano-2,2′-stilbenedisulfonic acid

The oval organ has been studied in both lobsters and crabs, but its characteristics are different in each species. In European or common lobsters, (*Homarus gammarus*), output from the sense organs takes the form of spiking action potentials, which are produced during both levation and depression of the bailer. In contrast, the output in shore crabs is non-spiking and the membrane potential is hyperpolarized with respect to the resting potential, as shown in Figure 22.12B. It is assumed that the information the oval organ sends to the central nervous system has a role in the regulation of ventilation, but the nature of its input to the central rhythm generator and its role in the production of the motor pattern are yet to be determined.

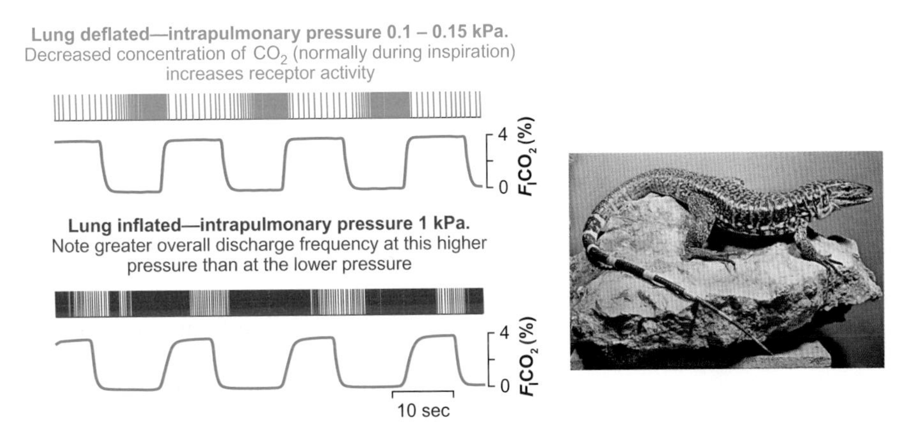

Figure 22.11 Carbon dioxide-sensitive mechanoreceptors in gold or common tegu lizards (*Tupinambis teguixin*, formerly *Tupinambis nigropunctatus*)

Responses of a single receptor to changes in intrapulmonary pressure and the percentage of carbon dioxide in the inspired air (F_ICO_2) during unidirectional ventilation of a lung of an anaesthetized tegu lizard. Both pressure in the lungs (degree of inflation) and the concentration of carbon dioxide in the air flowing through the lungs affect the activity of the receptor; inflation stimulates the receptor, whereas increase in CO_2 concentration inhibits it.

Source: modified from Fedde MR et al (1977). Intrapulmonary receptors in the tegu lizard: I. Sensitivity to CO_2. Respiration Physiology, 29: 35-48. Image: © John H. Gerard—The National Audubon Society Collection/Photo Researchers.

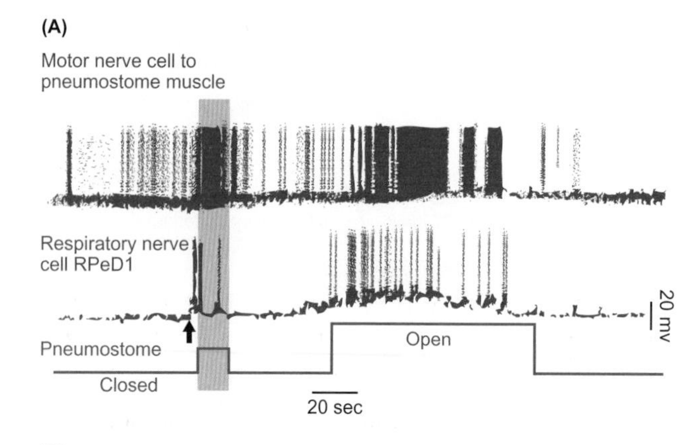

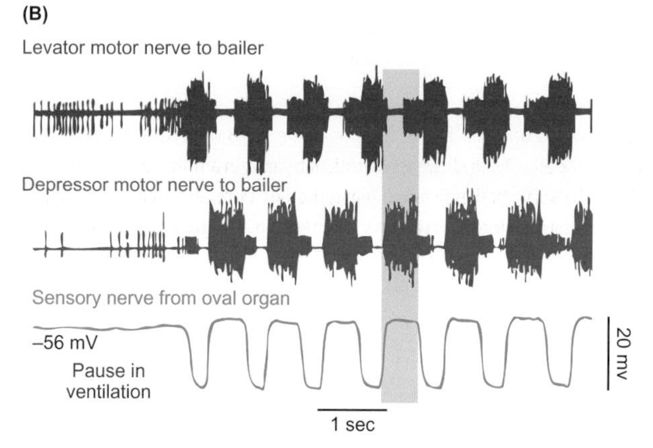

Figure 22.12 Mechanoreceptors in the respiratory systems of some species of invertebrates

(A) In great pond snails (*Lymnaea stagnalis*), the respiratory neuron, RPeD1 receives input from mechanoreceptors around the pneumostome as it breaks the air/water interface (at black arrow). There is an initial brief burst of action potentials in RPeD1 and partial opening of the pneumostome (highlighted by purple shading) followed by a cycle of complete opening and closing of the pneumostome. (B) In shore crabs or European green crabs (*Carcinus maenas*), the sensory nerve from the oval organ is periodically hyperpolarized by about 20 mV in phase with the oscillation of the bailer with which it is associated (highlighted by green shading) The resting potential, as measured during a pause in ventilation, is –56 mV.

Sources: A: modified from Haque Z et al (2006). An identified central pattern-generating neuron co-ordinates sensory-motor components of respiratory behavior in *Lymnaea*. European Journal of Neuroscience 23: 94-104. (B): DiCaprio RA (1999). Gating of afferent input by a central pattern generator. Journal of Neurophysiology 81: 950-953.

The situation is much the same in insects. There is evidence that mechanoreceptors are located in appropriate positions and activated during ventilatory movements of the abdomen, but their functional role is not known. In larvae of dragonflies (*Aeshna* spp.), segmentally arranged, paired mechanoreceptors called chordotonal organs[39] are located laterally in the abdomen and are stimulated during inspiration. In locusts, chordotonal organs span the folds between the dorsal and ventral plates of each segment and stretch receptors are attached to muscles that span the segments. The nerve fibres of these sense organs are active during ventilation, project to several ganglia and form the anatomical basis for the necessary connections required to ensure coordinated movements of the abdomen during ventilation.

22.2.3 Peripheral and central respiratory chemoreceptors

Peripheral and central chemoreceptors are involved in the control of ventilation with respect to the supply of adequate oxygen to the metabolizing tissues, and the removal of carbon dioxide. However, the differences in the capacitance coefficients of oxygen and carbon dioxide in water[40] mean that, aquatic animals must produce a relatively high flow rate of water across their gills in order to maintain a sufficiently high concentration and partial pressure of oxygen (PO_2) in the arterial blood to meet their oxygen demands. Such high flow rates of water over the gills lead to rapid excretion of carbon dioxide, which in turn leads to a low partial pressure of CO_2 (PCO_2) and a relatively alkaline pH in arterial blood.

In contrast, the relatively low ventilation rate of the lungs by air-breathing animals leads to high PCO_2 in arterial blood and a relatively acid pH, as illustrated in Figure 22.13.

[39] We discuss chordotonal organs in Section 17.4.1.

[40] Capacitance coefficients, partial pressures and concentrations of the respiratory gases in water are discussed in Section 11.1.2.

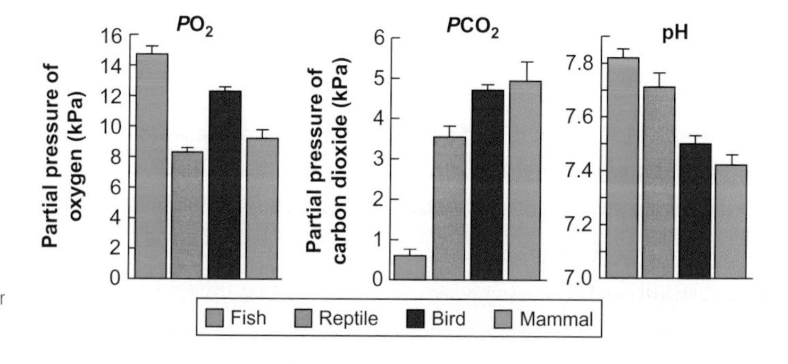

Figure 22.13 Partial pressures of oxygen (PO_2) and carbon dioxide (PCO_2) and pH in the arterial blood of a representative fish, reptile, bird and mammal

Note the progressive increase in partial pressure of carbon dioxide and decrease in pH in arterial blood throughout this phylogenetic progression. Mean values (+ SEM)

Source: modified from Milsom WK (1998). Phylogeny of respiratory chemoreceptor function in vertebrates. Zoology, 101, 316-332.

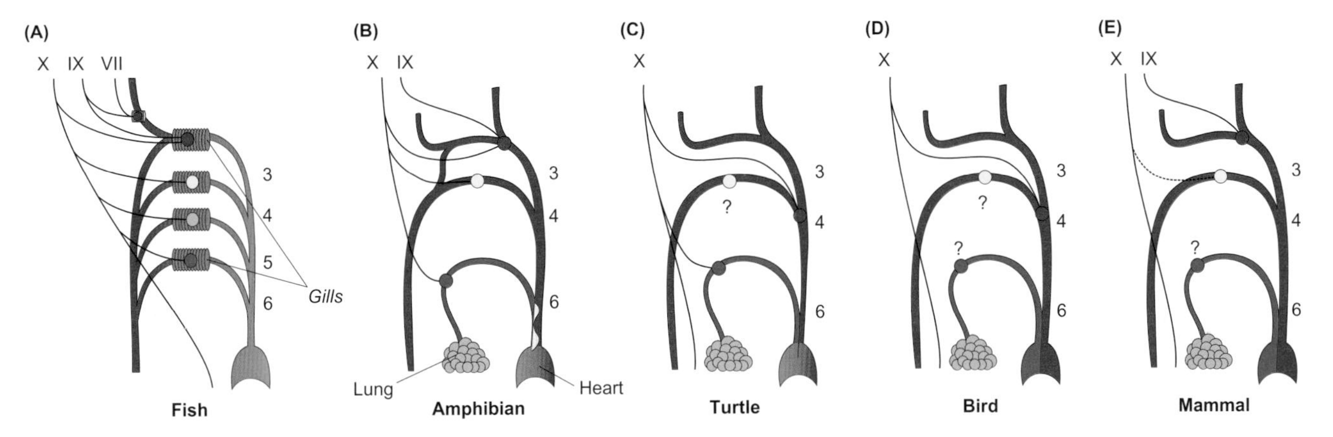

Figure 22.14 The distribution of the oxygen-sensitive chemoreceptors among the major anterior blood vessels in different groups of vertebrates

The cranial nerves are labelled VII (facial), IX (glossopharyngeal) and X (vagus). The remaining embryonic aortic arches are labelled 2–6. They are usually numbered with Roman numerals, II-VI, but the current labelling is used to avoid confusion with labelling of cranial nerves (look at Figure 14.43 for further details of aortic arches). The location of the chemoreceptors is colour-coded. Red represents those in aortic arch 3, which are the carotid labyrinth in amphibians and the carotid body in reptiles, birds and mammals. Yellow denotes those in aortic arch 4, which are the aortic bodies in amphibians, reptiles, birds and mammals. Green signifies those associated with aortic arch 5, which is only present in fish. Blue signifies receptors in aortic arch 6, which is the pulmocutaneous arch of amphibians and the pulmonary arch of reptiles, birds and mammals.

Source: reproduced from Milsom WK and Burleson ML (2007). Peripheral arterial chemoreceptors and the evolution of the carotid body. Respiratory Physiology & Neurobiology 157: 4-11.

As a result, air-breathing animals are more sensitive to CO_2/pH as the primary stimulus to ventilation compared with water-breathers. Differences in PO_2 in arterial blood between the different groups are less extreme.

Also related to the different capacitance coefficients of O_2 and CO_2 in water is the fact that a low PO_2 (hypoxia) in the surrounding water may be relatively common, whereas high PCO_2 (hypercapnia) is comparatively rare. Not surprisingly, therefore, oxygen rather than carbon dioxide exerts the dominant influence on ventilation in water-breathing animals.

Oxygen receptors that influence ventilation in vertebrates

In all vertebrates that have been studied, a reduction in environmental oxygen causes a reflex increase in ventilation (hyperventilation) as a result of increased respiratory tidal volume and/or respiratory frequency.

The location of the receptors responsible for this response can be determined by two methodologies:

(i) recording the effect of environmental hypoxia on ventilation before and after sectioning the nerves to a specific area or areas (denervation);

(ii) recording activity in the sensory nerve from a specific area in response to a hypoxic stimulus.

In fish, peripherally located oxygen-sensitive chemoreceptors responsible for changes in ventilation are found in virtually every part of the respiratory system, as shown in Figure 22.14A. Peripheral chemoreceptors occur in the walls of the mouth (or orobranchial cavity in elasmobranchs) and are innervated by cranial nerves V. Oxygen receptors also occur on gills in all species of fish, and are innervated by cranial nerves IX and X, as shown in Figure 22.14A. In teleost fish, the peripheral oxygen receptors seem to monitor PO_2 in the surrounding water and in the blood. In most fish that can breathe water or air (bimodal breathers), hypoxia in water stimulates air breathing.

In amphibians, reptiles, birds and mammals, the oxygen-sensitive chemoreceptors are associated with one or more of the remaining aortic arches[41], as shown in Figures 22.14B–E. These receptors monitor the level of oxygen in the blood. The most prominent location of these receptors in amphibians is within the carotid labyrinths at the base of the internal carotid arteries (3 in Figure 22.14B). Note in this diagram that the carotid labyrinths are innervated by branches from the IXth, as well as the Xth cranial nerve, whereas the systemic and pulmocutaneous arches are innervated by branches of the Xth cranial nerve only. Denervation of the carotid labyrinths in amphibians reduces ventilation in the resting animals but a similar proportional increase in ventilation occurs in response to hypoxia in intact and denervated animals. This suggests that the carotid labyrinths may contribute to, but are not completely responsible for, increasing ventilation during hypoxia.

Figure 22.14C shows that the situation is more complex in reptiles with Figures 22.14D and E, indicating that the location of peripheral chemoreceptors receptors differs between birds and mammals. In mammals, the major peripheral chemoreceptors involved in the control of ventilation are the **carotid bodies**, which are located where the common carotid arteries divide, high in the neck; these structures are innervated by a branch of the IXth cranial

[41] The aortic arches in vertebrates are discussed in, Section 14.4 and Figures 14.31, 14.32, 14.39 and 14.46.

nerve. More diffusely located aortic bodies are situated at the base of the aorta are innervated by the Xth cranial nerve in mammals. In contrast, in birds, the carotid bodies are low in the neck, at the base of the common carotid arteries, and are innervated by branches of the Xth cranial nerve. In both birds and mammals, denervation of the carotid bodies abolishes the ventilatory response to hypoxia.

Figure 22.15 shows that the response of the peripheral, oxygen-sensitive chemoreceptors to a hypoxic stimulus is qualitatively similar in all vertebrates; there is an increase in nervous activity as the partial pressure of oxygen falls. However, Figure 22.15 also shows that in mammals, the sensitivity of the aortic bodies to hypoxia is lower than that of the carotid bodies. This difference in sensitivity of the two groups of receptors seems to relate to the fact that the carotid bodies of mammals are responsible for the major part of the increase in ventilation in response to hypoxia. In contrast, and as we discuss later, the aortic bodies are thought to be more involved in control of the cardiovascular system.

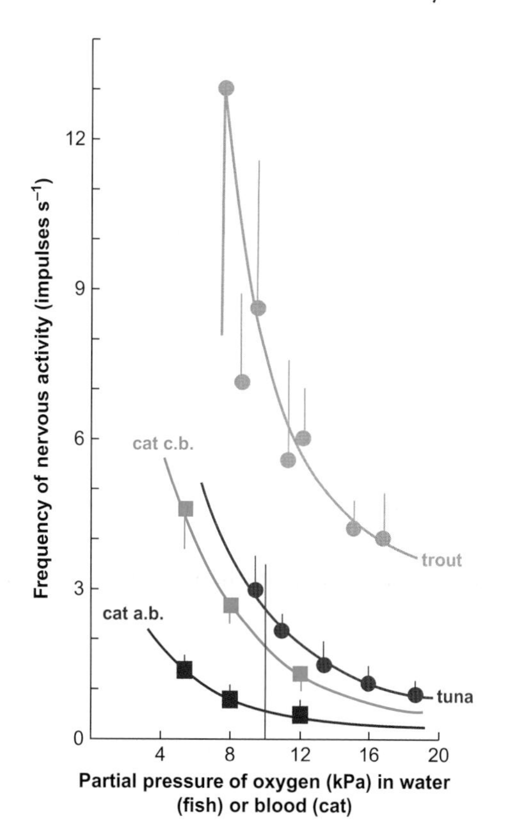

Figure 22.15 Responses of peripheral arterial oxygen receptors in vertebrates to reduced partial pressure of oxygen

Average (± standard error of the mean (SEM)) nervous activity in the sensory nerves from the peripheral oxygen-sensitive receptors of fish and mammals in response to decreasing partial pressure of oxygen (hypoxia). Note that the sensitivity of the aortic body in cat to a decrease in partial pressure of oxygen in arterial blood (P_aO_2) is less than that of the carotid body, which equates to a lower rate of nervous activity for a given P_aO_2, as indicated by the vertical red line. Cat c.b. = cat carotid body, cat a.b. = cat aortic body. n = 8 for trout, 11 for cat and 11 for tuna.

Source: modified from Burleson ML et al (1992). Afferent inputs associated with cardioventilatory control in fish. In: Fish Physiology, Vol XII, Part B. Hoar WS et al (eds.) pp 389-426. San Diego: Academic Press.

Oxygen receptors that influence ventilation in invertebrates

Oxygen-sensing cells are present in a number of groups of invertebrates, but their location varies between different groups. In great pond snails (*Lymnaea stagnalis*), a sense organ known as the osphradium, which is shown in Figure 22.16A, is located close to the pneumostome. Figure 22.16B shows intracellular recordings from nerve cells within the osphradium, which indicate that they respond to hypoxia with bursts of activity that cease when the organ becomes normoxic. Activity in these nerve cells triggers activity in one of the three cells in the respiratory network (RPeD1), which we illustrated in Figure 22.3B. However, the oxygen-sensitive cells in the osphradium are not essential for normal respiratory activity. Denervation of the osphradium reduces the response of whole animals to hypoxia but does not eliminate it; they just surface less frequently.

Among the arthropods, Danube crayfish (*Astacus leptodactylus*) have oxygen-sensitive receptors in their branchiocardiac veins.[42] These receptors are innervated by branches of the branchial nerves which themselves arise from the thoracic ganglia, as illustrated in Figure 22.16C. The branchiocardiac veins carry oxygen-rich haemolymph from the gills to the heart. A decrease in the partial pressure of oxygen in this haemolymph causes an increase in activity in the branchial nerves, which presumably stimulates ventilation.

Insects have both oxygen and carbon dioxide receptors located within their central nervous system, probably in the metathoracic ganglia.

Carbon dioxide receptors that influence ventilation in vertebrates

Although increased levels of carbon dioxide (**hypercapnia** or hypercarbia) are not that common in water, they do occur, especially in bodies of tropical fresh water which are covered at the surface. For example, a covering of plant material such as hyacinth mats can lead to PCO_2 in the water beneath increasing to up to 8 kPa. Even in seawater, stratification and the lack of photosynthesis can cause the partial pressure of carbon dioxide to be over 1 kPa at depths of 100–500 m, or in tidal pools. If carbon dioxide is unable to diffuse out of the water, oxygen will be unable to diffuse in, so environmental hypercapnia is invariably accompanied by environmental hypoxia—a phenomenon also experienced by terrestrial animals (for example, those that live in burrows).

[42] The circulatory system of crustaceans is discussed in Section 14.3.3.

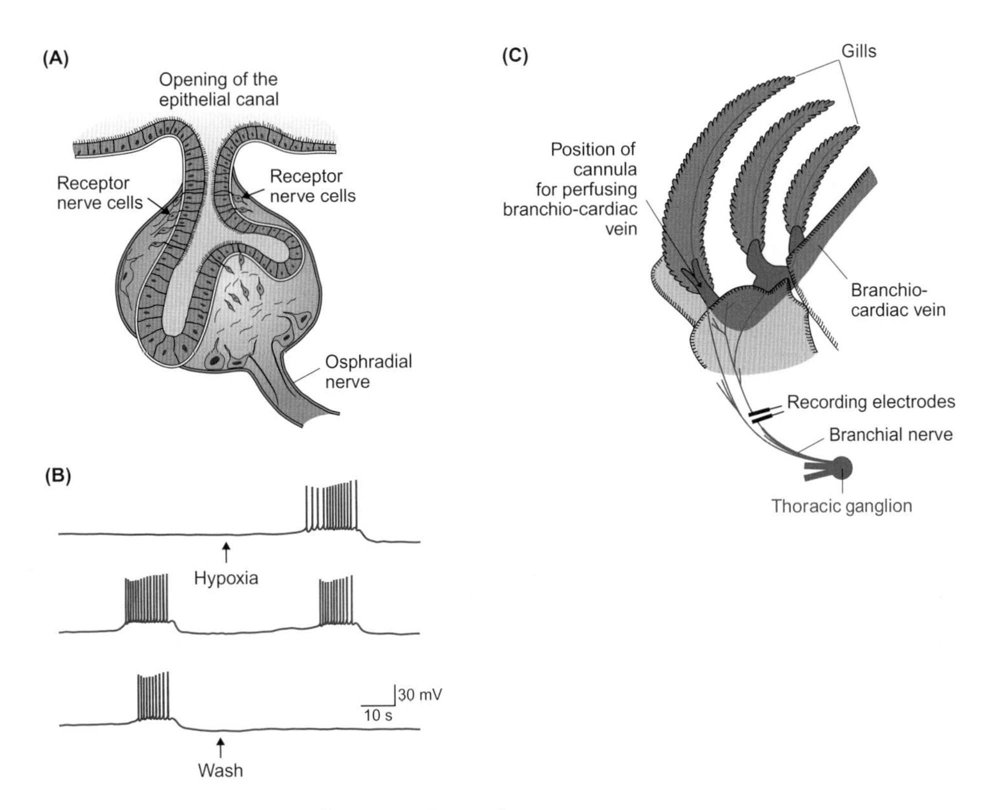

Figure 22.16 Location of oxygen-sensitive cells in some invertebrates

(A) Diagram of the epithelial tube within the osphradium of a great pond snail (*Lymnaea stagnalis*) showing the location of the sensory (receptor) neurons. (B) Intracellular recordings from a sensory neuron in the osphradium of a pond snail in response to exposure to hypoxic water followed by a return to normoxia (wash). The traces are continuous. (C) Location of sensory nerves in Danube crayfish (*Astacus leptodactylus*) which increase their activity when stimulated by perfusion of the brachio-cardiac vein with hypoxic saline.

Sources: (A) and (C): modified from Wedemeyer H and Schild D (1995). Chemosensitivity of the osphradium of the great pond snail *Lymnaea stagnalis*. Journal of Experimental Biology 198: 1743-1754; Ishii K et al (1989). Oxygen-sensitive chemoreceptors in the brachio-cardiac veins of the crayfish, *Astacus leptodactylus*. Respiration Physiology 78: 73-81. (B): reproduced from Bell HJ et al (2008). A peripheral oxygen sensor provides direct activation of an identified respiratory CPG neuron in *Lymnaea*. In: Integration in Respiratory Control: From Genes to Systems. Poulin MJ and Wilson RJA (eds.) pp 25-29. Springer, New York. Image (©) Alexander Mrkvicka

Peripheral carbon dioxide receptors

As well as possessing receptors sensitive to oxygen, vertebrates also have peripheral chemoreceptors that are sensitive to carbon dioxide/pH and which stimulate ventilation. In most fish that are bimodal breathers, however, exposure to hypercapnia in the water stimulates air breathing.

Experiments on rainbow trout suggest that carbon dioxide/pH receptors that stimulate ventilation are located on the first gill arch in teleost fish and respond exclusively to changes in carbon dioxide/pH in the surrounding water. However, it has yet to be determined whether or not the carbon dioxide/pH-sensitive receptors on the gills of fish are the same as those that are sensitive to changes in oxygen. That said, the peripheral carbon dioxide/pH sensitive receptors in birds and mammals *are* the same as those that are sensitive to oxygen, which are illustrated in Figure 22.14.

Figure 22.17A shows that, in mammals, the aortic bodies are less sensitive than the carotid bodies to an increase in carbon dioxide. This is similar to their respective responses to a decrease in PO_2(take a look at Figure 22.15 to remind yourself of this). Figure 22.17B shows that, at lower partial pressures of oxygen in arterial blood, the response to increasing partial pressure of carbon dioxide is proportionately greater. In other words, there is a positive interaction (**synergism**) between the two stimuli of hypoxia and hypercapnia.

Central carbon dioxide receptors

Denervation of the peripheral chemoreceptors in birds and mammals abolishes the ventilatory response to hypoxia, but only reduces the response to hypercapnia. These data tell us that there are carbon dioxide-sensitive receptors located elsewhere. There is now evidence that some species of air-breathing fish, such as long-nosed gar (*Lepisosteus osseus*), and all other air-breathing vertebrates have **centrally located carbon dioxide/pH receptors**, within the brainstem, which detect changes in PCO_2/pH in the cerebrospinal fluid. The presence and location of the central carbon dioxide/pH receptors have been determined in most groups of air-breathing vertebrates.

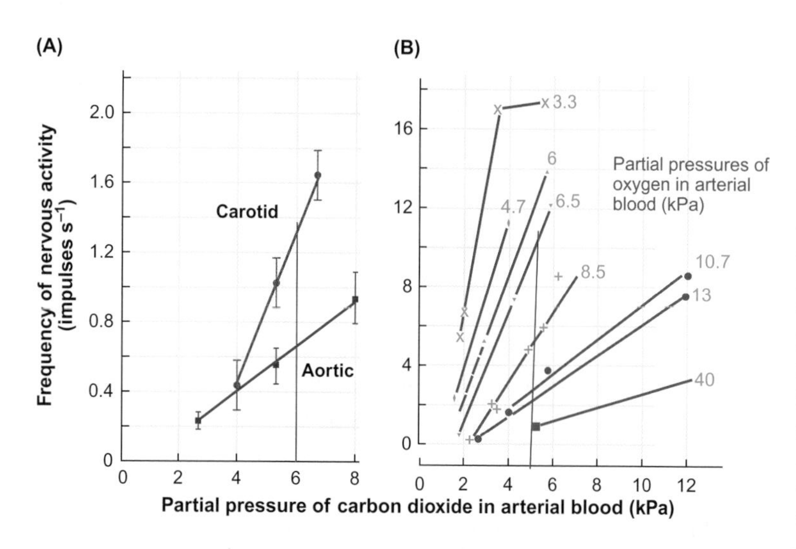

Figure 22.17 Characteristics of peripheral carbon dioxide receptors in mammals

(A) Average (± SEM) activity in the sensory nerves from the carotid and aortic bodies of anaesthetized cats in response to changes in the partial pressure of arterial carbon dioxide with a high (>50 kPa) partial pressure of oxygen. Note the greater sensitivity of the carotid body receptors, which equates to a higher rate of nervous activity for a similar partial pressure of CO_2 in arterial blood, as indicated by the vertical red line. (B) Activity in a single sensory nerve fibre from a carotid body of an anaesthetized cat in response to different partial pressures of arterial CO_2 and O_2 in arterial blood. Note the greater sensitivity to an increase in partial pressure of CO_2 at lower partial pressures of O_2, as indicated by the vertical red line.

Sources: (A): reproduced from Lahiri S et al (1981). Comparison of aortic and carotid chemoreceptor responses to hypercapnia and hypoxia. Journal of Applied Physiology 51: 55-61. (B) Lahiri S and DeLaney RG (1975). Stimulus interactions in the responses of carotid body chemoreceptor single afferent fibres. Respiration Physiology 24: 249-266.

In mammals, carbon dioxide/pH-sensitive neurons have been identified at various locations in the brain, including the nucleus of the solitary tract (NTS)[43], the medial portions of the reticular formation known as the raphe nuclei, and the retrotrapezoid nucleus (RTN) which is close to the ventral surface of the brainstem[44]. Figure 22.18 shows the location of the raphe nuclei and the RTN.

Cells in the raphe and RTN are not only sensitive to carbon dioxide/pH but also project to important respiratory neurons, including those in the ventral respiratory group and in the pre-Bötzinger complex, and to motor neurons which supply the respiratory muscles. Although they may both serve a similar purpose, the cells that are sensitive to carbon dioxide/pH in the raphe secrete serotonin as a neurotransmitter, while those in the RTN secrete glutamate[45].

The function of some of these sites varies according to whether the animal is awake or asleep. For example, it has been proposed that the cells in the RTN have their greatest effect during wakefulness, and those in the raphe

[43] The location of the nucleus of the solitary tract is shown in Figure 22.32.

[44] We discuss the RTN as part of the respiratory rhythm generator in Section 22.1.2 and Figure 22.8A,B.

[45] Neurotransmitters are discussed in Section 16.3.4.

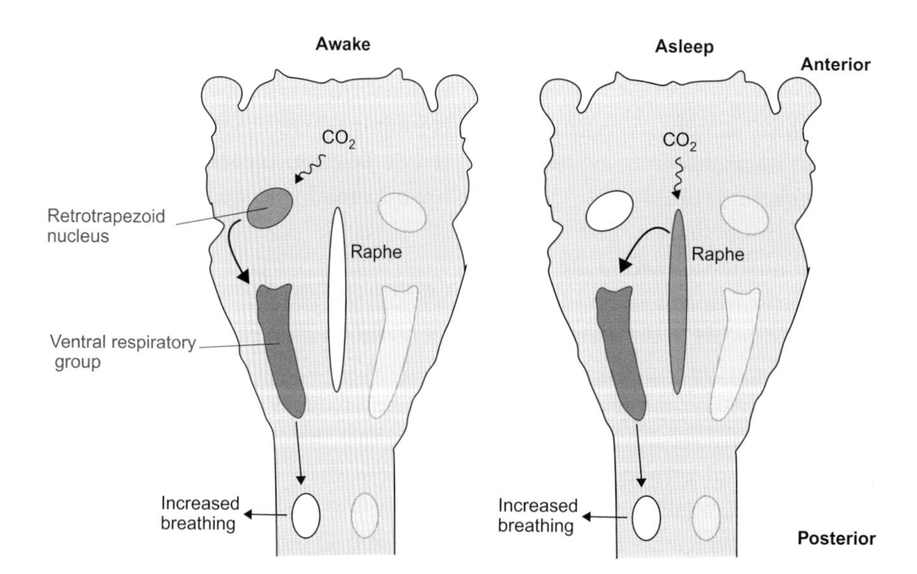

Figure 22.18 Location and possible functioning of major, central carbon dioxide/pH chemoreceptors in rats

When the animal is awake, nerve cells in the retrotrapezoid nucleus (RTN), when stimulated by increased carbon dioxide/decreased pH, may activate nerve cells in the ventral respiratory group. These may, in turn, activate respiratory motor nerve cells in the spinal column. When the animal is asleep, the contribution from the RTN may be reduced and the carbon dioxide/pH sensitive nerve cells in the raphe nuclei may take on a more prominent role.

Source: reproduced from Mitchell GS (2004). Back to the future: carbon dioxide chemoreceptors in the mammalian brain. Nature Neuroscience, 7: 1288-1290.

nucleus have their greatest effect during sleep. These proposals are illustrated in Figure 22.18. In addition, the carbon dioxide sensitive nerve cells of the RTN are activated by input from the carotid body chemoreceptors when stimulated by hypoxic blood. The input from the carotid body chemoreceptors is a direct pathway from the nucleus of the solitary tract[46].

Structure and sensory processes in peripheral chemoreceptor cells of vertebrates

Most of the studies on peripheral chemoreceptors have been performed on mammals, so we concentrate on this group. The carotid bodies are among the smallest organs in the body of mammals. In humans they weigh about 10 mg and are about 4 mm in length, while in rats they weigh only 40 μg and are 0.6 mm in length. The Belgian physiologist, Corneille Heymans, identified the carotid and aortic bodies as sites for sensing the levels of oxygen and carbon dioxide in the blood in the early 1930s. For this, he was awarded the Nobel prize in physiology or medicine in 1938. Basically, a carotid body consists of two types of cell, the **glomus** or **type I cells**, and the sustentacular or type II cells, as shown in Figure 22.19. In addition, there are both sensory (afferent) and motor (efferent) nerves and blood vessels. The glomus cells are thought to detect changes in oxygen and carbon dioxide/pH.

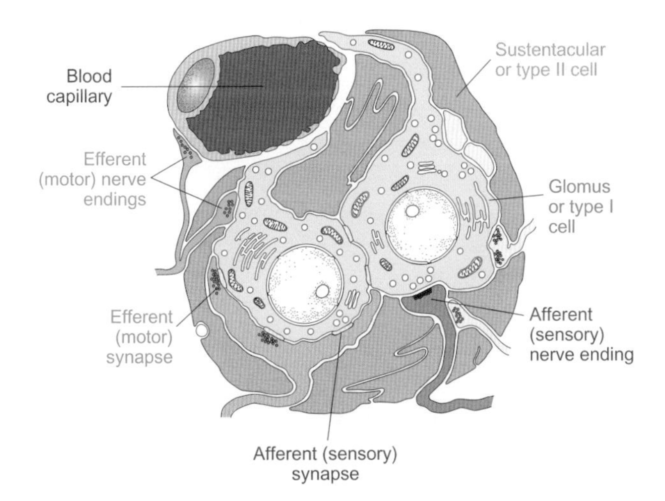

Figure 22.19 Structure of peripheral arterial chemoreceptors in mammals

Arrangement of the presumed chemoreceptive unit consisting of glomus (type I) cells, sustentacular (type II) cells and both afferent and efferent nerve endings. The chalice-like nerve endings have both efferent and afferent synapses. Efferent nerves from the sympathetic nervous system are shown innervating both a glomus cell and a capillary. Those to the glomus cells can modulate the output from the carotid body.

Source: modified from Jones DR and Milsom WK (1982). Peripheral receptors affecting breathing and cardiovascular function in non-mammalian vertebrates. Journal of Experimental Biology 100: 59-91.

Neurotransmitters are located in vesicles close to the cell membrane of the glomus cell. When released from the vesicles, the neurotransmitters stimulate the sensory nerve ending. Conversely, the vesicles located in the nerve endings of the motor (efferent) nerves transmit information from the brain to the glomus cells.

The carotid bodies are very well supplied with blood vessels, with 25–35 per cent of their tissue in cats being occupied by capillaries and venules. This proportion of blood vessels is 5–6 times greater than that of the brain. As a result, a carotid body has the highest blood flow rate, relative to its size, of any organ in the body at 15–20 mL g^{-1} min^{-1}; in humans this flow rate is almost 15 times greater than that in the brain. This rate exceeds that required for its metabolic requirements and ensures a sufficiently high supply of oxygen to the tissue, even during hypoxia.

The aortic bodies do not have a rate of blood flow as high as that of the carotid bodies. As a consequence, they are more sensitive to the rate of delivery of oxygen than the carotid bodies. In other words, the aortic bodies are more influenced by the rate of perfusion and the amount of oxygen in the blood, as indicated in Figure 22.20. So, when stimulated, the aortic bodies cause reflex changes in the circulatory system.

We do not know exactly how carotid bodies sense the change in the level of oxygen which causes a release of neurotransmitter(s) to stimulate the sensory nerve endings[47]. However, there are a number of possible processes and mechanisms. The ability to sense and respond to changes in the level of oxygen is a fundamental property of the cells of all metazoan animals. An important element in this process is a transcription factor known as **hypoxia-inducible factor** or **HIF-1**, which controls the expression of over 70 genes. Among the important physiological processes regulated by HIF-1 target genes are the production of red blood cells (erythropoiesis)[48], the growth of new blood vessels (angiogenesis)[49] and glycolysis, which are examples of responses to general, local tissue and intracellular hypoxia.

HIF-1 consists of two subunits, HIF-1α and HIF-1β. HIF-1β is expressed at a constant rate and does not respond to changes in the partial pressure of oxygen. In contrast, HIF-1α is rapidly degraded in cells at normal partial pressures of oxygen (normoxia) but accumulates in hypoxic

[46] We discuss the role of the nucleus of the tractus solitarius in the integration of the respiratory and cardiovascular systems in Section 22.5.3.

[47] The basic processes of chemoreception are discussed in Section 17.3.3.

[48] Erythropoiesis is discussed in Box 13.1.

[49] Angiogenesis is discussed in, Section 14.1.5 and Experimental Panel 14.1.

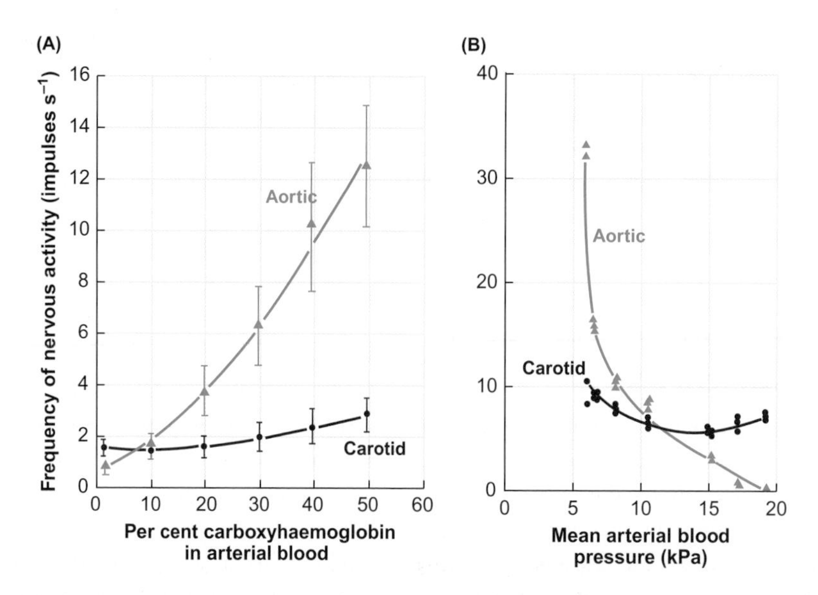

Figure 22.20 Responses of peripheral chemoreceptors in mammals to changes in rate of oxygen supply

(A) Responses of the carotid and aortic bodies of anaesthetized cats to a reduction in the oxygen carrying capacity of the blood (mean values ± SEM) by poisoning it with carbon monoxide, which irreversibly combines with haemoglobin to form carboxyhaemoglobin (discussed in Section 13.1.2), thus decreasing the oxygen carrying capacity of the blood. (B) Effect of reducing rate of perfusion of the organs with blood by reducing arterial blood pressure. In both (A) and (B), the partial pressure of oxygen in the blood was normal, at approximately 11.5 kPa. As either the carrying capacity of the blood or blood pressure (rate of perfusion) decreased, there was little effect on the output of the sensory nerve from the carotid body whereas there was a large increase in sensory activity from the aortic bodies. These results suggest that the aortic chemoreceptors monitor the rate of delivery of oxygen by the circulatory system.

Sources: (A): reproduced from Lahiri S et al (1981). Relative responses of aortic and carotid body chemoreceptors to carboxyhemoglobinemia. Journal of Applied Physiology 50: 580-586. (B): Lahiri S et al (1980). Relative responses of aortic and carotid chemoreceptors to hypotension. Journal of Applied Physiology 48: 781-788.

cells. This accumulation leads to the induction of HIF-1 target genes.

The role of HIF-1α in the functioning of the carotid body has been demonstrated using gene knock-out technology. While a complete deficiency of HIF-1α results in developmental defects, partial deficiency leads to responses to physiological stimuli being impaired. One example of this is the loss of oxygen sensing by the carotid body of mice in which a single allele at the locus encoding for HIF-1α has been deleted, leading to reduced production of HIF-1α. Although the carotid bodies in these mice have no obvious anatomical or histological abnormalities, there is no increase in activity in their sensory nerves in response to a hypoxic stimulus, and the increase in ventilation that does occur is mediated by the aortic bodies.

Another important feature of the oxygen sensing/neurotransmitter release process is that hypoxia causes rapid inhibition of potassium ion (K^+) channels in the glomus cells. This inhibition leads to the depolarization of the cells and the subsequent entry of calcium ions (Ca^{2+}) through voltage-dependent channels. The increase in intracellular concentration of Ca^{2+} then leads to the release of neurotransmitter, as shown in Figure 22.21. Increased levels of cyclic AMP may increase the sensitivity of the process by further inhibiting the potassium channels and stimulating the release of transmitter. Although the nature of the actual sensor is unknown, those that have been suggested include:

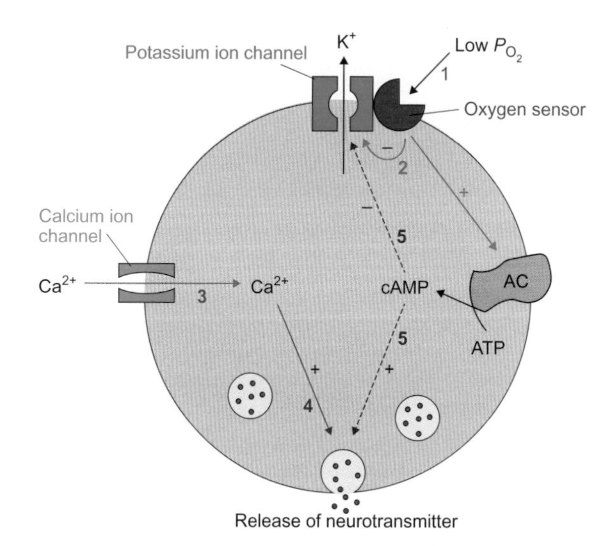

Figure 22.21 A model for sensory transduction in the chemoreceptor (glomus) cells of the carotid bodies

As the partial pressure of oxygen, PO_2, falls (**1**) a signal originates at the oxygen sensor which inhibits the potassium channels and activates adenylate cyclase, AC (**2**). The decrease in membrane potential that follows the inhibition of the potassium channels activates voltage-dependent calcium channels leading to the influx of calcium ions and an increase in their intracellular concentration (**3**). The increase in intracellular concentration of calcium ions then results in the release of neurotransmitter(s) by the process of exocytosis (**4**). Increased levels of cyclic AMP, cAMP, may then enhance the sensitivity of the processes to low PO_2 by further inhibiting the potassium channels and stimulating the release of neurotransmitter (**5**). + indicates excitatory influence, – indicates inhibitory influence

Source: reproduced from González C et al (1992). Oxygen and acid chemoreception in the carotid body chemoreceptors. Trends in Neurosciences, 15: 146-153.

- a haem-linked nicotinamide adenine dinucleotide phosphate-oxidase (NADPH-oxidase) located in the membrane of the glomus cells;
- the mitochondria;
- adenosine monophosphate (AMP)-activated protein kinase.

Both the mitochondria and AMP-activated protein kinase provide a link between oxygen sensing and metabolism, while NADPH oxidase raises the possibility that the haem-protein may have similar oxygen binding properties to those of haemoglobin. If this is the case, the oxygen binding properties of the receptors could be modulated in the same way as those of haemoglobin in response to changes in temperature, organophosphates, carbon dioxide/pH, etc[50]. This would maintain a tight coupling between the activation of the receptors, oxygen saturation of the blood and ventilation, despite any shift in the relationship between PO_2 and oxygen content in arterial blood. Even though the receptor cells respond to changes in local PO_2, an increase in sensory activity may only occur at the partial pressure at which the haemoglobin begins to desaturate.

[50] The effects of various modulators on the affinity of haemoglobin for oxygen are discussed in Section 13.1.2.

The glomus cells synthesize and store a variety of neurotransmitters, including acetylcholine, ATP, substance P, dopamine and serotonin (also known as 5-hydroxytryptamine, 5-HT). As such, the generation of the sensory discharge is extremely complex and, as yet, unresolved. In addition, sensory discharge can be altered by input from the central nervous system via the efferent (motor) fibres to the glomus cells.

It has been proposed that the glomus cells of the carotid body of mammals respond to a decrease in pH (acidosis) by inhibiting one or more types of K^+ channels followed by depolarization of the cells. This, in turn, causes an increase in the concentration of intracellular calcium ions via voltage-gated channels and the release of neurotransmitter. An alternative proposal is that acidosis directly activates acid-sensitive ion channels in the glomus cells.

We know much less about the oxygen-sensitive tissue in other vertebrates, although a number of studies have been performed on proposed oxygen-sensing cells in the gills of teleost fish. These cells are thought to be the neuroepithelial cells (NECs), which are characterized by the presence

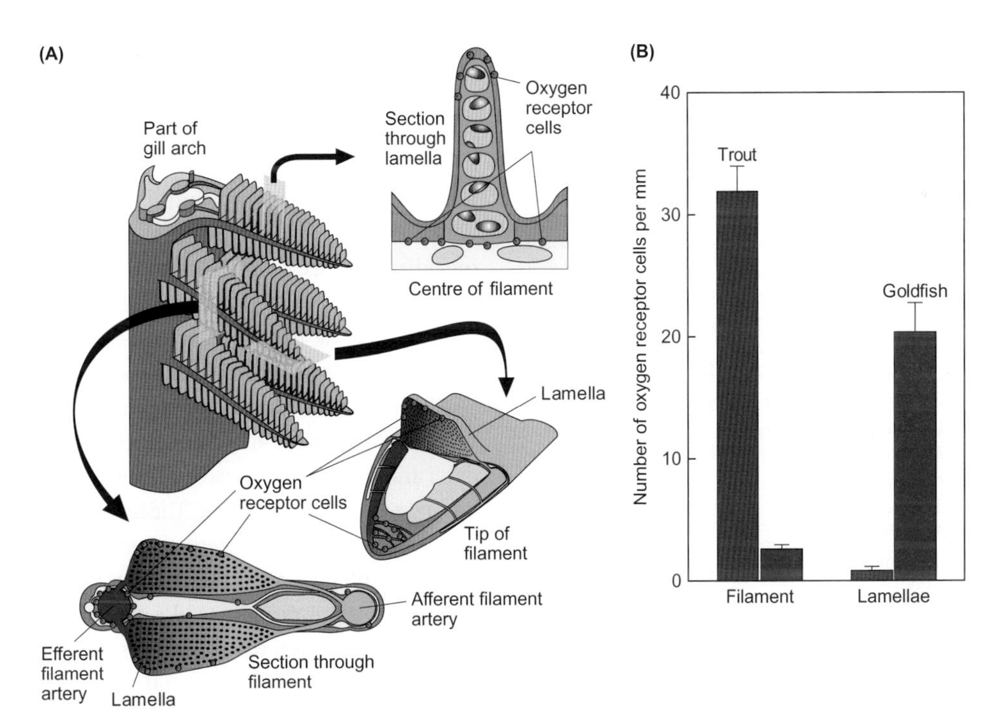

Figure 22.22 Location and density of proposed oxygen receptor cells in the gills of teleost fish

(A) Diagram showing all locations of neuroepithelial cells, as characterized by the presence of 5-hydroxytryptamine (serotonin). These cells are shown as green dots and are thought to be oxygen receptor cells. Details of the anatomy of the gills can also be seen in Fig 13.15. (B) Mean densities (+ SEM) of oxygen receptor cells on the gill filaments and lamellae of rainbow trout (*Onchorhynchus mykiss*) in blue, and goldfish (*Carassius auratus*) in red. The relative tolerance of these species to environmental hypoxia is indicated by the oxygen affinity of the haemoglobin in their blood (P_{50}). Trout are relatively intolerant of hypoxia and have low-affinity haemoglobin (P_{50} = 3 kPa), whereas goldfish are relatively tolerant of hypoxia and have high affinity haemoglobin (P_{50} = 0.3 kPa).

Source: Coolidge EH et al (2008). A comparative analysis of putative oxygen-sensing cells in the fish gill. Journal of Experimental Biology 211: 1231-1242.

of 5-HT in dense-cored vesicles. Neuroepithelial cells and their associated nerves have been located at the tips of the filaments of all species of fish that have been studied. Figure 22.22A shows the location of NECs. Those located at the tips of the filaments and in the lamellae could sense changes in partial pressure of oxygen (PO_2) in the water, whereas those within the filament surround the efferent filament artery where they could monitor the PO_2 in arterial blood.

The distribution of NECs seems to be related to the ability of the particular species to tolerate environmental hypoxia, as illustrated in Figure 22.22B. Rainbow trout (*Onchorhynchus mykiss*), which are less tolerant to hypoxia than many species, have the most NECs within their gill filaments (where they sense PO_2 of blood), with almost none in their lamellae. In contrast, goldfish (*Carassius auratus*), which are more tolerant to hypoxia, have most NECs in their lamellae (where they sense PO_2 of water), and almost none in their gill filaments. Like the glomus cells in the carotid body of mammals, neuroepithelial cells from the gills of fish also respond to hypoxia by inhibiting K^+ channels in their membranes leading to their subsequent depolarization.

› *Review articles*

Gonzalez C, Almaraz L, Obeso A, Rigual R (1994). Carotid body chemoreceptors: from natural stimuli to sensory discharges. Physiological Reviews 74: 829–898.

Kemp PJ (2006). Detecting acute changes in oxygen: will the real sensor please stand up? Experimental Physiology 91: 829–834.

Kumar P, Prabhakar NR (2012). Peripheral chemoreceptors: function and plasticity of the carotid body. Comprehensive Physiology 2: 141–219.

Lahiri S, Forster RE (2003). CO_2/H^+ sensing: peripheral and central chemoreception. International Journal of Biochemistry & Cell Biology 35: 1413–1435.

Milsom WK (1990). Mechanoreceptor modulation of endogenous respiratory rhythms in vertebrates. American Journal of Physiology 259: R898–R910.

Milsom WK (2012). New insights into gill chemoreceptors: Receptor distribution and roles in water and air breathing fish. Respiration Physiology & Neurobiology 184: 326–339.

Milsom WK, Burleson ML (2007). Peripheral arterial chemoreceptors and the evolution of the carotid body. Respiration Physiology & Neurobiology 157: 4–11.

Milsom WK, Abe AS, Andrade DV, Tattersall GJ (2004). Evolutionary trends in airway CO_2/H^+ chemoreception. Respiration Physiology & Neurobiology 144: 191–202.

Nattie E, Aihua L (2012). Central chemoreceptors: locations and functions. Comprehensive Physiology 2: 221–254.

Semenza GL (2004). Hydroxylation of HIF-1: oxygen sensing at the molecular level. Physiology 19: 176–182.

Widdicombe JG (1998). Afferent receptors in the airways and cough. Respiration Physiology 114: 5–15.

22.3 Generation of the cardiac rhythm

While the respiratory muscles rely on input from motor nerve fibres from the central nervous system (CNS) to cause them to contract, the hearts of all animals are able to continue contracting in the absence of any connection to the CNS. In most groups of animals, the cardiac rhythm is generated by cardiac muscle cells themselves. In vertebrates, some specialized cells spontaneously depolarize and act as localized pacemakers which then excite the rest of the heart. These hearts have **myogenic pacemakers** (myogenic: generated in muscles). In arthropods, except insects, the rhythm of the heart is set by a group of neurons, the cardiac ganglion, which is usually located on the dorsal wall of the heart. Hence, these hearts have neurogenic pacemakers (neurogenic: generated in nerves).

22.3.1 Myogenic pacemakers

Let us now investigate the basic principles of the functioning of myogenic cardiac pacemakers and associated events in vertebrates by studying those in the hearts of mammals, with occasional reference to those in other groups. A number of features distinguish heart muscle from skeletal muscle in vertebrates[51]. Of particular significance is the fact that isolated muscle cells of the heart, **cardiac myocytes**, will contract spontaneously in culture.

Figure 22.23 shows the basic structure of muscle fibres in the hearts of mammals. Fibres in the atria and ventricles are connected together by gap junctions, which act as electrical synapses to allow the rapid transmission of excitation to occur between the cells. As a result of this rapid transmission, all cells contract almost simultaneously so that the atria and ventricles can function as two distinct units. The gap junctions are particularly abundant at the intercalated discs, which consist of an undulating double membrane separating adjacent cells. There are also regions of tight mechanical contact in the intercalated discs called desmosomes. Desmosomes allow the force of contraction generated by the individual cells to be transmitted to neighbouring cells.

Although all cardiac myocytes have the ability to contract spontaneously, the rate of contraction of the intact heart is determined by those muscle cells with the fastest spontaneous rate of depolarization. These are the pacemaker cells: modified muscle cells with poorly developed contractile systems. The pacemaker cells do not have a stable resting potential[52], exhibiting instead a slow depolarization known as a **pacemaker potential**, which is shown in Figure 22.24A. When the pacemaker potential reaches the threshold, an action potential is generated, after which the membrane repolarizes and then slowly depolarizes until another action potential is generated.

[51] All types of muscles are discussed in Sections 18.1, 18.2, 18.3 and 18.4.

[52] Pacemaker cells are discussed in Section 16.2.10.

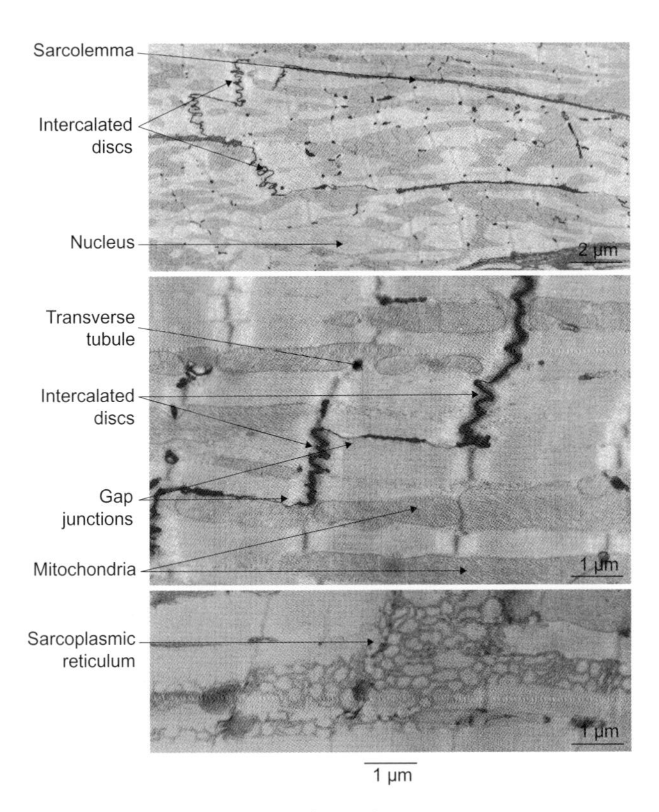

Figure 22.23 Heart muscle of vertebrates

Low-power electron micrographs of heart muscle from a mouse illustrating the major characteristics of the muscle cells (myocytes) of the hearts of mammals.

Source: reproduced from Berne RM and Levy MN (1998). Physiology, 4th edn. St Louis: Mosby.

Pacemaker cells are found in three locations:

- the sinus venous in fish, amphibians and reptiles (not shown in Figure 22.24B)[53], or its remnant in birds and mammals;
- the **sinoatrial (S-A) node**, where the superior vena cava joins the right atrium;
- the **atrioventricular (A-V) node** in the wall of the right atrium.

These locations are shown in Figure 22.24B. The S-A node is the dominant pacemaker and sets the pace of the heartbeat. From the S-A node, the excitation spreads throughout both atria along ordinary muscle fibres at a velocity of about 1 m s^{-1} and eventually reaches the atrioventricular node, as shown in Figure 22.24B.

The A-V node is connected to the **bundle of His** which is the only pathway for the transmission of excitation from the atria to the ventricles. The bundle of His divides into two branches that run either side of the interventricular septum. These branches eventually subdivide into a network of conducting fibres called Purkinje fibres, which are shown in Figure 22.24B. These fibres are much thicker (70–80 μm in diameter) than the other myocytes in the ventricles (10–15 μm in diameter) and their conduction velocity of 1–4 m s^{-1} is much greater than that from one ventricular myocyte to another. This enables the wave of excitation to reach the whole muscle mass of the two ventricles very rapidly, thus contributing to their uniform contraction. The impulses transmitted by the Purkinje fibres to the ventricular myocytes depolarize the cells and the muscles fibres contract[54].

There is a delay in the passage of the impulse from the atria to the ventricles as a result of the slow conduction velocity through the A-V node, which is as low as 0.05 m s^{-1}

[53] See Figure 14.33 for sinus venosus in the heart of fish.

[54] Contraction of cardiac muscle fibres is discussed in Section 18.3.1.

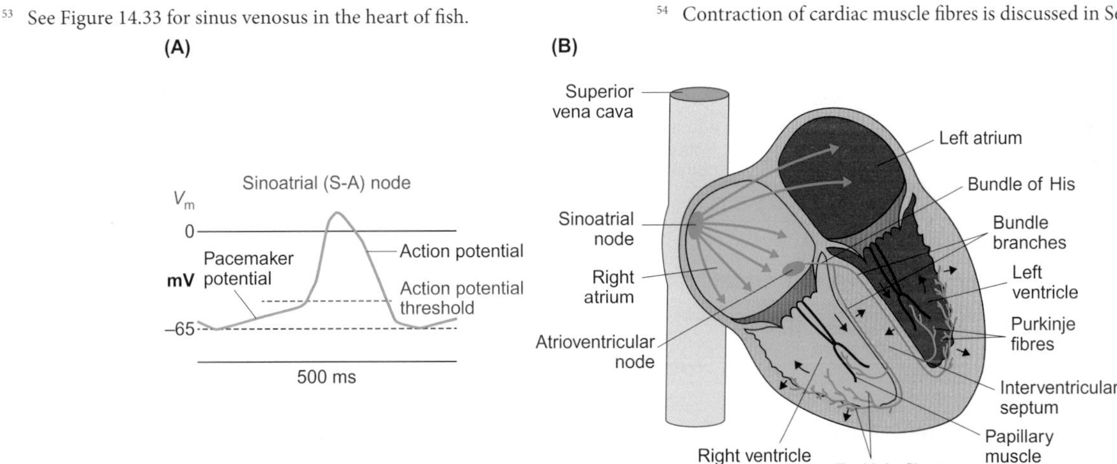

Figure 22.24 Sino-atrial pacemaker and the conducting system of the mammalian heart

(A) Intracellular recordings of the membrane potential (V_m) from the sinoatrial pacemaker region. (B) Major components of the excitation conducting system in the heart of mammals. The sinoatrial node is the dominant pacemaker. Electrical activity spreads throughout the atria to the atrioventricular node which is connected to the bundle of His. The bundle of His divides and runs either side of the interventricular septum, eventually subdividing into a network of conducting fibres, the Purkinje fibres.

Sources: (A) modified from Rushmer RF (1976). Cardiovascular dynamics, 4th edn. Philadelphia: W B Saunders Company. (B): reproduced from Berne RM and Levy MN (1998). Physiology, 4th edn. St Louis: Mosby.

through the middle portion of the node. This reduction in conduction velocity provides sufficient time for the ventricles to be filled adequately during contraction of the atria.

The electrical events in the pacemaker cells of the sinoatrial node are the result of movements of various ions across their cell membranes, most of which are controlled by time- or voltage-dependent ion channels[55]. Inward movements of cations (Na^+ and Ca^{2+}) cause membrane depolarization, while outward movement of cations (such as K^+) cause membrane hyperpolarization.[56] These events are shown in Figure 22.25A.

The pacemaker potential of myocytes in the S-A node is the result of the inward movement of sodium ions, Na^+, through background 'leak' channels and through the so-called I_f ('funny'[57]) channels that are activated near the end of repolarization following an action potential, when the membrane potential is more negative than −35 mV. The opening of T-type voltage-gated calcium channels (T for causing tiny or transient currents) that are activated after membrane repolarization also contribute to the pacemaker potential.

[55] Ion channels are discussed in Section 16.2.3 and 16.2.5.
[56] Hyperpolarization and depolarization are discussed in Section 16.2.2.
[57] The I_f current is called 'funny' because the discovery of an inward Na^+ current after repolarization of the pacemaker cell was unexpected.

When the pacemaker potential depolarizes to about −45 mV, the threshold condition for triggering an action potential is reached, as shown in Figure 22.24A. Unlike ventricular myocytes in which the primary channel responsible for the generation of the action potential is the voltage-gated Na^+-channel, the pacemaker cells in the S-A node have voltage-gated L-type Ca^{2+} channels (L for long lasting) which cause the upstroke of the action potential, as shown in Figure 22.25A.

The time course of the action potential in S-A pacemaker cells is much slower than the action potential in nerve fibres because the L-type Ca^{2+} channels activate and then inactivate at a much slower rate than the Na^+ channels. The action potential causes activation of voltage-dependent potassium channels, and the outward movement of K^+, the I_K current, repolarizes the membrane. After membrane repolarization, the K^+ channels gradually close, contributing to the slow membrane depolarization until the threshold is reached for triggering another action potential so that the cycle can begin again, as shown in Figure 22.25A.

There is also evidence that the release of calcium ions from the sarcoplasmic reticulum may contribute to the pacemaker potential. However, it appears that neither the funny current nor the spontaneous release of Ca^{2+} on its own is responsible for the activity of the pacemaker cells; there is always interaction between them.

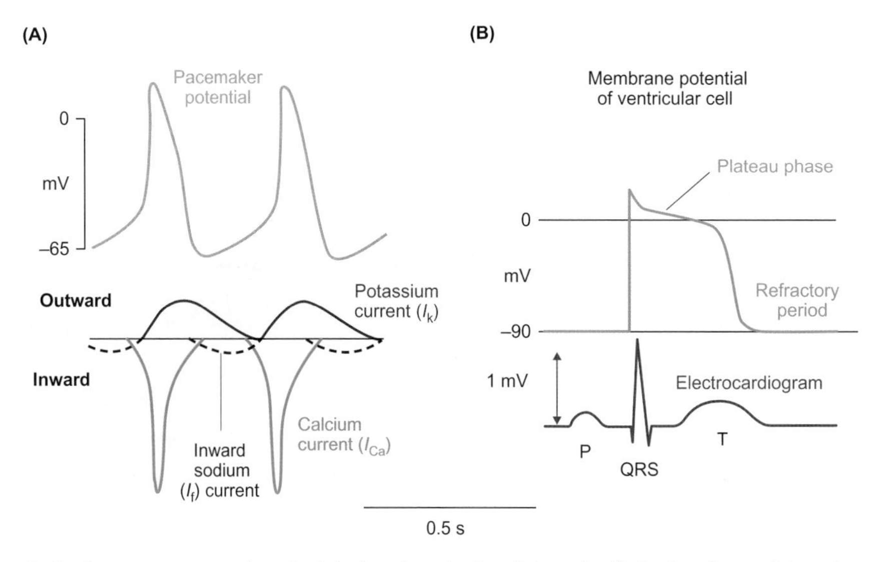

Figure 22.25 Changes in ionic currents associated with the electrical activity of cells in the sinoatrial node and the action potential of cells in the ventricle

(A) The changes in membrane potential that occur in the cells of the sinoatrial node are the result of three main currents: the inward movement of sodium ions which causes primarily the slow depolarization; an inward movement of calcium ions which causes the rapid depolarization (action potential); and an outward movement of potassium ions which causes repolarization. (B) The action potential in cells of the ventricle and the electrocardiogram associated with the contraction of the whole heart. The P-wave corresponds to the depolarization of the atria, the complex QRS wave to the depolarization of the ventricles and the T-wave to the repolarization of ventricles. The heart muscle cells will not respond to a stimulus (they won't contract) during the refractory period which lasts until shortly after the T-wave. Notice the much smaller amplitude of the electrocardiogram waves compared with the intracellular recording of an action potential.

Sources: (A): reproduced from Berne RM and Levy MN (1998). Physiology, 4th edn. St Louis: Mosby. (B): reproduced from Rushmer RF (1976). Cardiovascular dynamics, 4th edn. Philadelphia: W B Saunders Company.

The impulses transmitted by the Purkinje fibres to the ventricular myocytes depolarize the cells. The resting membrane potential of ventricular myocytes is about −90 mV and when the membrane potential reaches about −65 mV, the properties of the cell membrane change causing the characteristic action potential of a ventricular cell[58], which is shown in Figure 22.25B. Unlike the action potentials in skeletal muscle cells, there is a long plateau phase in ventricular cells. Also, during the relatively long refractory period, the muscle cells do not respond to stimulation, which means that they cannot be tetanized[59].

The overall process that convert the action potentials in the myocytes of the atria and ventricles into their contraction is known as **excitation–contraction coupling**[60]. Calcium ions play a crucial role in this process in the hearts of all vertebrates. What is more, the amount of calcium ions that are available to the contractile proteins also regulates the force generated by the contracting heart.

In birds and mammals, most of the calcium ions that activate contraction are released by the sarcoplasmic reticulum. However, the sarcoplasmic reticulum of cardiac muscle cells of most fish and amphibians occupies a much smaller volume (0.5 per cent in frogs) compared with those of mammals (7.3 per cent in rats). As such, the contribution of the intracellular cycling of calcium via the sarcoplasmic reticulum is minimal in most ectothermic vertebrates and most calcium necessary for activation of the contractile apparatus enters the cell via the sarcolemma.

In species with unusually high heart rates and blood pressures, such as tuna and varanid lizards[61], the sarcoplasmic reticulum does seem to have a role in the cycling of calcium during excitation–contraction coupling and, in this respect, these species appear to be intermediate between the majority of ectotherms and mammals.

The electrical activity of the heart can be recorded externally in humans by placing electrodes in specific locations on the surface of the body. The waveform obtained is called the **electrocardiogram** (ECG or EKG) and is shown at the bottom of Figure 22.25B. The ECG is an extremely useful tool for diagnosing clinical problems such as irregularity in the relative sizes of the chambers, the extent and location of ischaemic damage to the heart muscle (that is, resulting from reduced blood supply), and the effects of altered electrolyte concentrations. According to the original system devised by the Dutch physician and physiologist, Willem Einthoven, at the beginning of the 20th century, electrodes are located on both arms and the left leg. The electrical potentials recorded between the two arms (known as lead I), the right arm and left leg (lead II) and the left arm and left leg (lead III) give the typical ECG which consists of P, QRS complex and T waves, shown diagrammatically in Figure 22.25B. The wave (P) corresponds to the depolarization of the atria, the second wave (Q) to the depolarization of the ventricles and the third (T) to repolarization of the ventricles[62].

The delay in the passage of the impulse from the atria to the ventricles as a result of the slow conduction velocity through the A-V node, which we discuss above, accounts for the delay between the start of the P wave and the QRS complex of the electrocardiogram.

As well as being a diagnostic tool for clinicians, the ECG is also extremely useful for animal physiologists, as it is a relatively easy way to record heart rate. For this purpose, it is only necessary to have one pair of recording electrodes, preferably positioned across the heart, for a good signal.

22.3.2 Neurogenic pacemakers of crustaceans

The cardiac ganglion of crustaceans is located on the dorsal surface of the heart and is shown for a decapod, the American lobster (*Homarus americanus*), in Figure 22.26[63]. The ganglion consists of 9–16 neurons, depending on the species. The importance of this ganglion is demonstrated by the fact that the heart stops beating when the nerves between the ganglion and the heart muscle are cut. In the cardiac ganglion of the majority of decapods, the most posterior four of the neurons are smaller than the others and their axons are contained within the ganglion where they provide excitatory input to the anterior cells. These smaller neurons are considered to be the pacemaker neurons and one of these is consistently active. The first five anterior neurons are larger, and their axons leave the ganglion and innervate the heart muscle. These axons are motor nerves and are responsible for contraction of the heart muscle.

Some processes from the large cells, known as **dendrites**[64], do not innervate the muscles but ramify over the heart muscle near the cardiac ganglion. These dendrites are stimulated when the heart is stretched by the suspensory ligaments[65] during diastole and their stimulation has an

[58] The action potential of ventricular cells is discussed in Section 18.3.1.

[59] Absence of tetanus in cardiac muscle of vertebrates is discussed in Section 18.3.1.

[60] Excitation–contraction coupling in cardiac muscle is discussed in Section 18.3.1.

[61] Table 14.3 gives some values of blood pressure in tuna and varanid lizards.

[62] Depolarization and repolarization are discussed in Sections 16.2.2 and 16.2.4, respectively.

[63] The circulatory system of a decapod crustacean is shown in Figure 14.28.

[64] Dendrites are discussed in Section 3.3.3 and those of the crustacean cardiac pacemaker are shown in Figure 22.31.

[65] Suspensory ligaments of the crustacean heart are discussed in Section 14.3.3 and those in decapod crustaceans are shown in Figure 22.31.

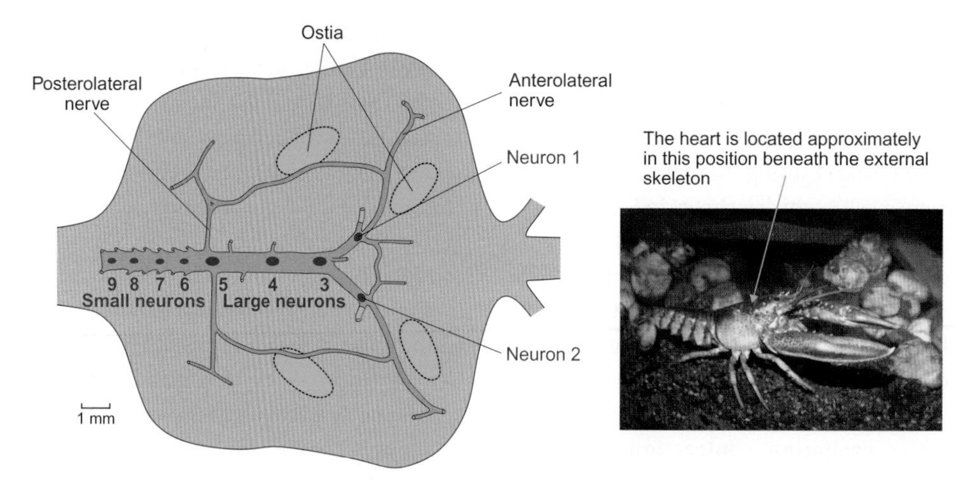

Figure 22.26 Neurogenic pacemaker of crustacean heart

Dorsal view of the cardiac ganglion of American lobsters (*Homarus americanus*). The ganglion is located on the inside of the dorsal surface and consists of nine neurons, five larger, anterior neurons, and four smaller, posterior neurons. The smaller neurons act as the pacemakers, while axons from the five anterior neurons leave the ganglion within the various nerves on the surface of the heart and innervate the muscle of the heart; they are motor neurons.

Source: modified from: Hartline DK (1967). The impulse identification and axon mapping of the nine neurons in the cardiac ganglion of the lobster *Homarus americanus*. Journal of Experimental Biology 47: 327-340. Image CaRMS Photogallery / Jim Cornall/Huntsman Marine Science Centre, 2012.

important influence on the heart rate and strength of contraction. Nonetheless, the isolated cardiac ganglion exhibits rhythmically occurring bursts of nervous activity. Figure 22.27A shows the slowly depolarizing pacemaker potential in a small posterior neuron of a a three spot swimming crab (*Portunus sanguinolentus*) and the excitatory postsynaptic potentials[66] in a larger anterior neuron. Repetitive impulses produce excitatory junction potentials which are facilitatory,

[66] Excitatory postsynaptic potentials are discussed in Section 16.3.3.

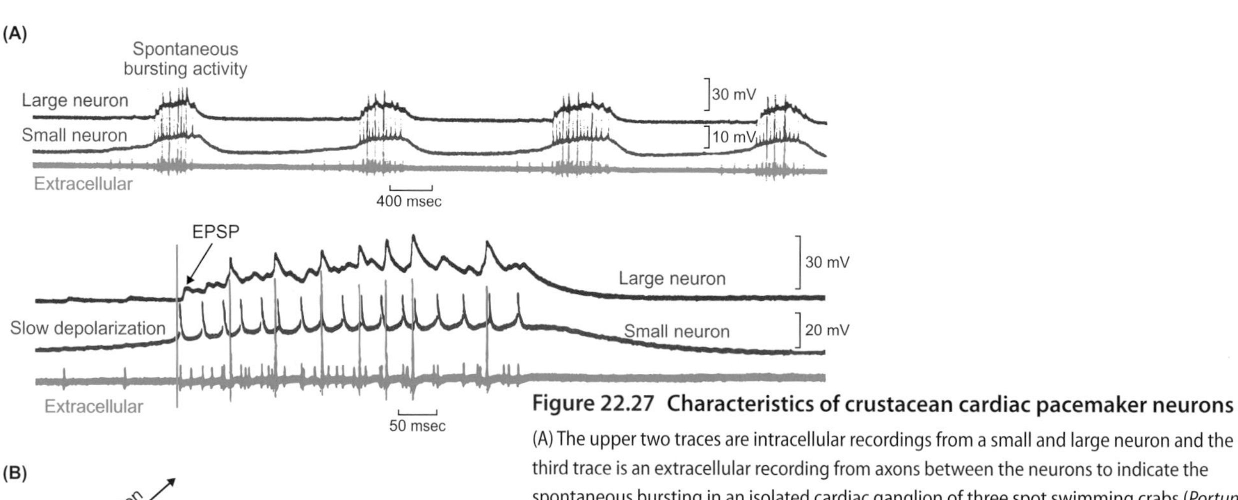

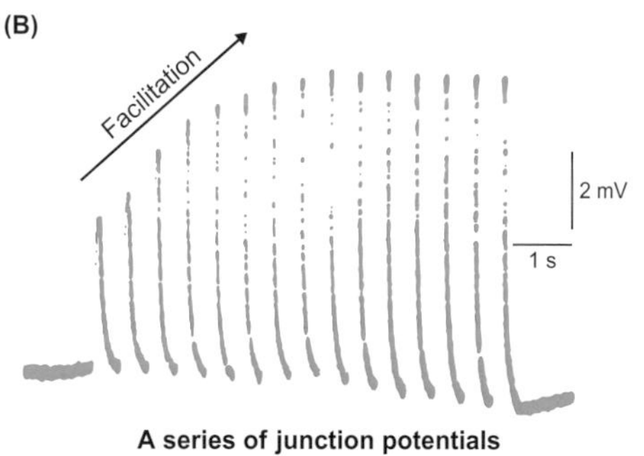

Figure 22.27 Characteristics of crustacean cardiac pacemaker neurons

(A) The upper two traces are intracellular recordings from a small and large neuron and the third trace is an extracellular recording from axons between the neurons to indicate the spontaneous bursting in an isolated cardiac ganglion of three spot swimming crabs (*Portunus sanguinolentus*). The bottom three traces show one of the bursts on an expanded timescale. Note the slowly depolarizing pacemaker potential in the small neuron which leads to a series of rapid depolarizations. At the same time as the first depolarization of the small neuron, indicated by the vertical blue line, there is a small impulse in the extracellular record which, after a short delay, is followed by an excitatory (depolarizing) post-synaptic potential (EPSP) in the large neuron. (B) Facilitation, as illustrated by the increasing size of excitatory junction potentials, in a heart muscle fibre of an American lobster in response to a series of electrical stimuli applied to the distal cut end of the posterolateral nerve, shown in Figure 22.26.

Sources: (A): modified from Cooke IM (2002). Reliable, responsive pacemaking and pattern generation with minimal cell numbers: the crustacean cardiac ganglion. Biological Bulletin 202: 108-136. (B): reproduced from Anderson M and Cooke IM (1971). Neural activation of the heart of the lobster *Homarus americanus*. Journal of Experimental Biology 55: 449-468.

as shown in Figure 22.27B. There is also a summation of responses to impulses arriving from different motor nerves.

The combined effect of facilitation and the summation of excitatory junction potentials can produce different levels of depolarization of a muscle cell with the extent of the depolarization determining the strength of contraction of the muscle cell, and of the heart as a whole. Heart rate and the strength of its contraction are, therefore, extremely sensitive to frequency and patterning of impulses from the motor nerves in the cardiac ganglion. The excitation–contraction coupling mechanism of the heart of decapod crustaceans is similar to that of the mammalian heart, except that the heart muscle can be tetanized[67].

We might wonder why, of all the groups of animals, arthropods alone, with the exception of insects, have a neurogenic cardiac pacemaker. However, it turns out that this is not always the case. Some crustaceans, such as tadpole shrimps (*Triops* spp.) have a diffuse myogenic cardiac pacemaker. On the other hand, the heartbeat of early juvenile sea slaters (*Ligia* sp.) is myogenic. A cardiac ganglion is present, but it displays no spontaneous activity. During development, however, the cardiac ganglion gradually becomes active and dominates the myogenic pacemaker so that, by adulthood, a myogenic rhythm is slower and subordinate to the neurogenic rhythm. It has been suggested, therefore, that the earliest crustaceans possessed myogenic hearts and that there was an evolutionary shift toward a neurogenic pacemaker. However, the evolutionary pressures that may have driven this process are unknown.

› *Review articles*

Burggren W, Farrell A, Lillywhite H (2011). Vertebrate cardiovascular systems. Comprehensive Physiology, Supplement 30: Handbook of Physiology, Comparative Physiology: 215-308. First published in print 1997.

Cooke IM (2002). Reliable, responsive pacemaking and pattern generation with minimal cell numbers: the crustacean cardiac ganglion. The Biological Bulletin 202: 108–136.

DiFrancesco D (2010). The role of the funny current in pacemaker activity. Circulation Research 106: 434–446.

Irisawa H, Brown HF, Giles W (1993). Cardiac pacemaking in the sinoatrial node. Physiological Reviews 73: 197–227.

Lakatta EG, Maltsev VA, Vinogradova TM (2010). A coupled SYSTEM of intracellular Ca^{2+} clocks and surface membrane voltage clocks controls the timekeeping mechanism of the heart's pacemaker. Circulation Research 106: 659–673.

McMahon BR, Wilkens JL, Smith PJS (1997). Invertebrate circulatory system. Comprehensive Physiology, Supplement 30: Handbook of Physiology, Comparative Physiology: 931-1008. First published in print 1997.

Maltsev VA, Lakatta EG (2007). Normal heart rhythm is initiated and regulated by an intracellular calcium clock within pacemaker cells. Heart, Lung and Circulation 16: 335–348.

Wilkens JL (1999). The control of cardiac rhythmicity and of blood distribution in crustaceans. Comparative Biochemistry and Physiology A 124: 531–538.

[67] Tetanus is discussed in Section 18.2.2

22.4 Control of the cardiovascular system

As with the respiratory system, the major groups of sense organs involved in the control of the cardiovascular system are the peripheral chemoreceptors and mechanoreceptors. In vertebrates, the chemoreceptors are the same as those involved in the control of ventilation. In contrast, the mechanoreceptors are different from those involved in control of the respiratory system and are located in the walls of the major arteries and veins, and in the various chambers of the heart.

The function of the mechanoreceptors is to monitor the degree of stretch of the wall of the blood vessel or chambers of the heart and to relay the information back to the brainstem. As the degree of stretch of a blood vessel is dependent on the compliance[68] of the wall of the vessel and the difference in pressure across the wall, these receptors are involved in the control of arterial blood pressure and are known as arterial pressure receptors or **baroreceptors** (*baros* is Greek for weight). On the venous side of the circulation, the mechanoreceptors are more involved in the control of central blood volume.

22.4.1 Peripheral chemoreceptors that influence the cardiovascular system

The chemoreceptors responsible for the cardiovascular changes in fish are restricted to the first pair of gill arches in many species, but can be found throughout the first three pairs in others. The most obvious cardiovascular response to being exposed to low oxygen in fish is a slowing in heart rate (**bradycardia**—from the Greek *brady* meaning slow and *cardia* meaning heart). In elasmobranchs (cartilaginous fish), this bradycardia is more than compensated for by an increase in cardiac stroke volume so that cardiac output increases slightly; in teleosts (bony fish), however, the response of cardiac stroke volume is more variable[69]. As well as a bradycardia in response to environmental hypoxia, the resistance to blood flow also changes, but this varies between elasmobranchs and teleosts. In elasmobranchs vascular resistance falls, as a result of the dilation of some peripheral blood vessels, but

[68] Compliance (C) is the change in volume (ΔV) of a blood vessel for a given pressure difference (ΔP) across the wall of the vessel: $C = \Delta V/\Delta P$

[69] Bradycardia in response to hypoxia in bony fish is discussed in Section 15.3.1.

increases (peripheral blood vessels constrict) or remains unchanged in teleosts.

While aquatic invertebrates show cardiovascular changes in response to environmental hypoxia, the location of the sense organs responsible for these changes is not clear. As in some species of fish, there is a slowing of the heart rate in crustaceans exposed to hypoxia which is compensated for by a similar increase in cardiac stroke volume so that cardiac output does not change. There is evidence to suggest that the receptors responsible for these responses are located somewhere in the gill cavities.

The cardiovascular systems of air-breathing vertebrates respond differently to hypoxia than those of fish. In mammals, such as dogs, spontaneously breathing, anaesthetized animals show an increase in heart rate (**tachycardia**—*tachy* is Greek for rapid) and a decrease in peripheral resistance (caused by dilation of peripheral blood vessels) when the carotid body chemoreceptors are stimulated by hypoxic blood. However, if ventilation is artificially prevented from increasing or the lungs are denervated, stimulation of the carotid bodies causes a bradycardia and an increase in peripheral vascular resistance. This has been called the primary cardiovascular response to stimulation of the carotid bodies. The increase in heart rate and dilation of peripheral blood vessels in spontaneously breathing, intact dogs is called the secondary cardiovascular response to carotid body stimulation.

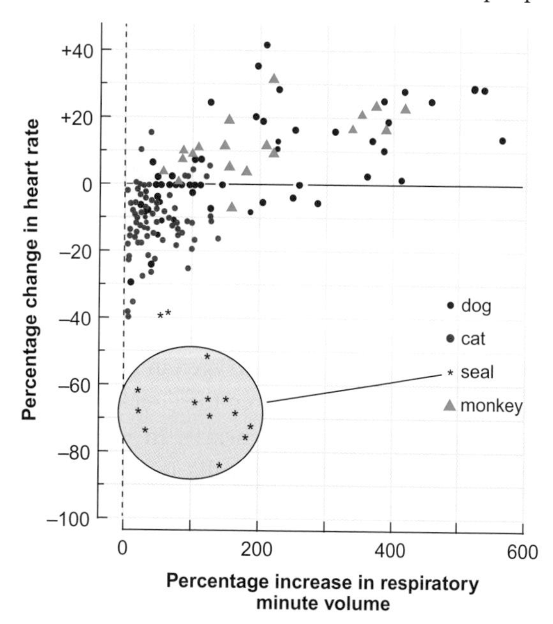

Figure 22.28 The relationship between increase in ventilation and change in heart rate resulting from stimulation of carotid body chemoreceptors by hypoxic blood in intact, spontaneously breathing mammals.

When the increase in ventilation is above approximately 200%, there is an accompanying increase in heart rate (tachycardia), but when the increase in ventilation is less than approximately 200%, there is no change or a reduction in heart rate (bradycardia). In seals, heart rate is lower for a given change in ventilation than in terrestrial species of mammals. This is related to the cardiovascular response during diving in marine mammals, which we discuss in section 15.2.3.

Source: modified from Marshall JM (1994). Peripheral chemoreceptors and cardiovascular regulation. Physiological Reviews 74: 543-594.

The primary response is similar to that seen in fish while the secondary response is caused at least partly by input from the slowly adapting receptors in the lungs which are stimulated during the increase in ventilation[70]. Similar responses to those seen in dogs are also seen in monkeys, but other species of mammals, such as cats, rabbits and rats, show less of a secondary response than dogs. This difference is probably because the increase in ventilation in response to carotid body stimulation is less in these species than in dogs, as illustrated in Figure 22.28.

Birds also show a slight secondary response (tachycardia and peripheral vascular dilation) which partly overrides the primary response (bradycardia and peripheral vascular constriction), but this is the result of decreased carbon dioxide in the lungs during the increased ventilation acting on the intrapulmonary chemoreceptors.

Other air-breathing vertebrates, such as lungfish, amphibians and most reptiles, have incompletely divided hearts and ventilate their lungs intermittently when at rest. Pronounced cardiovascular changes occur during the intermittent ventilation of many amphibians and reptiles, and it is difficult to distinguish the primary effects of a given stimulus on the cardiovascular system from the secondary effects arising from changes in ventilation. We do know, however, that the increase in heart rate and in blood flow along the pulmocutaneous arches which occurs during spontaneous lung inflation in African clawed frogs (*Xenopus laevis*), can be simulated by artificially inflating the lungs during a period of breath-hold. This observation suggests an important role for the lung stretch receptors in mediating the cardiovascular changes associated with lung ventilation in amphibians.

22.4.2 Mechanoreceptors (baroreceptors) in the cardiovascular system

In addition to chemoreceptors, mechanoreceptors (or baroreceptors) also help to control the cardiovascular system. The vast majority of our knowledge of mechanoreceptors in the cardiovascular system has been obtained from vertebrates, particularly mammals.

Baroreceptors of vertebrates

In mammals, the area of an artery where the baroreceptor endings are located is more compliant than the rest of

[70] Slowly adapting stretch receptors in the lungs of vertebrates are discussed in Section 22.2.1.

the vessel, so it stretches and relaxes more easily as blood pressure changes. This greater compliance is because there is a higher proportion of elastin and a lower proportion of smooth muscle in the region of the receptors than in the remainder of the blood vessel. For example, the proportion of elastin in the carotid sinus of dogs is greater (around 50 per cent) than that in the common carotid artery (approximately 30 per cent), while the smooth muscle content is less in the carotid sinus (around 15 per cent vs 35 per cent)[71].

In vertebrates, the arterial baroreceptors are located in much the same places as the peripheral chemoreceptors, as shown in Figure 22.29. They are present in the gill arches of fish, in the carotid labyrinths, systemic arches and pulmo-cutaneous arches of amphibians, in the systemic and pulmonary arches of reptiles, at the base of the aorta in birds and in the carotid sinus, aorta and pulmonary artery of mammals. One of their functions is to cause changes in cardiac output (mainly via heart rate) and peripheral vascular resistance via a reflex mediated by the nervous system. These changes maintain central arterial blood pressure at a given level—a set point. So, if for any reason there is an increase in central blood pressure, a reflex reduction in heart rate and dilation of peripheral blood vessels occur so that pressure returns toward its normal value. This response is shown in Figure 22.30. The opposite occurs if there is a fall in blood pressure.

These reflex responses are often called the **barostatic reflexes**, which don't necessarily involve all of the baroreceptor locations. For example, the receptors responsible for the barostatic reflex in toads and frogs are those located in the pulmo-cutaneous arch. The significance of this may be to protect the blood vessels in the lungs against the relatively high pressures generated in the incompletely divided heart of these animals. Baroreceptors are located in the gills of fish and may serve a similar function—mainly to protect the delicate blood vessels of the gas-exchange regions of the gills which are exposed to the relatively high pressure generated by the heart[72].

Baroreceptors and innervation of the heart in invertebrates

There have been very few studies of barostatic reflexes in invertebrates. However, there is evidence of functional baroreceptors in some species of crustaceans[73]. Experimental changes in volume of the haemolymph of land crabs, (*Cardisoma* spp.) cause opposite changes in heart rate, as in vertebrates. However, the location of the baroreceptors is unknown.

The nerves to the heart and associated structures originate from the thoracic ganglion in decapod crustaceans, as shown in Figure 22.31. There are two separate accel-

[71] The structure of blood vessels is discussed in Section 14.1.3.

[72] Pressures in the circulatory system of fish are discussed in Section 14.4.1 and Table 14.3.

[73] The cardiovascular system in crustaceans is discussed in Section 14.3.3.

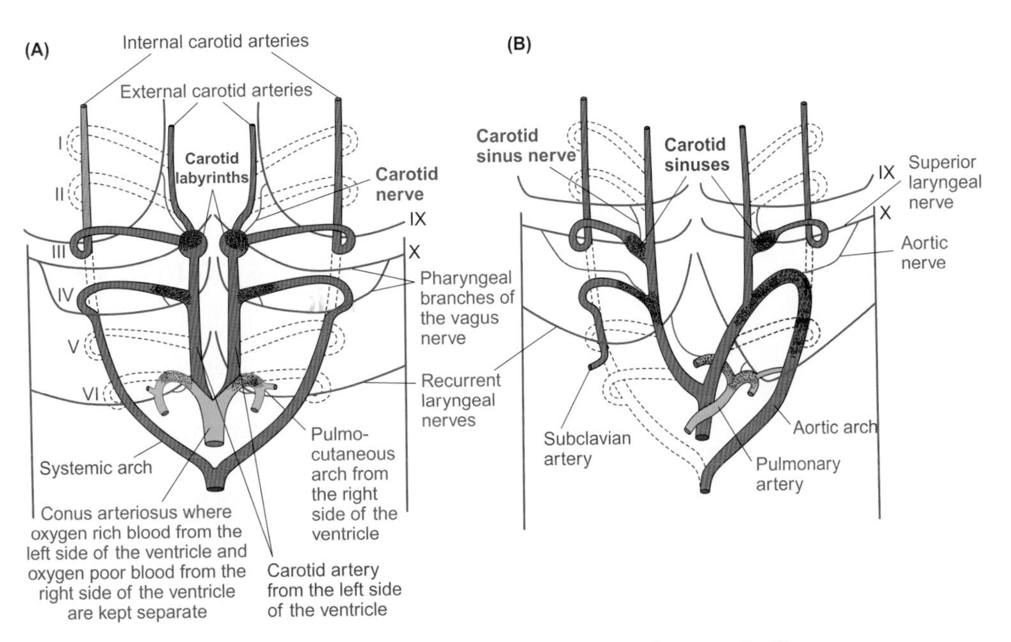

Figure 22.29 Location and innervation of baroreceptors in amphibians (A) and mammals (B)

The shaded areas indicate the location of the baroreceptors in the remaining aortic arches. The IXth (glossopharyngeal) and Xth (vagus) cranial nerves are indicated on the right of each diagram and the Roman numbers on the extreme left hand side indicate the ancestral aortic arches (look at Figure 14.43 for further details). The aortic arches that disappear during development are signified by dashed lines.

Source: modified from West NH and Van Vliet BN (1992). Sensory mechanisms regulating the cardiovascular and respiratory systems. In: Environmental Physiology of the Amphibians. Feder ME and Burggren WW (eds). pp 151-182. Chicago: University of Chicago Press.

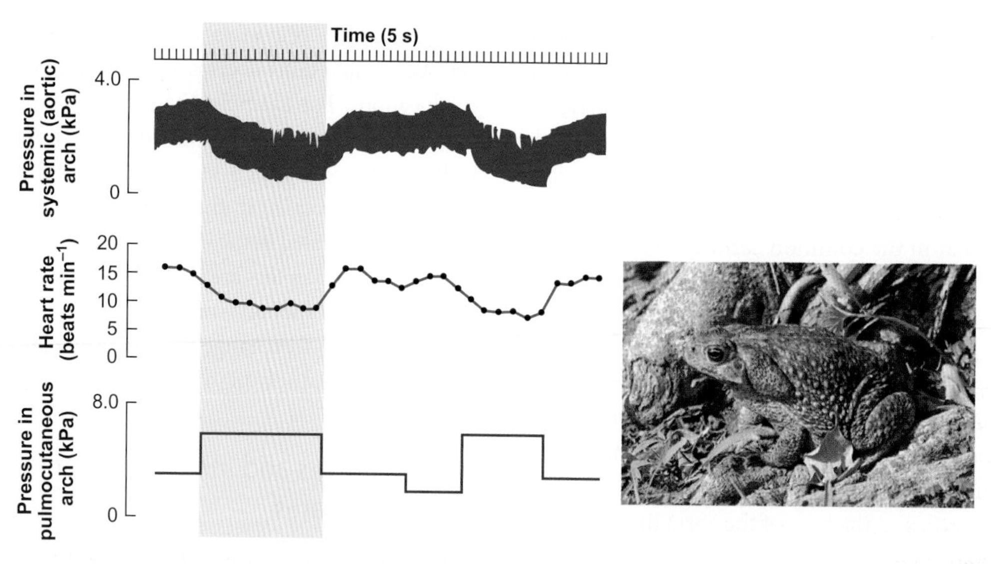

Figure 22.30 The baroreflex and baroreceptors in vertebrates

Changes in central (aortic) blood pressure and heart rate in toads (*Bufo marinus*) in response to the experimental raising and lowering of pressure in the pulmocutaneous arch. Whenever pressure in the pulmocutaneous arch is raised, both heart rate and central arterial pressure decline—indicated by area shaded. The opposite occurs when pressure in the pulmocutaneous arch is lowered.

Modified from West NH and van Vliet BN (1983). Open-loop analysis of the pulmocutaneous baroreflex in the toad, *Bufo marinus*. American Journal of Physiology 245: R642-R650. Image (©) Bill Love.

erator nerves[74] and one inhibitory nerve on each side. These not only innervate the cardiac ganglion, but also the pericardial organs and the suspensory ligaments. The neurotransmitter released from the accelerator nerves seems to be different in different taxa. For example, those in hermit crabs (*Aniculus aniculus*) release dopamine and those in Japanese mantis shrimps (*Oratosquilla oratoria*) release glutamate. In contrast, the transmitter released from the inhibitory nerve is thought to be γ-amino butyric acid (GABA). In addition, the pericardial organs release a number of hormones into the haemolymph which affect the performance of the heart. For example, serotonin and dopamine cause an increase in heart rate and in strength of contraction of the heart.

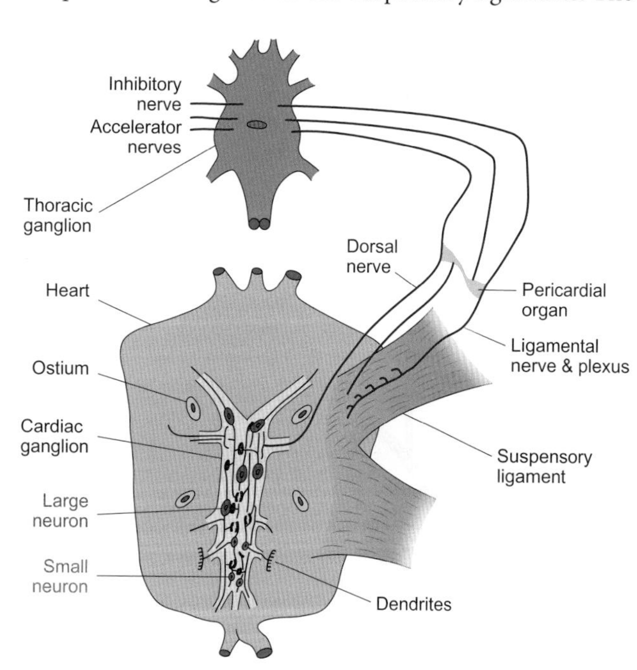

Figure 22.31 Innervation of the heart of decapod crustaceans

In decapod crustaceans, nerves innervating the heart originate in the thoracic ganglion. A pair of dorsal nerves to the heart contain one inhibitor and two accelerator nerves. The cardiac ganglion (also shown in Figure 22.26) is drawn disproportionately large.

Modified from Taylor EW (1982). Control and co-ordination of ventilation and circulation in crustaceans: responses to hypoxia and exercise. Journal of Experimental Biology 100: 289-319.

22.4.3 Nervous output to the cardiovascular system of vertebrates

The motor (efferent) limbs of the cardiovascular reflexes in vertebrates are components of the **autonomic nervous system**[75]. This part of the nervous system is composed entirely of efferent pathways that have a synapse within a ganglion outside of the central nervous system. There are two branches to the autonomic nervous system: the **parasympathetic** and **sympathetic**.

Parasympathetic nerve fibres are found in some of the cranial nerves[76]; as far as the cardiovascular system is

[74] The accelerator nerves in crustaceans are excitatory nerves, as we discuss in Section 18.2.1.

[75] The autonomic nervous system is discussed in Sections 16.1.3 and 16.3.7.

[76] Take a look at Box 16.2 and Figure 16.13 to remind yourself of the cranial nerves in vertebrates.

concerned, those in cranial nerve X, the vagus nerves, are the most important as they innervate the heart. The autonomic ganglia of the vagal motor pathways are usually embedded in the wall of the organ they innervate. Among the sense organs innervated by cranial nerves IX and/or X are those in the lungs and heart, the chemo- and baroreceptors.

The sympathetic nervous system consists of interconnected ganglia which form a chain on each side of the spinal cord[77]. One or more of the anterior sympathetic ganglia send nerve fibres to the heart (except in elasmobranchs where there is no sympathetic innervation to the heart) and the more posterior ganglia send fibres to the blood vessels. In most groups of vertebrates, some sympathetic nerve fibres are present in the vagus nerves which are therefore known as vago-sympathetic trunks.

The parasympathetic nerve fibres to the heart release **acetylcholine**, which causes the heart to slow down and, possibly, to reduce its force of contraction. In contrast, the sympathetic fibres release **adrenaline** or **noradrenaline** (also known as epinephrine or norepinephrine, respectively, and collectively together with dopamine, known as **catecholamines**[78]). Catecholamines cause the heart to speed up and to beat more strongly. The effect on the rate of contraction of the heart is known as a **chronotropic effect** (*chrono*—Greek for time, *tropic*—Greek for a turn), whereas the effect on the strength of contraction is known as an **inotropic effect** (*ino*—Greek for fibre). The two arms of the autonomic nervous system function in an antagonistic fashion, and both are continuously active, i.e. they have a 'tone'. This means that an increase in heart rate can be the result of either an increase in sympathetic activity, or tone, a decrease in vagal activity (vagal tone), or both.

The sympathetic fibres to the blood vessels control the degree of contraction of the smooth muscle in the vessels (the arterioles in particular) and, therefore, their diameter and resistance to the flow of blood. As with the sympathetic fibres to the heart, those to the blood vessels are also continuously active, and provide what is often called 'sympathetic tone'. This tone is important; without it, vasodilator mechanisms caused by a reduction in sympathetic tone, could not prevent central arterial pressure being elevated.

Resting sympathetic tone in mammals is determined by regions within the brainstem which include the rostral ventrolateral medulla (RVLM), shown in Figure 22.32. Intracellular recordings from sympathetic neurons within the RVLM have shown that the neurons have an irregular tonic activity and that they receive both excitatory and inhibitory synaptic inputs. Excitatory input originates from both within and above the level of the medulla oblongata. Inhibitory input originates from neurons within the

[77] Figure 16.14 shows the sympathetic nervous system of mammals.
[78] We discuss catecholamines in Section 16.3.4.

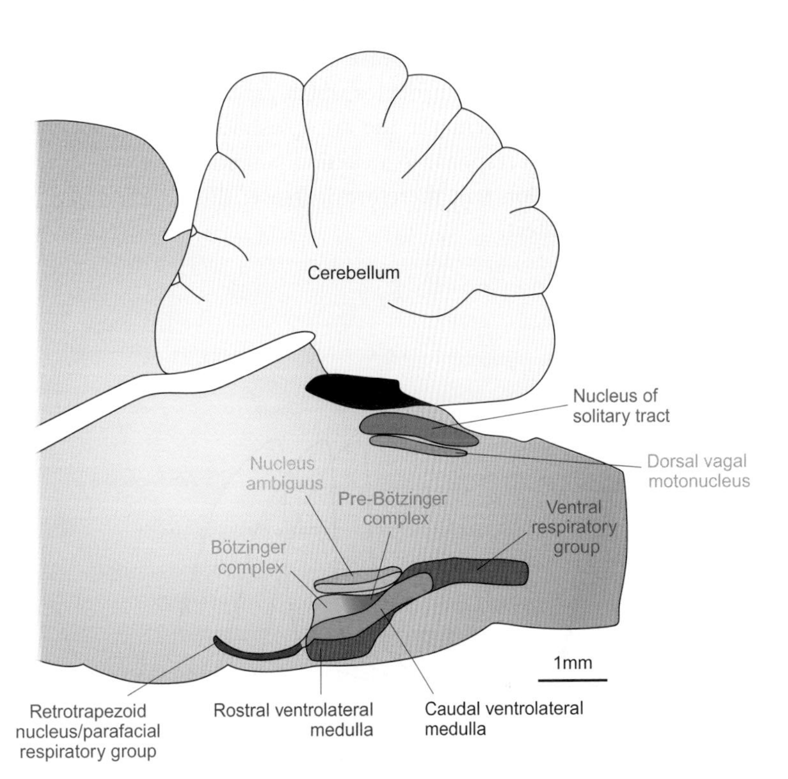

Figure 22.32 Anatomy of the cardiovascular control network in the hindbrain of a rat

Vertical longitudinal section through the middle of the hindbrain showing the locations of the main groups of neurons involved in the control of respiratory, sympathetic and parasympathetic activities in mammals.

Modified from Spyer KM and Gourine AV (2009). Chemosensory pathways in the brainstem controlling cardiorespiratory activity. Philosophical Transactions of the Royal Society B 364: 2603-2610.

caudal ventrolateral medulla, shown in Figure 22.32. However, there is still some tonic activity when both the excitatory and inhibitory inputs to the neurons in the RVLM are blocked. This means either that other types of inputs to the neurons exist or that their membranes have pacemaker-type properties.

Increasing evidence suggests that an increase in sympathetic tone to blood vessels can lead to a chronic increase in blood pressure (hypertension). The increase in sympathetic tone may be the result of a reduction in the level of activity from the arterial baroreceptors, perhaps as the blood vessels become less compliant with age, or may be generated directly by specific regions of the medulla within the brainstem.

In mammals, sympathetic nerve fibres also innervate the blood vessels of skeletal muscle, releasing acetylcholine and causing vasodilation. Another group of vasodilatory nerve fibres are those from the sacral part of the parasympathetic system. These cause dilatation of the erectile tissue of the genitalia, a function which has been described as 'of paramount importance to the preservation of the species'.

Acetylcholine and adrenaline or noradrenaline affect the rate of depolarization of the pacemaker cells of the sinoatrial node of the heart of vertebrates, as shown in Figure 22.33. The major effect of adrenaline and noradrenaline is to increase the permeability of the membrane to sodium ions. This increased permeability means that the membrane depolarizes more quickly to reach the threshold for generation of the action potentials. Consequently, heart rate increases. This is a positive chronotropic effect. Catecholamines also cause an increase in the rate of entry of calcium ions into cardiac myocytes which gradually increases the amount of calcium in the sarcoplasmic reticulum that is released in response to an action potential. The strength of contraction will therefore increase; this is a positive inotropic effect.

The major effect of acetylcholine on the pacemaker cells of the sinoatrial node is to decrease the rate of depolarization after an action potential. This decrease means that it takes longer to reach the threshold for the generation of the next action potential and so heart rate decreases (a negative chronotropic effect). Acetylcholine may also decrease the rate of influx of calcium ions into the muscle cells of the heart, thereby contributing to the reduced strength of contraction in response to increased activity of the vagus (a negative inotropic effect).

As well as acting as neurotransmitters, catecholamines are also released into the circulatory system either by the nerve endings and/or by special secretory cells, which are called **chromaffin cells**[79]. These circulating catecholamines can also have effects on the cardiovascular system, in addition to those released by nerves that directly innervate the heart and blood vessels.

The receptors for acetylcholine and catecholamines[80] have been the subjects of much study, not only because of their intrinsic interest, but also because of their pharmacological significance and clinical value. Receptors for adrenaline/noradrenaline are given the general name of **adrenoceptors**[81] (or adrenoreceptors). One group, the α-adrenoceptors, cause constriction of blood vessels, while the other group, the β-adrenoceptors, cause dilation of blood vessels and stimulation of the heart. In mammals at least, there is evidence of further divisions into α_1- and α_2-, and β_1- and β_2- adrenoceptors. The α_1-adrenoceptors are found in the smooth muscle of blood vessels.

The β_1-adrenoceptors on the heart respond to noradrenaline, whether it is released by nerve endings (as in mammals) or is circulating in the blood. By contrast, the β_2- adrenoceptors

79 Chromaffin tissue is discussed in Section 19.3 and Figure 19.22.

80 Synaptic receptors in the central nervous system of vertebrates are discussed in more detail in Section 16.3.4.

81 See Table 16.1 for a full list of adrenoceptor types in vertebrates.

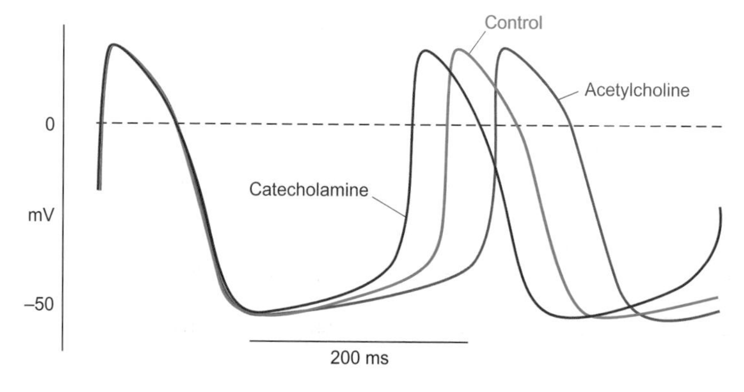

Figure 22.33 Effect of acetylcholine and a catecholamine such as noradrenaline (norepinephrine), on the spontaneous activity in a pacemaker cell in the sino-atrial node of a mammal

Note how acetylcholine slows the rate of depolarization while catecholamines have the opposite effect, thus causing a decrease or increase in heart rate, respectively.

Modified from DiFrancesco D (1993). Pacemaker mechanisms in cardiac tissue. Annual Review of Physiology 55: 455-472.

on the heart respond to adrenaline. Both cause increases in the rate and strength of contraction of the heart. β_2- adrenoceptors also cause relaxation (dilation) of blood vessels and of smooth muscle in the lungs. Thus, specific adrenoceptor blocking drugs (antagonists) or stimulating drugs (agonists) can be used to treat disorders of the circulatory and respiratory systems.

For example, atenolol and bisoprolol block the β_1-adrenoceptors of the heart and, therefore, reduce the rate and strength of contraction of the heart. They, or similar compounds, may be used in combination with other drugs such as α_1-adrenoceptor blockers, as a treatment for high blood pressure. Drugs which stimulate specific receptors (agonists) are also used clinically. For example, salbutamol stimulates β_2- adrenoceptors in the airways of the lungs causing them to dilate and is used in the treatment of asthma.

22.4.4 Long-term control of blood pressure in vertebrates

Although the arterial baroreceptors are important in the short-term control of arterial pressure, they may be only part of a complex system involved in long-term control. For example, Figure 22.34A shows that denervation of the baroreceptors of the major arteries of dogs does not lead to any significant change in the average long-term blood pressure. However, the daily fluctuations around these mean values are less in the control dogs (innervated baroreceptors) than in the denervated animals. These variations suggest that the baroreceptors play a significant role in the short-term control of blood pressure.

Why may arterial baroreceptors not be so important in the long-term control of blood pressure? A possible reason is that they adapt to an increase in stretch. This and other processes such as an increase in the resting diameter of the blood vessel at higher pressures, give rise to the situation where the average impulse frequency of the baroreceptors slowly declines for a given increase in mean arterial pressure, leading to a shift in the pressure–response curve to the right, as shown in Figure 22.34B. This rightward shift of the pressure–response curve is known as 'resetting'. Arterial blood pressure is now controlled by the baroreceptors at a higher level; there is a new set point.

The exact transduction mechanism of the baroreceptors is unknown[82], but current evidence points to the involvement of a group of proton-sensitive sodium channels (ASIC2);

[82] Transduction of stretch receptors is discussed in Section 17.4.4.

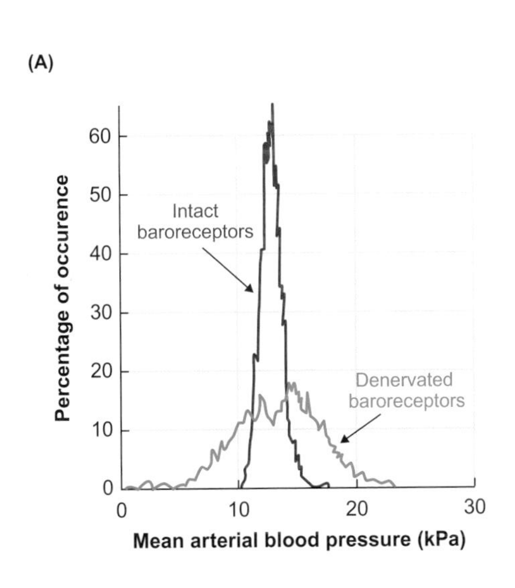

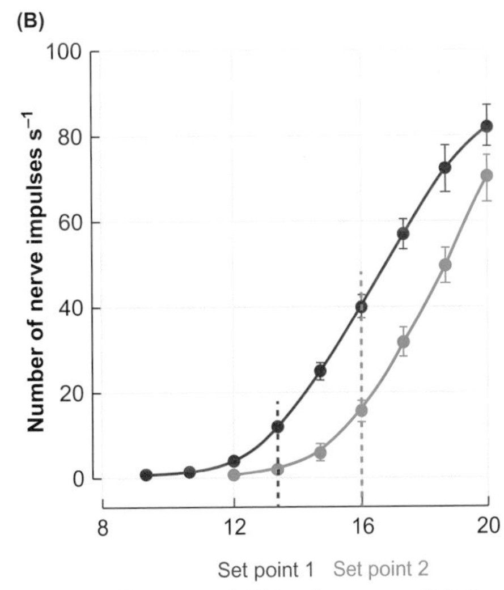

Figure 22.34 Arterial baroreceptors and the control of blood pressure

(A) Frequency distribution of average daily blood pressure in a dog with intact and denervated carotid and aortic baroreceptors. (B) Mean nervous activity (± SEM) of 16 aortic baroreceptors from six anaesthetized dogs when the set point was at a normal level of approximately 13 kPa (100 mmHg), i.e. the system was normotensive (●) and when the set point was raised to approximately 16 kPa (125 mmHg) and maintained there for 20 min, i.e. the system was hypertensive (●). The number of nerve impulses is similar at the two set points, which means that the number of nerve impulses at a given blood pressure is lower when the dogs are hypertensive, that is, the baroreceptors had reset.

Sources: (A): modified from Cowley AW et al (1973). Role of the baroreceptor reflex in daily control of arterial blood pressure and other variables in dogs. Circulation Research 32: 564-576. (B): modified from Coleridge HM et al (1984). Aortic wall properties and baroreceptor behaviour at normal arterial pressure and in acute hypertensive resetting in dogs. Journal of Physiology 350: 309-326.

deletion of the ASIC2 gene in mice leads to increased blood pressure and heart rate. Baroreceptor neurons from ASIC2-deficient mice are also less sensitive to stretch than those from wild-type mice. These and other data indicate a significant contribution of ASIC2 ion channels to the activation of arterial baroreceptors, and their reduced sensitivity could contribute to **hypertension** and other diseases of the cardiovascular system.

In addition to being the sensory side of the nervous reflexes which affect blood pressure, the arterial baroreceptors can also influence the release of vasoactive hormones which can have a profound influence on blood pressure. These vasoactive hormones include arginine vasotocin (AVT) in non-mammalian vertebrates and arginine vasopressin (AVP) in mammals. In addition, the stretching of the walls of the atria and/or ventricles can result in the release of **natriuretic peptides**, which can also influence blood pressure.

There is evidence that the vasoactive hormones are important in the long-term control of blood pressure. The hormones AVP, AVT and angiotensin II[83] have major roles in the maintenance of salt and water balance[84], and cause constriction of peripheral blood vessels. In dogs, the reflex secretion of AVP depends largely on input from the arterial baroreceptors (a decrease in blood pressure causes an increase in release of AVP and *vice versa* for a rise in blood pressure). However, there is no evidence that either arterial baroreceptors or receptors in the heart have a major role in activating the renin–angiotensin system in response to a fall in blood pressure. Instead, a baroreceptor (stretch-sensitive) mechanism within the kidney regulates renin secretion in response to a decrease in pressure in the renal blood vessel[85].

An increase in blood volume may not only cause a rise in blood pressure and stimulate the reflexes discussed above, but may also lead to the secretion of natriuretic peptides from the heart itself. Under normal conditions in mammals, the most important factor is the local stretching of the walls of the atria, which stimulates their myocytes to secrete natriuretic peptides. These peptides cause relaxation of peripheral blood vessels, and reductions in cardiac output and blood volume, all of which lead to a fall in blood pressure.

In the kidneys, natriuretic peptides increase the rate of filtration at the glomeruli and decrease the reabsorption of sodium. Thus, natriuretic peptides produce both diuresis (increased urine flow) and natriuresis (increased excretion of sodium)[86]. The opposite occurs when there is a reduction in blood volume and a decrease in the amount of stretching of the heart. In other words, changes in blood pressure cause directionally similar changes in fluid and sodium excretion.

The role of the kidneys in the long-term regulation of blood pressure

A major hypothesis for the long-term regulation of arterial blood pressure in mammals is that the kidneys adjust the rate of sodium and fluid excretion in order to maintain blood pressure. Figure 22.35A demonstrates how an increase in blood pressure leads to an increase in the rate of sodium excretion. So, although there is no known mechanism to detect total body water, extracellular fluid volume or

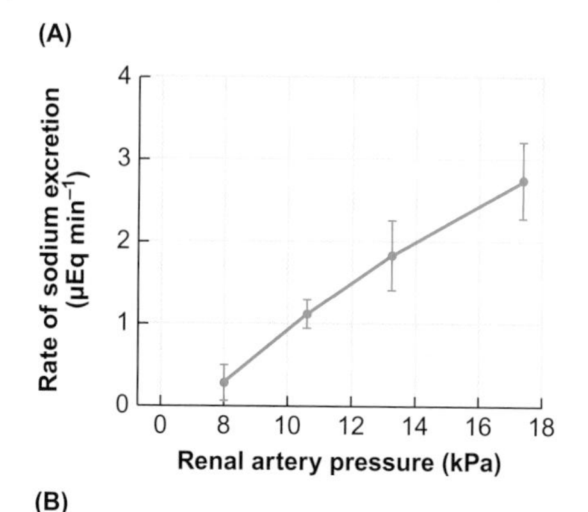

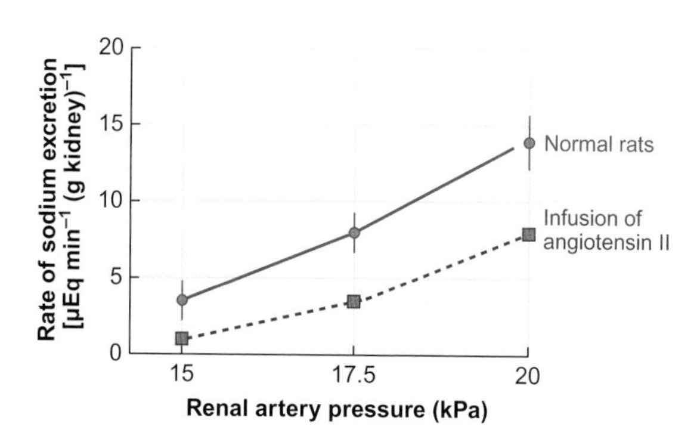

Figure 22.35 The relationship between blood pressure and rate of sodium excretion in mammals

(A) The effect of renal arterial pressure on the rate of sodium excretion in normal anaesthetized dogs. Mean values ± SEM from six dogs. (B) The effect of infusing angiotensin II (10 ng min^{-1}) into rats on the relationship between blood pressure and the rate of sodium excretion. Note that a higher blood pressure is required to maintain rate of sodium excretion after administration of angiotensin II. Mean values ± SEM from six rats.

Source: Brands MW (2012). Chronic blood pressure control. Comprehensive Physiology 2: 2481-2494.

[83] The renin–angiotensin system is discussed in Section 21.1.1.

[84] Control of water and salt balance and osmotic balance are discussed in Section 21.

[85] Factors affecting the release of renin are discussed in Section 21.1.1.

[86] Diuresis and natriuresis are discussed in Section 21.1.2.

blood volume, there are mechanisms which can detect changes in the degree of filling in various regions of the circulation, such as the atria, and can signal to the kidneys to change the rate of loss of sodium and water. In other words, the kidneys determine the set point for long-term pressure control in terms of the interrelationship between fluid loss and sodium balance; there is only one arterial pressure at which sodium balance is maintained. Any deviation of blood pressure away from the set point causes the loss or retention of sodium in order to restore pressure to the set point.

It must be borne in mind, however, that the cardiovascular system also provides oxygen and nutrients etc. to the cells of the body. These requirements have to be balanced against those of total fluid volume and its distribution.

Various factors can shift the position of the pressure/sodium excretion curve, which depends on the genetic make-up of the animal, the physiological state of the kidneys and the combined nervous and hormonal influences on the kidneys. Figure 22.35B illustrates the effect of elevated levels of angiotensin II on the position of the sodium excretion curve; an increase in blood pressure is required to maintain sodium and water balance. Hence, the inhibition of the action of angiotensin-converting enzyme (ACE, which converts angiotensin I to angiotensin II) with drugs called ACE inhibitors is a common way of decreasing blood pressure in hypertensive human patients.

Aquatic or semi-aquatic vertebrates, such as fish and amphibians, are liable to gain or lose water to a much greater extent than birds or mammals. Fish living in seawater tend to lose water and gain ions, while those living in fresh water tend to gain water and lose ions[87]. Similarly, amphibians will tend rapidly to lose water across their permeable skin when in air and gain water via the same route when in fresh water. This means that there is a tendency for blood volume to vary to a greater extent in these animals than in birds and mammals.

› *Review articles*

Barrett CJ, Malpas SC (2005). Problems, possibilities, and pitfalls in studying arterial baroreflexes' influence over long-term control of blood pressure. American Journal of Physiology 288: R837–R845.

Brands MW (2012). Chronic blood pressure control. Comprehensive Physiology 2: 2481–2494.

Cowley AW (1992). Long-term control of arterial blood pressure. Physiological Reviews 72: 231–300.

Dampney RAL, Horiuchi J, Tagawa T, Fontes MAP, Potts PD, Polson JW (2003). Medullary and supramedullary mechanisms regulating sympathetic vasomotor tone. Acta Physiological Scandinavica 177: 209–218.

Guyenet PG (2006). The sympathetic control of blood pressure. Nature Reviews Neuroscience 7: 335–346.

Marshall JM (1994). Peripheral chemoreceptors and cardiovascular regulation. Physiological Reviews 74: 543–594.

McMahon BR (1999). Intrinsic and extrinsic influences on cardiac rhythms in crustaceans. Comprehensive Biochemistry and Physiology A 124: 539–547.

Nilsson S (2011). Comparative anatomy of the autonomic nervous system. Autonomic Neurosciences 165: 3–9.

Thrasher TN (2005). Baroreceptors, baroreceptor unloading, and the long-term control of blood pressure. American Journal of Physiology 288: R819–R827.

22.5 Central terminations of respiratory and cardiovascular sense organs and their interactions in vertebrates

In mammals, the sensory inputs from the cardiovascular and respiratory systems to the central nervous system (CNS) all terminate in distinct regions of the nucleus of the solitary tract (NTS)[88]. The NTS is located within the dorsal respiratory group of nerve cells in the medulla oblongata, as illustrated back in Figure 22.8, so there is the potential here for interactions to take place between the respiratory and cardiovascular systems.

22.5.1 Location of terminations of the respiratory and cardiovascular sense organs in the central nervous system

One useful technique for determining where the sensory inputs from the respiratory and cardiovascular systems terminate within the CNS involves taking recordings from a sensory cell of identified function within the ganglion of the IXth or Xth cranial nerve (for example, a slowly adapting lung receptor in the ganglion of the Xth cranial nerve) and electrically stimulating different areas of the medulla with fine electrodes until activity is recorded in the sensory cell. This is known as antidromic mapping (*dromic*, Greek for course or direction), as the nerve impulse travels in the opposite direction to normal. The locations of the areas that cause such activity are noted so that the branches of the nerve fibre from the sensory cell can be mapped. Figure 22.36 shows two maps, one from a slowly adapting lung receptor which terminates in the medial region of the NTS in front of the region known as the obex (Figure 22.36A),

[87] Water and ion balance in fish are discussed in Chapter 5.

[88] NTS is, strictly speaking, the abbreviation for the nucleus of the tractus solitarius, but is also frequently used for the anglicised version: nucleus of the solitary tract.

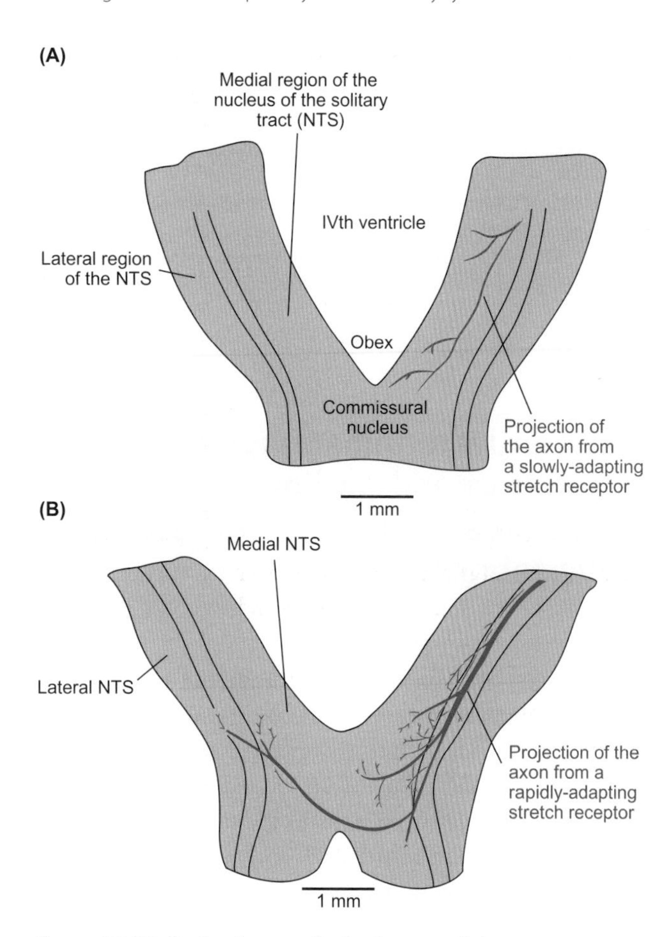

Figure 22.36 Projections to the brainstem of the sensory nerve fibres (axons) from stretch receptors in the lungs of mammals

(A) Nerve fibre from a slowly adapting stretch receptor in the lung of a rabbit. (B) Nerve fibre from a rapidly adapting stretch receptor in the lung of a cat.

Sources: (A) modified from Donoghue S et al (1982). The brain-stem projections of pulmonary stretch afferent neurones in cats and rabbits. Journal of Physiology 322: 353-363. (B): reproduced from Kubin L and Davies RO (1995). Central pathways of pulmonary and airway vagal afferents. In: Regulation of Breathing 2nd edition. Dempsey JA and Pack AI (eds.), pp 219-284. New York: Marcel Dekker, Inc.

and one from a rapidly adapting lung receptor that has more extensive terminations, including to the other side of the brain stem (Figure 22.36B).

The mapping method has also been used to determine the areas of termination of baroreceptor and chemoreceptor sense organs. Figure 22.37 summarizes data from a number of studies investigating the locations at which the various sense organs terminate in the NTS of cats. Data from amphibians and birds indicate that the fibres from sensory receptors in the lungs, peripheral chemoreceptors and baroreceptors also terminate in the NTS.

Figure 22.37 shows that the areas of termination of the sensory receptors from the respiratory and cardiovascular systems in mammals overlap widely. This arrangement provides the means by which various sensory inputs can be integrated into specific patterns of motor output to the heart and blood vessels.

22.5.2 Origins of motor outputs to heart and blood vessels

The vagal motor nerve cells that send branches to the heart are located in one or both of two regions within the medulla of the brains in vertebrates: the **nucleus ambiguus** (NA), which is within the location of the ventral group of respiratory nerve cells, as shown in Figure 22.32, and the **dorsal vagal motor nucleus (DVN)**, shown in Figures 22.32 and 22.38. In mammals, nervous connections have been identified from the NTS to both the DVN and the NA. These connections provide the basis for the reflex changes in vagal activity to the heart in response to input from the peripheral chemoreceptors, lung receptors and baroreceptors.

Sensory input from the baroreceptors enters the NTS and connections are made with the rostral and caudal ventrolateral medulla (RVLM and CVLM, respectively). As discussed in Section 22.4.3, the RVLM seems to be important in generating sympathetic tone, and a possible pathway from the RVLM to the heart and blood vessels is shown in Figure 22.38.

	Regions of the nucleus of the solitary tract		
Receptor type	Medial	Commissural	Lateral
Aortic baroreceptor	● ● ● ○	●	● ● ● ● (shaded)
Carotid baroreceptor	● ● ●	●	● ● ● ● (shaded)
Carotid chemoreceptor	● ● ● ● ● ● ○ (shaded)	● ● ● ● ● ○ ○ (shaded)	● ●
Slowly adapting lung stretch receptor	● ● ● ● ● ● (shaded)		● ●
Rapidly adapting lung stretch receptor	● ● ● ● ○ ○ ○ (shaded)	● ● ● ● ● ○ ○ ○ (shaded)	● ●

Figure 22.37 Major areas of termination within the nucleus of the solitary tract (NTS) of cats of nerve fibres from receptors in the cardiovascular and respiratory systems

The relative density of nerves terminating on the same side (●) and on the opposite side (●) of the NTS is denoted by the number of dots and the most extensive regions of termination are shaded green. See Figure 22.36 for locations of the medial, commissural and lateral regions of the NTS.

Source: modified from Jordan D (1995). CNS integration of cardiovascular regulation, In: Cardiovascular Regulation. Jordan D and Marshall J (eds.). London: Portland Press.

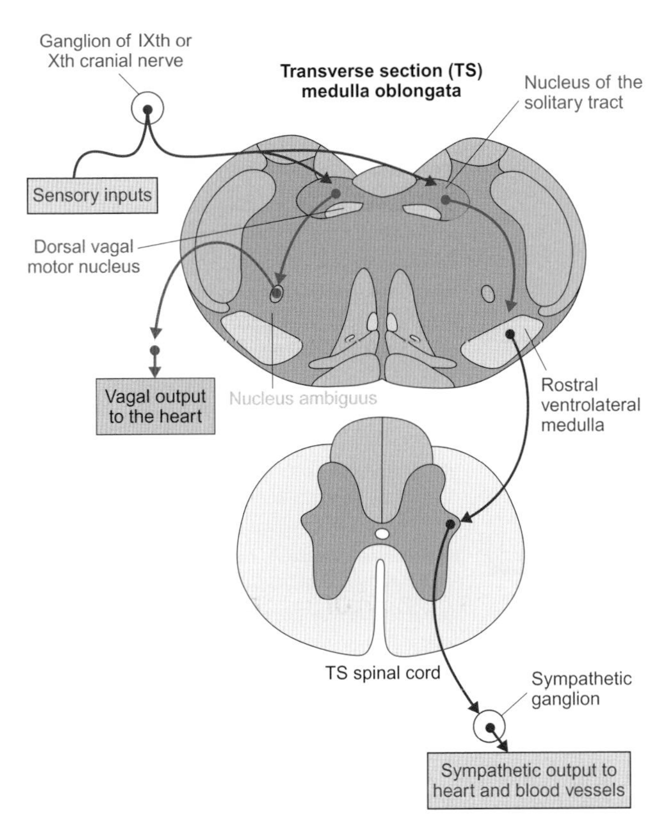

Figure 22.38 Nervous connections in the brain of mammals between the peripheral sense organs in the respiratory and circulatory systems and the autonomic output to the heart and blood vessels

The top panel shows the reflex arc for the vagal output to the heart. Sensory inputs from the peripheral chemoreceptors and baroreceptors terminate in the nucleus of the solitary tract (NTS). Vagal output to the heart is shown on the left side of the panel and is from either the dorsal vagal motor nucleus or the nucleus ambiguous (NA) both of which have connections from the NTS. Only the connection from the NTS to the NA is shown in the figure. The reflex arc for the sympathetic output to the heart and blood vessels is shown in both panels. Sensory inputs to the NTS connect to a region near the NA, the rostral ventrolateral region of the medulla (RVLM). Axons pass from the from the RVLM to the heart and blood vessels via the spinal cord and sympathetic ganglia, shown in the bottom panel.

Source: modified from Guyenet PG (1990). Role of the ventral medulla oblongata in blood pressure regulation. In: Central Regulation of Autonomic Functions. Loewy AD and Spyer KM (eds.), pp 145-187. Oxford: Oxford University Press.

22.5.3 Interactions between the respiratory and circulatory systems

Perhaps the most widely known interaction between the respiratory and circulatory systems is the change in heart rate associated with ventilation of the lung. In resting air-breathing vertebrates, heart rate increases during inspiration and falls during expiration, as shown in Figure 22.39A. This is known as **respiratory sinus arrhythmia**. Figure 22.39B shows that this arrhythmia is mainly the result of the inhibition of the activity of the vagal motor neurons to the heart during inspiration being removed during expiration. The variation in activity of the cardiac vagal motor neurons is due to both rhythmic information from the lungs via the slowly adapting stretch receptors and the direct influence of inspiratory neurons in the medulla.

Figure 22.40 shows that sympathetic nervous activity also changes during the respiratory cycle. Bursts of activity from the sympathetic nervous system (which are related to heartbeat) are inhibited during expiration, thus causing a decrease in heart rate. So, during a cycle of ventilation, there is no vagal activity to the heart during inspiration but there is some sympathetic activity. In contrast, there is some vagal activity to the heart during expiration but no sympathetic activity.

Respiratory sinus arrhythmia may serve to improve gas exchange in the lungs by reducing the fraction of physiological dead space and shunting of blood within the lungs[89]. The magnitude of the sinus arrhythmia indicates the degree of vagal tone on the heart.

The reflex effects of inputs from the peripheral arterial chemoreceptors and baroreceptors of mammals are also affected by the ventilation cycle. These receptors only cause a reduction in heart rate if they are stimulated during the expiratory phase of the ventilatory cycle but not if they are stimulated during inspiration, as shown in Figure 22.41A. This phenomenon is known as respiratory gating. Input from central inspiratory neurons within the medulla and stretch receptors from the lungs are responsible for this inhibition of chemoreceptor and baroreceptor input during inspiration, as shown in Figure 22.41B.

Despite the fact that the inputs from the various sense organs terminate in the nucleus of the NTS, the influence of ventilation (via the pulmonary stretch receptors) on the reflex responses of these sense organs occurs outside of the solitary tract, as indicated in Figure 22.41B. The inspiratory neurons in the ventral respiratory group, close to the NA, inhibit the vagal motor nerves to the heart, which are located in the NA, while input from the stretch receptors in the lungs inhibits pathways from the NTS to the NA.

When an animal suddenly experiences a threat, the physiological components of a defence-alerting response are apparent before the animal shows signs of retreat or defensive reactions. This response consists of increases in heart rate and cardiac output, ventilation, blood pressure and blood flow to the skeletal muscles. At the same time, blood flow to the kidneys, skin and gut is reduced. There is an increase in supply of blood (and therefore of oxygen) to the skeletal muscles in preparation for a hasty retreat or a

[89] Shunting of blood in gas-exchange organs is discussed in Box 12.1.

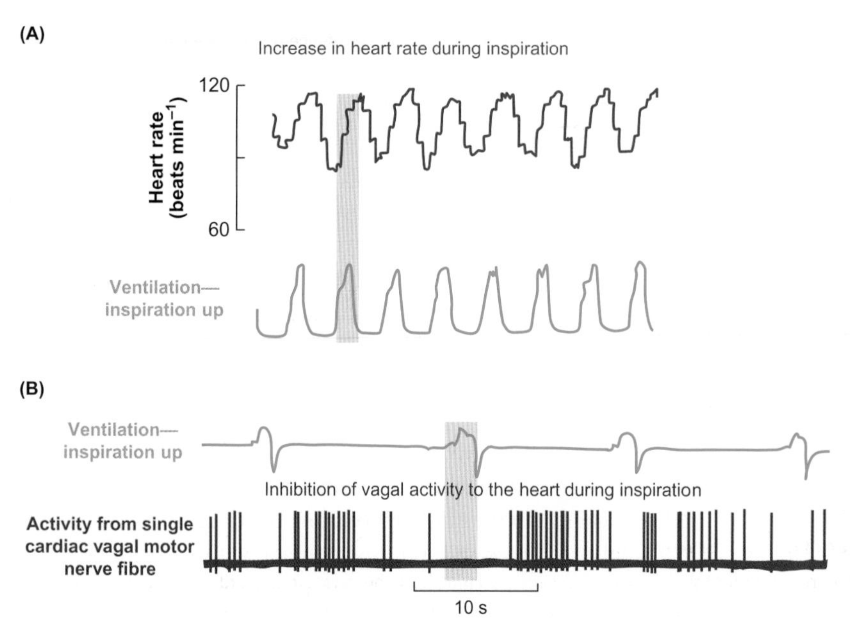

Figure 22.39 Respiratory-related output from the vagus nerve to the heart

(A) Variation in heart rate during ventilation in an anaesthetized dog. This variation in heart rate is known as respiratory sinus arrhythmia. Heart rate increases during inspiration as shown by the orange shading. (B) Inhibition of activity in the vagal motor nerve to the heart during inspiration of an anaesthetized dog, shown by the orange shading. This inhibition is the major cause of respiratory sinus arrhythmia.

Sources: (A): modified from Gandevia SC et al (1978). Inhibition of baroreceptor and chemoreceptor reflexes on the heart rate by afferents from the lungs. Journal of Physiology 276: 369-38. (B): modified from Davidson NS et al (1976). Respiratory modulation of baroreceptor and chemoreceptor reflexes affecting heart rate and cardiac vagal nerve activity. Journal of Physiology 259: 523-530.

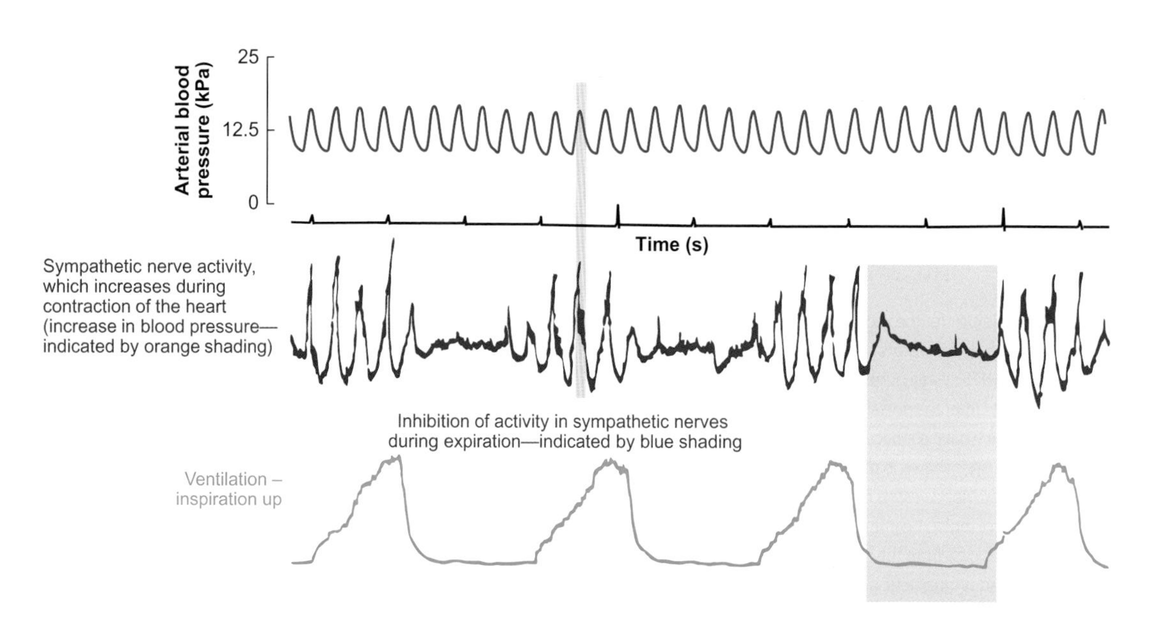

Figure 22.40 Respiratory-related output from the sympathetic nervous system to the heart

Inhibition of cardiac-related activity in a sympathetic nerve during the expiratory phase of ventilation in an anaesthetized cat.

Source: modified from Barman SM (1990). Brainstem control of cardiovascular function. In: Brainstem Mechanisms of Behaviour. Klemm WR and Vertes RP (eds.), pp 353-381 New York: Wiley and Sons.

vigorous defence of life and limb. Some of these responses are illustrated in Figure 22.42A. Figure 22.42B shows the regions of the hypothalamus, midbrain and medulla which, when stimulated, give rise to the defence-alerting response.

The full pattern of the alerting response can be produced by a variety of new or unexpected stimuli, as well as by stimuli that pose a threat or danger or create emotional stress. The fact that both cardiac output and blood pressure increase during the alerting response indicates that the baroreflex is inhibited under such circumstances. This is achieved by inhibition of neurons within the NTS which are normally activated by stimulation of the baroreceptors. This inhibition is indicated in Figure 22.41B.

Thus, the defence-alerting response is a coordinated set of behavioural and physiological responses to unexpected or threatening stimuli which modifies the more basic cardiovascular reflexes in order to meet the increasing energy demands that may be necessary for the animal's survival. As well as responding to an imminent, unexpected, threat by attacking or fleeing, animals exercise when pursuing prey, migrating or moving to and from a feeding ground. We explore the role of the nervous

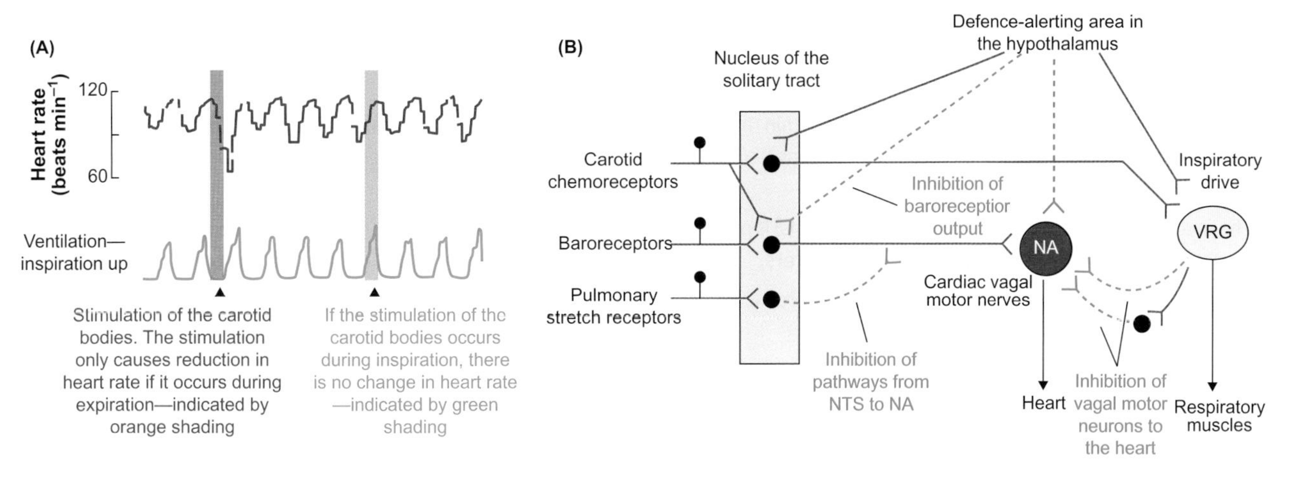

Figure 22.41 Influence of inputs from peripheral receptors, inspiratory nerve cells and the defence-arousal area of the hypothalamus on the activity of the cardiac vagal motor nerves of mammals

(A) Stimulation of the carotid body chemoreceptors and baroreceptors (not shown) only causes a reflex reduction in heart rate (bradycardia) during the expiratory phase of ventilation in anaesthetized dogs. (B) The control of the cardiac vagal motor nerves by inputs from baroreceptors, chemoreceptors, inspiratory nerve cells, stretch receptors in the lungs and the defence-alerting area in the hypothalamus. Continuous lines indicate excitatory pathways and dashed lines indicate inhibitory pathways. VRG, ventral respiratory group; NA, nucleus ambiguous.

Sources: (A): modified from Gandevia SC et al (1978). Inhibition of baroreceptor and chemoreceptor reflexes on the heart rate by afferents from the lungs. Journal of Physiology 276: 369-381. (B): modified from Daly MdeB (1995). Aspects of the integration of the respiratory and cardiovascular systems. In: Cardiovascular Regulation. Jordan D and Marshall J (eds.), pp 15–35 London: Portland Press.

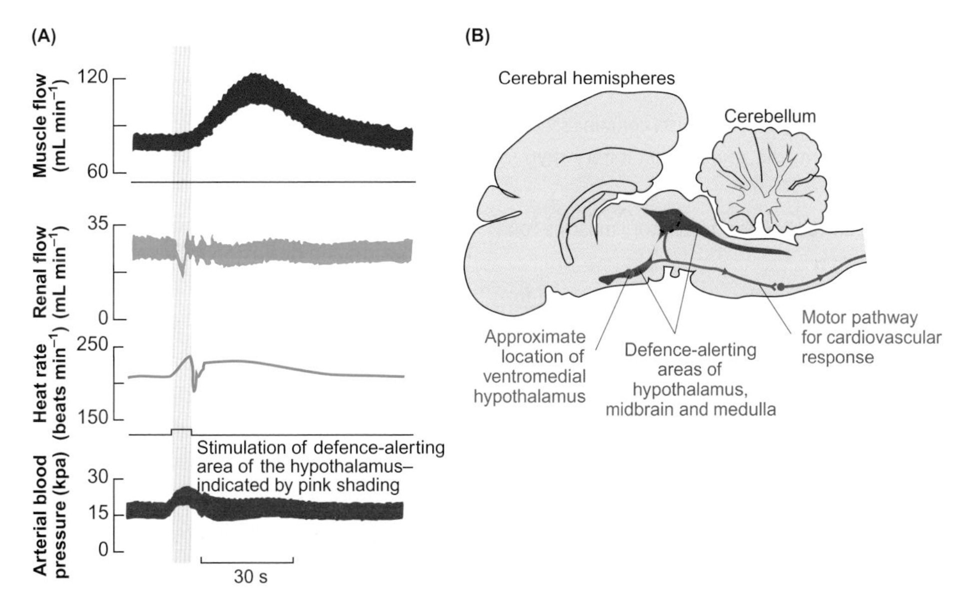

Figure 22.42 The defence-alerting regions of the brain of mammals and the effects of their stimulation

(A) Effects of stimulation of the defence-alerting areas of the brainstem of an anaesthetized cat on the cardiovascular system. Heart rate and blood pressure increase, while blood flow decreases, albeit temporarily, to the kidneys. There is a rise in blood flow to the skeletal muscles, but this occurs after the other responses. (B) Vertical longitudinal section through the middle of the brain of a cat showing the defence-alerting areas and the motor pathway to the cardiovascular system. The location of the ventromedial hypothalamus is also indicated as it has been suggested that it may serve as a common output for the defence-alerting response and the exercise response.

Source: modified from Marshall JM (1995). Cardiovascular changes associated with behavioural alerting. In: Cardiovascular Regulation. Jordan D and Marshall J (eds.), pp 37–59 London: Portland Press.

system in coordinating the responses of the respiratory and cardiovascular systems to exercise in mammals in Case Study 22.2, available online[90].

[90] The respiratory and cardiovascular responses to exercise in mammals are discussed in Section 15.2.3.

› *Review articles*

Spyer KM, Gourine AV (2009). Chemosensory pathways in the brainstem controlling cardiorespiratory activity. Philosophical Transactions of the Royal Society B 364: 2603–2610.

Taylor EW, Jordan D, Coote JH (1999). Central control of the cardiovascular and respiratory systems and their interactions in vertebrates. Physiological Reviews 79: 855–916.

Checklist of key concepts

- The activity of the respiratory pumps is produced within the central nervous system by a network of oscillating neurons—a **central rhythm (pattern) generator**.
- The rhythmic activity of the circulatory pumps (hearts) is produced in most groups of animals by **myogenic pacemakers**.
- In arthropods (except insects), the rhythm is produced by a **cardiac ganglion**, or **neurogenic pacemaker**.
- The control of the pumps and the diameters of peripheral blood vessels in many circulatory systems rely on sensory information from **chemoreceptors** and **mechanoreceptors**.
- The sensory organs send information to the central nervous system where it is integrated.
- The ensuing **motor output** is then sent to the central rhythm generators, pumps and blood vessels to ensure the adequate delivery of oxygen to the metabolizing cells.

Generation of the respiratory rhythm

- Relatively few cells are involved in generating the respiratory rhythm in invertebrates.
- Eight **oscillator neurons** in each side of the **sub-oesophageal ganglion** in decapod crustaceans alternately depolarize and hyperpolarize but do not produce action potentials.
- The first **fused abdominal ganglion** of locusts contains six neurons which are thought to form the primary respiratory rhythm generator.
  - When at **rest**, thoracic and abdominal respiratory muscles contract at the same frequency.
  - **During flight**, the thoracic muscles contract at the same frequency as the wings and faster than the abdominal muscles.
- Pond snails have a three-neuron respiratory rhythm generator which alternately opens and closes the **pneumostome** when the animal is at the water surface by the processes of **post-inhibitory rebound excitation** and **reciprocal inhibition**.
- One of the three cells in the rhythm generator is capable of remembering learned respiratory behaviour.
- Respiratory neurons are found in the two **reticular formations** of fish, which run alongside the motor nuclei of cranial nerves V, VII, IX and X.
- Nerves from these nuclei innervate the muscles that create the ventilatory current.
- Each half of the brain in amphibians features two **respiratory oscillators**: one for buccal ventilation and one for lung ventilation.
- The buccal oscillator seems to be based on a **network of interacting cells**, whereas the lung oscillator may rely on the activity of **pacemaker cells**.
- The **episodic pattern** of lung ventilation in vertebrates is an intrinsic property of the central respiratory control system and does not depend on feedback from oxygen and carbon dioxide sense organs.
- Mammals possess **dorsal** and **ventral groups of respiratory neurons**, but only the ventral group is essential for generation of the respiratory rhythm.
- **Self re-excitation** occurs within the inspiratory and expiratory groups of neurons (**I neurons** and **E neurons,** respectively), and **reciprocal inhibition** occurs between the two groups.
- Reciprocal inhibition results in the activity of one group of neurons being inhibited when the other is active.
- Inspiration is eventually terminated by an **inspiratory off switch**, which consists of input from stretch receptors in the lungs and of neurons located in the pons.
- The medulla of mammals has two distinct but coupled rhythm generators: the **pre-Bötzinger complex** and the **retrotrapezoid nucleus/parafacial respiratory group (RTN/pFRG)**.

Control of the respiratory system

- **Phasic** and **tonic mechanoreceptors** are found in various locations on the gills of fish.
  - Phasic receptors adapt rapidly and respond to deflection of the gills during inspiration and expiration.
  - Tonic receptors adapt slowly and provide an almost linear output related to the position of the gills.
- **Slowly** and **rapidly adapting (stretch) receptors** are also present in the air-breathing organs and lungs of vertebrates, except birds.
- Slowly adapting receptors are involved in the **Breuer–Hering inflation** and **deflation reflexes**.
- Lungfish, amphibians and some species of reptiles have lung mechanoreceptors that are also affected by the concentration of CO_2 in their lungs.
- The lungs of birds do not change volume during the respiratory cycle and possess **intrapulmonary chemoreceptors (IPCs)** that only respond to changes in the level of CO_2 through changes in intracellular pH.
- Mechanoreceptors are located around the pneumostome in pond snails and are thought to contribute to the initiation of the respiratory rhythm.
- The **oval organ** is located in the bailer of decapods crustaceans and **chordotonal organs** are in various locations in insects, but their exact role in the production and control of the respiratory rhythm is not known.
- **Peripheral chemoreceptors** in vertebrates monitor the level of respiratory gases in the in the blood and in the environment of aquatic animals and stimulate an increase in ventilation during hypoxia.
- **Peripheral oxygen-sensitive receptors** are located in the walls of the mouth and on the gills of fish.
- The peripheral oxygen-sensitive receptors in teleost fish monitor oxygen in the water and blood.

- In air-breathing vertebrates, oxygen-sensitive chemoreceptors are associated with one or more of the remaining aortic arches.
- Denervation of the carotid bodies in both birds and mammals abolishes the ventilatory response to hypoxia.
- The **osphradium** of pond snails, located close to the pneumostome, is sensitive to hypoxia.
- **CO_2-sensitive receptors** that stimulate ventilation are located on the first gill arch of teleost fish and respond exclusively to CO_2/pH.
- In birds and mammals, the same receptors that are sensitive to oxygen are also sensitive to increases levels of CO_2 (hypercapnia).
- These sense organs respond **synergistically** to hypoxia and hypercapnia combined.

Structure and sensory processes in peripheral chemoreceptor cells of vertebrates

- **Glomus** or **type I cells** are the oxygen-sensing cells of the carotid bodies.
- **HIF-1α** plays an important part in the response of the carotid body to hypoxia.
- Hypoxia also **inhibits potassium ion channels** in the glomus cells, which leads to depolarization of the cells and the **entry of calcium ions** via voltage-dependent channels.
- The nature of the actual sensor is unknown; those that have been suggested include:
  - a haem-linked nicotinamide adenine dinucleotide phosphate-oxidase (NADPH-oxidase) located in the membrane of the glomus cells,
  - the mitochondria,
  - adenosine monophosphate (AMP)-activated protein kinase.
- The glomus cells synthesize and store a variety of neurotransmitters.
- The oxygen-sensing cells on the gills of fish are thought to be **neuroepithelial cells (NECs)** which are characterized by the presence of the neurotransmitter 5-hydroxytryotamine (serotonin) in dense-cored vesicles.

Central carbon dioxide receptors

- **Centrally located carbon dioxide/pH receptors** are found in most groups of air-breathing vertebrates.
- In mammals, these receptors have been identified in a number of locations in the brain, including the **raphe nuclei** and in the **retrotrapezoid nucleus (RTN)**.
- The CO_2/pH-sensitive cells in the RTN are activated by input from the carotid body chemoreceptors when stimulated by hypoxia.

Generation of the cardiac rhythm

- Rate of excitation (and contraction) of the heart of vertebrates is determined by those **myocytes** with the fastest spontaneous rate of excitation—the **pacemaker cells**.
- Important pacemakers in the hearts of mammals are the **sino-atrial (S-A) node** and the **atrioventricular (A-V) node**.
- The A-V node is connected to the **bundles of His** which eventually subdivide into a network of conducting fibres, the **Purkinje fibres**.
- The **pacemaker potential** of myocytes in the S-A node is the result of the movements of various ions across their cell membranes, most of which are controlled by **time-** and/or **voltage-dependent ion channels**.
- The action potentials in the myocytes of the atria and ventricles are converted into the contraction of the muscles by the process of **excitation-contraction coupling**.
- Excitation and contraction of the myocytes in the hearts of arthropods, except insects, rely on the spontaneous activity of the **cardiac ganglion**.

Control of the cardiovascular system

- The **peripheral chemoreceptors** and **mechanoreceptors** (**baroreceptors**) are the major groups of sense organs involved in the control of the cardiovascular system:
  - The chemoreceptors are the same as those involved in the control of ventilation.
  - The baroreceptors are located in the walls of the major arteries and veins, and in various chambers of the heart.
- The **primary cardiovascular response** to the stimulation of the chemoreceptors by hypoxia causes a reduction in heart rate (bradycardia) in both fish and crustaceans, and a change in **resistance to blood flow** around the body in fish.
- Many species of air-breathing vertebrates exhibit the **secondary cardiovascular response** during which increased input from the lungs as ventilation increases in response to hypoxia, overrides the primary cardiovascular response and causes an increase in heart rate (**tachycardia**).
- The baroreceptors cause changes in the cardiovascular system that maintain central arterial blood pressure at a given level or **set point**. This is the **barostatic reflex**.
- Effects on the hearts of invertebrates by baroreflexes are mediated both by nerves and by circulating hormones.
- Nervous control of the cardiovascular system of vertebrates is mediated by the **parasympathetic** and **sympathetic** branches of the **autonomic nervous system (ANS)**.
- Parasympathetic fibres to the heart release **acetylcholine** which has **negative chronotropic** and **inotropic effects** on the heart.
- Sympathetic fibres release **catecholamines** which have **positive chronotropic** and **inotropic effects** on the heart and cause contraction of the smooth muscle in the walls of blood vessels.
- Both branches of the ANS are continuously active; they have a 'tone'—**vagal tone** and **sympathetic tone**.
- Resting sympathetic tone is determined by regions of the brainstem, including the **rostral ventrolateral medulla (RVLM)**.
- Sympathetic tone is important in the control of blood pressure.

Long-term control of blood pressure in vertebrates

- Reduced sensitivity of the baroreceptors may contribute to **hypertension**.

- The hormones **arginine vasopressin (AVP)**, **arginine vasotocin (AVT)** and **angiotensin II** have major roles in salt and water balance and cause constriction of peripheral blood vessels.
- The reflex secretion of AVP depends largely on input from the arterial baroreceptors, but stretch-sensitive receptors in the kidney regulate the secretion of **renin**, which is involved in the production of angiotensin II.
- A major hypothesis for the long-term regulation of arterial blood pressure in mammals is that the kidneys determine the set point in terms of the interrelationship between fluid loss and sodium balance; there is only one arterial pressure at which sodium balance is maintained.

Location of terminations of the respiratory and cardiovascular sense organs in the central nervous system

- The fibres from sensory receptors in the lungs, peripheral chemoreceptors and baroreceptors terminate in the **nucleus of the solitary tract (NTS)**.
- In mammals, nervous connections from the NTS to both the **dorsal vagal nucleus (DVN)** and the **nucleus ambiguus (NA)** provide the basis for the reflex changes in vagal activity to the heart in response to input from the peripheral chemoreceptors, lung receptors and baroreceptors.
- The NTS also makes connections with the rostral ventrolateral medulla, which is important in generating sympathetic tone.
- **Respiratory sinus arrhythmia** is mainly the result of the inhibition of the activity of the **vagal motor neurons** to the heart during inspiration causing an increase in heart rate.
- Bursts of activity from the sympathetic nervous system are inhibited during expiration, thus causing a decrease in heart rate.
- A reduction in heart rate only occurs if the peripheral arterial chemoreceptors and baroreceptors of mammals are stimulated during the expiratory phase of the ventilatory cycle, but not if they are stimulated during inspiration—a phenomenon known as **respiratory gating**.
- The baroreflex is inhibited when the **defence-alerting response** is activated.
- The respiratory and cardiovascular responses to exercise are controlled and coordinated by a combination of **central command** and **afferent feedback** from peripheral receptors.

Study questions

1. Briefly outline the major differences between respiratory rhythm generation in invertebrates and vertebrates. (Hint: Section 22.1.)
2. What role does reciprocal inhibition play in respiratory rhythm generation? (Hint: Section 22.1.)
3. Discuss the roles of self re-excitation and the inspiratory off switch in the respiratory rhythm generators of vertebrates. (Hint: Section 22.1.2.)
4. What is the relationship between episodic ventilation in amphibians and reptiles and continual ventilation in most species of fish, birds and mammals? (Hint: Section 22.1.2.)
5. Discuss the functional significance of the slowly adapting receptors in the lungs of mammals. (Hint: Section 22.2.1.)
6. Why is it thought that birds have carbon dioxide-sensitive receptors in their lungs instead of stretch receptors? How are these receptors able to perform a similar function to the stretch receptors in the lungs of mammals? (Hint: Section 22.2.1.)
7. What roles do the peripheral chemoreceptors play in the control of the respiratory and cardiovascular systems of vertebrates? (Hint: Sections 22.2.3 and 22.4.1.)
8. How are the carotid body chemoreceptors thought to detect changes in the partial pressure of oxygen? (Hint: Section 22.2.3.)
9. Describe the basic features of the myogenic cardiac pacemakers of vertebrates. (Hint: Section 22.3.1.)
10. To what extent are the arterial baroreceptors involved in the long-term control of blood pressure in vertebrates? (Hint: Section 22.4.4.)
11. What is the significance of the nucleus of the tractus solitarius (NTS) in the integration of the respiratory and cardiovascular systems in mammals? (Hint: Section 22.5.)
12. Discuss the characteristics of the defence-alerting response in mammals. (Hint: Section 22.5.3.)
13. What major factors ensure the respiratory and cardiovascular systems of mammals meet the increased demand for oxygen during exercise? (Hint: Case Study 22.2, available online.)

Bibliography

Berne RM, Levy MN (1998). Physiology, 4th Edition. Mosby, St Louis.

Appendix

The International System of Units (SI)

The SI system is based on a set of seven base units from which all other SI units are derived; some units are named after influential scientists. The names of these units are spelt in lower case, but the associated symbol is in upper case. Standard prefixes of SI units covering the range from 10^{-24} (yocto-) to 10^{24} (yotta-) are indicated in a different table below.

Quantity measured in a base SI unit	Base unit	Relationships among other units	Comment
Absolute temperature, T	kelvin, K	Centigrade scale, t (°C) 1°C = 1 K 0°C = 273 K (0 K = −273°C) T (K) = 273 + t (°C)	°C is commonly used in physiology; the centigrade scale is the official temperature scale for most countries in the world
		Fahrenheit scale, t (°F) 1°F = 1.8 K = 1.8°C 0°F = −32°C = 241 K t (°F) = 32 + 1.8 t (°C)	°F; the Fahrenheit scale is commonly used in some countries, including the USA
Amount of substance	mole, mol		1 mol contains 6.022×10^{23} molecules, ions or particles of substance (Avogadro's number)
Electric current	ampere, A	1 A = 1 amp	
Length	metre, m	1 m = 10^2 centimetres (cm) = 10^3 millimetres (mm)	
Luminous intensity	candela, cd		
Mass	kilogram, kg	1 g = 10^{-3} kg	
Time, t	second, s	1 min = 60 s 1 h = 60 min	

Quantity measured in derived SI units	Derived unit	Relationships among other units	Comment
Acceleration	m s^{-2}		gravitational acceleration = 9.81 m s^{-2}
Area	square metre (m^2)	1 m^2 = 10^4 square cm (cm^2) = 10^6 square mm (mm^2)	
Buffer value	slyke, (mmol L^{-1} (pH unit)$^{-1}$)		For example, a solution has a buffer value of 1 slyke when the addition of 1 mmol L^{-1} of strong acid or strong base changes the solution's pH by one unit
Capacitance	farad, F (kg^{-1} m^{-2} s^4 A^2)	1 F = 1 C V^{-1}	
Concentration	mol m^{-3}	1 mol m^{-3} = 10^{-3} mol L^{-1} = 0.001 mol L^{-1} 1 mol L^{-1} = 1 molar (M)	mol L^{-1} or M (molar) are more commonly used in physiology
Density	kg m^{-3}	1 kg m^{-3} = 10^{-3} g cm^{-3}	g cm^{-3} = g mL^{-1} is commonly used in physiology
Diffusion rate of substance	mol s^{-1}		
Electrical charge	coulomb, C (s A)		1 electron carries a charge of -1.6 10^{-19} C

Quantity measured in derived SI units	Derived unit	Relationships among other units	Comment
Electrical conductance	siemens, S ($kg^{-1}\ m^{-2}\ s^3\ A^2$)	$1\ S = 1\ \Omega^{-1}$ = mho (ohm spelt backward)	
Electrical resistance, impedance	ohm, Ω ($kg\ m^2\ s^{-3}\ A^{-2}$)	$1\ \Omega = 1\ V\ A^{-1}$	
Energy, heat, work	joule, J ($kg\ m^2\ s^{-2}$)	1 J = 0.239 calorie (cal) 1 cal = 4.182 J 1 kcal = 10^3 cal = 1 Cal = 4182 J	
Flux of substance	$mol\ m^{-2}\ s^{-1}$		
Force, weight	newton, N ($kg\ m\ s^{-2}$)	1 kg weight = 1 kg w = 9.81 N 1 g weight = 1 g w = 9.81 mN	weight = mass × gravitational acceleration
Frequency, **v** or **f**	hertz, Hz (s^{-1})	1 Hz = one cycle per second = 1 s^{-1}	
Osmotic concentration Osmolarity Osmolality	 $osmol\ m^{-3} = Osm\ m^{-3}$ Osm kg solvent^{-1}	 1 osmol $m^{-3} = 10^{-3}$ Osm L^{-1} 1 Osm L^{-1} = 1 Osmolar (OsM)	osmol = Osm = number of particles dissolved in solution expressed in mol Osmolarity (Osm L^{-1}) is commonly used in physiology, but osmometers measure osmolality (Osm kg solvent^{-1}), and the solvent is generally water
Power	watt, W ($kg\ m^2\ s^{-3}$)	1 W = 1 J s^{-1}	
Pressure	pascal, Pa ($kg\ m^{-1}\ s^{-2}$)	1 Pa = 1 N m^{-2} = 0.0075 mm of mercury (mmHg) 1 mmHg = 133.3 Pa 1 atmosphere (atm) = 101.3 kPa	atm and mmHg continue to be used in some branches of physiology. Human blood pressure is almost always given in mmHg
Velocity	$m\ s^{-1}$		
Voltage	volt, V ($kg\ m^2\ s^{-3}\ A^{-1}$)		
Volume	cubic metre (m^3)	1 m^3 = 10^3 litres (L) 1 L = 10^3 cm^3 1 millilitre (mL) = 1 cm^3 1 microlitre (μL) = 1 mm^3	
Planar angle	Radian, rad (dimensionless)	1 radian is the planar angle subtended by a circular arc that is equal in length to the radius of the arc; For example, a planar angle of 90° subtends an arc that is one quarter of the circle's length ($0.5\ \pi$ R) and therefore has 0.5π R/ R rad = 0.5π rad = 1.57 rad. Conversely, 1 rad = 57.3°.	The degree (°defined as one 360th part of a full circle) continues to be extensively used in Geometry, Trigonometry, Physiology and Earth Sciences.

Standard prefixes of SI units

Prefix	Symbol	Multiplicationfactor	Prefix	Symbol	Multiplication factor
deca-	**da**	10^1	deci-	**d**	10^{-1}
hecto-	**h**	10^2	centi-	**c**	10^{-2}
kilo-	**k**	10^3	milli-	**m**	10^{-3}
mega-	**M**	10^6	micro-	**μ**	10^{-6}
giga-	**G**	10^9	nano-	**n**	10^{-9}
tera-	**T**	10^{12}	pico-	**p**	10^{-12}
peta-	**P**	10^{15}	femto-	**f**	10^{-15}
exa-	**E**	10^{18}	atto-	**a**	10^{-18}
zetta-	**Z**	10^{21}	zepto-	**z**	10^{-21}
yotta-	**Y**	10^{24}	yocto-	**y**	10^{-24}

Index

Tables, figures, and boxes are indicated by an italic *t*, *f*, and *b* following the page number.

Many individual animal species are listed under animal groups such as frogs (species), toads, lungfish, slugs, cockroaches, locusts, teleost fish (species), and mosquitoes.

A

E

F

G

H

J

N

O

T

U

V